Volkswagen EuroVan

Repair Manual
1992, 1993, 1994, 1995, 1996, 1997, 1998, 1999

Gasoline, Diesel, TDI, 5-cylinder and VR6, including MultiVan and CV Camper

Selected Books and Repair Information From Bentley Publishers

Motorsports

Alex Zanardi: My Sweetest Victory
Alex Zanardi and Gianluca Gasparini
ISBN 978-0-8376-1249-2

The Unfair Advantage
Mark Donohue and Paul van Valkenburgh
ISBN 978-0-8376-0069-7

Equations of Motion - Adventure, Risk and Innovation *William F. Milliken*
ISBN 978-0-8376-1570-7

Engineering

Mechanical Ignition Handbook: The Hack Mechanic Guide to Vintage Ignition Systems
Rob Siegel ISBN 978-0-8376-1767-1

Hack Mechanic Guide to European Automotive Electrical Systems
Rob Siegel ISBN 978-0-8376-1751-0

Bosch Automotive Handbook *Robert Bosch GmbH* ISBN 978-0-8376-1732-9

Bosch Fuel Injection and Engine Management *Charles O. Probst, SAE*
ISBN 978-0-8376-0300-1

Maximum Boost: Designing, Testing, and Installing Turbocharger Systems
Corky Bell ISBN 978-0-8376-0160-1

Supercharged! Design, Testing and Installation of Supercharger Systems
Corky Bell ISBN 978-0-8376-0168-7

Scientific Design of Exhaust and Intake Systems *Phillip H. Smith & John C. Morrison* ISBN 978-0-8376-0309-4

Physics for Gearheads
Randy Beikmann ISBN 978-0-8376-1615-5

Audi

Audi A4 Service Manual: 2002-2008, 1.8L Turbo, 2.0L Turbo, 3.0L, 3.2L
Bentley Publishers
ISBN 978-0-8376-1574-5

Audi TT Service Manual: 2000-2006, 1.8L turbo, 3.2 L *Bentley Publishers*
ISBN 978-0-8376-1625-4

Audi A6 (C5 platform) Service Manual: 1998-2004 *Bentley Publishers*
ISBN 978-0-8376-1670-4

BMW

Memoirs of a Hack Mechanic
Rob Siegel ISBN 978-0-8376-1720-6

BMW 3 Series (F30, F31, F34) Service Manual: 2012-2015 *Bentley Publishers*
ISBN 978-0-8376-1752-7

BMW X3 (E83) Service Manual: 2004-2010 *Bentley Publishers*
ISBN 978-0-8376-1731-2

BMW X5 (53) Service Manual: 2000-2006 *Bentley Publishers*
ISBN 978-0-8376-1643-8

BMW 3 Series (E90, E91, E92, E93) Service Manual: 2006-2011
Bentley Publishers ISBN 978-0-8376-1723-7

BMW 3 Series (E46) Service Manual: 1999-2005 *Bentley Publishers*
ISBN 978-0-8376-1657-5

BMW 3 Series (E36) Service Manual: 1992-1998 *Bentley Publishers*
ISBN 978-0-8376-1709-1

BMW Z3 (E36/7) Service Manual: 1996-2002 *Bentley Publishers*
ISBN 978-0-8376-1617-9

BMW 5 Series (E60, E61) Service Manual: 2004-2010 *Bentley Publishers*
ISBN 978-0-8376-1689-6

BMW 5 Series (E39) Service Manual: 1997-2003 *Bentley Publishers*
ISBN 978-0-8376-1672-8

Corvette

Corvette: America's Star-Spangled Sports Car *Karl Ludvigsen*
ISBN 978-0-8376-1659-9

Corvette by the Numbers: The Essential Corvette Parts Reference 1955-1982 *Alan Covin*
ISBN 978-0-8376-0228-2

Zora Arkus-Duntov: The Legend Behind Corvette *Jerry Burton*
ISBN 978-08376-0858-7

Porsche

Porsche 911 (996) Service Manual: 1999-2005 *Bentley Publishers*
ISBN 978-0-8376-1710-7

Porsche 911 (993) Service Manual: 1995-1998 *Bentley Publishers*
ISBN 978-0-8376-1719-0

Porsche Boxster Service Manual: 1997-2004 *Bentley Publishers*
ISBN 978-0-8376-1645-2

Porsche 911 Carrera Service Manual: 1984-1989 *Bentley Publishers*
ISBN 978-0-8376-1696-4

Porsche 911 SC Service Manual: 1978-1983 *Bentley Publishers*
ISBN 978-0-8376-1705-3

Porsche: Excellence Was Expected
Karl Ludvigsen
ISBN 978-0-8376-0235-6

Porsche — Origin of the Species
Karl Ludvigsen
ISBN 978-0-8376-1331-4

Volkswagen

Volkswagen Rabbit, GTI Service Manual: 2006-2009 *Bentley Publishers*
ISBN 978-0-8376-1664-3

Volkswagen Jetta Service Manual: 2005-2010 *Bentley Publishers*
ISBN 978-0-8376-1616-2

Volkswagen Jetta, Golf, GTI Service Manual: 1999-2005 *Bentley Publishers*
ISBN 978-0-8376-1678-0

Volkswagen Jetta, Golf, GTI: 1993-1999, Cabrio: 1995-2002 Service Manual *Bentley Publishers*
ISBN 978-0-8376-1660-5

Volkswagen GTI, Golf, Jetta Service Manual: 1985-1992 *Bentley Publishers*
ISBN 978-0-8376-1637-7

Volkswagen Corrado Repair Manual: 1990-1994 *Bentley Publishers*
ISBN 978-0-8376-1699-5

Volkswagen Passat, Passat Wagon Service Manual: 1998-2005
Bentley Publishers
ISBN 978-0-8376-1669-8

MINI Repair Manuals

MINI Cooper Service Manual: 2007-2013 *Bentley Publishers*
ISBN 978-0-8376-1730-5

MINI Cooper Service Manual: 2002-2006 *Bentley Publishers*
ISBN 978-0-8376-1639-1

Mercedes-Benz

Mercedes-Benz C-Class (W202) Service Manual 1994-2000
Bentley Publishers
ISBN 978-0-8376-1692-6

Mercedes Benz E-Class (W124) Owner's Bible: 1986-1995
Bentley Publishers
ISBN 978-0-8376-0230-1

Mercedes-Benz Technical Companion
Staff of The Star and members of Mercedes-Benz Club of America
ISBN 978-0-8376-1033-7

BentleyPublishers®
Automotive Reference

Bentley Publishers has published service manuals and automobile books since 1950. For more information, please contact Bentley Publishers at 1734 Massachusetts Avenue, Cambridge, MA 02138 USA, or visit our web site at
BentleyPublishers.com

Volkswagen EuroVan

Repair Manual
1992, 1993, 1994, 1995, 1996, 1997, 1998, 1999

Gasoline, Diesel, TDI, 5-cylinder and VR6, including MultiVan and CV Camper

[B] BentleyPublishers® .com

[B] BENTLEY PUBLISHERS® | Automotive Reference™

Bentley Publishers, a division of Robert Bentley, Inc.
1734 Massachusetts Avenue
Cambridge, MA 02138 USA
800-423-4595 / 617-547-4170

Information that makes the difference®

BentleyPublishers®.com

Technical contact information

We welcome your feedback. Please submit corrections and additions to our Volkswagen technical discussion forum at:

`http://www.BentleyPublishers.com`

Updates and Corrections

We will evaluate submissions and post appropriate editorial changes online as updates or tech discussion. Appropriate updates and corrections will be added to the book in future printings. Check for updates and corrections for this book before beginning work on your vehicle. See the following web address for additional information:

`http://www.BentleyPublishers.com/updates/`

WARNING—Important Safety Notice

Do not use this manual unless you are familiar with automotive repair procedures and safe workshop practices. This manual illustrates workshop procedures required for service work; it is not a substitute for full and up-to-date information from the vehicle manufacturer or for proper training as an automotive technician. Note that it is not possible for us to anticipate all of the ways or conditions under which vehicles may be serviced or to provide cautions as to all of the possible hazards that may result.

We have endeavored to ensure the accuracy of the information in this manual. Please note, however, that considering the vast quantity and the complexity of the service information involved, we cannot warrant the accuracy or completeness of the information contained in this manual.

FOR THESE REASONS, NEITHER THE PUBLISHER NOR THE AUTHOR MAKES ANY WARRANTIES, EXPRESS OR IMPLIED, THAT THE INFORMATION IN THIS BOOK IS FREE OF ERRORS OR OMISSIONS AND WE EXPRESSLY DISCLAIM THE IMPLIED WARRANTIES OF MERCHANTABILITY AND OF FITNESS FOR A PARTICULAR PURPOSE, EVEN IF THE PUBLISHER OR AUTHOR HAVE BEEN ADVISED OF A PARTICULAR PURPOSE, AND EVEN IF A PARTICULAR PURPOSE IS INDICATED IN THE MANUAL. THE PUBLISHER AND AUTHOR ALSO DISCLAIM ALL LIABILITY FOR DIRECT, INDIRECT, INCIDENTAL OR CONSEQUENTIAL DAMAGES THAT RESULT FROM ANY USE OF THE EXAMPLES, INSTRUCTIONS OR OTHER INFORMATION IN THIS BOOK. IN NO EVENT SHALL OUR LIABILITY WHETHER IN TORT, CONTRACT OR OTHERWISE EXCEED THE COST OF THIS MANUAL.

Before attempting any work on your Volkswagen, read the **WARNINGS** and **CAUTIONS** on page vii and viii, and any **WARNING** and **CAUTION** that accompanies a procedure in the service manual. Review the **WARNINGS** and **CAUTIONS** on page vii and viii each time you prepare to work on your Volkswagen.

Your common sense and good judgment are crucial to safe and successful service work. Read procedures through before starting them. Think about whether the condition of your car, your level of mechanical skill, or your level of reading comprehension might result in or contribute in some way to an occurrence which might cause you injury, damage your car, or result in an unsafe repair. If you have doubts for these or other reasons about your ability to perform safe repair work on your car, have the work done at an authorized Volkswagen dealer or other qualified shop.

Part numbers listed in this manual are for identification purposes only, not for ordering. Always check with your authorized Volkswagen dealer to verify part numbers and availability before beginning service work that may require new parts.

Special tools required to perform certain service operations are identified in the manual and are recommended for use. Use of tools other than those recommended in this service manual may be detrimental to the car's safe operation as well as the safety of the person servicing the car.

The vehicle manufacturer will continue to issue service information updates and parts retrofits after the editorial closing of this manual. Some of these updates and retrofits will apply to procedures and specifications in this manual. We regret that we cannot supply updates to purchasers of this manual.

Copies of this manual may be purchased from authorized Volkswagen dealers, most automotive accessories and parts dealers specializing in Volkswagen automobiles, from selected booksellers, or directly from the publisher.

ISBN 978-0-8376-1681-0 (3-Volume Set)
ISBN 978-0-8376-1682-7 (Volume 1)
ISBN 978-0-8376-1683-4 (Volume 2)
ISBN 978-0-8376-1684-1 (Volume 3)

Bentley Stock No. VV99 **Job code: VV99-10** **VWoA, Inc. Part No. LPV 800 149**

Library of Congress Cataloging-in-Publication Data
Volkswagen EuroVan: official factory repair manual 1992,1993, 1994, 1995, 1996, 1997, 1998, 1999: gasoline, diesel, TDI, 5-cylinder, and VR6, including multivan and CV camper
p. cm.
ISBN 978-8376-0335-3 (alk. paper)
1. EuroVan van --Maintenance and repair--Handbooks, manuals, etc.
I. Robert Bentley, inc.
TL230.5.E92V55 1997
629.28'722--DC21
97-31787
CIP

 The paper used in this publication is acid free and meets the requirements of the National Standard for Information Sciences-Permanence of Paper for Printed Library Materials. ∞

Manufactured in the United States of America.
202510R003

Foreword

Service to Volkswagen owners is a top priority of the Volkswagen organization that has always included the continuing development and introduction of new and expanded services. In line with this purpose, Bentley Publishers in cooperation with Volkswagen of America, Inc. has introduced this Volkswagen Official Factory Repair Manual.

This Volkswagen EuroVan manual covers the model years 1992, 1993, 1994, 1995, 1996, 1997, 1998, and 1999. It includes Volkswagen EuroVan models equipped with gasoline and diesel engines. This manual was created specifically to cover only those models built for sale in the United States and Canada.

This manual has been prepared from the information that the Volkswagen organization provides to its own factory-trained technicians. It has been developed primarily with the Volkswagen dealership technician in mind. The aim throughout has been clarity and completeness, with step-by-step procedures and factory specifications.

For the automotive service professional, this manual includes the specifications and procedures that were available in an authorized Volkswagen dealer service department as this manual went to press. The Volkswagen owner who has no intention of working on his or her car will find that owning and referring to this manual will make it possible to be better informed and to more knowledgeably discuss repairs with a professional automotive service technician.

Disclaimer

We have endeavored to ensure the accuracy of the information in this manual. When the vast array of data presented in the manual is taken into account, however, no claim to infallibility can be made. We therefore cannot be responsible for the result of any errors that may have crept into the text. Please also read **WARNING–Important Safety Notice** on the copyright page.

Prior to starting a repair procedure, read the procedure, **Warnings and Cautions** on page vii and viii and the **warnings** and **cautions** that accompany the procedure. Reading a procedure before beginning work helps you determine in advance the need for specific skills, identify hazards, prepare for appropriate capture and handling of hazardous materials, and the need for particular tools and replacement parts such as gaskets.

For those intending to do maintenance and repair on their Eurovan, it is essential that safety equipment be used and safety precautions observed when working on your vehicle. A minimum safety equipment list includes hand protection, eye protection, and a fire extinguisher. A selection of good quality hand tools is also needed. This includes a torque wrench to ensure that fasteners are tightened in accordance with specifications.

Special tools that may be required for a particular repair procedure are usually identified by their Volkswagen special tool number and illustrated. A thorough review of each procedure is recommended to determine in advance the complexity of the repair work and the need for particular tools and replacement parts such as gaskets.

Some of the information in this manual applies only to cars of a particular model year or range of years. For example, “m.y. 1993” refers to the 1993 model year. The model year does not necessarily match the calender year in which the car was manufactured or sold. To be sure of the model year of a particular car, check the vehicle identification number (VIN) on the car. Technical changes made in production within a model year are identified in this manual by listing the VIN for the first car produced with this change.

The VIN is a unique sequence of 17 characters assigned by Volkswagen to identify each individual car. WV2MC0709PH016886 is an example. Each of the 17 letters and numbers indicates certain facts about the car and its manufacture. The characters in the 7th and 8th spaces are Volkswagen's model designation. The letter in the 10th space indicates model year. The last six digits are the unique sequential production number. References to VIN in this manual will often include only the last several characters, such as 709PH016886 or 70_P_016886 from the sample VIN above.

The Volkswagen VIN can be found on a plate mounted on the top of the instrument panel, on the driver's side where the number can be seen through the windshield, and on an identification label on the left door jamb. This manual covers Volkswagen EuroVan from model years 1992 through 1999, with these VINs:

position	1	2	3	4	5	6	7	8	9	10	11	12	13	14	15	16	17		1	2	3	4	5	6	7	8	9	10	11	12	13	14	15	16	17
1992:							7	0		N		0	0	0	0	0	1	to							7	0		N		9	9	9	9	9	9
1993:							7	0		P		0	0	0	0	0	1	to							7	0		P		9	9	9	9	9	9
1994:							7	0		R		0	0	0	0	0	1	to							7	0		V		9	9	9	9	9	9
1995:							7	0		S		0	0	0	0	0	1	to							7	0		S		9	9	9	9	9	9
1996:							7	0		T		0	0	0	0	0	1	to							7	0		T		9	9	9	9	9	9
1997:							7	0		V		0	0	0	0	0	1	to							7	0		V		9	9	9	9	9	9
1998:							7	0		W		0	0	0	0	0	1	to							7	0		W		9	9	9	9	9	9
1999:							7	0		X		0	0	0	0	0	1	to							7	0		X		9	9	9	9	9	9

Parts of this manual apply only to certain models or components. Notice especially that the Fuel Injection and Ignition, Engine, and Transmission Repair Group numbers are repeated, each Group covering a particular engine, fuel and ignition system, or transmission type. Most pages are clearly labeled to indicate the systems or components to which the information applies.

Bentley Publishers encourages comments from the readers of this manual with regard to errors, and/or suggestions for improvement of our product. These communications have been and will be carefully considered in the preparation of this and other manuals. If you identify inconsistencies in the manual, you may have found an error. Please contact the publisher and we will endeavor to post applicable corrections on our website. Posted updates and corrections should be reviewed before beginning work. Please see the following web address:

`http://www.BentleyPublishers.com/updates/`

Volkswagen continues to issue service information and parts retrofits after the editorial closing of this manual. Some of this updated information may apply to procedures and specifications in this manual. For the latest information, please see the following web address:

`http://www.ebahn.com/vw`

Volkswagen offers extensive warranties, especially on components of the fuel delivery and emission control systems. Before deciding to repair a car that may be covered wholly or in part by any warranties issued by Volkswagen of America, Inc., consult an authorized Volkswagen dealer. You may find that the dealer can make the repair either without charge or at minimal cost. Regardless of its age, or whether it is covered by warranty, a Volkswagen is an easy car to get serviced. If at any time a repair is too difficult, or requires special equipment that is not available, a trained Volkswagen dealer service technician is ready to do the job for you.

Bentley Publishers

Please read these WARNINGS and CAUTIONS before proceeding with maintenance and repair work.

WARNING—

• Some repairs may be beyond your capability. If you lack the skills, tools and equipment, or a suitable workplace for any procedure described in this manual, we suggest you leave such repairs to an authorized Volkswagen dealer service department, or other qualified shop.

• Volkswagen is constantly improving its cars. Sometimes these changes, both in parts and specifications, are made applicable to earlier models. Therefore, before starting any major jobs or repairs to components on which passenger safety may depend, consult an authorized Volkswagen dealer about Technical Bulletins that may have been issued since the editorial closing of this manual.

• Do not re-use any fasteners that are worn or deformed in normal use. Some fasteners are designed to be used only once, and are unreliable and may fail if used a second time. This includes, but is not limited to, nuts, bolts, washers, circlip and cotter pins. Always replace these fasteners with new parts.

• Never work under a lifted car unless it is solidly supported on stands designed for the purpose. Do not support a car on cinder blocks, hollow tiles or other props that may crumble under continuous load. Never work under a car that is supported solely by a jack. Never work under the car while the engine is running.

• If you are going to work under a car on the ground, make sure that the ground is level. Block the wheels to keep the car from rolling. Disconnect the battery negative terminal (ground strap) to prevent others from starting the car while you are under it.

• Never run the engine unless the work area is well ventilated. Carbon monoxide (CO) kills.

• Do not attempt to work on a vehicle if you do not feel well. You increase the danger of injury to yourself and others if you are tired, upset or have taken medication or any other substance that may impair you or keep you from being fully alert.

• Disconnect the battery negative (–) terminal (ground strap) whenever you work on the fuel system or the electrical system. Do not smoke or work near heaters or other fire hazards. Keep an approved fire extinguisher handy.

• Batteries produce explosive hydrogen gas. Keep sparks, lighted matches and open flame away from the top of the battery. If hydrogen gas escaping from the cap vents is ignited, it will ignite gas trapped in the cells and cause the battery to explode.

• Connect and disconnect battery cables, jumper cables or a battery charger only with the ignition switched off, to prevent sparks. Do not quick-charge the battery (for boost starting) for longer than one minute and do not allow charging voltage to exceed 16.5 volts. Wait at least one minute before boosting the battery a second time.

• Illuminate your work area adequately but safely. Use a portable safety light for working inside or under the car. Make sure the bulb is enclosed by a wire cage. The hot filament of an accidentally broken bulb can ignite spilled fuel or oil.

• Tie long hair behind your head. Do not wear a necktie, a scarf, loose clothing, or a necklace when you work near machine tools or running engines. If hair, clothing, or jewelry were to get caught in the machinery, severe injury could result.

• Finger rings, bracelets and other jewelry should be removed so that they cannot cause electrical shorts, get caught in running machinery, or be crushed by heavy parts.

• Catch draining fuel, oil, or brake fluid in suitable containers. Do not use food or beverage containers that might mislead someone into drinking from them. Store flammable fluids away from fire hazards. Wipe up spills at once, but do not store the oily rags, which can ignite and burn spontaneously.

• Always observe good workshop practices. Wear goggles when you operate machine tools or work with battery acid. Gloves or other protective clothing should be worn whenever the job requires working with harmful substances.

• The Air-Conditioning (A/C) system is filled with chemical refrigerant, which is hazardous. The A/C system should be serviced only by trained technicians using approved refrigerant recovery/recycling equipment, trained in related safety precautions, and familiar with regulations governing the discharging and disposal of automotive chemical refrigerants.

• Do not expose any part of the A/C system to high temperatures such as open flame. Excessive heat will increase system pressure and may cause the system to burst.

• Some aerosol tire inflators are highly flammable. Be extremely cautious when repairing a tire that may have been inflated using an aerosol tire inflator. Keep sparks, open flame or other sources of ignition away from the tire repair area. Inflate and deflate the tire at least four times before breaking the bead from the rim. Completely remove the tire from the rim before attempting any repair.

• Friction materials such as brake or clutch discs may contain asbestos fibers. Do not create dust by grinding, sanding, or by cleaning with compressed air. Avoid breathing asbestos fibers and asbestos dust. Breathing asbestos can cause serious diseases such as asbestosis or cancer, and may result in death.

• Greases, lubricants and other automotive chemicals contain toxic substances, many of which are absorbed directly through the skin. Read manufacturer's instructions and warnings carefully. Use hand and eye protection. Avoid direct skin contact.

• An airbag is an explosive device. Handled improperly or without adequate safeguards, it can be accidently activated and cause serious personal injury. To prevent personal injury or airbag system failure, **only trained Volkswagen dealer service technicians** should test, disassemble or service airbag system.

WARNINGS and CAUTIONS continued on next page ▶

Please read these WARNINGS and CAUTIONS before proceeding with maintenance and repair work.

WARNING (continued) —

• ALWAYS disconnect the power supply before working on the airbag system, or when doing repairs that require removing airbag system components. Disconnect the battery negative (–) terminal and cover the battery.

• The airbag unit should not be removed, nor should any work be performed on the airbag system, without first reviewing the warnings and cautions that accompany the airbag system repair information in this manual.

• Never apply stickers or any other coverings on the steering wheel. Do not let chemical cleaners, oil or grease come into contact with vinyl covering of the airbag unit.

• Never open or otherwise attempt to repair airbag system parts. Always use new parts. Never leave airbag parts or the partially disassembled airbag system unattended.

• Never use a test light or volt meter to conduct electrical tests on the airbag system. The airbag system must only be tested by trained Volkswagen Service technicians using special airbag tester equipment. The airbag system must only be tested while it is installed in the car.

• Do not expose an airbag unit to temperatures above 194°F (90°C), even for a brief period, while handling it during the repair process. Keep the unit clear of heat sources such as hot plates, soldering irons, heat lamps, welding equipment, or the like.

• Always replace airbag system components that have been mechanically damaged (bubbles, cracks, etc.).

• Always set down a removed airbag unit so that the horn pad is facing upward.

• The storage, transportation and disposal of airbag units are subject to the laws for flammable solids.

• When driving or riding in an airbag-equipped vehicle, never hold test equipment in your hands or lap while the vehicle is in motion. Objects between you and the airbag can increase the risk of injury in an accident.

CAUTION—

• If you lack the skills, tools and equipment, or a suitable workshop for any procedure described in this manual, we suggest you leave such repairs to an authorized Volkswagen dealer or other qualified shop. We especially urge you to consult an authorized Volkswagen dealer before beginning repairs on any car that may still be covered wholly or in part by any of the extensive warranties issued by Volkswagen of America, Inc.

• Volkswagen is constantly improving its cars and sometimes these changes, both in parts and specifications, are made applicable to earlier models. Therefore part numbers in this manual are listed for reference only. Always check with an authorized Volkswagen dealer parts department for the latest information.

• Before starting a job, make certain that you have all the necessary tools and parts on hand. Read all the instructions thoroughly, do not attempt shortcuts. Use tools that are appropriate to the work and use only replacement parts meeting Volkswagen specifications. Makeshift tools, parts and procedures will not make good repairs.

• Use pneumatic and electric tools only to loosen threaded parts and fasteners. Never use these tools to tighten fasteners, especially on light alloy parts. Always use a torque wrench to tighten fasteners to the tightening torque specification listed.

• Be mindful of the environment and ecology. Before you drain the crankcase, find out the proper way to dispose of the oil. Do not pour oil onto the ground, down a drain, or into a stream, pond or lake. Consult local ordinances that govern the disposal of wastes.

• When connecting cables or ground straps, all connections must be cleaned to bare metal. Be sure mounting bolts are tight.

• Do not replace fuse links with fuse links of other gauge or with standard wire. Always replace the entire length of the fuse link.

• Before doing any electrical welding on vehicles equipped with Anti-lock Brakes (ABS), disconnect the battery negative terminal (ground strap) and the ABS Control Module connector.

• On cars equipped with ABS, the ABS control unit cannot withstand temperatures from a paint-drying booth or a heat lamp in excess of 203°F (95°C) and should not be subjected to temperatures in excess of 185°F (85°C) for more than two hours.

• On cars equipped with anti-theft radios, make sure you know the correct radio activation code before disconnecting the battery or removing the radio. If the wrong code is entered into the radio when power is restored, that radio may lock up and be rendered inoperable, even if the correct code is then entered.

• When boost-charging the battery, first remove the fuses for the Engine Control Module (ECM), the Transmission Control Module (TCM), the ABS Control Module, and the On-Board computer. In cases where one or more of these components is not separately fused, disconnect the Control Module connector(s).

• Do not quick-charge the battery (for boost starting) for longer than one minute. Wait at least one minute before boosting the battery a second time.

◀ **WARNINGS and CAUTIONS continued from previous page**

You are now in ▶ **Volume 3**

Body–Exterior, Body–Interior

Section **Q**

WARNING!
Before proceeding with any work, please read the warnings and cautions immediately following this index. Also read any warning and caution that accompanies a procedure in the manual.

Please read these WARNINGS and CAUTIONS before proceeding with maintenance and repair work.

WARNING—

- Some repairs may be beyond your capability. If you lack the skills, tools and equipment, or a suitable workplace for any procedure described in this manual, we suggest you leave such repairs to an authorized Volkswagen dealer service department, or other qualified shop.

- Volkswagen is constantly improving its cars. Sometimes these changes, both in parts and specifications, are made applicable to earlier models. Therefore, before starting any major jobs or repairs to components on which passenger safety may depend, consult your authorized Volkswagen dealer about the Technical Bulletins that may have been issued since the editorial closing of this manual.

- Do not re-use any fasteners that are worn or deformed in normal use. Many fasteners are designed to be used only once and become unreliable and may fail when used a second time. This includes, but is not limited to, nuts, bolts, washers, self–locking nuts or bolts, circlips and cotter pins. Always replace these fasteners with new parts.

- Never work under a lifted car unless it is solidly supported on stands designed for the purpose. Do not support a car on cinder blocks, hollow tiles or other props that may crumble under continuous load. Never work under a vehicle that is supported solely by a jack. Never work under the car while the engine is running.

- If you are going to work under a car on the ground, make sure that the ground is level. Block the wheels to keep the car from rolling. Disconnect the battery negative terminal (ground strap) to prevent others from starting the vehicle while you are under it.

- Never run the engine unless the work area is well ventilated. Carbon monoxide (CO) kills.

- Do not attempt to work on your vehicle if you do not feel well. You increase the danger of injury to yourself and others if you are tired, upset or have taken medicine or any other substance that may impair you or keep you from being fully alert.

- Disconnect the battery negative (–) terminal (ground strap) whenever you work on the fuel system or the electrical system. Do not smoke or work near heaters or other fire hazards. Keep an approved fire extinguisher handy.

- Batteries produce explosive hydrogen gas. Keep sparks, lighted matches and open flame away from the top of the battery. If hydrogen gas escaping from the cap vents is ignited, it will ignite gas trapped in the cells and cause the battery to explode.

- Connect and disconnect battery cables, jumper cables or a battery charger only with the ignition switched off, to prevent sparks. Do not quick–charge the battery (for boost starting) for longer than one minute, and do not allow charging voltage to exceed 16.5 volts. Wait at least one minute before boosting the battery a second time.

- Illuminate your work area adequately but safely. Use a portable safety light for working inside or under the car. Make sure the bulb is enclosed by a wire cage. The hot filament of an accidentally broken bulb can ignite spilled fuel or oil.

- Tie long hair behind your head. Do not wear a necktie, a scarf, loose clothing, or a necklace when you work near machine tools or running engines. If your hair, clothing, or jewelry were to get caught in the machinery, severe injury could result.

- Finger rings, bracelets and other jewelry should be removed so that they cannot cause electrical shorts, get caught in running machinery, or be crushed by heavy parts.

- Catch draining fuel, oil, or brake fluid in suitable containers. Do not use food or beverage containers that might mislead someone into drinking from them. Store flammable fluids away from fire hazards. Wipe up spills at once, but do not store the oily rags, which can ignite and burn spontaneously.

- Always observe good workshop practices. Wear goggles when you operate machine tools or work with battery acid. Gloves or other protective clothing should be worn whenever the job requires working with harmful substances.

- The Air-Conditioning (A/C) system is filled with a chemical refrigerant, which is hazardous. The A/C system should be serviced only by trained technicians using approved refrigerant recovery/recycling equipment, trained in related safety precautions, and familiar with regulations governing the discharging and disposal of automotive chemical refrigerants.

- Do not expose any part of the A/C system to high temperatures such as an open flame. Excessive heat will increase system pressure and may cause the system to burst.

- Some aerosol tire inflators are highly flammable. Be extremely cautious when repairing a tire that may have been inflated using an aerosol tire inflator. Keep sparks, open flame or other sources of ignition away from the tire repair area. Inflate and deflate the tire at least four times before breaking the bead from the rim. Completely remove the tire from the rim before attempting any repair.

- Friction materials such as brake or clutch discs may contain asbestos fibers. Do not create dust by grinding, sanding, or by cleaning with compressed air. Avoid breathing asbestos fibers and asbestos dust. Breathing asbestos can cause serious diseases such as asbestosis or cancer, and may result in death.

- Greases, lubricants and other automotive chemicals contain toxic substances, many of which are absorbed directly through the skin. Read manufacturer's instructions and warnings carefully. Use hand and eye protection. Avoid direct skin contact.

- An airbag is operated by an explosive device. Handled improperly or without adequate safeguards, it can be accidently activated and cause serious personal injury. To prevent personal injury or airbag system failure, **only trained Volkswagen dealer service technicians** should test, disassemble or service the airbag system.

WARNINGS and CAUTIONS continued on next page ▶

Please read these WARNINGS and CAUTIONS before proceeding with maintenance and repair work.

WARNING (continued) —

- ALWAYS disconnect the airbag power supply before using a computer memory saver. Failure to disconnect the airbag power supply may result in accidental activation of the airbag.
- The airbag unit should not be removed, nor should any work be performed on the airbag system, without first reviewing the warnings and cautions that accompany the airbag system repair information in this manual.
- Never apply stickers or any other coverings on the steering wheel. Do not let chemical cleaners, oil or grease come into contact with the vinyl covering of the airbag unit.
- Never open or otherwise attempt to repair airbag system parts. Always use new parts. Never leave airbag parts or the partially disassembled airbag system unattended.
- Never use a test light or volt meter to conduct electrical tests on the airbag system. The airbag system must only be tested by trained Volkswagen Service technicians using special airbag tester equipment. The airbag system must only be tested while it is installed in the car.
- Do not expose an airbag unit to temperatures above 194°F (90°C) even for a brief period, while handling it during the repair process. Keep the unit clear of heat sources such as hot plates, soldering irons, heat lamps, welding equipment, or the like.
- Always replace airbag system components that have been mechanically damaged (bubbles, cracks, etc.).
- Always set down a removed airbag unit so that the horn pad is facing upward.
- When driving or riding in an airbag-equipped vehicle, never hold test equipment in your hands or lap while the vehicle is in motion. Objects between you and the airbag can increase the risk of injury in an accident.
- The storage, transportation and disposal of airbag units are subject to the laws for flammable solids.

CAUTION—

- If you lack the skills, tools and equipment, or a suitable workshop for any procedure described in this manual, we suggest you leave such repairs to an authorized Volkswagen dealer or other qualified shop. We especially urge you to consult an authorized Volkswagen dealer before beginning repairs on any car that may still be covered wholly or in part by any of the extensive warranties issued by Volkswagen of America, Inc.
- Volkswagen is constantly improving its cars and sometimes these changes, both in parts and specifications, are made applicable to earlier models. Therefore, part numbers listed in this manual are for reference only. Always check with your authorized Volkswagen dealer parts department for the latest information.
- Before starting a job, make certain that you have all the necessary tools and parts on hand. Read all the instructions thoroughly, do not attempt shortcuts. Use tools that are appropriate to the work and use only replacement parts meeting Volkswagen specifications. Makeshift tools, parts and procedures will not make good repairs.
- Use pneumatic and electric tools only to loosen threaded parts and fasteners. Never use these tools to tighten fasteners, especially on light alloy parts. Always use a torque wrench to tighten fasteners to the tightening torque specification listed.
- Be mindful of the environment and ecology. Before you drain the crankcase, find out the proper way to dispose of the oil. Do not pour oil onto the ground, down a drain, or into a stream, pond, or lake. Consult local ordinances that govern the disposal of wastes.
- When connecting cables or ground straps, all connections must be cleaned to bare metal. Be sure mounting bolts are tight.
- Do not replace fuse links with fuse links of other gauge or with standard wire. Always replace the entire length of the fuse link.
- Before doing any electrical welding on vehicles equipped with Anti-lock Brakes (ABS), disconnect the battery negative terminal (ground strap) and the ABS Control Module connector.
- On cars equipped with ABS, the ABS control unit cannot withstand temperatures from a paint–drying booth or a heat lamp in excess of 203°F (95°C) and should not be subjected to temperatures in excess of 185°F (85°C) for more than two hours.
- On cars equipped with anti-theft radios, make sure you know the correct radio activation code before disconnecting the battery or removing the radio. If the wrong code is entered when power is restored, the radio may lock up and be rendered inoperable, even if the correct code is then entered.
- When boost-charging the battery, first remove the fuses for the Engine Control Module (ECM), the Transmission Control Module (TCM), the ABS Control Module, and the On-Board computer. In cases where one or more of these components is not separately fused, disconnect the Control Module connector(s).
- Do not quick-charge the battery (for boost starting) for longer than one minute, and do not exceed 16.5 volts at the battery with the boosting cables attached. Wait at least one minute before boosting the battery a second time.

◀ **WARNINGS and CAUTIONS continued from previous page**

Driver and passenger airbag system On Board Diagnostic (OBD)

Function

The airbag control module -J234- is located behind the console on the tunnel. It is equipped with a Diagnostic Troouble Code (DTC) memory. The Data Link Connector (DLC) is located next to the instrument cluster on the lower right side.

The airbag control module -J234- detects malfunctions in the airbag system and stores them in a permanent memory.

After the ignition is switched on, the Airbag Malfunction Indicator Lamp (MIL) -K75- comes on for about 3 seconds and then goes out again.

♦ If the airbag MIL -K75- stays on, the voltage supply to the airbag control module -J234- has been interrupted. Check the DTC memory and correct malfunctions ⇒ page Q 01-11.

♦ If the airbag MIL comes on again, there is a malfunction. Check the DTC memory and correct malfunctions ⇒ page Q 01-11.

To begin troubleshooting, initiate the On Board Diagnostic (OBD) program and check for stored Diagnostic Trouble Code (DTC) information using the VAG 1551 Scan Tool.

The malfunction information displayed by the VAG 1551 Scan Tool is used to refer to the Diagnostic Trouble Code (DTC) table for notes on possible causes and directions for specific repair procedures ⇒ page Q 01-11.

> ***WARNING!***
>
> ♦ ***Check airbag system wiring by VISUAL INSPECTION only.***
>
> ♦ ***Do not perform electrical continuity tests or measurements to ignition circuits.***
>
> ♦ ***Check wiring routing only with igntion switched OFF.***

OBD requirements

♦ All fuses OK per wiring diagram

♦ Battery voltage OK (at least 9 V).

VAG 1551 Scan Tool, connecting and selecting function

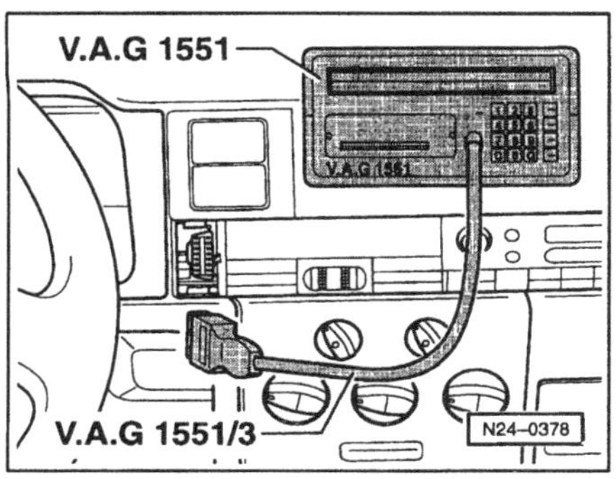

Note:

♦ *The Diagnostic Trouble Code (DTC) memory can only be erased after repairs have been performed and the DTC memory has been read.*

– Remove shelf in front of relay plate.

– Remove cover from Data Link Connector (DLC) socket (next to instrument cluster on lower right side).

◄ – With ignition switched OFF, connect VAG 1551/3 adapter harness between DLC and VAG 1551 Scan Tool (ST).

VAG On Board Diagnostic (OBD) HELP
1–Rapid data transfer*
2–Blink code output*

◄ Indicated on display:

* Is displayed alternately.

Notes:

- *If the display remains blank, check the VAG 1551 voltage supply using the wiring diagram.*
- *Additional operating instructions can be printed out by pressing the HELP button on the VAG 1551 Scan Tool.*
- *The → button advances the Scan Tool program sequence.*
- *The PRINT button switches the printer on(indicator lamp in button lights up).*

Checking control module version

– Switch on ignition.

– Press button -1- for "Rapid data transfer" mode.

Rapid data transfer HELP
Enter address word XX

◄ Indicated on display:

– Press buttons -1- and -5- (to insert address word 15 for "Airbag").

Rapid data transfer Q
15 Airbag

◄ Indicated on display:

– Confirm entry with Q button.

and then the display shows e.g.:

6N0 909 601 S AIRBAG VW2 V01 →
Coding 00083 WSC12345

◄ The control module identification number is displayed:

Rapid data transfer HELP control module does not answer!

Note:

- *If any other message appears in the display, the possible causes of the malfunction can be printed out by pressing the HELP button.*

Rapid data transfer HELP K wire not switching to positive!

(The ignition must be switched on!)

Rapid data transfer → No signal from control module!

- *Malfunctions have occurred at the start of or during the program (external interference).*

Rapid data transfer → Fault in communication build–up

- *Check diagnostic cables, voltage supply and ground connections.*
- After correcting possible causes of malfunction, enter address word 15 for "Airbag" again and confirm with Q button.

Rapid data transfer Tester sends the address word 15

Indicated in display after entering the address word 15:

and then the following appears in the display:

6N0 909 601 S AIRBAG VW2 V01 → Coding 00083 WSC12345

The control module identification number is displayed:

- Press → button.

Rapid data transfer HELP Select function XX

Indicated on display:

Note:

A list of the possible functions is printed out after pressing the HELP button:

List of functions	Page
01 – Checking control module version	Q 01-4
02 – Checking DTC memory	Q 01-6
05 – Erasing DTC memory	Q 01-8
06 – Ending output	Q 01-10
07 – Coding control module	Q 01-14
08 – Reading measured value block	Q 01-17

Only the functions listed can be used at present for On Board Diagnostic (OBD).

Diagnostic Trouble Code (DTC) memory, checking (VAG 1551 Scan Tool Function 02)

Rapid data transfer HELP Select function XX

Indicated on display:

- Press buttons -0- and -2- (to enter function 02 "Check Diagnostic Trouble Code (DTC) memory").

Rapid data transfer Q 02 – check DTC memory

Indicated on display:

- Press Print button
- Confirm entry with Q button.

X DTCs recognized!

◄ The number of stored DTCs appears in the display.

The stored DTCs are displayed and printed out one after the other.

– Refer to DTC table with displayed malfunction(s) and printed DTC listing ⇒ page Q 01-12.

No DTC recognized!	→

◄ If "No DTC recognized" is displayed the program will return to the start position after pressing the → button.

Rapid data transfer Select function XX	HELP

◄ Indicated on display:

– End output (function 06) ⇒ Page Q 01-10

– Switch off ignition and disconnect adapter harness VAG 1551/3.

Notes:

♦ *If a DTC was recognized:*

– 1. Correct malfunction.

– 2. Check DTC memory (function 02).

– 3. Erase DTC memory (function 05).

DTC memory, erasing (VAG 1551 Scan Tool Function 05)

Requirements:

● DTC(s) are corrected

– check DTC memory again

Erasing DTC memory

– Switch ignition on.

– Press button -1- for "Rapid data transfer" mode.

Rapid data transfer Enter address word XX	HELP

◄ Indicated on display:

– Press buttons -1- and -5- (to insert address word 15 for "Airbag").

Rapid data transfer 15 Airbag	Q

◄ Indicated on display:

– Confirm entry with Q button.

and then the display shows e.g.:

6N0 909 601 S AIRBAG VW2 V01 →	
Coding 00083 WSC12345	

◄ The control module identification number is displayed:

– – Press → button.

Rapid data transfer	HELP
Select function XX	

◄ Indicated on display:

– Press buttons -0- and -5- (to select "Erase DTC memory").

Rapid data transfer	Q
05 Erase DTC memory	

◄ Indicated on display:

– Confirm entry with Q button.

Rapid data transfer	→
DTC memory is erased!	

◄ Indicated on display:

– Press → button.

Rapid data transfer	HELP
Select function XX	

◄ Indicated on display:

Notes:

Important!	
DTC memory was not checked	

◄
- ♦ *If this appears in the display, the test sequence has not been followed.*
- ♦ *Follow the test sequence exactly; first check DTC memory (function 02), then erase memory (function 05).*

End output (VAG 1551 Scan Tool Function 06)

– Press buttons -0- and -6- to end output.

Rapid data transfer	Q
06 End output	

◄ Indicated on display:

– Confirm entry with Q button.

Rapid data transfer	Help
Enter address word XX	

◄ Indicated on display:

– Switch ignition off.

– Disconnect adapter harness VAG 1551/3.

– Switch ignition on.

- ● Airbag Malfunction Indicator Lamp -K75- must go out after approx. 3 seconds.

Diagnostic Trouble Code (DTC) table

Notes:

- *The table that follows lists all possible malfunctions that can be detected and stored by the airbag control module -J234- and printed out by the VAG 1551 Scan Tool according to the 5–digit Diagnostic Trouble Code.*
- *DTC code appears (in the "Rapid data transfer" mode) only on the printout.*

  Example:

  DTC code

  5 digit

  00277

- *In addition, the DTC type may also appear in the DTC table.*
- *Before replacing a component suspected as being faulty, check all appropriate wiring, component connections, harness connectors and ground connections according to wiring diagram.*
- *After repairs, the DTC memory must be checked again using the VAG 1551 Scan Tool, and then erased.*

DTC table

VAG 1551 printout	Possible cause of malfunction	Correcting malfunction
00000		
No DTC recognized	If "No DTC recognized" appears after completing repairs, the On Board Diagnostic (OBD) is ended.	
00532		
Supply voltage (B+)		
Signal too high	Generator (GEN) faulty	– Test Generator (GEN): ⇒ Electrical Wiring Diagrams
Signal too low	Wiring or connections to airbag control module -J234-	– Check wiring and connections to control module using wiring diagram
	control module faulty	– Replace control module
00588		
Airbag Igniter–Driver's Side -N95-	Faulty wiring or connections	– Check wiring and connections using wiring diagram
Resistance too high	Airbag Spiral Spring -F138- faulty	– Replace airbag spiral spring -F138- if necessary
Resistance too low	control module faulty	– Replace control module

VAG 1551 printout	Possible cause of malfunction	Correcting malfunction
00589		
Airbag Igniter 1–Passenger Side -N131-	Faulty wiring or connections	– Test wiring and connections using wiring diagram
Resistance value too high	Airbag Igniter 1–Passenger Side -N131- faulty	– Replace front passenger airbag unit
Resistance too low	control module faulty	– Replace control module
00651		
Wiring to ignition circuit	Faulty wiring or connections to airbag units	– Test wiring and connections using wiring diagram
	control module faulty	– Replace control module
65535		
Control module malfunction	Electrical interference from external sources, poor ground connections or positive connections to airbagcontrol module -J234-	– Check wiring and connections to control module using wiring diagram
	Airbag control module -J234- faulty	– Replace airbag control module -J234-

Airbag Control Module -J234-, coding using VAG 1551 Scan Tool (only when installing a new control module)

– Connect VAG 1551 Scan Tool ⇒ page Q 01-2

– Switch ignition on.

The Airbag Malfunction Indicator (MIL) Lamp -K75- lights up continuously.

– Press button -1- for "Rapid data transfer" mode.

```
Rapid data transfer                HELP
Enter address word XX
```

◄ Indicated on display:

– Press buttons -1- and -5- (to insert address word 15 for "Airbag").

```
Rapid data transfer                Q
15 Airbag
```

◄ Indicated on display:

– Confirm entry with Q button.

```
6N0 909 601     AIRBAG  VW2     V01
Coding 00000              WSC 12345
```

◄ Indicated on display:

– Press → button.

```
Rapid data transfer                HELP
Select function XX
```

◄ Indicated on display:

– Press buttons -0- and -7- (to insert function 07 "Code control module").

```
Rapid data transfer                Q
07 Code control module
```

◄ Indicated on display:

– Confirm entry with Q button.

Code control module	Q
Enter code number XXXXX	

◄ Indicated on display:

– Enter code number according to table:

Vehicle equipment	Index	Code number
Driver airbag only	N	00078
Driver/front passenger airbag	S	00083
US-Version	M	00077

– Confirm entry with Q button.

6N0 909 601 S AIRBAG VW2	V01
Coding 00083	WSC 12345

◄ The control module identification number with the relevant index letter, code number and workshop code is displayed.

If the contents of the display are as described then the coding is successful.

If the control module code number entered is not accepted then the following will be displayed:

6N0 909 601 AIRBAG VW2	V01
Coding 00000	WSC 12345

◄ Indicated on display:

In this case, the control module has not been correctly programmed with the relevant data for the vehicle. Coding is no longer possible. The airbag MIL -K75- will not go out and stays lit continuously.

– Check that airbag control module of correct part number has been installed (compare part number identification and index letter) or whether incorrect code number has been entered.

Ending function:

– Press → button.

Rapid data transfer	HELP
Select function XX	

◄ Indicated on display:

– Press buttons -0- and -6- to end output.

– Confirm entry with Q button.

Rapid data transfer	Q
06 End output	

◄ Indicated on display:

- The Airbag Malfunction Indicator Lamp -K75- must go out after approx. 3 seconds.

Measuring value block, reading (VAG 1551 Scan Tool function 08)

> ***WARNING!***
> - ***Check airbag system wiring by VISUAL INSPECTION only.***
> - ***Do not perform electrical continuity tests or measurements to ignition circuits.***
> - ***Check wiring routing only with igntion switched OFF.***

Rapid data transfer	HELP
Select function XX	

◄ Indicated on display (function selection):

- Press buttons -0- and -8- (08 initiates "Read measuring value block" function).

Rapid data transfer	Q
08 Read measuring value block	

◄ Indicated on display:

- Confirm entry with Q button.

Read measuring value block
Input display group number XX

◄ Indicated on display:

Table of selectable display group numbers

Display group number	Display field	Designation
01	1	Resistance of driver's side airbag igniter -N95-
	2	Resistance of passenger's side airbag igniter -N131-
02	1	Supply voltage (B+)

- Select relevant display group number
- Confirm entry with Q button.

Read measuring value block 1
1 2 3 4

◄ Indicated on display:

Notes:

- *For a key to the content of each display field ⇒ page Q 01-20*
- *With the printer switched on, the actual display content is printed out.*

If the displayed values in all display zones are OK:

– Press → button.

Rapid data transfer HELP
Select function XX

◄ Indicated on display (function selection):

Note:

Check DTC memory after completing the function "Read measuring value block" ⇒ Page Q 01-6

Measuring value block, display content

WARNING!

- ***Check airbag system wiring by VISUAL INSPECTION only.***
- ***Do not perform electrical continuity tests or measurements to ignition circuits.***
- ***Check wiring routing only with ignition switched OFF.***

Display group No.	Display field	Designation	Display contents	Definition of display	Correction
01	1	Driver's side airbag igniter -N95-	Correct Too large Too small To ground To positive	Resistance OK Resistance too high Resistance too low Short to ground Short to positive (B+)	– Visually check wiring – While watching display make sure circuit connections are correct and secure. – If display content changes to "Correct" (resistance OK), erase DTC memory. – If display content does not change to "Correct:", replace airbag igniter -N95- or -N131-
	2	Passenger's side airbag igniter -N131-	Correct Too large Too small To ground To positive	Resistance OK Resistance too high Resistance too low Short to ground Short to positive (B+)	
	3 4	Not applicable Not applicable	Correct Correct		
02	1	Voltage supply (B+)	Correct Too small	Voltage too low Voltage OK	– Check battery voltage – Check Generator (GEN) – Visually check wiring

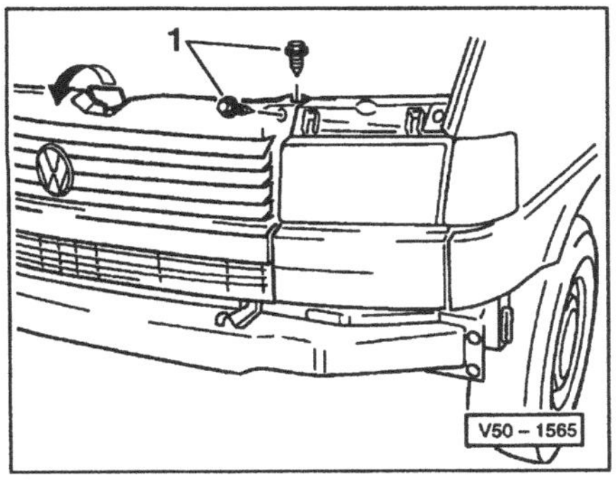

Cross member, removing/installing

- Remove dampening pan, ⇒ page Q 50-5.
- Remove front bumper.
- Remove left and right combi hex bolts -1-.
- Pivot radiator and lock carrier plate toward front -arrow-.

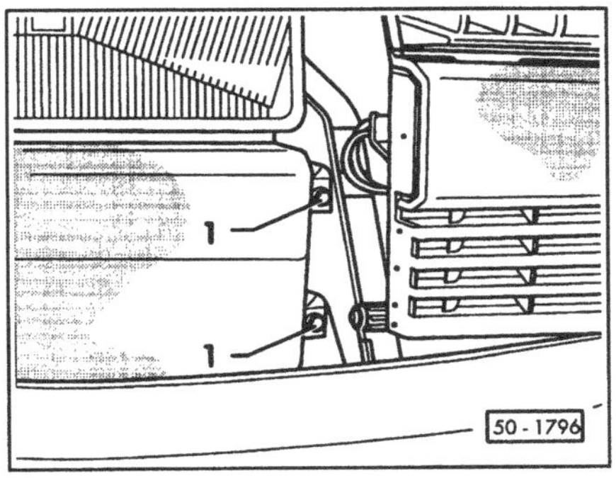

- Remove hex bolt and washer assemblies -1- from both sides.

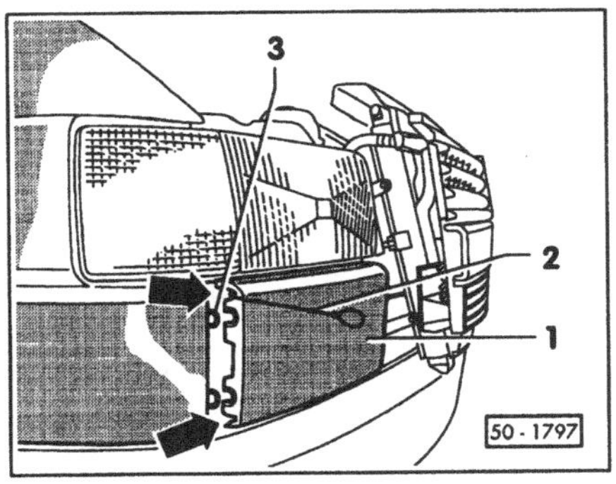

- Pull apron -1- off pins -3- using shop made hooks -2- (2.0 mm dia.), as shown.
- Insert hook over or under pins -3--arrows-.

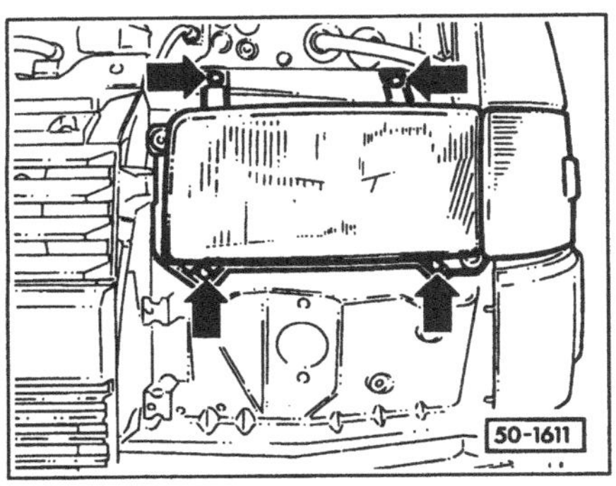

- Remove headlight and turn signal assemblies from both sides (four combi hex bolts -arrows-) and disconnect electrical connectors.

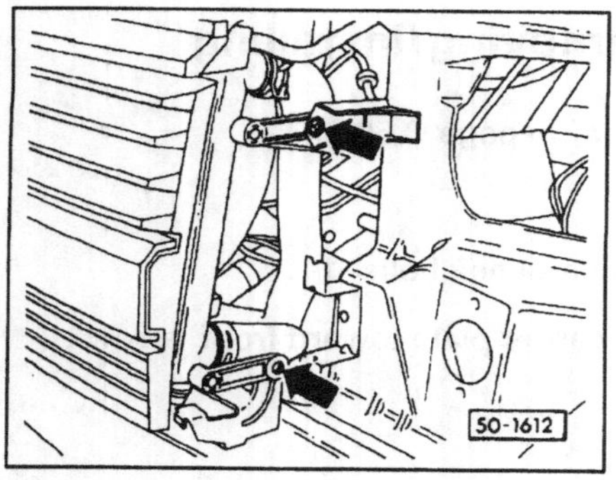

- Connect radiator to workshop crane (VAG 1202A or equivalent).
- Drive pins out of spreader clips -arrows- on both sides and remove spreader clips from radiator brackets.
- Lift radiator with workshop crane.

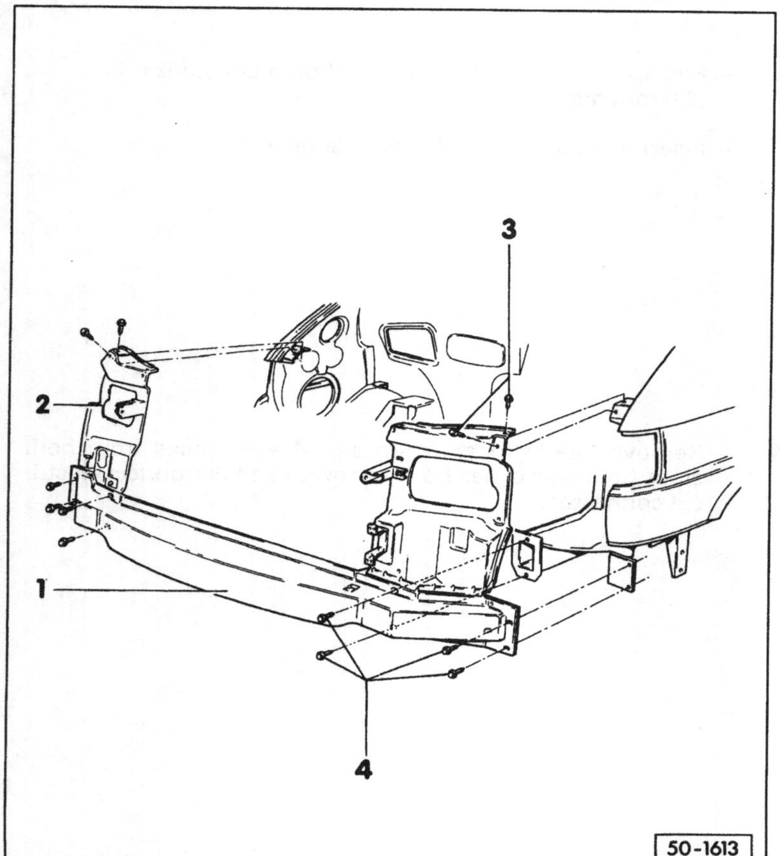

- Remove cross member -1- together with headlight brackets–2–.
  - ♦ Combi hex bolt-3- = 8.0 Nm (71 in. lb)
  - ♦ Combi hex bolt-4- = 8.0 Nm (71 in. lb)
- Unclip electrical connectors from cross member/headlight brackets.

Reinstall in reverse sequence.

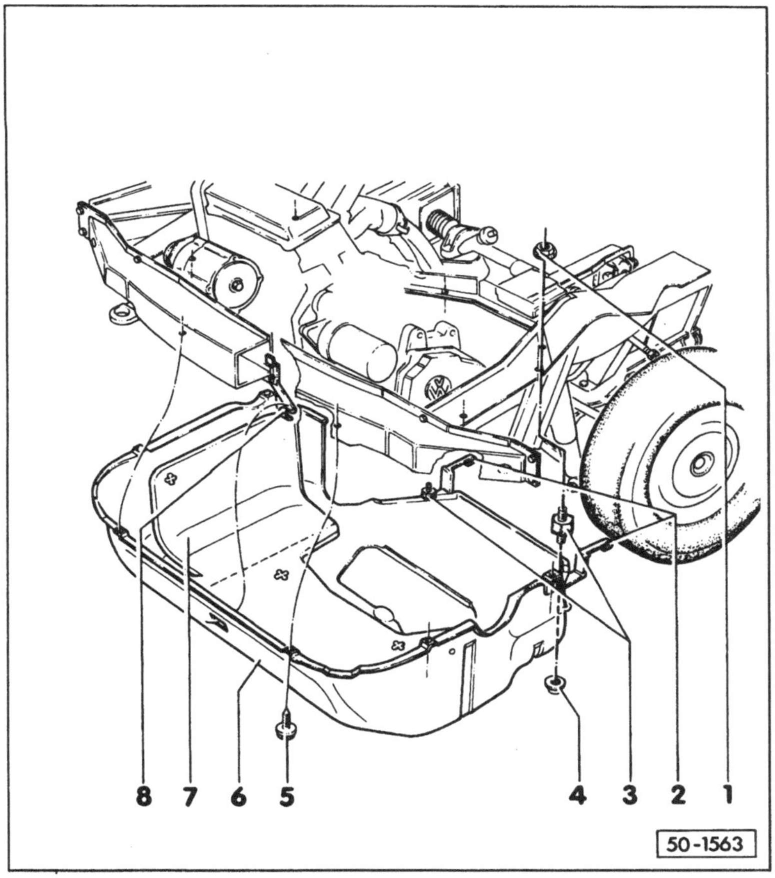

Sound dampening pan, components

1 - Hex nut
- ♦ 9 Nm (80 in. lb)

2 - Bracket

3 - Hex dampening element

4 - Hex nut
- ♦ 9 Nm (80 in. lb)

5 - Combi hex bolt
- ♦ 9 Nm (80 in. lb)

6 - Sound dampening pan
- ♦ Removing: Unscrew two Hex nuts -4- and six combi hex bolts -5-. Unhook locking lever -8-, swing down dampening pan and unhook at brackets -2-.

7 - Dampening material
- ♦ Replacing ⇒ Page Q 50-6

8 - Locking hook

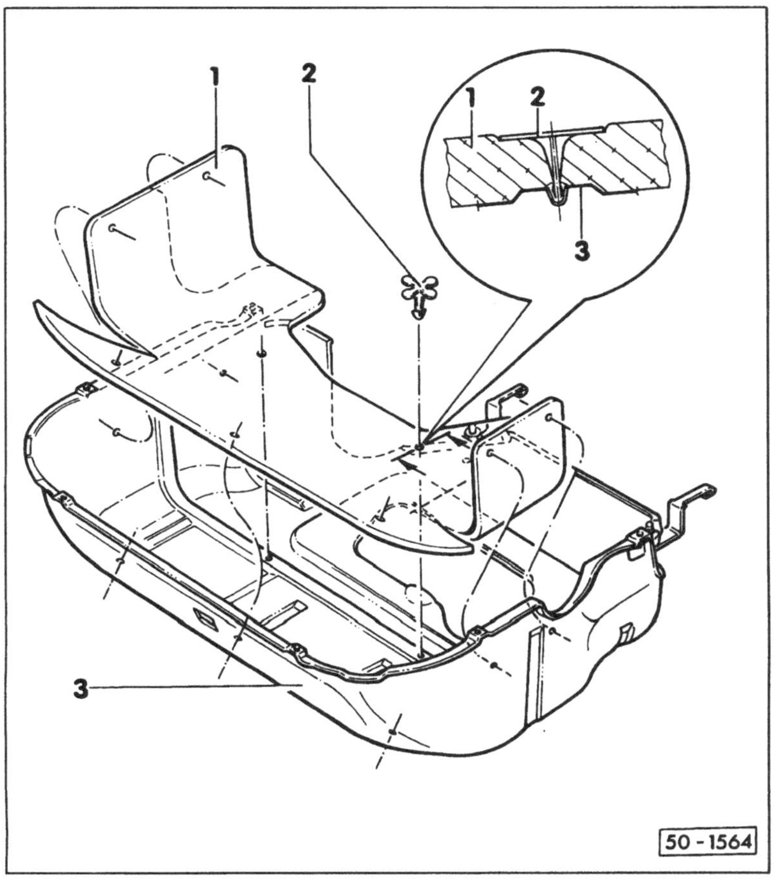

Sound dampening material, replacing

1 - Sound dampening material
- ♦ On vehicles with automatic transmission or air conditioner, dampening material must be removed according to perforation.

2 - Clip

3 - Sound dampening pan

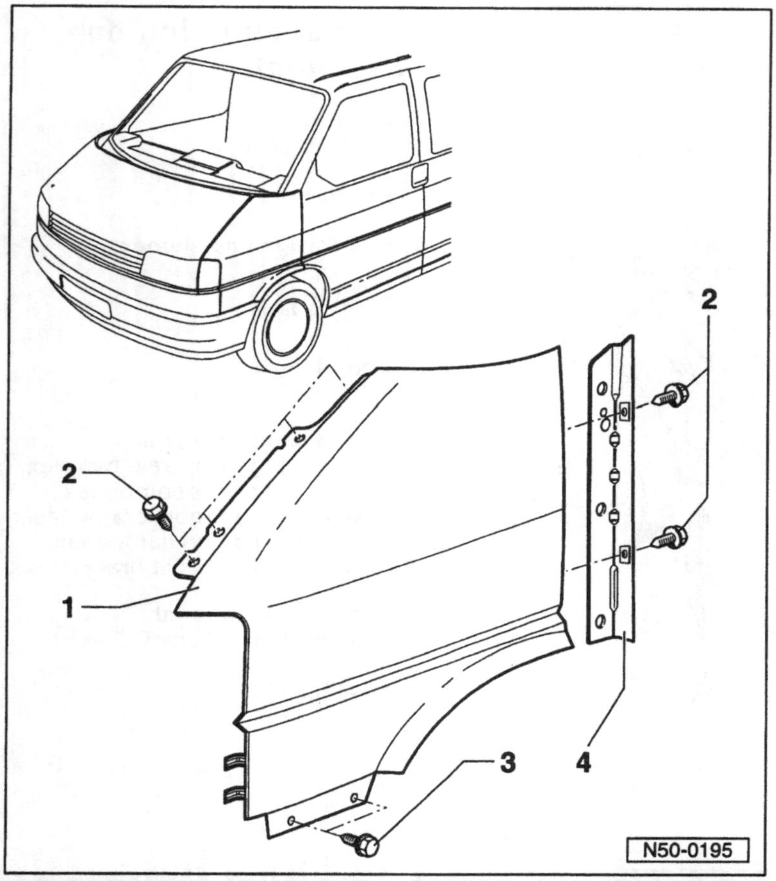

Front fenders, removing/installing 01.96 ➤

1 - Fender

- ♦ Front bumper, removing ⇒ page Q 63-1
- – Remove turn signals
- – Remove hex bolts -2- and -3-.
- – Heat fender and PVC with heat gun
- – Remove fender (separate PVC with utility knife, second technician needed)

Note:

Do not overheat PVC; it must not change color or form bubbles.

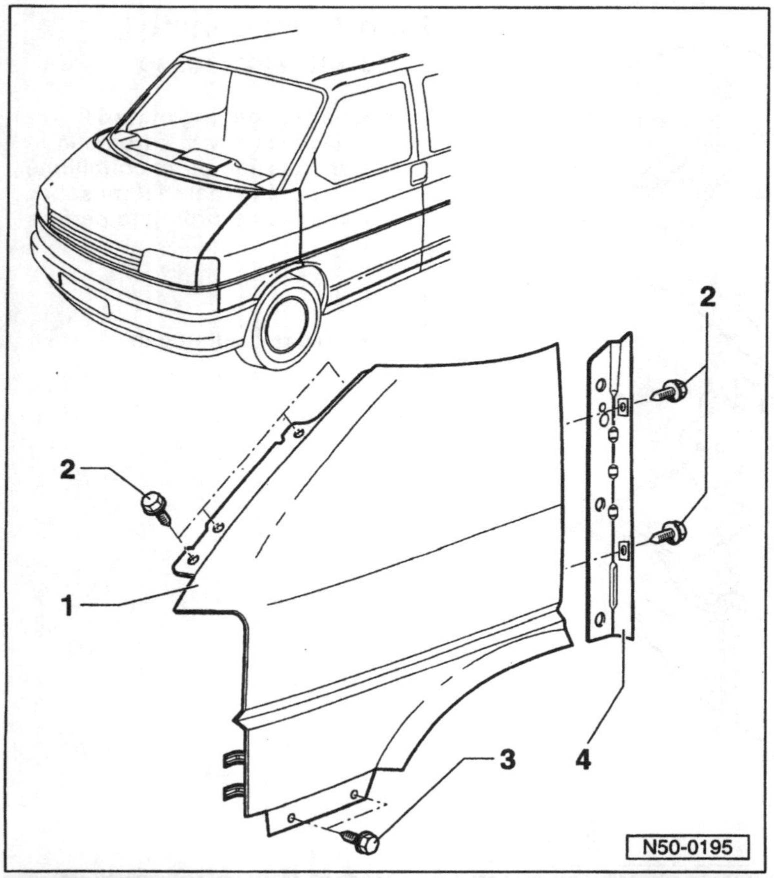

2 - Hex screw
- ♦ With small washer

3 - Hex screw
- ♦ With large washer

4 - Rib plate
- ♦ Provided with slots for adjustment between fender, front lid and front door.

Before installing fender, treat mating surfaces/-A- pillar with zinc AKL 381 035 50.

Hood, removing/installing/adjusting ➤12.95

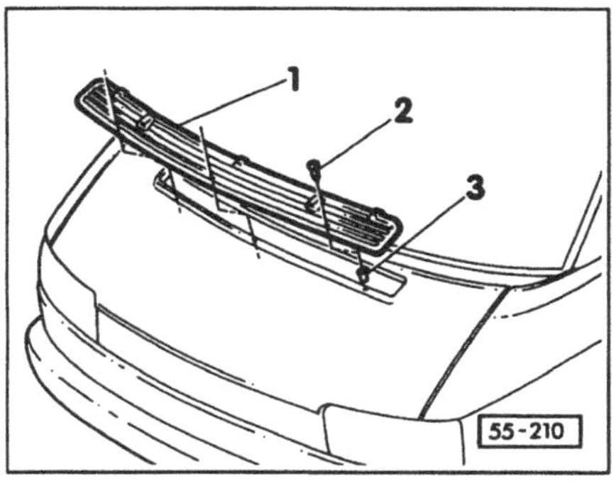

◄ – Remove three Phillips screws -2- (from expanding nuts -3-), and carefully pry out grille -1-.

– Pull hose off windshield washer nozzles, unclip and remove from hood.

– Disconnect plug connectors for heated washer nozzles, if applicable.

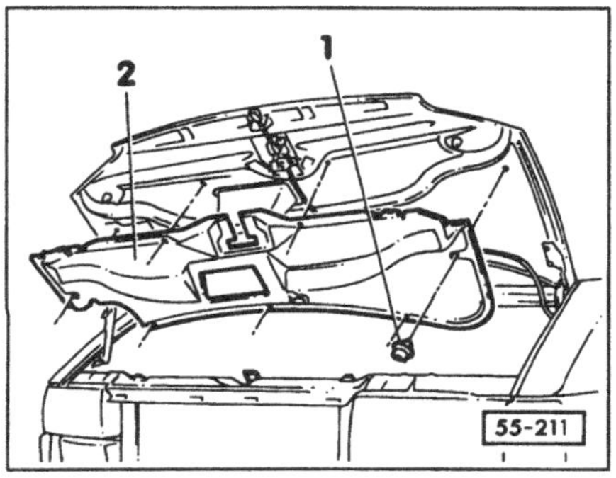

Vehicles with hood insulation

◄ – Turn clips -1- 60°, remove.

– Remove insulation material -2-.

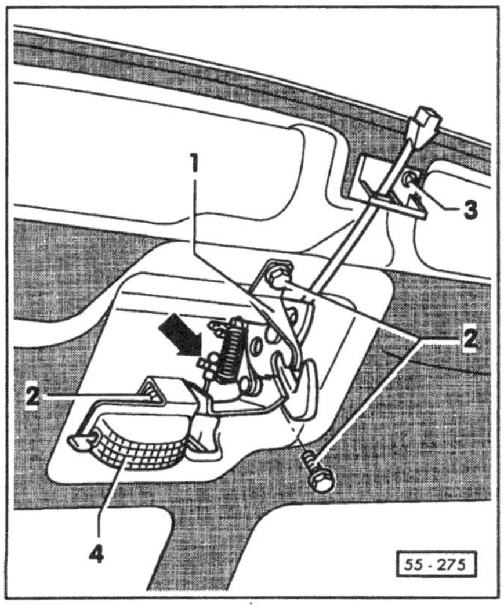

◄ – Remove Engine Compartment Light -W27- -4-, if applicable.

– Remove Phillips screw -3- (starting 9/92) clip and hex bolts -2-.

Tightening torque:

Hex bolts -2- = 8.0 Nm (71 in. lb)

– Remove hood latch -1-, disconnect hood latch cable - arrow

– Unclip hood latch cable at front of hood ⇒ Page Q 55-5.

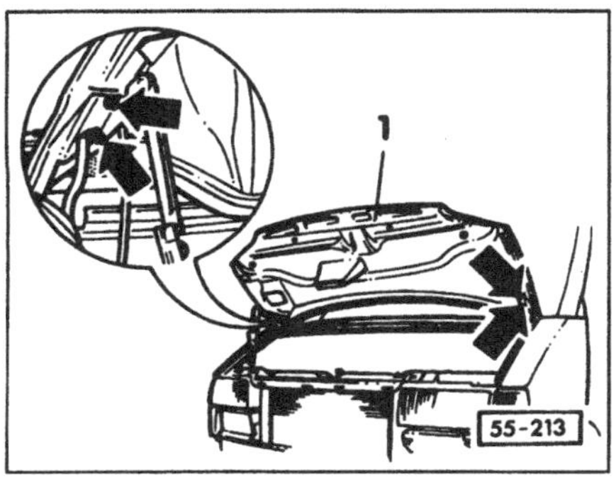

◄ – Remove hex bolts -1- -arrows- on both sides, remove hood.

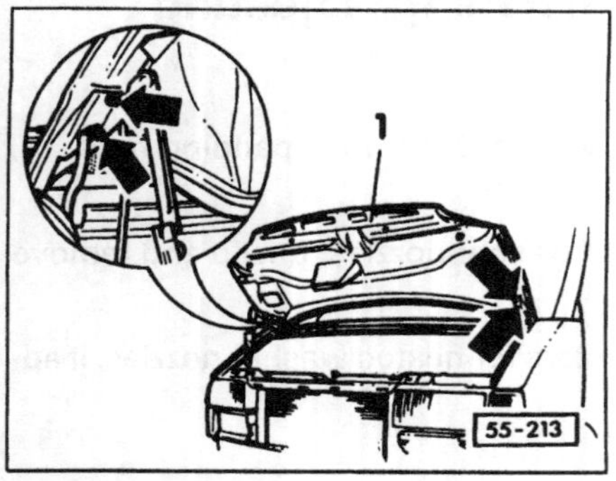

- Loosely mount hood -1- at hinges.
- Align hood and torque bolts.

  Tightening torque:

  Hex bolts = 8.0 Nm (71 in. lb)

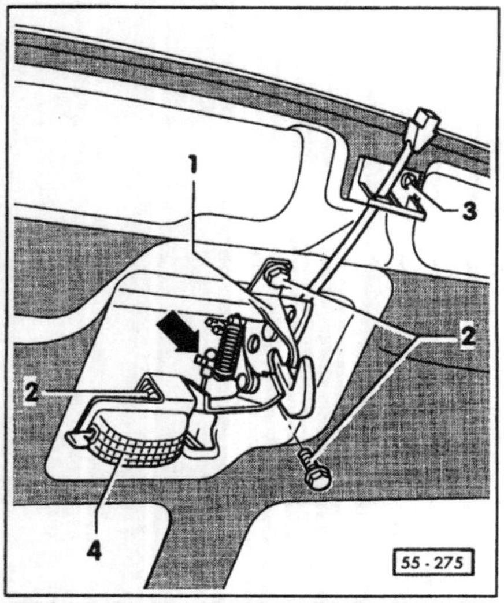

- Reinstall hood latch -1-, hood latch cable -arrow-, and hood ⇒ Page Q 55-5.

  Tightening torque:

  Hex bolts -2- = 8.0 Nm (71 in. lb)

- Reinstall Engine Compartment Light -W27- -4-.

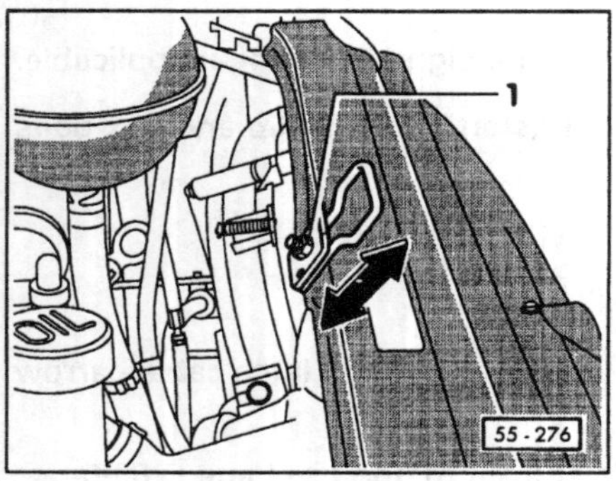

- Move latch on upper radiator support after loosening hex nut -arrow-.

  Tightening torque:

  Hex nut -1- 5.0 Nm (44 in. lb)

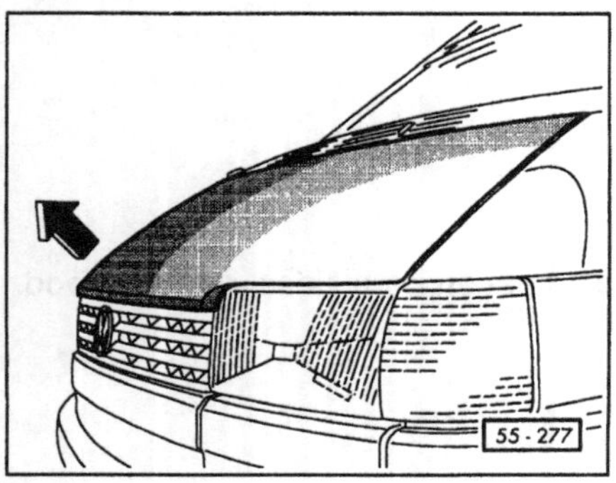

- Adjust latch so that hood latches completely after it is dropped from 30 cm (12 in.) without having to push down afterward, and:
  - ♦ Hood is aligned with body contour
  - ♦ Hood has no upward free play -arrow-

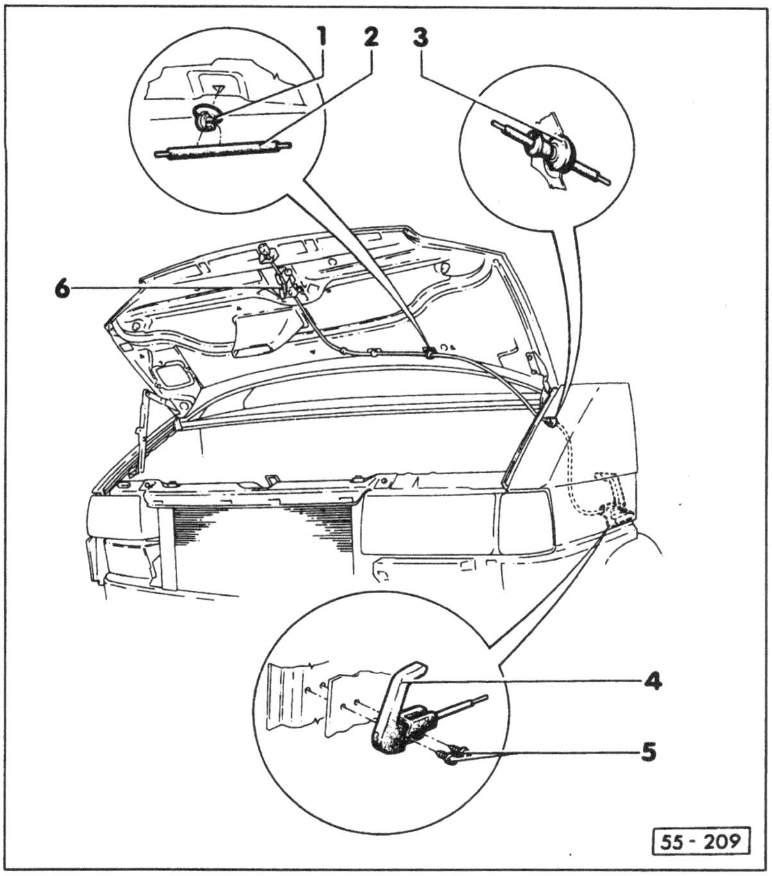

Hood release cable, components

1 - Clip
- ♦ Removing: turn 60° and pull out
- ♦ Installing: insert and turn 60°

2 - Hood release cable

3 - Cable grommet
- ♦ In front panel

4 - Release lever

5 - Self-tapping screw

6 - Hood latch
- ♦ Cable installed

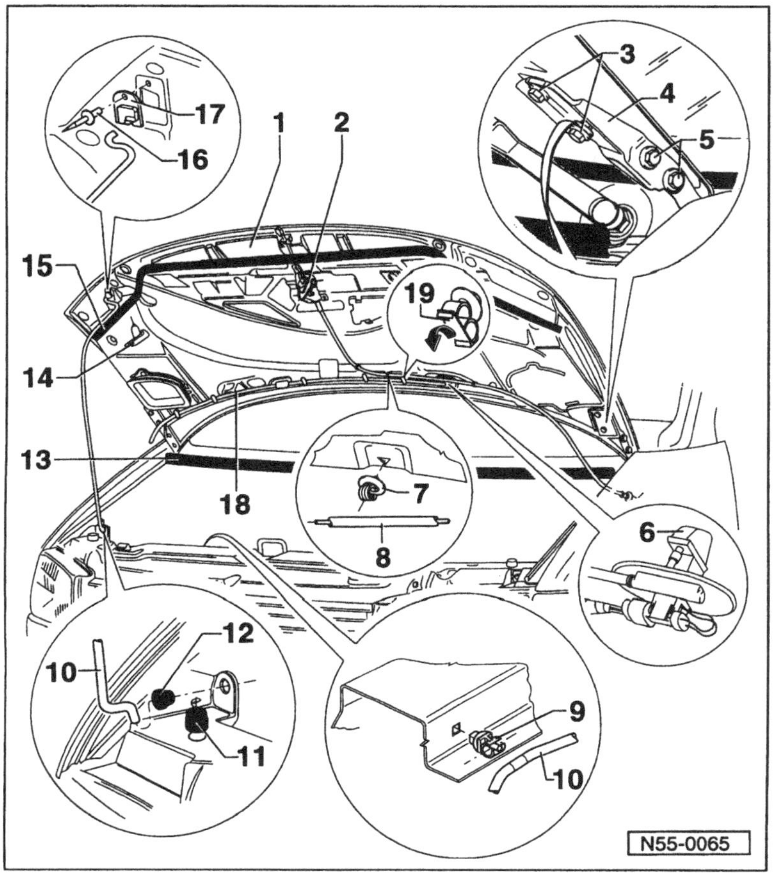

Hood, removing/ installing/adjusting 01.96 ➤

1 - Hood
- ♦ Removing ⇒ page Q 55-9
- ♦ Installing ⇒ page Q 55-11
- ♦ Adjusting ⇒ page Q 55-11
- – Apply corrosion protection to hinge and screws after installation or adjusting procedures.

2 - Hood lock
- – Remove insulation (if applicable).
- – Remove hood release cable bracket (press latch hook to left).
- – Remove screws and lock.
- – Disconnect cable.

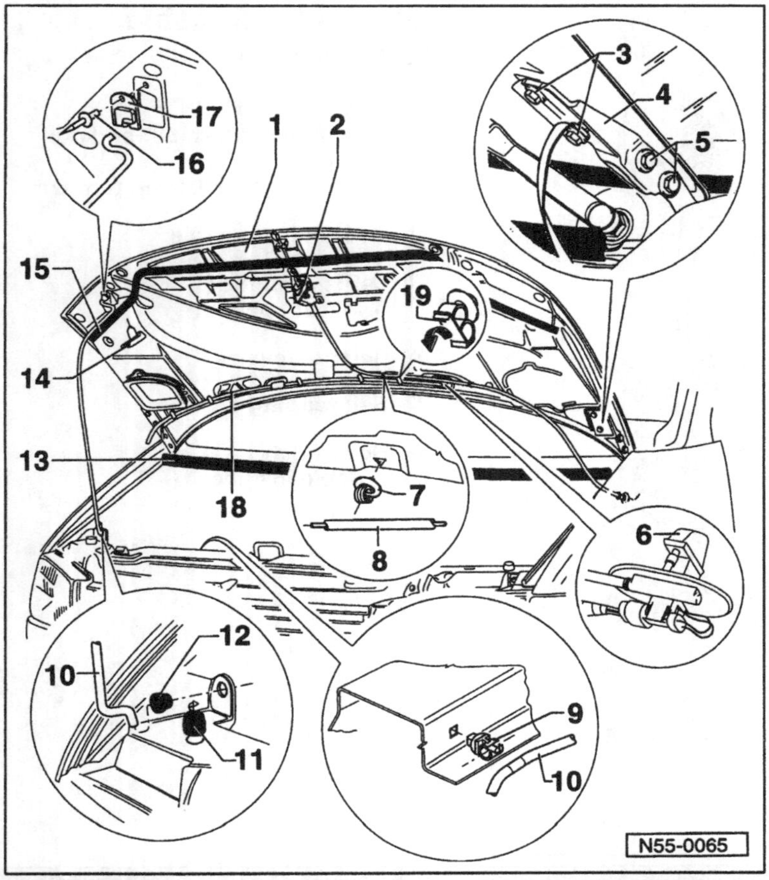

3 - Hex screw
- 23 Nm (17 ft lb)
- Fore/aft adjustment

4 - Hinge
- Two piece

5 - Hex bolt
- 23 Nm (17 ft lb)

6 - Nozzle
- Cowl grille must be removed to remove nozzles ⇒ page Q 55-9
- Electrical system

7 - Clip
- Rotate clip 180° to remove

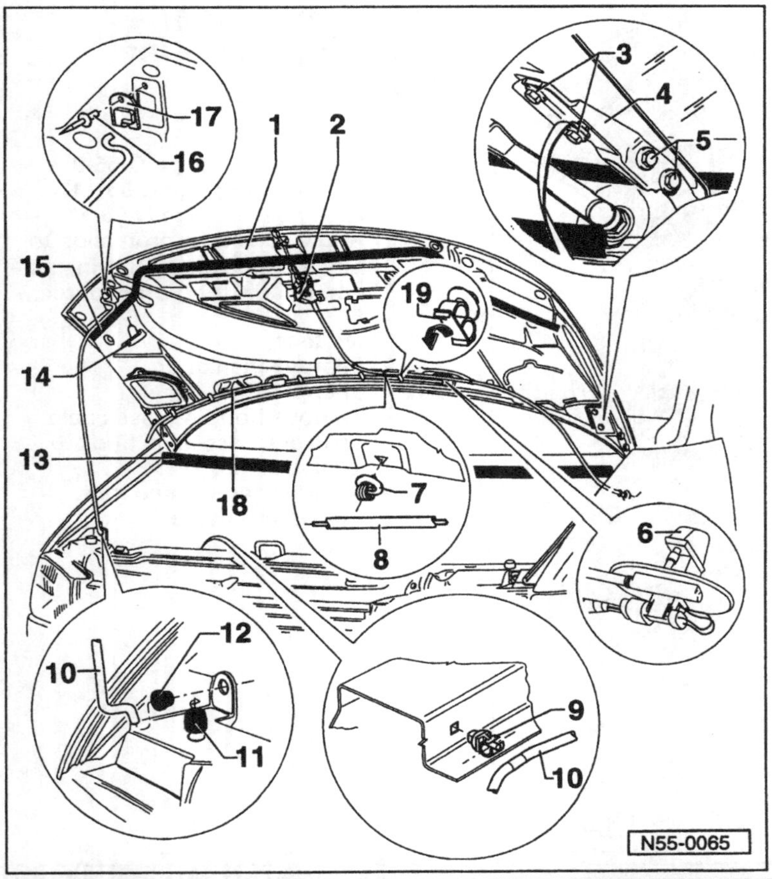

8 - Cable
- Cable connected

9 - Lid support

10 - Rubber bumper

11 - Bonded rubber bushing

12 - Cowl gasket
- Installed on flange

13 - Clip

14 - CW gasket

15 - Blind rivet

16 - Bracket, lid support

17 - Clip

18 - Hose

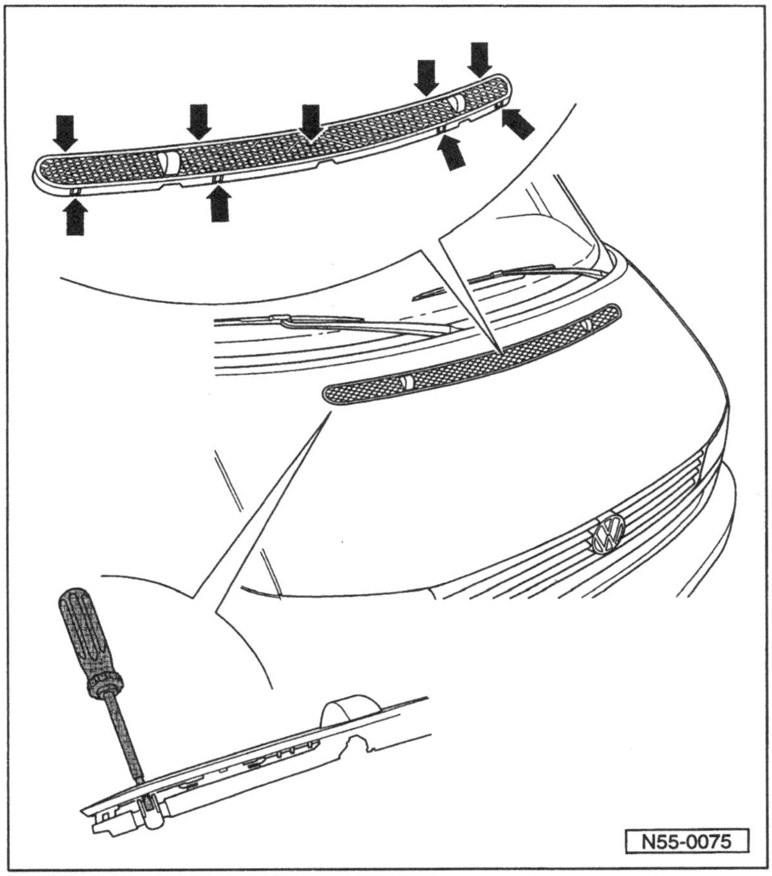

Hood, removing 01.96 ➤

Grille, removing/installing

– Disconnect grille -arrows-, unlatch with small screwdriver - starting at one end.

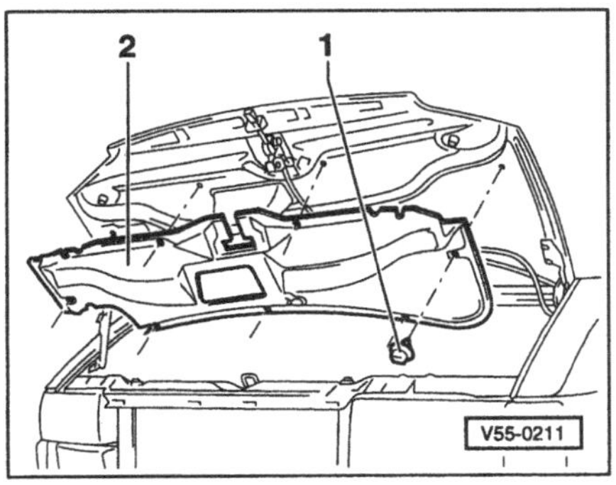

◄ **Hood with insulation**

– Rotate clips -1- 60 remove

– Remove insulation -2-.

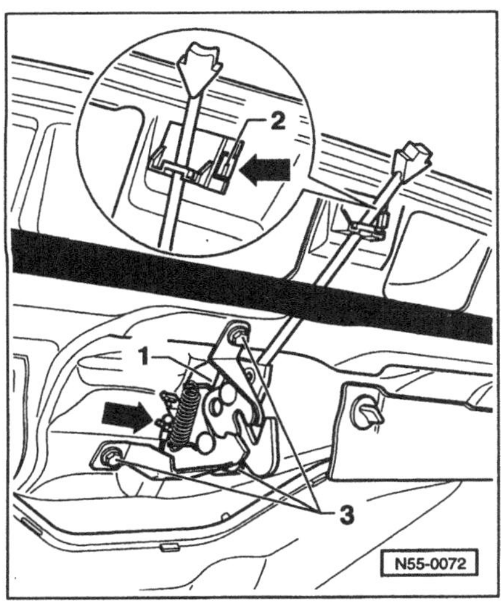

◄ – Press linkage locking tab -2- left (direction of arrow) and remove.

– Remove hex screws -3-.

Tightening torque:

Hex screws -3- = 8 Nm (71 in. lb)

– Disconnect cable and windshield washer hose, on hood (if applicable, heated nozzle wiring) ⇒ page Q 55-7, number 7.

– Disconnect latch cable -arrow-.

– Remove hood -1-.

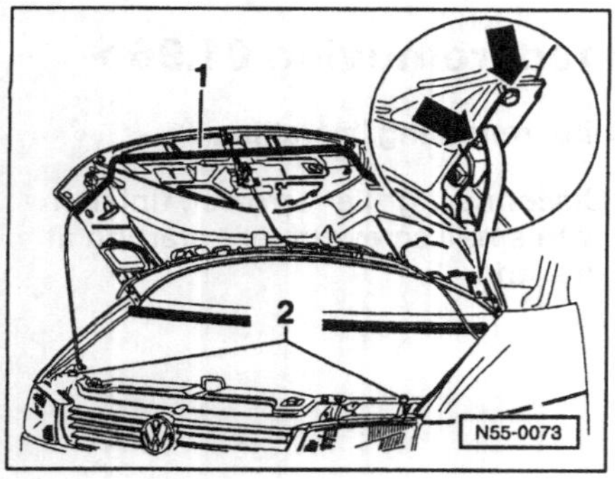

◄ – Remove hex bolts -arrows-, remove hood -1-.

Hood, installing/adjusting

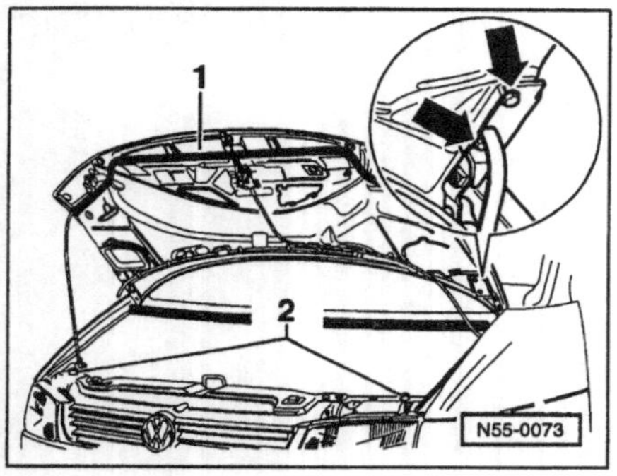

◄ – Loosely connect hood -1- to hinges.

Tightening torque = 8 Nm (71 in. lb)

– Align hood and torque bolts.

– Adjust rubber bumpers -2- so hood is flush with fenders.

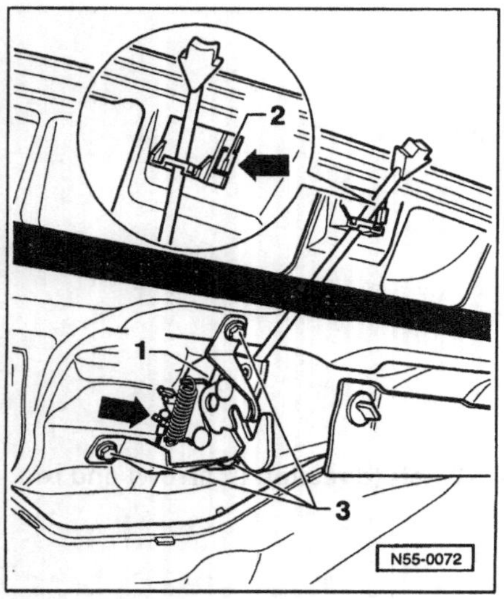

◄ – Install hood latch -1-, connect cable -arrow- and clip on hood ⇒ page Q 55-7, item 7. Torque: Hex bolt -2- = 8 Nm (71 in. lb)

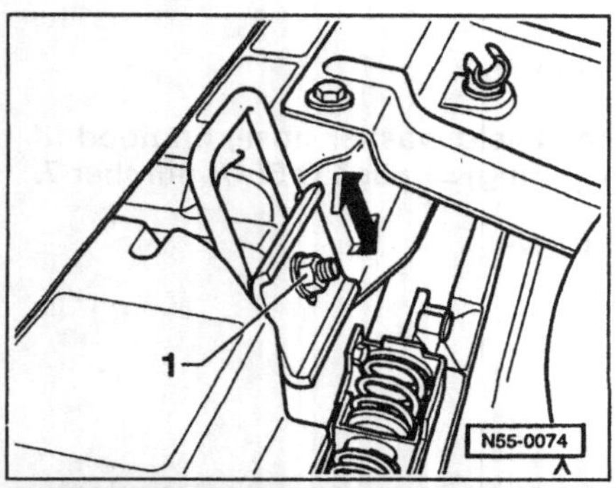

◄ – Loosen nuts -1- to adjust latch carrier -arrow-. Torque: Hex nut -1- = 5 Nm (44 in. lb)

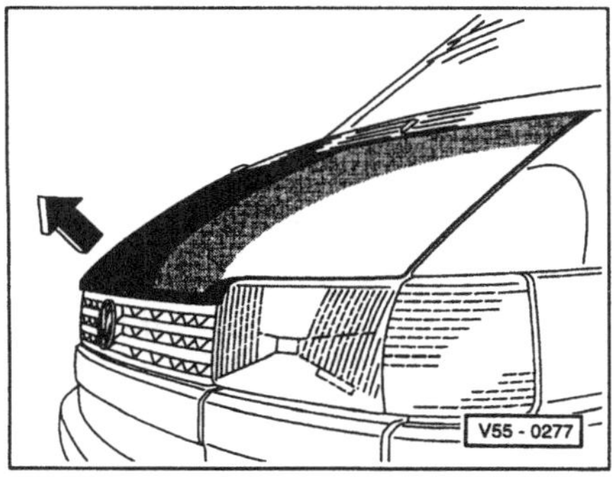

◄ – Adjust latch so that hood latches completely when dropped from 50 cm (20 in.) without additional assistance. The hood must be flush with fenders and have no free play -arrow-.

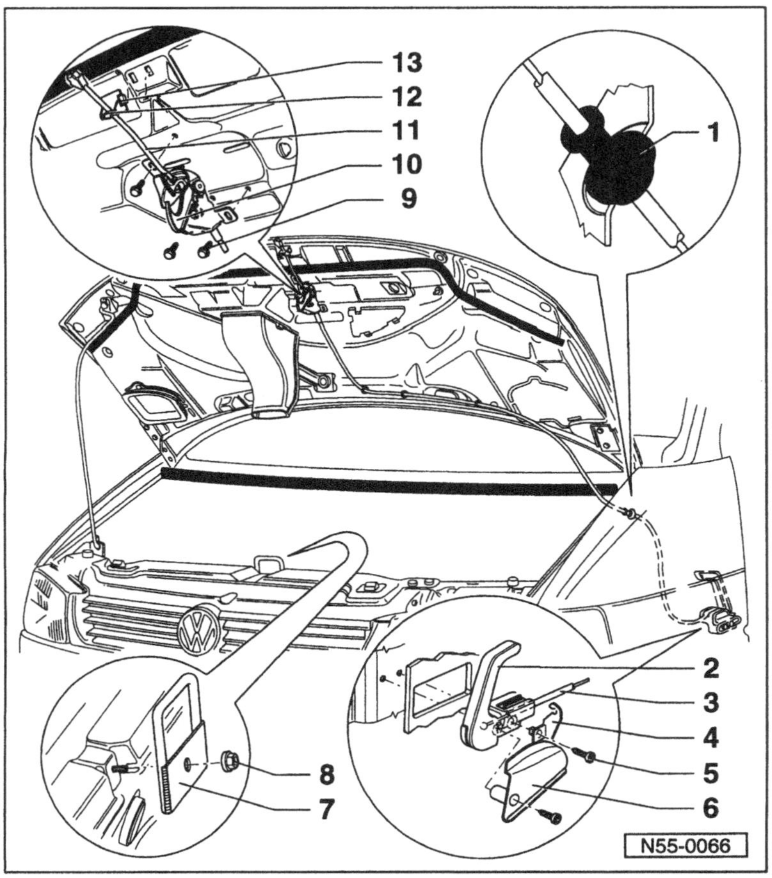

Hood release cable and latch, removing/installing 01.96 ➤

1 - Sleeve

2 - Hood release link
- ♦ Cable connected

3 - Cable
- ♦ Replacing - screw with linkage and latch, disconnect cable

4 - Cable bracket

5 - Combination screw and washer

6 - Cable bracket
- ♦ Only with carpet

7 - Hood latch hook

8 - Hex nut
- ♦ 5 Nm (44 in. lb)

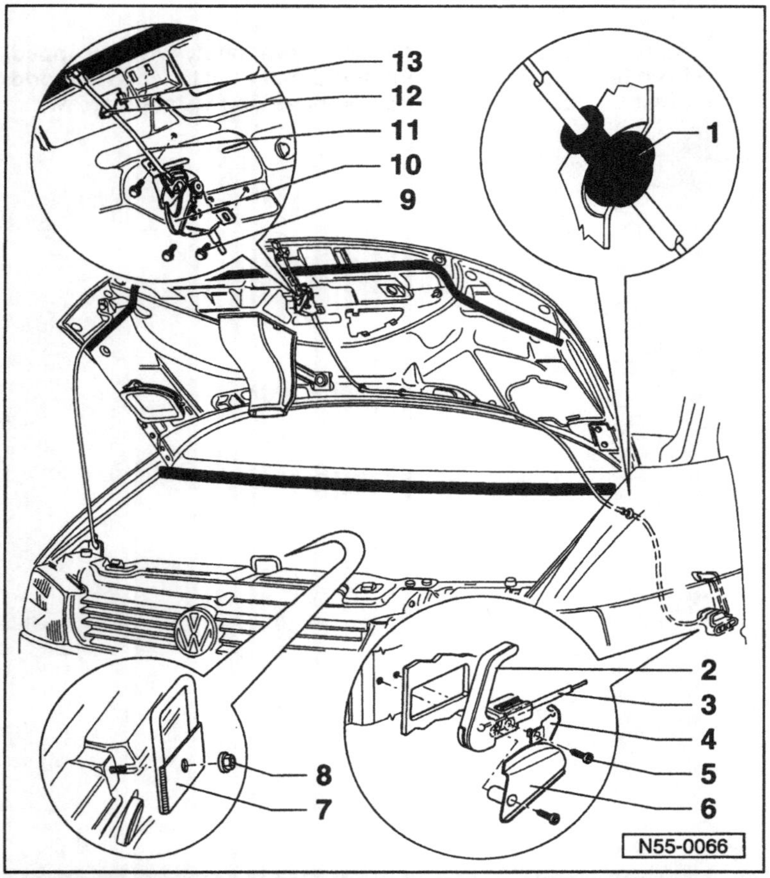

9 - **Hex bolt**
- 8 Nm (71 in. lb)

10 - **Hood latch**
- Cable connected
- Remove insulation (if applicable)
- Remove linkage bracket (press latching hooks to left)
- Remove latch
- Disconnect cable.

11 - **Link, hood release cable**

12 - **Bracket, hood release cable link**
- To remove, press hood latch bracket to left

13 - **Latch hook**

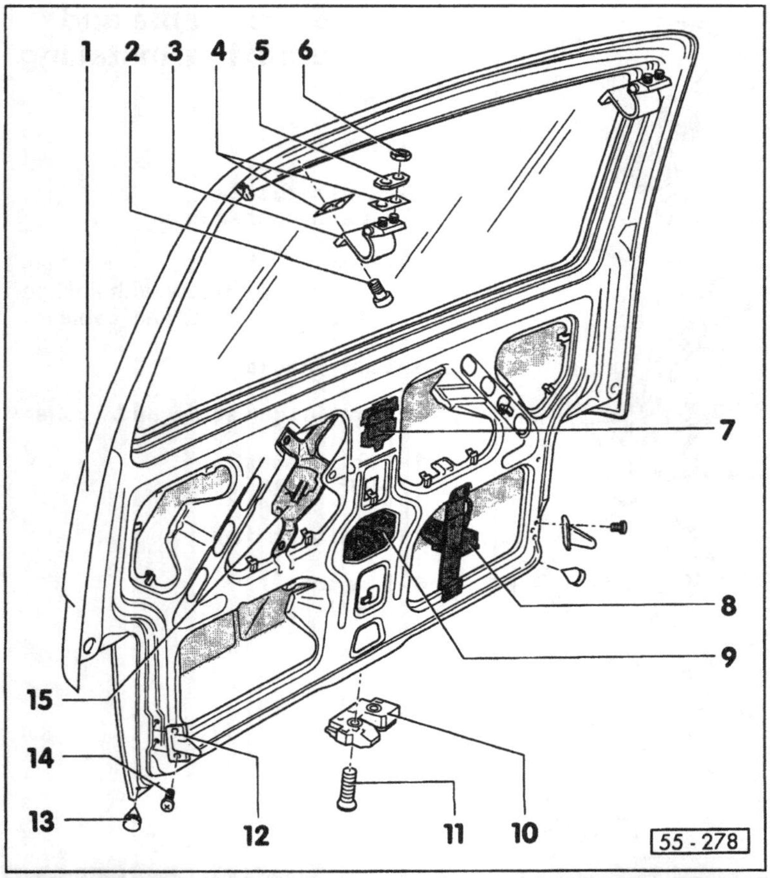

Tailgate, components

1 - **Tailgate**
- Removing and installing ⇒ Page Q 55-19
- Adjusting ⇒ Page Q 55-21

2 - **Phillips screw**
- 19 Nm (14 ft lb)

3 - **Hinge**

4 - **Sealing Plate**

5 - **Spacer plate**
- During removal of hinge -3-, do not let spacer plate fall into headliner

6 - **Hex nut**
- 30 Nm (22 ft lb)

7 - **Latch, with opening mechanism**

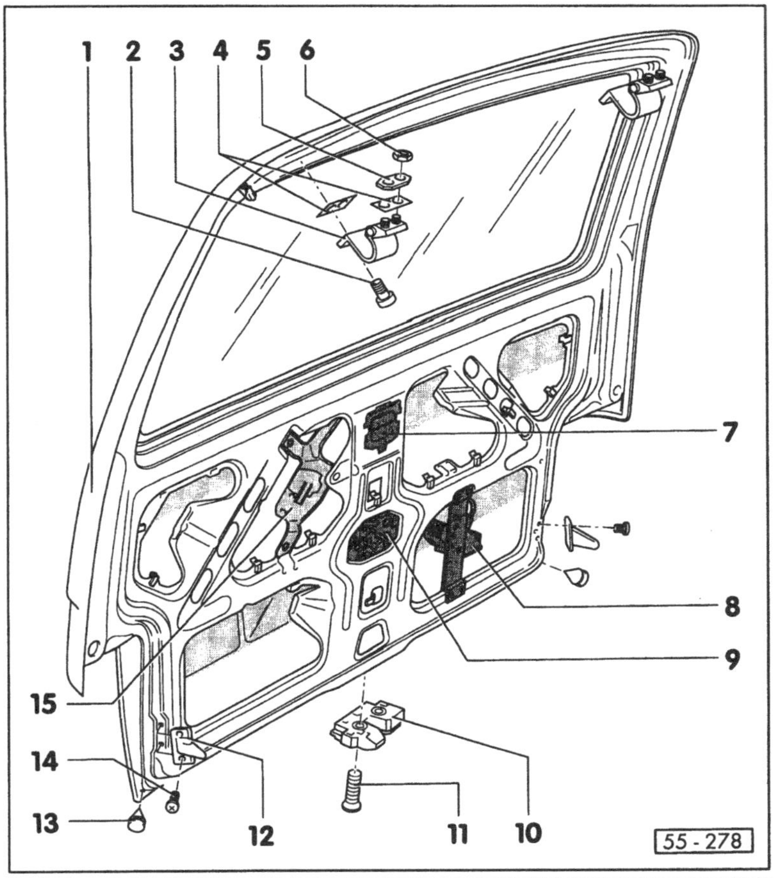

8 - Decklid Central Locking System Motor -V53-

9 - Lock mechanism
- ♦ Removing ⇒ Page Q 55-25
- ♦ Lock cylinder, removing ⇒ Page Q 55-26

10 - Lock

11 - Screw
- ♦ 19 Nm (14 ft lb)

12 - Guide pin
- ♦ Note identification for left -L- and right -R-
- ♦ Adjusting ⇒ Page Q 55-22

13 - Plug

14 - Phillips screw
- ♦ 8.0 Nm (71 in. lb)

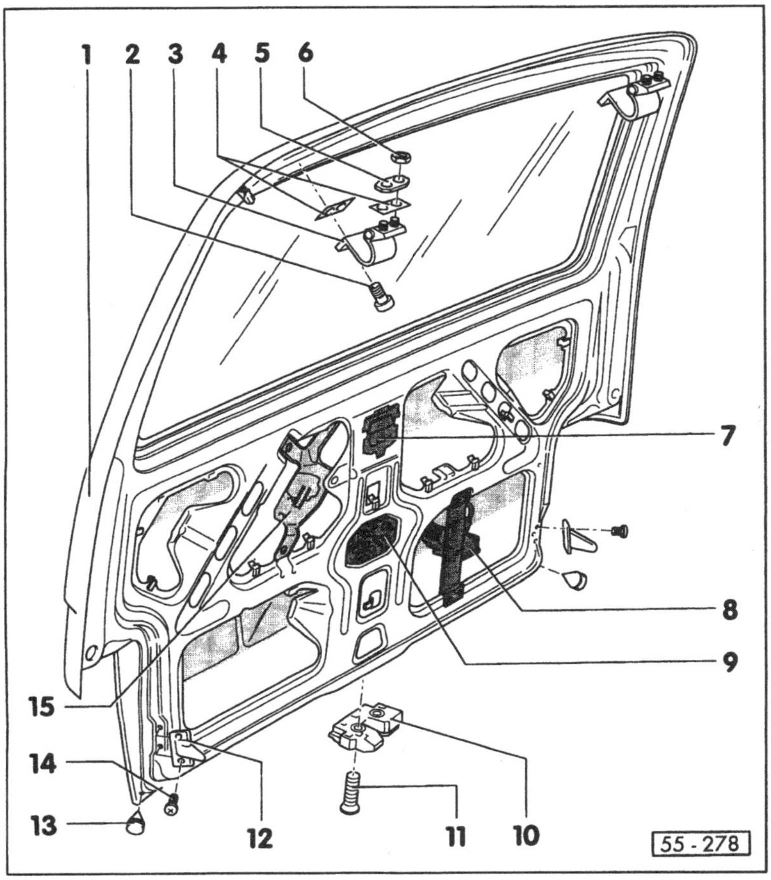

15 - Rear Window Wiper Motor -V12-
- ♦ Removing and installing:

⇒ Repair Group 92; Electrical system; Rear Window wiper and washer system for tailgate, repairing

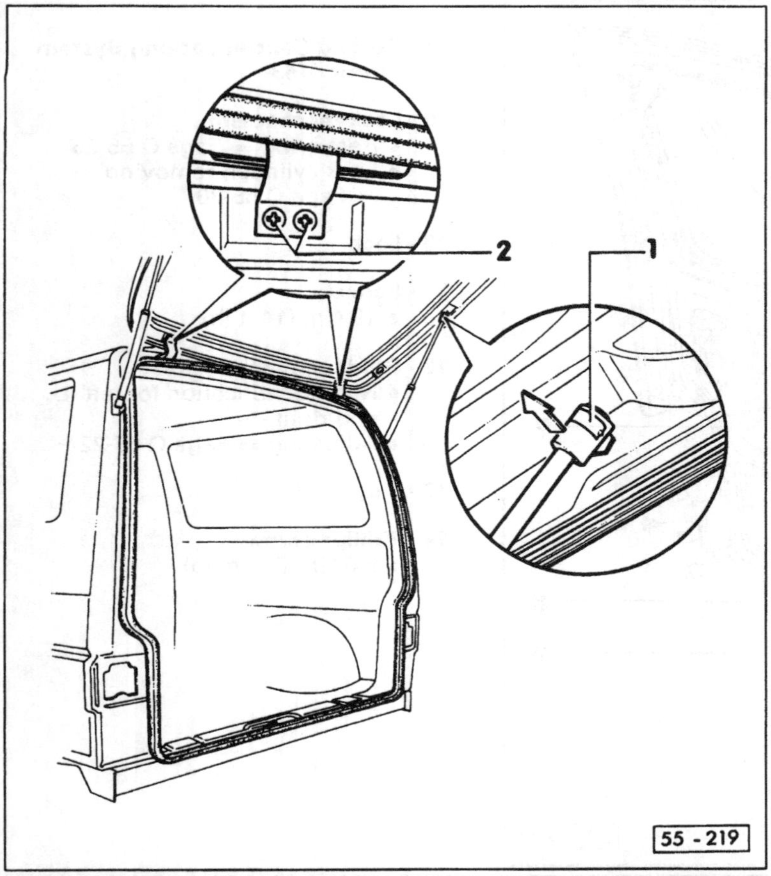

Tailgate, removing/installing

- Remove trim ⇒ Page Q 70-91 and/ or remove tailgate molded trim ⇒ Page Q 70-84.
- Disconnect connectors for rear window heating, license plate light, rear window wiper motor and central locking motor, if applicable, remove wiring from tailgate.
- Support tailgate.
- Lift locking clip -1- with screwdriver and pull gas-filled struts off ball pins on both sides.

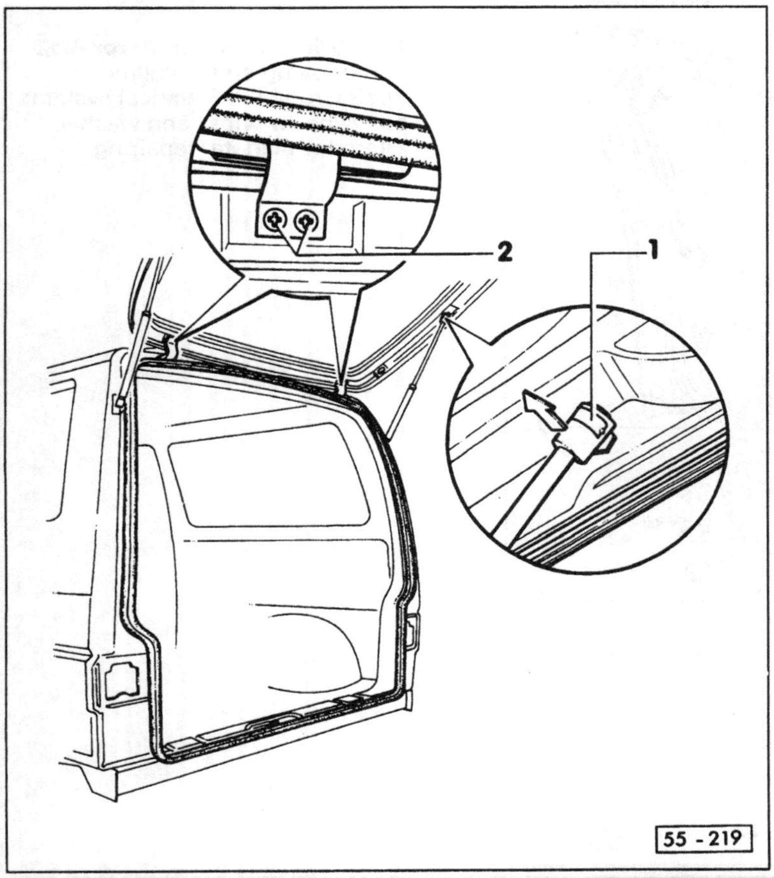

Note:

Do not pry locking clip -1- completely out of ball socket, otherwise clip will be damaged

- Remove Phillips screws -2- from hinges and remove tailgate.
- If a new tailgate is installed, it must be adjusted ⇒ Page Q 55-21.

  Tightening torque:

  Phillips screw -2- = 19 Nm (14 ft lb)

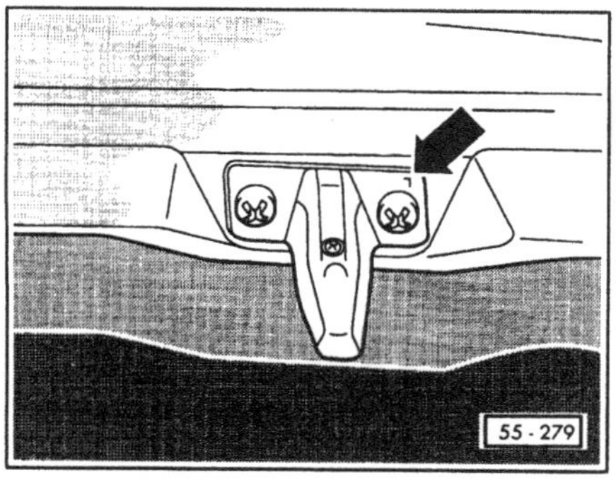

Tailgate, adjusting

Height adjustment

– Remove molded headliner from passenger compartment ⇒ Page Q 70-75.

– Remove tailgate gas struts ⇒ Page Q 55-27.

◄ – Remove guide pins, right and left (arrow shows L for left and R for right).

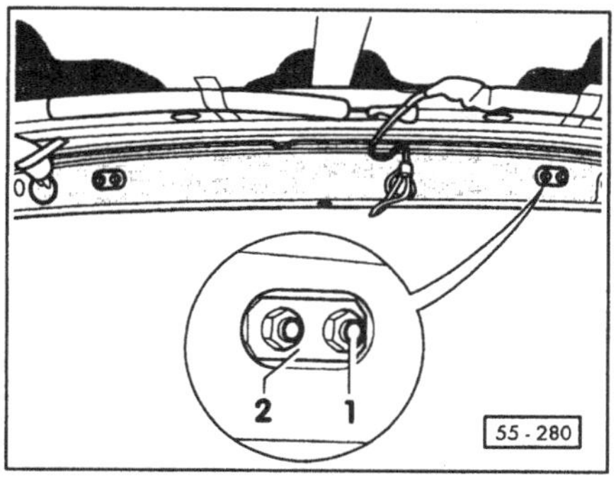

◄ – Loosen hex nuts -1- (do not remove hinge on rear roof beam)

Note:

If hex nuts -1- are completely removed, it is absolutely necessary to prevent hex nuts or spacer plates -2- from falling into headliner.

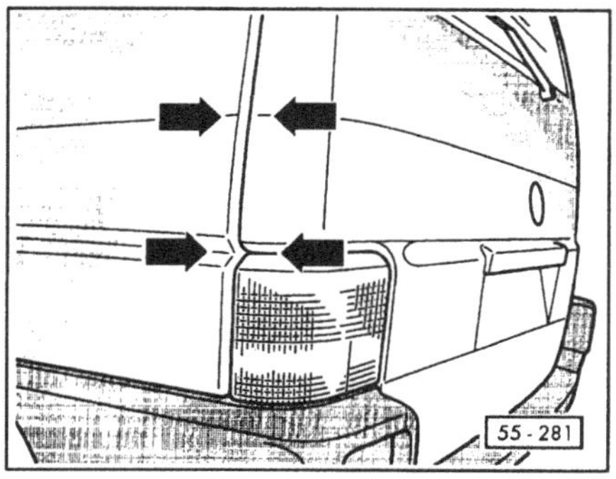

◄ – With second technician, position tailgate to match outer contour -arrows- and tighten hex nuts of hinges.

Tightening torque:

Hex nuts = 30 Nm (22 ft lb)

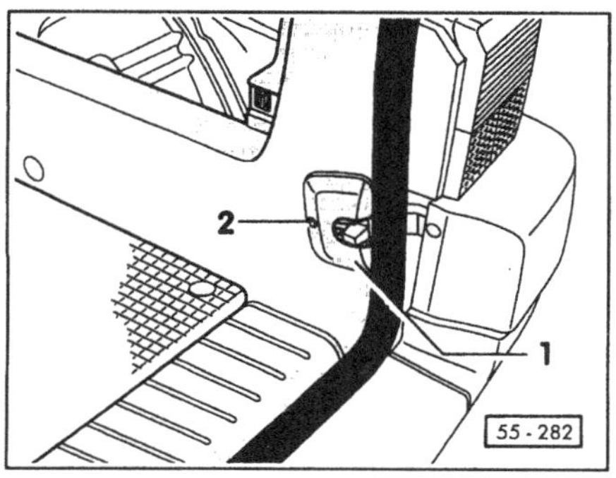

Guide pins, adjusting

◄ – Remove guide trim -1- after removing Phillips screws -2- from both sides.

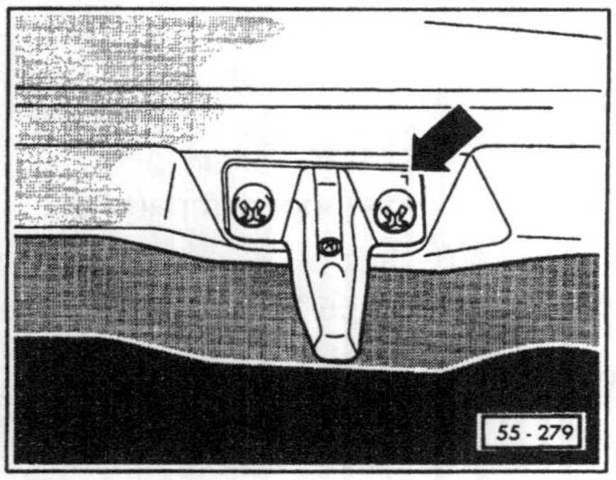

- Install left and right guide pins, note identification -arrow- (L for left side and R for right side) of tailgate.
- Install gas struts.

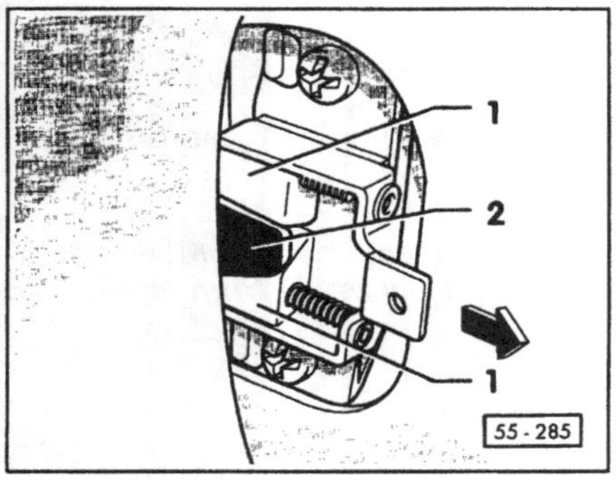

- Using a second technician, carefully close tailgate and observe from inside whether guide wedges -1- are pressed uniformly by the guide pins -2- approx. 10–15 mm (3/8 – 5/8 in.) in driving direction -arrow-.
- Make left and right adjustments by changing position of guide pin on tailgate and tighten guide pin Phillips screws.

  Tightening torque:

  Phillips screws = 8.0 Nm (71 in. lb)

Striker, adjusting

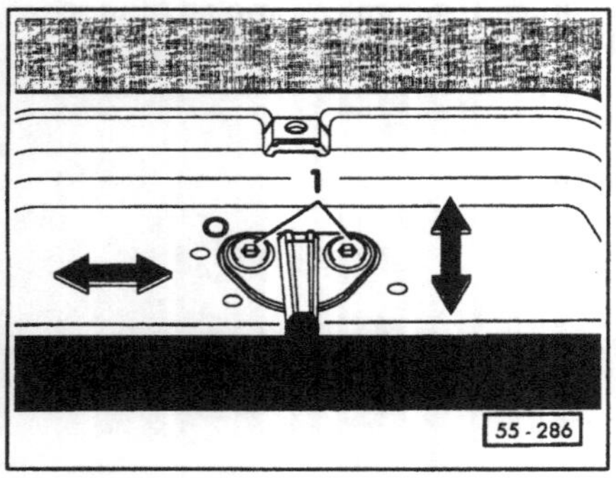

- Remove rear edge trim.
- After loosening mounting bolts -1-, striker can be moved in directions of arrows.

  Tightening torque:

  Mounting bolts -1- = 20 Nm (15 ft. lb)

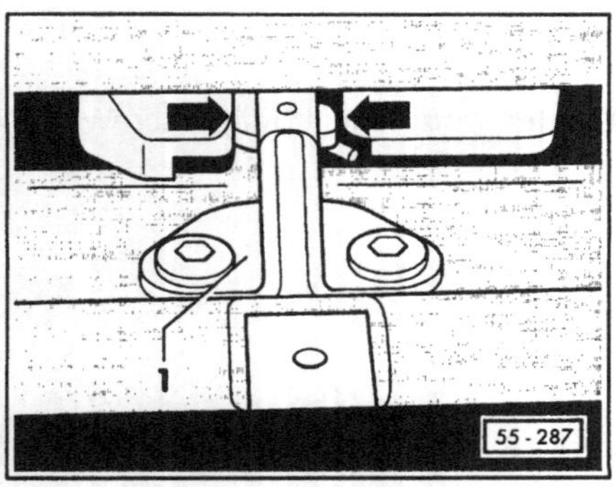

- Adjust striker -1- so that recess on door lock turn clip is concentric -arrows- to striker.
- From inside vehicle, check adjustment while slowly closing tailgate, adjust as needed.

Tailgate lock mechanism, removing

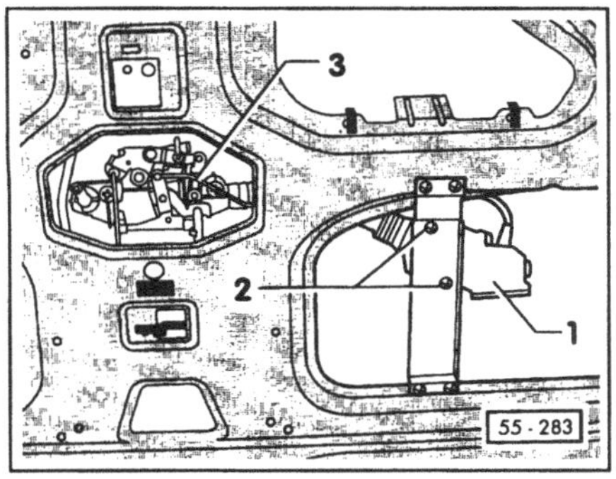

- Remove molded trim from tailgate ⇒ Page Q 70-92 and/or remove tailgate trim ⇒ Page Q 70-96.
- ◄ Loosen Decklid Central Locking Motor -V53- -1-,if applicable, by removing Phillips screws -2- and rod -3- at the lock mechanism.

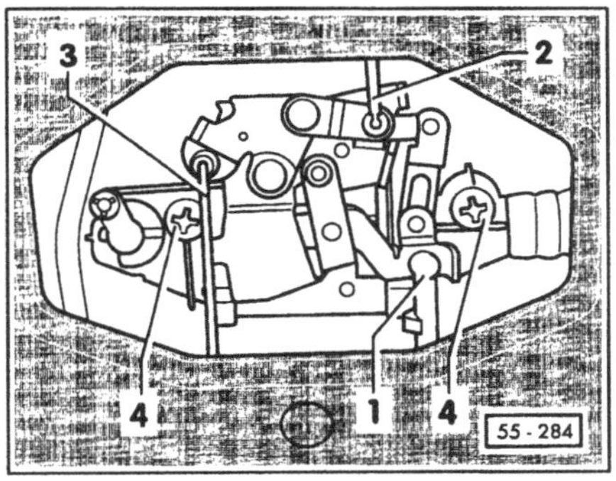

- ◄ Remove tailgate lock rod -1- for inner mechanism and child lock -3-.
- Remove Phillips screw -4- and tailgate lock mechanism.

Tailgate lock cylinder, removing

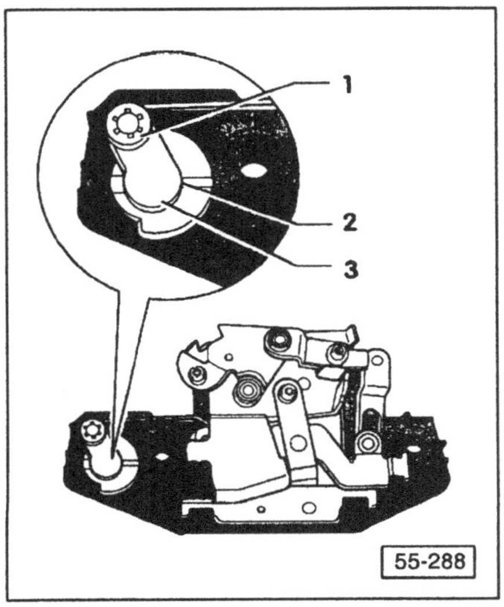

- Remove tailgate molded trim ⇒ Page Q 70-92 and/or remove tailgate trim ⇒ Page Q 70-96.
- Remove tailgate lock mechanism ⇒ Page Q 55-16.
- ◄ Remove lock washer -1- and rods,
- Drive out lock pin -2-.
- Remove casting -3- from four-cornered shaft .
- Remove spring.
- Remove lock cylinder with wrench.

Tailgate gas strut, removing

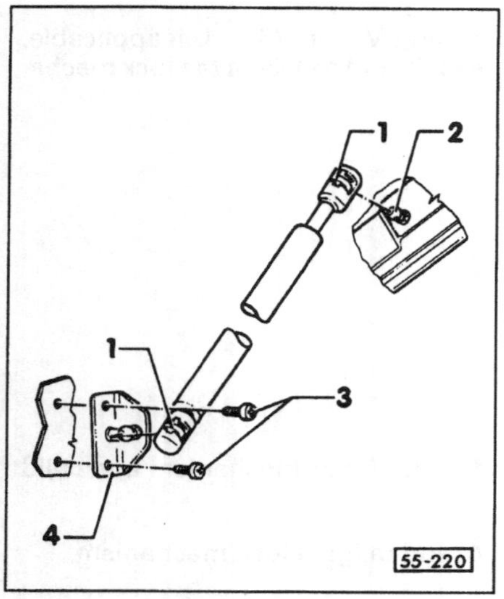

(Diagram shows left gas strut)

- Open and support tailgate.

- Lift locking clip -1- with screwdriver and pull gas strut off ball-head pin.

Note:

When reusing gas strut, locking clip -1- must not be completely removed from ball socket, otherwise clip will be damaged.

Tightening torque:

Ball-head pin -2- = 20 Nm (15 ft lb)

Screw -3- = 8 Nm (71 in. lb)

- Bracket with ball-head pin -4-.

Gas strut, discarding

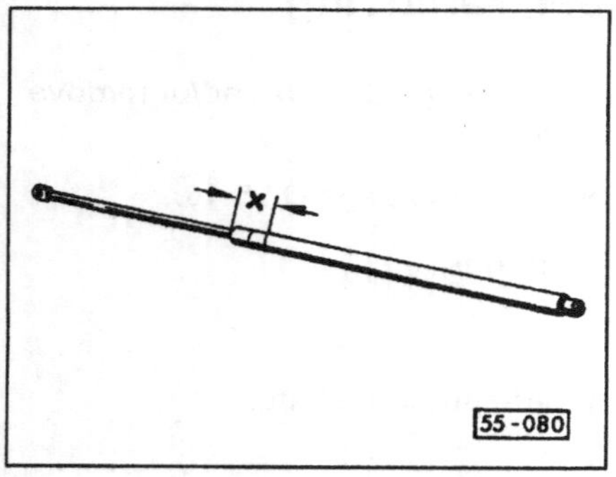

- Clamp gas strut into vice.

Note:

After replacing gas strut, gas should be released from removed part before discarding.

> ***WARNING!***
>
> - ***Never squeeze strut in vise in area other than shown. To do so can cause strut caps on either end to pop off and release spring, which is under high pressure.***
> - ***Use protective eye wear.***

- Cut strut cylinder open in first third of cylinder length, starting from reference edge on piston side.

- Cover area being cut with rag to catch oil.

- Dispose of oil and rags according to applicable regulations.

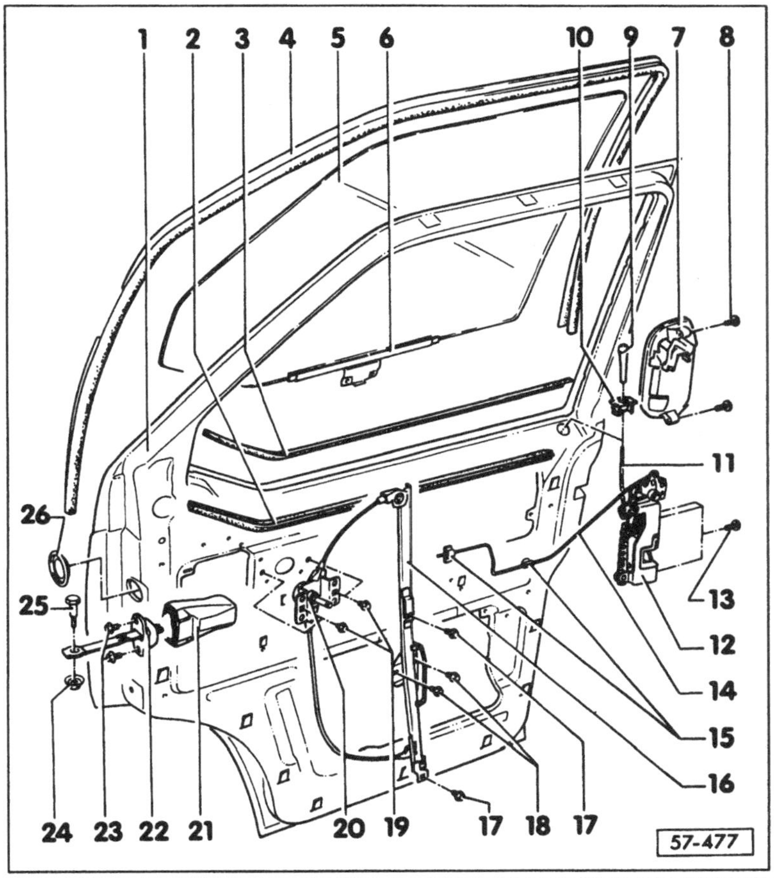

Front door (manual windows), components ➤12.95

1 - Front door

2 - Inner window slot seal
- Pressed into flange
- Remove together with door trim

⇒ Page Q 70-81

3 - Outer window slot seal
- Pressed into flange

4 - Window channel

5 - Door glass
- Removing

⇒ Page Q 57-20

- Installing

⇒ Page Q 57-20

6 - Window regulator rail

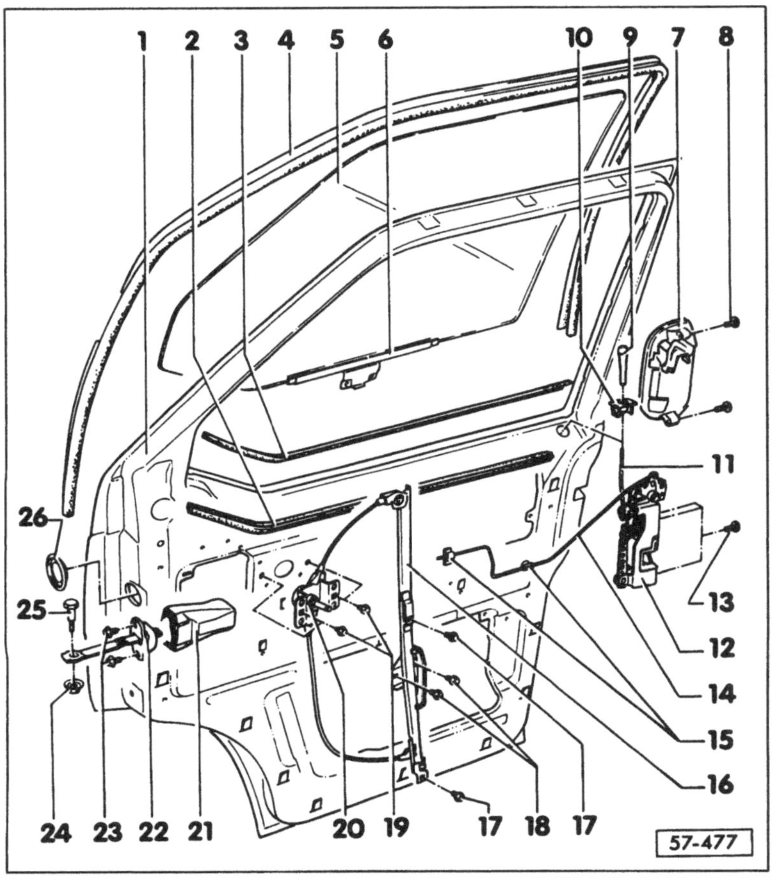

7 - Outside door handle
- Removing/installing

⇒ page Q 57-21

- Lock cylinder removing

⇒ page Q 57-22

8 - Screw
- 8.0 Nm (71 in. lb)

9 - Lock knob

10 - Trim plate

11 - Lock rod

12 - Door lock

13 - Screw
- 8.0 Nm (71 in. lb)

14 - Pull rod

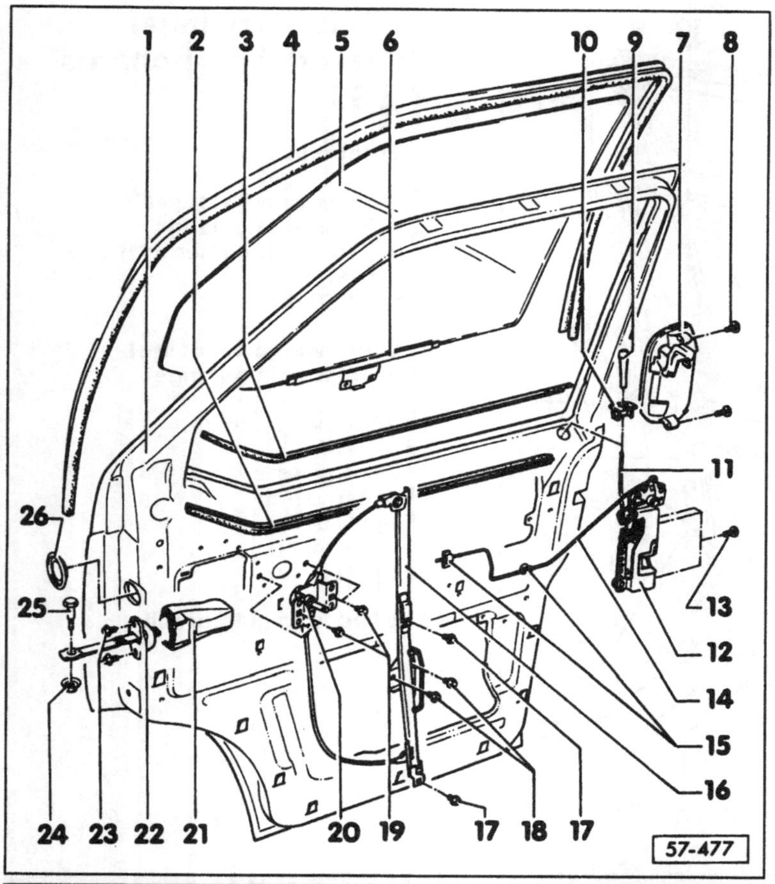

15 - Clip
- Secures and guides pull rod

16 - Window regulator
- Removing/installing

⇒ page Q 57-19

17 - Hex bolt
- 7.0 Nm (62 in. lb)

18 - Hex bolt
- 7.0 Nm (62 in. lb)
- Secures regulator rail to window regulator

19 - Hex bolt
- 7.0 Nm (62 in. lb)

20 - Window regulator crank mechanism
- Removing

⇒ page Q 57-19

21 - Cover
- Clipped on

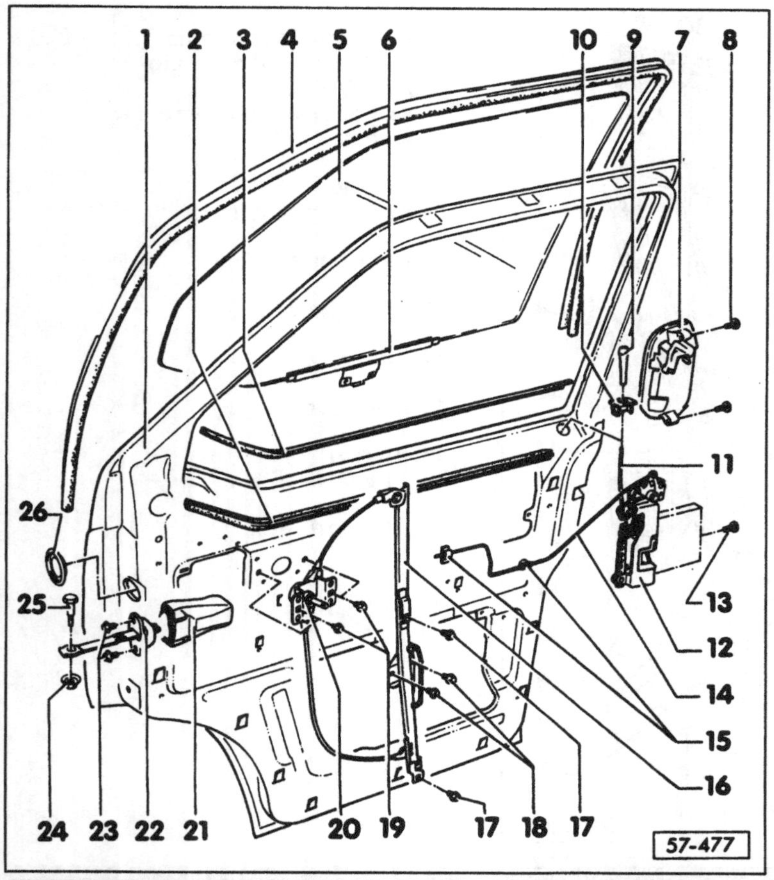

22 - Door check strap

23 - Hex bolt
- 7.0 Nm (62 in. lb)

24 - Lock nut
- 6.5 Nm (57 in. lb)

25 - Pin

26 - Trim ring

Front door, removing/installing/adjusting ➤ 12.95

Vehicles with central locking, power windows, heated outside mirrors and/or door speakers

– Remove front door trim
⇒ Page Q 70-81

– Locate wiring connectors, disconnect wiring and remove from door

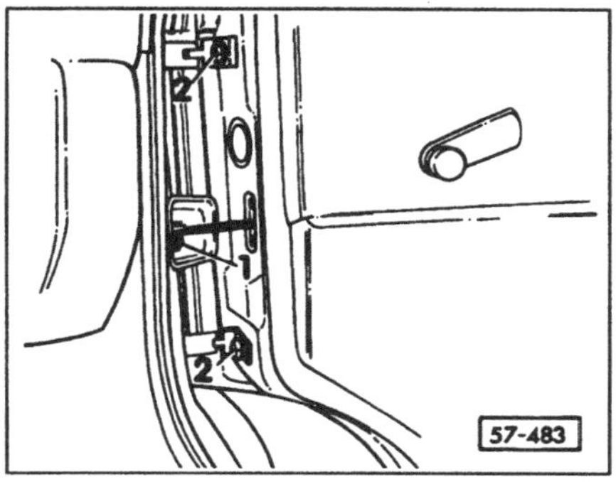

◄ – Remove nut -1- and pull out pin.

– Remove hex bolt -2- from upper and lower hinge.

– Remove door.

– Install hex bolts -2- into upper/lower door hinges.

– Adjust door by moving hinges in oversized holes.

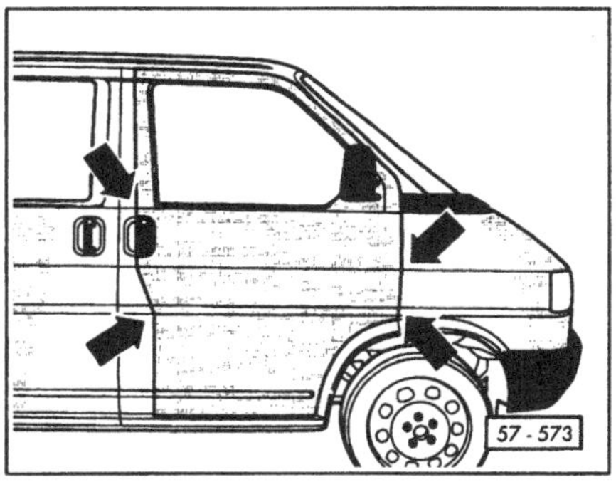

◄ • Door is correctly adjusted if, when closed, has uniform gap all around opening, is not sunken, does not protrude, matches contour -arrow-.
⇒ Page Q 57-25

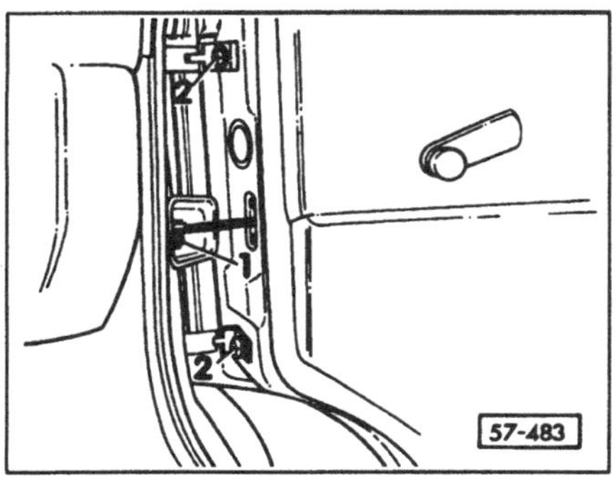

◄ – Tighten hex bolts -2-. After installation of bolt, tighten nut -1-.

Tightening torques:

Nut -1- = 6.5 Nm (57 in. lb)

Hex bolt -2- = 55 Nm (41 ft lb)

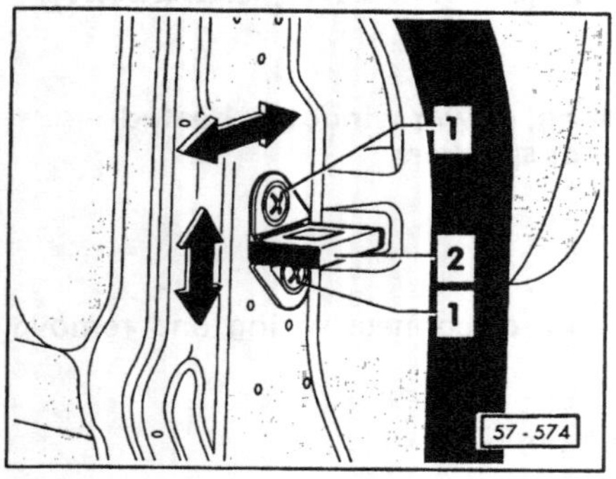

Striker, adjusting ➤12.95

◄ – After Phillips screws -1- are loosened, striker -2- can be moved horizontally and vertically -arrows-.

Tightening torque:

Phillips screws -1- 19 Nm (14 ft lb)

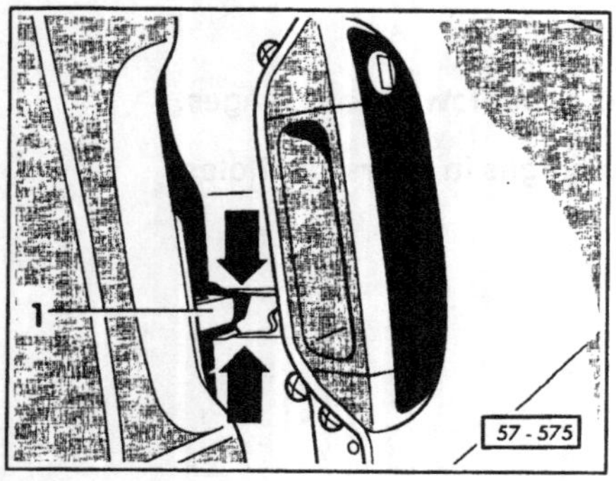

◄ – Adjust striker -1- so recess on door lock turn clip is concentric -arrows- to striker (door must not be raised or pressed down). Door must not move when closed.

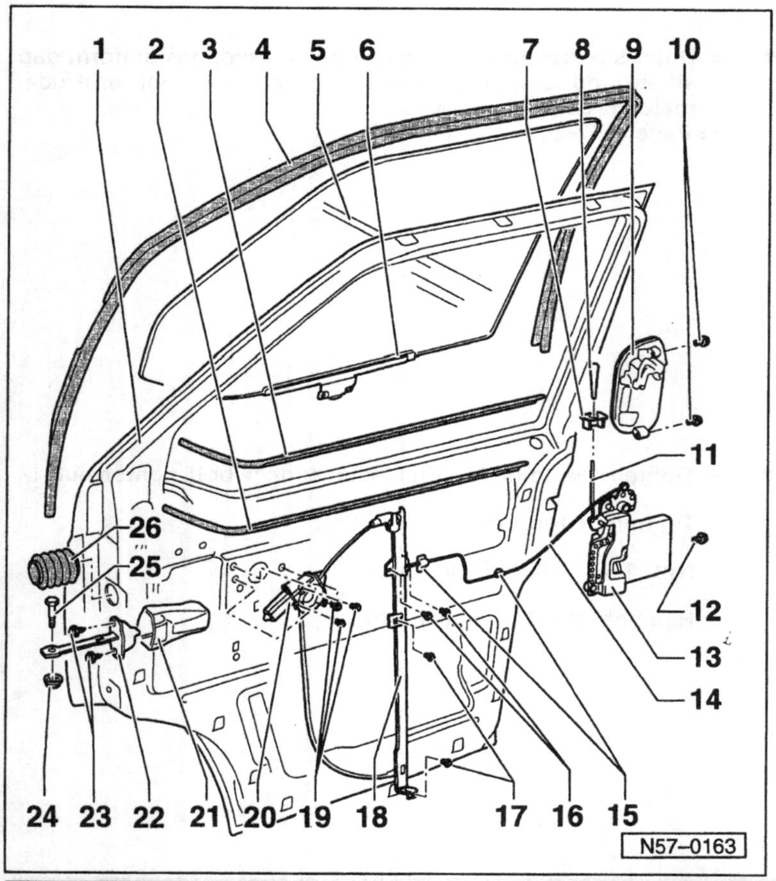

Front door (power windows), components

1 - Door, front
- ♦ removing installing/adjusting

⇒ Page Q 57-5

2 - Window channel seal, inner
- ♦ on flange
- ♦ removing: together with door trim

⇒ Page Q 57-15

3 - Window channel seal, outer
- ♦ on flange

4 - Window channel

5 - Door window glass
- ♦ removing/installing

⇒ Page Q 57-20

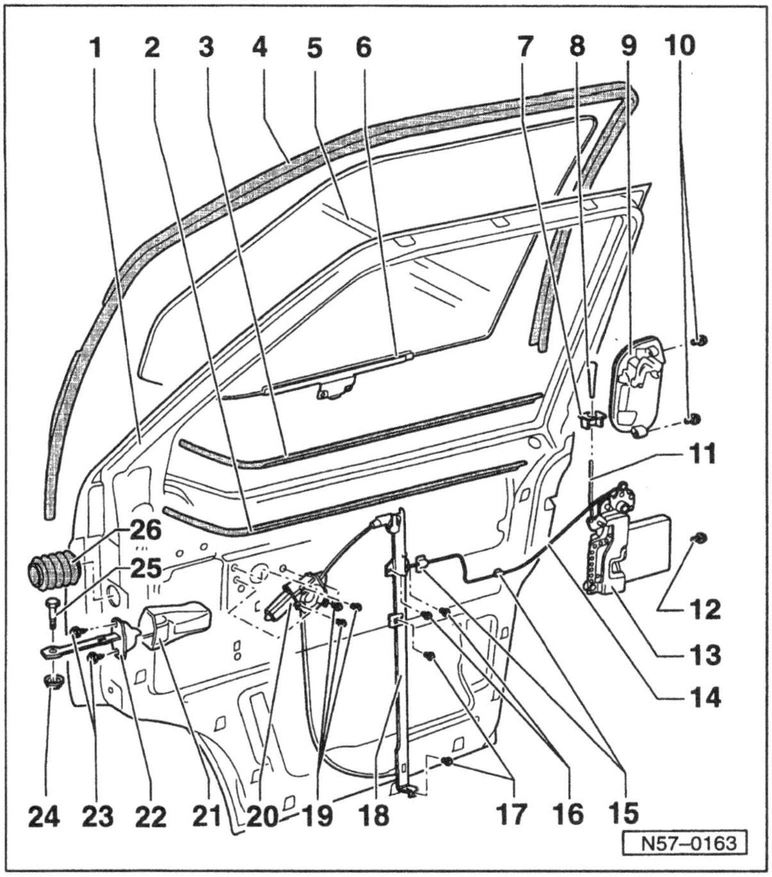

6 - Lift rail

7 - Escutcheon

8 - Safety Button
- screwed on

9 - Outside door handle
- removing

⇒ Page Q 57-22
- Remove lock cylinder

⇒ Page Q 57-23

10 - Oval head screw
- 8 Nm (71 in. lb)

11 - Lock Rod

12 - Oval head screw
- 8 Nm (71 in. lb)

13 - Door lock
- removing

⇒ Page Q 57-22

14 - Door latch release rod

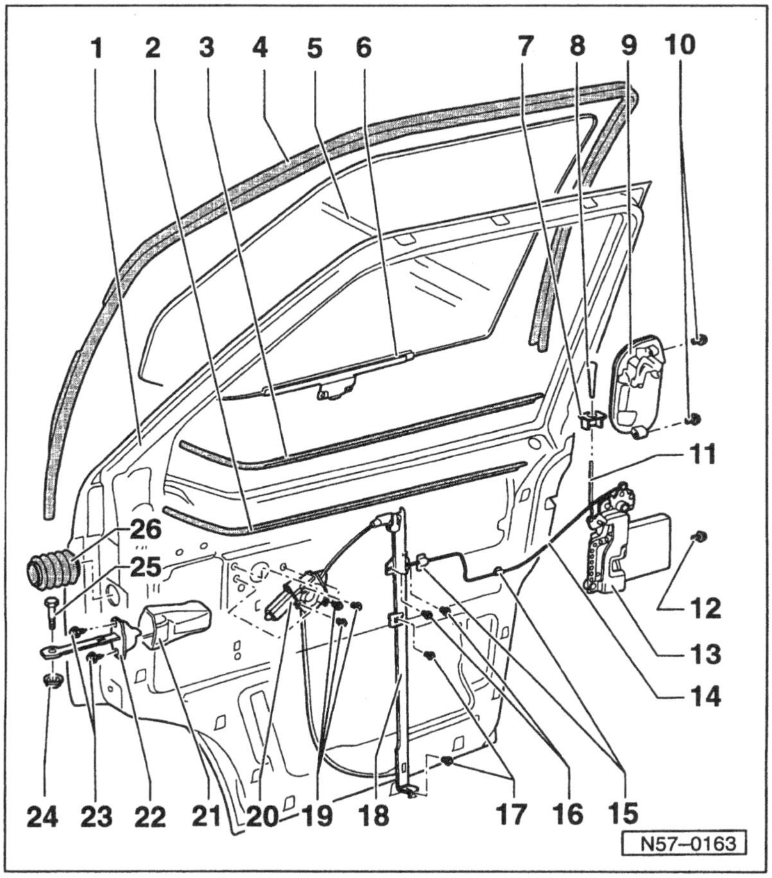

15 - Clip
- locates/guides pull rod

16 - Hex bolt
- 7 Nm (62 in. lb)
- fastens lift rail to window regulator

17 - Hex bolt
- 7 Nm (62 in. lb)

18 - Window regulator
- removing

⇒ Page Q 57-12

19 - Hex bolt
- 7 Nm (62 in. lb)

20 - Motor, window regulator
- remove together with window regulator

⇒ Page Q 57-12

21 - Sleeve
- inserted

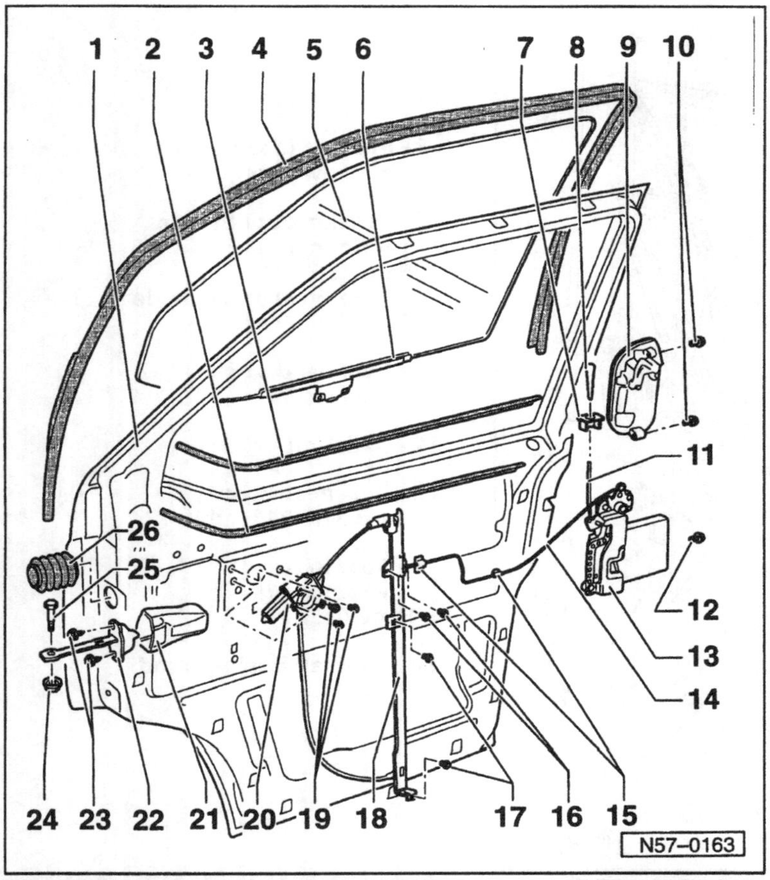

22 - **Door check**

23 - **Hex bolt**
- 7 Nm (62 in. lb)

24 - **Lock nut**
- 6.5 Nm (57 in. lb)

25 - **Bolt**

26 - **Boot**

Power window regulator, removing

– Remove door trim
⇒ Page Q 57-15

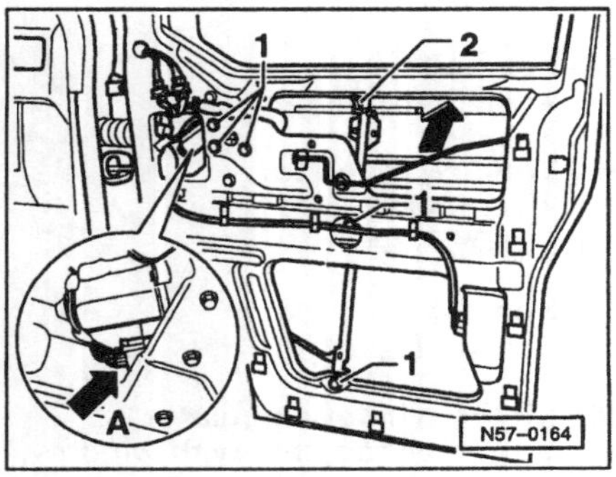

– Remove lift rail with door glass from window regulator.

– Slide door glass upward and secure (e.g. adhesive strip)

– Disconnect window regulator motor connector -arrow A- (V14 or V15)

– Remove hex bolts -1-.

– Unhook window regulator clip -2- on door opening.

– Remove window regulator with motor from upper opening -arrow-.

– Tightening torque: Hex bolt -1- = 7 Nm (62 in. lb)

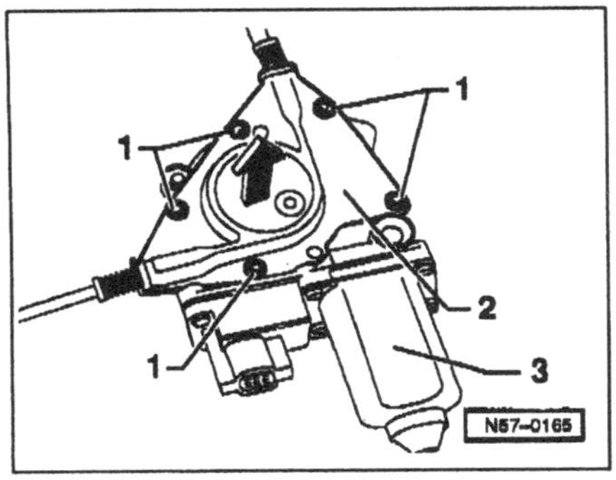

Power window regulator, repairing

– Remove door trim
⇒ page Q 57-15.

– Remove window regulator
⇒ page Q 57-12.

– Remove bolt -1-

– Remove window regulator -2- in direction of arrow from regulator motor -3-.

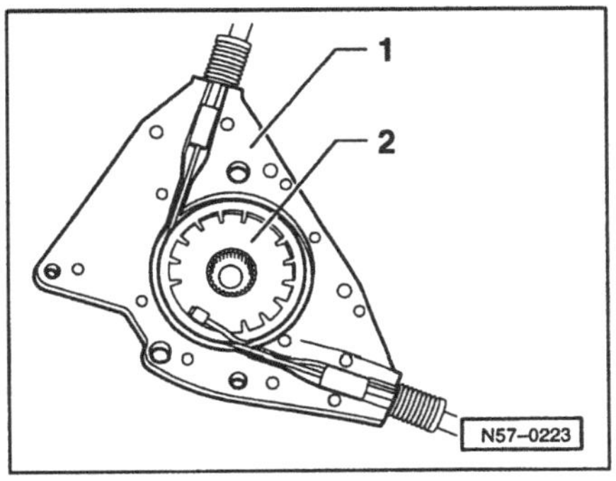

– During window regulator assembly or disassembly ensure cable drum -2-, with cables, does not come out of housing.

– Remove new regulator housing cover -1- from cable drum -2-.

Note:

Greased surfaces must not be contaminated with dust or dirt. Grease with original grease G 000 450 02.

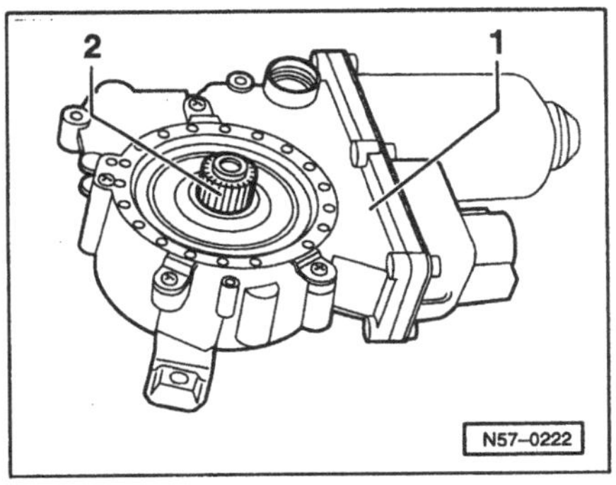

– Remove cover from new motor assembly -1- drive pinion.

Note:

Greased surfaces must not be contaminated with dust or dirt. Grease with original grease G 000 450 02.

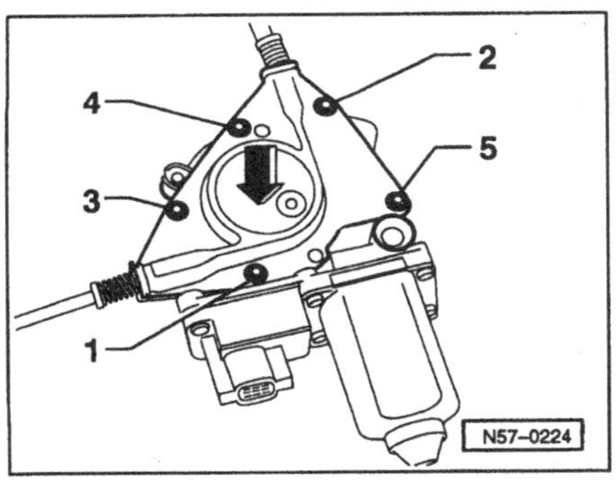

– Insert window regulator in direction of arrow with window regulator motor and tighten all bolts.

– Tighten bolts in specified sequence
Tightening torque: 3.5 Nm (2.6 in. lb)

Note:

Before removing window regulator, conduct functional check.

Trim, removing and installing

Removing/installing front door trim with power window regulator

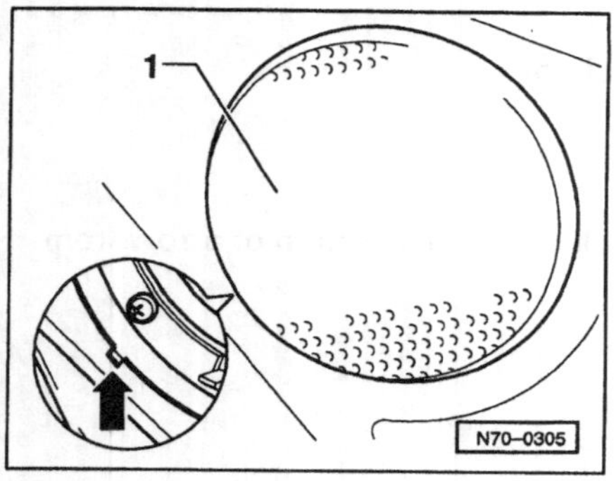

– Carefully remove speaker cover.

Note:

When installing, watch for opening in door trim -arrow- and corresponding tab on speaker cover.

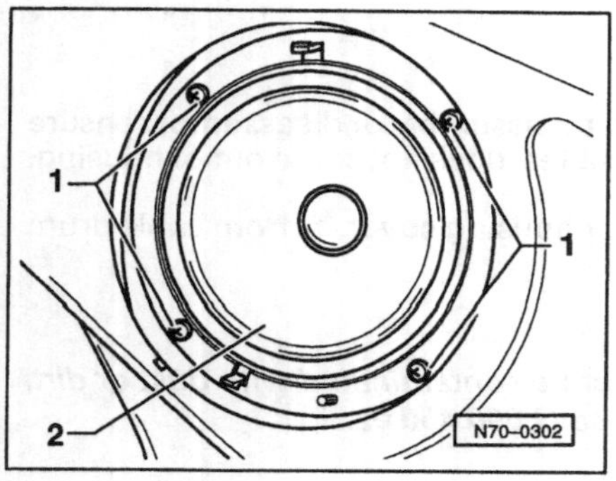

– Remove Phillips screws -1- from speaker-2-.

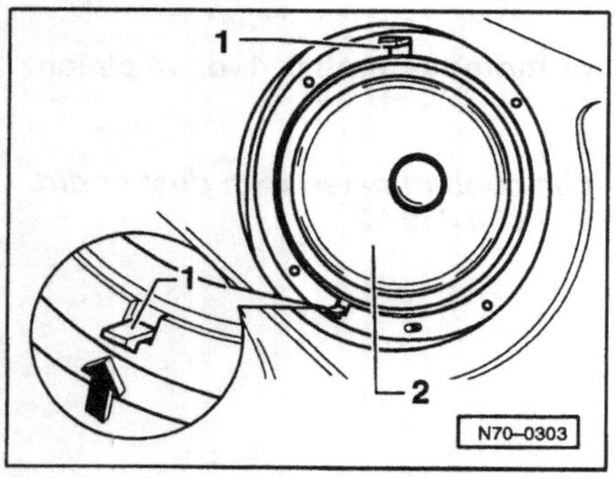

– Unhook tabs 1 -arrow- and remove speaker -2- from door trim.

– Remove speaker connector.

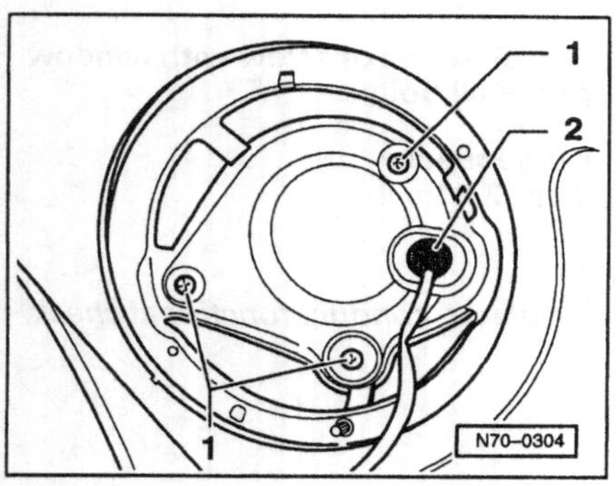

– Remove phillips screws -1-.

– Press speaker wire grommet -2- in door interior.

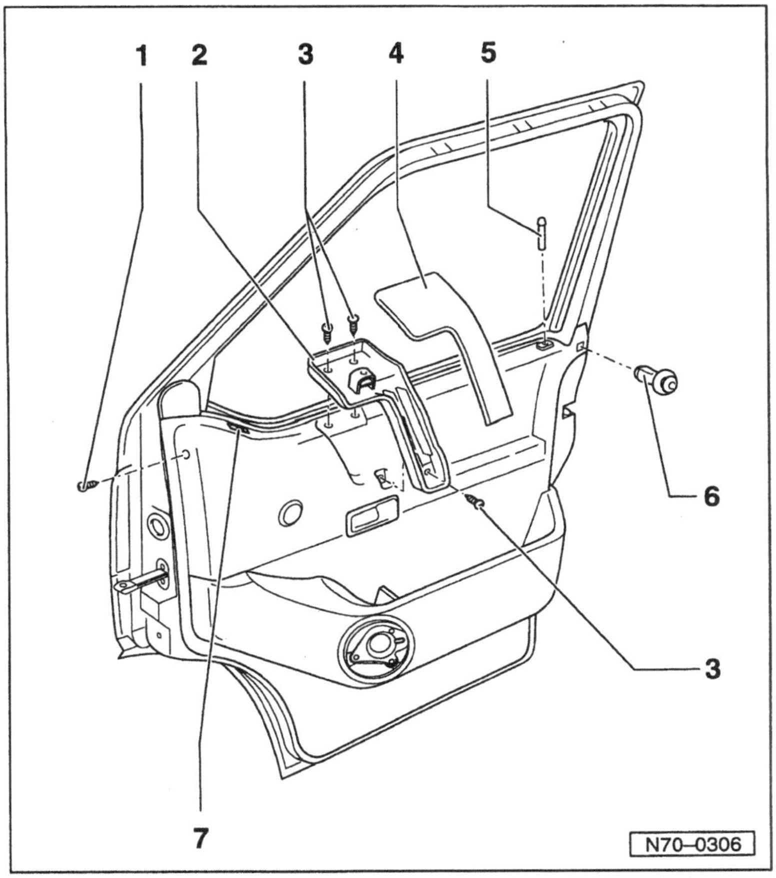

– Lower window.

– Carefully pry out window regulator switch -7- and disconnect connector.

– Remove lock button -5-.

– Remove inner part of clip -6- with socket wrench and remove clip from hole.

– Pry out trim -4-, remove oval head screw -3- and base/closing handle -1-.

– Remove Phillips screw -1-.

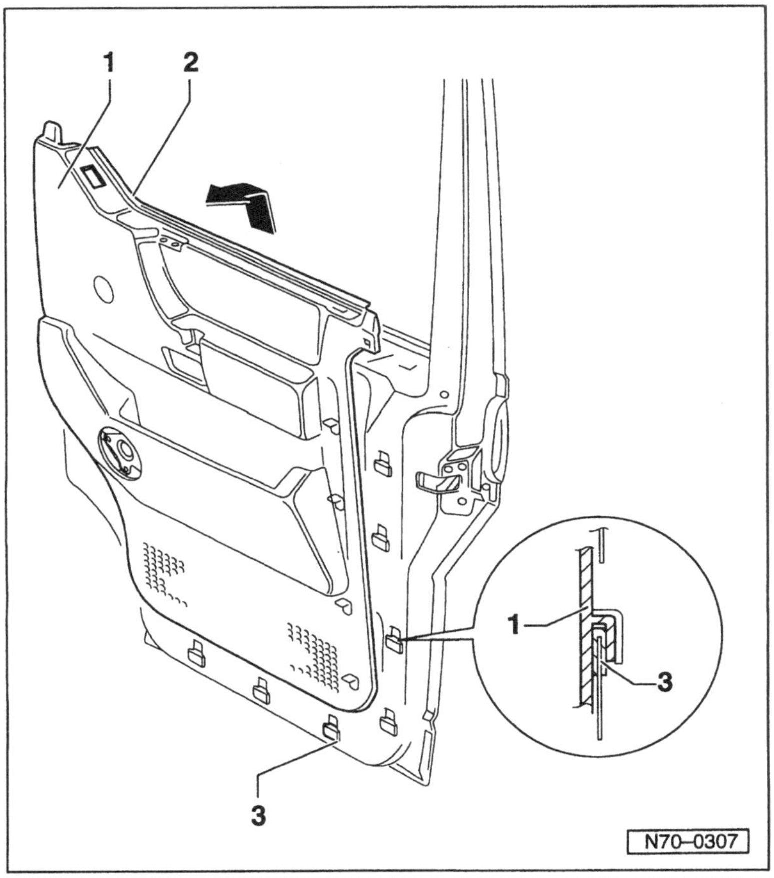

– Lift out door trim -1- together with inner window channel seal -2- in direction of arrow from mount holes.

– Guide speaker wires and window regulator/mirror adjustment switch through trim.

Note:

Before installing trim, insert clip -3- in mount holes. Guide speaker and switch wires for window regulator/ mirror adjustment through trim.

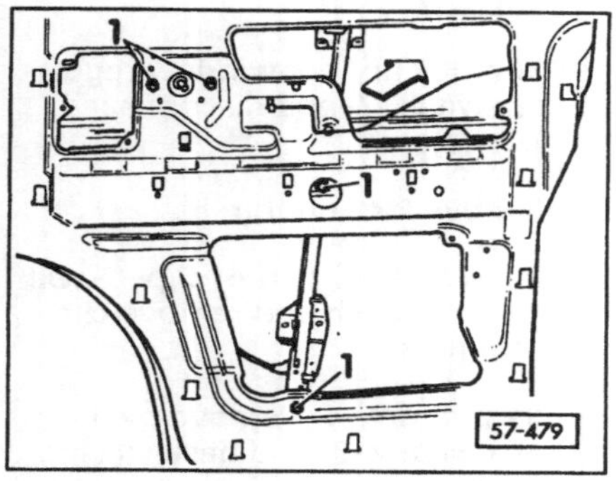

Window regulator, removing

– Remove front door trim
⇒ Page Q 70-81

– Remove door glass from window regulator.

– Slide door window upward and secure (e.g. with adhesive tape).

Vehicles with manual windows

◄ – Remove Hex bolts -1-.

– Carefully remove window regulator together with crank mechanism through upper opening -arrow-.

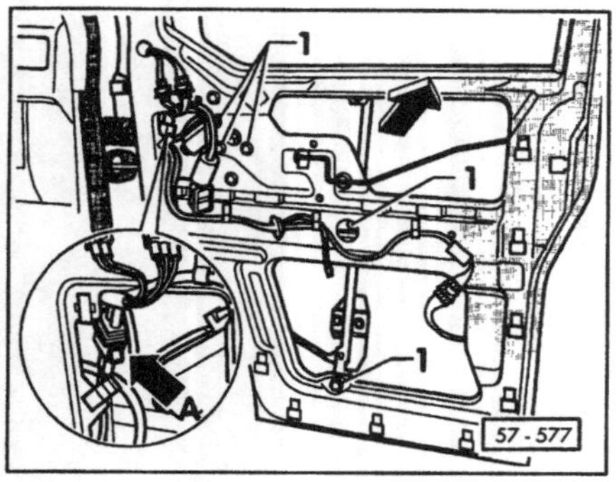

Vehicles with power windows

◄ – Disconnect connector -arrow A- of Window Motor -V14 and/or V15-.

– Remove hex bolts -1-.

– Remove window regulator with motor through upper opening -arrow-.

– Tightening torque:

– Hex bolts -1- = 8.0 Nm (71 in. lb)

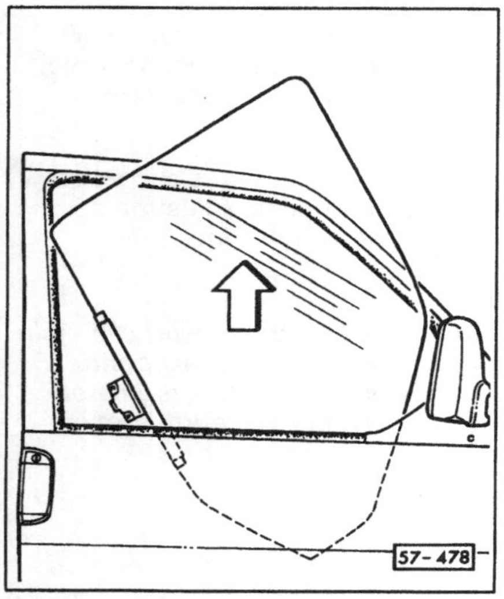

Door glass, removing/installing

– Remove front door trim
⇒ Page Q 70-81

– Open window completely and remove from regulator.

◄ – Tilt door glass toward front and lift out through window slot.

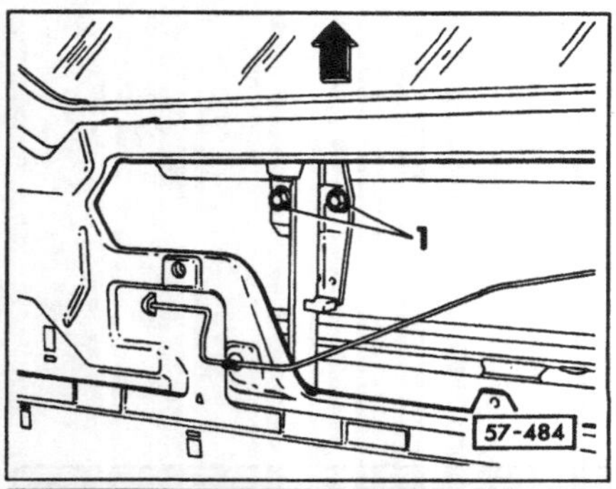

– Position window in door.

– Loosely screw regulator rail to regulator.

◄ – Close window completely with crank mechanism and tighten hex bolts -1- to 70 Nm (52 ft lb).

Tightening torque:

Hex bolts -1- = 7.0 Nm (62 in. lb)

Outside door handle, removing

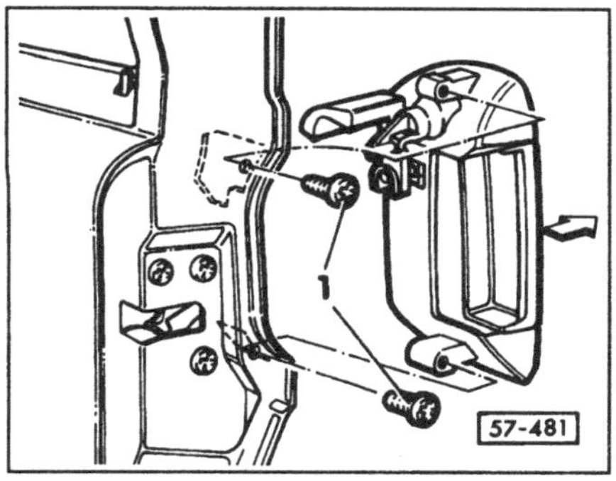

◄ – Remove screws -1-, remove outside door handle in direction of arrow.

Tightening torque:

Screws -1- = 8.0 Nm (71 in. lb)

Door lock, removing

– Remove front door trim ⇒ Page Q 70-81

– Remove outside door handle ⇒ Page Q 57-21.

– Close window completely.

– Unclip pull rod from retaining clips on inner door panel.

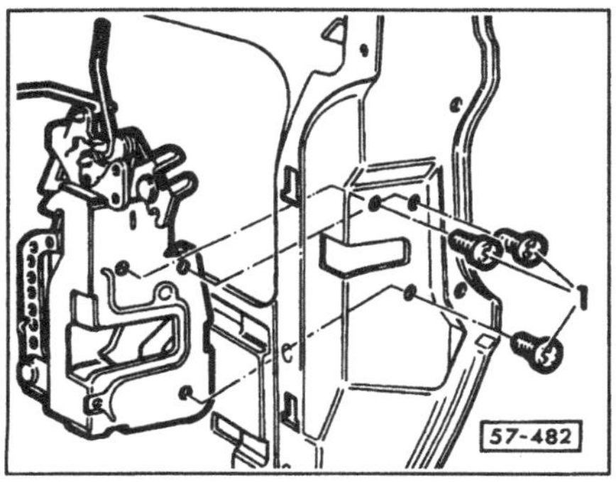

◄ – Remove screws -1-, remove door lock together with pull rod and lock rod.

Tightening torque:

Hex bolts -1- = 8.0 Nm (71 in. lb)

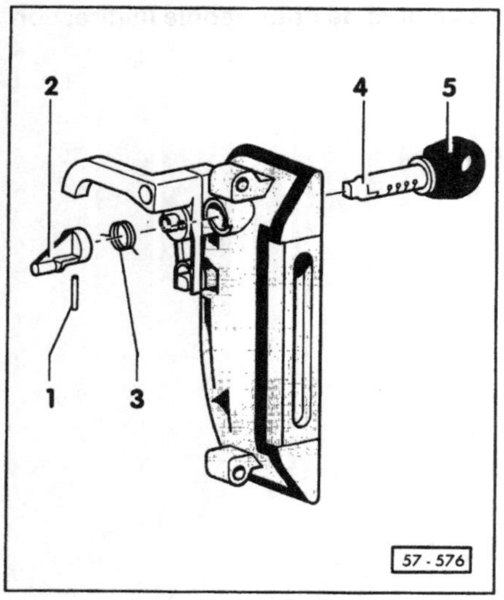

Lock cylinder, removing (double side door) 01.96 ➤

– Remove outside door handle
⇒ Page Q 57-21.

◄ – Drive out lock pin -1-.

– Remove casting -2- from four-sided shaft.

– Remove spring -3-.

– Remove lock cylinder -4- with key inserted -5-.

Front door, removing 01.96 ➤

Vehicles with central locking, power windows, heated outside mirrors and/or door speakers

♦ Door trim, removing
⇒ Page Q 70-81

– Move electrical wires aside, disconnect connections and remove harness from door.

– Remove nut -1-, take out bolt

◄ – Remove hex bolts -2- from hinges.

– Remove door.

Front door, installing/adjusting 01.96 ➤

Adjusting door to fender, to A-pillar / to B-pillar

♦ Door, removing
⇒ Page Q 57-5

◄ – Loosen hex bolts -3-, slide hinges and tighten hex bolts again

Tightening torques:

Nut -1- = 6.5 Nm (57 in. lb)

Hex bolt -3- = 55 Nm (41 ft lb)

– Reinstall door.

– Install screws -2- in upper and lower hinge.

– Adjust door flush to fender, to A-pillar and to B-pillar.

– Adjust door at hinge, flush to fender.

♦ Front door is correctly adjusted with uniform distance in closed position, does not extend too far inward or outward and contours -arrows- meet

◄ – Tighten hex bolts -2-, tighten nut -1- after bolt installation.

Tightening torques:

Nut -1- = 6.5 Nm (57 in. lb)

Hex bolt -2- = 55 Nm (40 ft lb)

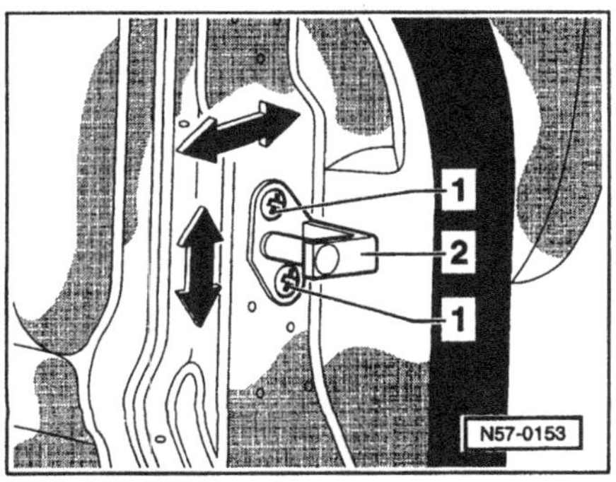

Lock striker, adjusting

◄ – After loosening hex bolts -1-, lock striker -2- can be moved in both directions -arrow-.

Tightening torques: Hex bolts -1- = 19 Nm (14 ft lb)

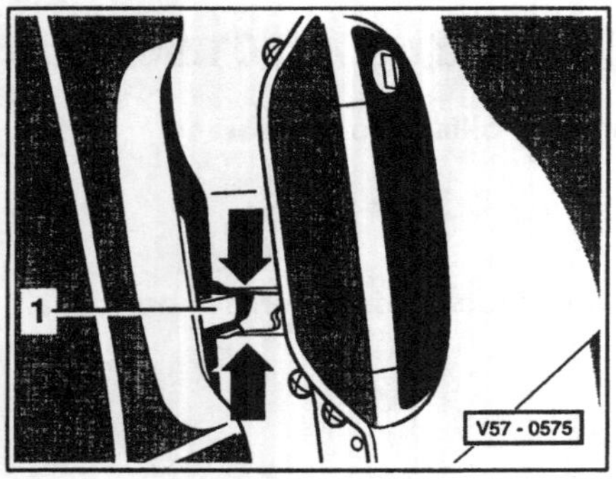

– Adjust striker -1- so recess on latch engages with catch in center of striker -arrows- (do not raise or press down on door).

- Door must not have free play when closed.

Sliding door, components ➤ 12.95

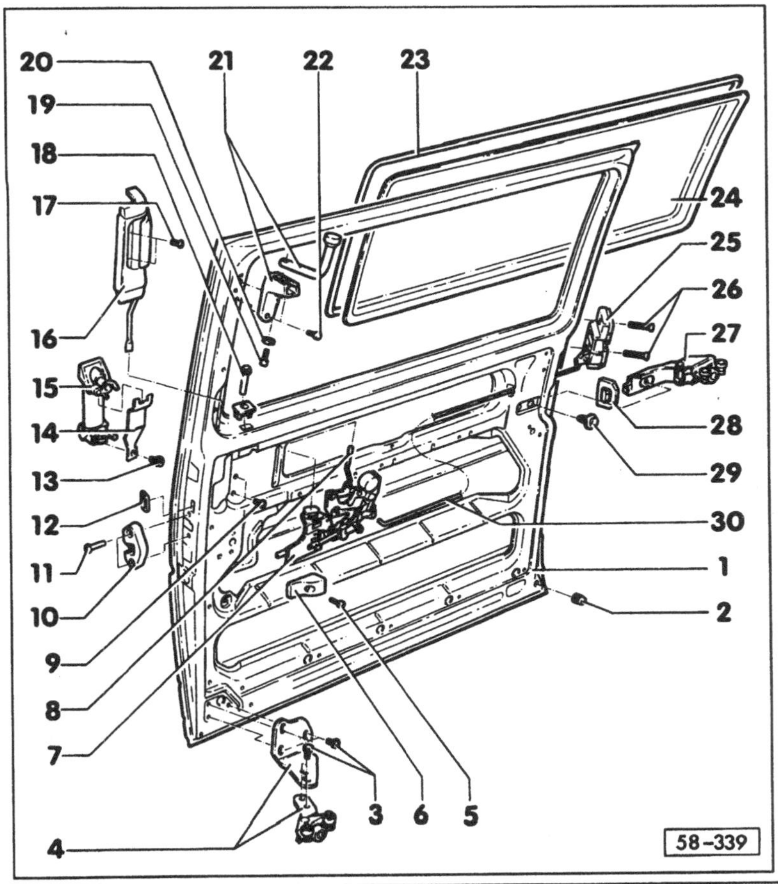

1 - Sliding door
- ♦ Removing ⇒ Page Q 58-5
- ♦ Adjusting ⇒ Page Q 58-8

2 - Rubber buffer

3 - Screw
- ♦ 19 Nm (14 ft lb)

4 - Lower roller guide assembly

5 - Screw
- ♦ 19 Nm (14 ft lb)

6 - Center stop

7 - Remote control

8 - Lock rod

9 - Screw
- ♦ 8.0 Nm (71 in. lb)

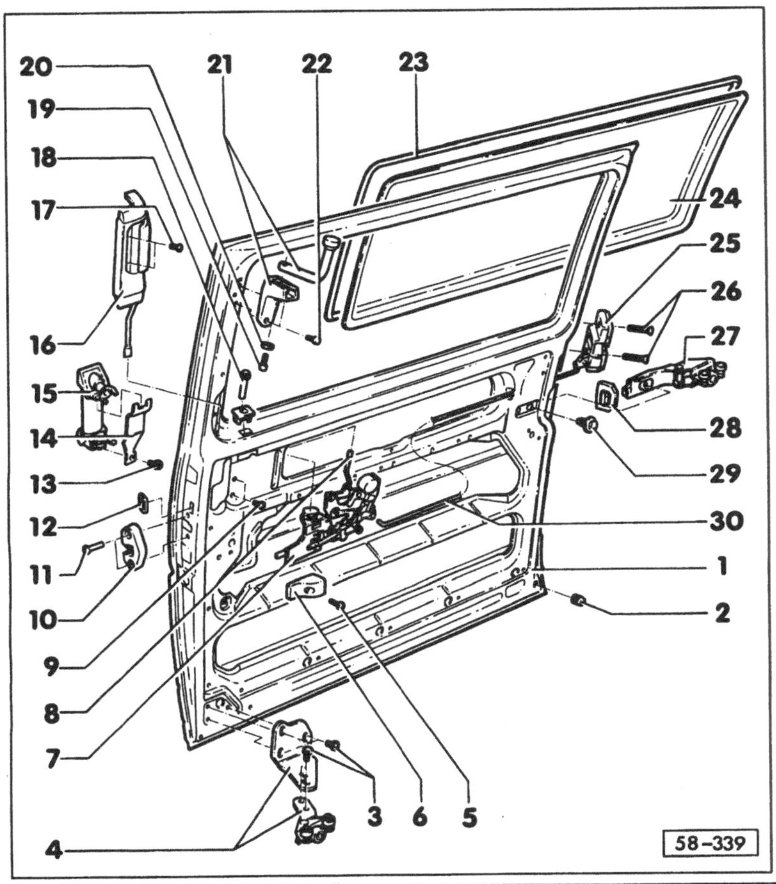

10 - Catch plate

11 - Phillips screw
- ♦ 19 Nm (14 ft lb)

12 - Trim plate

13 - Screw
- ♦ 19 Nm (14 ft lb)

14 - Tension spring

15 - Outside door handle
- ♦ Removing ⇒ Page Q 58-14

16 - Inside door handle

17 - Allen screw
- ♦ 8.0 Nm (71 in. lb)

18 - Lock knob
- ♦ Screwed into pull rod

19 - Screw
- ♦ 8.0 Nm (71 in. lb)

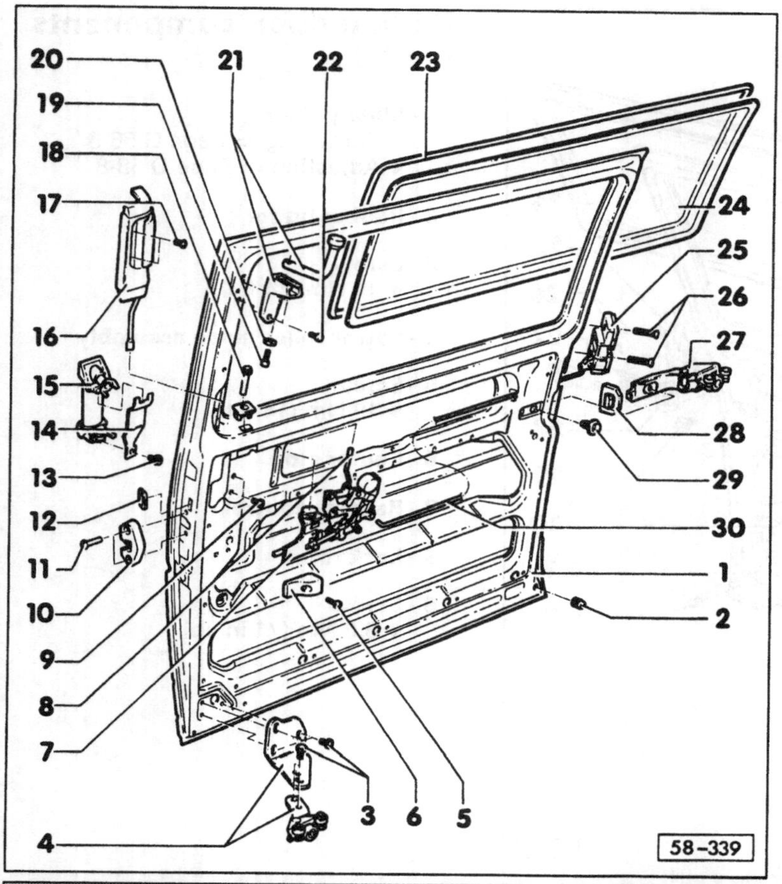

20 - **Washer**

21 - **Upper roller guide assembly**

22 - **Screw**
- 8.0 Nm (71 in. lb)

23 - **Window frames**
- Destroyed during removal

24 - **Sliding door window**
- Flush-bonded window ⇒ Q 64-7
- Sliding window, repairing ⇒ Page Q 64-27

25 - **Rotary switch**

26 - **Phillips screw**
- 19 Nm (14 ft lb)

27 - **Hinge link**

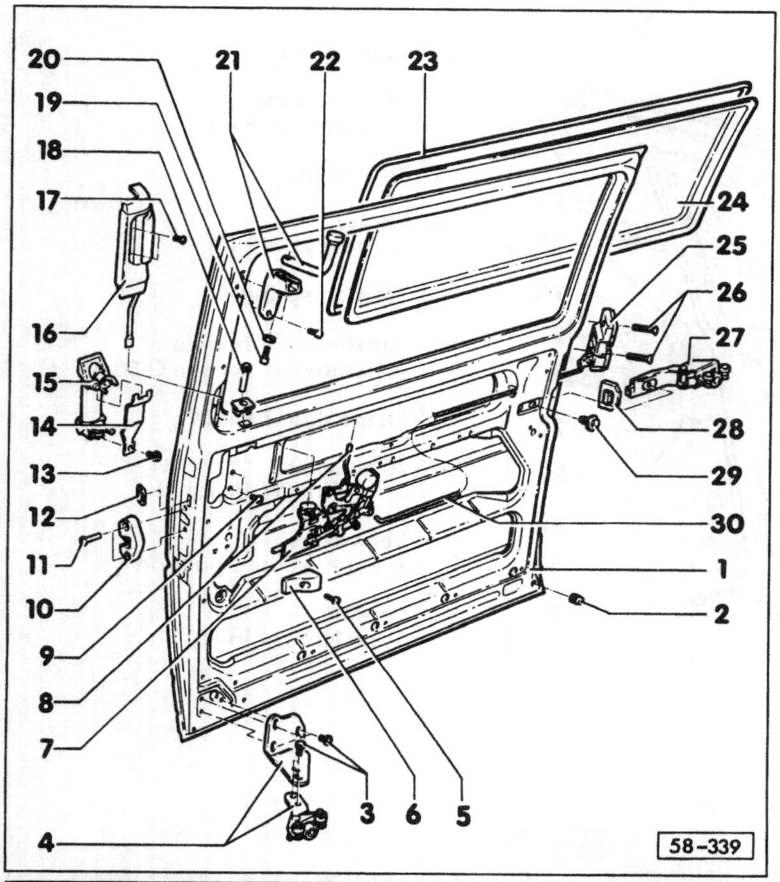

28 - **Seal**

29 - **Bolt w/washer**
- 54 Nm (40 ft lb)

30 - **Pull rod**
- Connects remote lock to rotary catch

Sliding door, removing ➤ 12.95

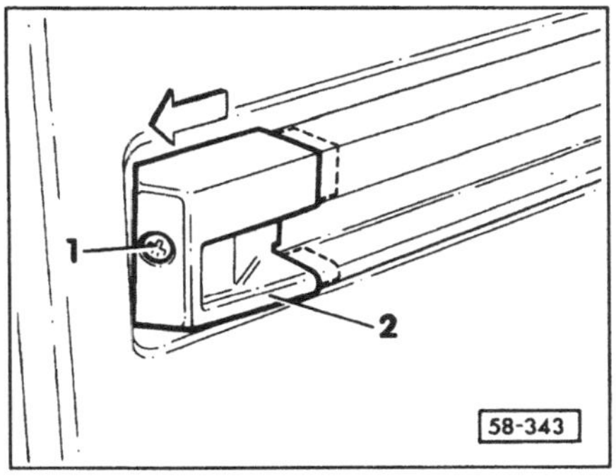

◄ – Remove Phillips screw -1-, pull end piece -2- off guide rail.

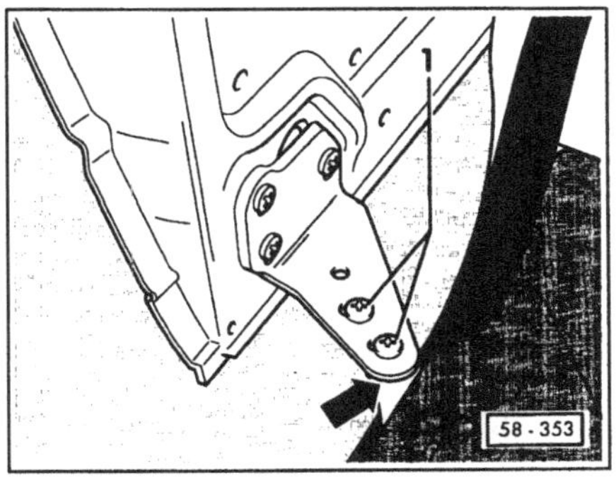

◄ – Mark position of lower roller guide.

– Remove Phillips screws -1-.

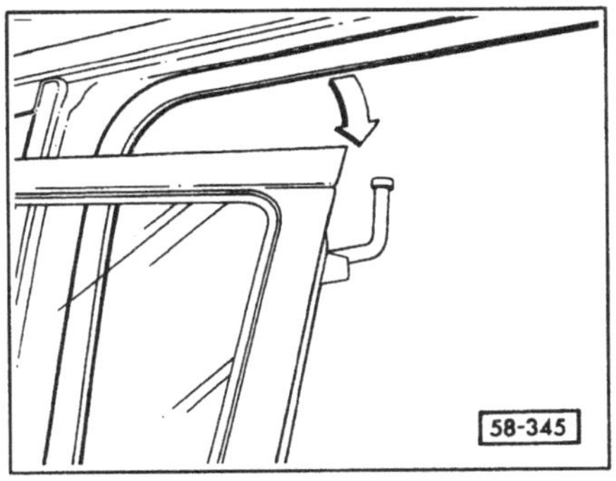

◄ – Lower sliding door at front until upper roller guide can be removed from guide rail.

◄ – With second technician, guide sliding door back -arrow- until hinge link -1- is out of guide rail -2-.

Door stop, removing

– If applicable, remove door step insert and Footwell Light with Delay Switch -W23- ⇒ Page Q 58-10.

– Place sliding door in half-opened position.

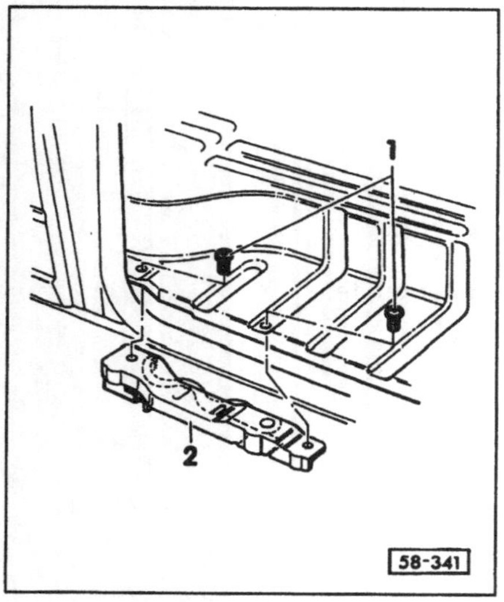

◄ – Take out Phillips screws -1- and pull door stop -2- downward out of the guide channel.

Tightening torque:

Phillips screws -1- = 19 Nm (14 ft lb)

Sliding door, adjusting

– Remove sliding door molded trim ⇒ Page Q 70-81 and/or remove sliding door trim ⇒ Page Q 70-85.

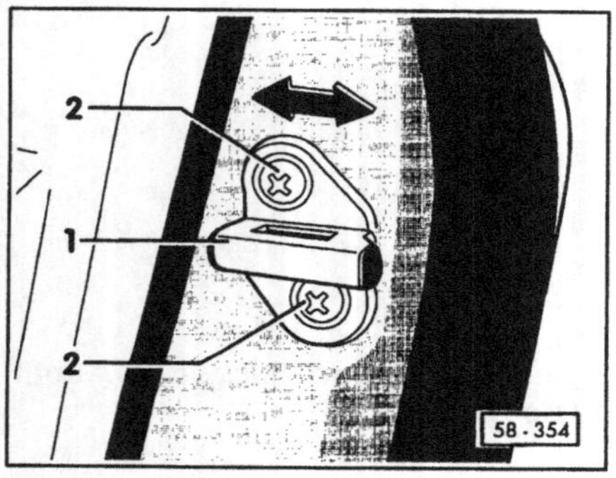

◄ – Loosen striker -1- by loosening Phillips screws -2-.

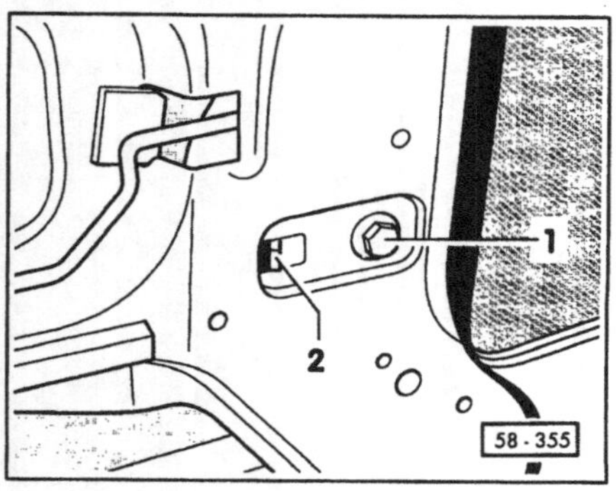

◄ – Close sliding door and loosen hex bolt -1- from inside vehicle.

– Have a second technician lift sliding door, push hinge fitting with hex bolts -1- and guide tab -2- as far forward as possible.

– Tighten hex bolt -1-:

Tightening torque:

Hex bolt -1- = 58 Nm (43 ft lb)

Lower roller guide, adjusting

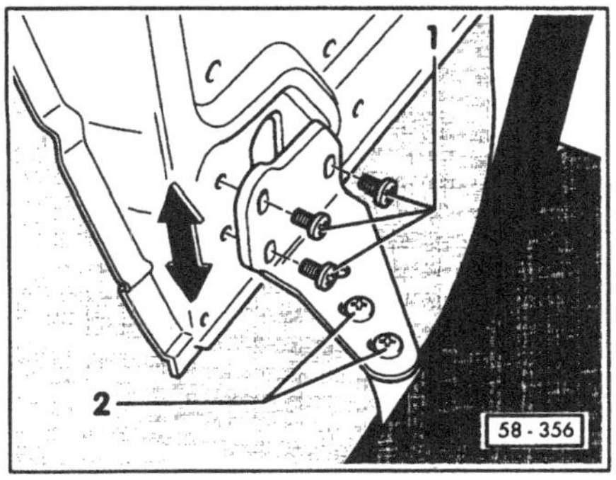

- Front Height of sliding door incorrect
- Sliding door does not match B-pillar outer contour

◄ – If sliding door height is incorrect, loosen Phillips screws -1-, align sliding door as required -arrow-.

Tightening torque:

Phillips screw -1- = 19 Nm (14 ft lb)

If sliding door does not match B-pillar outer contour adjust with Phillips screws -2- as follows:

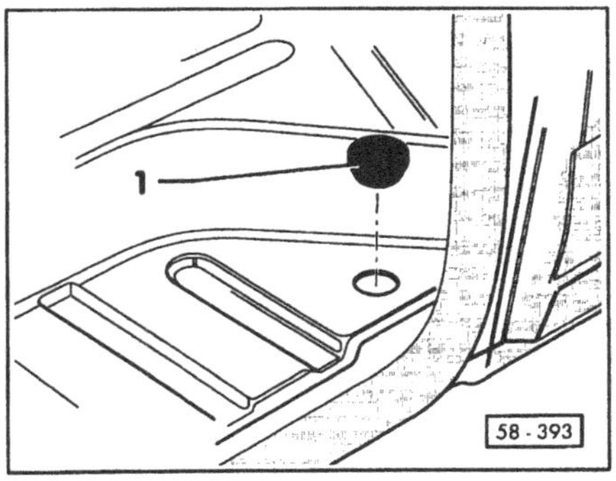

Vehicles without door step Insert

◄ – Press plastic cap -1- out of step from below.

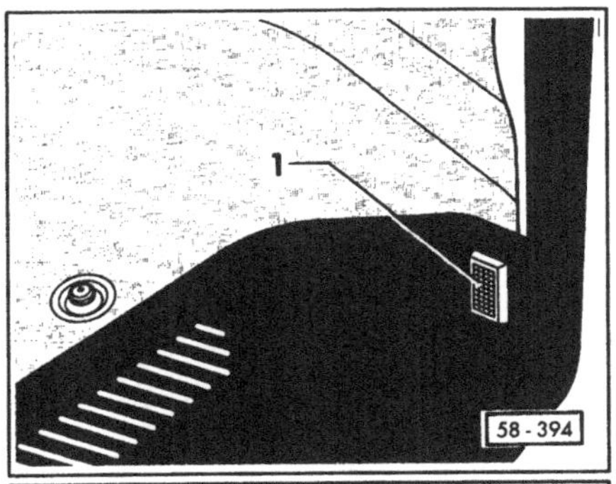

Vehicles with Footwell Light with Delay Switch -W23-

◄ – If applicable, pry out Footwell Light with Delay Switch -W23- -1- and disconnect connectors.

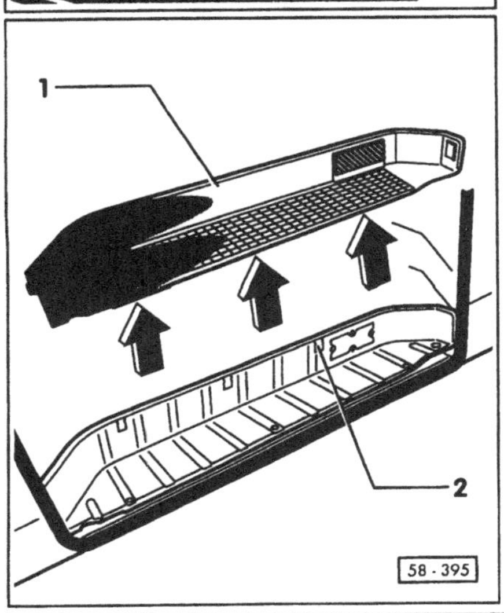

◄ – Unclip door step insert -1- from outside -arrows-, installation position, lift upward out of holding clips -2-.

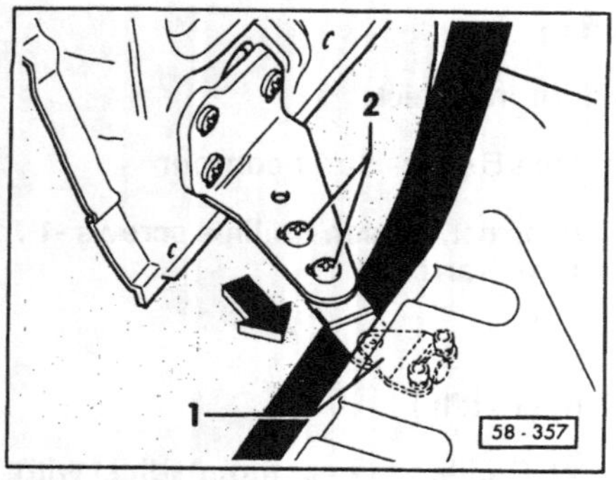

- Loosen Phillips screws -2-, press sliding door against roller guide -1- -arrow-, tighten only screw -2- on door side.
- Close sliding door carefully.
- Loosen screw -2- carefully while second technician outside indicates when sliding door is flush with B-pillar.
- Tighten both screws -2-.

Tightening torque:

Phillips screws -2- =19 Nm (14 ft lb)

Upper roller guide, adjusting

- Top front of sliding door does not match outer B-pillar contour.

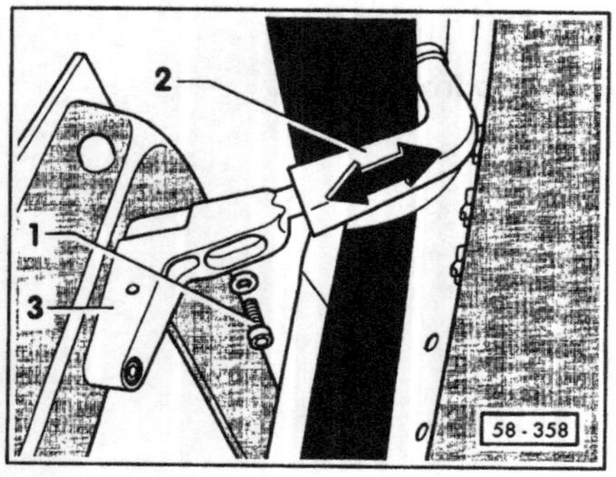

- Loosen hex socket bolt -1- and slide upper roller guide -2- on bracket -3- -arrows- until top front of sliding door matches outer B-pillar contour.

Tightening torque:

Hex socket bolt -1- = 8.0 Nm (71 in. lb)

Striker, adjusting

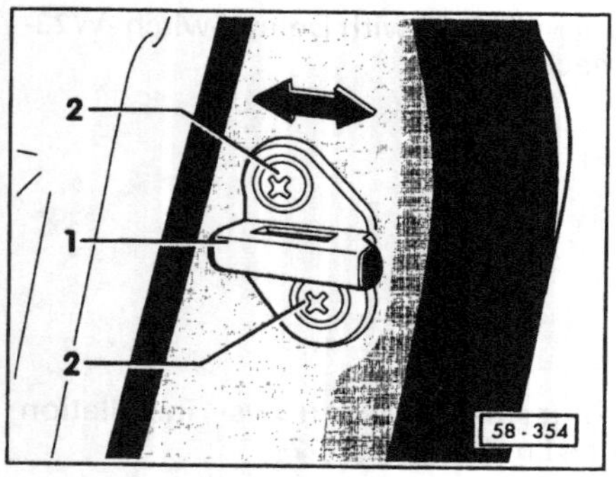

- Striker -1- can move horizontally -arrow- after loosening Phillips screw -2-.

Tightening torque:

Phillips screws -2- = 19 Nm (14 ft lb)

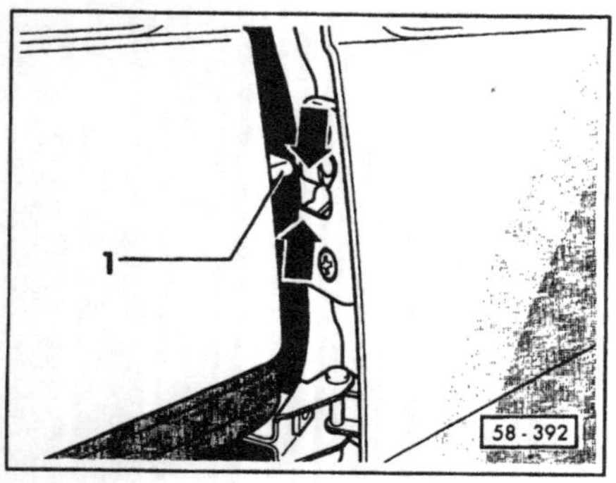

- Adjust striker -1- so recess on remote control with turn clip is concentric -arrows- to striker (door must not be raised or pressed down).
- Sliding door must have no free play when closed.

Remote lock, removing

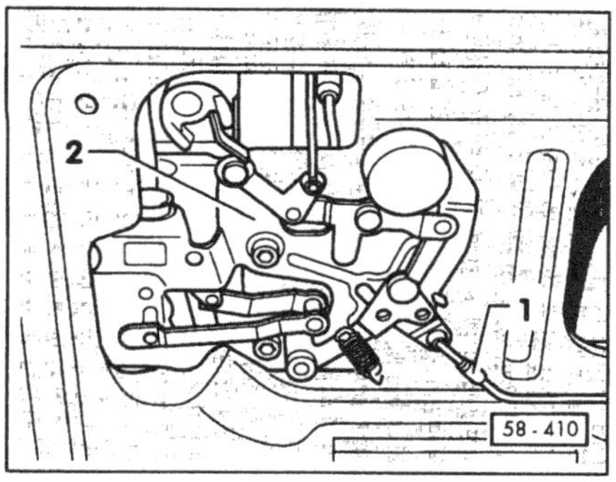

- Remove sliding door molded trim ⇒ Page Q 70-81 and/or remove sliding door trim ⇒ Page Q 70-85.
- Disconnect pull rod -1- from remote lock -2-.

- Remove remote lock -1- by removing Phillips screws and disconnecting pull rod -2- at remote control.

  Tightening torque:

  Phillips screw = 19 Nm (14 ft lb)

Outside door handle, removing

- Remove sliding door molded trim ⇒ Page Q 70-81 and/or remove sliding door trim ⇒ Page Q 70-85.

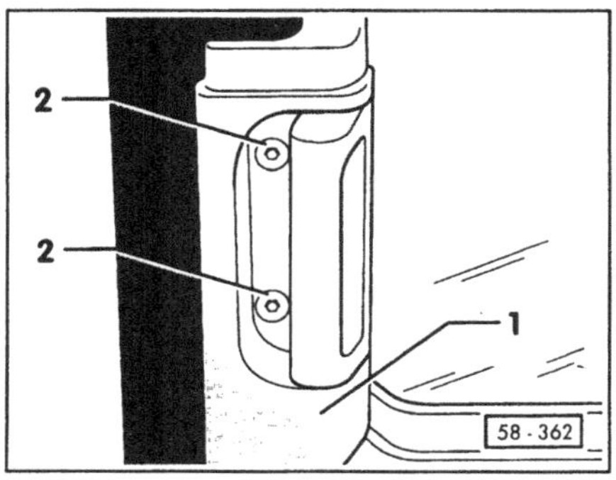

- Loosen inner mechanism by removing hex socket bolts -2-, remove by pulling upward.

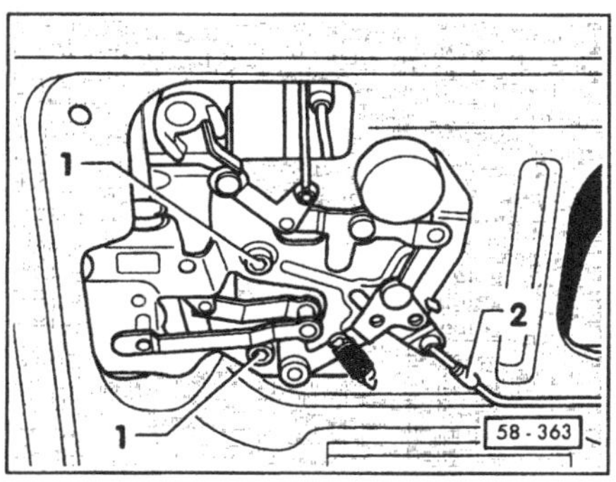

- Remove hex socket bolts and disconnect pull rod -2-. Push remote control to one side.

  Tightening torque:

  Hex socket bolts -1- = 8 Nm (71 in. lb)

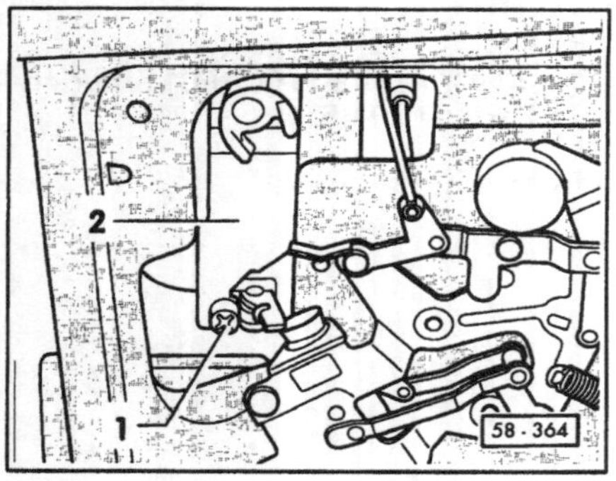

- Remove Phillips screw -1- and disconnect tension spring -2-.

  Tightening torque:

  Phillips screw -1- = 19 Nm (14 ft lb)

Lock cylinder, removing (sliding door)

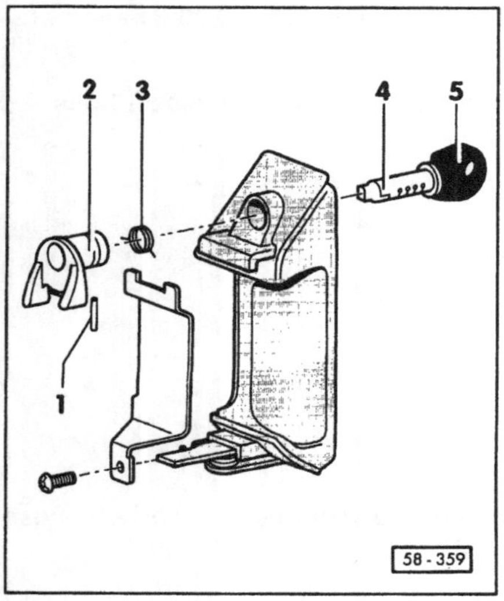

- Remove sliding door molded trim ⇒ Page Q 70-81 and/or remove sliding door trim ⇒ Page Q 70-85.
- Remove outer door handle ⇒ Page Q 58-14.
- Remove spring pin -1-.
- Remove casting -2- off four-sided shaft.
- Remove outer spring -3-.
- Remove lock cylinder -4- with key -5- inserted.

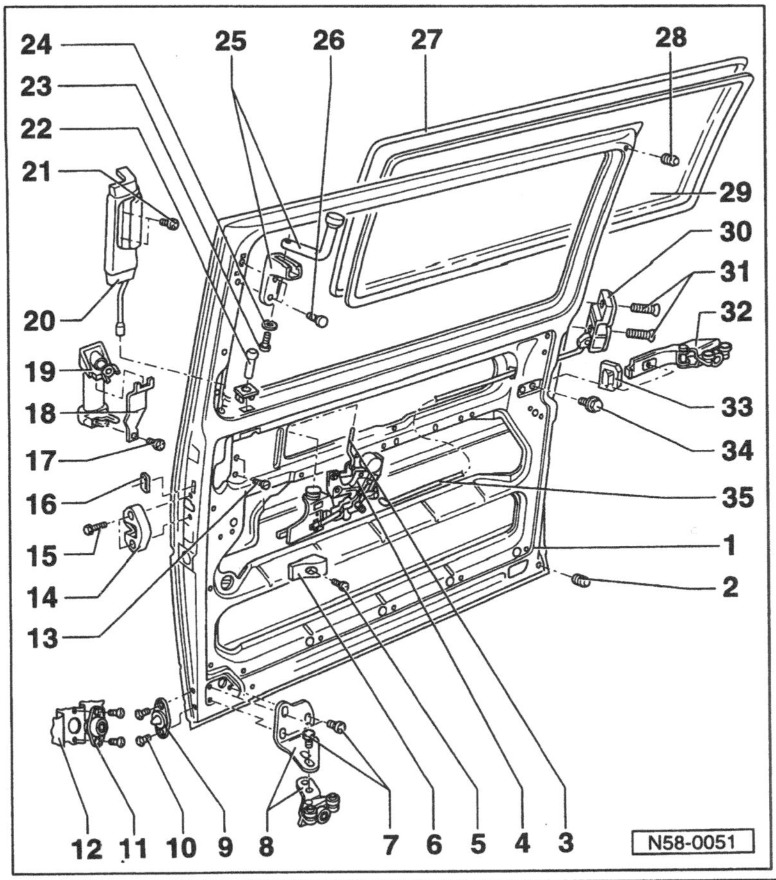

Sliding door, components 01.96 ➤

1 - Sliding door
- ♦ Removing ⇒ page Q 58-5
- ♦ Adjusting ⇒ page Q 58-21

2 - Rubber bumper

3 - Lock rod

4 - Remote control

5 - Phillips screw
- ♦ 19 Nm (14 ft lb)

6 - Center stop

7 - Oval head screw
- ♦ 19 Nm (14 ft lb)

8 - Lower roller guide

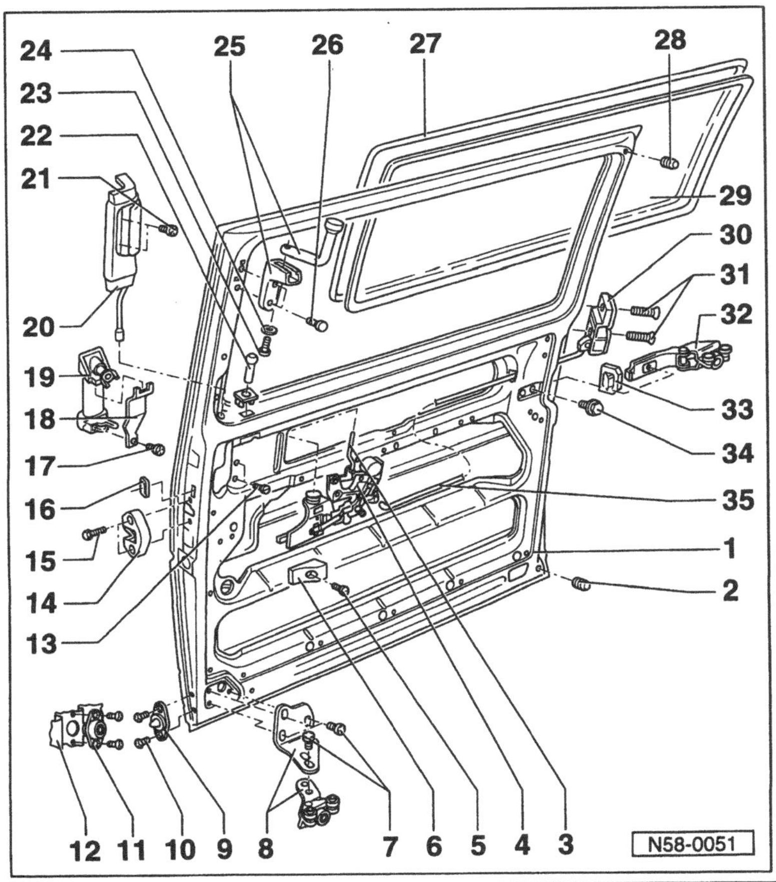

9 - Latch pin
- ♦ Adjusting ⇒ page Q 58-26

10 - Phillips screw
- ♦ 19 Nm (14 ft lb)

11 - Catch plate

12 - B-pillar

13 - Hex recess bolt
- ♦ 8 Nm (6 ft lb)

14 - Catch plate

15 - Phillips screw
- ♦ 19 Nm (14 ft lb)

16 - Anchor plate

17 - Phillips screw
- ♦ 19 Nm (14 ft lb)

18 - Tension spring

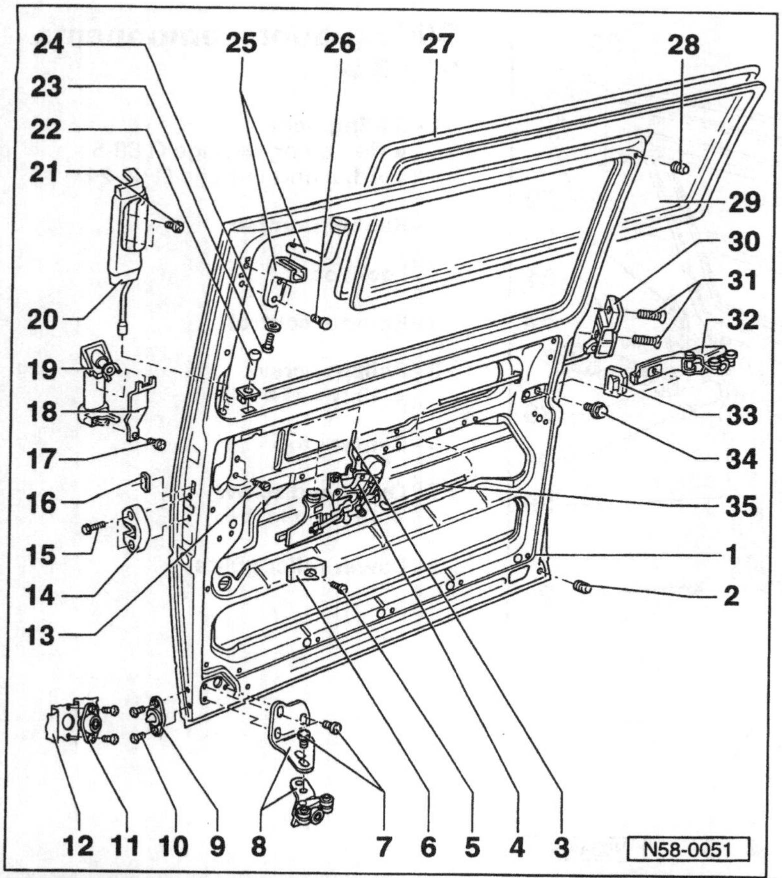

19 - Outer door handle
- Removing ⇒ page Q 58-14

20 - Inside door handle

21 - Hex recess bolt
- 8 Nm (71 in. lb)

22 - Lock knob
- Screw on to rod

23 - Hex recess bolt
- 12 Nm (9 ft lb)

24 - Washer

25 - Upper roller guide

26 - Hex recess bolt
- 12 Nm (9 ft lb)

27 - Window molding
- Destroyed during removal

28 - Rubber bumper

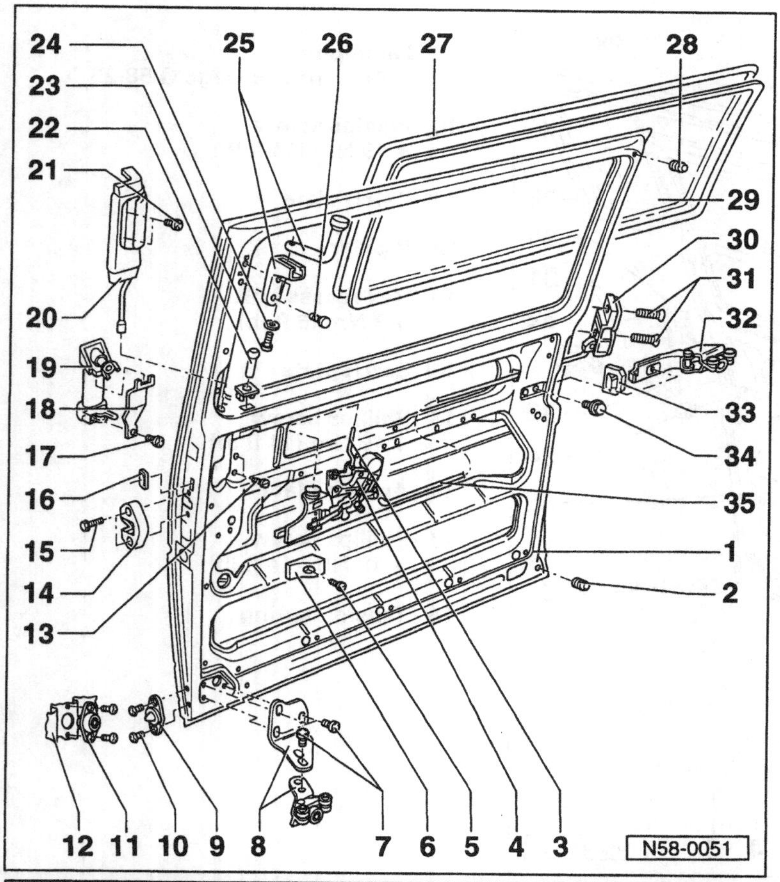

29 - Sliding door window
- Flush bonded window ⇒ page Q 64-20
- Sliding window, repairing ⇒ page Q 64-30

30 - Remote locking
- Removing⇒ page Q 58-13

31 - Phillips screw
- 19 Nm (14 ft lb)

32 - Hinge

33 - Seal

34 - Hex screw and washer assembly
- 54 Nm (40 ft lb)

35 - Latch actuator rod
- Connection from remote control to central lock

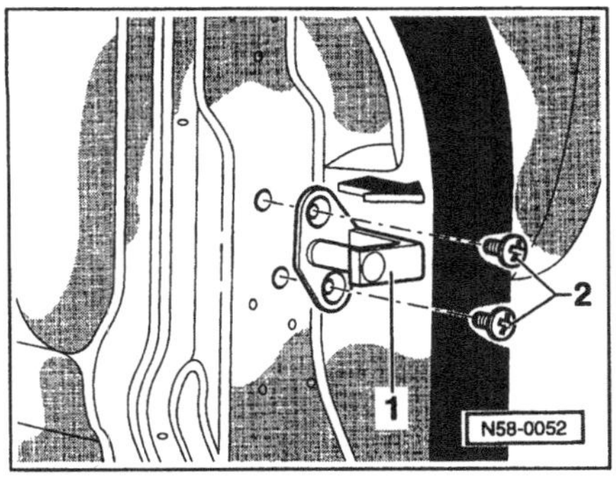

Sliding door, adjusting

– Remove sliding door molded trim ⇒ page Q 70-84 and/or remove lower door trim ⇒ Page Q 70-85.

Hinge, adjusting

◄ – Loosen latch striker -1- by loosening phillips screws -2-.

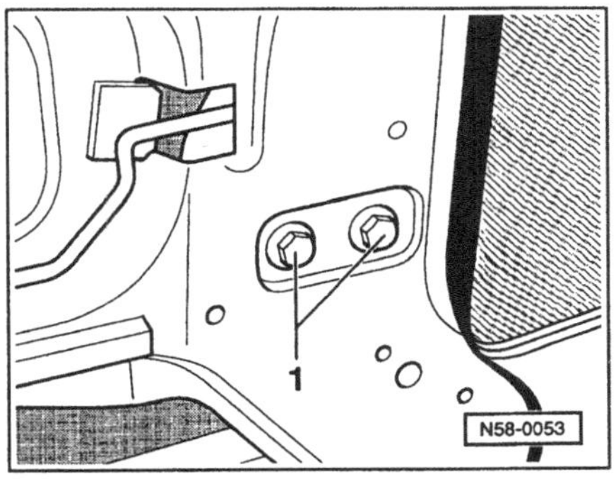

◄ – Close sliding door, loosen hex bolts -1- from inside vehicle

– Have second technician lift sliding door and push hinge as far forward as possible.

– Tighten hex bolts -1-.

Tightening torque: hex bolt -1- = 58 Nm (43 ft lb)

Lower roller guide, adjusting

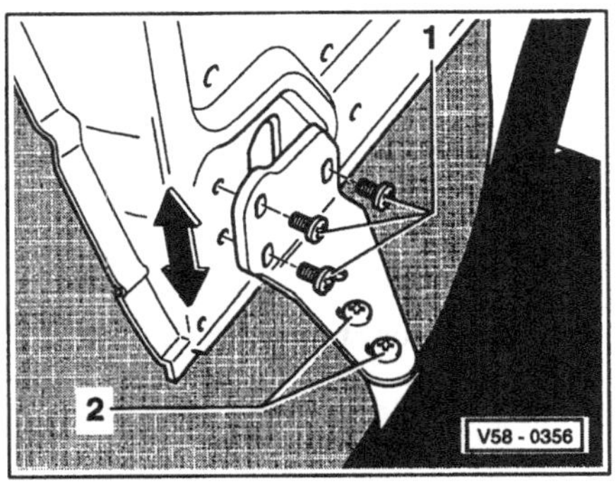

♦ Sliding door height not correct

♦ Sliding door outer contour does not match B-pillar

◄ – To adjust height, loosen Phillips screws -1- and align accordingly -arrow-.

Tightening torque: Phillips screw -1- = 19 Nm (14 ft lb)

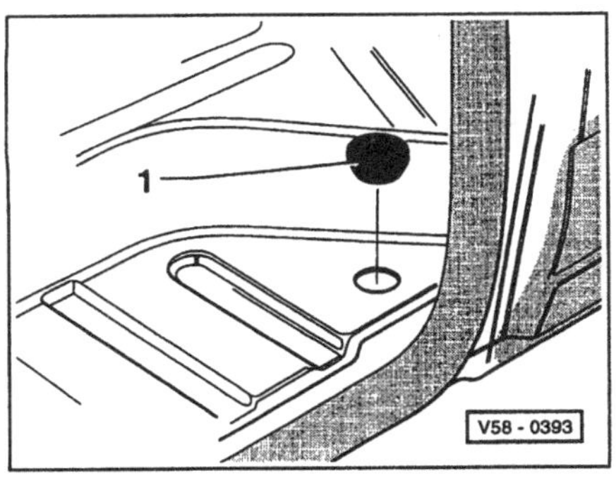

If sliding door contour does not match B-pillar, make adjustments with Phillips screws -2- as follows:

Without step trim

◄ – Press plastic cap -1- out of step from below.

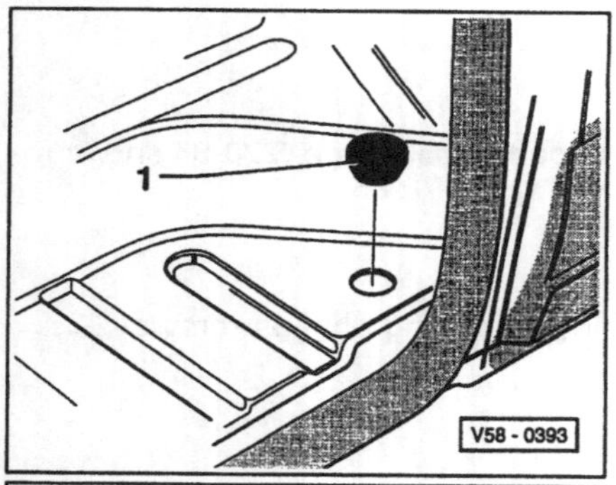

With step trim

◄ – Pry step light -W23- -1- out, if applicable, disconnect connector.

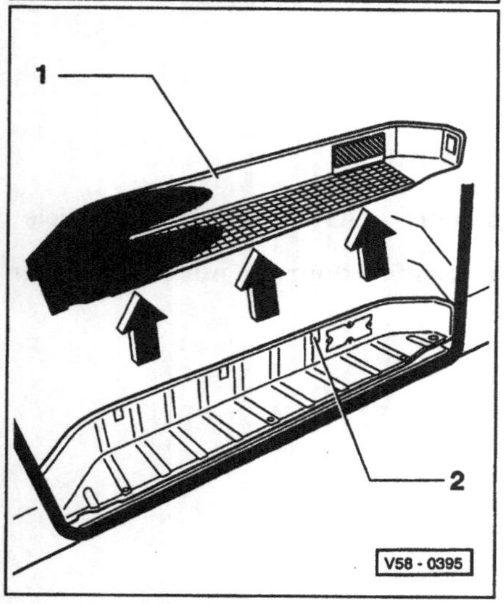

◄ – Unclip step trim -1- at outside -arrows-, lift up out of clips -2-.

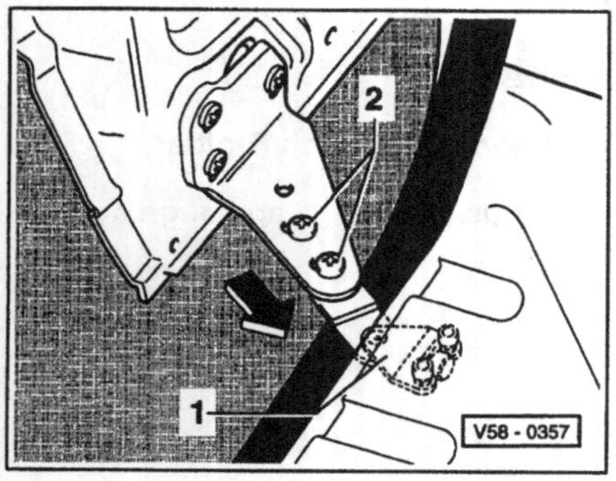

Continued, all vehicles

With help of second technician

– Loosen Phillips screws -2-, press sliding door against roller guide -1- -arrow-, tighten one screw -2- next to door.

– Carefully close door

– Loosen screw -2- enough to adjust contour with outer B-pillar, then tighten both screws -2-.

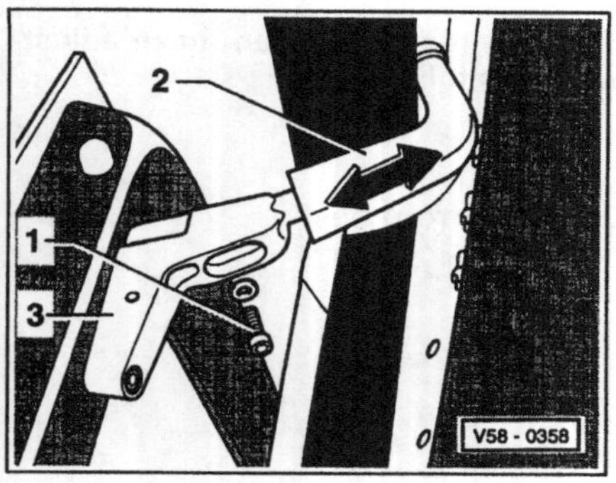

Upper roller guide, adjusting

♦ Sliding door top front does not match B-pillar contour

– Loosen hex recess bolt -1-, slide upper roller guide -2- on bracket -3- -arrows- until top front of sliding door matches outer B-pillar contour.

Tightening torque: Hex recess bolt -1- = 8 Nm (71 in. lb)

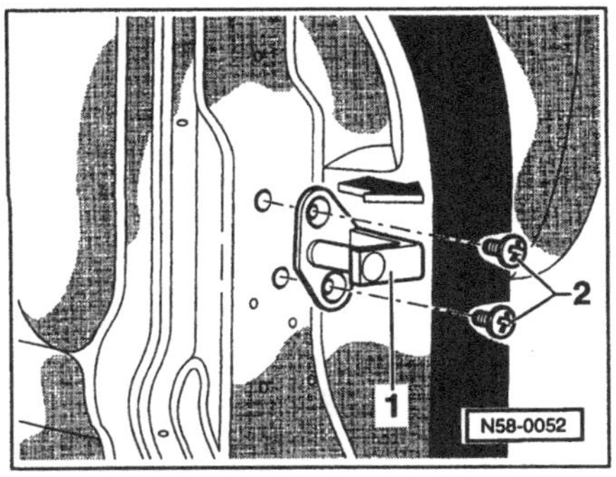

Latch striker, adjusting

◄ – Loosen Phillips screws -2-, striker can now slide horizontally -arrow-.

Tightening torque: Phillips screws -2- = 19 Nm (14 ft lb)

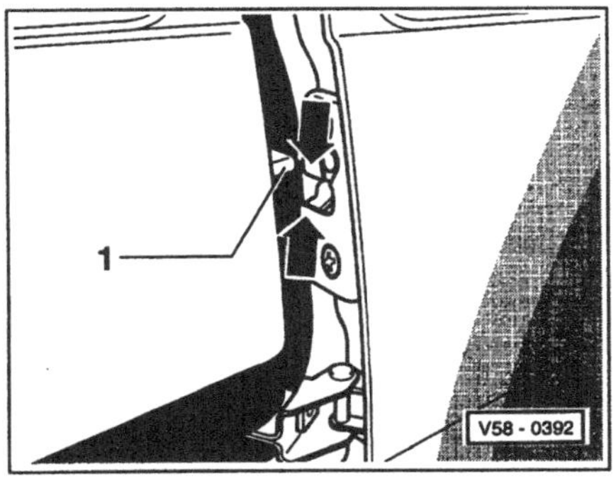

◄ – Adjust latch striker -1- so recess on central lock engages with catch in center of striker -arrow- (door must not be raised or pressed).

- Door must have no free play when closed.

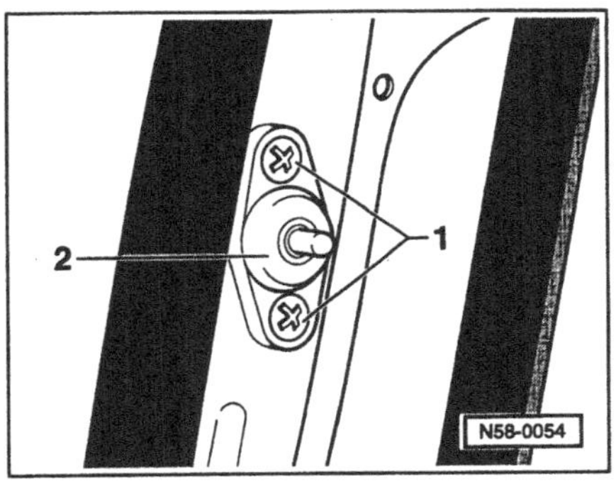

Latch pin, adjusting

♦ Latch pin must face inside of vehicle, when installing

◄ – Loosen screws -1- enough to move latch pin -2-.

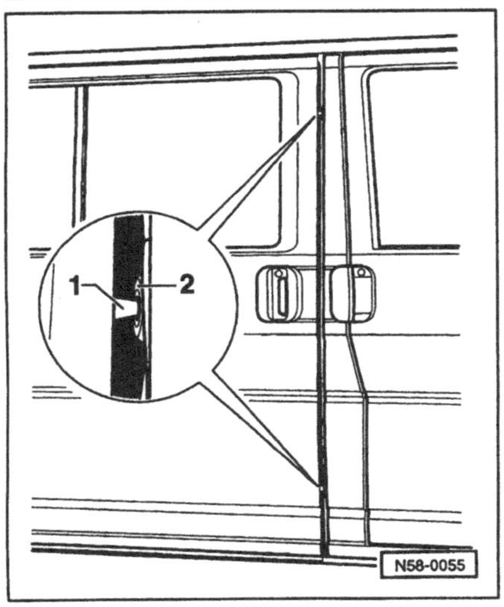

◄ – Close door slowly and check latch pin -1- position to catch plates -2-, latch pins must be parallel to catch plates.

– Close door completely to confirm adjustment.

– Open door and tighten screws.

Tightening torque: screws -1- = 19 Nm (14 ft lb)

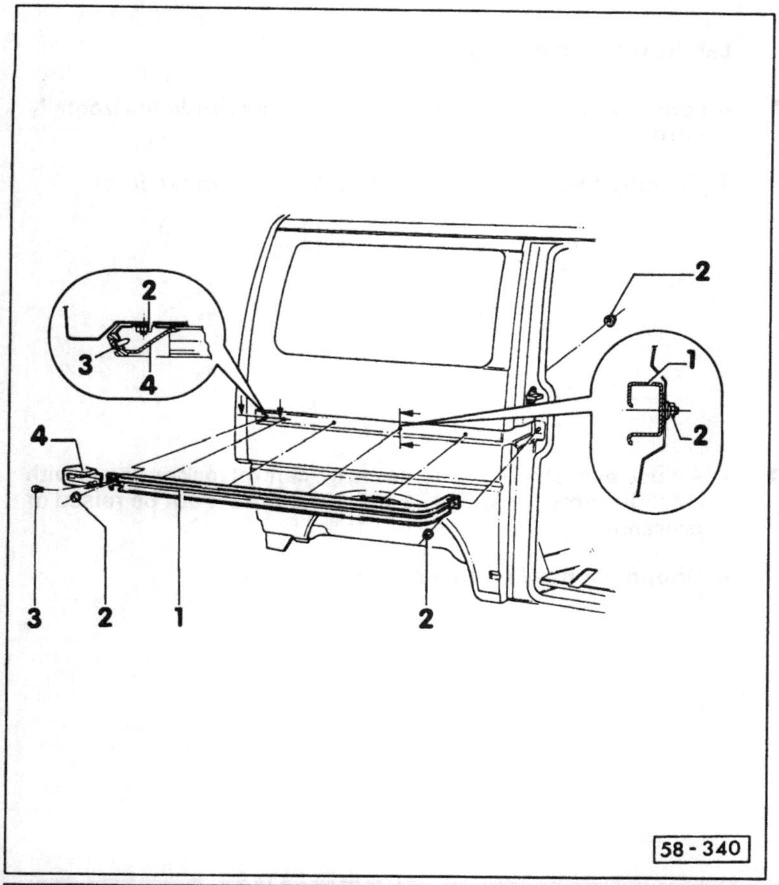

Guide rail, components

Note:

To remove guide rail: remove sliding door ⇒ Page Q 58-5, if necessary, molded trim on right rear side panel ⇒ Page Q 70-48 and/or ⇒ Page Q 70-61.

1 - **Guide rail**
- Removing: Remove end piece -4-, loosen hex nuts -2-

2 - **Hex nuts**
- 8.0 Nm (71 in. lb)

3 - **Phillips screws**

4 - **End piece**

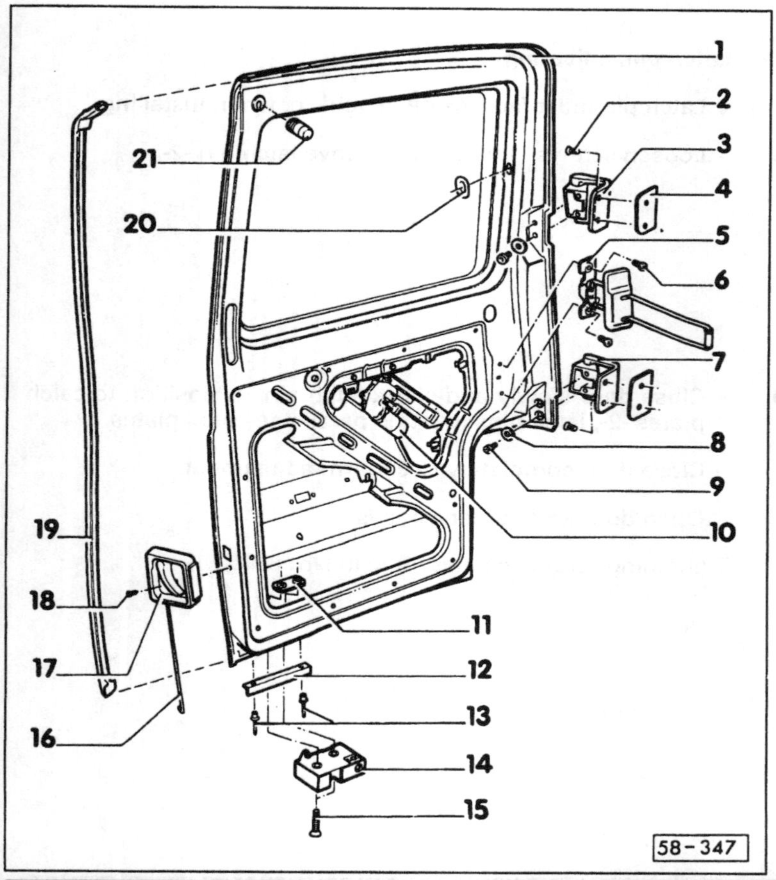

Double rear door (left), components

1 - **Left rear double door**
- Removing ⇒ Page Q 58-35
- Adjusting ⇒ Page Q 58-36

2 - **Countersunk screw**
- Secures hinges to D-pillar

3 - **Upper hinge**

4 - **Seal**

5 - **Door check strap**

6 - **Screw**
- 28 Nm (21 ft lb)

7 - **Lower hinge**

8 - **Washer**

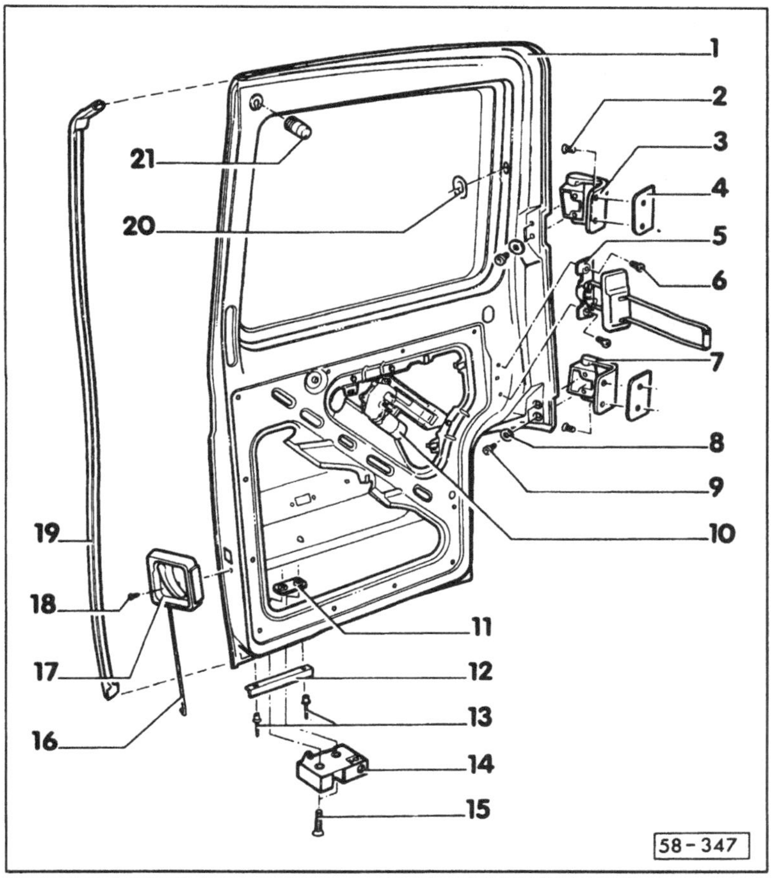

9 - Allen screw
- Secures door hinge

10 - Wiper motor
- Removing installing

⇒ Repair Group 92; electrical system; rear window wiper and washer system for double doors, repairing

11 - Threaded plate

12 - Stop strip

13 - Rivet

14 - Door lock
- Removing ⇒ Page Q 58-39

15 - Countersunk screw
- 19 Nm (14 ft lb)

16 - Operating rod

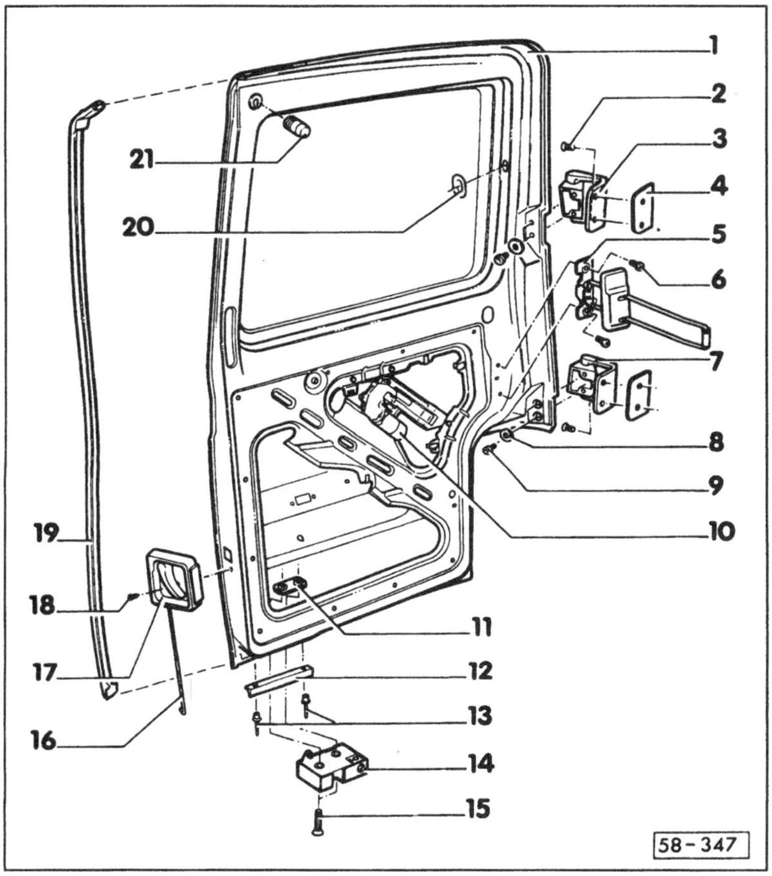

17 - Operating lever

18 - Phillips screw

19 - Seal

20 - Grommet

21 - Rubber stop

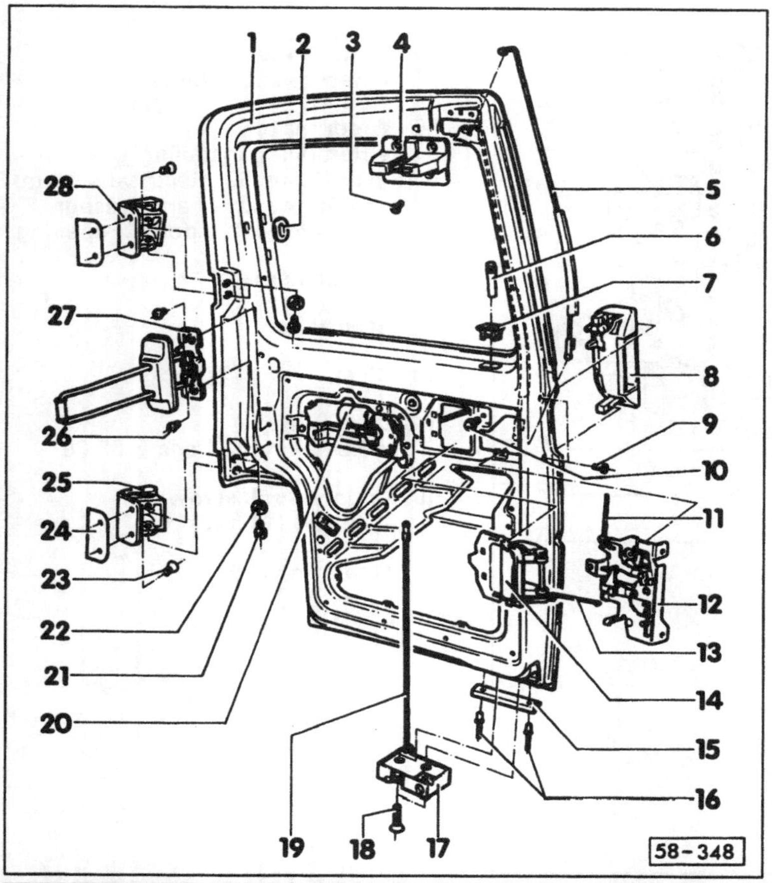

Double rear door (right), components

1 - **Right rear double door**
- ♦ Removing ⇒ Page Q 58-35
- ♦ Adjusting ⇒ Page Q 58-36

2 - **Grommet**

3 - **Countersunk screw**
- ♦ 8.0 Nm (71 in. lb)

4 - **Upper rotary catch**

5 - **Upper lock rod**
- ♦ with ball joint

6 - **Lock knob**

7 - **Trim piece**

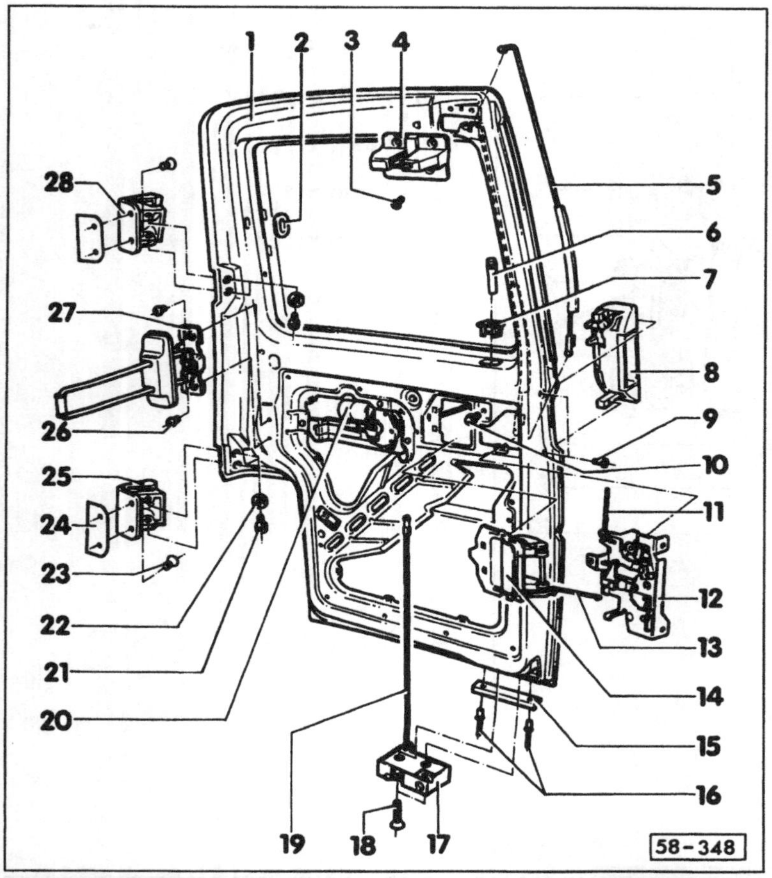

8 - **Outside door handle**
- ♦ Removing: remove pan-head tapping screws -9- and remove outside door handle outward

9 - **Screw**
- ♦ 8.0 Nm (71 in. lb)

10 - **Bolt w/washer**
- ♦ 8.0 Nm (71 in. lb)

11 - **Lock rod**

12 - **Lock mechanism**
- ♦ Removing ⇒ Page Q 58-42

13 - **Operating rod**

14 - **Inner door handle**
- ♦ Removing ⇒ Page Q 58-44

15 - **Stop strip**

16 - **Rivet**

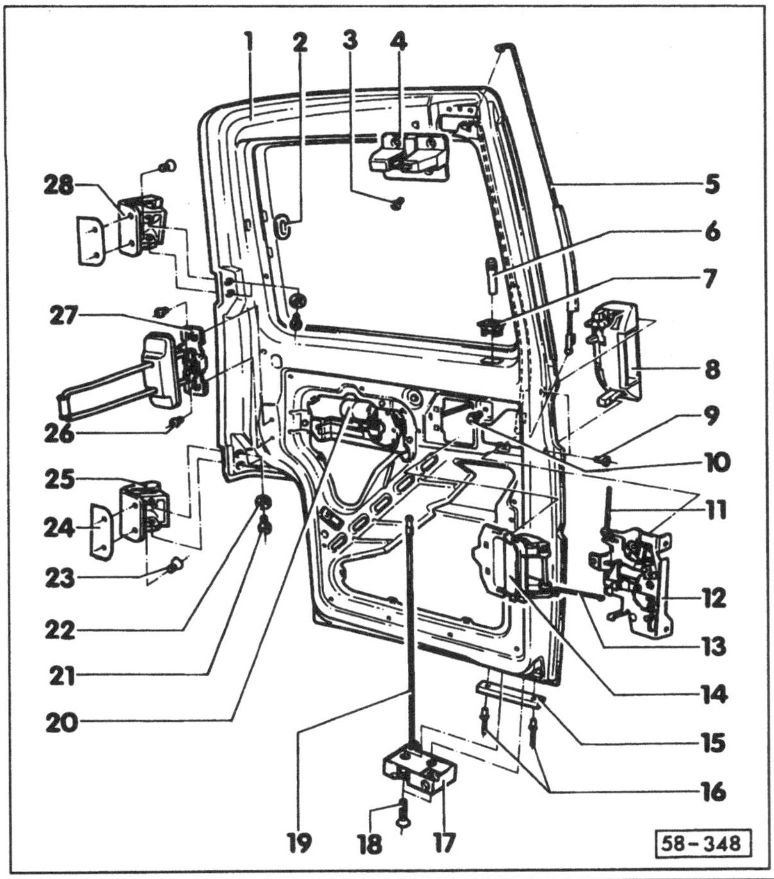

17 - Lower door lock
- Removing ⇒ Page Q 58-40

18 - Countersunk screw
- 19 Nm (14 ft lb)

19 - Lower operating rod
- with ball joint

20 - Wiper motor
- Removing and installing

⇒ Repair Group 92; electrical system; rear window wiper and washer system for double doors, repairing

21 - Allen bolt
- Secures door hinge

22 - Washer

23 - Countersunk screw
- Secures hinge to D-pillar

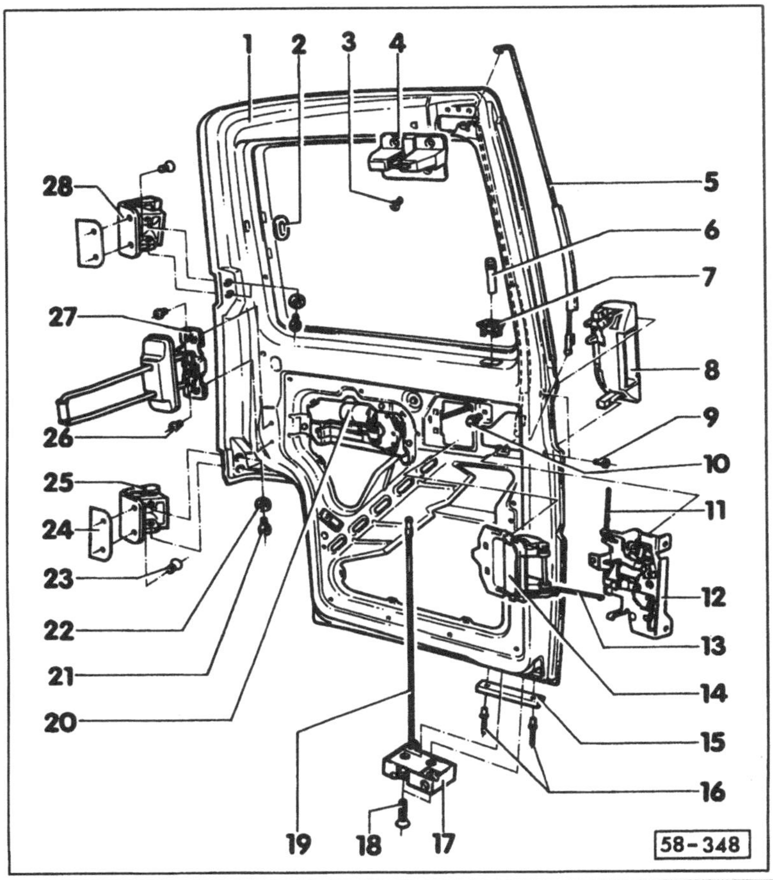

24 - Gasket

25 - Lower hinge

26 - Screw
- 28 Nm (21 ft lb)

27 - Door check strap

28 - Upper hinge

Double rear doors, removing

Note:

Description refers to right rear double door. Removal of left rear double door is mirror image.

– Remove trim ⇒ Page Q 70-101 and/or Page Q 70-103.

– If applicable, disconnect wiring for rear wiper, window heating and central locking and pull out of door.

– Remove door check trim, if applicable.

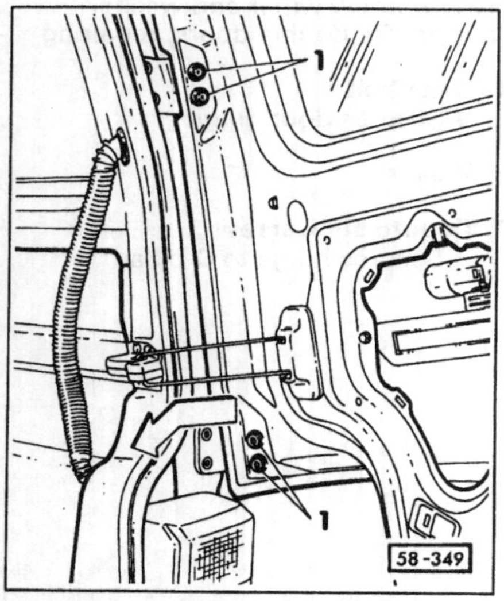

◄ – Remove Allen screw -1- from upper and lower hinge.

– Unhook door check strap -arrow- and remove door.

Double rear doors, adjusting

Lower strikers, adjusting

Note:

Adjustments should be made first on left and then right double door. A second technician is required.

– Remove rear end cover.

– Remove headliner trim ⇒ Page Q 70-37.

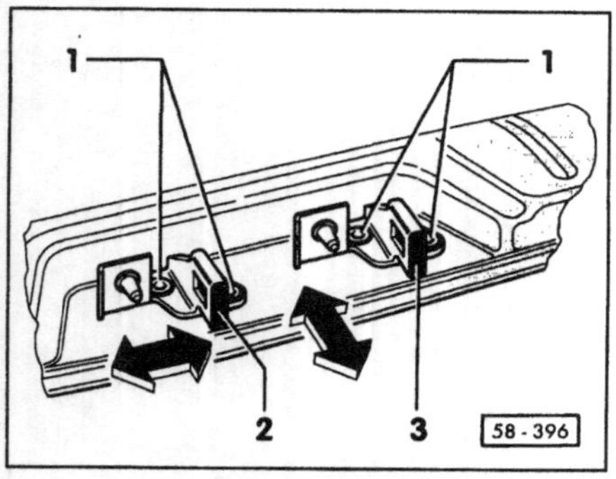

◄ – After loosening screws -1-, strikers -2, 3- can be moved -arrows-.

– Loosen left striker -2-.

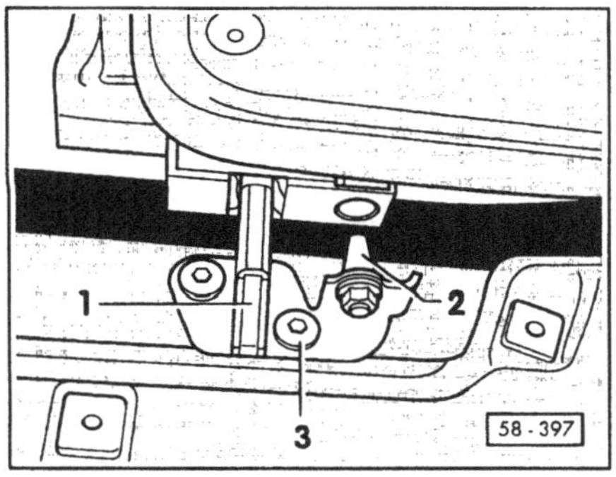

- Close left double door. In this process, the striker -1- is adjusted by the pins -2-.
- From outside, press double door against vehicle and tighten fastening screw -3-.
- Open double door and tighten second fastening screw.

  Tightening torque:

  Fastening screws -3- = 20 Nm (15 ft lb)

- Double door must latch completely when closing without using additional force and must not have play.

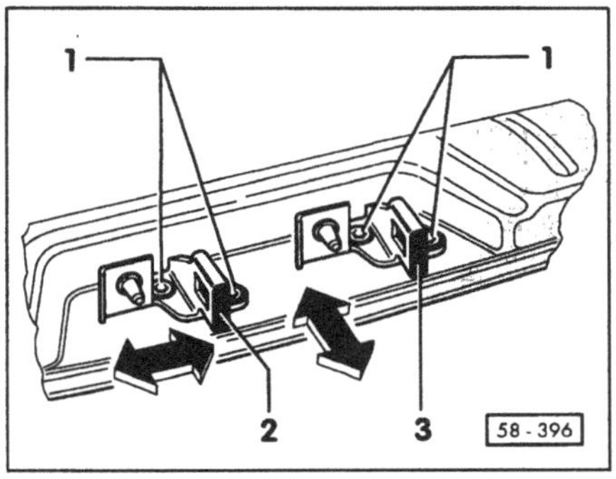

- Loosen right striker -3-.

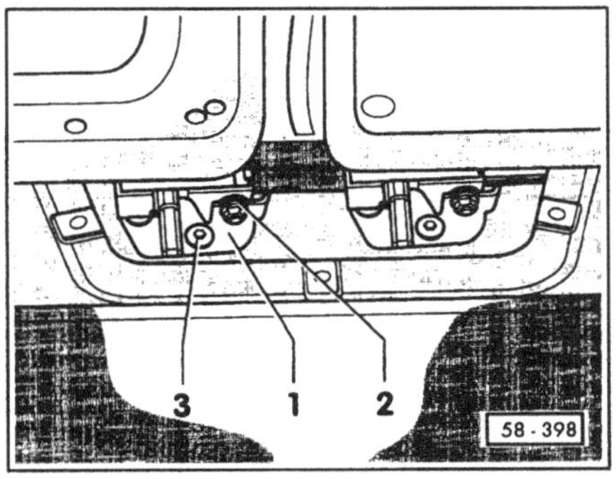

- With left double door closed, close right double door. In this process, striker -1- is adjusted by pin -2-.
- Push right double door from outside against the vehicle enough so that both double doors match the outside contour.
- Tighten fastening screw -3-.
- Open right double door and tighten second screw.

  Tightening torque:

  Screws -3- = 20 Nm (15 ft lb)

Upper striker, adjusting

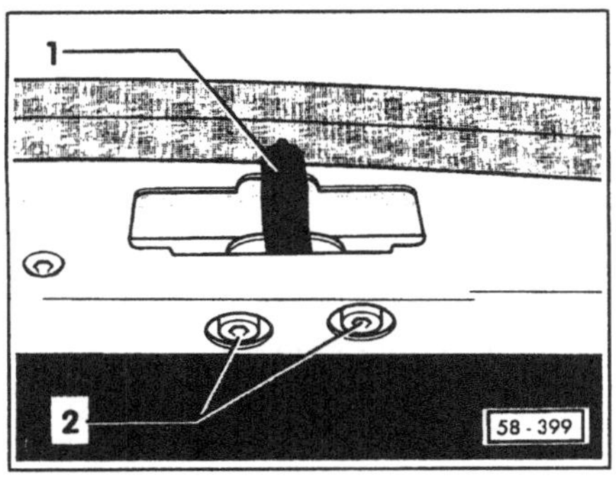

- Loosen upper striker -1- by loosening screws -2-.
- Close both double doors.
- Push right double door in upper area to outside or inside until both double doors match outer contour.
- Tighten fastening screws -2-.

  Tightening torque:

  Fastening screws -2- = 20 Nm (15 ft lb)

- When closing right double door (left double door closed), the upper and lower door lock must latch completely without use of extra force and must not have free play.

Door locks (rear double door), removing

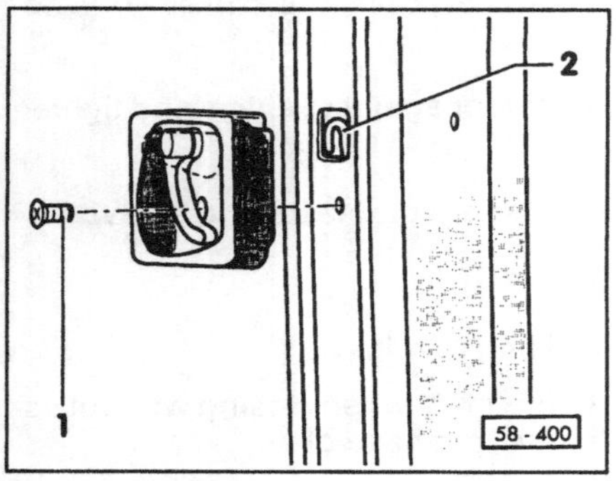

Left door

– Remove operating lever by removing Phillips screw -1-, disconnecting operating rod -2-.

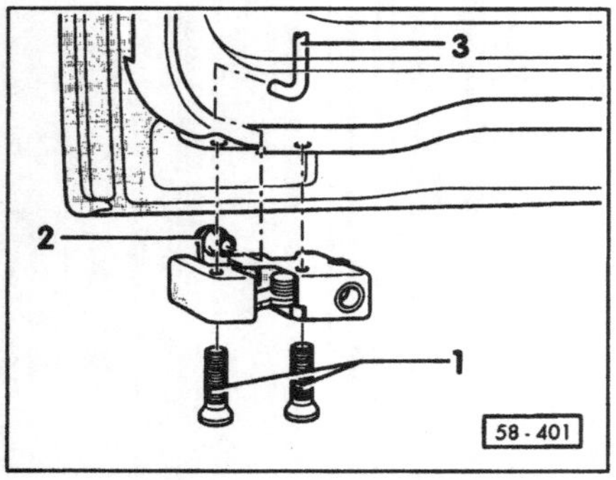

– Remove countersunk screws -1- and door lock -2- with operating rod -3-.

Tightening torque:

Fastening screws -1- = 19 Nm (14 ft lb)

Lower right double door lock

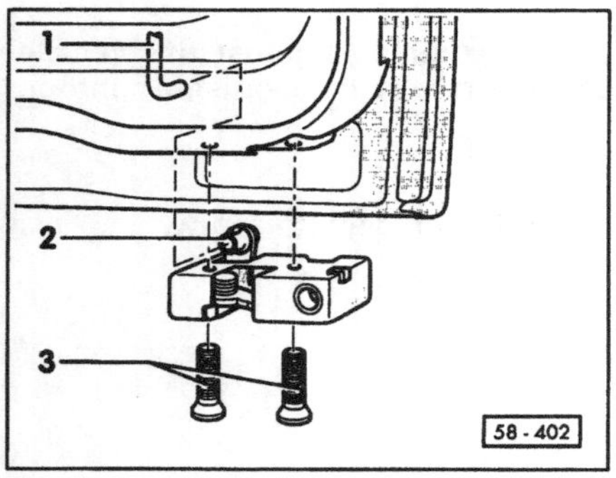

– Remove right rear double door trim ⇒ Page Q 70-103.

– Disconnect operating rod -1- on lower door lock -2-.

– Remove countersunk screws -3-.

– Remove lower door lock.

Tightening torque:

Countersunk screws -3- = 19 Nm (14 ft lb)

Note:

After installing door lock check lock operation, adjust if necessary ⇒ Page Q 58-36.

Upper right double door lock

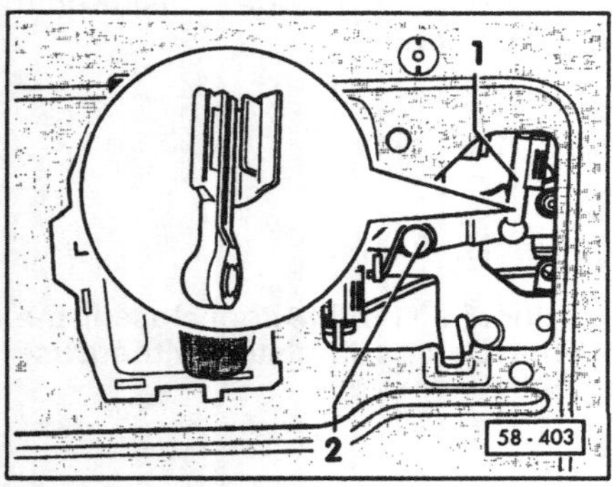

– Remove right rear double door trim ⇒ Page Q 70-103.

– Unclip ball joint -1- on lock mechanism -2-.

– Remove ball joint (with clip) from operating rod.

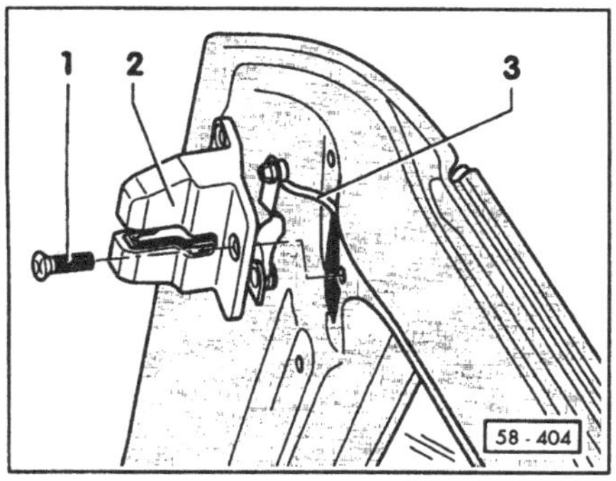

◄ – Remove countersunk screws -1-, remove door lock -2- with operating rod -3- by pulling upward.

Tightening torque:

Countersunk screws -1- = 8 Nm (71 in. lb)

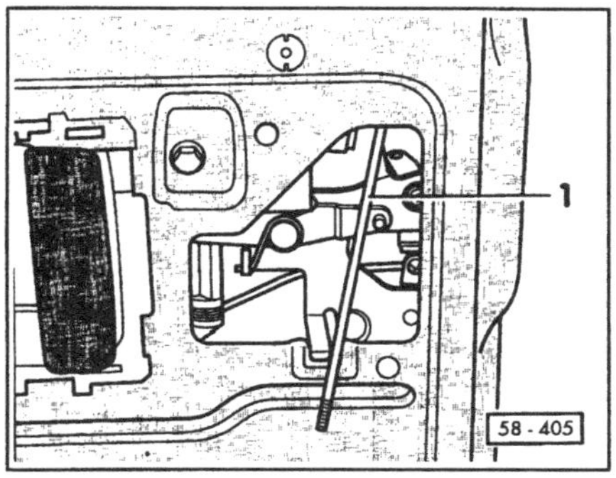

Note:

◄ ♦ *When installing operating rod -1-, push it through window frame from below until operating rod can be clipped on door lock at top. Then screw on upper door lock.*

♦ *After installing door lock, check lock operation and adjust if necessary ⇒ Page Q 58-43.*

Lock mechanism, removing

– Remove right rear double door trim ⇒ Page Q 70-103.

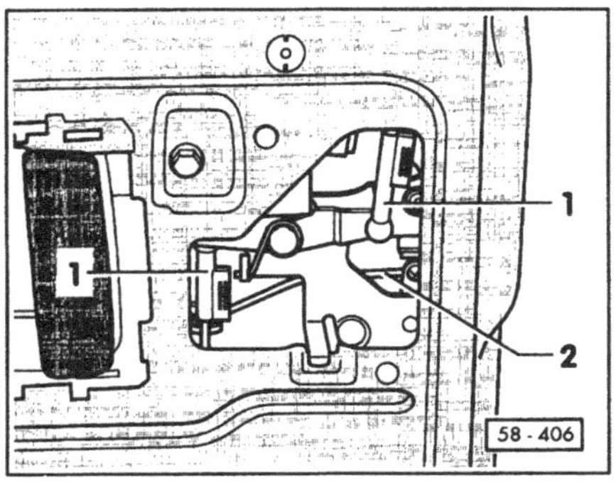

◄ – Remove ball joints -1- of operating rods, unclip internal mechanism operating rod -2-.

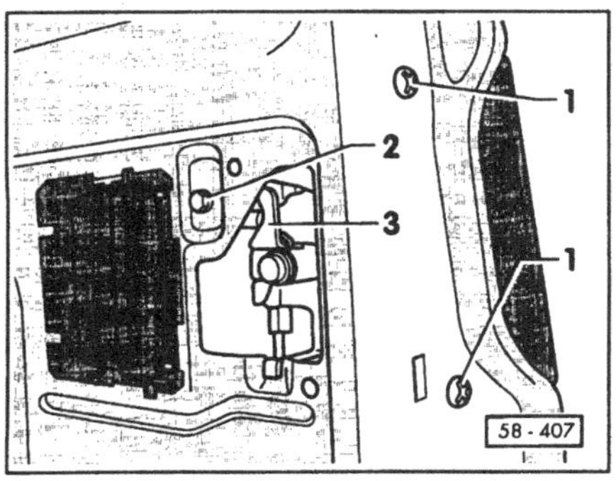

◄ – Remove Phillips screws -1-, remove outside door handle.

– Remove hex screw -2-, remove lock mechanism from double door.

Tightening torque:

Phillips screws -1- = 8.0 Nm (71 in. lb)

Hex screw -2- = 8.0 Nm (71 in. lb)

Lock mechanism, adjusting

Note:

Adjustment is made on right rear double door. Top and bottom door locks must latch completely.

- When activating lock from outer door handle, both door locks must unlock at the same time. If door locks unlock audibly one after the other, perform adjustment described below.

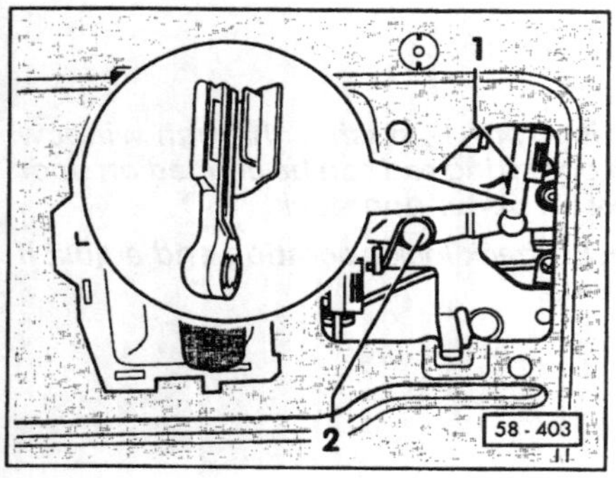

– Remove right rear double door trim ⇒ Page Q 70-103.

◄ – Unclip ball joint -1- (figure shows ball joint for upper door lock) at lock mechanism -2-.

– Turn ball joint -1- on threaded operating rod to shorten or lengthen rod.

- Lengthening operating rod causes respective door lock to unlock "later"; shortening it, makes lock unlock "sooner".

– Adjust length of both operating rods so that both door locks can be unlocked individually and simultaneously.

Inside door handle, removing

– Remove right rear double door trim ⇒ Page Q 70-103.

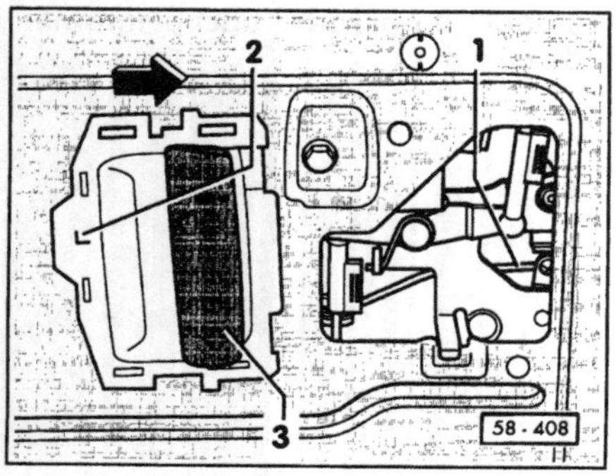

◄ – Disconnect operating rod -1- from lock mechanism.

– Using screwdriver, pry out holding tab -2- and slide outer door handle -3- out of bracket in double door -arrow-.

Lock cylinder, removing (rear double door)

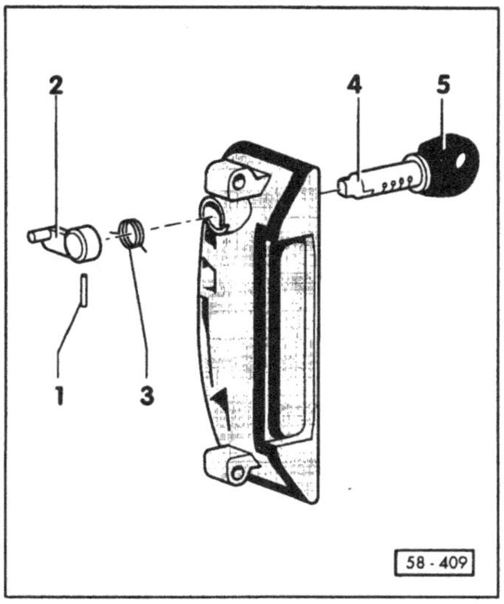

- Remove outside door handle ⇒ Page Q 58-46.
- ◄ Drive out spring pin -1-.
- Remove casting -2- from four-sided shaft.
- Remove spring -3-.
- Remove lock cylinder -4- with key -5- inserted.

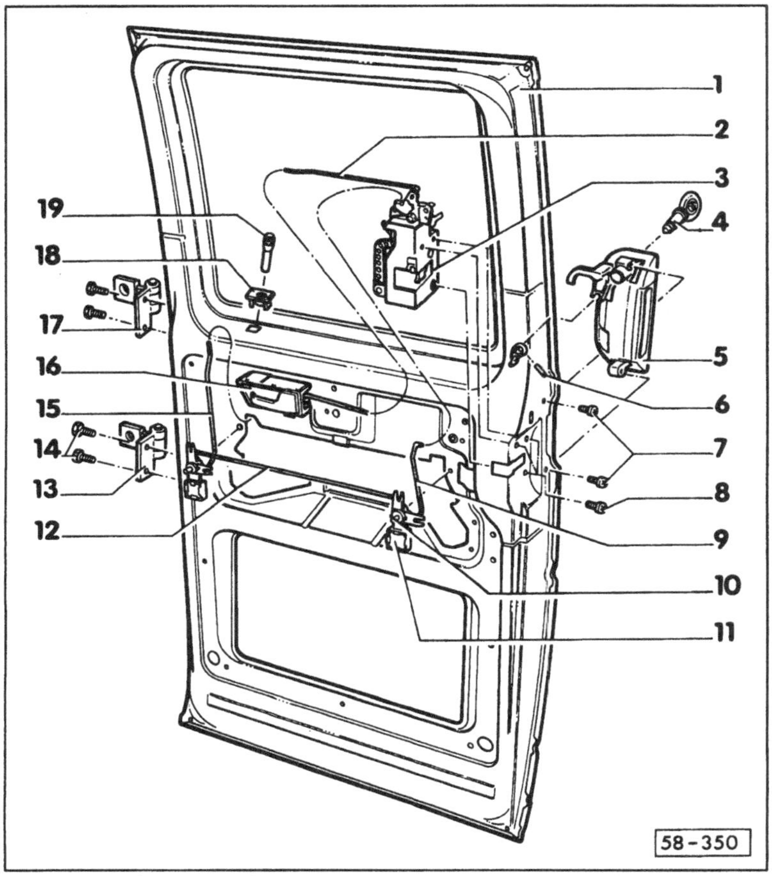

Double cab door (rear), components

1 - **Double cab door**
- ♦ Removing ⇒ Page Q 58-49
- ♦ Installing and adjusting ⇒ Page Q 58-50

2 - **Operating rod**

3 - **Door lock**
- ♦ Removing ⇒ Page Q 58-39

4 - **Lock cylinder**
- ♦ Removing: Remove outside door handle -5-, drive out locking pin -6-, remove lock cylinder with key inserted ⇒ Page Q 58-45.

5 - **Outside door handle**
- ♦ Removing: Remove screws -7-, remove door handle.

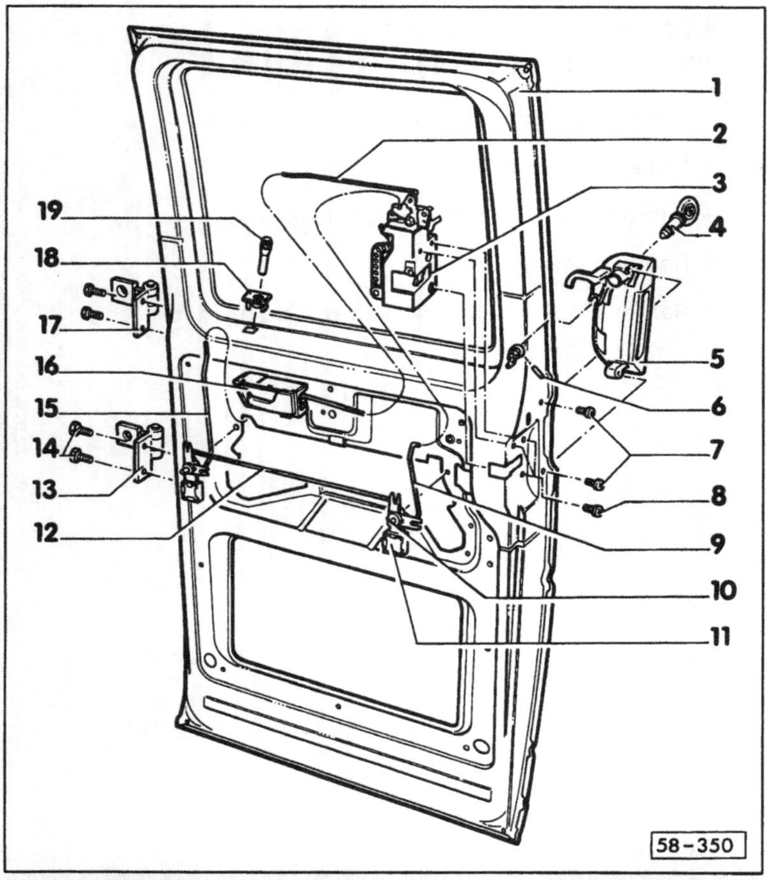

6 - **Lock pin**

7 - **Phillips screw**
- ♦ 8.0 Nm (71 in. lb)

8 - **Phillips screw**
- ♦ 8.0 Nm (71 in. lb)

9 - **Operating rod**

10 - **Expanding clip**
- ♦ Removing: Push pin out through expanding clip
- ♦ Installing: Press pin in far enough to make it flush with upper edge of expanding clip

11 - **Bell lever**

12 - **Operating rod**

13 - **Lower hinge**

14 - **Hex bolt**

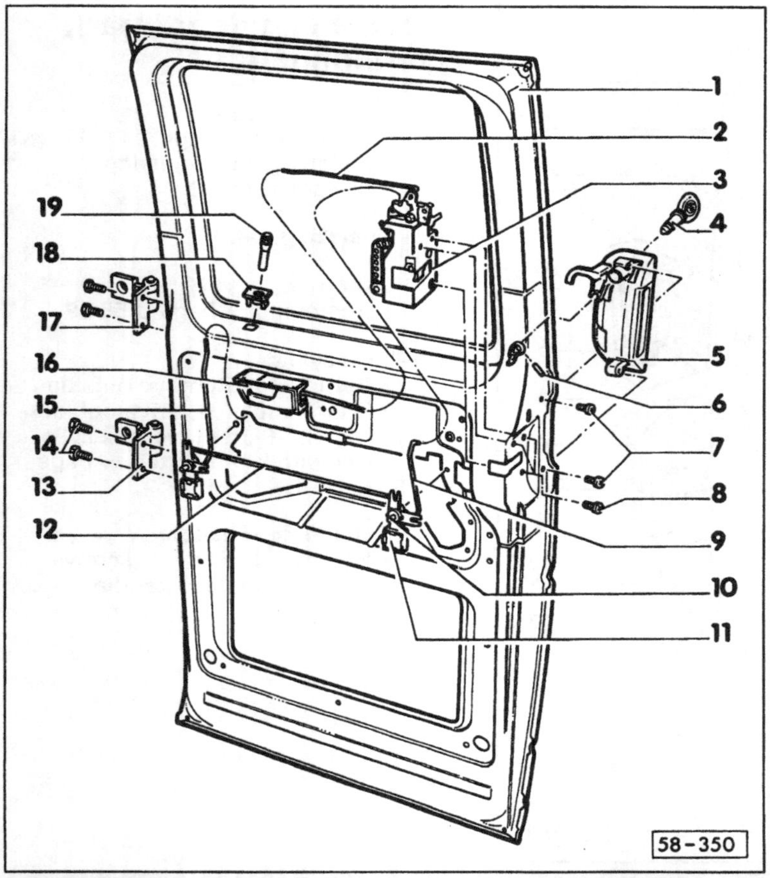

15 - **Lock rod**

16 - **Inner door opening mechanism**

17 - **Upper hinge**

18 - **Trim piece**

19 - **Lock knob**

Double cab door (rear), removing

With central locking

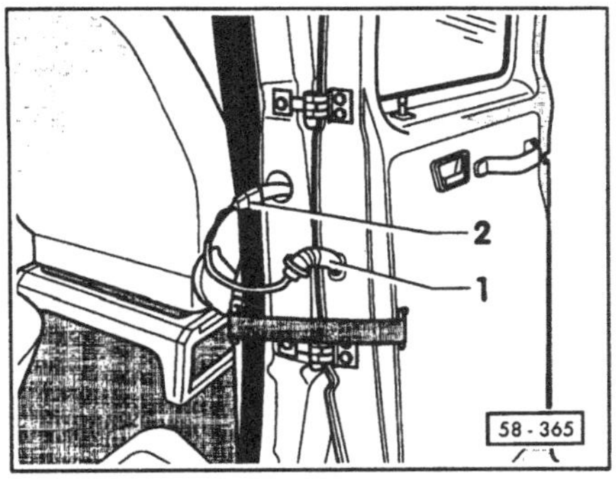

- Remove cable grommet -1- from B-pillar, pull out plug connector -2- from B-pillar and disconnect.
- Remove hinge covers from outside B-pillar.

Continuation, all vehicles

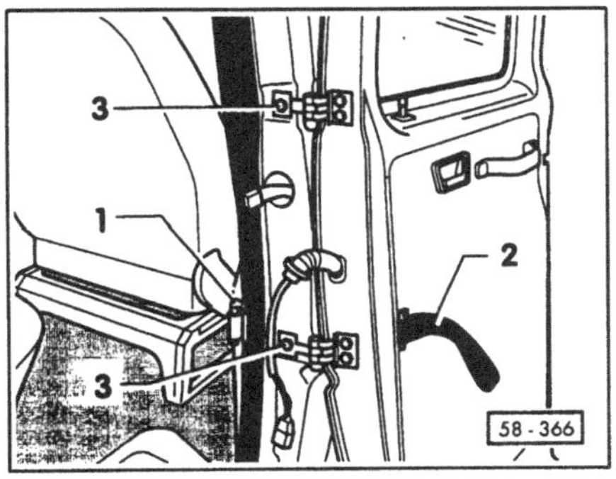

- Lift up holding pin -1- and disconnect door strap -2-.
- Remove hex bolt -3- from upper and lower hinges.
- Remove door.

Double cab door (rear), installing/adjusting

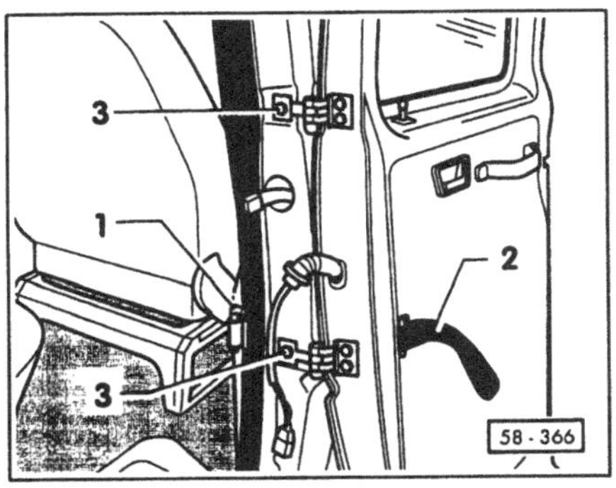

- Install hex bolts -3- in upper and lower hinges.
- Adjust double door through oversized holes in hinge.
- Adjust striker.

- Double cab door (rear) is correctly adjusted if, when closed, has a uniform gap around door opening, is not too far in or out, matches contour -arrows-.

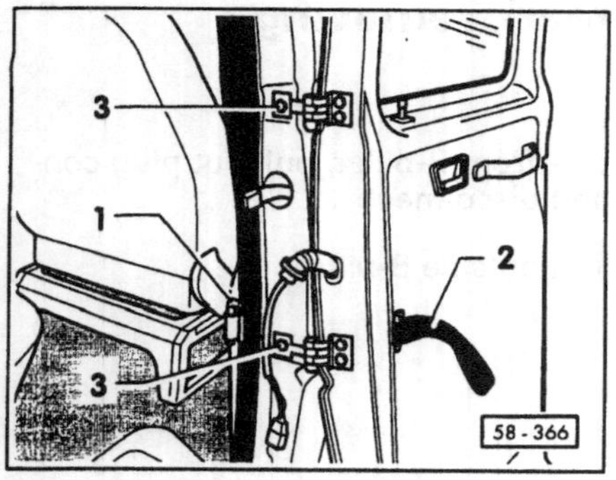

- Tighten hex bolts -3-.

  Tightening torque:

  Hex bolt -3- = 55 Nm (41 ft lb)

- Connect connector and push into B-pillar, install cable grommet (vehicles with central locking).

- Lift holding pin -1- and install door strap -2-.

Striker, adjusting

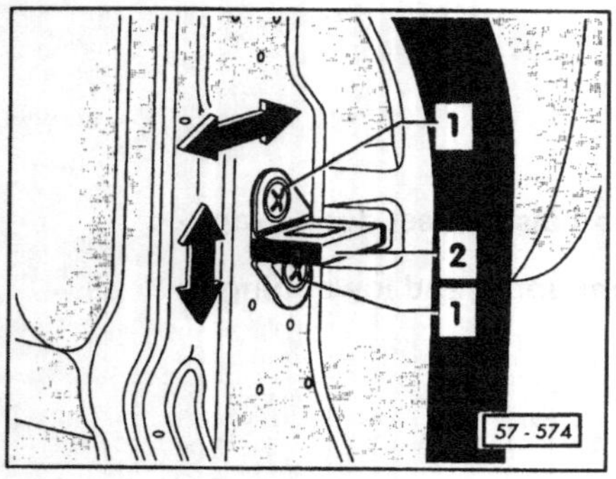

- After loosening Phillips screws -1-, striker -2- can be moved horizontally and vertically -arrows-.

  Tightening torque:

  Phillips screws -1- = 19 Nm (14 ft lb)

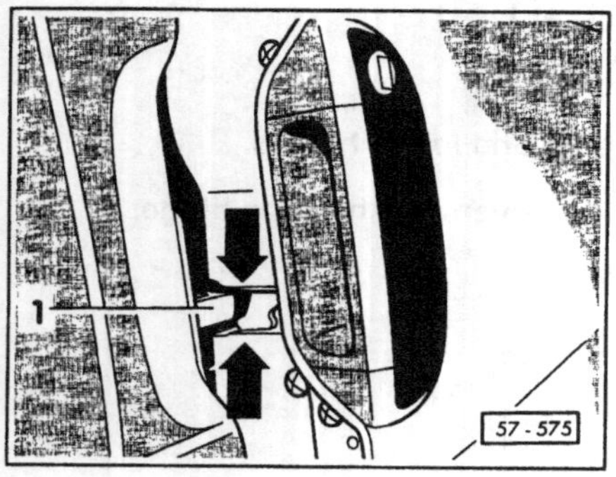

- Adjust striker -1- so that recess on lock with turn clip is concentric -arrows- to striker (door must not be raised or pressed down).

• Door must have no play when closed.

Door lock (double cab door rear), removing

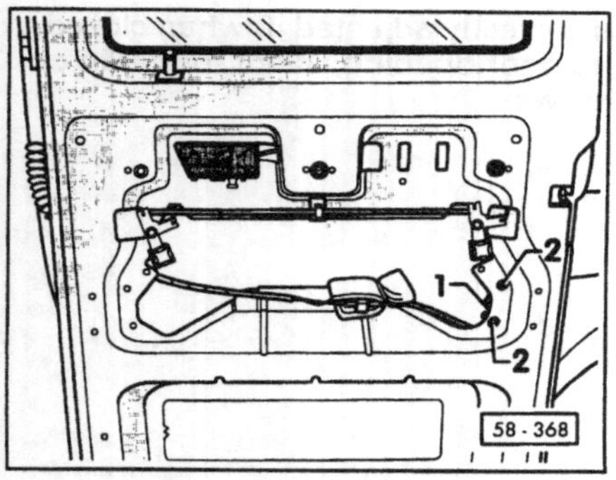

- Remove side double door trim ⇒ Page Q 70-89.

- Remove outside door handle ⇒ Page Q 57-21.

- If applicable, loosen Double Cab Right Rear Door Central Locking System Motor -V97- by taking out the Phillips screws -2- and disconnect operating rod on motor.

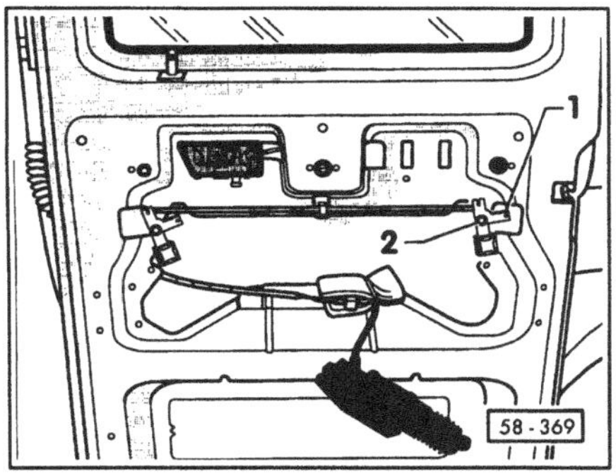

◄ – Loosen expanding clips -2- (push pin through) and disconnect interior lock operating rod -1-.

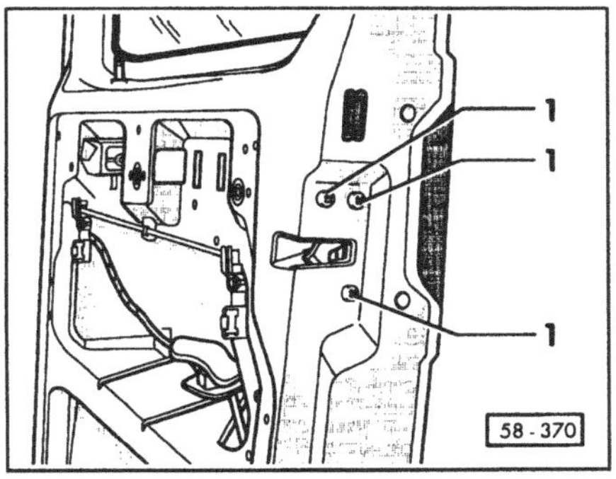

◄ – Remove Phillips screws -1-, remove door lock out downward.

Tightening torque:

Phillips screws -1- = 19 Nm (14 ft lb)

– Disconnect door lock rod.

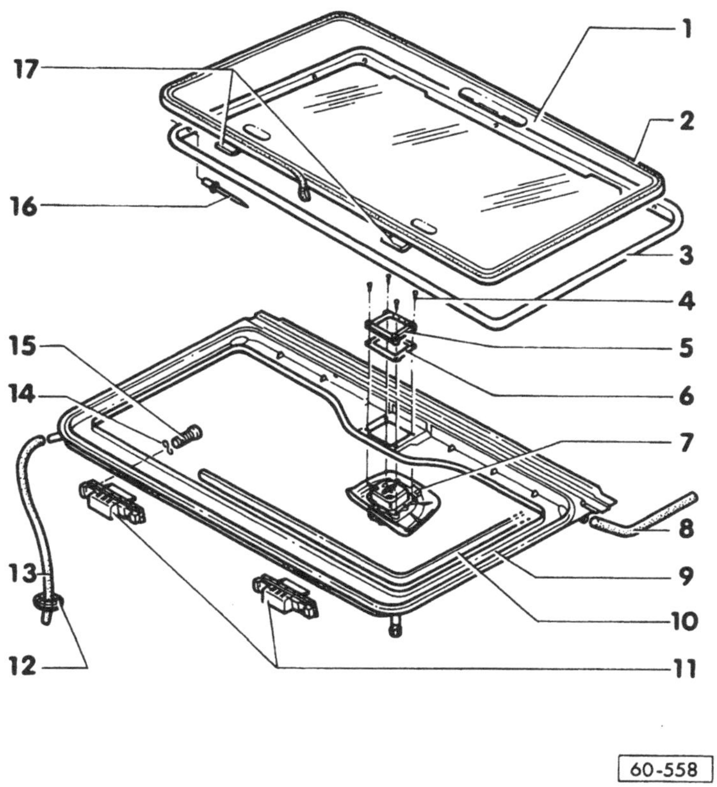

Sunroof, components

1 - **Sunroof panel**
- ♦ Removing ⇒ Page Q 60-3
- ♦ Installing ⇒ Page Q 60-4

2 - **Gasket**

3 - **Trim**
- ♦ riveted

4 - **Phillips screw**

5 - **Mounting ring**

6 - **Gasket**

7 - **Crank handle**

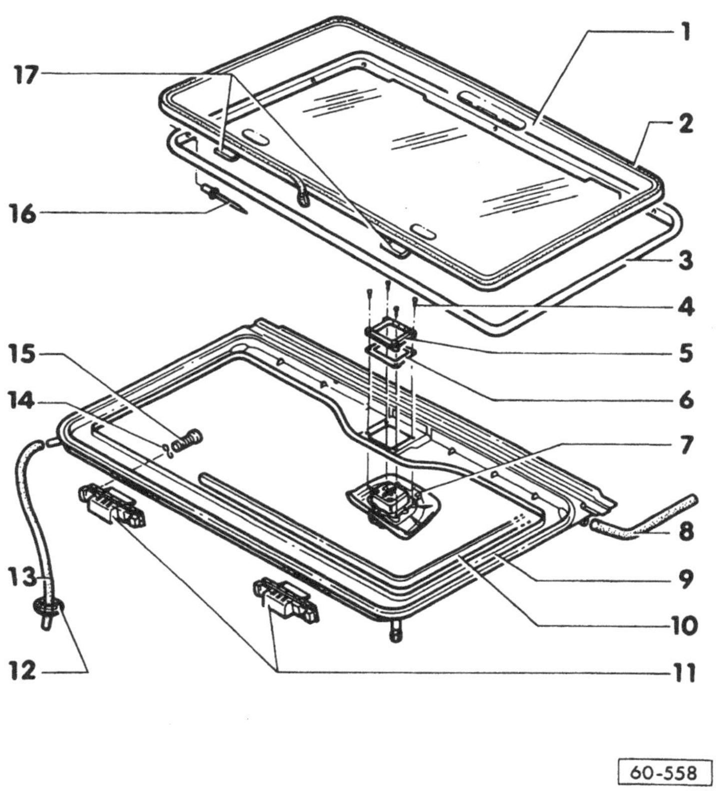

8 - **Rear drain hose**
- ♦ Routing and cleaning ⇒ Page Q 60-5

9 - **Frame**

10 - **Gasket**

11 - **Bracket**

12 - **Grommet**

13 - **Front drain hose**
- ♦ Routing and cleaning ⇒ Page Q 60-5

14 - **Washer**

15 - **Screw**

16 - **Rivet**

17 - **Hinge**
- ♦ Hooked into bracket -11-

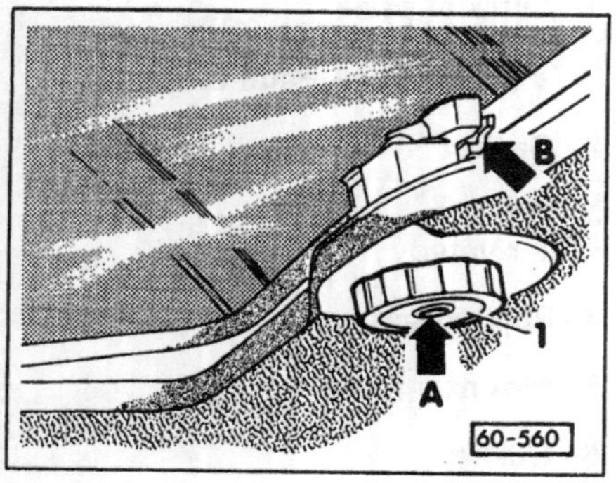

Sunroof panel, removing

◄ – Turn unlocking screws -arrow -A- in center of rotary handle -1- (sunroof panel closed) with screwdriver approx. 1/4 turn to right or left until sunroof panel disengages from latch.

– Turn rotary handle -1- right until sunroof is raised. Also, raise sunroof panel slightly so it does not engage in latch.

– Press lock lever -arrow -B- upward.

– Unlatch sunroof panel and pull with hinges, out of brackets.

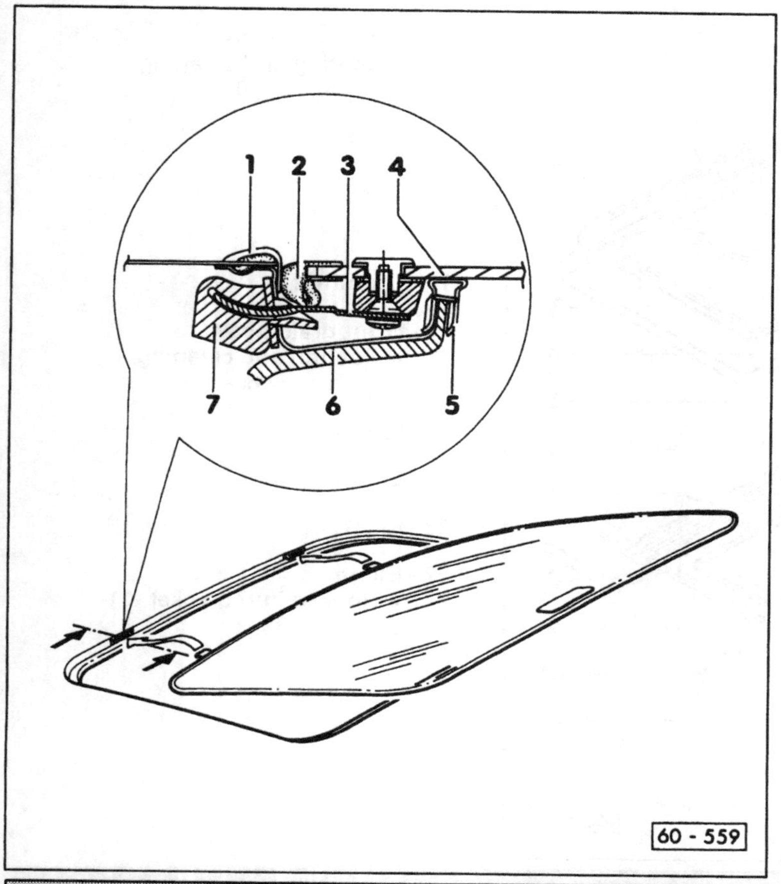

Sunroof panel, installing

1 - Trim frame

2 - Gasket

3 - Hinge

4 - Sunroof panel
- ♦ Installing: Hook sunroof panel with hinges -3- into brackets -7- and let drop into latch

5 - Gasket

6 - Roof trim

7 - Bracket

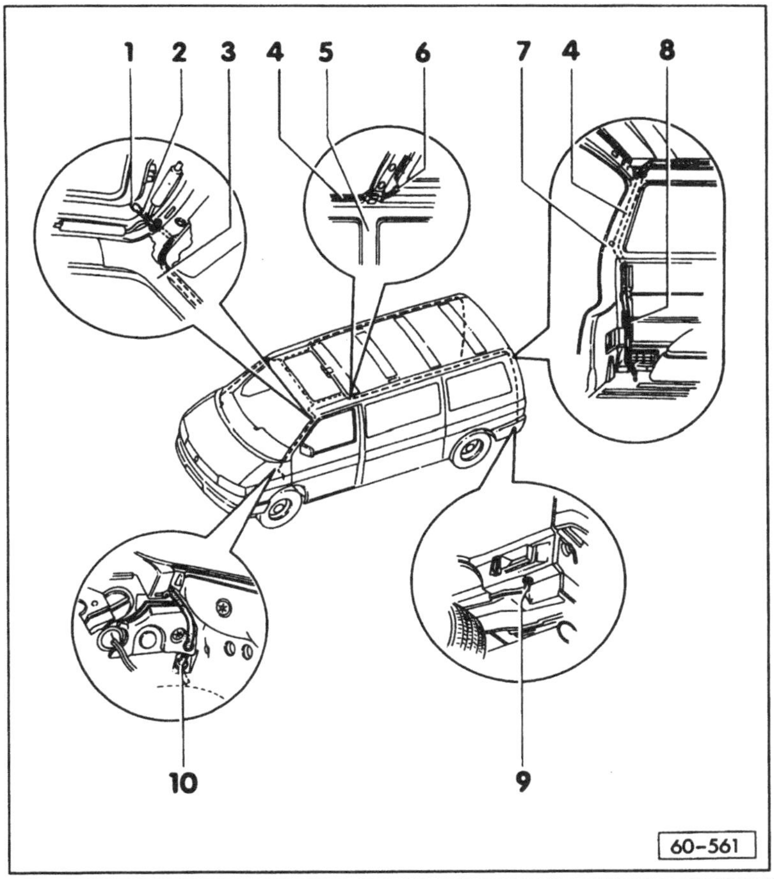

Water drain hoses, routing and cleaning

Panel truck / van with sunroof in driver's compartment

1 - **Front frame drain**

2 - **Front drain hose**
- Blow through front frame drain -1- with compressed air to clean.

3 - **A-pillar**

4 - **Rear hose**
- Covered with a foam tube in D-pillar area.
- To clean blow through with compressed air from rear frame drain -6-.

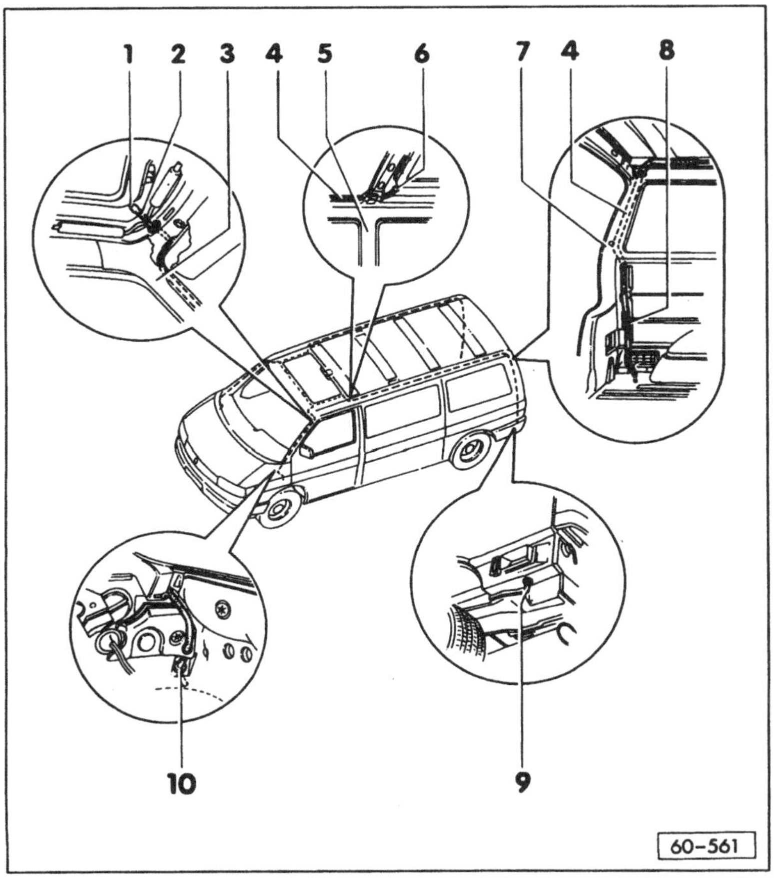

5 - **B-pillar**

6 - **Rear frame drain**

7 - **D-pillar**

8 - **Fastening clip**

9 - **Grommet**
- Rear water drain located behind bumper cover.

10 - **A-pillar cover**

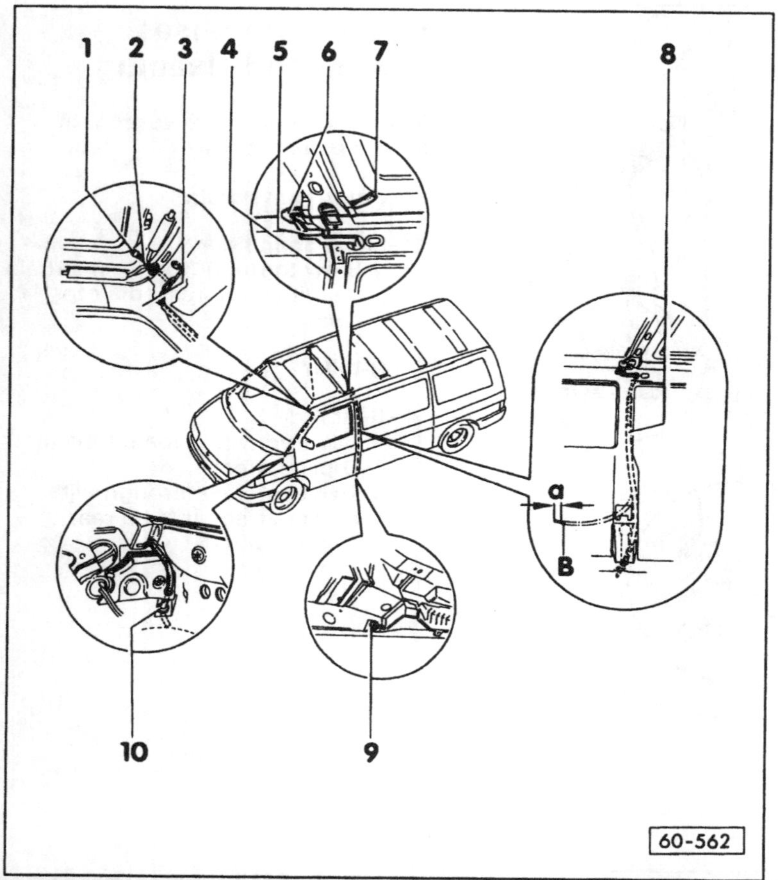

Double cab with sunroof in driver's compartment

1 - **Front frame drain**

2 - **Front drain hose**
- To clean, blow through front frame drain -1- with compressed air.

3 - **A-pillar**

4 - **B-pillar**

5 - **Adapter**

6 - **Molded hose**

7 - **Rear frame drain**

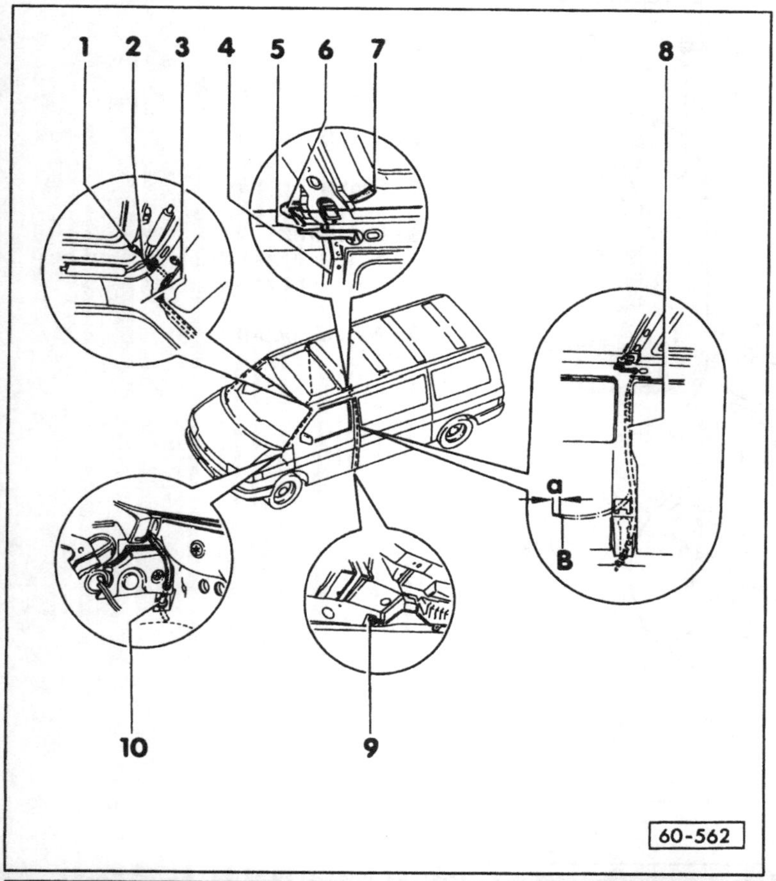

8 - **Rear drain hose**
- Covered with a foam tube in D-pillar area
- Water drain hose and foam tube are taped together with fabric tape in lower area -B-
a = 200 mm (approx. 8.0 in.)
- To clean, blow through with compressed air from rear frame drain -7-.

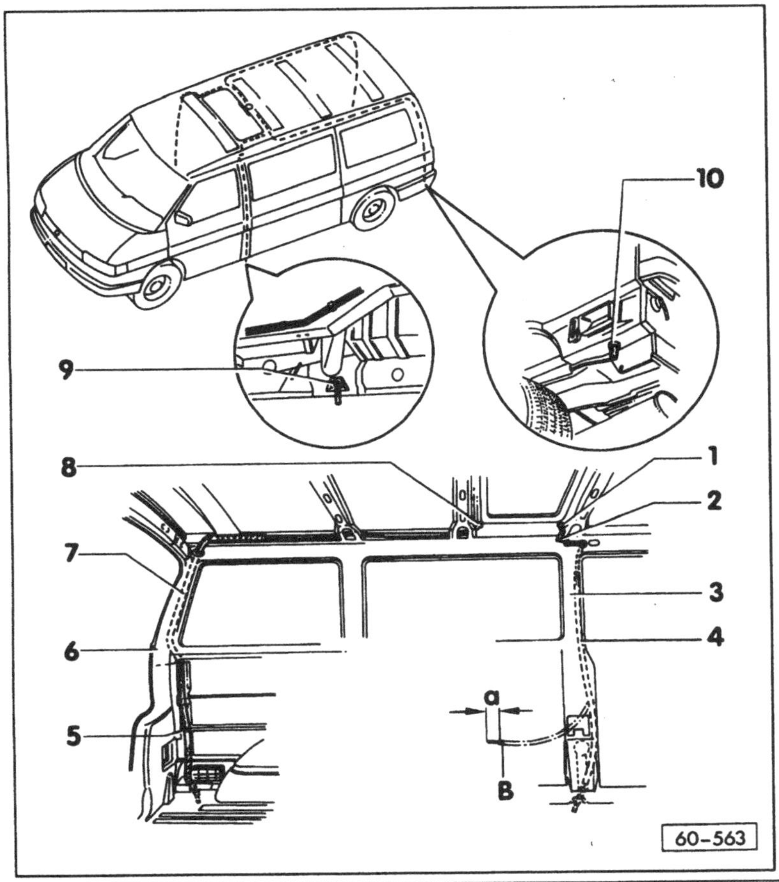

Vehicles with sunroof in passenger compartment

1 - **Front frame drain**

2 - **Adapter**

3 - **B-pillar**

4 - **Front drain hose**
- Covered with a foam tube.
- Water drain hose and foam tube are taped together with fabric tape in lower area -B- a = 200 mm (approx. 8.0 in.).
- Blow through front frame drain -1- with compressed air to clean.

5 - **Fastening clip**

6 - **D-pillar**

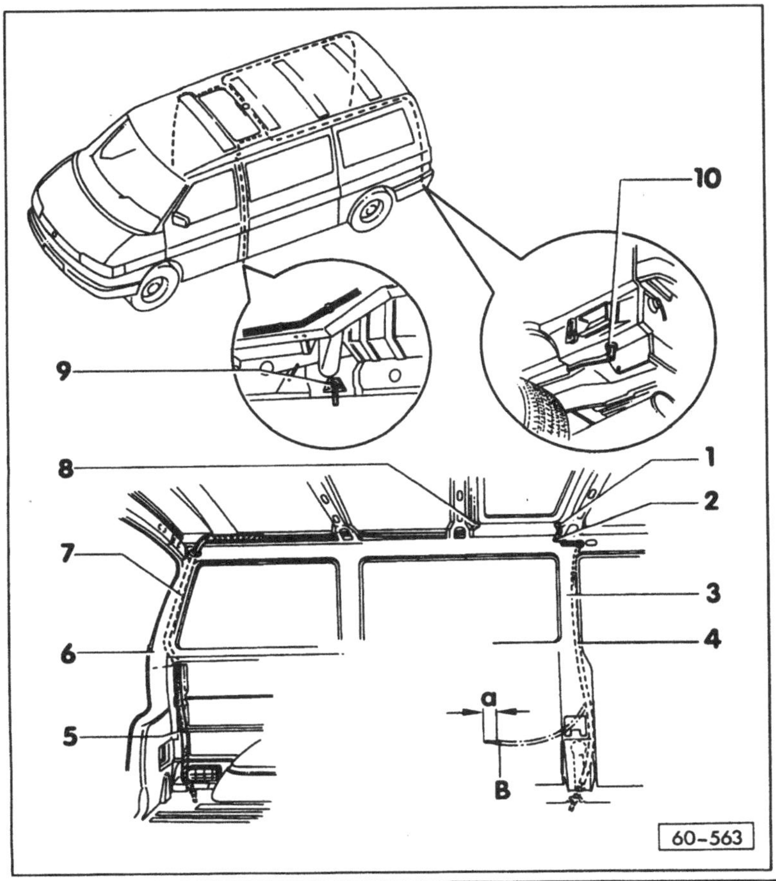

7 - **Rear drain hose**
- Covered with a foam tube in D-pillar area.
- To clean, blow through with compressed air from rear frame drain -8-.

8 - **Rear frame drain**

9 - **B-pillar cover**

10 - **Grommet**
- Rear water drain located behind bumper cover.

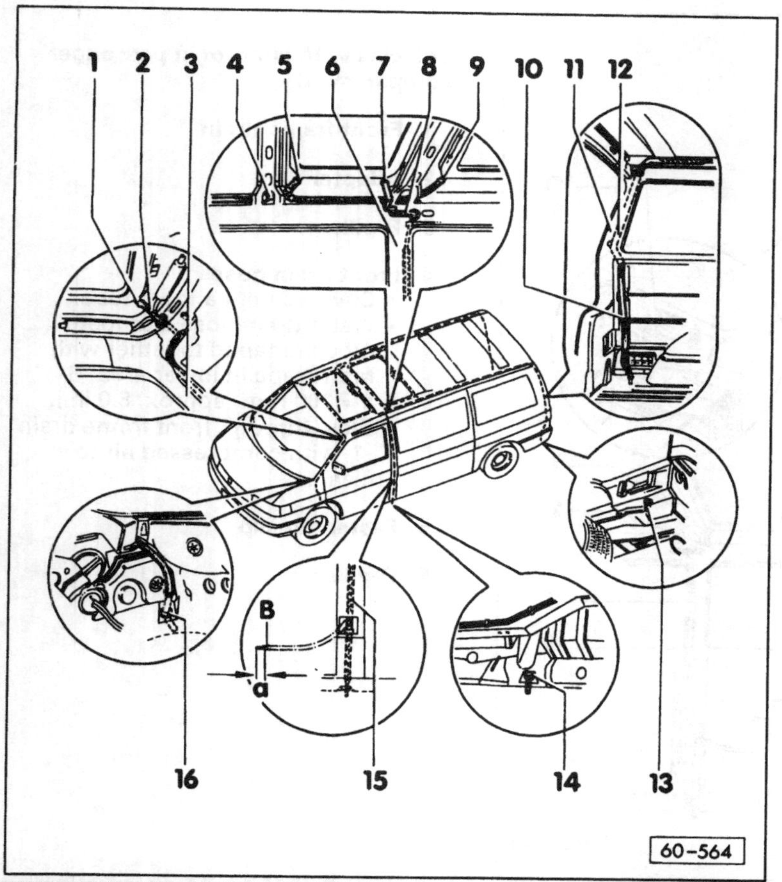

Vehicles with sunroof in driver and passenger compartments

1 - **Front frame drain**

2 - **Front drain hose**
- ♦ To clean, blow through front frame drain -1- with compressed air

3 - **A-pillar**

4 - **Y-Connector**

5 - **Rear frame drain**

6 - **B-pillar**

7 - **Front frame drain**

8 - **Adapter**

9 - **Rear frame drain**

10 - **Fastening Clip**

11 - **D-pillar**

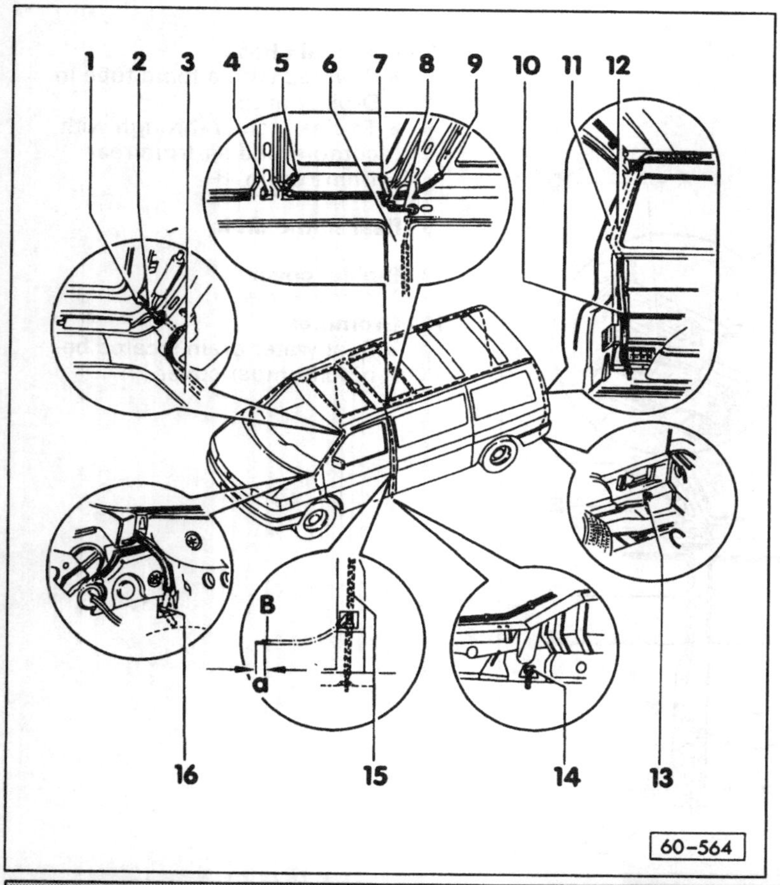

12 - **Rear drain hose**
- ♦ Covered with a foam tube in D-pillar area.
- ♦ To clean, blow through with compressed air from rear frame drains -5- and -9-.

13 - **Sleeve**
- ♦ Rear drain located behind bumper cover.

14 - **B-pillar cover**

15 - **Front drain hose**
- ♦ Covered with foam tube.
- ♦ Water drain hose and foam tube are taped together with fabric tape in lower area -B- a = 200 mm (approx. 8.0 in.).
- ♦ To clean, blow through with compressed air from front frame drain -7-.

16 - **A–pillar cover**

Tarpaulin Frame and Tarpaulin, components

Note:

- *Support pipes are supplied as replacement parts only in a length of 2864 mm (approx. 113 in.), which corresponds to flat bed trucks with the long wheel base.*
- *For short wheel base flat bed trucks, support pipes must be cut to 2464 mm (approx. 97 in.).*
- *On double cab vehicles, support pipes must be cut to 2094 mm (approx. 82 in.).*
- *Corrosion protection must be applied to joints and holes of modified pipes.*

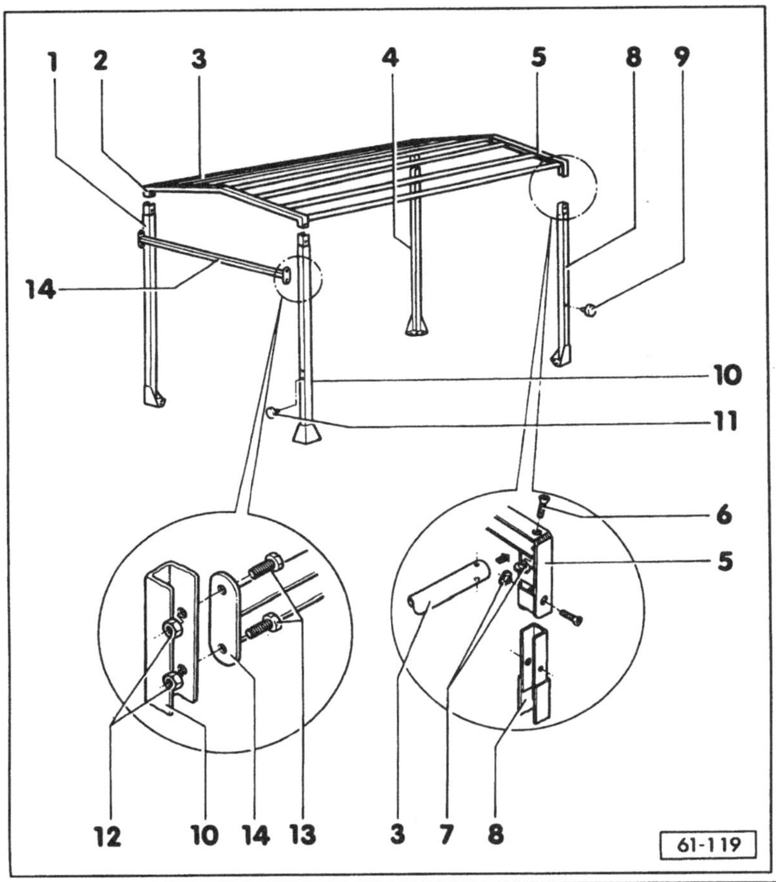

1 - Right front rail

2 - Upper front rail
- Bolt together with left front rail -10- and right front rail -1-

3 - Support pipe
- Bolt together with upper front rail -2- and upper rear rail -5-.

4 - Right rear rail

5 - Upper rear rail

6 - Countersunk screw M6 x 30

7 - Screw M6

8 - Left rear rail

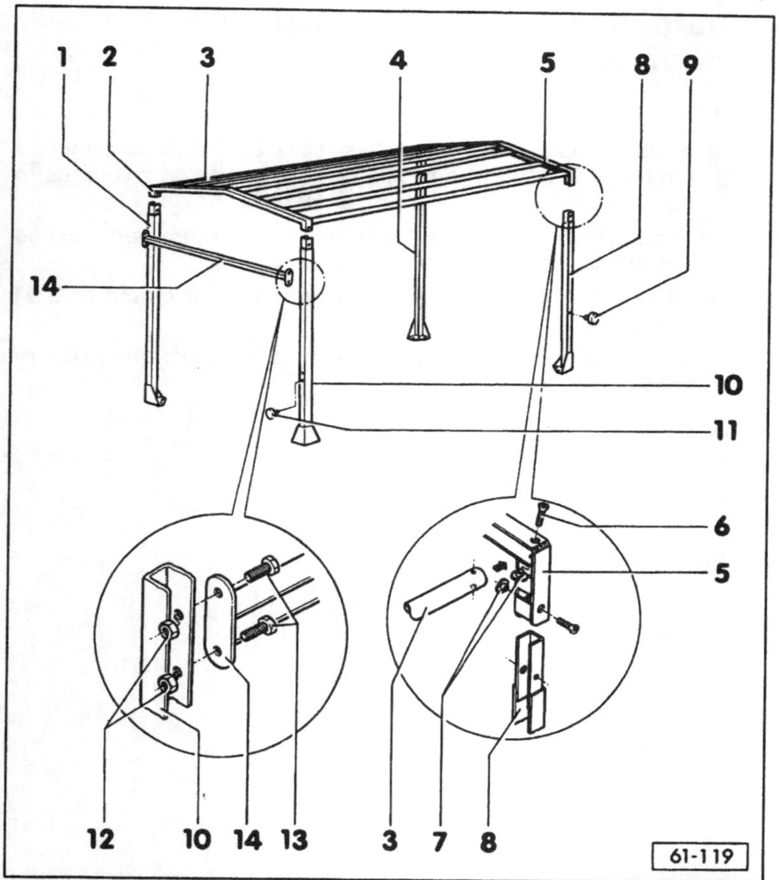

9 - **Rubber bumper**
- Press into hole on left -8-, right -4- rear rails

10 - **Left front rail**

11 - **Rubber bumper**
- Press into hole on left -10-, right -1- front rails

12 - **Hex nut M8**

13 - **Hex bolt M8 x 22**

14 - **Brace**
- Bolted together with left front -10-, right front -1- rail

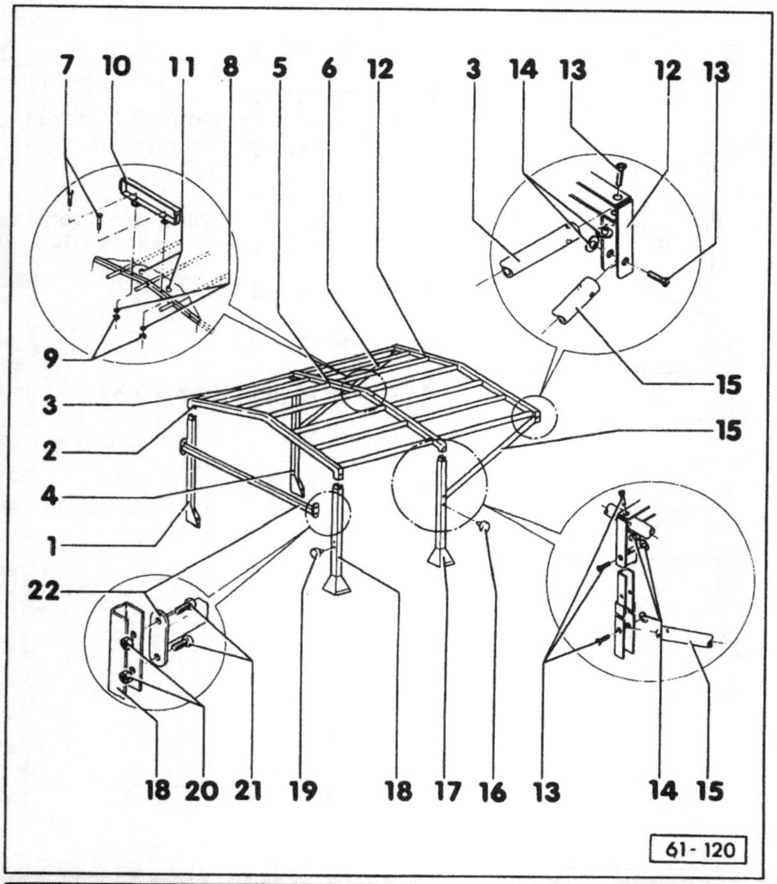

Tarpaulin assembly (double cab)

1 - **Right front rail**

2 - **Rail, upper front**
- Connect and bolt together with left front rail -18-, right front rail -1-.

3 - **Support pipe**
- Length = 2094 mm (approx. 82 in.)

4 - **Right center rail**

5 - **Rail, upper center**
- Bolt together with left center rail -17-, right center rail -4-

6 - **Right brace pipe**
- Bolt together with right center rail -4-, rail rear upper part -12-.

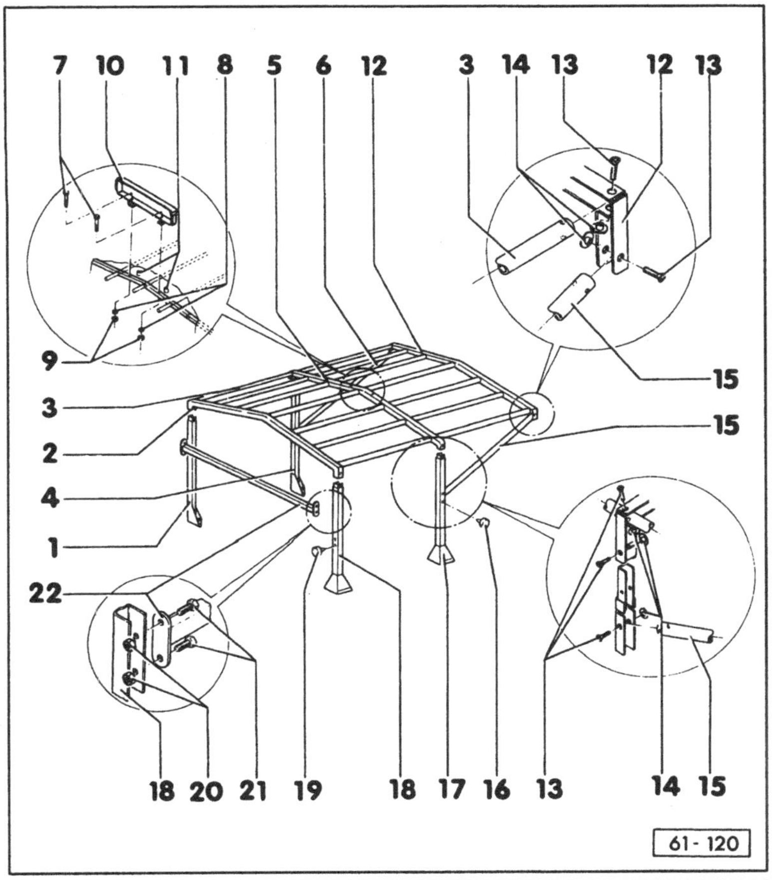

7 - **Hex bolt M6 x 45**

8 - **Spring ring B6**

9 - **Hex nut M6**

10 - **Ladder support**
- ♦ Not for USA

11 - **Spacer Pipe**

12 - **Rail, rear upper**

13 - **Countersunk screw M6 x 30**

14 - **Tapping screw M6**

15 - **Support pipe, left**
- ♦ Bolted together with center left rail -17-, rail rear upper part -12-.

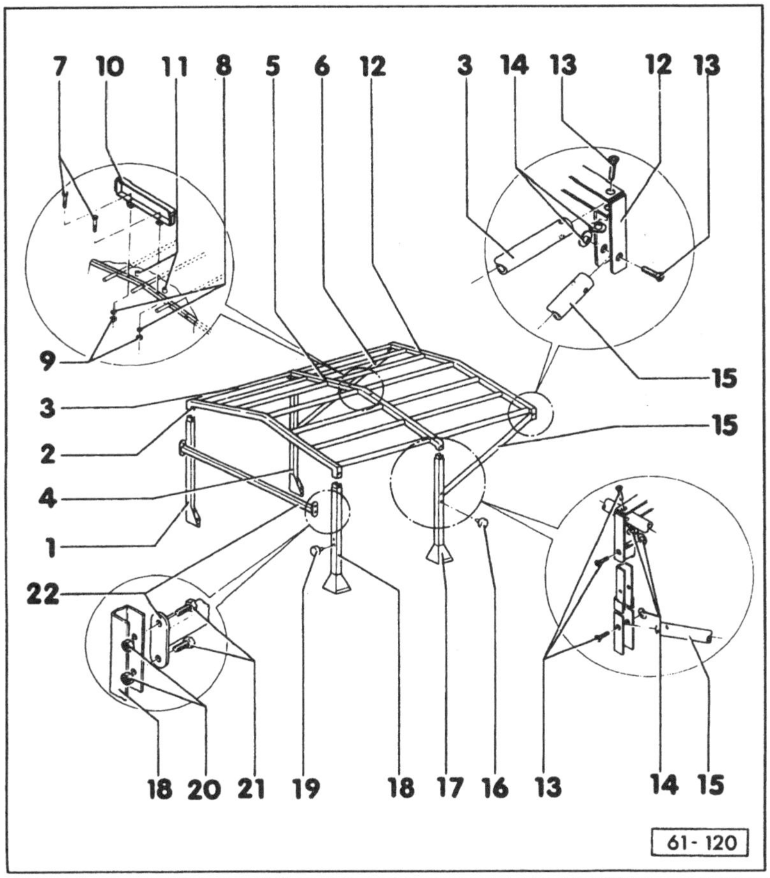

16 - **Rubber buffer**
- ♦ Press into hole on center left rail -17-, center right rail -4-.

17 - **Center left rail**

18 - **Left front rail**

19 - **Rubber buffer**
- ♦ Press into hole on left rail -18-, right front rail -1-.

20 - **Hex nut M8**

21 - **Hex bolts M8 x 22**

22 - **Brace**
- ♦ Bolted together with left front rail -18-, right front rail -1-.

Tarpaulin frame/tarpaulin, installing

– Place tarpaulin frame on bed (requires four technicians).

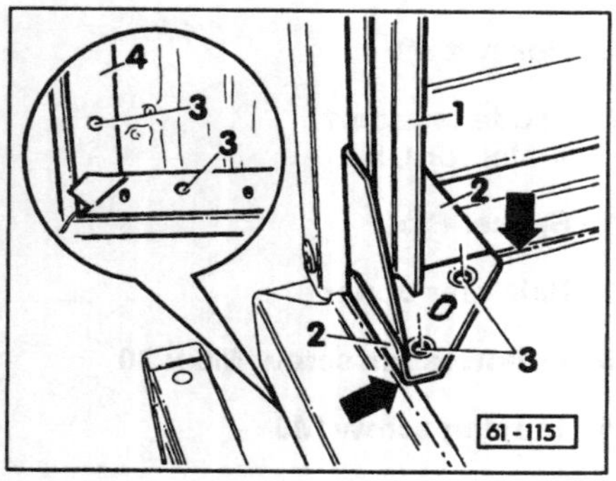

◄ – Mount rail -1- in corners of bed so outer foot edges -2- are flush with center of profile -arrows-.

– Mark holes -3- (9.0 mm dia.) and drill.

– Deburr holes in sideboard flange -4-.

Note:

Touch-up drilled holes, apply corrosion protection as required.

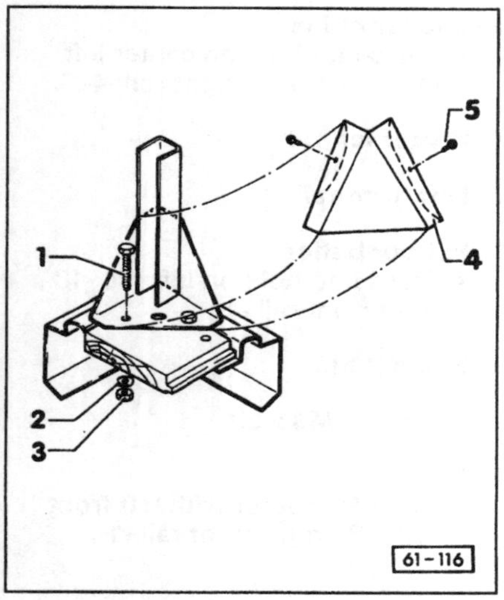

◄ – Bolt tarpaulin frame to bed.

1 – hex bolt M8 x 40

2 – washer A8.4

3 – hex nut M8

4 – deflector plate

5 – self-tapping screw

Note:

Cover two rear rail feet with deflector plate -4-, due to safety considerations.

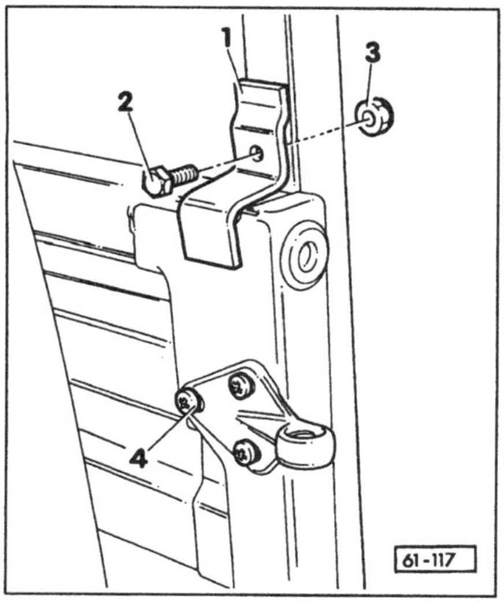

- Fasten left and right front rails to sideboard with mounting bracket -1-, hex bolt (M8 x 22) -2- and hex nut (M8) -3-.
- Remove Phillips screws -4- from both sides.

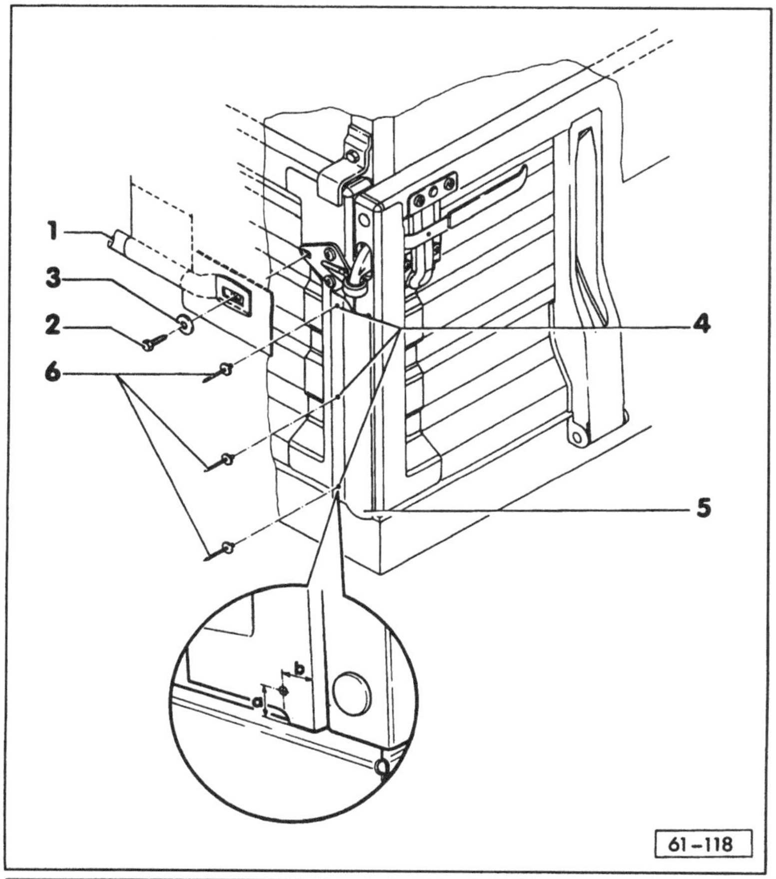

- Pull tarpaulin over tarpaulin frame and secure.
- Secure holding pipe for tarpaulin -1- in place with six hex bolts (M6 x 26) -2-, washer (B6.5 x 21 x2) -3- on left and right front.
- Mark holes (4.1 mm dia.) -4- for water deflector plate -5- on left and right front, drill and deburr.
  - ♦ Lower hole

    a=15 mm (approx. 19/32 in.)

    b = 18 mm (approx. 45/64 in.)
  - ♦ Water deflector plate -5- serves as a template for center and upper holes.

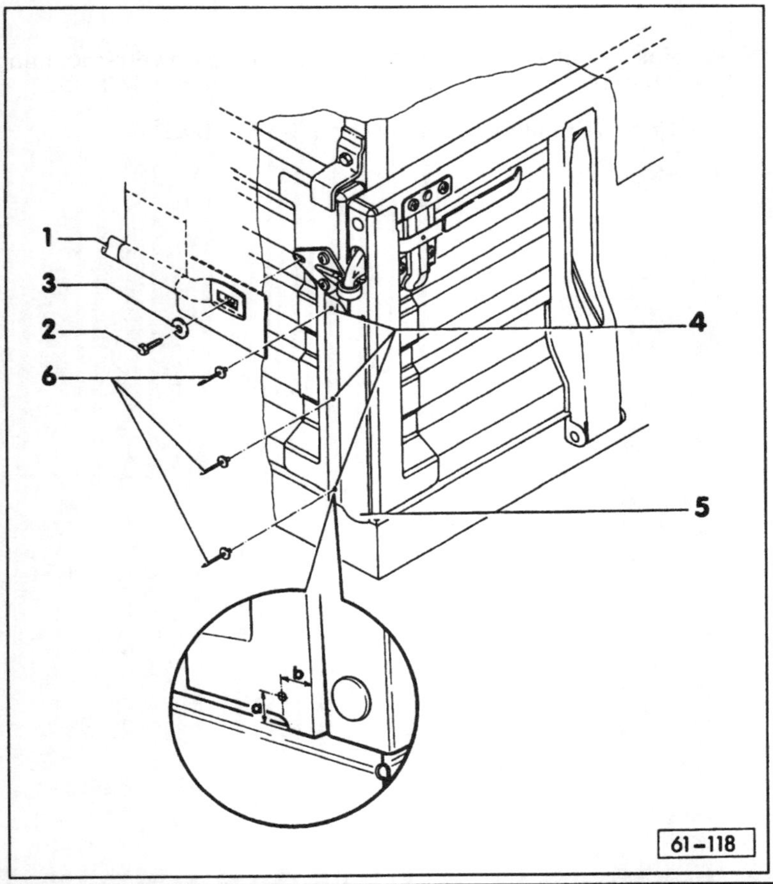

- Repair paint damage

> ***CAUTION!***
>
> ***Part numbers are for reference only. Always check with your Parts Department for latest parts information.***

- Fasten water deflector plate -5-, left and right front, with tubular rivets (N 900 978 01).

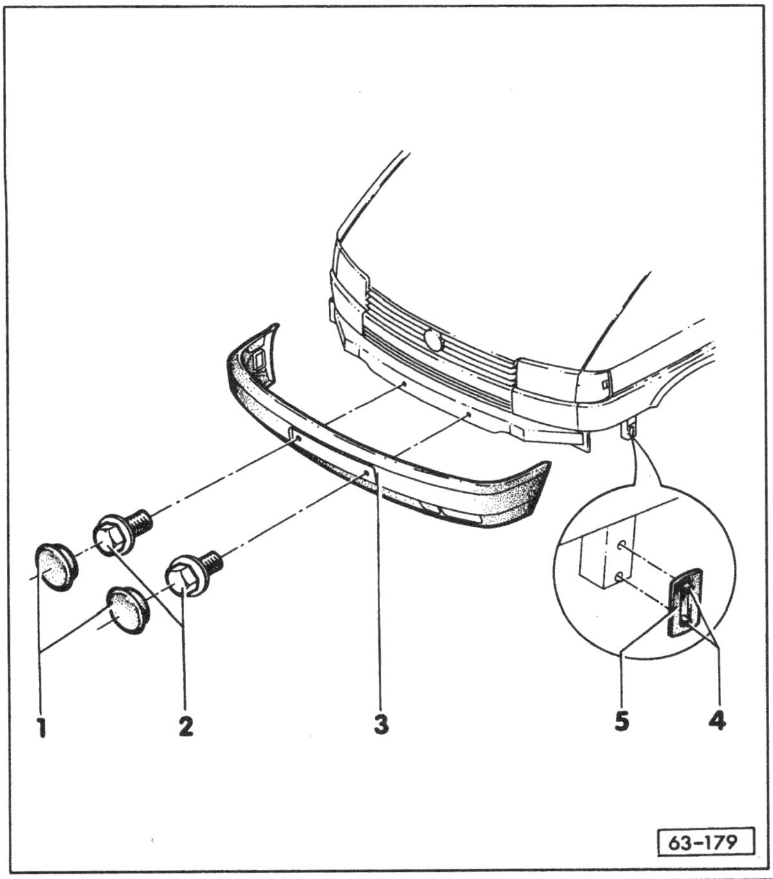

Front bumper ➤12.95, components

1 - Cover

2 - Bolt w/washer
- ♦ 10 Nm (7 ft lb)

3 - Front bumper
- ♦ Removing: Pull/drive pins -4- out and pull bumper parallel out of guides -5-
- ♦ Installing: Slide bumper parallel onto guides -5-, secure by driving in pins -4- flush

4 - Pin

5 - Guide

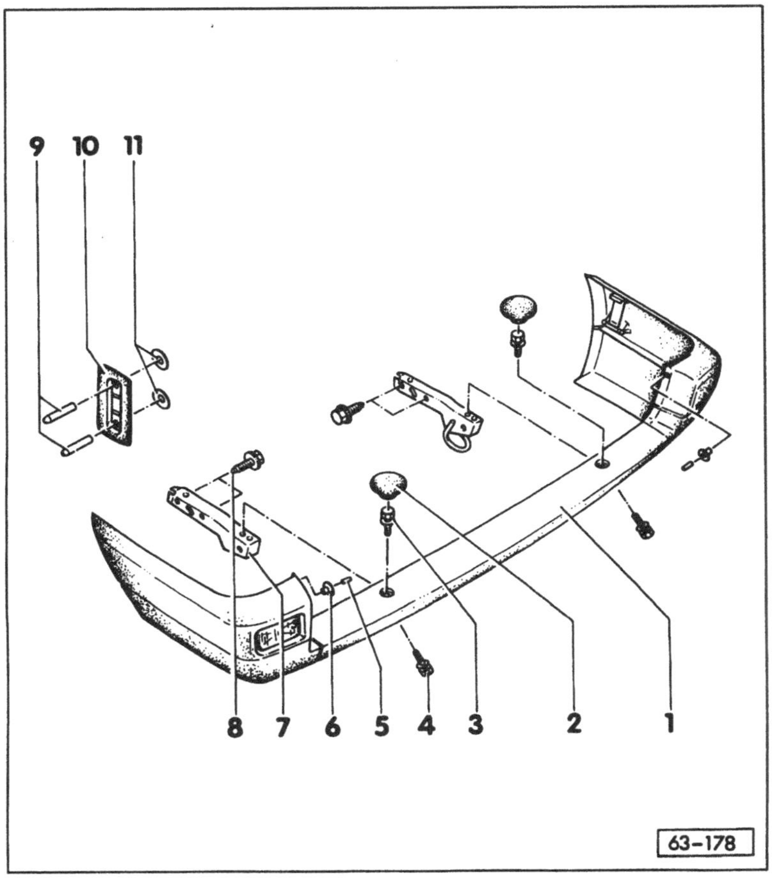

Rear bumper ➤12.95, components

1 - Rear bumper

2 - Cover

3 - Bolt w/washer
- ♦ 15 Nm (11 ft lb)

4 - Bolt w/washer
- ♦ 15 Nm (11 ft lb)

5 - Pin
- ♦ Replacement part includes -5- and -6-

6 - Expanding clip
- ♦ Replacement part includes -5- and -6-

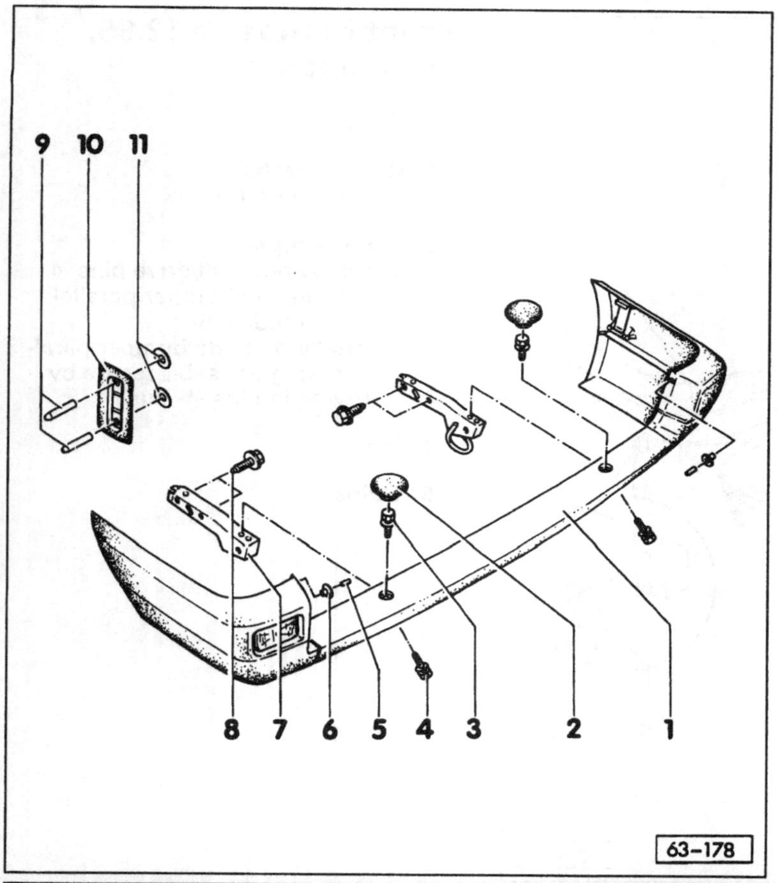

7 - **Bumper bracket**

8 - **Bolt w/washer**
- ♦ 110 Nm (81 ft lb)

9 - **Pin**

10 - **Guide**
- ♦ To remove and install, push in or slide out bumper parallel from guide (left and right)

11 - **Sealing washer**

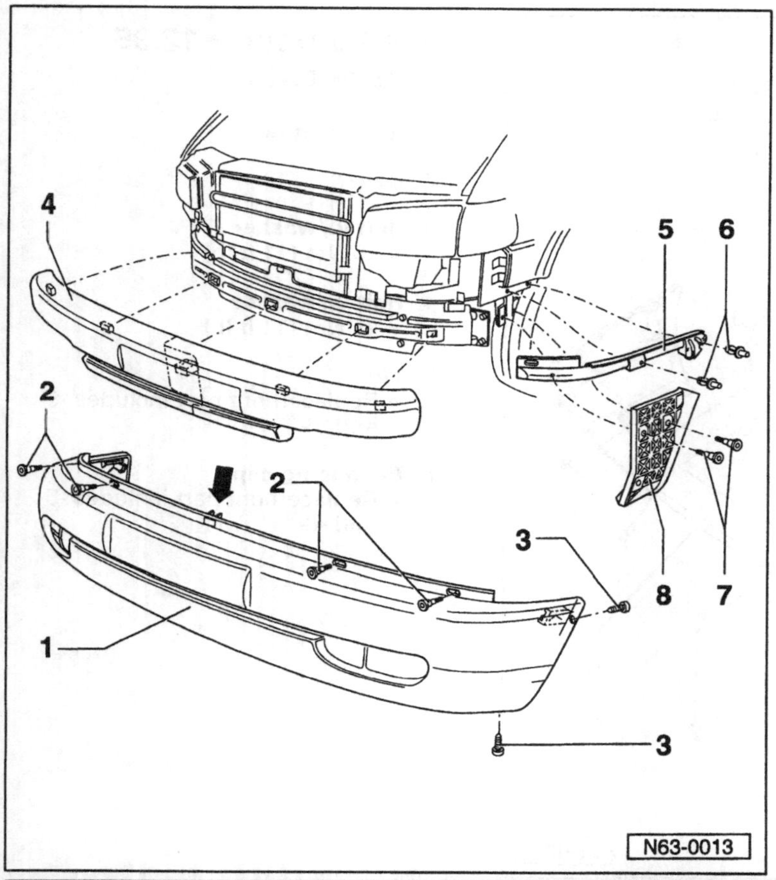

Front bumper 01.96 ➤ removing/installing (long front end)

1 - **Cover**
- – Remove radiator grille with trim frame ⇒ page Q 66-2.
- – Remove torx screws -2- and -3-.
- – Pull cover forward out of guides.
- ♦ Note center faster -arrow-

2 - **Torx screw T 30 (4x)**
- ♦ 6 Nm (53 in. lb)

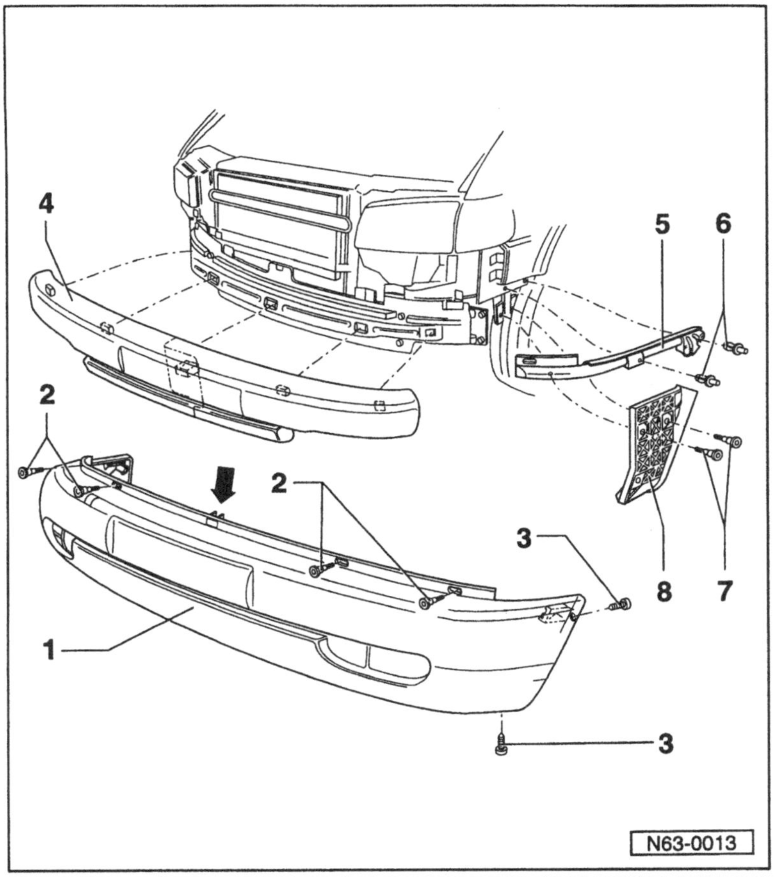

3 - **Torx screw T25 (2x)**
- 1.2 Nm (11 in. lb)

4 - **Bumper beam**
- Disconnect right side then left

5 - **Guide bracket**
- To remove/install bumper beam, push/pull beam straight (parallel) out of guide bracket

6 - **Rivet**

7 - **Hex screw (4x)**
- 1.5 Nm (13 in. lb)

8 - **Cover bracket**

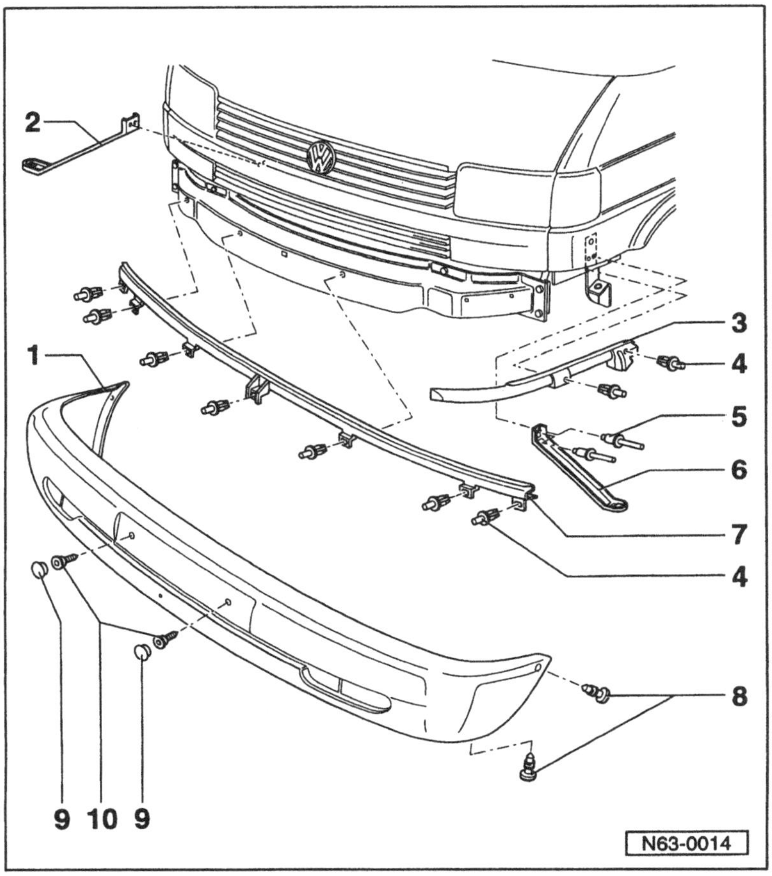

Front bumper 01.96 ➤ removing/installing (short front end)

1 - **Cover**
- Pry out caps -9-, remove combination hex screw and washer assemblies -10-, Torx screws -8-.

2 - **Bumper bracket**

3 - **Bumper guide bracket**
- To remove/install bumper beam, push/pull beam straight (parallel) out of guide bracket

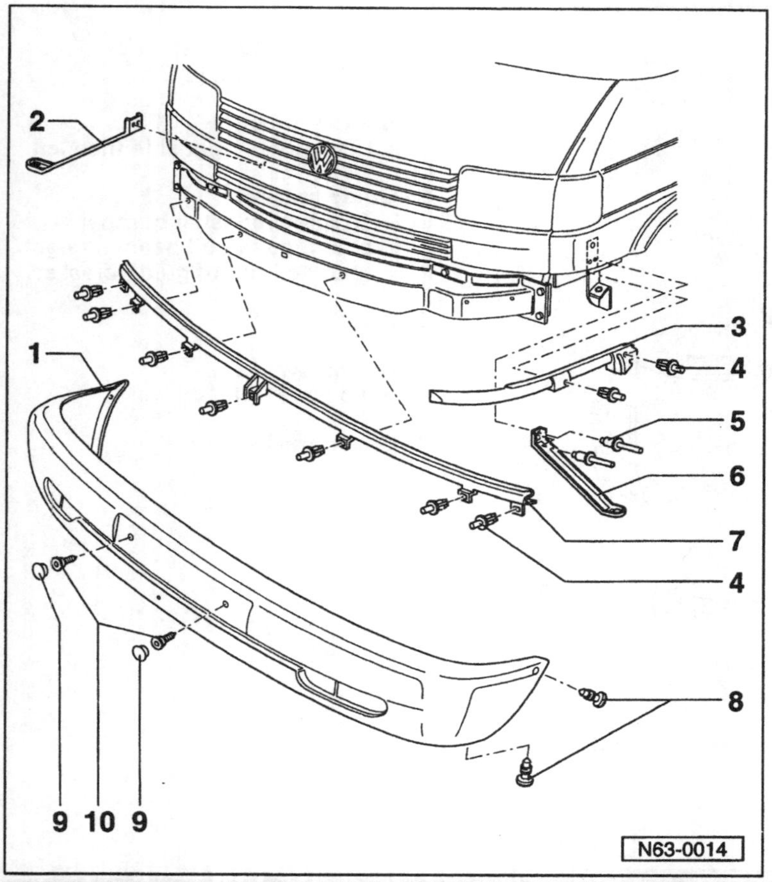

4 - **Rivet**

5 - **Blind rivet**

6 - **Bumper bracket**

7 - **Cover guide bracket**

8 - **Torx screw T25 (2x)**
- ♦ 1.2 Nm (11 in. lb)

9 - **Cap, hex screw**

10 - **Screw / washer assembly**
- ♦ 6.5 Nm (58 in. lb)

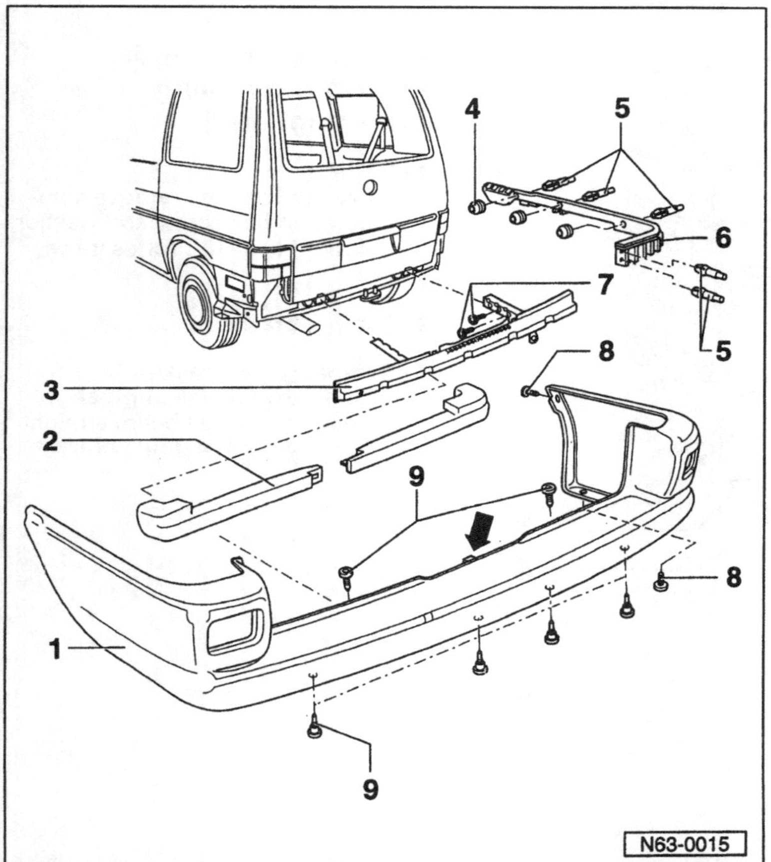

Rear bumper 01.96 ➤ removing/installing

1 - **Cover**
- – Remove torx screws -8-, hex screws -7-.
- – Remove bumper with rail -3- from guides -6-.
- – Remove torx screws -9-.
- – Pull cover from rail with brackets -3-.
- ♦ Note center fastening point

2 - **Bumper beam**
- ♦ When removing, disconnect right side then left

3 - **Rail with bracket**

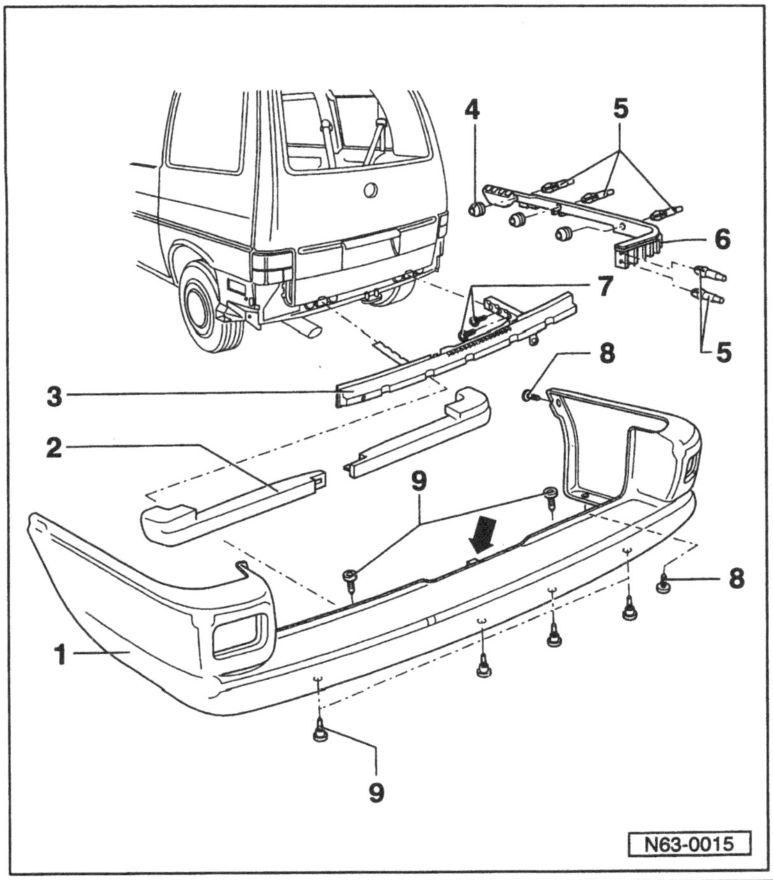

4 - Sleeve

5 - Expansion clip

6 - Bumper guide bracket
- ♦ To remove/install bumper beam, push/pull beam straight (parallel) out of guide bracket.

7 - Hex bolt with washer
- ♦ 15 Nm (11 ft lb)

8 - Torx screw T25 (2x)
- ♦ 1.2 Nm (11 in. lb)

9 - Torx screw T30
- ♦ 6 Nm (53 in. lb)

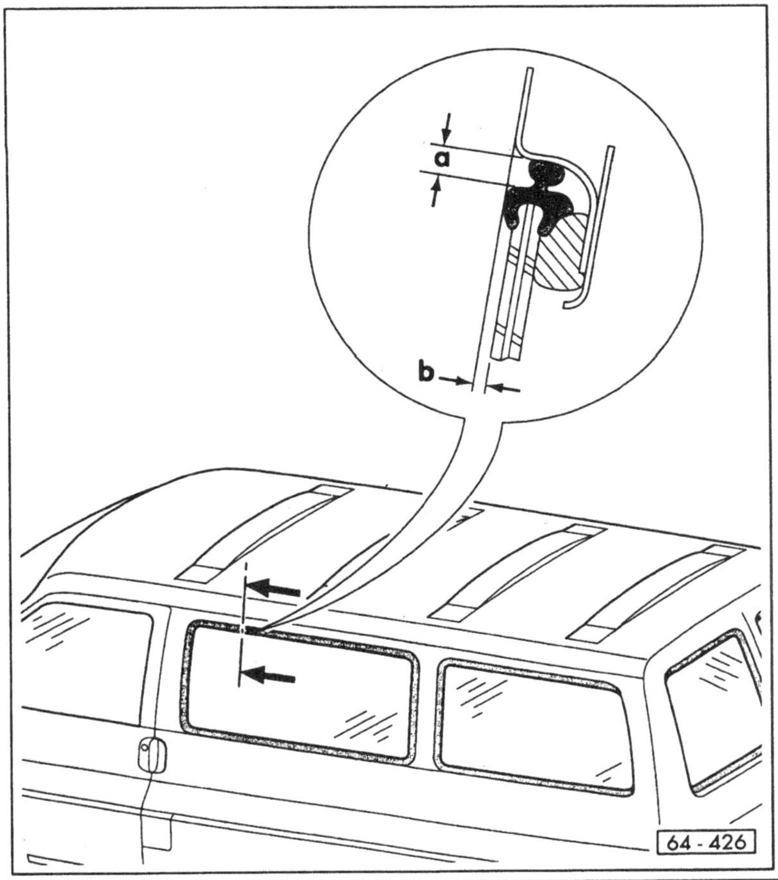

Flush bonded windows, removing/installing

Note:

Windshield, rear, and side windows are flush-bonded in window opening with 2K PUR adhesive.

1 - Windshield
- Removing and installing ⇒ Page Q 64-6

2 - Front side panel/sliding door window
- Removing and installing ⇒ Page Q 64-7

3 - Rear side panel window
- Removing and installing ⇒ Page Q 64-7

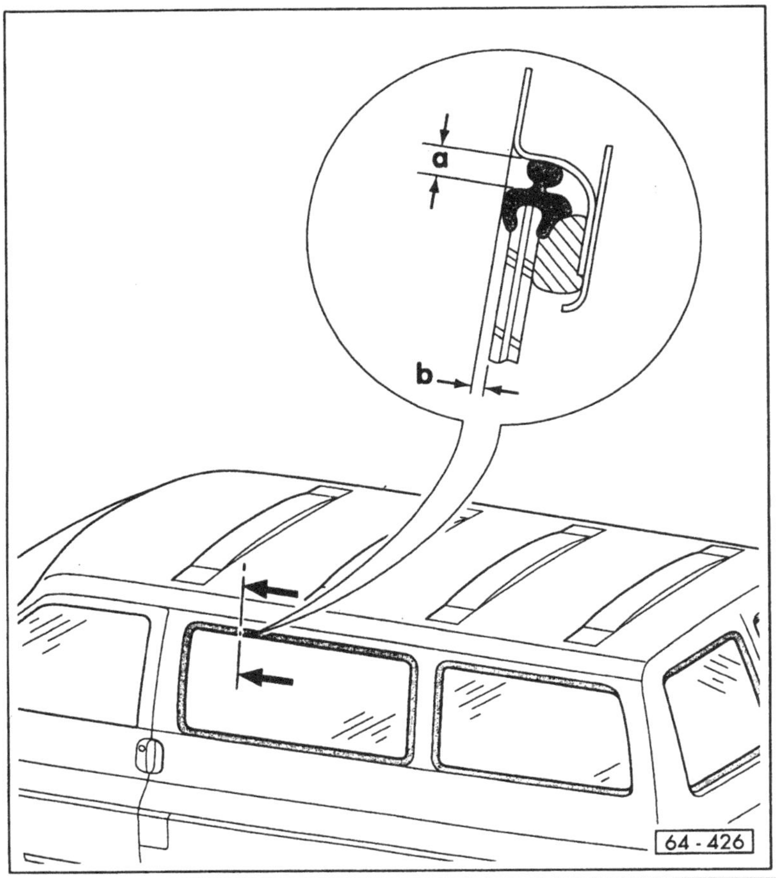

4 - Rear double door window
- Removing and installing ⇒ Page Q 64-7

5 - Tailgate window
- Removing and installing ⇒ Page Q 64-7

6 - Cab rear window
- Removing and installing ⇒ Page Q 64-7

7 - Side panel/double cab window
- Removing and installing ⇒ Page Q 64-7

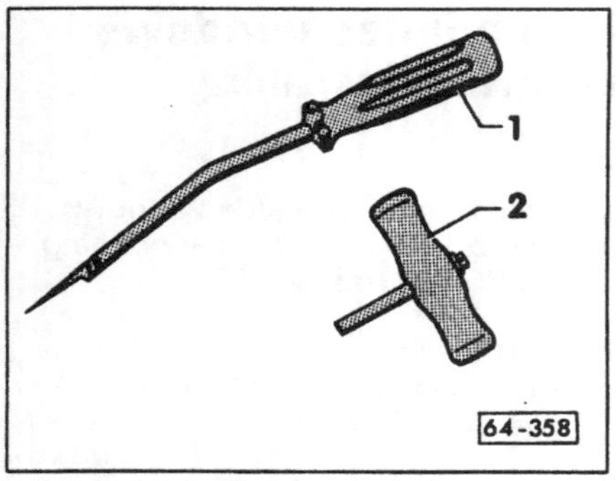

Tools

◄ ♦ Tool Kit VAG 1351 (handle -1- and pulling handle -2-) to remove windshield or back and side windows that are not broken

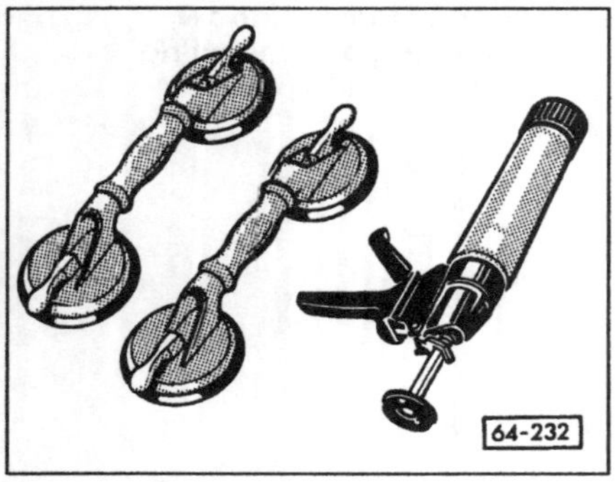

◄ ♦ Caulk gun VAG 1628 or equivalent

♦ Double suction pad VAG 1344 or equivalent

Not shown

♦ Protective goggles, gloves (leather)

♦ Utility Knife

♦ Awl (from Tool Kit VAG 1474) or equivalent

♦ Fabric or cloth tape

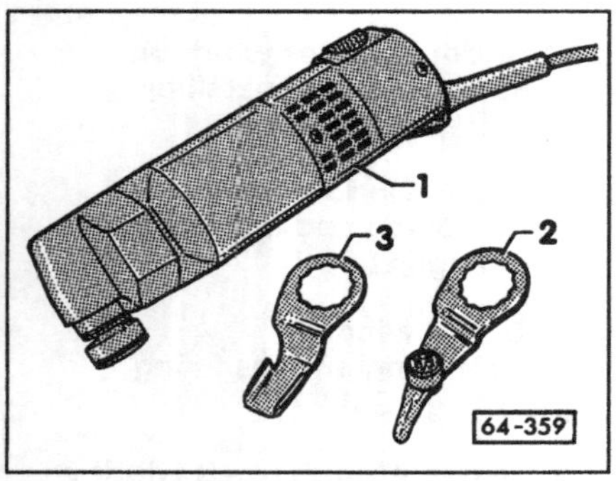

◄ ♦ Electric cutting tool VAG 1561 -1- or equivalent

♦ Knife blade with stop roller VAG 1561/9 -2- for removal of broken side and rear windows

♦ Scraping knife VAG 1561 -3- to cut adhesive sealing material on window and/or on body flange

> ***CAUTION!***
>
> ***Tape around window opening with cloth or fabric tape to prevent damage to paint when removing/installing glass with electric cutter.***

CAUTION!

Part numbers are for reference only. Always check with your Parts Department for latest Information.

Materials

- 2K adhesive sealant D 004 300 05 or equivalent
- Activator AMV 181 800 02 or equivalent
- Glass/paint primer D 009 200 02 or equivalent
- Cleaning solution D 009 401 04 or equivalent
- Primer/applicator (felt ball with wire) or equivalent
- Adhesive remover D 002 00010 or equivalent
- Cutting wire 357 845 955

Note:

If 300 ml cartridge is not sufficient, small (110 ml) adhesive sealant cartridge D 009 100 03 or equivalent is recommended.

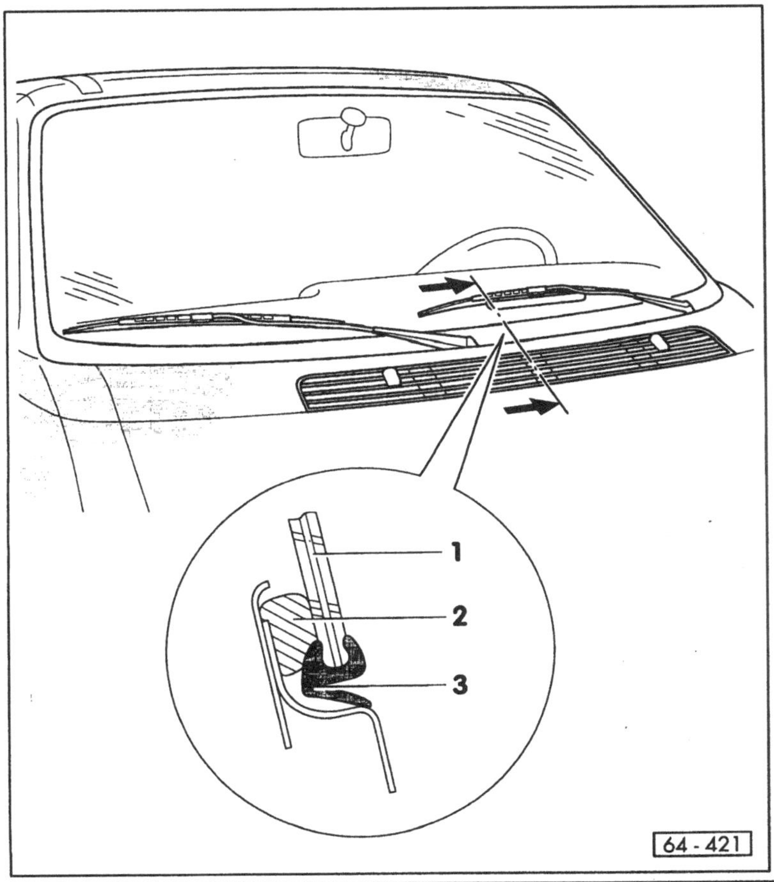

Windshield, removing/installing

1 - Windshield
- Removing ⇒ Page Q 64-11
- Preparations for installation ⇒ Page Q 64-20
- Installing ⇒ Page Q 64-15
- Re-sealing ⇒ Page Q 64-24

2 - PUR adhesive sealant
- Bead cross section
  Width = 8 mm (1/3 in.)
  Height = 13 mm (1/2 in.)
- Curing time ⇒ Q 64-22

3 - Window molding
- One-piece destroyed during removal
- Mount first before applying adhesive bead

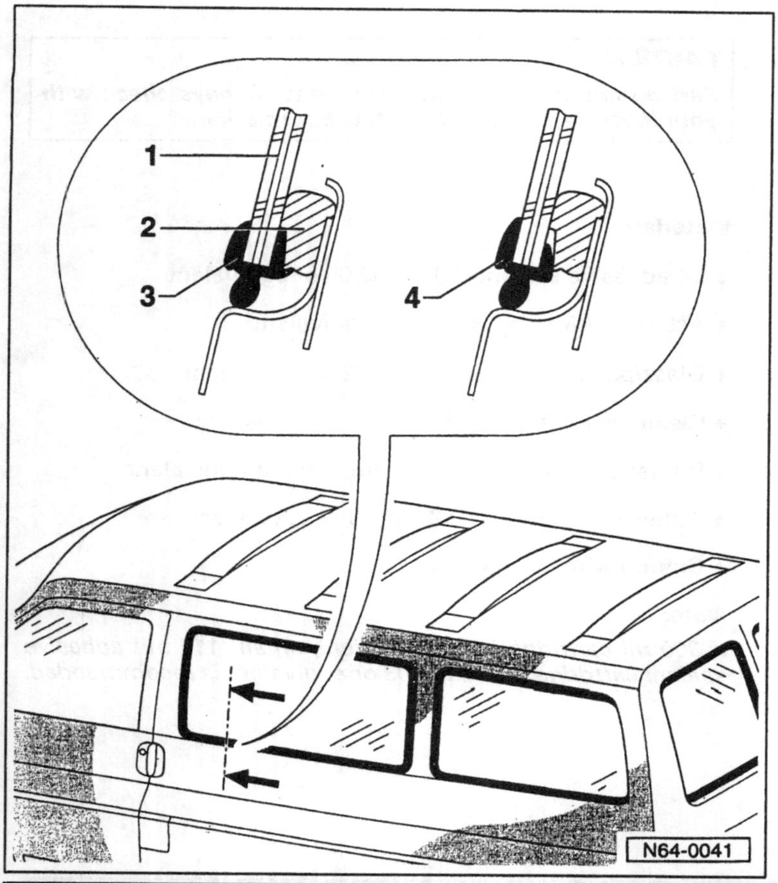

Side and rear window, removing/installing

1 - Side/rear window
- Removing:
  Unbroken ⇒ Page Q 64-11
  Broken ⇒ Page Q 64-14
- Preparations for installation ⇒ Page Q 64-20
- Installing ⇒ Page Q 64-11
- Re-sealing ⇒ Page Q 64-24

2 - PUR adhesive sealant
- Bead cross section
  Width = 8 mm (1/3 in.)
  Height =13 mm (1/2 in.)
- Curing time ⇒ Page Q 64-22

3 - Window molding
- One-piece, destroyed during removal
- Mount first before applying adhesive bead
- See item 4 for re-repair profile. If glass is replaced use production seal.

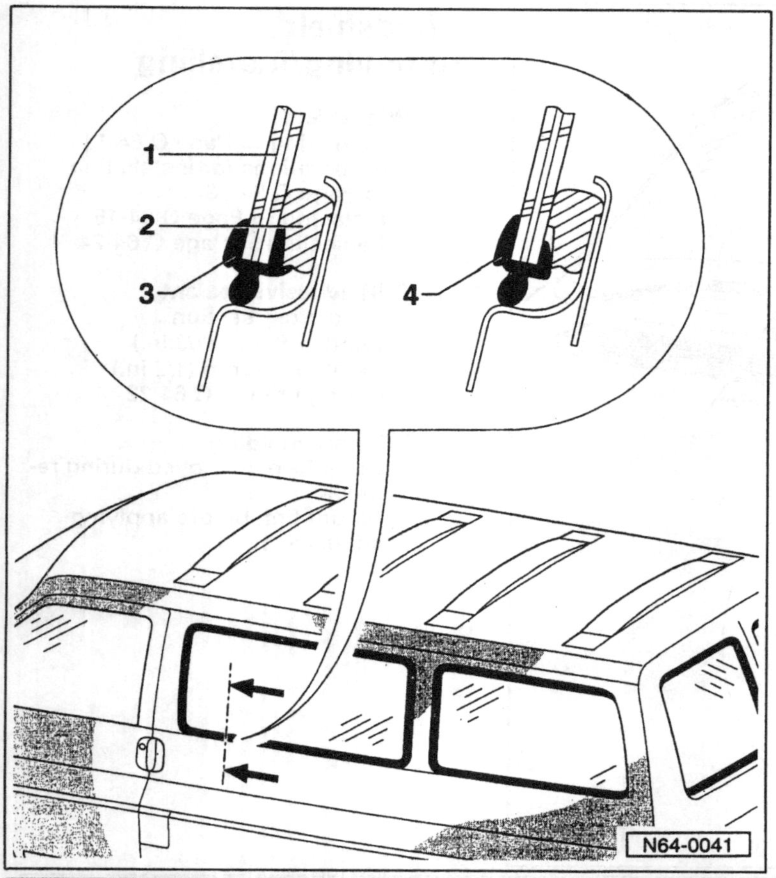

4 - Molding (Repair profile)
- Beginning 10.95
- Size to fit
- To be used during painting, without removing glass.
- See Parts Department for correct part number
- Installing ⇒ page Q 64-9

Molding (repair profile), installing

The production molding is removed.

- Spray glass edge with D 009 401 04 lubricant.
- ◄ Place repair molding in middle lower area -arrow A- and fasten along window cutout -arrow B-.
- Cut repair molding approximately 4 mm (0.15 in.) longer.

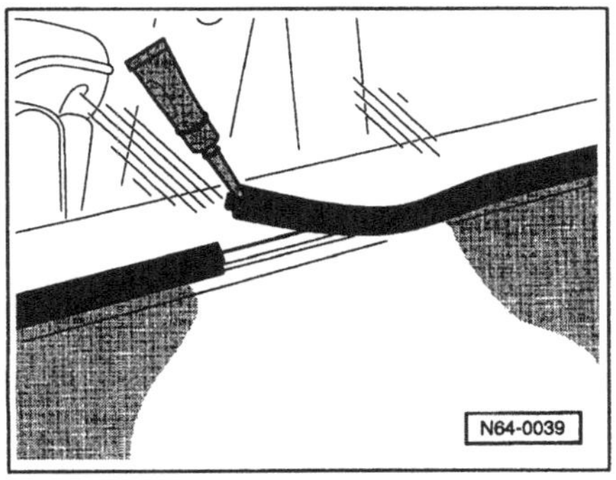

- ◄ Apply Cyan acrylic adhesive AMV 195 KD1 to repair molding end.

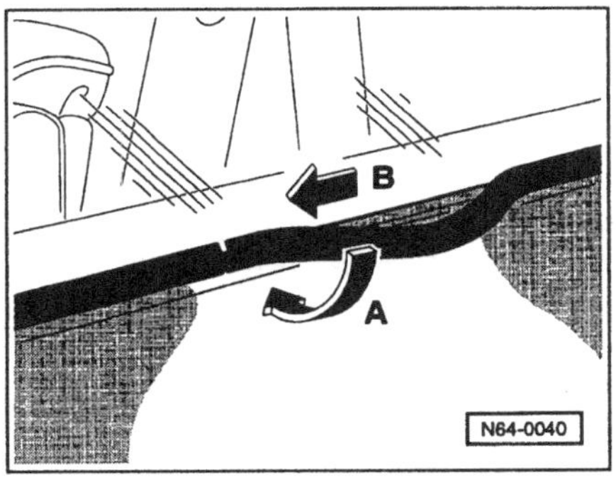

- ◄ Repair molding end should also be attached -arrow A-.
- Press repair molding -arrow B- together and hold for five seconds.
- Install remainder of seal in window cut out.

Windshield, unbroken side windows and unbroken rear windows, removing

> ***CAUTION!***
>
> ***Always wear protective goggles and leather gloves when removing/installing glass.***

Windshield

– Remove windshield wipers.

– Disconnect window antenna.

Rear window

– Disconnect rear window heating element.

– Pull window frame completely off window.

– Mask window opening.

– Push awl (from VAG 1474) through sealing bead from inside vehicle.

Note:

For windshield, push awl through at lower left or lower right corner.

– Adjust cutting wire to 350 - 400 mm (approx. 14 to 15 3/4 in.) and pull through sealing compound from outside with awl.

Notes:

- *Two technicians required to remove windshield.*
- *Specified method for cutting out windshield must be used.*
- *Do not cut out windshield using sawing motions. This will heat wire and cause it to break. Wire can also break if it becomes twisted or bent.*

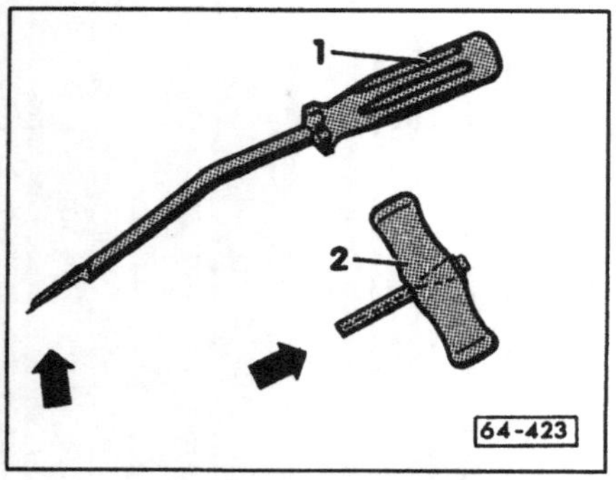

◄ – Fasten cutting wire with wire ends on bracket -1- (inside vehicle) and handle -2- (outside vehicle).

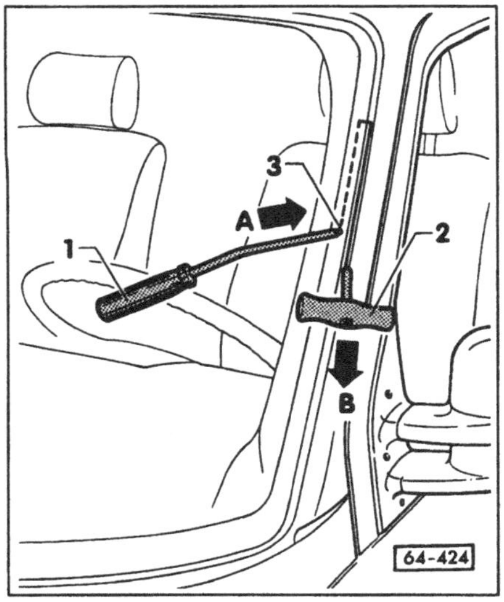

- Insert handle -1- into adhesive sealing component and hold in direction of -arrow A-.
- Use pulling handle -2-, pull wire evenly in direction of -arrow B- to stopping point -3-.
- Pull wire with handle -1- approx. 100 – 500 mm (approx. 4 – 20 in.) toward the inside, insert handle -1- on adhesive material again and hold. Cut to next stopping point.
- Cut out window completely.

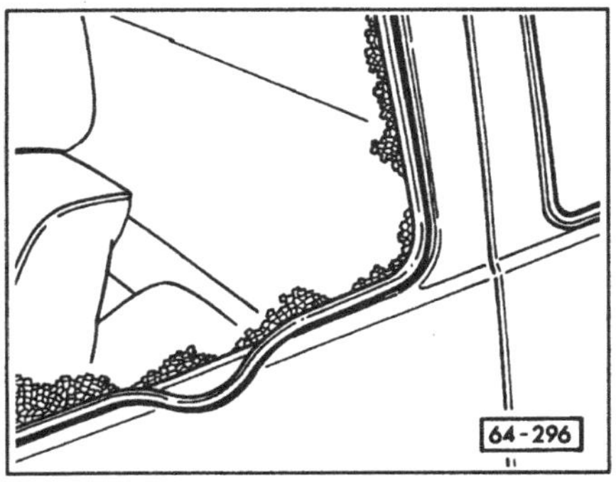

Broken side and rear window, removing

WARNING!

Always wear protective goggles and leather gloves when removing glass.

CAUTION!

- ***Tape around window opening with cloth or fabric tape to prevent damage to paint when removing/ installing glass with electric cutter.***
- ***Protect body and interior from glass splinters.***

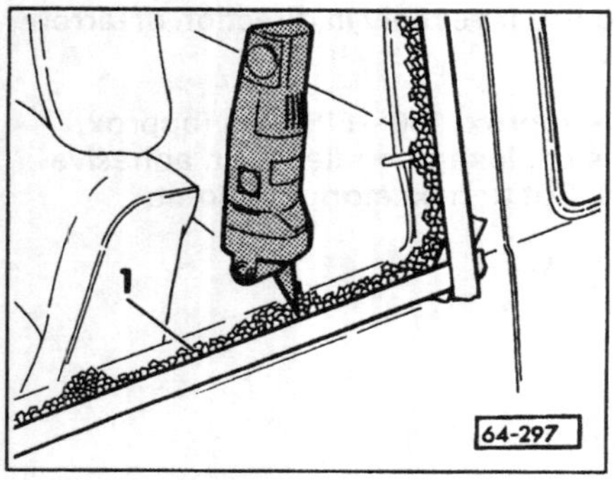

- Pull molding out all around.
- Remove glass pieces down to adhesive sealing compound.
- Cut off adhesive sealing compound -1- (with glass residue) using electric cutting tool VAG 1561 ,or equivalent, and knife blade VAG 1561/9 (with stop roller).
- If necessary, separate adhesive sealing material in corners with utility knife.

Windshield, side and rear window, installing

- Prepare window for installation ⇒ Page Q 64-20.

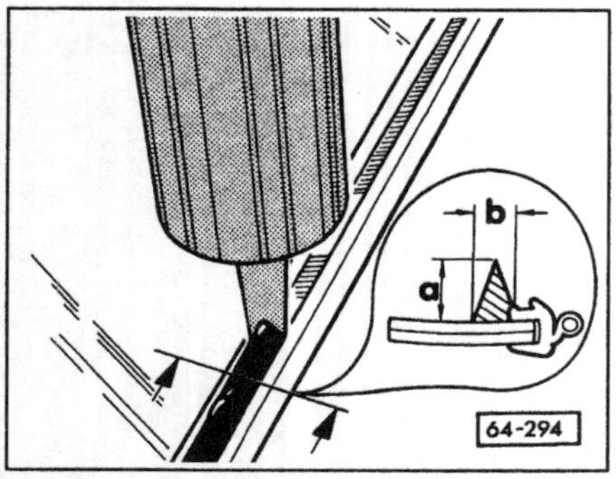

- Apply adhesive sealing material around window:

♦ a = 13 mm (1/2 in.)

♦ b = 8 mm (1/3 in.)

Note:

- *To install rear double door and sliding door windows, adhesive sealing compound is applied to window flange on door.*
- *Height of sealing bead must be reduced by height of remaining material on window body flange.*

CAUTION!

- ***2K adhesive compound must be mixed according to supplied instructions.***
- ***Window must be installed within 15 minutes otherwise adhesion of adhesive sealing material will be reduced.***

- Install window in opening using two double suction pads (VAG 1344).
- Align window and press into window opening as follows:

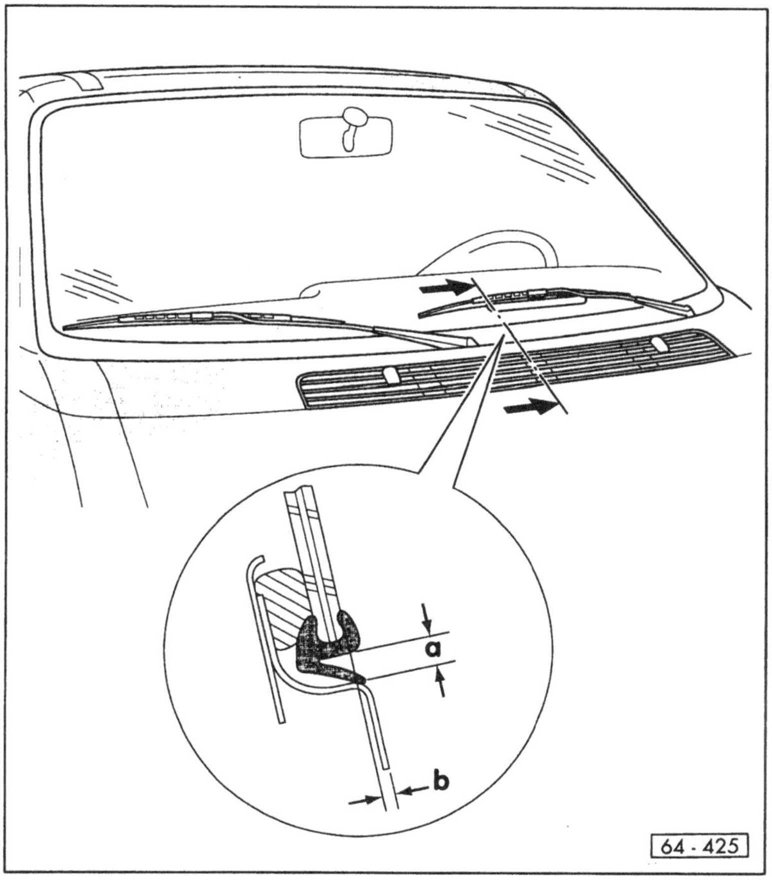

Windshield

- a = 3 – 4 mm (1/8 to 5/32 in.) all around
- b = 1 – 2 mm (1/32 to 5/64 in.) all around

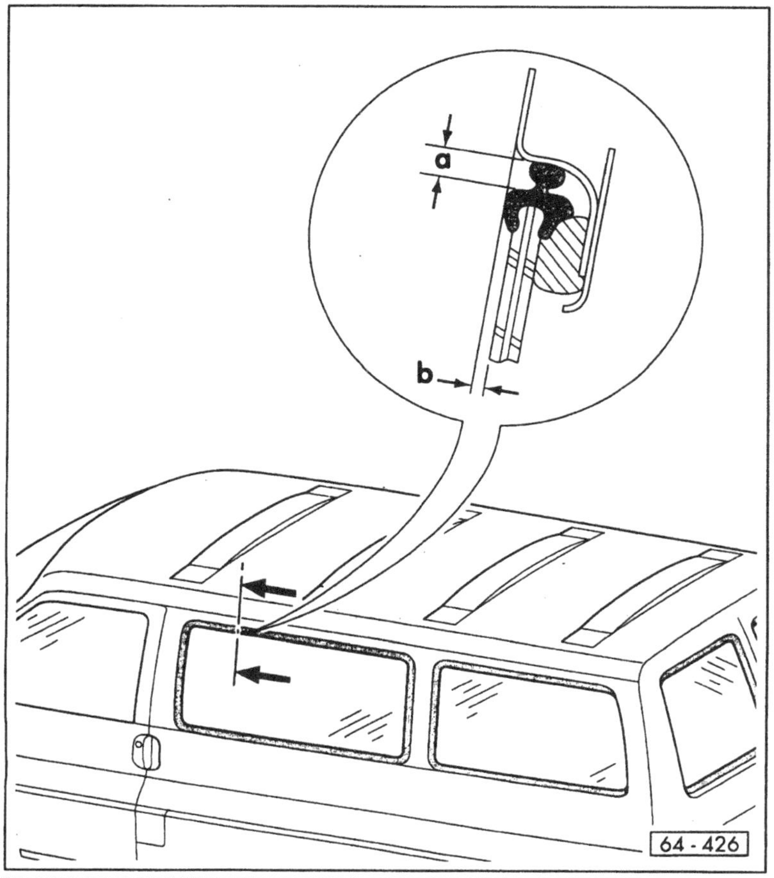

Side and rear window

- a = 3 – 4 mm (1/8 to 5/32 in.) all around
- b = 1 – 2 mm (1/32 to 5/64 in.) all around

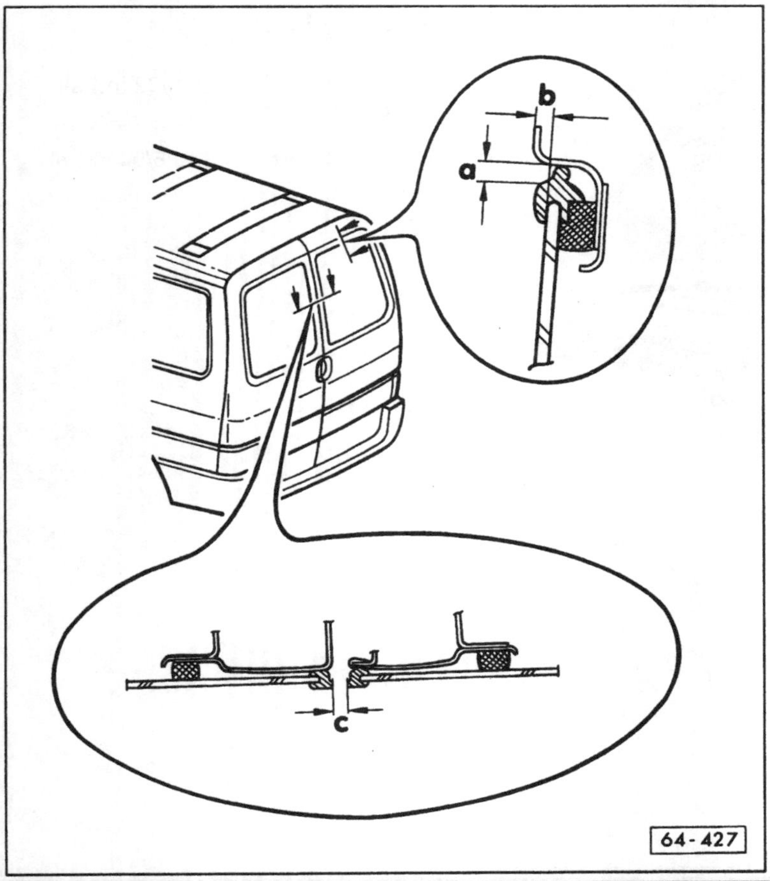

Rear double door window

- a= 3 – 4 mm (1/8 to 5/32 in.) all around
- b = 1 – 2 mm (1/32 to 5/64 in.) all around
- c = 7 – 8 mm (9/32 to 5/16 in.) all around

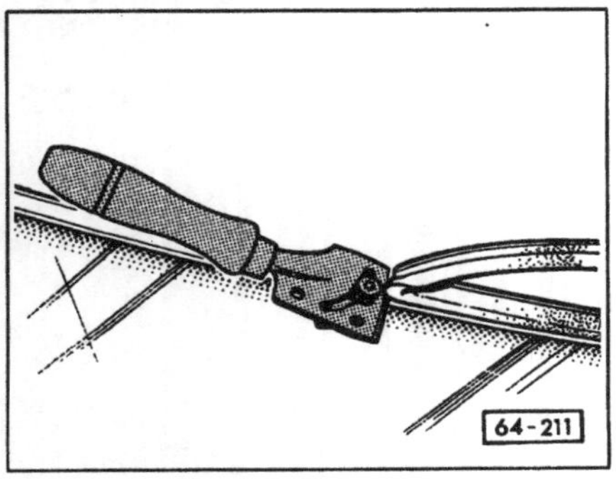

Window installation, preparations

Unbroken window

Notes:

◄

- *When an unbroken window is reused, cut back remaining adhesive material to 1-2 mm (1/32 to 5/64) in.) before re-gluing (with electric cutting tool and scraping knife VAG 1561/8).*
- *Remaining material serves as adhesive base for new adhesive*

> ***CAUTION!***
>
> ***Do not prime or activate adhesive surface or treat it with cleaning solution. Keep adhesive surface dirt and grease free.***

Exception:

If gluing will not take place immediately after cutting back, remaining residual material must be activated with activator AMV 181 800 02 or equivalent.

– Install molding on window.

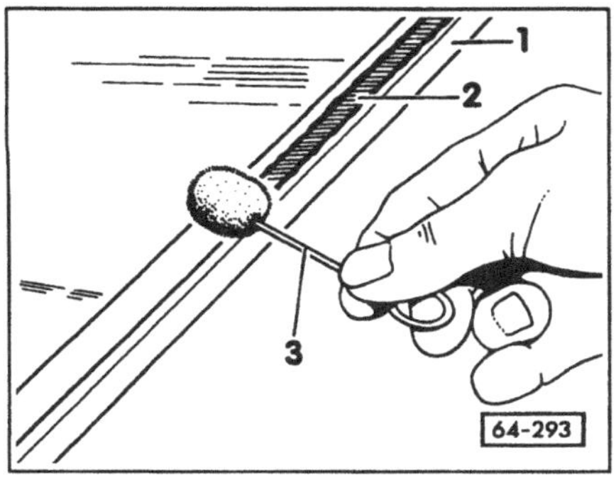

New window installation, preparing

- Clean edge of window around to width of about 30 mm (approx. 1.2 in.) with cleaning solution D 009 401 04 or equivalent and wipe dry with lint-free cloth.
- Install molding -1- window.

CAUTION!
Ceramic application on window Is not primer. Priming must be done before applying adhesive sealing compound Use only primer D 009 200 02 or equivalent.

- Using applicator D 009 500 25 or equivalent -3-, apply primer -2- evenly in one continuous motion.

♦ Drying time approx. 10 minutes

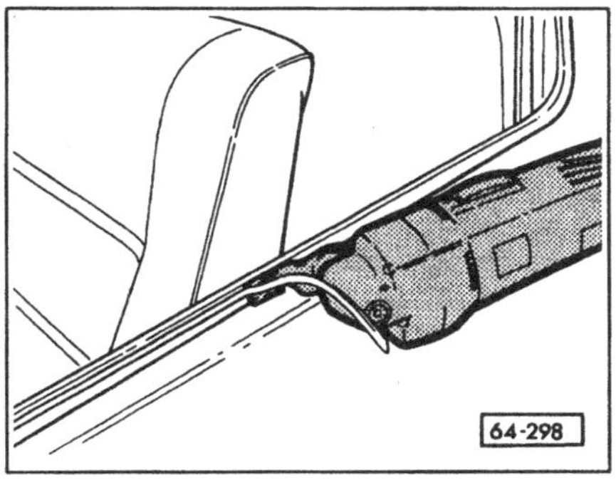

Body flange and/or window flange on door, preparing

- Cut back remaining adhesive material on body flange to 1-2 mm (1/32 to 5/64 in.) height with electric cutting tool VAG 1561 and scraping knife VAG 1561/8.

Note:

Do not remove adhesive completely. It is used as a base for new adhesive material.

CAUTION!
Do not prime or activate adhesive surface or treat it with cleaning solution. Keep adhesive surface dirt and grease free.

Exception:

If gluing will not take place immediately after cutting back, remaining residual material must be activated with activator AMV 181 800 02 or equivalent.

♦ If body flange at area of glass is being worked on or replaced flange must be cleaned and primed after painting.

Curing time of adhesive sealing material

Hardening time: from gluing in window until vehicle is placed in service. Vehicle must stand on a level surface for 1 hour at room temperature before it can be moved.

Sealing leaks

– Water test window areas.

– Visually inspect from inside.

– Blow unsealed area dry and clean.

- Windshield can only be sealed from inside in area of A-pillars and roof.
- Rear and side windows can be sealed from inside.

– Let adhesive sealing material air dry.

Note:

Small adhesive sealant cartridge D 009 100 03 (110 ml) or equivalent is recommended for resealing.

Paint damage, repairing

Refer to EuroVan Body-Collision Repair Manual for paint repair and corrosion protection.

For paint damage in non-visible areas, following method is suitable:

> ***CAUTION!***
>
> ***Part numbers are for reference only. Always check with your Parts Department for latest information.***

♦ Paint twice (wet on wet) with glass/paint primer D 009 200 01 or equivalent - drying time approx. 10 minutes.

Surplus bonding/sealing compound, removing

CAUTION!

- ***When using any cleaning agent, follow all cautions and warnings listed on the containers.***
- ***While adhesive sealant is hardening, avoid putting pressure on glass from inside car when cleaning glass.***
- ***Part numbers are for reference only. Always check with your Parts Department for latest information.***

Adhesive remover D 002 000 10 or equivalent is recommended for use as a cleaning agent. When using adhesive remover, all safety instructions must be observed.

– First clean painted areas roughly with a dry cloth. Remove remaining sealant with cleaning agent D 002 000 10 or equivalent.

Plastic trim, cleaning

– Allow bonding/sealing material to cure (approx. 1 hour) and then pull off.

Sliding window, repairing

Notes:

- *There are different types of sliding windows, depending on vehicle equipment.*
- *Sliding window in side panel, sliding door, and side double door are bolted on. Sliding window in dividing panel is clamped on with a frame.*
- *If fixed window is broken, sliding window must be replaced as a complete unit.*
- *Following descriptions relate to sliding window in sliding door. Repair to other sliding windows is identical.*

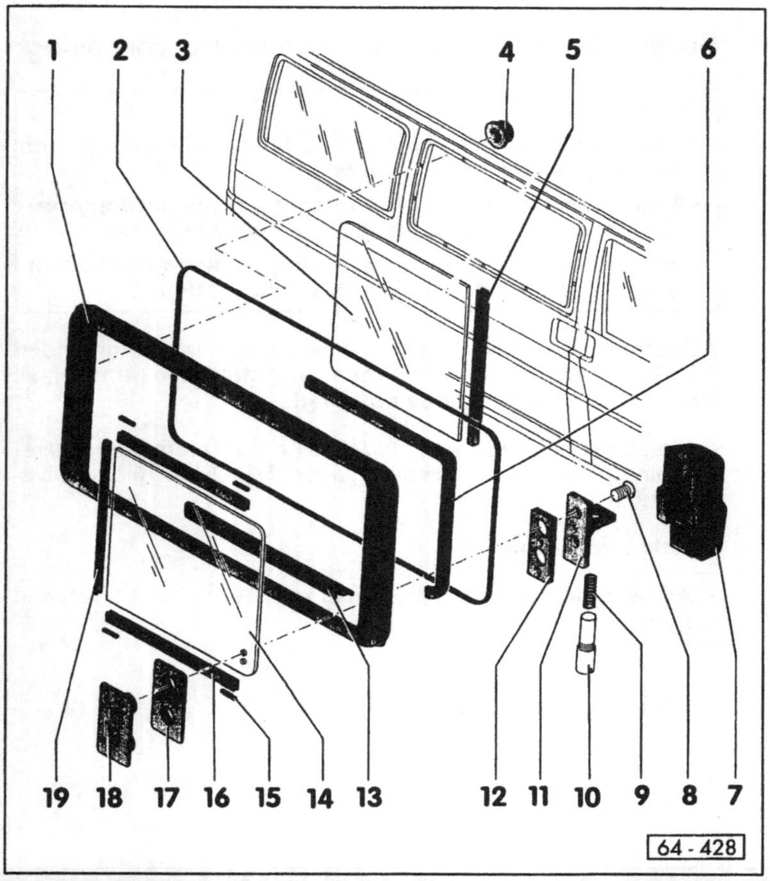

Sliding window (sliding door), components

> ***CAUTION!***
>
> ***Part numbers are for reference only. Always check with your Parts Department for latest information.***

Note:

To remove sliding window (sliding door), first remove molded trim on sliding door *⇒ Page Q 70-84.*

1 - Frame
- With studs

2 - Gasket
- Glued on with frame

3 - Fixed window
- If broken, replace sliding window unit

4 - Hex lock nut
- 1.5 Nm (13 in. lb)
- Cap nut for vehicles without molded trim

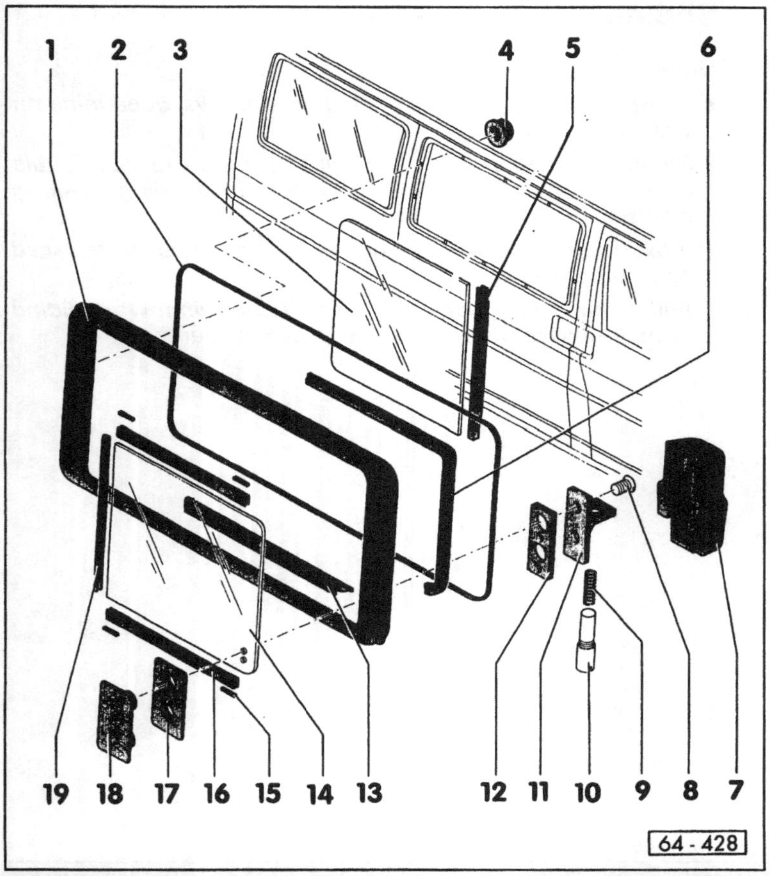

5 - Divider post
- Glued with silicone adhesive sealing compound (black) AMV 176 000 05 or equivalent
- Replacing ⇒ Page Q 64-37

6 - Sealing frame
- If necessary, shorten new sealing frame
- Glued with silicone adhesive sealing compound (black) AMV 176 000 05 or equivalent
- Replacing ⇒ Page Q 64-36

7 - Handle

8 - Countersunk screw

9 - Spring

10 - Bolt

11 - Bracket

12 - Base plate

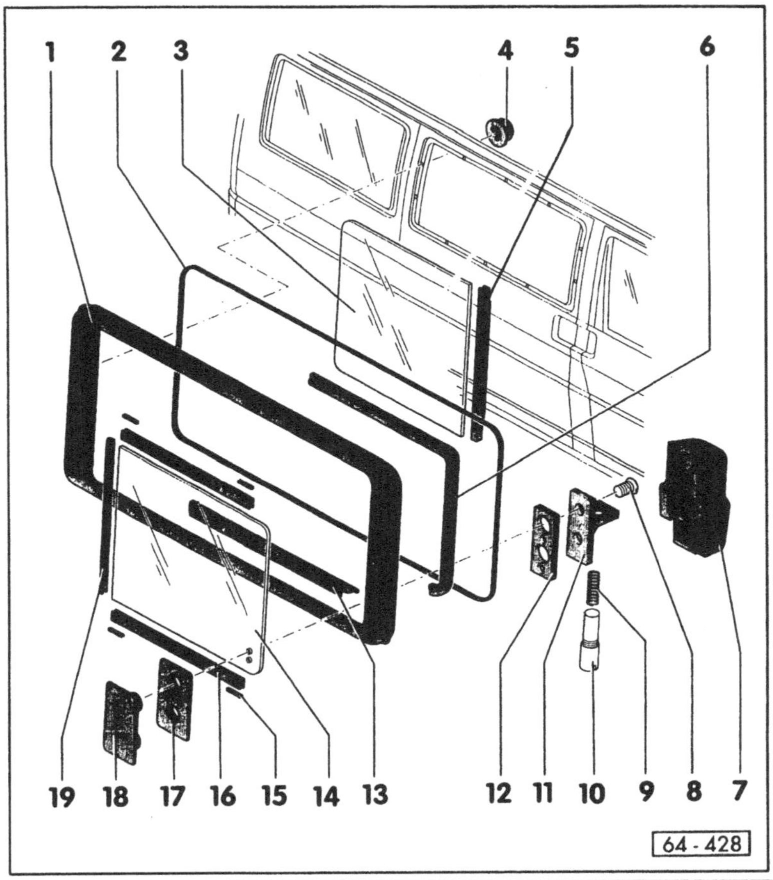

13 - Stop rail
- If necessary, shorten new stop rail
- Glued in with silicone adhesive sealing material (black) AMV 176 000 05 or equivalent
- Replacing ⇒ Page Q 64-34

14 - Movable window
- Replacing ⇒ Page Q 64-38

15 - Guide piece
- For left and right
- Replace in pairs

16 - Guide molding
- Shorten new guide moldings if required.

17 - Gasket

18 - Cover plate

19 - Clamping frame

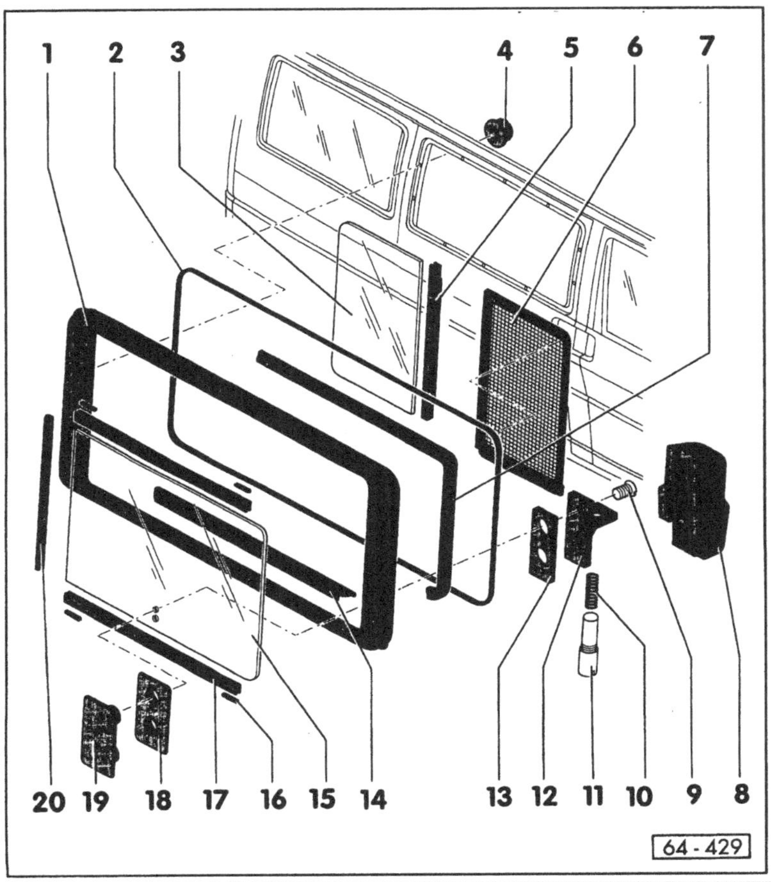

Sliding window with screen (sliding door), components

> ***Caution!***
>
> ***Part numbers are for reference only. Always check with your Parts Department for latest information.***

Note:

To remove the sliding window (sliding door) first remove the sliding door molded trim ⇒ Page Q 70-84.

1 - Frame
- With studs

2 - Gasket
- Glued to frame

3 - Fixed window
- If broken, replace complete sliding window unit

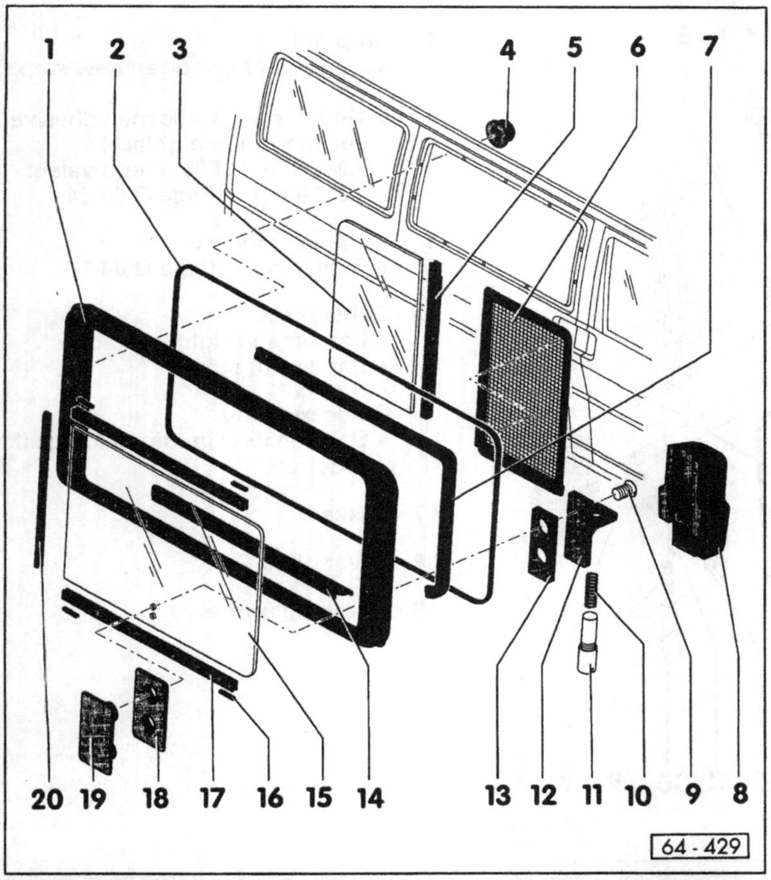

4 - Hex nut
- ♦ 1.5 Nm (13 in. lb)
- ♦ Cap nut in vehicles without molded trim

5 - Divider post
- ♦ Glued with silicone adhesive sealing compound (black) AMV 176 000 05 or equivalent
- ♦ Replacing ⇒ Page Q 64-37

6 - Screen
- ♦ Clipped on from inside

7 - Sealing frame
- ♦ If necessary, cut to fit
- ♦ Glued with silicone adhesive sealing compound (black) AMV 176 000 05 or equivalent
- ♦ Replacing ⇒ Page Q 64-36

8 - Handle

9 - Countersunk screw

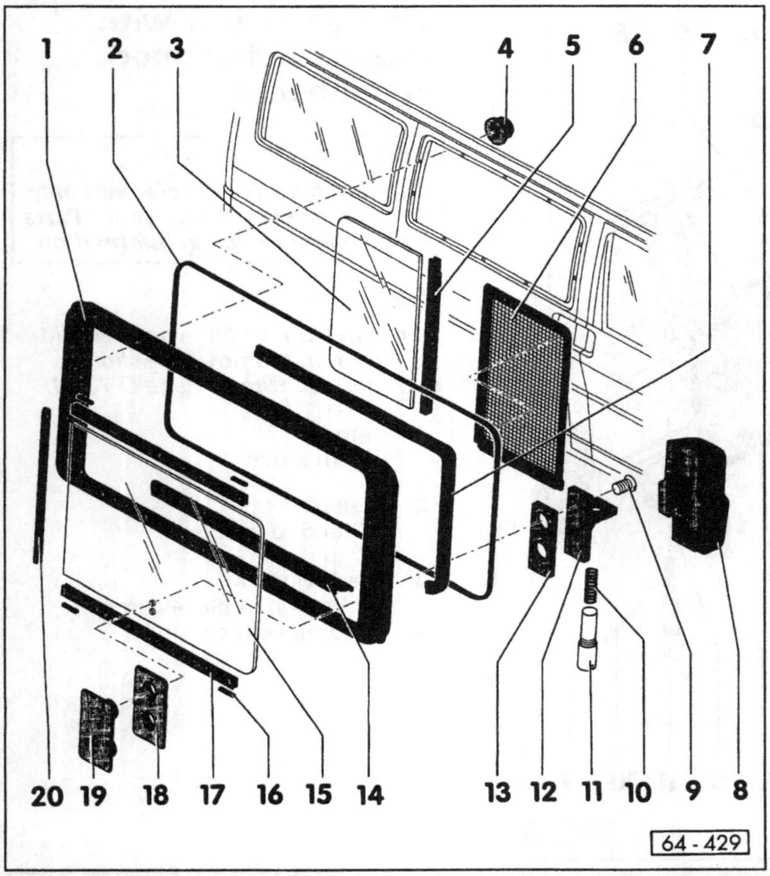

10 - Spring

11 - Bolt

12 - Bracket

13 - Base plate

14 - Stop rail
- ♦ If necessary, cut to fit
- ♦ Glued in with silicone adhesive sealing material (black) AMV 176 000 05 or equivalent
- ♦ Replacing ⇒ Page Q 64-34

15 - Movable window
- ♦ Replacing ⇒ Page Q 64-38

16 - Guide piece
- ♦ For left and right
- ♦ Replace in pairs

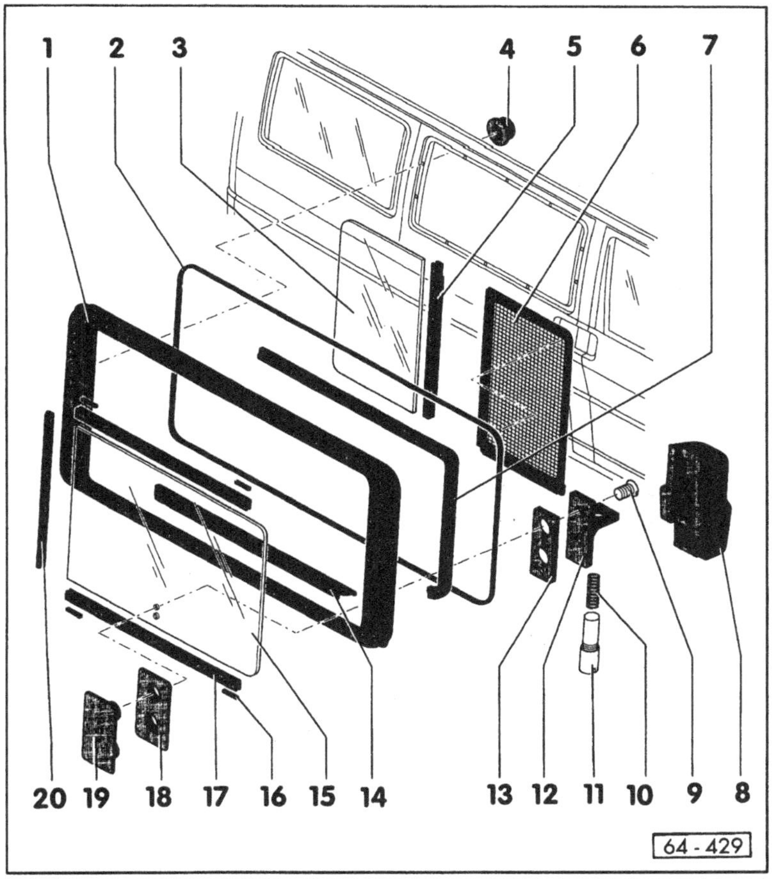

17 - **Guide molding**
- ♦ If necessary, cut to fit

18 - **Gasket**

19 - **Cover plate**

20 - **Clamping frame**

Stop rail, replacing

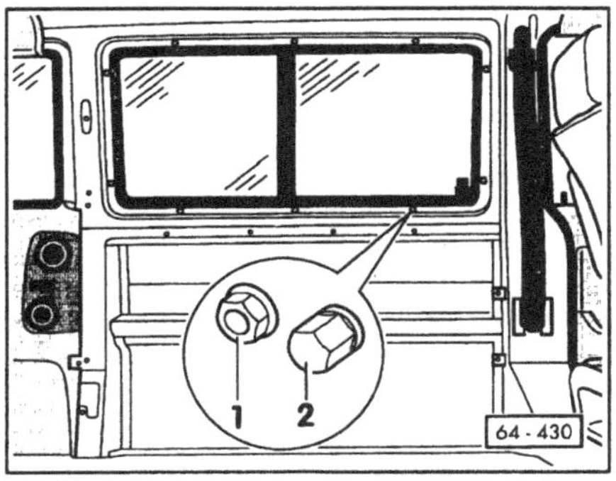

Stop rail, removing

- Depending on equipment, remove trim and seats or seat benches ⇒ Page Q 70-11 and Page Q 72-7.
- Remove hex lock nuts -1- and/or cap nuts -2-, remove sliding window from vehicle.
- Open sliding window completely.

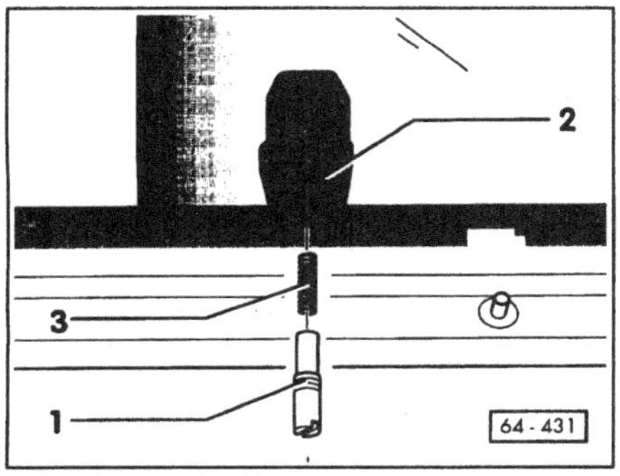

- Remove bolts -1- from handle -2-, remove spring -3-.
- Remove handle.
- Remove exposed countersunk screws, take bracket and cover plate off movable window.

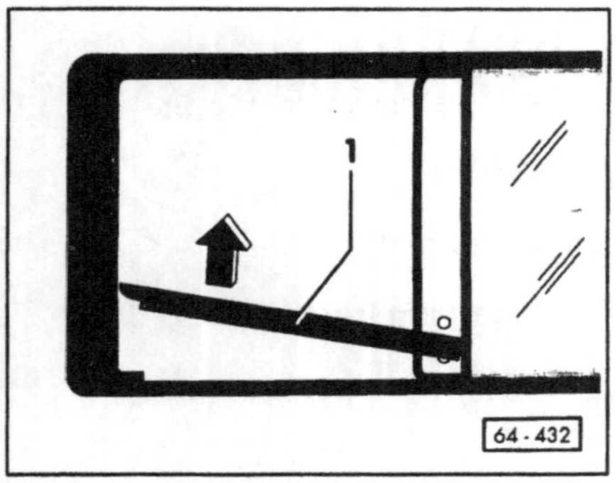

- Loosen stop rail -1- by tapping or prying out of frame -arrow- and remove.

Stop rail, Installing

- Clean groove in sliding window molding.
- If necessary, shorten new stop rail.

Note:

Refer to EuroVan Body-Collision Repair Manual for corrosion protection and paint repair.

- Clean stop rail with cleaning solution D 009 401 04 or equivalent and wipe dry with lint-free cloth.
- Install stop rail with silicone adhesive sealing compound (black) AMV 176 000 05 or equivalent and hold in place with fabric tape while curing.

- Curing time approx. 8 hours.

- Install latch, check sliding window for proper function and re-install in vehicle.

Sealing frame, replacing

Sealing frame, removing

- Depending on equipment, remove trim and seats or seat benches ⇒ Page Q 70-11 and Page Q 72-7.
- Remove stop rail ⇒ Page Q 64-34.

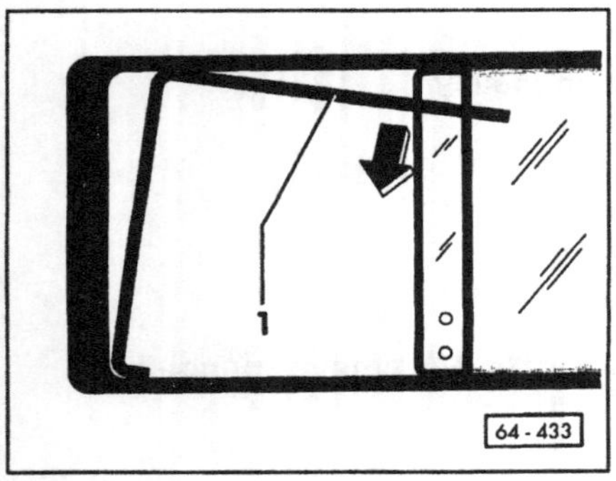

- Pry sealing frame -1- out of sliding window molding, or use pliers.

Sealing frame, installing

- Clean groove in sliding window molding.
- If necessary, shorten sealing frame.

Note:

Refer to EuroVan Body-Collision Repair Manual for corrosion protection and paint repair.

- Clean sealing frame with cleaning solution D 009 401 04 or equivalent and wipe dry with lint-free cloth.
- Install sealing frame with silicone adhesive compound (black) AMV 176 000 05 or equivalent and let cure under tension.

- Curing time approx. 8 hours.

Divider post, replacing

Divider post, removing

- Depending on equipment version, remove trim and seats or seat benches ⇒ Page Q 70-11 and Page Q 72-7.
- Remove stop rail ⇒ Page Q 64-34.
- Remove sealing frame ⇒ Page Q 64-36.

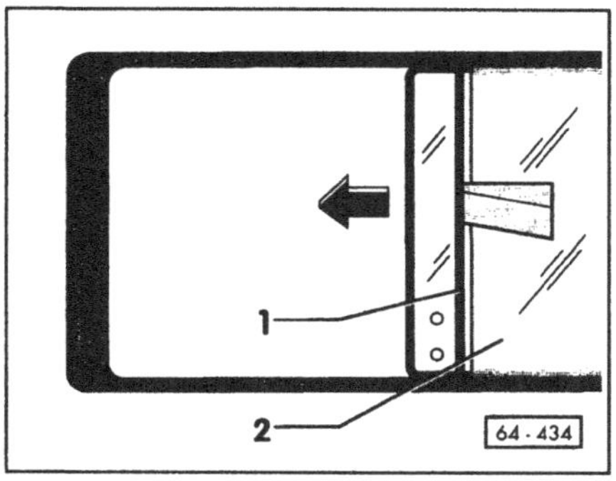

◄ - Loosen divider post -1- from fixed window -2- by tapping with a piece of wood or plastic wedge.

Divider post, installing

- Clean fixed window with cleaning solution D 009 401 04 or equivalent and wipe dry with lint-free cloth.
- Install new divider post with silicone adhesive sealing compound (black) AMV 176 000 05 or equivalent.

● Curing time approx. 8 hours.

Movable window, replacing

Movable window, removing

- Depending on equipment version, remove trim and seats or seat benches ⇒ Page Q 70-11 and Page Q 72-7.
- Remove stop rail ⇒ Page Q 64-34.
- Remove sealing frame ⇒ Page Q 64-36.

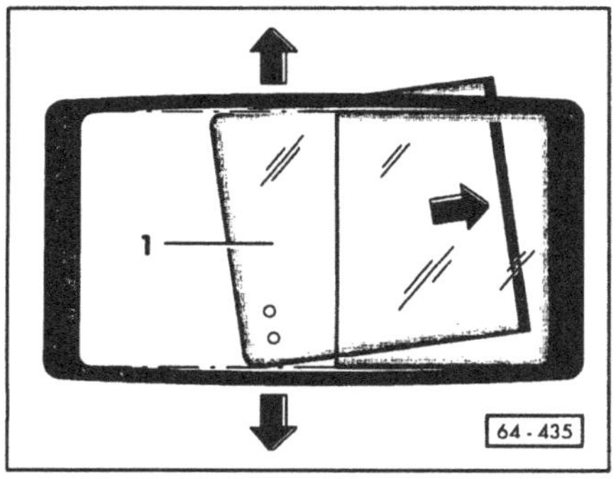

◄ - Spread open sliding window molding -arrows- and slide out movable window -1-.

Movable window, installing

- Clean adhesive surface of movable window and frame with cleaning solution D 009 401 04 or equivalent and wipe dry with lint-free cloth.
- Glue frame and movable window with silicone adhesive sealing compound (black) AMV 176 000 05 or equivalent.

● Curing time approx. 8 hours.

- Adapt guide molding, cut to fit.

Note:

Refer to EuroVan Body-Collision Repair Manual for paint repair and corrosion protection.

– Glue guide pieces in guide moldings.

– Clean window, frame guide mouldings, and adhesive surfaces with cleaning solution D 009 401 04 or equivalent and wipe dry with lint-free cloth.

– Glue frame and movable window with silicone adhesive sealing compound (black) AMV 176 000 05 or equivalent.

● Curing time approx. 8 hours.

– Install sealing frame ⇒ Page Q 64-36 and install stop rail ⇒ Page Q 64-34, seal fixed window with silicone sealing compound (black) AMV 176 000 05 or equivalent.

Sliding window, retrofitting

Notes:

♦ *In some versions, sliding windows can be installed instead of flush-bonded windows in the sliding door, side panel of transporter and/or double cab, and side door of double cab.*

♦ *The installation description is given for only one side. Directions for other side are mirror image.*

♦ *Depending on equipment, remove trim and seats or seat benches ⇒ Page Q 70-11 and Page Q 72-7.*

– Remove flush bonded window ⇒ Page Q 64-11.

– Cut back remainder of adhesive sealing material.

– Transferring appropriate measurements to window molding and setting corner points:

– Sliding door ⇒ Page Q 64-42

– Side panel ⇒ Page Q 64-43

– Side door (double cab) ⇒ Page Q 64-44

– Side panel (double cab) ⇒ Page Q 64-45

– Drill holes 7.5 mm dia (0.30 in.) and deburr.

– If necessary, rework holes with a round file to fit the sliding window.

Note:

Refer to EuroVan Body-Collision Repair Manual for paint repair and corrosion protection.

– After completion of the painting/corrosion protection, install sliding window.

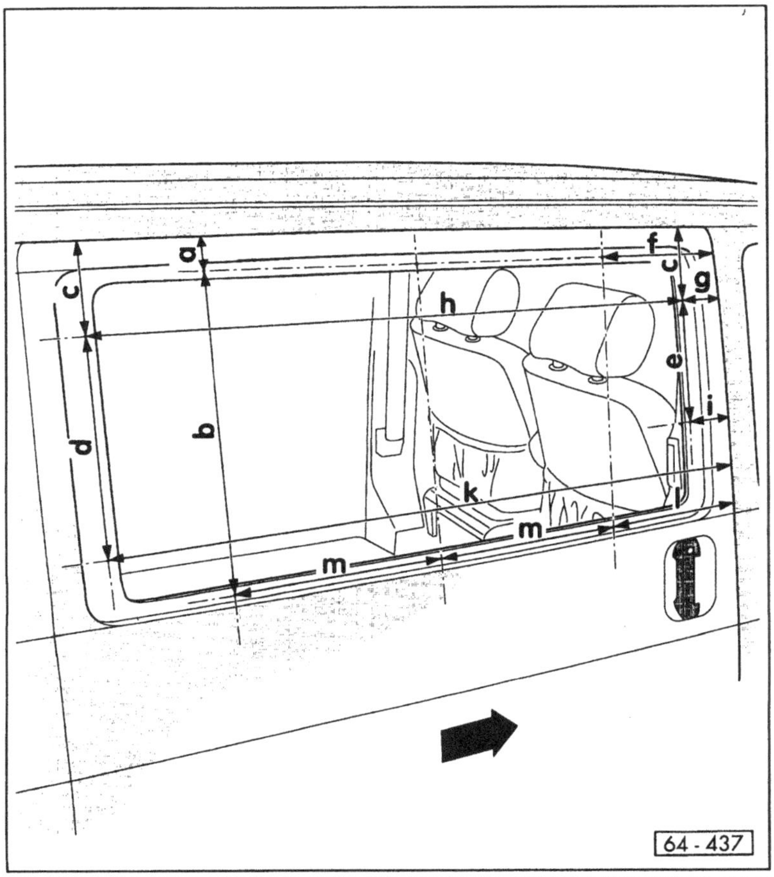

Sliding door, right

Note:

Front outer edge and upper edge of sliding door are used as reference points. Arrow points in driving direction.

Dimensions:

a = 55 mm (2.16 in.)

b = 510 mm (20.07 in.)

c = 140 mm (5.51 in.)

d = 337 mm (13.26 in.)

e = 237 mm (9.33 in.)

f = 260 mm (10.23 in.)

9 = 80 mm (3.14 in.)

h = 1060 mm (41.73 in.)

i = 82 mm (3.14 in.)

k = 1146 mm (45.11 in.)

l = 267 mm (10.51 in.)

m = 350 mm (13.77 in.)

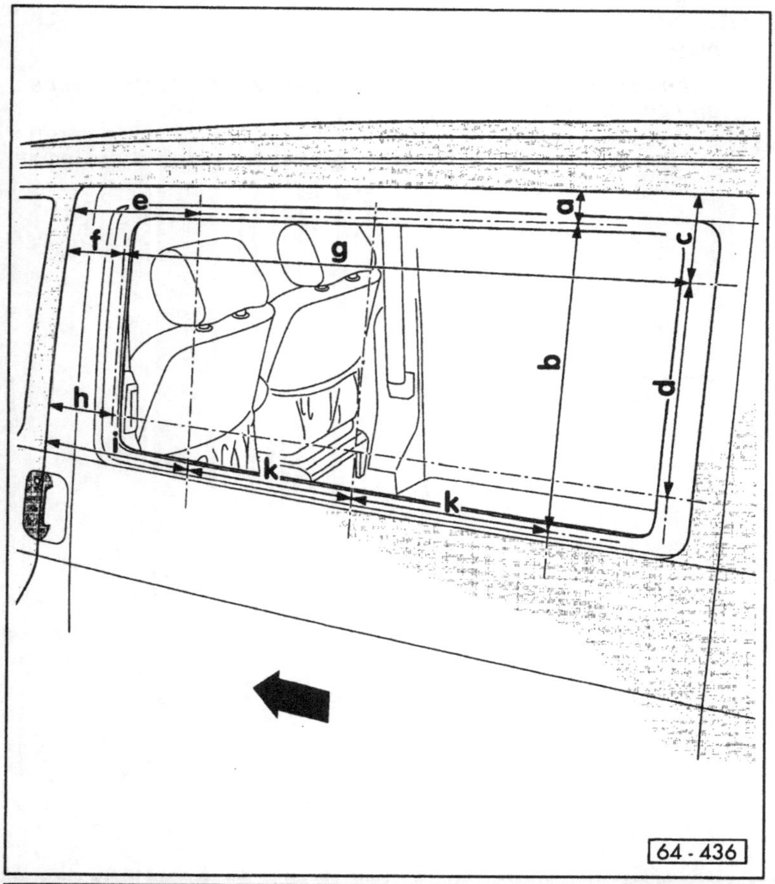

Side panel, left

Note:

Front edge of B-pillar and roof edge are used as reference points. Arrow points in driving direction.

Dimensions:

a = 52 mm (2.04 in.)

b = 510 mm (20.07 in.)

c = 135 mm (5.31 in.)

d = 340 mm (13.38 in.)

e = 310 mm (12.20 in.)

f = 107 mm (4.21 in.)

9 = 1118 mm (44.01 in.)

h = 122 mm (4.80 in.)

i = 335 mm (13.18 in.)

k = 350 mm (13.77 in.)

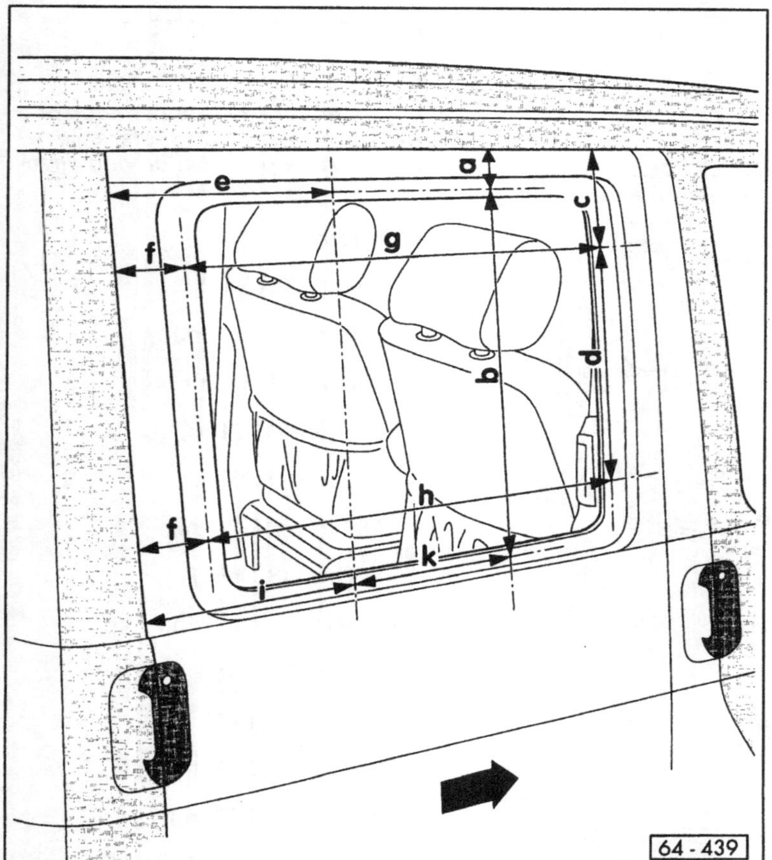

Double cab door, right

Note:

Rear outer edge and upper edge of hinged door are used as reference points. Arrow points in driving direction.

Dimensions:

a = 52 mm (2.04 in.)

b = 510 mm (20.07 in.)

c = 138 mm (5.43 in.)

d = 340 mm (13.38 in.)

e = 265 mm (10.43 in.)

f = 85 mm (3.34 in.)

g = 560 mm (22.04 in.)

h = 565 mm (22.24 in.)

i = 267 mm (10.51 in.)

k = 200 mm (7.87 in.)

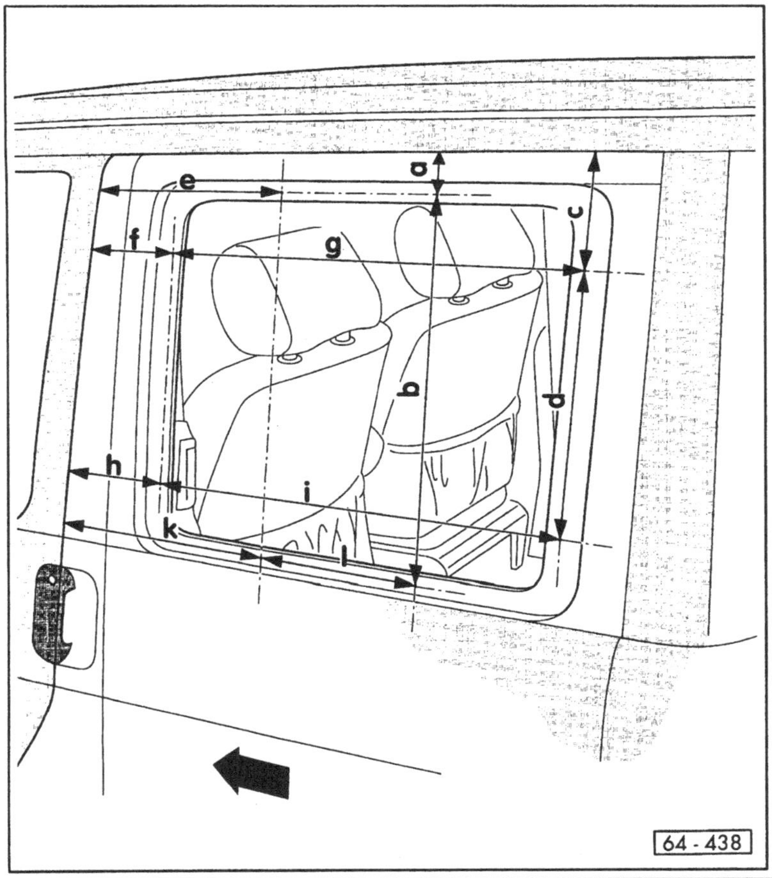

Double cab side panel, left

Note:

Front edge of B-pillar and roof edge are used as reference points. Arrow points in driving direction.

Dimensions:

a = 55 mm (2.16 in.)

b = 510 mm (20.07 in.)

c = 143 mm (5.62 in.)

d = 340 mm (13.38 in.)

e = 310 mm (12.20 in.)

f = 110 mm (4.33 in.)

9 = 615 mm (24.21 in.)

h = 122 mm (4.80 in.)

i = 620 mm (24.40 in.)

k = 335 mm (13.81 in.)

l = 200 mm (7.87 in.)

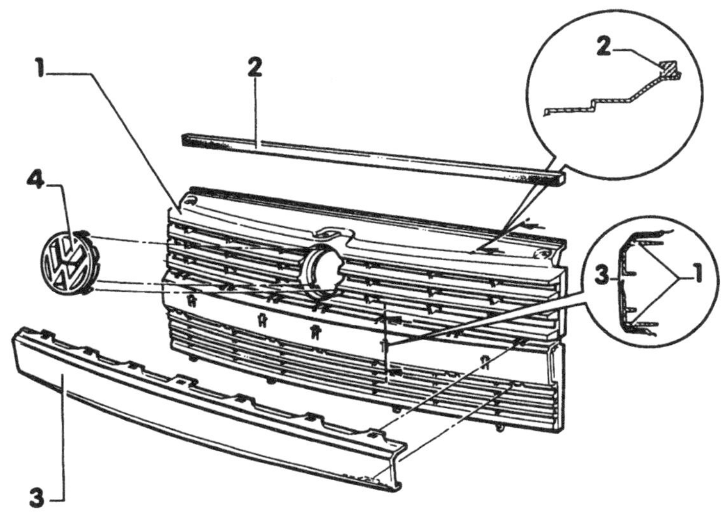

66 - 351

Radiator grille, components ➤ 12.95

1 - **Radiator grille**
- ♦ Removing ⇒ Page Q 66-2

2 - **Seal**
- ♦ Self-adhesive

3 - **Trim**
- ♦ Removing: remove radiator grille ⇒ Page Q 66-2 and unclip from back side.

4 - **Emblem**
- ♦ Fastened with clip

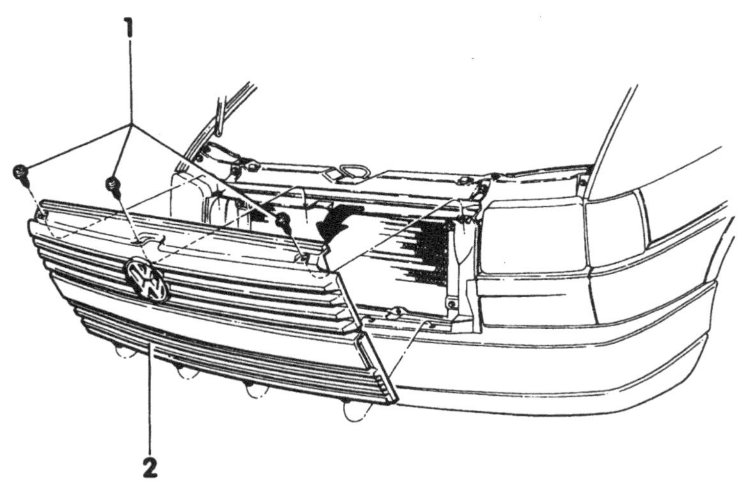

66 - 352

Radiator grille, removing ➤12.95

– Remove Phillips screws -1-.

– Tip radiator grille -2- forward -arrow- and unhook from mounting holes.

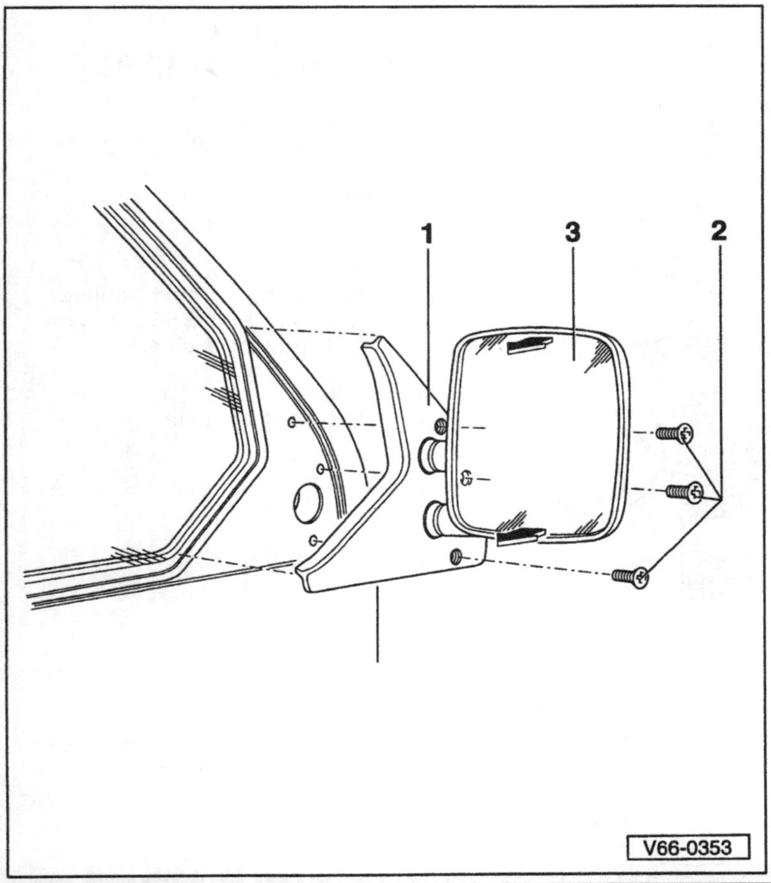

Outside rear view mirror, components

> *WARNING!*
>
> ***Always wear protective gloves when removing and installing mirror glass.***

1 - Mirror housing
- Remove Phillips screws -2-, remove mirror, complete.

2 - Phillips screws

3 - Mirror glass
- Removing: remove rear view mirror (unscrew). Push mirror glass from bottom, then at top with lever 80-200.
- Installing: Place mirror housing on work surface. Clip mirror glass in by pushing vertically in center with uniform force.

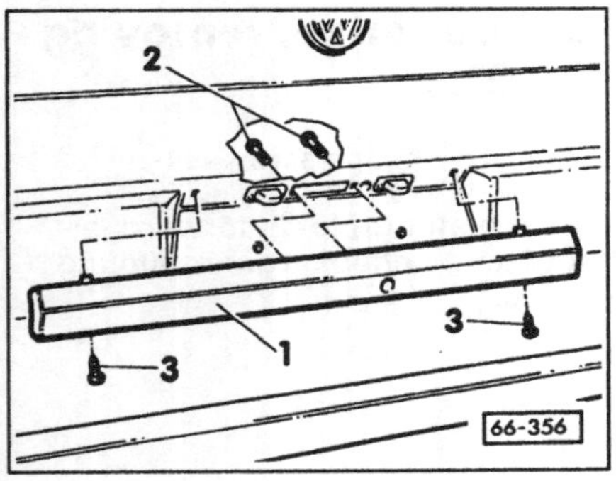

License plate light trim strip (tailgate), removing

- Remove tailgate trim ⇒ Page Q 70-96 and/or tailgate molded trim ⇒ Page Q 70-91.
- Remove screws -2-.

Note:

◄ *Screws -2- are also used to fasten the lid lock mechanism ⇒ Page Q 55-15.*

- Remove Phillips screws -3-, unclip trim -1-.

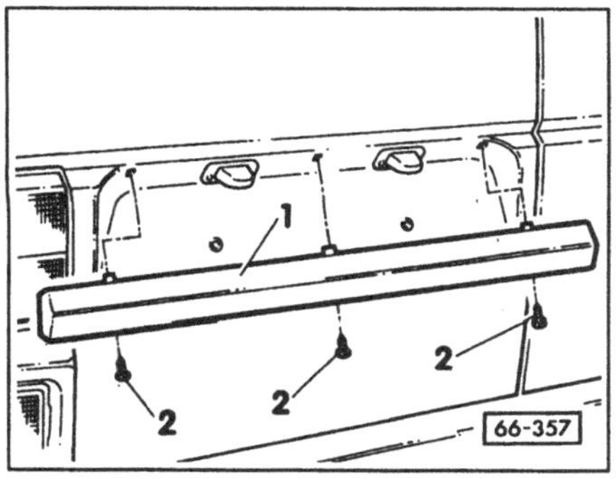

License plate light trim strip (double doors), removing

◄ – Remove Phillips screws.

– Unclip trim strip -1-.

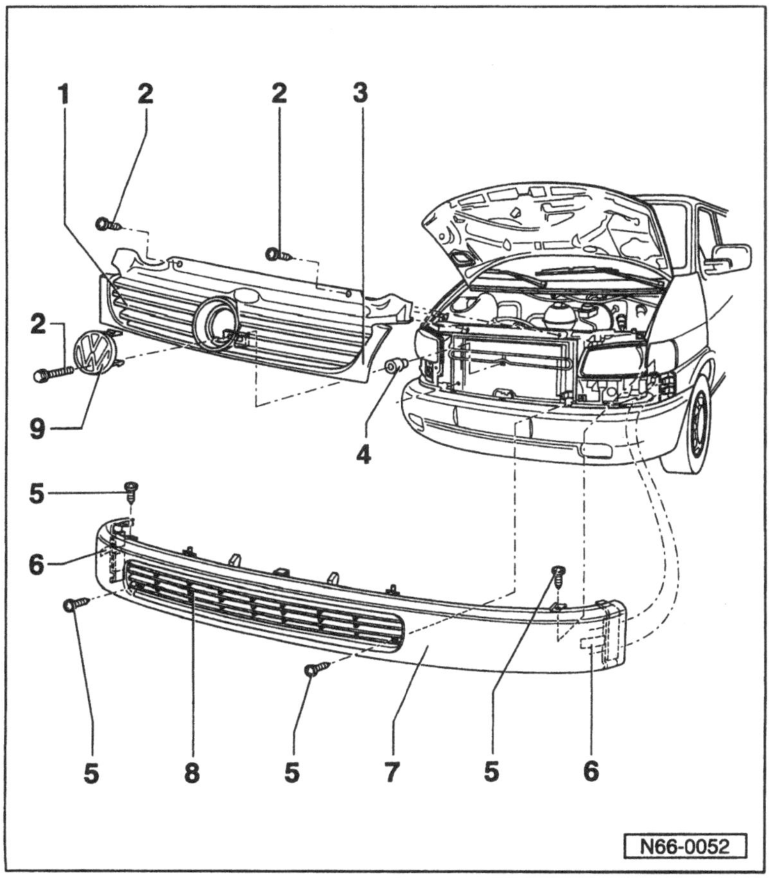

Radiator grille, removing/installing 01.96 ➤

1 - Radiator grille, upper
- ♦ Grille is clipped together with trim frame
- ♦ Grille with trim frame, removing
- – Remove screw/washer assemblies -2-, pull upper radiator grille upward with trim frame.

2 - Screw/washer assembly

3 - Trim frame, upper
- ♦ Remove with upper radiator grille

4 - Bonded rubber bushing

5 - Screw/washer assembly

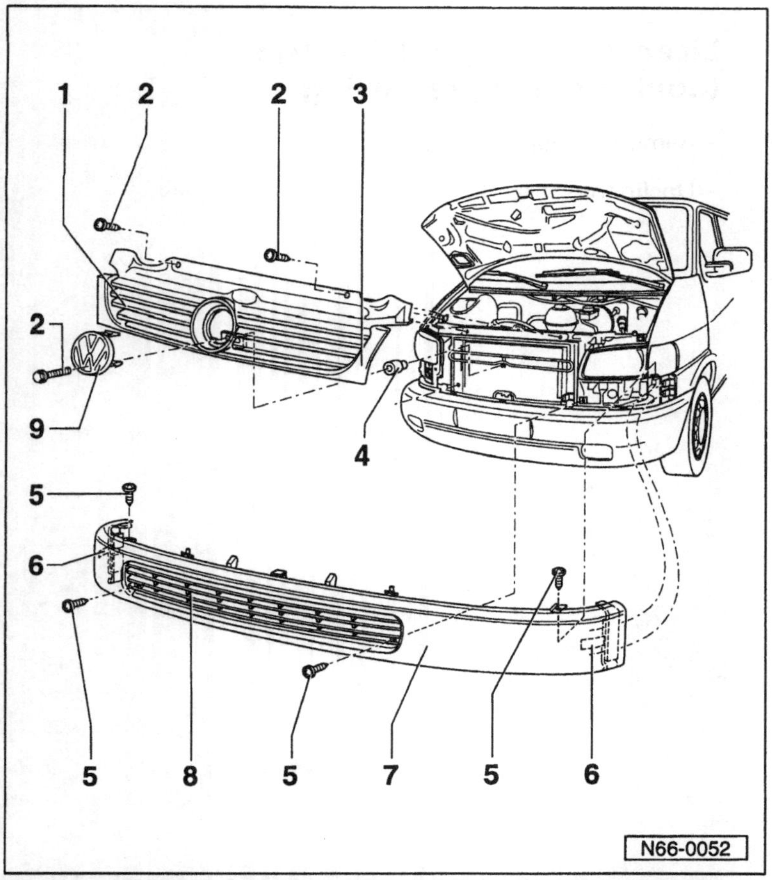

6 - Unlocking tab

- Unlocks corners

7 - Trim frame, lower

- Remove with lower radiator grille

8 - Radiator grille, lower

- Lower grille is clipped together with trim frame
- Radiator grille with trim frame, removing:
  - Remove screw/washer assemblies -5-, remove turn signals, loosen side attachments with unlocking tabs -6-.
  - Remove radiator grille toward front with trim frame.

9 - Logo

- Clipped in radiator grille

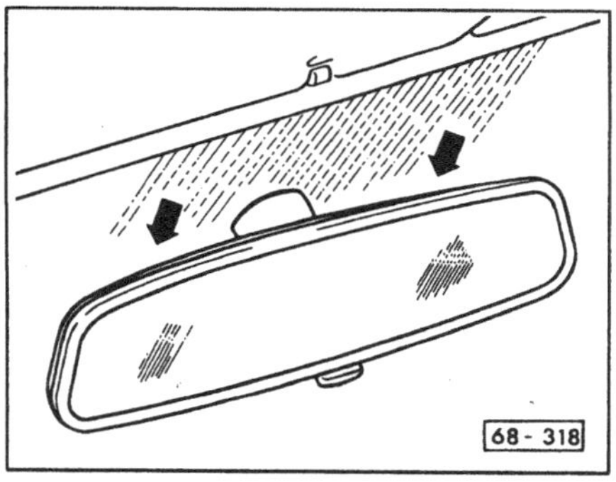

Inside rear view mirror, removing/installing

◄ – Press interior rear view mirror downward at an angle, off retaining plate (spring clip in mirror base).

◄ – Push mirror on 60°- 90° angle to mounted position, rotate until locking spring clicks in place.

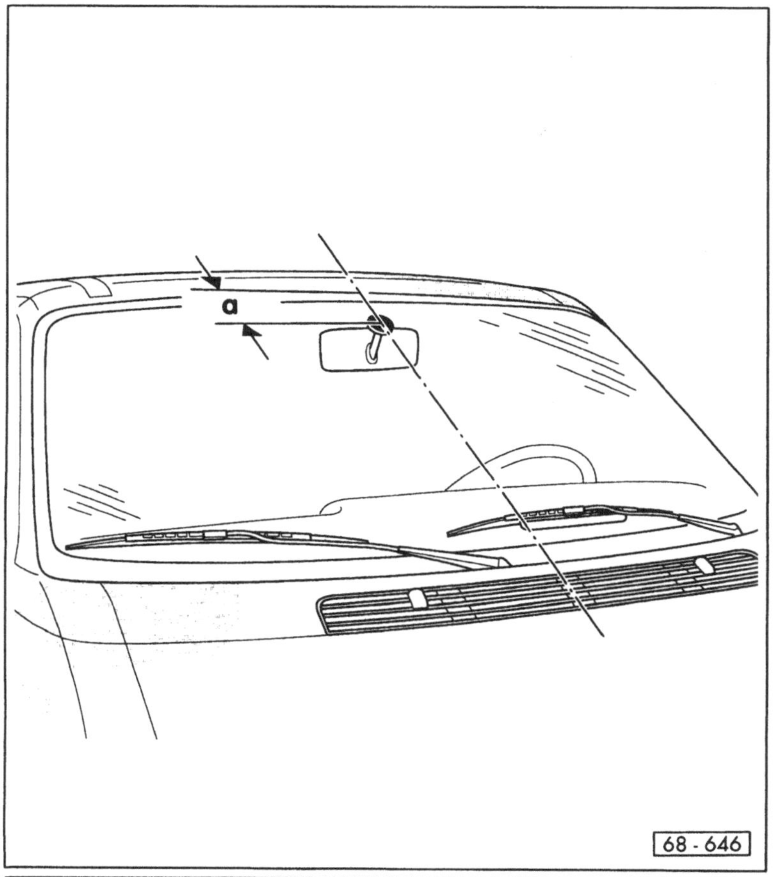

Inside rear view mirror mounting plate, installing

Material:

Activator D 000 701

Metal adhesive D 000 702

> ***CAUTION!***
>
> - ***Handle adhesive material at room temperature.***
> - ***Read and observe all cautions and warnings on adhesive compound labels.***
> - ***Adhesive is completely hardened after several hours.***

> ***CAUTION!***
>
> ***Part numbers are for reference only. Always check with your Parts Department for latest information.***

- Dimension for rear view mirror:

a = 100 mm (approx. 4.0 in).

– Clean bonding surfaces.

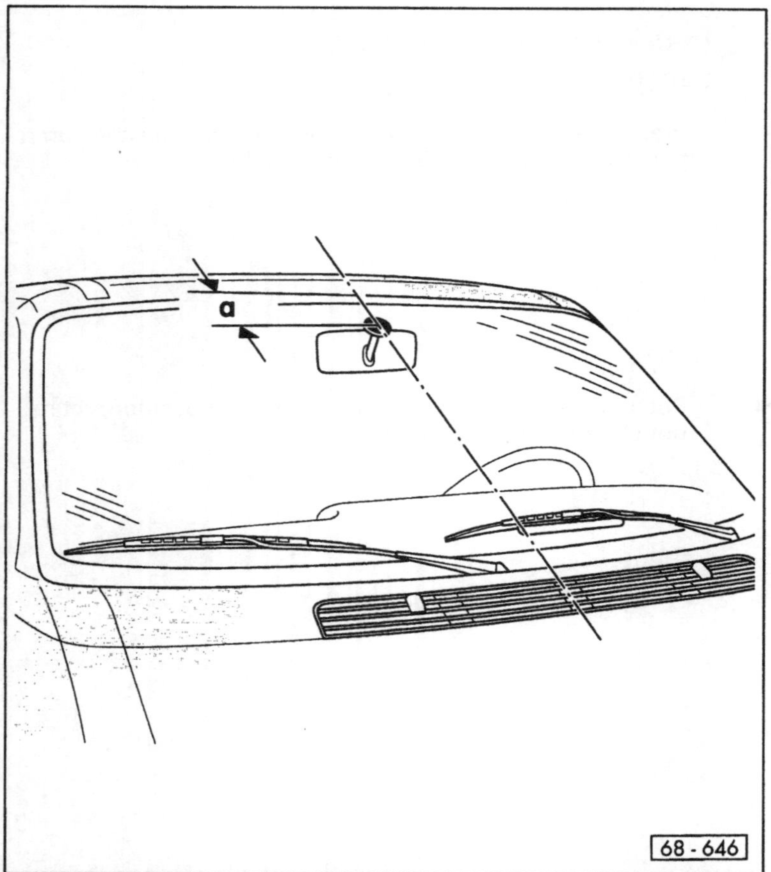

- Apply (spray) activator, wipe dry.
- Spray cleaned windshield surface with activator again and let dry 1 to 2 minutes
- Apply one drop of adhesive to center of cleaned retaining plate surface.
- Press and hold retaining plate on adhesive surface for 30 seconds.
- Install mirror.
- Wipe off excess activator.

Note:

Residual adhesive must be sanded off bonding surface of retaining plates.

Seat belts, removing/installing

Notes:

- *Seat belt version depends on how vehicle is equipped.*
- *Following descriptions include basic seat belt versions. They serve as a basis for repair.*
- *Descriptions are given for only one side, other side is mirror image.*

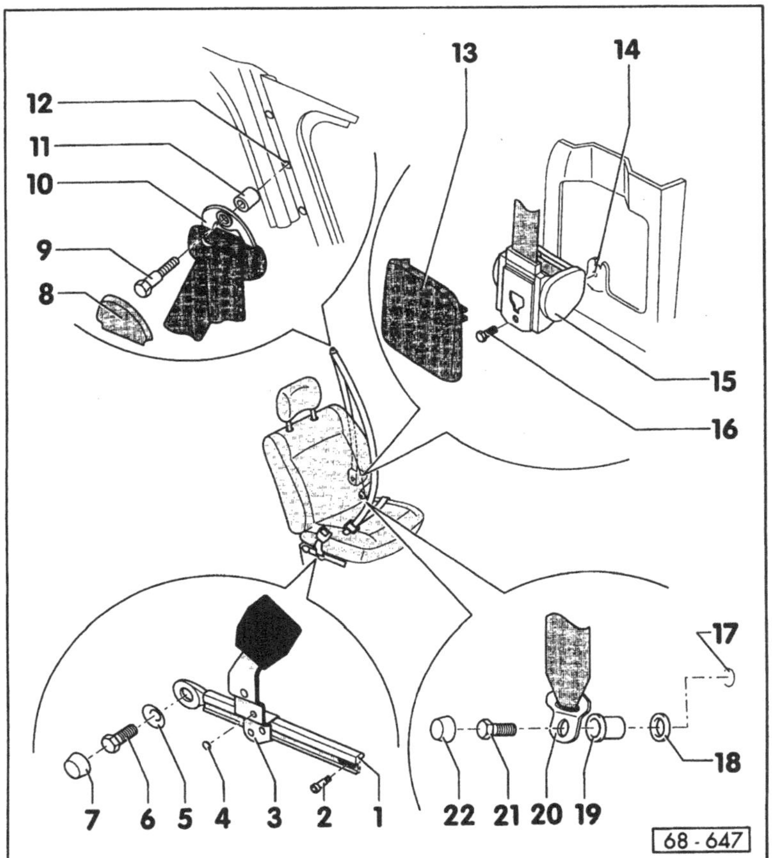

Driver/passenger seat belt (without height adjustment), components

Note:

Depending on how vehicle is equipped, remove trim and seats and/or seat benches ⇒ Page Q 70-11 and Q 72-7.

1 - Rail

2 - Mounting bolt
- 10 Nm (7 ft lb)

3 - Belt buckle with steel mounting stalk
- Trim removing ⇒ Page Q 69-9

4 - Clamp cap
- Pry off and/or press on

5 - Beveled spring washer

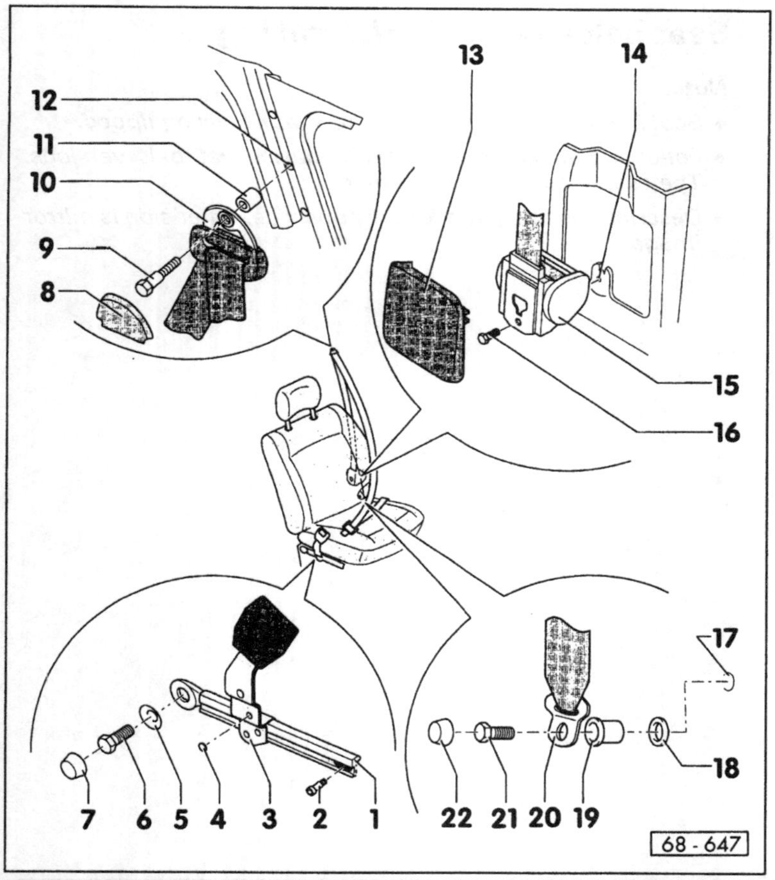

6 - **Hex bolt**
- 40 Nm (30 ft lb)

7 - **Cover**

8 - **O-ring cover**
- Without B-pillar trim

9 - **Hex bolt**
- 40 Nm (30 ft lb)

10 - **O-ring**

11 - **Spacer**

12 - **Mounting point, B-pillar**

13 - **Automatic tensioning seat belt retractor cover**
- Without B-pillar trim

14 - **Automatic tensioning seatbelt retractor mounting point**
- With retaining lug

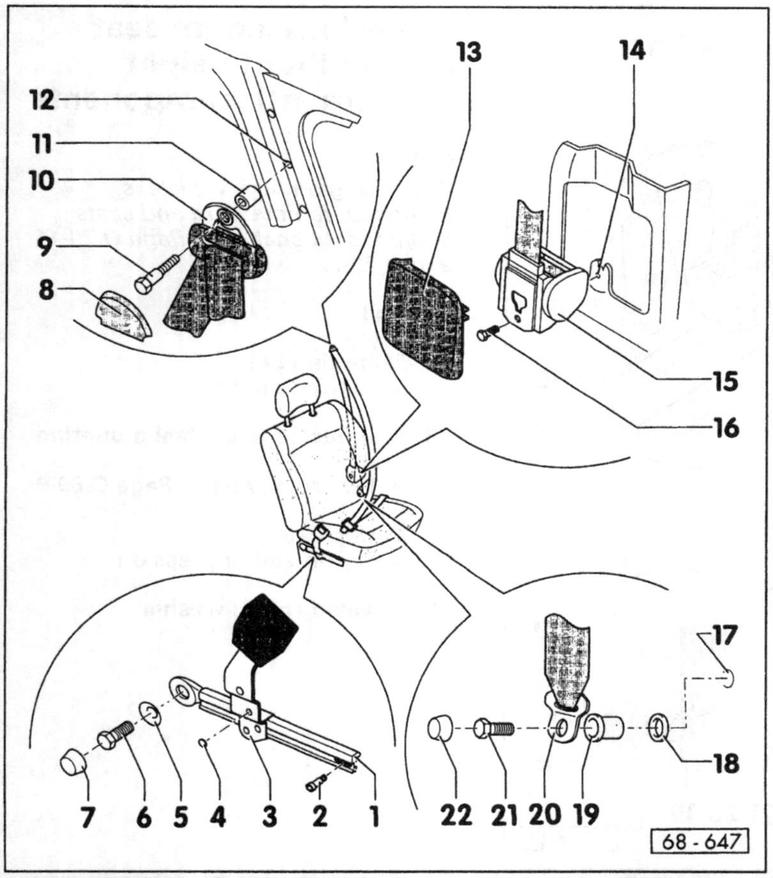

15 - **Automatic tensioning seat belt retractor**

16 - **Hex bolt**
- 40 Nm (30 ft lb)

17 - **B-pillar mounting point**

18 - **Spacer ring**
- Without B-pillar trim

19 - **Spacer bushing**
- With B-pillar trim

20 - **Seat belt anchor**

21 - **Hex bolt**
- 40 Nm (30 ft lb)

22 - **Cover**

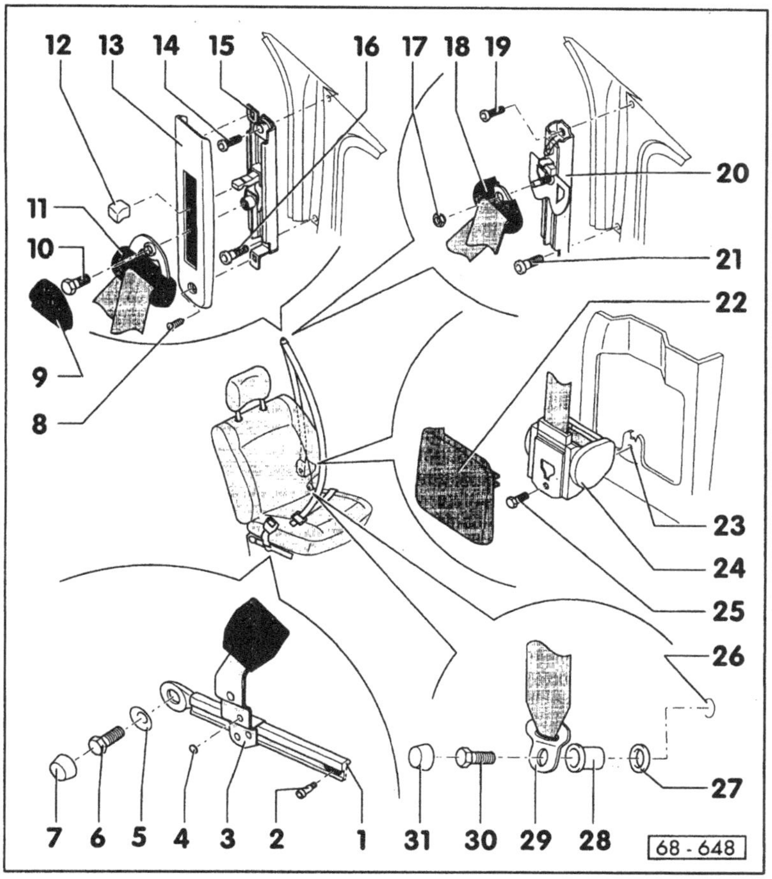

Driver/passenger seat belt (with height adjustment), components

Note:

Depending on how vehicle is equipped, remove trim and seats and/or seat benches ⇒ Page Q 70-11 and Q 72-7.

1 - **Rail**

2 - **Mounting bolt**
- ♦ 10 Nm (7 ft lb)

3 - **Belt buckle with steel mounting stalk**
- ♦ Trim removing ⇒ Page Q 69-9

4 - **Clamp cap**
- ♦ Pry off/press on

5 - **Beveled spring washer**

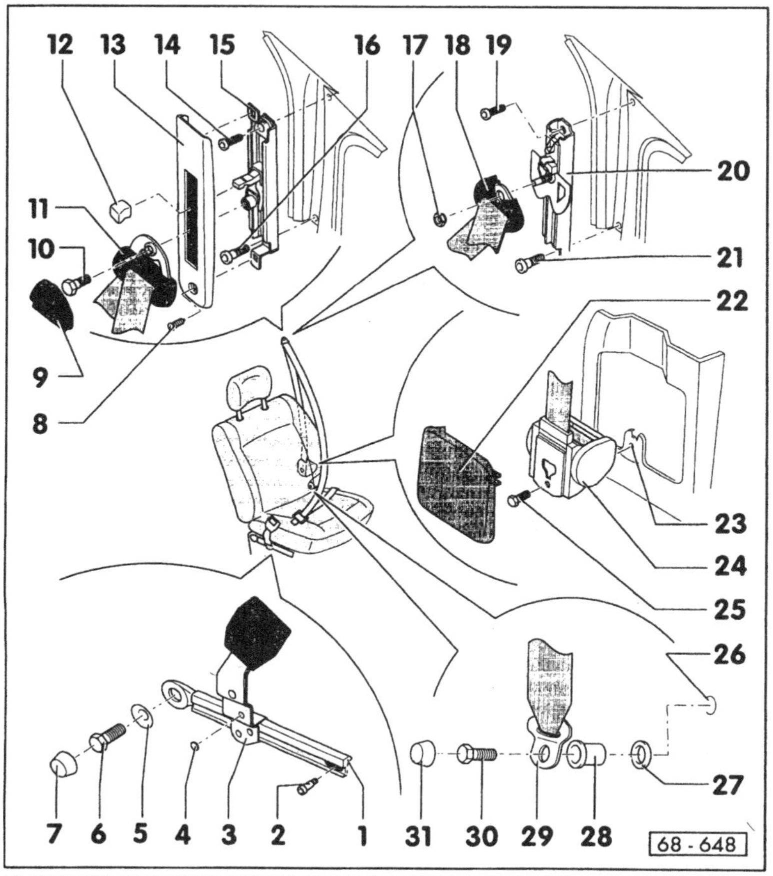

6 - **Hex bolt**
- ♦ 40 Nm (30 ft lb)

7 - **Cap**

Without B-pillar trim:

8 - **Self-tapping screw**

9 - **D-ring cover**

10 - **Hex bolt**
- ♦ 40 Nm (30 ft lb)

11 - **D-ring**

12 - **Height-adjustment knob**

13 - **Height adjustment mechanism trim cover**

14 - **Allen screw**
- ♦ 18 Nm (13 ft lb)

15 - **Height adjustment mechanism**

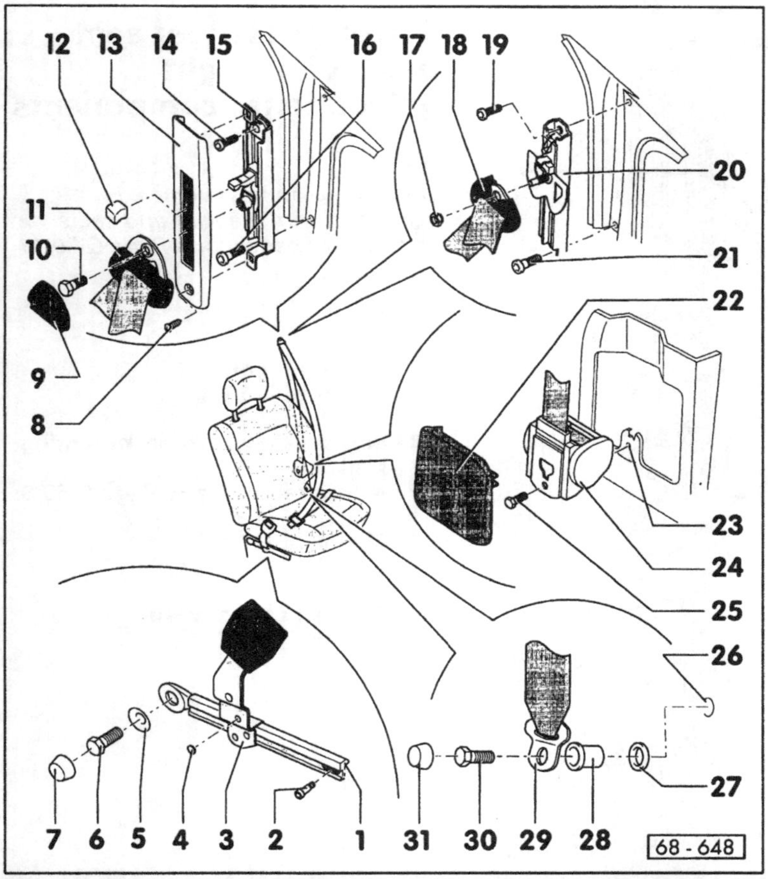

16 - Allen screw
- 40 Nm (30 ft lb)

With B-pillar trim:

17 - Shoulder nut
- 55 Nm (41 ft lb)

18 - D-ring

19 - Allen screw
- 18 Nm (13 ft lb)

20 - Height adjustment mechanism

21 - Allen screw
- 40 Nm (30 ft lb)

Continuation, all vehicles:

22 - Automatic tensioning seat belt retractor cover
- Without B-pillar trim

23 - Automatic tensioning seat belt retractor mounting point
- With retaining lug

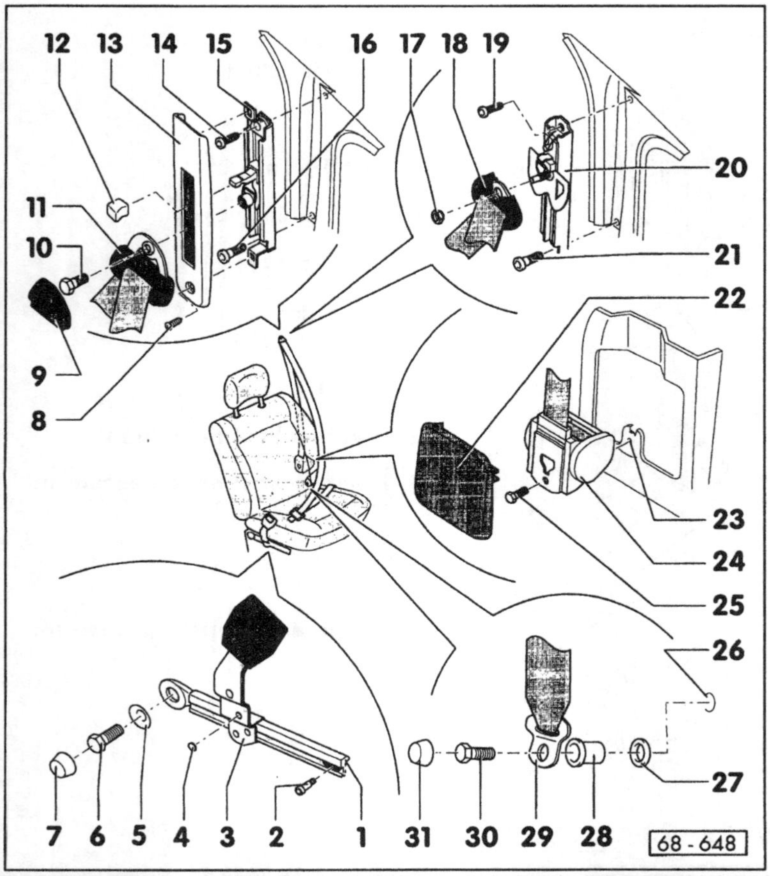

24 - Automatic tensioning seat belt retractor

25 - Hex bolt
- 40 Nm (30 ft lb)

26 - B-pillar mounting point

27 - Spacer
- Without B-pillar trim

28 - Spacer bushing
- With B-pillar trim

29 - Seat belt anchor

30 - Hex bolt
- 40 Nm (30 ft lb)

31 - Cover

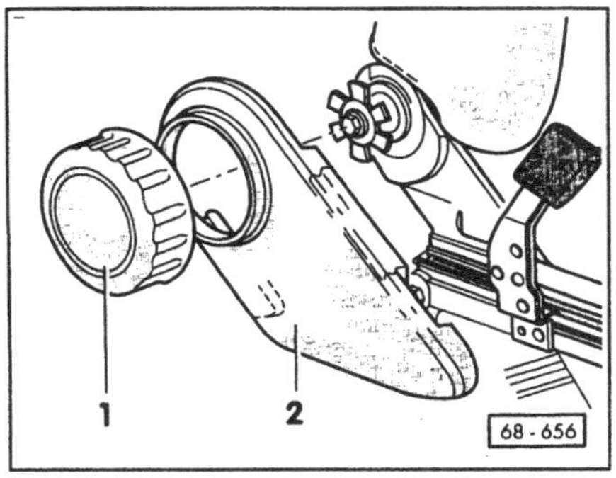

Trim, removing

◄ – To remove belt buckle from driver and/or passenger seat, unclip seat back adjusting knob -1- and inside cover -2-.

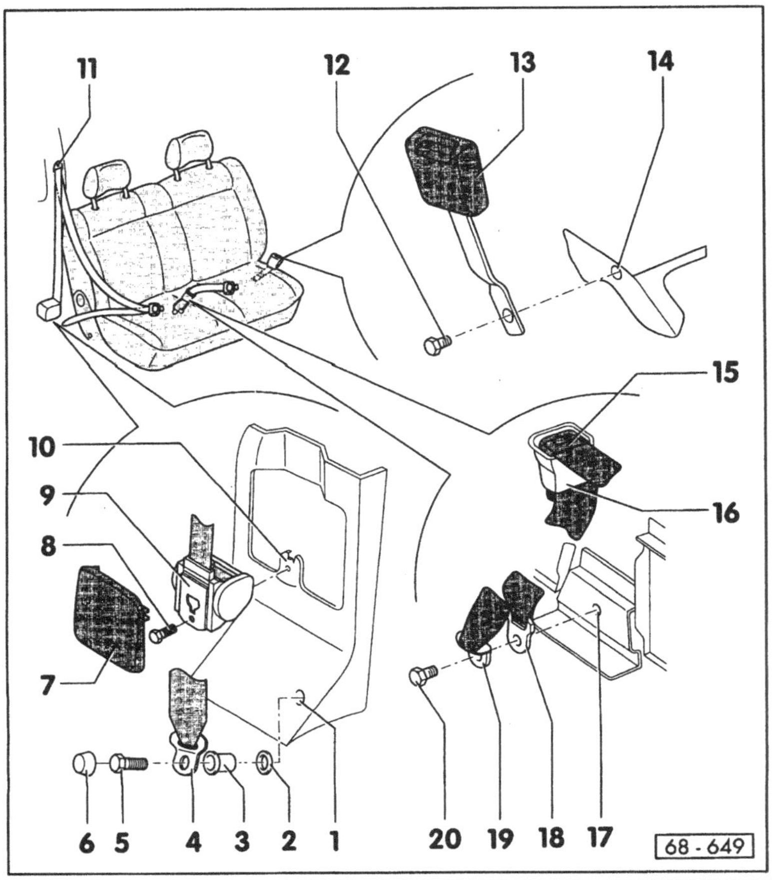

Two-place seat, seat belt components

Note:

Depending on how vehicle is equipped, remove trim and seats and/or seat benches *⇒ Page Q 70-11 and Q 72-7.*

1 - B-pillar mounting point

2 - Spacer
- ♦ Without B-pillar trim

3 - Spacer
- ♦ With B-pillar trim

4 - Seat belt anchor

5 - Hex bolt
- ♦ 40 Nm (30 ft lb)

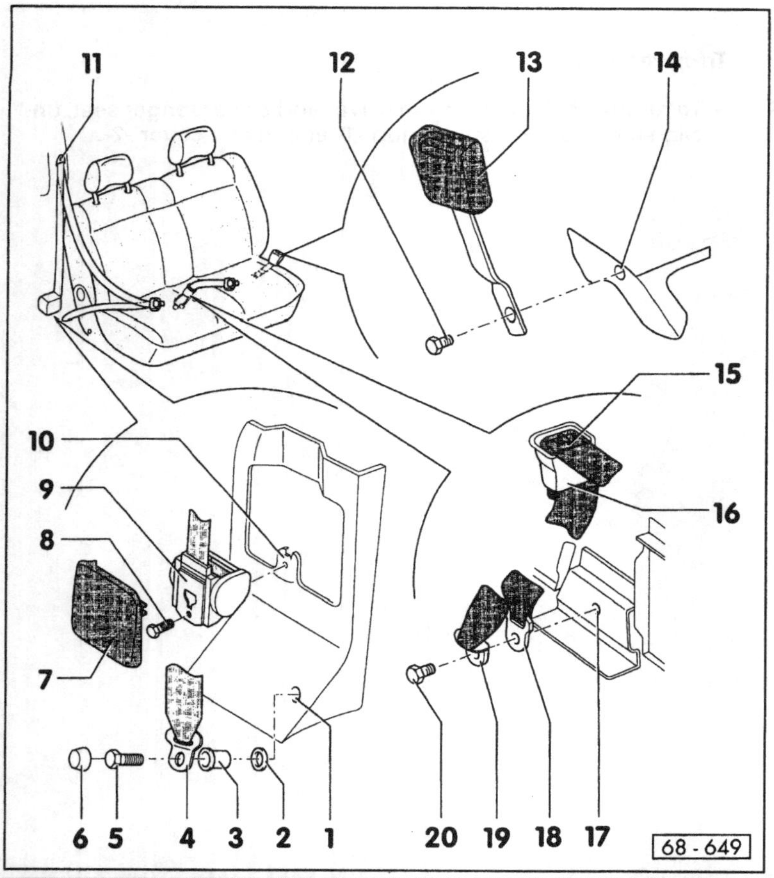

6 - **Cover**

7 - **Automatic tensioning seat belt retractor cover**
- ♦ Without B-pillar trim

8 - **Hex bolt**
- ♦ 40 Nm (30 ft lb)

9 - **Automatic tensioning seat belt retractor**

10 - **Automatic tensioning seat belt retractor mounting point**
- ♦ With retaining lug

11 - **Upper belt anchor**
- ♦ without height adjustment ⇒ Page Q 69-2
- ♦ with height adjustment ⇒ Page Q 69-5

12 - **Hex bolt**
- ♦ 40 Nm (30 ft lb)

13 - **Belt buckle with mounting stalk**

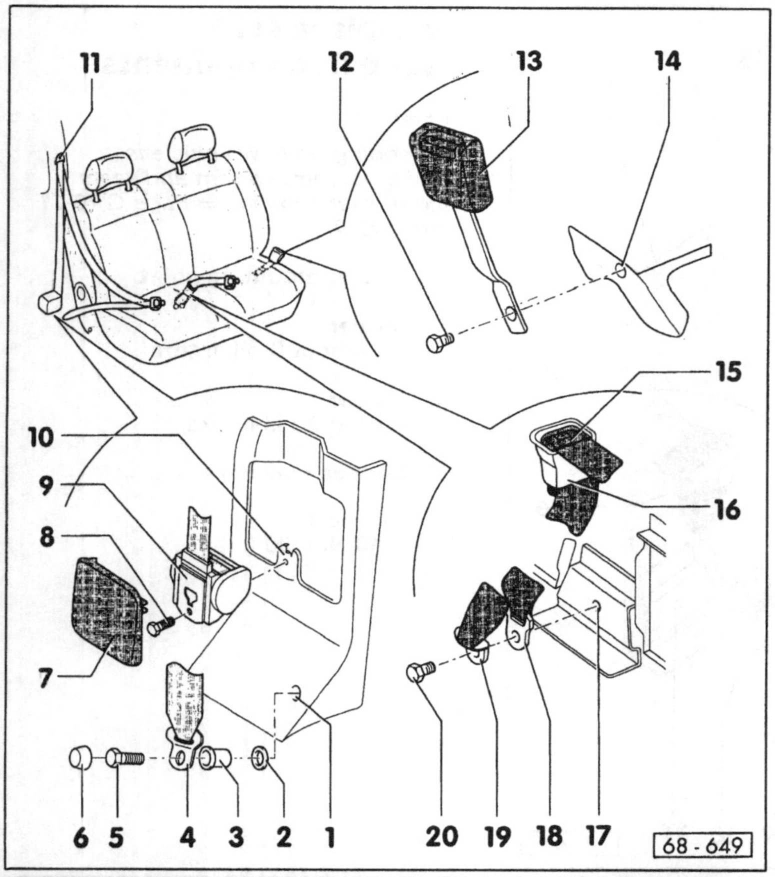

14 - **Seat belt mounting point on seat frame**

15 - **Seat belt buckle with belt**

16 - **Belt buckle guide**

17 - **Seat belt mounting point on seat frame**

18 - **Seat belt anchor for belt buckle**

19 - **Belt anchor for lap belt**

20 - **Hex bolt**
- ♦ 40 Nm (30 ft lb)

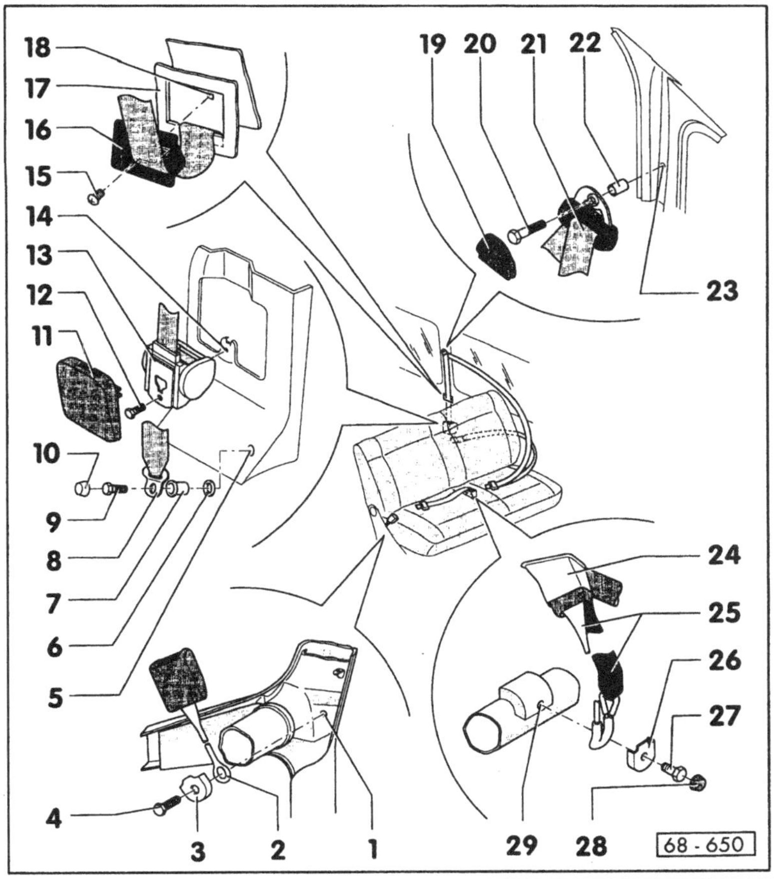

Two-place seat with 3-point belt, components

Note:

Depending on vehicle equipment, remove trim and seats and/or seat benches *⇒ Page Q 70-11 and Q 72-7.*

1 - B-pillar mounting point

2 - Belt buckle with cable

3 - Cap

4 - Hex bolt
- ♦ 40 Nm (30 ft lb)

5 - C-pillar mounting point

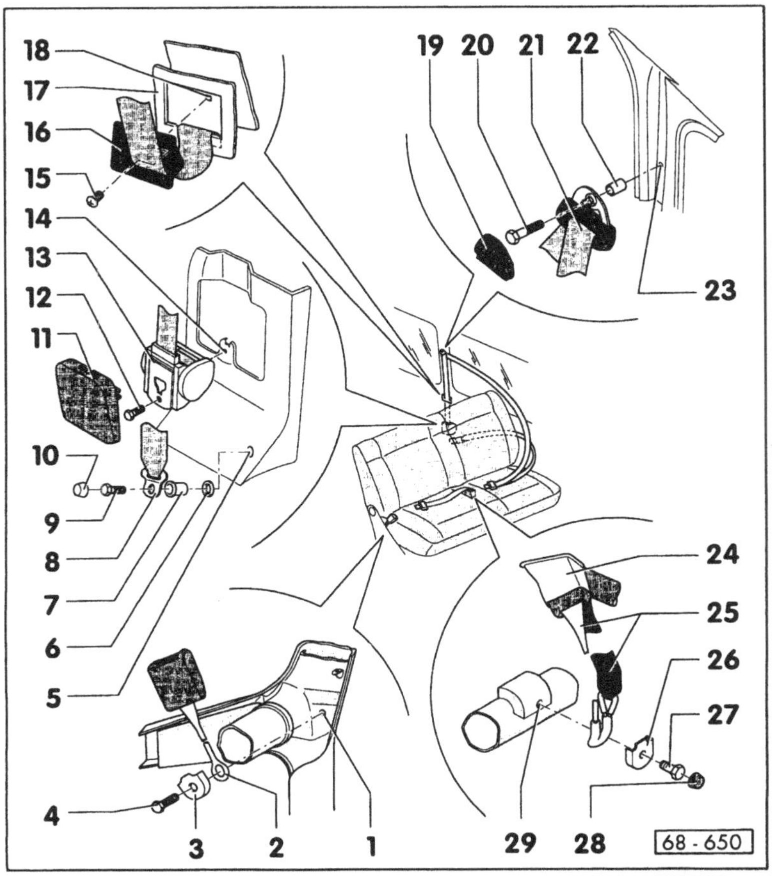

6 - Spacer
- ♦ Without C-pillar trim

7 - Spacer
- ♦ With C-pillar trim

8 - Belt anchor

9 - Hex bolt
- ♦ 40 Nm (30 ft lb)

10 - Cover

11 - Automatic tensioning seat belt retractor cover
- ♦ Without C-pillar trim

12 - Hex bolt
- ♦ 40 Nm (30 ft lb)

13 - Automatic tensioning seat belt retractor

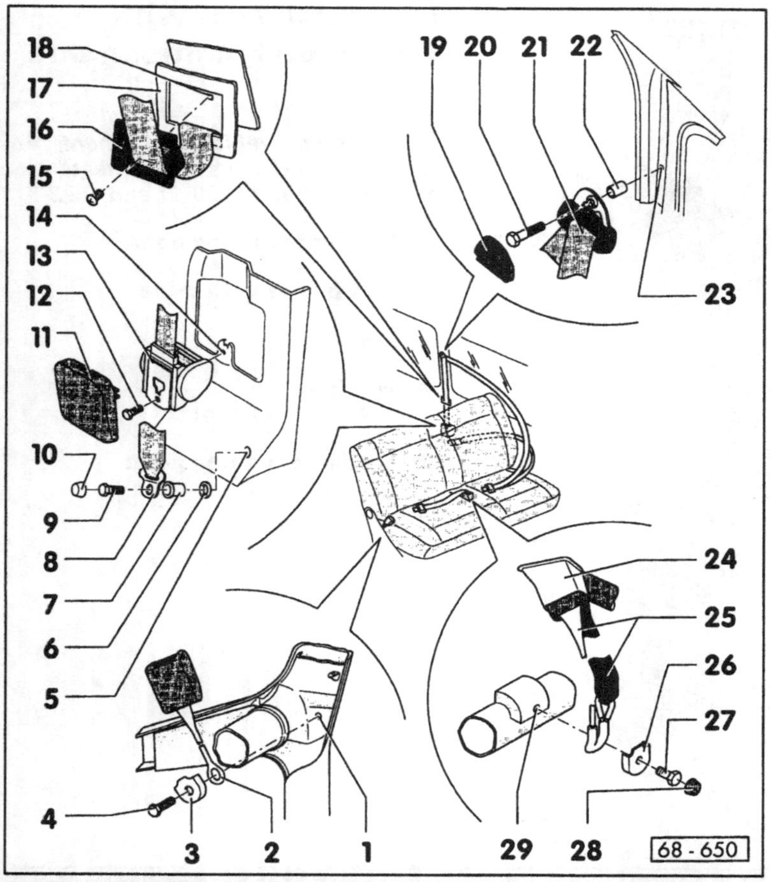

14 - Automatic tensioning seat belt retractor mounting point
- With retaining lug

15 - Phillips screw

16 - Webbing guide
- Can not be removed from webbing
- If replaced, destroy old webbing guide and cut new webbing guide

17 - Webbing guide cover

18 - Webbing guide mounting point

19 - D-ring cover

20 - Hex bolt
- 40 Nm (30 ft lb)

21 - D-ring

22 - Spacer

23 - C-pillar mounting point

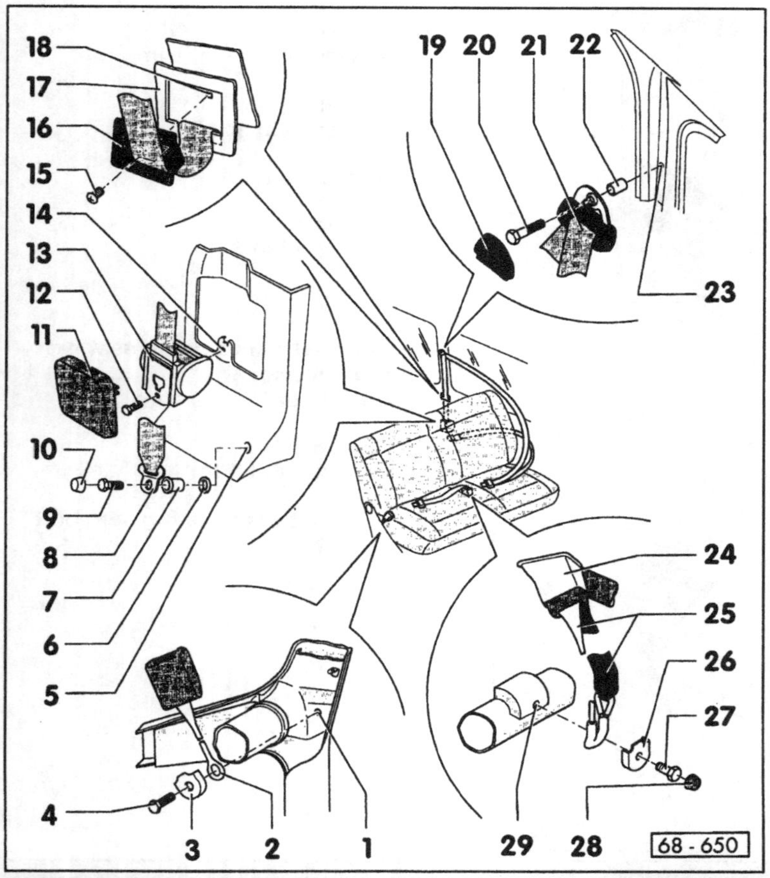

24 - Belt buckle guide

25 - Lap belt with buckle

26 - Cover

27 - Hex bolt
- 40 Nm (30 ft lb)

28 - Cover

29 - Seat frame mounting point

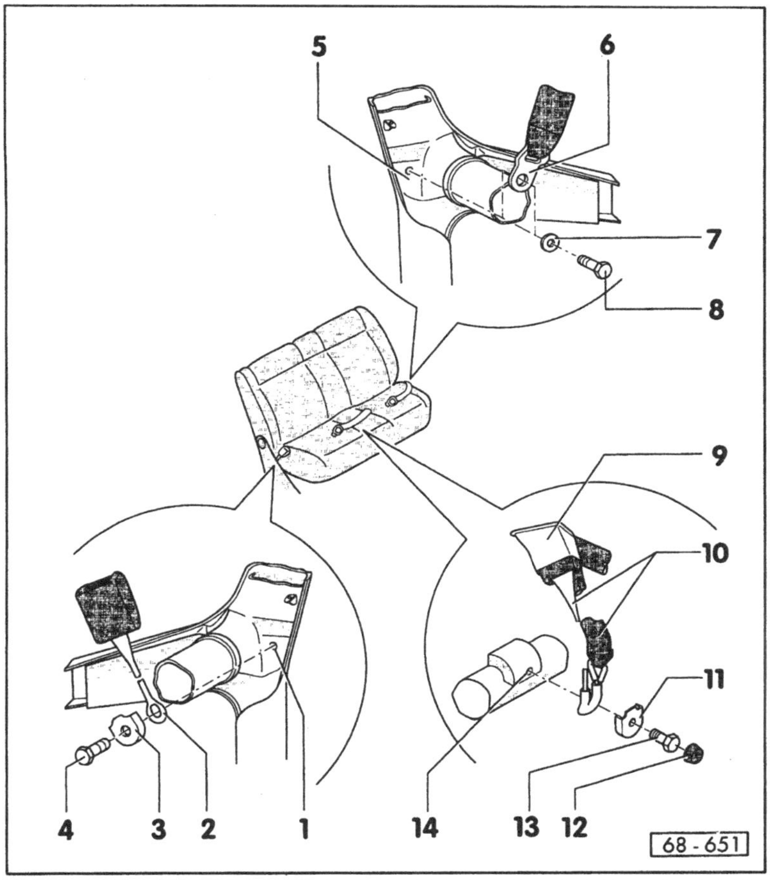

Two-place seat with lap belt, components

1 - Mounting point on seat frame

2 - Belt buckle with cable

3 - Cap

4 - Hex bolt
- ♦ 40 Nm (30 ft lb)

5 - Mounting point on seat frame

6 - Belt anchor

7 - Washer

8 - Hex bolt
- ♦ 40 Nm (30 ft lb)

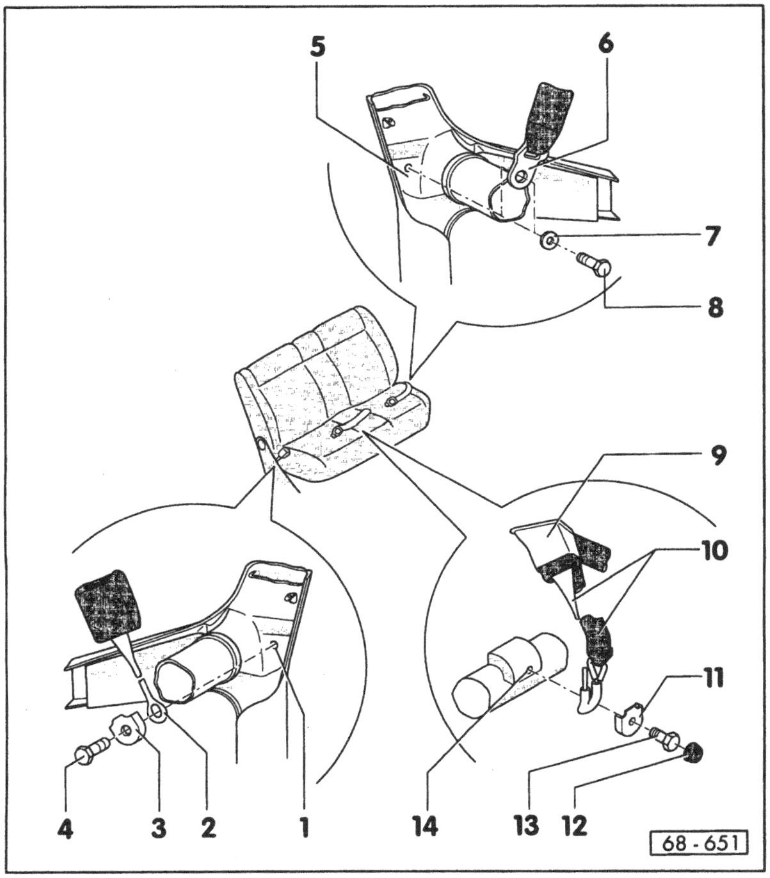

9 - Belt buckle guide

10 - Lap belt with buckle

11 - Cover

12 - Cap

13 - Hex bolt
- ♦ 40 Nm (30 ft lb)

14 - Mounting point on seat frame

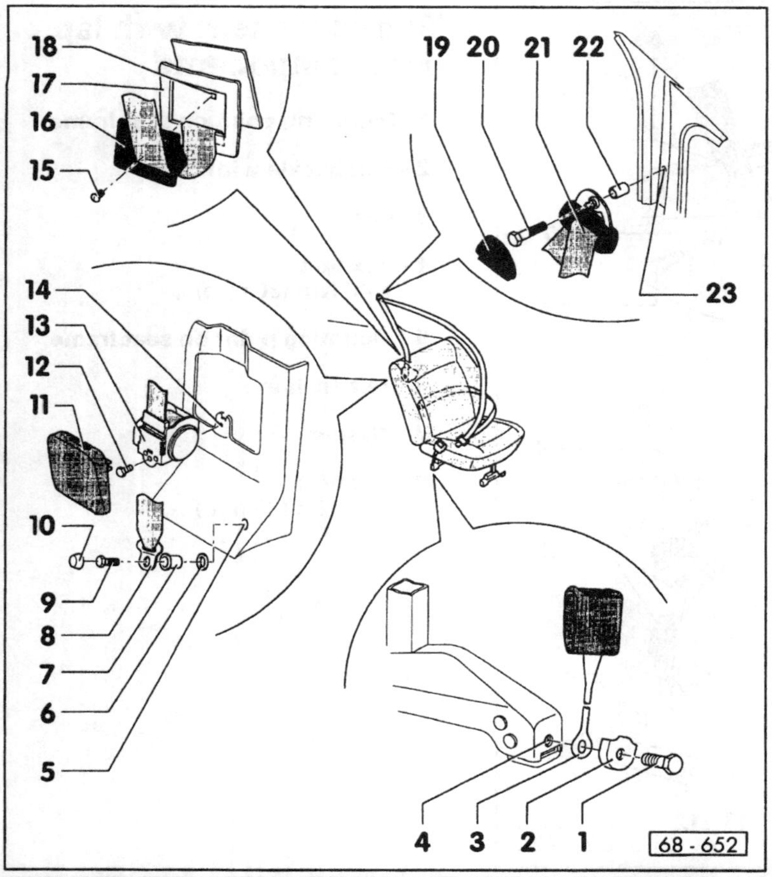

Folding seat, seat belt, components

Note:

Depending on vehicle equipment, remove trim and seats and/or seat benches ⇒ Page Q 70-11 and Q 72-7.

1 - Hex bolt
- 40 Nm (30 ft lb)

2 - Cover

3 - Belt buckle with cable

4 - Mounting point

5 - Mounting point on C-pillar

6 - Spacer
- Without C-pillar trim

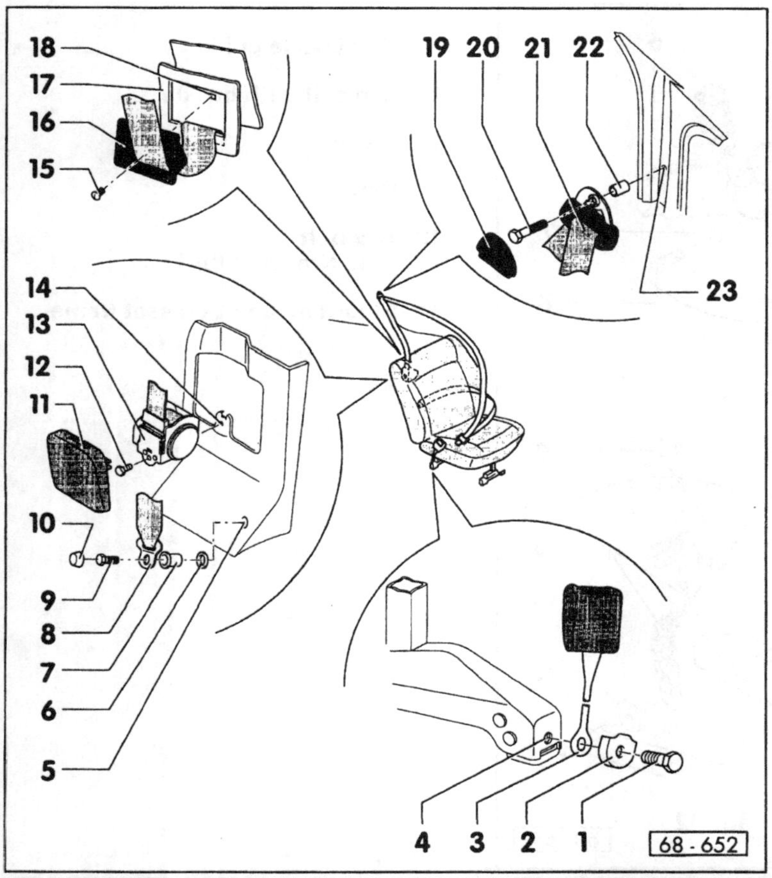

7 - Spacer
- With C-pillar trim

8 - Belt anchor

9 - Hex bolt
- 40 Nm (30 ft lb)

10 - Cap

11 - Automatic tensioning seat belt retractor cover
- Without C-pillar trim

12 - Hex bolt
- 40 Nm (30 ft lb)

13 - Automatic tensioning seat belt retractor

14 - Automatic tensioning seat belt retractor mounting point
- With retaining lug

15 - Phillips screw

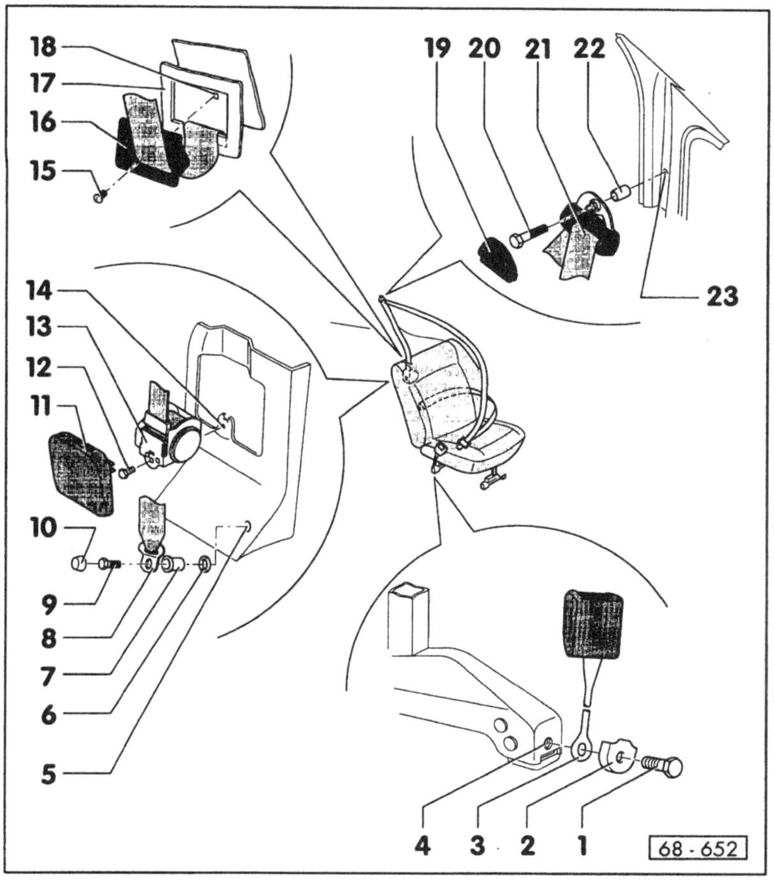

16 - **Webbing guide**
- ♦ Can not be removed from webbing
- ♦ If replaced, destroy webbing guide and cut new webbing guide

17 - **Webbing guide cover**

18 - **Webbing guide mounting point**

19 - **D-ring cover**

20 - **Hex bolt**
- ♦ 40 Nm (30 ft lb)

21 - **D-ring**

22 - **Spacer**

23 - **C-pillar mounting point**

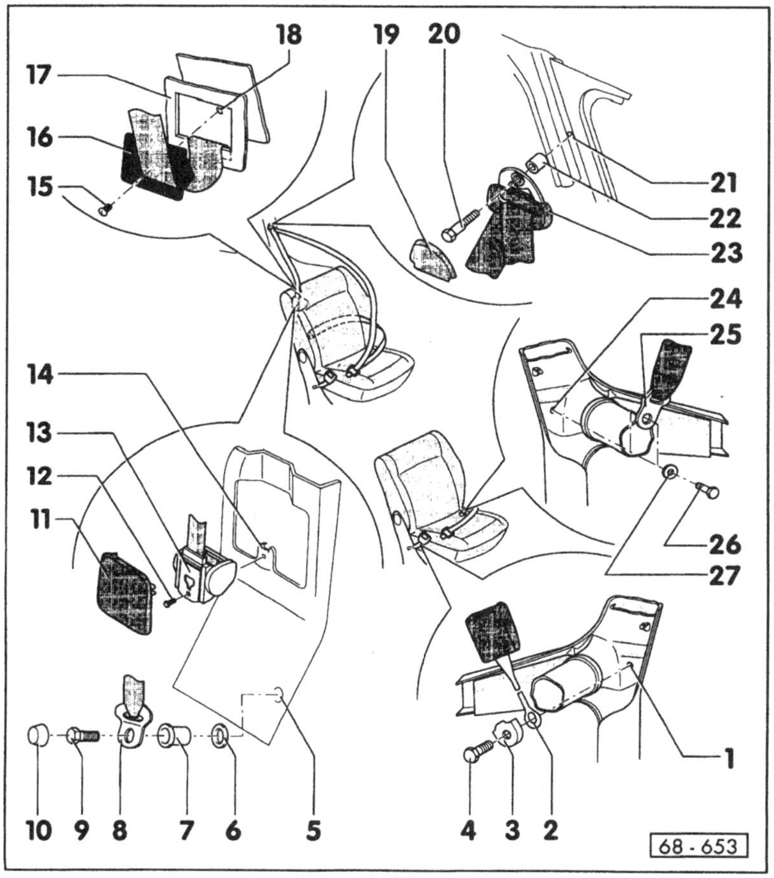

Single seat belt, components

Note:

Depending on vehicle equipment, remove trim and seats and/or seat benches ⇒ Page Q 70-11 and Q 72-7.

1 - **Mounting point on seat frame**

2 - **Belt buckle with cable**

3 - **Cover**

4 - **Hex bolt**
- ♦ 40 Nm (30 ft lb)

5 - **C-pillar mounting point**

6 - **Spacer**
- ♦ Without C-pillar trim

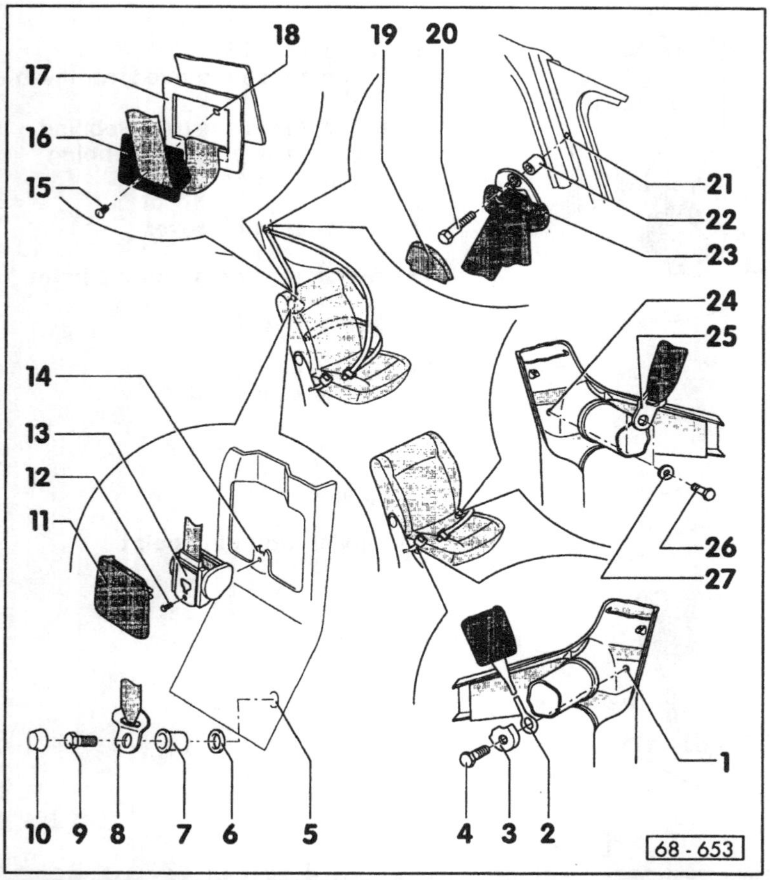

7 - Spacer
- With C-pillar trim

8 - 3-point belt anchor

9 - Hex bolt
- 40 Nm (30 ft lb)

10 - Cap

11 - Automatic tensioning seat belt retractor cover

12 - Hex bolt
- 40 Nm (30 ft lb)

13 - Automatic tensioning seat belt retractor

14 - Automatic tensioning seat belt retractor mounting point
- With retaining lug

15 - Phillips screw

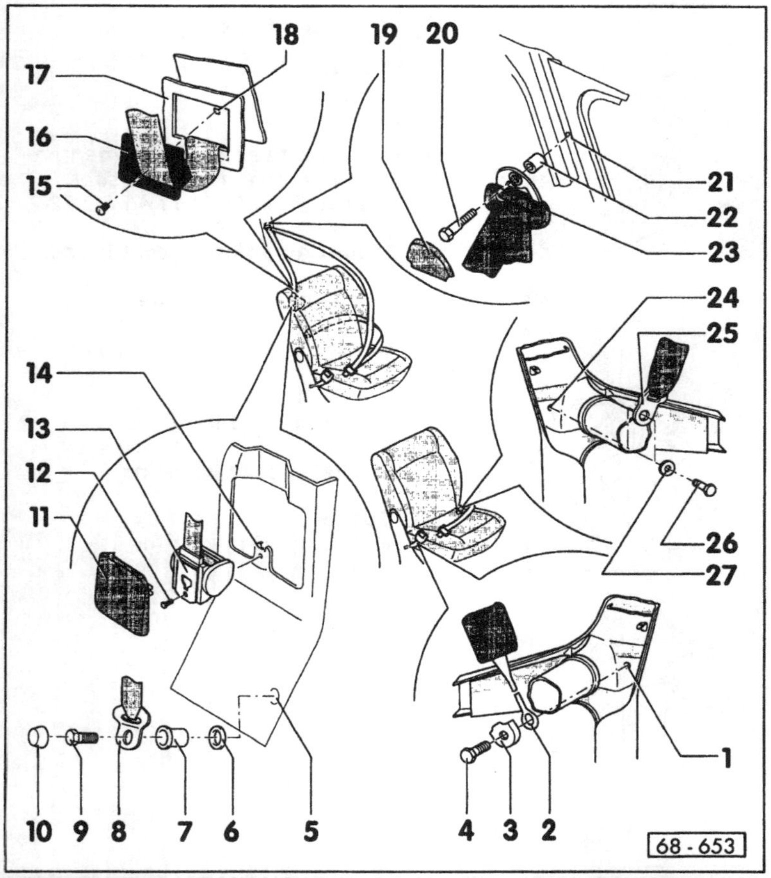

16 - Webbing guide
- Can not be removed from webbing
- If replaced, destroy old webbing guide and cut new webbing guide

17 - Webbing guide cover

18 - Webbing guide mounting point

19 - D-ring cover

20 - Hex bolt
- 40 Nm (30 ft lb)

21 - C-pillar mounting point

22 - Spacer

23 - D-ring

24 - Mounting point on seat frame

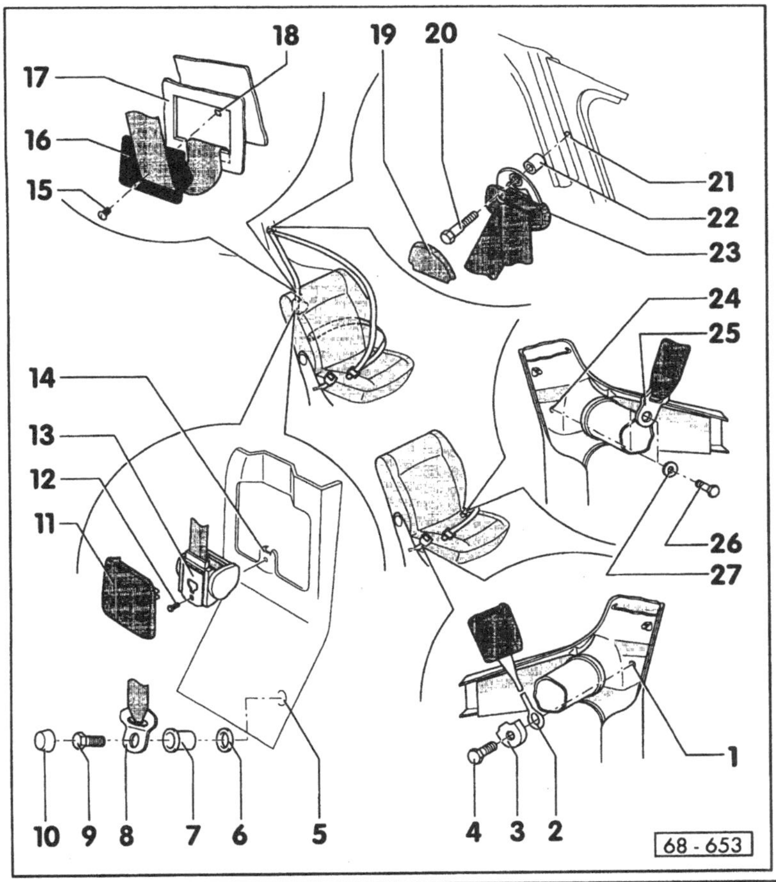

25 - **Lap belt anchor**

26 - **Hex bolt**
- ♦ 40 Nm (30 ft lb)

27 - **Washer**

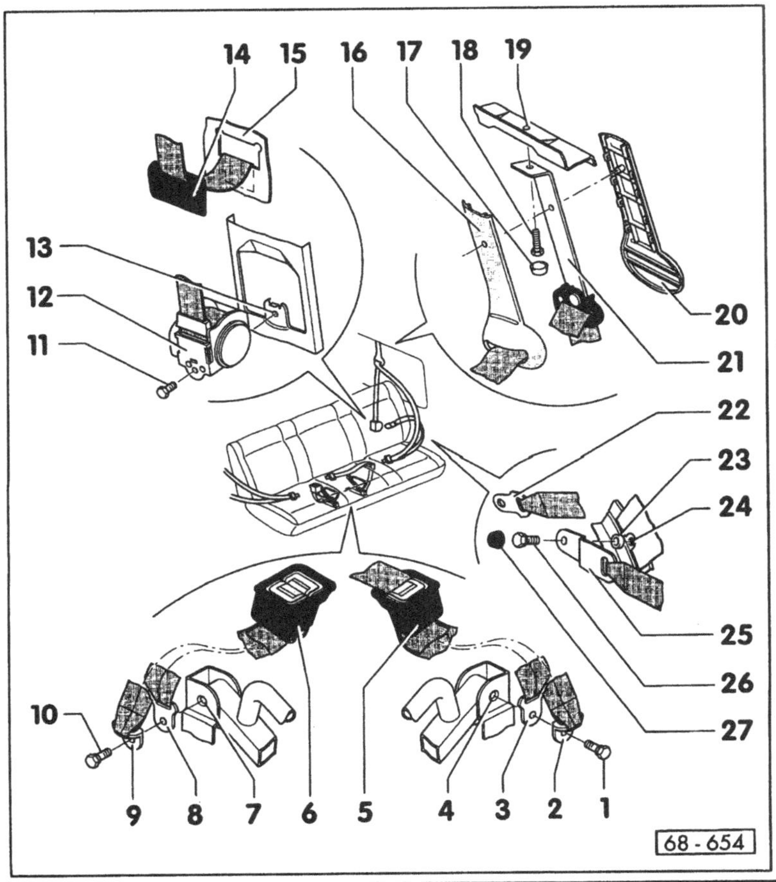

Three-place seat, seat belt components

Note:

Depending on vehicle equipment, remove trim and seats and/or seat benches ⇒ Page Q 70-11 and Q 72-7.

1 - **Hex bolt**
- ♦ 40 Nm (30 ft lb)

2 - **Belt buckle anchor**

3 - **Lap belt anchor**

4 - **Mounting point on seat frame**

5 - **Belt buckle guide**

6 - **Guide for belt buckles**

7 - **Mounting point on seat frame**

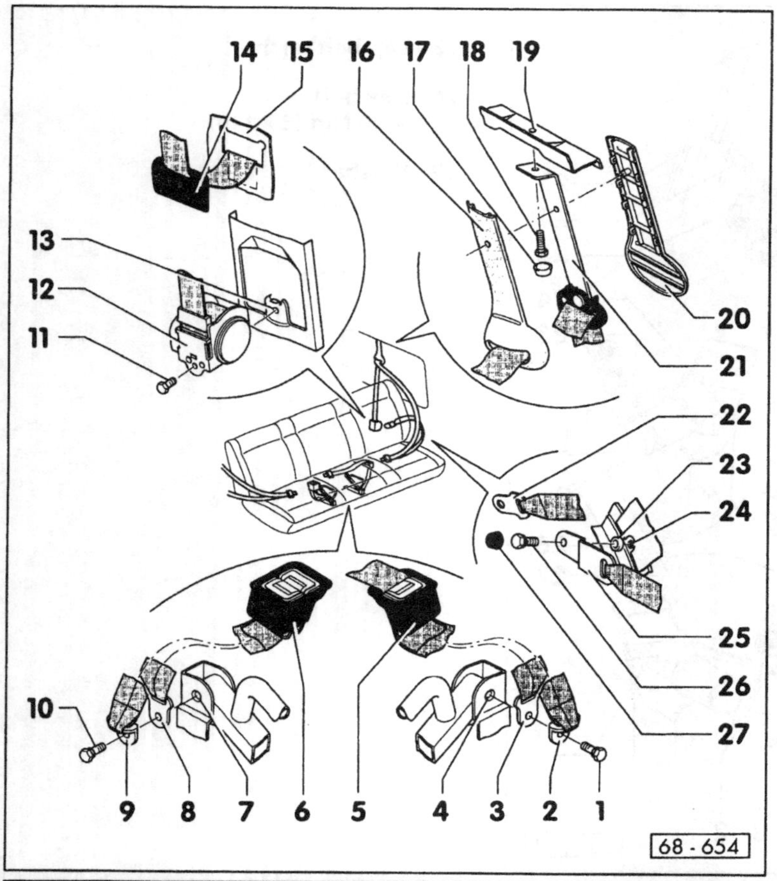

8 - **Belt buckle anchor**

9 - **Belt buckle anchor**

10 - **Hex bolt**
- ♦ 40 Nm (30 ft lb)

11 - **Hex bolt**

12 - **Automatic tensioning seat belt retractor**

13 - **Automatic tensioning seat belt retractor mounting point**
- ♦ With retaining lug

14 - **Webbing guide**

15 - **Webbing guide cover**

16 - **D-ring trim**

17 - **Cap**

18 - **Hex bolt**
- ♦ 40 Nm (30 ft lb)

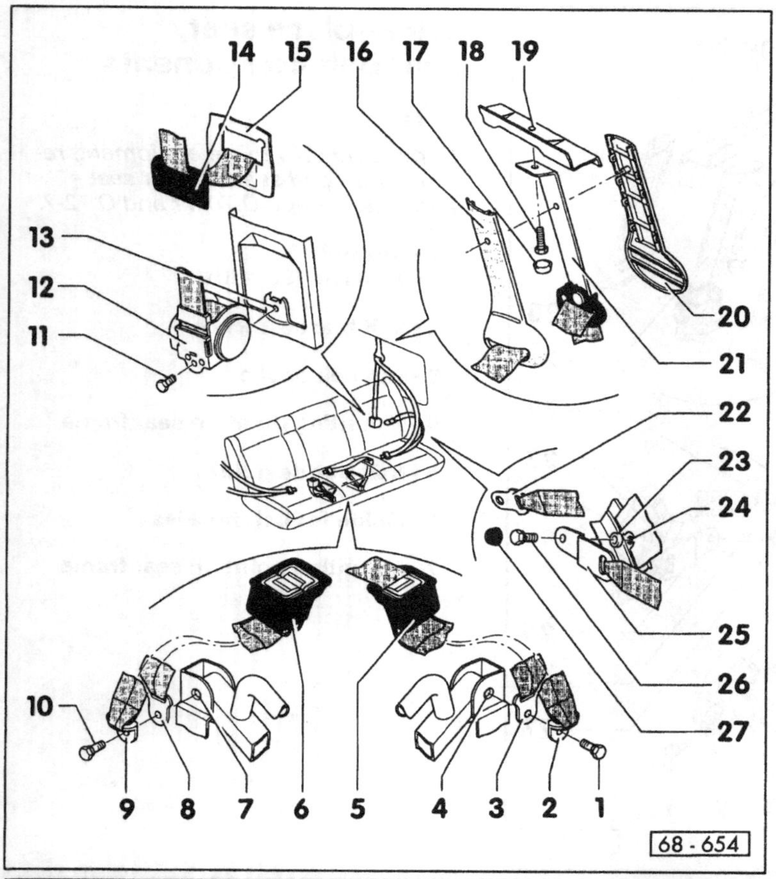

19 - **Mounting point at headliner**

20 - **D-ring trim**

21 - **3-point seat belt D-ring**

22 - **Lap belt anchor**

23 - **Spacer**

24 - **Mounting point on side panel**

25 - **3-point seat belt anchor**

26 - **Hex bolt**
- ♦ 40 Nm (30 ft lb)

27 - **Cap**

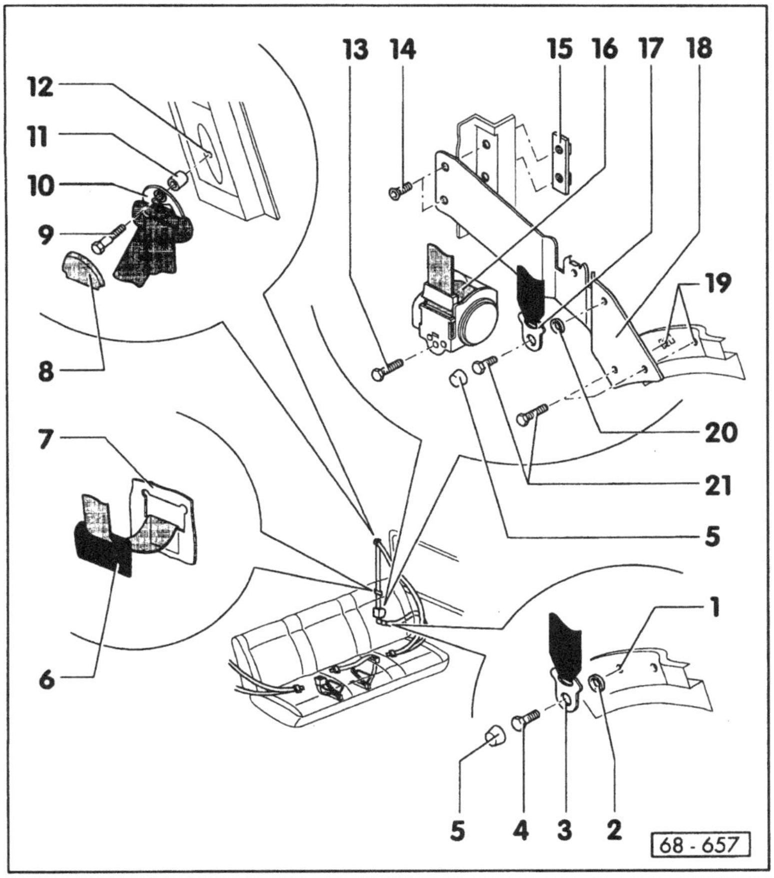

Rear seat, seat belt, components (long wheel base)

Note:

- *Depending on vehicle equipment, remove trim and seats and/or seat benches ⇒ Page Q 70-11 and Q 72-7.*
- *Lap belt and belt buckles ⇒ Page Q 69-26*

1 - Lap belt mounting point

2 - Spacer

3 - Lap belt anchor

4 - Hex bolt
- 40 Nm (30 ft lb)

5 - Cap

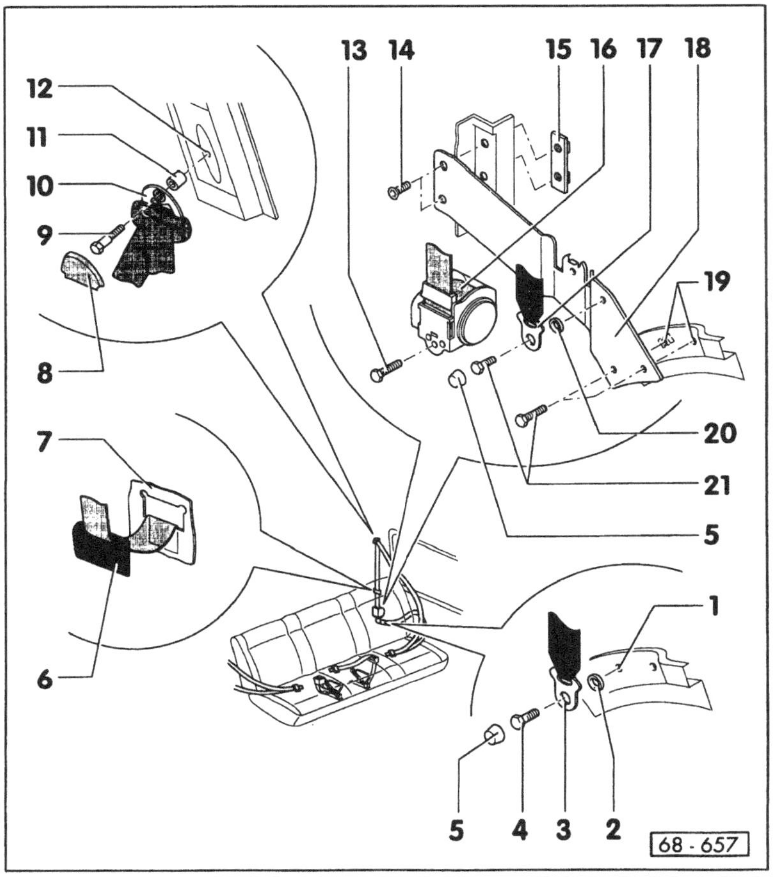

6 - Webbing guide

7 - Webbing guide cover

8 - D-ring cap

9 - Hex bolt
- 40 Nm (30 ft lb)

10 - D-ring

11 - Spacer

12 - D-pillar mounting point

13 - Hex bolt
- 40 Nm (30 ft lb)

14 - Countersunk screw
- 20 Nm (15 ft lb)

15 - Threaded plate

16 - Automatic tensioning seat belt retractor

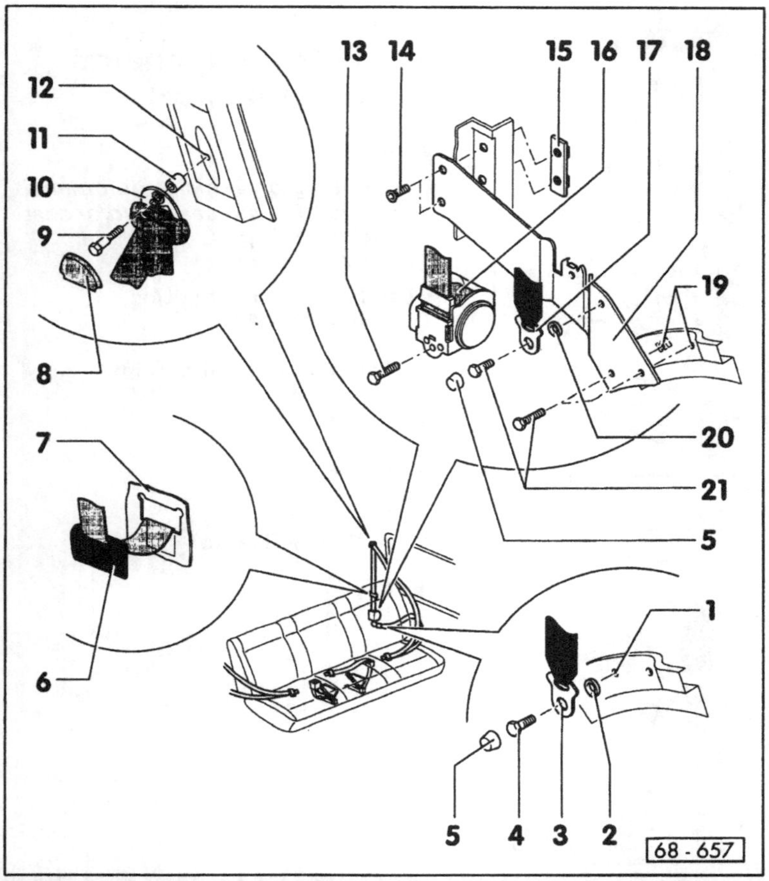

17 - **3-point belt anchor**

18 - **Bracket with mounting point for 3-point belt**

19 - **Mounting points for brackets**

20 - **Spacer**

21 - **Hex bolt**
- ♦ 40 Nm (30 ft lb)

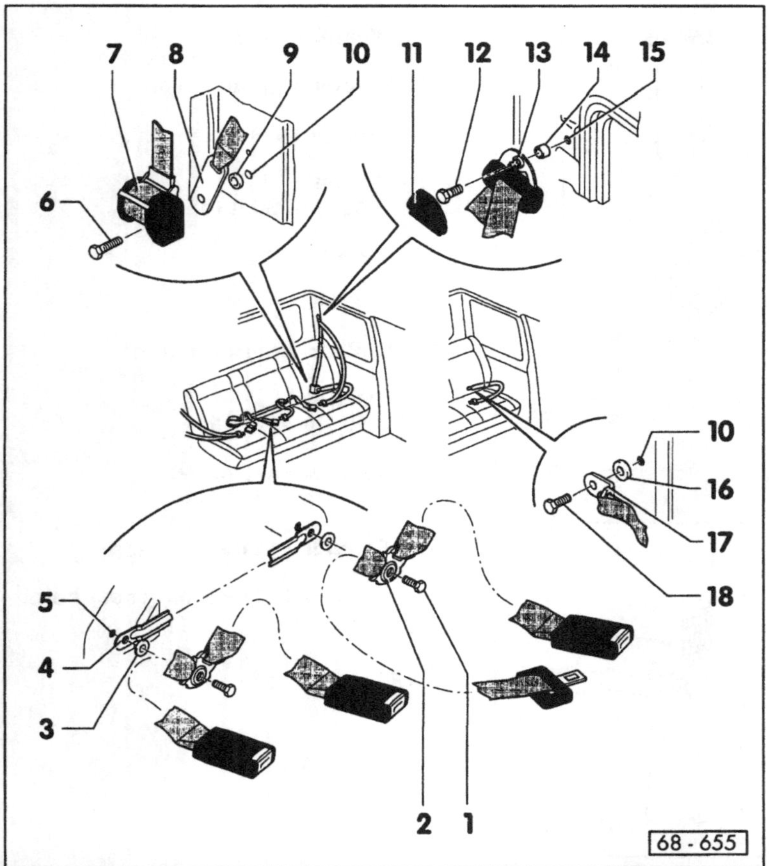

Rear seat, seat belt, components (double cab)

Note:

If necessary, remove rear seat bench
⇒ Page Q 72-10

1 - **Hex bolt**
- ♦ 40 Nm (30 ft lb)

2 - **Belt fitting**
- ♦ Right: with two belt buckles
- ♦ Left: with belt buckle and lap belt

3 - **Gasket**

4 - **Pull strap**

5 - **Cab rear panel mounting point**

6 - **Hex bolt**
- ♦ 40 Nm (30 ft lb)

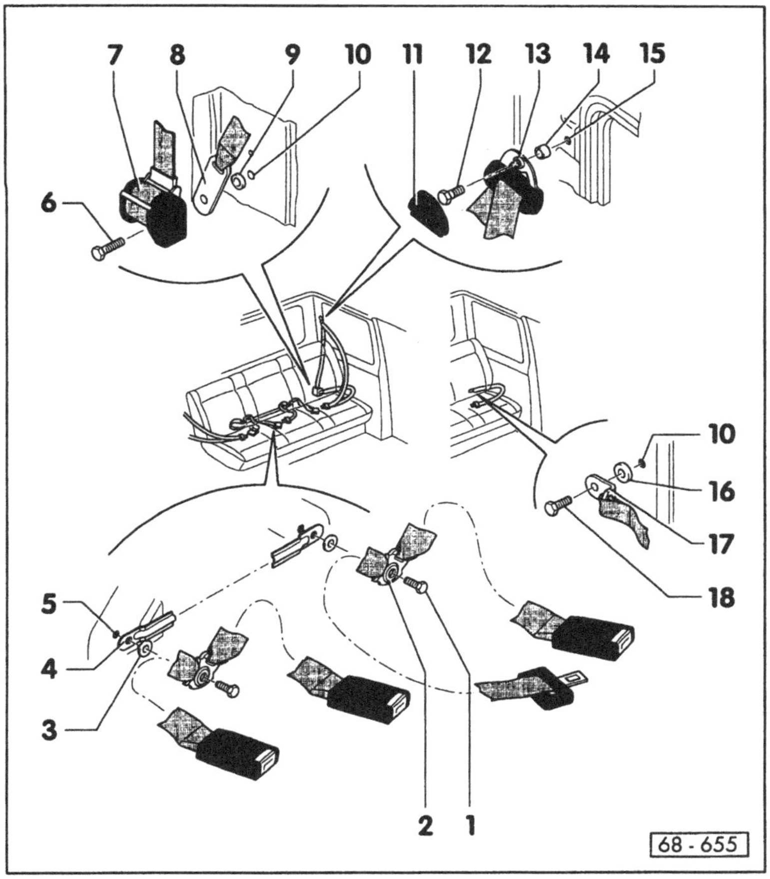

7 - **Automatic seat belt retractor**

8 - **3-point belt anchor**

9 - **Spacer**

10 - **Side panel mounting point**

11 - **D-ring cover**

12 - **Hex bolt**
- 40 Nm (30 ft lb)

13 - **3-point belt D-ring**

14 - **Spacer**

15 - **C-pillar mounting point**

16 - **Spacer**

17 - **Lap belt anchor**

18 - **Hex bolt**
- 40 Nm (30 ft lb)

Seat belt system, inspection

CAUTION!

Seat belt assemblies must be replaced after being subjected to loading by occupants in collision. Replace components as necessary.

Test points:

- Check belt webbing
- Check automatic seat belt lock retractor
- Belt lock visual inspection
- Belt lock function test
- Check D-rings and lock tongues
- Check mounting components and mounting points
- Check lap belt automatic lock retractor

Note:

If customer refuses to have a damaged seat belt or belt component replaced, a note to this effect must be made.

Seat belt webbing

– Fully extend webbing from retractor.

– If dirty, clean belt webbing only with mild soap solution recommended for cleaning upholstery or carpets. Follow instructions provided with cleaner.

● If following types of damage (1 and 2) is found in a vehicle that has been in an accident, replace seat belt complete with belt lock.

● If damage 1, 2, or 3 is found in a vehicle that has not been in an accident, replace just damaged seat belt.

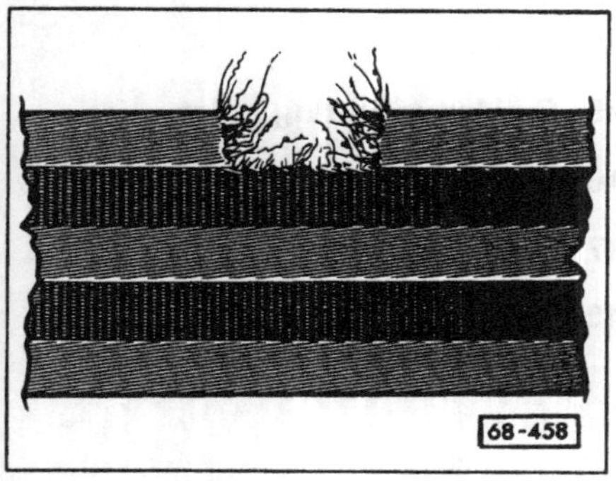

◄ 1 - Cut or damaged webbing

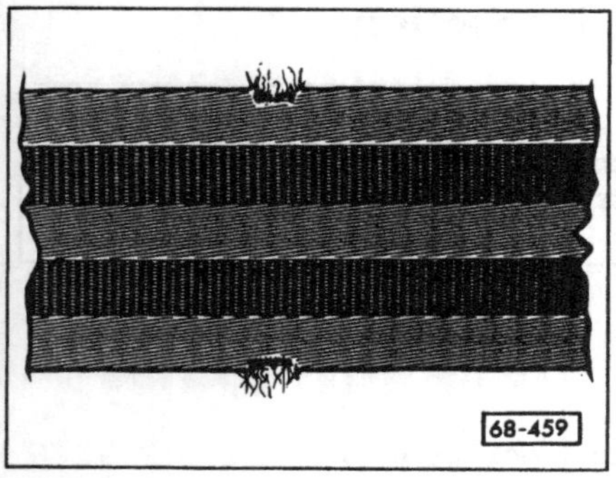

◄ 2 - Cut loops at belt edge

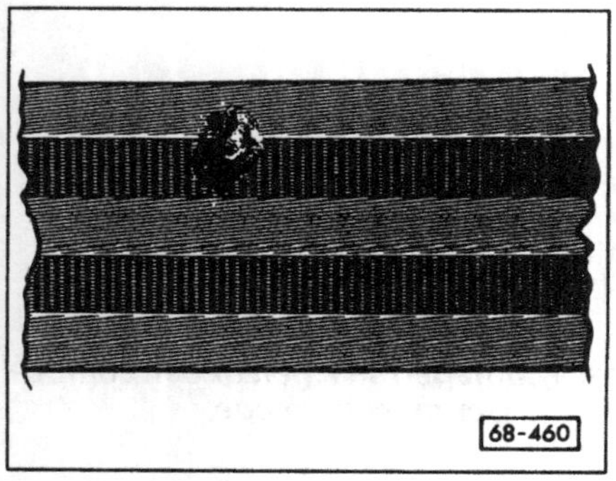

◄ 3 - Burnt spots from cigarettes, etc.

Seat belt retractor inspection

Automatic retractors are dual sensing and will lock up in two modes:

♦ Retractor will lock up as belt is being withdrawn if inertia increases dramatically

– While pulling webbing from retractor, give belt a fast jerk.

- No locking action - replace complete seat belt assembly with buckle
- Seat belt retractor malfunction: check for change in retractor position

– Put seat belt on.

– Accelerate vehicle to 20 km/h (approx. 12 mph) and quickly apply footbrake.

- If retractor does not lock up under these conditions, replace seat belt assembly

> ***CAUTION!***
> ***For safety reasons, drive vehicle in open area away from other vehicles when performing driving test***

Buckle, visual inspection

– Check buckle for cracks and chipping.

- If damaged, replace complete seat belt assembly

Buckle, function test

– Insert seat belt tongue in buckle until audible click is heard. Pull back on webbing quickly to assure buckle is latched properly.

- If after at least 5 tests belt tongue did not engage in buckle even once, replace complete seat belt assembly with buckle

Unlocking, checking:

– Depress button on buckle to release belt.

If belt webbing is under tension, the lock tongue must spring out by itself.

- Perform test at least 5 times. If lock tongue fails to spring out once, replace complete seat belt assembly with buckle

CAUTION!
Do not lubricate belt buckles to eliminate noise or difficulty in operating buttons.

D-rings and lock tongue, checking

Check for scoring. If belt was worn during accident, plastic covered D-rings will show fine parallel grooves indicating stress. D-rings worn from normal usage will be smooth.

– Check plastic for deformation, chipping and cracking.

- If there is scoring and/or damage, replace complete seat belt assembly with buckle.

Mounting parts and mounting points, checking

◆ Lock strap deformed (stretched)

◆ Height adjustment mechanism does not work

◆ Mounting points (seat, column, vehicle floor) warped or threads damaged

– If damage is found on parts, replace complete seat belt assembly with belt buckle.

– Replace mounting points.

Note:

If there is damage that did not occur as a result of an accident, e.g. wear, replace only damaged parts.

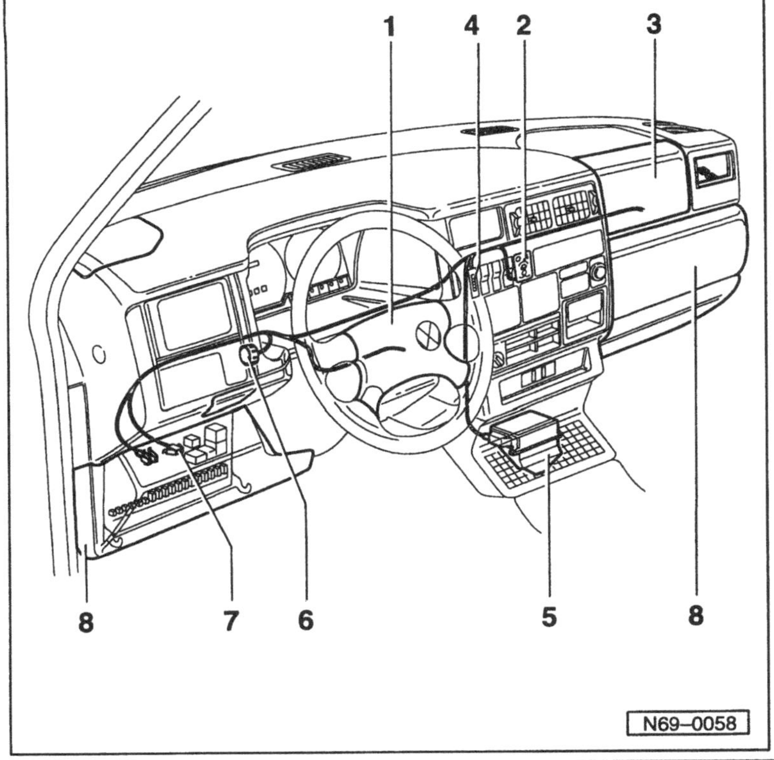

Airbag system, components

1 - **Airbag unit, driver's side**
- ♦ Removing and installing ⇒ Page Q 69-46

2 - **Airbag Malfunction Indicator (MIL) Lamp -K75-**

3 - **Airbag unit, passenger's side**
- ♦ Removing and installing ⇒ Page Q 69-50

4 - **Data Link Connectors (DLC)**

5 - **Airbag Control Module -J234-**
- ♦ Removing and installing ⇒ Page Q 69-53

6 - **Central Ground connection**

7 - **Relay plate (terminal 15)**

8 - **Knee padding**
- ♦ Removing and installing ⇒ Page Q 69-58

Airbag, safety precautions

CAUTION!

- ♦ ***Testing, removing, installing and repair work may only be performed by trained Service department personnel.***
- ♦ ***Always disconnect the battery ground strap (–) before working on airbag related systems.***
- ♦ ***No waiting time is necessary after disconnecting the battery.***
- ♦ ***When connecting the airbag system to a voltage source, there must be no person inside the vehicle.***
- ♦ ***Any time the battery has been disconnected it will be necessary to reestablish basic settings for the Engine Control Module (ECM) and, on vehicles with automatic transmission, for the Transmission Control Module (TCM), using VAG 1551 Scan Tool (ST).***
- ♦ ***On vehicles equipped with anti-theft radios, make sure you know the radio activation code before you disconnect the battery.***
- ♦ ***The airbag units should be installed immediately after removing them from their packaging.***
- ♦ ***If the work on the airbag system is interrupted, the airbag unit should again be placed in its packaging. Do not leave the airbag unit unattended!***
- ♦ ***If the airbag unit or the airbag control module has been dropped from a height of 0.5 meter (19.7 in.) or more, DO NOT install.***

CAUTION!

- ***Airbag units which have been removed should be stored so that the padded side is facing up.***
- ***Always replace any airbag system component that has been mechanically damaged (example: dented or cracked).***
- ***Airbag units which have NOT been activated should be marked and returned to the manufacturer for disposal (use airbag unit transport container).***

Replacing airbag units following an accident

Accident involving activation of airbag

- The following components must always be replaced:
  Driver airbag unit
  Front passenger airbag unit
  Airbag control module with holder
  Coil connector with slip ring
  Driver seat belt
  Front passenger seat belt (if used)
- In addition, the following must also be replaced if necessary (visual check):
  All damaged components

Accident not involving activation of airbag

- If the Airbag Malfunction Indicator Lamp -K75- does not indicate a malfunction, it is not necessary to replace the airbag components. The seat belts should be checked separately.

After replacing the airbag units or the control module, affix stickers (tear-off strips) to the registration card and return the registration card to the relevant Importer for registration.

Note:

For vehicles produced from 09.98 ⇒ Page N 48-51 to remove airbag unit from steering wheel.

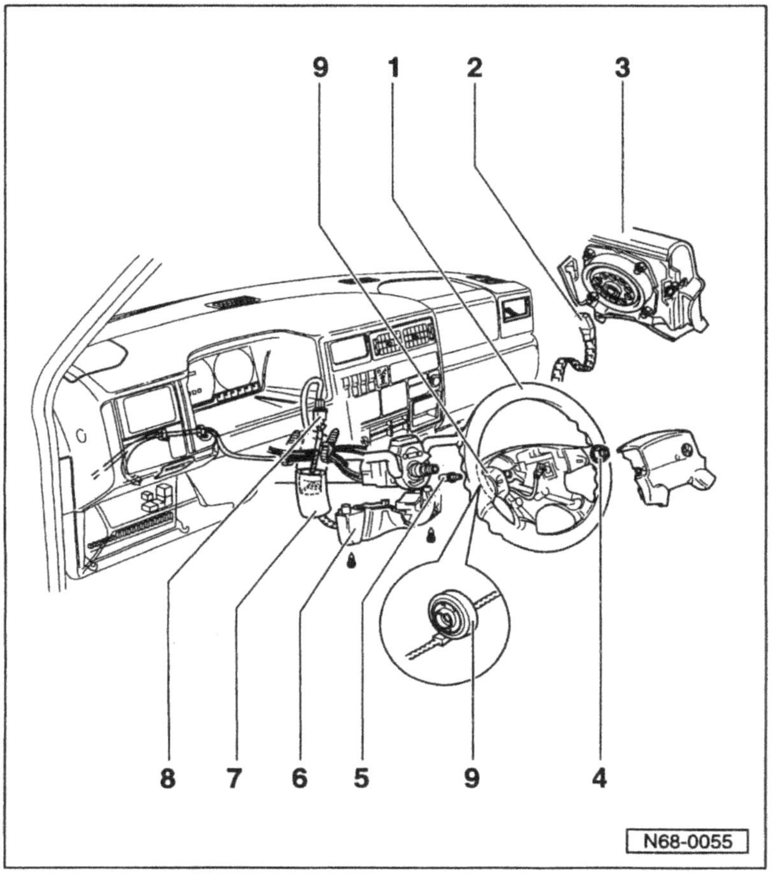

Driver's airbag unit, removing and installing

1 - **Steering wheel**
   - Removing ⇒ Page Q 69-48

2 - **Airbag unit connector**

3 - **Airbag unit**

4 - **Hex nut**
   - 67 Nm (49 ft lb)

5 - **Torx screw (T 30)**
   - 6.5 Nm (58 in. lb)
   - Always replace

6 - **Lower trim**

7 - **Foam tube**

8 - **Connector**

9 - **Airbag spiral spring -F138-**
   - Removing ⇒ Page Q 69-49

> *CAUTION!*
>
> ***On cars equipped with anti-theft radios, make sure you know the radio activation code before you disconnect the battery.***

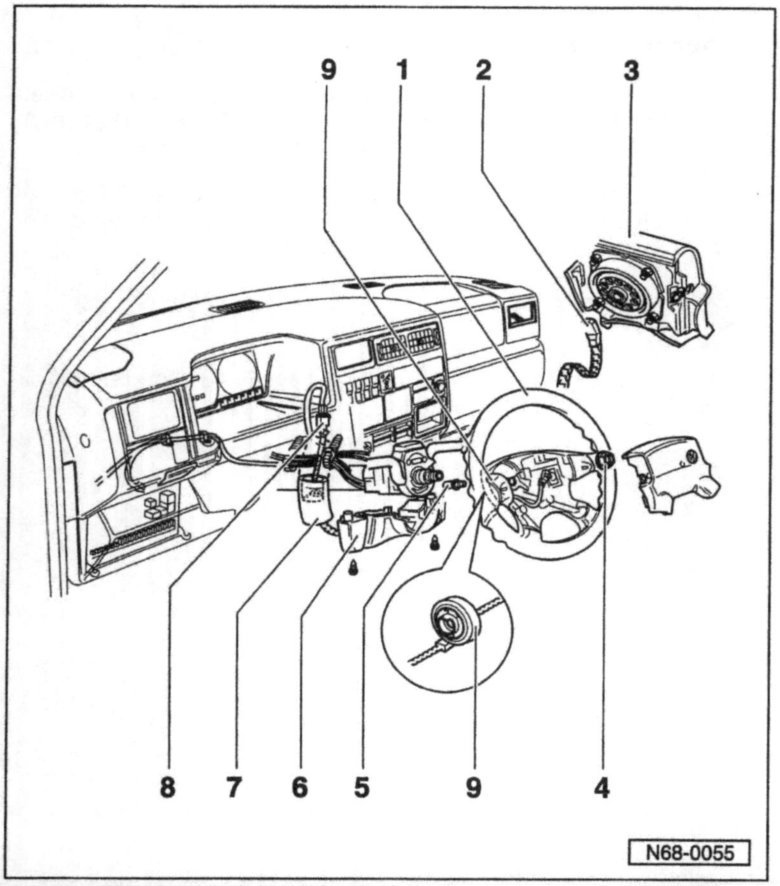

Removing

- Disconnect battery ground strap.
- Turn steering wheel -1- to center position (wheels straight-ahead).
- Remove Torx socket-head screws (T 30) -5- (through 08.98 production).
- Remove airbag unit -3- from steering wheel and disconnect red connector -2- from airbag unit.

Installing

- Install in reverse order or removal.

Tightening torque (bolts -5-): 6.5 Nm (58 in. lb)

> ***WARNING!***
> ***Ensure that there is no person inside the vehicle during this operation.***

After all driver's side airbag components are installed:

- Switch ignition ON.
- Connect battery ground strap (-).

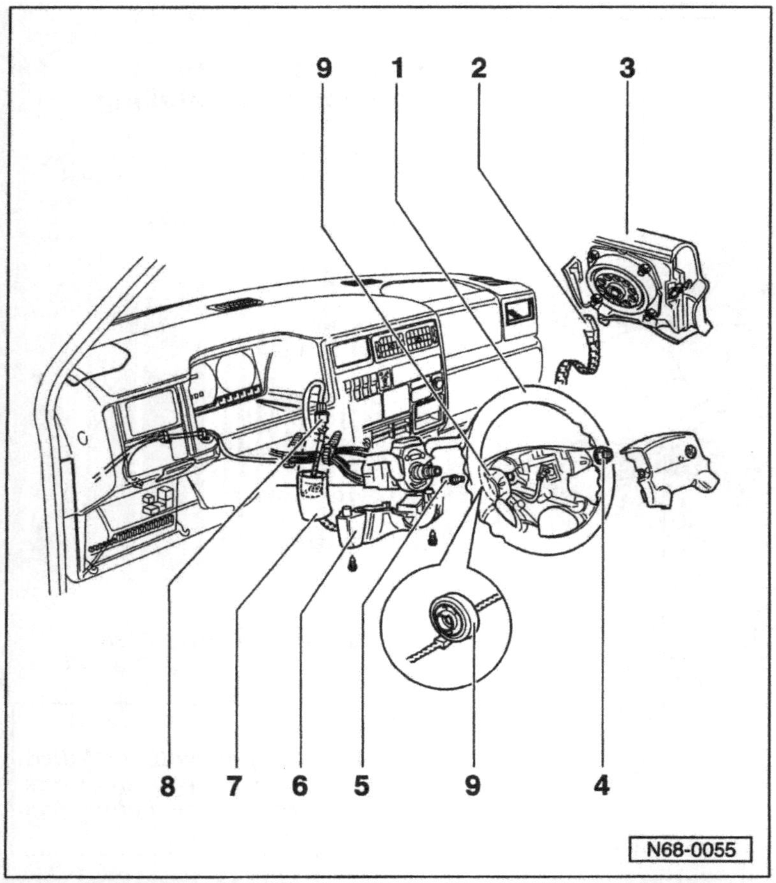

Steering wheel with airbag, removing and installing

Requirement

- ♦ Driver's side airbag unit removed ⇒ Page Q 69-46 (through 08.98 prod.) ⇒ Page N 48-51 (from 09.98 prod.)

- Remove lower trim -6- from steering column switch.
- Pull foam tube -7- off connector -8- and separate connector.
- Remove hex nut -4- and remove steering wheel from steering column.

  Tightening torque (nut -4-): 67 Nm (49 ft lb)

Note:

Airbag spiral spring -F138- -9- remains on the steering wheel.

Airbag spiral spring -F138-, removing and installing

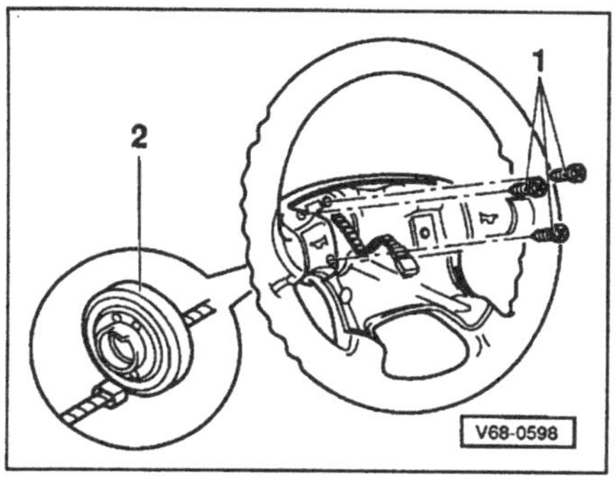

Requirements

- ♦ Driver's side airbag unit removed ⇒ page Q 69-46
- ♦ Steering wheel removed ⇒ page Q 69-48
- – Separate single connector for horn.
- – Remove Phillips screws -1- and remove airbag spiral spring -2- from steering wheel.

Note:

- ♦ *The steering must be in the center position (wheels straight-ahead) when the airbag spiral spring -F138- is removed and installed.*
- ♦ *Replacement airbag spiral springs are secured in the center position with a cable tie.*

Checking center position

- – Turn airbag spiral spring (clockwise) onto stop and then turn back 3.5 turns.

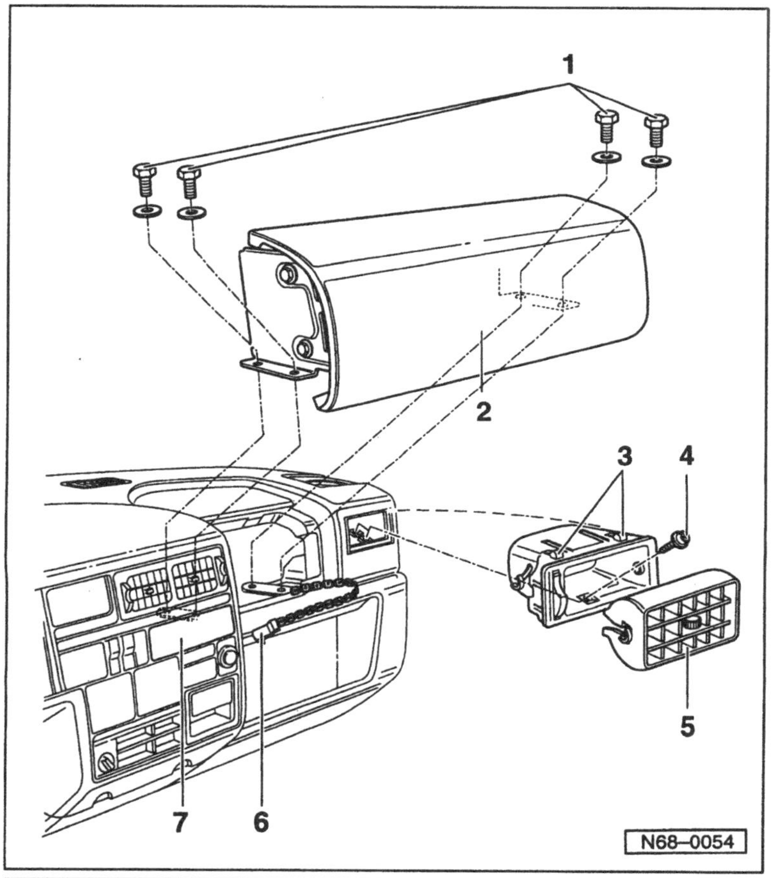

Passenger's side airbag unit, removing and installing

1 - **Hex bolt**
  - ♦ 10 Nm (7 ft lb)

2 - **Airbag unit**
  - ♦ Removing ⇒ page Q 69-51
  - ♦ Installing ⇒ page Q 69-52

3 - **Retaining tabs on air vent housing**

4 - **Screw**

5 - **Air vent grille**

6 - **Airbag unit connector**

7 - **Trim panel**

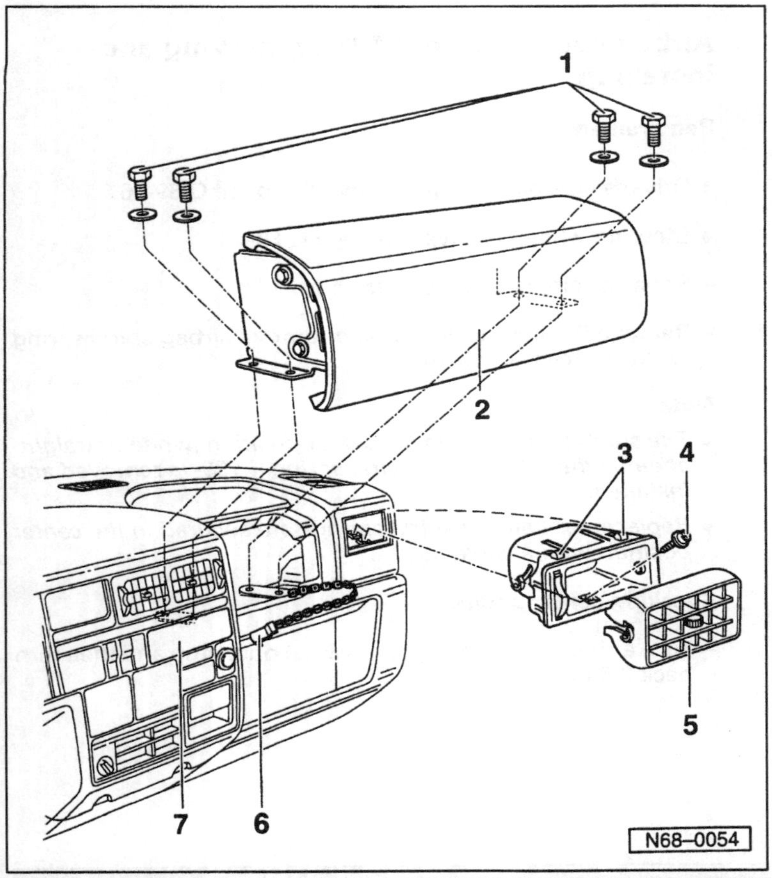

Removing

> ***CAUTION!***
> ***On cars equipped with anti-theft radios, make sure you know the radio activation code before you disconnect the battery.***

- Disconnect battery ground strap (–).
- Carefully pull out air vent grille -5- with long-nosed pliers.
- Remove screw -4-.
- Release retaining tabs -3- and pull out air vent housing for air vent grille -5-.
- Remove trim panel -7- or radio.
- Remove hex bolts -1-.

Note:

The bolts are accessible via openings for the radio and air vent in the dash panel.

- Remove airbag unit -2- from dash panel.
- Separate airbag unit connector -6-.

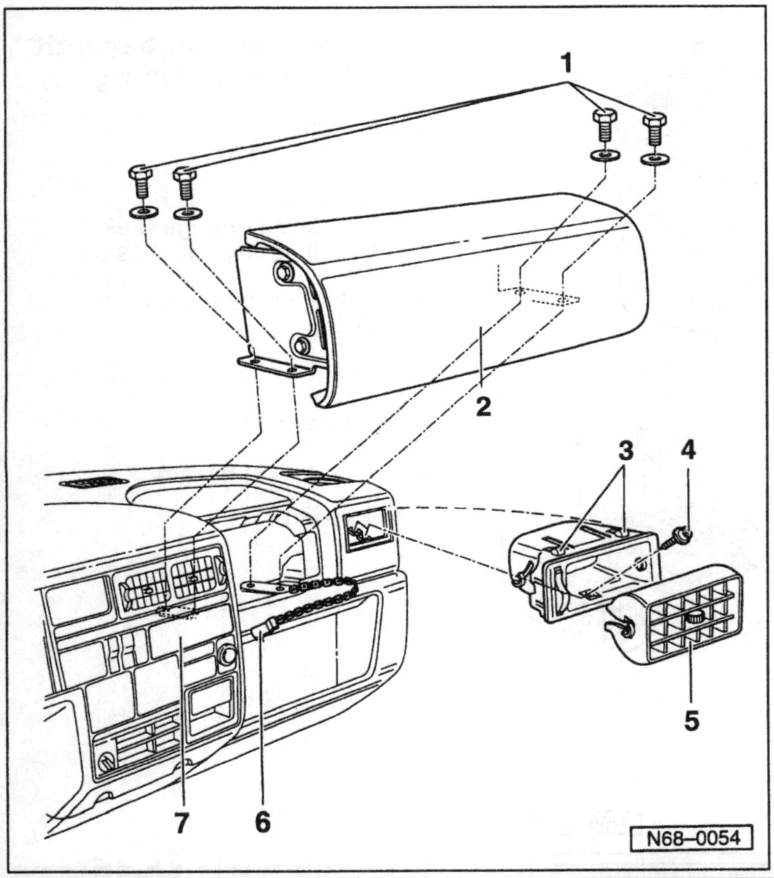

Installing

◆ Battery ground strap disconnected

- Connect airbag unit connector -6- to passenger's side airbag unit -2- and install airbag unit with four hex bolts -1-.

  Tightening torque (bolts -1-):
  10 Nm (7 ft lb)

- Install air vent housing by engaging retaining tabs -3- into dash panel.
- Install screw -4-.
- Install air vent grille -5-.
- Switch on ignition.
- Connect battery ground strap.

> ***WARNING!***
> ***Ensure that there is no person inside the vehicle during this operation.***

Airbag control module -J234-, removing and installing

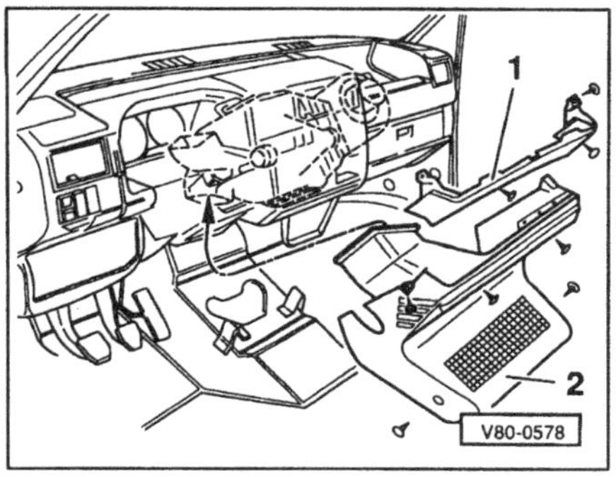

- Remove trim panel -1-.
- Remove footwell air vent housing -2-.

> ***WARNING!***
> ***Disconnect the battery ground strap (–) before disconnecting the airbag control module.***

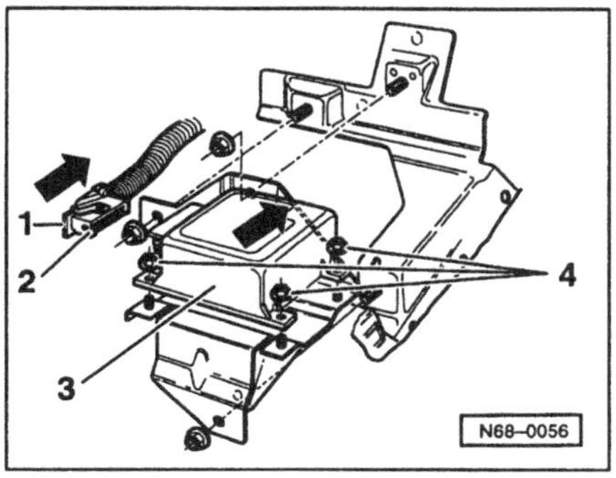

◄

- Move bar -1- in opposite direction of arrow and disconnect connector -2- from control module -3-.
- Remove nuts -4- and remove control module from studs (nuts -4- are sealed with paint).

  Tightening torque (nuts -4-): 10 Nm (7 ft lb)

Installation is the reverse order of removal.

Airbag unit, disposal

For vehicles to be returned to service

> ***CAUTION!***
> - ***The storage, transportation and disposal of airbag units are subject to the laws for flammable solids.***
>
> ***Disposal of airbag units:***
> - ***Active airbag units (gas generators): remove an active (not deployed) airbag unit carefully from the vehicle and return to the Warranty Parts Test Center for proper disposal. DO NOT deploy airbag in vehicle!***
> - ***Deployed airbag units (gas generators): deployed airbag units DO NOT have to be disposed of as hazardous waste but can be discarded in the normal manner. However, it is recommended to have the unit recycled.***

For vehicles being scrapped

> ***WARNING!***
>
> ***Do not ignite airbag unit(s) on vehicles that are to be returned to service. When replacing non-deployed airbag unit(s) return old airbag unit(s) to the Warranty Parts Test Center.***

To conform with accident prevention regulations, airbag units must be rendered unusable by igniting them electrically before scrapping. This is necessary because the airbag unit is a pyrotechnic object that may cause injuries if activated in an improper manner (Example: scrapping methods using cutting torches).

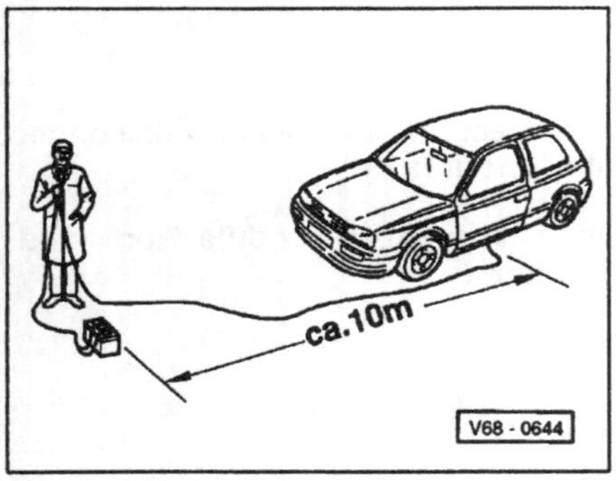

Driver's side airbag unit, igniting

Airbag units must be ignited from the outside with the airbag inside the vehicle and the doors closed.

◄ • Minimum distance from vehicle: 10 meters (33 ft)

– Remove airbag unit ⇒ page Q 69-46.

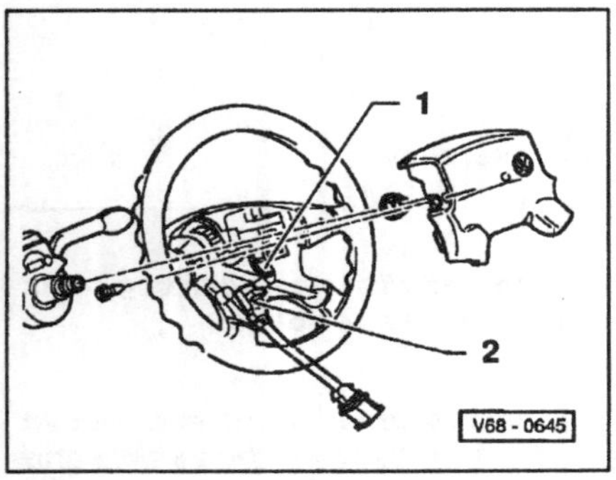

◄ – Disconnect red connector -1- from airbag unit.

– Connect activation cable -2- (Part No. 357 971 419) to airbag unit and firmly install airbag again. Activation cable is positioned between airbag unit and steering wheel.

Note:

The activation cable must be placed between the airbag unit and the steering wheel.

> ***WARNING!***
>
> ***Make sure that there is no person inside the vehicle during this operation.***

– Route airbag igniter cables through door gap and connect to external battery.

– Connect ignition device from VAG 1821 and cable connector from adapter set VAG 1594 A to two-pin connector of activation cable.

– Activate airbag unit.

– Dispose of airbag unit as scrap.

Front passenger's side airbag unit, igniting

> ***WARNING!***
>
> ***Make sure that there is no person inside the vehicle during this operation.***

– Remove shelf on front passenger's side.

– Unclip and separate connector.

– Cut off airbag unit connector and connect to VAG 1821 airbag igniter.

– Route airbag igniter cables through door gap and in front of vehicle and connect to external battery.

● Minimum distance from vehicle: 10 meters (33 ft)

– Activate airbag unit.

– Dispose of airbag unit as scrap.

Knee padding, removing and installing

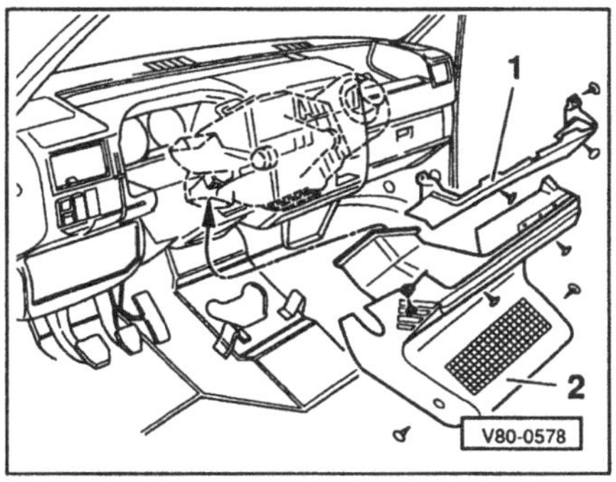

◄ – Remove trim panel -1-.

– Remove footwell air vent housing -2-.

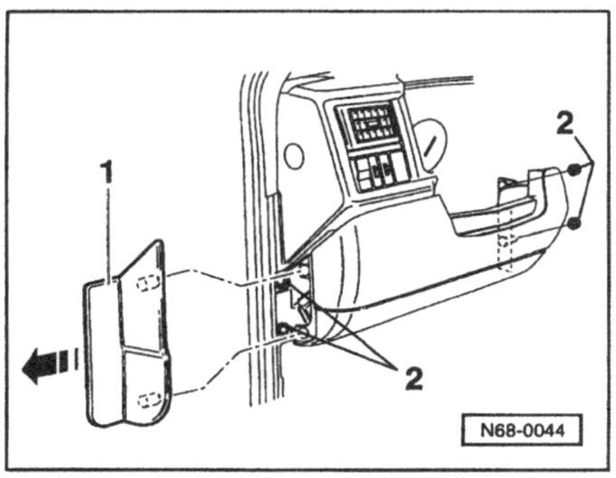

◄ – Remove trim panel -1- on left and right.

– Remove nuts and bolts -2- (2 each side) and remove driver's knee pad.

Tightening torque (nuts and bolts -2-): 21 Nm (15 ft lb)

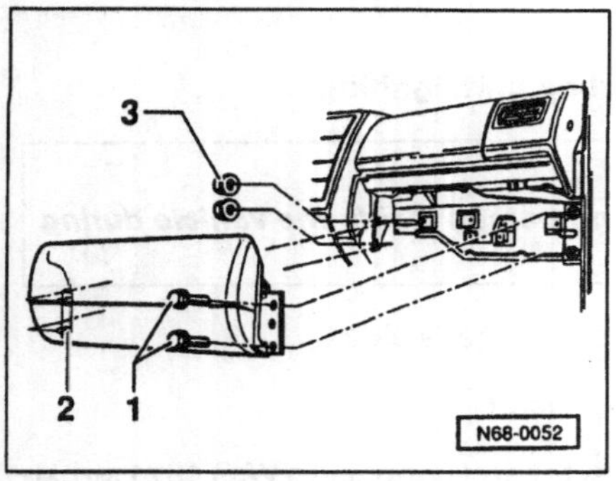

◄ – Remove bolts -1- and nuts -3- and remove front passenger's knee pad toward door.

-2- = Guide pins

Tightening torque (bolts -1- and nuts -3-): 21 Nm (15 ft lb)

Instrument panel, removing

> ***CAUTION!***
>
> ***Before working on any part of the electrical system, obtain radio security code and disconnect battery Ground (GND) strap.***

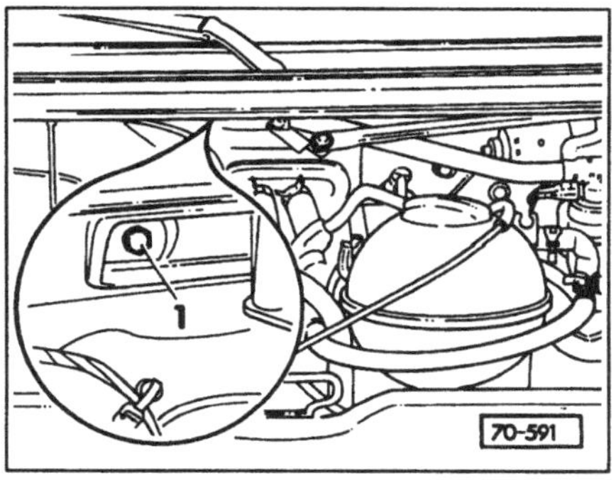

- Remove hex bolt -1- from air duct and cross panel.
- Remove steering wheel ⇒ Page Q 70-2 (without airbag)
  Page Q 69-48 (with airbag)
- Remove steering column switch and trim ⇒ Page Q 70-3.
- Remove instrument panel trim ⇒ Page Q 70-3.
- Remove left heat/fresh air vent ⇒ Page Q 70-4.
- Remove left switch panel ⇒ Page Q 70-4.
- Remove Cold Start Accelerator (CSA) handle -Diesel- ⇒ Page Q 70-5.
- Remove storage tray ⇒ Page Q 70-6.
- Take out center heat/fresh air vent ⇒ Page Q 70-7.
- Remove instruments and controls in center part of instrument panel ⇒ Page Q 70-8.

- Remove glove compartment ⇒ Page Q 70-9.
- Remove right heat/fresh air vent ⇒ Page Q 70-10.
- Remove instrument panel.

Steering wheel, removing

(without airbag)

For steering wheel with airbag ⇒ Page Q 69-48.

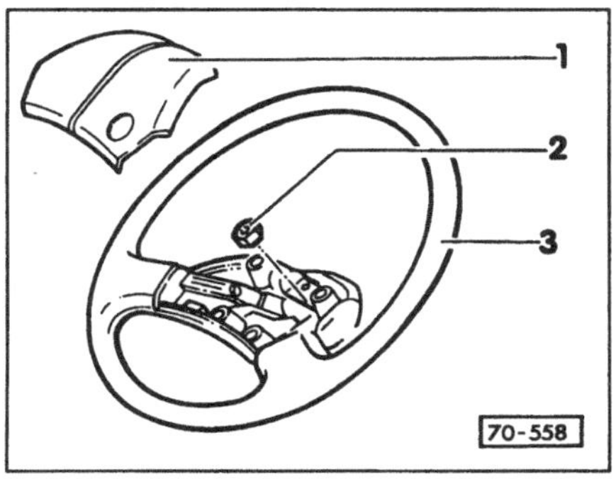

- Unclip horn pad -1-.
- Remove nut -2-.
- ♦ When installing, tighten to 70 Nm (52 ft lb)
- Mark position of steering wheel -3- for reference during installation.
- Remove steering wheel -3-.

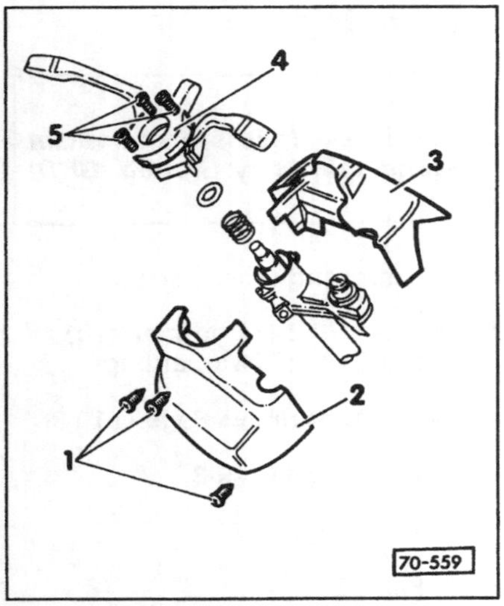

Steering column switch and trim, removing

◄ – Remove screws -1-, trim -2-.

– Disconnect electrical connectors.

– Remove screws-5, switch -4- .

– Remove trim -4-.

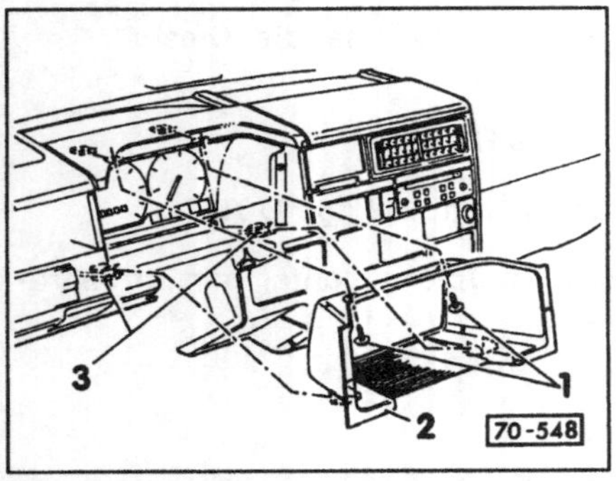

Instrument panel trim, removing

◄ – Remove screws -1-.

– Release trim -2- from clips -3-.

– Disconnect speedometer cable and coupling from back of instrument panel insert.

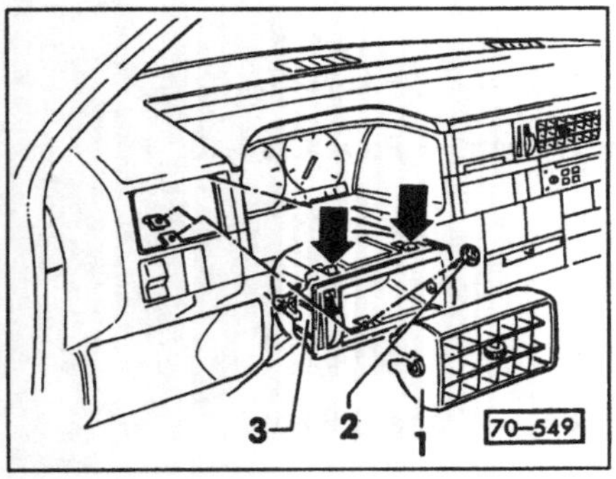

Heat/fresh air vent, removing

◄ – Unclip vent -1-.

– Remove screw -2-.

– Press tabs -arrow- and pry out air outlet -4-.

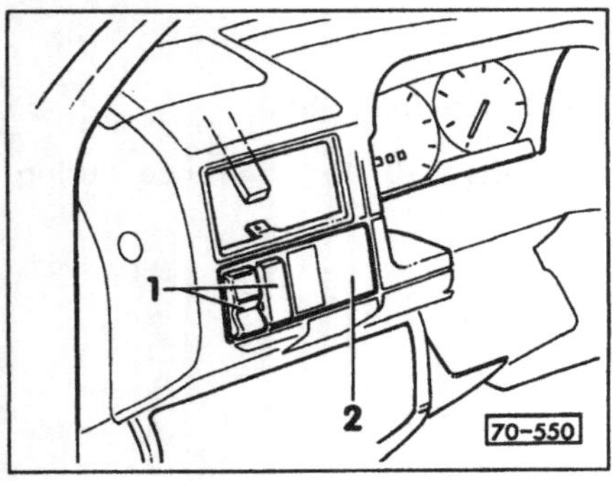

Left switch panel, removing

◄ – Pry out light switch -1-, disconnect electrical connector and remove trim -2-.

– Disconnect electrical connector for left speaker, if applicable (only when removing complete instrument panel).

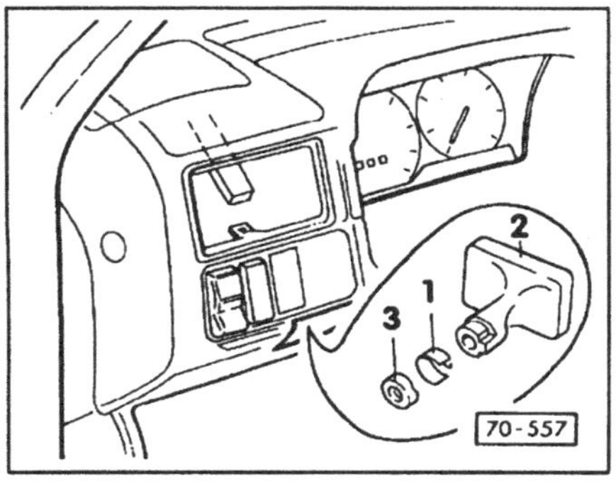

Cold start accelerator (CSA) handle, removing

With Diesel engine

◄ – Remove clamp -1-.

– Turn and remove handle -2- from Cold Start Accelerator (CSA) cable.

– Remove nut -3-.

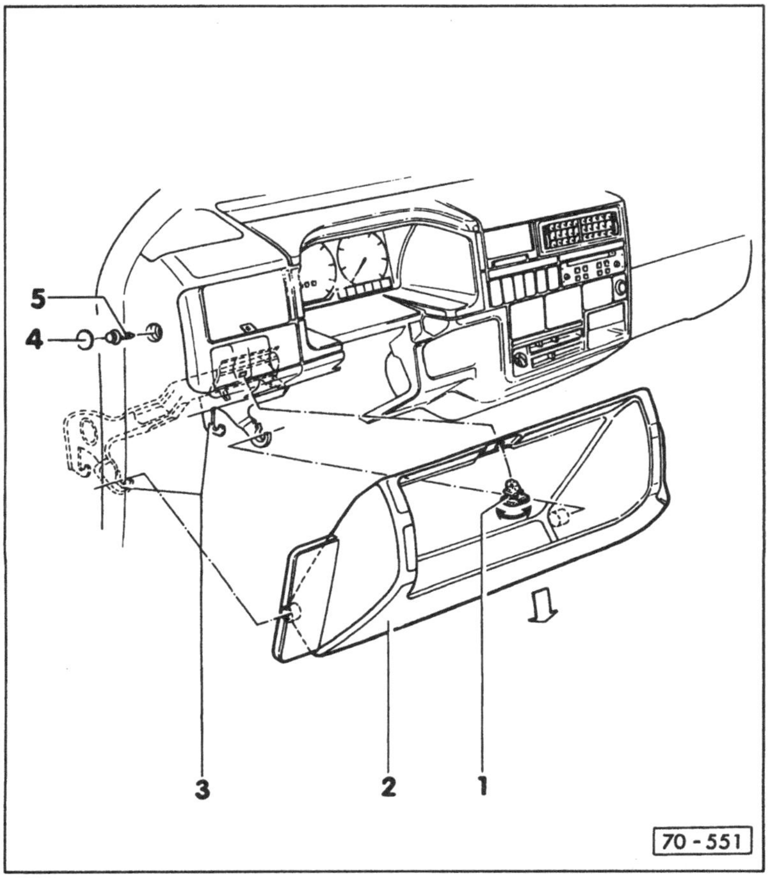

Storage tray, removing

Note:

Storage tray also is fuse/relay panel cover.

– Turn fitting -1- (90°).

– Fold down storage tray -2- and pull from brackets -3-.

– Remove cap -4-, screw, and washer assembly -5- (only when removing complete instrument panel).

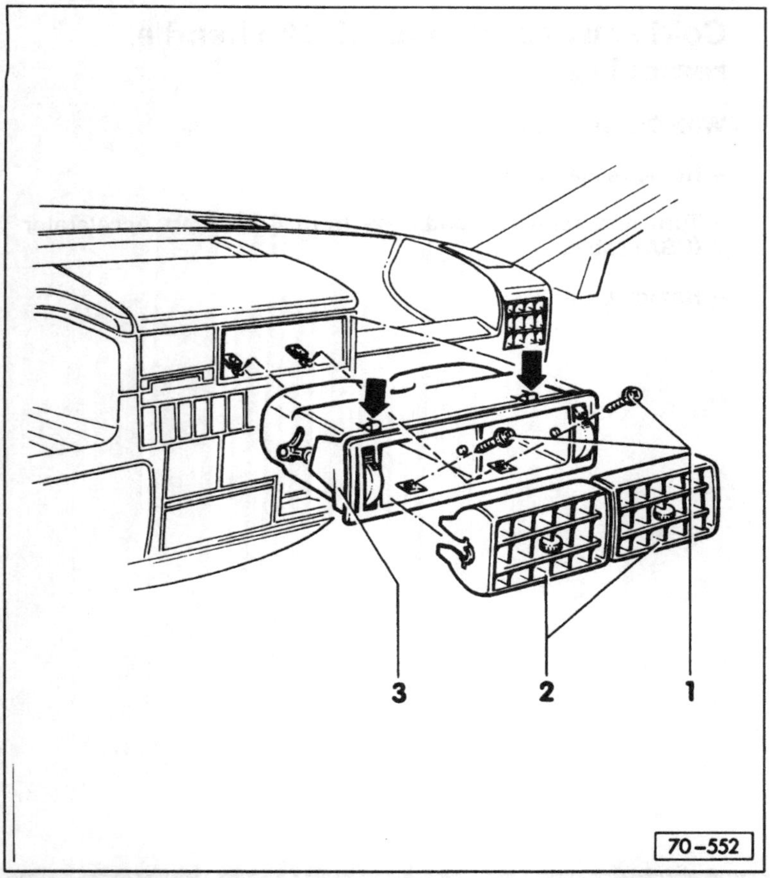

Center heat/fresh air vent, removing

- Unclip vents -2-.
- Remove Phillips screws -1-.
- Press down holding tabs -arrows- and pry out air outlet -3-.

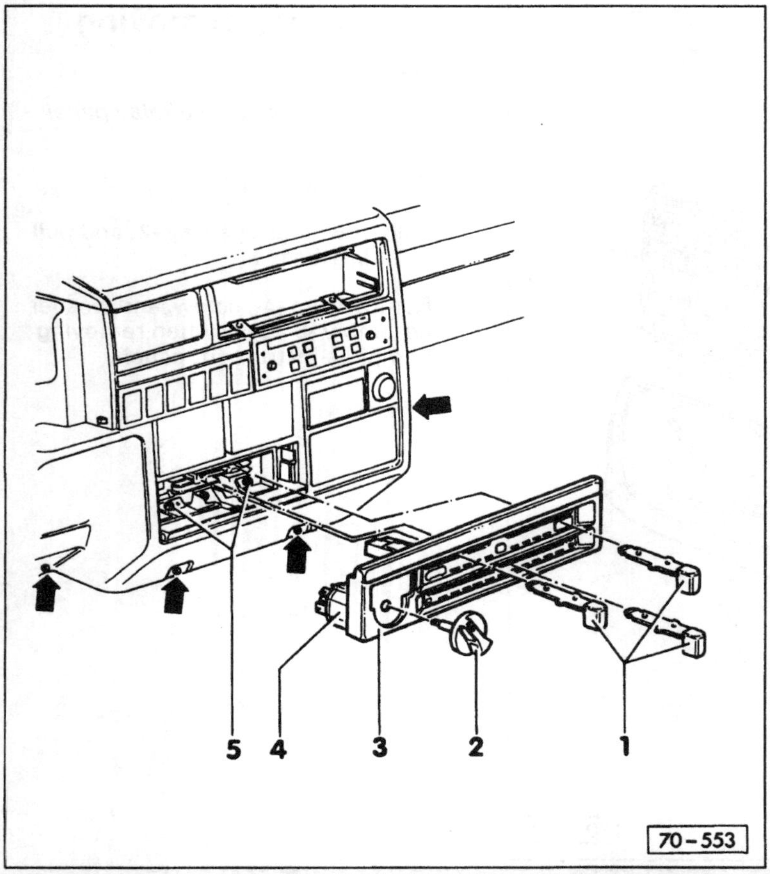

Instrument panel switches (center), removing

- Remove levers -1- and knob -2-.
- Unclip trim -3-, disconnect fan switch -4- electrical connector.
- Remove Phillips screws -5-.
- Remove radio, if applicable.
- Disconnect antenna and electrical connectors, pull electrical connectors off all switches in center of instrument panel and from cigarette lighter.
- Remove mounting screws -arrows- (only when removing complete instrument panel).

Glove compartment, removing

(Vehicles without passenger's side airbag)

- Remove light -arrow- (only in glove compartment with door) and disconnect electrical connector.
- Remove seven Phillips screws and glove compartment.

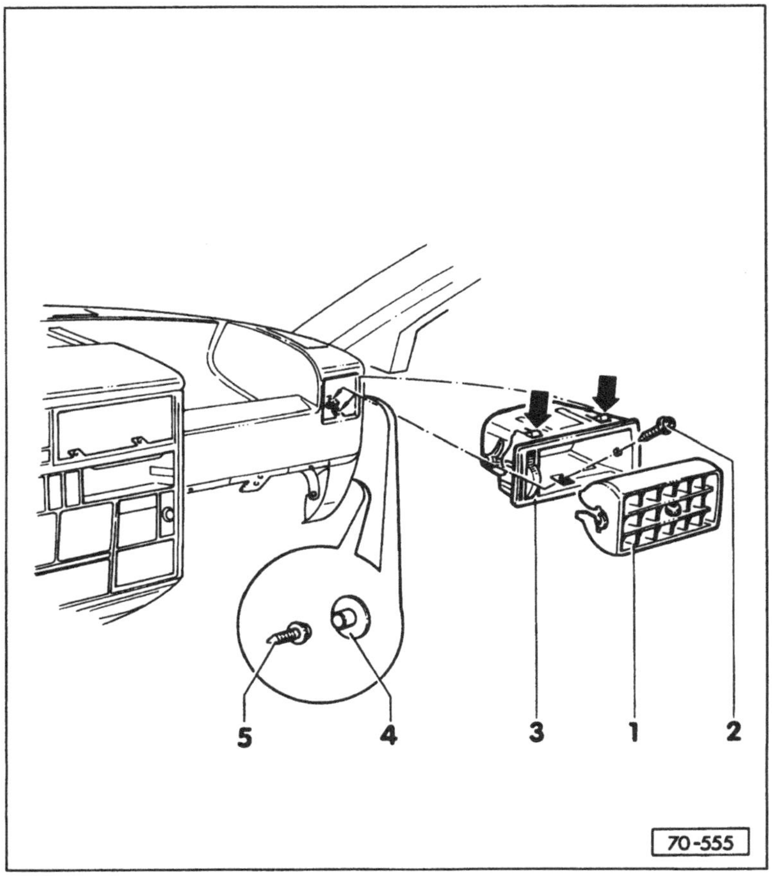

Heat/fresh air vent (right), removing

- Unclip vent -1-.
- Remove screw -2-.
- Press tabs -arrows- and remove air outlet -3-.
- Disconnect speaker electrical connector, if applicable, remove cover caps -4- and remove combi screw -5- (only when removing complete instrument panel).

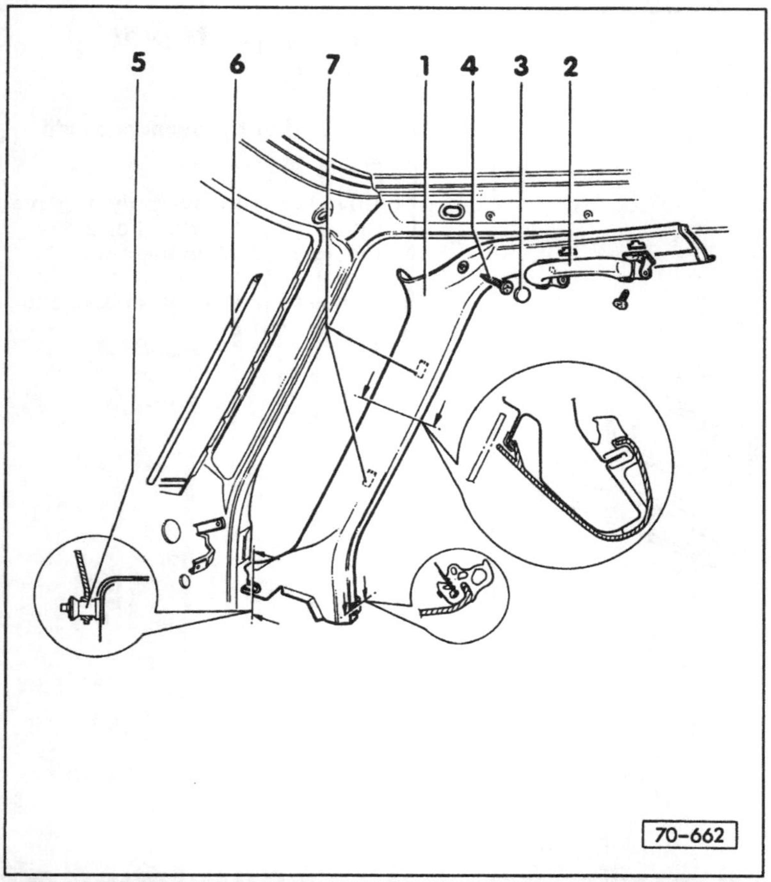

A-pillar trim, removing/installing

Notes:

- *Instructions are shown for one side, directions for other side are opposite.*
- *In order to remove A-pillar trim, upper B-pillar trim must first be removed*
  ⇒ Page Q 70-15 and/or Page Q 70-19.

1 - Trim

- Carefully unclip from headliner and A-Pillar
- It is hooked into bonded rubber bushing -5- on lower side behind instrument panel.

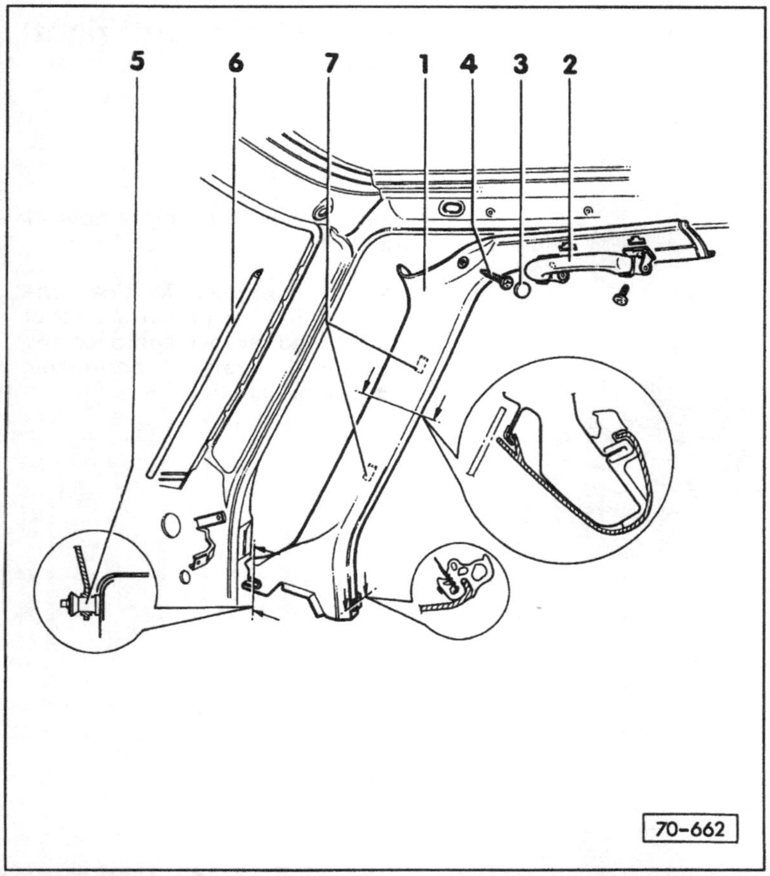

2 - Assist handle

- Passenger side only
- Removing ⇒ Page Q 70-13

3 - Cover cap

- Pry out

4 - Phillips screw

5 - Bonded rubber bushing

6 - Trim strip

- Ensure correct positioning before reinstalling trim

7 - Foam pad

- Two pieces, glued on back of trim

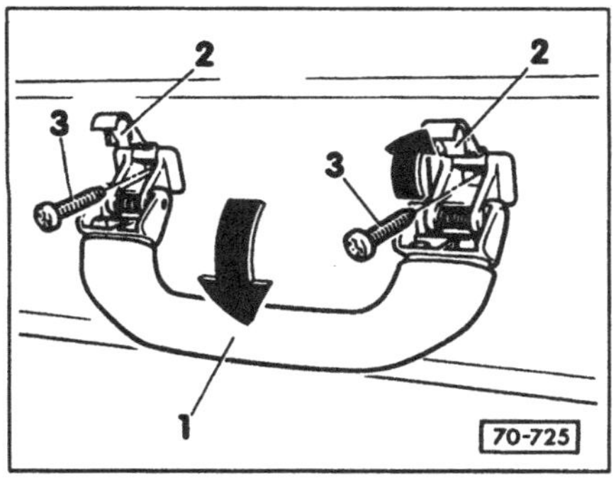

Assist handle, removing

◄

- Pull assist handle -1- downward -arrow-.
- Pry up caps -2- with screwdriver.
- Remove Phillips screws -3-, assist handle -1-.

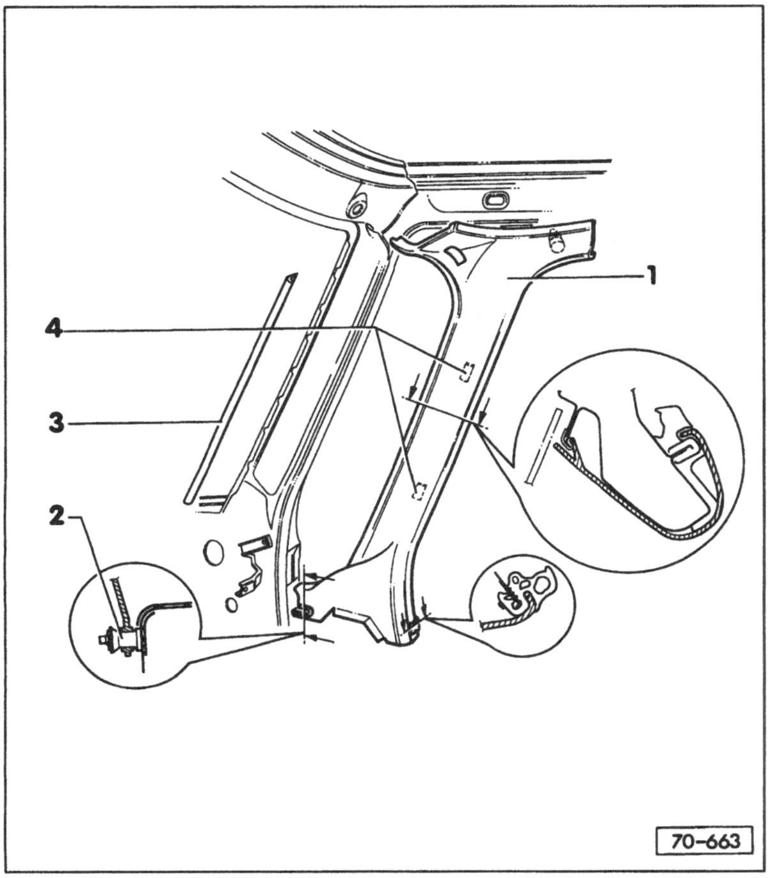

A-pillar trim, components

(Van, Panel Truck, Flat Bed, Double Cab)

Note:

Description is given for one side only, opposite side is mirror image.

1 - Trim
- ♦ Carefully unclip from headliner and A-pillar.
- ♦ It is hooked into bonded rubber bushing -2- on lower side behind the instrument panel.

2 - Bonded rubber bushing

3 - Trim strip
- ♦ Ensure correct positioning before reinstalling trim -1-.

4 - Foam pad
- ♦ Two pieces, glued on back of trim -1-.

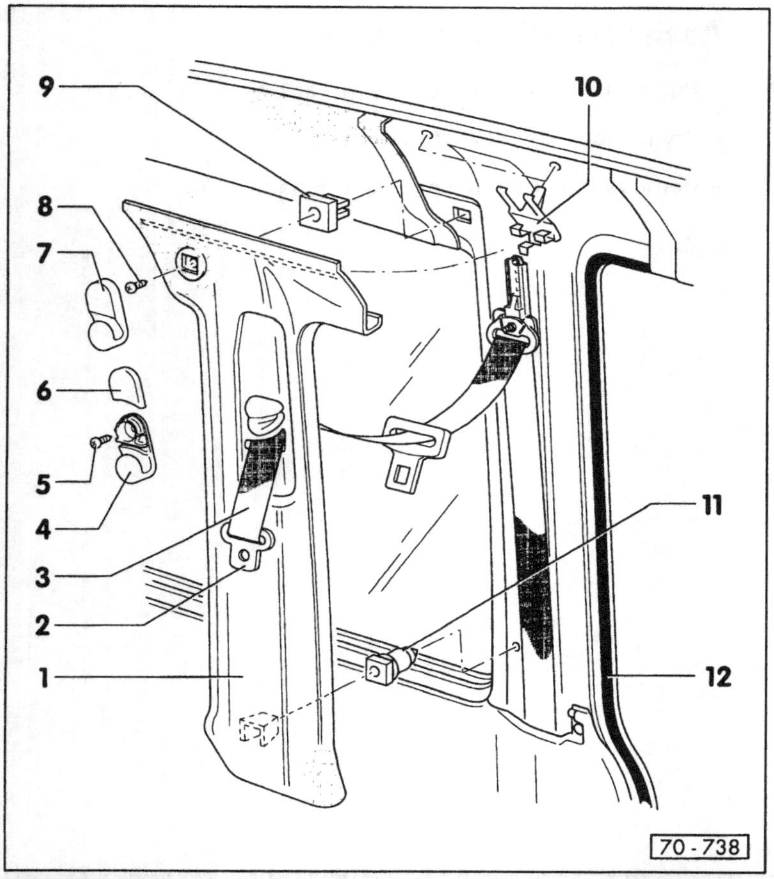

Upper B-pillar trim (left), components

1 - B-pillar upper trim
- ♦ Removing ⇒ Page Q 70-17
- ♦ Installing ⇒ Page Q 70-18

2 - Belt Tongue
- ♦ Bolted at bottom of B-pillar.
- ♦ Tightening torque: 40 Nm (30 ft lb)

3 - Seat belt

4 - Coat hook
- ♦ From 03/92

5 - Phillips screw

6 - Cover

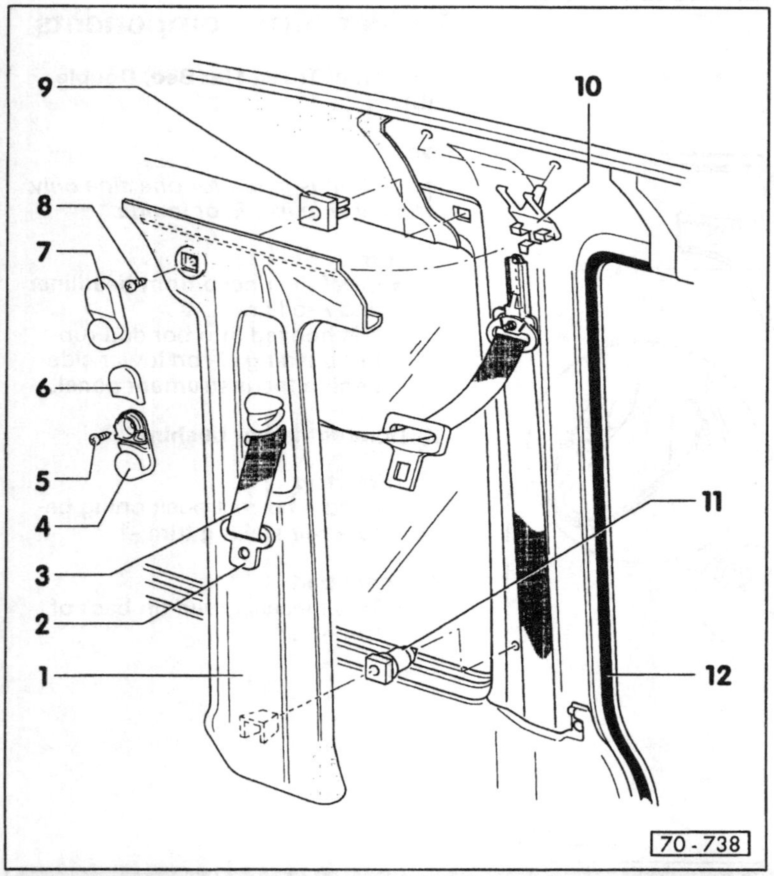

7 - Coat hook
- ♦ Up to 02/92

8 - Phillips screw

9 - Expanding nut

10 - Spring clip

11 - Clip
- ♦ Two pieces
- ♦ Install sleeve in B-pillar hole, install pin in trim.

12 - Gasket
- ♦ After installing trim, ensure correct positioning of gasket

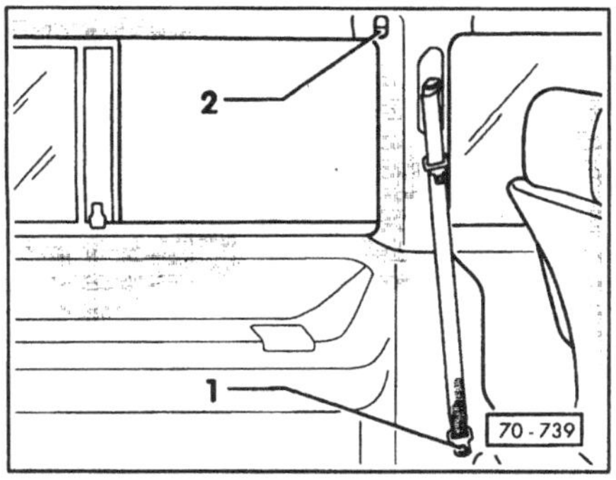

Upper B-pillar trim (left), removing

◄ – Remove cap -1-, remove anchor screw.

– Remove cap -2- and/or coat hook, remove coat hook and/or Phillips screw underneath.

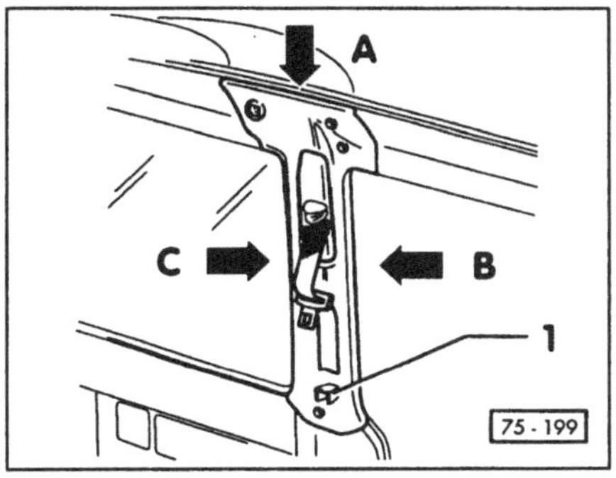

◄ – Unclip and remove trim in following sequence:

- ♦ Along molded headliner -arrow A-
- ♦ Right side -arrow B-

– Pull trim from B-pillar starting at height of clip -1-.

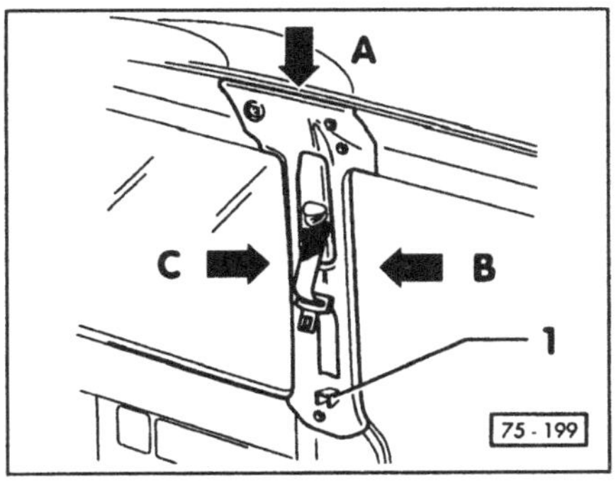

Upper B-pillar trim (left), installing

◄ – Clip trim on B-pillar and into spring clips on headliner. Ensure correct installation of clips -1- and belt height adjustment mechanism.

– After installing trim, ensure correct positioning of gasket -arrow B-.

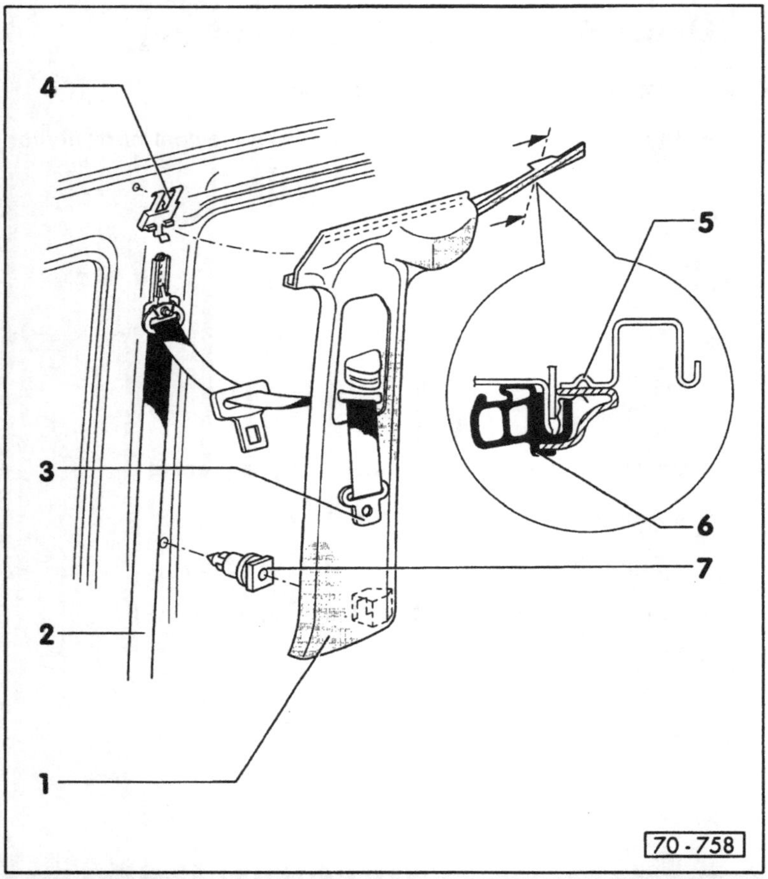

Upper B-pillar trim (right), components

1 - Upper B-pillar trim
- ♦ Removing ⇒ Page Q 70-21
- ♦ Installing ⇒ Page Q 70-21

2 - Seat belt

3 - Belt tongue
- ♦ Bolted at bottom of B-pillar.
- ♦ Tightening torque: 40 Nm (30 ft lb).

4 - Spring clip

5 - Locating tab in roller guide
- ♦ Clamp must be centered
- ♦ Incorrect installation of spring clip can cause damage to upper B-pillar trim by sliding door.

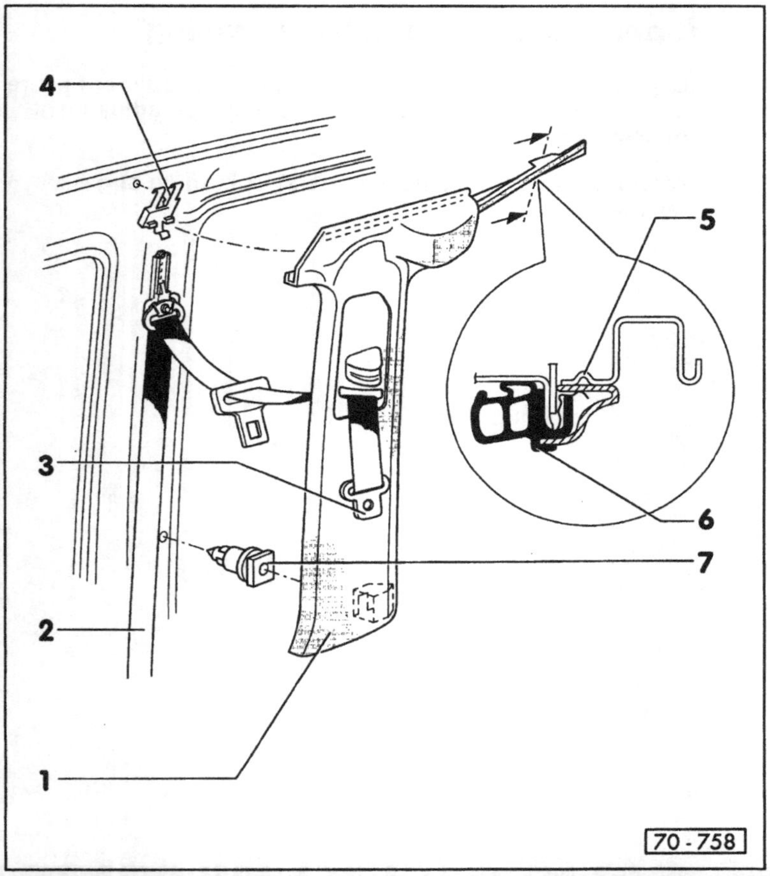

6 - Sliding door gasket

7 - Clip
- ♦ Two-piece
- ♦ Install sleeve in B-pillar hole, install pin in trim.

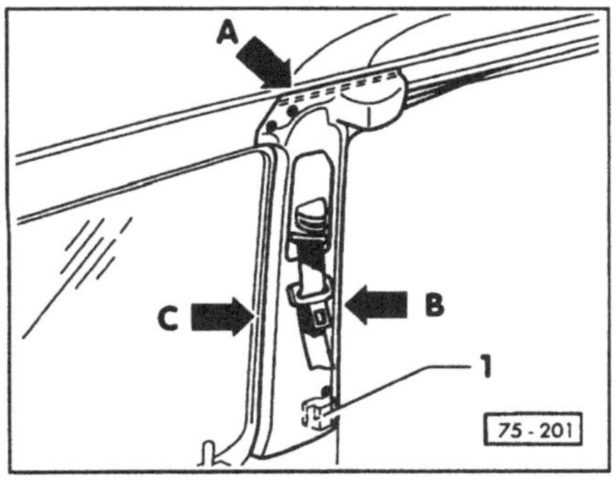

Upper B-pillar trim (right), removing/installing

◄ – Remove seat belt anchor cap and hex screw at bottom of B-pillar.

– Unclip and remove trim in following sequence:

- ♦ Along molded headliner -arrow A-.
- ♦ Right -arrow B-.
- ♦ Left -arrow C-.

– Remove B-pillar trim starting at clip -1- position.

– Clip trim onto B-pillar and into spring clips on headliner, in installed position. Ensure correct installation of clip -1- and belt height adjustment mechanism.

– After installing trim, ensure correct positioning of gaskets -arrows B and C-.

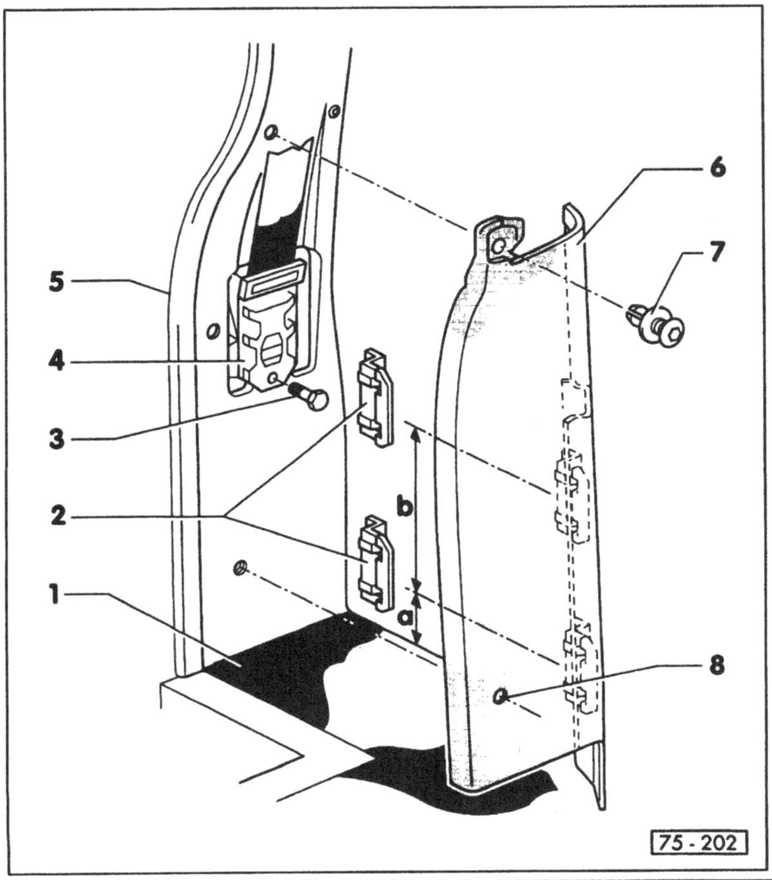

Lower B-pillar trim, components

Note:

To remove lower B-pillar trim, first remove upper B-pillar trim, right side *⇒ Page Q 70-21.*

1 - Cross member piece

2 - Spring clips
- ♦ Distance of clips from cross member piece -1-.

a = 120 mm (approx. 4.6 in.)

b = 200 mm (approx. 7.8 in.)

3 - Hex bolt
- ♦ 40 Nm (30 ft lb)

4 - Automatic tensioning seat belt retractor

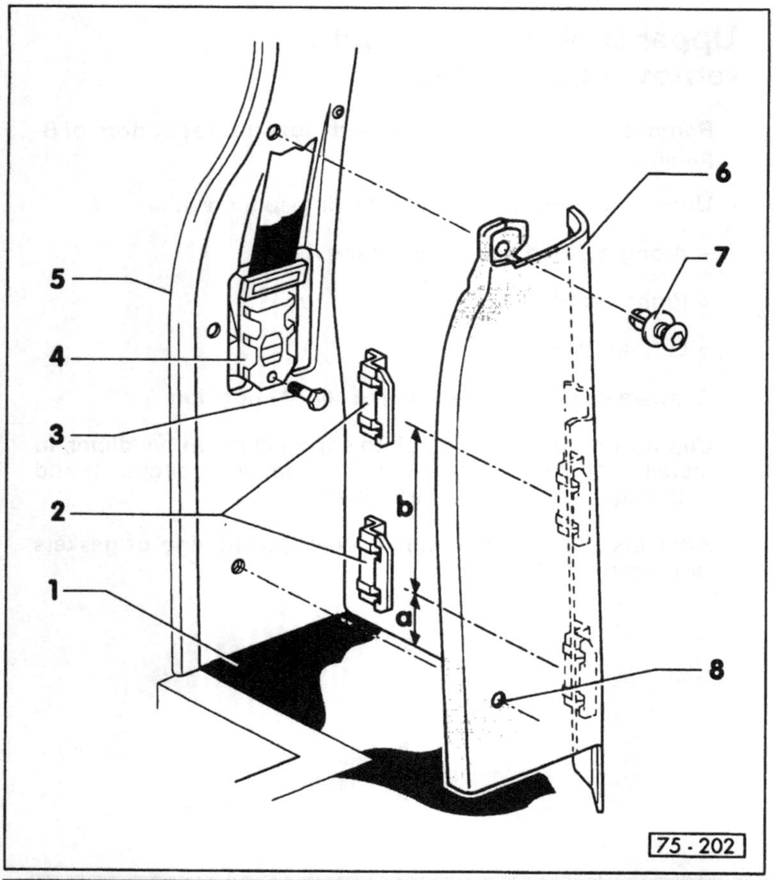

5 - Gasket
- After installing lower B-pillar trim, ensure correct positioning of gasket.

6 - Lower B-pillar trim

7 - Clip
- Two-piece
- Removing: Use Allen wrench
- Installing: Press outer and inner parts into hole one after the other

8 - Anchor point for seat belt

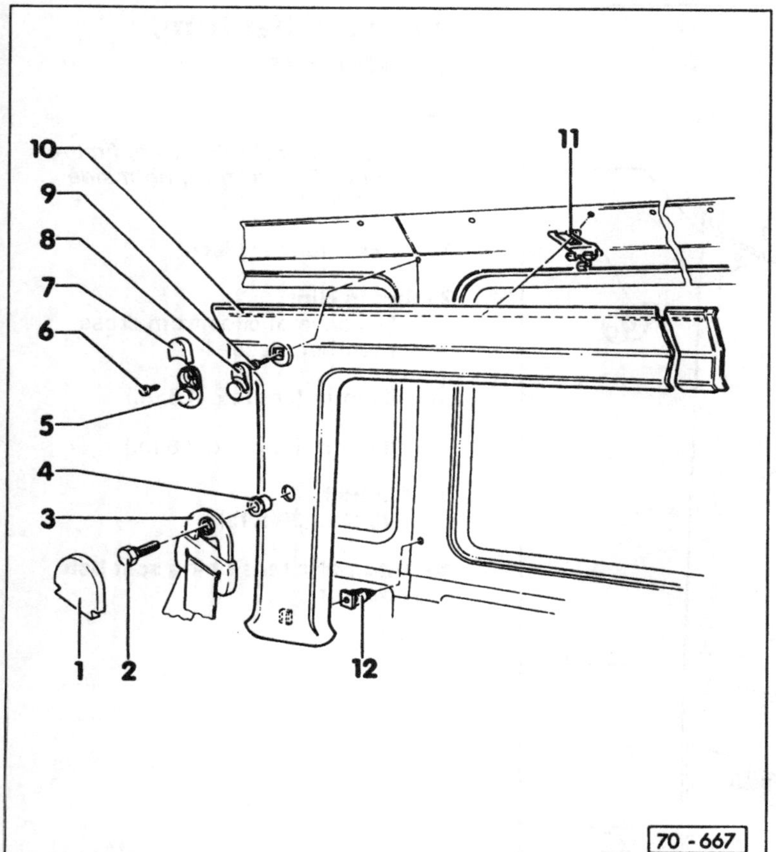

C-pillar trim (left), components

Note:
- *Before removing C-pillar trim, any accessories (fluorescent light, heating controls, etc.) must be removed first.*
- *To remove left C-pillar trim first remove upper B-pillar trim ⇒ Page Q 70-15.*

1 - Cover

2 - Hex bolt
- 40 Nm (30 ft lb)

3 - D-Ring

4 - Bushing

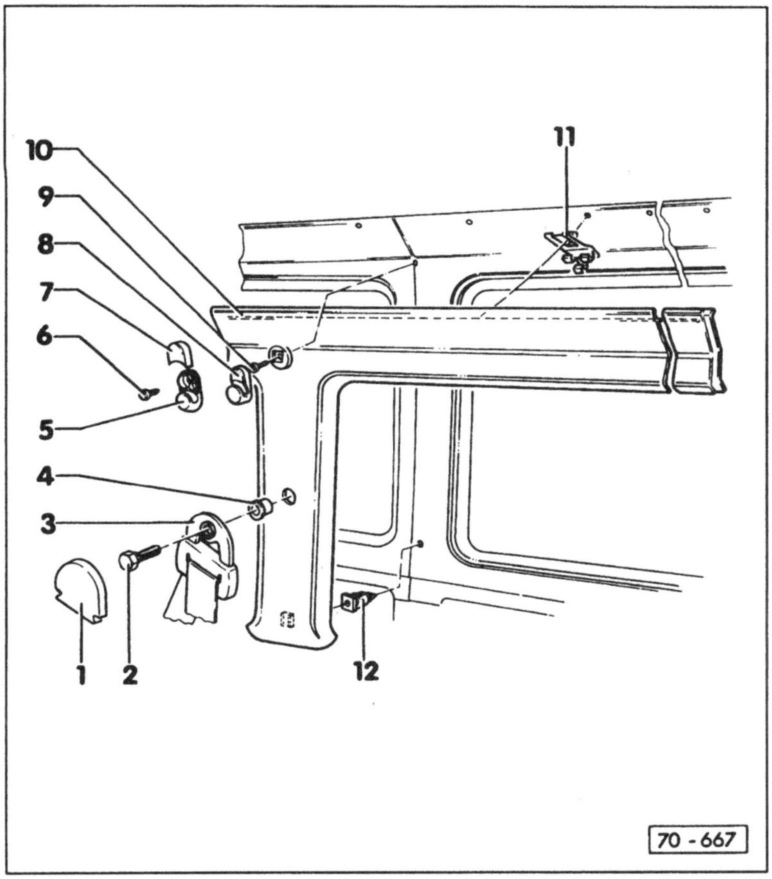

5 - **Coat hook**
  - From 03/92

6 - **Phillips screw**

7 - **Cap**

8 - **Coat hook**
  - Up to 02/92

9 - **Phillips screw**

10 - **C-pillar trim (left)**
  - Removing ⇒ Page Q 70-26
  - Installing ⇒ Page Q 70-27

11 - **Spring clip**

12 - **Clip**
  - Two-piece
  - Install C-pillar sleeve in hole, then install pin in trim.

C-pillar trim (left), removing

– Remove upper B-pillar trim ⇒ Page Q 70-15. Remove additional equipment if present.

– Remove coat hook cap and/or pry out coat hook.

– Remove coat hook and/or Phillips screw underneath.

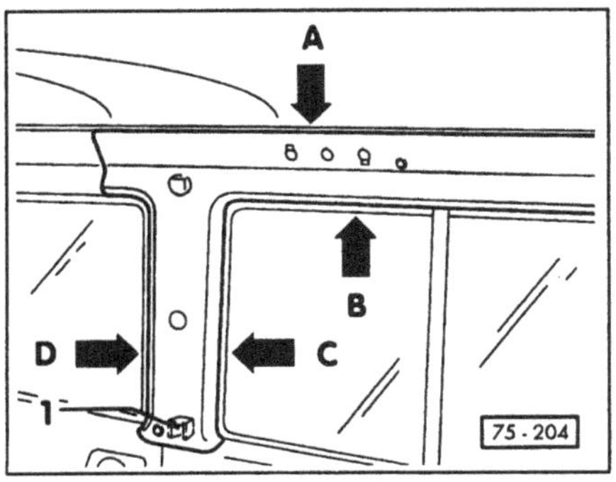

◄ – Unclip and remove trim in following sequence:

- Along molded headliner -arrow A-
- Window beam -arrow B-
- Front C-pillar -arrow C-
- Rear C-pillar -arrow D-

– Pull trim from C-pillar at clips -1- as much as possible.

Note:

When gluing trim in window areas -arrows B, C and D-, unclip a little at a time, while supporting trim from back with your hand or lever tool/ 80–200.

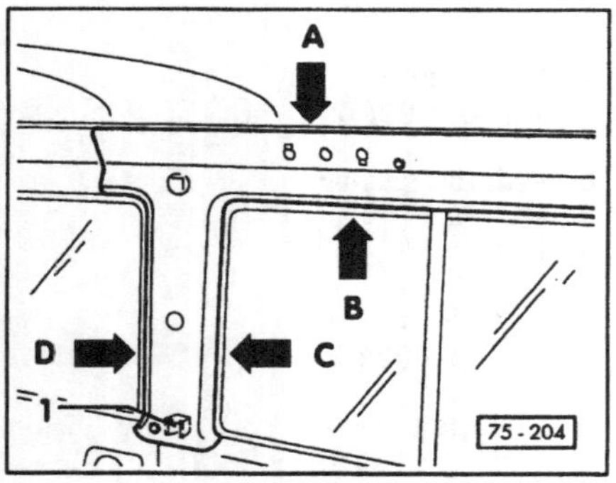

C-pillar trim (left), installing

◄

- Place trim in installed position, simultaneously ensure correct clip positioning.
- Clip trim in window areas -arrows B, C and D- and into spring clips on headliner -arrow A-.

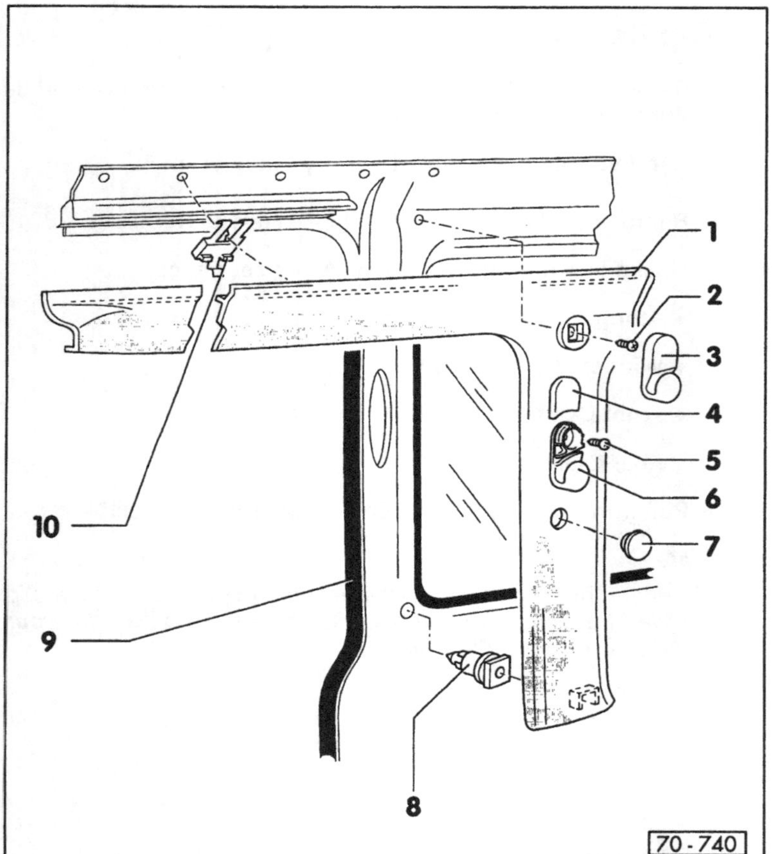

C-pillar trim (right), components

Note:

To remove right C-pillar trim, first remove B-pillar trim ⇒ Page Q 70-19.

1 - C–pillar trim (right)
- ♦ Removing ⇒ Page Q 70-30
- ♦ Installing ⇒ Q 70-31

2 - Phillips screw

3 - Coat hook
- ♦ Up to 0292

4 - Cap

5 - Phillips screw

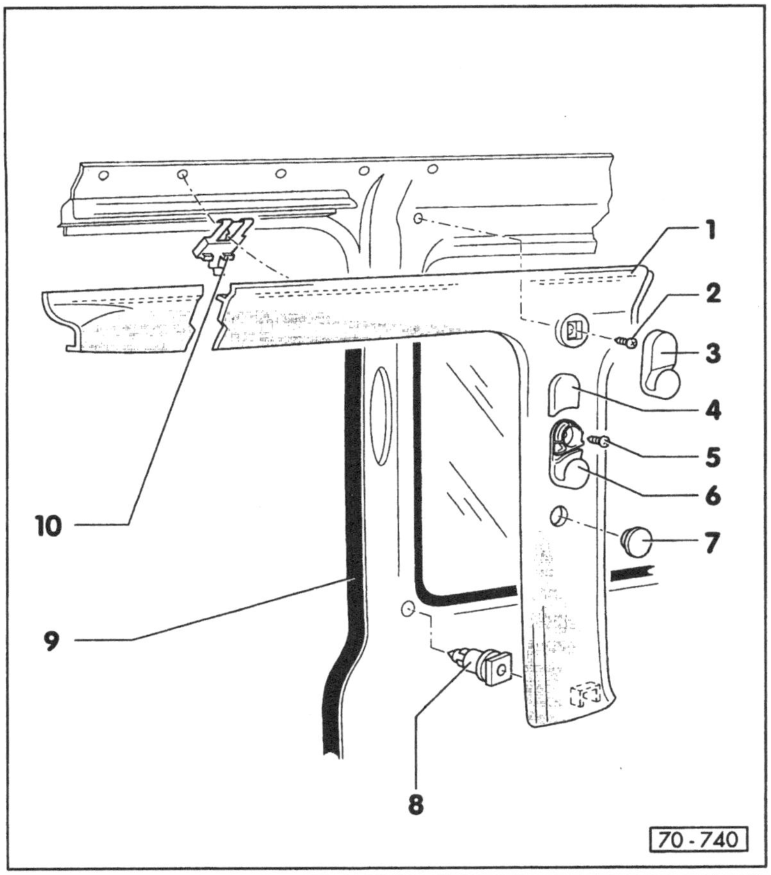

6 - **Coat hook**
- From 03/92

7 - **Cap**
- Not included with automatic 3-point seat belts in the rear passenger compartment.

8 - **Clip**
- Two-piece
- When installing, insert sleeve in C-pillar hole and install pin in trim.

9 - **Gasket**
- After installing trim, ensure correct positioning of gasket

10 - **Spring Clip**

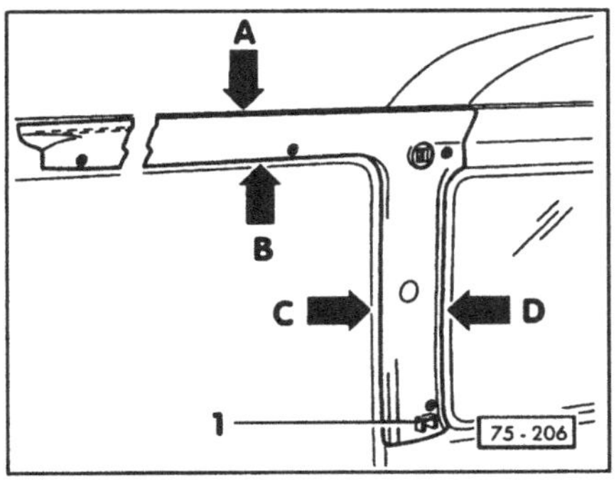

C-pillar trim right, removing

– Remove right upper B-pillar trim ⇒ Page Q 70-19.

– Remove coat hook cap and/or Phillips screw underneath.

– Remove coat hook.

– Unclip trim in following sequence:

- Along molded headliner -arrow A-
- Along door guide rail -arrow B-
- Center bottom C-pillar, clip-1-
- Right side, below -arrow C-
- Left side, above -arrow D-

Note:

When gluing trim in window areas -arrow D- unclip a little at a time while supporting trim from the back with your hand or lever tool 80–200.

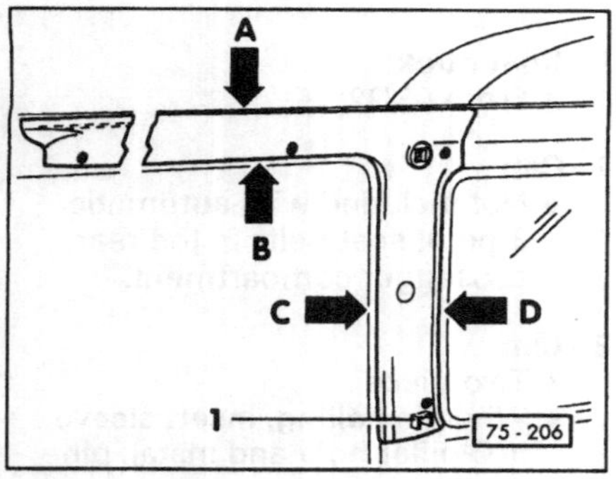

C-pillar trim (right), installing

◄

- Place trim in installed position while ensuring correct positioning of clips -1-.
- Clip trim onto C-pillar in window areas -arrow D- and into spring clips on headliner-arrow A-.
- Ensure correct gasket positioning -arrow C-.

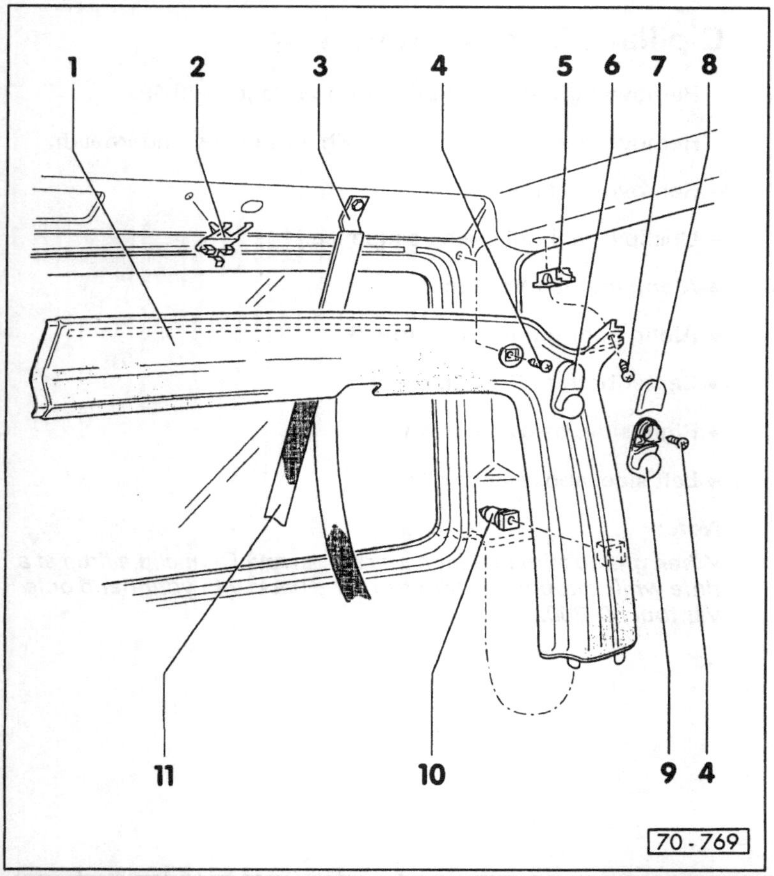

D-pillar trim, components

Notes:

- *Removal description is given for only one side, other side is mirror image.*
- *To remove D-pillar trim, first remove rear headliner trim ⇒ Page Q 70-37*
- *Left upper B-pillar trim ⇒ Page Q 70-15*
- *Right upper B-pillar trim ⇒ Page Q 70-19*
- *Left C-pillar trim ⇒ Page Q 70-24*
- *Right C-pillar trim ⇒ Page Q 70-28*

1 - D-pillar trim
- Removing ⇒ Page Q 70-35
- Installing ⇒ Page Q 70-36

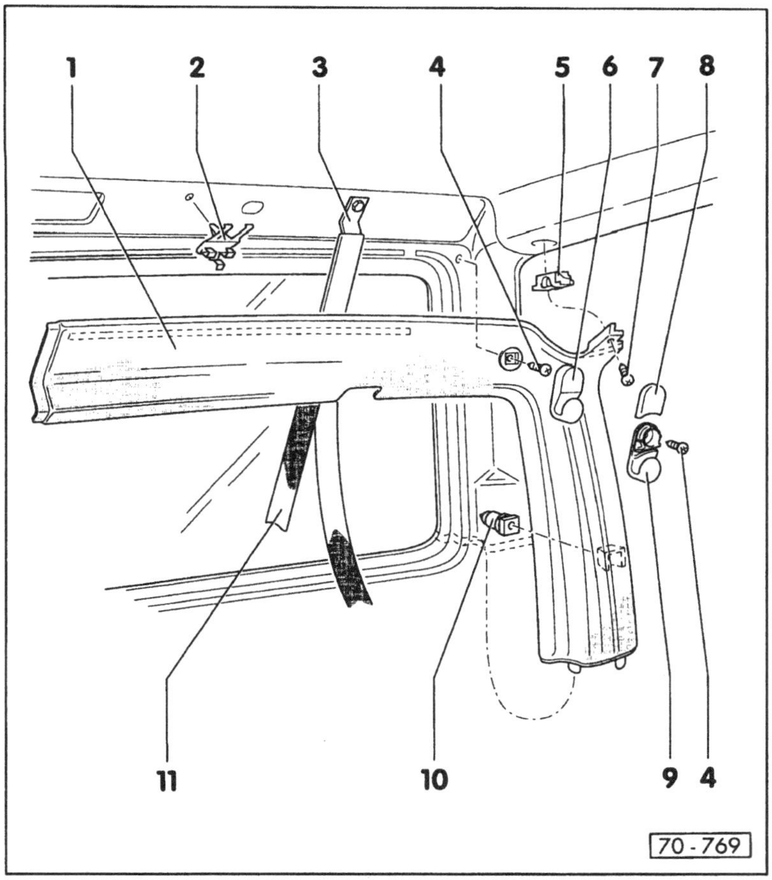

2 - **Spring clip**

3 - **D-ring for seat belt**
- Bolted to window frame with hex bolt
- 40 Nm (30 ft lb)

4 - **Phillips screw**

5 - **Clip**

6 - **Coat hook**
- Up to 02/92

7 - **Phillips screw**

8 - **Cover**

9 - **Coat hook**
- From 03/92

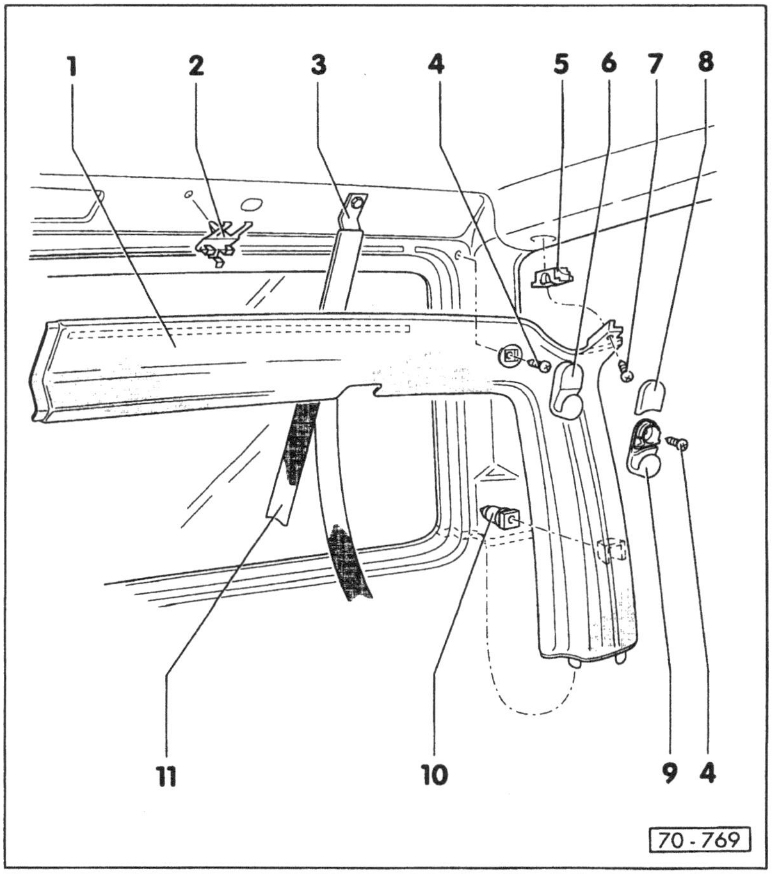

10 - **Clip**
- Two-piece
- Insert sleeve in D-pillar hole, then install pin in trim.

11 - **Seat belt**

D-pillar trim, removing

Note:

- *Remove rear headliner trim ⇒ Page Q 70-37, left upper B-pillar trim ⇒ Page Q 70-15, right upper B-pillar trim ⇒ Page Q 70-19, left C-pillar trim ⇒ Page Q 70-24, right C-pillar trim ⇒ Page Q 70-28.*

– Remove coat hook cover to remove Phillips screw, remove coat hook.

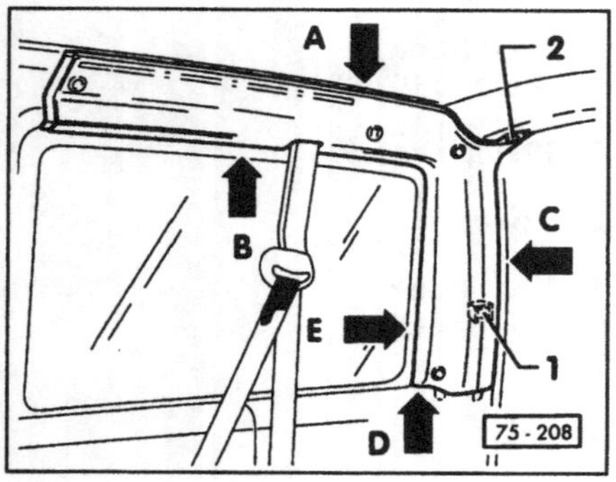

– Remove Phillips screw -2-.

– Unclip trim in following sequence:

- Along molded headliner-arrow A-
- Window frame -arrow B-
- Rear D-pillar-arrow C-
- Loosen clip -1-
- Remove D-pillar trim from bottom -arrow -
- Front D-pillar -arrow E-

Note:

When gluing trim in window areas -arrows B, E- unclip short sections while supporting trim from the back by hand or lever tool 80–200.

D-pillar trim, installing

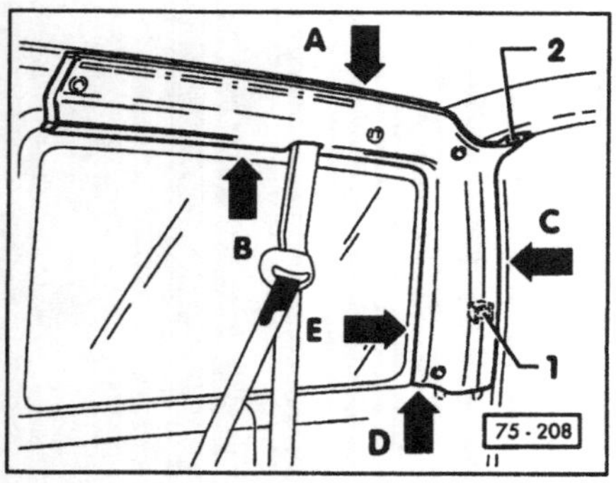

– Install trim on bottom -arrow D-, place in installation position and ensure correct positioning of clip -1- simultaneously.

– Clip D-pillar trim into clamps on headliner-arrow A- and in window area -arrows B and E-.

– Ensure correct positioning of gasket-arrow C-.

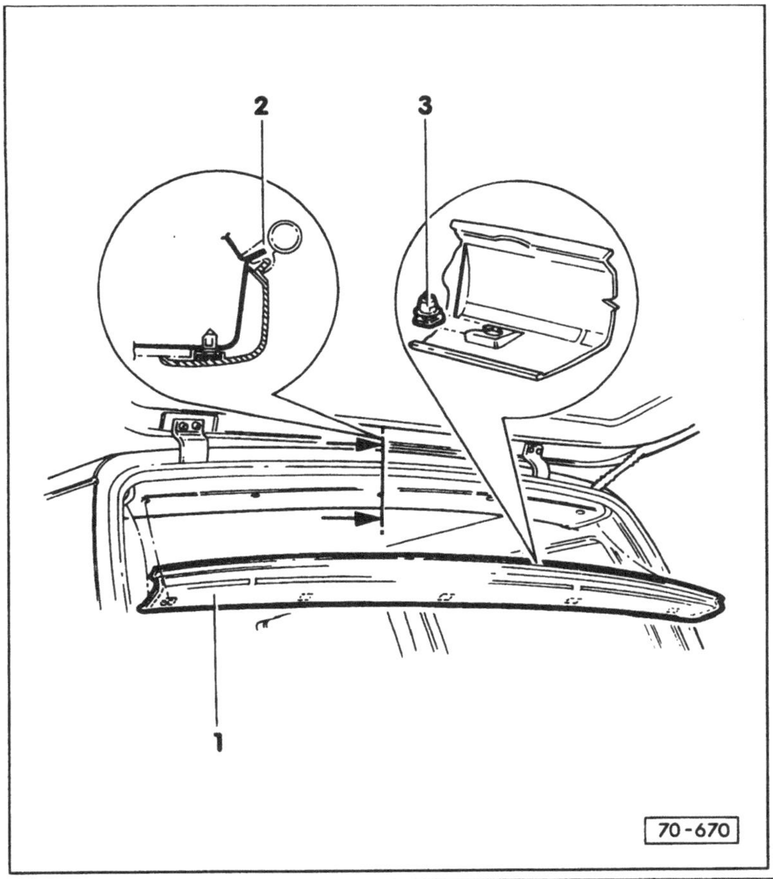

Headliner trim (rear), components

1 - Rear headliner trim
- Removing: Carefully unclip and remove
- Installing: Press clips -3- into holes and ensure correct gasket fit -2-

2 - Gasket

3 - Clip
- Two-piece
- Installing: insert sleeve in headliner hole, install pin in headliner through trim

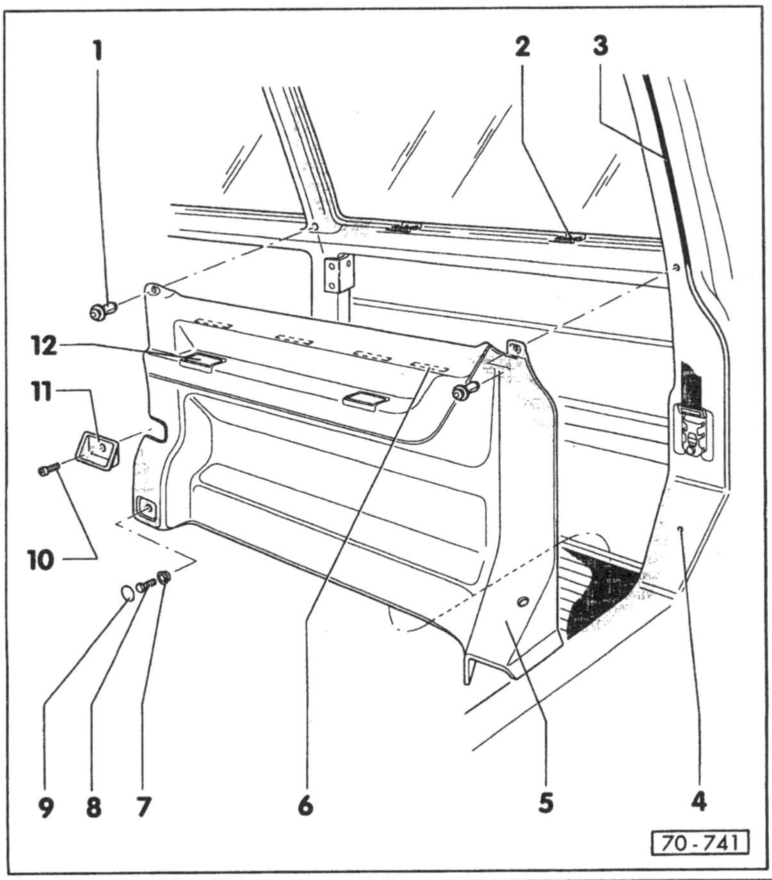

Front side panel molded trim, components

Note:

- *Removal description is given for only one side, directions for the other side are mirror image.*
- *To remove front side panel molded trim, remove left upper B-pillar trim ⇒ Page Q 70-15, left C-pillar trim ⇒ Page Q 70-24.*
- *Left rear side panel molded trim (short wheel base) ⇒ Page Q 70-48, and/or right side panel molded trim (long wheel base) ⇒ Page Q 70-55, loosen only where it overlaps front side panel molded trim.*
- *Two-place bench and/or individual seat must be removed ⇒ Page Q 72-7.*

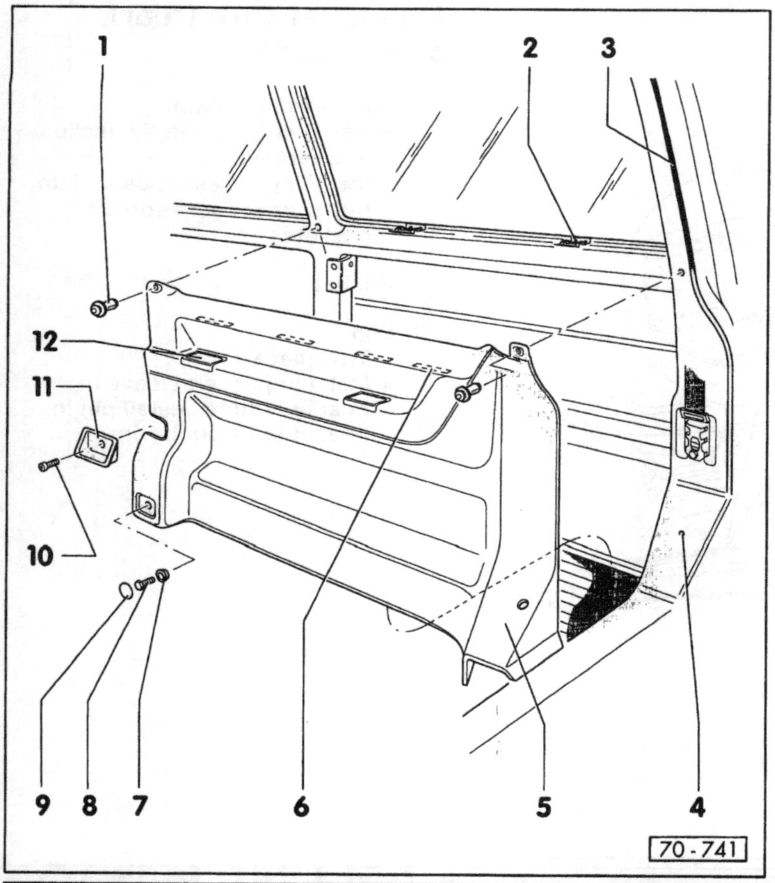

1 - Clip
- Two-piece
- Removing: Remove inner part with Allen wrench
- Installing: Press outer and inner part into hole simultaneously

2 - Clip

3 - Gasket
- After installing front side panel molded trim, ensure correct gasket positioning.

4 - Seat belt anchor point

5 - Front side panel molded trim
- Removing ⇒ Page Q 70-41
- Installing ⇒ Page Q 70-43

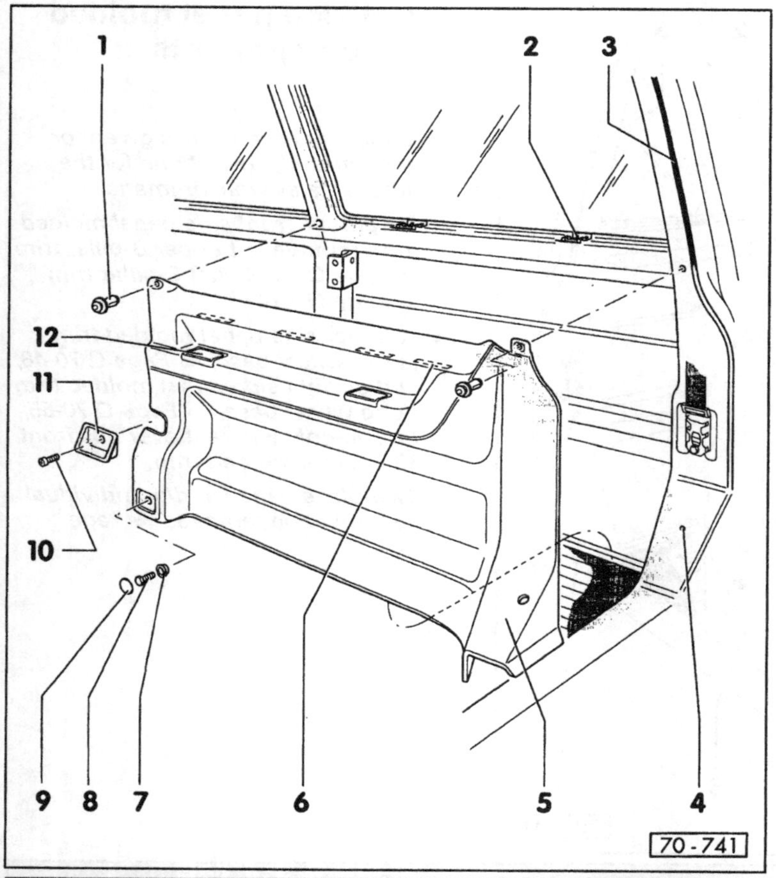

6 - Clip
- On back
- Several along window edge

7 - Spacer bushing

8 - Hex bolt
- 40 Nm (30 ft lb)

9 - Cap

10 - Phillips screw

11 - Belt guide with closure strips

12 - Ash tray
- Removing: Pull upward and out of molded trim
- Installing: Press into molded trim until spring engages audibly

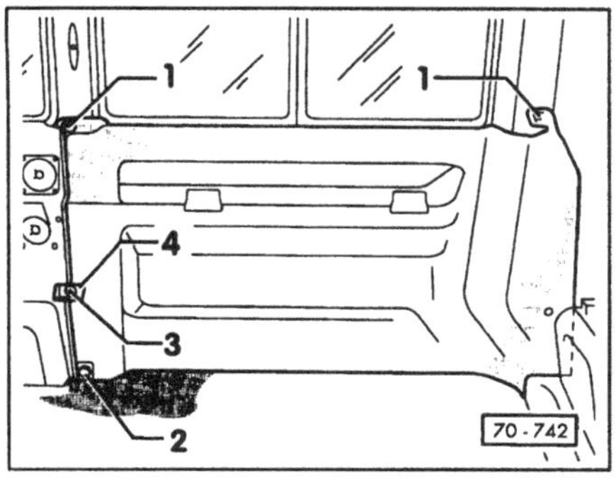

Front side panel molded trim, removing

- Remove left upper B-pillar trim ⇒ Page Q 70-15, left C-pillar trim ⇒ Page Q 70-24.
- Remove two-place bench and/or individual seat ⇒ Page Q 72-7.
- Remove inner part of clips with Allen wrench.
- Remove hex bolt under cap -2- with spacer bushing and, if necessary, belt fitting.
- Remove Phillips screw -3- and carefully pry out belt guide with closure strips -4- and/or seat belt.
- Loosen left rear side panel molded trim (short wheel base) ⇒ Page Q 70-43, and/or left rear side panel molded trim (long wheel base) ⇒ Page Q 70-55 at overlap of front side wall molded trim.

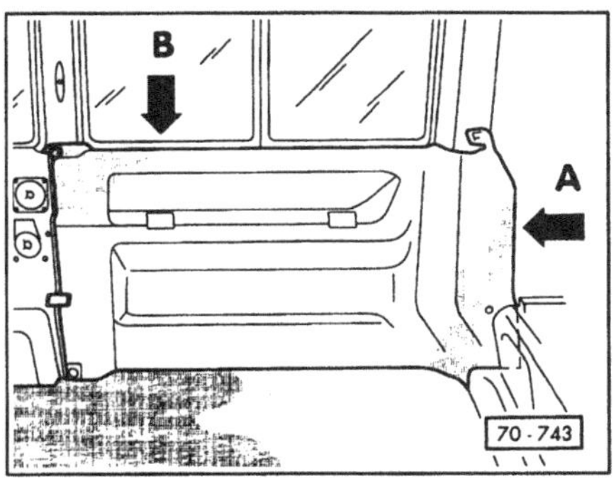

- Unclip front side panel molded trim in the following sequence:
  - B-pillar -arrow A-
  - Window frame -arrow B-

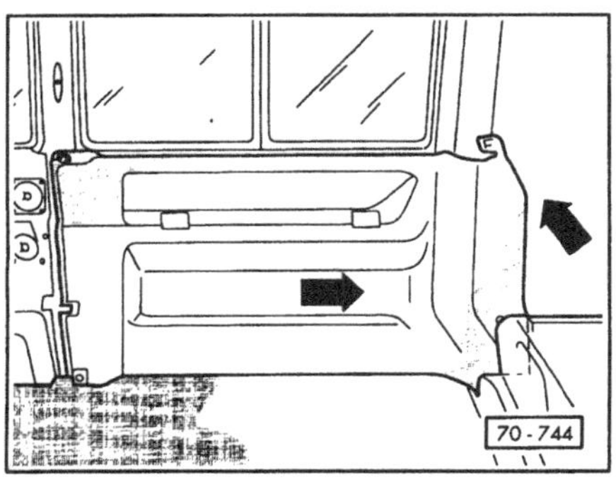

- Remove front side wall molded trim at area of B-pillar and push forward as far as possible.

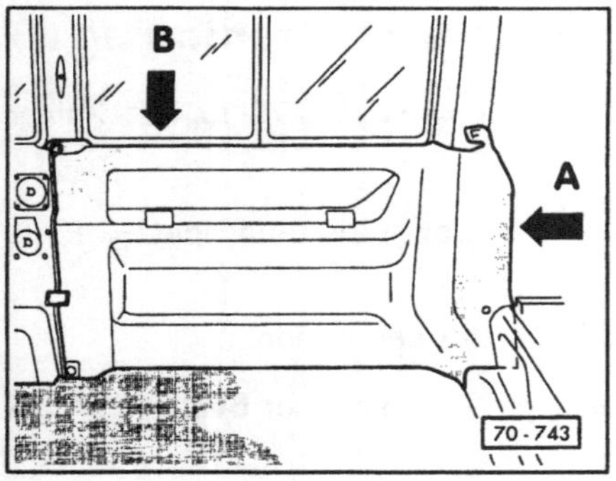

Front side molded trim, installing

◄

– Install in reverse sequence. After clipping in the front side panel molded trim, ensure correct positioning of gasket -arrow A-.

Q 70-105

Rear side molded trim, components (left side, short wheel base)

Note:

- *Removal description is given for one side, opposite side is mirror image.*
- *Depending on equipment installed, remove seats and/or seat benches ⇒ Page Q 72-7, for storage area ⇒ Page Q 70-105.*
- *To remove rear side panel molded trim, remove rear headliner trim, ⇒ Page Q 70-37, left upper B-pillar trim, ⇒ Page Q 70-15, left C-pillar trim ⇒ Page Q 70-24, D-pillar trim ⇒ Page Q 70-32, loosen rear edge cover on both sides.*

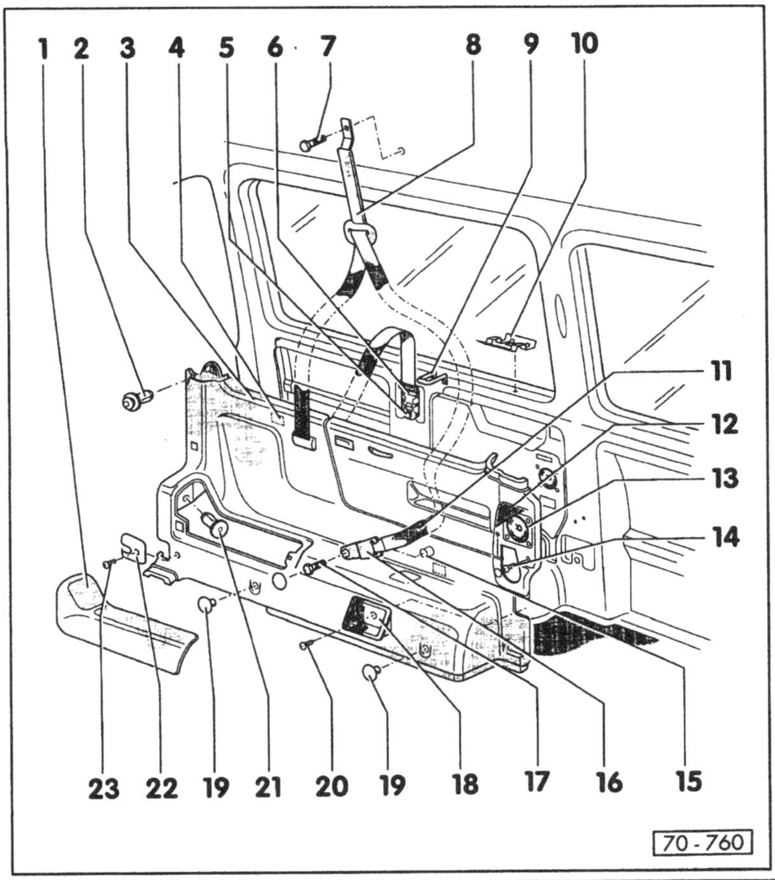

1 - Cover

2 - Clip
- Two-piece
- Removing: inner section requires an Allen wrench.
- Installing: Press outer and inner part into hole simultaneously.

3 - Rear side panel molded trim
- Removing and installing ⇒ Page Q 70-48.

4 - Clip
- On back
- Several located along the window edge

5 - Hex bolt
- 40 Nm (30 ft lb)

6 - Automatic tensioning seat belt retractor

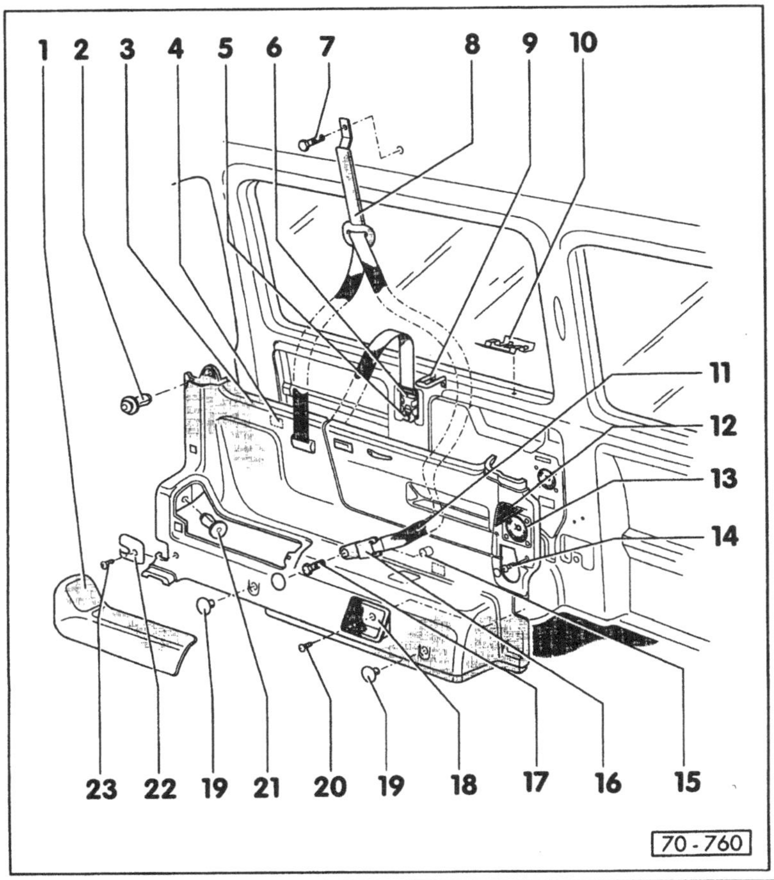

7 - Hex bolt
- 40 Nm (30 ft lb)

8 - D-ring for seat belt
- Can be taken off with two-piece trim

9 - Three-place bench seat belt anchor point

10 - Spring clip

11 - Seat belt

12 - Speaker cover

13 - Speaker with connector

14 - Phillips screw
- With washer and cap

15 - Spacer

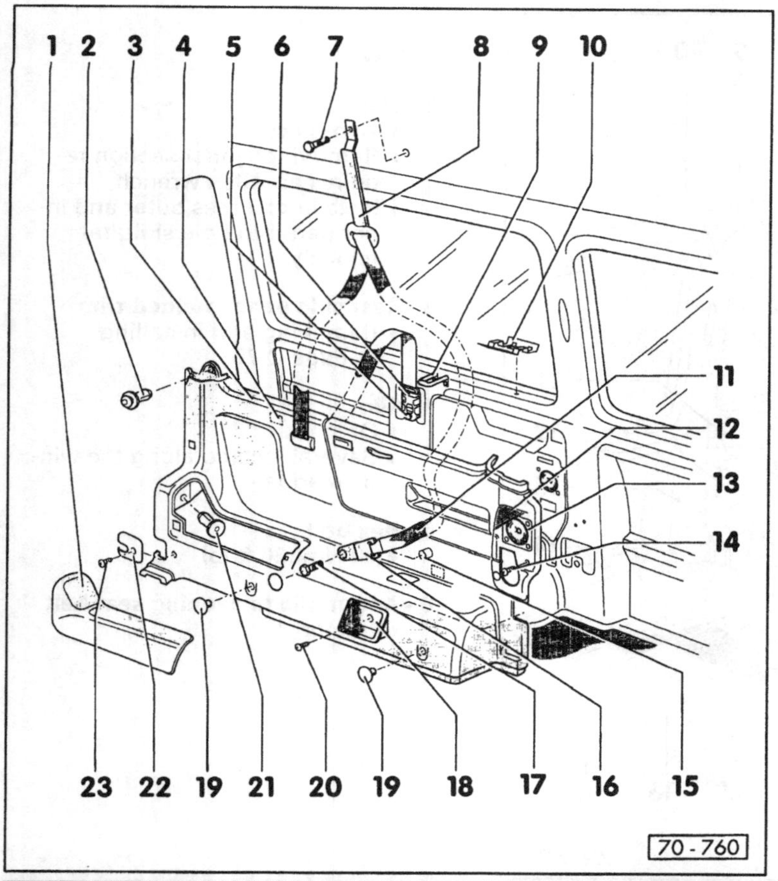

16 - **Belt anchor**

17 - **Hex bolt**
- 40 Nm (30 ft lb)
- With cap

18 - **Belt guide with closure strips**

19 - **Trim plug**

20 - **Phillips screw**

21 - **Trim plug**

22 - **Guide cover**

23 - **Phillips screw**

Rear side molded trim, removing/installing (left side, short wheel base)

Note:

- *Depending on equipment level remove seats and/or seat benches ⇒ Page Q 72-7, storage area ⇒ Page Q 70-105.*
- *Remove rear headliner trim ⇒ Page Q 70-37, left B-pillar upper trim ⇒ Page Q 70-15, left C-pillar trim ⇒ Page Q 70-24, D-pillar trim ⇒ Page Q 70-32.*

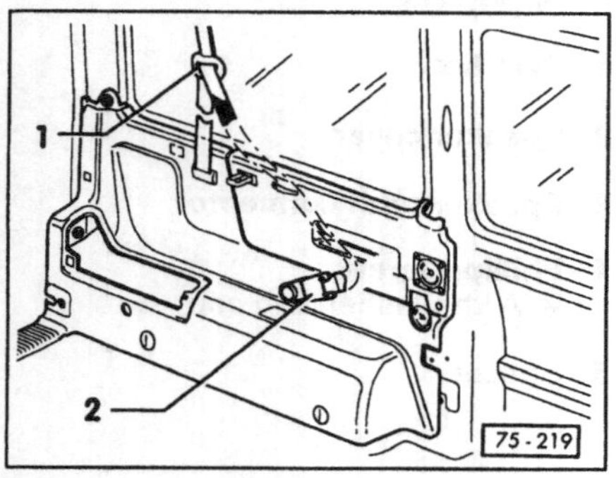

◄ – Loosen cover on left and/or right rear edge.

– Remove D-ring -1- belt anchor -2-, and/or only mounting bolt with spacer.

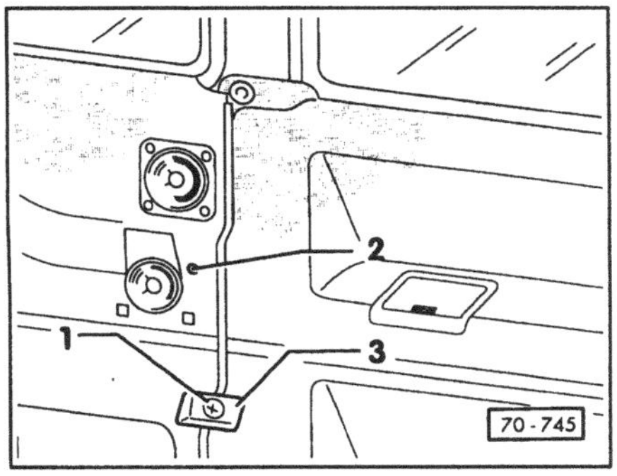

◄ – Remove Phillips screw -1-, -2- (with cap) and carefully pry out belt guide with closure strips -3- and/or seat belt.

Note:

Depending on version, Phillips screw can be located under speaker cover; remove this screw first.

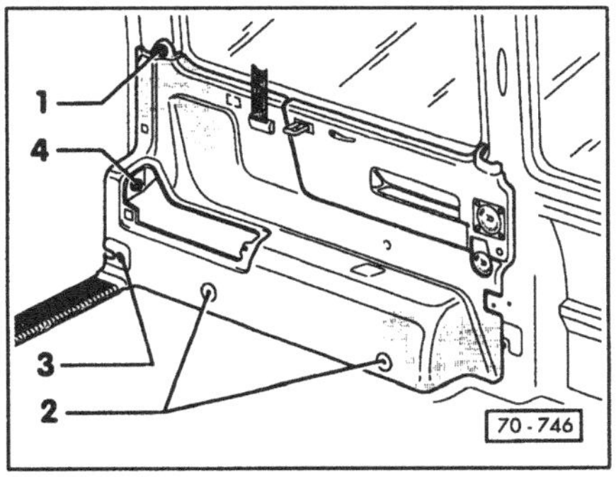

◄ – Remove inner part of clip -1- with Allen wrench.

– Carefully remove trim plug -2- with screwdriver.

– Remove guide cover -3-.

– Remove trim plug -4-.

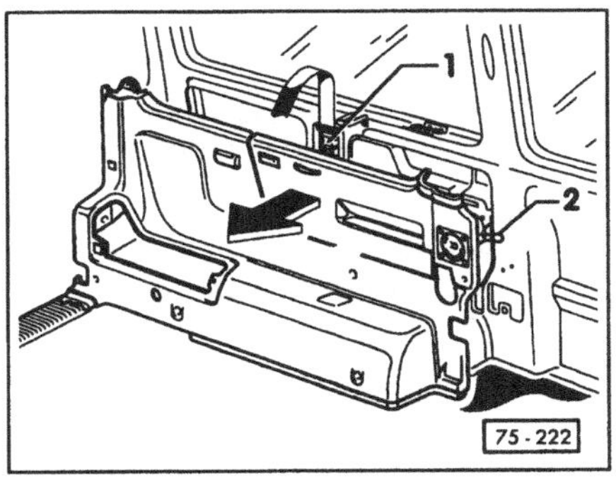

◄ – Remove molded trim panel from window frame -arrow-.

– Remove seat belt retractor -1- and, if necessary, disconnect speaker connector -2-.

– Install in reverse sequence.

– After installing rear side panel molded trim, ensure correct positioning of sliding door gasket.

Rear side molded trim, components (right side, short wheel base)

Note:

- *Depending on equipment level remove seats and/or seat benches ⇒ Page Q 72-7, storage area ⇒ Page Q 70-105.*
- *To remove rear side panel molded trim, first remove rear headliner trim ⇒ Page Q 70-37, right upper B-pillar trim ⇒ Page Q 70-19, right C-pillar trim ⇒ Page Q 70-28, D-pillar trim ⇒ Page Q 70-32, loosen the rear edge cover, left and right.*

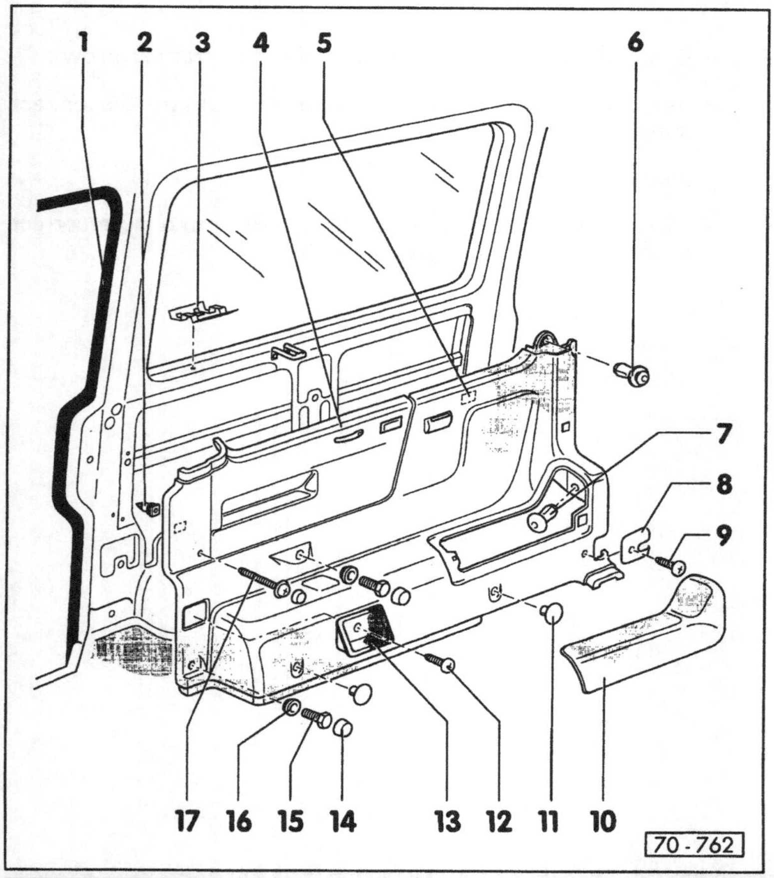

1 - Gasket
- Ensure correct positioning of rear side panel molded trim.

2 - Clip
- Two-piece
- Insert sleeve in C-pillar hole, install pin in trim.

3 - Spring clamp

4 - Rear side panel molded trim
- Removing ⇒ Page Q 70-55

5 - Clip
- On back
- Several located along window edge.

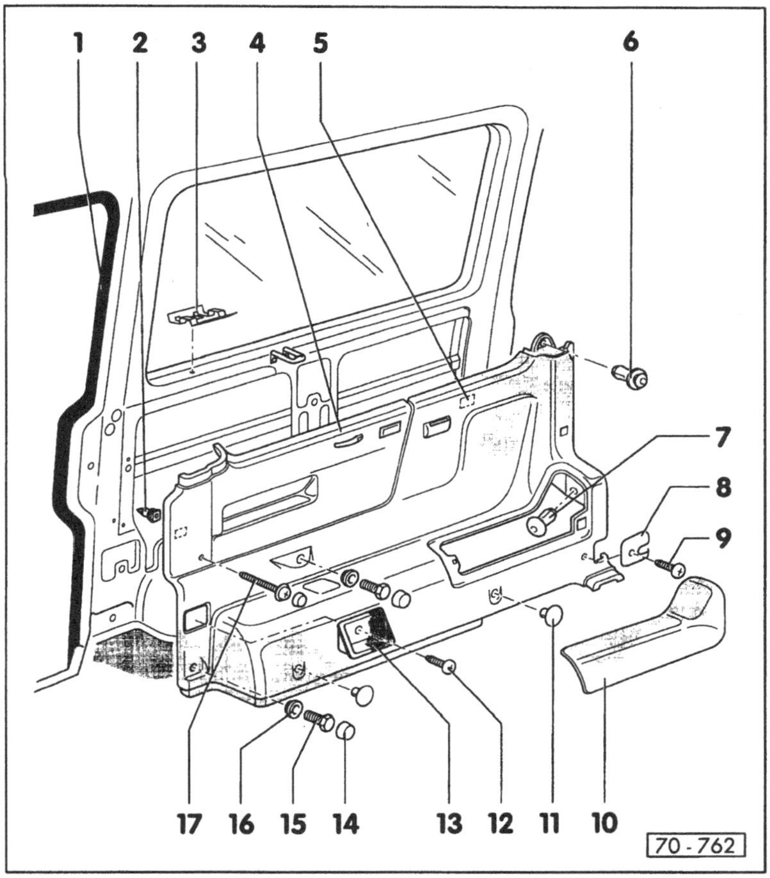

6 - Clip
- Two-piece
- Removing: Remove inner part with Allen wrench
- *Installing:* Press outer part and inner part into hole simultaneously

7 - Trim plug

8 - Guide cover

9 - Phillips screw

10 - Cover

11 - Plug

12 - Phillips screw

13 - Belt guide with closure strip

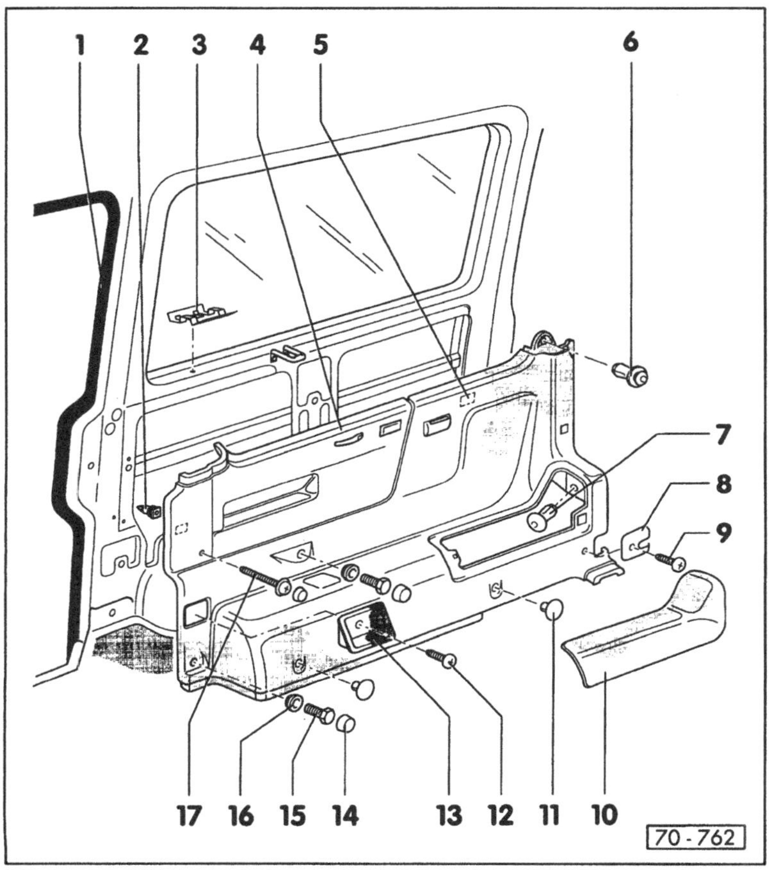

14 - Cap

15 - Hex bolt
- 40 Nm (30 ft lb)

16 - Spacer

17 - Phillips screw
- With washer and cap

Rear side molded trim, removing/installing (right side, short wheel base)

Note:

- *Depending on equipment level remove seats and/or seat benches ⇒ Page Q 72-7, storage area ⇒ Page Q 70-105.*
- *Remove rear headliner trim ⇒ Page Q 70-37, right upper B-pillar trim ⇒ Page Q 70-19, right C-pillar trim ⇒ Page Q 70-28, D-pillar trim ⇒ Page Q 70-32.*

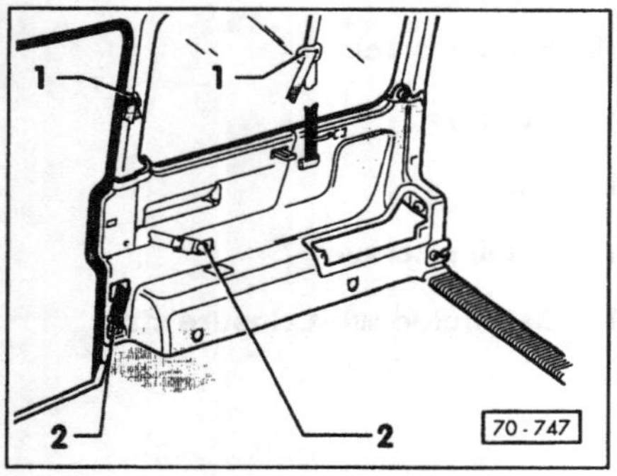

◄

- Loosen rear edge cover on left and right.
- Remove D-ring -1- and belt anchor -2- or only mounting bolts with spacer rings.

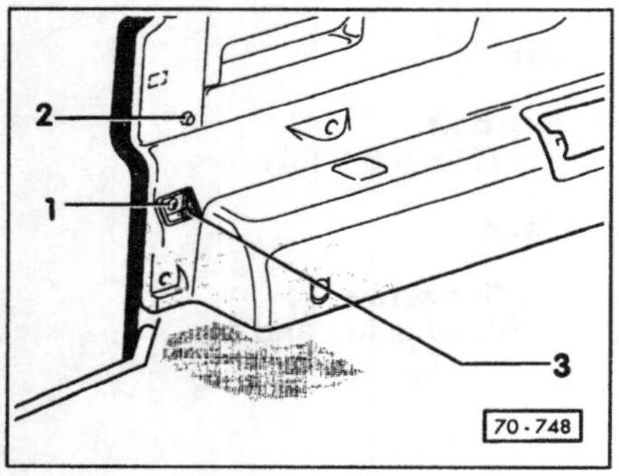

◄

- Remove Phillips screws -1-, -2- (with cap), and carefully pry out belt guide with closure strip and/or seat belt.
- Remove inner part of clip with Allen wrench.

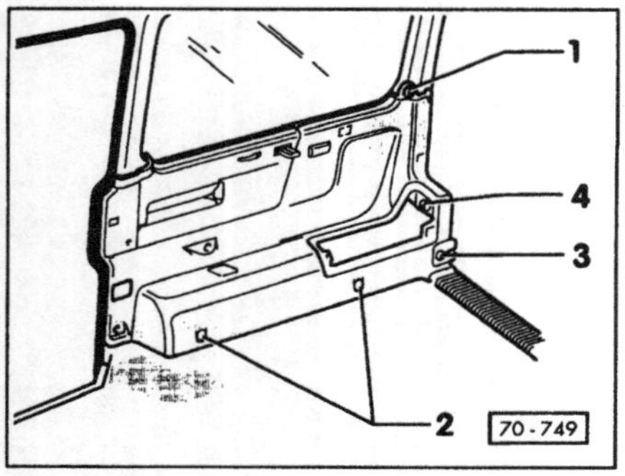

◄

- Remove trim plug -2- with screwdriver.
- Remove cover guide -3-.
- Remove trim plug -4-.

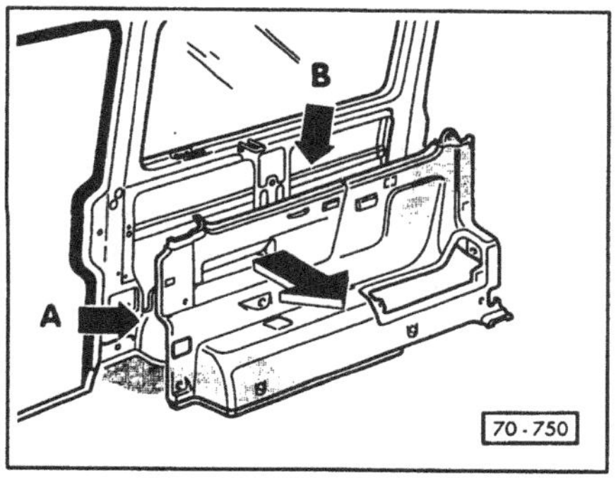

◄ – Unclip right molded trim -arrow A- and along the window frame edge -arrow B-.

– Remove molded trim from window edge -arrow-.

– If applicable, remove automatic tensioning seat belts and disconnect speaker connection.

– Install in reverse sequence. After installing rear side panel molded trim, ensure correct rear lid gasket positioning .

Rear side molded trim, components (left side, long wheel base)

Notes:

- *Removal description is given for only one side, other side is mirror image.*
- *Depending on equipment installed, remove seats and/or seat benches ⇒ Page Q 72-7*
- *To remove rear side panel molded trim, remove rear headliner trim ⇒ Page Q 70-37, left upper B-pillar trim ⇒ Page Q 70-15, left C-pillar trim ⇒ Page Q 70-24, D-pillar trim ⇒ Page Q 70-32, right and/or left rear edge cover.*

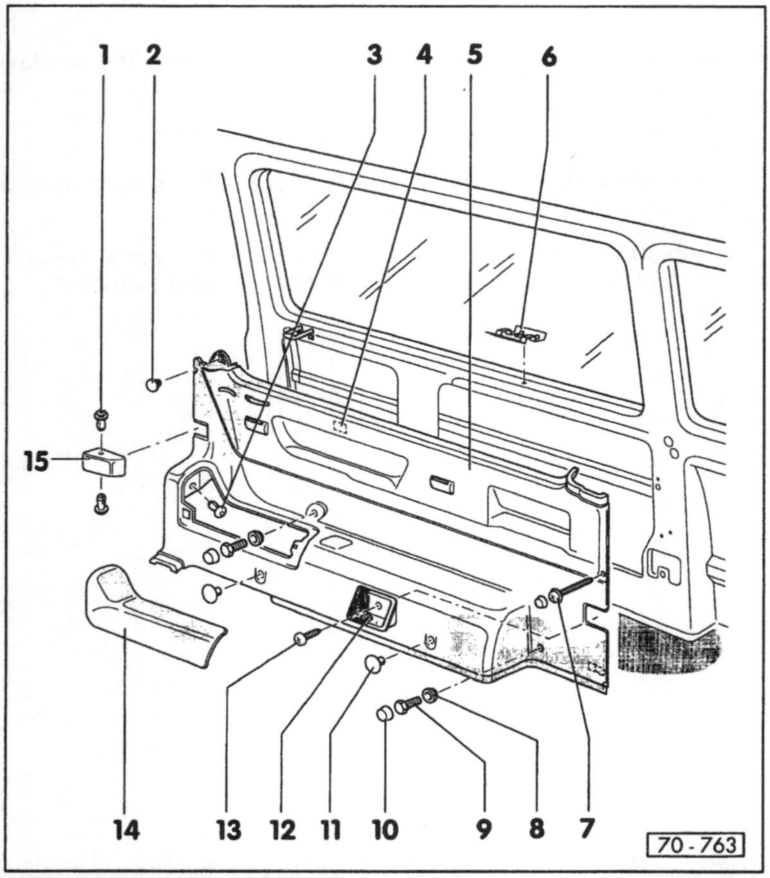

1 - Clip
- Two-piece
- Removing: Remove inner part with Allen wrench
- Installing: Press outer part and inner part into hole simultaneously

2 - Cap
- From 12/91
- Up to11/91, clip, item 1

3 - Trim plug
- From 12/91
- Up to 11/91, clip, item 1

4 - Clip
- On back
- Several located along window edge

5 - Rear side panel molded trim
- Removing and installing ⇒ Page Q 70-61

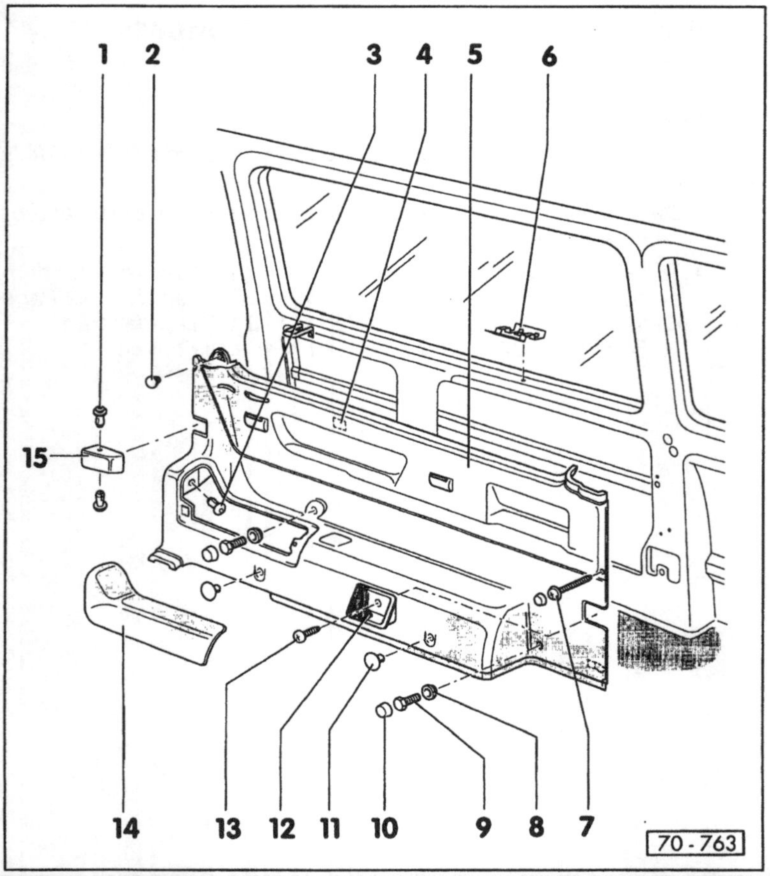

6 - Spring clip

7 - Phillips screw
- with cap

8 - Spacer

9 - Hex bolt
- 40 Nm (30 ft lb)

10 - Cover

11 - Trim plug

12 - Belt guide with closure strip

13 - Phillips screw

14 - Cover

15 - Door stop cover

Rear side molded trim, removing/installing (left side, long wheel base)

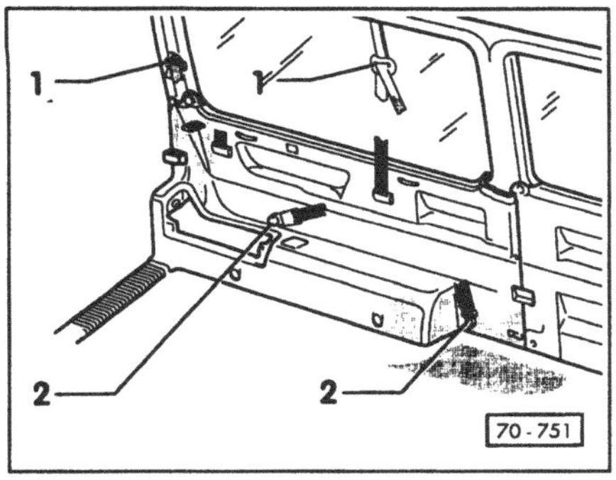

Note:

- *Depending on equipment level remove seats and/or seat benches ⇒ Page Q 72-7.*
- *Remove rear headliner trim ⇒ Page Q 70-37, left B-pillar upper trim ⇒ Page Q 70-15, left C-pillar trim ⇒ Page Q 70-24, D-pillar trim ⇒ Page Q 70-32.*

◄ – Loosen rear edge cover, left and right.

– Remove D-ring -1- and belt anchor -2- or only mounting bolts with spacer rings.

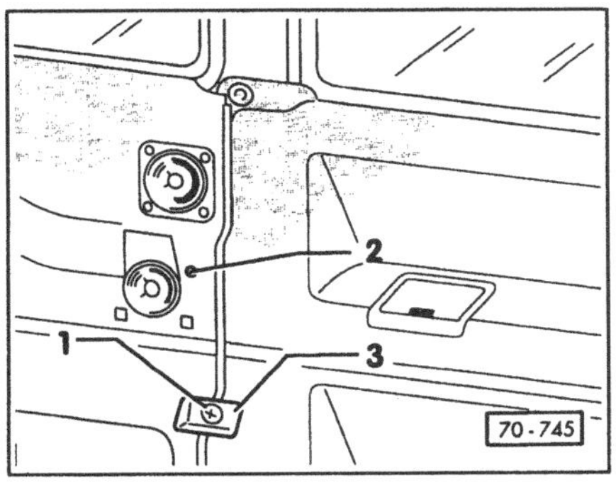

◄ – Remove Phillips screws -1-, -2- (with cap), carefully pry out belt guide with closure strip -3- and/or seat belt.

Note:

Depending on version, Phillips screw -2- may be under speaker trim; remove Phillips screw first.

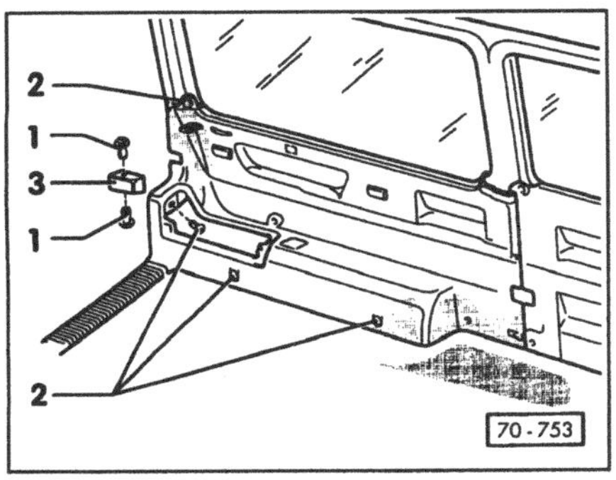

◄ – Remove inner part of clips -1- with Allen wrench.

– Turn or pry out plug -2- with screwdriver.

– Remove door stop cover -3-.

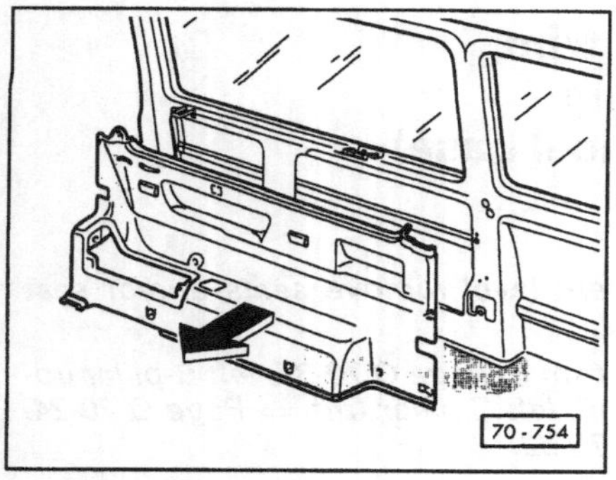

◄ – Remove molded trim from window frame -arrow-.

– If necessary, remove automatic tensioning seat belt retractor and disconnect speaker connector.

– Install in reverse sequence. After installing rear side panel molded trim, ensure correct positioning of sliding door gasket.

Rear side molded trim, components (right side, long wheel base)

Notes:

- *Removal description is given for only one side, other side is mirror image*
- *Depending on equipment installed, remove seats and/or seat benches ⇒ Page Q 72-7*
- *To remove rear side panel molded trim, first remove rear headliner trim ⇒ Page Q 70-37, right upper B-pillar trim ⇒ Page Q 70-19, right C-pillar trim ⇒ Page Q 70-28, D-pillar trim ⇒ Page Q 70-32.*

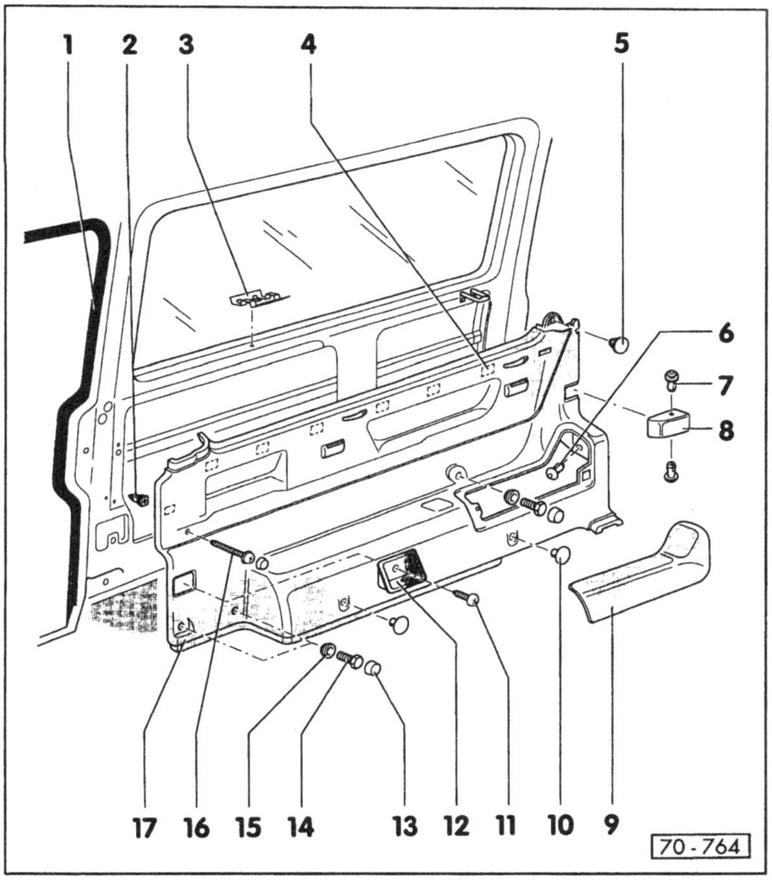

1 - Gasket
- After installing rear side panel molded trim, ensure correct positioning.

2 - Clip
- Two-piece
- Place sleeve in C-pillar hole and install pin.

3 - Spring clip

4 - Clip
- On back
- Several located along window edge.

5 - Cap
- From 12/91
- Up to 11/91, clip, item 7

6 - Plug
- From 12/91
- Up to 11/91, clip, item 7

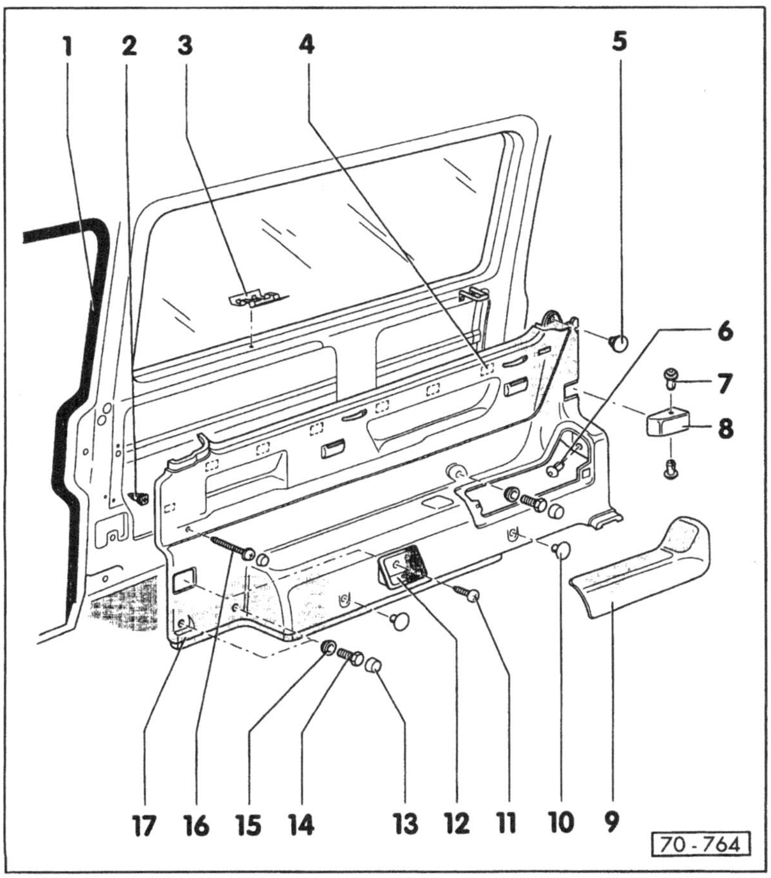

7 - Clip
- Removing: Remove inner part with Allen wrench
- Installing: Press outer part and inner part into hole simultaneously

8 - Door stop cover

9 - Cover

10 - Trim plug

11 - Phillips screw

12 - Belt guide with closure strip

13 - Cover

14 - Hex bolt
- 40 Nm (30 ft lb)

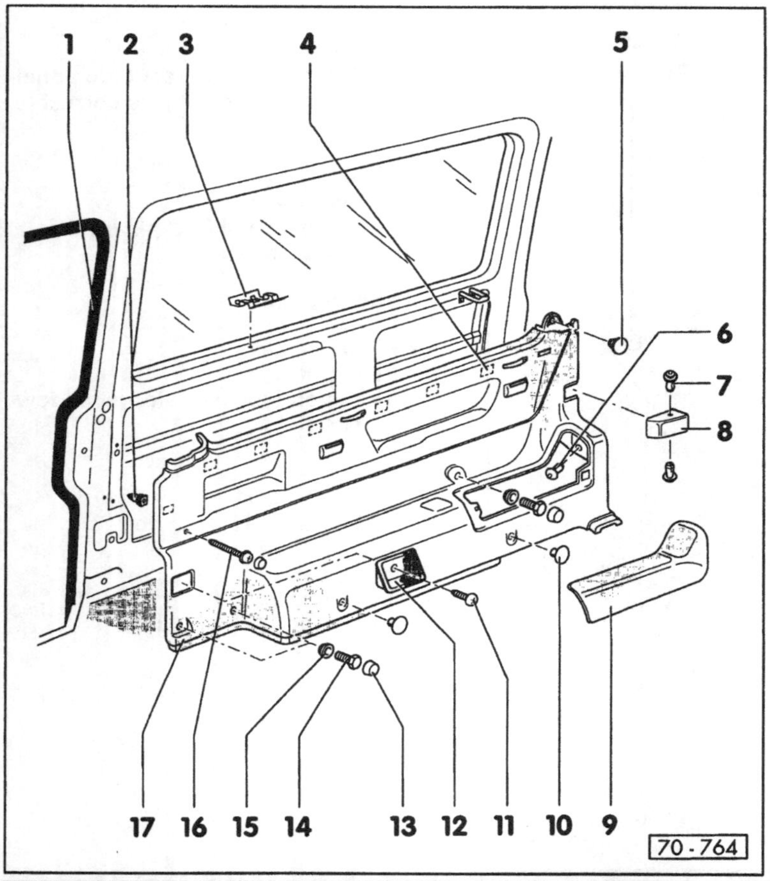

15 - Spacer

16 - Phillips screw

- with washer and cap

17 - Rear side panel molded trim

- Removing and installing ⇒ Page Q 70-68

Rear side molded trim, removing/installing (left side, long wheel base)

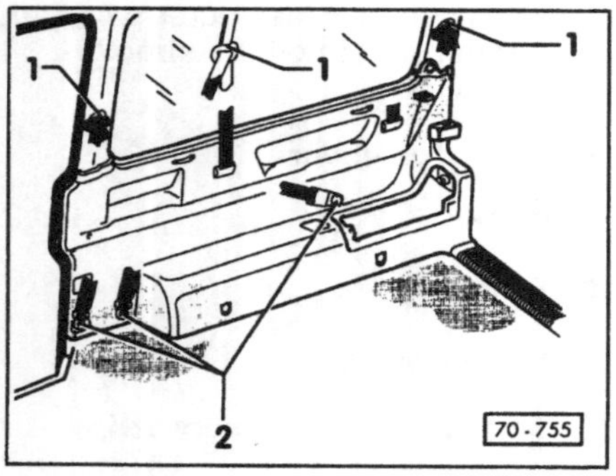

Note:

- *Depending on equipment level, remove seats and/or seat benches ⇒ Page Q 72-7.*
- *Remove rear headliner trim ⇒ Page Q 70-37, left B-pillar upper trim ⇒ Page Q 70-15, left C-pillar trim ⇒ Page Q 70-24, D-pillar trim ⇒ Page Q 70-32.*

◄ – Loosen rear end cover, left and right.

– Remove D-ring -1-, belt anchor -2- and/or mounting bolts with spacer.

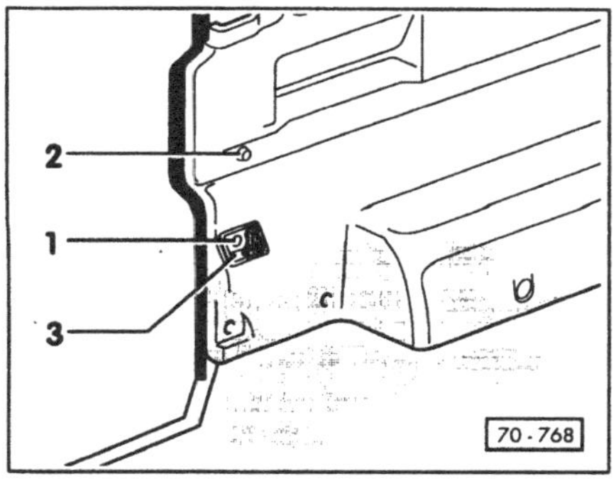

◄ – Remove Phillips screws -1-, -2- (with cap), and carefully pry out belt guide with closure strip -3- and/or seat belt.

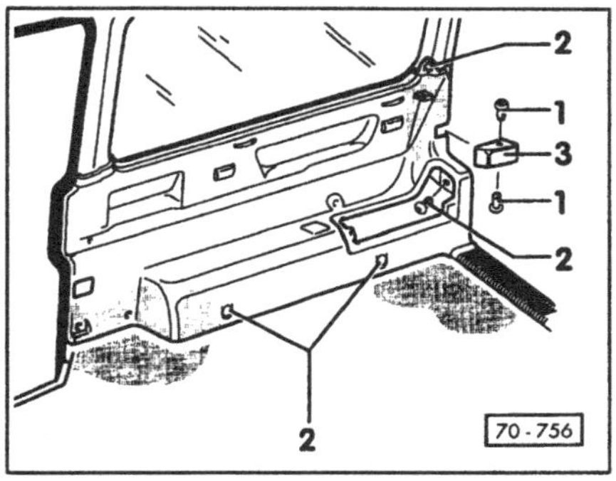

◄ – Remove inner parts of clips -1- with Allen wrench.

– Turn or pry plug -2- out with screwdriver.

– Remove door stop cover -3-.

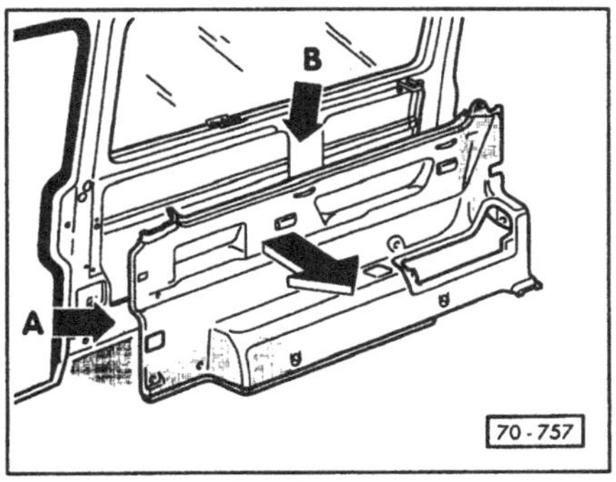

◄ – Unclip right molded trim -arrow A- and along window -arrow B-.

– Pull molded trim off window frame in direction of arrow.

– If necessary, remove automatic tensioning seat belts and disconnect speaker connector.

– Install in reverse sequence. After installing rear side panel molded trim, ensure correct positioning of rear door and sliding door gaskets.

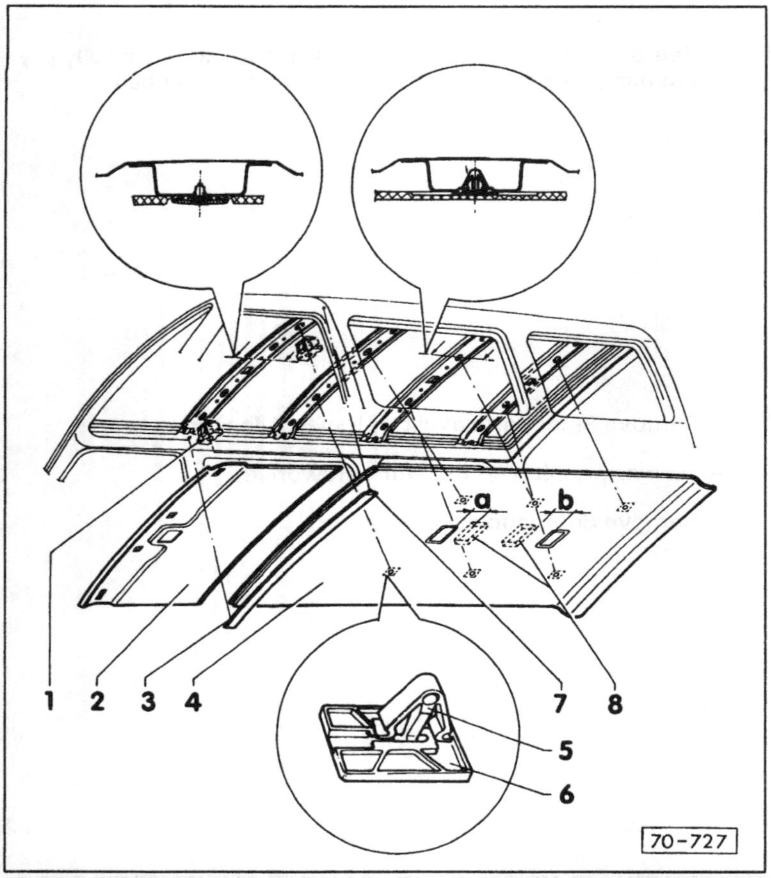

Molded headliner, components

Note:

- *To remove molded headliner in front passenger compartment, remove upper B-pillar trim ⇒ Page Q 70-15 and Page Q 70-19, A-pillar trim ⇒ Page Q 70-11*
- *To remove rear molded headliner, remove upper B-pillar trim ⇒ Page Q 70-15 and Page Q 70-19, C-pillar trim ⇒ Page Q 70-24 and Page Q 70-28, D-pillar trim ⇒ Page Q 70-32, rear headliner trim ⇒ Page Q 70-37*

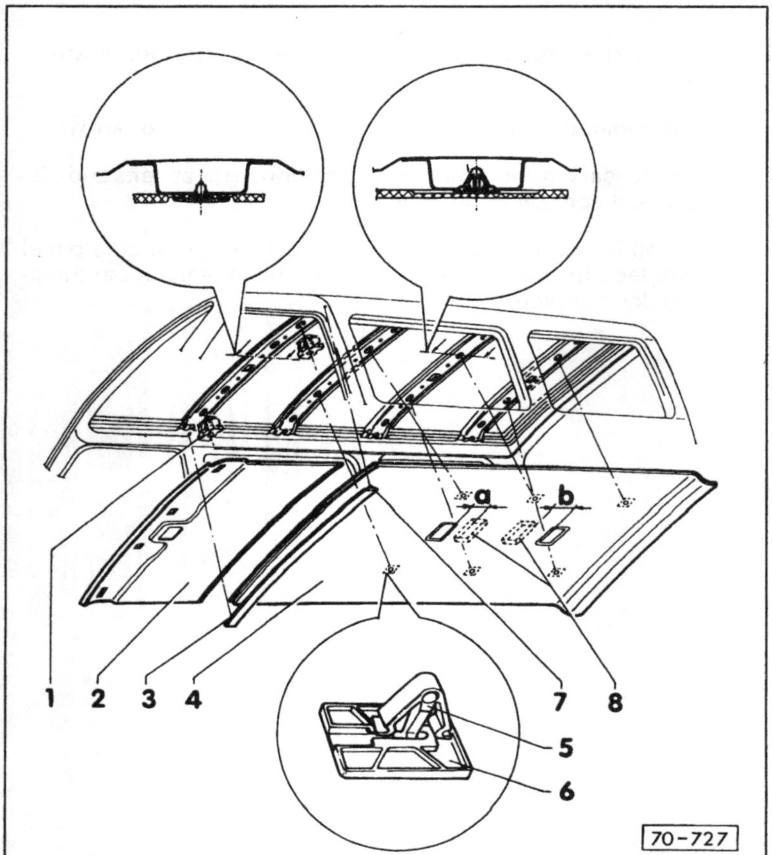

1 - Mounting clip
- Two-piece

2 - Front passenger compartment molded headliner
- Removing ⇒ Page Q 70-74

3 - Cover strip

4 - Rear passenger compartment molded headliner
- Removing ⇒ Page Q 70-75
- Installing ⇒ Page Q 70-77

5 - Clip
- Six pieces

6 - Clip holder
- Six pieces
- Self–adhesive

7 - Clip
- Eight pieces

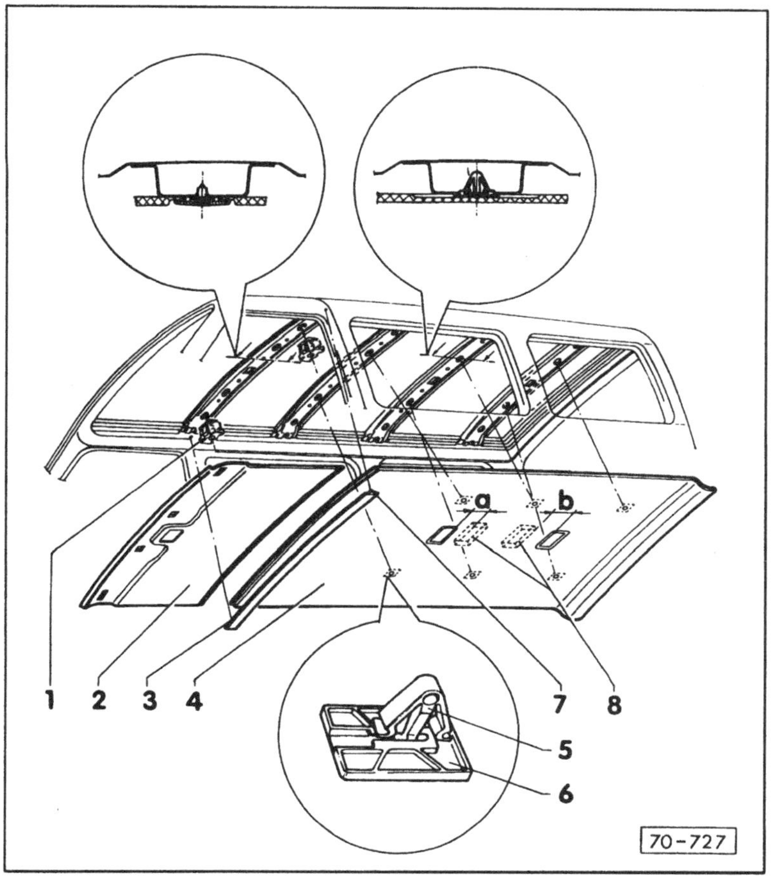

8 - **Spacer block**
- Only on vehicles w/o A/C
- Two-piece
- Self-adhesive, glued to back of molded headliner
- a = 60 mm (approx. 2.4 in.) (back edge cut out for front interior light to front edge of spacer block)
- b = 60 mm (approx. 2.4 in.) (front edge cut out for rear interior light to rear edge of spacer block)

Front passenger compartment molded headliner, removing

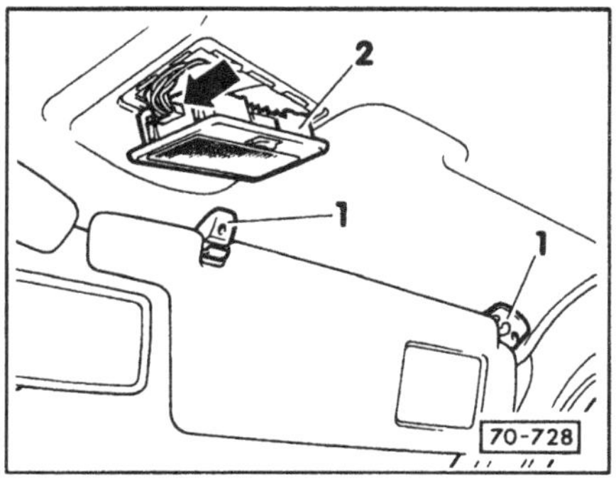

- Remove upper B-pillar trim ⇒ Page Q 70-17 and Page Q 70-21.
- Remove A-pillar trim ⇒ Page Q 70-11.
- Remove bracket -1- for sun visors and disconnect electrical connector for illuminated vanity mirror.
- Remove Front Interior Light -2- and disconnect electrical connector-arrow-.

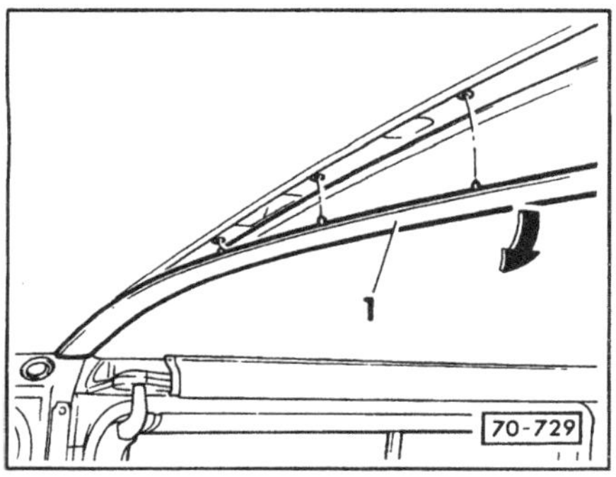

- Unclip cover strip -1-.
- Remove front passenger compartment molded headliner by pulling downward.

Rear passenger compartment molded headliner, removing

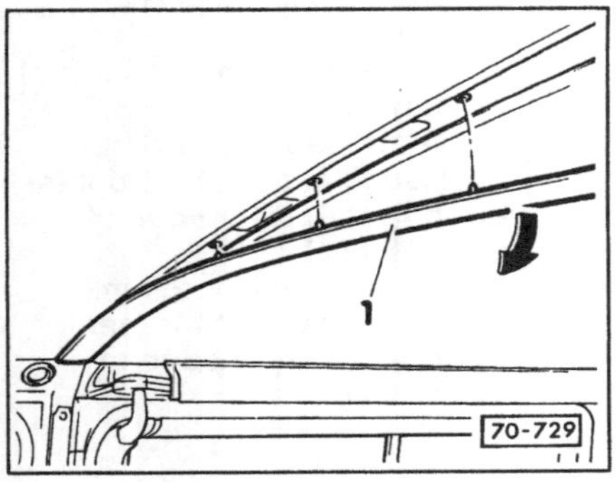

- Remove upper B-pillar trim ⇒ Page Q 70-17 and Page Q 70-21.
- Remove C-pillar trim ⇒ Page Q 70-26 and Page Q 70-30.
- Remove rear headliner trim ⇒ Page Q 70-37.
- Remove D-pillar trim ⇒ Page Q 70-35.
- ◄ Unclip cover strip -1-.

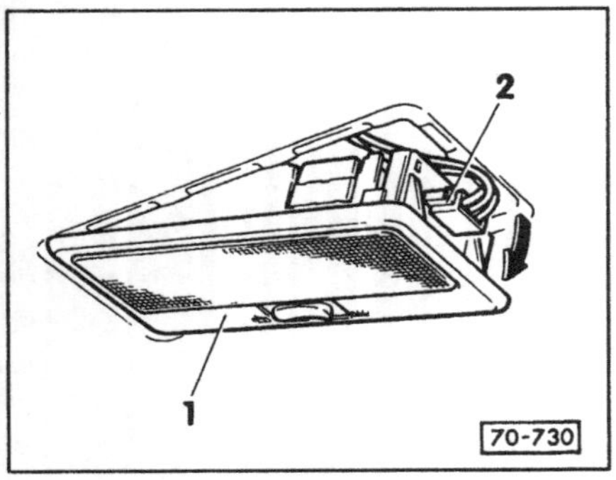

- ◄ Remove Rear/Luggage Compartment Interior Light -1- and Center Interior Light -1- in rear passenger compartment and disconnect electrical connectors -2-.

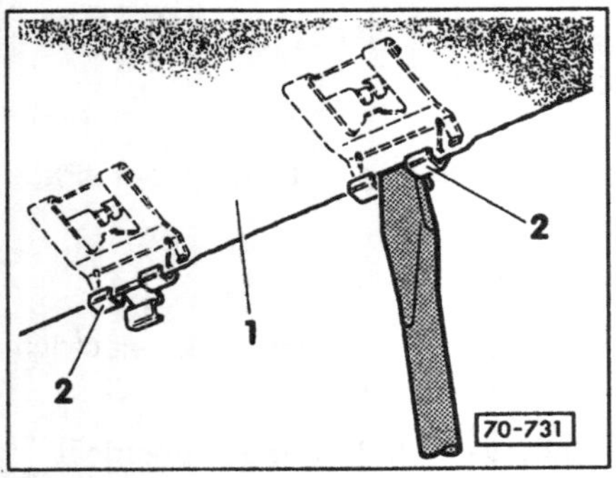

- ◄ Carefully pry molded headliner -1- out of clips on headliner.

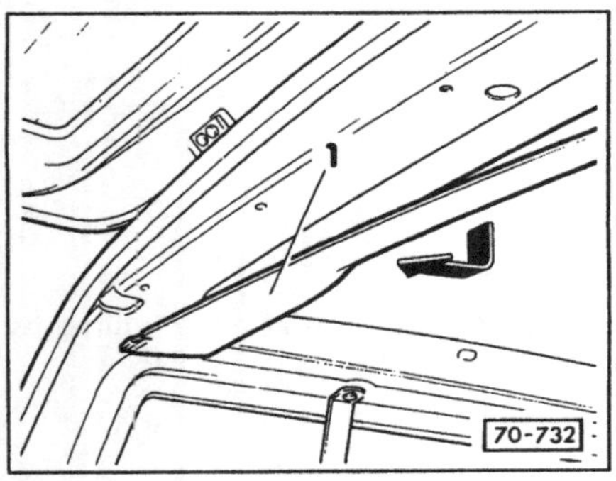

- ◄ Slightly lower molded headliner-1- in rear area and pull back out of clips, out of vehicle -arrow-.

Passenger compartment molded headliner, installing

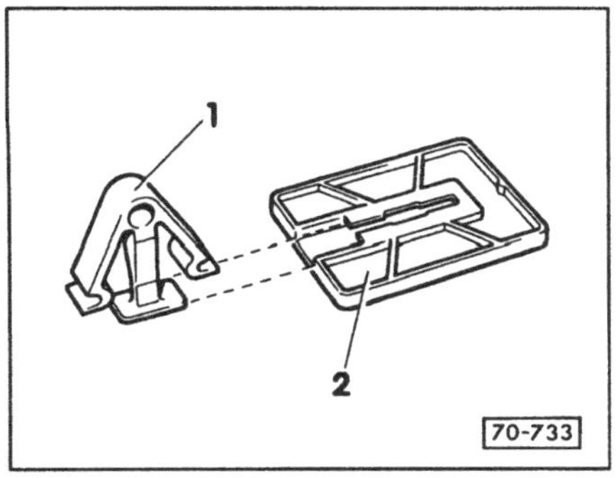

◄ – Turn clips -1- 90°, remove from slots on roof bow and install in clip holder (on back side of molded headliner).

– Mount molded trim into position and clip under headliner.

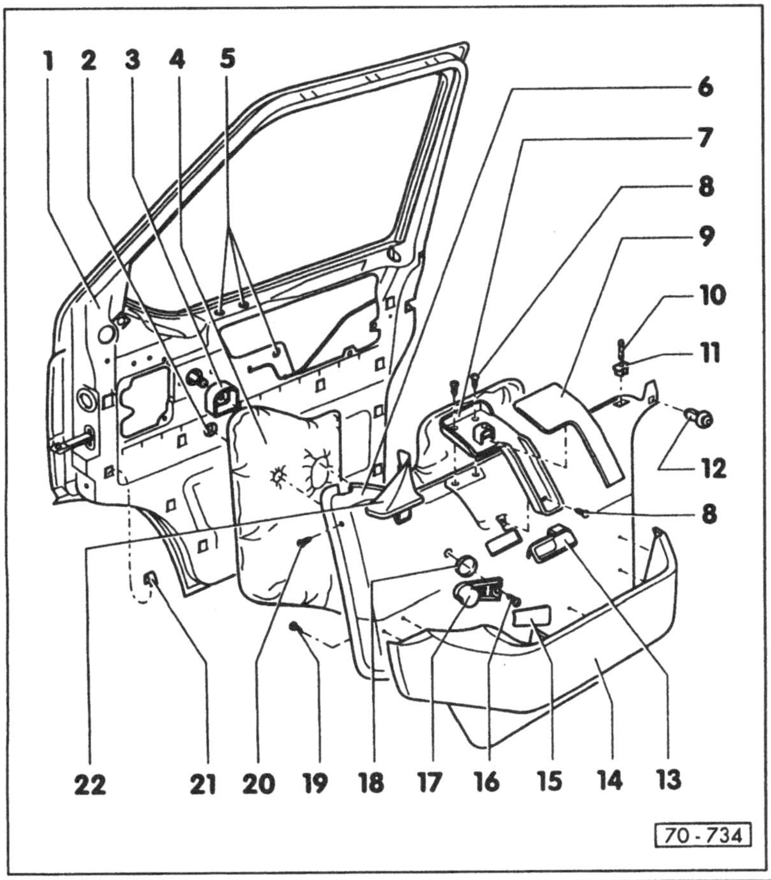

Front door trim, components

1 - **Front door**

2 - **Retaining plate**
  ♦ Secures insulation on door trim -6-.

3 - **Seal**

4 - **Insulation**

5 - **Blind rivet nut**
  ♦ Install with blind rivet tool e.g. VAG 1618

6 - **Door trim**
  ♦ Removing ⇒ Page Q 70-81

7 - **Handle**

8 - **Fillister head screws**

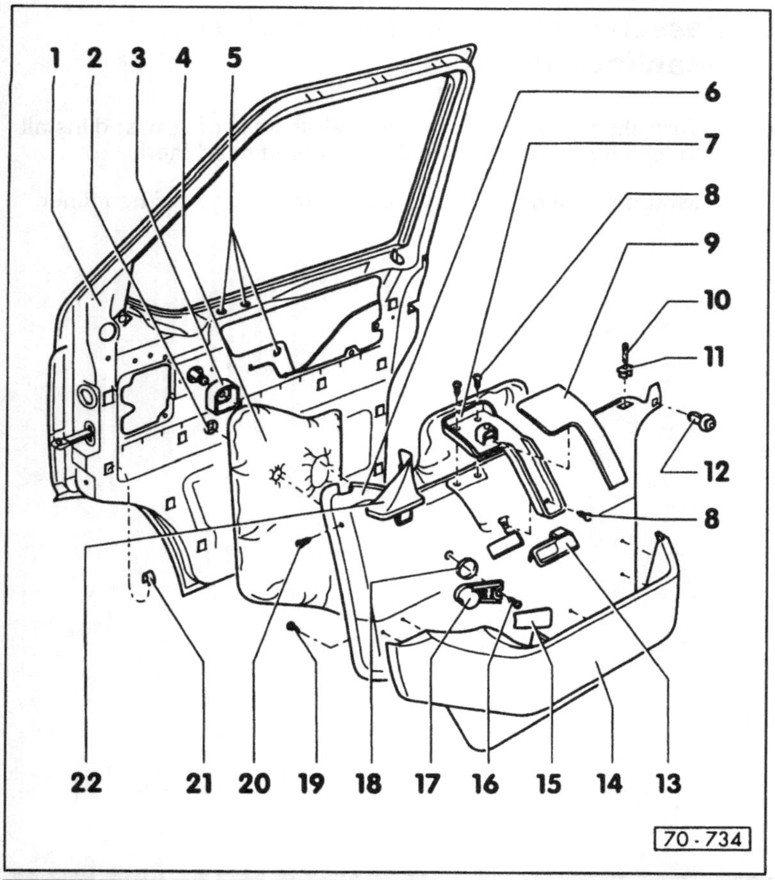

9 - **Trim**

10 - **Locking knob**

11 - **Trim**

12 - **Clip**
- Two-piece
- Removing: Remove inner part with Allen wrench
- Installing: Press outer and inner part into hole simultaneously

13 - **Inner door mechanism**
- Unclip with door trim removed

14 - **Door storage compartment**
- Fastened to door trim with screws

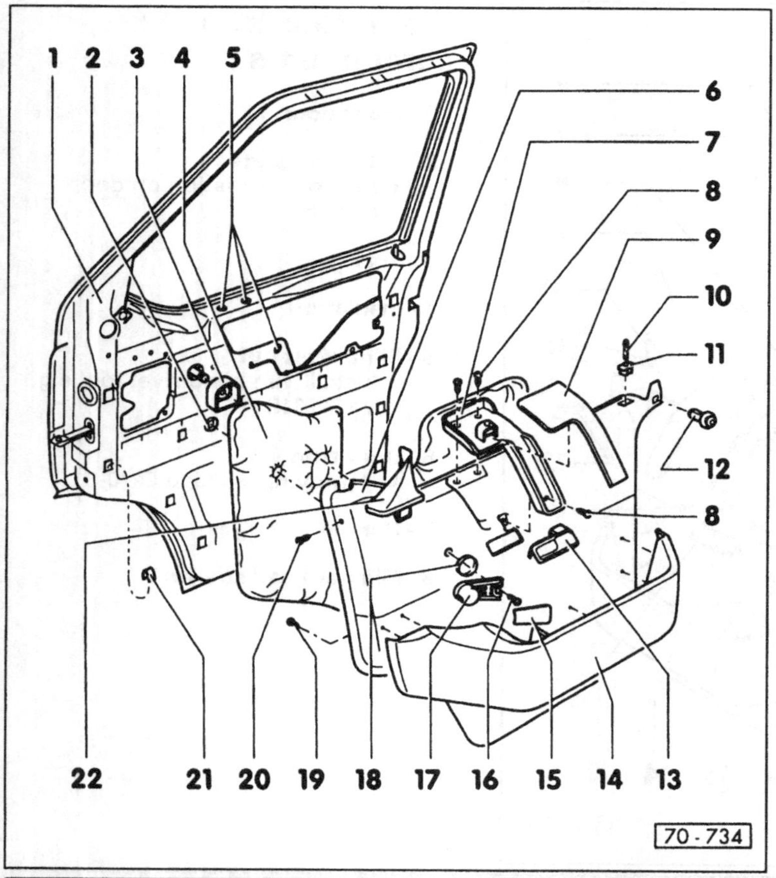

15 - **Cover**

16 - **Phillips screw**
- 4.0 Nm (35 in. lb)

17 - **Window crank**
- Removing ⇒ Page Q 70-83
- Installing ⇒ Page Q 70-83

18 - **Crank shim**

19 - **Phillips screw**
- Secures door storage compartment -1- to door trim -6-.

20 - **Phillips screw**

21 - **Hook**

22 - **Trim**

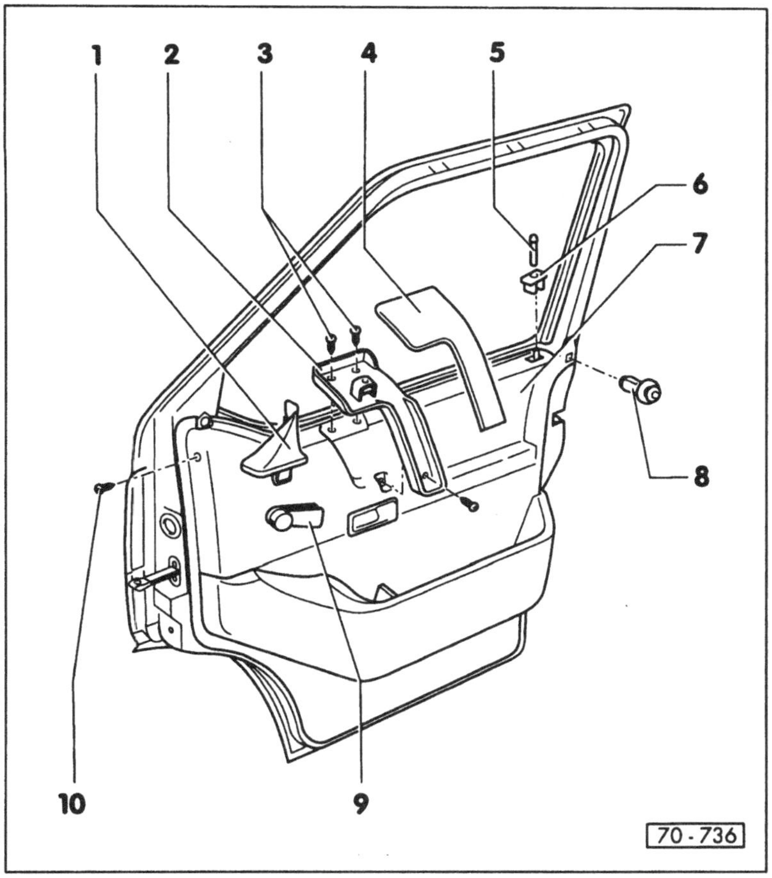

Door trim, removing

- Crank window down.
- Remove window crank -9- ⇒ Page Q 70-83.
- Remove locking knob -5- and pry off escutcheon trim -6-.
- Turn inner part of clip -8- out with Allen wrench.
- Remove trim -4-, remove fillister-head screws -3-, remove handle -2-.
- Remove Phillips screw -10-.
- Remove trim -1-.

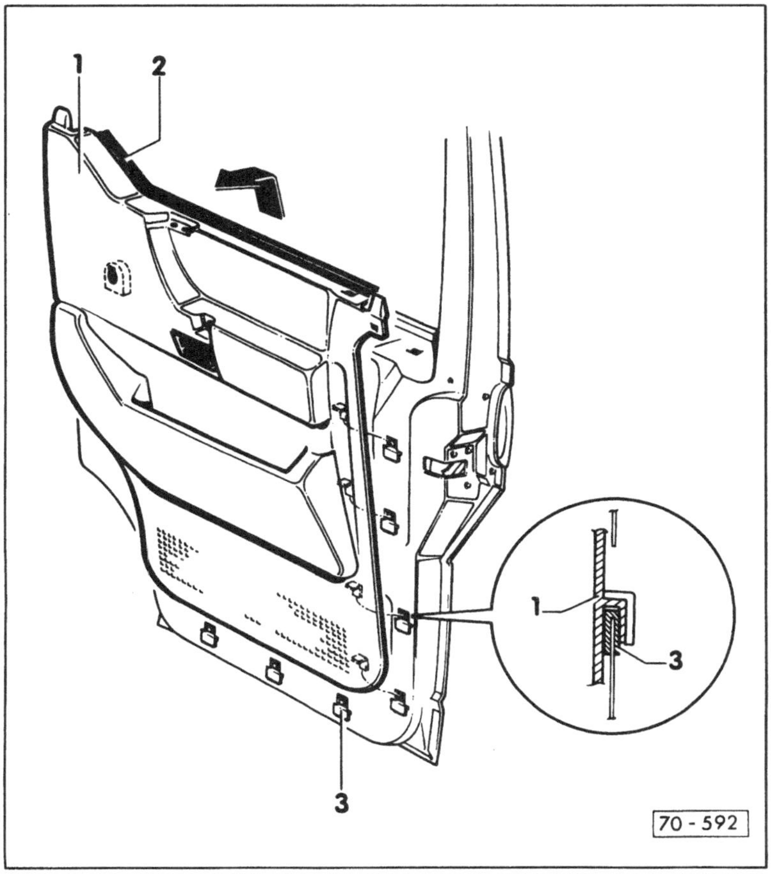

- Lift door trim -1- with inside window slot seal -2- out of mounting holes in direction of arrow.

Note:

Before installing trim, install hook -3- into mounting holes.

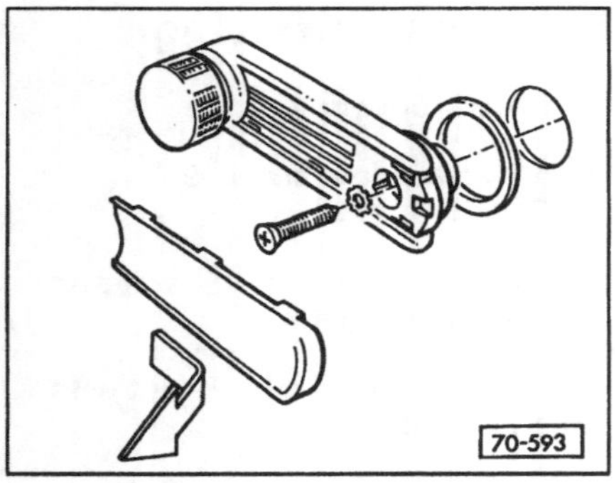

Window crank, removing

◄ – Unclip cover in direction of arrow.

– Remove Phillips screw with toothed washer.

– Remove window crank with crank shim.

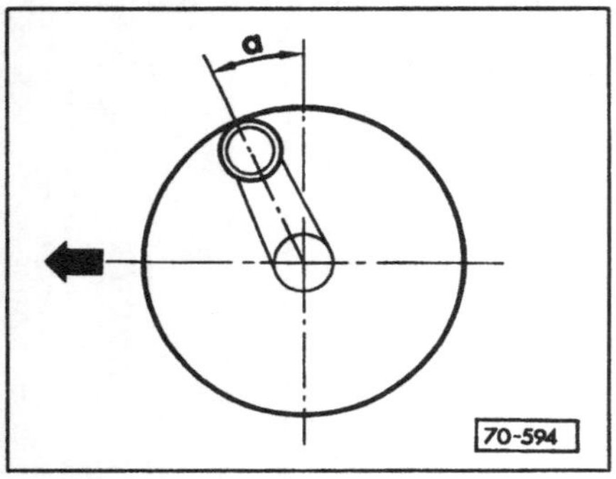

Window crank, installing

◄ After installation the window crank must be in the position shown when the window is closed.

Arrow = driving direction

Angle -a- = 25° + 12°

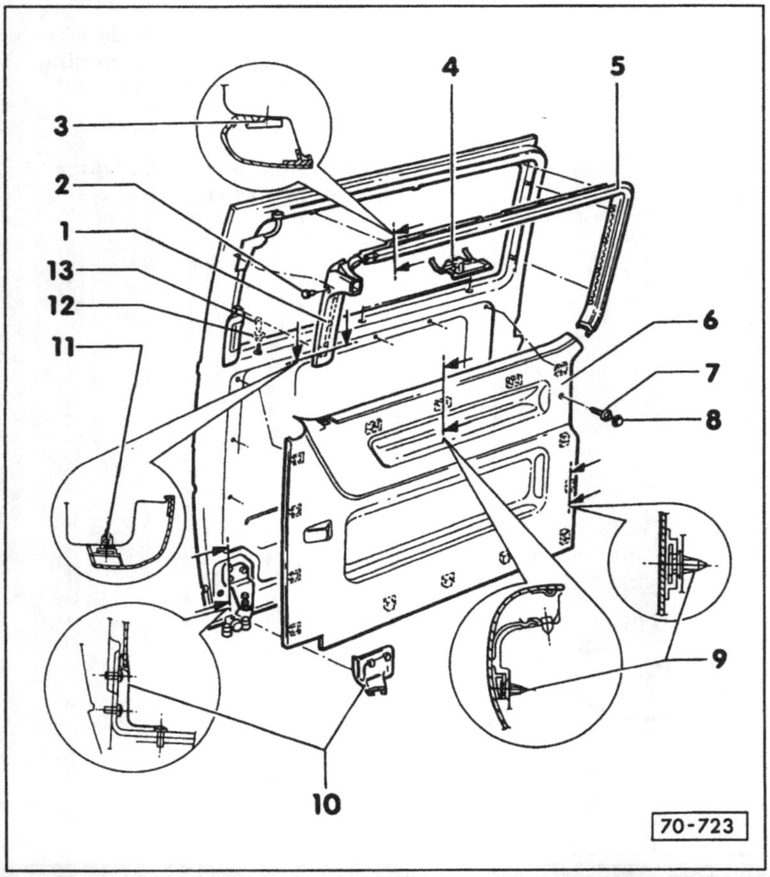

Sliding door molded trim, components

1 - Frame trim, front
- Removing: Pry out plug -2-.

2 - Plug

3 - Clamp
- On window frame, top

4 - Spring clip
- On window frame, lower

5 - Frame trim, rear
- Unclip

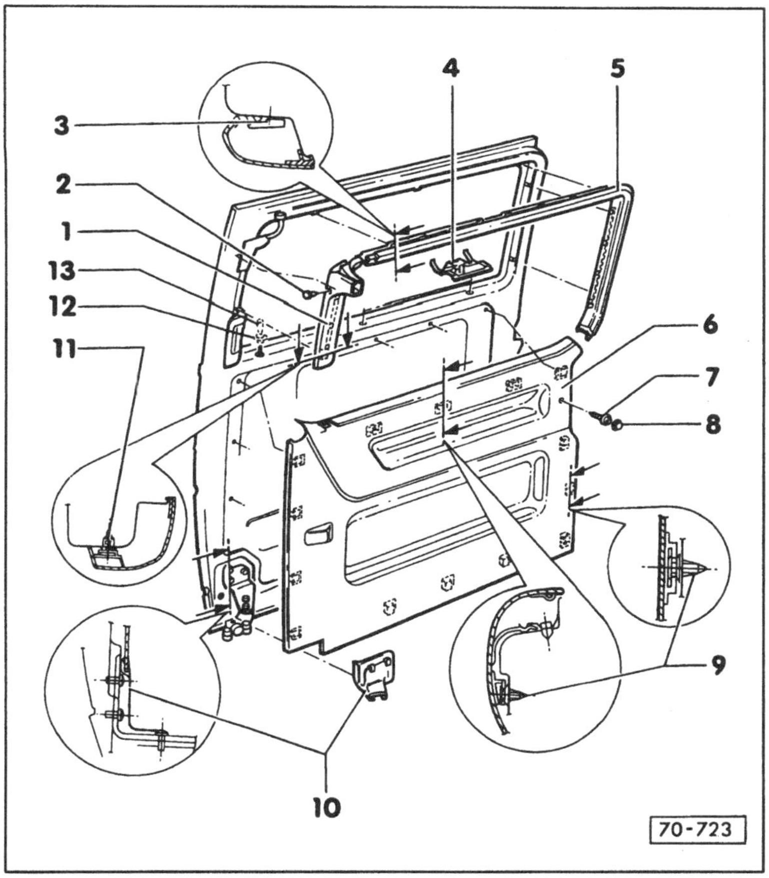

6 - Lower sliding door molded trim
- ♦ Removing:
  - Remove locking knob -13-, pry off trim -12-.
  - Pry off cap -8-, remove Phillips screw -7-.
  - Pry out plug -2-, unclip front frame trim -1- and rear frame trim -5-.
  - Unclip lower sliding door molded trim -6-.

7 - Phillips screw

8 - Cover

9 - Clip

10 - Roller guide trim, bottom
- ♦ Clipped

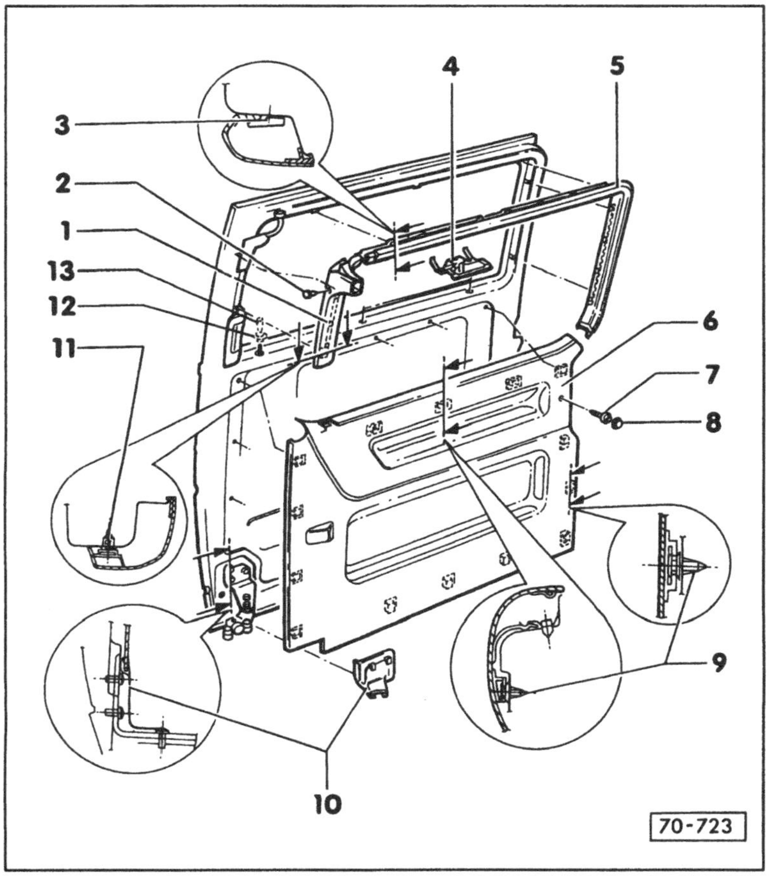

11 - Clip

12 - Trim

13 - Locking knob

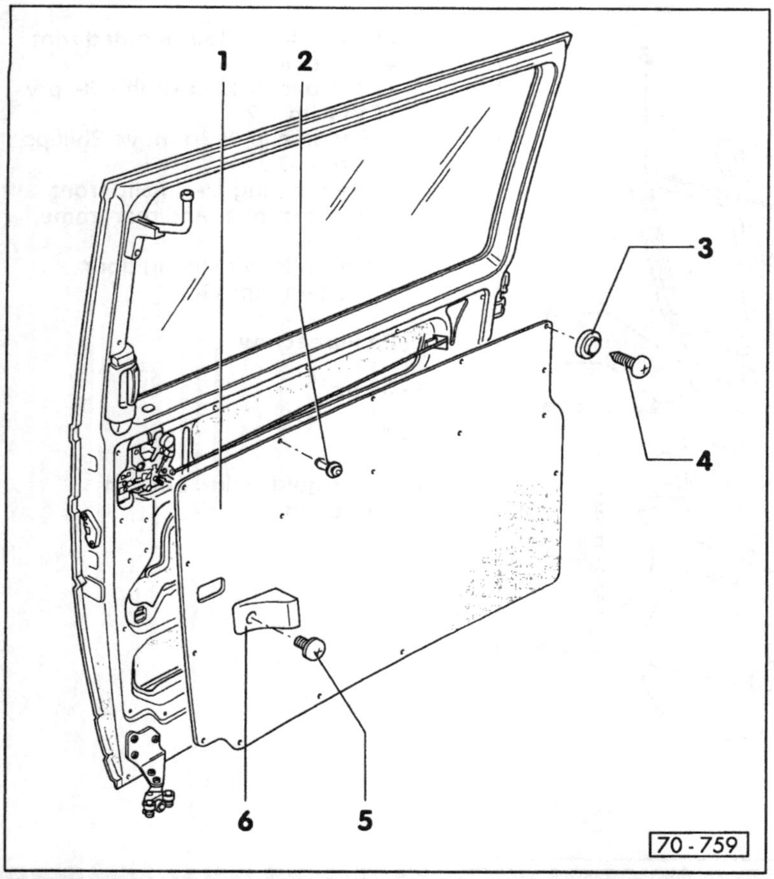

Lower sliding door trim panel, components

1 - Trim panel
- Removing:
  - Remove screw -5- and center stop -6-.
  - Remove Phillips screw -4-.
  - Remove clips -2- with Allen wrench and remove trim.

2 - Clip
- Two-piece
- Removing: Remove inner part with Allen wrench.
- Installing: Press outer and inner parts into hole simultaneously.

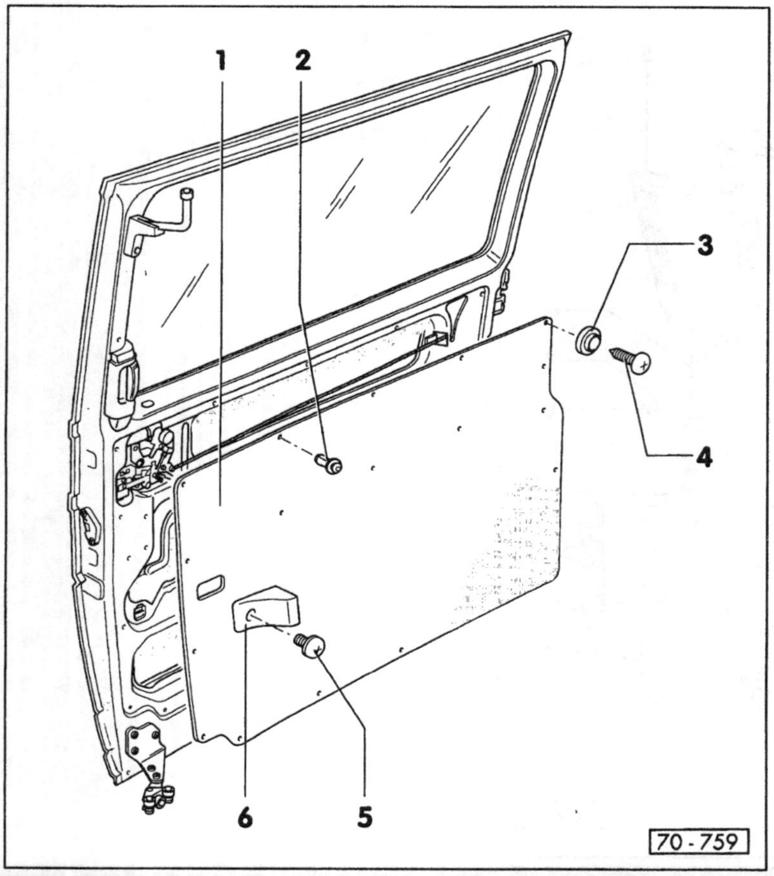

3 - Washer

4 - Phillips screw

5 - Screw
- 19 Nm (14 ft lb)

6 - Center stop

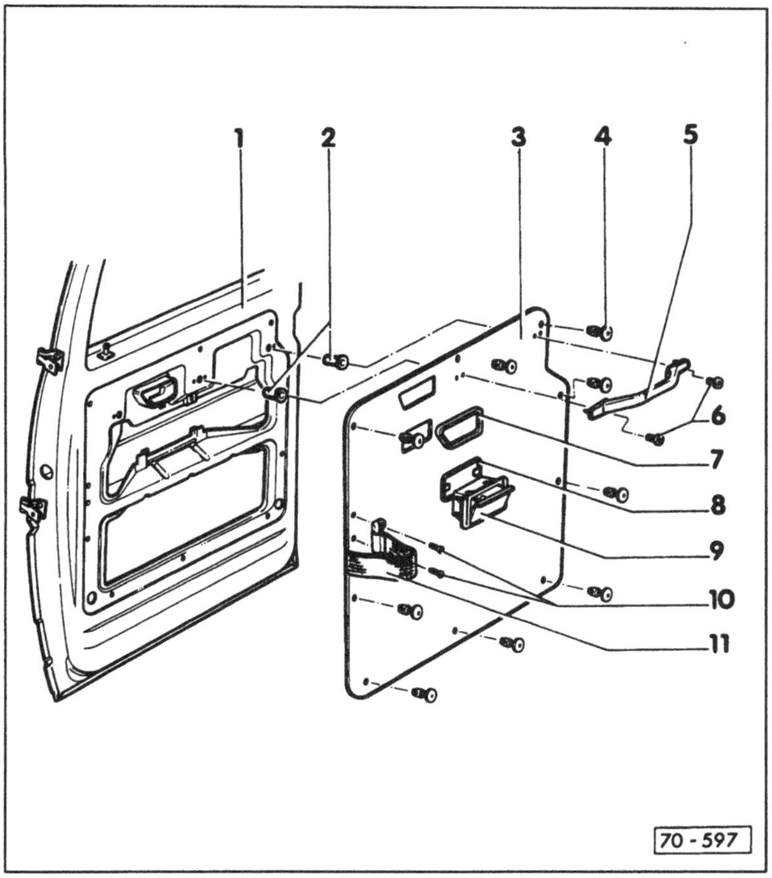

Double door trim, components

1 - Side hinged door

2 - Blind rivet nut
- ♦ Install with blind rivet tool e.g. VAG 1618 or equivalent

3 - Trim
- ♦ Removing: Remove door check strap -11-, handle -5-, unclip trim -7-, remove clips -4- and trim

4 - Clip
- ♦ Two-piece
- ♦ Removing: Remove inner part with Allen wrench
- ♦ Installing: Press outer and inner parts into hole simultaneously

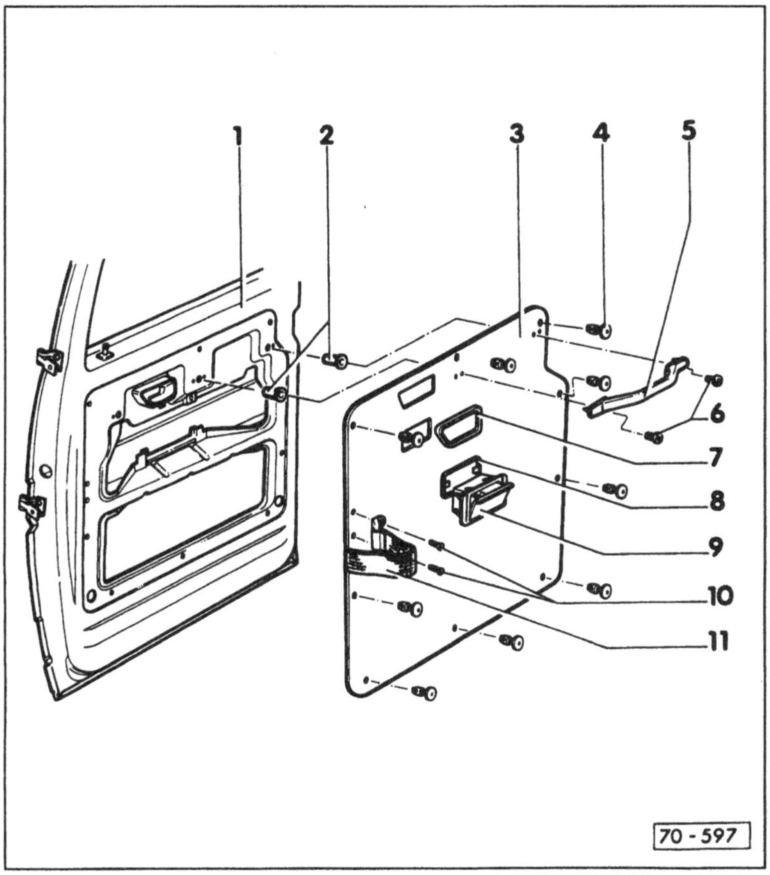

5 - Handle

6 - Screw

7 - Trim

8 - Washer
- ♦ Only for left double door

9 - Ash tray
- ♦ Only for left double door

10 - Phillips screw
- ♦ 3.5 Nm (31 in. lb)

11 - Door check strap

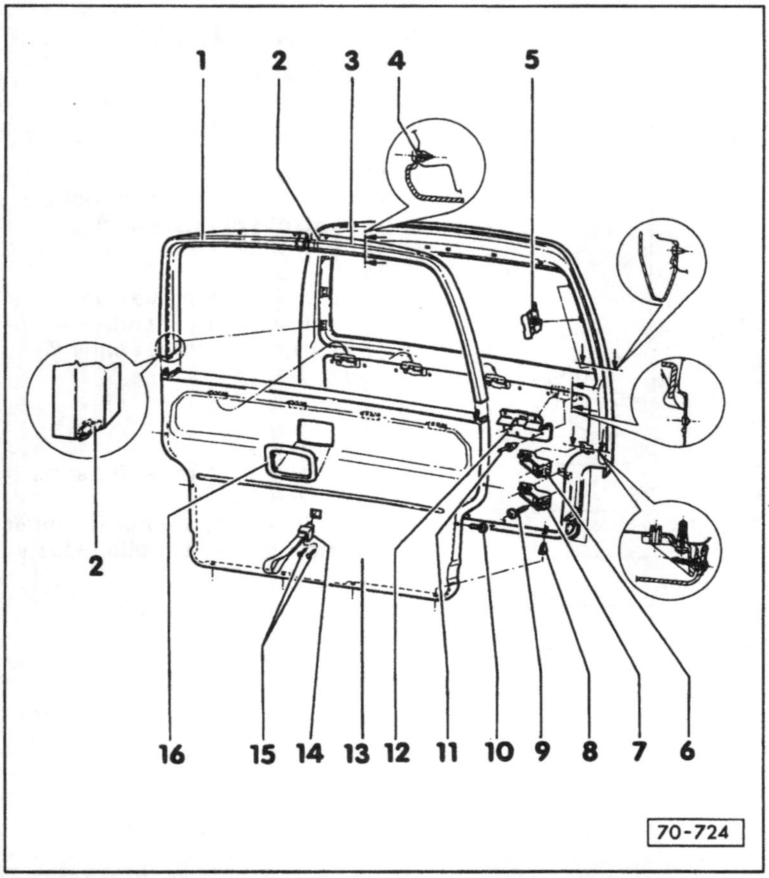

Tailgate molded trim, components

1 - **Right frame trim**

2 - **Felt strip**
- ♦ Self-adhesive - glued on back of trim

3 - **Left frame trim**

4 - **Expanding pins**
- ♦ Removing: pry off
- ♦ Installing: press in

5 - **Spring clip**

6 - **Clip**
- ♦ left and right

7 - **Clip**
- ♦ left and right

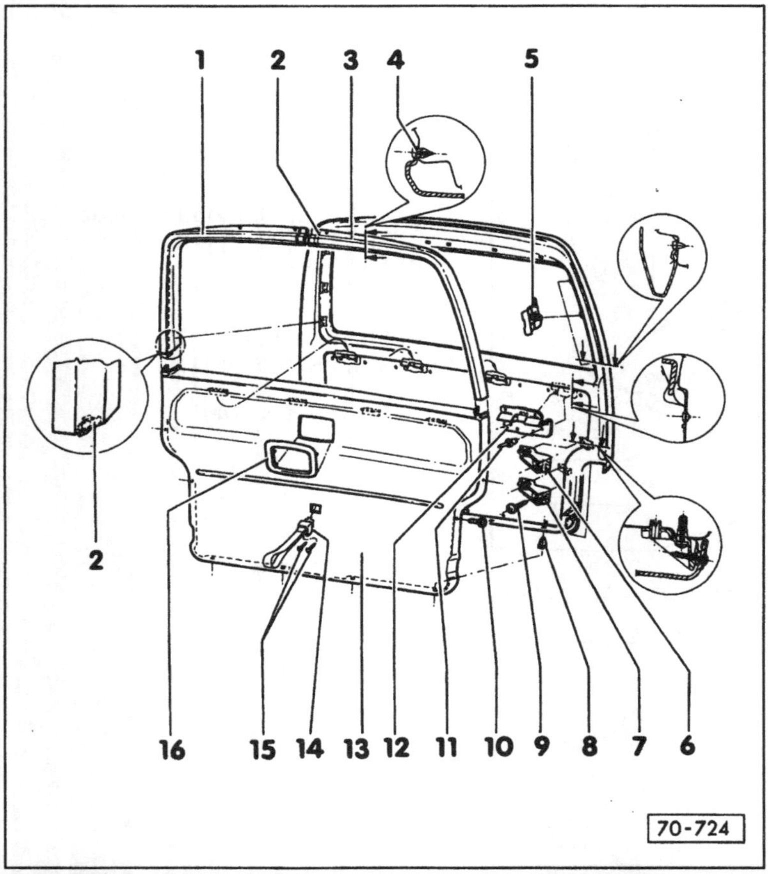

8 - **Expanding nut**
- ♦ Remains in tailgate

9 - **Phillips screw**
- ♦ Secures clips -6- and -7-

10 - **Phillips screw**
- ♦ Secures molded trim -13-

11 - **Banjo bolt**

12 - **Clamp**
- ♦ Expand to install the molded trim ⇒ Page Q 70-95

13 - **Molded trim panel**
- ♦ Removing:
- – Press trim -16- up and unhook
- – Remove Phillips screws -15- and unhook strap

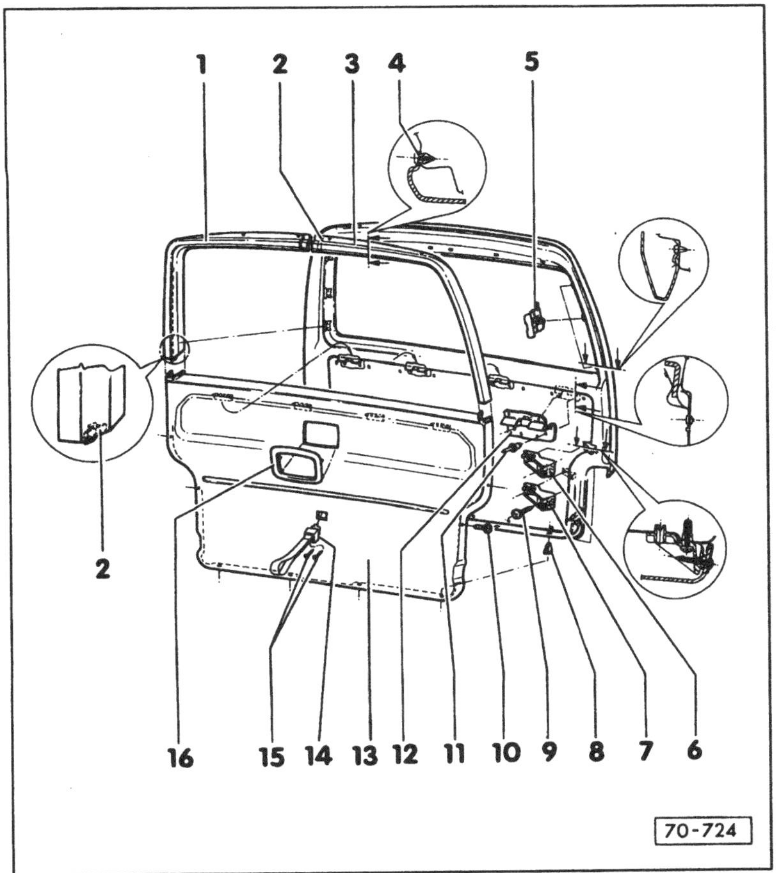

– Pry out expanding pins -4-, unclip left frame trim -3-, right frame trim -1-.

– Remove screws -10-.

– Pull molded trim out slightly from tailgate in lower area and press upward out of clamps -12-.

– Expand clamps -12- ⇒ Page Q 70-95.

– After installing molded trim, pull auxiliary wedges out of clamp -12-.

14 - Strap
- Removing: Remove Phillips screws -15-, unhook strap from mounting hole.
- Retrofitting ⇒ Page Q 70-97.

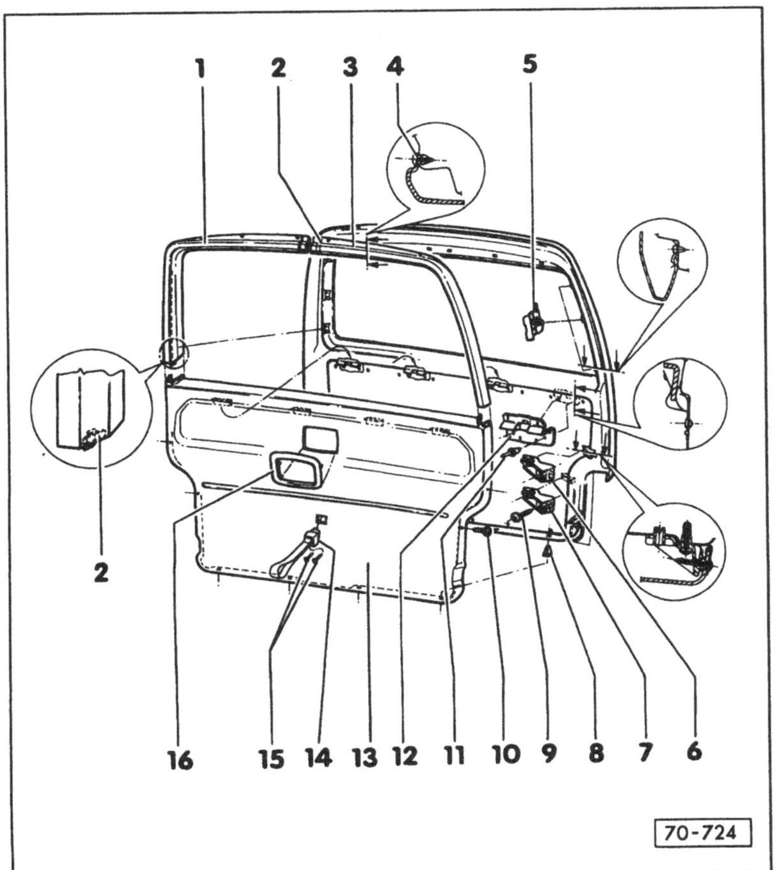

15 - Phillips screw
- Two-piece

16 - Trim
- Press upward, unhook to remove

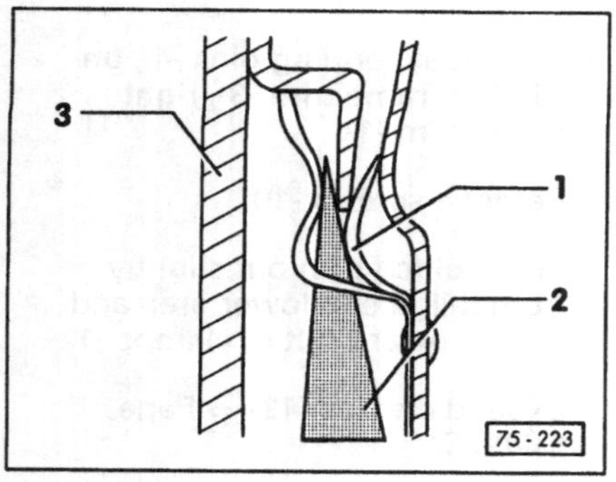

Clamps, expanding

◄

- Expand clamps -1- with shop made auxiliary wedges -2- (wood or plastic).
- Install molded trim -3-.
- Pull out auxiliary wedges -2- or, when molded trim is installed, auxiliary wedges will fall out by themselves.

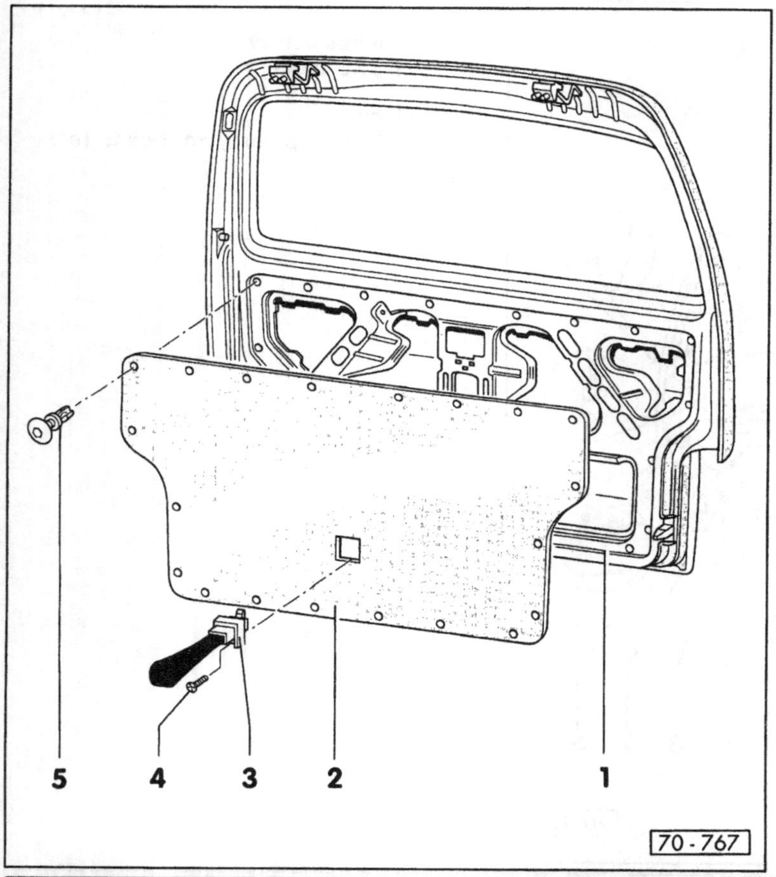

Tailgate fiber board trim, components

1 - Tailgate

2 - Trim panel (fiber board)

3 - Strap

4 - Phillips screw
- ♦ Retrofitting ⇒ Page Q 70-97

5 - Clip
- ♦ Two parts
- ♦ Removing: remove inner part with Allen wrench
- ♦ Installing: Press outer and inner parts into hole simultaneously

Tailgate strap, retrofitting

> ***CAUTION!***
> ***Part numbers are for reference only. Always check with your Parts Department for latest Information.***

Parts required

Vehicle with trim panel

Strap - Part No.701 829 621 01 C

Double Clip - Part No. 701 829 623

2 fillister head screws (B 4.8 x 22) N 012 971 3

Vehicle with molded trim

Strap Part No.:

703 829 621 01C

703 829 621 3KX

703 829 621 9UW

703 829 621 9XK

Double clip - Part No. 701 829 623

2 fillister head screws (B 4.8 x 38), Part No. N 013 938 3

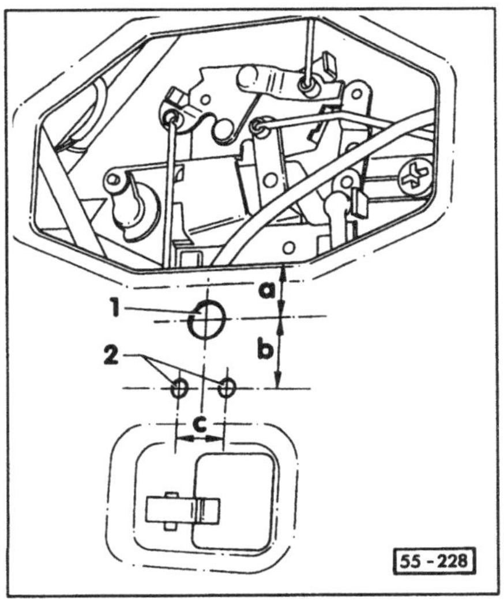

◄ – Mark holes -1- and -2-, drill:

a = 22 mm (0.86 in.)

b= 29 mm (1.14 in.)

c = 21 mm (0.82 in.)

1 = 16 mm dia. (0.62 in.)

2 = 8.1 mm dia. (0.31 In.)

– Deburr holes and apply corrosion protection as necessary.

Note:

Following method is suitable for paint damage in non-visible areas: coat twice (wet on wet) with glass/paint primer D 009 200 01 or equivalent, drying time approx. 10 minutes.

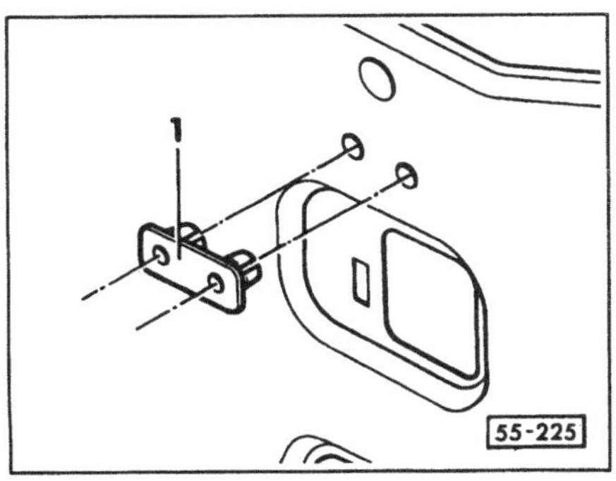

◄ – Clip double clip -1- in two bottom holes.

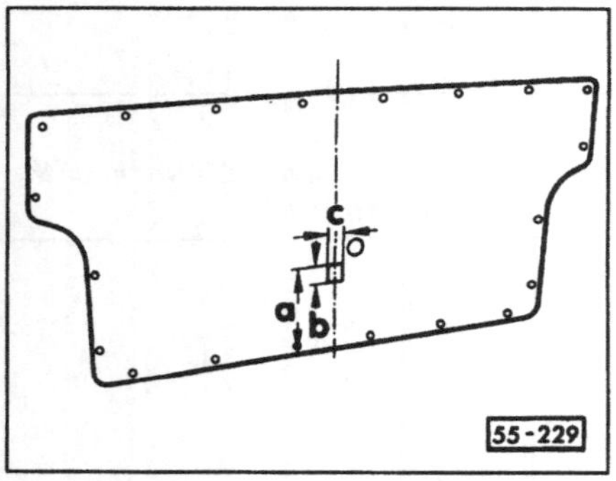

Trim panel (single)

- Draw cut out on trim according to these dimensions:

a = 220 mm (8.66 in.)

b = 42 mm (1.65 in.)

c = 35 mm (1.37 in.)

Drill through in the corner points with drill (8.0 mm dia.), cut out (e.g. with body saw VAG 1523 or equivalent).

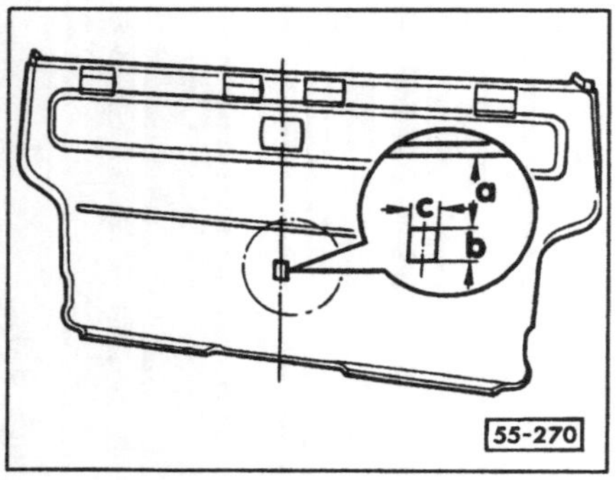

Molded trim panel

- Draw cut out on the back side of trim according to these dimensions:

a = 90 mm (3.54 in.)

b = 42 mm (1.65 in.)

c = 30 mm (1.18 in.)

Drill through in corner points with drill (8.0 mm dia.), cut out (e.g. with body saw VAG 1523 or equivalent)

- Mount trim panel or molded trim panel on tailgate.

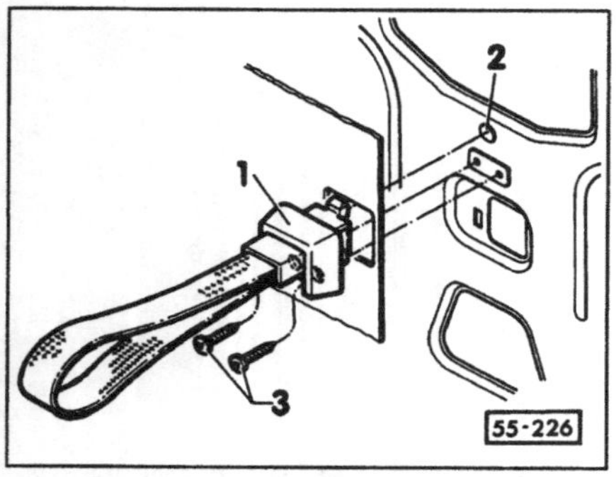

- Hook strap -1- with locking tab in upper hole -2-, fasten with fillister screws -2-.

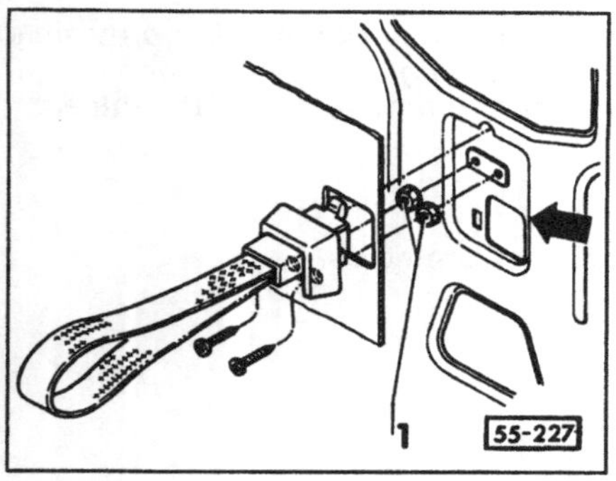

Note:

On tailgates with large inner sheet metal recesses -arrow-, two additional spacers, 5.0 mm (0.19 in.) height (e.g. nuts M6 -1-) are placed underneath for height adjustment

Rear double door trim (left), components

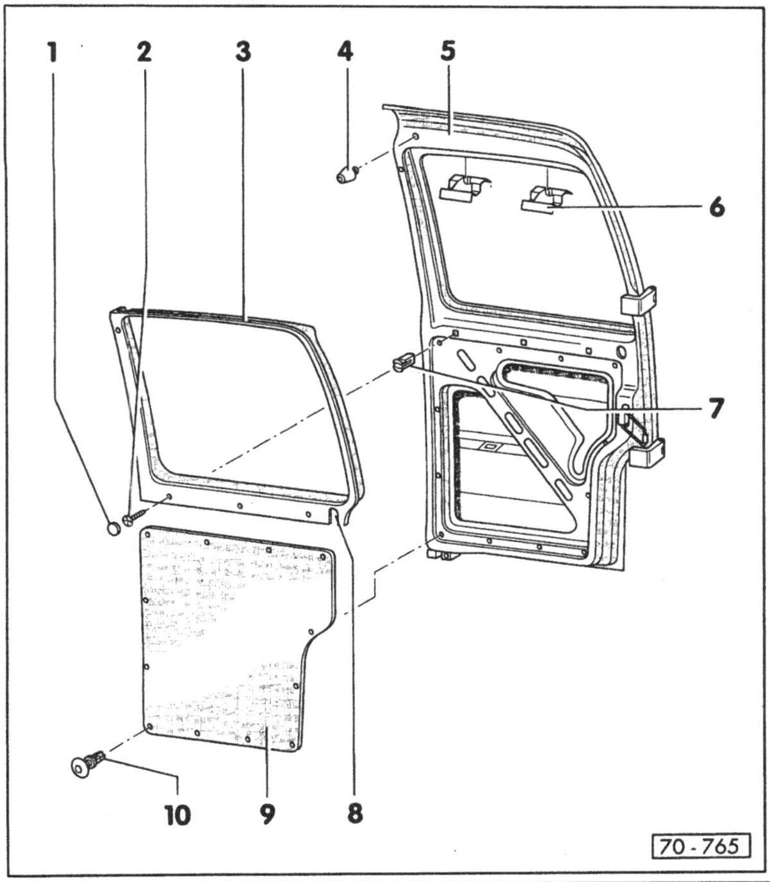

1 - Cap

2 - Phillips screw

3 - Window frame trim

4 - Door stop

5 - Rear double door (left)

6 - Spring clip

7 - Expanding nut

8 - Cutout for electrical connectors

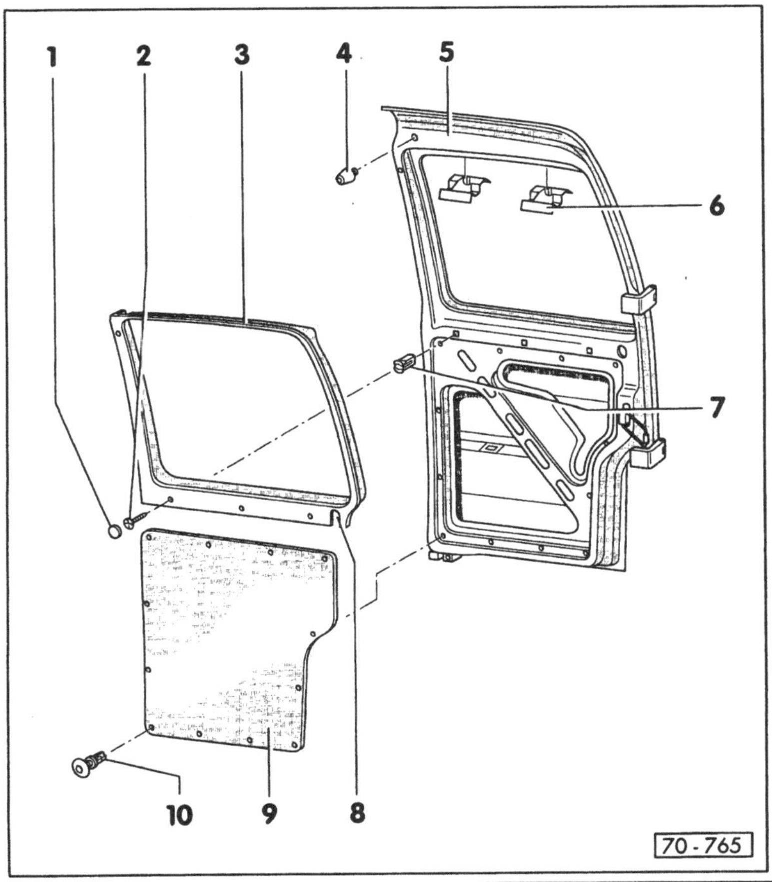

9 - Trim panel

10 - Clip
- Two-piece
- Removing: Remove inner part with Allen wrench
- Installing: Press outer and inner parts in holes on trim panel -9- simultaneously

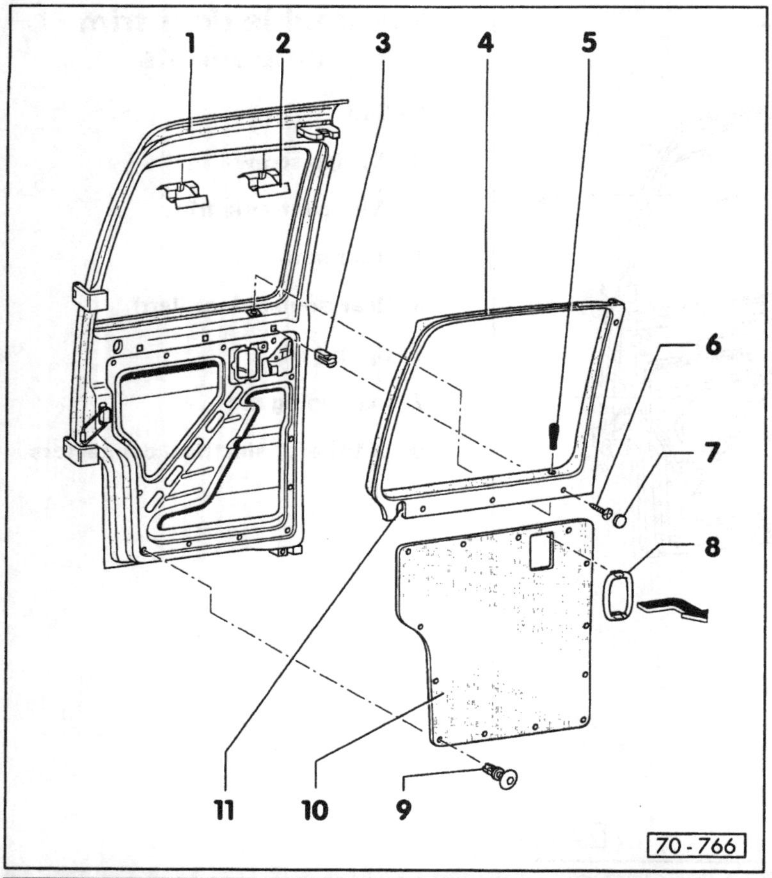

Rear double door trim (right), components

1 - **Rear double door (right)**

2 - **Spring clip**

3 - **Expanding nut**

4 - **Window frame trim**

5 - **Lock knob**

6 - **Phillips screw**

7 - **Cover**

8 - **Trim**
- ♦ Removing: slide to right and remove -arrow-

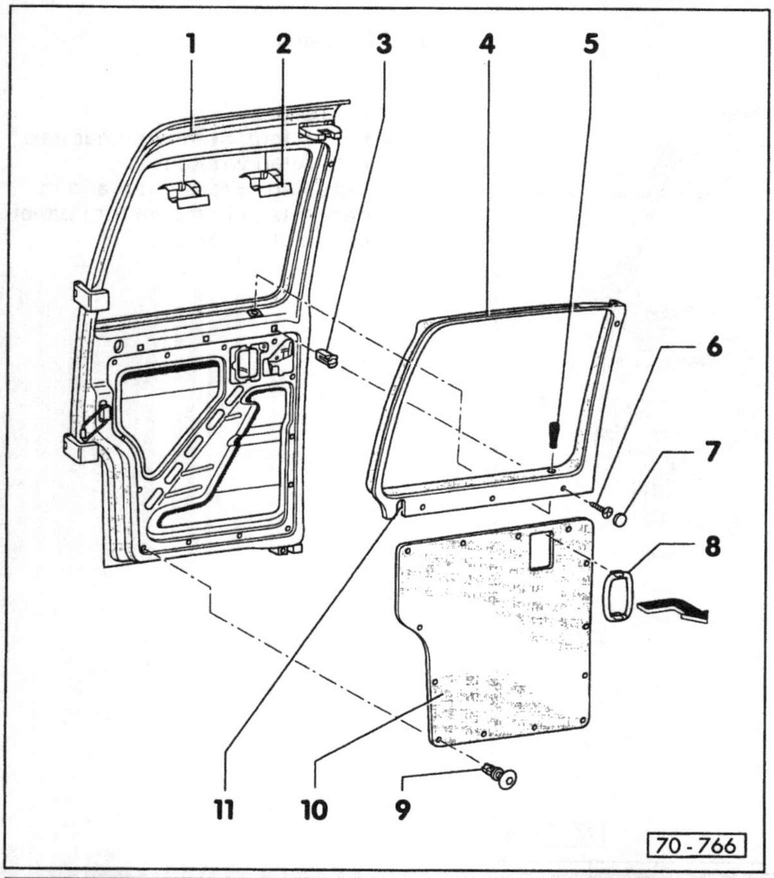

9 - **Clip**
- ♦ Two-piece
- ♦ Removing: Remove inner part with Allen wrench
- ♦ Installing: Press outer and inner parts in holes on trim panel -10- simultaneously

10 - **Trim panel**

11 - **Cutout for electrical connectors**

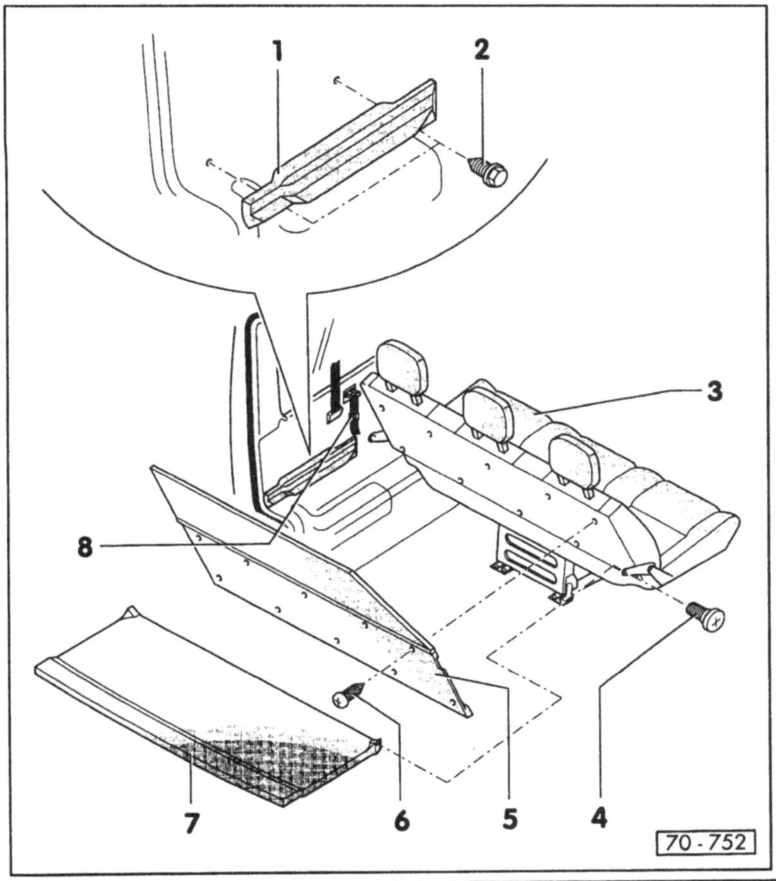

Storage area, removing

1 - Base
- Removing: Loosen hex screws -2-, if necessary, disconnect electrical connector for storage area lights

2 - Hex screw
- 18 Nm (13 ft lb)

3 - Bench seat (three place)

4 - Fillister head tapping screw
- 14 Nm (10 ft lb)

5 - Front part of storage area
- Removing: remove Phillips screws -6-

6 - Phillips screws

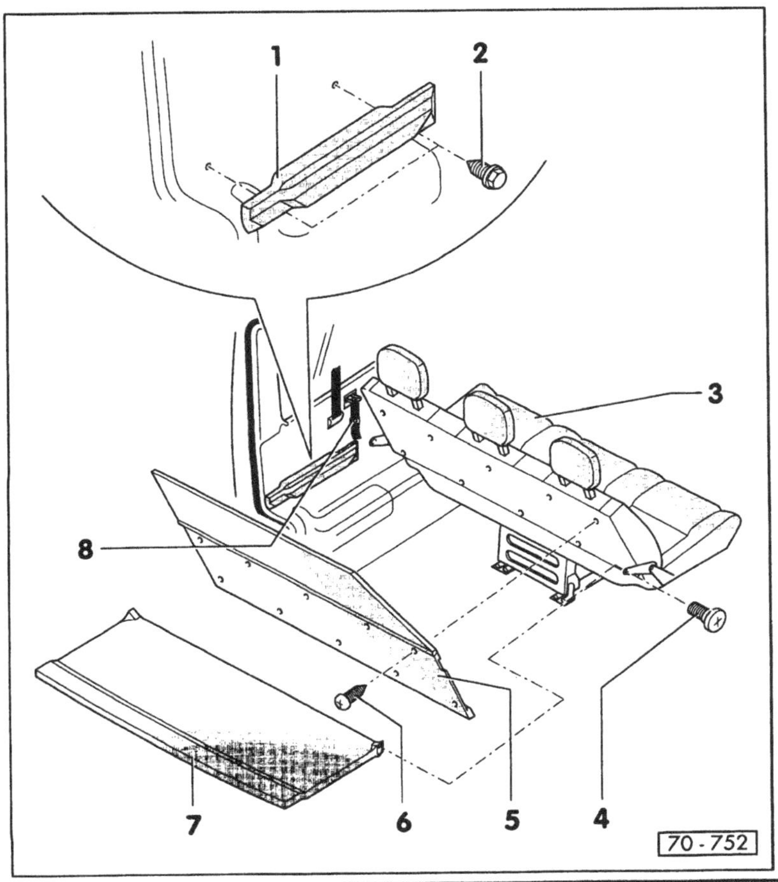

7 - Rear storage area
- Removing:
  Remove full bench seat / bed ⇒ Page Q 72-9. Remove screws -4-.

8 - Stop strap
- Only on left mounting point for back of full bench seat

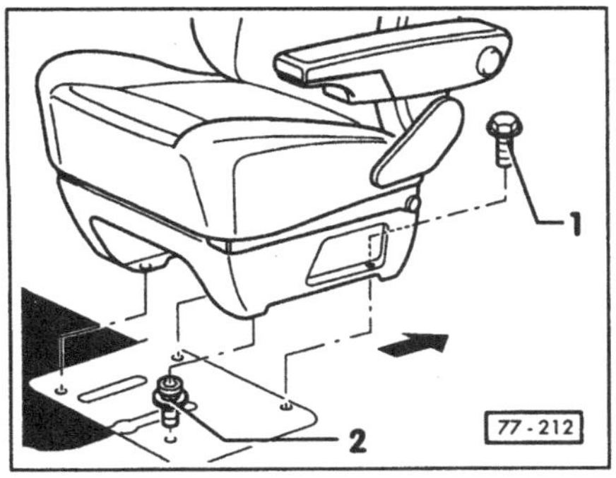

Front seat and frame, removing

- Remove hex bolts -1-.
- Move seat rearward -arrow-, off bolts -2-, remove from vehicle.

  Tightening torques:

  Hex bolts -1- = 44 Nm (32 ft lb)

  Bolts -2- = 30 Nm (22 ft lb)

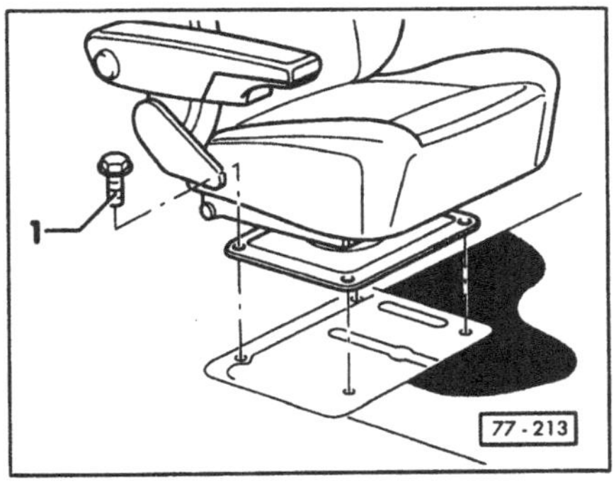

Swivel seat

- Remove hex bolts -1-, remove swivel seat.

  Tightening torque:

  Hex bolts -1- = 44 Nm (32 ft lb)

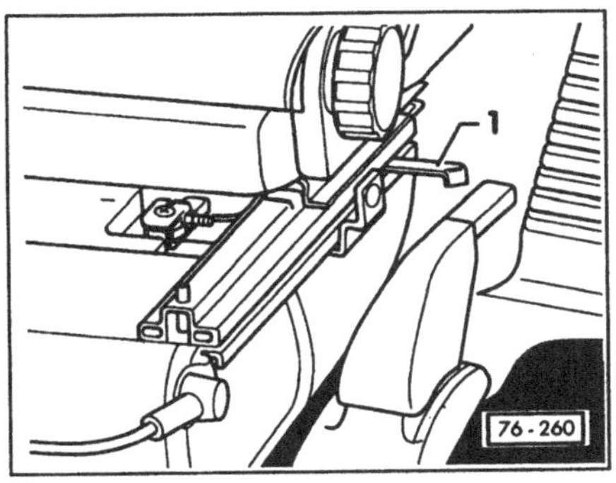

Driver seat with auxiliary battery -A1-

- Move seat forward on seat track.
- Pull locking lever -1- toward inside of vehicle and push seat further forward.
- Disconnect auxiliary battery -A1-.

> ***CAUTION!***
> ***Insulate the positive terminal of the auxiliary battery if the engine is to be started; a danger of a short circuit exists because of charging current.***

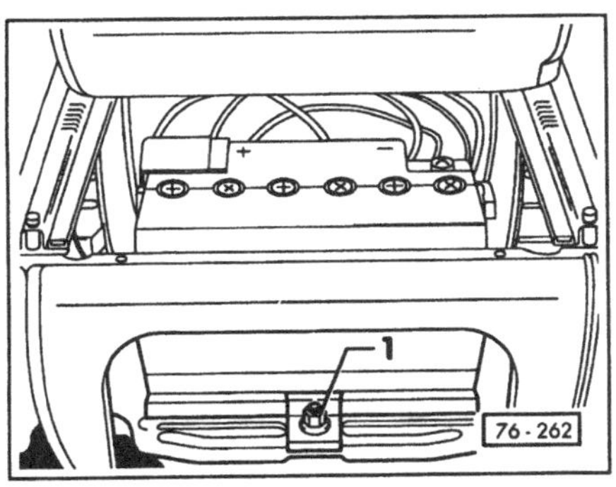

- Unclip and remove rear trim on seat frame.
- Loosen bracket -1- mounting nuts, remove auxiliary battery -A1- from seat frame.

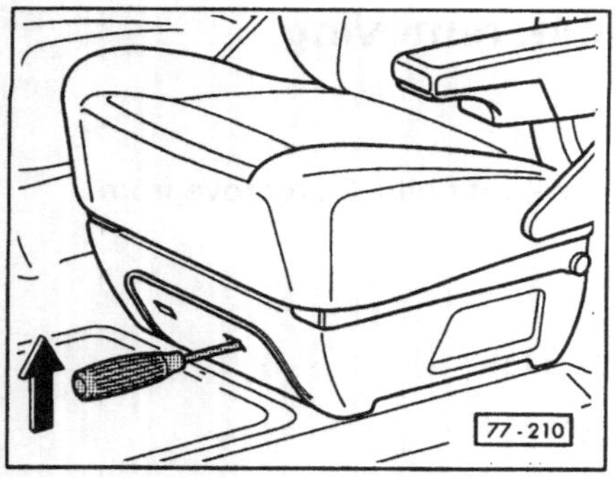

- Open seat frame front trim with screwdriver, as shown.

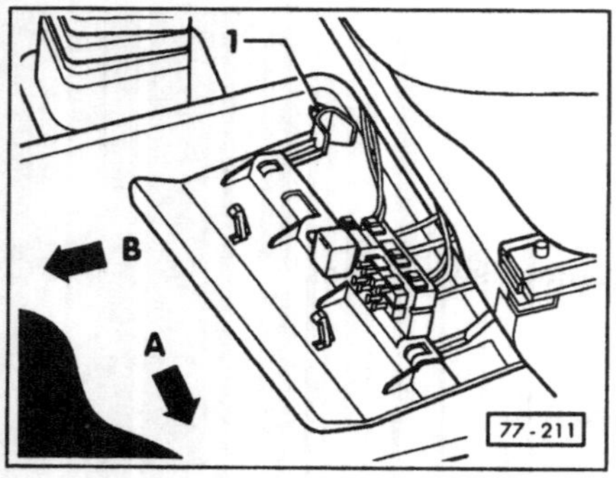

- Push trim to left -arrow A-, unhook joint pins -1- -arrow B-.
- Place trim on vehicle floor.

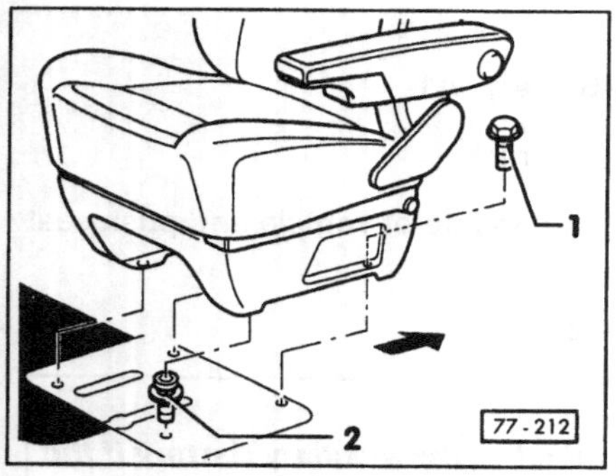

- Remove hex bolts-1-.
- Pull seat off pins -2- direction of arrow, remove from vehicle.

  Tightening torques:

  Hex bolts -1- = 44 Nm (32 ft lb)

  Bolts -2- = 30 Nm (22 ft lb)

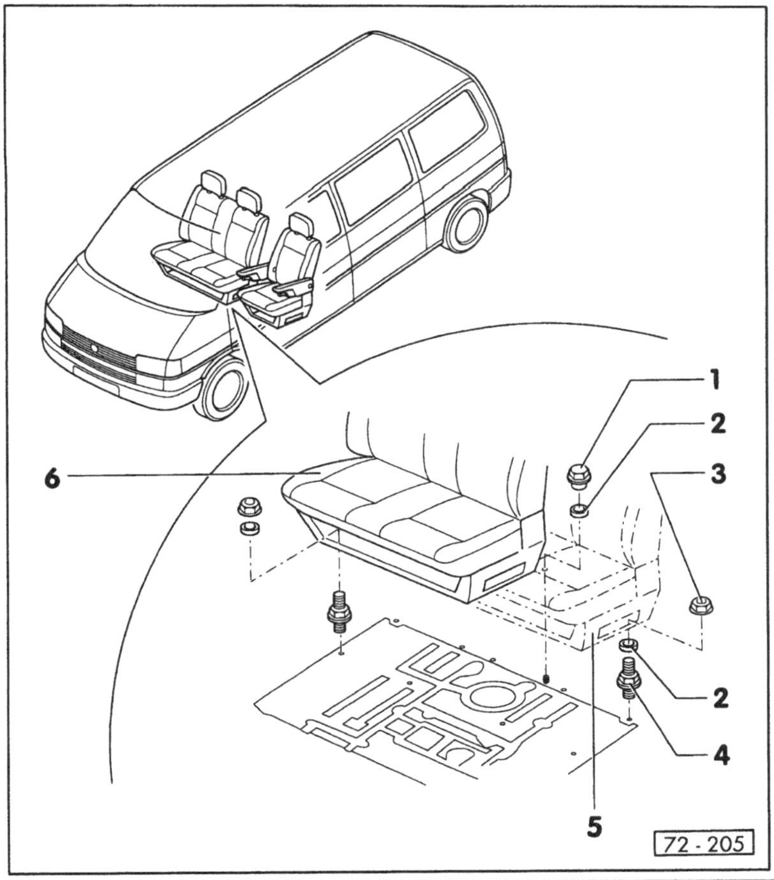

Double passenger seat and seat frame, removing

1 - **Spacer nut**
- ♦ Only right-hand drive vehicles
- ♦ 25 Nm (18 ft lb)
- ♦ Bolting point: mounting bolts for parking brake cable and seat frame

2 - **Spacer washer**
- ♦ Only in vehicles with partition

3 - **Hex nut**
- ♦ 44 Nm (32 ft lb)

4 - **Center bolt w/washer**
- ♦ 44 Nm (32 ft lb)

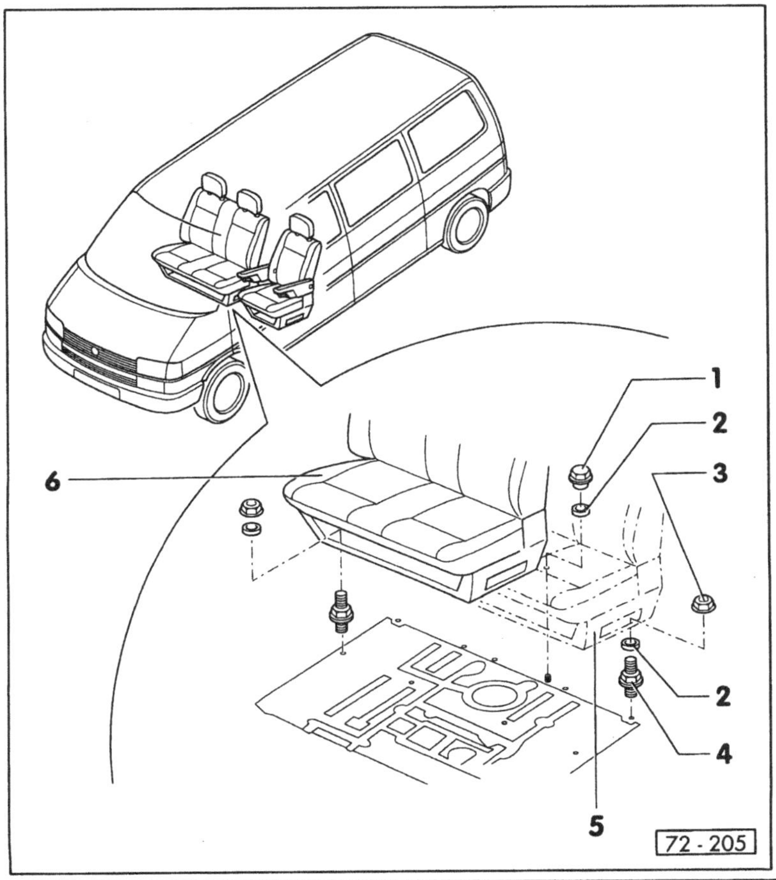

5 - **Installation position for dual passenger seat, right-hand drive vehicles**

6 - **Dual passenger seat**
- ♦ Removing: Remove hex nuts -3-, remove seat from vehicle.

Center seats, components

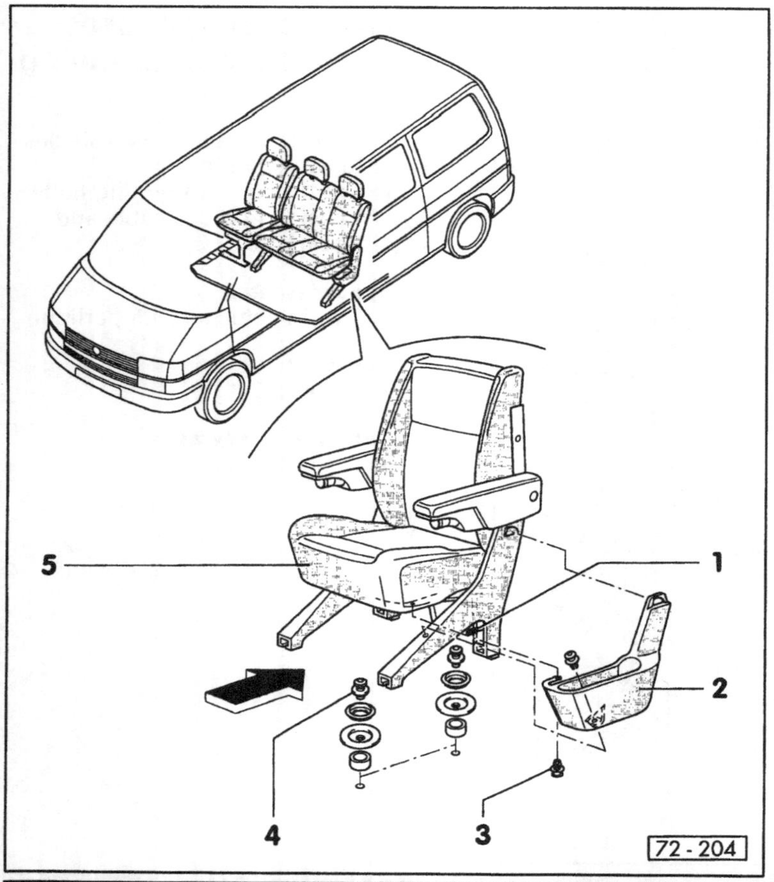

1 - Locking lever
- ♦ Left and/or right on seat frame

2 - Map pocket
- ♦ Removing: remove clips -3-, detach map pocket

3 - Clip
- ♦ Two-piece
- ♦ Removing: Remove inner part with Allen wrench
- ♦ Installing: Press outer and inner parts in hole simultaneously

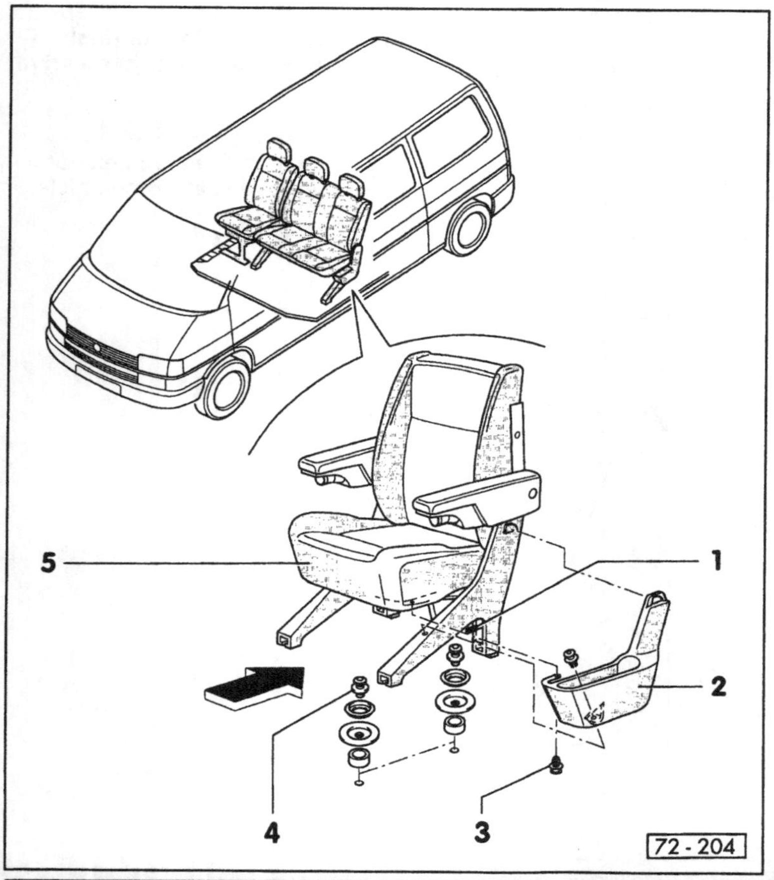

4 - Bolt
- ♦ 30 Nm (22 ft lb)

5 - Single seat or dual bench seat
- ♦ Removing: Pull locking lever -1- upward, pull seat or bench seat off bolts direction of -arrow-.

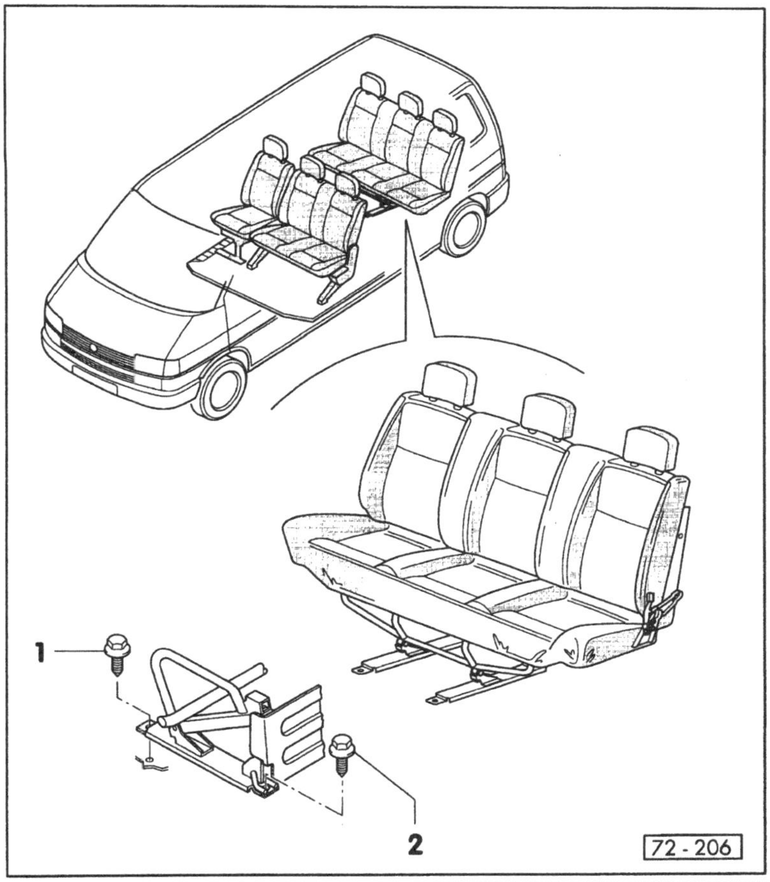

Rear bench seat, removing

- Remove front hex bolts -1-.
- Unlock seat back and fold forward.
- Unlock seat base, tip forward (hex bolts will be visible).
- Remove rear hex bolts -2-.
- Remove seat bench from vehicle.

Tightening torque:

Hex bolts -1-, -2- = 35 Nm (26 ft lb)

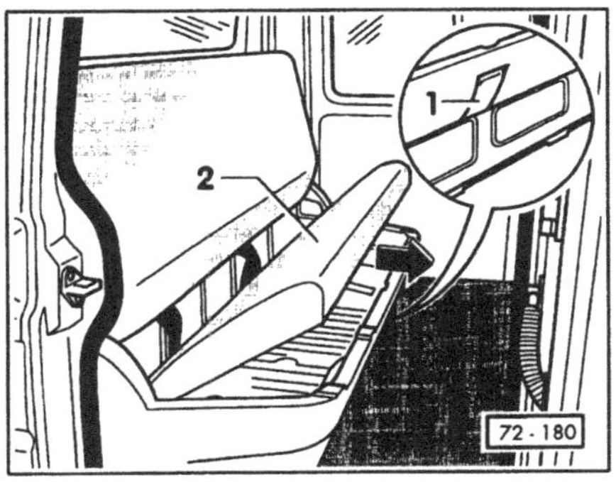

Rear bench seat (double cab), removing

◄ – Unlatch seat cushion with locking lever -1-, lift up.

– Pull seat cushion -2- forward to stop -arrow-.

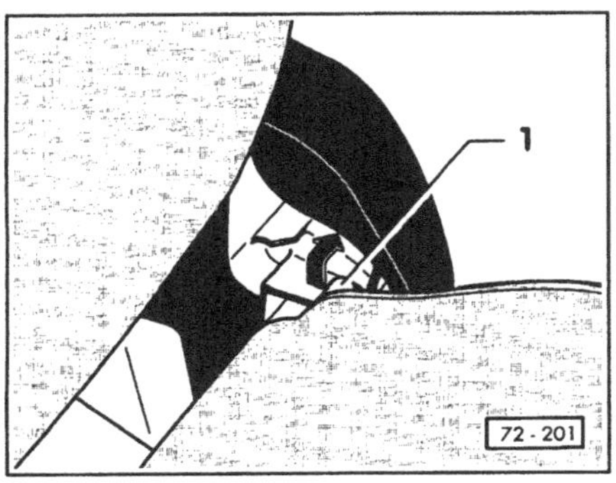

◄ – Lift seat cushion on one side (left) until guide pins -1- can be slid into upper slot -arrow- (in the process, seat cushions will be pushed to outside of vehicle).

– Lift opposite side of seat cushion out of seat bracket slot, pull seat cushion diagonally out of bracket.

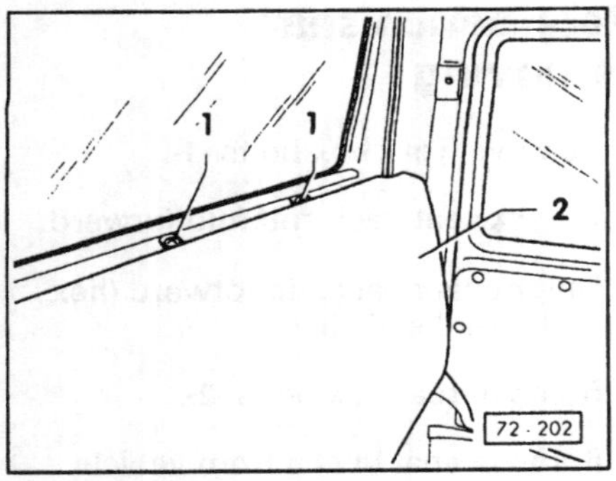

◄ – Remove Phillips screws -1-, back cushion -2- on seat frame and remove from vehicle.

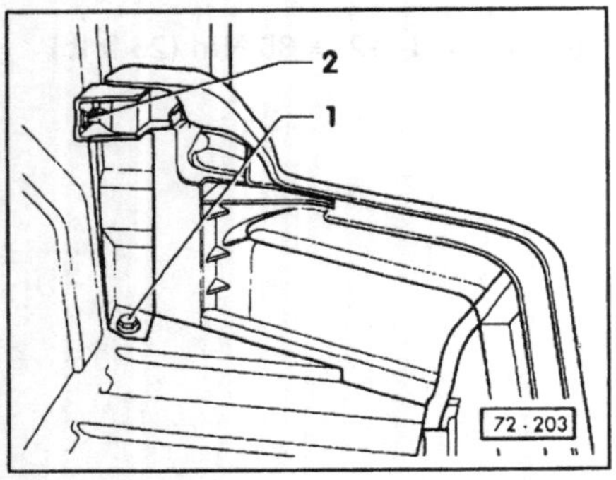

◄ – Remove four hex bolts -1- from cab floor, remove nuts -2- from cab rear panel.

– Remove seat box from vehicle.

Tightening torques:

Hex bolts -1- = 22 Nm (16 ft lb)

Nuts -2- = 10 Nm (7 ft lb)

Single seat upholstery, removing/installing

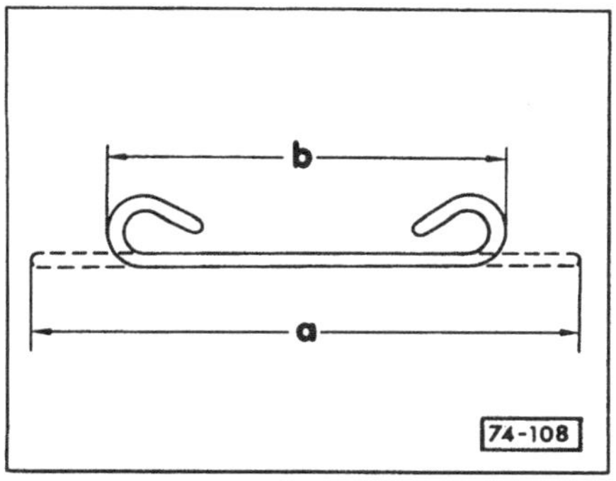

- Make up two upholstery wires (2.0 mm or approx. 5/64 in. dia.) to following dimensions:

  a = 320 mm (approx. 12.5 in.)

  b = 280 mm (approx.11.0 in.)

- Make up one upholstery wire (2.0 mm or approx. 5/64 in. dia.) to following dimensions:

  a = 360 mm (approx.14.0 in.)

  b = 320 mm (approx.12.5 in)

- Make up four upholstery wires (2.8 mm or approx. 1/8 in. dia.) to following dimensions:

  a = 360 mm (approx.14.0 in.)

  b = 320 mm (approx.12.5 in.)

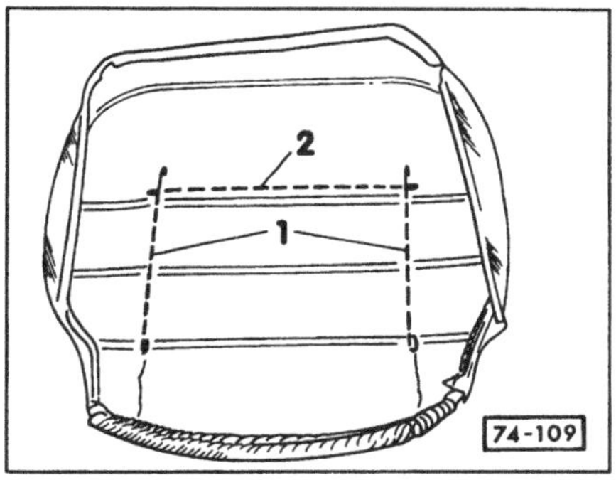

- Insert upholstery wires (2.0 mm or approx. 5/64 in. dia.) in seat cover straps as shown.

  1 = 280 mm (approx.11.0 in.)

  2 = 320 mm (approx.12.5 in)

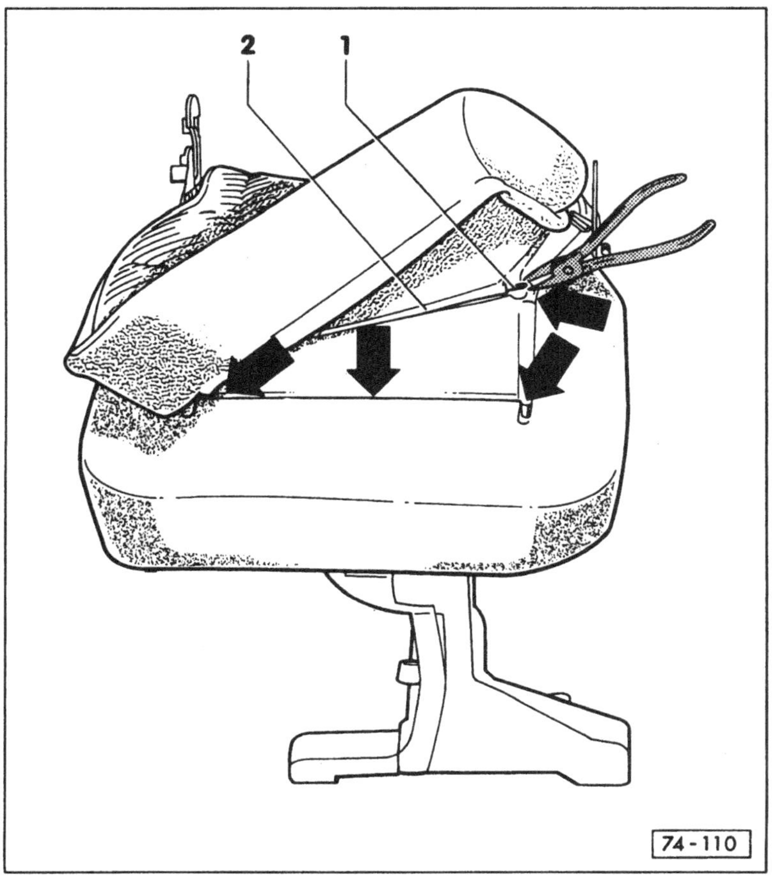

- Place seat cover on cushion.
- Clamp upholstery wires -1- with loop bands -2- on wire in seat cushion -arrows-, use upholstery pliers VAG 1634 or equivalent.

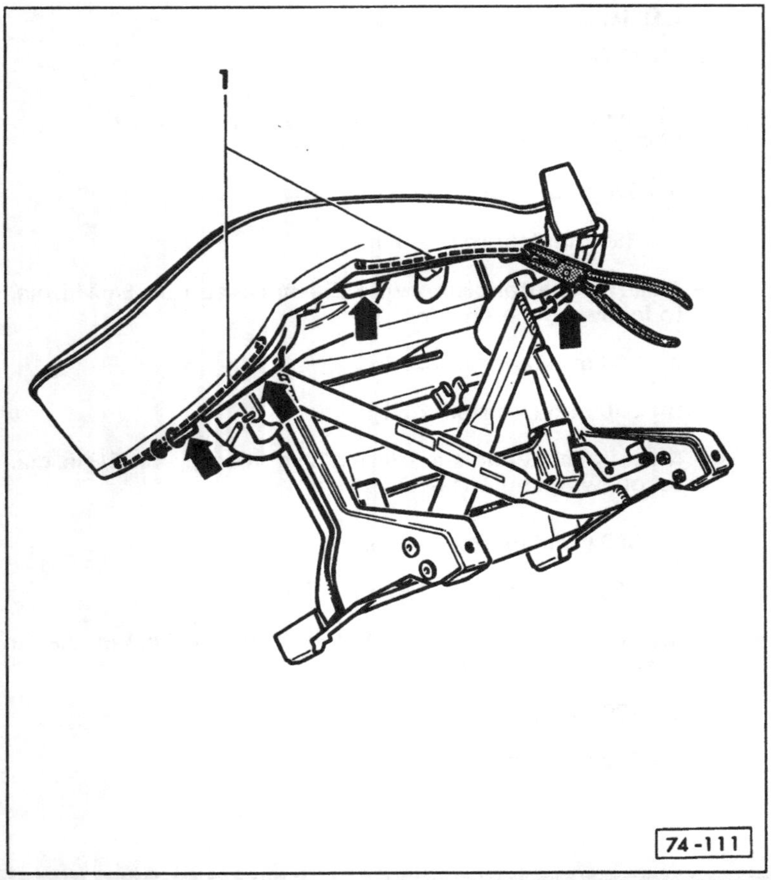

– Insert four 2.8 mm (approx. 1/8 in. dia.) upholstery wires in seat cover -1- edge loops, fasten to retainer -arrows- with upholstery clips, use upholstery pliers VAG 1634 or equivalent.

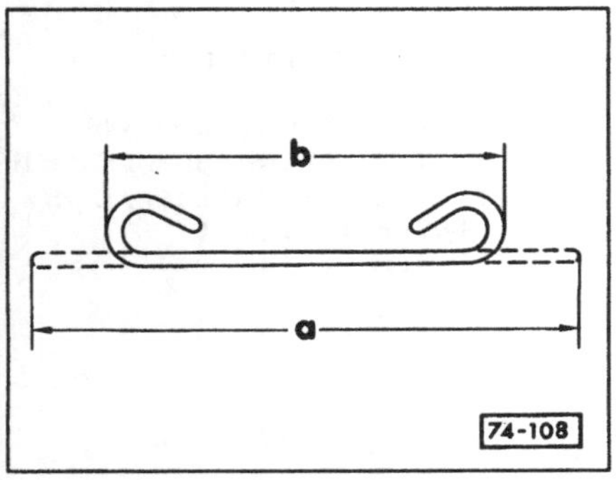

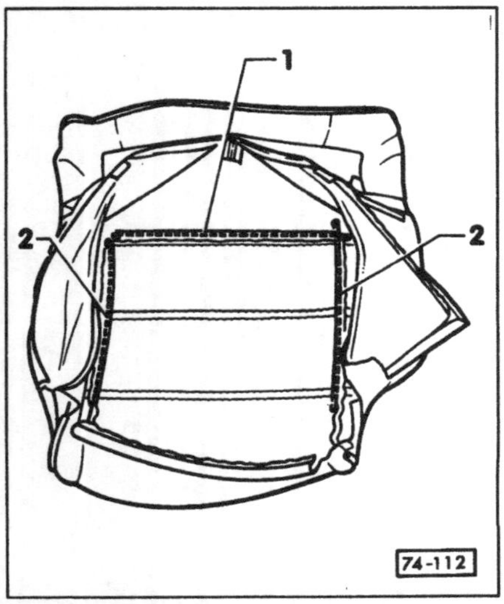

Individual folding seat, installing seat back cover

◄ – Make up four upholstery wires (2.0 mm or approx. 5/64 in. dia.) to following dimensions:

a = 360 mm (approx.14.0 in.)

b = 320 mm (approx.12.5 in.)

– Make up one upholstery wire (2.0 mm or approx. 5/64 in. dia.) to following dimensions:

a = 320 mm (approx.12.5 in.)

b = 280 mm (approx.11.0 in.)

◄ – Insert upholstery wires (2.0 mm or approx. 5/64 in. dia.) in seat cover loop straps as shown.

1 = 280 mm (approx.11.0 in.)

2 = 320 mm (approx.12.5 in.)

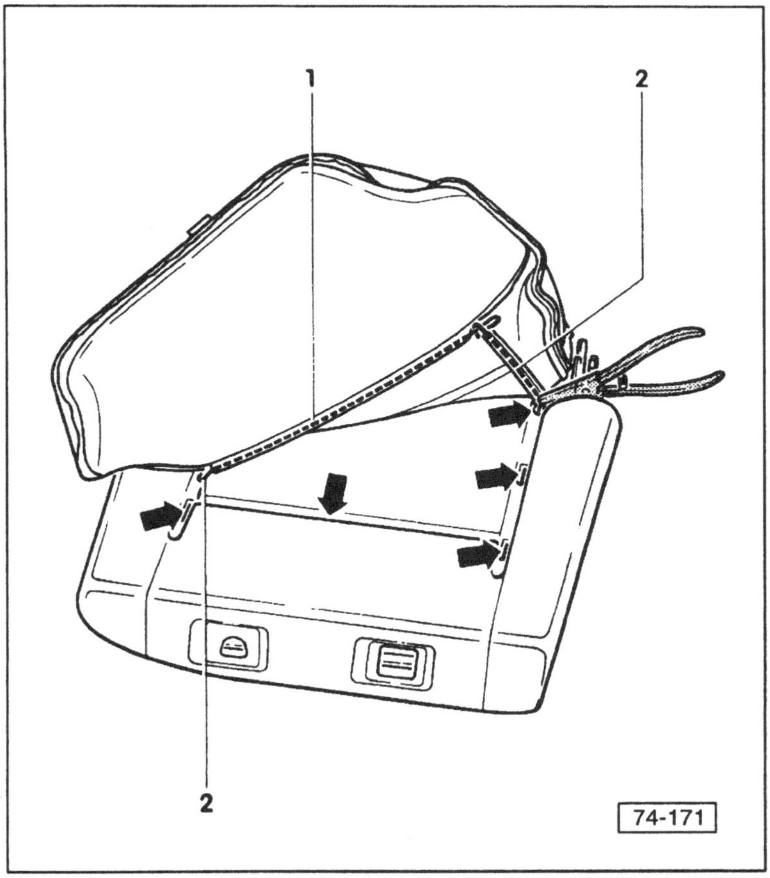

- Place seatback cover on cushion.
- Clamp upholstery wires -1- and -2- on wire in seatback cushion -arrows-, use upholstery pliers VAG 1634 or equivalent.

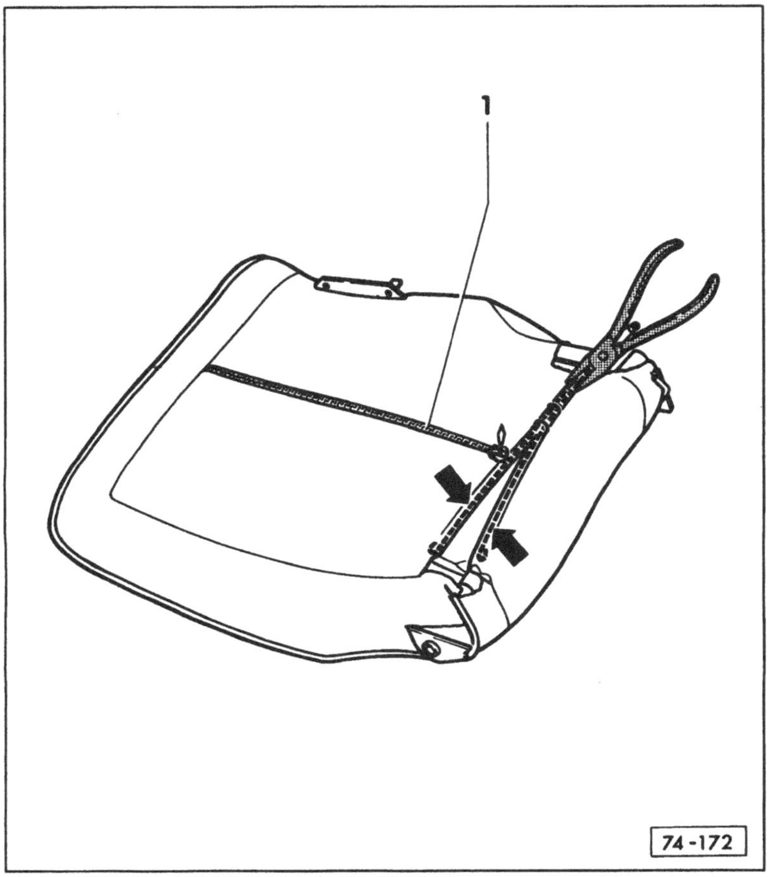

- Pull seatback cover over cushion.
- Close zipper -1-.
- Insert two prepared upholstery wires in edge loops -arrows-, staple together with upholstery clips using upholstery pliers VAG 1634 or equivalent.

Body – CV Camper and MultiVan

Section R

> ***WARNING!***
> ***Before proceeding with any work, please read the warnings and cautions immediately following this index. Also read any warning and caution that accompanies a procedure in the manual.***

Please read these WARNINGS and CAUTIONS before proceeding with maintenance and repair work.

WARNING—

- Some repairs may be beyond your capability. If you lack the skills, tools and equipment, or a suitable workplace for any procedure described in this manual, we suggest you leave such repairs to an authorized Volkswagen dealer service department, or other qualified shop.

- Volkswagen is constantly improving its cars. Sometimes these changes, both in parts and specifications, are made applicable to earlier models. Therefore, before starting any major jobs or repairs to components on which passenger safety may depend, consult your authorized Volkswagen dealer about the Technical Bulletins that may have been issued since the editorial closing of this manual.

- Do not re-use any fasteners that are worn or deformed in normal use. Many fasteners are designed to be used only once and become unreliable and may fail when used a second time. This includes, but is not limited to, nuts, bolts, washers, self–locking nuts or bolts, circlips and cotter pins. Always replace these fasteners with new parts.

- Never work under a lifted car unless it is solidly supported on stands designed for the purpose. Do not support a car on cinder blocks, hollow tiles or other props that may crumble under continuous load. Never work under a vehicle that is supported solely by a jack. Never work under the car while the engine is running.

- If you are going to work under a car on the ground, make sure that the ground is level. Block the wheels to keep the car from rolling. Disconnect the battery negative terminal (ground strap) to prevent others from starting the vehicle while you are under it.

- Never run the engine unless the work area is well ventilated. Carbon monoxide (CO) kills.

- Do not attempt to work on your vehicle if you do not feel well. You increase the danger of injury to yourself and others if you are tired, upset or have taken medicine or any other substance that may impair you or keep you from being fully alert.

- Disconnect the battery negative (–) terminal (ground strap) whenever you work on the fuel system or the electrical system. Do not smoke or work near heaters or other fire hazards. Keep an approved fire extinguisher handy.

- Batteries produce explosive hydrogen gas. Keep sparks, lighted matches and open flame away from the top of the battery. If hydrogen gas escaping from the cap vents is ignited, it will ignite gas trapped in the cells and cause the battery to explode.

- Connect and disconnect battery cables, jumper cables or a battery charger only with the ignition switched off, to prevent sparks. Do not quick–charge the battery (for boost starting) for longer than one minute, and do not allow charging voltage to exceed 16.5 volts. Wait at least one minute before boosting the battery a second time.

- Illuminate your work area adequately but safely. Use a portable safety light for working inside or under the car. Make sure the bulb is enclosed by a wire cage. The hot filament of an accidentally broken bulb can ignite spilled fuel or oil.

- Tie long hair behind your head. Do not wear a necktie, a scarf, loose clothing, or a necklace when you work near machine tools or running engines. If your hair, clothing, or jewelry were to get caught in the machinery, severe injury could result.

- Finger rings, bracelets and other jewelry should be removed so that they cannot cause electrical shorts, get caught in running machinery, or be crushed by heavy parts.

- Catch draining fuel, oil, or brake fluid in suitable containers. Do not use food or beverage containers that might mislead someone into drinking from them. Store flammable fluids away from fire hazards. Wipe up spills at once, but do not store the oily rags, which can ignite and burn spontaneously.

- Always observe good workshop practices. Wear goggles when you operate machine tools or work with battery acid. Gloves or other protective clothing should be worn whenever the job requires working with harmful substances.

- The Air-Conditioning (A/C) system is filled with a chemical refrigerant, which is hazardous. The A/C system should be serviced only by trained technicians using approved refrigerant recovery/recycling equipment, trained in related safety precautions, and familiar with regulations governing the discharging and disposal of automotive chemical refrigerants.

- Do not expose any part of the A/C system to high temperatures such as an open flame. Excessive heat will increase system pressure and may cause the system to burst.

- Some aerosol tire inflators are highly flammable. Be extremely cautious when repairing a tire that may have been inflated using an aerosol tire inflator. Keep sparks, open flame or other sources of ignition away from the tire repair area. Inflate and deflate the tire at least four times before breaking the bead from the rim. Completely remove the tire from the rim before attempting any repair.

- Friction materials such as brake or clutch discs may contain asbestos fibers. Do not create dust by grinding, sanding, or by cleaning with compressed air. Avoid breathing asbestos fibers and asbestos dust. Breathing asbestos can cause serious diseases such as asbestosis or cancer, and may result in death.

- Greases, lubricants and other automotive chemicals contain toxic substances, many of which are absorbed directly through the skin. Read manufacturer's instructions and warnings carefully. Use hand and eye protection. Avoid direct skin contact.

- An airbag is operated by an explosive device. Handled improperly or without adequate safeguards, it can be accidently activated and cause serious personal injury. To prevent personal injury or airbag system failure, **only trained Volkswagen dealer service technicians** should test, disassemble or service the airbag system.

WARNINGS and CAUTIONS continued on next page ▶

Please read these WARNINGS and CAUTIONS before proceeding with maintenance and repair work.

WARNING (continued) —

- ALWAYS disconnect the airbag power supply before using a computer memory saver. Failure to disconnect the airbag power supply may result in accidental activation of the airbag.
- The airbag unit should not be removed, nor should any work be performed on the airbag system, without first reviewing the warnings and cautions that accompany the airbag system repair information in this manual.
- Never apply stickers or any other coverings on the steering wheel. Do not let chemical cleaners, oil or grease come into contact with the vinyl covering of the airbag unit.
- Never open or otherwise attempt to repair airbag system parts. Always use new parts. Never leave airbag parts or the partially disassembled airbag system unattended.
- Never use a test light or volt meter to conduct electrical tests on the airbag system. The airbag system must only be tested by trained Volkswagen Service technicians using special airbag tester equipment. The airbag system must only be tested while it is installed in the car.
- Do not expose an airbag unit to temperatures above 194°F (90°C) even for a brief period, while handling it during the repair process. Keep the unit clear of heat sources such as hot plates, soldering irons, heat lamps, welding equipment, or the like.
- Always replace airbag system components that have been mechanically damaged (bubbles, cracks, etc.).
- Always set down a removed airbag unit so that the horn pad is facing upward.
- When driving or riding in an airbag-equipped vehicle, never hold test equipment in your hands or lap while the vehicle is in motion. Objects between you and the airbag can increase the risk of injury in an accident.
- The storage, transportation and disposal of airbag units are subject to the laws for flammable solids.

CAUTION—

- If you lack the skills, tools and equipment, or a suitable workshop for any procedure described in this manual, we suggest you leave such repairs to an authorized Volkswagen dealer or other qualified shop. We especially urge you to consult an authorized Volkswagen dealer before beginning repairs on any car that may still be covered wholly or in part by any of the extensive warranties issued by Volkswagen of America, Inc.
- Volkswagen is constantly improving its cars and sometimes these changes, both in parts and specifications, are made applicable to earlier models. Therefore, part numbers listed in this manual are for reference only. Always check with your authorized Volkswagen dealer parts department for the latest information.
- Before starting a job, make certain that you have all the necessary tools and parts on hand. Read all the instructions thoroughly, do not attempt shortcuts. Use tools that are appropriate to the work and use only replacement parts meeting Volkswagen specifications. Makeshift tools, parts and procedures will not make good repairs.
- Use pneumatic and electric tools only to loosen threaded parts and fasteners. Never use these tools to tighten fasteners, especially on light alloy parts. Always use a torque wrench to tighten fasteners to the tightening torque specification listed.
- Be mindful of the environment and ecology. Before you drain the crankcase, find out the proper way to dispose of the oil. Do not pour oil onto the ground, down a drain, or into a stream, pond, or lake. Consult local ordinances that govern the disposal of wastes.
- When connecting cables or ground straps, all connections must be cleaned to bare metal. Be sure mounting bolts are tight.
- Do not replace fuse links with fuse links of other gauge or with standard wire. Always replace the entire length of the fuse link.
- Before doing any electrical welding on vehicles equipped with Anti-lock Brakes (ABS), disconnect the battery negative terminal (ground strap) and the ABS Control Module connector.
- On cars equipped with ABS, the ABS control unit cannot withstand temperatures from a paint–drying booth or a heat lamp in excess of 203°F (95°C) and should not be subjected to temperatures in excess of 185°F (85°C) for more than two hours.
- On cars equipped with anti-theft radios, make sure you know the correct radio activation code before disconnecting the battery or removing the radio. If the wrong code is entered when power is restored, the radio may lock up and be rendered inoperable, even if the correct code is then entered.
- When boost-charging the battery, first remove the fuses for the Engine Control Module (ECM), the Transmission Control Module (TCM), the ABS Control Module, and the On-Board computer. In cases where one or more of these components is not separately fused, disconnect the Control Module connector(s).
- Do not quick-charge the battery (for boost starting) for longer than one minute, and do not exceed 16.5 volts at the battery with the boosting cables attached. Wait at least one minute before boosting the battery a second time.

◀ **WARNINGS and CAUTIONS continued from previous page**

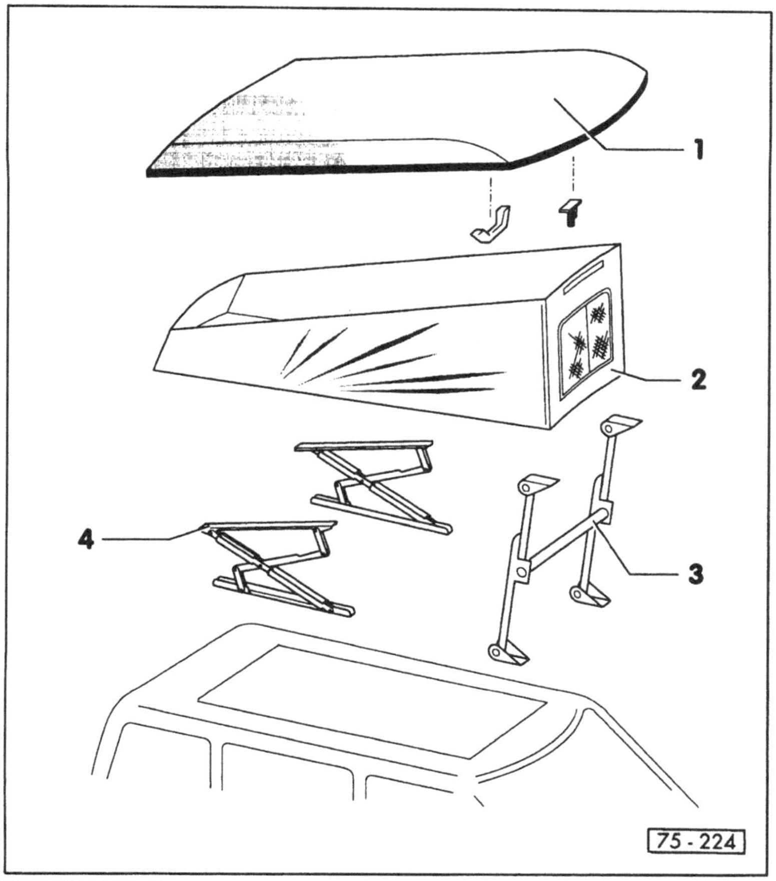

Pop-up (camper) roof, component layout

1 - Pop-up roof
- Removing ⇒ page R 75-2
- Installing ⇒ page R 75-4

2 - Pop-up roof canvas
- Removing and installing ⇒ page R 75-6

3 - Pop-up roof support struts

4 - Pop-up roof opening bracket
- Removing and installing ⇒ page R 75-5

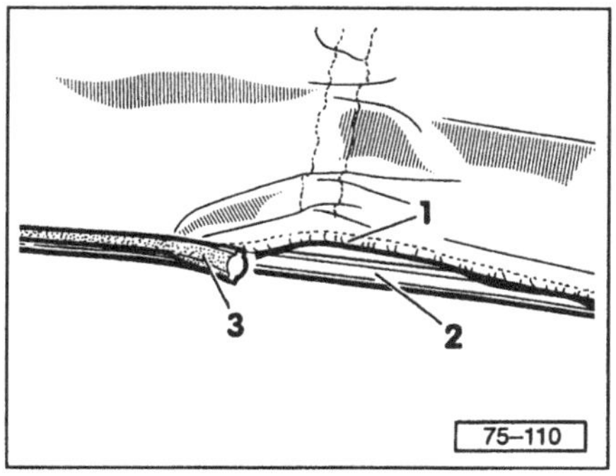

Pop-up (camper) roof, removing

- Open pop-up (camper) roof, remove front bedboard and both cushions.
- Pull piping -3- and canvas cord -1- out of lower rail -2-.

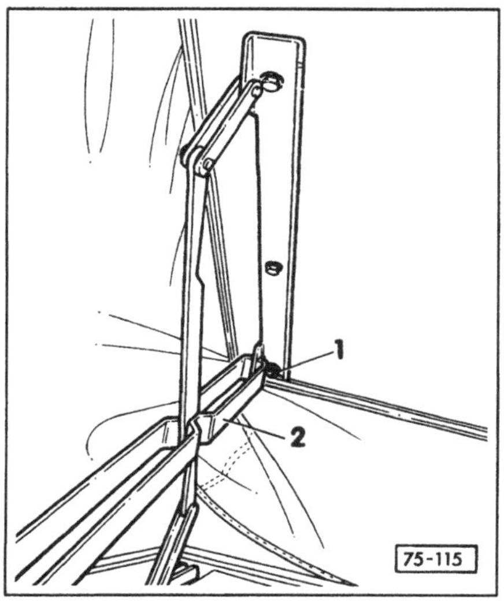

- Remove rear hex bolt -1- from -2- left and right.

- Remove hex bolts -1- from support struts -2- left and right.
- Place support struts -2- on roof frame.
- Lower front of camper roof to approx. 300 mm (11.8 in.) and support.

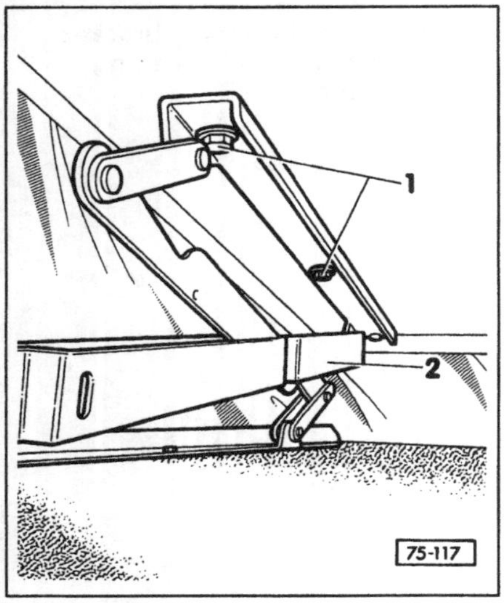

- Remove hex bolts -1- from camper roof opening bracket, left and right.

Notes:

- *Mark position of shims.*
- *Carefully remove camper roof with three technicians.*

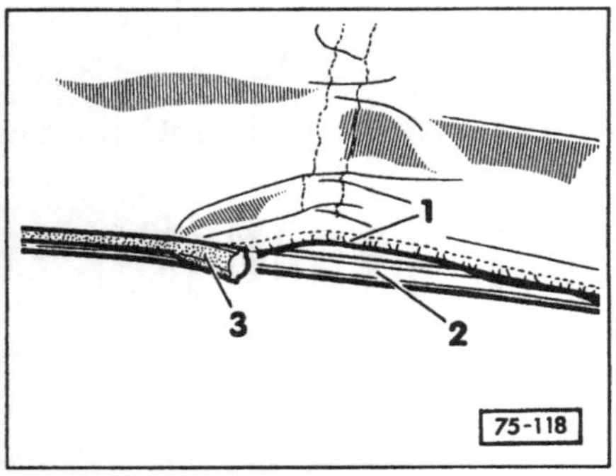

Pop-up (camper) roof, installing

- Install in reverse sequence, note the following:
- Insert canvas cord -1- in rail -2-, use auxiliary tool VAG 1474/5 to press piping -3- into lower rail.

Opening bracket, removing/installing

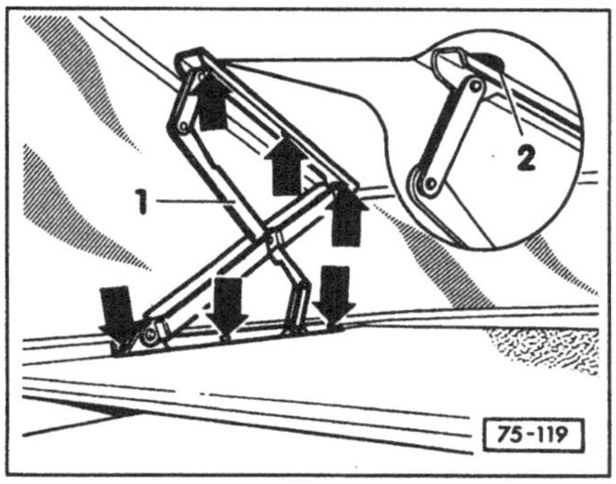

Removing

– Open camper roof, remove front bedboard and both cushions.

– Support camper roof at rear.

◄ – Remove hex bolts -arrows- and remove opening bracket -1-.

Note:
Mark position of shims.

Installing

Install in reverse sequence, note the following:

– Install opening bracket to camper roof first, then to vehicle roof.

Pop-up roof (camper) canvas, removing/installing

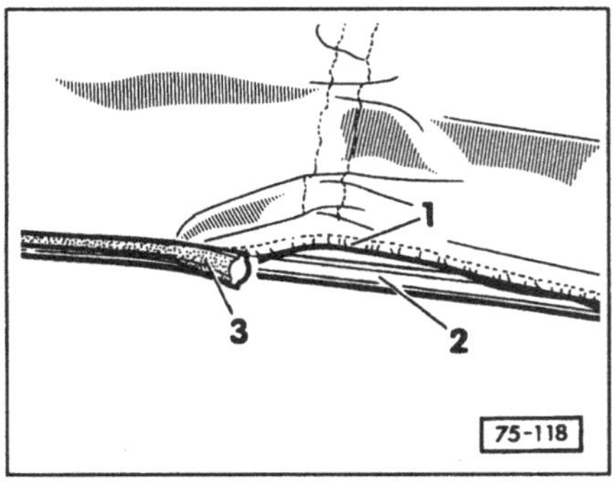

Removing

– Open camper roof, remove front bedboard and both cushions.

◄ – Pull piping -3- and cord of canvas -1- out of lower rail -2-.

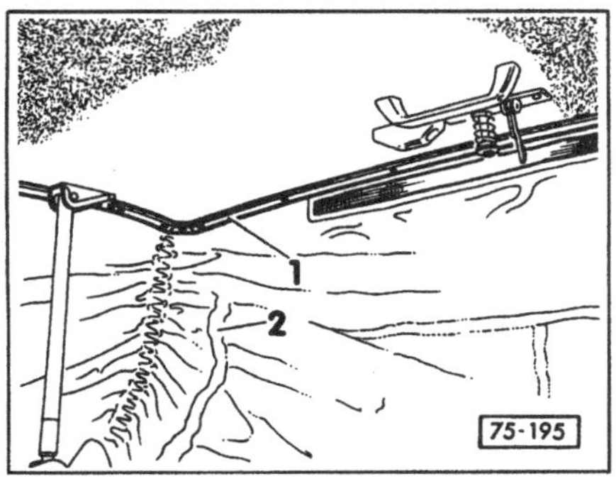

◄ – Remove upper rails -1- from camper roof around perimeter.

– Remove canvas -2- in an upward direction over camper roof.

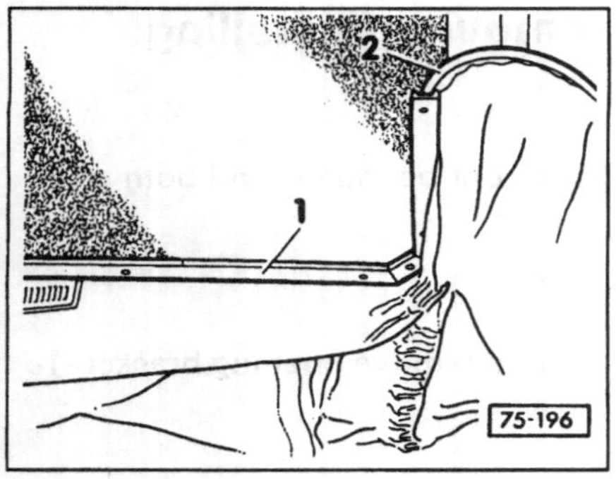

Installing

- Fasten upper rails -1- at front and rear loosely to pop-up roof.
- Pull canvas over camper roof and position.

Note:

Ensure canvas does not become dirty when pulling it over pop-up roof.

- Lay upper canvas cord -2- underneath, loosely screwed–in rails -1-.
- Fasten upper rails -1- at front and rear tightly.
- Fasten upper rails to long sides.
- Lower pop-up roof approx. 100 mm (4 in.), provide support.
- Insert cord canvas -1- into lower rail -2- and press piping -3- into lower rail with help of auxiliary tool VAG 1474/5.

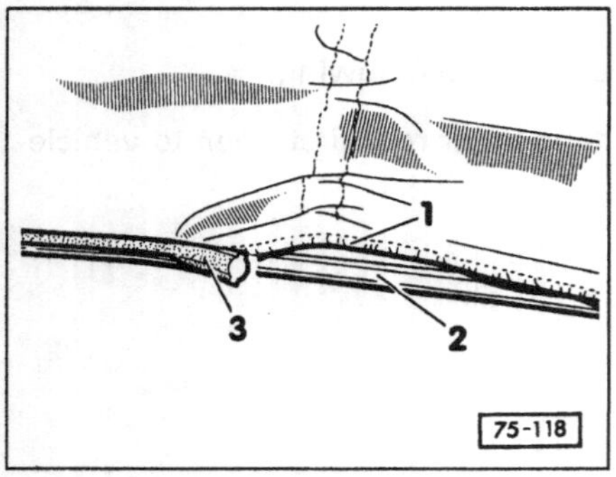

Sunroof, removing

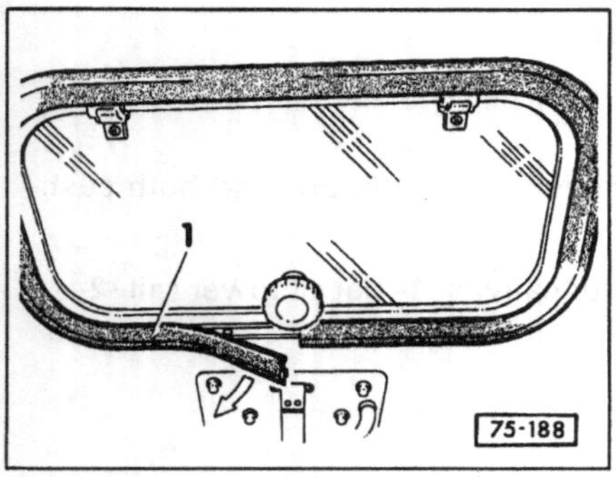

- Remove trim strip -1-.

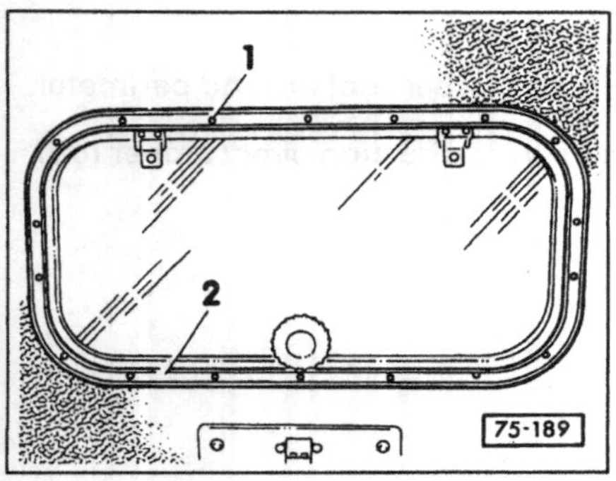

- Remove recessed screws -1-, remove lower frame -2- and press sunroof out, upward direction.

High roof, removing/installing

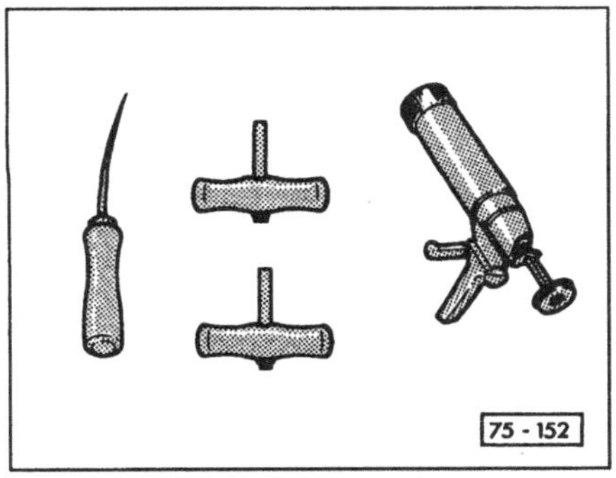

75 - 152

Tools:

- ◄ Awl VAG 1474/2
- For feeding cutting wire
- Handles VAG 1351/1
- Caulking gun VAG 1628 or equivalent

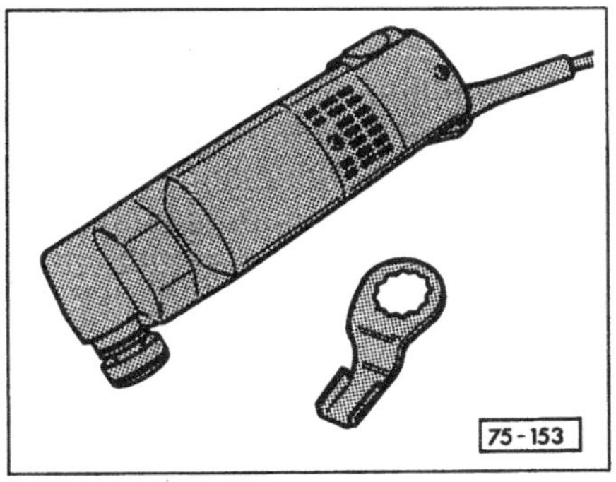

75 - 153

- ◄ Electric cutting tool VAG 1561 or equivalent
- Blade VAG 1561/8 or equivalent. For cutting bonding/sealing material.

> ***CAUTION!***
> ***Part numbers are for reference only. Always check with your Parts Department for latest Information.***

Materials:

- Cutting wire part no. 357 845 955
- Cleaning fluid D 009 401 04

  To clean the bonding surfaces before priming
- Glass/paint primer D 009 200 02

> ***CAUTION!***
> ***Do not prime cut surface material***

- Applicator D 009 500 25 or equivalent
- Bonding/sealing material D 009 100, six cartridges (310 ml)
- Bonding material remover D 002 000 10 or equivalent

Removing

– Remove bedboards (high roof) ⇒ page R 75-21.

– Remove roof storage boxes (high roof) ⇒ page R 77-32.

– Remove reading lamp (high roof) ⇒ page R 76-21.

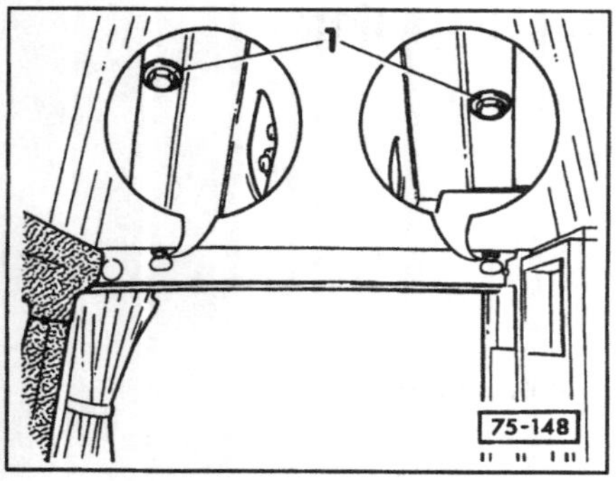

- Remove bolts -1- at left and right rear.

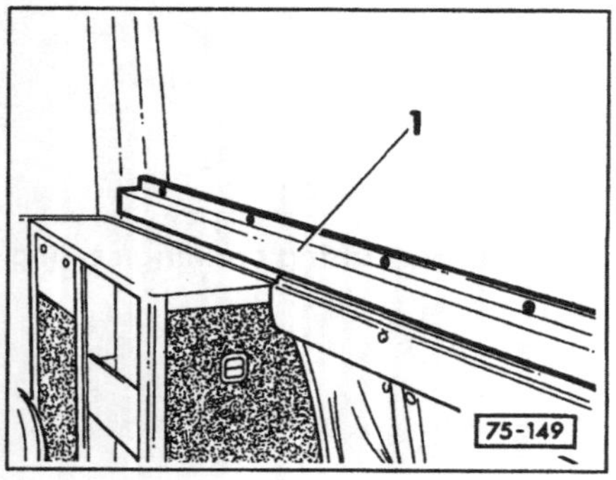

- Remove retainer for guide rail -1- left and right (five recessed screws each side).

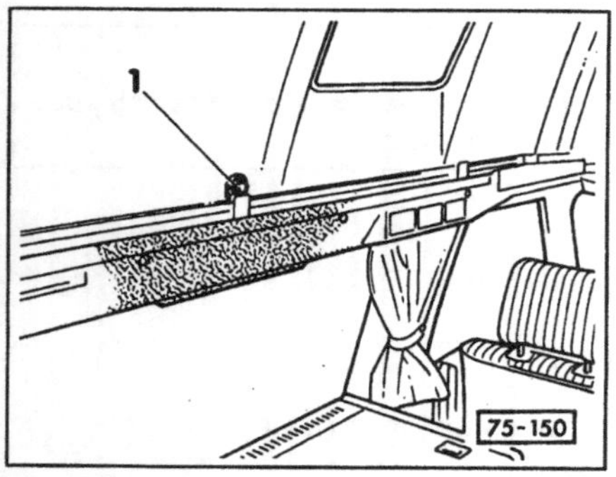

- Remove hex bolts -1- from brackets, left and right.

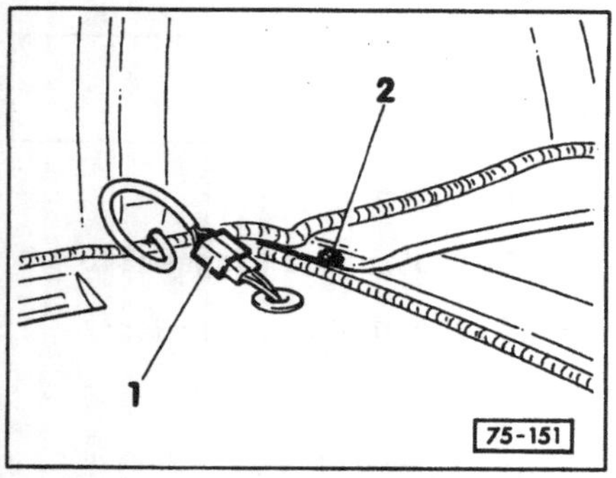

- Remove connector -1- for reading lamp, front roof area.
- Drill out rivet 4.8 mm dia. (0.19 in.) -2-, left and right.

- Remove seal -arrows-, around high roof perimeter.

Note:

To prevent paint damage, mask off roof frame outside with fabric adhesive tape.

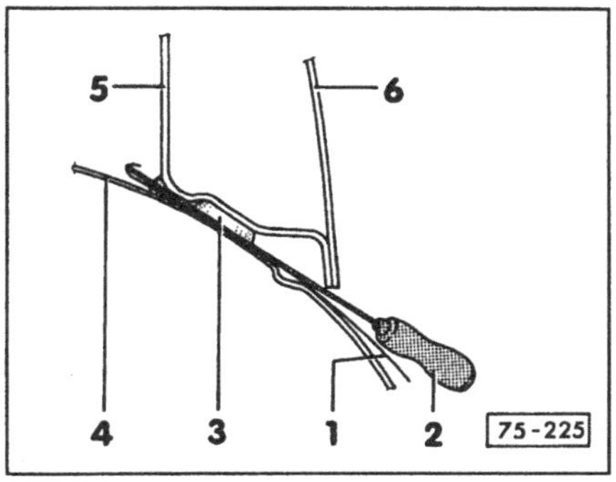

- Cut off approx. 2 meters (6.5 ft) of cutting wire (P/N 357 845 955) -1-, use awl (VAG 1474/2) -2-, guide wire through bonding/sealing material -3- from outside.

  Vehicle roof frame -4-

  High roof – inner shell -5-

  High roof – outer shell -6-

- Secure handle (VAG 1351/1) to each end of wire, cut high roof away from roof frame.

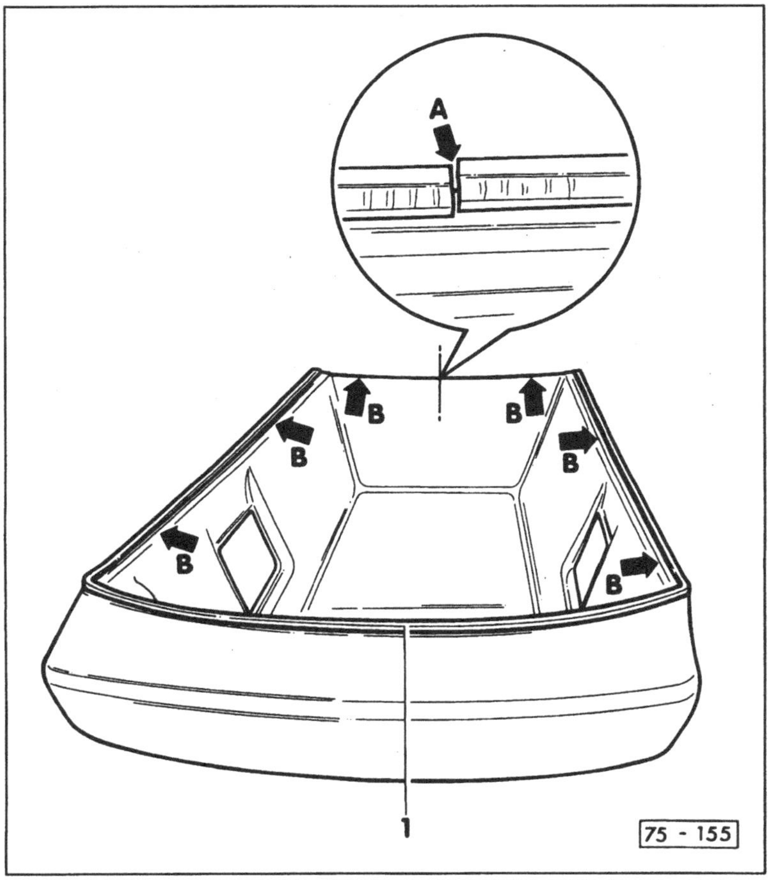

Installing

- Press seal -1- on all around

  Installation position: Seal butt joint is centered, at rear -arrow A-.

- Clean bonding area (only on new high roof) -arrows B- with cleaning fluid D 009 401 04 and prime with glass/paint primer D 009 200 02. Note flash-off time.

- Cut back excess bonding material from roof frame (and on high roof if reusing roof which has been removed) to approx. 1 mm (0.039 in.)

Notes:

- *Completely remove loose excess bonding material and prime corresponding area with glass/paint primer D 009 200 02.*
- *Do not prime cut-back bonding material remaining on roof frame.*

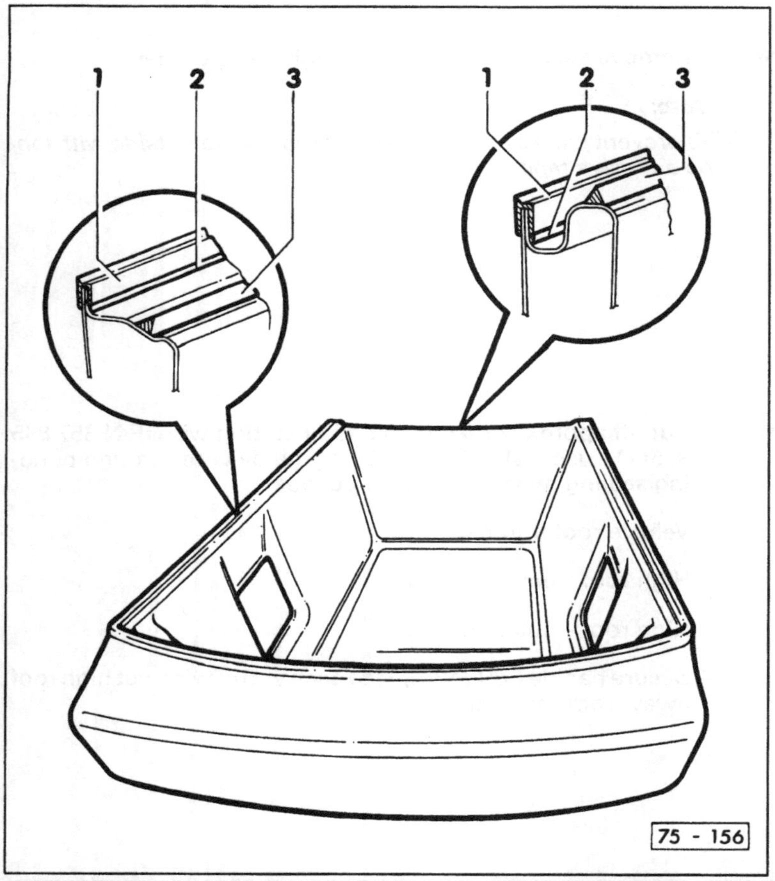

- Bond seal -1- to high roof around perimeter with bonding/sealing agent D 009 100 -2-.

  Bead thickness 5 mm dia (0.20 in.)

- Position spacers (Part No. 165 845 125A) in area of bolts at rear left and right.

- Apply bonding/sealing agent -3- (Part No. D 009100) on both long sides and rear area of high roof.

  Bead thickness:

  Width = 10 mm (0.39 in.)

  Height = 10 mm (0.39 in.)

- Lift high roof on vehicle with four technicians.
- Tighten high roof loosely, first at rear center flap hinge, then center of long side.
- ◄ Align outside high roof to vehicle roof edge -arrows-.
- Tighten bolts at rear on long sides.

- Apply bonding/sealing material D 009 100 at front -arrows- and wipe.

Notes:

- *Select bead thickness according to gap.*
- *To wipe, a cloth soaked with liquid soap is suitable.*

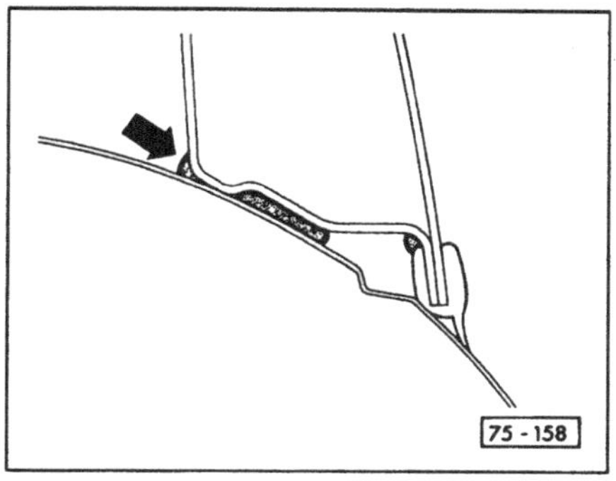

- Apply bonding/sealing agent D 009 100 from inside -arrow- on long sides and rear, then wipe.

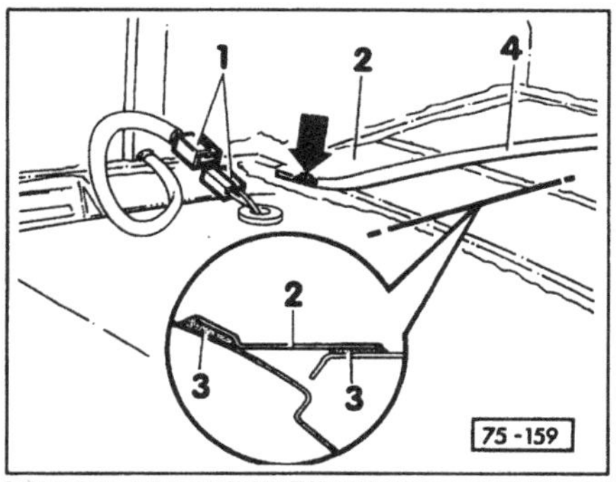

- Reconnect reading lamp connector -1-.
- Secure trim strip (GRP) -2- with bonding/sealing agent -3- (part no. D 009 100).
- Secure metal straps -4- left and right with rivets, 4.8 mm dia. (0.20 in.).

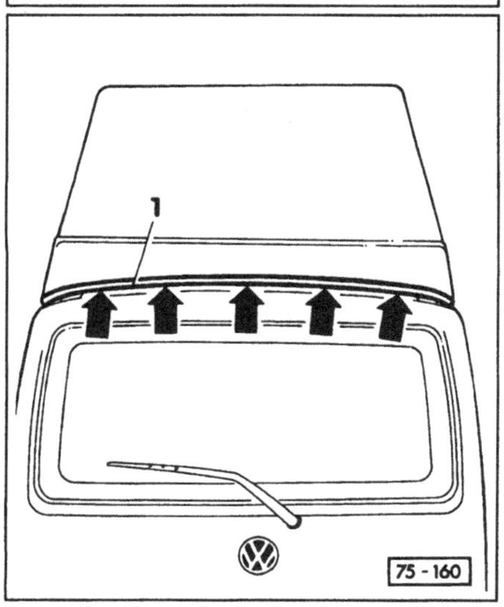

- Apply bonding/sealing agent D 009100 between -1- and vehicle roof at rear -arrows-, then wipe ⇒ Notes, page R 75-17.

Paint damage, repairing

Refer to the EuroVan Body-Collision Repair Manual for application of corrosion protection.

For paint damage in non–visible areas the following repair method is suitable:

> ***CAUTION!***
>
> ***Part numbers are for reference only. Always check with your Parts Department for latest information.***

– Paint twice (wet on wet) with glass/paint primer D 009 200 02 -flash-off time approx. 10 minutes.

Surplus bonding/sealing compound, removing

> ***CAUTION!***
>
> ***Part numbers are for reference only. Always check with your Parts Department for latest information.***

> ***CAUTION!***
>
> ***When using any cleaning agent follow all cautions and warnings listed on the containers.***

The adhesive remover D 002 000 10 or equivalent is recommended for use as a cleaning agent. When using adhesive remover, all safety instructions must be observed.

– First clean painted areas with a dry cloth. Remove the remaining sealant with cleaning agent D 002 000 10.

Cleaning plastic trim:

– Allow bonding/sealing agent to harden (approx. 1 hour) and then pull off.

Bedboards (high roof), removing

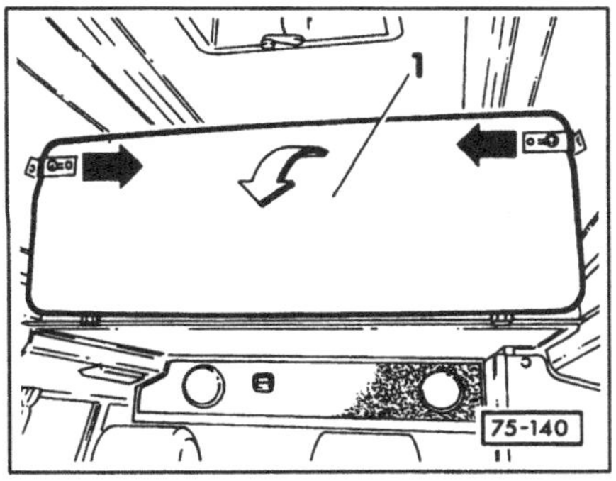

◄ – Unlatch center bedboard -1- and swing downward.

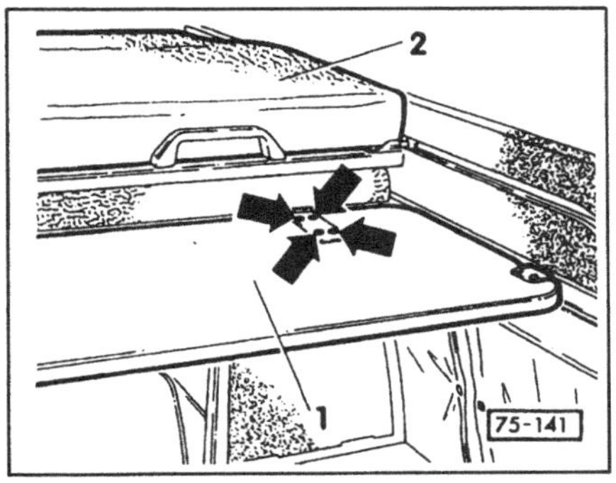

◄ – Remove cushion padding -2-.

– Remove recessed screws -arrows- at left and right.

– Remove center bedboard -1-.

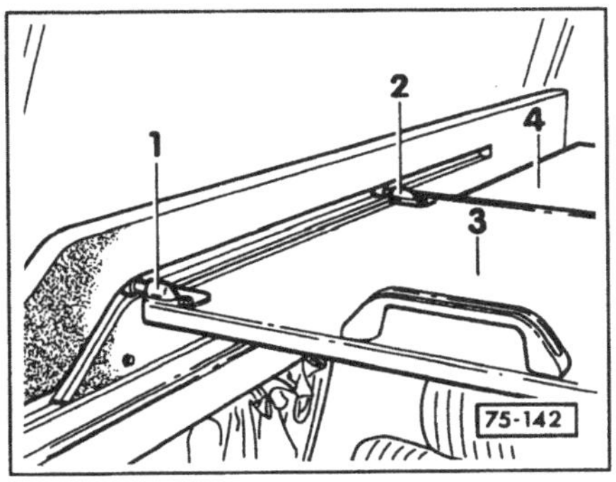

◄ – Remove front bedboard guide rollers -1- and -2-, -3- on left and right side.

– Remove front bedboard -3- and rear padding -4-.

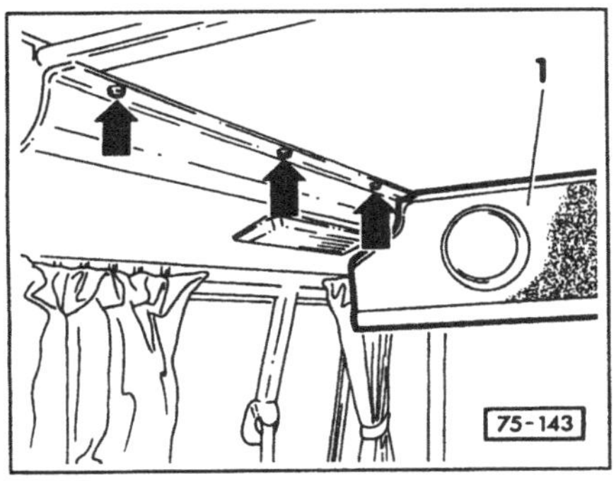

◄ – Remove wall cupboard -1- ⇒ page R 77-17.

– Pry off three caps -arrows- and remove recessed screws.

- Pry out two caps -arrows- and remove recessed screws.

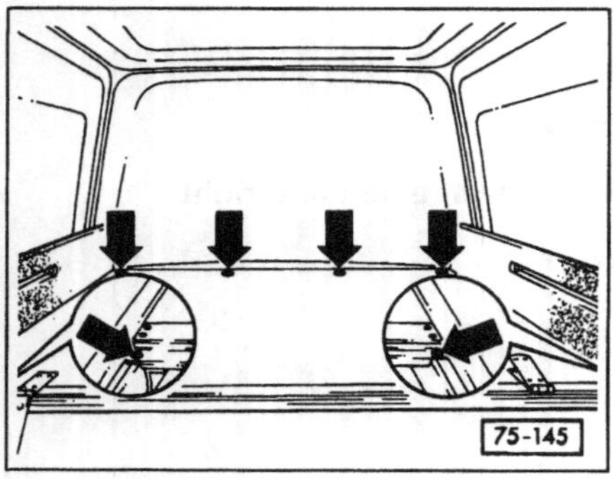

- Drill out rivets 5 mm dia. (0.20 in.) -arrows-, remove rear bedboard.

Roof rails, removing

Note:

Following description relates to removal of left roof rail (right side – mirror image).

- Remove bedboards (high roof) ⇒ page R 75-21.

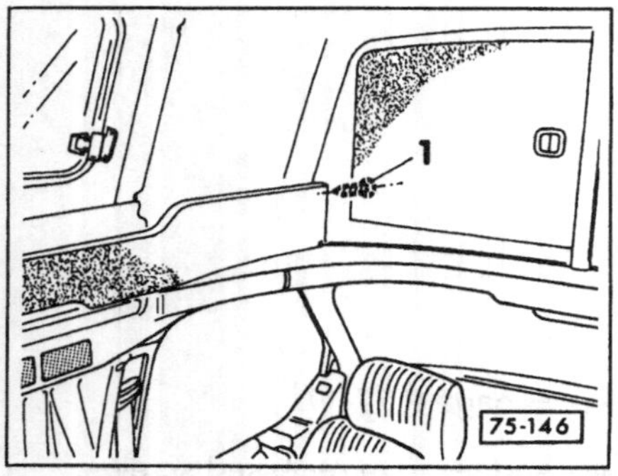

- Remove recessed screw in roof storage box from front face ofroof rail -1-.

- Pry off caps -arrows A-, remove three recessed screws.
- Remove eight recessed screws -arrows B- from retainer behind roof rail.
- Remove roof rail.

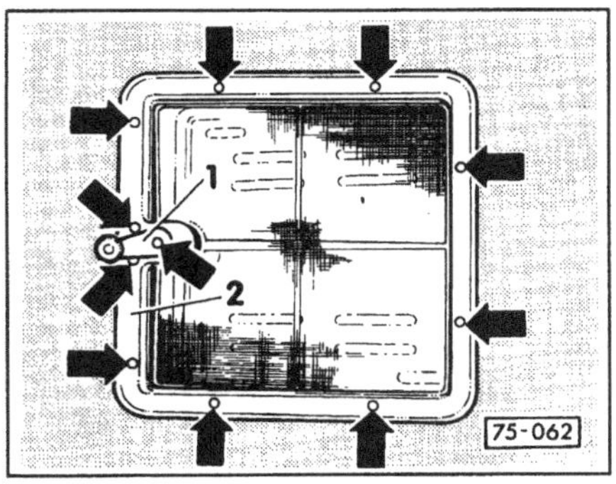

Roof vent, removing/installing

Removing

Note:

The upper frame is damaged when removing roof vent.

◄ – Pry off caps -arrows- and remove eleven recessed screws.

– Remove crank -1- and lower frame -2-.

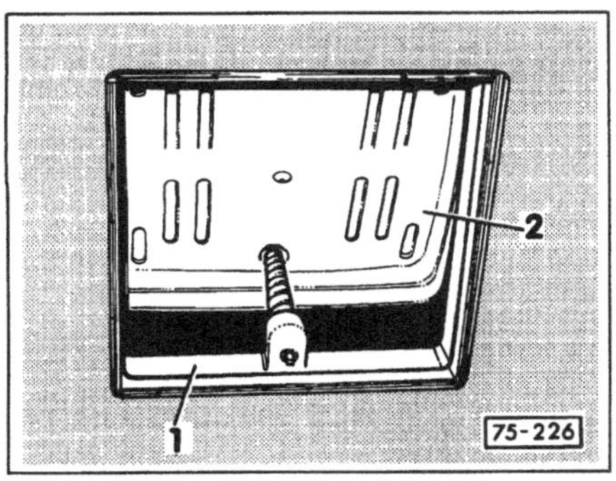

◄ – Press out upper frame -1- and vent cover -2- forcefully in an upward direction.

Installing

> ***CAUTION!***
>
> ***Part numbers are for reference only. Always check with your Parts Department for latest information.***

- Cut off excess bonding material to approx. 1.0 mm (0.039 in.) on roof outer surface where upper frame rests .

Notes:

- *Remove excess bonding material, prime corresponding area with primer D 009 600.*

> ***CAUTION!***
>
> ***Do not prime cut surface material***

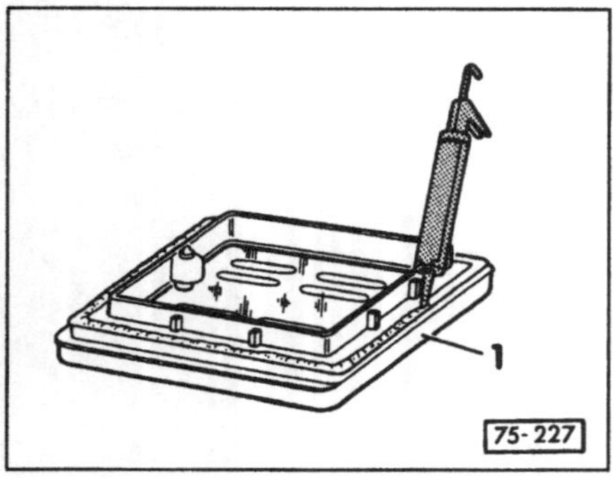

◄ - Clean upper frame (new part) -1- in contact area.

- Apply primer D 009 600 to cleaned surface, allow at least 10 minutes flash-off time. Apply bonding material AKD 476 KD5 05.

- Insert upper frame in roof cut-out.

- Hold lower frame with drill holes in wood frame and in upper frame flush under roof trim and fasten all parts.

- Install crank, caps and screen netting.

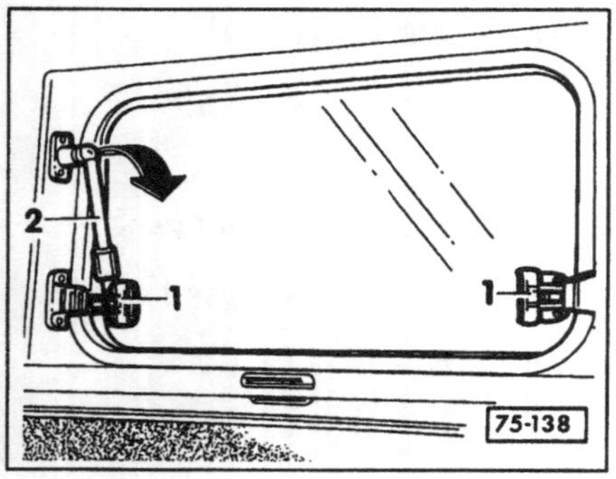

Roof window, removing

◄ - Open window catches -1-.

- Unclip tilt support -2-.

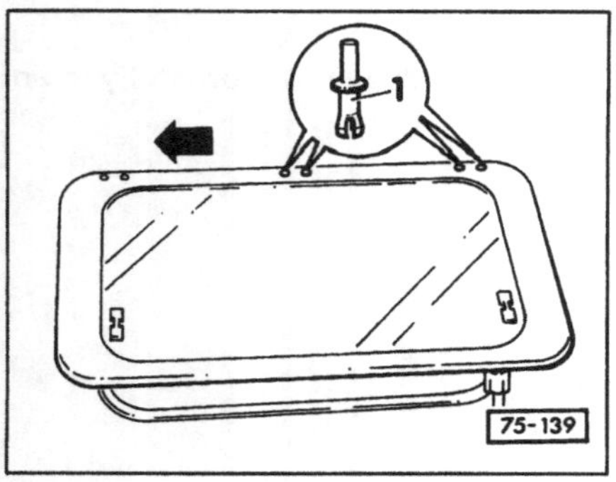

◄ - Push out four spreader clips -1-.

- Remove roof window in direction of arrow together with front hinge.

Trim with screen and insect protection blind, removing

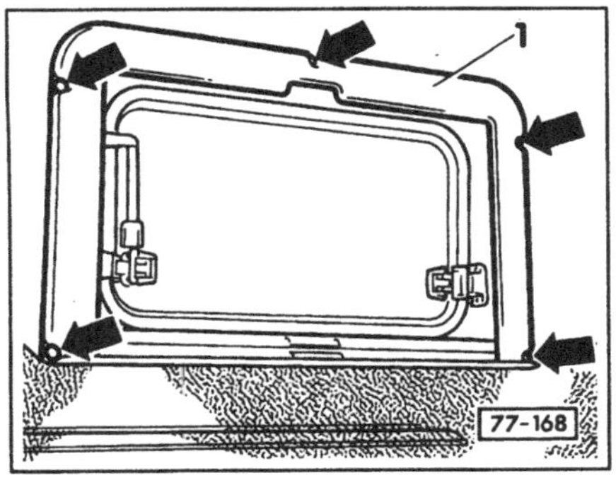

◄ – Pry off caps -arrows- and remove recessed screws.

– Pull trim -1- upward and out together with screen and insect protection blind.

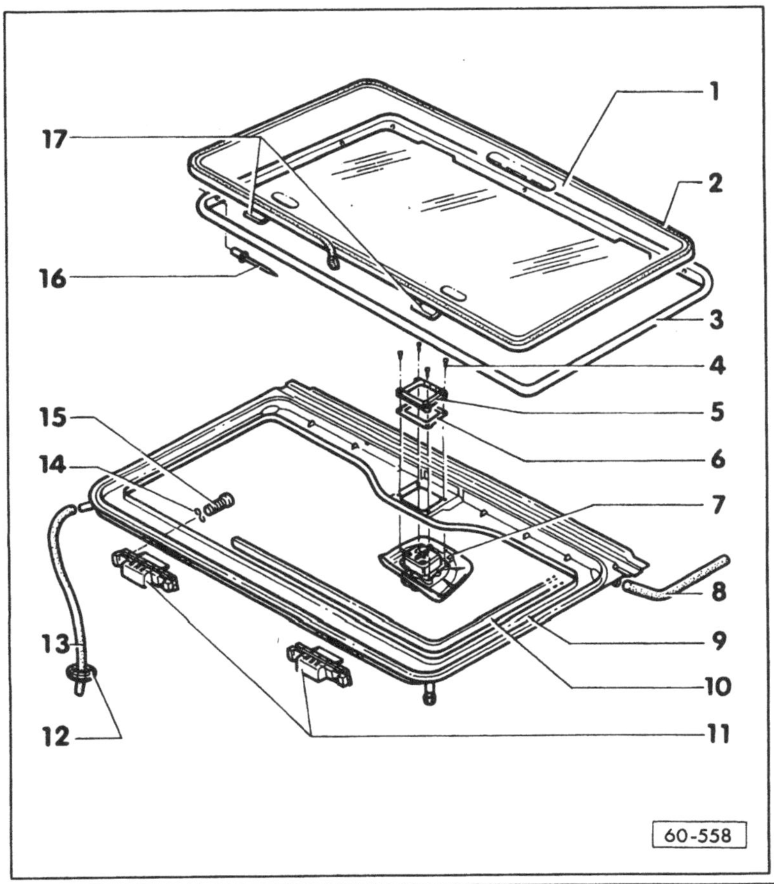

Glass roof vent, component layout (Multivan with standard roof)

1 - **Glass top**
  - ♦ Removal ⇒ page R 75-32
  - ♦ Installation ⇒ R 75-33

2 - **Seal**

3 - **Profile frame**
  - ♦ riveted

4 - **Recessed screw**

5 - **Securing ring**

6 - **Seal**

7 - **Twist grip**

8 - **Water drain hose, rear**
  - ♦ Location/cleaning ⇒ page R 75-34

9 - **Frame**

10 - **Seal**

11 - **Mount**

12 - **Grommet**

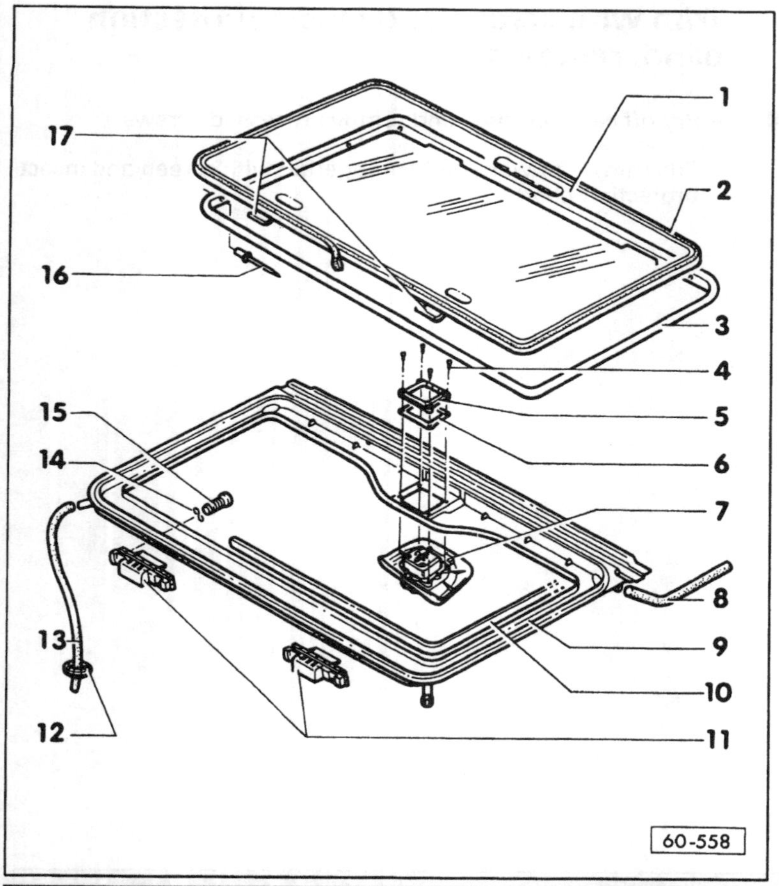

13 - **Water drain hose, front**
- ♦ Location/cleaning ⇒ page R 75-34

14 - **Washer**

15 - **Fillister-head screw**

16 - **Rivet**

17 - **Hinge**
- ♦ hooked into mount -11-

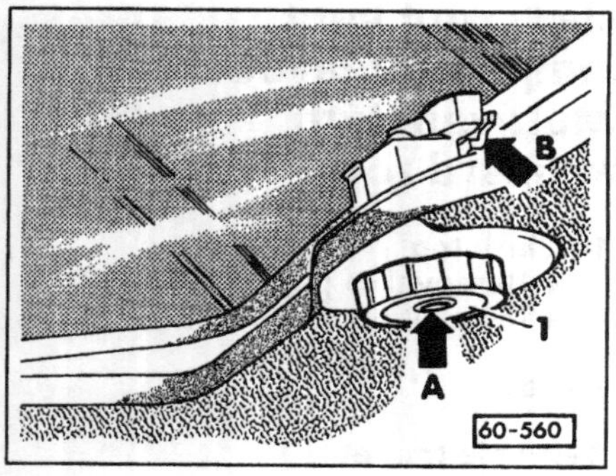

Glass top, removing

◄

- Turn unlocking screw -arrow A- in center of twist grip -1- (with glass top closed) with a screwdriver approx. 1/4 turn left or right until glass top is unlatched.
- Turn twist grip -1- right until glass roof vent is released; simultaneously, lift glass slightly to prevent it from relatching.
- Press locking lever -arrow B- upward.
- Unlatch glass top and remove from mounts with hinges.

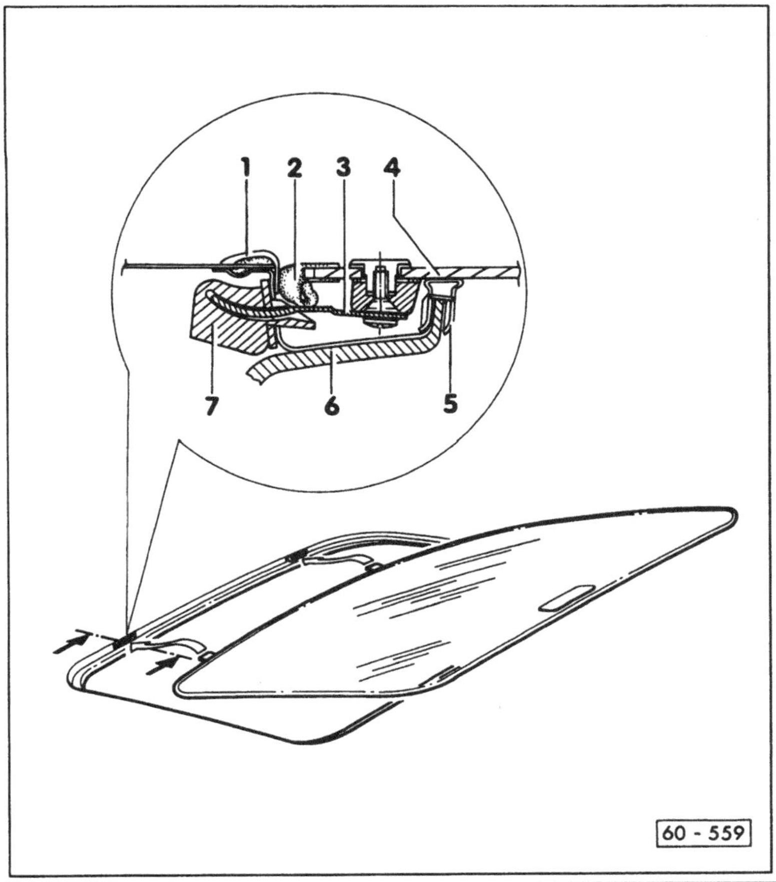

Glass top, installing

1 - **Profile frame**

2 - **Seal**

3 - **Hinge**

4 - **Glass top**
- ♦ Installation: hook glass with hinges -3- to mount -7- and drop them into locking position.

5 - **Seal**

6 - **Roof trim**

7 - **Mount**

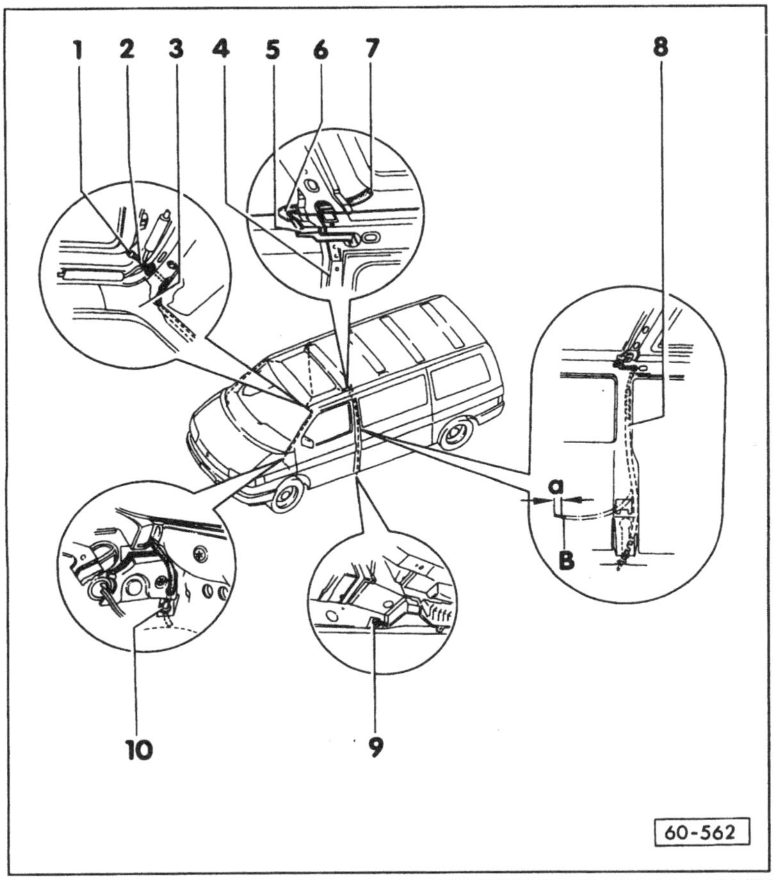

Water drain hoses, location/cleaning

With glass roof vent in driver's compartment

1 - **Frame drain, front**

2 - **Water drain hose, front**
- ♦ use compressed air to clean, blow from frame drain -1-.

3 - **A-pillar**

4 - **B-pillar**

5 - **Adaptor**

6 - **Molded hose**

7 - **Frame drain, rear**

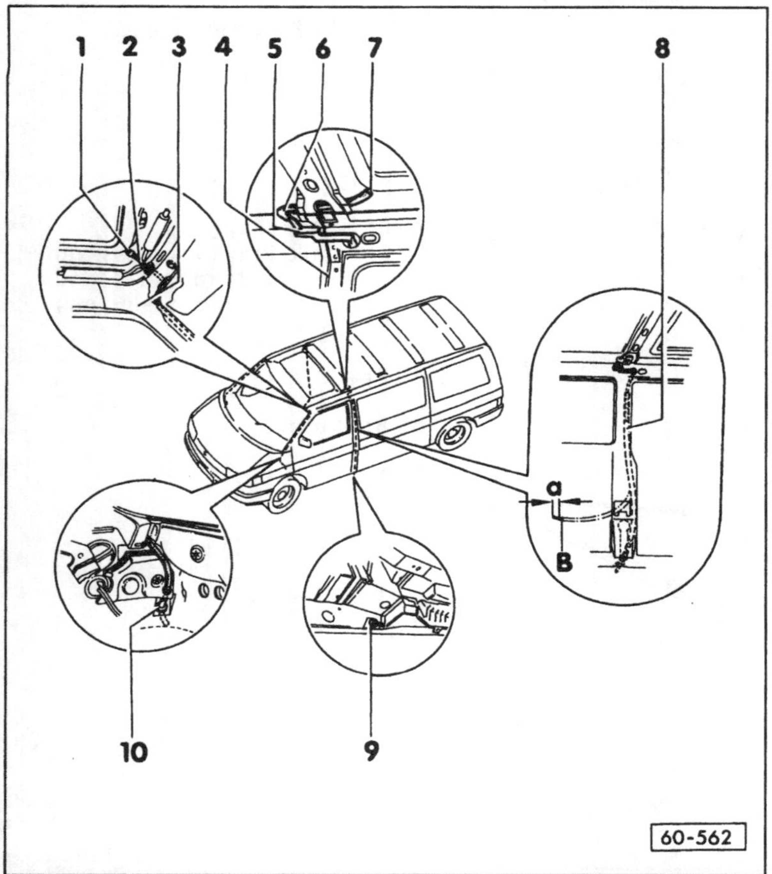

8 - **Water drain hose, rear**
- covered with foam tube
- water drain hose and foam tube are fastened together with fabric adhesive tape in lower area -B- to prevent slipping -a = 200 mm (7.9 in.)-
- use compressed air to clean, from rear frame drain -7-.

9 - **Cap, B-pillar**

10 - **Cap, A-pillar**

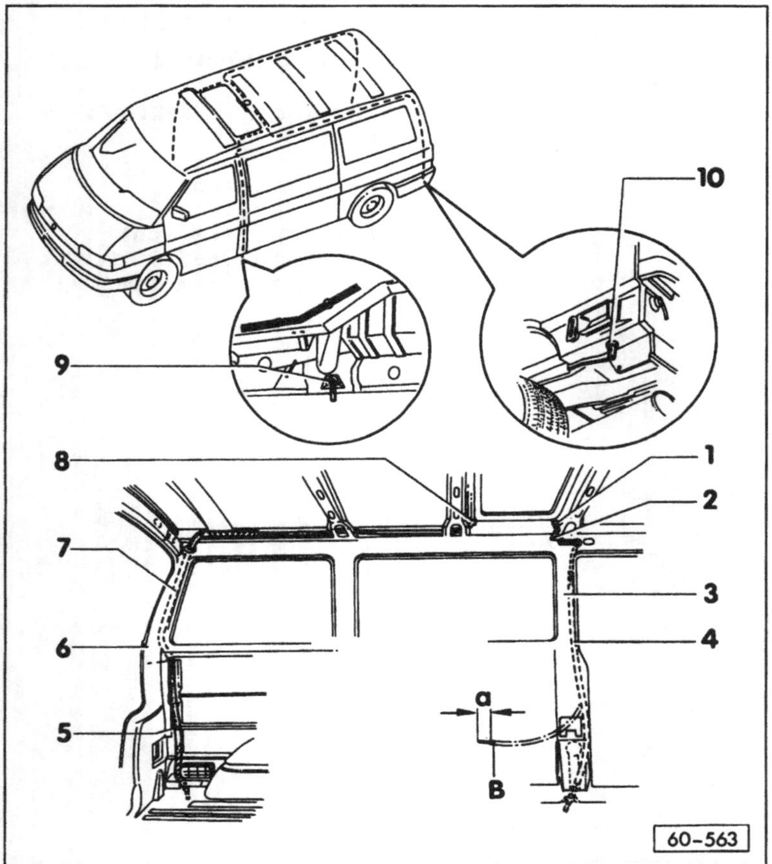

Glass roof vent

Passenger compartment

1 - **Frame drain, front**

2 - **Adaptor**

3 - **B-pillar**

4 - **Water drain hose, front**
- covered with foam tube
- water drain hose and foam tube are fastened together with fabric adhesive tape in lower area -B- to prevent slipping -a = 200 mm (7.9 in.)-
- use compressed air to clean, blow from rear frame drain -7-

5 - **Clip**

6 - **D-pillar**

7 - **Water drain hose, rear**
- covered with foam tube in D-pillar area
- use compressed air to clean, blow from rear frame drain –8–

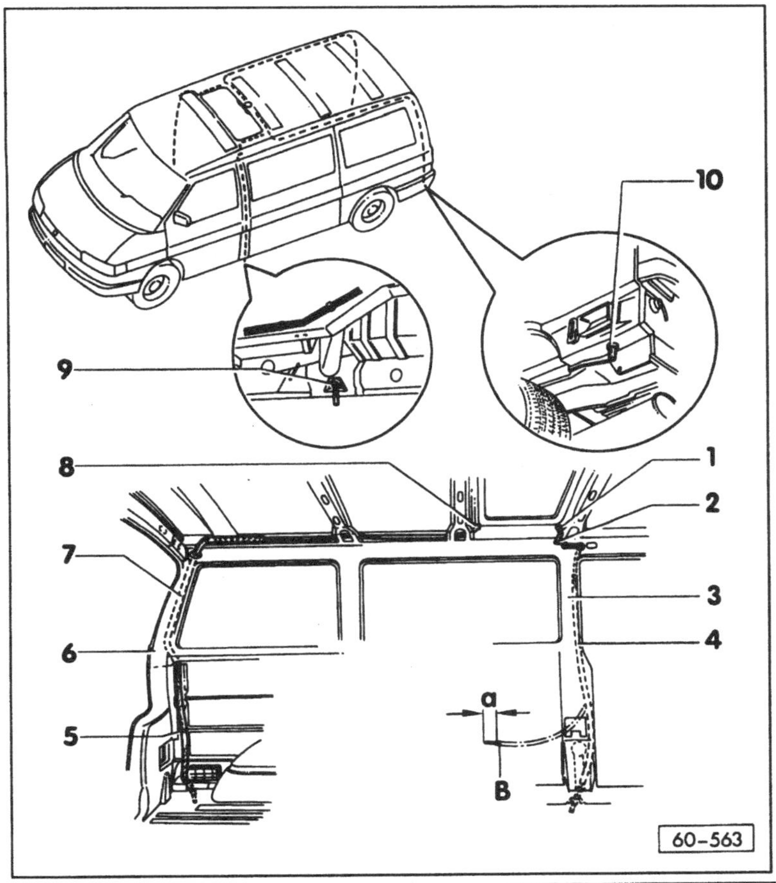

8 - **Frame drain, rear**

9 - **Cap, B-pillar**

10 - **Grommet**
- rear water drain is behind bumper cover

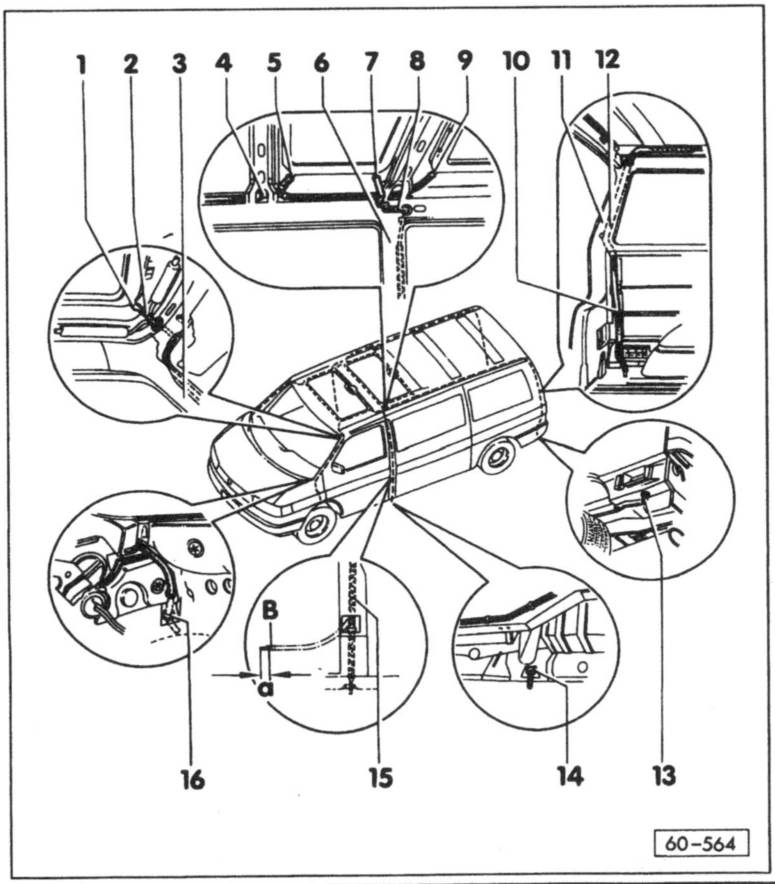

Glass roof vent

Driver and passenger compartments

1 - **Frame drain, front**

2 - **Water drain hose, front**
- use compressed air to clean, blow from front frame drain -1-

3 - **A-pillar**

4 - **Y-fitting**

5 - **Frame drain, rear**

6 - **B-pillar**

7 - **Frame drain, front**

8 - **Adaptor**

9 - **Frame drain, rear**

10 - **Clip**

11 - **D-pillar**

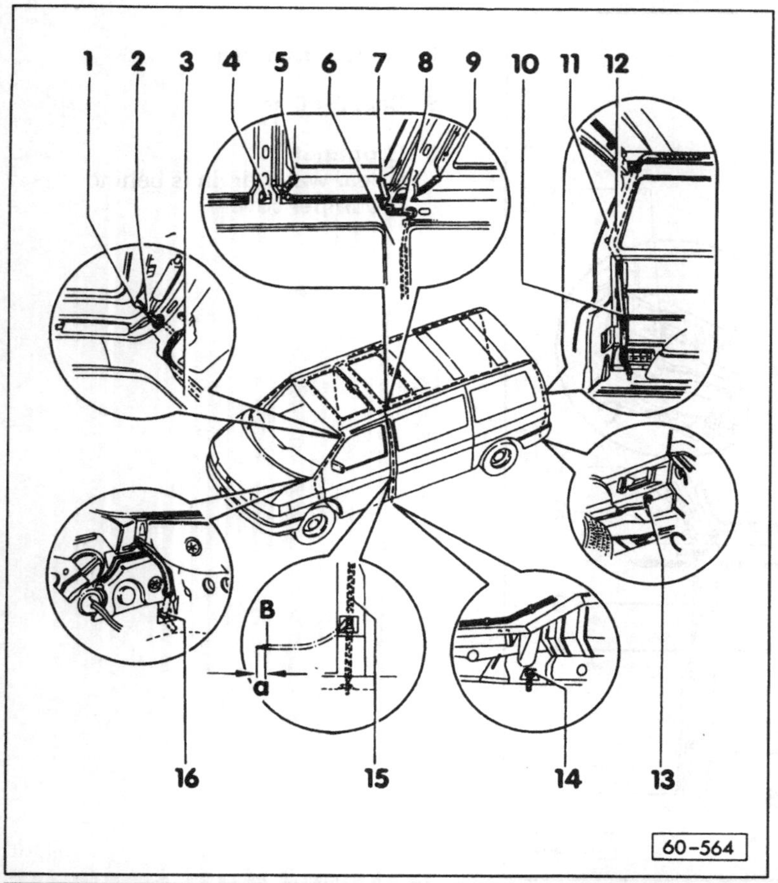

12 - **Water drain hose, rear**
- covered with foam in D-pillar area
- use compressed air to clean, blow from rear frame drains -5- and -9-

13 - **Grommet**
- rear water drain is behind bumper cover

14 - **Cap, B-pillar**

15 - **Water drain hose, front**
- covered with foam
- water drain hose and foam are fastened together with fabric adhesive tape in lower area -B- to prevent slipping
  a = 200 mm (7.8 in.)
- use compressed air to clean, blow from front frame drain -7-

16 - **Cap, A-pillar**

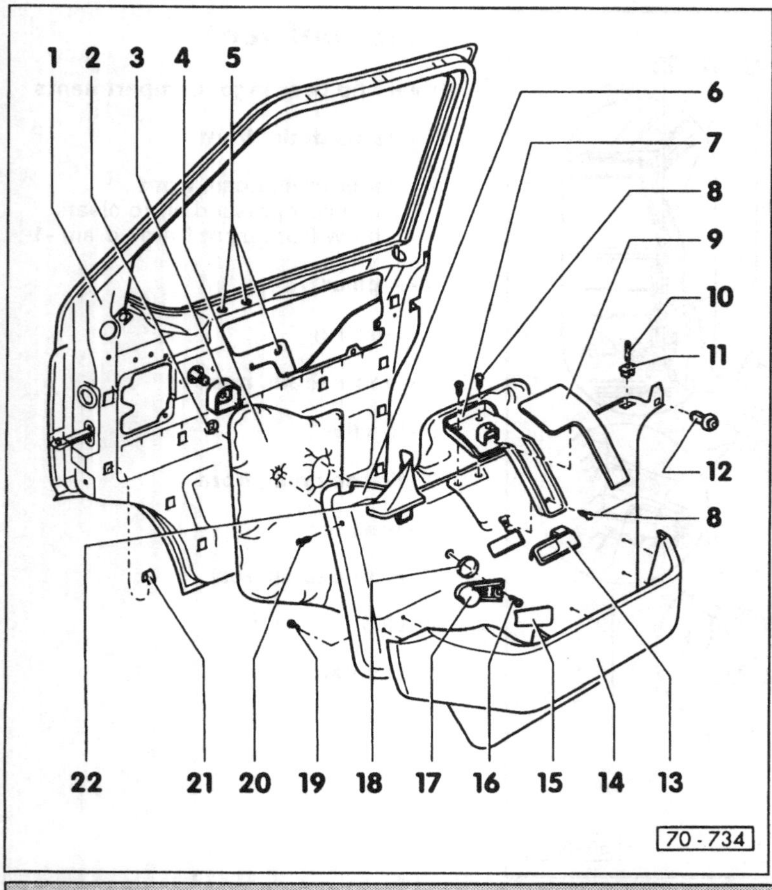

Front door trim, component layout

1 - **Door, front**

2 - **Retaining plate**
- secures damper to trim

3 - **Seal**

4 - **Damper**

5 - **Blind rivet nut**
- insert with blind rivet tool e.g., VAG 1618

6 - **Door trim**
- removal ⇒ fig. 3 and fig. 4

7 - **Handle**

8 - **Fillister-head screws**

9 - **Trim**

10 - **Locking knob**
- screwed on

11 - **Trim**

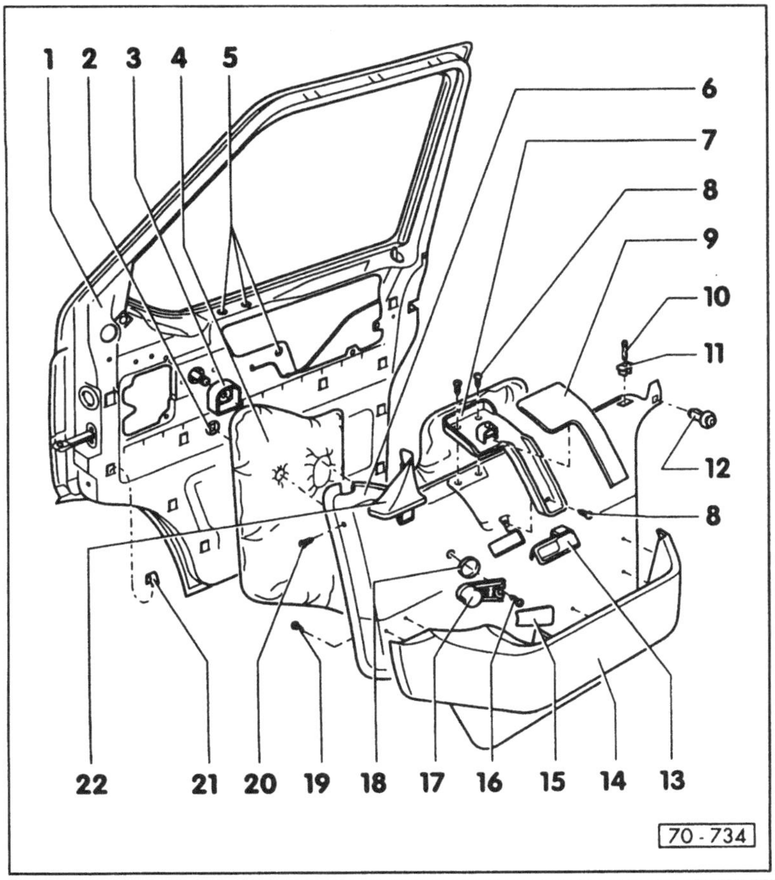

12 - Clip
- 2-part
- removal: unscrew inner section with Allen wrench
- installation: press outer section and inner section simultaneously into mounting hole

13 - Inside door mechanism
- remove door trim, then unclip

14 - Door storage compartment
- screwed onto door trim

15 - Cover

16 - Recessed screw
- 4 Nm (35 in. lb)

17 - Window crank
- removal ⇒ fig. 1
- installation ⇒ fig. 2

18 - Crank shim

19 - Recessed screw
- secures door storage compartment to door trim

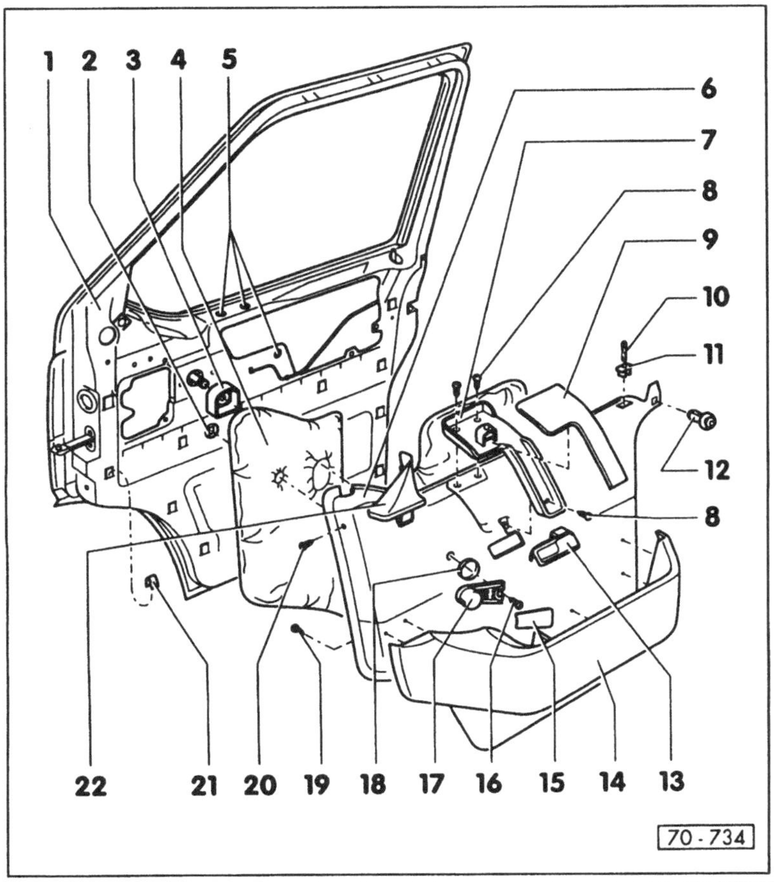

20 - Recessed screw

21 - Hook

22 - Trim

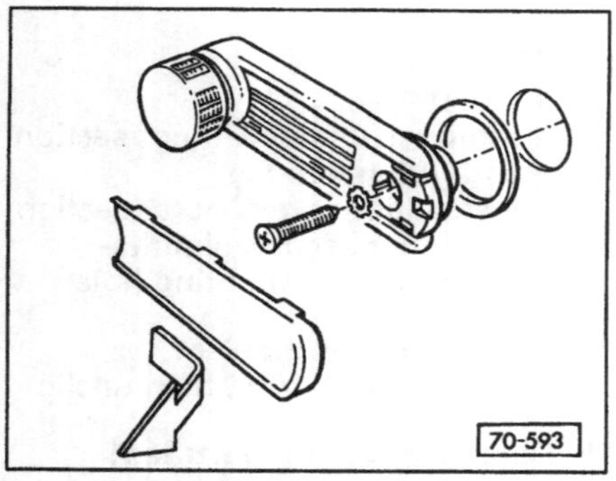

Fig. 1 Window crank, removing

- Unclip cover in direction of arrow.
- Remove recessed screw with star washer.
- Pull off window crank with crank shim.

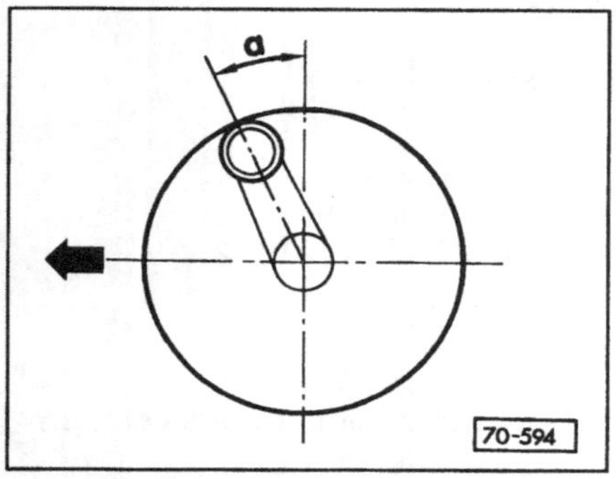

Fig. 2 Window crank, installing

- After installation window crank must be in closed position, shown when window is closed.

  -Arrow- = direction of travel

  Angle -a- = 25° ± 12°

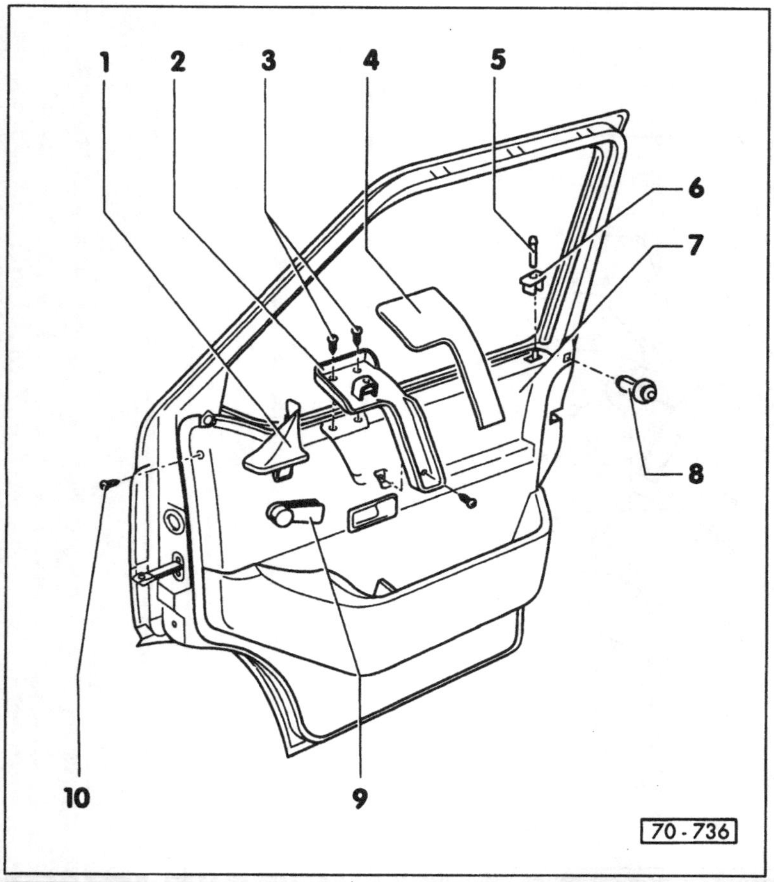

Fig. 3 Door trim, removing

- Crank down window.
- Remove window crank -9- ⇒ fig. 1
- Remove lock knob -5- and pry out trim -6-.
- Remove inner section of clip -8- with Allen wrench.
- Pry off trim -4-, unscrew Fillister-head screws -3-, remove handle -2-.
- Remove recessed screw -10-.
- Remove trim -1-.
- Remove door trim -7- ⇒ fig. 4.

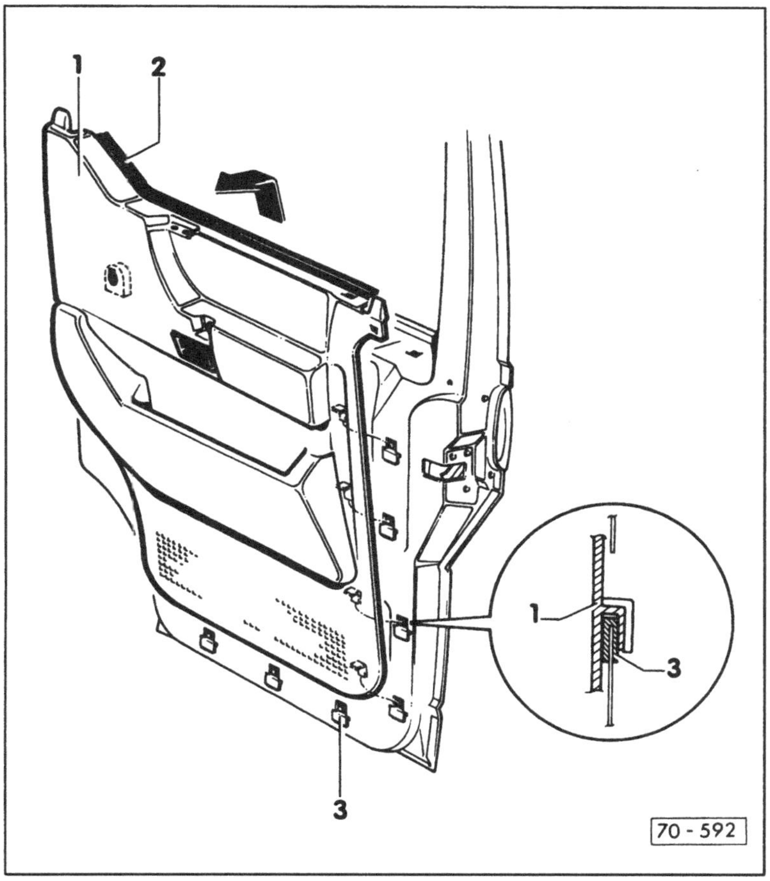

Fig. 4 Door trim, removing

– Lift door trim -1- together with inside window seal -2- from door, in direction of arrow.

Note:

Before installing trim, place clip -3- into mounting holes.

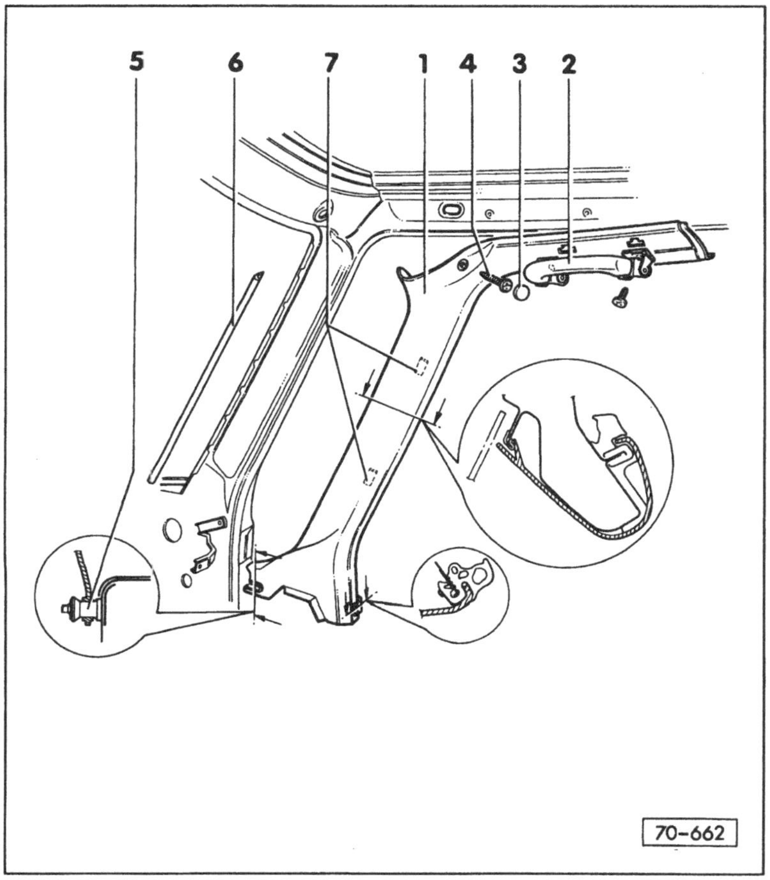

A-pillar trim, removing

Note:

- *Procedure is shown for one side only; procedure for other side is a mirror–image of this illustration.*
- *Remove B-pillar trim before removing A-pillar trim ⇒ page R 75-56, page R 75-58 and page R 75-74.*
- *Loosen headliner (EuroVan CV Camper) near A-pillar trim ⇒ page R 75-66 and page R 75-68.*

1 - Trim
- Unclip carefully from headliner and A-pillar
- Is hooked sideways onto bonded rubber mount behind instrument panel -5-

2 - Handle
- Only front passenger side
- Removal ⇒ fig. 1

3 - Cap
- pry out

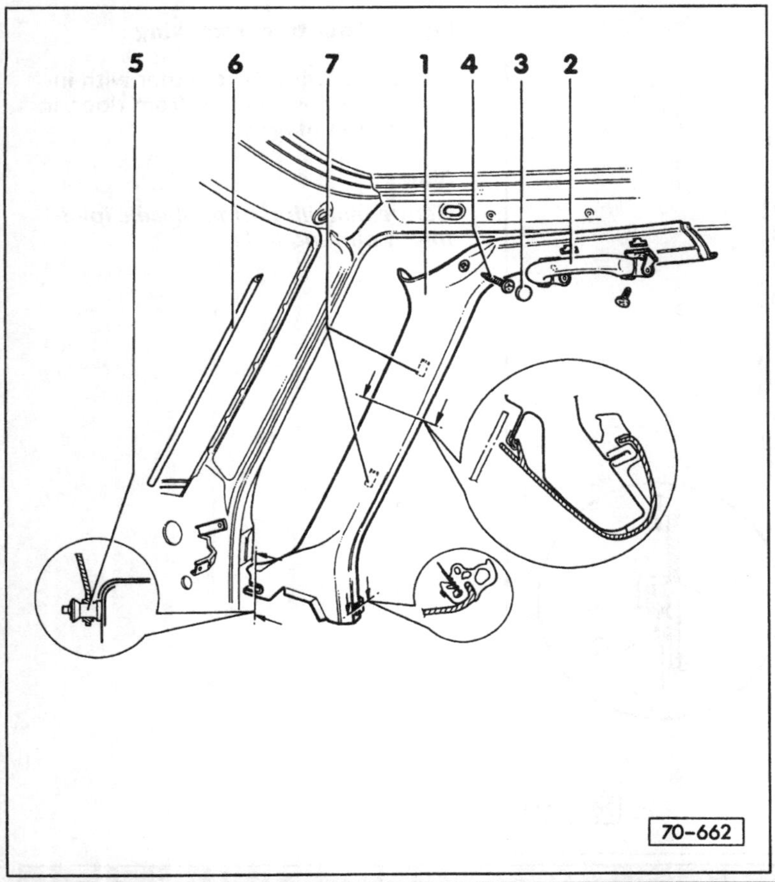

4 - **Recessed screw**

5 - **Bonded rubber mount**

6 - **Weatherstrip**

7 - **Foam pad**

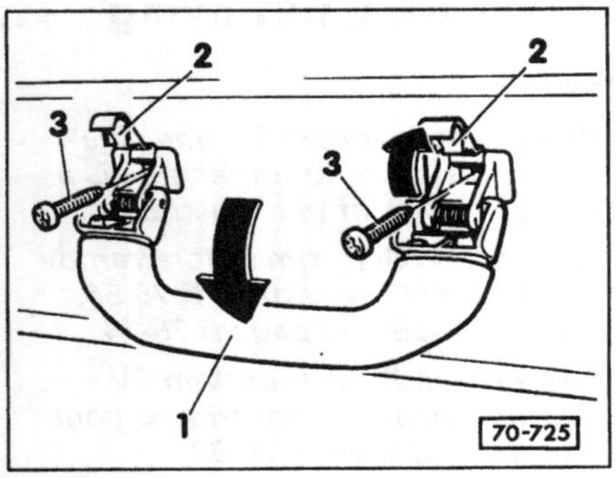

◄ **Fig. 1 Handle, removing**

– Push handle -1- downward.

– Pry up covers -2- with screwdriver and open.

– Remove recessed screws -3-.

– Remove handle -1-.

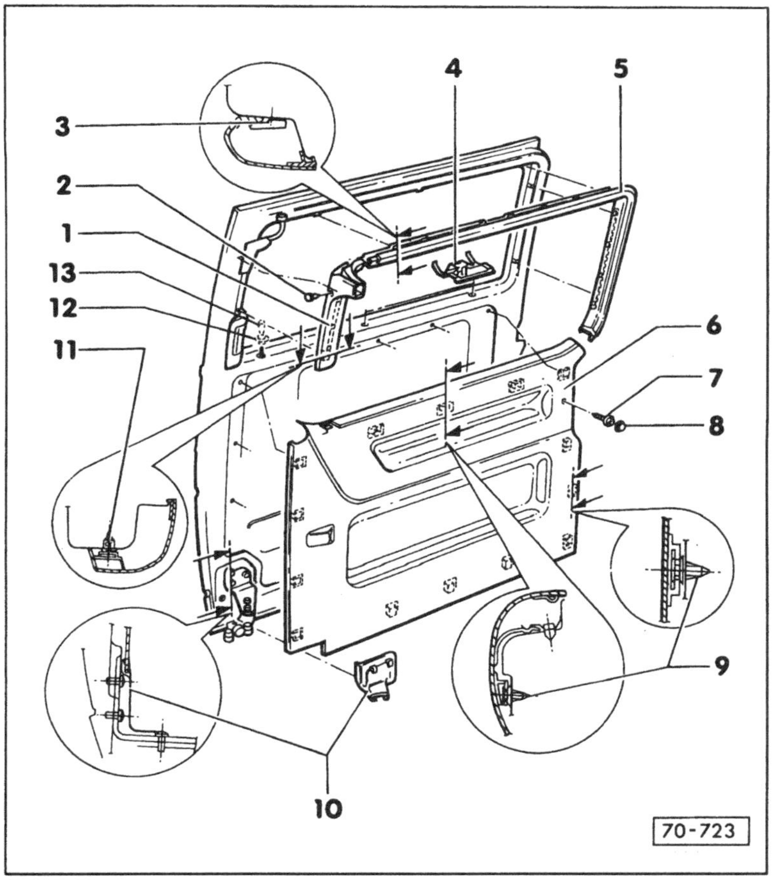

Sliding door trim, component layout

1 - **Trim, front**
   - ♦ Removing: pry out plugs -2- and unclip

2 - **Plugs**

3 - **Bracket**
   - ♦ On window frame, top

4 - **Bracket**
   - ♦ On window frame, bottom

5 - **Trim, rear**
   - ♦ Unclip

6 - **Sliding door molded trim, bottom**
   - ♦ Removal:

– Remove lock knob -13- and pry out trim -12-.

– Pry off cap -8- and remove recessed screw -7-.

– Pry out plugs -2-, unclip trim at front -1- and trim -5- at rear.

– Unclip sliding door molded trim at bottom -6-

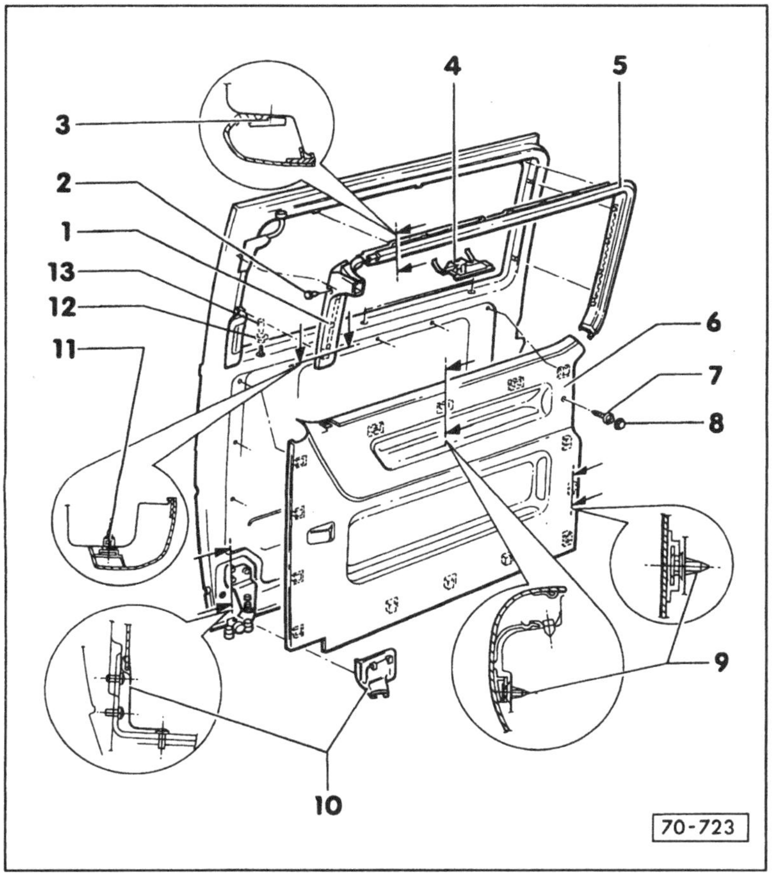

7 - **Recessed screw**

8 - **Cap**

9 - **Clip**

10 - **Roller guide trim, bottom**
   - ♦ Clipped

11 - **Clip**

12 - **Trim**

13 - **Lock knob**
   - ♦ Threads on

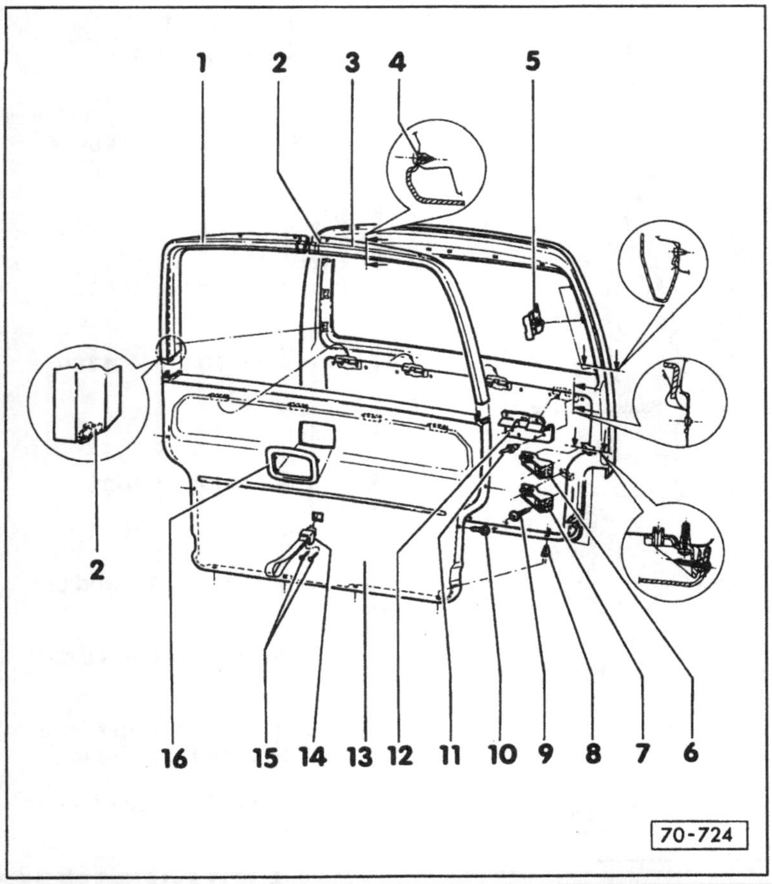

Tailgate trim, component layout

1 - Trim, right

2 - Felt strip
- ♦ Self-adhesive - back of trim

3 - Trim, left

4 - Spreader pins
- ♦ Removal: pry out
- ♦ Installation: press in

5 - Clamp

6 - Clip
- ♦ Left and right

7 - Clip
- ♦ left and right

8 - Spreader nut
- ♦ Remains in tailgate

9 - Recessed screw
- ♦ For clips -6- and -7-

10 - Recessed screw
- ♦ For securing molded trim -13-

11 - Rivet

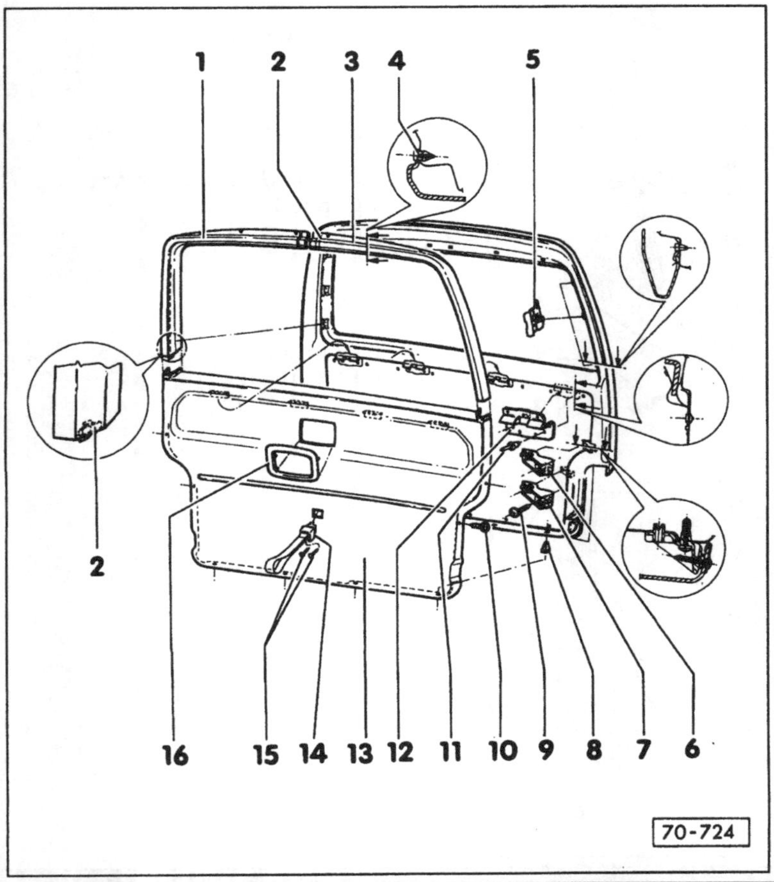

12 - Bracket
- ♦ Spread bracket, to install molded trim ⇒ fig. 1

13 - Molded trim
- ♦ Removal:
- – Press trim -16- upward and unlatch.
- – Remove recessed screws -15- and unhook strap.
- – Pry out spreader pins -4-, unclip trim left side -3- and frame trim right side -1-.
- – Remove recessed screws -10-.
- – Pull molded trim away from tailgate slightly in lower area and push upward out of brackets -12-.
- ♦ Installation:
- – Spread brackets -12- ⇒ fig. 1
- – After mounting molded trim, pull wedge aids out of bracket -12-.

14 - Strap
- ♦ Removal:
- – Remove recessed screws -15- and unhook strap from mounting hole.

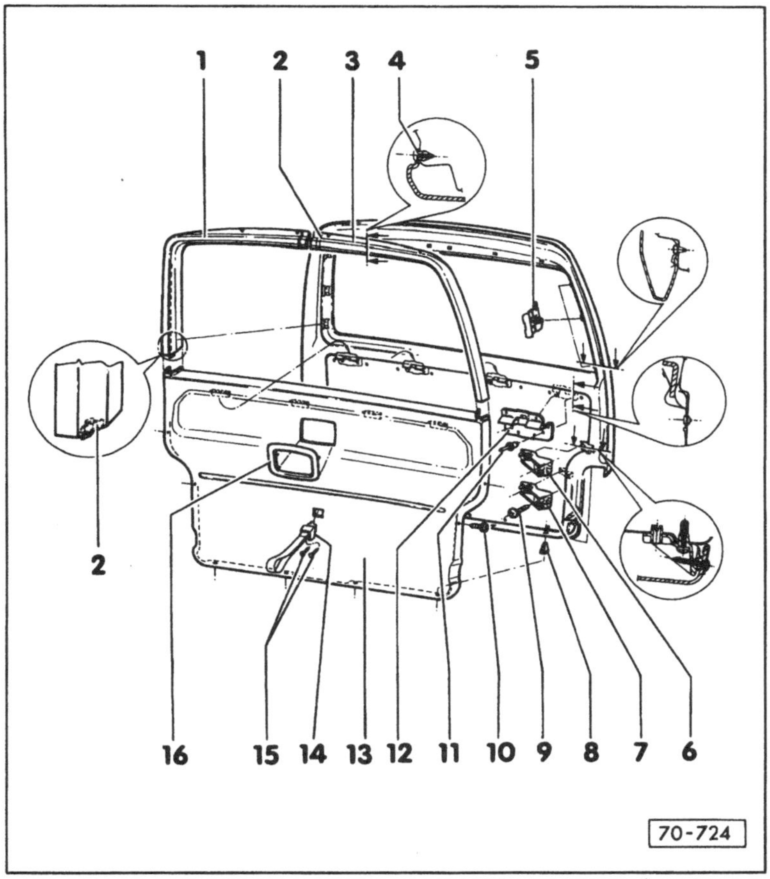

15 - Recessed screw
- Two, one on each side

16 - Trim
- Removing: push upward and unlatch

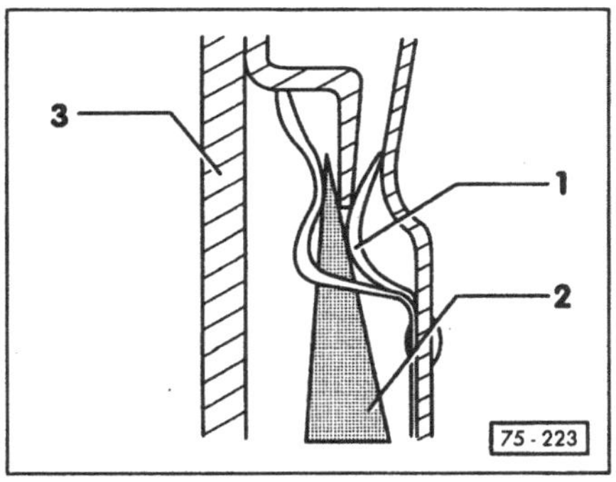

◄ **Fig. 1 Spreading brackets**

– Spread brackets -1- with self-made wedges -2- (wood or plastic).

– Mount molded trim -3-.

– Remove wedges -2-, or allow them to drop out when molded trim is mounted.

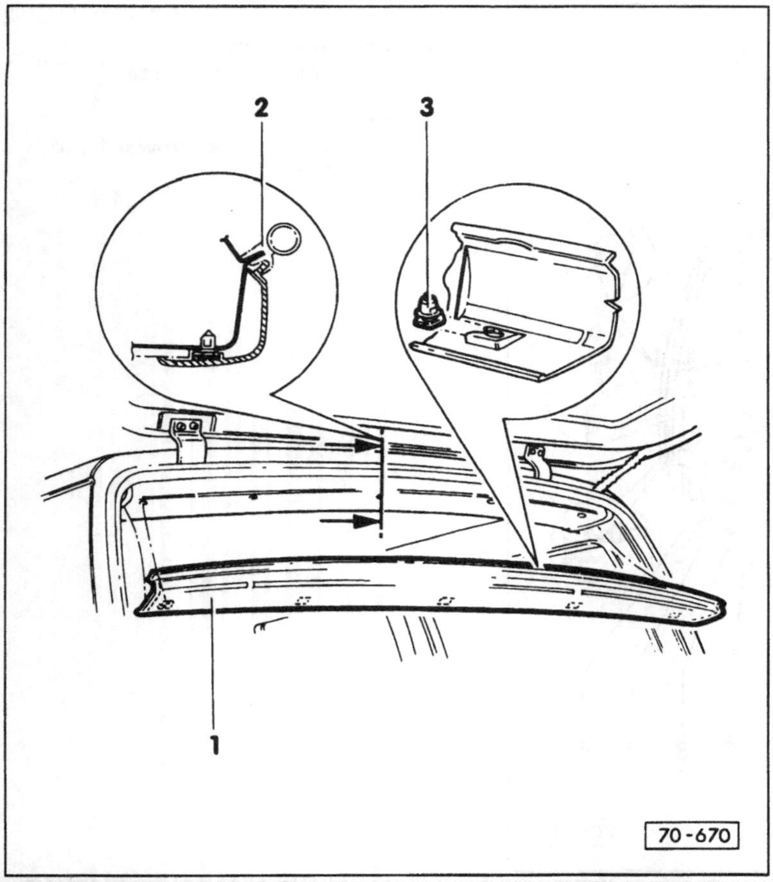

Headliner trim rear, removing/installing

Notes:

- *Before removing rear headliner trim, (EuroVan CV Camper), remove wall cupboard ⇒ page R 77-17, Before installation, remove Auxiliary Batteries -A1- ⇒ page R 76-5, clothes closet ⇒ page R 77-27.*
- *Multivan with hinged roof, headliner trim at rear has additional screw attachments.*

1 - Headliner trim, rear
- Removal: carefully unclip downward from headliner.
- Installation: press clip pins into mounting holes and ensure seal -2- makes proper contact.

2 - Seal

3 - Clip
- 2-part
- When installing, insert sleeve into headliner mounting hole and pin into headliner trim at rear.

Trim, removing/installing (EuroVan CV Camper)

Left side B-pillar trim, removing

CAUTION!

Before working on any part of electrical system obtain radio security code. Disconnect battery (GND) strap.

CAUTION!

Note safety measures ⇒ page R 76-1!

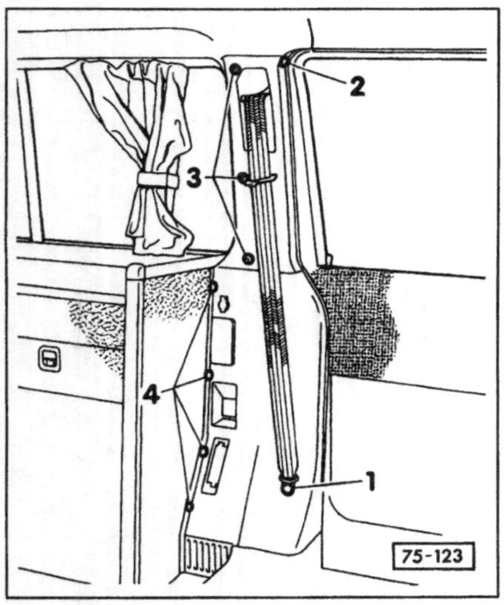

- Remove driver's seat ⇒ page R 77-1.
- Unfasten left B-pillar curtain.
- ◄ Pry off cap -1-, remove hex bolt
- ♦ 40 Nm (30 ft lb).
- Pry off cap -2-, remove recessed screw.
- Remove recessed screws with press-stud -3- and unclip upper B-pillar trim left side.

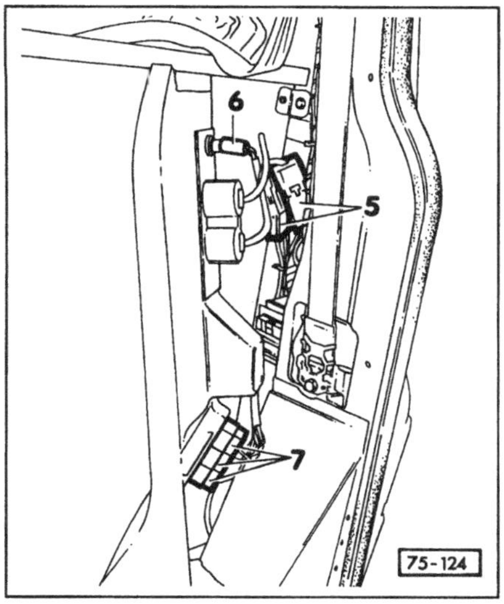

- Pull seat belt out of trim and remove upper B-pillar trim left side.
- Remove lower B-pillar trim left side.
- Remove recessed screws.
- Remove 220 V connectors -5-.
- Remove connector -6- from 12 V Socket -U 5-.
- Remove connectors -7- from fuse block.

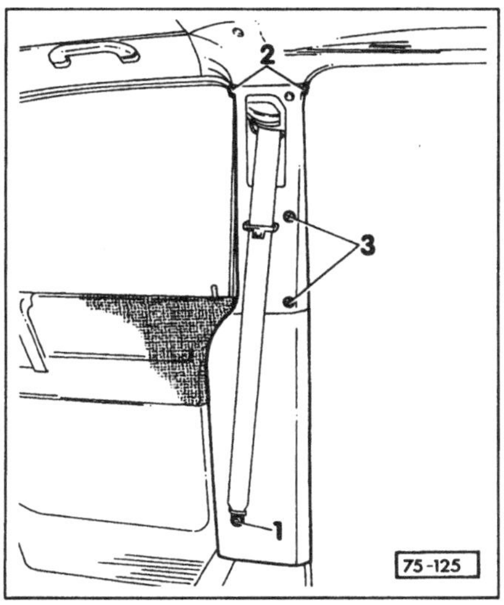

B-pillar trim upper right, removing

- Pry off cap -1-, remove hex bolt

♦ 40 Nm (30 ft lb)

- Pry off caps -2-, remove two recessed screws.
- Remove recessed screws with press-studs -3-.
- Unclip upper B-pillar trim right side, pull seat belt out of trim.

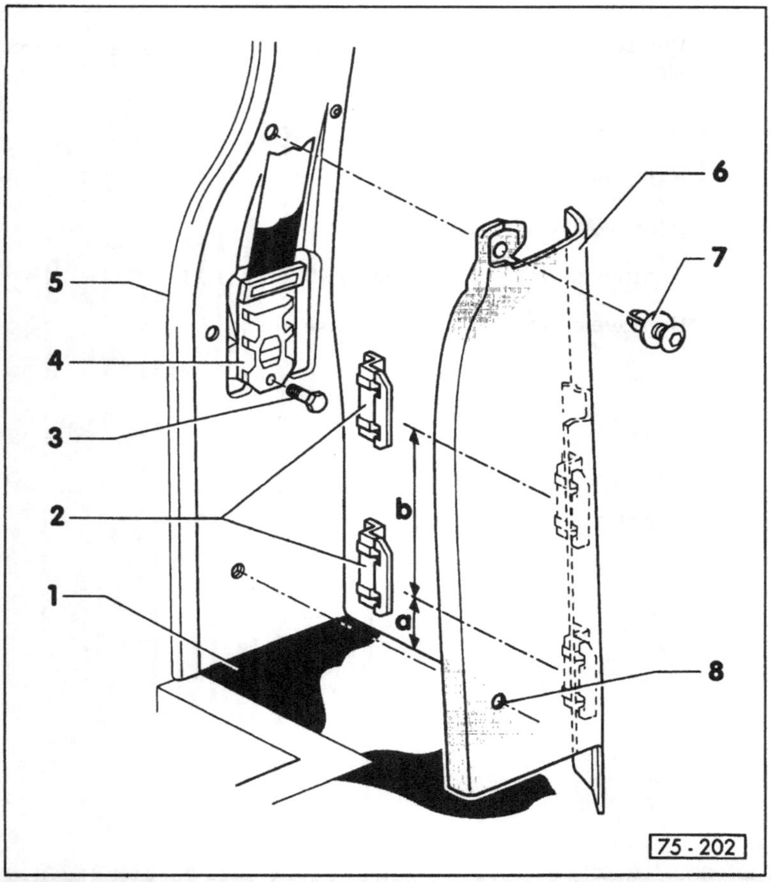

B-pillar trim lower right, component layout

Note:

To remove lower right side B-pillar trim, remove upper B-pillar trim first *⇒ page R 75-56.*

1 - Cross member, seal

2 - Bracket
- Distance of brackets from cross member, sill -1-.
  a = 120 mm (4.7 in)
  b = 200 mm (7.8 in)

3 - Bolt
- 40 Nm (30 ft lb)

4 - Belt retractor

5 - Seal
- After installing trim, ensure seal sits correctly.

6 - B-pillar trim, lower

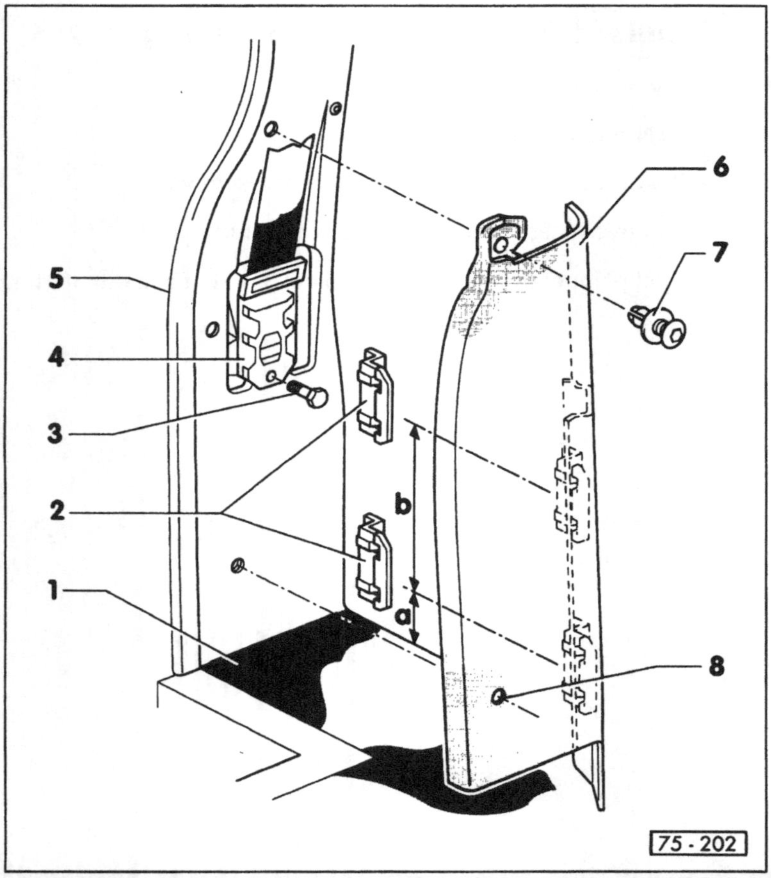

7 - Clip
- 2-part
- Removal: remove inner section with Allen wrench
- Installation: press outer and inner sections in mounting hole simultaneously

8 - Seat belt securing point

C-pillar trim, removing

- Unfasten curtains from C-pillar.

◄ – Remove recessed screws with press-studs -arrows- and remove C-pillar trim.

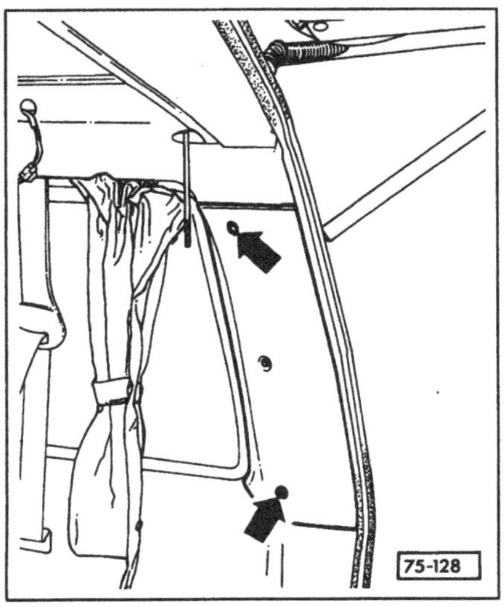

D-pillar trim right, removing

- Remove wall cupboard ⇒ page R 77-17.

◄ – Remove curtain from D-pillar.

- Remove recessed screws with press-studs -arrows-.
- Remove D-pillar trim.

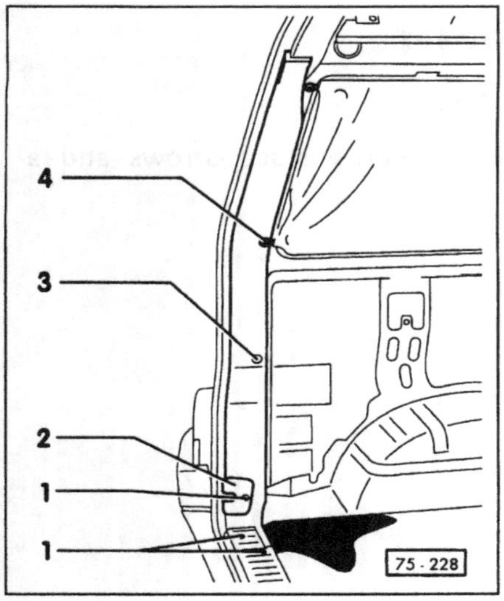

D-pillar trim left, removing

- Remove Auxiliary Batteries -A1- ⇒ page R 76-5.
- Remove wall cupboard ⇒ page R 77-17.
- Remove clothes closet ⇒ page R 77-27.
- ◄ Remove recessed screws -1-.
- Guide blocks -2-, remove cover.
- Remove clip -3- inner section with Allen wrench.
- Remove curtain-holder -4-.
- Remove trim.

Headliner trim, removing

- Remove upper B-pillar trim ⇒ page R 75-56 and page R 75-58.
- Loosen headliner trim in area of A-pillar trim ⇒ page R 75-66 and page R 75-68.
- Remove A-pillar trim ⇒ page R 75-46.
- Remove Front Interior Light -W28- ⇒ page R 76-12.
- Remove Camper Central Electronic Display -E153- ⇒ page R 76-8.

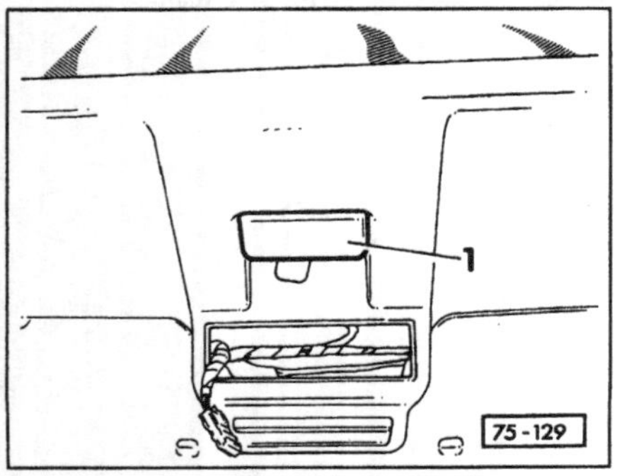

Vehicles with hinged roof

- ◄ Pry out trim -1-.

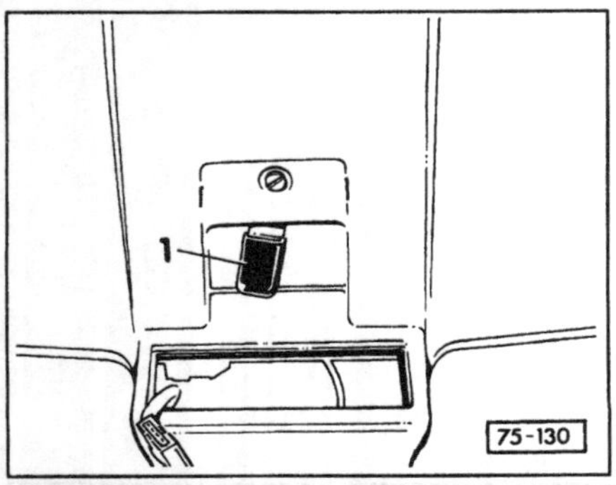

- ◄ Remove locking lever –1 –.

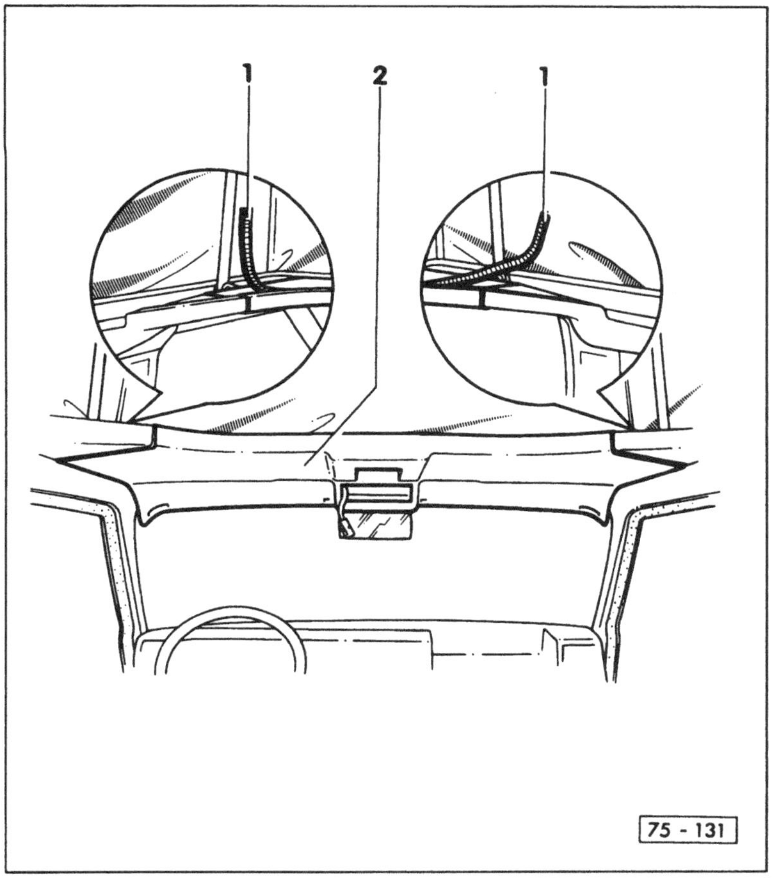

Continuation for all vehicles

- Remove seals -1- left and right from roof trim.
- Unclip roof trim -2-.

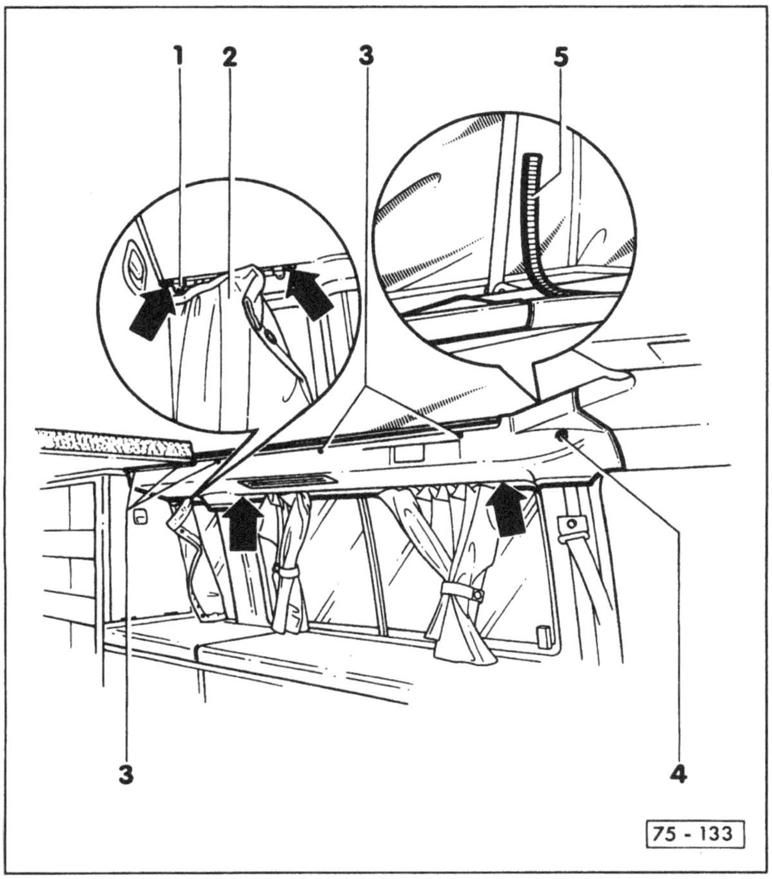

Headliner trim left, removing

- Remove upper B-pillar trim left side ⇒ page R 75-56
- Remove C-pillar trim left side ⇒ page R 75-61
- Pry out stop -1-, pull out rear curtain -2- from curtain rail.
- Remove recessed screws -arrows-
- Pry off caps -3-, remove recessed screws.
- Remove recessed screw with press-stud -4-.
- Pull off seal -5- in area of roof frame trim.
- Carefully pry trim away from roof frame.

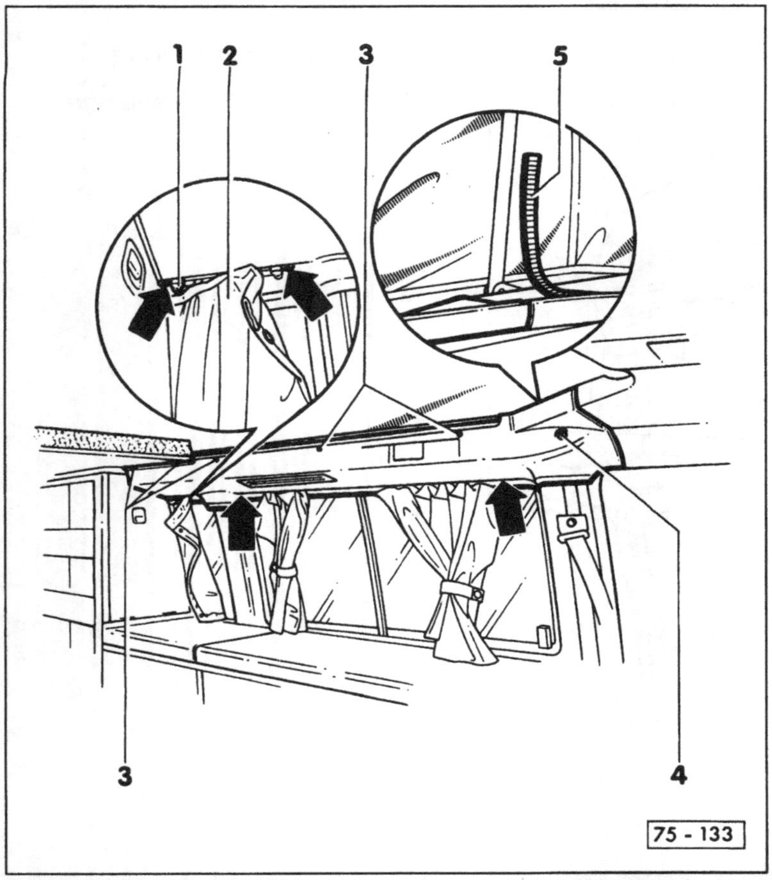

Note:

Trim is also fastened with double-sided adhesive tape.

– Pull connectors from Heater timer/ Temperature Control Switch -E13- and Heater/Heat Output Switch -E16- if equipped, and from Left Fluorescent Light -W8-.

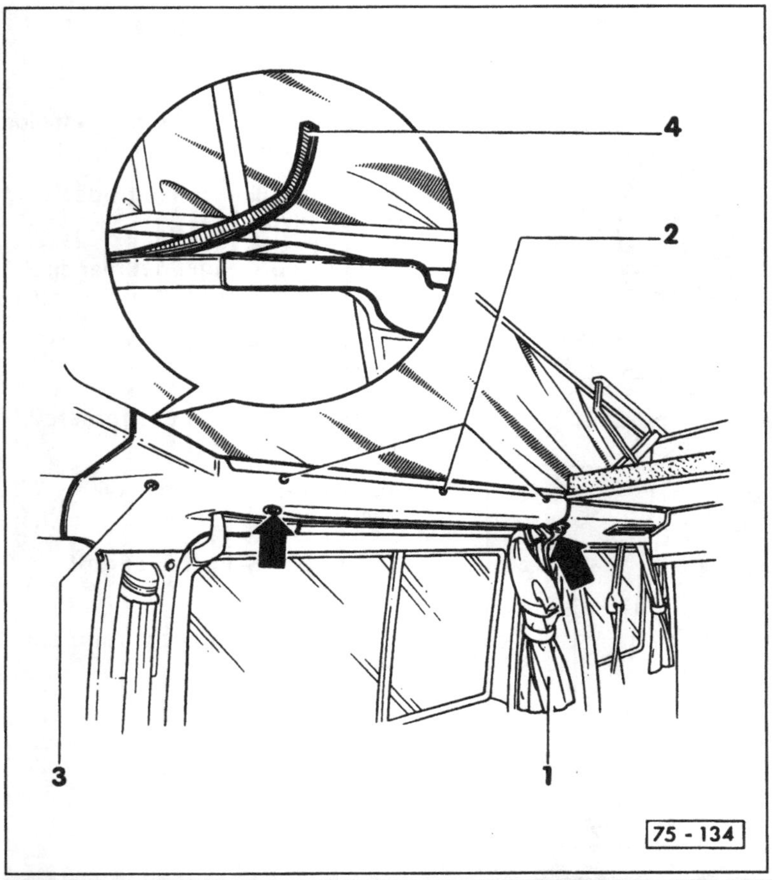

Headliner trim right, removing

– Removing upper B-pillar trim right side ⇒ page R 75-58.

– Unfasten curtain -1- from C-pillar.

– Pry off caps -2- and remove three recessed screws.

– Remove recessed screw with press-stud -3- and two recessed screws -arrows-.

– Remove seal -4- in area of roof frame trim.

– Carefully pry trim away from roof frame.

Note:

Trim is also fastened with double-sided adhesive tape.

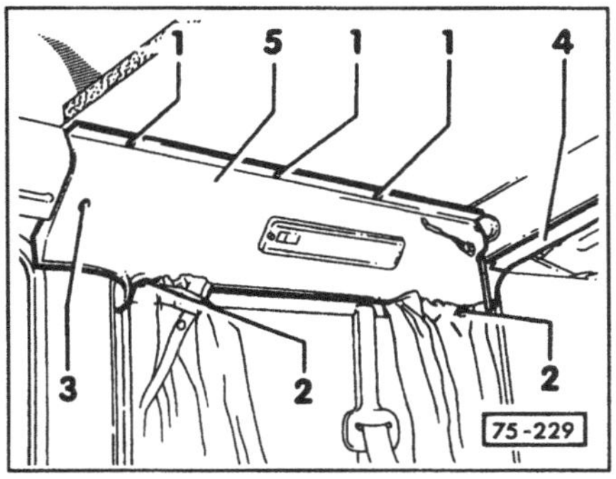

Headliner trim right rear, removing

- Remove wall cupboard ⇒ page R 77-17.
- Remove C-pillar trim right side ⇒ R 75-61 and remove D-pillar trim right side ⇒ page R 75-62.
- Pry off caps -1- and remove recessed screws.
- Remove stop -2- from curtain rail.
- Remove recessed screw with press-stud -3- and two recessed screws (from curtain rail).
- Unclip roof frame trim at rear -4- on right side.
- Remove roof frame trim right rear -5-, disconnect connections from Right Fluorescent Light -W25-.

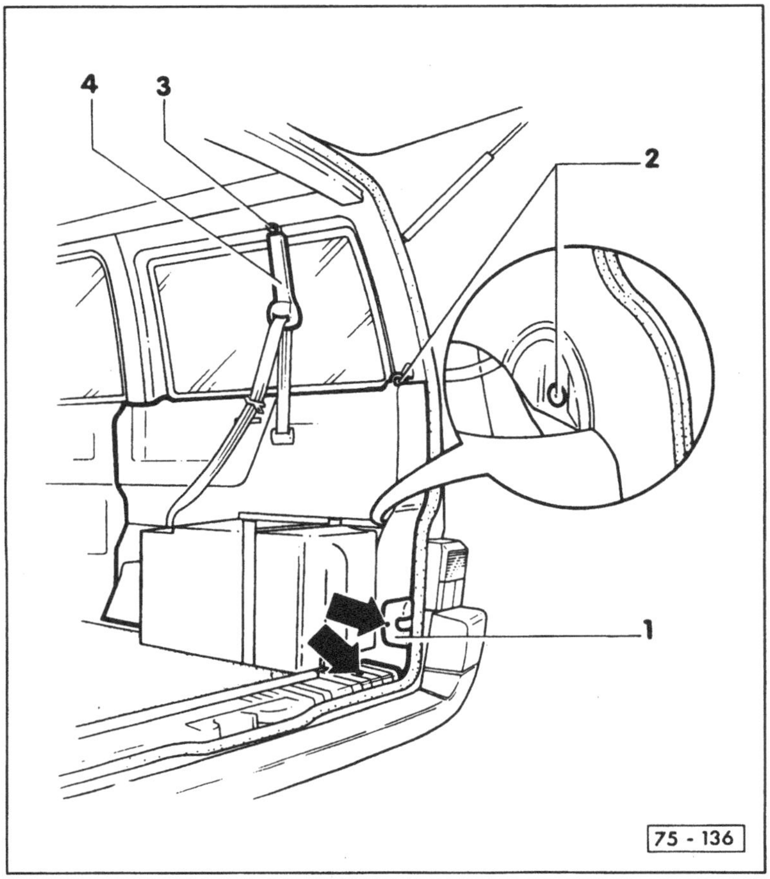

Trim right rear, removing (MultiVan)

- Remove lockable storage compartment ⇒ page R 77-16.
- Remove wall cupboard ⇒ page R 77-17.
- Remove C-pillar trim ⇒ page R 75-61 and remove D-pillar trim right side ⇒ page R 75-62.
- Remove roof frame trim right rear side ⇒ page R 75-69, remove water tank trim ⇒ page R 76-32.
- Pull bench forward.
- Remove recessed screws -arrows-.
- Remove cover -1-.
- Remove clips -2- with Allen wrench.
- Remove hex bolt -3- and unclip cover -4- from seat belt holder.

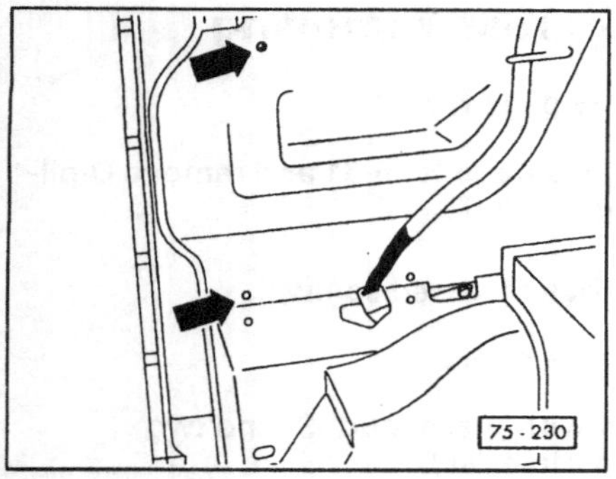

◄ – Remove recessed screws -arrows- near C-pillar (one screw with cap).

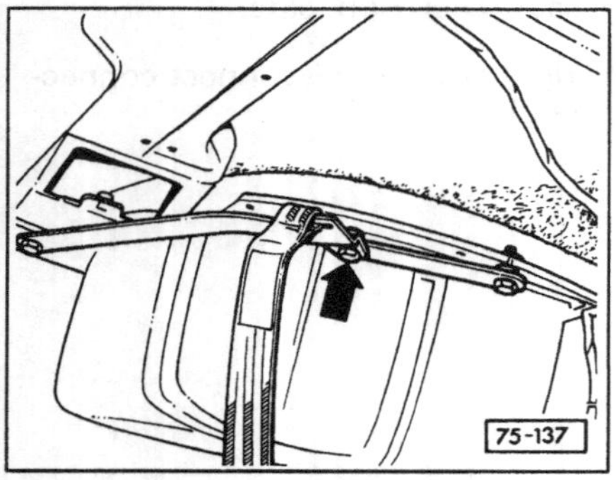

◄ – Unclip side trim and pull away from body.

– Remove hex bolt -arrow-, pull seat belt out of trim.

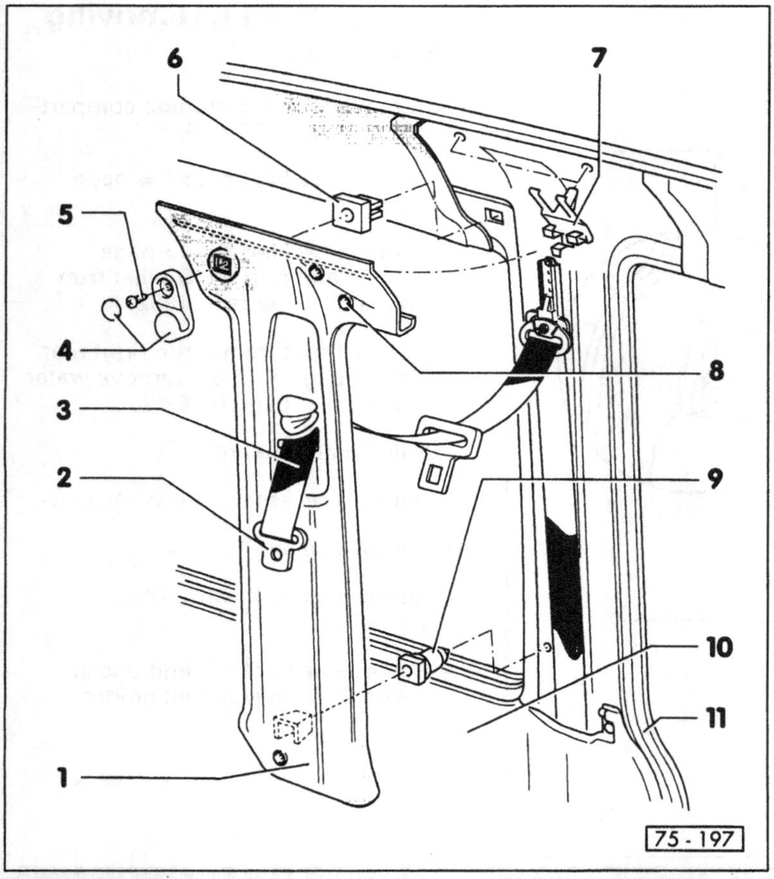

Trim, removing/installing (Multivan)

Note:

Multivan with camper roof, trim above window edge is to be removed and installed according to EuroVan CV Camper instructions

Upper B-pillar trim, component layout

Left side

1 - **Upper B-pillar trim**
- ♦ Removal ⇒ page R 75-74.
- ♦ Installation ⇒ page R 75-75.

2 - **Seat belt assembly**
- ♦ On B-pillar at bottom with hex bolt
- ♦ 48 Nm (35 ft lb)

3 - **Seat belt**

4 - **Coat hook**
- ♦ With cap for recessed screw

5 - **Recessed screw**

6 - **Spreader nut**

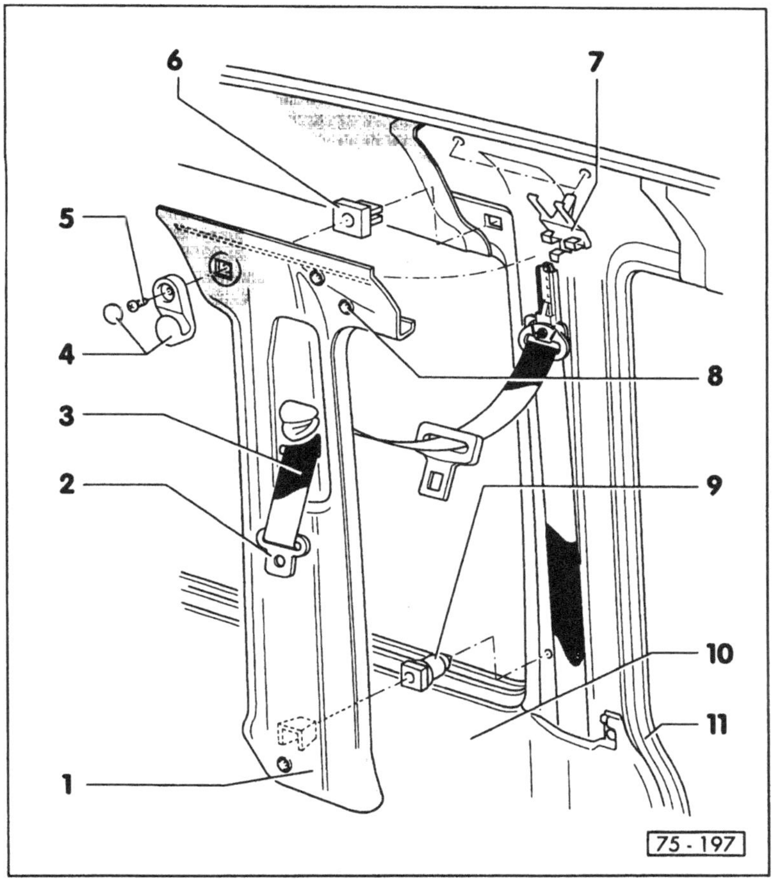

7 - **Bracket**

8 - **Press–stud**
- Only screwed onto trim

9 - **Clip**
- 2-part
- When mounting, insert sleeve into mounting hole in B-pillar and pin into trim.

10 - **Side panel molded trim, front**

11 - **Seal**
- After mounting trim, ensure that seal is positioned correctly.

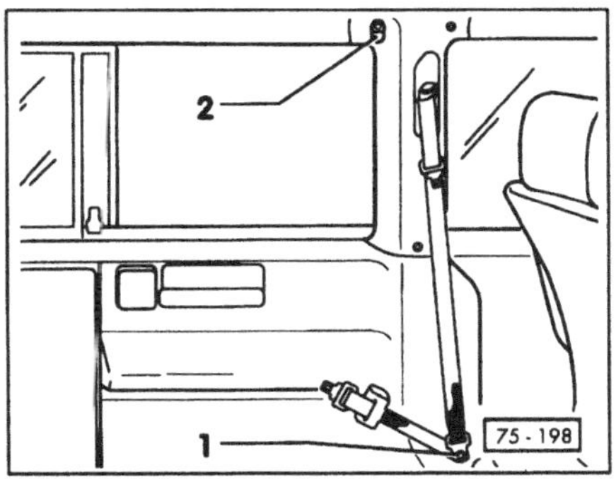

Upper B-pillar trim left, removing

◄ – Pry off cap -1-, remove belt assembly bolt.

– Pry out cap -2-, remove coat hook.

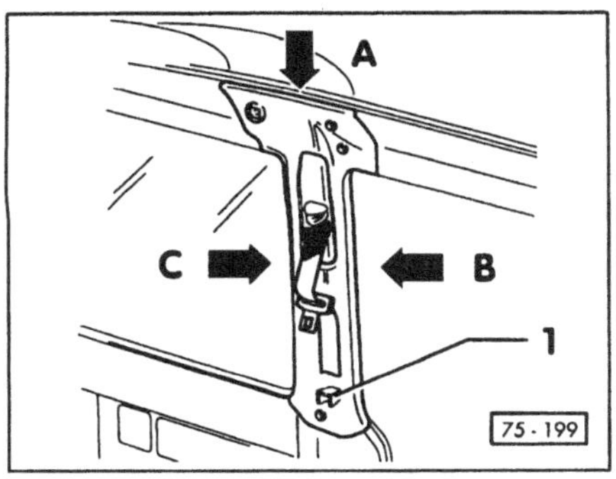

◄ – Unclip trim in following order:

- Along molded headliner -arrow A-
- Rear side of B-pillar -arrow B-
- Front side of B-pillar -arrow C-

– At level of clip -1-, remove trim in installation position of B-pillar.

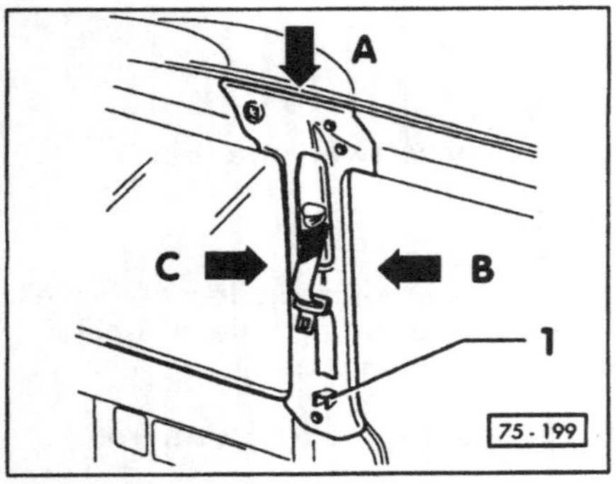

Upper B-pillar trim left, installing

◄

– Clip trim in position on B-pillar and into brackets; ensure clip -1- and seat belt height adjuster are mounted correctly.

– After installing trim, ensure seal -arrow B- is positioned correctly.

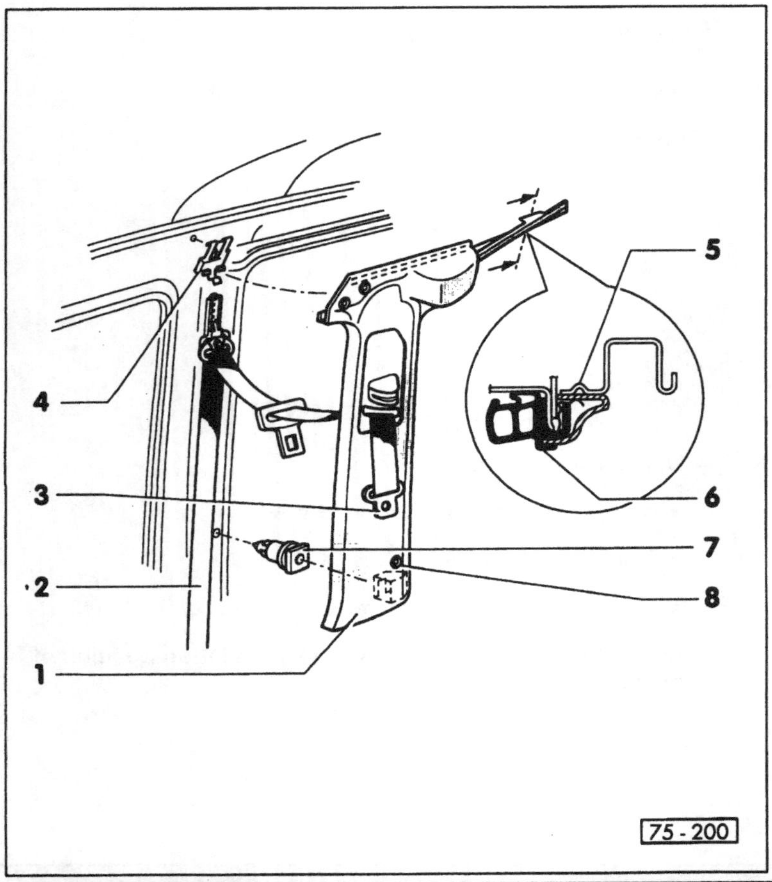

Right side

1 - Upper B-pillar trim
- ♦ Removing/installing ⇒ page R 75-77.

2 - Seat belt

3 - Belt assembly
- ♦ Fastened to B-pillar with hex bolt.
- ♦ 40 Nm (30 ft lb)

4 - Bracket

5 - Positioning buckle in roller guide
- ♦ Bracket must be centered
- ♦ Incorrectly positioning bracket may damage upper B-pillar trim, caused by sliding door.

6 - Seal, sliding door

7 - Clip
- ♦ 2-part
- ♦ When mounting, insert sleeve in B-pillar mounting hole, insert pin in trim.

8 - Press-stud
- ♦ Screwed onto trim.

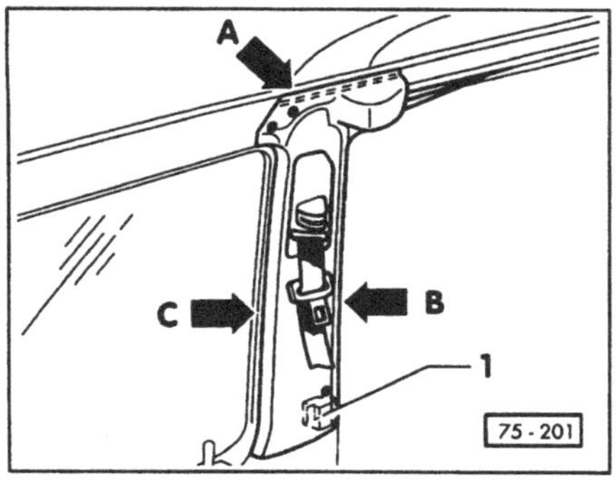

Upper B-pillar trim right, removing

– Pry off cap, remove seat belt bolt at bottom of B-pillar.

◄ – Unclip trim in following order:

- ♦ Along molded headliner -arrow A-
- ♦ Rear side of B-pillar -arrow B-
- ♦ Front side of B-pillar -arrow C-

– At level of clip -1- remove trim.

Upper B-pillar trim right, installing

– Set trim in installation position on B-pillar, clip in brackets on roof frame -arrow A-, ensure clip -1- and belt height adjuster are mounted correctly.

– After mounting trim, ensure seals -arrows B and C- are positioned correctly.

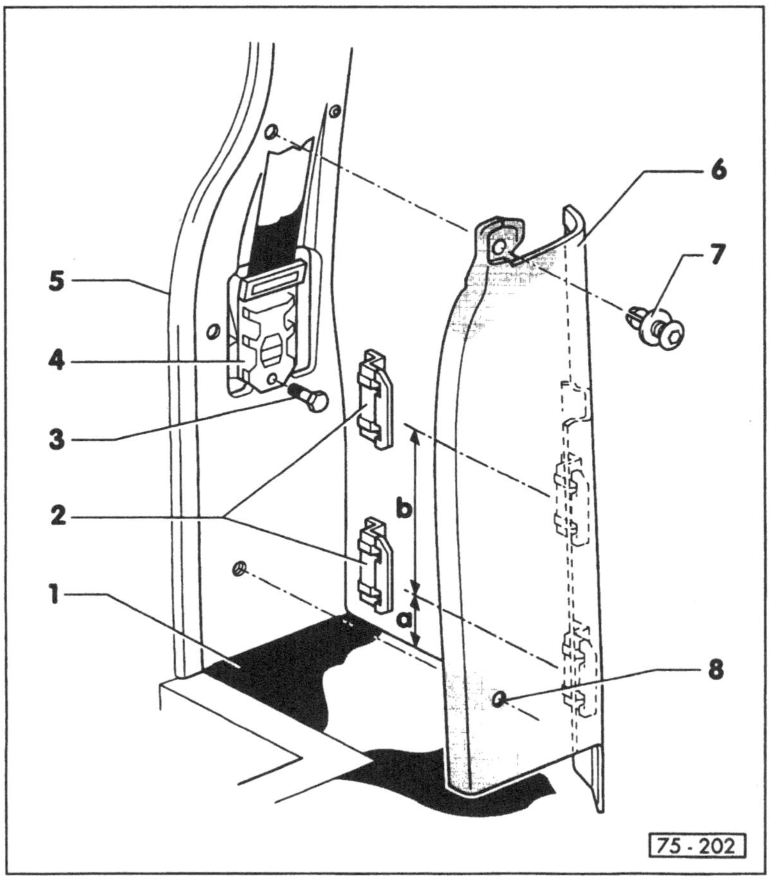

B-pillar trim lower, component layout

Note:

To remove lower B-pillar trim, upper B-pillar trim must be removed first *⇒ page R 75-77.*

1 - Cross member sill

2 - Brackets
- ♦ Distance of brackets from cross member, sill -2-.

a = 120 mm (4.7 in)
b = 200 mm (7.8 in)

3 - Bolt
- ♦ 40 Nm (30 ft lb)

4 - Belt retractor

5 - Seal
- ♦ After mounting lower B-pillar trim, ensure seal is positioned correctly.

6 - Lower B-pillar trim

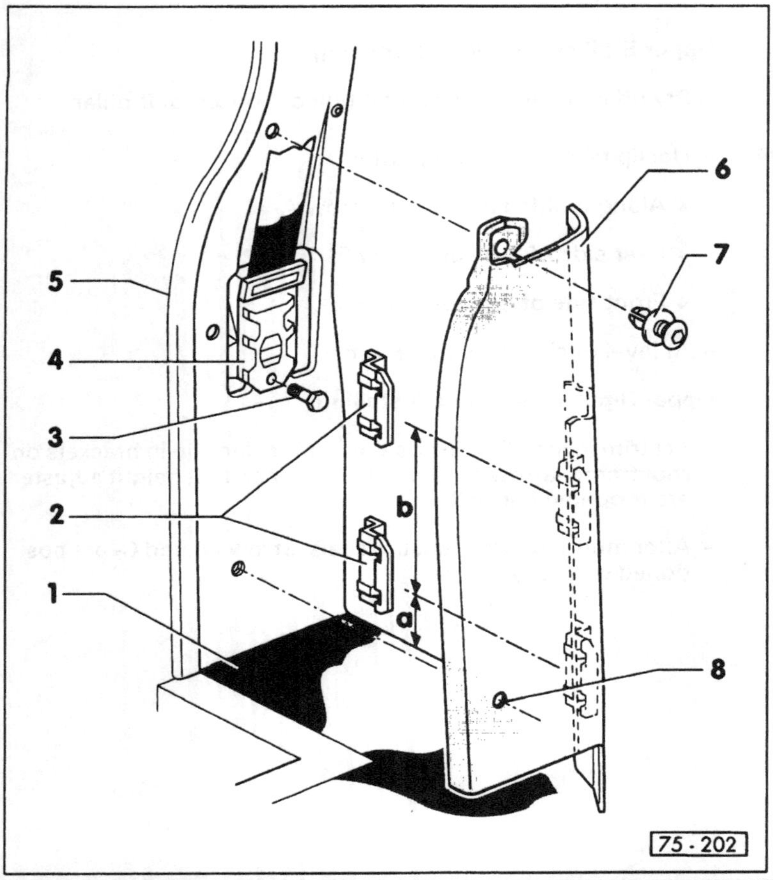

7 - **Clip**
- 2-part
- Removal: unscrew inner section with Allen wrench
- Installation: press outer section and inner section simultaneously into mounting hole.

8 - **Securing point for seat belt**

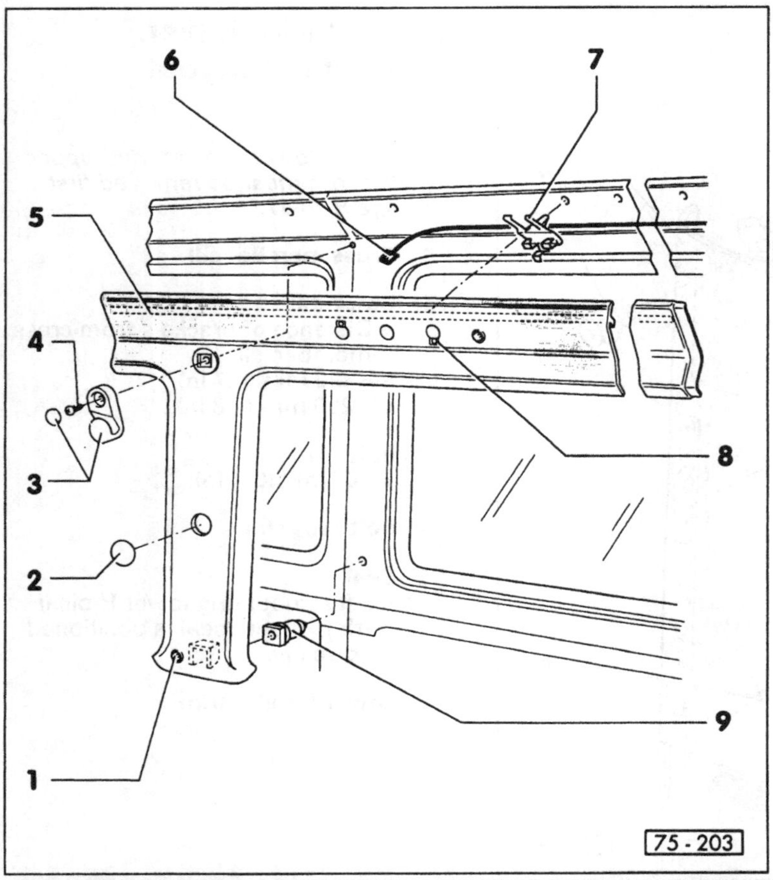

C-pillar trim, component layout

Left side

Note:

To remove C-pillar trim, remove upper B-pillar trim first ⇒ page R 75-74 and remove Left Fluorescent Light -W8- ⇒ page R 76-24.

1 - **Press stud**
- Screwed onto trim

2 - **Cap**
- Securing point, seat belt

3 - **Coat hook**
- With cap for recessed screw.

4 - **Recessed screw**

5 - **C-pillar trim**
- Removal ⇒ page R 75-82
- Installation ⇒ page R 75-83

6 - **Plug, Left Fluorescent Light -W8-**

7 - **Bracket**

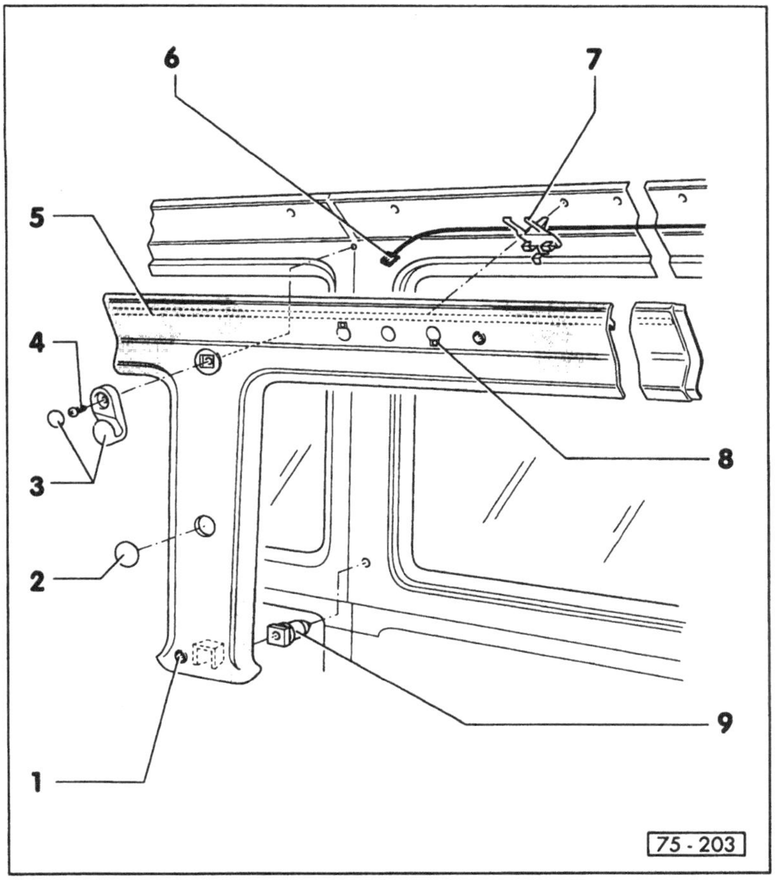

8 - **Speed nut**
- Secures Left Fluorescent Light -W8-

9 - **Clip**
- 2-part
- Installing: insert sleeve in C-pillar mounting hole and pin in trim.

C-pillar trim left, removing

– Remove upper B-pillar trim ⇒ page R 75-74 and Left Fluorescent Light -W8- ⇒ page R 76-24.

– Pry out coat hook cap, remove coat hook.

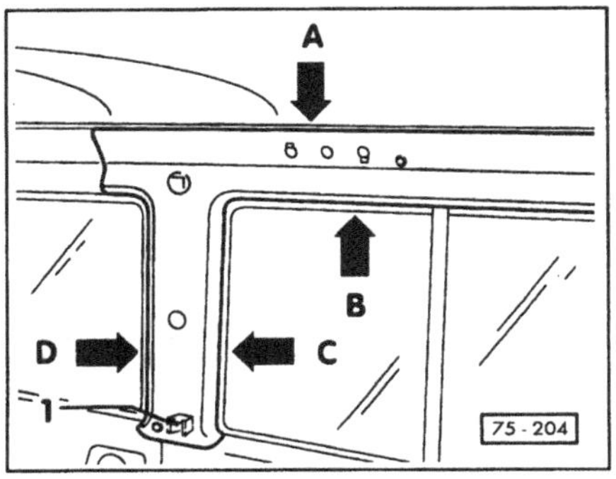

◄ – Unclip trim in following order:

- Along molded headliner -arrow A-
- Window pillar -arrow B-
- C-pillar front -arrow C-
- C-pillar rear -arrow D-

– At level of clip -1-, pull off trim as close to C-pillar as possible.

Note:

When removing trim in window areas -arrows B, C and D- unclip trim in small sections, supporting from back by hand or with tool 80–200.

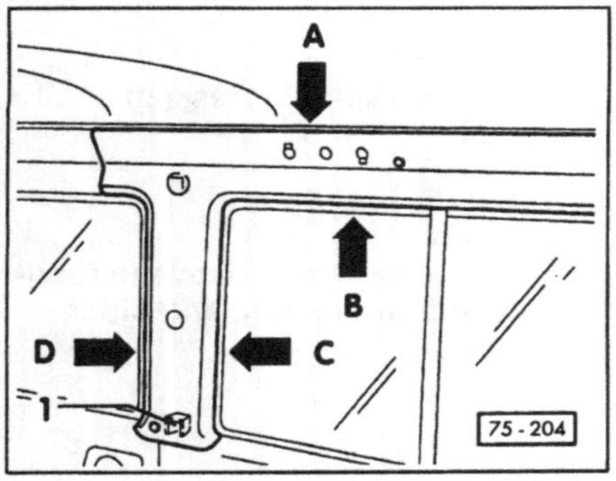

C-pillar trim left, installing

◄

– Position trim and ensure clip -1- is positioned correctly, insert connector for Left Fluorescent Light -W8-.

– Clip trim in window areas -arrows B, C and D- and in brackets on roof frame -arrow A-.

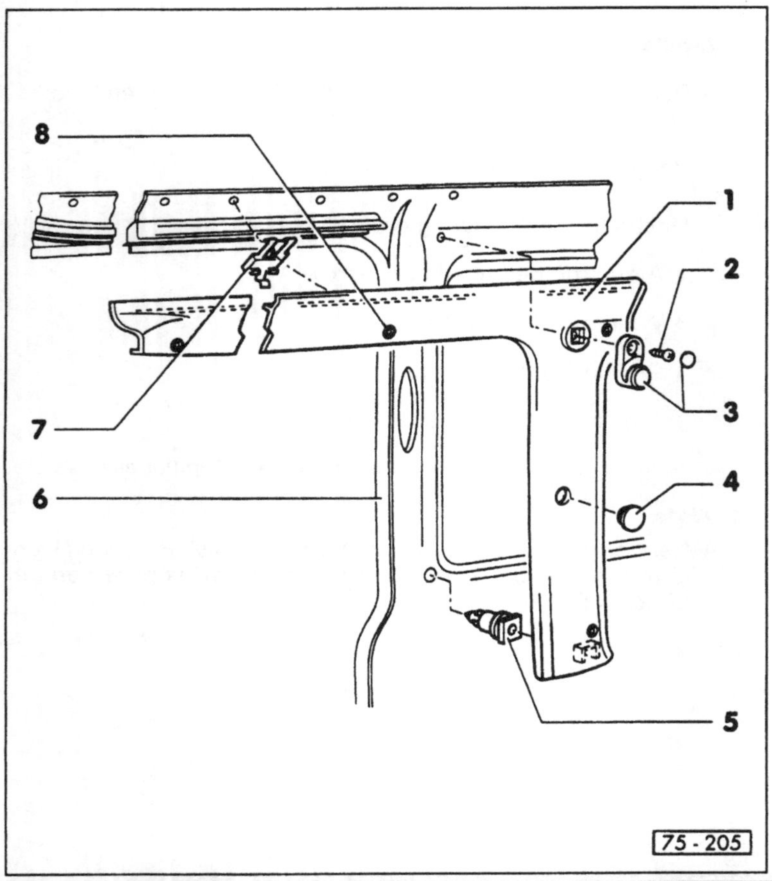

Right side

Note:

To remove C-pillar trim, remove upper B-pillar trim first. *⇒ page R 75-77.*

1 - **C-pillar trim**
- ♦ Removing/installing ⇒ page R 75-86

2 - **Recessed screw**

3 - **Coat hook**
- ♦ With cap for recessed screw.

4 - **Cap**
- ♦ Securing point, seat belt.

5 - **Clip**
- ♦ 2-part
- ♦ When mounting, insert sleeve in C-pillar mounting hole, and pin in trim.

6 - **Seal**
- ♦ After mounting trim, ensure that seal is positioned correctly.

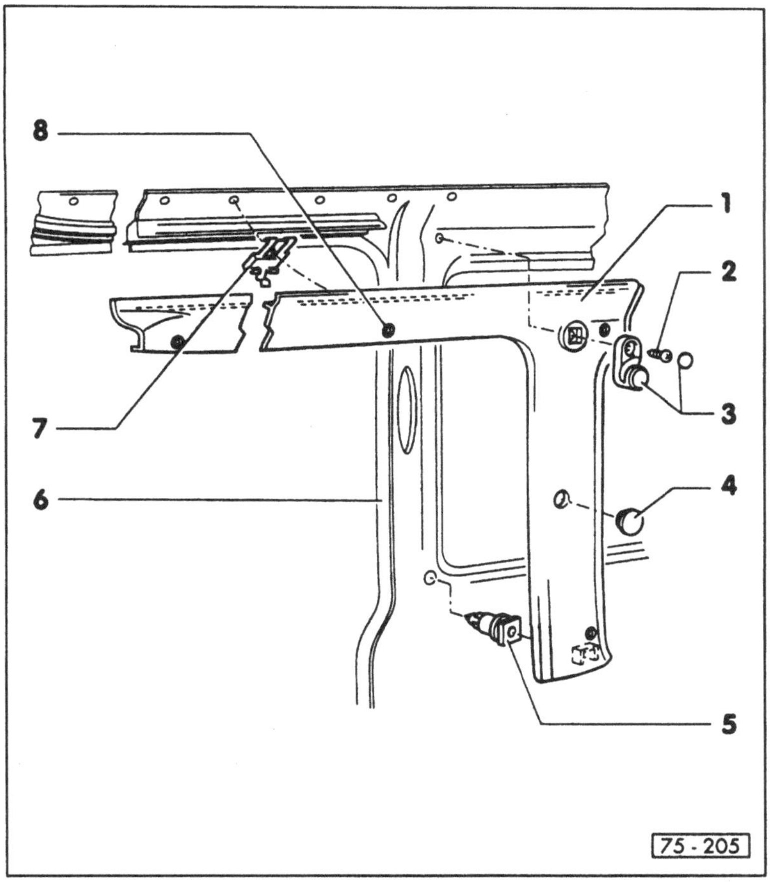

7 - Bracket

8 - Press-stud
- Screws onto trim.

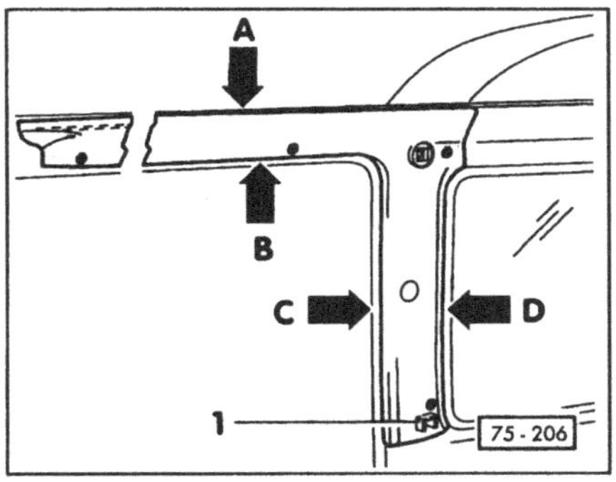

C-pillar trim right, removing

– Remove upper B-pillar trim ⇒ page R 75-77.

– Pry out coat hook cap, remove coat hook.

◄ – Unclip trim in following order:

- Along molded headliner -arrow A-
- Along door runner -arrow B-
- C-pillar center bottom, clip -1-
- Right side bottom -arrow C-
- Window side top -arrow D-

Note:

When removing trim in window area -arrow D- unclip small sections, support from back by hand or with tool 80–200.

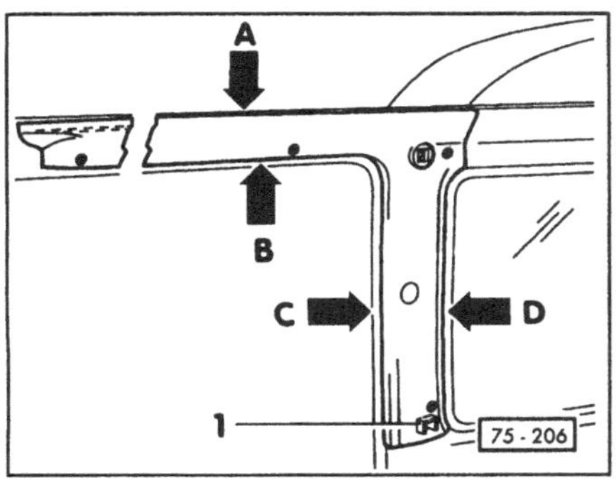

C-pillar trim right, installing

◄ – Position trim, and ensure clip -1- is positioned correctly.

– Clip trim on C-pillar in window area -arrow C-, and in brackets on roof frame -arrow A-.

– Ensure seal -arrows B and C- is positioned correctly.

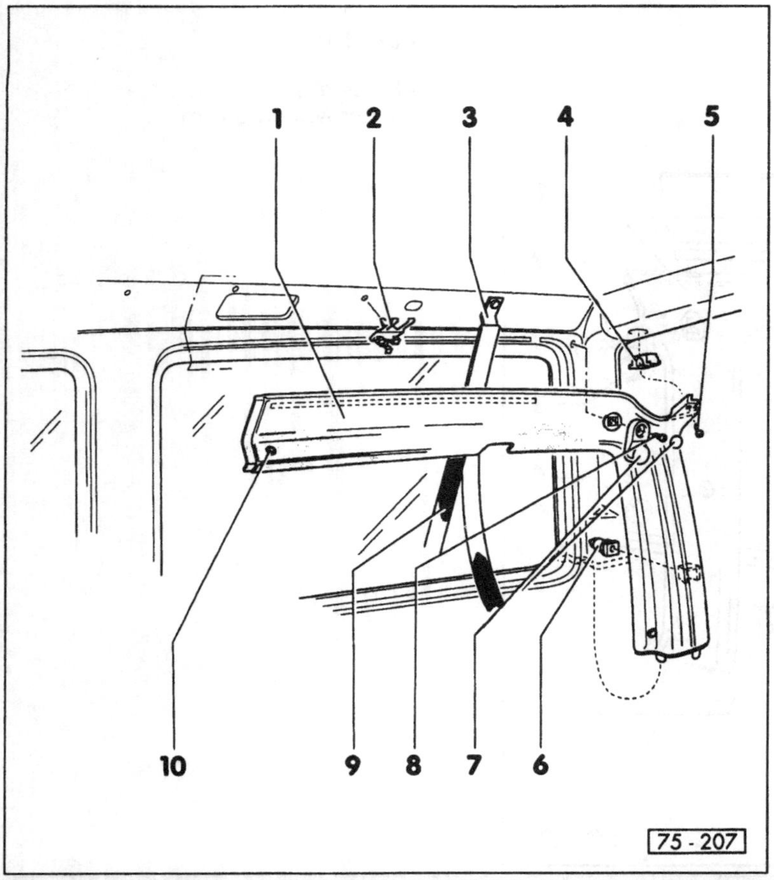

D-pillar trim, component layout

Notes:

- *Procedure is for one side, opposite side is mirror-image.*
- *To remove D-pillar trim, remove rear roof frame trim first, ⇒ page R 75-55, upper B-pillar trim ⇒ page R 75-77 and C-pillar trim ⇒ page R 75-82.*

1 - D-pillar trim
- Removal ⇒ page R 75-89
- Installation ⇒ page R 75-90

2 - Bracket

3 - Anchor, seat belt
- Bolted on window pillar with hex bolt
- 40 Nm (30 ft lb)

4 - Bracket

5 - Recessed screw

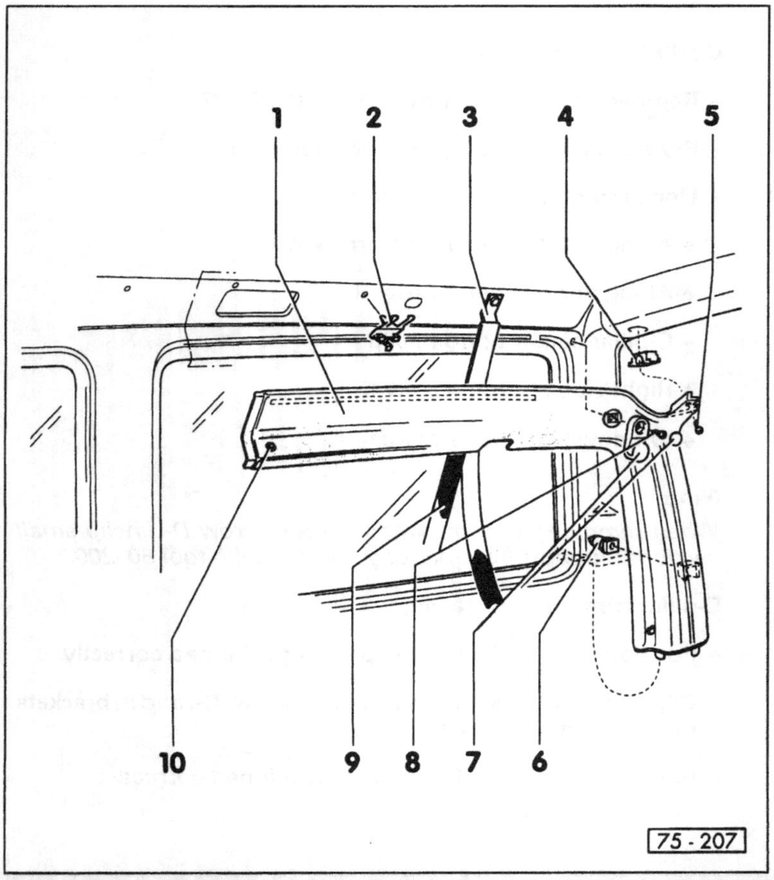

6 - Clip
- 2-part
- When mounting, insert sleeve in D-pillar mounting hole and pin in trim.

7 - Coat hook
- With cap for recessed screw

8 - Recessed screw

9 - Seat belt

10 - Press-stud
- Screws onto trim.

D-pillar trim, removing

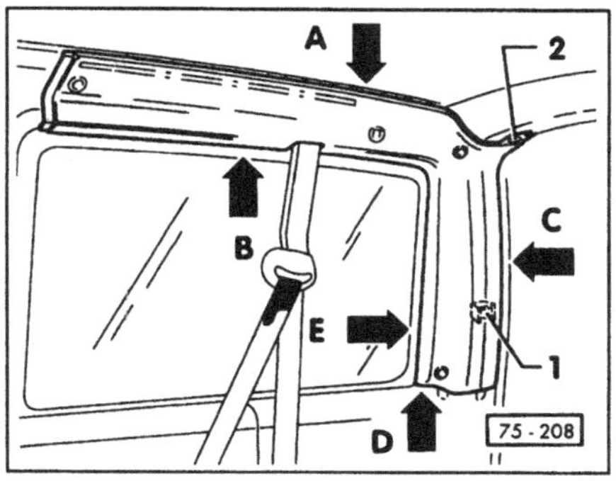

- Remove rear roof frame trim ⇒ page R 75-55, upper B-pillar trim ⇒ page R 75-77 and C-pillar trim ⇒ page R 75-82.
- Pry out coat hook cap, remove coat hook.
- Remove recessed screw -2-.
- Unclip trim in the order indicated:
  - ♦ Along molded headliner -arrow A-
  - ♦ Window pillar -arrow B-
  - ♦ D-pillar at rear -arrow C-
  - ♦ Loosen clip -1-
  - ♦ Carefully detach underside -arrow D-
  - ♦ D-pillar at front -arrow E-

Note:

Removing trim in window area -arrows B and E-, unclip in short sections, support from back by hand or with tool 80–200.

D-pillar trim, installing

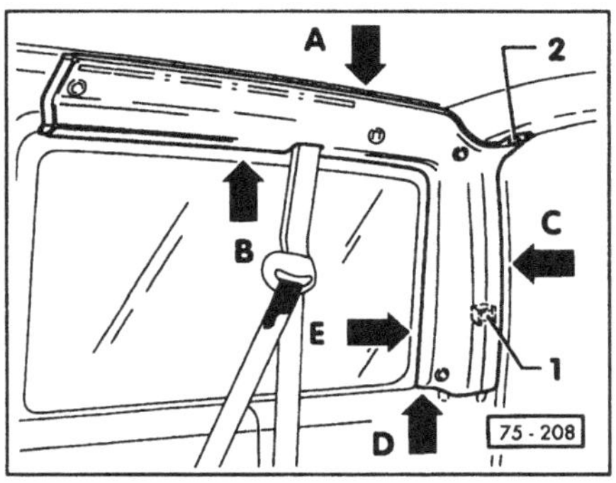

- Latch trim on underside -arrow D–, position trim and ensure clip -1- is positioned correctly.
- Clip trim on D-pillar, on roof frame brackets -arrow A- and window area -arrows B and E-
- Ensure seal -arrow C- is positioned correctly.

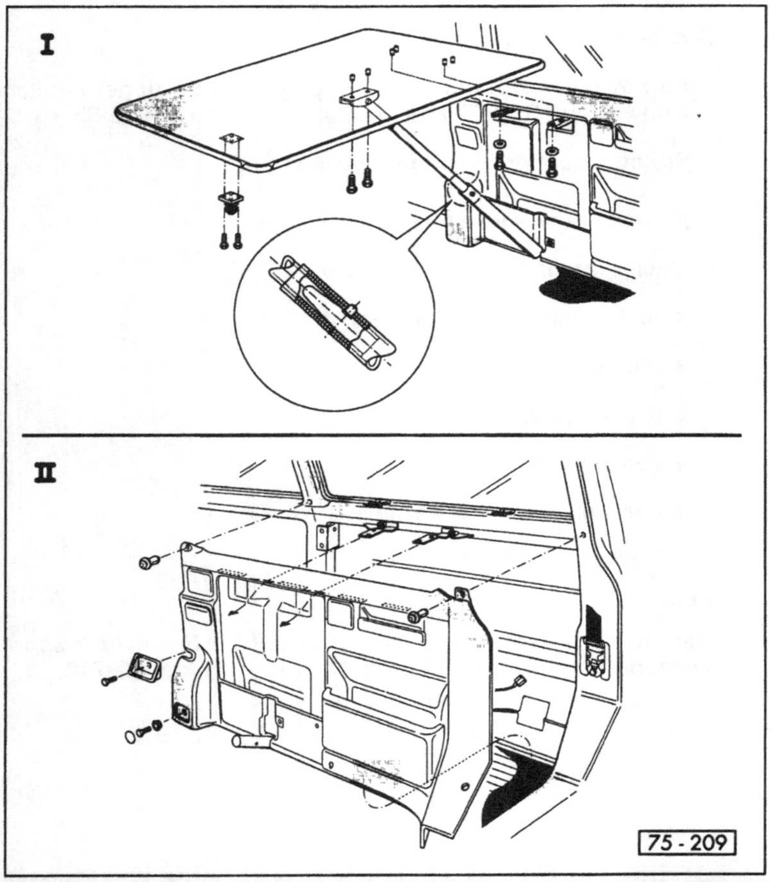

Front side panel molded trim, component layout

Part I ⇒ page R 75-92

Part II ⇒ page R 75-93

Notes:

- *To remove front molded trim, remove upper B-pillar trim: page R 75-77 and C-pillar trim ⇒ page R 75-82.*
- *Only loosen side panel molded trim at rear ⇒ page R 75-99.*
- *Single seat (passenger compartment) is to be removed ⇒ page R 77-12.*

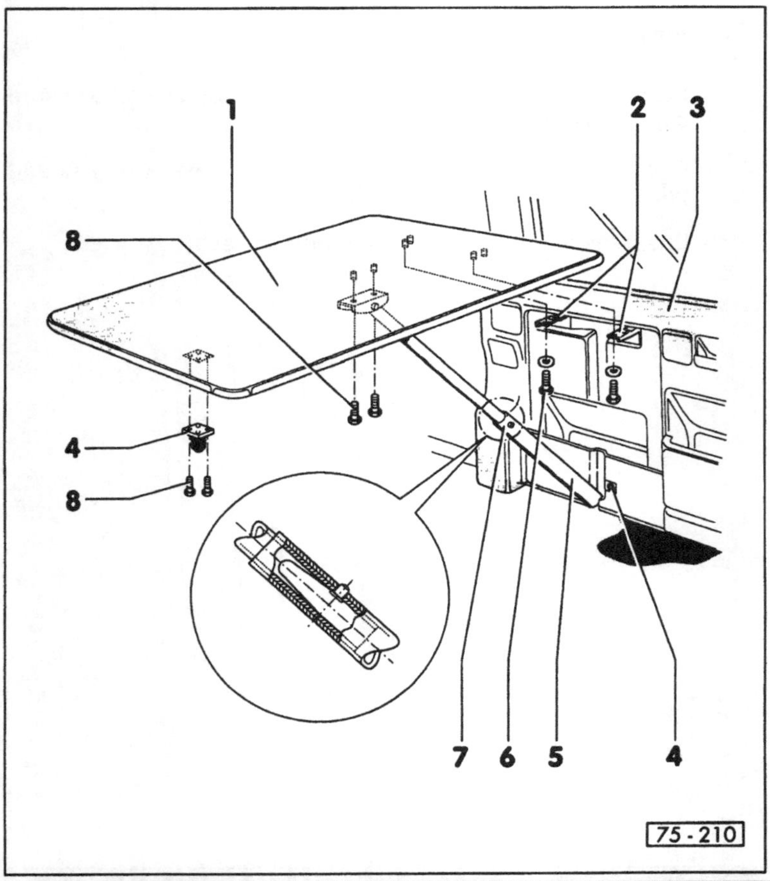

Part I

1 - Table top

2 - Mount
- Screws on body

3 - Side panel molded trim, front

4 - Catch
- Table folded down

5 - Support

6 - Bolt
- With washer

7 - Catch
- To remove table, disconnect support

8 - Bolt

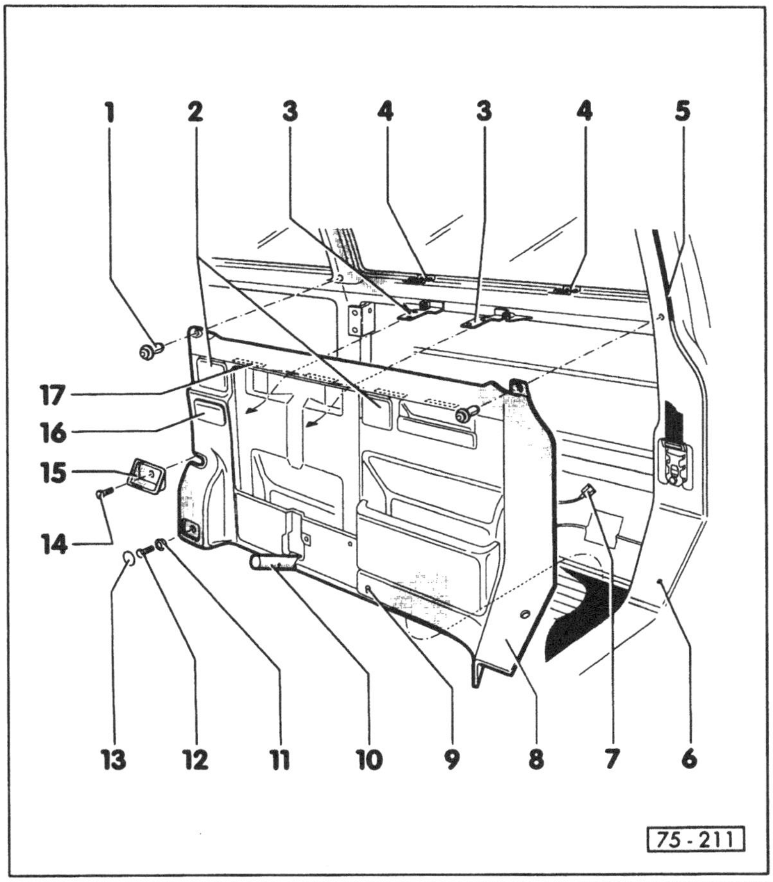

Part II

1 - Clip
- 2-part
- Removal: Remove inner section with Allen wrench.
- Installation: press in outer section and inner section simultaneously into mounting hole.

2 - Cup holder
- Removal ⇒ fig. 1

3 - Mount
- Fastened onto body
- For table top

4 - Bracket

5 - Seal
- After mounting side panel molded trim front, ensure seal makes proper contact.

6 - Securing point for seat belt

7 - Connector, 12 V Socket -U5-

8 - Side panel molded trim, front
- Removal ⇒ page R 75-96
- Installation ⇒ page R 75-98

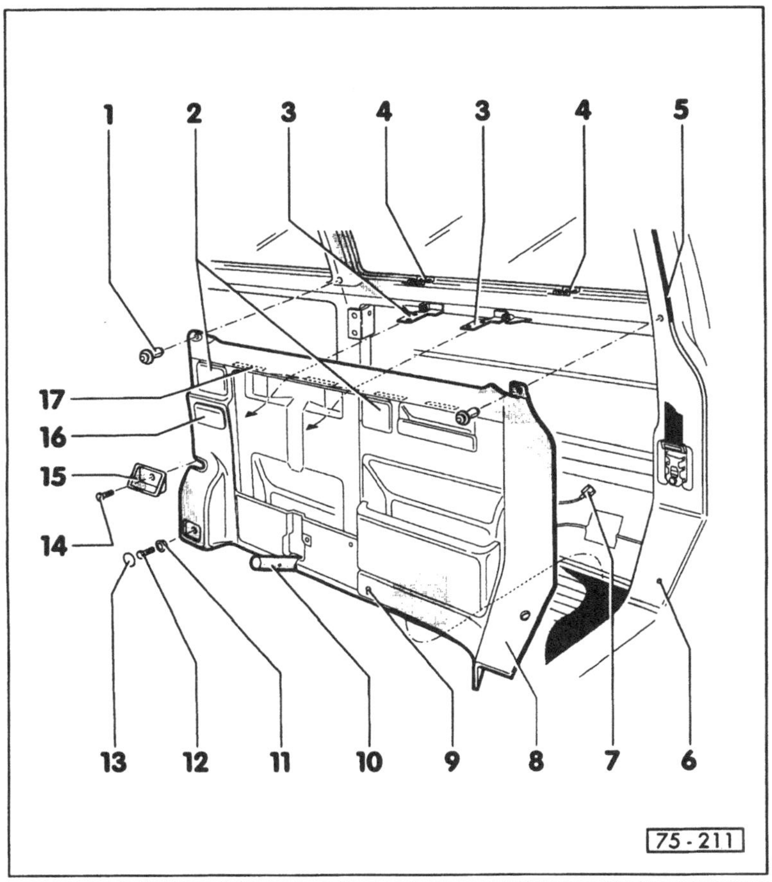

9 - 12 V Socket -U5-

10 - Support, bottom section
- Fastened to body

11 - Washer

12 - Hex bolt

13 - Cap

14 - Recessed screw

15 - Belt guide with locking panel

16 - Ash tray
- Removal ⇒ fig. 2

17 - Clip

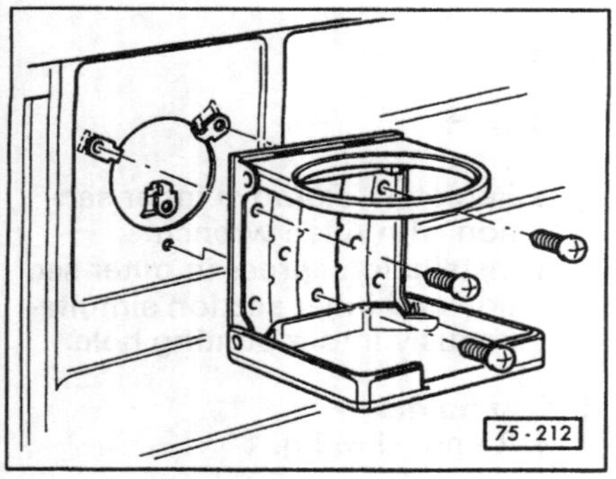

Fig. 1 Cup holder, removing

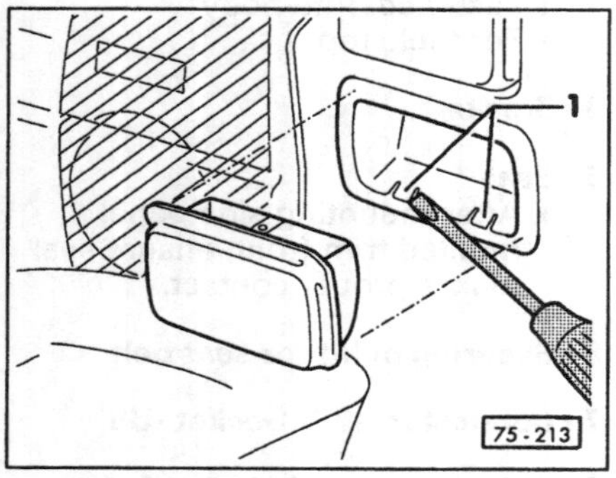

Fig. 2 Ash tray, removing

- Remove ash tray (inner section).
- Pry retaining lugs -1- over edge of molded trim recess with screwdriver.

Front molded side panel trim, removing

- Remove upper B-pillar trim ⇒ page R 75-77 and C-pillar trim ⇒ page R 75-82.
- Remove single seat (passenger compartment) ⇒ page R 77-12.
- Unlatch table by pressing inward, and fold out.
- Loosen bolts -1- on table top mount and disconnect support at catch -2-.
- Remove table top with top section of support.

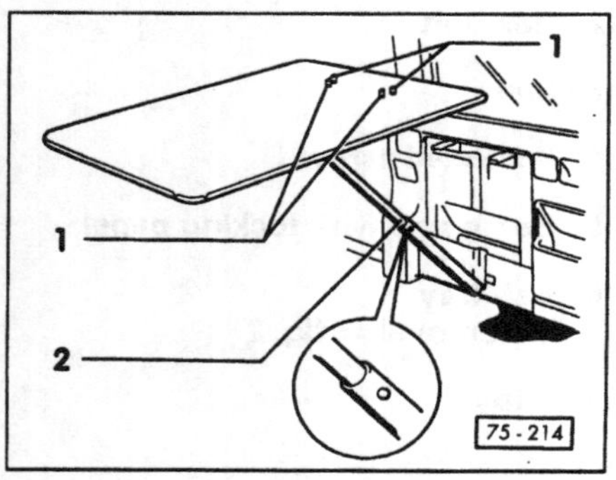

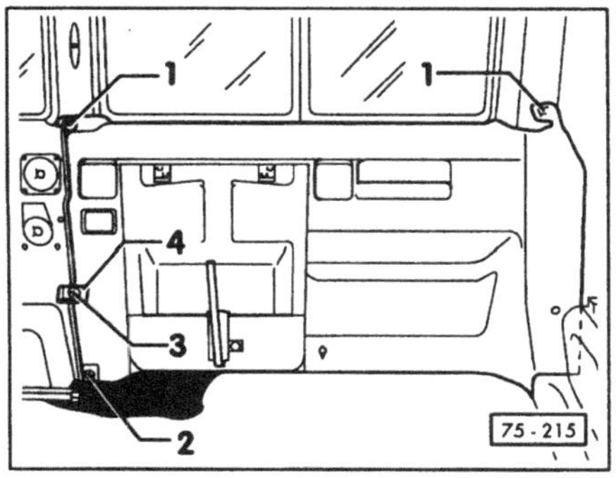

- Remove inner sections of clips -1- with Allen wrench.
- Pry off cap -2-, remove hex bolt with washer.
- Remove recessed screw -3-, carefully pry out belt guide with locking panel -4-.
- Loosen molded trim at rear in area of overlap with side panel molded trim front ⇒ page R 75-96.

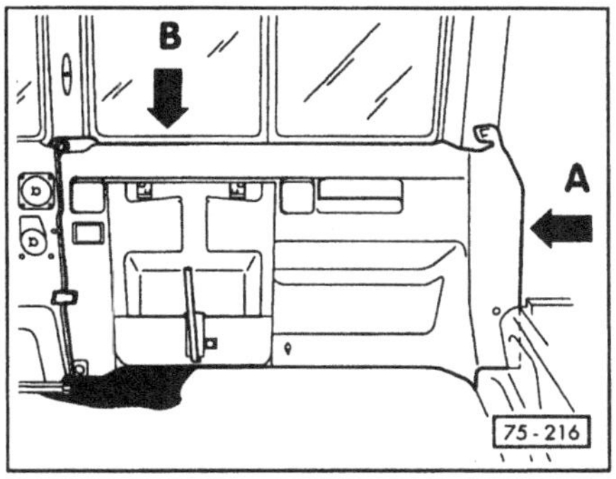

- Unclip side panel molded trim front, in following order:
  - ♦ B-pillar -arrow A-
  - ♦ Window pillar -arrow B-

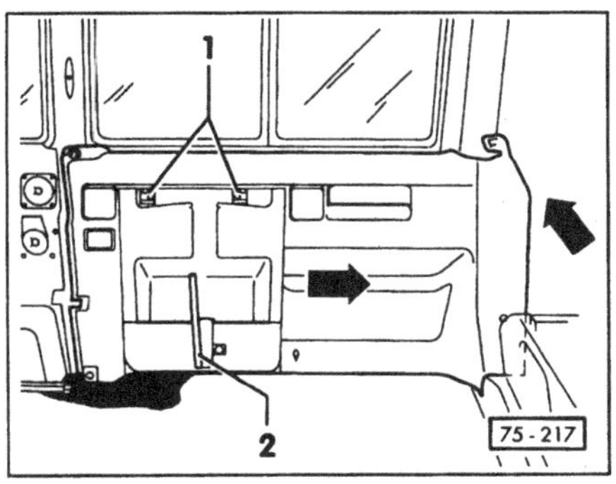

- Pull off front of side panel molded trim, near B-pillar, push as far as possible in direction of travel.
- Disconnect connector for 12 V Socket -U5-.
- Pull off molded trim -top- near table -1- mount, fold down mount and lift molded trim over bottom support -2-.

Front molded side panel trim, installing

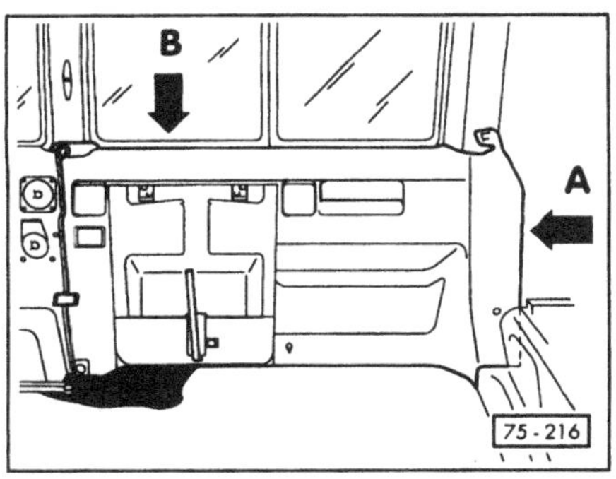

- Install in reverse sequence and note the following:
- Clip side panel molded trim in at front, ensure seal -arrow A- is positioned correctly.

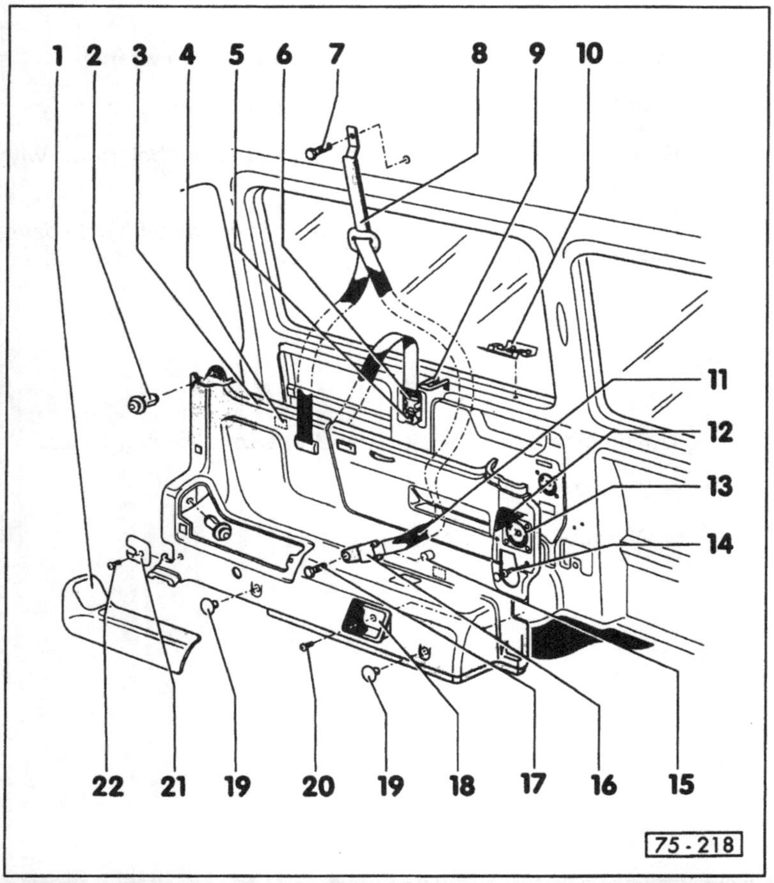

Rear side panel molded trim, component layout

Notes:

- *Procedure is for one side, opposite side is mirror-image.*
- *Single seat (passenger compartment) ⇒ page R 77-12, bench ⇒ page R 77-13, bedboard ⇒ page R 77-15, remove together with left or right guide rail and supporting frame.*
- *To remove molded trim at rear remove roof frame trim at rear ⇒ page R 75-55, upper B-pillar trim ⇒ page R 75-77, C-pillar trim ⇒ page R 75-82 and D-pillar trim ⇒ page R 75-89, and loosen left and right side sill covers.*

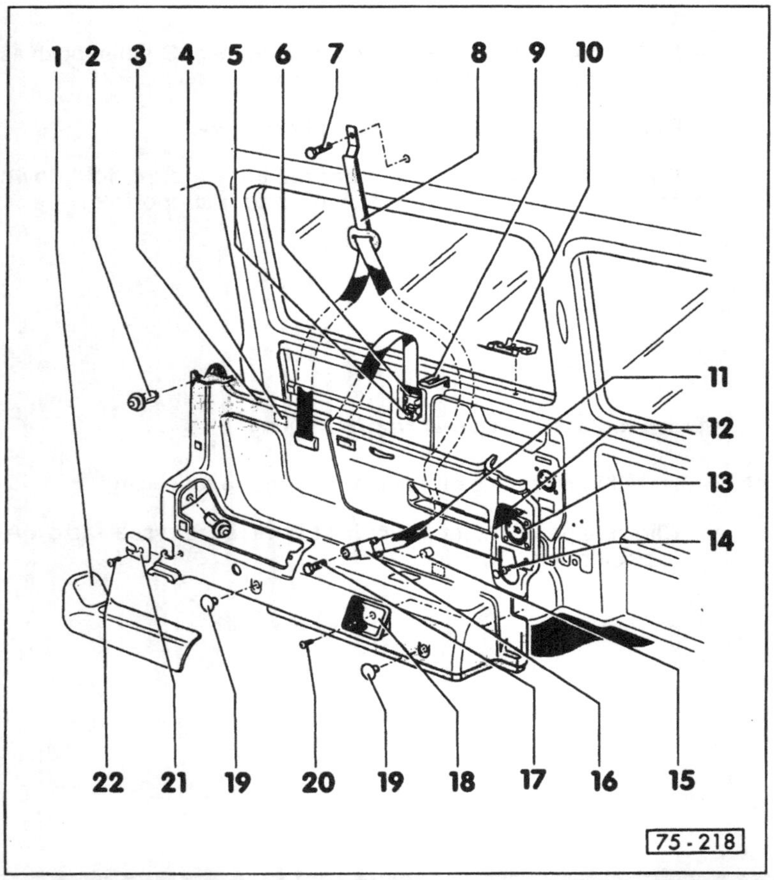

1 - Cover

2 - Clip
- 2-part
- Removal: Remove inner section with Allen wrench.
- Installation: Press outer section and inner section simultaneously in mounting hole.

3 - Side panel molded trim, rear
- Removal ⇒ page R 75-102
- Installation ⇒ page R 75-103

4 - Clip

5 - Bolt
- 40 Nm (30 ft lb)

6 - Inertia reel

7 - Bolt
- 40 Nm (30 ft lb)

8 - Relay fitting, seat belt
- Removable with 2–part trim

9 - Securing point, bench backrest

10 - Bracket

11 - Seat belt

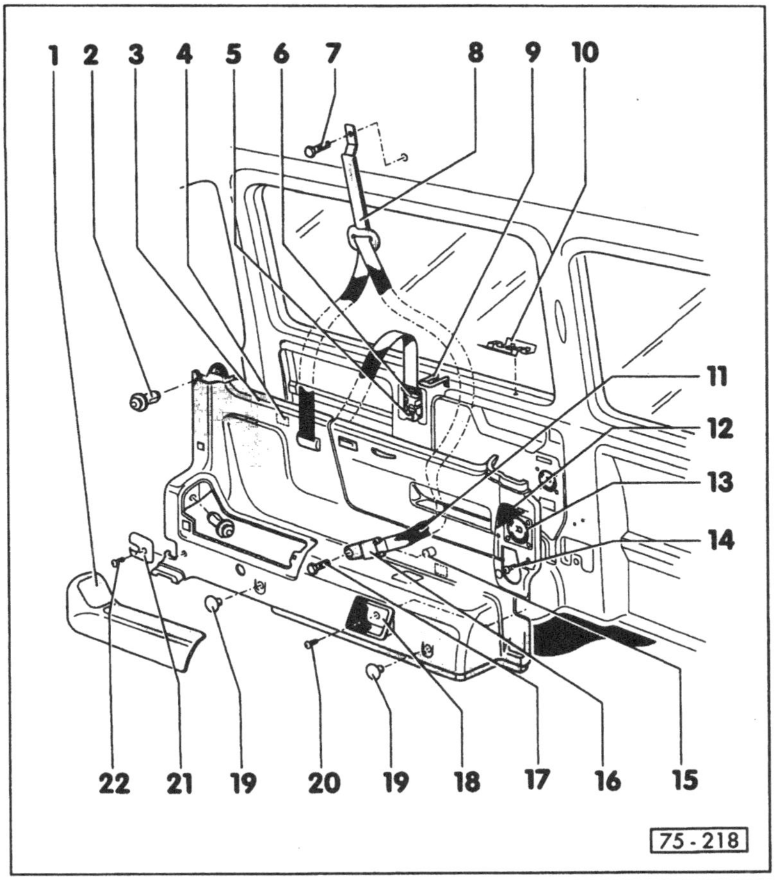

12 - **Speaker cover**

13 - **Speaker with connector**

14 - **Recessed screws**
- With washer and cap

15 - **Spacer sleeve**

16 - **Belt fitting**

17 - **Bolt**
- 40 Nm (30 ft lb)
- With cap

18 - **Belt guide with locking panel**

19 - **Plug**

20 - **Recessed screw**

21 - **Cover, guide blocks**

22 - **Recessed screw**

Rear molded side panel trim, removing

– Remove single seat (passenger compartment) ⇒ page R 77-12, bench ⇒ page R 77-13, bedboard ⇒ page R 77-15 with guide rail or supporting frame on corresponding side.

– Remove roof frame trim at rear ⇒ page R 75-55, upper B-pillar trim ⇒ page R 75-77, C-pillar trim ⇒ page R 75-82 and D-pillar trim ⇒ page R 75-89.

– Loosen sill covers left/right side.

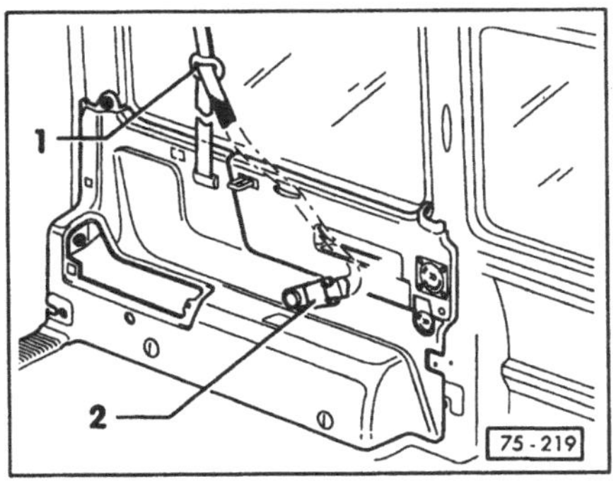

◄ – Remove anchor -1- and belt anchor -2-.

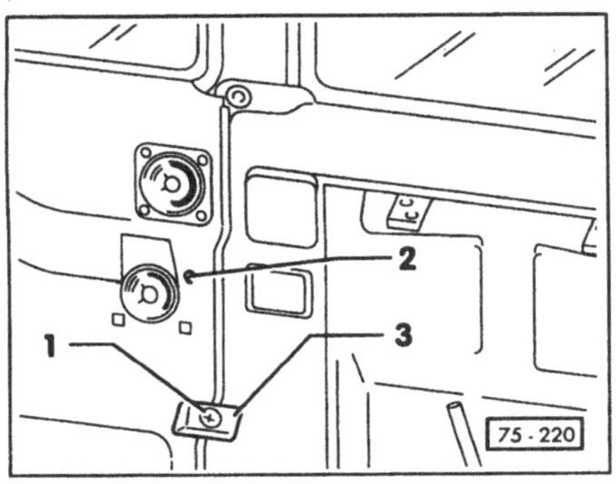

◄ – Remove Phillips screws -1- and -2- (with cap), carefully lift out belt guide with locking panel -3-.

Note:

Depending on design, Phillips screw -2- may be located under speaker cover; if so, remove it first.

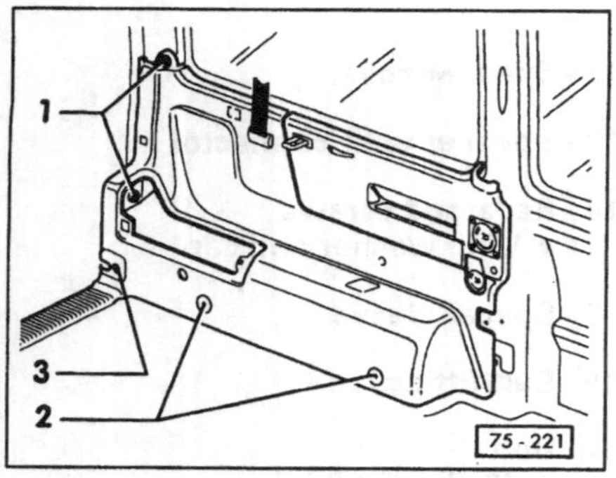

- Remove inner sections of clips -1- with hex socket wrench.
- Remove or pry out plugs -2- with screwdriver.
- Remove cover of guide blocks -3-.

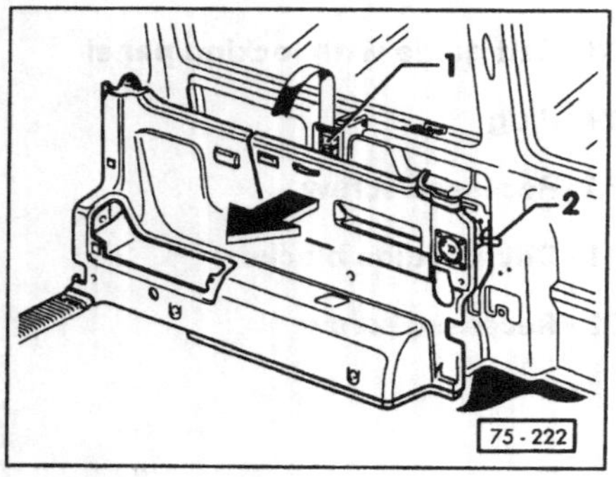

- Remove molded trim from window pillar in direction of arrow -arrow-.
- Remove inertia reel -1- and if necessary disconnect connector for speaker -2-.

Rear molded side panel trim, installing

- Install in reverse sequence, note the following:
- After installing rear molded side panel trim, ensure tailgate seal is positioned correctly.

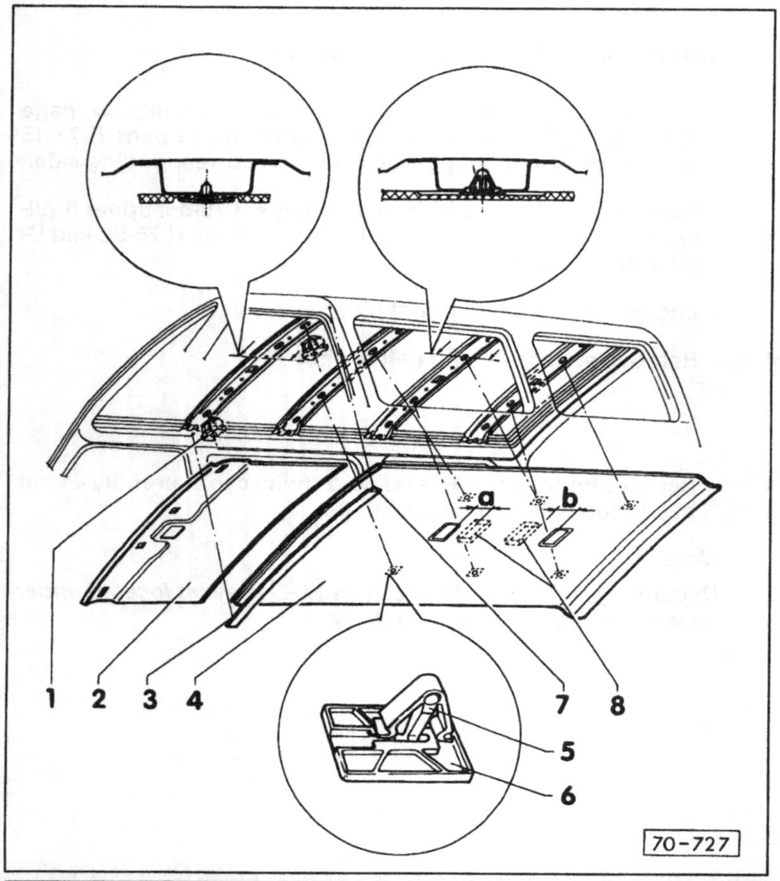

Molded headliner, component layout

Notes:

- *To remove molded headliner in driver's compartment remove upper B-pillar trim ⇒ page R 75-77, A-pillar trim ⇒ page R 75-46.*
- *To remove molded headliner in passenger compartment first remove upper B-pillar trim ⇒ page R 75-77, C-pillar trim page R 75-82 and D-pillar trim page R 75-89 as well as roof frame at rear ⇒ page R 75-55.*

1 - Bracket
- two, one on each side

2 - Molded headliner, driver's compartment
- Removal ⇒ page R 75-106

3 - Masking strip

4 - Molded headliner, passenger compartment
- Removal ⇒ page R 75-107
- Installation ⇒ page R 75-109

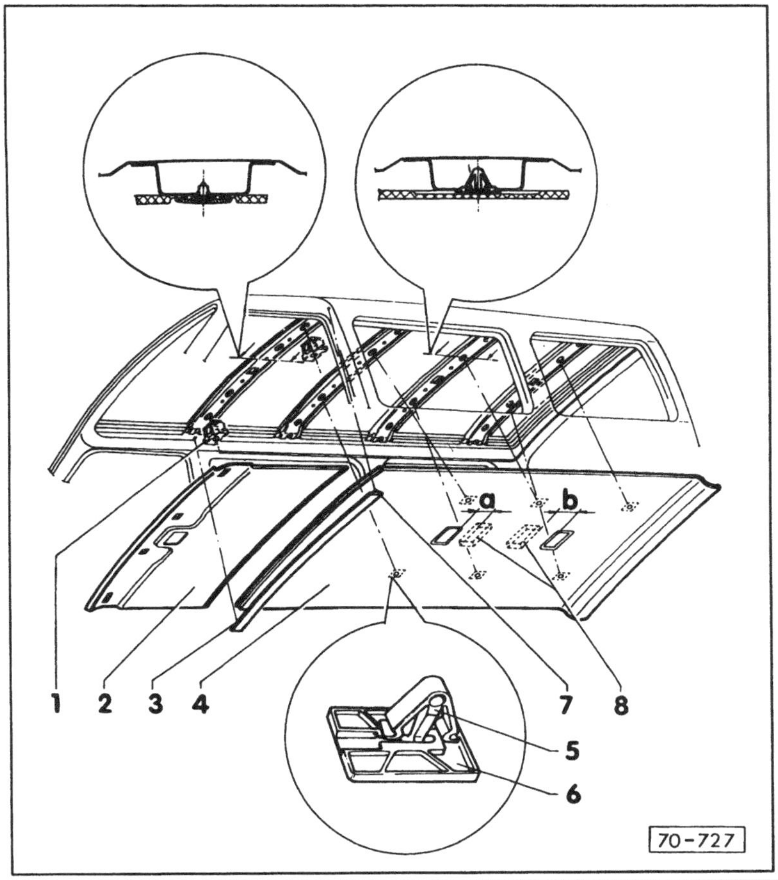

5 - Clip
- six

6 - Clip holder
- six
- self-adhesive

7 - Clip
- eight

8 - Spacer pad
- Not applicable with air-conditioner
- Self-adhesive, is stuck on back of molded headliner in passenger compartment.
- a = 60 mm (2.4 in)
(Rear edge of cut-out for front interior light to front edge of spacer pad.)
- b = 60 mm (2.4 in)
(Front edge of cut-out for rear interior light to rear edge of spacer pad.)
- two

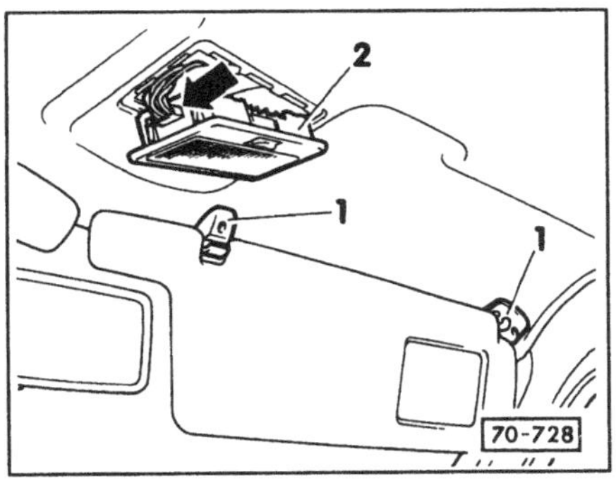

Molded headliner in driver's compartment, removing

- Remove upper B-pillar trim ⇒ page R 75-77.
- Remove A-pillar trim ⇒ page R 75-46.
- Remove holders -1- for sun visors and disconnect connectors for mirror lighting.
- Pry out interior light -2- and disconnect connector -arrow-.

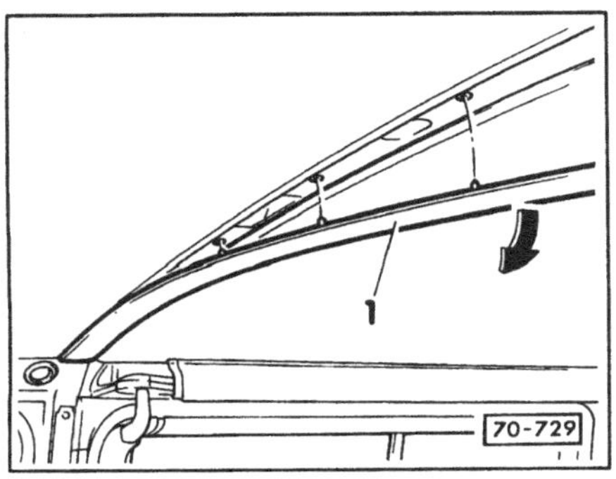

- Unclip masking strip -1-.
- Remove driver's compartment molded headliner downward.

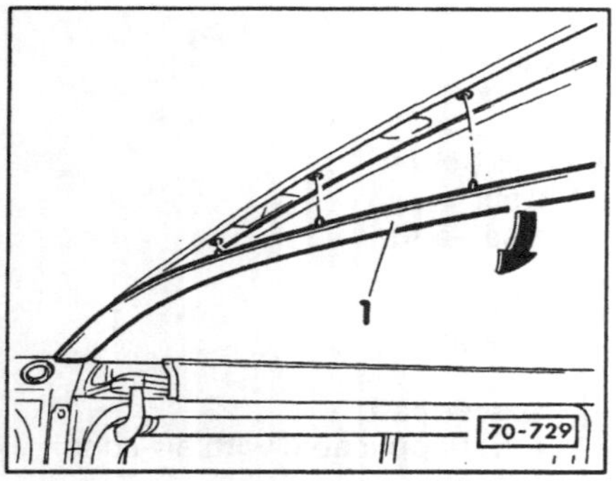

Molded headliner in passenger compartment, removing

– Remove upper B-pillar trim ⇒ page R 75-77.

– Remove C-pillar trim ⇒ page R 75-82.

– Remove D-pillar trim ⇒ page R 75-89.

– Remove roof frame trim at rear ⇒ page R 75-55.

◄ – Unclip masking strip -1-.

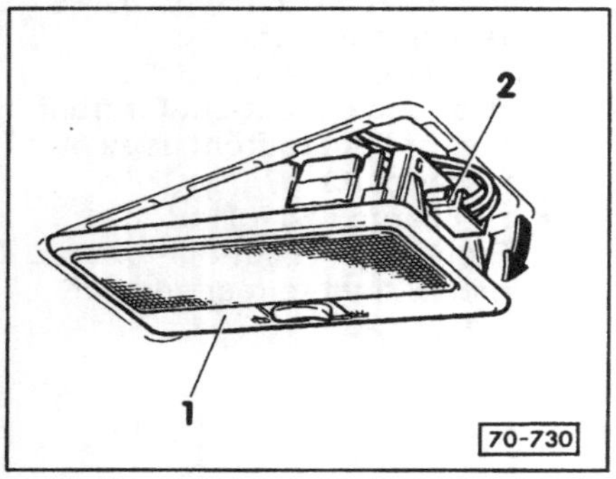

◄ – Pry out interior lights -1- in passenger compartment, disconnect connectors -2-.

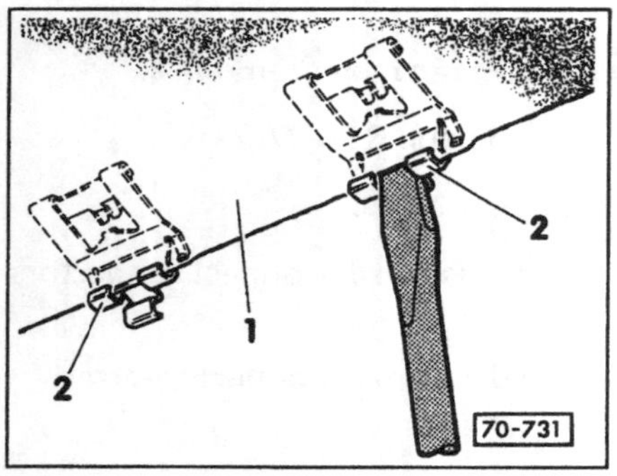

◄ – Carefully pry out molded headliner -1- from brackets -2- on roof frame.

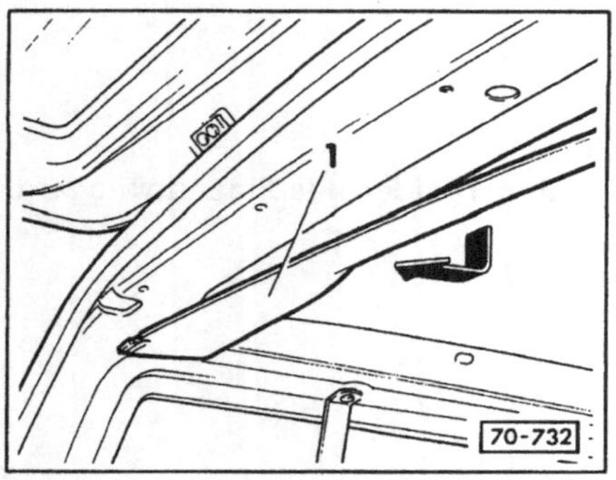

◄ – Lower molded headliner -1- slightly in rear area and pull out of clips backward out of vehicle.

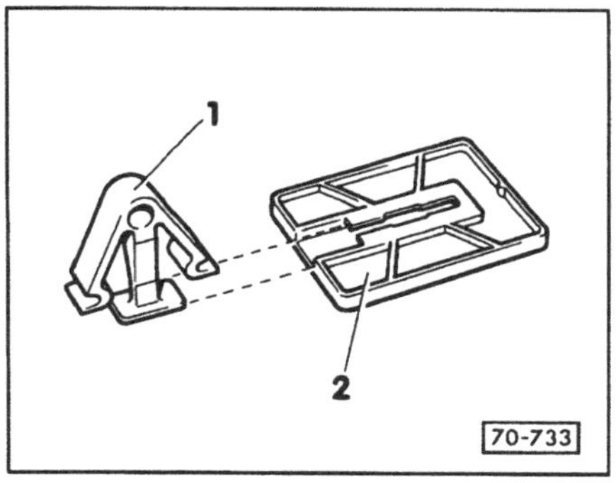

Molded headliner in passenger compartment, installing

◄ – Rotate clips -1- 90°, remove from slits in roof bow and insert into clip holders (on back of molded headliner) -2-.

– Set molded trim in position and clip into place under roof.

Safety Measures

- During repair work on any part of electrical system, the 110/220 V power source must be disconnected.
- 110/220 V repairs must be carried out either by an electrician or with the supervision of an electrician.
- If propane gas lines are opened during repair work, a pressurized leak test must be performed, prior to returning vehicle to service. Use soapy water or equivalent to check fittings.
- Disconnect battery Ground (GND) strap before working on electrical system.
- Note all appropriate regulations!

Electrical equipment, removing (EuroVan CV Camper)

Removing Battery Charger -A10-

> ***CAUTION!***
> ***Before working on any part of electrical system obtain radio code. Disconnect battery Ground (GND) strap.***

◄ – Remove stop -1- from rail.

> ***CAUTION!***
> ***Note safety measures ⇒ page R 76-1!***

– Release catch -2- and pull bench forward in direction of travel.

◄ – Pry off caps -arrows- and remove recessed screws from lower screen cover -1-.

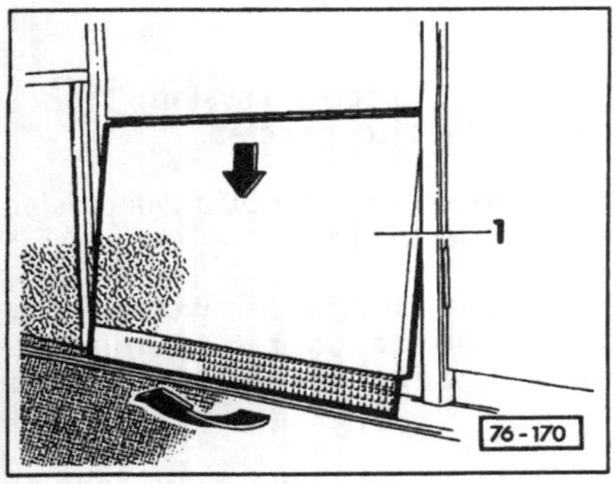

◄ – Press screen cover -1- downward and pull out toward center of vehicle.

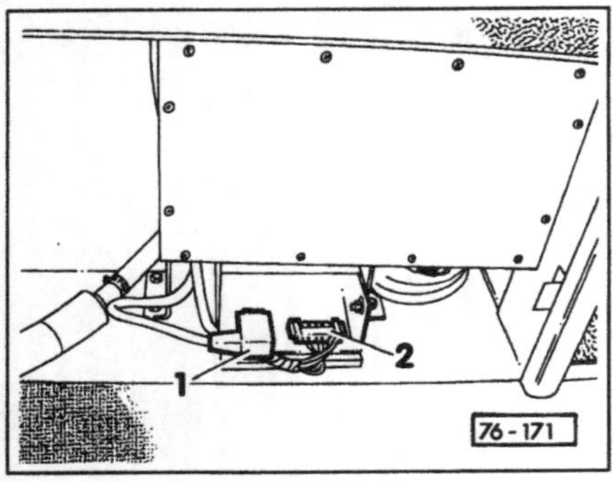

◄ – Pull plug -1- and -2- off Battery Charger -A10-.

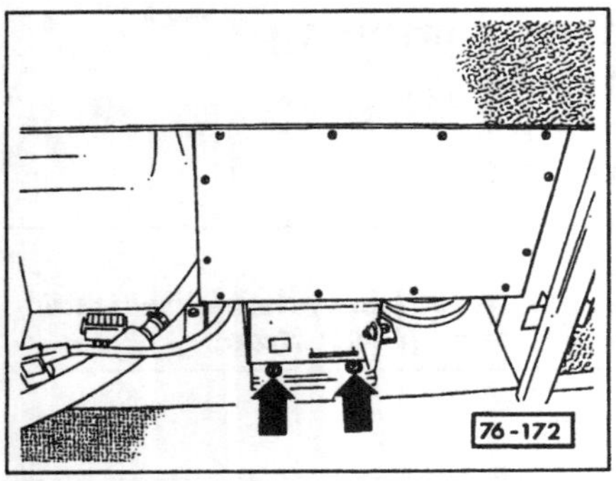

◄ – Remove recessed screws -arrows- and remove Battery Charger -A10-.

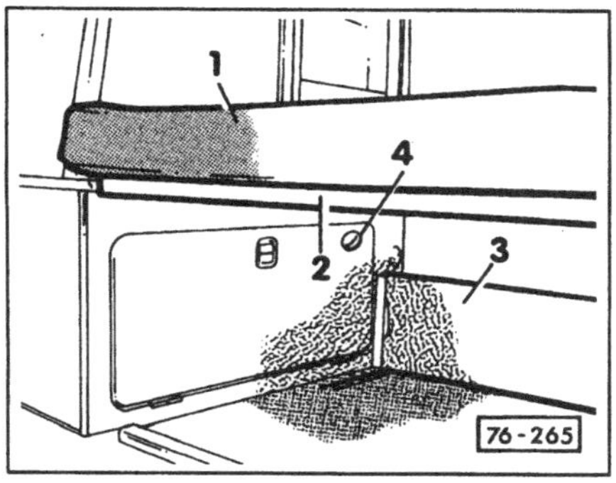

Auxiliary Batteries, removing -A1-

> ***CAUTION!***
> ***Note safety measures ⇒ page R 76-1!***

- Pull bench forward in direction of travel.
- ◄ Remove upholstered top -1-, panel -2- and insert panel -3-.
- Remove screw -4-.

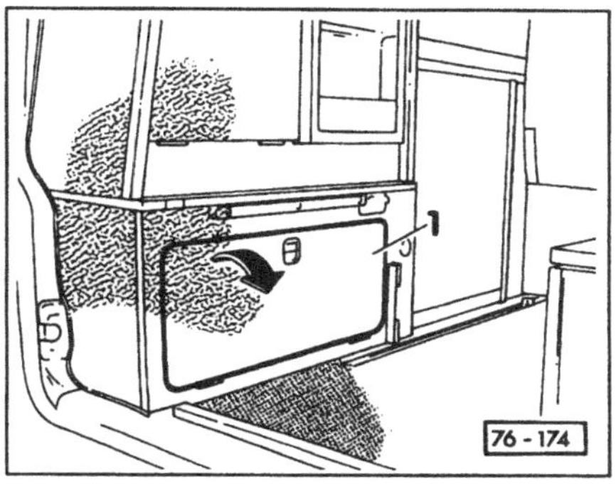

- ◄ Open flap -1- and remove

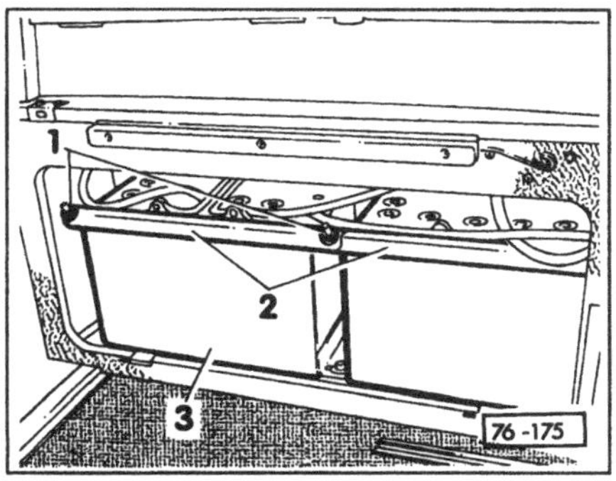

- ◄ Remove hex nuts -1- and remove brackets -2-.
- Lift out rear Auxiliary Battery -A1 - -3-, disconnect connections and insulate positive cable (B+).

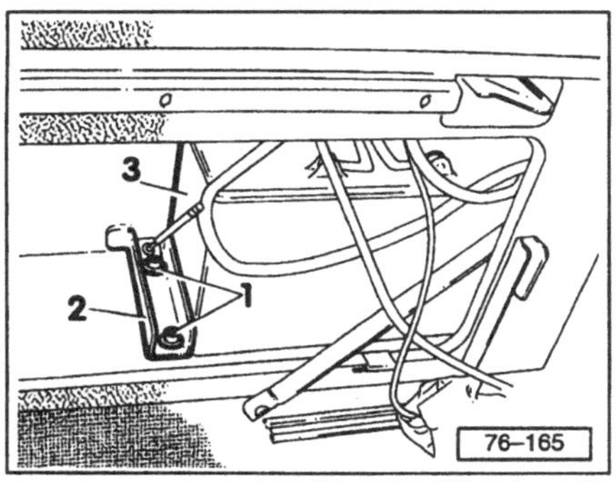

- ◄ Remove hex bolts -1- and take off angle bracket -2-.
- Lift out front Auxiliary Battery -A1- -3-, disconnect connections and insulate positive cable (B+).

110/220 V Exterior Socket -U8-, removing

> ***CAUTION!***
> ***Note safety measures ⇒ page R 76-1!***

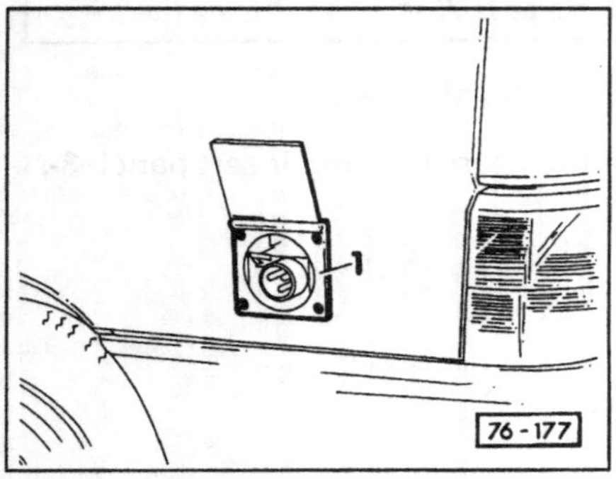

◄ – Remove recessed screws and pull out 110/220 V Exterior Socket -U8- -1-.

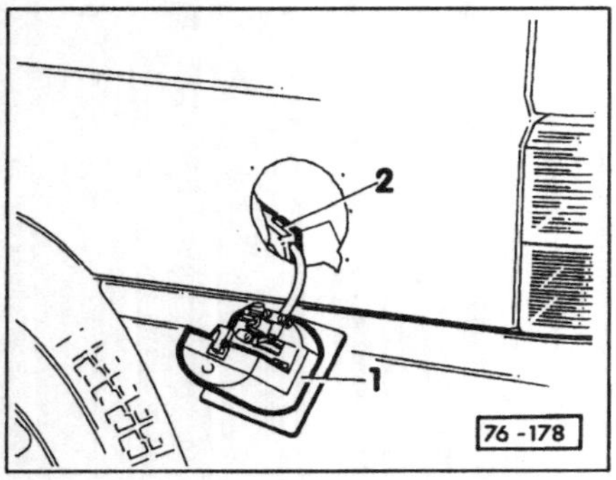

◄ – Disconnect 110/220 V connector -2- and remove 110/220 Exterior Socket -U8- -1-.

Central Electronic Display -E153-, removing

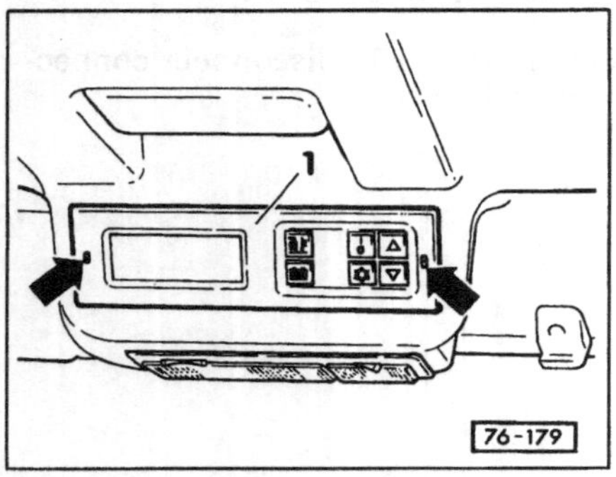

◄ – Press auxiliary tool (e.g. dia. 2.5 mm mandrel) into mounting holes -arrows- to a depth of approx. 20 mm and pull out Camper Central Electronic Display -1-.

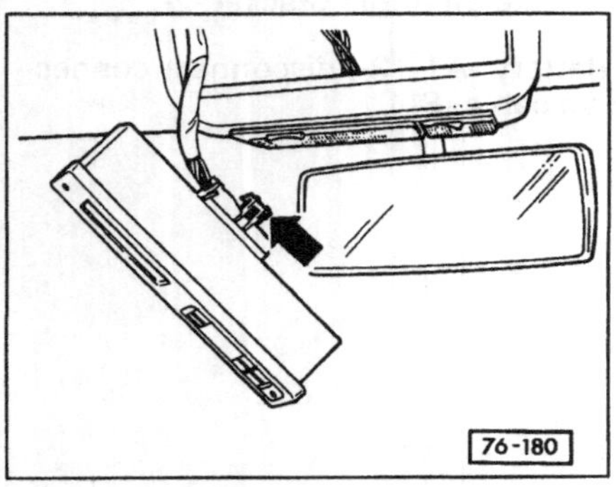

◄ – Pry out securing clip -arrow- and pull off connector.

Central Electronic Display -E153-, replacing

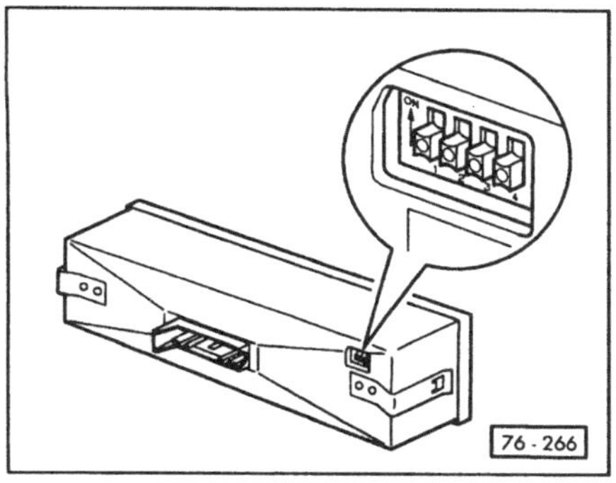

◄ – When installing replacement unit, adjust DIP switches on back to equipment on vehicle.

– Adjust with suitable implement (e.g. a ball–point pen) according to the following table:

DIP switch no.	Switch "OFF"	Switch "ON"
1	one auxiliary battery	two auxiliary batteries
2	battery type Sonnenschein	battery type Deta
3	cooler	refrigerator
4	Degrees Celsius (C°)	Degrees Fahrenheit (F°)

Fluorescent Light right -W8- / Fluorescent Light left -W25-, removing

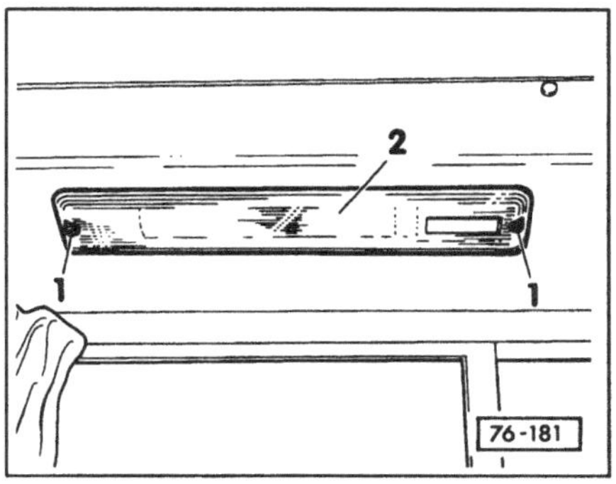

◄ – Remove plastic nuts -1-.

– Remove panel -2-.

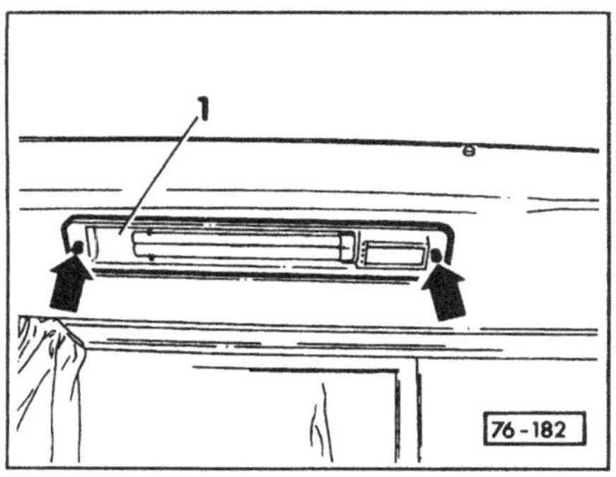

◄ – Remove recessed screws -arrows-.

– Remove lamp housing -1-.

– Disconnect 12 V connectors.

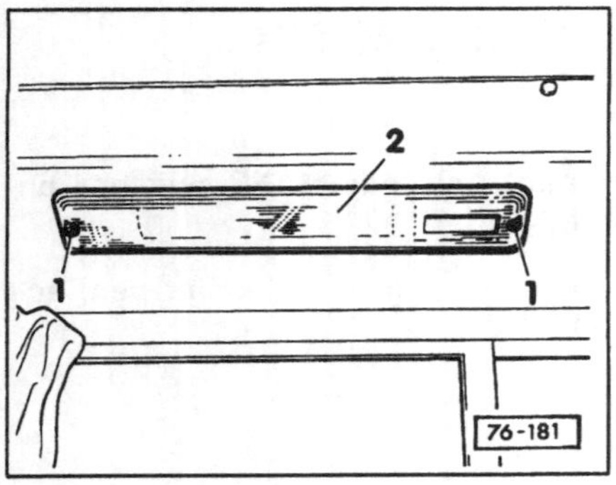

Fluorescent tubes, removing

◄ – Remove plastic nuts -1-.

– Remove panel -2-.

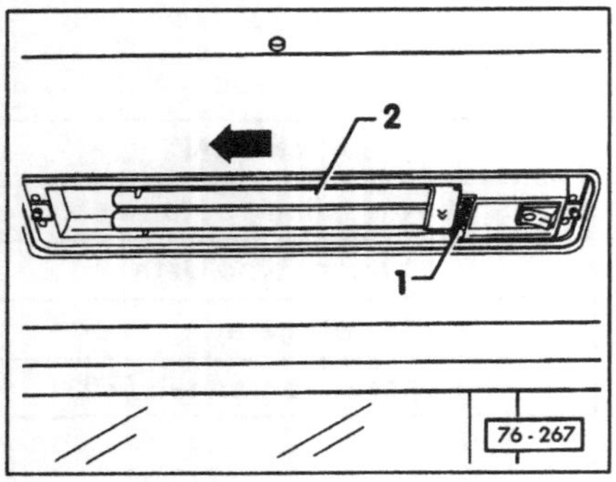

◄ – Loosen four group screws -1- and pull out fluorescent tubes -2- in direction of arrow.

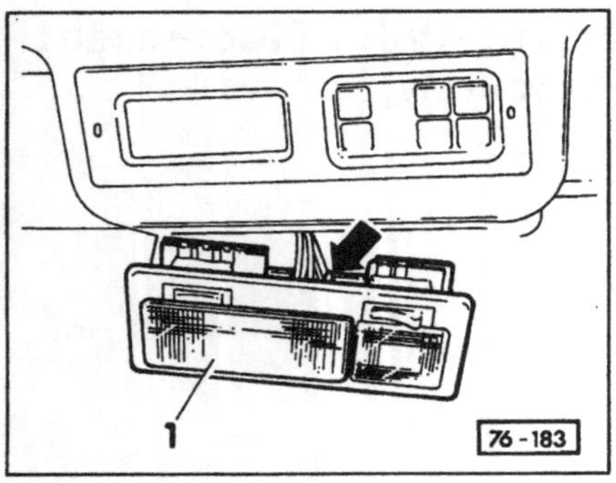

Front Interior Light -W28-, removing

◄ – Pry out Front Interior Light -W28- -1- and disconnect connector -arrow-.

Reading lamp, removing

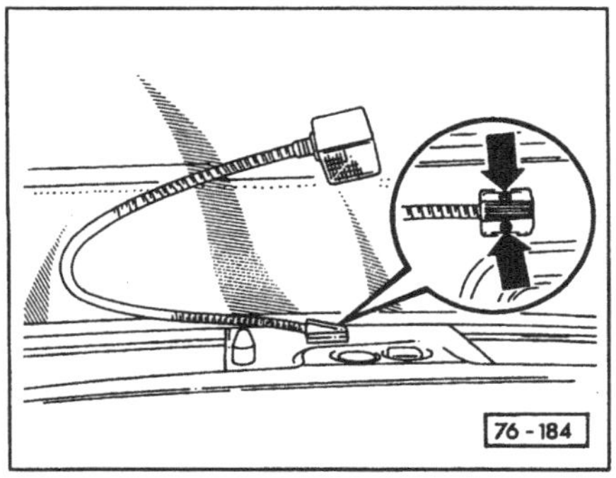

– Remove recessed screws -arrows-.

– Pull cable and grommet out of hole until connectors are visible.

– Disconnect connectors -arrows-.

Refrigerator -I-, removing

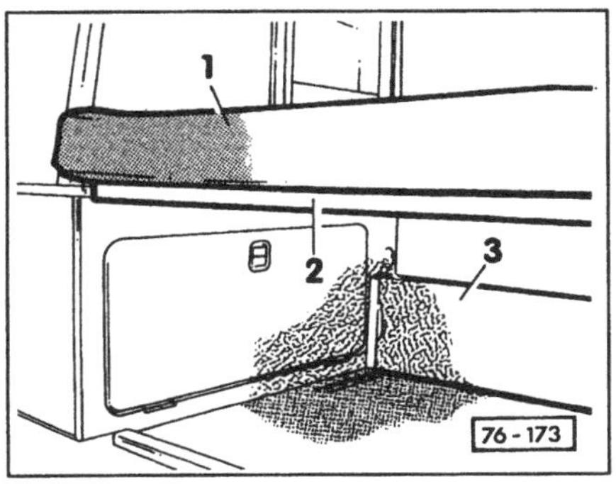

– Remove Battery Charger -A10- ⇒ page R 76-2.

– Remove upholstered top -1-, panel -2- and insert panel -3-.

– Lift kitchenette cover -1-.

– Remove guide rail -2-.

– Open door of left side storage compartment -3- and remove.

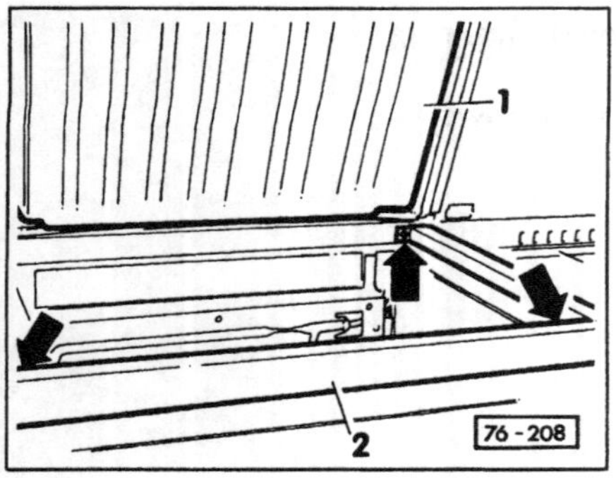

◄ – Raise refrigerator -I- lid -1-.

– Remove frame -2- (three recessed screws -arrows-).

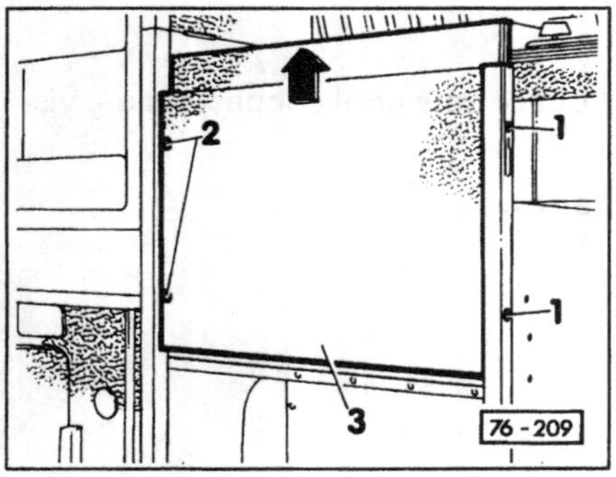

◄ – Remove recessed screws -1-.

– Pry off caps -2-, remove recessed screws and lift out screen cover -3- upward.

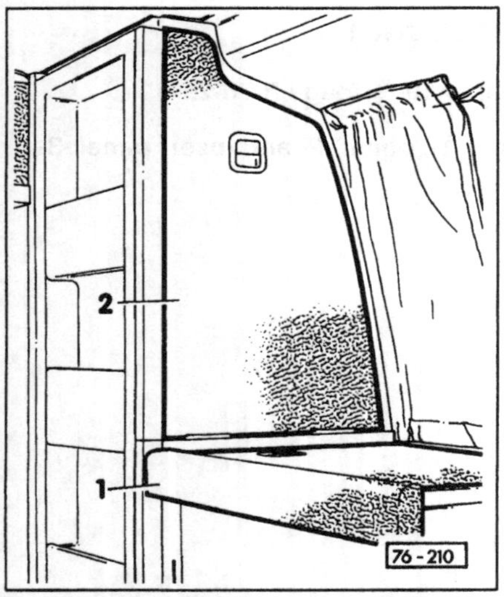

◄ – Close refrigerator -I- -1- lid.

– Open clothes closet -2- flap and remove.

– Open refrigerator -1- lid.

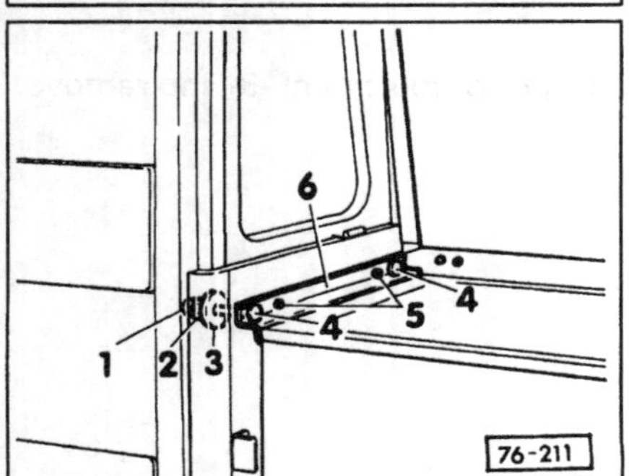

◄ – From inside clothes closet, pry off caps -1- (if applicable-, remove hex nuts -2-, remove washers -3- and press out saucerhead screws -4-.

– Remove recessed screws -5-, if necessary remove refrigerator lift support screws.

– Remove cover -6-.

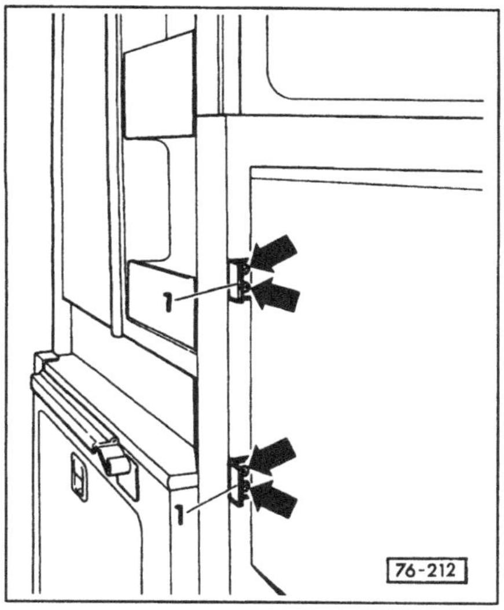

◄ – Remove recessed screws -arrows- and remove angle bracket -1-.

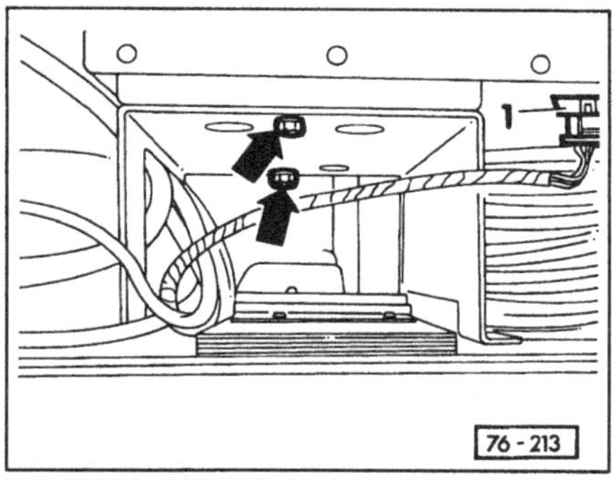

◄ – Disconnect plug -1- on underside of refrigerator -I-.

– Remove hex bolts -arrows- from bottom of refrigerator -I-.

◄ – Remove false bottom -1- from kitchenette left-side storage compartment.

– Remove cover -2- (six recessed screws -arrows-).

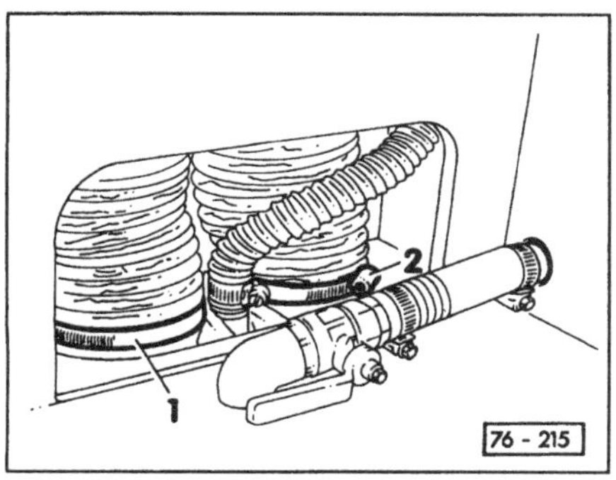

◄ – Loosen screw clamps -1- and -2- and disconnect vent hoses from base connections.

- Pull out cover for cooking stove.
- Remove recessed screws -arrows-.
- Lift out refrigerator -l-.

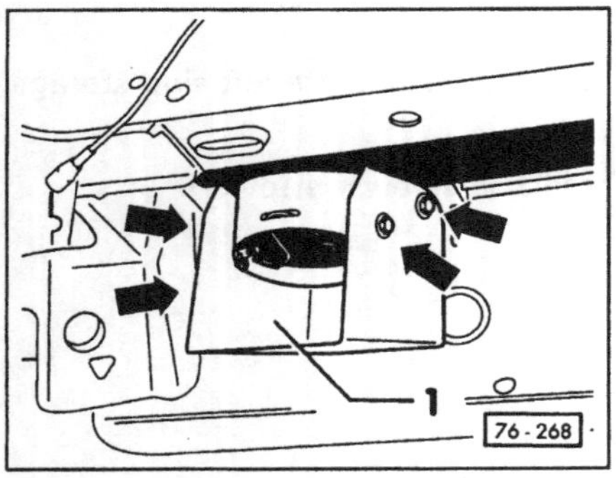

Compressor/refrigerator -l- filter, removing

- Remove securing bolts -arrows- under vehicle floor.
- Remove cover -1-.

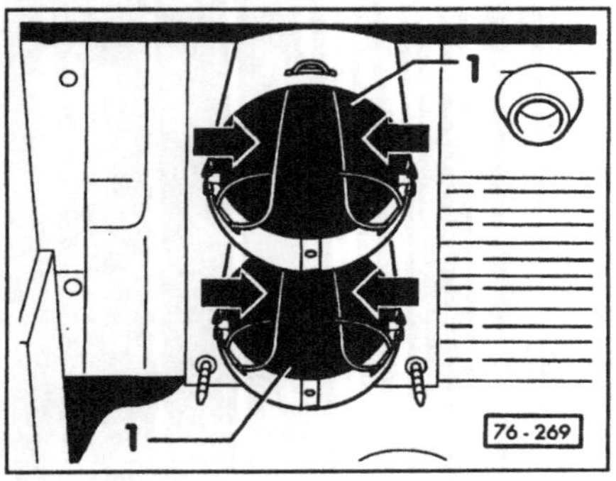

- Press springs together in direction of arrow, and remove.
- Remove filter -1-.

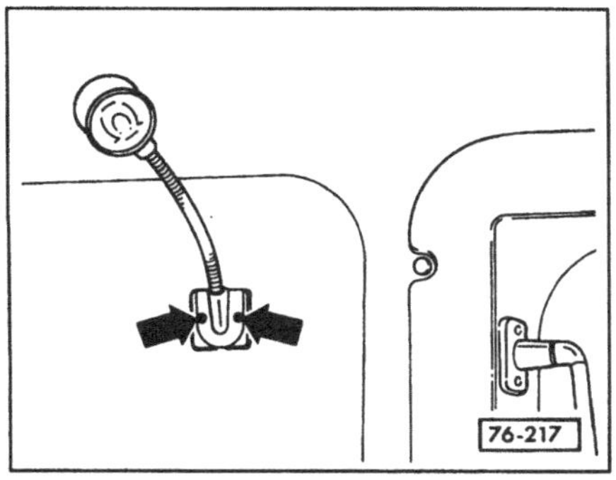

Reading lamp (high roof), removing

◄ – Remove recessed screws -arrows-.

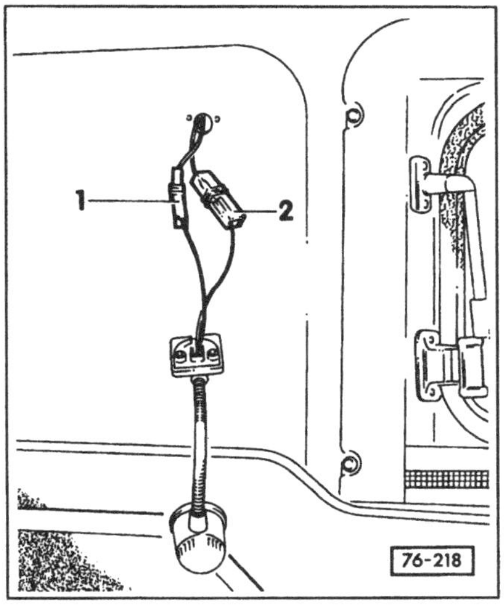

◄ – Pull connector cable with connectors -1- and -2- carefully out of drill hole.

– Disconnect connectors -1- and -2-.

Electrical equipment, removing (Multivan)

Auxiliary Battery -A1-, removing

> ***CAUTION!***
> ***Before working on any part of electrical system obtain radio code. Disconnect battery Ground (GND) strap.***

– Auxiliary Battery -A1- is in driver's seat frame.

– Remove single seat (passenger compartment) ⇒ page R 77-12.

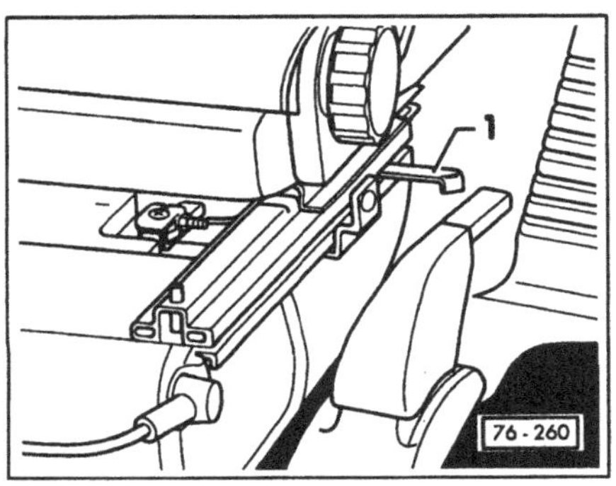

– Push driver's seat to full forward position.

◄ – Pull on locking lever -1- (interior side of seat frame), push seat further forward.

– Disconnect auxiliary batteries.

> ***CAUTION!***
> ***Insulate auxiliary batteries positive (B+) connection, if engine is to be started. Danger of a short-circuit is result of charging current.***

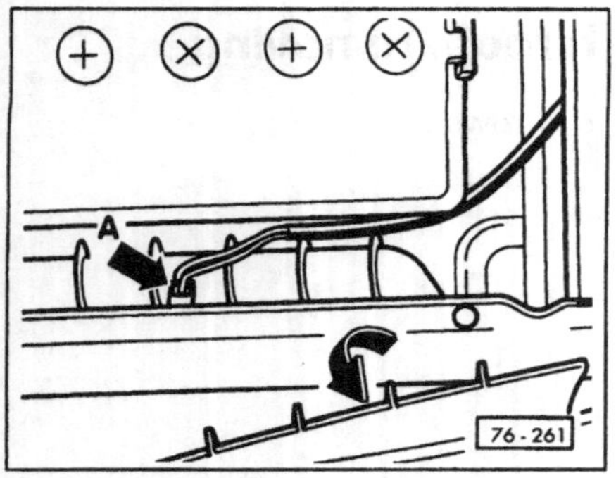

- Bend back rear seat frame trim upper section and disconnect 110/220V Refrigerator connector -U12- -arrow A-.

- Unclip rear seat frame trim, remove from seat frame.

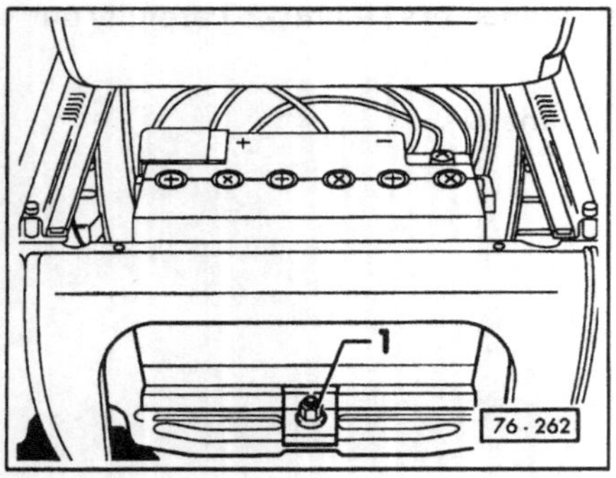

- Loosen nut of bracket -1-, remove Auxiliary Battery -A1- from seat frame.

Fluorescent Light -W8- left, removing

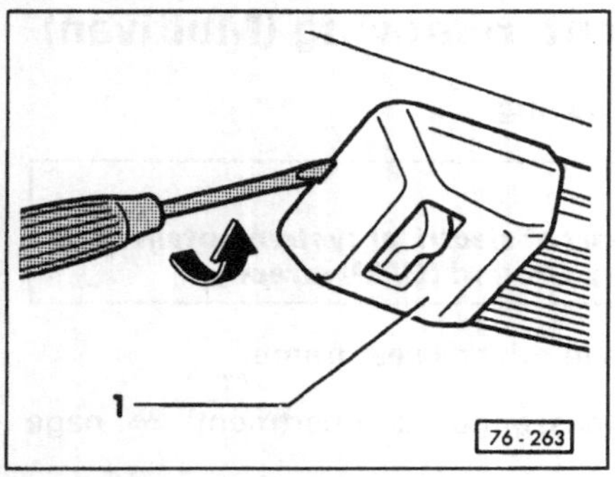

- Pry off cover -1- with screwdriver.

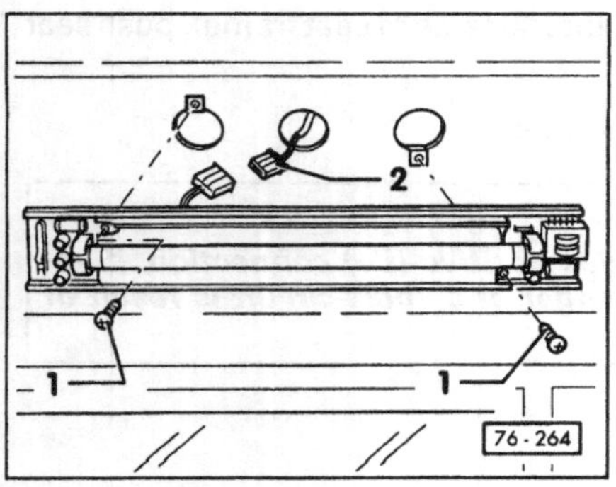

- Remove recessed screws -1-, remove left Fluorescent Light -W8- and disconnect connector -2-.

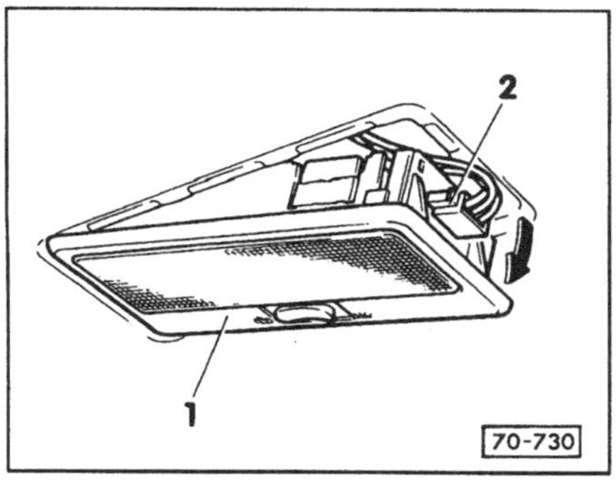

Interior Light -W7- center , removing

- Pry out center Interior Light -W7- -1-, pull off connector -2-.

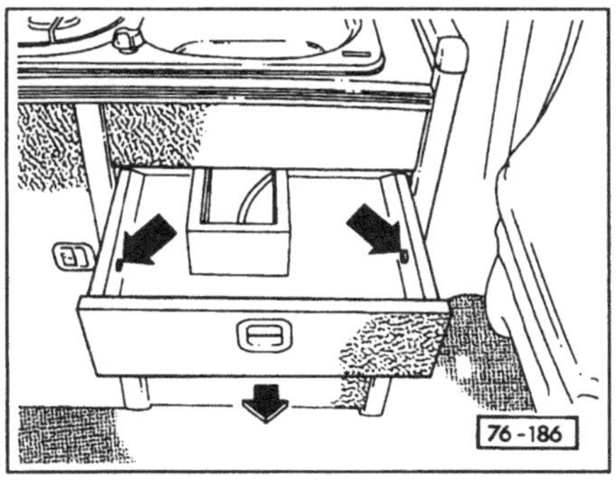

Water supply/drain, removing/installing (EuroVan CV Camper)

Water faucet & Integrated Water Pump Switch -E80-, removing

- Lift kitchenette cover.
- Open drawer, remove recessed screws -arrows- and remove drawer.
- Open and remove right side storage compartment door.

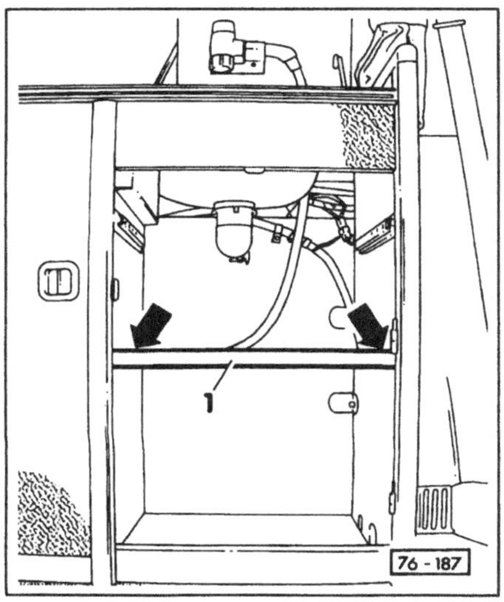

- Pull out supports -arrows- (if applicable), remove false bottom -1-.

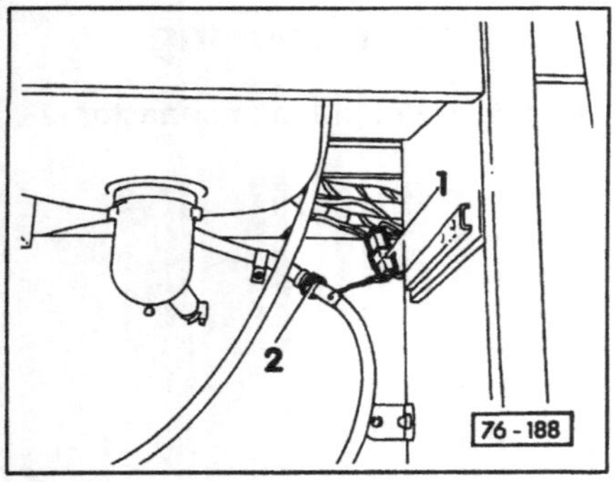

- Remove connector -1-.
- Loosen screw connection -2- and pull out water faucet with hose upward out of kitchenette.

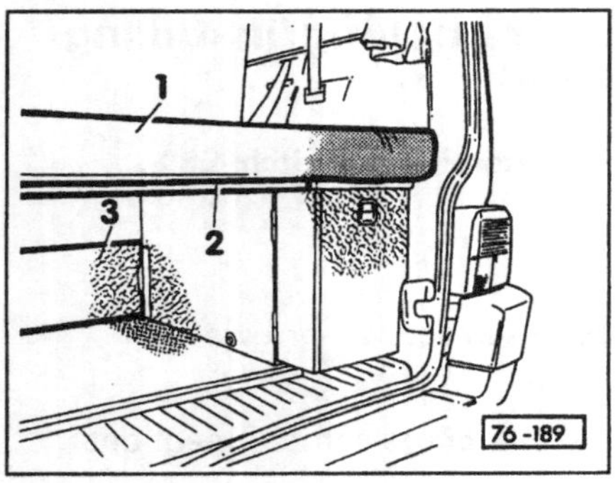

Water Pump -V36-, removing/installing

Removing

- Pull bench forward in direction of travel.
- Remove upholstered top -1-, panel -2- and insert panel -3-.

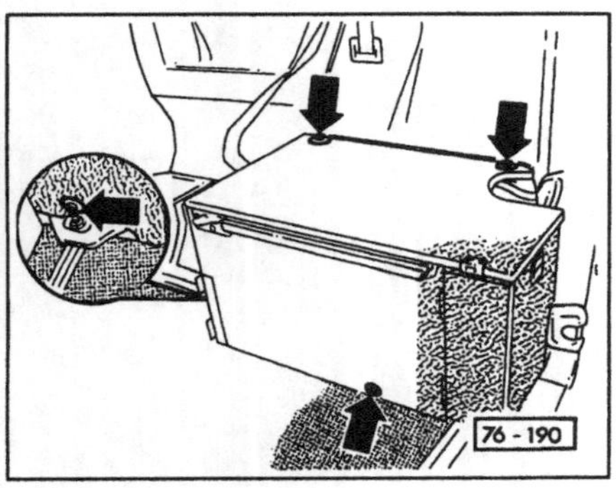

- Remove four bolts -arrows-, lift out water tank trim.

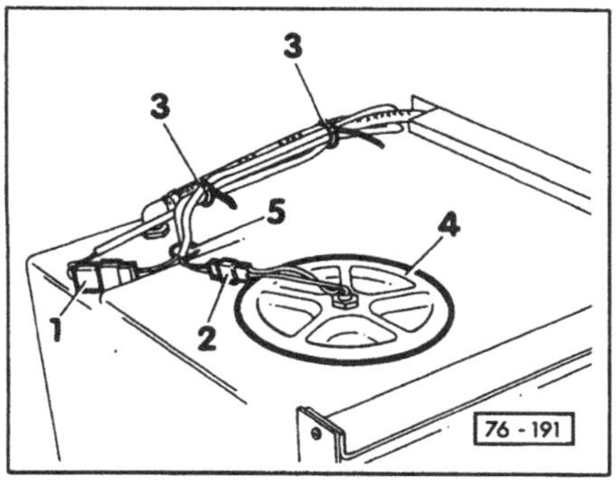

- Remove connectors -1-, -2-.
- Loosen cable tie -3-.
- Remove inspection cover -4-, then remove together with Water Level Sensor -G33-.
- Disconnect Water pump -V36- connections from connector -1-, pull inward into water tank through rubber grommet -5-.

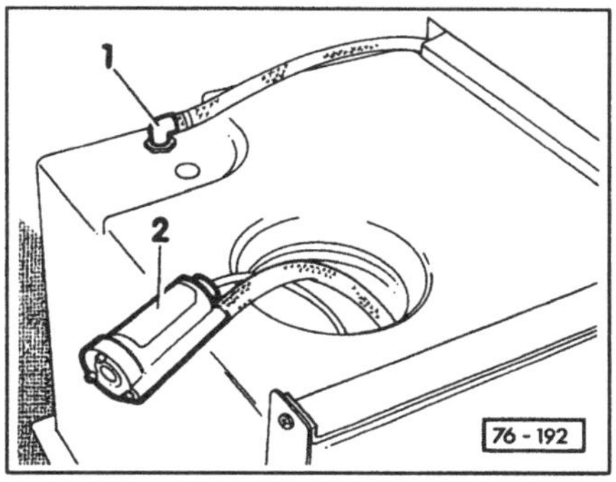

- Disconnect hose inside water tank from connection -1-, remove Water Pump -V36- -2- from tank.

Installing

Note:

Do not mix up connections (12 V)! Water Pump -V36- only delivers correct amount in one direction (see arrow on base of water pump).

Water Level Sensor -G33-, removing

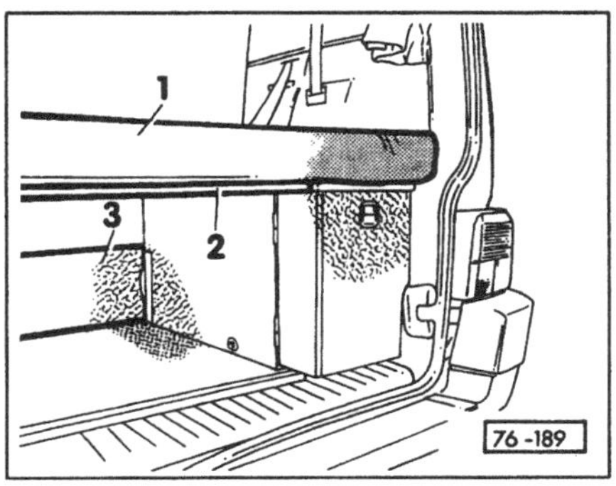

- Pull bench forward in direction of travel.
- Remove upholstered top -1-, panel -2- and insert panel -3-.

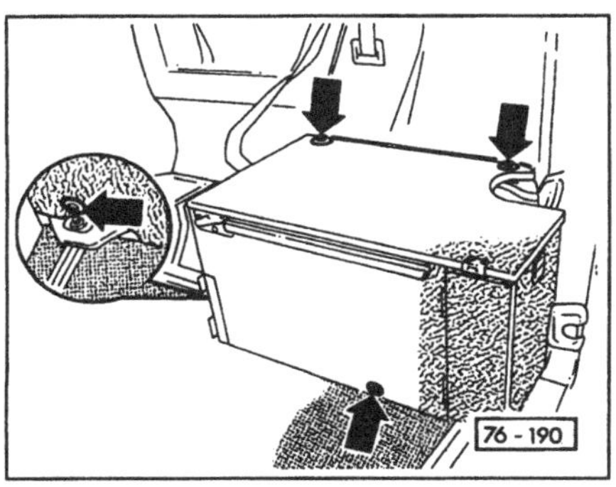

- Remove four bolts -arrows-, lift out water tank trim.

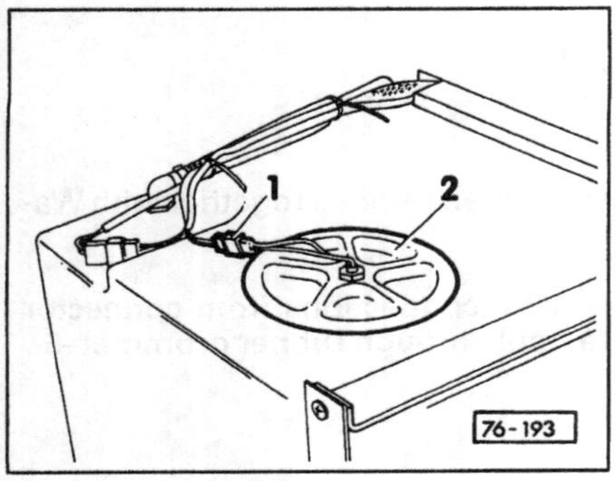

- Disconnect connector -1-.
- Remove inspection cover -2-, remove with Water-Level Sensor -G33-.
- Remove Water-Level Sensor -G33- from inspection cover.

Water tank, removing

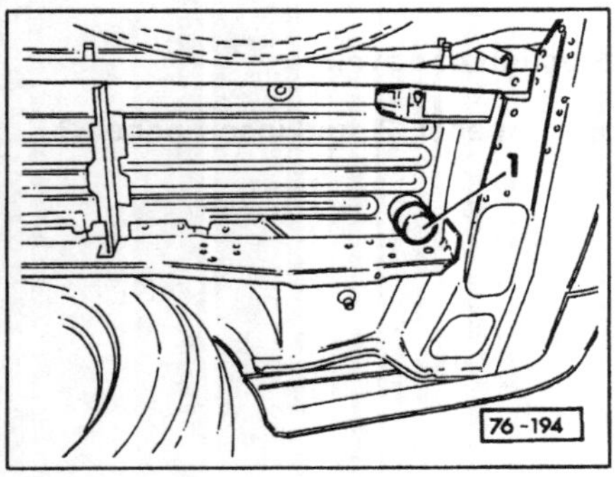

- Remove tank filler cap nut -1- beneath right rear floor plate and drain off any water in tank.
- Pull bench forward in direction of travel.

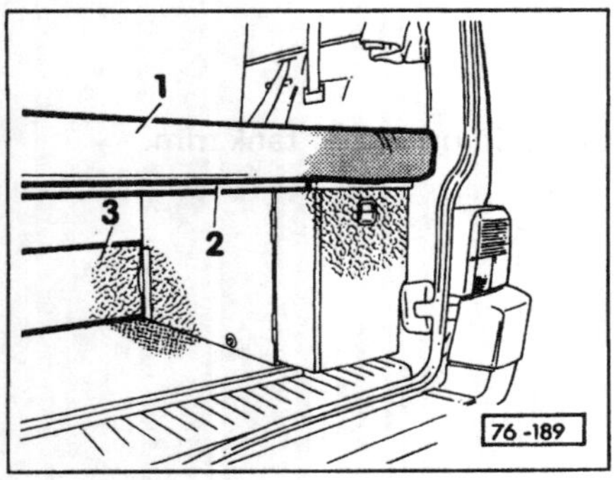

- Remove upholstered top -1-, panel -2- and insert panel -3-.

◄ – Remove four bolts -arrows-, lift out water tank trim.

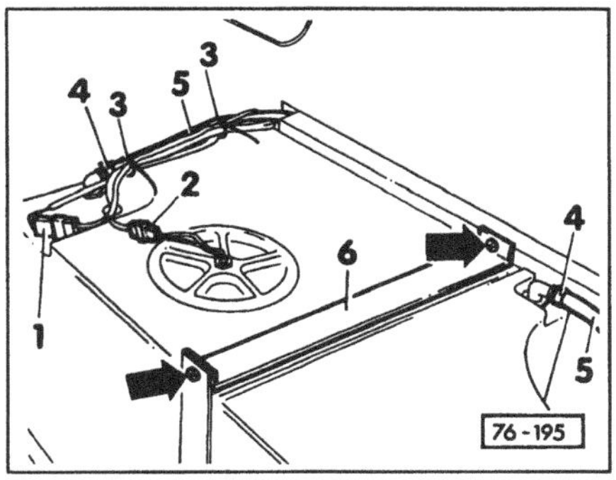

◄ – Disconnect connectors -1-, -2-.

– Loosen cable tie -3-.

– Cut clips -4- and detach hoses -5-.

– Remove recessed screws -arrows-, remove upper retaining strap -6-.

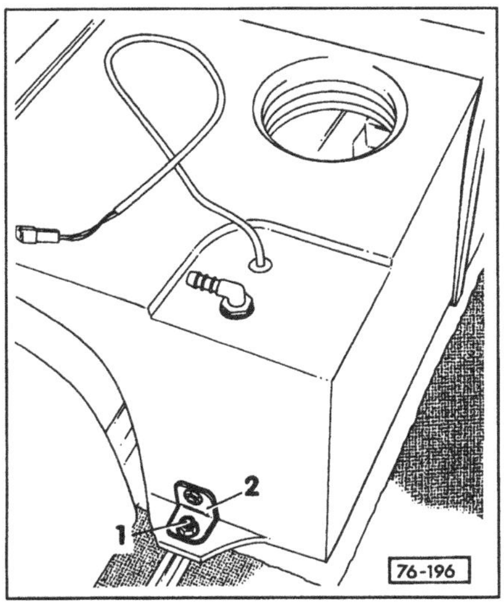

◄ – Remove hex nut -1-, remove angle bracket -2-.

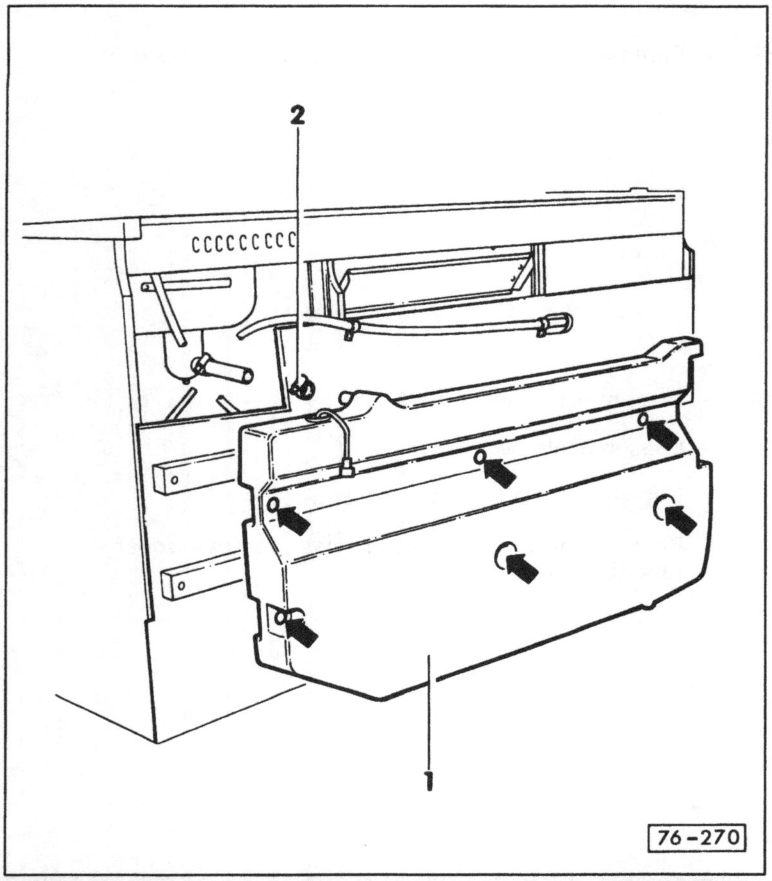

Waste water tank, removing

- Remove drain valve for waste water tank ⇒ page R 76-36 and kitchenette ⇒ page R 77-19.
- Loosen screw clamp -2- and disconnect hose from waste water tank.
- Remove hex nuts from inside kitchenette and press out saucer-head screws -arrows-.
- Remove waste water tank -1-.

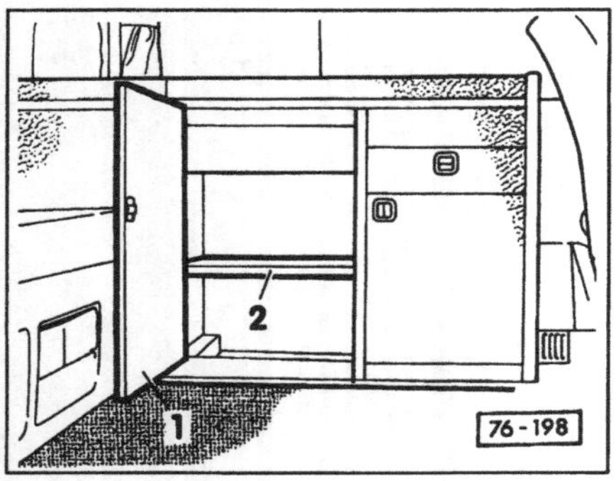

Waste water tank drain valve, removing

Note:

Drain waste water tank prior to removal.

- Open kitchenette -1- left side storage compartment door.
- Remove false bottom -2-.

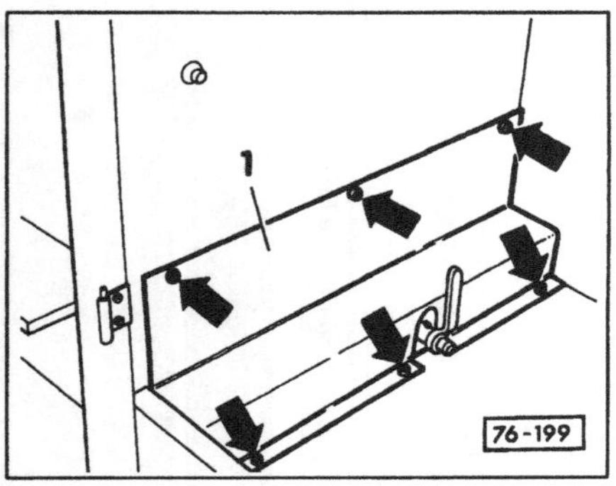

- Remove six recessed screws -arrows- and remove cover -1-.

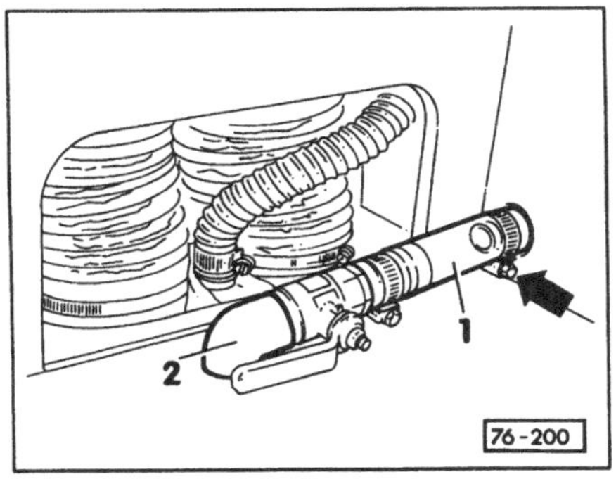

- Loosen screw clamp -arrow-, disconnect hose -1-.
- Pull drain cock -2- upward out of base.

Discharge valve, removing

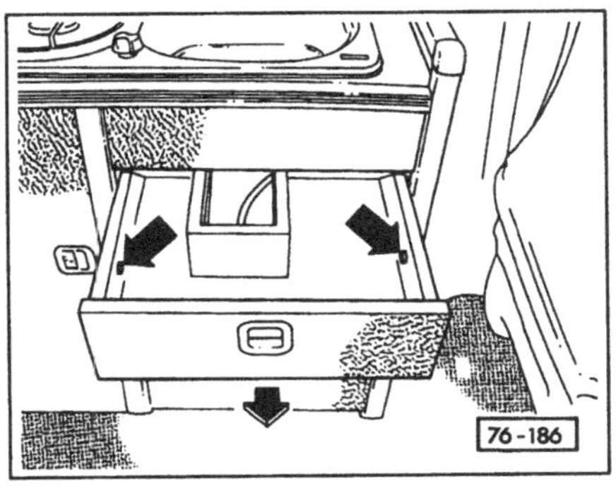

- Lift kitchenette cover.
- Open drawer, remove recessed screws -arrows- and remove drawer.
- Open and remove drawer from right side storage compartment.

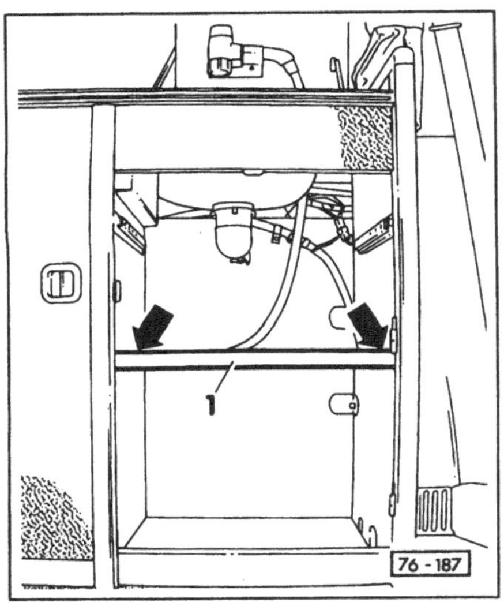

- Pull out supports -arrows-, (if applicable), and remove false bottom -1-.

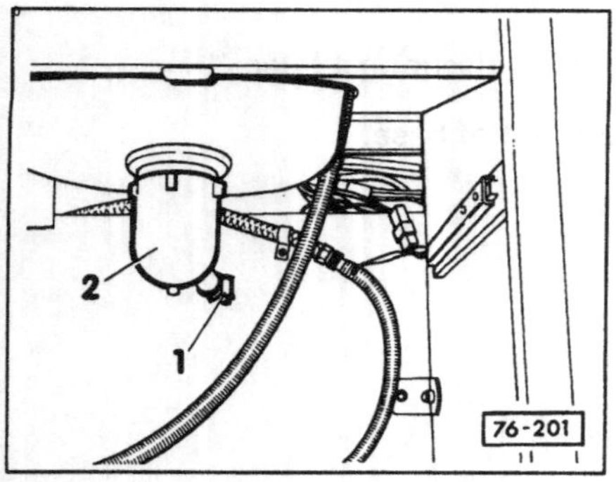

- Loosen screw clamps -1-, disconnect hose.
- Remove discharge valve -2-.

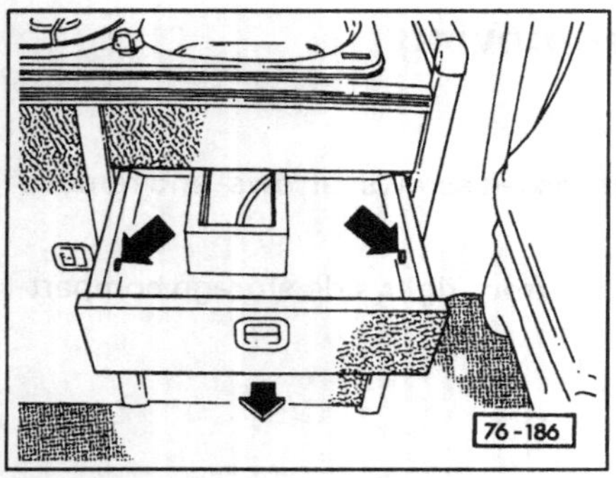

Waste Water Level Sensor -G126-, removing

- Lift kitchenette cover.
- Open drawer, remove recessed screws -arrows- and pull out drawer.
- Open and remove drawer from right side storage compartment.

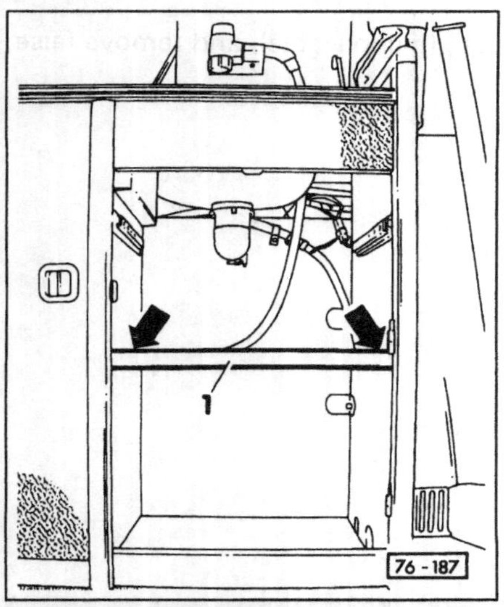

- Pull out supports -arrows-, (if applicable), remove false bottom -1-.

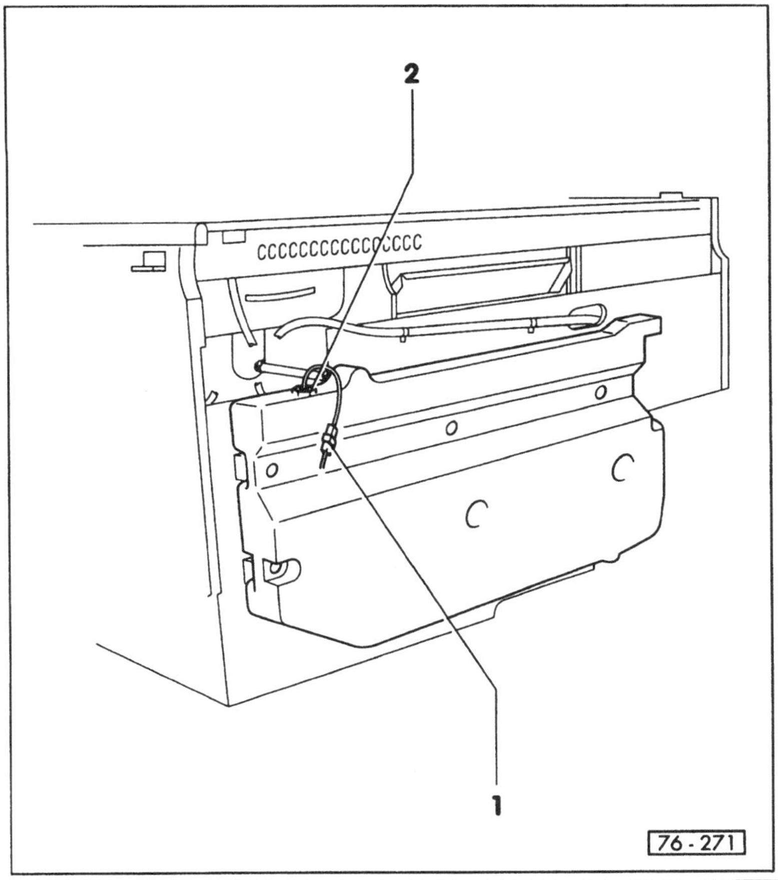

– Remove connector -1-.

– Remove Waste Water Level Sensor -G126- -2-.

Note:

Connector -1- and Waste Water Level Sensor -G126- -2- are accessible through kitchenette right side storage compartment. Removal of kitchenette is not necessary.

Propane Gas tank, removing/installing

Removing

WARNING!
Fire hazard! Do not smoke or have anything in area that could ignite propane fuel. Ensure adequate ventilation.

CAUTION!
Note safety measures ⇒ page R 76-1!

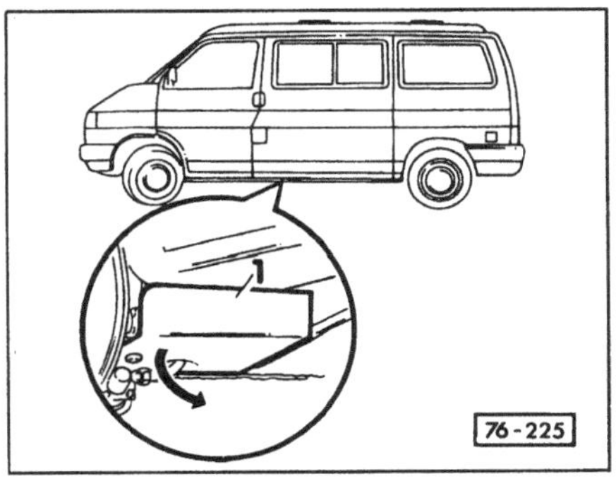

◄ – Open cover underneath left side center floor plate.

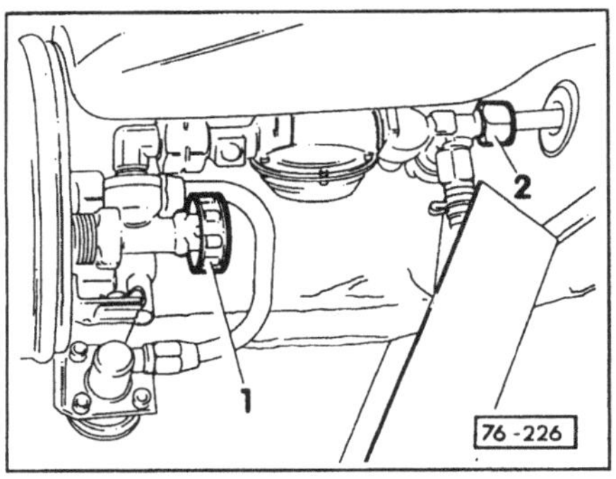

◄ – Close bleeder valve -1-.

– Disconnect union -2-.

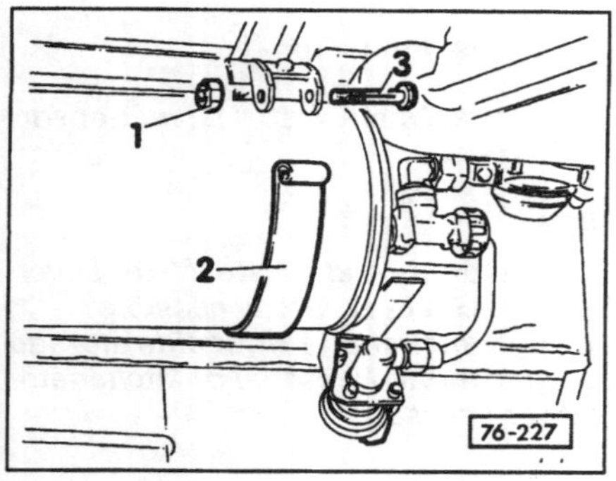

- Remove two hex nuts -1-.
- Release tension on retaining straps -2- by rotating hex bolts -3-.
- Pull out hex bolts -3-.

Installing

- Install in reverse sequence.

> ***CAUTION!***
> ***Note safety measures ⇒ page R 76-1!***

Propane gas stove with sink, removing/installing

Removing

> ***WARNING!***
> ***Fire hazard! Do not smoke or have anything in area that could ignite propane fuel. Ensure adequate ventilation.***

> ***CAUTION!***
> ***Note safety measures ⇒ page R 76-1!***

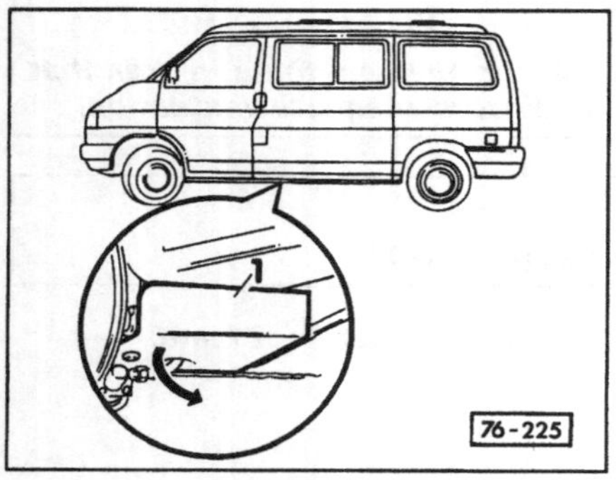

- Open cover -1- underneath left center floor plate.

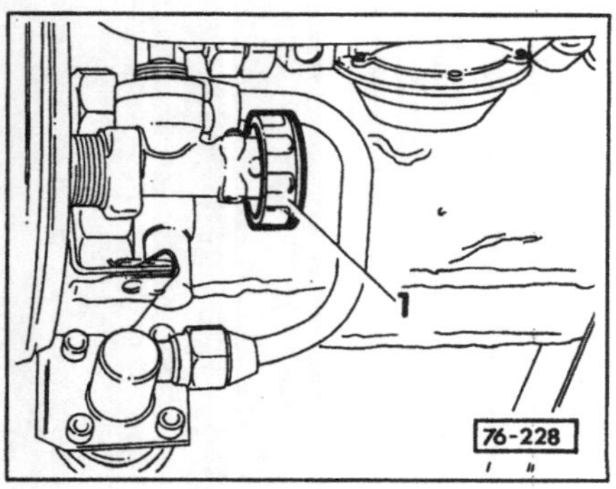

- Close bleeder valve -1-.

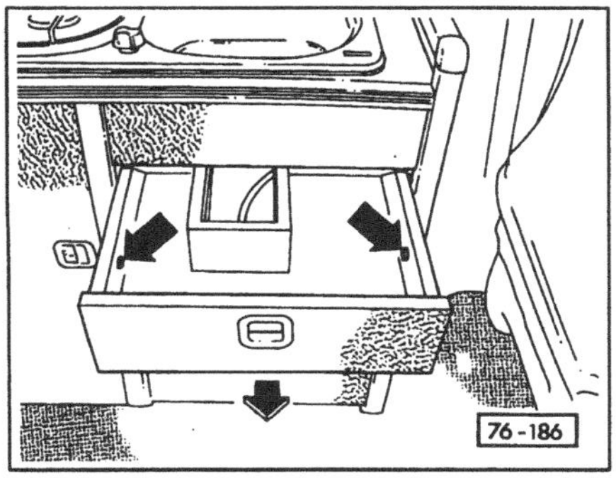

- Raise kitchenette -1- cover.
- Open drawer, remove recessed screws -arrows-, remove drawer.
- Open and remove door from right side storage compartment.

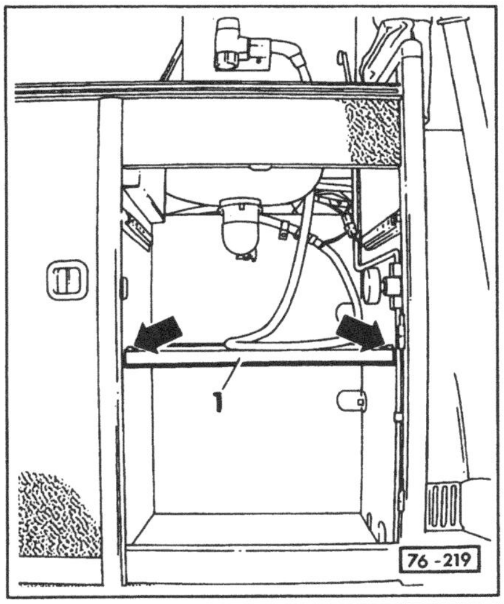

- Pull out supports -arrows- (if applicable), remove false bottom -1-.

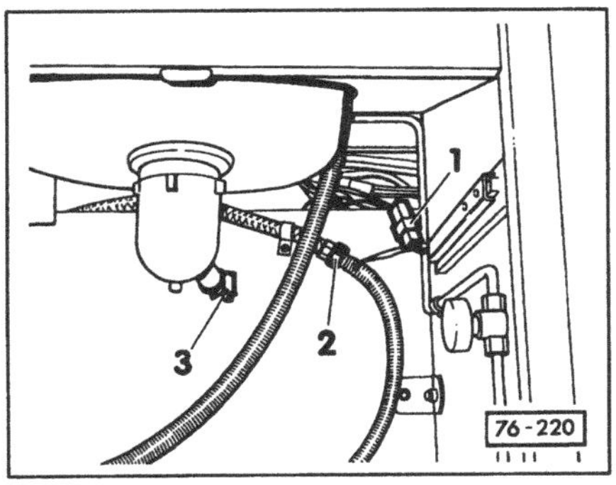

- Disconnect connector -1-.
- Loosen union -2-, pull drain valve with hose upward out of kitchenette.

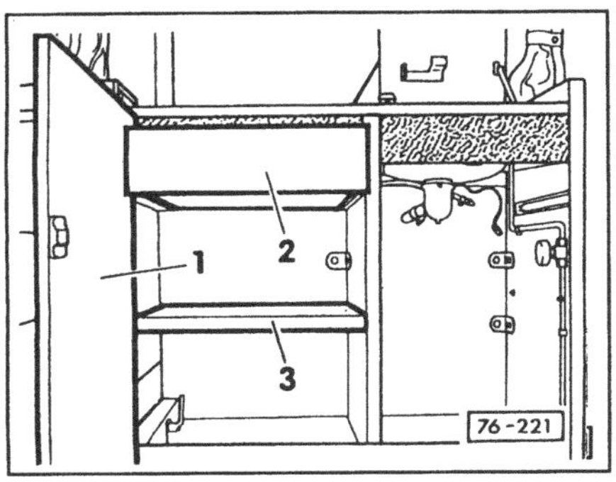

- Open left storage compartment -1- door.
- Pull out cover -2-.
- Remove false bottom -3-.

◄ – Disconnect union -1-.

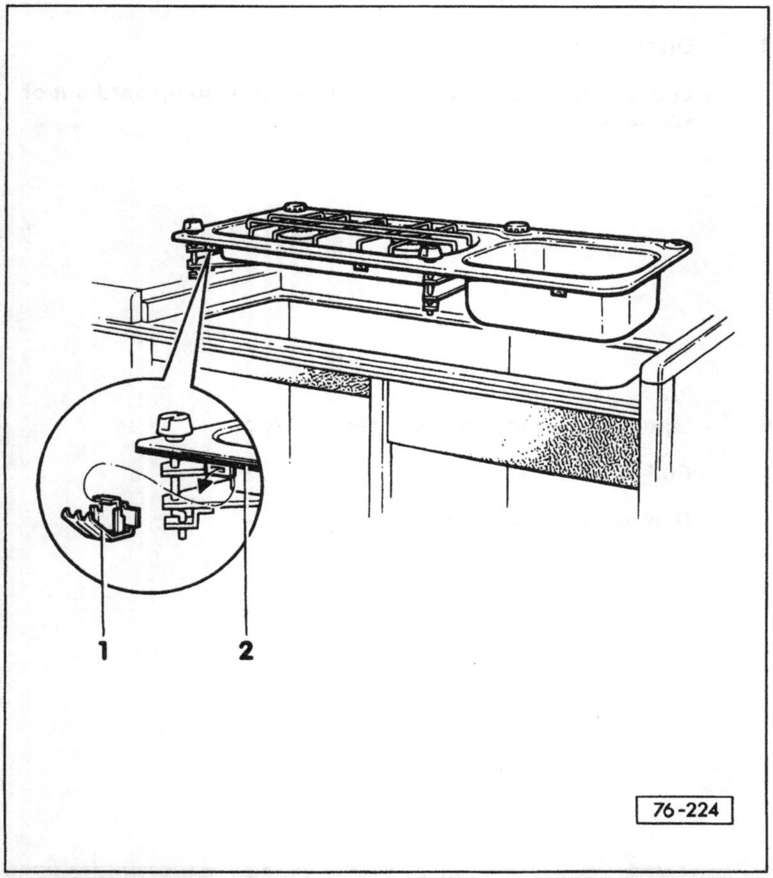

– Loosen six retaining clips -1- and rotate 90°.

– Remove gas stove with sink from kitchenette.

Installing

Note:

Before mounting, ensure that seal -2- makes proper contact.

> ***CAUTION!***
>
> ***Note safety measures ⇒ page R 76-1!***

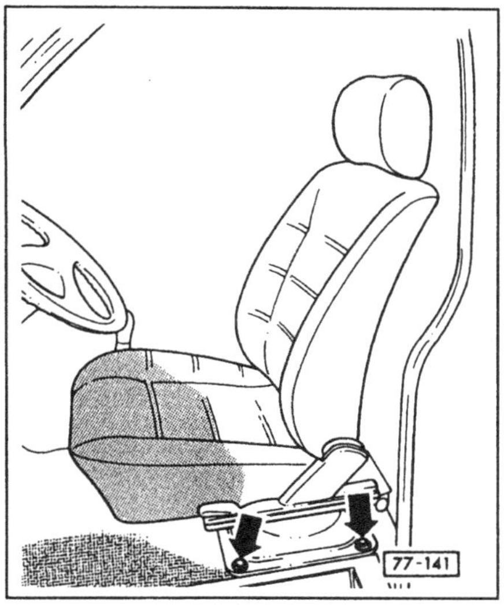

Seats, (EuroVan CV Camper)

Driver's/front passenger seat, removing

◄ – Remove four hex bolts -arrows- from floor.

Tightening torque:

Hex bolt = 44 Nm (32 ft lb)

– Lift out driver's/front passenger seat.

Bench seat rear, removing/installing (EuroVan CV Camper)

– Pry out plastic caps from both rail ends.

– Remove stop -1-.

– Release catch -2-, pull bench forward in direction of travel and lift out of rails.

Note:

When mounting bench into rails catch -2- must be released.

Seat cushion, removing

- Unlatch seat cushion -1- and lift up.
- Remove recessed screws-arrows-.
- Pull seat cushion -1- upward out of fittings.

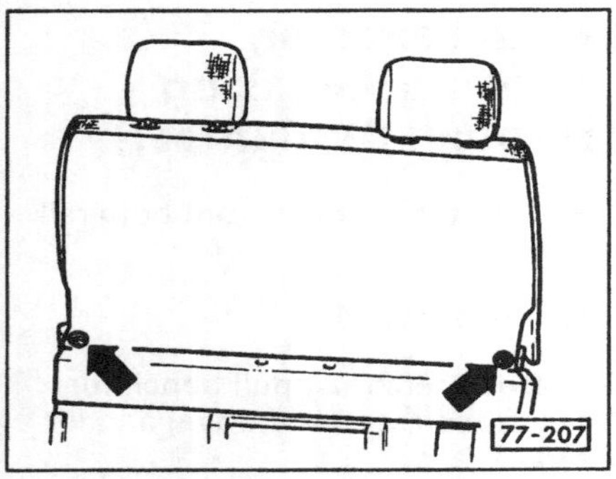

Backrest cushion, removing

◄ – Remove recessed screws -arrows-.

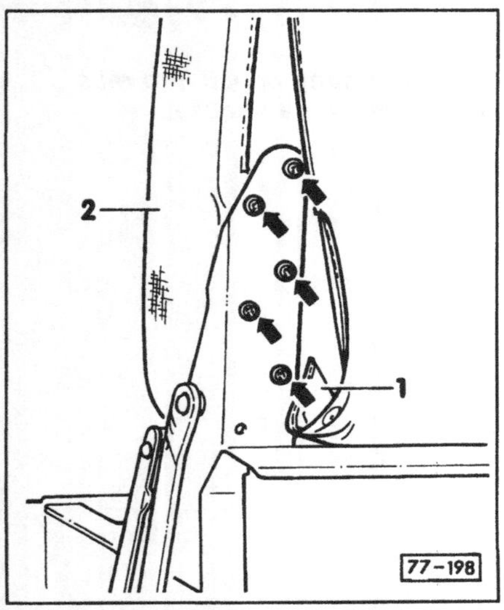

◄ – Pull off backrest cover -1- from tabs left and right side.

– Remove recessed screws -arrows- left and right side.

– Pull backrest cushion -2- upward out of fittings.

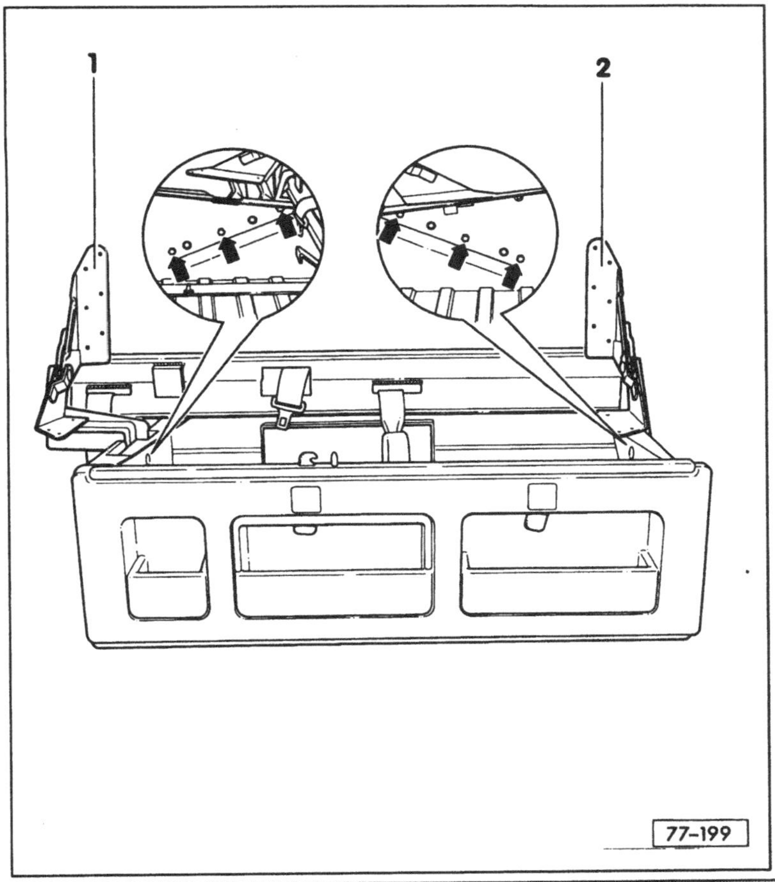

Bench hardware, removing

- Remove seat cushion ⇒ page R 77-3
- Remove backrest cushion ⇒ page R 77-4
- Remove hex bolts -arrows- remove fittings -1- and -2-.

Bench seat front trim, removing

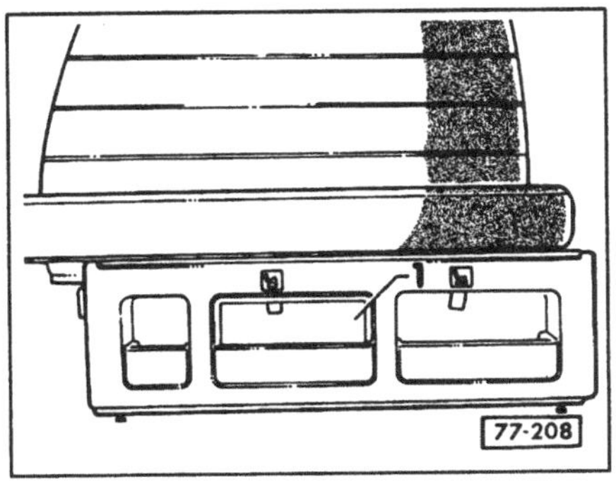

◄ – Remove bench ⇒ page R 77-2.

– Remove center storage tray -1-.

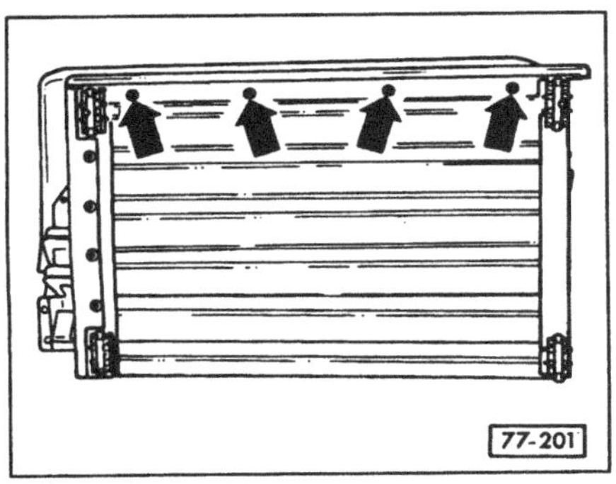

◄ – Remove recessed screws -arrows- underneath bench and remove front trim.

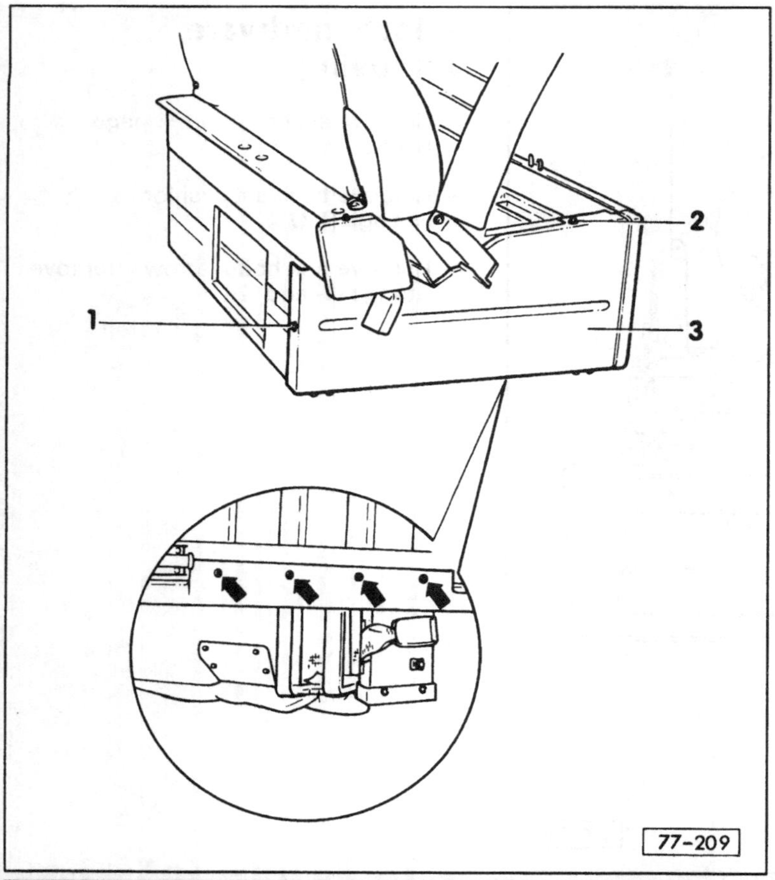

Bench seat side trim, removing

– Remove bench ⇒ page R 77-2

– Pry off cap -1-, remove recessed screw.

– Lift seat cushion, pry off cap -2-, and remove recessed screw.

– Remove recessed screws -arrows- underneath bench and remove side trim -3-.

Bench seat release cable, adjusting/removing

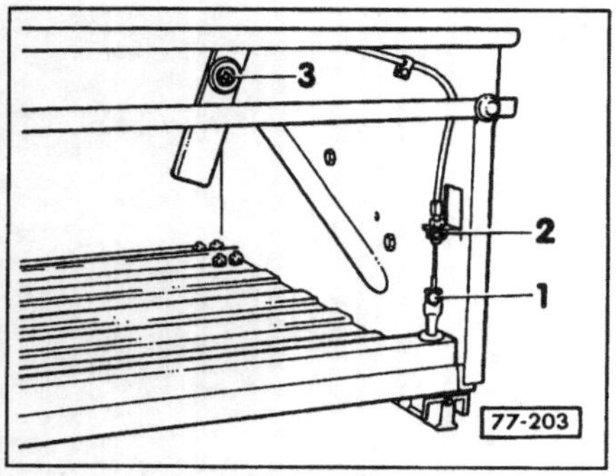

– Remove bench ⇒ page R 77-2

– Remove bench front trim ⇒ page R 77-6

◄ – Disconnect union -1-.

– Remove hex nut -2-.

– Remove recessed screw -3-, remove release cable.

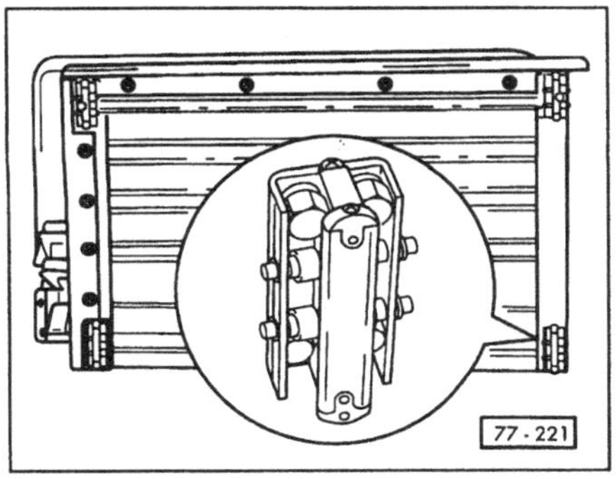

Bench seat sliding blocks, replacing

- Remove bench ⇒ page R 77-2
- Replace sliding blocks on underside of bench, assemble loosely.
- Install bench, move back and forth several times.
- Tighten screws in bench storage compartment.

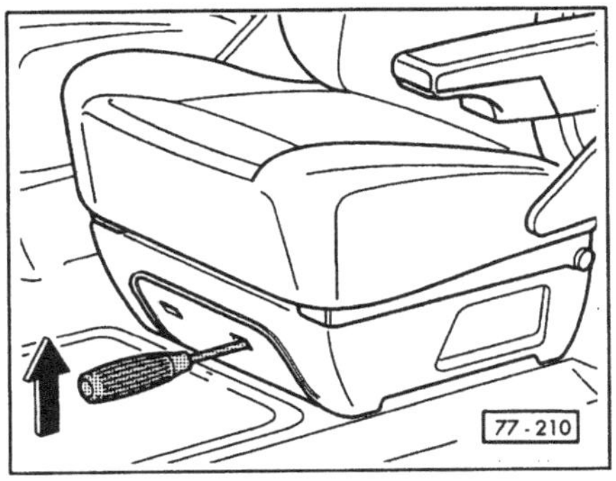

Seats, (Multivan)

Driver's/front passenger seat, removing

Driver's seat

- Remove auxiliary battery -A1- ⇒ page R 76-22.
- Open front seat frame trim with screw driver.

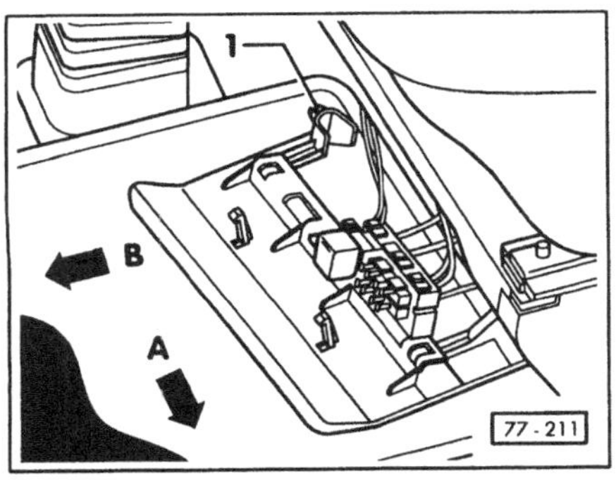

- Push trim to left -A-, disconnect pivot pin -1- -arrow B-.
- Place trim on vehicle floor.

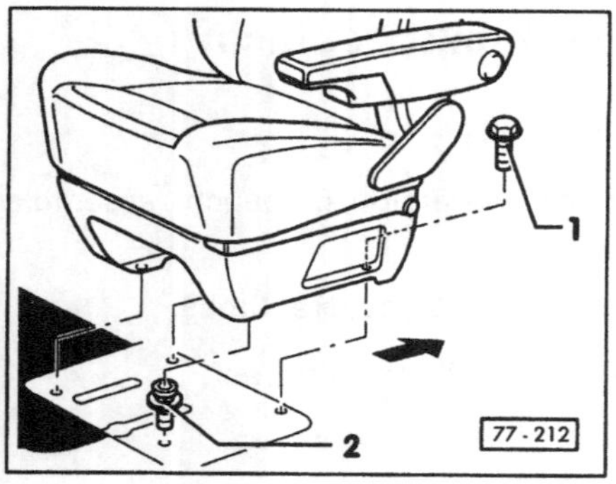

– Remove rear bolts -1-, pull seat out of bolts -2- toward rear and lift out of vehicle.

Tightening torques:

Hex bolts -1- = 44 Nm (32 ft lb)

Bolts -2- = 30 Nm (22 ft lb)

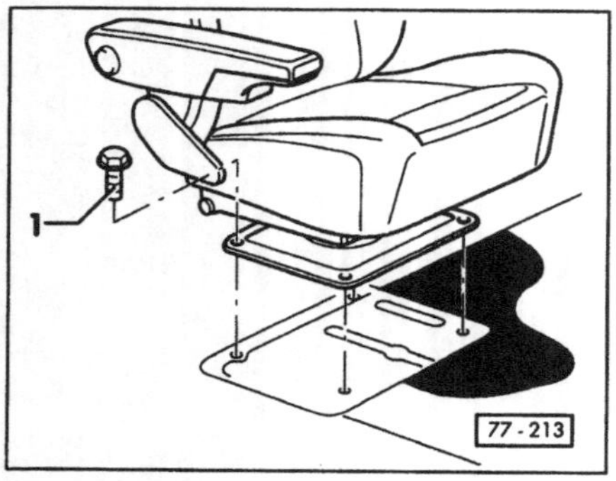

Front passenger seat

– Remove securing bolts -1- and remove front passenger seat.

Tightening torque:

Hex bolts -1- = 44 Nm (32 ft lb)

Single seat (passenger compartment), removing

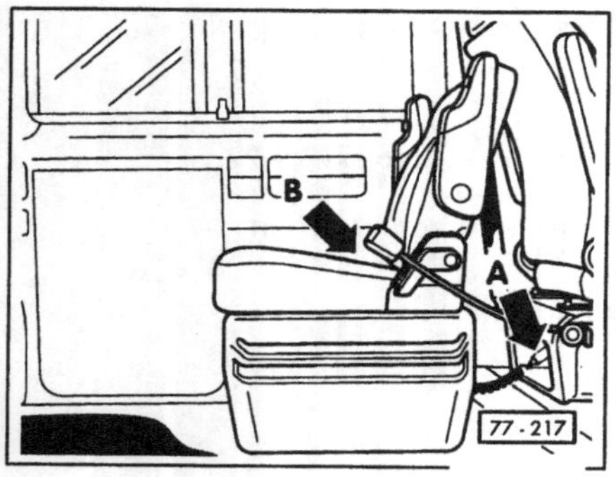

– Pull cooler 12 V plug out of 110/220 V Refrigerator Socket -U12-, -arrow A-.

– Raise seat cushion; depending on model, pull out knob or pull
lever upward -arrow B-.

– Remove cooler.

– Remove securing bolts -arrows-.

Tightening torque:

Hex bolt = 44 Nm (32 ft lb)

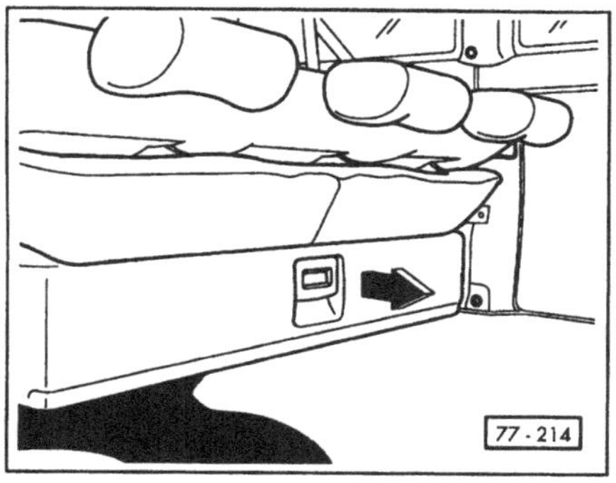

Bench seat, removing

- Remove single seat (passenger compartment) ⇒ page R 77-12.
- Unlatch bench at backrest and fold backrest forward.
- ◄ Pull bench forward to stop with handle, push back slightly.

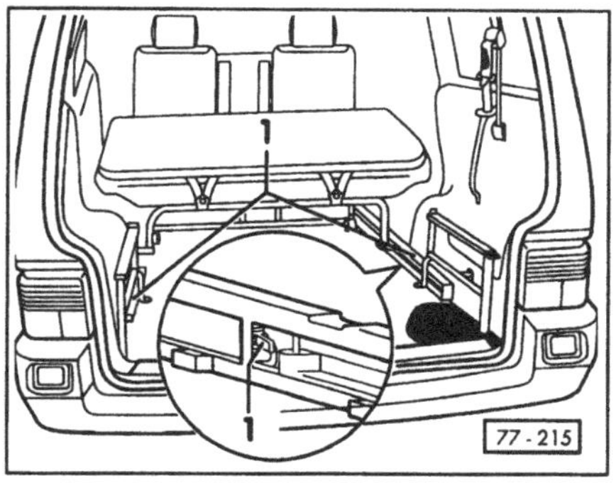

- ◄ Press retaining hook -1- in guide rails simultaneously on left and right, push bench forward out of guide rails.
- Remove bench from vehicle.

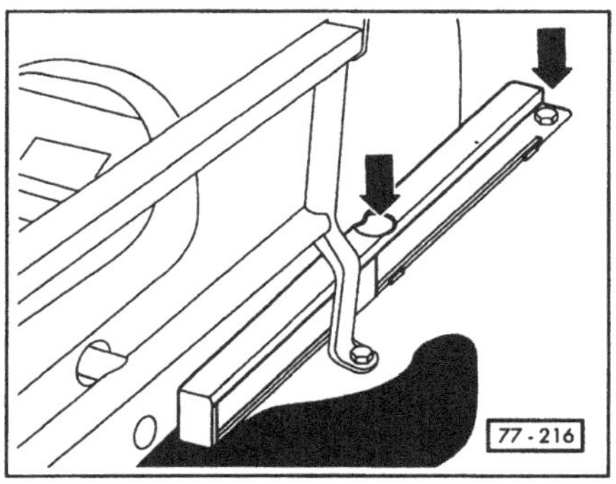

Guide rails, removing

- ◄ Remove guide rail securing bolts -arrows-.

  Tightening torque:

  Hex bolt = 44 Nm (32 ft lb)

Bedboard and support frame, removing

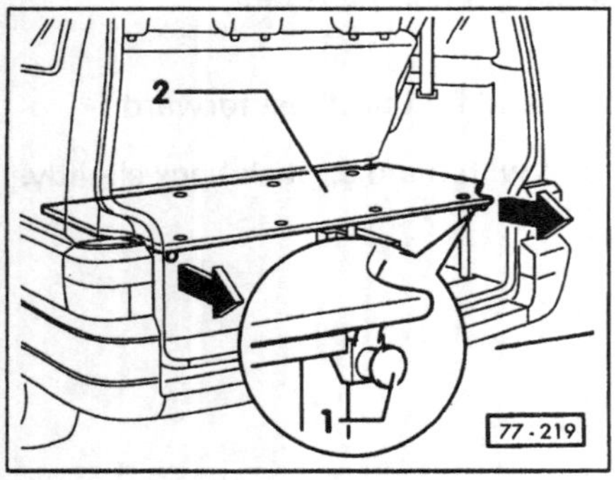

- Remove cushion padding.
- ◄ Pull back catch -1- -arrows- and take bedboard -2- out of vehicle.
- Remove side panel molded trim at rear left/right.
- Remove tool kit.

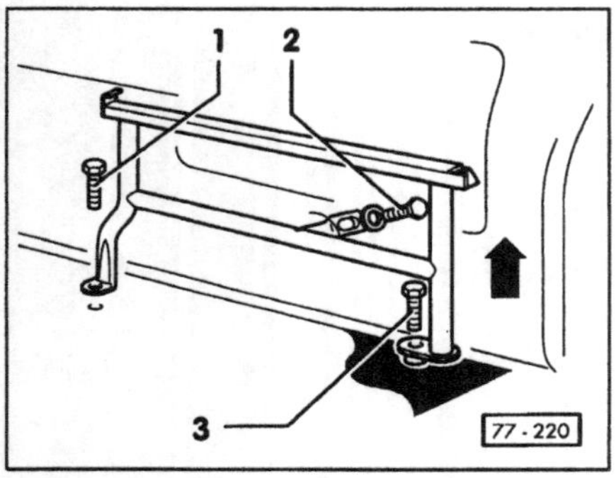

- ◄ Remove support frame bolts -1-, -2- and -3-.
- Lift support frame at back -arrow- and remove left or right.

Tightening torques:

Hex bolt -1- = 22 Nm (16 ft lb)

Hex bolt -2- = 20 Nm (15 ft lb)

Hex bolt -3- = 40 Nm (30 ft lb)

Cupboards and storage compartments, removing/installing (EuroVan CV Camper)

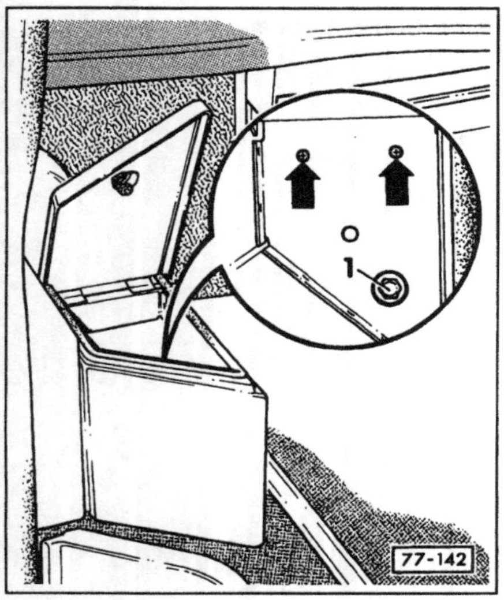

Lockable storage compartment, removing

- ◄ Open storage compartment lid, remove two recessed screws -arrows- and one hex bolt -1- from bottom of storage compartment.

Wall cupboard, removing

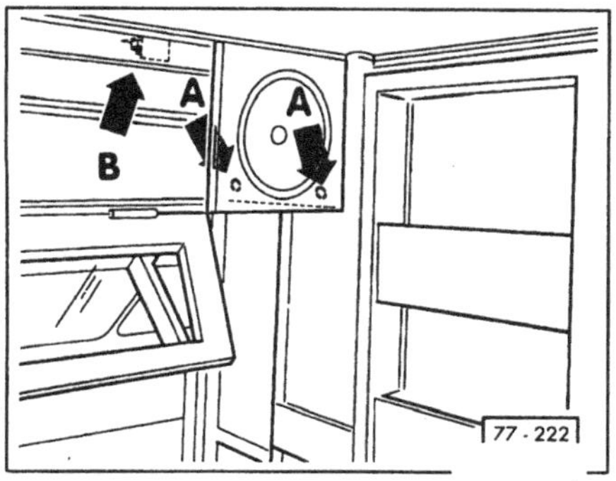

◄ – From inside clothes closet, pry off two caps (if applicable), remove hex nuts and press saucer-head screws -arrows- through into wall cupboard.

– Loosen screw connection to bedboard -arrow B-.

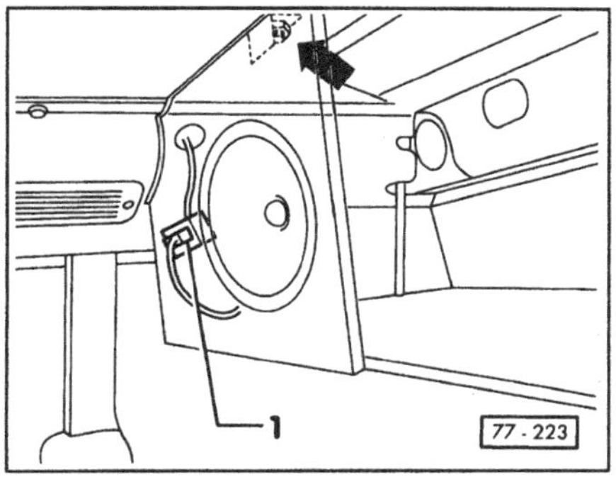

◄ – Disconnect speaker cable connectors -1-.

– Loosen screw -arrow- to bedboard.

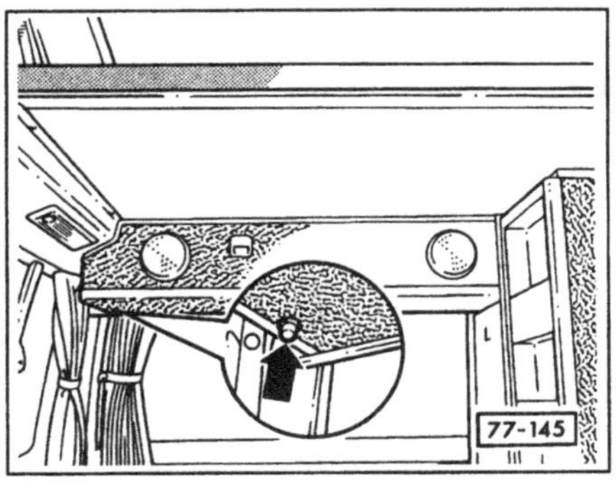

◄ – Pry off cap -arrow-, remove hex nut and remove wall cupboard.

Kitchenette, removing

> ***CAUTION!***
> ***Note safety measures ⇒ page R 76-1!***

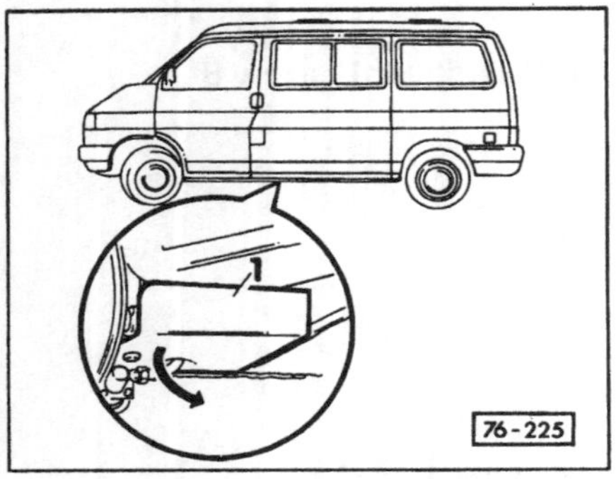

- Remove driver's seat ⇒ page R 77-1
- Remove bench ⇒ page R 77-2.
- Remove drain cock for waste water tank ⇒ page R 76-36.

Vehicles with gas installation

◄ – Open cover -1- underneath left center floor plate.

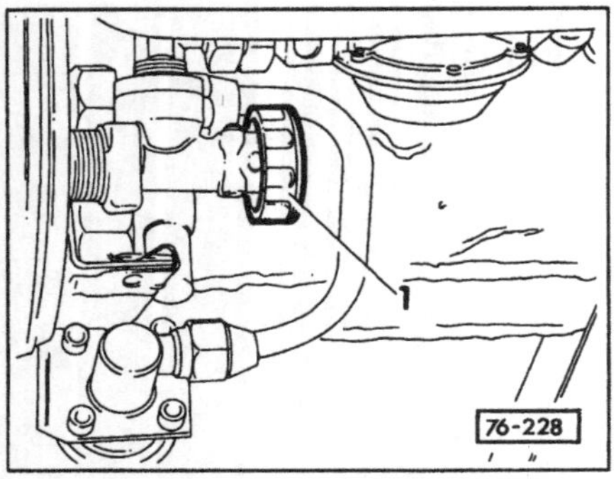

◄ – Close bleeder valve -1-.

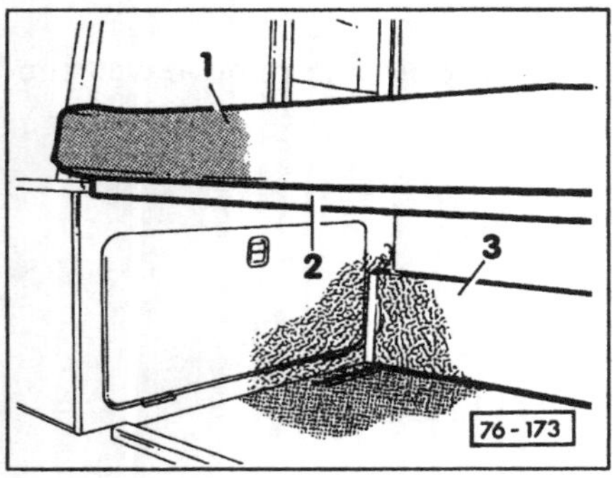

Continuation for all vehicles

◄ – Remove upholstered top -1- panel -2- and insert panel -3-.

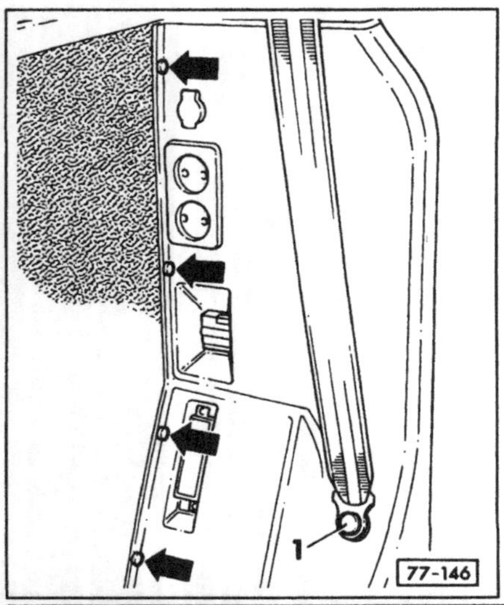

◄ – Pry off caps -arrows- and remove recessed screws.

– Pry off cap -1- and remove hex bolt.

◆ 40 Nm (30 ft lb)

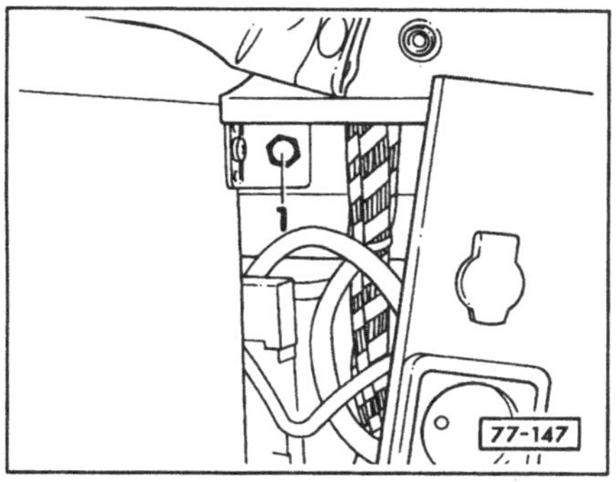

- Pull off lower B-pillar trim and remove hex bolts -1-.

Vehicles with gas installation

> ***WARNING!***
>
> ***Fire hazard! Do not smoke or have anything in area that could ignite propane fuel. Ensure adequate ventilation.***

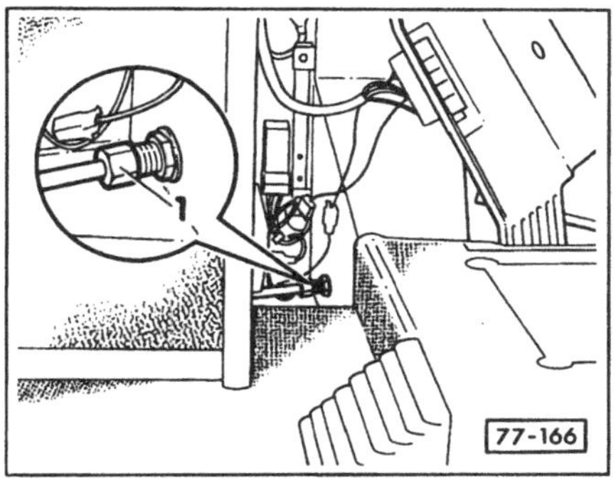

- Loosen screw connection -1-.

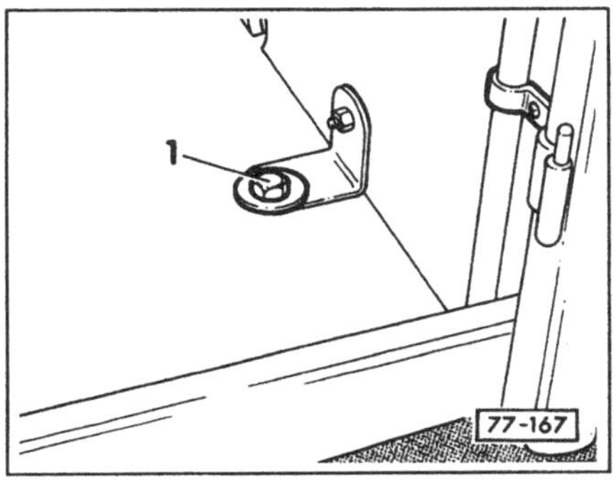

Continuation for all vehicles

- Open and remove kitchenette right side storage compartment door.
- Remove hex bolt -1- from bottom.

- From left side storage compartment of kitchenette, loosen screw clamp -1-, -2- and -3-
- Disconnect vent hoses from base connections.

◄ – Pry off caps -arrows-, remove screws from lower screen cover.

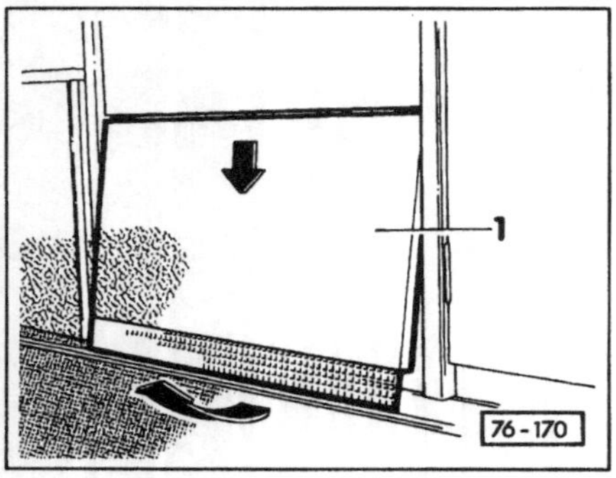

◄ – Press screen cover -1- downward, pull out toward center of vehicle.

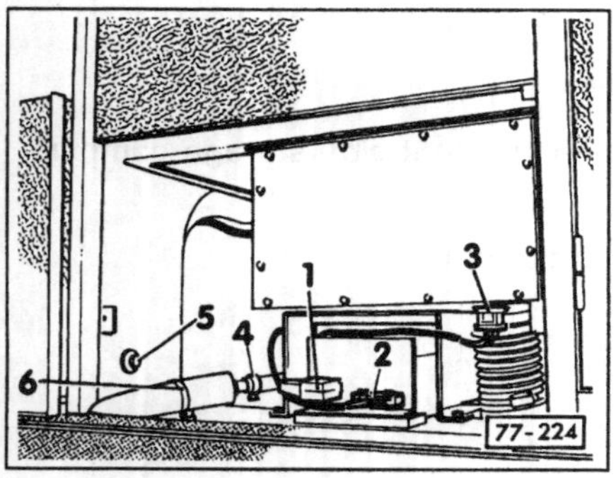

◄ – Disconnect connectors -1-, -2- and -3-.

– Loosen screw clamps -4-, remove hose.

– Remove hex bolt -5-.

– Remove clamp -6-.

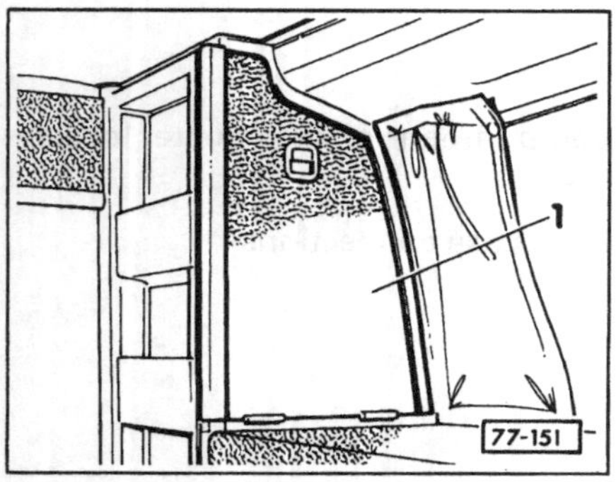

◄ – Open flap for clothes closet -1- and remove.

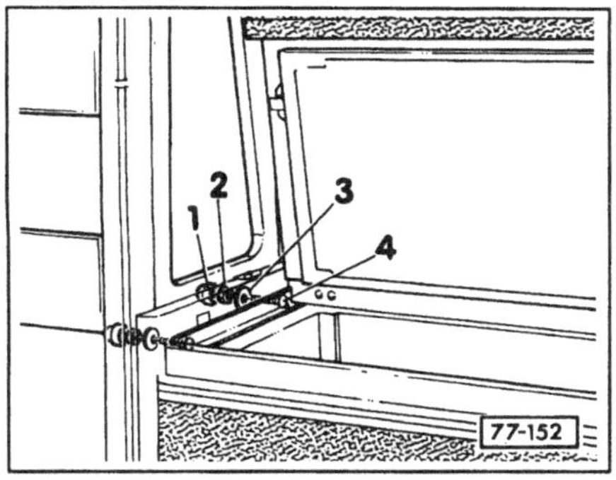

◄ – Open refrigerator lid.

– From inside clothes closet, pry off caps -1-, (if applicable), remove hex nuts -2-, remove washers -3- and press off saucerhead screws -4-.

– Close refrigerator lid.

– Unfasten curtains at C-pillar.

◄ – Remove recessed screws with press-studs -arrows- and remove C-pillar trim.

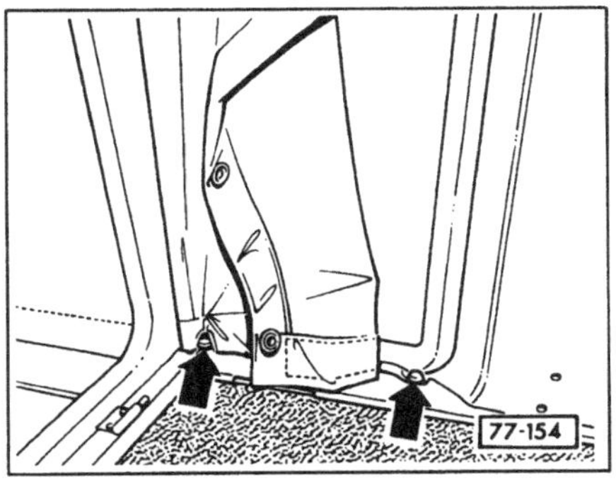

◄ – Pry off caps -arrows- and remove recessed screws (1992 ➤, only one screw).

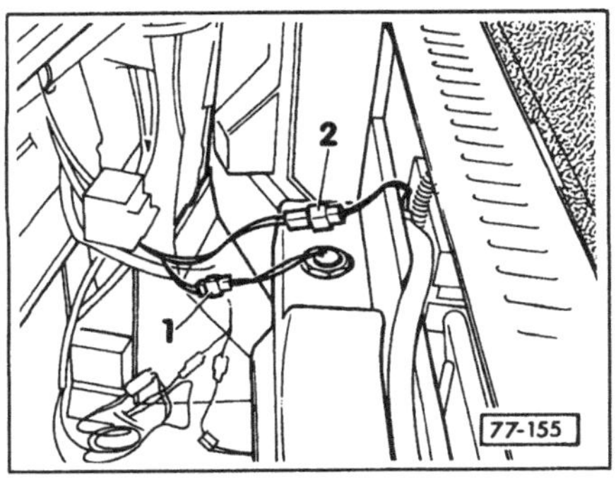

◄ – Lift kitchenette off outside wall toward center of vehicle.

– Disconnect connectors -1- and -2-.

– Lift kitchenette out of vehicle.

Clothes closet, removing

- Remove Auxiliary Batteries -A1- ⇒ page R 76-5.
- Remove wall cupboard ⇒ page R 77-17.
- ◄ Pry off caps -arrows- and remove recessed screws from lower screen cover.

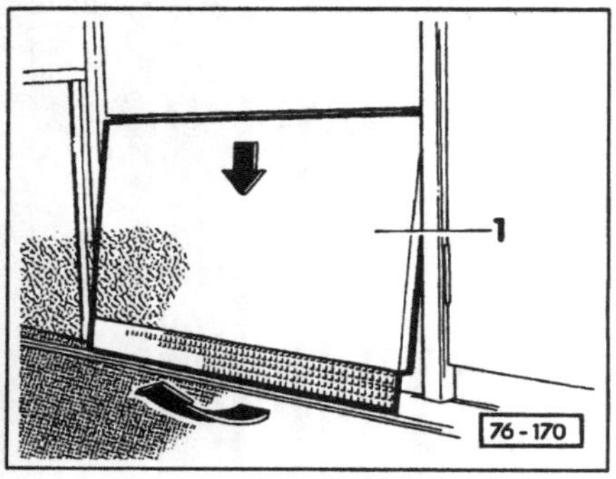

- ◄ Press screen cover -1- downward and pull out toward center of vehicle.

- ◄ Remove hex bolt -arrow-.

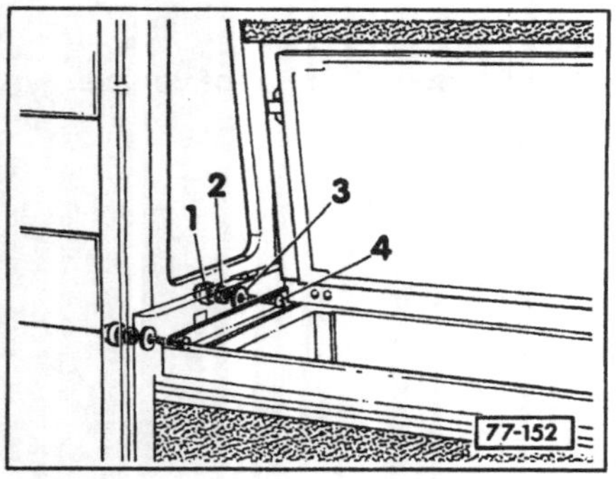

- ◄ From inside clothes closet, pry off caps -1- (if applicable), remove hex nuts -2-, remove washers -3- and press out saucerhead screws -4-.

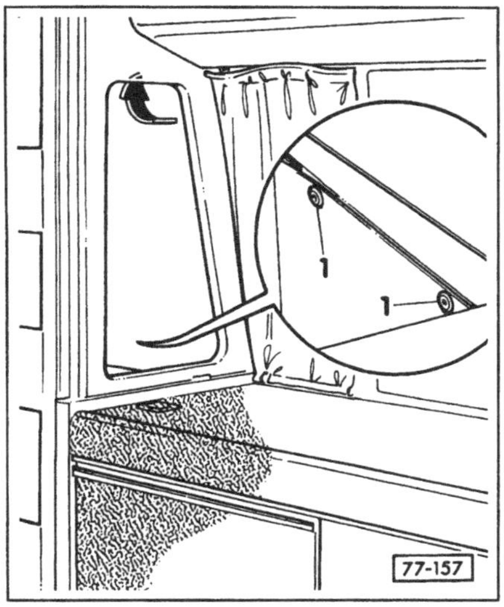

◄ – Remove hex bolt with washer.

– Remove clips -1- with Allen wrench.

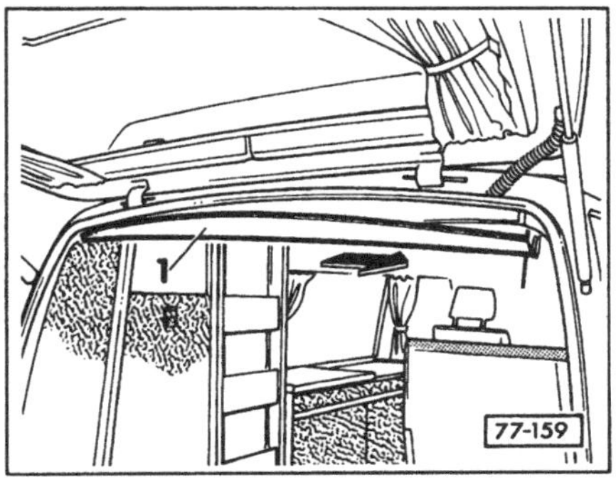

◄ – Unclip roof frame trim -1- and pull out in direction of arrow.

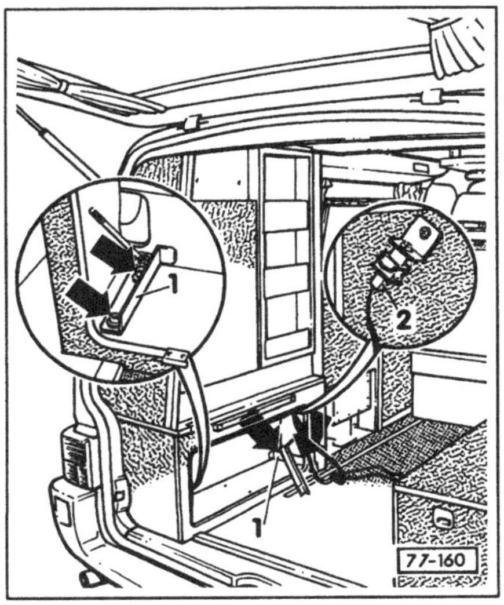

◄ – Remove bolts -arrows- and remove battery -1-.

Note:
Mark position of battery brackets before loosening bolts.

– Pull connector off Temperature Sensor -G18- -2-.

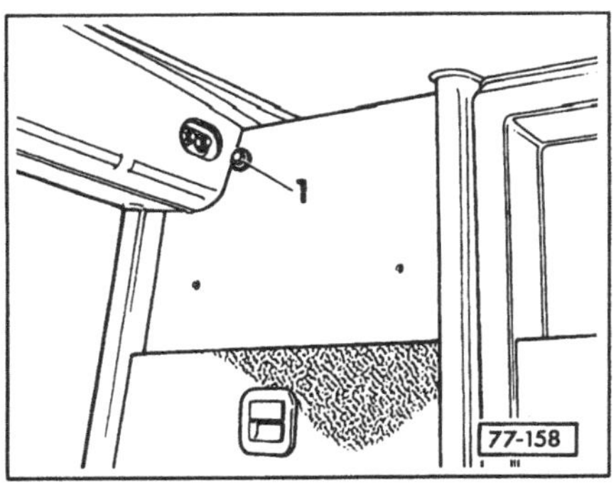

◄ – Remove hex bolt -1-.

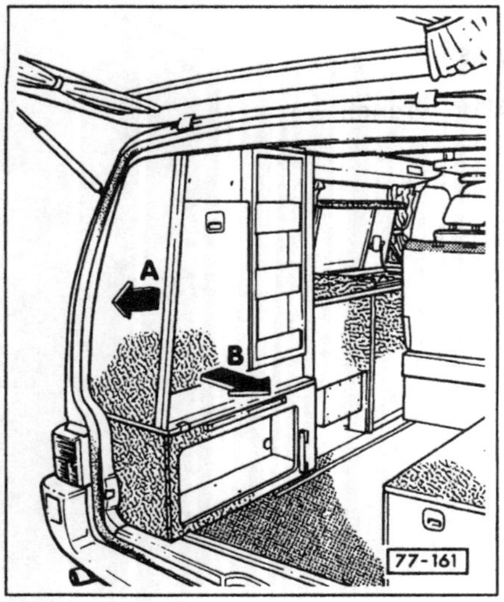

- Remove clothes closet from kitchenette -arrow A- and lift toward center of vehicle -arrow B-.
- Pull battery cable out of battery box and lift clothes closet out of vehicle.

Roof storage box (high roof), removing

- Remove bedboards (high roof) ⇒ page R 75-21
- Remove roof rails ⇒ page R 75-24
- Remove trim with screen and insect protection blind ⇒ page R 75-29

- Remove front retaining brackets for roof rails –1– left and right side (three recessed screws in each case –arrows–).

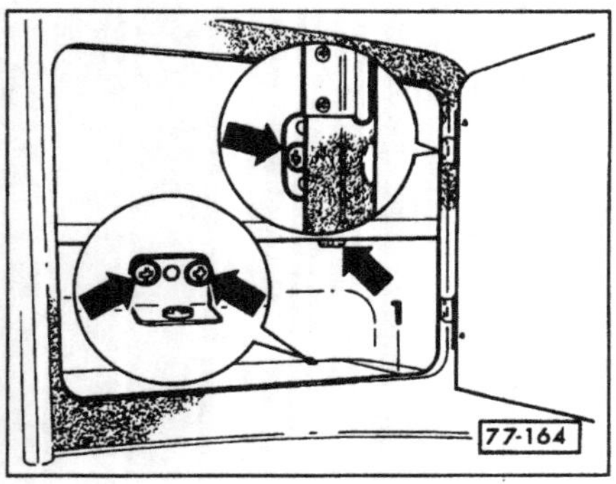

- Open roof storage box right door .
- Remove trim –1–.
- Remove recessed screws –arrows– from brackets.

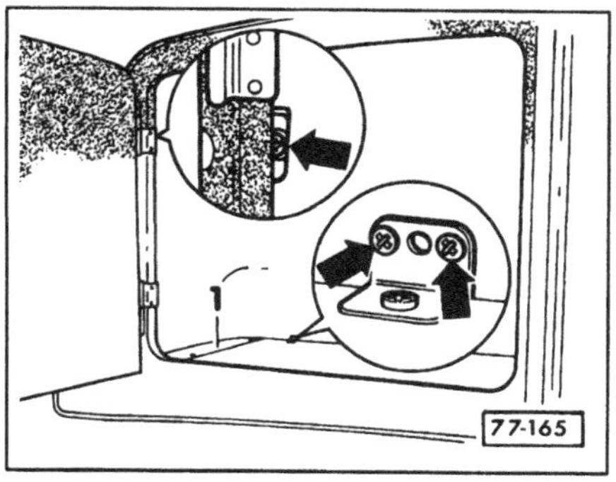

◄ – Open roof storage left door of box.

– Remove trim –1–.

– Remove recessed screws –arrows– from angle brackets.

Heating and Air Conditioning (1992 ➤ 1996)

Section **S**

WARNING!
Before proceeding with any work, please read the warnings and cautions immediately following this index. Also read any warning and caution that accompanies a procedure in the manual.

87 Air conditioning

Please read these WARNINGS and CAUTIONS before proceeding with maintenance and repair work.

WARNING—

- Some repairs may be beyond your capability. If you lack the skills, tools and equipment, or a suitable workplace for any procedure described in this manual, we suggest you leave such repairs to an authorized Volkswagen dealer service department, or other qualified shop.

- Volkswagen is constantly improving its cars. Sometimes these changes, both in parts and specifications, are made applicable to earlier models. Therefore, before starting any major jobs or repairs to components on which passenger safety may depend, consult your authorized Volkswagen dealer about the Technical Bulletins that may have been issued since the editorial closing of this manual.

- Do not re-use any fasteners that are worn or deformed in normal use. Many fasteners are designed to be used only once and become unreliable and may fail when used a second time. This includes, but is not limited to, nuts, bolts, washers, self–locking nuts or bolts, circlips and cotter pins. Always replace these fasteners with new parts.

- Never work under a lifted car unless it is solidly supported on stands designed for the purpose. Do not support a car on cinder blocks, hollow tiles or other props that may crumble under continuous load. Never work under a vehicle that is supported solely by a jack. Never work under the car while the engine is running.

- If you are going to work under a car on the ground, make sure that the ground is level. Block the wheels to keep the car from rolling. Disconnect the battery negative terminal (ground strap) to prevent others from starting the vehicle while you are under it.

- Never run the engine unless the work area is well ventilated. Carbon monoxide (CO) kills.

- Do not attempt to work on your vehicle if you do not feel well. You increase the danger of injury to yourself and others if you are tired, upset or have taken medicine or any other substance that may impair you or keep you from being fully alert.

- Disconnect the battery negative (–) terminal (ground strap) whenever you work on the fuel system or the electrical system. Do not smoke or work near heaters or other fire hazards. Keep an approved fire extinguisher handy.

- Batteries produce explosive hydrogen gas. Keep sparks, lighted matches and open flame away from the top of the battery. If hydrogen gas escaping from the cap vents is ignited, it will ignite gas trapped in the cells and cause the battery to explode.

- Connect and disconnect battery cables, jumper cables or a battery charger only with the ignition switched off, to prevent sparks. Do not quick–charge the battery (for boost starting) for longer than one minute, and do not allow charging voltage to exceed 16.5 volts. Wait at least one minute before boosting the battery a second time.

- Illuminate your work area adequately but safely. Use a portable safety light for working inside or under the car. Make sure the bulb is enclosed by a wire cage. The hot filament of an accidentally broken bulb can ignite spilled fuel or oil.

- Tie long hair behind your head. Do not wear a necktie, a scarf, loose clothing, or a necklace when you work near machine tools or running engines. If your hair, clothing, or jewelry were to get caught in the machinery, severe injury could result.

- Finger rings, bracelets and other jewelry should be removed so that they cannot cause electrical shorts, get caught in running machinery, or be crushed by heavy parts.

- Catch draining fuel, oil, or brake fluid in suitable containers. Do not use food or beverage containers that might mislead someone into drinking from them. Store flammable fluids away from fire hazards. Wipe up spills at once, but do not store the oily rags, which can ignite and burn spontaneously.

- Always observe good workshop practices. Wear goggles when you operate machine tools or work with battery acid. Gloves or other protective clothing should be worn whenever the job requires working with harmful substances.

- The Air-Conditioning (A/C) system is filled with a chemical refrigerant, which is hazardous. The A/C system should be serviced only by trained technicians using approved refrigerant recovery/recycling equipment, trained in related safety precautions, and familiar with regulations governing the discharging and disposal of automotive chemical refrigerants.

- Do not expose any part of the A/C system to high temperatures such as an open flame. Excessive heat will increase system pressure and may cause the system to burst.

- Some aerosol tire inflators are highly flammable. Be extremely cautious when repairing a tire that may have been inflated using an aerosol tire inflator. Keep sparks, open flame or other sources of ignition away from the tire repair area. Inflate and deflate the tire at least four times before breaking the bead from the rim. Completely remove the tire from the rim before attempting any repair.

- Friction materials such as brake or clutch discs may contain asbestos fibers. Do not create dust by grinding, sanding, or by cleaning with compressed air. Avoid breathing asbestos fibers and asbestos dust. Breathing asbestos can cause serious diseases such as asbestosis or cancer, and may result in death.

- Greases, lubricants and other automotive chemicals contain toxic substances, many of which are absorbed directly through the skin. Read manufacturer's instructions and warnings carefully. Use hand and eye protection. Avoid direct skin contact.

- An airbag is operated by an explosive device. Handled improperly or without adequate safeguards, it can be accidently activated and cause serious personal injury. To prevent personal injury or airbag system failure, **only trained Volkswagen dealer service technicians** should test, disassemble or service the airbag system.

WARNINGS and CAUTIONS continued on next page ▶

Please read these WARNINGS and CAUTIONS before proceeding with maintenance and repair work.

WARNING (continued) —

- ALWAYS disconnect the airbag power supply before using a computer memory saver. Failure to disconnect the airbag power supply may result in accidental activation of the airbag.
- The airbag unit should not be removed, nor should any work be performed on the airbag system, without first reviewing the warnings and cautions that accompany the airbag system repair information in this manual.
- Never apply stickers or any other coverings on the steering wheel. Do not let chemical cleaners, oil or grease come into contact with the vinyl covering of the airbag unit.
- Never open or otherwise attempt to repair airbag system parts. Always use new parts. Never leave airbag parts or the partially disassembled airbag system unattended.
- Never use a test light or volt meter to conduct electrical tests on the airbag system. The airbag system must only be tested by trained Volkswagen Service technicians using special airbag tester equipment. The airbag system must only be tested while it is installed in the car.
- Do not expose an airbag unit to temperatures above 194°F (90°C) even for a brief period, while handling it during the repair process. Keep the unit clear of heat sources such as hot plates, soldering irons, heat lamps, welding equipment, or the like.
- Always replace airbag system components that have been mechanically damaged (bubbles, cracks, etc.).
- Always set down a removed airbag unit so that the horn pad is facing upward.
- When driving or riding in an airbag-equipped vehicle, never hold test equipment in your hands or lap while the vehicle is in motion. Objects between you and the airbag can increase the risk of injury in an accident.
- The storage, transportation and disposal of airbag units are subject to the laws for flammable solids.

CAUTION—

- If you lack the skills, tools and equipment, or a suitable workshop for any procedure described in this manual, we suggest you leave such repairs to an authorized Volkswagen dealer or other qualified shop. We especially urge you to consult an authorized Volkswagen dealer before beginning repairs on any car that may still be covered wholly or in part by any of the extensive warranties issued by Volkswagen of America, Inc.
- Volkswagen is constantly improving its cars and sometimes these changes, both in parts and specifications, are made applicable to earlier models. Therefore, part numbers listed in this manual are for reference only. Always check with your authorized Volkswagen dealer parts department for the latest information.
- Before starting a job, make certain that you have all the necessary tools and parts on hand. Read all the instructions thoroughly, do not attempt shortcuts. Use tools that are appropriate to the work and use only replacement parts meeting Volkswagen specifications. Makeshift tools, parts and procedures will not make good repairs.
- Use pneumatic and electric tools only to loosen threaded parts and fasteners. Never use these tools to tighten fasteners, especially on light alloy parts. Always use a torque wrench to tighten fasteners to the tightening torque specification listed.
- Be mindful of the environment and ecology. Before you drain the crankcase, find out the proper way to dispose of the oil. Do not pour oil onto the ground, down a drain, or into a stream, pond, or lake. Consult local ordinances that govern the disposal of wastes.
- When connecting cables or ground straps, all connections must be cleaned to bare metal. Be sure mounting bolts are tight.
- Do not replace fuse links with fuse links of other gauge or with standard wire. Always replace the entire length of the fuse link.
- Before doing any electrical welding on vehicles equipped with Anti-lock Brakes (ABS), disconnect the battery negative terminal (ground strap) and the ABS Control Module connector.
- On cars equipped with ABS, the ABS control unit cannot withstand temperatures from a paint–drying booth or a heat lamp in excess of 203°F (95°C) and should not be subjected to temperatures in excess of 185°F (85°C) for more than two hours.
- On cars equipped with anti-theft radios, make sure you know the correct radio activation code before disconnecting the battery or removing the radio. If the wrong code is entered when power is restored, the radio may lock up and be rendered inoperable, even if the correct code is then entered.
- When boost-charging the battery, first remove the fuses for the Engine Control Module (ECM), the Transmission Control Module (TCM), the ABS Control Module, and the On-Board computer. In cases where one or more of these components is not separately fused, disconnect the Control Module connector(s).
- Do not quick-charge the battery (for boost starting) for longer than one minute, and do not exceed 16.5 volts at the battery with the boosting cables attached. Wait at least one minute before boosting the battery a second time.

◀ **WARNINGS and CAUTIONS** continued from previous page

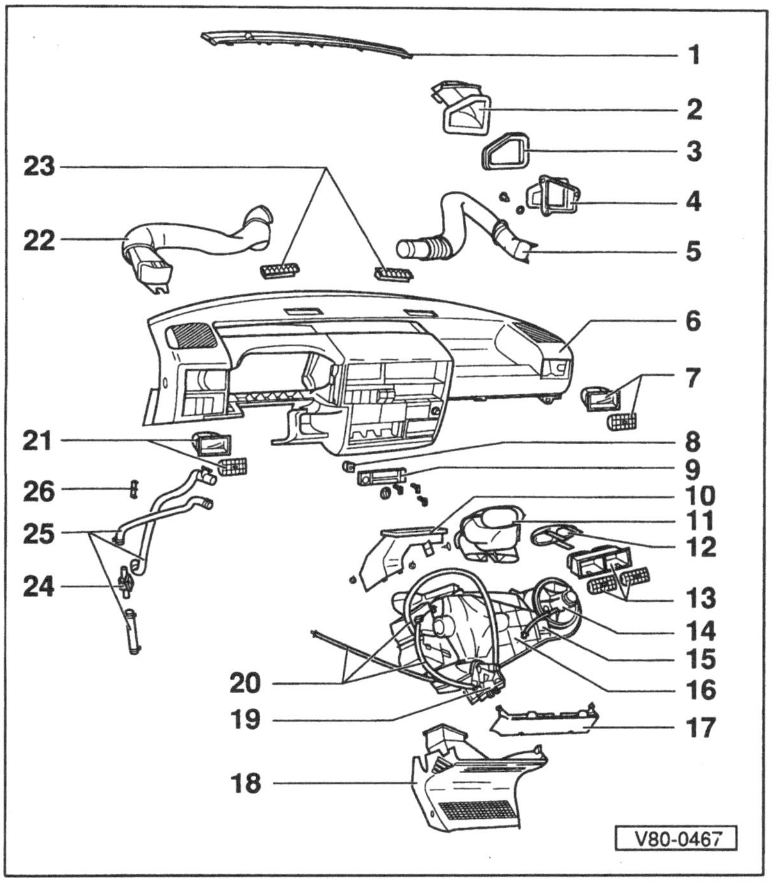

Heating and ventilation system, servicing

> ***CAUTION!***
>
> ***Before working on the electrical system, obtain radio security code and disconnect battery Ground (GND) strap.***

1 - Air intake grill

2 - Air intake duct, front (under hood)
- ♦ Removing/installing ⇒ Fig. 1

3 - Seal

4 - Air intake duct, rear (on bulkhead)
- ♦ Removing/installing ⇒ Fig. 1

5 - Right air duct
- ♦ Removing, first remove glove box

6 - Instrument panel

7 - Right air outlet
- ♦ Removing/installing ⇒ Repair Group 70

8 - Fresh air blower switch -E9-

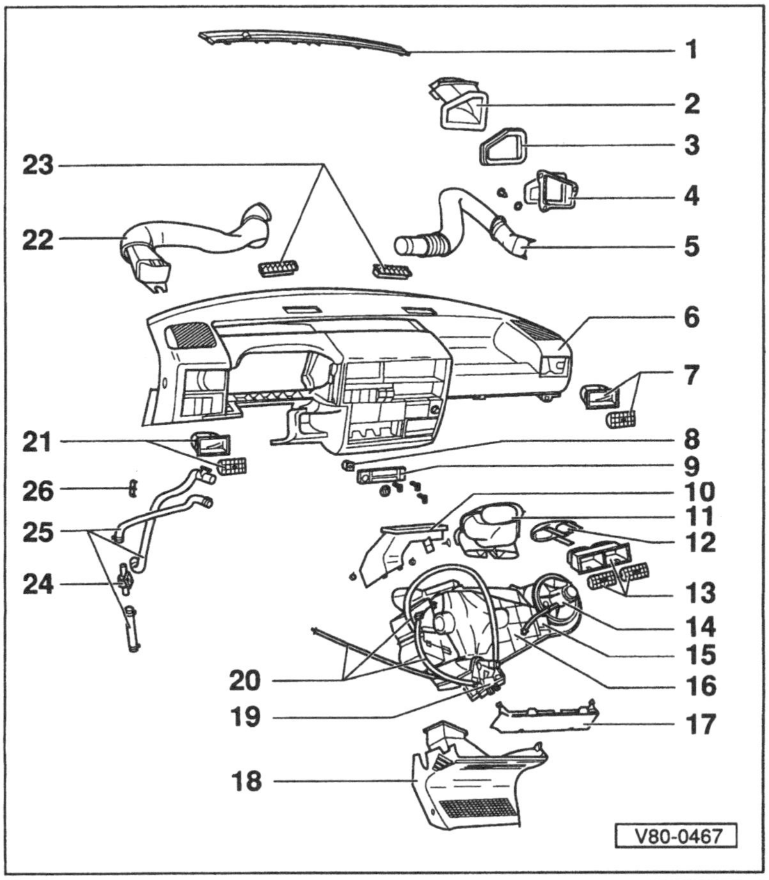

9 - Heater/fresh air control plate
- ♦ Removing/installing (vehicles without A/C) ⇒ Fig. 3

10 - Defroster duct
- ♦ To remove, first remove heater box/evaporator housing

11 - Center air duct
- ♦ To remove, first remove heater box/evaporator housing

12 - Center air outlet adaptor
- ♦ Removing: remove radio, remove center air outlet

13 - Center air outlet
- ♦ Removing ⇒ Fig. 4

14 - Fresh air blower -V2-
- ♦ Removing ⇒ Fig. 5

15 - Fresh air blower series resistance with fuse -N24-
- ♦ Removing, first remove glove box
- ♦ Removing ⇒ Fig. 5

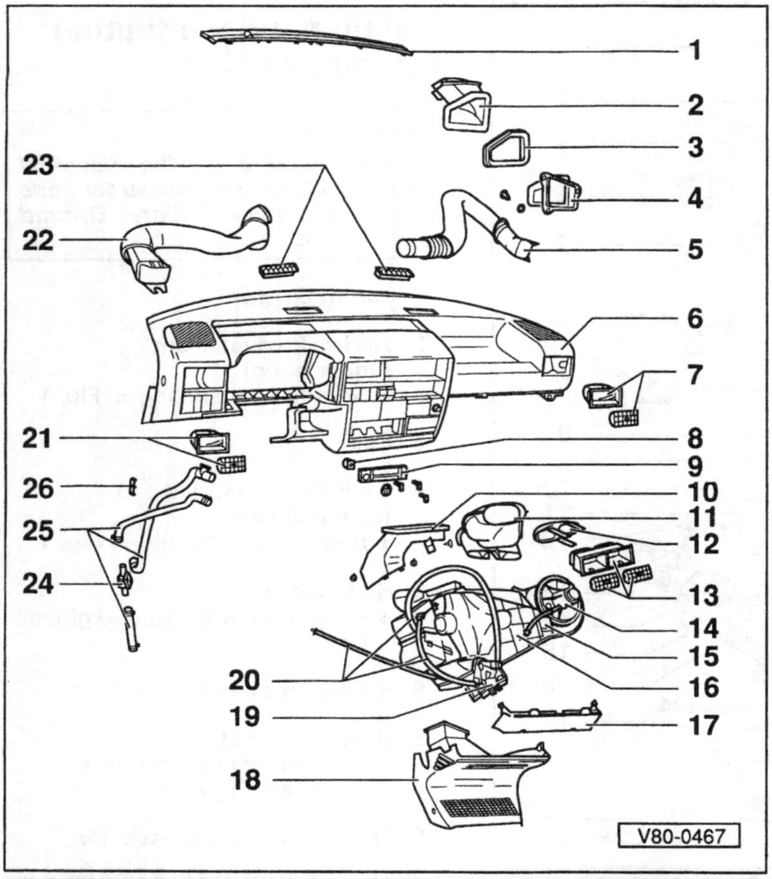

16 - Heater/evaporator housing
- ♦ Removing ⇒ Fig. 8
- ♦ Heater assembly ⇒ page S 80-15
- ♦ Installing evaporator water drain valve ⇒ Repair Group 87

17 - Cover

18 - Footwell air outlet console
- ♦ Removing ⇒ Fig. 6

19 - Heater / fresh air control assembly
- ♦ Removing/installing ⇒ Fig. 3
- ♦ Installing/adjusting cables ⇒ page S 80-6

20 - Cables
- ♦ Installing/adjusting ⇒ page S 80-11

21 - Left air outlet
- ♦ Removing ⇒ Fig. 2

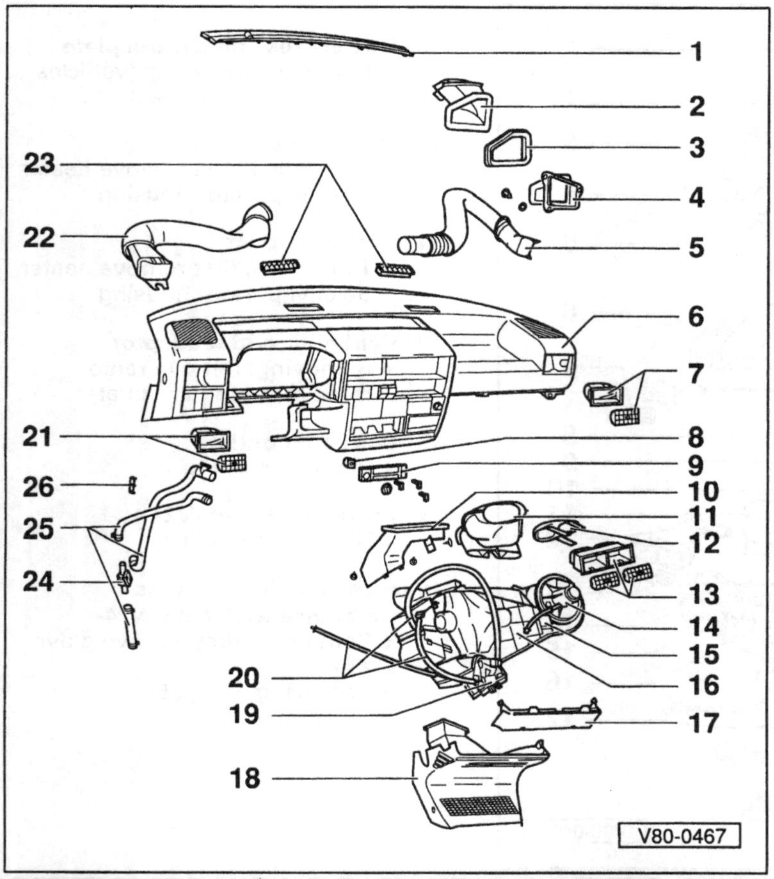

22 - Left air duct
- ♦ Removing

– Remove fuse/relay panel cover

– Remove footwell air outlet cover and console

– Separate and remove air duct

23 - Defroster outlet

– Carefully pry out with screwdriver

24 - Heater control valve
- ♦ Removing ⇒ Fig. 7

25 - Heater (coolant) hoses
- ♦ For hose routing ⇒ Repair Group 19

26 - Clip

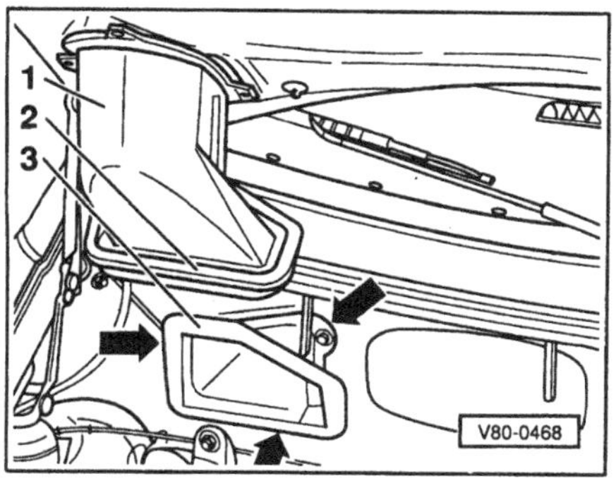

Fig. 1 Air intake ducts, front and rear, removing / installing

1 - Air intake duct, front (under hood)

2 - Seal

3 - Air intake duct, rear (on bulkhead)

Removing

– Remove screws –arrows– and remove rear duct

– Remove screws from hood and remove front duct

– Install in reverse order of removal

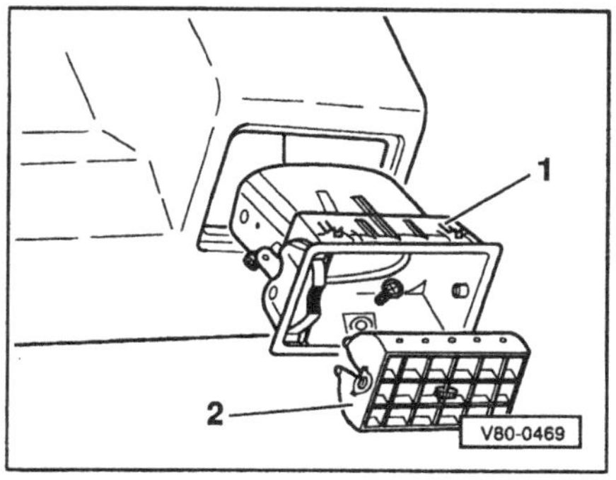

Fig. 2 Air outlet, removing

– Using a screwdriver, carefully pry out pivoting vent -2-

– Remove screw from center of outlet

– Remove air outlet -1-

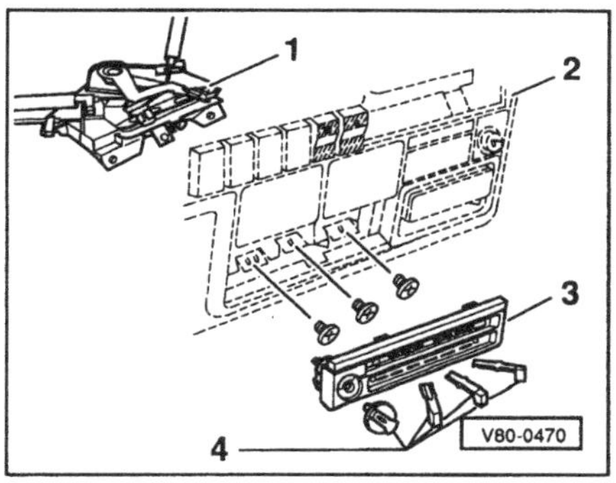

Fig. 3 Heater/fresh air control plate and control assembly, removing/installing (vehicles without A/C)

Removing

– Remove levers -4-

– Remove control plate -3-

– Remove screws from instrument panel -2-

– Push control assembly -1- (cables attached) forward and down to bottom of instrument panel

– Remove control cables and remove heater/fresh air control

Installing

– Attach cables to heater/fresh air control assembly

– Install in reverse order of removal

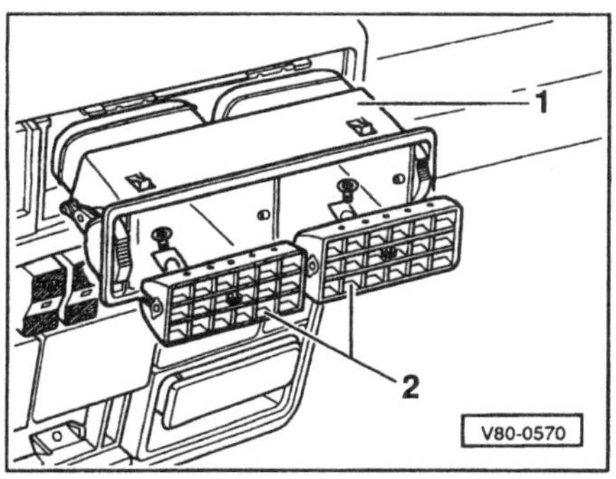

Fig. 4 Center air outlet, removing

– Using a screwdriver, carefully pry out pivoting vents -2-

– Remove screws from center of each vent opening

– Remove center air outlet -1-

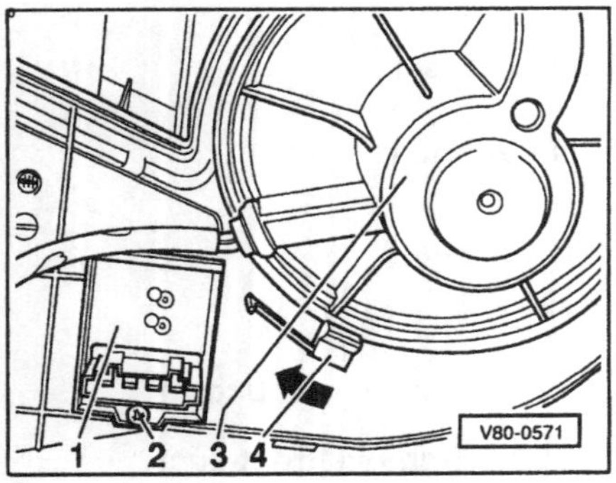

◄ **Fig. 5 Fresh air blower series resistor with fuse -N24- and fresh air blower -V2-, removing**

> ***CAUTION!***
>
> ***Before working on the electrical system, obtain radio security code and disconnect battery Ground (GND) strap.***

Removing series resistor -1-

– Remove harness connector from series resistor

– Remove screw -2- and remove series resistor

Removing fresh air blower -3- (with heater box removed)

– Disengage retaining tab -4– and turn fresh air blower in direction of –arrow–

– Removing fresh air blower -3- (with instrument panel and heater box installed)

– Remove glove box

– Remove harness connector from series resistor

– Remove right air duct

– Disengage retaining tab -4-, turn fresh air blower in direction of–arrow– and remove from housing

– Move fresh air blower to center of instrument panel and remove

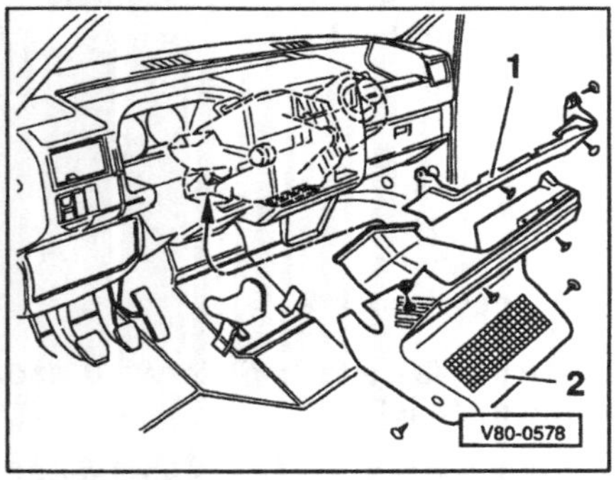

◄ **Fig. 6 Footwell air outlet console, removing**

– Remove cover -1-

– Detach and remove footwell air outlet console -2-

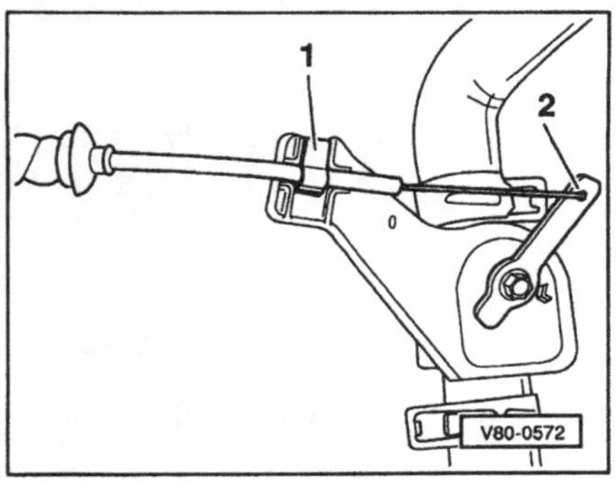

◄ **Fig. 7 Heater control valve (vehicles without A/C), removing**

– Remove clip -1- and remove cable end -2- from lever

– Clamp coolant hoses near valve

– Remove spring clamps from coolant hoses and remove coolant hoses from valve

– Remove valve

Note

♦ *Release coolant system pressure and drain coolant before beginning repair.*

♦ *Refill with coolant after installing/replacing control valve.*

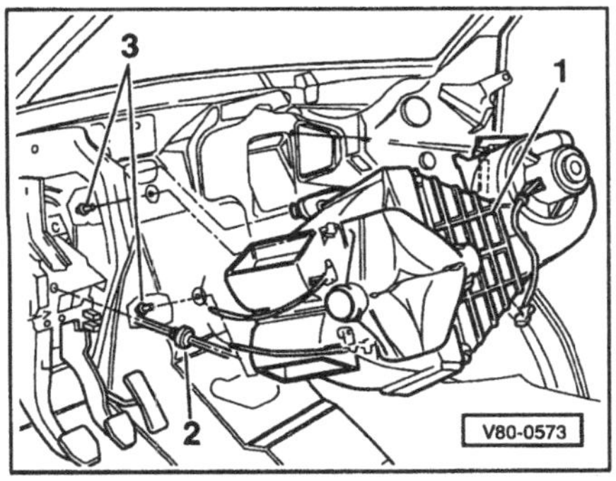

◄ **Fig. 8 Heater box/evaporator housing, removing**

Note:

For illustration purposes, heater box/evaporator housing is shown with the instrument panel removed.

1 - Heater box/evaporator housing

2 - Rubber grommet

3 - Screws

Note:

Plug heater core outlets to prevent coolant spills in the passenger compartment during heater box/evaporator housing removal.

Removing

– Remove glove box

– Remove right air duct

– Remove air intake duct, rear (on bulkhead)

– Remove center air outlet and center air duct (if equipped)

– Loosen and remove heater/fresh air control cables from housing

– Remove harness connector from fresh air blower series resistor

– Remove footwell air outlet console and cover

– Remove coolant hoses from heater core and plug connections

– Loosen and remove cable at heater control valve

– Remove screws -3-

Note

The lower screw is located on the bulkhead beneath the engine compartment trim.

– Remove heater box/evaporator housing 1

Heater/fresh air control cables (vehicles without A/C), installing/adjusting

Note:

- *Install cables on control valve first*
- *Place cable housing at stop and fasten*
- *Fasten cables to the levers*
- *Defroster and floor duct cables are self adjusting – slide lever –5– and –6– from stop to stop to adjust.*
- *With temperature control valve in the installed position adjust cable at the valve.*

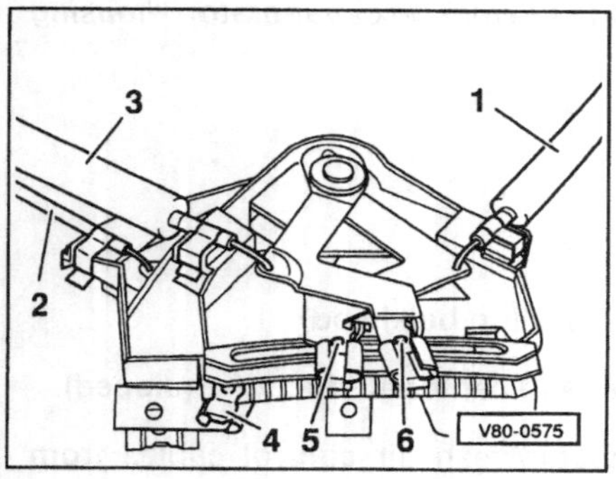

◄ **Cables, installing**

1 - Footwell flap cable

– Slide lever -5- to right stop when adjusting cable

– Adjust cable at floor outlet ⇒ Fig. 1

2 - Temperature control cable (to heater control valve)

– Adjust cable at control valve ⇒ Fig. 2

– Slide lever -4- to left stop when adjusting cable

3 - Defroster flap cable

– Slide lever -6- to left stop when adjusting cable

– Adjust cable at defroster flap ⇒ Fig. 3

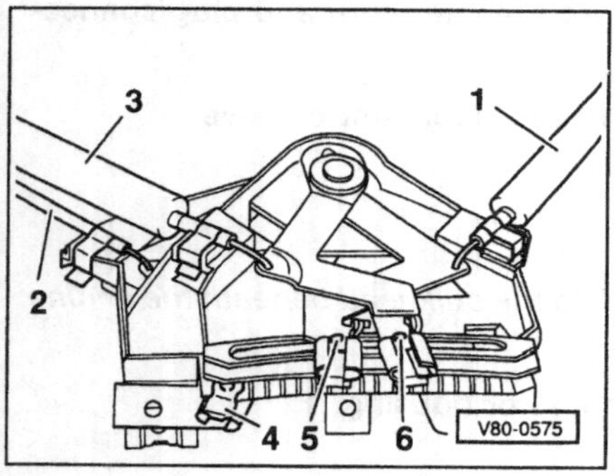

◄ **Cable for defoster flap**

– Slide lever -6- to the left

– Adjust cable at defroster flap ⇒ Fig. 3

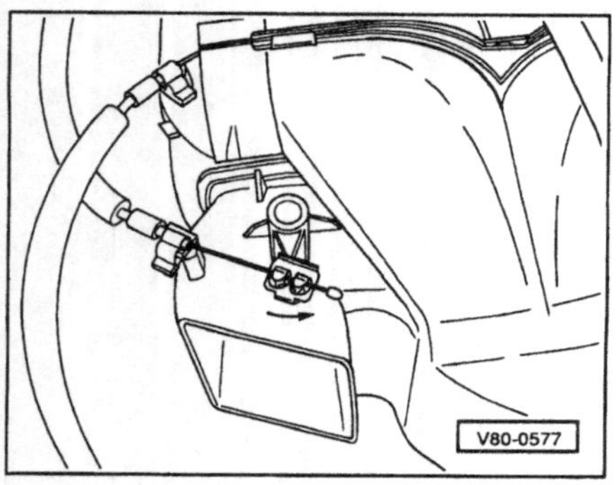

◄ **Fig. 1 Adjusting cable at floor outlet flap**

– Fasten cable to floor outlet flap by simultaneously pressing the cable in direction of arrow.

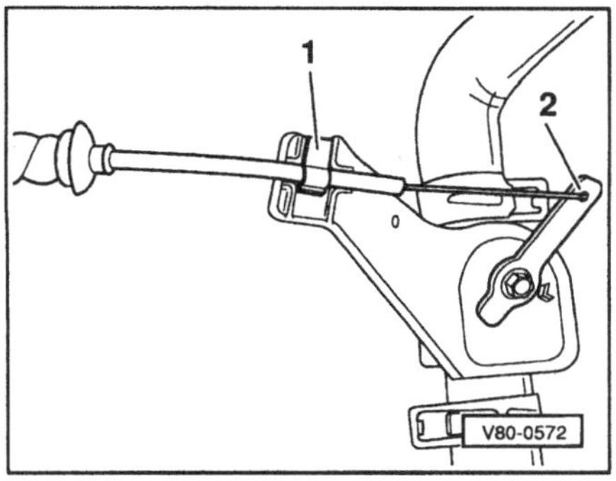

◄ **Fig. 2 Fastening cable to control valve**

– Direct cable to control valve through bulkhead in engine compartment

– Postion lever -2- to the left up to the stop (closed/cool)

– Fasten cable to control valve with clip -1-

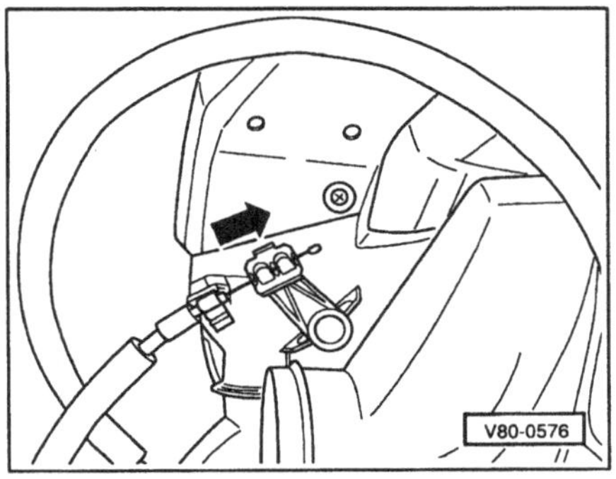

◄ **Fig. 3 Adjusting cable at defroster flap**

– Fasten cable to defroster flap by simultaneously pushing lever in direction of arrow.

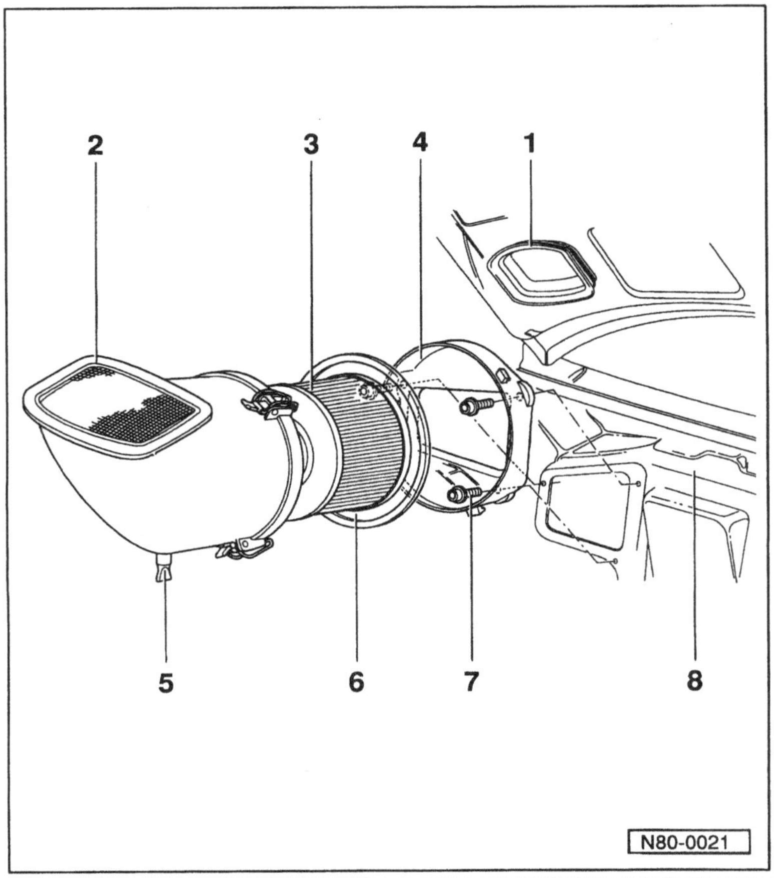

Dust and pollen filter, removing and installing

Note:

- *Retrofitting dust and pollen filter in fresh air intake is only possible on vehicles from January 1995.*

1 - Hood seal

2 - Filter housing

3 - Dust and pollen filter
- The passenger compartment dust and pollen filter is identical

4 - Adaptor

5 - Valve

6 - Filter retainer

7 - Hex bolt
- 3 Nm (26 in. lb)

8 - Bulk head

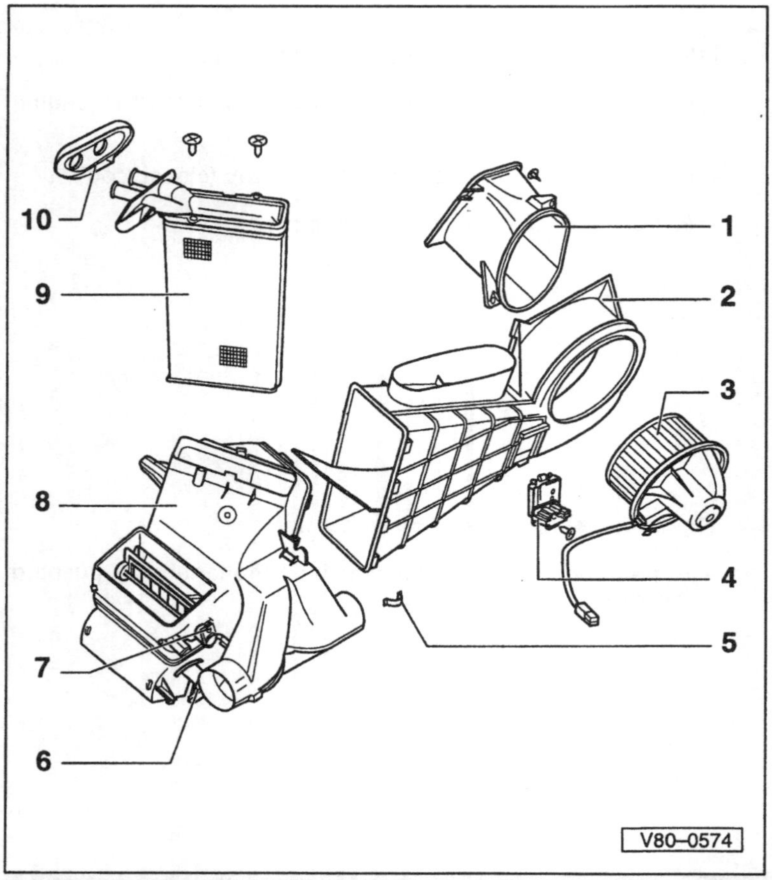

Heater box, assembly

1 - **Air intake duct**

2 - **Fresh air blower duct**

3 - **Fresh air blower -V2-**

4 - **Fresh air blower series resistor with fuse -N24-**

5 - **Clip**

6 - **Footwell flap lever**

– Removing footwell flap, unclip lever using a screwdriver

7 - **Defroster flap lever**

– To remove defroster flap, unclip lever using a screwdriver

8 - **Air distribution housing**

9 - **Heater core**
- ♦ Install seals around entire circumference of heater core

10 - **Gasket**

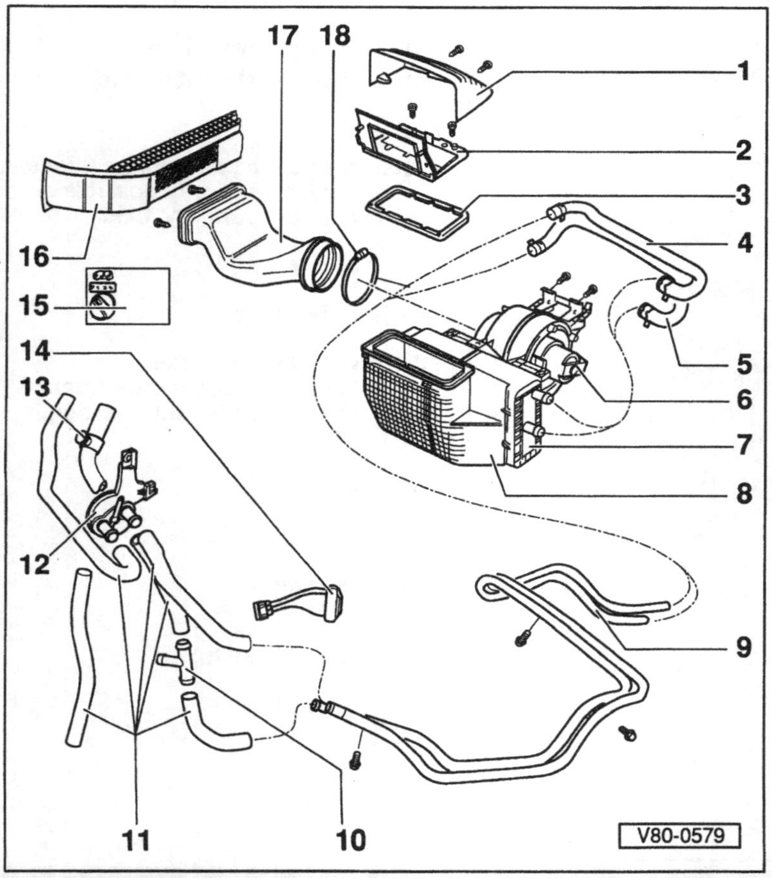

Rear heater, components

EuroVan MV (USA), EuroVan CV (Canada), located in loading area

1 - **Upper air outlet vent (in passenger compartment)**
- ♦ Removing ⇒ Fig. 2
- ♦ In loading area

2 - **Lower air outlet vent (in passenger compartment)**
- ♦ Removing ⇒ Fig. 2
- ♦ In loading area

3 - **Seal**

4 - **Coolant return hose**

5 - **Coolant supply hose**

6 - **Warm air blower -V47-**
- ♦ Removing ⇒ page S 80-26

7 - **Heater core**
- ♦ Removing coolant hoses ⇒ page S 80-21
- ♦ All seals must be bonded all around and gap free

Note:

Always change coolant.

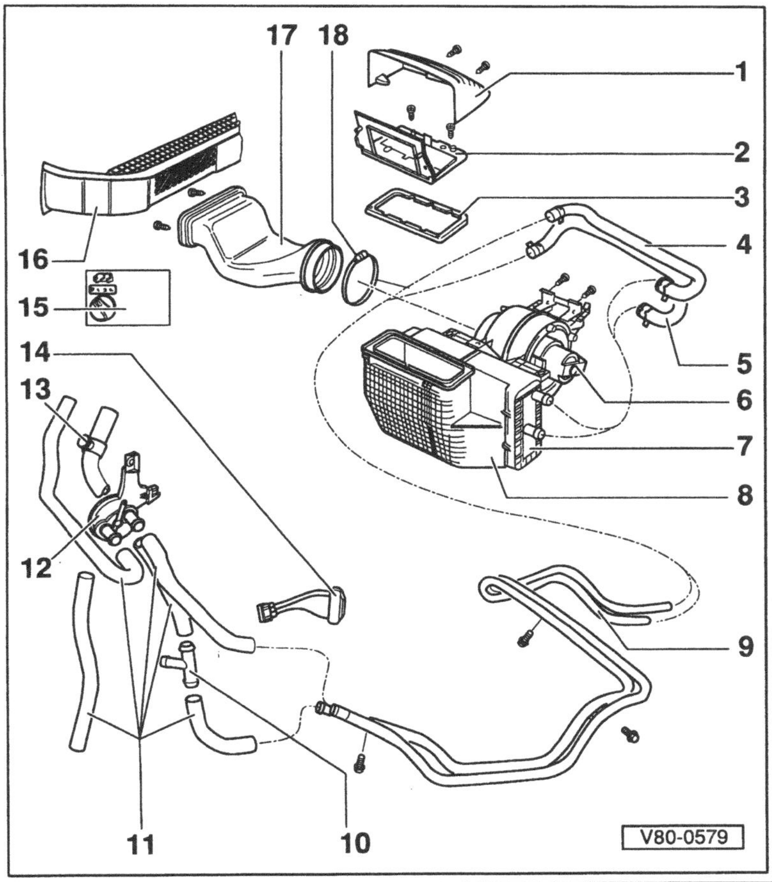

8 - Air distribution housing
- ♦ Assembly ⇒ page S 80-26
- ♦ Removing ⇒ Fig. 1; rear heater, removing ⇒ page S 80-29

9 - Coolant pipe
- ♦ Removing/installing ⇒ Fig. 4

10 - T–adaptor

11 - Coolant hoses in engine compartment
- ♦ Connections ⇒ page S 80-24

12 - Control valve (vehicles without A/C)
- ♦ Removing ⇒ Fig. 6

13 - Air bleeder screw

14 - Heater blower series resistance -N72-
- ♦ Located in left side of engine compartment

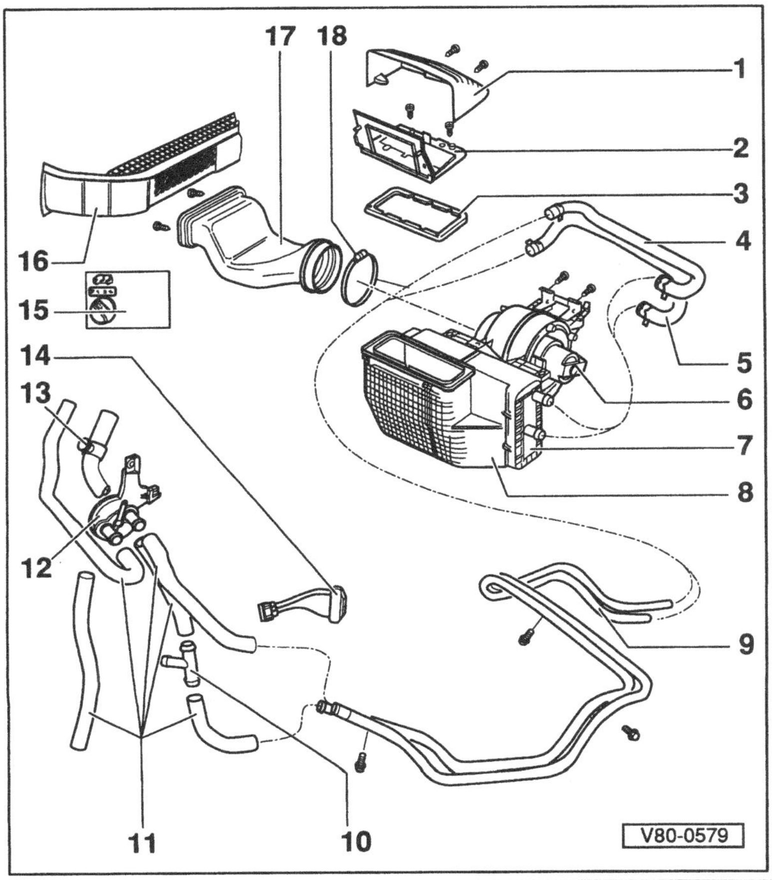

15 - Warm air blower switch -E100-
- ♦ Removing ⇒ Fig. 5

16 - Sliding door footwell insert
- ♦ Removing ⇒ Repair Group 68

17 - Air intake duct
- ♦ Removing ⇒ Fig. 3

18 - Clamp

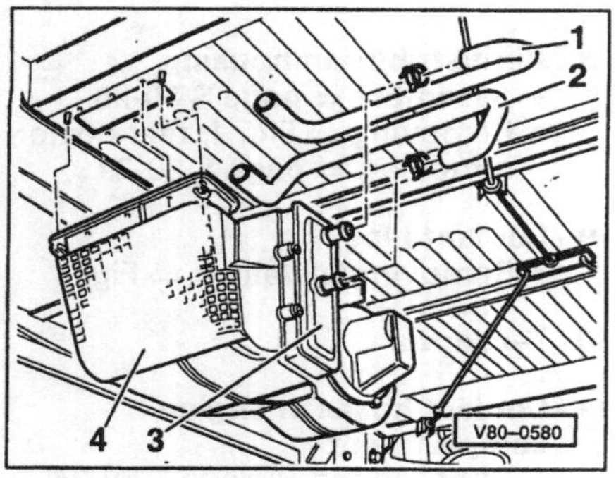

◄ **Fig. 1 Rear heater, removing**

1 - Coolant return hose
2 - Coolant supply hose
3 - Heater core
4 - Rear heater

Removing

– Disconnect wire harness connector from rear heater

– Remove coolant hoses -1- and -2- from heater core -3-

– Remove air outlet vent ⇒ Fig. 2, page S 80-20

– Remove screws and remove rear heater

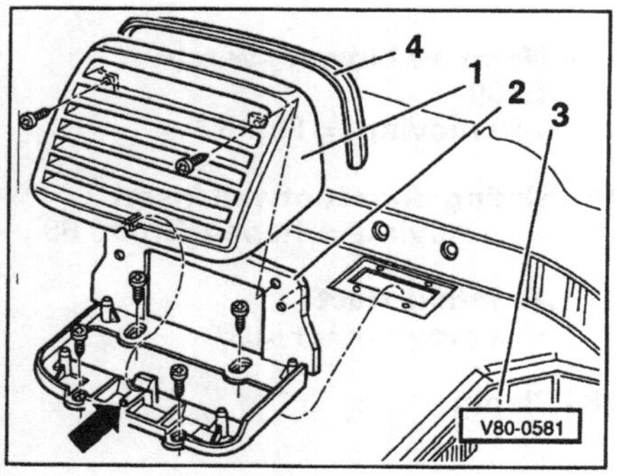

◄ **Fig. 2 Air outlet vent, removing**

1 - Upper air outlet vent
2 - Lower air outlet vent
3 - Sliding door footwell insert
4 - Seal (only on vehicles with partition)

Removing

– Remove upper air outlet vent screws

– Using a screwdriver, unlock latch –arrow– and remove upper air outlet vent

– Remove lower air outlet vent screws

– Remove lower air outlet vent

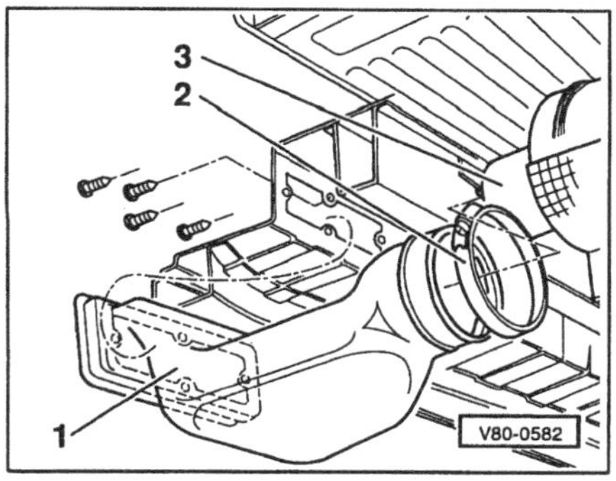

◄ **Fig. 3 Air intake duct, removing**

1 - Air intake duct

2 - Clamp

3 - Rear heater

Removing

– Remove sliding door footwell insert

– Remove air duct screws

– Loosen clamp -2- and remove air duct -1-

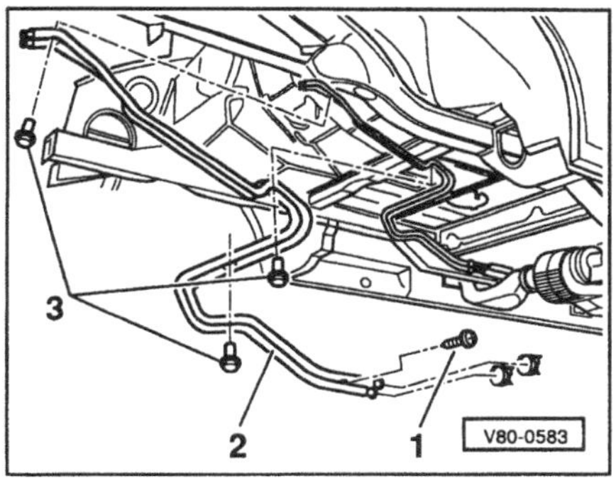

◄ **Fig. 4 Coolant pipe, removing/installing**

1 - Sheet metal screw

2 - Coolant pipe

3 - Hexagon nut

Removing

– Remove fuel tank ⇒ Repair Group 20

– Remove torsion bar ⇒ Repair Group 40

– Remove nuts -3- and screw -1-

– Remove coolant pipe -2-

– Reinstall in reverse sequence

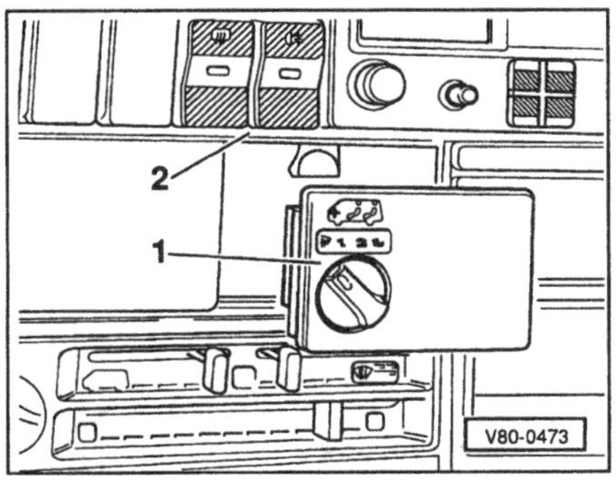

◄ **Fig 5. Warm air blower switch -E100-, removing**

– Carefully pry out warm air blower switch -1- from instrument panel -2- with screw driver

– Remove connector

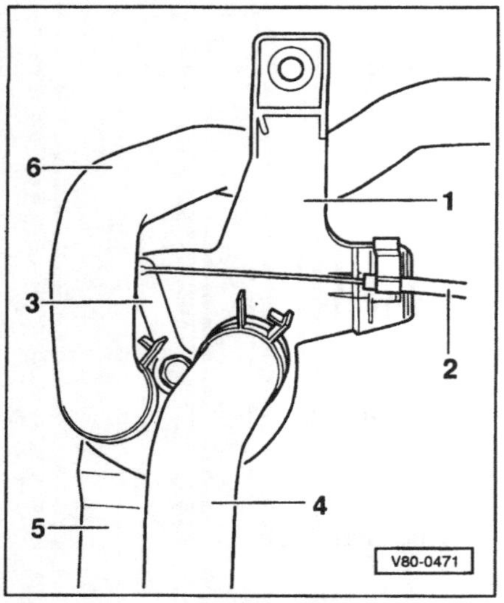

◄ **Fig. 6 Control valve, removing**

1 - Control valve

2 - Cable

3 - Lever

4 - Coolant return hose, (from coolant pipe - Auxiliary heater core to control valve)

5 - Coolant return hose, (from control valve to motor)

6 - Coolant return hose, (from main heater core to control valve)

Removing

– Loosen clip and remove cable

– Loosen spring clamps and remove coolant hoses

– Remove control valve from over flow reservoir bracket

Note:

- *Release coolant system pressure and drain coolant before beginning repair.*
- *Refill with coolant after installing/replacing control valve.*

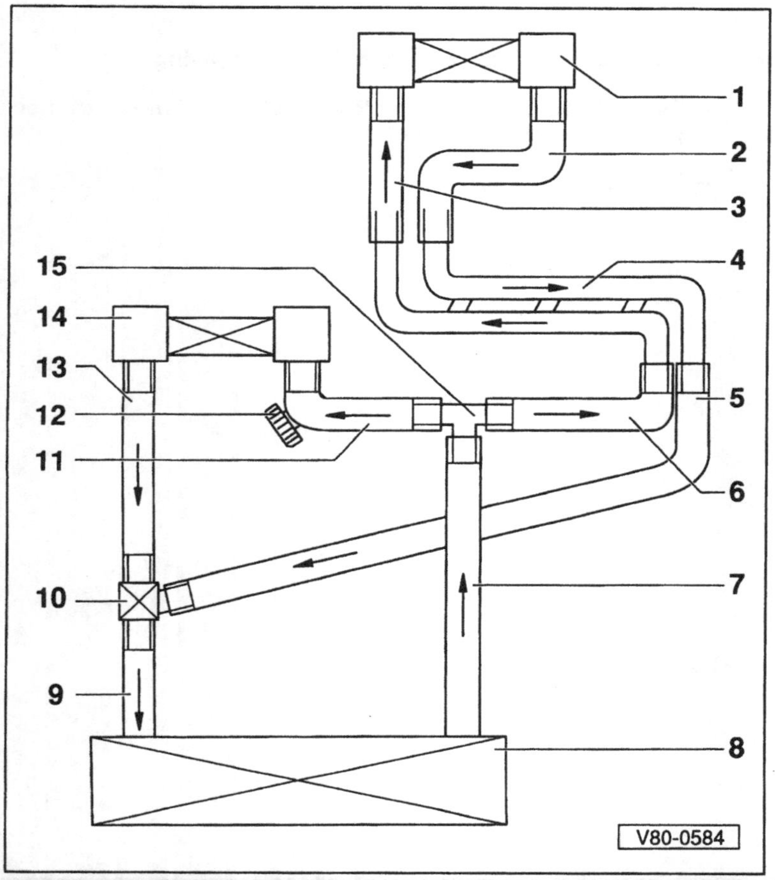

Coolant hoses, connections

Note:

For routing of all hoses
⇒ Repair Group 19

1 - Rear heater (heater core)

2 - Coolant return hose
- From rear heater core to coolant pipe

3 - Coolant supply hose
- From coolant pipe to rear heater core

4 - Coolant pipe

5 - Coolant return hose
- From coolant pipe to control valve

6 - Coolant supply hose
- From T–adaptor to coolant pipe

7 - Coolant supply hose
- From engine to T–adaptor

8 - Engine

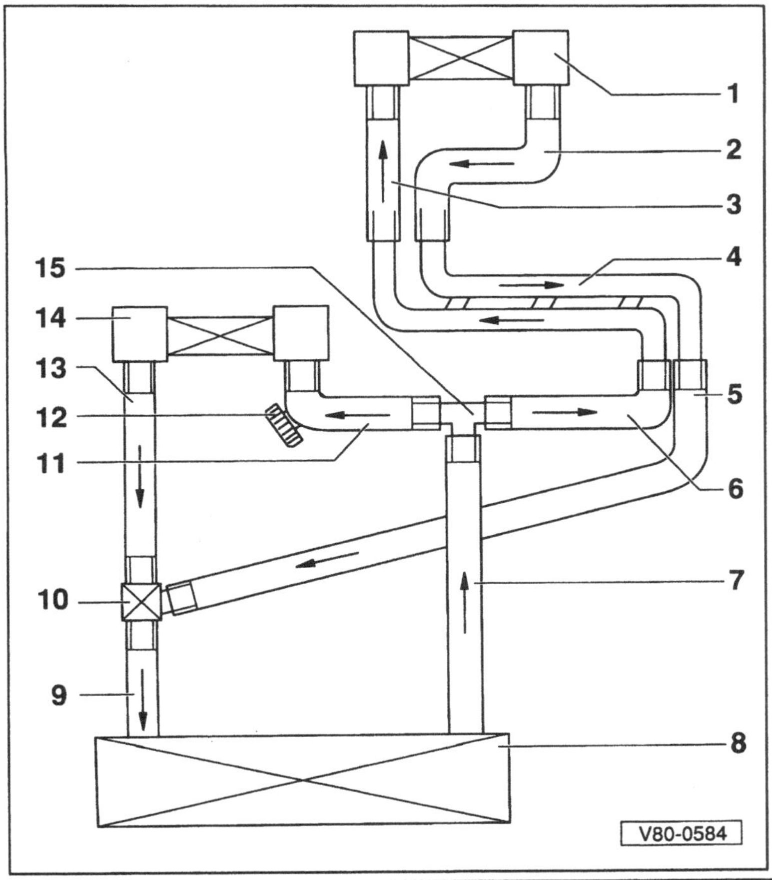

9 - **Coolant return hose**
- From control valve to engine

10 - **Control valve (vehicles without A/C), coolant cut–off valve (vehicles with A/C)**

11 - **Coolant supply hose**
- From T–adaptor to front heater core

12 - **Air bleeder screw**

13 - **Coolant return hose**
- From front heater core to control valve

14 - **Front heater core**

15 - **T–adaptor**

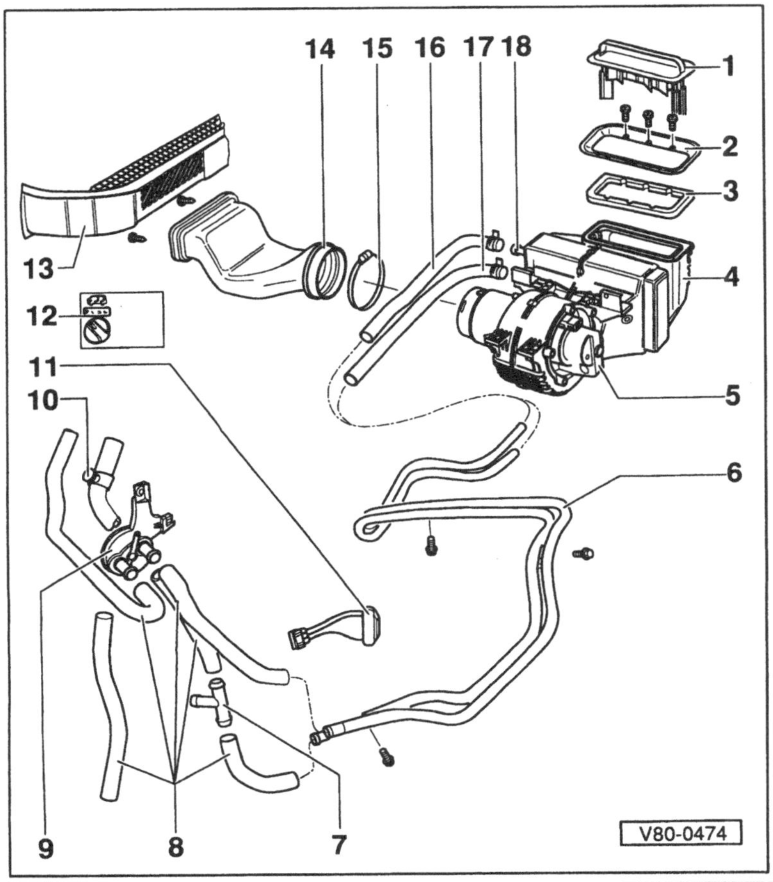

Rear heater air distribution housing, assembly

1 - **Gasket**

2 - **Air distribution housing**

3 - **Bracket**

4 - **Clip**
- Snaps into position on cross member

5 - **Rear warm air fan -V47-**

6 - **Heater core**
- Install seals around entire circumference of heater core so there are no gaps

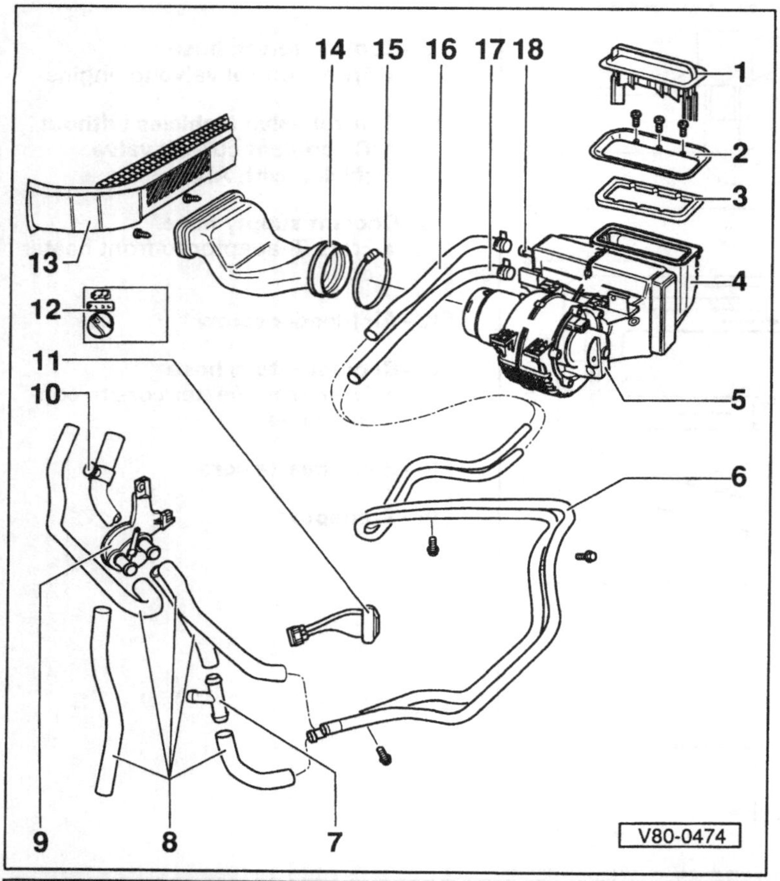

Rear heater, components

EuroVan and Transporter models

1 - Air outlet vent
- ♦ Removing ⇒ Fig. 2

2 - Cover plate

3 - Seal

4 - Air distribution housing
- ♦ Removing ⇒ Fig. 1

5 - Rear warm air fan -V47-

6 - Coolant pipe
- ♦ Removing/installing ⇒ Fig. 4

7 - T–adaptor

8 - Coolant hoses in engine compartment
- ♦ Connections ⇒ Page S 80-24

9 - Control valve (vehicles without A/C)
- ♦ Removing ⇒ Fig. 6

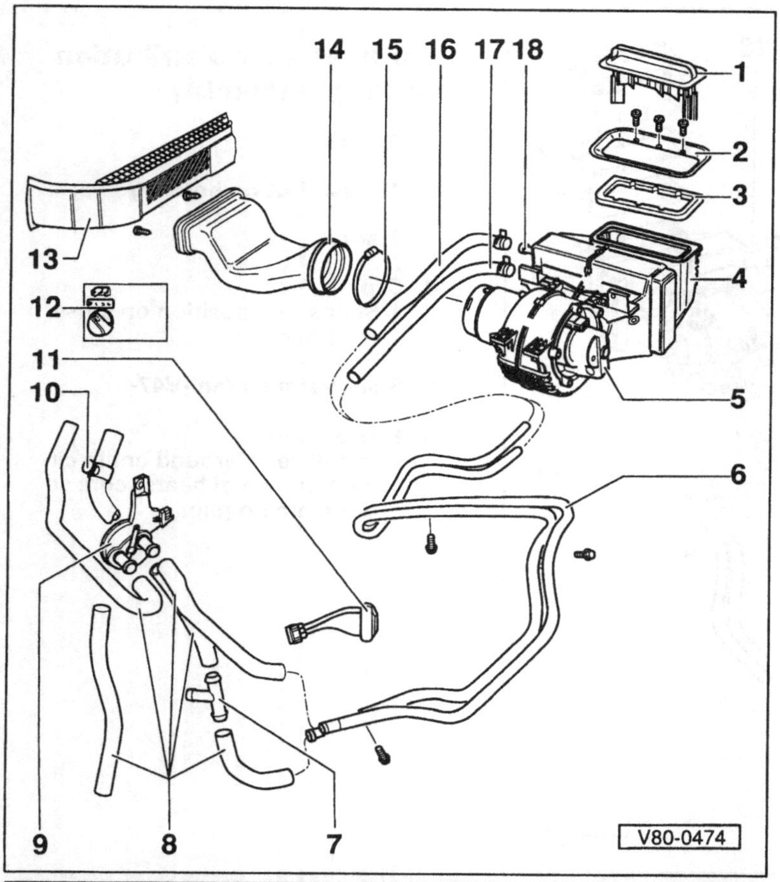

10 - Air bleeder screw

11 - Heater blower series resistance -N72-
- ♦ Located in left side of engine compartment

12 - Warm air blower switch -E100–
- ♦ Removing ⇒ Fig. 7

13 - Sliding door footwell insert
- ♦ Removing ⇒ Repair Group 68

14 - Air intake duct
- ♦ Removing ⇒ Fig. 3

15 - Clamp

16 - Coolant return hose

17 - Coolant supply hose

18 - Heater core
- ♦ Removing ⇒ Fig. 1
- ♦ Ensure seal is installed flat without folds or gaps

Note:

Always change coolant.

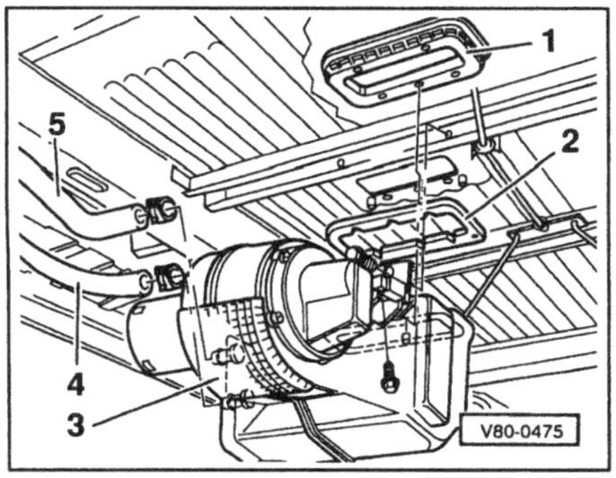

◄ **Fig. 1 Rear heater, removing**

1 - Cover plate (in passenger compartment)
2 - Seal
3 - Rear heater (fan housing)
4 - Coolant supply hose
5 - Coolant return hose

Removing

– Disconnect wire harness connector from rear heater

– Clamp coolant hoses near heater core

– Remove coolant hoses -4- and -5- from heater core

– Remove air outlet vent ⇒ see Fig. 2

– Remove cover plate -1-

– Remove screws and remove rear heater

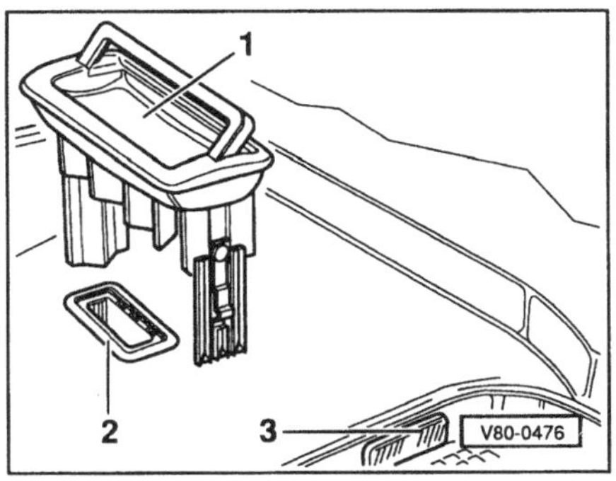

◄ **Fig. 2 Air outlet vent, removing**

1 - Air outlet vent
2 - Cover plate
3 - Sliding door footwell insert

Removing

– Pull out air outlet vent -1-

– Remove cover plate -2-

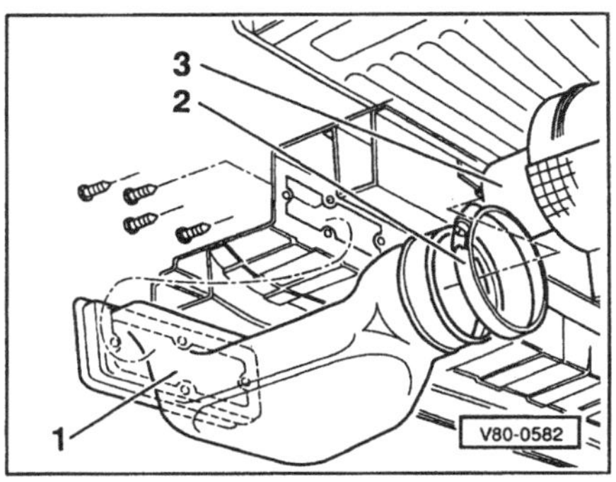

◄ **Fig. 3 Air intake duct, removing**

1 - Air intake duct
2 - Clamp
3 - Rear heater

Removing

– Remove sliding door footwell insert

– Remove air duct screws

– Loosen clamp -2- and remove air duct -1-

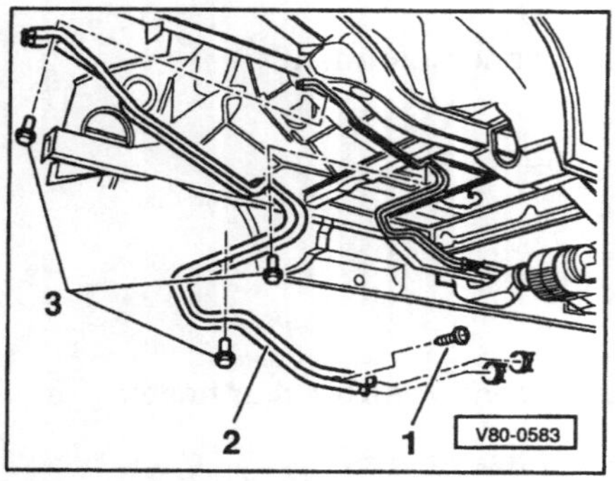

Fig. 4 Coolant pipe, removing/installing

1 - Sheet metal screw

2 - Coolant pipe

3 - Hexagon nut

Removing

– Remove fuel tank ⇒ Repair Group 20

– Remove torsion bar ⇒ Repair Group 40

– Remove coolant pipe -2-

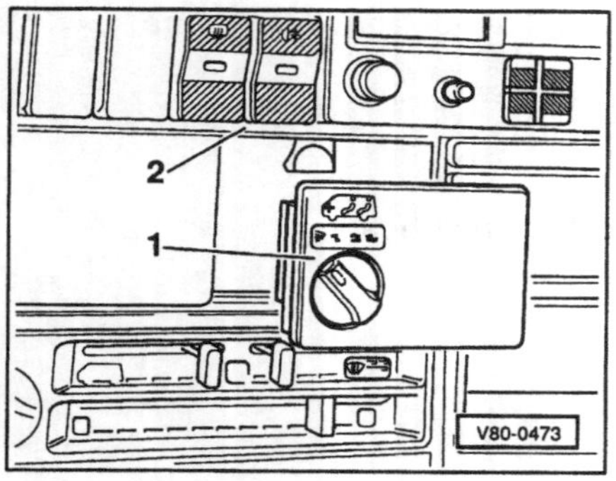

Fig. 5 Warm air blower switch -E100-, removing

– Carefully pry warm air blower switch -1- from instrument panel -2- with a screw driver

– Remove connector

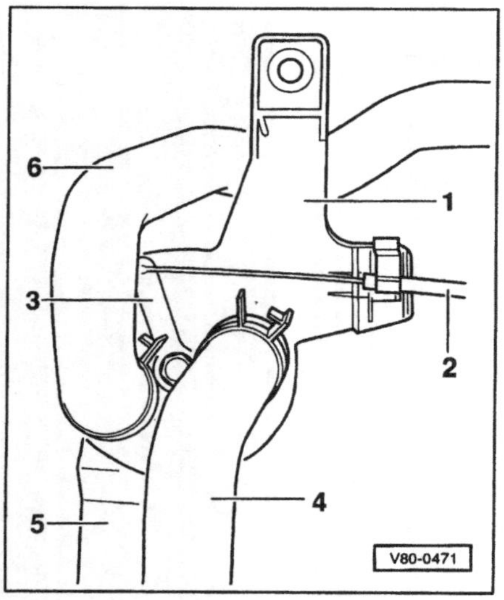

Fig. 6 Control valve (vehicles without A/C), removing

1 - Control valve

2 - Cable

3 - Lever

4 - Coolant return hose

♦ From coolant pipe (rear heater core) to control valve

5 - Coolant return hose

♦ From control valve to engine

6 - Coolant return hose

♦ From front heater core to control valve

Removing

– Detach clip and remove cable -2-

– Remove coolant hoses

– Remove control valve from over flow bracket

Note:

If coolant is in the system, clamp coolant hoses before removing.

♦ Release coolant system pressure and drain coolant before beginning repair.

♦ Refill with coolant after installing/replacing control valve.

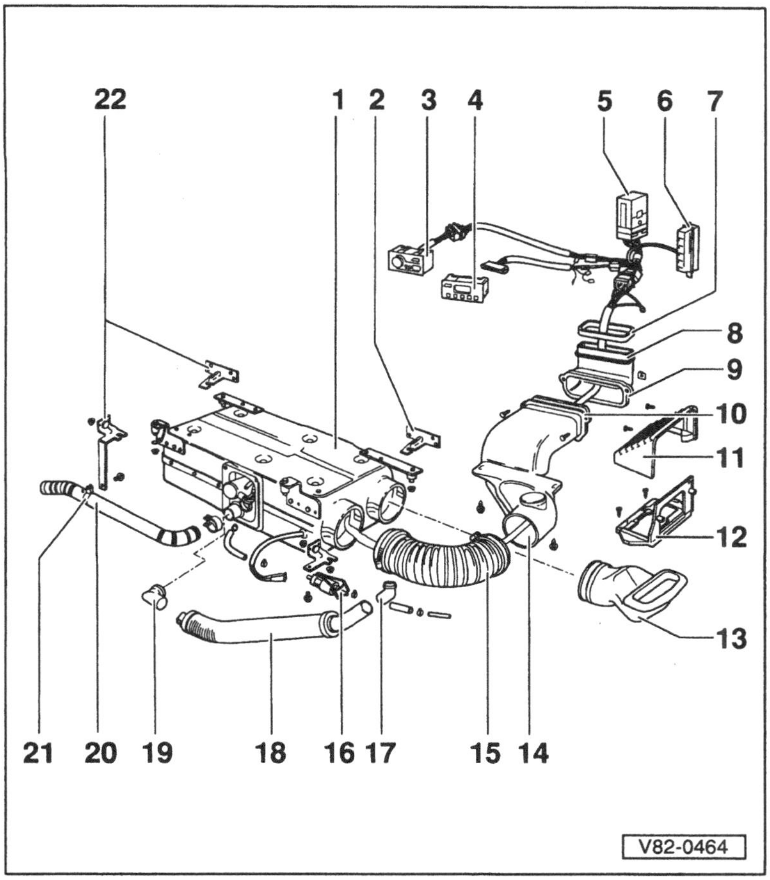

Auxiliary heater B2L (gasoline), servicing (Canada only)

CAUTION!

Before working on the electrical system, obtain radio security code and disconnect battery Ground (GND) strap.

Note:

Refer to auxiliary heater identification plate ⇒ Fig. 1 for information on type of heater installed in vehicle.

1 - **Auxiliary heater B2L**
- ♦ Removing/installing ⇒ Fig. 3
- ♦ Disassembling/reassembling ⇒ page S 82-21

2 - **Bracket**

3 - **Heater-time temperature control switch -E13-**
- ♦ Removing/installing ⇒ Fig. 4
- ♦ Operation/function ⇒ Fig. 5

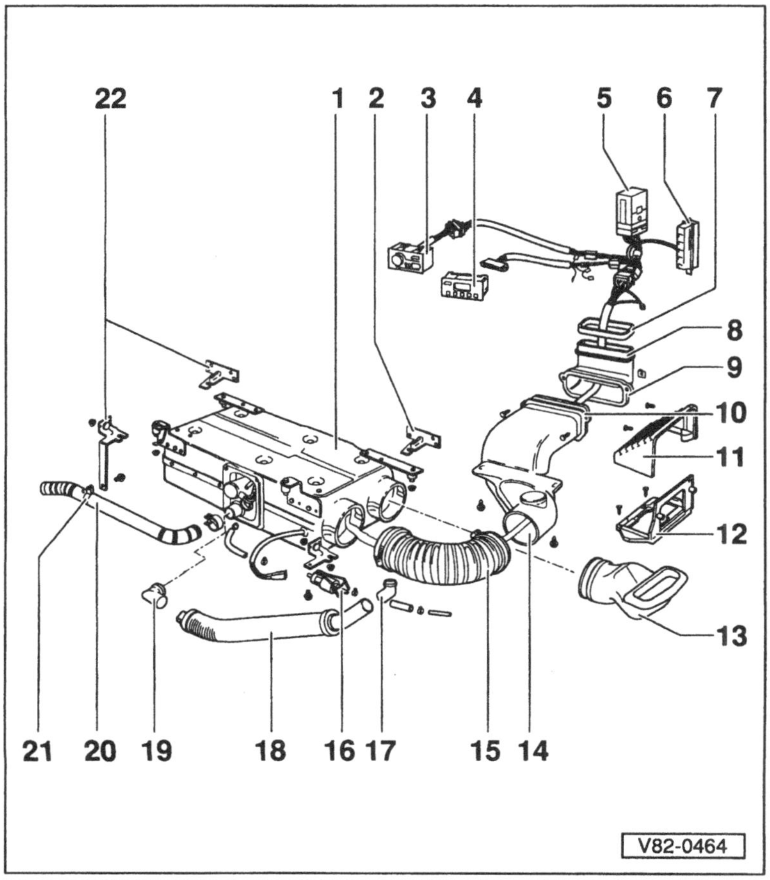

4 - **Programmable timer -E111-**
- ♦ Removing/installing ⇒ Fig. 6
- ♦ Operation/function ⇒ page S 82-16
- ♦ Timer is optional equipment in certain models

5 - **Heater control module -J65-**
- ♦ Removing/installing ⇒ Fig. 7
- ♦ Replacing fuse -S23-

6 - **Fuse holder**
- ♦ Replacing heater fuse -S62-
- ♦ For vehicles with rotating seat – on left lower B-pillar
- ♦ Without rotating seat – located under driver's seat

7 - **Gasket**

8 - **Elbow, lower**

9 - **Gasket**

10 - **Gasket**

11 - **Vent, upper**
- ♦ ⇒ Fig. 10 – removing

12 - **Vent, lower**
- ♦ ⇒ Fig. 10 – removing

13 - **Exhaust duct**
- ♦ Remove/installing ⇒ Fig. 11

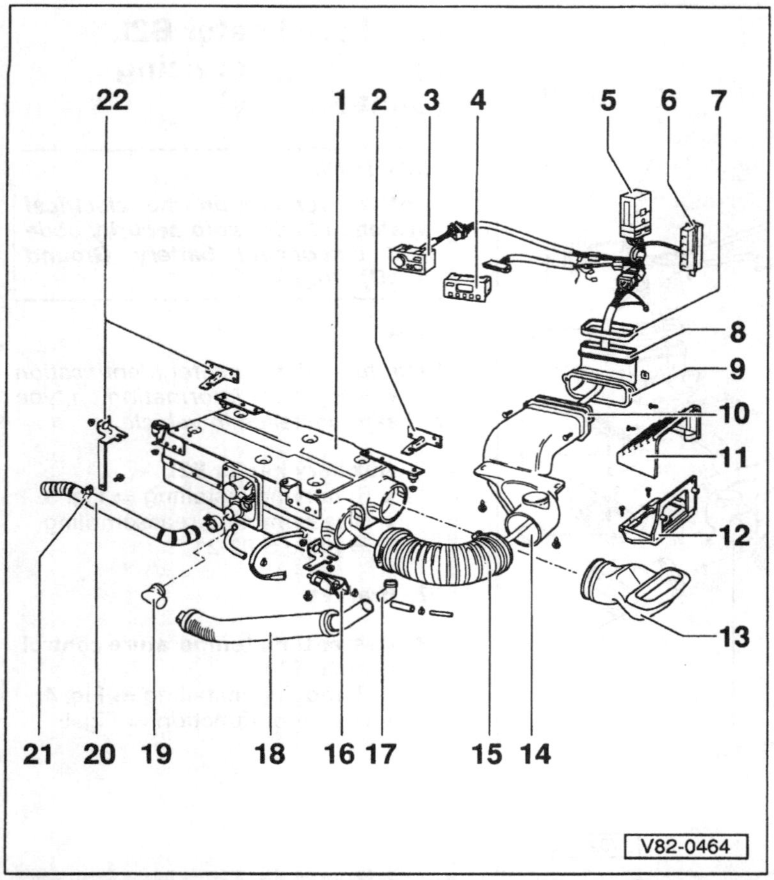

14 - Intake duct
- Removing/installing ⇒ Fig 11

15 - Intake hose
- Removing ⇒ Fig 11

16 - Transfer fuel pump -G23-
- Assembly ⇒ Fig 13
- Installed position ⇒ Fig 13, page S 82-13
- Check flow ⇒ Fig 15

17 - Combustion air intake adaptor
- Removing/installing ⇒ Fig 11

18 - Combustion air hose
- Removing/installing ⇒ Fig 11

19 - Combustion air pipe adaptor
- Removing/installing ⇒ Fig 11

20 - Exhaust pipe
From NH 057 927 a 2-part exhaust pipe is installed.
- Removing/installing ⇒ Fig. 12, page S 82-12

21 - Clamp

22 - Bracket

◄ **Fig. 1 Identification plate**

Note:

- *Factory only installs version 20 168 901 for auxiliary heater B2L in EuroVan models from 1992 m.y.*
- *Year heater is first placed into service is indicated on lower portion of plate*
- *The heater core must be replaced after 10 years*

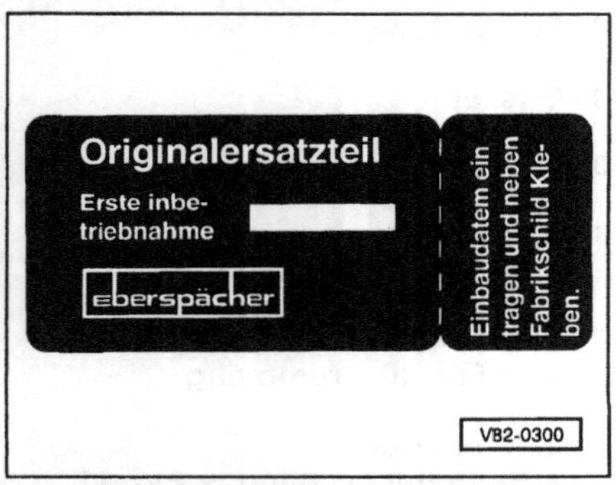

◄ **Fig. 2 Original replacement part identification plate**

Note:

- *After installing a new heater core, identification plate must be replaced*
- *Install replacement identification plate in same postion as previous plate, and enter year of heater core installation*

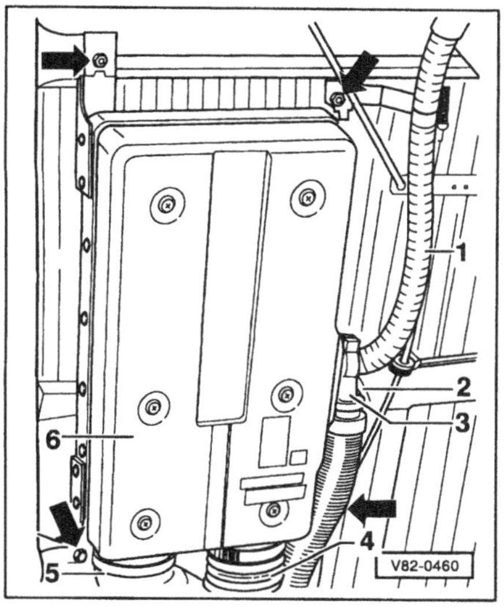

Fig. 3 Auxiliary heater, removing/installing

- Remove lower left B-pillar trim ⇒ Repair Group 75
- Disconnect plugs T10 and T1a located behind lower left B-pillar trim
- Disconnect Ground (GND) connection located at point -34- behind lower left B-pillar trim panel
- Disconnect intake hose -4- on heater -6-
- Pull wiring harness out of intake hose
- Pull combustion air inlet adaptor -3- off heater
- Pull off fuel hoses -2- between fuel pump and heater
- Remove exhaust pipe -1-
- Remove screws -arrows- slide heater in drive direction and remove

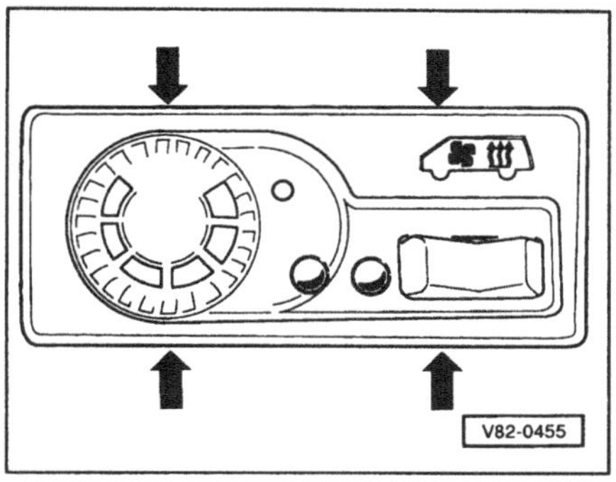

Fig. 4 Heater-time temperature control switch -E13-, removing/installing

- Carefully pry switch at indicated points -arrows- from trim panel with a screwdriver
- Reinstall in reverse sequence

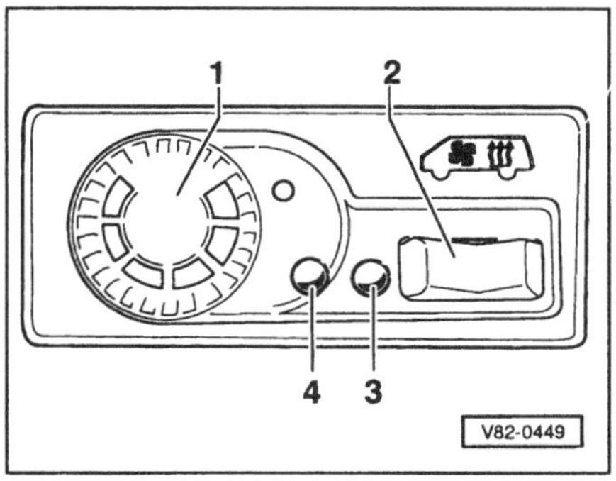

Fig. 5 Heater-time temperature control switch -E13- operation/function

1 - Heater-time temperature control switch -E13-

Note:

- *When auxiliary heater is started, it operates at full heating capacity*
- *After the glow plug shuts OFF, heating is regulated according to setting of temperature control switch*
- *If inside temperature is lower than the selected temperature, heater will operate at 100% capacity*
- *If inside temperature exceeds the selected temperature, transfer fuel pump -G23- is switched OFF*
- *Once inside temperature has been cooled below the selected temperature, the fuel pump -G23- is switched ON again*

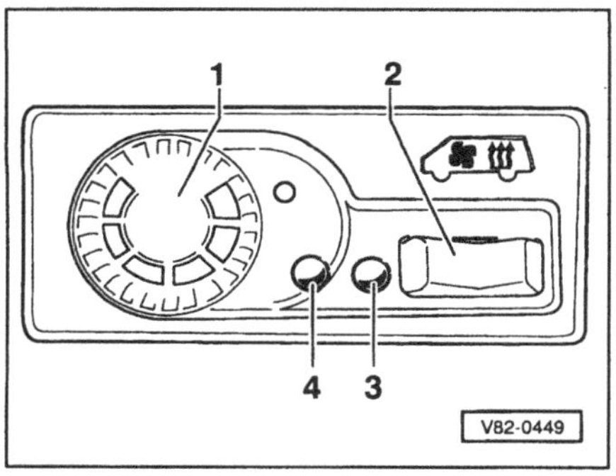

2 - Heater/heat output switch -E16-

Switch position	Function
Left	Fresh air
Middle	Off
Right	Heated air

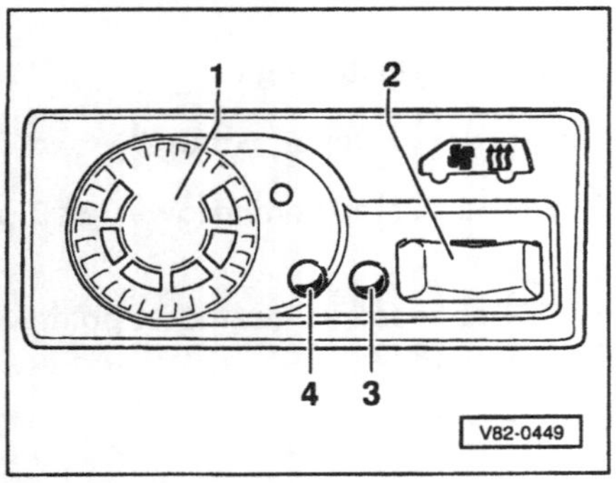

3 - Heater ON light -K89-, green

- ♦ Indicates heater is ON
- ♦ Goes out during after run
- ♦ Varies in brightness:
  Bright - full heating power
  Faint - low heating power, fresh air mode

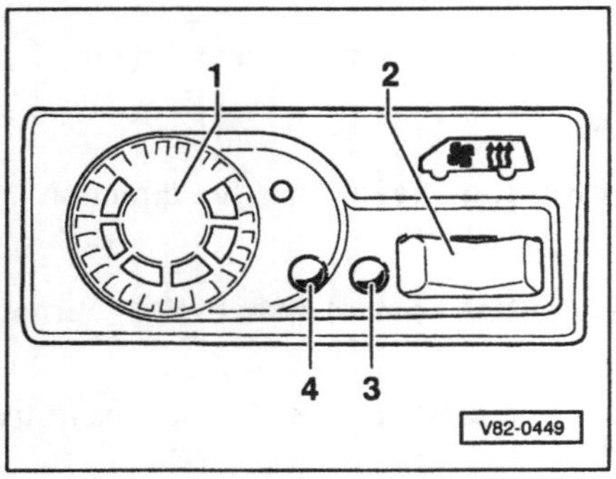

4 - Heater malfunction light -K88-, red

- ♦ Flashes to indicate that the maximum heated air temperature has been exceeded, and the heater will switch OFF
- ♦ Heater can be switched back ON again (after malfunction corrected) by briefly switching the heater blower switch -E16- OFF and ON again

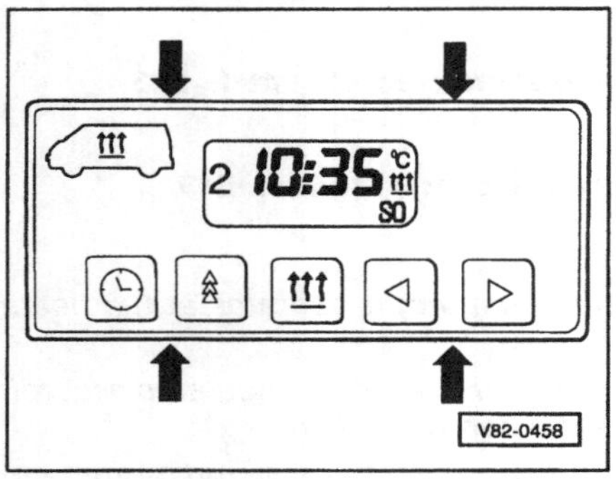

◄ **Fig. 6 Programmable timer -E111-, removing/installing**

- Carefully pry timer at indicated points -arrows- from trim panel with screw driver
- Disconnect plug

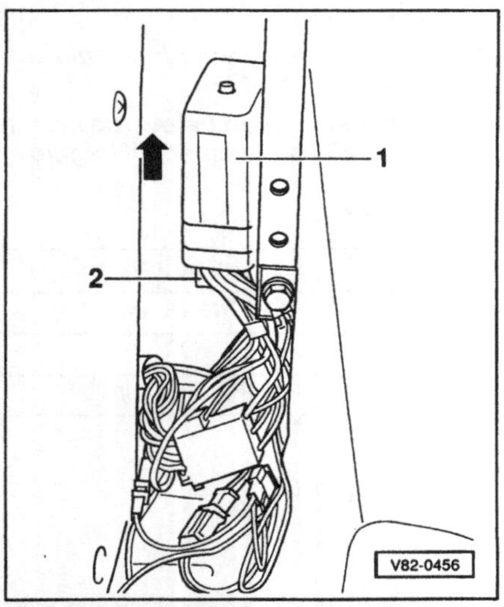

◄ **Fig. 7 Heater control module -J65-, removing/installing**

- Remove trim from bottom left side B-pillar trim ⇒ Repair Group 75
- Pull control module -1- up in direction of arrow
- Disconnect plugs -2-
- Reinstall in reverse sequence

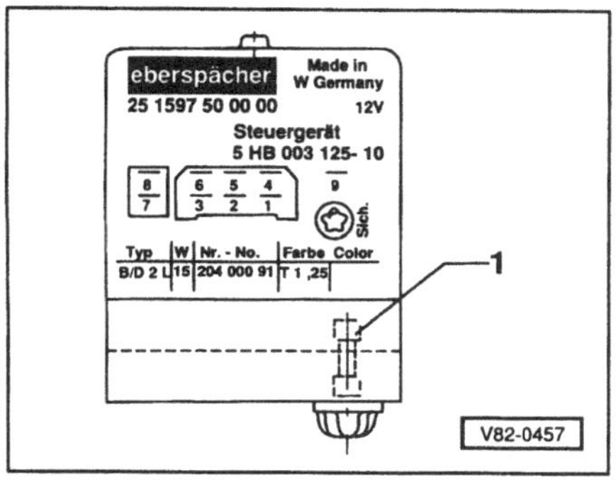

Fig. 8 Heater fuse -S23-

– Fuse 1 for fresh air/heater blower -V81-

Note:

If the voltage supplied to the control module -J65- drops below 10.2 Volts for longer than 20 seconds, or goes above 15 Volts, a voltage protection circuit in the control module will switch the heater off.

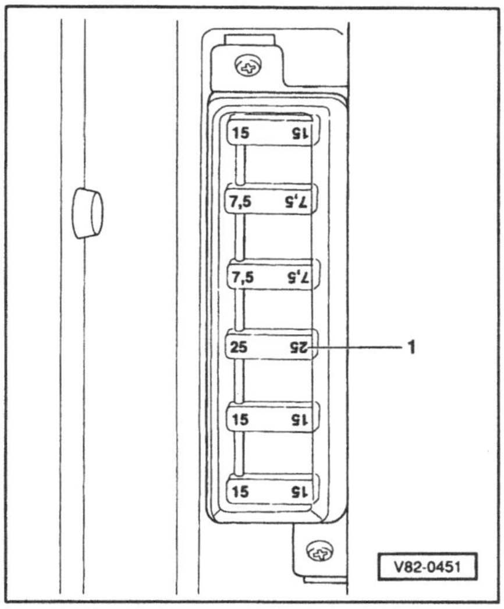

Fig. 9 Auxiliary heater fuse -S62-, replacing

– Fuse -1- -S62- is located in fuse holder at bottom of left side B-pillar

– Use 25A (ampere) fuse only

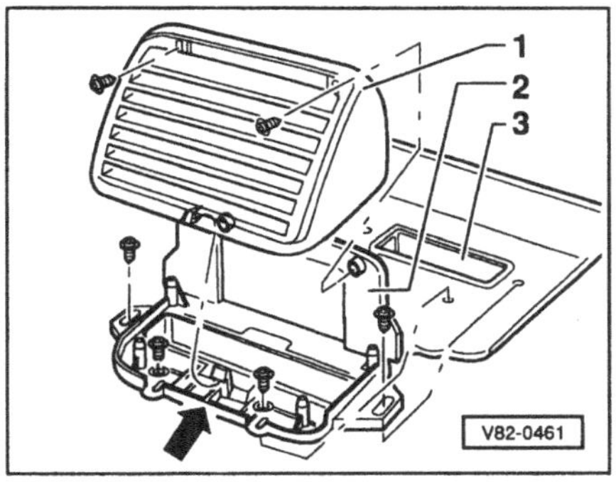

Fig. 10 Vent, removing/installing

1 - Vent, upper

2 - Vent, lower

3 - Outlet channel

– Remove screws for upper vent section

– Release tab -arrow- with screwdriver, and remove upper vent

– Remove lower vent screws then remove vent

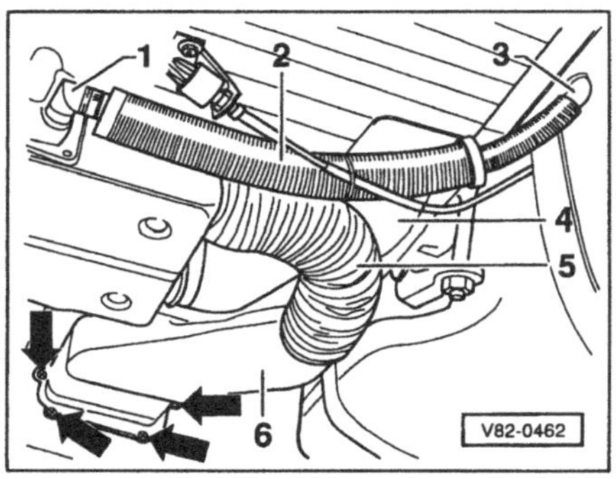

Fig. 11 Intake duct, exhaust duct, intake hose, combustion air hose, removing/installing

1 - Combustion air pipe adapter

– Pull off

2 - Combustion air hose

3 - Combustion air intake adaptor

– Remove cross member

– Remove adaptor from combustion air hose

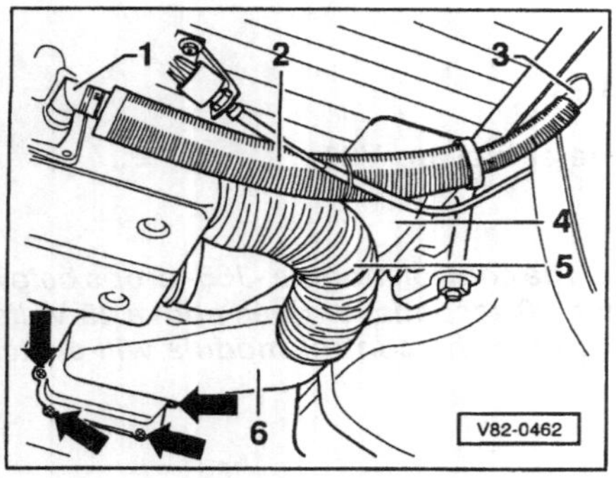

4 - Exhaust duct

- Remove floor plate
- Remove upper/lower outlets in passenger compartment
- Pull duct off heater

5 - Intake hose

- Disconnect plugs T10 and T1a located behind lower left B-pillar trim panel
- Disconnect Ground (GND) connection at point -34- behind lower left B-pillar trim panel
- Loosen clamps and remove hose
- Pull out wiring harness from intake hose and intake duct

6 - Intake duct

- Disconnect plugs T10 and T1a located behind lower left B-pillar trim panel
- Disconnect Ground (GND) connection at point -34- behind lower left B-pillar trim panel
- Loosen clamp and remove intake hose
- Pull out wiring harness from intake duct
- Remove screws -arrows- and take out intake duct

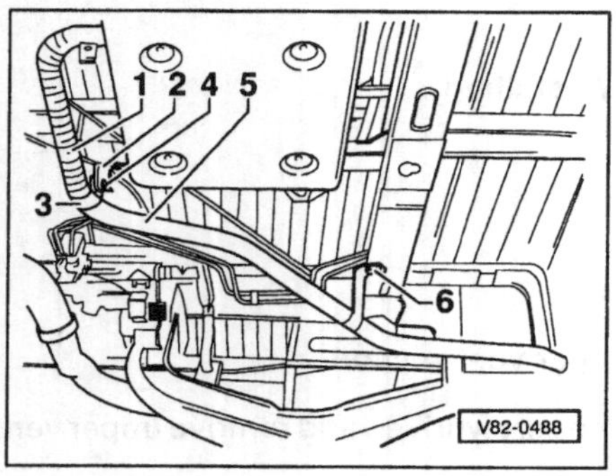

◄ **Fig. 12 Exhaust pipe, removing/installing**

1 - Exhaust pipe, front

2 - Bracket

3 - Clamp

4 - Screw and washer assembly M6 x 15

5 - Exhaust pipe, rear

6 - Tapping screws

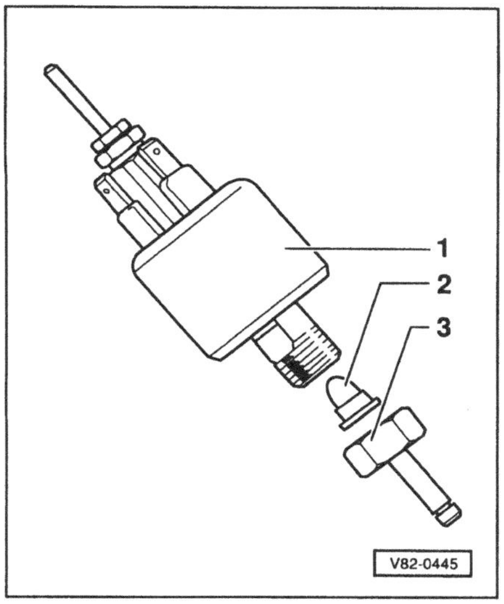

Fig. 13 Transfer fuel pump -G23- assembly

1 - Fuel pump

2 - Strainer

3 - Hose adaptor

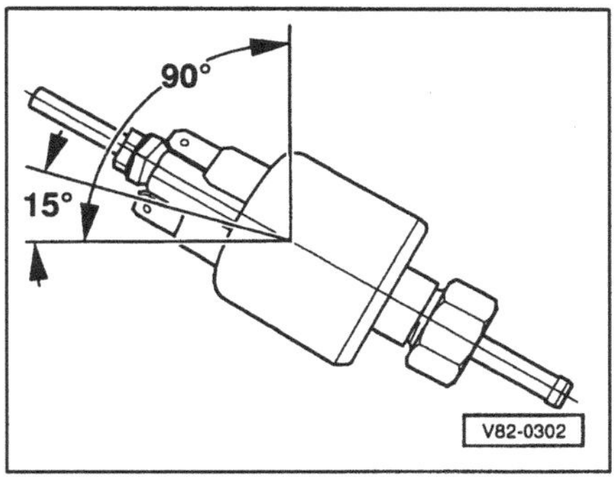

Fig. 14 Transfer fuel pump -G23-, installed position

– Inclined 15° to 90°

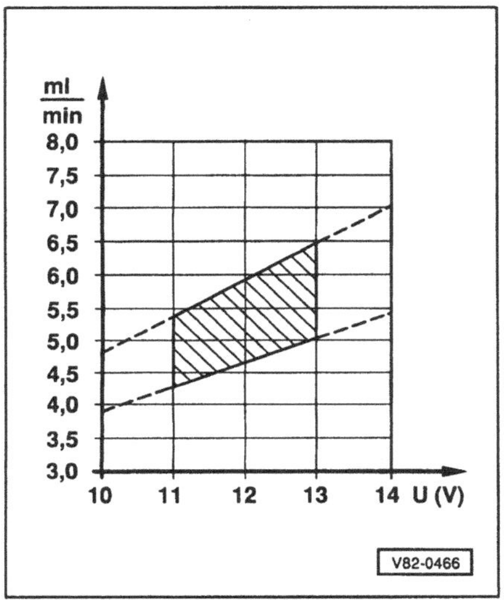

Fig. 15 Fuel flow rate, checking

Test requirements:

♦ Intake and outlet channel are not plugged

♦ Combustion air lines and exhaust lines are not plugged.

♦ Between terminal -3- and -4- of control module -J65- there must be a voltage of between 11V and 13V.

Note:

Hold measuring glass at height of glow/spark plug during checking.

Checking:

– Set test range selector switch on Fluke 83 to direct voltage measurement, test range 20V

– Connect test points to contact 3 (–) and contact 4 (+) of control module -J65-

– Remove connector on the glow/spark plug -Q5-

– Place fuel line into measuring glass (10 ml–25 ml)

– Switch on heater

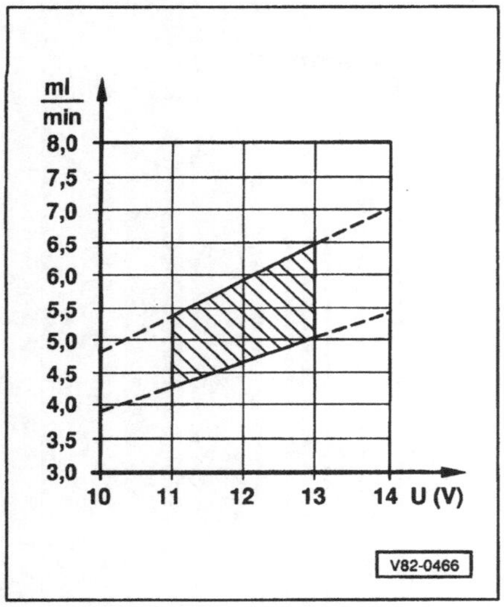

- Bleed fuel hose
- Switch off heater and drain measuring glass
- Switch heater on again
- As soon as fuel flows, activate stop watch
- Read voltage on Fluke 83
- After 1 minute switch off heater
- Read amount of fuel flow and compare measuring value with diagram
- The fuel flow amount is OK if the value read is in the highlighted area of the diagram

Notes:

- *Replace fuel transfer pump -G23- if fuel flow is out of specification.*
- *If fuel flow is in specification, fresh air/heater blower -V81- may be turning too slowly. Replace.*

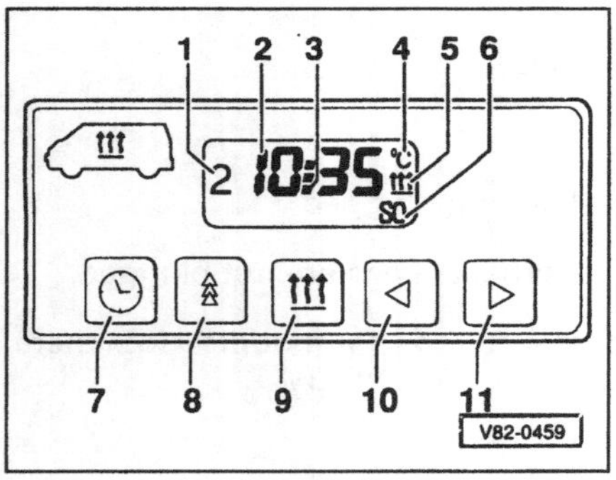

Programmable timer -E111-, operation/function

Note:

- *Starting time programmable up to one week*
- *Maximum operating time with timer is 120 minutes (with ignition OFF)*
- *Ignition switch OFF, heater is still operated another 5 minutes*
- *Up to 3 different times can be programmed*

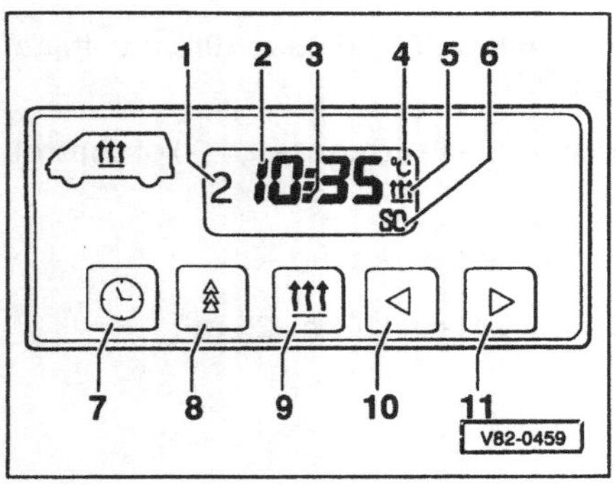

Display functions – symbols

1 - Programmable display, up to three times

2 - Time of day, programmed time or time remaining

3 - Malfunction indicator display

- Optional

4 - Temperature display

- Optional

5 - Heating mode

6 - Day of week

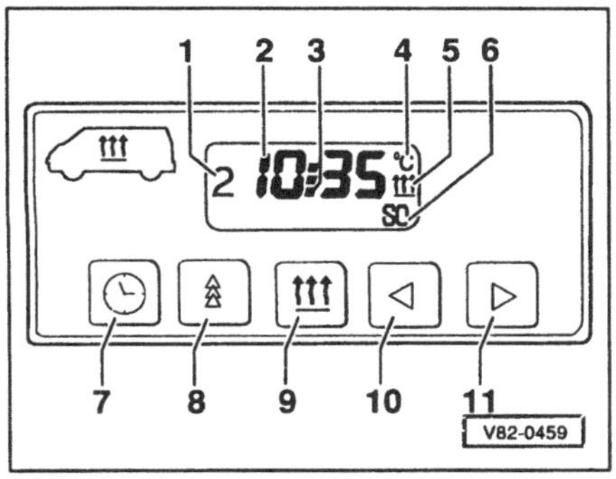

◄ **Control functions – buttons**

7 - Setting, displaying time of day

8 - Displaying, activating programmed time. Setting actual day of week

9 - Heater ON/OFF

10 - Moving time of day and programmed time backward displaying time remaining

11 - Moving time of day and programmed time forward. Setting programmed day. Displaying temperature (optional)

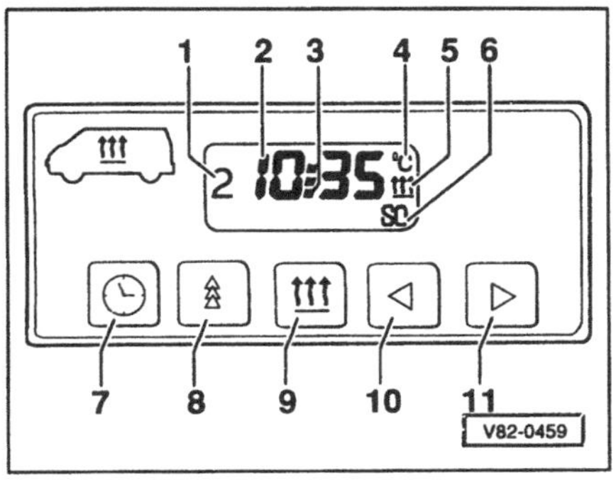

◄ **Setting clock time**

– Press and hold button -7-

– Set clock by pressing button -10- or -11-

Setting day of week

– Press and hold button -7-

– Set day of week by pressing button -8-

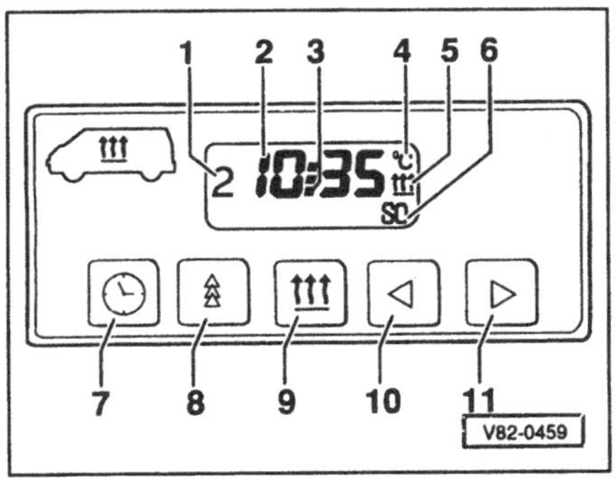

Switching heater ON

– Set heater/heat output switch -E16- in heater-time temperature control switch -E13- to middle position

– Press button -9-

  ♦ Symbol "heater ON" -5- appears in display

Switching heater OFF

– Press button -9-

  ♦ Symbol "heater OFF" -5- will disappear from display

Note:

When the heater is switched OFF, the blower motor will continue to operate for approx. 3 minutes to cool down the unit. The thermostat will switch OFF the afterrun function automatically.

Displaying or shortening remaining operation time

– Only possible during heating mode

Displaying remaining time

– Briefly press button -10-

Shortening remaining time

– Press and hold button -10- until desired remaining operating time appears in the display

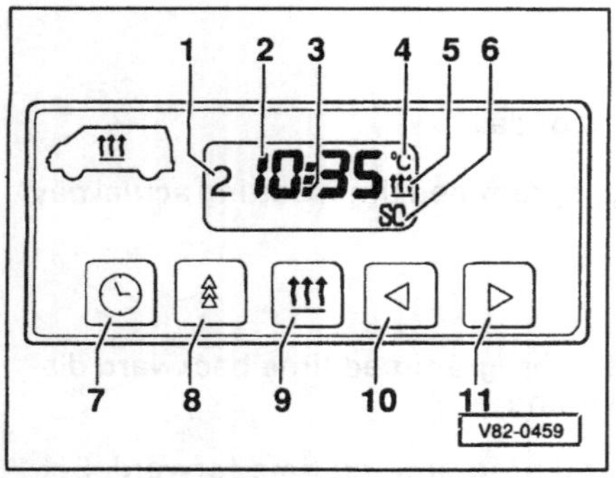

Programmed times, setting/activating

– Press button -8- to display, in sequence, -3- different preselected times, and the clock time

– The desired preselected time can be set at any level by pressing button -10- or -11-

– After several seconds weekday display -6- will flash, and day of week can then be set by pressing button -11-

Note:

- *If no other buttons are pressed entries will be stored after approx. 15 seconds*
- *Display will switch to current clock time after switching ignition ON*
- *Display disappears after switching ignition OFF*
- *Numbers 1, 2, or 3 for preselected times remain in display 1 to indicate that a preselected time is activated*

Cancelling activation

– Press and hold button -8- until number in display -1- disappears

Switching temperature display

The programmable timer -E111- does not have a temperature sensor. Therefore, value "–52°C" appears in display

In order to use the temperature display, a temperature sensor must be ordered from heater manufacturer Eberspacher, order No. 201 689 800 700

Connect sensor to terminals -1- and -2- of timer -E111-

– Press button -11-, with ignition switch OFF, the temperature display will appear for approx. 10 seconds

– Press button -11-, with ignition switched ON, the temperature display will appear continuously

> ***Caution!***
> ***Part numbers are for reference only. Always check with your Parts Department for latest information.***

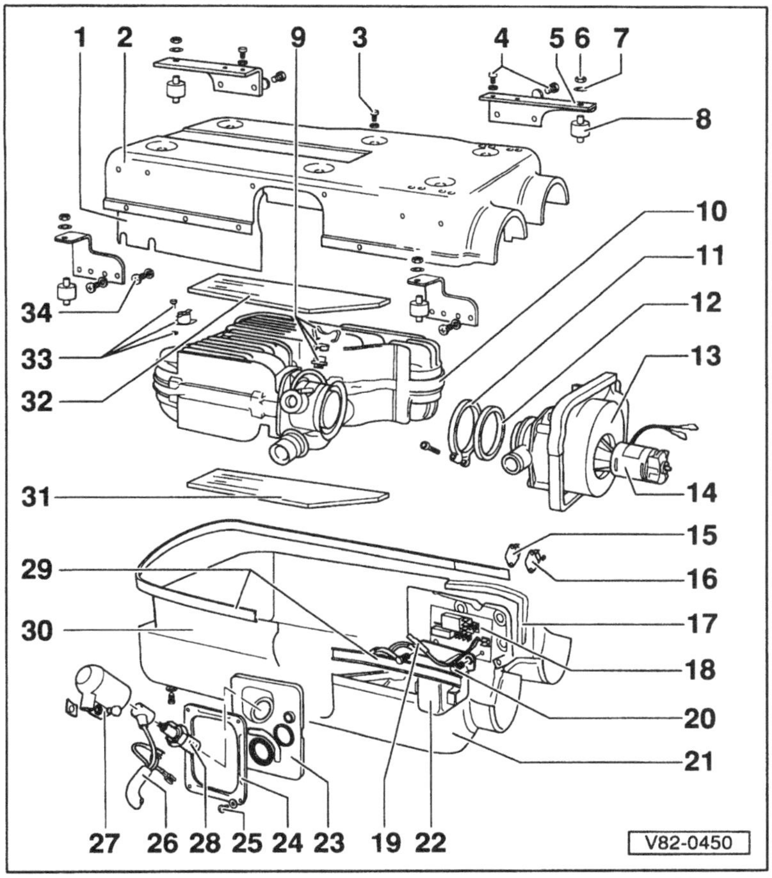

Auxiliary heater B2L, assembly

1 - **Heat shield, upper**

2 - **Outside cover, upper**
- ♦ Removing:
- – remove support -24-and screws -34- on both sides of outside covers
- – remove outside cover
- – pull off seal -29-
- ♦ Installing:
- – reinstall seal -29-
- – screw both outside cover together
- – check for proper sealing of covers

3 - **Screw - B4, 2 x 9, 5**
- ♦ Use silicone washer

4 - **Screw - M5 x 8**
- ♦ Use silicone washer

5 - **Bracket, support**

6 - **Nut - M6**

7 - **Spring washer**

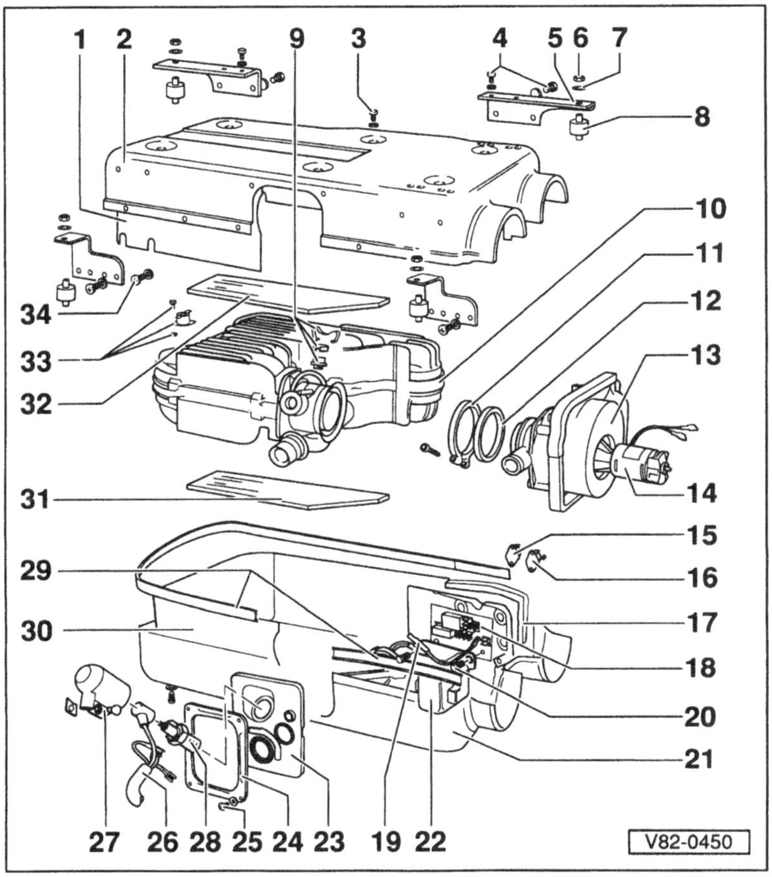

8 - **Bonded rubber bushing**

9 - **Heater flame sensor -F16-**
- ♦ The flame sensor shuts off glow/spark plug -28- after stable combustion is achieved. It also ends the after run cycle when the heater has cooled sufficiently after shut–off.
- ♦ Opens at temperatures over 100°C ± 3°C (212° ± 2°F)
- ♦ Closes at temperatures under 90°C ± 4°C (194° ± 2°F)
- ♦ Replacing ⇒ Fig. 1

10 - **Heat exchanger**
- ♦ Do not separate combustion chamber from heat exchanger

11 - **Clamp**

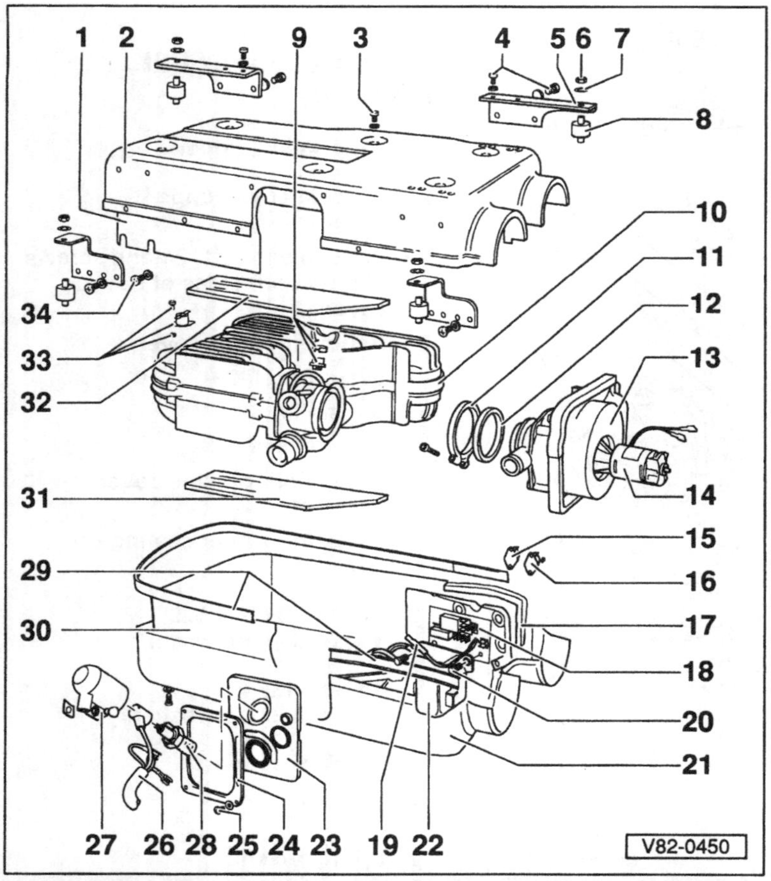

12 - Seal
- Always replace

13 - Blower motor assembly
- Removing/installing ⇒ Fig. 4
- Assembly ⇒ page S 82-31

14 - Motor for combustion blower, fresh and hot air blower -V81-
- Removing/installing ⇒ Fig. 4
- Assembly ⇒ page S 82-31

15 - Temperature limit switch
- The temperature limit switch shuts off the fuel pump at temperatures over 125°C (257°F)
- Replacing ⇒ Fig. 2

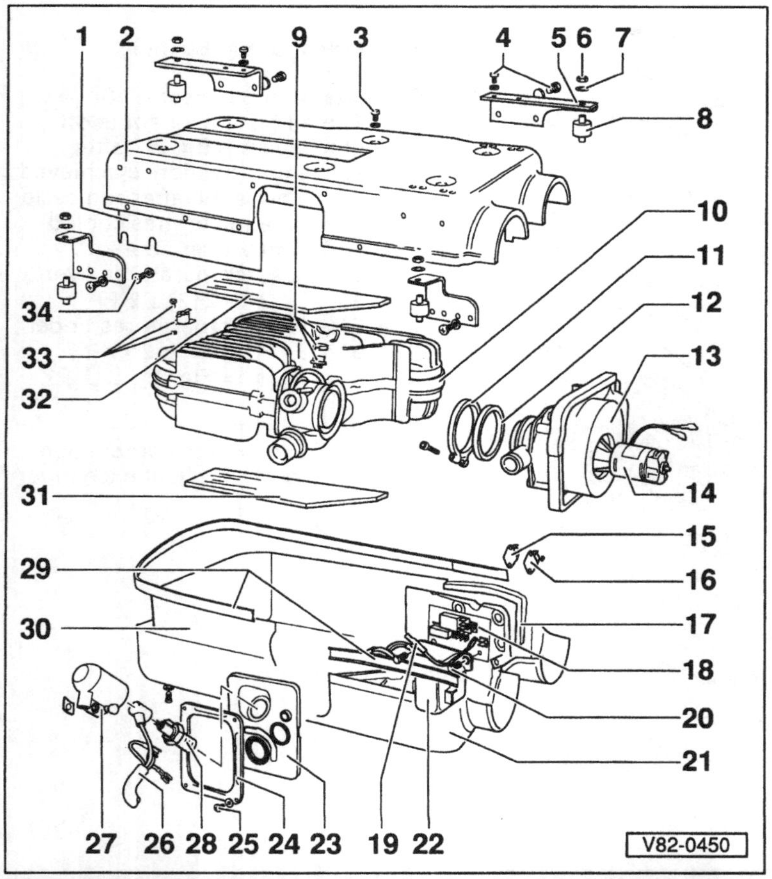

16 - Glow plug thermoswitch -F153-
- The thermoswitch for glow plug shuts off the glow/spark plug -28-, if during normal operation the ignition sparks are sufficient for flame formation
- Opens at temperatures over 45°C ± 3°C (114° ± 2°F)
- Closes at temperatures below 25°C ± 3°C (77° ± 2°F)
- Replacing ⇒ Fig. 2

17 - Insulation tape

18 - Junction box (T)
- The wire routing for fuel -electrical heater is identified with a white "B"
- Removing/installing ⇒ Fig. 5
- Connector routing: Electrical Wiring Diagrams

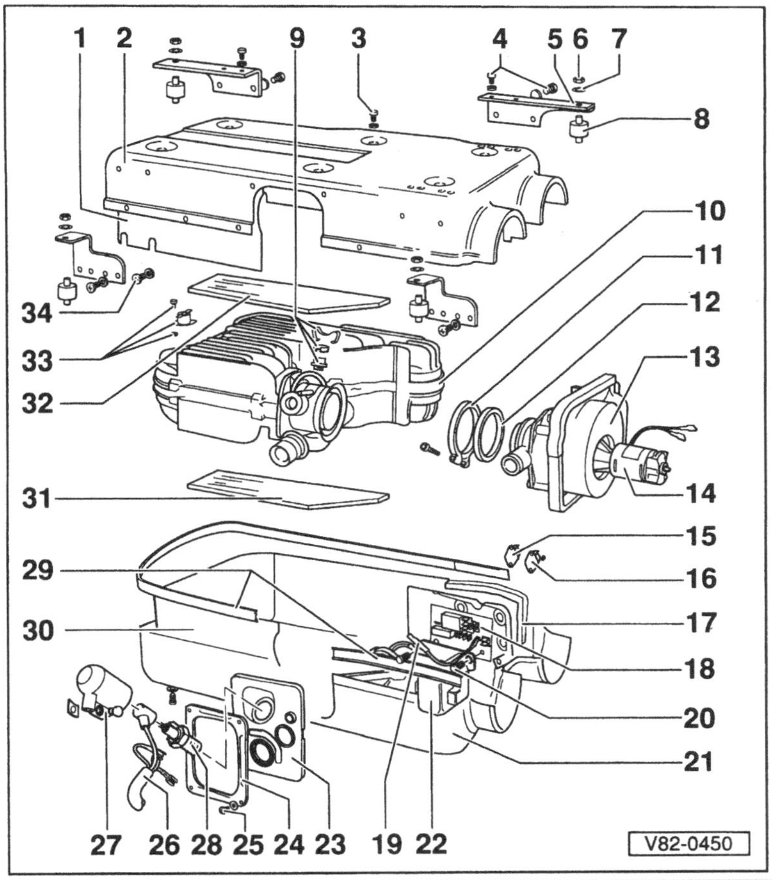

19 - Auxiliary heater temperature sensor -G18-

20 - Screw - M5 x 8

21 - Outside cover, lower
- ♦ Keep dirt out of 2.0 mm dia. drain hole in cover

22 - Ignition coil -N152-
- ♦ Removing/installing ⇒ Fig. 5

23 - Gasket, black
- ♦ Ensure gasket is seated correctly

24 - Support plate

25 - Screw B4, 2 x 13
- ♦ Use silicone washer

26 - Glow plug wire

27 - Protective cap, black

28 - Glow plug, 17 Nm (12 ft lb)

29 - Seal
- ♦ To install:
  - – coat with petroleum jelly
  - – install onto lower outside cover
  - – ensure correct seating

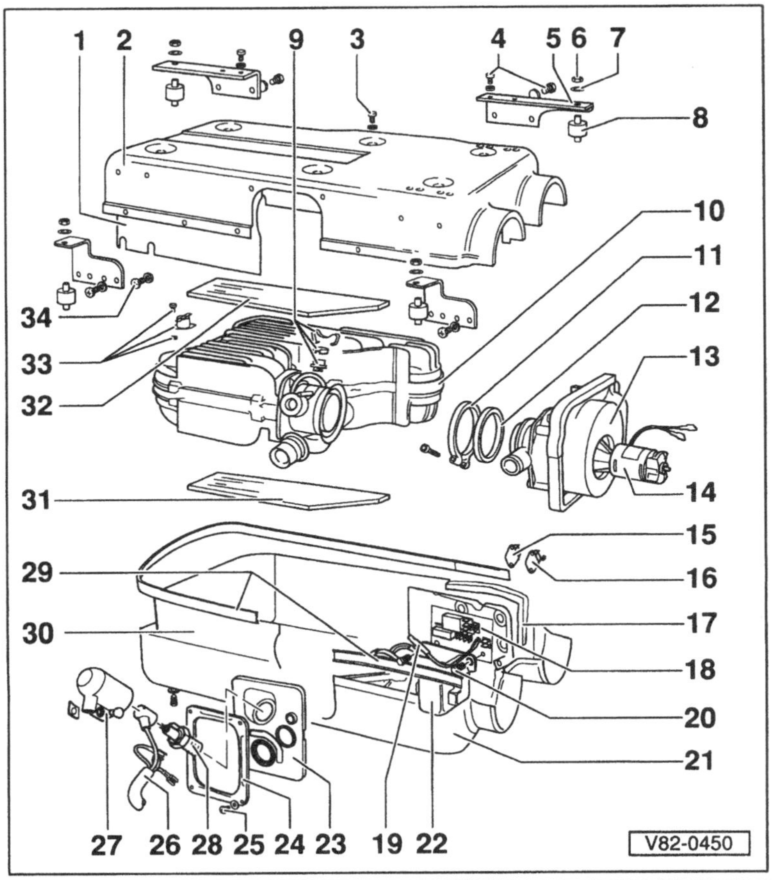

30 - Heat shield, lower

31 - Sealing plate, lower
- ♦ To install:

align plate with old seat and glue to lower heat shield with a silicone cement sealing compound

32 - Sealing plate, upper
- ♦ to install:

align plate with old seat and glue to upper heat shield with a silicone cement sealing compound

33 - Heater overheat control switch -F17-
- ♦ The overheat switch shuts off the fuel feed when hot air exceeds maximum temperature. Red warning light -K88- in temperature control switch also flashes
- ♦ Opens at temperture over 110°C ± 4°C (230° ± 2°F)
- ♦ Closes at temperatures below 95°C ± 4°C (205° ± 2°F)
- ♦ Replacing ⇒ Fig. 3

34 - Screw B4, 2 x 9.5
- ♦ Use silicone washer

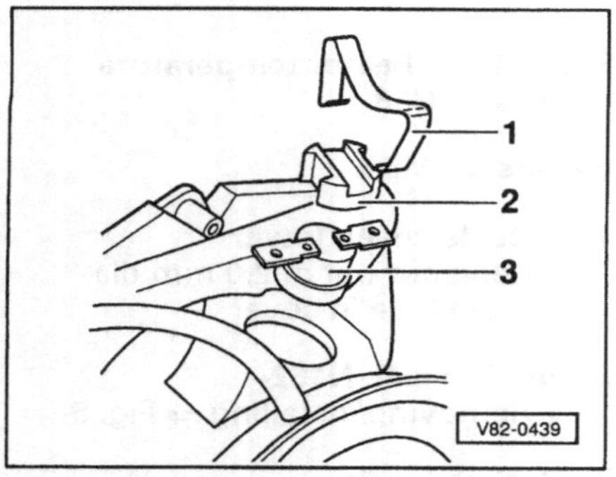

Fig. 1 Heater flame sensor -F16-, removing/installing

- Remove wire connector
- Unclip retainer -1- using screwdriver
- Remove connector -2-, flame sensor -3-

Reinstall in reverse sequence, noting the following:

- Do not pinch wires between sides of heat exchanger and heat shield

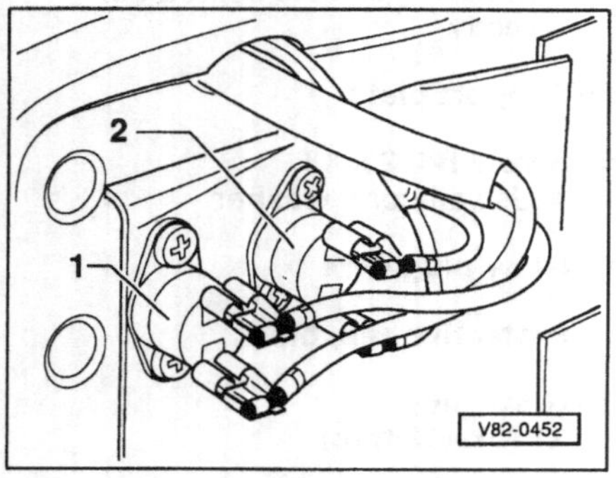

Fig. 2 Glow plug thermoswitch -F153- and temperature limiting switch -F82-, replacing

1 - Glow plug thermoswitch -F153- (white wires)

2 - Temperature limiting switch -F82- (gray wires)

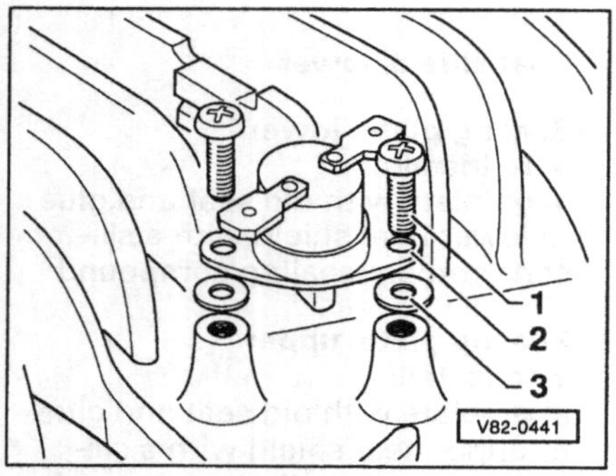

Fig. 3 Heater overheat control switch –F17-, replacing

1 - Screw M3 x 10

2 - Overheating switch

3 - Insulator

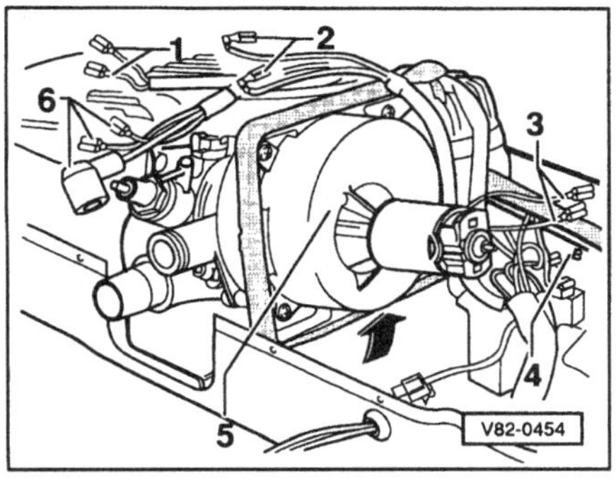

- Fig. 4 Blower motor assembly, removing/installing
- Pull off plugs -1-, of overheating switch
- Pull off plugs -2-, of flame sensor
- Pull blower motor wire -3- off junction box -4-
- Disconnect plugs -6- on glow plug
- Pull auxiliary heater temperature sensor -G18- from holder on blower motor housing -5-
- Combustion air, fresh air, and hot air blower with heater core is raised in direction of arrow, until clamp is accessible
- Loosen clamp, remove blower and seal
- Coat outside cover plates in sealing areas with petroleum jelly
- Install new seal between blower and combustion chamber
- Place blower motor assembly on combustion chamber, mount and lightly tighten clamp
- Press heater into lower outside cover plate
- Fully tighten clamp
- Remount temperature sensor on blower motor housing
- Reconnect plugs

Fig. 5 Ignition coil -N152-, junction box -T-, removing/installing

- Pull plugs off glow plug
- Remove screws -arrows-
- Remove ignition transformer -2-, junction box -1-

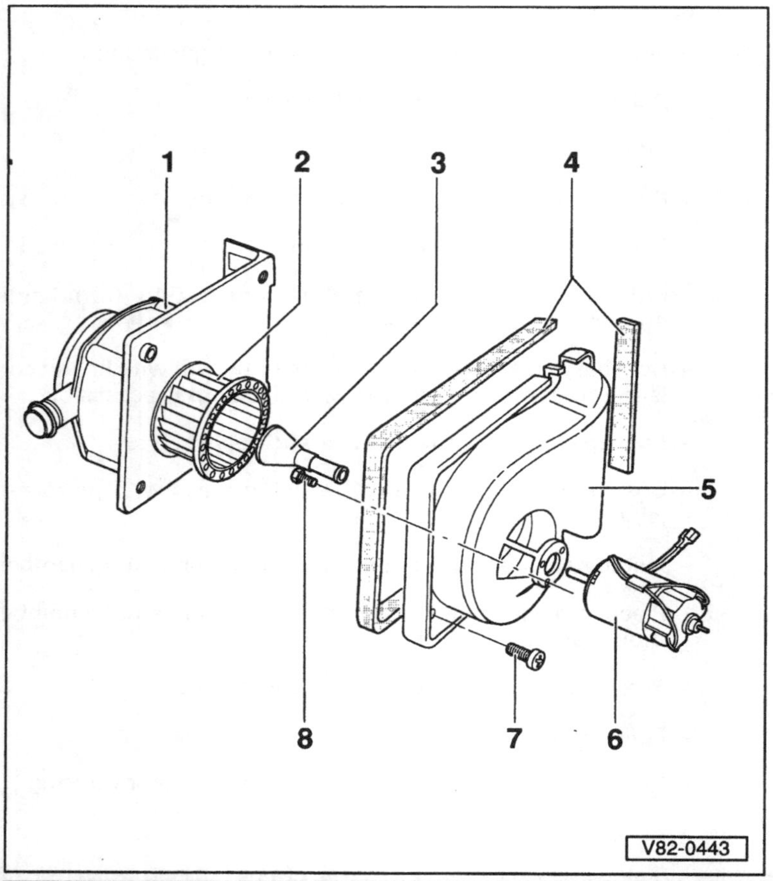

Blower motor, assembly

1 - Combustion air blower
- Do not disassemble

2 - Blower wheel (fresh/heated air)
- Checking installed position ⇒ Fig. 1

3 - Connector
- To install:
  – place on shaft of motor
- Checking installed position ⇒ Fig. 1

4 - Insulation tape
- Always replace

5 - Blower housing (fresh/heated air)

6 - Fresh air/heater blower -V81-

7 - Screw M5 x 16

8 - Screw M4 x 8

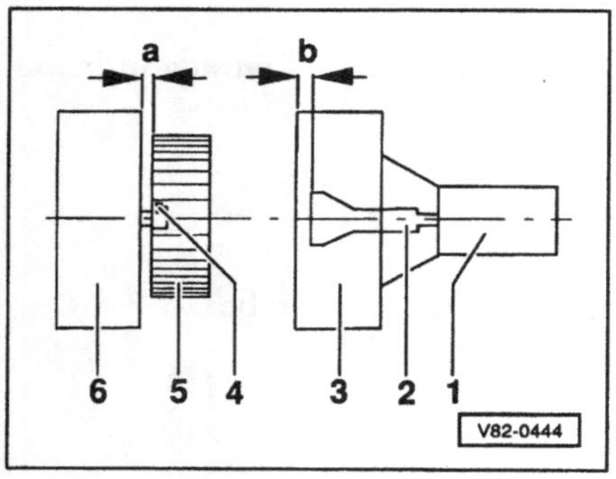

◄ **Fig. 1 Blower wheel/connector, installed poistion**

1 - Blower motor

2 - Connector

3 - Blower housing

4 - Stud

- 120 ± 30 Ncm (10.6 ± 2.7 in lb)

5 - Fan

6 - Combustion air blower

Installed dimensions

-a- = 1.4 ± 0.3 mm (0.055 ± 0.012 in)

-b- = 4.0 ± 1.0 mm (0.156 ± 0.039 in)

Auxiliary heater B2L (gasoline), troubleshooting

Tool requirements:

- Multimeter Fluke 83 (US 1119) or equivalent
- LED voltage tester VAG 1527B, US 1115 or equivalent
- Adapter cable set VAG 1594
- Applicable wiring diagram
- Stopwatch
- Measuring glass (10 ml – 25ml)

Heater does not come on

(no audible sound from blower within 5 seconds after switching on)

Test requirements

- Battery voltage OK
- Fuse -S62- OK

Open circuit in wiring between auxiliary battery (30+) and heater–time temperature control switch -E13-

– Remove temperature control switch -E13- from B pillar trim
– Connect LED voltage tester between ground and -E13- terminal T12/11

LED lights up	**LED does not light up**
→ A	→ **Check circuit using wiring diagram**

A

Heater–time temperature control switch -E13- faulty

– Connect LED voltage tester between ground and -E13- terminal -T12/12-
– Turn heater/heat output switch -E16- to "heat" position

LED lights up	**LED does not light up**
→ (continue below)	→ **Replace heater–time temperature control switch -E13-**

Open circuit in wiring between -E13- connector -T12/12- and -T12/6-

– Connect LED voltage tester between ground and -E13- terminal -T12/6-

LED lights up	**LED does not light up**
→ B	→ **Check circuit using wiring diagram**

(B)

Heater–time temperature control switch -E13- faulty

– Connect LED voltage tester between ground and -E13- terminal T12/8
– Heater/heat output switch -E16- in "heat" position

LED does not light up	LED lights up

LED does not light up:

– Connect LED voltage tester between ground and -E13- terminal T12/9

LED lights up	LED does not light up
Open circuit in wiring between -E13- connector -T12/9- and junction box -T- – Check circuit using electrical wiring diagrams	– Replace heater–time temperature control switch -E13-

LED lights up:

Open circuit in wiring between -E13- terminal -T12/8- and heater conrtol module -J65- terminal 7

– Connect LED voltage tester between ground and -J65- terminal 7

LED lights up	LED does not light up
(C)	Check circuit using wiring diagram

(C)

Heater control module -J65- faulty

Prerequisite: Heater fuse -S23- OK

– Connect LED voltage tester between ground and -J65- terminal 2

LED lights up	LED does not light up
(see below)	Heater control module -J65- faulty

LED lights up:

Junction box -T- (on heater unit) faulty

– Turn heater/heat output switch -E16- from "heat" position to "ventilation" position

Fresh air/heater blower -V81- starts	Fresh air/heater blower -V81- does not start
Replace junction box -T-	(D)

D

Open circuit or faulty solder connection on junction box -T-

– Visual check, if necessary check continuity using wiring diagram

- **Open circuit found**
  - Replace junction box -T-
- **Open circuit not found**
  - **Motor for fresh air/heater blower -V81- faulty**
    - – Remove intake hose at heater unit
    - – Rotate blower motor output shaft
    - **Output shaft rotates**
      - Replace motor for fresh air/heater blower -V81-
    - **Output shaft does not rotate**
      - Repair/replace fresh air/heater blower -V81- motor

Heater shuts off after approx. 10 seconds

Test requirements

♦ Heater switched off

- **From April 1993 production**
  - **Open circuit between heater control module -J65- terminal 5 and ground point (34) located under driver's seat**
    - – Check circuit using wiring diagram
    - **Open circuit not found**
      - Replace heater control module -J65-
    - **Open circuit found**
      - Repair open circuit
- **Up to April 1993 production**
  - **Heater overheat control switch -F17- does not close**
    - **Switching points:**
      - ♦ Opens above 110°C (230°F)
      - ♦ Closes below 99°C (211°F)
    - – Connect LED voltage tester between positive (30+) and heater control module -J65- terminal 5
    - **LED lights up**
      - Replace heater control module -J65-
    - **LED does not light up**
      - Replace overheat control switch -F17-

Heater shuts off after 3 minutes, fuel pump clicks

- Transfer fuel pump -G23- clicks after appox. 30 seconds

– Turn on heater, set heater/heat output switch -E16- to "heat" position

- Combustion takes place, but heater switches off by itself after approx. 3 minutes
- Combustion/fresh air/heater blower runs on

→ Glow/spark plug -Q5- faulty or covered with carbon deposits

- Combustion does not take place, heater switches off by itself after approx. 3 minutes
- Combustion/fresh air/heater blower does not run on

→ **Replace heater flame sensor -F16-**

Glow/spark plug -Q5- faulty or covered with carbon deposits

– Remove glow/spark plug -Q5-

– Perform visual inspection and check for excessive carbon deposits

Glow/spark plug -Q5- OK	**Glow/spark plug -Q5- not OK**
→ A	→ **Replace glow/spark plug -Q5-**

A

A

WARNING!
When checking for ignition spark as described below, ensure that no fuel, rags or other combustible materials are in the area where the test is performed.

Glow plug thermal switch -F153- faulty

Switching points:

- Opens above 45°C (113°F)
- Closes below 25°C (77°F)

Glow plug thermal switch -F153- OK	**Glow plug thermal switch -F153- not OK**
→ Ignition coil -N152- faulty	→ **Replace glow pug thermal switch -F153-** → Ignition coil -N152- faulty

Ignition coil -N152- faulty

– Disconnect wires to transfer fuel pump -G23-

– Turn heater/heat output switch -E16- to "heat" position

– Remove spark plug cable from glow/spark plug and hold approx. 5 mm from ground

Ignition spark visible	**Ignition spark not visible**
→ B	→ **Replace ignition coil -N152-**

B

(B)

Strainer in transfer fuel pump -G23- dirty or blocked

Disassemble transfer fuel pump and inspect strainer

Strainer OK	**Strainer dirty or blocked**
Fuel delivery rate insufficient **– Replace fuel transfer pump -G23-**	**Clean or replace strainer**

Heater shuts off after 3 minutes, fuel pump does not click

♦ Transfer fuel pump -G23- does not click after appox. 30 seconds

Transfer fuel pump -G23- inoperative

– Turn heater/heat output switch -E16- to "heat" position
– Connect LED voltage tester across terminals on -G23-, wait 30 seconds

LED does not flash	**LED flashes**
→ see below	**Replace transfer fuel pump -G23-**

From April 1993 production	**Up to April 1993 production**
Heater overheat control switch -F17- faulty (see below)	→ (B)

Heater overheat control switch -F17- faulty

Switching points:

♦ Opens above 110°C (230°F)
♦ Closes below 99°C (211°F)

Overheat control switch -F17- OK	**Overheat control switch -F17- not OK**
→ (A)	**Replace overheat control switch -F17-**

(A) (B)

Check and test circuit from heater control module -J65- to transfer fuel pump -G23- using wiring diagram

No open circuit found	Open circuit found
Replace heater control module -J65-	Repair open circuit

Heater malfunction light -K88- flashes, heater emits soot or smoke

Test requirements

- ♦ Intake and outlet channels are not blocked
- ♦ Combustion air and exhaust pipes are not blocked
- ♦ Voltage between heater control module -J65- terminals 3 and 4 must be between 11V and 13V

Possible cause:

- ♦ Excessive fuel delivery

– Perform fuel flow rate test ⇒ page S 82-14

Heater run-on cycle exceeds 4 minutes after switching heater off

Heater flame sensor -F16- faulty

- Switch off heater, remove heater fuse -S62- if necessary
- Set range switch on multimeter to test continuity
- Connect multimeter to heater control module -J65- terminals 1 and 8

Continuity	No continuity
↓	↓
Replace heater control module -J65-	Replace heater flame sensor -F16-

Heater does not run-on after switching off and "Ventilation" function inoperative

Possible cause:

- ♦ Auxiliary heater blower relay -J233- on junction box -T- sticking

Correction:

- Replace junction box -T- (relay and junction box are integrated)

Heater does not reduce output

Short in wiring to temperature sensor II -G18-

- Remove heater–time temperature control switch -E13-
- Set range switch on multimeter to resistance test, 2 kΩ range
- Connect multimeter to connector terminals T12/2 and T12/3

Reading between 800Ω and 1.15 kΩ	Reading less than 800Ω
↓	↓
Replace heater–time temperature control switch -E13-	Check wiring to temperature sensor II -G18- using wiring diagram

Heater output inadequate or heater always running at low output

Test requirement

- outlet channel not damaged

Open circuit in wiring to temperature sensor II -G18-

- Remove heater–time temperature control switch -E13-
- Set range switch on multimeter to resistance test, 2 kΩ range
- Connect multimeter to connector terminals T12/2 and T12/3

Reading between 800Ω and 1.15 kΩ	Reading greater than 1.15 kΩ
↓	↓
Replace heater–time temperature control switch -E13-	Check for open circuit in wiring to temperature sensor II -G18- using wiring diagram

Check for open circuit in wiring to temperature sensor II -G18- using wiring diagram

Open circuit found	No open circuit found
↓	↓
Repair open circuit	A

A

↓

Temperature limiting switch -F82- faulty

- Set range switch on multimeter to test continuity
- Connect to -F82-terminals at junction box -T-

Continuity	No continuity
↓	↓
Replace junction box -T- (temperature sensor -G18- integrated)	Replace temperature limiting switch -F82-

Heater does not cycle (combustion does not resume) after selecting lower temperature

Possible cause:

- Setting heater–time temperature control switch -E13- to low temperatures over long periods may lead to excessive carbon deposit build–up on the glow–ignition plug. Combustion will not resume automatically when thermostatic control starts fuel delivery again.

Correction:

– Remove ignition–glow plug and inspect for carbon deposits

– Clean or replace glow–ignition plug as necessary.

– Check strainer in heater unit for excessive carbon build–up. If necessary, replace heat exchanger.

– Run heater at maximum output.

Note:

On vehicles up to chassis No. 70P H 030000 check for marking L 125 C on temperature limiting switch -F82-. If previous marking L110 C is present, replace -F82-.

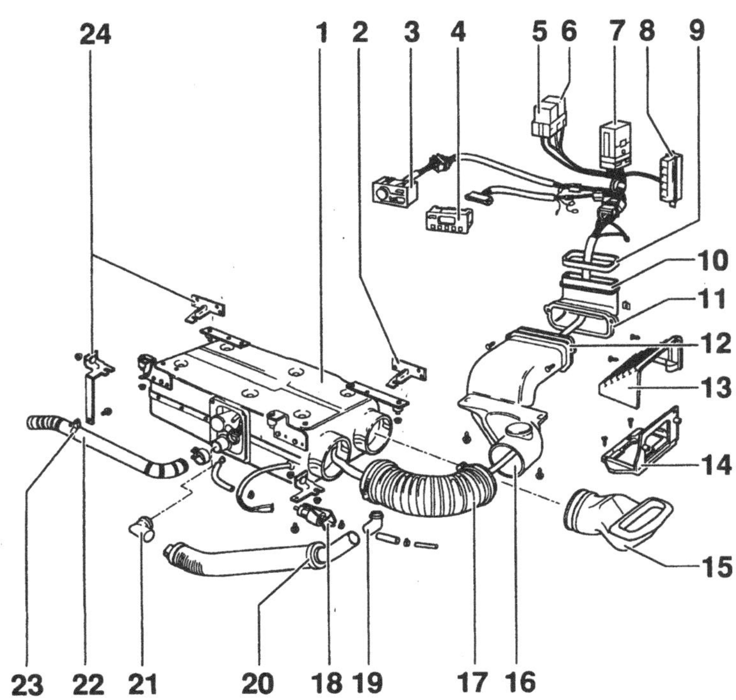

V82-0465

Auxiliary heater D2L (Diesel), servicing, (Canada only)

CAUTION!

Before working on the electrical system, obtain radio security code and disconnect battery Ground (GND) strap.

Note:

Refer to auxiliary heater ID plate ⇒ Fig. 1 for information on type of heater installed in vehicle.

1 - **Auxiliary heater D2L**
- Removing/installing ⇒ Fig. 3
- Disassembling/reassembling ⇒ page S 82-66

2 - **Bracket**

3 - **Heater–time temperature control switch -E13-**
- Removing/installing ⇒ Fig. 4
- Operation/function ⇒ Fig. 5

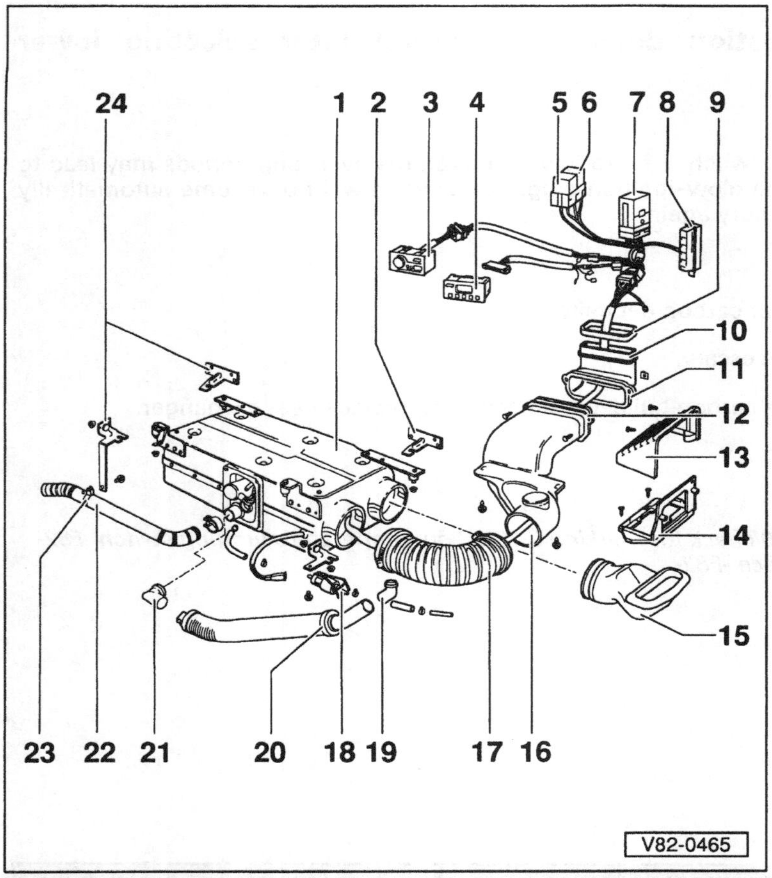

4 - Programmable timer -E111-

- Removing ⇒ Fig. 6
- Operation/function ⇒ page S 82-16
- Timer optional in certain models

5 - Resistance wire relay -J116-

- For vehicles without rotating dirver's seat
- For vehicles with rotating seat, behind left B-pillar trim lower

6 - Relay – glow plug monitoring

- For vehicles wilthout rotating seat
- For vehicles with rotating seat, behind left B-pillar trim lower

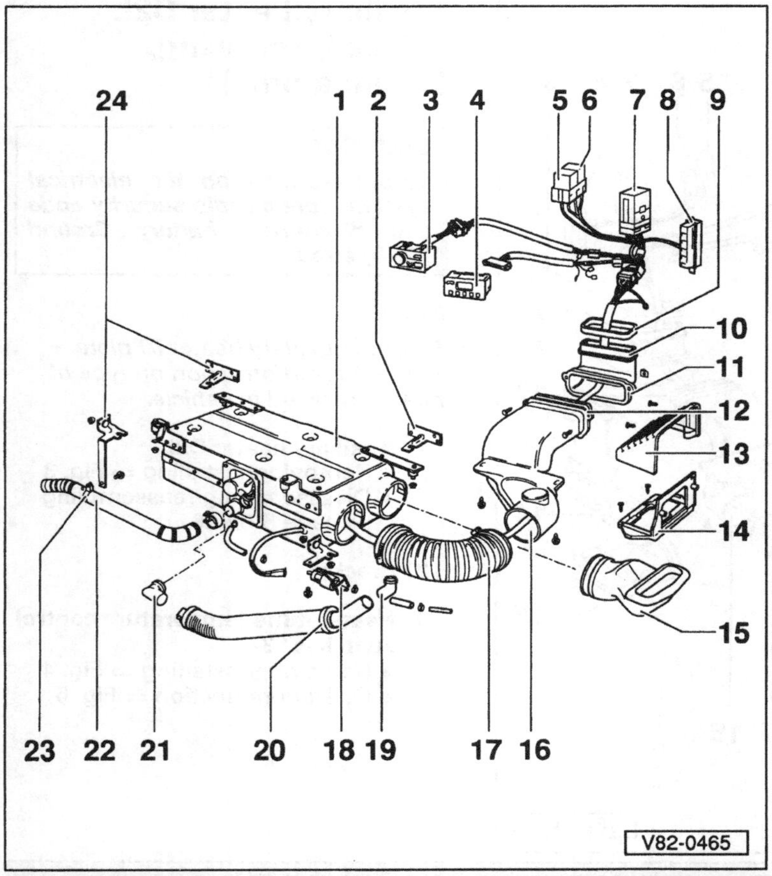

7 - Heater control module -J65-

- For vehicles without rotating driver's seat
- For vehicles with rotating driver's seat behind left B-pillar trim lower ⇒ Fig. 7
- Replace fuse -S23- ⇒ Fig. 8

8 - Fuse Holder

- For vehicles with rotating seat, at left B-pillar lower
- Heater fuse -S62- replace
- For vehicles without rotating seat, located under driver's seat

9 - Gasket

10 - Elbow, lower

11 - Gasket

12 - Gasket

13 - Vent, upper

- Removing ⇒ Fig. 10

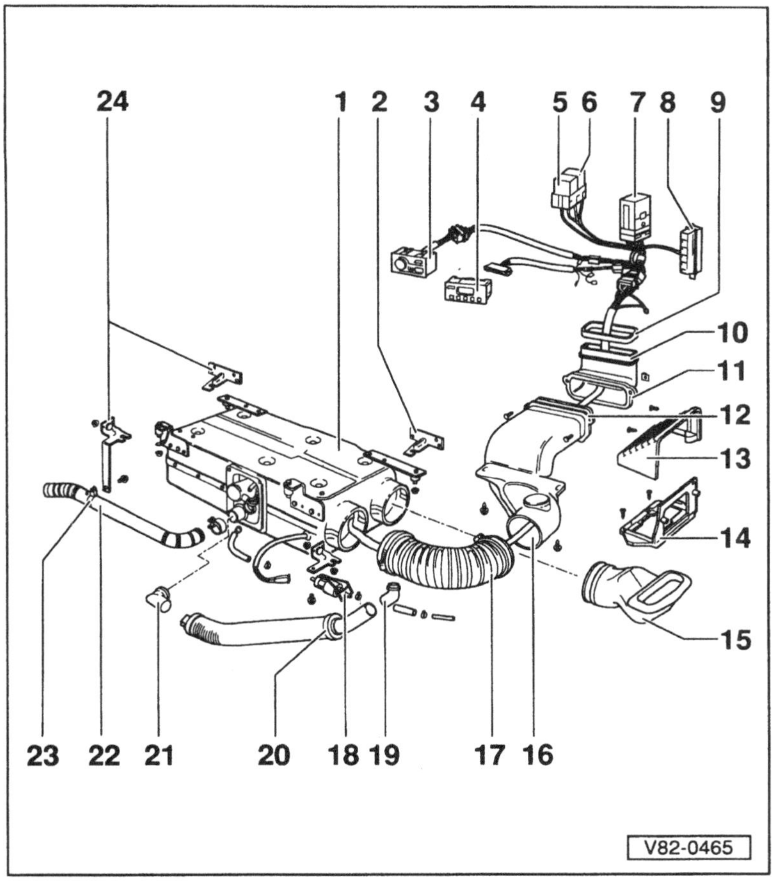

14 - Vent, lower
- Removing ⇒ Fig. 10

15 - Exhaust duct
- Removing ⇒ Fig. 11

16 - Intake duct
- Removing ⇒ Fig. 11

17 - Intake hose ⇒ Fig. 11

18 - Transfer fuel pump -G23-
- Assembly ⇒ Fig. 13
- Installed position ⇒ Fig. 14
- Check flow amount ⇒ Fig. 15

19 - Combustion air intake adaptor
- Removing/installing ⇒ Fig. 11

20 - Combustion air hose
- Removing ⇒ Fig. 11

21 - Combustion air pipe adaptor
- Removing ⇒ Fig. 11

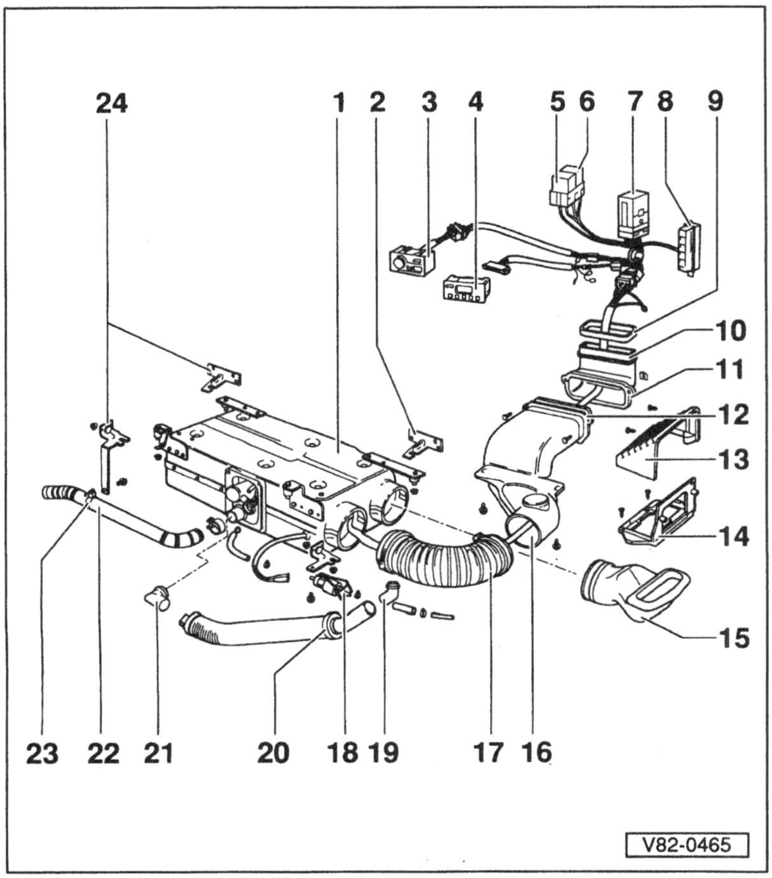

22 - Exhaust pipe
From HN 057 927, a 2 part exhaust pipe is installed
- Removing/installing ⇒ Fig. 12

23 - Clamp

24 - Bracket

Fig. 1 Identification plate

Note:

- *Factory only installs version 25 172 501 for auxiliary heater D2L-1 in diesel EuroVan/Transporter models from m.y. 1992.*
- *The year heater is first placed into service is indicated on lower portion of plate*
- *The heater core must be replaced after 10 years ⇒ Fig. 2*

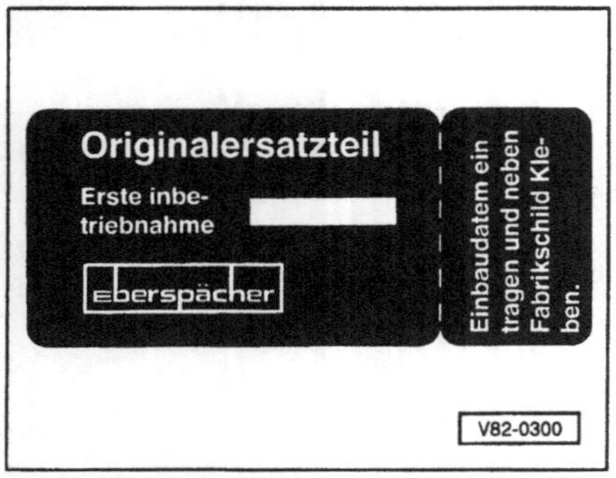

Fig. 2 Original replacement part identification plate

Note:

- *After installing a new heater core, identification plate must be replaced*
- *Install replacement indentification plate in same position as previous plate, and enter year of heater core installation*

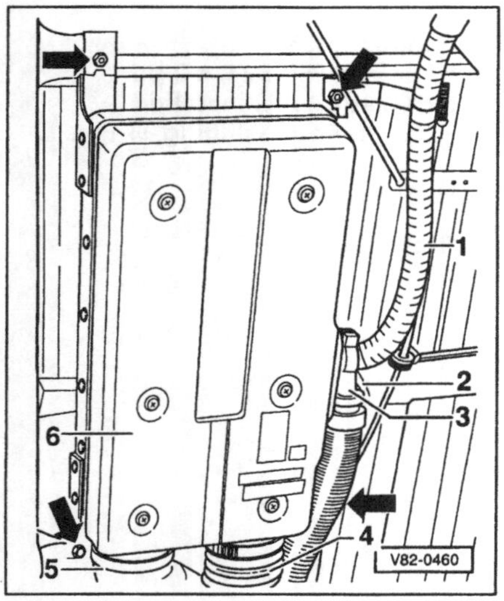

Fig. 3 Auxiliary heater, removing/installing

- Remove lower left B-pillar trim ⇒ Repair Group 75
- Disconnect plugs T10 and T1a located behind lower left B-pillar trim
- Disconnect Ground (GND) connection located at point -34- behind lower left B-pillar trim

Disconnect intake hose -4- on heater -6-

- Pull wiring harness out of intake hose
- Disconnect exhaust duct -5- on heater
- Pull combustion air inlet adaptor -3- off heater
- Pull off fuel hose -2- between fuel pump and heater
- Remove exhaust pipe -1- (twist off to remove)
- Remove screws -arrows- slide heater in drive direction and remove

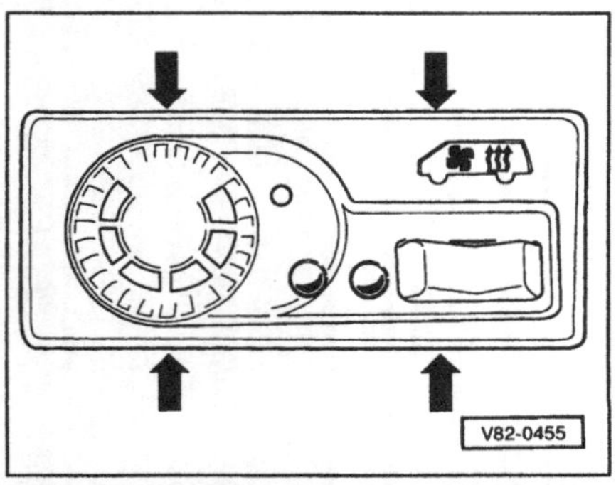

Fig. 4 Heater-time temperature control switch -E13-, removing/installing

Carefully pry switch from trim panel -arrows- with screwdriver

- Disconnect plug

Reinstall in reverse sequence

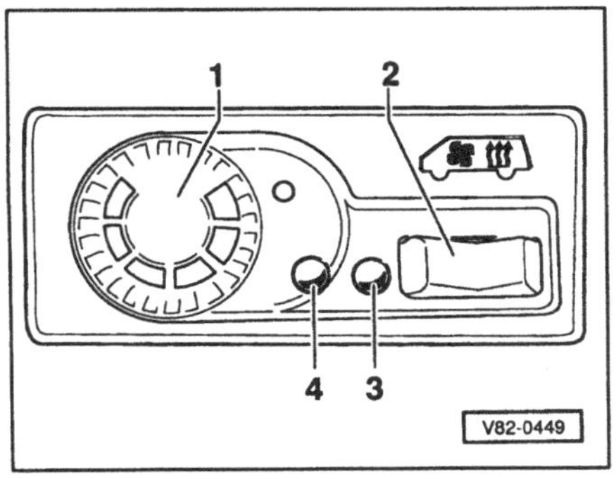

◄ **Fig. 5 Heater-time temperature control switch -E13-, operation/function**

1 - Heater-time temperature control switch -E13-

Note:

- *When auxiliary heater is started, it operates at full heating capacity*
- *When stable combustion is achieved, the glow plug shuts OFF and heating is regulated according to setting of temperature control switch*
- *If inside temperature is lower than the selected temperature, heater will operate at 100% capacity*
- *If inside temperature exceeds the selected temperature, fuel pump -G23- is switched OFF*

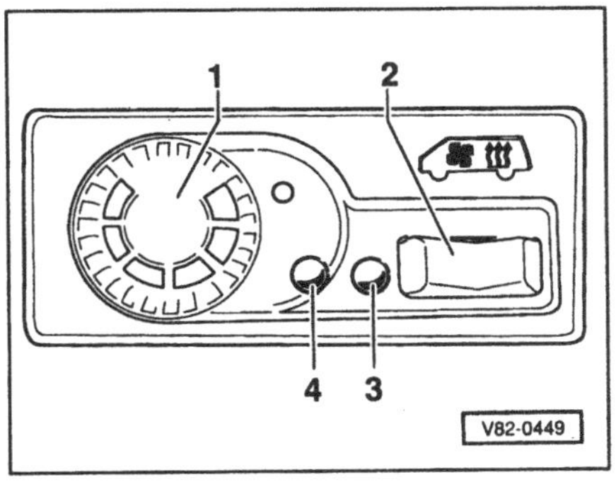

2 - Heater/heat output switch -E16-

Switch Position	Function
Left	Fresh air
Middle	OFF
Right	Heated air

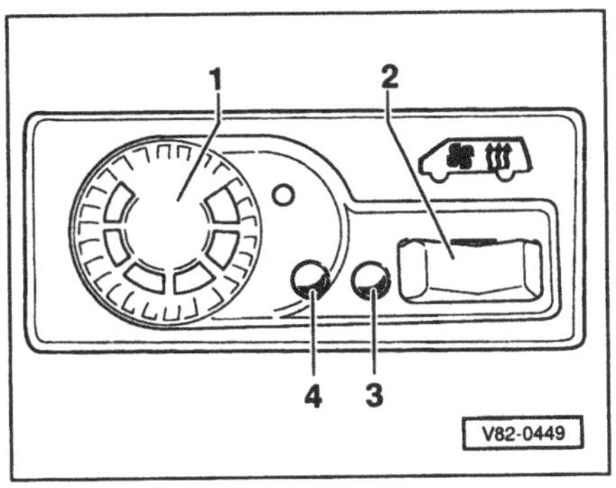

3 - Heater ON light -K89-, green

- Indicates heater is ON
- Goes out during after run
- Varies in brightness:
  Bright - full heating power
  Faint - low heating power, fresh air mode

4 - Heater malfunction light -K88-, red

Flashes during interference (overheating), switches off after fault eliminated by switching off and switching on again

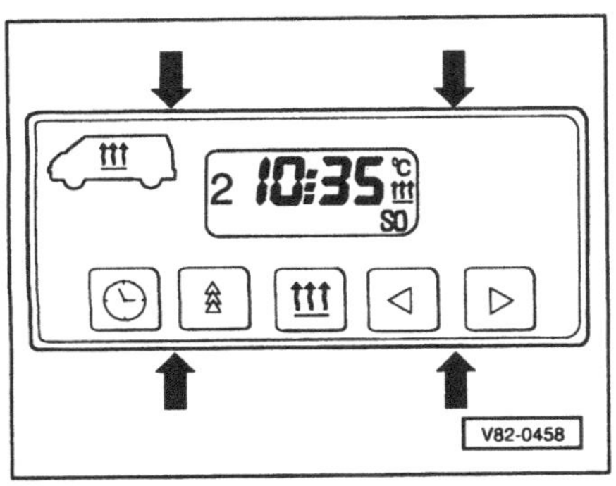

◄ **Fig. 6 Removing programmable timer -E111-**

– Carefully pry out timer with screw driver from trim -arrows-

– Remove connector

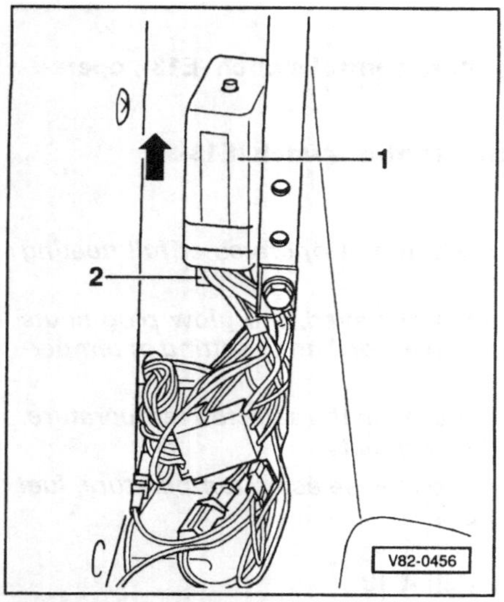

Fig. 7 Heater control module -J65-, removing/installing

– Remove trim from bottom left side B-pillar ⇒ Repair Group 75 (camping equipment)

Pull control module -1- up in direction of arrow

– Disconnect plugs -2-

Reinstall in reverse sequence

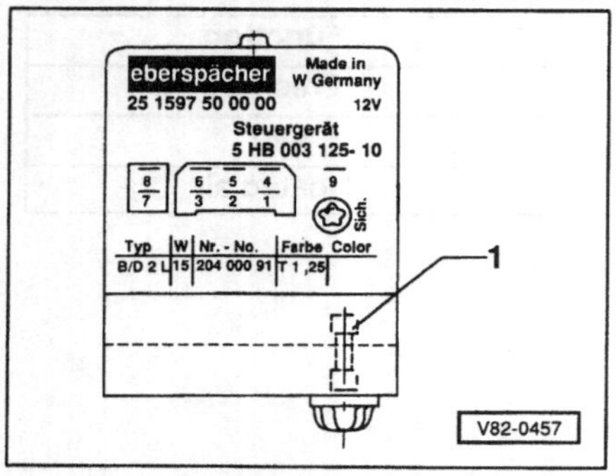

Fig. 8 Fuse -S23-, replacing

– Fuse -1- for fresh air/heater blower -V81-

Note:

If the voltage supplied to the control module -J65- drops below 10.2 Volts for longer than 20 seconds, or goes above 15 Volts, a voltage protection circuit in the control module will switch the heater off.

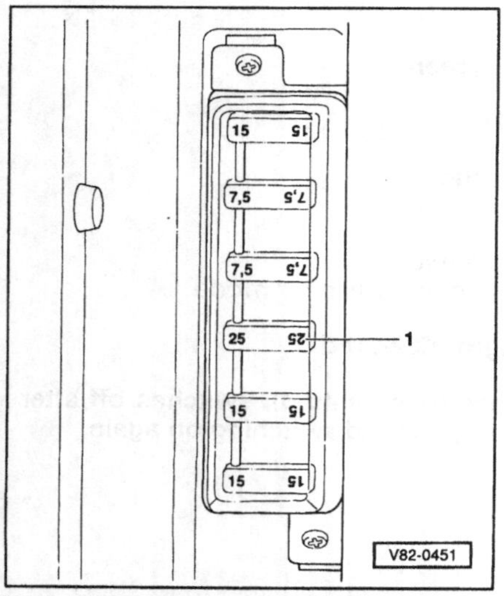

Fig. 9 Heater fuse -S62-, replacing

– Fuse -1-, -S62- is located in fuse holder at bottom of left B-pillar

– Use 25A (ampere) fuse only

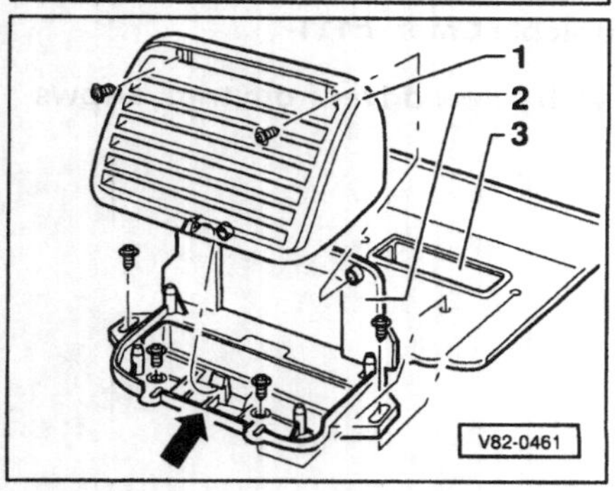

Fig. 10 Vent, removing/installing

1 - Vent, upper

2 - Vent, lower

3 - Outlet channel

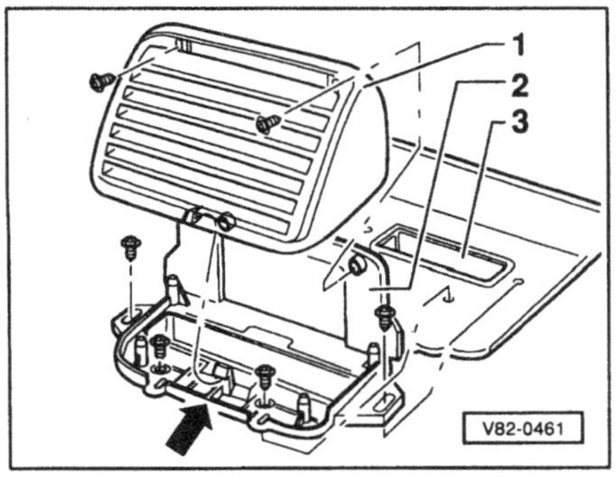

- Remove screws – upper vent section
- Release tab -arrow- with screwdriver, and remove upper vent
- Remove lower vent (remove screws)

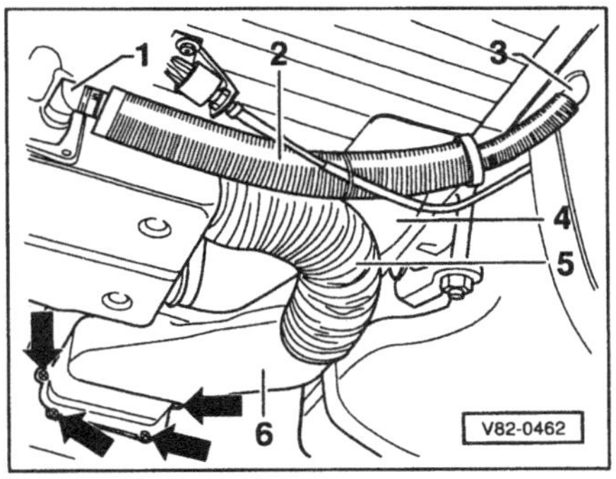

◄ **Fig. 11 Intake duct, exhaust duct, intake hose, combustion air hose, removing/installing**

1 - Combustion air pipe adapter

- Pull off

2 - Combustion air hose

3 - Combustion air intake adaptor

- Take out cross member
- Remove adaptor from combustion air hose (remove screws)

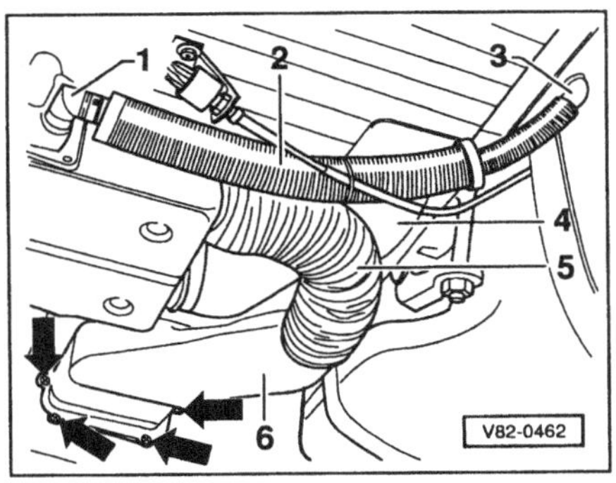

4 - Exhaust duct

- Remove floor plate
- Remove upper/lower outlets in passenger compartment
- Pull duct off heater

5 - Intake hose

- Disconnect plugs T10 and T1a located behind lower left B-pillar trim
- Disconnect Ground (GND) connection at point -34- behind lower left B-pillar trim
- Loosen clamps and remove hose
- Pull out wiring harness from intake hose and intake duct

6 - Intake duct

- Disconnect plugs T10 and T1a located behind lower left B-pillar trim
- Disconnect Ground (GND) connection at point -34- benind lower left B-pillar trim
- Loosen clamp and remove intake hose
- Pull out wiring harness from intake duct
- Remove screws -arrows- then remove intake duct

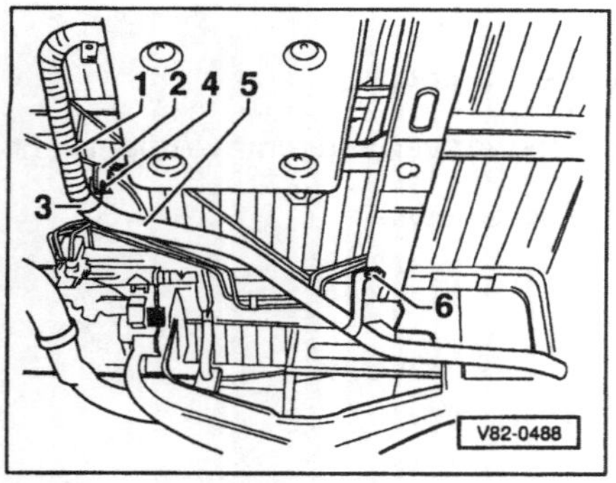

Fig. 12 Exhaust pipe, removing/installing

1 - Exhaust pipe, front

2 - Bracket

3 - Clamp

4 - Screw and washer M6 x15

5 - Exhaust pipe, rear

6 - Tapping screw

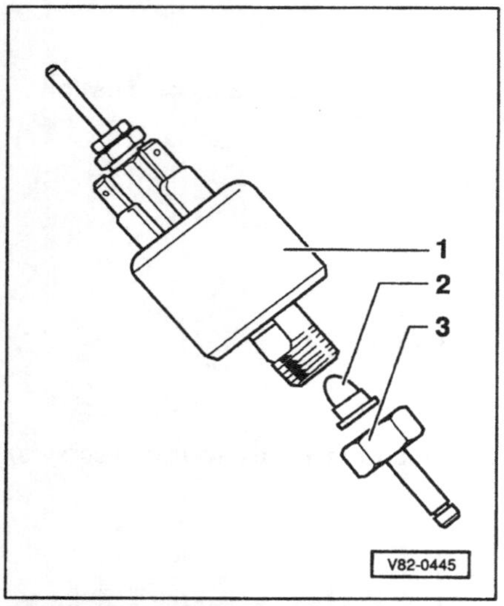

Fig. 13 Transfer fuel pump -G23-, assembly

1 - Fuel pump

2 - Strainer

3 - Hose adaptor

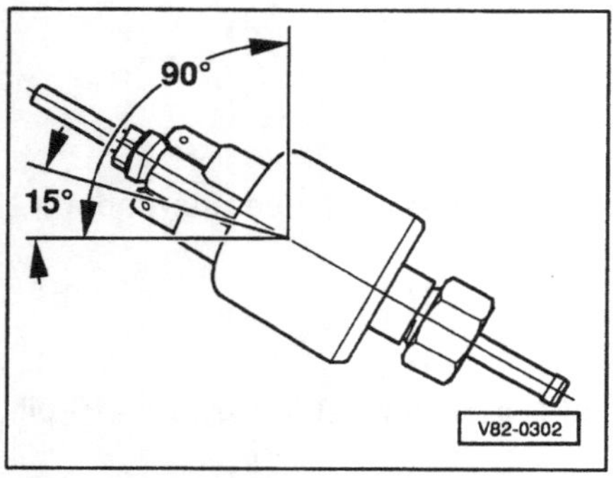

Fig. 14 Transfer fuel pump -G 23-, installed position

– Inclined 15° to 90°

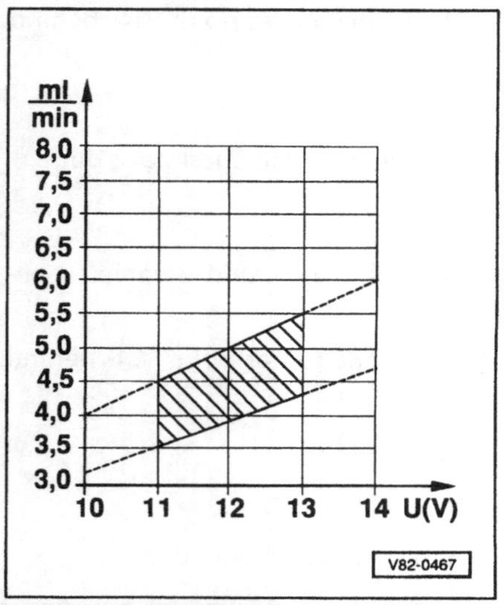

Fig. 15 Fuel flow rate, checking

Test requirements:

- Intake and outlet channel are not plugged
- Combustion air lines and exhaust lines are not plugged
- Bridging terminals -3- and -4- of control module -J65-, there must be a voltage from 11 to 13 Volts

Note:

Hold measuring glass at height of glow/spark plug during testing

Checking:

– Set Fluke 83 to direct voltage measurement, range 20V

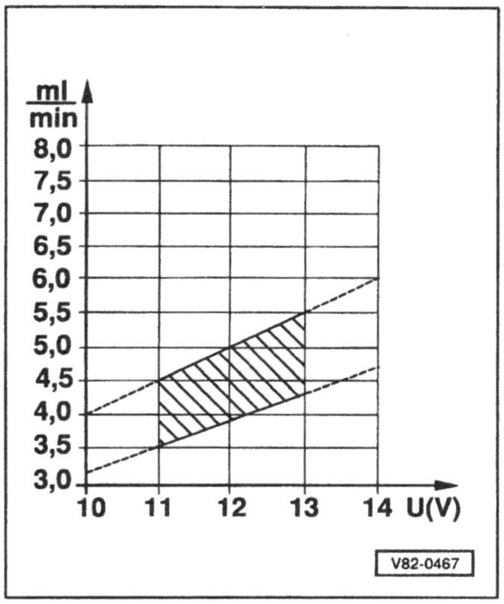

- Connect test points to contact 3 (–) and contact 4 (+) of heater contrtol module -J65-
- Remove connector on the glow/spark plug
- Place fuel line into measuring glass (10 ml – 25 ml)
- Switch on heater
- Bleed fuel hose
- Switch off heater and drain measuring glass
- Switch heater on again
- As soon as fuel flows, activate stop watch
- Read voltage value on Fluke 83
- After 1 minute switch off heater
- ◄ Read amount of fuel flow and compare measuring value with diagram
- The fuel flow amount is OK if the value read is in the highlighted area of the diagram

Notes:

- ♦ *Replace fuel transfer pump -G23- if fuel flow is out of specification.*
- ♦ *If fuel flow is in specification, fresh air/heater blower -V81- may be turning too slowly. Replace.*

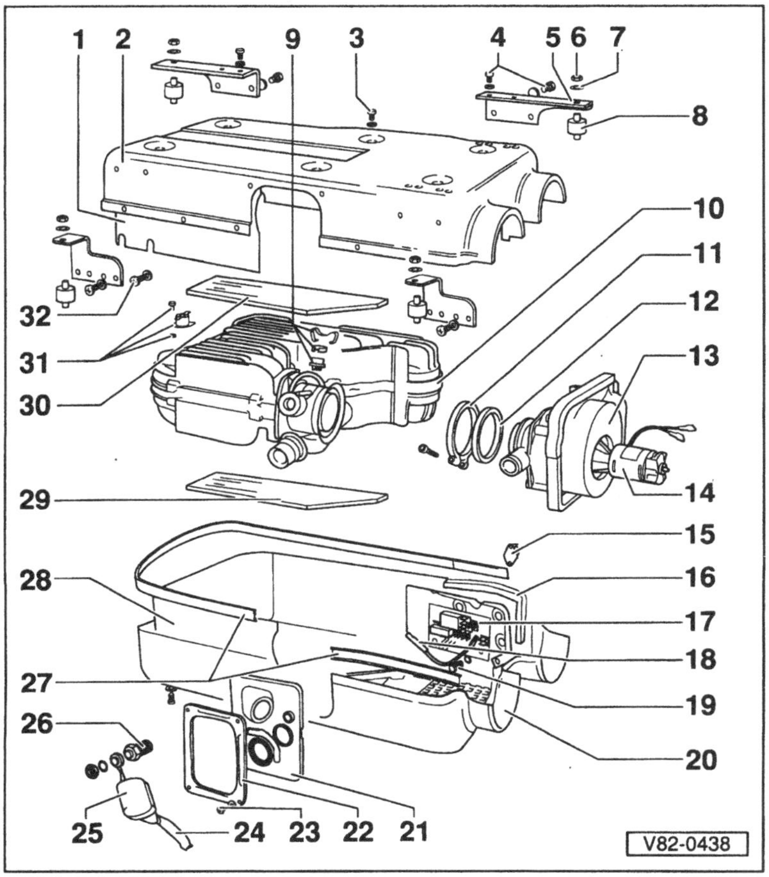

Auxiliary heater D2L, assembly

1 - Heatshield, upper

2 - Outside cover, upper
♦ Removing:

- Remove screws from support panel -22-, remove screws -32- on both sides from outside cover
- Remove outside cover
- Pull off seal -27-
  ♦ Installing
- Reinstall seal -27-
- Assemble both outside covers
- Check for proper seal of covers

3 - Screw B4.2 x 9.5
♦ Use silicone washer

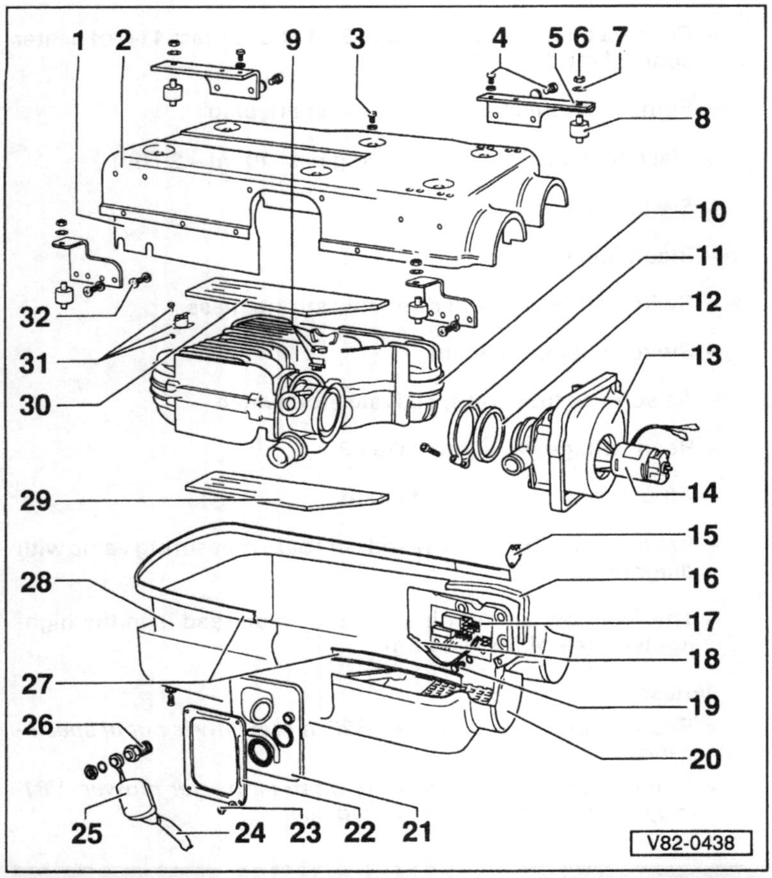

4 - **Screw M5 x 8**
- ♦ Use silicone washer

5 - **Bracket support**

6 - **Hex nut M6**

7 - **Spring washer**

8 - **Bonded rubber bushing**

9 - **Heater flame sensor -F16-**
- ♦ The flame sensor shuts off the glow plug -26-, after a stable flame has been formed and erases the safety time. In addition, it ends the afterrun when heater has cooled down
- ♦ Opens at temperatures over 100°C ±3°C (212° ± 2°F)
- ♦ Closes at temperatures below 90°C ±4°C (194° ± 2°F)
- ♦ Replacing ⇒ Fig. 1

10 - **Heat exchanger**
- ♦ Do not remove (unscrew) combustion chamber from exchanger

11 - **Clamp**

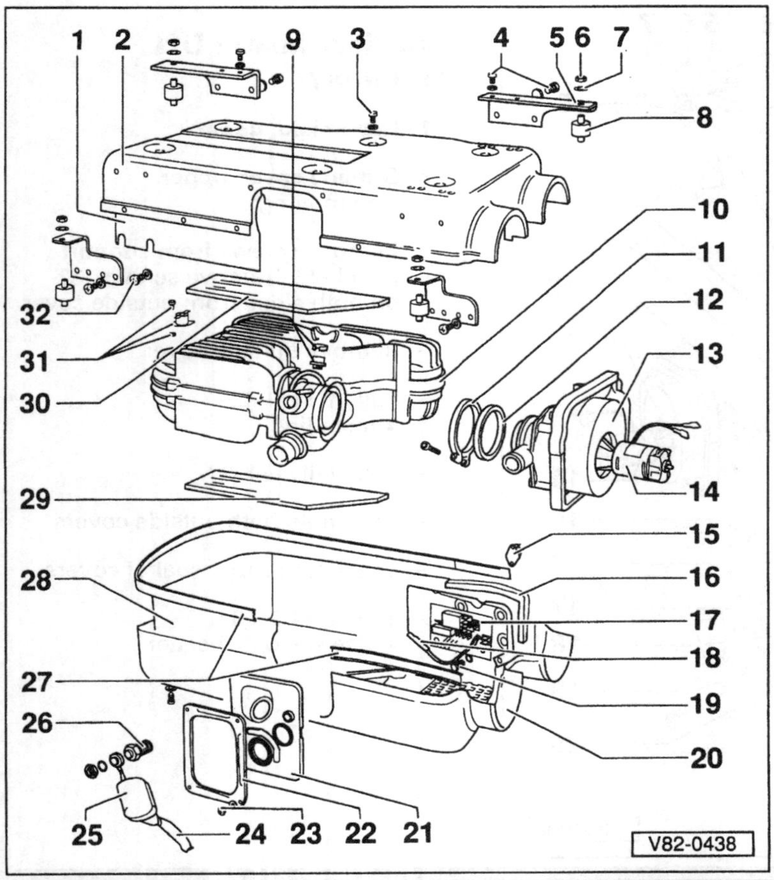

12 - **Seal**
- ♦ Always replace

13 - **Blower motor assembly**
- ♦ Removing/installing ⇒ Fig. 4
- ♦ Assembly ⇒ page S 82-75

14 - **Motor for combustion blower, fresh and hot air blower -V81-**
- ♦ Removing/installing ⇒ Fig. 4
- ♦ Assembly ⇒ page S 82-75

15 - **Temperature limiting switch -F82-**
- ♦ The temperature limiting switch shuts off the heater at temperatures over 125°C (257°F)
- ♦ Replacing ⇒ Fig. 2

16 - **Insulation tape**

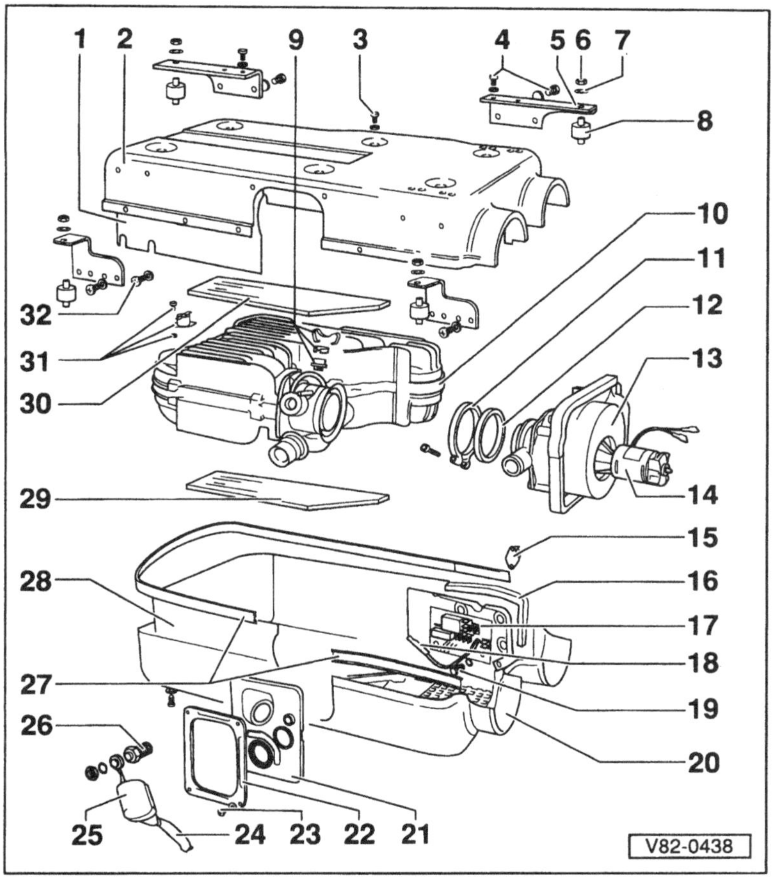

17 - Junction box (T)
- Junction box for diesel electrical heater is identified with a white "D"
- Connector routing: ⇒ Electrical Wiring Diagrams

18 - Temperature sensor -G18-
- Depending on the setting of the temperature control switch, -G18- regulates the heat output by measuring the outlet temperature
- -G18- is fastened to the combustion fresh air and hot air blower

19 - Bolt M5 x 8

20 - Outer cover, lower
- Note: 2mm - hole in the lower outer cover must be kept clean

21 - Seal, red
- Installing

– check for proper fit

22 - Support panel

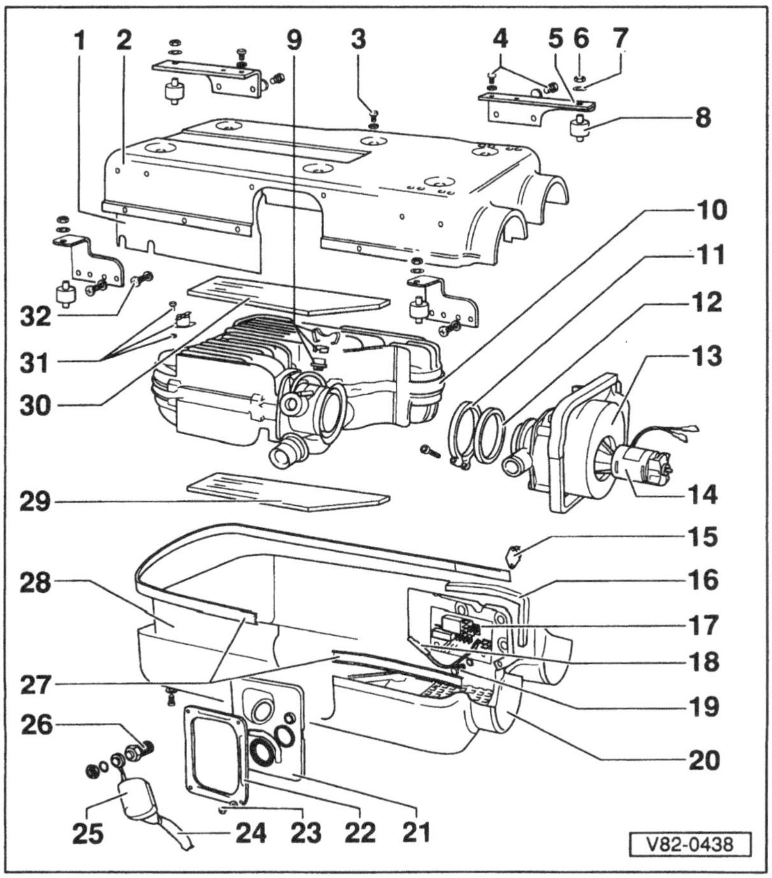

23 - Screw B4 2 x 13
- Use silicone washer

24 - Glow plug wire

25 - Cap, red

26 - Glow plug
- 17 Nm (12 ft lb)

27 - Seal
- Installing

– Coat with petroleum jelly

– Install onto lower outer cover

– Ensure cover is seated correctly

28 - Heatshield, lower

29 - Sealing plate, lower
- Installing

– Align plate with old seal and glue to heatshield with silicone cement

30 - Sealing plate, upper
- Installing

– Align plate with old seal and glue to heatshield with silicone cement sealing compound

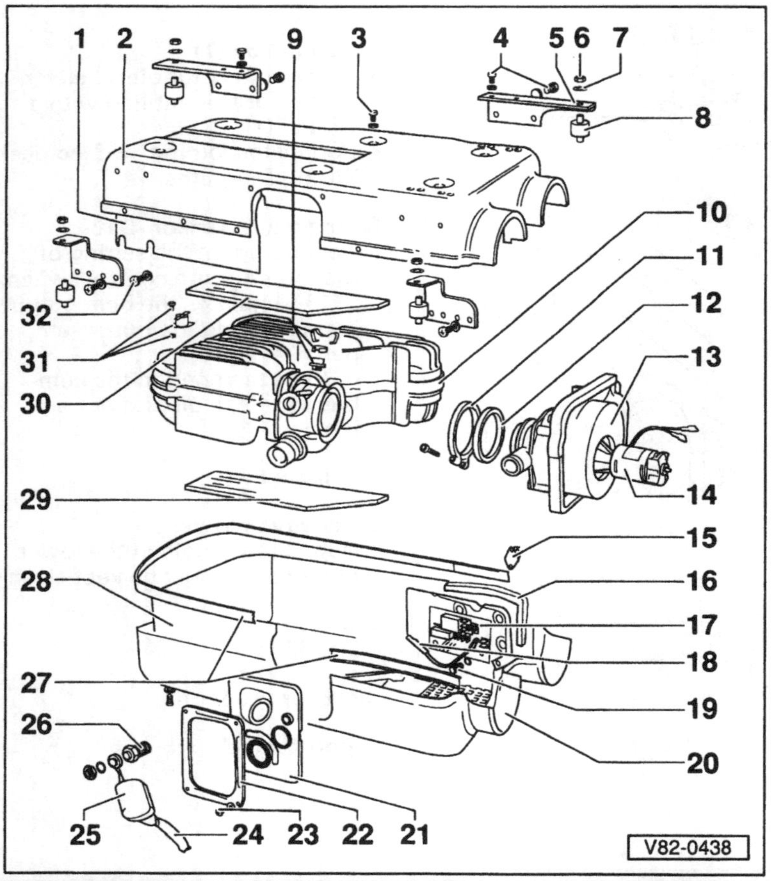

31 - **Heater overheat control switch -F17-**
- The overheat control switch shuts off the fuel feed when hot air exceeds maximum temperature. Red warning light -K81- in temperature control switch will also flash
- Opens at temperatures over 110°C ± 4°C (230°± 2°F)
- Closes at 95°C ±4°C (203° ± 2°F)
- Replacing ⇒ Fig. 3

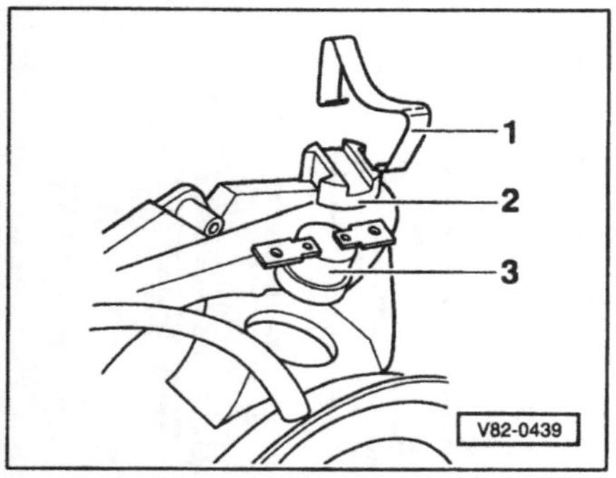

◄ **Fig. 1 Heater flame sensor -F16-, removing/installing**

– Remove wire connector

– Using screwdriver, unclip retainer -1-

– Remove connector -2-, flame sensor 3--

Reinstall in reverse sequence

Note:

Do not pinch wires between sides of heat exchanger and heat shield

– Replace flame sensor with retaining spring

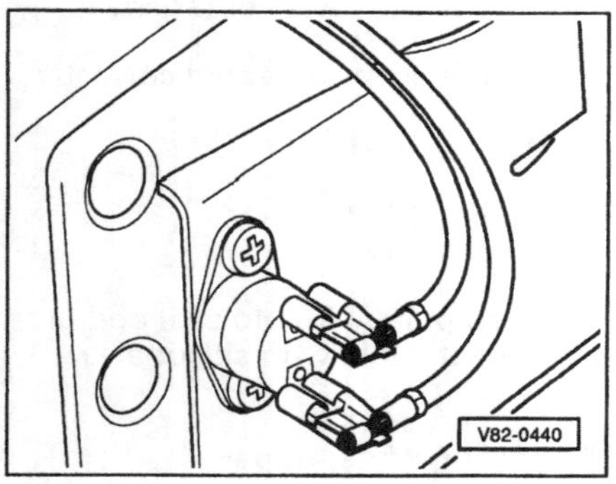

◄ **Fig. 2 Temperature limiting switch -F82-, replacing**

– Remove connector

– Remove temperature sensor switch (unscrew) from heat-shield

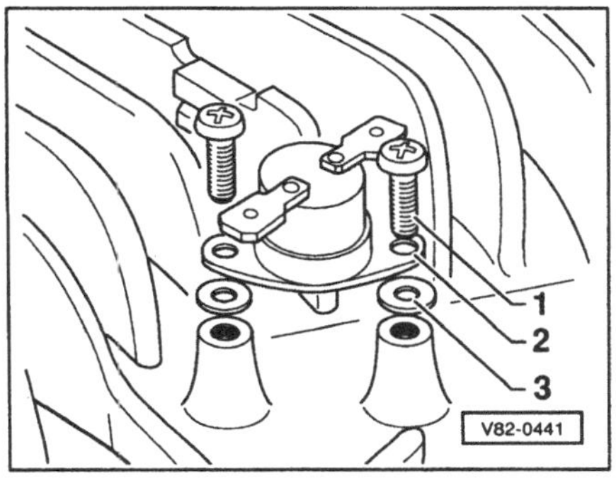

Fig. 3 Heater overheat control switch -F17-, replacing

1 - Screw M3 x 10

2 - Heater overheat control switch -F17-

3 - Insulating washer

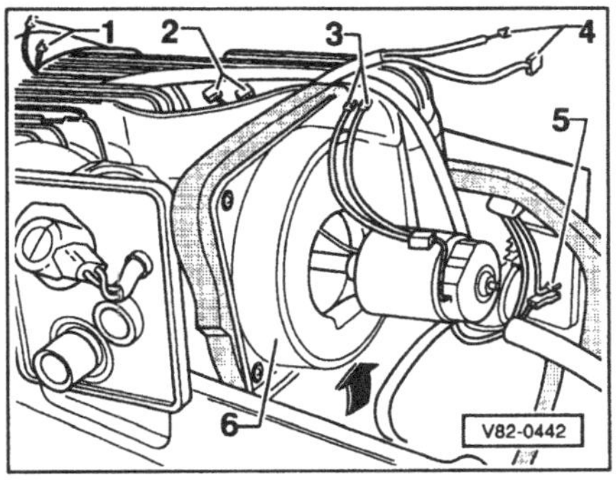

Fig. 4 Combustion, fresh air, and hot air blower, removing/ installing

Removing

– Pull off plugs -1- of overheating switch

– Pull off plugs -2- of flame sensor

– Pull blower motor wires -3- off junction box -5-

– Remove glow plug wires -4- from wire distributor -5-

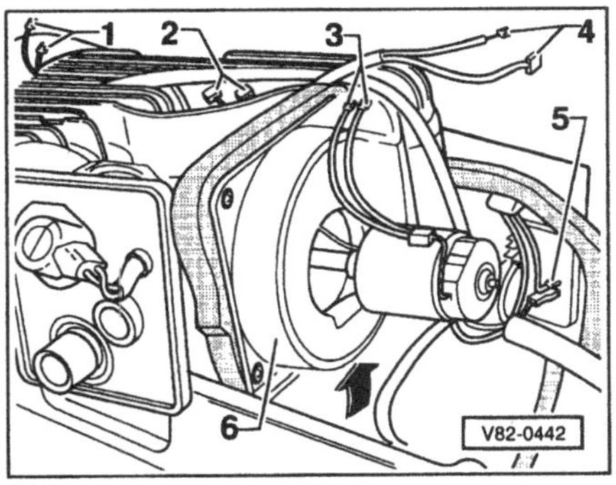

– Remove temperature sensor -G18- from retainer on spiral housing (6)

– Raise blower with heater core in direction of arrow until clamp is accessible

– Loosen clamp, remove blower and seal

Installing

– Coat outer cover in seal area with petroleum jelly

– Place new seal between blower and combustion chamber

– Place blower on combustion chamber, install clamp and tighten slightly

– Press heater in lower outer cover

– Tighten clamp

– Fasten temperature sensor to spiral housing

– Reconnect conectors

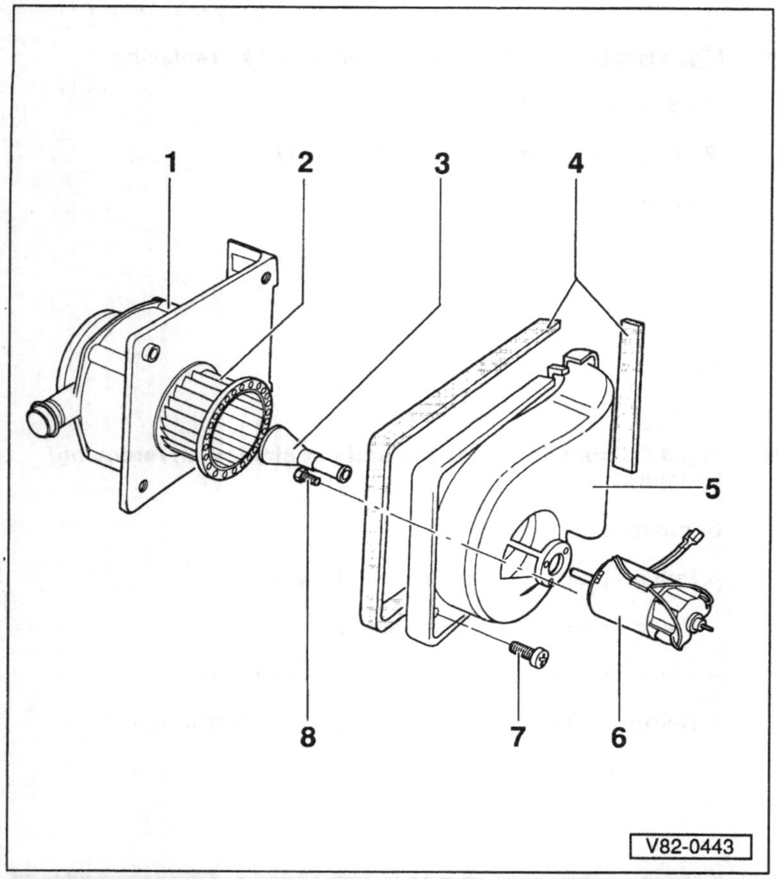

Blower motor, assembly

1 - **Combustion air blower**
  - ♦ Do not disassemble

2 - **Blower wheel (fresh/heated air)**
  - ♦ Checking installed position ⇒ Fig. 1

3 - **Connector**
  - ♦ To install: place on shaft of motor
  - ♦ Checking installed position ⇒ Fig. 1

4 - **Insulation tape**
  - ♦ Always replace

5 - **Blower motor (fresh/heated air)**

6 - **Fresh air/heater blower -V81-**

7 - **Screw M5 x 16**

8 - **Screw M4 x 8**

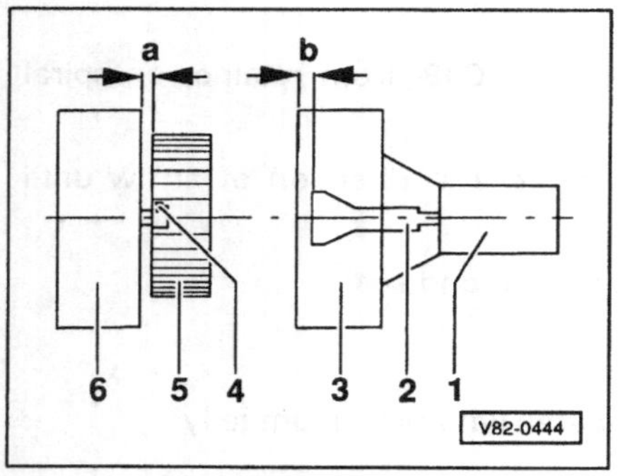

◄ **Fig. 1 Blower wheel/connector, installed position**

1 - Blower motor

2 - Connector

3 - Blower housing

4 - Stud

  - ♦ 120 ± 30 Ncm (10.6 ± 2.7 in. lb)

5 - Blower wheel

6 - Combustion air blower

Installed dimension
a = 1.4 ± 0.3 mm (0.055 ± 0.012 in.)
b = 4.0 ± 0.3 mm (0.156 ± 0.039 in.)

Auxiliary heater D2L (Diesel), troubleshooting

Tool requirements:

- Multimeter Fluke 83 (US 1119) or equivalent
- LED voltage tester VAG 1527B, US 1115 or equivalent
- Adapter cable set VAG 1594
- Applicable wiring diagram
- Stopwatch
- Measuring glass (10 ml – 25ml)

Heater does not come on

(no audible sound from blower within 5 seconds after switching on)

Test requirements

- Battery voltage OK
- Fuse -S62- OK

Glow plug -Q6- faulty or covered with carbon deposits

– Remove glow plug
– Inspect for carbon deposits

Glow plug OK	**Glow plug not OK**
→ A	→ **Replace glow plug**

A

Open circuit in wiring between auxiliary battery (30+) and heater–time temperature control switch -E13-

– Remove heater–time temperature control switch -E13-from B pillar trim
– Connect LED voltage tester between ground and -E13- terminal -T12a/11-

LED lights up	**LED does not light up**
↓	→ **Check circuit using wiring diagram**

Heater–time temperature control switch -E13- faulty

– Connect LED voltage tester between ground and -E13- terminal -T12a/12-
– Turn heater/heat output switch -E16- to "heat" position

LED lights up	**LED does not light up**
→ B	→ **Replace heater–time temperature control switch -E13-**

(B)

↓

Open circuit in wiring between -E13- connector -T12a/12- and -heater control module -J65- terminal 7

– Connect LED voltage tester between ground and -J65- terminal 7

LED lights up	LED does not light up
↓	↓
	Check circuit using wiring diagram

Heater control module -J65- faulty

Prerequisite: Heater fuse -S23- OK
– Connect LED voltage tester between ground and -J65- terminal 2

LED lights up	LED does not light up
↓	↓
(C)	**Heater control module -J65- faulty**

(C)

↓

Junction box -T- (in heater unit) faulty

– Turn heater/heat output switch -E16- from "heat" position to "ventilation" position

Fresh air/heater blower -V81- starts	Fresh air/heater blower -V81- does not start
↓	↓
Replace junction box -T-	

Motor for fresh air/heater blower -V81- faulty

– Remove intake hose at heater unit
– Rotate blower motor output shaft

Output shaft rotates	Output shaft does not rotate
↓	↓
Replace motor for fresh air/heater blower -V81-	**Repair/replace fresh air/heater blower -V81- motor**

Heater shuts off after approx. 10 seconds

Test requirements

- Heater switched off

Heater overheat control switch -F17- does not close

– Separate terminal connector -T12b-
– Set range switch on multimeter to test continuity
– Connect test leads to terminals -T12b/9 and -T12b/10 at terminal housing/harness to heater unit

Continuity	**No continuity**
Replace heater control module -J65-	**Replace overheat control switch -F17-**

Heater shuts off after 3 minutes, fuel pump clicks

- Transfer fuel pump -G23- clicks after appox. 30 seconds

– Turn on heater, set heater/heat output switch -E16- to "heat" position

♦ Combustion takes place, but heater switches off by itself after approx. 3 minutes ♦ Combustion/fresh air/heater blower runs on	♦ Combustion does not take place, heater switches off by itself after approx. 3 minutes ♦ Combustion/fresh air/heater blower doesn't run on
↓ (to strainer check below)	**Replace heater flame sensor -F16-**

Strainer in transfer fuel pump -G23- dirty or blocked

Disassemble transfer fuel pump and inspect strainer

Strainer OK	**Strainer dirty or blocked**
Fuel delivery rate insufficient **– Replace fuel transfer pump -G23-**	**Clean or replace strainer**

Heater shuts off after 3 minutes, fuel pump does not click

- Transfer fuel pump -G23- does not click after appox. 30 seconds

Transfer fuel pump -G23- inoperative

– Turn heater/heat output switch -E16- to "heat" position
– Connect LED voltage tester acrross terminals on -G23-, wait approx. 30 seconds

- **LED does not flash** → go to "Heater control module -J65- faulty"
- **LED flashes** → **Replace transfer fuel pump -G23-**

Heater control module -J65- faulty (does not provide impulse)

– Remove heater–time temperature control switch -E13-
– Connect LED voltage tester to ground and -E13- terminal -T12a/8

- **LED flashes** → go to "Heater–time temperature control switch -E13- faulty"
- **LED does not flash** → **Replace heater control module -J65-**

Heater–time temperature control switch -E13- faulty

– Connect LED voltage tester between ground and -E13- terminal -T12a/9-

- **LED does not flash** → **Replace heater–time temperature control switch -E13-**
- **LED flashes** → **Check for open circuit between -T12a/9- and -G23- using wiring diagram**

Heater malfunction light -K88- flashes, heater emits soot or smoke

Test requirements

- Intake and outlet channels are not blocked
- Combustion air and exhaust pies are not blocked
- Voltage between heater control module -J65- terminals 3 and 4 must be between 11V and 13V

Possible cause:

- Excessive fuel delivery

Correction:

– Perform fuel flow rate test ⇒ page S 82-14

Heater run-on cycle exceeds 4 minutes after switching heater off

Heater flame sensor -F16- faulty

- Switch off heater, remove heater fuse -S62- if necessary
- Set range switch on multimeter to test continuity
- Connect multimeter to heater control module -J65- terminals 1 and 8

Continuity	No continuity
↓	↓
	Replace heater flame sensor -F16-

Glow plug control relay -J195- inoperative

- Set range switch on multimeter to check continuity
- Connect multimeter to -J195- terminals -30- and -87-

Continuity	No Continuity
↓	↓
Replace glow plug control relay -J195-	Replace heater control module -J65-

Heater does not run-on after switching off and "Ventilation" function inoperative

Possible cause:

- ♦ Auxiliary heater blower relay -J233- on junction box -T- sticking

Correction:

- Replace junction box -T- (relay and junction box are integrated)

Heater does not reduce output

Short in wiring to temperature sensor II -G18-

- Remove heater–time temperature control switch -E13-
- Set range switch on multimeter to resistance test, 2 kΩ range
- Connect multimeter to connector terminals T12a/2 and T12a/3

Reading between 800Ω and 1.15 kΩ	Reading less than 800Ω
↓	↓
Replace heater–time temperature control switch -E13-	Check wiring to temperature sensor II -G18- using wiring diagram

Heater output inadequate or heater always running at low output

Test requirement

- ♦ outlet channel not damaged

Open circuit in wiring to temperature sensor II -G18-

- Remove heater–time temperature control switch -E13-
- Set range switch on multimeter to resistance test, 2 kΩ range
- Connect multimeter to connector terminals T12a/2 and T12a/3

Reading between 800Ω and 1.15 kΩ → **Replace heater–time temperature control switch -E13-**

Reading greater than 1.15 kΩ →

Check for open circuit in wiring to temperature sensor II -G18- using wiring diagram

Open circuit found → **Repair open circuit**

No open circuit found → **(A)**

(A) →

Temperature limiting switch -F82- faulty

- Set range switch on multimeter to test continuity
- Connect to -F82-terminals at junction box -T-

Continuity → **Replace junction box -T- (temperature sensor -G18- integrated)**

No continuity → **Replace temperature limiting switch -F82-**

A/C refrigerant system, identification

1992 m.y. EuroVan and Transporter models (Canada only) contain A/C systems charged with refrigerant R-12.

Starting with 1993 m.y. (USA and Canada), A/C systems are charged with refrigerant R-134a.

Labels specifying refrigerant type are located on the compressor or radiator support. Before proceeding with refrigerant system servicing or repairs, confirm refrigerant type used.

Certain equipment, specifications and procedures vary from R-12 to R-134a systems. Always refer to the appropriate refrigerant system section in this manual when servicing A/C systems.

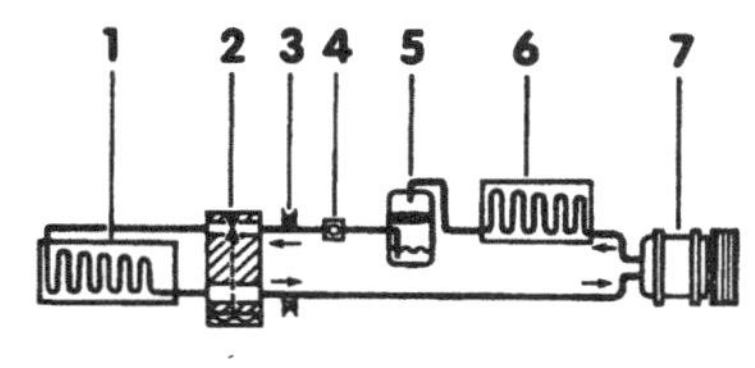

87-1437

◄ **Basic A/C refrigerant circuit**

1 - Evaporator
2 - Expansion valve
3 - High-pressure service valve
4 - Sight glass (if equipped)
5 - Receiver drier
6 - Condenser
7 - Compressor

Note:
Arrows indicate flow direction of refrigerant.

A/C refrigerant system, component functions

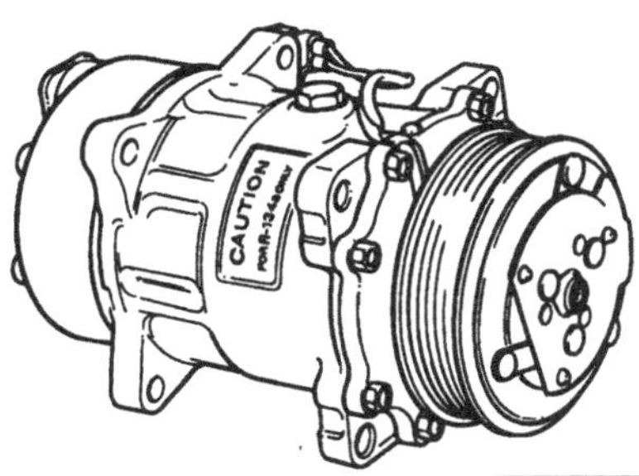

87-1447

◄ **Compressor**

The compressor is driven via a belt on the engine when the A/C clutch engages (A/C ON).

Low-pressure refrigerant gas from the evaporator is compressed by the compressor. After compression, the refrigerant gas (now high-pressure) flows to the condenser.

Notes:

- *The compressor contains refrigerant oil that is mixable under all temperatures with the refrigerant.*
- *A label on the compressor indicates that compressor is for R-134a systems only.*

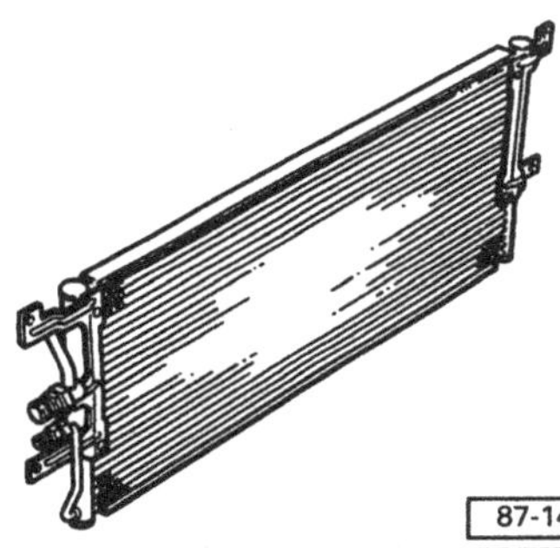
87-1450

◄ **Condenser**

The condenser transfers heat from the compressed refrigerant gas to the outside air which causes the refrigerant to change state from a gas to a liquid.

Note:
The condenser for the R-134a refrigerant system is identified with a green label.

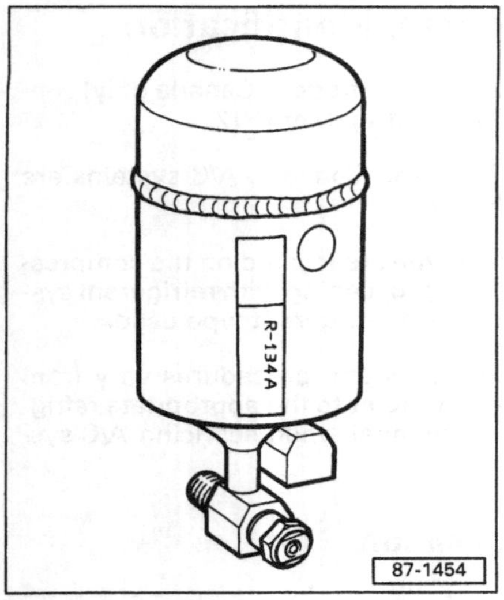

◄ **Receiver drier**

The receiver drier acts as a refrigerant reservoir for the system. Any moisture in the system is absorbed in the drier desiccant.

> ***CAUTION!***
>
> ***The R-134a receiver drier is identified with a green label and MUST NOT be used in R-12 systems (drier desiccant is only compatible with R-134a refrigerant). Likewise, NEVER use an R-12 receiver drier in an R-134a system.***

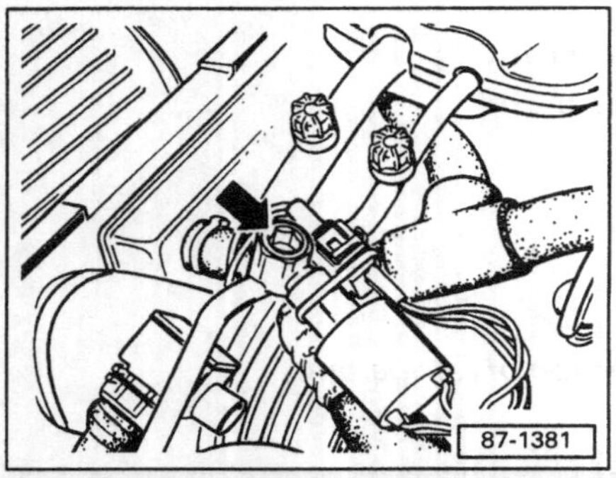

◄ **Sight glass (if equipped)**

With a properly charged A/C system and A/C clutch engaged, the sight glass (arrow) should remain clear and bubble-free.

Note:

The mixture of R-134a refrigerant and refrigerant oil (PAG oil) may appear milky in the sight glass.

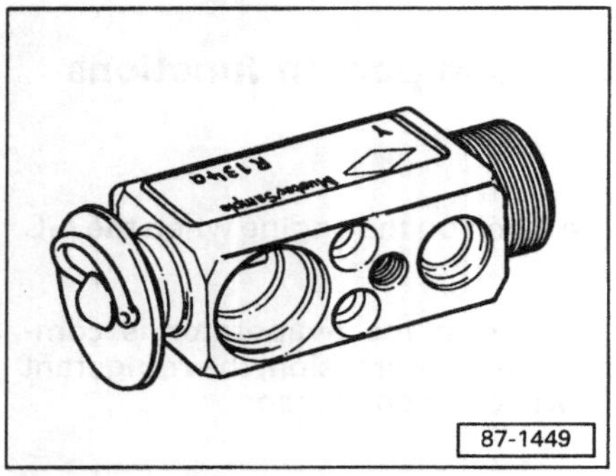

◄ **Expansion valve**

The expansion valve restricts and controls refrigerant flow to the evaporator thus lowering refrigerant temperature and pressure.

Note:

The R-134a expansion valve is identified with a green adhesive label.

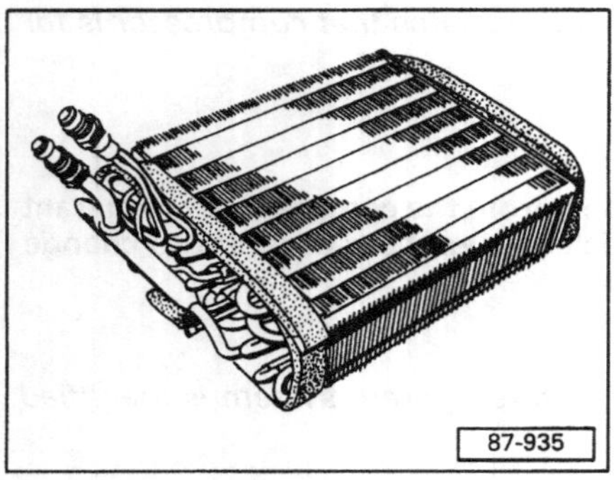

◄ **Evaporator**

Liquid refrigerant entering the evaporator absorbs heat from air passing through the evaporator fins and cools the air. As the refrigerant absorbs heat it turns to vapor and then is suctioned by the compressor.

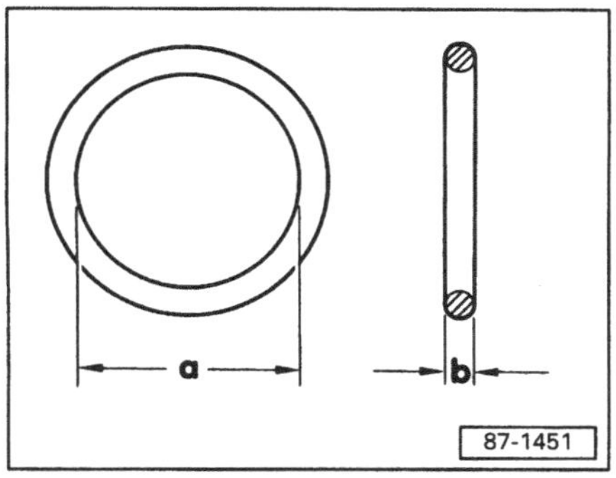

O-rings

O-rings seal connections between A/C system components.

Notes:

- *Always use correct size O-rings (dimensions -a- and -b-).*
- *Do not reuse O-rings, always replace. Use only new O-rings that are compatible with R-134a refrigerant and refrigerant (PAG) oil on R-134a systems.*
- *Lubricate O-rings with the appropriate refrigerant oil before installing (use PAG oil on R-134a systems).*

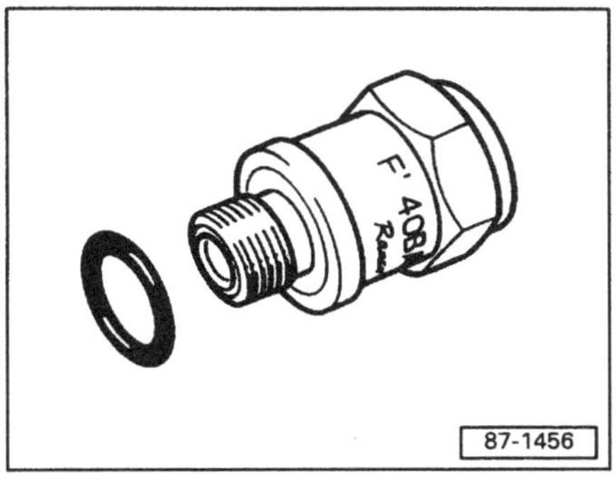

Pressure relief valve

The pressure relief valve is mounted on the compressor or the receiver drier. At pressures above 40 bar (580 psi), the pressure relief valve opens to vent excessive pressure. When the system pressure is reduced, the valve closes to prevent total refrigerant loss.

A cap on the pressure relief valve will pop out if the valve has opened.

A/C system hoses and lines

The mixture of refrigerant oil (PAG oil) and refrigerant R-134a attacks some metals and alloys (for example, copper) and breaks down certain hose material. Use only hoses and lines that are identified with a green mark (stripe) or the lettering "R-134a".

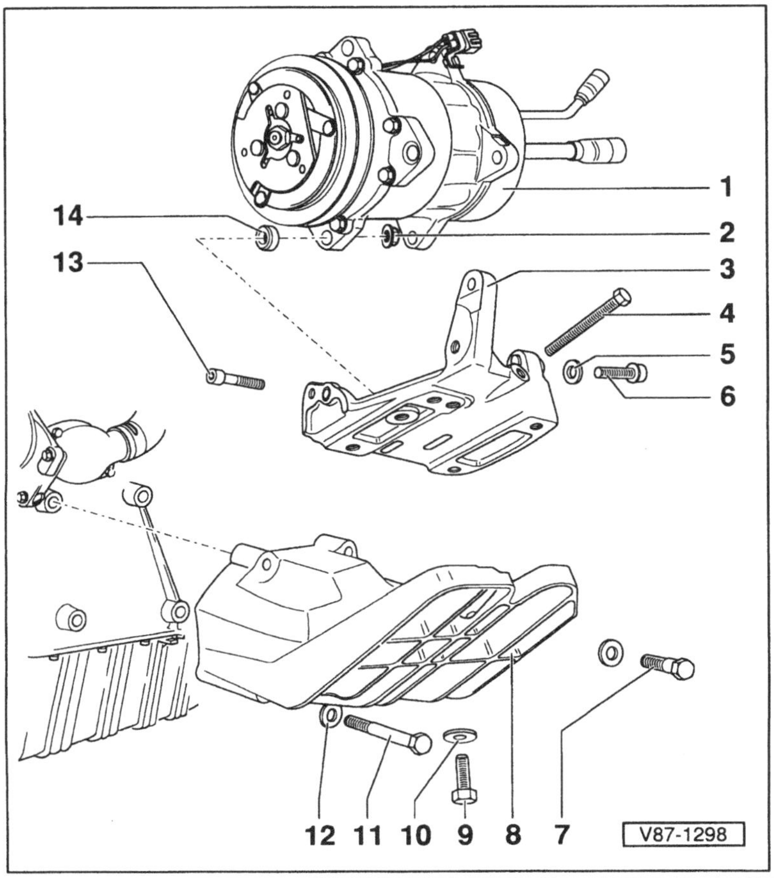

Compressor and bracket assembly

Compressor bracket, removing/installing 5 cyl gas engine ➤ 08.91

Note:

The compressor mounting bracket can be removed and installed without discharging the A/C refrigerant system.

1 - Compressor
- Compressor mounts to bracket so oil filler/drain plug faces up

2 - Nut
- M 8, 25 Nm (18.4 ft lb)

3 - Mounting bracket

4 - Adjusting bolt
- M 8
- Used to tension V–belt
- V–belt, installing ⇒ Fig. 1

5 - Washer
- 10.5 x 18 x 1.6 mm

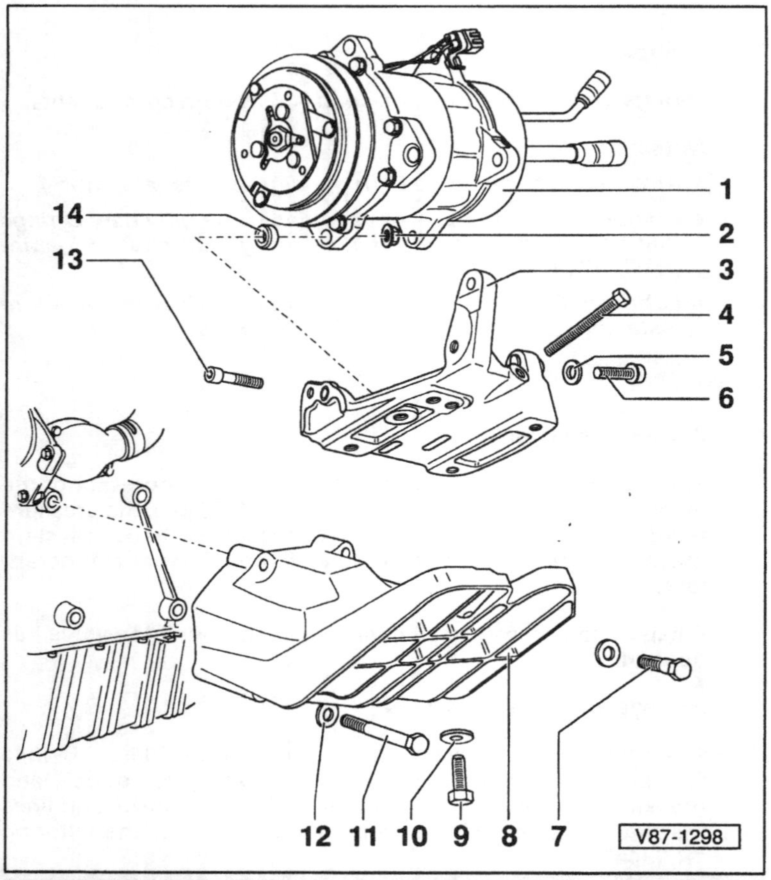

6 - Bolt (2x)
- M 10 x 35 mm
- installed through bushings
- 45 Nm (33.2 ft lb)

7 - Bolt (3x)
- M 10 x 45 mm
- 45 Nm (33.2 ft lb)

8 - Accessory drive bracket

9 - Hex bolt (4x)
- M 10 x 30
- 45 Nm (33.2 ft lb)

10 - Washer
- 10.5 x 2.6 x 4 mm

11 - Bolt (2x)
- M 10 x 80
- 45 Nm (33.2 ft lb)

12 - Washer
- 10.5 x 21 x 2 mm

13 - Bolt
- M 8 x 45 mm
- 25 Nm (18.4 ft lb)

14 - Bushing

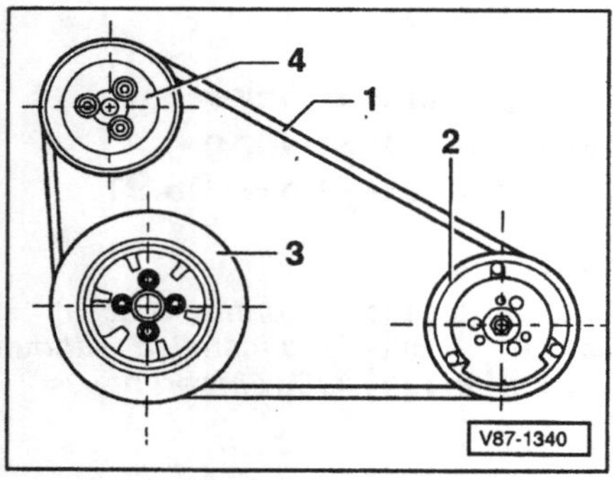

Fig. 1 V–belt and pulley arrangement

1 - V–belt, 13 mm x 1330 mm

2 - A/C compressor

3 - Engine crankshaft pulley

4 - Power steering pump

Note:

V–belt is correctly installed if it can be pushed with thumb pressure 5mm to 10 mm (0.2 in. to 0.4 in.)

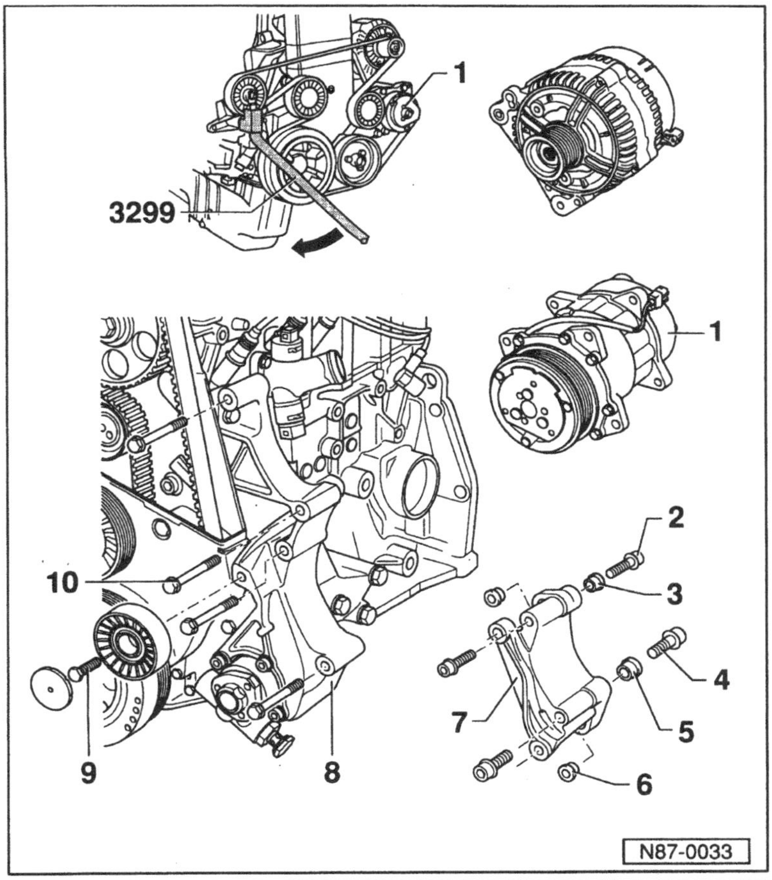

Compressor bracket, removing/installing 5 cyl. gas engine 08.91 ➤

Note:

The compressor mounting bracket can be removed and installed without discharging the A/C refrigerant system.

1 - Compressor
- ♦ Compressor mounts to bracket so oil filler/drain plug faces up

2 - Hex socket combination screw
- ♦ 2 part – 25 Nm (18 ft lb)

3 - Bushing
- ♦ M8 thread

4 - Hex socket combination screw (M10 x 35)
- ♦ 2 part 45 Nm (33 ft lb)

5 - Bushing
- ♦ M10 thread

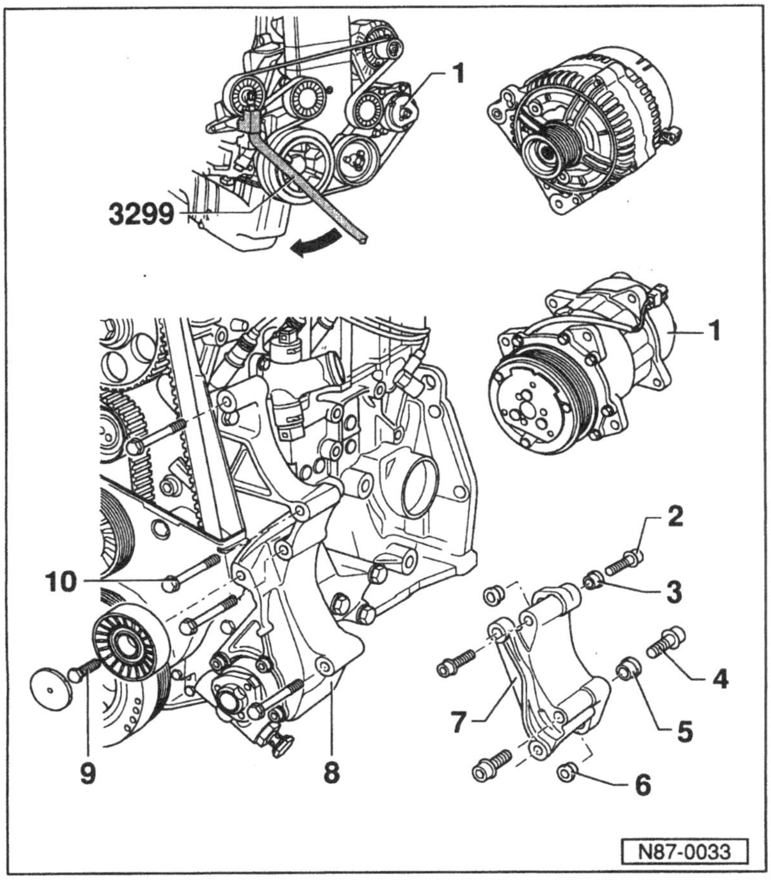

6 - Bushing
- ♦ 2 part – M8 thread

7 - Compressor bracket

8 - Generator (GEN) bracket

9 - Hex bolt M8 x 40
- ♦ 10 Nm (0.4 in. lb)

10 - Hex bolt M8 x 85 (4x)
- ♦ 20 Nm (0.8 in. lb)

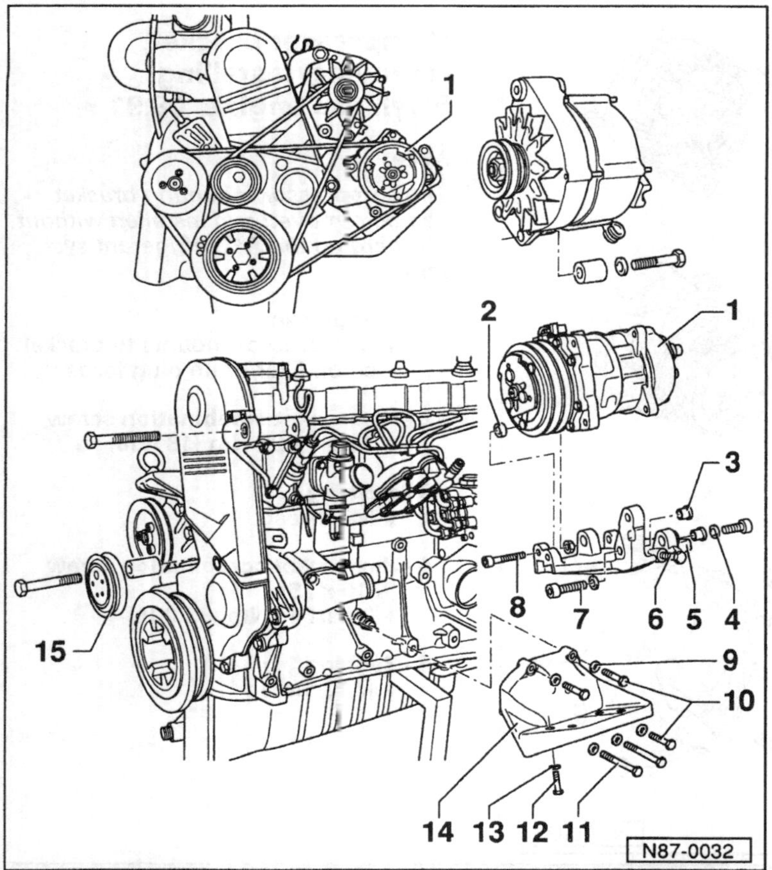

Compressor bracket, removing/installing, 5 cyl. Diesel

Note:

The compressor mounting bracket can be removed and installed without discharging the A/C refrigerant system.

1 - Compressor
- ♦ Compressor mounts to bracket so that the oil filler/drain plug faces up

2 - Bushing

3 - Bushing (2x)

4 - Washer 10.5 x 18 x 1.6 mm (3x)

5 - Compressor bracket

6 - Set screw M8 (for tensioning V–belt)

Note:

V–belt is correctly installed if it can be pressed with thumb pressure 5mm to 10mm (0.2 in.to 0.4 in.)

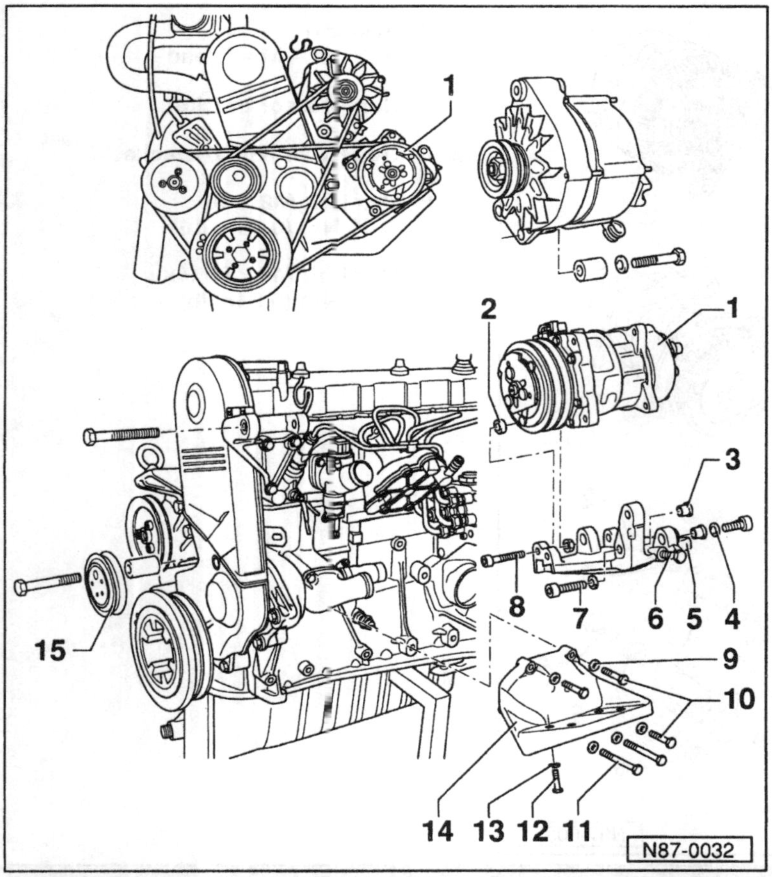

7 - Allen screw M10 x 35 (3x)
- ♦ 40 Nm (30 ft lb)

8 - Allen screw M8 x 45
- ♦ 20 Nm (15 ft lb)

9 - Washer 10.5 x 21 x 2 (6x)

10 - Hex bolt M10 x 45 (3x)
- ♦ 40 Nm (30 ft lb)

11 - Hex bolt M10 x 80 (2x)
- ♦ 40 Nm (30 ft lb)

12 - Hex bolt M10 x 30 (4x)
- ♦ 40 Nm (30 ft lb)

13 - Washer 10.5 x 26 x 4

14 - Accessory bracket

15 - Tension roller

A/C refrigerant system R-12, (Canada only, 1992 m.y.)

Safety measures

The subassemblies and lines of the air conditioner are filled with difluorodichloromethane (CF_2Cl_2) or dichlorodifluoromethane (CCL_2F_2) as a refrigerant. These refrigerants are known under the commercial name of Frigen R-12 and Freon R-12.

> ***WARNING!***
>
> ***Automotive refrigerant containing CFC's is hazardous to the earth's atmosphere. To protect our environment, use an Underwriter's Laboratory (UL) approved refrigerant recovery/recycling unit such as Kent–Moore ACR^3, or equivalent, whenever discharging an A/C system***
>
> ***- As of January 1, 1992 any person who services a motor vehicle air conditioner MUST, by law (in USA), be properly trained and certified and use approved refrigerant recycling equipment. Technicians must complete an EPA approved recycling course to be certified***
>
> ***- State and local governments may have additional requirements regarding air conditioning servicing. Always comply with state and local laws.***
>
> ***- A/C system is filled with R-12 refrigerant gas which is under pressure***
>
> ***- Always be careful that R-12 refrigerant does not come in contact with your skin***

> ***WARNING!***
>
> ***- If liquid R-12 refrigerant has come in contact with your skin or eyes:***
>
> - ***Do not rub skin or eyes***
> - ***Immediately flush with cool water for 15 minutes***
> - ***Rush to a doctor or hospital***
> - ***Do not attempt to treat yourself***
>
> ***- Always wear hand and eye protection (gloves and goggles) when working around the A/C system***
>
> ***- Work in a well ventilated area because R-12 refrigerant gas evaporates quickly and breathing may be difficult due to lack of oxygen***
>
> ***- R-12 refrigerant gas is heavier than air and will gather first in low places, like under the car***
>
> ***- Keep R-12 refrigerant containers stored below 50° C (122° F) and never drop from high places***
>
> ***- Keep R-12 refrigerant away from open flames because poisonous gas will be produced if it burns. Do not smoke when R-12 refrigerant gases are present for the same reason***
>
> ***-Electric welding near refrigerant hoses causes R-12 refrigerant to decompose from ultraviolet light. Discharge system before electric welding ⇒ page S 87-80***

WARNING!

- Do not expose any component of the A/C system to high temperatures (for example, open flames). Excessive heat will cause a pressure increase which could burst the system

- Do not steam clean condensers or evaporators. Use only cold water or compressed air

CAUTION!

R-12 and R-134a refrigerant are NOT compatible. Never add R-12 refrigerant to an R-134a system or R-134a refrigerant to an R-12 system. If the refrigerants are mixed, total system contamination will occur and compressor failure may result

- Refrigerant oils used for the R-134a system and R-12 systems are NOT compatible. Use only the specified synthetic oil (Polyalkylene Glycol/PAG) for the R-134a refrigerant system. DO NOT use R-12 system oil in an R-134a system or R-134a system oil in an R-12 system. If the refrigerant oils are mixed, system contamination will occur and compressor failure may result

-Always use separate refrigerant recovery/recycling/recharging servicing equipment for R-12 and R-134a systems. DO NOT use one piece of equipment for both R-12 and R-134a systems. The residual traces of refrigerant will contaminate and damage the equipment. Servicing equipment includes recovery/recycling/recharging unit, charging station, vacuum pump, manifold gauges, etc.

CAUTION!

DO NOT use R-12 servicing equipment on R-134a systems or R-134a equipment on R-12 systems or damage to both the vehicle A/C system and servicing equipment may result. Use only equipment designed to meet Society of Automotive Engineers (SAE) standards

R-134a and R-12 systems use different size service fittings. NEVER use adaptors to convert an R-12 fitting to R-134a size or R-134a fitting to R-12 size

R-134a and R-12 A/C components including compressor, hoses, O–rings, evaporator, condenser, receiver–drier, etc. are NOT interchangeable. Components of the R-134a system are identified by lettering (R-134a) or by a green label (or stripe). In addition a label on the upper radiator support identifies which type refrigerant is used. Use only the correct system component for each refrigerant type

- Always replace damaged and/or leaking A/C system components. Do not attempt repair by soldering or welding

- Immediately plug open connections on A/C components to prevent dirt and moisture contamination. Likewise, DO NOT remove new components from packaging until ready to install. Immediately tighten component connections after installation

CAUTION!

- DO NOT leave A/C system open for a long period of time. Prolonged exposure to the air will contaminate the system and will require the replacement of the receiver drier and possibly other system components

- Work area must be extremely clean when working on A/C system components

- Discharge A/C system before removing any A/C system component

- Keep A/C refrigerant oil container sealed to prevent moisture contamination

- Always reinstall caps over A/C service valves

- Always replace O-rings, DO NOT reuse. Lubricate new O-rings with refrigerant oil before installing. Replace O-rings only with the correct size and type

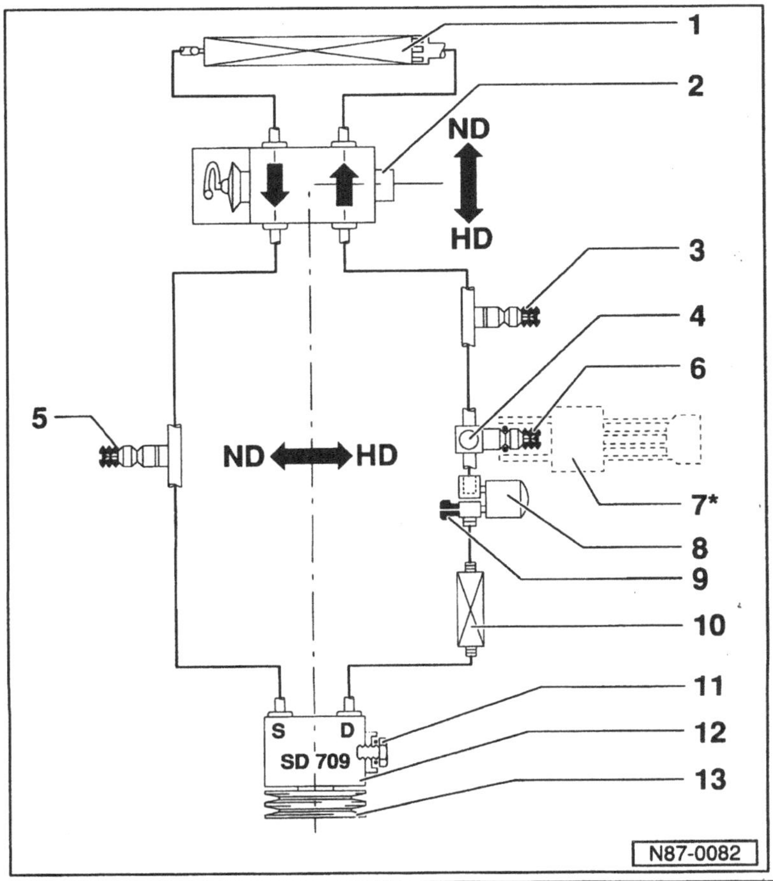

A/C refrigerant system, components

Note:

- *Illustration shows refrigerant circuit without second (rear) evaporator*

ND = Low pressure

HD = High pressure

1 - Evaporator

2 - Expansion valve
- In engine compartment, right

3 - High pressure service valve
- 3/8" x 24 UNF

4 - Sight glass

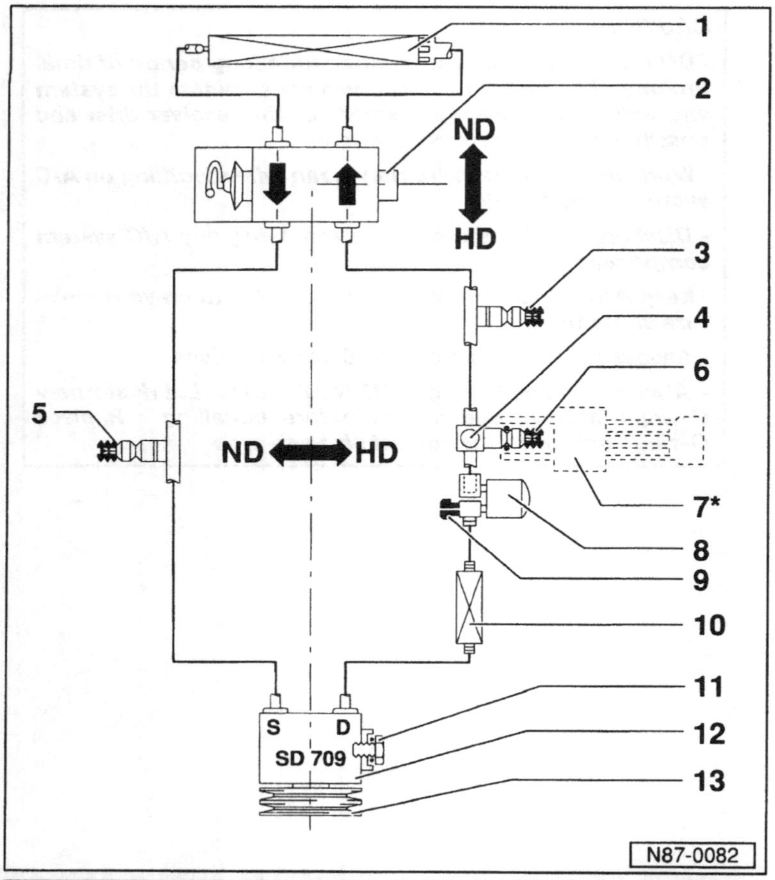

5 - **Low pressure service valve**
- 7/16" x 20 UNF

6 - **Connector/valve for A/C pressure switch -F129-**

7 - **A/C pressure switch -F129- (shown as dotted line)**
- Can be removed/installed without discharging A/C system

8 - **Receiver drier**

9 - **Pressure relief valve**

10 - **Condenser**

11 - **Oil filler/drain plug**

12 - **A/C compressor**

13 - **A/C clutch -N25-**

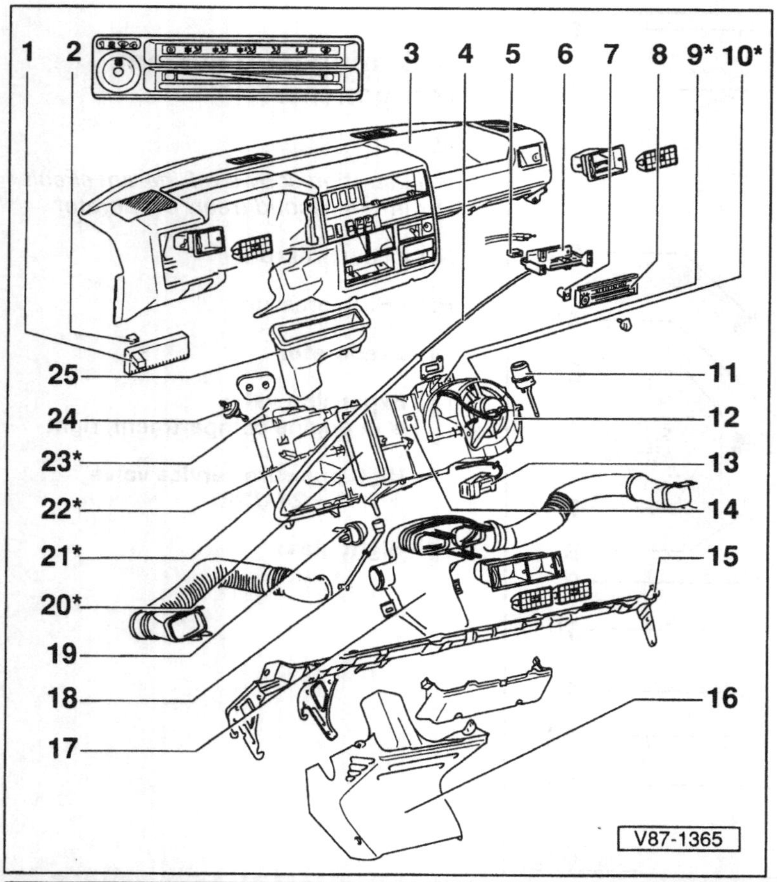

A/C controls and mechanical components, servicing

Heater/air conditioner air ducts, assembly

Note:

*System components identified with an * must only be serviced or replaced after discharging refrigerant system. Use recovery/recycling unit Kent Moore ACR³ or equivalent.*

1 - **A/C relay -J32-**

2 - **A/C system and heater fuse S23 30 Amps**

3 - **Instrument panel**
- Removing/installing ⇒ Repair Group 70

4 - **Temperature flap cable**
- Installing/adjusting ⇒ page S 87-68

5 - **Heater switch -F102-**
- For auxiliary heater B7W or D7W - not applicable to USA/ Canada
- Removing/installing ⇒ Fig. 1

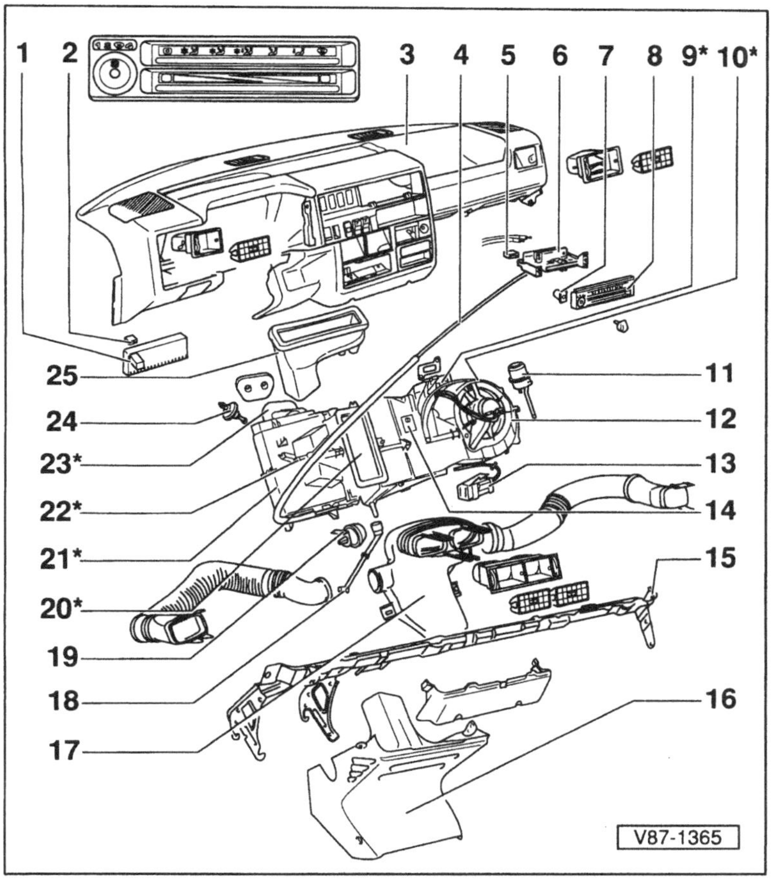

6 - A/C control head
- ♦ Includes A/C switch -E35-
- ♦ Temperature flap cable, installing/adjusting ⇒ page S 87-68
- ♦ Vacuum hose layout ⇒ page S 87-69
- ♦ removing/installing ⇒ Fig. 2

7 - Fresh air blower switch -E9-

8 - Trim plate – A/C control head -6-

9 - Expansion valve*
- ♦ Located in engine compartment
- ♦ Ensure valve is protected from rain and splash water
- ♦ Checking insulator ⇒ Fig. 3

10 - Air intake duct with fresh air/recirculation flap*

11 - Fresh air/recirculating flap vacuum servo
- ♦ Vaccum hose layout ⇒ page S 87-69
- ♦ Servo is in fresh air mode with no vacuum applied
- ♦ Removing ⇒ Fig. 3

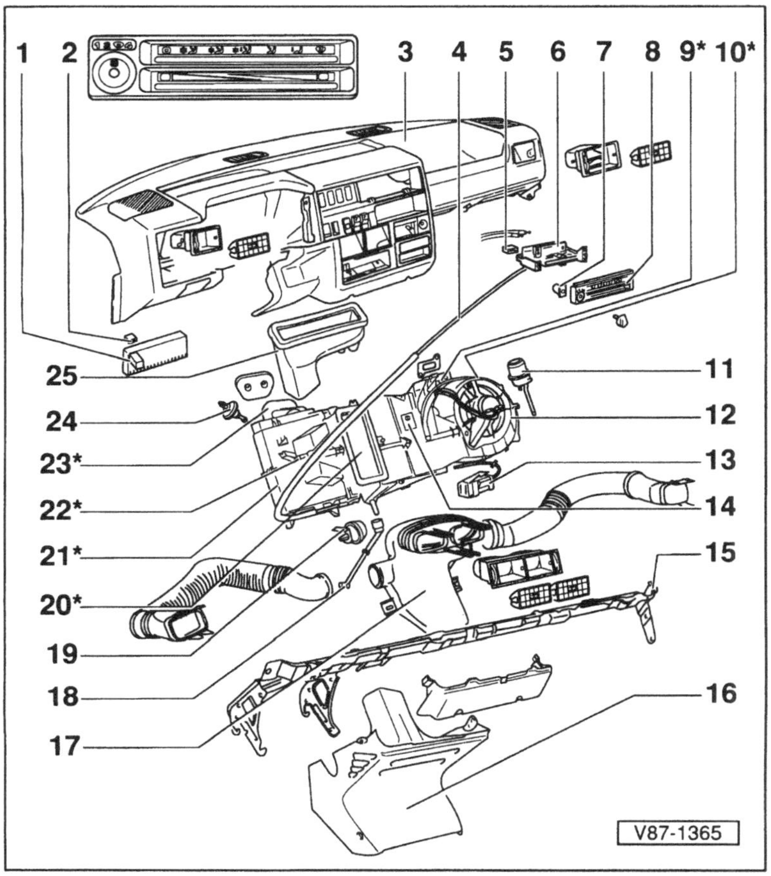

12 - Fresh air blower -V2-
- ♦ To replace, remove instrument panel and loosen right side of bracket -15-
- ♦ Apply silicone rubber sealant (locally available) to sealing surfaces

13 - Evaporator temperature switch -E33-
- ♦ Prevents evaporator from freezing
- – To replace, first remove glove box
- – Insert sensor tube of temperature switch into evaporator 330 mm (13.0 in.)
- ♦ Removing/installing ⇒ Fig. 6

14 - Fresh air blower series resistance with overheat fuse -N24-
- ♦ Replacing ⇒ Fig. 4

15 - Bracket

16 - Footwell air outlet console

17 - Center air duct
- – To remove, first remove instrument panel -3- and bracket -15-

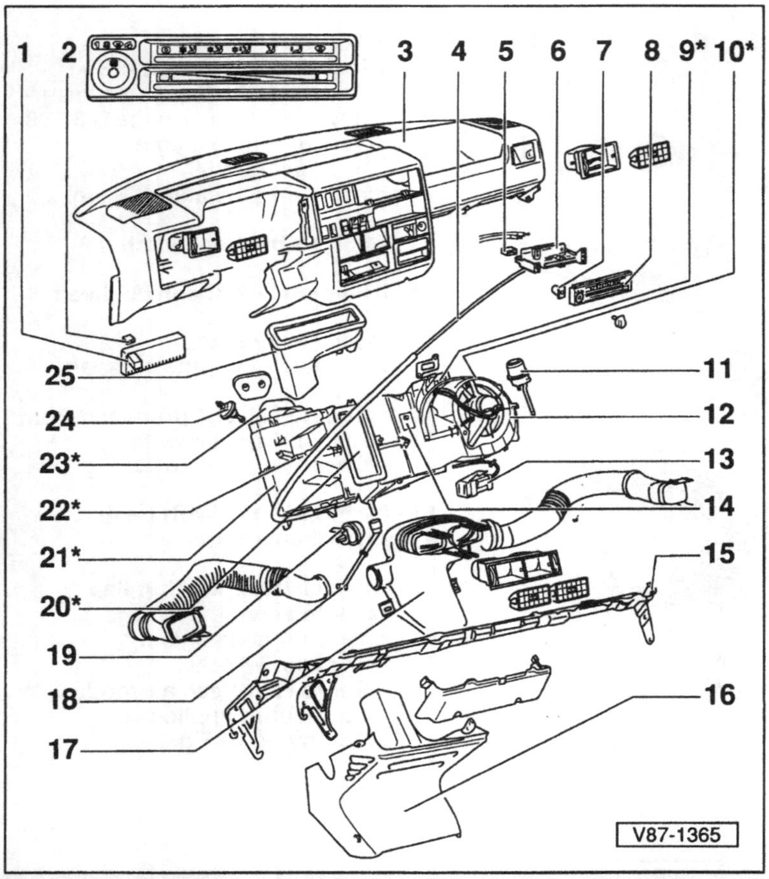

18 - Evaporator water drain hose
- Installing in passenger compartment ⇒ Fig. 5
- Installing in engine compartment ⇒ page S 87-64; item -5-

19 - Central flap vacuum servo
- Air should not flow from instrument panel vents with no vacuum applied
- Vacuum hose layout ⇒ page S 87-69

20 - Central flap*
- Not available separately as a replacement part

21 - Heater/evaporator housing*

22 - Footwell/defroster flap*
- Not available separately as a replacement part

23 - Heater core*
- Seals must be bonded all around, free of gaps
- After removing, replace coolant completely

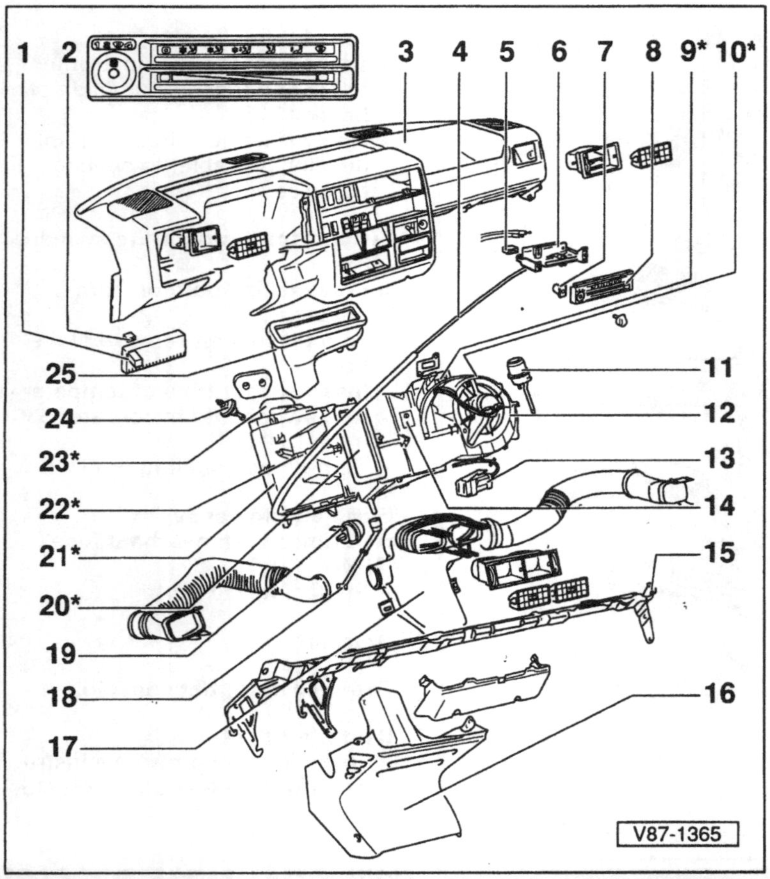

24 - Footwell/defroster flap vacuum servo
- Air flows to defroster vents with no vacuum applied

25 - Defroster duct
- To replace, first remove instrument panel ⇒ Repair Group 70

Note:

*System components identified with an * must only be serviced or replaced after discharging refrigerant system. Use recovery/recycling unit Kent Moore ACR[3] or equivalent.*

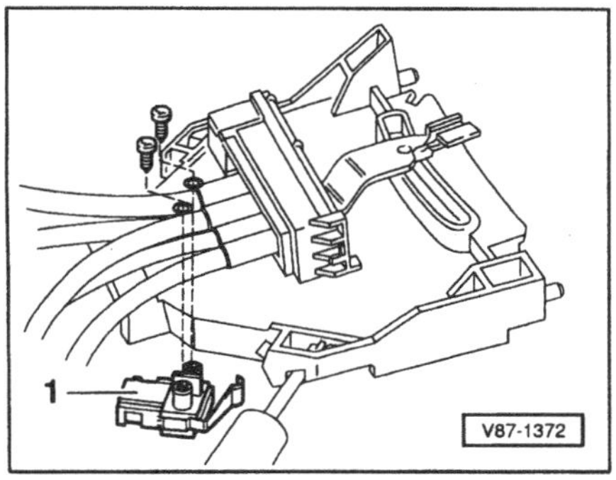

Fig. 1 Auxiliary heater switch -F102-, removing/installing

Note:

This switch is used for auxiliary heaters B7W (gas) or D7W (diesel) which are not currently available on USA/Canada vehicles.

1 - Switch for heater on -F102-

– remove switch screws

– remove switch -1-

– Reinstall in reverse sequence

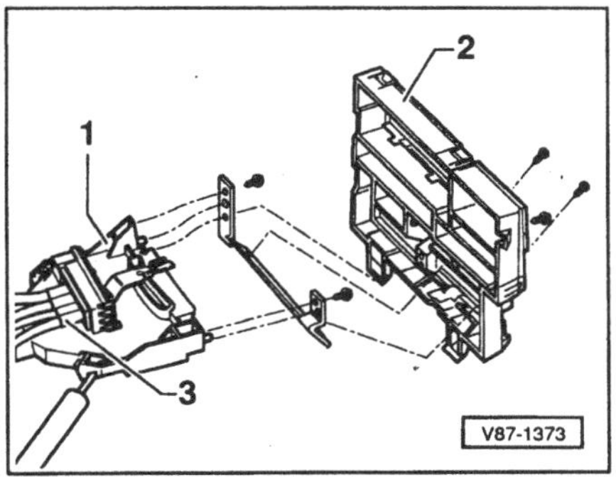

Fig. 2 A/C control head, removing/installing

1 - A/C control head

2 - Instrument panel center section (do not remove)

3 - Multi-point vacuum hose connector

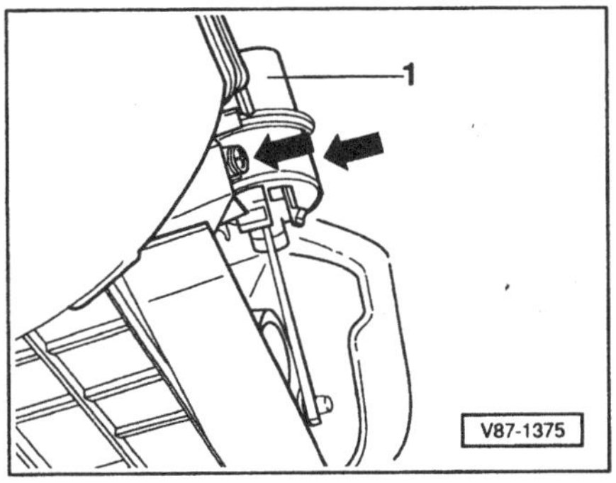

Fig. 3 Fresh air/recirculating flap vacuum servo, removing/installing

– Remove glove box

– Remove screws -arrows-

– Rotate vacuum servo -1- and remove arm from lever

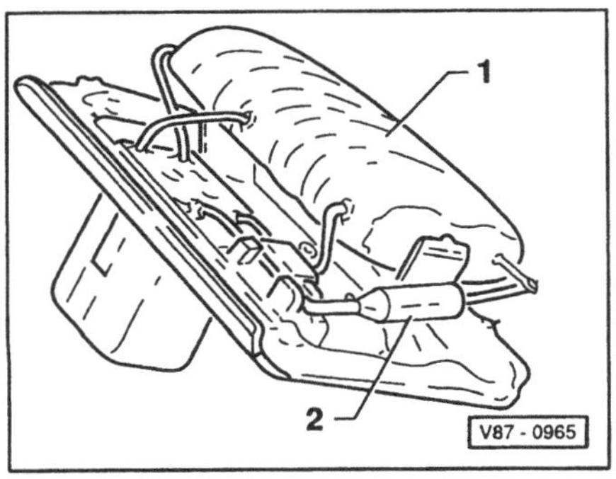

Fig. 4 Fresh air blower series resistance with overheat fuse -N24-, removing/installing

– Remove glove box

– Remove series resistance -1- with overheat fuse -2-

1 - Series resistance

2 - Overheat fuse

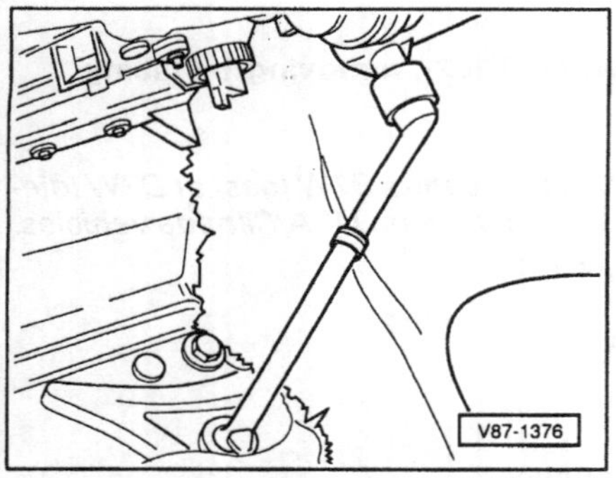

◄ **Fig. 5 Evaportator water drain hose, installing in passenger compartment**

Note:

For installing water drain hose in engine compartment, ⇒ Fig. 2

Hose must fit as shown in illustration or water will leak into passenger compartment

– Install hose so that grommet in bulkhead seals tight around hose

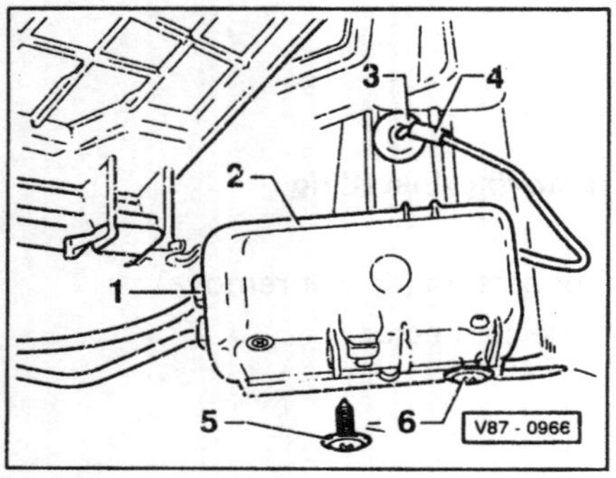

◄ **Fig. 6 Evaporator temperature switch -E33-, removing/installing**

1 - Evaporator temperature switch

- ♦ Do not bend sensor tube

2 - Temperature switch with housing

3 - Grommet

4 - Tape

- ♦ Install tape on sensor tube 330 mm (13.0 in.) from end and install sensor tube into evaporator up to tape

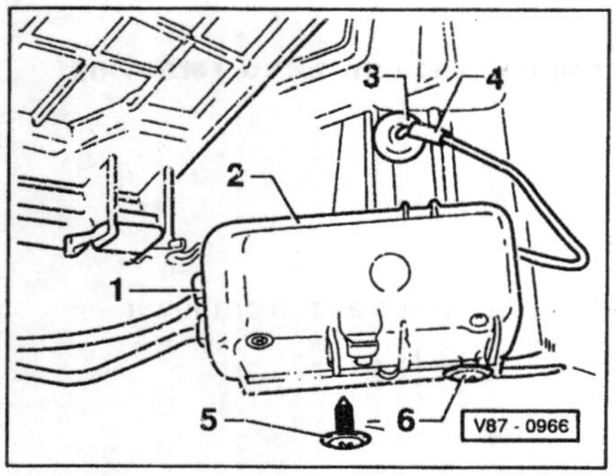

5 - Screw

- ♦ Remove screw to remove temperature switch from evaporator housing

6 - Screw

- ♦ Remove screw to remove cover from temperature switch

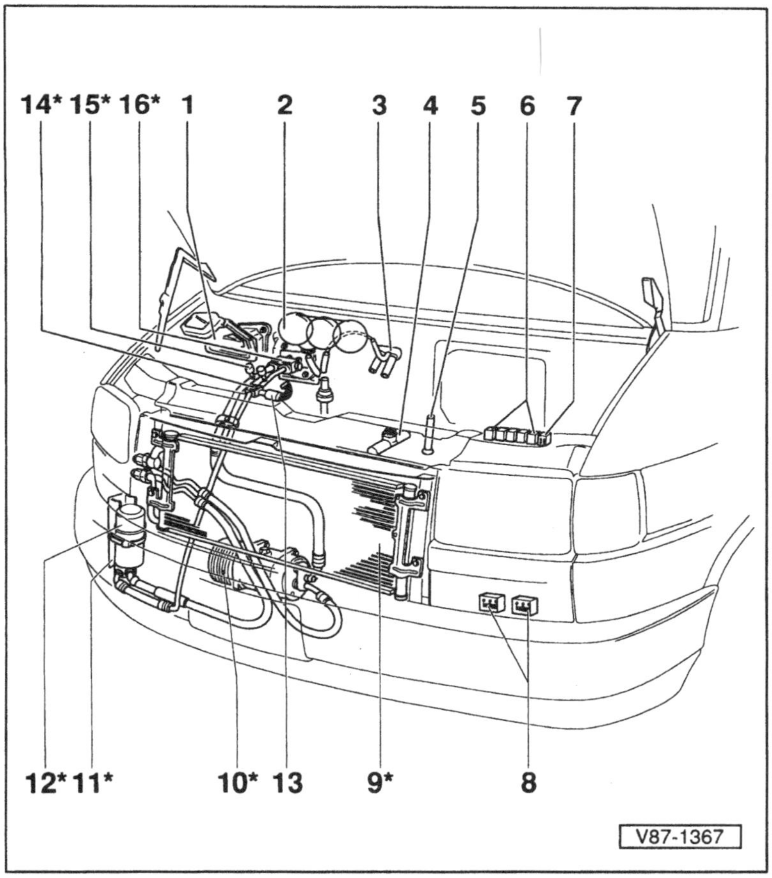

A/C refrigerant and heating system, components in engine compartment

Note:

*System components identified with an * must only be serviced or replaced after discharging refrigerant system. Use recovery/recycling unit Kent Moore ACR^3 or equivalent.*

1 - Air intake duct, rear
- Removing/installing ⇒ Repair Group 80

2 - Vacuum reservoir
- Insert vacuum hose 30 mm (1 3/16 in.) into reservoir
- Vacuum hose layout ⇒ page S 87-69

3 - Bulkhead opening for heater core and vacuum hose

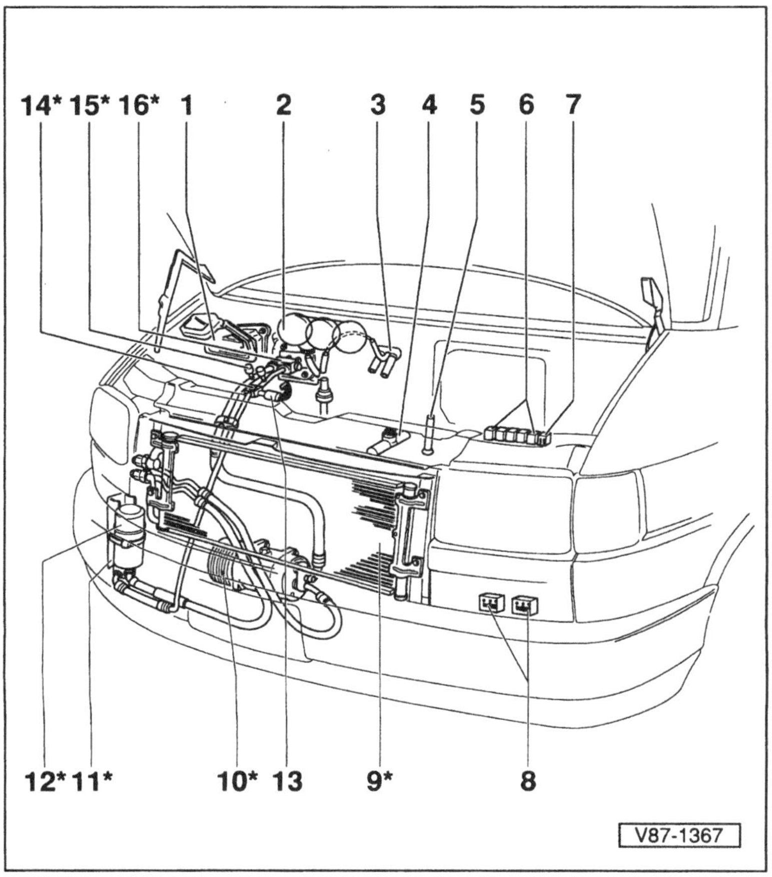

4 - A/C thermoswitch -F45-
- Switches coolant fan third speed ON at temperatures above 112°C (234°F)
- Switches coolant fan third speed OFF at temperatures below 108°C (226°F)
- Switching occurs between terminals -2- and -3- of thermoswitch
- Switches A/C clutch -N25- OFF at temperatures above 120°C (248°F)
- Switches: A/C clutch ON at temperatures below 112°C (234°F)
- Switching occurs between terminals -1- and -4- of thermoswitch
- Removing/installing, ⇒ Fig. 4

5 - Evaporator water drain hose with valve

- installing ⇒ Fig. 2
- clean any dirt, wax or undercoating from valve flap, valve flap must close

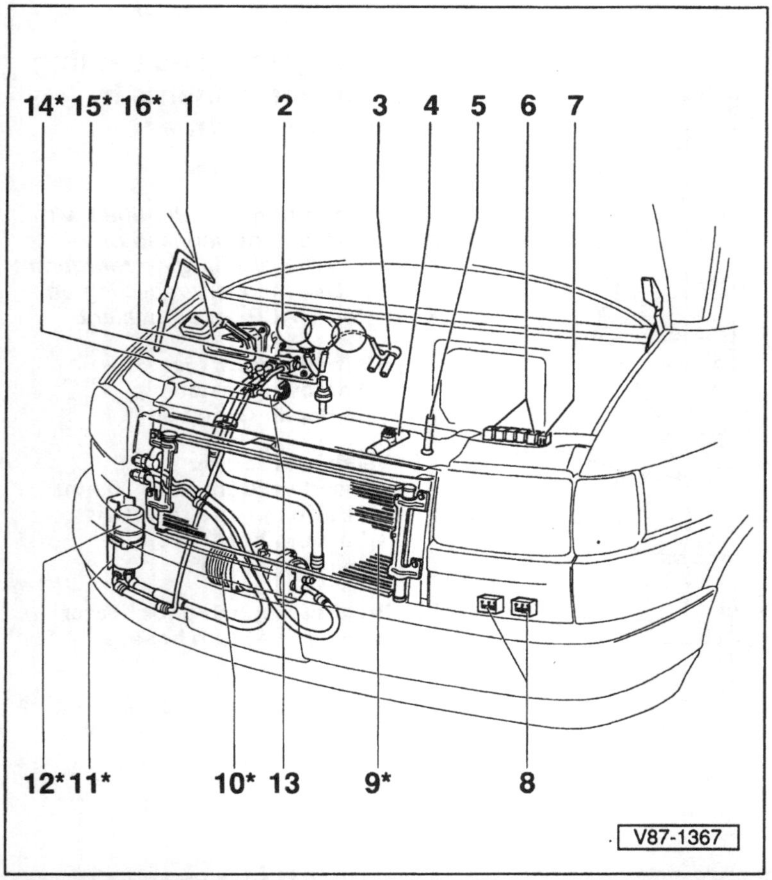

6 - Relays
- For exact relay location and wire routing, see Electrical Wiring Diagrams, Troubleshooting and Component Locations binder

7 - Individual fuses (2x) for coolant fan -S24-
- 30 amps
- 2 piece

8 - Coolant fan control (FC) series resistance -N39-

9 - Condenser*

10 - A/C clutch -N25-*

11 - Pressure relief valve*
- Protects system against excessive pressure
- If pressure relief valve has opened, check system for cause of excessive pressure, repair as necessary
- Checking ⇒ Fig. 5

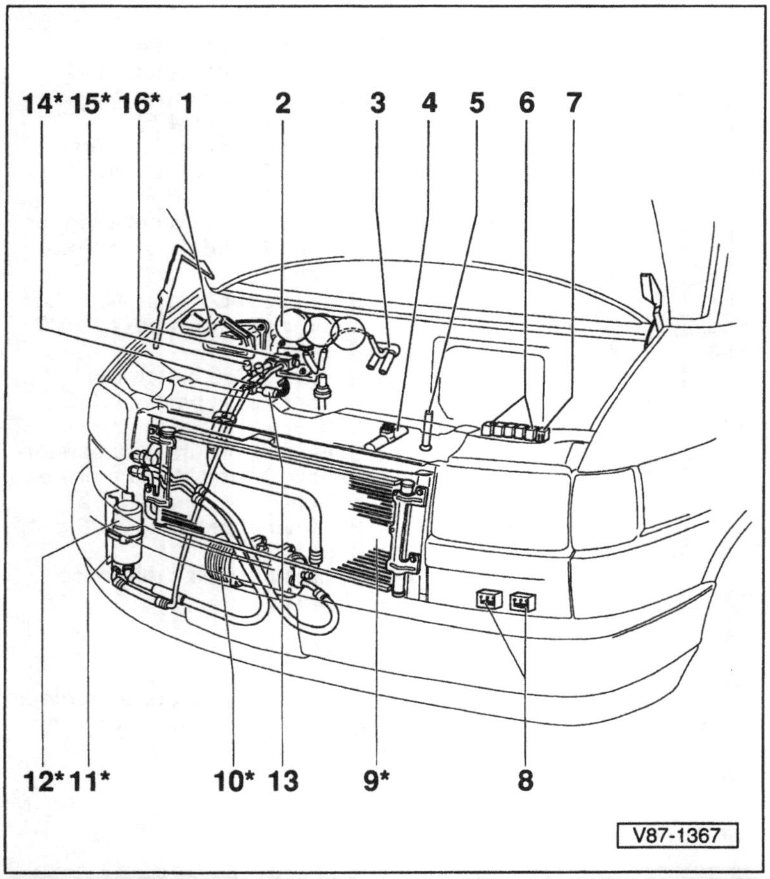

12 - Receiver drier*

13 - A/C pressure switch -F129-
- Checking ⇒ Fig. 3
- Removing/installing
  - Tightening torque: 8 Nm (70 in. lb)
  - Replace O–ring seal (note part number)

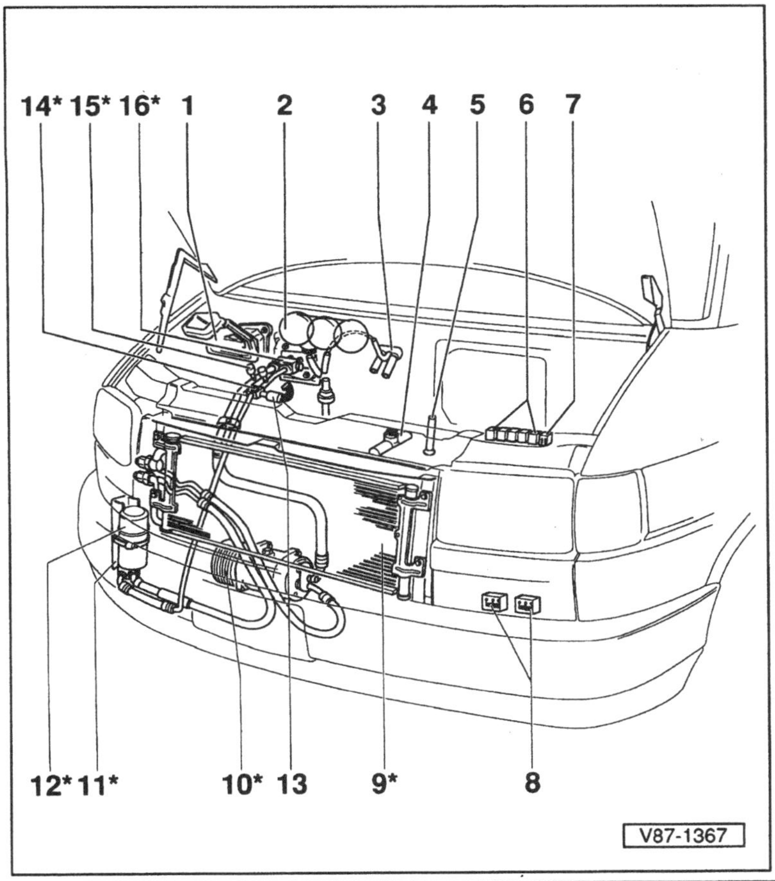

14 - Sight glass*
- To check refrigerant charge,
- Start engine
- Switch A/C ON
- Check sight glass for bubbles after A/C clutch switches ON
- ♦ If bubbles are present continuously with A/C clutch engaged, system is low on refrigerant

15 - Service valves

16 - Expansion valve*
- ♦ Ensure valve is protected from rain and splash water
- ♦ Ensure insulator is in good condition ⇒ Fig. 1

Note:

*System components identified with an * must only be serviced or replaced after discharging refrigerant system. Use Kent Moore ACR³ or equivalent.*

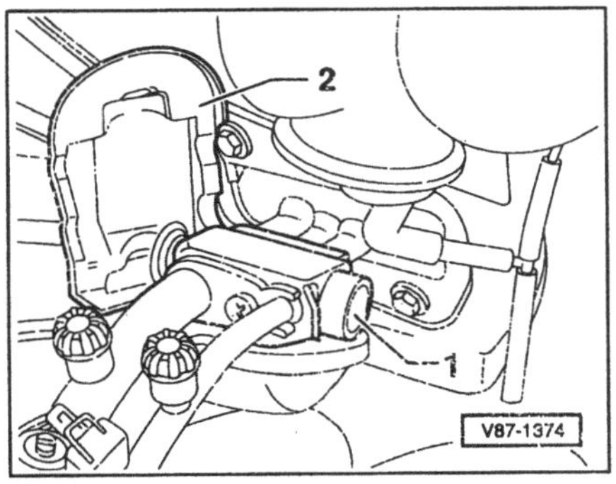

◄ **Fig. 1 Expansion valve insulator, checking**

1 - Expansion valve

2 - Insulator

The insulator protects the expansion valve against high engine compartment temperatures to prevent a decrease in A/C system cooling performance.

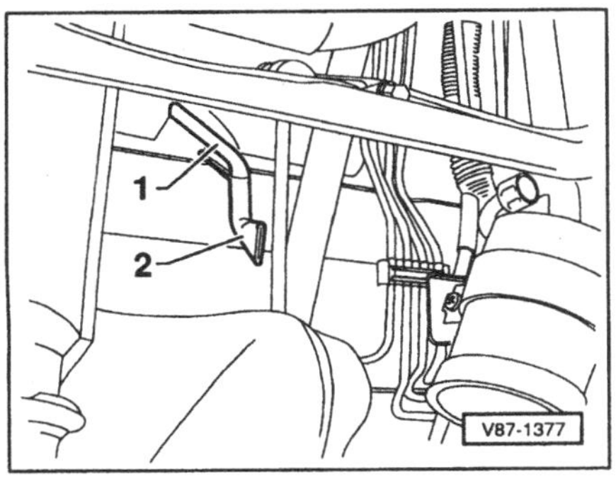

◄ **Fig. 2 Evaporator water drain hose and valve, removing/ installing**

1 - Replace water drain hose, remove engine dampener for bulkhead, glove box and console for floor outlet

2 - Push valve on by hand

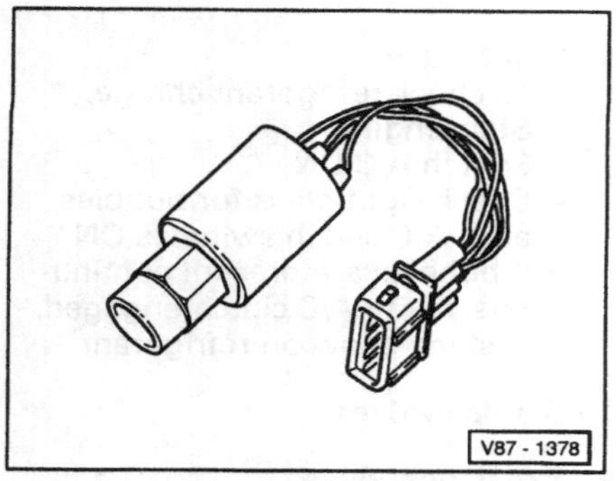

◄ **Fig. 3 A/C pressure switch -F129-, checking**

– Ensure O–ring (10.8 mm x 1.8 mm) fits properly in groove (color: black)

Functions

♦ Switches A/C clutch -N25- OFF if system pressure is too low (low refrigerant charge) – switching occurs between blue wires

♦ Switch opens (compressor OFF): below 2.0 bar (29 psi)

♦ Switch closes (compressor ON): above 3.0 bar (43 psi)

♦ Switches A/C clutch -N25- OFF if system pressure is too high – switching occurs between blue wires

♦ Switch opens (compressor OFF): above 32.0 bar (464 psi)

♦ Switch closes (compressor ON): below 24.0 bar (348 psi)

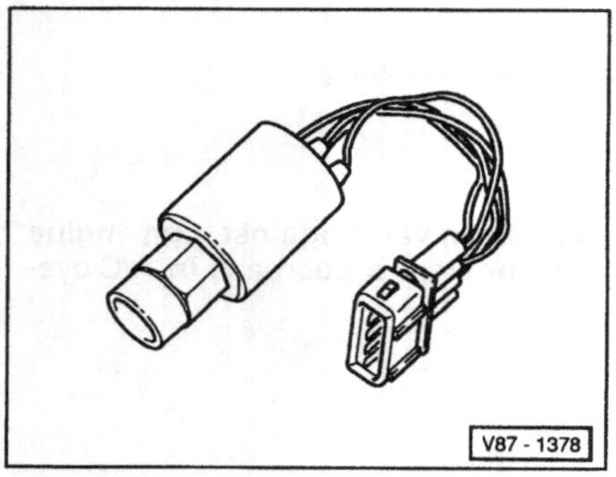

♦ Switches coolant fan -V7- to second speed when A/C system pressure increases (switching occurs between red and black wires)

♦ Switch closes (2nd speed ON): above 16.0 bar (232 psi)

♦ Switch opens (2nd speed OFF): below 12.5 bar (181 psi)

♦ To remove, remove switch from Schrader valve tightening torque: 8 Nm (71 in lb)

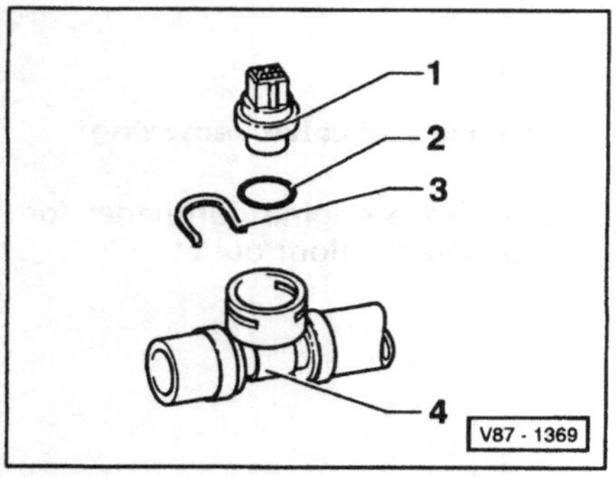

◄ **Fig. 4 A/C thermostwitch -F45-, removing/installing**

> ***CAUTION!***
> ***Relieve cooling system pressure before removing A/C thermoswitch.***

1 - A/C thermoswitch -F45-

2 - O–ring

3 - Clip

4 - Coolant hose

– Relieve cooling system pressure and drain cooling system

– Remove clip -3- and remove thermoswitch from coolant hose

– Reinstall in reverse sequence

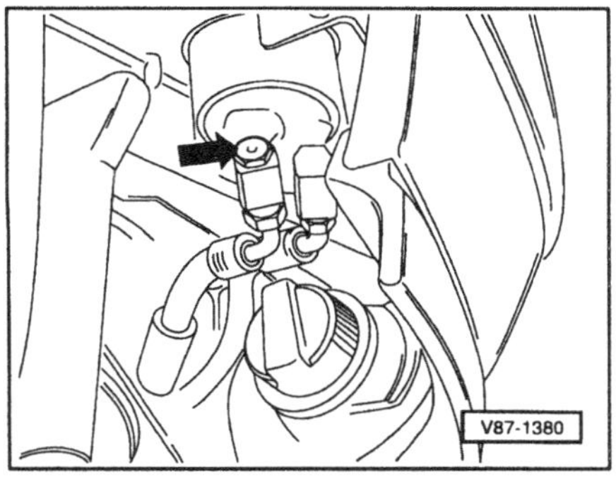

Fig. 5 A/C refrigerant system R-12, safety devices

Pressure relief valve

- ♦ Protects A/C system against excessive pressure

If pressure relief valve -arrow- has opened,

- Check system for cause of excessive pressure, repair as necessary

Temperature flap cable, installing/adjusting

- Hook temperature flap cable end into lower control lever of A/C control head
  - ♦ Temperature flap cable sleeve is blue

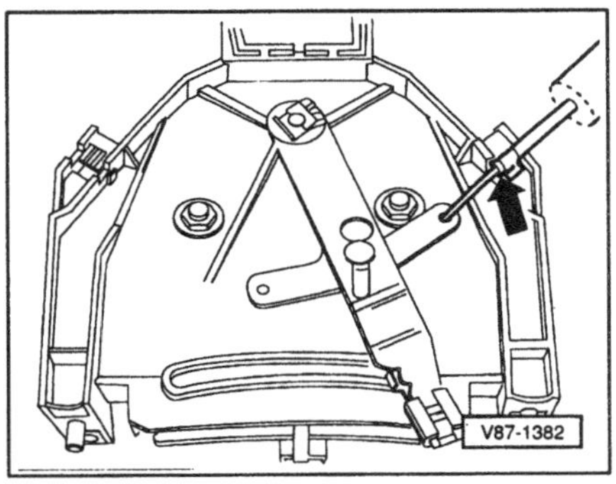

Position cable sleeve on stop of A/C control head -arrow- and secure cable with clip

- Install A/C control head
- Slide temperature control lever fully left to stop (OFF/cool)
- Connect other end of temperature flap cable to temperature flap lever
  - ♦ Temperature flap must audibly contact stops

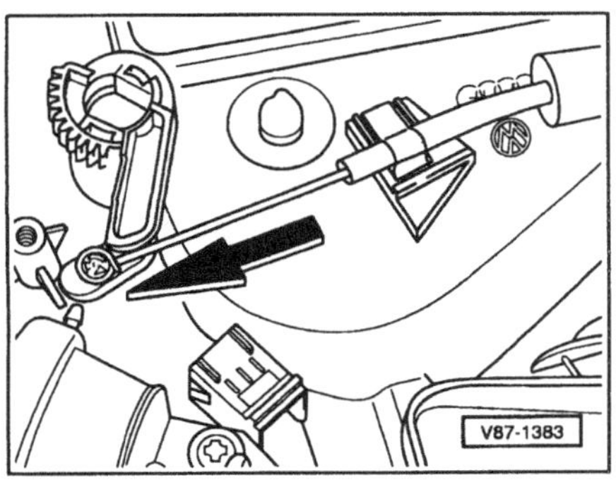

Push lever in direction of -arrow- to stop

- Hold lever in this position and install cable retaining clip

To check proper adjustment,

- Slide temperature control lever back and forth from stop to stop
  - ♦ Temperature flap must audibly contact stops

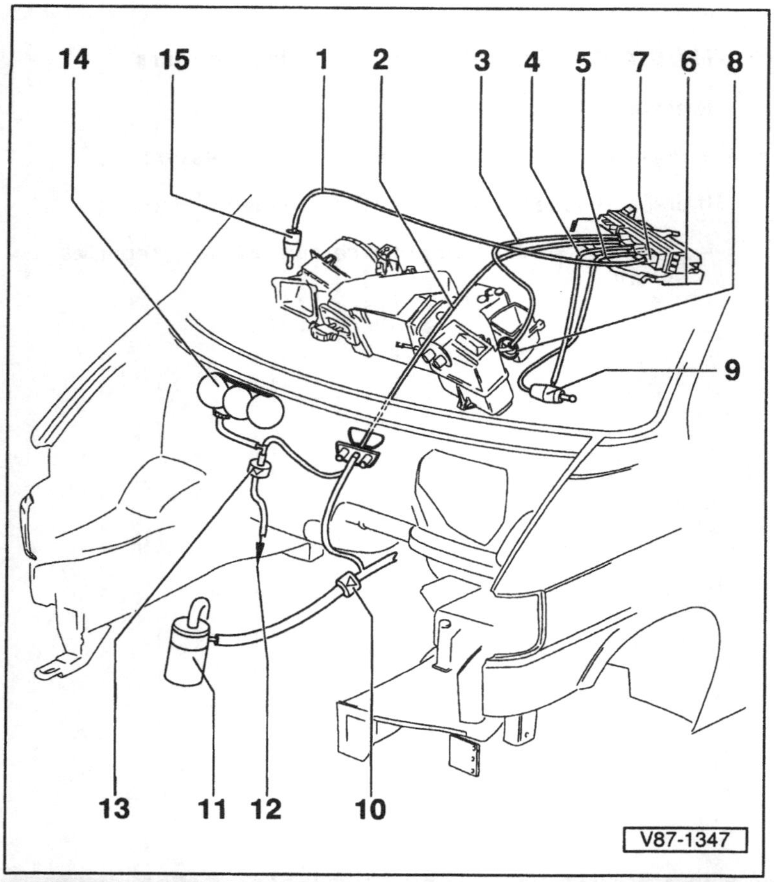

Vacuum hose, layout

Note:

Insert vacuum hose into vacuum reservoir 30 mm (1 3/16 in.). Insert vacuum hoses onto plastic connectors 15 mm (19/32 in.)

Vacuum hose colors

1 - White
- From multi point connector -7- to vacuum canister (for fresh air recirculation door)

2 - Black
- From vacuum reservoir in engine compartment to multi-point connector -7-

3 - Red
- From multi-point connector -7- to vacuum cansiter (for floor outlet/defrost

4 - Green
- From multi-point connector -7- to vacuum canister (for central flap)

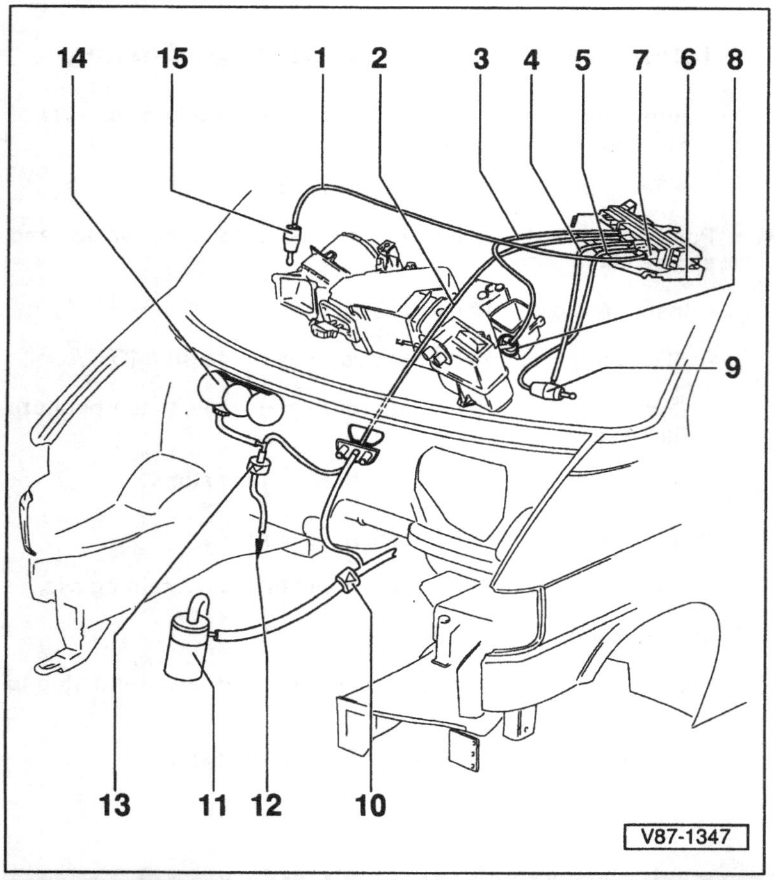

5 - Yellow
- From multi-point connector -7- to vacuum canister for central flap

Components

6 - A/C control head

7 - Vacuum hose multi–point connector

8 - Footwell/defroster flap vacuum servo

9 - Central flap vacuum servo

10 - Check valve (diesel engine only)

11 - Vacuum pump (diesel engine only)

12 - To engine vacuum supply

13 - Check valve (arrow points to vacuum source)

14 - Vacuum reservoir

15 - Fresh air/recirculating flap vacuum servo

A/C refrigerant system R-134a, (USA & Canada from 1993 m.y.)

Safety measures

The subassemblies and lines of the air conditioner are filled with Tetrafluoroethane also known as refrigerant R-134a.

refrigerant R-134a replaces the refrigerant R-12 which is reported to have harmful effects on the earth's ozone layer.

R-134a and R-12 systems are similar in design but there are very important differences in the refrigerant, lubricating oils, service equipment and A/C system components.

WARNING!

- Always use an Underwriter's Laboratory (UL) approved refrigerant recovery/recycling/recharging unit such as Kent–Moore ACR^4, or equivalent, whenever discharging an R-134a A/C system

- As of January, 1992 any person who services a motor vehicle air conditioner MUST, by law, be properly trained and certified and use approved refrigerant recycling equipment. Technicians must complete an EPA approved recycling course to be certified

- State and local governments may have additional requirements regarding air conditioning servicing. Always comply with state and local laws

WARNING!

- A/C system is filled with refrigerant gas which is under pressure

- Always be careful that refrigerant does not come in contact with your skin

- If liquid refrigerant has come in contact with your skin or eyes:

- *Do not rub skin or eyes*
- *Immediately flush with cool water for 15 minutes*
- *Rush to a doctor or hospital*
- *Do not attempt to treat yourself*

- Work in a well ventilated area because refrigerant gases are heavier than air, displace oxygen and may cause suffocation in areas of poor air circulation, like under the car

- Avoid breathing refrigerant vapors. Exposure may irritate eyes, nose and throat

- Always wear hand and eye protection (gloves and goggles) when working around the A/C system

- Keep refrigerant containers stored below 50°C (122°F) and never drop from high places. DO NOT warm refrigerant containers with an open flame. If refrigerant needs to be warmed, place bottom of tank in warm water.

- Keep refrigerant away from open flames because poisonous gas will be produced if it burns. Do not smoke when refrigerant gases are present for the same reason

WARNING!

- Electric welding near refrigerant hoses causes R-134a to decompose from ultraviolet light. Discharge system before electric welding ⇒ S 87-80

- Pressurized R-134a refrigerant in the presence of oxygen may form a combustible mixture. Never introduce compressed air into any R-134a container (full or empty), A/C component or piece of service equipment

- Do not expose any component of the A/C system to high temperatures (for example, open flames). Excessive heat will cause a pressure increase which could burst the system

- Do not steam clean condensers or evaporators. Use only cold water or compressed air

CAUTION!

- R-12 and R-134a refrigerant are NOT compatible. Never add R-12 refrigerant to an R-134a system or R-134a refrigerant to an R-12 system. If the refrigerants are mixed, total system contamination will occur and compressor failure may result

- Refrigerant oils used for the R-134a system and R-12 system are NOT compatible. Use only the specified synthetic oil (Polyalkylene Glycol/PAG) for the R-134a refrigerant system. DO NOT use R-12 system oil in an R-134a system or R-134a system oil in an R-12 system. If the refrigerant oils are mixed, system contamination will occur and compressor failure may result

- R-134a refrigerant system oil (PAG oil) absorbs moisture very rapidly. Moisture combines with the refrigerant to form acids which will damage the system. Use only the specified oil from a sealed container and ALWAYS reseal oil container immediately after use. DO NOT use oil if it has become contaminated with moisture

- Immediately plug open connections on A/C components to prevent dirt and moisture contamination. Likewise, DO NOT remove new components from packaging until ready to install. Immediately tighten component connections after installation

CAUTION!

- Always use separate refrigerant recovery/recycling/recharging servicing equipment for R-12 and R-134a systems. DO NOT use one piece of equipment for both R-12 and R-134a systems. The residual traces of refrigerant will contaminate and damage the equipment. Servicing equipment includes recovery/recycling/recharging unit, charging station, vacuum pump, manifold gauges, etc.

- DO NOT use R-12 servicing equipment on R-134a systems or R-134a equipment on R-12 systems or damage to both the vehicle A/C system and servicing equipment may result. Use only equipment designed to meet Society of Automotive Engineers (SAE) standards

- R-134a and R-12 systems use different size service fittings. NEVER use adaptors to convert an R-12 fitting to R0134a size or R-134a fitting to R-12 size

- R-134a and R-12 A/C components including compressor, hoses, O–rings, evaporator, condenser, receiver–drier, etc. are NOT interchangeable. Components of the R-134a system are identified by lettering (R-134a) or by a green label (or stripe). In addition a label on the upper radiator support identifies which type refrigerant is used. Use only the correct system component for each refrigerant type

CAUTION!

- Always replace damaged and/or leaking A/C system components. Do not attempt repair by soldering or welding

- Work area must be extremely clean when working on A/C system components

discharge A/C system before removing any A/C system component

- Always reinstall caps over A/C service valves

- Always replace O–rings, DO NOT reuse. Use only the correct size and type of O–rings specified for use with R-134a refrigerant. Lubricate O–ring with refrigerant oil before installing

A/C refrigerant (R-134a), properties

Commercial designation

The following designations are used and appear on component labels and factor tags:

- R-134a
- Tetrafluoroethane
- $CH_2F\ CF_3$
- H–KW 134a
- SUVA® 134a
- ARCTON® 134a

Color

refrigerant R-134a is colorless and is invisible as a gas. R-134a when viewed through the sight glass may appear milky due to the mixture of refrigerant and lubricating oil (PAG oil).

Temperature/pressure relationship

Similar to R-12 refrigerant, R-134a in an enclosed container will have a specific temperature/pressure relationship.

Temperature in °C (°F)		Pressure in bar (psi)	
–30	(–22)	0.0	(0.0)
–20	(–4)	0.3	(4.4)
–10	(14)	1.0	(14.5)
0	(32)	1.9	(27.5)
10	(50)	3.1	(45.0)
20	(68)	4.7	(68.2)
30	(86)	6.7	(97.2)
40	(104)	9.1	(132.0)
50	(122)	12.2	(177.0)
60	(140)	15.8	(229.2)
70	(158)	20.2	(293.0)

Refrigerant oil

A special Polyalkylene Glycol (PAG) synthetic oil is used in R-134a systems. This oil is NOT compatible with mineral based oils used in R-12 systems. See safety measures ⇒ page S 87-1, for Warnings and Cautions regarding proper oil usage.

Airborne properties

Escaped refrigerant gases are heavier than air and will gather first in low places, like under the car. R-134a refrigerant gas displaces oxygen and may cause suffocation in low areas of poor air circulation.

refrigerant R-134a is not poisonous in any state (liquid or gas) and is safe when used properly.

Affects on plastic

refrigerant R-134a will deteriorate some plastics. Therefore, when making system repairs, use only genuine VW replacement parts which are specified for use with R-134a refrigerant.

Affects on metal

In its pure state, refrigerant R-134a is chemically stable and will not attack iron, copper, brass or aluminum. However, the mixture of R-134a and PAG oil may deteriorate certain metals (copper). Therefore, when making system repairs, use only genuine VW replacement parts which are specified for use with R-134a refrigerant.

Water solubility

Liquid R-134a refrigerant will absorb only very minute quantities of moisture. However, R-134a vapor can absorb large amounts of moisture.

Flammability

R-134a refrigerant is not flammable.

Refrigerant containers

DO NOT exceed maximum rated capacity of refrigerant container.

Leak detection

Use halogen leak detector Hitec HI400A–TEL or equivalent to check for R-134a system leaks. This tool can also be used to detect leaks in R-12 systems. Many currently available R-12 leak detectors cannot detect R-134a refrigerant leaks.

R-134a refrigerant oil

A special Polyalkylene Glycol (PAG) synthetic oil is used in R-134a systems. This oil is NOT compatible with mineral based oils used in R-12 systems.

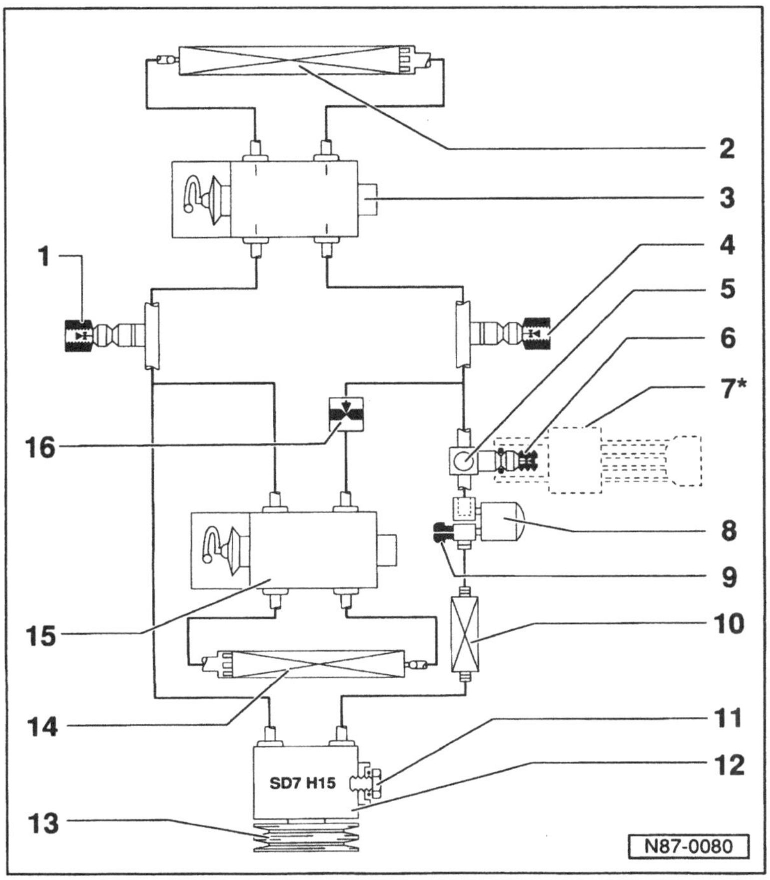

A/C refrigerant system (R-134a), components

1 - **Low pressure service valve**
   - For charging/discharging

2 - **Evaporator**

3 - **Expansion valve**
   - In engine compartment, right

4 - **High pressure service valve**
   - For charging/discharging

5 - **Sight glass**

6 - **Connector and valve for A/C pressure switch -F129-**

7 - **A/C pressure switch -F129-**
   - Can be removed/installed without discharging system

8 - **Receiver dryer**

9 - **Pressure relief valve**

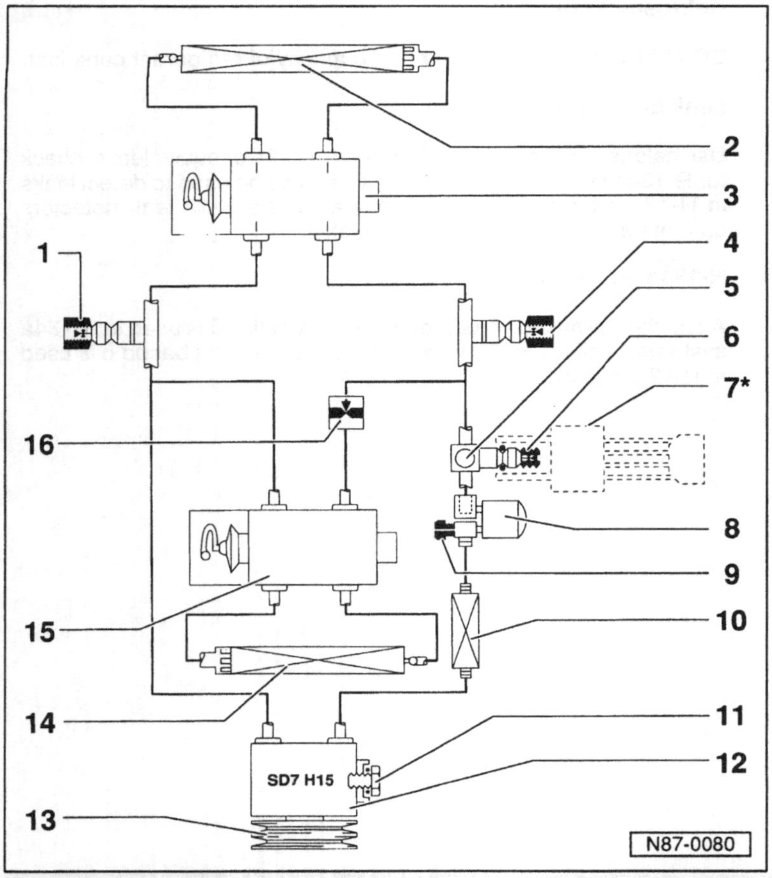

10 - Condersor

11 - Oil drain plug

12 - A/C compressor

13 - A/C clutch -N25-

14 - Second evaporator (rear)
- ♦ Not installed on all vehicles

15 - Expansion valve for second evaporator
- ♦ Not installed on all vehicles

16 - A/C refrigerant shut-off valve -N43-
- ♦ For second evaporator
- ♦ Not installed on all vehicles

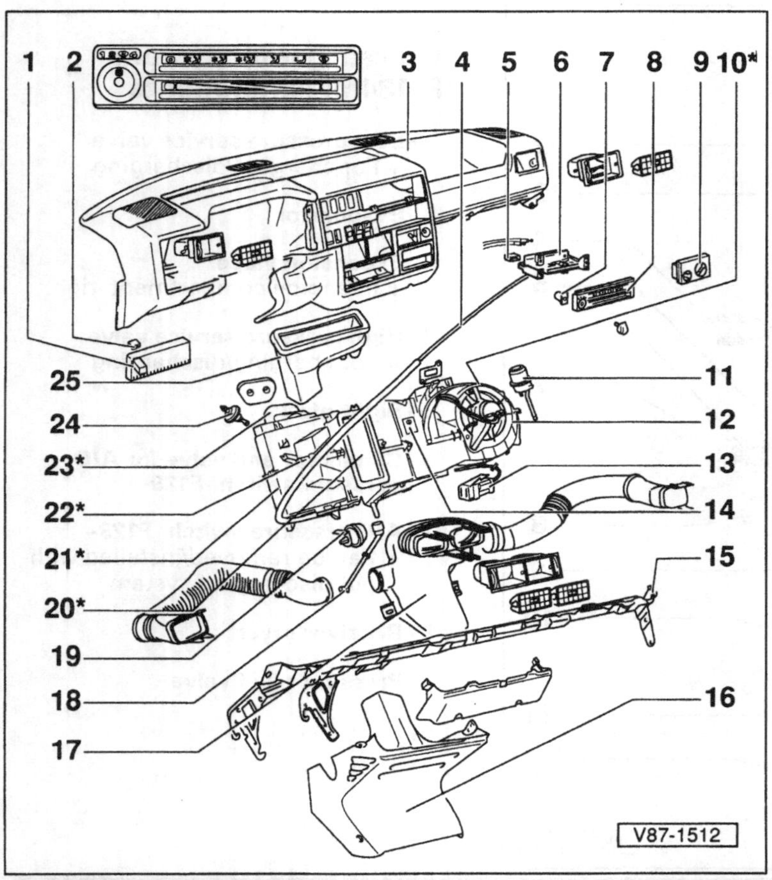

A/C controls and mechanical components, servicing

Heater/air conditioner air ducts, assembly

Note:

*System components identified with an * must only be serviced or replaced after discharging refrigerant system. Use Kent Moore ACR4 or equivalent.*

1 - A/C relay -J32-

2 - A/C system and heater fuse, -S23-
- ♦ 30 amps

3 - Instrument panel
- ♦ Removing/installing ⇒ Repair Group 70

4 - Temperature flap cable
- ♦ Installing/adjusting ⇒ page S 87-68

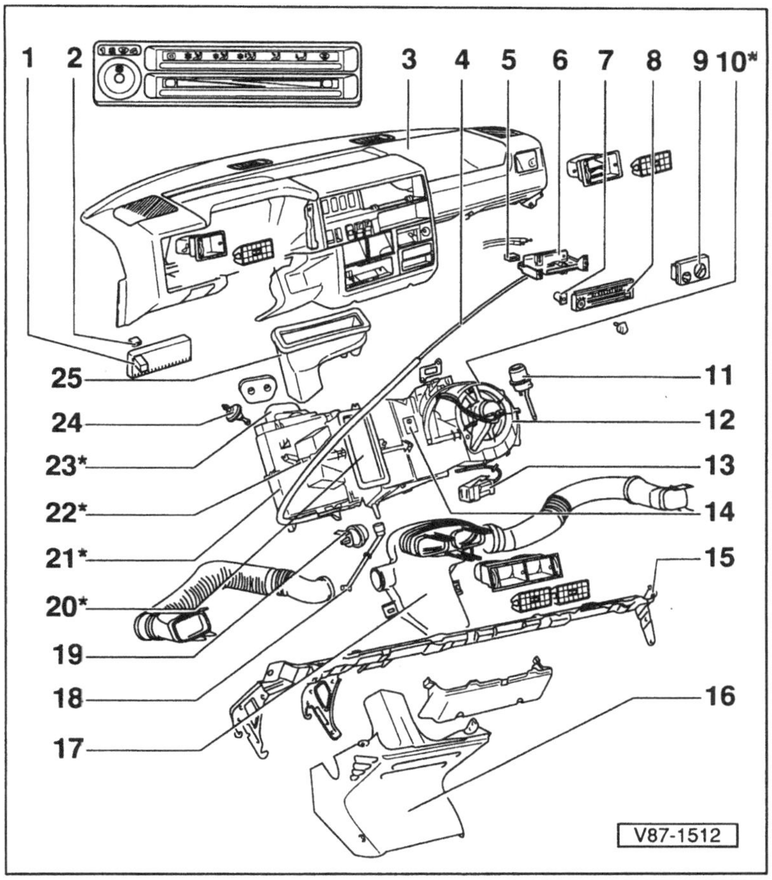

5 - **Heater switch -F102-**
- For auxiliary heater B7W or D7W – not currently applicable to USA/Canada
- Removing and installing ⇒ Fig. 1

6 - **A/C control head**
- Includes A/C switch -E35-
- Temperature flap cable, installing/adjusting ⇒ page S 87-68
- Vacuum hose layout ⇒ page S 87-69
- removing/installing ⇒ Fig. 2

7 - **Fresh air blower switch -E9-**

8 - **Trim plate**

9 - **Second evaporator blower switch -E30-**
- Only on vehicles with second (rear) evaporator
- To remove, carefully pry off with screwdriver

10 - **Air intake duct with fresh air/recirculation flap***

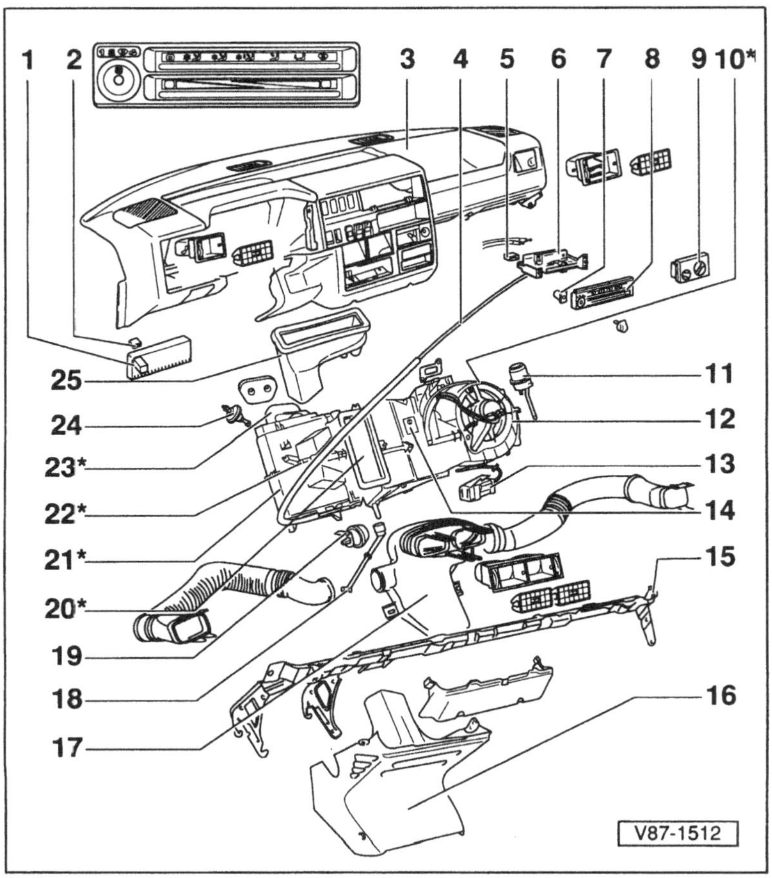

11 - **Fresh air/recirculating flap vacuum servo***
- Removing ⇒ Fig. 3
- Fresh air vent operation
- Vacuum hose layout ⇒ page S 87-69

12 - **Fresh air blower -V2-**
- To replace, remove instrument panel and loosen right side of bracket -15-
- Apply silicone rubber sealant to sealing surfaces

13 - **Evaporator temperature switch -E33-**
- Prevents evaporator from freezing
- Opens at temperatures beginning at –2°C (28°F)
- Insert sensor tube of temperature switch into evaporator 330 mm (13.0 in.)
- Removing/installing ⇒ Fig. 6

14 - **Fresh air blower series resistance with overheat fuse -N24-**
- Removing/installing ⇒ Fig. 4

15 - **Bracket**

16 - **Footwell air outlet console**

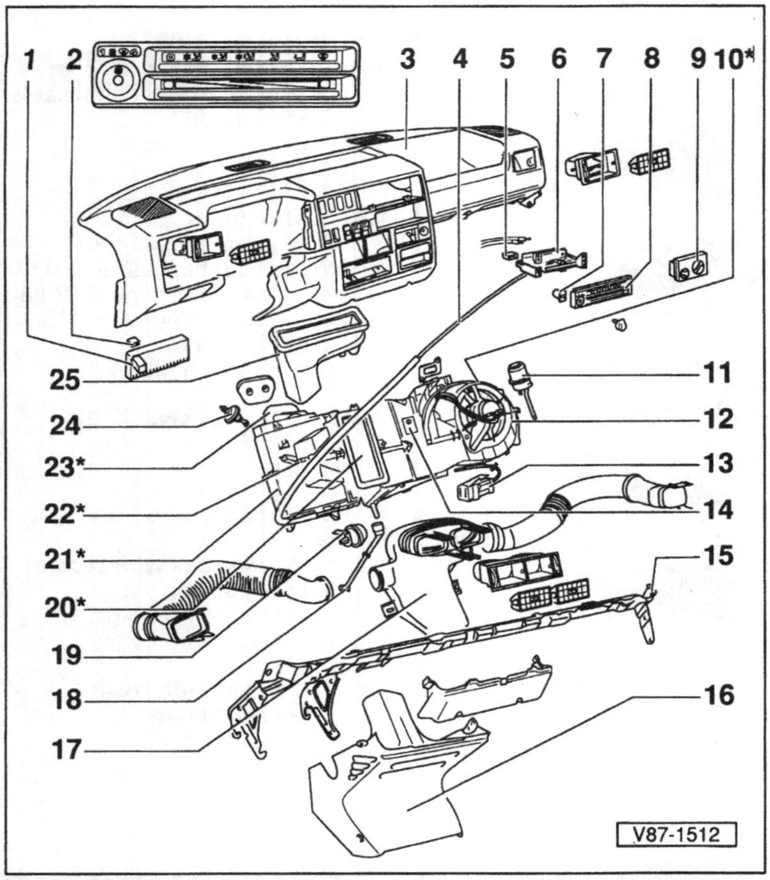

17 - Center air duct

- To remove, first remove instrument panel -3- and bracket -15-

18 - Evaporator water drain hose

- Installing in passenger compartment ⇒ page S 87-59
- Installing in engine compartment ⇒ page S 87-64

19 - Central flap vacuum servo

- Vacuum hose layout ⇒ page S 87-69
- Directs air flow to outlets of closed instrument panel

20 - Central flap*

- Not available separately as a replacement part

21 - Heater/evaporator housing*

22 - Footwell/defroster flap *

- Not available separately as a replacement part

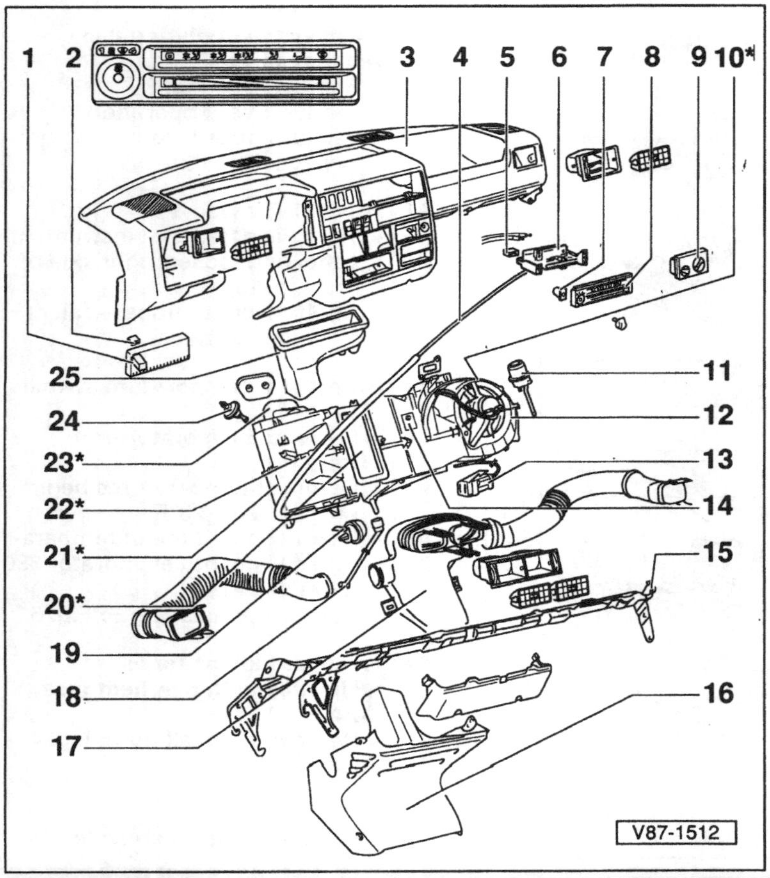

23 - Heater core*

- Seals must be bonded all around without gaps

24 - Footwell/defroster flap vacuum servo

- Duct: guides airflow to windshield

25 - Defroster duct

- To replace, first remove instrument panel

Note:

*System components identified with an * must only be serviced or replaced after discharging refrigerant system. Use Kent Moore ACR4 or equivalent.*

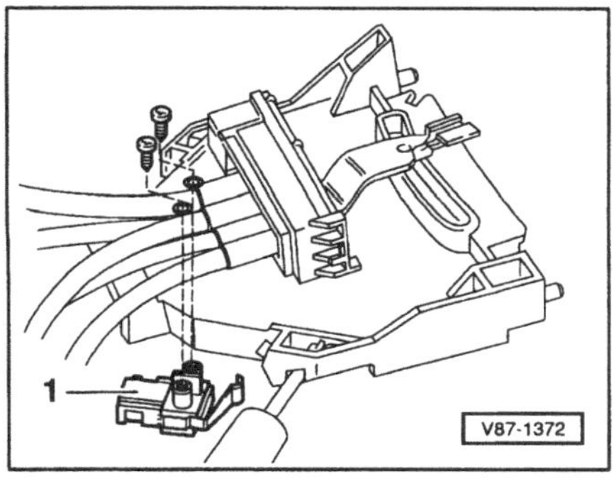

Fig. 1 Auxiliary heater switch -F102-, removing/installing

Note:

This switch is used for auxiliary heaters B7W (gas) or D7W (diesel) which are not currently available on USA/Canada vehicles.

1 - Switch for heater on -F102-

- remove switch screws
- remove switch -1-
- Reinstall in reverse sequence

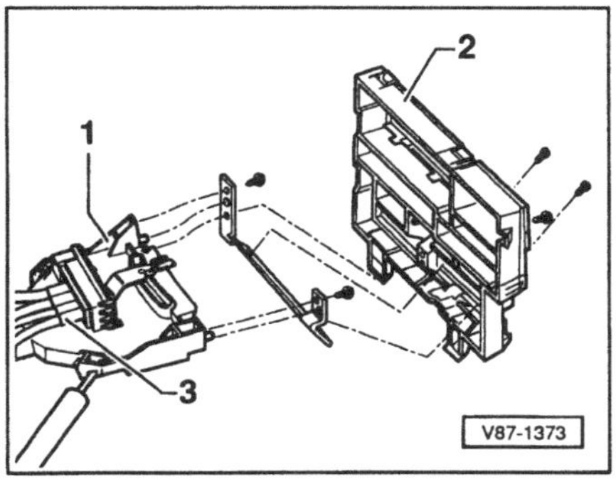

Fig. 2 A/C control head, removing/installing

1 - A/C control head

2 - Instrument panel center section (do not remove)

3 - Multi–point vacuum hose connector

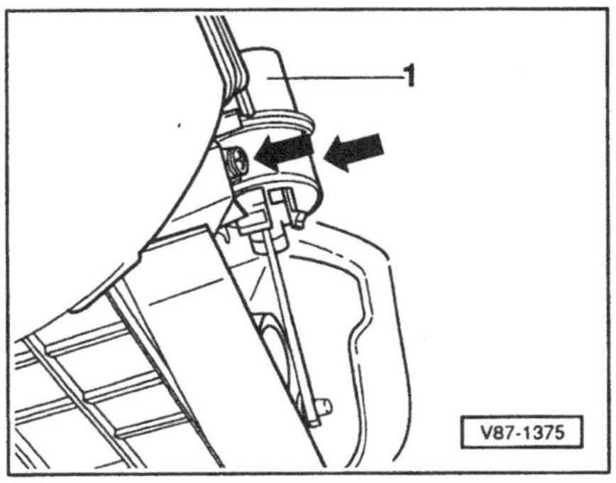

Fig. 3 Fresh air/recirculating flap vacuum servo, removing/installing

- remove glove box
- remove screws -arrows-
- rotate vacuum servo -1- and remove arm from lever

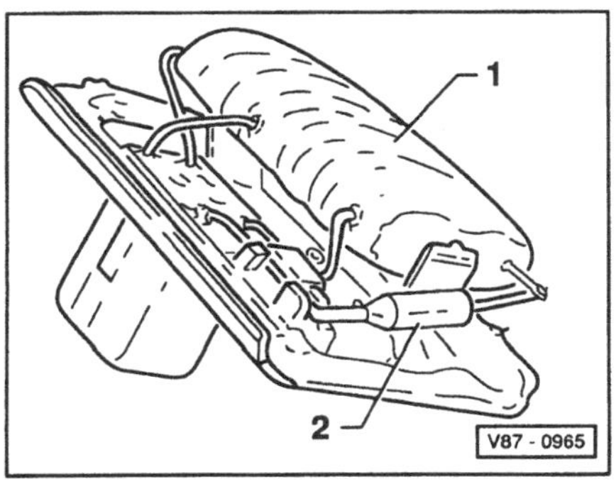

Fig. 4 Fresh air blower series resistance with overheat fuse -N24-, removing/installing

1 - Series resistor

2 - Overheat fuse

- remove glove box

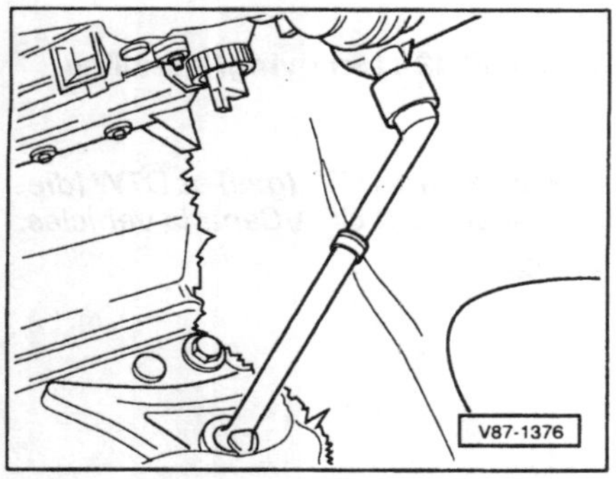

Fig. 5 Evaporator water drain hose, installing in passenger compartment

Note:

Installing water drain hose in engine compartment
⇒ page S 87-64

Hose must fit as shown in illustration or water will leak into passenger compartment.

– install hose so that grommet seals tightly around hose

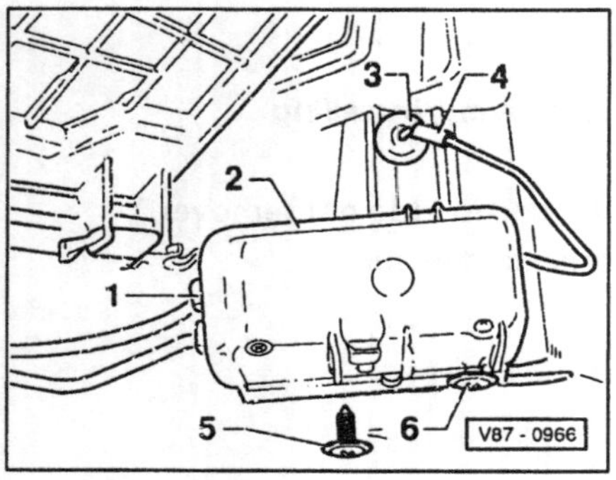

Fig. 6 Evaporator temp. switch -E33-, removing/installing

1 - Evaporator temperature switch

– do not bend sensor tube

2 - Cover

3 - Grommet

4 - Tape

– install tape on sensor tube 330 mm (13.0 in.) from end and sensor tube into evaporator up to tape

5 - Screw

6 - Screw

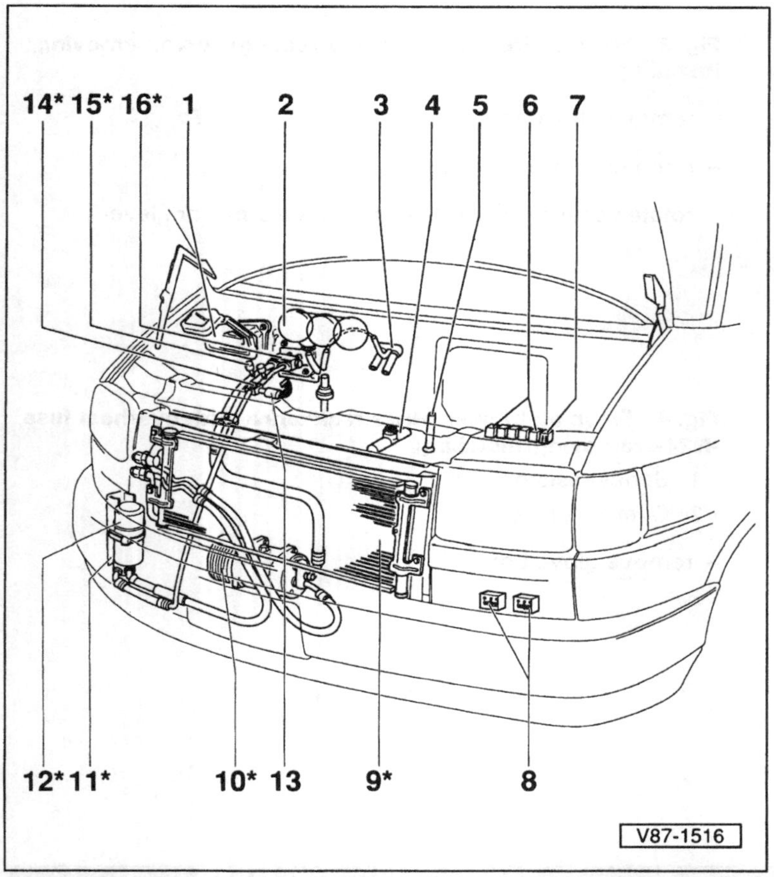

A/C refrigerant (R-134a) and heating system, components in engine compartment

Note:

*System components identified with an * must only be serviced or replaced after discharging refrigerant system. Use Kent Moore ACR4 or equivalent.*

1 - Air intake duct, rear
- removing/installing
  ⇒ Repair Group 80 item -4-

2 - Vacuum reservoir
- Insert vacuum hose 30 mm (1 3/16 in.) into reservoir
- Vacuum hose layout
  ⇒ page S 87-69

3 - Bulkhead opening for heater core and vacuum hose

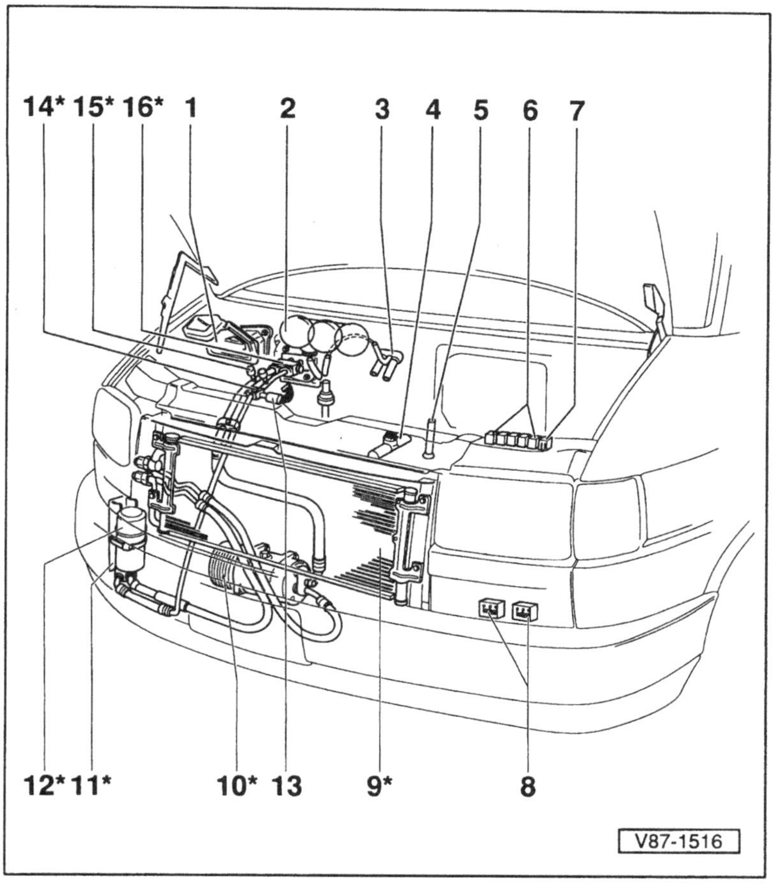

4 - **A/C thermoswitch -F45-**
- ♦ Switching occurs between terminals -2- and -3- of thermoswitch
- ♦ Thermoswitch switches the coolant fan when coolant temperature increases, to third speed (on at 112°C (234°F), off at 108°C (226°F)
- ♦ Switching occurs between terminals -2- and -3- of thermoswitch
- ♦ Thermoswitch also switches the A/C clutch -N25- to a higher coolant temperature (off at 120°C (248°F), on at 112°C (234°F)
- ♦ removing/installing ⇒ Fig. 4

5 - **Evaporator water drain hose with valve**
- ♦ Removing/installing ⇒ Fig. 2

6 - **Relays**
- ♦ For exact relay location and wire routing, see Electrical Wiring Diagrams

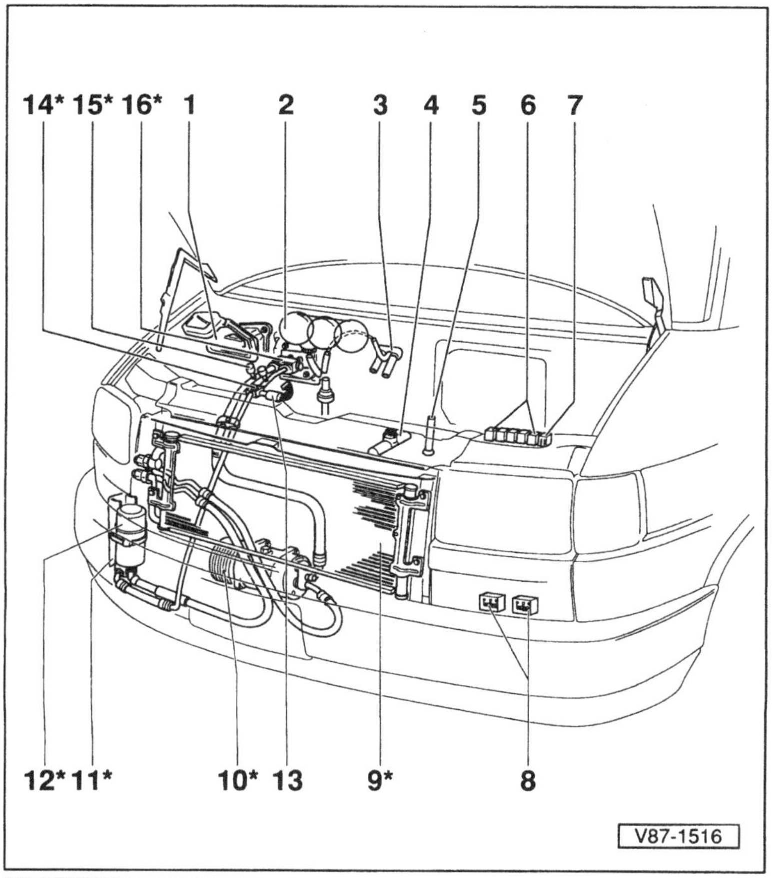

7 - **Individual fuses for coolant fan -S42- (2x)**

8 - **Coolant Fan Control (FC) series resistance -N39-**

9 - **Condenser ***

10 - **A/C clutch -N25-***

11 - **Pressure relief valve***
- ♦ Protects refrigerant circuit from excessive pressure
- ♦ Checking ⇒ Fig. 5

12 - **Receiver drier***

13 - **A/C pressure switch -F129-**
- ♦ Checking ⇒ Fig. 3
- ♦ Removing/installing:
  - – Tightening torque 8 Nm
  - – Replace o–ring seal (note part mumber)

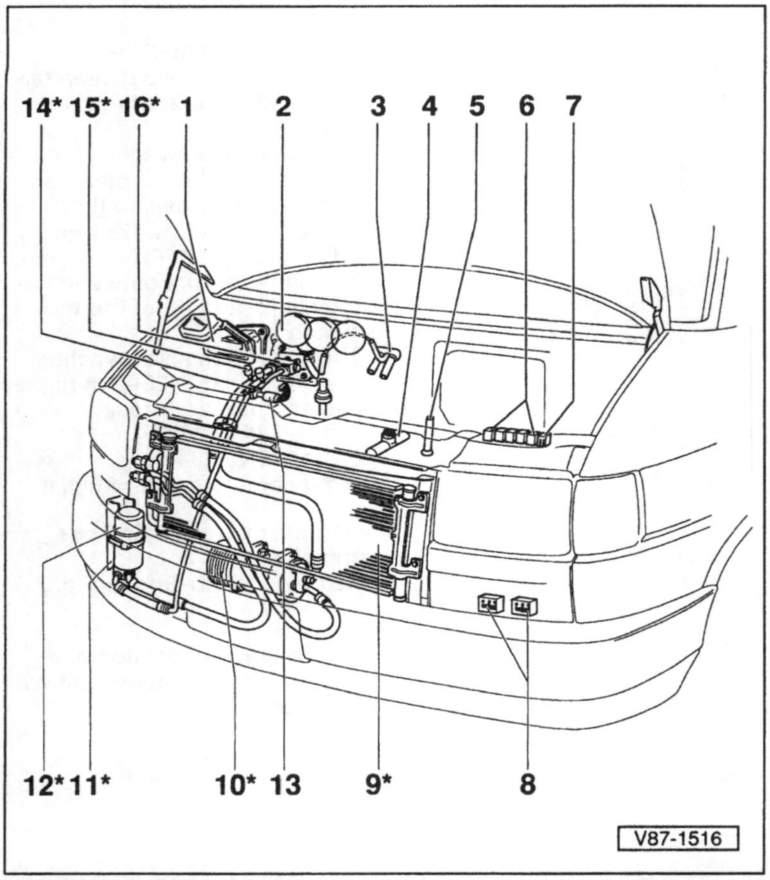

14 - Sight glass*

15 - Service valves*

16 - Expansion valve*

- Ensure valve is protected from rain and splash water
- Ensure insulator is in good condition ⇒ Fig.1

Note:

*System components identified with an * must only be serviced or replaced after discharging refrigerant system. Use Kent Moore ACR4 or equivalent.*

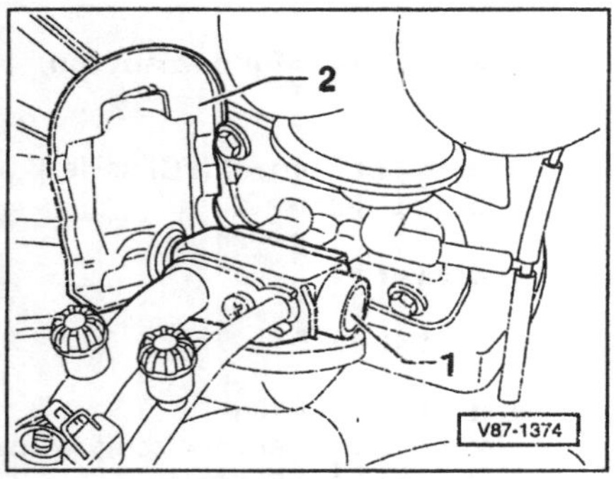

◄ **Fig. 1 Expansion valve insulator, checking**

1 - Expansion valve

2 - Insulator

The insulator protects the expansion valve against high engine compartment temperatures to prevent a decrease in A/C system cooling performance.

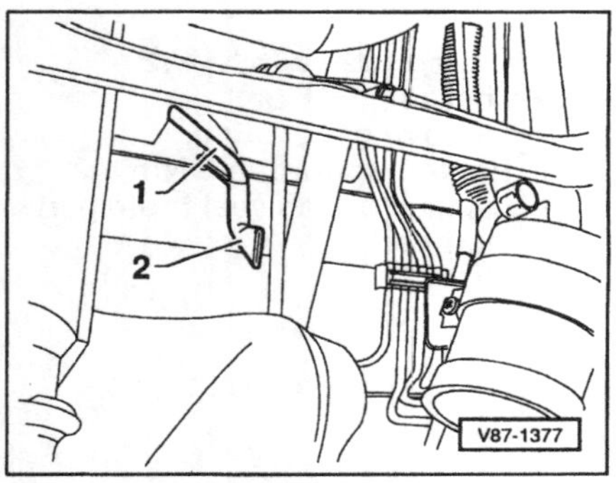

◄ **Fig. 2 Evaporator water drain hose and valve, removing/ installing**

1 - Hose

2 - Valve

- Remove engine, ⇒ Repair Group 10
- Remove sound dampener from bulkhead
- Remove glove compartment
- Remove floor outlet console
- Valve -2- is pushed on hose -1-

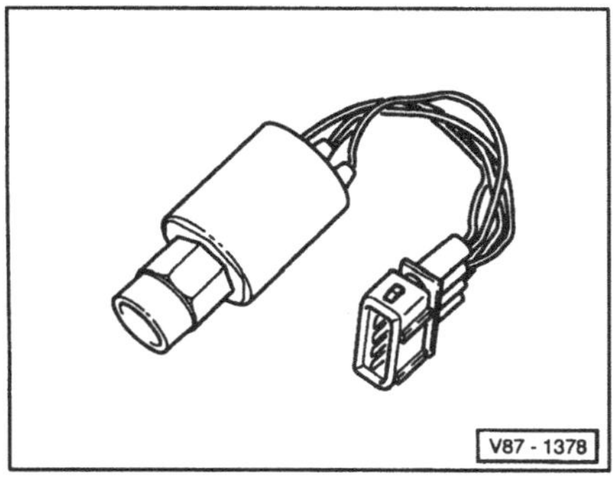

Fig. 3 A/C pressure switch -F129-, checking

– Ensure O–ring (10.8 mm x 1.8 mm) fits properly in groove

Functions

Switch between the blue wires switches the A/C clutch -N25- when pressure is too high in refrigerant circuit or if system not charged properly:

- ♦ Switch opens (compressor OFF): below 2.0 bar (29.0 psi)
- ♦ Switch closes (compressor ON): above 3.0 bar (43.5 psi)

Checking switch:

– Briefly bridge connections from terminal 1 and 2 of connector with engine running. If A/C clutch switches on -N25-, refrigenant circuit is empty.

– Closes when over 16 bar (232 psi) gauge pressure and opens again when under 12.5 bar (181 psi) gauge pressure (threshhold voltage)

R134a pressure switch is identified with a green label or lettering “For R134a”

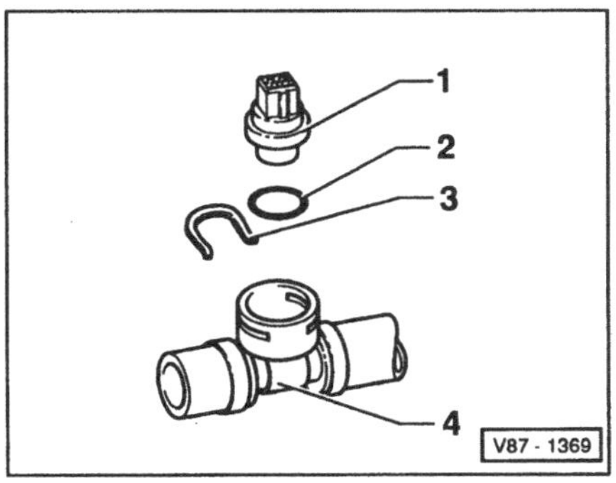

Fig. 4 A/C thermoswitch -F45-, removing/installing

> ***CAUTION!***
> ***Relieve cooling system pressure before removing A/C thermoswitch.***

- ♦ Refill coolant after thermoswitch installation

1 - A/C thermoswitch -F45-

2 - O–ring

3 - Clip

4 - Coolant hose

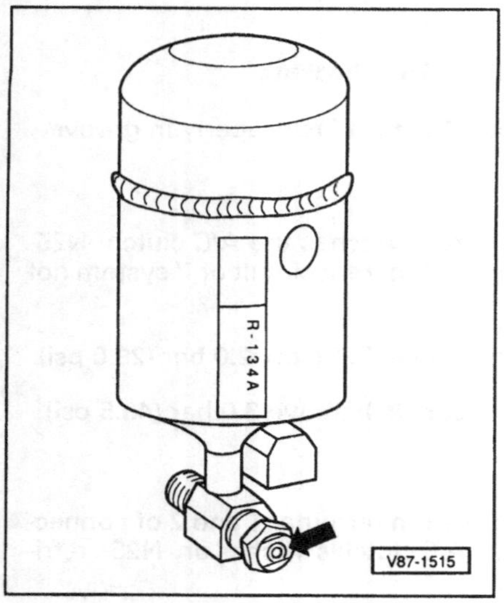

Fig. 5 Pressure relief valve on compressor, checking

- Protects against excessive system pressure
- At approximatley 40 bar (580 psi) the valve opens to relieve system pressure then closes again when pressure has dropped. The refrigerant system does not discharge completely
- Located on receiver drier

The plastic washer -arrow- will break if the pressure relief valve releases (opens)

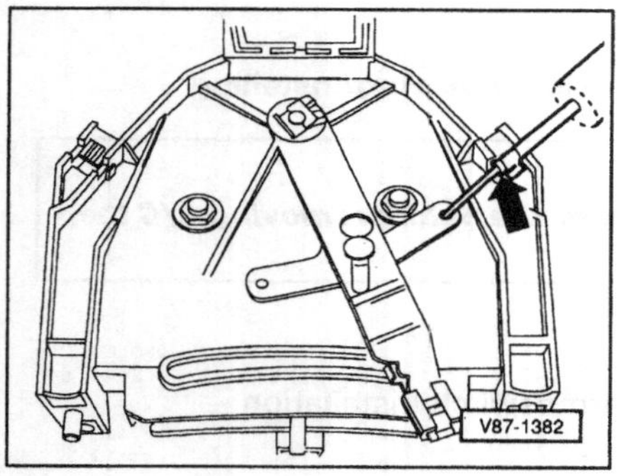

Temperature flap cable (vehicles with A/C), installing/adjusting

- Connect temperature flap cable end into lower control lever of A/C control head
  - Temperature flap cable sleeve is blue

Position cable sleeve on stop of A/C control head -arrow- and secure cable with clip

- Install A/C control head
- Slide temperature control lever fully left to stop (OFF/cool)

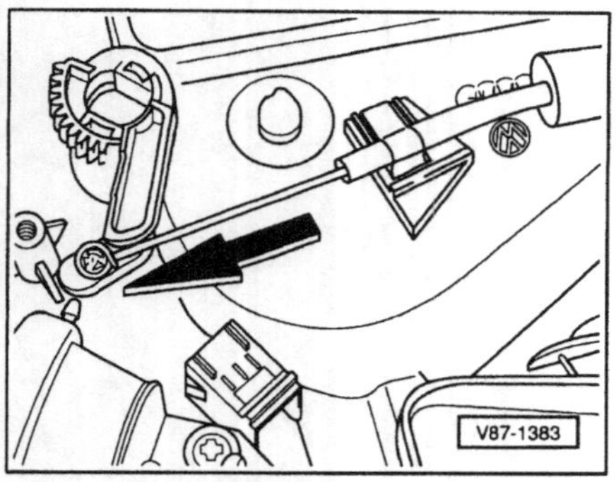

Push lever in direction of -arrow- to stop

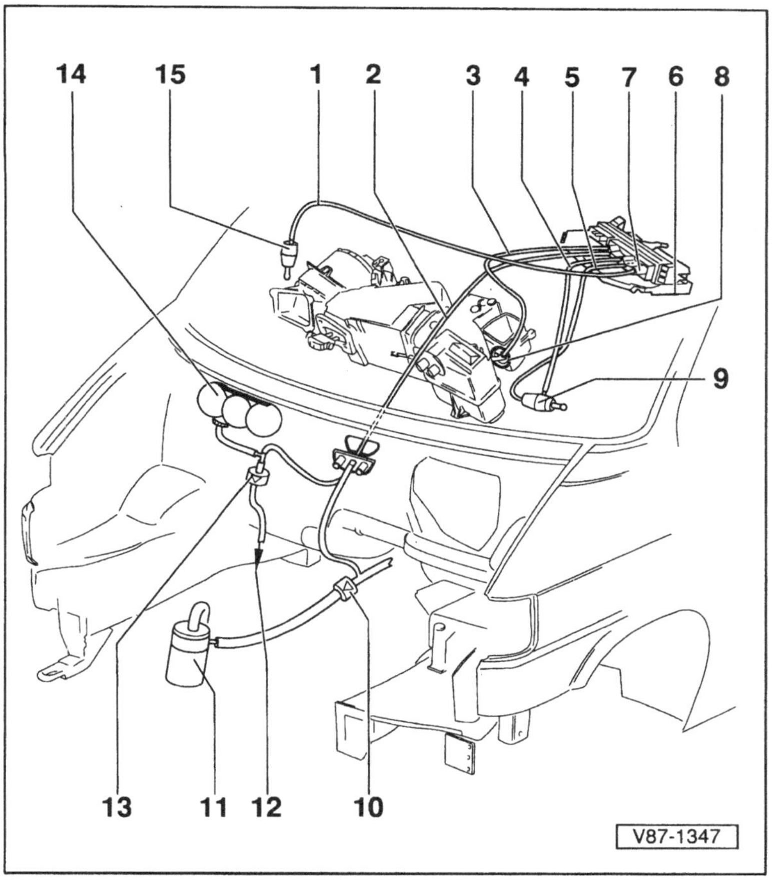

Vacuum hose, layout

Note:

Insert vacuum hose into vacuum reservoir 30 mm (1 3/16 in.). Insert vacuum hoses onto plastic connectors 15 mm (19/32 in.)

Vacuum hose colors

1 - White

- From multi-point conector -7- to vacuum canister for fresh air/recirculation door

2 - Black

- From vacuum reservoir in engine compartment to multi-point connector -7-

3 - Red

- From multi-point connector -7- to vacuum canister for flap floor outlet/defroster

4 - Green

- From multi-point connector -7- to vacuum canister for central flap

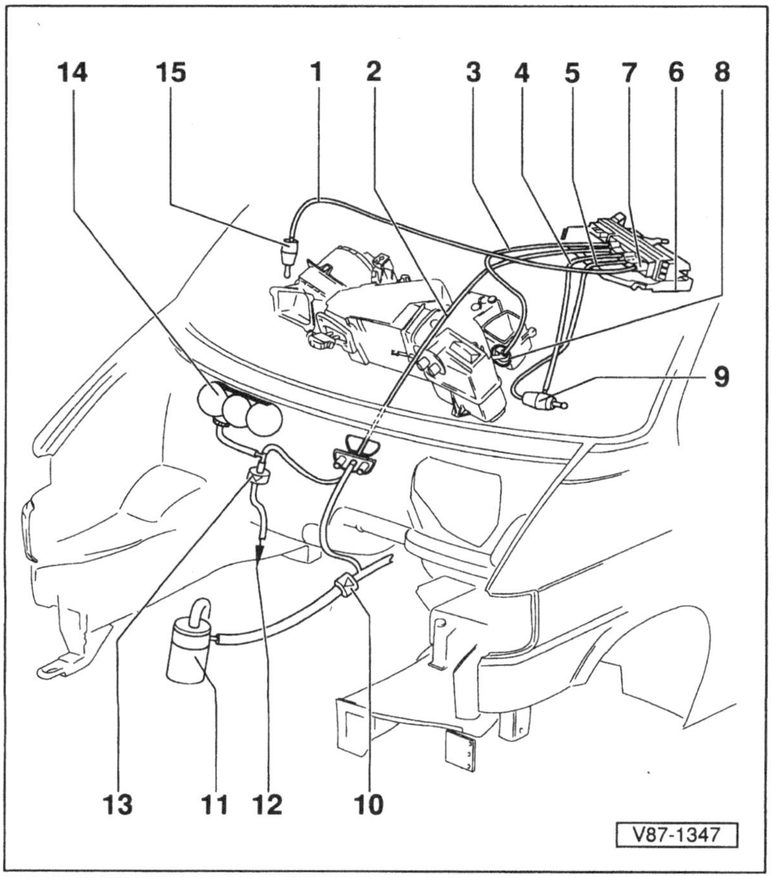

5 - Yellow

- From multi-point connector -7- to vacuum canister for central flap

Components

6 - A/C control head

7 - Vacuum hose multi-point connector

8 - Footwell/defroster flap vacuum servo

9 - Central flap vacuum servo

10 - Check valve (diesel engine only)

11 - Vacuum pump (diesel engine only)

12 - To engine vacuum supply

13 - Check valve (arrow points to vacuum source)

14 - Vacuum reservoir

15 - Fresh air/recirculating flap vacuum servo

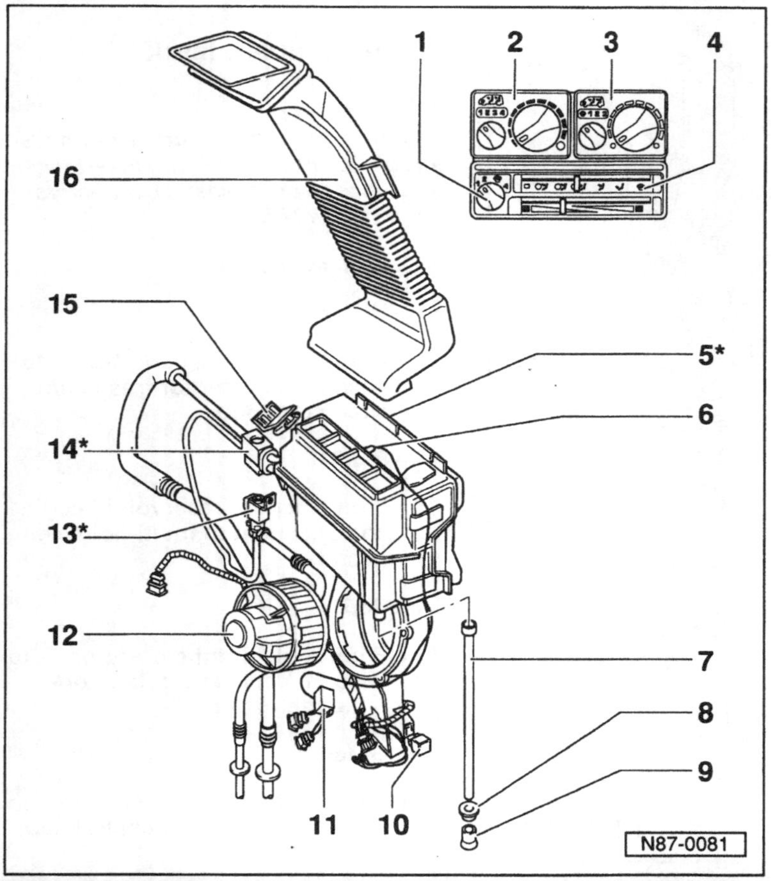

Heating and A/C, second evaporator (where equipped)

Note:

*System components identified with an * must only be serviced or replaced after discharging refrigerant system. Use Kent Moore ACR4 or equivalent.*

1 - Switch – fresh air blower -E9-

2 - Switch – evaporator blower -E110-, temperature control for 2nd evaporator

3 - Switch – warm air blower -E100-

4 - Control – warm and fresh air
- ♦ Including A/C switch

5 - Evaporator

6 - Temperature sensor -G18-

7 - Water drain tube

8 - Grommet

9 - Evaporator water drain valve

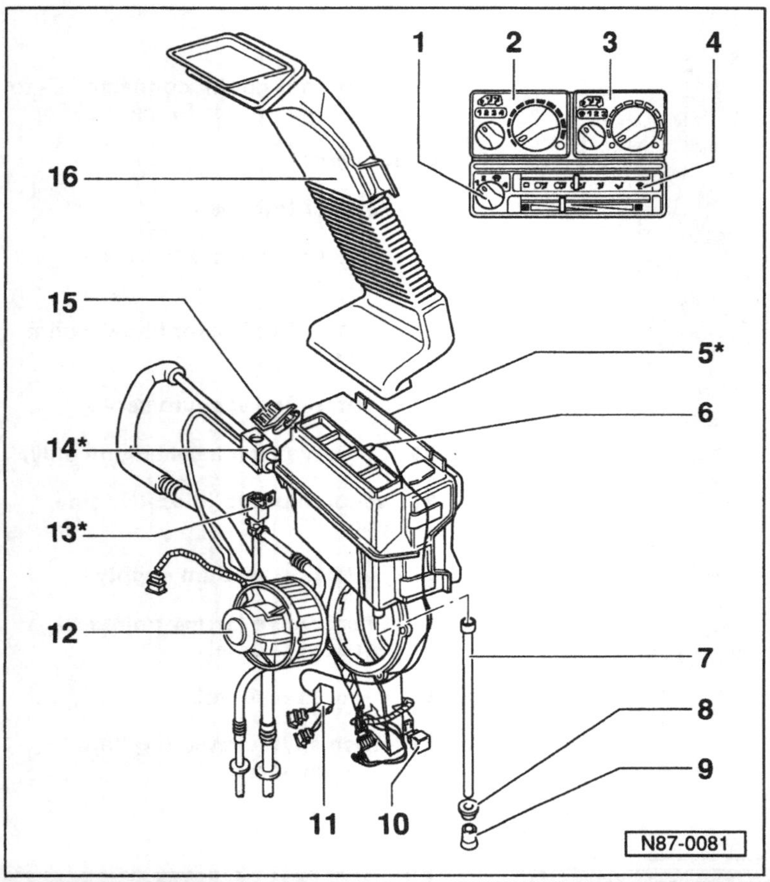

10 - Relay – fresh air blower 2nd evaporator

11 - A/C programmer (control and regulating unit) for second evaporator -J127-
- ♦ For wire routing – terminal connections: ⇒ Electrical Wiring Diagrams

12 - Fresh air blower for second evaporator

13 - A/C refrigerant shut–off valve -N43-*

14 - Expansion valve for second evaporator*

15 - Fresh air blower series resistance for second evaporator with overheat fuse -N6-
- ♦ Air duct

16 - Air duct
- ♦ To remove, first remove D–pillar trim

A/C refrigerant system, servicing

A/C refrigerant system, replacing components

Replacement A/C compressors, evaporators and condensers supplied by the Parts Dept. are filled with Nitrogen. If gas (Nitrogen) does not escape when component is first opened, the component may be faulty (leaking), do not install.

Replacement A/C compressors are filled with the total refrigerant oil quantity needed for the entire refrigerant system. When an additional, rear evaporator is present, additional compressor oil must be added after replacing the compressor.

Always replace the receiver drier whenever the refrigerant system has been opened. Install receiver drier immediately after opening to prevent moisture contamination of drier desiccant. Moreover, keep refrigerant system and other replacement components sealed for as long as possible to minimize the chance of dirt and moisture contamination.

Always plug open refrigerant line connections to prevent dirt and moisture contamination.

If the system has been discharged due to a damaged or leaking component (refrigerant hose/line, compressor, evaporator, condenser, etc.), flush the refrigerant system first with compressed air, then with nitrogen (available locally) and collect the oil that runs out. This will remove the refrigerant oil which may be saturated with moisture.

CAUTION!

DO NOT flush R-134a refrigerant system with R-11. R-11 is not compatible with refrigerant R-134a and PAG oil and will cause total system contamination.

If the compressor is not replaced after flushing the system, fill compressor with the correct type and quantity of refrigerant oil as specified under A/C system capacities ⇒ page S 87-98. Fill compressor with the total system oil capacity.

If the compressor is replaced, do not add any additional oil as the total amount of oil required is already in the compressor.

Dispose of contaminated refrigerant oil following laws governing hazardous waste disposal. Do not combine any refrigerant oil with any other old oils such as engine oil or transmission fluid.

Follow all applicable Warnings and Cautions when working on the A/C system.

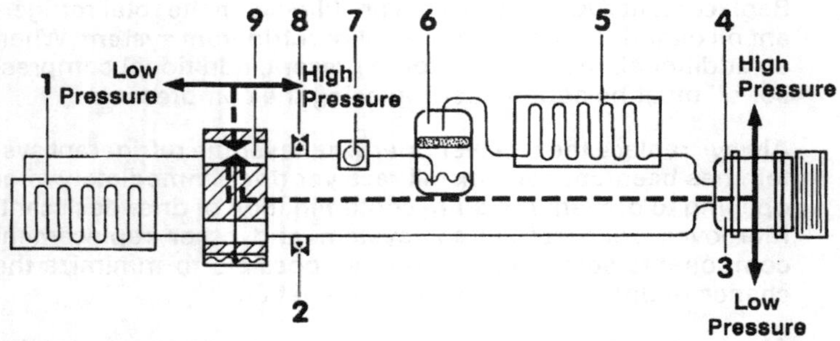

V87-1457

A/C refrigerant system (R-134a) pressures and temperatures, checking

The pressures and temperatures in the A/C system will vary depending on engine speed (RPM), coolant fan speed, engine coolant temperature, A/C clutch engagement, outside temperature, humidity, etc.

Pressure and temperature specifications are based on the following:

- Engine speed (RPM) at 1500
- Fresh air blower on high speed
- A/C switch adjusted to Max. cooling

1 - Evaporator

2 - Low-pressure service valve

3 - Compressor, low pressure side

4 - Compressor, high pressure side

5 - Condenser

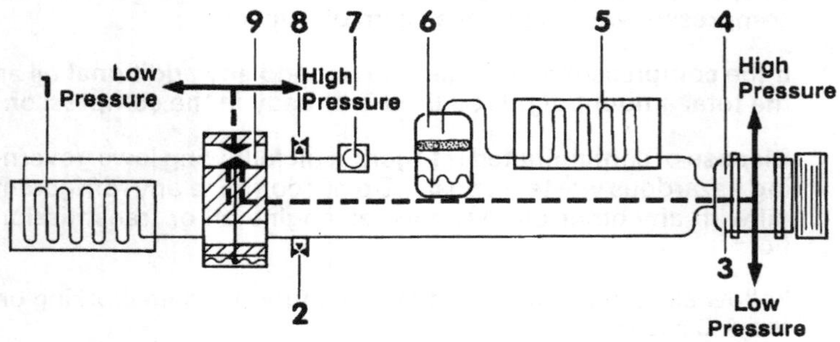

V87-1457

6 - Receiver drier

7 - Sight glass
- ♦ Not applicable to all vehicles

8 - High-pressure service valve

9 - Expansion valve

Pressure and temperature specifications

Component identification ⇒ pages S 87-75 and S 87-76

Component	Refrigerant state	Approximate pressure (bar/psi)	Approximate temperature
1 – Evaporator, inlet to outlet	Vapor	1.2 bar (17.4 psi) [1)] 1.8 bar (26.1 psi) [3)]	−7°C (19°F) [2)] −1°C (30°F) [3)]
2, 3 – Compressor, low-pressure side	Gas	1.2 bar (17.4 psi) 1.8 bar (26.1 psi) [3)]	−1°C (30°F) +1°C (34°F) [3)]
4 – Compressor, high-pressure side	Gas	14 bar (203 psi)	65°C (149°F)
5 – Condenser	Gas Vapor Liquid	14 bar (203 psi)	55°C (131°F) at outlet
6 – Receiver drier	Liquid	14 bar (203 psi)	55°C (131°F)
7 – Sight glass (if equipped)	Liquid	14 bar (203 psi)	55°C (131°F)
8 – Service valve, high-pressure	Liquid	14 bar (203 psi)	55°C (131°F)
9 – Expansion valve	Liquid, turns to vapor	inlet: 14 bar (203 psi) outlet: 1.2 bar(17.4 psi)	inlet: 55°C (131°F) outlet: −7°C (19°F)

1) This pressure is maintained in the refrigerant system by the variable displacement compressor despite changing conditions and varying engine speeds (RPM).

2) This temperature is maintained in the refrigerant system by the variable displacement compressor despite changing conditions and varying engine speeds (RPM).

3) Test values for vehicles equipped with second (rear) evaporator.

A/C refrigerant system, testing with pressure gauges

Due to the constant temperature/pressure relationship of R-12 and R-134a, approximate high-side system temperature can be determined based on system pressure.

Pressure gauges may have one or more temperature scales in addition to the pressure scale. Since various refrigerants have different temperature/pressure relationships, each gauge is specific for a particular refrigerant.

Measuring pressure and temperature in a "switched-on" refrigerant system

Use the A/C refrigerant high-pressure gauge to measure high pressure between the compressor and restrictor (including condenser) with the A/C switched on.

Use the A/C refrigerant low-pressure gauge to measure low pressure between the expansion valve and the compressor inlet (including evaporator) with the A/C switched on.

The temperature displays of the low- and high-pressure gauges apply only to a portion of the low- and high-pressure sides. The low-pressure gauge displays approximate temperature between the expansion valve and evaporator outlet. The high-pressure gauge displays approximate temperature between the compressor outlet and restrictor.

Note:

The temperature/pressure relationship only holds true when the refrigerant system contains liquid and vapor, but not when it contains only gas. In the gaseous state, the temperature is approx. 10-30°C (18-54°F) higher than indicated on the pressure gauge.

Measuring pressure and temperature in a closed container or "switched-off" refrigerant system

Due to the constant temperature/pressure relationship of R-12 and R-134a, approximate temperature in a closed container or in a switched-off A/C system can be determined based on its pressure.

If the temperature displayed on a gauge is lower than the actual temperature of the refrigerant, the refrigerant container or system is empty (discharged).

Note:

The temperature/pressure relationship only holds true when the refrigerant system contains liquid and vapor, but not when it contains only gas. In the gaseous state, the temperature is approx. 10-30°C (18-54°F) higher than indicated on the pressure gauge.

A/C refrigerant system, discharging

WARNING!

- ***Always use an Underwriter's Laboratory (UL) approved refrigerant recovery/recycling/recharging unit such as Kent-Moore ACR[4], or equivalent, whenever discharging an R-134a A/C system.***
- ***Any person who services a motor vehicle air conditioner MUST, by law, be properly trained and certified and use approved refrigerant recycling equipment. Technicians must complete an EPA approved recycling course to be certified.***
- ***State and local governments may have additional requirements regarding air conditioning servicing. Always comply with state and local laws.***
- ***Always wear safety goggles when charging or discharging system.***

CAUTION!

- ***Always use separate refrigerant recovery/recycling/recharging servicing equipment for R-12 and R-134a systems. DO NOT use one piece of equipment for both R-12 and R-134a systems. The residual traces of refrigerant inside the equipment will contaminate and damage the equipment. Servicing equipment includes recovery/recycling/recharging unit, charging station, vacuum pump, manifold gauges, etc.***
- ***DO NOT use R-12 servicing equipment on R-134a systems or R-134a equipment on R-12 systems or damage to both the vehicle A/C system and servicing equipment may result. Use only equipment designed to meet Society of Automotive Engineers (SAE) standards.***
- ***R-134a and R-12 systems use different size service fittings. NEVER use adaptors to convert an R-12 fitting to R-134a size or R-134a fitting to R-12 size.***

Note:

Refer to R-12 safety measures starting on page S 87-13 and R-134a safety measures starting on page S 87-41 prior to discharging or charging A/C refrigerant system.

Note:

Make sure that initial set-up of the refrigerant recovery/recycling/recharging unit has been completed before discharging the A/C system.

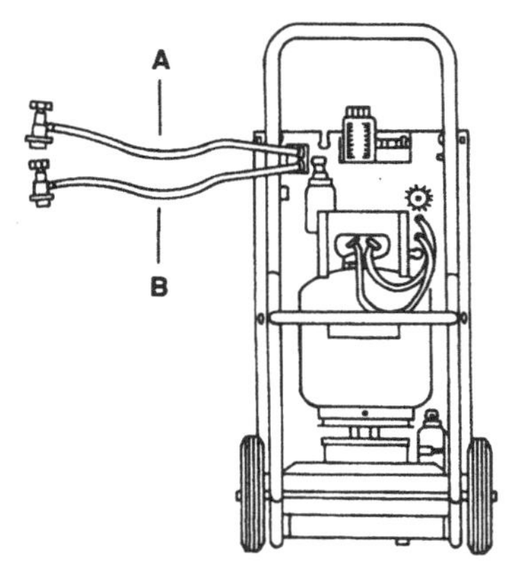

87-A197

◄ – Connect red high-pressure hose -A- of refrigerant recovery/recycling/recharging unit to high side fitting on vehicle and open coupler valve.

– Connect blue low-pressure hose -B- of refrigerant recovery/recycling/recharging unit to low side fitting on vehicle and open coupler valve.

CAUTION!

Always follow manufacturer's instructions when using a refrigerant recovery/recycling/recharging unit.

– Following refrigerant recovery/recycling/recharging unit manufacturer's instructions, discharge A/C system into refrigerant recovery/recycling/recharging unit.

– Disconnect power supply from A/C clutch to prevent accidental compressor operation with A/C system discharged.

A/C refrigerant system, flushing with compressed air and nitrogen

> ***CAUTION!***
>
> - ***When using compressed nitrogen always use a pressure regulator and the proper adaptor hoses and fittings (available locally). During flushing, use existing exhaust/ventilation systems to draw off the gas mixture escaping from the A/C system.***
> - ***DO NOT flush R-134a refrigerant system with R-11. R-11 is not compatible with R-134a refrigerant and PAG oil and will cause total system contamination.***

- Use compressed air and nitrogen (available locally) to remove moisture, impurities and old refrigerant oil from A/C refrigerant system.
- First blow out old refrigerant oil and dirt with compressed air, then dry components with nitrogen.
- DO NOT blow compressed air and nitrogen through the compressor or expansion valve. Only blow compressed air and nitrogen through disconnected, free flowing components (i.e. disconnected hose, condenser, evaporator, etc.)
- DO NOT blow compressed air and nitrogen into a capped off A/C component. Pressurized R-134a refrigerant in the presence of oxygen may form a combustible mixture.

- Always flush components in opposite direction of refrigerant flow.
- Flush evaporator through the low-pressure line with the high-pressure line removed.
- If any component has dark thick deposits that cannot be removed with compressed air, replace component.
- Thin light gray deposits in refrigerant lines and hoses are normal and do not impair the function of the system.
- Always replace receiver drier and expansion valve after flushing.
- Dispose of contaminated refrigerant (PAG) oil following laws governing hazardous waste disposal. Do not combine PAG oil with any other old oils such as engine oil or transmission fluid.

Flush refrigerant system with compressed air and nitrogen if:

- Refrigerant oil is dark and viscous (thick)
- Too much refrigerant oil is in the system following compressor replacement
- Unclear or do not know how much refrigerant oil is in the system
- Moisture, dirt or other impurities have entered the refrigerant system (i.e. following an accident)
- Unable to pull a constant vacuum during evacuation of a leak-free system due to excessive moisture in the system
- Refrigerant system has been open longer than the time required for normal repairs (i.e. following an accident)
- Based on temperature and pressure measurements, system is diagnosed with moisture contamination
- Compressor is replaced due to noises or internal damage
- Flushing is required after replacing certain components in certain situations ⇒ page S 87-73

A/C refrigerant system, evacuating and recharging

WARNING!

- ***Always use an Underwriter's Laboratory (UL) approved refrigerant recovery/ recycling/recharging unit such as*** **Kent-Moore ACR4**, ***or equivalent, when evacuating and recharging an R-134a A/C system.***
- ***Any person who services a motor vehicle air conditioner MUST, by law, be properly trained and certified and use approved refrigerant recycling equipment. Technicians must complete an EPA approved recycling course to be certified.***
- ***State and local governments may have additional requirements regarding air conditioning servicing. Always comply with state and local laws.***
- ***Always wear safety goggles when discharging, evacuating and recharging an A/C system.***

CAUTION!

Always follow manufacturer's instructions when using a refrigerant recovery/recycling/recharging unit.

Notes:

- *Refer to R-12 safety measures starting on page S 87-13 and R-134a safety measures starting on page S 87-41 prior to discharging or charging A/C refrigerant system.*
- *Follow refrigerant recovery/recycling/recharging unit manufacturer's instructions for evacuating and recharging A/C system.*
- *Evacuate refrigerant system for a minimum of 30 minutes.*
- *When recharging A/C system, add correct amount of refrigerant and PAG oil to system ⇒ page S 87-98.*
- *After system recharge, manually rotate A/C compressor approx. 10 turns before starting engine. Start engine with A/C OFF. After idle speed has stabilized, switch A/C ON and let engine idle (compressor running) for a minimum of two minutes before raising engine speed.*

A/C refrigerant system, troubleshooting

Requirements

- Electrical systems OK
- Vacuum control systems OK
- Air flow distribution systems OK

Correct and repair as necessary before proceeding.

A/C cooling performance

If any one of the following system conditions exists, check A/C system pressures ⇒ page S 87-91.

- A/C does not cool (cooling has stopped completely)
- Insufficient cooling during all driving speeds or engine speeds (RPM)
- No cooling or insufficient cooling after driving several miles

A/C compressor noisy

- Tighten A/C compressor and compressor bracket mounting bolts using torque wrench.
- Check routing of A/C refrigerant hoses and lines; hoses and lines must not contact other components and must be installed without tension or kinks, reposition or realign as necessary.

A/C compressor noisy immediately after switching A/C ON and/or when driving around corners or braking (refrigerant shock)—system overcharged

- Discharge A/C refrigerant system ⇒ page S 87-80, then evacuate and recharge ⇒ page S 87-86.

With an otherwise properly operating A/C system, water sprays out of instrument panel vents

- Check evaporator water drain hose for proper routing (should not be kinked or pinched).
- Check evaporator water drain valve.
  - Must not be plugged with wax, dirt or undercoating (clean if necessary)
  - Valve flap must open/close properly
- Check plenum tray.

  Must be properly installed and not damaged

A/C refrigerant system pressures, checking

Check A/C refrigerant system pressures using these three tests:

1 - Air temperature drop from center instrument panel vent (A/C system cooling performance)

2 - A/C system high pressure

3 - A/C system low pressure

The combined results of all three steps determine which part of the A/C system is causing the problem, see table ⇒ page S 87-95.

Test conditions

- A/C refrigerant system fully charged; discharge, evacuate and recharge system if necessary
- Condenser and radiator clean and free of obstructions (spray clean if necessary)
- Air distribution can be adjusted correctly using control knobs (all air distribution flaps reach end positions)
- No wiring problems as per wiring diagram
- Outside (ambient) air temperature between 20–30°C (68–86°F)
- Drive belts for A/C compressor and Generator in good condition and properly tensioned

Test 1: Checking air temperature drop from center instrument panel vent (A/C system cooling performance)

– Start engine.

– Adjust air distribution so air flows from instrument panel vents.

– Set temperature control knob to full "cold" position (against stop).

– Set fresh air blower to second speed (position 2).

– Insert thermometer into center instrument panel vent.

– Switch A/C on by pressing A/C "Norm" button.

– Raise engine speed to 2000 RPM.

Specified result

With humidity normal and outside (ambient) temperature between 20–25°C (68–77°F), system is cooling satisfactorily if air temperature from center instrument panel vent drops below 10°C (50°F) within 1 minute.

For higher ambient temperatures and/or higher humidity, specified air temperature from center instrument panel vent can be slightly higher.

If specified reading is not obtained, perform tests 2 and 3, then compare results of all three tests ⇒ table, page S 87-95.

Test 2: Checking A/C system high pressure

- Connect high- and low-pressure hoses of refrigerant recovery/ recycling/recharging unit Kent-Moore ACR^4 or equivalent, to high- and low-pressure service valves.
- Disconnect coolant fan connector.
- Adjust air distribution knob to footwell outlets.
- Adjust temperature control knob to full hot.
- Set fresh air blower to high speed (position 4).
- Start engine.
- Switch Max. A/C on by pressing A/C Max. button.
- Raise engine speed to 1500 RPM.

Specified result

System high pressure is normal if high-pressure gauge reads 232 psi (16 bar) within 30 seconds.

If specified reading is not obtained, also perform test 3 and compare results of all three tests ⇒ table, page S 87-95.

Test 3: Checking A/C system low pressure

- Connect high- and low-pressure hoses of refrigerant recovery/recycling/recharging unit Kent-Moore ACR^4, or equivalent, to high- and low-pressure service valves.
- Adjust air distribution knob to instrument panel outlets.
- Adjust temperature control knob to full cold.
- Set fresh air blower to low speed (position 1).
- Start engine.
- Switch A/C on by pressing A/C Norm button.
- Raise engine speed to 1500 RPM.

Specified result

System low pressure is normal if low-pressure gauge reads 22–36 psi (1.5–2.5 bar) within 30 seconds.

If specified reading is not obtained, compare results of all three tests ⇒ table, page S 87-95.

Test 1 Temperature from center air vent[1]	Test 2 High pressure[2]	Test 3 Low pressure[2]	Possible causes of incorrect readings	Corrective measures
Normal	Normal	Normal	None	---
Too high	Normal	Normal	Temperature flap position incorrect	Adjust temperature flap cable
Too high	Too low	Normal	Compressor	Replace compressor
Normal	Too low	Normal	Compressor	Replace compressor
Normal	Normal	Too high or too low	Expansion valve or compressor	Replace expansion valve or compressor
Too high	Normal	Too high or too low	Expansion valve or compressor	Replace expansion valve or compressor
Normal	Too high or too low	Too high or too low	Expansion valve or compressor	Replace expansion valve or compressor

[1] Normal air outlet temperature approx. 6°C (43°F).

[2] For normal system temperatures and pressures ⇒ page S 87-77.

A/C system cooling performance, checking

Note:

A/C evaporator temperature switch -E33- (if equipped) switches A/C compressor OFF only in special instances so evaporator does not ice up; variable displacement A/C compressor maintains temperature of 0°C (32°F) in evaporator.

Test conditions

- Condenser and radiator are clean and free of obstructions
- Air distribution can be adjusted properly with ventilation control knobs
- No wiring problems detected using wiring diagram
- Outside (ambient) temperature is 20°C–30°C (68°F–86°F)

Checking

- Start engine.
- Set fresh air blower to second speed (position 2).
- Set temperature control knob to full "cold" position (against stop).
- Adjust air distribution so air flows from instrument panel vents.
- Switch A/C on by pressing "A/C" button.
- Insert thermometer into center instrument panel outlet.
- Raise engine speed to 2000 RPM.

Specified result

With normal humidity and outside (ambient) temperature between 20°–25°C (68°–77°F), system is sufficiently charged if air temperature from center instrument panel vent drops below 10°C (50°F) within one minute.

For higher ambient temperatures and/or higher humidity, specified air temperature from center instrument panel vent can be slightly higher.

A/C refrigerant system, capacities

Refrigerant R-12 (Canada only, 1992 m.y.)

Model	Quantity
EuroVan, Transporter	1050 g + 50 g (37 oz + 1.8 oz)

Refrigerant oil, R-12 (Canada only, 1992 m.y.)

Model	Total system capacity
EuroVan, Transporter	135 cc (4.6 fl. oz.)

Note:

Replacement A/C compressors supplied by the Parts Department are filled with 135 cc (4.6 fl. oz) R-12 compatible refrigerant oil. This is the total A/C system refrigerant oil capacity.

Refrigerant R-134a

Model	Quantity
EuroVan, Transporter with one evaporator (front)	950 g + 50 g (33.5 oz + 1.8 oz)
EuroVan with two evaporators (front & rear)	1350 g + 50 g (47.6 oz + 1.8 oz)

Obtain R-134a refrigerant from a local A/C supplier under one of the following names:

- R-134a
- Tetrafluoroethane
- $CH_2F\ CF_3$
- H-FKW 134a
- SUVA® TRANS A/C
- ARCTON® 134a

Note:

R-134a refrigerant is packaged in different containers. Some are used only for commercial applications which are sold in cylinders using a 1/4" flare fitting. This does not connect to the vehicle fittings and servicing equipment. Use only R-134a which come in containers having the correct type of service fitting.

Refrigerant (PAG) oil, R-134a

Model	Total system capacity
EuroVan, Transporter with one evaporator (front)	135 cc (4.6 fl. oz.)
EuroVan with two evaporators (front & rear)	240 cc (8.1 fl. oz.)

Use only the correct type of refrigerant oil (PAG oil) specified for use with R-134a.

Note:

Replacement A/C compressors supplied by the Parts Department are filled with 135 cc (4.6 fl. oz) of refrigerant oil. This is the total A/C system refrigerant oil capacity for systems with one evaporator. Systems with two evaporators require an additional 105 cc (3.6 fl. oz.) be added to the system after compressor replacement.

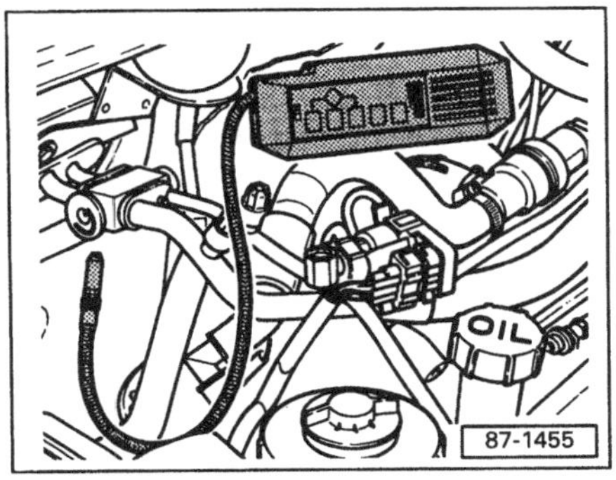

A/C refrigerant system, checking for leaks

◄ Use halogen leak detector, Hitec HI400A-TEL or equivalent, to check for refrigerant leaks, following the leak detector manufacturer's instructions.

Refrigerant gas dissipates very quickly. To make the job easier, avoid drafty or windy areas when checking for leaks.

If the refrigerant system is discharged (empty), re-charge the system with approx. 100 g (3.5 oz) of refrigerant in order to check for leaks.

Heating and Air Conditioning (1997 ➤)

Section **T**

WARNING!
Before proceeding with any work, please read the warnings and cautions immediately following this index. Also read any warning and caution that accompanies a procedure in the manual.

87 Air conditioning

Please read these WARNINGS and CAUTIONS before proceeding with maintenance and repair work.

WARNING—

- Some repairs may be beyond your capability. If you lack the skills, tools and equipment, or a suitable workplace for any procedure described in this manual, we suggest you leave such repairs to an authorized Volkswagen dealer service department, or other qualified shop.

- Volkswagen is constantly improving its cars. Sometimes these changes, both in parts and specifications, are made applicable to earlier models. Therefore, before starting any major jobs or repairs to components on which passenger safety may depend, consult your authorized Volkswagen dealer about the Technical Bulletins that may have been issued since the editorial closing of this manual.

- Do not re-use any fasteners that are worn or deformed in normal use. Many fasteners are designed to be used only once and become unreliable and may fail when used a second time. This includes, but is not limited to, nuts, bolts, washers, self–locking nuts or bolts, circlips and cotter pins. Always replace these fasteners with new parts.

- Never work under a lifted car unless it is solidly supported on stands designed for the purpose. Do not support a car on cinder blocks, hollow tiles or other props that may crumble under continuous load. Never work under a vehicle that is supported solely by a jack. Never work under the car while the engine is running.

- If you are going to work under a car on the ground, make sure that the ground is level. Block the wheels to keep the car from rolling. Disconnect the battery negative terminal (ground strap) to prevent others from starting the vehicle while you are under it.

- Never run the engine unless the work area is well ventilated. Carbon monoxide (CO) kills.

- Do not attempt to work on your vehicle if you do not feel well. You increase the danger of injury to yourself and others if you are tired, upset or have taken medicine or any other substance that may impair you or keep you from being fully alert.

- Disconnect the battery negative (–) terminal (ground strap) whenever you work on the fuel system or the electrical system. Do not smoke or work near heaters or other fire hazards. Keep an approved fire extinguisher handy.

- Batteries produce explosive hydrogen gas. Keep sparks, lighted matches and open flame away from the top of the battery. If hydrogen gas escaping from the cap vents is ignited, it will ignite gas trapped in the cells and cause the battery to explode.

- Connect and disconnect battery cables, jumper cables or a battery charger only with the ignition switched off, to prevent sparks. Do not quick–charge the battery (for boost starting) for longer than one minute, and do not allow charging voltage to exceed 16.5 volts. Wait at least one minute before boosting the battery a second time.

- Illuminate your work area adequately but safely. Use a portable safety light for working inside or under the car. Make sure the bulb is enclosed by a wire cage. The hot filament of an accidentally broken bulb can ignite spilled fuel or oil.

- Tie long hair behind your head. Do not wear a necktie, a scarf, loose clothing, or a necklace when you work near machine tools or running engines. If your hair, clothing, or jewelry were to get caught in the machinery, severe injury could result.

- Finger rings, bracelets and other jewelry should be removed so that they cannot cause electrical shorts, get caught in running machinery, or be crushed by heavy parts.

- Catch draining fuel, oil, or brake fluid in suitable containers. Do not use food or beverage containers that might mislead someone into drinking from them. Store flammable fluids away from fire hazards. Wipe up spills at once, but do not store the oily rags, which can ignite and burn spontaneously.

- Always observe good workshop practices. Wear goggles when you operate machine tools or work with battery acid. Gloves or other protective clothing should be worn whenever the job requires working with harmful substances.

- The Air-Conditioning (A/C) system is filled with a chemical refrigerant, which is hazardous. The A/C system should be serviced only by trained technicians using approved refrigerant recovery/recycling equipment, trained in related safety precautions, and familiar with regulations governing the discharging and disposal of automotive chemical refrigerants.

- Do not expose any part of the A/C system to high temperatures such as an open flame. Excessive heat will increase system pressure and may cause the system to burst.

- Some aerosol tire inflators are highly flammable. Be extremely cautious when repairing a tire that may have been inflated using an aerosol tire inflator. Keep sparks, open flame or other sources of ignition away from the tire repair area. Inflate and deflate the tire at least four times before breaking the bead from the rim. Completely remove the tire from the rim before attempting any repair.

- Friction materials such as brake or clutch discs may contain asbestos fibers. Do not create dust by grinding, sanding, or by cleaning with compressed air. Avoid breathing asbestos fibers and asbestos dust. Breathing asbestos can cause serious diseases such as asbestosis or cancer, and may result in death.

- Greases, lubricants and other automotive chemicals contain toxic substances, many of which are absorbed directly through the skin. Read manufacturer's instructions and warnings carefully. Use hand and eye protection. Avoid direct skin contact.

- An airbag is operated by an explosive device. Handled improperly or without adequate safeguards, it can be accidently activated and cause serious personal injury. To prevent personal injury or airbag system failure, **only trained Volkswagen dealer service technicians** should test, disassemble or service the airbag system.

WARNINGS and CAUTIONS continued on next page ►

Please read these WARNINGS and CAUTIONS before proceeding with maintenance and repair work.

WARNING (continued) —

- ALWAYS disconnect the airbag power supply before using a computer memory saver. Failure to disconnect the airbag power supply may result in accidental activation of the airbag.
- The airbag unit should not be removed, nor should any work be performed on the airbag system, without first reviewing the warnings and cautions that accompany the airbag system repair information in this manual.
- Never apply stickers or any other coverings on the steering wheel. Do not let chemical cleaners, oil or grease come into contact with the vinyl covering of the airbag unit.
- Never open or otherwise attempt to repair airbag system parts. Always use new parts. Never leave airbag parts or the partially disassembled airbag system unattended.
- Never use a test light or volt meter to conduct electrical tests on the airbag system. The airbag system must only be tested by trained Volkswagen Service technicians using special airbag tester equipment. The airbag system must only be tested while it is installed in the car.
- Do not expose an airbag unit to temperatures above 194°F (90°C) even for a brief period, while handling it during the repair process. Keep the unit clear of heat sources such as hot plates, soldering irons, heat lamps, welding equipment, or the like.
- Always replace airbag system components that have been mechanically damaged (bubbles, cracks, etc.).
- Always set down a removed airbag unit so that the horn pad is facing upward.
- When driving or riding in an airbag-equipped vehicle, never hold test equipment in your hands or lap while the vehicle is in motion. Objects between you and the airbag can increase the risk of injury in an accident.
- The storage, transportation and disposal of airbag units are subject to the laws for flammable solids.

CAUTION—

- If you lack the skills, tools and equipment, or a suitable workshop for any procedure described in this manual, we suggest you leave such repairs to an authorized Volkswagen dealer or other qualified shop. We especially urge you to consult an authorized Volkswagen dealer before beginning repairs on any car that may still be covered wholly or in part by any of the extensive warranties issued by Volkswagen of America, Inc.
- Volkswagen is constantly improving its cars and sometimes these changes, both in parts and specifications, are made applicable to earlier models. Therefore, part numbers listed in this manual are for reference only. Always check with your authorized Volkswagen dealer parts department for the latest information.
- Before starting a job, make certain that you have all the necessary tools and parts on hand. Read all the instructions thoroughly, do not attempt shortcuts. Use tools that are appropriate to the work and use only replacement parts meeting Volkswagen specifications. Makeshift tools, parts and procedures will not make good repairs.
- Use pneumatic and electric tools only to loosen threaded parts and fasteners. Never use these tools to tighten fasteners, especially on light alloy parts. Always use a torque wrench to tighten fasteners to the tightening torque specification listed.
- Be mindful of the environment and ecology. Before you drain the crankcase, find out the proper way to dispose of the oil. Do not pour oil onto the ground, down a drain, or into a stream, pond, or lake. Consult local ordinances that govern the disposal of wastes.
- When connecting cables or ground straps, all connections must be cleaned to bare metal. Be sure mounting bolts are tight.
- Do not replace fuse links with fuse links of other gauge or with standard wire. Always replace the entire length of the fuse link.
- Before doing any electrical welding on vehicles equipped with Anti-lock Brakes (ABS), disconnect the battery negative terminal (ground strap) and the ABS Control Module connector.
- On cars equipped with ABS, the ABS control unit cannot withstand temperatures from a paint–drying booth or a heat lamp in excess of 203°F (95°C) and should not be subjected to temperatures in excess of 185°F (85°C) for more than two hours.
- On cars equipped with anti-theft radios, make sure you know the correct radio activation code before disconnecting the battery or removing the radio. If the wrong code is entered when power is restored, the radio may lock up and be rendered inoperable, even if the correct code is then entered.
- When boost-charging the battery, first remove the fuses for the Engine Control Module (ECM), the Transmission Control Module (TCM), the ABS Control Module, and the On-Board computer. In cases where one or more of these components is not separately fused, disconnect the Control Module connector(s).
- Do not quick-charge the battery (for boost starting) for longer than one minute, and do not exceed 16.5 volts at the battery with the boosting cables attached. Wait at least one minute before boosting the battery a second time.

◀ **WARNINGS and CAUTIONS continued from previous page**

Digital Climate Control (Climatronic) On Board Diagnostic (OBD)

Function

The Climatronic control module -J255- receives input information from various electrical and electronic components (sensors). This information is processed by the control module in accordance with specified values and then provides corresponding output signals that control electrical components (actuators).

The Climatronic control module -J255- is controlled by an internal microprocessor with On Board Diagnostic (OBD) capability. If malfunctions, short or open circuits occur on monitored components, Diagnostic Trouble Codes (DTC) are stored in memory.

Malfunctions recognized by OBD and stored as coded DTCs include an indication of the malfunction type. Tables containing all DTC codes and corresponding repair information guide technicians through the OBD program.

Stored DTCs are checked with VAG 1551 or VAG 1552 Scan Tool (ST).

When DTCs are stored, the Climatronic system will switch off either partially or entirely depending on the severity of the fault.

Component malfunctions on monitored components having a permanent adverse effect on the operation of the Climatronic will cause the A/C control head -E87- display panel to flash for approx. 15 seconds when the ignition is switched on. DTCs are also stored.

Component malfunctions on monitored components not having a permanent adverse effect on the operation of the Climatronic (-E87- display does not flash) will result in the Climatronic operating in emergency mode at default settings. Substitute values are assumed for temperature sensors and flap motor positions.

Some components and functions of the Climatronic are not monitored by OBD. If a verified symptom is present and no DTCs are stored in memory, carry out the following tests:

- Output Diagnostic Test Mode (DTM) (scan tool function 03) ⇒ Page T 01-29
- Read Measuring Value Block (scan tool function 08) ⇒ Page T 01-37.
- Climatronic system, troubleshooting ⇒ Page T 87-111

Additional information:

- Climatronic system, General/Function ⇒ from Page T 87-56
- Climatronic system, servicing ⇒ from Page T 87-66
- A/C refrigerant system, servicing ⇒ from Page T 87-118

VAG 1551/1552 Scan Tool (ST), connecting and selecting functions

Notes:

- *Print-out is not possible with VAG 1552 Scan Tool (ST).*
- *The following text display examples apply to VAG 1551 only. VAG 1552 operating sequence and displays may be different.*
- *Always ensure most recent VAG 1551/1552 program card version is used.*

Requirements

- All fuses OK per wiring diagram.
- Battery voltage (B+) minimum 9 Volts.

– Remove cover for Data Link Connector (DLC) next to instrument cluster.

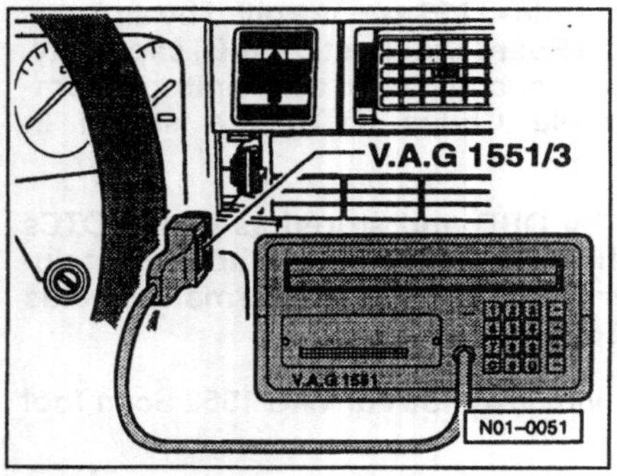

◄ – With ignition switched off, connect VAG 1551/1552 to Data Link Connector (DLC) using adapter cable VAG 1551/3.

Notes:

- *Additional operating information can be displayed by pressing the HELP button on VAG 1551 Scan Tool.*
- *The → button is used for moving forward within program.*
- *Pressing PRINT button switches on the printer (indicator lamp in button comes on).*

– Switch on ignition.

Rapid data transfer	HELP
Insert address word XX	

◄ Indicated on display

– Press buttons -0- and -8- to select "AC/Heating Electronics" address word 08.

Rapid data transfer	Q
08 AC/Heating Electronics	

◄ Indicated on display

– Press -Q- button to confirm input.

Rapid data transfer
Tester sends address word 08

◄ Indicated on display

701 907 044 A CLIMATRONIC VXX →
Coding 00001 WSC XXXXX

◄ Indicated on display after selecting address word 08

- 701 906 044 = control module identification (for latest version see parts catalog)
- Climatronic = System designation
- VXX = Software version of control module
- Coding:

  00001 = with dust and pollen filter

  00000 = without dust and pollen filter

  Coding Climatronic control module ⇒ Page T 01-35
- WSC = Workshop code

Note:

Improper control module coding results in incorrect blower output.

	List of Scan Tool Functions	Page
01	Check Control Module Version	T 01-3
02	Check DTC Memory	T 01-9
03	Output Diagnostic Test Mode	T 01-29
05	Erase DTC Memory	T 01-33
06	End Output	T 01-33
07	Code Control Module	T 01-35
08	Read Measuring Value Block	T 01-37

Note:

Only the functions listed here can be used for AC/Heating On Board Diagnostic (OBD) program

After pressing the → button, the OBD program returns to the starting position.

Rapid data transfer HELP
Select function XX

◄ Indicated on display (function selection, e.g. 02 – Checking DTC memory)

Rapid data transfer HELP Control module does not answer!

◄ If one of the following messages are displayed, the possible causes can be printed out by pressing the HELP button.

Rapid data transfer HELP K wire not switching to B+!

◄ The ignition must be switched on!

Rapid data transfer → No signal from control module!

◄ Check VAG 1551 adapter cables and vehicle wiring to Data Link Connector (DLC) using wiring diagram.

Rapid data transfer → Error in communication link

◄ If this display appears at the beginning of or during the OBD program, malfunctions have occurred and data transfer is no longer possible between the VAG 1551 Scan tool and A/C control module.

– After repairing the possible causes of the malfunction, once again select address word 08 for "AC/Heating electronics" and confirm entry with -Q- button.

Rapid data transfer Q 08 AC/Heating electronics

◄ Indicated on display after entering the address word 08:

– Press -Q- button to confirm input.

Rapid data transfer Tester sends address word 08

◄ Indicated on display

701 907 044 A CLIMATRONIC VXX → Coding 00001 WSC XXXXX

◄ Control module identification is displayed

– Press → button.

Rapid data transfer HELP Select function XX

◄ Indicated on display (function selection, e.g. 02 – Checking DTC memory)

Note:

A list of available functions is printed out after pressing the HELP button

Diagnostic Trouble Code (DTC) memory, checking (scan tool function 02)

– Connect VAG 1551 Scan Tool (ST), input address word 08 "AC/Heating electronics" and press → until "Select function XX" is displayed ⇒ Page T 01-3.

– Switch printer on by pressing the PRINT button (indicator lamp in button comes on).

Rapid data transfer	HELP
Select function XX	

◄ Indicated on display (function selection):

– Press buttons -0- and -2- to select "Check DTC memory" function 02.

Rapid data transfer	Q
02 – Check DTC memory	

◄ Indicated on display

– Press -Q- button to confirm input.

X faults recognized!

◄ Number of stored DTCs or "No DTC recognized" is displayed.

No DTC recognized!	→

– Press → button.

Stored DTCs are displayed and printed out one after another.

When all DTCs are displayed/printed out, eliminate malfunctions as described in Diagnostic Trouble Code (DTC) table ⇒ Page T 01-11.

If "No DTC recognized" is displayed, program returns to the starting position after pressing → button.

Rapid data transfer	HELP
Select function XX	

◄ Indicated on display (function selection)

– End output (function 06) ⇒ Page T 01-34.

– Switch off ignition and disconnect Data Link Connector (DLC) adapter harness.

Notes:

♦ *If DTC(s) are recognized:*

– *Refer to DTC table ⇒ Page T 01-11.*

– *Rectify malfunction(s).*

– *Check DTC memory again (scan tool function 02) ⇒ Page T 01-9.*

– *Erase DTC memory (scan tool function 05) ⇒ Page T 01-33.*

♦ *If no DTC(s) are recognized:*

– *Verify customer's complaint.*

– *Carry out "Output Diagnostic Test Mode (DTM)" function 03 ⇒ Page T 01-29.*

– *Carry out "Read Measuring Value Block" function 08 ⇒ Page T 01-37.*

– *Perform Climatronic system troubleshooting ⇒ Page T 87-111.*

Diagnostic Trouble Code (DTC) table

Notes:

- *The following table lists all DTCs recognized by the Climatronic control module -J255- and displayed/printed out by VAG 1551 Scan Tool (ST).*
- *If malfunctions occur only occasionally or if the DTC memory was not erased after repairs, these DTCs are displayed as sporadic, ("SP" appears in the display).*
- *DTCs appear only on print-out.*
- *Where applicable, an indication of possible malfunction types may also appear below the DTC code and identified component or signal description.*
- *After performing repairs, DTC memory should always be checked again and erased with VAG 1551 Scan Tool.*
- *Before replacing components, always test the relevant positive and Ground connections as well as all harness connectors and wiring between the component indicated to be malfunctioning and the Climatronic control module. Refer to the wiring diagram.*
- *If a customer complaint is verified and no DTCs are stored in memory:*
  - *select "Output Diagnostic Test Mode (DTM)" (scan tool function 03)* *⇒ Page T 01-29*
  - *select "Read Measuring Value Block" (scan tool function 08)* *⇒ Page T 01-37*
  - *perform Climatronic troubleshooting* *⇒ Page T 87-111*

VAG 1551 Scan Tool printout	Possible cause	Corrective steps
No DTC recognized	If "No DTC recognized" appears after performing repairs, On Board Diagnostic (OBD) program is ended.	
00532 Supply Voltage (B+)		
Signal too high	♦ Generator (GEN) malfunction	– Check voltage supply using "Read Measuring Value Block" (function 08) ⇒ Page T 01-37
Signal too low	♦ Contact resistance in wiring or connectors from fuse/relay panel terminal Q/4 (15 circuit) to Climatronic control module -J255-, terminal -T5e/2- or from -T5e/1- to Ground	– Check wiring and connectors to control module wiring using diagram – Check Generator (GEN) ⇒ Electrical Wiring Diagrams

VAG 1551 Scan Tool printout	Possible cause	Corrective steps
00538 Reference Voltage	Voltage at Climatronic control module -J255- terminals -T28a/9- or -T28a/13- more than 5.5 V or less than 4.4 V	
Signal too high	♦ Short or open circuit in wiring or connectors to -J255- terminals -T28a/9- or -T28a/13-	– Check wiring and connectors to -J255- using wiring diagram
Signal too low	♦ Sunlight photo sensor -G107- malfunction ♦ Temperature regulator flap motor position sensor -G92- malfunction ♦ Central flap motor position sensor -G112- malfunction ♦ Footwell/defrost flap motor position sensor -G114- malfunction	– Disconnect electrical connectors from components (listed at left) one at a time and follow the sequence below for each component: – Erase DTC memory ⇒ Page T 01-33 – Check DTC memory again ⇒ Page T 01-9 • If "Reference voltage" DTC is no longer recognized, replace the applicable component.
	♦ -J255- malfunction	– Replace -J255- if necessary

VAG 1551 Scan Tool printout	Possible cause	Corrective steps
00603 Footwell/Defroster Flap Motor -V85-[1]	♦ Short or open circuit in wiring or connectors to footwell/defroster flap motor -V85-	– Check wiring and connectors to -V85- using wiring diagram
	♦ -V85- incorrectly adjusted	– Check footwell/defrost flap position using "Read Measuring Value Block" (function 08) ⇒ Page T 01-37 – Carry out Output Diagnostic Test Mode (function 03) ⇒ Page T 01-29 – Check and adjust end position of -V85- ⇒ Page T 87-82
	♦ -V85- malfunction	– Replace -V85-
00658 Operating Module Instrument Panel Temperature Sensor -G56-	♦ Short or open circuit in wiring or connectors to A/C control head -E87-	– Check segments of -E87-: carry out Output Diagnostic Test Mode (DTM) (function 03) ⇒ Page T 01-29
	♦ -G56- malfunction	– Check -G56- using "Read Measuring Value Block" (function 08) ⇒ Page T 01-37
	♦ -E87- malfunction	– Replace -E87-
	♦ Climatronic control module -J255- malfunction	– Replace -J255- if necessary

[1] DTC is recognized after approx. 2 seconds.

VAG 1551 Scan Tool printout	Possible cause	Corrective steps
00739 Coolant Pump -V50-		
Short circuit to Ground	♦ Short circuit in wiring or connectors from Climatronic control module -J255- terminal -T5e/3- to Coolant Pump -V50-	– Check -V50- using "Read Measuring Value Block" (function 08) ⇒ Page T 01-37 – Check wiring and connectors to -V50- using wiring diagram
	♦ -V50- malfunction	– Replace -V50-
00766 Front Vent Temperature Sensor -G152-		
Open/Short circuit to B+	♦ Short or open circuit in wiring or connectors to -G152-	– Check -G152- using "Read Measuring Value Block" (function 08) ⇒ Page T 01-37 ♦ If DTCs for the outside air temperature sensor -G17-, sunlight photo sensor -G107- and the A/C engine coolant temperature sensor -G110- are also present, first check wiring and connections from Climatronic control module -J255- terminal -T28a/14- to internal splice junction (L56) for sensors in wiring harness for Climatronic.
Short circuit to Ground	♦ -G152- malfunction	– Check wiring and connectors using wiring diagram – Replace -G152-

VAG 1551 Scan Tool printout	Possible cause	Corrective steps
00767 Rear Evaporator Temperature Sensor -G153-		
Open/Short circuit to B+	♦ Short/open circuit in wiring or connections from Climatronic control module -J255- terminal -T28b/15- via rear evaporator temperature sensor -G153- to -J255- terminal -T28b/3-	– Check -G153- using "Read Measuring Value Block" (function 08) ⇒ Page T 01-37 ♦ If a DTC for the rear heater core temperature sensor -G154- is also present, first check wiring and connections from Climatronic control module -J255- terminal -T28b/3- to internal splice junction (244) for Ground connections in wiring harness for Climatronic.
Short circuit to Ground	♦ -G153- malfunction	– Check wiring and connectors using wiring diagram – Replace -G153-

VAG 1551 Scan Tool printout	Possible cause	Corrective steps
00768 Rear Heater Core Temperature Sensor -G154-		
Open/Short circuit to B+	♦ Short/open circuit in wiring or connections from Climatronic control module -J255- terminal -T28b/17- via rear heater core temperature sensor -G154- to -J255- terminal -T28b/3-	– Check -G154- using "Read Measuring Value Block" ⇒ Page T 01-37 ♦ If a DTC for the rear evaporator sensor -G153- is also present, first check wiring and connections from Climatronic control module -J255- terminal -T28b/3- to internal splice junction (244) for Ground connections in wiring harness for Climatronic.
Short circuit to Ground	♦ -G154- malfunction	– Check wiring and connectors using wiring diagram – Replace -G154-

VAG 1551 Scan Tool printout	Possible cause	Corrective steps
00779 Outside Air Temperature Sensor -G17-		
Open/Short circuit to B+	♦ Short or open circuit in wiring or connectors to outside air temperature sensor -G17-	– Check -G17- using "Read Measuring Value Block" (function 08) ⇒ Page T 01-37 ♦ If DTCs for the sunlight photo sensor -G107- A/C engine coolant temperature sensor -G110- and the front vent temperature sensor -G152- are also present, first check wiring and connections from Climatronic control module -J255- terminal -T28a/14- to internal splice junction (L56) for sensors in wiring harness for Climatronic.
Short circuit to Ground	♦ -G17- malfunction	– Check wiring and connectors using wiring diagram – Replace -G17-

VAG 1551 Scan Tool printout	Possible cause	Corrective steps
00791 A/C Evaporator Temperature Switch -E33-[2]	◆ Short or open circuit in wiring or connectors to A/C evaporator temperature switch -E33-	– Check -E33- with "Read Measuring Value Block" (function 08) ⇒ Page T 01-37
	◆ -E33- malfunction	– Replace -E33-
No check possible at this time	◆ This message appears only when Output Diagnostic Test Mode (function 03) was carried out before checking DTC Memory and the temperature switch could not be checked (e.g. if outside air temperature measured was below 12°C / 54°F). This DTC will be erased from memory when ignition is switched off.	

[2] This DTC is not loaded if the outside air temperature is below 12°C (54°F) or if outside air temperature sensor -G17- malfunctions. Condition required for loading of DTC: A/C readiness "On", and -E33- has been opened for at least 10 minutes or no voltage at Climatronic control module -J255- terminal -T28a/3-.

VAG 1551 Scan Tool printout	Possible cause	Corrective steps
00792 A/C Pressure Switch -F129-[3]	◆ Short or open circuit in wiring or connectors to A/C pressure switch -F129-	– Check wiring and connectors to -F129- using wiring diagram
	◆ Refrigerant system malfunctions: overcharged insufficient charge condensor or evaporator blocked	– Check operation of -F129- using "Read Measuring Value Block" (function 08) ⇒ Page T 01-37
	◆ -F129- malfunction	– Replace -F129-
No check possible at this time	◆ This message appears only when Output Diagnostic Test Mode (function 03) was carried out before checking DTC Memory and the pressure switch could not be checked (e.g. if outside air temperature measured was below 12°C / 54°F). This DTC will be erased from memory when ignition is switched off.	
00796 Fan for Interior Temperature Sensor -V42-	◆ Air intake for -V42- blocked or restricted	– Carry out visual check; if necessary, clean air intake and -V42-
	◆ Short or open circuit in wiring or connectors to A/C Control head -E87- or to temperature sensor blower -V42-	– Check wiring and connectors to -E87- and -V42- using wiring diagram
	◆ -V42- malfunction	– Replace -V42-

[3] This DTC is not loaded if the outside air temperature is below 12°C (54°F) or if outside air temperature sensor -G17- malfunctions.

VAG 1551 Scan Tool printout	Possible cause	Corrective steps
00797 Sunlight Photo Sensor -G107-		
Open/Short circuit to B+	♦ Open or short circuit circuit in wiring or connectors to sunlight photo sensor -G107-	– Check -G107- using "Read Measuring Value Block" (function 08) ⇒ Page T 01-37 ♦ If DTCs for the outside air temperature sensor -G17-, A/C engine coolant temperature sensor -G110- and the front vent temperature sensor -G152- are also present, first check wiring and connections from Climatronic control module -J255- terminal -T28a/14- to internal splice junction (L56) for sensors in wiring harness for Climatronic.
	♦ -G107- malfunction	– Check wiring and connectors using wiring diagram – Replace -G107-

VAG 1551 Scan Tool printout	Possible cause	Corrective steps
00799 Engine Coolant Temperature Sensor -G110-		
Open/Short circuit to B+	♦ Open or short circuit circuit in wiring or connectors to engine coolant temperature sensor -G110-	– Check -G110- using "Read Measuring Value Block" (function) ⇒ Page T 01-37 ♦ If DTCs for the outside air temperature sensor -G17-, sunlight photo sensor -G107- and the front vent temperature sensor -G152- are also present, first check wiring and connections from Climatronic control module -J255- terminal -T28a/14- to internal splice junction (L56) for sensors in wiring harness for Climatronic.
Short circuit to Ground	♦ -G110- malfunction	– Check wiring and connectors using wiring diagram – Replace -G110-

VAG 1551 Scan Tool printout	Possible cause	Corrective steps
01026 Rear Evaporator Fan -V20-	♦ Short or open circuit in wiring or connectors to rear evaporator fan -V20-	– Check wiring and connectors to -V20- using wiring diagram – Check voltage at -V20- using "Read Measuring Value Block" (function 08) ⇒ Page T 01-37 – Carry out Output Diagnostic Test Mode (function 03) ⇒ Page T 01-29
	♦ Control module for blower for evaporator -J349- malfunction	– Replace -J349-
	♦ -V20- malfunction	– Replace -V20-
01027 Rear Warm Air Fan -V47-	♦ Short or open circuit in wiring or connectors to rear warm air fan -V47-	– Check wiring and connectors to -V47- using wiring diagram – Check voltage at -V47- using "Read Measuring Value Block" (function 08) ⇒ Page T 01-37 – Carry out Output Diagnostic Test Mode (function 03) ⇒ Page T 01-29
	♦ Control module for warm air blower -J350- malfunction	– Replace -J350-
	♦ -V47- malfunction	– Replace -V47-

VAG 1551 Scan Tool printout	Possible cause	Corrective steps
01028 Fan Relay -J323-	Short or open circuit in wiring or connectors from Climatronic control module -J255- terminal -T28b/11- to blower relay -J323- or from -J255- terminal -T28a/25- to relay for Climatronic-J254-	– Check -J323- using "Read Measuring Value Block" (function 08) ⇒ Page T 01-37 – Check wiring and connectors to -J323- using wiring diagram
	-J323- malfunction	– Replace -J323-

VAG 1551 Scan Tool printout	Possible cause	Corrective steps
01086		
Speedometer Vehicle Speed Sensor -G22-[4]	♦ Short or open circuit in wiring or connectors from terminal -T28/7- (instrument cluster -K-) to Climatronic control module -J255- terminal -T28b/28-	– Check signal from -G22- or -G54-[5] using "Read Measuring Value Block" (function 08) ⇒ Page T 01-37 – Check wiring and connectors to -G22- or -G54- using wiring diagram[5]
	♦ Vehicle speed sensor -G54- malfunction	– Replace -G54-[5] ⇒ Repair Manual, *Electrical Equipment*, Repair Group 90
	♦ Speedometer vehicle speed sensor -G22-, on transmission malfunction (only in conjunction with inoperative Speedometer -G21-)	– Replace -G22-[5] ⇒ Repair Manual, *Electrical Equipment*, Repair Group 90
No check possible at this time	This message appears under the following conditions:[5] ♦ Vehicle speed lower than 10 km/h (6.25 mph). or ♦ Output Diagnostic Test Mode (function 03) was performed before checking DTC memory. This display does not effect OBD program. display is erased from DTC memory after switching ignition off.	

4) DTC is loaded for speedometer vehicle speed sensor -G22- (on vehicles with electronic speedometer) or vehicle speed sensor -G54- (on vehicles with mechanical speedometer).

5) Supplementary check after repair/replacement:
- Erase DTC memory, drive vehicle for at least 2 minutes in excess of 15 Km/h (9.5 mph), switch engine off.
- Repeat sequence 3 times. If the malfunction has not been rectified the DTC will again load into memory

VAG 1551 Scan Tool printout	Possible cause	Corrective steps
01205		
Climatronic Relay -J254-		
Short circuit to Ground	♦ Short or open circuit in wiring from Climatronic control module -J255- terminal -T28b/14- via -E33-, -F129-, -F163- to Climatronic relay -J254-	– Check wiring and connectors to -J254- using wiring diagram
	♦ -V254- malfunction	– Replace -J254-
No check possible at this time	♦ This message appears only when Output Diagnostic Test Mode (function 03) was carried out before checking DTC Memory and the Climatronic relay -J254- could not be checked (e.g. if outside air temperature measured was below 12°C / 54°F). This DTC will be erased from memory when ignition is switched off.	

VAG 1551 Scan Tool printout	Possible cause	Corrective steps
01271 Temperature Regulator Flap Motor -V68-[6]	◆ Short or open circuit in wiring or connectors to temperature regulator flap motor -V68-	– Check wiring and connectors to -V68- using wiring diagram
	◆ -V68- incorrectly adjusted	– Check position of temperature flap using "Read Measuring Value Block" (function 08) ⇒ Page T 01-37 – Check and adjust end position of -V68- ⇒ Page T 87-81
	◆ -V68- malfunction	– Carry out Output Diagnostic Test Mode (function 03) ⇒ Page T 01-29 – Replace -V68-
01272 Central Air Flap Motor -V70-[6]	◆ Short or open circuit in wiring or connectors to central air flap motor -V70-	– Check wiring and connectors to -V70- using wiring diagram
	◆ -V70- incorrectly adjusted	– Check central flap position using "Read Measuring Value Block" (function 08) ⇒ Page T 01-37
	◆ -V70- malfunction	– Carry out Output Diagnostic Test Mode (function 03) ⇒ Page T 01-29 – Replace -V70-

6) DTC is recognized after approx. 2 seconds.

VAG 1551 Scan Tool printout	Possible cause	Corrective steps
01273 Fresh Air Blower -V2-	◆ Short or open circuit in wiring or connectors to fresh air blower -V2-	– Check wiring and connectors to -V2- using wiring diagram – Carry out Output Diagnostic Test Mode (function 03) ⇒ Page T 01-29
	◆ Control module for fresh air blower -J126- malfunction	– Replace -J126-
	◆ -V2- malfunction	– Replace -V2-
65535 Control Module Malfunctioning	◆ Outside electrical interference caused by poor Ground or power supply connections to Climatronic control module -J255-	– Check all Ground and power supply wiring and connectors to -J255- using wiring diagram
	◆ Sunlight photo sensor -G107- malfunction ◆ Temperature regulator flap motor position sensor -G92- malfunction ◆ Central flap motor position sensor -G112- malfunction ◆ Footwell/defrost flap motor pos. sensor -G114- malfunction	– Disconnect electrical connectors from components (listed at left) one at a time and follow the sequence below for each component: – Erase DTC memory ⇒ Page T 01-33 – Check DTC memory again ⇒ Page T 01-9 • If the "Control Module Malfunctioning" DTC is no longer recognized, replace the applicable component.
	◆ -J255- malfunction	– Replace -J255- if necessary

Output Diagnostic Test Mode (DTM) (scan tool function 03)

Notes:

- *Output DTM must only be carried out with the engine not running.*
- *For definitive results, output DTM must be carried out at an ambient temperature of at least 12°C (54°F) as displayed on A/C control head Page T 87-59.*
- *Component locations ⇒ Repair Group 87.*
- *During output DTM, both coolant fans (-V7- and -V35-) are activated.*
- *Climatronic does not function or automatically regulate during output DTM.*
- *Output DTM can be repeated several times If necessary.*

Initiating Output DTM

– Connect VAG 1551 Scan Tool (ST) and select AC/Heating electronics (address word 08) ⇒ Page T 01-3.

Rapid data transfer	HELP
Select function XX	

◄ Indicated on display (function selection):

– Press buttons -0- and -3- to select "Output Diagnostic Test Mode" function 03.

Rapid data transfer	Q
03 – Output diagnostic test mode	

◄ Indicated on display

– Press Q button to confirm input.

Output diagnostic test mode

◄ Indicated on display

The following functional tests are carried out in sequence:

- A/C Control head -E87- display ⇒ Page T 01-31
- Flap motors Page T 01-31
- Electrical circuit to the following blower motors:
  - Fresh air blower -V2-
  - Rear evaporator fan -V20-
  - Rear warm air fan -V47- ⇒ Page T 01-32
- Electrical circuit to A/C clutch -N25- Page T 01-32

Function is not recognized or cannot	→
be carried out at the moment!	

◄ Indicated on display

– Output DTM is concluded.

– Press → button.

Note:

After completing output DTM, always check DTC memory again and rectify any malfunctions that may have been recognized during output DTM.

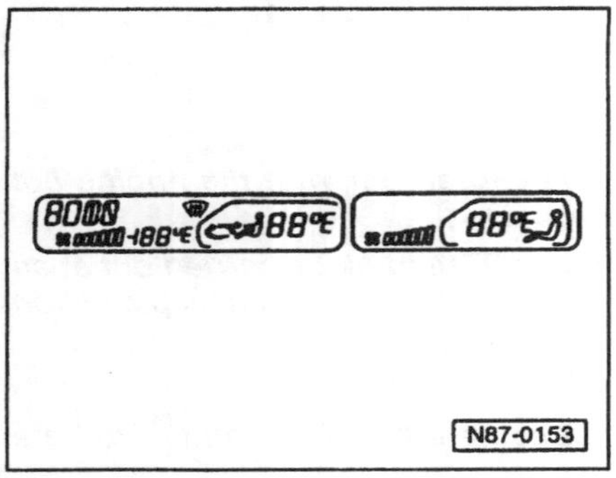

N87-0153

A/C Control head -E87- display test

◄ All display segments must appear fully in display. If not, replace -E87-.

Flap motor test

During output DTM:

- Temperature regulator flap motor -V68-
- Central air flap motor -V70-
- Footwell/defrost flap motor -V85-

are activated in sequence and move over their complete operating range. The first half of the motor movement range will be continuously activated, the remainder will be driven in stages.

The movement will be completed in approx. 30 seconds and the flap motor adopts the position selected on the A/C control head.

Note:

If after subsequent checking of DTC memory, no DTC is recognized (even though the selected positions were not attained), check the appropriate flap motor or air flap for mechanical damage.

Electrical circuit tests

Blower/fan motors

During output DTM, the following are activated in sequence with 2 different voltages:

- Fresh air blower -V2-
- Rear evaporator fan -V20-
- Rear warm air fan -V47-

Should a malfunction be detected in the electrical circuit to one of the motors listed, it will be stored in DTC memory.

A/C clutch -N25-

During output DTM, the electrical circuit is tested from connector T28b/14 via A/C evaporator temperature switch -E33- to after A/C pressure switch -F129-. Should a malfunction be detected, it will be stored in the DTC memory.

Note:

If after subsequent checking of DTC memory, no DTC is detected (even though the A/C clutch is NOT activated during cooling operation), then check wiring and connections from A/C pressure switch -F129- via -F163- and -J32- to A/C clutch -N25- using wiring diagram.

DTC Memory, erasing (scan tool function 05)

Prerequisite:

- DTC Memory has been checked and malfunctions corrected.

Erasing DTC Memory

– Press → button.

Rapid data transfer	HELP
Select function XX	

◄ Indicated on display (function selection):

– Press buttons -0- and -5- to select the "Erase DTC memory" function 05.

Rapid data transfer	Q
05 Erase DTC memory	

◄ Indicated on display

– Press -Q- button to confirm input.

Rapid data transfer	→
DTC memory is erased!	

◄ Indicated on display

– Press → button.

Rapid data transfer	HELP
Select function XX	

◄ Indicated on display (function selection):

Notes:

Warning!
DTC memory was not checked

◄ ♦ *If this appears in the display, the test sequence is faulty.*

♦ *Follow test sequence precisely: first check DTC memory and if correct malfunctions.*

Ending output (scan tool function 06)

– Press buttons -0- and -6- to end the OBD program.

Rapid data transfer	Q
06 End output	

◄ Indicated on display

– Press -Q- button to confirm input.

Rapid data transfer	Help
Enter address word XX	

◄ Indicated on display

– Switch off ignition.

– Disconnect VAG 1551 Scan Tool (ST) from Data Link Connector (DLC).

Climatronic Control Module -J255-, coding (scan tool function 07)

If the appropriate code is not displayed, or if the control module has been replaced, the control module must be coded as follows:

– Connect VAG 1551 Scan Tool (ST), input address word 08 "AC/Heating electronics" and press → until "Select function XX" appears in display ⇒ Page T 01-3.

Rapid data transfer	HELP
Select function XX	

◄ Indicated on display (function selection):

– Press buttons -0- and -7- to select the "Coding control module" function 07.

Coding control module	Q
Enter code number	XXXXX (0-32000)

◄ Indicated on display

– Press -Q- button to confirm input.

– Using VAG 1551 keypad, enter code number and confirm with Q button.

Coding	For vehicles
00001	with dust and pollen filter

	→
Coding 00001	WSC XXXXX

◄ Indicated on display after entering code.

Notes:

- *The ignition must be switched off in order for the desired coding to be loaded into the control module. Always "End Output" (function 06) after coding control module and switch ignition off before proceeding with any additional VAG 1551 tests.*
- *If an unauthorized code number is entered, coding 00000 (for vehicles without dust and pollen filter) is automatically entered by default.*
- *If no code number is entered after control module replacement, coding 00000 (for vehicles without dust and pollen filter) is automatically entered by default.*
- *All vehicles for the USA and Canada are equipped with the dust and pollen filter. Always ensure that coding 00001 is entered.*
- *Improper coding results in incorrect blower output.*

– Press → button.

Rapid data transfer	HELP
Select function XX	

◄ Indicated on display (function selection):

– End output (function 06) ⇒ Page T 01-33.

– Switch off ignition.

– Disconnect VAG 1551 Scan Tool (ST) from Data Link Connector (DLC).

Read Measuring Value Block (scan tool function 08)

Notes:

- *During the function "Read Measuring Value Block" all Climatronic functions and automatic regulation take place.*
- *Provision is made for 7 display groups (measuring value blocks) each containing 4 display fields (measuring values).*

Initiating Read Measuring Value Block

– Connect Scan Tool (ST) VAG 1551 and input address word "08 AC/Heating electronics" and then press → button, until "Select function XX" is shown in display ⇒ Page T 01-3.

Rapid data transfer	HELP
Select function XX	

◄ Indicated on display (function selection):

– Press buttons -0- and -8- to select the "Read Measuring Value Block" function 08.

Rapid data transfer	Q
08 – Read Measuring Value Block	

◄ Indicated on display

– Press -Q- button to confirm input.

Read Measuring Value Block
Input display group number XX

◄ Indicated on display

Table of selectable display group numbers

Display Group	Display Field	Designation
001	1	Outside air temperature sensor -G17-
	2	Front vent temperature sensor -G152-
	3	Instrument panel interior temperature sensor -G56-
	4	A/C engine coolant temperature sensor -G110-
002	1	Speedometer vehicle speed sensor -G22- or vehicle speed sensor -G54-
	2	Sunlight photo sensor -G107-
	3	Reference voltage supply for sensors -G92-, -G107-, -G112- and -G114-
	4	Voltage supply (terminal 15)
003	1	A/C Evaporator temperature switch -E33-
	2	A/C pressure switch -F129-
	3	Climatronic readiness/engine start recognition
	4	Climatronic cut-off by kick-down switch

Display Group	Display Field	Designation
004	1 2 3 4	Temperature regulator flap position Central air flap position Footwell/defrost flap position Vacuum valve rail -N53-
005	1 2 3 4	Auxiliary water heater Fan relay -J323- Coolant pump -V50- Voltage signal (terminal X)
006	1 2 3 4	Rear evaporator temperature sensor -G153- A/C refrigerant shut-off valve -N43- Rear heater core temperature sensor -G154- Auxiliary heater valve -N172-
007	1 2 3 4	Fresh air blower -V2- Rear warm air fan -V47- Rear evaporator fan -V20- Not applicable

Read Measuring Value Block			1 →
1	2	3	4

◄ Indicated on display

Notes:

- *Decoding display zones readouts ⇒ Test table, Page T 01-41.*
- *With the printer switched on the information on the display will be printed-out.*
- *If the specifications are attained in all display zones:*

– Press → button.

Rapid data transfer	HELP
Select function XX	

◄ Indicated on display (function selection):

Note:

After completing the "Read Measuring Value Block" function check DTC Memory ⇒ Page T 01-9

Test table

Display Group	Display field	Designation	Test conditions	Specified readout on VAG 1551 display	Corrective steps
001	1	Test result of outside air temperature sensor -G17-	Displayed value equates to actual air temperature measured at sensor	... °C	– Check wiring and connectors to sensor using wiring diagram – Replace -G17-
	2	Test result of front vent temperature sensor -G152-	Displayed value equates to actual air temperature measured at sensor	... °C	– Check wiring and connectors to sensor using wiring diagram – Replace -G152-
	3	Test result of instrument panel interior temperature sensor -G56-	Displayed value equates to actual air temperature measured at sensor	... °C	– Check wiring and connectors to sensor using wiring diagram – Replace -G56-
	4	Test result of A/C engine coolant temperature sensor -G110-	Displayed value equates to actual coolant temperature measured at sensor	... °C	– Check wiring and connectors to sensor using wiring diagram – Replace -G110-

Display group	Display field	Designation	Test conditions	Specified readout on VAG 1551 display	Corrective steps
002	1	Speed signal from vehicle road speed sensor -G22- or from vehicle speed sensor -G54-	Displayed value equates to actual road speed measured by sensor[1)]	... km/h	– Check wiring and connectors from sensor using wiring diagram – Replace -G22- (only when the speedometer -G21- displays no speed) ⇒ Repair Manual, *Electrical Equipment,* Repair Group 90 – Replace -G54- ⇒ Repair Manual, *Electrical Equipment,* Repair Group 90
	2	Measurement value of sunlight photo sensor -G107-	Displayed value depends on light intensity	0..100%	– Check wiring and connectors to sensor using wiring diagram – Replace -G107-
			dark	0%	
Cont.			light	100%	

1) Displayed value dependant on transmission and tire options. A functional hall sensor is indicated by speed signal display.

Display group	Display field	Designation	Test conditions	Specified readout on VAG 1551 display	Corrective steps
002	3	Reference voltage supply for sensors -G92-, -G107-, -G112- and -G114-)	Voltage range: 4.4 to 5.5V	..V	– Check wiring and connectors for voltage supply using wiring diagram
	4	Supply voltage at Climatronic control module -J255- terminal -T5e/2-	Displayed value equates to battery voltage	..V	– Check wiring and connectors for voltage supply using wiring diagram – Check Generator (GEN). For troubleshooting procedure, ⇒ Electrical Wiring Diagrams

Display group	Display field	Designation	Test conditions		Specified readout on VAG 1551 display	Corrective steps
003	1	A/C evaporator temperature switch -E33- switched status	Switch	closed	1	
				open	0	
Cont.					–	If "–" appears in display, test is not possible because: • Climatronic is not switched on (not ready) • Outside air temperature is below 12°C (54°F) • Outside air temperature sensor -G17- malfunction.

Display group	Display field	Designation	Test conditions	Specified readout on VAG 1551 display	Corrective steps
003	2	A/C pressure switch -F129- switched status	No malfunctions in refrigerant circuit Condenser clean and has sufficient air flow (cooling) 2 to 32 bar switch component closed 16 bar switch component open	1 0[2)]	If "0 0" appears in display • 2 to 32 bar & 16 bar switch components open, indicating that the refrigerant circuit is empty. – Check refrigerant pressures ⇒ Page T 87-120
Cont.			No malfunctions in refrigerant circuit 2nd speed radiator blower switched on 2 to 32 bar switch component closed 16 bar switch component closed	1 1[2)]	If "0 1" appears in display • 2 to 32 bar switch component open & 16 bar switch component closed indicates that the condenser cooling is insufficient (e.g.: dirt/debris). If condensor OK, refrigerant circuit is overcharged. – Check refrigerant pressures ⇒ Page Page T 87-120

[2)] Coded value indicating switch component status. First figure indicates 2 to 32 bar switch component status, the second figure indicates 16 bar switch component status (in both cases, 1 = closed, 0 = open).

Display group	Display field	Designation	Test conditions	Specified readout on VAG 1551 display	Corrective steps
003 Cont.	2				If "– 0" appears in display • Checking the 2 to 32 bar switch component is not possible at the moment because the Climatronic system is not switched on (not ready) or the A/C evaporator temperature switch -E33- is open. Condenser cooling is sufficient. If "– 1" appears in display • Checking the 2 to 32 bar switch component is not possible at the moment because the Climatronic system is not switched on (not ready) or the A/C evaporator temperature switch -E33- is open. Radiator fan 2nd speed has switched on.

Display group	Display field	Designation	Test conditions	Specified readout on VAG 1551 display	Corrective steps
	3	Climatronic system readiness[3)]/engine start recognized	System readiness recognized and engine started[4)]	1 1[5)]	If the engine start is not recognized:
			System readiness recognized and engine not started	1 0[5)]	– Check wiring and connections from terminal X to Climatronic control module -J255- terminal -T28a/1-
			System readiness not recognized and engine not started	0 0[5)]	
Cont.			System readiness not recognized and engine started	0 1[5)]	

[3)] System readiness recognized if voltage is established at connector -T28a/3-

[4)] A/C clutch -N25- will only be switched on when engine start is recognized.

[5)] Coded value indicating both the Climatronic readiness and starting status. First figure indicates: 1 = on, 0 = off.
The second figure indicates the engine starting status: 1 = started, 0 = not started.

Display group	Display field	Designation	Test conditions	Specified readout on VAG 1551 display	Corrective steps
003	4	Climatronic system cut-out by kick-down[6)]	Kick-down switch operated (full throttle)	1 1[7)]	If "1 0" appears in display – Check wiring and connections from Automatic Transmission Control Module (TCM) -J217- to Climatronic control module-J255- – Replace control module -J255- if necessary
			Kick-down switch not operated	0 1[7)]	If "0 0" appears in display, test is not possible because: • Climatronic is not switched on (not ready) • Outside air temperature is below 12°C (54°F) • Outside air temperature sensor -G17- malfunction.

[6)] Check possible only when vehicle is driven.

[7)] Coded value indicating both the kickdown and transmission status. The first figure indicates the kickdown switch status: 1 = closed, 0 = open. The second figure indicates the transmission type: 1 = automatic, 0 = Manual. If a transmission other than indicated is installed, check wiring to -J225- using wiring diagram.

Display group	Display field	Designation	Test conditions	Specified readout on VAG 1551 display	Corrective steps
Cont.	1	Position of temperature regulator flap	Maximum heater output "HI" selected with front passenger "warmer" button Minimum heater output "LO" selected with front passenger "cooler" button	90% ..100% 0% ..10%	– Check wiring and connections from Climatronic control module -J255- to temperature regulator flap motor position sensor -G92- – Check end position of temperature regulator flap motor -V68-, adjust if necessary ⇒ Page T 87-81 – Replace -V68-

Display group	Display field	Designation	Test conditions	Specified readout on VAG 1551 display	Corrective steps
Cont.	2	Position of central air flap	Air distribution to defrost selected[9] with "defrost" button Air distribution to footwell selected[9] with front passenger "footwell" button Air distribution to vent selected[9] with front passenger "vent" button	0% ..10% 0% ..10% 90% ..100%	– Check wiring and connections from Climatronic control module -J255- to central flap motor position sensor -G112- – Check end position of central air flap motor -V70-, adjust if necessary ⇒ Page T 87-81 – Replace central air flap motor -V70-

[8] For valid test results, air distribution must be set ONLY to level indicated, not in conjunction with other levels.

Display group	Display field	Designation	Test conditions	Specified readout on VAG 1551 display	Corrective steps
Cont.	3	Position of footwell/ defrost flap	Air distribution to defrost selected[9] with front passenger "defrost" button Air distribution to footwell selected[9] with front passenger "footwell" button	90% ..100% 0% ..10%	– Check wiring and connections from Climatronic control module -J255- to footwell/defrost flap motor position sensor -G114- – Replace footwell/defroster flap motor -V85-

[9] For valid test results, air distribution must be set ONLY to level indicated, not in conjunction with other levels.

Display group	Display field	Designation	Test conditions	Specified readout on VAG 1551 display	Corrective steps
004	4	Vacuum valve rail -N53- switch status (fresh air/recirculating air flap and center vent)	Recirculating air button not pressed and button for air flow to vent not pressed	0 0	– Check wiring and connections from Climatronic control module -J255- to vacuum valve rail -N53-
			Recirculating air button not pressed and button for air flow to vent pressed	0 1	– Replace -N53-
			Recirculating air button pressed and button for air flow to vent not pressed	1 0	
			Recirculating air button pressed and button for air flow to vent pressed	1 1	

Display group	Display field	Designation	Test conditions	Specified readout on VAG 1551 display	Corrective steps
005	1	Auxiliary water heater switched status[10]	Auxiliary water heater operating	1	– Check wiring and connections from terminal X to Climatronic control module -J255- terminal -T28a/1-.
			Auxiliary water heater not operating	0	
	2	Fan relay -J323- switched status	Climatronic switched on and auxiliary water heater not operating[11]	1	– Check wiring and connections from Climatronic control module -J255- terminal -T28b/11- via relay -J323- to Ground.
Cont.			Climatronic switched off and auxiliary water heater operating[11]	0	

[10] If equipped. (This display field indicates whether the water heater is switched on).

[11] If equipped. Switching condition appears in display group 05, field 1.

Display group	Display field	Designation	Test conditions	Specified readout on VAG 1551 display	Corrective steps
005	3	Coolant pump -V50- switched status	Maximum heater output "HI" selected with front passenger "warmer" button	1	– Check wiring and connections from Climatronic control module -J255- terminal -T5e/3- via coolant pump -V50- to Ground.
			Minimum heater output "LO" selected with front passenger "cooler" button	0	– Replace -V50-
	4	Voltage signal (terminal X)	Ignition switched on	1	– Check wiring and connections from terminal X to Climatronic control module -J255- terminal -T28a/1-
Cont.			Auxiliary water heater operating[11]	0	

[11] If equipped. Switching condition appears in display group 05, field 1.

Display group	Display field	Designation	Test conditions	Specified readout on VAG 1551 display	Corrective steps
006	1	Test result of rear evaporator temperature sensor -G153-	Displayed value equates to actual air temperature measured at sensor	... °C	– Check wiring and connections to rear evaporator temperature sensor -G153- using wiring diagram – Replace -G153-
	2	A/C refrigerant shut-off valve valve -N43- switched status	Engine running Air temperature[1] at least 12°C. Minimum heater output "LO" selected with front passenger "cooler" button	1	– Check wiring and connections from terminal 15 via A/C refrigerant shut-off valve valve -N43- to Climatronic control module -J255- terminal -T5e/5- – Replace -N43-
Cont.			Maximum heater output "HI" selected with front passenger "warmer" button	0	– Check wiring and connections from terminal 15 via -N43- to -J255 terminal -T5e/5- – Replace -N43-

[12] For valid results, A/C control head -E87- must display an outside air temperature of at least 12°C (54°F).

Display group	Display field	Designation	Test conditions	Specified readout on VAG 1551 display	Corrective steps
006	3	Test result of rear heater core temperature sensor -G154-	Displayed value equates to actual air temperature measured at sensor	... °C	– Check wiring and connections to rear heater core temperature sensor -G154- using wiring diagram. – Replace -G154-
	4	Auxiliary heater valve -N172- switched status	Minimum heater output "LO" selected with front passenger "cooler" button	1	– Check wiring and connections from terminal 15 via valve -N172- to Climatronic control module -J255- terminal -T5e/4-.
			Maximum heater output "HI" selected with front passenger "warmer" button	0	– Replace -N172-

Display group	Display field	Designation	Test conditions	Specified readout on VAG 1551 display	Corrective steps
Cont.	1	Voltage at fresh air blower -V2-	Engine running. A/C control head -E87- front and rear passenger temperature displays indicate maximum heater output "HI" selected with front passenger "warmer" button Maximum output of Fresh air blower -V2- selected with front passenger "increase blower speed" button	approx. 12 V	– Check wiring and connections to fresh air blower -V2- using wiring diagram

Display group	Display field	Designation	Test conditions	Specified readout on VAG 1551 display	Corrective steps
	2	Voltage at rear warm air fan -V47-	Engine running A/C control head -E87- front and rear passenger temperature displays indicate maximum heater output "HI" selected with front passenger "warmer" button Maximum output of rear warm air fan -V47- with rear passenger "increase blower speed" button	approx. 12 V	– Check wiring and connections to rear warm air fan -V47- using wiring diagram
Cont.			A/C control head -E87- front and rear passenger temperature displays indicate minimum heater output "LO" selected with front passenger "cooler" button Maximum output of rear warm air fan -V47- with rear passenger "increase blower speed" button	approx. 0 V	– Check wiring and connections to rear warm air fan -V47- using wiring diagram

Display group	Display field	Designation	Test conditions	Specified readout on VAG 1551 display	Corrective steps
Cont.	3	Voltage at rear evaporator fan -V20-	Engine running. A/C control head -E87- front and rear passenger temperature displays indicate minimum heater output "LO" selected with front passenger "cooler" button Maximum output of rear evaporator blower -V20- with rear passenger "increase blower speed" button	approx. 12 V	– Check wiring and connections to rear evaporator fan -V20- using wiring diagram

Display group	Display field	Designation	Test conditions	Specified readout on VAG 1551 display	Corrective steps
007	3		A/C control head -E87- front and rear passenger temperature displays indicate maximum heater output "HI" selected with front passenger "warmer" button Maximum output of rear evaporator fan -V20- with rear passenger "increase blower speed" button	approx. 0 V	– Check wiring and connections to rear evaporator fan -V20- using wiring diagram
	4	Not assigned			

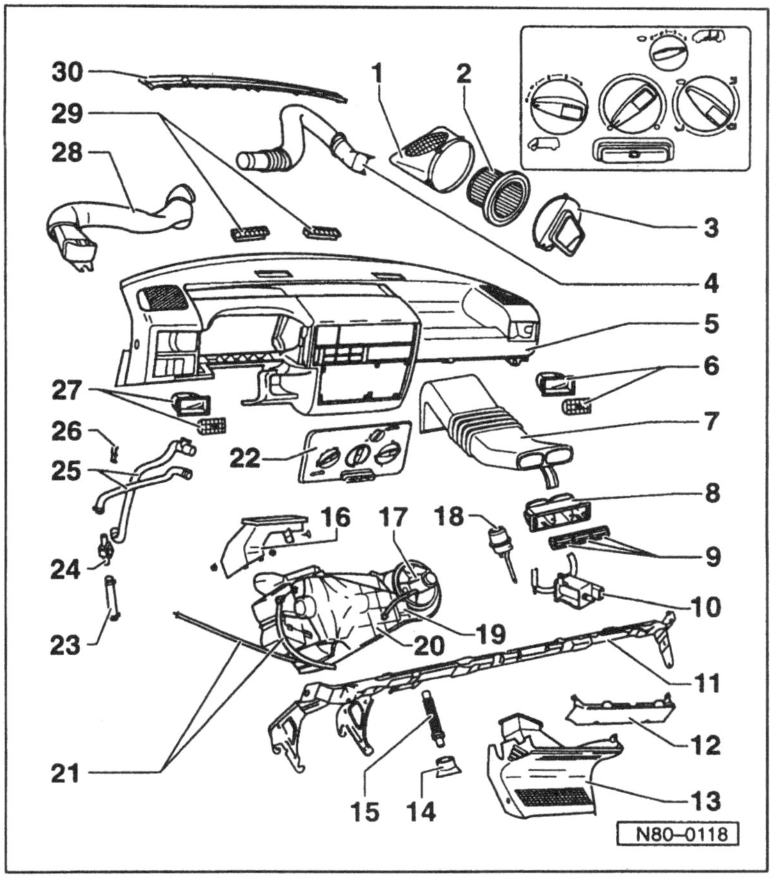

Heating & ventilation system, servicing (without A/C)

> ***CAUTION!***
>
> ***Before working on electrical system obtain radio security code and disconnect battery Ground (GND) strap.***

1 - Filter housing

2 - Dust and pollen filter

3 - Air intake duct
- ♦ Removing and installing

⇒ Fig. 1

4 - Right air duct
- ♦ Replacing
- – Remove glove compartment or passenger's side airbag unit.

⇒ Repair Manual, *Body - Exterior, Body - Interior,* Repair Group 68

5 - Instrument panel
- ♦ Removing and installing

⇒ Repair Manual, *Body - Exterior, Body - Interior,* Repair Group 70

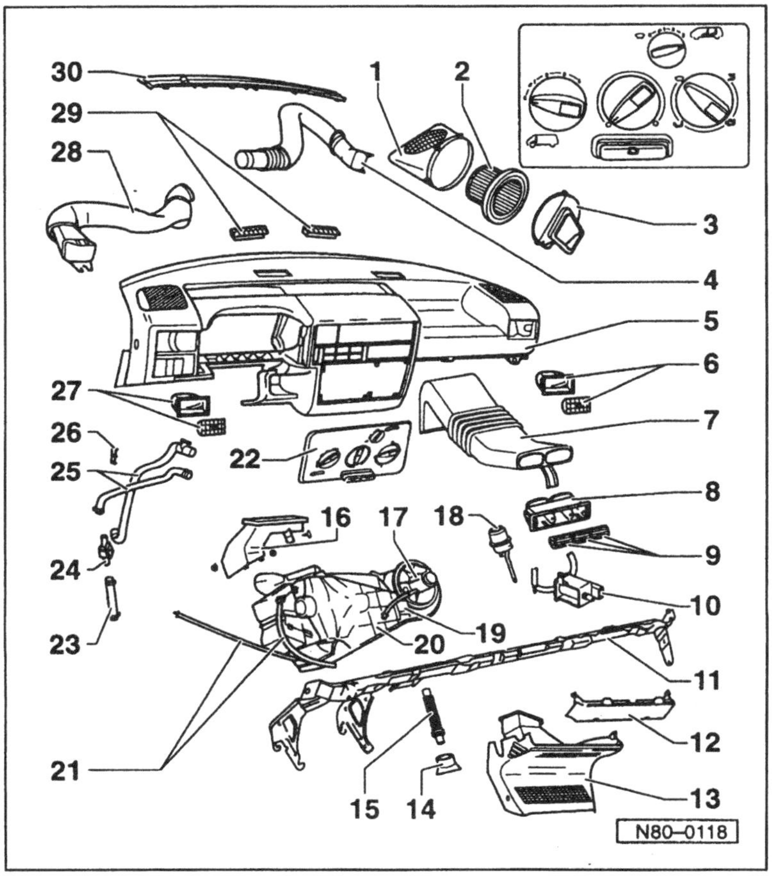

6 - Right air outlet
- ♦ Removing and installing

⇒ Fig. 2

7 - Center air duct
- ♦ Replacing
- – Remove heating unit -20-.

8 - Center air outlet housing
- ♦ Removing

⇒ Fig. 4

9 - Air guide
- ♦ Removing

⇒ Fig. 4

10 - Fresh air/recirculating flap two-way valve -N63-
- ♦ Cap must be left vented
- ♦ Vacuum hose layout

⇒ Page T 80-34
- ♦ Replacing
- – Remove glove compartment, knee padding or passenger side airbag unit.

⇒ Repair Manual, *Body - Exterior, Body - Interior,* Repair Group 69

11 - Carrier

12 - Trim panel

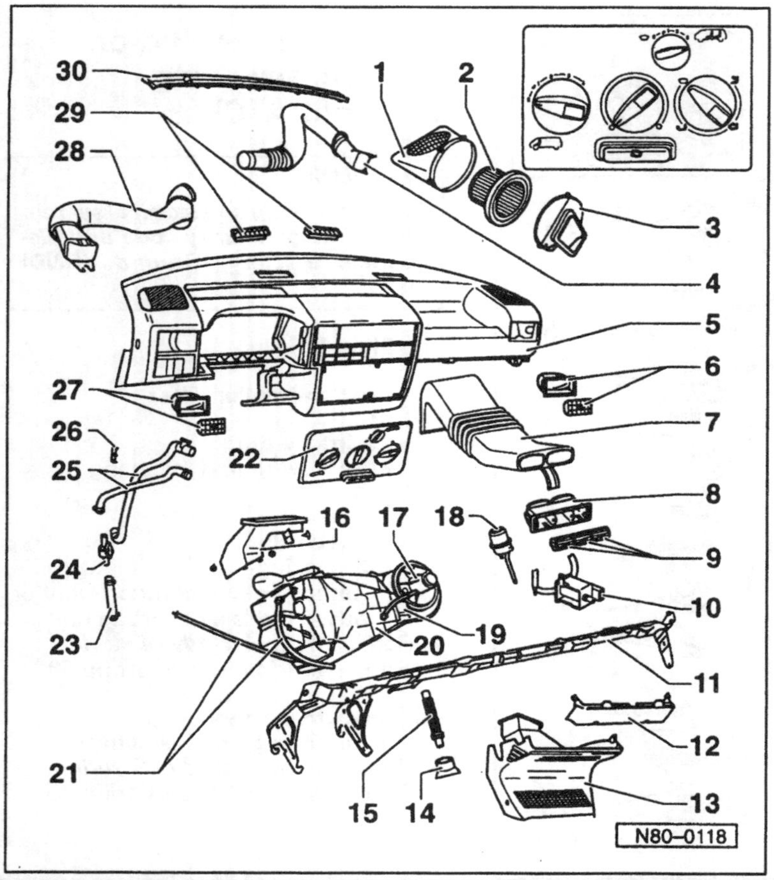

13 - Footwell air outlet console
- ♦ Removing ⇒ Fig. 7
- ♦ Replacing
- – Remove radio.
- – Remove center vent -9-.

14 - Water drain valve
- ♦ Water leak at heater unit housing may result if valve blocked

15 - Water drain pipe

16 - Defroster duct
- ♦ Replacing
- – Remove heating unit -20-.

17 - Fresh air blower -V2-
- ♦ Removing ⇒ Fig. 6

18 - Fresh air/recirculating flap vacuum unit
- ♦ Flap in fresh air position when no vacuum is present
- ♦ Flap in recirculated air position when vacuum is present
- ♦ Vacuum hose layout ⇒ Page T 80-34
- ♦ Removing ⇒ Fig. 5

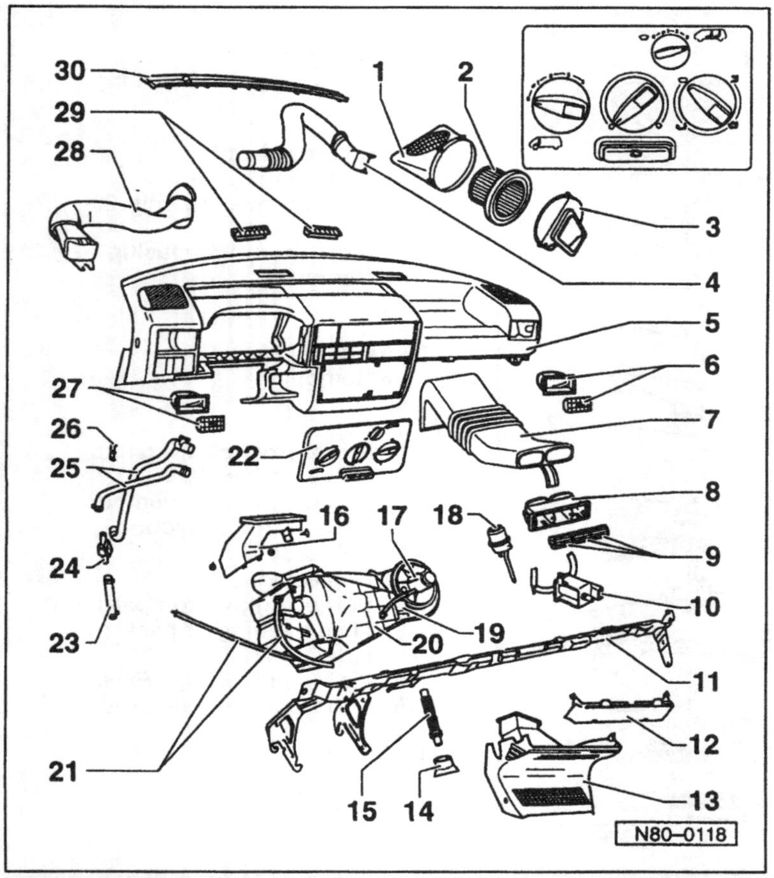

19 - Fresh air blower series resistance and fuse -N24-
- ♦ Removing ⇒ Fig. 6

20 - Heating unit
- ♦ Removing ⇒ Page T 80-12
- ♦ Assembly ⇒ Page T 80-15

21 - Cables
- ♦ Installing and adjusting ⇒ Page T 80-19

22 - Heating and ventilation controls
With:
- ♦ Fresh air blower switch -E9-
- ♦ Fresh air/recirculating flap switch -E159-
- ♦ Removing and installing ⇒ Fig. 3
- ♦ Disassembling and assembling ⇒ Page T 80-17
- ♦ Cables, installing and adjusting ⇒ Page T 80-19

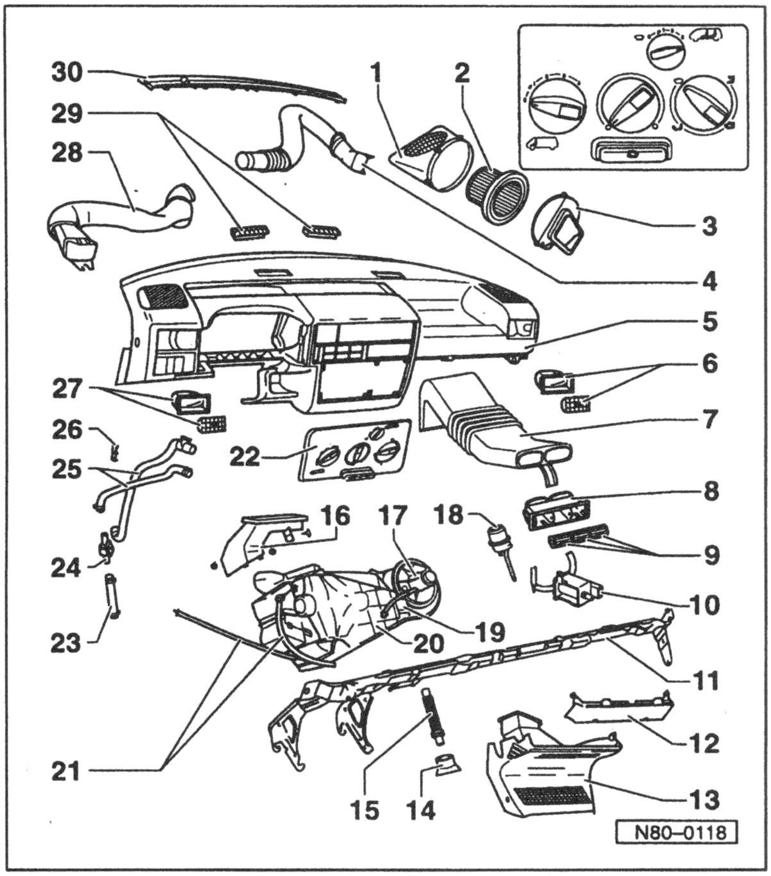

23 - Coolant hose
- ♦ Hose routing diagram

⇒ Repair Manual, *General, Engine,* Repair Group 19

24 - Heater control valve
- ♦ Removing

⇒ Fig. 8

25 - Coolant hoses
- ♦ Hose routing diagram
- – Repair Manual, *General, Engine,* Repair Group 19

26 - Clip

27 - Left air outlet
- ♦ Removing and installing

⇒ Fig. 2

28 - Left air duct
- ♦ Replacing
- – Remove relay plate cover.
- – Remove footwell vent console and trim.
- – Remove air duct.

29 - Defroster vents
- ♦ Removing
- – Carefully pry out with screwdriver.

30 - Air intake grille

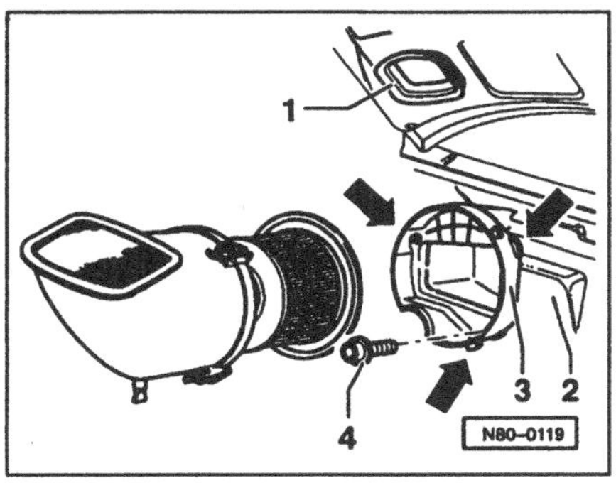

◄ **Fig. 1 Air intake duct, removing and installing**

1 - Hood insert

2 - Bulkhead

3 - Duct

4 - Socket head combination bolt, 3 Nm (26 in. lb)

Removing:

– Remove dust and pollen filter.
⇒ Page T 80-14

– Remove screws -arrows-.

Installing:

Reassemble in reverse order. Ensure that duct -3- is not inserted into bulkhead at an angle.

◄ **Fig. 2 Air outlets, removing and installing**

Removing:

– Carefully pull grill -2- out with needle nose pliers.

– Remove screw.

– With a small screwdriver, release locking tabs -arrows- from grille opening while gently pulling on housing -1-.

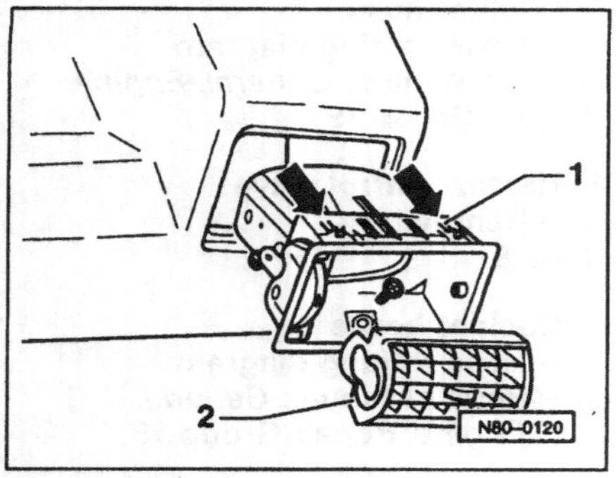

Installing:

Install in reverse order, noting the following:

- First locate grille -2- on the left guide lug of housing -1-.
- Carefully press in vent until locating tabs engage.

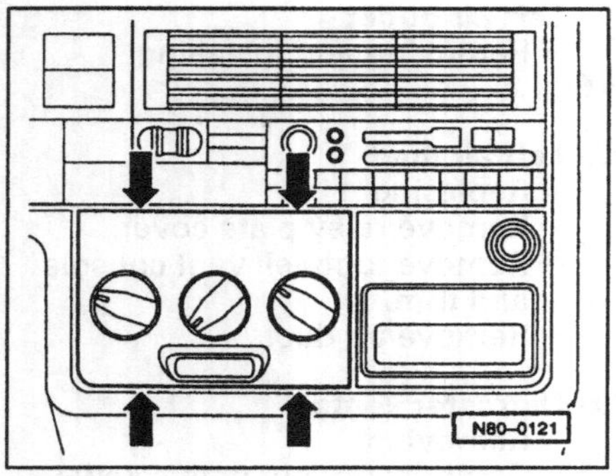

Fig. 3 Heating/ventilation controls and trim panel, removing and installing

Removing:

- Carefully remove trim panel at locating lugs -arrows- with a screwdriver. Take care not to damage instrument panel.
- Remove ashtray/trim panel.
- Remove footwell vent console.
⇒ Fig. 7

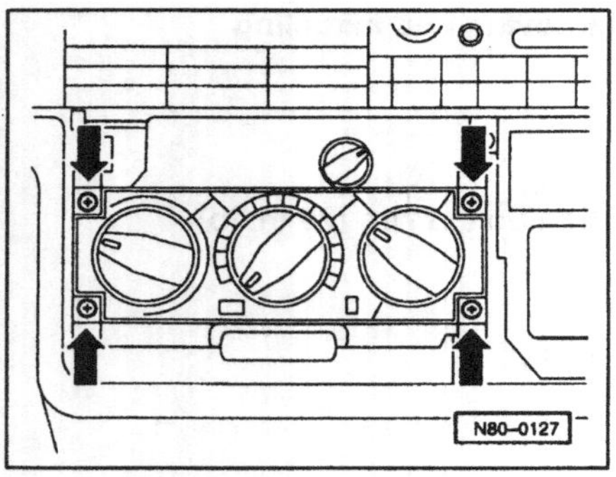

Removing (continued)

- Remove screws from instrument trim panel -arrows-.
- Push controls with cables attached below instrument panel.
- Disconnect electrical connectors.
- Detach cables from controls.

Installing:

Install in reverse order, noting the following:

- Before installing controls, attach cables first.
⇒ Page T 80-19

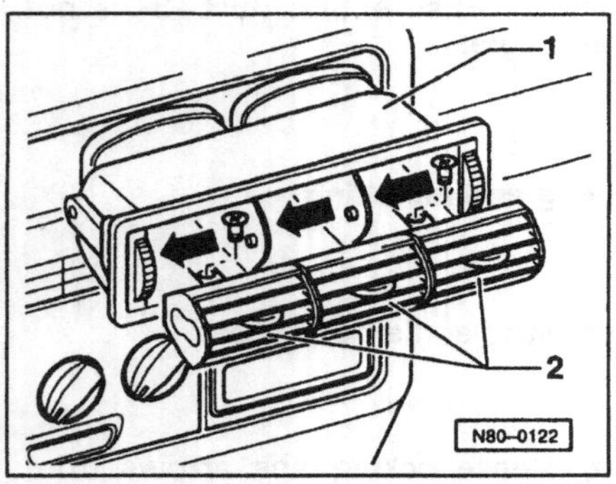

Fig. 4 Center air outlet, removing and installing

Removing:

- Carefully pull grille -2- out with needle nose pliers.
- Remove screws.
- Remove vent housing -1-.

Installing

Install in reverse order, noting the following:

- First locate grille -2- on the left guide lug of housing -1-.
- Carefully press in vent until locating tabs engage.

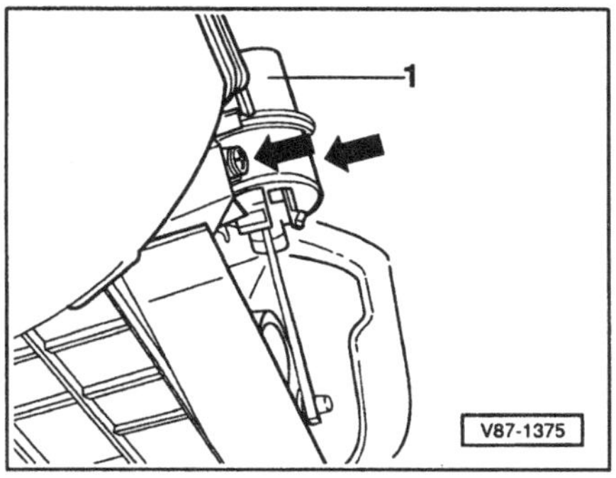

Fig. 5 Fresh air/recirculating flap vacuum unit, removing and installing

- Remove right air outlet.
⇒ Page T 80-6, Fig. 2
- Remove glove box, knee padding or passenger side airbag unit.
⇒ Repair Manual, *Body - Exterior, Body - Interior,* Repair Group 69
- Remove screws -arrows-.
- Rotate and lift vacuum unit -1- from lever.

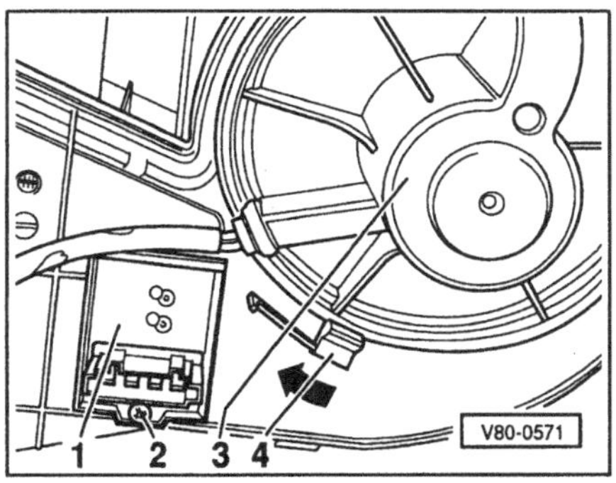

Fig. 6 Fresh air blower series resistance with fuse -N24- and fresh air blower -V2- removing

Series resistance -N24-, removing

- Remove screw -2-.
- Remove series resistance module -1-.

Fresh air blower -3-, removing (heater unit installed)

- Remove glove box, knee padding or passenger side airbag unit.
⇒ Repair Manual, *Body - Exterior, Body - Interior,* Repair Group 69
- Remove electrical connector from series resistance -1-.
- Remove right air duct.
- Press retaining tab -4-, turn fresh air blower in direction of -arrow- and remove from housing.
- Guide blower to center of instrument panel and remove.

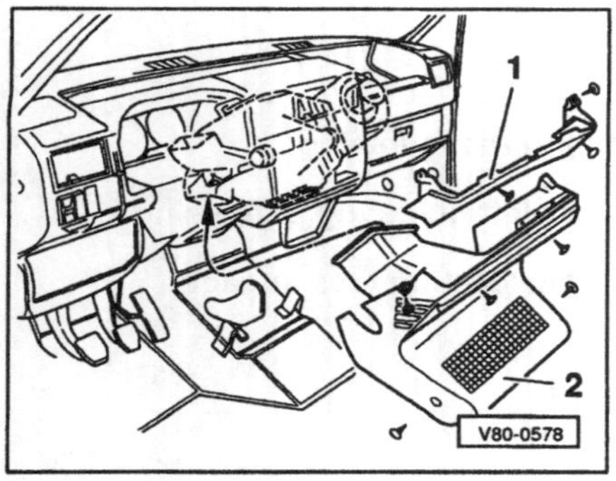

◄ **Fig. 7 Footwell air outlet console, removing**

– Remove trim panel -1-.

– Remove footwell vent console -2-.

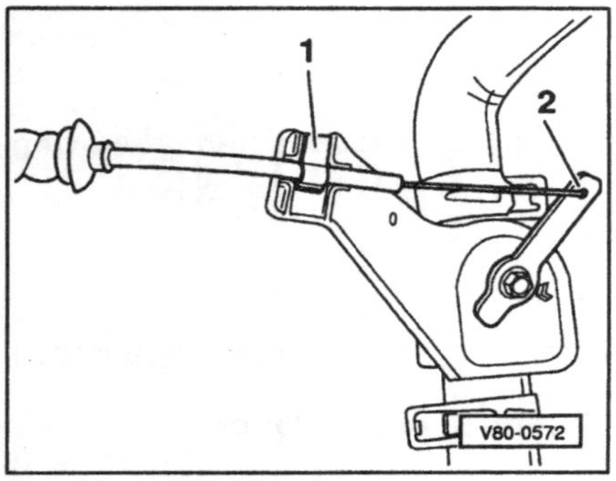

◄ **Fig. 8 Heater control valve, removing**

– Remove clip -1- and remove cable end -2- from lever.

– Clamp coolant hoses near valve.

> ***WARNING!***
> ***The cooling system is pressurized when the engine is warm. Wear gloves and other appropriate protection, and slowly and carefully release system pressure, if necessary, before performing repairs.***

– Remove spring clamps from coolant hoses and and remove coolant hoses from valve.

– Remove valve.

Heating unit, removing

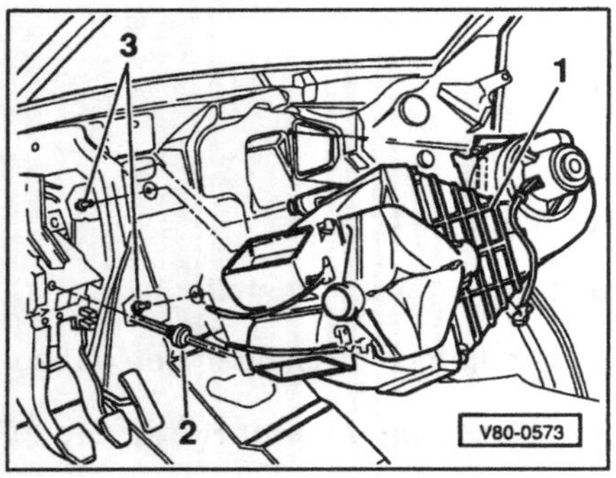

Heating unit removal shown with instrument panel removed.

◄ 1 - Heating unit

2 - Rubber grommet

3 - Combination screw

Removing:

– Remove dust and pollen filter
⇒ Page T 80-14

– Remove air intake duct (on bulkhead)

> ***WARNING!***
> ***The cooling system is pressurized when the engine is warm. Wear gloves and other appropriate protection, and slowly and carefully release system pressure, if necessary, before performing repairs.***

⇒ Page T 80-6

– Clamp coolant (heater) hoses near bulkhead.

– Remove right air duct.

– Remove complete instrument panel, passenger side airbag and associated trim
⇒ Repair Manual, *Body - Exterior, Body - Interior,* Repair Groups 69 – 70.

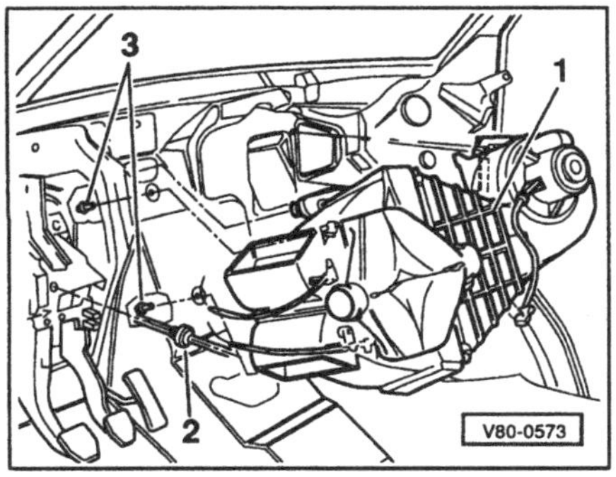

◄ – Remove bolts -3-.

Note:

Lower bolt is located on bulkhead under engine compartment cladding.

– Remove coolant hoses from heater core and plug heater core connections.

– Remove heater unit.

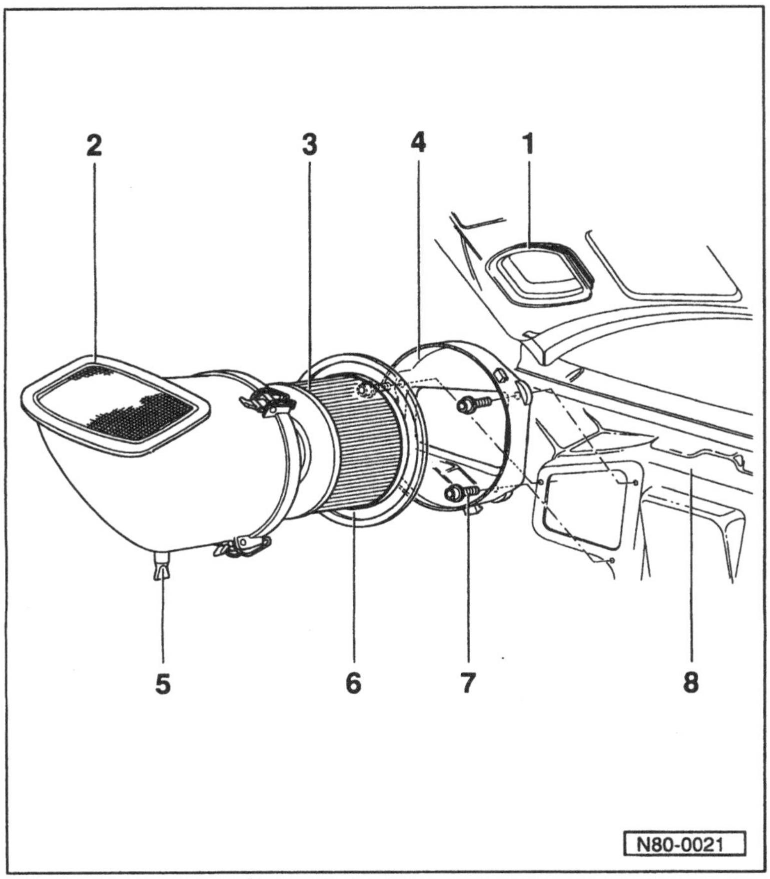

Dust and pollen filter, removing and installing

1 - Hood seal

2 - Filter housing

3 - Dust and pollen filter

4 - Air intake duct

5 - Valve

6 - Filter retainer

7 - Hex bolt
- ♦ 3 Nm (26 in. lb)

8 - Bulkhead

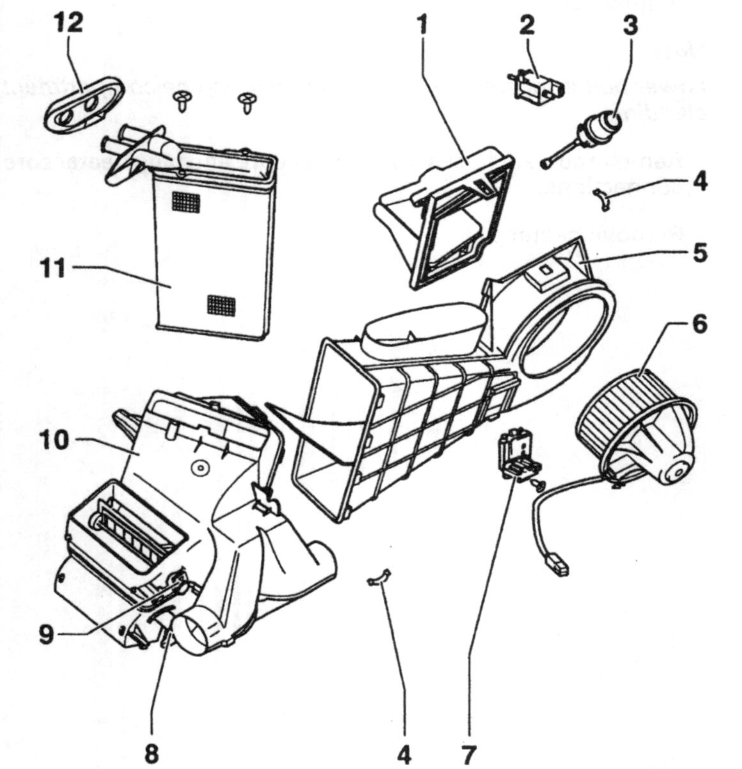

N80–0123

Heater unit, assembly

> ***CAUTION!***
>
> ***Before working on electrical system obtain radio security code and disconnect battery Ground (GND) strap.***

Note:

Heater unit, removing ⇒ Page T 80-12

1 - **Air intake duct**
- ♦ With recirculating air flap

2 - **Fresh air/recirculating flap two-way valve -N63-**

3 - **Fresh air/recirculating flap vacuum unit**
- ♦ Vacuum hose, layout

⇒ Page T 80-34
- ♦ Removing

⇒ Fig. 5

4 - **Clip**

5 - **Air duct**

6 - **Fresh air blower -V2-**

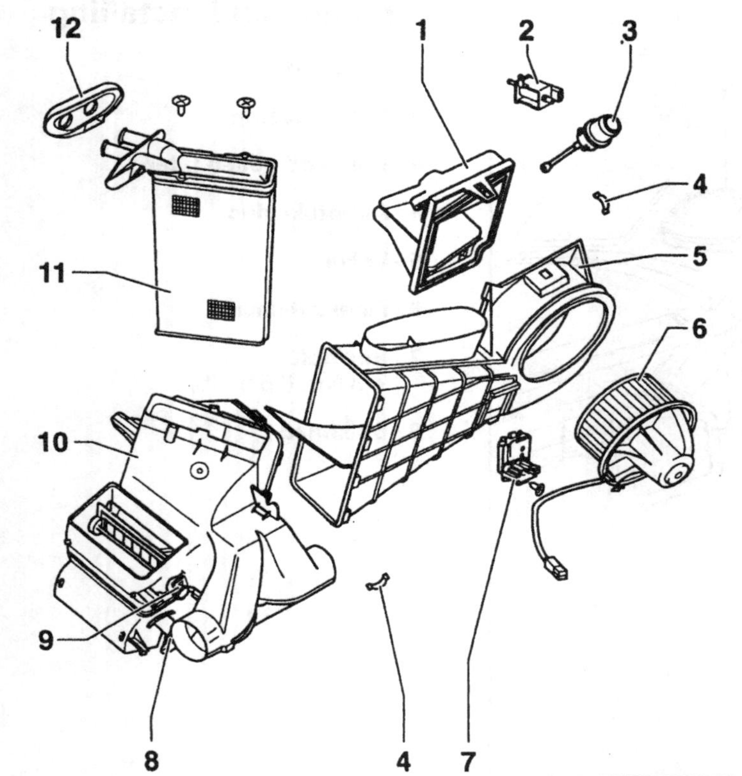

N80–0123

7 - **Fresh air blower series resistance with fuse -N24-**

8 - **Footwell flap lever**
- ♦ Removing:
  - – Disconnect lever with a screwdriver.

9 - **Defroster flap lever**
- ♦ Removing:
  - – Disconnect lever with a screwdriver.

10 - **Air distribution housing**

11 - **Heater core**
- ♦ Install seals around entire circumference of heater core

12 - **Gasket**

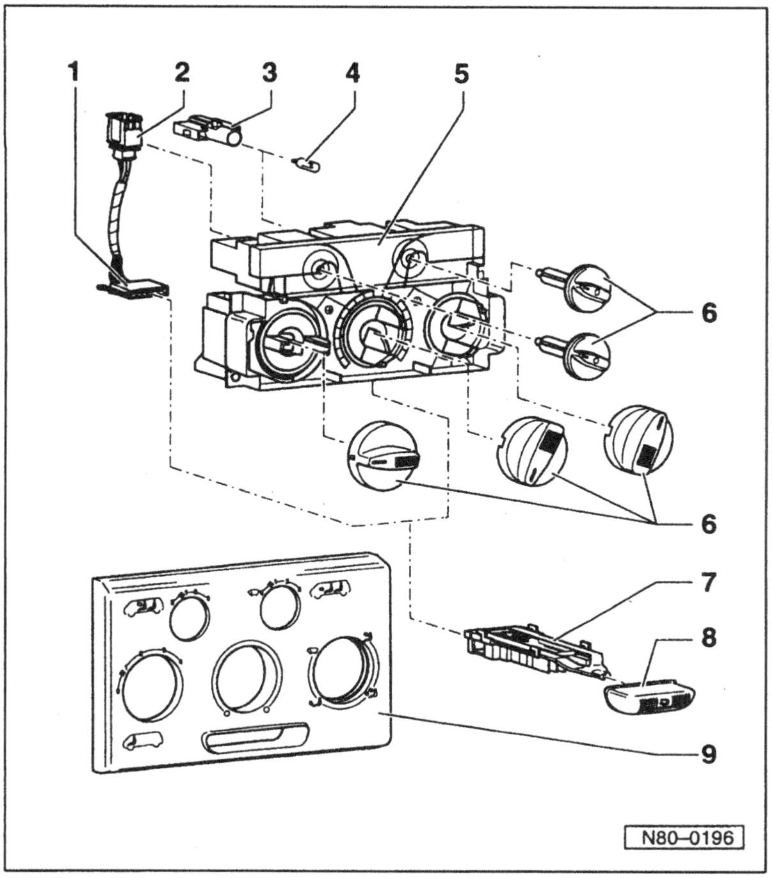

Heating and ventilation controls, assembly

CAUTION!

Before working on electrical system, obtain radio code and disconnect battery Ground (GND) strap.

Note:

- *Illustrated control from vehicle with additional (rear) heat exchanger and (rear) passenger compartment ventilation.*

1 - Connector, 6 pin -T6-

2 - Connector, 6 pin -T6c-

3 - Illumination filter

4 - Illumination bulb -L16-

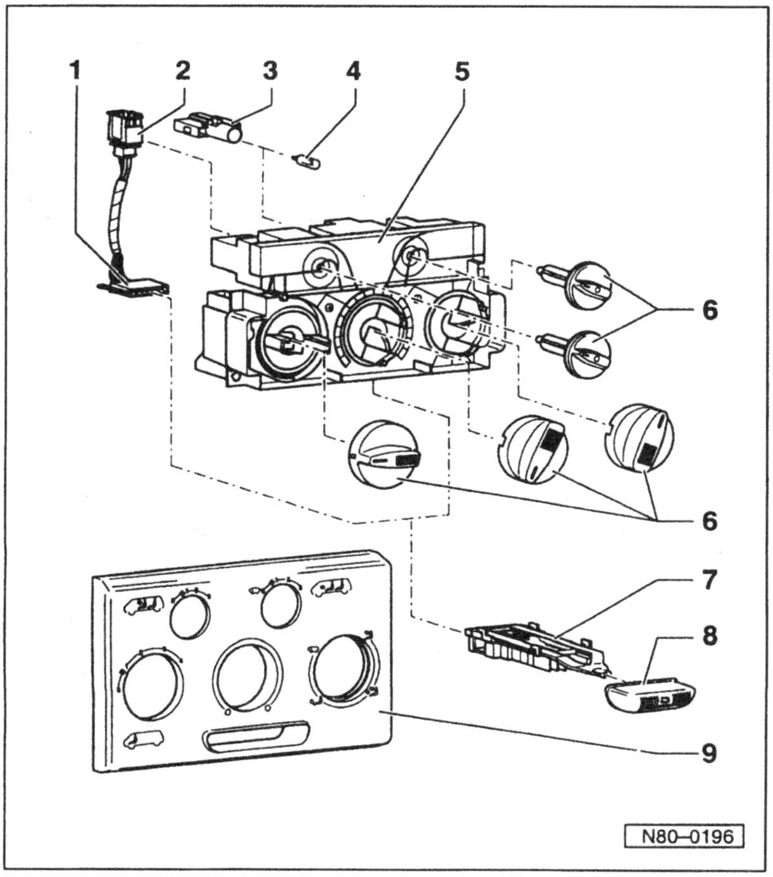

5 - Control assembly
With:
- Fresh air blower switch -E9-
- Warm air blower switch -E100-
- Fresh air blower switch -E179-
- Removing and installing

⇒ Page T 80-7
- Cables, installing and adjusting

⇒ Page T 80-19

6 - Rotary knobs

7 - Fresh air/recirculating flap switch -E159-
- With illumination bulb -L43-
- Removing:
- – Remove assembly -5-.
- – Release retainers between push buttons and controls.
- – Press switch in direction of push button.

8 - Push button

9 - Control panel trim

Heating and ventilation cables, installing and adjusting

Notes:

- *Attach cables before installing controls.*
- *Place outer cable against stops on controls and secure.*
- *Defroster and footwell flap cables are self-adjusting. Adjust by turning air distribution knob fully from stop to stop.*
- *Adjust cable on regulating valve with controls installed.*

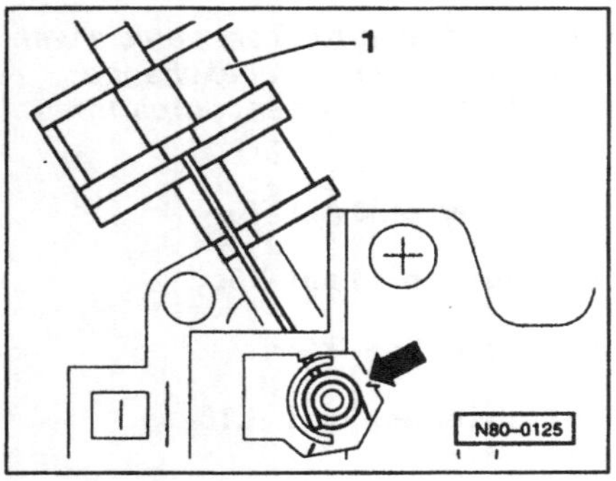

Cables, installing

◄ **Heater control valve cable**

(Length 754 mm (29.68 in), identification red/white)

- Turn heater control valve knob fully to left stop.
- Connect heater control valve cable -arrow-.
- Secure cable to controls -1- with clip.
- Heater control valve cable, adjusting ⇒ Fig. 3

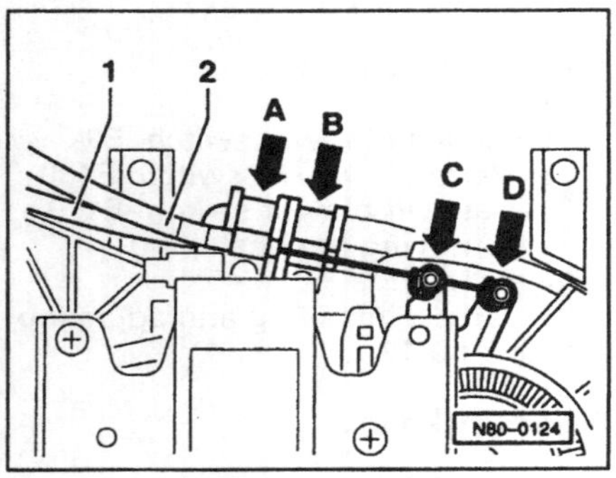

◄ **Footwell flap cable**

(Length 543 mm (21.38 in), identification yellow/white)

- Turn air distribution knob fully to left stop.
- Attach footwell flap cable -arrow -D-.
- Secure cable to controls with clip -arrow -B-.
- Adjusting cable ⇒ Fig. 1

Defroster flap cable

(Length 458 mm (18.03 in), identification red/green)

- Turn air distribution knob fully to left stop.
- Attach defroster flap cable -arrow -C-.
- Secure cable to controls with clip -arrow -A-.
- Adjusting cable ⇒ Fig. 2

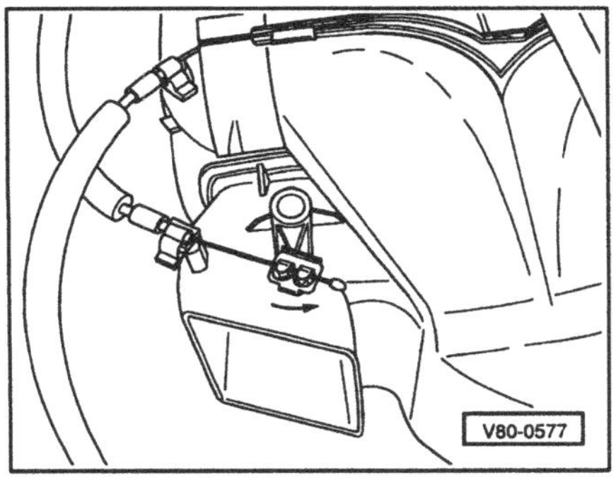

◄ **Fig. 1 Footwell flap cable, adjusting**

- Secure cable to footwell flap while simultaneously pressing lever in direction of arrow.
- Turn control knob fully from stop to stop.

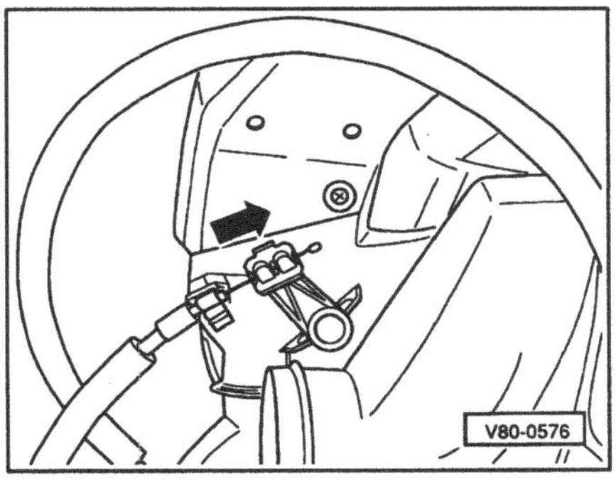

◄ **Fig. 2 Defroster flap cable, adjusting**

- Secure cable to defroster flap while simultaneously pressing lever in direction of arrow.
- Turn control knob fully from stop to stop.

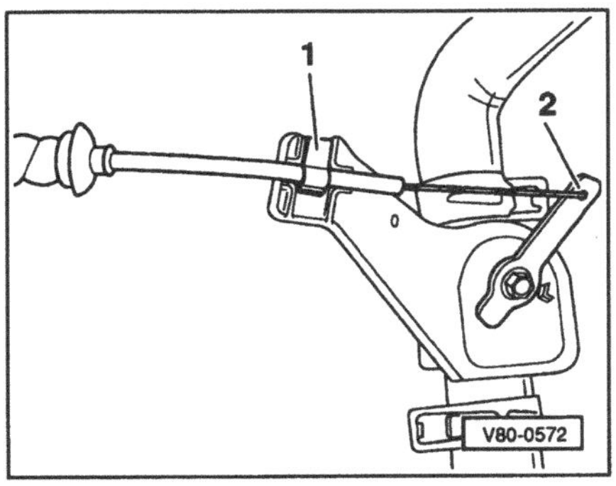

◄ **Fig. 3 Heater control valve cable, adjusting**

- Guide cable through engine compartment bulkhead to valve lever -2- and connect.
- Slide lever -2- left to stop (off/cool).
- Secure cable housing to valve with clip -1-.

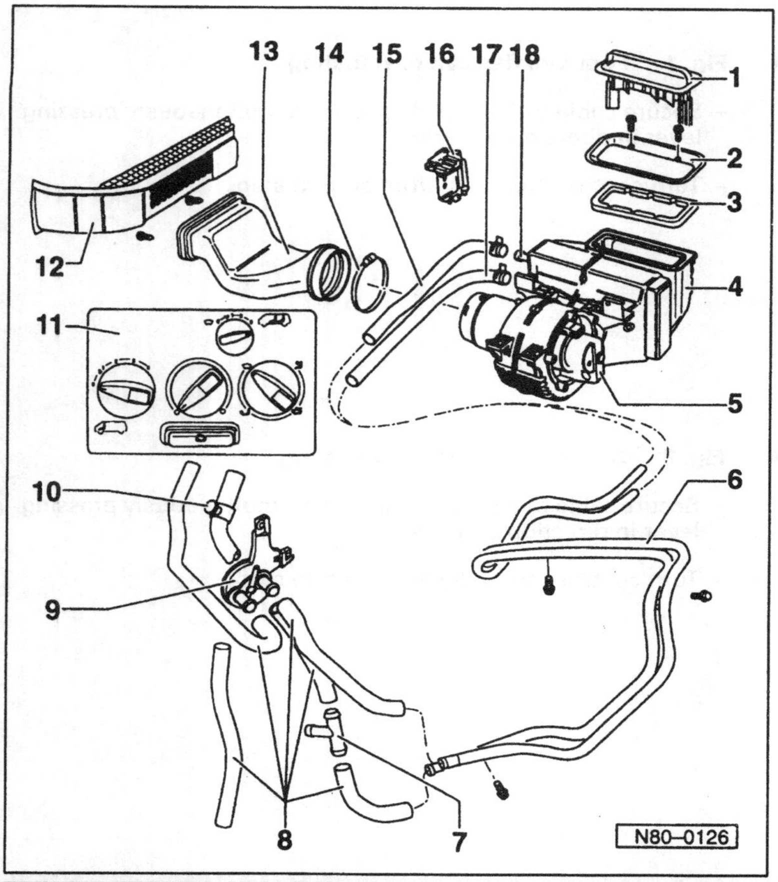

Rear heater, servicing

> ***CAUTION!***
>
> ***Before working on electrical system, obtain radio code and disconnect battery Ground (GND) strap.***

1 - Vent, in rear passenger compartment or cargo area
- ♦ Removing

⇒ Fig. 2

2 - Cover plate

3 - Gasket

4 - Air distribution housing
- ♦ Rear heater core, removing

⇒ Fig. 1
- ♦ Assembly

⇒ Page T 80-33

5 - Rear warm air fan -V47-
- ♦ Removing

⇒ Page T 80-33

6 - Coolant pipe
- ♦ Removing

⇒ Fig. 5

7 - T fitting

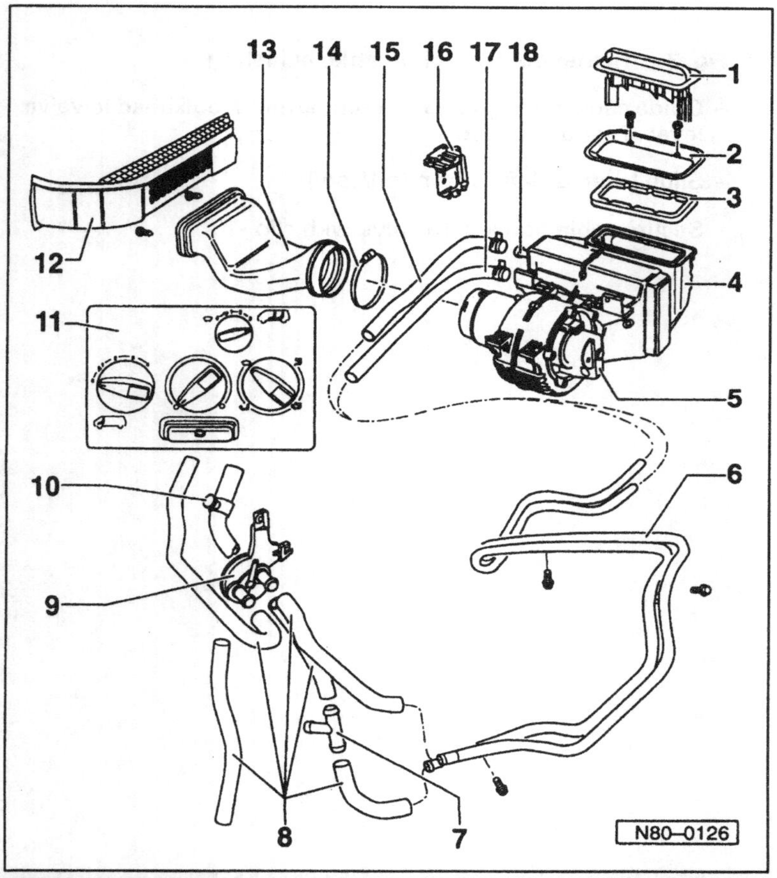

8 - Coolant hoses in engine compartment
- ♦ Coolant hose connections

⇒ Page T 80-31

9 - Rear heater control valve
- ♦ Removing

⇒ Page T 80-30

10 - Bleeder screw

11 - Control assembly
With:
- ♦ Warm air blower switch -E100-
- ♦ Fresh air blower switch -E9-
- ♦ Fresh air/recirculating flap switch -E159-
- ♦ Disassembling and assembling

⇒ Page T 80-17

12 - Sliding door footwell insert
- ♦ Removing
- – Repair Manual, *Body - Exterior, Body - Interior,* Repair Group 68

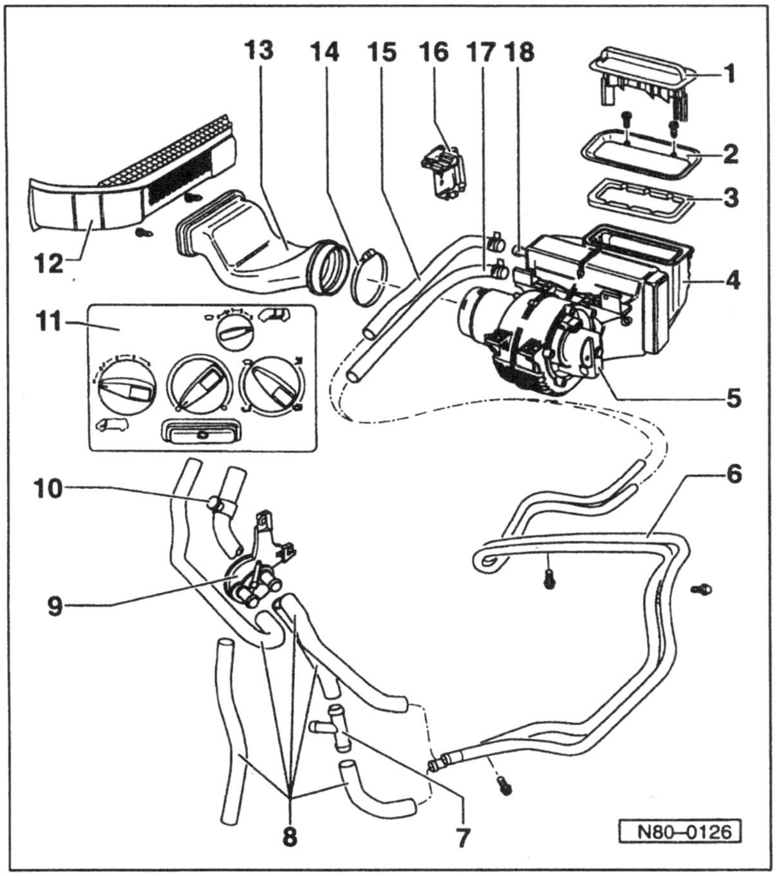

13 - Air intake duct
- Kombi models*, removing ⇒ Fig. 3
- EuroVan & Transporter models, removing ⇒ Fig. 4

14 - Hose clamp

15 - Coolant hose, return

16 - Heater blower series resistance -N72-

17 - Coolant hose, supply

18 - Heater core
- Install seals around entire circumference of heater core
- Removing

⇒ Page T 80-33

Note:

Always change coolant after repairs.

* Not applicable for U.S. models

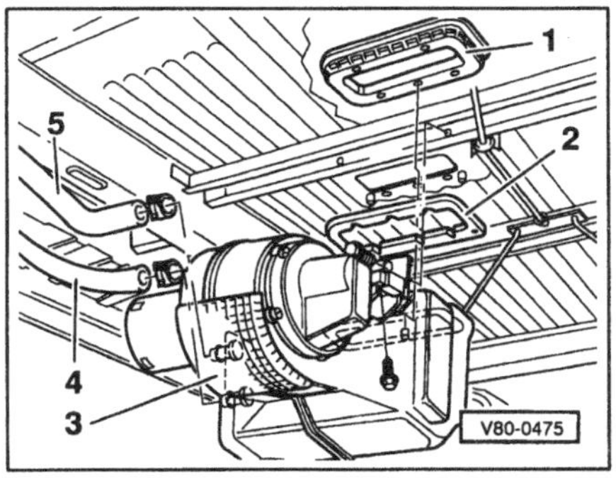

◄ **Fig. 1 Rear heater, removing**

1 - Cover plate, in passenger compartment
2 - Gasket
3 - Rear heater air distribution housing
4 - Coolant supply hose
5 - Coolant return hose

Removing:

– Disconnect electrical connectors.

> ***WARNING!***
> ***The cooling system is pressurized when the engine is warm. Wear gloves and other appropriate protection, and slowly and carefully release system pressure, if necessary, before performing repairs.***

– Clamp coolant hoses as close as possible to heater core.

– Loosen hose clamps and remove coolant hoses. Plug heater core outlet/inlet.

– Remove vent ⇒ Fig. 2.

– Remove cover plate.

– Remove bolts.

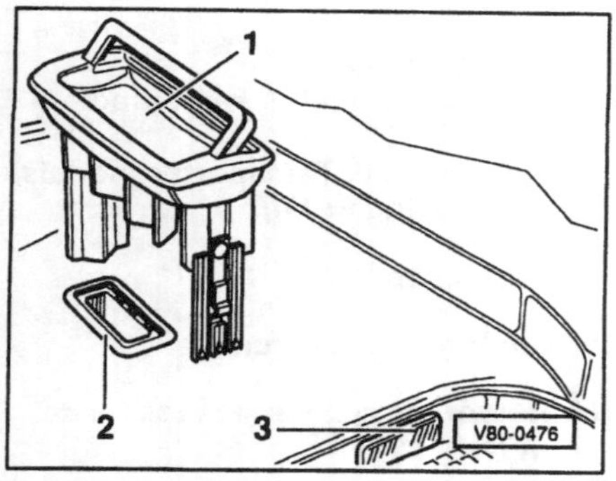

Fig. 2 Vent, removing

1 - Vent
2 - Cover plate
3 - Sliding door footwell insert

Removing:

– Pull vent out of floor plate.

– Remove cover plate

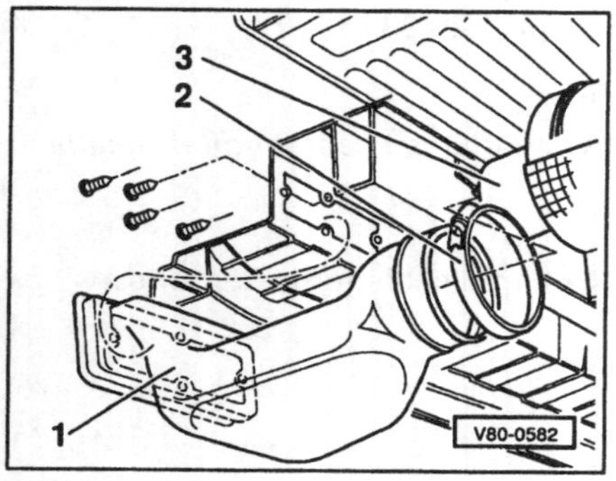

Fig. 3 Air intake duct (Kombi), removing

1 - Air intake duct
2 - Hose clamp
3 - Rear heater air distribution housing

Removing:

– Remove footwell insert.

– Remove bolts.

– Loosen hose clamp.

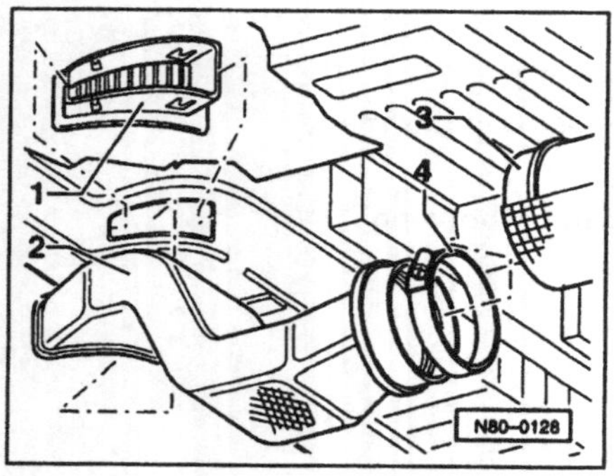

Fig. 4 Intake duct (EuroVan/Transporter), removing

1 - Intake trim
2 - Intake duct
3 - Rear heater core
4 - Hose clamp

Removing:

– Carefully remove intake trim -1- with screwdriver.

– Loosen hose clamp.

– Remove intake duct.

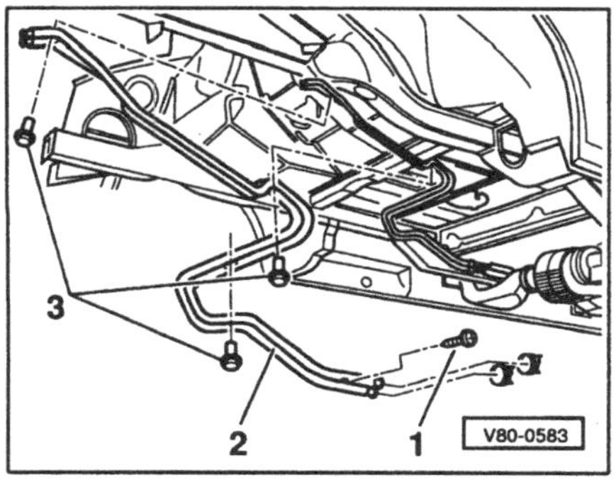

Fig. 5 Coolant pipe, removing

1 - Self-tapping screw

2 - Coolant pipe

3 - Hex nut

Removing:

> ***WARNING!***
> ***The cooling system is pressurized when the engine is warm. Wear gloves and other appropriate protection, and slowly and carefully release system pressure, if necessary, before performing repairs.***

– Remove fuel tank
⇒ Repair Manual, *Engine, General*, Repair Group 20

– Remove torsion bar
⇒ Repair Manual, *Suspension, Brakes, Wheels, Steering*, Repair Group 40

– Remove screws -3- and -1-.

– Remove coolant pipe -2-.

Heater control valve, removing

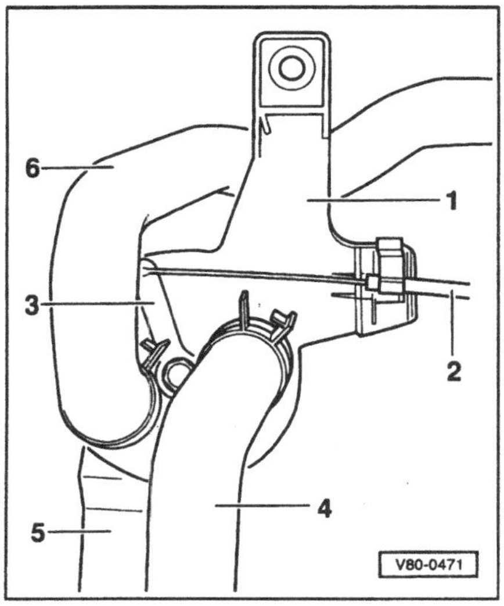

1 - Control valve

2 - Cable

3 - Lever

4 - Coolant hose, return
(from rear heater core coolant pipe to control valve)

5 - Coolant hose, return
(from control valve to engine)

6 - Coolant hose, return
(from front heater core to control valve)

Removing:

– Remove expansion tank.

– Loosen clip and remove cable.

> ***WARNING!***
> ***The cooling system is pressurized when the engine is warm. Wear gloves and other appropriate protection, and slowly and carefully release system pressure, if necessary, before performing repairs.***

– Clamp coolant hoses as close as possible to valve.

– Loosen spring type hose clamps and pull coolant hoses off.

– Remove regulating valve from expansion tank console.

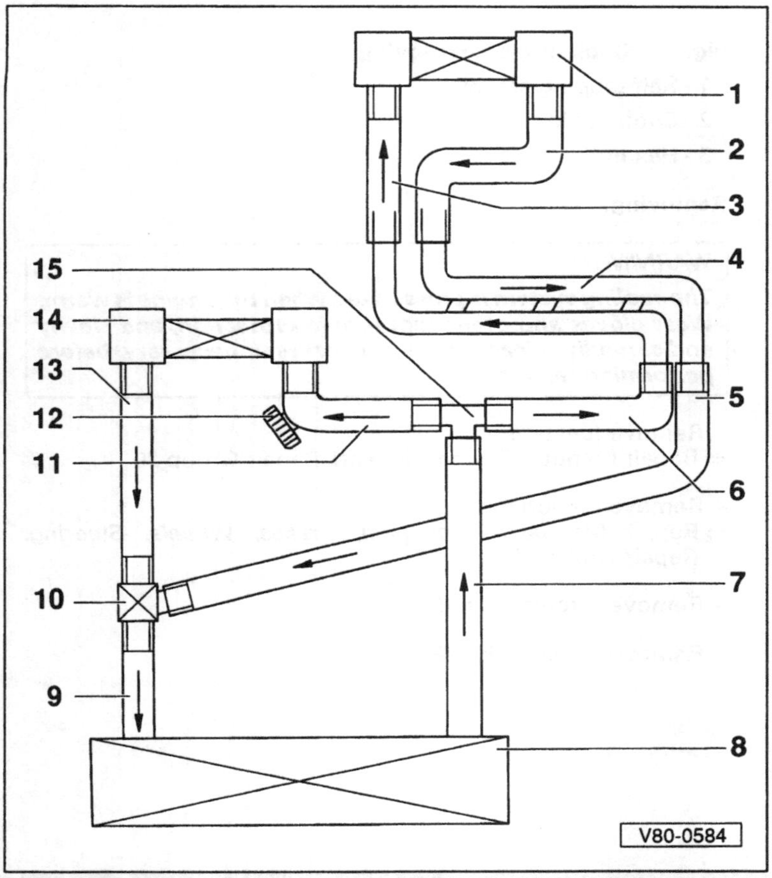

Coolant hoses, layout

Note:

For routing of all hoses
⇒ Repair Manual, *Engine, General,* Repair Group 19

1 - **Rear heater core**

2 - **Coolant return hose**
(from rear heater core to coolant pipe)

3 - **Coolant supply hose**
(from coolant pipe to rear heater core)

4 - **Coolant pipe**

5 - **Coolant return hose**
(from coolant pipe to rear heater control valve)

6 - **Coolant supply hose**
(from T-piece to coolant pipe)

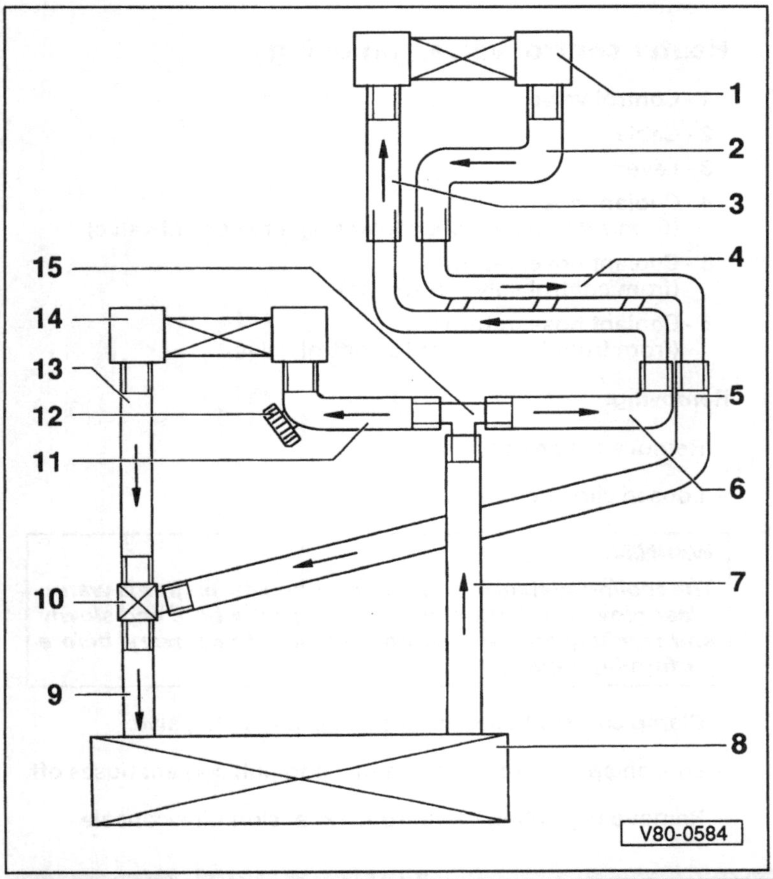

7 - **Coolant supply hose**
(from engine to T-piece)

8 - **Engine**

9 - **Coolant return hose**
(from control valve to engine)

10 - **Control valve**

11 - **Coolant supply hose**
(from T-fitting to front heater core)

12 - **Bleeder screw**

13 - **Coolant return hose**
(from front heater core to control valve)

14 - **Front heater core**

15 - **T-piece**

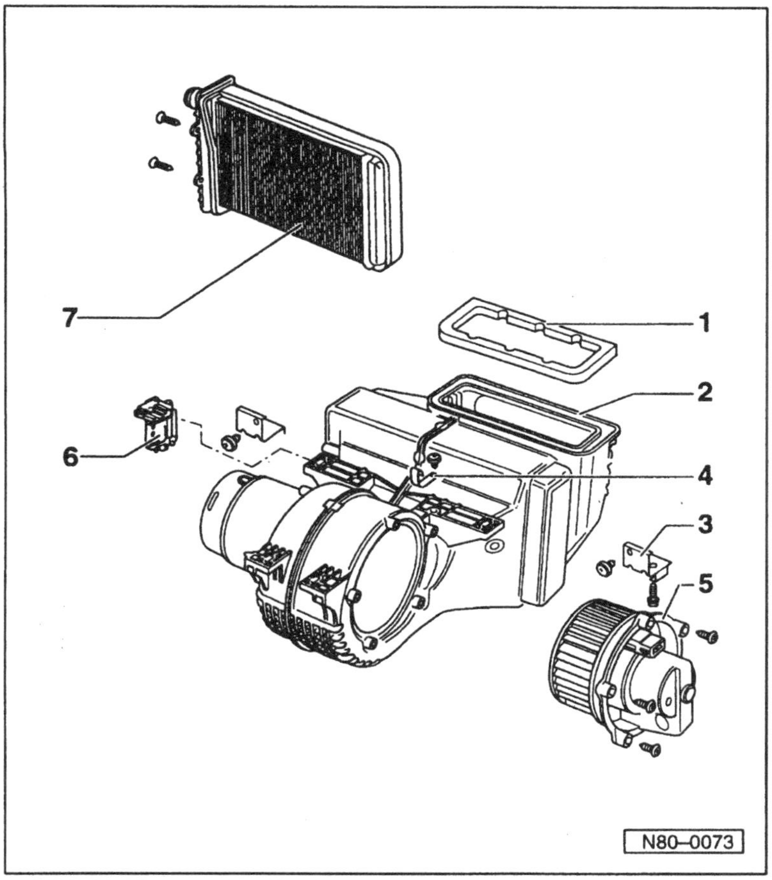

Rear heater air distribution housing, assembly

1 - **Gasket**

2 - **Air distribution housing**

3 - **Bracket**

4 - **Clip**
- ♦ Snaps into position on cross member

5 - **Rear warm air fan -V47-**
- ♦ Note installation position

6 - **Warm air blower series resistor -N72-**

7 - **Heater core**
- ♦ Install seals around entire circumference of heater core

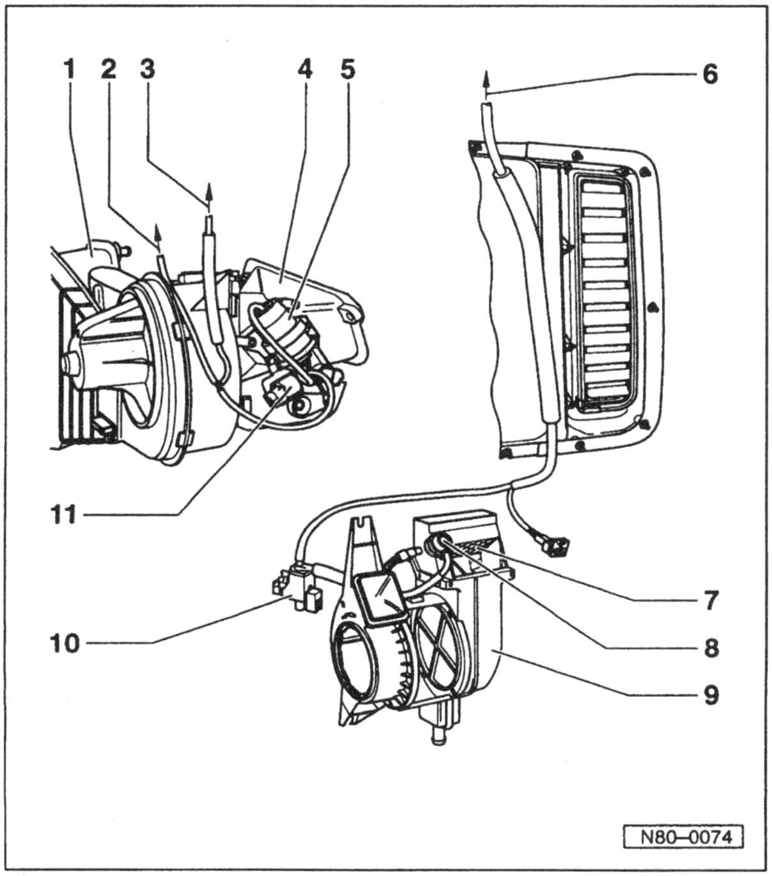

Vacuum hose, layout

1 - **Heater unit**

2 - **To brake servo or vacuum pump**

3 - **Vacuum hose routing**
- ♦ Routed through loops to rear fresh and recirculating air flap two-way valve -N45-.

4 - **Air intake housing**
- ♦ With recirculating air flap

5 - **Vacuum unit for fresh air/recirculating flap**
- ♦ Flap in fresh air position when no vacuum is present
- ♦ Flap in recirculated air position when vacuum is present

6 - **Vacuum hose routing**
- ♦ Routed through loops up to instrument panel.

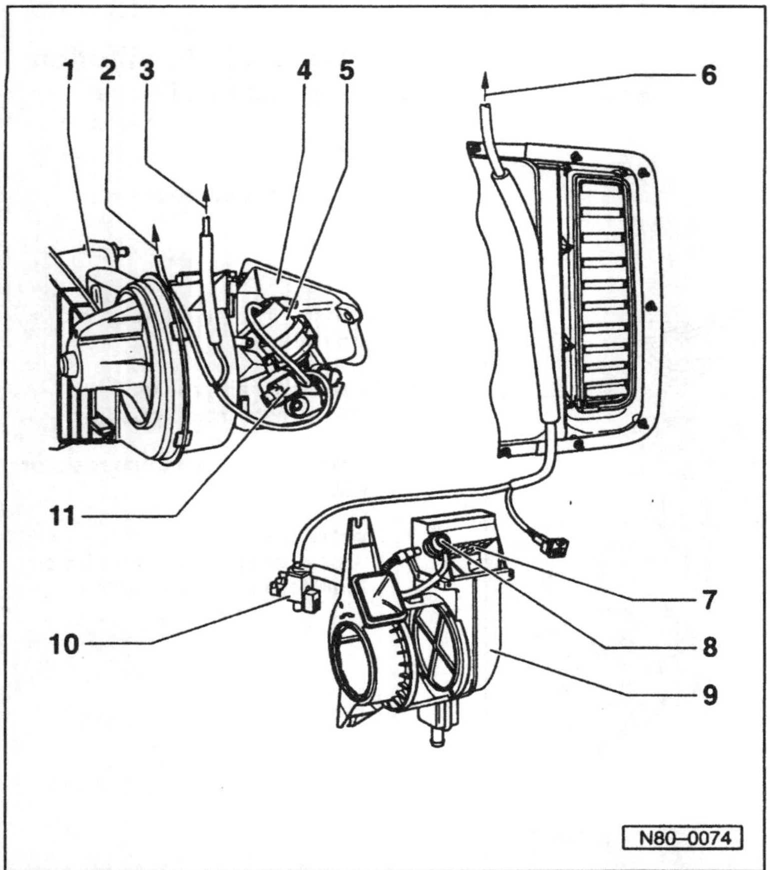

7 - Fresh air /recirculating flap

8 - Vacuum unit

- When switched off vacuum unit is vented.

9 - Air intake

10 - Rear fresh air/recirculating flap two-way valve -N45-

- Cap must be vented

11 - Rear fresh air/recirculating flap two-way valve -N63-

- Replacing

–

– Remove glove compartment, knee padding or passenger side airbag unit.

⇒ Repair Manual, *Body - Exterior, Body - Interior,* Repair Group 69

A/C refrigerant system, identification

A/C refrigerant systems on EuroVan and Transporter models from m.y. 1993 are charged with refrigerant R-134a.

Labels specifying refrigerant type are located on the compressor or radiator support. Before proceeding with refrigerant system servicing or repairs, always confirm refrigerant type used.

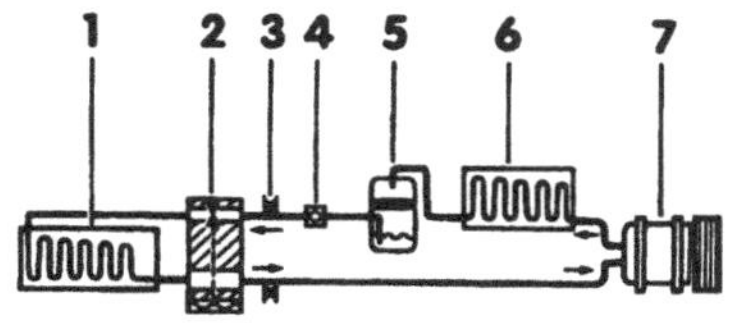

87-1437

Manual A/C

◄ **A/C refrigerant circuit with one evaporator**

1 - Evaporator
2 - Expansion valve
3 - High-pressure service valve
4 - Sight glass (if equipped)
5 - Receiver drier
6 - Condenser
7 - Compressor

Note:

Arrows indicate direction of refrigerant flow.

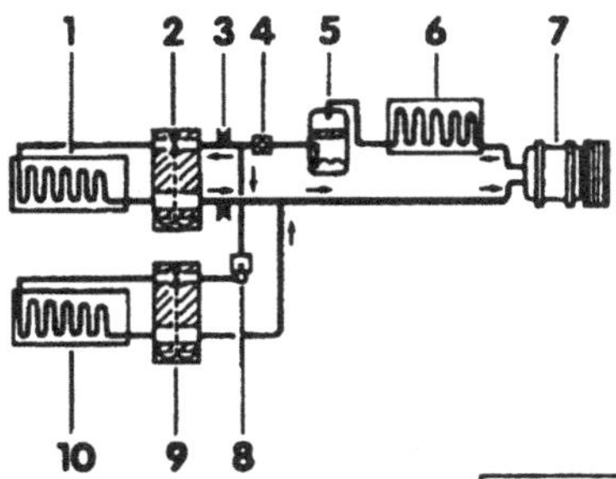

87-1438

Climatronic

◄ **A/C refrigerant circuit with two evaporators**

1 - Evaporator
2 - Expansion valve
3 - High-pressure service valve
4 - Sight glass (if equipped)
5 - Receiver drier
6 - Condenser
7 - Compressor
8 - Solenoid valve for refrigerant circuit
9 - Expansion valve
10 - Second evaporator

Note:

Arrows indicate direction of refrigerant flow.

A/C refrigerant system, component overview

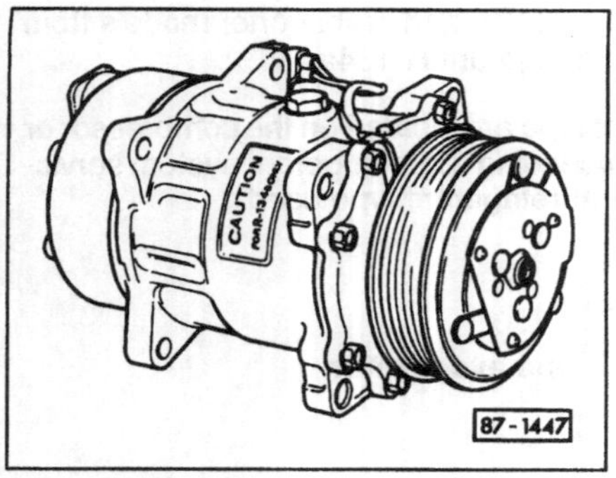

◄ **Compressor**

The compressor is driven via a belt on the engine when the A/C clutch engages (A/C ON).

Low-pressure refrigerant gas from the evaporator is compressed by the compressor. After compression, the refrigerant gas (now high-pressure) flows to the condenser.

Notes:

- *The compressor contains refrigerant oil that is mixable under all temperatures with the refrigerant.*
- *A label on the compressor indicates that compressor is for R-134a systems only.*

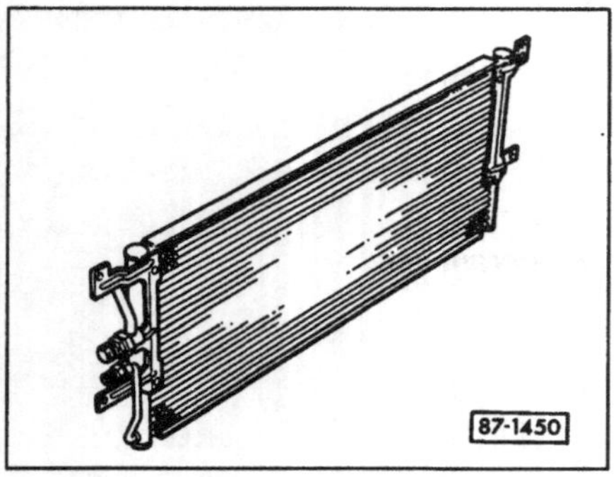

◄ **Condenser**

The condenser transfers heat from the compressed refrigerant gas to the outside air which causes the refrigerant to change state from a gas to a liquid.

Note:

The condenser for the R-134a refrigerant system is identified with a green label.

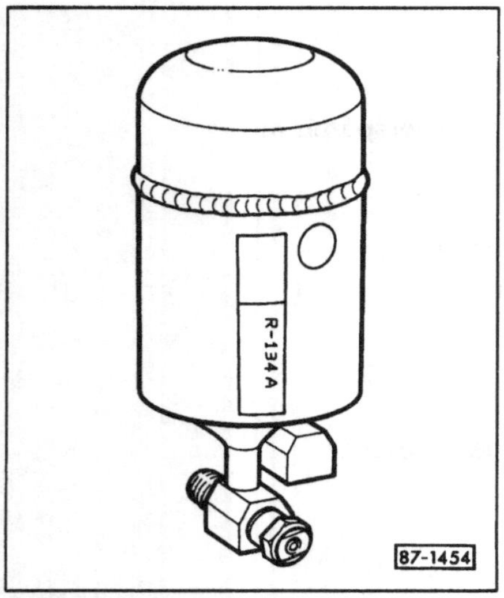

◄ **Receiver drier**

The receiver drier acts as a refrigerant reservoir for the system. Any moisture in the system is absorbed in the drier desiccant.

> ***CAUTION!***
>
> ***The R-134a receiver drier is identified with a green label and MUST NOT be used in R-12 systems (drier desiccant is only compatible with R-134a refrigerant). Likewise, NEVER use an R-12 receiver drier in an R-134a system.***

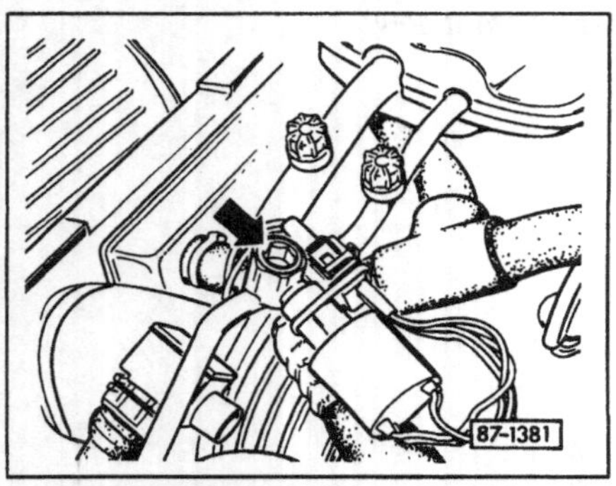

◄ **Sight glass (if equipped)**

With a properly charged A/C system and A/C clutch engaged, the sight glass (arrow) should remain clear and bubble-free.

Note:

The mixture of R-134a refrigerant and refrigerant oil (PAG oil) may appear milky in the sight glass.

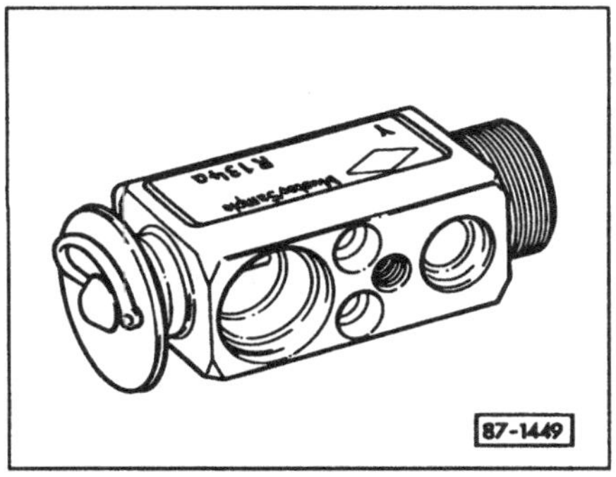

Expansion valve

The expansion valve restricts and controls refrigerant flow to the evaporator thus lowering refrigerant temperature and pressure.

Note:

The R-134a expansion valve is identified with a green adhesive label.

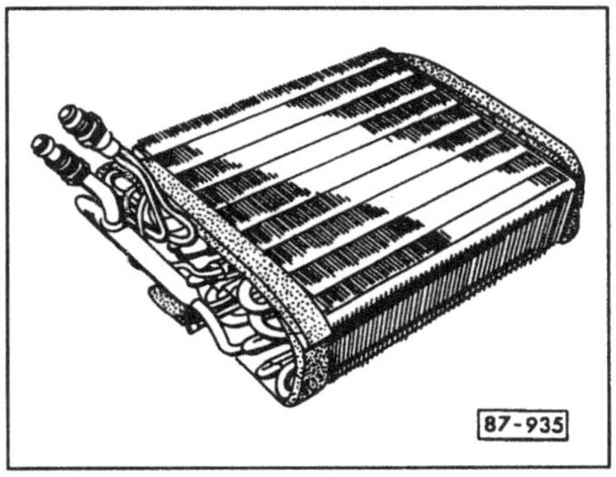

Evaporator

Liquid refrigerant entering the evaporator absorbs heat from air passing through the evaporator fins and cools the air. As the refrigerant absorbs heat it turns to vapor and then is suctioned by the compressor.

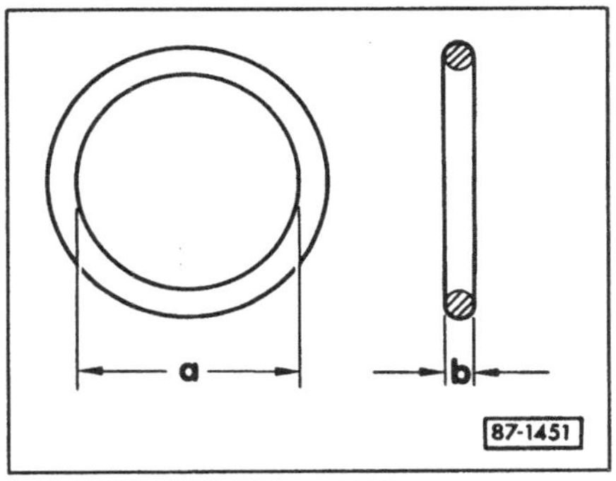

O-rings

O-rings seal connections between A/C system components.

Notes:

- *Always use correct size O-rings (dimensions -a- and -b-).*
- *Do not reuse O-rings, always replace. Use only new O-rings that are compatible with R-134a refrigerant and refrigerant (PAG) oil on R-134a systems.*
- *Lubricate O-rings with the appropriate refrigerant oil before installing (use PAG oil on R-134a systems).*

Pressure relief valve

The pressure relief valve is mounted on the compressor or the receiver drier. At pressures above 40 bar (580 psi), the pressure relief valve opens to vent excessive pressure. When the system pressure is reduced, the valve closes to prevent total refrigerant loss.

On some models, a cap on the pressure relief valve will pop out if the valve has opened.

A/C system hoses and lines

The mixture of refrigerant oil (PAG oil) and refrigerant R-134a attacks some metals and alloys (for example, copper) and breaks down certain hose material. Use only hoses and lines that are identified with a green mark (stripe) or the lettering “R-134a”.

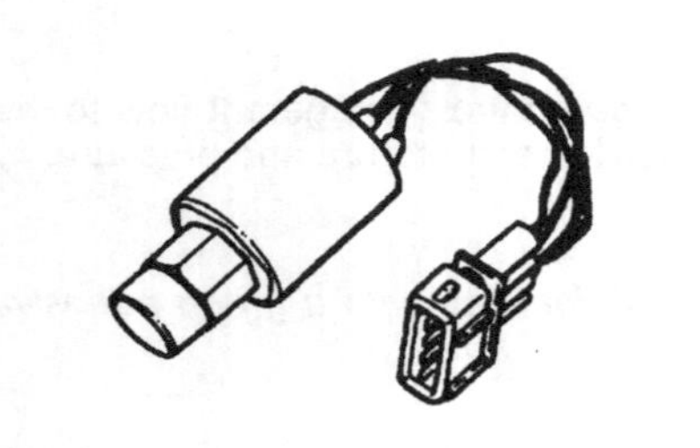

87-1378

◄ **A/C pressure switch -F129-**

Switch has three functions:

- Switches the A/C clutch -N25- off when excessive refrigerant circuit pressure is present (e.g.: insufficient air flow over condensor or when overcharged).
- Switches the A/C clutch -N25- off when insufficient refrigerant circuit pressure is present (e.g.: when refrigerant has leaked out)
- Switches on the second speed coolant fan -V7- when refrigerant circuit pressure increases.

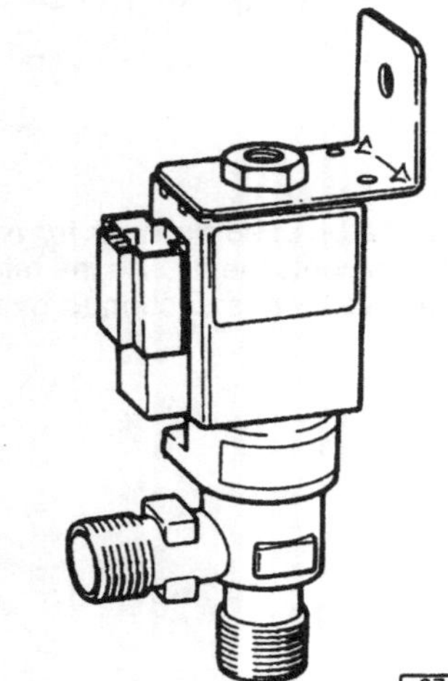

87-453A

◄ **A/C refrigerant shut-off valve -N43-**

The flow of refrigerant to the second evaporator (where applicable) is controlled by valve -N43-. The valve opens (refrigerant flow) when voltage is applied by A/C control head -E87-.

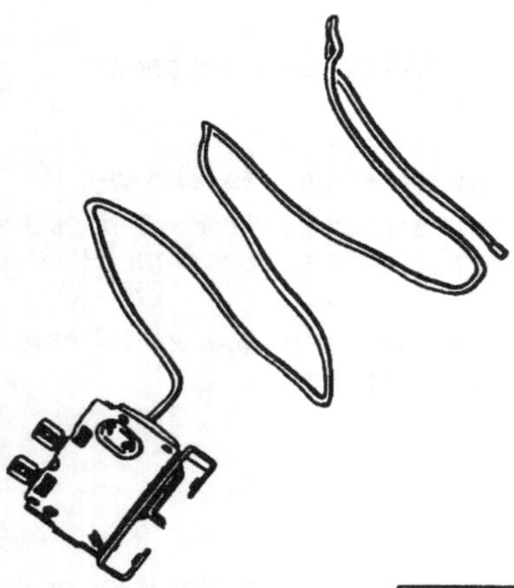

87-422

◄ **A/C evaporator temperature switch -E33-**

The temperature switch monitors the temperature between the evaporator fins to prevent evaporator icing. When the evaporator temperature drops to below 0°C (32°F), the A/C clutch -N25- is switched off.

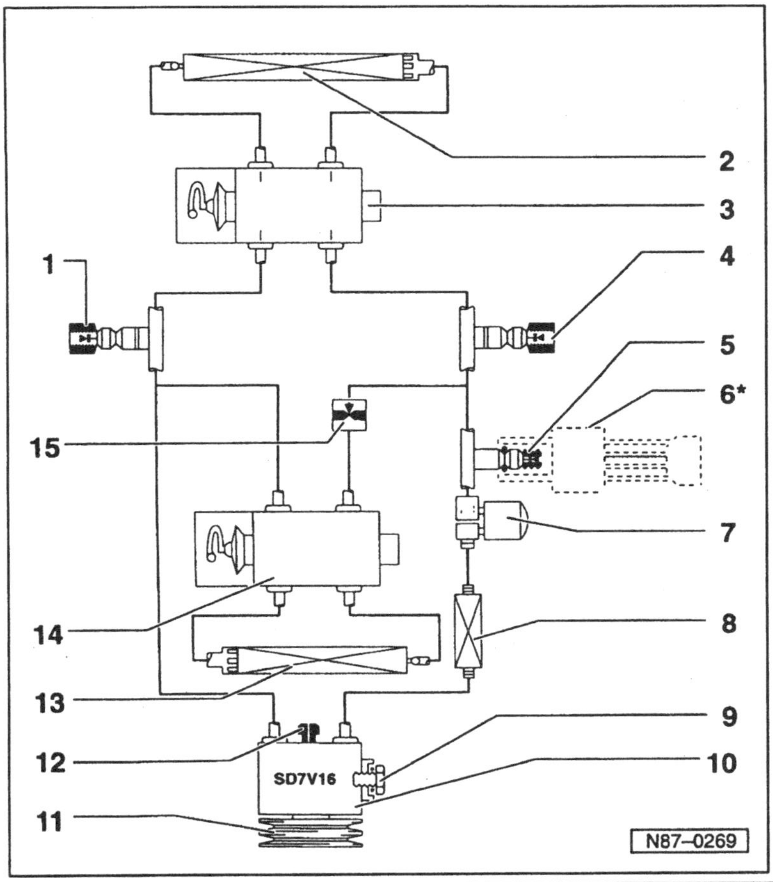

A/C refrigerant circuit

Note:

Illustration shows Climatronic refrigerant circuit with second evaporator.

1 - Low pressure service valve
- For charging/discharging

2 - Evaporator

3 - Expansion valve
- In engine compartment, right

4 - High pressure service valve
- For charging/discharging

5 - Connector with valve
- For A/C pressure switch -F129-

6 - A/C pressure switch -F129-
- Can be removed/installed without discharging system

7 - Receiver dryer

8 - Condenser

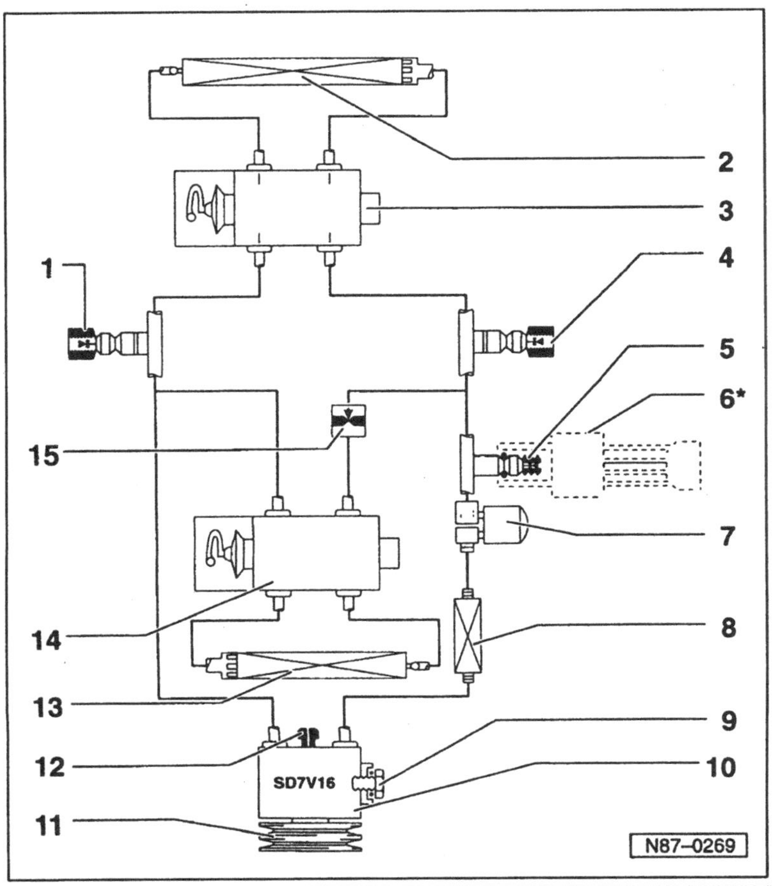

9 - Oil fill/drain plug

10 - A/C compressor

11 - A/C clutch -N25-

12 - Pressure relief valve

13 - Second evaporator (rear)*

14 - Expansion valve for second evaporator*

15 - A/C refrigerant shut-off valve -N43-*

* Where applicable

A/C refrigerant system R-134a

Safety measures

The subassemblies and lines of the air conditioner are filled with Tetrafluoroethane also known as refrigerant R-134a.

Refrigerant R-134a replaces the refrigerant R-12 which is reported to have harmful effects on the earth's ozone layer.

R-134a and R-12 systems are similar in design but there are very important differences in the refrigerant, lubricating oils, service equipment and A/C system components.

> ***WARNING!***
>
> ***- Always use an Underwriter's Laboratory (UL) approved refrigerant recovery/recycling/recharging unit such as Kent–Moore ACR^4, or equivalent, whenever discharging an R-134a A/C system***
>
> ***- As of January, 1992 any person who services a motor vehicle air conditioner MUST, by law, be properly trained and certified and use approved refrigerant recycling equipment. Technicians must complete an EPA approved recycling course to be certified***
>
> ***- State and local governments may have additional requirements regarding air conditioning servicing. Always comply with state and local laws***

> ***WARNING!***
>
> ***- A/C system is filled with refrigerant gas which is under pressure***
>
> ***- Always be careful that refrigerant does not come in contact with your skin***
>
> ***- If liquid refrigerant has come in contact with your skin or eyes:***
>
> ***♦ Do not rub skin or eyes***
>
> ***♦ Immediately flush with cool water for 15 minutes***
>
> ***♦ Rush to a doctor or hospital***
>
> ***♦ Do not attempt to treat yourself***
>
> ***- Work in a well ventilated area because refrigerant gases are heavier than air, displace oxygen and may cause suffocation in areas of poor air circulation, like under the car***
>
> ***- Avoid breathing refrigerant vapors. Exposure may irritate eyes, nose and throat***
>
> ***- Always wear hand and eye protection (gloves and goggles) when working around the A/C system***
>
> ***- Keep refrigerant containers stored below 50°C (122°F) and never drop from high places. DO NOT warm refrigerant containers with an open flame. If refrigerant needs to be warmed, place bottom of tank in warm water.***
>
> ***- Keep refrigerant away from open flames because poisonous gas will be produced if it burns. Do not smoke when refrigerant gases are present for the same reason***

WARNING!

- Electric welding near refrigerant hoses causes R-134a to decompose from ultraviolet light. Discharge system before electric welding

- Pressurized R-134a refrigerant in the presence of oxygen may form a combustible mixture. Never introduce compressed air into any R-134a container (full or empty), A/C component or piece of service equipment

- Do not expose any component of the A/C system to high temperatures (for example, open flames). Excessive heat will cause a pressure increase which could burst the system

- Do not steam clean condensers or evaporators. Use only cold water or compressed air

CAUTION!

- R-12 and R-134a refrigerant are NOT compatible. Never add R-12 refrigerant to an R-134a system or R-134a refrigerant to an R-12 system. If the refrigerants are mixed, total system contamination will occur and compressor failure may result

- Refrigerant oils used for the R-134a system and R-12 system are NOT compatible. Use only the specified synthetic oil (Polyalkylene Glycol/PAG) for the R-134a refrigerant system. DO NOT use R-12 system oil in an R-134a system or R-134a system oil in an R-12 system. If the refrigerant oils are mixed, system contamination will occur and compressor failure may result

- R-134a refrigerant system oil (PAG oil) absorbs moisture very rapidly. Moisture combines with the refrigerant to form acids which will damage the system. Use only the specified oil from a sealed container and ALWAYS reseal oil container immediately after use. DO NOT use oil if it has become contaminated with moisture

- Immediately plug open connections on A/C components to prevent dirt and moisture contamination. Likewise, DO NOT remove new components from packaging until ready to install. Immediately tighten component connections after installation

CAUTION!

- Always use separate refrigerant recovery/recycling/recharging servicing equipment for R-12 and R-134a systems. DO NOT use one piece of equipment for both R-12 and R-134a systems. The residual traces of refrigerant will contaminate and damage the equipment. Servicing equipment includes recovery/recycling/recharging unit, charging station, vacuum pump, manifold gauges, etc.

- DO NOT use R-12 servicing equipment on R-134a systems or R-134a equipment on R-12 systems or damage to both the vehicle A/C system and servicing equipment may result. Use only equipment designed to meet Society of Automotive Engineers (SAE) standards

- R-134a and R-12 systems use different size service fittings. NEVER use adaptors to convert an R-12 fitting to R0134a size or R-134a fitting to R-12 size

- R-134a and R-12 A/C components including compressor, hoses, O–rings, evaporator, condenser, receiver–drier, etc. are NOT interchangeable. Components of the R-134a system are identified by lettering (R-134a) or by a green label (or stripe). In addition a label on the upper radiator support identifies which type refrigerant is used. Use only the correct system component for each refrigerant type

CAUTION!

- Always replace damaged and/or leaking A/C system components. Do not attempt repair by soldering or welding

- Work area must be extremely clean when working on A/C system components

- Discharge A/C system before removing any A/C system component

- Always reinstall caps over A/C service valves

- Always replace O–rings, DO NOT reuse. Use only the correct size and type of O–rings specified for use with R-134a refrigerant. Lubricate O–ring with refrigerant oil before installing

A/C refrigerant (R-134a), properties

Commercial designation

The following designations are used and appear on component labels and factor tags:

- R-134a
- Tetrafluoroethane
- $CH_2F\ CF_3$
- H–KW 134a
- SUVA® 134a
- ARCTON® 134a

Color

Refrigerant R-134a is colorless and is invisible as a gas. R-134a when viewed through the sight glass may appear milky due to the mixture of refrigerant and lubricating oil (PAG oil).

Temperature/pressure relationship

Similar to R-12 refrigerant, R-134a in an enclosed container will have a specific temperature/pressure relationship.

Temperature in °C (°F)	Pressure in bar (psi)
–30 (–22)	0.0 (0.0)
–20 (–4)	0.3 (4.4)
–10 (14)	1.0 (14.5)
0 (32)	1.9 (27.5)
10 (50)	3.1 (45.0)
20 (68)	4.7 (68.2)
30 (86)	6.7 (97.2)
40 (104)	9.1 (132.0)
50 (122)	12.2 (177.0)
60 (140)	15.8 (229.2)
70 (158)	20.2 (293.0)

Refrigerant oil

A special Polyalkylene Glycol (PAG) synthetic oil is used in R-134a systems. This oil is NOT compatible with mineral based oils used in R-12 systems. See safety measures ⇒ page T 87-1, for Warnings and Cautions regarding proper oil usage.

Airborne properties

Escaped refrigerant gases are heavier than air and will gather first in low places, like under the car. R-134a refrigerant gas displaces oxygen and may cause suffocation in low areas of poor air circulation.

Refrigerant R-134a is not poisonous in any state (liquid or gas) and is safe when used properly.

Affects on plastic

Refrigerant R-134a will deteriorate some plastics. Therefore, when making system repairs, use only genuine VW replacement parts which are specified for use with R-134a refrigerant.

Affects on metal

In its pure state, refrigerant R-134a is chemically stable and will not attack iron, copper, brass or aluminum. However, the mixture of R-134a and PAG oil may deteriorate certain metals (copper). Therefore, when making system repairs, use only genuine VW replacement parts which are specified for use with R-134a refrigerant.

Water solubility

Liquid R-134a refrigerant will absorb only very minute quantities of moisture. However, R-134a vapor can absorb large amounts of moisture.

Flammability

R-134a refrigerant is not flammable.

Refrigerant containers

DO NOT exceed maximum rated capacity of refrigerant container.

Leak detection

Use halogen leak detector Hitec HI400A–TEL or equivalent to check for R-134a system leaks. This tool can also be used to detect leaks in R-12 systems. Many currently available R-12 leak detectors cannot detect R-134a refrigerant leaks.

R-134a refrigerant oil

A special Polyalkylene Glycol (PAG) synthetic oil is used in R-134a systems. This oil is NOT compatible with mineral based oils used in R-12 systems.

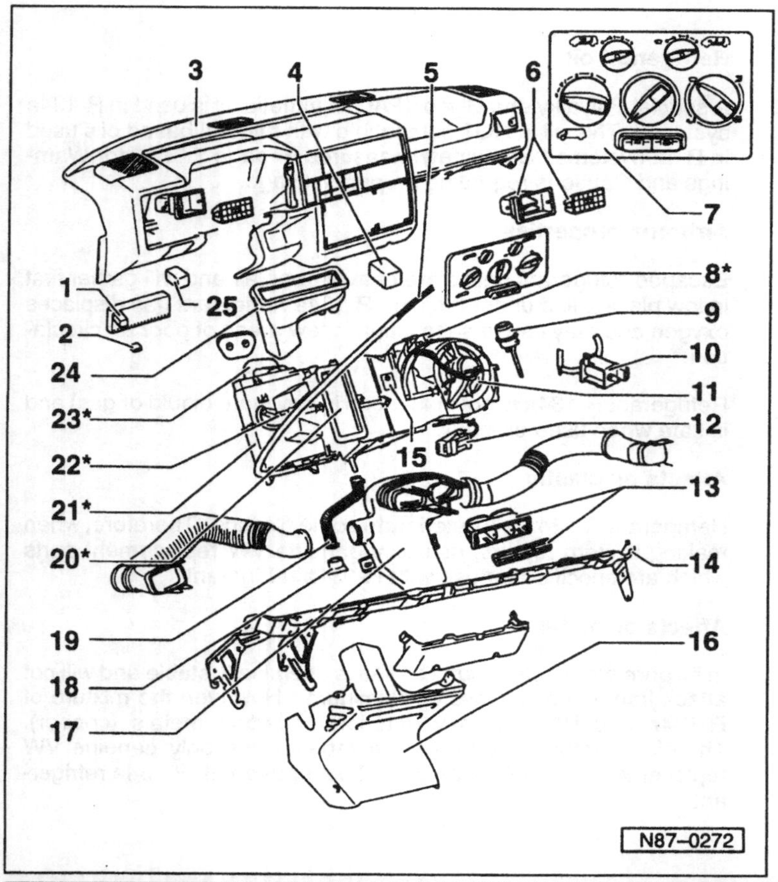

Manual A/C

Controls and mechanical components, servicing

> ***CAUTION!***
>
> ***Before working on electrical system, obtain radio code and disconnect battery Ground (GND) strap.***

Note:

*System components identified with an * must only be serviced or replaced after discharging refrigerant system. Use Kent Moore ACR4 or equivalent.*

1 - A/C relay -J32-
- ♦ Control number 140

2 - A/C system and Heating fuse -S97-
- ♦ Below relay plate

3 - Instrument panel
- ♦ Removing and installing

⇒ Repair Manual, *Body - Exterior, Body - Interior*, Repair Group 70

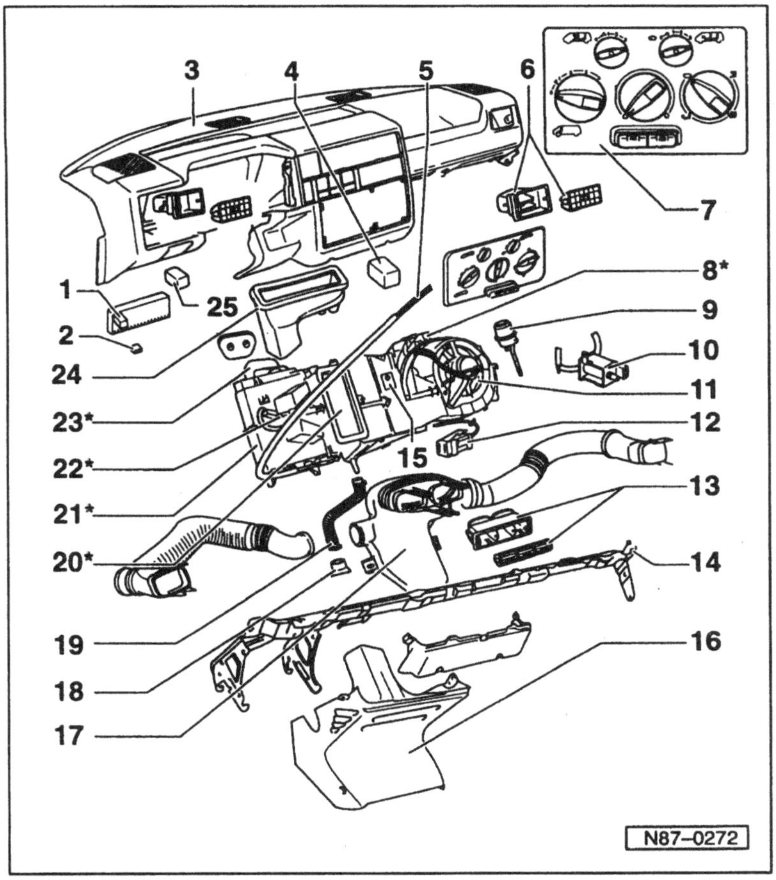

4 - Relay for fresh air and recirculating air flap rear -J353-
- Behind center instrument panel
- Control number 94

5 - Air temperature & distribution cables
- Installing and adjusting temperature flap cable

⇒ Page T 87-37

- Installing and adjusting footwell/ defrost flap cable

⇒ Page T 87-34

- Installing and adjusting central flap cable

⇒ Page T 87-34

6 - Right air outlet
- Removing and installing

⇒ Page T 80-6

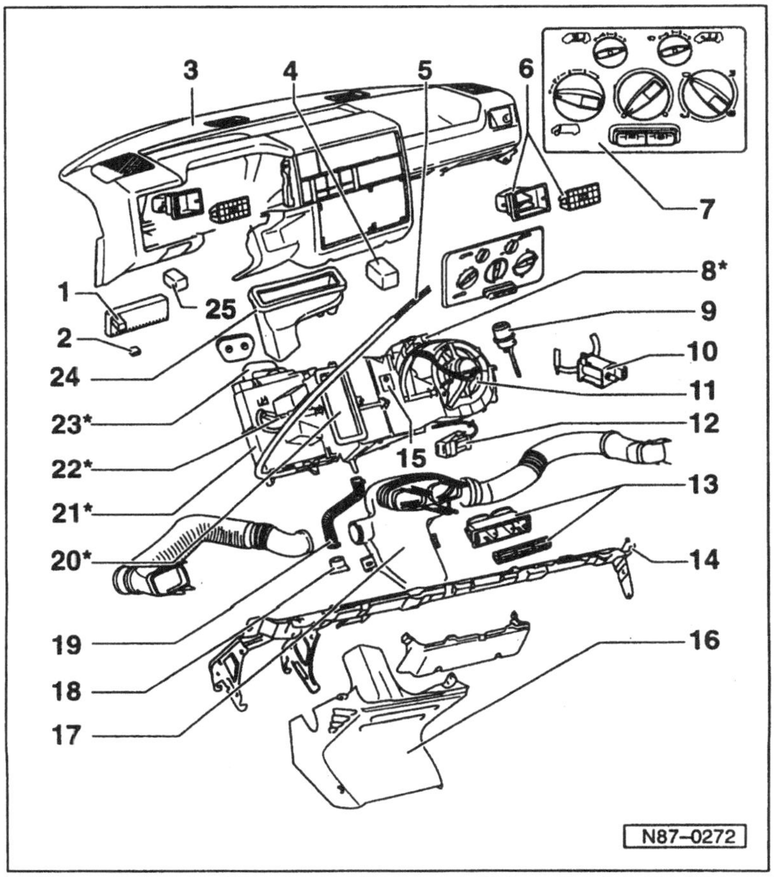

7 - Heating and A/C controls

With:
- A/C switch -E35-
- Fresh air blower switch -E9-
- Warm air blower switch -E100-
- Switch for fresh air blower switch rear -E179-
- Fresh air/recirculating flap switch -E159-
- Controls, removing

⇒ Page T 87-30

- Disassembling

⇒ Page T 87-31

- Installing and adjusting temperature flap cable

⇒ Page T 87-37

- Installing and adjusting footwell/ defrost flap cable

⇒ Page T 87-34

- Installing and adjusting central flap cable

⇒ Page T 87-34

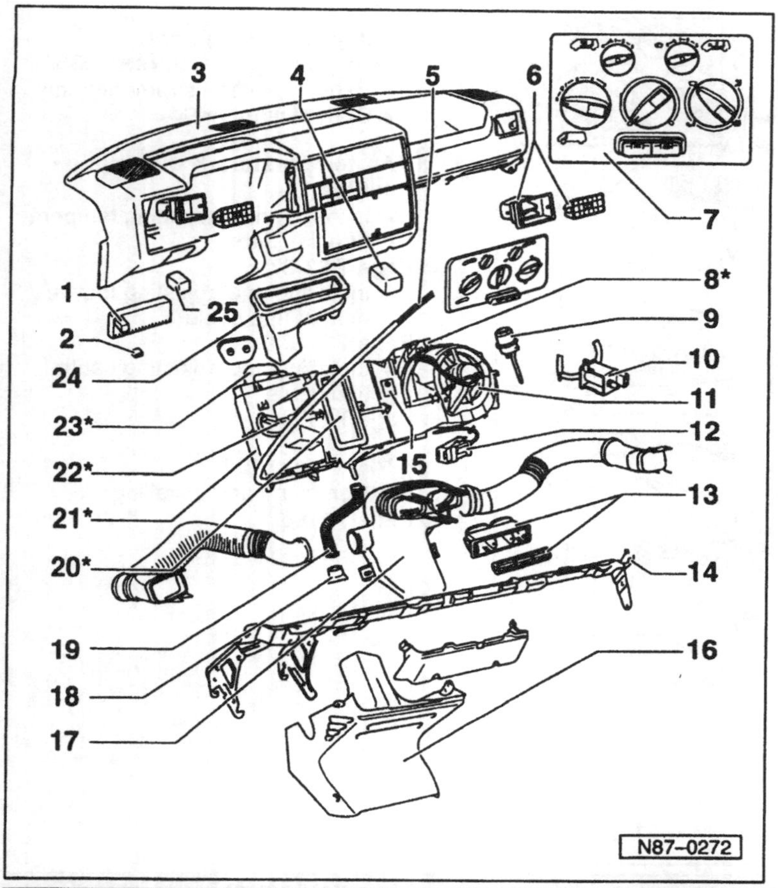

8 - Air intake housing*
- ♦ With fresh air/recirculating flap

9 - Fresh air/recirculating flap vacuum unit
- ♦ Flap in fresh air position when no vacuum is present
- ♦ Flap in recirculated air position when vacuum is present
- ♦ Vacuum hose routing

⇒ Page T 87-45
- ♦ Removing

⇒ Fig. 1

10 - Fresh air/recirculating flap two-way valve -N63-
- ♦ Checking

⇒ Page T 87-45
- ♦ Replacing:
- – Remove glove compartment, knee padding or passenger side airbag unit.

⇒ Repair Manual, *Body - Exterior, Body - Interior,* Repair Group 68

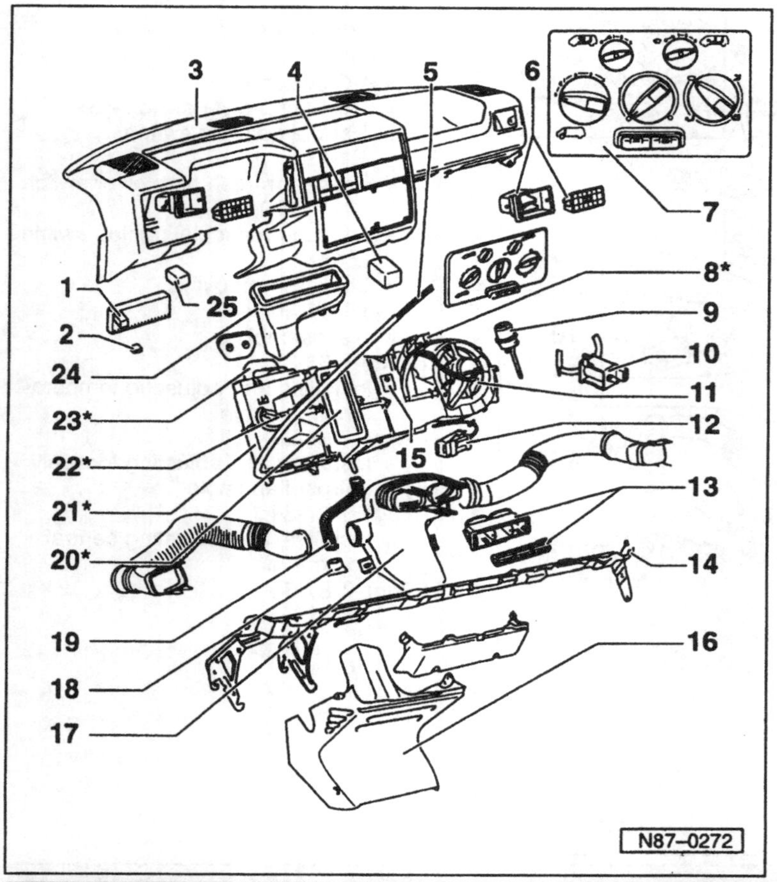

11 - Fresh air blower -V2-
- ♦ Replacing:
- – Remove instrument panel -7- and loosen center air duct -17- on right-hand side
- ♦ Before installing coat contact surfaces with AMV 176 000 05

12 - A/C evaporator temperature switch -E33-
- ♦ Prevents evaporator icing
- ♦ Insertion depth of sensor tube: 330 mm (13 in)
- ♦ Replacing

⇒ Fig. 6
- ♦ Switching temperatures: opens below –2 °C (30°F)

13 - Center air outlet
- ♦ Removing

⇒ Page T 80-8

14 - Carrier

15 - Fresh air blower series resistance -N24-
- ♦ Replacing

⇒ Fig. 2

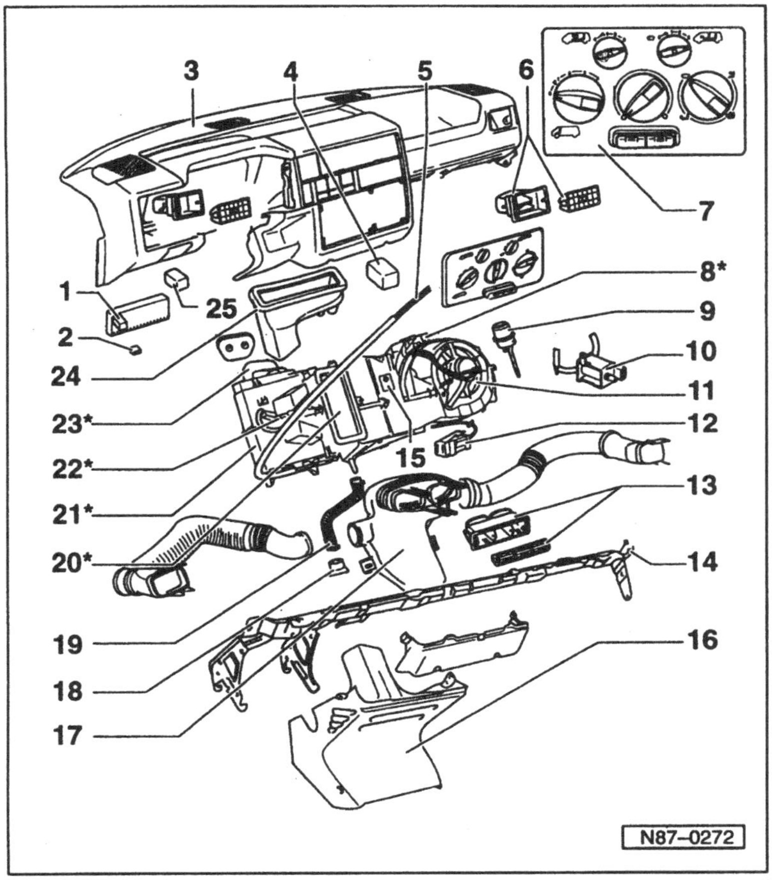

16 - Footwell air outlet console
- Removing

⇒ Fig. 5

17 - Center air duct
- Replacing
  - Remove instrument panel -7- and carrier -14-.

18 - Evaporator water drain valve
- Water leak at housing may result if valve is blocked

19 - Evaporator drain pipe
- Removing and installing

⇒ Fig. 3

20 - Central flap*
- Not a replaceable part

21 - Heater/evaporator unit*

22 - Footwell/defrost flap*
- Not a replaceable part

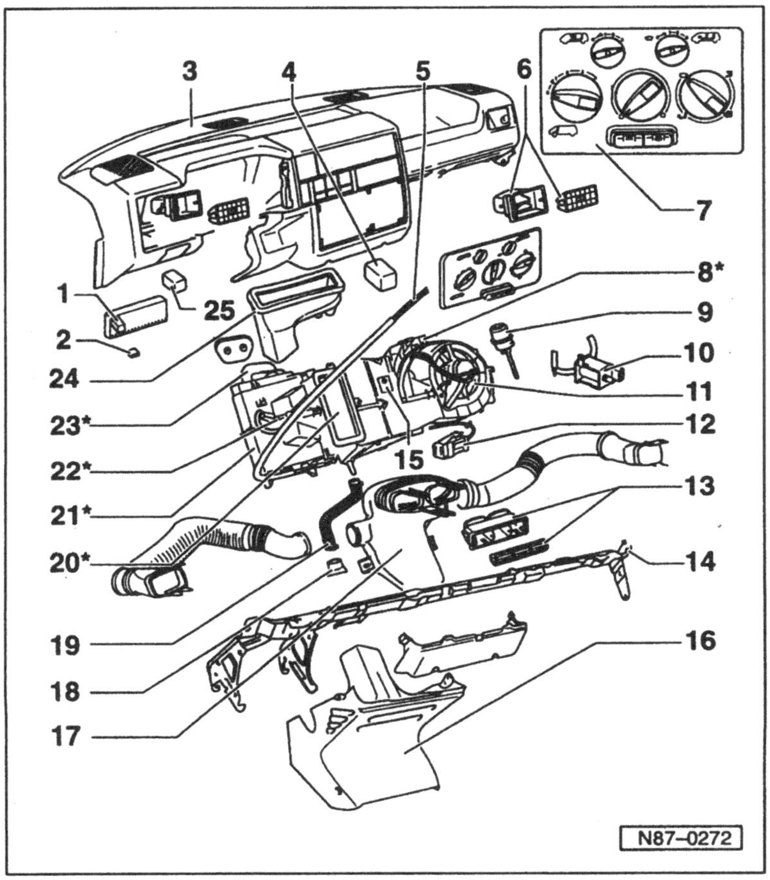

23 - Heater core*
- Install seals around entire circumference of Heating core
- Always replace coolant

24 - Defroster duct
- Replacing:
  - Remove instrument panel -7-

25 - Relay for A/C shut-off -J365-
- Above relay plate
- Only on vehicles with engine codes AES and ACV
- Control number 147

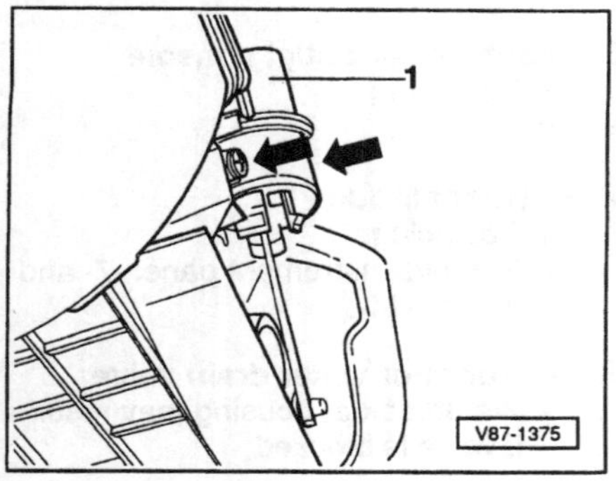

◄ **Fig. 1 Fresh air/recirculating flap vacuum unit, removing and installing**

Removing:

– Remove glove box, knee padding or passenger side airbag ⇒ Repair Manual, *Body - Exterior, Body - Interior,* Repair Group 69

– Remove screws -arrows-.

– Twist and lift vacuum unit -1- from lever.

Installing:

Install in reverse order.

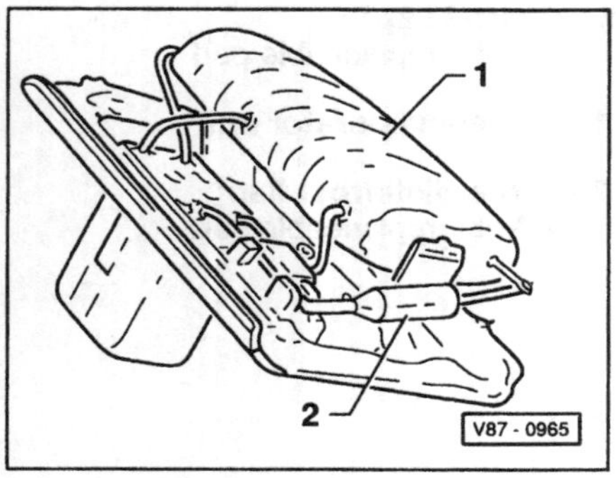

◄ **Fig. 2 Fresh air blower series resistance with fuse -N24-, removing and installing**

CAUTION!

Before working on the electrical system, obtain radio security code and disconnect battery Ground (GND) strap.

– Remove glove box, knee padding or passenger side airbag ⇒ Repair Manual, *Body - Exterior, Body - Interior,* Repair Group 69

1 - Series resistance

2 - Fuse

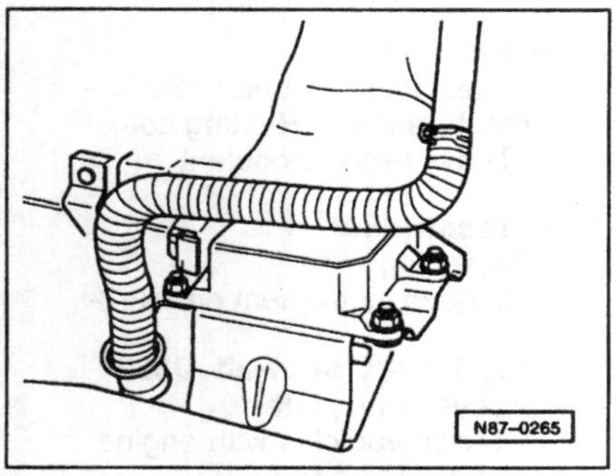

◄ **Fig. 3 Evaporator drain pipe, installing**

Notes:

- *Water drain hose must be pushed into distributor box onto stop.*
- *Water may leak into passenger compartment if hose is not located on the drain valve as shown.*
- *Drain valve must not be blocked.*

– Remove footwell vent console ⇒ Fig. 5.

Install hose so that drain valve grommet seals tightly around hose.

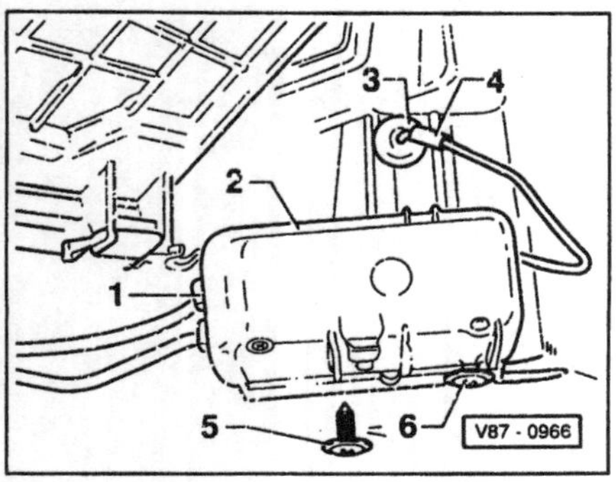

◄ **Fig. 4 Evaporator temperature switch -E33-, removing and installing**

1 - Connector

2 - Temperature switch with housing

3 - Grommet

4 - Sensor tube with insertion depth marking at 330mm (13 in)

5 - Screw for evaporator housing cover

6 - Screw

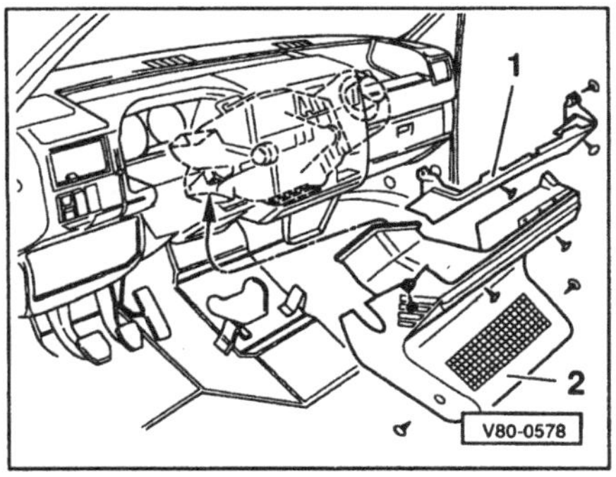

Fig. 5 Removing footwell vent console

– Remove trim -1-.

– Remove footwell vent console -2-.

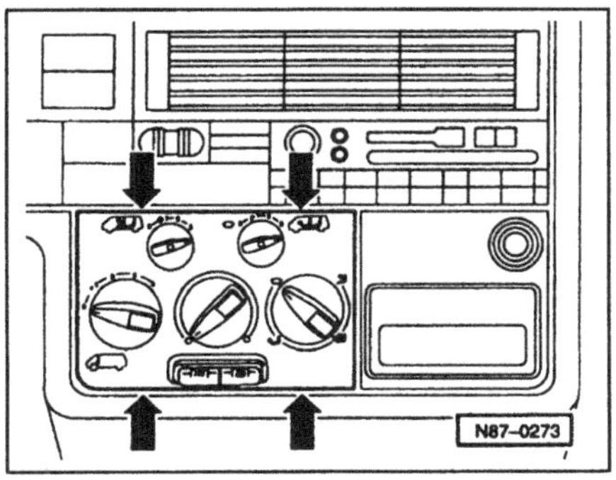

Heating & A/C controls, removing and installing

Removing:

– Carefully remove trim panel locating lugs -arrows- with a screwdriver. Take care not to damage instrument panel.

– Remove ashtray/trim panel.

– Remove footwell vent console
⇒ Page T 87-29, Fig. 5.

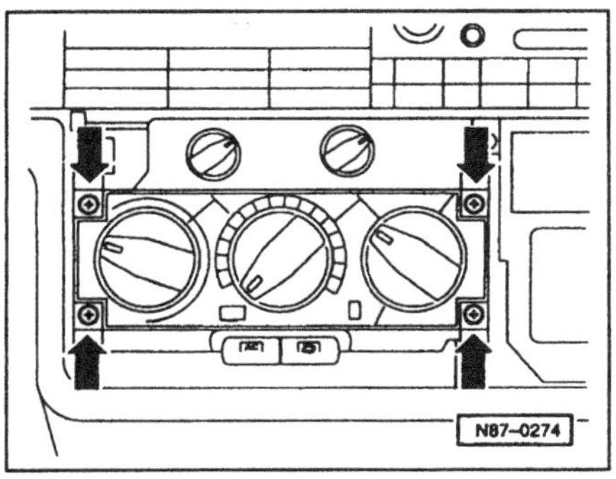

– Remove screws from instrument trim panel -arrows-.

– Push controls with cables attached below the instrument panel.

– Disconnect electrical connectors.

– Detach cables from controls.

Installing:

Install in reverse sequence, noting the following:

– Before installing controls, attach cables first.
⇒ Page T 87-34, Cables, installing and adjusting
⇒ Page T 87-37, Temperature flap cable, installing and adjusting

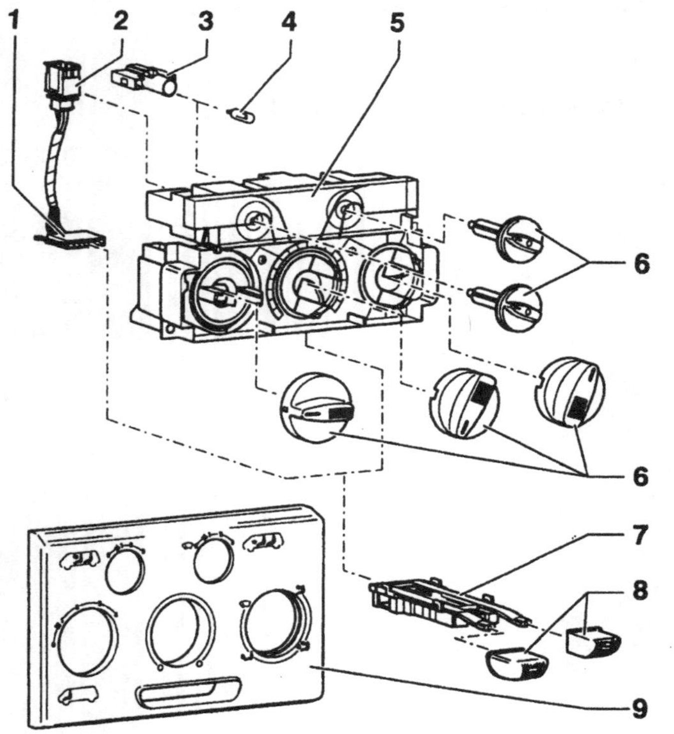

N87-0283

Heating & A/C controls, assembly

CAUTION!

Before working on electrical system, obtain radio code and disconnect battery Ground (GND) strap.

Note:

- *Illustrated control from vehicle with additional (rear) heat exchanger and (rear) passenger compartment ventilation.*

1 - Connector, 6 pin -T6-

2 - Connector, 6 pin -T6c-

3 - Illumination filter

4 - Illumination bulb -L16-

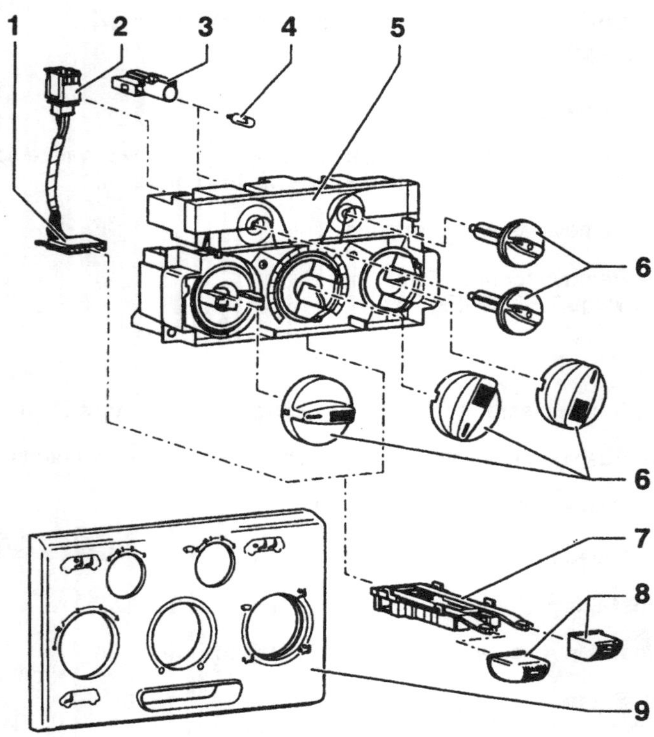

N87-0283

5 - Control assembly
With:
- Fresh air blower switch -E9-
- Warm air blower switch -E100-
- Fresh air blower switch -E179-
- Removing and installing

⇒ Page T 87-30

- Temperature flap cable, installing and adjusting

⇒ Page T 87-37

- Footwell/defrost flap cable, installing and adjusting

⇒ Page T 87-34

- Central flap cable, installing and adjusting

⇒ Page T 87-34

6 - Rotary knobs

7 - A/C switch
With:
- A/C switch -E35-
- Fresh air/recirculating flap switch -E159-
- Illumination bulb -L43-
- Removing

⇒ Fig. 1

8 - Push button

9 - Control panel trim

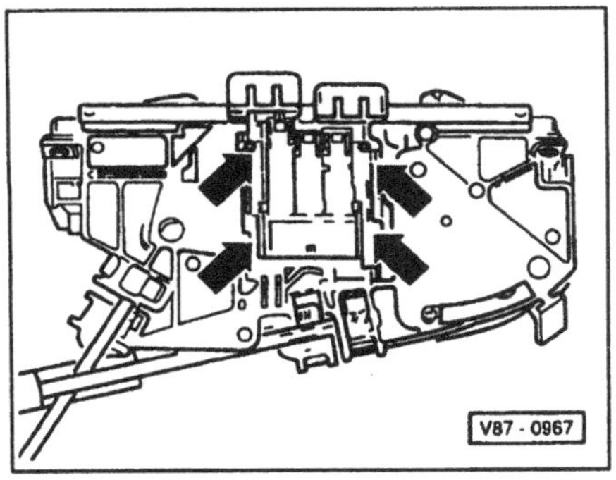

Fig. 1 A/C switch -E35- and -E159-, removing

– Remove Heating & A/C controls ⇒ Page T 87-30.

– Disconnect electrical connector.

– Press switch at the four locating points -arrows- against the controls and in direction of buttons simultaneously.

Heating & A/C cables, installing and adjusting

Notes:

- *Attach cables first before installing controls.*
- *Place outer cable against stops on controls and secure.*

Cables, installing

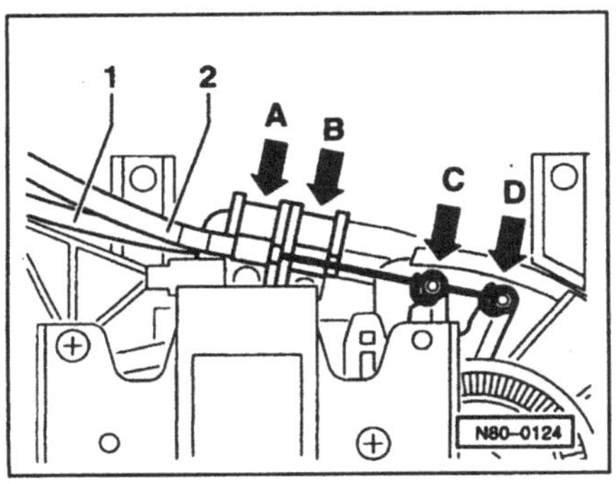

1 - Central flap cable

(Length 640 mm (25.20 in), identification yellow/green)

– Turn air distribution rotary knob fully to left stop.

– Connect central flap cable -arrow D-.

– Secure cable to controls with clip -arrow B-.

– Central flap cable, adjusting ⇒ Fig. 1

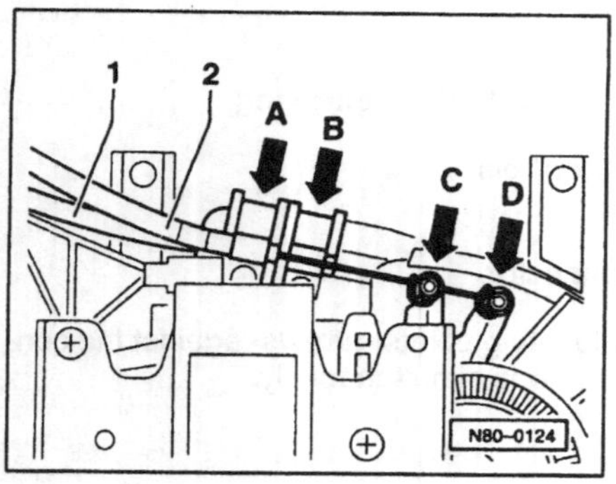

2 - Footwell/defroster flap cable

(Length 715 mm (28.14 in), identification yellow/blue)

- Turn air distribution rotary knob fully to left stop.
- Connect footwell and defrost flap cable -arrow C-.
- Secure cable to control with clip -arrow A-.
- Adjust footwell/defroster flap cable ⇒ Fig. 2

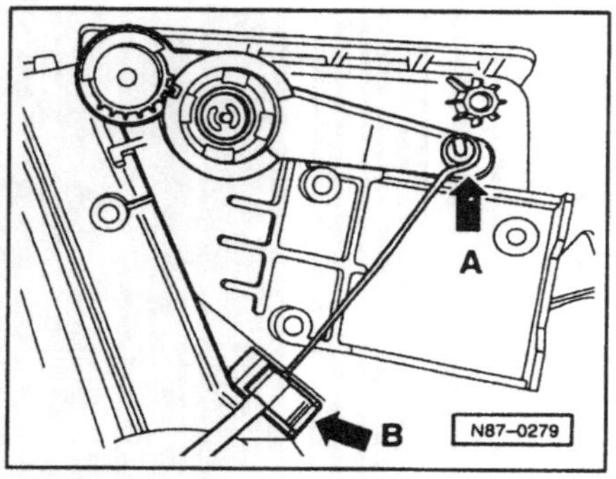

Fig. 1 Central flap cable, adjusting

- Secure cable to central flap -arrow B- while simultaneously pressing lever -arrow A- in direction of arrow.

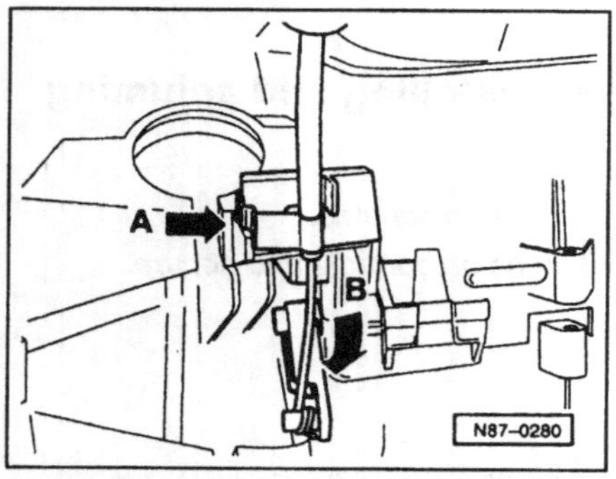

Fig. 2 Footwell/defroster flap cable, adjusting

- Secure cable to footwell flap -arrow A- while simultaneously pressing lever -arrow B- in direction of arrow.

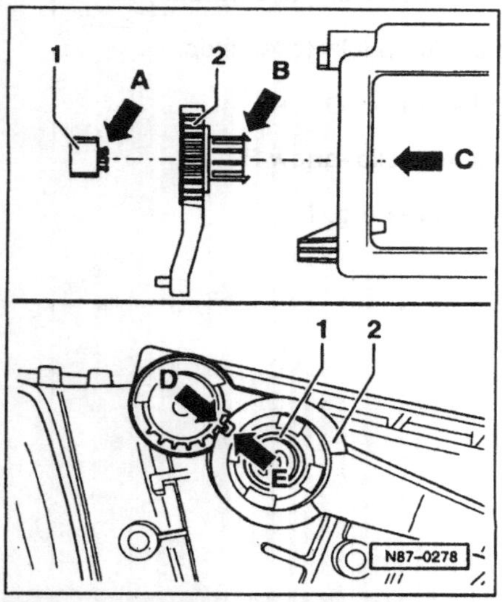

Central flap lever, removing and installing

Removing:

- Carefully depress tabs -arrow A- from cap -1- with a screwdriver at housing -arrow C-.
- Carefully depress tabs -arrow B- at housing -C-.
- Remove central flap lever -2-.

Installing:

Install in reverse order ensuring correct lever alignment (-arrows-, -D- and -E-)

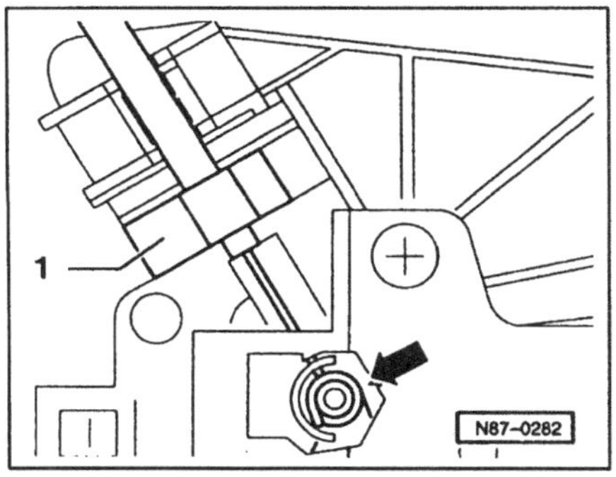

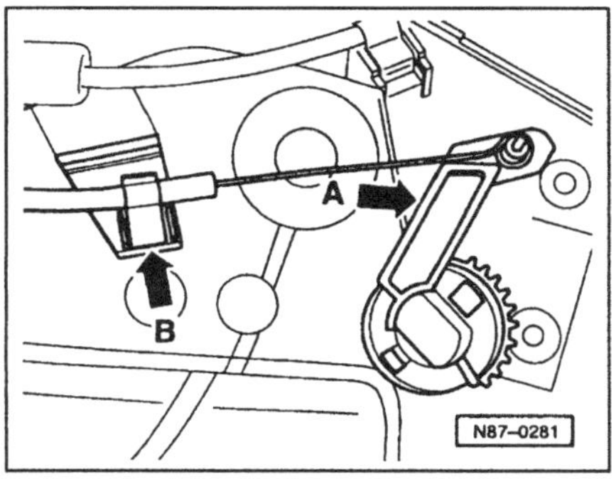

Temperature flap cable, installing and adjusting

(Length 585 mm (23.03 in), identification red/blue)

Note:

The temperature flap must audibly make contact when turning the rotary knob fully from stop to stop.

◄ – Turn temperature control rotary knob fully to left stop.

– Connect temperature flap cable -arrow-.

– Secure cable with clip to controls -1-.

◄ – Press temperature flap lever -arrow A- onto stop.

– Secure cable to control with clip -arrow B-.

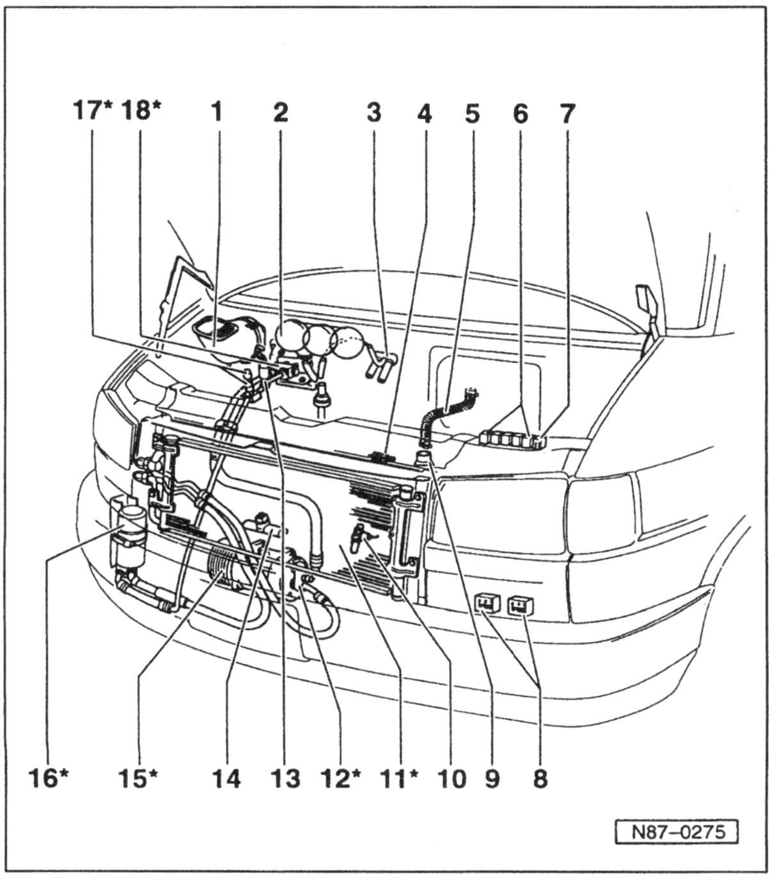

Heating & A/C components in engine compartment, servicing

CAUTION!

Before working on electrical system, obtain radio code and disconnect battery Ground (GND) strap.

Notes:

- *System components identified with an * must only be serviced or replaced after discharging refrigerant system. Use Kent Moore ACR4 or equivalent.*
- *Illustration shows vehicle with engine code AES.*

1 - Air intake duct

- Dust and pollen filter, removing and installing

⇒ Page T 80-14

2 - Vacuum reservoir

- Insert vacuum hose: 30 mm (1 3/16 in) into reservoir
- Vacuum layout

⇒ Page T 87-45

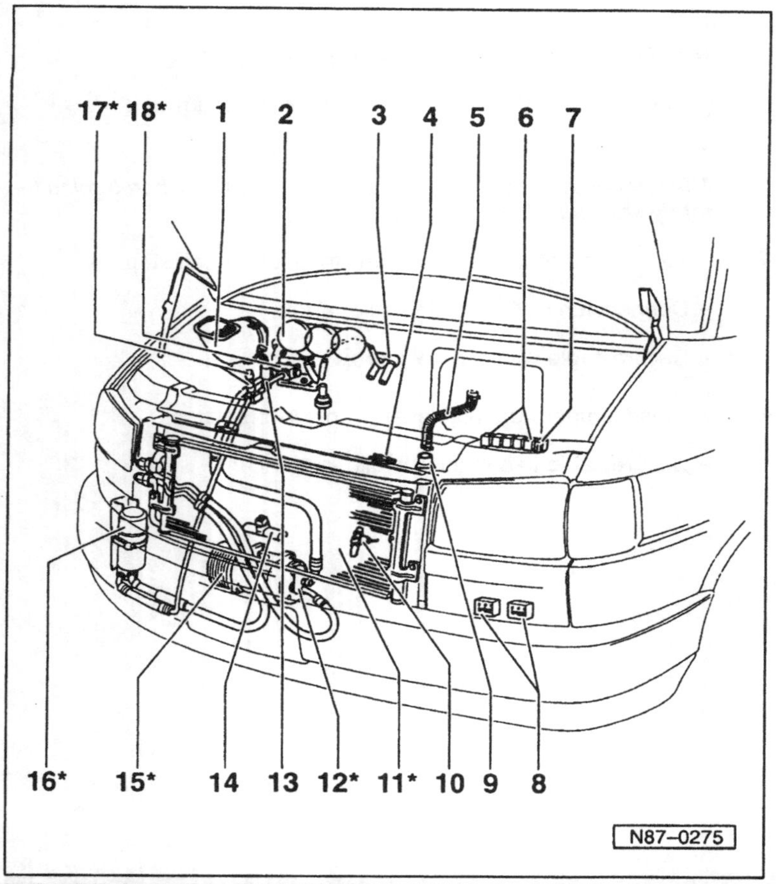

3 - Heater core and vacuum hose guide at bulkhead

4 - Engine coolant two-way vacuum valve -N147-

- ♦ Checking

⇒ Page T 87-45, vacuum hose layout

5 - Evaporator drain pipe

- ♦ Removing and installing

⇒ Page T 87-28

6 - Relays

Locations and wiring

- – Electrical Wiring Diagrams

7 - Individual fuses for coolant fan -S42- (2x)

8 - Coolant Fan Control (FC) series resistance -N39-

9 - Evaporator water drain valve

- ♦ Water leak at housing may result if valve blocked.

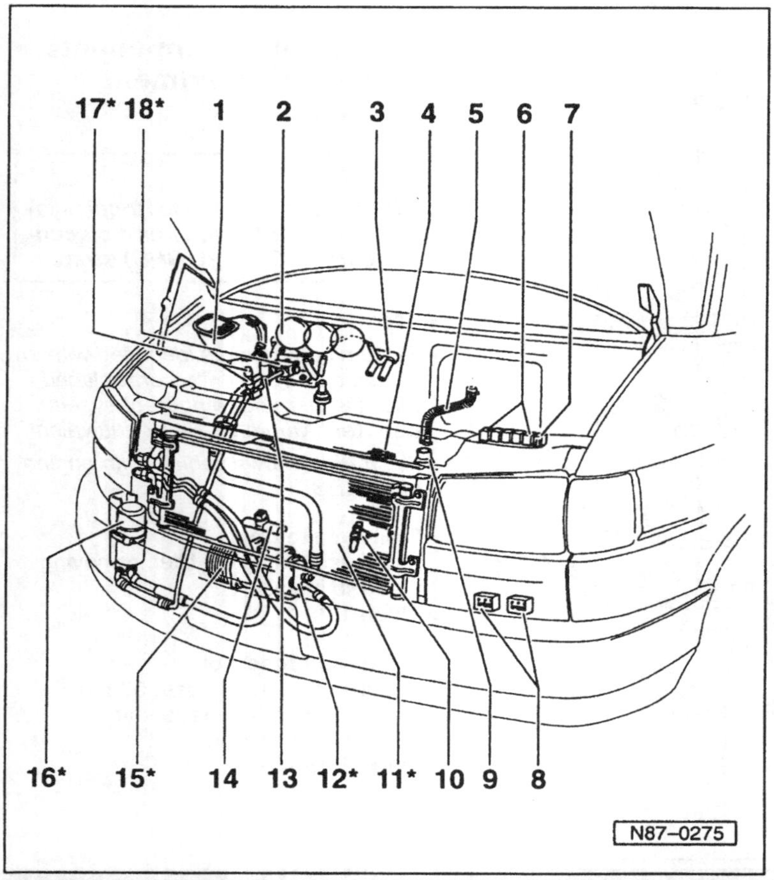

10 - Coolant shut-off valve

Function:

- ♦ Engine off: Vacuum supply to coolant shut-off valve is controlled by engine coolant two-way vacuum valve -N147-. Valve allows coolant to circulate from engine to radiator when after-run coolant pump operates.

11 - Condenser*

12 - Pressure relief valve*

- ♦ Protects refrigerant circuit against over-pressure
- ♦ Checking

⇒ Fig. 4

13 - A/C pressure switch -F129-

- ♦ Checking

⇒ Fig. 3

- ♦ Removing and installing:
- – Tightening torque 8 Nm
- – Replace O ring (note Part No.).

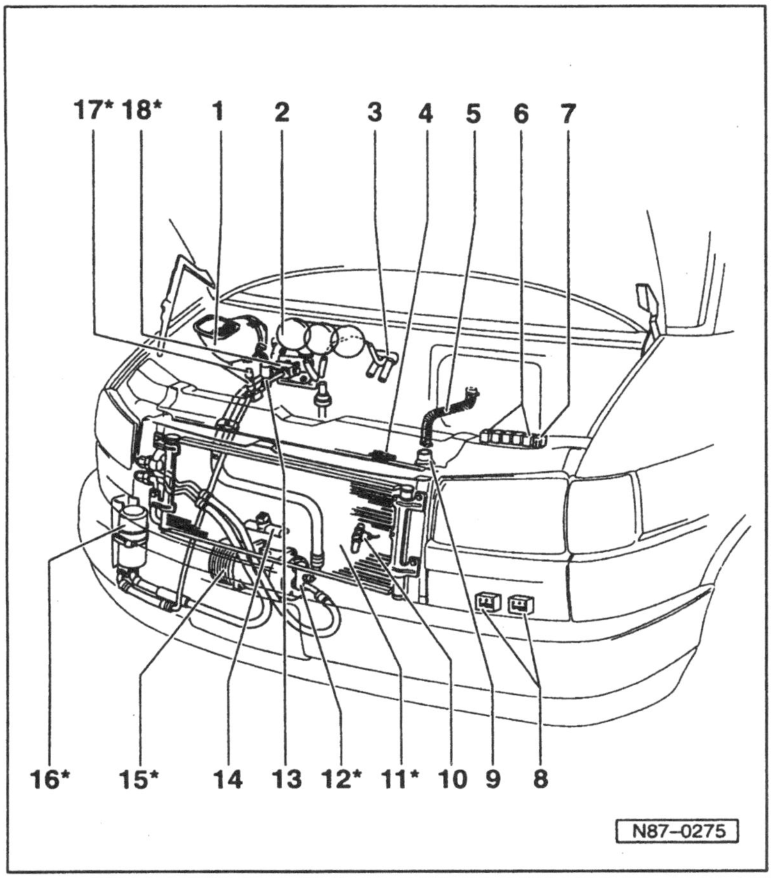

14 - A/C cut-out thermal switch -F163- and Third Speed Coolant Fan Control (FC) Thermal Switch -F165-

- In engine compartment, front

Function:

- -F165- switches-on coolant fan 3rd speed (on at 112°C / 234°F, off at 108°C / 226°F).
- -F163- switches off A/C clutch -N25- at excessive coolant temperatures (off at 119°C / 246°F, on at 112°C / 234°F).
- Removing

⇒ Fig. 2

15 - A/C clutch -N25-*

16 - Receiver drier*

17 - Service valves*

18 - Expansion valve*

- Opening at bulkhead must be sealed to prevent water ingress
- Expansion valve insulation ⇒ Fig. 1

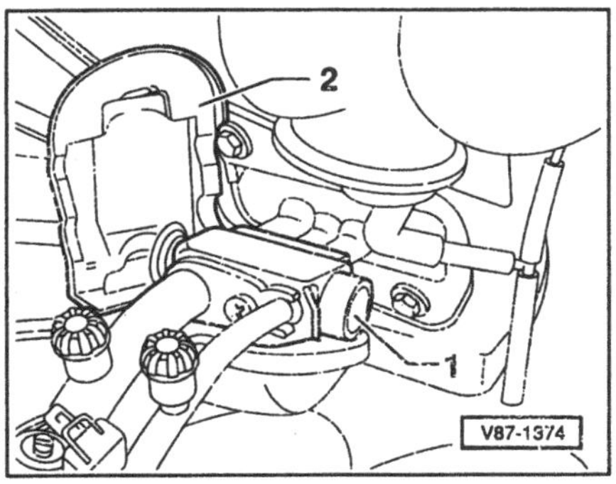

◄ **Fig. 1 Expansion valve insulation**

1 - Expansion valve

2 - Insulator

- The insulator protects the expansion valve against high engine compartment temperatures to prevent a decrease in A/C system cooling performance.

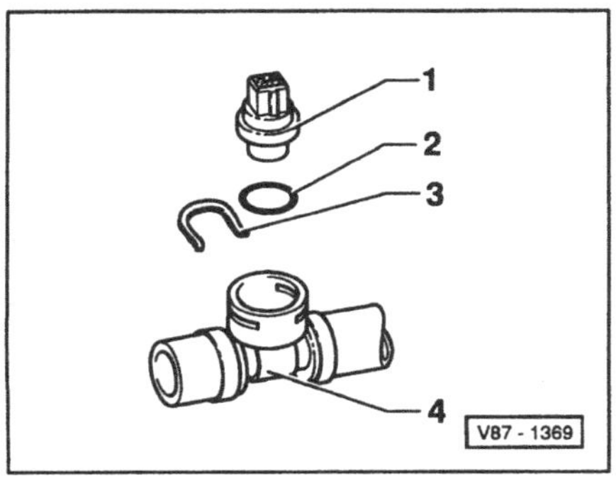

◄ **Fig. 2 A/C cut-out thermal switch -F163- and Third Speed Coolant Fan Control (FC) Thermal Switch -F165-, removing and installing**

> ***WARNING!***
>
> ***The cooling system is pressurized when the engine is warm. Wear gloves and other appropriate protection, and slowly and carefully release system pressure, if necessary, before performing repairs.***

- Refill with coolant after replacing switch.

1 - Thermal switch -F163- / -F165-

2 - O ring

3 - Clip

4 - Coolant hose

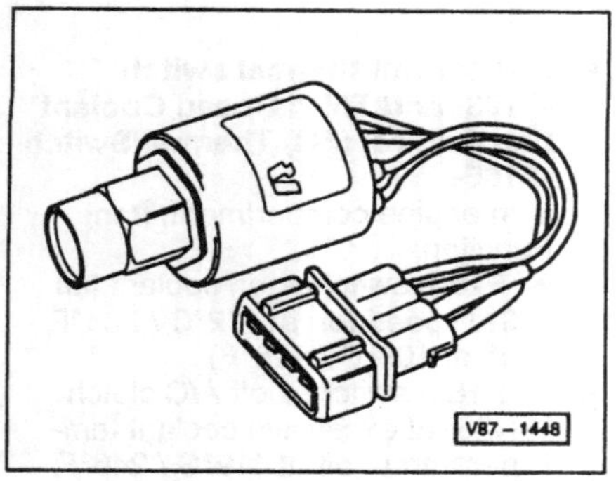

Fig. 3 A/C pressure switch -F129-, checking

♦ Ensure O-ring (color: black) 10.8 mm x 1.8 mm fits properly in groove.

Functions

Pressure switch component between blue wires switches A/C clutch -N25- off when too little or excess refrigerant system pressure is present.

♦ Switch opens (compressor off) below 2 bar (29.4 psi) and closes (compressor on) above 3 bar (43.5 psi).

♦ Switch opens (compressor off) above 32 bar (464 psi) and closes (compressor on) below 24 bar (348 psi).

Switch, checking

– Briefly bridge connections from terminal 1 and 2 with engine running.

● If A/C clutch -N25- switches on, refrigerant circuit is empty.

Pressure switch component between red/black wiring switches coolant fan -V7- to 2nd speed when refrigerant pressure rises.

♦ Switch closes (fan on) above 16 bar (232 psi) and opens again (fan off) below 12.5 bar (181.25 psi).

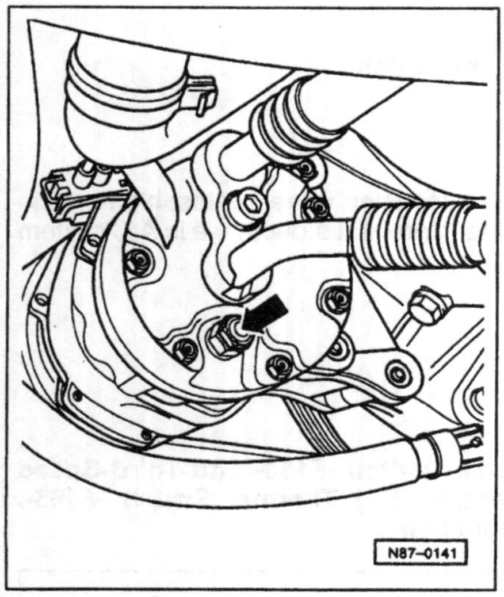

Fig. 4 Pressure relief valve, checking

♦ Function: Protects refrigerant circuit against over–pressure

♦ The pressure relief valve plug indicates if valve has opened to release pressure -arrow-. On some types, an adhesive plate is pushed out.

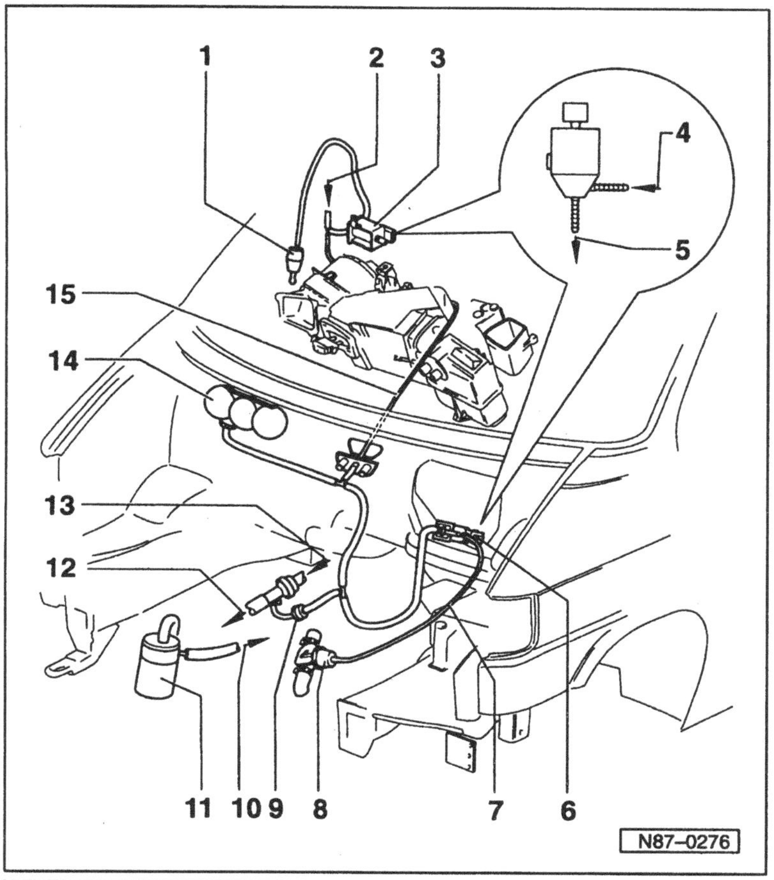

Vacuum hose, layout

Notes:

- *Arrows indicate vacuum.*
- *Insert vacuum hose into vacuum reservoir 30 mm (1 3/16 in). Insert vacuum hoses onto plastic connectors 15 mm (19/32 in).*
- *Illustration shows vehicle with engine, code AES*

1 - Fresh air/recirculating flap vacuum unit

2 - Vacuum supply to rear passenger compartment ventilation unit.*

* Not applicable

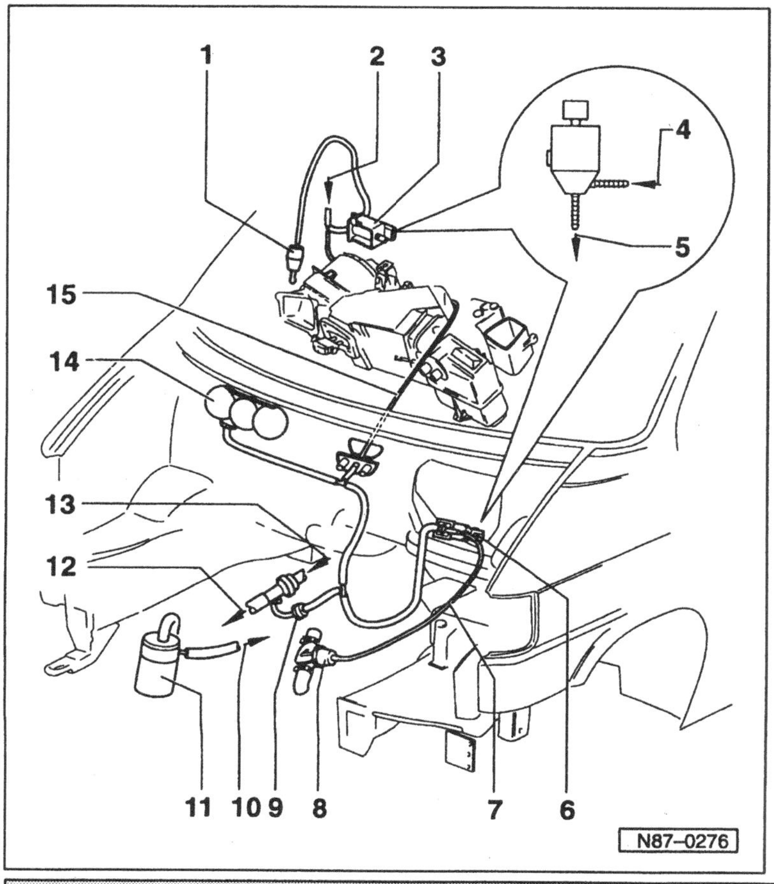

3 - Fresh air/recirculating flap two-way valve -N63-

- Controls vacuum flow to fresh air/recirculating flap vacuum unit.
- Replacing:

–

– Remove glove compartment, knee padding or passenger side airbag unit.

⇒ Repair Manual, *Body - Exterior, Body - Interior*, Repair Group 69

– After attaching vacuum hoses, check:

- The fresh air/recirculating flap vacuum unit must pull in if vacuum and voltage are present at -N63-.
- The fresh air/recirculating flap vacuum unit is vented if vacuum is present and voltage is not (switched off).

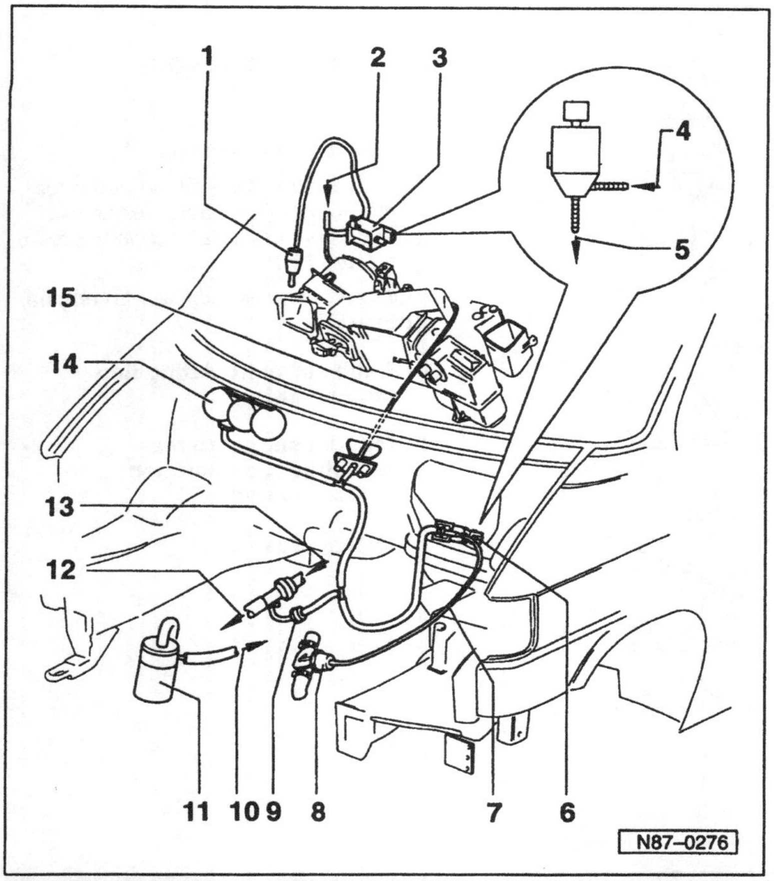

4 - Vacuum unit connection

5 - Intake manifold connection

6 - Engine coolant two-way vacuum valve -N147-
- ♦ Shuts off vacuum flow when no voltage is present.
- – After attaching vacuum hoses, check:
- ♦ When connected the vacuum unit must pull in if voltage and vacuum are present.
- ♦ When connected the vacuum unit is vented if vacuum is present and voltage is not.

7 - Vacuum hose
- ♦ From -N147- to coolant cut-off valve

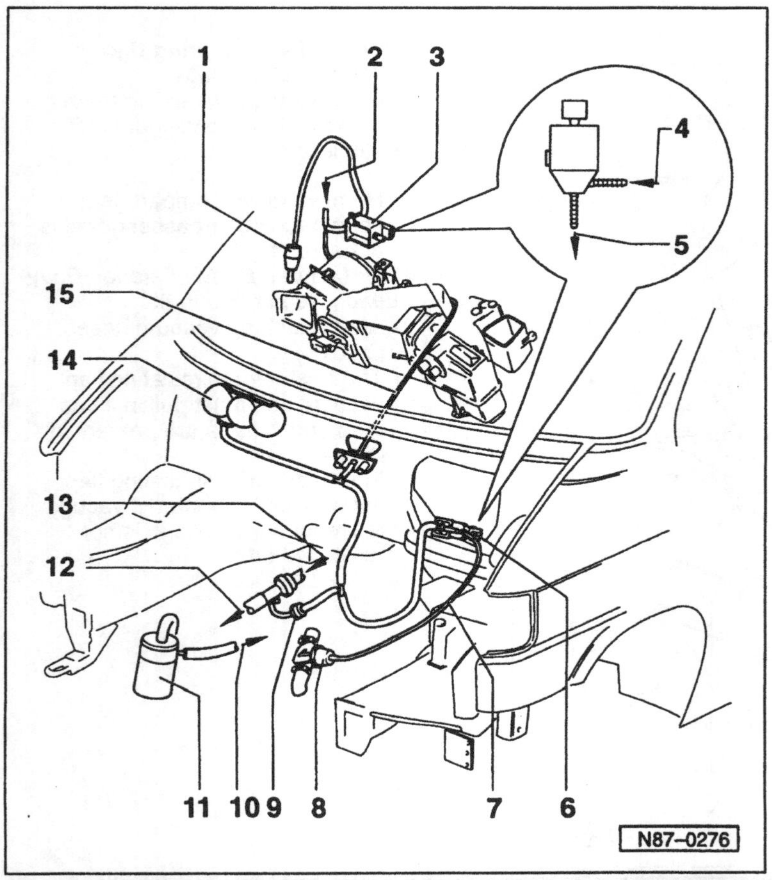

8 - Coolant cut-off valve
- ♦ Switched by vacuum from two-way valve -N147- when engine is running

9 - One-way valve

10 - From brake booster vacuum hose connection
- ♦ Only on engine, code AAB

11 - Vacuum unit
- ♦ Only on engine, code AAB

12 - To intake manifold

13 - To brake booster

14 - Vacuum reservoir

15 - Vacuum hose
- ♦ From vacuum reservoir in engine compartment to fresh air/recirculating flap vacuum unit

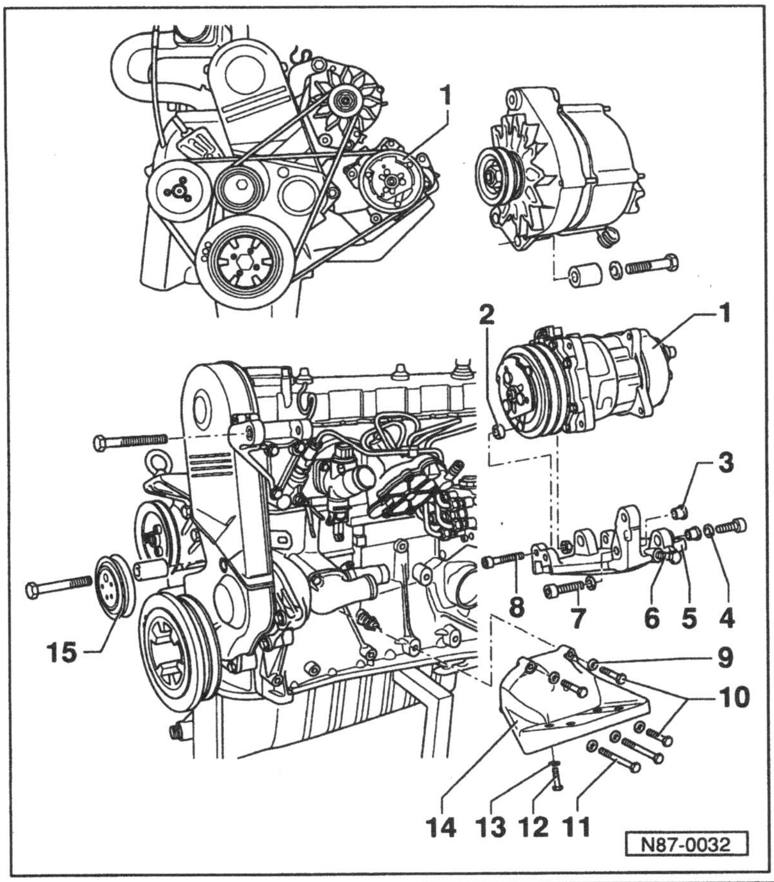

Compressor and bracket, assembly

Compressor bracket, removing and installing 2.4L diesel engine (AAB)

Required:

- Torque wrench 5 - 50 Nm (44 in lb - 40 ft lb)

Note:
The compressor bracket and related parts can be removed and installed without discharging the A/C refrigerant system.

1 - Compressor
- Must be installed so that the oil filler/drain plug is on top

2 - Bushing

3 - Bushing (2x)

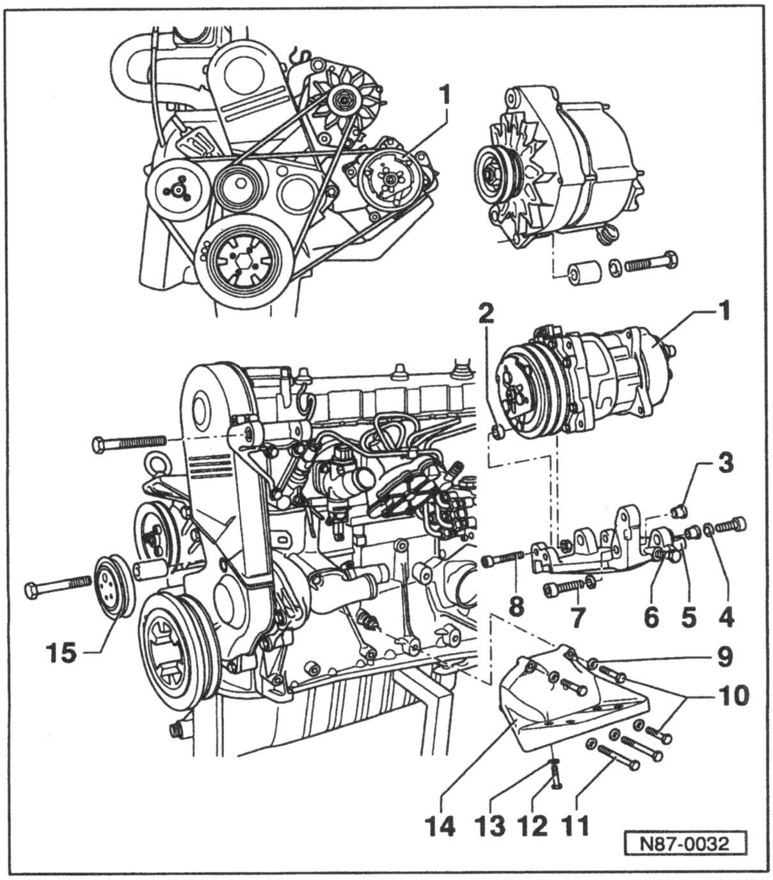

4 - Washer (3x)
- 10.5 x18 x 1.6

5 - Compressor bracket

6 - Adjustment bolt
- M8
- Use to tension V-belt

Note:

V-belts are correctly installed if they can be pressed with thumb pressure 5 mm to 10 mm (0.2 in to 0.4 in)

7 - Socket head bolt (3x)
- M10 x 35
- 40 Nm (30 ft lb)

8 - Socket head bolt
- M8 x 45
- 20 Nm (15 ft lb)

9 - Washer (6x)
- 10.5 x 21 x 2

10 - Hex bolt (3x)
- M10 x 45
- 40 Nm (30 ft lb)

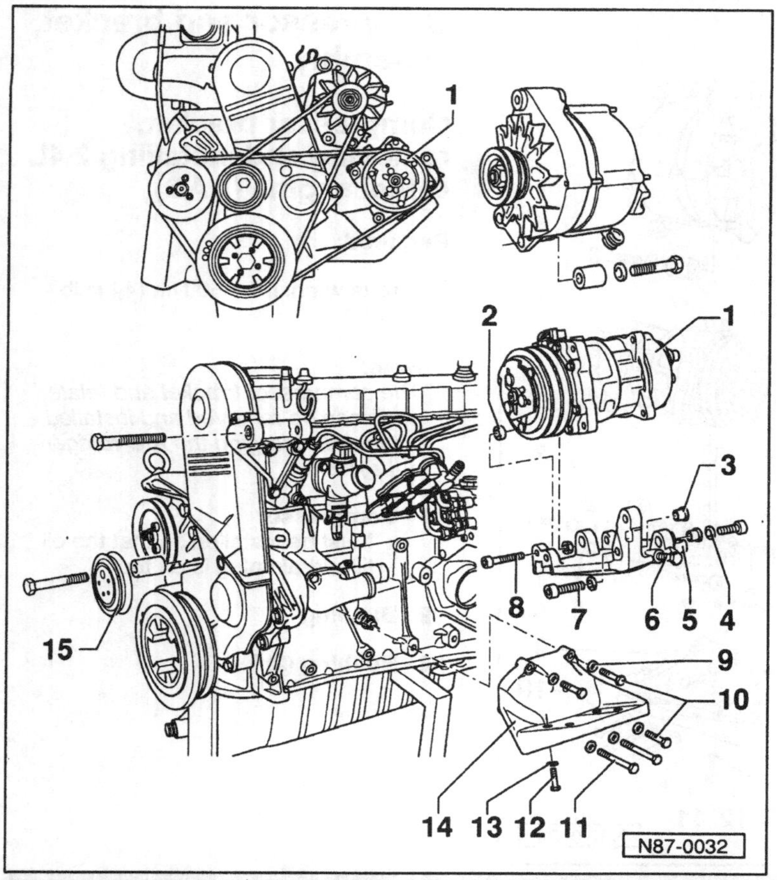

11 - **Hex bolt (2x)**
- M10 x 80
- 40 Nm (30 ft lb)

12 - **Hex bolt (4x)**
- M10 x 30
- 40 Nm (30 ft lb)

13 - **Washer (4x)**
- 10.5 x 26 x 4

14 - **Accessory drive bracket**

15 - **Tension roller**

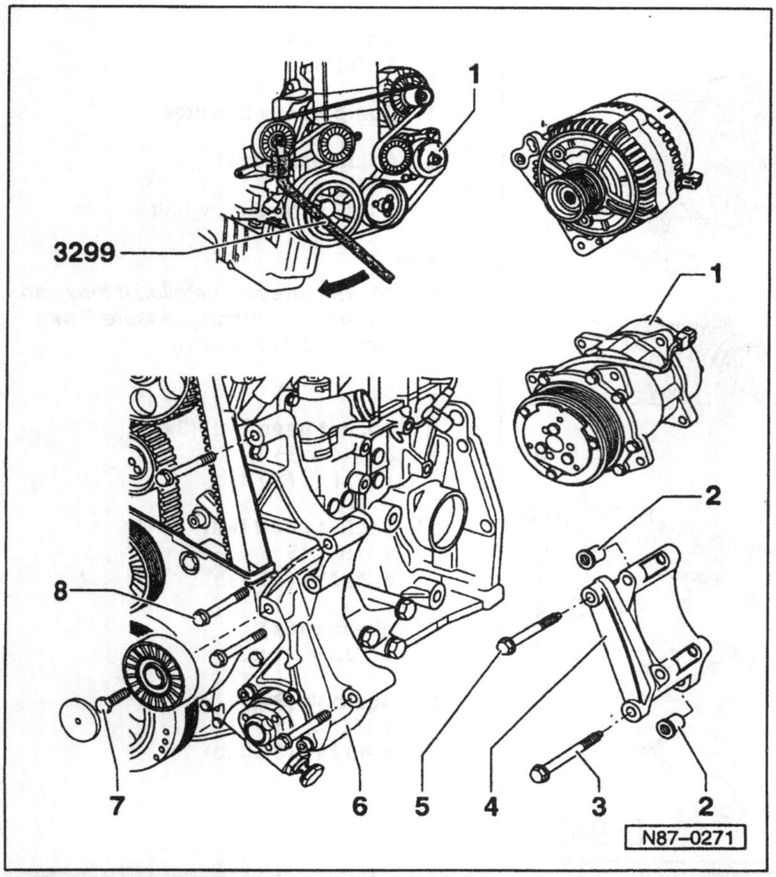

Compressor bracket, removing and installing 2.4L diesel engine (ACV)

Required:

- Torque wrench 5 - 50 Nm (44 in lb - 40 ft lb.)

Note:
The compressor bracket and related parts can be removed and installed without discharging the A/C refrigerant system.

1 - **Compressor**

2 - **Threaded bushing (2x)**
- M8

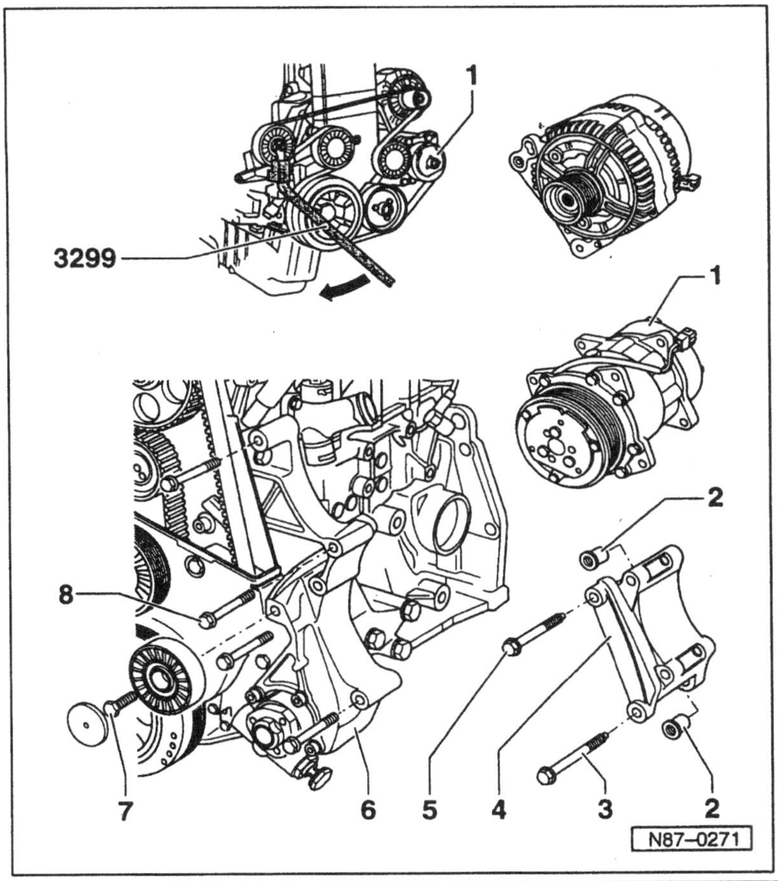

3 - **Hex bolt**
- M10 x 110
- 45 Nm (33 ft lb)

4 - **Compressor bracket**

5 - **Hex bolt**
- M8 x 85
- 20 Nm (15 ft lb)

6 - **Generator (GEN) bracket**

7 - **Hex bolt**
- M8 x 12
- 10 Nm (7 ft lb)
- With washer 8.4 x 16 x 1.0

8 - **Hex bolt**
- M10
- 45 Nm (33 ft lb)

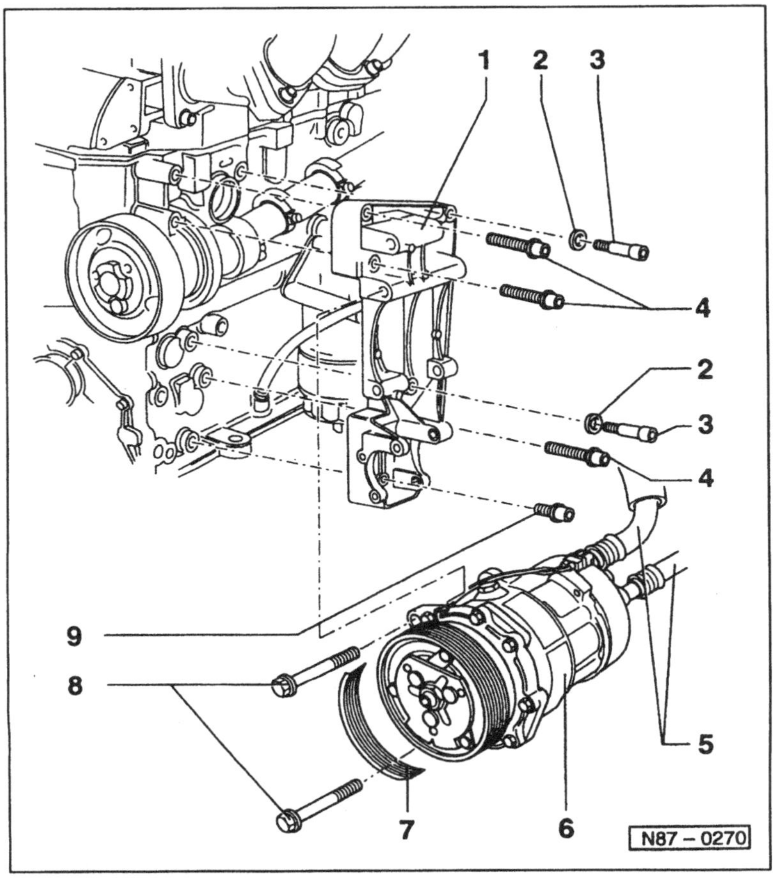

Compressor bracket, removing and installing 2.8L VR6 engine (AES)

Required:

- Torque wrench 5 - 50 Nm (44 in lb - 40 ft lb).
  Note: *The compressor bracket and related parts can be removed and installed without discharging the A/C refrigerant system.*

1 - **Generator (GEN), compressor and power steering pump bracket**

- Removing:

–

– Remove Generator (GEN)

– Remove compressor

– Remove power steering pump

⇒ Repair Manual, *Suspension, Wheels, Brakes, Steering (1997 ➤)*, Repair Group 48.

- Installing:

– First install guide bolts -3-

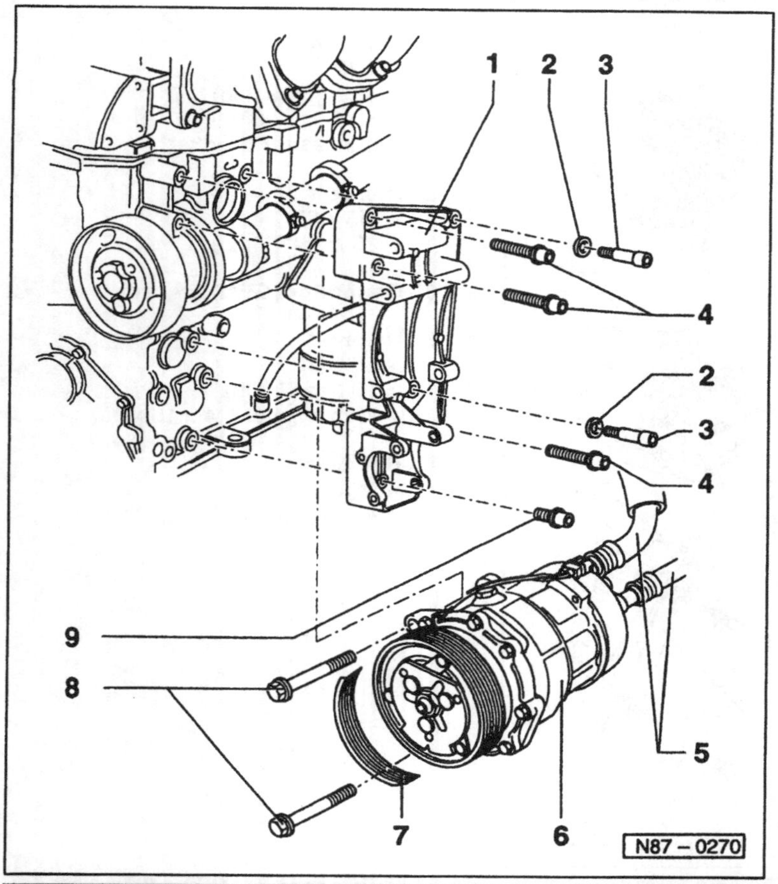

2 - Washer
- 8.4 x 16 x 1.6

3 - Guide bolt (2x)
- M8 x 60
- 25 Nm (18 ft lb)

4 - Socket head bolt
- M8 x 60
- 25 Nm (18 ft lb)

5 - Refrigerant hoses

6 - Compressor
- Removing:

–

– Loosen bolts -8-, knock-back clamping sleeves on compressor, then remove bolts completely.

7 - Poly ribbed drive belts
- Removal and installing

⇒ Repair Manual, *2.8L General, Engine*, Repair Group 13

8 - Hex bolt
- M10 x 112
- 45 Nm (33 ft lb)

9 - Socket head combination bolt
- M8 x 25
- 25 Nm (18 ft lb)

Digital Climate Control (Climatronic)

General

Climatronic is a dual zone, fully automatic climate control system. The temperature of the interior air supply, fresh air blower speeds and passenger compartment air distribution are regulated automatically. The system features fully digital electronic circuitry and On Board Diagnostic (OBD) capability.

In addition to fully automatic function, air distribution settings and fresh air blower speeds can be selected manually to suit individual needs.

The Climatronic control module -J255- is the heart of the system. It receives input information from the A/C control head -E87- and various electronic components (sensors and switches). This information is processed by the control module in accordance with specified values and provides corresponding output signals that control electrical components (actuators). Component functions, overview ⇒ Page T 87-61.

The Climatronic control module -J255- is controlled by an internal microprocessor with On Board Diagnostic (OBD) capability. If malfunctions, short or open circuits occur on monitored components, Diagnostic Trouble Codes (DTC) are stored in memory ⇒ Page T 01-1.

The system controls interior air temperature and distribution for front and rear passenger zones. The system continually samples and compares the selected interior air temperature with the actual air temperature measured by the system. Blower speeds, air temperature and distribution are then adjusted and regulated accordingly.

System functions, overview

Front passenger zone, climate control options

- Automatic temperature and fan speed mode
- Manual temperature and fan speed mode
- Manual selection of air recirculation mode
- Manual selection of air flow to instrument panel vents
- Manual selection of air flow to footwell vents
- Manual selection of air flow to windshield defroster

Rear passenger zone, climate control options

- Automatic temperature and fan speed mode
- Manual selection of temperature and fan speed

The A/C control head -E87- contains all controls and relevant displays needed to operate the Climatronic system. The system's functional status (mode selection, temperature and air distribution) is displayed in two separate LCD readouts. The left display is for the front passenger zone, the right display is for the rear passenger zone ⇒ Page T 87-59.

All automatic or manual function selections are input into the A/C control head via buttons located below both displays.

When AUTO mode is selected, fresh air blower speed and air distribution signals are provided by the A/C control head -E87-. The fresh air blower speed is high when there is a large difference between the selected temperature and the actual interior temperature, and low when the difference is small or the same.

The interior temperature setting, operation modes, approximate outside temperature and blower speed are indicated on the A/C control head display at all times. The defrost, vent and footwell air distribution symbols only appear in the display when air distribution is selected manually (AUTO function override). During manual air distribution, the "AUTO" display remains off until such time the "AUTO" button is pressed.

The system can be switched off completely by pressing the front passenger "decrease fan speed" button until "OF" appears in the display.

The Climatronic control module can be replaced as a separate part. Four screws locate the module behind the center instrument panel trim.

The A/C control head can be replaced as a separate part. Four plastic tabs locate the module in front of the center instrument panel trim.

A ribbon harness is used to electrically connect the A/C control head to the Climatronic control module for processing of selected functions and system regulation.

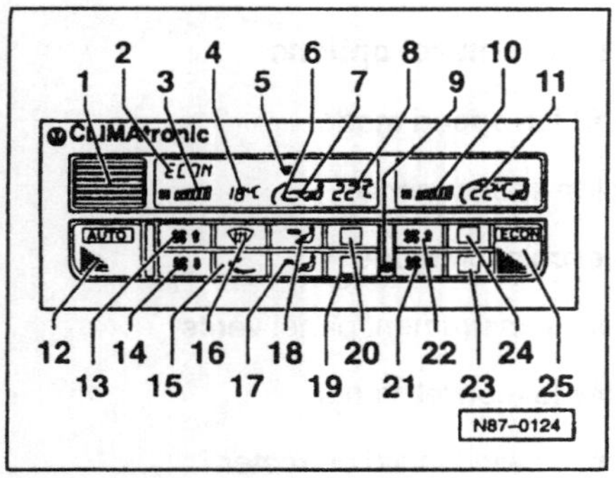

N87-0124

A/C control head -E87-, display

1 - Instrument panel interior temperature sensor -G56-

2 - "AUTO" (automatic operation) or "ECON" (compressor off) display.

3 - Front passenger "blower speed" display

4 - Ambient (outside) temperature display

Notes:

- *Temperature is measured by outside air temperature sensor -G17-. In event of malfunction "- - -" appears in display.*
- *If road speed drops below 15 km/h (approx. 9 mph) and the coolant temperature is above 70° C (158° F), the displayed value will not change. Under these conditions, the actual measured value is not displayed because engine heat may adversely affect reading.*

5 - Windshield defrost display

6 - Front passenger "fresh/recirculating air" display

7 - Front passenger "air distribution" display

8 - Front passenger "selected temperature" display

9 - Celsius/Fahrenheit selection button

10 - Rear passenger "blower speed" display

11 - Rear passenger "temperature selection" display

12 - "Automatic operation" button

13 - Front passenger "increase blower speed" button

14 - Front passenger "decrease blower speed" button (this button also switches system off)

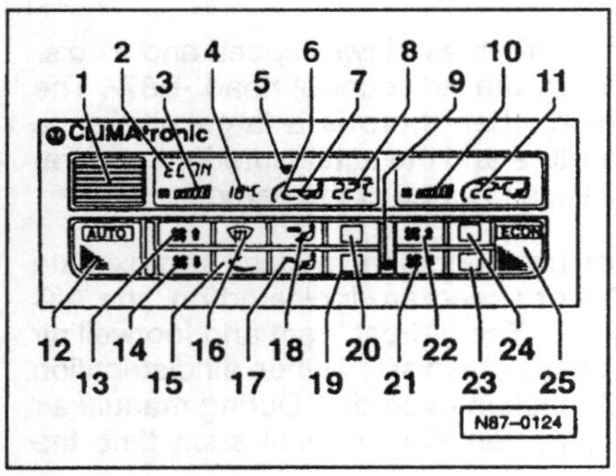

N87-0124

15 - Front passenger "fresh/recirculating air" button

16 - "Windshield defrost" button

17 - Front passenger "air flow to foot well" button

18 - Front passenger "air flow to upper body" button

19 - Front passenger "cooler" button

20 - Front passenger "warmer" button

21 - Rear passenger "decrease blower speed" button

22 - Rear passenger "increase blower speed" button

23 - Rear passenger "cooler" button

24 - Rear passenger "warmer" button

25 - "ECON" (compressor off) button

Notes:

- *Manual override of fully automatic control is possible with buttons 13 to 18, 21 and 22.*
- *Climatronic can only be completely switched off with button -14-. To do this press button until "OF" appears in the display. Because in this function the flaps should remain in the last set position, this setting should only be used in exceptional circumstances e.g. in case of a malfunction.*
- *A maximum temperature difference of 3° C (5° F) can be selected between front and rear passenger zones.*
- *When "ECON" operation is selected, only the compressor is switched off. Automatic heating and ventilation still operates.*
- *When blower speeds and air distribution have been selected manually, pressing "AUTO" button -12- will cancel all manually selected functions.*

Component functions, overview

Inputs

Sensors & Switches	Function
Instrument Panel Interior Temperature Sensor -G56-	Interior air temperature measurement (resistance value) for front interior air temperature regulation
Sunlight Photo Sensor -G107-	Sunlight intensity measurement (resistance value) for interior air temperature/fan speed regulation
Front Vent Temperature Sensor -G152-	Heater/evaporator output measurement (resistance value) for front interior air temperature and fan speed regulation
Rear Evaporator Temperature Sensor -G153-	Rear evaporator temperature measurement (resistance value) for rear interior air temperature and fan speed regulation
A/C Evaporator Temperature Switch -E33-	Rear evaporator temperature monitoring – -J255- switches off A/C clutch -N25- in the event of evaporator icing –
Outside Air Temperature Sensor -G17-	Outside air temperature measurement (resistance value) for interior air temperature/fan speed regulation and control head display
Rear Heater Core Temperature Sensor -G154-	Rear heater output measurement (resistance value) for rear interior air temperature regulation
A/C Engine Coolant Temperature Sensor -G110-	Engine coolant temperature measurement (resistance value) for front interior air temperature and fan speed regulation

→ Climatronic Control Module -J255-

Continued

Sensors & Switches	Function
Temperature Regulator Flap Motor Postion Sensor -G92-	Temperature regulator flap position indication (potentiometer resistance value) for front heater output regulation
Central Air Flap Motor Position Sensor -G112-	Central air flap position indication (potentiometer resistance value) for front air distribution control – instrument panel vents and footwell vents–
Footwell/Defroster Flap Motor Position Sensor -G114-	Footwell/defroster flap position indication (potentiometer resistance value) for front air distribution control – defroster vents and footwell vents–
A/C Pressure Switch -F129-	Refrigerant circuit high/low pressure indication (voltage signal) – -J255- switches off A/C clutch -N25- in the event of excess pressure and controls second speed of coolant fan -V7- –
Speedometer Vehicle Speed Sensor -G22-	Vehicle speed measurement (hall signal) for fresh air fan speed control

→ Climatronic Control Module -J255-

Continued

Outputs

Climatronic Control Module -J255-

Actuators and Relays	Function
Interior Temperature Sensor Fan -V42-	Draws continuous air sample over instrument panel interior air temperature sensor -G56- for front interior air temperature regulation
Vacuum Valve Rail -N53-	Controls vacuum supply to actuate vacuum units controlling air distribution – fresh air / recirculating flap and footwell defrost flap –
Relay for Climatronic -J254-	Activates (voltage supply): – Climatronic control module -J255- (functions) – Fresh air blower -V2- (interior air supply) – A/C clutch -N25- (refrigerant circuit operation)
Control Module for Fresh Air Blower -J126-	Controls speed of fresh air blower -V2- (variable ground signal) for interior air supply
Fan Relay -J323-	Activates blower for rear evaporator -V20- and rear warm air fan -V47- (voltage supply) for interior air supply
Control Module for Blower for Evaporator -J349- (rear)	Controls speed of blower for rear evaporator -V20- (variable ground signal) for interior air supply
Control Module for Warm air Blower -J350- (rear)	Controls blower speed of rear warm air fan -V47- (variable ground signal) for interior air supply
Relay for A/C Shut-Off -J365-	Activates relay for Climatronic -J254- (voltage supply) – dependant on switch status of A/C cut-out thermal switch -F163-

Continued

Climatronic Control Module -J255-

Actuators & Relays	Function
Temperature Regulator Flap Motor -V68-	Motor adjusts position of temperature regulating flap (voltage and ground supply) for front interior air temperature regulation
Central Air Flap Motor -V70-	Motor adjusts position of central air flap (voltage and ground supply) for front air distribution control – instrument panel vents and footwell vents –
Footwell/Defroster Flap Motor -V85-	Motor adjusts position of footwell/defroster flap (voltage and ground supply) for front air distribution control – defroster vents and footwell vents –
Engine Coolant Two-Way Vacuum Valve -N147-	Controls vaccum supply to actuate coolant cut-off valve – shut-offs coolant flow to heater core –
Auxiliary Heater Valve -N172-	Controls coolant flow through rear heater core (pulsed voltage supply) for rear heater output regulation
A/C Refrigerant Shut-Off Valve -N43-	Shuts off refrigerant flow to rear evaporator
A/C Cut-Out Thermo Switch -F163-	Activates relay for Climatronic -J254- through relay for A/C shut-off (voltage supply) dependant on switch status of A/C cut-out thermal switch -F163-
Coolant Fan Control Relay -J26-	Activates coolant fans -V7- and -V35- for refrigerant circuit pressure control
Coolant Pump -V50-	Ciculates engine coolant to rear heater core

Continued

Climatronic Control Module -J255-

Actuators & Relays	Function
Fresh Air Blower -V2-	Supplies fresh air for front passenger zone interior air heating, cooling and recirculation
Rear Evaporator Fan -V20-	Supplies fresh air for rear passenger zone interior air cooling
Rear Warm Air Fan -V47-	Supplies fresh air for rear passenger zone interior air heating
A/C Clutch -N25-	Engages A/C compressor to poly-ribbed drive belt – circulates refrigerant through refrigerant circuit for interior air cooling

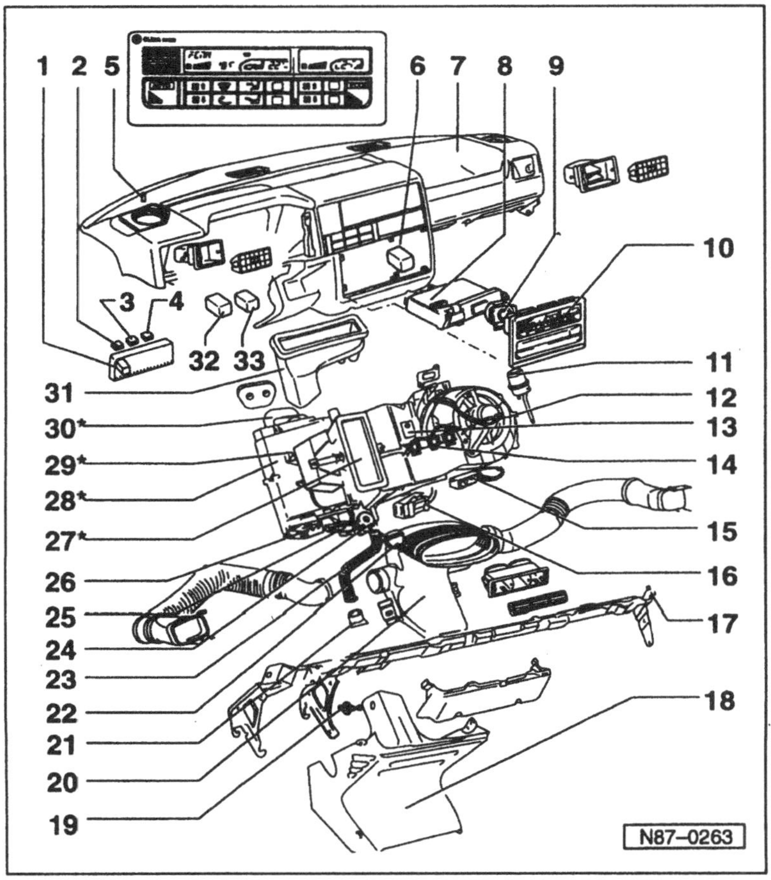

Climatronic

Controls and mechanical components, servicing

> ***CAUTION!***
>
> ***Before working on electrical system, obtain radio code and disconnect battery Ground (GND) strap.***

Note:

*System components identified with an * must only be serviced or replaced after discharging refrigerant system. Use Kent Moore ACR4 or equivalent.*

1 - **Climatronic relay -J254-**
- Control number 140
- Checking

⇒ Repair Group 01

2 - **Fuse for Climatronic front -S140-**
- Above relay plate

3 - **Fuse for blower for rear evaporator -S120-**
- Above fuse-relay plate

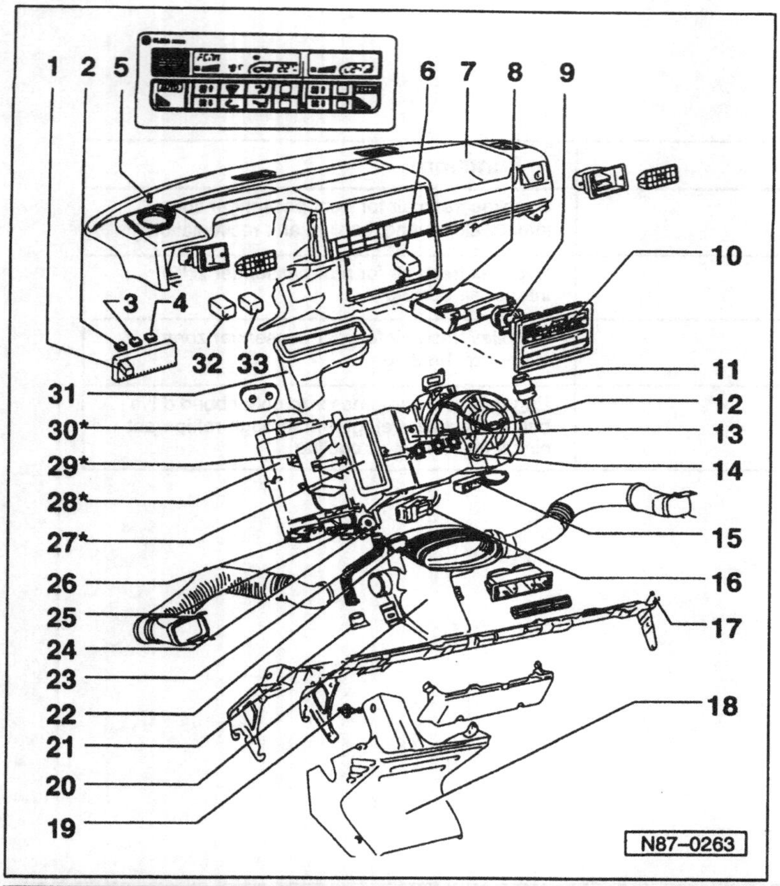

4 - Warm air blower fuse -S118-
- Above fuse-relay plate

5 - Sunlight photo sensor -G107-
- Regulates temperature flap position and fresh air blower depending on light intensity
- In event of malfunction, Climatronic control module -J255- assumes fixed value
- Checking

⇒ Repair Group 01
- Removing
- – Remove cover for left speaker, disconnect harness connector, press sensor out from inside outwards.

6 - Fan relay -J323-
- Control number 114
- Behind center instrument panel trim
- Controls blower for rear evaporator -V20- and rear warm air fan -V47-
- Checking

⇒ Repair Group 01

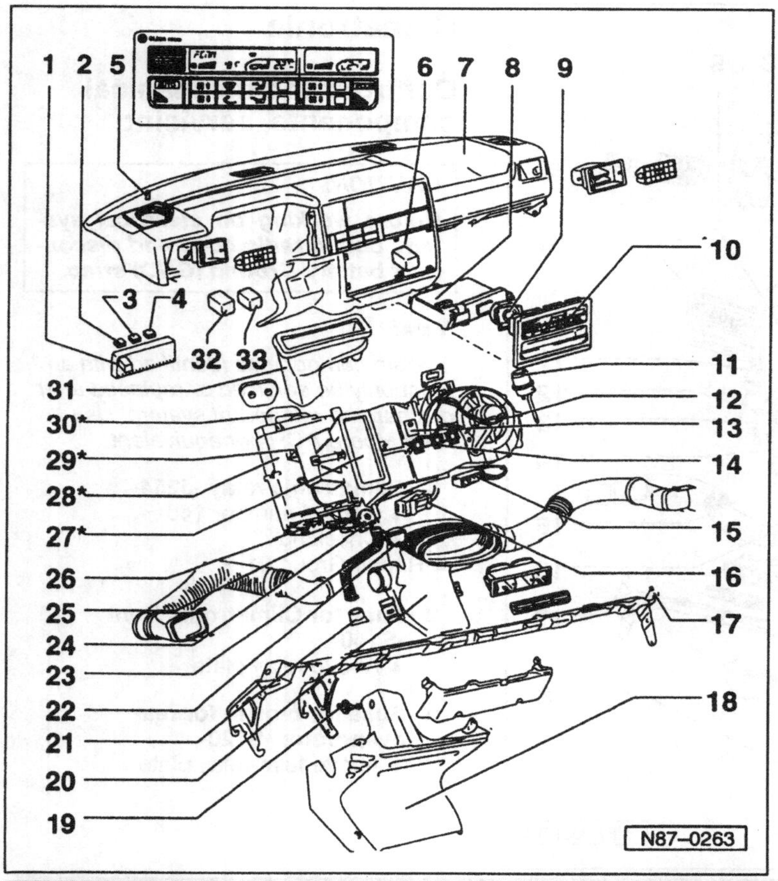

7 - Instrument panel
- Removing and installing
- – Repair Manual, *Body - Exterior, Body - Interior,* Repair Group 70

8 - Climatronic control module -J255-
- Checking

⇒ Repair Group 01
- Removing

⇒ Fig. 1
- Always check control unit coding when installing

⇒ Page T 01-35

9 - Interior temperature sensor fan -V42-
- Mounted to A/C control head -E87-
- Checking

⇒ Repair Group 01
- Removing
- – Carefully pry out with a screwdriver.

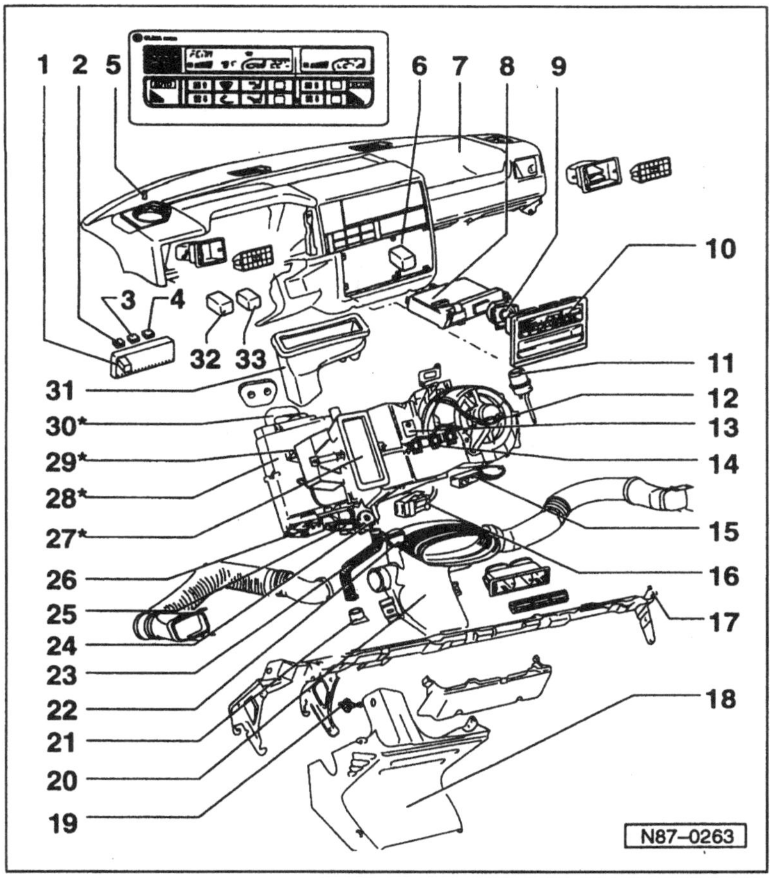

10 - A/C control head -E87-
- ♦ With instrument panel interior temperature sensor -G56-
- ♦ Operation/function

⇒ Page T 87-59
- ♦ Checking

⇒ Repair Group 01
- ♦ Removing
- – Carefully pry out -E87- with a screwdriver and pull out of center instrument panel trim.

11 - Fresh air/recirculating flap vacuum unit
- ♦ Flap in fresh air position when no vacuum is present
- ♦ Flap in recirculated air position when vacuum is present
- ♦ Vacuum hose layout

⇒ Page T 87-108
- ♦ Removing

⇒ Fig. 2

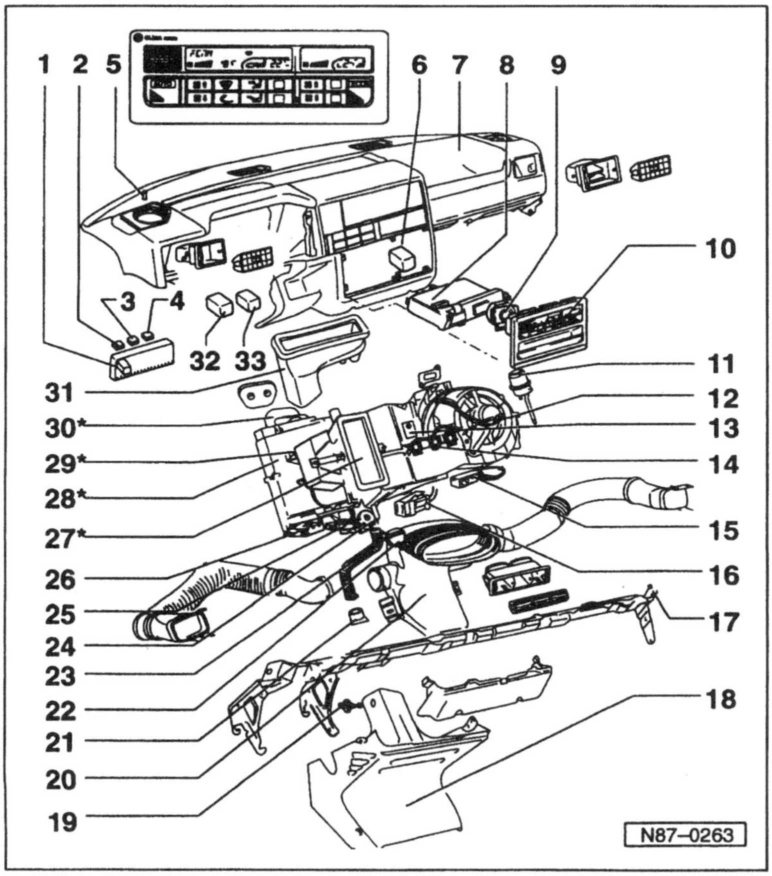

12 - Fresh air blower -V2-
- ♦ Checking

⇒ Repair Group 01
- ♦ Replacing
- – Remove Instrument panel -7- and loosen carrier -17- on right side

⇒ Repair Manual, *Body - Exterior*, *Body - Interior*, Repair Group 70
- ♦ Before installing coat contact surfaces with AMV 176 000 05 or equivalent

13 - Control module for fresh air blower -J126-
- ♦ Controls -J126- with variable voltage signal
- ♦ Checking

⇒ Repair Group 01
- ♦ Removing

⇒ Fig. 4

14 - 5-Pin connectors for flap motors
- ♦ Identification

⇒ Fig. 9

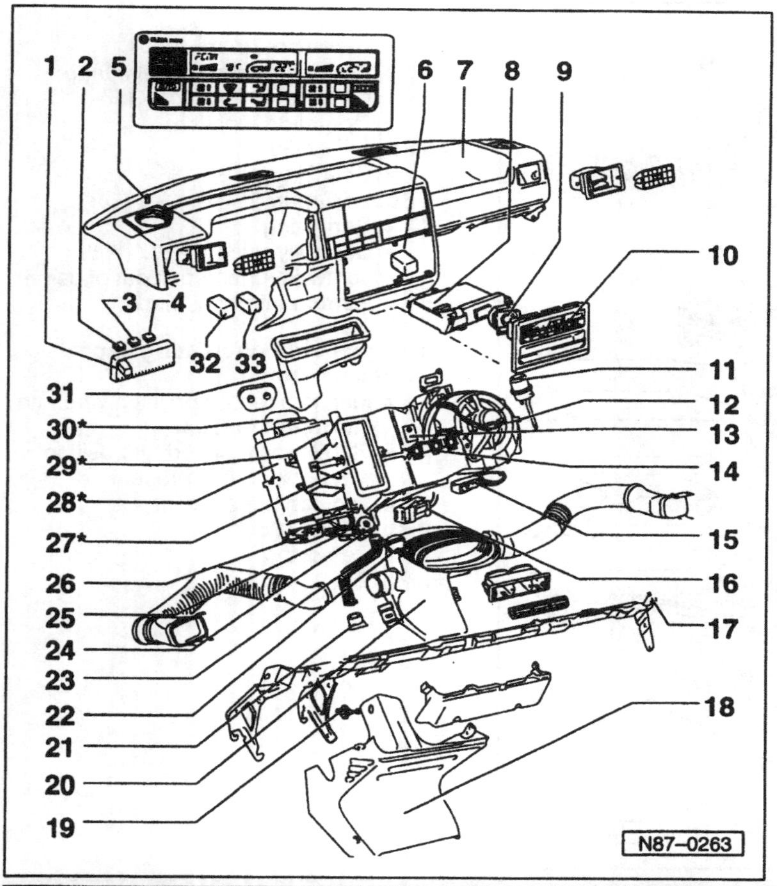

15 - Vacuum valve rail -N53-

- Power supply (15) fuse -S16-
- Climatronic control module switches ground to -N53- to control vacuum supply to fresh air/recirculating flap and footwell/defrost flap
- Checking

⇒ Repair Group 01

- Removing and installing

⇒ Fig. 8

- Vacuum hose, layout

⇒ Page T 87-108

16 - Evaporator temperature switch -E33-

- Prevents evaporator icing
- Checking
- – Repair Group 01
- Sensor tube depth: 330 mm (13 in)
- Replacing
- – Fig. 6
- Switching temperatures: opens below -2° C (28° F)

17 - Carrier

18 - Footwell vent console

- Insert front vent temperature sensor -G152- first before installing vent console (rotate -G152- 90° to lock).

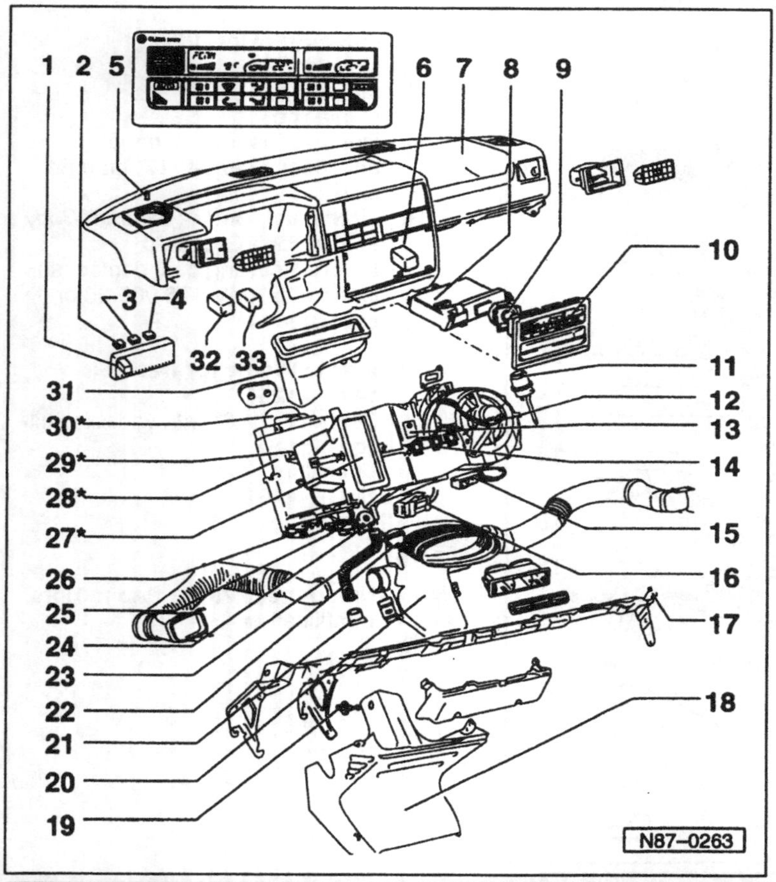

19 - Front vent temperature sensor -G152-

- If sensor -G152- malfunctions, default value of 42°C (108°F) is assumed
- Checking

⇒ Repair Group 01

20 - Center air duct

- Replacing
- – Remove instrument panel -7- and carrier -17-.

⇒ Repair Manual, *Body - Exterior, Body - Interior,* Repair Group 70

21 - Water drain valve

- Water leak at housing may result if valve is blocked

22 - Center vent flap vacuum unit

- Center vent closed when no vacuum is present
- Vacuum hose, layout

⇒ Page T 87-108

- Removing

⇒ Fig. 3

23 - Evaporator drain pipe

- Removing and installing

⇒ Fig. 5

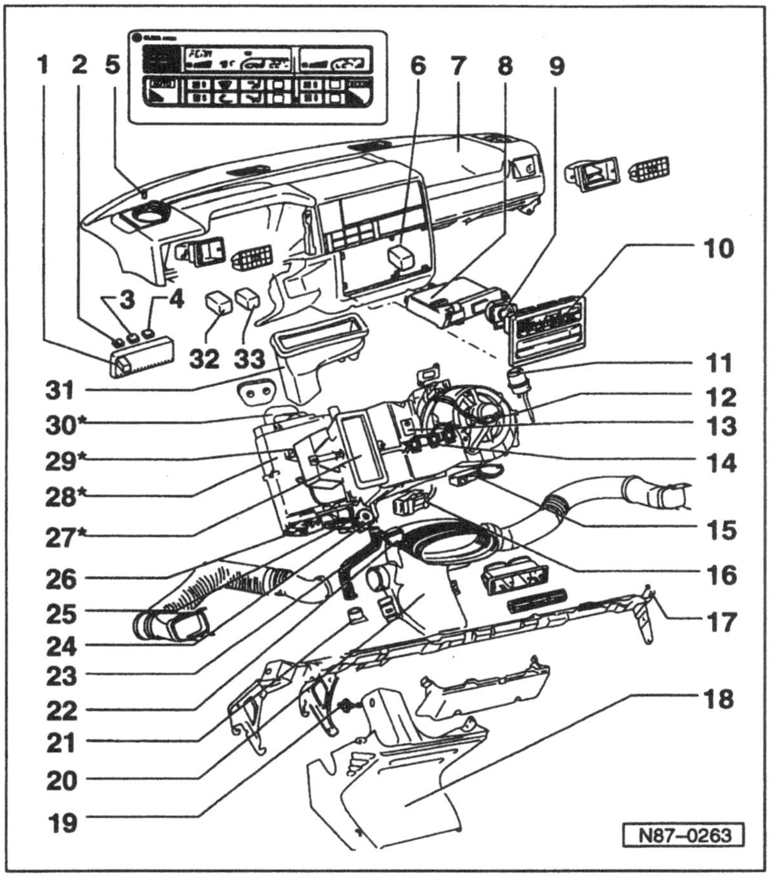

24 - Temperature regulator flap motor -V68-
- Checking

⇒ Repair Group 01
- End position, checking and adjusting

⇒ Page T 87-81
- Removing
  – Remove passenger side footwell vent console and footwell trim.

25 - Central air flap motor -V70-
- Checking

⇒ Repair Group 01
- Removing and installing

⇒ Fig. 7

26 - Footwell/defroster flap motor -V85-
- Checking

⇒ Repair Group 01
- End position, checking and adjusting

⇒ Page T 87-82
- Removing
  – Remove passenger side footwell vent console and footwell trim.

27 - Central flap*
- Not a replaceable part

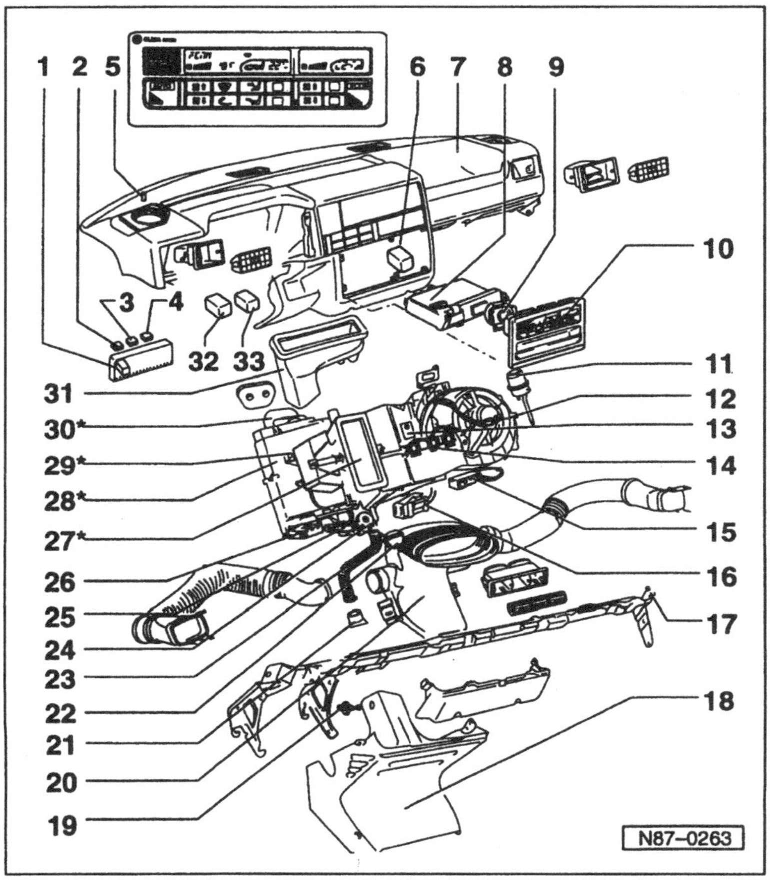

28 - Heater/evaporator housing*

29 - Footwell/defrost flap*
- Not replaceable part

30 - Heater core*
- Install seals around entire circumference
- Always replace coolant

31 - Defroster duct
- Replacing
  – Remove instrument panel -7-

⇒ Repair Manual, *Body - Exterior, Body - Interior,* Repair Group 70

32 - Relay for A/C shut-off -J365-
- Above fuse-relay plate

33 - Not applicable

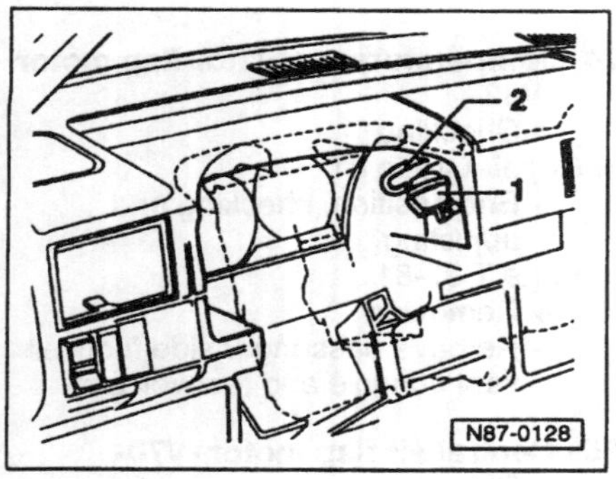

Fig. 1 **Climatronic control module -J255-, removing**

– Remove lower trim -1-.

– Remove screws -2-.

– Disconnect electrical connector and remove unit -3- downwards.

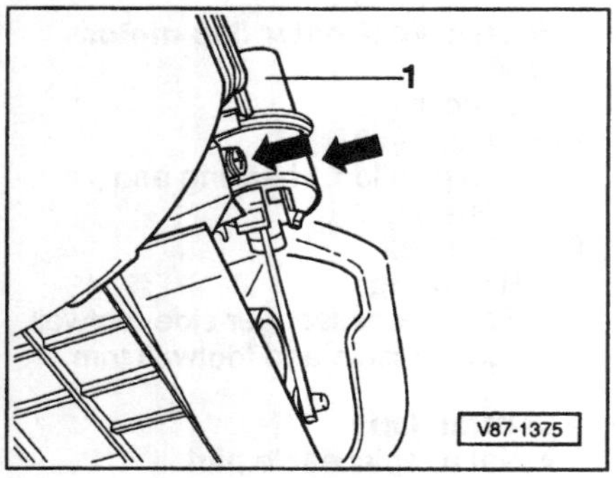

Fig. 2 **Fresh air/recirculating flap vacuum unit, removing and installing**

– Remove right air outlet ⇒ Page T 80-6, Fig. 2.

– Remove glove box, knee padding or passenger side airbag unit ⇒ Repair Manual, *Body - Exterior, Body - Interior,* Repair Group 69

– Remove screws -arrows-.

– Twist and lift vacuum unit -1- from lever.

– Install in reverse order

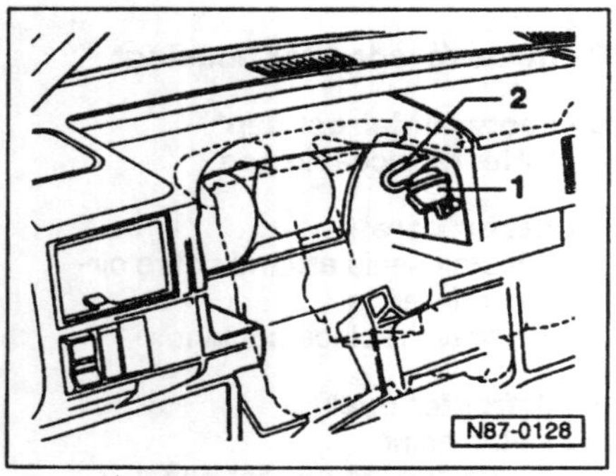

Fig. 3 **Center vent flap vacuum unit, removing and installing**

– Remove Instrument panel insert trim.

– Remove center vent ⇒ Page T 80-8, Fig. 4.

– Pull vacuum hose -2- from vacuum unit.

– Twist and lift vacuum unit -1- from lever.

– Install in reverse order.

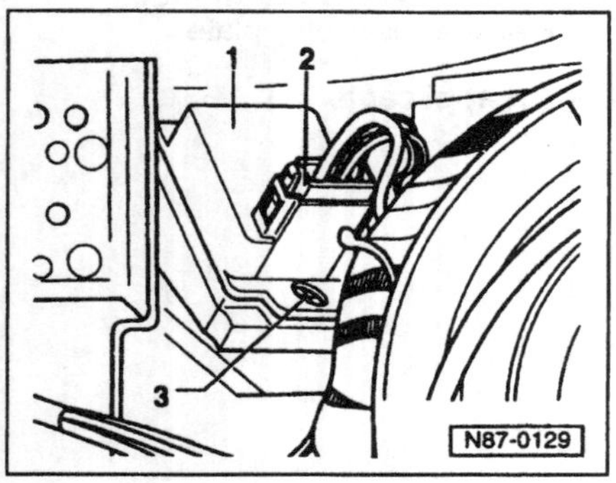

Fig. 4 **Control module for fresh air blower -J126-, replacing**

– Remove glove box, knee padding or passenger side airbag unit ⇒ Repair Manual, *Body - Exterior, Body - Interior,* Repair Group 69

– Disconnect electrical connector -2- and remove screw -3-.

– Seal area between resistor -1- and air duct with AMV 176 000 05 (or equivalent) before fitting.

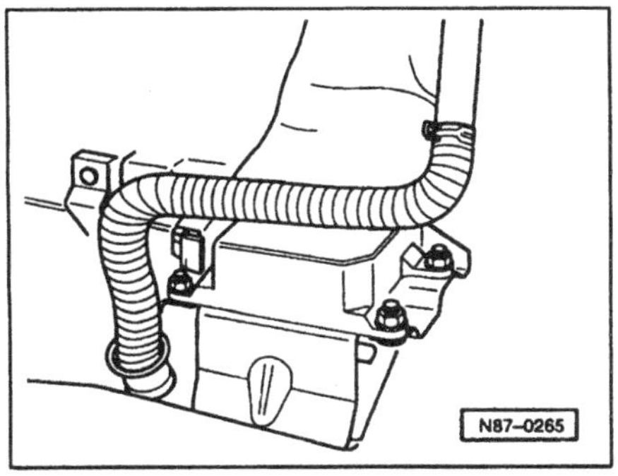

◄ **Fig. 5 Evaporator drain pipe, installing**

Notes:

- *Water drain pipe must be pushed into heater/evaporator housing onto stop.*
- *Water may leak into passenger compartment if hose is not located on the drain valve as shown.*
- *Drain valve must not be blocked.*

– Remove footwell vent console ⇒ Page T 87-29

Install hose so that drain valve grommet seals tightly around hose.

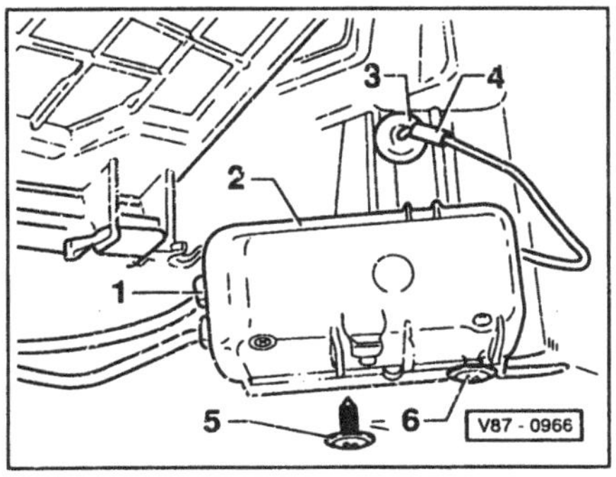

◄ **Fig. 6 Evaporator temperature switch -E33-, replacing**

1 - Connector

2 - Temperature switch with housing

3 - Grommet

4 - Sensor pipe with insertion depth marking at 330mm (13in)

5 - Screw for evaporator housing cover

6 - Screw

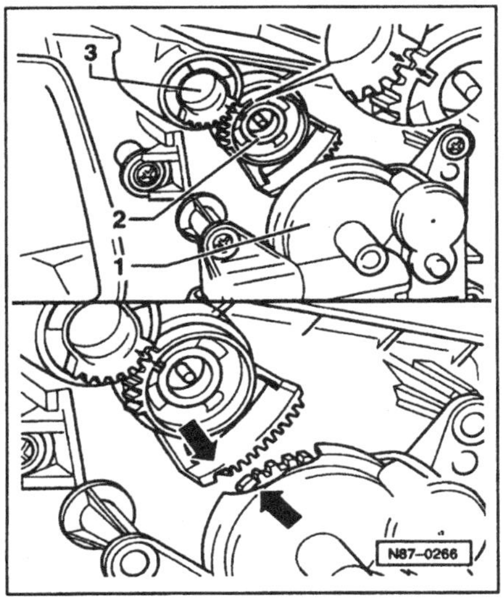

◄ **Fig. 7 Central air flap motor -V70-, removing and installing**

1 - Central air flap motor

2 - Intermediate gear

3 - Central flap bearing

Removing

– Remove footwell vent console ⇒ Page T 87-29.

– Remove central flap motor -V70- from heater/evaporator housing.

– Disconnect 5-pin connector T5b (for flap motor) at heater/evaporator housing ⇒ Fig. 9.

Installing

Install in reverse order, noting the following:

– Turn central flap lever until the recess aligns with lug on flap motor -arrows-

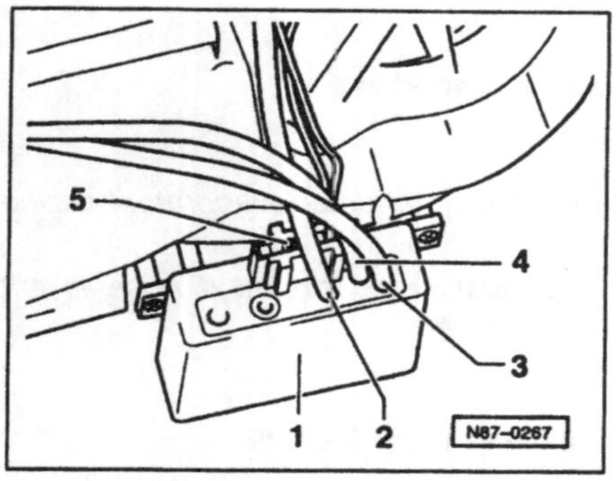

Fig. 8 Vacuum valve rail -N53-, removing and installing

1 - Vacuum valve rail
2 - Vacuum hose, green
3 - Vacuum hose, yellow
4 - Vacuum hose, black
5 - Connector

Removing

– Remove lower trim on passenger side

– Pull off vacuum hoses and electrical connections.

– Remove mounting screws.

Installing

Install in reverse order, noting the following:

– Vacuum hoses connected correctly (according to color coding)

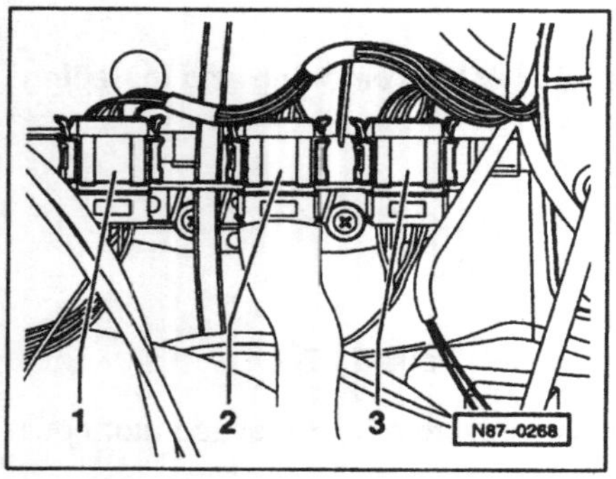

Fig. 9 5-Pin connectors for flap motors, identification

1 - Connector T5b
For central flap positioning motor -V70- (red connector)

2 - Connector T5
For temperature flap positioning motor -V68- (red connector)

3 - Connector T5a
For footwell/defrost flap positioning motor -V85- (yellow connector)

Note:

Individual terminal assignments ⇒ electrical wiring diagrams.

Temperature regulator flap motor -V68- end position, checking and adjusting

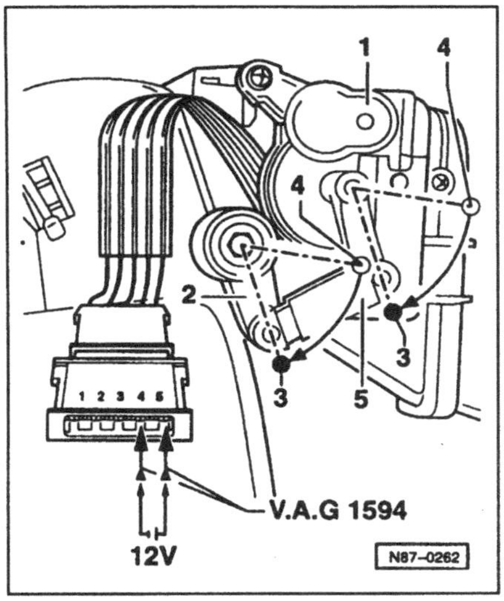

◄ – Remove footwell vent console and footwell trim ⇒ Page T 87-29

– Disconnect 5-pin connector T5 for -V68- at heater/evaporator housing ⇒ Page T 87-80.

– Connect appropriate cables from adapter kit VAG 1594 to terminals 4 and 5.

– Connect to 12V supply. Direction of flap motor rotation can be changed by reversing polarity of voltage supply.

– Operate temperature flap -2- with flap motor -1- from heater end position -4- to fresh air end position -3-.

– Check if temperature flap lever -2- reaches stop at end position. If necessary, adjust length of connecting rod -5- with temperature flap -2- on stop.

Checking:

By reversing polarity of 12V supply, flap motor must move fully to both end positions.

Footwell/defroster flap motor -V85- end position, checking and adjusting

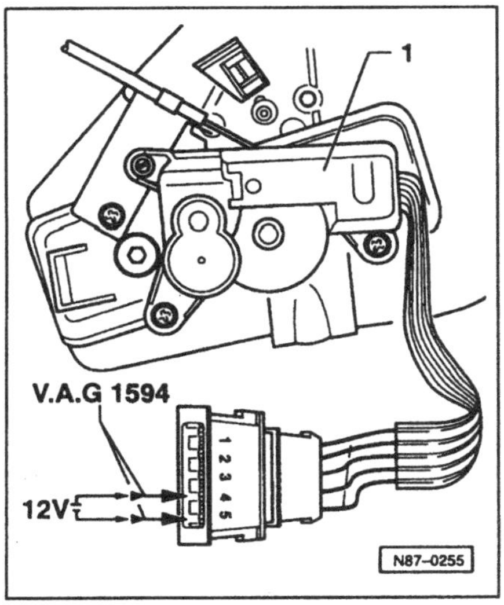

◄ – Remove footwell vent console and footwell trim ⇒ Page T 87-29.

– Disconnect 5 pin connector T5a for -V85- at heater/evaporator housing ⇒ Page T 87-80.

– Connect appropriate cables from adapter kit VAG 1594 to terminals 4 and 5.

– Connect to 12V supply. Direction of flap motor rotation can be changed by reversing polarity of voltage supply.

– Operate flap motor -1- from one end position to the other.

– Check if footwell/defrost flap lever reaches stop at end position. If necessary, adjust cable.

Checking:

By reversing polarity of 12V supply, positioning motor must move fully to both end positions.

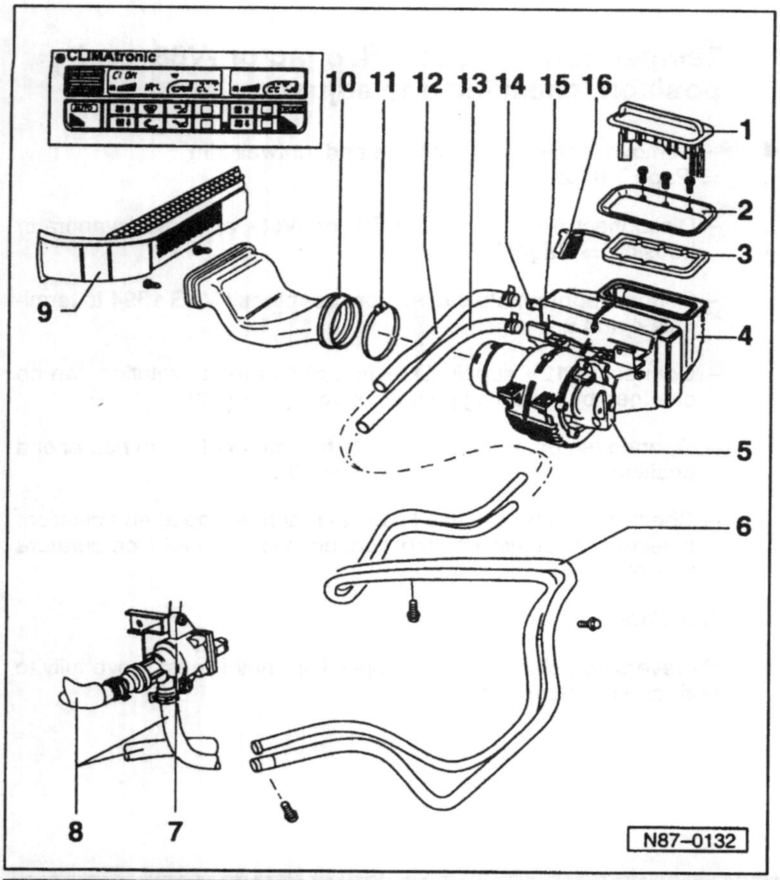

Rear heater, servicing

> ***CAUTION!***
>
> ***Before working on electrical system, obtain radio code and disconnect battery Ground (GND) strap.***

1 - Vent, for rear passenger zone
- ♦ Removing

⇒ Fig. 2

2 - Cover plate

3 - Gasket

4 - Air distribution housing
- ♦ Removing

⇒ Fig. 1
- ♦ Assembly

⇒ Page T 80-33

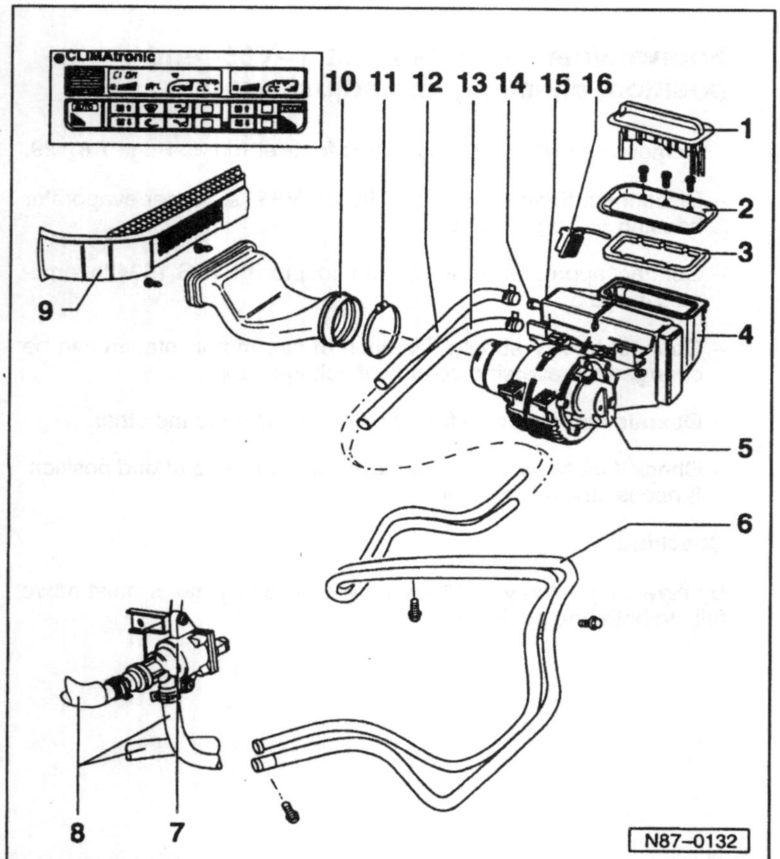

5 - Rear warm air fan -V47-
- ♦ Protected by fuse -S118

⇒ Page T 87-67, key No. 4
- ♦ Controlled by fan relay -J323- and control module for warm air blower -J350-
- ♦ Checking

⇒ Repair Group 01
- ♦ Removing

⇒ Page T 80-33

6 - Coolant pipe
- ♦ Removing

⇒ Fig. 5

7 - Auxiliary heater valve -N172-
- ♦ Removing

⇒ Fig. 6

8 - Coolant hose
- ♦ From engine compartment connection diagram
- ♦ Coolant hose connection diagram

⇒ Page T 87-105

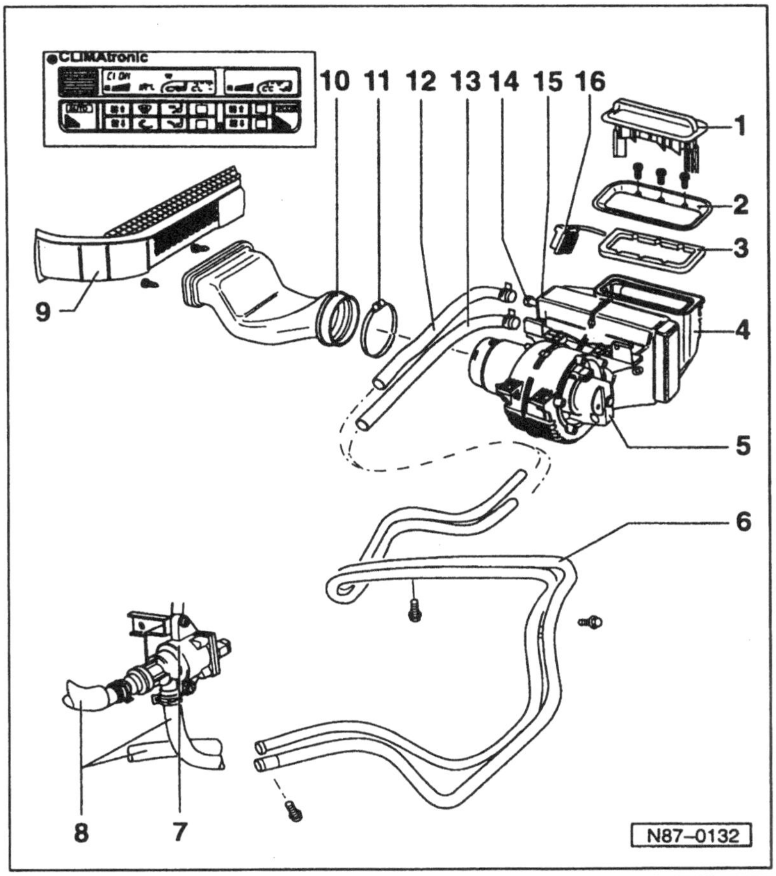

9 - **Sliding door footwell insert**
- ◆ Removing

⇒ Repair Manual, *Body - Exterior, Body - Interior,* Repair Group 68

10 - **Air intake duct**
- ◆ Kombi models*

⇒ Fig. 3
- ◆ EuroVan & Transporter models

⇒ Fig. 4

11 - **Hose clamp**

12 - **Coolant hose, return**

13 - **Coolant hose, supply**

14 - **Heater core**
- ◆ Install seals around entire circumference of heater core
- ◆ Removing

⇒ Page T 80-33

Note:

Always change coolant after repairs.

* Not applicable

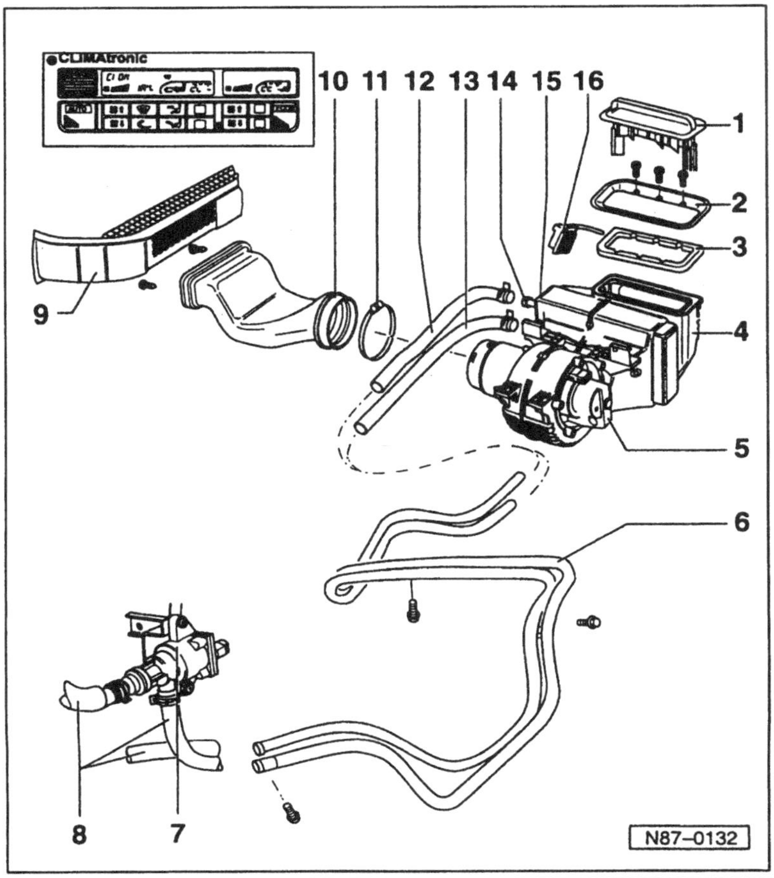

15 - **Rear heater core temperature sensor -G154-**
- ◆ Inserted into heater core
- ◆ If sensor -G56- malfunctions, default value of 42°C (108°F) is assumed
- ◆ Checking

⇒ Repair Group 01
- ◆ Removing
- – Remove rear heater core first

⇒ Page T 80-33

16 - **Control module for warm air blower -J350- (rear)**
- ◆ Function: Controls blower speed of rear warm air fan -V47- with variable ground signal
- ◆ Checking

⇒ Repair Group 01
- ◆ Removing

⇒ Fig. 7

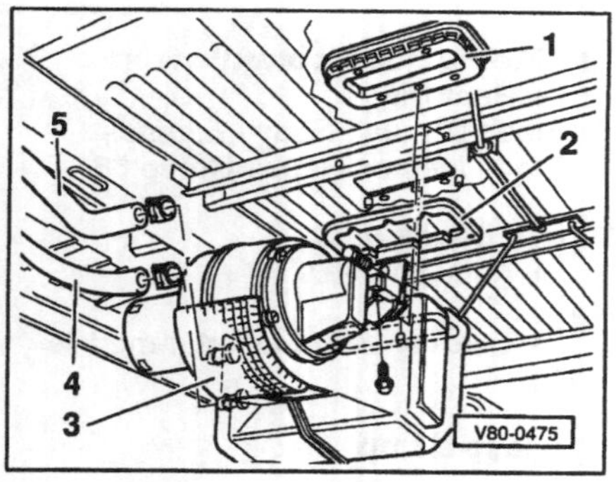

Fig. 1 Rear heater, removing

1 - Cover plate, in passenger compartment
2 - Gasket
3 - Rear heater air distribution housing
4 - Coolant supply hose
5 - Coolant return hose

Removing

– Disconnect electrical connectors.

> ***WARNING!***
> ***The cooling system is pressurized when the engine is warm. Wear gloves and other appropriate protection, and slowly and carefully release system pressure, if necessary, before performing repairs.***

– Clamp coolant hoses as close as possible to heater core.

– Loosen hose clamps and remove coolant hoses. Plug heater core outlet/inlet.

– Remove vent ⇒ Fig. 2.

– Remove cover plate.

– Remove bolts.

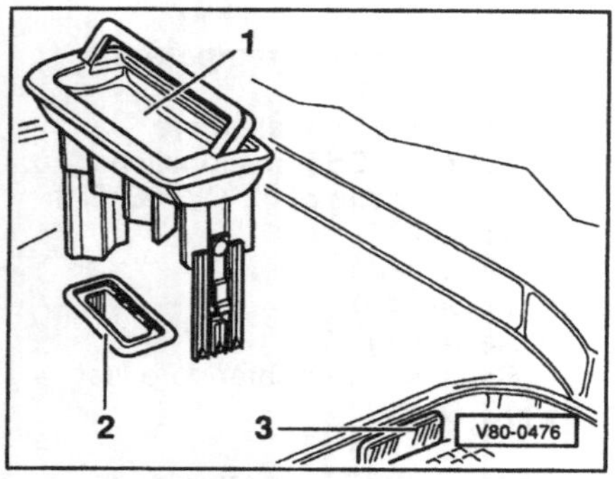

Fig. 2 Vent, removing

1 - Vent
2 - Cover plate
3 - Sliding door footwell insert

Removing

– Pull vent out of floor plate.

– Remove cover plate

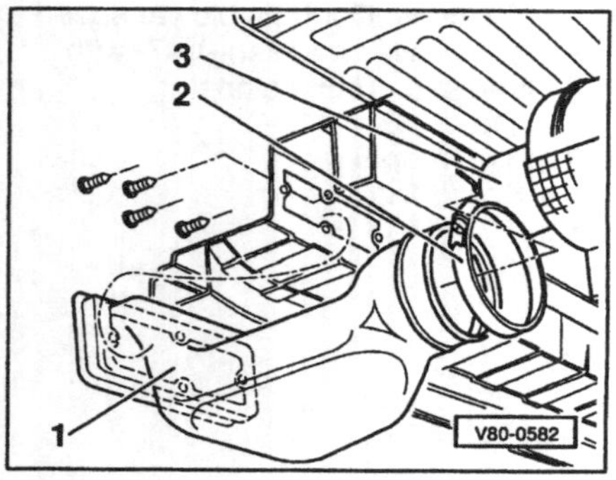

Fig. 3 Air intake duct (Kombi), removing

1 - Air Intake duct
2 - Hose clamp
3 - Rear heater air distribution housing

Removing

– Remove footwell insert.

– Remove bolts.

– Loosen hose clamp.

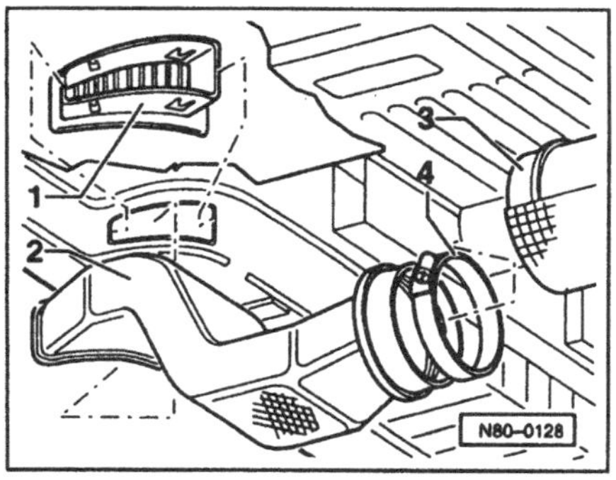

Fig. 4 Air intake duct (EuroVan/Transporter), removing

1 - Intake trim
2 - Intake duct
3 - Rear heater core
4 - Hose clamp

Removing

– Carefully remove intake trim -1- with screwdriver.

– Loosen hose clamp.

– Remove intake duct.

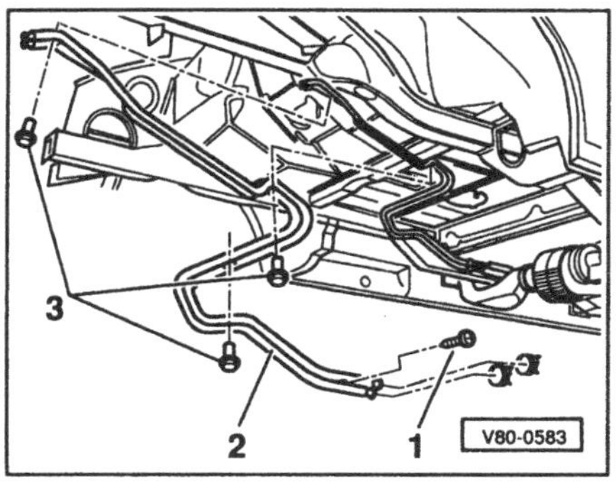

Fig. 5 Coolant pipe, removing

1 - Self-tapping screw
2 - Coolant pipe
3 - Hexagon nut

Removing:

> ***WARNING!***
> ***The cooling system is pressurized when the engine is warm. Wear gloves and other appropriate protection, and slowly and carefully release system pressure, if necessary, before performing repairs.***

– Remove fuel tank
⇒ Repair Manual, *Engine, General,* Repair Group 20

– Remove torsion bar
⇒ Repair Manual, *Suspension, Brakes, Wheels, Steering,* Repair Group 40

– Remove screws -3- and -1-.

– Remove coolant pipe -2-.

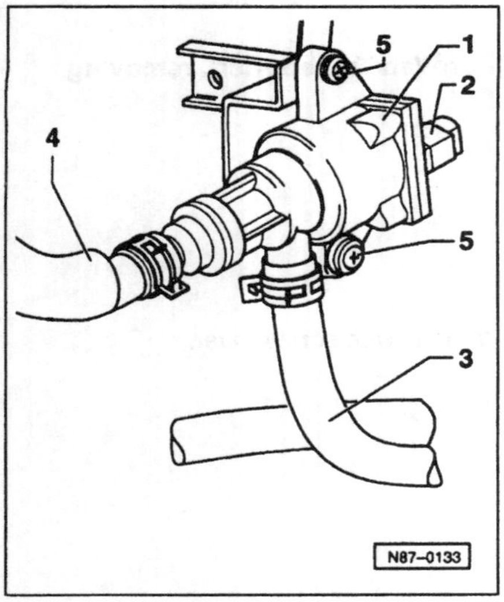

◄ **Fig. 6 Auxiliary heater valve -N172- (rear heater), removing**

1 - Auxiliary heater valve -N172-

2 - Electrical connector

3 - Coolant return hose,
(from rear heater core coolant pipe to valve)

4 - Coolant return hose
(from valve to engine)

5 - Securing screw

Removing

> ***WARNING!***
>
> ***The cooling system is pressurized when the engine is warm. Wear gloves and other appropriate protection, and slowly and carefully release system pressure, if necessary, before performing repairs.***

– Disconnect electrical connector -2- from valve.

– Clamp coolant hoses as close as possible to valve.

– Loosen spring type hose clamps and pull coolant hoses off.

– Unbolt valve from bracket and remove.

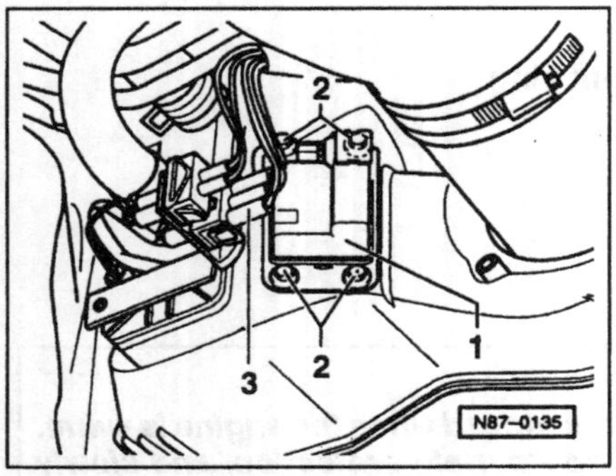

◄ **Fig. 7 Control module for warm air blower -J350- (rear), replacing**

1 - Control module for warm air blower -J350-

2 - Bolts (Qty. 4)

3 - Connector

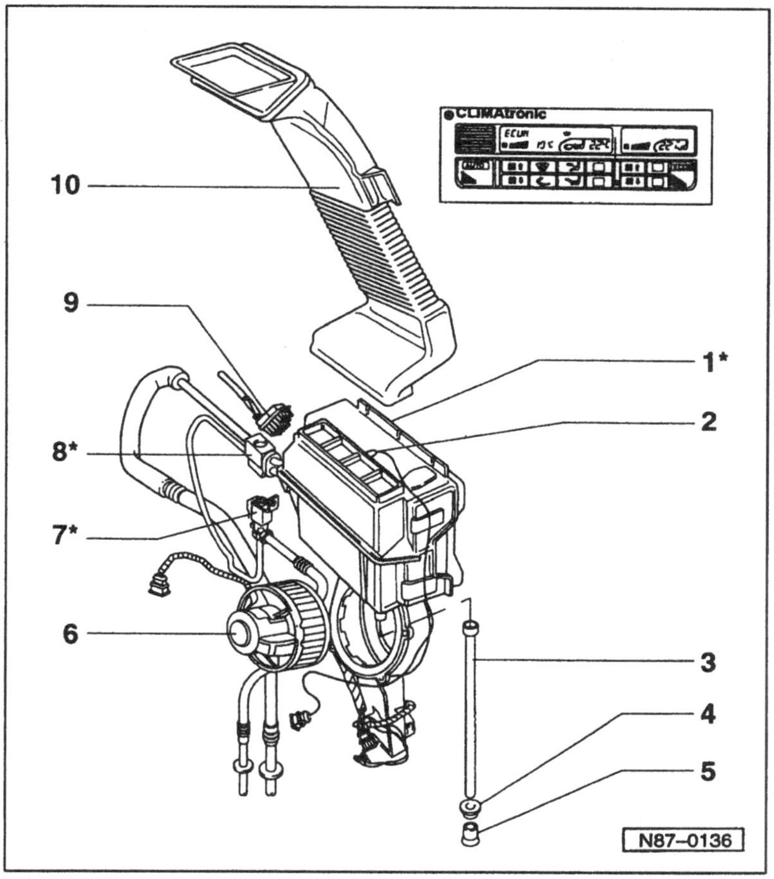

Rear (2nd) evaporator, servicing

> **CAUTION!**
>
> ***Before working on electrical system, obtain radio code and disconnect battery Ground (GND) strap.***

Note:

*System components identified with an * must only be serviced or replaced after discharging refrigerant system. Use Kent Moore ACR4 or equivalent.*

1 - Rear evaporator*

2 - Rear evaporator temperature sensor -G153-

- ♦ Inserted into rear evaporator
- ♦ If sensor-G153- malfunctions, default value of 10°C (50°F) is assumed
- ♦ Checking
- – Repair Group 01

3 - Evaporator drain pipe

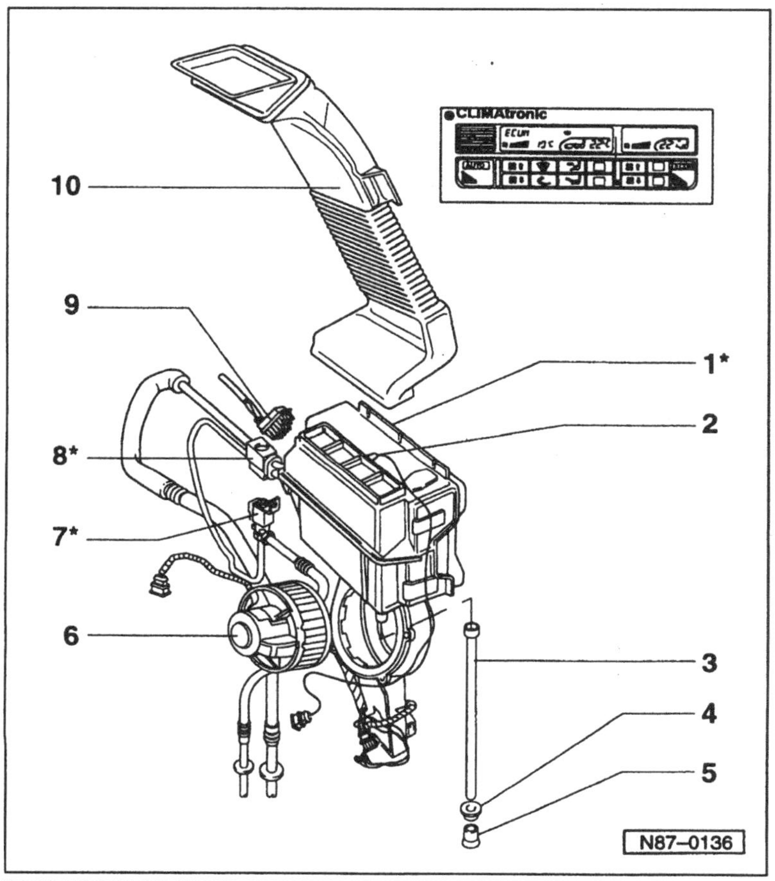

4 - Grommet

5 - Evaporator water drain valve

6 - Rear evaporator fan -V20-

- ♦ Protected by fuse -S120-

⇒ Page T 87-66, key No. 3

- ♦ Controlled by fan relay -J323- and control module for blower for evaporator -J349-.
- ♦ Checking

⇒ Repair Group 01

- ♦ Removing
- – Remove right "D" pillar trim first.

⇒ Repair Manual, *Body - Exterior, Body - Interior,* Repair Group 70

7 - A/C refrigerant shut-off valve -N43-*

- ♦ Checking

⇒ Repair Group 01

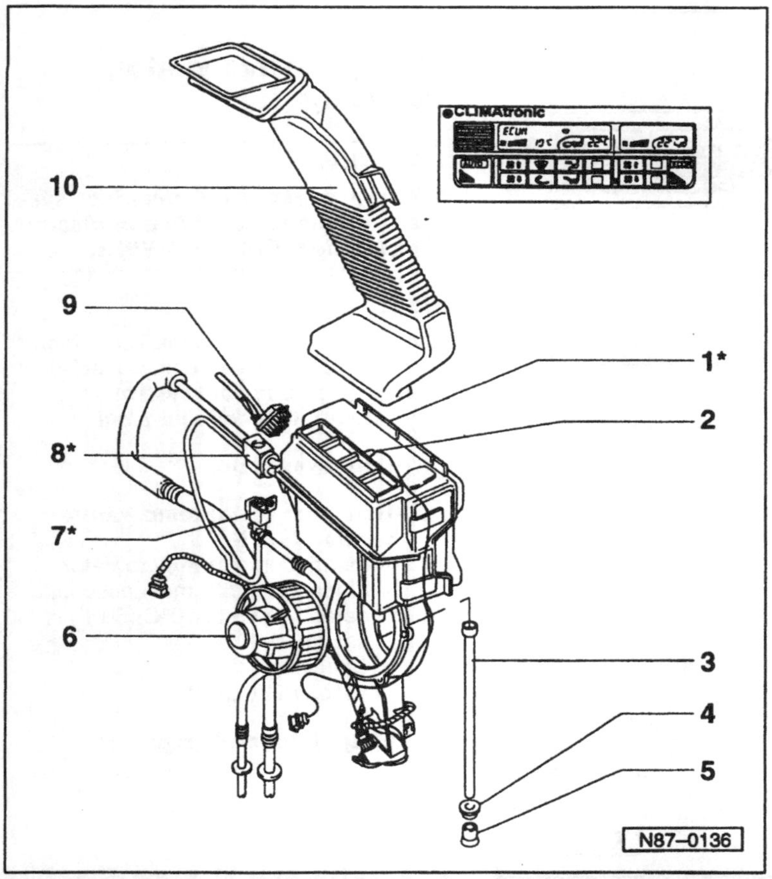

8 - **Expansion valve for rear evaporator***

9 - **Blower control unit for 2nd evaporator -J349-**
- ♦ Function: Controls blower speed of rear evaporator fan -V20- with variable ground signal

10 - **Air intake duct**
- ♦ Removing
- – Remove right "D" pillar trim first.

⇒ Repair Manual, *Body - Exterior, Body - Interior,* Repair Group 70

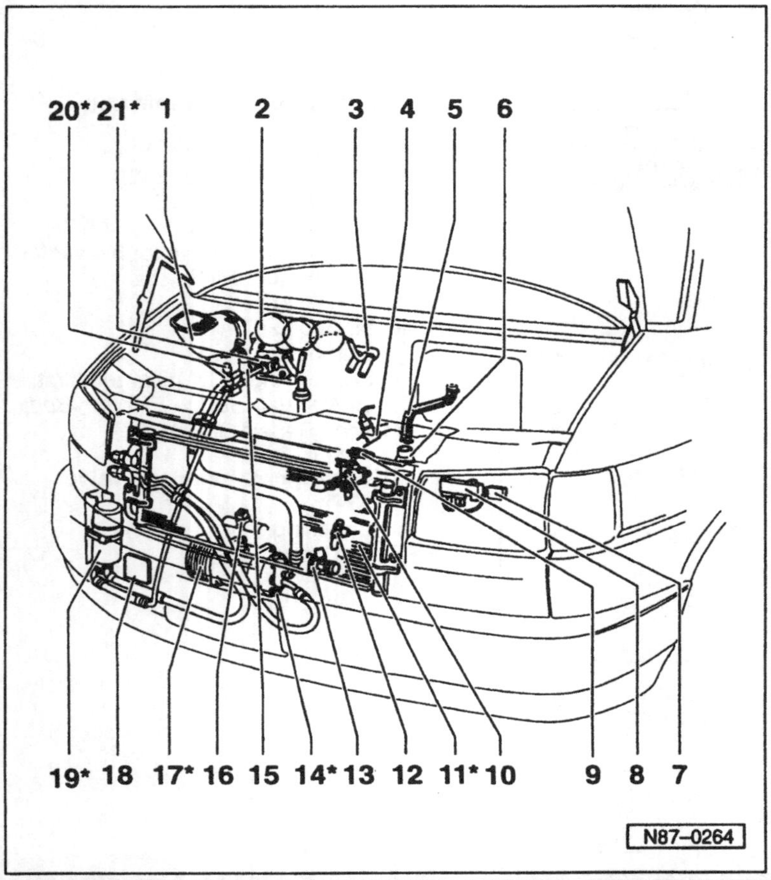

Components in engine compartment, servicing

> ***CAUTION!***
>
> ***Before working on electrical system, obtain radio code and disconnect battery Ground (GND) strap.***

Note:

*System components identified with an * must only be serviced or replaced after discharging refrigerant system. Use Kent Moore ACR4 or equivalent.*

1 - **Air intake duct**
- ♦ Dust and pollen filter, removing and installing

⇒ Page T 80-14

2 - **Vacuum reservoir**
- ♦ Insert vacuum hose: 30 mm (1 3/16 in) into reservoir
- ♦ Vacuum hose layout

⇒ Page T 87-108

3 - **Heater core and vacuum hose guide at bulkhead**

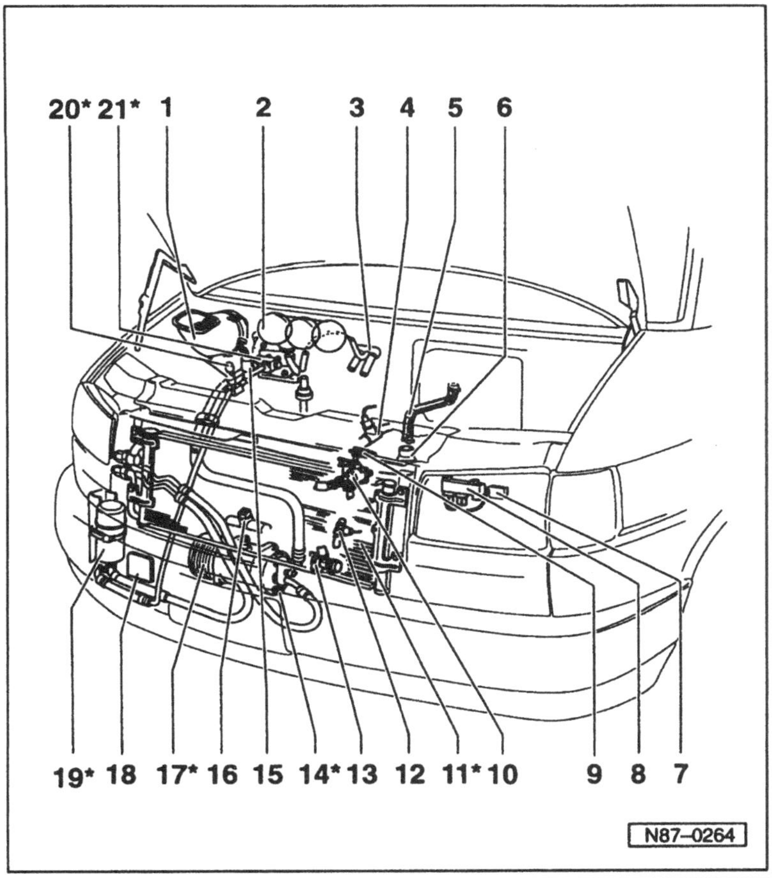

4 - A/C engine coolant temperature sensor -G110-
- Function: Provides engine coolant temperature measurement signal to Climatronic control module -J255- for defrost/footwell air distribution and fresh air blower control
- If sensor -G110- malfunctions, default value of 80°C (176°F) is assumed
- Checking

⇒ Repair Group 01

5 - Evaporator drain pipe
- Removing and installing

⇒ Fig. 4

6 - Evaporator water drain valve
- Water leak at housing may result if valve blocked

7 - Individual fuses for coolant fan -S42- (2x)
- In engine compartment, front left

8 - Coolant fan control (FC) relay -J26-
- In engine compartment, front left
- Checking

⇒ Repair Group 01

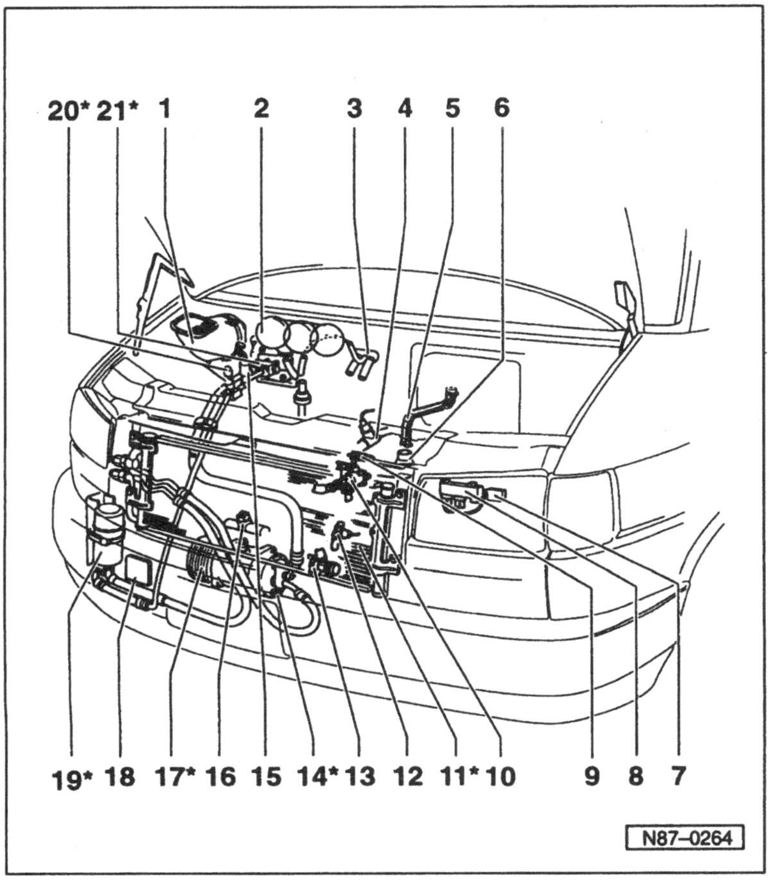

9 - Engine coolant two-way vacuum valve -N147-
- Checking

⇒ Page T 87-109, vacuum hose layout

10 - Auxiliary heater valve -N172- (rear heater)
- Function: Controls coolant flow through rear heater core
- Checking

⇒ Repair Group 01
- Removing

⇒ Page T 87-91, Fig. 6

11 - Condenser*

12 - Coolant shut-off valve
Function:
- Engine off: Vacuum supply to coolant shut-off valve is controlled by engine coolant two-way vacuum valve -N147-. Valve allows coolant to circulate from engine to radiator when after-run coolant pump operates.

13 - Coolant pump -V50-
- Checking

⇒ Repair Group 01

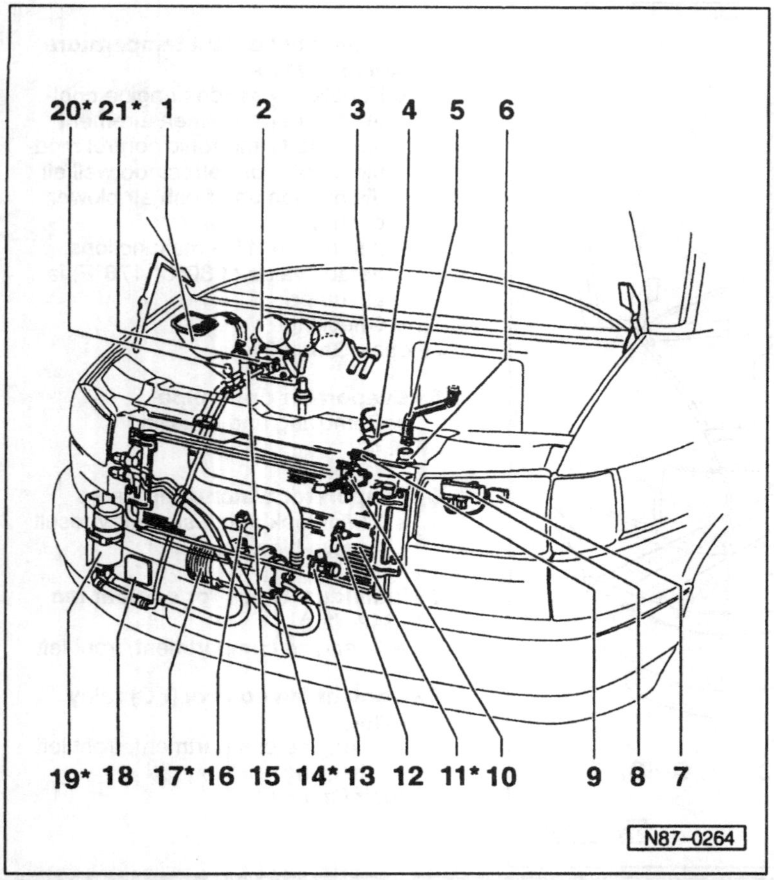

14 - Pressure relief valve*
- Protects refrigerant circuit against over-pressure
- Checking

⇒ Fig. 3

15 - A/C pressure switch -F129-
- Checking

⇒ Page T 87-43, Fig. 3
- Removing and installing:
  – Tightening torque 8 Nm (71 in lb)
  – Replace O ring (note Part No.).

16 - A/C cut-out thermal switch -F163- and Third Speed Coolant Fan Control (FC) Thermal Switch -F165-
- In engine compartment, front

Function:
- -F165- switches-on coolant fan 3rd speed (on at 112°C / 234°F, off at 108°C / 226°F).
- -F163- switches off A/C clutch -N25- at excessive coolant temperatures (off at 119°C / 246°F, on at 112°C / 234°F).
- Removing

⇒ Fig. 1
- Removing and installing, engine code AES

⇒ Fig. 2

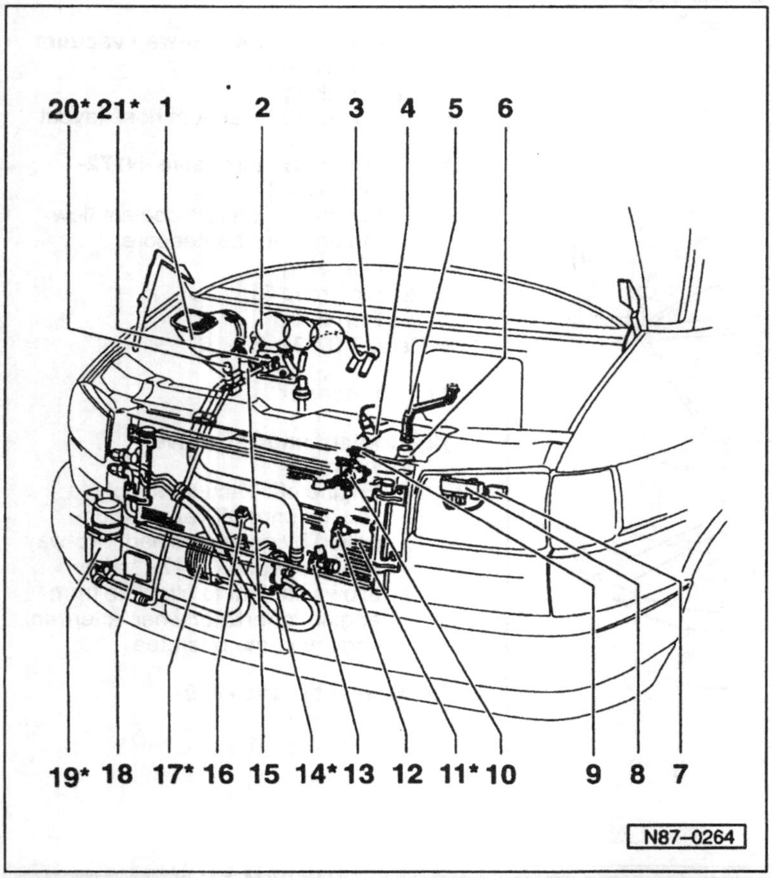

17 - A/C clutch -N25-*

18 - Outside air temperature sensor -G17-
- Function: Provides outside (ambient) temperature measurement signal to Climatronic control module -J255- for temperature flap and fresh air blower control
- If sensor -G17- malfunctions, default value of 10°C (50°F) is assumed
- Checking

⇒ Repair Group 01
- Replacing

⇒ Page T 87-104

19 - Receiver drier*

20 - Service valves*

21 - Expansion valve*
- Opening at bulkhead must be sealed to prevent water ingress
- Expansion valve insulation ⇒ Page T 87-42, Fig. 1

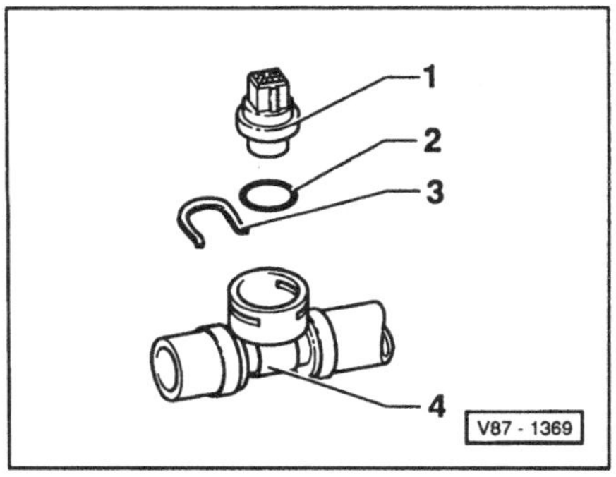

Fig. 1 A/C cut-out thermal switch -F163- and Third Speed Coolant Fan Control (FC) Thermal Switch -F165-, removing and installing

> ***WARNING!***
>
> ***The cooling system is pressurized when the engine is warm. Wear gloves and other appropriate protection, and slowly and carefully release system pressure, if necessary, before performing repairs.***

♦ Refill with coolant after replacing switch.

1 - Thermo switch -F163-/-F165-

2 - O ring

3 - Clip

4 - Coolant hose

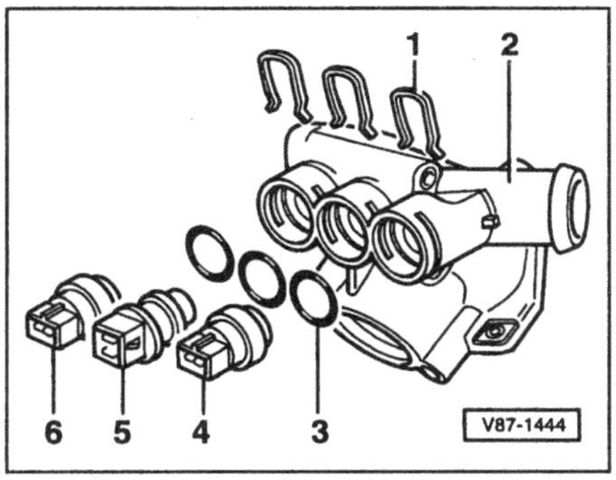

Fig. 2 A/C cut-out thermal switch -F163- and Third Speed Coolant Fan Control (FC) Thermal Switch -F165- engine code AES, removing and installing

> ***WARNING!***
>
> ***The cooling system is pressurized when the engine is warm. Wear gloves and other appropriate protection, and slowly and carefully release system pressure, if necessary, before performing repairs.***

♦ Refill with coolant after replacing switch.

1 - Clip

2 - Thermostat housing

3 - O ring

4 - A/C cut-out thermal switch -F163- & third speed coolant fan control (FC) thermal switch-F165-

5 - Engine coolant temperature (ECT) sensor -G2- & after-run coolant fan control (FC) -F87-.

6 - Engine coolant temperature (ECT) sensor -G62-

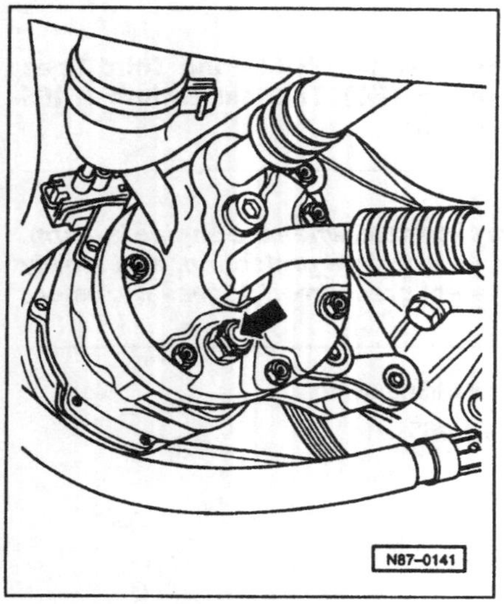

Fig. 3 Pressure relief valve, checking

- Function: Protects refrigerant circuit against over–pressure
- The pressure relief valve plug indicates if valve has opened to release pressure -arrow-. On some types, an adhesive plate is pushed out.

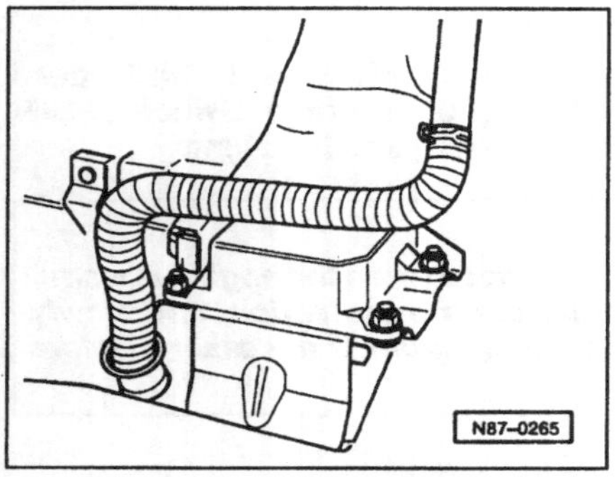

Fig. 4 Evaporator drain pipe, installing

Notes:

- *Water drain hose must be pushed into distributor box onto stop.*
- *Water may leak into passenger compartment if hose is not located on the drain valve as shown.*
- *Drain valve must not be blocked.*

– Remove footwell vent console ⇒ Page T 87-29, Fig. 5

Install hose so that drain valve grommet seals tightly around hose.

Outside air temperature sensor -G17-, replacing

Note:

- *Prior to installing new sensors, the locating tabs must be shortened to ensure proper seating.*

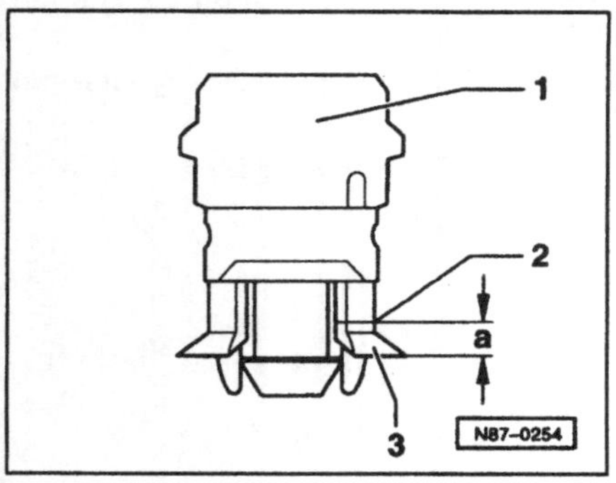

1 - Outside air temperature sensor -G17-
2 - Area to be cut
3 - Locating tabs (3x)

– Mark cutting area, dimension -a- = 3.5 mm (0.13 in) on all three locating tabs.

– Carefully cut off locating lugs with knife.

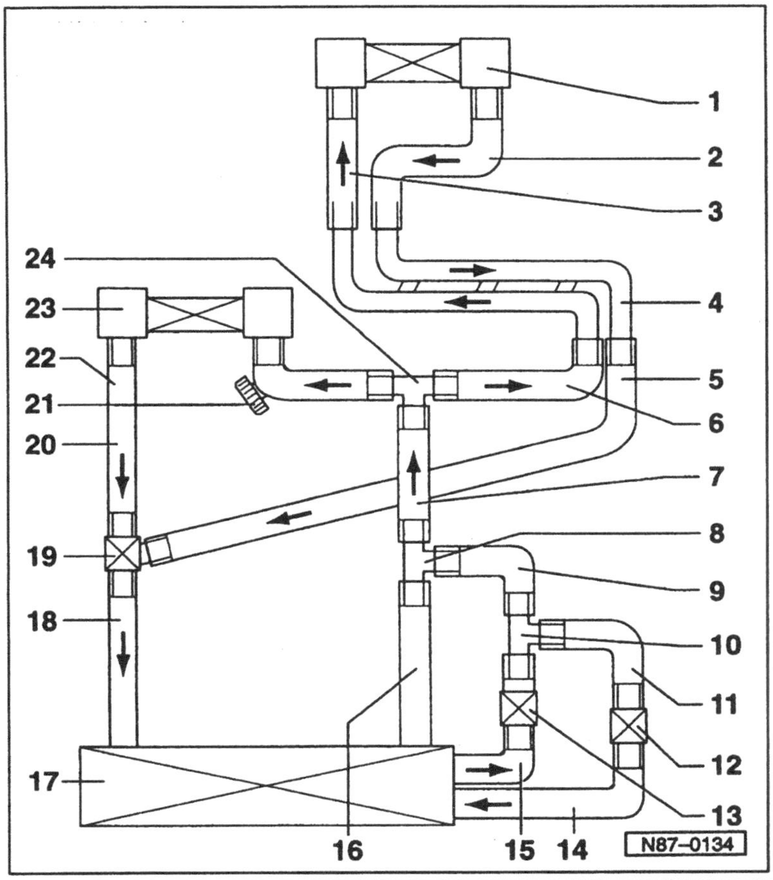

Coolant hoses, layout

Note:

For routing of all hoses
⇒ Repair Manual, *Engine, General,* Repair Group 19

1 - Rear heater core

2 - Coolant return hose
(from rear heater core to coolant pipe)

3 - Coolant supply hose
(from coolant pipe to rear heater core)

4 - Coolant pipe

5 - Coolant return hose
(from coolant pipe to auxiliary heater valve -N172-)

6 - Coolant supply hose
(from T piece to coolant pipe)

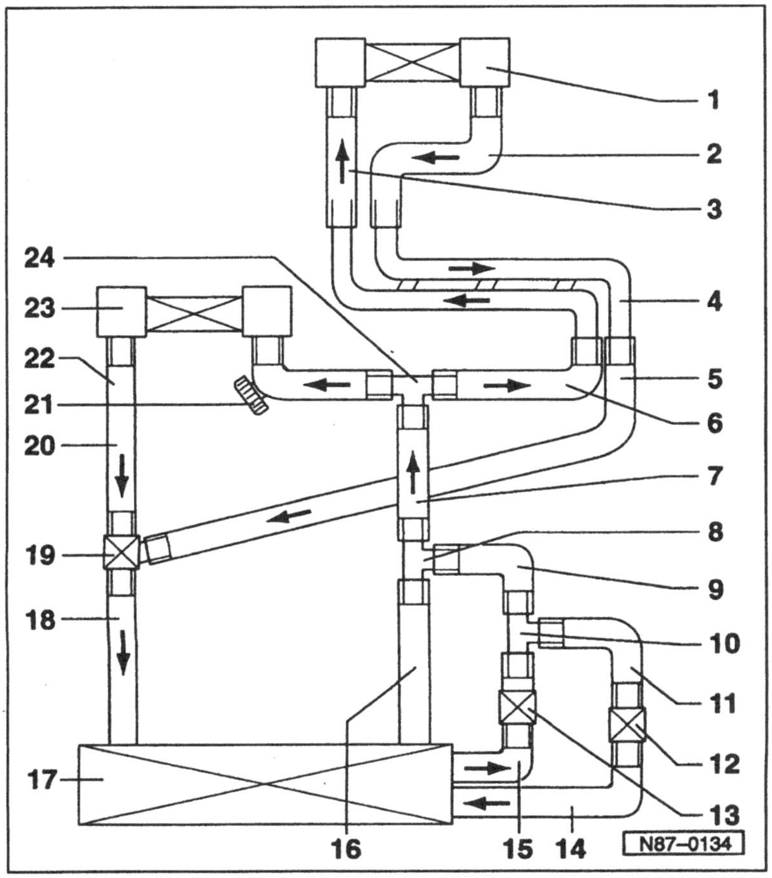

7 - Coolant supply hose
(from T piece to T piece)

8 - T piece

9 - Coolant supply hose
(from T piece to T piece)

10 - T piece

11 - Coolant return hose
(from T piece to coolant cut-off valve)

12 - Coolant cut-off valve

13 - Coolant pump -V50- (for after-run)

14 - Coolant return hose
(from coolant cut-off valve to engine)

15 - Coolant supply hose
(from engine to coolant pump -V50-)

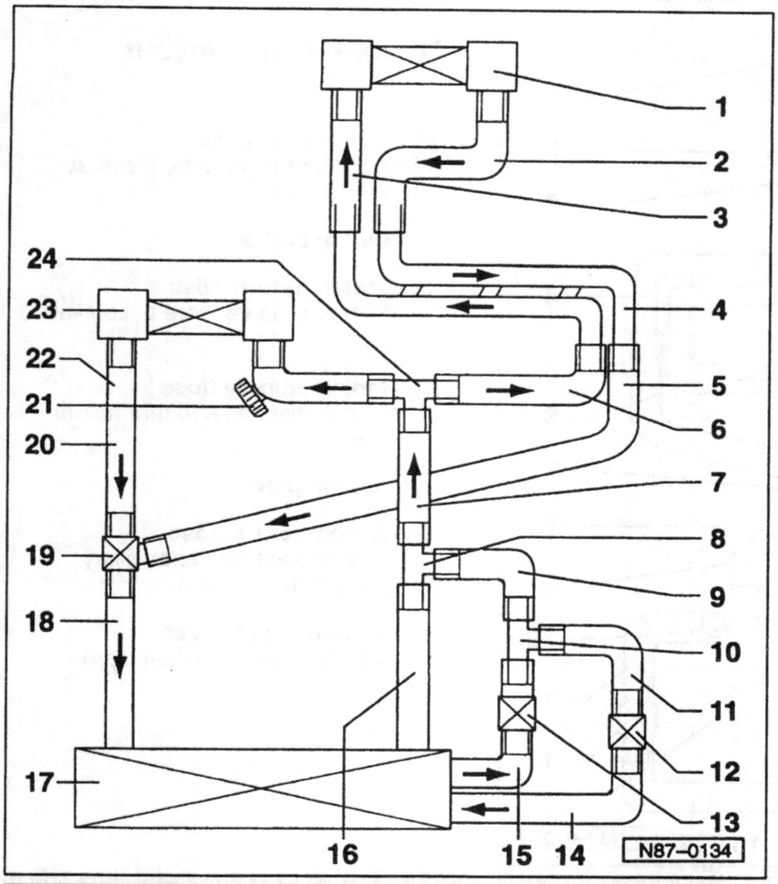

16 - Coolant supply hose
(from engine to T piece)

17 - Engine

18 - Coolant return hose
(from auxiliary heater valve -N172- to engine)

19 - Auxiliary heater valve -N172- (for rear heater)

20 - Coolant return hose
(from T piece to front heater core)

21 - Bleeder screw

22 - Coolant return hose
(from front heater core to auxiliary heater valve -N172-)

23 - Front heater core

24 - T piece

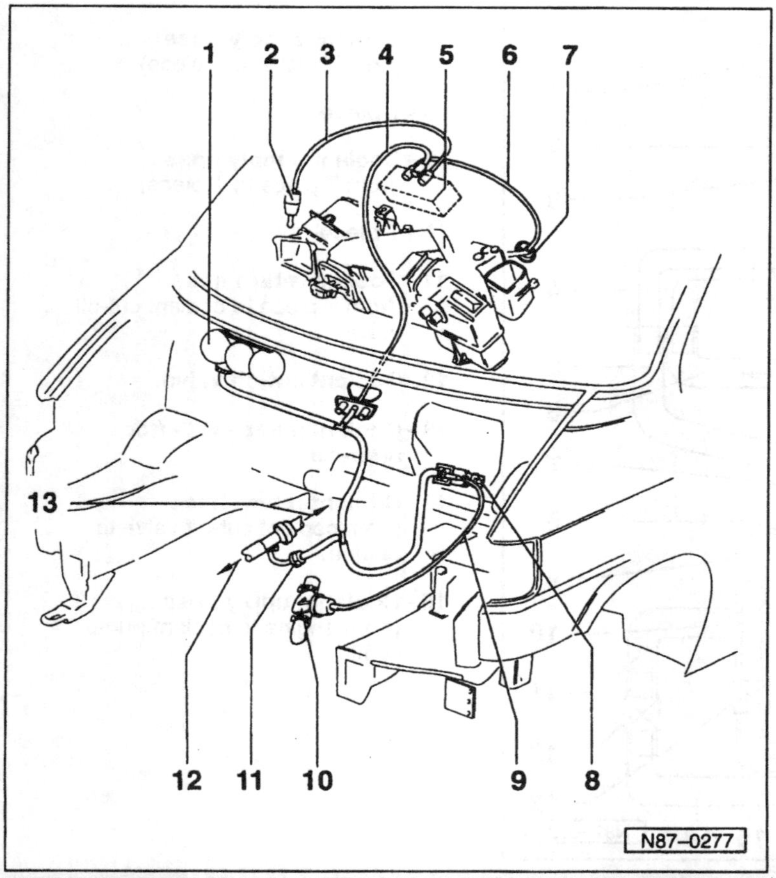

Vacuum hoses, layout

Notes:

- *Arrows indicate vacuum.*
- *Insert vacuum hose into vacuum reservoir 30 mm (1 3/16 in). Insert vacuum hoses onto plastic connectors 15 mm (19/32 in).*

1 - Vacuum reservoir

2 - Fresh air/recirculating flap vacuum unit

3 - Vacuum hose
- From vacuum valve rail -N53- to fresh air/recirculating flap vacuum unit
- Vacuum valve rail connection with green marking

4 - Vacuum hose
- From vacuum reservoir in engine compartment to vacuum valve rail -N53-
- Vacuum valve rail connection with black marking

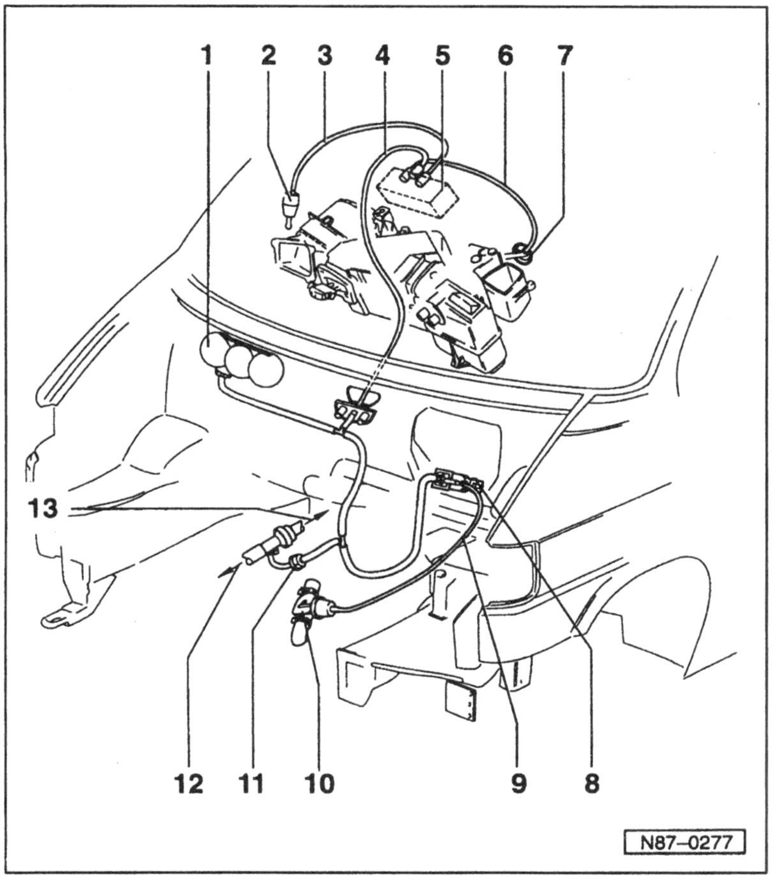

5 - Vacuum valve rail -N53-
- Connections for vacuum hoses have colored markings

6 - Vacuum hose
- From Vacuum valve rail -N53- to center vent vacuum unit
- Vacuum valve rail connection with yellow marking

7 - Central flap vacuum unit

8 - Engine coolant two-way vacuum valve -N147-
- Shuts off vacuum flow when no voltage is present.
- – After attaching vacuum hoses, check:
- When connected the vacuum unit must pull in if voltage and vacuum are present.
- When connected the vacuum unit is vented if vacuum is present and voltage is not.

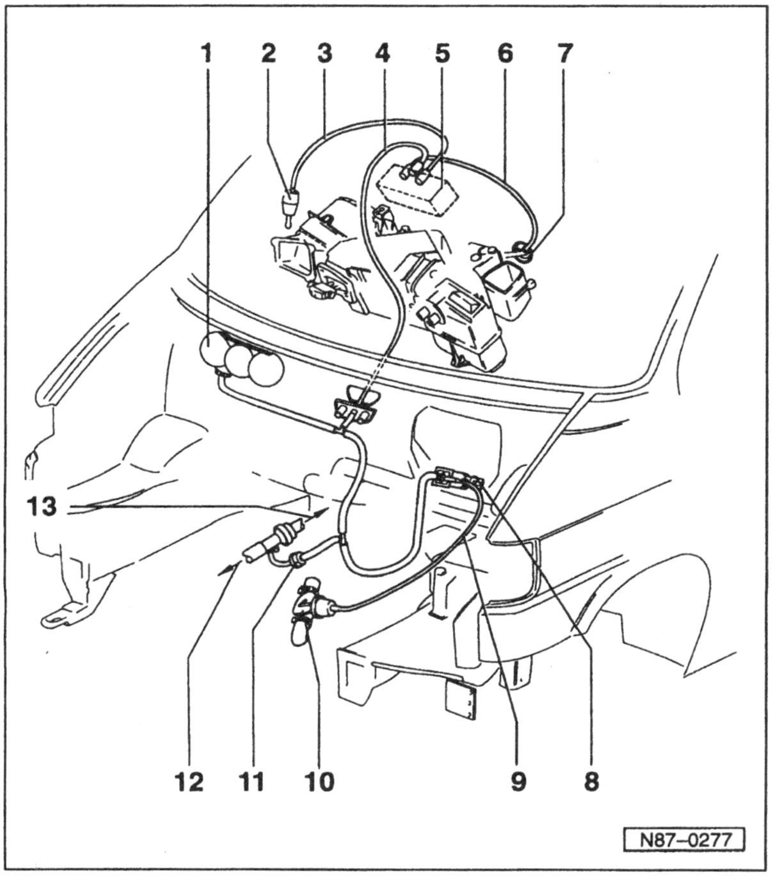

9 - Vacuum hose
- From -N147- to coolant cut-off valve

10 - Coolant shut-off valve
- Switched by vacuum from two-way valve -N147- when engine is running

11 - One-way valve

12 - To intake manifold

13 - To brake booster

Climatronic, troubleshooting

Notes:

- *The Climatronic is equipped with On Board Diagnostic (OBD) capabilities. Always check DTC memory with VAG 1551 Scan Tool (ST) function 02 and rectify all indicated malfunctions ⇒ Page T 01-9.*
- *In addition, carry out "Output Diagnostic Test Mode (DTM)"⇒ Page T 01-29 and "Read Measuring Value Block" function 08 ⇒ Page T 01-37.*
- *The following information applies only when the customer's symptom description has been verified and no DTCs are stored in memory.*
- *Before replacing components, always test the relevant positive and Ground connections as well as all harness connectors and wiring between the component indicated to be malfunction and the Climatronic control module. Refer to the wiring diagram.*
- *After performing repairs, DTC memory should always be checked again and if necessary, erased with VAG 1551 Scan Tool.*
- *Climatronic, servicing ⇒ Page T 87-66*
- *Refrigerant system, servicing ⇒ Page T 87-118*
- *Wiring diagrams ⇒ Electrical Wiring Diagrams.*

Fresh air mode cannot be switched off, no recirculating air function

Test requirements:

- Fuse No. S140, S6 and S16 OK
- Check DTC memory with VAG 1551 Scan Tool (ST), no DTCs recognized.
- Engine running
- Vacuum being built-up

Test sequence:

– Select "recirculating air mode" at A/C control head -E87- ("recirculating air flow" indicated on A/C control head display)

– Check components and specified functions listed below.

Function check	Corrective steps
Fresh air intake blocked	– Check wiring and connections to vacuum valve rail -N53- using wiring diagram
Fresh/recirculating air flap vacuum unit has pulled-in	– Check vacuum hoses ⇒ Page T 87-108

Air flow to center vents cannot be adjusted

Test requirements:

- Fuse No. S140, S6 and S16 OK
- Check DTC memory with VAG 1551 Scan Tool (ST), no DTCs recognized.
- Engine running
- Vacuum being built-up

Test sequence:

– Adjust air flow to instrument panel outlets (air flow direction indicated on A/C control head display)

– Check components and specified functions listed below.

Function check	Corrective steps
Air flow to center vents	– Check wiring and connections to vacuum valve rail -N53- using wiring diagram
Double vacuum unit has pulled-in	– Check vacuum hoses ⇒ Page T 87-108

Insufficient or no heating

Test requirements:

- Fuse No. S140, S6 and S16 OK
- Check DTC memory with VAG 1551 Scan Tool (ST), no DTCs recognized.
- Engine running and at operating temperature.
- Vacuum being built-up

Test sequence:

– Select "maximum temperature" mode at A/C control head -E87- (air flow direction indicated on A/C control head display)

– Check components and specified functions listed below.

Function check	Corrective steps
Coolant pipes to heater core are warm	– If not warm, check thermostat ⇒ Repair Manual, *General, Engine,* Repair Group 19
Coolant shut-off valve open (vacuum unit has pulled in)	– Check wiring and connections to engine coolant two-way vacuum valve -N147- ⇒ wiring diagram – Check vacuum connections at engine coolant two-way vacuum valve -N147- ⇒ Page T 87-108 – Check coolant shut-off valve ⇒ Page T 87-98 – Check vacuum hoses ⇒ Page T 87-108
No warm air from vents	– Check and adjust temperature regulator flap motor -V68- ⇒ Page T 87-81

Insufficient cooling

Test requirements:

- Fuse No. S140, S6 and S16 OK
- Check DTC memory with VAG 1551 Scan Tool (ST); no DTCs recognized
- Climatronic control module coding correct: must be 00001 (model with dust and pollen filter)
- Vacuum system and hoses OK, air distribution can be adjusted properly
- Wiring OK as per wiring diagram
- A/C refrigerant system fully charged
- A/C clutch and compressor function OK
- Condensor and evaporator clean
- Outside air temperature 20°C to 30°C (68°F to 86°F)

Test sequence:

- Start engine
- Set temperature control to "LO" (maximum cooling) on A/C control head
- Select minimum blower speed (manual override) on A/C control head
- Adjust air distribution to instrument panel outlets
- Insert thermometer into center instrument panel outlet
- Raise engine speed to approximately 1500 rpm

- System is OK if outlet temperature drops below 10°C (50°F) within one minute
- If not OK, check A/C refrigerant system ⇒ Page T 87-118

Note:

If drops of water spray out center vents, check evap. water drain valve (behind bulkhead insulation) in eng. compartment.

Fresh air fan does not run after starting engine or "ZU HZ" appears in outside temperature display

Test requirements:

- Fuse No. S6 OK
- Check DTC memory with VAG 1551 Scan Tool (ST), no DTCs recognized.

Possible cause:

♦ Open circuit in wiring (X circuit) to Climatronic control module -J255-.

Remedy:

- Check wiring and connections from fuse/relay panel (X circuit) to terminal T28a/1 at Climatronic control module -J255-.

Manually selected functions do not remain stored in memory

Function:
Any manual selections of fan speed and air distribution ("AUTO" display off) remain stored in memory for approximately one hour. Fully automatic function resumes after the ignition is switched on again after one hour. Temperature settings input during the previous automatic function are then resumed along with automatic air distribution determined by Climatronic control module -J255-.

Test requirements:

- Fuse No. S6 OK
- Check DTC memory with VAG 1551 Scan Tool (ST), no DTCs recognized.

Possible cause:

- ♦ Open circuit in wiring (30 circuit) to Climatronic control module -J255-.

Remedy:

– Check wiring and connections from fuse/relay panel (30 circuit) to terminal T28a/1 at Climatronic control module -J255-.

A/C refrigerant system, servicing

A/C refrigerant system, replacing components

Replacement A/C compressors, evaporators and condensers supplied by the Parts Dept. are filled with Nitrogen. If gas (Nitrogen) does not escape when component is first opened, the component may be faulty (leaking), do not install.

Replacement A/C compressors are filled with the total refrigerant oil quantity needed for the entire refrigerant system. When an additional, rear evaporator is present, additional compressor oil must be added after replacing the compressor.

Always replace the receiver drier whenever the refrigerant system has been opened. Install receiver drier immediately after opening to prevent moisture contamination of drier desiccant. Moreover, keep refrigerant system and other replacement components sealed for as long as possible to minimize the chance of dirt and moisture contamination.

Always plug open refrigerant line connections to prevent dirt and moisture contamination.

If the system has been discharged due to a damaged or leaking component (refrigerant hose/line, compressor, evaporator, condenser, etc.), flush the refrigerant system first with compressed air, then with nitrogen (available locally) and collect the oil that runs out. This will remove the refrigerant oil which may be saturated with moisture.

> ***CAUTION!***
> ***DO NOT flush R-134a refrigerant system with R-11. R-11 is not compatible with refrigerant R-134a and PAG oil and will cause total system contamination.***

If the compressor is not replaced after flushing the system, fill compressor with the correct type and quantity of refrigerant oil as specified under A/C system capacities ⇒ page T 87-143. Fill compressor with the total system oil capacity.

If the compressor is replaced, do not add any additional oil as the total amount of oil required is already in the compressor.

Dispose of contaminated refrigerant oil following laws governing hazardous waste disposal. Do not combine any refrigerant oil with any other old oils such as engine oil or transmission fluid.

Follow all applicable Warnings and Cautions when working on the A/C system.

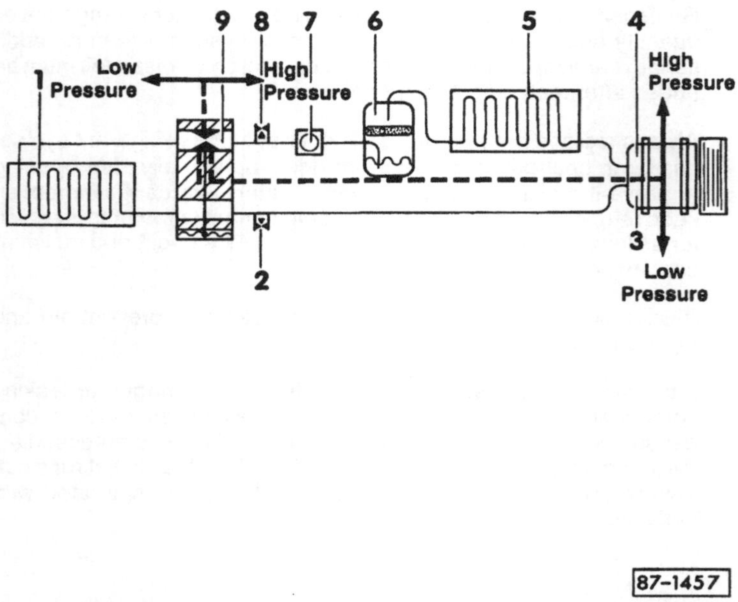

A/C refrigerant system pressures and temperatures, checking

The pressures and temperatures in the A/C system will vary depending on engine speed (RPM), coolant fan speed, engine coolant temperature, A/C clutch engagement, outside temperature, humidity, etc.

Pressure and temperature specifications are based on the following:

- Engine speed (RPM) at 1500
- Fresh air blower on high speed
- A/C switch adjusted to Max. cooling

1 - Evaporator

2 - Low-pressure service valve

3 - Compressor, low pressure side

4 - Compressor, high pressure side

5 - Condenser

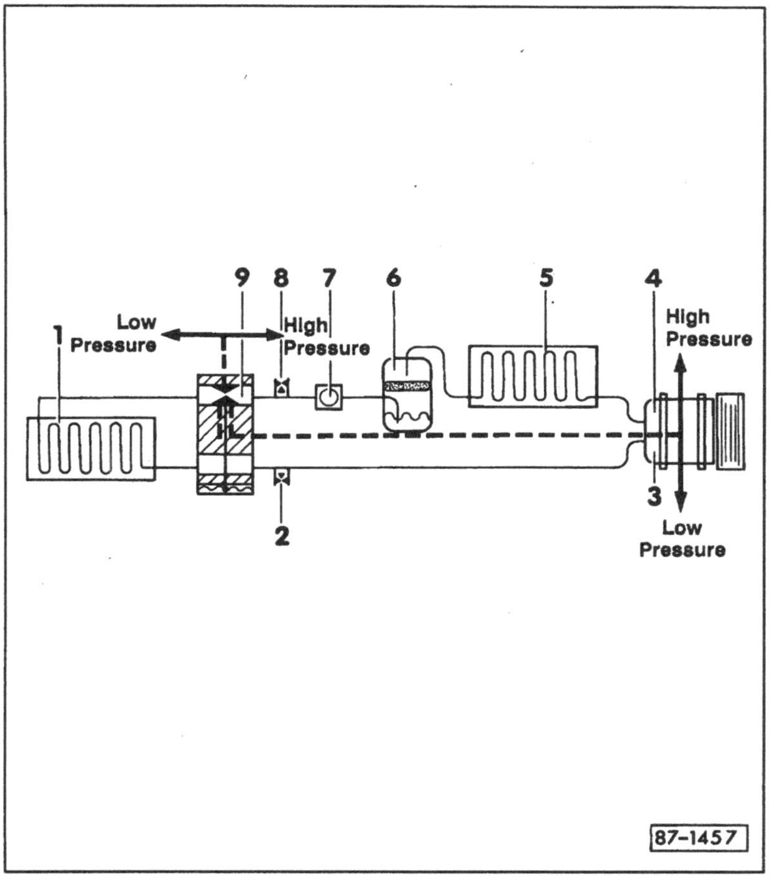

6 - Receiver drier

7 - Sight glass
- Not applicable to all vehicles

8 - High-pressure service valve

9 - Expansion valve

Pressure and temperature specifications

Refer to pages T 87-120 and T 87-121 for component identification.

Component	Refrigerant state	Approximate pressure (bar/psi)	Approximate temperature
1 – Evaporator, inlet to outlet	Vapor	1.2 bar (17.4 psi) 1) 1.8 bar (26.1 psi) 3)	–7°C (19°F) 2) –1°C (30°F) 3)
2, 3 – Compressor, low-pressure side	Gas	1.2 bar (17.4 psi) 1.8 bar (26.1 psi) 3)	–1°C (30°F) +1°C (34°F) 3)
4 – Compressor, high-pressure side	Gas	14 bar (203 psi)	65°C (149°F)
5 – Condenser	Gas Vapor Liquid	14 bar (203 psi)	55°C (131°F) at outlet
6 – Receiver drier	Liquid	14 bar (203 psi)	55°C (131°F)
7 – Sight glass (if equipped)	Liquid	14 bar (203 psi)	55°C (131°F)
8 – Service valve, high-pressure	Liquid	14 bar (203 psi)	55°C (131°F)

Component	Refrigerant state	Approximate pressure (bar/psi)	Approximate temperature
9 – Expansion valve	Liquid, turns to vapor	inlet: 14 bar (203 psi)	inlet: 55°C (131°F)
		outlet: 1.2 bar(17.4 psi)	outlet: –7°C (19°F)

1) This pressure is maintained in the refrigerant system by the variable displacement compressor despite changing conditions and varying engine speeds (RPM).

2) This temperature is maintained in the refrigerant system by the variable displacement compressor despite changing conditions and varying engine speeds (RPM).

3) Test values for vehicles equipped with second (rear) evaporator.

A/C refrigerant system, testing with pressure gauges

Due to the constant temperature/pressure relationship of R-12 and R-134a, approximate high-side system temperature can be determined based on system pressure.

Pressure gauges may have one or more temperature scales in addition to the pressure scale. Since various refrigerants have different temperature/pressure relationships, each gauge is specific for a particular refrigerant.

Measuring pressure and temperature in a "switched-on" refrigerant system

Use the A/C refrigerant high-pressure gauge to measure high pressure between the compressor and restrictor (including condenser) with the A/C switched on.

Use the A/C refrigerant low-pressure gauge to measure low pressure between the expansion valve and the compressor inlet (including evaporator) with the A/C switched on.

The temperature displays of the low- and high-pressure gauges apply only to a portion of the low- and high-pressure sides. The low-pressure gauge displays approximate temperature between the expansion valve and evaporator outlet. The high-pressure gauge displays approximate temperature between the compressor outlet and restrictor.

Note:

The temperature/pressure relationship only holds true when the refrigerant system contains liquid and vapor, but not when it contains only gas. In the gaseous state, the temperature is approx. 10-30°C (18-54°F) higher than indicated on the pressure gauge.

Measuring pressure and temperature in a closed container or "switched-off" refrigerant system

Due to the constant temperature/pressure relationship of R-12 and R-134a, approximate temperature in a closed container or in a switched-off A/C system can be determined based on its pressure.

If the temperature displayed on a gauge is lower than the actual temperature of the refrigerant, the refrigerant container or system is empty (discharged).

Note:

The temperature/pressure relationship only holds true when the refrigerant system contains liquid and vapor, but not when it contains only gas. In the gaseous state, the temperature is approx. 10-30°C (18-54°F) higher than indicated on the pressure gauge.

A/C refrigerant system, discharging

WARNING!

- ***Always use an Underwriter's Laboratory (UL) approved refrigerant recovery/recycling/recharging unit such as Kent-Moore ACR[4], or equivalent, whenever discharging an R-134a A/C system.***
- ***Any person who services a motor vehicle air conditioner MUST, by law, be properly trained and certified and use approved refrigerant recycling equipment. Technicians must complete an EPA approved recycling course to be certified.***
- ***State and local governments may have additional requirements regarding air conditioning servicing. Always comply with state and local laws.***
- ***Always wear safety goggles when charging or discharging system.***

CAUTION!

- ***Always use separate refrigerant recovery/recycling/recharging servicing equipment for R-12 and R-134a systems. DO NOT use one piece of equipment for both R-12 and R-134a systems. The residual traces of refrigerant inside the equipment will contaminate and damage the equipment. Servicing equipment includes recovery/recycling/recharging unit, charging station, vacuum pump, manifold gauges, etc.***
- ***DO NOT use R-12 servicing equipment on R-134a systems or R-134a equipment on R-12 systems or damage to both the vehicle A/C system and servicing equipment may result. Use only equipment designed to meet Society of Automotive Engineers (SAE) standards.***
- ***R-134a and R-12 systems use different size service fittings. NEVER use adaptors to convert an R-12 fitting to R-134a size or R-134a fitting to R-12 size.***

Note:

Refer to safety measures starting on page T 87-41 prior to discharging or charging A/C refrigerant system.

Note:

Make sure that initial set-up of the refrigerant recovery/recycling/recharging unit has been completed before discharging the A/C system.

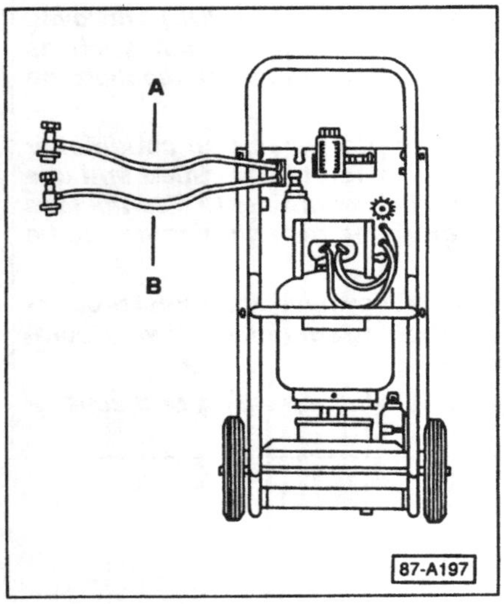

◄

- Connect red high-pressure hose -A- of refrigerant recovery/recycling/recharging unit to high side fitting on vehicle and open coupler valve.
- Connect blue low-pressure hose -B- of refrigerant recovery/recycling/recharging unit to low side fitting on vehicle and open coupler valve.

CAUTION!

Always follow manufacturer's instructions when using a refrigerant recovery/recycling/recharging unit.

- Following refrigerant recovery/recycling/recharging unit manufacturer's instructions, discharge A/C system into refrigerant recovery/recycling/recharging unit.
- Disconnect power supply from A/C clutch to prevent accidental compressor operation with A/C system discharged.

A/C refrigerant system, flushing with compressed air and nitrogen

CAUTION!

- ***When using compressed nitrogen always use a pressure regulator and the proper adaptor hoses and fittings (available locally). During flushing, use existing exhaust/ventilation systems to draw off the gas mixture escaping from the A/C system.***
- ***DO NOT flush R-134a refrigerant system with R-11. R-11 is not compatible with R-134a refrigerant and PAG oil and will cause total system contamination.***

- Use compressed air and nitrogen (available locally) to remove moisture, impurities and old refrigerant oil from A/C refrigerant system.
- First blow out old refrigerant oil and dirt with compressed air, then dry components with nitrogen.
- DO NOT blow compressed air and nitrogen through the compressor or expansion valve. Only blow compressed air and nitrogen through disconnected, free flowing components (i.e. disconnected hose, condenser, evaporator, etc.)
- DO NOT blow compressed air and nitrogen into a capped off A/C component. Pressurized R-134a refrigerant in the presence of oxygen may form a combustible mixture.

- Always flush components in opposite direction of refrigerant flow.
- Flush evaporator through the low-pressure line with the high-pressure line removed.
- If any component has dark thick deposits that cannot be removed with compressed air, replace component.
- Thin light gray deposits in refrigerant lines and hoses are normal and do not impair the function of the system.
- Always replace receiver drier and expansion valve after flushing.
- Dispose of contaminated refrigerant (PAG) oil following laws governing hazardous waste disposal. Do not combine PAG oil with any other old oils such as engine oil or transmission fluid.

Flush refrigerant system with compressed air and nitrogen if:

- Refrigerant oil is dark and viscous (thick)
- Too much refrigerant oil is in the system following compressor replacement
- Unclear or do not know how much refrigerant oil is in the system
- Moisture, dirt or other impurities have entered the refrigerant system (i.e. following an accident)
- Unable to pull a constant vacuum during evacuation of a leak-free system due to excessive moisture in the system
- Refrigerant system has been open longer than the time required for normal repairs (i.e. following an accident)
- Based on temperature and pressure measurements, system is diagnosed with moisture contamination
- Compressor is replaced due to noises or internal damage
- Flushing is required after replacing certain components in certain situations ⇒ page T 87-118

A/C refrigerant system, evacuating and recharging

WARNING!

- ***Always use an Underwriter's Laboratory (UL) approved refrigerant recovery/ recycling/recharging unit such as Kent-Moore ACR4, or equivalent, when evacuating and recharging an R-134a A/C system.***
- ***Any person who services a motor vehicle air conditioner MUST, by law, be properly trained and certified and use approved refrigerant recycling equipment. Technicians must complete an EPA approved recycling course to be certified.***
- ***State and local governments may have additional requirements regarding air conditioning servicing. Always comply with state and local laws.***
- ***Always wear safety goggles when discharging, evacuating and recharging an A/C system.***

CAUTION!

Always follow manufacturer's instructions when using a refrigerant recovery/recycling/recharging unit.

Notes:

- *Refer to R-134a safety measures starting on page T 87-11 prior to discharging or charging A/C refrigerant system.*
- *Follow refrigerant recovery/recycling/recharging unit manufacturer's instructions for evacuating and recharging A/C system.*
- *Evacuate refrigerant system for a minimum of 30 minutes.*
- *When recharging A/C system, add correct amount of refrigerant and PAG oil to system ⇒ page T 87-143.*
- *After system recharge, manually rotate A/C compressor approx. 10 turns before starting engine. Start engine with A/C OFF. After idle speed has stabilized, switch A/C ON and let engine idle (compressor running) for a minimum of two minutes before raising engine speed.*

A/C refrigerant system, troubleshooting

Requirements

- Electrical systems OK
- Vacuum hoses and control systems OK
- Air flow distribution systems OK

Correct and repair as necessary before proceeding.

If any one of the following system conditions exists, check A/C refrigerant system pressures ⇒ page T 87-136.

- A/C does not cool (cooling has stopped completely)
- Insufficient cooling during all driving speeds or engine speeds (RPM)
- No cooling or insufficient cooling after driving several miles

A/C compressor noisy

– Tighten A/C compressor and compressor bracket mounting bolts using torque wrench.

– Check routing of A/C refrigerant hoses and lines; hoses and lines must not contact other components and must be installed without tension or kinks, reposition or realign as necessary.

A/C compressor noisy immediately after switching A/C ON and/or when driving around corners or braking (refrigerant shock)—system overcharged

– Discharge A/C refrigerant system ⇒ page T 87-126, then evacuate and recharge ⇒ page T 87-132.

With an otherwise properly operating A/C system, water sprays out of instrument panel vents

– Check evaporator drain pipe for proper routing (should not be kinked or pinched).

– Check evaporator water drain valve.

- Must not be plugged with wax, dirt or undercoating (clean if necessary)
- Valve flap must open/close properly

– Check plenum tray.

- Must be properly installed and not damaged

A/C refrigerant system pressures, checking

Check A/C refrigerant system pressures using these three tests:

1 - Air temperature drop from center instrument panel outlet (A/C system cooling performance)

2 - A/C system high pressure

3 - A/C system low pressure

The combined results of all three steps determine which part of the A/C system is causing the problem, see table ⇒ page T 87-140.

Test conditions

- ♦ A/C refrigerant system fully charged; discharge, evacuate and recharge system if necessary
- ♦ Condenser and radiator clean and free of obstructions (spray clean if necessary)
- ♦ Air distribution can be adjusted correctly using control knobs (all air distribution flaps reach end positions)
- ♦ Wiring OK per wiring diagram
- ♦ Outside (ambient) air temperature between 20–30°C (68–86°F)
- ♦ Drive belts for A/C compressor and Generator in good condition and properly tensioned

Test 1: Air temperature drop from center instrument panel outlet, checking (A/C system cooling performance)

– Start engine.

– Set temperature control to maximum "cold" (if Manual A/C) or "LO" (if Climatronic).

– Manual A/C: press "A/C" button and select second blower speed.

– Climatronic: Select minimum blower speed (manual override) by pressing "decrease blower speed" button (observe front "blower speed" display in control head) .

– Adjust air distribution to instrument panel outlets.

– Insert thermometer into center instrument panel outlet and raise engine speed to approximately 1500 RPM.

Specified result

With humidity normal and outside (ambient) temperature between 20–25°C (68–77°F), system is cooling satisfactorily if air temperature from center instrument panel vent drops below 10°C (50°F) within 1 minute.

For higher ambient temperatures and/or higher humidity, specified air temperature from center instrument panel vent can be slightly higher.

If specified reading is not obtained, perform tests 2 and 3, then compare results of all three tests ⇒ table, page T 87-140.

Test 2: A/C system high pressure, checking

– Connect high- and low-pressure hoses of refrigerant recovery/ recycling/recharging unit Kent-Moore ACR^4 or equivalent, to high- and low-pressure service valves.

– Disconnect electrical connector from coolant fan.

– Start engine.

– Set temperature control to maximum "hot" (if Manual A/C) or "HI" (if Climatronic).

– Manual A/C: press "A/C" button and select highest blower speed.

– Adjust air distribution to footwell outlets.

– Raise engine speed to approximately 1500 RPM.

Specified result

System high pressure is normal if high-pressure gauge reads 232 psi (16 bar) within 30 seconds.

If specified reading is not obtained, also perform test 3 and compare results of all three tests ⇒ table, page T 87-140.

Test 3: A/C system low pressure, checking

– Connect high- and low-pressure hoses of refrigerant recovery/recycling/recharging unit Kent-Moore ACR[4], or equivalent, to high- and low-pressure service valves.

– Start engine.

– Set temperature control to maximum "cold" (if Manual A/C) or "LO" (if Climatronic).

– Manual A/C: press "A/C" button and select first blower speed.

– Climatronic: Reduce blower speed to lowest setting (manual override) by pressing "decrease blower speed" button (observe front "blower speed" display in control head).

– Adjust air distribution to instrument panel outlets.

– Raise engine speed to 1500 RPM.

Specified result

System low pressure is normal if low-pressure gauge reads 22–36 psi (1.5–2.5 bar) within 30 seconds.

If specified reading is not obtained, compare results of all three tests ⇒ table, page T 87-140.

Test 1 Temperature from center air vent[1]	Test 2 High pressure[2]	Test 3 Low pressure[2]	Possible causes of incorrect readings	Corrective measures
Normal	Normal	Normal	None	---
Too high	Normal	Normal	Temperature flap position incorrect	Adjust temperature flap cable
Too high	Too low	Normal	Compressor	Replace compressor
Normal	Too low	Normal	Compressor	Replace compressor
Normal	Normal	Too high or too low	Expansion valve or compressor	Replace expansion valve or compressor
Too high	Normal	Too high or too low	Expansion valve or compressor	Replace expansion valve or compressor
Normal	Too high or too low	Too high or too low	Expansion valve or compressor	Replace expansion valve or compressor

[1] Normal air outlet temperature approx. 43°F (6°C).

[2] For normal system temperatures and pressures ⇒ page T 87-122.

A/C system cooling performance, checking

Note:

A/C evaporator temperature switch -E33- (if equipped) switches A/C compressor OFF only in special instances so evaporator does not ice up; variable displacement A/C compressor maintains temperature of 0°C (32°F) in evaporator.

Test conditions

- A/C refrigerant system fully charged
- A/C clutch -N25- and compressor function OK
- Condenser and radiator clean and free of obstructions
- Air distribution can be adjusted properly
- Wiring OK as per wiring diagram
- Outside (ambient) temperature 20°C–30°C (68°F–86°F)

Checking

– Start engine.

– Set temperature control to maximum "cold" (if Manual A/C) or "LO" (if Climatronic).

– Manual A/C: press A/C button and select second blower speed.

– Climatronic: Select minimum blower speed (manual override) by pressing "decrease blower speed" button (observe front "blower speed" display in control head) .

– Adjust air distribution to instrument panel outlets.

– Insert thermometer into center instrument panel outlet.

– Raise engine speed to approximately 1500 RPM.

Specified result

With normal humidity and outside (ambient) temperature between 20°–25°C (68°–77°F), system is sufficiently charged if air temperature from center instrument panel vent drops below 10°C (50°F) within one minute.

For higher ambient temperatures and/or higher humidity, specified air temperature from center instrument panel vent can be slightly higher.

A/C refrigerant system, capacities

Refrigerant R-134a

Model	Quantity
EuroVan & Transporter with Manual A/C (one evaporator)	950 g + 50 g (33.5 oz + 1.8 oz)
EuroVan with Climatronic (two evaporators, front & rear)	1350 g + 50 g (47.6 oz + 1.8 oz)

Obtain R-134a refrigerant from a local A/C supplier under one of the following names:

- R-134a
- Tetrafluoroethane
- $CH_2F\ CF_3$
- H-FKW 134a
- SUVA® TRANS A/C
- ARCTON® 134a

Note:

R-134a refrigerant is packaged in different containers. Some are used only for commercial applications which are sold in cylinders using a 1/4" flare fitting. This does not connect to the vehicle fittings and servicing equipment. Use only R-134a which come in containers having the correct type of service fitting.

Refrigerant (PAG) oil, R-134a

Model	Total system capacity
EuroVan & Transporter with Manual A/C (one evaporator)	135 cc (4.6 fl. oz.)
EuroVan with Climatronic (two evaporators, front & rear)	240 cc (8.1 fl. oz.)

Use only the correct type of refrigerant oil (PAG oil) specified for use with R-134a.

Note:

Replacement A/C compressors supplied by the Parts Department are filled with 135 cc (4.6 fl. oz) of refrigerant oil. This is the total A/C system refrigerant oil capacity for systems with one evaporator. Systems with two evaporators require an additional 105 cc (3.6 fl. oz.) be added to the system after compressor replacement.

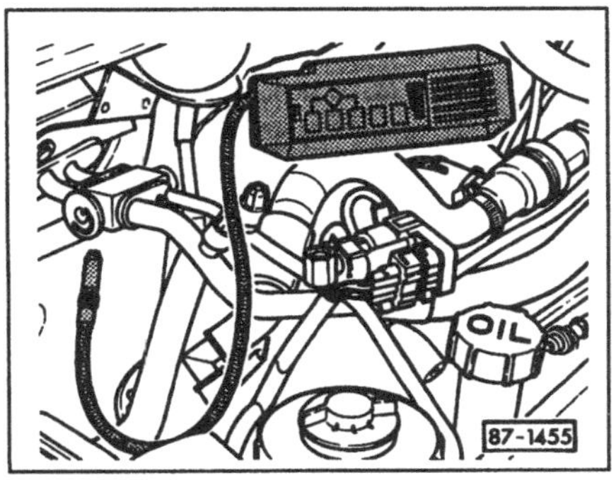

A/C refrigerant system, checking for leaks

◄ Use halogen leak detector, Hitec HI400A-TEL or equivalent, to check for refrigerant leaks, following the leak detector manufacturer's instructions.

Refrigerant gas dissipates very quickly. To make the job easier, avoid drafty or windy areas when checking for leaks.

If the refrigerant system is discharged (empty), re-charge the system with approx. 100 g (3.5 oz) of refrigerant in order to check for leaks.

Electrical Equipment

Section V

WARNING!
Before proceeding with any work, please read the warnings and cautions immediately following this index. Also read any warning and caution that accompanies a procedure in the manual.

Please read these WARNINGS and CAUTIONS before proceeding with maintenance and repair work.

WARNING—

- Some repairs may be beyond your capability. If you lack the skills, tools and equipment, or a suitable workplace for any procedure described in this manual, we suggest you leave such repairs to an authorized Volkswagen dealer service department, or other qualified shop.

- Volkswagen is constantly improving its cars. Sometimes these changes, both in parts and specifications, are made applicable to earlier models. Therefore, before starting any major jobs or repairs to components on which passenger safety may depend, consult your authorized Volkswagen dealer about the Technical Bulletins that may have been issued since the editorial closing of this manual.

- Do not re-use any fasteners that are worn or deformed in normal use. Many fasteners are designed to be used only once and become unreliable and may fail when used a second time. This includes, but is not limited to, nuts, bolts, washers, self–locking nuts or bolts, circlips and cotter pins. Always replace these fasteners with new parts.

- Never work under a lifted car unless it is solidly supported on stands designed for the purpose. Do not support a car on cinder blocks, hollow tiles or other props that may crumble under continuous load. Never work under a vehicle that is supported solely by a jack. Never work under the car while the engine is running.

- If you are going to work under a car on the ground, make sure that the ground is level. Block the wheels to keep the car from rolling. Disconnect the battery negative terminal (ground strap) to prevent others from starting the vehicle while you are under it.

- Never run the engine unless the work area is well ventilated. Carbon monoxide (CO) kills.

- Do not attempt to work on your vehicle if you do not feel well. You increase the danger of injury to yourself and others if you are tired, upset or have taken medicine or any other substance that may impair you or keep you from being fully alert.

- Disconnect the battery negative (–) terminal (ground strap) whenever you work on the fuel system or the electrical system. Do not smoke or work near heaters or other fire hazards. Keep an approved fire extinguisher handy.

- Batteries produce explosive hydrogen gas. Keep sparks, lighted matches and open flame away from the top of the battery. If hydrogen gas escaping from the cap vents is ignited, it will ignite gas trapped in the cells and cause the battery to explode.

- Connect and disconnect battery cables, jumper cables or a battery charger only with the ignition switched off, to prevent sparks. Do not quick–charge the battery (for boost starting) for longer than one minute, and do not allow charging voltage to exceed 16.5 volts. Wait at least one minute before boosting the battery a second time.

- Illuminate your work area adequately but safely. Use a portable safety light for working inside or under the car. Make sure the bulb is enclosed by a wire cage. The hot filament of an accidentally broken bulb can ignite spilled fuel or oil.

- Tie long hair behind your head. Do not wear a necktie, a scarf, loose clothing, or a necklace when you work near machine tools or running engines. If your hair, clothing, or jewelry were to get caught in the machinery, severe injury could result.

- Finger rings, bracelets and other jewelry should be removed so that they cannot cause electrical shorts, get caught in running machinery, or be crushed by heavy parts.

- Catch draining fuel, oil, or brake fluid in suitable containers. Do not use food or beverage containers that might mislead someone into drinking from them. Store flammable fluids away from fire hazards. Wipe up spills at once, but do not store the oily rags, which can ignite and burn spontaneously.

- Always observe good workshop practices. Wear goggles when you operate machine tools or work with battery acid. Gloves or other protective clothing should be worn whenever the job requires working with harmful substances.

- The Air-Conditioning (A/C) system is filled with a chemical refrigerant, which is hazardous. The A/C system should be serviced only by trained technicians using approved refrigerant recovery/recycling equipment, trained in related safety precautions, and familiar with regulations governing the discharging and disposal of automotive chemical refrigerants.

- Do not expose any part of the A/C system to high temperatures such as an open flame. Excessive heat will increase system pressure and may cause the system to burst.

- Some aerosol tire inflators are highly flammable. Be extremely cautious when repairing a tire that may have been inflated using an aerosol tire inflator. Keep sparks, open flame or other sources of ignition away from the tire repair area. Inflate and deflate the tire at least four times before breaking the bead from the rim. Completely remove the tire from the rim before attempting any repair.

- Friction materials such as brake or clutch discs may contain asbestos fibers. Do not create dust by grinding, sanding, or by cleaning with compressed air. Avoid breathing asbestos fibers and asbestos dust. Breathing asbestos can cause serious diseases such as asbestosis or cancer, and may result in death.

- Greases, lubricants and other automotive chemicals contain toxic substances, many of which are absorbed directly through the skin. Read manufacturer's instructions and warnings carefully. Use hand and eye protection. Avoid direct skin contact.

- An airbag is operated by an explosive device. Handled improperly or without adequate safeguards, it can be accidently activated and cause serious personal injury. To prevent personal injury or airbag system failure, **only trained Volkswagen dealer service technicians** should test, disassemble or service the airbag system.

WARNINGS and CAUTIONS continued on next page ►

Please read these WARNINGS and CAUTIONS before proceeding with maintenance and repair work.

WARNING (continued) —

- ALWAYS disconnect the airbag power supply before using a computer memory saver. Failure to disconnect the airbag power supply may result in accidental activation of the airbag.
- The airbag unit should not be removed, nor should any work be performed on the airbag system, without first reviewing the warnings and cautions that accompany the airbag system repair information in this manual.
- Never apply stickers or any other coverings on the steering wheel. Do not let chemical cleaners, oil or grease come into contact with the vinyl covering of the airbag unit.
- Never open or otherwise attempt to repair airbag system parts. Always use new parts. Never leave airbag parts or the partially disassembled airbag system unattended.
- Never use a test light or volt meter to conduct electrical tests on the airbag system. The airbag system must only be tested by trained Volkswagen Service technicians using special airbag tester equipment. The airbag system must only be tested while it is installed in the car.
- Do not expose an airbag unit to temperatures above 194°F (90°C) even for a brief period, while handling it during the repair process. Keep the unit clear of heat sources such as hot plates, soldering irons, heat lamps, welding equipment, or the like.
- Always replace airbag system components that have been mechanically damaged (bubbles, cracks, etc.).
- Always set down a removed airbag unit so that the horn pad is facing upward.
- When driving or riding in an airbag-equipped vehicle, never hold test equipment in your hands or lap while the vehicle is in motion. Objects between you and the airbag can increase the risk of injury in an accident.
- The storage, transportation and disposal of airbag units are subject to the laws for flammable solids.

CAUTION—

- If you lack the skills, tools and equipment, or a suitable workshop for any procedure described in this manual, we suggest you leave such repairs to an authorized Volkswagen dealer or other qualified shop. We especially urge you to consult an authorized Volkswagen dealer before beginning repairs on any car that may still be covered wholly or in part by any of the extensive warranties issued by Volkswagen of America, Inc.
- Volkswagen is constantly improving its cars and sometimes these changes, both in parts and specifications, are made applicable to earlier models. Therefore, part numbers listed in this manual are for reference only. Always check with your authorized Volkswagen dealer parts department for the latest information.
- Before starting a job, make certain that you have all the necessary tools and parts on hand. Read all the instructions thoroughly, do not attempt shortcuts. Use tools that are appropriate to the work and use only replacement parts meeting Volkswagen specifications. Makeshift tools, parts and procedures will not make good repairs.
- Use pneumatic and electric tools only to loosen threaded parts and fasteners. Never use these tools to tighten fasteners, especially on light alloy parts. Always use a torque wrench to tighten fasteners to the tightening torque specification listed.
- Be mindful of the environment and ecology. Before you drain the crankcase, find out the proper way to dispose of the oil. Do not pour oil onto the ground, down a drain, or into a stream, pond, or lake. Consult local ordinances that govern the disposal of wastes.
- When connecting cables or ground straps, all connections must be cleaned to bare metal. Be sure mounting bolts are tight.
- Do not replace fuse links with fuse links of other gauge or with standard wire. Always replace the entire length of the fuse link.
- Before doing any electrical welding on vehicles equipped with Anti-lock Brakes (ABS), disconnect the battery negative terminal (ground strap) and the ABS Control Module connector.
- On cars equipped with ABS, the ABS control unit cannot withstand temperatures from a paint–drying booth or a heat lamp in excess of 203°F (95°C) and should not be subjected to temperatures in excess of 185°F (85°C) for more than two hours.
- On cars equipped with anti-theft radios, make sure you know the correct radio activation code before disconnecting the battery or removing the radio. If the wrong code is entered when power is restored, the radio may lock up and be rendered inoperable, even if the correct code is then entered.
- When boost-charging the battery, first remove the fuses for the Engine Control Module (ECM), the Transmission Control Module (TCM), the ABS Control Module, and the On-Board computer. In cases where one or more of these components is not separately fused, disconnect the Control Module connector(s).
- Do not quick-charge the battery (for boost starting) for longer than one minute, and do not exceed 16.5 volts at the battery with the boosting cables attached. Wait at least one minute before boosting the battery a second time.

◀ **WARNINGS and CAUTIONS** continued from previous page

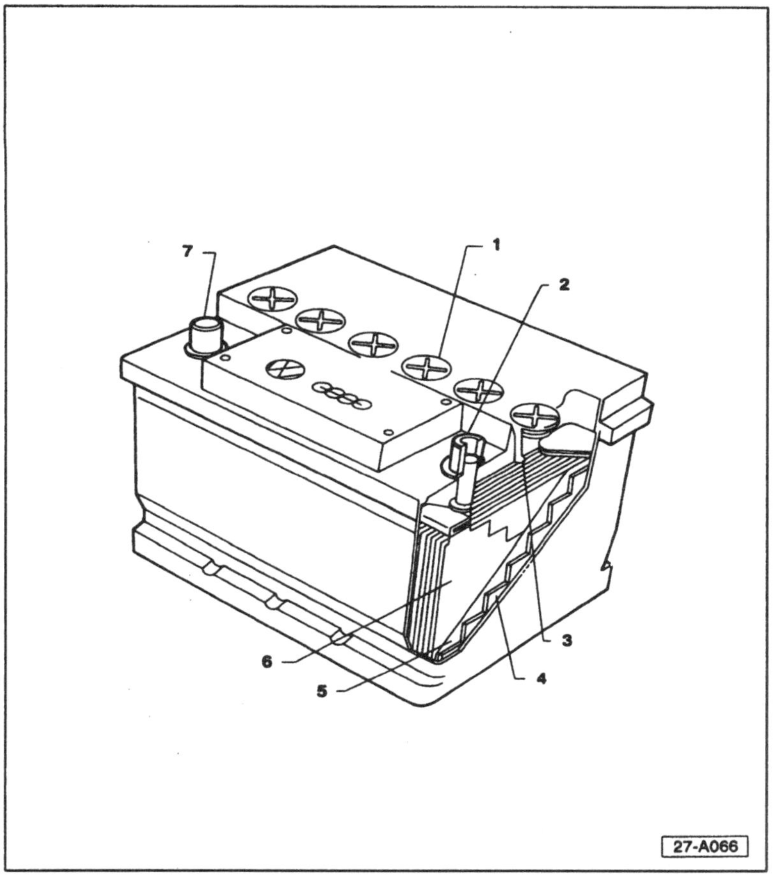

Battery, servicing

> *CAUTION!*
>
> ***Before working on the electrical system, obtain radio security code, turn off ignition, and disconnect battery Ground (GND) strap***

> *CAUTION!*
>
> - ***Disconnecting the battery may cause the basic setting of the Engine Control Module (ECM) to be erased, causing driveability problems.***
> - ***Whenever the battery has been disconnected or replaced during a repair, the Engine Control Module (ECM) "Basic Setting" must be reestablished using the VAG 1551 Scan Tool (ST),*** ***⇒ Repair Manual Fuel Injection & Ignition, Repair Group 24.***

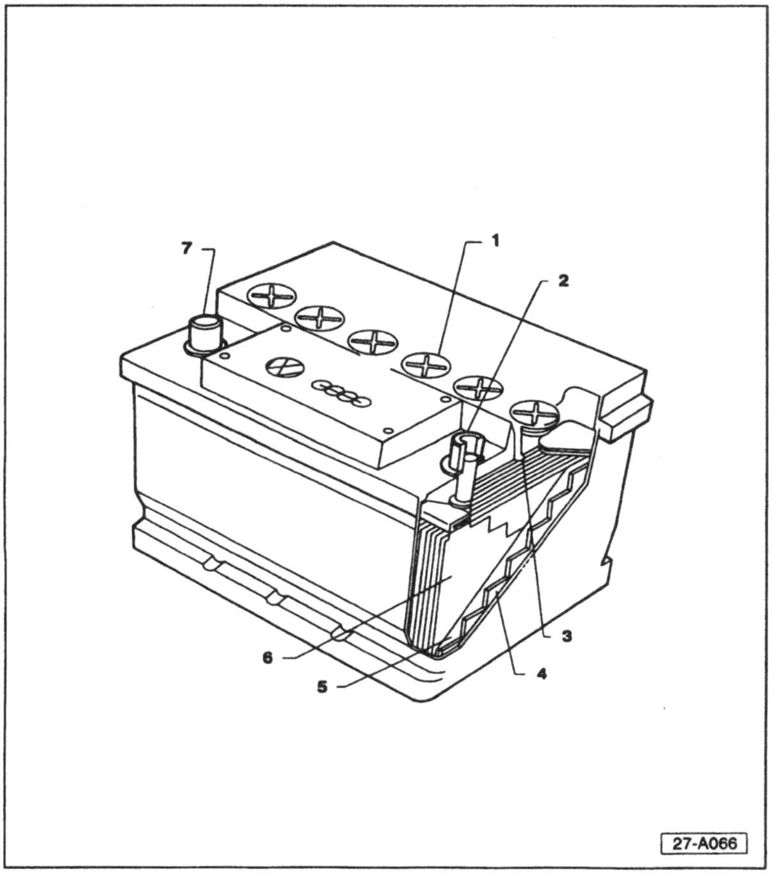

1 - Filler cap

2 - Positive (B+) terminal
- Larger diameter than negative (–) terminal.
- Clean off any oxidation.

3 - Electrolyte level mark
- If necessary, fill to mark with distilled water.

4 - Negative plate
- Color: metallic gray.

5 - Separator
- Insulator

6 - Positive plate
- Color: dark brown.

7 - Negative (–) terminal
- Smaller diameter than positive (B+) terminal.
- Clean off any oxidation.

Battery, checking

A weak battery can be caused by:

- Generator belt malfunctioning or slipping (replace belt if glazed).
- Ground straps corroded, loose or broken.
- Voltage Regulator (VR) malfunctioning.
- Generator Warning Light -K2- bulb failed.
- Poor Ground connection between Generator, Warning Light socket, and circuit board
- Battery and cable terminals should be cleaned and coated with acid resistant grease to prevent corrosion.

Visual checks

– **Check for:**

  - Cracked or leaking case.
  - Loose or corroded terminals.
  - Low electrolyte level (below MIN mark). If electrolyte is low, refill with distilled water and charge battery for several minutes after cells begin gassing (boiling). Charging battery, ⇒ Page V 27-10.

> ***CAUTION!***
>
> ***DO NOT overfill. Overfilled batteries can boil over. Too little electrolyte reduces the service life of the battery.***

Load testing

Requirements:

- Battery at least 75% charged.
- Battery Ground strap disconnected.
- Battery load tester connected per manufacturer's instructions (Christie Automotive BT-900 or equivalent).

Note:

The load current and the minimum voltage varies, depending upon the capacity of the battery, and should be taken from the sticker on the heavy discharge tester, or taken from the following chart:

Load current and minimum voltage

Capacity	Cold Cranking Amps	Load Current (DIN)*	Mininum Voltage
60 Amp-hr	280A	200A	9.2V
92 Amp-hr	450A	300A	9.5V
110 Amp-hr	420A	300A	9.5V

*Alternate method: 1/2 of Cold Cranking Amps for 15 seconds.

Notes:

- *If the minimum voltage is too low after a load period of 5-10 seconds, the battery is discharged or faulty.*
  Check specific gravity, *⇒ Page V 27-6.*
- *If the battery is colder than room temperature, voltage under load will be lower than the specified value.*

If battery must be tested cold:

– Use chart below as a guide:

Approximate electrolyte temp. °F (°C)	Minimum acceptable voltage under load
60 (16)	9.5 Volts
50 (10)	9.4 Volts
40 (4)	9.3 Volts
30 (–1)	9.1 Volts
20 (–7)	8.9 Volts
10 (–12)	8.7 Volts
0 (–18)	8.5 Volts

Specific gravity, checking

Notes:

- *The specific gravity test, in addition to a voltage test under load (described previously), can accurately determine the battery's state of charge.*
- *Hydrometers are used for testing specific gravity. Check the hydrometer calibration frequently to ensure accuracy.*

> ***CAUTION!***
>
> ***DO NOT expose a fully discharged battery to freezing temperatures; the battery will be damaged.***

– Test electrolyte in all cells.

♦ Average specific gravity should be at least 1.225

– Correct for electrolyte temperature if necessary:

♦ For every 10°F above 80°F, add 0.004 to hydrometer reading

♦ For every 10°F below 80°F, subtract 0.004 from hydrometer reading

Specific gravity	% of charge
1.265	100%
1.225	75%
1.190	50%
1.155	25%
1.120	0%

If specific gravity is above 1.225:

– Load test battery, ⇒ Page V 27-4.

If specific gravity is below 1.225:

– Charge battery, ⇒ Page V 27-10.

If specific gravity varies by more than 0.040 between cells:

– Replace battery.

In-stock batteries, checking

Note:

Batteries that have not been used for a lengthy period of time eventually discharge themselves, and may become sulphated. If these batteries are boost-charged using a conventional charger, they may not accept the charging current, or may only take on a "surface charge" and then prematurely test "fully charged."

Although a boost-charged battery appears to be malfunctioning, check it first, as follows:

– Check specific gravity in every cell, ⇒ Page V 27-6.

– Readings must not deviate by more than 0.040 from each other (example: When the highest specific gravity reading is 1.150, none of the other cell readings should fall below 1.110).

If cell readings are within 0.040 of each other:

– Perform complete battery charging procedure, ⇒ Page V 27-10.

– Perform load test to determine if battery is actually malfunctioning, ⇒ Page V 27-4.

If reading in one or two adjacent cells is noticeably lower (for example, five cells are 1.160, one cell is 1.080):

♦ Battery has a short circuit and is malfunctioning.

Voltage test, engine not running (for sealed batteries only)

– Turn headlights on high beam for one minute to remove surface charge.

– Disconnect battery Ground (GND) strap.

– Check Battery Positive Voltage (B+), using SUN VAT 40 or equivalent.

- 12.5 Volts minimum.

Voltage	% charge
12.6 or more	100%
12.5	75%
12.2	50%
12.0	25%
11.7 or less	0%

If above the minimum voltage:

– Load test battery, ⇒ Page V 27-4.

If below the minimum voltage:

– Charge battery (12.5 Volts minimum), ⇒ page V 27-10.

Battery, charging

Requirement

- Temperature of battery must not be less than 10°C (50°F).

Tools required

- Battery charger: Christie Automotive Model 400 QDD or equivalent

Notes:

- *Follow battery charger manufacturer's instructions*
- *Before testing a battery that has been charged, load battery with 15 amps for one minute to remove surface charge.*

WARNING!

- ***Gases given off during charging are explosive. DO NOT smoke in rooms containing a charging battery, or have bare light bulbs or any other source of sparks or flame.***
- ***Battery charger MUST be turned off when connecting or disconnecting cables on battery.***
- ***"Boosting" a sulfated battery at a high charging rate can cause an explosion!***

Battery with removable caps

Remove cell caps while charging. If water is added, fast-charge battery for a few minutes after battery begins gassing (boiling).

Slow-charging rate should be approximately 10% of battery capacity (for example, with a 45 Amp-hour battery, charging current should be about 4.5 A).

Fast-charging rate should be approximately 80-90% of battery capacity.

> ***CAUTION!***
>
> - ***Batteries may be damaged by fast-charging. It should be used only as a last alternative, when slow charging is not possible.***
> - ***DO NOT allow battery charging voltage to exceed 17 Volts.***
> - ***If battery begins gassing (boiling) violently when charging, REDUCE charging rate.***
> - ***DO NOT disconnect the battery while the engine is running.***

– Charge battery according to following table:

Specific gravity	Fast charge up to
1.150 or less	1 hour
1.150 to 1.175	3/4 hour
1.175 to 1.200	1/2 hour
1.200 to 1.225	1/4 hour
above 1.225	slow charge ONLY to 1.250-1.280

Sealed batteries

Only slow-charge sealed batteries.

Sealed batteries, often called "maintenance-free" batteries, will not accept a high rate of charge. It may be necessary to charge them for up to twice as long as batteries with removable caps.

Voltage will NOT increase as rapidly as when charging batteries with removable caps.

Battery, jump starting

> ***CAUTION!***
>
> ***Car with good battery must NOT be running when connecting jumper cables.***

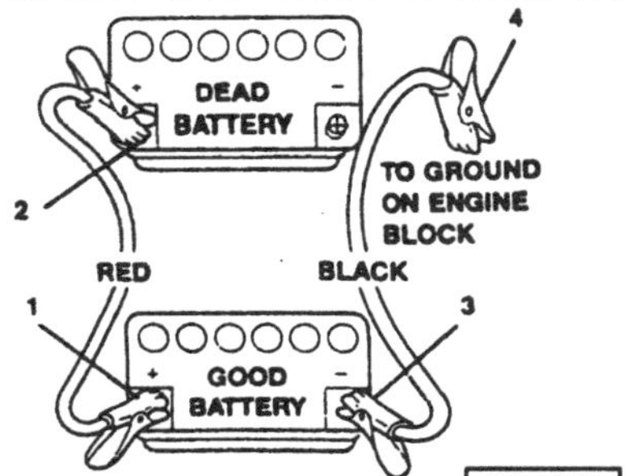

27-A027

- Connect jumper cables in following order:

1 –One end of positive cable -red- to (+) post -1- of good battery

2 – Other end of positive cable -red- to (+) post -2- of dead battery

3 – One end of negative cable -black- to (–) post -3- of good battery

4 – Other end of negative cable -black- to engine block -4- of car with dead battery

- Start car with good battery first.
- Next start car with dead battery.
- Disconnect jumper cables in reverse order.

Battery, removing/installing

> ***WARNING!***
>
> - ***Battery acid can cause severe burns. When working with electrolyte, always wear goggles, rubber gloves, and apron. If electrolyte is spilled on skin or clothing, flush at once with large quantities of water. If it gets into eyes, immediately flush with large quantities of water for several minutes and call a doctor.***
> - ***Batteries produce explosive gases. Keep flames and sparks away from batteries. DO NOT smoke near batteries.***

Notes:

- *Some early production vehicles are equipped with batteries 190 mm (7.5 in.) in height. Later production vehicles are equipped with batteries 175 mm (6.9 in.) in height.*
- *Batteries with a height of 190 mm (7.5 in.) are no longer available. If battery replacement is necessary, replace the 190 mm (7.5 in.) battery with the 175 mm (6.9 in) battery.*

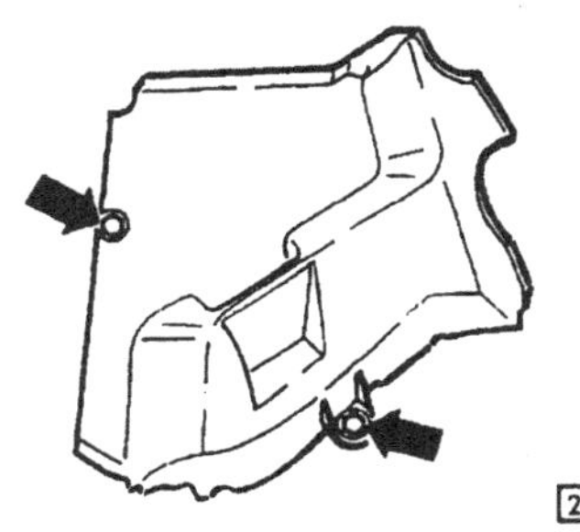
27-652

◄ – Remove bolts M6x15 10 Nm (88 in lb) -arrows- and remove battery cover.

Removing

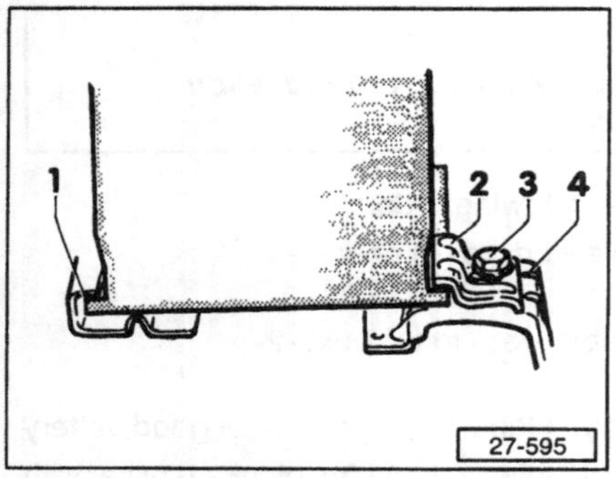

- Disconnect battery cable at negative (-) terminal.
- Disconnect battery cable at positive (+) terminal.
- ◄ Remove battery hold-down nut -3- and clamp -2-.
- Remove battery.

Installing

- Position battery.
  - ♦ Seat lower lip -1- in clamping strip, as shown.
- Retain other lip of battery at battery securing pedestal -4- using hold-down -2- and M8 x 25 hex bolt -3-.
- Tighten hex bolt -3- (tightening torque: 20 Nm [15 ft lb]).
- Connect battery positive (+) cable and negative (–) cable or Ground strap. (tightening torque : 5 Nm [44 in lb]).

> ***CAUTION!***
> ***If the battery is not secured properly, normal shock and vibration can damage the battery plates, which result in premature battery failure.***

Auxiliary battery under driver's seat, removing and installing

Removing

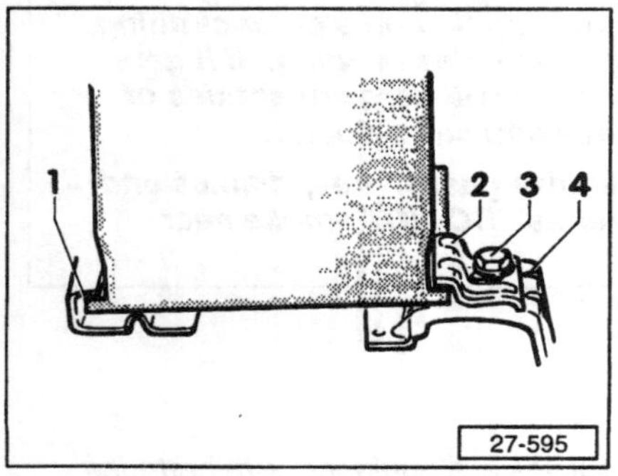

- Disconnect battery cable at negative (-) terminal.
- Disconnect battery cable at positive (+) terminal.
- ◄ Remove battery hold-down nut -3- and clamp -2-.
- Remove battery.

Installing

- Position battery.
  - ♦ Seat lower lip -1- in clamping strip, as shown.
- Retain other lip of battery at battery securing pedestal -4- using hold-down -2- and M8 x 25 hex bolt -3-.
- Tighten hex bolt -3- (tightening torque: 20 Nm [15 ft lb]).
- Connect battery positive (+) cable and negative (–) cable or Ground strap (tightening torque : 5 Nm [44 in lb]).

> ***CAUTION!***
> ***DO NOT exchange or replace a 175 mm (6.89 in.) height battery, fitted below the driver's seat, with a 190 mm (7.48 in.)height battery.***

Starter, removing and installing

Starter application chart:

- 2.5 ltr. Gas (12V – 1.8kW)
- 2.4 ltr. Diesel (12 V - 2.4 kW)
- 2.8 ltr. VR6 (12 V - 2.4 kW)

Removing:

– Disconnect Ground (GND) cable at battery.

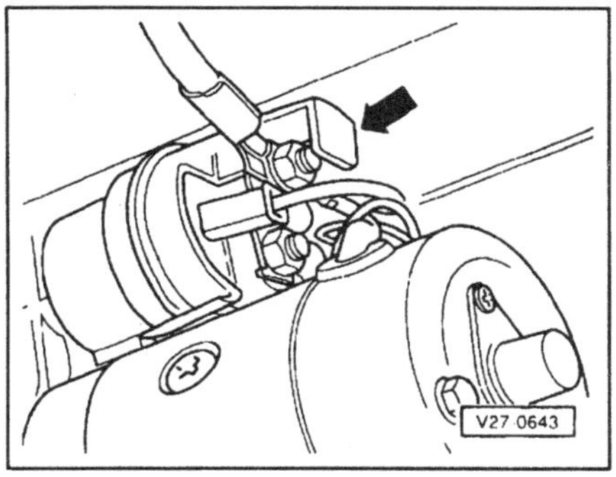

– Remove plastic cap -arrow- from solenoid switch.

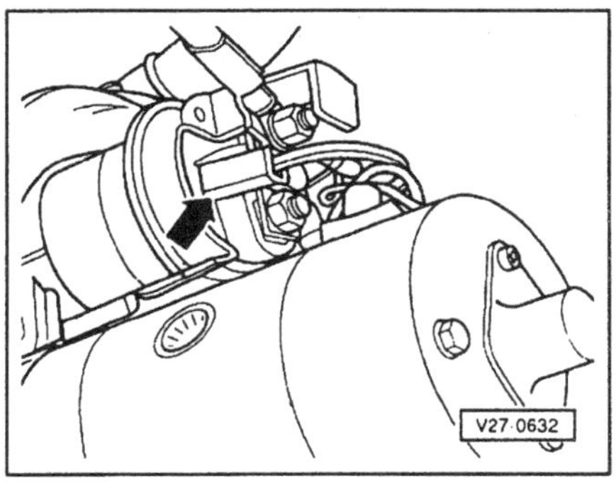

– Release push-on terminal -50- connector -arrow- and disconnect.

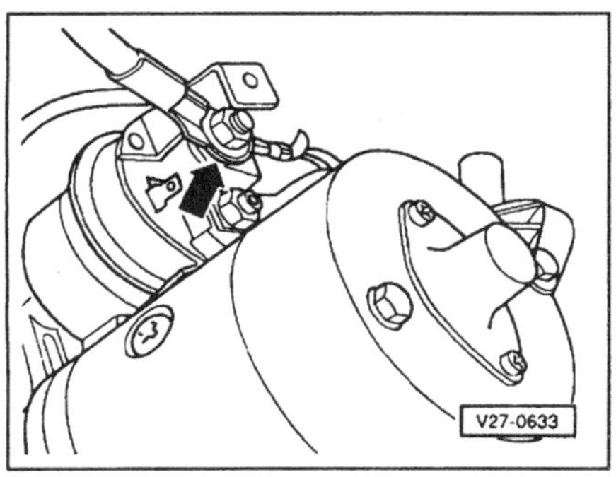

– Remove nut securing terminal -30- -arrow- 13 Nm (10 ft lb) on solenoid switch and remove wiring.

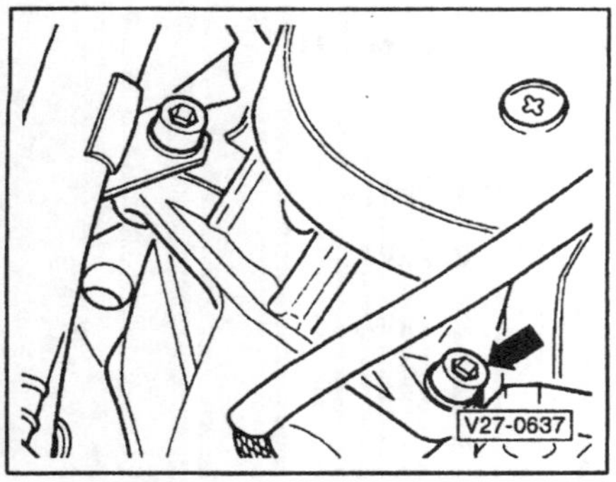

◄ – Remove securing bolt – 20 Nm (15 ft lb) -arrow-.

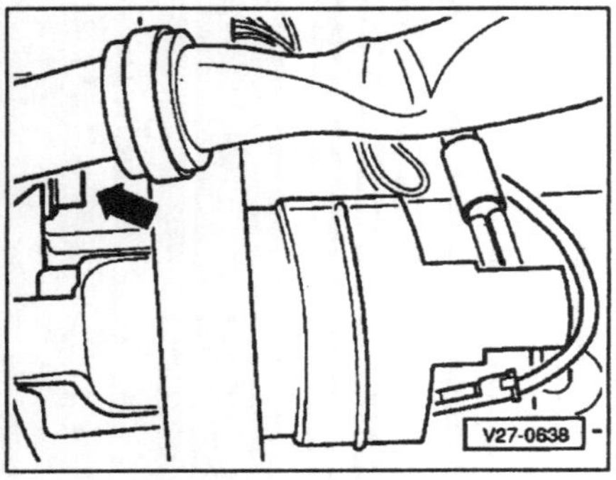

◄ – Remove securing bolt – 20 Nm (15 ft lb) -arrow-.

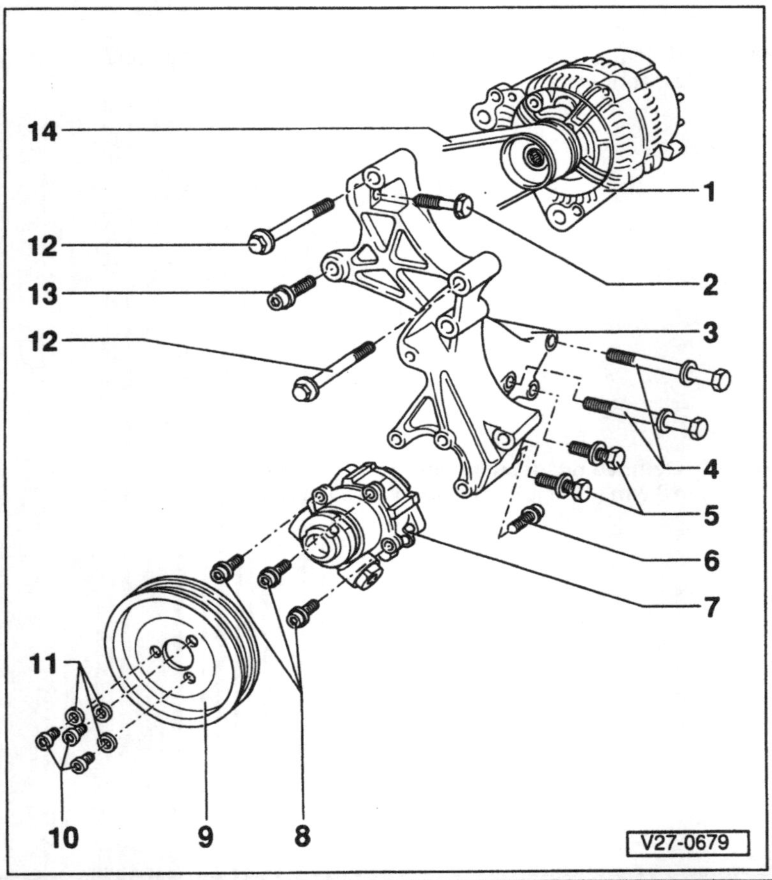

Generator (GEN), component layout

5-Cylinder gas engine

1 - Generator (GEN)
- assembly:

⇒ Page V 27-25.

2 - Hex bolt M8 x 45
- 40 Nm (29.5 ft lb)

3 - Mounting bracket

4 - Hex bolts M10 x 120
- 40 Nm (29.5 ft lb)
- With washer

5 - Hex bolts M10 x 45
- 40 Nm (29.5 ft lb)
- With washer

6 - Allen bolt M8 x 35
- 20 Nm (14.8 ft lb)

7 - Power steering pump

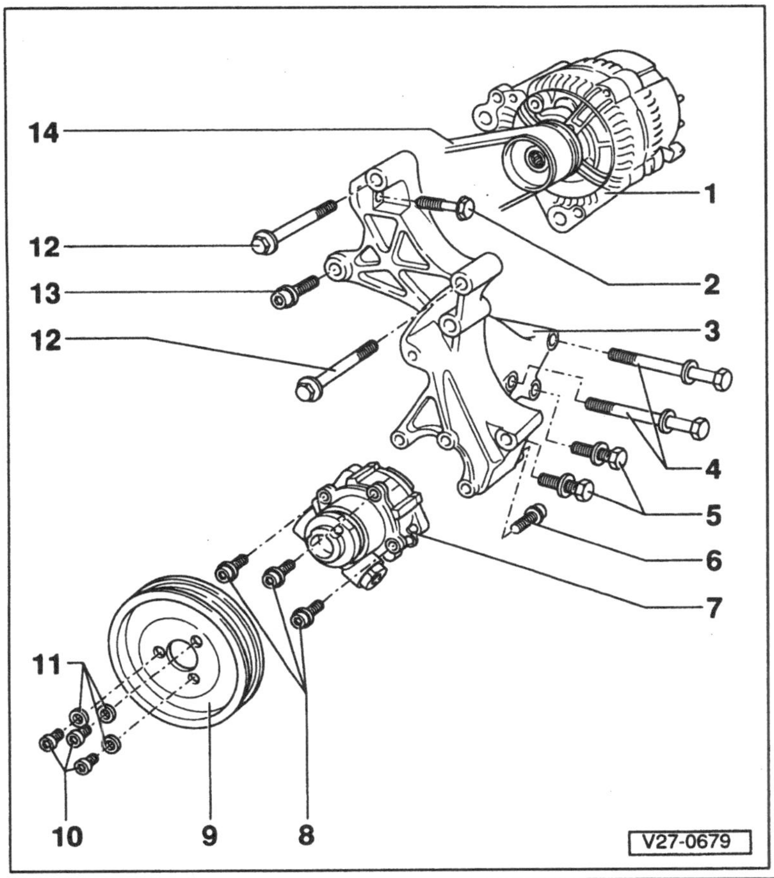

8 - Allen bolt M8 x 30
- 20 Nm (14.8 ft lb)

9 - Power steering pump pulley
- Install so that Part No. stamped on pulley faces out. Pulley will be misaligned if installed backward.

10 - Allen bolt M8 x 12
- 20 Nm (14.8 ft lb)

11 - Washers

12 - Bolt M8x85
- 25 Nm (18.4 ft lb)

13 - Bolt/washer assembly M10 x 38
- 40 Nm (29.5 ft lb)

14 - Poly-ribbed drive belt
- Checking/adjusting:

⇒ page V 27-24.
- Belt routing:

⇒ Figs. 1, 2 and 3.

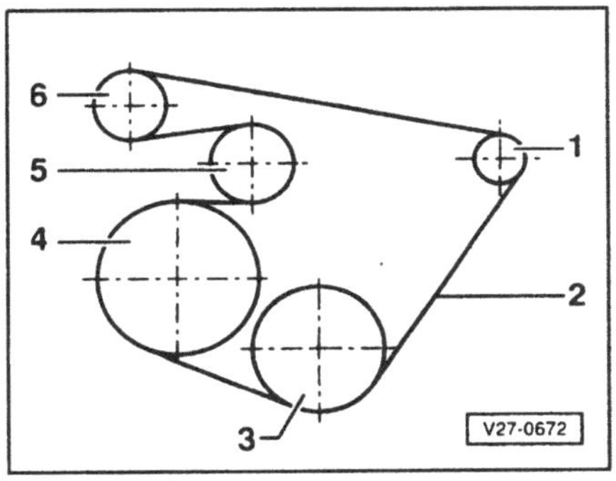

◄ **Fig. 1 Ribbed drive belt and pulley arrangement , Generator and power steering pump only**

1 - Generator pulley
2 - Ribbed drive belt
3 - Power steering pump pulley
4 - Crankshaft pulley
5 - Idler pulley
6 - Belt tensioner

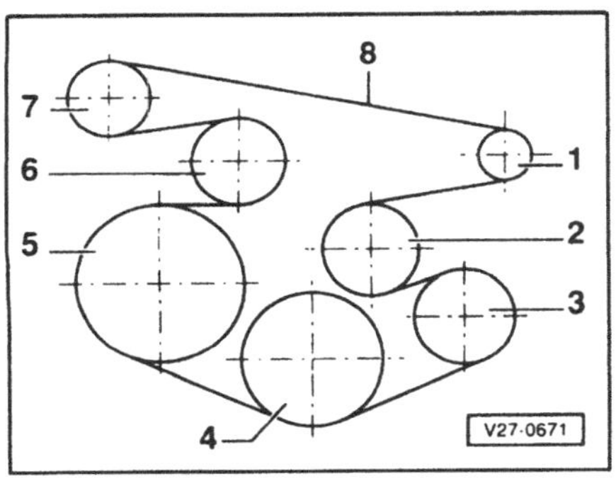

◄ **Fig. 2 Ribbed drive belt and pulley arrangement , Generator, power steering pump and A/C compressor**

1 - Generator,pulley
2 - Idler pulley
3 - A/C compressor pulley
4 - Power steering pump pulley
5 - Crankshaft pulley
6 - Idler pulley
7 - Belt tensioner
8 - Ribbed drive belt

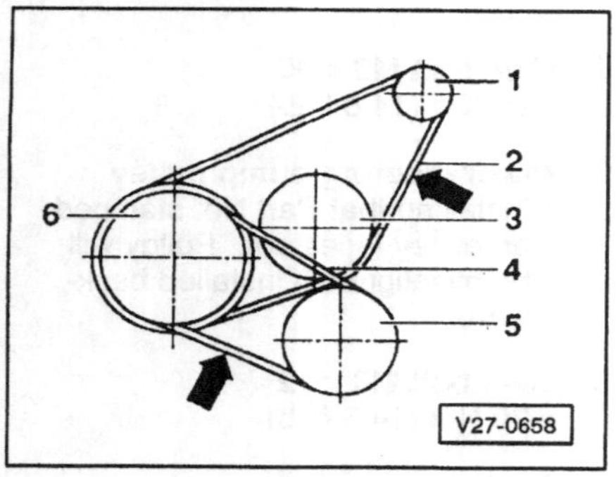

◄ **Fig. 3 Ribbed/V drive belts and pulley arrangement , Generator and Power Steering Pump (Diesel engine)**

1 - Generator pulley
2 - Ribbed drive belt
3 - Coolant pump pulley
4 - V-belt
5 - Power steering pump pulley
6 - Crankshaft pulley

Poly-ribbed drive belt, checking and adjusting

Checking

– Before removing drive belt, mark running direction for installation reference. Reversing belt running direction can lead to belt damage

– Inspect drive belt for any damage. Always replace a damaged drive belt to avoid breakage or malfunction.

Adjusting

Periodic adjustment of the ribbed drive belt is not required. Proper tension is maintained by a spring-loaded tensioner.

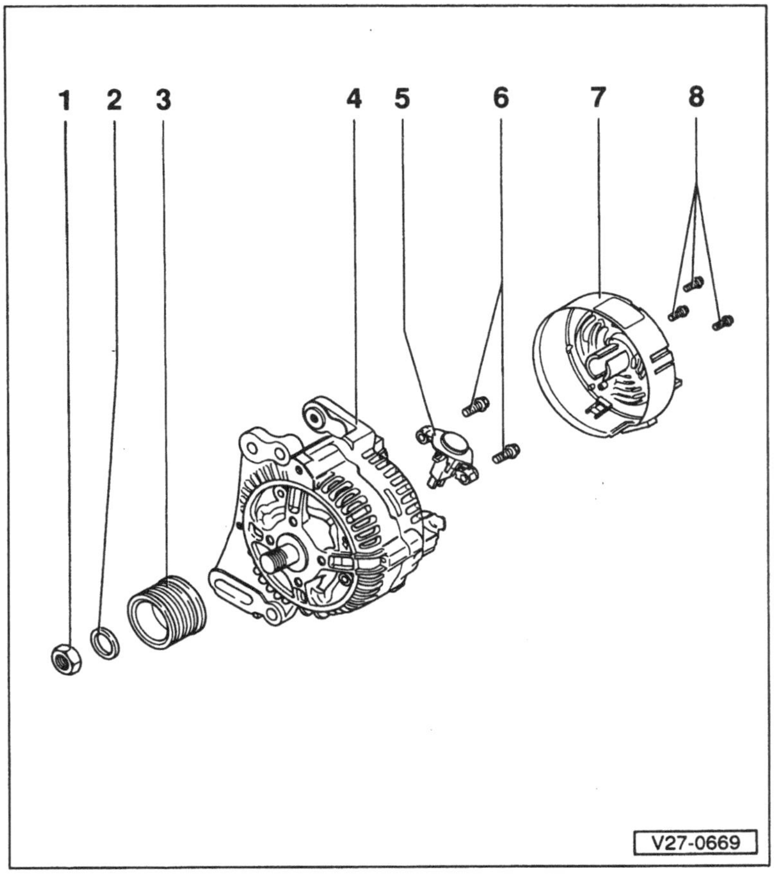

Generator (GEN), assembly

90A and 120A Generator

Note:

For checking Generator, see Electrical Wiring Diagrams.

1 - Hex nut M16 x 1.5
- 65 Nm (48 ft lb)

2 - Lock washer

3 - Pulley
- removing/installing:

⇒ Fig. 1.

4 - Generator (GEN)

5 - Voltage Regulator (VR)
- to remove, first remove bolts -8-, cover -7-, and screws -6-
- checking carbon brushes:

⇒ Fig. 2.

6 - Phillips head screws M4x25

7 - Cover

8 - Hex bolts M3x18

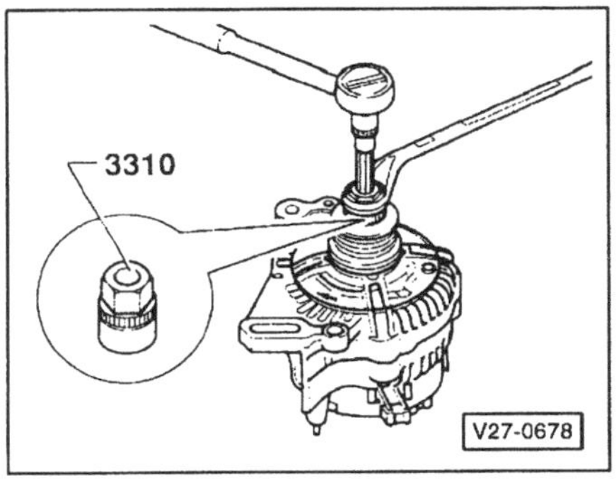

◄ **Fig. 1 Generator pulley, removing and installing**

– **Use special tool 3310 to remove/install Generator pulley hex nut**

- Hex nut: M16 x 1.5
- Torque: 65 Nm (48 ft lb)

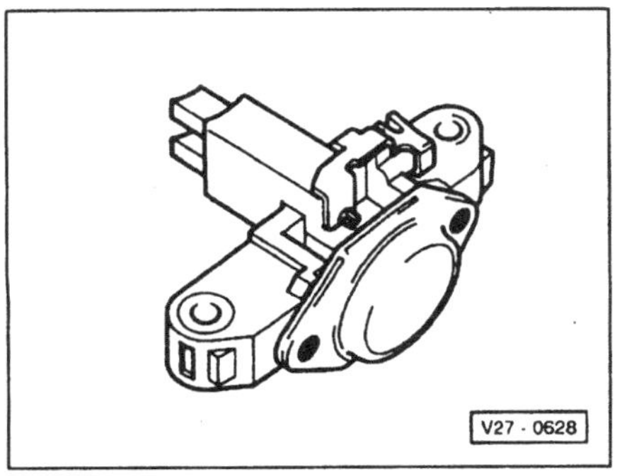

◄ **Fig. 2 Voltage Regulator (VR) brushes, checking**

– **Measure length of carbon brushes, must be:**

- New: 12 mm (15/32 in.)
- Service limit: not under 5 mm (3/16 in.)
- Tolerance: +1 mm (approx. 3/64 in.)

Instrument cluster ➤ 05.96,

Removing and installing

> ***CAUTION!***
>
> ***Before working on the electrical system, obtain radio security code, turn off ignition, and disconnect battery Ground (GND) strap.***

Notes:

- *Troubleshooting, ⇒ Electrical Wiring Diagrams.*
- *The multi-pin connection and speedometer cable can be disconnected without removing the housing.*
- *The instruments, warning lights with push-in holders, and the bulbs for illumination, along with the voltage regulator can also be removed without removing the housing.*
- *The following can only be removed after removing the housing, bulb for multi-function indicator/digital clock illumination, distance display or selector lever position display; contact plate with speedometer sender -G54- and multi-function indicator pressure sender.*

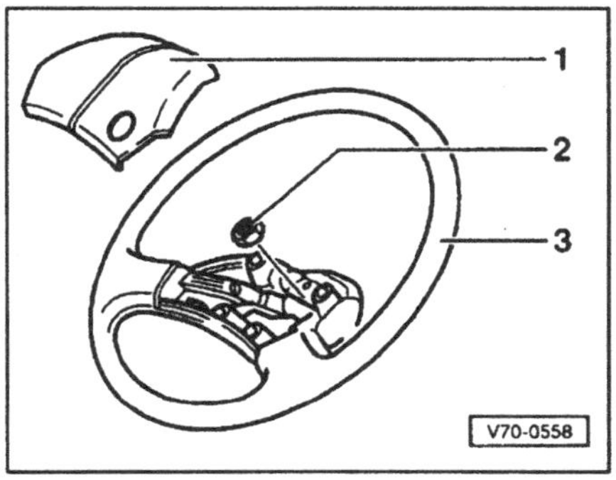

Steering wheel, removing

◄ – Unclip cover plate -1-.

– Remove hex nut -2- 70 Nm (52 ft lb).

– Mark steering wheel position and remove steering wheel -3-.

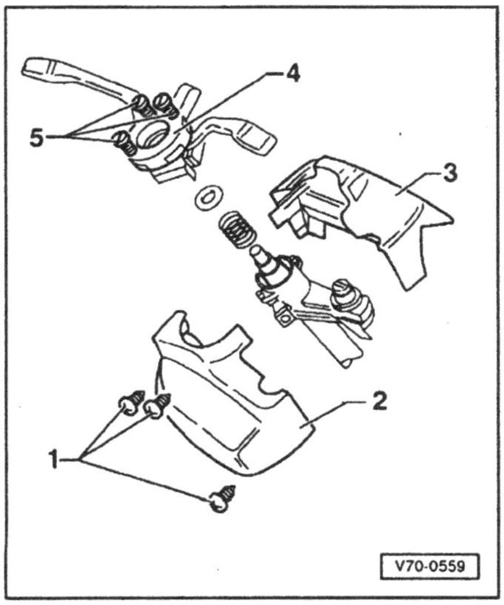

Steering column switch and trim, removing

◄ – Remove phillip screws -1- and take off trim -2-.

– Pull connectors off steering column switch -4-.

– Remove screws -5- and pull off steering column switch -4-.

– Take off trim -3-.

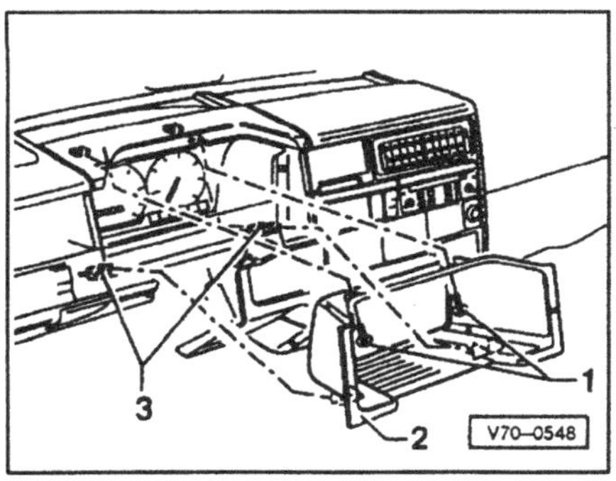

Instrument cluster trim, removing

◄ – Remove two self-tapping screws -1- and pull instrument cluster trim -2- out of the spring clips -3-.

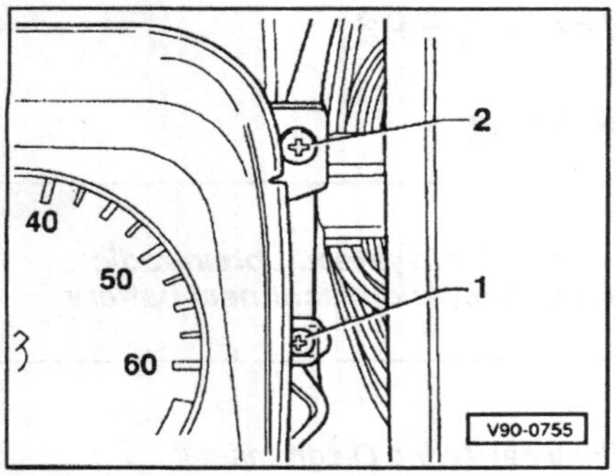

Instrument cluster complete, removing

- Remove securing screws left and right -2-.
- Tip cluster out at top and lift slightly, pull off multi-pin connector, hose for multi-function indicator pressure sender and speedometer drive cable from the rear of the instrument cluster.
- Remove instrument cluster from instrument panel.
- Remove transparent cover securing screws -1-.
- Remove trip recorder zeroing button and pull transparent cover off instrument panel insert.

Instrument cluster 05.96 ➤

Removing and installing

> ***CAUTION!***
>
> ***Before working on the electrical system, obtain radio security code, turn off ignition, and disconnect battery Ground (GND) strap.***

Notes:

- *From 05.96 production, do not disassemble or attempt repairs on components within the instrument cluster. Only replace as a complete unit.*
- *Troubleshooting, ⇒ Electrical Wiring Diagrams.*
- *The multi-pin connection and speedometer cable can be disconnected without removing the housing.*

Steering wheel, removing

> ***CAUTION!***
>
> ***Special safety precautions apply to vehicles equipped with airbags, ⇒ Repair Manual* Body Exterior/Interior, Repair Group 69 *for airbag removal and handling instructions.***

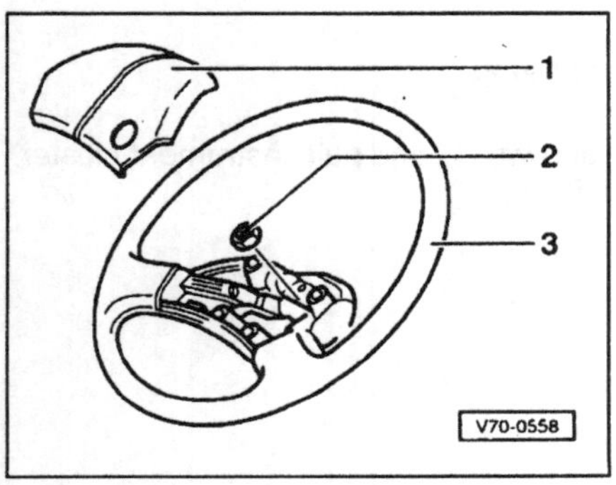

- Remove hex nut -2- 70 Nm (52 ft lb).
- Mark steering wheel position and remove steering wheel -3-.

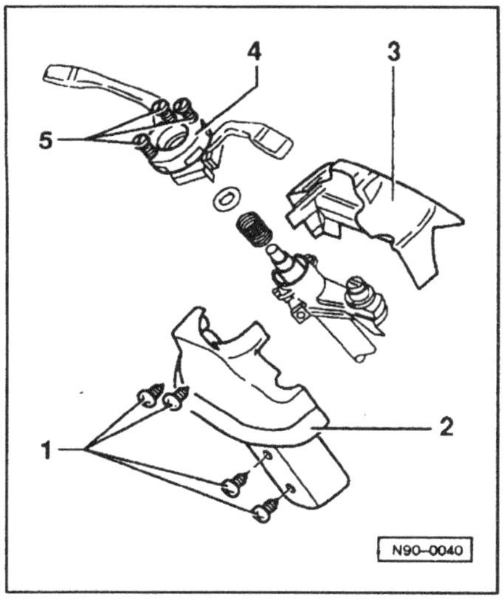

Steering column switch and trim, removing

- Remove phillip screws -1- and take off trim -2-.
- Pull connectors off steering column switch -4-.
- Remove screws -5- and pull off steering column switch -4-.
- Take off trim -3-.

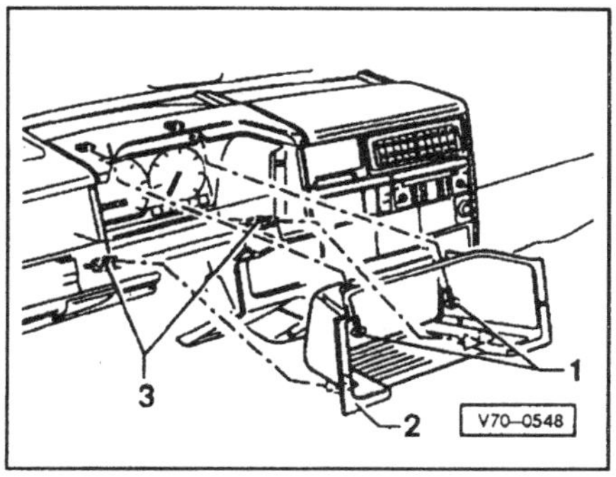

Instrument cluster trim, removing

- Remove two self-tapping screws -1- and pull instrument cluster trim -2- out of the spring clips -3-.

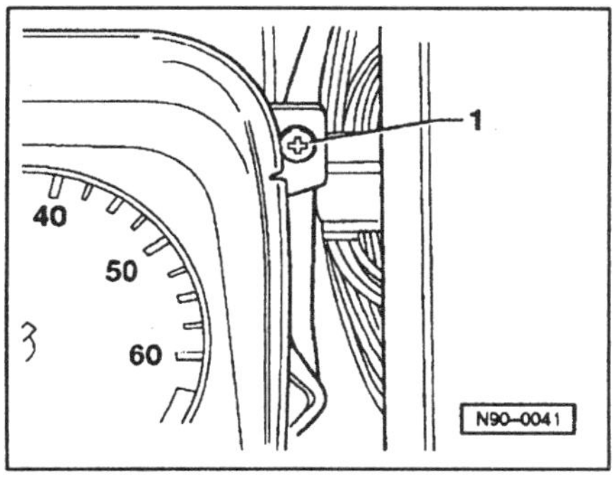

Instrument cluster complete, removing

- Remove securing screws left and right -1-.
- Tip cluster out at top and lift slightly, pull off both multi-pin connectors, hose for multi-function indicator, pressure sender, and speedometer drive cable from the rear of the instrument cluster.
- Remove instrument cluster from instrument panel

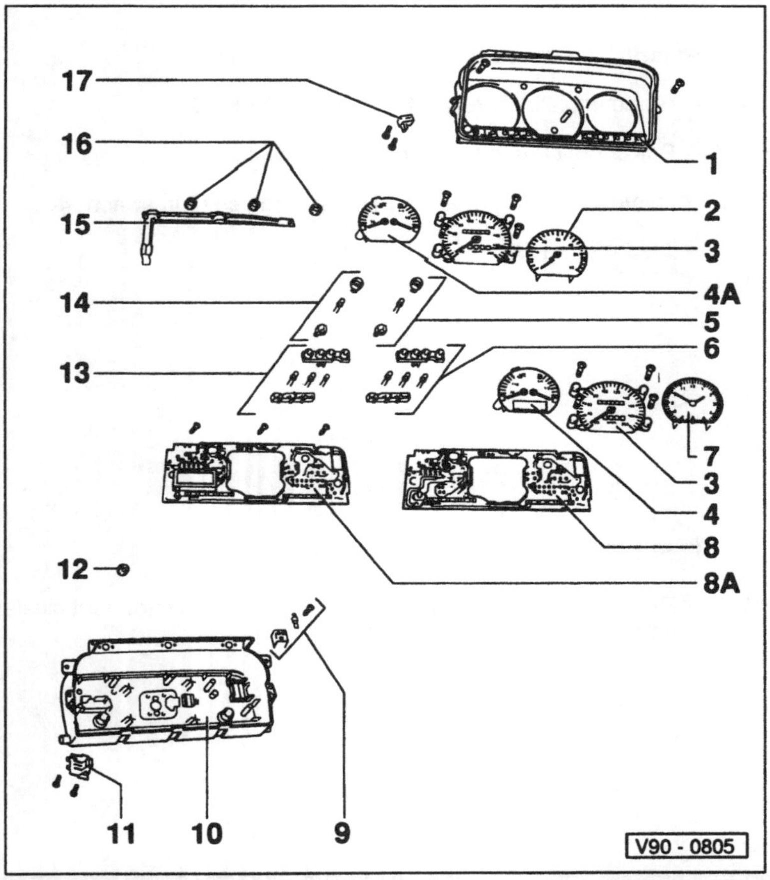

Instrument cluster ➤ 05.96 only, servicing

Notes:

- *Troubleshooting,*

⇒ Electrical Wiring Diagrams.

- *Checking voltages, LEDs and bulbs with VAG 1598,*

⇒ *Page V 90-23.*

1 - Trim cover

- Removing and installing instrument cluster,

⇒ Page V 90-1.

2 - Tachometer

- Note various versions.
- Removing and installing,

⇒ Fig. 1.

3 - Speedometer, mechanical

- Removing and installing instruments,

⇒ Fig. 1.

- Removing and installing drive cable,

⇒ Fig. 12.

- Installing Vehicle Speed Sensor (VSS) -G54-,

⇒ Page V 90-20.

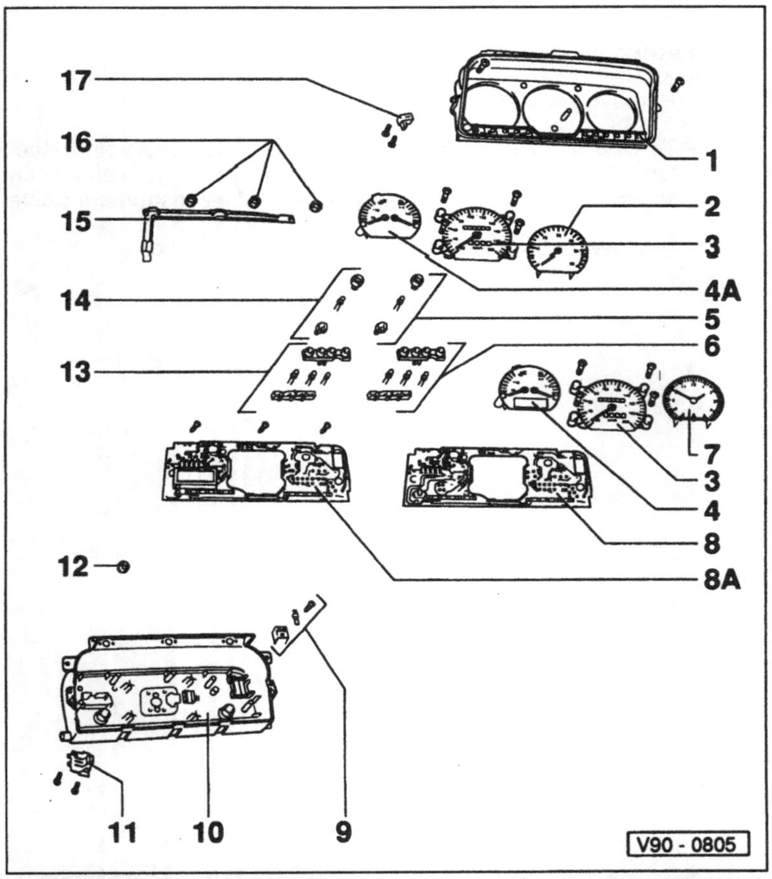

4 - Coolant temperature gauge and fuel gauge with cut-out for digital clock, or
4 A - Coolant temperature gauge and fuel gauge

- Removing and installing instruments,

⇒ Fig. 1.

- Checking,

⇒ Electrical Wiring Diagrams.

5 - Turn signal warning light, right

- Checking LED,

⇒ Page V 90-19.

- Installing,

⇒ Fig. 5.

6 - Warning lights, right

- Installation position,

⇒ Fig. 3.

- Checking LED,

⇒ Page V 90-19.

- Installing,

⇒ Fig. 4.

7 - Analog clock

- Removing and installing instruments,

⇒ Fig. 1.

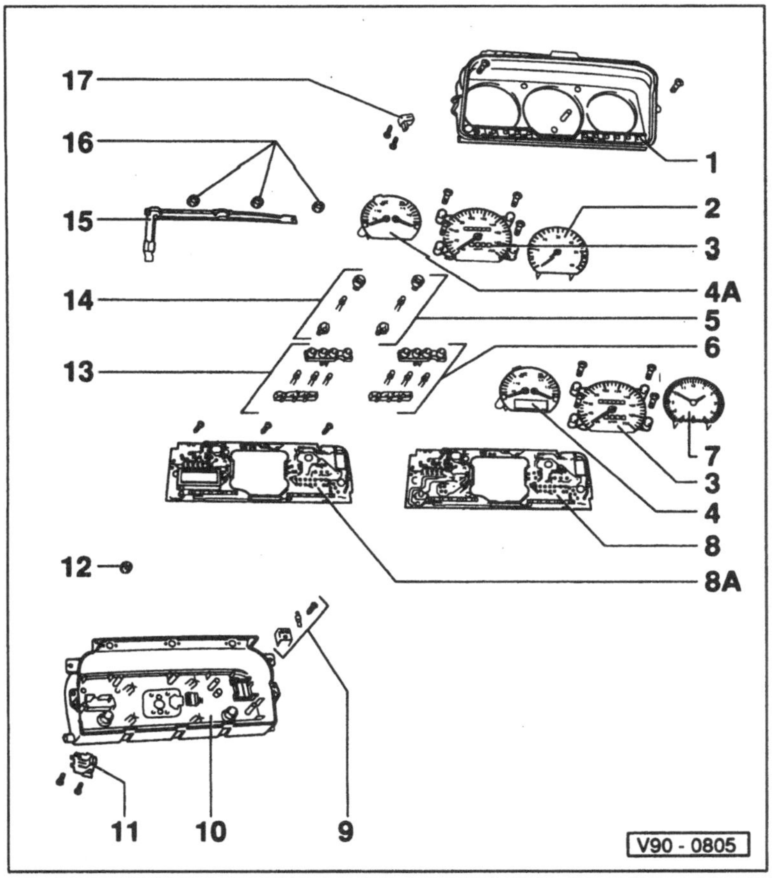

8 - Analog clock printed circuit, or
8 A - Digital clock printed circuit
- Only replace printed circuit together with housing.
- Components on printed circuit,

⇒ Figs. 8, 9.

9 - Voltage stabilizer with heat sink
- Checking,

⇒ Fig. 6.
- Removing and installing,

⇒ Fig. 7.

10 - Housing
- Only replace housing together with printed circuit.

11 - Speed Sensor (VSS) -G54- contact plate
- Installing speed sensor,

⇒ Page V 90-20.

12 - Bulb for digital clock
- Black holder
- 12 V / 1.2 W
- Installation position,

⇒ Fig. 10.
- Removing and installing,

⇒ Fig. 11.

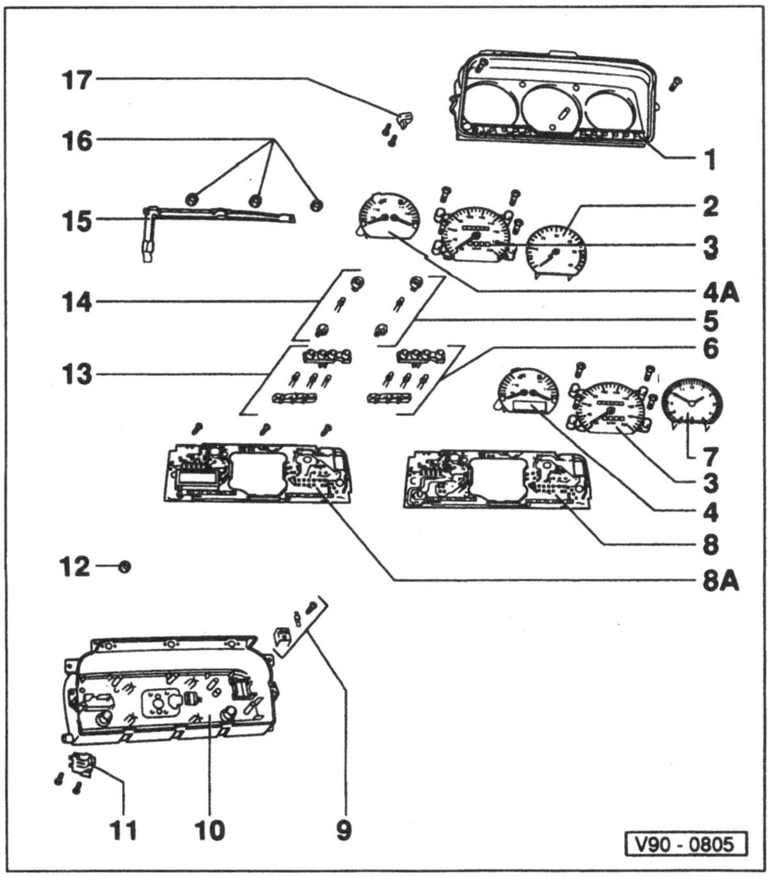

13 - Warning lights, left
- Installation position,

⇒ Fig. 2.
- Checking LED,

⇒ Page V 90-19.
- Installing,

⇒ Fig. 4.

14 - Turn signal warning light, left
- Checking LED,

⇒ Page V 90-19.
- Installing,

⇒ Fig. 5.

15 - Instrument cluster lighting conductor
- Connection to printed circuit,

⇒ Fig. 9.

16 - Instrument cluster light bulb
- Blue holder
- 12 V / 1.2 W

17 - Speed Sensor (VSS) -G54-
- Only with mechanical speedometer.
- Installing,

⇒ Page V 90-20.

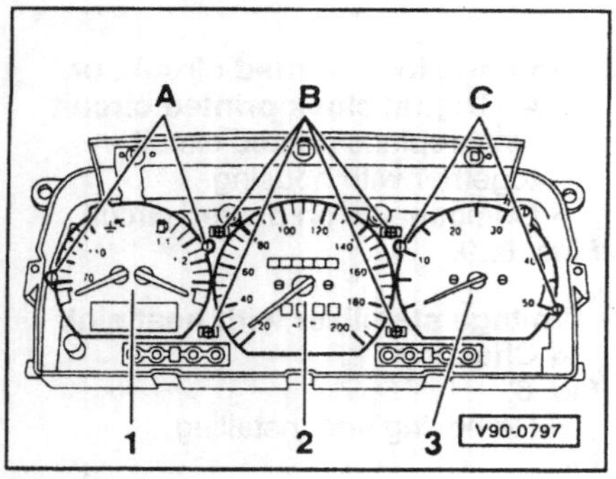

Fig. 1 Instruments, removing and installing

> ***CAUTION!***
> ***Before working on the electrical system, obtain radio security code, turn off ignition, and disconnect battery Ground (GND) strap.***

1 - Fuel gauge and coolant temperature gauge
- additionally with digital clock or multi function indicator (MFI)

2 - Mechanical speedometer or electronic speedometer -G21-

3 - Analog clock or tachometer
- additionally with selector lever position display

– Pull instruments -1 and 3- forward out of housing, and also electronic speedometer (instrument -2-), if equipped.

– For mechanical speedometers -2-, remove securing screws -B-, and pull instrument out forward.

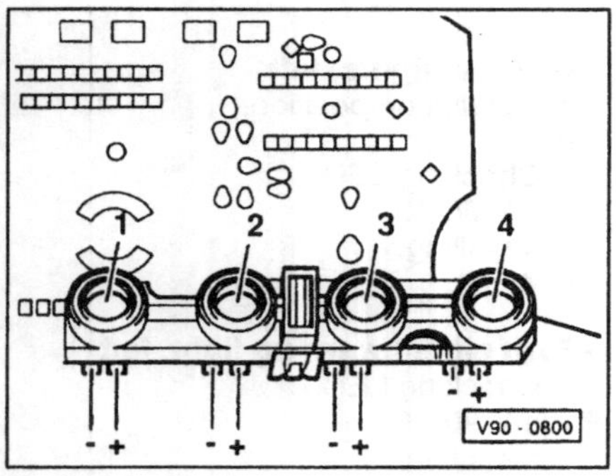

Fig. 2 Warning lamps installed position, left

1 - Warning lamp (red), coolant temperature gauge/coolant level indicator, LED

2 - Vacant

3 - Warning lamp (red) for oil pressure, LED

4 - Warning lamp (blue) for main beam, bulb

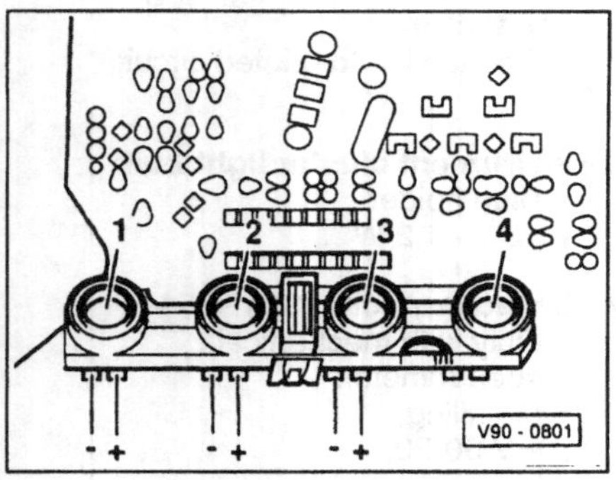

Fig. 3 Warning lamps installed position, right

1 - Warning lamp (red) for dual circuit and handbrake warning, LED
- Warning lamp (red) for seat belt warning system, LED

2 - Warning lamp (red) for Generator, LED

3 - Warning lamp (yellow) for glow plugs, LED
- Warning lamp (red) for catalyst monitoring, LED

4 - Vacant

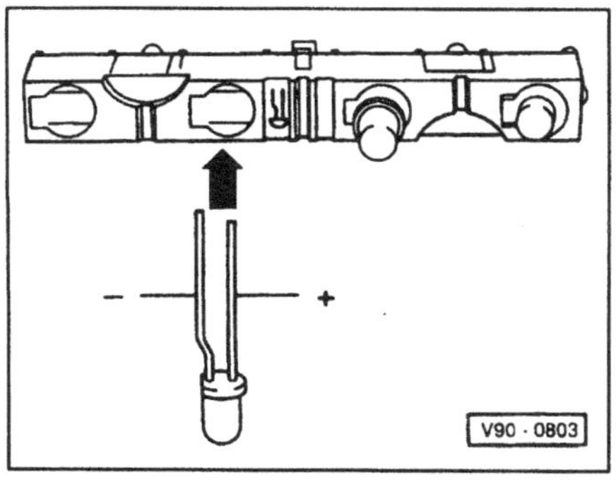

Fig. 4 Warning lamps, installing left/right

The warning lamps can be exchanged after pulling off the upper part of the bulb holder, Qty. 4.

Note:

When installing LEDs ensure the polarity as illustrated on the upper bulb holder part is observed.

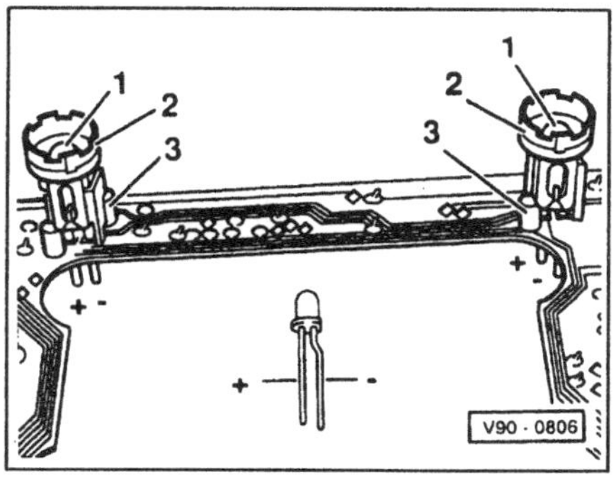

Fig. 5 Turn signal warning lamps, installing

The turn signal warning lamps are to be installed per illustration.

1 - LED

2 - Bulb holder upper part with locking device

3 - Bulb holder lower part

Fig. 6 Voltage stabilizer, checking

Note:

Only check when fuel and coolant temperature gauges both indicate false readings. Do not disconnect battery ground strap; instrument cluster multi-pin connector remains connected.

Voltage supply, checking

– Remove right-hand instrument.

– Switch on ignition.

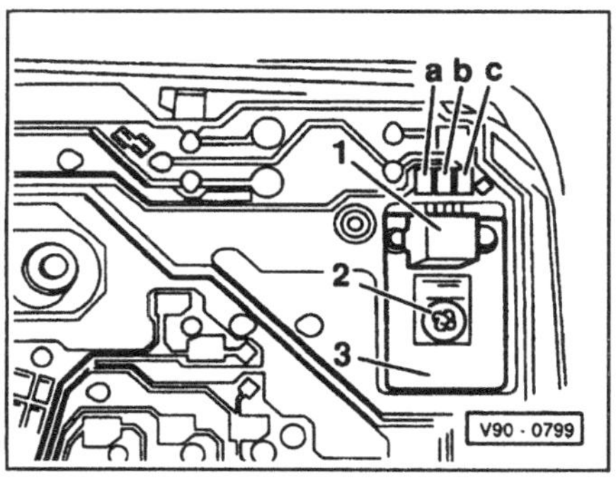

– Connect voltmeter between terminal 15 -a- and ground -b-.
  Specification: approximate battery voltage.

– If specification not achieved, check wiring according to wiring diagram.

Output voltage, checking

– Connect voltmeter between outlet -c- and ground -b-.
  Specification: 9.5–10.5 V.

– If specifications not achieved, voltage stabilizer defective; replace.

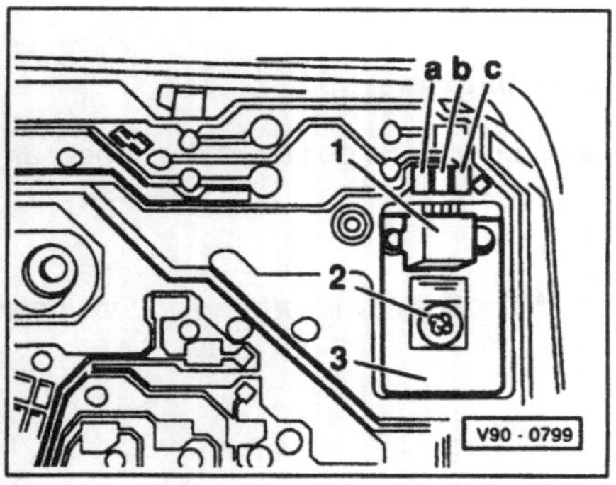

◄ **Fig. 7 Voltage stabilizer, removing and installing (only on mechanical speedometers)**

– Remove right instrument.

– Remove securing screw -2- and pull voltage stabilizer -1- downwards and out of contacts -a, b, c-.

Note:

When installing, ensure that the 3 connections are inserted in contacts -a, b, c-, and that the stabilizer is screwed tightly to heat sink -3- (heat dissipation away from housing).

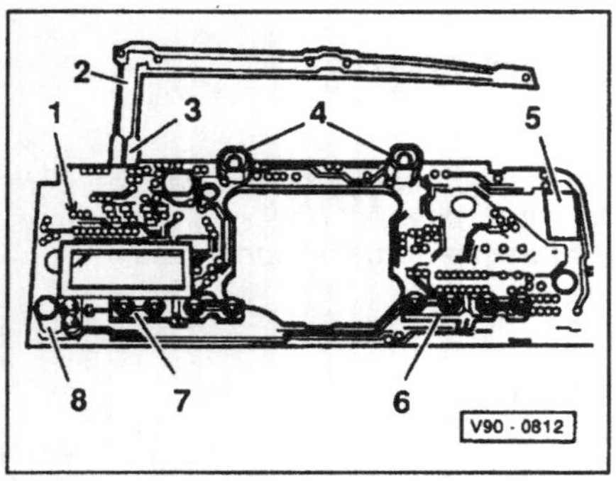

◄ **Fig. 8 Components on front of printed circuit, with digital clock**

1 - Printed circuit

2 - Instrument cluster lighting conductor

3 - Conductor foil connection

4 - Left/right turn signal warning light

5 - Cut-out for voltage stabilizer, which with a heat sink is screwed to instrument cluster housing

6 - Warning lights right

7 - Warning lights left

8 - Digital clock, cannot be replaced separately

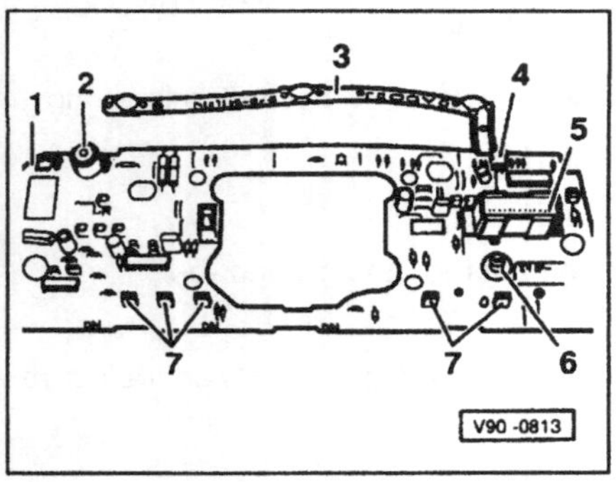

◄ **Fig. 9 Components on rear of printed circuit, with digital clock**

1 - Printed circuit

2 - Oil pressure warning buzzer, cannot be replaced separately

3 - Instrument cluster lighting conductor

4 - Conductor foil connection

5 - 28-pin connector to instrument cluster wiring loom

6 - Digital clock lighting

7 - Holders for warning lights

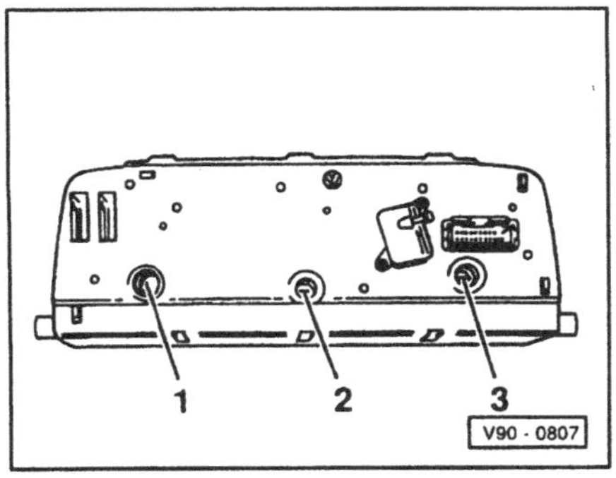

◄ **Fig. 10 Bulbs for odometer, multi-function indicator/digital clock and selector lever position display**

1 - Bulb for selector lever position display

2 - Bulb for odometer

3 - Bulb for MFI/digital clock

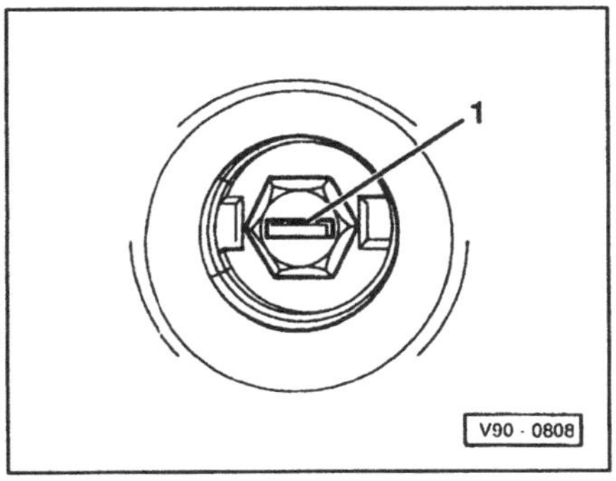

◄ **Fig. 11 Bulbs for odometer, multi-function indicator/digital clock, and selector lever position display, removing and installing**

– Turn holder until the slot -1- is vertical (unlocked) and remove bulb.

– Insert new bulb and turn holder until slot -1- is horizontal (locked).

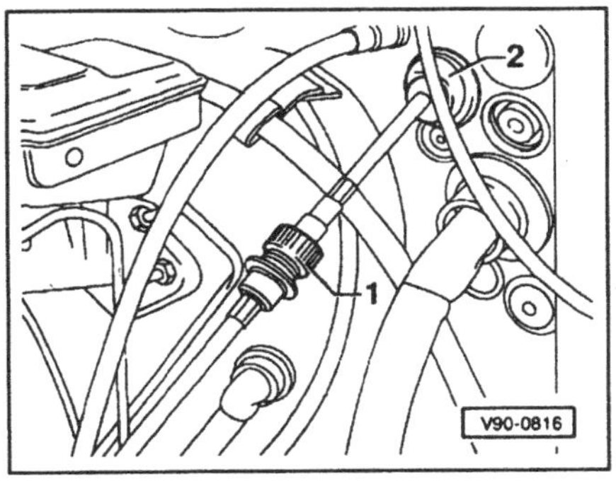

◄ **Fig. 12 Speedometer drive cable, removing from instrument cluster**

Note:

The drive cable is split in the engine compartment and engages the instrument cluster.

– Release drive cable plastic coupling on instrument cluster by pressing lugs together, then pull drive cable out of housing.

– Unscrew drive cable coupling -1- and pull drive cable with rubber grommet -2- out of the hole in the dash panel which leads to the passenger compartment.

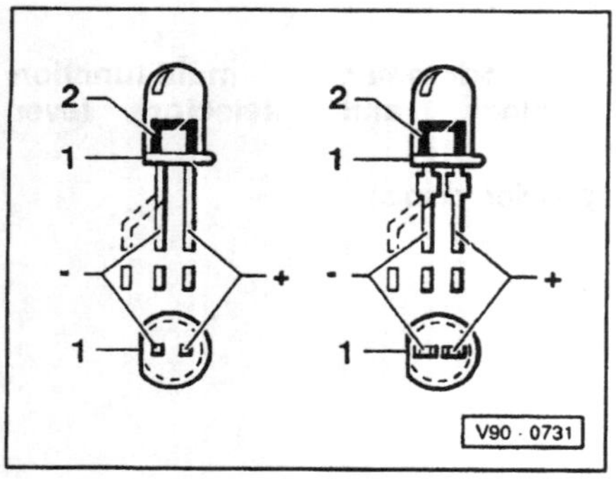

Light emitting diode (LED), checking

The negative pole is indicated by:

1 - Flat on diode housing

2 - Larger pole in diode housing

On some diodes the negative pole connector is additionally angled.

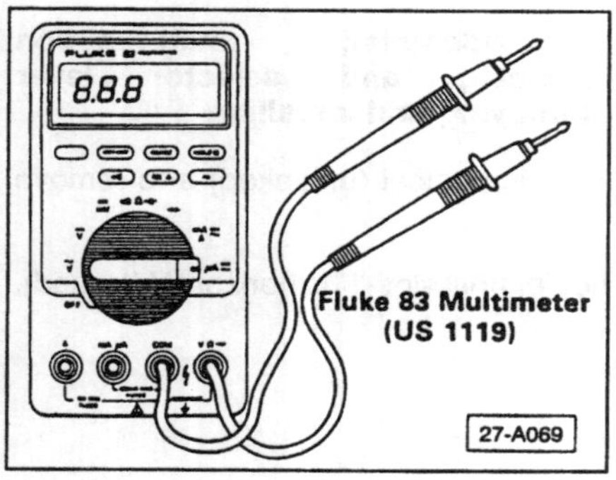

– Digital multimeter VAG 1315A (or Fluke 83), press the resistance and voltage buttons simultaneously -arrows-.

– Connect red terminal (+) to LED (+) .

– Connect black terminal (–) to LED (–).

LED must light.

Vehicle Speed Sensor (VSS) -G54-, installing

> ***CAUTION!***
>
> ***Before working on electrical system, obtain radio security code, turn off ignition, and disconnect battery Ground (GND) strap.***

Note:

Some systems in the vehicle require a speed signal (instrument cluster, 28 pin multi-pin connector, contact 7).

On vehicles with a mechanical speedometer the speed sensor is bolted to the rear side of the instrument and delivers 4 impulses per drive cable revolution. An inductive sensor (2-pin) or crankshaft position sensor (3-pin) is used for radios with Gala 1994 ➤.

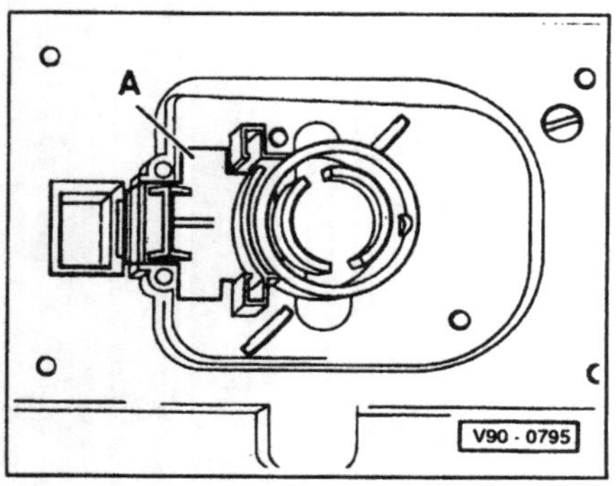

– On reverse side of instrument cluster, carefully break part -A- off along perforated line.

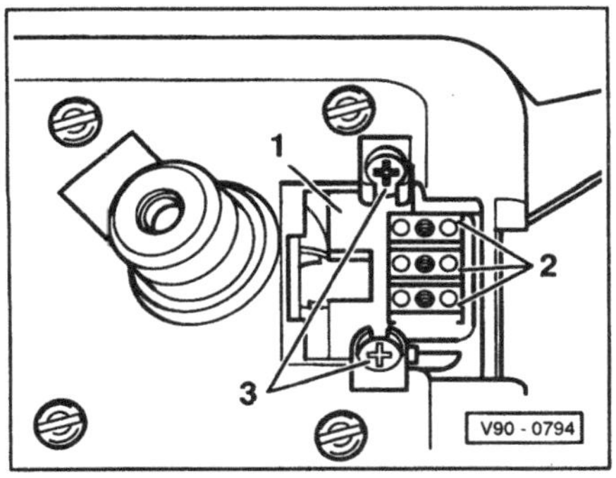

◄ – Insert the speed sensor -1- into the resulting recess and secure with phillips screws -3-.

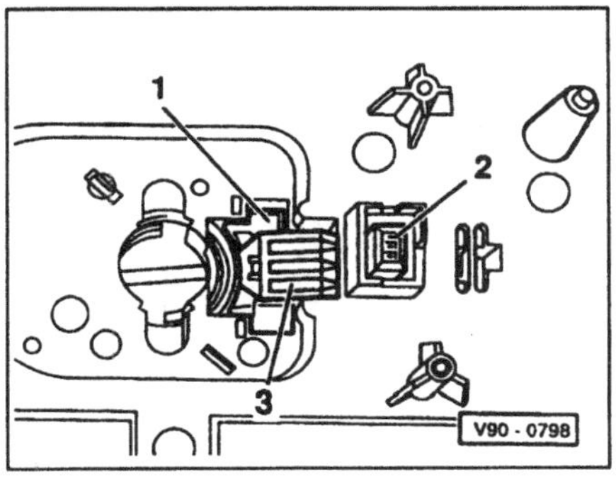

◄ – Ensure that the conductive strip -3- lies crease-free on the sensor spring contacts.

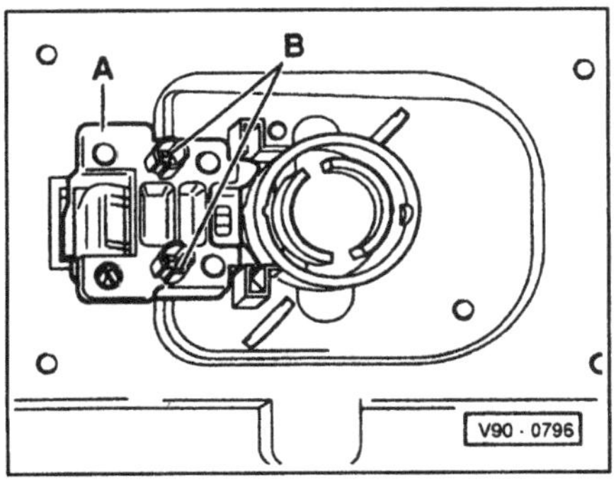

◄ – Secure contact plate -A- with screws -B-.

Instrument cluster LEDs and bulbs, electrical testing using test box VAG 1598

Notes:

- *Tools required:*
  - *Multimeter US 1119 (Fluke 83 or equivalent)*
  - *Test light VAG 1527 or equivalent*
  - *Test box VAG 1598 with adapter cable VAG 1598/8*
- *Test box VAG 1598 connector pin assignment is identical to the instrument cluster contact pin assignment.*
- *If specifications are achieved, check printed circuit foil with connections and components.*
- *If specifications are not achieved, check wiring according to wiring diagram,*

⇒ Electrical Wiring Diagrams.

> ***CAUTION!***
>
> ***If wiring-to-instrument cluster is damaged, replace the complete harness. DO NOT repair.***

Test prerequisites:

- Battery voltage OK.
- Relevant fuses OK.

⇒ Electrical Wiring Diagrams.

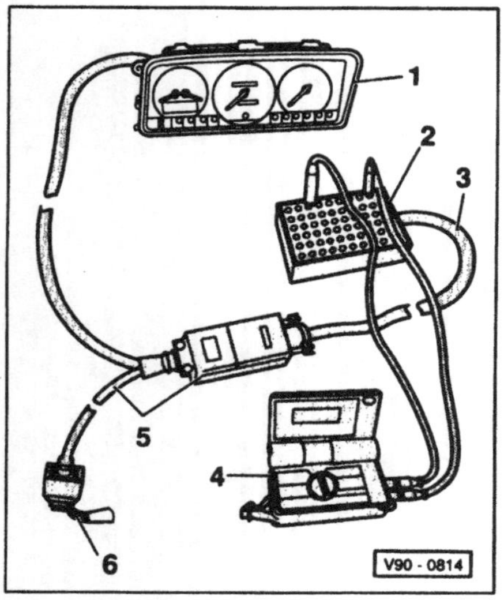

1 - Instrument cluster

2 - Test box VAG 1598 with connecting lead -3-

4 - Multimeter US 1119 (Fluke 83 or equivalent)

5 - Adapter cable VAG 1598/8

6 - instrument cluster wiring harness

- Fold down relay plate cover.
- Pull multi-pin connector off from rear of instrument cluster -1-.
- Connect test box VAG 1598 with adapter cable VAG 1598/8 to instrument cluster and detached wiring harness (plug must be felt to engage).

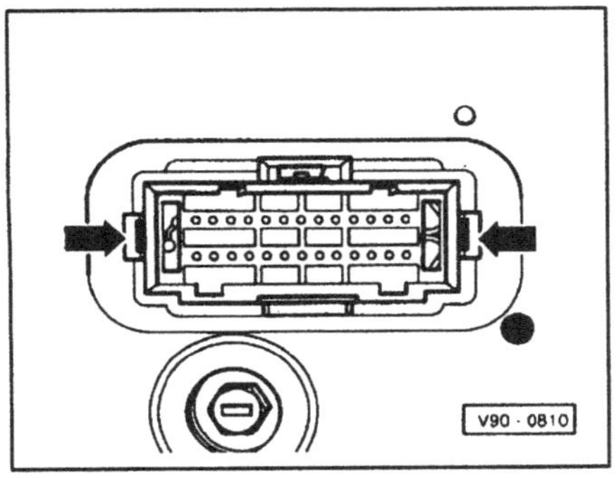

◄ **Instrument cluster ➤ 05.96, multi-pin connector terminal assignments**

1 - To coolant level indicator sensor
2 - Terminal 31, ground
3 - Terminal 31, ground
4 - Signal from speed sensor on instrument cluster -G54-
5 - From oil pressure switch 1.8 / 1.4 or 0.9 bar
6 - Oil pressure switch 0.3 bar
7 - Terminal 1 / Terminal W or RPM signal from Digifant Engine Control Module
8 - Terminal 30, battery positive
9 - Terminal 58b, illumination
10 - Terminal 15
11 - Vacant
12 - Generator warning light, terminal 61

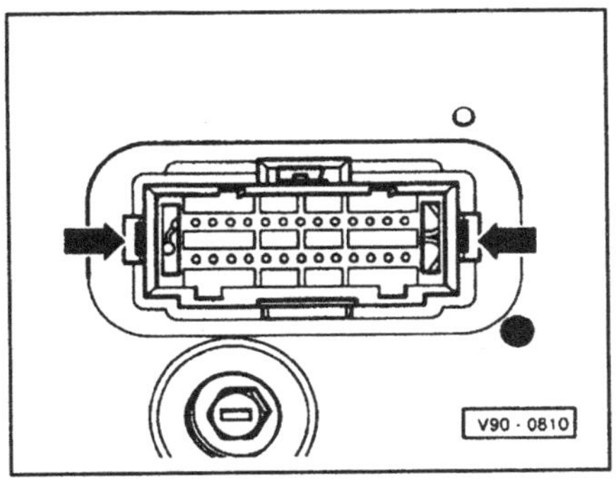

13 Brake, handbrake/seat belt warning light
14 -Glow period warning light
15 - Fuel gauge
16 - Turn signal warning light, left
17 - Coolant temperature gauge
18 - Turn signal warning light, right
19 - High beam warning light
20 - Automatic transmission selector range display

Instrument cluster ➤ 05.96, electrical testing using VAG 1598

– Measuring range: Switch to voltage measurement -V-.				
Test step	**VAG 1598 socket**	**Voltage supply to be tested**	**• Test conditions – additional work**	**Specifications**
1	11 + 3	Terminal 30/Terminal 31	• Ignition off	approx. battery voltage
2	13 + 3	Terminal 15/Terminal 31	– Switch ignition on	approx. battery voltage
3 a)	22 + 3	Turn signal warning light, left	• Ignition on – Connect test light. – Switch left turn signal on.	Test light flashes
4 a)	24 + 3	Turn signal warning light, right	• Ignition on – Connect test light. – Switch right turn signal on.	Test light flashes
5	12 + 3	instrument cluster and digital display lighting	• Ignition on – Switch lights on.	approximately 2.5 V up to battery voltage depending on regulator E 20 position
6	25 + 3	Main beam warning light	• Ignition on – Switch on main beam.	approximately battery voltage

a) Use LED voltage tester VAG 1527 or equivalent

– Measuring range: Switch to voltage measurement -V-.				
Test step	**VAG 1598 socket**	**Voltage supply will be tested**	**• Test conditions – additional work**	**Specifications**
7	18 + 3	Dual circuit and handbrake system warning light	• Ignition on – Pull handbrake on. – Release handbrake.	approximately battery voltage approximately 0 V approximately battery voltage
8	16 + 3	Generator warning light	• Ignition on – Start engine.	approximately 0 V approximately battery voltage
9 b)	13 + 20	Glow plug warning light	• Engine cold (under glow temperature) – Switch on ignition.	approximately battery voltage until glow temperature is reached

b) Not possible with engine at operating temperture.

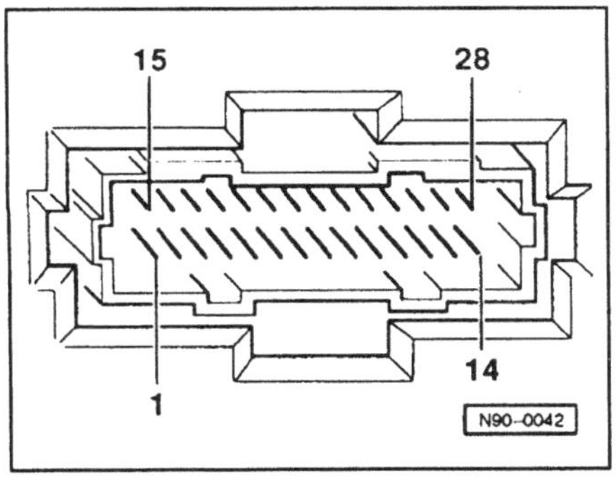

◄ **Instrument cluster 05.96➤, multi-pin connector terminal assignments**

1 - Outside air temperature sensor, ground*
2 - Coolant warning light
3 - Terminal 31, ground
4 - Multi-Function Indicator (reset)*
5 - Terminal 31b, ground
6 - Multi-Function Indicator (memory)*
7 - Road speed signal
8 - From oil pressure switch 1.8 bar
9 - Oil pressure switch 0.3 bar
10 - Terminal 1 / Terminal W
11 - Terminal 30, battery positive
12 - Terminal 58b, instrument illumination
13 - Terminal 15
14 - Fog light indicator (if equipped)

* only on vehicles with Multi-Function Indicator (MFI)

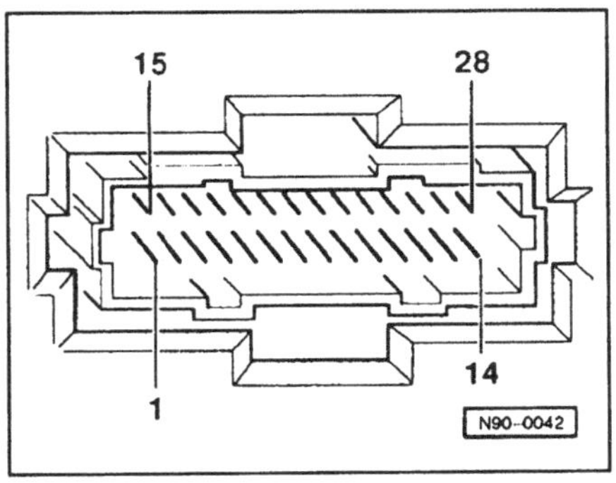

15 - Multi-Function Indicator selector switch (mode)*
16 - Generator warning light, terminal 61
17 - Oil temperature sender
18 - Brake, handbrake/seat belt warning light
19 - Outside air temperature sensor, signal*
20 - Malfunction Indicator Lamp (MIL)
21 - Fuel gauge
22 - Turn signal warning light, left
23 - Coolant temperature gauge
24 - Turn signal warning light, right
25 - High beam control light
26 - Fuel consumption signal*
27 - Speedometer Vehicle Speed Sensor (VSS) -G22-, signal
28 - Automatic transmission selector range display

* only on vehicles with Multi-Function Indicator (MFI)

Radio, removing and installing 1994 ➤

> ***CAUTION!***
>
> ***Before working on the electrical system, obtain radio security code, turn off ignition, and disconnect the battery Ground (GND) strap.***

Note:

Before removing the radio, obtain the code number from the customer. If the radio is replaced, always activate the anti-theft coding. Inform the customer of the new code number.

Removing:

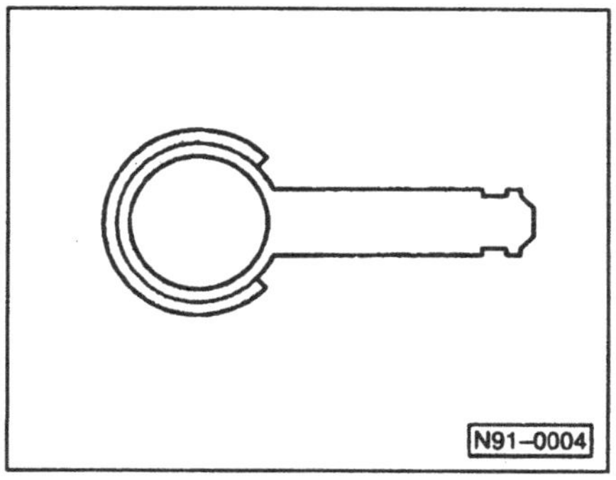

Use radio release tool 3316, to remove radio.

Radio release tool 3316 consists of two identical parts.

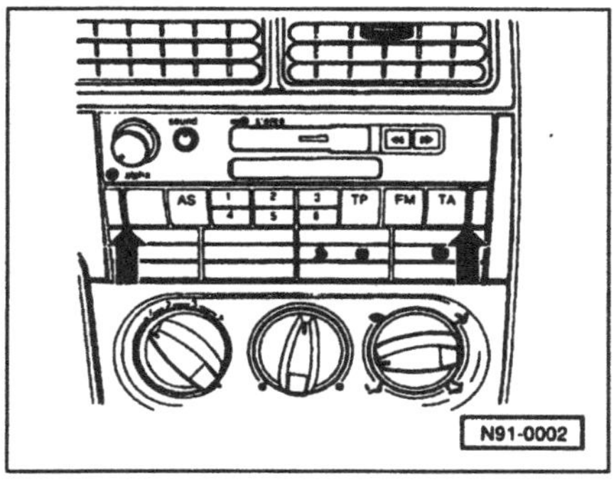

– Slide radio release tool 3316 into the release slots -arrows- until they locate.

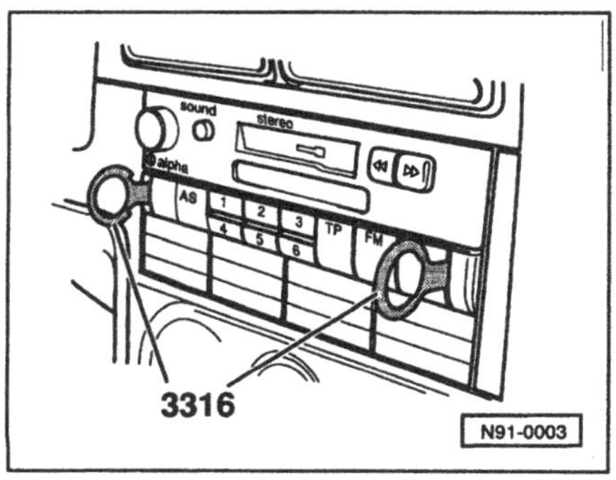

– Pull radio out of the instrument panel using the grip rings of the release tools.

Notes:

- *When using the radio release tool 3316, do not push to the side or tilt.*
- *To remove the release tools from the radio press inward on the locating lugs on the side of the radio.*

– Remove connectors from radio.

Installing:

– Connect plug to radio.

– Slide radio level into the instrument panel, until it locates in the assembly frame.

Fixed or windshield antenna, installing

Notes:

- *After turning the radio on, battery voltage is supplied to the antenna connector and therefore the antenna cable; this serves, together with the ground connection on the amplifier housing, as the voltage supply for the amplifier (remote supply via antenna cable).*
- *This voltage is not required for fixed antennas.*

Fixed Antennas

– Antenna cable. Only install the prescribed antenna cable with integrated condenser; the condenser blocks the voltage supply to the fixed antenna.

Windshield antenna

– Removing and installing flush-mounted windshield ⇒ Repair Manual, Body Exterior/Interior, Repair Group 64

Notes:

- *The wire on the radio 8 pin multi-pin connector, chamber 5 (voltage supply) is not required if the voltage supply is via the antenna cable.*
- *Only install the prescribed antenna cable to the antenna amplifier (without integrated condenser).*

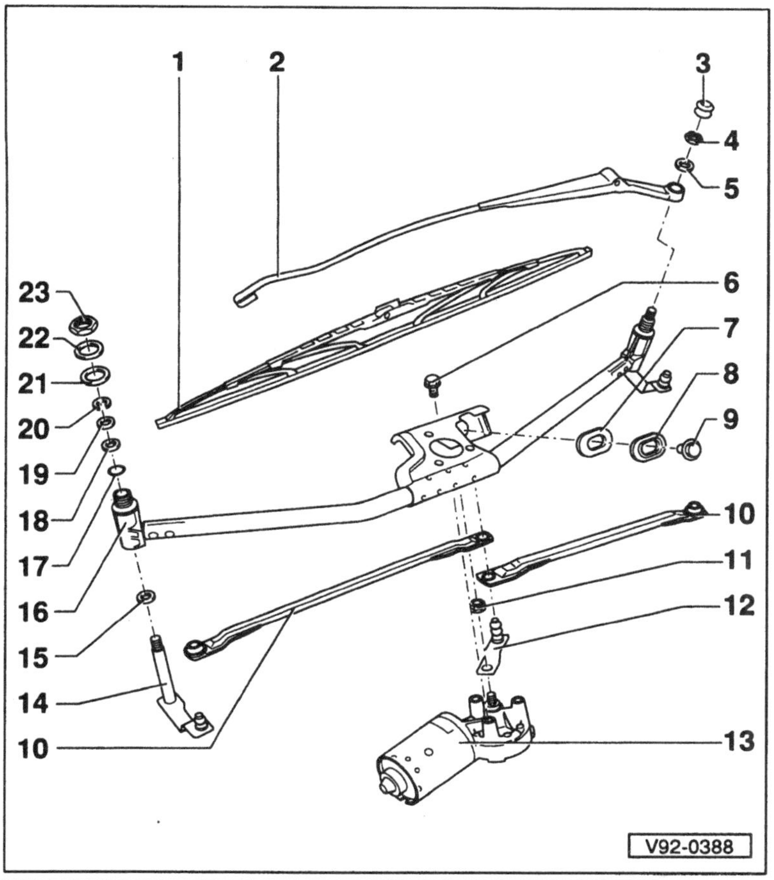

Windshield wiper system, servicing

> ***CAUTION!***
> ***Before working on the electrical system, obtain radio security code, turn off ignition, and disconnect battery Ground (GND) strap.***

1 - Wiper blade
- ♦ Adjusting park position ,

⇒ Fig. 2.

- ♦ Wiper blade inserts removing and installing,

⇒ Page V 92-5.

2 - Wiper arm

3 - Cap

4 - Hex nut M8
- ♦ 19 Nm (14 ft lb)

5 - Washer

6 - Hex bolt M6
- ♦ 5 Nm (4 ft lb)

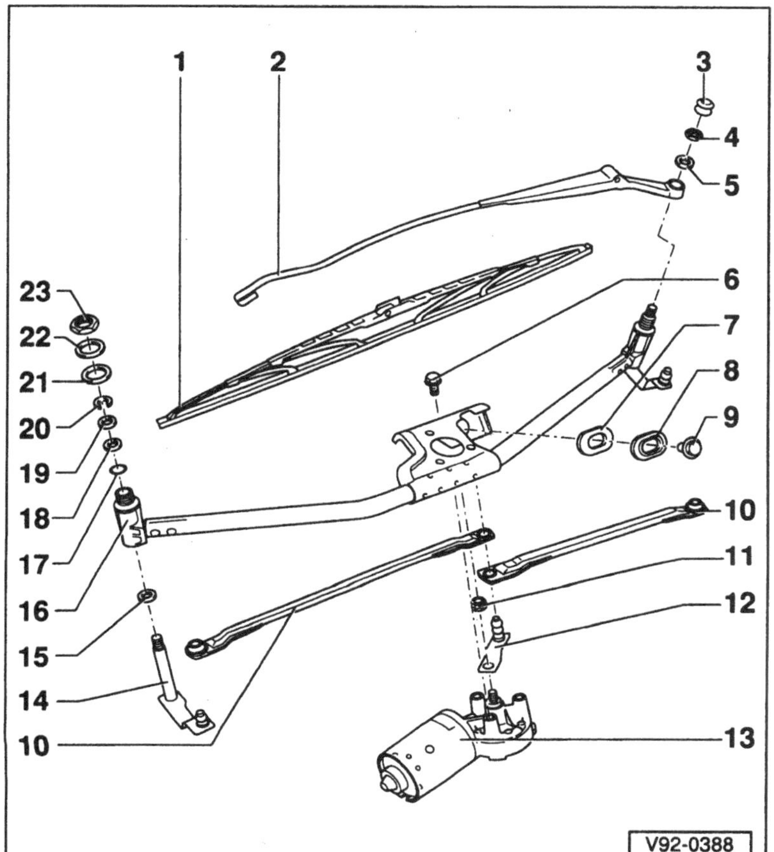

7 - Rubber grommet

8 - Spacer

9 - Hex nut M8
- ♦ 25 Nm (18 ft lb)

10 - Connecting rod

11 - Hex nut M8
- ♦ Self-locking

12 - Crank
- ♦ Setting park position,

⇒ Fig. 1.

13 - Wiper motor
- ♦ Remove wiper frame first.

14 - Connecting lever

15 - Spring washer

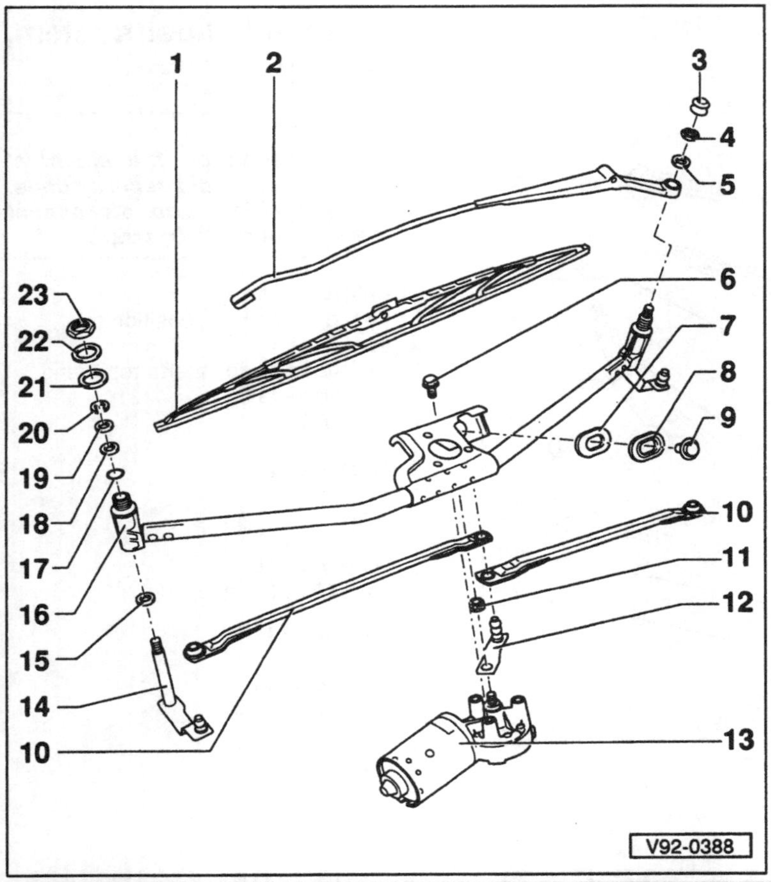

16 - **Wiper frame**

17 - **O ring**

18 - **Spacer washer**

19 - **Washer**

20 - **Snap ring**

21 - **Plastic washer**

22 - **Spacer ring**

23 - **Hex nut M16**
- ♦ 10 Nm (7 ft lb)

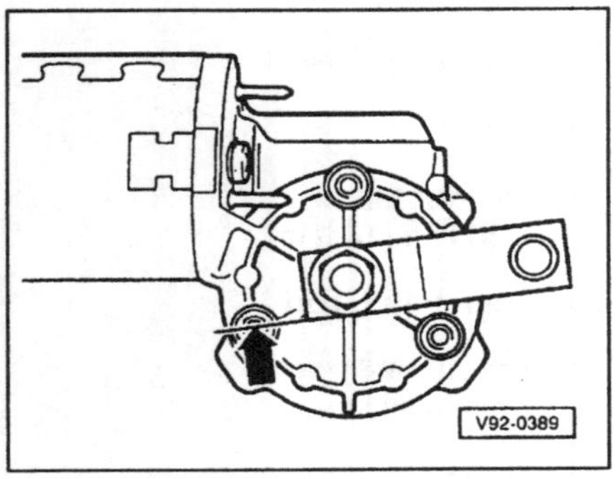

◄ **Fig. 1 Front wiper motor crank park position, adjusting**

– Let wiper motor run to end position.

– Adjust so that the upper edge aligns with the drilling -arrow-.

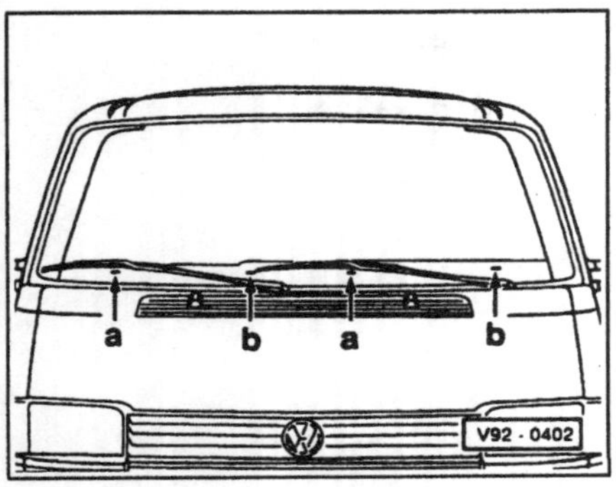

◄ **Fig. 2 Front wiper blade park position, adjusting**

– Adjust wiper blades to the markings on the windshield.

a - Markings for LHD vehicles

b - Markings for RHD vehicles

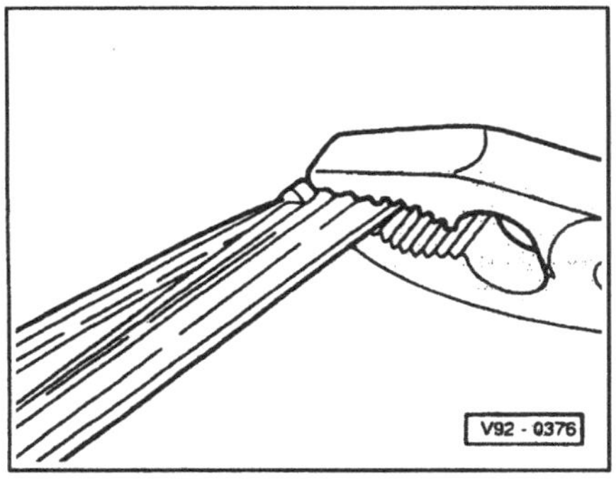

Wiper blade inserts, removing and installing

◄ – Using a pair of pliers press both steel strips together on the enclosed side of the rubber, slide sideways out of the upper clip and remove insert complete with strips from the other clips of the wiper blades.

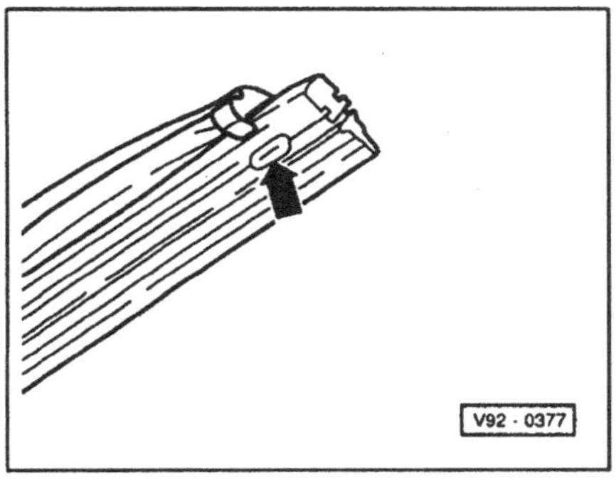

◄ – Install new insert into lower clips of wiper blade.

– Install both strips in the first groove of the rubber so that the recesses in the strips face the rubber and engage in the rubber lugs of the groove.

– Using pliers press both steel strips and rubber together and install in the upper clips so that the lugs of the clips on both sides engage in the retaining slots –arrow– of the wiper blade insert.

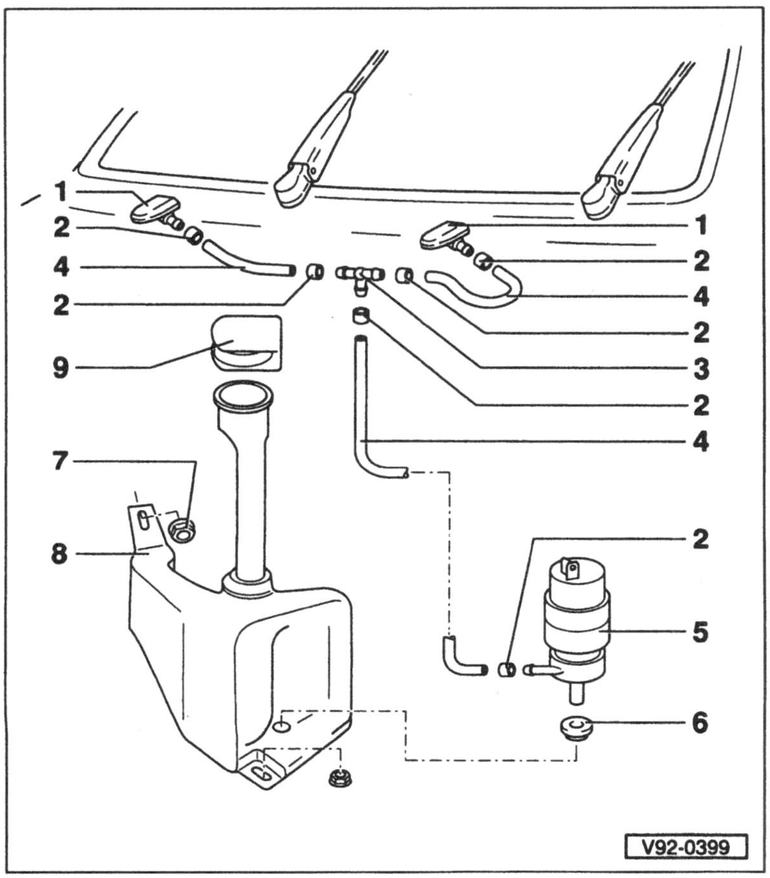

Windshield washer system ➤ 05.96, servicing

1 - **Nozzles**
♦ Removing,
⇒ Fig. 1.
♦ Adjusting,
⇒ Fig. 2.

2 - **Hose clip**

3 - **T-fitting**

4 - **Hose**

5 - **Windshield washer pump**
♦ Connections,
⇒ Fig. 3.

6 - **Seal**

7 - **Plastic nut**

8 - **Container**
♦ Capacity 5.0 L (5.3 Qts.).

9 - **Container cap**

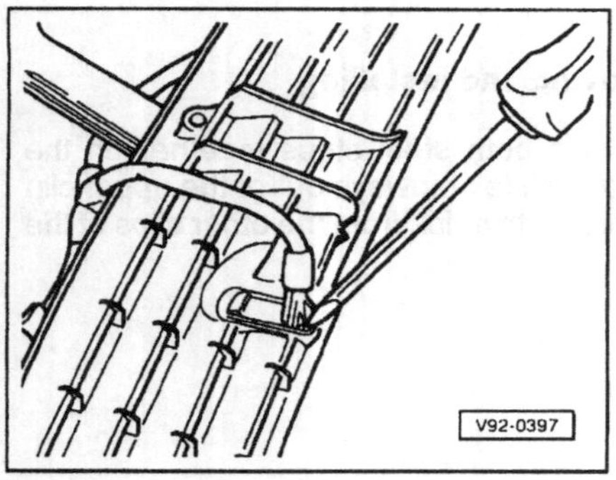

Fig. 1 Washer nozzle, removing

– Remove screws and take off air intake grille.

– Using a screwdriver press nozzle locking device to side and remove nozzle.

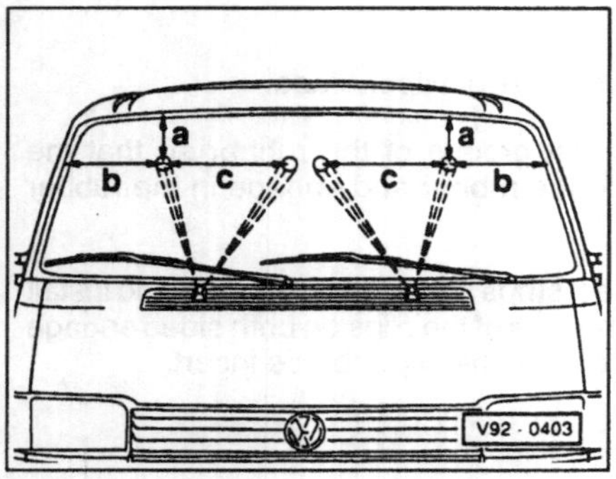

Fig. 2 Washer nozzle, adjusting

– Use special tool 3125.

a - 160 mm (6 in.)

b - 320 mm (12 in.)

c - 400 mm (15.8 in.)

Note:

Only check nozzle adjustment on vehicles with preset nozzles (white spot on under side of nozzle). Adjustment not possible.

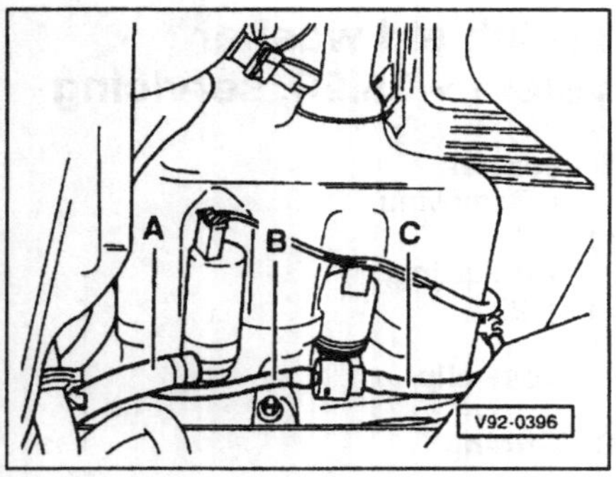

Fig. 3 Windshield washer pump connections

A - Line to headlight washer system (if equipped)

B - Line to windshield washer system

C - Line to rear window washer system

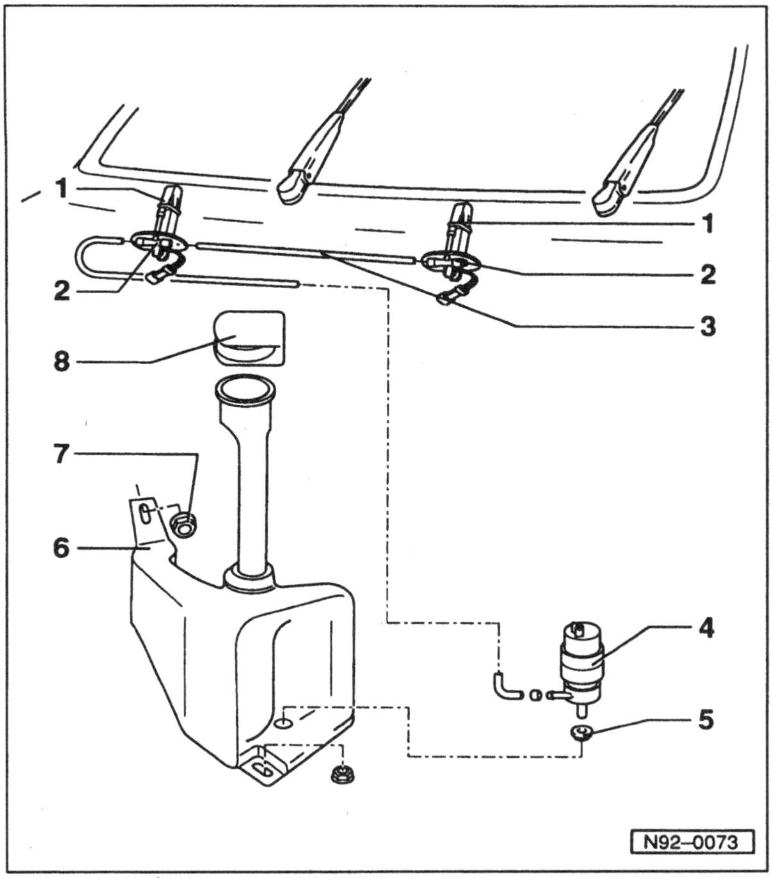

Windshield washer system 05.96 ➤, servicing

1 - **Nozzles**
- ♦ Removing

⇒ Fig. 1

2 - **Hose junction**

3 - **Hose**

4 - **Windshield washer pump**
- ♦ Connections

⇒ Fig. 2

5 - **Seal**

6 - **Container**

7 - **Plastic nut**

8 - **Container cap**

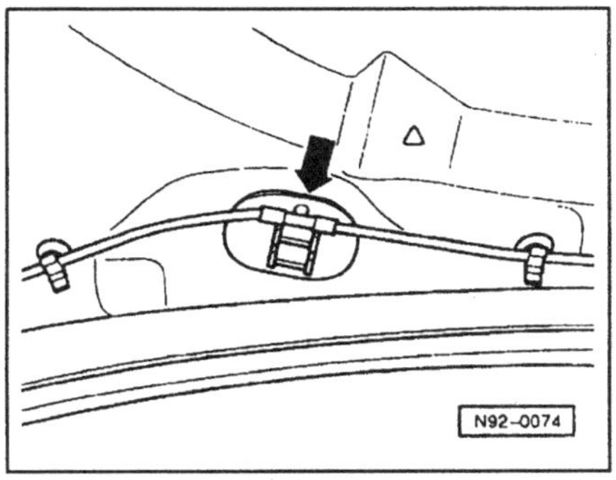

◄ **Fig. 1 Washer nozzle, removing**

– Unclip and remove cover from inside of front hood.

– Loosen rubber seal at nozzle -arrow-, unclip and pull down on nozzle to remove.

Washer nozzle, checking

Notes:

- ♦ *Washer nozzle spray pattern and position can only be checked. Adjustment is not possible.*
- ♦ *Nozzle spray pattern correct when contacting the upper third of the windshield.*

◄ **Fig. 2 Windshield washer pump connections**

A - Line to headlight washer system (if equipped)

B - Line to windshield washer system

C - Line to rear window washer system

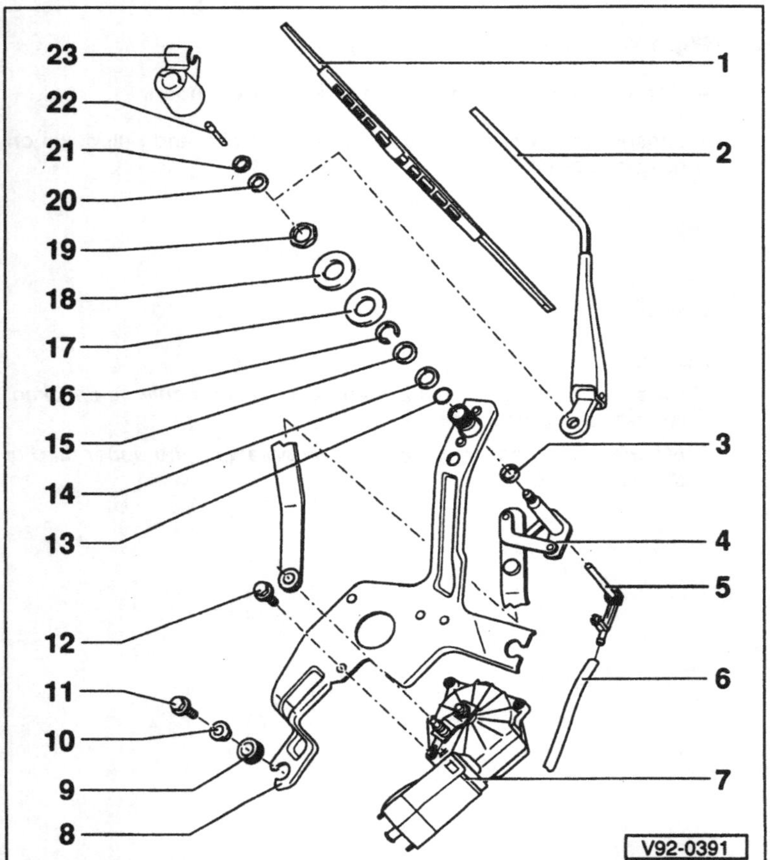

Rear window wiper/ washer system, repairing

Vehicles with hatch door

1 - Wiper blade
- Park position, adjusting,

⇒ Fig. 2.
- Wiper blade insert, removing and installing,

⇒ Page V 92-5.

2 - Wiper arm

3 - Spring washer

4 - Connecting lever with rods

5 - Nozzle connecting pipe

6 - Hose

7 - Wiper motor
- Park position, adjusting,

⇒ Fig. 1.

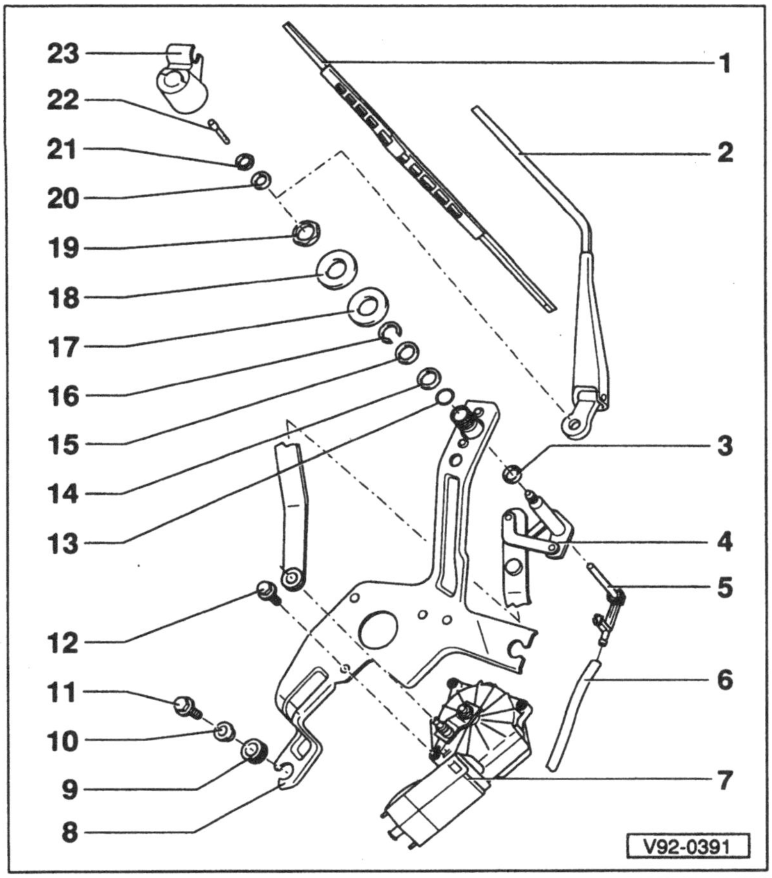

8 - **Bracket, wiper motor**

9 - **Rubber grommet**

10 - **Spacer**

11 - **Hex bolt M6**
- ♦ 5 Nm (44 in. lb)

12 - **Hex bolt M6**
- ♦ 5 Nm (44 in. lb)

13 - **O–ring**

14 - **Spacer washer**

15 - **Washer**

16 - **Snap ring**

17 - **Plastic washer**

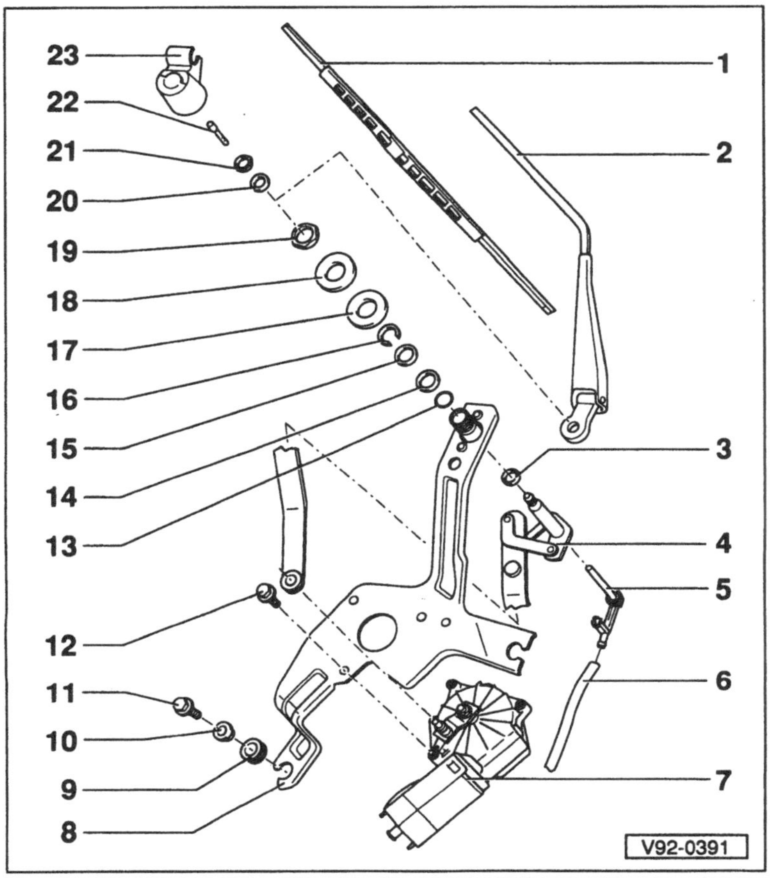

18 - **Spacer ring**

19 - **Hex nut M16**
- ♦ 10 Nm (7 ft lb)

20 - **Washer**

21 - **Hex nut M8**
- ♦ 16 Nm (12 ft lb)

22 - **Nozzle**

23 - **Cap**

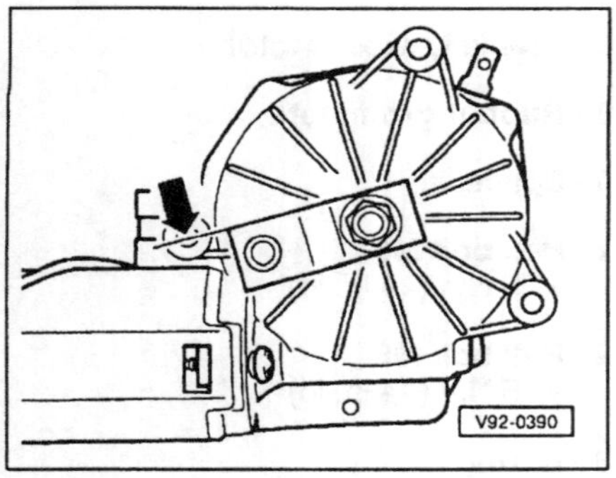

Fig. 1 Rear wiper motor crank park position, adjusting

– Let wiper motor run to end position.

– Install crank arm and align so that the upper edge aligns with the drilling -arrow-.

Fig. 2 Rear window wiper park position, adjusting

a - 45 mm, (1.77 in.)

Adjusting nozzles

Center in wiped area.

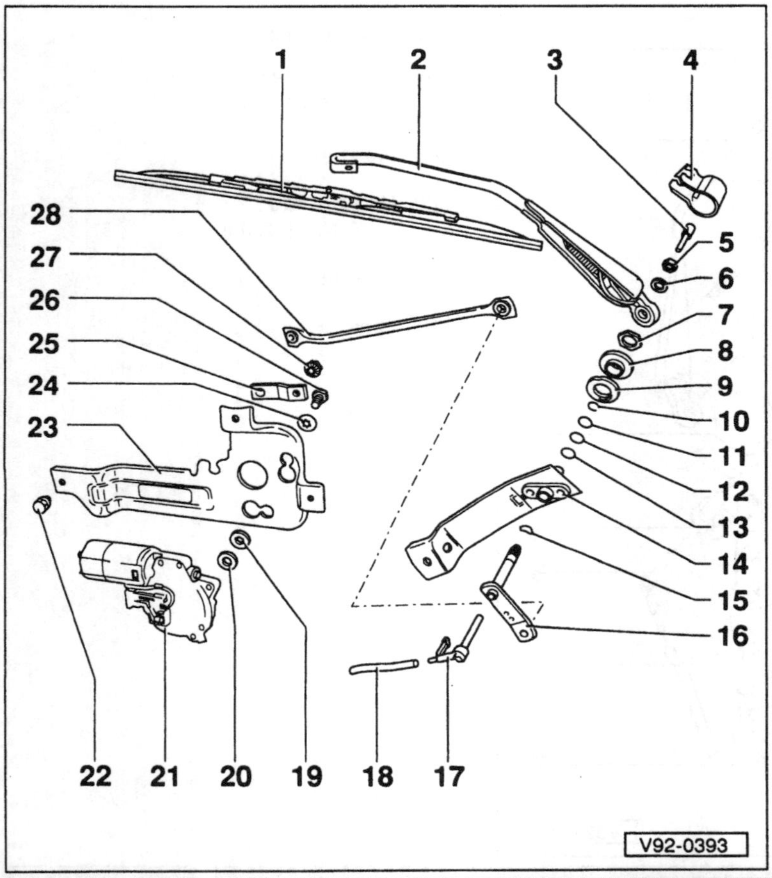

Vehicles with rear double doors

Note:

The components for the left hand and right hand doors are the same,
⇒ Figs. 3, 4.

1 - Wiper blade
- Park position, adjusting,

⇒ Fig. 2.
- Wiper blade insert, removing and installing,

⇒ Page V 92-5.

2 - Wiper arm

3 - Nozzle

4 - Cap

5 - Hex nut M8
- 16 Nm (12 ft lb)

6 - Washer

7 - Hex nut M16
- 10 Nm (7 ft lb)

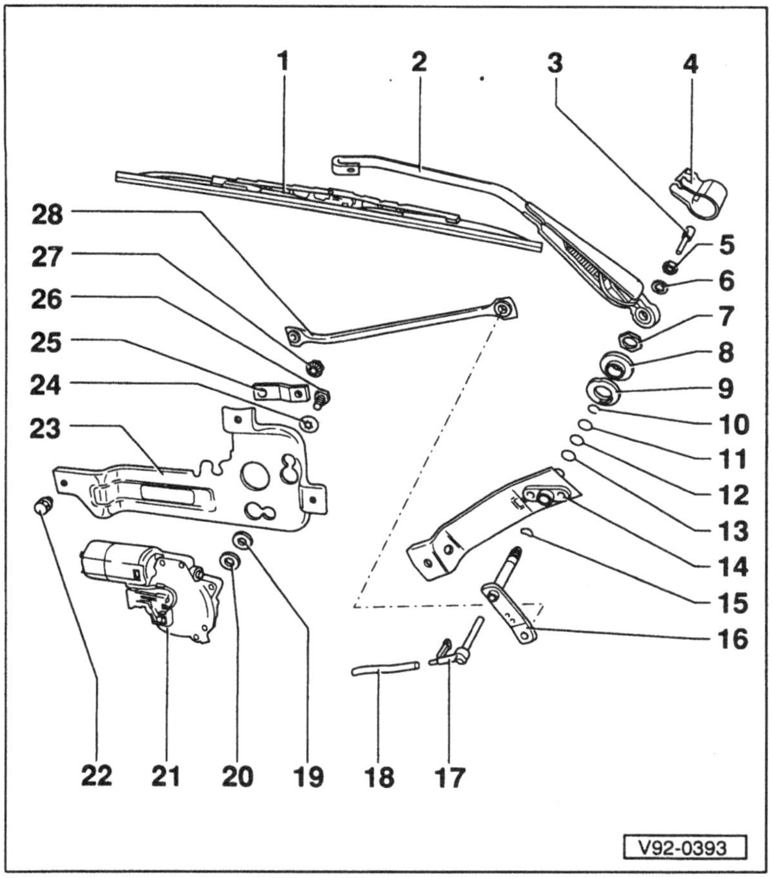

8 - Spacer ring

9 - Plastic washer

10 - Snap ring

11 - Washer

12 - Spacer washer

13 - O–ring

14 - Wiper shaft holder

15 - Spring washer

16 - Connecting lever

17 - Nozzle connecting pipe

18 - Hose

19 - Rubber grommet

20 - Spacer

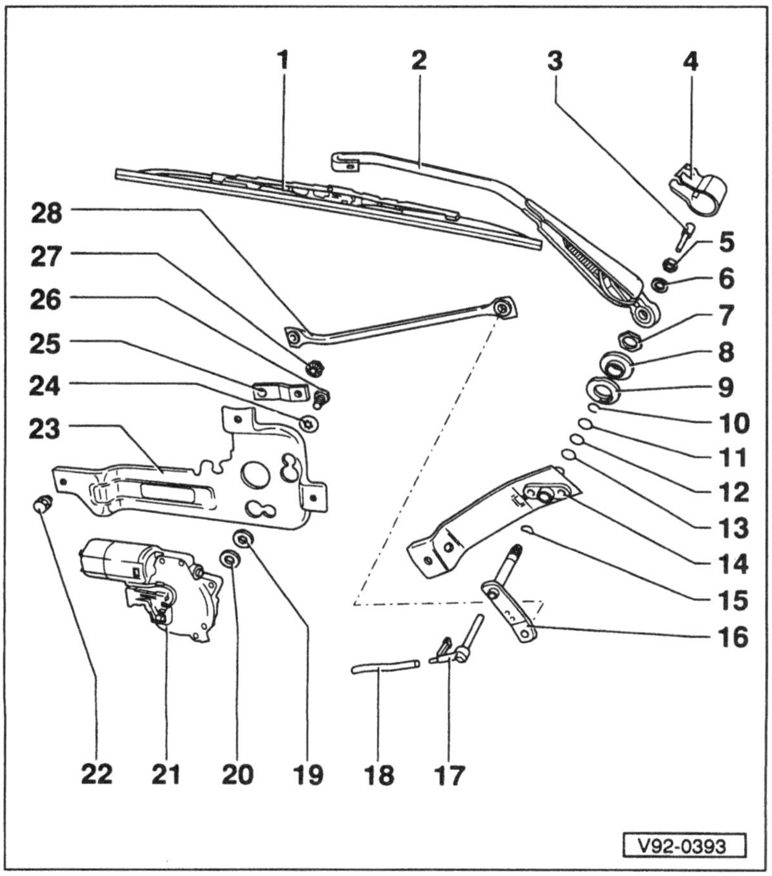

21 - Wiper motor

22 - Hex bolt M6
- ♦ 5 Nm, (44 in lb)

23 - Wiper motor bracket

24 - Toothed spring washer

25 - Crank
- ♦ Adjusting park position, ⇒ Fig. 1.

26 - Hex nut M6
- ♦ 5 Nm (44 in lb)

27 - Hex nut M8
- ♦ Self-locking

28 - Connecting rod

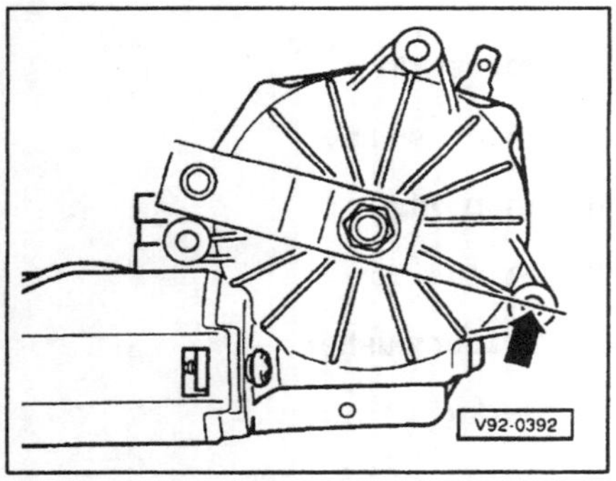

Fig. 1 Rear wiper motor crank park position, adjusting

- Let wiper motor run to end position.
- Adjust linkage arm so that the lower edge aligns with the hole centerline -arrow-.

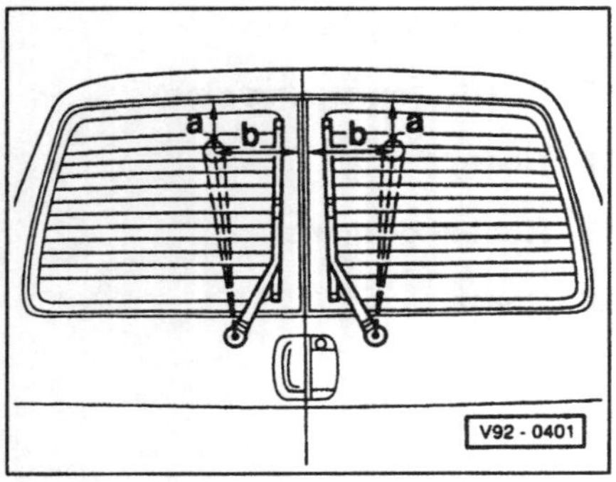

Fig. 2 Rear window wiper park position, adjusting

- Adjust the wiper arm so that the wiper blade is vertical.

Adjusting nozzles

a - 150 mm, (6 in.)

b - 250 mm, (9.8 in.)

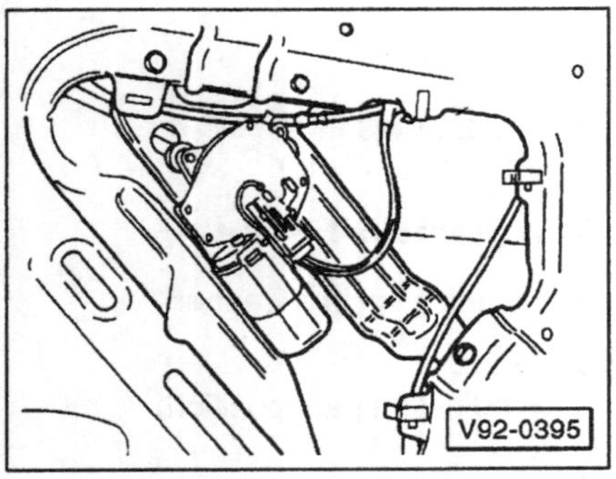

Fig. 3 Wiper motor installation position, left rear door

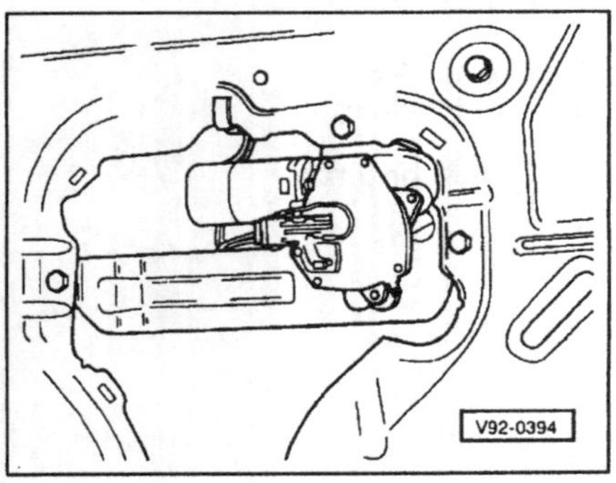

Fig. 4 Wiper motor installation position, right rear door

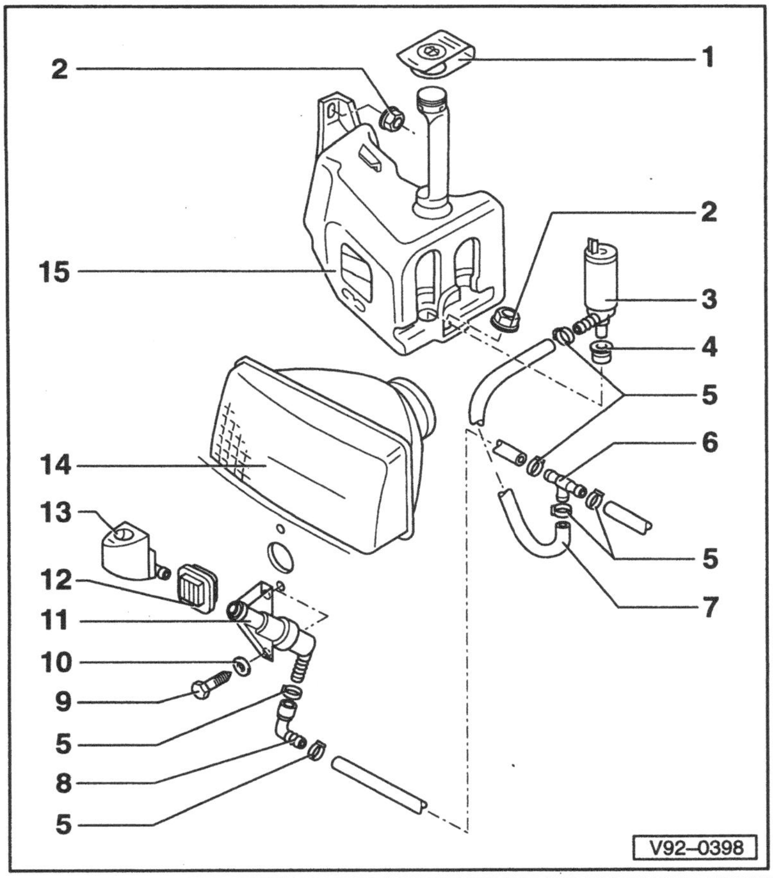

Headlight washer system, servicing

1 - **Container cap**

2 - **Hex nut M6**

3 - **Headlight washer pump**

4 - **Seal**

5 - **Hose clamp**

6 - **Distributor**
 - ♦ Note hose connections.

7 - **Hose 10 x 3**

8 - **Angled connector**

9 - **Hex bolt M6**

10 - **Washer**

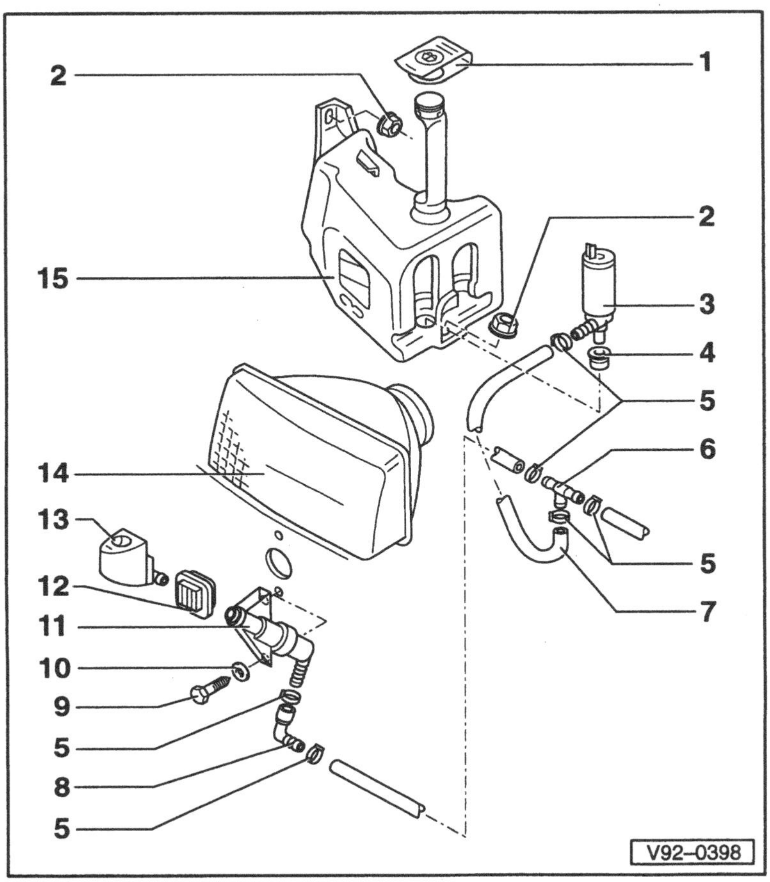

11 - **Lift cylinder**

12 - **Rubber grommet**

13 - **Nozzles**
 - ♦ Adjusting ⇒ V 92-23

14 - **Headlight**

15 - **Container**
 - ♦ Capacity approx. 7 litre (7.4 Qt).

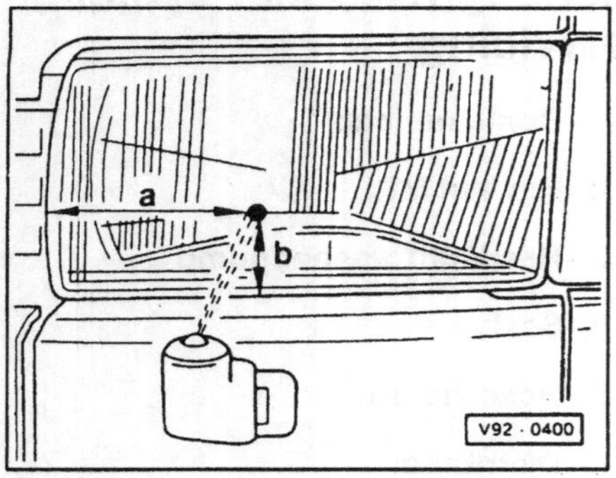

Fig. 1 Nozzles, adjusting

– Use special tool 3019 A.

– Pull nozzles out onto stop and secure with suitable piece of wood.

Left headlight:

a - 120 mm, (4.7 in)
b - 50 mm, (2 in.)

Nozzle dimensions for the right headlight are a mirror image.

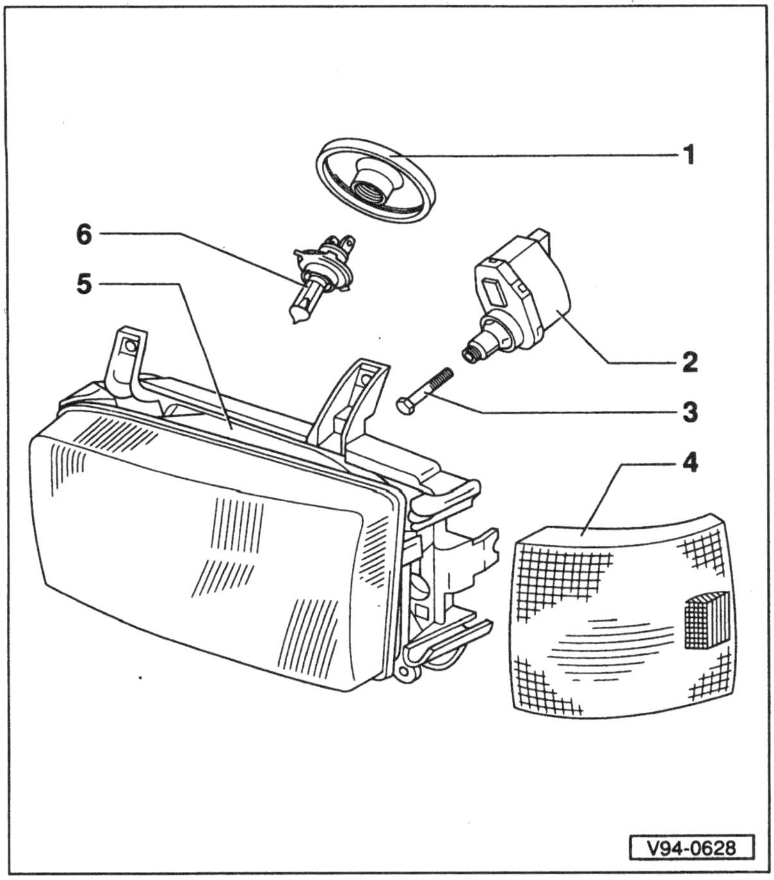

Headlight and front turn signal ➤ 05.96, servicing

CAUTION!

Before starting work on the electrical system, obtain radio security code, turn off ignition, and disconnect battery Ground (GND) strap.

Note:

After any work which could affect the headlight setting, the headlight must be adjusted.

1 - Cap

2 - Positioning motor (if equipped)

3 - Adjuster for positioning motor (if equipped)

4 - Left-hand turn signal housing
- Removing,

⇒ Fig. 4.
- Removing and installing turn signal bulb,

⇒ Fig. 5.

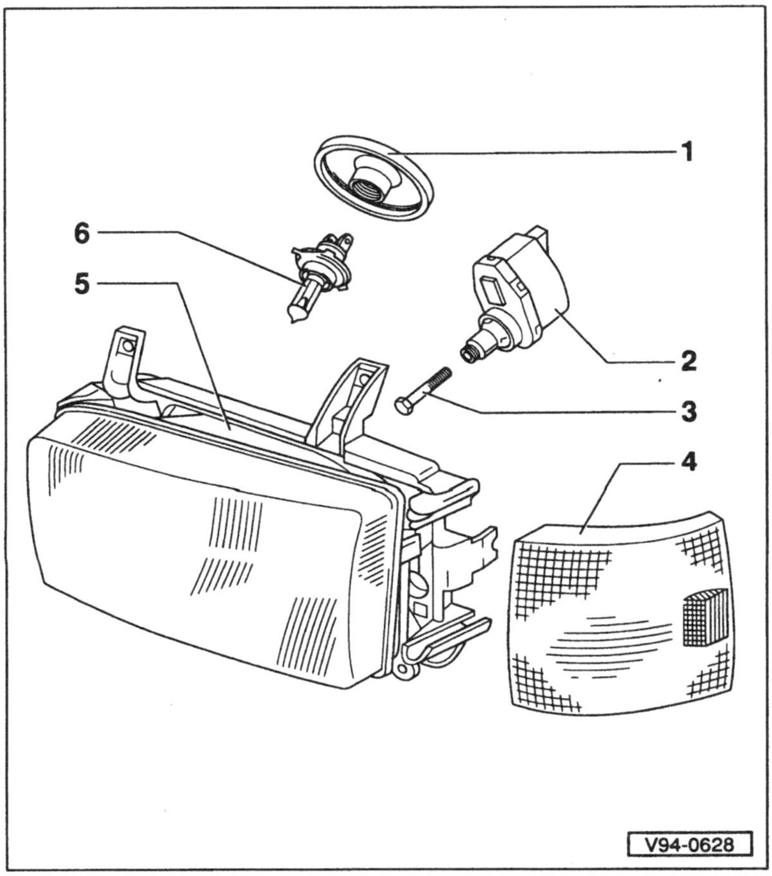

5 - Headlight housing
- Removing and installing,

⇒ Fig. 1.
- Adjusting headlight,

⇒ Fig. 2.

6 - Low and high beam twin filament bulb
- Halogen bulb H4 12 V, 60 W/55 W,
- Removing and installing,

⇒ Fig. 3.

Fig. 1 Headlights, removing and installing

Note:

After installing headlight housing, check headlight and adjust, if necessary.

- Remove cross panel below headlight,
⇒ Repair Manual, Body Exterior/Interior, Repair Group 50.
- Remove securing screws -arrows- and pull headlight out forward.
- Disconnect headlight and turn signal multi-pin connector.

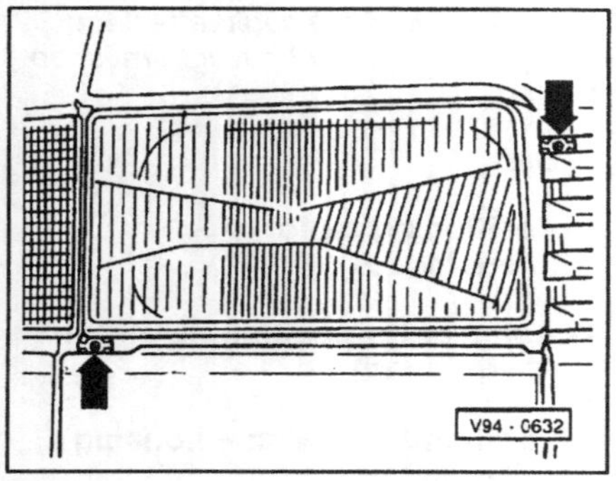

Fig. 2 Headlights, adjusting

Note:

For headlight checking and adjustment specifications,
⇒ Repair Manual, Maintenance.

- Headlight range control adjuster in basic setting.
- Adjust right-hand headlight using a phillip screwdriver at adjustment screws -arrows-.

Upper adjustment screw – lateral adjustment

Lower adjustment screw – height adjustment

Note:

Both left headlight adjustment screws are located in mirror image.

Fig. 3 Main and high beam twin filament bulb, removing and installing

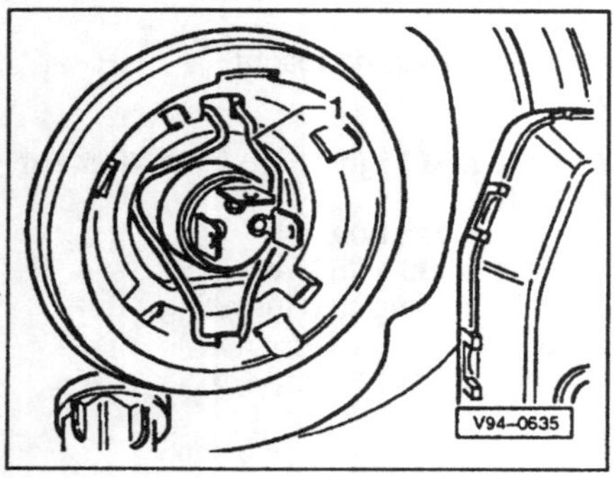

- Pull connector off twin filament bulb.
- Remove cap on rear of headlight insert.
- Release spring retainer -1- and fold down.
- Take bulb out of reflector.

Installing

CAUTION!   ***DO NOT touch glass portion of bulb.***

- Insert new bulb so that the lugs on the laminated plate align with the grooves in the reflector.
- After installing a new bulb check headlight adjustment.

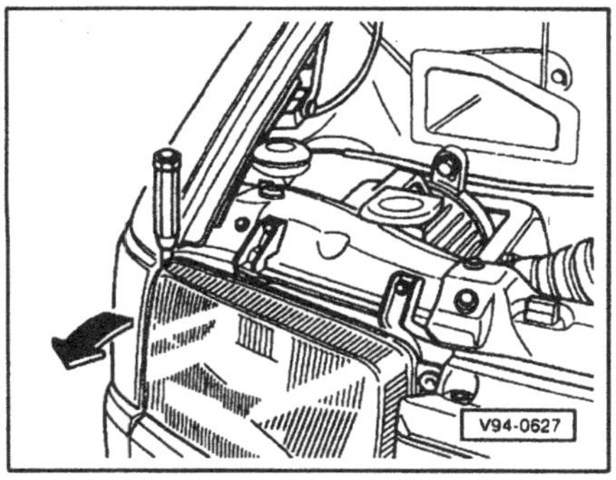

Fig. 4 Turn signal housing, removing

– Press locking lug to side and pull turn signal housing out forward from headlight housing in direction of arrow.

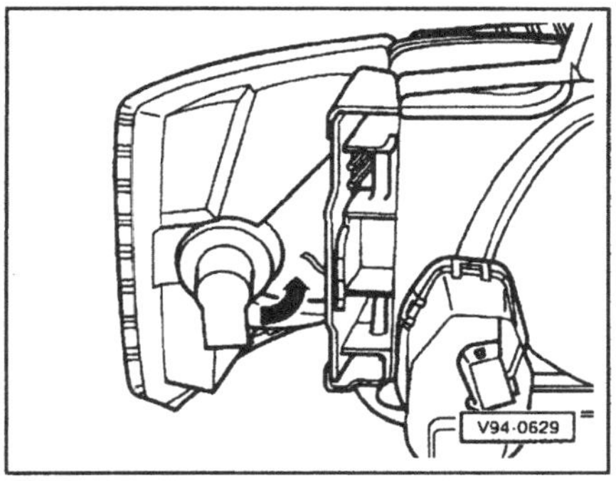

Fig. 5 Turn signal bulb, removing and installing

– Turn coupling inward in direction of arrow and pull out of housing.

– Remove bulb from holder.

– When installing, the coupling must be turned (opposite to direction of arrow) until it engages.

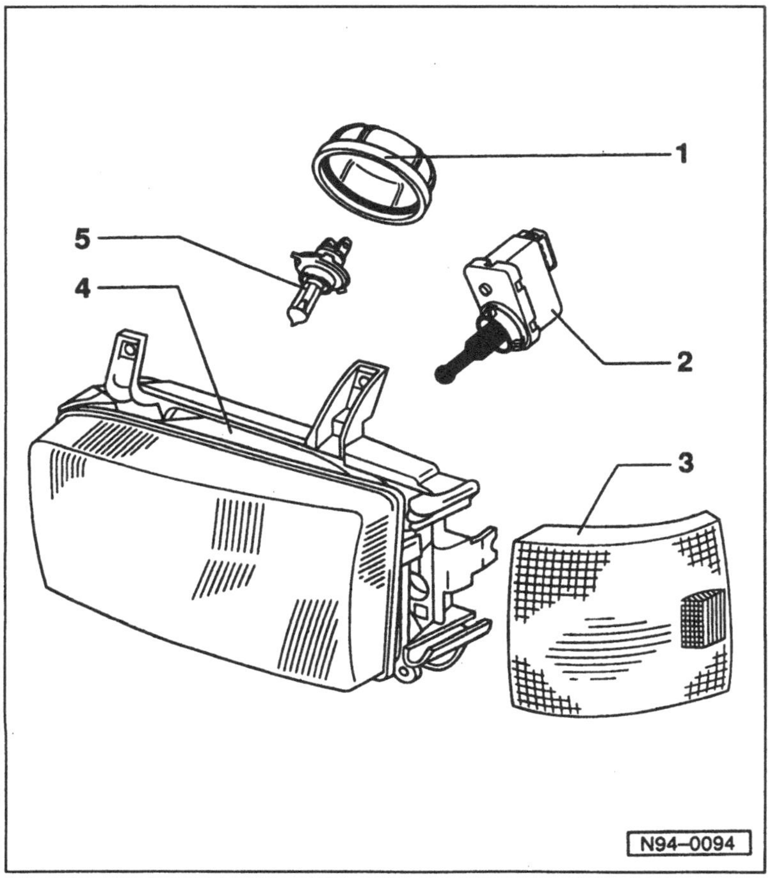

Headlight and front turn signal 05.96 ➤, servicing

> ***CAUTION!***
> ***Before starting work on the electrical system, obtain radio security code, turn off ignition, and disconnect battery Ground (GND) strap.***

Note:

After any work which could affect the headlight setting, the headlight must be adjusted.

1 - Cap

2 - Positioning motor (if equipped)

3 - Left-hand turn signal housing

- Removing,

⇒ Fig. 4.

- Removing and installing turn signal bulb,

⇒ Fig. 5.

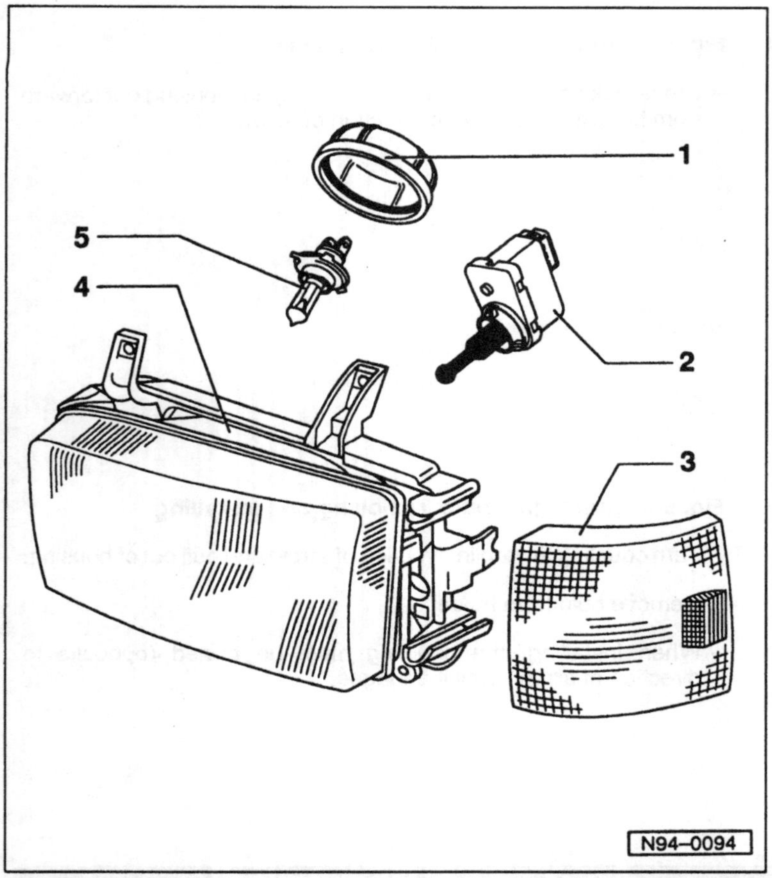

4 - Headlight housing

- Removing and installing,

⇒ Fig. 1.

- Adjusting headlight,

⇒ Fig. 2.

5 - Low and high beam twin filament bulb

- Halogen bulb H4
  12 V, 60 W/55 W,
- Removing and installing,

⇒ Fig. 3.

◄ **Fig. 1 Headlights, removing and installing**

Note:

After installing headlight housing, check headlight and adjust if necessary.

– Remove cross panel below headlight,
⇒ Repair Manual, Body Exterior/Interior, Repair Group 50.

– Remove securing screws -arrows- and pull headlight out forward.

– Disconnect headlight and turn signal multi-pin connector.

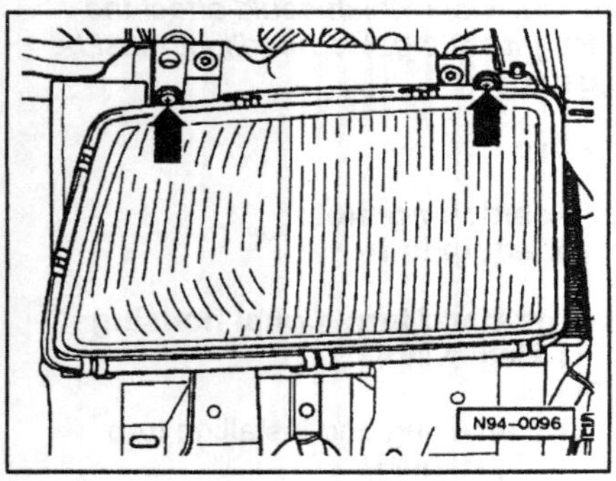

◄ **Fig. 2 Headlights, adjusting**

Note:

For headlight checking and adjustment specifications,
⇒ Repair Manual, Maintenance.

– Headlight range control adjuster in basic setting

– Adjust right-hand headlight using a phillip screwdriver at adjustment screws -arrows-.

Left adjustment screw – lateral adjustment.

Right adjustment screw – height adjustment.

Note:

Both left headlight adjustment screws are located in mirror image.

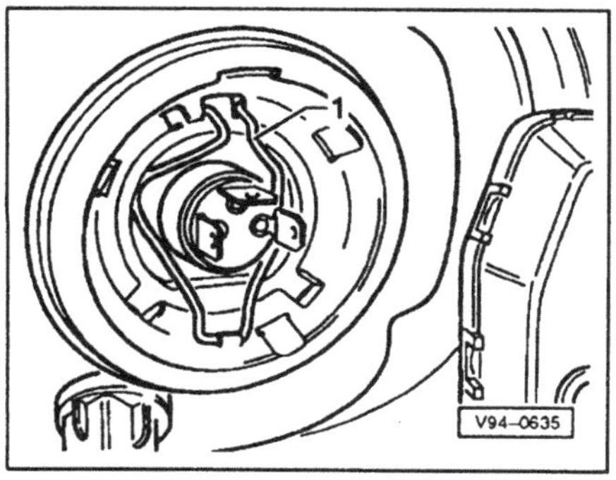

Fig. 3 Main and high beam twin filament bulb, removing and installing

◄ **Removing**

– Pull connector off twin filament bulb.

– Remove cap on rear of headlight insert.

– Release spring retainer -1- and fold down.

– Take bulb out of reflector.

Installing

CAUTION!
DO NOT touch glass portion of bulb.

– Insert new bulb so that the lugs on the laminated plate align with the grooves in the reflector.

– After installing a new bulb check headlight adjustment.

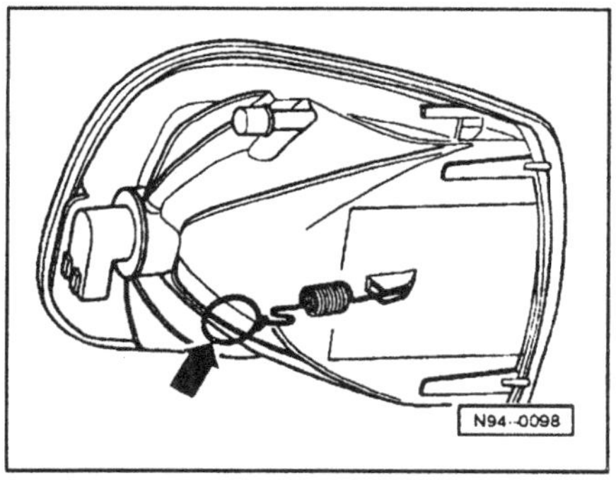

◄ **Fig. 4 Turn signal housing, removing**

– Release spring retainer -arrow- on inner side of housing.

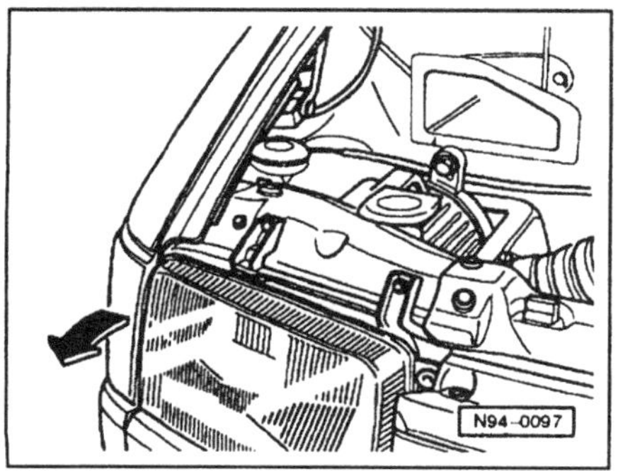

◄ – Remove from headlight housing in direction of arrow.

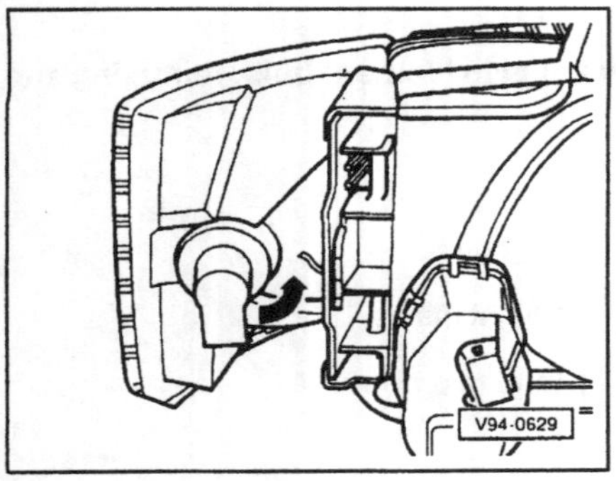

◄ **Fig. 5 Turn signal bulb, removing and installing**

- Turn coupling inward in direction of arrow and pull out of housing.
- Remove bulb from holder.
- When installing, the coupling must be turned (opposite to direction of arrow) until it engages.

Instrument panel switches 05.96 ➤, removing and installing

◄ **Switches, removing**

- Carefully insert screwdriver between switch trim insert and switch housing.
- Release switch by pulling forward slightly.
- Remove switch from trim and instrument panel.
- Pull off connector.

Installing

- Re-install connector.
- Insert light switch into instrument panel opening/trim and gently push in until it engages.

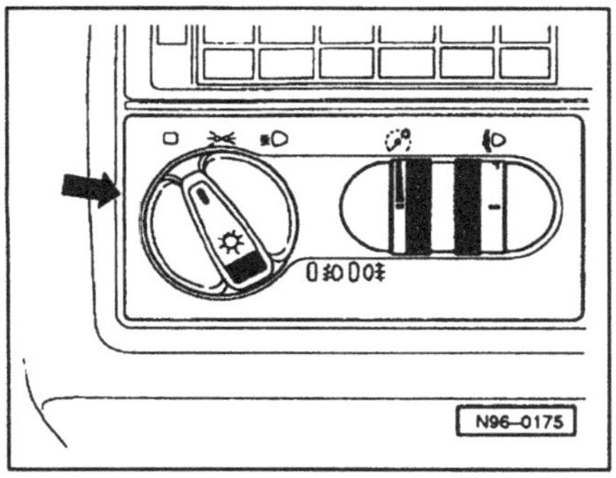

Light switch, removing

– Carefully insert screwdriver between instrument panel and the left of light switch housing -arrow-.

– Release switch by pulling forward slightly.

– Remove light switch from instrument panel.

– Pull off connector

Installing

– Re-install connector

– Insert light switch into instrument panel opening and gently push switch until it engages.

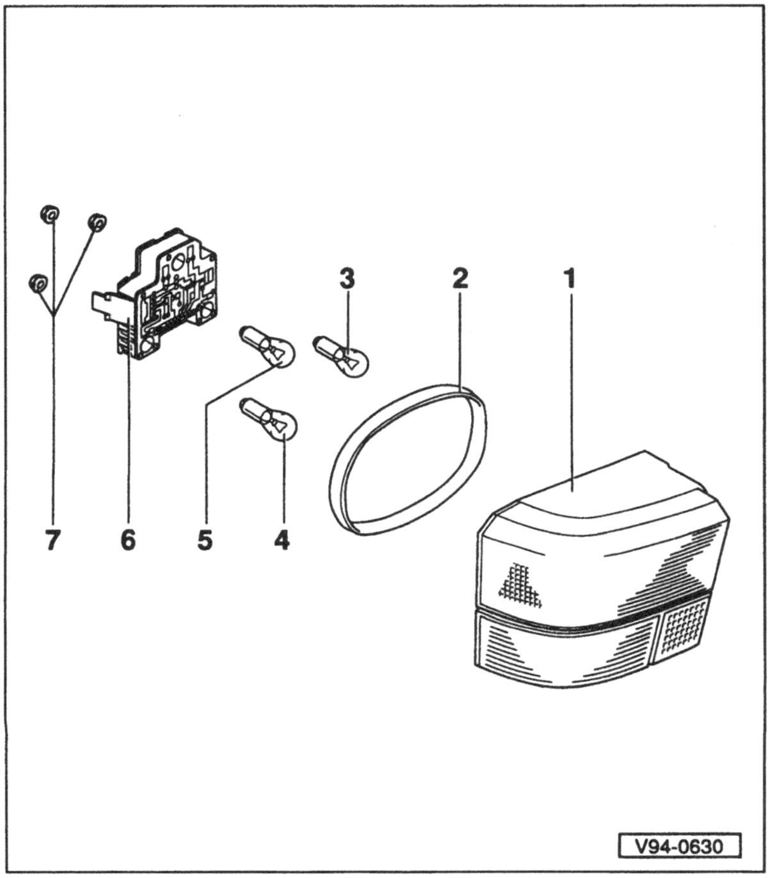

Tail lights, removing and installing

Brake/tail and turn signals

1 - Lens housing
♦ Removing,
⇒ Fig. 1.

2 - Gasket

3 - Backup light bulb - left/right
12V / 21W

4 - Turn signal bulb - left/right
♦ 12V / 21W

5 - Brake and tail light bulb - left/right
♦ 12 V, 21 W/5 W

6 - Bulb holder

7 - Hex nut M5 with washer

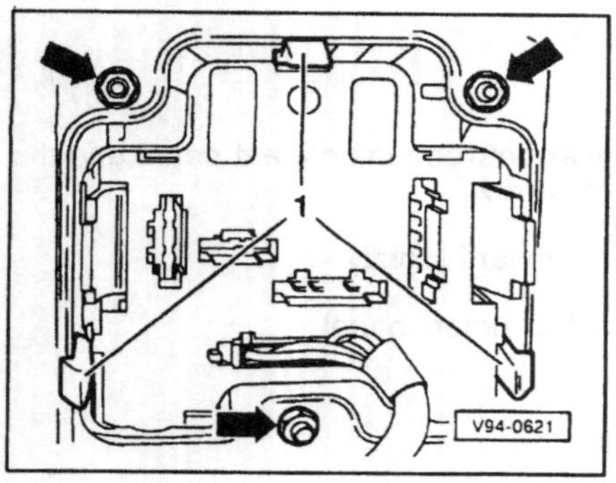

Fig. 1 Lens housing, removing

- Remove connector from bulb carrier.
- Remove hex nuts -arrows-.
- Release lugs -1- and remove housing by turning.
- Reassemble in reverse sequence.

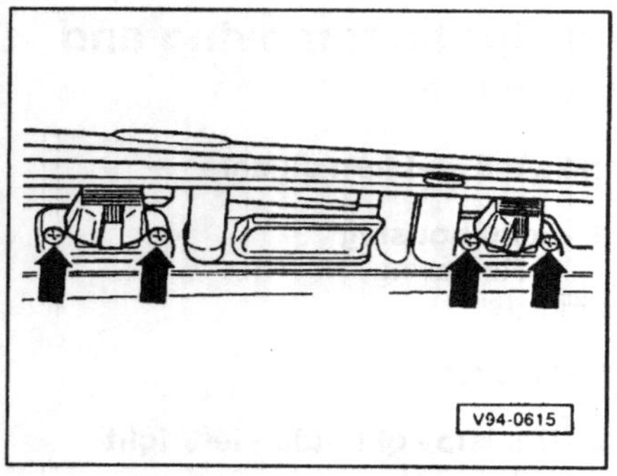

License plate lights, removing

- Remove screws -arrows- from the respective license plate light and take out light.

Note:

- *The light upper section engages in the base plate, and by tilting both parts slightly, can be removed to facilitate bulb changing.*

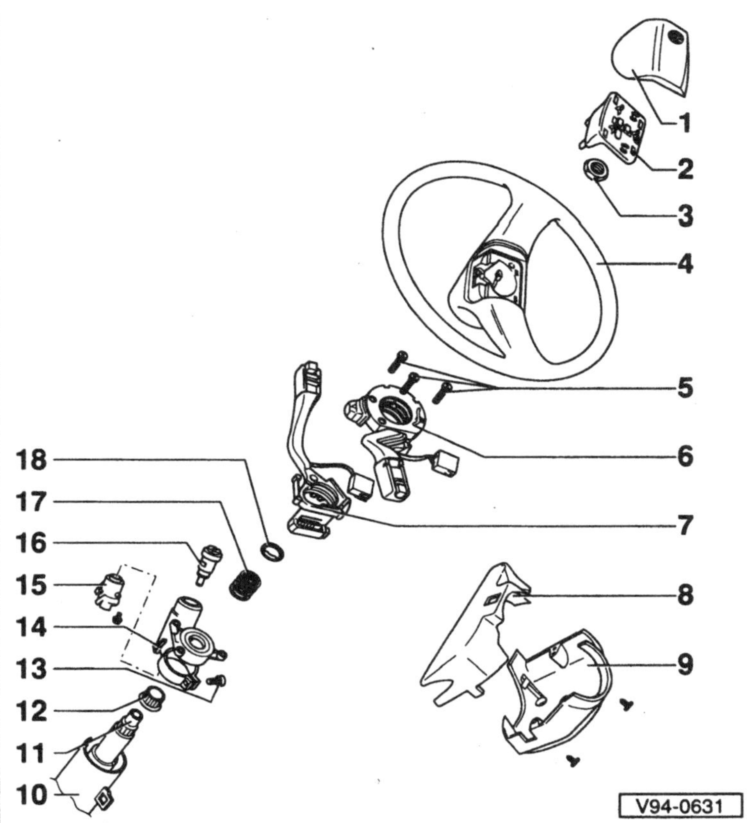

Steering column switches, servicing

> ***CAUTION!***
> ***Before working on the electrical system, obtain radio security code, turn off ignition, and disconnect battery Ground (GND) strap.***

Note:

For further information on installing the steering column,

⇒ Repair Manual, Suspension, Brakes, Wheels, Steering, Repair Group 48.

1 - Horn pad,
- Removing,

⇒ Fig. 1.

2 - Horn plate
- Removing contact plate,

⇒ Fig. 2.

3 - Hex nut
- 70 Nm, (51 ft lb)

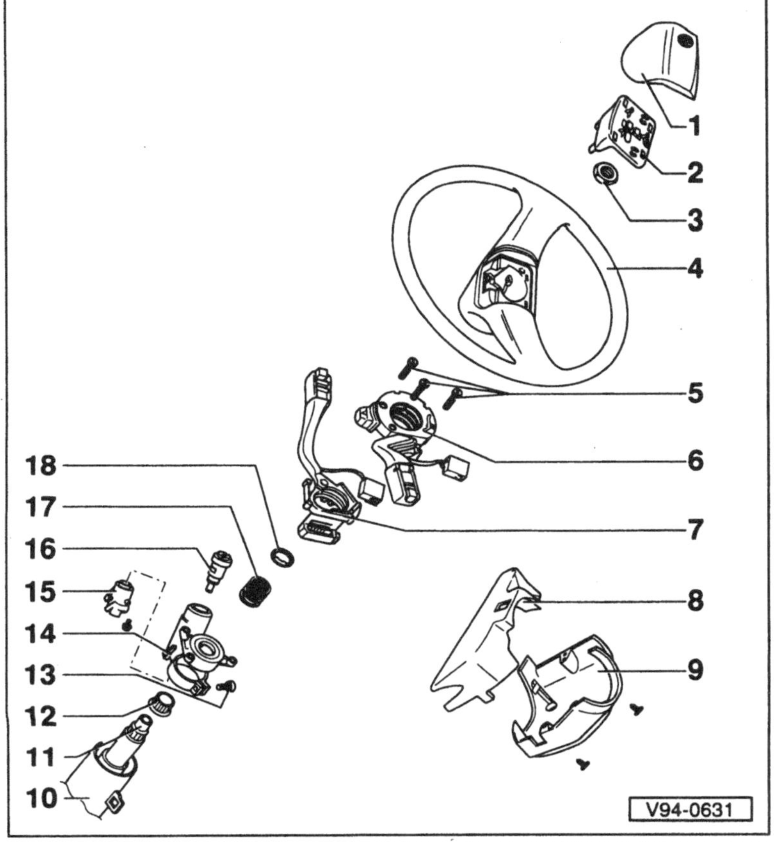

4 - Steering wheel
- Horn operation contact ring,

⇒ Fig. 3.

5 - Steering column switch securing screws

6 - Turn signal/ cruise control switch, high and low beam, headlight flasher, and emergency light switch
- Removing,

⇒ Fig. 4.

7 - Windshield wiper, wipe/wash operation, and multi-function indicator switch
- Removing,

⇒ Fig. 4.

8 - Upper steering column trim

9 - Lower steering column trim

10 - Column Tube

11 - Steering column

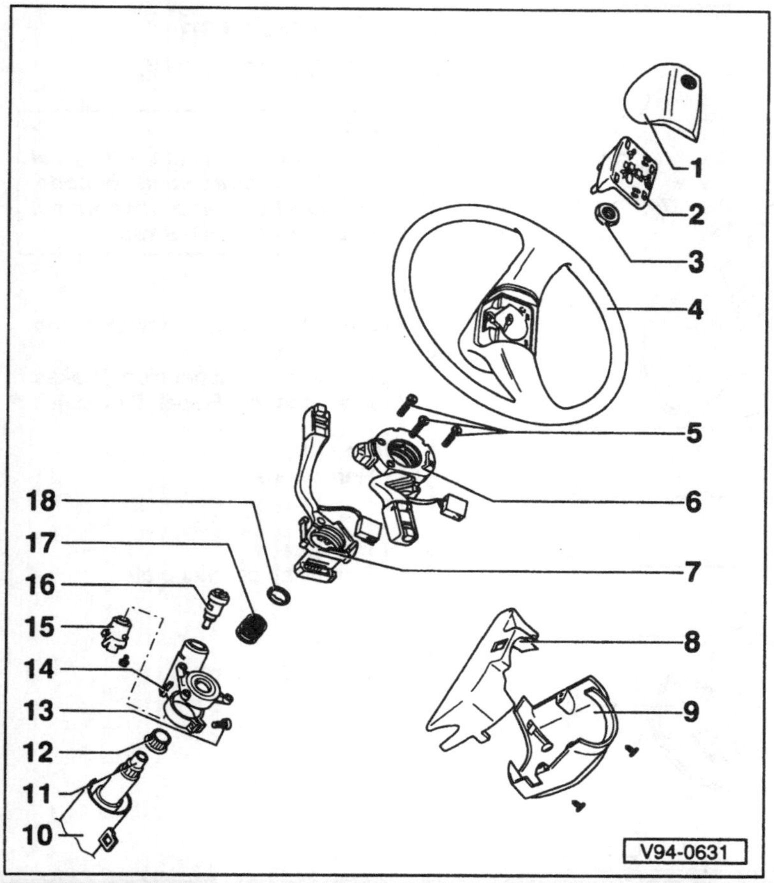

12 - Support ring

13 - Steering column lock housing securing bolt
- ♦ 10 Nm, (7 ft lb)

14 - Steering lock housing
- ♦ Removing and installing,
⇒ Page V 94-23.

15 - Ignition switch–remove steering lock housing, then remove securing screws

16 - Lock cylinder
- ♦ Removing and installing,
⇒ Fig. 5.

17 - Spring

18 - lock washer

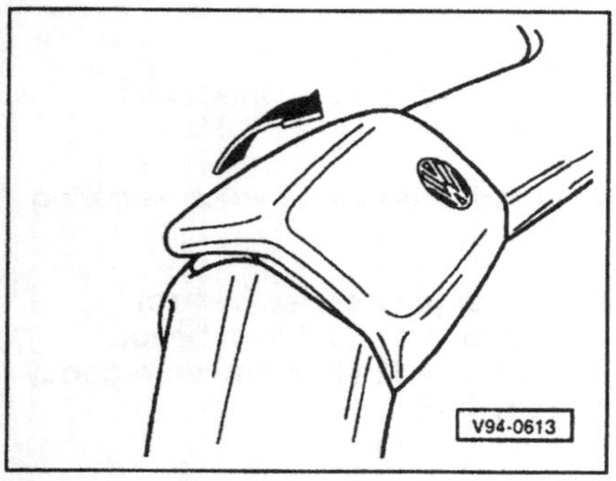

◄ **Fig. 1 Horn pad, removing**

> ***CAUTION!***
>
> ***Special safety precautions apply to vehicles equipped with airbags ⇒ Repair Manual* Body Exterior/Interior, Repair Group 69 *for airbag removal and handling instructions.***

– Lift top of cover and pull off steering wheel in direction of arrow.

Note:

Under the cover there are three springs, which keep the cover and the contact plate separated,
⇒ Fig. 2.

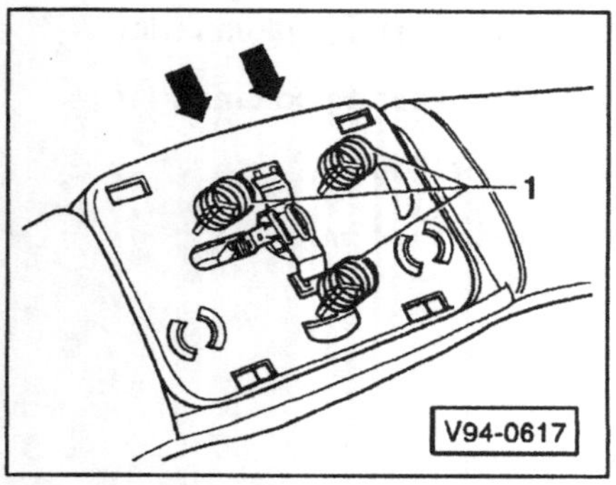

◄ **Fig. 2 Horn plate, removing**

– Take springs -1- off contact plate.

– Press contact plate in direction of arrow and pull off steering wheel, upward.

– Pull off wiring connectors.

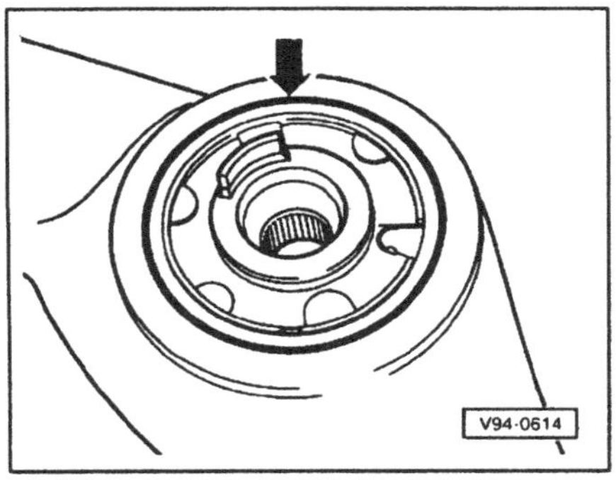

Fig. 3 Horn contact ring, operation

The positive voltage for operating the horn is supplied via a spring contact on the turn signal switch and a slip ring -arrow- under the steering wheel.

Grease with G 000 400 or equivalent.

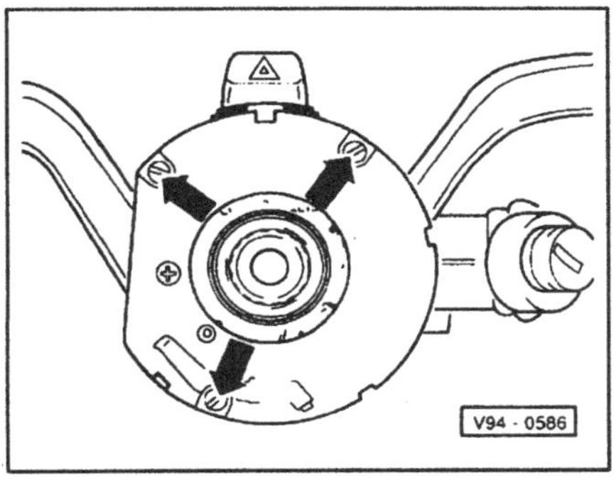

Fig. 4 Steering column switches, removing

> ***CAUTION!***
>
> ***Special safety precautions apply to vehicles equipped with airbags ⇒ Repair Manual* Body Exterior/Interior, Repair Group 69 *for airbag removal and handling instructions.***

– Remove steering wheel.

– Remove trim lower part.

– Remove securing screws -arrows-.

– Separate connectors to switches.

– Pull switches together with upper trim from steering column.

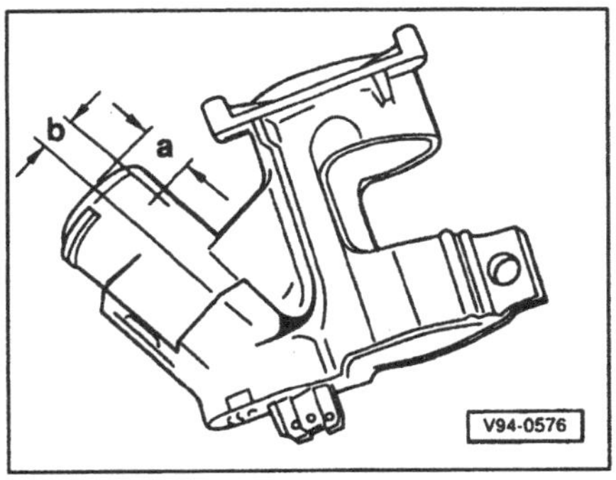

Fig. 5 Lock cylinder, removing and installing

– Removing steering lock housing, ⇒ Page V 94-23.

– Mark steering lock housing for drilling

a - 12 mm, (0.5 in.)

b - 10 mm, (0.4 in.)

– Drill steering lock housing (approx. 3 mm, [0.12 in.]), until the lock cylinder spring is evident (hole depth approx. > 3 mm, [0.12 in.]).

– Using opposite end of drill, compress lock spring and pull out lock cylinder.

Installing

> ***CAUTION!***
>
> ***The original steering lock housing must not be reinstalled.***

– Push lock cylinder into steering lock housing.

– By slightly turning key in lock cylinder it can be pushed onto the stop.

Steering lock housing, removing and installing

– Remove steering column switches.

– Drill out steering lock housing securing bolts.

– Pull lock washer and spring off steering column.

Note:

The steering column is pulled down and away , after removing the lock washer.

– Remove steering lock housing from column tube and steering column.

Installing

Note:

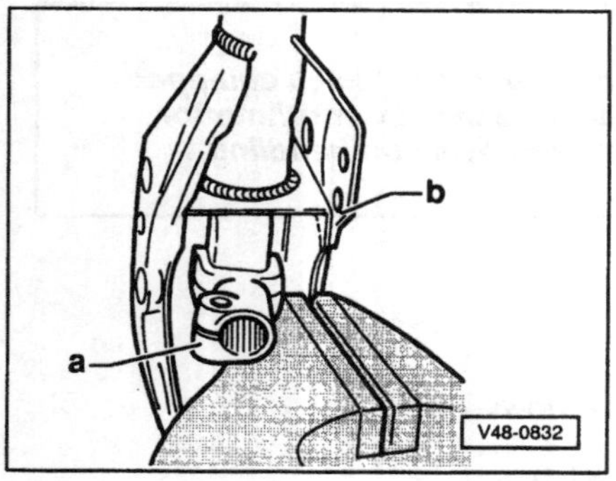

◄ *Installation of the steering lock housing can only be undertaken when the steering column has been removed*
⇒ Repair Manual Suspension, Wheels, Brakes, Steering, Repair Group 48

– Place column tube in vice ensuring that the universal joint -a- is supported on the vice.

– Secure column in vice.

– Place steering lock housing on column tube.

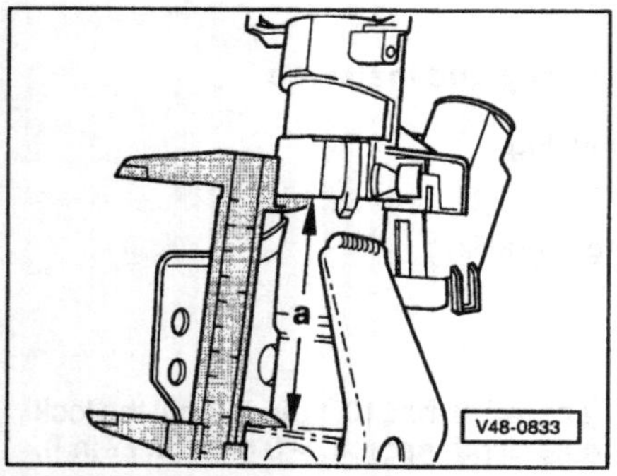

◄ – Push steering lock housing onto the column tube until dimension -a- is attained.

a = 95 mm (3.7 in.)

– Tighten steering lock housing shear bolt until the head shears off.

– Slide spring and lock washer on steering column.

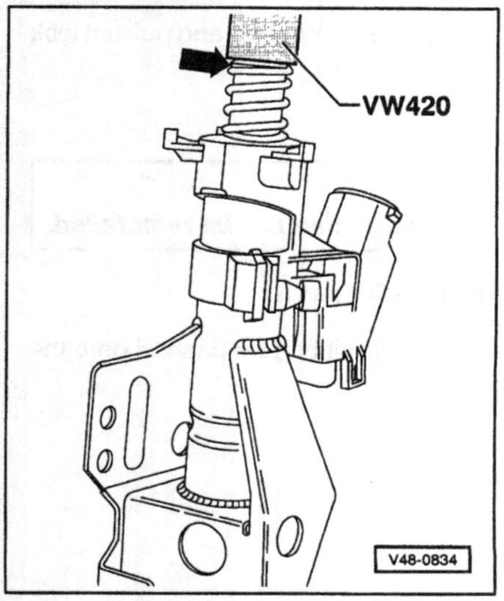

◄ – Drive lock washer onto steering column using special tool VW 420 until the installation position is reached.

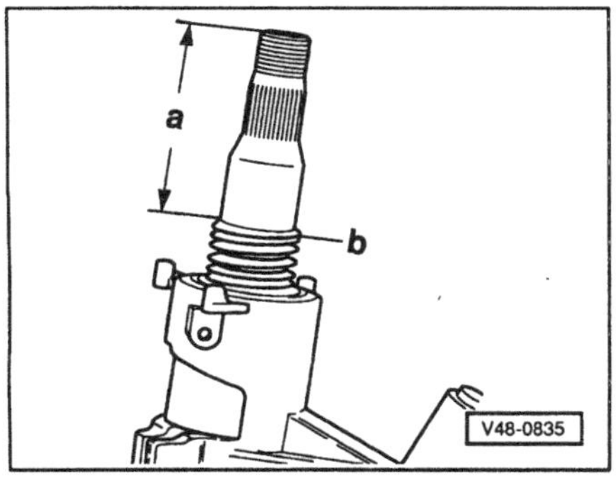

◄ Lock washer installation position

a = 65 mm (2.5 in.)

b = lock washer position

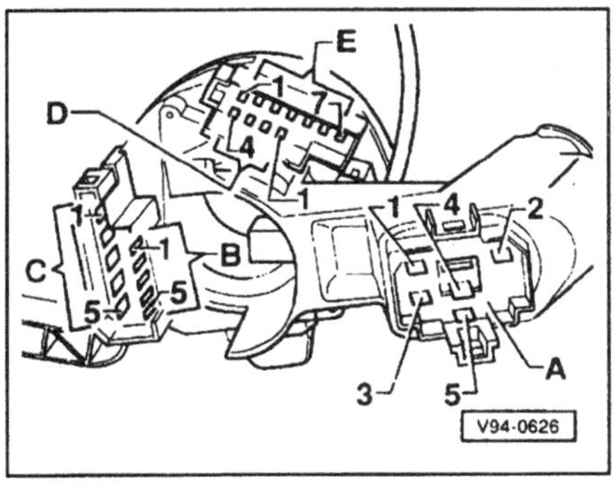

Wiring connections on steering column switch

◄ **A – Ignition/starter switch**

1 - Ignition/starter switch – Terminal 15
2 - Ignition/starter switch – Terminal X
3 - Ignition/starter switch – Terminal 50
4 - Ignition/starter switch – Terminal 30
5 - Ignition/starter switch – Terminal SU
6 - Ignition/starter switch – Terminal P

B – Connector, 5–pin

1 - Windshield wiper switch – Terminal 53b
2 - Windshield wiper switch – Terminal 53a
3 - Windshield wiper switch – Terminal 53
4 - Windshield wiper switch – Terminal 53e
5 - Windshield wiper switch – Terminal J

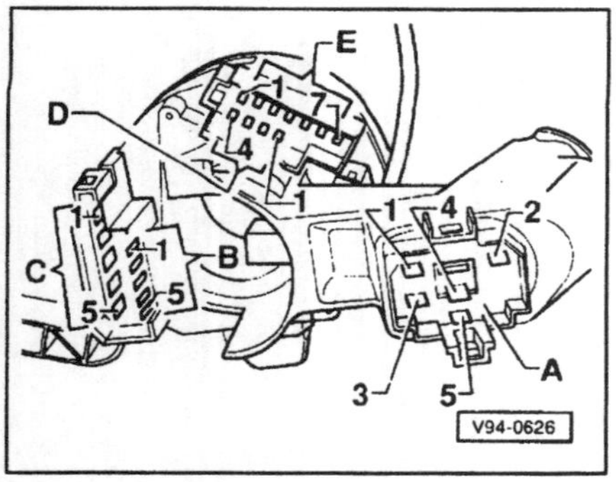

C – Connector, 5–pin

1 - Hazard light switch – Terminal 49
2 - High beam/flasher switch– Terminal 56b
3 - High beam/flasher switch– Terminal 56
4 - High beam/flasher switch– Terminal 56a
5 - High beam/flasher switch– Terminal 30

D – Connector, 4–pin

1 - Windshield wiper switch – Terminal L
2 - Windshield wiper switch – Terminal T
3 - Windshield wiper switch – Terminal 31
4 - Hazard light switch – Terminal 15

E – Connector, 7–pin

1 - Horn operation – Terminal 71
2 - Hazard light switch – Terminal 49a
3 - Turn signal switch – Terminal L
4 - Parking light switch – Terminal PL
5 - Parking light switch – Terminal P
6 - Parking light switch – Terminal PR
7 - Turn signal switch – Terminal R

Multi–function indicator (MFI) and cruise control system (CCS) (on steering lock housing), connector pin assignment

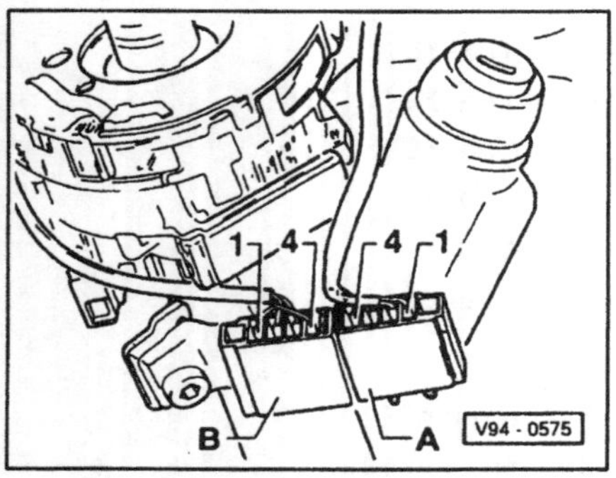

◄ **A - Connector, 4-pin MFI switch**

1 - Terminal 31
2 - Function (Mode)
3 - Memory selection switch (Memory)
4 - Cancel (Reset)

B - Connector, 4-pin, cruise control switch

1 - On
2 - Resume
3 - Memory (FIX)
4 - Terminal 15

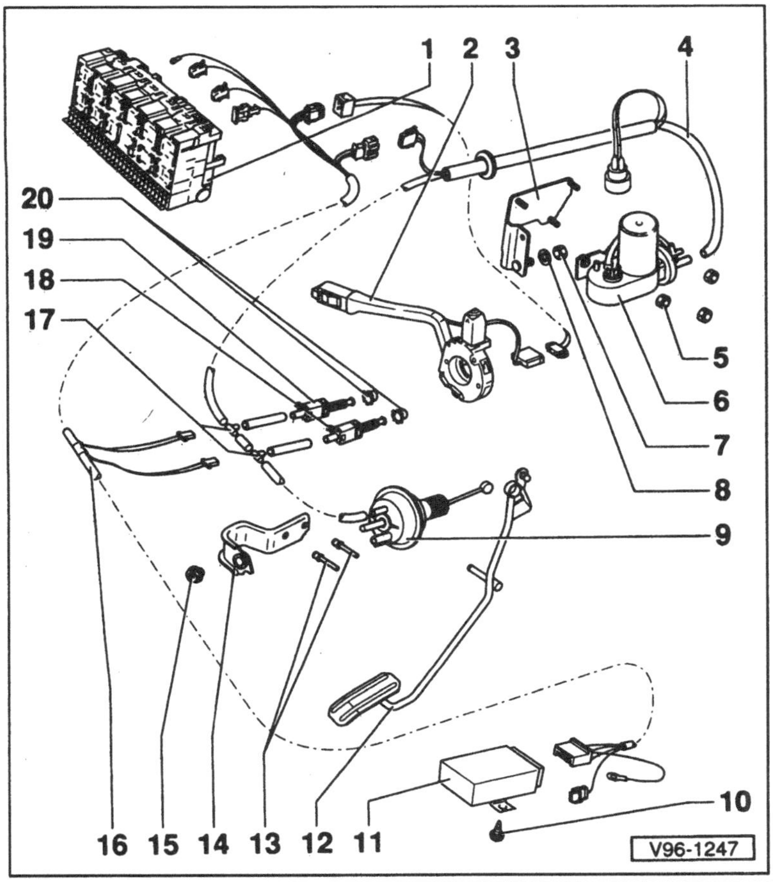

Cruise control system (CCS), servicing

> ***CAUTION!***
> ***Before working on the electrical system, obtain radio security code, turn off ignition, and disconnect battery Ground (GND) strap.***

Note:

- *Troubleshooting,*

⇒ Electrical Wiring Diagrams.

1 - Relay plate
- Removing and installing,

⇒ Page V 97-1.

2 - CCS switch -E45-
- Removing and installing,

⇒ Page V 94-21.

3 - Vacuum pump bracket

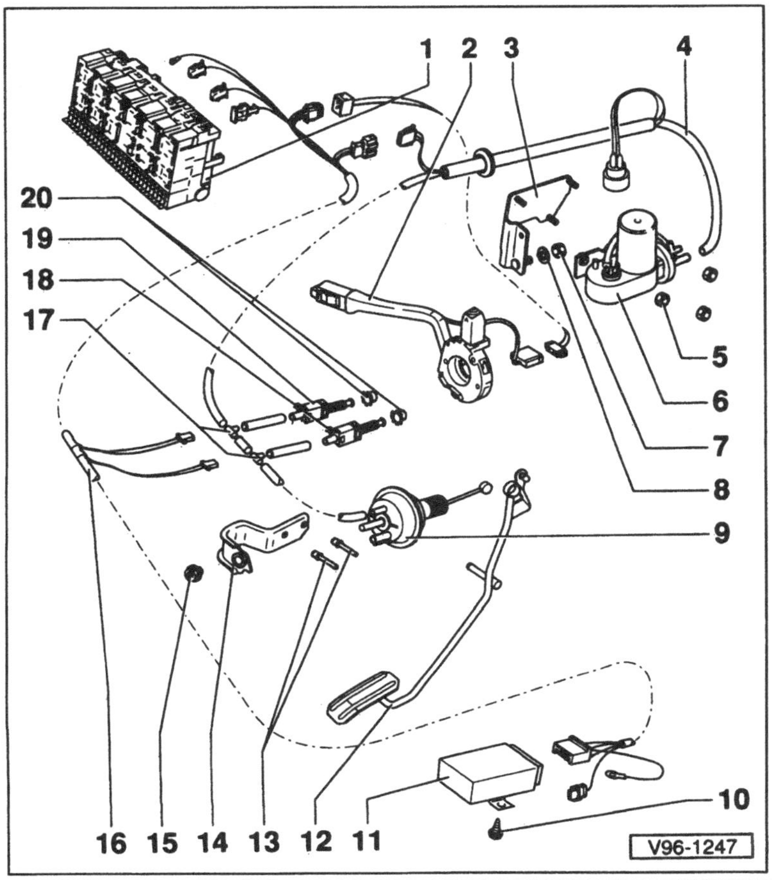

4 - Vacuum hose
- Checking for leaks,

⇒ Page V 96-7.

5 - Hex nut M5
- 3 Nm, (27 in lb)

6 - CCS vacuum pump -V18-
- Removing,

⇒ Fig. 4.

7 - Hex nut M8
- 16 Nm, (12 ft lb)

8 - Washer

9 - Positioning element on accelerator pedal
- Removing,

⇒ Fig. 1.
- Adjusting,

⇒ Fig. 2.

10 - Fillister head self-tapping screw B4.8x9.5

11 - CCS control unit
- Behind instrument panel, right.
- Removing,

⇒ Fig. 5.

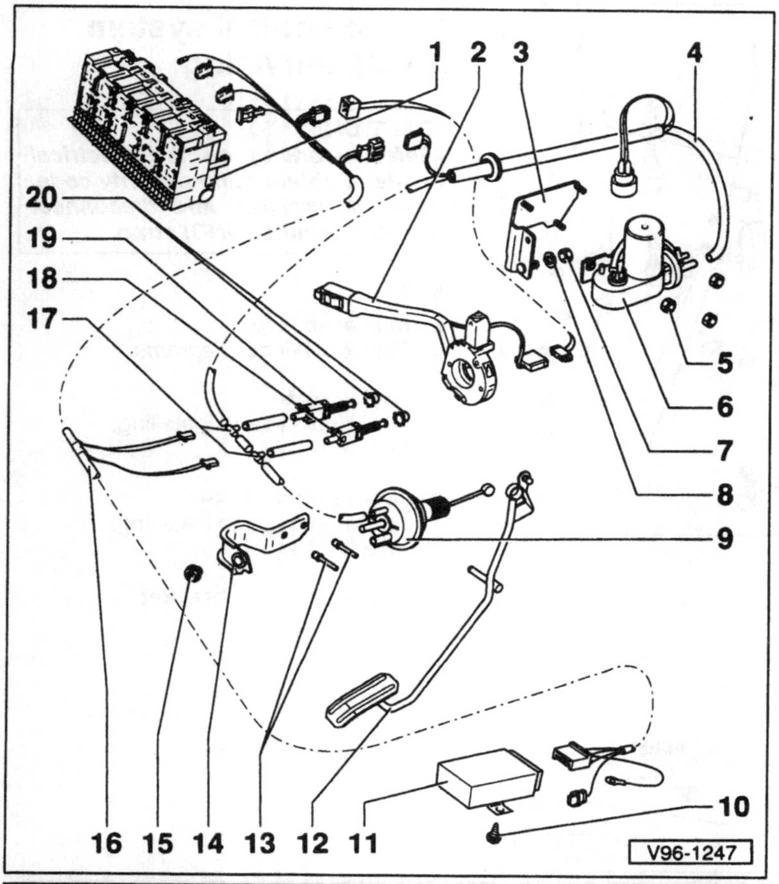

12 - **Accelerator pedal**

13 - **Pop-rivet**

14 - **Positioning element bracket**
- Riveted to pedal cluster.

15 - **Plastic nut**
- 3 Nm, (27 in lb)

16 - **CCS wiring harness**

17 - **Y-fitting for vacuum hose**

18 - **Vent valve with clutch pedal switch -F36-**
- Removing and installing,

⇒ Fig. 3.

19 - **Vent valve with brake pedal switch -F47-**
- Removing and installing,

⇒ Fig. 3.

20 - **Clip**

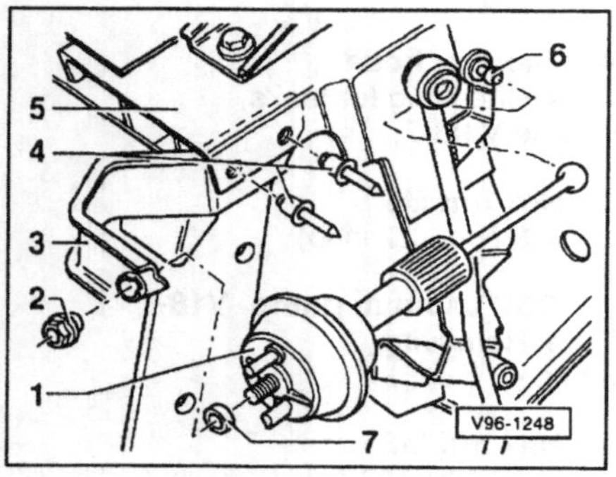

◄ **Fig. 1 Positioning element, removing**

1 - Positioning element

2 - Plastic nut

3 - Positioning element bracket

4 - Pop-rivet

5 - Pedal cluster

6 - Ball pin, on accelerator pedal

7 - Spacer - Only vehicles with automatic transmission

– Pull vacuum hose off positioning element -1-.

– Unclip positioning element pull rod from ball pin -6-.

– Remove plastic nut -2- and take out positioning element -1-.

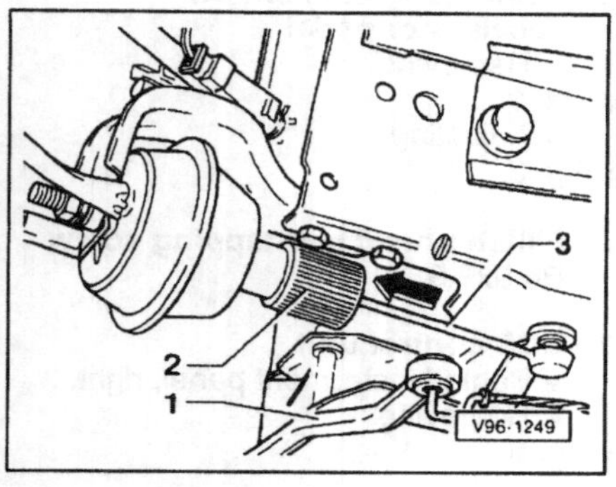

◄ **Fig. 2 Positioning element, adjusting**

– Hold accelerator pedal -1- in idle position.

– Pull adjusting sleeve -2- together with pull rod -3- in direction of arrow until resistance is felt.

– Lock adjusting sleeve by turning.

– Check accelerator pedal for freedom of movement.

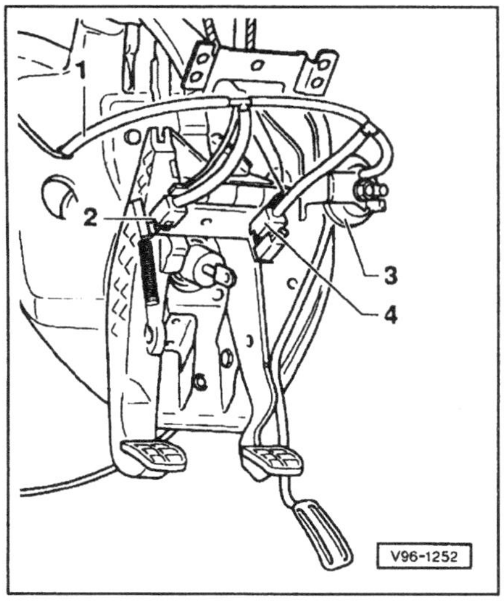

Fig. 3 Vent valves, removing and installing

1 - Vacuum hose

2 - Vent valve on clutch pedal

3 - Positioning element

4 - Vent valve on brake pedal

Removing:

– Pull connector and vacuum hose off.

– Remove vent valve from bracket.

Installing:

– Screw vent valve into bracket onto stop. When doing this, the pedal must be pulled fully to the rear.

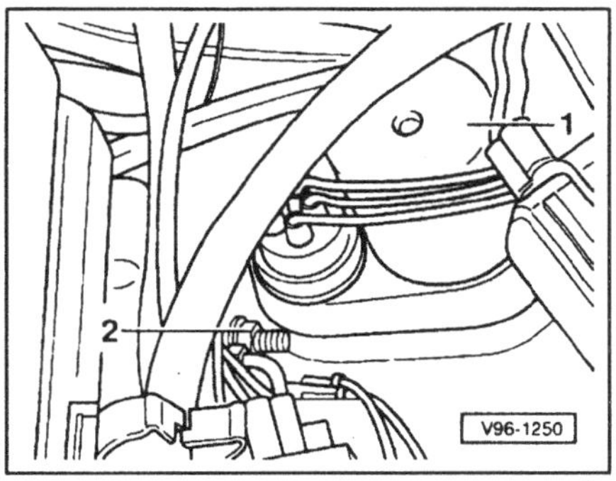

Fig. 4 CCS vacuum pump -V18-, removing

– Pull connector and vacuum hose from vacuum pump -1-.

– Remove hex nut -2-.

– Take out vacuum pump -1- complete with bracket.

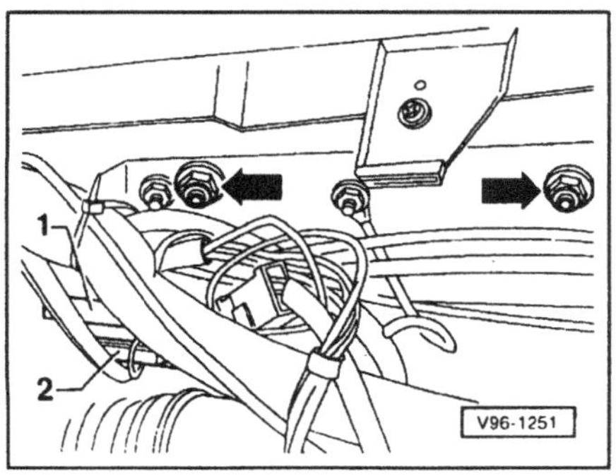

Fig. 5 CCS control unit -J213-, removing

– Remove glove compartment.

– Unbolt mounting plate with control unit -1- and relay -arrows-.

– Pull connector -2- off control unit -1-.

– Take out control unit with mounting plate.

Vacuum system, checking for leaks

- Pull vacuum hose off vacuum pump and seal.
- Push positioning element diaphragm in. To do this, press brake pedal down by hand to pressure point.
- Release brake pedal.
- Diaphragm must not move.

Central electric and relay panel, removing and installing

> ***CAUTION!***
>
> ***Before working on the electrical system, obtain radio security code, turn off ignition, and disconnect battery Ground (GND) strap.***

Notes:

- *The relay plate with fuse box is located under the instrument panel on the left-hand side.*
- *Removing the instrument panel trim,*

⇒ Repair Manual, Body Exterior/Interior, Repair Group 70.

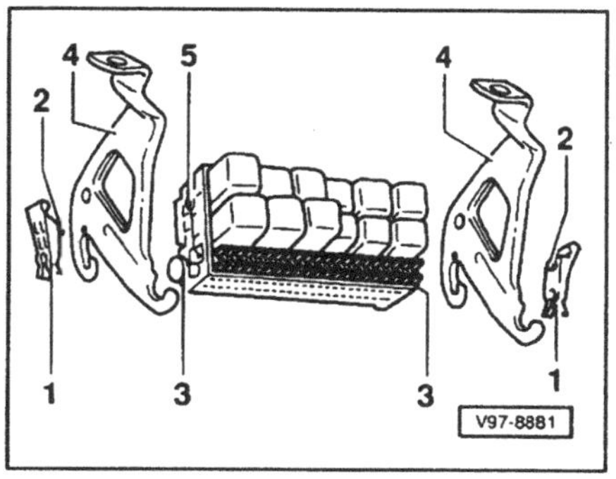

◄ – Press securing clips -1- on left- and right-hand side of relay plate in direction of arrow, until the retaining pins -2- on the clips can be pulled out of the recesses on the retainer -4-.

– Swing securing clips -1- forward and pull off relay plate pins -3-.

– Press left-hand retainer -4- outward until pin -5- is no longer engaged in the retainer and then unhook relay plate.

Note:

When installing, pin-5- and retaining pins -2- must engage in retainer recesses.

Multi–pin plug behind relay panel, removing and installing

Note:

Multi-pin connector pin assignment,
⇒ Electrical Wiring Diagrams.

Multi-pin plug, removing

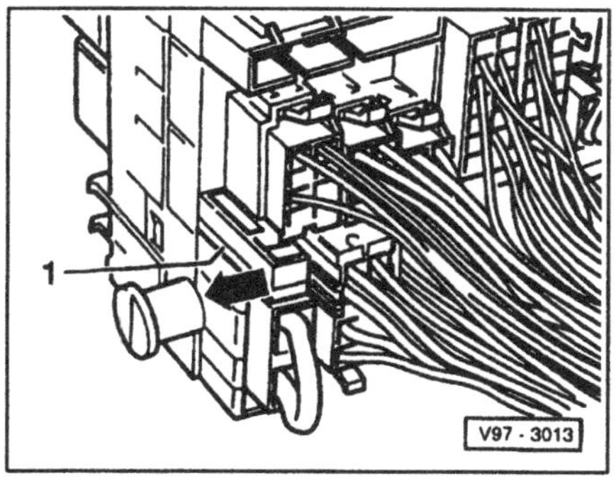

◄ – Pull the retaining slide -1-, on rear of relay plate, out of housing -arrow-, as far as possible (about 5 mm [0.2 in.]).

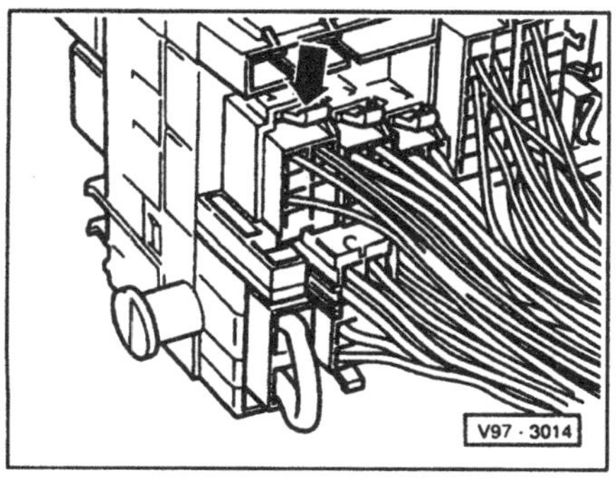

◄ – Press retaining lug on multi-pin connector -arrow- and pull out of relay plate.

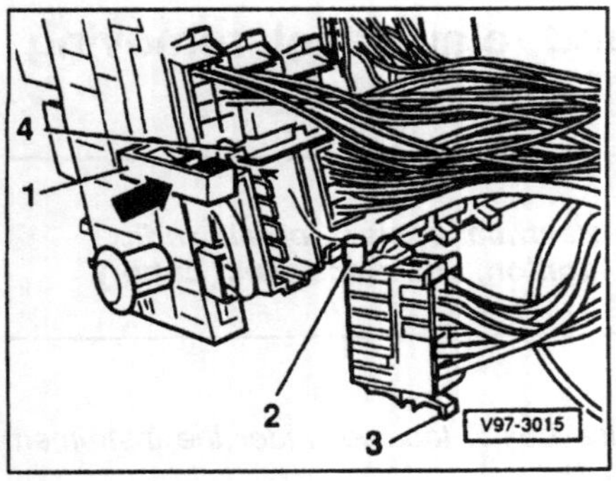

◄ **Multi-pin connector, installing**

Note:

Insert connector so that locking pin -2- faces toward locking slide -1-.

– Push multi-pin connector into appropriate location to stop, with locking pin -2- located in recess -4- until retaining tab -3- engages.

– Push locking slide -1- against stop -arrow-; only possible when all multi-pin connectors are pushed into retaining tab -3- engagement.

Electrical Wiring Diagrams

Section
X

__WARNING!__
Before proceeding with any work, read the warnings and cautions immediately following this index. Also read any warning and caution that accompanies a procedure in the manual.

Please read these WARNINGS and CAUTIONS before proceeding with maintenance and repair work.

WARNING—

• Some repairs may be beyond your capability. If you lack the skills, tools and equipment, or a suitable workplace for any procedure described in this manual, we suggest you leave such repairs to an authorized Volkswagen dealer service department, or other qualified shop.

• Volkswagen is constantly improving its cars. Sometimes these changes, both in parts and specifications, are made applicable to earlier models. Therefore, before starting any major jobs or repairs to components on which passenger safety may depend, consult your authorized Volkswagen dealer about Technical Bulletins that may have been issued since the editorial closing of this manual.

• Do not re-use any fasteners that are worn or deformed in normal use. Some fasteners are designed to be used only once, and are unreliable and may fail if used a second time. This includes, but is not limited to, nuts, bolts, washers, circlips and cotter pins. Always replace these fasteners with new parts.

• Never work under a lifted car unless it is solidly supported on stands designed for the purpose. Do not support a car on cinder blocks, hollow tiles or other props that may crumble under continuous load. Never work under a car that is supported solely by a jack. Never work under the car while the engine is running.

• If you are going to work under a car on the ground, make sure that the ground is level. Block the wheels to keep the car from rolling. Disconnect the battery negative terminal (ground strap) to prevent others from starting the car while you are under it.

• Never run the engine unless the work area is well ventilated. Carbon monoxide (CO) kills.

• Do not attempt to work on your vehicle if you do not feel well. You increase the danger of injury to yourself and others if you are tired, upset or have taken medication or any other substance that may impair you or keep you from being fully alert.

• Disconnect the battery negative (–) terminal (ground strap) whenever you work on the fuel system or the electrical system. Do not smoke or work near heaters or other fire hazards. Keep an approved fire extinguisher handy.

• Batteries produce explosive hydrogen gas. Keep sparks, lighted matches and open flame away from the top of the battery. If hydrogen gas escaping from the cap vents is ignited, it will ignite gas trapped in the cells and cause the battery to explode.

• Connect and disconnect battery cables, jumper cables or a battery charger only with the ignition switched off, to prevent sparks. Do not quick-charge the battery (for boost starting) for longer than one minute and do not allow charging voltage to exceed 16.5 volts. Wait at least one minute before boosting the battery a second time.

• Illuminate your work area adequately but safely. Use a portable safety light for working inside or under the car. Make sure the bulb is enclosed by a wire cage. The hot filament of an accidentally broken bulb can ignite spilled fuel or oil.

• Tie long hair behind your head. Do not wear a necktie, a scarf, loose clothing, or a necklace when you work near machine tools or running engines. If your hair, clothing, or jewelry were to get caught in the machinery, severe injury could result.

• Finger rings, bracelets and other jewelry should be removed so that they cannot cause electrical shorts, get caught in running machinery, or be crushed by heavy parts.

• Catch draining fuel, oil, or brake fluid in suitable containers. Do not use food or beverage containers that might mislead someone into drinking from them. Store flammable fluids away from fire hazards. Wipe up spills at once, but do not store the oily rags, which can ignite and burn spontaneously.

• Always observe good workshop practices. Wear goggles when you operate machine tools or work with battery acid. Gloves or other protective clothing should be worn whenever the job requires working with harmful substances.

• The Air-Conditioning (A/C) system is filled with chemical refrigerant, which is hazardous. The A/C system should be serviced only by trained technicians using approved refrigerant recovery/recycling equipment, trained in related safety precautions, and familiar with regulations governing the discharging and disposal of automotive chemical refrigerants.

• Do not expose any part of the A/C system to high temperatures such as an open flame. Excessive heat will increase system pressure and may cause the system to burst.

• Some aerosol tire inflators are highly flammable. Be extremely cautious when repairing a tire that may have been inflated using an aerosol tire inflator. Keep sparks, open flame or other sources of ignition away from the tire repair area. Inflate and deflate the tire at least four times before breaking the bead from the rim. Completely remove the tire from the rim before attempting any repair.

• Friction materials such as brake or clutch discs may contain asbestos fibers. Do not create dust by grinding, sanding, or by cleaning with compressed air. Avoid breathing asbestos fibers and asbestos dust. Breathing asbestos can cause serious diseases such as asbestosis or cancer, and may result in death.

• Greases, lubricants and other automotive chemicals contain toxic substances, many of which are absorbed directly through the skin. Read manufacturer's instructions and warnings carefully. Use hand and eye protection. Avoid direct skin contact.

• An airbag is an explosive device. Handled improperly or without adequate safeguards, it can be accidently activated and cause serious personal injury. To prevent personal injury or airbag system failure, **only trained Volkswagen dealer service technicians** should test, disassemble or service the airbag system.

WARNINGS and CAUTIONS continued on next page ▶

Please read these WARNINGS and CAUTIONS before proceeding with maintenance and repair work.

WARNING (continued) —

• ALWAYS disconnect the power supply before working on the airbag system, or when doing repairs that require removing airbag system components. Disconnect the battery negative (–) terminal and cover the battery.

• The airbag unit should not be removed, nor should any work be performed on the airbag system, without first reviewing the warnings and cautions that accompany the airbag system repair information in this manual.

• Never apply stickers or any other coverings on the steering wheel. Do not let chemical cleaners, oil or grease come into contact with vinyl covering of the airbag unit.

• Never open or otherwise attempt to repair airbag system parts. Always use new parts. Never leave airbag parts or the partially disassembled airbag system unattended.

• Never use a test light or volt meter to conduct electrical tests on the airbag system. The airbag system must only be tested by trained Volkswagen Service technicians using special airbag tester equipment. The airbag system must only be tested while it is installed in the car.

• Do not expose an airbag unit to temperature above 194°F (90°C), even for a brief period, while handling it during the repair process. Keep the unit clear of heat sources such as hot plates, soldering irons, heat lamps, welding equipment, or the like.

• Always replace airbag system components that have been mechanically damaged (bubbles, cracks, etc.).

• Always set down a removed airbag unit so that the horn pad is facing upward.

• When driving or riding in an airbag-equipped vehicle, never hold test equipment in your hands or lap while the vehicle is in motion. Objects between you and the airbag can increase the risk of injury in an accident.

• The storage, transportation and disposal of airbag units are subject to the laws for flammable solids.

CAUTION—

• If you lack the skills, tools and equipment, or a suitable workshop for any procedure described in this manual, we suggest you leave such repairs to an authorized Volkswagen dealer or other qualified shop. We especially urge you to consult an authorized Volkswagen dealer before beginning repairs on any car that may still be covered wholly or in part by any of the extensive warranties issued by Volkswagen of America, Inc.

• Volkswagen is constantly improving its cars and soemtimes these changes, both in parts and specifications, are made applicable to earlier models. Therefore part numbers in this manual are liset for reference only. Always check with your authorized Volkswagen dealer parts department for the latest information.

• Before starting a job, make certain that you have all the necessary tools and parts on hand. Read all the instructions thoroughly, do not attempt shortcuts. Use tools that are appropriate to the work and use only replacement parts meeting Volkswagen specifications. Makeshift tools, parts and procedures will not make good repairs.

• Use pneumatic and electric tools only to loosen threaded parts and fasteners. Never use these tools to tighten fasteners, especially on light alloy parts. Always use a torque wrench to tighten fasteners to the tightening torque specification listed.

• Be mindful of the environment and ecology. Before you drain the crankcase, find out the proper way to dispose of the oil. Do not pour oil onto the ground, down a drain, or into a stream, pond or lake. Consult local ordinances that govern the disposal of wastes.

• When connecting cables or ground straps, all connections must be cleaned to bare metal. Be sure mounting bolts are tight.

• Do not replace fuse links with fuse links of other gauge or with standard sire. Always replace the entire length of the fuse link.

• Before doing any electrical welding on vehicles equipped with Anti-lock Brakes (ABS), disconnect the battery negative termanial (ground strap) and the ABS Control Module connector.

• On cars equipped with ABS, the ABS control unit cannot withstand temperatures form a paint-drying booth or a heat lamp in excess of 203°F (95°C) and should not be subjected to temperatures in excess of 185°F (85°C) for more than two hours.

• On cars equipped with anti-theft radios, make sure you know the correct radio activation code before disconnecting the battery or removing the radio. If the wrong code is entered into the radio when power is restored, that radio may lock up and be rendered inoperable, even if the correct code is then entered.

• When boost-charging the battery, first remove the fuses for the Engine Control Module (ECM), the Transmission Control Module (TCM), the ABS Control Module, and the On-Board computer. In cases where one or more of these components is not seperately fused, disconnect the Control Module connector(s).

• Do not quick-charge the battery (for boost starting) for longer than one minute, and do not exceed 16.5 volts at the battery with the boosting cables attached. Wait at least one minute before boosting the battery a second time.

◀ **WARNINGS and CAUTIONS continued from previous page**

How to use the electrical wiring diagram section

This electrical wiring diagram section has been organized based on the way the electrical wiring diagrams were originally distributed. Each wiring diagram page appears in its original and complete form, including original page numbers or technical bulletin numbers. In addition, every page contains a unique Repair Manual page number to facilitate indexing and easy access to individual circuits.

The electrical wiring diagram section consists of a general information group, a component locations section showing relay, fuse, connector, and control module locations, and comprehensive indexes designated by model year and identified by black page tabs. These indexes list the various components and electrical systems on a particular vehicle model. The actual wiring diagrams themselves follow the indexes.

The diagram below shows important wiring diagram information.

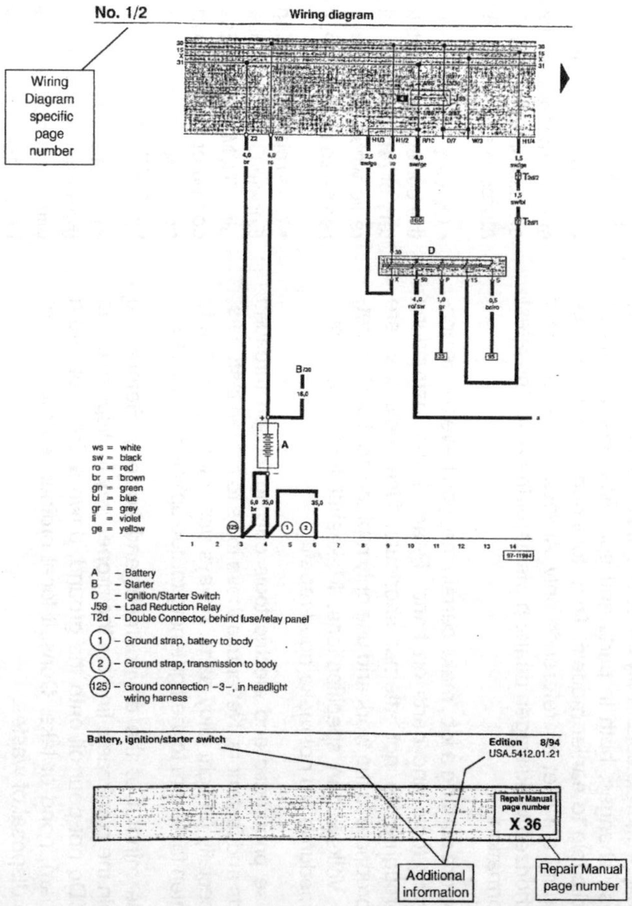

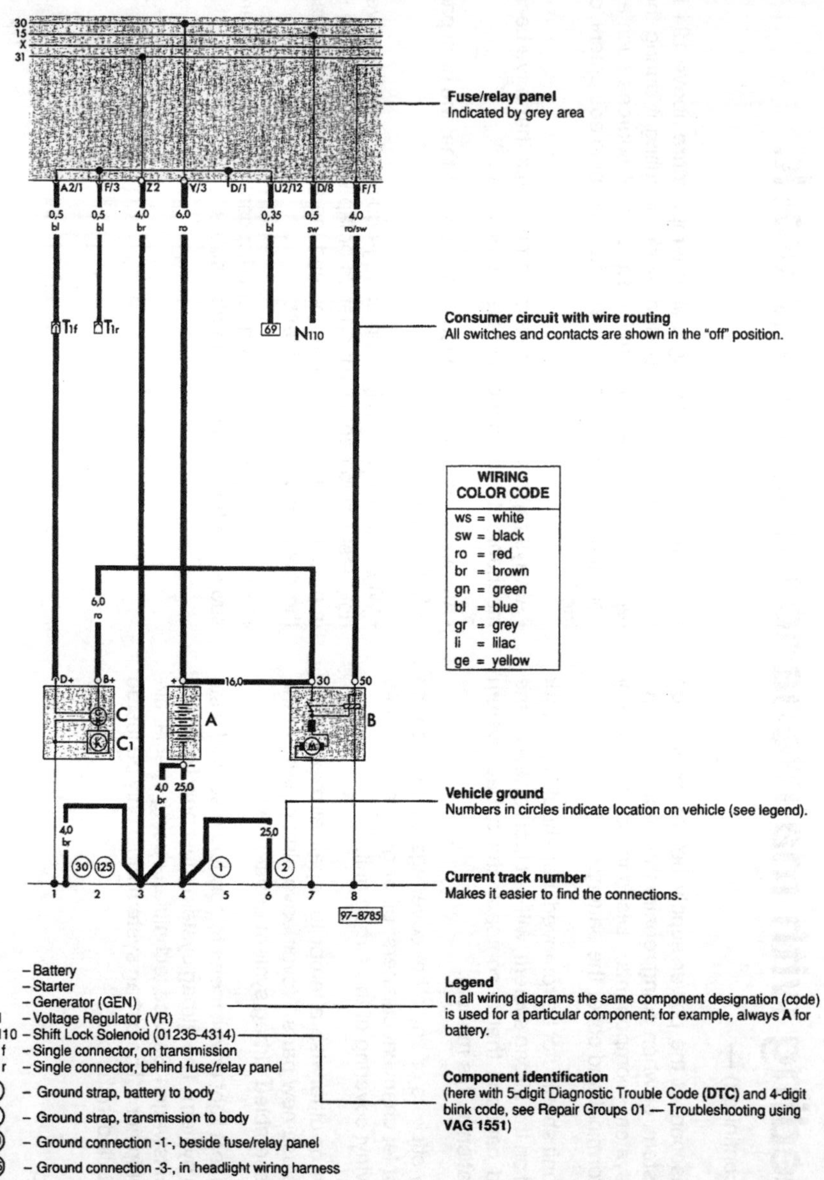

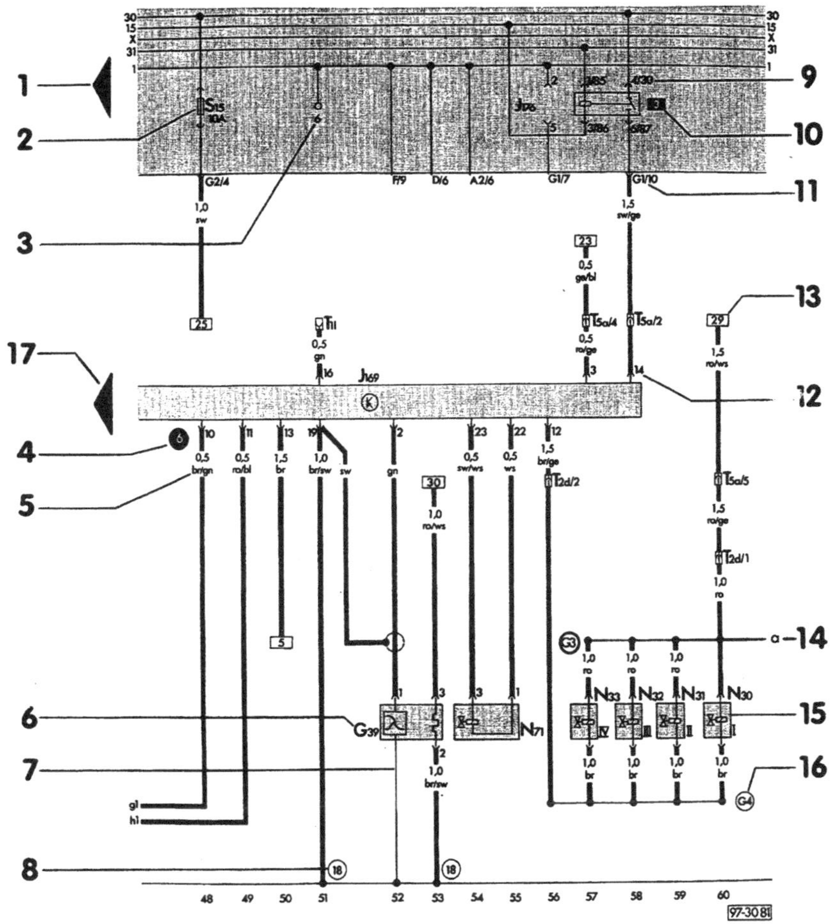

1 – **Arrow**
Indicates wiring circuit is continued on the previous and/or next page.

2 – **Fuse designation**
For example: **S15** = Fuse number **15**, 10 amps, in fuse/relay panel.

3 – **Test points on fuse/relay panel**

4 – **Test points used during troubleshooting**
Numbers in black circles may be found in a regular wiring diagram or in a wiring diagram included with troubleshooting.

5 – **Wire cross-section size (in mm²) and wire colors**
See chart at right for wire size conversion to American Wire Gauge (AWG). Wire color abbreviations are explained in color code chart on each page of the wiring diagram.

6 – **Component designation (code)**
Use legend at bottom of page to identify the component code.

7 – **Internal connections (thin lines)**
These connections are not always wires. Internal connections are current carrying and are listed to allow tracing of current flow inside components and wiring harnesses.

8 – **Wire harness ground connection**
Locations of ground connections are indicated in the legend.

9 – **Relay/control module connectors on fuse/relay panel**
Shows the individual terminals in a multi-point connector.
Example: 4/30
4 = contact **4** at location **3** on relay panel or relay carrier
30 = terminal **30 (B+)** on relay/control module

10 – **Relay location number**
Indicates relay location on fuse/relay panel.

11 – **Fuse/relay panel connectors**
Shows wiring of multi-point or single connectors on fuse/relay panel.
Example: G1/10 = multi-point connector **G1**, terminal **10**

12 – **Terminal designation**
Designation which appears on actual component and/or terminal number of a multi-point connector.

13 – **Wire continuation (current track number)**
Number in frame indicates current track where wire is continued.

14 – **Continuation of internal connections**
Letters indicate where connection continues on the previous and/or next page.

15 – **Component symbols (see page IV and V)**

16 – **Wiring harness wire connection**
Locations of wire connections are indicated in the legend.

17 – **Arrow**
Indicates component is continued on the previous and/or next page.

Wire Size

Wiring Diagrams identify wires by the metric wire size. Metric wire sizes indicate cross-sectional area in square millimeters (mm²). The chart below lists metric wire sizes and their approximate equivalents in American Wire Gauge (AWG) sizes.

Metric size Cross-section (mm²)	American Wire Gauge size (AWG)
0.35	22
0.50	20
0.75	18
1.00	16
1.50	14
2.50	12
4.00	10
6.00	8
16.00	4
25.00	2
35.00	2

Wiring diagram terminal (circuit) identification

Several wiring circuits in the vehicle's electrical system are identified with a number or letter designation. These circuits are identified the same in all wiring diagrams and are most commonly shown near the top of each page in the fuse/relay panel portion of the wiring diagram. The circuit designations may also be used to identify switch connector terminals (switch circuits). Following are the most common numbered/lettered circuits:

Terminal (circuit) 1-	Ignition coil/ignition distributor low voltage (typically used as an Engine Speed (RPM) signal for the tachometer)
Terminal (circuit) 15-	Switched Battery Positive Voltage (B+) from ignition/starter switch
Terminal (circuit) 30-	Battery Positive Voltage (B+), hot at all times
Terminal (circuit) 31-	Ground (GND)
Terminal (circuit) 50-	Starter control; switched B+ from ignition/starter switch
Terminal (circuit) 56-	Switched headlight B+ from light switch
Terminal (circuit) 58-	Switched parking light, taillight, illumination B+ from light switch
Terminal (circuit) S (SU)-	Key in ignition circuit; switched B+ from ignition/starter switch
Terminal (circuit) X-	Load reduction circuit; switched B+ from load reduction relay

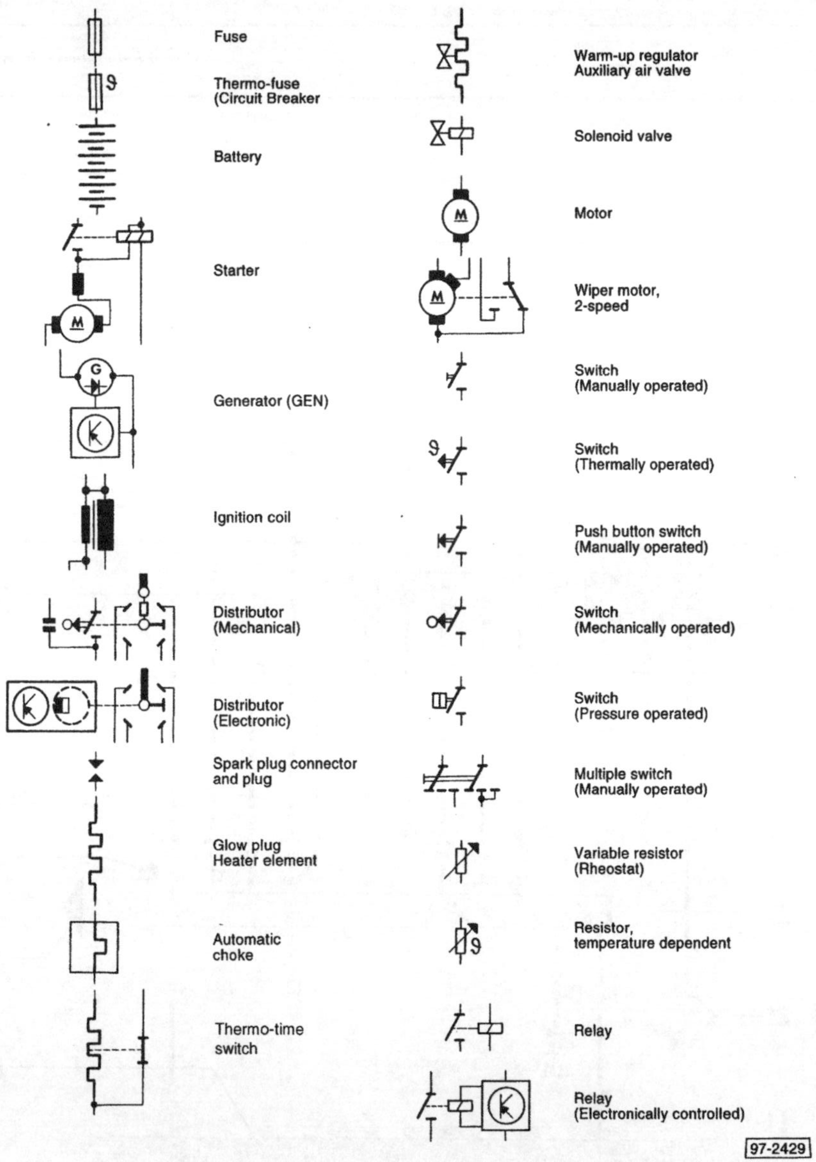
Fuse
Thermo-fuse (Circuit Breaker
Battery
Starter
Generator (GEN)
Ignition coil
Distributor (Mechanical)
Distributor (Electronic)
Spark plug connector and plug
Glow plug Heater element
Automatic choke
Thermo-time switch
Warm-up regulator Auxiliary air valve
Solenoid valve
Motor
Wiper motor, 2-speed
Switch (Manually operated)
Switch (Thermally operated)
Push button switch (Manually operated)
Switch (Mechanically operated)
Switch (Pressure operated)
Multiple switch (Manually operated)
Variable resistor (Rheostat)
Resistor, temperature dependent
Relay
Relay (Electronically controlled)
97-2429

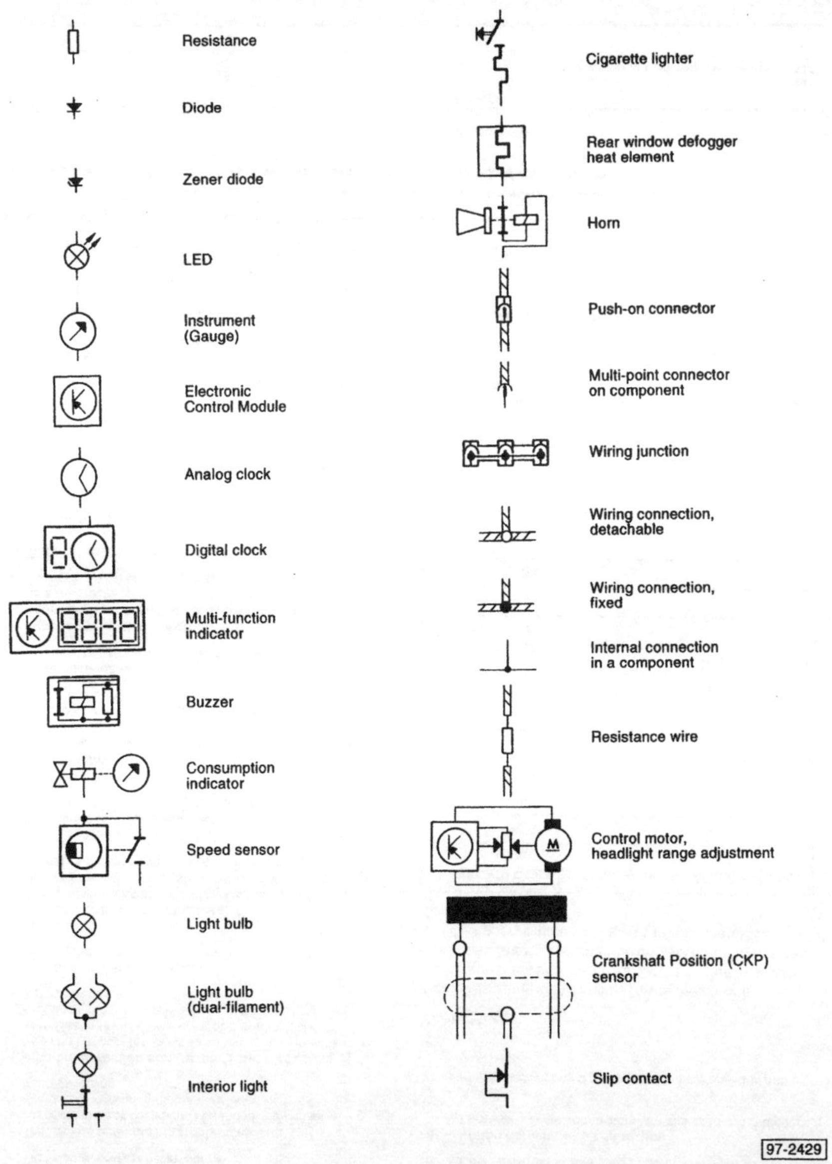
Resistance
Diode
Zener diode
LED
Instrument (Gauge)
Electronic Control Module
Analog clock
Digital clock
Multi-function indicator
Buzzer
Consumption indicator
Speed sensor
Light bulb
Light bulb (dual-filament)
Interior light
Cigarette lighter
Rear window defogger heat element
Horn
Push-on connector
Multi-point connector on component
Wiring junction
Wiring connection, detachable
Wiring connection, fixed
Internal connection in a component
Resistance wire
Control motor, headlight range adjustment
Crankshaft Position (CKP) sensor
Slip contact
97-2429

2. Test Equipment

Special Tools

The most basic electrical troubleshooting calls for specialized electrical test equipment. Dealer Service Department technicians do this work using Special Tools, many of which are designed specifically for working on electrical systems. Proper tools are essential to ensure accurate test results and to prevent damage to sensitive components.

The Special Tools shown in this manual are identified by the order numbers that can be used to special-order the tools through your authorized dealer Parts Department.

2.1 LED Test Light

A test light is an inexpensive tool used to perform many simple electrical tests that would otherwise require a multimeter. It lights to indicate when there is Voltage potential between any two test-points in a circuit.

With one probe contacting the battery negative (–) terminal or a known ground (GND) connection, the other probe can be used to check for Voltage at other points in the circuit. The test light will light when the test probe contacts a Voltage source, and current flows from the Voltage source to ground (GND).

With one probe contacting a known source of Voltage, the other probe can be used to check for continuity to ground (GND) at other points in the circuit. Current will flow and the test light will light when the test-point provides the known Voltage source with a path to ground (GND).

Caution

A common test light (with incandescent bulb) may no longer be used. The bulb's high current draw will damage sensitive electronic components.

LED Test Light (US 1115)

Used to check for Voltage reaching components or when searching for malfunctions in a circuit.

Voltage range: 3 to 48 VDC

CAUTION

A common test light (with incandescent bulb) may no longer be used. The bulb's high current draw will damage sensitive electronic components.

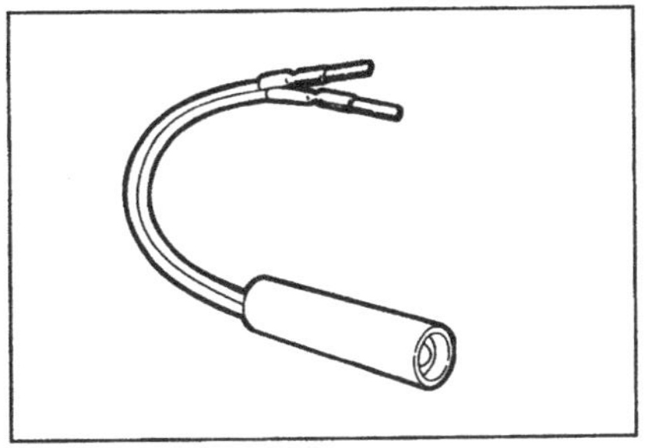

Note

This tool has been superseded by Special Tool VAG 1527B, LED Tester. (US 1115 is still considered an acceptable equivalent to VAG 1527B).

LED Tester (VAG 1527B)
Order Number: TAG 152 7B0 28 ZEL

Two-pole Voltage tester with two LEDs (light emitting diodes), suitable for measuring from 3 to 48 Volts DC or AC.

When used to test DC circuits, the positive (+) or negative (–) LED will light, depending on polarity. When used to test AC circuits, both LEDs will light.

Supplied with banana clip and detachable probe-can be used directly with VAG 1598 Test Box.

For troubleshooting all common electrical and electronic components in automobiles. Safe for testing electronic components and circuits because of extremely low current consumption (1.5 mA maximum).

Voltage range: 3 to 48 Volts (AC or DC)

CAUTION

A common test light (with incandescent bulb) may no longer be used. The bulb's high current draw will damage sensitive electronic components.

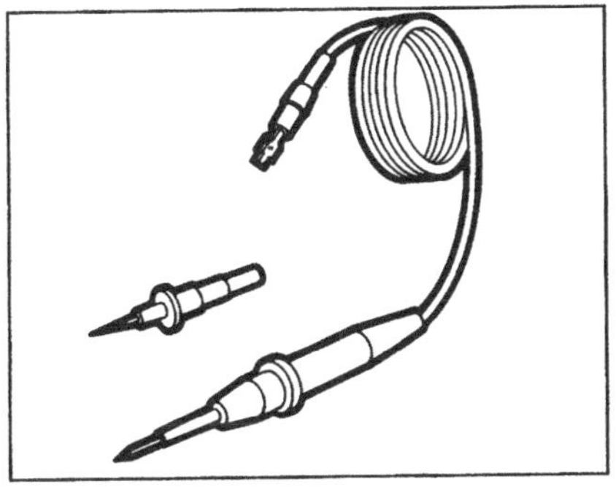

Note

This tool supersedes Special Tool US 1115 LED Test Light.

Making an LED Test Light

A lower-cost alternative to buying an LED Test Light such as US 1115 or VAG 1527B is to make one, using parts available from most any electronics supply outlet.

Assemble the components as shown. Use needle nose pliers to hold the parts and to act as a heat dam while soldering. Insulate the connections with heat-shrinkable tubing or electrical tape.

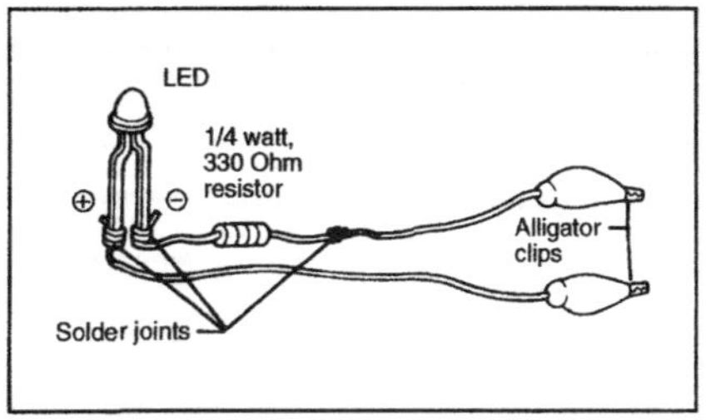

Do-it-yourself LED Test Light for safely testing sensitive solid-state circuits.

Parts:

1. LED (1)
2. 1/4 watt, 330 Ω resistor (1)
3. Alligator clips (2)
4. Wire, Solder and Soldering Iron
5. Heat-shrinkable tubing or electrical tape

2.2 Multimeter or Volt/Ohm Meter (VOM)

A multimeter or Volt-Ohm meter (VOM) is used to measure Voltage, resistance (Ohms) or current (amps or millamps). Two types of meters are in common use. The analog or swing-needle Volt-Ohm meter displays test values according to the position of a needle on the meter face. A digital multimeter displays test values as numbers. The meters shown here and recommended are digital multimeters.

Accuracy is an important consideration when choosing a meter. Analog or swing-needle Voltmeters are generally rated for accuracy as a percentage of full-scale on the meter face. A typical analog meter may be rated as accurate to ± 3% of full scale. Some analog meters offer 10-Volt and 50-Volt scales, with no range in between. For 12-Volt automotive electrical systems, this means testing on the 50-Volt scale, with a corresponding decrease in accuracy. On a 50-Volt scale, ± 3% accuracy is equal to ± 1.5 Volts. While still useful for testing circuits such as lights and horns, analog meters may not be accurate enough for use where precise measurements are required.

Digital multimeters are preferred for precise measurement and for electronics work because most digital multimeters are more accurate then most analog meters. Another advantage of digital multimeters is that they are less likely to be misread, since there is no needle position to be misinterpreted or distorted by reading at an angle.

Perhaps the most important consideration is input impedance-how much of a load the meter places on the circuit being tested. Meters with low input impedance place a greater load on the circuit because they allow more current flow through the meter. This can exceed design limits and be damaging to sensitive electronic components.

Most digital multimeters have very high input impedance-10 megohms (10,000,000 Ω) or higher-and make their measurements while allowing very little current flow through the meter. The meter itself is not a load in the circuit, and does not induce damaging current flow.

Multimeter Kit - digital (US 1119, VAG 1526)
Order Number: TU1 119 000 00 KTM

Used for monitoring electronic engine controls (Example: setting/checking the idle speed control system) and for general electrical troubleshooting.

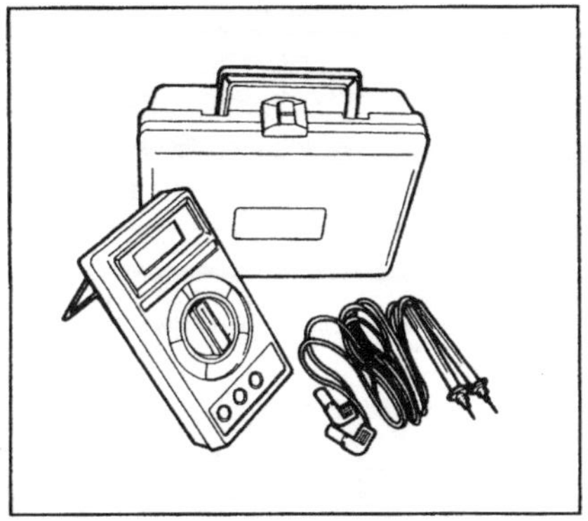

> ***Note***
>
> *This tool has been superseded by the Fluke 83® Digital Multimeter. (US 1119 or VAG 1526 are still considered acceptable equivalents).*

Fluke 83® Digital Multimeter
Order Number: TFL UKE 830 00 RSE

Used for monitoring electronic engine controls (Example: setting/checking the idle speed control system) and for general electrical troubleshooting.

- 10 Megohm input impedance
- ± 0.3% DC Voltage accuracy
- fully overload-protected

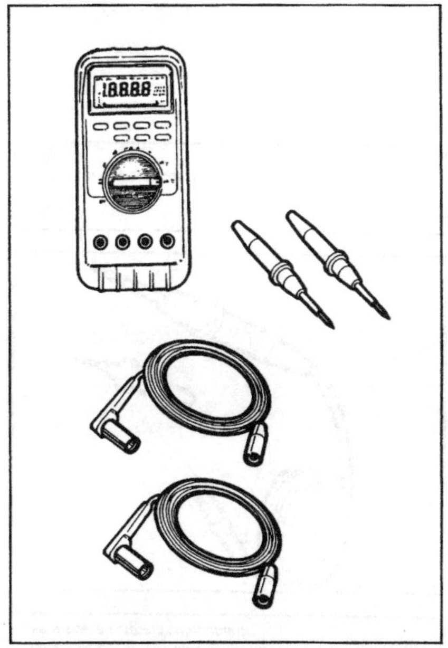

> ***Note***
>
> *This tool supersedes Special Tool US 1119 Digital Multimeter Kit.*

2.3 Test Kits

Adaptor Kit (VW 1594)
Order Number: TV1 594 400 97 KTM

Used to connect electrical measuring equipment to various harness connectors in the electrical system.

> **CAUTION**
>
> *Test connections made carelessly or without proper tools can damage harness connectors, causing poor connections and future problems.*

Test Box (VAG 1598)
Order Number: WAG 159 800 00 VOA

Used to perform static and dynamic tests of electrical and electronic systems without damaging miniature contacts, especially where limited access to connectors makes testing difficult.

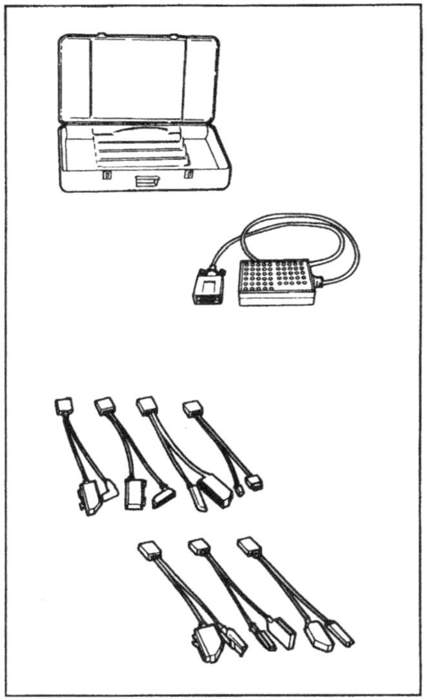

Tester (VAG 1466)

Used to systematically troubleshoot electrical circuits.

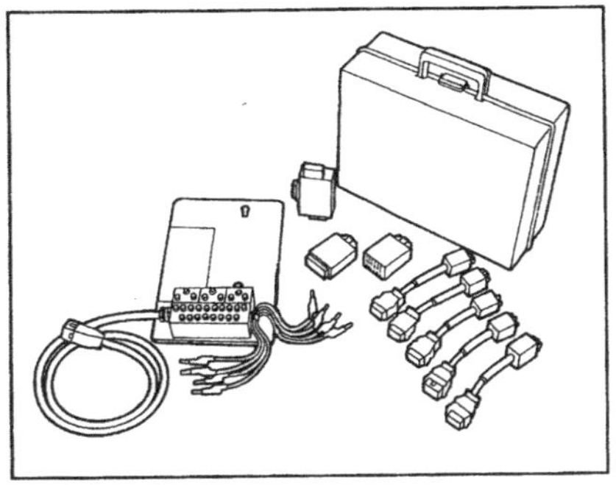

2.4 Computer Memory Saver

Used on vehicles with theft-protected radios to prevent the radio from electronically locking when battery power is disconnected. Use of this tool eliminates the need to re-activate the radio (reenter the correct code) after reconnecting battery power.

> **WARNING**
>
> *Always separate the airbag Voltage supply connector before using a computer memory saver. Failure to do so may result in accidental activation of the airbag. For more information, see the Circuit Section in this manual entitled Airbag, or see the Repair Manual.*

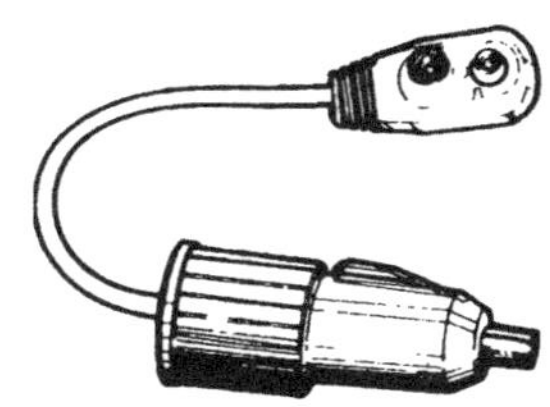

Snap-On
Computer Memory Saver
Part No. YA960

91-A051

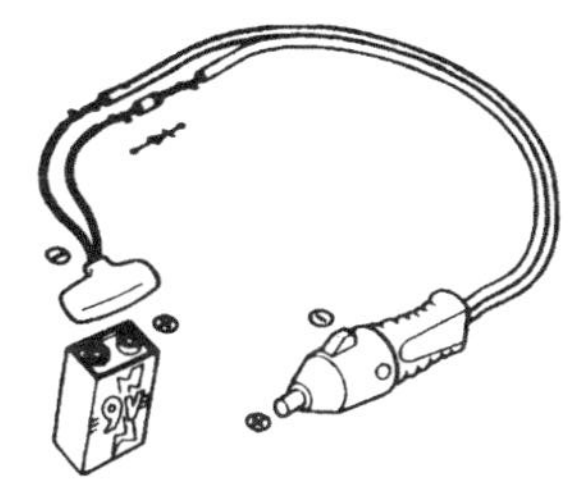

Home-made computer memory saver

91-A052

3. Handling Components and Connectors

Harness Connectors

The harness connectors used throughout the vehicle are designed to positively lock into place. Press on wire lock as shown to release the connector.

> ***CAUTION***
>
> *To disconnect, pull only on the connector body. Never pull on the wires themselves.*

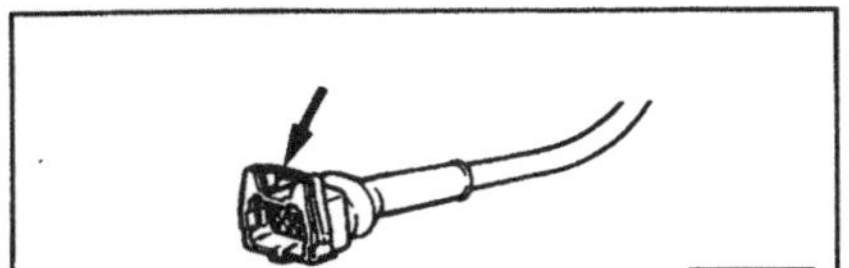
91-A051

Typical wiring harness connector. Push on wire lock (arrow) and gently pull on connector to release.

Making Test Connections

Many electrical troubleshooting tests will require hooking up to wiring harness connectors, or socket connectors on electrical components. Test connections made carelessly or without proper tools can easily damage the connectors, causing poor connections and future problems.

To avoid connector damage, test connections to wiring harness connectors must be made using small, flat-blade terminals which will mate properly with the connectors. The Special Tools recommended include special adapters which can be used to make test connections safely and prevent connector damage. See **2. Test Equipment.**

All test connections to harness connectors should be made using the proper adapter or correct size flat connectors that will not deform the connector cavities.

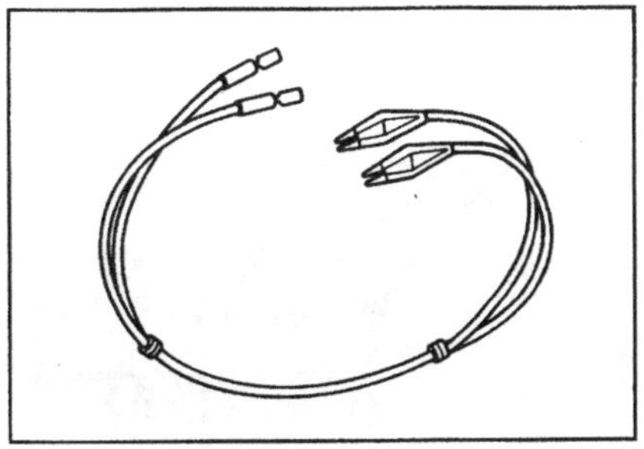

Alligator-clip test leads with flat connectors (Part Number: N 17 457 2)

Wiring Harness Repairs

When repairing wiring harnesses, use only high-quality electrical connectors suitable for use with electronic components.

To make connections, use crimping pliers that make a "W"-type crimp. Only this type of crimp provides the necessary mechanical strength.

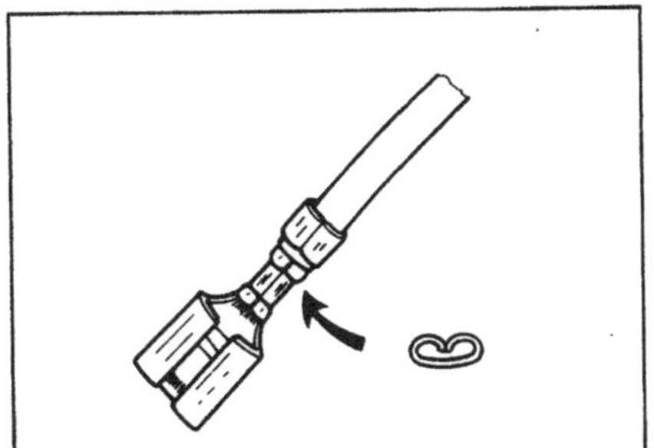
97-A400

Typical "W"-type crimp connection

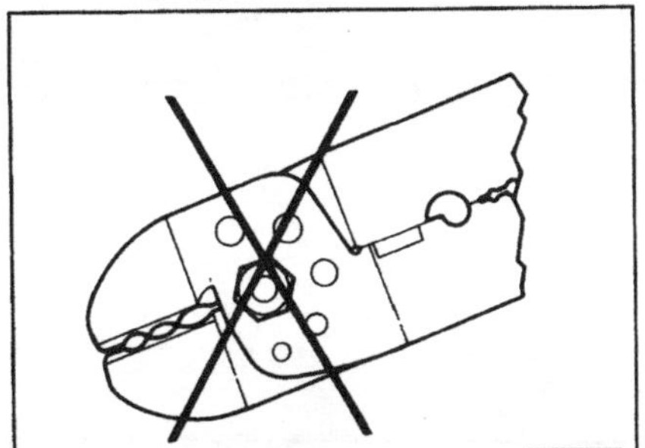
97-A401

Typical wire crimping pliers, not suitable for making "W"-type crimps.

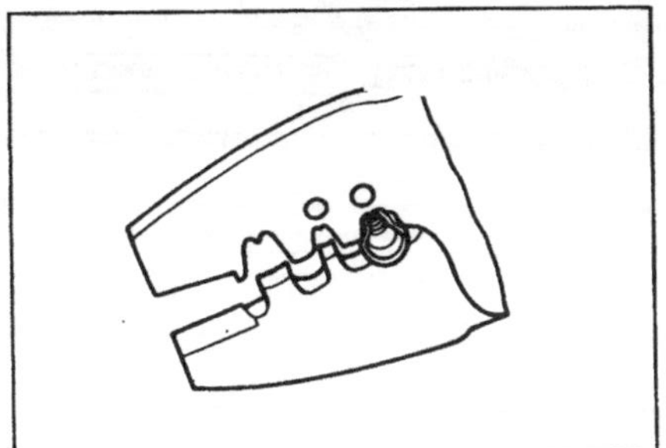
97-A402

Wire crimping pliers, used to make "W"-type crimps.

Connector repair kits, connectors and the correct crimping pliers are available through your authorized dealer Parts Department.

4. Troubleshooting

4.1 Basic Electricity

Electricity is defined by three basic elements: Voltage, Current and Resistance.

Voltage

Voltage is a measure of electromotive force, sometimes referred to as electrical "pressure". It can be described as the difference in potential (potential for the flow of electricity) between any two points in a circuit.

A typical automobile battery, for example, has a difference in potential of about 12 Volts between the positive (+) terminal and the negative (–) terminal.

The basic units of electrical potential are **Volts** (V). Very low Voltages are expressed as **millivolts** (mV).
1 V = 1000 mV; 1 mV = .001 V

Current

Current is the term describing the flow of electricity through a conductor. In a complete circuit, potential (Voltage) will cause current to flow from positive (+) to negative (–).

The basic units of current flow are amperes or **amps** (A). Small amounts of current flow are often measured in **milliamps** (mA).
1 A = 1000 mA; 1 mA = .001 A

Resistance

Resistance resists or opposes the flow of electricity. Conductors are made from materials of low resistance that allow electricity to flow easily. Insulators are materials of very high resistance that inhibit the low of electricity.

The basic unit of resistance is the **Ohm Ω**. High resistance values are often expressed as Kilohms (KΩ).
1 KΩ = 1000Ω

Resistance vs. Current Flow

The basic rule of electricity (Ohm's Law) states that one unit of force (1 Volt) is required for one unit of current (1 am) to flow against one unit of resistance (1 Ohm). From Ohm's Law, we also know that:

Voltage = Current x Resistance

When Voltage is approximately constant, as in an automobile electrical system, current and resistance affect each other. As resistance increases, there will be less current flow. And lower resistance will permit higher current flow.

Higher resistance = lower current flow
Example: Corrosion on a headlight connector (higher resistance) causes the light to be dim (lower current flow).

Lower resistance = higher current flow
Example: A damaged wire shorted to ground (GND) (lower resistance) overloads circuit capacity (higher current flow) and blows a fuse.

Definition of a Circuit

Four things are required for current to flow in any electrical circuit, and for that circuit to function as intended:

- **Power Source** (Voltage)
- **Conductors** (wires, printed circuits, etc.)
- **Load or Consumer** (a user of electrical power)
- **Complete Circuit** (a connection to ground [GND])

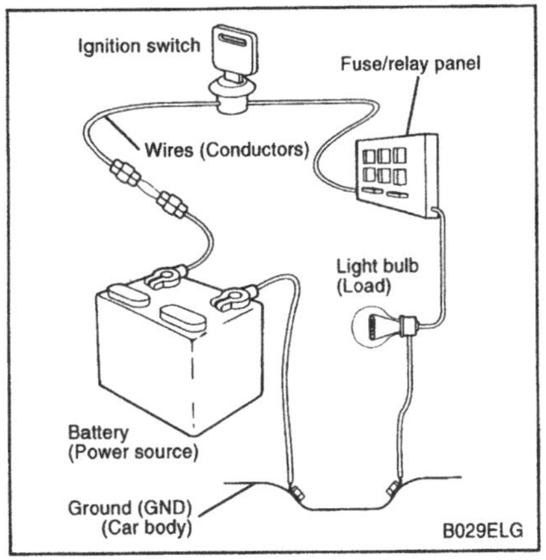

A complete circuit

Open Circuits

An open circuit is an incomplete circuit. An open circuit occurs when some kind of malfunction interrupts the circuit path and prevents current flow. Some common causes of open circuits are:

- broken wire
- loose or disconnected connector
- loose or damaged connector terminal
- corrosion
- malfunctioning fuse or component

Test for an open circuit by checking continuity using an Ohmmeter (multimeter), or by checking for Voltage at various points of the circuit using a test light or Voltmeter (multimeter). See **4.4 Checking Wiring and Components.**

Short Circuits

A short circuit is an unintended complete circuit. A short circuit occurs when some kind of malfunction causes current flow to follow the wrong path.

A short circuit to ground (GND) (grounded circuit) may prevent Voltage from reaching a component. If Voltage is shorted directly to ground (GND), bypassing any load, the unrestricted current flow will damage fuses wires or components. Some common causes of short circuits are:

- damaged wire or wiring harness
- malfunctioning insulation
- internally damaged component
- incorrect connection

Test for a short circuit to ground (GND) using a multimeter or a test light to indicate circuit malfunctions and abnormal current flow paths. See **4.5 Checking for Short Circuit to Ground (GND).**

Troubleshooting Procedure

Verify the complaint – Check the complaint. Try to understand the problem. If possible, let the driver show you what happens. Check all functions of the system and note the symptoms before starting any testing or disassembly.

Analyze the problem – Identify the part of the electrical system that is most likely to be causing the problem. Find the Circuit Section in the manual that applies to that part of the system. Find the wiring diagram that applies to the vehicle. By following the circuit from a ground (GND) back to the power source, get an understanding of how the circuit works.

Find the problem – You will find the problem if you follow a simple and logical step-by-step procedure. Test portions of the circuit one at a time, starting with he area or component most likely to be malfunctioning. Test first at points that you can reach most easily.

Repair the problem – When you find the cause of the problem, make the repair. Use appropriate tools and procedures.

Check the results – Be sure it works. Check the functions of all parts of the circuit that you worked on.

Working on the Electrical System

A test light or a multimeter can be very helpful for testing circuits. See **2. Test Equipment** for more information on test equipment.

Current flow is logical, always moving from the highest potential at the Voltage source (+) toward the lowest potential at ground (–). Using a wiring diagram to trace a circuit, you should start with the ground (GND) and then follow the wires back to the source of power.

To troubleshoot a circuit:

1. Inspect all connections, especially grounds (GND). Make sure they are clean, tight and corrosion-free.
2. Check the fuses.

> ***Note***
>
> *Repeated fuse failures are the sign of a malfunctioning wire, a failed component, or a short to ground (GND) somewhere in the circuit.*

3. Check for Voltage reaching particular components or points in the circuit.
4. Check continuity between points to look for breaks in the circuit (open circuit).
5. Check Voltage drop at connections, especially ground (GND) connections.

4.2 Checking for voltage

Checking for Voltage confirms that the circuit is uninterrupted between the Voltage source and the test point. The example illustrates troubleshooting the high-speed circuit for the radiator cooling fan.

Voltage: If the test light or multimeter indicates Voltage potential, then the circuit between the Voltage source and terminal 1 of the fan connector is OK.

NO Voltage: Power is not reaching the fan connector. The fan is probably OK. Look for a malfunction somewhere between the Voltage source and the fan connector. (Example: Check for Voltage reaching terminal E87La of the fuse/relay panel)

> ***CAUTION***
>
> *Direct contact with meter probes at the connector terminals can easily damage the small contacts, causing poor connections and risking future intermittent malfunctions.*
>
> *Special Tools include adapters for making test connections safely and preventing connector damage. See **2. Test Equipment.***

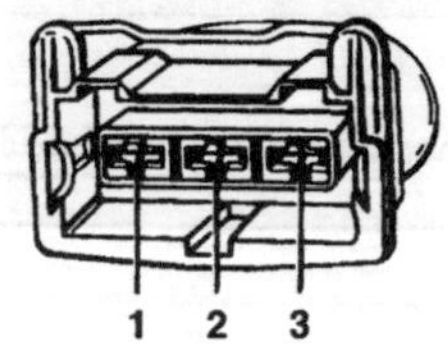

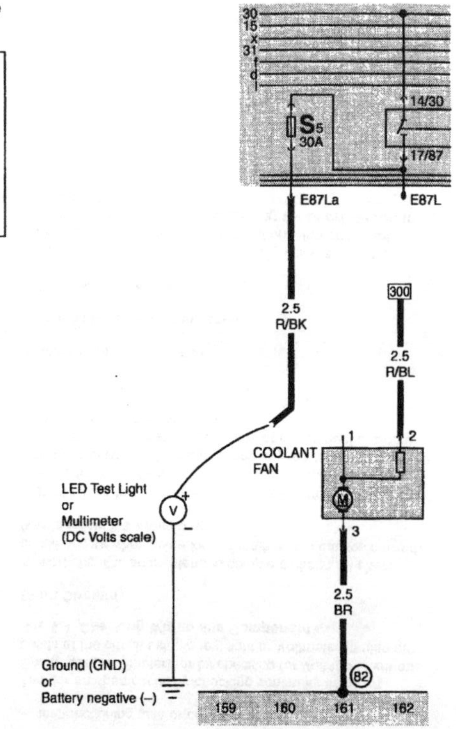

4.3 Checking Ground (GND) Connections

Checking ground (GND) connections as shown confirms that the circuit is complete - that the necessary path to ground (GND) is uninterrupted and current can flow in the circuit. The example illustrates troubleshooting the high-speed circuit for the radiator cooling fan.

Voltage: If the test light or multimeter indicates Voltage, then there is potential for current flow between the known Voltage source and ground (GND) at the test point. The ground (GND) side of the circuit, between terminal 3 of the fan connector and battery negative (–), is OK.

NO Voltage: The test point is not providing a path that completes the circuit to ground (GND). The fan is probably OK. Look for a malfunction somewhere in the wiring between the fan connector and chassis ground (GND). Also check the mechanical ground connection at the chassis (body).

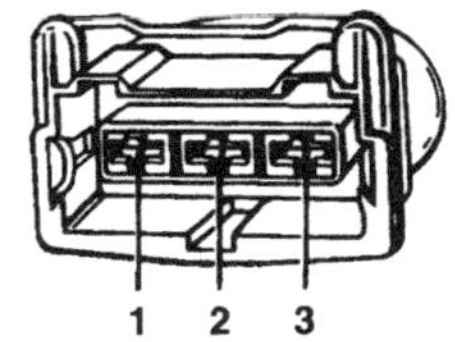

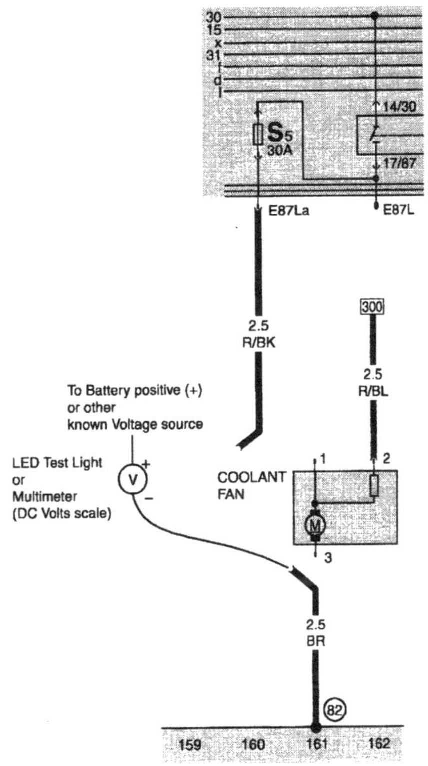

Continuity (approximately 0 Ω): Little or no resistance indicates that there is a continuous conductive path between the two test points - the circuit's ground (GND) path between terminal 3 and battery negative (–) is OK.

No Continuity: There is resistance to current flow in the ground (GND) side of the circuit. The fan is probably OK. Look for a malfunction somewhere in the wiring between the fan connector and chassis ground (GND). Also check the mechanical ground (GND) connection at the chassis (body).

> ***CAUTION***
>
> *direct contact with meter probes at the connector terminals can easily damage the small contacts, causing poor connections and risking future intermittent malfunctions.*
>
> *Special Tools include adapters for making test connections safely and preventing connector damage. See **2. Test Equipment.***

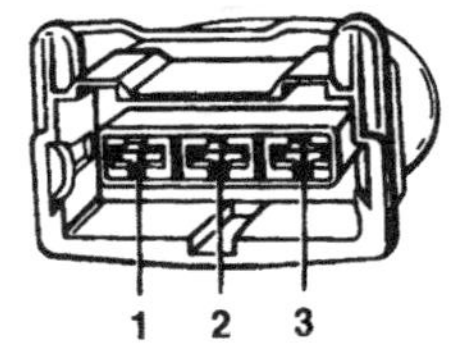

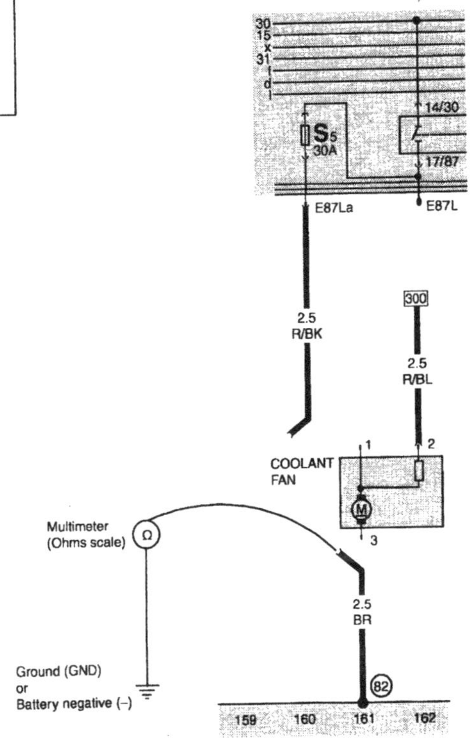

4.4 Checking Wiring and Components

Resistance or Continuity

Checking a portion of the wiring harness or a component as shown indicates whether or not there is a continuous conductive path - whether current can flow between the two test points. The example illustrates troubleshooting the Close Throttle Position switch and the Wide Open Throttle Position switch in the fuel injection system.

> **CAUTION**
>
> *Resistance measurements and continuity checks must always be made with all power to the circuit or component switched OFF. When testing continuity in a circuit that is always powered (fuse/relay panel "30" circuit for example) disconnect the battery before testing.*
>
> *Always use a digital (low current) meter. An ohmmeter, or the Ohms scale of a multimeter, measures resistance by passing a small amount of current through the circuit or component being checked.*
>
> *Improper testing may damage sensitive electronic components.*

Continuity (approximately 0 Ω): Little or no resistance indicates that there is a continuous conductive path between the two test points. As shown, this is correct for the normally Closed Throttle Position switch in the Closed Throttle Position (switch not actuated). Also check that the switch opens (no continuity) when actuated by the throttle.

No Continuity: There is resistance to current flow through the switch. In the example, this indicates that the normally Closed Throttle Position switch or the wires between it and the connector (t3) are malfunctioning.

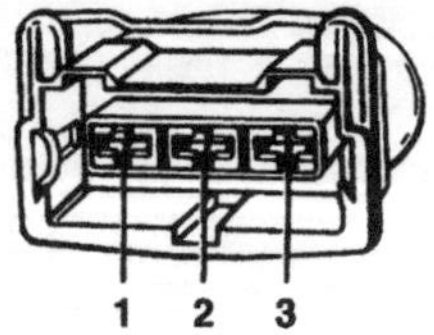

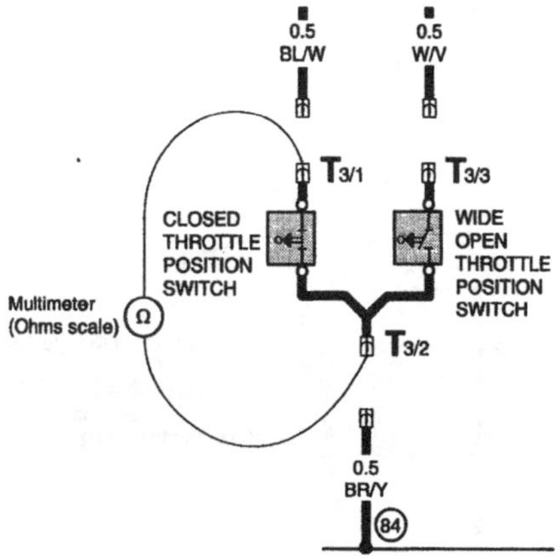

> **CAUTION**
>
> *Direct contact with meter probes at the connector terminals can easily damage the small contacts, causing poor connections and risking future intermittent malfunctions.*
>
> *Special Tools include adapters for making test connections safely and preventing connector damage. See **2. Test Equipment**.*

Continuity (approximately 0 Ω): Little or no resistance indicates that there is a continuous conductive path between the two test points. In the example, this indicates that the normally Wide Open Throttle Position switch is malfunctioning.

No Continuity: There is no connection-an open circuit. As shown, this is correct for the normally Wide Open Throttle Position switch. Also check that the switch closes, completing the circuit, when actuated by the Wide Open Throttle Postion switch.

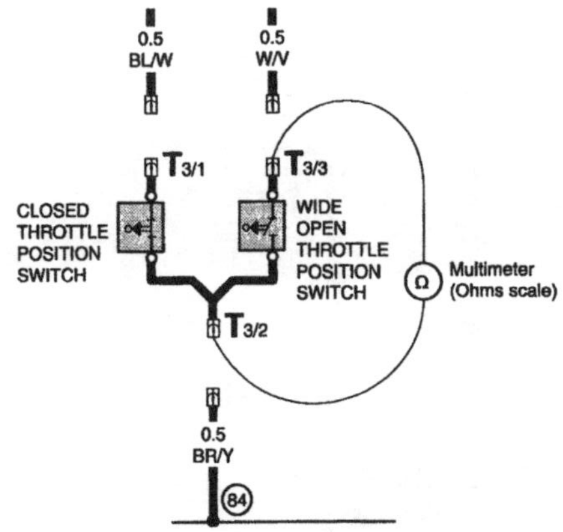

Voltage Drop

Checking Voltage drop across connections or components as shown will indicate whether there is abnormal resistance creating an additional load in the circuit - consuming power and dropping the Voltage available to other parts of the circuit.

> **Note**
>
> *Voltage drop measurements can only be made when the circuit is powered and there is normal current flow.*

The example illustrates troubleshooting the back-up light switch. The switch is in the circuit to switch power to the back-up lights On and Off. When Reverse gear is selected and the switch is closed, it should have very little resistance and not be a consumer.

If dirt or corrosion on the switch contacts creates resistance, some of the available battery Voltage goes into overcoming that resistance. Less Voltage is available to light the back-up lights, and they will not be as bright.

> **Note**
>
> *An ohmmeter can measure resistance (or check continuity) only when the circuit is not powered, i.e. when there is almost no load.*

> ***CAUTION***
>
> *Direct contact with meter probes at the connector terminals can easily damage the small contacts, causing poor connections and risking future intermittent malfunctions.*
>
> *Special Tools include adapters for making test connections safely and preventing connector damage.* *See 2.* ***Test Equipment.***

Low Voltage: A low Voltage reading across the two switch contacts indicates almost no difference in potential. Resistance across the switch is low - most of the battery Voltage is passing through the switch and is still available to power the lights.

High Voltage: Any significant Voltage reading indicates a difference in potential across the switch contacts. Excessive resistance is loading the circuit, causing a Voltage drop - Voltage is consumed overcoming the resistance of the switch, and less is left to power the lights.

Maximum allowable Voltage drops recommended by the Society of Automotive Engineers are:

- 0.0 Volt for small wire connections
- 0.1 Volt for high current connections (example: fuel pump, headlights)
- 0.1 Volt for ground (GND) connections
- 0.2 Volt for high-current cables (example: battery/starter cable)
- 0.3 Volt for switch or relay contacts

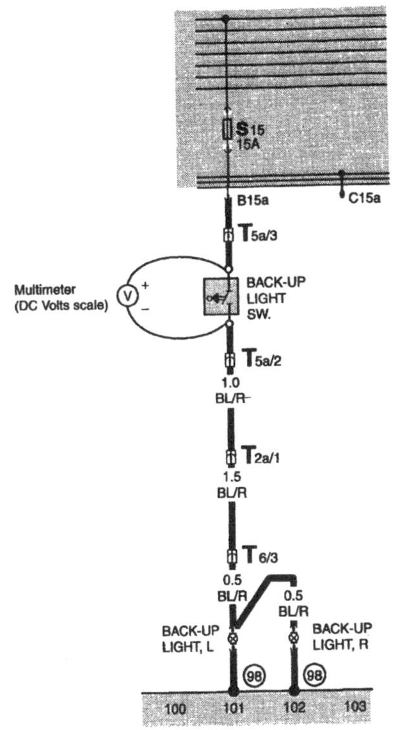

4.5 Checking for Short Circuit to Ground (GND)

Checking the circuit as shown will detect circuit malfunctions that are providing an unintended current flow path to ground (GND). The examples illustrate two methods of troubleshooting a short to ground (GND) that is causing a blown fuse in the circuit powering the license plate lights and the glove compartment light.

Using an Ohmmeter (Multimeter)

Step 1 - Remove the fuse

Step 2 - Disconnect the load (powered components) to eliminate the circuit's normal path to ground (GND)

> ***CAUTION***
>
> *Always use a digital (low current) meter. An ohmmeter, or the Ohms scale of a multimeter, measures resistance by passing a small amount of current through the circuit or component being checked. Improper testing may damage sensitive electronic components.*

Continuity (approximately 0 Ω): Little or no resistance indicates that there is a continuous conductive path between the isolated circuit and ground (GND), even though all the circuit's normal ground (GND) paths are eliminated. There is a short - an unintentional connection to ground (GND) - somewhere in the circuit.

No Continuity: The circuit's normal ground (GND) paths have been disconnected, and there is no other connection between the isolated circuit and ground (GND) - no short has been detected.

> ***CAUTION***
>
> *Direct contact with meter probes at the connector terminals can easily damage the small contacts, causing poor connections and risking future intermittent malfunctions.*
>
> *Special Tools include adapters for making test connections safely and preventing connector damage. See **2. Test Equipment.***

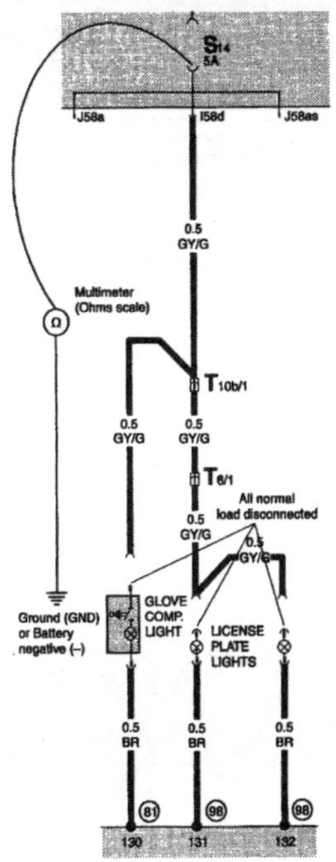

Using an LED Test Light or Voltmeter (Multimeter)

Step 1 - Remove the fuse

Step 2 - Disconnect the load (powered components) to eliminate the circuit's normal path to ground (GND)

Voltage: If the test light or multimeter indicates Voltage, then there is a complete circuit - a connection to ground (GND) even though all the circuit's Normal ground (GND) paths are eliminated. There is a short - an unintentional connection to ground (GND) - somewhere in the circuit.

NO Voltage: There is not a complete circuit. The circuit's normal ground (GND) paths have been disconnected, and there is no other connection between the isolated circuit and ground (GND) - no short has been detected.

> ***CAUTION***
>
> *Direct contact with meter probes at the connector terminals can easily damage the small contacts, causing poor connections and risking future intermittent malfunctions.*
>
> *Special Tools include adapters for making test connections safely and preventing connector damage. See **2. Test Equipment.***

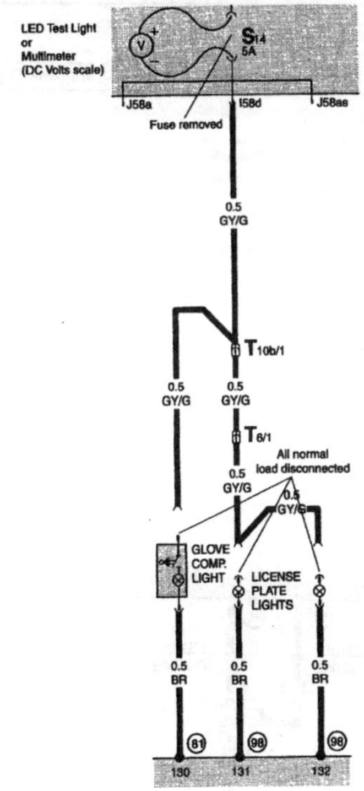

Relay panel with fuse holder

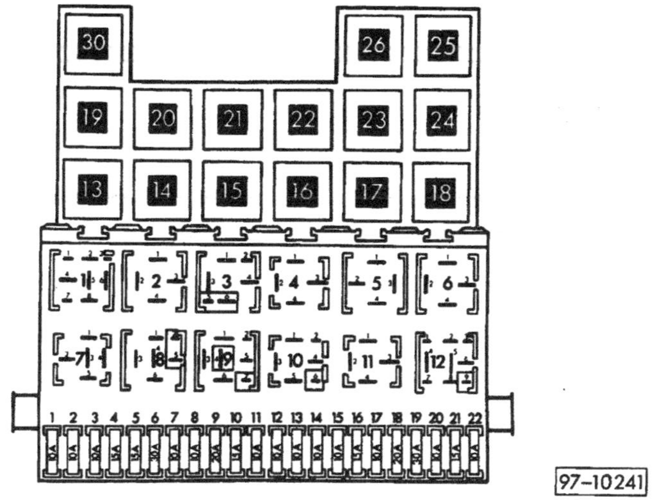

Warning!

The negative battery cable must be disconnected before working on the electrical system.

Notes:

- For other relay locations and fuse arrangements see respective wiring diagrams.
- Number in parentheses indicates production control number stamped on relay housing.

Relay location:

1 – A/C Relay (13) or
– A/C Control Module (140)
2 – Rear Window Wiper/Washer Relay (72)
3 – Digifant Engine Control Module (ECM) Power Supply Relay (30)
4 – Load Reduction Relay (18)
5 – Open
6 – Emergency Flasher Relay (21) or
– Trailer Operation Hazard Flasher Relay (22)
7 – Headlight Washer System Relay (95)
8 – Washer/Wiper Intermittent Relay (19)
9 – Seat Belt Warning System Relay (4)
10 – Fog Light Relay (53)
11 – Dual Horn Relay (53) or
– Busbar for Horn
12 – Fuel Pump (FP) Relay (80) or
– Glow Plug Relay (60) or
– Glow Plug Relay (102)

Seperate relays above fuse/relay panel:

Notes:

- Number in parentheses indicates production control number stamped on relay housing.
- A different assignment is possible above the relay panel depending on the option.

13 – Heater Control Module (152), A/C

14 – Engine Coolant Pump Relay (After-run) (38), 5-cylinder-engine

15 – Idle Air Control (IAC) Control Module (155), 5-cylinder-diesel-engine, with A/C and automatic transmission

16 – 5-cylinder-diesel-engine-relay (125) with A/C and automatic transmission

17 – Trailer Towing Rear Fog Light–Cut-Out Relay – flatbed (151)

18 – Roof Ventilator Fuse
– Terminal 58B Wire Connector

19 – 2nd Evaporator Relay (24), not on A/C
– 2nd Evaporator Cut-Off Relay (94), A/C

20 – Park/Neutral Position (PNP) Relay (150), automatic transmission
– Engine Coolant Pump Circuit Breaker, 5-cylinder-engine

21 – Back-Up Lights Fuse, automatic transmission
– Cruise Control Fuse, automatic transmission

22 – A/C Fuse
– 2nd Evaporator Fuse

23 – Heated Mirror and Rear Window Defogger Relay (94), not on flatbed

24 – Central Locking System Fuse
– Power Window Fuse

25 – ABS Fuse

26 – Heated Mirror and Rear Window Defogger Fuse, not on flatbed

30 – A/C Clutch Cut-Off Relay (94) – A/C, automatic transmission

Fuse arrangement

The fuse protection for the options which are not listed are shown where necessary with the respective wiring diagram.

1 – Left headlight, low beam (56b) ...... 10 A
2 – Right headlight, low beam (56b) ..... 10 A
3 – Instrument panel lights, license plate light (58) .... 10 A
4 – Rear window wiper, heated seats(X) . 15 A
5 – Windshield wiper/washer system, heated washer nozzles (X) .... 15 A
6 – Fresh air blower, A/C (X) .... 30 A
7 – Right parklight and taillight (58R), engine compartment light .... 10 A
8 – Left parklight and taillight (58L) ...... 10 A
9 – Rear window defogger, heated mirrors (X) .... 20 A
10 – Fog lights, rear fog light .... 15 A
11 – Left headlight, high beam (56a) ..... 10 A
12 – Right Headlight, high beam (56a) 10A
13 – Dual horns, coolant fan (15) .... 10 A
14 – Back-up lights, power mirrors (15) ... 10 A
15 – Engine electrics (15) .... 10 A
16 – Instrument cluster, ABS warning light, glove compartment light (15) .... 15 A
17 – Emergency flasher system (15) ..... 10 A
18 – Fuel pump (FP), heated oxygen sensor (HO2S) .... 20 A
19 – Coolant pump, A/C relay (30) .... 30 A
20 – Brake lights (30) .... 10 A
21 – Interior lights, clock, radio, central locking system, B+(battery positive voltage) for on board diagnostic (OBD) (30) ..... 15 A
22 – Cigarette lighter .... 10 A

Multiple connector arrangement

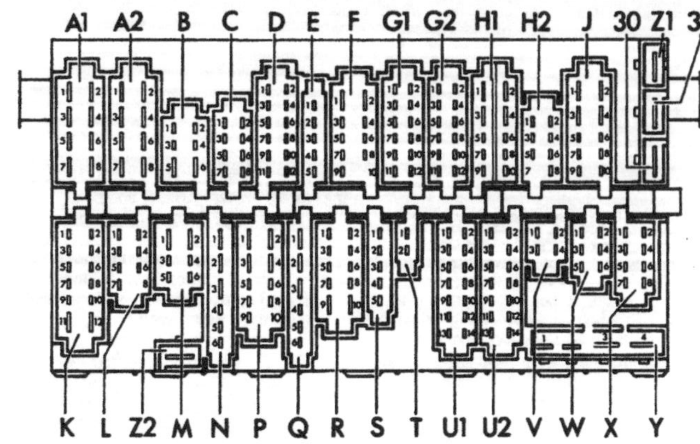

A1 – Headlight wiring harness (yellow), 8-Pin

A2 – Headlight wiring harness (yellow), 8-Pin

B – Headlight washer system wiring harness (green), 6-Pin

C – Wiper motor wiring harness (yellow), 8-Pin

D – Optional equipment wiring harness (green), 12-Pin

E – Brake light wiring harness (green), 5-Pin

F – Digifant engine control module (ECM), engine compartment (white), 12-Pin

G1 – Digifant engine control module (ECM), engine compartment (white), 12-Pin

G2 – Digifant engine control module (ECM), engine compartment (white), 12-Pin

H1 – Steering column switch wiring harness (red), 10-Pin

H2 – Steering column switch wiring harness (red), 8-Pin

J – Steering column switch wiring harness (red),10-Pin

K – Rear wiring harness (black), 12-Pin

L – Parking brake wiring harness (black), 7-Pin

M – Fuel tank wiring harness (black), 6-Pin

N – A/C wiring harness (green), 6-Pin

P – Rear window defogger/fog light switch wiring harness (blue), 9-Pin

Q – Fresh air blower wiring harness (blue), 6-Pin

R – Light switch wiring harness (blue), 10-Pin

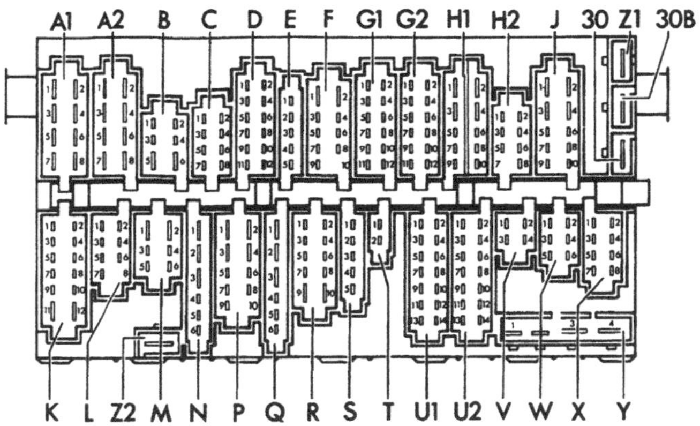

S – Wiper motor wiring harness (white), 5-Pin

T – Double Connector (green)

U1 – Instrument cluster wiring harness (blue), 14-Pin

U2 – Instrument cluster wiring harness (blue), 14-Pin

V – Steering column wiring harness (green), 4-Pin

W – Optional equipment wiring harness (green), 6-Pin

X – Warning lamps (trailer towing, anti-lock brake system) wiring harness (green), 8-Pin

Y – Single Connector, terminal 30

Z1 – Single Connector

Z2 – Single Connector, terminal 31, battery-minus

30 – Single Connector, terminal 30*

30B – Single Connector*

* See No. 1/6, Terminal (B+) Wire Connector (TV2)

Terminal 30 (B+) Wire Connector (TV2)

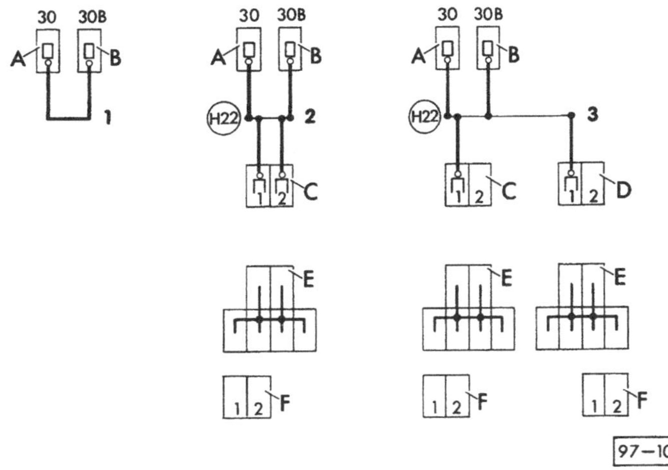

Terminals "30" and "30B" are located on the back of relay panel depending on the electrical equipment of the vehicle.
The terminals are connected with each other using either
- a wire connection -1- or
- a wire connector TV2 -2- or -3-

The wire connectors make it possible to connect additional components to terminal 30. If TV2 does not exist, it must be connected by hand.

1 – Simple wire connection between terminal "30" and "30B"
2 – Terminal 30 (B+) Wire Connector (TV2) with 4 outputs
3 – Terminal 30 (B+) Wire Connector (TV2) with 8 outputs
A – Female Single Connector
B – Female Single Connector
C – Female Double Connector
D – Female Double Connector
E – Flat-Pin Connector, 6-Pin
F – Flat-Pin Double Connector

(H22) – Plus connection (30), in terminal 30/30B wiring harness

Notes:

- Additional connected components must be protected by a fuse depending on their current consumption.
- Open connections on connector -E- must be covered with an empty female connector to avoid short circuits.
- Wire diameter of all wires in wire connector TV2: 6,0mm; Wire color: red.

Relay panel with fuse holder and auxiliary relay panel

from January 1994

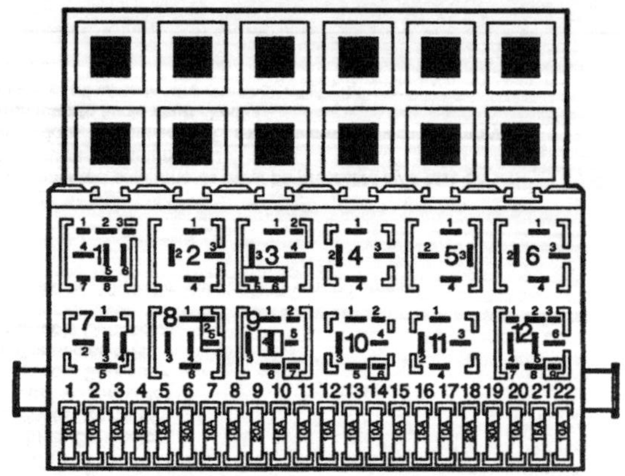

Fuse colors

30 A – green
20 A – yellow
15 A – blue
10 A – red

Fuse arrangement

The fuse protection for the options which are not listed are shown where necessary with the respective wiring diagram.

1 – Left headlight, low beam (56b) ...... 10 A

2 – Right headlight, low beam (56b) ..... 10 A

3 – Instrument panel lights, license plate light (58) .... 10 A

4 – Rear window wiper, heated seats (X) . 15 A

5 – Windshield wiper/washer system, heated washer nozzles (X) .... 15 A

6 – Fresh air blower, A/C (X) .... 30 A

7 – Right parklight and tail light (58R) ... 10 A

8 – Left parklight and tail light (58L) ..... 10 A

9 – Rear window defogger, heated mirrors (X) .... 20 A

10 – Fog lights .... 15 A

11 – Left headlight, high beam (56a) ..... 10 A

12 – Right headlight, high beam (56a) .... 10 A

13 – Horn, dual horns (15) .... 10 A

14 – Back-up light, sunroof, cruise control, power mirrors (15) .... 10 A

15 – Engine electric .... 10 A

16 – Instrument cluster, ABS warning light, glove compartment light (15) ... 15 A

17 – Emergency flasher system (15) ..... 10 A

18 – Fuel pump (FP), heated oxygen sensor (HO2S) .... 20 A

19 – Coolant fan, A/C relay (30) .... 30 A

20 – Brake light (30) .... 10 A

21 – Interior lights, digital clock, radio, central locking system, B+(battery positive voltage) for on-board diagnostic (OBD) system (30) .... 15 A

22 – Cigarette lighter .... 10 A

Relay arrangement

Notes:

- ◆ For other relay locations and fuse arrangements see respective wiring diagrams.
- ◆ Number in parentheses indicates production control number stamped on relay housing.

Relay locations on fuse/relay panel:

1 – A/C Control Module (140)

2 – Rear Window Wiper/Washer Relay (72)

3 – Digifant Engine Control Module (ECM) Power Supply Relay (30)

4 – Load Reduction Relay (18)

5 – Not used

6 – Emergency Flasher Relay (21)

7 – Headlight Washer System Relay (33)

8 – Washer/Wiper Intermittent Relay (19)
– Programmable Washer/Wiper Intermittent Relay (99)

9 – Seat Belt Warning System Relay (4),

10 – Fog Light Relay (53)

11 – Dual Horn Relay (53)
– Busbar for Horn

12 – Fuel Pump (FP) Relay (80)
– Glow Plug Relay (102)

Continued on next page!

Relay locations above fuse/relay panel:

- – Engine Coolant Pump Fuse, 5-Cyl. Engine (38)
- – Secondary Air Injection (AIR) Control Module (111)
- – Idle Air Control (IAC) Control Module (155)
- – Idle Boost Relay (125)
- – Vehicle Speed Warning Buzzer (4-wheel drive with ABS)
- – Data Link Connector (DLC) Wire Connector for on-board diagnostic (OBD) system
- – Coolant Pump Fuse
- – Secondary Air Injection (AIR) Fuse
- – ABS Fuse
- – Fuse for Automatic Transmission with Cruise Control System
- – Auxiliary Heating Fuse
- – Climatronic Fuse (planned)
- – 2nd Heat Exchanger Fuse
- – A/C Fuse
- – 2nd Evaporater Fuse
- – Passenger Compartment Ventilation Fuse
- – Power Mirrors and Rear Window Defogger Fuse
- – Power Seats Fuse
- – Power Windows Fuse
- – Central Locking Fuse
- – Terminal 58b Wire Connector
- – Terminal 15 Wire Connector
- – Terminal X Wire Connector, protected
- – Terminal 15 Wire Connector, protected

Auxiliary relay panel behind instrument panel – center:

- – Passenger Compartment Ventilation Relay (53) and (38)
- – Load Reduction Relay II (53)
- – 2nd Heat Exchanger Relay, Blower (53)
- – 2nd Heat Exchanger, A/C (94)
- – 2nd Heat Exchanger and 2nd Evaporator Relay (114)
- – Heater Control Module, Auxiliary Heat Exchanger (152)
- – Park/Neutral Position (PNP) Relay (150)
- – Freewheel Lock Control Module, Syncro with ABS (87)

Multiple connector arrangement:

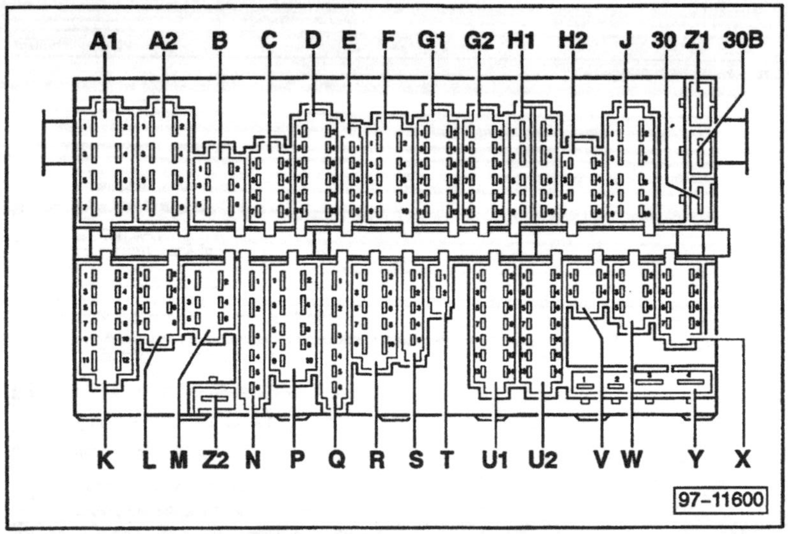

A1 – Headlight wiring harness (yellow), 8-Pin

A2 – Headlight wiring harness (yellow), 8-Pin

B – Headlight washer system wiring harness (green), 6-Pin

C – Wiper motor wiring harness (yellow), 8-Pin

D – Optional equipment wiring harness (green), 12-Pin

E – Brake light switch wiring harness (green), 5-Pin

F – Engine compartment wiring harness (white), 10-Pin

G1 – Engine compartment wiring harness (white), 12-Pin

G2 – Engine compartment wiring harness (white), 12-Pin

H1 – Steering column switch wiring harness (red), 10-Pin

H2 – Steering column switch wiring harness (red), 8-Pin

J – Steering column switch wiring harness (red),10-Pin

K – Rear wiring harness (black), 12-Pin

L – Parking brake wiring harness (black), 7-Pin

M – Fuel tank wiring harness (black), 6-Pin

N – A/C wiring harness (green), 6-Pin

Continued on next page!

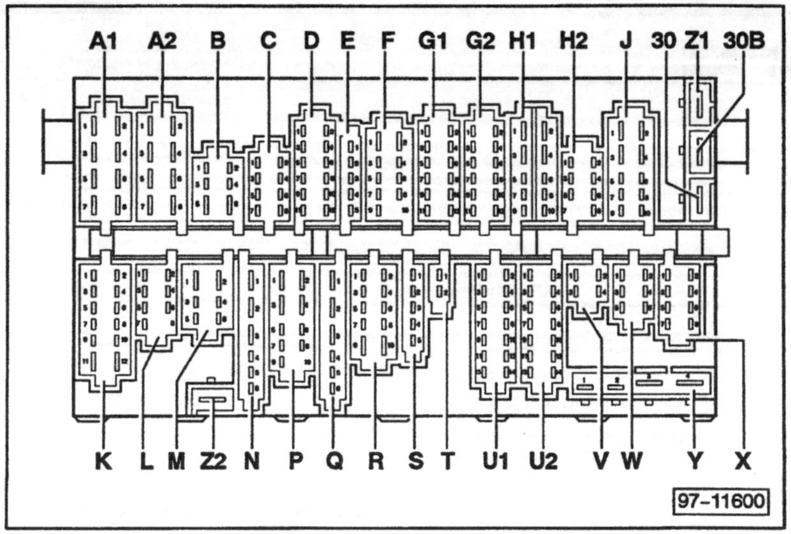

P – Rear window defogger switch and fog light switch wiring harness (blue), 9-Pin

Q – Fresh air blower wiring harness (blue), 6-Pin

R – Light switch wiring harness (blue), 10-Pin

S – Wiper motor wiring harness (white), 5-Pin

T – Double Connector (green)

U1 – Instrument cluster wiring harness (blue), 14-Pin

U2 – Instrument cluster wiring harness (blue), 14-Pin

V – Steering column switch wiring harness -multi-function display- (green), 4-Pin

W – Optional equipment wiring harness (green), 6-Pin

X – Warning lamps (trailer towing, anti-lock brake system) wiring harness (green), 8-Pin

Y – Single Connector, terminal 30

Z1 – Single Connector

Z2 – Single Connector, terminal 31, battery-minus

30 – Single Connector, terminal 30

30B – Single Connector

Relay panel with fuse holder and auxiliary relay panel

Index

Fuse arrangement

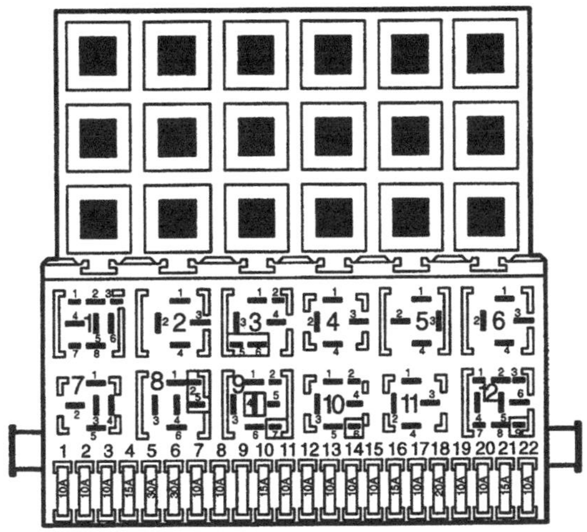

The fuse protection for the options which are not listed are shown where necessary with the respective wiring diagram.

1	– Left headlight, low beam, left headlight beam adjusting (56b)	10 A
2	– Right headlight, low beam, right headlight beam adjusting (56b)	10 A
3	– License plate light (58b)	10 A
4	– Rear window wiper, optional equipment (X)	15 A
5	– Windshield wiper/washer system, heated washer nozzles (X)	30 A
6	– Fresh air blower, airconditioning (X)	30 A
7	– Right parking and tail light (58R)	10 A
8	– Left parking and tail light (58L)	10 A
9	– Open	
10	– Fog lights, rear fog light	15 A
11	– Left headlight, high beam (56a)	10 A
12	– Right headlight, high beam (56a)	10 A
13	– Dual horns (15)	10 A
14	– Back-up light, optional equipment (15)	10 A
15	– Engine electric (code AES), Fuel Cut-off Valve (code AAB)	10 A
16	– Turn signal light system, ABS warning light, airbag malfunction indicator lamp, glove compartment light (15)	15 A
17	– Open	10 A
18	– Fuel pump (FP), heated oxygen sensor (HO2S), Coolant pump after-run	20 A
19	– After-Run Coolant Pump (30)	10 A
20	– Brake light (30)	10 A
21	– Instrument cluster, interior lights, radio, B+ (battery positiv voltage) for on board diagnostic (OBD) (30)	15 A
22	– Cigarette lighter (30)	10 A

Fuse colors

30 A	– green		10 A	–	red
25 A	– white		7,5 A	–	brown
20 A	– yellow		5 A	–	beige
15 A	– blue		3 A	–	violet

Relay arrangement

Notes:

Number in parntheses indicates production control number stamped on relay housing.

Relay locations on fuse/relay panel:

1 – Fresh Air Blower Relay (105),
– A/C Relay (140),
– Relay for Climatronic (140)

2 – Rear Window Wiper/Washer Relay (72)(174)

3 – Not used

4 – Load Reduction Relay (18)

5 – Not used

6 – Emergency Flasher Relay (21)

7 – Headlight Washer System Relay (95)

8 – Programmable Washer/Wiper Intermittent Relay (99)

9 – Warning Buzzer for Parking Light (36)

10 – Busbar,
– Fog Light Relay (53)

11 – Busbar for Horn,
– Dual Horn Relay (53)

12 – Fuel Pump (FP) Relay (167),
– Glow Plug Relay (102)

Auxiliary relay panel behind instrument cluster, center:

Indication

- Heater Air Blower Relay (53)
- Relay for Fresh-air and recirculating air flap - rear (38), (94)
- Heater Control Module (152)
- Fan Relay (114)
- Park/Neutral Position (PNP) Relay (175)

Auxiliary relay panel below driver's seat:

Indication

- Driver's Seat Heater Control Module (59)
- Passenger's Seat Heater Control Module (59)
- Fuse for Auxiliary Heater B3L/D3L, (Glow Plug) 25A
- Fuse for Auxiliary Heater B3L/D3L, (Heater Control Module) 5A

Relay locations above fuse/relay panel:

Indication	Color code
– Seat Belt Warning System Relay (49)	
– Idle Boost Relay (125)	
– Relay for A/C shutoff (147)	
– Idle Air Control (IAC) Control Module (155)	
– Load Reduction Relay II (53)	
– Coolant Circulation Pump Relay (53)	
– Secondary Air Injection (AIR) Pump Relay (53)	
– Power Window Circuit Breaker 30A	
– Fuse for secondary air pump 30A, vehicles before September 1997	
– Fuse for ignition coil terminal 15 15A, vehicles from September 1997	
– Fuel system control fuse 15A, vehicles from September 1997	
– Data Link Connector (DLC) Wire Connector for on-board diagnostic (OBD) system	
– Wire junction -2- for terminal 30, 4-Pin	red
– Distributor for terminal Xa, 4-Pin	yellow
– Terminal 15 Wire Connector, 8-Pin	black
– Wire junction -2- for terminal 15a, 8-Pin	green
– Terminal 15a Wire Connector, 4-Pin, – Terminal 58b Wire Connector, 4-Pin	green grey
– Wire junction -2- for terminal 58b, 8-Pin	grey
– Terminal 54 Wire Connector, 4-Pin	orange
– Vehicle Speed Signal Wire Connector, 8-Pin	blue

Multiple fuse-adapter below fuse/relay panel:

Indication	Fuse-placement
– Fuse for 12V outlets 15A	1
– Fuse for Swivel Seat lock 10A	2
– Power Sunroof Fuse 20A	3
– Fuse for auxiliary heater 10A	4
– Fuse 1 for control module - ABS 10A	5
– Open	6
– Open	7
– Open	8
– Open	9
– Central Locking System Fuse 15A	10
– Heated Seats Fuse 15A	11
– Power Mirrors and Rear Window Defogger Fuse 30A	12
– Fuse for Rear Evaporator Fan (Climatronic) 30A	13
– Fuse for Fresh air blower rear 30A, – Fuse for Fresh air blower rear (A/C) 30A, – Fuse for Fresh Air Blower (Climatronic) 30A	14
– Fuse for Rear Warm Air Fan 20A, – Fuse for Rear Warm Air Fan (A/C) 30A, – Fuse for Rear Warm Air Fan (Climatronic) 30A	15

Multiple connector arrangement:

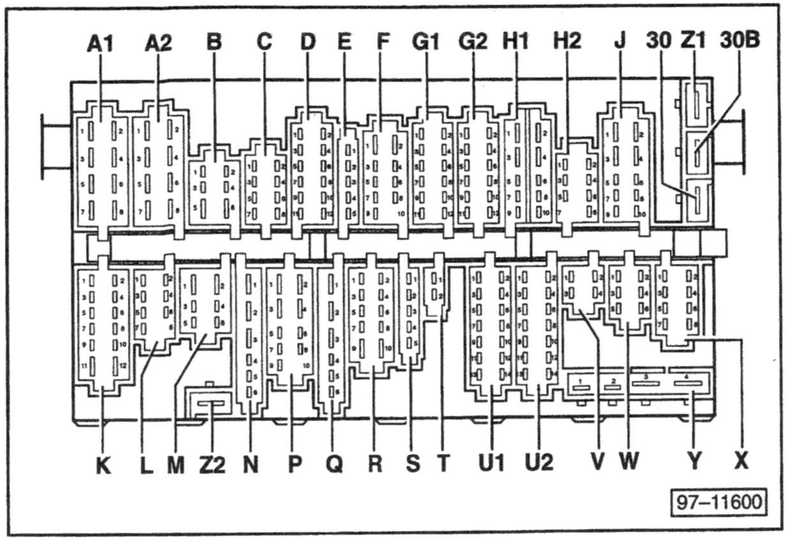

A1 – Headlight wiring harness (yellow), 8-Pin

A2 – Headlight wiring harness (yellow), 8-Pin

B – Headlight washer system wiring harness (green), 6-Pin

C – Wiper motor wiring harness (yellow), 8-Pin

D – Optional equipment wiring harness (green), 12-Pin

E – Brake light switch wiring harness (green), 5-Pin

F – Engine compartment wiring harness (white), 10-Pin

G1 – Engine compartment wiring harness (white), 12-Pin

G2 – Engine compartment wiring harness (white), 12-Pin

H1 – Steering column switch wiring harness (red), 10-Pin

H2 – Steering column switch wiring harness (red), 8-Pin

J – Steering column switch wiring harness (red),10-Pin

K – Rear wiring harness (black), 12-Pin

L – Parking brake wiring harness (black), 8-Pin

M – Fuel tank wiring harness (black), 6-Pin

N – A/C wiring harness (green), 6-Pin

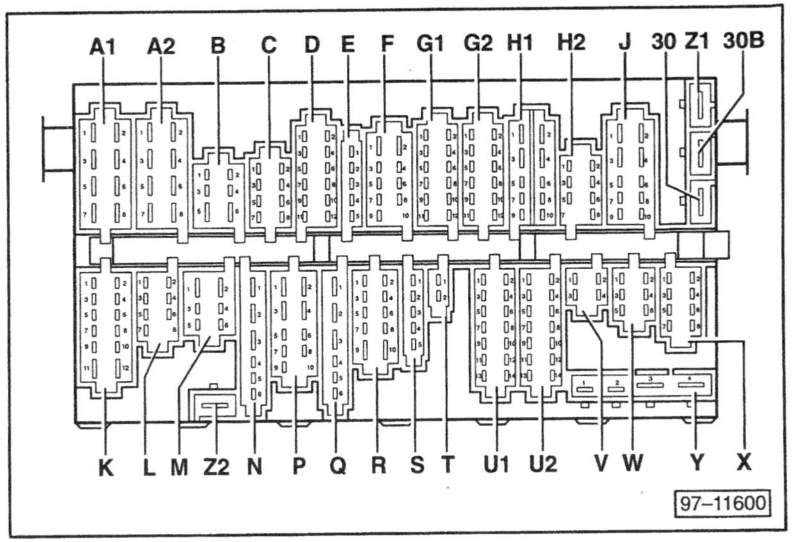

P – Rear window defogger switch and fog light switch wiring harness (blue), 10-Pin

Q – Fresh air blower wiring harness (blue), 6-Pin

R – Light switch wiring harness (blue), 10-Pin

S – Wiper motor wiring harness (white), 5-Pin

T – Double Connector (green)

U1 – Instrument cluster wiring harness (blue), 14-Pin

U2 – Instrument cluster wiring harness (blue), 14-Pin

V – Steering column switch wiring harness multi-function display- (green), 4-Pin

W – Optional equipment wiring harness (green), 6-Pin

X – Warning lamps (trailer towing, anti-lock brake system) wiring harness (green), 8-Pin

Y – Single Connector, terminal 30

Z1 – Single Connector

Z2 – Single Connector, terminal 31, battery minus

30 – Single Connector, terminal 30

30B – Single Connector

Electrical Wiring Diagram Index

2.5 Liter engine (code AAF) from September 1991 to April 1992

2.5 Liter engine (code AAF) from April 1992 to August 1992

1992 EuroVan, additional electrical wiring diagrams

2.5 Liter engine (code AAF) from August 1992 to Dec. 1992

Electrical Wiring Diagram Index

2.5 Liter engine (code AAF) from January 1993

2.5 Liter engine(code AAF, CA only) from April 1992 to Dec 1992

2.5 Liter engine (code AAF, CA only) from January 1993

	Wiring Diagram page number	Repair Manual page number
• **EuroVan (2.5L-Engine – Digifant Multiport Fuel Injection (MFI)/81kW), code AAF, California only, from January 1993**	33/1	X 265
– Generator (GEN), battery, starter, ignition/starter switch	33/2	X 266
– Ignition system, digifant engine control module (ECM)	33/3	X 267
– Digifant engine control module (ECM), on board diagnostic (OBD) system, malfunction indicator lamp (MIL)	33/4	X 268
– Digifant engine control module (ECM), heated oxygen sensor (HO2S)	33/5	X 269
– Digifant engine control module (ECM), injectors, fuel pump (FP)	33/6	X 270
– Transmission control module (TCM), park/neutral position (PNP) relay	27/6	X 218
– Transmission control module (TCM), solenoid valves	27/7	X 219
– Transmission control module (TCM), multi-function transmission range (TR) switch	27/8	X 220
– Transmission control module (TCM), transmission range (TR) program switch	27/9	X 221
– Fuel supply, coolant pump after-run	27/10	X 222
– Instrument cluster, oil pressure and engine coolant level/temperature (ECL/ECT) warning lights	27/11	X 223
– Instrument cluster, tachometer, clock, indicator lights	27/12	X 224
– Instrument cluster, transmission range (TR) selector lever display, seat belt warning system	27/13	X 225
– Brake and parking brake warning light, glove compartment light	27/14	X 226
– Radio connection, cigarette lighter, engine compartment light	27/15	X 227
– Interior lights	27/16	X 228
– Headlights, front turn signals	27/17	X 229
– Back-up lights, headlight dimmer/flasher switch	27/18	X 230
– Turn signal and emergency flasher system	27/19	X 231
– Rear turn signals, brake lights, taillights	27/20	X 232
– Taillights (flatbed and double cabin), from April 1992	18/1	X 173
– Light switch, license plate light (tailgate)	27/21	X 233
– Coolant fans (2 x 200W or 2 x 350W)	27/22	X 234
– Coolant fans (2 x 450W), from April 1992	9/1	X 141
– Fresh air blower	27/23	X 235
– Rear window defogger (tailgate), dual horns	27/24	X 236
– Windshield wiper/washer system	27/25	X 237
– Rear window wiper/washer system (tailgate), heated washer nozzles	27/26	X 238

2.4 Liter Diesel (code AAB) from August 1992

	Wiring Diagram page number	Repair Manual page number
• **EuroVan (2.4L-Diesel Engine/57kW), code AAB, from August 1992**	26/1	X 193
– Generator (GEN), battery, starter, ignition/starter switch	26/2	X 194
– Preheating system, fuel cut-off valve	26/3	X 195
– Coolant pump after-run	26/4	X 196
– Instrument cluster, oil press. and engine cool. level/temp. (ECL/ECT) warning lights	26/5	X 197
– Instrument cluster, tachometer, clock, indicator lights	26/6	X 198
– Brake and parking brake warning light, seat belt warning system	26/7	X 199
– Radio connection, cigarette lighter, engine compartment light	26/8	X 200
– Interior lights	26/9	X 201
– Headlights, front turn signals	26/10	X 202
– Back-up lights, headlight dimmer/flasher switch	26/11	X 203
– Turn signal and emergency flasher system	26/12	X 204
– Rear turn signals, brake lights, taillights	26/13	X 205
– Taillights (flatbed and double cabin), from April 1992	18/1	X 173
– Light switch, license plate light (tailgate)	26/14	X 206
– Coolant fans (2 x 200W or 2 x 350W)	26/15	X 207
– Coolant fans (2 x 450W), from April 1992	9/1	X 141
– Fresh air blower	26/16	X 208
– Rear window defogger (tailgate), dual horns	26/17	X 209
– Windshield wiper/washer system	26/18	X 210
– Rear window wiper/washer system (tailgate), heated washer nozzles	26/19	X 211

1993 EuroVan additional electrical wiring diagrams

1994 Eurovan

Model Year 1994

see Model Year 1993

2.5 Liter engine (code ACU) from September 1994

	Wiring Diagram page number	Repair Manual page number
• **EuroVan (2.5L-Engine – Digifant Multiport Fuel Injection (MFI)/81kW), code ACU, from September 1994**	35/1	X 277
Note: This wiring diagram is also used for Winnebago Camper, from January 1994		
– Generator (GEN), battery, starter	35/2	X 278
– Ignition/starter switch	35/3	X 279
– Ignition system, digifant engine control module (ECM)	35/4	X 280
– Digifant engine control module (ECM), on board diagnostic (OBD) system	35/5	X 281
– Digifant engine control module (ECM), knock sensors	35/6	X 282
– Digifant engine control module (ECM), heated oxygen sensor (HO2S), injectors	35/7	X 283
– Park/Neutral position (PNP) relay, transmission control module (TCM)	35/8	X 284
– Transmission control module (TCM), solenoid valves	35/9	X 285
– Transmission control module (TCM)	35/10	X 286
– Transmission control module (TCM), multi-function transmission range (TR) switch	35/11	X 287
– Fuel supply, coolant after-run, secondary air injection (AIR) system	35/12	X 288
– Instrument cluster, oil pressure and engine coolant level/temperature (ECL/ECT) warning lights	35/13	X 289
– Instrument cluster, clock, vehicle speed sensor (VSS), indicator lights	35/14	X 290
– Instrument cluster, tachometer, transmission range (TR) selector lever display	35/15	X 291
– Seat belt warning system, brake system sensor, glove compartment light	35/16	X 292
– Radio connection, cigarette lighter	35/17	X 293
– Interior lights	35/18	X 294
– Interior lights	35/19	X 295
– Headlights, front turn signal light	35/20	X 296
– Tail light, headlight dimmer/flasher switch	35/21	X 297
– Emergency flasher system	35/22	X 298
– Tail light, brake light, rear turn signal	35/23	X 299
– Light switch, license plate light	35/24	X 300
– Coolant fans (2 x 200W or 2 x 350W)	35/25	X 301
– Coolant fans (2 x 450W), from April 1992	9/1	X 141
– Fresh air blower	35/26	X 302
– Rear window defogger/heat element, dual horns	35/27	X 303
– Windshield wiper/washer system	35/28	X 304
– Rear window wiper washer system, heated washer nozzles	35/29	X 305

2.4 Liter diesel (code AAB) from August 1992

	Wiring Diagram page number	Repair Manual page number
• **EuroVan (2.4L-Diesel Engine/57kW), code AAB, from August 1992**	26/1	X 193
– Generator (GEN), battery, starter, ignition/starter switch	26/2	X 194
– Preheating system, fuel cut-off valve	26/3	X 195
– Coolant pump after-run	26/4	X 196
– Instrument cluster, oil press. and engine cool. level/temp. (ECL/ECT) warning lights	26/5	X 197
– Instrument cluster, tachometer, clock, indicator lights	26/6	X 198
– Brake and parking brake warning light, seat belt warning system	26/7	X 199
– Radio connection, cigarette lighter, engine compartment light	26/8	X 200
– Interior lights	26/9	X 201
– Headlights, front turn signals	26/10	X 202
– Back-up lights, headlight dimmer/flasher switch	26/11	X 203
– Turn signal and emergency flasher system	26/12	X 204
– Rear turn signals, brake lights, taillights	26/13	X 205
– Taillights (flatbed and double cabin), from April 1992	18/1	X 173
– Light switch, license plate light (tailgate)	26/14	X 206
– Coolant fans (2 x 200W or 2 x 350W)	26/15	X 207
– Coolant fans (2 x 450W), from April 1992	9/1	X 141
– Fresh air blower	26/16	X 208
– Rear window defogger (tailgate), dual horns	26/17	X 209
– Windshield wiper/washer system	26/18	X 210
– Rear window wiper/washer system (tailgate), heated washer nozzles	26/19	X 211

1995 EuroVan additional electrical wiring diagrams

1996 EuroVan

Model Year 1996

see Model Year 1995

2.8 Liter engine (code AES) from May 1996

	Wiring Diagram page number	Repair Manual page number
● EuroVan (2.8L-Engine – Motronic Multiport Fuel Injection (MFI)/103 kW), code AES, from May 1996	41/1	X 343
– Battery, load reduction relay II, data link connector (DLC) wire connector	39/2	X 324
– Ignition/starter switch	39/3	X 325
– Fresh air blower, fresh air blower switch, glove compartment light	39/4	X 326
– Generator (GEN), starter, fuel pump, fuel level sensor, coolant after-run	41/2	X 344
– Motronic engine control module (ECM), fuel injectors	41/3	X 345
– Motronic engine control module (ECM), ignition system, throttel valve control module	41/4	X 346
– Motronic engine control module (ECM), heated oxygen sensor, camshaft position (CMP) sensor, engine speed (RPM) sensor	41/5	X 347
– Motronic engine control module, knock sensors, EGR temperature sensor, intake air temperature (IAT) sensor	41/6	X 348
– Motronic engine control module (ECM), (MAF) sensor, evapurator emission (EVAP) canister purge regulator valve, secondary air injection (AIR) solenoid valve, secondary air injection (AIR) pump motor, EGR vacuum regulator solenoid valve	41/7	X 349
– Motronic engine control module, speedometer vehicle speed sensor (VSS), heated oxygen sensor, oil pressure switch	41/8	X 350
– Instrument cluster, visual acoustical pressure oil control, engine coolant level (ECL) sensor, engine coolant tempreature (ECT) sensor	41/9	X 351
– Coolant pump after-run	42/2	X 355
– Instrument cluster, fuel gauge, engine coolant level/temperature (ECL/ECT) gauge, speedometer, odometer display	41/10	X 352
– Instrument cluster, multi-function indicator (MFA)	41/11	X 353
– Instrument cluster, tachometer, odometer display	39/6	X 328
– Instrument cluster, digital clock, parking brake warning light switch	39/7	X 329
– Brake and parking brake warning light, ABS warning light, airbag malfunction indicator lamp (MIL), radio connection	39/8	X 330
– Interior lights	39/9	X 331
– Headlights, parking light, front turn signal lights	39/10	X 332
– Emergency flasher system	39/11	X 333
– Turn signal switch, headlight dimmer/flasher switch	39/12	X 334
– Rear turn signal light, brake lights, tail light	39/13	X 335
– Light switch	39/14	X 336
– Cigarette lighter, brake light switch, license plate light, back up light	39/15	X 337
– Rear window defogger heat element, dual horns	39/16	X 338
– Windshield wiper/washer system	39/17	X 339
– Rear window wiper/washer system (rear lid), heated windshield washer nozzles	39/18	X 340

2.4 Liter diesel (code AAB) from May 1996

	Wiring Diagram page number	Repair Manual page number
● EuroVan (2.4L-Diesel Engine/57 kW), code AAB, from May 1996	43/1	X 356
– Battery, load reduction relay II, data link connector (DLC) wire connector	39/2	X 324
– Ignition/starter switch	39/3	X 325
– Fresh air blower, fresh air blower switch, glove compartment light	39/4	X 326
– Generator (GEN), starter, coolant after-run	43/2	X 357
– Preheating system, fuel level sensor, sensor for engine temperature, engine coolant level (ECL) sensor	43/3	X 358
– Instrument cluster, oil pressure switch, speedometer vehicle speed sensor, fuel cut off valve, heating element	43/4	X 359
– Instrument cluster, fuel gauge, engine coolant temperature (ECT) gauge, glow plug indicator light, speedometer, odometer display	43/5	X 360
– Instrument cluster, tachometer, odometer display	39/6	X 328
– Instrument cluster, digital clock, parking brake warning light switch	39/7	X 329
– Brake and parking brake warning light, ABS warning light, airbag malfunction indicator lamp (MIL), radio connection	39/8	X 330
– Interior lights	39/9	X 331
– Headlights, parking light, front turn signal lights	39/10	X 332
– Emergency flasher system	39/11	X 333
– Turn signal switch, headlight dimmer/flasher switch	39/12	X 334
– Rear turn signal light, brake lights, tail light	39/13	X 335
– Light switch	39/14	X 336
– Cigarette lighter, brake light switch, license plate light, back up light	39/15	X 337
– Rear window defogger heat element, dual horns	39/16	X 338
– Windshield wiper/washer system	39/17	X 339
– Rear window wiper/washer system (rear lid), heated windshield washer nozzles	39/18	X 340

Electrical Wiring Diagram Index

1997 EuroVan additional electrical wiring diagrams

1998-99 EuroVan electrical wiring diagrams

Model Year 1998

see Repair Manual page number X 467

Model Year 1999

see supplement in back of book

2.5 L-Engine – Digifant Multiport Fuel Injection (MFI)/81kW, code AAF,

from April 1992 to August 1992

Relay location
For relay location and multiple connectors see also section "component locations".

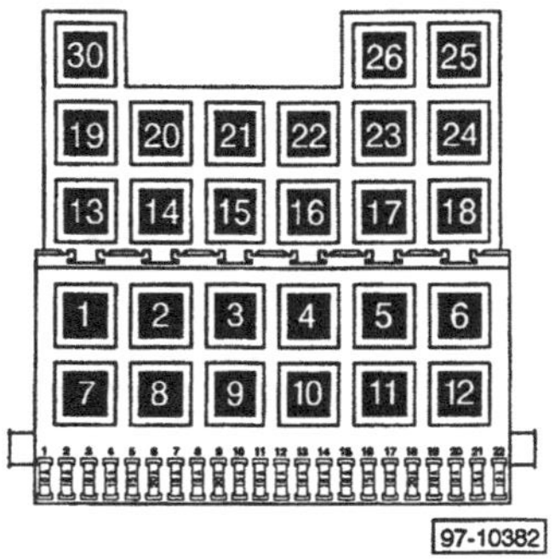

- 2 – Rear Window Wiper/Washer Relay (72)
- 3 – Digifant Engine Control Module (ECM) Power Supply Relay (30)
- 4 – Load Reduction Relay (18)
- 6 – Emergency Flasher Relay (21)
- 8 – Washer/Wiper Intermittent Relay (19)
- 9 – Seat Belt Warning System Relay (4)
- 11 – Dual Horn Relay (53)
- 12 – Fuel Pump (FP) Relay (80)
- 14 – Engine Coolant Pump Relay (38)
- 20 – Park/Neutral Position (PNP) Relay (150)
- 21 – Automatic Transmission Fuse

Fuse colors

30A – green
25A – white
20A – yellow
15A – blue
10A – red

Note: Number in parentheses indicates production control number stamped on relay housing.

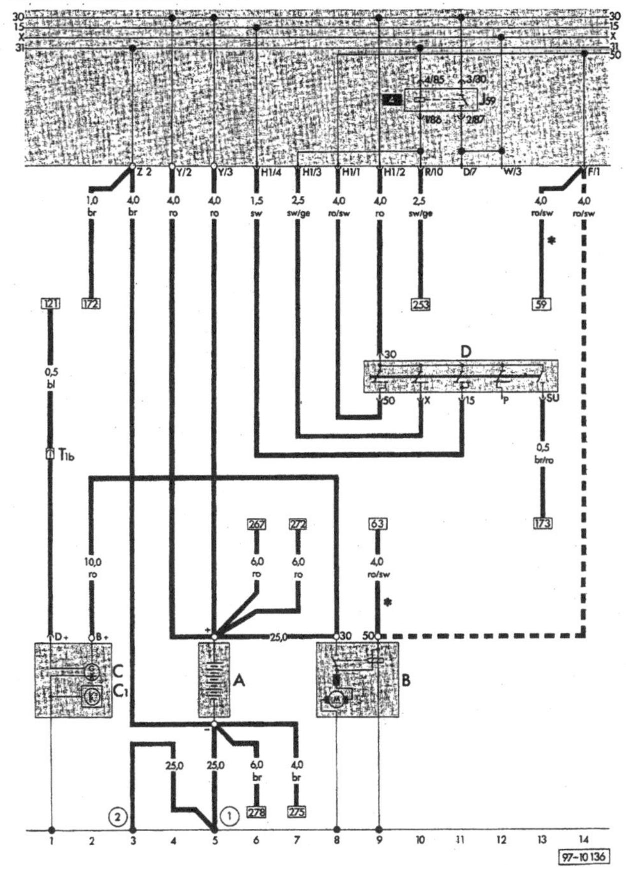

ws = white
sw = black
ro = red
br = brown
gn = green
bl = blue
gr = grey
li = violet
ge = yellow

A – Battery
B – Starter
C – Generator (GEN)
C1 – Voltage Regulator (VR)
D – Ignition/Starter Switch
J59 – Load Reduction Relay
T1b – Single Connector, near starter

(1) – Ground strap, battery to body

(2) – Ground strap, transmission to body

* – Automatic transmission only
▬ ▬ – Manual transmission only

Generator (GEN), battery, starter, ignition/starter switch

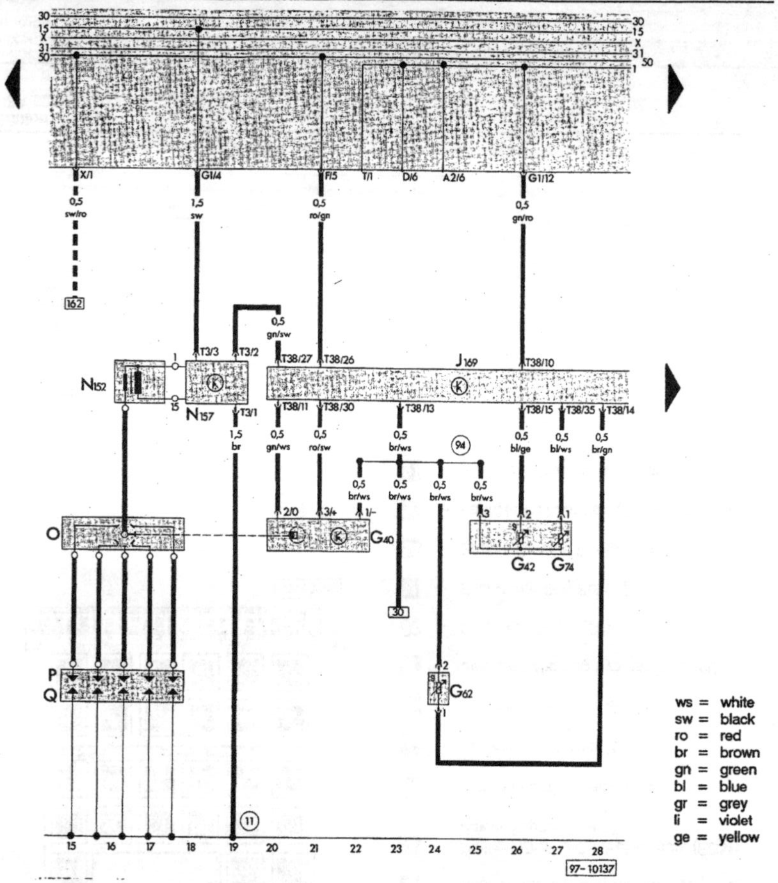

ws = white
sw = black
ro = red
br = brown
gn = green
bl = blue
gr = grey
li = violet
ge = yellow

G40 – Camshaft Position (CMP) Sensor
G42 – Intake Air Temperature (IAT) Sensor (00523–2322)
G62 – Engine Coolant Temperature (ECT) Sensor (0522–2312)
G74 – CO Fuel Trim (FT) Potentiometer (00521–2242)
J169 – Digifant Engine Control Module (ECM), near battery
N152 – Ignition Coil
N157 – Ignition Coil Power Output Stage
O – Ignition Distributor
P – Spark Plug Connectors
Q – Spark Plugs
T3 – 3-Pin Connector
T38 – 38-Pin Connector

(11) – Ground connection, in battery box

(94) – Ground connection -1-, in Digifant Multiport Fuel Injection (MFI) wiring harness

▬ ▬ – Manual transmission only

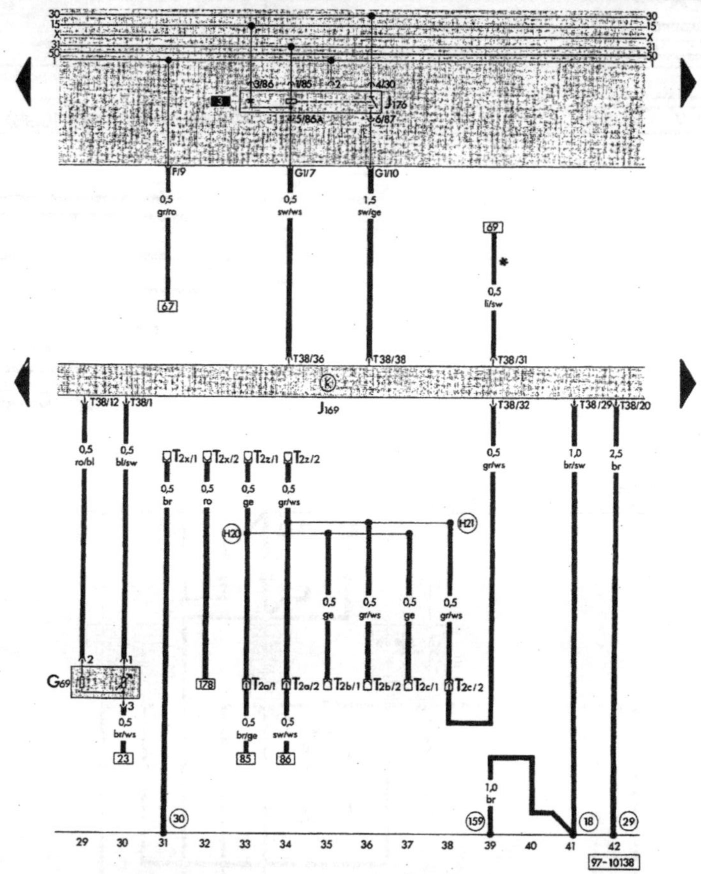

ws = white
sw = black
ro = red
br = brown
gn = green
bl = blue
gr = grey
li = violet
ge = yellow

G69 – Throttle Position (TP) Sensor (00518–2212)
J169 – Digifant Engine Control Module (ECM), near battery
J176 – Digifant Engine Control Module (ECM) Power Supply Relay
T2a – Double Connector, behind instrument panel, left
T2b – Double Connector, behind instrument panel, left*
T2c – Double Connector behind instrument panel, left
T2x – Double Connector, –black–, behind fuse/relay panel cover (OBD)
T2z – Double Connector, –white–, behind fuse/relay panel cover (OBD)
T38 – 38-Pin Connector (T38/31 is connected with automatic transmission only)

(18) – Ground connection, on engine block
(29) – Ground connection, near battery
(30) – Ground connection, -1-, beside fuse/relay panel
(159) – Ground connection -2-, in Digifant Multiport Fuel Injection (MFI) wiring harness
(H20) – Wire connection -1-, in On Board Diagnostic (OBD) wiring harness
(H21) – Wire connection -2-, in On Board Diagnostic (OBD) wiring harness
* Automatic transmission only
** With further On Board Diagnostic (OBD) system only

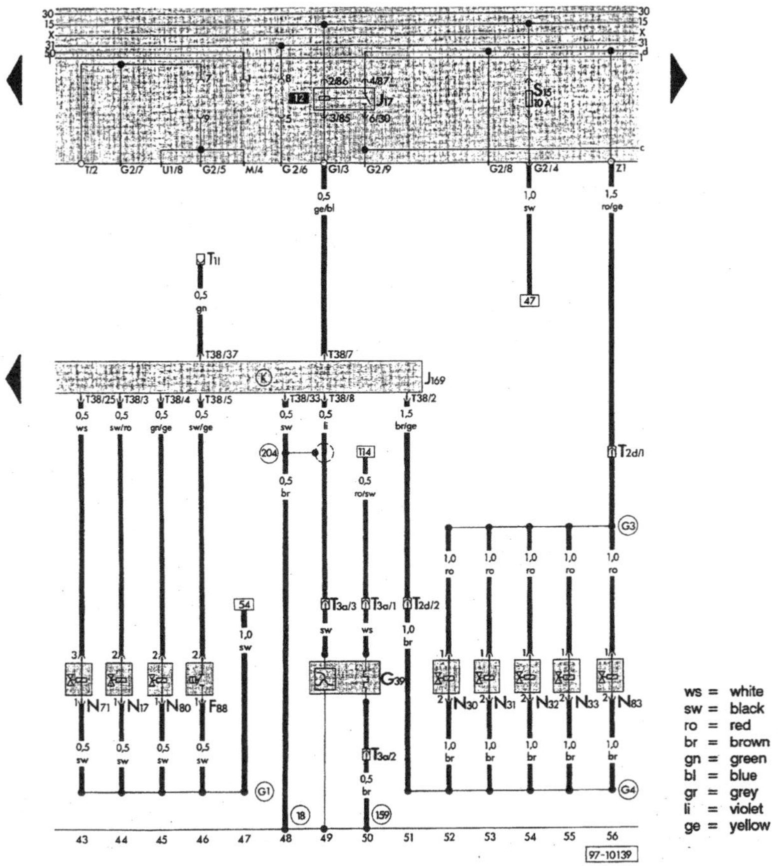

ws = white
sw = black
ro = red
br = brown
gn = green
bl = blue
gr = grey
li = violet
ge = yellow

F88 – Power Steering Pressure (PSP) Switch
G39 – Heated Oxygen Sensor (HO2S) (00525–2342)
J17 – Fuel Pump (FP) Relay
J169 – Digifant Engine Control Module (ECM), near battery
N17 – Cold Start Injector
N30 – Cylinder 1 Fuel Injector
N31 – Cylinder 2 Fuel Injector
N32 – Cylinder 3 Fuel Injector
N33 – Cylinder 4 Fuel Injector
N71 – Idle Air Control (IAC) Valve
N80 – Evaporative Emission (EVAP)Canister Purge Regulator Valve
N83 – Cylinder 5 Fuel Injector
T1l – Single Connector, behind fuse/relay panel (for A/C)
T2d – Double Connector, in engine compartment, right
T3a – 3-Pin Connector, near exhaust manifold
T38 – 38-Pin Connector

(18) – Ground connection, on engine block

(159) – Ground connection -2-, in Digifant Multiport Fuel Injection (MFI) wiring harness

(204) – Ground connection -3-, in Digifant Multiport Fuel Injection (MFI) wiring harness

(G1) – Plus connection, in Digifant Multiport Fuel Injection (MFI) wiring harness

(G3) – Plus connection, in injector wiring harness

(G4) – Wire connection, in injector wiring harness

Digifant engine control module (ECM), heated oxygen sensor (HO2S), injectors

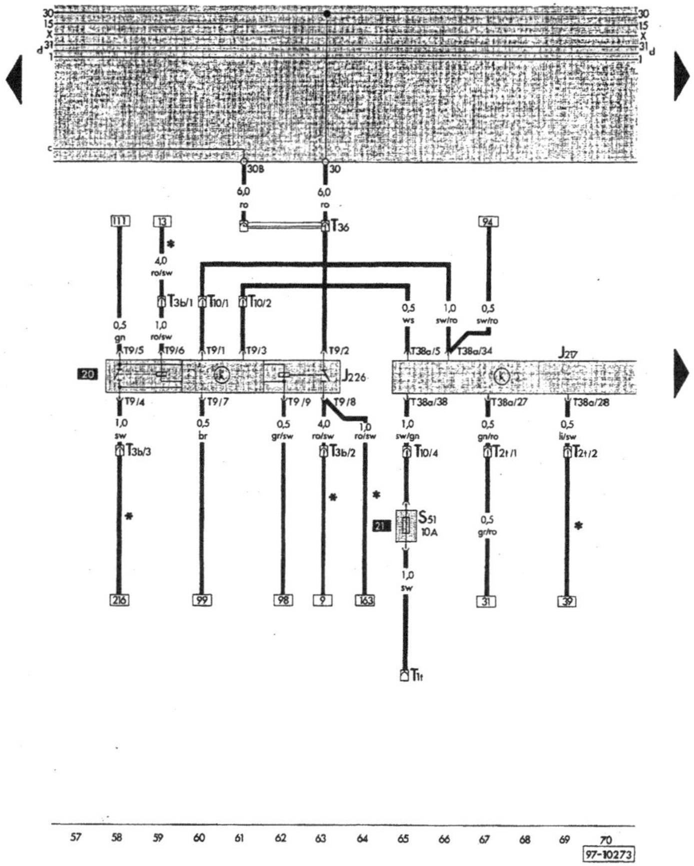

ws = white
sw = black
ro = red
br = brown
gn = green
bl = blue
gr = grey
li = violet
ge = yellow

J217 – Transmission Control Module (TCM) near A-pillar, right
J226 – Park/Neutral Position (PNP) Relay*
T1t – Single Connector, behind fuse/relay panel (cruise control)
T2t – Double Connector, in engine compartment, near transmission
T3b – 3-Pin Connector, behind fuse/relay panel
T9 – 9-Pin Connector
T10 – 10-Pin Connector, behind instrument panel, right
T36 – Wire Connector, Terminal 30 (B+), behind fuse/relay panel
T38a – 38-Pin Connector
S51 – Fuse

* – Automatic transmission only

Transmission control module (TCM), park/neutral position (PNP) relay

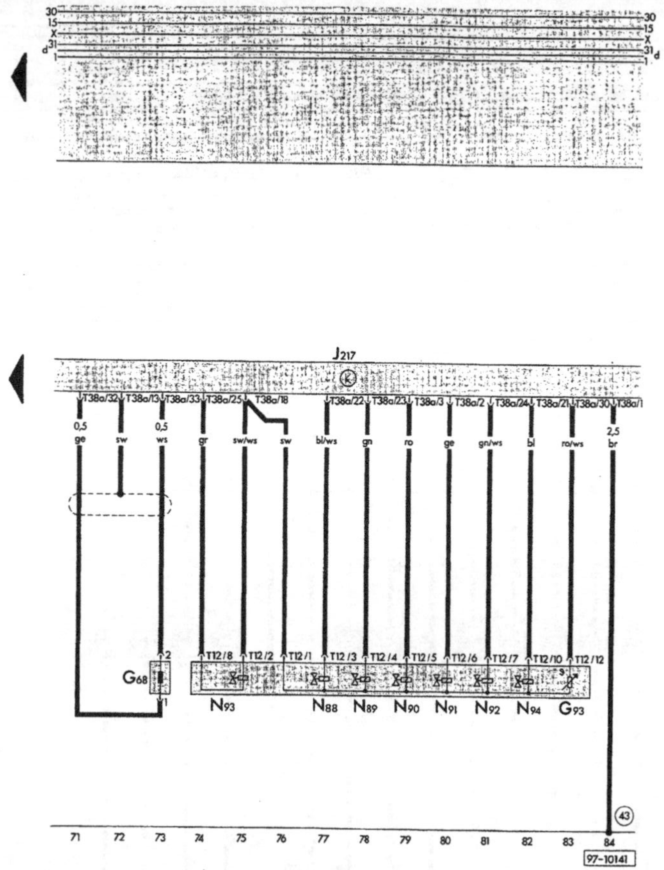

ws = white
sw = black
ro = red
br = brown
gn = green
bl = blue
gr = grey
li = violet
ge = yellow

G68 – Vehicle Speed Sensor (VSS) (00281–1231)
G93 – Transmission Fluid Temperature Sensor (00300–1333)
J217 – Transmission Control Module (TCM), near A-pillar, right
N88 – Solenoid Valve 1 (00258–1113)
N89 – Solenoid Valve 2 (00260–1121)
N90 – Solenoid Valve 3 (00262–1123)
N91 – Solenoid Valve 4 (00264–1131)
N92 – Solenoid Valve 5 (00266–1133)
N93 – Solenoid Valve 6 (00268–1141)
N94 – Solenoid Valve 7 (00270–1143)
T12 – 12-Pin Connector, on valve body
T38a – 38-Pin Connector

(43) – Ground connection, on right A–pillar, lower part

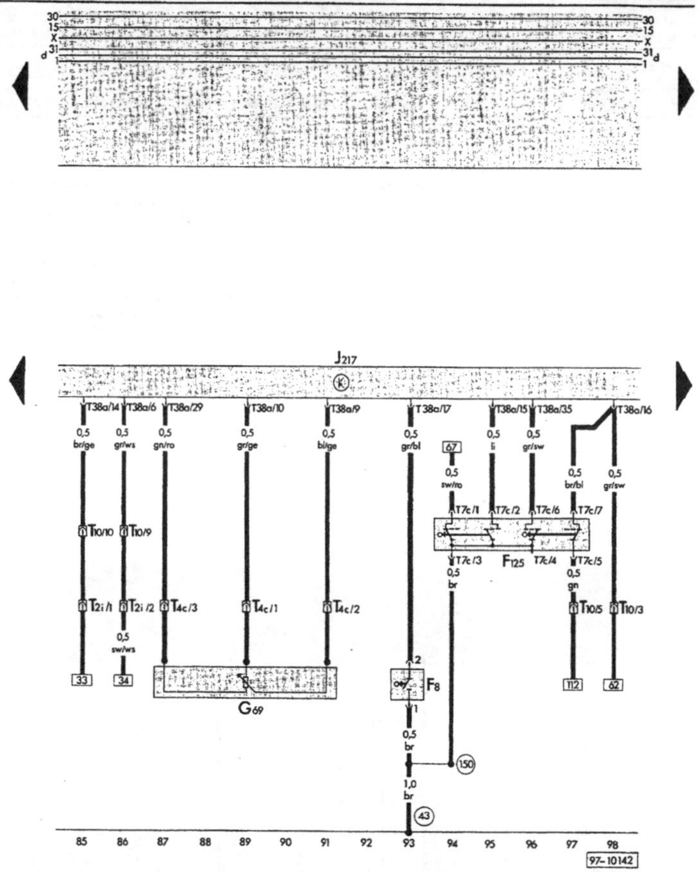

ws = white
sw = black
ro = red
br = brown
gn = green
bl = blue
gr = grey
li = violet
ge = yellow

F8 – Kick Down Switch
F125 – Multi-Function Transmission Range (TR) Switch (00293–1314)
G69 – Throttle Position (TP) Sensor (00518–2212)
J217 – Transmission Control Module (TCM), near A-pillar, right
T2i – Double Connector, behind instrument panel, left
T4c – 4-Pin Connector, near G69
T7c – 7-Pin Connector
T10 – 10-Pin Connector, behind instrument panel, right
T38a – 38-Pin Connector

(43) – Ground connection, on right A–pillar, lower part

(150) – Ground connection, in automatic transmission wiring harness

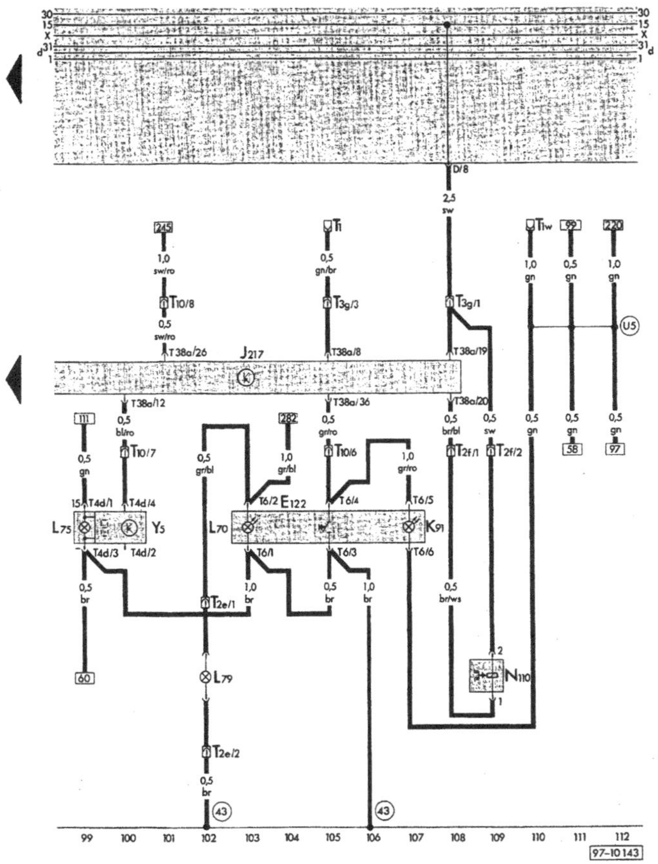

ws = white
sw = black
ro = red
br = brown
gn = green
bl = blue
gr = grey
li = violet
ge = yellow

E122 – Transmission Range (TR) Program Switch (00299–1332)
J217 – Transmission Control Module (TCM), near A-pillar, right
K 91 – Automatic Transmission (A/T) Sport Mode Indicator Light
L70 – Program Switch Illumination
L75 – Digital Display Light
L79 – Selector Lever Light
N110 – Shift Lock Solenoid (01236–4314)
T1 – Single Connector, near fuse/ relay panel (for A/C)
T1w – Single Connector, near fuse/ relay panel
T2e – Double Connector, near selector lever
T2f – Double Connector, behind instrument panel, right
T3g – 3-Pin Connector
T4d – 4-Pin Connector
T6 – 6-Pin Connector
T10 – 10-Pin Connector
T38a – 38-Pin Connector
Y5 – Transmission Range (TR) Selector Lever Display

(43) – Ground connection, on right A-pillar, lower part

(U5) – Wire connection -2- (15), in automatic transmission wiring harness

Transmission control module (TCM), transmission range (TR) selector lever display

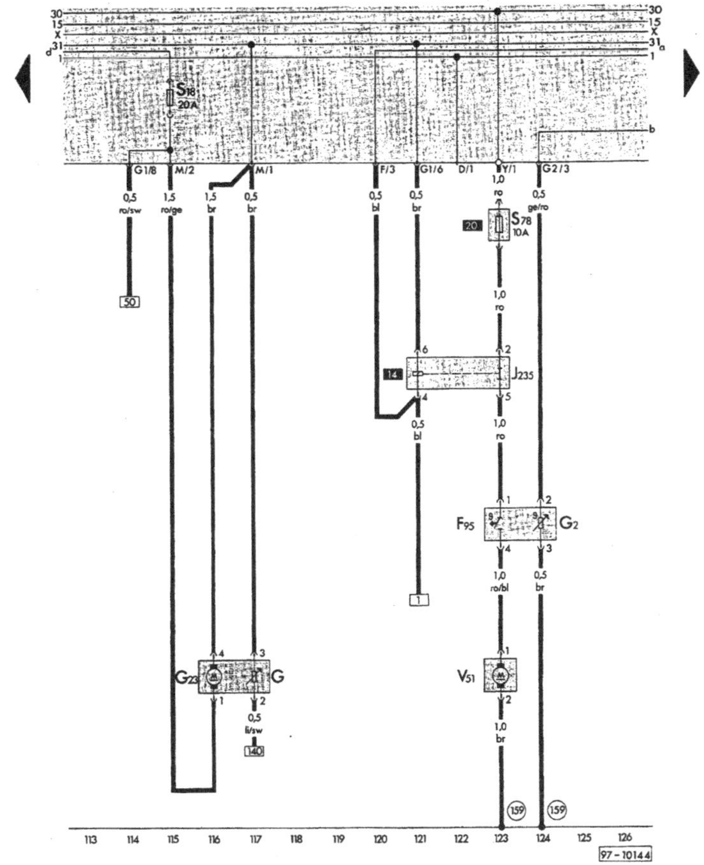

ws = white
sw = black
ro = red
br = brown
gn = green
bl = blue
gr = grey
li = violet
ge = yellow

F95 – After-Run Coolant Pump Thermal Switch *
G – Fuel Level Sensor
G2 – Engine Coolant Temperature (ECT) Sensor
G23 – Transfer Fuel Pump (FP)
J235 – Engine Coolant Pump Relay
S78 – Engine Coolant Pump Circuit Breaker
V51 – After-Run Coolant Pump

(159) – Ground connection -2-, in Digifant Multiport Fuel Injection (MFI) wiring harness

* – Switch value, see engine repair manual

Fuel supply, coolant pump after-run

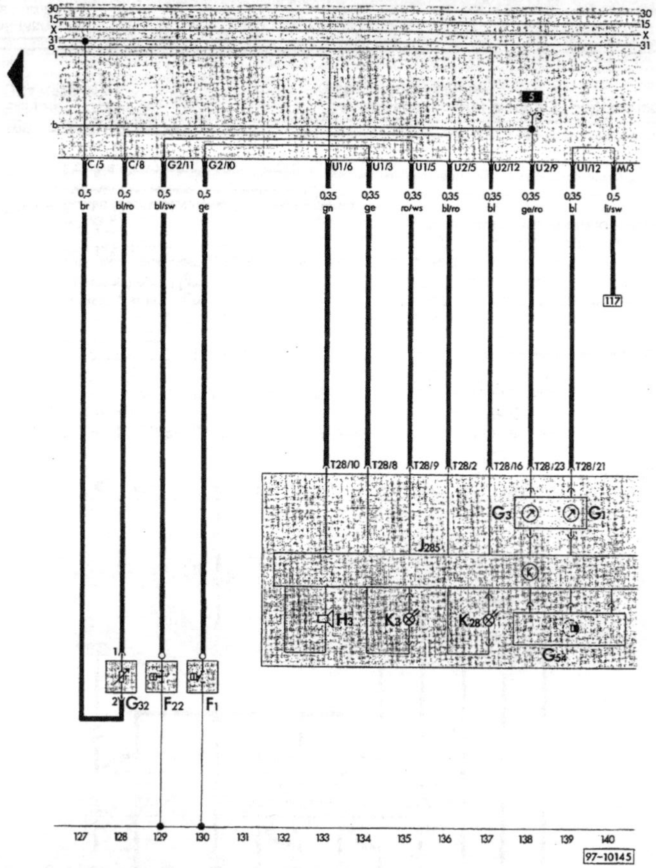

ws = white
sw = black
ro = red
br = brown
gn = green
bl = blue
gr = grey
li = violet
ge = yellow

F1 – Oil Pressure Switch *
F22 – 0.3bar Oil Pressure Switch
G1 – Fuel Gauge
G3 – Engine Coolant Temperature (ECT) Gauge
G32 – Engine Coolant Level (ECL) Sensor
G54 – Vehicle Speed Sensor (VSS), (in instrument cluster)
H3 – Warning Buzzer
J285 – Control Module With Display Unit In Instrument Cluster
K3 – Oil Pressure Warning Light
K28 – Engine Coolant Level/Temperature (ECL/ECT) Warning Light
T28 – 28-Pin Connector, on instrument cluster

* – Switch value, see engine repair manual

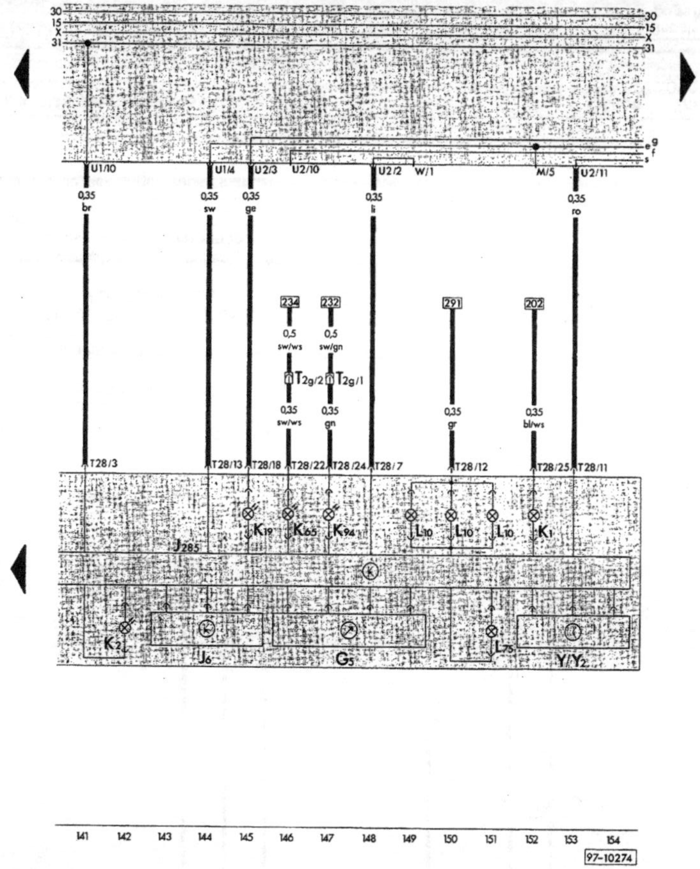

ws = white
sw = black
ro = red
br = brown
gn = green
bl = blue
gr = grey
li = violet
ge = yellow

G5 – Tachometer
J6 – Voltage Stabilizer
J285 – Control Module With Display Unit In Instrument Cluster
K1 – Headlight High Beam Indicator Light
K2 – Generator (GEN) Warning Light
K19 – Seat Belt Warning Light
K65 – Left Turn Signal Indicator Light
K94 – Right Turn Signal Indicator Light
L10 – Instrument Cluster Light
L75 – Digital Display Light
T2g – Double Connector, behind fuse/relay panel
T28 – 28-Pin Connector, on instrument cluster
Y – Analog Clock
Y2 – Digital Clock

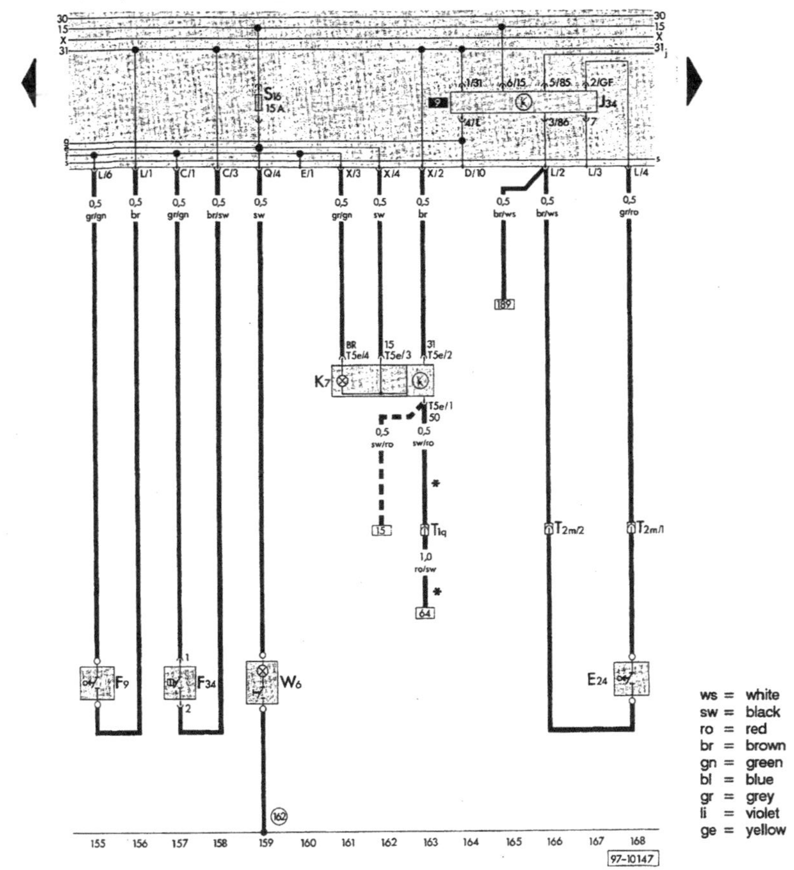

ws = white
sw = black
ro = red
br = brown
gn = green
bl = blue
gr = grey
li = violet
ge = yellow

E24 – Left Seat Belt Switch
F9 – Parking Brake Warning Light Switch
F34 – Brake Fluid Level Warning Switch
J34 – Seat Belt Warning System Relay
K7 – Brake and Parking Brake Warning Light
T1g – Single Connector, behind fuse/relay panel
T2m – Double Connector, near seat belt switch
T5e – 5-Pin Connector
W6 – Glove Compartment Light

(162) – Ground connection, in blower motor wiring harness

* – Automatic transmission only
▬ ▬ – Manual transmission only

Brake and parking brake warning light, seat belt warning system

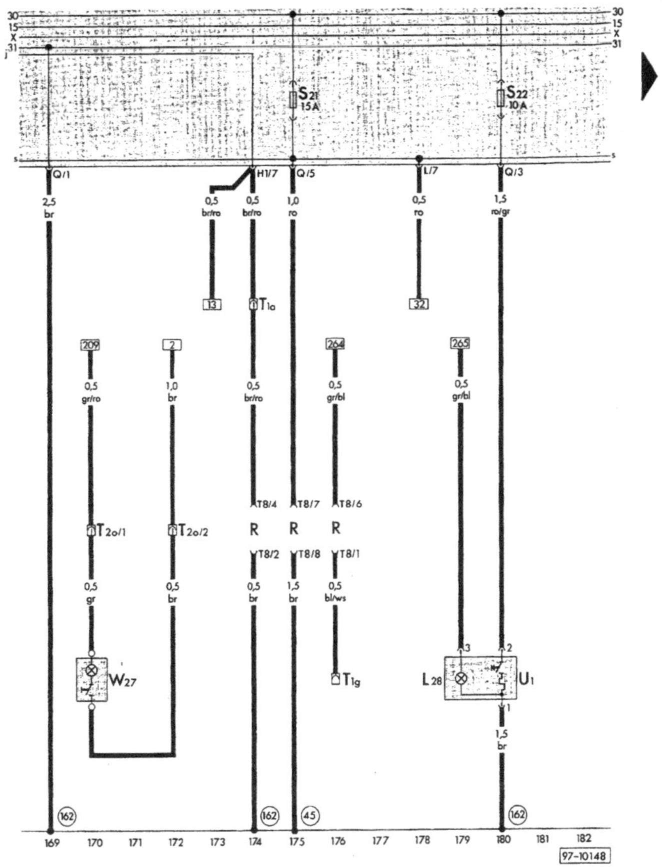

ws = white
sw = black
ro = red
br = brown
gn = green
bl = blue
gr = grey
li = violet
ge = yellow

L28 – Cigarette Lighter Light
R – Radio
T1a – Single Connector, behind instrument panel, left
T1g – Single Connector, behind fuse/relay panel
T2o – Double Connector, near battery
T8 – 8-Pin Connector, on radio
U1 – Cigarette Lighter
W27 – Engine Compartment Light

(45) – Ground connection, behind instrument panel, center
(162) – Ground connection, in blower motor wiring harness

Radio connection, cigarette lighter, engine compartment light

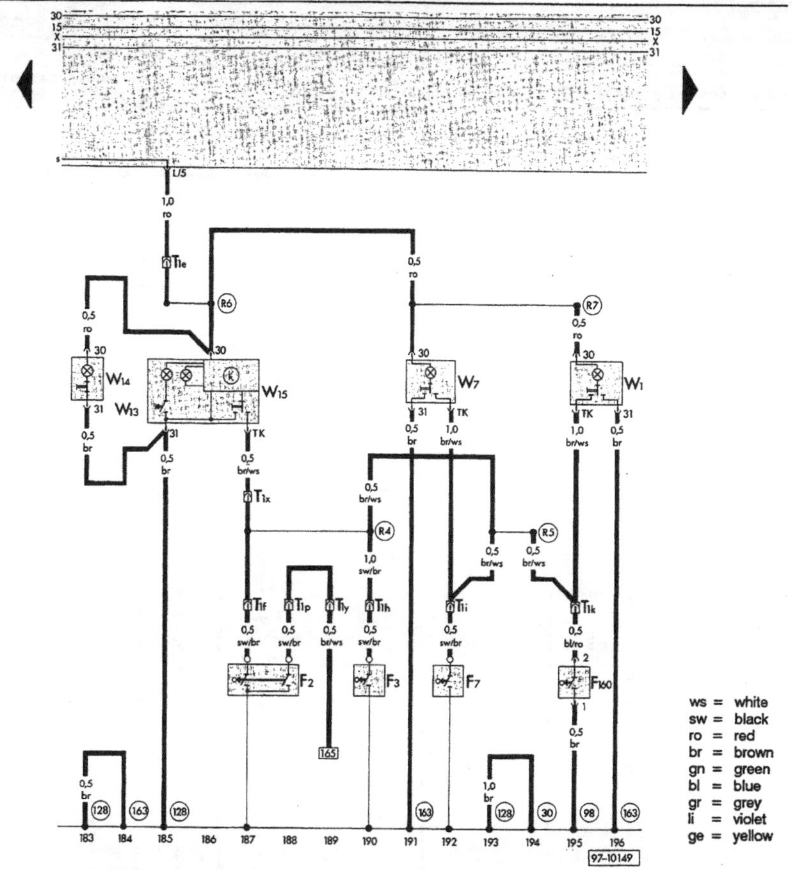

F2 – Left Front Door Contact Switch
F3 – Right Front Door Contact Switch
F7 – Right Rear Side/Sliding Door ContactSwitch
F160 – Rear Lid Contact Switch
T1e – Single Connector, behind fuse/relay panel
T1f – Single Connector, behind fuse/relay panel
T1h – Single Connector, behind instrument panel, right
T1i – Single Connector, behind fuse/relay panel
T1k – Single Connector, near taillight, right
T1p – Single Connector, behind fuse/relay panel
T1x – Single Connector, behind fuse/relay panel
T1y – Single Connector, behind fuse/relay panel
W1 – Rear/Luggage Compartment Interior Light
W7 – Center Interior Light
W13 – Right Front Map/Reading Light
W14 – Right Make-Up Mirror Light
W15 – Interior Light With Delay Switch
(30) – Ground connection, -1-, beside fuse/relay panel
(98) – Ground connection, in rear lid wiring harness
(128) – Ground connection -1-, in interior light wiring harness
(163) – Ground connection -2-, in interior light wiring harness
(R4) – Wire connection -1-, in interior light/door contact switch wiring harness
(R5) – Wire connection -2-, in interior light wiring harness
(R6) – Plus connection -1-, in interior light wiring harness
(R7) – Plus connection -2-, in interior light wiring harness

ws = white
sw = black
ro = red
br = brown
gn = green
bl = blue
gr = grey
li = violet
ge = yellow

Interior lights

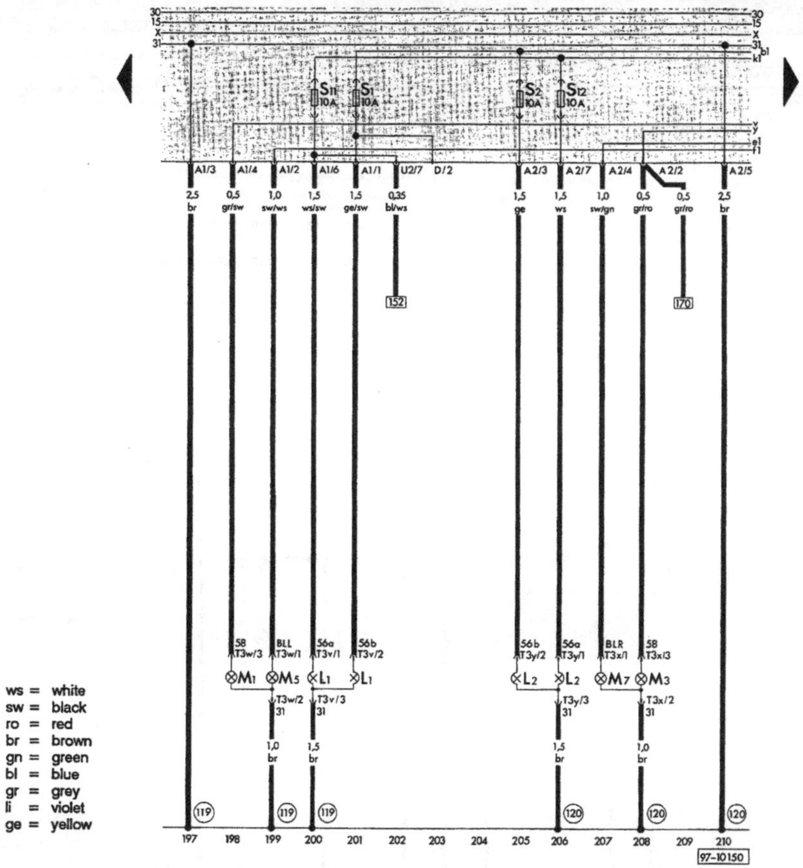

ws = white
sw = black
ro = red
br = brown
gn = green
bl = blue
gr = grey
li = violet
ge = yellow

L1 – Left Headlight
L2 – Right Headlight
M1 – Left Parking Light
M3 – Right Parking Light
M5 – Left Front Turn Signal Light
M7 – Right Front Turn Signal Light
T3v – 3-Pin Connector, on Left Headlight
T3w – 3-Pin Connector, on Left Headlight
T3x – 3-Pin Connector, on Right Headlight
T3y – 3-Pin Connector, on Right Headlight
(119) – Ground connection –1–, in headlight wiring harness
(120) – Ground connection –2–, in headlight wiring harness

Headlights, front turn signals

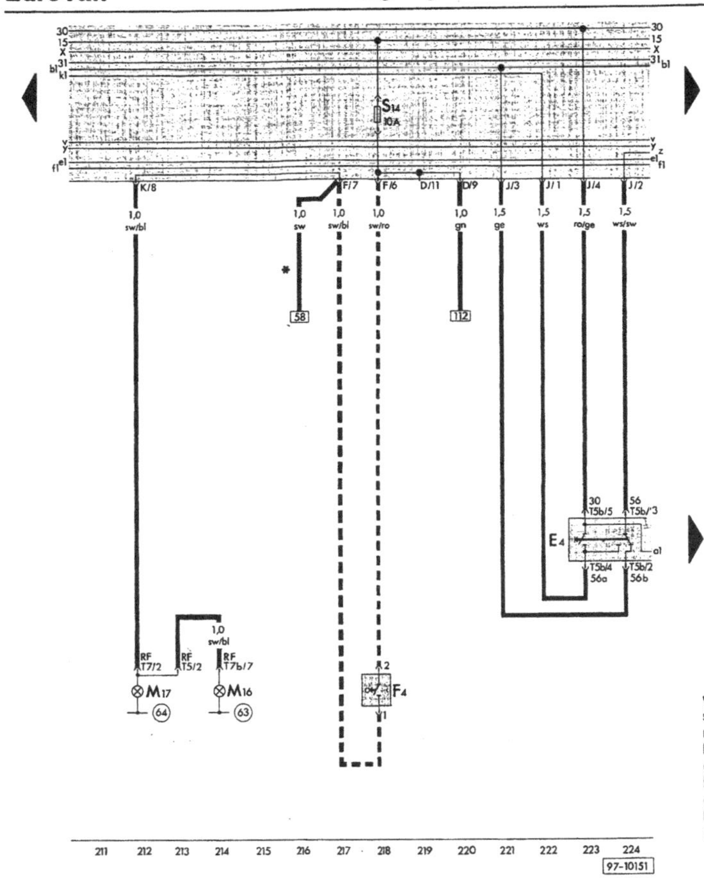

ws = white
sw = black
ro = red
br = brown
gn = green
bl = blue
gr = grey
li = violet
ge = yellow

E4 – Headlight Dimmer/Flasher Switch
F4 – Back-Up Light Switch (manual transmission)
M16 – Left Back-Up Light
M17 – Right Back-Up Light
T5 – 5-Pin Connector, on Right Back-Up light
T5b – 5-Pin Connector, behind steering column switch cover
T7 – 7-Pin Connector, on Right Taillight
T7b – 7-pin Connector, on Left Taillight

* – Automatic transmission only
▬ ▬ – Manual transmission only

(63) – Ground connection, on left taillight bulb holder
(64) – Ground connection, on right taillight bulb holder

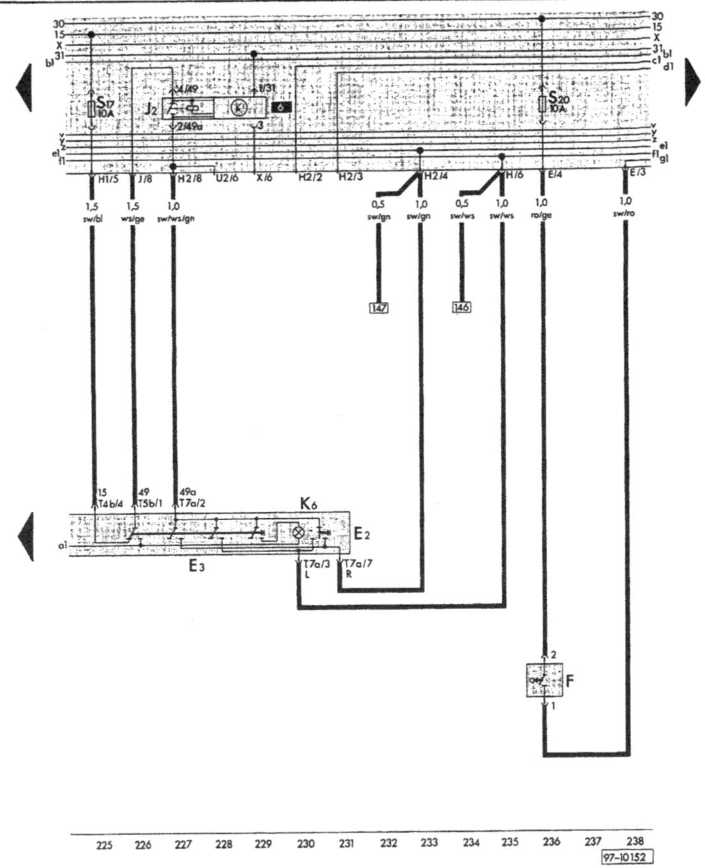

ws = white
sw = black
ro = red
br = brown
gn = green
bl = blue
gr = grey
li = violet
ge = yellow

E2 – Turn Signal Switch
E3 – Emergency Flasher Switch
F – Brake Light Switch (00526–2131)
J2 – Emergency Flasher Relay
K6 – Emergency Flasher Warning Light
T4b – 4-Pin Connector, behind steering column switch cover
T5b – 5-Pin Connector, behind steering column switch cover
T7a – 7-Pin Connector, behind steering column switch cover

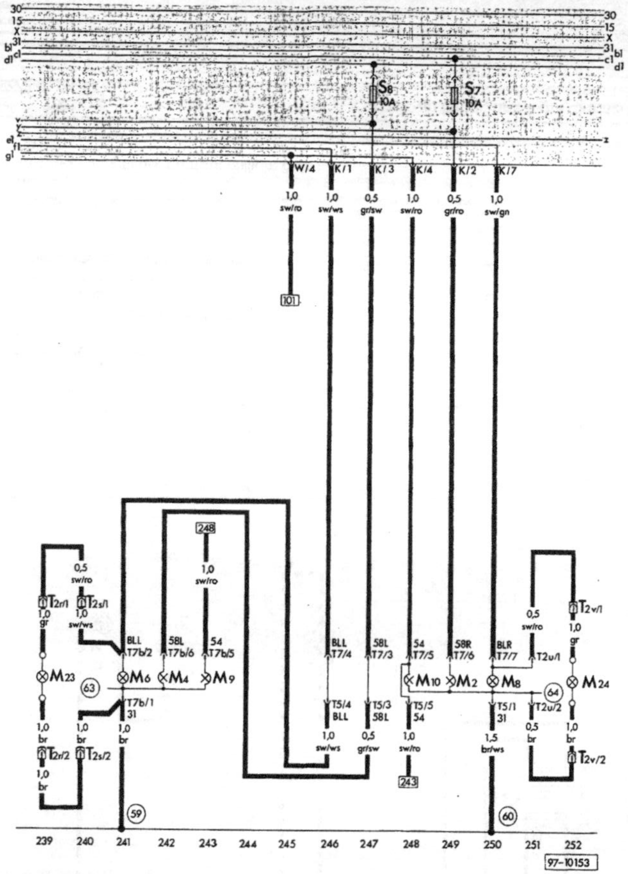

ws = white
sw = black
ro = red
br = brown
gn = green
bl = blue
gr = grey
li = violet
ge = yellow

M2 – Right Taillight
M4 – Left Taillight
M6 – Left Rear Turn Signal Light
M8 – Right Rear Turn Signal Light
M9 – Left Brake Light
M10 – Right Brake Light
M23 – Left Roof Turn Signal Light **
M24 – Right Roof Turn Signal Light **
T2r – Double Connector, near left roof turn signal
T2s – Double Connector, near left taillight
T2u – Double Connector, on Right Taillight
T2v – Double Connector, near right roof turn signal
T5 – 5-Pin Connector, on Right Taillight
T7 – 7-Pin Connector, on Right Taillight
T7b – 7-Pin Connector, on Left Taillight

(59) – Ground connection, near left taillight
(60) – Ground connection, near right taillight
(63) – Ground connection, on left taillight bulb holder
(64) – Ground connection, on right taillight bulb holder

* – Different wire routing for flatbed models
** – Roof turn signals not installed on all vehicles

Edition 7/93
USA.5272.01.21

Rear turn signals*, brake lights*, taillights*

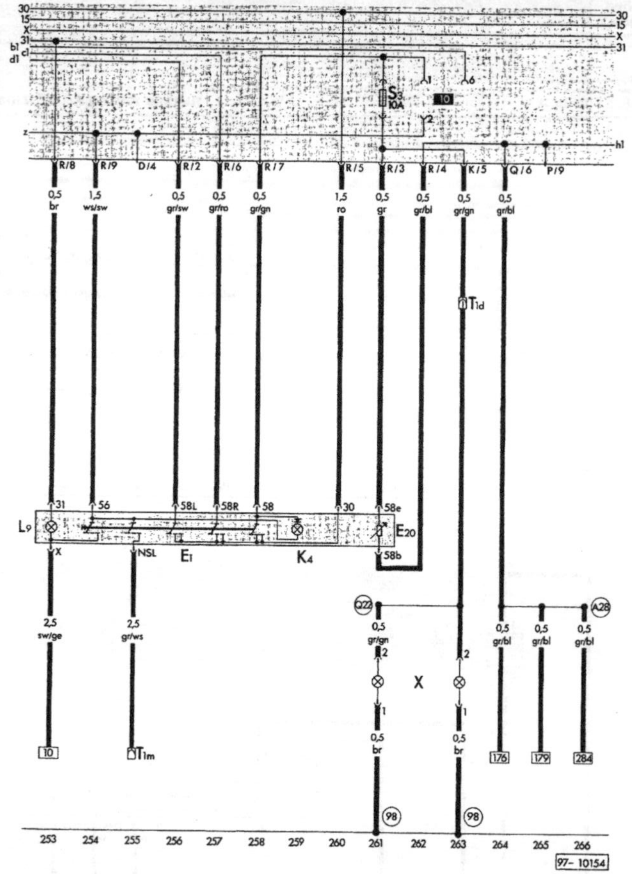

ws = white
sw = black
ro = red
br = brown
gn = green
bl = blue
gr = grey
li = violet
ge = yellow

E1 – Light Switch
E20 – Instrument Panel Light Dimmer Switch
K4 – Park Light Indicator Light
L9 – Headlight Switch Light
T1d – Single Connector, near right taillight
T1m – Single Connector, behind fuse/relay panel
X – License Plate Light

(98) – Ground connection, in rear lid wiring harness
(A28) – Plus connection (58b), in fresh air blower wiring harness
(Q22) – Wire connection, in rear lid wiring harness

Light switch, license plate light (tailgate)

Edition 7/93
USA.5272.01.21

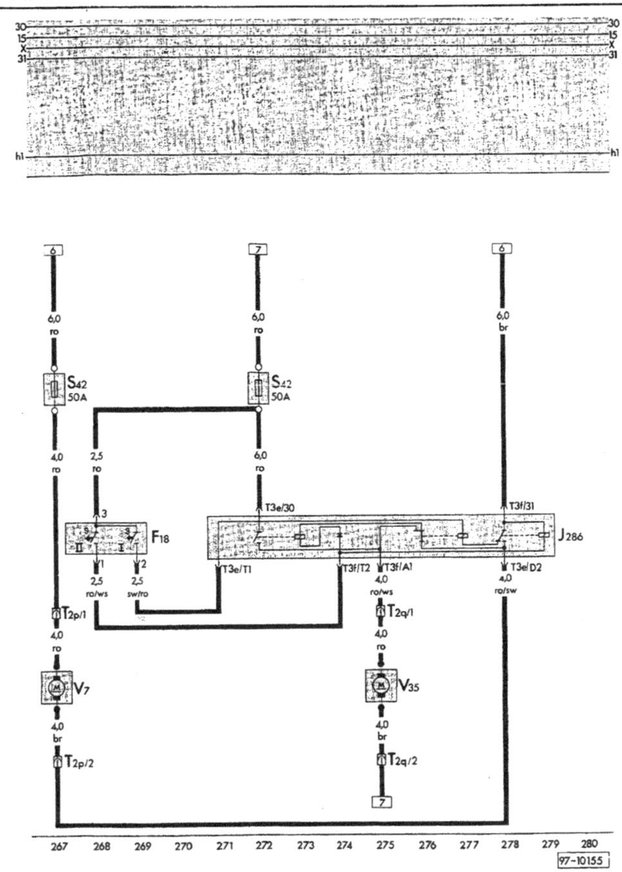

ws = white
sw = black
ro = red
br = brown
gn = green
bl = blue
gr = grey
li = violet
ge = yellow

F18 – Coolant Fan Control (FC) Thermal Switch
J286 – Series/Parallel Coolant Fan Control (FC) Relay (V7, V35)
S42 – Coolant Fan Fuse
T2p – Double Connector, in engine compartment, front
T2g – Double Connector, in engine compartment, front
T3e – 3-Pin Connector
T3f – 3-Pin Connector
V7 – Coolant Fan
V35 – Right Coolant Fan

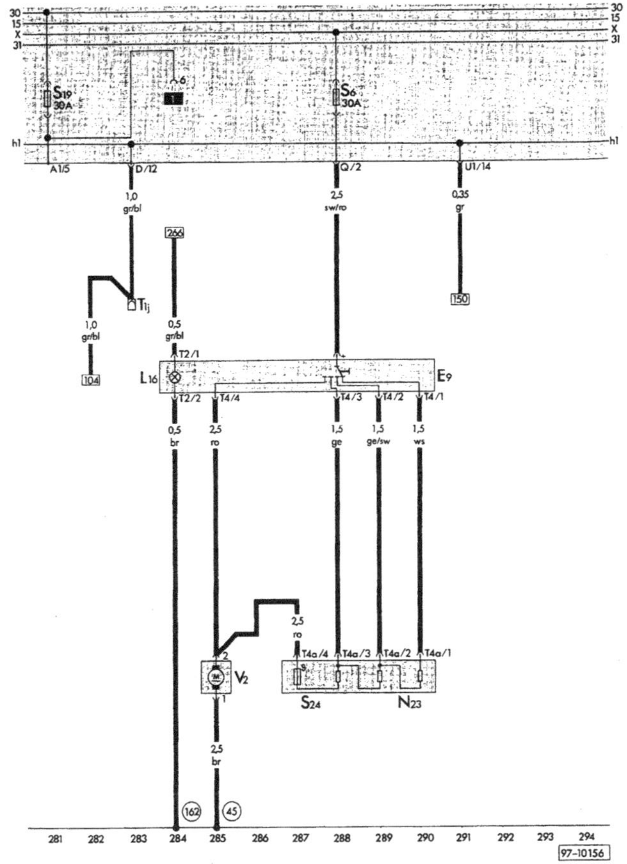

ws = white
sw = black
ro = red
br = brown
gn = green
bl = blue
gr = grey
li = violet
ge = yellow

E9 – Fresh Air Blower Switch
L16 – Fresh Air Control Lever Light
N23 – Fresh Air Blower Series Resistance
S24 – Heater, A/C Clutch Fuse
T1j – Single Connector, behind instrument panel, left
T2 – Double Connector
T4 – 4-Pin Connector
T4a – 4-Pin Connector
V2 – Fresh Air Blower

(45) – Ground connection, behind instrument panel, center
(162) – Ground connection, in blower motor wiring harness

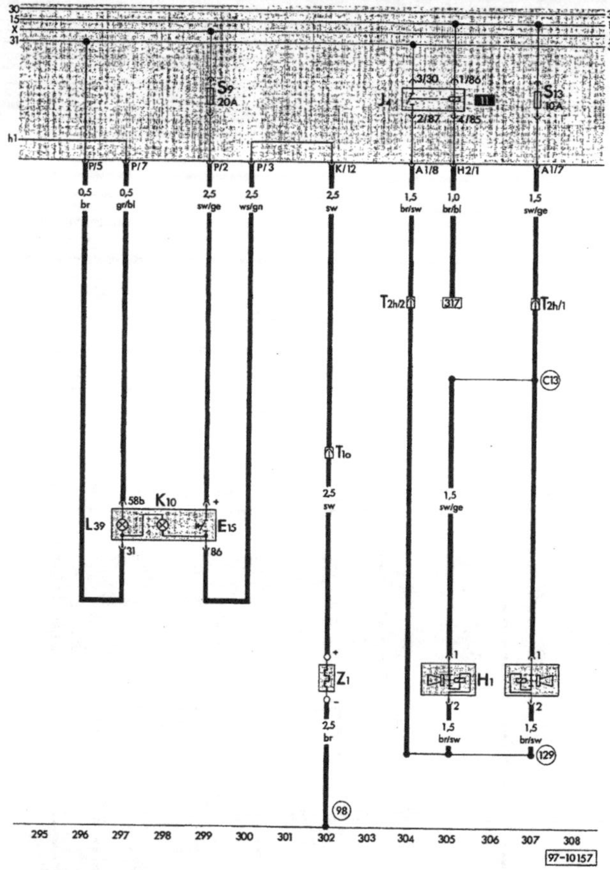

ws = white
sw = black
ro = red
br = brown
gn = green
bl = blue
gr = grey
li = violet
ge = yellow

E15 – Rear Window Defogger Switch
H1 – Dual Horns
J4 – Dual Horn Relay
K10 – Rear Window Defogger Indicator Light
L39 – Rear Window Defogger Switch Light
T1o – Single Connector, near right taillight
T2h – Double Connector, in engine compartment, left front
Z1 – Rear Window Defogger/Heat Element

(98) – Ground connection, in rear lid wiring harness
(129) – Ground connection, in dual tone horn wiring harness
(C13) – Plus connection, in dual tone horn wiring harness

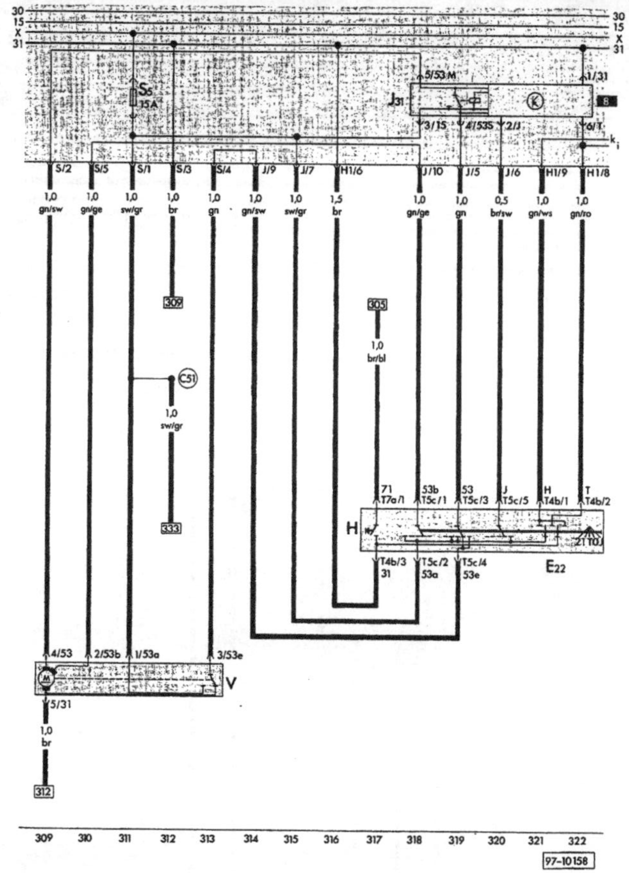

ws = white
sw = black
ro = red
br = brown
gn = green
bl = blue
gr = grey
li = violet
ge = yellow

E22 – Windshield Wiper/Washer Switch
H – Horn Button
J31 – Washer/Wiper Intermittent Relay
T4b – 4-Pin connector, behind steering column switch cover
T5c – 5-Pin connector, behind steering column switch cover
T7a – 7-Pin connector, behind steering column switch cover

(C51) – Plus connection (x), in wiper motor wiring harness

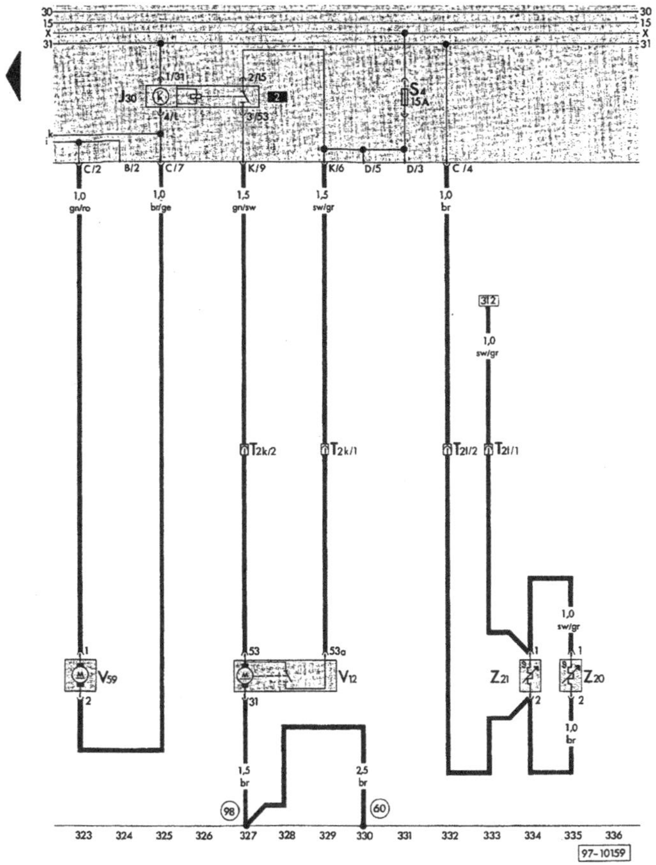

ws = white
sw = black
ro = red
br = brown
gn = green
bl = blue
gr = grey
li = violet
ge = yellow

J30 – Rear Window Wiper/Washer Relay
T2k – Double Connector, near right taillight
T2l – Double Connector, near right firewall
V12 – Rear Window Wiper Motor
V59 – Windshield and Rear Window Washer Pump
Z20 – Left Washer Nozzle Heater
Z21 – Right Washer Nozzle Heater

(60) – Ground connection, near right taillight

(98) – Ground connection, in rear lid wiring harness

THIS PAGE INTENTIONALLY LEFT
BLANK

2.5 L-Engine–Digifant Multiport Fuel Injection (MFI)/81kW, code AAF, California only,

from April 1992

Relay location

For relay location and multiple connectors see also section "component locations".

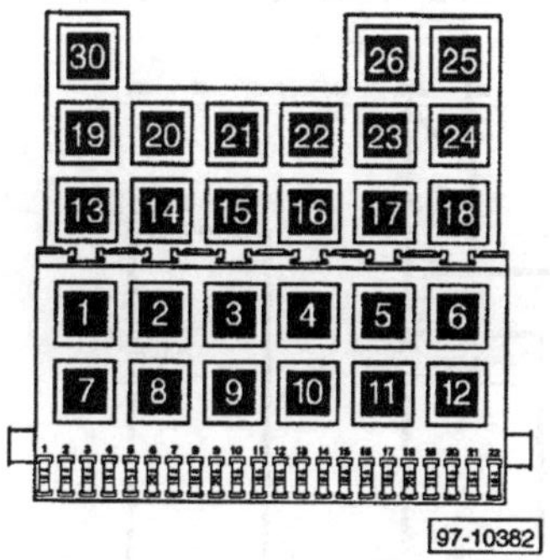

- **3** – Digifant Engine Control Module (ECM) Power Supply Relay (30)
- **4** – Load Reduction Relay (18)
- **9** – Seat Belt Warning System Relay (4)
- **12** – Fuel Pump (FP) Relay (80)

Note: Number in parentheses indicates production control number stamped on relay housing.

Fuse colors

30A – green
25A – white
20A – yellow
15A – blue
10A – red

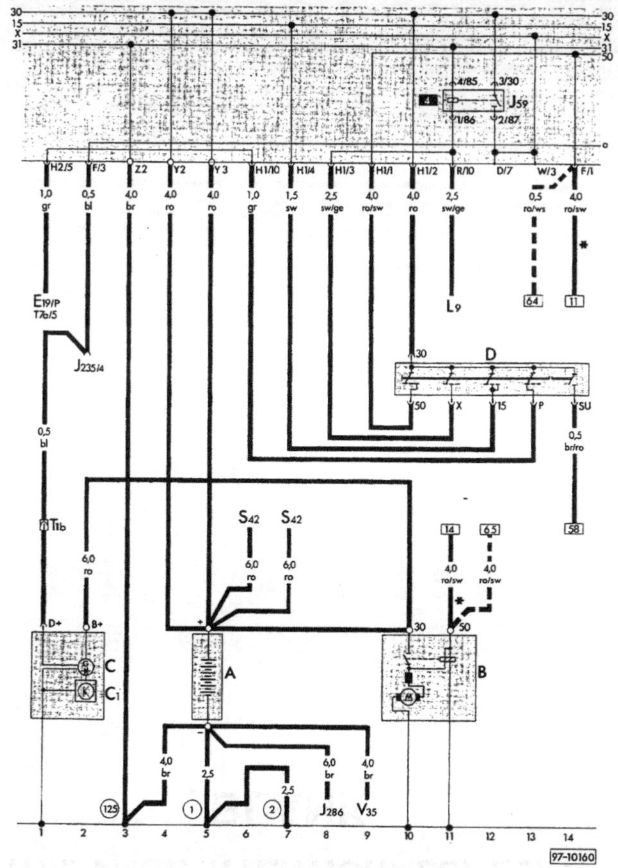

ws = white
sw = black
ro = red
br = brown
gn = green
bl = blue
gr = grey
li = violet
ge = yellow

A – Battery
B – Starter
C – Generator (GEN)
C1 – Voltage Regulator (VR)
D – Ignition/Starter Switch
E19 – Park Light Switch
J59 – Load Reduction Relay
J235 – Engine Coolant Pump Relay
J286 – Series/Parallel Coolant Fan Control (FC) Relay (V7,V35)
L9 – Headlight Switch Light
S42 – Coolant Fan Fuse
T1b – Single Connector, near starter
V35 – Right Coolant Fan

(1) – Ground strap, battery to body
(2) – Ground strap, transmission to body
(125) – Ground connection -3-, in headlight wiring harness
* – Manual transmission only
▬ ▬ – Automatic transmission only

Generator (GEN), battery, starter, ignition/starter switch

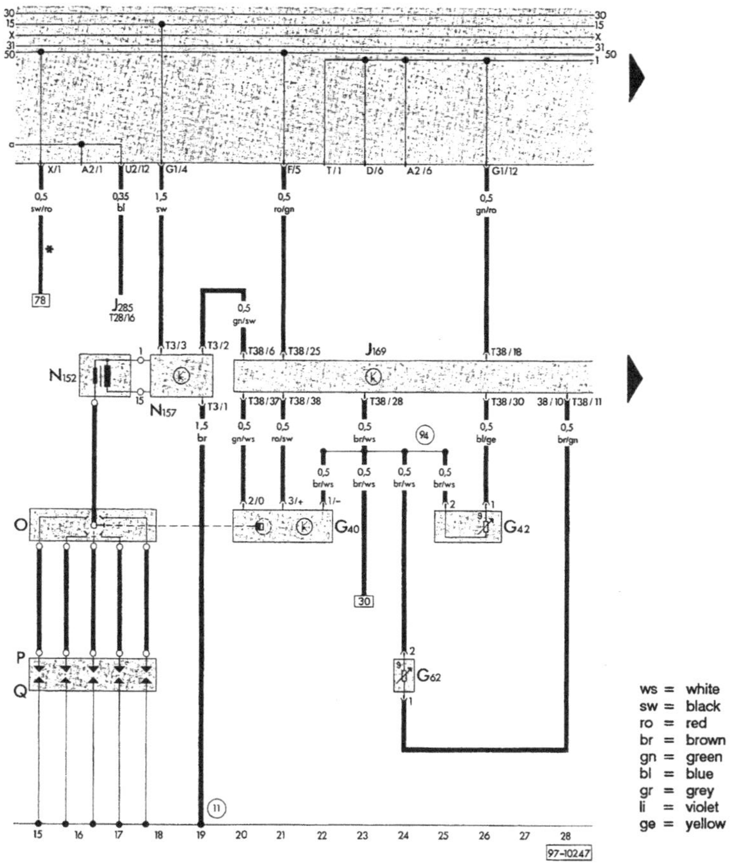

ws = white
sw = black
ro = red
br = brown
gn = green
bl = blue
gr = grey
li = violet
ge = yellow

G40 – Camshaft Position (CMP) Sensor
G42 – Intake Air Temperature (IAT) Sensor (00523–2322)
G62 – Engine Coolant Temperature (ECT) Sensor (00522–2312)
J169 – Digifant Engine Control Module (ECM), near battery
J285 – Control Module With Display Unit In Instrument Cluster
N152 – Ignition Coil
O – Ignition Distributor
P – Spark Plug Connectors
Q – Spark Plugs
T3 – 3-Pin Connector
T 28 – 28-Pin Connector, on instrument cluster
T 38 – 38-Pin Connector

(11) – Ground connection, in battery box
(94) – Ground connection -1-, in Digifant Multiport Fuel Injection (MFI) wiring harness
* – Manual transmission only

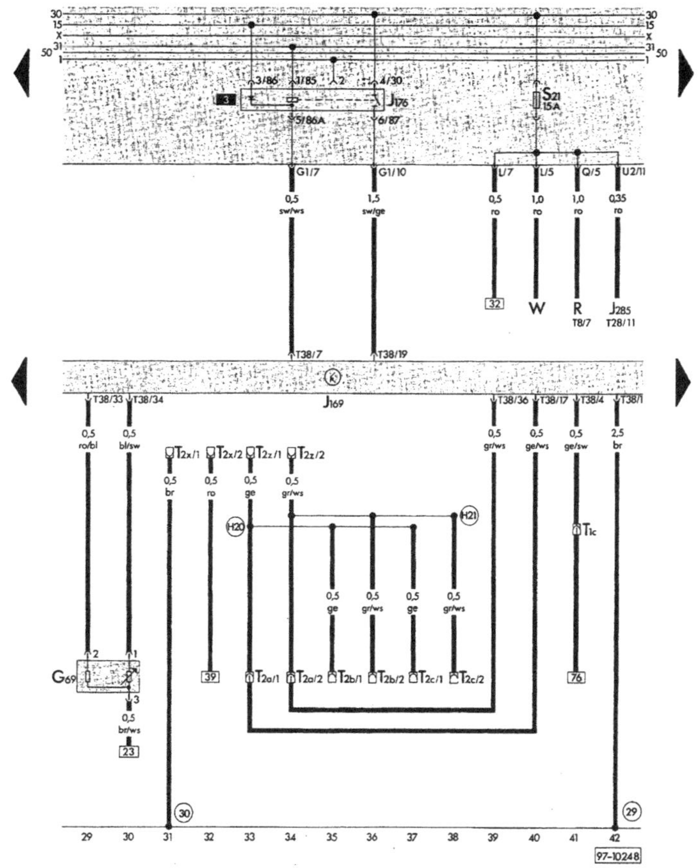

ws = white
sw = black
ro = red
br = brown
gn = green
bl = blue
gr = grey
li = violet
ge = yellow

G69 – Throttle Position (TP) Sensor (00518–2212)
J176 – Digifant Engine Control Module (ECM) Power Sypply Relay
J285 – Control Module With Display Unit In Instrument Cluster
R – Radio
T1c – Single Connector, behind fuse/relay panel
T2a – Double Connector, behind instrument panel, left
T2b – Double Connector, behind instrument panel, left*
T2c – Double Connector, behind instrument panel, left*
T2x – Double Connector, –black–, behind fuse/relay panel cover (OBD)
T2z – Double Connector, –white–, behind fuse/relay panel cover (OBD)
T8 – 8-Pin Connector, on radio
T28 – 28-Pin Connector, on instrument cluster
T38 – 38-Pin Connector

(29) – Ground connection, near battery
(30) – Ground connection, -1-, beside fuse/relay panel
(H20) – Wire connection -1-, in On Board Diagnostic (OBD) wiring harness
(H21) – Wire connection -2-, in On Board Diagnostic (OBD) wiring harness
* With further On Board Diagnostic (OBD) systems only

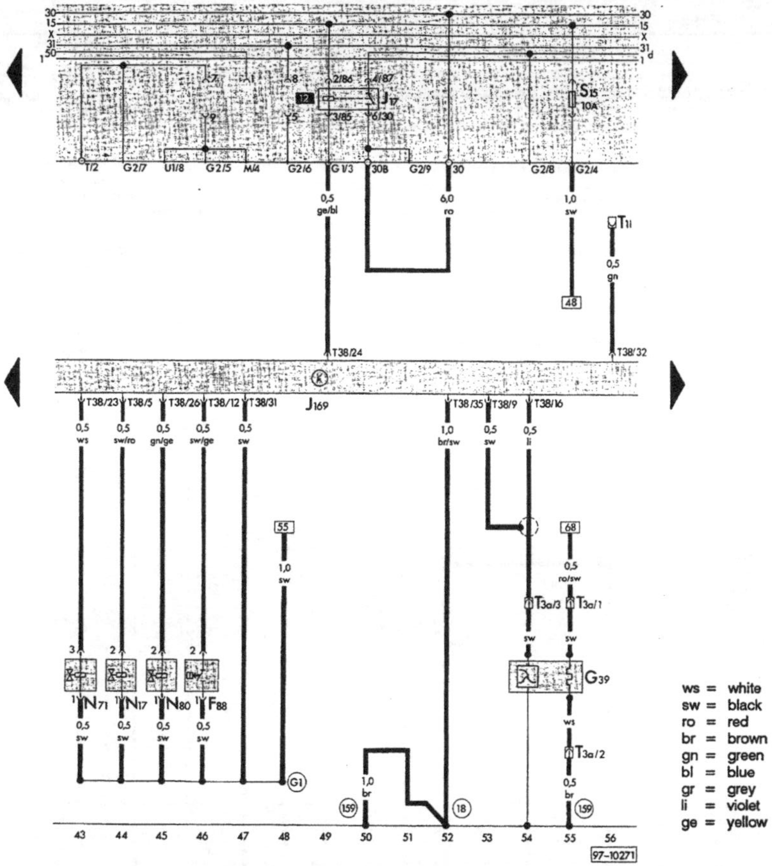

ws = white
sw = black
ro = red
br = brown
gn = green
bl = blue
gr = grey
li = violet
ge = yellow

F88 – Power Steering Pressure (PSP) Switch
G39 – Heated Oxygen Sensor (HO2S) (00525–2342)
J17 – Fuel Pump (FP) Relay
J169 – Digifant Engine Control Module (ECM), near battery
N17 – Cold Start Injector
N71 – Idle Air Control (IAC) Valve
N80 – Evaporative Emission (EVAP) Canister Purge Regulator Valve
T1I – Single Connector, behind fuse/relay panel (for A/C)
T3a – 3-Pin Connector, near exhaust manifold
T38 – 38-Pin Connector

(18) – Ground connection, on engine block

(159) – Ground connection -2-, in Digifant Multiport Fuel Injection (MFI) wiring harness
(G1) – Plus connection, in Digifant Multiport Fuel Injection (MFI) wiring harness

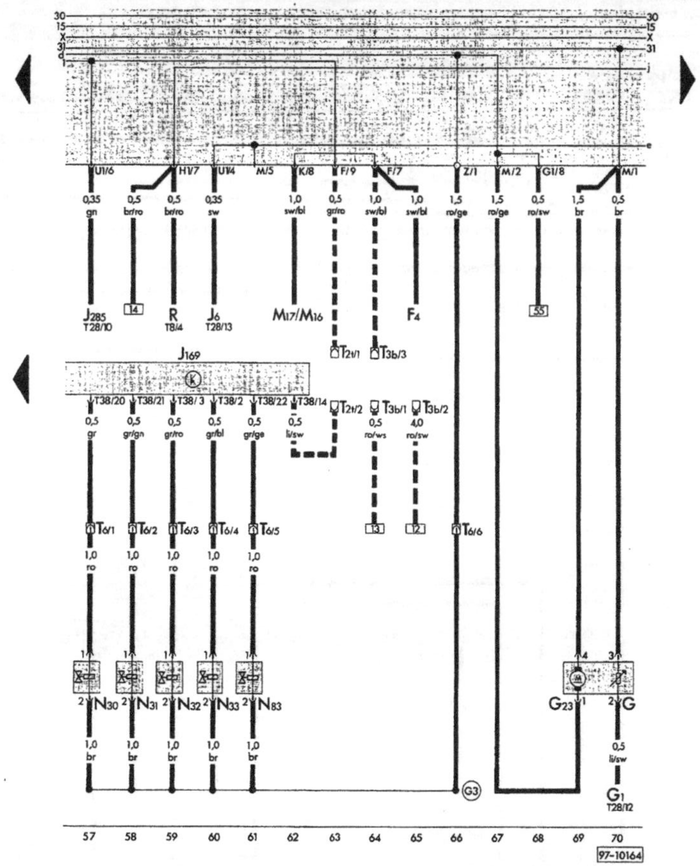

ws = white
sw = black
ro = red
br = brown
gn = green
bl = blue
gr = grey
li = violet
ge = yellow

F4 – Back-Up Light Switch
G – Fuel Level Sensor
G1 – Fuel Gauge
G23 – Transfer Fuel Pump (FP)
J6 – Voltage Stabilizer
J169 – Digifant Engine Control Module (ECM), near battery
J285 – Control Module With Display Unit In Instrument Cluster
M16 – Left Back-Up Light
M17 – Right Back-Up Light
N30 – Cylinder 1 Fuel Injector
N31 – Cylinder 2 Fuel Injector
N32 – Cylinder 3 Fuel Injector
N33 – Cylinder 4 Fuel injector
N83 – Cylinder 5 Fuel Injector
R – Radio
T2t – Double Connector, behind instrument panel, left
T3b – 3-Pin Connector, behind instrument panel, left
T6 – 6-Pin Connector, near injectors
T8 – 8-Pin Connector, on radio
T28 – 28-Pin Connector, on instrument cluster
T38 – 38-Pin Connector

(G3) – Plus connection, in injector wiring harness

▬ ▬ – Automatic transmission only

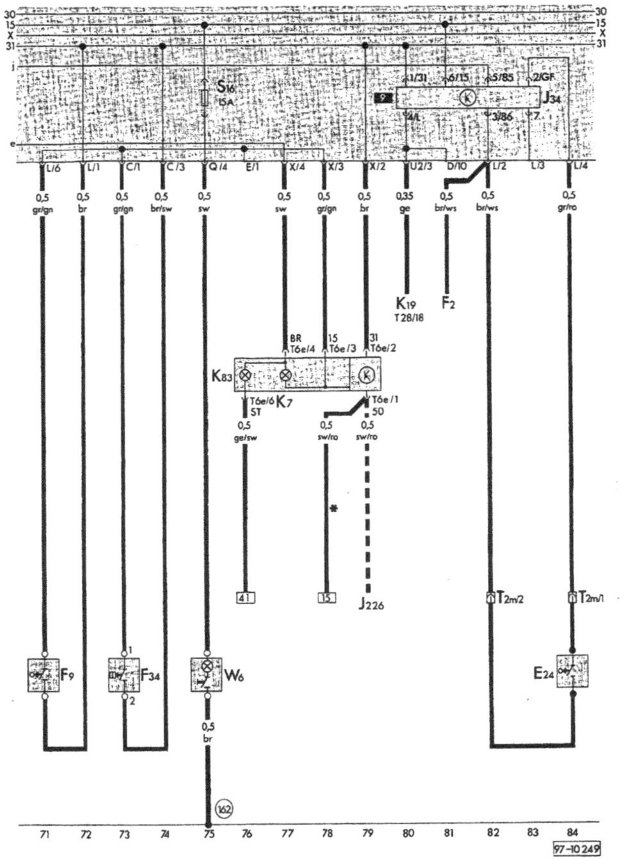

ws = white
sw = black
ro = red
br = brown
gn = green
bl = blue
gr = grey
li = violet
ge = yellow

- E 24 – Left Seat Belt Switch
- F2 – Left Front Door Contact Switch
- F9 – Parking Brake Warning Light Switch
- F34 – Brake Fluid Level Warning Switch
- J34 – Seat Belt Warning System Relay
- J226 – Park/NeutralPosition (PNP) Relay
- K7 – Brake and Parking Brake Warning Light
- K19 – Seat Belt Warning Light
- K83 – Malfunction Indicator Lamp (MIL)
- T2m – Double Connector, near seat belt switch
- T6e – 6-Pin Connector
- W6 – Glove Compartment Light
- (162) – Ground connection, in blower motor wiring harness

- ▬ ▬ – Automatic transmission only
- * – Manual transmission only

THIS PAGE INTENTIONALLY LEFT

BLANK

- **Rear window defogger (rear swinging doors),**
- **License plate light (rear swinging doors),**
- **Rear window wiper/washer system (rear swinging doors),**

from April 1992,

Relay location
For relay location and multiple connectors see also section "component locations".

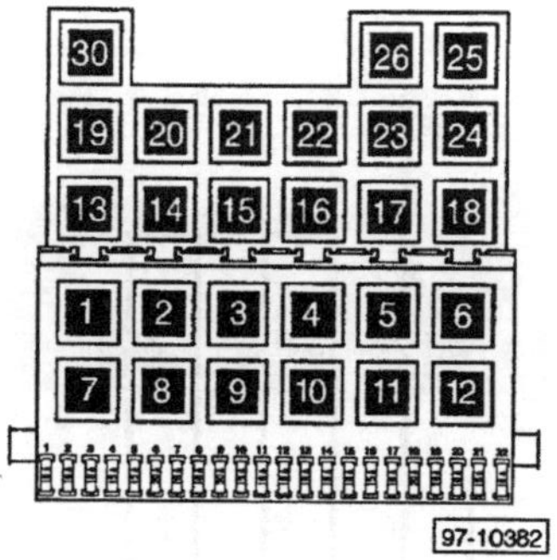

2 – Rear Window Wiper/Washer Relay (72)

8 – Washer/Wiper Intermittent Relay (19)

Note: Number in parentheses indicates production control number stamped on relay housing.

Fuse colors

30A – green
25A – white
20A – yellow
15A – blue
10A – red

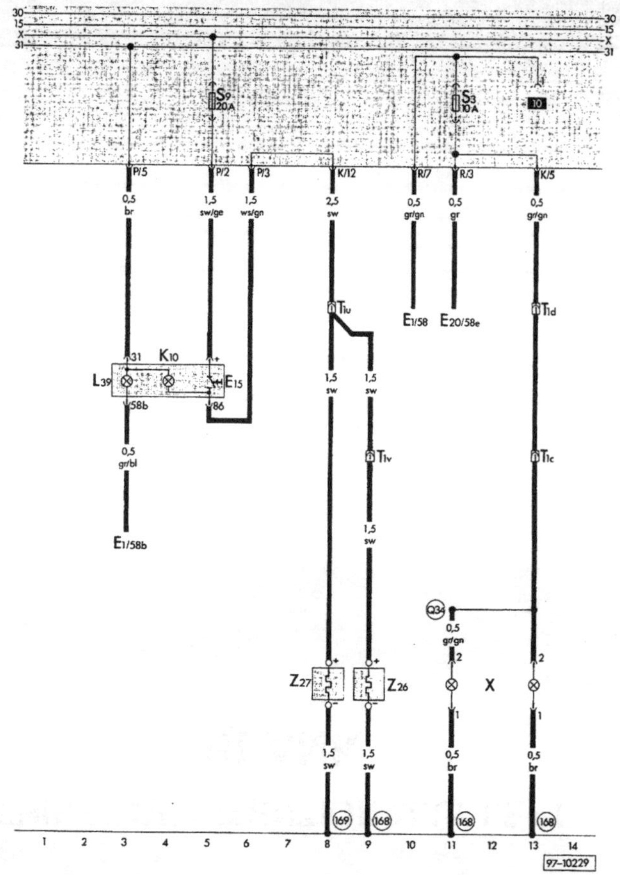

ws = white
sw = black
ro = red
br = brown
gn = green
bl = blue
gr = grey
li = violet
ge = yellow

E1 – Light Switch
E15 – Rear Window Defogger Switch
E20 – Instrument Panel Light Dimmer Switch
K10 – Rear Window Defogger Indicator Light
L39 – Rear Window Defogger Switch Light
T1c – Single Connector, on Left Taillight
T1d – Single Connector, on Right Taillight
T1u – Single Connector, on Right Taillight
T1v – Single Connector, on Left Taillight
X – License Plate Light
Z26 – Left Swinging Door Rear Window Defogger
Z27 – Right Swinging Door Rear Window Defogger
(168) – Ground connection, in left swinging door wiring harness
(169) – Ground connection, in right swinging door wiring harness
(Q34) – Wire connection (58e), in left rear swinging door wiring harness

Rear window defogger, license plate light

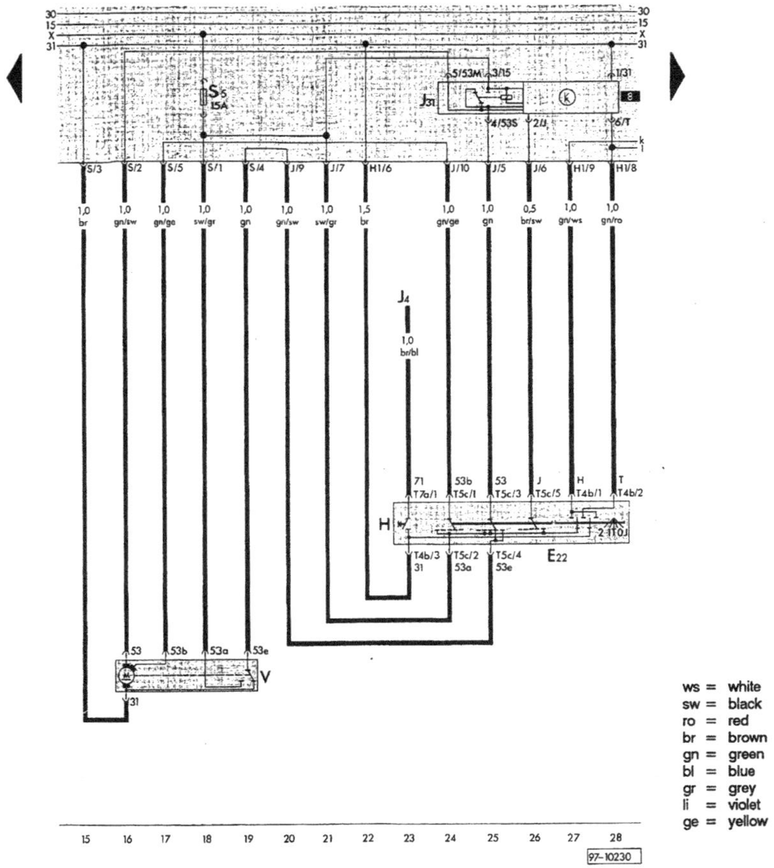

ws = white
sw = black
ro = red
br = brown
gn = green
bl = blue
gr = grey
li = violet
ge = yellow

E22 – Windshield Wiper Intermittent Switch
H – Horn Button
J4 – Dual Horn Relay
J31 – Washer/Wiper Intermittent Relay
T4b – 4-Pin Connector, behind steering column switch cover
T5c – 5-Pin Connector, behind steering column switch cover
T7a – 7-Pin Connector, behind steering column switch cover
V – Windshield Wiper Motor

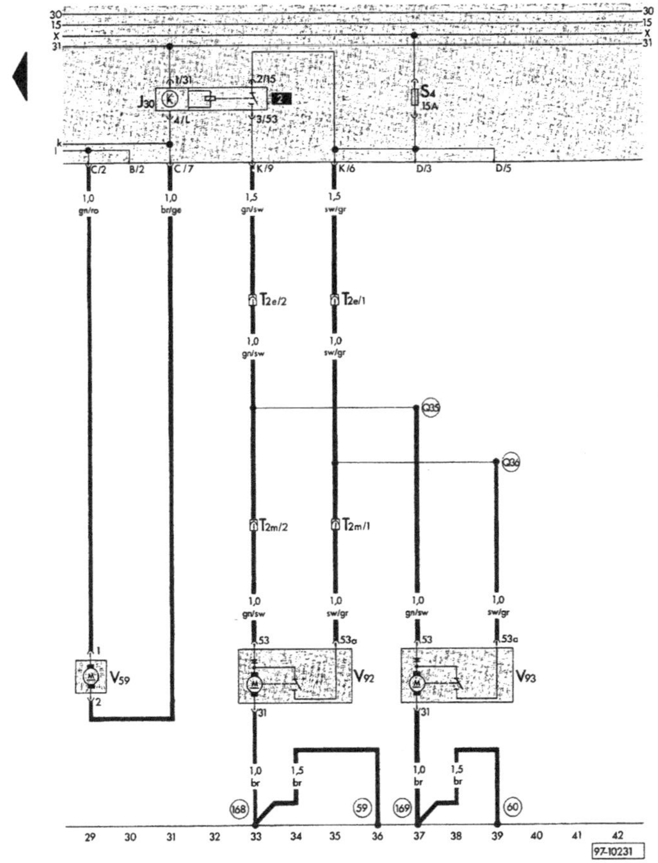

ws = white
sw = black
ro = red
br = brown
gn = green
bl = blue
gr = grey
li = violet
ge = yellow

J30 – Rear Window Wiper/Washer Relay
T2e – Double Connector, on Right Taillight
T2m – Double Connector, on Left Taillight
V59 – Windshield and Rear Window Washer Pump
V92 – Left Swinging Door Rear Window Wiper Motor
V93 – Right Swinging Door Rear Window Wiper Motor

(59) – Ground connection, near left taillight

(60) – Ground connection, near right taillight

(168) – Ground connection, in left swinging door wiring harness

(169) – Ground connection, in right swinging door wiring harness

(Q35) – Wire connection (53), in right rear swinging door wiring harness

(Q36) – Wire connection (53a), in right rear swinging door wiring harness

Stereo radio with 6 speakers,

from April 1992

Relay location

For relay location and multiple connectors see also section "component locations".

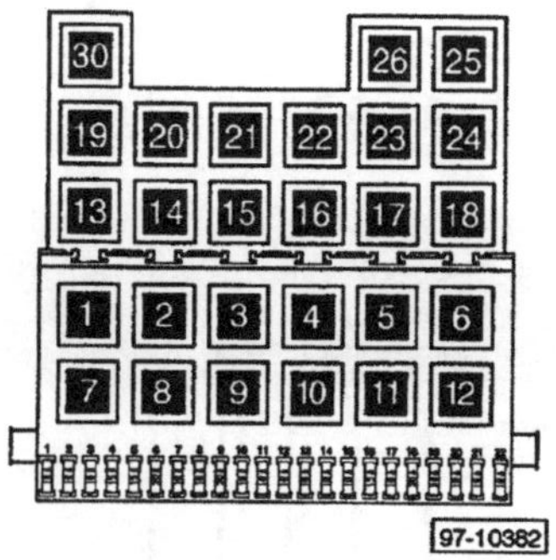

Fuse colors

30A – green
25A – white
20A – yellow
15A – blue
10A – red

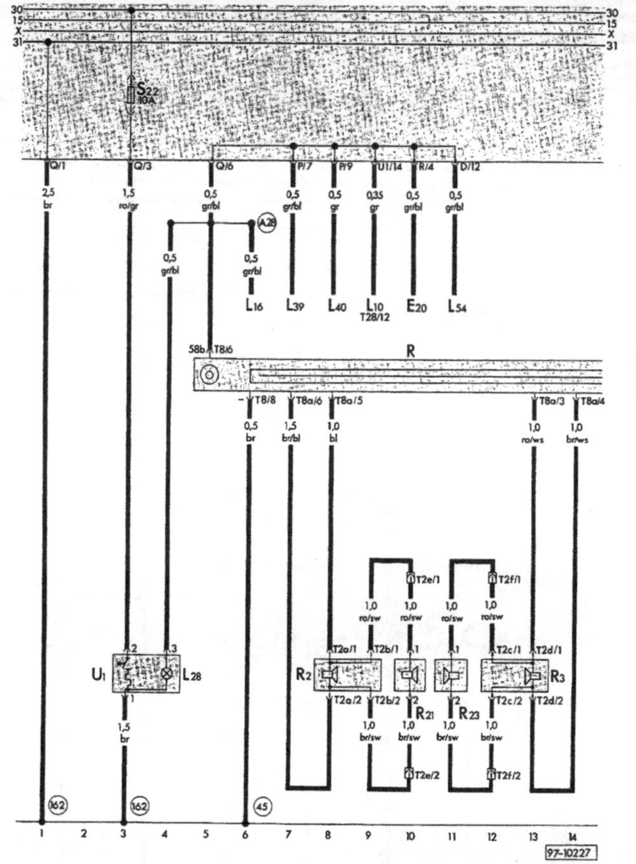

ws = white
sw = black
ro = red
br = brown
gn = green
bl = blue
gr = grey
li = violet
ge = yellow

E20 – Instrument Panel Light Dimmer Switch
L10 – Instrument Cluster Light
L16 – Fresh Air Control Lever Light
L28 – Cigarette Lighter Light
L39 – Rear Window Defogger Switch Light
L40 – Front And Rear Fog Light Switch Light
L54 – Headlight Adjuster Control Light
R – Radio
R2 – Left Front Speaker
R3 – Right Front Speaker
R21 – Left Front Woofer
R23 – Right Front Woofer
T2a – Double Connector
T2b – Double Connector
T2c – Double Connector
T2d – Double Connector
T2e – Double Connector, in driver's door
T2f – Double Connector, in passenger's door
T8 – 8-Pin Connector
T8a – 8-Pin Connector
T28 – 28-Pin Connector, on instrument cluster
U1 – Cigarette Lighter

(45) – Ground connection, behind instrument panel, center
(162) – Ground connection, in blower motor wiring harness
(A28) – Plus connection (58b), in fresh air blower wiring harness

Stereo radio, front speaker

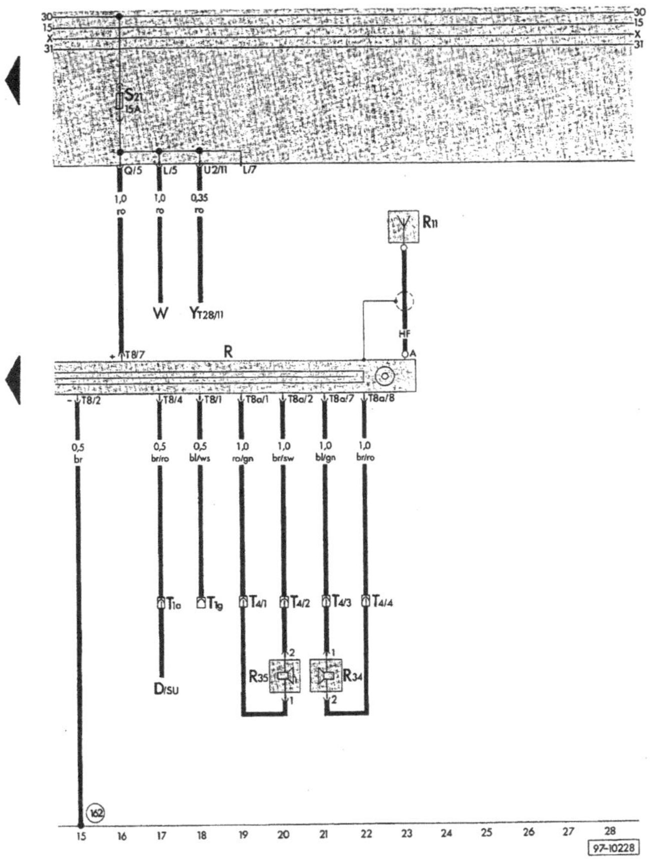

ws = white
sw = black
ro = red
br = brown
gn = green
bl = blue
gr = grey
li = violet
ge = yellow

D –Ignition/Starter Switch
R –Radio
R11 –Antenna
R34 –Left Rear Midrange Speaker
R35 –Right Rear Midrange Speaker
T1a –Single Connector, behind instrument panel, left
T1g –Single Connector, behind instrument panel, left
T4 –4-Pin Connector, behind fuse/relay panel
T8 –8-Pin Connector
T8a –8-Pin Connector
T28 –28-Pin Connector, on instrument cluster
Y –Analog Clock

(162) – Ground connection, in blower motor wiring harness

THIS PAGE INTENTIONALLY LEFT
BLANK

Daytime running lights (DRL) and headlight washer system, Canada only,

from April 1992

Relay location
For relay location and multiple connectors see also section "component locations".

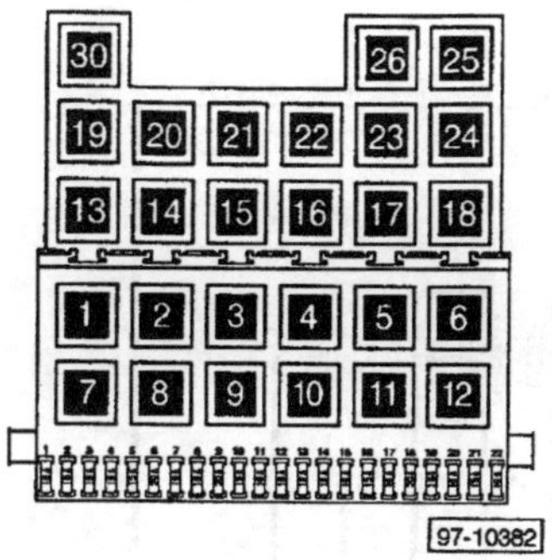

- **4** – Load Reduction Relay (18)
- **7** – Headlight Washer System Relay (95)
- **8** – Washer/Wiper Intermittent Relay (19)

Note: Number in parentheses indicates production control number stamped on relay housing.

Fuse colors

30A – green
25A – white
20A – yellow
15A – blue
10A – red

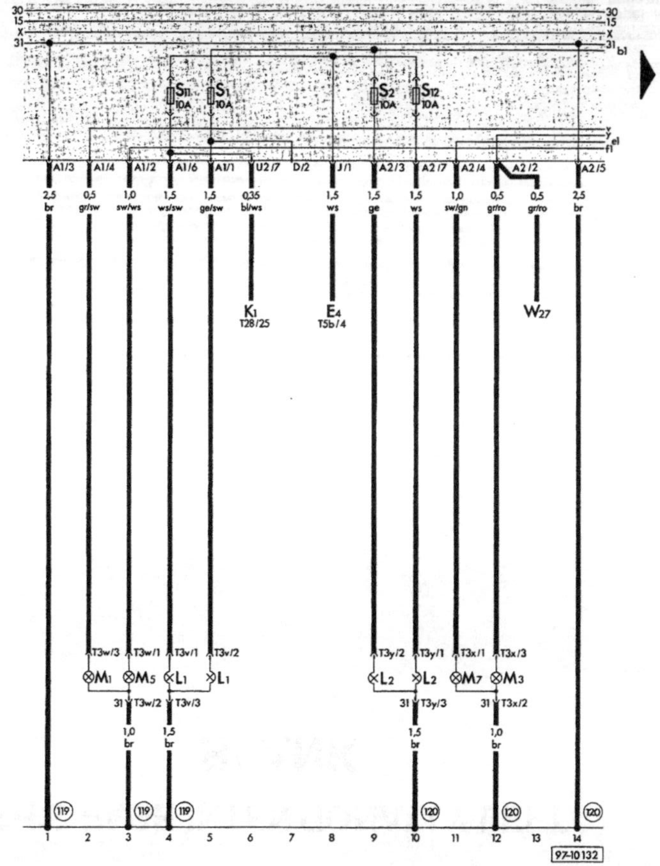

ws = white
sw = black
ro = red
br = brown
gn = green
bl = blue
gr = grey
li = violet
ge = yellow

E4 – Headlight Dimmer/Flasher Switch
K1 – Headlight High Beam Indicator Light
L1 – Left Headlight
L2 – Right Headlight
M1 – Left Parking Light
M3 – Right Parking Light
M5 – Left Front Turn Signal Light
M7 – Right Front Turn Signal Light
T3v – 3-Pin Connector, on Left Headlight
T3w – 3-Pin Connector, on Left Headlight
T3x – 3-Pin Connector, on Right Headlight
T3y – 3-Pin Connector, on Right Headlight
T5b – 5-Pin Connector, behind steering column switch cover
T28 – 28-Pin Connector, on instrument cluster
W27 – Engine Compartment Light

(119) – Ground connection -1-, in headlight wiring harness
(120) – Ground connection -2-, in headlight wiring harness

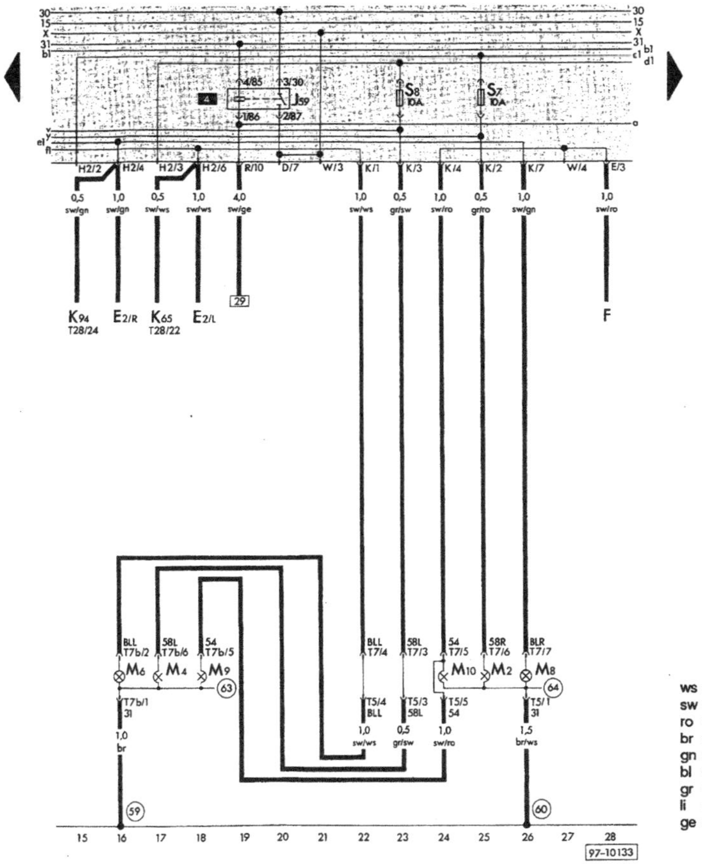

ws = white
sw = black
ro = red
br = brown
gn = green
bl = blue
gr = grey
li = violet
ge = yellow

E2 – Turn Signal Switch
F – Brake Light Switch
J59 – Load Reduction Relay
K65 – Left Turn Signal Indicator Light
K94 – Right Turn Signal Indicator Light
M2 – Right Taillight
M4 – Left Taillight
M6 – Left Rear Turn Signal Light
M8 – Right Rear Turn Signal Light
M9 – Left Brake Light
M10 – Right Brake Light
T5 – 5-Pin Connector, on Right Tallight
T7 – 7-Pin Connector, on Right Taillight
T7b – 7-Pin Connector, on Left Taillight
T28 – 28-Pin Connector, on instrument cluster

(59) – Ground connection, near left taillight
(60) – Ground connection, near right taillight
(63) – Ground connection, on left taillight bulbholder
(64) – Ground connection, on right taillight bulb holder

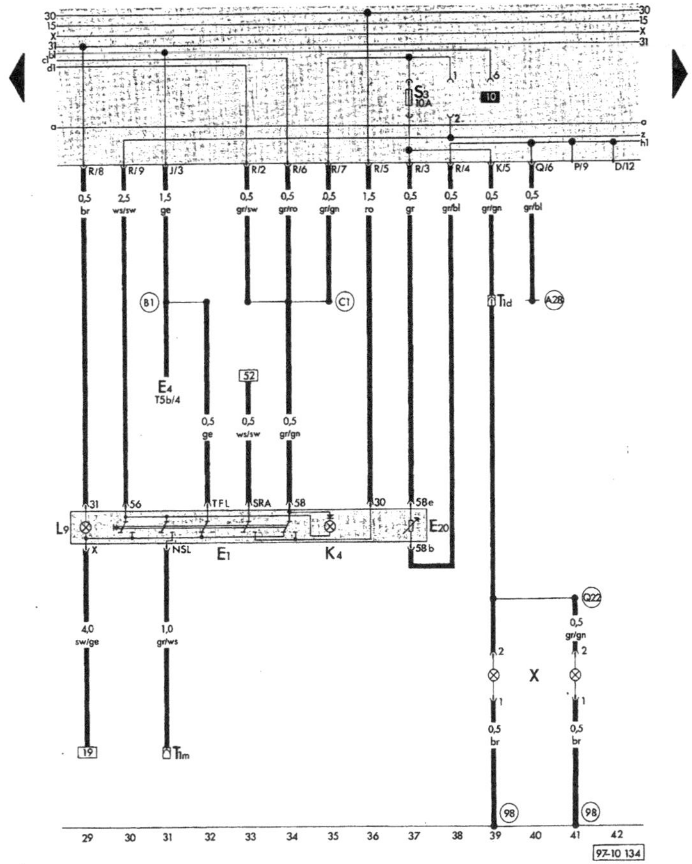

ws = white
sw = black
ro = red
br = brown
gn = green
bl = blue
gr = grey
li = violet
ge = yellow

E1 – Light Switch
E4 – Headlight Dimmer/Flasher Switch
E20 – Instrument Panel Light Dimmer Switch
K4 – Park Light Indicator Light
L9 – Headlight Switch Light
T1d – Single Connector, behind fuse/relay panel
T1m – Single Connector, behind fuse/relay panel
T5b – 5-Pin Connector, behind steering column switch cover
X – License Plate Light

(98) – Ground connection, in rear lid wiring harness
(A28) – Plus connection (58b), in fresh air blower wiring harness

(B1) – Wire connection (56 b), in steering column switch wiring harness
(C1) – Plus connection (58), in left front wiring harness
(Q22) – Wire connection, in rear lid wiring harness

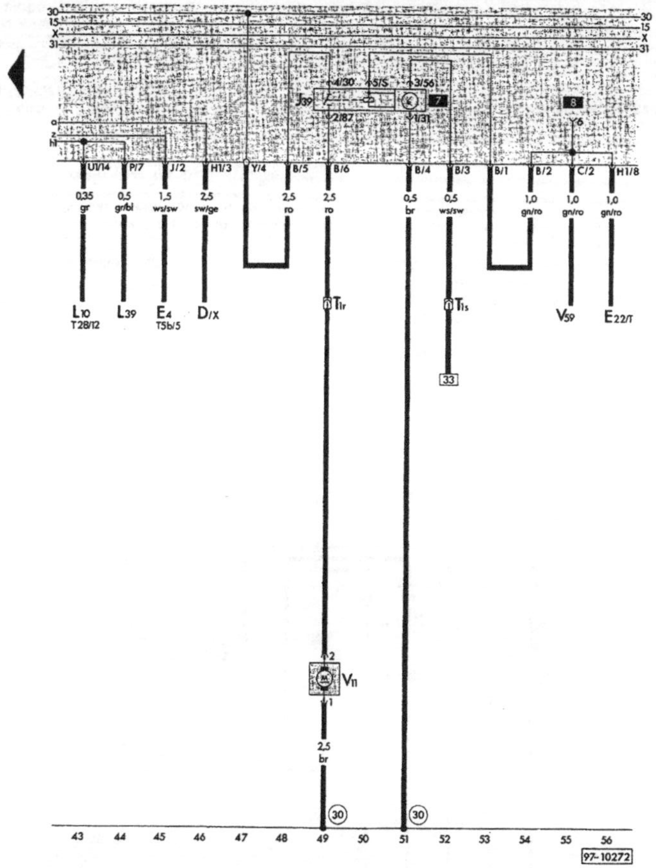

ws = white
sw = black
ro = red
br = brown
gn = green
bl = blue
gr = grey
li = violet
ge = yellow

D – Ignition/Starter Switch
E4 – Headlight Dimmer/Flasher Switch
E22 – Windshield Wiper Intermittent Switch
J39 – Headlight Washer System Relay
L10 – Instrument Cluster Light
L39 – Rear Window Defogger Switch Light
T1r – Single Connector, behind fuse/relay panel
T1s – Single Connector, behind fuse/relay panel
T5b – 5-Pin Connector, behind steering column switch cover
T28 – 28-Pin Connector, on instrument cluster
V11 – Headlight Washer Pump
V59 – Windshield and Rear Window Washer Pump

(30) – Ground connection, –1–, beside fuse/relay panel

Air conditioning with one evaporator,
• 2.5 L-Engine – Digifant Multiport Fuel Injection (MFI)/81 kW, code AAF,

to December 1991

Relay location
For relay location and multiple connectors see also section "component locations".

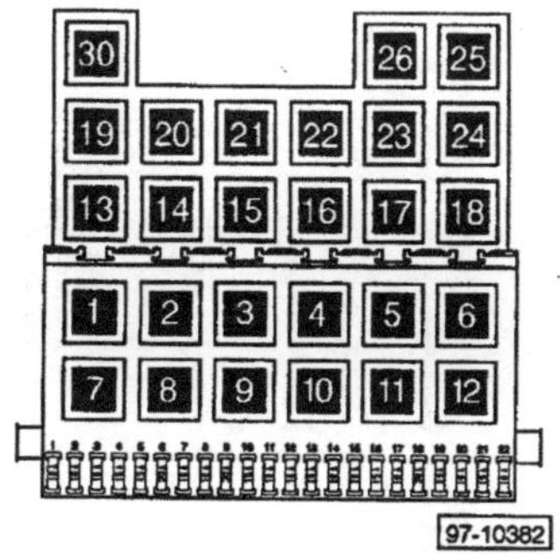

1 – A/C Relay (13)

Note: Number in parentheses indicates production control number stamped on relay housing.

Fuse colors

30A – green
25A – white
20A – yellow
15A – blue
10A – red

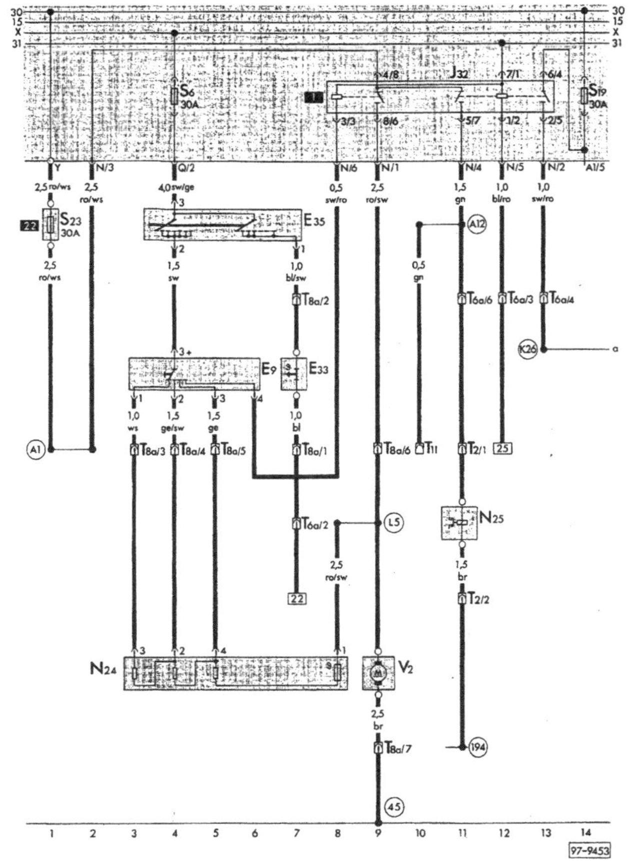

ws = white
sw = black
ro = red
br = brown
gn = green
bl = blue
gr = grey
li = violet
ge = yellow

E9 – Fresh Air Blower Switch
E33 – A/C Evaporator Temperature Switch
E35 – A/C Switch
J32 – A/C Relay
N24 – Fresh Air Blower Series Resistance With Fuse
N25 – A/C Clutch
S23 – Heater Fuse
T1l – Single Connector, behind fuse/relay panel, to Digifant Engine Control Module (ECM)–J169/37
T2 – Double Connector, on A/C Clutch
T6a – 6-Pin Connector, behind instrument panel, left
T8 – 8-Pin Connectorbehind instrument panel, center

V2 – Fresh Air Blower
(45) – Ground connection, behind instrument panel, center
(194) – Ground connection –2–, in fan wiring harness
(A1) – Plus connection (30a), in instrument panel wiring harness
(A12) – Plus connection in instrument panel wiring harness
(K26) – Wire connection –5–, in fan wiring harness
(L5) – Wire connection –1–, in evaporator housing wiring harness

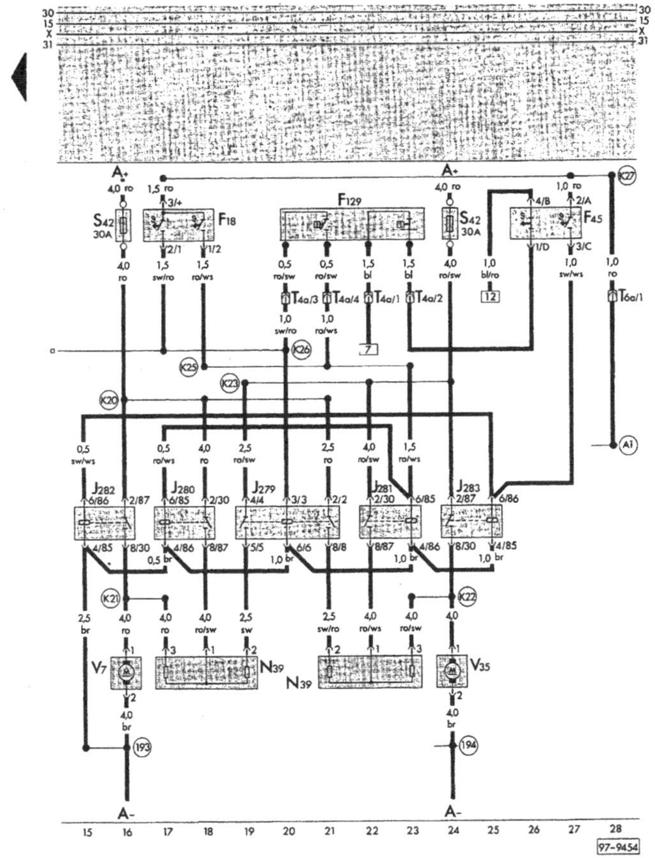

ws = white
sw = black
ro = red
br = brown
gn = green
bl = blue
gr = grey
li = violet
ge = yellow

A – Battery
F18 – Coolant Fan Control (FC) Thermal Switch, right radiator
F45 – A/C Thermal Switch, in coolant hose
F129 – A/C Pressure Switch
J279 – First Speed Coolant Fan Controlo (FC) Relay (V7,V35), (114)
J280 – Second Speed Coolant Fan Control (FC) Relay (V7), (98)
J281 – Second Speed Coolant Fan Control (FC) Relay (V35), (98)
J282 – Third Speed Coolant Fan Control (FC) Relay (V7), (98)
J283 – Third Speed Coolant Fan Control (FC) Relay (V35), (11)
N39 – Series Resistance, behind left headlight
S42 – Coolant Fan Fuse, in engine compartment, left front
T4a – 4-Pin Connector, on A/C Pressure Switch
T6a – 6-Pin Connector, behind instrument panel, left
V7 – Coolant Fan
V35 – Right Coolant Fan
(193) – Ground connection -1-, in fan wiring harness

(194) – Ground connection -2-, in fan wiring harness
(A1) – Plus connection (30a), in instrument panel wiring harness
(K20) – Plus connection -1-(30), in fan wiring harness
(K21) – Wire connection -1-, in fan wiring harness
(K22) – Wire connection -2-, in fan wiring harness
(K23) – Plus connection -2-(30), in fan wiring harness
(K25) – Wire connection -4-, in fan wiring harness
(K26) – Wire connection -5-, in fan wiring harness
(K27) – Wire connection -6-, in fan wiring harness

Auxiliary heater B2L*, with programmable timer,

from April 1992

Relay location
For relay location and multiple connectors see also section "component locations".

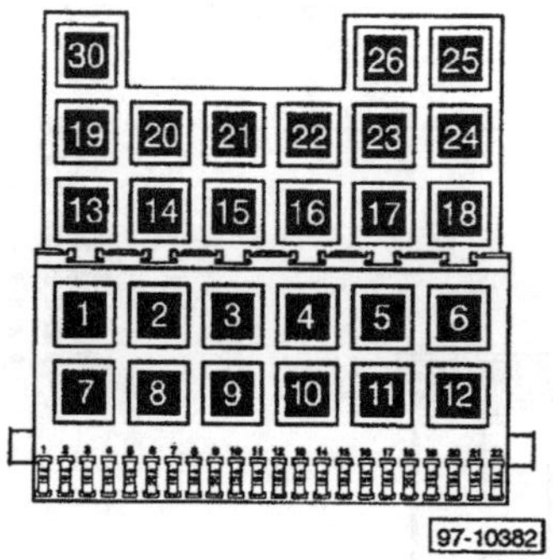

Fuse colors

30A – green
25A – white
20A – yellow
15A – blue
10A – red

* Type 20168901, see identification plate

Edition 7/93
USA.5272.01.21

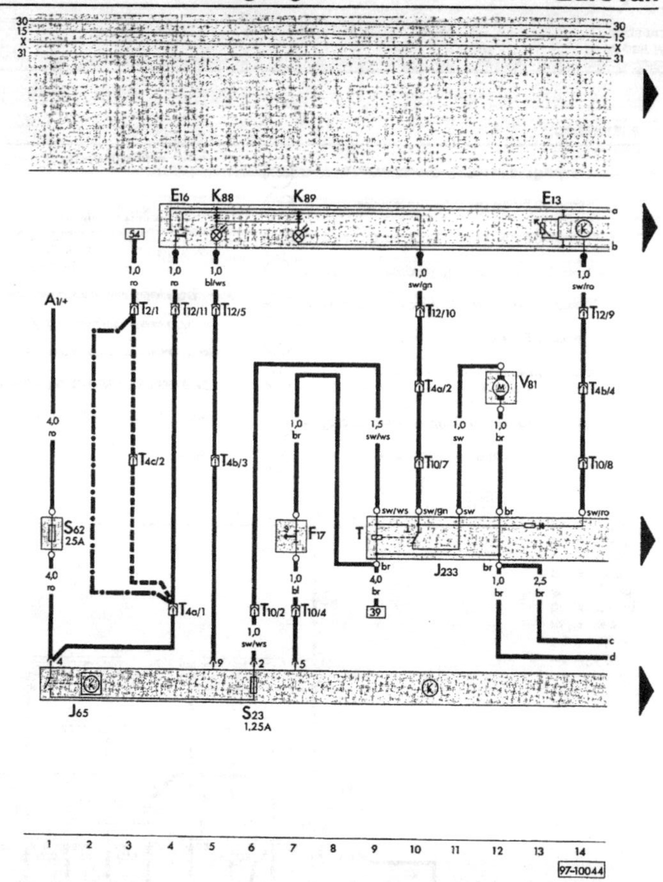

ws = white
sw = black
ro = red
br = brown
gn = green
bl = blue
gr = grey
li = violet
ge = yellow

A1 – Auxiliary Battery
E13 – Heater-Time Temperature Control Switch, on left B-pillar, upper part
E16 – Heater/Heat Output Switch
F17 – Haeter Overheat Control Switch
J65 – Heater Control Module, below driver's seat*
J233 – Auxiliary Heater Blower Relay
K88 – Water Heater Malfunction Light (red)
K89 – Water Heater On Light (green)
S23 – Heater Fuse
S62 – Auxiliary Heater Fuse, below driver's seat**
T – Junction Box
T2 – Double Connector, below driver's seat*
T4a – 4-Pin Connector, below driver's seat*
T4b – 4-Pin Connector, below driver's seat*
T4c – 4-Pin Connector, below driver's seat*
T10 – 10-Pin Connector, below driver's seat*
T12 – 12-Pin Connector, behind left B-pillar trim, upper part
V81 – Fresh Air/Heater Blower
* – Located behind lower part of left B-pillar trim on vehicles with swivel seat
** – Located on lower part of left B-pillar on vehicles with swivel seat
▬•▬ – "Camper" only
▬ ▬ – "Standard" only

Heater-time temperature control switch, combustion/fresh air heater blower, heater control module

Edition 7/93
USA.5272.01.21

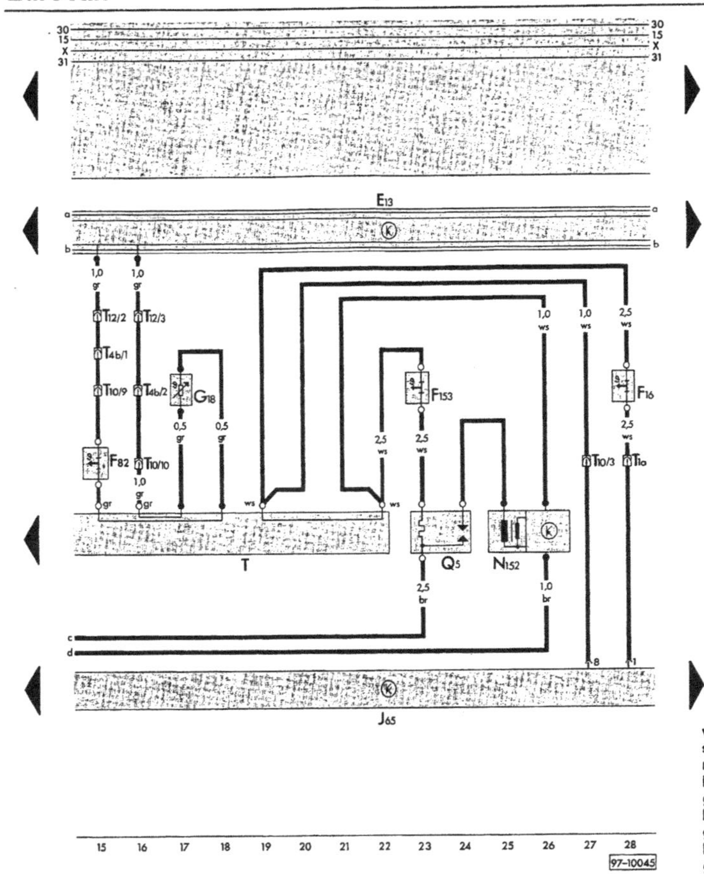

ws = white
sw = black
ro = red
br = brown
gn = green
bl = blue
gr = grey
li = violet
ge = yellow

E13 – Heater-Time Temperature Control Switch, on left B-pillar, upper part
F16 – Haeter Flame Sensor
F82 – Temperature Limiting Switch, in heater
F153 – Glow Plug Thermal Switch
G18 – Temperature Sensor II
J65 – Heater Control Module, below driver's seat*
N152 – Ignition Coil
Q5 – Glow/Spark Plug
T – Junction Box
T1a – Single Connector, below driver's seat*
T4b – 4-Pin Connector, below driver's seat*
T10 – 10-Pin Connector, below driver's seat*
T12 – 12-Pin Connector, behind left B-pillar trim, upper part

* – Located behind lower part of left B-pillar trim on vehicles with swivel seat

Heater-time temperature control switch, glow/spark plug, ignition coil, heater control module

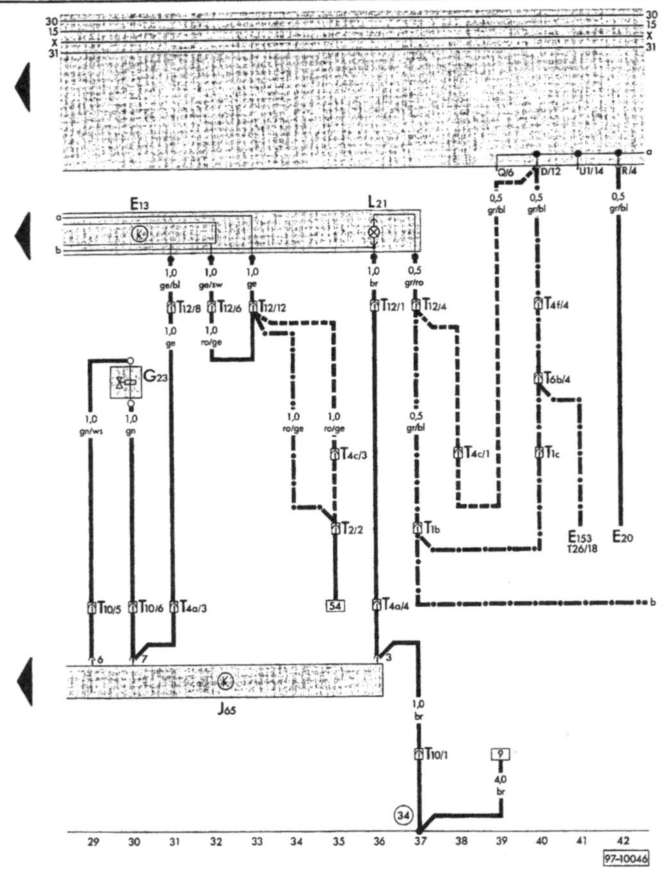

ws = white
sw = black
ro = red
br = brown
gn = green
bl = blue
gr = grey
li = violet
ge = yellow

E13 – Heater-TimeTemperature Control Switch, on left B-pillar, upper part
E20 – Instrument Panel Light Dimmer Switch
E153 – Camper Central Electronic Display (Camping Equipment), near front interior light
G23 – Transfer Fuel Pump (FP)
J65 – Heater Control Module, below driver's seat*
L21 – Heater Control Lever Light
T1b – Single Connector, below driver's seat
T1c – Single Connector, behind left B-pillar trim, lower part
T2 – Double Connector, below driver's seat*
T4a – 4-Pin Connector, below driver's seat*
T4c – 4-Pin Connector, below driver's seat*
T4f – 4-Pin Connector, behind fuse/relay panel
T6b – 6-Pin Connector, on left A-pillar, upper part
T10 – 10-Pin Connector, below driver's seat*
T12 – 12-Pin Connector, behind left B-pillar trim, upper part

(34) – Ground connection, below driver's seat*

* – Located behind lower part of left B-pillar trim on vehicles with swivel seat
▬•▬ – "Camper" only
▬ ▬ – "Standard" only

Heater-time temperature control switch, heater control module, fuel supply

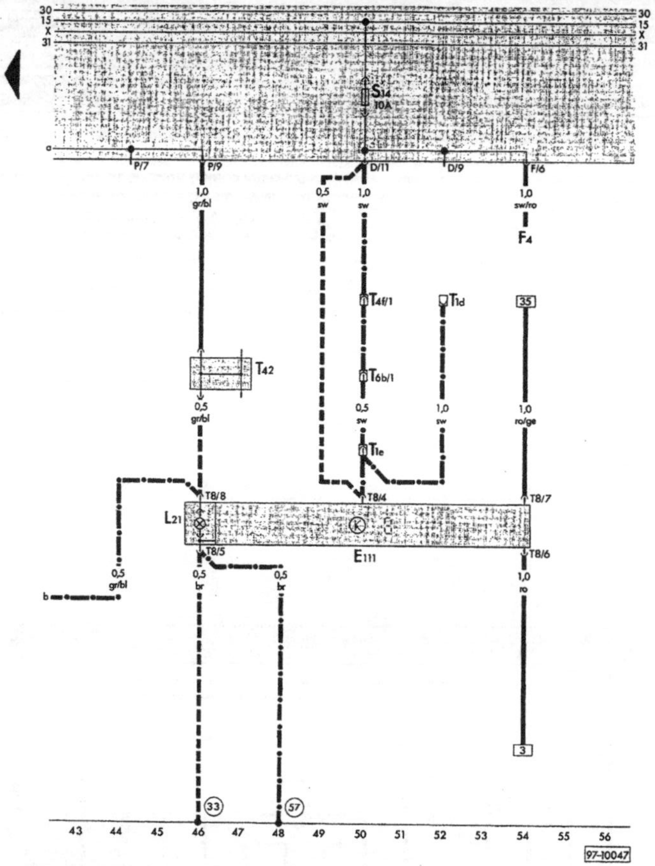

ws = white
sw = black
ro = red
br = brown
gn = green
bl = blue
gr = grey
li = violet
ge = yellow

E111 – Programmable Timer, in instrument panel, center
F4 – Back-Up Light Switch
L21 – Heater control Lever Light
T1d – Single Connector, on B-pillar, left
T1e – Single Connector, on B-pillar, left
T4f – 4-Pin Connector, behind fuse/relay panel
T6b – 6-Pin Connector, on left A-pillar, upper part
T8 – 8-Pin Connector
T42 – Wire Connector, Terminal 58B, above fuse/relay panel

(33) – Ground connection, behind instrument panel, right
(57) – Ground connection, on left rear pillar

–•– – "Camper" only
– – – "Standard" only

THIS PAGE INTENTIONALLY LEFT
BLANK

2.5 L-Engine – Digifant Multiport Fuel Injection (MFI)/81kW, code AAF, Canada only,

from September 1991 to April 1992

Relay location
For relay location and multiple connectors see also section "component locations".

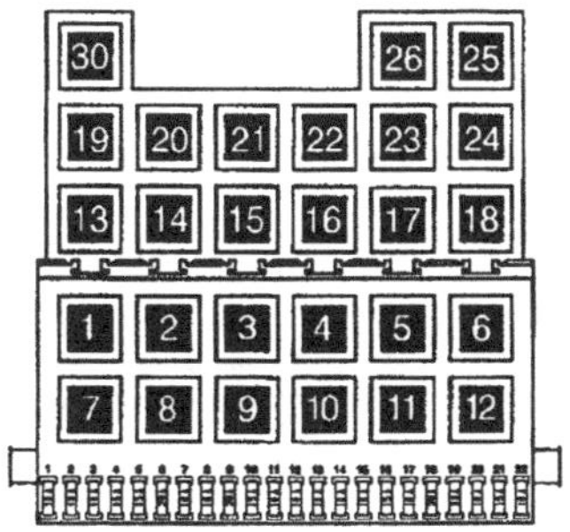

97-10382

- 2 – Rear Window Wiper/Washer Relay (72)
- 3 – Digifant Engine Control Module (ECM) Power Supply Relay (30)
- 4 – Load Reduction Relay (18)
- 6 – Emergency Flasher Relay (21)
- 7 – Headlight Washer System Relay (33)
- 8 – Washer/Wiper Intermittent Relay (19)
- 9 – Seat Belt Warning System Relay (4)
- 11 – Dual Horn Relay (53)
- 12 – Fuel Pump (FP) Relay (80)
- 14 – Engine Coolant Pump Relay (38)
- 20 – Engine Coolant Pump Circuit Breaker

Note: Number in parentheses indicates production control number stamped on relay housing.

Fuse colors

30A – green
25A – white
20A – yellow
15A – blue
10A – red

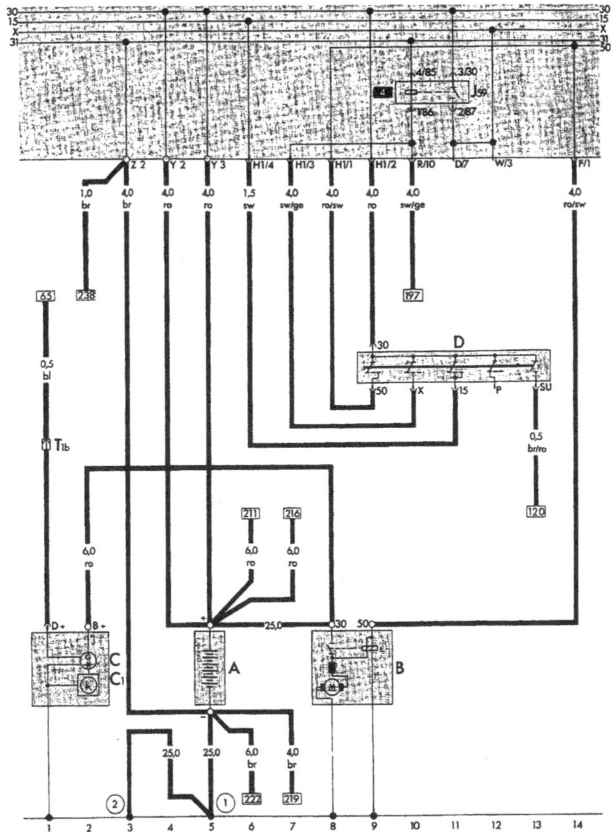

97-9820

ws = white
sw = black
ro = red
br = brown
gn = green
bl = blue
gr = grey
li = violet
ge = yellow

A – Battery
B – Starter
C – Generator (GEN)
C1 – Voltage Regulator (VR)
D – Ignition/Starter Switch
J59 – Load Reduction Relay
T1b – Single Connector, near starter

(1) – Ground strap, battery to body

(2) – Ground strap, transmission to body

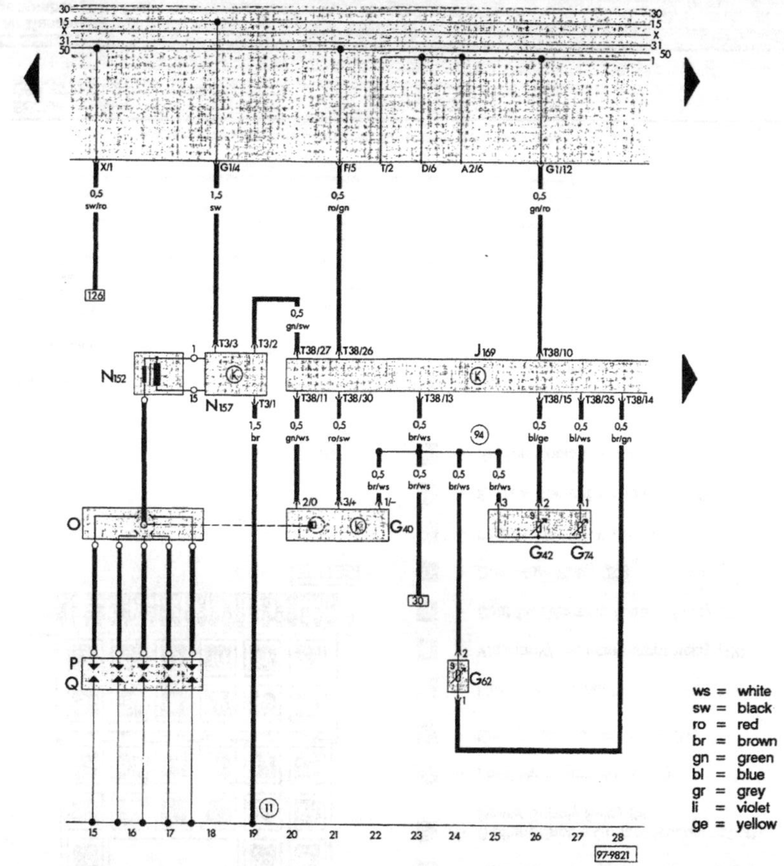

ws = white
sw = black
ro = red
br = brown
gn = green
bl = blue
gr = grey
li = violet
ge = yellow

G40 – Camshaft Positioin (CMP) Sensor
G42 – Intake Air Temperature (IAT) Sensor (00523-2322)
G62 – Engine Coolant Temperature (ECT) Sensor (0522–2312)
G74 – CO Fuel Trim (FT) Potentiometer (00521–2242)
J169 – Digifant Engine Control Module (ECM)
N152 – Ignition Coil
N157 – Ignition Coil Power Output Stage
O – Ignition Distributor
P – Spark Plug Connectors
Q – Spark Plugs
T3 – 3-Pin Connector, on Ignition Coil Power Output Stage
T38 – 38-Pin Connector

(11) – Ground connection, in battery box
(94) – Ground connection -1-, in Digifant Multiport Fuel Injection (MFI) wiring harness

Ignition system, digifant engine control module (ECM)

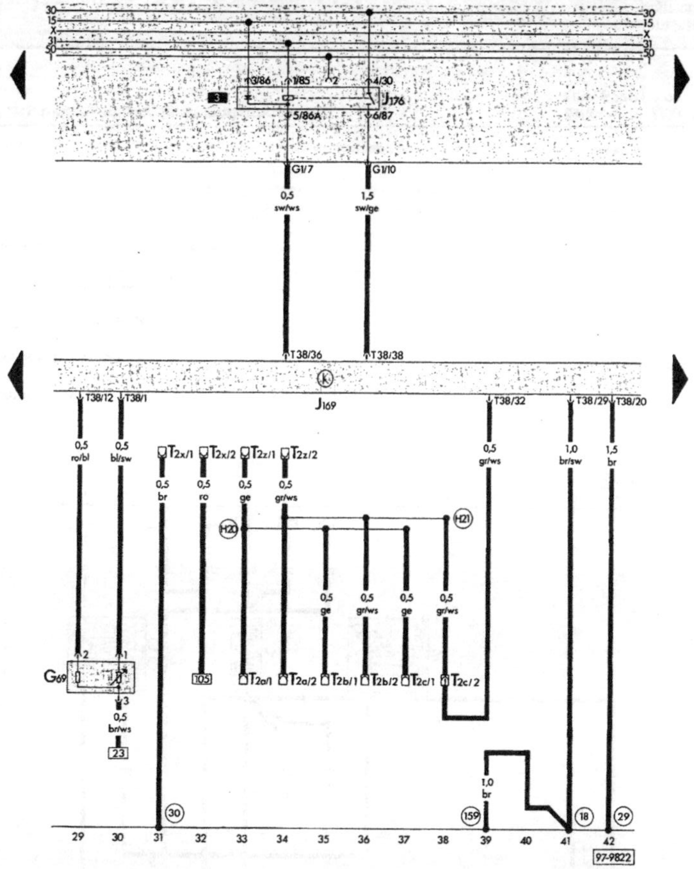

ws = white
sw = black
ro = red
br = brown
gn = green
bl = blue
gr = grey
li = violet
ge = yellow

G69 – Throttle Position (TP) Sensor (00518–2212)
J169 – Digifant Engine Control Module (ECM)
J176 – Digifant Engine Control Module (ECM) Power Supply Relay
T2a – Double Connector, behind instrument panel, left*
T2b – Double Connector, behind instrument panel, left *
T2c – Double Connector behind instrument panel, left
T2x – Double Connector, –black–, behind fuse/relaypanel cover (OBD)
T2z – Double Connector, –white–, behind fuse/relay panel cover (OBD)
T38 – 38-Pin Connector (T38/31 is connected with automatic transmission only)

(18) – Ground connection, on engine block
(29) – Ground connection, near battery
(30) – Ground connection, -1-, beside fuse/relay panel
(159) – Ground connection -2-, in Digifant Multiport Fuel Injection (MFI) wiring harness
(H20) – Wire connection -1-, in On Board Diagnostic (OBD) wiring harness
(H21) – Wire connection -2-, in On Board Diagnostic (OBD) wiring harness
* – With further On Board Diagnostic (OBD) systems only

Digifant engine control module (ECM), on board diagnostic (OBD) system

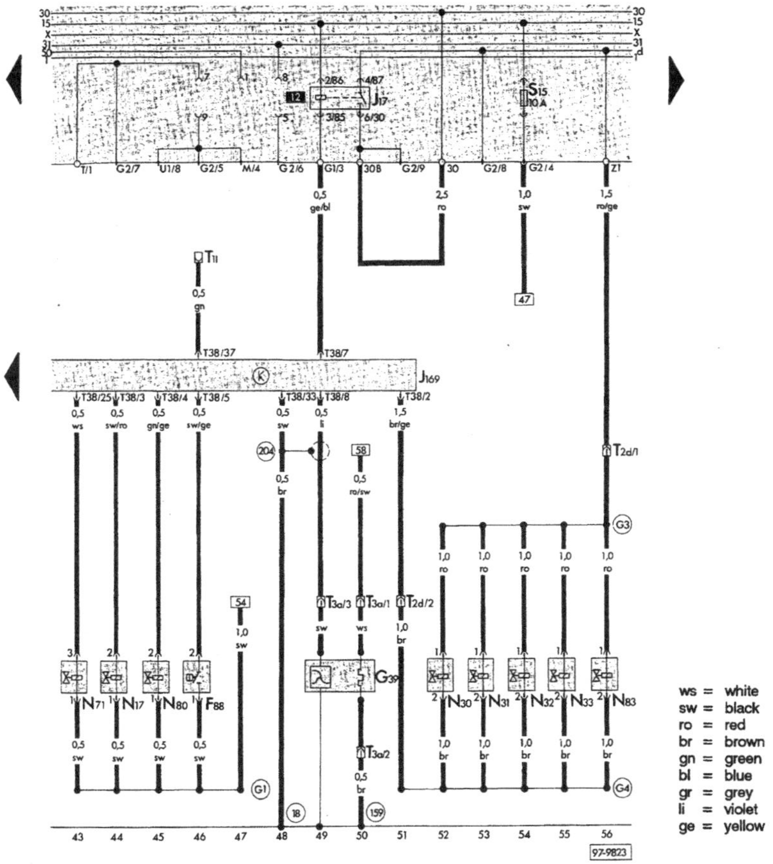

ws = white
sw = black
ro = red
br = brown
gn = green
bl = blue
gr = grey
li = violet
ge = yellow

F88 – Power Steering Pressure (PSP) Switch
G39 – Heated Oxygen Sensor (HO2S) (00525–2342)
J17 – Fuel Pump (FP) Relay
J169 – Digifant Engine Control Module (ECM)
N17 – Cold Start Injector
N30 – Cyiinder 1 Fuel Injector
N31 – Cyiinder 2 Fuel Injector
N32 – Cyiinder 3 Fuel Injector
N33 – Cyiinder 4 Fuel Injector
N71 – Idle Air Control (IAC) Valve
N80 – Evaporative Emission (EVAP) Canister Purge Regulator Valve
N83 – Cylinder 5 Fuel injector
T1l – Single Connector, behind fuse/relay panel
T2d – Double Connector, in engine compartment, right
T3a – 3-Pin Connector, near exhaust manifold
T38* – 38-Pin Connector

(18) – Ground connection, on engine block
(159) – Ground connection -2-, in Digifant Multiport Fuel Injection (MFI) wiring harness
(204) – Ground connection -3-, in Digifant Multiport Fuel Injection (MFI) wiring harness
(G1) – Plus connection, in Digifant Multiport Fuel Injection (MFI) wiring harness
(G3) – Plus connection, in injector wiring harness
(G4) – Wire connection, in injector wiring harness
* – T38/31 is used with automatic transmission only

Digifant engine control module (ECM), heated oxygen sensor (HO2S), injectors

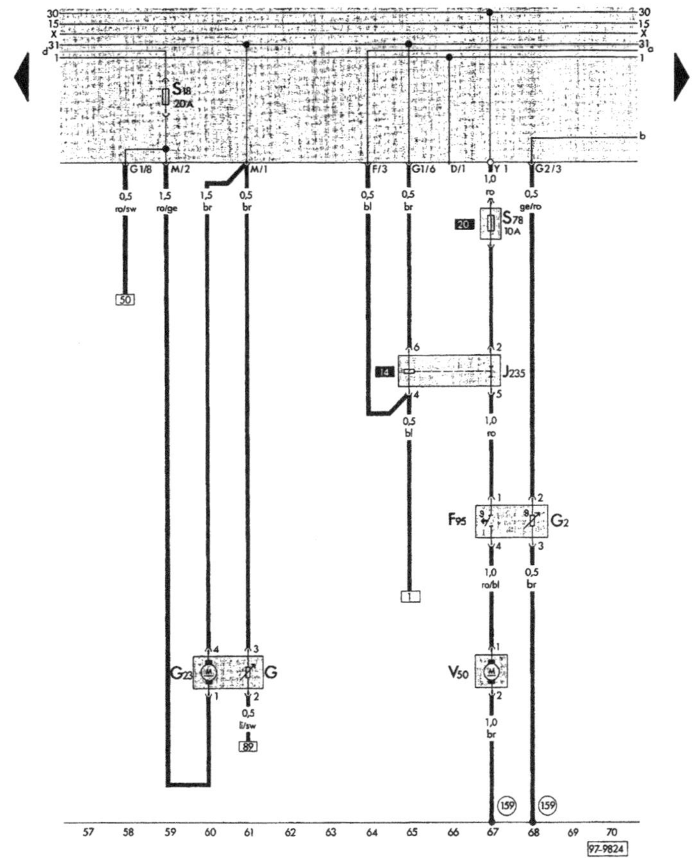

ws = white
sw = black
ro = red
br = brown
gn = green
bl = blue
gr = grey
li = violet
ge = yellow

F95 – After-Run Coolant Pump Thermal Switch *
G – Fuel Level Sensor
G2 – Engine Coolant Temperature (ECT) Sensor
G23 – Transfer Fuel Pump (FP)
J235 – Engine Coolant Pump Relay
S78 – Engine Coolant Pump Circuit Breaker
V50 – Water Pump

(159) – Ground connection -2-, in Digifant Multiport Fuel Injection (MFI) wiring harness

* – Switch value, see engine repair manual

Fuel supply, coolant pump after-run

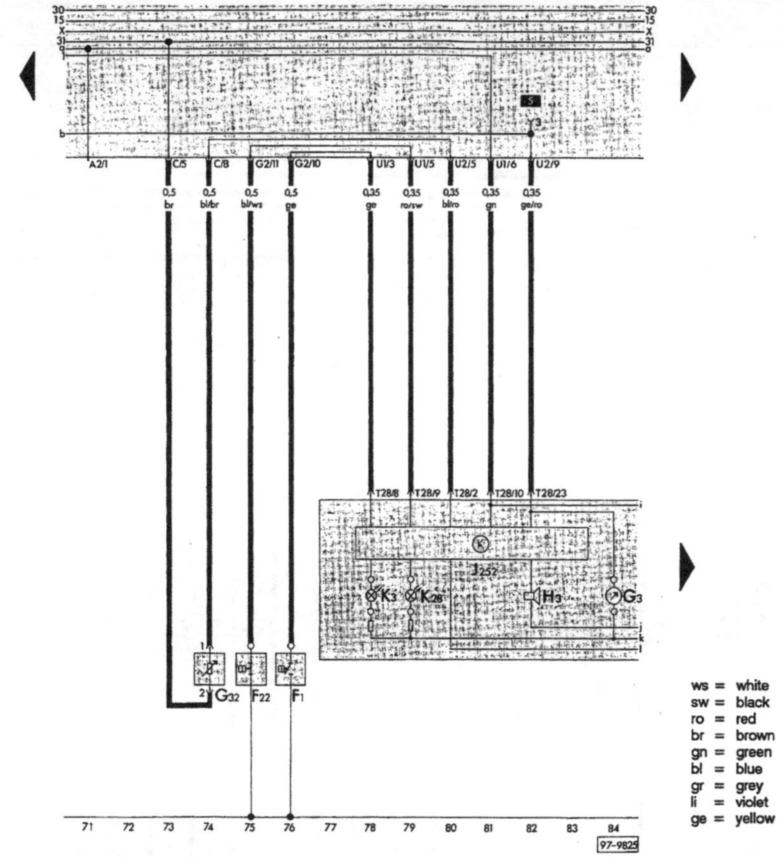

ws = white
sw = black
ro = red
br = brown
gn = green
bl = blue
gr = grey
li = violet
ge = yellow

F1 – Oil Pressure Switch *
F22 – 0.3bar Oil Pressure Switch
G3 – Engine Coolant Temperature (ECT) Gauge
G32 – Engine Coolant Level (ECL) Sensor
H3 – Warning Buzzer
J252 – Oil Pressure and Engine Coolant Level/Temperature (ECL/ECT) Warning Control Module
K3 – Oil Pressure Warning Light
K28 – Engine Coolant Level/Temperature (ECL/ECT) Warning Light
T28 – 28-Pin Connector, on instrument cluster

* – Switch value, see engine repair manual

Instrument cluster, oil pressure- and engine coolant level/temperature (ECL/ECT) warning lights

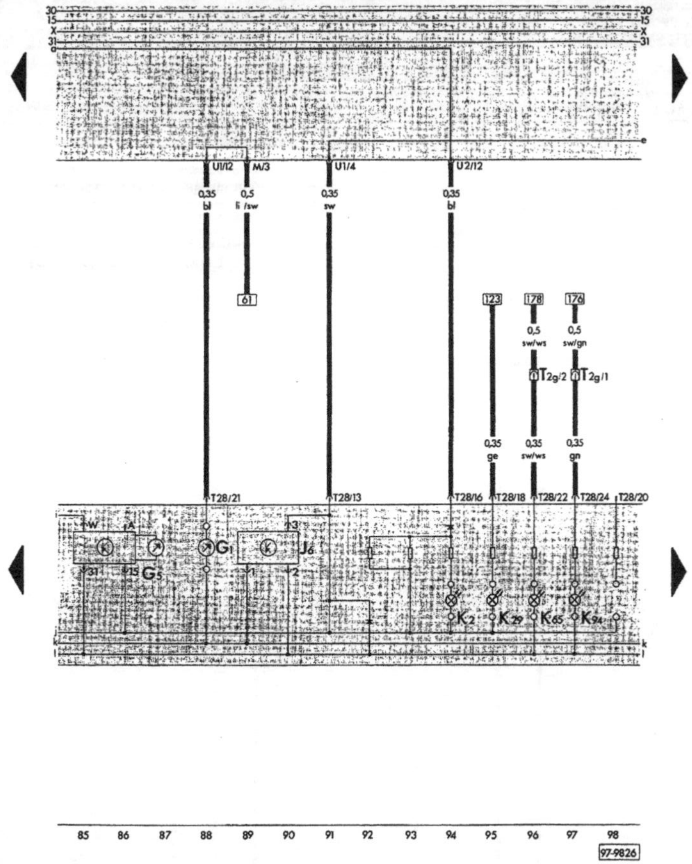

ws = white
sw = black
ro = red
br = brown
gn = green
bl = blue
gr = grey
li = violet
ge = yellow

G1 – Fuel Gauge
G5 – Tachometer
J6 – Voltage Stabilizer
K2 – Generator (GEN) Warning Light
K29 – Glow Plug Indicator Light
K65 – Left Turn Signal Indicator Light
K94 – Right Turn Signal Indicator Light
T2g – Double Connector, behind fuse/relay panel
T28 – 28-Pin Connector, on instrument cluster

Instrument cluster, tachometer, fuel gauge, indicator lights

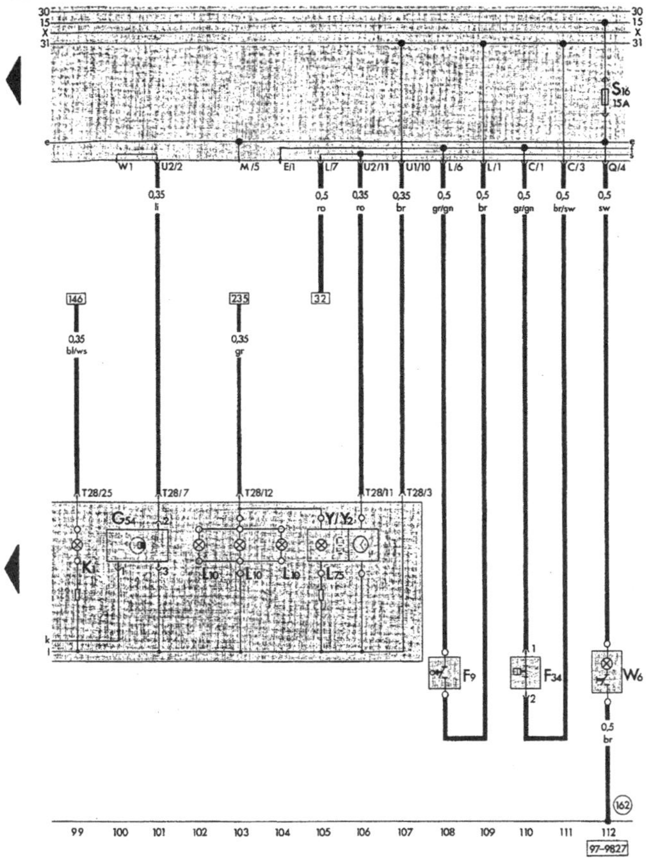

ws = white
sw = black
ro = red
br = brown
gn = green
bl = blue
gr = grey
li = violet
ge = yellow

97-9827

F9 – Parking Brake Warning Light Switch
F34 – Brake Fluid Level Warning Switch
G54 – Vehicle Speed Sensor (VSS)
K1 – Headlight High Beam Indicator Light
L10 – Instrument Cluster Light
L75 – Digital Display Light
T28 – 28-Pin Connector, on instrument cluster
W6 – Glove Compartment Light
Y – Analog Clock
Y2 – Digital Clock

(162) – Ground connection, in blower motor wiring harness

Instrument cluster, vehicle speed sensor (VSS), clock, glove compartment light

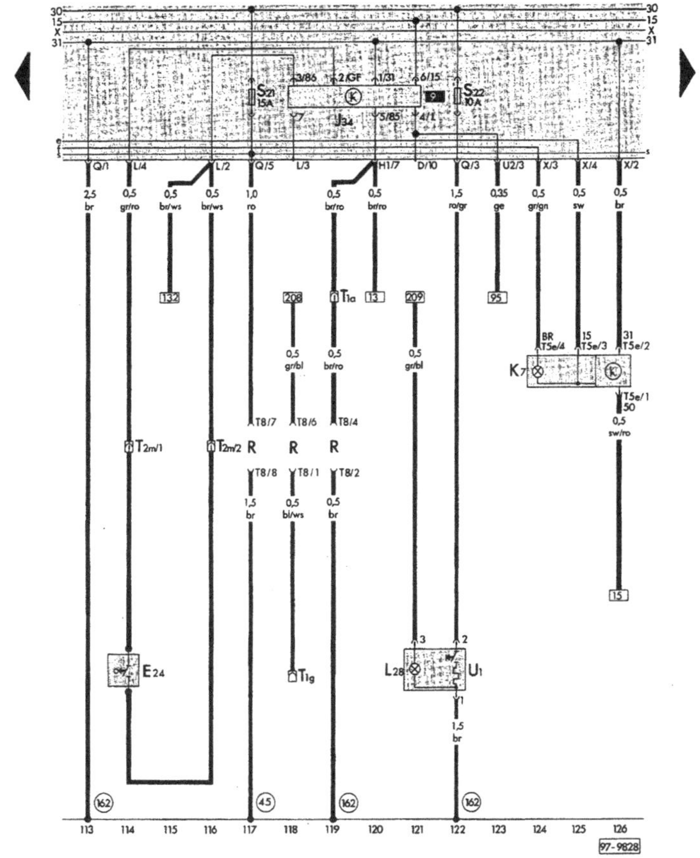

ws = white
sw = black
ro = red
br = brown
gn = green
bl = blue
gr = grey
li = violet
ge = yellow

97-9828

E24 – Left Seat Belt Switch
J34 – Seat Belt Warning System Relay
K7 – Brake and Parking Warning Light
L28 – Cigarette Lighter Light
R – Radio
T1a – Single Connector, behind instrument panel, left
T1g – Single Connector, behind fuse/relay panel
T2m – Double Connector, near left seat belt switch
T5e – 5-Pin Connector
T8 – 8-Pin Connector, on radio
U1 – Cigarette Lighter

(45) – Ground connection, behind instrument panel, center

(162) – Ground connection, in blower motor wiring harness

Radio connection, cigarette lighter, brake and parking brake indicator light, seat belt warning system

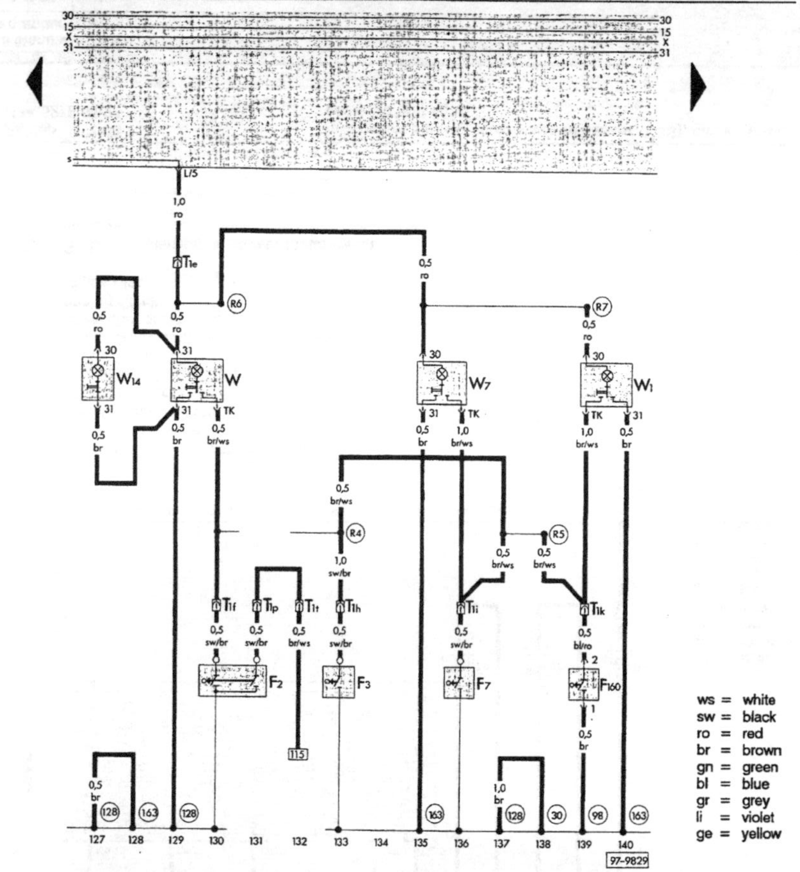

ws = white
sw = black
ro = red
br = brown
gn = green
bl = blue
gr = grey
li = violet
ge = yellow

F2 – Left Front Door Contact Switch
F3 – Right Front Door Contact Switch
F7 – Right Rear Side/Sliding Door Contact Switch
F160 – Rear Lid Contact Switch
T1e – Single Connector, behind fuse/relay panel
T1f – Single Connector, behind fuse/relay panel
T1h – Single Connector, behind instrument panel, right
T1i – Single Connector, behind fuse/relay panel
T1k – Single Connector, near taillight, right
T1p – Single Connector, behind fuse/relay panel
T1t – Single Connector, on B-pillar, right
W – Front Interior Light
W1 – Rear/Luggage Compartment Interior Light
W7 – Center Interior Light
W14 – Right Make-Up Mirror Light

(30) – Ground connection, -1-, beside fuse/relay panel
(98) – Ground connection, in rear lid wiring harness
(128) – Ground connection -1-, in interior light wiring harness
(163) – Ground connection -2-, in interior light wiring harness
(R4) – Wire connection -1-, in interior light/door contact switch wiring harness
(R5) – Wire connection -2-, in interior light wiring harness
(R6) – Plus connection -1-, in interior light wiring harness
(R7) – Plus connection -2-, in interior light wiring harness

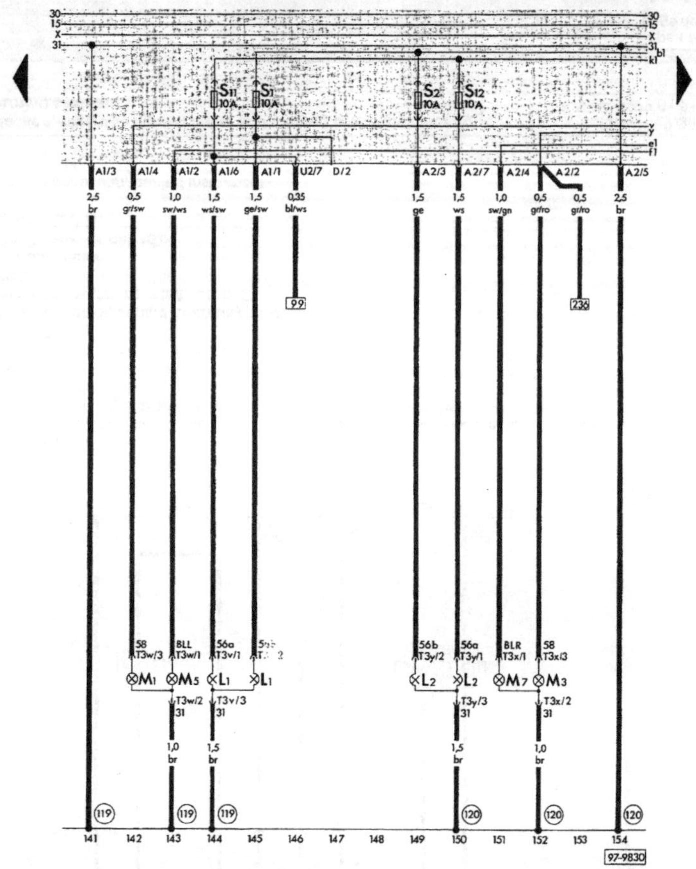

ws = white
sw = black
ro = red
br = brown
gn = green
bl = blue
gr = grey
li = violet
ge = yellow

L1 – Left Headlight
L2 – Right Headlight
M1 – Left Parking Light
M3 – Right Parking Light
M5 – Left Front Turn Signal Light
M7 – Right Front Turn Signal Light
T3v – 3-Pin Connector, on Left Headlight
T3w – 3-Pin Connector, on Left Front Turn Signal Light
T3x – 3-Pin Connector, on Right Front Turn Signal Light
T3y – 3-Pin Connector, on Right Headlight

(119) – Ground connection -1-, in headlight wiring harness
(120) – Ground connection -2-, in headlight wiring harness

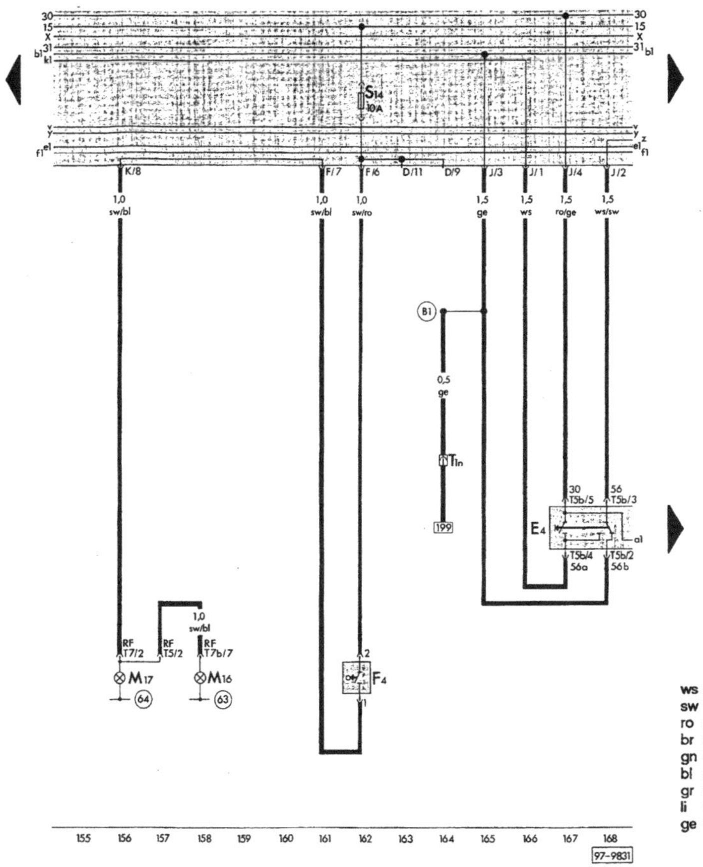

ws = white
sw = black
ro = red
br = brown
gn = green
bl = blue
gr = grey
li = violet
ge = yellow

E4 – Headlight Dimmer/Flasher Switch
F4 – Back-Up Light Switch
M16 – Left Back-Up Light
M17 – Right Back-Up Light
T1h – Single Connector, behind instrument panel, left
T5 – 5-Pin Connector, on Right Back-up Light
T5b – 5-Pin Connector, behind steering column switch cover
T7 – 7-Pin Connector, on Right Taillight
T7b – 7-Pin Connector, on Left Taillight

(63) – Ground connection, on left taillight bulb holder
(64) – Ground connection, on right taillight bulb holder

(B1) – Wire connection (56b), in steering column switch wiring harness

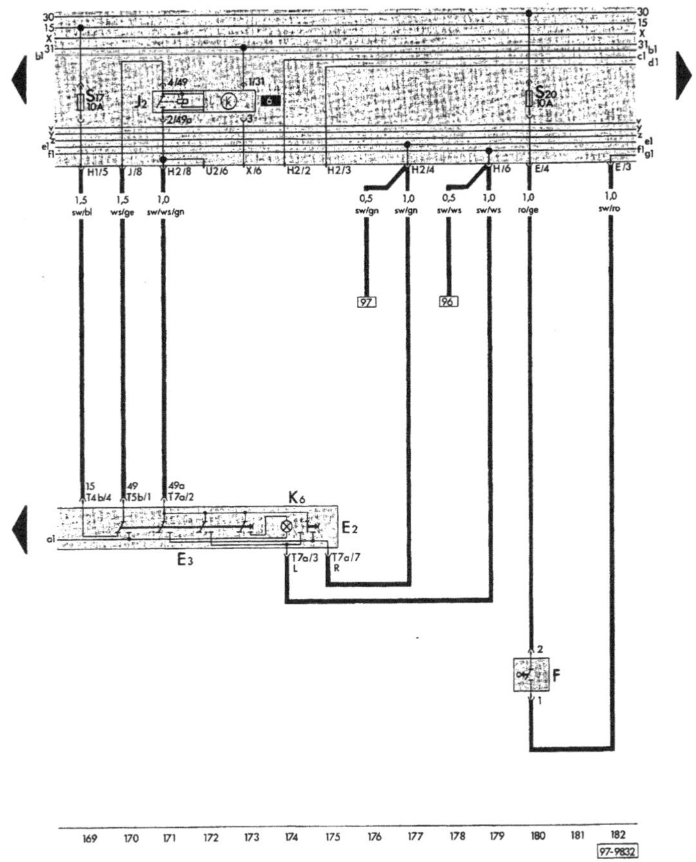

ws = white
sw = black
ro = red
br = brown
gn = green
bl = blue
gr = grey
li = violet
ge = yellow

E2 – Turn Signal Switch
E3 – Emergency Flasher Switch
F – Brake Light Switch
J2 – Emergency Flasher Relay
K6 – Emergency Flasher Warning Light
T4b – 4-Pin Connector, behind steering column switch cover
T5b – 5-Pin Connector, behind steering column switch cover
T7a – 7-Pin Connector, behind steering column switch cover

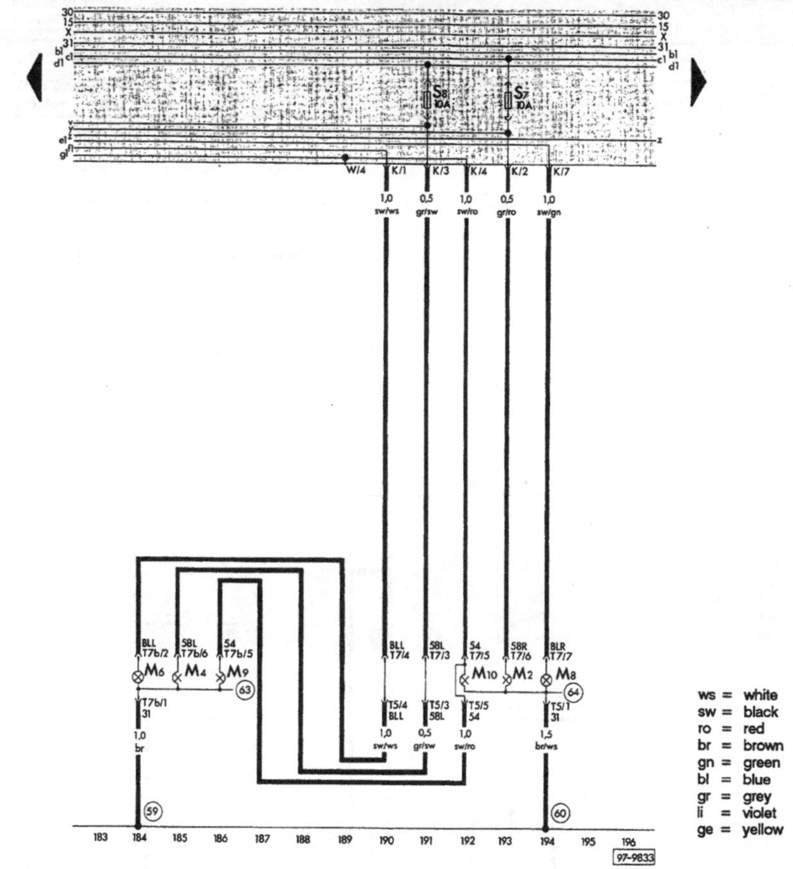

ws = white
sw = black
ro = red
br = brown
gn = green
bl = blue
gr = grey
li = violet
ge = yellow

M2 – Right Taillight
M4 – Left Taillight
M6 – Left Rear Turn Signal Light
M8 – Right Rear Turn Signal Light
M9 – Left Brake Light
M10 – Right Brake Light
T5 – 5-Pin Connector, on Right Taillight
T7 – 7-Pin Connector, on Right Taillight
T7b – 7-Pin Connector, on Left Taillight

(59) – Ground connection, near left taillight

(60) – Ground connection, near right taillight

(63) – Ground connection, on left taillight bulb holder

(64) – Ground connection, on right taillight bulb holder

* – Different wire routing for flatbed vehicles

Rear turn signals*, brake lights*, taillights*

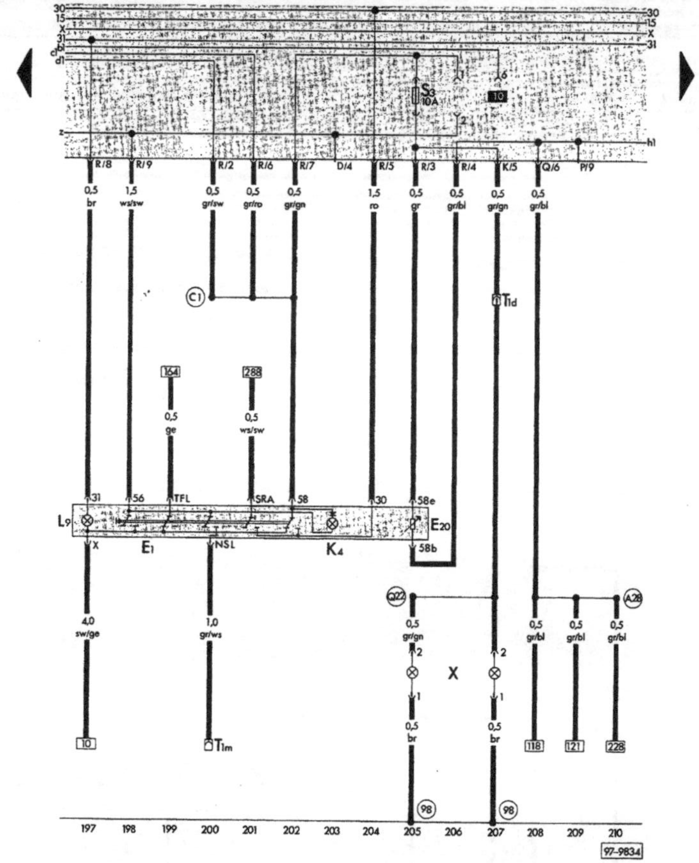

ws = white
sw = black
ro = red
br = brown
gn = green
bl = blue
gr = grey
li = violet
ge = yellow

E1 – Light Switch
E20 – Instrument Panel Light Dimmer Switch
K4 – Park Light Indicator Light
L9 – Headlight Switch Light
T1d – Single Connector, behind fuse relay panel
T1m – Single Connector, behind fuse/relay panel
X – License Plate Light

(98) – Ground connection, in rear lid wiring harness
(A28) – Plus connection (58b), in fresh air blower wiring harness
(C1) – Plus connection (58L), in left front wiring harness
(Q22) – Wire connection, in rear lid wiring harness

* – Different wire routing on vehicles with rear swinging doors

Light switch, license plate light (tailgate)*

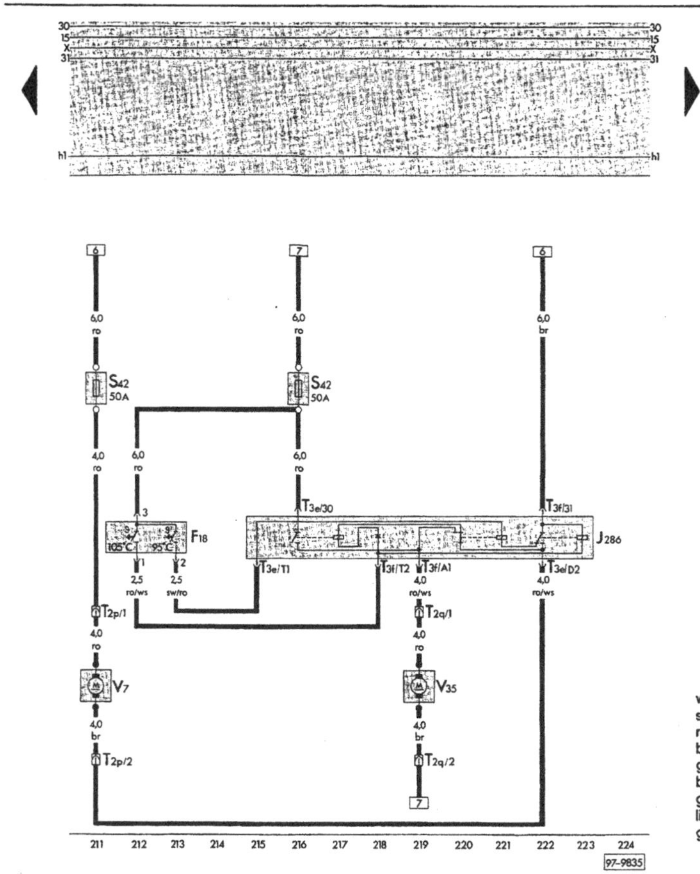

ws = white
sw = black
ro = red
br = brown
gn = green
bl = blue
gr = grey
li = violet
ge = yellow

F18 – Coolant Fan Control (FC) Thermal Switch
J286 – Series/Parallel Coolant Fan Control (FC) Relay (V7, V35)
S42 – Coolant Fan Fuse
T2p – Double Connector, in engine compartment, front
T2q – Double Connector, in engine compartment, front
T3e – 3-Pin Connector
T3f – 3-Pin Connector
V7 – Coolant Fan
V35 – Right Coolant Fan

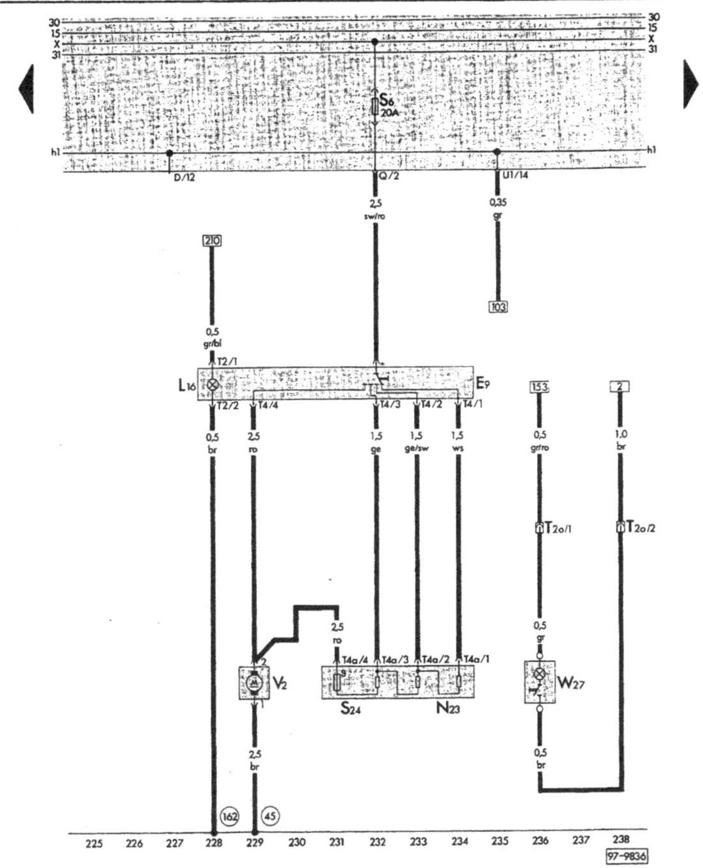

ws = white
sw = black
ro = red
br = brown
gn = green
bl = blue
gr = grey
li = violet
ge = yellow

E9 – Fresh Air Blower Switch
L16 – Fresh Air Control Lever Light
N23 – Fresh Air Blower Series Resistance
S24 – Heater, A/C Clutch Fuse
T2 – Double Connector
T2o – Double Connector, near battery
T4 – 4-Pin Connector
T4a – 4-Pin Connector
V2 – Fresh Air Blower
W27 – Engine Compartment Light

(45) – Ground connection, behind instrument panel, center
(162) – Ground connection, in blower motor wiring harness

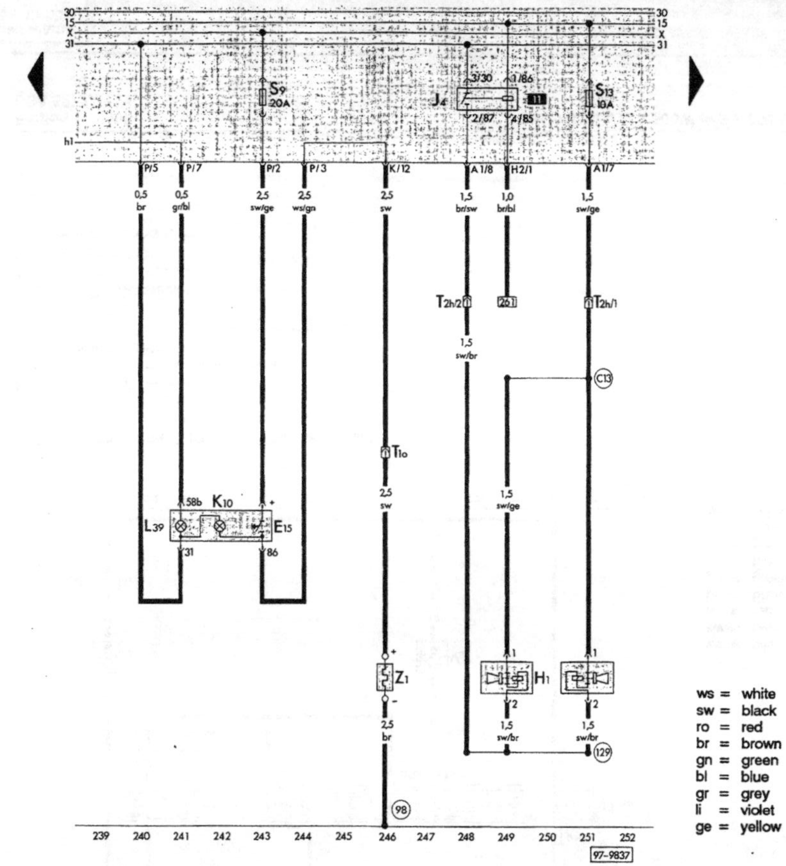

ws = white
sw = black
ro = red
br = brown
gn = green
bl = blue
gr = grey
li = violet
ge = yellow

E15 – Rear Window Defogger Switch
H1 – Dual Horns
J4 – Dual Horn Relay
K10 – Rear Window Defogger Indicator Light
L39 – Rear Window Defogger Switch Light
T1o – Single Connector, near right taillight
T2h – Double Connector, in engine compartment, left front
Z1 – Rear Window Defogger/Heat Element

(98) – Ground connection, in rear lid wiring harness
(129) – Ground connection, in dual tone horn wiring harness
(C13) – Plus connection, in dual tone horn wiring harness

* – Different wire routing on vehicles with rear swinging doors

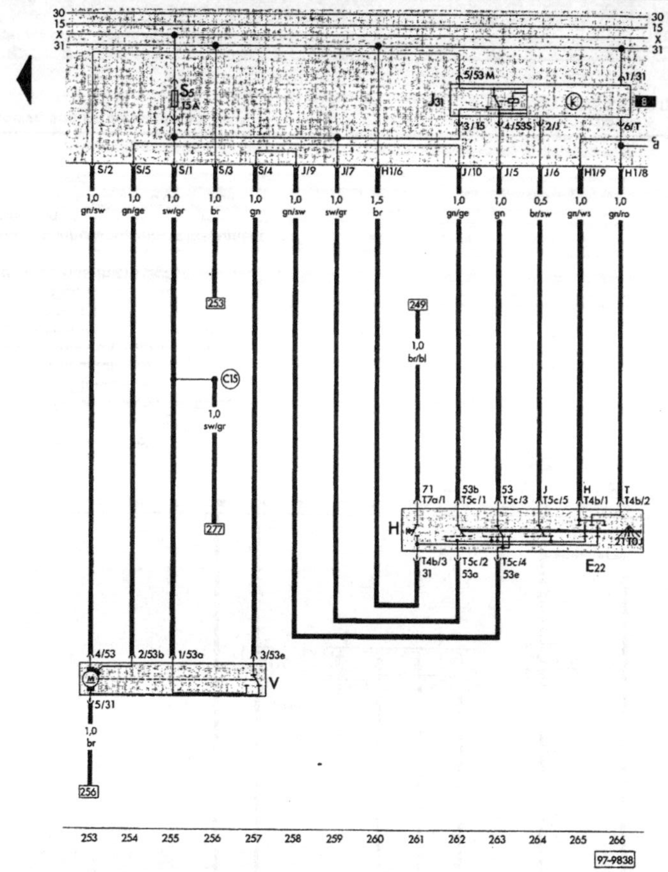

ws = white
sw = black
ro = red
br = brown
gn = green
bl = blue
gr = grey
li = violet
ge = yellow

E22 – Windshield Wiper/Washer Switch
H – Horn Button
J31 – Washer/Wiper Intermittent Relay
T4b – 4-Pin Connector, behind steering column switch cover
T5c – 5-Pin Connector, behind steering column switch cover
T7a – 7-Pin Connector, behind steering column switch cover
V – Windshield Wiper Motor

(C51) – Plus connection (x), in wiper motor wiring harness

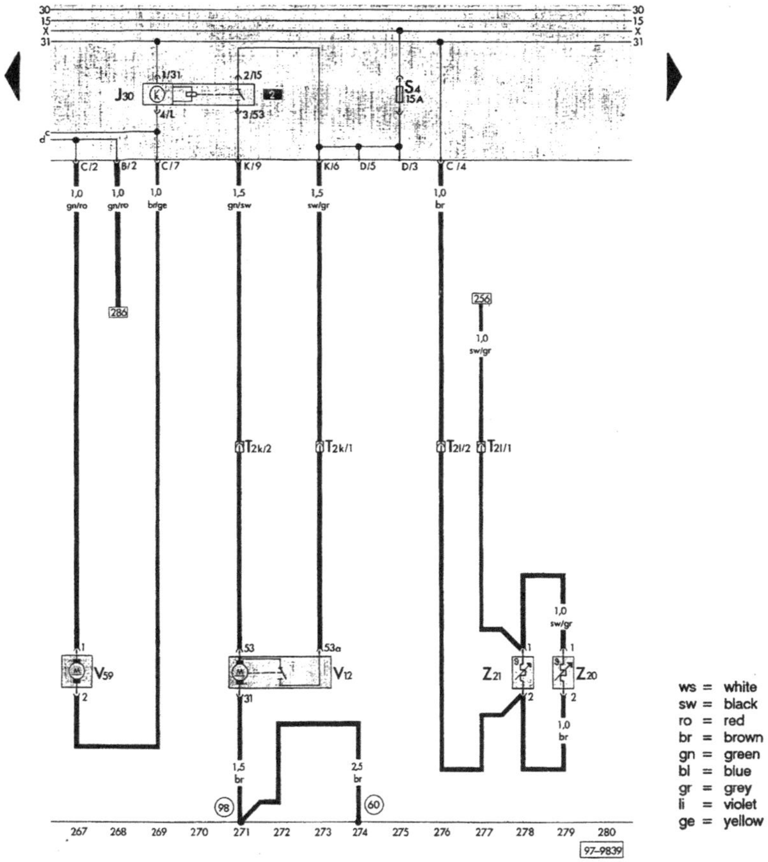

ws = white
sw = black
ro = red
br = brown
gn = green
bl = blue
gr = grey
li = violet
ge = yellow

J30 – Rear Window Wiper/Washer Relay
T2k – Double Connector, near right taillight
T2l – Double Connector, near right firewall
V12 – Rear Window Wiper Motor
V59 – Windshield and Rear Window Washer Pump
Z20 – Left Washer Nozzle Heater
Z21 – Right Washer Nozzle Heater

(60) – Ground connection, near right taillight

(98) – Ground connection, in rear lid wiring harness

* – Different wire routing on vehicles with rear swinging doors

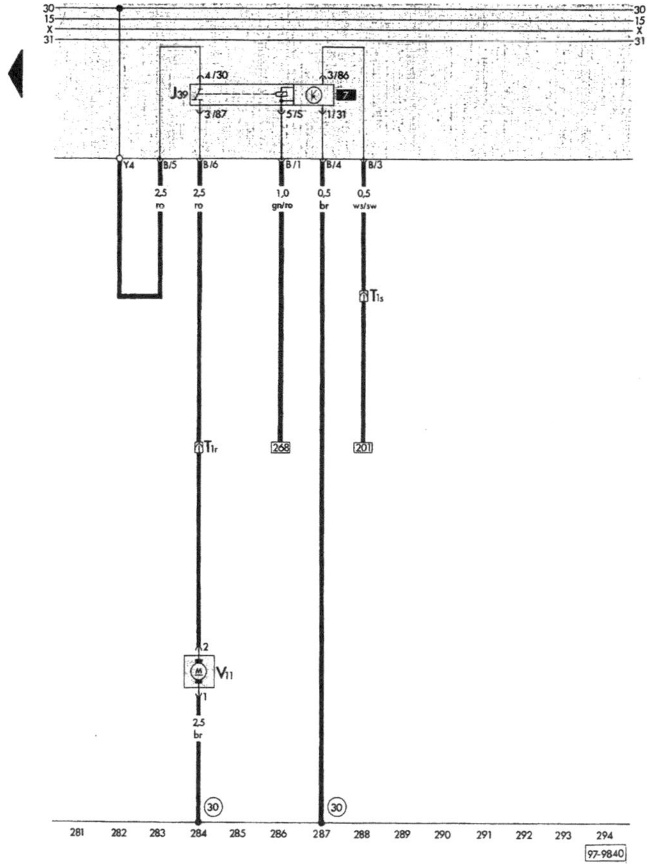

ws = white
sw = black
ro = red
br = brown
gn = green
bl = blue
gr = grey
li = violet
ge = yellow

J39 – Headlight Washer System Relay
T1r – Single Connector, behind fuse relay panel
T1s – Single Connector, behind fuse relay panel
V11 – Headlight Washer Pump

(30) – Ground connection, -1-, beside fuse relay panel

Coolant fans (2 x 450W),

from April 1992

Relay location
For relay location and multiple connectors see also section "component locations".

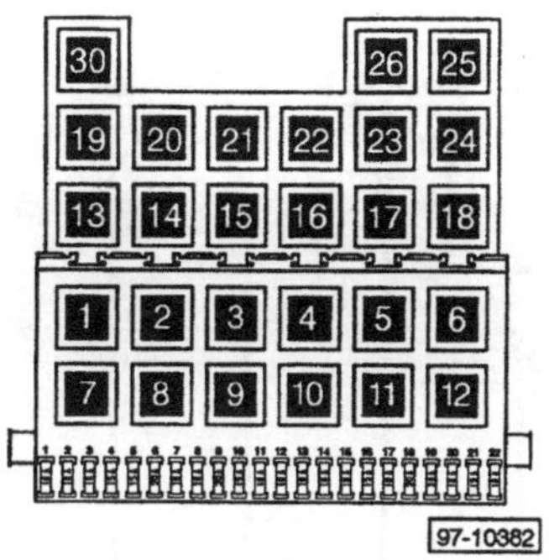

Fuse colors

30A – green
25A – white
20A – yellow
15A – blue
10A – red

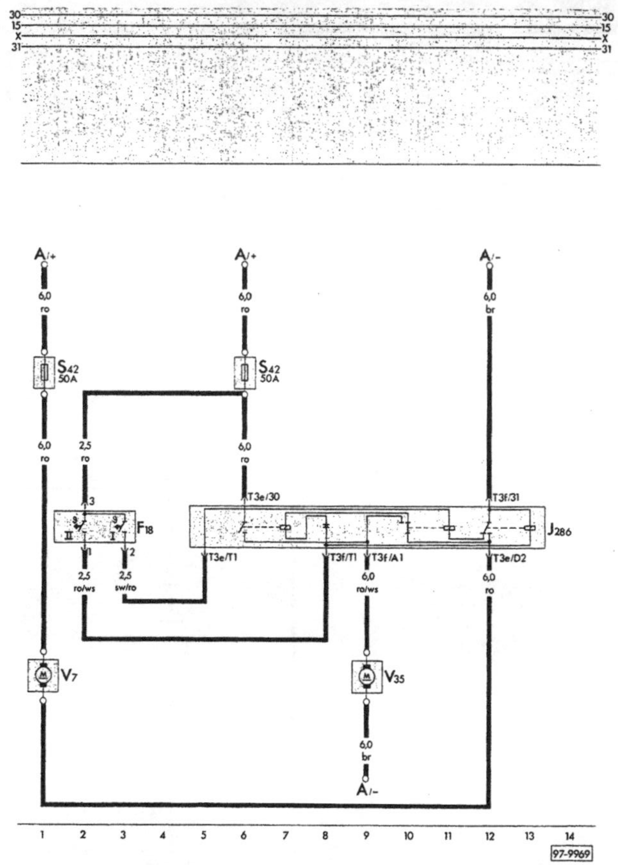

ws = white
sw = black
ro = red
br = brown
gn = green
bl = blue
gr = grey
li = violet
ge = yellow

A – Battery
F18 – Coolant Fan Control (FC) Thermal Switch
J286 – Series/Parallel Coolant Fan Control (FC) Relay (V7, V35)
S42 – Coolant Fan Fuse
T3 – 3-Pin Connector
T3f – 3-Pin Connector
V7 – Coolant Fan
V35 – Right Coolant Fan

Air conditioning with manual control, two evaporators and auxiliary heater core

- **2.5 L-Engine–Digifant Multiport Fuel Injection (MFI)/81 kW, code AAF,**

from April 1992

Relay location
For relay location and multiple connectors see also section "component locations".

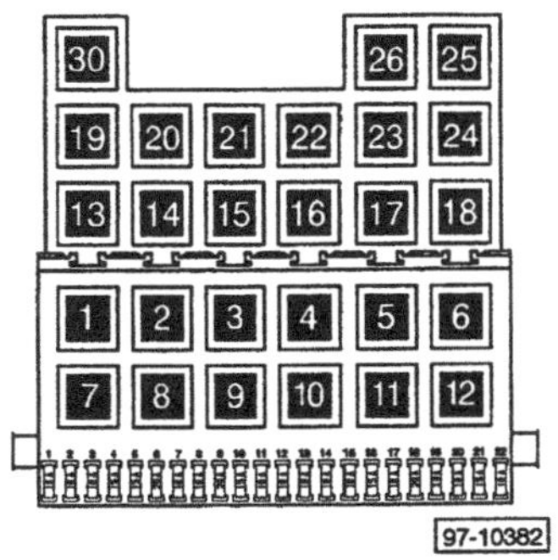

- 1 – A/C Control Module (140)
- 13 – Heater Control Module (152)
- 19 – Heater Air Blower Relay (94)
- 22 – Heater Fuse
- 30 – A/C Clutch Cut-Off Relay (94)

Note: Number in parentheses indicates production control number stamped on relay housing.

Fuse colors

30A – green
25A – white
20A – yellow
15A – blue
10A – red

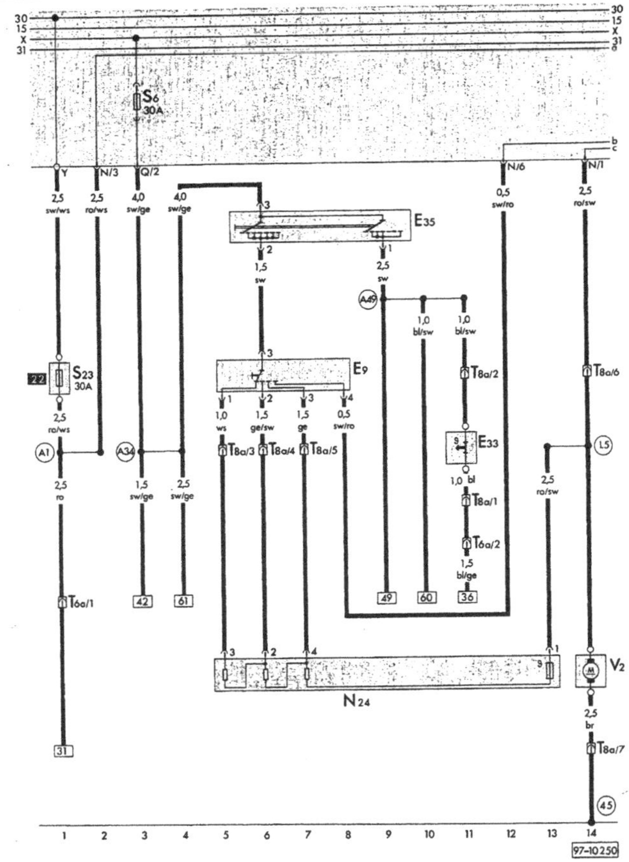

ws = white
sw = black
ro = red
br = brown
gn = green
bl = blue
gr = grey
li = violet
ge = yellow

E9 – Fresh Air Blower Switch
E33 – A/C Evaporator Temperature Switch
E35 – A/C Switch
N24 – Fresh Air Blower Series Resistance With Fuse
S23 – Heater Fuse
T6a – 6-Pin Connector, behind instrument panel, left
T8a – 8-Pin Connector, behind instrument panel, center
V2 – Fresh air blower

(45) – Ground connection, behind instrument panel, center
(A1) – Plus connection (30a), in instrument panel wiring harness

(A34) – Wire connection (75x), in instrument panel wiring harness
(A49) – Wire connection -1-, in instrument panel wiring harness
(L5) – Wire connection -1-, in evaporator housing wiring harness

A/C switch, first evaporator fresh air blower

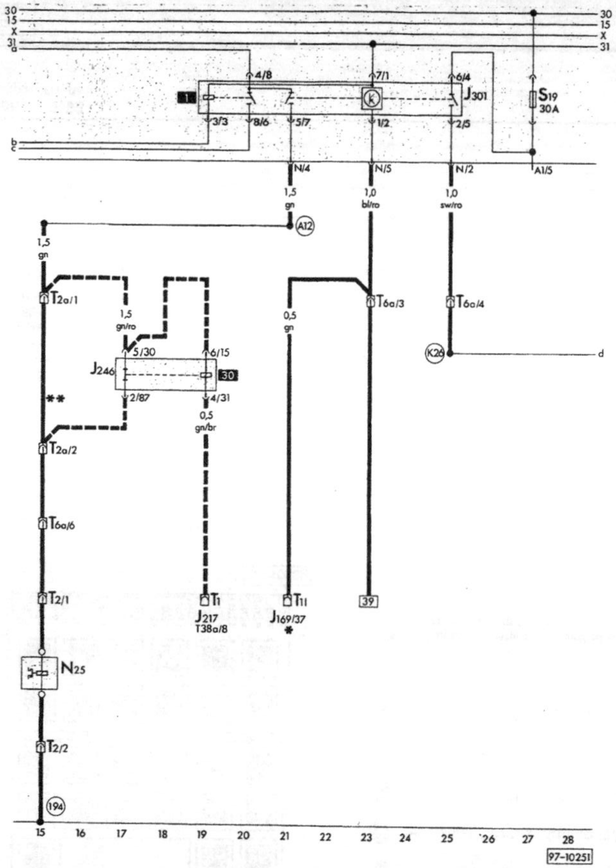

ws = white
sw = black
ro = red
br = brown
gn = green
bl = blue
gr = grey
li = violet
ge = yellow

J169 – Digifant Engine Control Module (ECM)*
J217 – Transmission Control Module (TCM)
J246 – A/C Clutch Cut-Off Relay
J301 – A/C Control Module
N25 – A/C Clutch
T1 – Single Connector, behind fuse/relay panel
T1l – Single Connector, behind instrument panel, left, for idle speed boost
T2 – Double Connector, on A/C Compressor Clutch
T2a – Double Connector,behind instrument panel, left
T6a – 6-Pin Connector, behind instrument panel, left

(194) – Ground connection -2-, in coolant fan wiring harness
(A12) – Plus connection (75aw) in instrument panel wiring harness
(K26) – Wire connection -5-, in coolant fan wiring harness

* – For California vehicles (CARB) this wire connected to terminal T38/32 of digifant engine control module (ECM)
** – Manual transmission only
▬ ▬ – Automatic transmission only

Edition 7/93
USA.5272.01.21

A/C clutch, Idle speed boost, kick-down cut-off

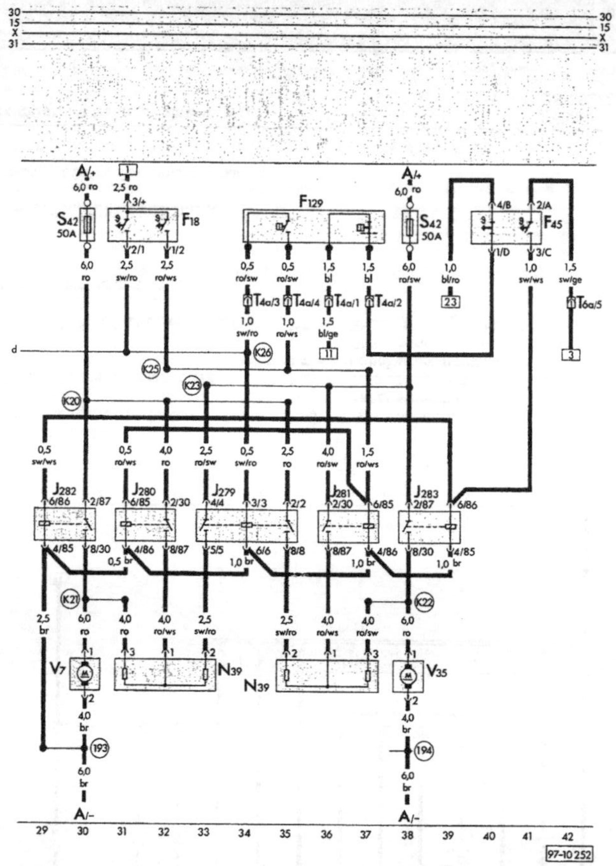

ws = white
sw = black
ro = red
br = brown
gn = green
bl = blue
gr = grey
li = violet
ge = yellow

A – Battery
F18 – Coolant Fan Control (FC) Thermal Switch
F45 – A/C Thermal Switch, in coolant hose
F129 – A/C Pressure Switch
J279 – First Speed Coolant Fan Control (FC) Relay (V7, V35), in engine compartment, left
J280 – Second Speed Coolant Fan Control (FC) Relay (V7)
J281 – Second Speed Coolant Fan Control (FC) Relay (V35)
J282 – Third Speed Coolant Fan Control (FC) Relay (V7)
J283 – Third Speed Coolant Fan Control (FC) Relay (V35)
N39 – Coolant Fan Control (FC) Series Resistance, behind headlight, left
S42 – Coolant Fan Fuse, in engine compartment, left front
T4a – 4-Pin Connector, on A/C Pressure Switch
T6a – 6-Pin Connector, behind instrument panel,left
V7 – Coolant Fan
V35 – Right Coolant Fan

(193) – Ground connection -1-, in coolant fan wiring harness
(194) – Ground connection -2-, in coolant fan wiring harness
(K20) – Plus connection -1-(30), in coolant fan wiring harness
(K21) – Wire connection -1-, in coolant fan wiring harness
(K22) – Wire connection -2-, in coolant fan wiring harness
(K23) – Plus connection -2-(30), in coolant fan wiring harness
(K25) – Wire connection -4-, in coolant fan wiring harness
(K26) – Wire connection -5-, in coolant fan wiring harness

Coolant fans

Edition 7/93
USA.5272.01.21

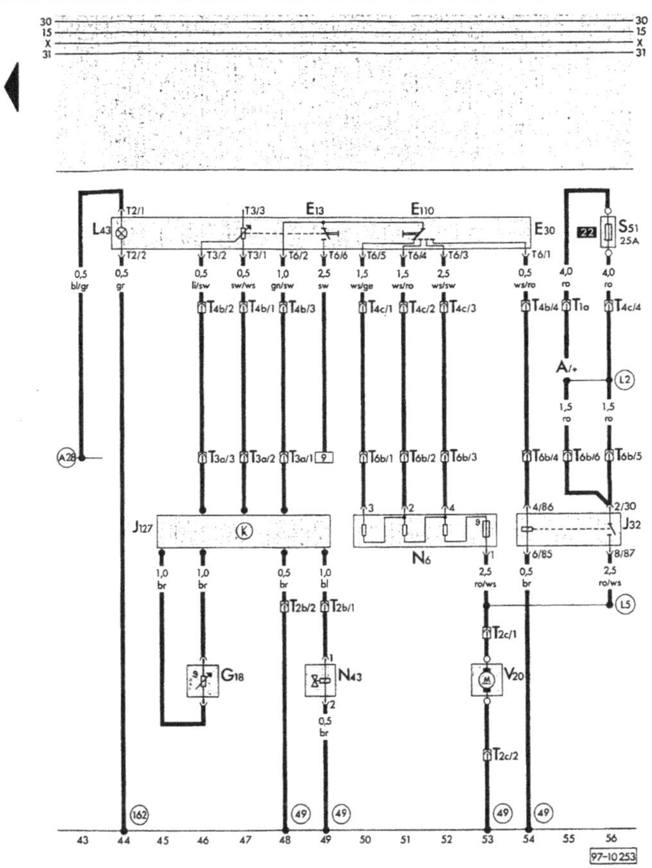

ws	=	white
sw	=	black
ro	=	red
br	=	brown
gn	=	green
bl	=	blue
gr	=	grey
li	=	violet
ge	=	yellow

A – Battery
E13 – Heater-Time Temperature Control Switch
E30 – A/C Switch
E110 – Evaporator Blower Switch
G18 – Temperature Sensor II
J127 – A/C Programmer (Control and Regulating Unit) for 2nd evaporator
J163 – Fresh Air Blower Relay, production contol number (24)
L43 – A/C Switch Illumination Light
N6 – Series Resistance for 2nd evaporator with over heating fuse
N43 – A/C Refrigerant Shut-Off Valve
S51 – Fuse
T1a – Single Connector, behind fuse/relay panel
T2b – Double Connector, on rear evaporator
T2c – Double Connector, on Rear Evaporator Fan
T3a – 3-Pin Connector, on rear evaporator
T4b – 4-Pin Connector, behind instrument panel, center
T4c – 4-Pin Connector, behind instrument panel, center
T6b – 6-Pin Connector, on rear evaporator
V20 – Rear Evaporator Fan

(49) – Ground connection, near ..
(162) – Ground connection, in blower motor wiring harness
(A28) – Plus connection (58b), in fresh air blower wiring harness
(L2) – Wire connection, in A/C wiring harness
(L5) – Wire connection -1-, in evaporator housing wiring harness

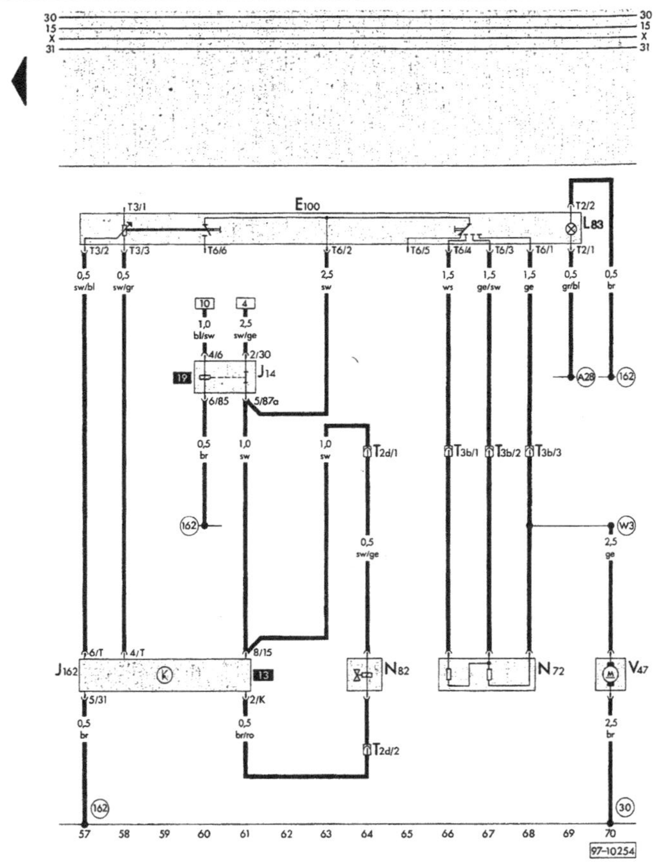

ws	=	white
sw	=	black
ro	=	red
br	=	brown
gn	=	green
bl	=	blue
gr	=	grey
li	=	violet
ge	=	yellow

E100 – Warm Air Blower Switch, in instrument panel
J14 – Heater Air Blower Relay
J162 – Heater Control Module/auxiliary heater core
L83 – Warm Air Blower Switch Light/auxiliary heater core
N72 – Heater Blower Series Resistance, near battery
N82 – Coolant Cut-off Valve, in engine compartment, rear
T2d – Double Connector, behind instrument panel
T3b – 3-pin Connector, behind instrument panel, left
V47 – Rear Warm Air Fan, passenger compartment, lower part

(30) – Ground connection, -1-, beside fuse/relay panel
(162) – Ground connection, in blower motor wiring harness
(A28) – Plus connection (58b), in fresh air blower wiring harness
(W3) – Wire connection, in rear wiring harness

Heated seats,

from April 1992

Relay location
For relay location and multiple connectors location see also section "component locations".

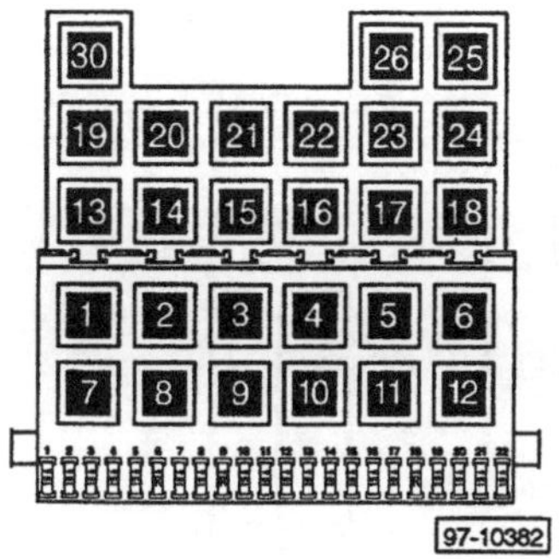

2 – Rear Window Wiper/Washer Relay (72)

Note: Number in parentheses indicates production control number stamped on relay housing.

Fuse colors

30A – green
25A – white
20A – yellow
15A – blue
10A – red

ws = white
sw = black
ro = red
br = brown
gn = green
bl = blue
gr = grey
li = violet
ge = yellow

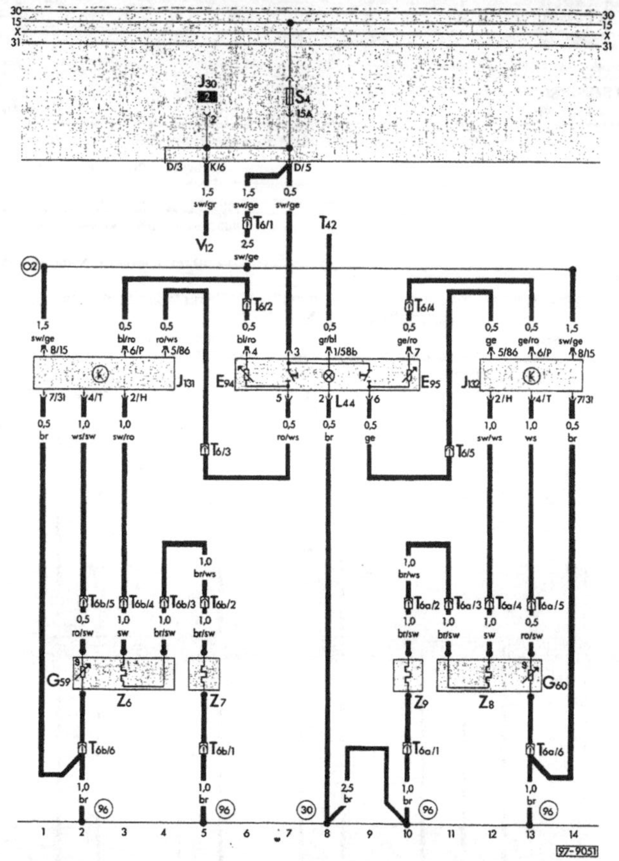

E94 – Driver's Seat Heat Regulating Switch
E95 – Passenger's Seat Heat Regulating Switch
G59 – Driver Seat Heater Temperature Sensor
G60 – Passenger Seat Heater Temperature Sensor
J30 – Rear Window Wiper/Washer Relay
J131 – Driver's Seat Heater Control Module
J132 – Passenger's Seat Heater Control Module
L44 – Seat Heater Switch Illumination Light
T6 – 6-Pin Connector, behind fuse/relay panel
T6a – 6-Pin Connector, below passenger's seat
T6b – 6-Pin Connector, below driver's seat
T42 – Wire Connector, Terminal 58 B, above fuse/relay panel
V12 – Rear Window Wiper Motor
Z6 – Driver's Seat Heat Element
Z7 – Driver's Backrest Heat Element
Z8 – Passenger's Seat Heat Element
Z9 – Passenger's Backrest Heat Element

(30) – Ground connection, beside fuse/relay panel
(96) – Ground connection -1-, in heated seats wiring harness
(O2) – Plus connection (15), in heated seats wiring harness

Heated seats (driver's/passenger's seat)

Front power windows,

from April 1992

Relay location
For relay location and multiple connectors see also section "component locations".

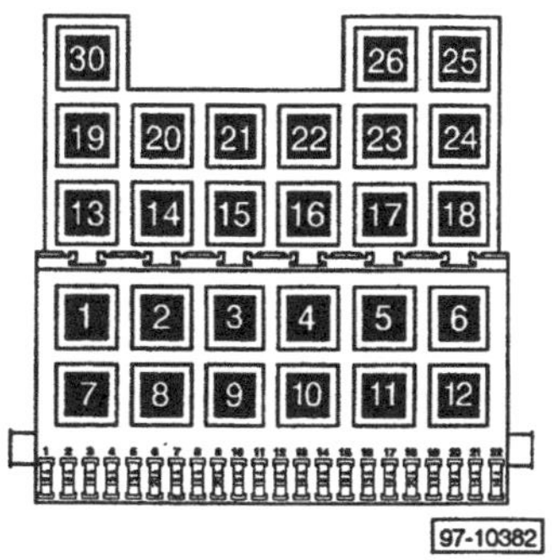

24 – Power Window Circuit Breaker

Note: Number in parentheses indicates production control number stamped on relay housing.

Fuse colors

30A – green
25A – white
20A – yellow
15A – blue
10A – red

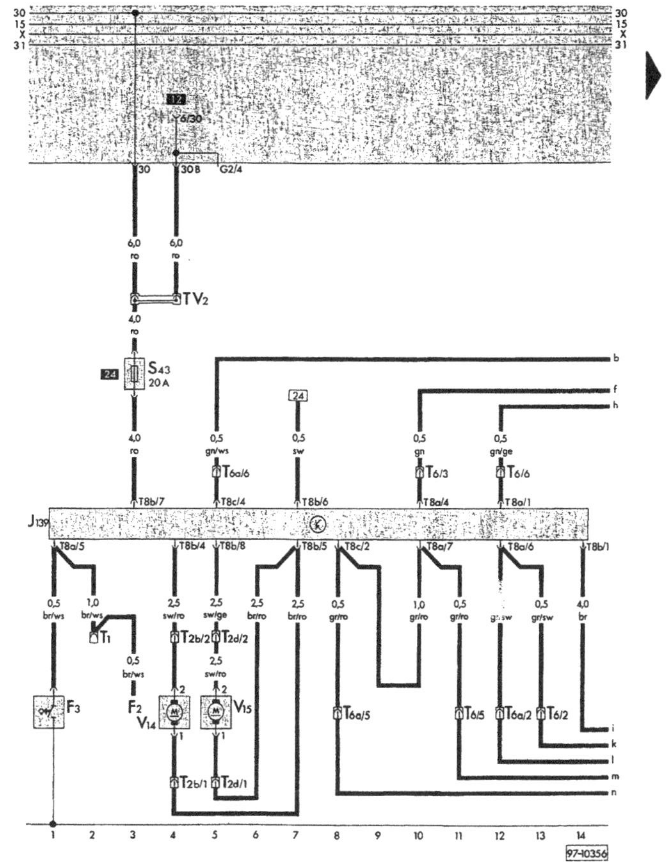

ws = white
sw = black
ro = red
br = brown
gn = green
bl = blue
gr = grey
li = violet
ge = yellow

F2 – Left Front Door Contact Switch
F3 – Right Front Door Contact Switch
J139 – Power Window/Sunroof Control Module, behind instrument panel, right
S43 – Power Window Circuit Breaker
T1 – Single Connector, behind instrument panel, left
T2b – Double Connector, behind instrument panel, left
T2d – Double Connector, behind instrument panel, right
T6 – 6-Pin Connector, behind instrument panel, left
T6a – 6-Pin Connector, behind instrument panel, right
T8a – 8-Pin Connector
T8b – 8-Pin Connector
T8c – 8-Pin Connector
TV2 – Terminal 30 (B+) Wire Connector, behind fuse/relay panel
V14 – Left Window Motor
V15 – Right Window Motor

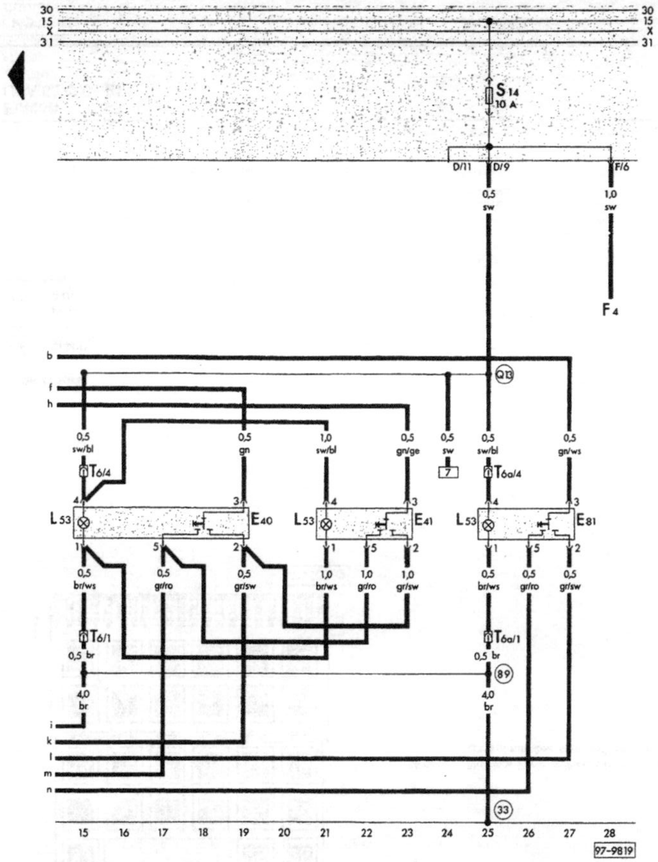

ws = white
sw = black
ro = red
br = brown
gn = green
bl = blue
gr = grey
li = violet
ge = yellow

E40 – Left Front Window Switch
E41 – Right Front Window Switch
E81 – Right Front Window Switch, (In RF Door)
F4 – Back-Up Light Switch
L53 – Power Window Switch Light
T6 – 6-Pin Connector, behind instrument panel, left
T6a – 6-Pin Connector, behind instrument panel, right

(33) – Ground connection, behind instrument panel, right
(89) – Ground connection -1-, in power window wiring harness

(Q13) – Plus connection (75 as/15), in power window/central locking system and door contact switch wiring harness

Heated power mirrors,

from April 1992

Relay location
For relay location and multiple connectors see also section "component locations".

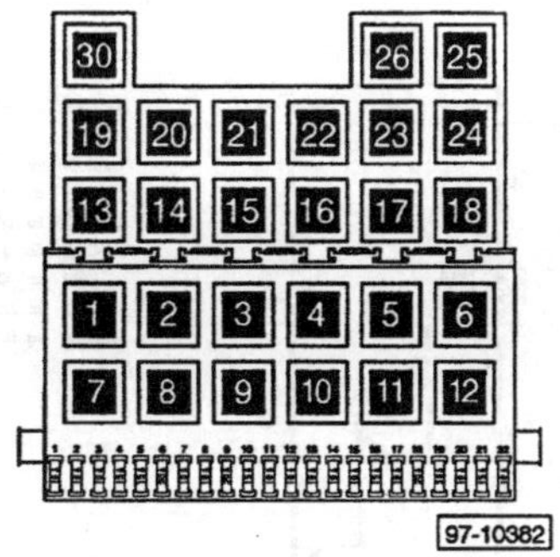

23 – Heated Mirror Relay (94)
26 – Radio Fuse

Note: Number in parentheses indicates production control number stamped on relay housing.

Fuse colors

30A – green
25A – white
20A – yellow
15A – blue
10A – red

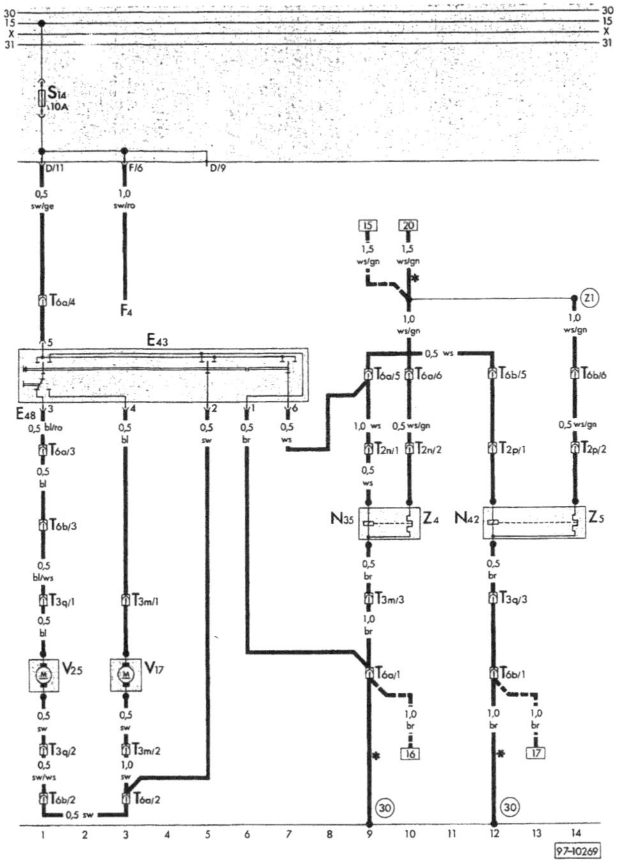

ws = white
sw = black
ro = red
br = brown
gn = green
bl = blue
gr = grey
li = violet
ge = yellow

E43 – Mirror Adjustment Switch
E48 – Mirror Selector Switch
F4 – Back-Up Light Switch
N35 – Left Mirror Magnetic Clutch
N42 – Right Mirror Magnetic Clutch
T2n – Double Connector, in driver's door
T2p – Double Connector, in passenger's door
T3m – 3-Pin Connector, in driver's door
T3q – 3-Pin Connector, in passenger's door
T6a – 6-Pin Connector, on A-pillar,left
T6b – 6-Pin Connector, on A-pillar, right
V17 – Driver's Side Mirror Adjustment Motor
V25 – Passenger's Side Mirror Adjustment Motor
Z4 – Driver's Side Heated Mirror
Z5 – Passenger's Side Heated Mirror

(30) – Ground connection, beside fuse/relay panel
(Z1) – Wire connection, in power heated mirror wiring harness

▬ ▬ – On vehicles with heated mirrors and rear window defogger only
* – On vehicles with heated mirrors but without rear window defogger only

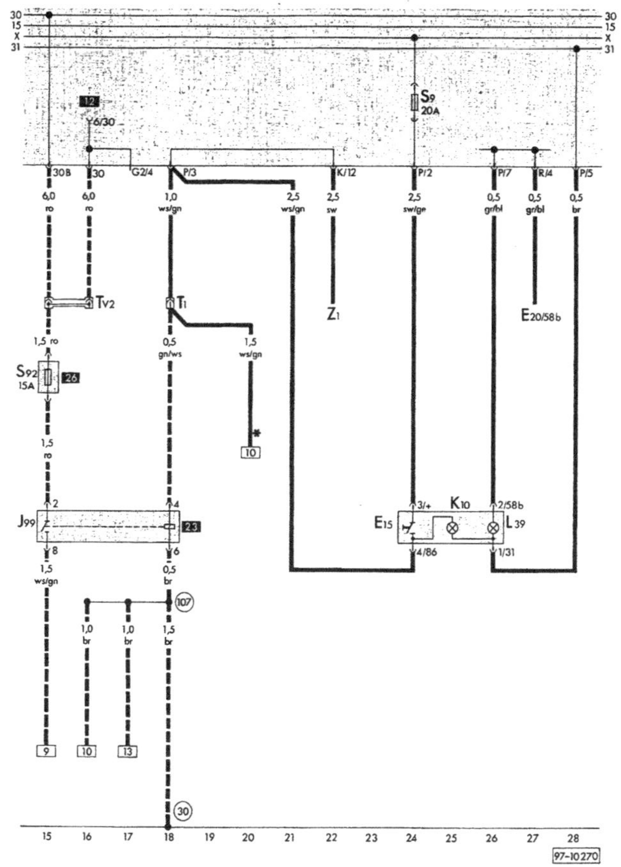

ws = white
sw = black
ro = red
br = brown
gn = green
bl = blue
gr = grey
li = violet
ge = yellow

E15 –Rear Window Defogger Switch
E20 –Instrument Panel Light Dimmer Switch
J99 –Heated Mirror Relay
K10 –Rear Window Defogger Indicator Light
L39 –Rear Window Defogger Switch Light
S92 –Radio Fuse
T1 –Single Connector, behind fuse/relay panel
TV2 –Terminal 30 (B+) Wire Connector
Z1 –Rear Window Defogger/Heat Element

(30) –Ground connection, -1-, beside fuse/relay panel
(107) –Ground connection, in outside mirrors wiring harness

▬ ▬ – On vehicles with heated mirrors and rear window defogger only
* – On vehicles with heated mirrors but without rear window defogger only

Central locking system for vehicles with right sliding door and tailgate or rear swinging doors,

from April 1992

Relay location

For relay location and multiple connectors location see also section "component locations".

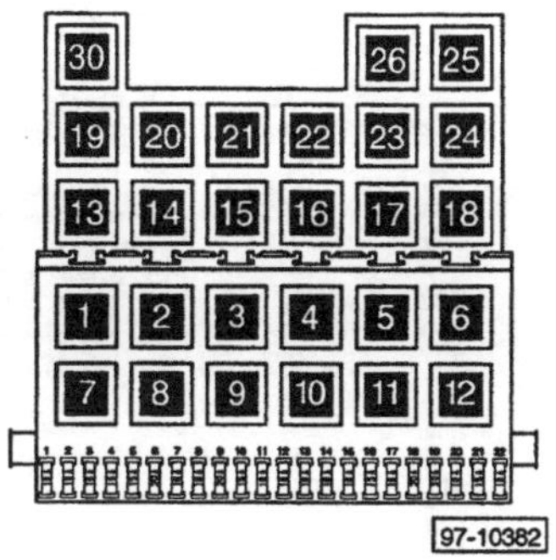

Fuse colors

30A – green
25A – white
20A – yellow
15A – blue
10A – red

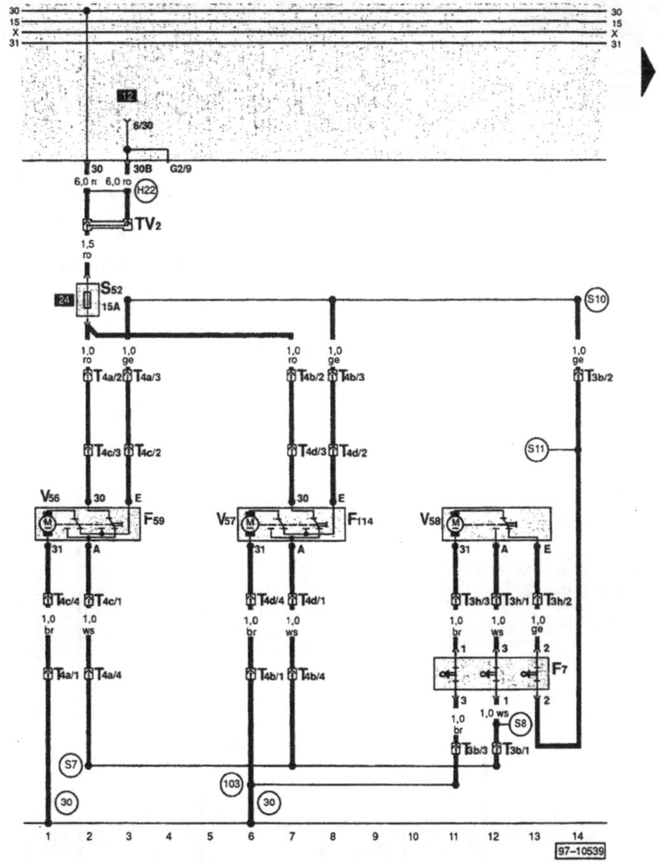

ws = white
sw = black
ro = red
br = brown
gn = green
bl = blue
gr = grey
li = violet
ge = yellow

F7 – Right Rear Side/Sliding Door Contact Switch
F59 – Driver's Door Central Locking System Switch
F114 – Passenger's Door Central Locking SystemSwitch
S52 – Central Locking System Fuse
T3b – 3-Pin Connector, behind instrument panel, right
T3h – 3-Pin Connector, in sliding door
T4a – 4-Pin Connector, behind fuse/relay panel
T4b – 4-Pin Connector, behind instrument panel, right
T4c – 4-Pin Connector, in driver's door
T4d – 4-Pin Connector, in passenger's door
TV2 – Terminal 30 (B+) Wire Connector
V56 – Driver's Door Central Locking System Motor
V57 – Passenger's Door Central Locking System Motor
V58 – Sliding Door Central Locking System Motor

(30) – Ground connection, beside fuse/relay panel
(103) – Ground connection -2-, in central locking system wiring harness
(H22) – Plus connection (30), in terminal 30/30B wiring harness
(S7) – Wire connection (A) -2-, in central locking system wiring harness
(S8) – Wire connection (A) -3-, in central locking system passenger compartment wiring harness
(S10) – Wire connection (E) -2-, in central locking system wiring harness
(S11) – Wire connection (E) -3-, in central locking system passenger compartment wiring harness

Central locking system for driver's/passenger's door and right sliding door

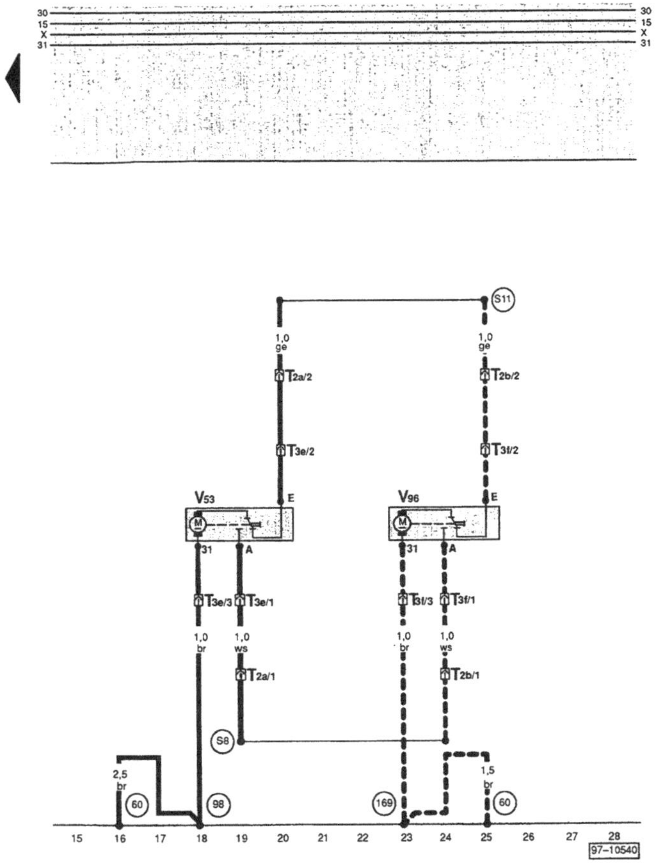

ws = white
sw = black
ro = red
br = brown
gn = green
bl = blue
gr = grey
li = violet
ge = yellow

T2a – Double Connector, on right D-pillar, upper part
T2b – Double Connector, on right D-pillar, upper part
T3e – 3-Pin Connector, in tailgate
T3f – 3-Pin Connector, in right rear swinging door
V53 – Decklid Central Locking System Motor
V96 – Left Rear Door Central Locking System Motor
(60) – Ground connection, near right taillight
(98) – Ground connection, in rear lid wiring harness
(169) – Ground connection, in right swinging door wiring harness
(S8) – Wire connection (A) -3-, in central locking system/passenger compartment wiring harness
(S11) – Wire connection (E) -3-, in central locking system/passenger compartment wiring harness
▬ ▬ – On models with rear swinging doors only

THIS PAGE INTENTIONALLY LEFT

BLANK

Rear passenger heating,

from April 1992

Relay location

For relay location and multiple connectors see also section "component locations".

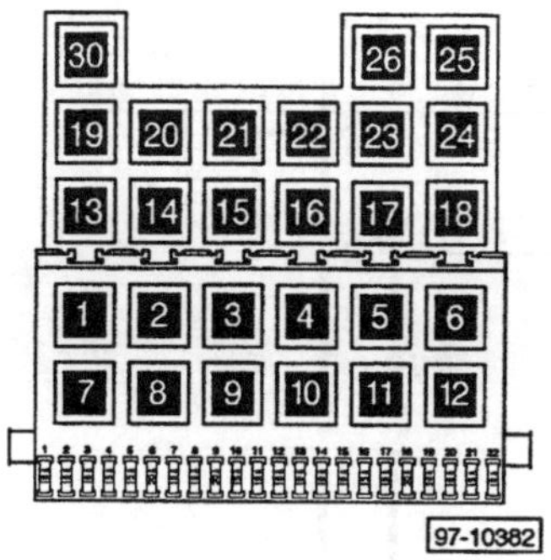

19 – Heater Air Blower Relay (24)

22 – Heater Air Blower Fuse

Note: Number in parentheses indicates production control number stamped on relay housing.

Fuse colors

30A – green
25A – white
20A – yellow
15A – blue
10A – red

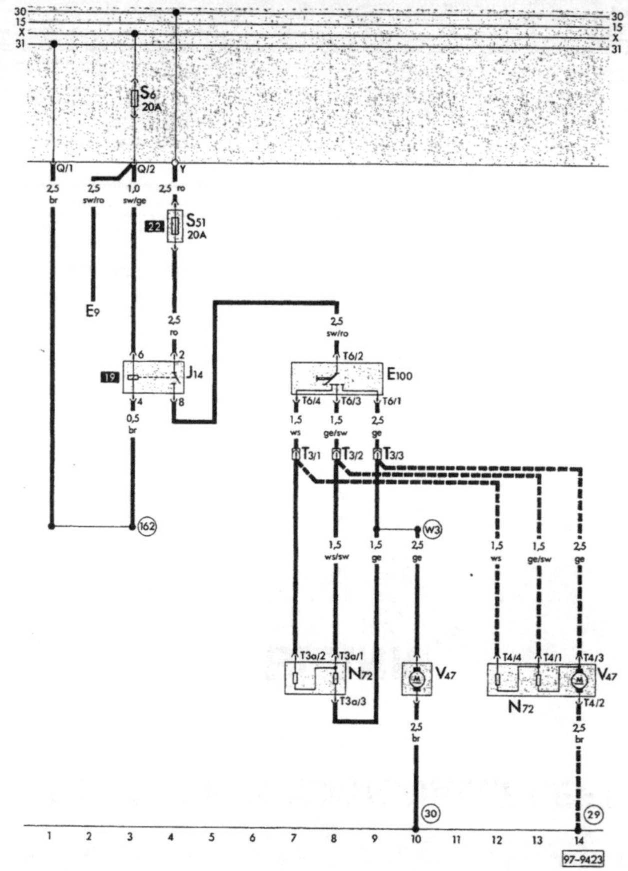

ws = white
sw = black
ro = red
br = brown
gn = green
bl = blue
gr = grey
li = violet
ge = yellow

E9 – Fresh Air Blower Switch
E100 – Warm Air Blower Switch, in instrument panel
J14 – Heater Air Blower Relay
N72 – Heater Blower Series Resistance
S51 – Fuse
T3 – 3-Pin Connector, behind instrument panel, left
T3a – 3-Pin Connector
T4 – 4-Pin Connector
T6 – 6-Pin Connector
V47 – Rear Warm Air Fan

(29) – Ground connection, near Rear Warm Air Fan

(30) – Ground connection, beside fuse/relay panel
(162) – Ground connection, in blower motor wiring harness
(W3) – Wire connection, in rear wiring harness
▬ ▬ – Cargo floor models only

Rear passenger heater core

Anti-lock brake system (ABS),

from April 1992

Relay location

For relay location and multiple connectors see also section "component locations".

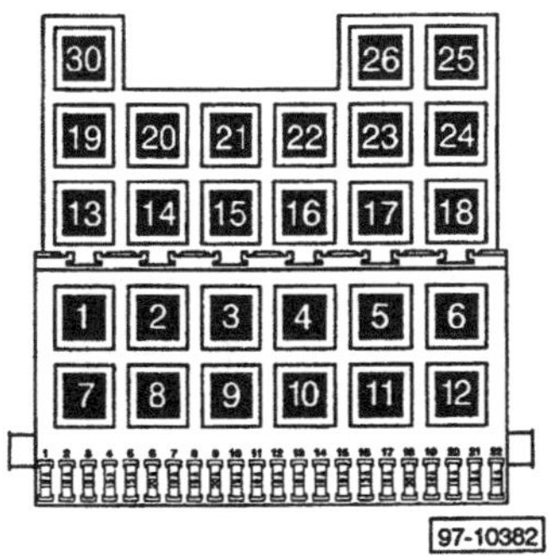

25 – ABS Fuse

Fuse colors

30A – green
25A – white
20A – yellow
15A – blue
10A – red

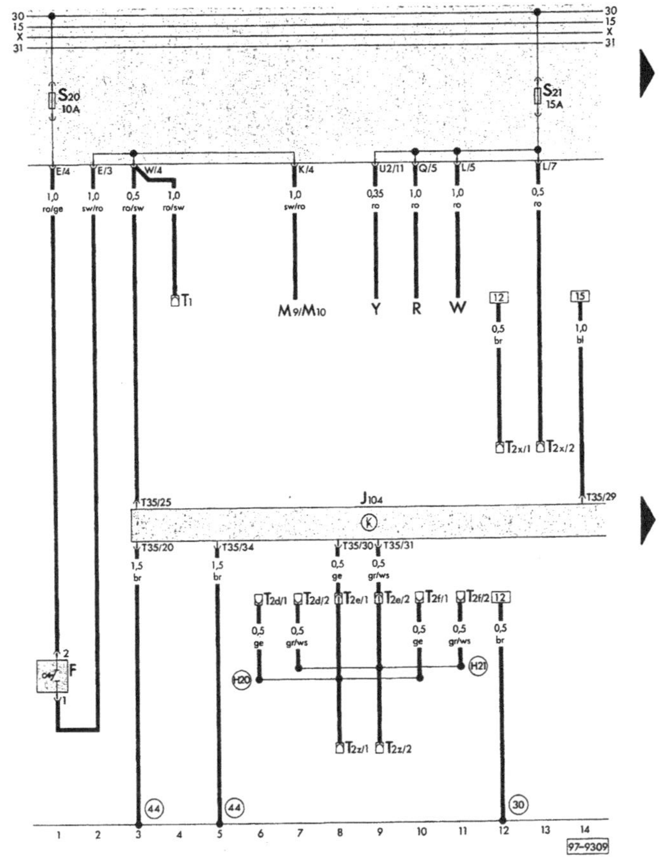

ws = white
sw = black
ro = red
br = brown
gn = green
bl = blue
gr = grey
li = violet
ge = yellow

F – Brake Light Switch (00526–2131)
J104 – ABS Control Module (w/EDL), on left A-pillar, lower part
M9 – Left Brake Light
M10 – Right Brake Light
R – Radio
T1 – Single Connector, behind fuse/relay panel
T2d – Double Connector, behind fuse/relay panel*
T2e – Double Connector, behind fuse/relay panel
T2f – Double Connector, behind fuse/relay panel*
T2x – Double Connector, –black–, behind fuse/relay panel cover (OBD)
T2z – Double Connector, –white–, behind fuse/relay panel cover (OBD)
T35 – 35-Pin Connector
W – Front Interior Light
Y – Analog Clock
(30) – Ground connection, -1-, beside fuse/relay panel
(44) – Ground connection, on left A-pillar, lower part
(H20) – Wire connection -1-, in On Board Diagnostic (OBD) wiring harness
(H21) – Wire connection -2-, in On Board Diagnostic (OBD) wiring harness
* – With further On Board Diagnostic (OBD) systems only

ABS control module, on board diagnostic (OBD) connection

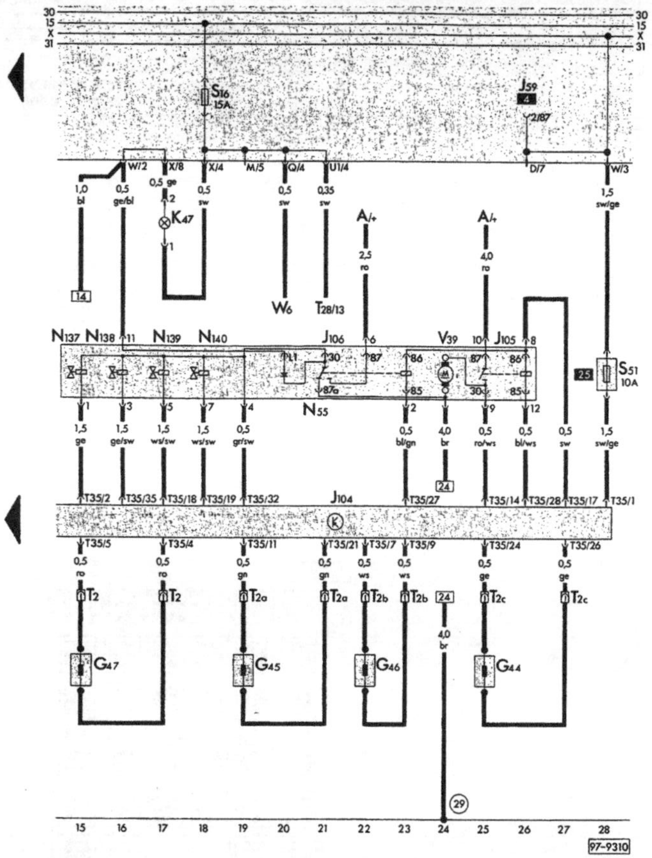

ws = white
sw = black
ro = red
br = brown
gn = green
bl = blue
gr = grey
li = violet
ge = yellow

A – Battery
G44 – Right Rear ABS Wheel Speed Sensor, (00287-1243)
G45 – Right Front ABS Wheel Speed Sensor (00285-1241)
G46 – Left Rear ABS Wheel Speed Sensor (00290-1311)
G47 – Left Front ABS Wheel Speed Sensor (00283-1233)
J59 – Load Reduction Relay
J104 – ABS Control Module (w/EDL)
J105 – ABS Return Flow Pump Relay
J106 – ABS Solenoid Valve Relay (00302-1341)
K47 – ABS Warning Light
N55 – ABS Hydraulic Unit
N137 – Left Front ABS Inlet/Outlet Valve (00277-1221)
N138 – Right Front ABS Inlet/Outlet Valve (00284-1234)
N139 – Left Rear ABS Inlet/Outlet Valve (00286-1242)
N140 – Right Rear ABS Inlet/Outlet Valve (00289-1244)
S51 – Fuse
T2 – Double Connector, in left front wheel housing
T2a – Double Connector, in right front wheel housing
T2b – Double Connector, on left side in front of rear cross member
T2c – Double Connector, on right side in front of rear cross member
T28 – 28-Pin Connector, on instrument cluster
T35 – 35-Pin Connector
V39 – ABS Return Flow Pump (00301-1334)
W6 – Glove Compartment Light
(29) – Ground connection, near ABS Hydraulic Unit

THIS PAGE INTENTIONALLY LEFT

BLANK

Camper equipment,

from April 1992

Relay location
For relay location and multiple connectors see also section "component locations".

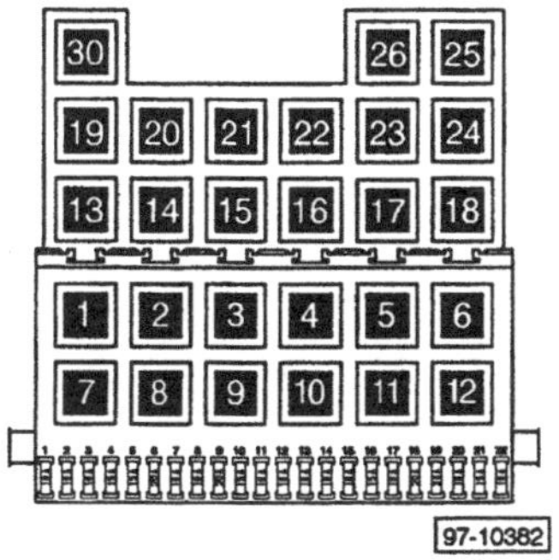

Warning!

- Do not connect a 110/220V voltage supply to the exterior socket during repair work.
- All repair work to the 110/220V system should only be performed by an electrician or under the guidance and supervision of an electrician.
- Observe all local regulations.

Fuse colors

30A – green
25A – white
20A – yellow
15A – blue
10A – red

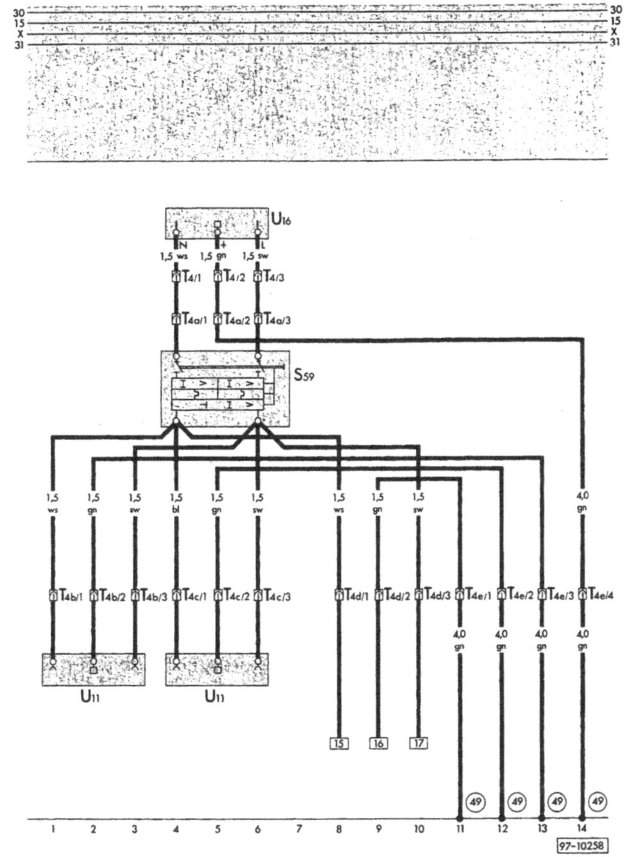

ws = white
sw = black
ro = red
br = brown
gn = green
bl = blue
gr = grey
li = violet
ge = yellow

S59 – Fuse, Automatic (110 V), behind driver's seat
T4 – 4-Pin Connector, near B-pillar, left
T4a – 4-Pin Connector, near B-pillar, left
T4b – 4-Pin Connector, near B-pillar, left
T4c – 4-Pin Connector, near B-pillar, left
T4d – 4-Pin Connector, near B-pillar, left
T4e – 4-Pin Connector, near B-pillar, left
U11 – 110 V/220 V Interior Socket, behind driver's seat
U16 – 110 V Connection Plug

(49) – Ground connection, near B-pillar, left

Connector for 110V-socket, fuse automatic (110V), interior socket (110V)

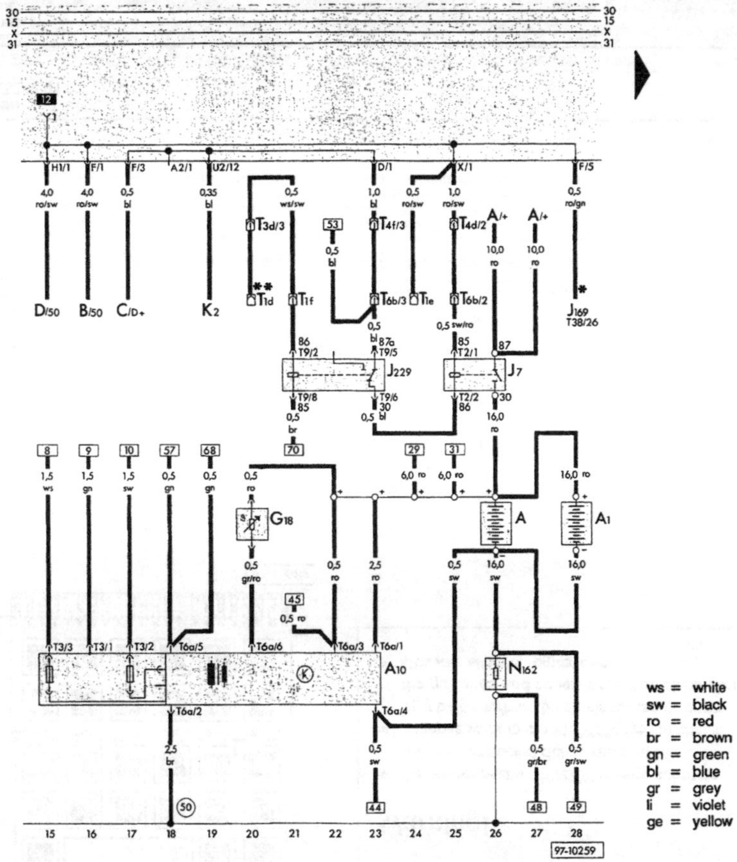

ws = white
sw = black
ro = red
br = brown
gn = green
bl = blue
gr = grey
li = violet
ge = yellow

A – Battery
A1 – Auxiliary Battery, left rear
A10 – Battery Charger (220V, 110V or 12 Volt)
B – Starter
C – Generator (GEN)
D – Ignition/Starter Switch
G18 – Temperature Sensor II, near auxiliary battery
J7 – Battery Cut Out Relay, near auxiliary battery
J169 – Digifant Engine Control Module (ECM), in engine compartment, left
J229 – First or Third Speed Coolant Fan Control (FC) Cut-off Relay (currently not available)
K2 – Generator (GEN) Warning Light, in instrument cluster
N162 – Battery Charging Control Resistance
T1d – Single Connector, behind fuse/relay panel
T1e – Single Connector, behind fuse/relay panel
T1f – Single Connector, on left A-pillar, upper part
T2 – Double Connector
T3 – 3-Pin Connector
T3d – 3-Pin Connector, behind fuse/relay panel
T4f – 4-Pin Connector, behind fuse/relay panel
T6a – 6-Pin Connector
T6b – 6-Pin Connector, on left A-pillar, upper part
T9 – 9-Pin Connector
T38 – 38-Pin Connector, on Digifant Engine Control Module (ECM)
(50) – Ground connection, in luggage compartment, left
* – With Digifant Engine Control Module (ECM) only
** – Fan control (FC) cut-off, currently not available
*** – With gel-electrolyte; maintenance free

Battery charger, auxiliary battery, battery cut-out relay

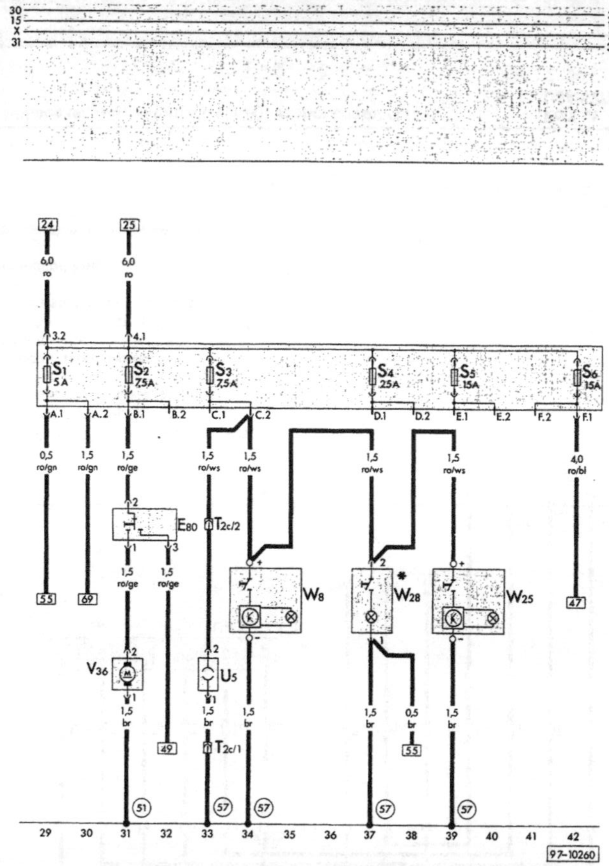

ws = white
sw = black
ro = red
br = brown
gn = green
bl = blue
gr = grey
li = violet
ge = yellow

E80 – Water Pump Switch
S1 – Fuse, in fuse bracket, on B-pillar, left
S2 – Fuse, in fuse bracket, on B-pillar, left, changed in 3 Amps.
S3 – Fuse, in fuse bracket, on B-pillar,left
S4 – Fuse, in fuse bracket, on B-pillar,left
S5 – Fuse, in fuse bracket, on B-pillar,left
S6 – Fuse, in fuse bracket, on B-pillar,left
T2 – Double Connector, on B-pillar, left
U5 – 12 V Socket, behind driver's seat
V36 – Water Pump
W8 – Left Flourescent Light, over kitchen
W25 – Right Flourescent Light, near sliding door
W28 – Front Interior Light, (High Roof)

(51) – Ground connection, in luggage copartment, right
(57) – Ground connection, on left rear pillar

Fuses, water supply, interior lights

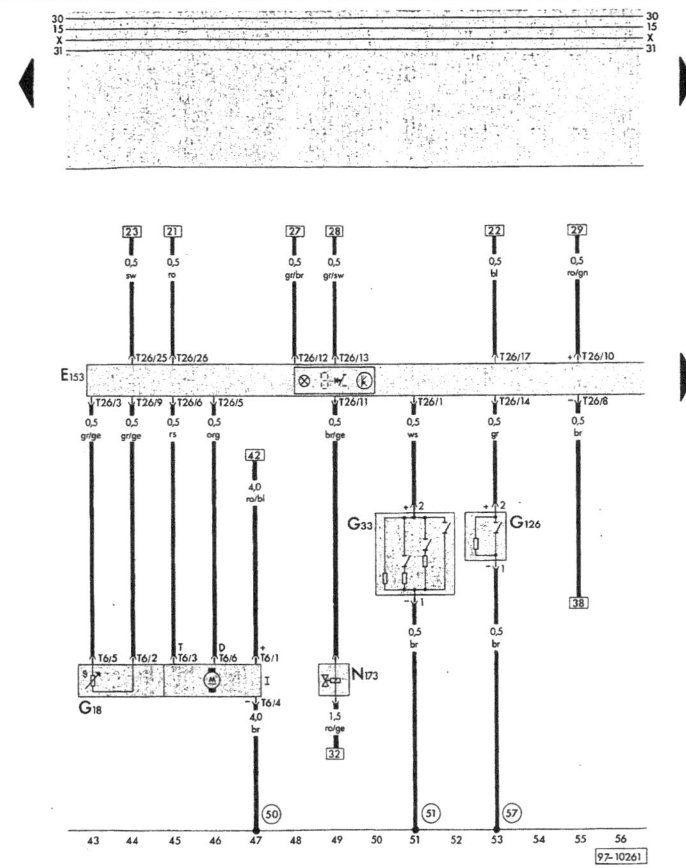

ws = white
sw = black
ro = red
br = brown
gn = green
bl = blue
gr = grey
li = violet
ge = yellow

E153 – Camper Central Electronic Displ, near front interior light
G18 – Temperature Sensor II, in refrigerator I
G33 – Windshield Washer Fluid Level Sensor
G126 – Waste Water Level Sensor
I – Refrigerator
N173 – External Water Supply Fresh Water Valve
T6 – 6-Pin Connector
T26 – 26-Pin Connector

(50) – Ground connection, in luggage compartment, left

(51) – Ground connection, in luggage compartment, right

(57) – Ground connection, left rear pillar

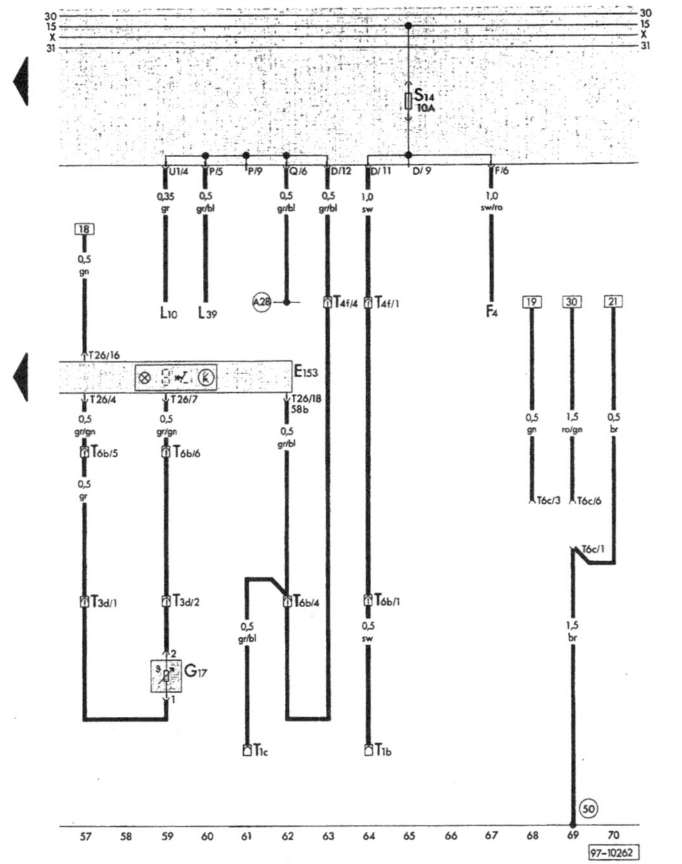

ws = white
sw = black
ro = red
br = brown
gn = green
bl = blue
gr = grey
li = violet
ge = yellow

E153 – Camper Central Electronic Display, near front interior light
F4 – Back-Up Light Switch
G17 – Outside Air Temperature Sensor, on front bumper
L10 – Instrument Cluster Light
L39 – Rear Window Defogger Switch Light
T1b – Single Connector, on B-pillar, left (auxiliary heater)
T1c – Single Connector, on B-pillar, left (auxiliary heater)
T3d – 3-Pin Connector, behind fuse/relay panel
T4f – 4-Pin Connector, behind fuse/relay panel
T6b – 6-Pin Connector, on left A-pillar, upper part
T6c – 6-Pin Connector, near auxiliary battery
T26 – 26-Pin Connector

(50) – Ground connection, in lugggage compartment, left

(A28) – Plus connection (58b), in fresh air blower wiring harness

Taillights (flatbed and double cabin),

from April 1992

Relay location
For relay location and multiple connectors see also section "component locations".

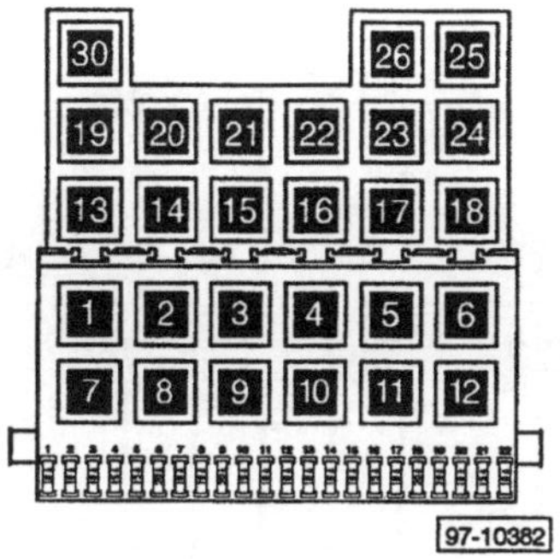

Fuse colors

30A – green
25A – white
20A – yellow
15A – blue
10A – red

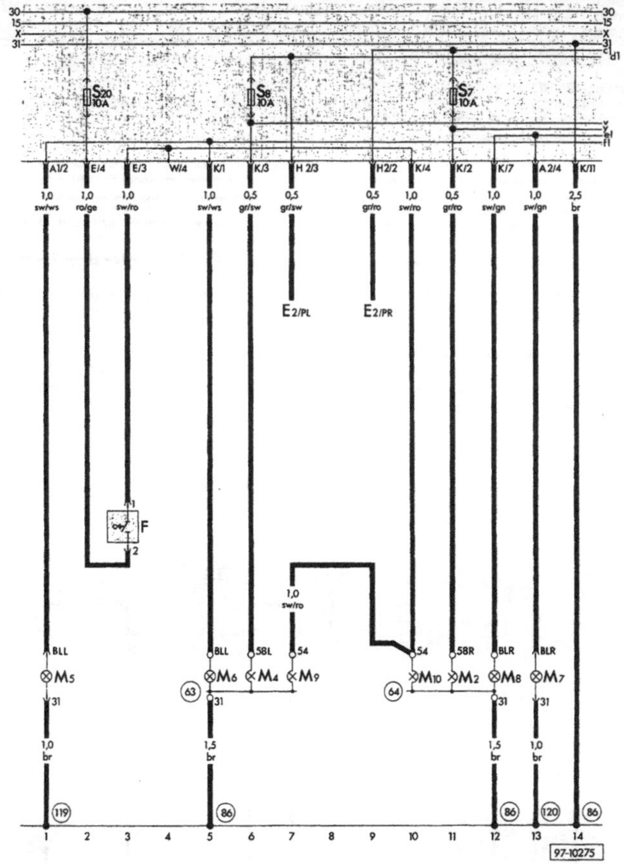

ws = white
sw = black
ro = red
br = brown
gn = green
bl = blue
gr = grey
li = violet
ge = yellow

E2 – Turn Signal Switch
F – Brake Light Switch
M2 – Right Taillight
M4 – Left Taillight
M5 – Left Front Turn Signal Light
M6 – Left Rear Turn Signal Light
M7 – Right Front Turn Signal Light
M8 – Right Rear Turn Signal Light
M9 – Left Brake Light
M10 – Right Brake Light

(63) – Ground connection, on left taillight bulb-holder
(64) – Ground connection, on right taillight bulb-holder

(86) – Ground connection -1-, in rear wiring harness

(119) – Ground connection -1-, in headlight wiring harness
(120) – Ground connection -2-, in headlight wiring harness

Rear turn signal lights, brake lights, taillights

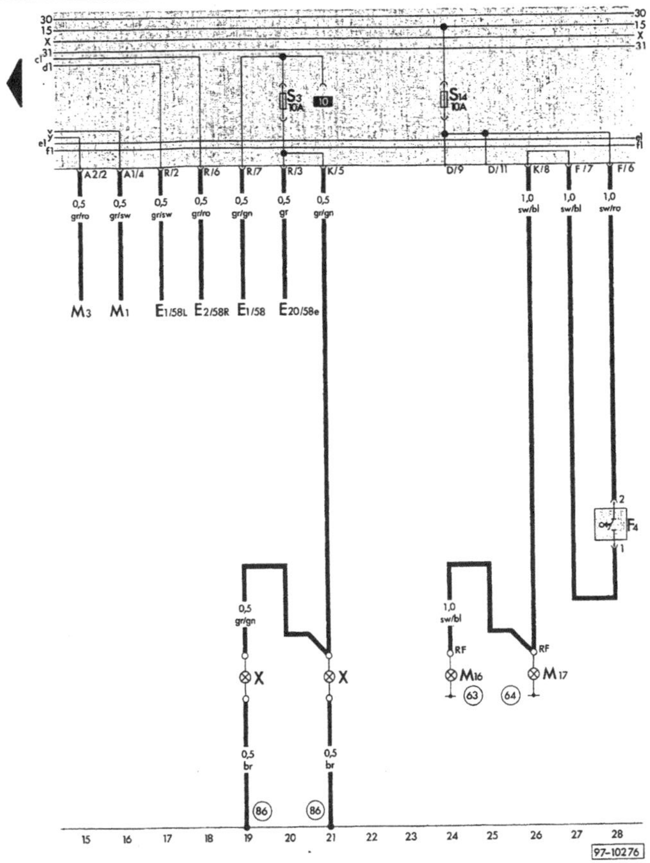

ws = white
sw = black
ro = red
br = brown
gn = green
bl = blue
gr = grey
li = violet
ge = yellow

E1 – Light Switch
E2 – Turn Signal Switch
E20 – Instrument Panel Light Dimmer Switch
F4 – Back-Up Light Switch
M1 – Left Parking Light
M3 – Right Parking Light
M16 – Left Back-Up Light
M17 – Right Back-Up Light
X – License Plate Light

(63) – Ground connection, on left taillight bulb-holder
(64) – Ground connection, on right taillight bulb-holder
(86) – Ground connection –1–, in rear wiring harness

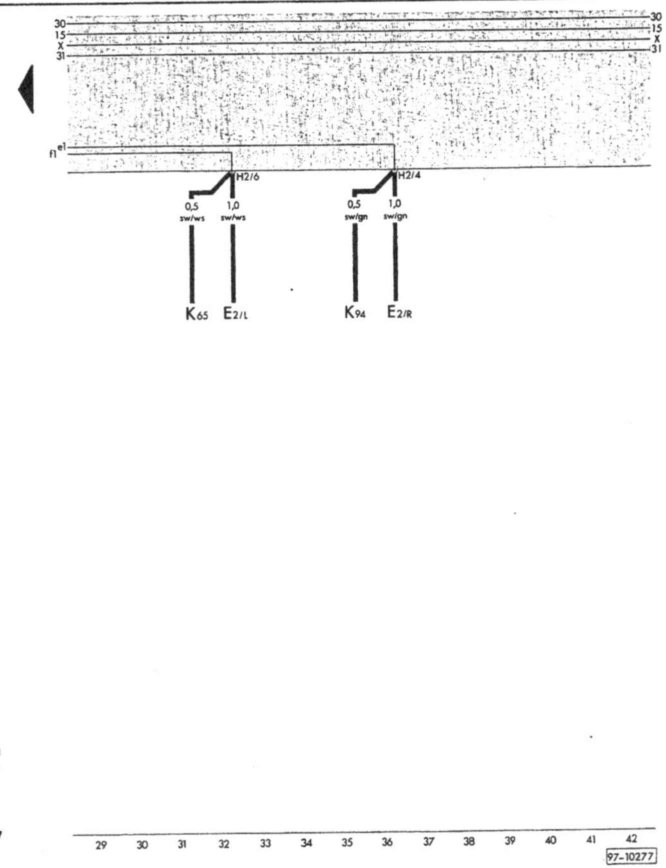

ws = white
sw = black
ro = red
br = brown
gn = green
bl = blue
gr = grey
li = violet
ge = yellow

E2 – Turn Signal Switch
K65 – Left Turn Signal Indicator Light
K94 – Right Turn Signal Indicator Light

Auxiliary battery with cut-out relay,

from April 1992

Relay location
For relay location and multiple connectors see also section "component locations".

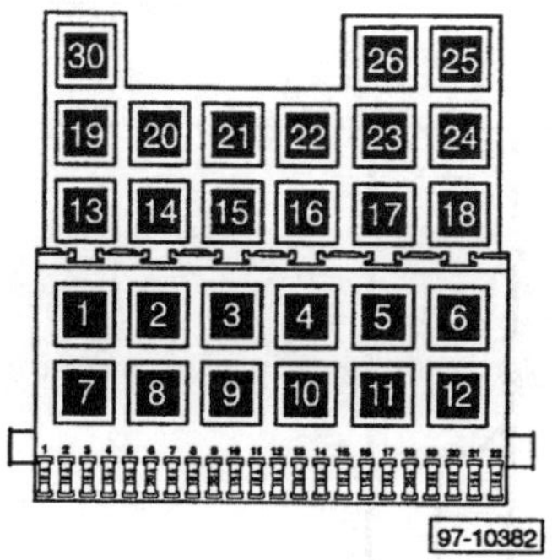

Fuse colors

30A – green
25A – white
20A – yellow
15A – blue
10A – red

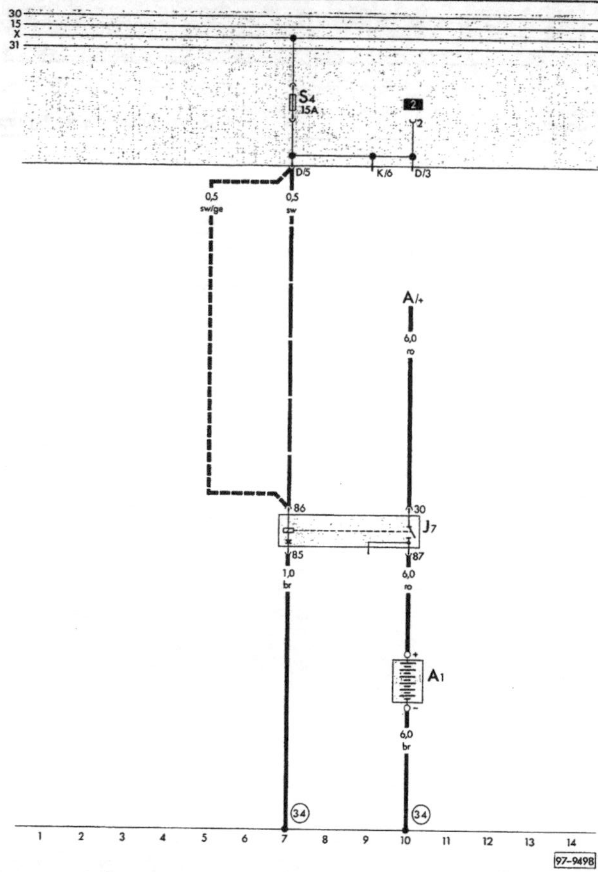

ws = white
sw = black
ro = red
br = brown
gn = green
bl = blue
gr = grey
li = violet
ge = yellow

A – Battery, in engine compartment
A1 – Auxiliary Battery, below driver's seat
J7 – Battery Cut Out Relay, below driver's seat

(34) – Ground connection, below driver's seat

▪▪ – Vehicles with auxiliary heater only
▬ ▬ – Vehicles without auxiliary heater only

Cruise control,

from April 1992

Relay location

For relay location and multiple connectors see also section "component locations".

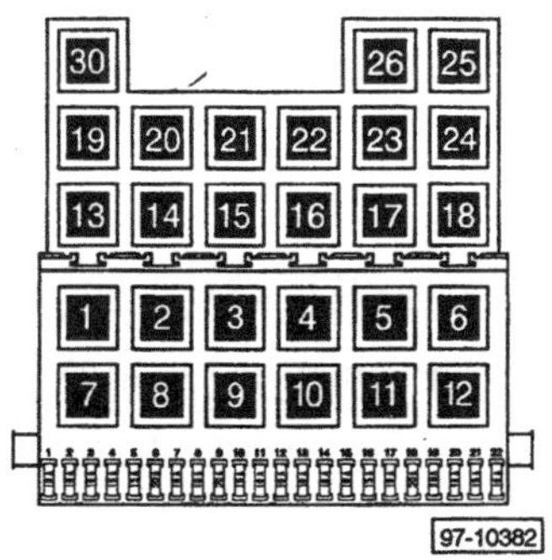

21 – Cruise Control Fuse

Fuse colors

30A – green
25A – white
20A – yellow
15A – blue
10A – red

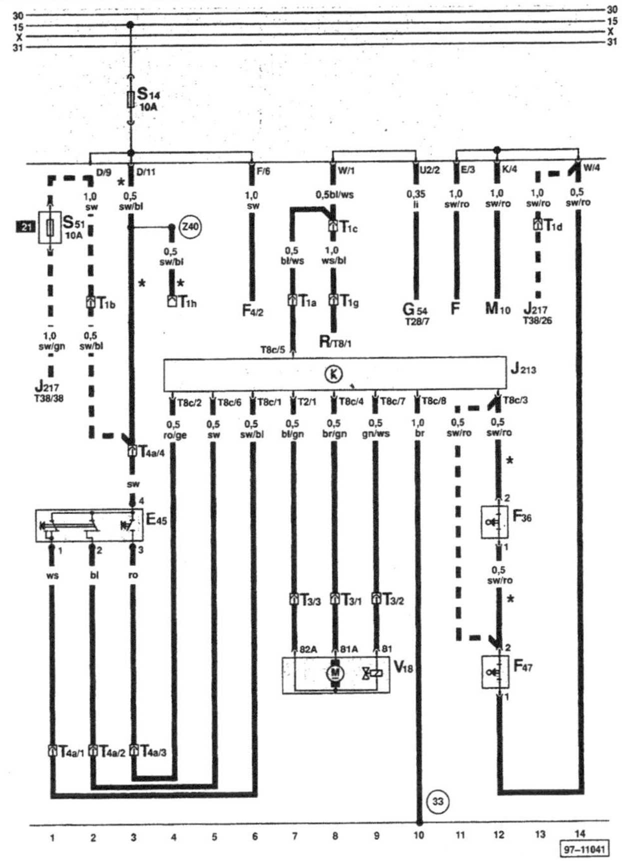

ws = white
sw = black
ro = red
br = brown
gn = green
bl = blue
gr = grey
li = violet
ge = yellow

E45 – Cruise Control Switch
F – Brake Light Switch
F4 – Back-Up Light Switch
F36 – Clutch Vacuum Vent Valve
F47 – Brake Vacuum Vent Valve
G54 – Vehicle Speed Sensor (VSS)
J213 – Cruise Control, Control Module
J217 – Transmission Control Module (TCM)
M10 – Right Brake Light
R – Radio
S51 – Fuse
T1a – Single Connector, behind fuse/relay panel
T1b – Single Connector, behind fuse/relay panel
T1c – Single Connector, behind fuse/relay panel
T1d – Single Connector, behind fuse/relay panel
T1g – Single Connector, behind fuse/relay panel
T1h – Single Connector, behind fuse/relay panel

T2 – Double Connector
T3 – 3-Pin Connector, behind fuse/relay panel
T4a – 4-Pin Connector, behind steering column switch cover
T8c – 8-Pin Connector
T28 – 28-Pin Connector, on instrument cluster
V18 – Cruise Control Vacuum Pump

(33) – Ground connection, behind instrument panel, right
(Z40) – Plus connection (15), in cruise control wiring harness

* – Manual transmission only
▬ ▬ – Automatic transmission only

- **Make-up mirror lights,**
- **Interior light with delay switch and front map/reading light,**
- **Running board lights,**

from April 1992

Relay location
For relay location and multiple connectors see also section "component locations".

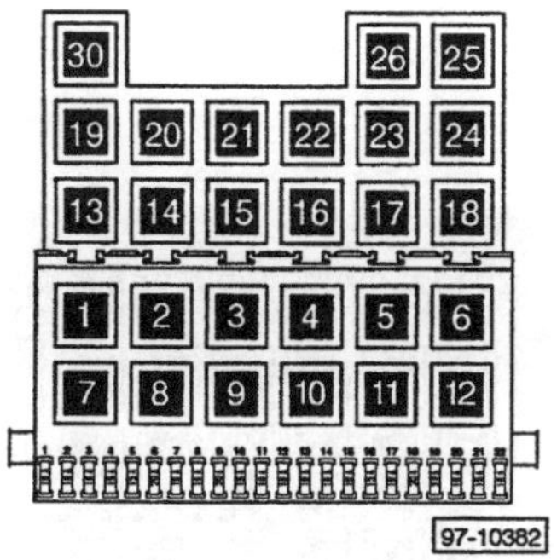

Fuse colors

30A – green
25A – white
20A – yellow
15A – blue
10A – red

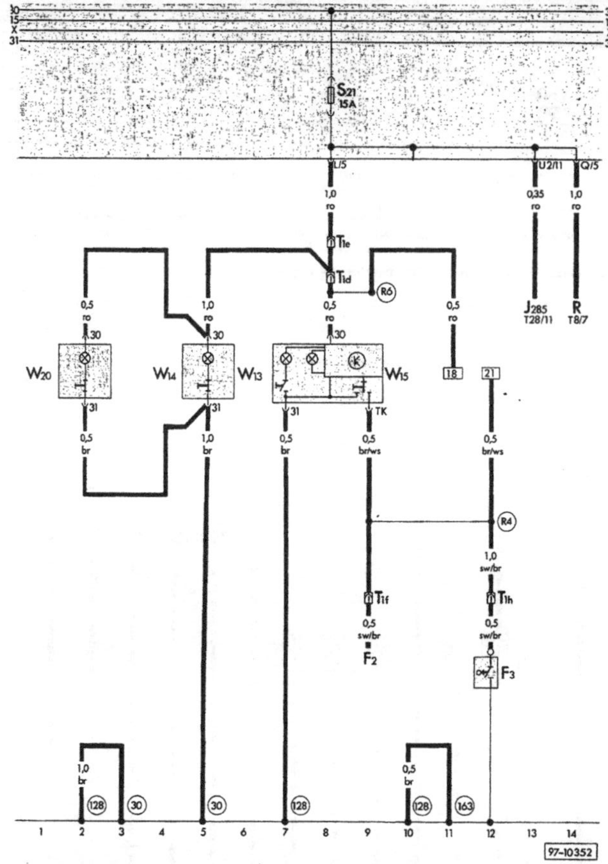

ws = white
sw = black
ro = red
br = brown
gn = green
bl = blue
gr = grey
li = violet
ge = yellow

F2 – Left Front Door Contact Switch
F3 – Right Front Door Contact Switch
J285 – Control Module With Display Unit In Instrument Cluster
R – Radio
T1d – Single Connector, behind fuse/relay panel
T1e – Single Connector, behind fuse/relay panel
T1f – Single Connector, behind fuse/relay panel
T1h – Single Connector, behind instrument panel right
T8 – 8-Pin Connector, on radio
W13 – Right Front Map/Reading Light
W14 – Right Make-Up Mirror Light
W15 – Interior Light With Delay Switch
W20 – Left Make-Up Mirror Light

(30) – Ground connection, beside fuse/relay panel
(128) – Ground connection -1-, in interior light wiring harness
(163) – Ground connection -2-, in interior light wiring harness
(R4) – Wire connection -1-, in interior light/door contact switch wiring harness
(R6) – Plus connection -1-, in interior light wiring harness

Illumination (driver's cabin), front door contact switch

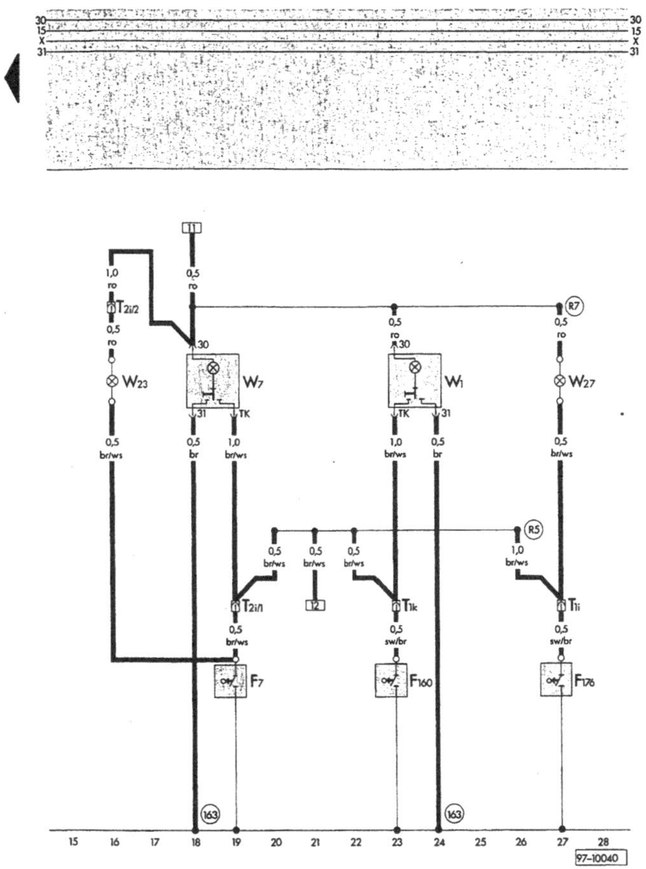

ws = white
sw = black
ro = red
br = brown
gn = green
bl = blue
gr = grey
li = violet
ge = yellow

F7 – Right Rear Side/Sliding Door Contact Switch
F160 – Rear Lid Contact Switch
F176 – Left Rear Sliding Door Contact Switch
T1i – Single Connector, near running board light with delay
T1k – Single Connector, near rear/luggage compartment interior light
T2i – Double Connector, near center interior light
W1 – Rear/Luggage Compartment Interior Light
W7 – Center Interior Light
W23 – Footwell Light With Delay Switch
W29 – Running Board Light With Delay

(163) – Ground connection -2-, in interior light wiring harness
(R5) – Wire connection -2-, in interior light wiring harness
(R7) – Plus connection -2-, in interior light wiring harness

THIS PAGE INTENTIONALLY LEFT

BLANK

Fresh air and recirculating flap,

from April 1992

Relay location
For relay location and multiple connectors see also section "component locations".

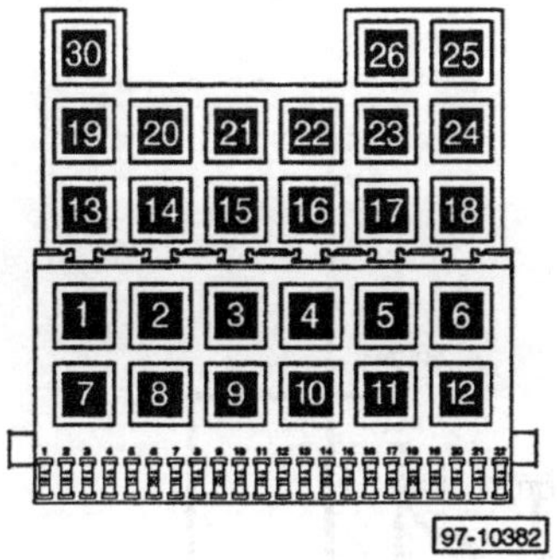

Fuse colors

30A – green
25A – white
20A – yellow
15A – blue
10A – red

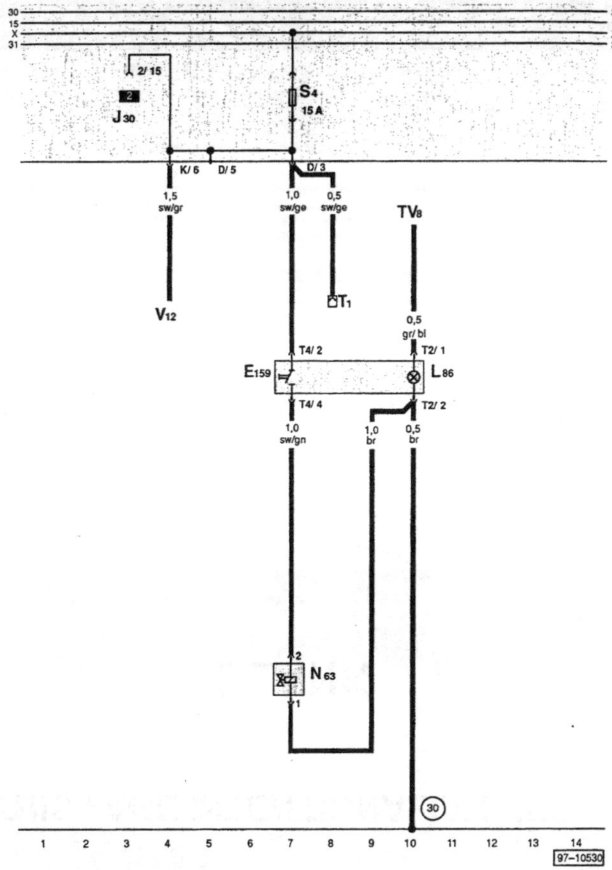

ws = white
sw = black
ro = red
br = brown
gn = green
bl = blue
gr = grey
li = violet
ge = yellow

E159 – Fresh Air-/Recirculating Flap Switch
J30 – Rear Window Wiper/Washer Relay
L86 – Fresh Air-/Recirculating Flap Switch Light
N63 – Fresh Air-/Recirculating Flap Two-Way Valve
T1 – Single Connector, behind fuse/relay panel
T2 – Double Connector
T4 – 4-Pin Connector
TV8 – Terminal 58b Wire Connector
V12 – Rear Window Wiper Motor

(30) – Ground connection, beside fuse/relay panel

Multivan,

from April 1992

Relay location

For relay location and multiple connectors see also section "component locations".

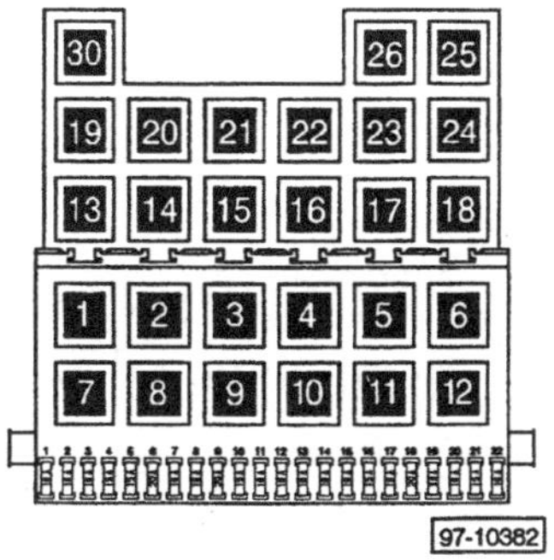

Fuse colors

30A – green
25A – white
20A – yellow
15A – blue
10A – red

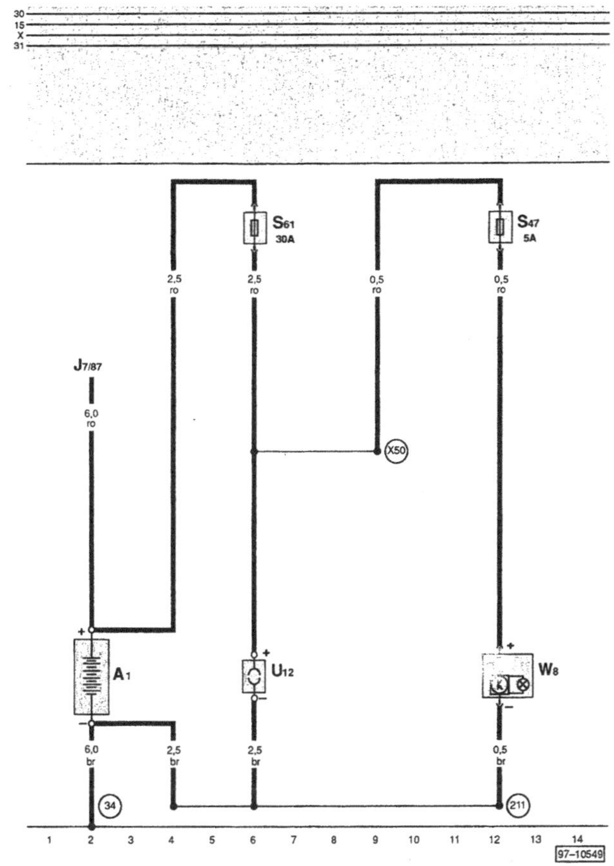

ws = white
sw = black
ro = red
br = brown
gn = green
bl = blue
gr = grey
li = violet
ge = yellow

A1 – Auxiliary Battery, below driver's seat
J7 – Battery Cut Out Relay, below driver's seat
S47 – Reading Interior Light Fuse, below driver's seat
S61 – Refrigerator/Water Pump/Socket Fuse, below driver's seat
U12 – 110 V/220 V Refrigerator Socket (12 V), below driver's seat
W8 – Left Flourescent Light, above folding table

(34) – Ground connection, below driver's seat

(211) – Ground connection, in Multivan wiring harness

(X50) – Plus connection (30), in Multivan wiring harness

Central locking system for vehicles with two sliding doors,

from April 1992

Relay location

For relay location and multiple connectors see also section "component locations".

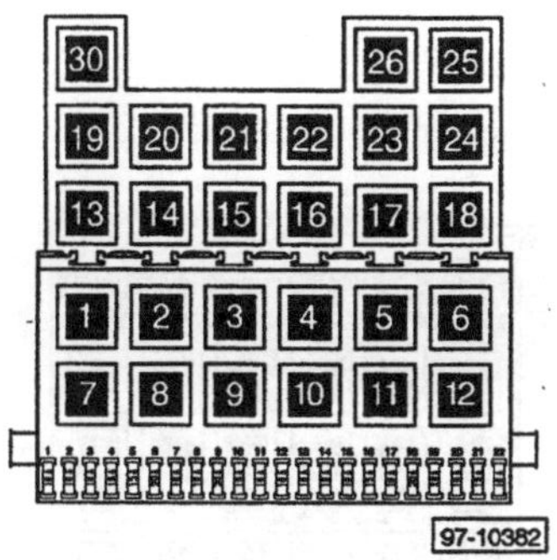

Fuse colors

30A – green
25A – white
20A – yellow
15A – blue
10A – red

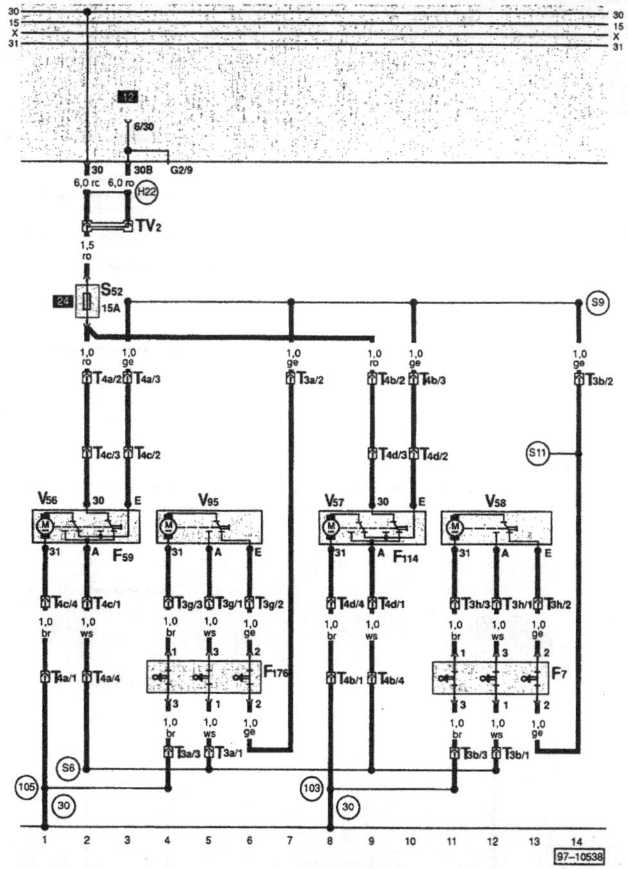

ws = white
sw = black
ro = red
br = brown
gn = green
bl = blue
gr = grey
li = violet
ge = yellow

F7 – Right Rear Side/Sliding Door Contact Switch
F59 – Driver's Door Central Locking System Switch
F114 – Passenger's Door Central Locking System Switch
F176 – Left Rear Side/Sliding Door Contact Switch
S52 – Central Locking System Fuse
T3a – 3-Pin Connector, behind fuse/relay panel
T3b – 3-Pin Connector, behind instrument panel, right
T3g – 3-Pin Connector, in sliding door, left
T3h – 3-Pin Connector, in sliding door, right
T4a – 4-Pin Connector, behind fuse/relay panel
T4b – 4-Pin Connector, behind instr. panel, right
T4c – 4-Pin Connector, in driver's door
T4d – 4-Pin Connector, in passenger's door
TV2 – Terminal 30 (B+) Wire Connector
V56 – Driver's Door Central Locking System Motor
V57 – Passenger's Door Central Locking System Motor
V58 – Sliding Door Central Locking System Motor
V95 – Left Sliding Door Central Locking System Motor
(30) – Ground connection, beside fuse/relay panel
(103) – Ground connection, in central locking system wiring harness
(105) – Ground connection -1-, in central locking system wiring harness
(H22) – Plus connection (30), in terminal 30/30B wiring harness
(S6) – Wire connection (A) -1-, in central locking system wiring harness
(S9) – Wire connection (E) -1-, in central locking system wiring harness
(S11) – Wire connection (E) -3-, in central locking system passenger compartment wiring harness

Central locking system for front doors and sliding doors

Central locking system for double cabin,

from April 1992

Relay location
For relay location and multiple connectors see also section "component locations".

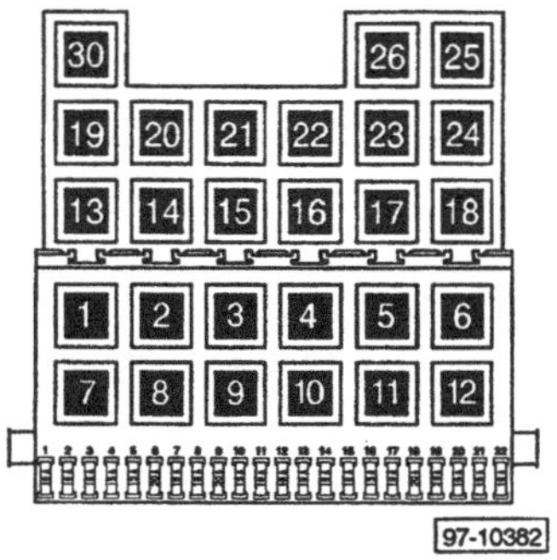

Fuse colors

30A – green
25A – white
20A – yellow
15A – blue
10A – red

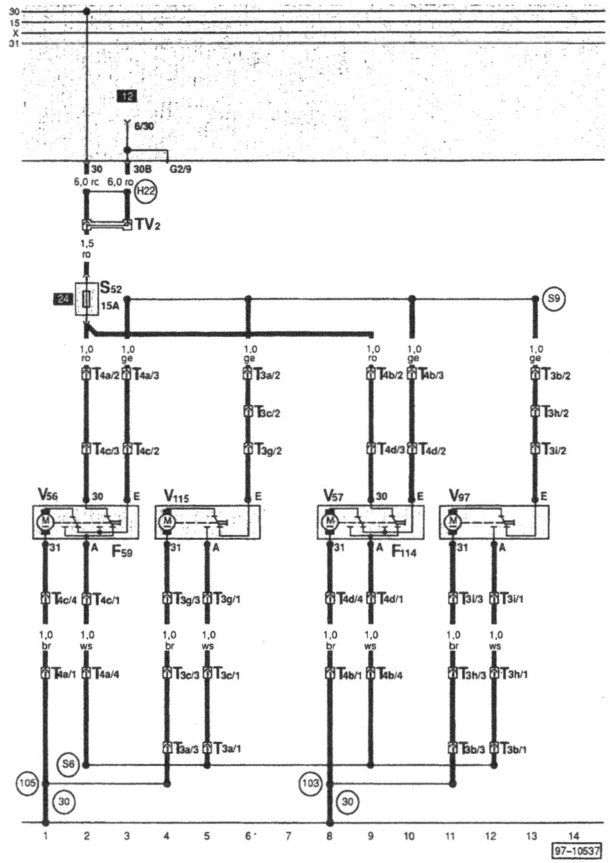

ws = white
sw = black
ro = red
br = brown
gn = green
bl = blue
gr = grey
li = violet
ge = yellow

F59 – Driver's Door Central Locking System Switch
F114 – Passenger's Door Central Locking System Switch
S52 – Central Locking System Fuse
T3a – 3-Pin Connector, behind fuse/relay panel
T3b – 3-Pin Connector, behind instrument panel, right
T3c – 3-Pin Connector, on B–pillar, left
T3g – 3-Pin Connector, in left rear door
T3h – 3-Pin Connector, on B–pillar, right
T3i – 3-Pin Connector, in right rear door
T4a – 4-Pin Connector, behind fuse/relay panel
T4b – 4-Pin Connector, behind instrument panel, right
T4c – 4-Pin Connector, in driver's door
T4d – 4-Pin Connector, in passenger's door
TV2 – Terminal 30 (B+) Wire Connector
V56 – Driver's Door Central Locking System Motor
V57 – Passenger's Door Central Locking System Motor
V97 – Double Cab Right Rear Door Central System System Motor
V115 – Left Rear Central Locking System Motor
(30) – Ground connection -1-, beside fuse/relay panel
(103) – Ground connection -2-, in power window/central locking system/door contact switch wiring harness
(105) – Ground connection -1-, in central locking system wiring harness
(H22) – Plus connection (30), in terminal 30/30B wiring harness
(S6) – Wire connection (A) -1-, in central locking system wiring harness
(S9) – Wire connection (E) -1-, in central locking system wiring harness

2.4 L-Diesel Engine/57 kW, code AAB,

from October 1992

Relay location
For relay location and multiple connectors see also section "component locations".

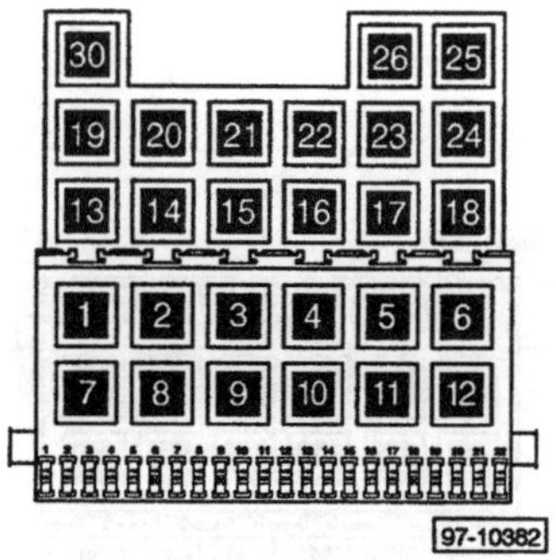

- 2 – Rear Window Wiper/Washer Relay (72)
- 4 – Load Reduction Relay (18)
- 6 – Emergency Flasher Relay (21)
- 8 – Washer/Wiper Intermittent Relay (19)
- 9 – Seat Belt Warning System Relay (4)
- 11 – Dual Horn Relay (53)
- 12 – Glow Plug Relay (102)
- 14 – Engine Coolant Pump Relay (38)
- 20 – Engine Coolant Pump Fuse

Note: Number in parentheses indicates production control number stamped on relay housing.

Fuse colors

30A – green
25A – white
20A – yellow
15A – blue
10A – red

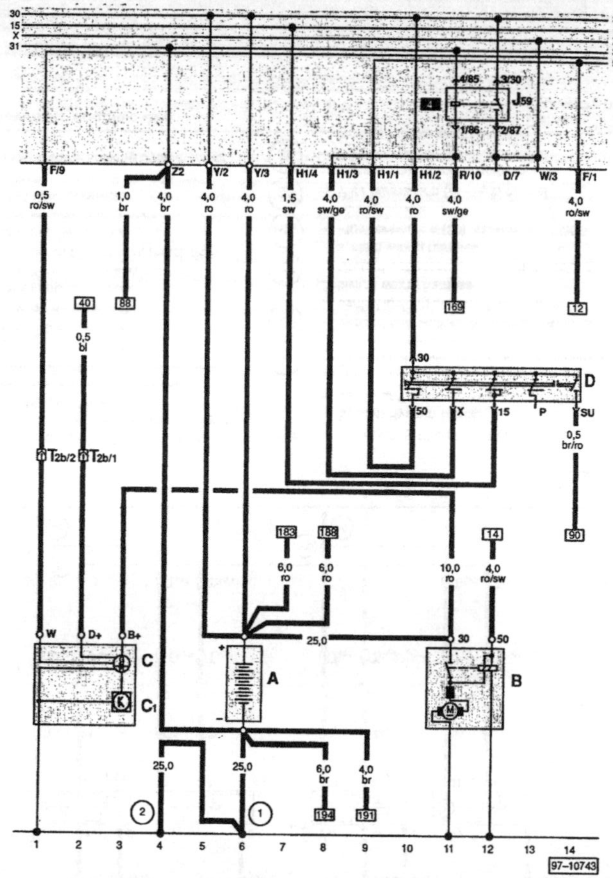

ws = white
sw = black
ro = red
br = brown
gn = green
bl = blue
gr = grey
li = violet
ge = yellow

- A – Battery
- B – Starter
- C – Generator (GEN)
- C1 – Voltage Regulator (VR)
- D – Ignition/Starter Switch
- J59 – Load Reduction Relay
- T2b – Single Connector, near starter
- (1) – Ground strap, battery to body
- (2) – Ground strap, transmission to body

Generator (GEN), battery, starter, ignition/starter switch

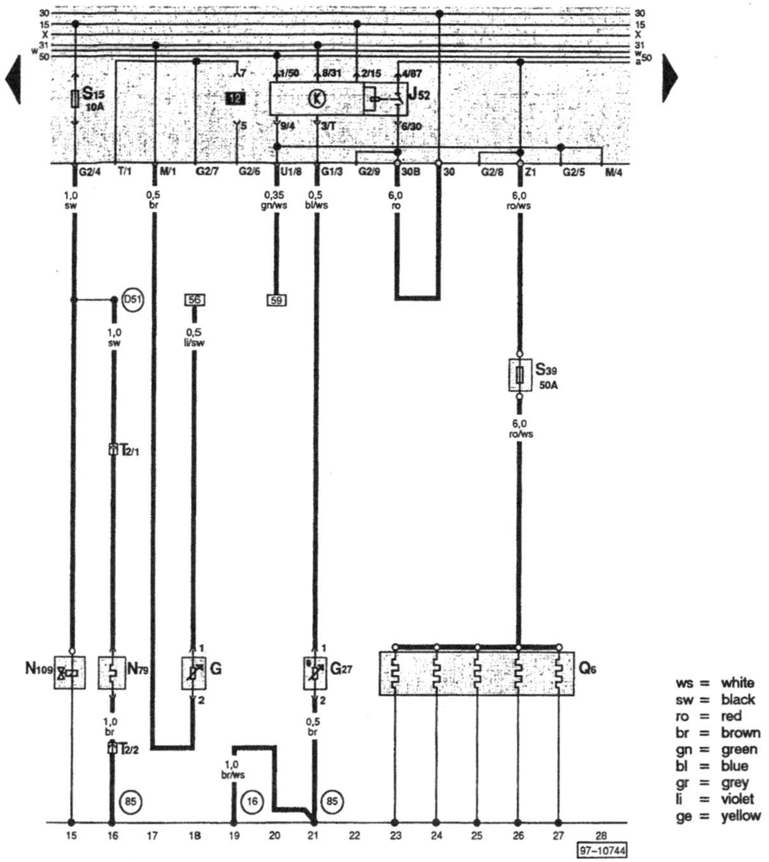

ws	=	white
sw	=	black
ro	=	red
br	=	brown
gn	=	green
bl	=	blue
gr	=	grey
li	=	violet
ge	=	yellow

G – Fuel Level Sensor
G27 – Engine Temperature Sensor
J52 – Glow Plug Relay
N79 – Positive Crankcase Ventilation (PCV) Heating Element
N109 – Fuel Cut-off Valve
Q6 – Engine Glow Plug
S39 – Glow Plug 5oA Fuse Strip
T2 – Double Connector, near heating element

(16) – Ground connection, -1-, on valve cover

(85) – Ground connection -1-, in engine compartment wiring harness

(D51) – Plus connection (15), in engine compartment wiring harness

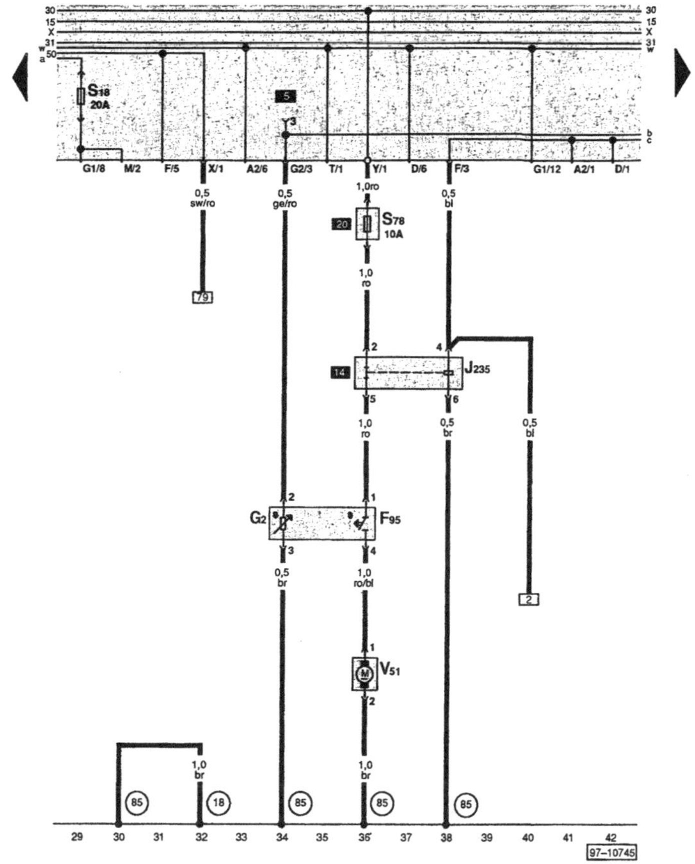

ws	=	white
sw	=	black
ro	=	red
br	=	brown
gn	=	green
bl	=	blue
gr	=	grey
li	=	violet
ge	=	yellow

F95 – After-Run Coolant Pump Thermal Switch*
G2 – Engine Coolant Temperature (ECT) Sensor
J235 – Engine Coolant Pump Relay
S78 – Engine Coolant Pump Circuit Breaker
V51 – After-Run Coolant Pump

(18) – Ground connection, on engine block

(85) – Ground connection -1-, in engine compartment wiring harness

* – Switch value, see engine repair manual

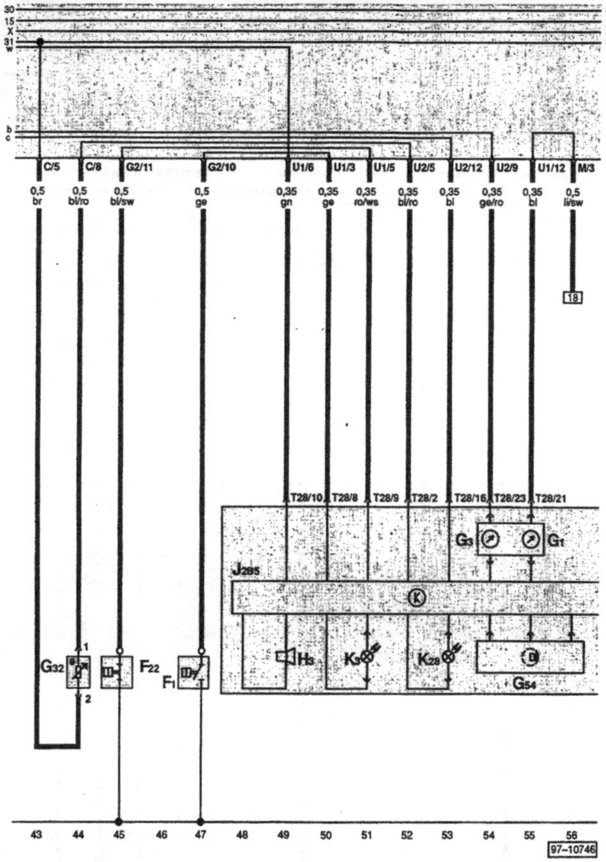

ws = white
sw = black
ro = red
br = brown
gn = green
bl = blue
gr = grey
li = violet
ge = yellow

F1 – Oil Pressure Switch*
F22 – 0.3bar Oil Pressure Switch
G1 – Fuel Gauge
G3 – Engine Coolant Temperature (ECT) Gauge
G32 – Engine Coolant Level (ECL) Sensor
G54 – Vehicle Speed Sensor (VSS)
H3 – Warning Buzzer
J285 – Control Module With Display Unit In Instrument Cluster
K3 – Oil Pressure Warning Light
K28 – Engine Coolant Level/Temperature (ECL/ECT) Warning Light
T28 – 28-Pin Connector

* – Switch value, see engine repair manual

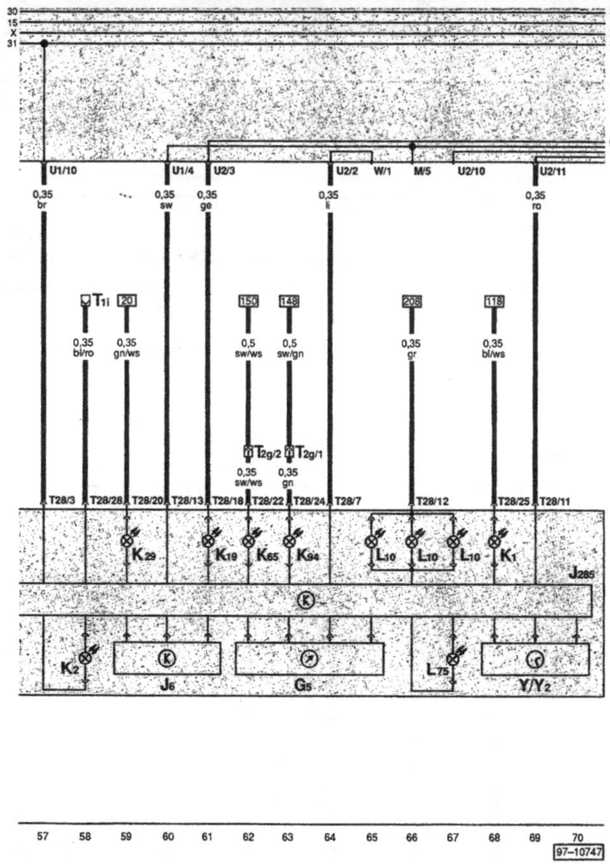

ws = white
sw = black
ro = red
br = brown
gn = green
bl = blue
gr = grey
li = violet
ge = yellow

G5 – Tachometer
J6 – Voltage Stabilizer
J285 – Control Module With Display Unit In Instrument Cluster
K1 – Headlight High Beam Indicator Light
K2 – Generator (GEN) Warning Light
K19 – Seat Belt Warning Light
K29 – Glow Plug Indicator Light
K65 – Left Turn Signal Indicator Light
K94 – Right Turn Signal Indicator Light
L10 – Instrument Cluster Light
L75 – Digital Display Light
T1i – Single Connector, near fuse/relay panel
T2g – Double Connector, behind fuse/relay panel
T28 – 28-Pin Connector

Y – Analog Clock
Y2 – Digital clock

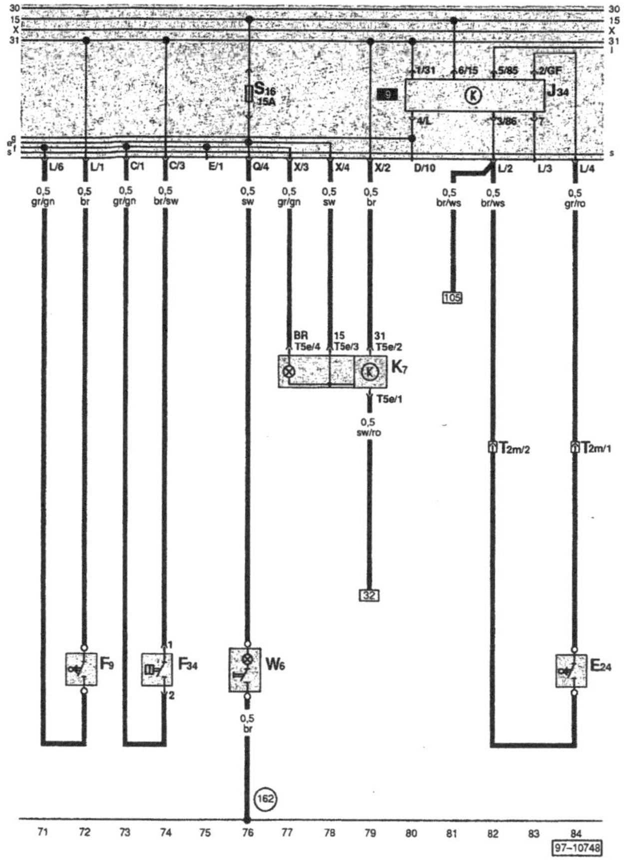

ws = white
sw = black
ro = red
br = brown
gn = green
bl = blue
gr = grey
li = violet
ge = yellow

E24 – Left Seat Belt Switch
F9 – Parking Brake Warning Light Switch
F34 – Brake Fluid Level Warning Switch
J34 – Seat Belt Warning System Relay
K7 – Brake and Parking Brake Warning Light
T2m – Double Connector, near seat belt switch
T5e – 5-Pin Connector
W6 – Glove Compartment Light

(162) – Ground connection, in blower motor wiring harness

Brake and parking brake warning light, seat belt warning system

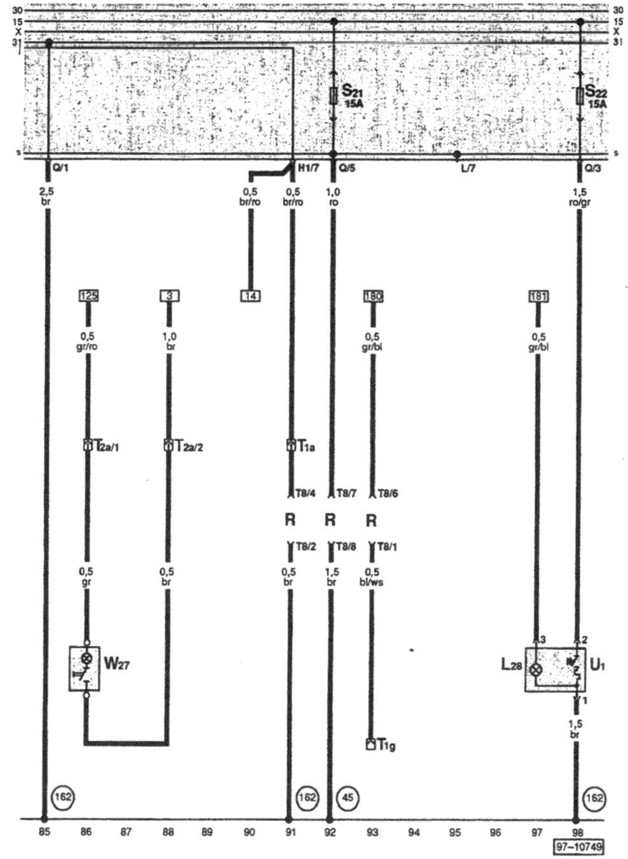

ws = white
sw = black
ro = red
br = brown
gn = green
bl = blue
gr = grey
li = violet
ge = yellow

L28 – Cigarette Lighter Light
R – Radio
T1a – Single Connector, behind fuse/relay panel
T1g – Single Connector, behind fuse/relay panel
T2o – Double Connector, near battery
T8 – 8-Pin Connector
U1 – Cigarette Lighter
W27 – Engine Compartment Light

(45) – Ground connection, behind instrument panel, center
(162) – Ground connection, in blower motor wiring harness

Radio connection, cigarette lighter, engine compartment light

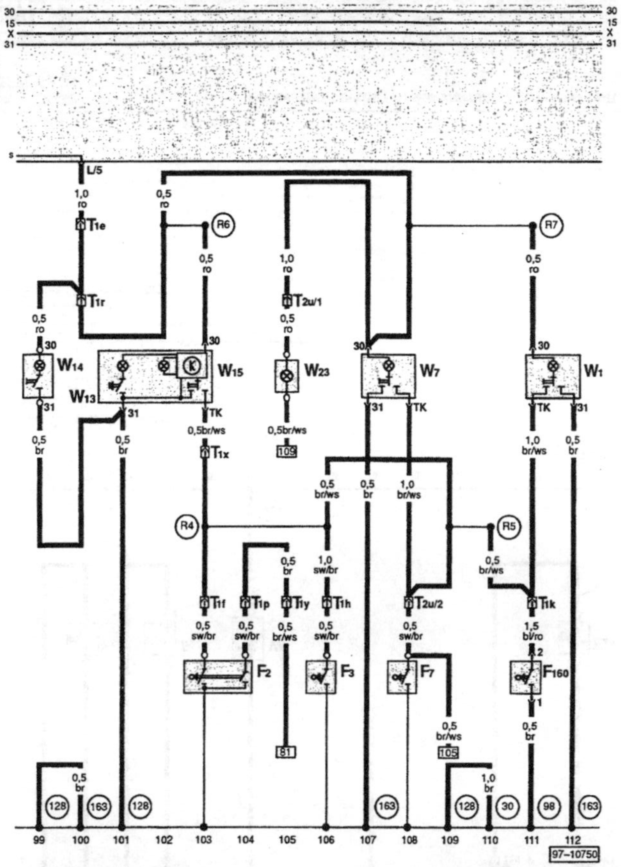

ws = white
sw = black
ro = red
br = brown
gn = green
bl = blue
gr = grey
li = violet
ge = yellow

F2 – Left Front Door Contact Switch
F3 – Right Front Door Contact Switch
F7 – Right Rear Side/Sliding Door Contact Switch
F160 – Rear Lid Contact Switch
T1e – Single Connector, behind fuse/relay panel
T1f – Single Connector, behind fuse/relay panel
T1h – Single Connector, behind instrument panel, right
T1k – Single Connector, near taillight, right
T1p – Single Connector, behind fuse/relay panel
T1r – Single Connector, behind fuse/relay panel
T1x – Single Connector, behind fuse/relay panel
T1y – Single Connector, behind fuse/relay panel
T2u – Double Connector, near center interior light
W1 – Rear/Luggage Compartment Interior Light
W7 – Center Interior Light
W13 – Right Front Map/Reading Light
W14 – Right Make-Up Mirror Light
W15 – Interior Light With Delay Switch
W23 – Footwell Light With Delay Switch
(30) – Ground connection, -1-, beside fuse/relay panel
(98) – Ground connection, in rear lid wiring harness
(128) – Ground connection -1-, in interior light wiring harness
(163) – Ground connection -2-, in interior light wiring harness
(R4) – Wire connection -1-, in interior light/door contact switch wiring harness
(R5) – Wire connection -2-, in interior light wiring harness
(R6) – Plus connection -1-, in interior light wiring harness
(R7) – Plus connection -2-, in interior light wiring harness

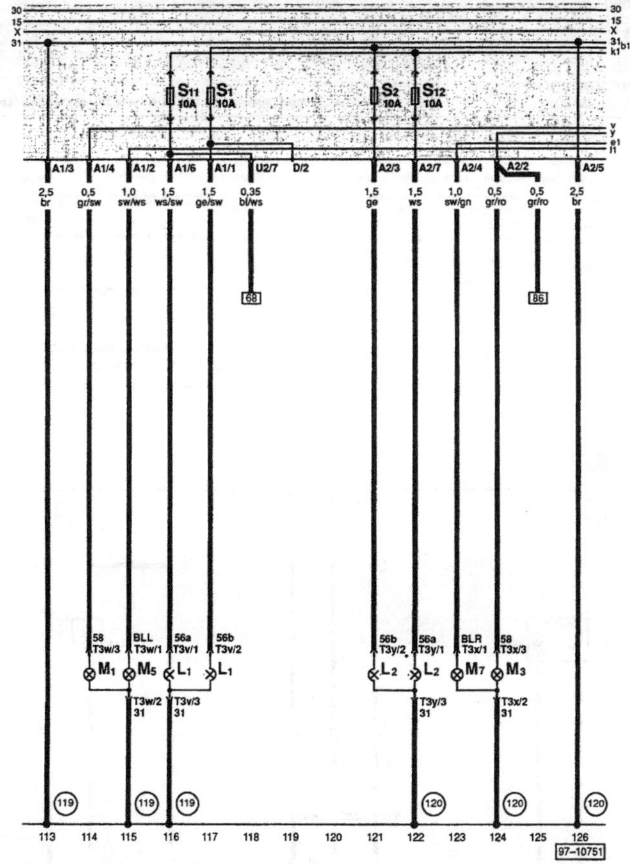

ws = white
sw = black
ro = red
br = brown
gn = green
bl = blue
gr = grey
li = violet
ge = yellow

L1 – Left Headlight
L2 – Right Headlight
M1 – Left Parking Light
M3 – Right Parking Light
M5 – Left Front Turn Signal Light
M7 – Right Front Turn Signal Light
T3v – 3-Pin Connector
T3w – 3-Pin Connector
T3x – 3-Pin Connector
T3y – 3-Pin Connector
(119) – Ground connection -1-, in headlight wiring harness
(120) – Ground connection -2-, in headlight wiring harness

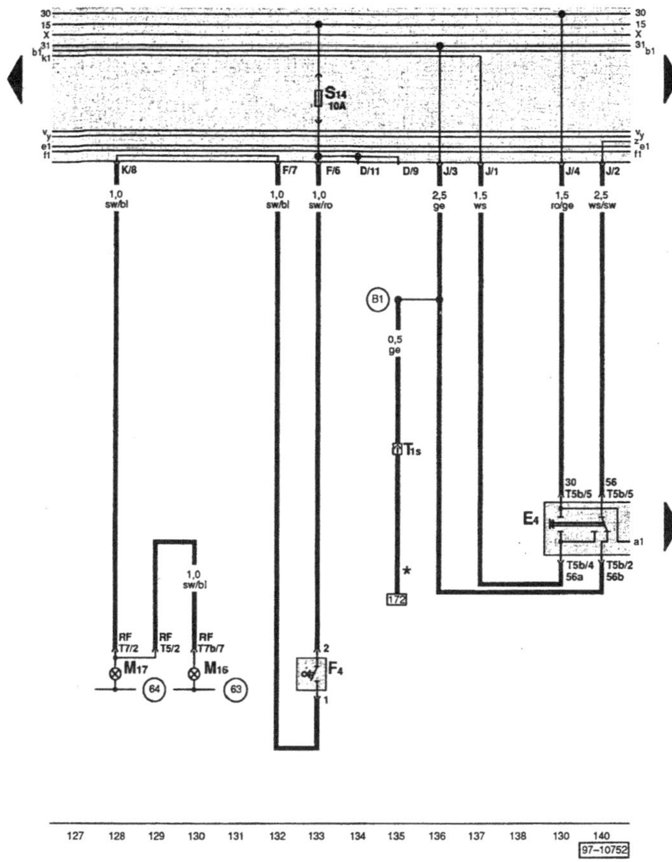

ws = white
sw = black
ro = red
br = brown
gn = green
bl = blue
gr = grey
li = violet
ge = yellow

E4 – Headlight Dimmer/Flasher Switch
F4 – Back-Up Light Switch (manual transmission)
M16 – Left Back-Up Light
M17 – Right Back-Up Light
T1s – Single Connector, behind fuse/relay panel
T5 – 5-Pin Connector
T5b – 5-Pin Connector, behind steering column switch cover
T7 – 7-Pin Connector
T7b – 7-Pin Connector

(63) – Ground connection, on left taillight bulb holder
(64) – Ground connection, on right taillight bulb holder

(B1) – Wire connection (56b), in steering column switch wiring harness

* – Resistance wire

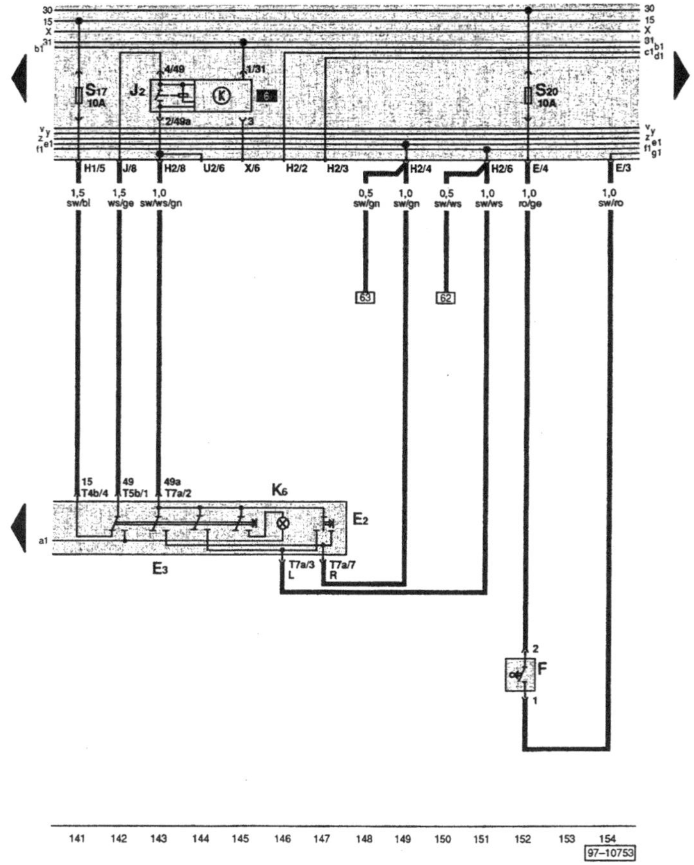

ws = white
sw = black
ro = red
br = brown
gn = green
bl = blue
gr = grey
li = violet
ge = yellow

E2 – Turn Signal Switch
E3 – Emergency Flasher Switch
F – Brake Light Switch
J2 – Emergency Flasher Relay
K6 – Emergency Flasher Warning Light
T4b – 4-Pin Connector, behind steering column switch cover
T5b – 5-Pin Connector, behind steering column switch cover
T7a – 7-Pin Connector, behind steering column switch cover

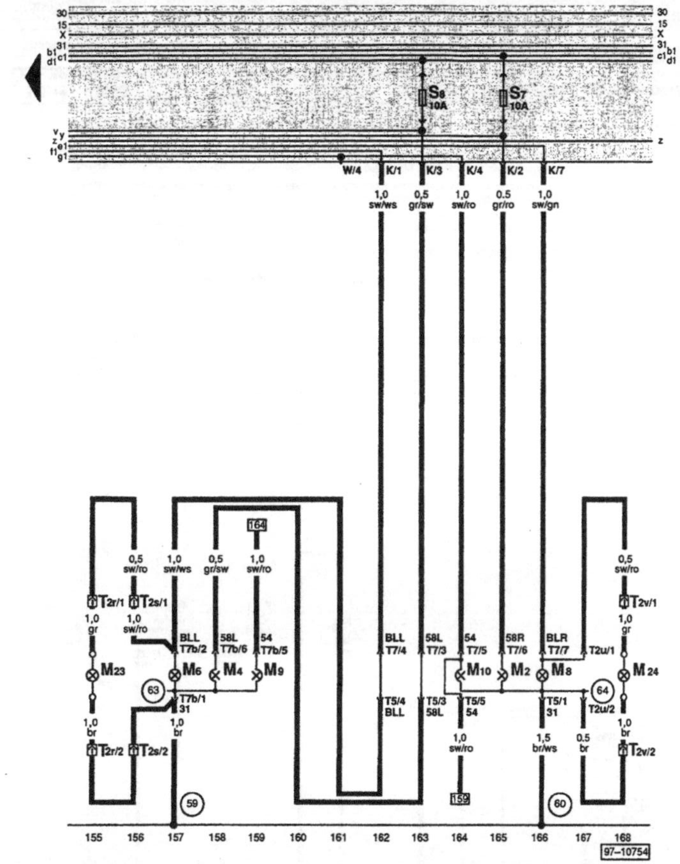

ws	=	white
sw	=	black
ro	=	red
br	=	brown
gn	=	green
bl	=	blue
gr	=	grey
li	=	violet
ge	=	yellow

M2 – Right Taillight
M4 – Left Taillight
M6 – Left Rear Turn Signal Light
M8 – Right Rear Turn Signal Light
M9 – Left Brake Light
M10 – Right Brake Light
M23 – Left Roof Turn Signal Light
M24 – Right Roof Turn Signal Light
T2r – Double Connector, near left roof turn signal
T2s – Double Connector, near left taillight
T2u – Double Connector, on Right Taillight
T2v – Double Connector, near right roof turn signal
T5 – 5-Pin Connector, on Right Taillight
T7 – 7-Pin Connector, on Right Taillight
T7b – 7-Pin Connector, on Left Taillight
(59) – Ground connection, near taillight, left
(60) – Ground connection, near taillight, right
(63) – Ground connection, on left taillight bulb holder
(64) – Ground connection, on right taillight bulb holder
* – Different wire routing for flatbed models

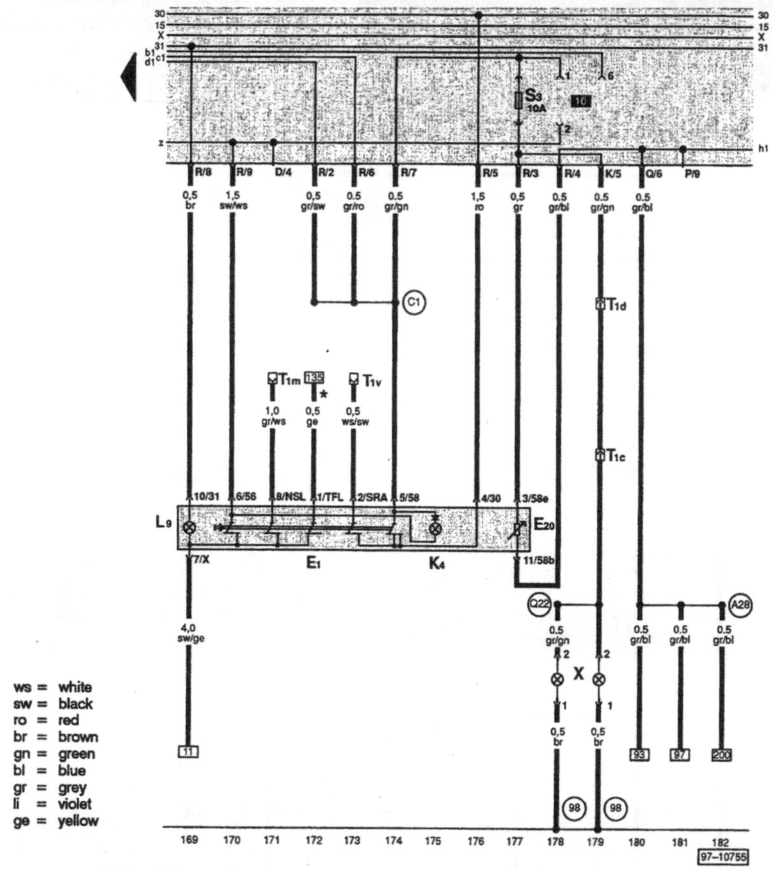

ws	=	white
sw	=	black
ro	=	red
br	=	brown
gn	=	green
bl	=	blue
gr	=	grey
li	=	violet
ge	=	yellow

E1 – Light Switch
E20 – Instrument Panel Light Dimmer Switch
K4 – Park Light Indicator Light
L9 – Headlight Switch Light
T1c – Single Connector, on Left Taillight
T1d – Single Connector, near right taillight
T1m – Single Connector, on Right Taillight
T1v – Single Connector, behind fuse/relay panel
X – License Plate Light
(98) – Ground connection, in rear lid wiring harness
(A28) – Plus connection (58b), in fresh air blower wiring harness
(C1) – Plus connection (58L), in left front wiring harness
(Q22) – Wire connection -1-, in rear lid wiring harness
* – Resistance wire

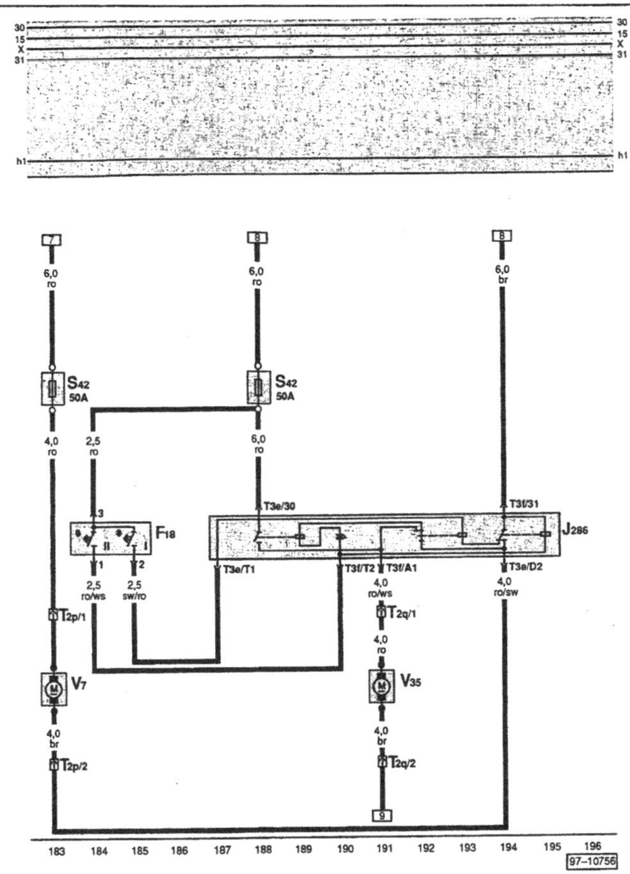

ws = white
sw = black
ro = red
br = brown
gn = green
bl = blue
gr = grey
li = violet
ge = yellow

F18 – Coolant Fan Control (FC) Thermal Switch
J286 – Series/Parallel Coolant Fan Control (FC) Relay (V7, V35)
S42 – Coolant Fan Fuse
T2p – Double Connector, in engine compartment, front
T2q – Double Connector, in engine compartment, front
T3e – 3-Pin Connector
T3f – 3-Pin Connector
V7 – Coolant Fan
V35 – Right Coolant Fan

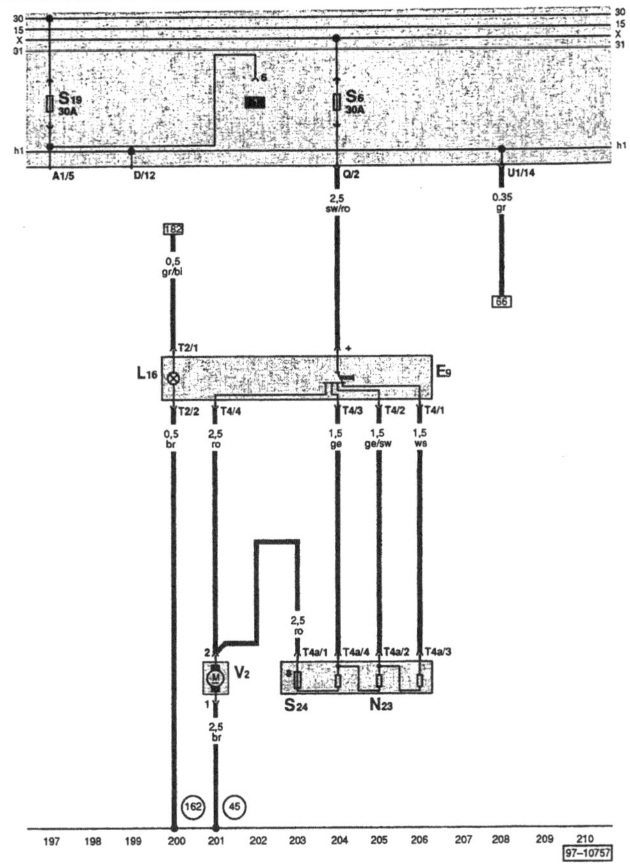

ws = white
sw = black
ro = red
br = brown
gn = green
bl = blue
gr = grey
li = violet
ge = yellow

E9 – Fresh Air Blower Switch
L16 – Fresh Air Control Lever Light
N23 – Fresh Air Blower Series Resistance
S24 – Heater, A/C Clutch Fuse
T2 – Double Connector
T4 – 4-Pin Connector
T4a – 4-Pin Connector
V2 – Fresh Air Blower

(45) – Ground connection, behind instrument panel, center
(162) – Ground connection, in blower motor wiring harness

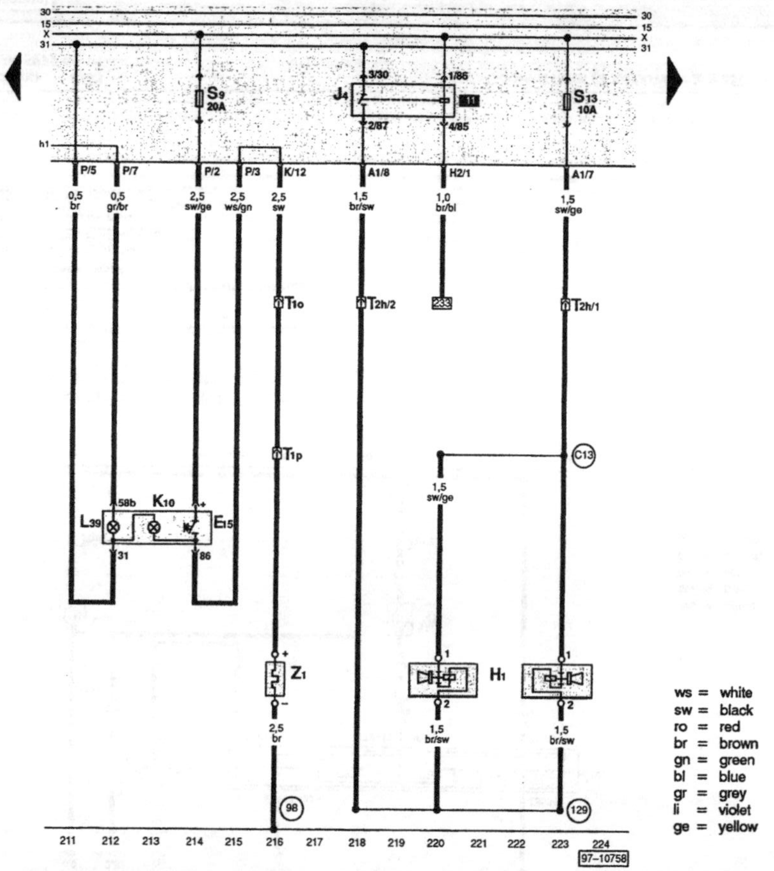

E15 – Rear Window Defogger Switch
H1 – Dual Horns
J4 – Dual Horn Relay
K10 – Rear Window Defogger Indicator Light
L39 – Rear Window Defogger Switch Light
T1o – Single Connector, near right taillight
T1p – Single Connector (deleted)
T2h – Double Connector, in engine compartment, left front
Z1 – Rear Window Defogger/Heat Element

(98) – Ground connection, in rear lid wiring harness
(129) – Ground connection, in dual tone horn wiring harness

(C13) – Plus connection, in dual tone horn wiring harness

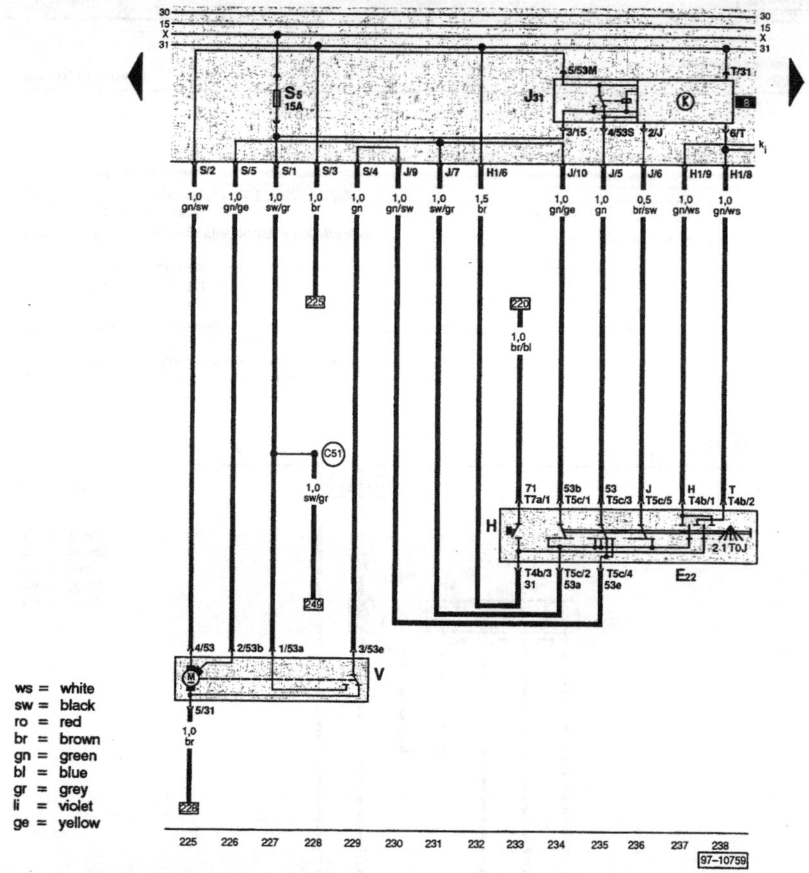

E22 – Windshield Wiper/Washer Switch
H – Horn Button
J31 – Washer/Wiper Intermittent Relay
T4b – 4-Pin Connector
T5c – 5-Pin Connector
T7a – 7-Pin Connector
V – Windshield Wiper Motor

(C51) – Plus connection (x), in wiper motor wiring harness

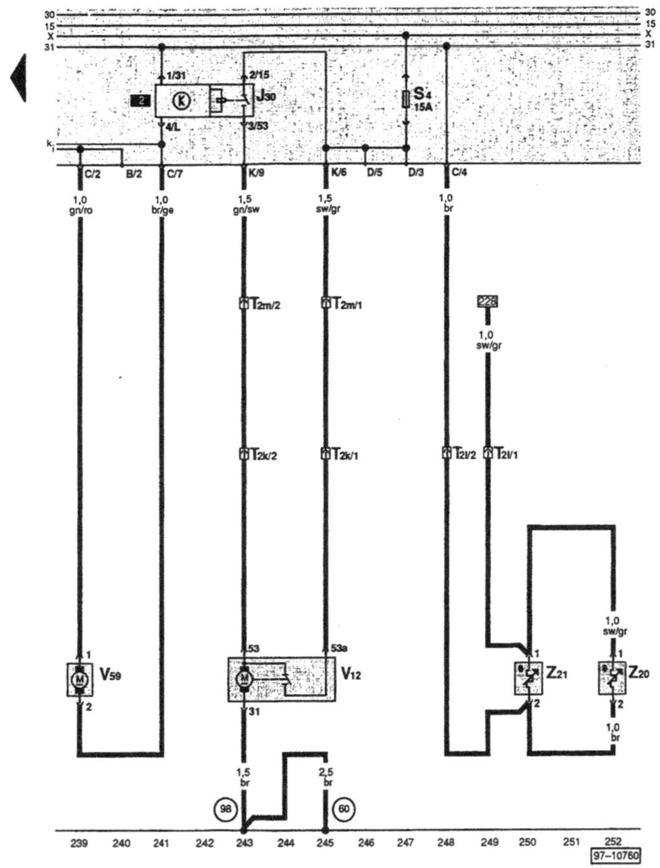

ws = white
sw = black
ro = red
br = brown
gn = green
bl = blue
gr = grey
li = violet
ge = yellow

J30 – Rear Window Wiper/Washer Relay
T2k – Double Connector, near right taillight
T2l – Double Connector, near right firewall
T2m – Double Connector (deleted)
V12 – Rear Window Wiper Motor
V59 – Windshield and Rear Window Washer Pump
Z20 – Left Washer Nozzle Heater
Z21 – Right Washer Nozzle Heater

(60) – Ground connection, near right taillight

(98) – Ground connection, in rear lid wiring harness

THIS PAGE INTENTIONALLY LEFT

BLANK

2.5 L-Engine – Digifant Multiport Fuel Injection (MFI)/81kW, code AAF,

from August 1992

Relay location
For relay location and multiple connectors see also section "component locations".

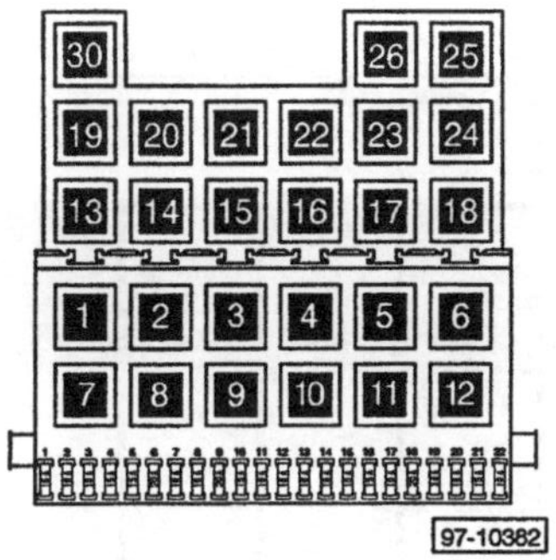

- 2 – Rear Window Wiper/Washer Relay (72)
- 3 – Digifant Engine Control Module (ECM) Power Supply Relay (30)
- 4 – Load Reduction Relay (18)
- 6 – Emergency Flasher Relay (21)
- 8 – Washer/Wiper Intermittent Relay (19)
- 9 – Seat Belt Warning System Relay (4)
- 11 – Dual Horn Relay (53)
- 12 – Fuel Pump (FP) Relay (80)
- 14 – Engine Coolant Pump Relay (38)
- 20 – Park/Neutral Position (PNP) Relay (150)
- 21 – Automatic Transmission Fuse

Note: Number in parentheses indicates production control number stamped on relay housing.

Fuse colors

30A – green
25A – white
20A – yellow
15A – blue
10A – red

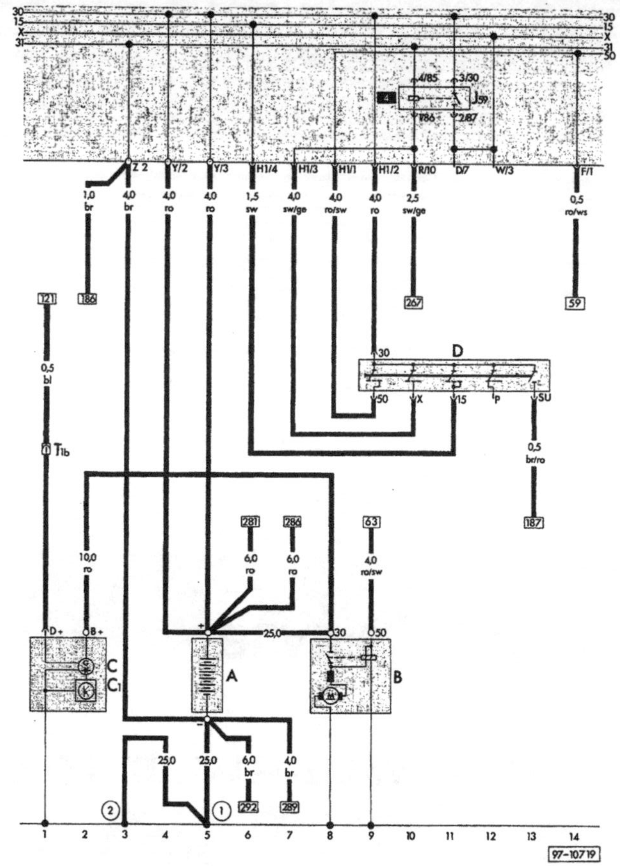

ws = white
sw = black
ro = red
br = brown
gn = green
bl = blue
gr = grey
li = violet
ge = yellow

- A – Battery
- B – Starter
- C – Generator (GEN)
- C1 – Voltage Regulator (VR)
- D – Ignition/Starter Switch
- J59 – Load Reduction Relay
- T1b – Single Connector, near starter
- (1) – Ground strap, battery to body
- (2) – Ground strap, transmission to body

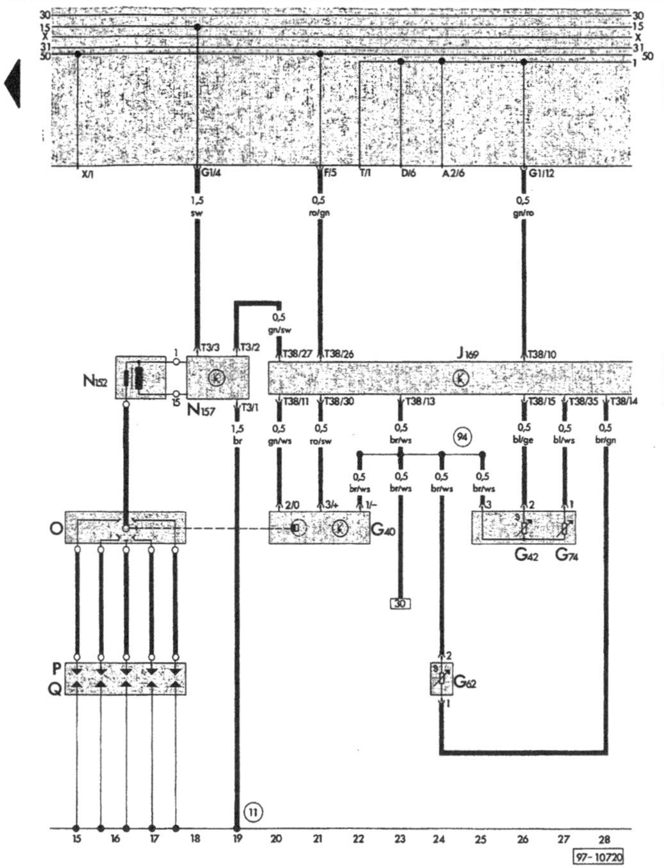

ws = white
sw = black
ro = red
br = brown
gn = green
bl = blue
gr = grey
li = violet
ge = yellow

G40 – Camshaft Position (CMP) Sensor
G42 – Intake Air Temperature (IAT) Sensor (00523–2322)
G62 – Engine Coolant Temperature (ECT) Sensor (0522–2312)
G74 – CO Fuel Trim (FT) Potentiometer (00521–2242)
J169 – Digifant Engine Control Module (ECM), near battery
N152 – Ignition Coil
N157 – Ignition Coil Power Output Stage
O – Ignition Distributor
P – Spark Plug Connectors
Q – Spark Plugs
T3 – 3-Pin Connector
T38 – 38-Pin Connector

(11) – Ground connection, in battery box
(94) – Ground connection -1-, in Digifant Multiport Fuel Injection (MFI) wiring harness

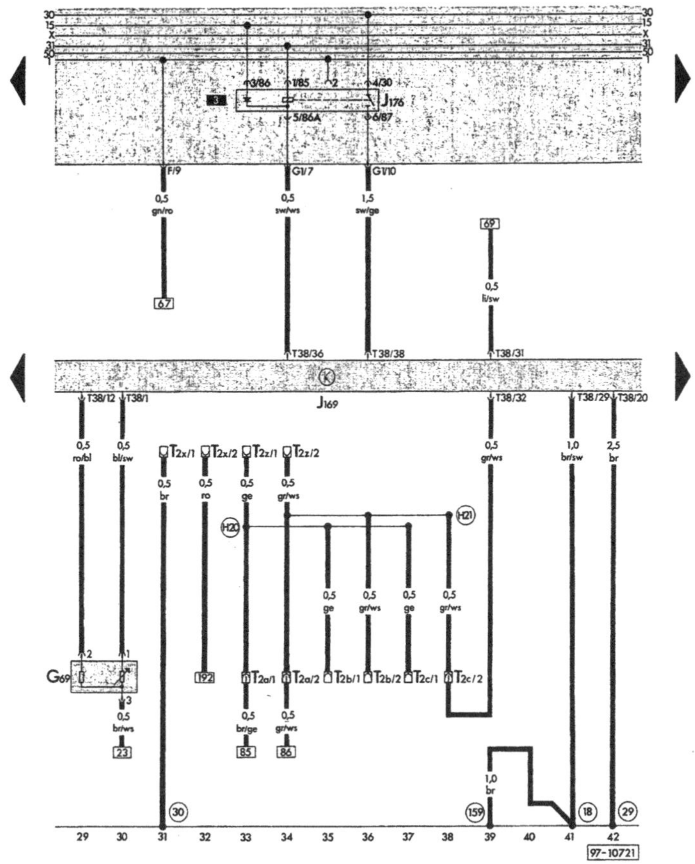

ws = white
sw = black
ro = red
br = brown
gn = green
bl = blue
gr = grey
li = violet
ge = yellow

G69 – Throttle Position (TP) Sensor (00518–2212)
J169 – Digifant Engine Control Module (ECM), near battery
J176 – Digifant Engine Control Module (ECM) Power Supply Relay
T2a – Double Connector, behind fuse/relay panel
T2b – Double Connector, behind fuse/relay panel*
T2c – Double Connector, behind fuse/relay panel
T2x – Double Connector, –black–, behind fuse/relay panel cover (OBD)
T2z – Double Connector, –white–, behind fuse/relay panel cover (OBD)
T38 – 38-Pin Connector

(18) – Ground connection, on engine block
(29) – Ground connection, near battery
(30) – Ground connection, -1-, beside fuse/relay panel
(159) – Ground connection -2-, in Digifant Multiport Fuel Injection (MFI) wiring harness
(H20) – Wire connection -1-, in On Board Diagnostic (OBD) wiring harness
(H21) – Wire connection -2-, in On Board Diagnostic (OBD) wiring harness
* With further On Board Diagnostic (OBD) system only

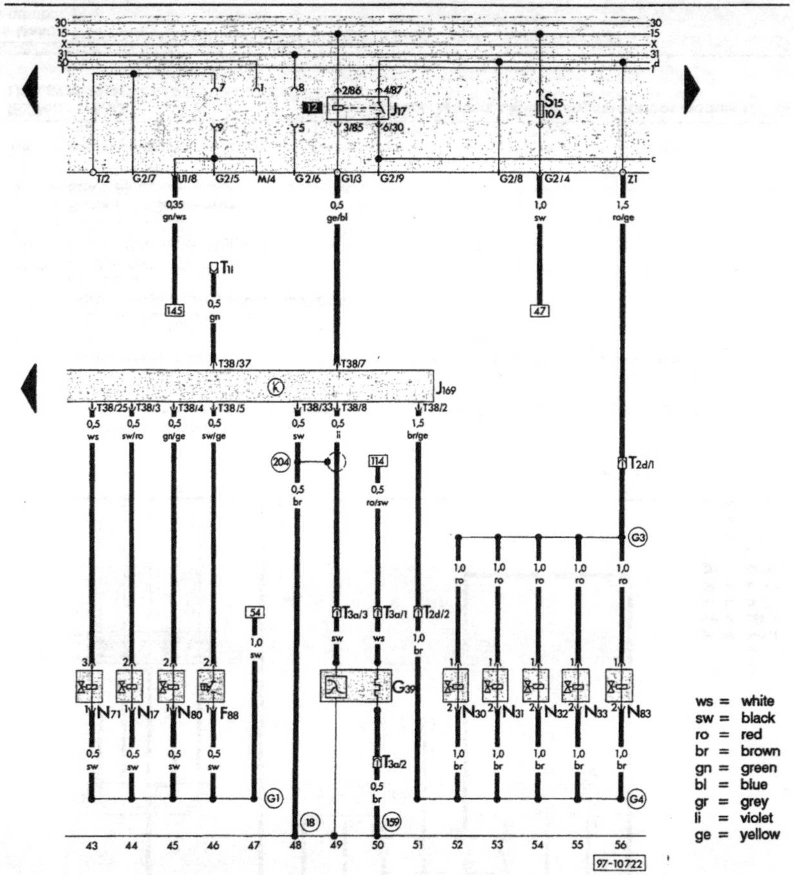

ws = white
sw = black
ro = red
br = brown
gn = green
bl = blue
gr = grey
li = violet
ge = yellow

F88 – Power Steering Pressure (PSP) Switch
G39 – Heated Oxygen Sensor (HO2S) (00525–2342)
J17 – Fuel Pump (FP) Relay
J169 – Digifant Engine Control Module (ECM), near battery
N17 – Cold Start Injector
N30 – Cylinder 1 Fuel Injector
N31 – Cylinder 2 Fuel Injector
N32 – Cylinder 3 Fuel Injector
N33 – Cylinder 4 Fuel Injector
N71 – Idle Air Control (IAC) Valve
N80 – Evaporative Emission (EVAP) Canister Purge Regulator Valve
N83 – Cylinder 5 Fuel Injector
T1l – Single Connector, behind fuse/relay panel (for A/C)
T2d – Double Connector, in engine compartment, right
T3a – 3-Pin Connector, near exhaust manifold
T38 – 38-Pin Connector

(18) – Ground connection, on engine block
(159) – Ground connection -2-, in Digifant Multiport Fuel Injection (MFI) wiring harness
(204) – Ground connection -3-, in Digifant Multiport Fuel Injection (MFI) wiring harness
(G1) – Plus connection, in Digifant Multiport Fuel Injection (MFI) wiring harness
(G3) – Plus connection, in injector wiring harness
(G4) – Wire connection, in injector wiring harness

Digifant engine control module (ECM), heated oxygen sensor (HO2S), injectors

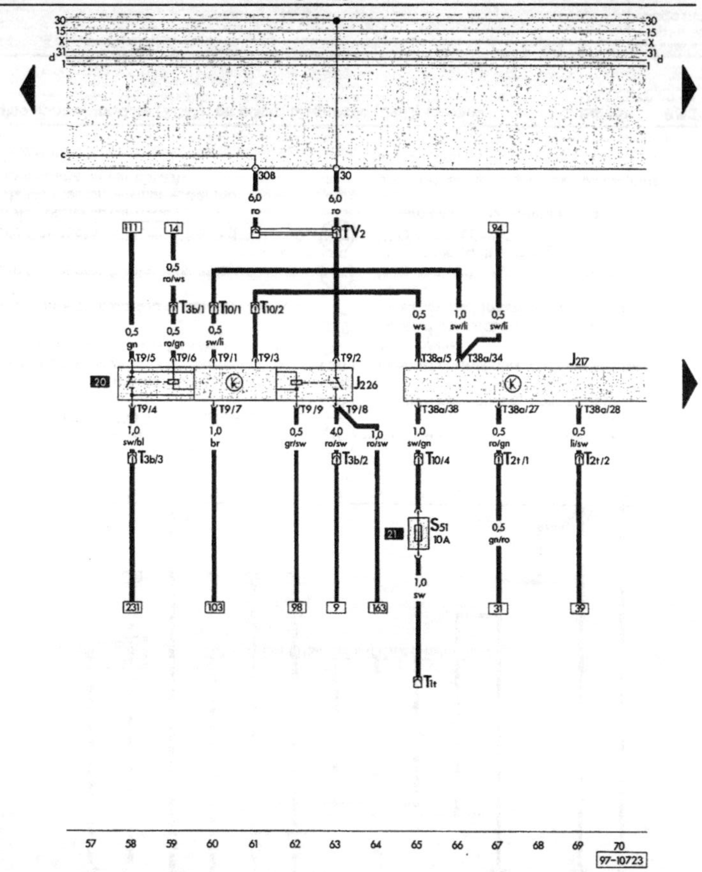

ws = white
sw = black
ro = red
br = brown
gn = green
bl = blue
gr = grey
li = violet
ge = yellow

J217 – Transmission Control Module (TCM), near A-pillar, right
J226 – Park/Neutral Position (PNP) Relay*
S51 – Fuse
T1t – Single Connector, behind fuse/relay panel (cruise control system)
T2t – Double Connector, in engine compartment, near transmission
T3b – 3-Pin Connector, behind fuse/relay panel
T9 – 9-Pin Connector
T10 – 10-Pin Connector, behind instrument panel, right
T38a – 38-Pin Connector
TV2 – Terminal 30 (B+) Wire Connector, behind – fuse/relay panel

* – Automatic transmission only

Transmission control module (TCM), park/neutral position (PNP) relay

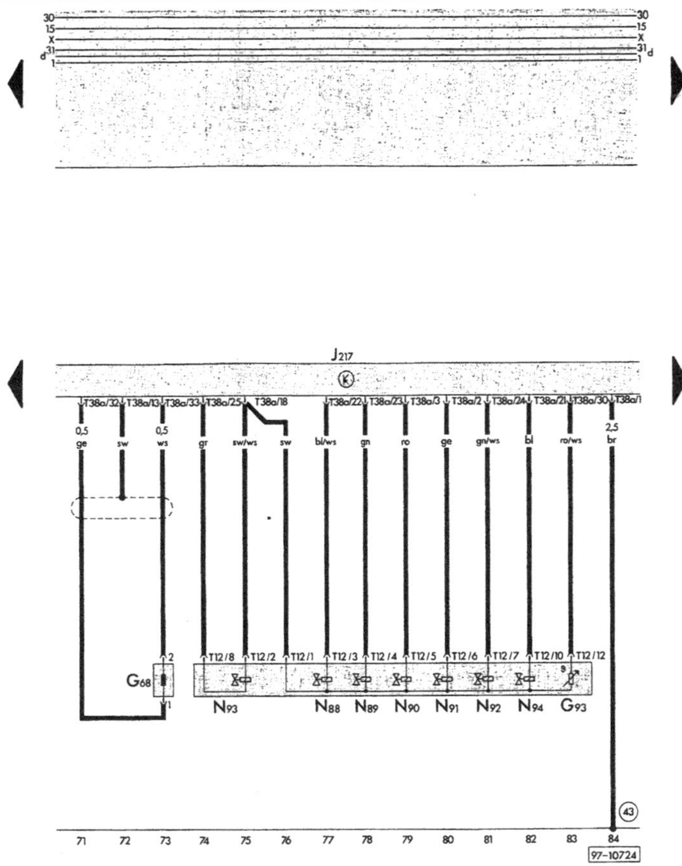

ws = white
sw = black
ro = red
br = brown
gn = green
bl = blue
gr = grey
li = violet
ge = yellow

G68 – Vehicle Speed Sensor (VSS) (00281–1231)
G93 – Transmission Fluid Temperature Sensor (00300–1333)
J217 – Transmission Control Module (TCM), near A-pillar, right
N88 – Solenoid Valve 1 (00258–1113)
N89 – Solenoid Valve 2 (00260–1121)
N90 – Solenoid Valve 3 (00262–1123)
N91 – Solenoid Valve 4 (00264–1131)
N92 – Solenoid Valve 5 (00266–1133)
N93 – Solenoid Valve 6 (00268–1141)
N94 – Solenoid Valve 7 (00270–1143)
T12 – 12-Pin Connector, on valve body
T38a – 38-Pin Connector

(43) – Ground connection, on right A-pillar, lower part

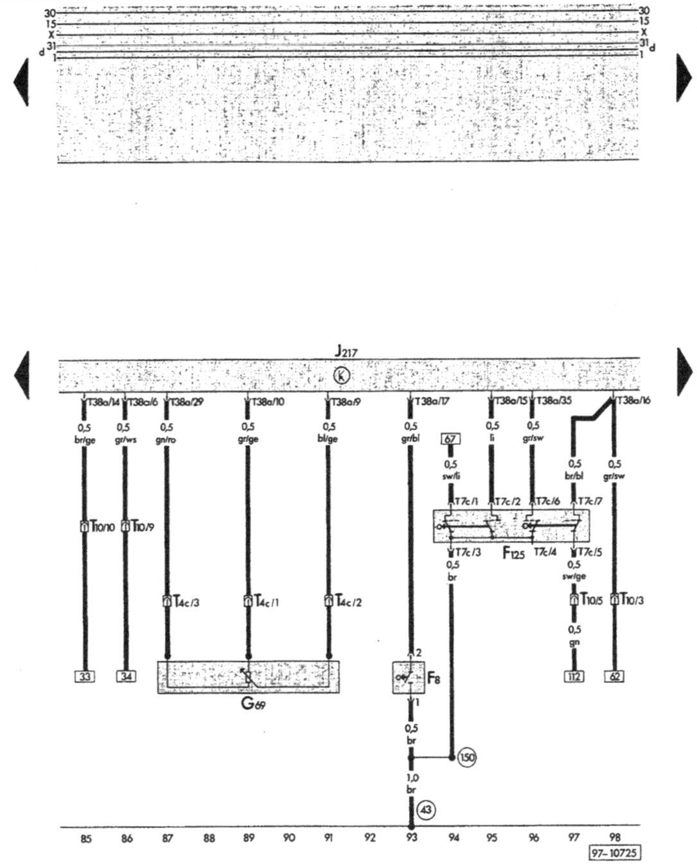

ws = white
sw = black
ro = red
br = brown
gn = green
bl = blue
gr = grey
li = violet
ge = yellow

F8 – Kick Down Switch (00296–1323)
F125 – Multi-Function Transmission Range (TR) Switch (00293–1314)
G69 – Throttle Position (TP) Sensor (00518–2212)
J217 – Transmission Control Module (TCM), near A-pillar, right
T4c – 4-Pin Connector
T7c – 7-Pin Connector
T10 – 10-Pin Connector, behind instrument panel, right
T38a – 38-Pin Connector

(43) – Ground connection, on right A-pillar, lower part
(150) – Ground connection, in automatic transmission wiring harness

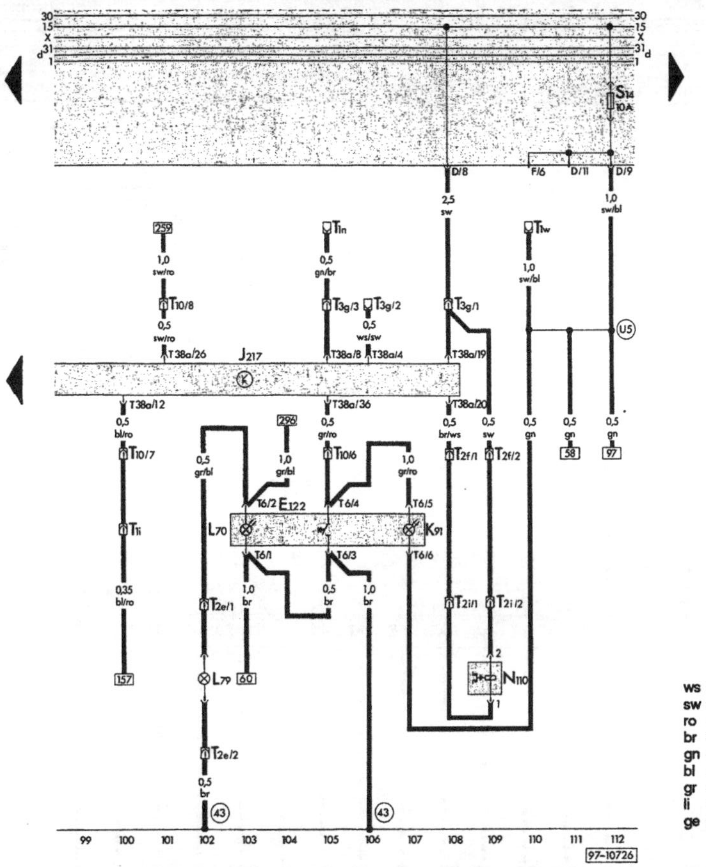

ws	=	white
sw	=	black
ro	=	red
br	=	brown
gn	=	green
bl	=	blue
gr	=	grey
li	=	violet
ge	=	yellow

E122 – Transmission Range (TR) Program Switch (00299–1332)
J217 – Transmission Control Module (TCM), near A-pillar, right
K 91 – Sport Mode Indicator Light (Automatic Transmission)
L70 – Program Switch Illumination
L79 – Selector Lever Light
N110 – Shift Lock Solenoid (01236–4314)
T1i – Single Connector, near fuse/relay panel
T1n – Single Connector, near fuse/relay panel (for A/C)
T1w – Single Connector, near fuse/relay panel
T2e – Double Connector, near selector lever
T2f – Double Connector, behind instrument panel, right
T2i – Double Connector, near selector lever
T3g – 3-Pin Connector, behind instrument panel, right
T6 – 6-Pin Connector
T10 – 10-Pin Connector, behind fuse/relay panel
T38a – 38-Pin Connector

(43) – Ground connection, on right A-pillar, lower part

(U5) – Wire connection -2- (15), in automatic transmission wiring harness

Transmission control module (TCM), transmission range (TR) program switch

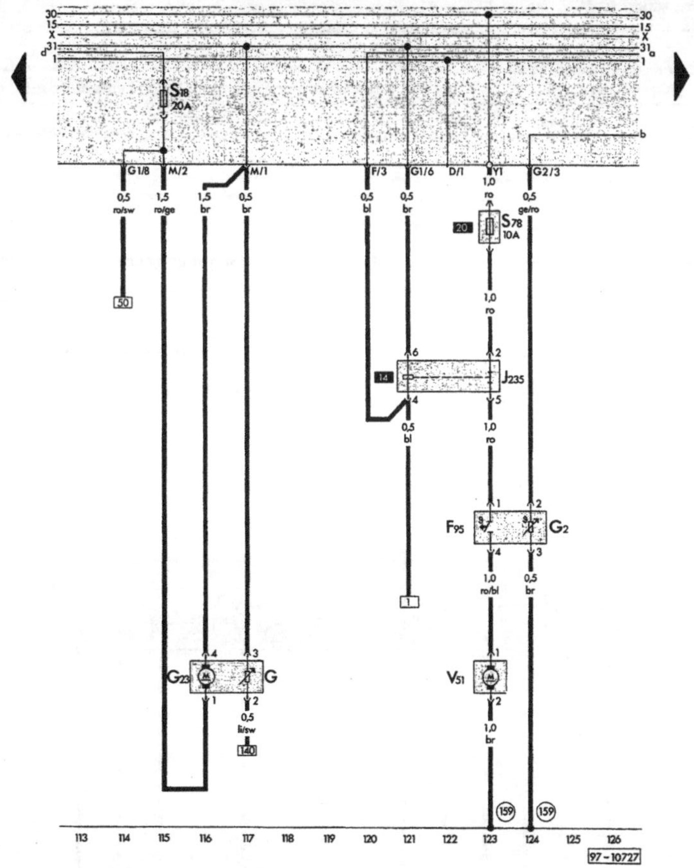

ws	=	white
sw	=	black
ro	=	red
br	=	brown
gn	=	green
bl	=	blue
gr	=	grey
li	=	violet
ge	=	yellow

F95 – After-Run Coolant Pump Thermal Switch*
G – Fuel Level Sensor
G2 – Engine Coolant Temperature (ECT) Sensor
G23 – Transfer Fuel Pump (FP)
J235 – Engine Coolant Pump Relay
S78 – Engine Coolant Pump Circuit Breaker
V51 – After-Run Coolant Pump

(159) – Ground connection -2-, in Digifant Multiport Fuel Injection (MFI) wiring harness

* – Switch value, see engine repair manual

Fuel supply, coolant after-run

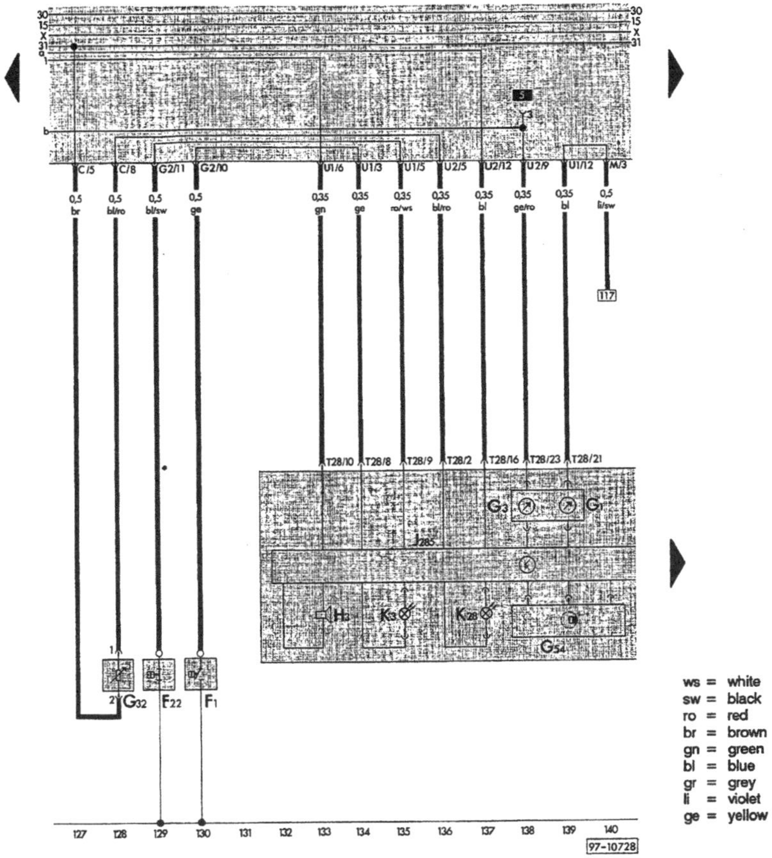

ws = white
sw = black
ro = red
br = brown
gn = green
bl = blue
gr = grey
li = violet
ge = yellow

F1 – Oil Pressure Switch*
F22 – 0.3bar Oil Pressure Switch
G1 – Fuel Gauge
G3 – Engine Coolant Temperature (ECT) Gauge
G32 – Engine Coolant Level (ECL) Sensor
G54 – Vehicle Speed Sensor (VSS), in instrument cluster
H3 – Warning Buzzer
J285 – Control Module With Display Unit In Instrument Cluster
K3 – Oil Pressure Warning Light
K28 – Engine Coolant Level/Temperature (ECL/ECT) Warning Light
T28 – 28-Pin Connector

* – Switch value, see engine repair manual

Instrument cluster, oil pressure- and engine coolant level/temperature (ECL/ECT) warning lights

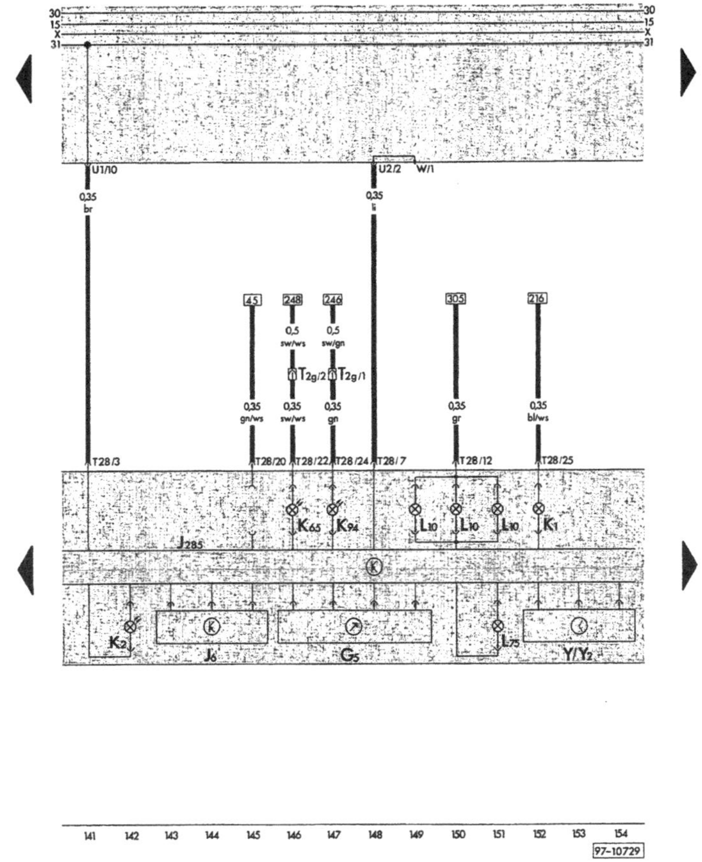

ws = white
sw = black
ro = red
br = brown
gn = green
bl = blue
gr = grey
li = violet
ge = yellow

G5 – Tachometer
J6 – Voltage Stabilizer
J285 – Control Module With Display Unit In Instrument Cluster
K1 – Headlight High Beam Indicator Light
K2 – Generator (GEN) Warning Light
K65 – Left Turn Signal Indicator Light
K94 – Right Turn Signal Indicator Light
L10 – Instrument Cluster Light
L75 – Digital Display Light
T2g – Double Connector, behind fuse/relay panel
T28 – 28-Pin Connector
Y – Analog Clock
Y2 – Digital Clock

Instrument cluster, tachometer, clock, indicator lights

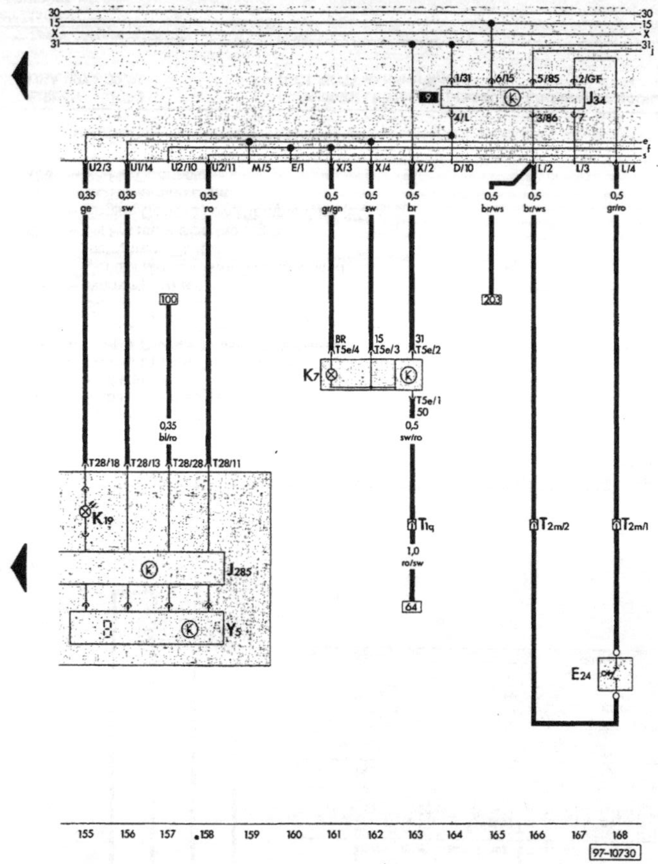

ws = white
sw = black
ro = red
br = brown
gn = green
bl = blue
gr = grey
li = violet
ge = yellow

E24 – Left Seat Belt Switch
J34 – Seat Belt Warning System Relay
J285 – Control Module With Display Unit In Instrument Cluster
K7 – Brake and Parking Brake Warning Light
K19 – Seat Belt Warning Light
T1g – Single Connector, behind fuse/relay panel
T2m – Double Connector, near seat belt switch
T5e – 5-Pin Connector
T28 – 28-Pin Connector
Y5 – Transmission Range (TR) Selector Lever Display

Instrument cluster, transmission range (TR) selector lever display, seat belt warning system

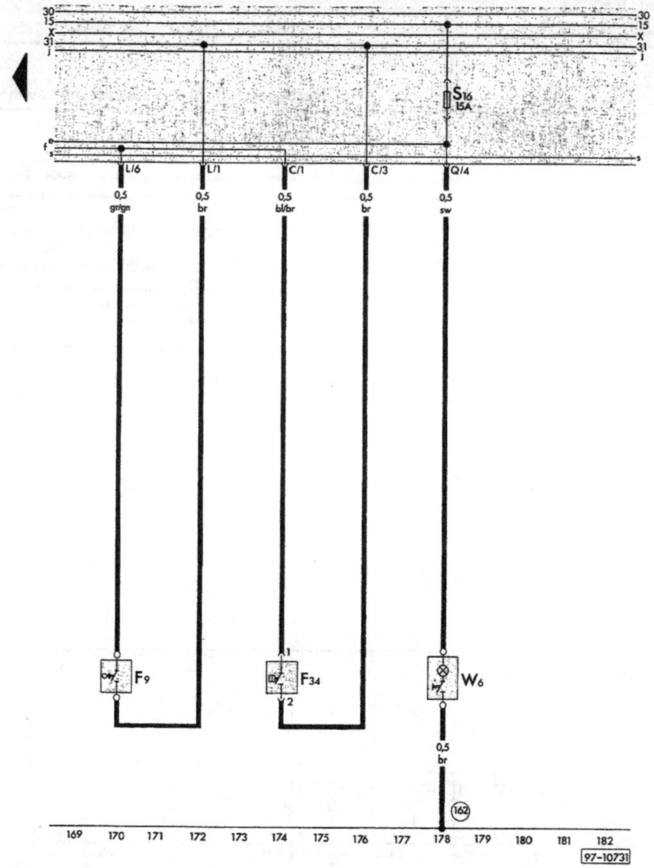

ws = white
sw = black
ro = red
br = brown
gn = green
bl = blue
gr = grey
li = violet
ge = yellow

F9 – Parking Brake Warning Light Switch
F34 – Brake Fluid Level Warning Switch
W6 – Glove Compartment Light

(162) – Ground connection, in blower motor wiring harness

Brake and parking brake indicator light, glove compartment light

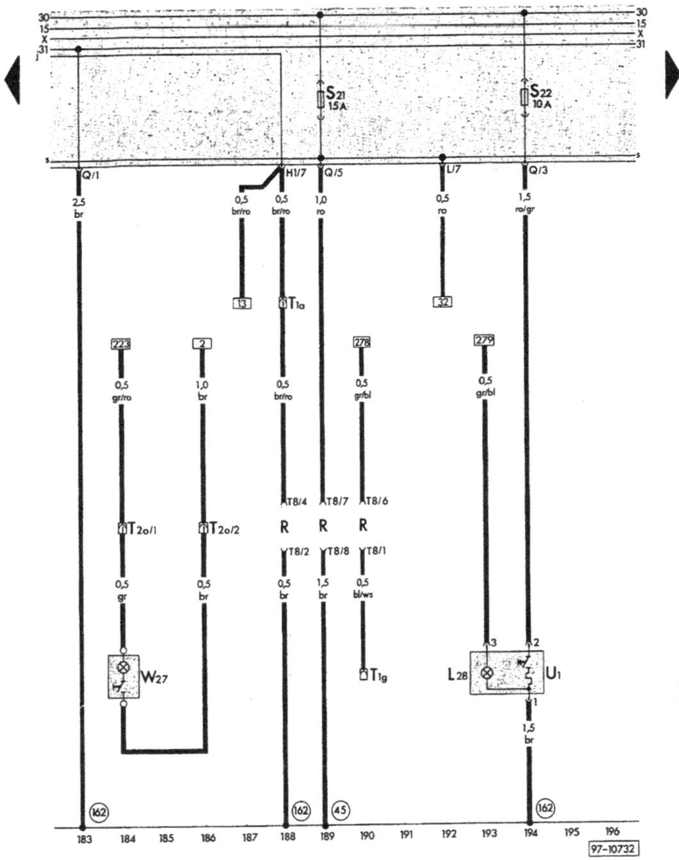

ws	=	white	
sw	=	black	
ro	=	red	
br	=	brown	
gn	=	green	
bl	=	blue	
gr	=	grey	
li	=	violet	
ge	=	yellow	

- L28 – Cigarette Lighter Light
- R – Radio
- T1a – Single Connector, behind fuse/relay panel
- T1g – Single Connector, behind fuse/relay panel
- T2o – Double Connector, near battery
- T8 – 8-Pin Connector
- U1 – Cigarette Lighter
- W27 – Engine Compartment Light

- (45) – Ground connection, behind instrument panel, center
- (162) – Ground connection, in blower motor wiring harness

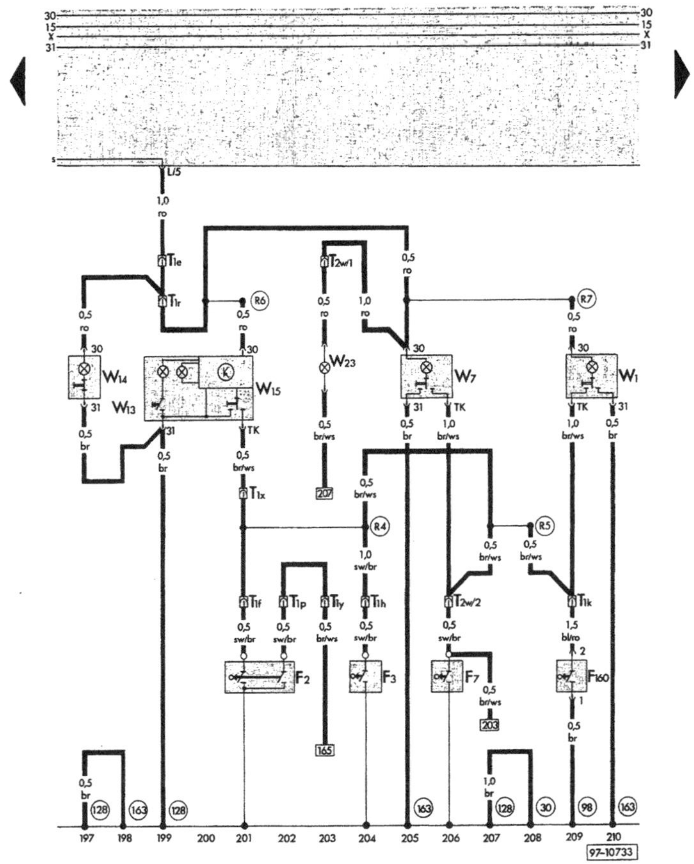

ws	=	white	
sw	=	black	
ro	=	red	
br	=	brown	
gn	=	green	
bl	=	blue	
gr	=	grey	
li	=	violet	
ge	=	yellow	

- F2 – Left Front Door Contact Switch
- F3 – Right Front Door Contact Switch
- F7 – Right Rear Side/Sliding Door Contact Switch
- F160 – Rear Lid Contact Switch
- T1e – Single Connector, behind fuse/relay panel
- T1f – Single Connector, behind fuse/relay panel
- T1h – Single Connector, behind instrument panel, right
- T1k – Single Connector, near taillight, right
- T1p – Single Connector, behind fuse/relay panel
- T1v – Single Connector, behind fuse/relay panel
- T1x – Single Connector, behind fuse/relay panel
- T1y – Single Connector, behind fuse/relay panel
- T2w – Double Connector, near center interior light
- W1 – Rear/Luggage Compartment Interior Light
- W7 – Center Interior Light
- W13 – Right Front Map/Reading Light
- W14 – Right Make-Up Mirror Light
- W15 – Interior Light With Delay Switch
- W23 – Footwell Light With Delay Switch

- (30) – Ground connection, -1-, beside fuse/relay panel
- (98) – Ground connection, in rear lid wiring harness
- (128) – Ground connection -1-, in interior light wiring harness
- (163) – Ground connection -2-, in interior light wiring harness
- (R4) – Wire connection -1-, in interior light/door contact switch wiring harness
- (R5) – Wire connection -2-, in interior light wiring harness
- (R6) – Plus connection -1-, in interior light wiring harness
- (R7) – Plus connection -2-, in interior light wiring harness

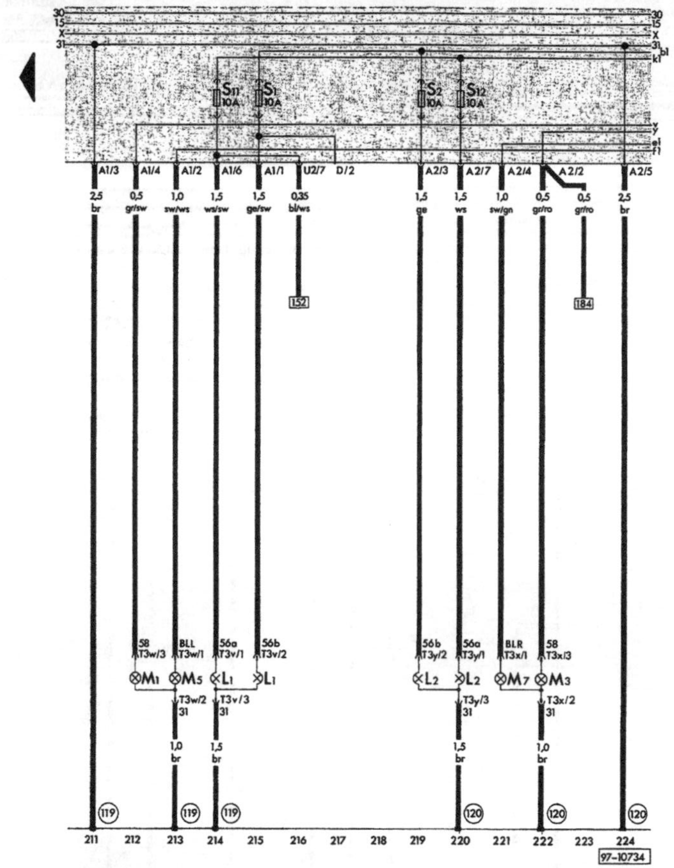

ws = white
sw = black
ro = red
br = brown
gn = green
bl = blue
gr = grey
li = violet
ge = yellow

- L1 – Left Headlight
- L2 – Right Headlight
- M1 – Left Parking Light
- M3 – Right Parking Light
- M5 – Left Front Turn Signal Light
- M7 – Right Front Turn Signal Light
- T3v – 3-Pin Connector
- T3w – 3-Pin Connector
- T3x – 3-Pin Connector
- T3y – 3-Pin Connector
- (119) – Ground connection -1-, in headlight wiring harness
- (120) – Ground connection -2-, in headlight wiring harness

ws = white
sw = black
ro = red
br = brown
gn = green
bl = blue
gr = grey
li = violet
ge = yellow

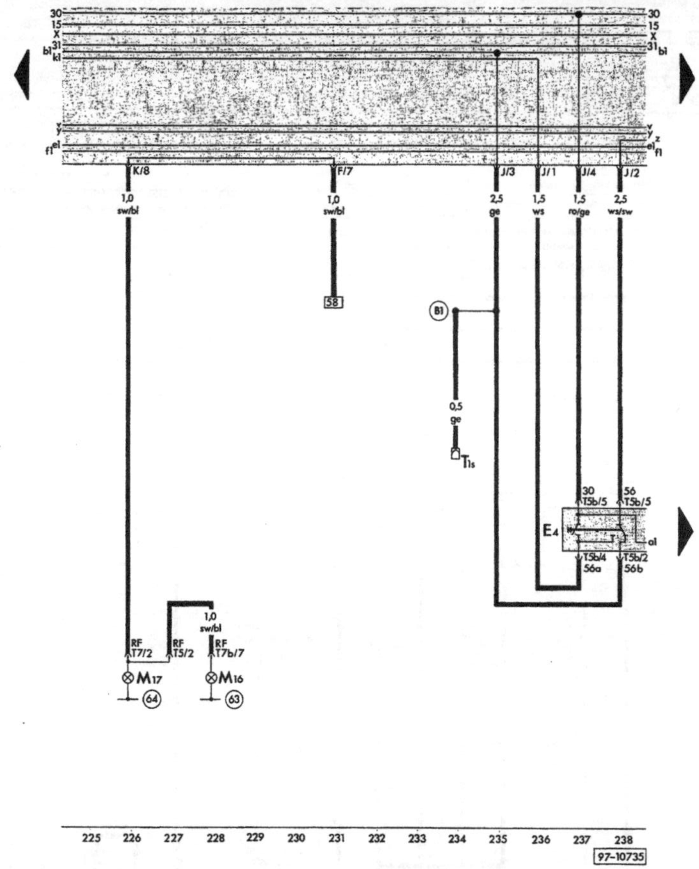

- E4 – Headlight Dimmer/Flasher Switch
- M16 – Left Back-Up Light
- M17 – Right Back-Up Light
- T1s – Single Connector, behind fuse/relay panel (daytime running light)
- T5 – 5-Pin Connector, on Right Tailight
- T5b – 5-Pin Connector, behind steering column switch cover
- T7 – 7-Pin Connector, on Right Taillight
- T7b – 7-Pin Connector, on Left Taillight
- (63) – Ground connection, on left taillight bulb holder
- (64) – Ground connection, on right taillight bulb holder
- (B1) – Wire connection (56b), in steering column switch wiring harness

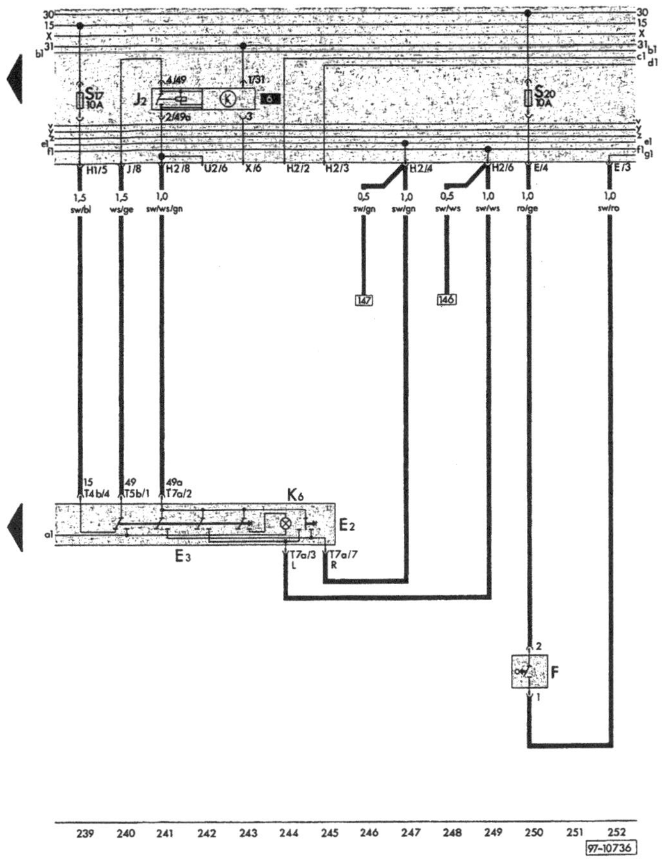

ws	=	white
sw	=	black
ro	=	red
br	=	brown
gn	=	green
bl	=	blue
gr	=	grey
li	=	violet
ge	=	yellow

- E2 – Turn Signal Switch
- E3 – Emergency Flasher Switch
- F – Brake Light Switch
- J2 – Emergency Flasher Relay
- K6 – Emergency Flasher Warning Light
- T4b – 4-Pin Connector, behind steering column switch cover
- T5b – 5-Pin Connector, behind steering column switch cover
- T7a – 7-Pin Connector, behind steering column switch cover

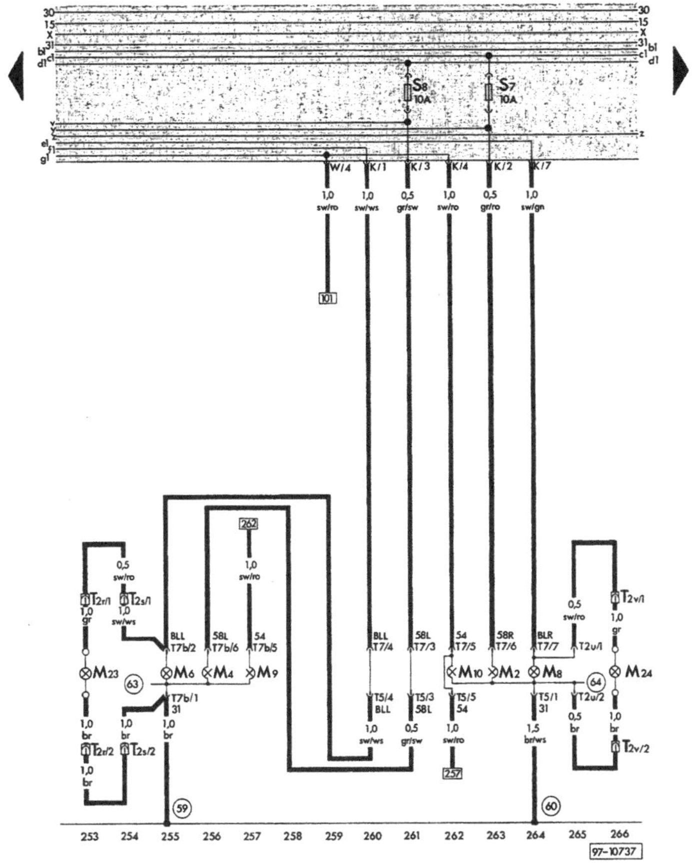

ws	=	white
sw	=	black
ro	=	red
br	=	brown
gn	=	green
bl	=	blue
gr	=	grey
li	=	violet
ge	=	yellow

- M2 – Right Taillight
- M4 – Left Taillight
- M6 – Left Rear Turn Signal Light
- M8 – Right Rear Turn Signal Light
- M9 – Left Brake Light
- M10 – Right Brake Light
- M23 – Left Roof Turn Signal Light
- M24 – Right Roof Turn Signal Light
- T2r – Double Connector, near left roof turn signal
- T2s – Double Connector, near left taillight
- T2u – Double Connector, on Right Taillight
- T2v – Double Connector, near right roof turn signal
- T5 – 5-Pin Connector, on Right Taillight
- T7 – 7-Pin Connector, on Right Taillight
- T7b – 7-Pin Connector, on Left Taillight
- (59) – Ground connection, near left taillight
- (60) – Ground connection, near right taillight
- (63) – Ground connection, on left taillight bulb holder
- (64) – Ground connection, on right taillight bulb holder
- * – Different wire routing for flatbed vehicles

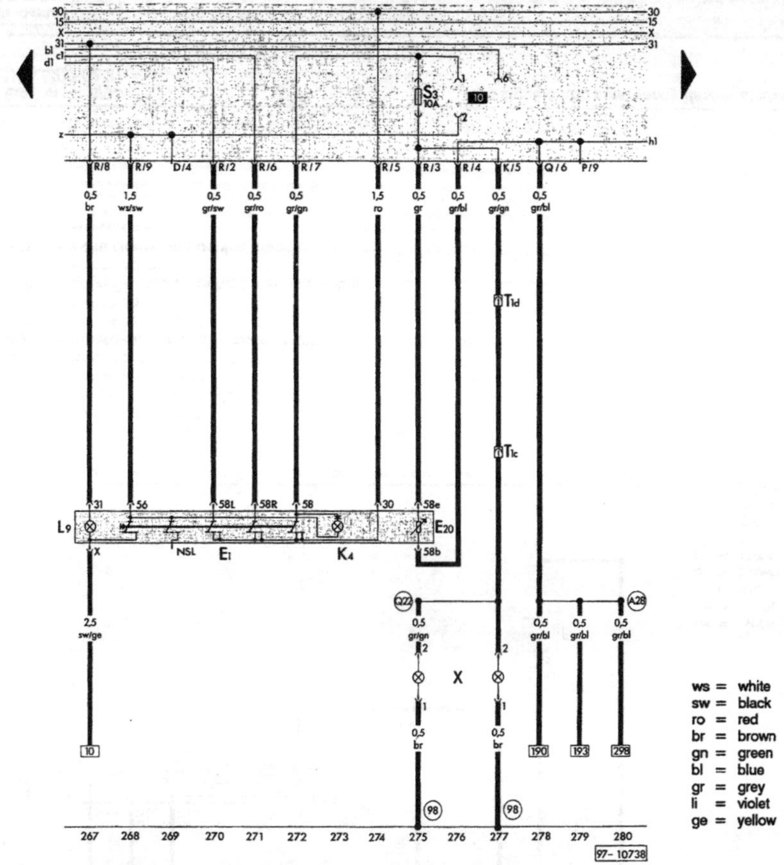

ws = white
sw = black
ro = red
br = brown
gn = green
bl = blue
gr = grey
li = violet
ge = yellow

- E1 – Light Switch
- E20 – Instrument Panel Light Dimmer Switch
- K4 – Park Light Indicator Light
- L9 – Headlight Switch Light
- T1c – Single Connector, taillight, left
- T1d – Single Connector, taillight, right
- X – License Plate Light
- (98) – Ground connection, in rear lid wiring harness
- (A28) – Plus connection (58b), in fresh air blower wiring harness
- (Q22) – Wire connection -1-, in rear lid wiring harness

Light switch, license plate light (tailgate)

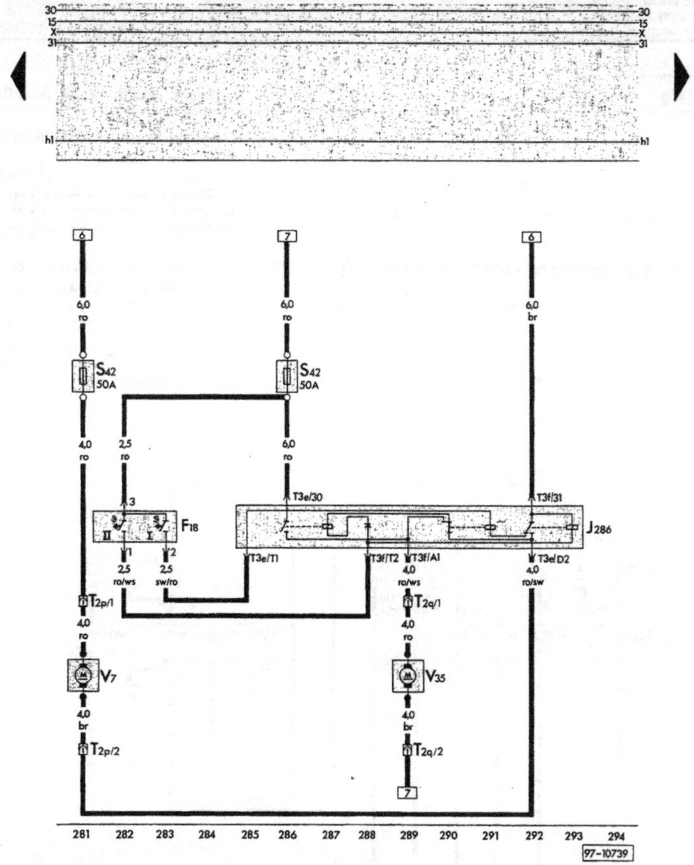

ws = white
sw = black
ro = red
br = brown
gn = green
bl = blue
gr = grey
li = violet
ge = yellow

- F18 – Coolant Fan Control (FC) Thermal Switch
- J286 – Series/Parallel Coolant Fan Control (FC) Relay (V7, V35)
- S42 – Coolant Fan Fuse
- T2p – Double Connector, in engine compartment, front
- T2q – Double Connector, in engine compartment, front
- T3e – 3-Pin Connector
- T3f – 3-Pin Connector
- V7 – Coolant Fan
- V35 – Right Coolant Fan

Coolant fans (2 x 200W or 2 x 350W)

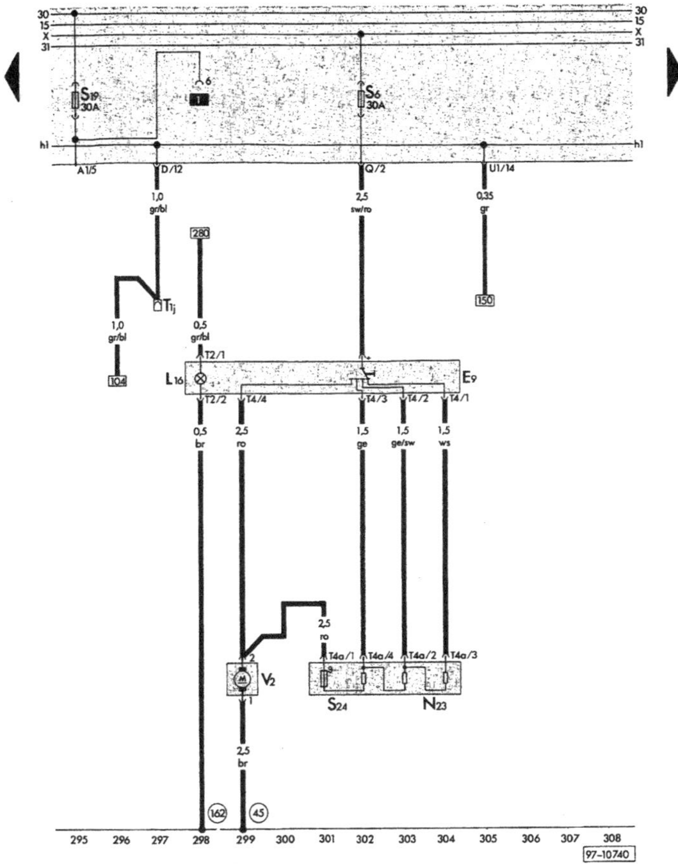

ws = white
sw = black
ro = red
br = brown
gn = green
bl = blue
gr = grey
li = violet
ge = yellow

E9 – Fresh Air Blower Switch
L16 – Fresh Air Control Lever Light
N23 – Fresh Air Blower Series Resistance
S24 – Heater, A/C Clutch Fuse
T1j – Single Connector, behind fuse/relay panel
T2 – Double Connector
T4 – 4-Pin Connector
T4a – 4-Pin Connector
V2 – Fresh Air Blower

(45) – Ground connection, behind instrument panel, center
(162) – Ground connection, in blower motor wiring harness

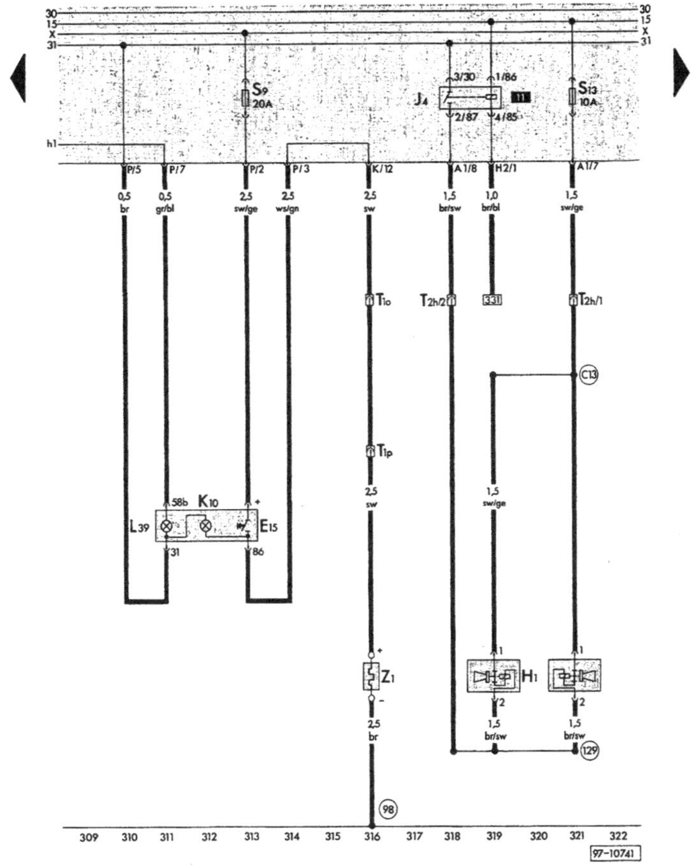

ws = white
sw = black
ro = red
br = brown
gn = green
bl = blue
gr = grey
li = violet
ge = yellow

E15 – Rear Window Defogger Switch
H1 – Dual Horns
J4 – Dual Horn Relay
K10 – Rear Window Defogger Indicator Light
L39 – Rear Window Defogger Switch Light
T1o – Single Connector, near taillight, right
T1p – Single Connector (deleted)
T2h – Double Connector, in engine compartment, left front
Z1 – Rear Window Defogger/Heat Element

(98) – Ground connection, in rear lid wiring harness

(129) – Ground connection, in dual-tone horn wiring harness

(C13) – Plus connection, in dual-tone horn wiring harness

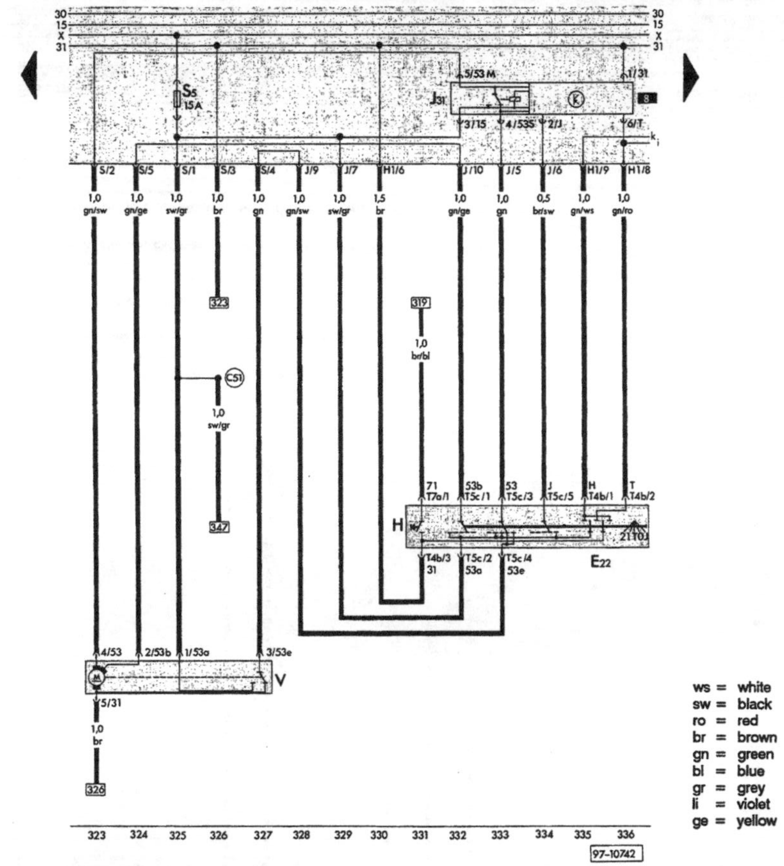

ws = white
sw = black
ro = red
br = brown
gn = green
bl = blue
gr = grey
li = violet
ge = yellow

- E22 – Windshield Wiper/Washer Switch
- H – Horn Button
- J31 – Washer/Wiper Intermittent Relay
- T4b – 4-Pin Connector, behind steering column switch cover
- T5c – 5-Pin Connector, behind steering column switch cover
- T7a – 7-Pin Connector, behind steering column switch cover
- V – Windshield Wiper Motor
- (C51) – Plus connection (x), in wiper motor wiring harness

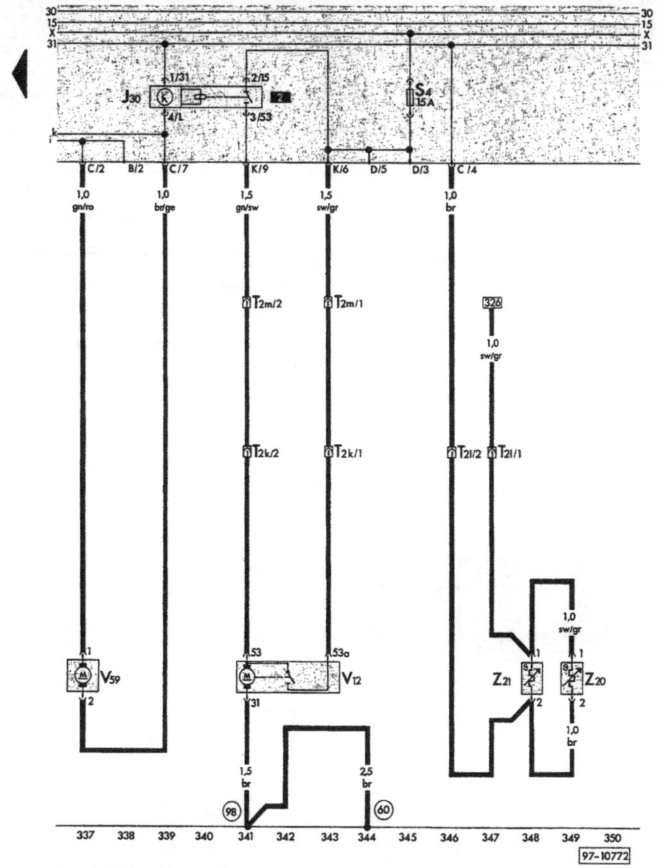

ws = white
sw = black
ro = red
br = brown
gn = green
bl = blue
gr = grey
li = violet
ge = yellow

- J30 – Rear Window Wiper/Washer Relay
- T2k – Double Connector, near taillight, right
- T2l – Double Connector, near firewall, right
- T2m – Double Connector (deleted)
- V12 – Rear Window Wiper Motor
- V59 – Windshield and Rear Window Washer Pump
- Z20 – Left Washer Nozzle Heater
- Z21 – Right Washer Nozzle Heater
- (60) – Ground connection, near taillight, right
- (98) – Ground connection, in rear lid wiring harness

- **Taillights (flatbed and double cabin),**
- **Rear window defogger (flatbed and double cabin),**

from August 1992

Relay location
For relay location and multiple connectors see also section "component locations".

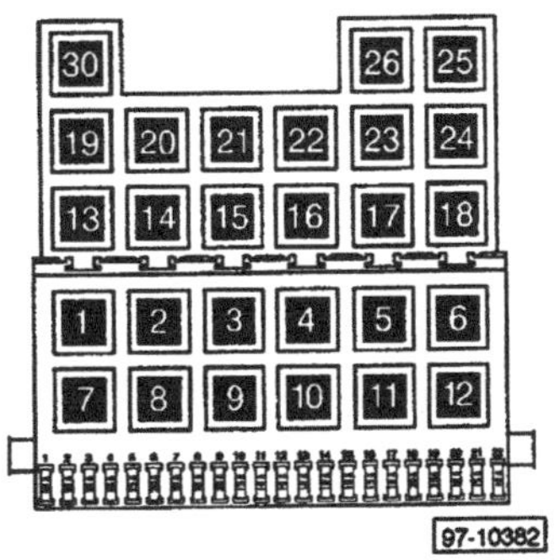

4 – Load Reduction Relay (18)

6 – Emergency Flasher Relay (21)

Note: Number in parentheses indicates production control number stamped on relay housing.

Fuse colors

30A – green
25A – white
20A – yellow
15A – blue
10A – red

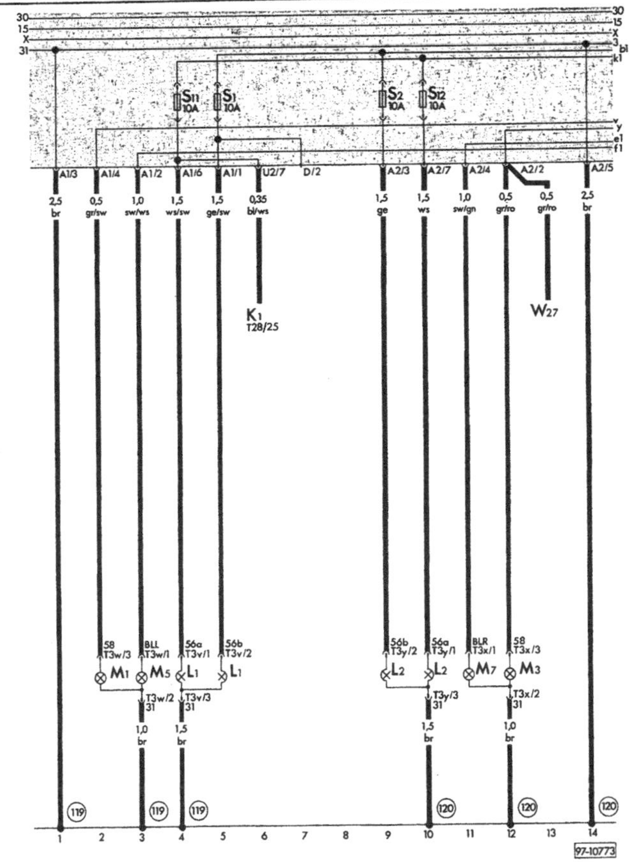

ws = white
sw = black
ro = red
br = brown
gn = green
bl = blue
gr = grey
li = violet
ge = yellow

K1 – Headlight High Beam Indicator Light
L1 – Left Headlight
L2 – Right Headlight
M1 – Left Parking Light
M3 – Right Parking Light
M5 – Left Front Turn Signal Light
M7 – Right Front Turn Signal Light
T3v – 3-Pin Connector
T3w – 3-Pin Connector
T3x – 3-Pin Connector
T3y – 3-Pin Connector
T28 – 28-Pin Connector, on instrument cluster
W27 – Engine Compartment Light

(119) – Ground connection -1-, in headlight wiring harness

(120) – Ground connection -2-, in headlight wiring harness

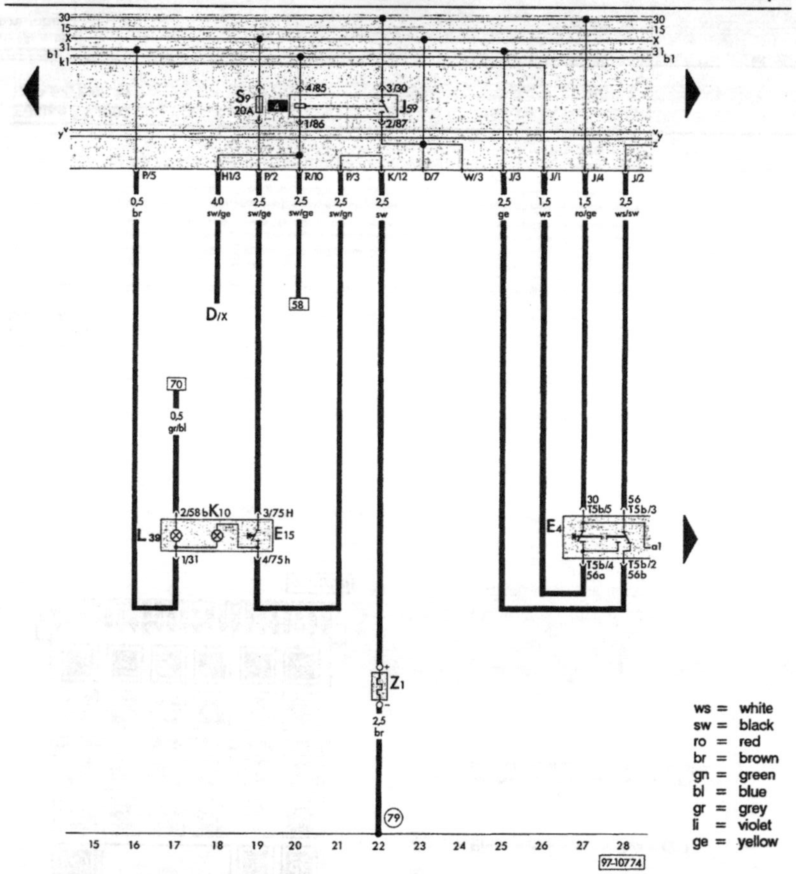

ws = white
sw = black
ro = red
br = brown
gn = green
bl = blue
gr = grey
li = violet
ge = yellow

D – Ignition/Starter Switch
E4 – Headlight Dimmer/Flasher Switch
E15 – Rear Window Defogger Switch
J59 – Load Reduction Relay
K10 – Rear Window Defogger Indicator Light
L39 – Rear Window Defogger Switch Light
T5b – 5-Pin Connector
Z1 – Rear Window Defogger/Heat Element

(79) – Ground connection, near rear window, right

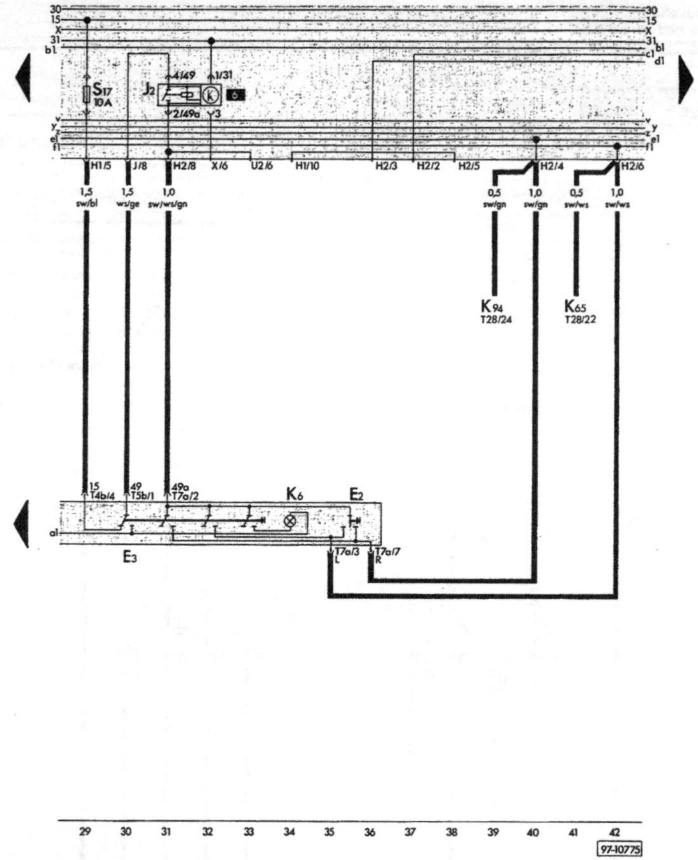

ws = white
sw = black
ro = red
br = brown
gn = green
bl = blue
gr = grey
li = violet
ge = yellow

E2 – Turn Signal Switch
E3 – Emergency Flasher Switch
J2 – Emergency Flasher Relay
K6 – Emergency Flasher Warning Light
K65 – Left Turn Signal Indicator Light
K94 – Right Turn Signal Indicator Light
T4b – 4-Pin Connector, behind steering column switch cover
T5b – 5-Pin Connector, behind steering column switch cover
T7a – 7-Pin Connector, behind steering column switch cover
T28 – 28-Pin Connector, on instrument cluster

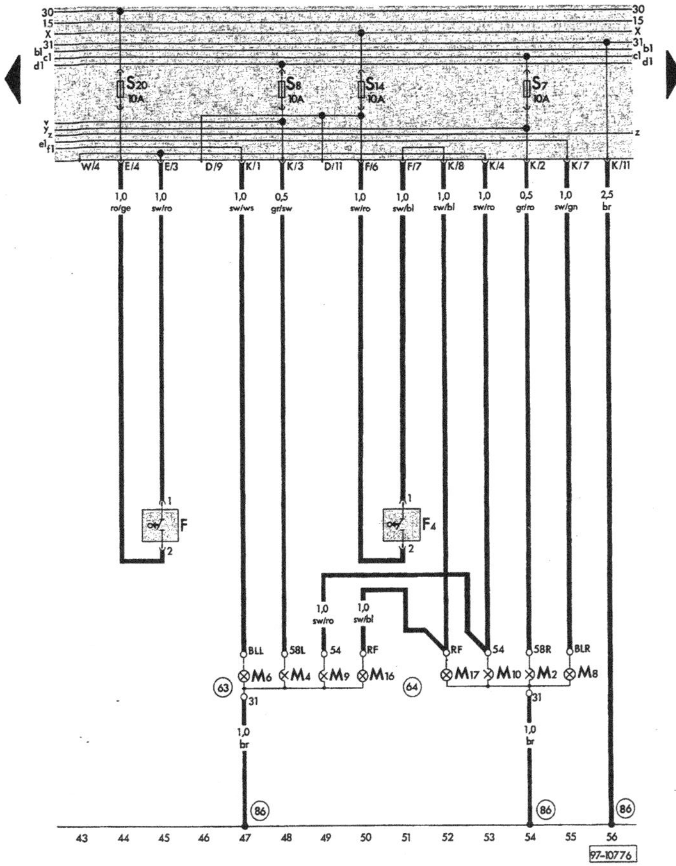

ws = white
sw = black
ro = red
br = brown
gn = green
bl = blue
gr = grey
li = violet
ge = yellow

F –Brake Light Switch
F4 –Back-Up Light Switch
M2 –Right Taillight
M4 –Left Taillight
M6 –Left Rear Turn Signal Light
M8 –Right Rear Turn Signal Light
M9 –Left Brake Light
M10 –Right Brake Light
M16 –Left Back-Up Light
M17 –Right Back-Up Light

(63) –Ground connection, on left taillight bulb holder
(64) –Ground connection, on right taillight bulb holder

(86) – Ground connection -1-, in rear wiring harness

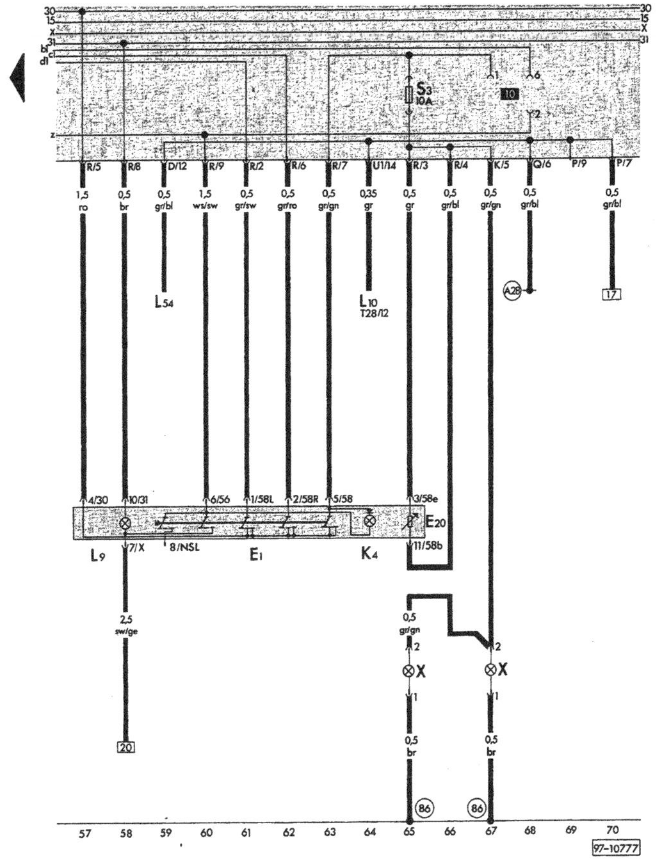

ws = white
sw = black
ro = red
br = brown
gn = green
bl = blue
gr = grey
li = violet
ge = yellow

E1 – Light Switch
E20 – Instrument Panel Light Dimmer Switch
K4 – Park Light Indicator Light
L9 – Headlight Switch Light
L10 – Instrument Cluster Light
L54 – Headlight Adjuster Control Light (not for USA, Canada)
T28 – 28-Pin Connector, on instrument cluster
X – License Plate Light

(86) – Ground connection -1-, in rear wiring harness
(A28) – Plus connection (58b), in fresh air blower wiring harness

Auxiliary heater B2L*, without programmable timer,

from April 1992

Relay location
For relay location and multiple connectors location see also section "component locations".

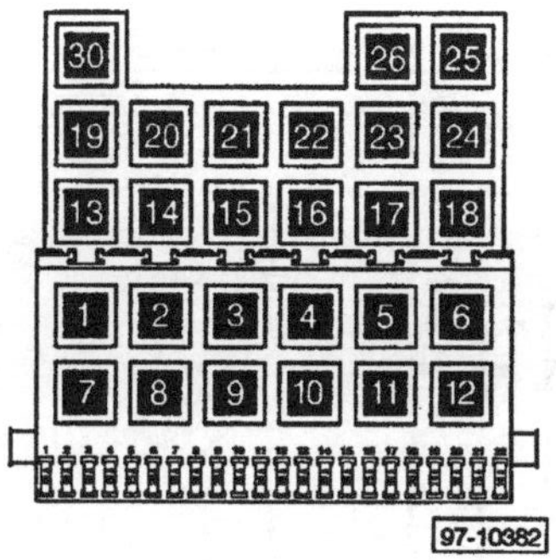

Fuse colors

30A – green
25A – white
20A – yellow
15A – blue
10A – red

* – Type 20168901, see identification plate

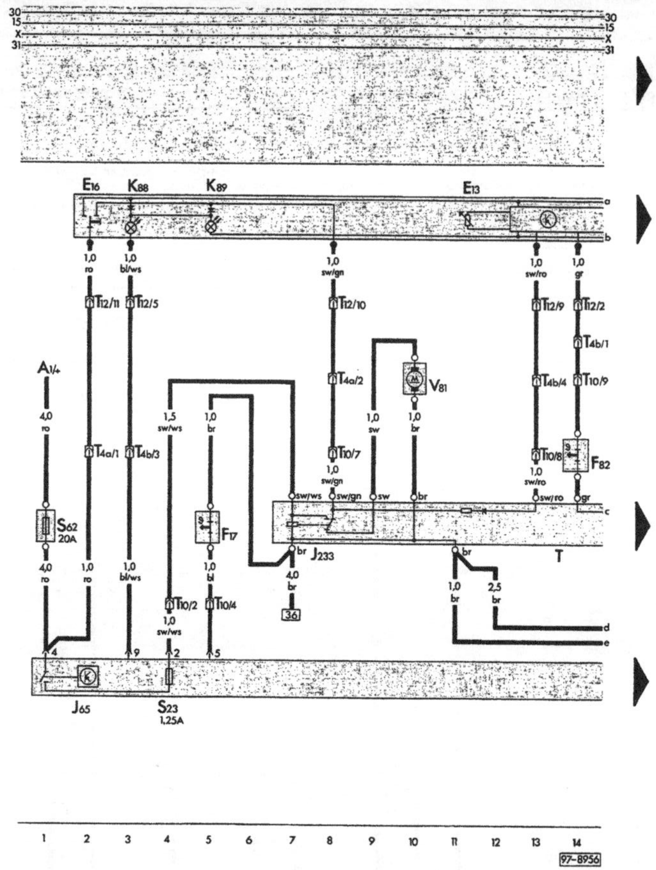

ws = white
sw = black
ro = red
br = brown
gn = green
bl = blue
gr = grey
li = violet
ge = yellow

A1 – Auxiliary Battery
E13 – Heater-Time Temperature Control Switch, on left B-pillar, upper part
E16 – Heater Switch
F17 – Heater Overheat Control Switch
F82 – Temperature Limiting Switch, in heater
J65 – Heater Control Module, below driver's seat*
J233 – Auxiliary Heater Blower Relay
K88 – Malfunction Light (red)
K89 – On Light (green)
S23 – Heater Fuse
S62 – Auxiliary Heater Fuse**
T – Junction Box (printed circuit board)
T4a – 4-Pin Connector, below driver's seat*
T4b – 4-Pin Connector, below driver's seat*
T10 – 10-Pin Connector, below driver's seat*
T12 – 12-Pin Connector, behind left B-pillar trim, upper part
V81 – Fresh Air/Heater Blower

* – Located behind lower part of left B-pillar trim on vehicles with swivel seat
** – Located on lower part of left B-pillar on vehicles with swivel seat

Heater-time temperature control switch, combustion/fresh air heater blower fan, heater control module

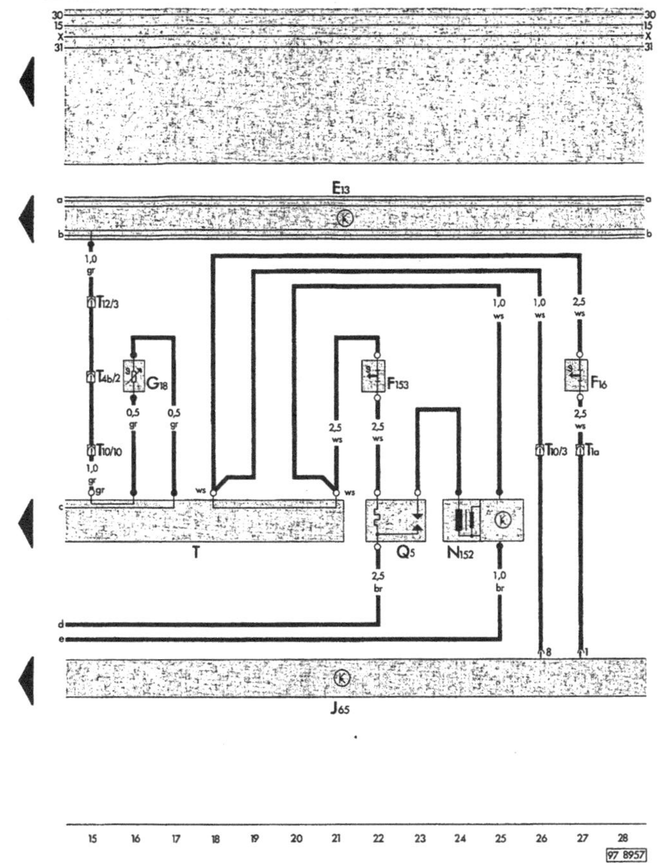

ws = white
sw = black
ro = red
br = brown
gn = green
bl = blue
gr = grey
li = violet
ge = yellow

E13 – Heater-Time Temperature Control Switch, on left B-pillar, upper part
F16 – Heater Flame Sensor
F153 – Glow Plug Thermoswitch
G18 – Temperature Sensor II
J65 – Heater Control Module, below driver's seat*
N152 – Ignition Coil
Q5 – Glow/Spark Plug
T – Junction Box (printed circuit board)
T1a – Single Connector, below driver's seat*
T4b – 4-Pin Connector, below driver's seat*
T10 – 10-Pin Connector, below driver's seat*
T12 – 12-Pin Connector, behind left B-pillar trim, upper part
* – Located behind lower part of left B-pillar trim on vehicles with swivel seat

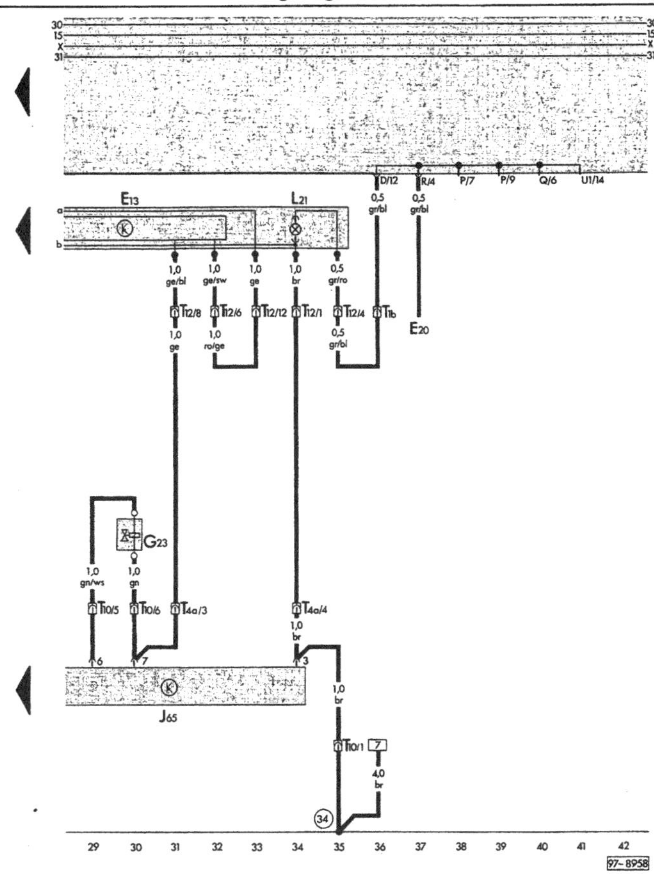

ws = white
sw = black
ro = red
br = brown
gn = green
bl = blue
gr = grey
li = violet
ge = yellow

E13 – Heater-Time Temperature Control Switch, on left B-pillar, upper part
E20 – Instrument Panel Light Dimmer Switch
G23 – Transfer Fuel Pump (FP)
J65 – Heater Control Module, below driver's seat*
L21 – Heater Control Lever Light
T1b – Single Connector, below driver's seat*
T4a – 4-Pin Connector, below driver's seat*
T10 – 10-Pin Connector, below driver's seat*
T12 – 12-Pin Connector, behind left B-pillar trim, upper part

(34) – Ground connection, below driver's seat*

* – Located behind lower part of B-pillar trim on vehicles with swivel seat

Auxiliary heater D2L*, without programmable timer,

from August 1992

Relay location

For relay location and multiple connectors location see also section "component locations".

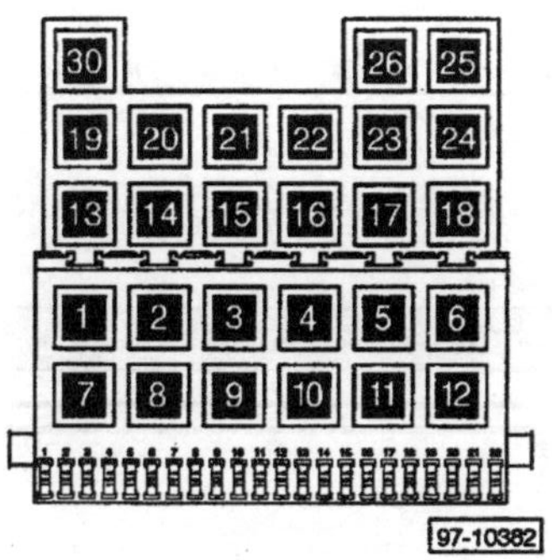

Fuse colors

30A – green
25A – white
20A – yellow
15A – blue
10A – red

* – Type 25172501, see identification plate

Edition 8/93
USA.5272.02.21

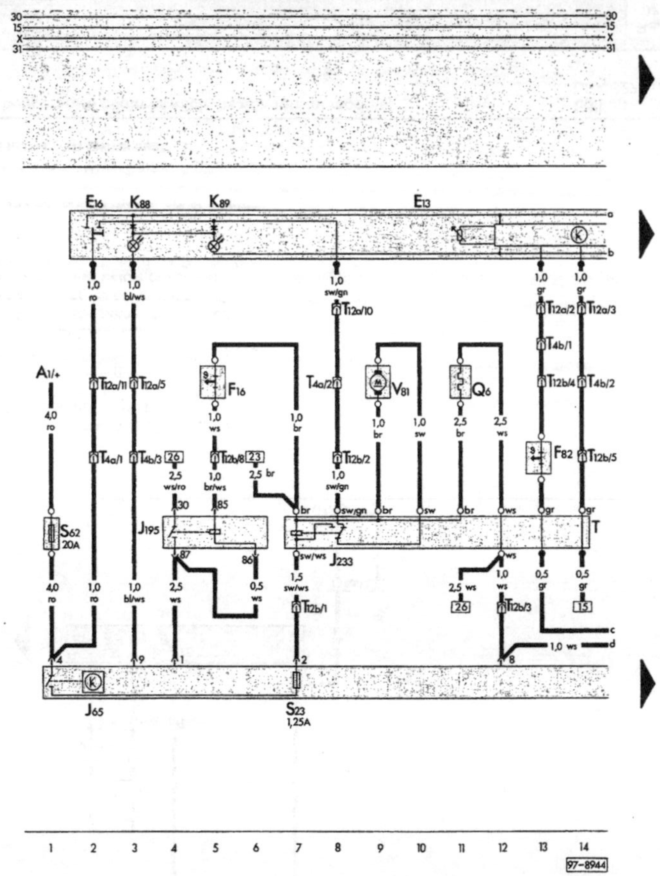

ws = white
sw = black
ro = red
br = brown
gn = green
bl = blue
gr = grey
li = violet
ge = yellow

A1 – Auxiliary Battery
E13 – Heater-Timer Temperature Control Switch, on left B-pillar, upper part
E16 – Heater/Heat Output Switch
F16 – Heater Flame Sensor
F82 – Temperature Limiting Switch, in heater
J65 – Heater Control Module, below driver's seat*
J195 – Glow Plug Control Relay, below driver's seat*
J233 – Auxiliary Heater Blower Relay
K88 – Water Heater Malfunction Light (red)
K89 – Water Heater On Light (green)
Q6 – Engine Glow Plug
S23 – Heater Fuse
S62 – Auxiliary Heater Fuse, below driver's seat**

T – Junction Box (printed circuit board)
T4a – 4-Pin Connector, below driver's seat*
T4b – 4-Pin Connector, below driver's seat*
T12a – 12-Pin Connector, behind left B-pillar trim, upper part
T12b – 12-Pin Connector, below driver's seat*
V81 – Fresh Air/Heater Blower

* – Located behind lower part of left B-pillar trim on vehicles with swivel seat
** – Located on lower part of left B-pillar on vehicles with swivel seat

Heater-time temperature control switch, combustion/fresh air heater blower fan, engine glow plug, heater control module, glow plug control

Edition 8/93
USA.5272.02.21

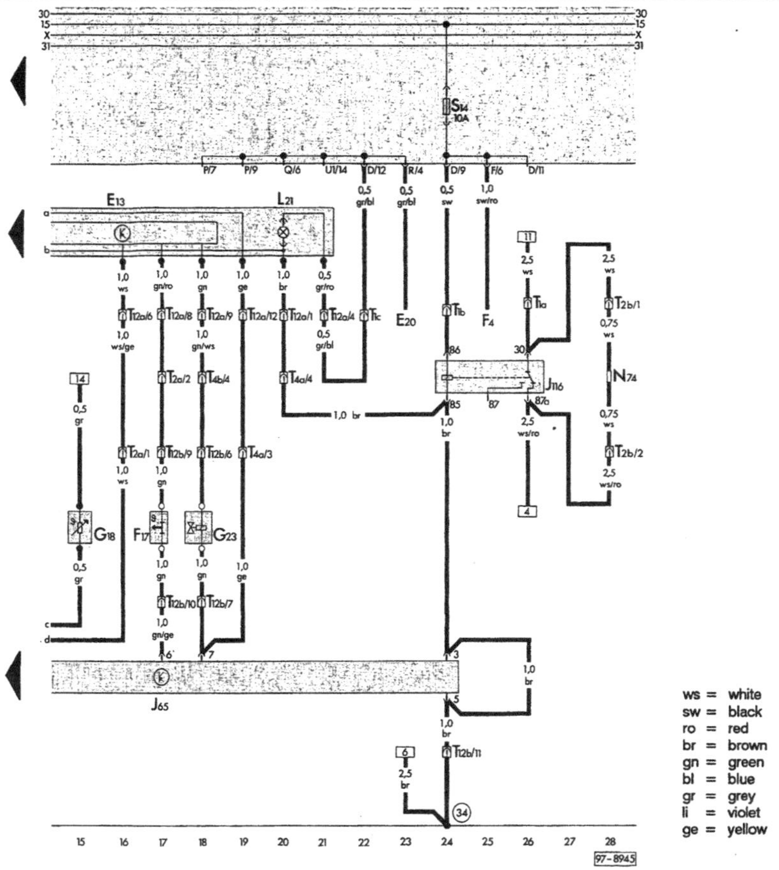

ws = white
sw = black
ro = red
br = brown
gn = green
bl = blue
gr = grey
li = violet
ge = yellow

E13 – Heater-Time Temperature Control Switch, on left B-pillar, upper part
E20 – Instrument Panel Light Dimmer Switch
F4 – Back-Up Light Switch
F17 – Heater Overheat Control Switch
G18 – Temperature Sensor II
G23 – Transfer Fuel Pump (FP)
J65 – Heater Control Module, below driver's seat*
J116 – Resistance Wire Relay
L21 – Heater Control Lever Light
N74 – Series Resistance Wiring
T1a – Single Connector, below driver's seat*
T1b – Single Connector, below driver's seat*
T1c – Single Connector, on B-pillar, left
T2a – Double Connector, below driver's seat*
T2b – Double Connector, below driver's seat*
T4a – 4-Pin Connector, below driver's seat*
T4b – 4-Pin Connector, below driver's seat*
T12a – 12-Pin Connector, behind left B-pillar trim, upper part
T12b – 12-Pin Connector, below driver's seat*

(34) – Ground connection, below driver's seat*

* – Located behind lower part of B-pillar trim on vehicles with swivel seat

Heater-time temperature control switch, heater control module, fuel supply

THIS PAGE INTENTIONALLY LEFT

BLANK

Auxiliary heater D2L*, without programmable timer,

from August 1992

Relay location
For relay location and multiple connectors location see also section "component locations".

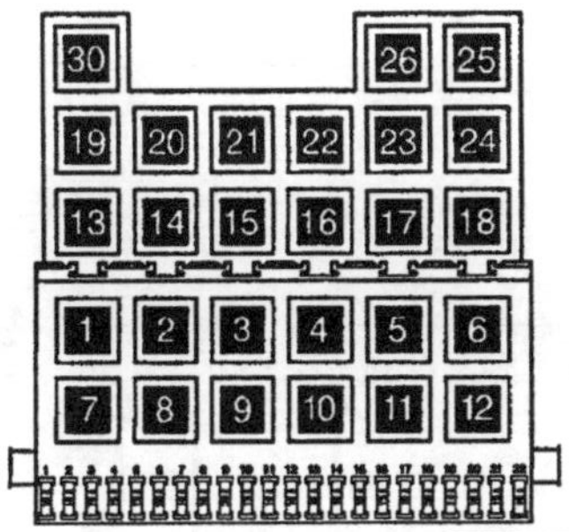

97-10382

Fuse colors

30A – green
25A – white
20A – yellow
15A – blue
10A – red

* – Type 25172501, see identification plate

Edition 8/93
USA.5272.02.21

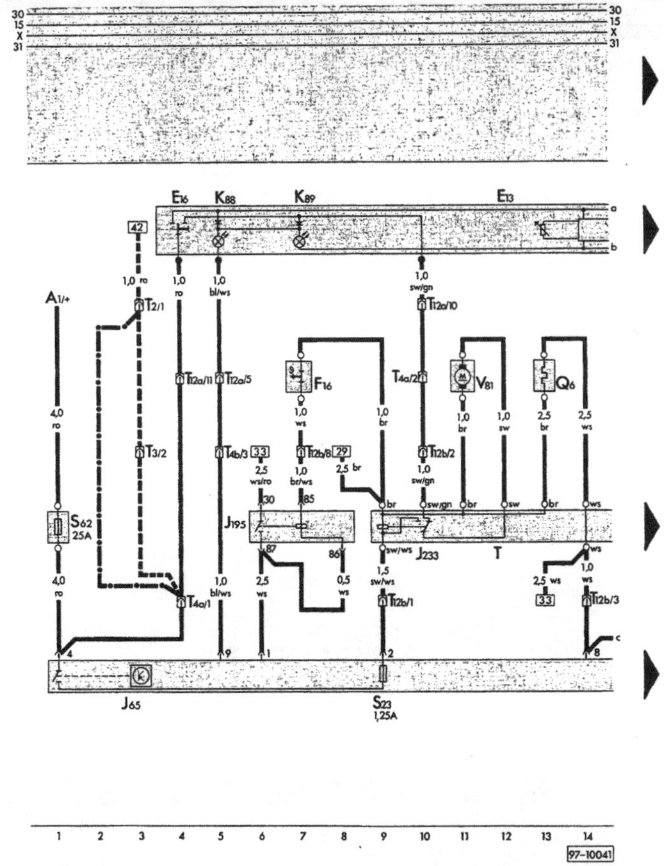

ws = white
sw = black
ro = red
br = brown
gn = green
bl = blue
gr = grey
li = violet
ge = yellow

A1 – Auxiliary Battery
E13 – Heater-Time Temperature Control Switch, on left B-pillar, upper part
E16 – Heater/Heat Output Switch
F16 – Heater Flame Sensor
J65 – Heater Control Module, below driver's seat*
J195 – Glow Plug Control Relay, below driver's seat*
J233 – Auxiliary Heater Blower Relay
K88 – Water Heater Malfunction Light (red)
K89 – Water Heater On Light (green)
Q6 – Engine Glow Plug
S23 – Heater Fuse
S62 – Auxiliary Heater Fuse, below driver's seat**
T – Junction Box (printed circuit board)

T2 – Double Connector, below driver's seat*
T3 – 3-Pin Connector, below driver's seat*
T4a – 4-Pin Connector, below driver's seat*
T4b – 4-Pin Connector, below driver's seat*
T12a – 12-Pin Connector, behind left B-pillar trim, upper part
T12b – 12-Pin Connector, below driver's seat*
V81 – Fresh Air/Heater Blower

* – Located behind lower part of left B-pillar trim on vehicles with swivel seat
** – Located on lower part of left B-pillar on vehicles with swivel seat
▪•▪ – "Camper" only
▪ ▪ – Not on "Camper"

Heater-time temperature control switch, combustion/fresh air heater blower fan, engine glow plug, heater control module, glow plug control

Edition 8/93
USA.5272.02.21

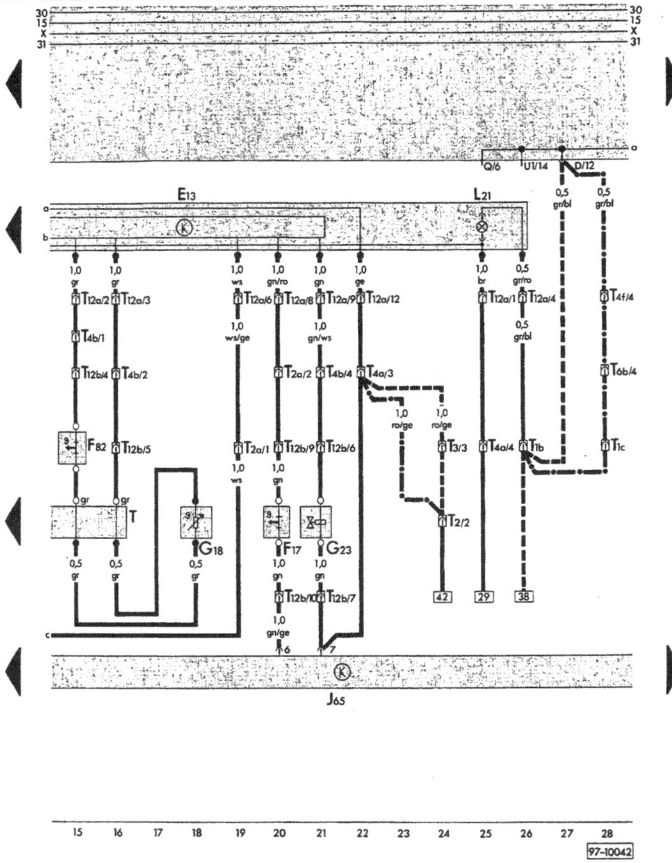

ws = white
sw = black
ro = red
br = brown
gn = green
bl = blue
gr = grey
li = violet
ge = yellow

E13 – Heater-Time Temperature Control Switch, on left B-pillar, upper part
E20 – Instrument Panel Light Dimmer Switch
F17 – Heater Overheat Control Switch
F 82 – Temperature Limiting Switch
G18 – Temperature Sensor
G23 – Transfer Fuel Pump (FP)
J65 – Heater Control Module, below driver's seat*
L21 – Heater Control Lever Light
T – Junction Box (printed circuit board)
T1b – Single Connector, below driver's seat*
T1c – Single Connector, on B-pillar, left
T2 – Double Connector, below driver's seat*
T2a – Double Connector, below driver's seat*
T3 – 3-Pin Connector, below driver's seat*
T4a – 4-Pin Connector, below driver's seat*
T4b – 4-Pin Connector, below driver's seat*
T4f – 4-Pin Connector, behind fuse/relay panel
T6b – 6-Pin Connector, on left A-pillar, upper part
T12a – 12-Pin Connector, behind left B-pillar trim, upper part
T12b – 12-Pin Connector, below driver's seat*

* – Located behind lower part of B-pillar trim on vehicles with swivel seat

▬•▬ – "Camper" only
▬ ▬ – Not on "Camper"

Heater-time temperature control switch, heater control module, fuel supply

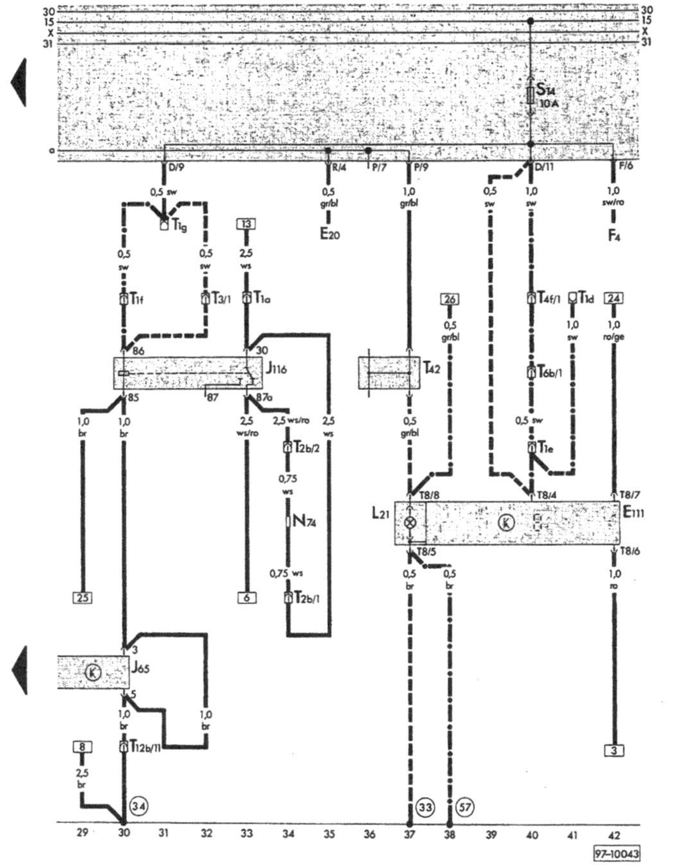

ws = white
sw = black
ro = red
br = brown
gn = green
bl = blue
gr = grey
li = violet
ge = yellow

E111 – Programmable Timer, in instrument panel, center
F4 – Back-Up Light Switch
J65 – Heater Control Module, below driver's seat*
J116 – Resistance Wire Relay, below driver's seat*
L21 – Heater Control Lever Light
N74 – Series Resistance Wiring
T1a – Single Connector, below driver's seat*
T1d – Single Connector, on B-pillar, left
T1e – Single Connector, on B-pillar, left
T1f – Single Connector, below driver's seat*
T1g – Single Connector, behind fuse/relay panel
T2b – Double Connector, below driver's seat*
T3 – 3-Pin Connector, below driver's seat
T4f – 4-Pin Connector, behind fuse/relay panel
T6b – 6-Pin Connector, on left A-pillar, upper part
T8 – 8-Pin Connector
T12b – 12-Pin Connector, below driver's seat*
T42 – Wire Connector, Terminal 58B

(33) – Ground connection, behind instrument panel, right
(34) – Ground connection, below driver's seat*
(57) – Ground connection, on left rear B-pillar

* – Located behind lower part of left B-pillar trim on vehicles with swivel seat

▬•▬ – "Camper" only
▬ ▬ – Not on "Camper"

Programmable timer

2.5 L-Engine – Digifant Multiport Fuel Injection (MFI)/81kW, code AAF,

from January 1993

Relay location
For relay location and multiple connectors see also section "component locations".

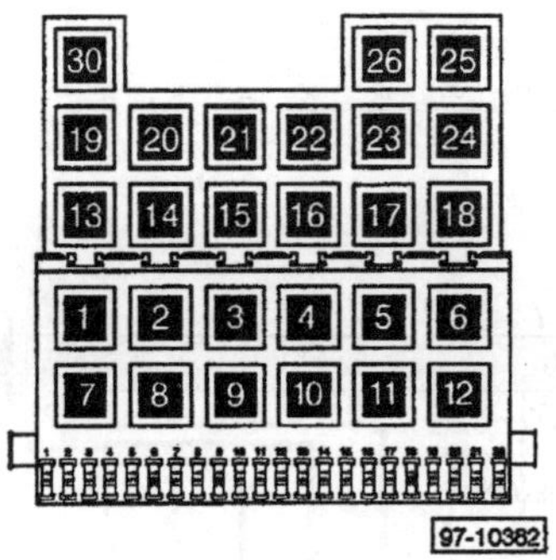

- 3 – Digifant Engine Control Module (ECM) Power Supply Relay (30)
- 4 – Load Reduction Relay (18)
- 9 – Seat Belt Warning System Relay (4)
- 12 – Fuel Pump (FP) Relay (80)
- 14 – Engine Coolant Pump Relay (38)

Note: Number in parentheses indicates production control number stamped on relay housing.

Fuse colors

30A – green
25A – white
20A – yellow
15A – blue
10A – red

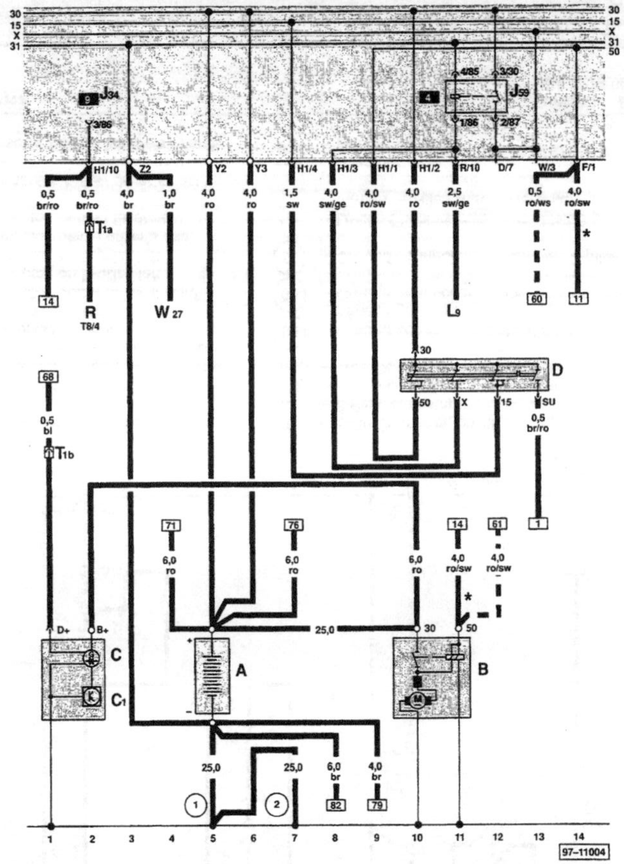

ws = white
sw = black
ro = red
br = brown
gn = green
bl = blue
gr = grey
li = violet
ge = yellow

A – Battery
B – Starter
C – Generator (GEN)
C1 – Voltage Regulator (VR)
D – Ignition/Starter Switch
J34 – Seat Belt Warning System Relay
J59 – Load Reduction Relay
L9 – Headlight Switch Light
R – Radio
T1a – Single Connector, behind instrument panel, left
T1b – Single Connector, near starter
T8 – 8-Pin Connector
W27 – Engine Compartment Light

(1) – Ground strap, battery to body
(2) – Ground strap, transmission to body
* – Manual transmission only
▬ ▬ – Automatic transmission only

Generator (GEN), battery, starter, ignition/starter switch

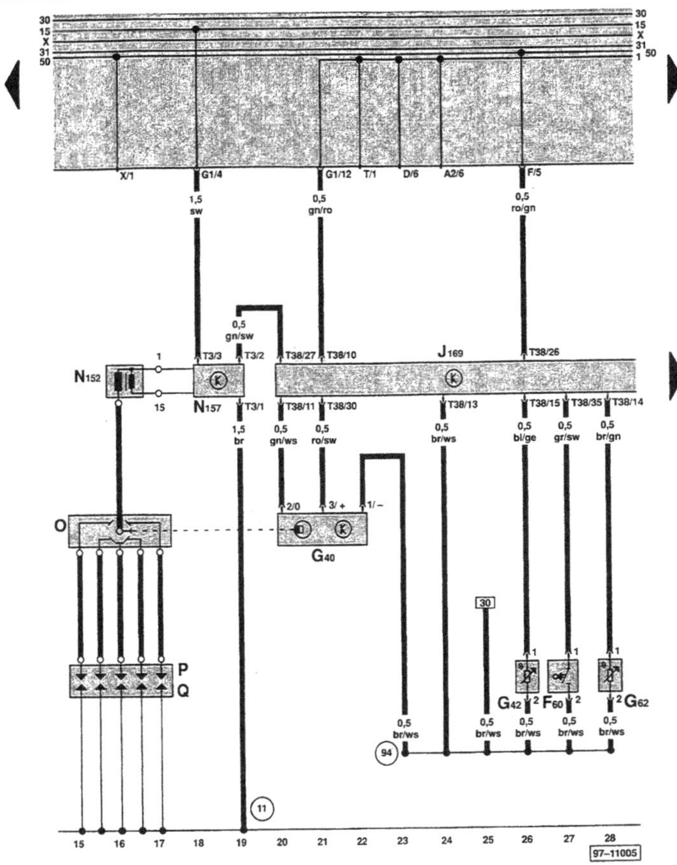

ws = white
sw = black
ro = red
br = brown
gn = green
bl = blue
gr = grey
li = violet
ge = yellow

F60 – Closed Throttle Position (CTP) Switch
G40 – Camshaft Position (CMP) Sensor
G42 – Intake Air Temperature (IAT) Sensor (00523–2322)
G62 – Engine Coolant Temperature (ECT) Sensor (0522–2312)
J169 – Digifant Engine Control Module (ECM), near battery
N152 – Ignition Coil
N157 – Ignition Coil Power Output Stage
O – Ignition Distributor
P – Spark Plug Connectors
Q – Spark Plugs
T3 – 3-Pin Connector
T38 – 38-Pin Connector

(11) – Ground connection, in battery box

(94) – Ground connection -1-, in Digifant Multiport Fuel Injection (MFI) wiring harness

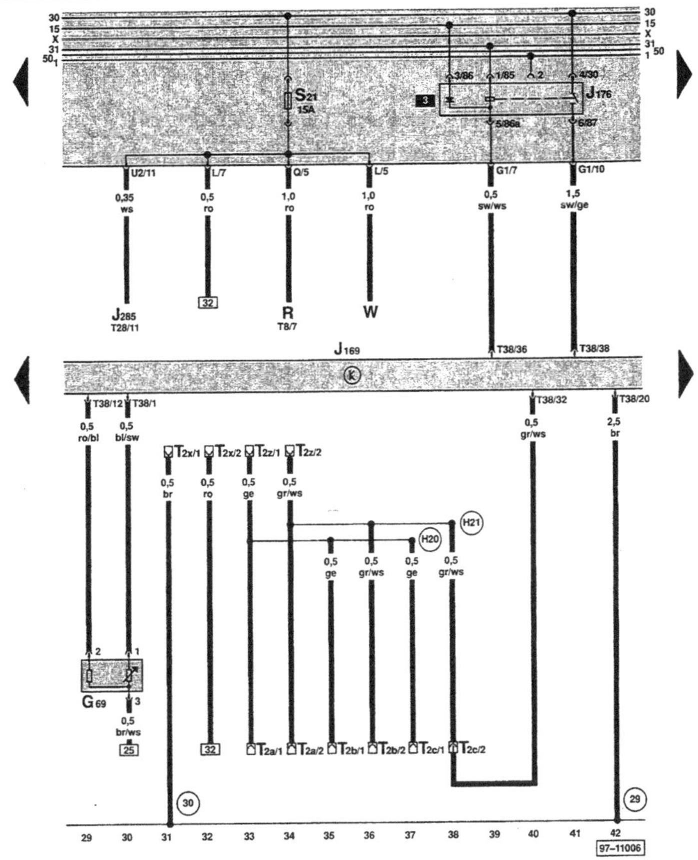

ws = white
sw = black
ro = red
br = brown
gn = green
bl = blue
gr = grey
li = violet
ge = yellow

G69 – Throttle Position (TP) Sensor (00518–2212)
J169 – Digifant Engine Control Module (ECM), near battery
J176 – Digifant Engine Control Module (ECM) Power Supply Relay
T2a – Double Connector, behind fuse/relay panel*
T2b – Double Connector, behind fuse/relay panel*
T2c – Double Connector, behind fuse/relay panel
T2x – Double Connector, –black–, behind fuse/relay panel cover (OBD)
T2z – Double Connector, –white–, behind fuse/relay panel cover (OBD)
T38 – 38-Pin Connector

(29) – Ground connection, near battery

(30) – Ground connection, -1-, beside fuse/relay panel

(H20) – Wire connection -1-, in On Board Diagnostic (OBD) wiring harness

(H21) – Wire connection -2-, in On Board Diagnostic (OBD) wiring harness

* – With further On Board Diagnostic (OBD) system only

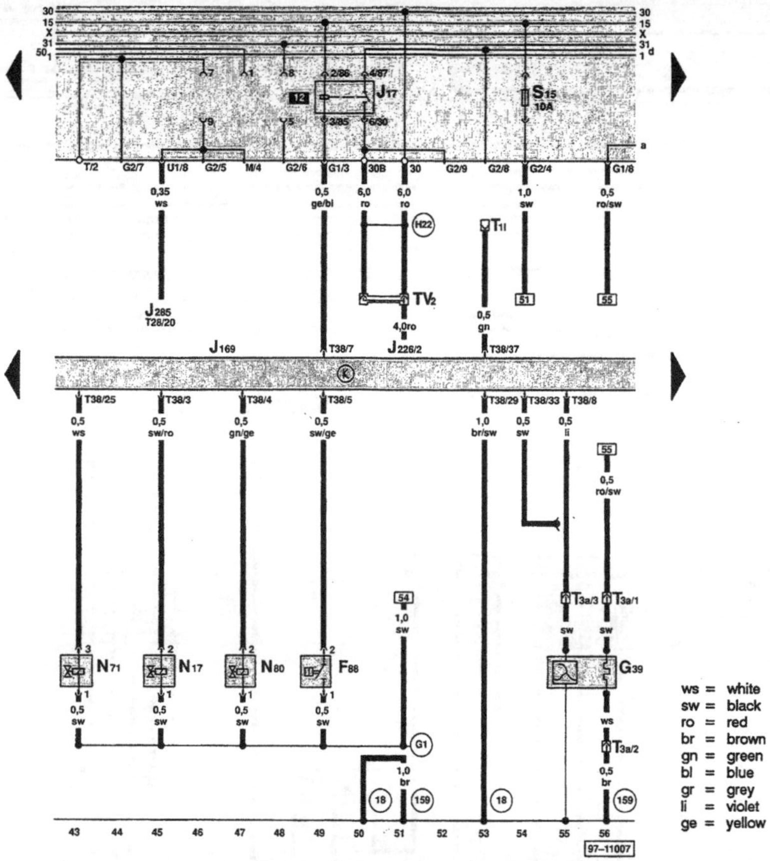

ws = white
sw = black
ro = red
br = brown
gn = green
bl = blue
gr = grey
li = violet
ge = yellow

F88 – Power Steering Pressure (PSP) Switch
G39 – Heated Oxygen Sensor (HO2S) (00525–2342)
J17 – Fuel Pump (FP) Relay
J169 – Digifant Engine Control Module (ECM), near battery
J226 – Park/Neutral Position (PNP) Relay
J285 – Control Module With Display Unit In Instrument Cluster
N17 – Cold Start Injector
N71 – Idle Air Control (IAC) Valve
N80 – Evaporative Emission (EVAP) Canister Purge Regulator Valve
T1I – Single Connector, behind fuse/relay panel (for A/C)
T3a – 3-Pin Connector, near exhaust manifold
T28 – 28-Pin Connector
T38 – 38-Pin Connector
TV2 – Terminal 30 (B+) Wire Connector
(18) – Ground connection, on engine block
(159) – Ground connection -2-, in Digifant Multiport Fuel Injection (MFI) wiring harness
(G1) – Plus connection, in Digifant Multiport Fuel Injection (MFI) wiring harness
(H22) – Plus connection (30), in terminal 30/30B wiring harness

Digifant engine control module (ECM), heated oxygen sensor (HO2S)

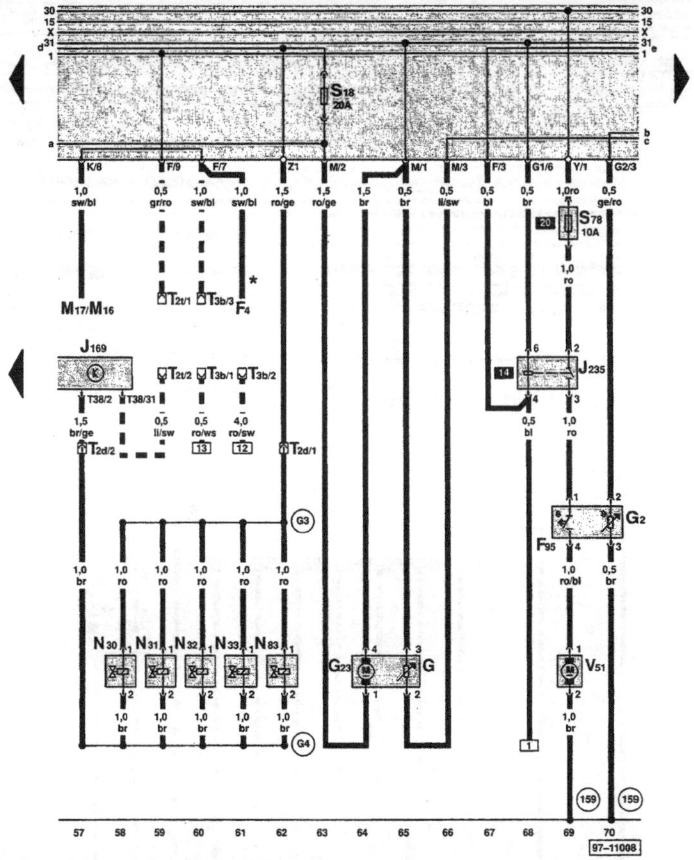

ws = white
sw = black
ro = red
br = brown
gn = green
bl = blue
gr = grey
li = violet
ge = yellow

F4 – Back-Up Light Switch
F95 – After-Run Coolant Pump Thermal Switch
G – Fuel Level Sensor (00771–3113)
G2 – Engine Coolant Temperature (ECT) Sensor
G23 – Transfer Fuel Pump (FP)
J169 – Digifant Engine Control Module (ECM)
J235 – Engine Coolant Pump Relay
M16 – Left Back-Up Light
M17 – Right Back-Up Light
N30 – Cylinder 1 Fuel Injector
N31 – Cylinder 2 Fuel Injector
N32 – Cylinder 3 Fuel Injector
N33 – Cylinder 4 Fuel Injector
N83 – Cylinder 5 Fuel Injector
S78 – Engine Coolant Pump Circuit Breaker
T2d – Double Connector, in engine comparment, right
T2t – Double Connector, behind instrument panel, left
T3b – 3-Pin Connector, behind instrument panel, left
T38 – 38-Pin Connector
V51 – After-Run Coolant Pump
(G3) – Plus connection, in injector wiring harness
(G4) – Wire connection, in injector wiring harness
(159) – Ground connection -2-, in Digifant Multiport Fuel Injection (MFI) wiring harness
* – Manual transmission only
▬ ▬ – Automatic transmission only

Digifant engine control module (ECM), injectors, coolant after-run

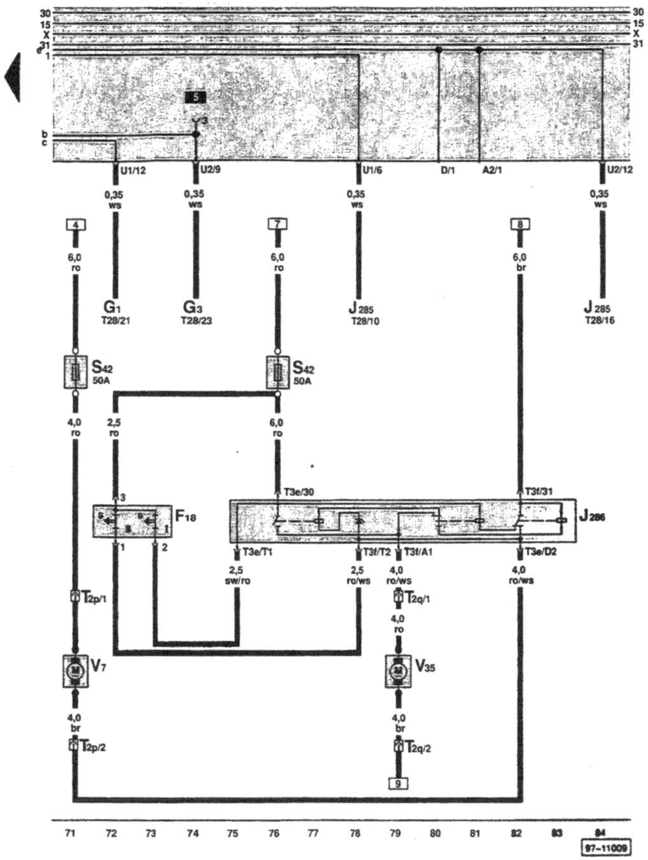

ws = white
sw = black
ro = red
br = brown
gn = green
bl = blue
gr = grey
li = violet
ge = yellow

F18 – Coolant Fan Control (FC) Thermal Switch
G1 – Fuel Gauge
G3 – Engine Coolant Temperature (ECT) Gauge
J285 – Control Module With Display Unit In Instrument Cluster
J286 – Series/Parallel Coolant Fan Control (FC) Relay (V7, V35)
S42 – Coolant Fan Fuse
T2p – Double Connector, in engine compartment, front
T2q – Double Connector, in engine compartment, front
T3e – 3-Pin Connector
T3f – 3-Pin Connector
V7 – Coolant Fan
V35 – Right Coolant Fan

THIS PAGE INTENTIONALLY LEFT

BLANK

2.5 L-Engine–Digifant Multiport Fuel Injection (MFI)/81kW, code AAF, California only,

from January 1993

Relay location
For relay location and multiple connectors see also section "component locations".

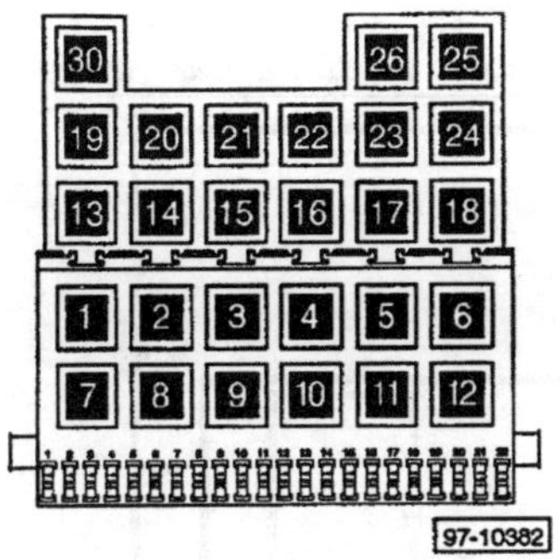

- 3 – Digifant Engine Control Module (ECM) Power Supply Relay (30)
- 4 – Load Reduction Relay (18)
- 12 – Fuel Pump (FP) Relay (80)

Note: Number in parentheses indicates production control number stamped on relay housing.

Fuse colors

30A – green
25A – white
20A – yellow
15A – blue
10A – red

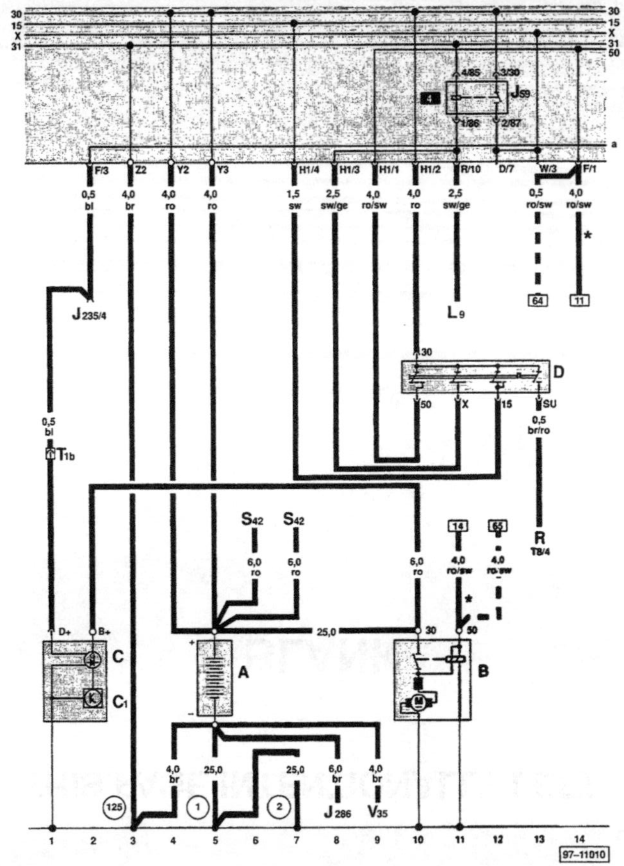

ws = white
sw = black
ro = red
br = brown
gn = green
bl = blue
gr = grey
li = violet
ge = yellow

A – Battery
B – Starter
C – Generator (GEN)
C1 – Voltage Regulator (VR)
D – Ignition/Starter Switch
J59 – Load Reduction Relay
J235 – Engine Coolant Pump Relay
J286 – Series/Parallel Coolant Fan Control (FC) Relay (V7,V35)
L9 – Headlight Switch Light
R – Radio
S42 – Coolant Fan Fuse
T1b – Single Connector, near starter
T8 – 8-Pin Connector
V35 – Right Coolant Fan

(1) – Ground strap, battery to body
(2) – Ground strap, transmission to body
(125) – Ground connection -3-, in headlight wiring harness

* – Manual transmission only
▬ ▬ – Automatic transmission only

Generator (GEN), battery, starter, Ignition/starter switch

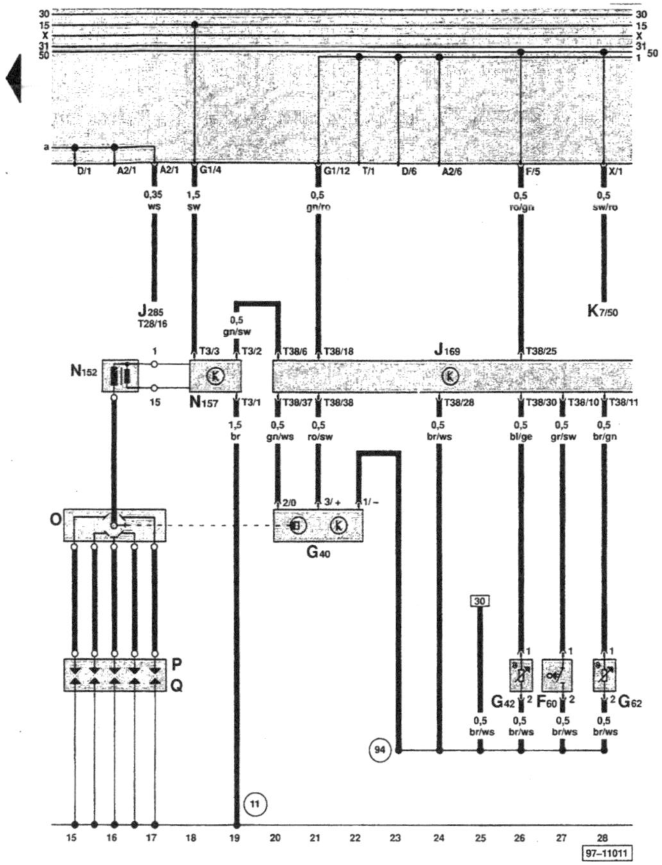

ws	= white
sw	= black
ro	= red
br	= brown
gn	= green
bl	= blue
gr	= grey
li	= violet
ge	= yellow

- F60 – Closed Throttle Position (CTP) Switch
- G40 – Camshaft Position (CMP) Sensor
- G42 – Intake Air Temperature (IAT) Sensor (00523–2322)
- G62 – Engine Coolant Temperature (ECT) Sensor (00522–2312)
- J169 – Digifant Engine Control Module (ECM), near battery
- J285 – Control Module With Display Unit In Instrument Cluster
- K7 – Brake and Parking Brake Warning Light
- N152 – Ignition Coil
- N157 – Ignition Coil Power Output Stage
- O – Ignition Distributor
- P – Spark Plug Connectors
- Q – Spark Plugs
- T3 – 3-Pin Connector
- T6e – 6-Pin Connector
- T28 – 28-Pin Connector, on instrument cluster
- T38 – 38-Pin Connector
- (11) – Ground connection, in battery box
- (94) – Ground connection -1-, in Digifant Multiport Fuel Injection (MFI) wiring harness

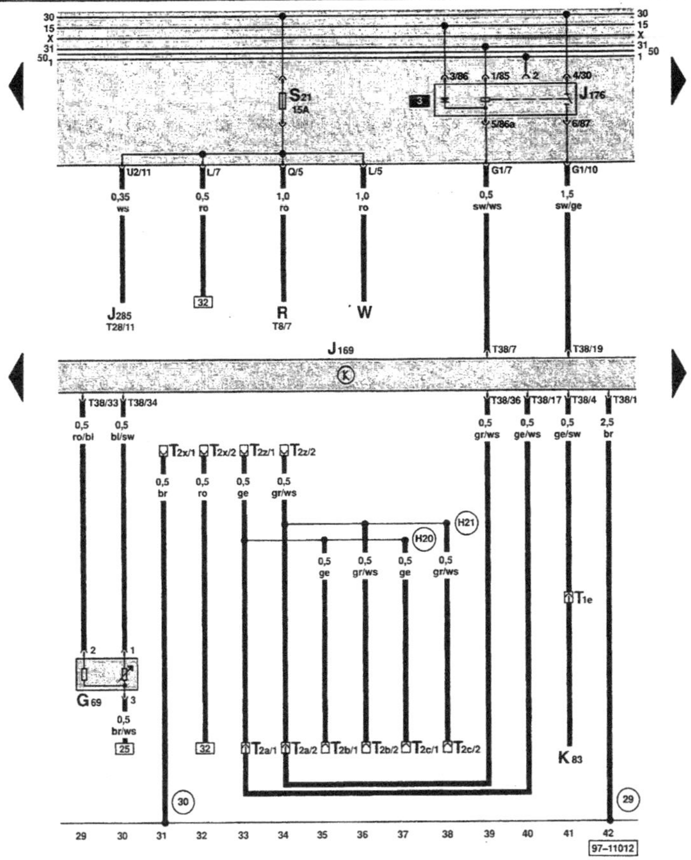

ws	= white
sw	= black
ro	= red
br	= brown
gn	= green
bl	= blue
gr	= grey
li	= violet
ge	= yellow

- G69 – Throttle Position (TP) Sensor (00518–2212)
- J176 – Digifant Engine Control Module (ECM) Power Supply Relay
- J285 – Control Module With Display Unit In Instrument Cluster
- K83 – Malfunction Indicator Lamp (MIL)
- R – Radio
- T1c – Single Connector, behind fuse/relay panel
- T2a – Double Connector, behind instrument panel, left
- T2b – Double Connector, behind instrument panel, left*
- T2c – Double Connector, behind instrument panel, left*
- T2x – Double Connector, –black–, behind fuse/relay panel cover (OBD)
- T2z – Double Connector, –white–, behind fuse/relay panel cover (OBD)
- T8 – 8-Pin Connector, on radio
- T28 – 28-Pin Connector, on instrument cluster
- T38 – 38-Pin Connector
- W – Front Interior Light
- (29) – Ground connection, near battery
- (30) – Ground connection, -1-, beside fuse/relay panel
- (H20) – Wire connection -1-, in On Board Diagnostic (OBD) wiring harness
- (H21) – Wire connection -2-, in On Board Diagnostic (OBD) wiring harness
- * – With further On Board Diagnostic (OBD) systems only

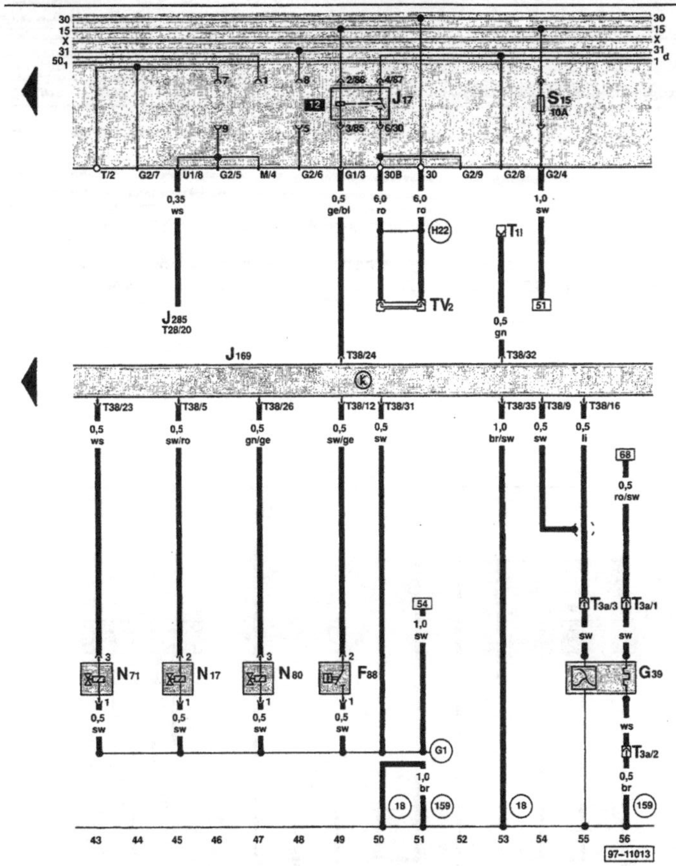

ws = white
sw = black
ro = red
br = brown
gn = green
bl = blue
gr = grey
li = violet
ge = yellow

F88 – Power Steering Pressure (PSP) Switch
G39 – Heated Oxygen Sensor (HO2S) (00525–2342)
J17 – Fuel Pump (FP) Relay
J169 – Digifant Engine Control Module (ECM), near battery
N17 – Cold Start Injector
N71 – Idle Air Control (IAC) Valve
N80 – Evaporative Emission (EVAP) Canister Purge Regulator Valve
T1l – Single Connector, behind fuse/relay panel (for A/C)
T3a – 3-Pin Connector, near exhaust manifold
T38 – 38-Pin Connector
TV2 – Terminal 30 (B+) Wire Connector

(18) – Ground connection, on engine block
(159) – Ground connection -2-, in Digifant Multiport Fuel Injection (MFI) wiring harness
(G1) – Plus connection, in Digifant Multiport Fuel Injection (MFI) wiring harness
(H22) – Plus connection (30), in terminal 30/30B wiring harness

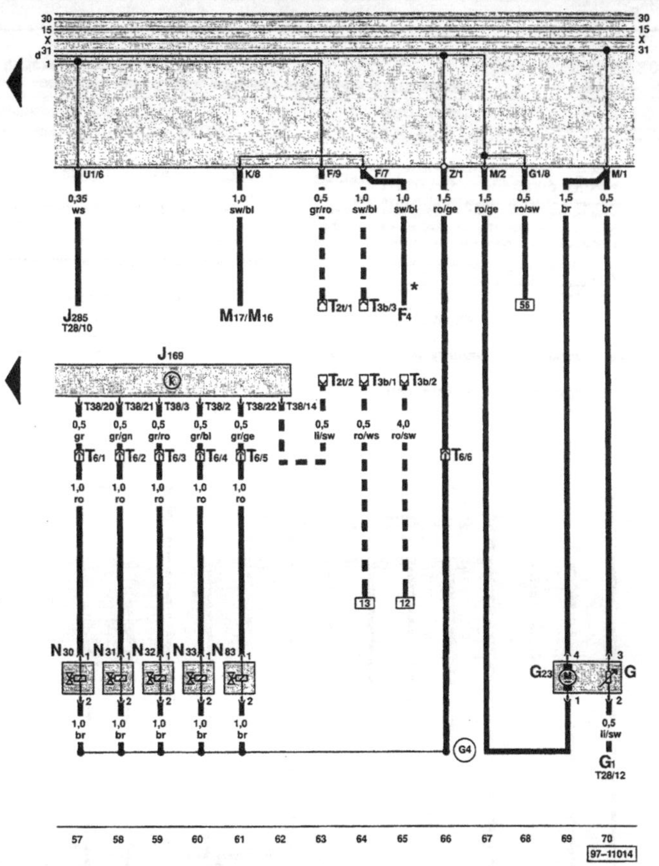

ws = white
sw = black
ro = red
br = brown
gn = green
bl = blue
gr = grey
li = violet
ge = yellow

F4 – Back-Up Light Switch
G – Fuel Level Sensor
G1 – Fuel Gauge
G23 – Transfer Fuel Pump (FP)
J169 – Digifant Engine Control Module (ECM), near battery
J285 – Control Module With Display Unit In Instrument Cluster
M16 – Left Back-Up Light
M17 – Right Back-Up Light
N30 – Cylinder 1 Fuel Injector
N31 – Cylinder 2 Fuel Injector
N32 – Cylinder 3 Fuel Injector
N33 – Cylinder 4 Fuel Injector
N83 – Cylinder 5 Fuel Injector

T2t – Double Connector, behind instrument panel, left
T3b – 3-Pin Connector, behind instrument panel, left
T6 – 6-Pin Connector, near injectors
T28 – 28-Pin Connector, on instrument cluster
T38 – 38-Pin Connector

(G3) – Plus connection, in injector wiring harness

▬ ▬ – Automatic transmission only

Automatic Transmission,
• 2.5L-Engine–Digifant Multiport Fuel Injection (MFI)/81kW, code AAF,

from January 1993

Relay location
For relay location and multiple connectors see also section "component locations".

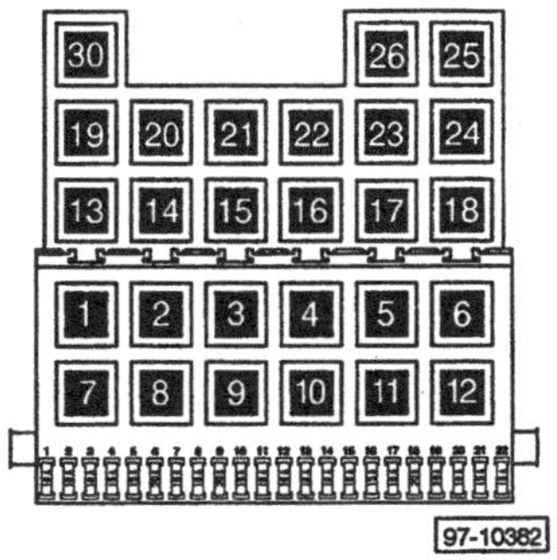

20 – Park/Neutral Position (PNP) Relay (150)

21 – Cruise Control Fuse

Note: Number in parentheses indicates production control number stamped on relay housing.

Fuse colors

30A – green
25A – white
20A – yellow
15A – blue
10A – red

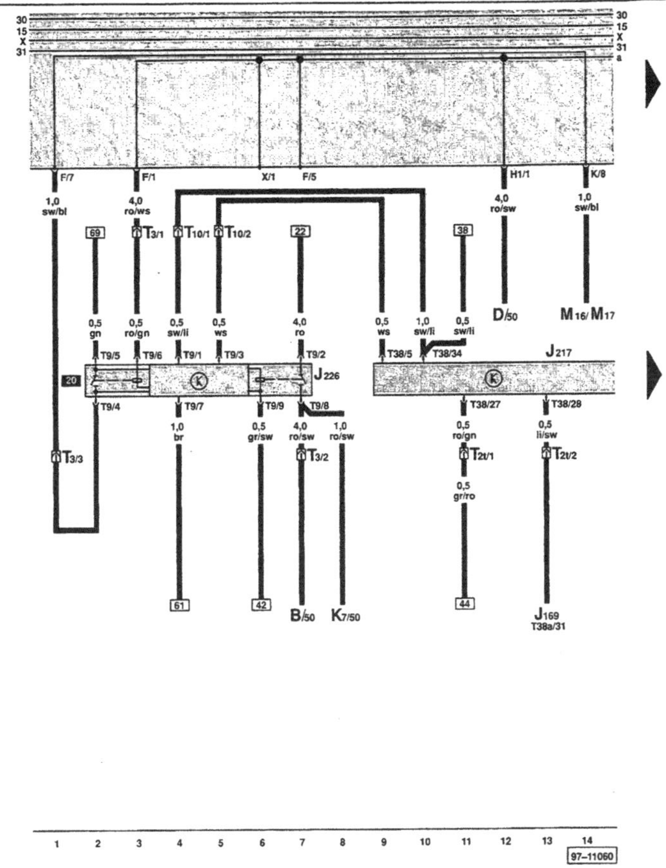

ws = white
sw = black
ro = red
br = brown
gn = green
bl = blue
gr = grey
li = violet
ge = yellow

B – Starter
D – Ignition/Starter Switch
J169 – Digifant Engine Control Module (ECM), near battery
J217 – Transmission Control Module (TCM), below instrument panel, right
J226 – Park/Neutral Position (PNP) Relay
K7 – Brake and Parking Brake Warning Light (01047)
M16 – Left Back-Up Light
M17 – Right Back-Up Light
T2t – Double Connector, engine compartment, near transmission
T3 – 3-Pin Connector, behind fuse/relay panel
T9 – 9-Pin Connector
T10 – 10-Pin Connector, behind instrument panel, right
T38 – 38-Pin Connector
T38a – 38-Pin Connector, on Digifant Engine Control Module (ECM)

Park/neutral position (PNP) relay, transmission control module (TCM)

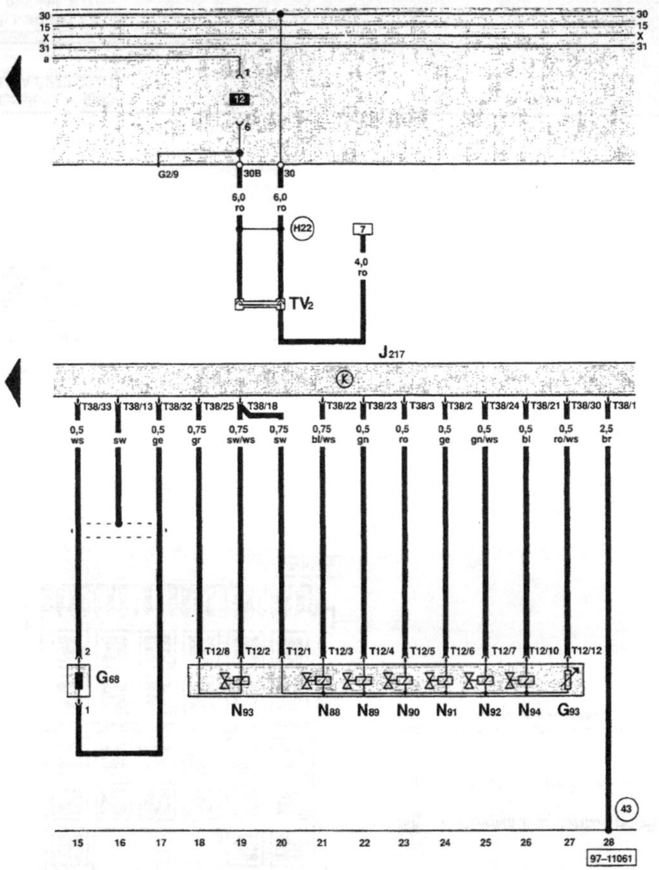

ws = white
sw = black
ro = red
br = brown
gn = green
bl = blue
gr = grey
li = violet
ge = yellow

G68 – Vehicle Speed Sensor (VSS) (00281–1231)
G93 – Transmission Fluid Temperature Sensor (00300–1333)
J217 – Transmission Control Module (TCM), below instrument panel, right
N88 – Solenoid Valve 1 (00258–1113)
N89 – Solenoid Valve 2 (00260–1121)
N90 – Solenoid Valve 3 (00262–1123)
N91 – Solenoid Valve 4 (00264–1131)
N92 – Solenoid Valve 5 (00266–1133)
N93 – Solenoid Valve 6 (00268–1141)
N94 – Solenoid Valve 7 (00270–1143)
T12 – 12-Pin Connector, on valve body
T38 – 38-Pin Connector
TV2 – Terminal 30 (B+) Wire Connector

(43) – Ground connection, on right A-pillar, lower part
(H22) – Plus connection (30), in terminal 30/30B wiring harness

Transmission control module (TCM), vehicle speed sensor (VSS), solenoid valves

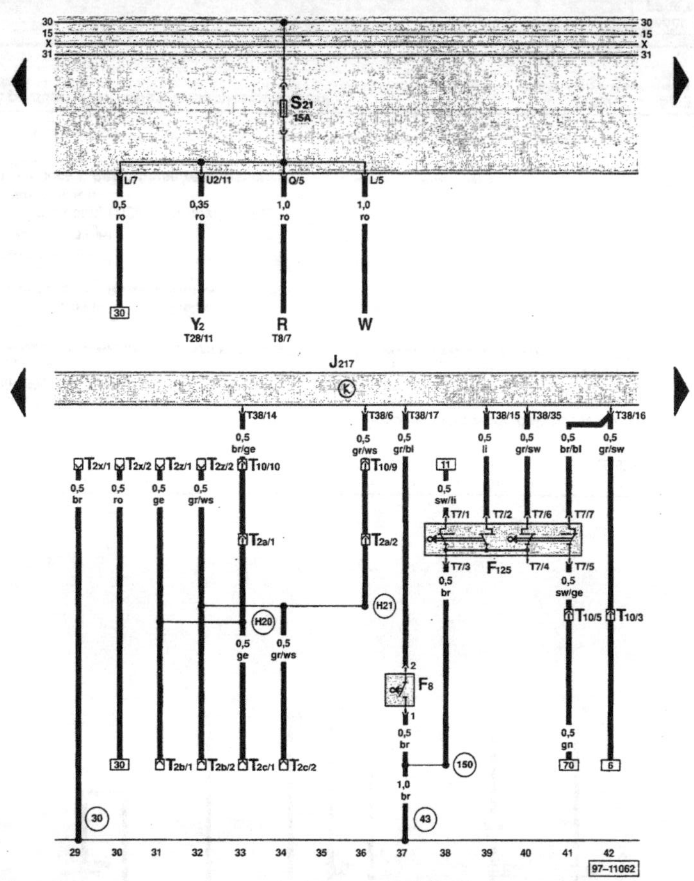

ws = white
sw = black
ro = red
br = brown
gn = green
bl = blue
gr = grey
li = violet
ge = yellow

F8 – Kick Down Switch
F125 – Multi-Function Transmission Range (TR) Switch (00293–1314)
J217 – Transmission Control Module (TCM), below instrument panel, right
R – Radio
T2a – Double Connector, behind fuse/relay panel
T2b – Double Connector, behind fuse/relay panel*
T2c – Double Connector, behind fuse/relay panel*
T2x – Double Connector, –black–, behind fuse/relay panel cover (OBD)
T2z – Double Connector, –white–, behind fuse/relay panel cover (OBD)
T7 – 7-Pin Connector
T8 – 8-Pin Connector, on radio
T10 – 10-Pin Connector, behind instrument panel, right
T28 – 28-Pin Connector, on instrument cluster
T38 – 38-Pin Connector
W – Front Interior Light
Y2 – Digital clock

(30) – Ground connection, -1-, beside fuse/relay panel
(43) – Ground connection, on right A-pillar, lower part
(150) – Ground connection, in automatic transmission (Digimat/AG4) wiring harness
(H20) – Wire connection -1-, in On Board Diagnostic (OBD) wiring harness
(H21) – Wire connection -2-, in On Board Diagnostic (OBD) wiring harness
* – With further On Board Diagnostic (OBD) systems only

Transmission control module (TCM), kick down switch, multi-function transmission range (TR) switch, on board diagnostic (OBD) system

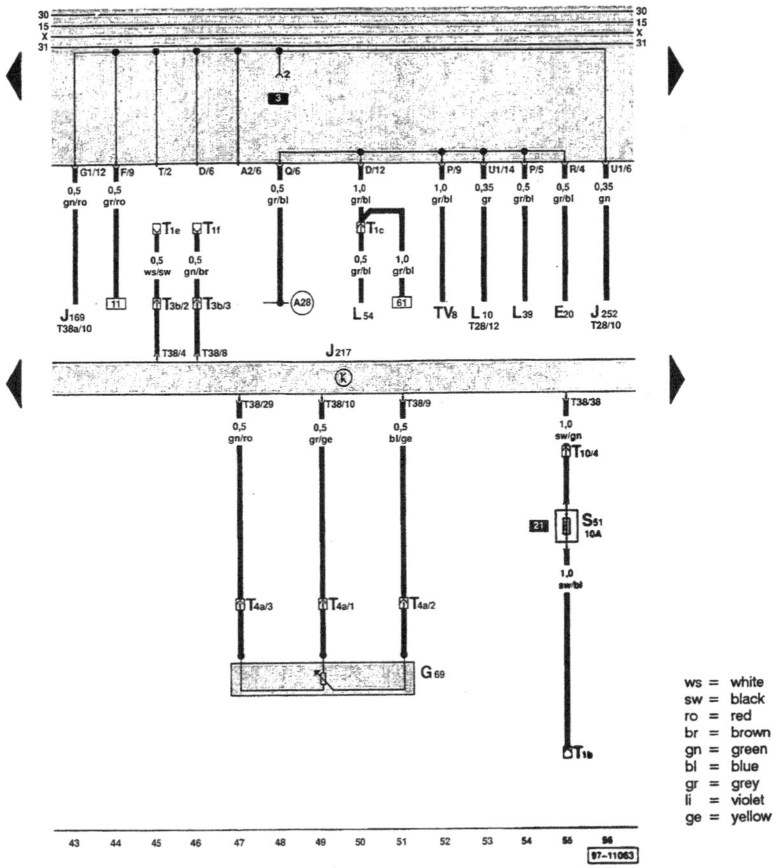

ws = white
sw = black
ro = red
br = brown
gn = green
bl = blue
gr = grey
li = violet
ge = yellow

- E20 – Instrument Panel Light Dimmer Switch
- G69 – Throttle Position (TP) Sensor (00518–2212)
- J169 – Digifant Engine Control Module (ECM), near battery
- J217 – Transmission Control Module (TCM), below instrument panel, right
- J252 – Oil Pressure and Engine Coolant Level/Temperature (ECL/ECT) Warning Control Module
- L10 – Instrument Cluster Light
- L39 – Rear Window Defogger Switch Light
- L54 – Headlight Adjuster ControlLight
- S51 – Fuse
- T1b – Single Connector, behind fuse/relay panel (cruise control system)
- T1c – Single Connector, behind fuse/relay panel
- T1e – Single Connector, behind fuse/relay panel
- T1f – Single Connector, behind fuse/relay panel (for A/C)
- T3b – 3-Pin Connector, behind instrument panel, right
- T4a – 4-Pin Connector, near throttle position (TP) sensor
- T10 – 10-Pin Connector, behind instrument panel, right
- T28 – 28-Pin Connector, on instrument cluster
- T38 – 38-Pin Connector
- T38a – 38-Pin Connector, on Digifant Engine Control Module (ECM)
- TV8 – Terminal 58b Wire Connector
- (A28) – Plus connection (58b), in fresh air blower wiring harness

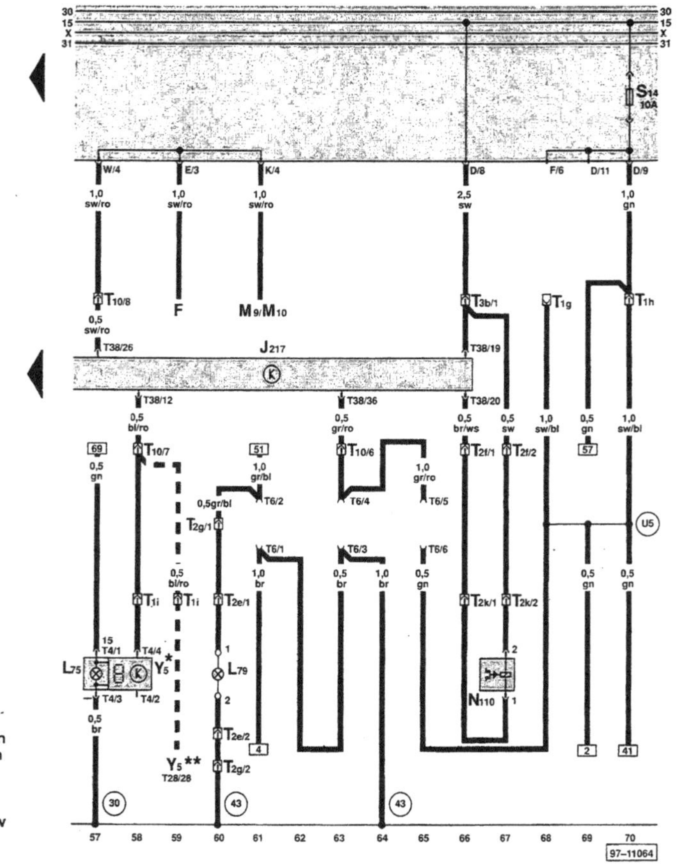

ws = white
sw = black
ro = red
br = brown
gn = green
bl = blue
gr = grey
li = violet
ge = yellow

- F – Brake Light Switch (00526–2131)
- J217 – Transmission Control Module (TCM), below instrument panel, right
- L75 – Digital Display Light
- L79 – Selector Lever Light
- M9 – Left Brake Light
- M10 – Right Brake Light
- N110 – Shift Lock Solenoid (01236–4314)
- T1g – Single Connector, behind fuse/relay panel
- T1h – Single Connector, behind fuse/relay panel
- T1i – Single Connector, behind fuse/relay panel
- T2e – Double Connector, near selector lever
- T2f – Double Connector, behind instrument panel, right
- T2g – Double Connector, behind instrument panel, right
- T2k – Double Connector, near shift lock solenoid
- T3b – 3-Pin Connector, behind instrument panel, right
- T4 – 4-Pin Connector
- T6 – 6-Pin Connector, for Transmission Range (TR) Program Switch (not used for transmission-codes CFZ, CGA, CGH and CGK beginning 01/93)
- T10 – 10-Pin Connector, behind instrument panel, right
- T38 – 38-Pin Connector
- Y5 – Transmission Range (TR) Selector Lever Display
- * – in instrument panel
- ** – in instrument cluster (instrument cluster without analog clock beginning August 1992)
- (30) – Ground connection, -1-, beside fuse/relay panel
- (43) – Ground connection, on right A-pillar, lower part
- (U5) – Wire connection -2- (15), in automatic transmission wiring harness

2.5 L-Engine – Digifant Multiport Fuel Injection (MFI)/81kW, code ACU,

from September 1994

Relay location
For relay location and multiple connectors see also section "component locations".

- 2 – Rear Window Wiper/Washer Relay (72)
- 3 – Digifant Engine Control Module (ECM) Power Supply Relay (30)
- 4 – Load Reduction Relay (18)
- 6 – Emergency Flasher Relay (21)
- 8 – Washer/Wiper Intermittent Relay (19)
- 11 – Dual Horn Relay (53)
- 12 – Fuel Pump (FP) Relay (80)

Note: Number in parentheses indicates production control number stamped on relay housing.

Fuse colors

30A – green
25A – white
20A – yellow
15A – blue
10A – red

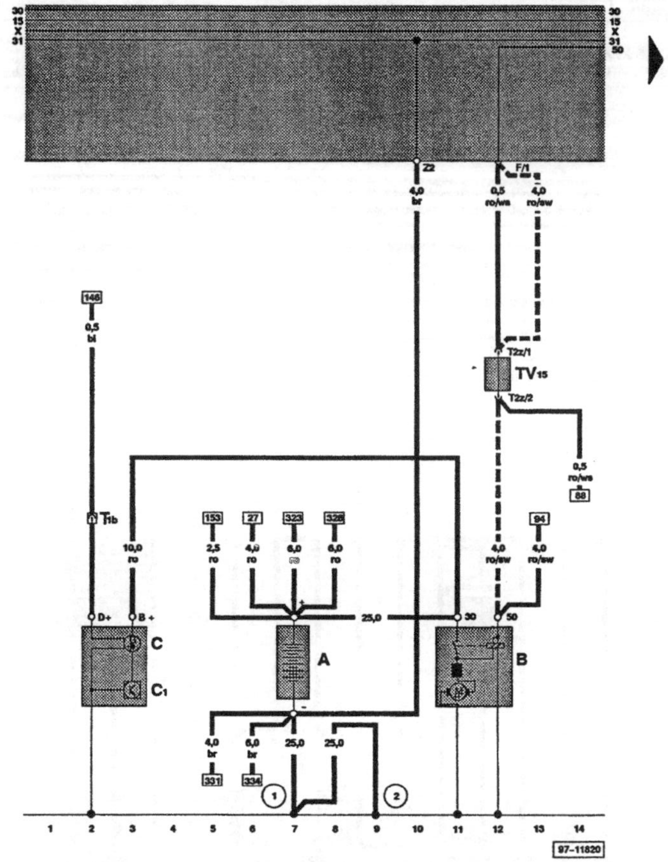

ws = white
sw = black
ro = red
br = brown
gn = green
bl = blue
gr = grey
li = violet
ge = yellow

A – Battery
B – Starter
C – Generator (GEN)
C1 – Voltage Regulator (VR)
T1b – Single Connector, near starter
T2z – Double Connector, behind instrument panel ,left
TV15 – Wire Connector, behind fuse/relay panel (cancelled)

(1) – Ground strap, battery to body

(2) – Ground strap, transmission to body

▬ ▬ – Manual transmission only

Generator (GEN), battery, starter

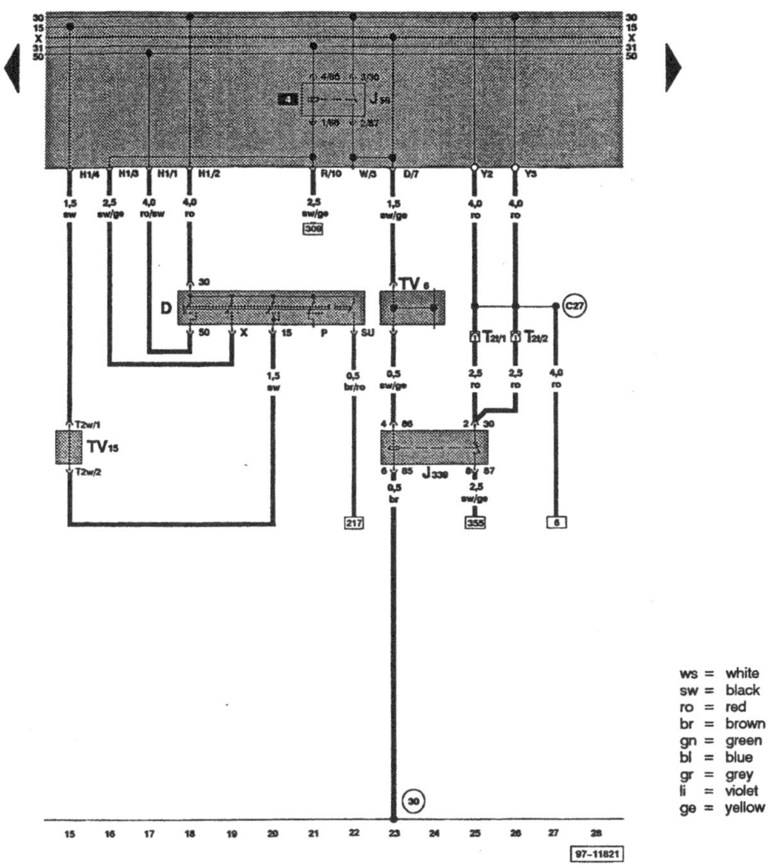

ws = white
sw = black
ro = red
br = brown
gn = green
bl = blue
gr = grey
li = violet
ge = yellow

D – Ignition/Starter Switch
J59 – Load Reduction Relay
J339 – Load Reduction Relay II, behinf instrument panel, center
T2t – Double Connector, behind fuse/relay panel
TV6 – Terminal X Wire Connector, above fuse/relay panel
TV15 – Wire Connector, behind fuse/relay panel (cancelled)

(30) – Ground connection, –1–, beside fuse/relay panel

(C27) – Plus connection –2– (30), in headlight wiring harness

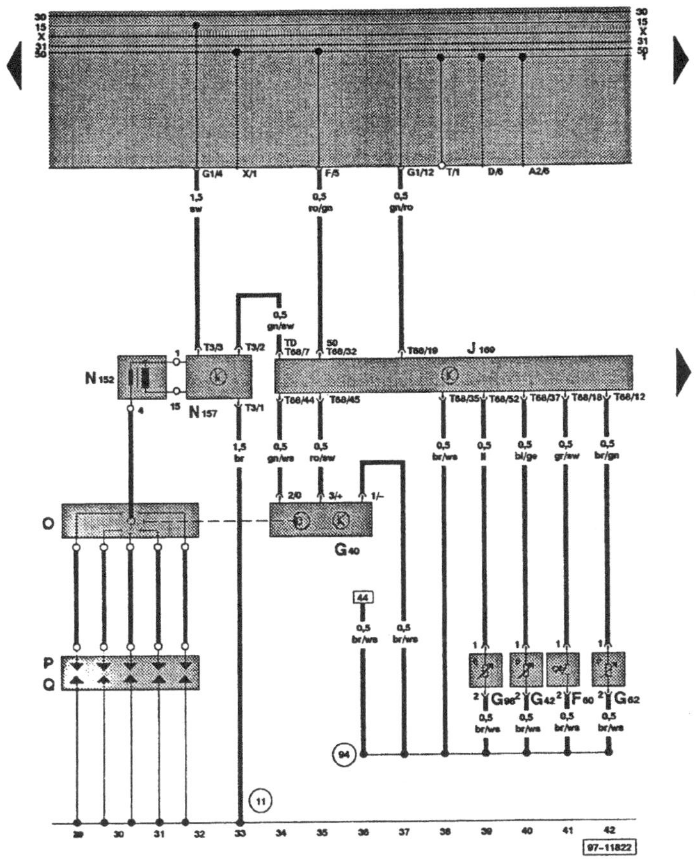

ws = white
sw = black
ro = red
br = brown
gn = green
bl = blue
gr = grey
li = violet
ge = yellow

F60 – Closed Thottle Position (CTP) Switch
G40 – Camshaft Position (CMP) Sensor (00515–2113)
G42 – Intake Air Temperature (IAT) Sensor (00523–2322)
G62 – Engine Coolant Temperature (ECT) Sensor (00522–2312)
G98 – EGR Temperature Sensor
J169 – Digifant Engine Control Module (ECM), near battery
N152 – Ignition Coil
N157 – Ignition Coil Power Output Stage
O – Ignition Distributor
P – Spark Plug Connectors
Q – Spark Plugs
T3 – 3-Pin Connector
T68 – 68-Pin Connector

(11) – Ground connection, in battery box

(94) – Ground connection –1–, in Digifant Multiport Fuel Injection (MFI) wiring harness

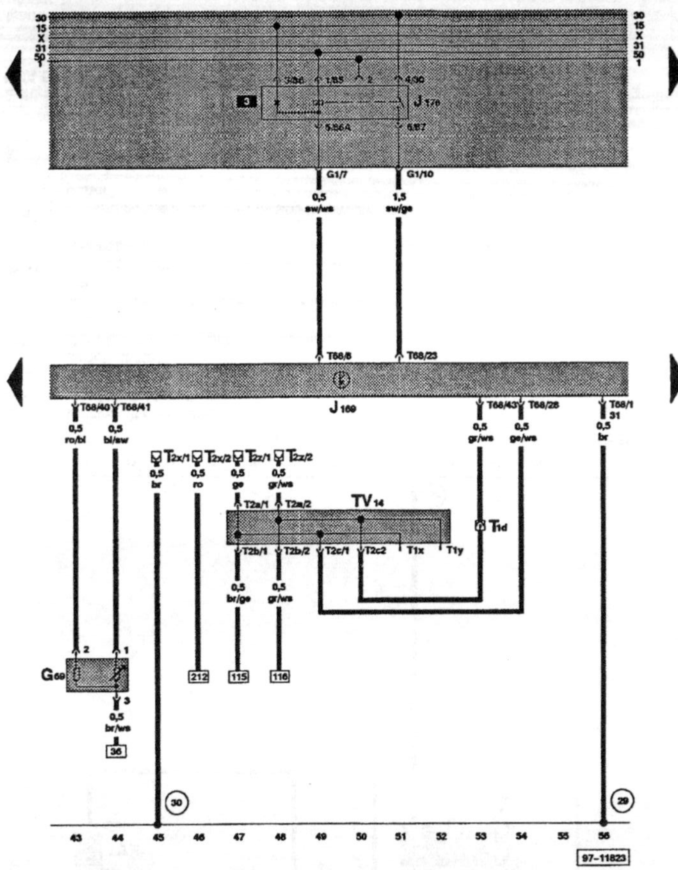

ws = white
sw = black
ro = red
br = brown
gn = green
bl = blue
gr = grey
li = violet
ge = yellow

G69 – Throttle Position (TP) Sensor (00518–2212)
J169 – Digifant Engine Control Module (ECM), near battery
J176 – Digifant Engine Control Module (ECM) Power Sypply Relay
T1d – Single Connector, behind fuse relay/panel
T1x – Single Connector
T1y – Single Connector
T2a – Double Connector, above fuse/relay panel
T2b – Double Connector, above fuse/relay panel
T2c – Double Connector
T2x – Double Connector, black, behind instrument panel, center (OBD)
T2z – Double Connector, white, behind instrument panel, center (OBD)
T68 – 68-Pin Connector

TV14 – Data Link Connector (DLC) Wire Connector for on-board diagnostic (OBD)

(29) – Ground connection, near battery

(30) – Ground connection, –1–, beside fuse/relay panel

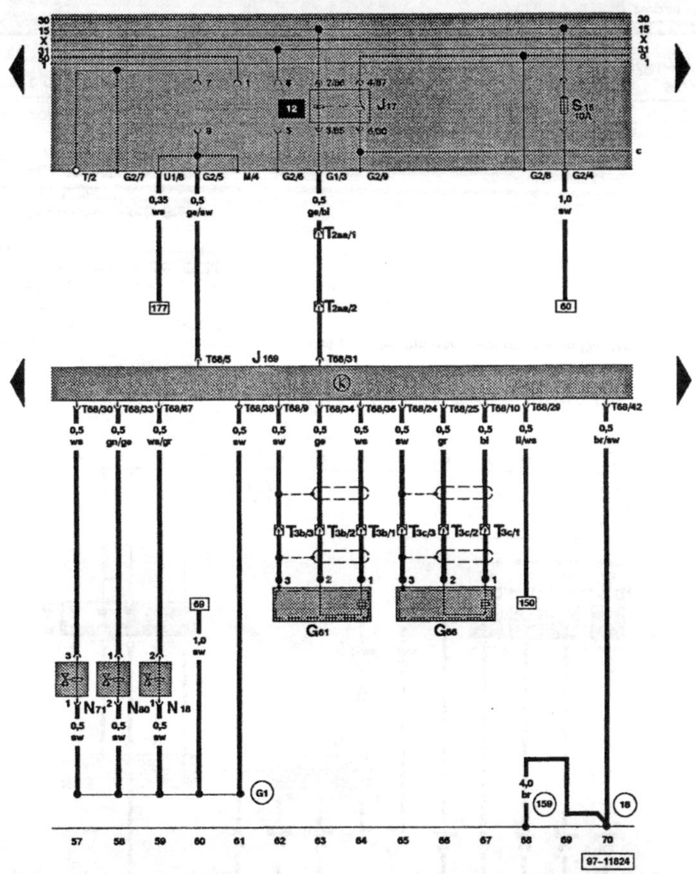

ws = white
sw = black
ro = red
br = brown
gn = green
bl = blue
gr = grey
li = violet
ge = yellow

G61 – Knock Sensor (KS) 1 (00524–2142)
G66 – Knock Sensor (KS) II (00540–2144)
J17 – Fuel Pump (FP) Relay
J169 – Digifant Engine Control Module (ECM), near battery
N18 – EGR Vacuum Regulator Solenoid Valve
N71 – Idle Air Control (IAC) Valve
N80 – Evaporative Emission (EVAP) Canister Purge Regulator Valve
T2aa – Double Connector (was not used)
T3b – 3-Pin Connector, near exhaust manifold
T3c – 3-Pin Connector, near exhaust manifold
T68 – 68-Pin Connector

(18) – Ground connection, on engine block

(159) – Ground connection –2–, in Digifant Multiport Fuel Injection (MFI) wiring harness

(G1) – Plus connection, in Digifant Multiport Fuel Injection (MFI) wiring harness

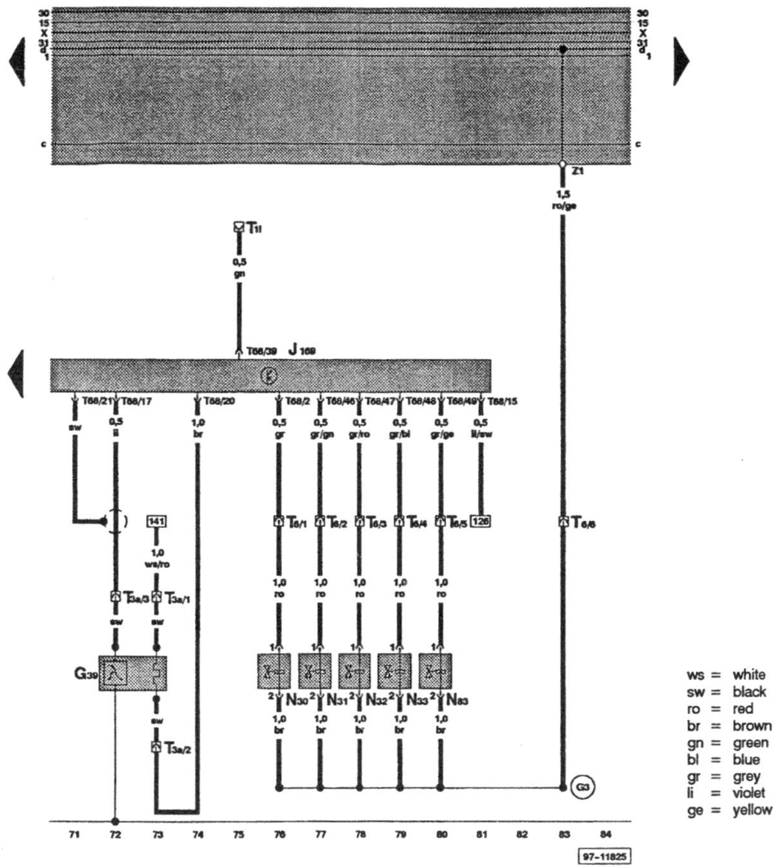

ws = white
sw = black
ro = red
br = brown
gn = green
bl = blue
gr = grey
li = violet
ge = yellow

G39 – Heated Oxygen Sensor (HO2S) (00525–2342)
J169 – Digifant Engine Control Module (ECM), near battery
N30 – Cylinder 1 Fuel Injector (01249–4411)
N31 – Cylinder 2 Fuel Injector
N32 – Cylinder 3 Fuel Injector
N33 – Cylinder 4 Fuel Injector
N83 – Cylinder 5 Fuel Injector
T1l – Single Connector, behind fuse/relay panel, (for A/C)
T3a – 3-Pin Connector, near exhaust pipe
T6 – 6-Pin Connector, in engine compartment, right
T68 – 68-Pin Connector

(G3) – Plus connection, in injector wiring harness

Digifant engine control module (ECM), heated oxygen sensor (HO2S), injectors

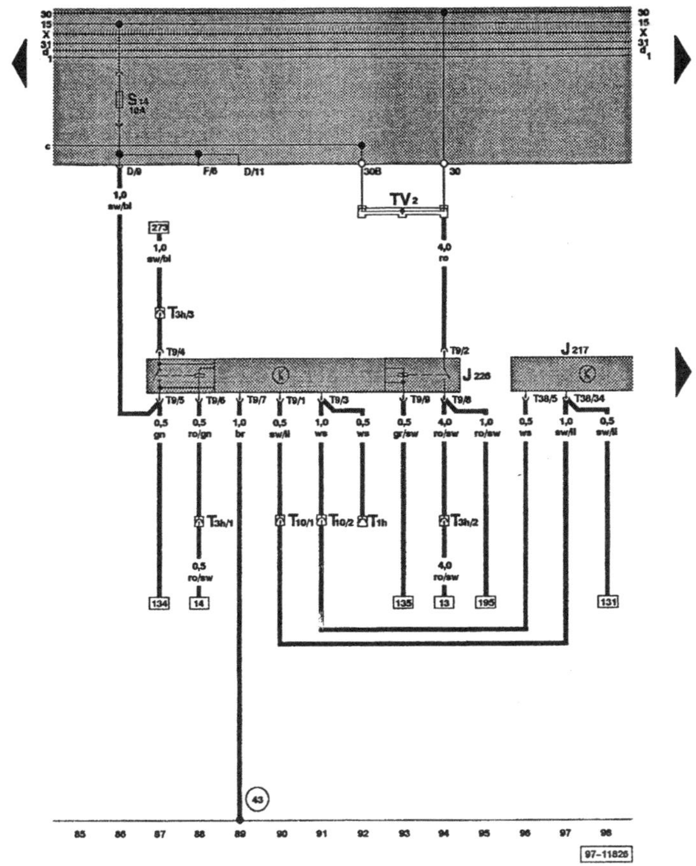

ws = white
sw = black
ro = red
br = brown
gn = green
bl = blue
gr = grey
li = violet
ge = yellow

J217 – Transmission Control Module (TCM), on right A-pillar, lower part
J226 – Park/Neutral Position (PNP) Relay, behind instrument panel, center
T1h – Single Connector, behind fuse/relay panel
T3h – 3-Pin Connector, behind fuse/relay panel
T9 – 9-Pin Connector
T10 – 10-Pin Connector, behind instrument panel, right
T38 – 38-Pin Connector
TV2 – Terminal 30 (B+) Wire Connector

(43) – Ground connection, on right A-pillar, lower part

Park/Neutral position (PNP) relay, transmission control module (TCM)

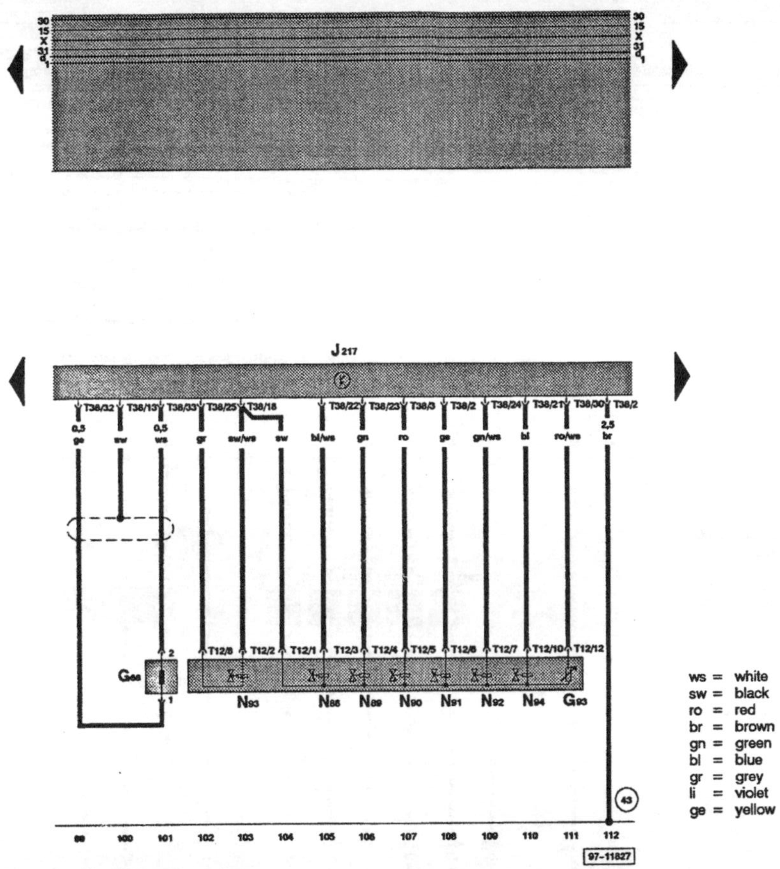

ws = white
sw = black
ro = red
br = brown
gn = green
bl = blue
gr = grey
li = violet
ge = yellow

G68 – Vehicle Speed Sensor (VSS) (00281–1231)
G93 – Transmission Fluid Temperature Sensor (00300–1333)
J217 – Transmission Control Module (TCM), on right A-pillar, lower part
N88 – Solenoid valve 1 (00258–1113)
N89 – Solenoid valve 2 (00260–1121)
N90 – Solenoid valve 3 (00262–1123)
N91 – Solenoid valve 4 (00264–1131)
N92 – Solenoid valve 5 (00266–1133)
N93 – Solenoid valve 6 (00268–1141)
N94 – Solenoid valve 7 (00270–1143)
T12 – 12-Pin Connector
T38 – 38-Pin Connector

(43) – Ground connection, on right A-pillar, lower part

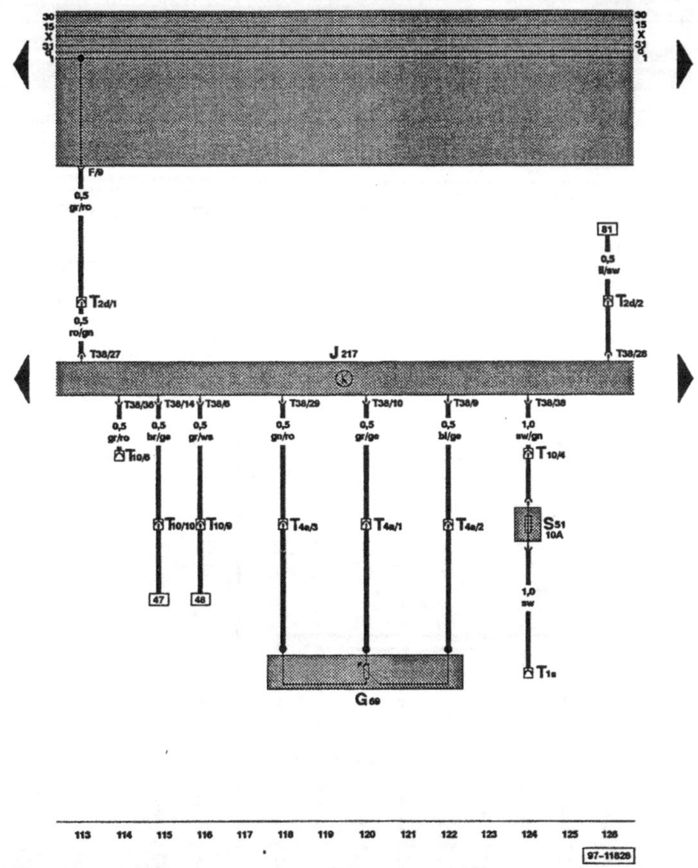

ws = white
sw = black
ro = red
br = brown
gn = green
bl = blue
gr = grey
li = violet
ge = yellow

G69 – Throttle Position (TP) Sensor (00518–2212)
J217 – Transmission Control Module (TCM), on right A-pillar, lower part
S51 – Fuse, above fuse relay panel
T1s – Single Connector, behind fuse/relay panel (cruise control system)
T2d – Double Connector
T4a – 4-Pin Connector
T10 – 10-Pin Connector, behind instrument panel, right
T38 – 38-Pin Connector

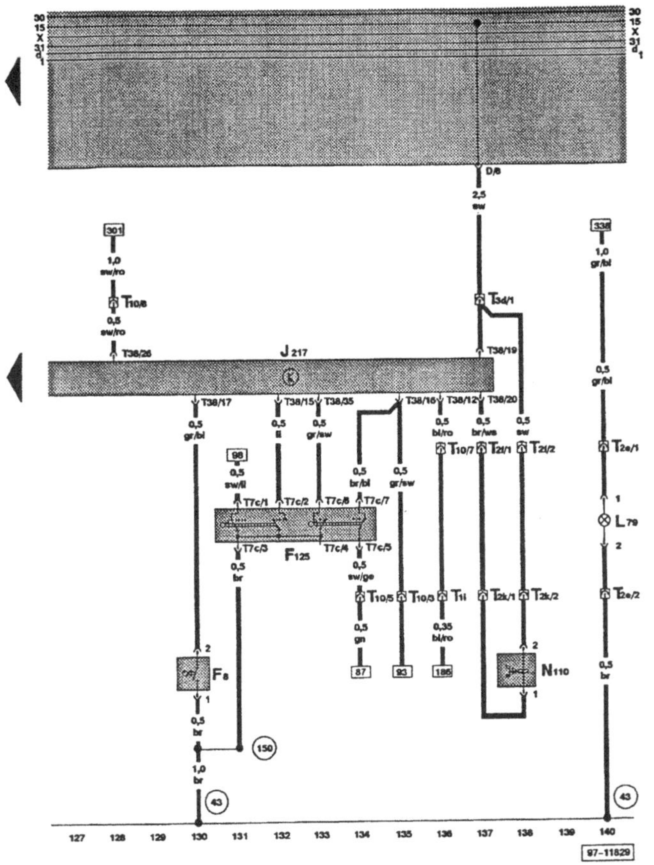

ws =	white
sw =	black
ro =	red
br =	brown
gn =	green
bl =	blue
gr =	grey
li =	violet
ge =	yellow

F8 – Kick Down Switch
F125 – Multi–Function Transmission Range (TR) Switch (00293–1314)
J217 – Transmission Control Module (TCM), on right A-pillar, lower part
L79 – Selector Lever Light
N110 – Shift Lock Solenoid (01236–4314)
T1i – Single Connector, behind fuse/relay panel
T2e – Double Connector, near selector lever
T2f – Double Connector, behind instrument panel, right
T2k – Double Connector, near selector lever
T3d – 3-Pin Connector, behind instrument panel, right
T7c – 7-Pin Connector
T10 – 10-Pin Connector, behind instrument panel, right
T38 – 38-Pin Connector

(43) – Ground connection, on right A-pillar, lower part

(150) – Ground connection, in automatic transmission wiring harness

Transmission control module (TCM), multi-function transmission range (TR) switch

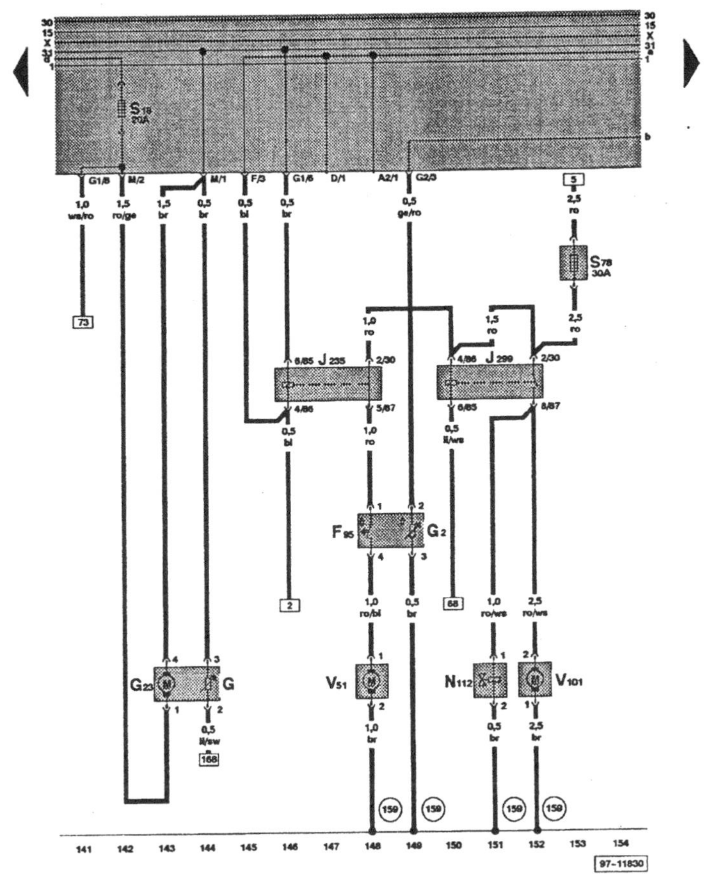

ws =	white
sw =	black
ro =	red
br =	brown
gn =	green
bl =	blue
gr =	grey
li =	violet
ge =	yellow

F95 – After-Run Coolant Pump Thermal Switch
G – Fuel Level Sensor
G2 – Engine Coolant Temperature (ECT) Sensor
G23 – Transfer Fuel Pump (FP)
J235 – Engine Coolant Pump Relay, behind fuse/relay panel
J299 – Secondary Air Injection (AIR) Pump Relay
N112 – Secondary Air Injection (AIR) Solenoid Valve
S78 – Engine Coolant Pump and Secondary Air Injection (AIR) Pump Motor Circuit Breaker, behind fuse/relay panel
V51 – After-Run Coolant Pump
V101 – Secondary Air Injection (AIR) Pump Motor

(159) – Ground connection –2–, in Digifant Multi-port Fuel Injection (MFI) wiring harness

Fuel supply, coolant after-run, secondary air injection (AIR) system

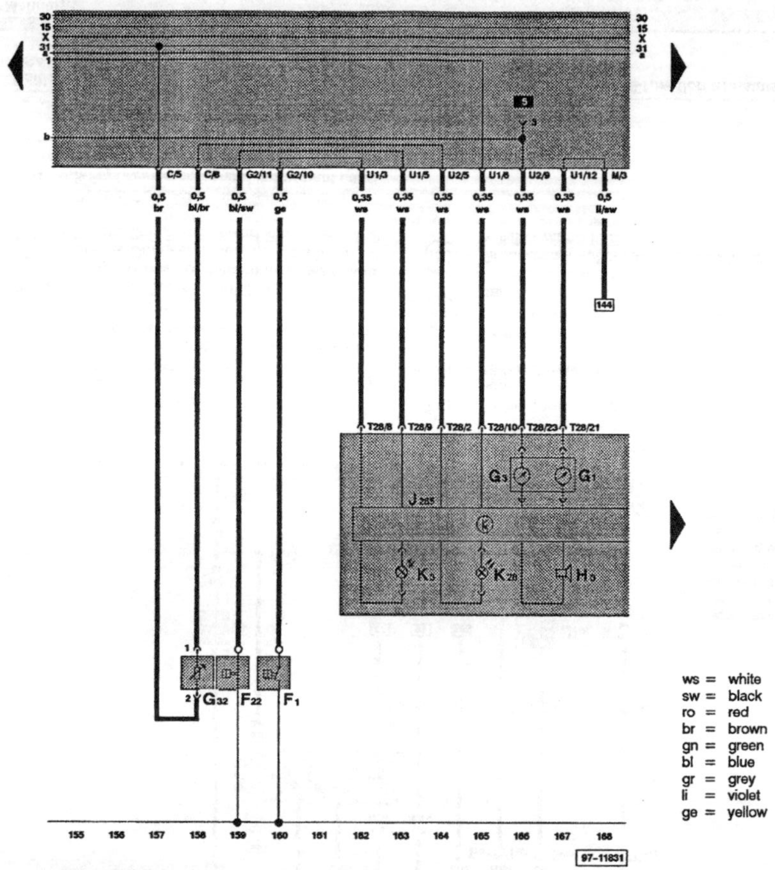

ws = white
sw = black
ro = red
br = brown
gn = green
bl = blue
gr = grey
li = violet
ge = yellow

F1 – Oil Pressure Switch
F22 – 0.3 Bar Oil Pressure Switch
G1 – Fuel Gauge
G3 – Engine Coolant Temperature (ECT) Gauge
G32 – Engine Coolant Level (ECL) Sensor
H3 – Warning Buzzer
J285 – Control Module With Display Unit In Instrument Cluster
K3 – Oil Pressure Warning Light
K28 – Engine Coolant Level/Temperature (ECL/ECT) Warning Light, on instrument cluster
T28 – 28-Pin Connector

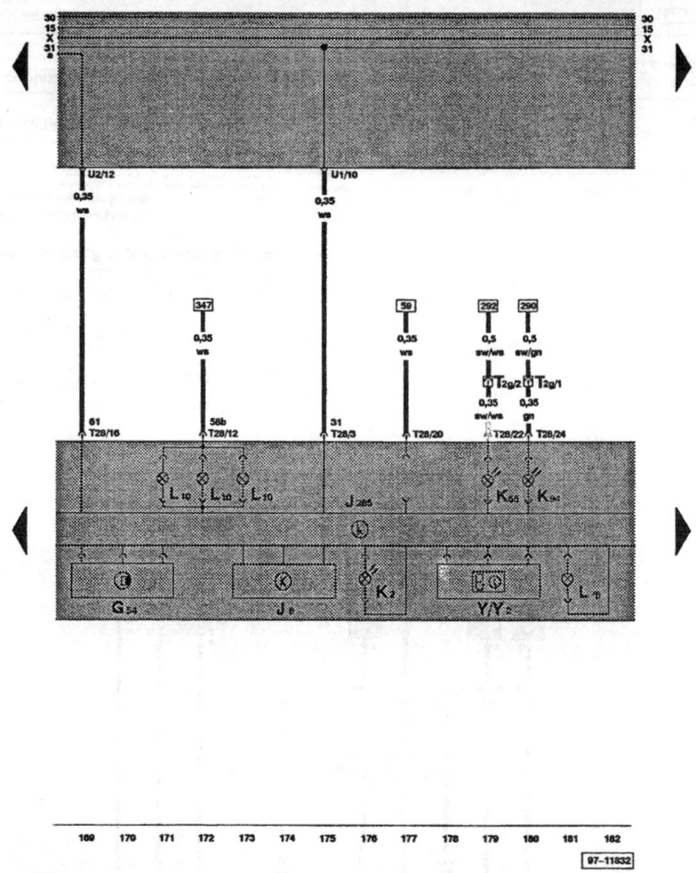

ws = white
sw = black
ro = red
br = brown
gn = green
bl = blue
gr = grey
li = violet
ge = yellow

G54 – Vehicle Speed Sensor (VSS)
J6 – Voltage Stabilizer
J285 – Control Module With Display Unit In Instrument Cluster
K2 – Generator (GEN) Warning Light
K65 – Left Turn Signal Indicator Light
K94 – Right Turn Signal Indicator Light
L10 – Instrument Cluster Light
L75 – Digital Display Light
T2g – Double Connector, behind fuse/relay panel
T28 – 28-Pin Connector, on instrument cluster
Y – Analog Clock
Y2 – Digital Clock

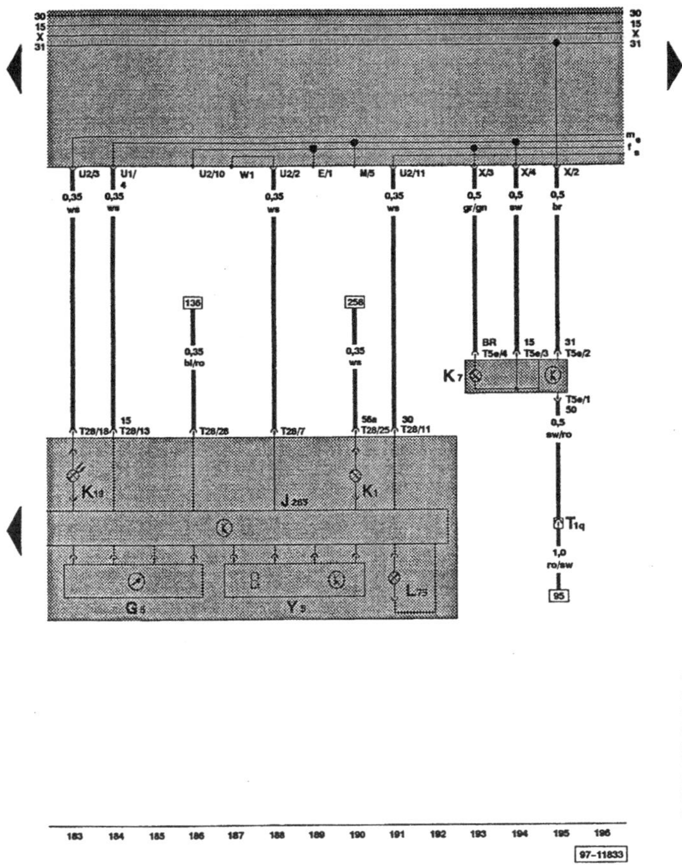

ws = white
sw = black
ro = red
br = brown
gn = green
bl = blue
gr = grey
li = violet
ge = yellow

G5 – Tachometer
J285 – Control Module With Display Unit In Instrument Cluster
K1 – Headlight High Beam Indicator Light
K7 – Brake and Parking Brake Warning Light
K19 – Seat Belt Warning Light
L75 – Digital Display Light
T1q – Single Connector, behind fuse/relay panel
T5e – 5-Pin Connector
T28 – 28-Pin Connector, on instrument cluster
Y5 – Transmission Range (TR) Selector Lever Display

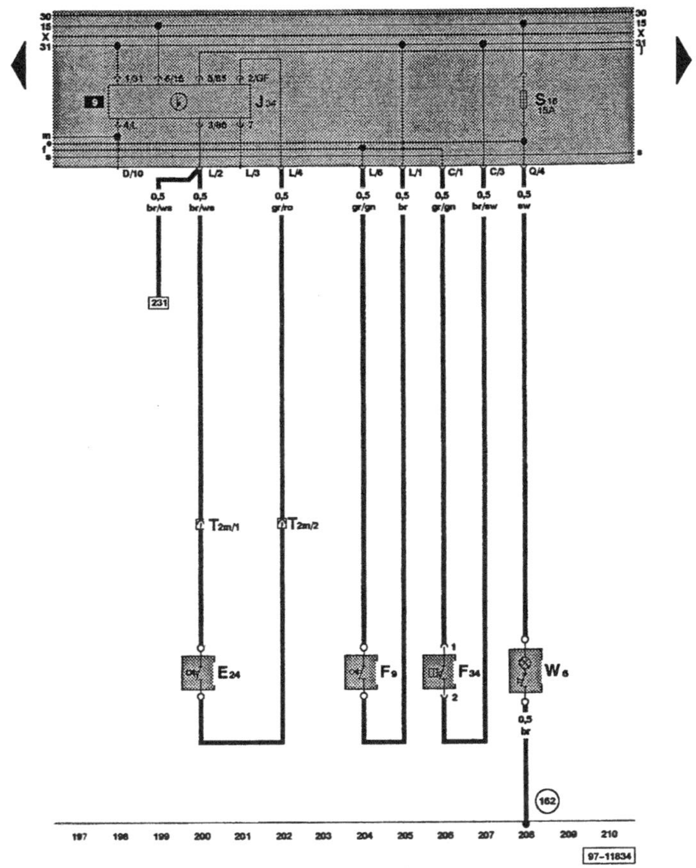

ws = white
sw = black
ro = red
br = brown
gn = green
bl = blue
gr = grey
li = violet
ge = yellow

E24 – Left Seat Belt Switch
F9 – Parking Brake Warning Light Switch
F34 – Brake Fluid Level Warning Switch
J34 – Seat Belt Warning System Relay
T2m – Double Connector, near seat belt switch
W6 – Glove Compartment Light

(162) – Ground connection, in blower motor wiring harness

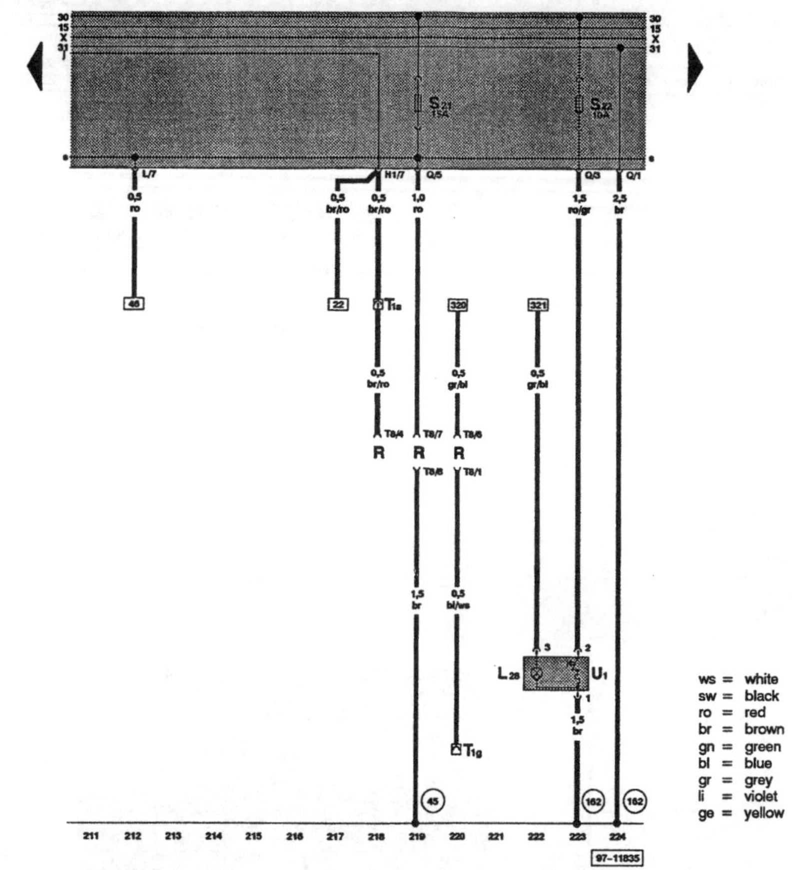

ws = white
sw = black
ro = red
br = brown
gn = green
bl = blue
gr = grey
li = violet
ge = yellow

L28 – Cigarette Lighter Light
R – Radio
T1a – Single Connector, behind fuse/relay panel
T1g – Single connector, behind fuse/relay panel (Radio with GALA)
T8 – 8-Pin Connector
U1 – Cigarette Lighter

(45) – Ground connection, behind instrument panel, center

(162) – Ground connection, in blower motor wiring harness

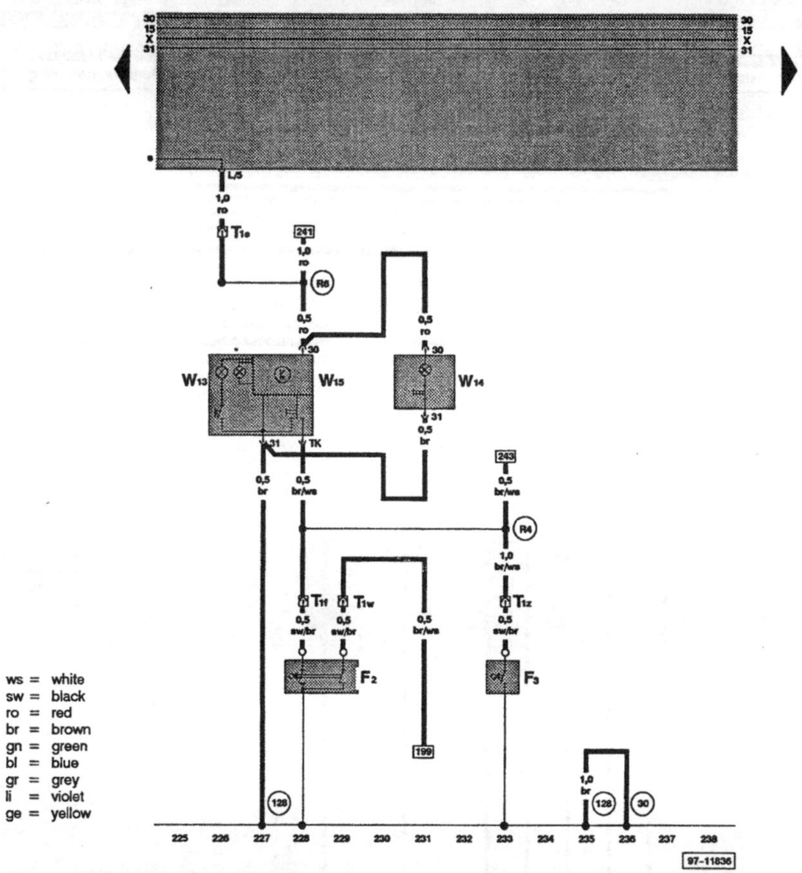

ws = white
sw = black
ro = red
br = brown
gn = green
bl = blue
gr = grey
li = violet
ge = yellow

F2 – Left Front Door Contact Switch
F3 – Right Front Door Contact Switch
T1e – Single Connector, behind fuse/relay panel
T1f – Single Connector, behind fuse/relay panel
T1w – Single Connector, behind fuse/relay panel
T1z – Single Connector, behind instrument panel, right
W13 – Right Front Map/Reading Light
W14 – Right Make-up Mirror Light
W15 – Interior Light With Delay Switch

(30) – Ground connection, –1–, beside fuse/relay panel

(128) – Ground connection –1–, in interior light wiring harness

(R4) – Wire connection –1–, in interior light/door contact switch wiring harness

(R6) – Plus connection –1–, in interior light wiring harness

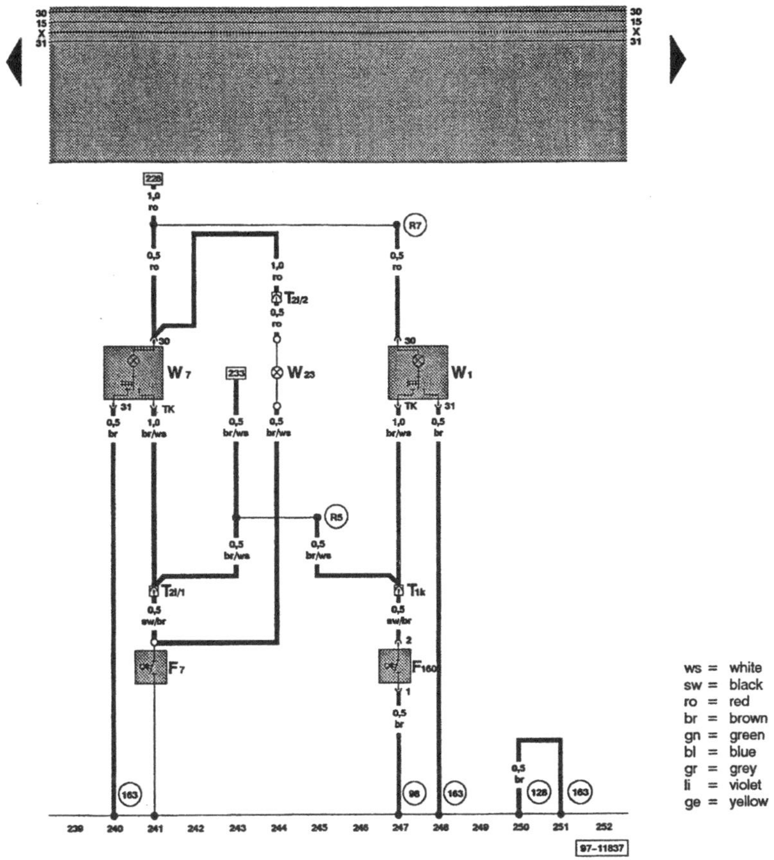

ws = white
sw = black
ro = red
br = brown
gn = green
bl = blue
gr = grey
li = violet
ge = yellow

F7 – Right Rear Side/Sliding Door Contact Switch
F160 – Rear Lid Contact Switch
T1k – Single Connector, near tail light, right
T2i – Double Connector, near interior light, center
W1 – Rear/Luggage Compartment Interior Light
W7 – Center Interior Light
W23 – Footwell Light With Delay Switch

(98) – Ground connection, in rear lid wiring harness

(128) – Ground connection –1–, in interior light wiring harness

(163) – Ground connection –2–, in interior light wiring harness

(R5) – Wire connection –2–, in interior light wiring harness

(R7) – Plus connection –2–, in interior light wiring harness

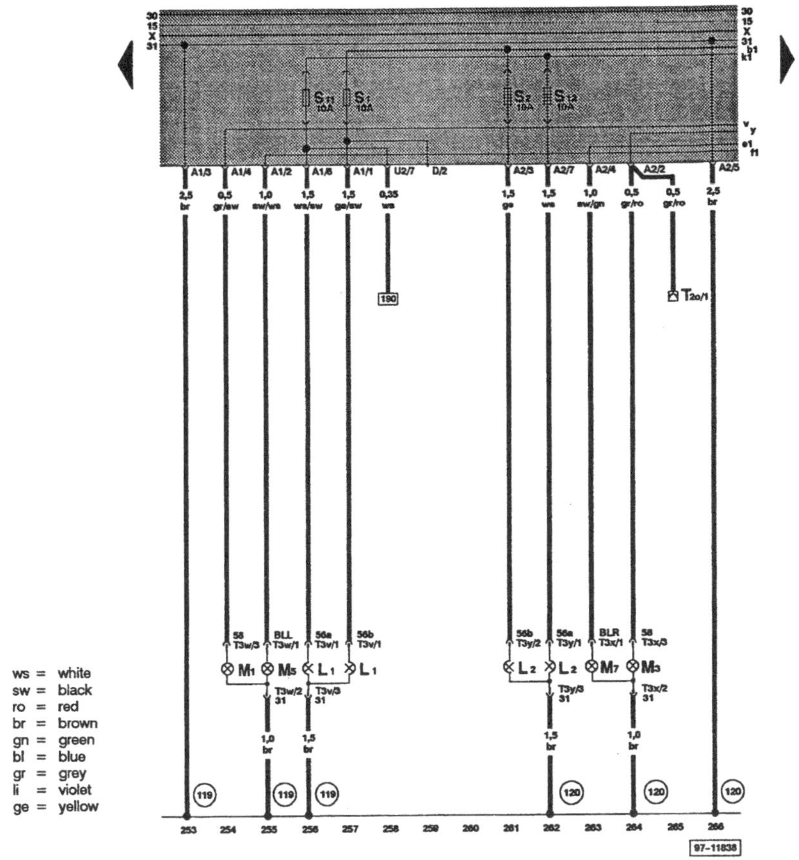

ws = white
sw = black
ro = red
br = brown
gn = green
bl = blue
gr = grey
li = violet
ge = yellow

L1 – Left Headlight
L2 – Right Headlight
M1 – Left Parking Light
M3 – Right Parking Light
M5 – Left Front Turn Signal Light
M7 – Right Front Turn Signal Light
T2o – Double Connector, near battery
T3v – 3-Pin Connector
T3w – 3-Pin Connector
T3x – 3-Pin Connector
T3y – 3-Pin Connector

(119) – Ground connection –1–, in headlight wiring harness

(120) – Ground connection –2–, in headlight wiring harness

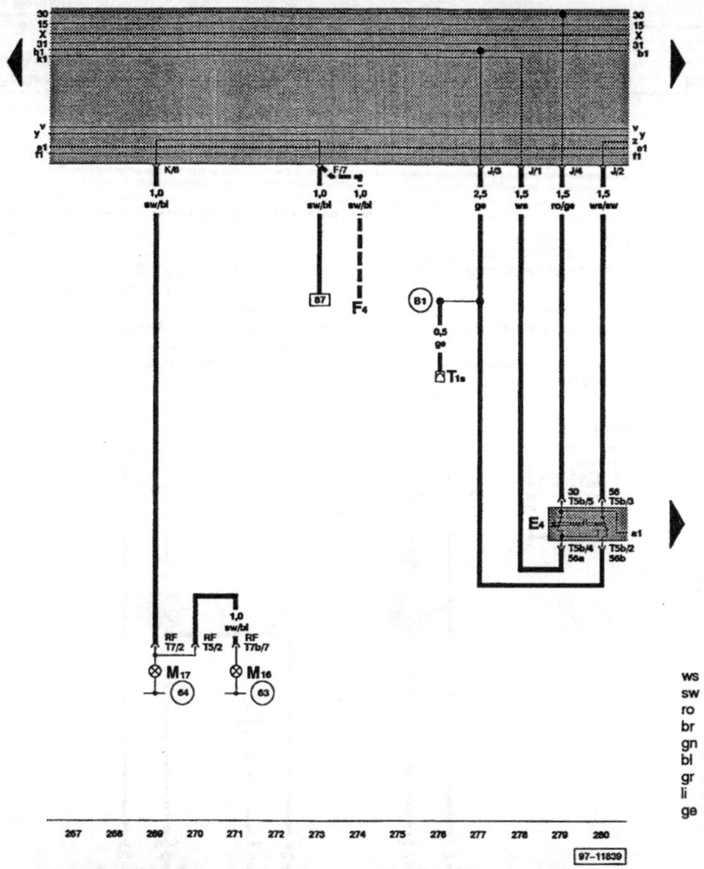

ws = white
sw = black
ro = red
br = brown
gn = green
bl = blue
gr = grey
li = violet
ge = yellow

E4 – Headlight Dimmer/Flasher Switch
F4 – Back-Up Light Switch
M16 – Left Back-Up Light
M17 – Right Back-Up Light
T1s – Single Connector, behind fuse/relay panel
T5 – 5-Pin Connector, on tail light, right
T5b – 5-Pin Connector, behind steering column switch cover
T7 – 7-Pin Connector, on tail light, right
T7b – 7-Pin Connector, on tail light, left

(63) – Ground connection, on left tail light bulb holder

(64) – Ground connection, on right tail light bulb holder

(B1) – Wire connection (56 b), in steering column switch wiring harness

▬ ▬ – Manual transmission only

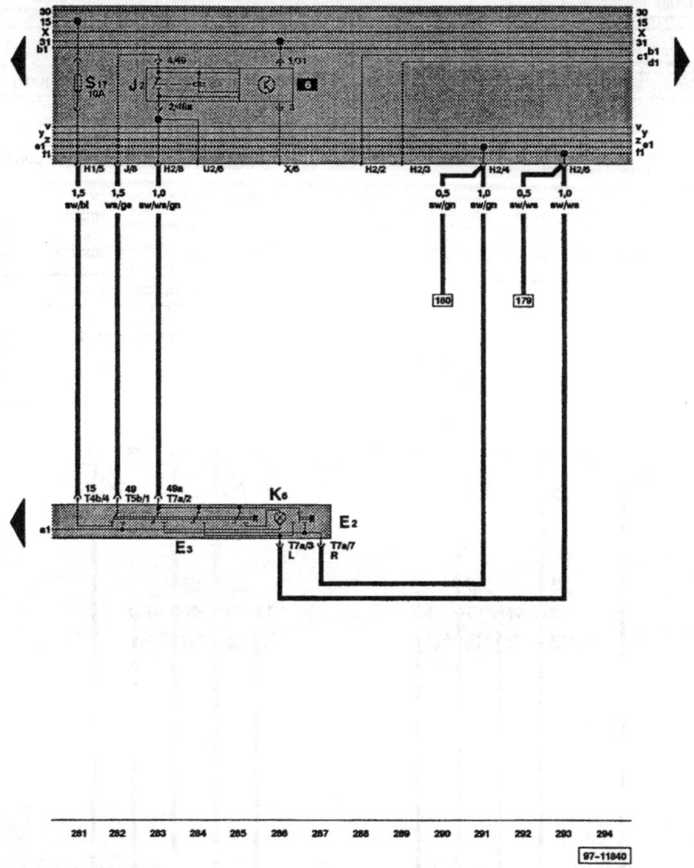

ws = white
sw = black
ro = red
br = brown
gn = green
bl = blue
gr = grey
li = violet
ge = yellow

E2 – Turn Signal Switch
E3 – Emergency Flasher Switch
J2 – Emergency Flasher Relay
K6 – Emergency Flasher Warning Light
T4b – 4-Pin Connector, behind steering column switch cover
T5b – 5-Pin Connector, behind steering column switch cover
T7a – 7-Pin Connector, behind steering column switch cover

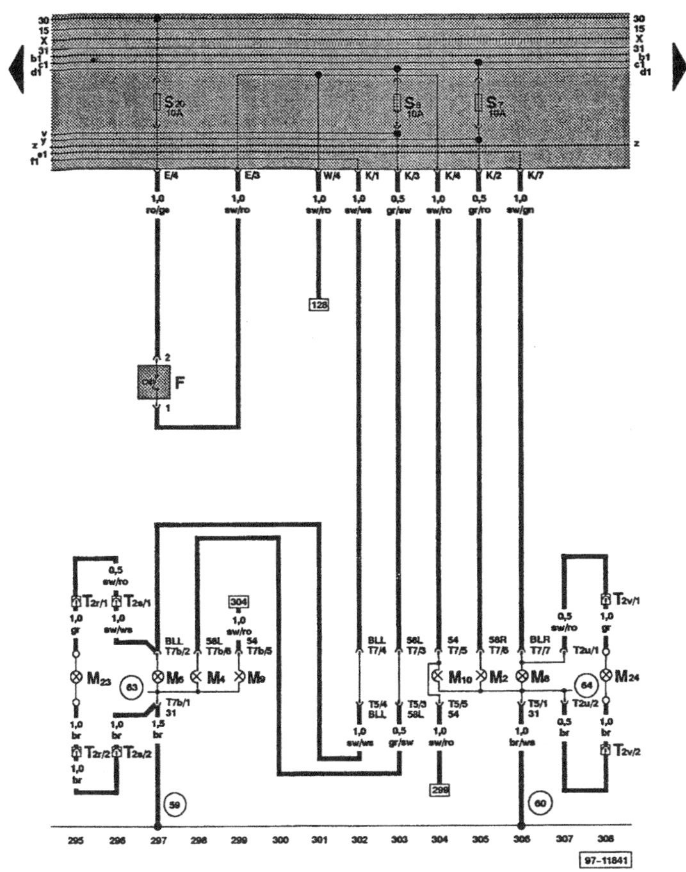

ws = white
sw = black
ro = red
br = brown
gn = green
bl = blue
gr = grey
li = violet
ge = yellow

F – Brake Light Switch
M2 – Right Tail Light
M4 – Left Tail Light
M6 – Left Rear Turn Signal Light
M8 – Right Rear Turn Signal Light
M9 – Left Brake Light
M10 – Right Brake Light
M23 – Left Roof Turn Signal Light
M24 – Right Roof Turn Signal Light
T2r – Double Connector, near roof turn signal, left
T2s – Double Connector, near tail light, left
T2u – Double Connector, near tail light, right
T2v – Double Connector, near roof turn signal, right
T5 – 5-Pin Connector
T7 – 7-Pin Connector
T7b – 7-Pin connector

(59) – Ground connection, near tail light, left

(60) – Ground connection, near tail light, right

(63) – Ground connection, on left tail light bulb-holder

(64) – Ground connection, on right tail light bulb holder

* – Different wire routing for flatbed vehicles

Edition 9/94
USA.5272.04.21

Tail light*, brake light*, rear turn signal*

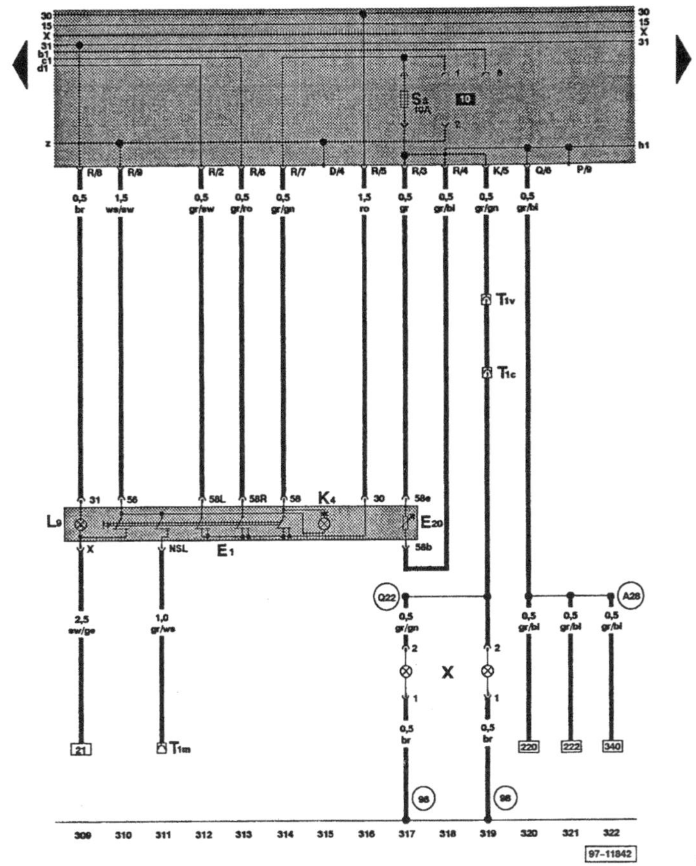

ws = white
sw = black
ro = red
br = brown
gn = green
bl = blue
gr = grey
li = violet
ge = yellow

E1 – Light Switch
E20 – Instrument Panel Light Dimmer Switch
K4 – Park Light Indicator Light
L9 – Headlight Switch Light
T1c – Single Connector, near tail light, right
T1m – Single Connector, behind fuse/relay panel
T1v – Single Connector, near tail light, left
X – License Plate Light

(98) – Ground connection, in rear lid wiring harness

(A28) – Plus connection (58b), in fresh air blower wiring harness

(Q22) – Wire connection –1–, in rear lid wiring

Light switch, license plate light

Edition 9/94
USA.5272.04.21

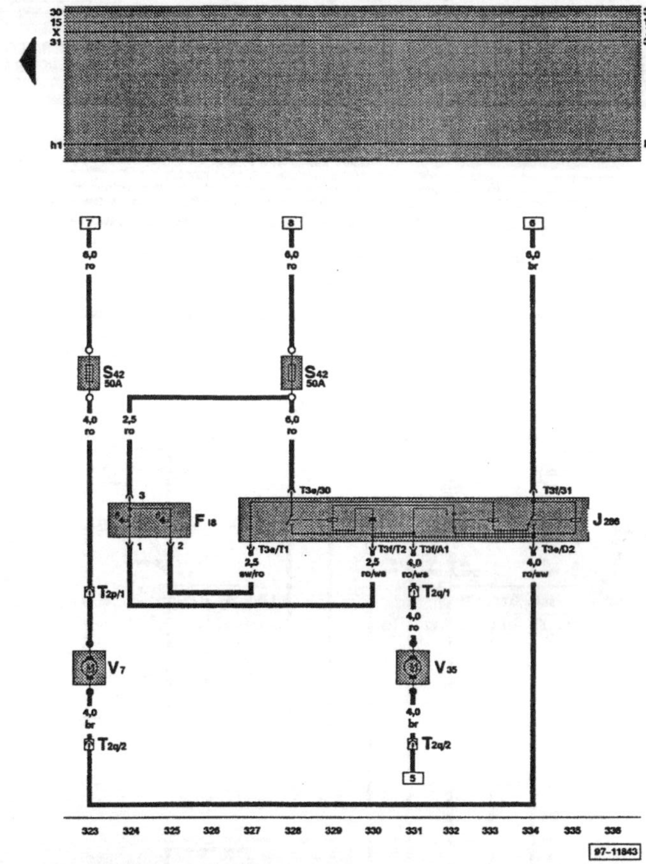

ws = white
sw = black
ro = red
br = brown
gn = green
bl = blue
gr = grey
li = violet
ge = yellow

F18 – Coolant Fan Control (FC) Thermal Switch
J286 – Series/Parallel Coolant Fan Control (FC) Relay (V7, V35)
S42 – Coolant Fan Fuse, near fan
T2p – Double Connector, in engine compartment, front
T2g – Double Connector, in engine compartment, front
T3e – 3-Pin Connector
T3f – 3-Pin Connector
V7 – Coolant Fan
V35 – Right Coolant Fan

ws = white
sw = black
ro = red
br = brown
gn = green
bl = blue
gr = grey
li = violet
ge = yellow

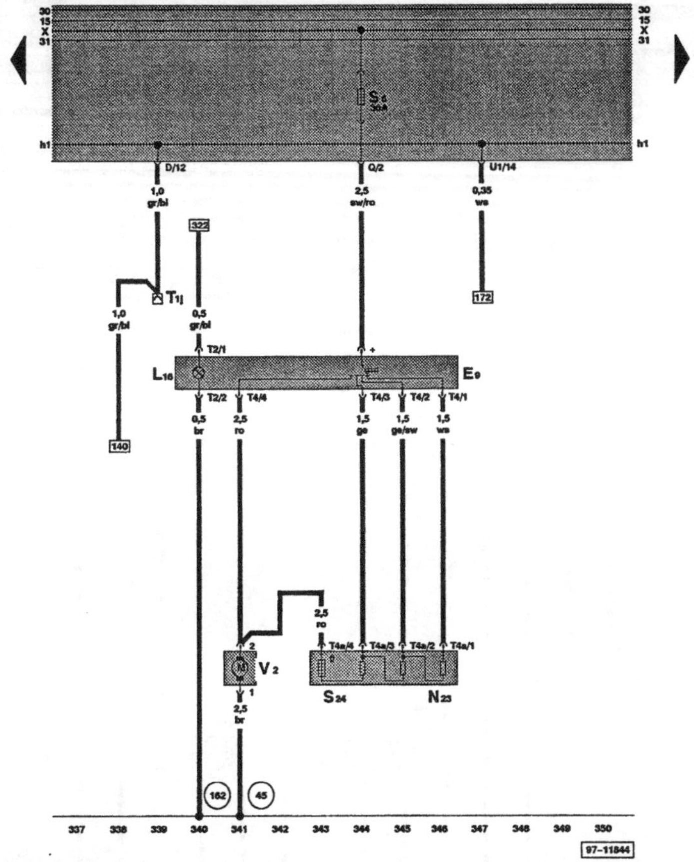

E9 – Fresh Air Blower Switch
L16 – Fresh Air Control Lever Light
N23 – Fresh Air Blower Series Resistance
S24 – Heater, A/C Clutch Fuse
T1j – Single Connector, behind fuse/relay panel
T2 – Double Connector
T4 – 4-Pin Connector
T4a – 4-Pin Connector
V2 – Fresh Air Blower

(45) – Ground connection, behind instrument panel, center

(162) – Ground connection, in blower motor wiring harness

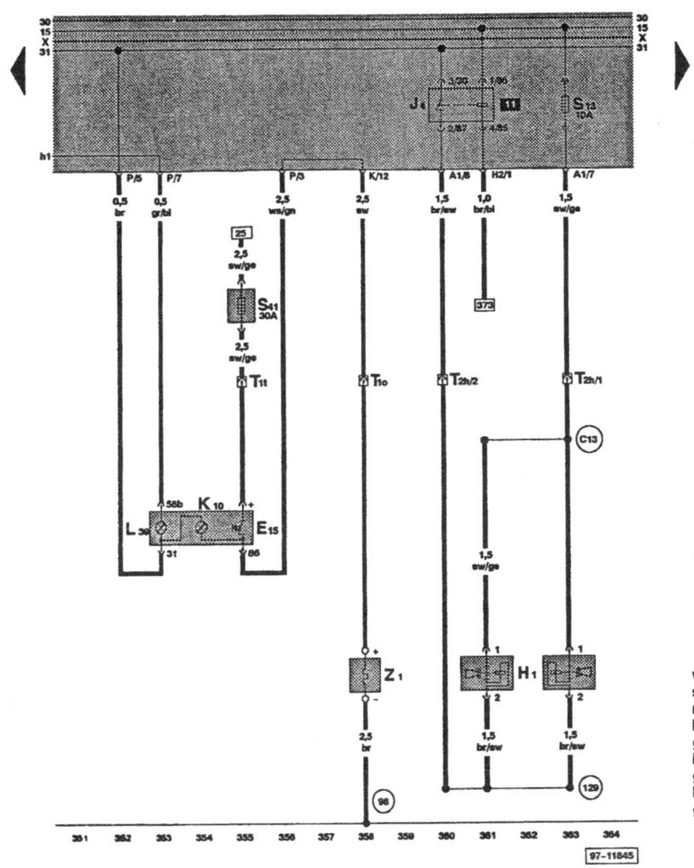

ws = white
sw = black
ro = red
br = brown
gn = green
bl = blue
gr = grey
li = violet
ge = yellow

E15 – Rear Window Defogger Switch
H1 – Dual Horns
J4 – Dual Horn Relay
K10 – Rear Window Defogger Indicator Light
L39 – Rear Window Defogger Switch Light
S41 – Rear Window Defogger Fuse, above fuse relay panel
T1o – Single Connector, near tail light, right
T1t – Single Connector, behind fuse/relay panel
T2h – Double Connector, in engine compartment, left front
Z1 – Rear Window Defogger/Heat Element

(98) – Ground connection, in rear lid wiring harness

(129) – Ground connection, in dual-tone horn wiring harness

(C13) – Plus connection, in dual-tone horn wiring harness

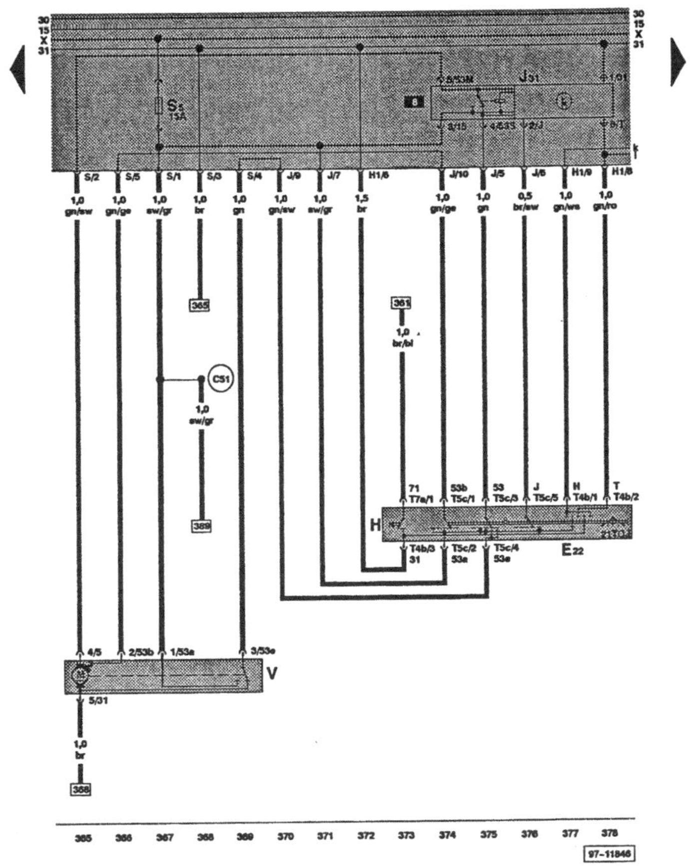

ws = white
sw = black
ro = red
br = brown
gn = green
bl = blue
gr = grey
li = violet
ge = yellow

E22 – Windshield Wiper/Washer Switch
H – Horn Button
J31 – Wiper/Washer Intermittent Relay
T4b – 4-Pin Connector
T5c – 5-Pin Connector
T7a – 7-Pin Connector
V – Windshield Wiper Motor

(C51) – Plus connection (x), in wiper motor wiring harness

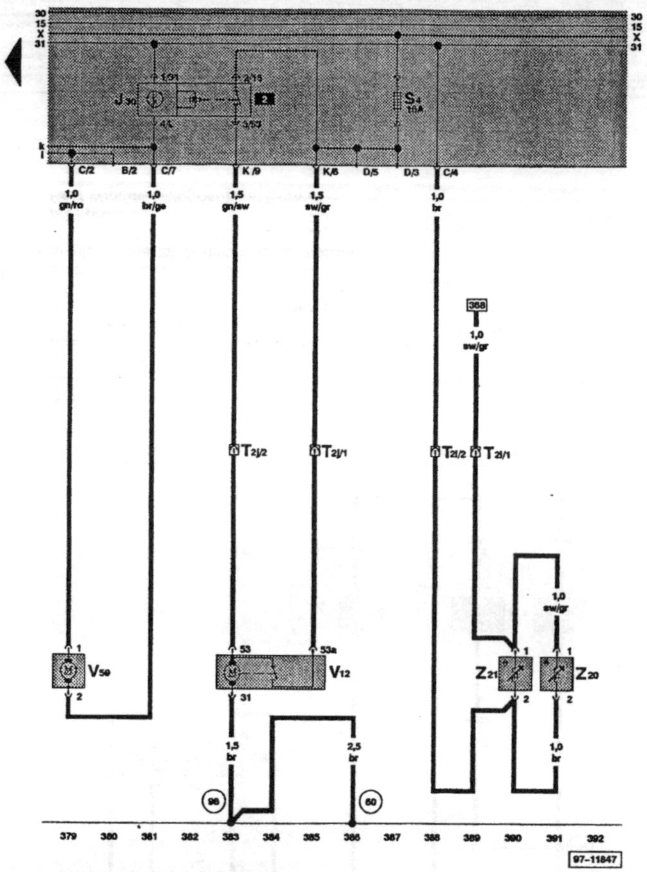

ws = white
sw = black
ro = red
br = brown
gn = green
bl = blue
gr = grey
li = violet
ge = yellow

J30 – Rear Window Wiper/Washer Relay
T2j – Double Connector, near tail light, right
T2l – Double Connector, near firewall, right
V12 – Rear Window Wiper Motor
V59 – Windshield and Rear Window Washer Pump
Z20 – Left Washer Nozzle Heater
Z21 – Right Washer Nozzle Heater

(60) – Ground connection, near tail light, right

(98) – Ground connection, in rear lid wiring harness

THIS PAGE INTENTIONALLY LEFT

BLANK

Air conditioning,
- **2.5 L-Engine – Digifant Multiport Fuel Injection (MFI)/81kW, code ACU,**

from September 1994

Relay location
For relay location and multiple connectors see also section "component locations".

1 – A/C Control Module (140)

Note: Number in parentheses indicates production control number stamped on relay housing.

Fuse colors

30A – green
25A – white
20A – yellow
15A – blue
10A – red

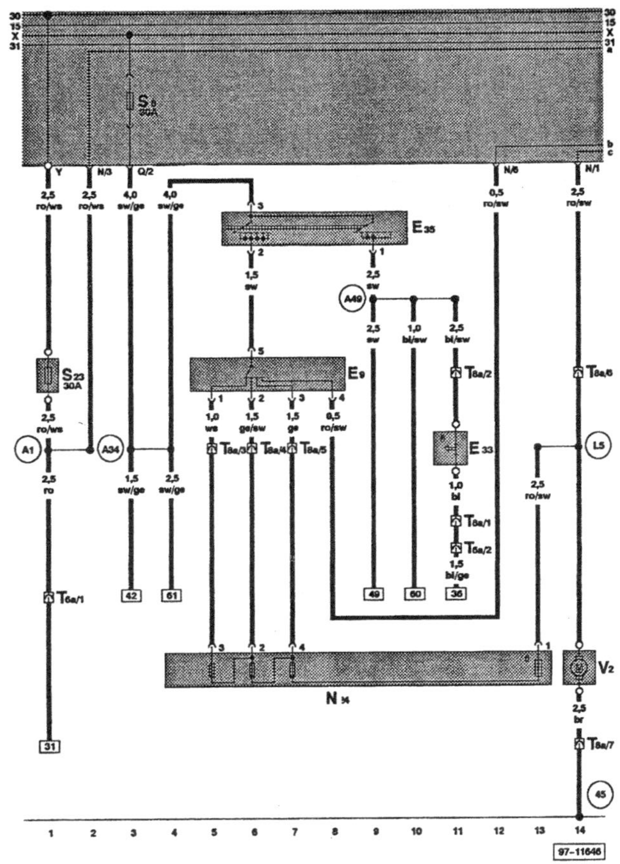

ws = white
sw = black
ro = red
br = brown
gn = green
bl = blue
gr = grey
li = violet
ge = yellow

E9 – Fresh Air Blower Switch
E33 – A/C Evaporator Temperature Switch
E35 – A/C Switch
N24 – Fresh Air Blower Series Resistance With Fuse
S23 – Heater and A/C Fuse
T6a – 6-Pin Connector, behind instrument panel, left
T8a – 8-Pin Connector, behind instrument panel, center
V2 – Fresh Air Blower

(45) – Ground connection, behind instrument panel, center

(A1) – Plus connection (30a), in instrument panel wiring harness
(A34) – Wire connection (x), in instrument panel wiring harness
(A49) – Wire connection –1–, in instrument panel wiring harness
(L5) – Wire connection –1–, in evaporator housing wiring harness

A/C switch, first evaporator fresh air blower

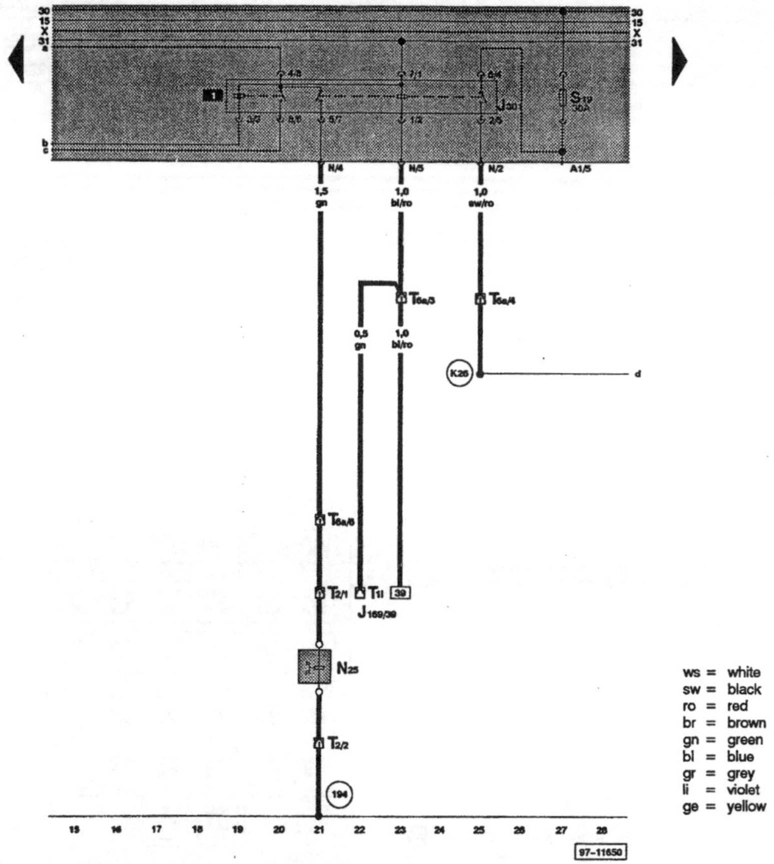

ws = white
sw = black
ro = red
br = brown
gn = green
bl = blue
gr = grey
li = violet
ge = yellow

J169 – Digitant Engine Control Module (ECM)
J301 – A/C Control Module, production control number (140), with 3 sec. switch delay
N25 – A/C Clutch
T1l – Single Connector, behind fuse/relay panel, for idle speed boost
T2 – Double Connector, on compressor
T6a – 6-Pin Connector, behind instrument panel, left

(194) – Ground connection –2–, in coolant fan wiring harness

(K26) – Wire connection –5–, in coolant fan wiring harness

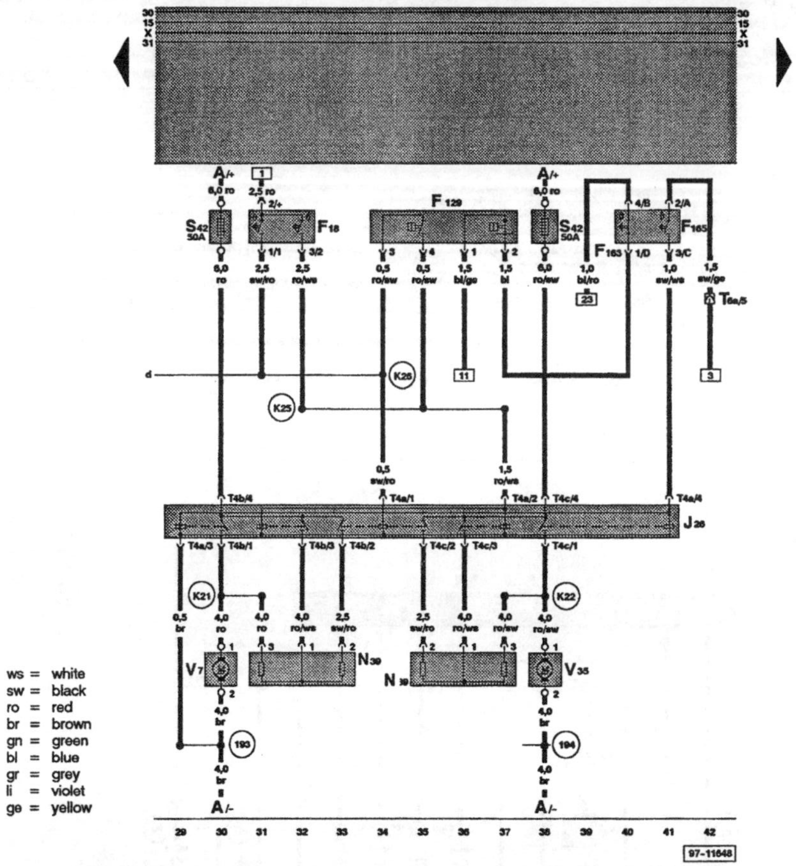

ws = white
sw = black
ro = red
br = brown
gn = green
bl = blue
gr = grey
li = violet
ge = yellow

A – Battery
F18 – Coolant Fan Control (FC) Thermal Switch
F129 – A/C Pressure Switch
F163 – A/C Cut-Out Thermal Switch
F165 – Third Speed Coolant Fan Control (FC) Thermal Switch
J26 – Coolant Fan control (FC) Relay
N39 – Coolant Fan Control (FC) Series Resistance, behind headlight, left
S42 – Coolant Fan Fuse, in engine compartment, left front
T4a – 4-Pin Connector
T4b – 4-Pin Connector
T4c – 4-Pin Connector
T6a – 6-Pin Connector, behind instrument panel, left
V7 – Left Coolant Fan
V35 – Right Coolant Fan

(193) – Ground connection –1–, in coolant fan wiring harness

(194) – Ground connection –2–, in coolant fan wiring harness

(K21) – Wire connection –1–, in coolant fan wiring harness

(K22) – Wire connection –2–, in coolant fan wiring harness

(K25) – Wire connection –4–, in coolant fan wiring harness

(K26) – Wire connection –5–, in coolant fan wiring harness

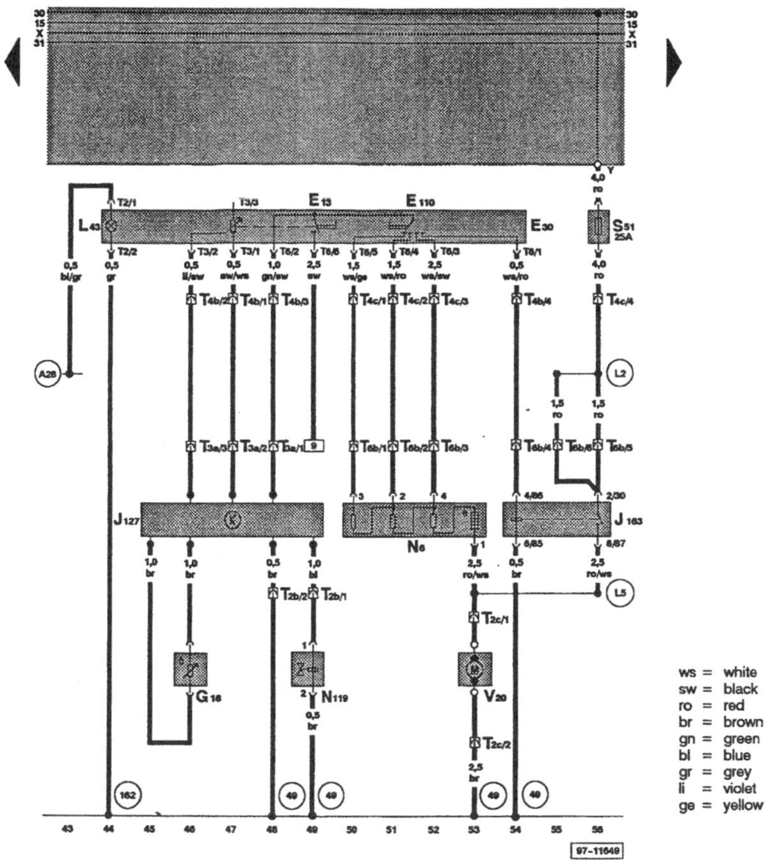

ws = white
sw = black
ro = red
br = brown
gn = green
bl = blue
gr = grey
li = violet
ge = yellow

E13 – Heater-Time Temperature Control Switch
E110 – Evaporator Blower Switch
G18 – Temperature Sensor
J127 – A/C Programmer for 2nd evaporator
J163 – Fresh Air Blower Relay for 2nd evaporator, production control number (24)
L43 – A/C Switch Illumination Light
N6 – Series Resistance for 2nd evaporator with overheating fuse
N81 – Evaporator Blower Series Resistance, rear
N119 – Refrigerant Circulation Solenoid Valve
S51 – Blower Fuse for 2nd evaporator
T2b – Double Connector, on 2nd evaporator
T2c – Double Connector, on 2nd Evaporator Fan
T3a – 3-Pin Connector, on 2nd evaporator
T4b – 4-Pin Connector, behind instrument panel
T4c – 4-Pin Connector, behind instrument panel
T6b – 6-Pin Connector, on 2nd evaporator
V20 – 2nd Evaporator Fan

(49) – Ground connection, near 2nd evaporator
(162) – Ground connection, in blower motor wiring harness
(A28) – Plus connection (58b), in fresh air blower wiring harness
(L2) – Wire connection, in A/C wiring harness
(L5) – Wire connection –1–, in A/C wiring harness

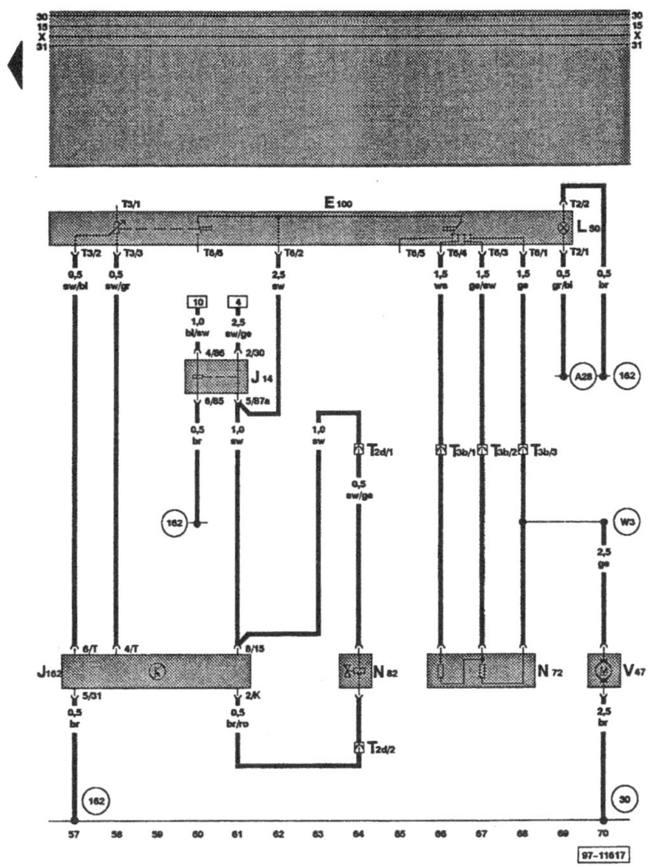

ws = white
sw = black
ro = red
br = brown
gn = green
bl = blue
gr = grey
li = violet
ge = yellow

E100 – Warm Air Blower Switch, in instrument panel
J14 – Heater Air Blower Relay, behind instrument panel, center, production control number (94)
J162 – Heater Control Module, auxiliary heater core, behind instrument panel, center, production control number (152)
L50 – Warm Air Blower Switch, auxiliary heater core
N72 – Heater Blower Series Resistance, near battery
N82 – Coolant Cut-off Valve, on firewall, center
T2d – Double Connector, behind instrument panel, left
T3b – 3-Pin Connector, behind instrument panel, center
V47 – Rear Warm Air Fan, below passenger compartment

(30) – Ground connection, –1–, beside fuse/relay panel
(162) – Ground connection, in blower motor wiring harness
(A28) – Plus connection (58b), in fresh air blower wiring harness
(W3) – Wire connection, in rear wiring harness

Passenger compartment ventilation,

from September 1994

Relay location
For relay location and multiple connectors see also section "component locations".

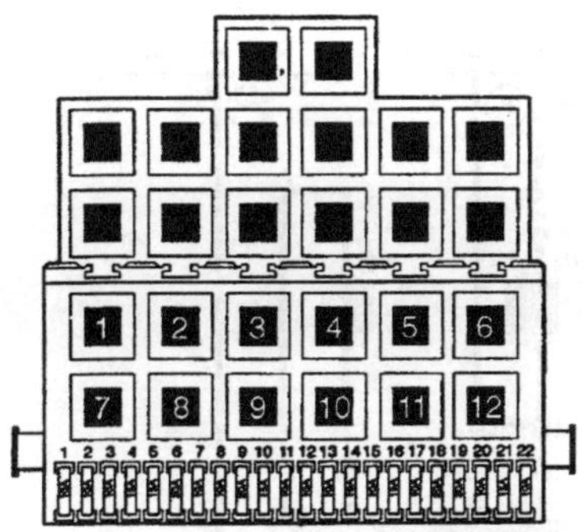

97–11715

Fuse colors

30A – green
25A – white
20A – yellow
15A – blue
10A – red

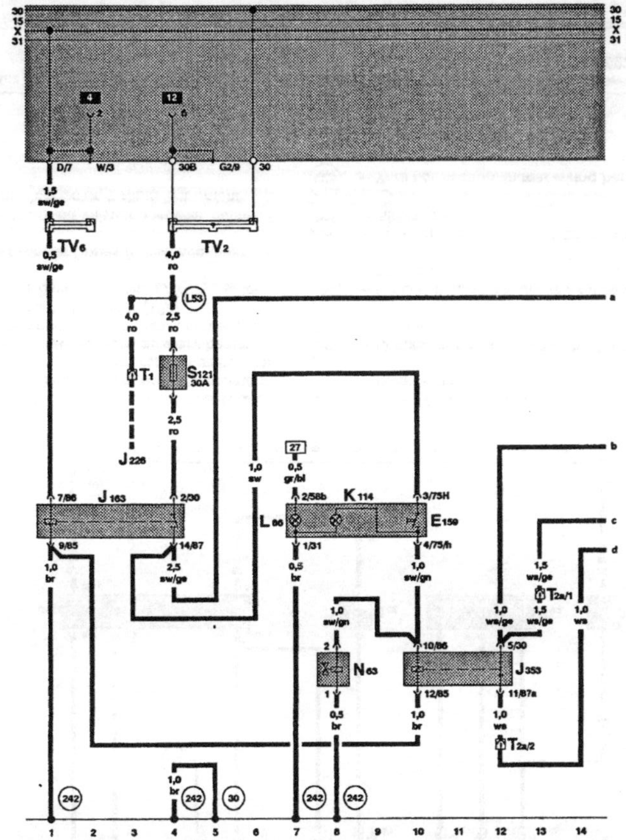

97-11932

ws = white
sw = black
ro = red
br = brown
gn = green
bl = blue
gr = grey
li = violet
ge = yellow

E159 – Fresh Air/Recirculating Flap Switch
J163 – Fresh Air Blower Relay, behind instrument panel, center
J226 – Park/Neutral Position (PNP) Relay
J353 – Relay for Fresh Air and Recirculating Air Flap-Rear, behind instrument panel, center
K114 – Fresh Air and Recirculating Air Mode Indicator Light
L86 – Fresh Air/Recirculating Flap Switch Light
N63 – Fresh Air/Recirculating Flap Two-Way Valve-Front, behind instrument panel, right
S121 – Fuse for Fresh Air Blower-Rear, above fuse/relay panel
T1 – Single Connector, above fuse/relay panel
T2a – Double Connector, behind instrument panel, right

TV2 – Terminal 30 (B+) Wire Connector, above fuse/relay panel
TV6 – Terminal X Wire Connector, above fuse/relay panel

(30) – Ground connection, –1–, beside fuse/relay panel
(242) – Ground connection, in wiring harness passenger ventilation
(L53) – Plus connection (30), in wiring harness passenger ventilation
▬▬ – Automatic transmission only

Voltage supply, protection, fresh air recirculating flap control-front

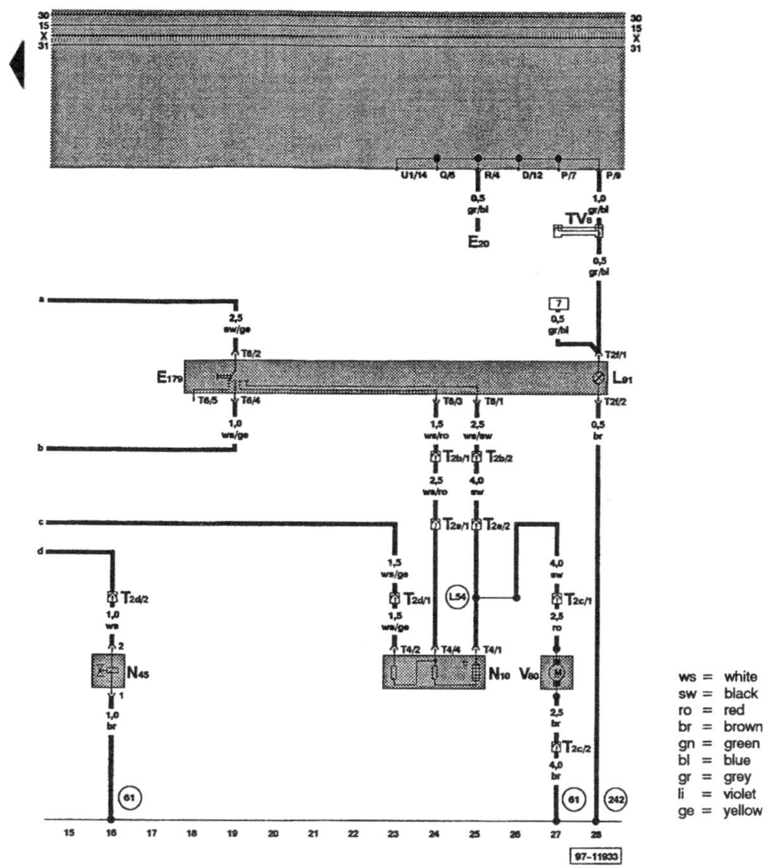

ws = white
sw = black
ro = red
br = brown
gn = green
bl = blue
gr = grey
li = violet
ge = yellow

E20 – Instrument Panel Light Dimmer Switch
E179 – Switch for Fresh Air Blower-Rear
L91 – Illumination for Switch for Fresh Air Blower-Rear
N10 – Series Resistance for Fresh Air Blower-Rear,behind right D-pillar, lower part
N45 – Two-Way Valve for Fresh and Recirculation Air Flap-Rear, behind right D-pillar, lower part
T2b – Double Connector, behind instrument panel, right
T2c – Double Connector, behind right D-pillar, lower part
T2d – Double Connector, behind right D-pillar, lower part
T2e – Double Connector, behind right D-pillar, lower part
T2f – Double Connector, on Switch for Fresh Air Blower-Rear
T4 – 4-Pin Connector, on Series Resistance for Fresh Air Blower-Rear
T6 – 6-Pin Connector, on Switch for Fresh Air Blower-Rear
TV8 – Terminal 58b Wire Connector, above fuse/relay panel
V80 – Auxiliary Fresh Air Blower, behind right D-pillar, lower part

(61) – Ground connection –2–, near tail light, right

(242) – Ground connection, in wiring harness passenger ventilation

(L54) – Plus connection (blower stage 3), in wiring harness blower-passenger ventilsation

THIS PAGE INTENTIONALLY LEFT
BLANK

Automatic transmission, 01P

- **2.5 L-Engine – Digifant Multiport Fuel Injection (MFI)/81kW, code ACU,**

from January 1995

Relay location
For relay location and multiple connectors see also section "component locations".

12 – Fuel Pump (FP) Relay (80)

Note: Number in parentheses indicates production control number stamped on relay housing.

Fuse colors

30A – green
25A – white
20A – yellow
15A – blue
10A – red

VA 308

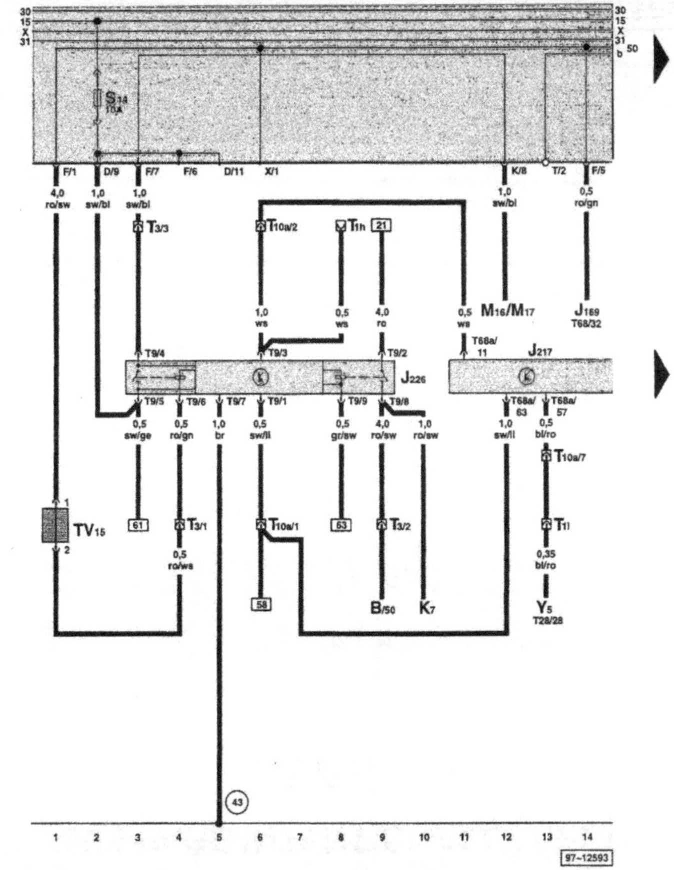

ws = white
sw = black
ro = red
br = brown
gn = green
bl = blue
gr = grey
li = violet
ge = yellow

B – Starter
J169 – Digifant Engine Control Module (ECM), near battery
J217 – Transmission Control Module (TCM), behind instrument panel, right
J226 – Park/Neutral Position (PNP) Relay, above fuse/relay panel
K7 – Brake and Parking Brake Warning Light
M16 – Left Back-Up Light
M17 – Right Back-Up Light
T1h – Single Connector, behind instrument panel, left
T1i – Single Connector, behind fuse/relay panel
T3 – 3-Pin Connector, behind fuse/relay panel
T9 – 9-Pin Connector, on Park/Neutral Position (PNP) Relay
T10a – 10-Pin Connector, behind instrument panel, right
T28 – 28-Pin Connector, on instrument cluster
T68 – 68-Pin Connector, on Digifant Engine Control Module (ECM)
T68a – 68-Pin Connector, on Transmission Control Module (TCM)
TV15 – Wire Connector, behind fuse/relay panel
Y5 – Transmission Range (TR) Selector Lever Display, in instrument cluster

(43) – Ground connection, on right A-pillar, lower part

Park/neutral position (PNP) relay, transmission control module (TCM)

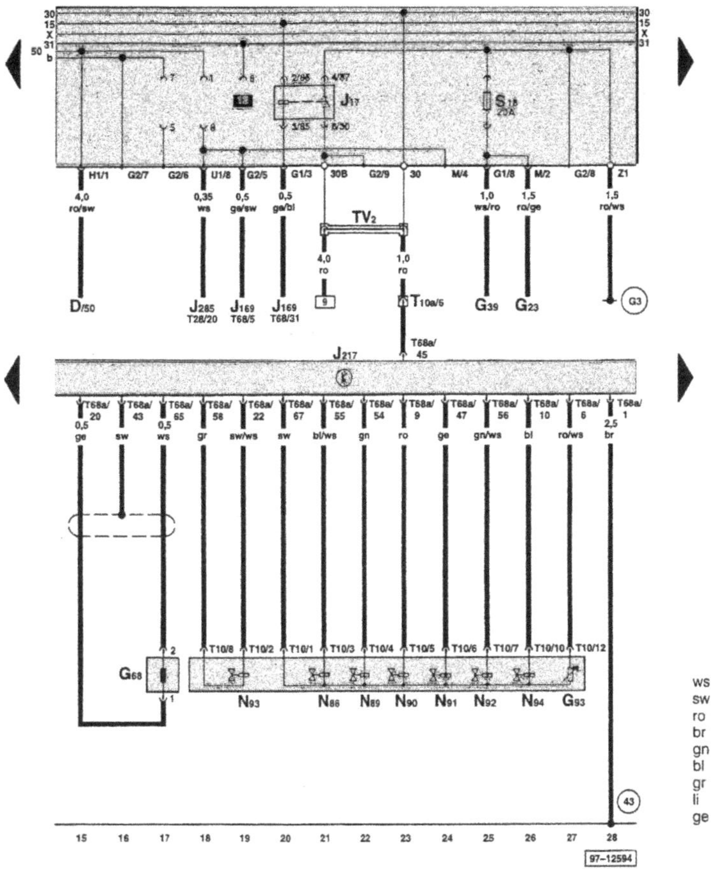

ws = white
sw = black
ro = red
br = brown
gn = green
bl = blue
gr = grey
li = violet
ge = yellow

D – Ignition/Starter Switch
G23 – Transfer Fuel Pump (FP)
G39 – Heated Oxygen Sensor (HO2S) (00525)
G68 – Vehicle Speed Sensor (VSS) (00281)
G93 – Transmission Fluid Temperature Sensor (00300)
J17 – Fuel Pump (FP) Relay
J169 – Digifant Engine Control Module (ECM), near battery
J217 – Transmission Control Module (TCM), behind instrument panel, right
J285 – Control Module With Display Unit In Instrument Cluster, in instrument cluster
N88 – Solenoid valve 1 (00258)
N89 – Solenoid valve 2 (00260)
N90 – Solenoid valve 3 (00262)
N91 – Solenoid valve 4 (00264)
N92 – Solenoid valve 5 (00266)
N93 – Solenoid valve 6 (00268)
N94 – Solenoid valve 7 (00270)
T10 – 10-Pin Connector
T10a – 10-Pin Connector, behind instrument panel, right
T28 – 28-Pin Connector, on instrument cluster
T68 – 68-Pin Connector, on Digifant Engine Control Module (ECM)
T68a – 68-Pin Connector, on Transmission Control Module (TCM)
TV2 – Terminal 30 (B+) Wire Connector
(43) – Ground connection, on right A-pillar, lower part
(G3) – Plus connection, in injector wiring harness

Transmission control module (TCM), solenoid valves

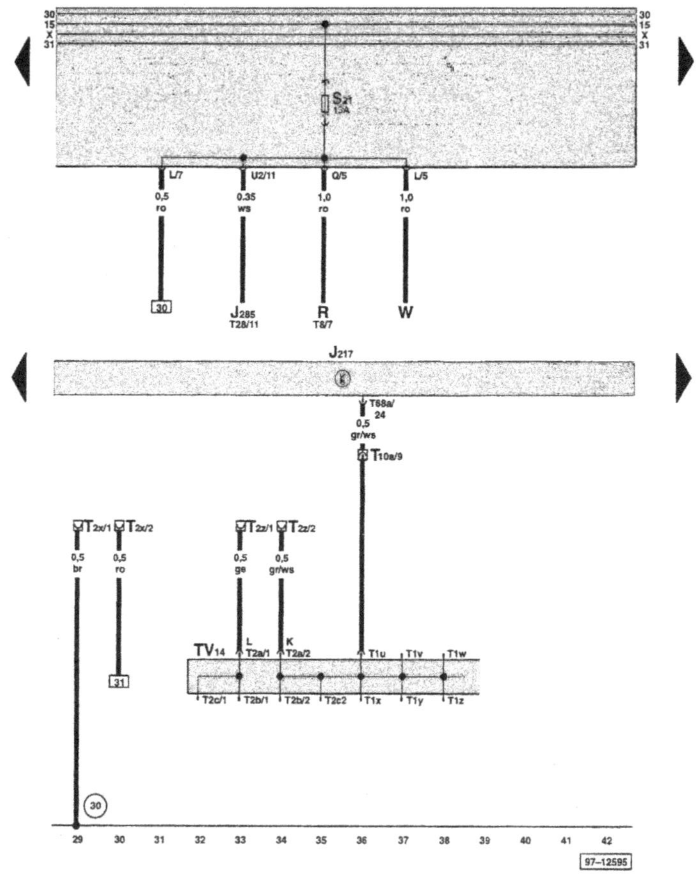

ws = white
sw = black
ro = red
br = brown
gn = green
bl = blue
gr = grey
li = violet
ge = yellow

J217 – Transmission Control Module (TCM), behind instrument panel, right
J285 – Control Module With Display Unit In Instrument Cluster
R – Radio
T1u – Single Connector
T1v – Single Connector
T1w – Single Connector
T1x – Single Connector
T1y – Single Connector
T1z – Single Connector
T2a – Double Connector
T2b – Double Connector
T2c – Double Connector
T2x – Double Connector, black, above fuse/relay panel, Data Link Connector (DLC)
T2z – Double Connector, white, above fuse/relay panel,Data Link Connector (DLC)
T8 – 8-Pin Connector, on radio
T10a – 10-Pin Connector, behind instrument panel, right
T28 – 28-Pin Connector, on instrument cluster
T68a – 68-Pin Connector, on Transmission Control Module (TCM)
TV14 – Data Link Connector (DLC) Wire Connector, above fuse/relay panel
W – Front Interior Light
(30) – Ground connection, –1–, beside fuse/relay panel

Transmission control module (TCM), data link connector (DLC) wire connector

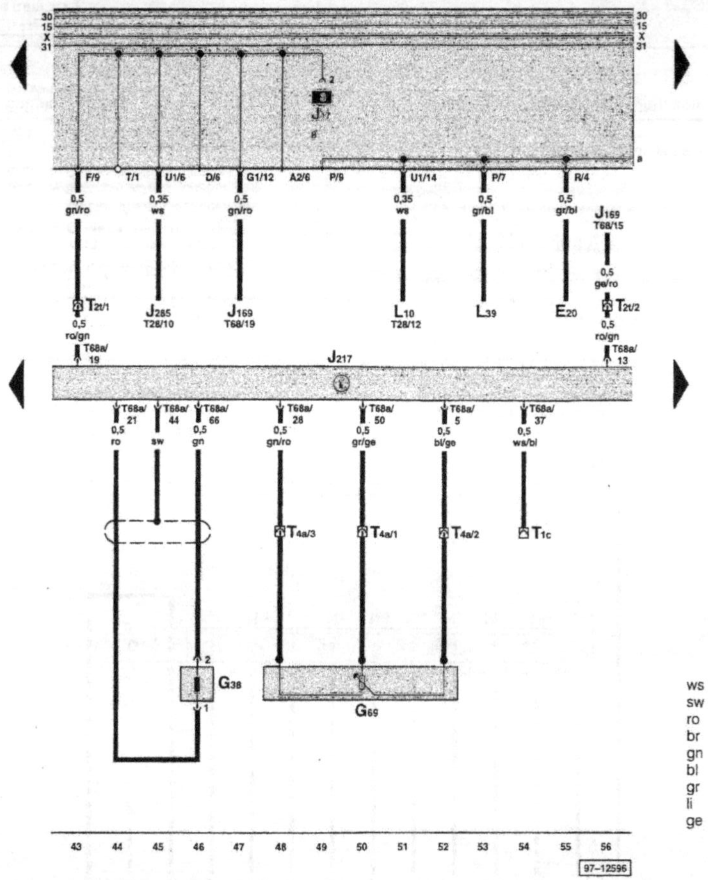

ws = white
sw = black
ro = red
br = brown
gn = green
bl = blue
gr = grey
li = violet
ge = yellow

97–12596

- E20 – Instrument Panel Light Dimmer Switch
- G38 – Transmission Vehicle Speed Sensor (VSS) (00297)
- G69 – Throttle Position (TP) Sensor (00518)
- J169 – Digifant Engine Control Module (ECM), near battery
- J176 – Digifant Engine Control Module (ECM) Power Sypply Relay
- J217 – Transmission Control Module (TCM), behind instrument panel, right
- J285 – Control Module With Display Unit In Instrument Cluster
- L10 – Instrument Cluster Light
- L39 – Rear Window Defogger Switch Light
- T1c – Single Connector, behind instrument panel, right
- T2t – Double Connector, engine compartment, near transmission
- T4a – 4-Pin Connector, near throttle position (TP) sensor
- T28 – 28-Pin Connector, on instrument cluster
- T68 – 68-Pin Connector, on Digifant Engine Control Module (ECM)
- T68a – 68-Pin Connector, on Transmission Control Module (TCM)

Transmission control module (TCM), throttle position (TP) sensor

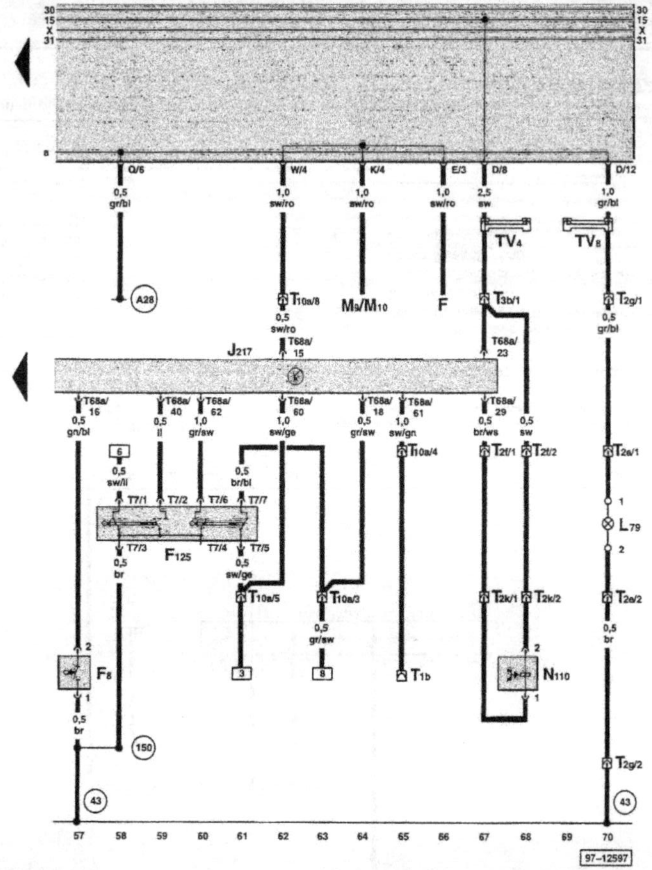

ws = white
sw = black
ro = red
br = brown
gn = green
bl = blue
gr = grey
li = violet
ge = yellow

97–12597

- F – Brake Light Switch
- F8 – Kick Down Switch
- F125 – Multi-Function Transmission Range (TR) Switch (00293)
- J217 – Transmission Control Module (TCM), behind instrument panel, right
- L79 – Selector Lever Light
- M9 – Left Brake Light
- M10 – Right Brake Light
- N110 – Shift Lock Solenoid
- T1b – Single Connector, behind fuse/relay panel (cruise control)
- T2e – Double Connector, near selector lever
- T2f – Double Connector, behind instrument panel, right
- T2g – Double Connector, behind fuse/relay panel
- T2k – Double Connector, near selector lever
- T3b – 3-Pin Connector, behind instrument, right
- T7 – 7-Pin Connector, on Multi-Function Transmission Range (TR) Switch
- T10a – 10-Pin Connector, behind instrument panel, right
- T68 – 68-Pin Connector, on Transmission Control Module (TCM)
- TV4 – Terminal 15 Wire Connector
- TV8 – Terminal 58b Wire Connector
- (43) – Ground connection, on right A-pillar, lower part
- (150) – Ground connection, in automatic transmission wiring harness
- (A28) – Plus connection (58b), in fresh air blower wiring harness

Transmission control module (TCM), multi-function transmission range (TR) switch

EuroVan - Standard Equipment,

from May 1996

Deviate relay location and fuseplacements as well as the locations of multiple connectors-see section "component locations".

- 2 – Rear Window Wiper/Washer Relay (72)
- 4 – Load Reduction Relay (18)
- 6 – Emergency Flasher Relay (21)
- 8 – Program. Washer/Wiper Intermittent Relay (99)
- 9 – Parking Light Warning Buzzer (36)
- 11 – Dual Horn Relay (53)

Note:

Number in parentheses indicates production control number stamped on relay housing.

Fuse colors

30 A – green
25 A – white
20 A – yellow
15 A – blue
10 A – red
7,5 A – brown
5 A – beige
3 A – violet

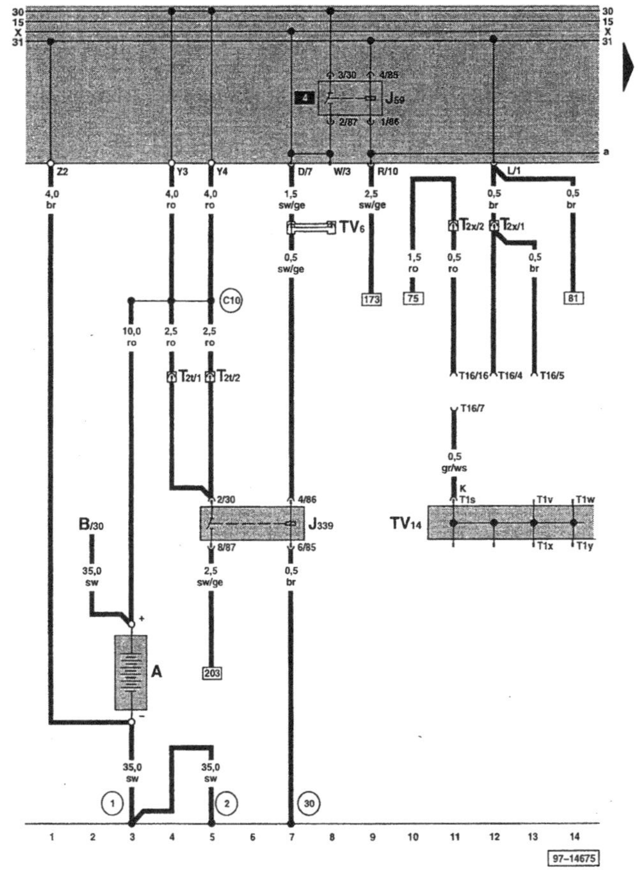

ws = white
sw = black
ro = red
br = brown
gn = green
bl = blue
gr = grey
li = violet
ge = yellow

A – Battery
B – Starter
J59 – Load Reduction Relay
J339 – Load Reduction Relay II, above fuse/relay panel
T1s – Single Connector
T1v – Single Connector
T1w – Single Connector
T1x – Single Connector
T1y – Single Connector
T2t – Double Connector, behind fuse/relay panel
T2x – Double Connector, behind fuse/relay panel
T16 – 16-Pin Connector, On Board Diagnostic, in instrument panel, center
TV6 – Terminal X Wire Connector, above fuse/relay panel
TV14 – Data Link Connector (DLC) wire Connector, above fuse/relay panel

(1) – Ground strap, battery to body
(2) – Ground strap, transmission to body
(30) – Ground connection, -1-, next to fuse/relay panel
(C10) – Plus connection -1- (30), in headlight wiring harness

Battery, load reduction relay II, data link connector (DLC) wire connector

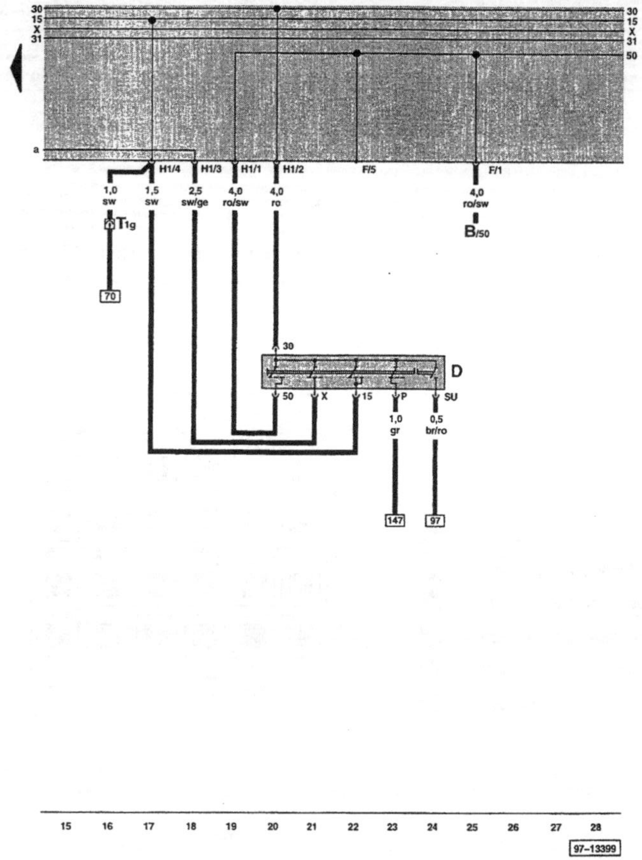

ws = white
sw = black
ro = red
br = brown
gn = green
bl = blue
gr = grey
li = violet
ge = yellow

- B – Starter
- D – Ignition/Starter Switch
- T1 – Single Connector, behind fuse/relay panel

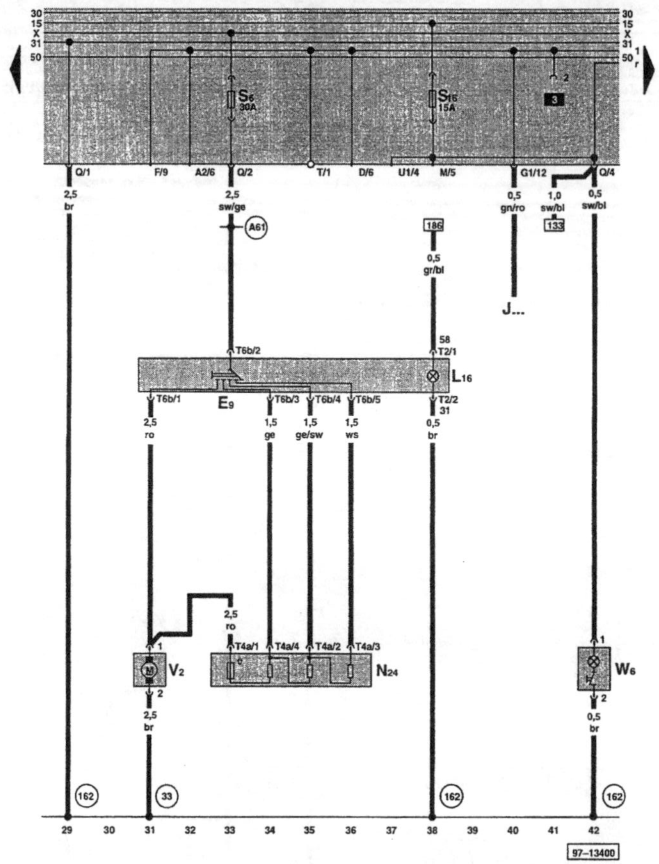

ws = white
sw = black
ro = red
br = brown
gn = green
bl = blue
gr = grey
li = violet
ge = yellow

- E9 – Fresh Air Blower Switch
- J... – Engine Control Modules
- L16 – Fresh Air Control Lever Light
- N24 – Fresh Air Blower Series Resistance with Fuse
- T2 – Double Connector
- T4a – 4-Pin Connector
- T6b – 6-Pin Connector
- V2 – Fresh Air Blower
- W6 – Glove Compartment Light

- (33) – Ground connection, behind instrument panel, right
- (162) – Ground connection, in blower motor wiring harness
- (A61) – Plus Connection (X), in Instrument cluster wiring harness

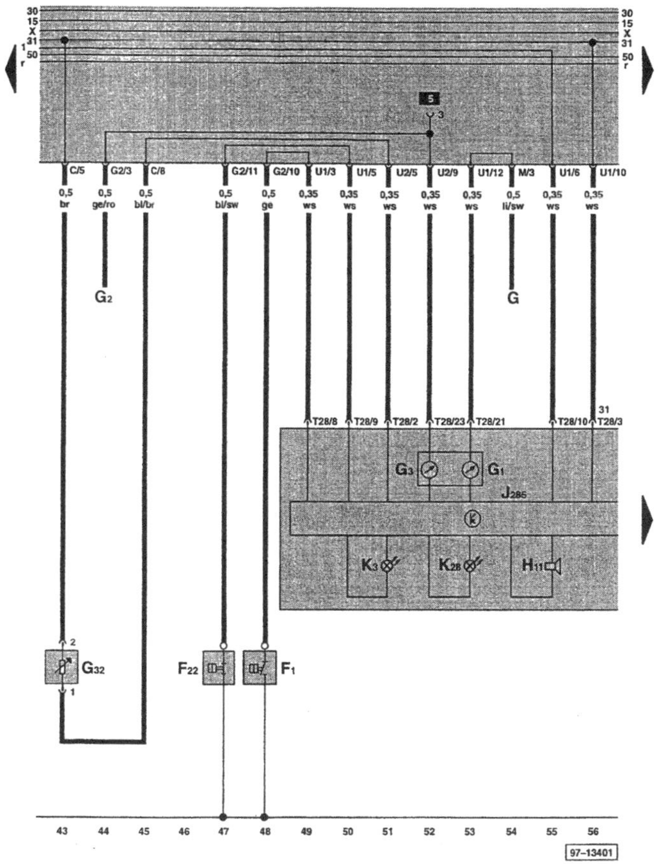

ws = white
sw = black
ro = red
br = brown
gn = green
bl = blue
gr = grey
li = violet
ge = yellow

F1 – Oil Pressure Switch
F22 – 0.3bar Oil Pressure Switch
G – Fuel Level Sensor
G1 – Fuel Gauge
G2 – Engine Coolant Temperature (ECT) Sensor
G3 – Engine Coolant Temperature (ECT) Gauge
G32 – Engine Coolant Level (ECL) Sensor
H11 – Oil Pressure Warning Buzzer
J285 – Control module with indicator unit in instrument panel insert
K3 – Oil Pressure Warning Light
K28 – Engine Coolant Level/Temperature (ECL/ECT) Warning Light
T28 – 28-Pin Connector

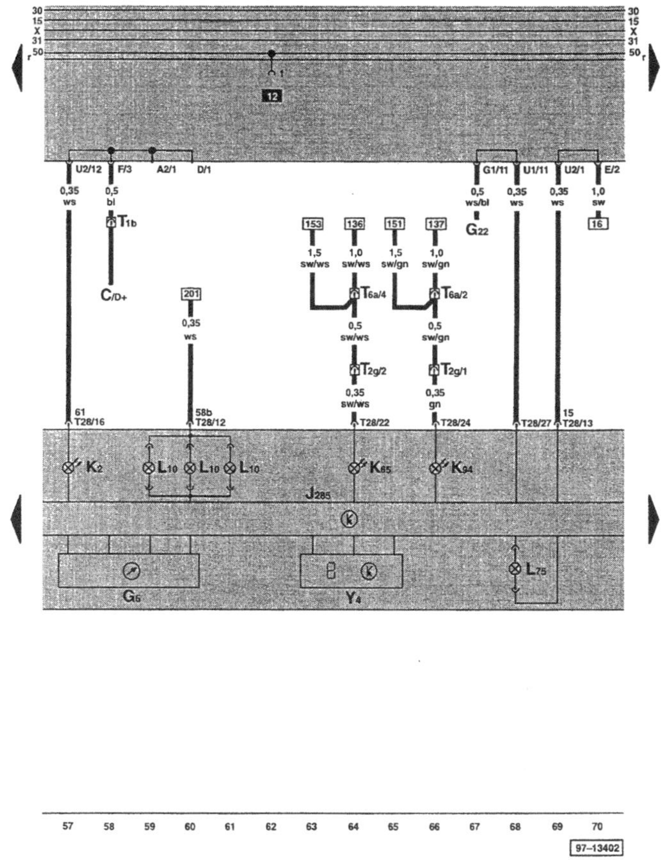

ws = white
sw = black
ro = red
br = brown
gn = green
bl = blue
gr = grey
li = violet
ge = yellow

C – Generator (GEN)
G5 – Tachometer
G22 – Speedometer Vehicle Speed Sensor (VSS)
J285 – Control module with indicator unit in instrument panel insert
K2 – Generator (GEN) Warning Light
K65 – Left Turn Signal Indicator Light
K94 – Right Turn Signal Indicator Light
L10 – Instrument Cluster Light
L75 – Digital Display Light
T1b – Single Connector, near starter
T2g – Double Connector, behind fuse/relay panel
T6a – 6-Pin Connector, behind fuse/relay panel
T28 – 28-Pin Connector
Y4 – Odometer Display

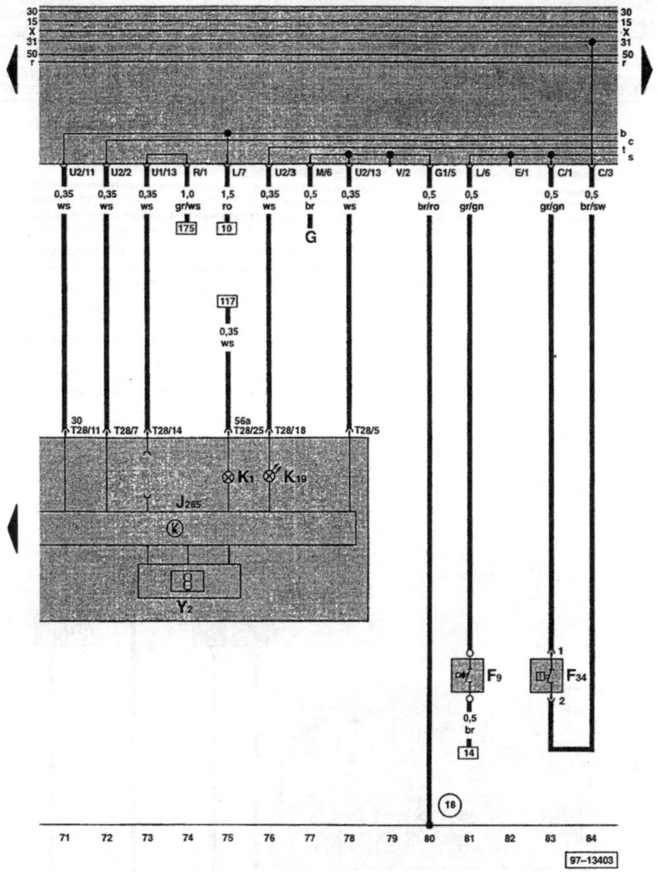

ws = white
sw = black
ro = red
br = brown
gn = green
bl = blue
gr = grey
li = violet
ge = yellow

F9 – Parking Brake Warning Light Switch
F34 – Brake Fluid Level Warning Switch
G – Fuel Level Sensor
J285 – Control module with indicator unit in instrument panel insert
K1 – Headlight High Beam Indicator Light
K19 – Seat Belt Warning Light
T28 – 28-Pin Connector
Y2 – Digital Clock
(18) – Ground connection, on engine block

Edition 12/96
USA.5272.08.21

Instrument cluster, digital clock, parking brake warning light switch

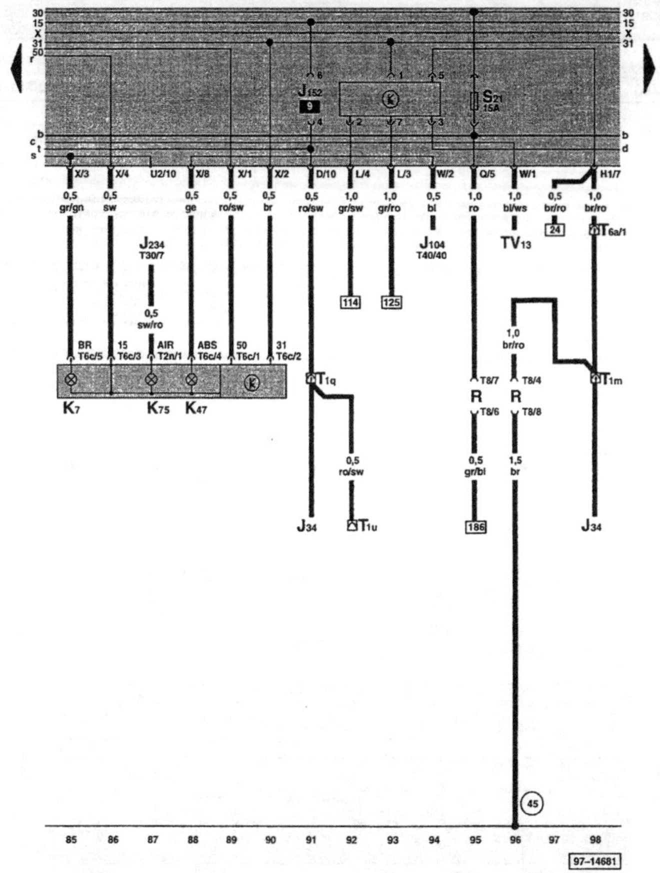

ws = white
sw = black
ro = red
br = brown
gn = green
bl = blue
gr = grey
li = violet
ge = yellow

J34 – Seat Belt Warning System Relay
J104 – ABS Control Module (W/EDL)
J152 – Parking Light Warning Buzzer
J234 – Airbag Control Module
K7 – Brake and Parking Brake Warning Light
K47 – ABS Warning Light
K75 – Airbag Malfunction Indicator Lamp (MIL)
R – Radio
T1m – Single Connector, behind fuse/relay panel
T1q – Single Connector, behind fuse/relay panel
T1u – Single Connector, behind fuse/relay panel
T2n – Double Connector
T6a – 6-Pin Connector, behind fuse/relay panel
T6c – 6-Pin Connector
T8 – 8-Pin Connector, on Radio
T30 – 30-Pin Connector, on Airbag Control Module
T40 – 40-Pin Connector, on ABS Control Module (w/EDL)
TV13 – Vehicle Speed Signal Wire Connector, above fuse/relay panel
(45) – Ground connection, behind instrument panel, center

Brake and parking brake warning light, ABS warning light, airbag malfunction indicator lamp (MIL), radio connection

Edition 12/96
USA.5272.08.21

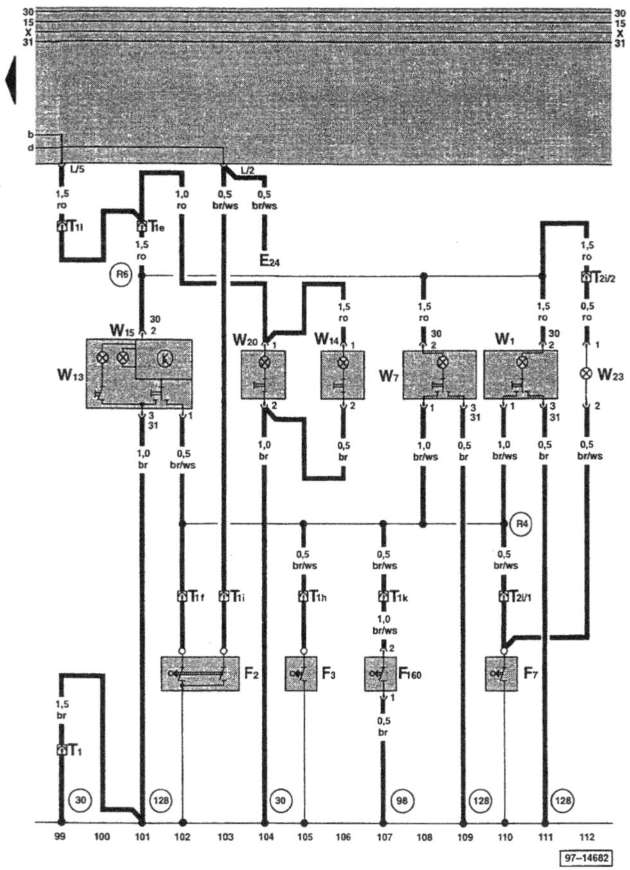

ws = white
sw = black
ro = red
br = brown
gn = green
bl = blue
gr = grey
li = violet
ge = yellow

E24 – Left Seat Belt Switch
F2 – Door contact Switch - Driver's side
F3 – Door contact switch-Passenger side
F7 – Right Rear Side/Sliding Door Contact Switch
F160 – Rear Lid Contact Switch
T1 – Single Connector
T1e – Single Connector, behind fuse/relay panel
T1f – Single Connector, behind fuse/relay panel
T1h – Single Connector, behind instrument panel, right
T1i – Single Connector, behind fuse/relay panel
T1k – Single Connector, near taillight, left
T1l – Single Connector, behind fuse/relay panel
T2i – Double Connector, near Center Interior Light
W1 – Interior light, front
W7 – Center Interior Light
W13 – Right Front Map/Reading Light
W14 – Right Make-up Mirror Light
W15 – Interior Light With Delay Switch
W20 – Left Make-up Mirror Light
W23 – Footwell Light With Delay Switch

(30) – Ground connection, -1-, next to fuse/relay panel
(98) – Ground connection, in rear lid wiring harness
(128) – Ground connection -1-, in interior light wiring harness
(R4) – Wire connection -1-, in interior light/door contact switch wiring harness
(R6) – Plus connection -1-, in interior light wiring harness

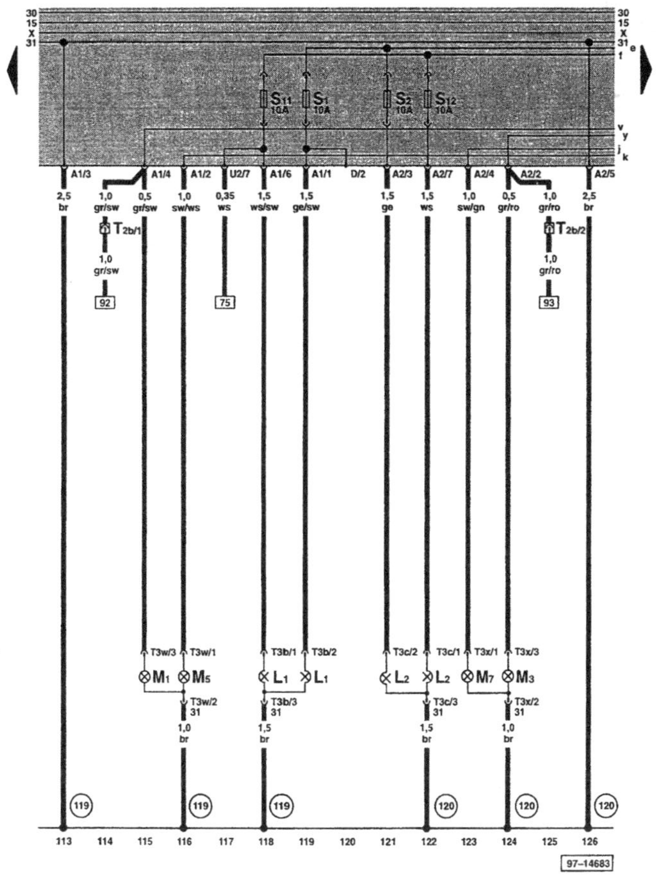

ws = white
sw = black
ro = red
br = brown
gn = green
bl = blue
gr = grey
li = violet
ge = yellow

L1 – Left Headlight
L2 – Right Headlight
M1 – Left Parking Light
M3 – Right Parking Light
M5 – Left Front Turn Signal Light
M7 – Right Front Turn Signal Light
T2b – Double Connector, behind instrument panel, left
T3b – 3-Pin Connector, behind headlight, left
T3c – 3-Pin Connector, behind headlight, right
T3w – 3-Pin Connector, behind headlight, left
T3x – 3-Pin Connector, behind headlight, right

(119) – Ground connection -1-, in headlight wiring harness
(120) – Ground connection -2-, in headlight wiring harness

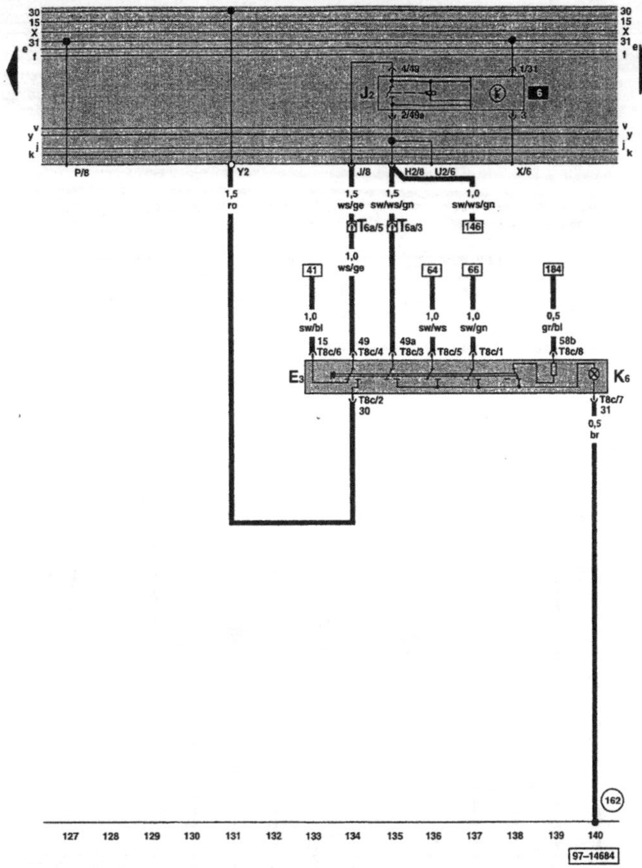

ws = white
sw = black
ro = red
br = brown
gn = green
bl = blue
gr = grey
li = violet
ge = yellow

E3 – Emergency Flasher Switch
J2 – Emergency Flasher Relay
K6 – Emergency Flasher Warning Light
T6a – 6-Pin Connector, behind fuse/relay panel
T8c – 8-Pin Connector

(162) – Ground connection, in blower motor wiring harness

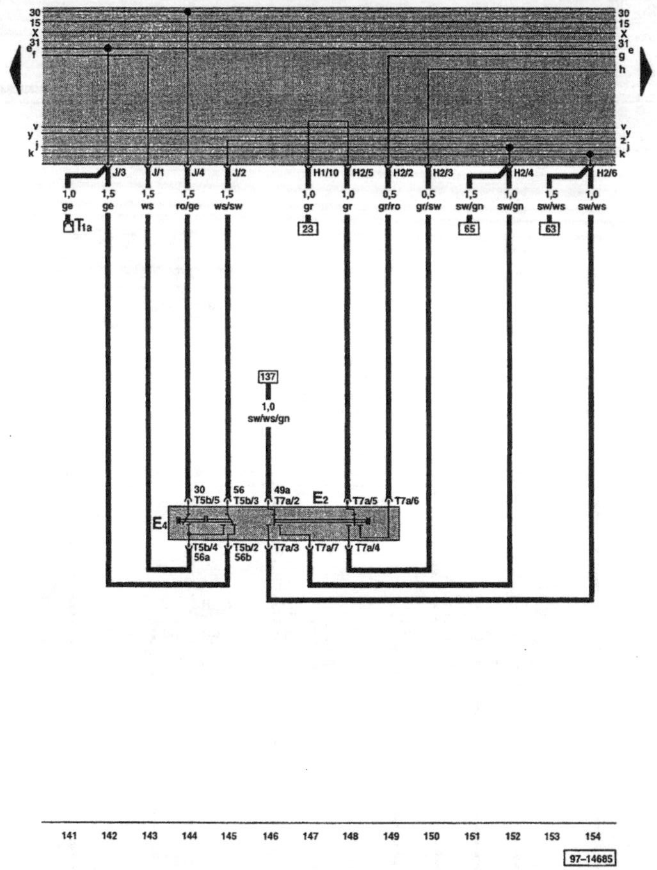

ws = white
sw = black
ro = red
br = brown
gn = green
bl = blue
gr = grey
li = violet
ge = yellow

E2 – Turn Signal Switch
E4 – Headlight Dimmer/Flasher Switch
T1a – Single Connector, behind instrument panel, left
T5b – 5-Pin Connector, behind steering column switch cover
T7a – 7-Pin Connector, behind steering column switch cover

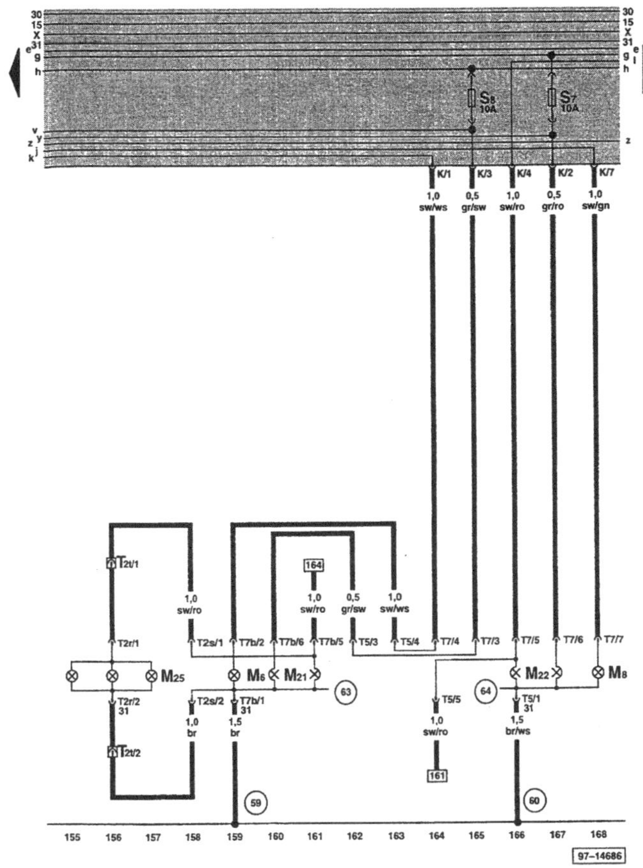

ws = white
sw = black
ro = red
br = brown
gn = green
bl = blue
gr = grey
li = violet
ge = yellow

M6 – Left Rear Turn Signal Light
M8 – Right Rear Turn Signal Light
M21 – Left Brake/Tail Light
M22 – Right Brake/Tail Light
M25 – High-mount Brake Light (rear lid)
T2r – Double Connector
T2s – Double Connector, on taillight, left
T2t – Double Connector, in rear lid
T5 – 5-Pin Connector, on taillight, right
T7 – 7-Pin Connector, on taillight, right
T7b – 7-Pin Connector, on taillight, left

(63) – Ground connection, on left taillight bulb holder

(64) – Ground connection, on right taillight bulb holder

(59) – Ground connection, near taillight, left

(60) – Ground point -2- near taillamp assembly, right

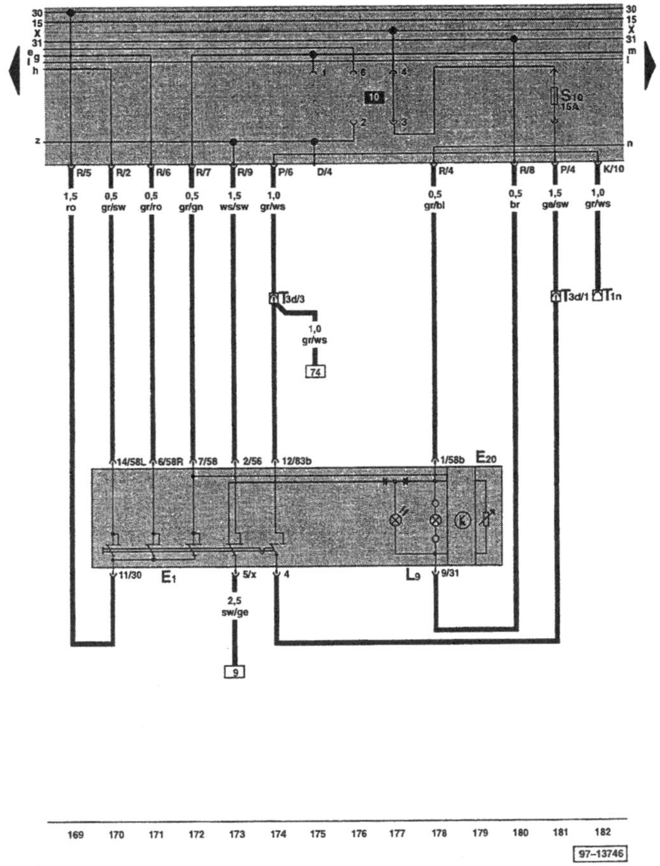

ws = white
sw = black
ro = red
br = brown
gn = green
bl = blue
gr = grey
li = violet
ge = yellow

E1 – Light switch
E20 – Instrument Panel Light Dimmer Switch
L9 – Headlight Switch Light
T1n – Single Connector, on taillight, right
T3d – 3-Pin Connector, behind fuse/relay panel

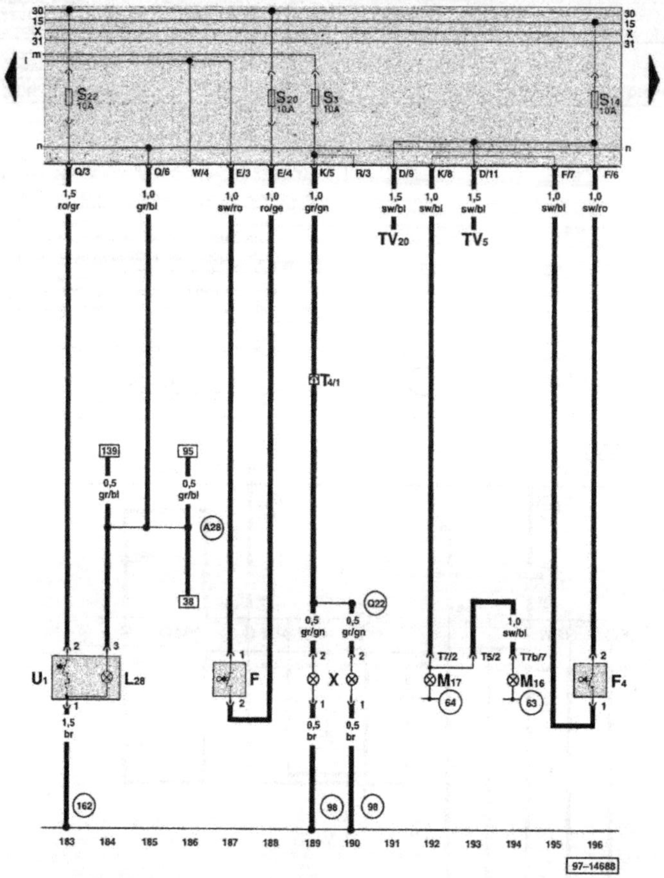

ws = white
sw = black
ro = red
br = brown
gn = green
bl = blue
gr = grey
li = violet
ge = yellow

- F – Brake Light Switch
- F4 – Back-Up Light Switch
- L28 – Cigarette Lighter Light
- M16 – Left Back-Up Light
- M17 – Right Back-Up Light
- T4 – 4-Pin Connector, near tail lamp assembly
- T5 – 5-Pin Connector, on taillight, right
- T7 – 7-Pin Connector, on taillight, right
- T7b – 7-Pin Connector, on taillight, left
- TV5 – Terminal 15a Wire Connector, above fuse/relay panel
- TV20 – Wire junction -2- for terminal 15a, above fuse/relay panel
- U1 – Cigarette Lighter
- X – License Plate Light

- (63) – Ground connection, on left taillight bulb holder
- (64) – Ground connection, on right taillight bulb holder
- (98) – Ground connection, in rear lid wiring harness
- (162) – Ground connection, in blower motor wiring harness
- (A28) – Plus connection (58b), in fresh air blower wiring harness
- (Q22) – Wire connection -1-, in rear lid wiring harness

Cigarette lighter, brake light switch, license plate light, back up light

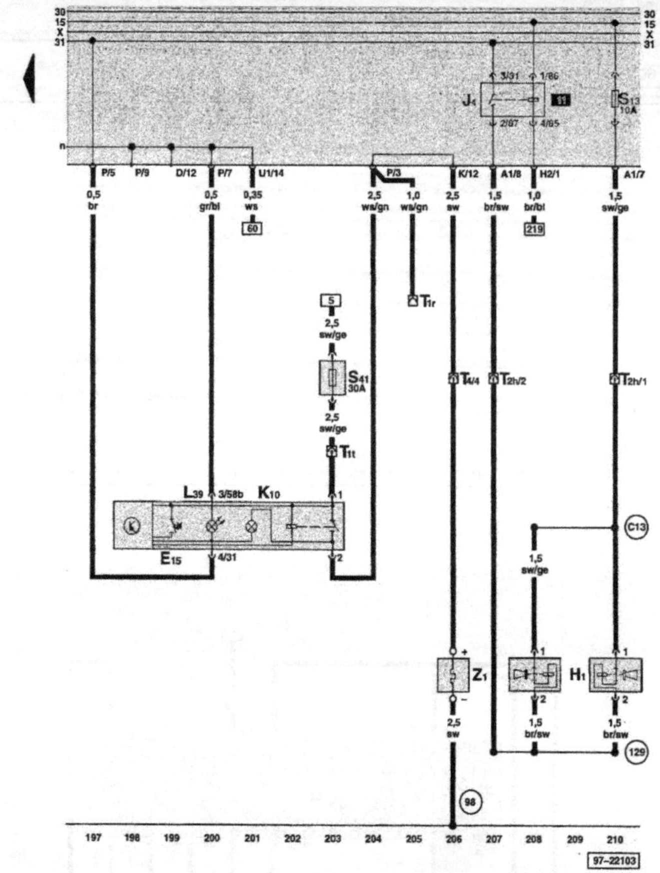

ws = white
sw = black
ro = red
br = brown
gn = green
bl = blue
gr = grey
li = violet
ge = yellow

- E15 – Rear Window Defogger Switch
- H1 – Dual Horns
- J4 – Dual Horn Relay
- K10 – Rear Window Defogger Indicator Light
- L39 – Rear Window Defogger Switch Light
- S41 – Rear Window Defogger Fuse, below fuse/relay panel
- T1r – Single Connector, behind fuse/relay panel
- T1t – Single Connector, behind fuse/relay panel
- T2h – Double Connector, in engine compartment, left front
- T4 – 4-Pin Connector, near tail lamp assembly
- Z1 – Rear Window Defogger/Heat Element

- (98) – Ground connection, in rear lid wiring harness
- (129) – Ground connection, in dual-tone horn wiring harness
- (C13) – Plus connection, in dual tone horn wiring harness

Rear window defogger heat element, dual horns

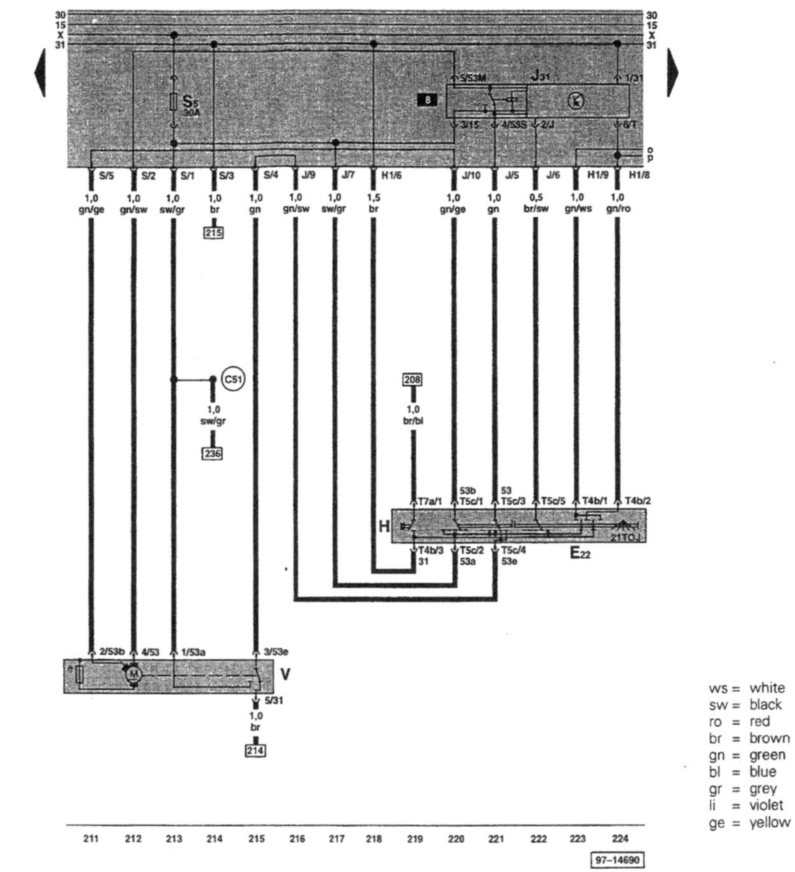

ws = white
sw = black
ro = red
br = brown
gn = green
bl = blue
gr = grey
li = violet
ge = yellow

E22 – Windshield Wiper/Washer Switch
H – Horn Button
J31 – Wiper/Washer Intermittent Relay
T4b – 4-Pin Connector, behind steering column switch cover
T5c – 5-Pin Connector, behind steering column switch cover
T7a – 7-Pin Connector, behind steering column switch cover
V – Windshield Wiper Motor

(C51) – Plus connection (x), in wiper motor wiring harness

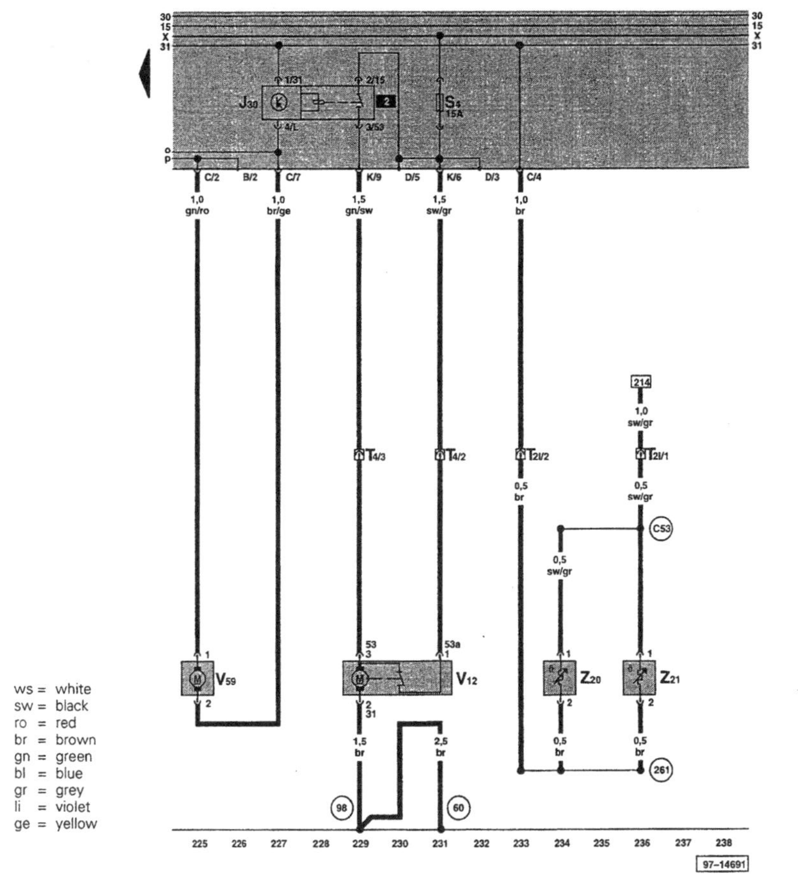

ws = white
sw = black
ro = red
br = brown
gn = green
bl = blue
gr = grey
li = violet
ge = yellow

J30 – Rear Window Wiper/Washer Relay
T2l – Double Connector, on firewall, left
T4 – 4-Pin Connector, near tail lamp assembly
V12 – Rear Window Wiper Motor
V59 – Windshield and Rear Window Washer Pump
Z20 – Left Washer Nozzle Heater
Z21 – Right Washer Nozzle Heater

(60) – Ground point -2- near tail lamp assembly, right

(98) – Ground connection, in rear lid wiring harness

(261) – Ground connection, in washer nozzles heater wiring harness

(C53) – Plus connector, in wiring harness washer nozzles heater

Seat Belt Warning System,

from May 1996

Deviate relay location and fuseplacements as well as the locations of multiple connectors-see section "component locations".

Fuse colors

30 A - green
25 A - white
20 A - yellow
15 A - blue
10 A - red
7,5 A - brown
5 A - beige
3 A - violet

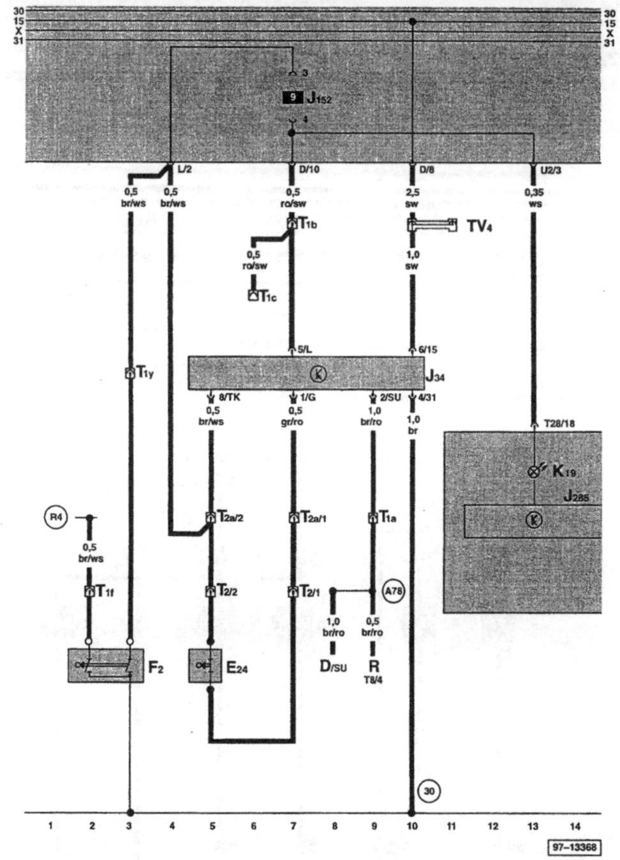

ws = white
sw = black
ro = red
br = brown
gn = green
bl = blue
gr = grey
li = violet
ge = yellow

D - Ignition/Starter Switch
E24 - Left Seat Belt Switch
F2 - Door contact Switch - Driver's side
J34 - Seat Belt Warning System Relay, above fuse/relay panel
J152 - Radio/Parking Light Warning Buzzer
J285 - Control module with indicator unit in instrument panel insert
K19 - Seat Belt Warning Light
R - Radio
T1a - Single Connector, behind fuse/relay panel
T1b - Single Connector, behind fuse/relay panel
T1c - Single Connector, behind fuse/relay panel
T1f - Single Connector, behind fuse/relay panel
T1y - Single Connector, behind fuse/relay panel
T2 - Double Connector, on Seat Belt Switch
T2a - Double Connector, behind fuse/relay panel
T8 - 8-Pin Connector, on Radio
T28 - 28-Pin Connector
TV4 - Terminal 15 Wire Connector, above fuse/relay panel

(30) - Ground connection, -1-, next to fuse/relay panel
(A78) - Connector (SU), in instrument panel wiring harness
(R4) - Wire connection -1-, in interior light/door contact switch wiring harness

2.8 L- Engine - Motronic Multiport Fuel Injection (MFI)/103 kW, code AES,

from May 1996

Devite relay location and fuseplacements as well as locations of multiple connectors - see section "component locations".

12 Fuel Pump (FP) Relay (167)

Note: Number in parentheses indicates production control number stamped on relay housing.

Fuse colors

30 A – green
25 A – white
20 A – yellow
15 A – blue
10 A – red
7,5 A – brown
5 A – beige
3 A – violet

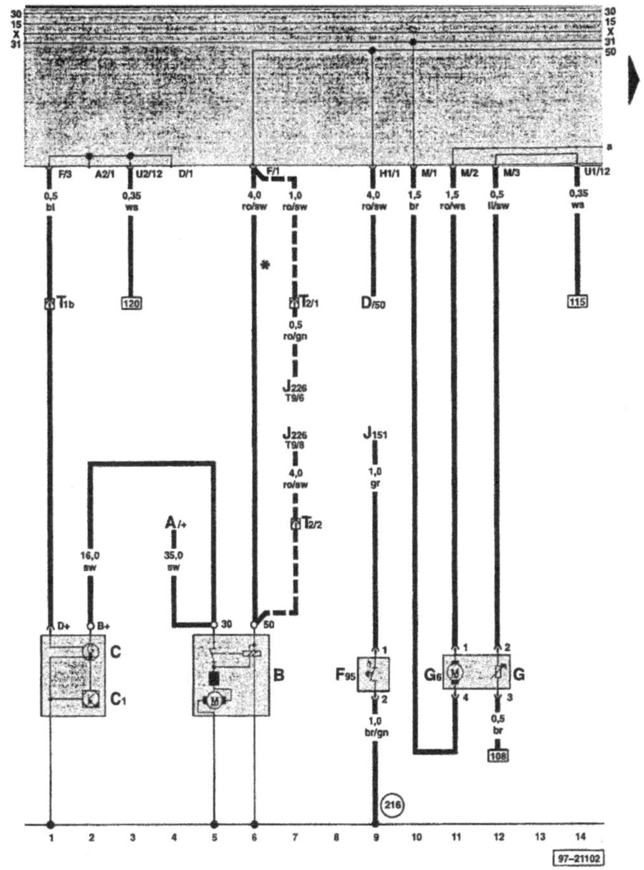

ws = white
sw = black
ro = red
br = brown
gn = green
bl = blue
gr = grey
li = violet
ge = yellow

- A – Battery
- B – Starter
- C – Generator (GEN)
- C1 – Voltage Regulator (VR)
- D – Ignition/Starter Switch
- F95 – After-Run Coolant Thermal Switch
- G – Fuel Level Sensor
- G6 – Fuel Pump (FP)
- J151 – Coolant Circulation Pump Relay
- J226 – Park/Neutral Position (PNP) Relay
- T1b – Single Connector, near starter
- T2 – Double Connector
- T9 – 9-Pin Connector
- (216) – Ground connection -2-, in Motronic Multiport Fuel Injection (MFI) wiring harness

- - - automatic transmission only
* – manual transmission only

Generator (GEN), starter, fuel pump, fuel level sensor, coolant after-run

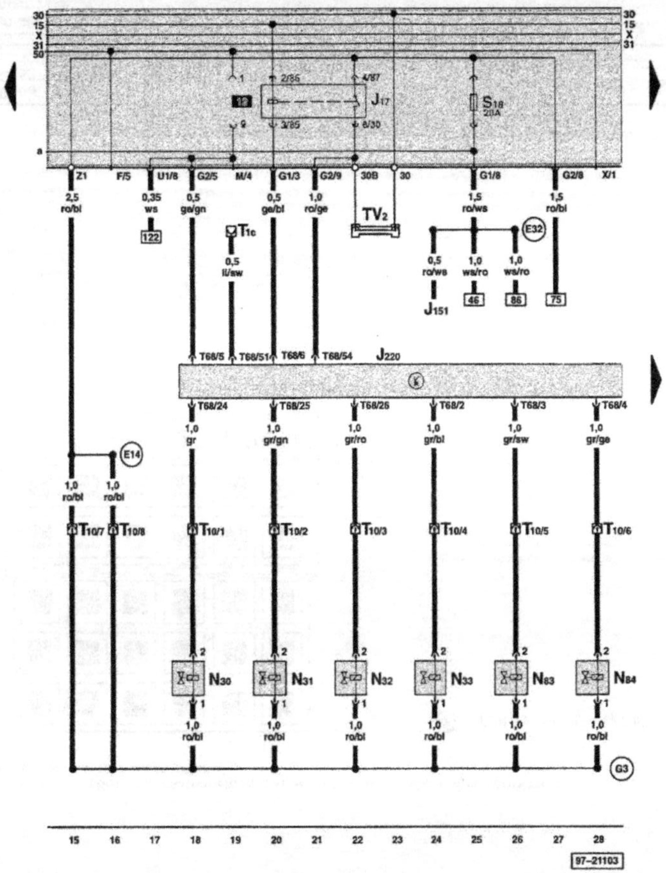

ws = white
sw = black
ro = red
br = brown
gn = green
bl = blue
gr = grey
li = violet
ge = yellow

- J17 – Fuel Pump (FP) Relay
- J220 – Motronic Engine Control Module (ECM), in engine compartment, left
- J151 – Coolant Circulation Pump Relay
- N30 – Cylinder 1 Fuel Injector
- N31 – Cylinder 2 Fuel Injector
- N32 – Cylinder 3 Fuel Injector
- N33 – Cylinder 4 Fuel Injector
- N83 – Cylinder 5 Fuel Injector
- N84 – Cylinder 6 Fuel Injector
- T1c – Single Connector, behind fuse/relay panel
- T10 – 10-Pin Connector, near cylinder head cover
- T68 – 68-Pin Connector
- TV2 – Terminal 30 (B+) Wire Connector
- (E14) – Wire connection -1-, in Motronic Multiport Fuel Injection (MFI) wiring harness
- (E32) – Connector (87), in wiring harness Motronic
- (G3) – Plus connection, in injector wiring harness

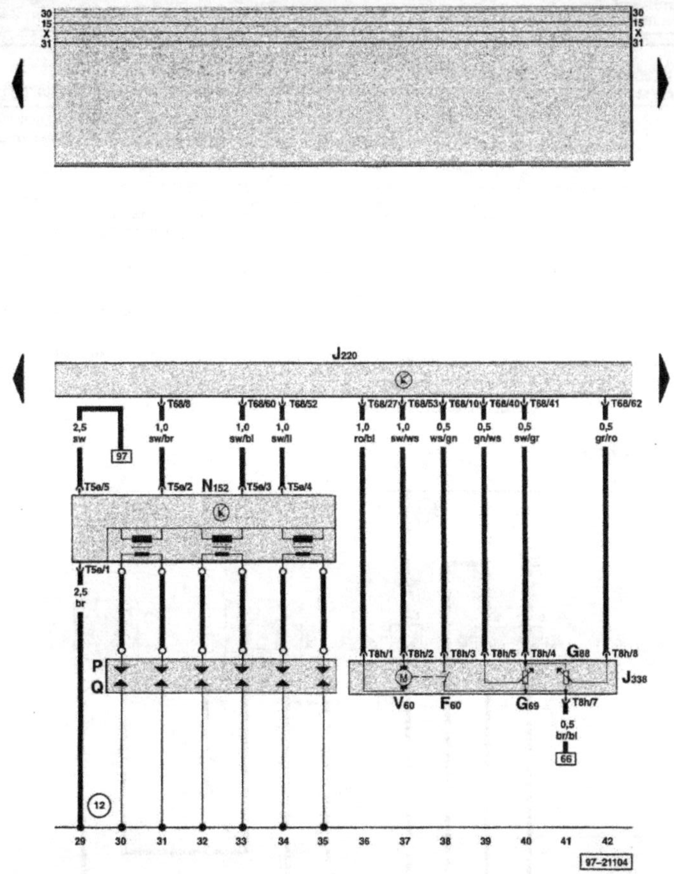

ws = white
sw = black
ro = red
br = brown
gn = green
bl = blue
gr = grey
li = violet
ge = yellow

- F60 – Closed Throttle Position (CTP) Switch
- G69 – Throttle Position (TP) Sensor
- G88 – Throttle Position (TP) Sensor
- J220 – Motronic Engine Control Module (ECM), in engine compartment, left
- J338 – Throttle Valve Control Module
- N152 – Ignition Coil
- P – Spark Plug Connectors
- Q – Spark Plugs
- T5e – 5-Pin Connector
- T8h – 8-Pin Connector
- T68 – 68-Pin Connector
- V60 – Throttle Position (TP) Actuator
- (12) – Ground connection, in engine compartment, left

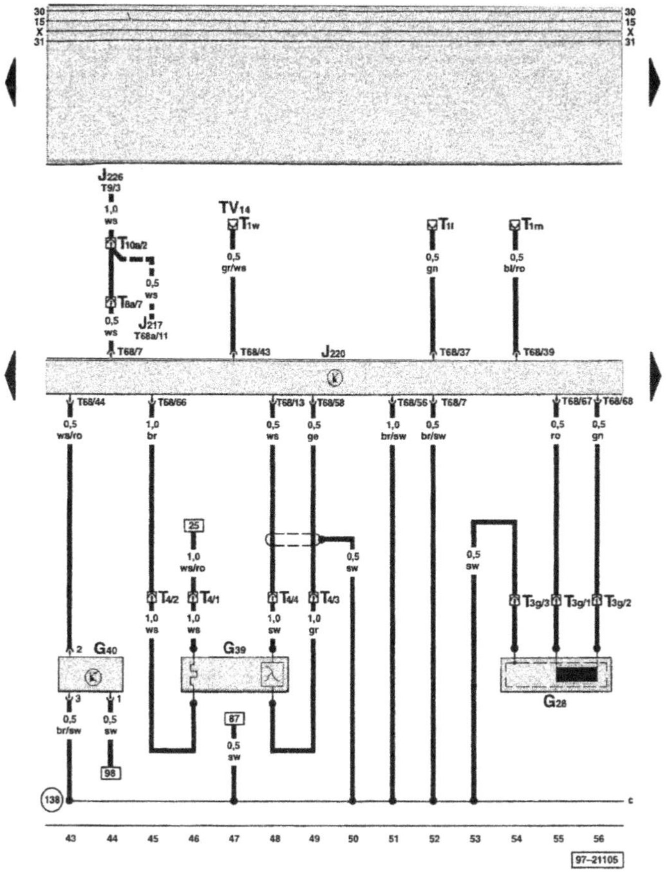

ws = white
sw = black
ro = red
br = brown
gn = green
bl = blue
gr = grey
li = violet
ge = yellow

G28 – Engine Speed (RPM) Sensor
G39 – Heated Oxygen Sensor (HO2S)
G40 – Camshaft Position (CMP) Sensor
J217 – Transmission Control Module (TCM)
J220 – Motronic Engine Control Module (ECM), in engine compartment, left
J226 – Park/Neutral Position (PNP) Relay
T1l – Single Connector
T1m – Single Connector
T1w – Single Connector
T3g – 3-Pin Connector, near starter
T4 – 4-Pin Connector, near Heated Oxygen Sensor (HO2S)
T8a – 8-Pin Connector
T9 – 9-Pin Connector
T10a – 10-Pin Connector
T68 – 68-Pin Connector
T68a – 68-Pin Connector
TV14 – Data Link Connector (DLC) wire Connector
(138) – Ground connection (control module), in Motronic Multiport Fuel Injection (MFI) wiring harness
- - - automatic transmission only

Motronic engine control module (ECM), heated oxygen sensor, camshaft position (CMP) sensor, engine speed (RPM) sensor

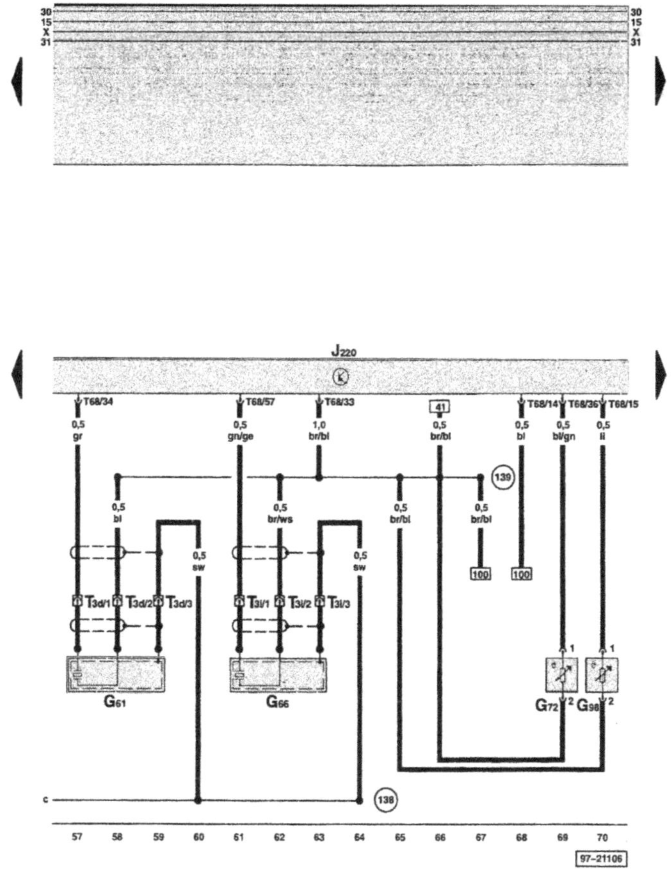

ws = white
sw = black
ro = red
br = brown
gn = green
bl = blue
gr = grey
li = violet
ge = yellow

G61 – Knock Sensor (KS) 1
G66 – Knock Sensor (KS) 2
G72 – Intake Air Temperature (IAT) Sensor
G98 – EGR Temperature Sensor
J220 – Motronic Engine Control Module (ECM)
T3d – 3-Pin Connector, near cylinder head cover
T3i – 3-Pin Connector, near cylinder head cover
T68 – 68-Pin Connector
(100) – Ground connection (control module), in Motronic Multiport Fuel Injection (MFI) wiring harness
(139) – Ground connection (sensor Ground), in Motronic Multiport Fuel Injection (MFI) wiring harness

Motronic engine control module, knock sensors, EGR Temperature Sensor, intake air temperature (IAT) sensor

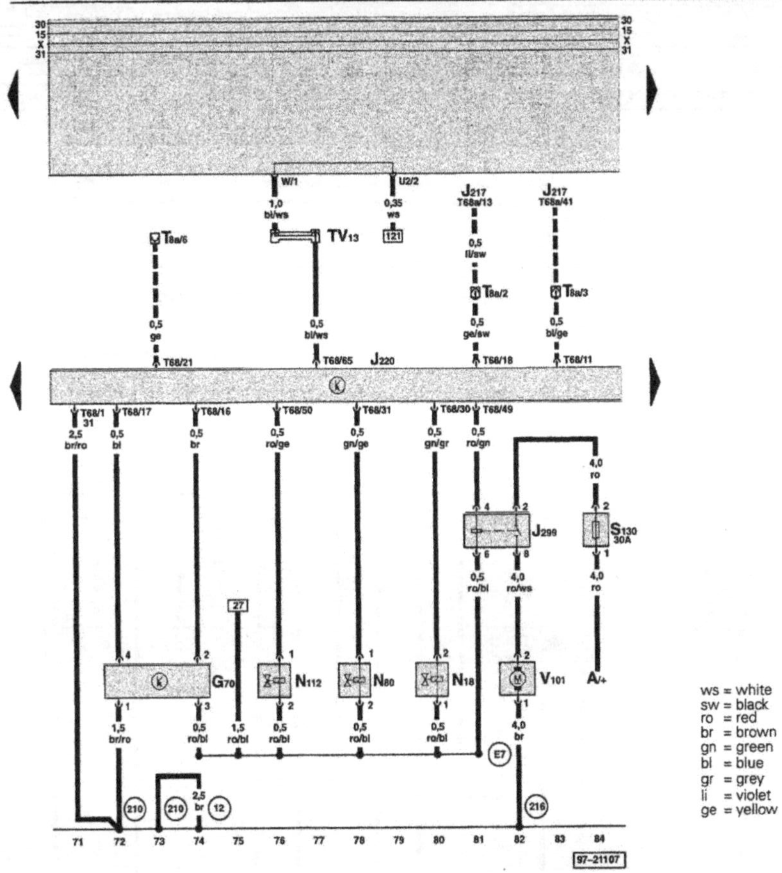

ws = white
sw = black
ro = red
br = brown
gn = green
bl = blue
gr = grey
li = violet
ge = yellow

- A – Battery
- G70 – Mass Air Flow (MAF) Sensor
- J217 – Transmission Control Module (TCM)
- J220 – Motronic Engine Control Module (ECM)
- J299 – Secondary Air Injection (AIR) Pump Relay, above fuse/relay panel
- N18 – EGR Vacuum Regulator Solenoid Valve
- N80 – Evaporative Emission (EVAP) Canister Purge Regulator Valve
- N112 – Secondary Air Injection (AIR) Solenoid Valve
- S130 – Fuse for secondary air pump, above fuse/relay panel
- T8a – 8-Pin Connector
- T68 – 68-Pin Connector
- T68a – 68-Pin Connector
- TV13 – Vehicle Speed Signal Wire Connector, above fuse/relay panel
- V101 – Secondary Air Injection (AIR) Pump Motor
- (12) – Ground connection, in engine compartment, left
- (210) – Ground connection -1-, in Motronic Multiport Fuel Injection (MFI) wiring harness
- (216) – Ground connection -2-, in Motronic Multiport Fuel Injection (MFI) wiring harness
- (E7) – Plus connection (87a), in Motronic Multiport Fuel Injection (MFI) wiring harness
- - - - automatic transmission only

Edition 11/97 USA.5272.09.21 **Motronic engine control module (ECM), (MAF) sensor, evaporative emission (EVAP) canister purge regulator valve, secondary air injection (AIR) solenoid valve, secondary air injection (AIR) pump motor, EGR vacuum regulator solenoid valve**

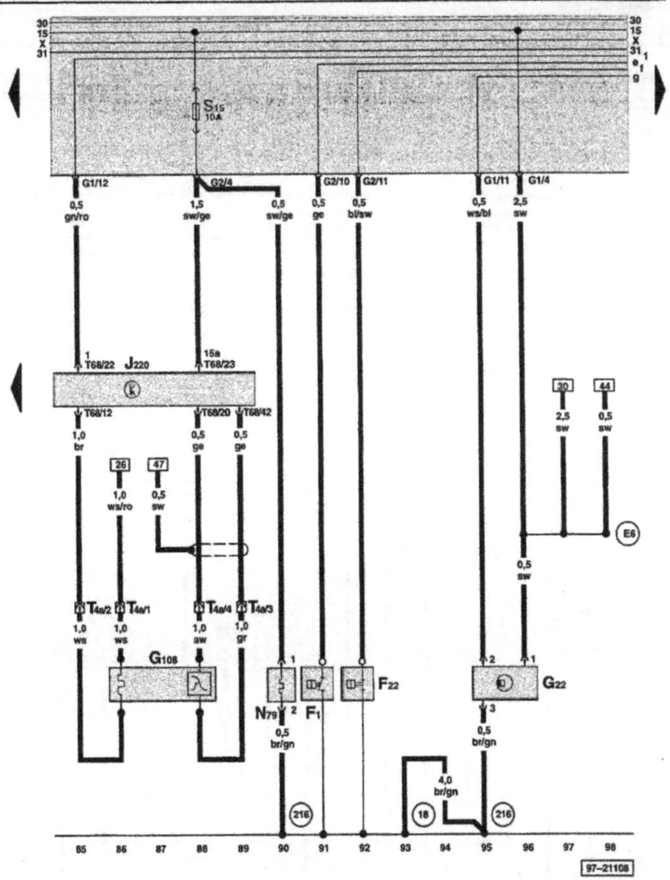

ws = white
sw = black
ro = red
br = brown
gn = brown
bl = blue
gr = grey
li = violet
ge = yellow

- F1 – Oil Pressure Switch
- F22 – 0.3 bar Oil Pressure Switch
- G22 – Speedometer Vehicle Speed Sensor (VSS)
- G108 – Heated Oxygen Sensor (HO2S) 2
- J220 – Motronic Engine Control Module (ECM), in engine compartment, left
- N79 – Positive Crankcase Ventilation (PCV) Heating Element
- T4a – 4-Pin Connector, near Heated Oxygen Sensor (HO2S)
- T68 – 68-Pin Connector
- (18) – Ground connection, on engine block
- (216) – Ground connection -2-, in Motronic Multiport Fuel Injection (MFI) wiring harness
- (E6) – Plus connection -1- (15), in Motronic Multiport Fuel Injection (MFI) wiring harness

Motronic engine control module, speedometer vehicle speed sensor (VSS), heated oxygen sensor, oil pressure switch **Edition 11/97** USA.5272.09.21

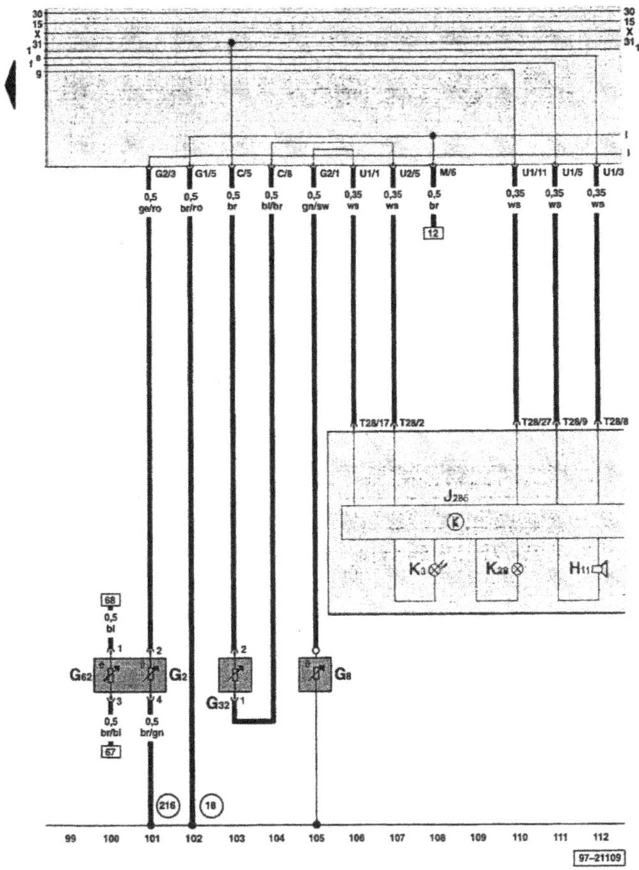

ws = white
sw = black
ro = red
br = brown
gn = green
bl = blue
gr = grey
li = violet
ge = yellow

G2 – Engine Coolant Temperature (ECT) Sensor
G8 – Engine Oil Temperature Sensor
G32 – Engine Coolant Level (ECL) Sensor
G62 – Engine Coolant Temperature (ECT) Sensor
H11 – Oil Pressure Warning Buzzer
J285 – Control module with indicator unit in instrument panel insert
K3 – Oil Pressure Warning Light
K28 – Engine Coolant Level/Temperature (ECL/ECT) Warning Light
T28 – 28-Pin Connector

(18) – Ground connection, on engine block

(216) – Ground connection -2-, in Motronic Multiport Fuel Injection (MFI) wiring harness

Instrument cluster, visual acoustical pressure oil control, engine coolant level (ECL) sensor, engine coolant temperature (ECT) sensor

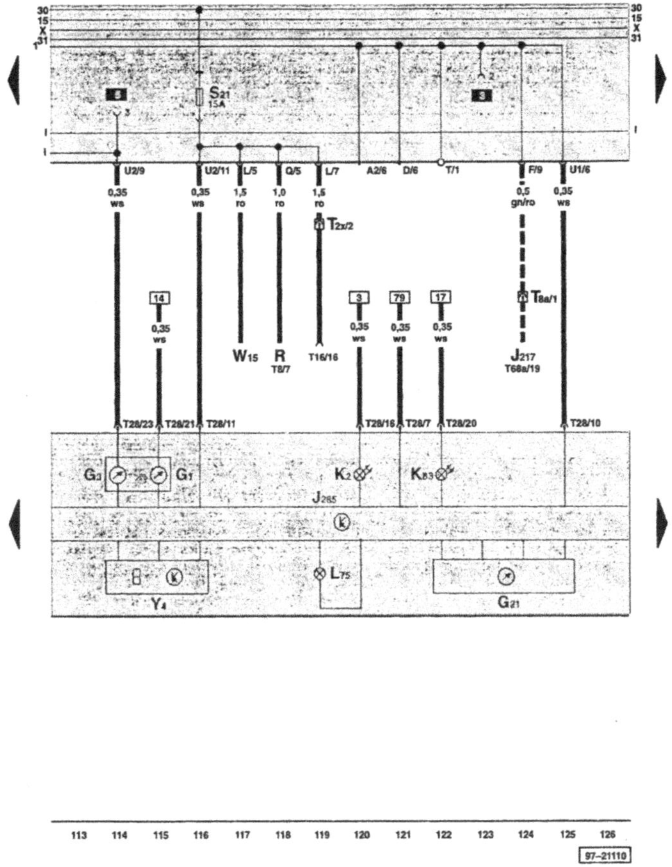

ws = white
sw = black
ro = red
br = brown
gn = green
bl = blue
gr = grey
li = violet
ge = yellow

G1 – Fuel Gauge
G3 – Engine Coolant Temperature (ECT) Gauge
G21 – Speedometer
J217 – Transmission Control Module (TCM)
J285 – Control module with indicator unit in instrument panel insert
K2 – Generator (GEN) Warning Light
K83 – Malfunction Indicator Lamp (MIL)
L75 – Digital Display Light
R – Radio
T2x – Double Connector, behind fuse/relay panel
T8 – 8-Pin Connector, on Radio
T8a – 8-Pin Connector
T16 – 16-Pin Connector, On Board Diagnostic, in instrument panel, center
T28 – 28-Pin Connector
T68a – 68-Pin Connector
W15 – Interior Light with Delay Switch
Y4 – Odometer Display

--- automatic transmission only

Instrument cluster, fuel gauge, engine coolant temperature (ECT) gauge, speedometer, odometer display

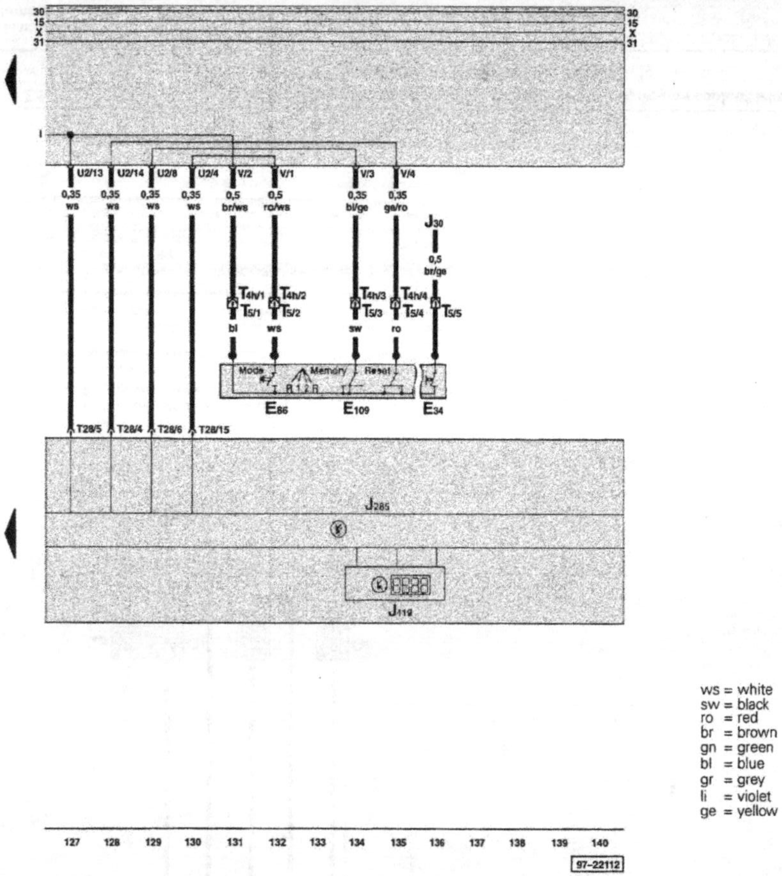

ws = white
sw = black
ro = red
br = brown
gn = green
bl = blue
gr = grey
li = violet
ge = yellow

- E34 – Rear Window Wiper Switch (interval)*
- E86 – Multi-Function Indicator Mode Select Switch
- E109 – Multi-Function Indicator Memory Switch
- J30 – Rear Window Wiper/Washer Relay
- J119 – Multi-Function Indicator (MFA)
- J285 – Control module with indicator unit in instrument panel insert
- T4h – 4-Pin Connector, behind steering column switch cover
- T5 – 5-Pin Connector, behind steering column switch cover
- T28 – 28-Pin Connector

* – from September 1997 with new rear window wiper switch (interval)

Coolant Pump After-Run,

from May 1996

- **2.8 L-Engine – Motronic Multiport Fuel Injection (MFI)/103 kW, code AES**

Deviate relay location and fuseplacements as well as the locations of multiple connectors- see section "component locations".

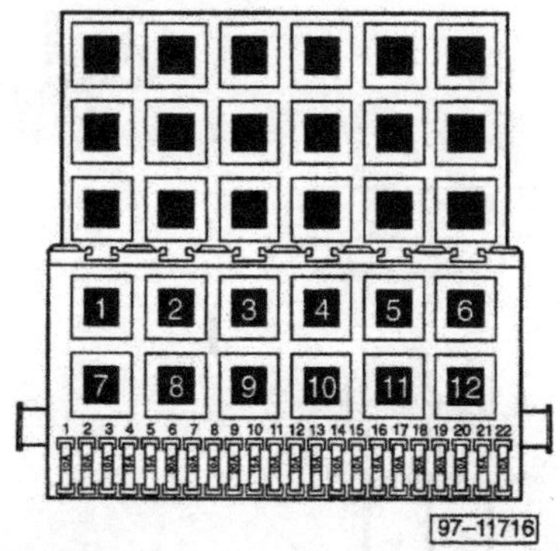

Fuse colors

- 30 A – green
- 25 A – white
- 20 A – yellow
- 15 A – blue
- 10 A – red
- 7,5 A – brown
- 5 A – beige
- 3 A – violet

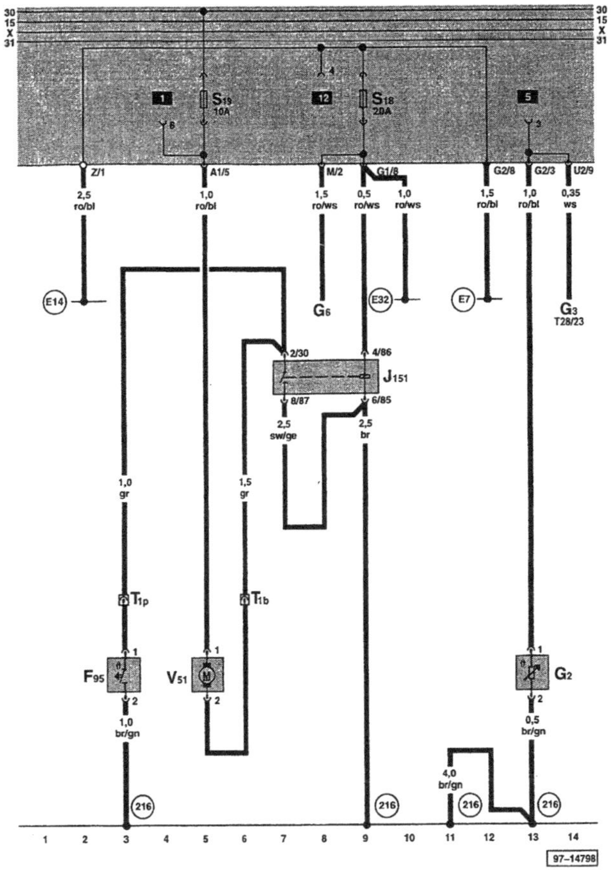

ws = white
sw = black
ro = red
br = brown
gn = green
bl = blue
gr = grey
li = violet
ge = yellow

F95 – After-Run Coolant Thermal Switch*
G2 – Engine Coolant Temperature (ECT) Sensor*
G3 – Engine Coolant Temperature (ECT) Gauge
G6 – Fuel Pump (FP)
J151 – Coolant Circulation Pump Relay
T1b – Single Connector, behind fuse/relay panel
T1p – Single Connector, behind fuse/relay panel
T28 – 28-Pin Connector, on instrument cluster
V51 – After-Run Coolant Pump

(216) – Ground connection -2-, in Motronic Multiport

(E7) – Plus connection (87a), in Motronic Multiport Fuel Injection (MFI) wiring harness

(E14) – Wire connection -1-, in Motronic Multiport Fuel Injection (MFI) wiring harness

(E32) – Connector (87), in wiring harness Motronic

* – F95 and G2 by some Modells in one housing

Coolant pump after-run

2.4 L-Diesel Engine/57 kW, code AAB,

from May 1996

Deviate relay location and fuseplacements as well as the locations of multiple connectors- see section "component locations".

12 – Glow Plug Relay (102)

Note:

Number in parentheses indicates production control number stamped on relay housing.

Fuse colors

30 A – green
25 A – white
20 A – yellow
15 A – blue
10 A – red
7,5 A – brown
5 A – beige
3 A – violet

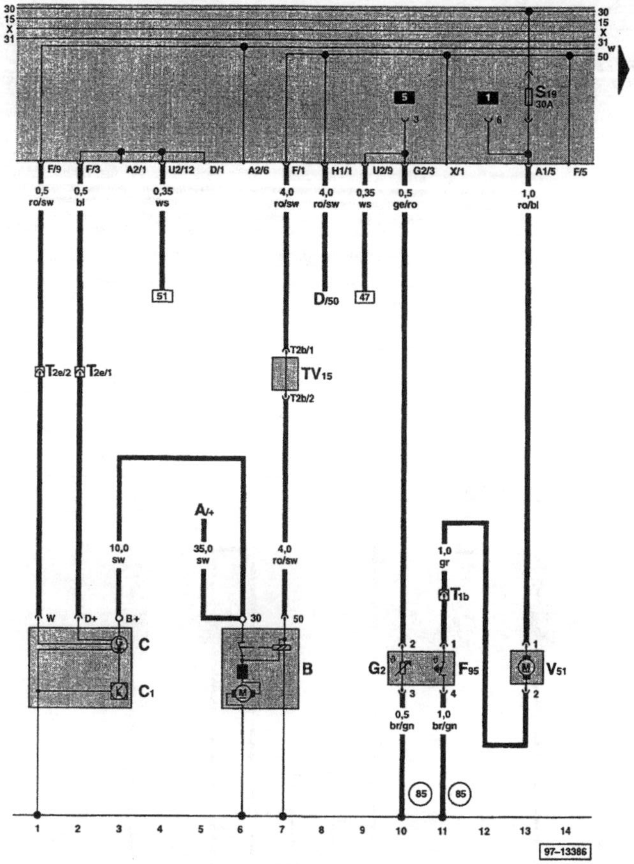

ws = white
sw = black
ro = red
br = brown
gn = green
bl = blue
gr = grey
li = violet
ge = yellow

A – Battery
B – Starter
C – Generator (GEN)
C1 – Voltage Regulator (VR)
D – Ignition/Starter Switch
F95 – After-Run Coolant Thermal Switch
G2 – Engine Coolant Temperature (ECT) Sensor
T1b – Single Connector, behind fuse/relay panel
T2b – Double Connector, behind fuse/relay panel
T2e – Double Connector, near starter
TV15 – Wire Connector, behind fuse/relay panel
V51 – After-Run Coolant Pump

(85) – Ground connection -1-, in engine compartment wiring harness

Generator (GEN), starter, coolant after-run

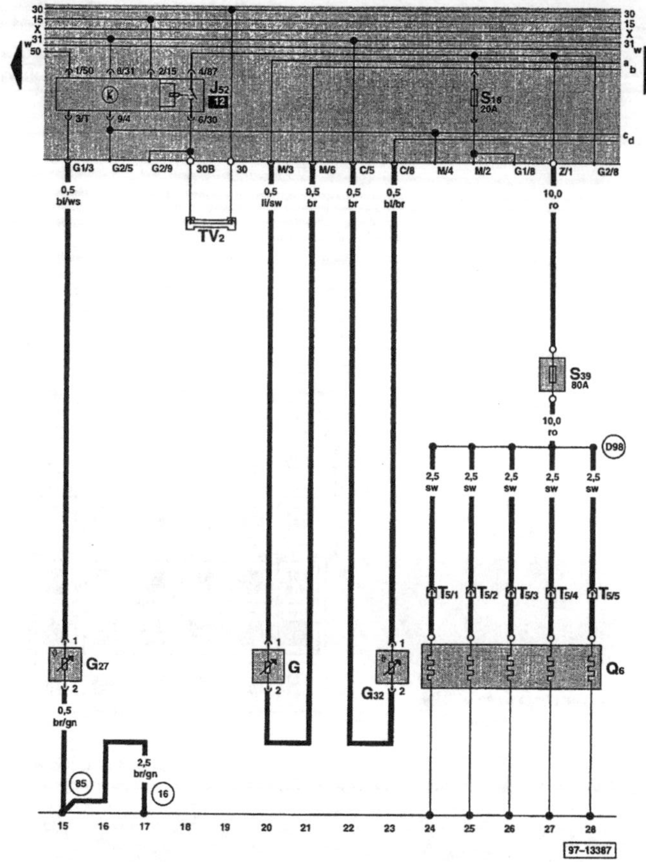

G – Fuel Level Sensor
G27 – Sensor for engine temperature
G32 – Engine Coolant Level (ECL) Sensor
J52 – Glow Plug Relay
Q6 – Spark plugs (engine)
S39 – Engine Glow Plug Fuse Strip
T5 – 5-Pin Connector, near cylinder head cover
TV2 – Terminal 30 (B+) Wire Connector

(16) – Ground connection, -1-, on valve cover

(85) – Ground connection -1-, in engine compartment wiring harness

(D98) – Wire connection (glow plugs), in engine compartment wiring harness

ws = white
sw = black
ro = red
br = brown
gn = green
bl = blue
gr = grey
li = violet
ge = yellow

Preheating system, fuel level sensor, sensor for engine temperature, engine coolant level (ECL) sensor

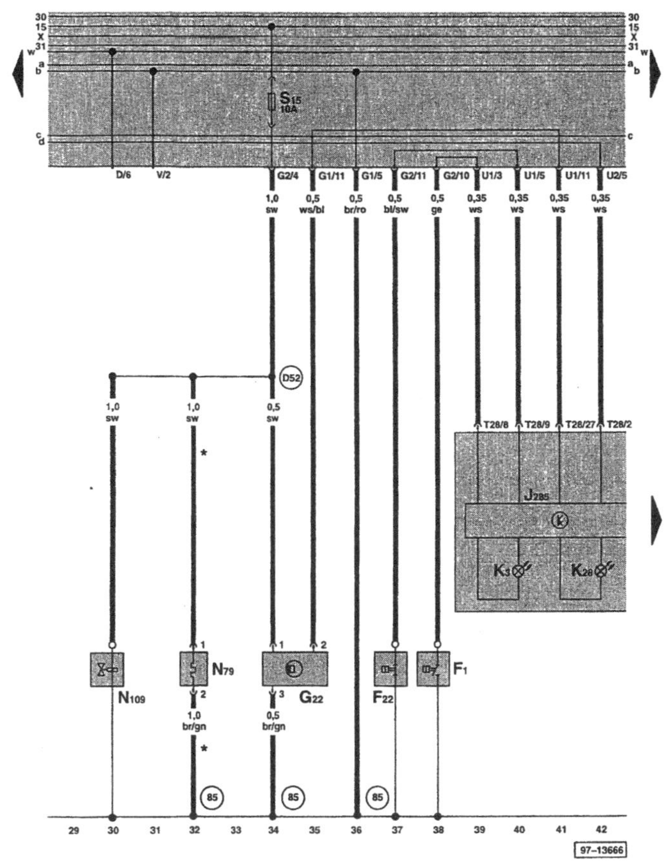

ws = white
sw = black
ro = red
br = brown
gn = green
bl = blue
gr = grey
li = violet
ge = yellow

F1 – Oil Pressure Switch
F22 – 0.3bar Oil Pressure Switch
G22 – Speedometer Vehicle Speed Sensor (VSS)
J285 – Control module with indicator unit in instrument panel insert
K3 – Oil Pressure Warning Light
K28 – Engine Coolant Level/Temperature (ECL/ECT) Warning Light
N79 – Positive Crankcase Ventilation (PCV) Heating Element
N109 – Fuel Cut-off Valve
T28 – 28-Pin Connector

(85) – Ground connection -1-, in engine compartment wiring harness
(D52) – Plus connection (15a), in engine compartment wiring harness

* – only for cold climatic zones

Instrument cluster, oil pressure switch, speedometer vehicle speed sensor, fuel cut off valve, heating element

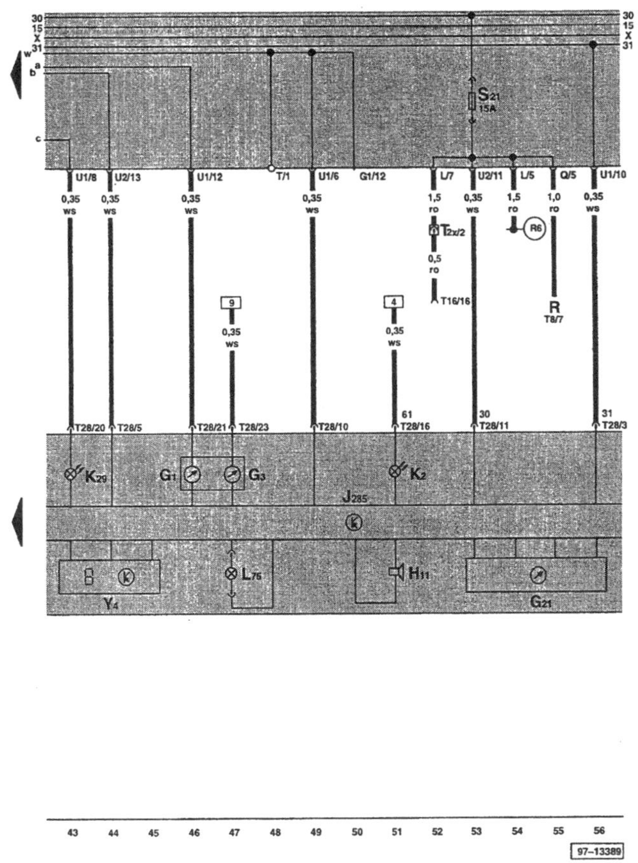

ws = white
sw = black
ro = red
br = brown
gn = green
bl = blue
gr = grey
li = violet
ge = yellow

G1 – Fuel Gauge
G3 – Engine Coolant Temperature (ECT) Gauge
G21 – Speedometer
H11 – Oil Pressure Warning Buzzer
J285 – Control module with indicator unit in instrument panel insert
K2 – Generator (GEN) Warning Light
K29 – Glow Plug Indicator Light
L75 – Digital Display Light
R – Radio
T2x – Double Connector, behind fuse/relay panel
T8 – 8-Pin Connector, on Radio
T16 – 16-Pin Connector, On Board Diagnostic, in instrument panel, center
T28 – 28-Pin Connector
Y4 – Odometer Display

(R6) – Plus connection -1-, in interior light wiring harness

Instrument cluster, fuel gauge, engine coolant temperature (ECT) gauge, glow plug indicator light, speedometer, odometer display

Airbag,

from May 1996

Deviate relay location and fuseplacements as well as the locations of multiple connectors- see section "component locations".

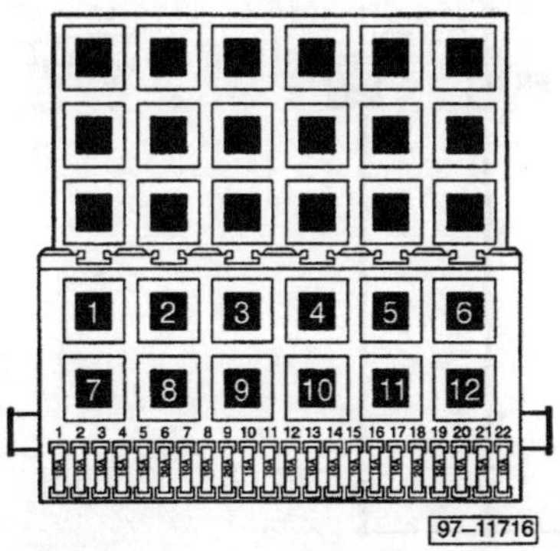

Fuse colors

30 A - green
25 A - white
20 A - yellow
15 A - blue
10 A - red
7,5 A - brown
5 A - beige
3 A - violet

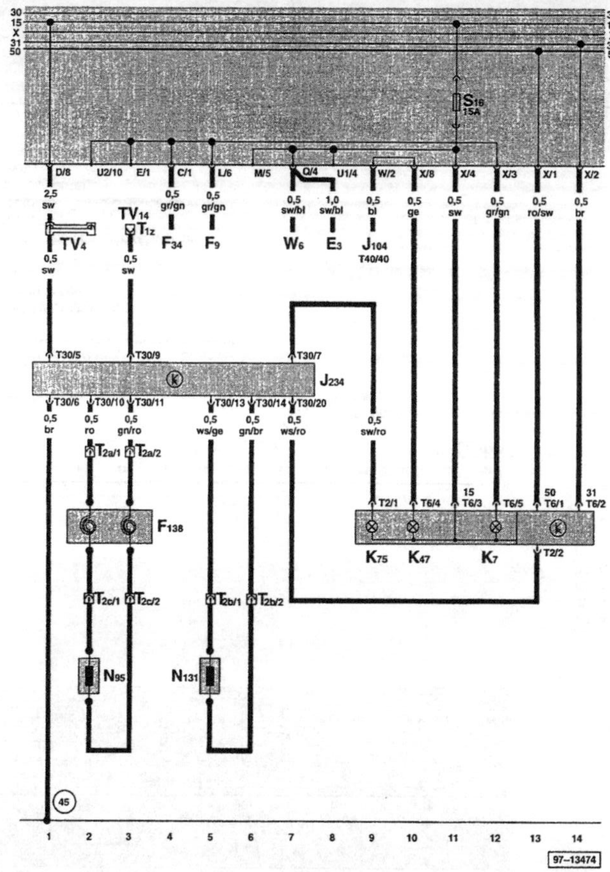

ws = white
sw = black
ro = red
br = brown
gn = green
bl = blue
gr = grey
li = violet
ge = yellow

E3 - Emergency Flasher Switch
F9 - Parking Brake Warning Light Switch
F34 - Brake Fluid Level Warning Switch
F138 - Airbag Spiral Spring/Return Spring With Slip Ring
J104 - ABS Control Module (w/EDL)
J234 - Airbag Control Module, behind instrument panel, center
K7 - Brake and Parking Brake Warning Light
K47 - ABS Warning Light
K75 - Airbag Malfunction Indicator Lamp (MIL)
N95 - Driver's Side Airbag Igniter
N131 - Passenger's Side Airbag Igniter 1
T1z - Single Connector, on Data Link Connector (DLC) wire Connector
T2 - Double Connector
T2a - Double Connector, behind steering column switch cover
T2b - Double Connector, behind instrument panel, right
T2c - Double Connector, in steering wheel
T6 - 6-Pin Connector
T30 - 30-Pin Connector
T40 - 40-Pin Connector, on ABS Control Module (w/EDL)
TV4 - Terminal 15 Wire Connector, above fuse/relay panel
TV14 - Data Link Connector (DLC) wire Connector, above fuse/relay panel
W6 - Glove Compartment Light
(45) - Ground connection, behind instrument panel, center

ABS control module (w/EDL), airbag malfunction indicator lamp (MIL), airbag igniter, airbag spiral spring/return spring with slip ring

Anti-lock brake system (ABS) with electronic differential lock (EDL),

from May 1996

Deviate relay location and fuseplacements as well as the locations of multiple connectors- see section "component locations".

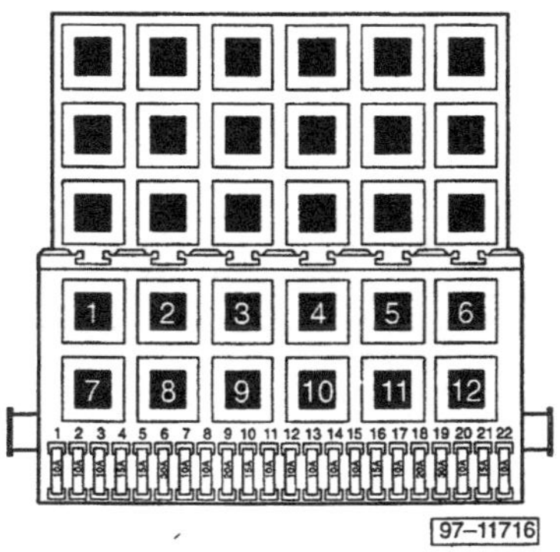

4 – Load Reduction Relay (18)

Note:

Number in parentheses indicates production control number stamped on relay housing.

Fuse colors

30 A – green
25 A – white
20 A – yellow
15 A – blue
10 A – red
7,5 A – brown
5 A – beige
3 A – violet

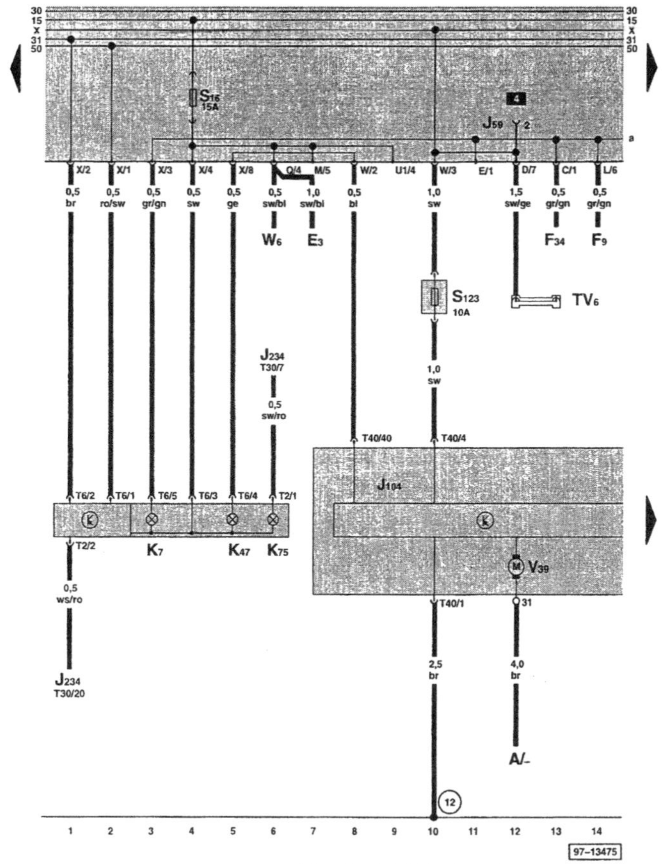

ws = white
sw = black
ro = red
br = brown
gn = green
bl = blue
gr = grey
li = violet
ge = yellow

A – Battery
E3 – Emergency Flasher Switch
F9 – Parking Brake Warning Light Switch
F34 – Brake Fluid Level Warning Switch
J59 – Load Reduction Relay
J104 – ABS Control Module (w/EDL), in engine compartment, left
J234 – Airbag Control Module
K7 – Brake and Parking Brake Warning Light
K47 – ABS Warning Light
K75 – Airbag Malfunction Indicator Lamp (MIL)
S123 – Fuse 1 for control module - ABS
T2 – Double Connector
T6 – 6-Pin Connector
T30 – 30-Pin Connector, on Airbag Control Module
T40 – 40-Pin Connector
TV6 – Terminal X Wire Connector
V39 – ABS Return Flow Pump
W6 – Glove Compartment Light

(12) – Ground connection, in engine compartment, left

ABS control module (w/EDL), ABS return flow pump, ABS warning light

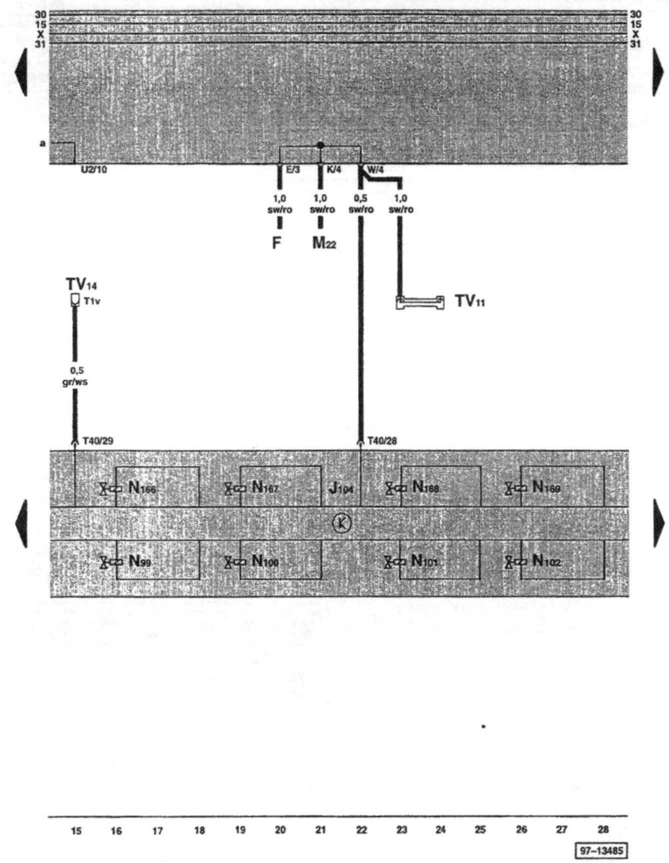

ws = white
sw = black
ro = red
br = brown
gn = green
bl = blue
gr = grey
li = violet
ge = yellow

F – Brake Light Switch
J104 – ABS Control Module (w/EDL), in engine compartment, left
M22. – Right Brake/Tail Light
N100 – Right Front ABS Outlet Valve
N101 – Left Front ABS Inlet Valve
N102 – Left Front ABS Outlet Valve
N166 – Right Front Traction Control Switch-Over Valve
N167 – Right Front Traction Control Outlet Valve
N168 – Left Front Traction Control Switch-Over Valve
N169 – Left Front Traction Control Outlet Valve
T1v – Single Connector, on Data Link Connector (DLC) wire Connector
T40 – 40-Pin Connector
TV11 – Terminal 54 Wire Connector, above fuse/relay panel
TV14 – Data Link Connector (DLC) wire Connector, above fuse/relay panel

ABS control module (w/EDL), valves for ABS front, valves for differential lock

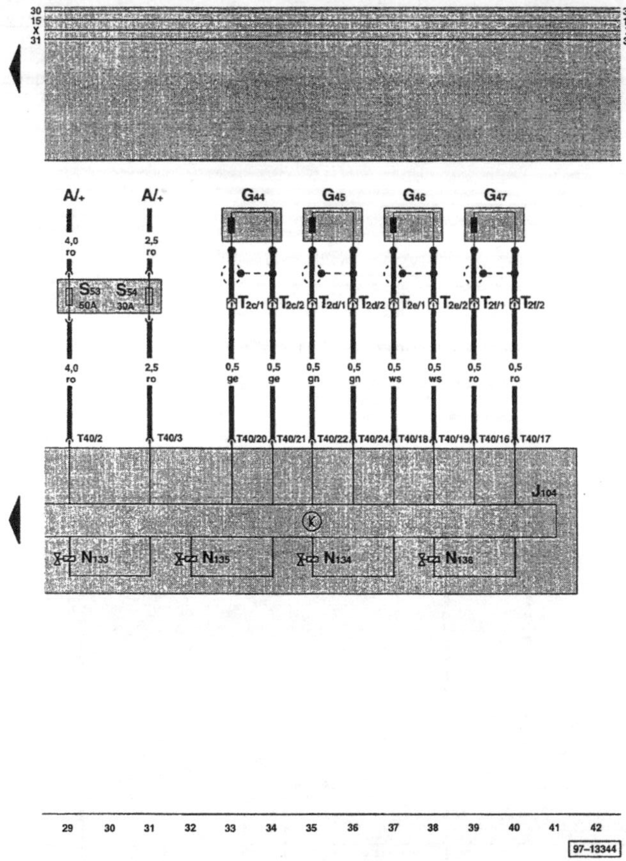

ws = white
sw = black
ro = red
br = brown
gn = green
bl = blue
gr = grey
li = violet
ge = yellow

A – Battery
G44 – Right Rear ABS Wheel Speed Sensor
G45 – Right Front ABS Wheel Speed Sensor
G46 – Left Rear ABS Wheel Speed Sensor)
G47 – Left Front ABS Wheel Speed Sensor
J104 – ABS Control Module (w/EDL), in engine compartment, left
N133 – Right Rear ABS Inlet Valve
N134 – Left Rear ABS Inlet Valve
N135 – Right Rear ABS Outlet Valve
N136 – Left Rear ABS Outlet Valve
S53 – ABS Hydraulic Pump Fuse
S54 – ABS Valves Fuse, in engine compartment, left
T2c – Double Connector
T2d – Double Connector, in front wheel casing, right
T2e – Double Connector
T2f – Double Connector, in front wheel casing, left
T40 – 40-Pin Connector

ABS control module (w/EDL), valves for ABS rear

Coolant Fans (2 x 200 W or 2 x 350 W),

from May 1996

Deviate relay location and fuseplacements as well as the locations of multiple connectors- see section "component locations".

Fuse colors

30 A – green
25 A – white
20 A – yellow
15 A – blue
10 A – red
7,5 A – brown
5 A – beige
3 A – violet

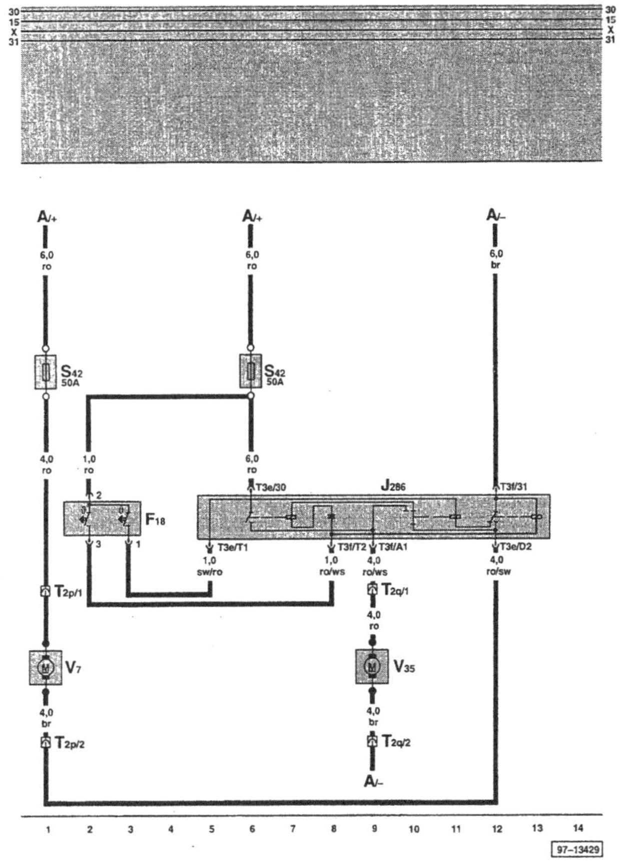

ws = white
sw = black
ro = red
br = brown
gn = green
bl = blue
gr = grey
li = violet
ge = yellow

A – Battery
F18 – Coolant Fan Control (FC) Thermal Switch
J286 – Series/Parallel Coolant Fan Control (FC) Relay (V7, V35)
S42 – Coolant Fan Fuse
T2p – Double Connector, in engine compartment, front
T2q – Double Connector, in engine compartment, front
T3e – 3-Pin Connector
T3f – 3-Pin Connector
V7 – Coolant Fan
V35 – Right Coolant Fan

Coolant Fans (2 x 450 W),

from May 1996

Deviate relay location and fuseplacements as well as the locations of multiple connectors- see section "component locations".

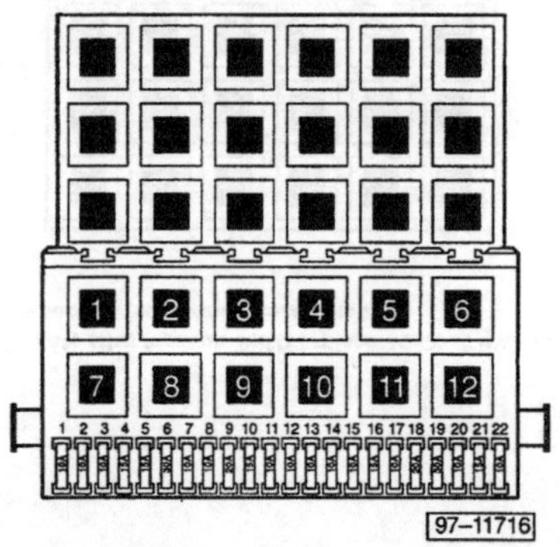

Fuse colors

30 A – green
25 A – white
20 A – yellow
15 A – blue
10 A – red
7,5 A – brown
5 A – beige
3 A – violet

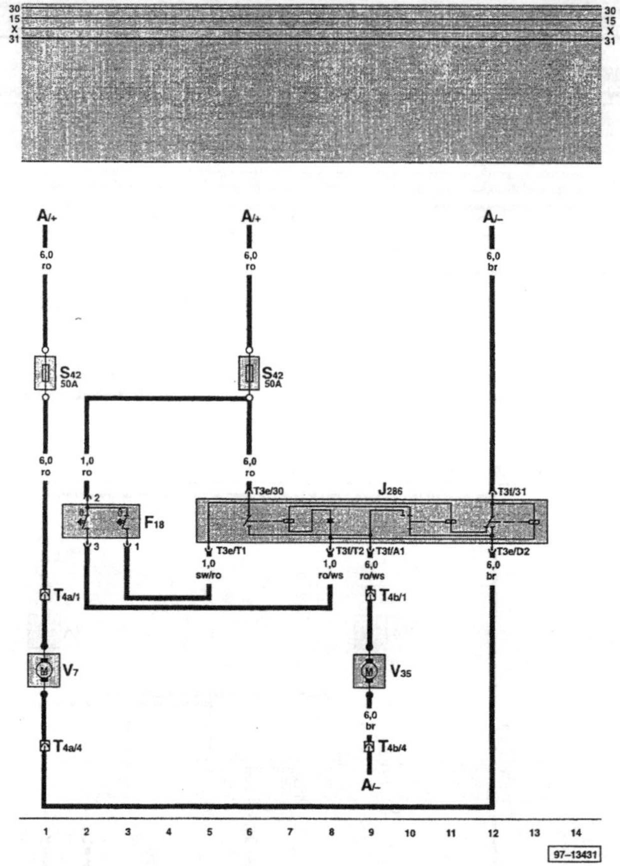

ws = white
sw = black
ro = red
br = brown
gn = green
bl = blue
gr = grey
li = violet
ge = yellow

A – Battery
F18 – Coolant Fan Control (FC) Thermal Switch
J286 – Series/Parallel Coolant Fan Control (FC) Relay (V7, V35)
S42 – Coolant Fan Fuse
T3e – 3-Pin Connector, in engine compartment, front
T3f – 3-Pin Connector, in engine compartment, front
T4a – 4-Pin Connector
T4b – 4-Pin Connector
V7 – Coolant Fan
V35 – Right Coolant Fan

Central locking system for vehicles with right sliding door and tailgate or rear swinging doors,

from May 1996

Deviate relay location and fuseplacements as well as the locations of multiple connectors- see section "component locations".

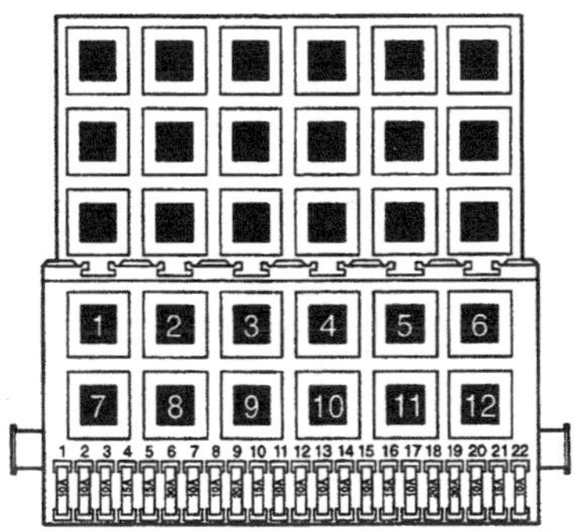

97-11716

Fuse colors

30 A – green
25 A – white
20 A – yellow
15 A – blue
10 A – red
7,5 A – brown
5 A – beige
3 A – violet

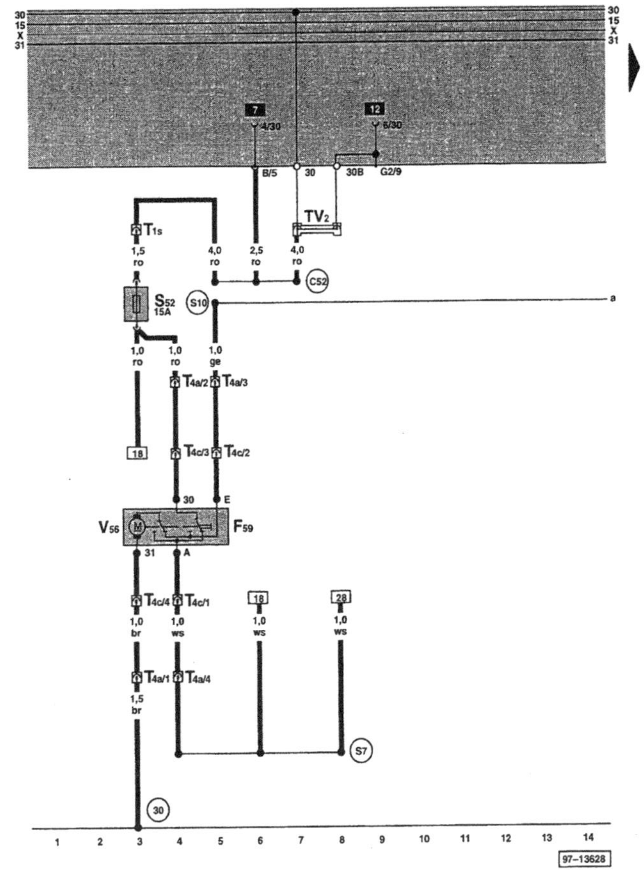

97-13628

ws = white
sw = black
ro = red
br = brown
gn = green
bl = blue
gr = grey
li = violet
ge = yellow

F59 – Driver's Door Central Locking System Switch
S52 – Central Locking System Fuse, below fuse/relay panel
T1s – Single Connector, behind fuse/relay panel
T4a – 4-Pin Connector, behind fuse/relay panel
T4c – 4-Pin Connector, in driver's door
TV2 – Terminal 30 (B+) Wire Connector
V56 – Driver's Door Central Locking System Motor

(30) – Ground connection, -1-, next to fuse/relay panel

(C52) – Plus connector (30), in wiring harness Head washer assembly
(S7) – Wire connection (A) -2-, in central locking system wiring harness
(S10) – Wire connection (E) -2-, in central locking system wiring harness

Central locking system for driver's door

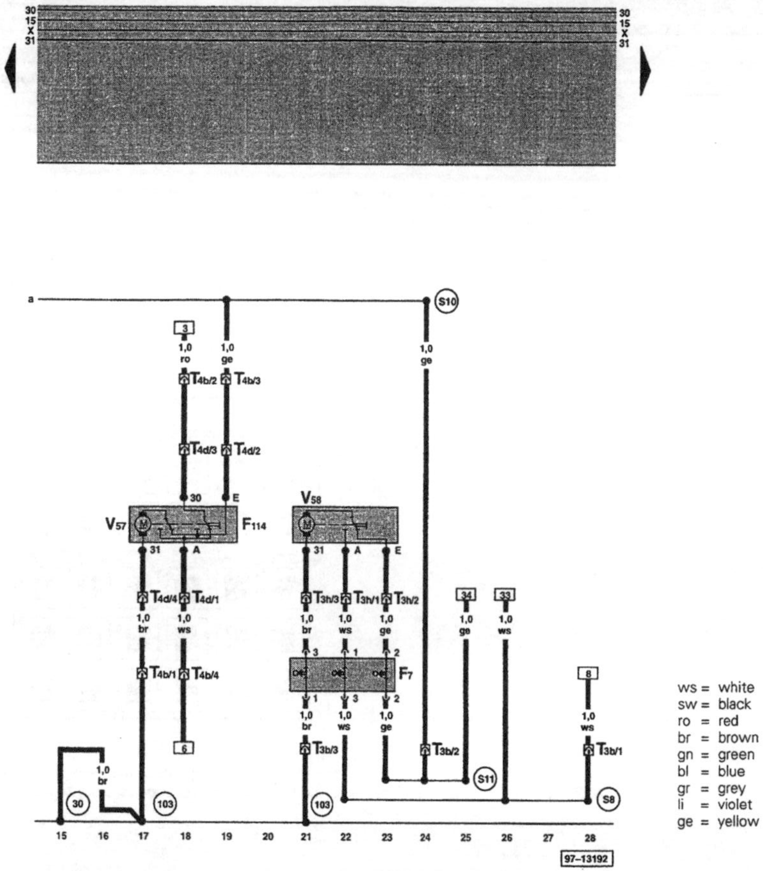

ws = white
sw = black
ro = red
br = brown
gn = green
bl = blue
gr = grey
li = violet
ge = yellow

F7 – Right Rear Side/Sliding Door Contact Switch*
F114 – Passenger's Door Central Locking System Switch
T3b – 3-Pin Connector, behind instrument panel, right
T3h – 3-Pin Connector, in sliding door, right
T4b – 4-Pin Connector, behind instrument panel, right
T4d – 4-Pin Connector, n front passenger's door
V57 – Passenger's Door Central Locking System Motor
V58 – Sliding Door Central Locking System Motor

(30) – Ground connection, -1-, next to fuse/relay panel

(103) – Ground connection -2-, in central locking system wiring harness

(S8) – Wire connection (A) -3-, in central locking system/passenger compartment wiring harness

(S10) – Wire connection (E) -2-, in central locking system wiring harness

(S11) – Wire connection (E) -3-, in central locking system passenger compartment wiring harness

Central locking system for passenger's door and sliding door right

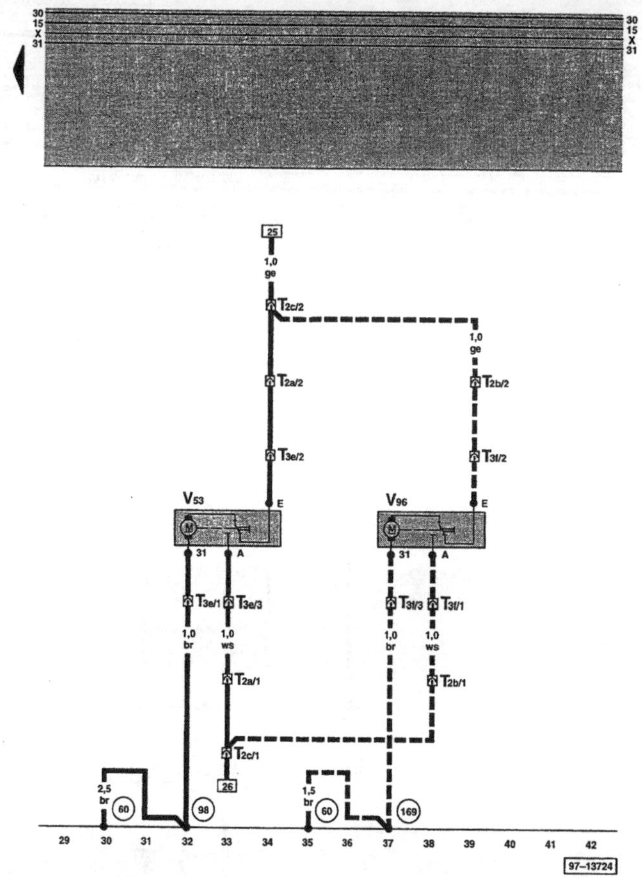

ws = white
sw = black
ro = red
br = brown
gn = green
bl = blue
gr = grey
li = violet
ge = yellow

T2a – Double Connector, on D-pillar, upper part
T2b – Double Connector, on D-pillar, upper part
T2c – Double Connector, on D-pillar, upper part
T3e – 3-Pin Connector, in rear lid
T3f – 3-Pin Connector, in right rear swinging door
V53 – Decklid Central Locking System Motor
V96 – Left Rear Door Central Locking System Motor

(60) – Ground point -2- near taillamp assembly, right

(98) – Ground connection, in rear lid wiring harness

(169) – Ground connection, in right swinging door wiring harness

- - - – Only Vehicles with rear swinging doors

Central locking system for rear lid or rear swinging doors

Heated Power Mirrows,

from May 1996

Deviate relay location and fuseplacements as well as the locations of multiple connectors-see section "component locations".

Fuse colors

30 A – green
25 A – white
20 A – yellow
15 A – blue
10 A – red
7,5 A – brown
5 A – beige
3 A – violet

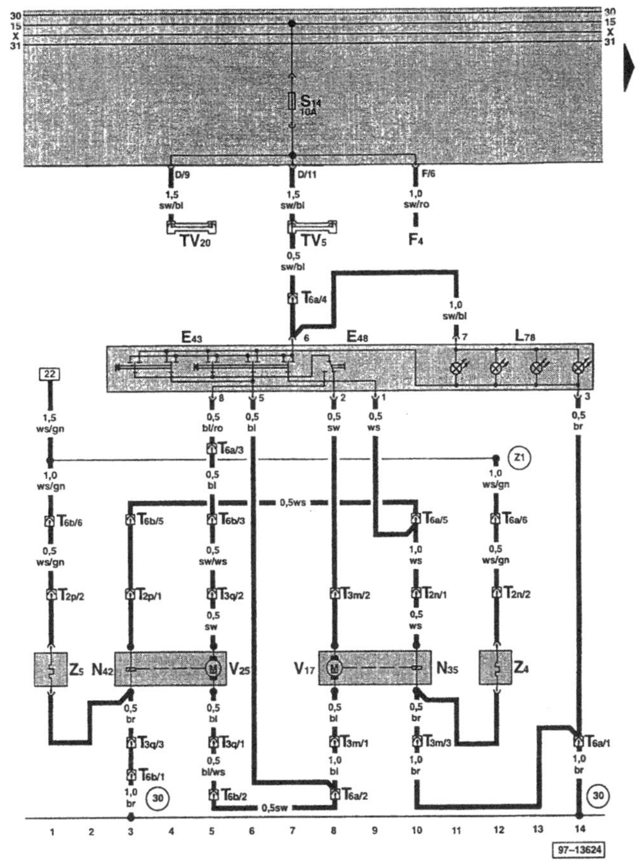

ws = white
sw = black
ro = red
br = brown
gn = green
bl = blue
gr = grey
li = violet
ge = yellow

E43 – Mirror Adjustment Switch
E48 – Mirror Selector Switch
F4 – Back-Up Light Switch
L78 – Mirror Adjusting Switch Light
N35 – Left Mirror Magnetic Clutch
N42 – Right Mirror Magnetic Clutch
T2n – Double Connector, in driver's door
T2p – Double Connector, in front passenger's door
T3m – 3-Pin Connector, in driver's door
T3q 3 Pin Connector, in front passenger's door
T6a – 6-Pin Connector, on left A-pillar
T6b – 6-Pin Connector, on right A-pillar
TV5 – Terminal 15a Wire Connector, above fuse/relay panel
TV20 – Terminal 15a Wire Connector -2-, above fuse/relay panel

V17 – Driver's Side Mirror Adjustment Motor
V25 – Passenger's Side Mirror Adjustment Motor
Z4 – Heated outside mirror, driver side
Z5 – Heated outside mirror, passenger side

(30) – Ground connection, -1-, next to fuse/relay panel
(Z1) – Wire connection -1-, in power heated mirror wiring harness

Mirror adjustment switch, left mirror magnetic clutch, heated outside mirror, mirror adjustment motor

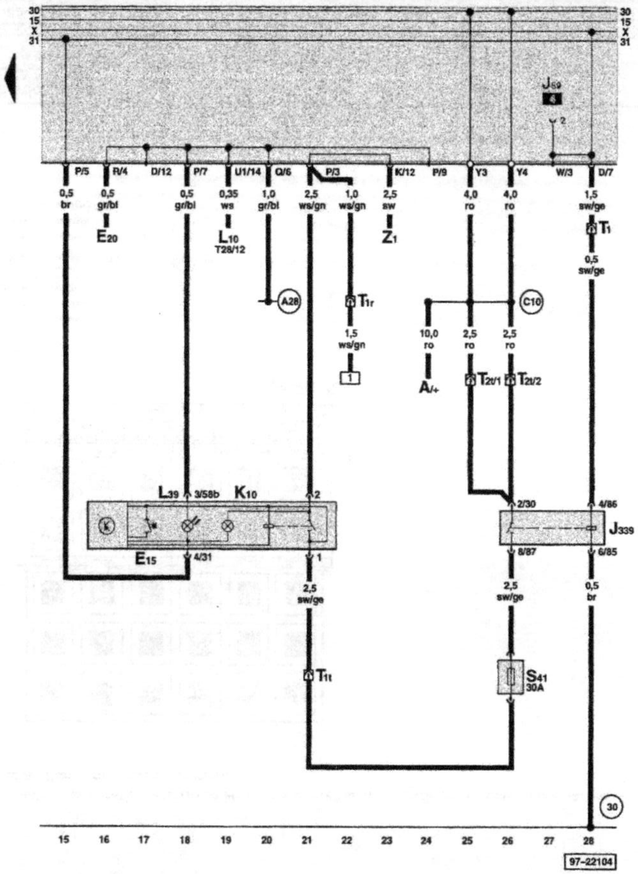

ws = white
sw = black
ro = red
br = brown
gn = green
bl = blue
gr = grey
li = violet
ge = yellow

A - Battery
E15 - Rear Window Defogger Switch
E20 - Instrument Panel Light Dimmer Switch
J59 - Load Reduction Relay
J339 - Load Reduction Relay II
K10 - Rear Window Defogger Indicator Light
L10 - Instrument Cluster Light
L39 - Rear Window Defogger Switch Light
S41 - Rear Window Defogger Fuse, below fuse/relay panel
T1 - Single Connector, behind fuse/relay panel
T1r - Single Connector, behind fuse/relay panel
T1t - Single Connector, behind fuse/relay panel
T2t - Double Connector, behind fuse/relay panel
T28 - 28-Pin Connector, on instrument cluster
Z1 - Rear Window Defogger/Heat Element

(30) - Ground connection, -1-, next to fuse/relay panel
(A28) - Plus connection (58b), in fresh air blower wiring harness
(C10) - Plus connection -1- (30), in headlight wiring harness

THIS PAGE INTENTIONALLY LEFT
BLANK

Cruise Control (manual transmission),

from May 1996

Deviate relay location and fuseplacements as well as the locations of multiple connectors-see section "component locations".

Fuse colors

30 A – green
25 A – white
20 A – yellow
15 A – blue
10 A – red
7,5 A – brown
5 A – beige
3 A – violet

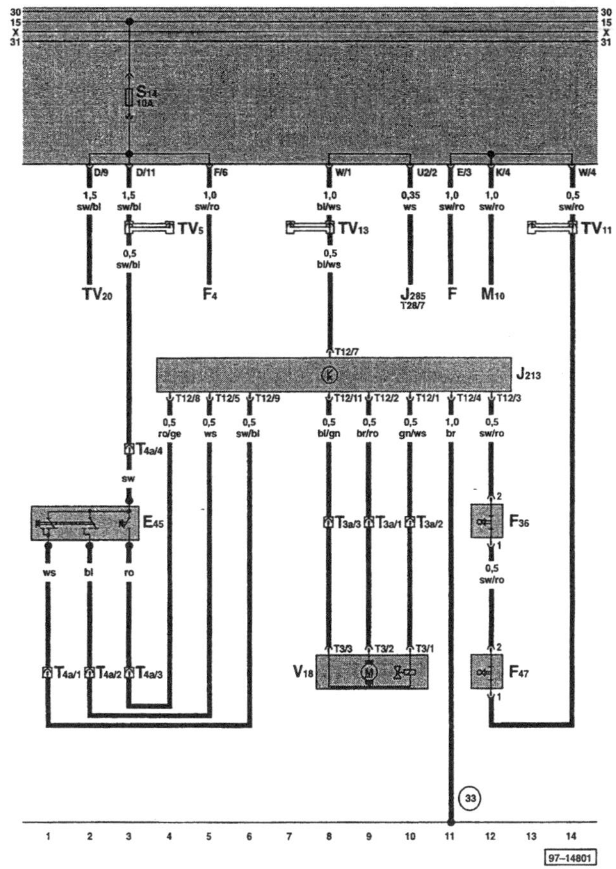

ws = white
sw = black
ro = red
br = brown
gn = green
bl = blue
gr = grey
li = violet
ge = yellow

E45 – Cruise Control Switch
F – Brake Light Switch
F4 – Back-Up Light Switch
F36 – Clutch Vacuum Vent Valve Switch
F47 – Brake Vacuum Vent Valve Switch for cruise control
J213 – Cruise Control, Control Module, behind instrument panel, right
J285 – Control module with indicator unit in instrument panel insert
M10 – Right Brake Light
T3 – 3-Pin Connector
T3a – 3-Pin Connector, behind fuse/relay panel
T4a – 4-Pin Connector, behind steering column switch cover
T12 – 12-Pin Connector
T28 – 28-Pin Connector, on instrument cluster
TV5 – Terminal 15a Wire Connector, above fuse/relay panel
TV11 – Terminal 54 Wire Connector, above fuse/relay panel
TV13 – Vehicle Speed Signal Wire Connector, above fuse/relay panel
TV20 – Wire junction -2- for terminal 15a, above fuse/relay panel
V18 – Cruise Control Vacuum Pump
(33) – Ground connection, behind instrument panel, right

Front Power Windows,

from May 1996

Deviate relay location and fuseplacements as well as the locations of multiple connectors- see section "component locations".

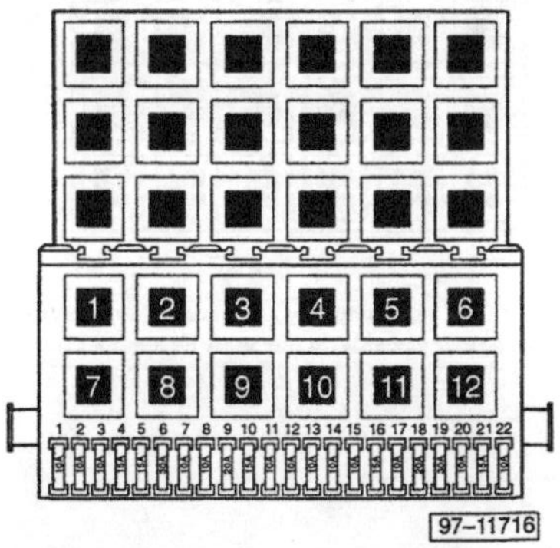

Fuse colors

30 A - green
25 A - white
20 A - yellow
15 A - blue
10 A - red
7,5 A - brown
5 A - beige
3 A - violet

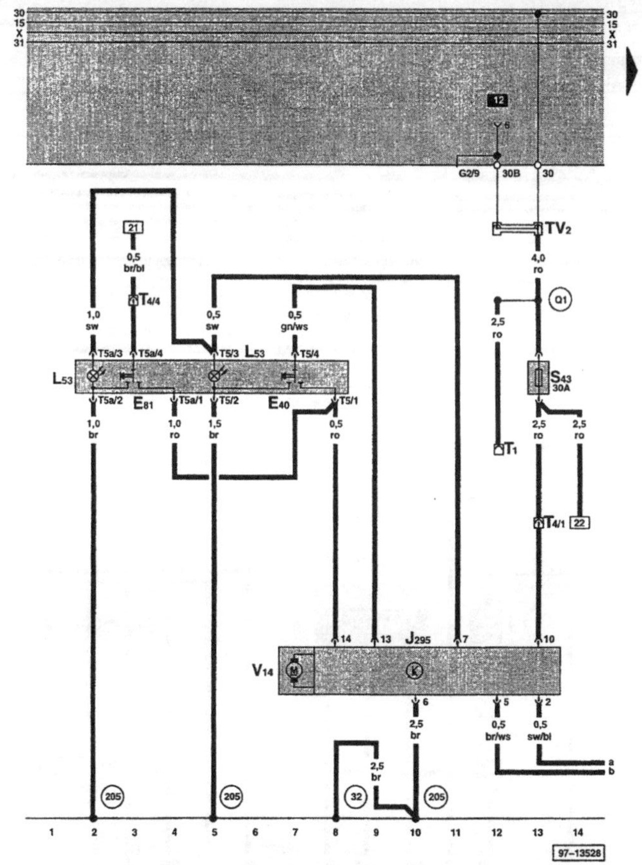

ws = white
sw = black
ro = red
br = brown
gn = green
bl = blue
gr = grey
li = violet
ge = yellow

E40 - Left Front Window Switch
E81 - Right Front Window Switch (In RF Door)
J295 - Left Front Power Window Control Module
L53 - Power Window Switch Light
S43 - Power Window Circuit Breaker, above fuse/relay panel
T1 - Single Connector
T4 - 4-Pin Connector, behind instrument panel, left
T5 - 5-Pin Connector, on Left Front Window Switch
T5a - 5-Pin Connector, on Right Front Window Switch
TV2 - Terminal 30 (B+) Wire Connector, above fuse/relay panel
V14 - Left Window Motor, in driver's door

(32) - Ground connection, behind instrument panel, left
(205) - Ground connection, in driver's door wiring harness
(Q1) - Plus connection (30), in power window wiring harness

Front power window control module, front window switches

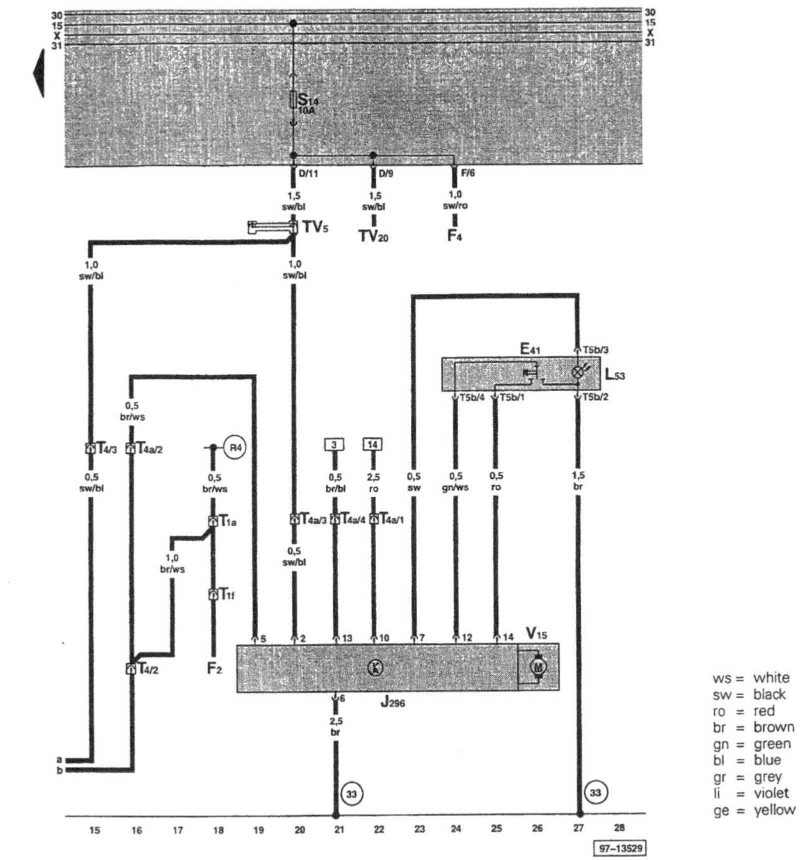

ws = white
sw = black
ro = red
br = brown
gn = green
bl = blue
gr = grey
li = violet
ge = yellow

E41 – Right Front Window Switch, in front passenger's door
F2 – Door contact Switch - Driver's side
F4 – Back-Up Light Switch
J296 – Right Front Power Window Control Module, in front passenger's door
L53 – Power Window Switch Light
T1a – Single Connector, behind fuse/relay panel
T1f – Single Connector, behind fuse/relay panel
T4 – 4-Pin Connector, behind instrument panel, left
T4a – 4-Pin Connector, behind instrument panel, right
T5b – 5-Pin Connector, on Right Front Window Switch

TV5 – Terminal 15a Wire Connector, above fuse/relay panel

TV20 – Terminal 15a Wire Connector -2-
V15 – Right Window Motor, in front passenger's door

(33) – Ground connection, behind instrument panel, right
(R4) – Wire connection -1-, in interior light/door contact switch wiring harness

Right front power window control module, right front window switch in front passenger's door

THIS PAGE INTENTIONALLY LEFT
BLANK

Air-Conditioning,

from May1996

- **2.8 L- Engine - Motronic Multiport Fuel Injection (MFI)/103 kW, code AES, (manual transmission)**

Deviate relay location and fuseplacements as well as the locations of multiple connectors- see section "component locations".

1 – A/C Relay (140)

Note:

Number in parentheses indicates production control number stamped on relay housing.

Fuse colors

30 A – green
25 A – white
20 A – yellow
15 A – blue
10 A – red
7,5 A – brown
5 A – beige
3 A – violet

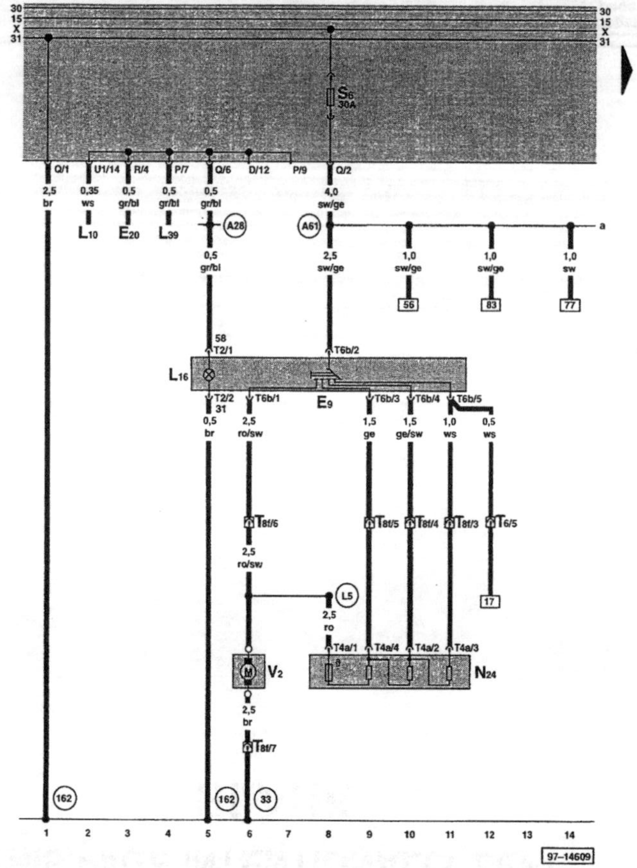

ws = white
sw = black
ro = red
br = brown
gn = green
bl = blue
gr = grey
li = violet
ge = yellow

E9 – Fresh Air Blower Switch
E20 – Instrument Panel Light Dimmer Switch
L10 – Instrument Cluster Light
L16 – Fresh Air Control Lever Light
L39 – Rear Window Defogger Switch Light
N24 – Fresh Air Blower Series Resistance with Fuse
T2 – Double Connector, on Fresh Air Blower Switch
T4a – 4-Pin Connector, on Resistance
T6 – 6-Pin Connector, behind instrument panel, center
T6b – 6-Pin Connector
T8f – 8-Pin Connector, behind instrument panel, right
V2 – Fresh Air Blower

(33) – Ground connection, behind instrument panel, right
(162) – Ground connection, in blower motor wiring harness
(A28) – Plus connection (58b), in fresh air blower Wiring harness
(A61) – Plus Connection (X), in Instrument cluster Wiring harness
(L5) – Wire connection -1-, in evaporator housing wiring harness

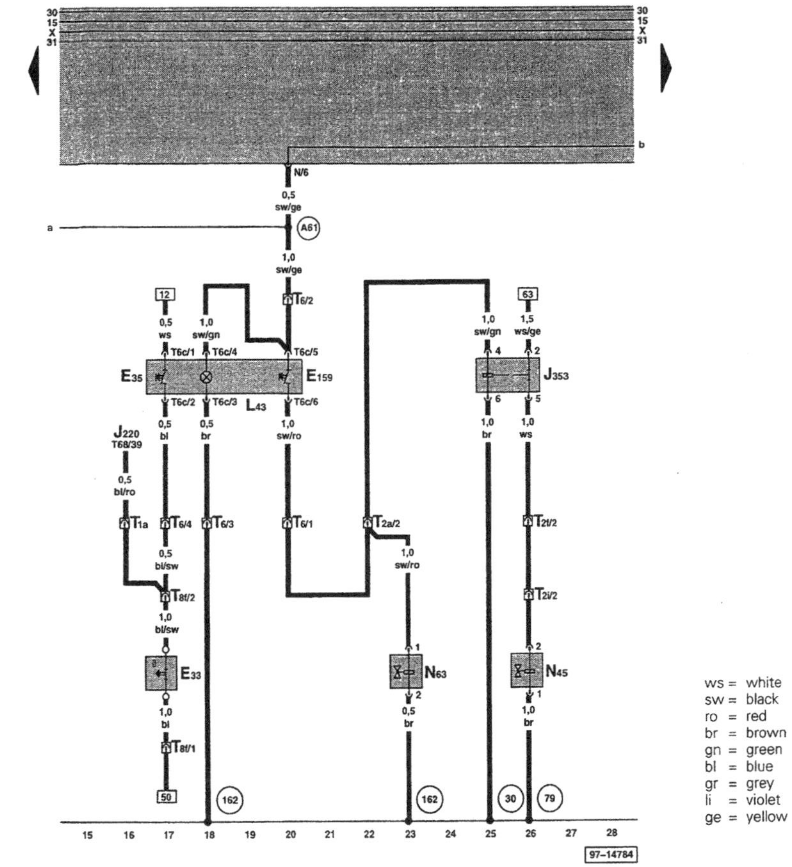

ws = white
sw = black
ro = red
br = brown
gn = green
bl = blue
gr = grey
li = violet
ge = yellow

E33 – A/C Evaporator Temperature Switch
E35 – A/C Switch
E159 – Fresh Air/Recirculating Flap Switch
J220 – Motronic Engine Control Module (ECM), ("Stand by")-signal
J353 – Relay for Fresh-air and recirculating air flap-rear, control number (94), behind instrument panel, center
L43 – A/C Switch Illumination Light
N45 – Two-way valve for Fresh-and recirculation air flap-rear
N63 – Fresh Air/Recirculating Flap Two-Way Valve
T1a – Single Connector, behind fuse/relay panel
T2a – Double Connector, behind fuse/relay panel
T2f – Double Connector
T2i – Double Connector
T6 – 6-Pin Connector, behind instrument panel, center
T6c – 6-Pin Connector
T8f – 8-Pin Connector, behind instrument panel, right
T68 – 68-Pin Connector, on Motronic Engine Control Module (ECM)

(30) – Ground connection, -1-, next to fuse/relay panel

(79) – Ground connection, near D-pillar, lower part

(162) – Ground connection, in blower motor wiring harness

(A61) – Plus Connection (X), in Instrument cluster wiring harness

Control for recirculating flap front and rear

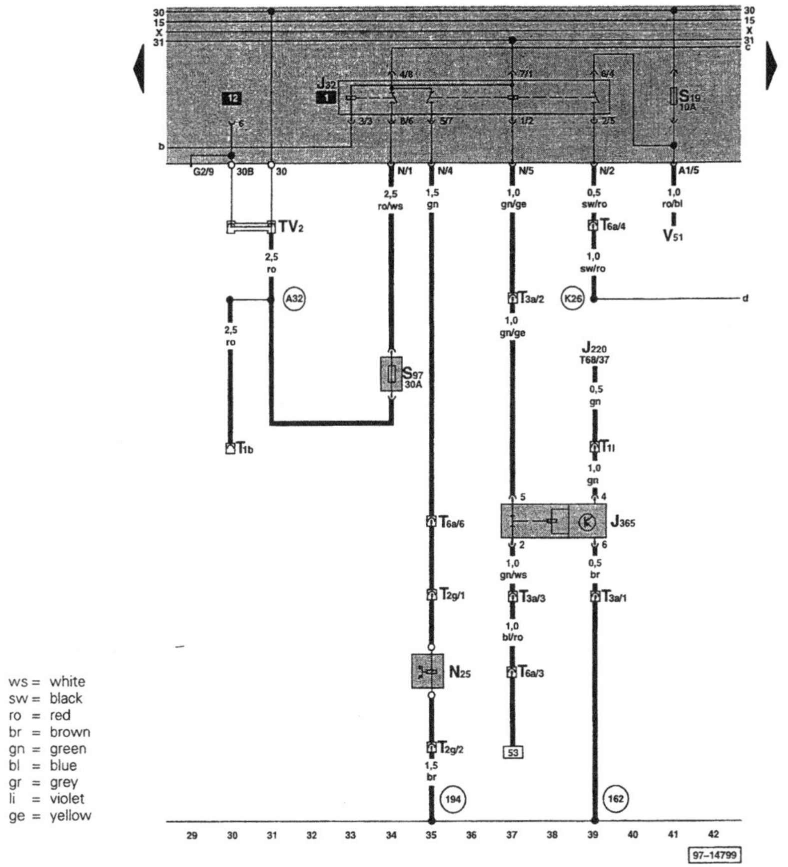

ws = white
sw = black
ro = red
br = brown
gn = green
bl = blue
gr = grey
li = violet
ge = yellow

J32 – A/C Relay, control number (140), switch delay time 1/2 second
J220 – Motronic Engine Control Module (ECM), ("on/off",-signal)
J365 – Relay for A/C shutoff, control number (147), above fuse/relay panel
N25 – A/C Clutch
S97 – Fresh Air Blower Fuse 1, below fuse/relay panel
T1b – Single Connector, behind fuse/relay panel
T1l – Single Connector, behind fuse/relay panel
T2g – Double Connector, near kompressor
T3a – 3-Pin Connector, behind fuse/relay panel
T6a – 6-Pin Connector, behind instrument panel, left
T68 – 68-Pin Connector, on Engine Control Module (ECM)
TV2 – Terminal 30 (B+) Wire Connector
V51 – After-Run Coolant Pump

(162) – Ground connection, in blower motor wiring harness

(194) – Ground connection -2-, in coolant fan wiring harness

(A32) – Plus connection (30), in instrument panel wiring harness

(K26) – Wire connection -5-, in coolant fan wiring harness

Magnetic clutch, A/C shut off

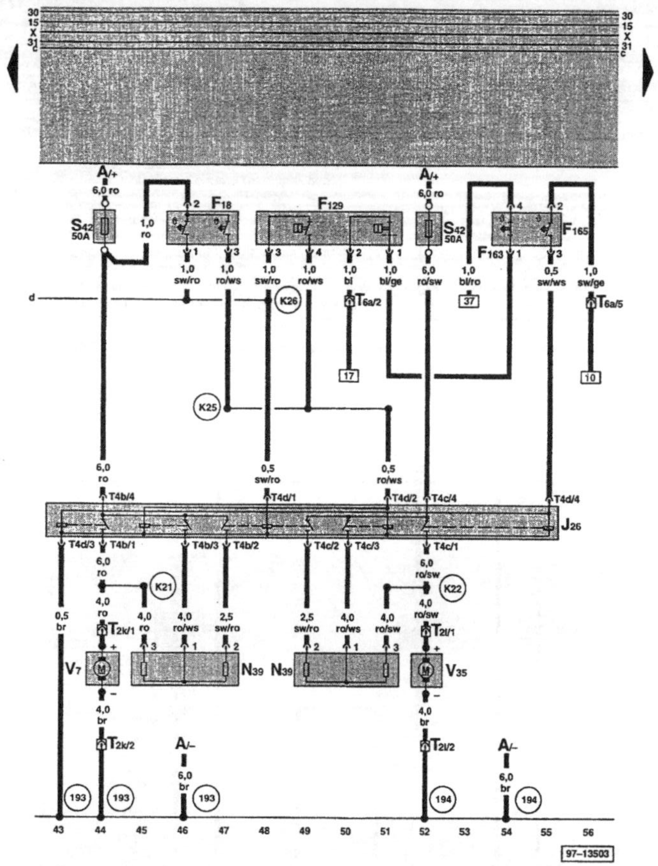

ws	=	white
sw	=	black
ro	=	red
br	=	brown
gn	=	green
bl	=	blue
gr	=	grey
li	=	violet
ge	=	yellow

A – Battery
F18 – Coolant Fan Control (FC) Thermal Switch
F129 – A/C Pressure Switch
F163 – A/C Cut-Out Thermal Switch
F165 – Third Speed Coolant Fan Control (FC) Thermal Switch
J26 – Coolant Fan control (FC) Relay, in engine compartment, left front
N39 – Coolant Fan Control (FC) Series Resistance
S42 – Coolant Fan Fuse, in engine compartment, left front
T2k – Double Connector
T2l – Double Connector
T4b – 4-Pin Connector
T4c – 4-Pin Connector
T4d – 4-Pin Connector
T6a – 6-Pin Connector, behind instrument panel, left

V7 – Coolant Fan (350 W)
V35 – Right Coolant Fan (350 W)

(193) – Ground connection -1-, in coolant fan wiring harness
(194) – Ground connection -2-, in coolant fan wiring harness
(K21) – Wire connection -1-, in coolant fan wiring harness
(K22) – Wire connection -2-, in coolant fan wiring harness
(K25) – Wire connection -4-, in coolant fan wiring harness
(K26) – Wire connection -5-, in coolant fan wiring harness

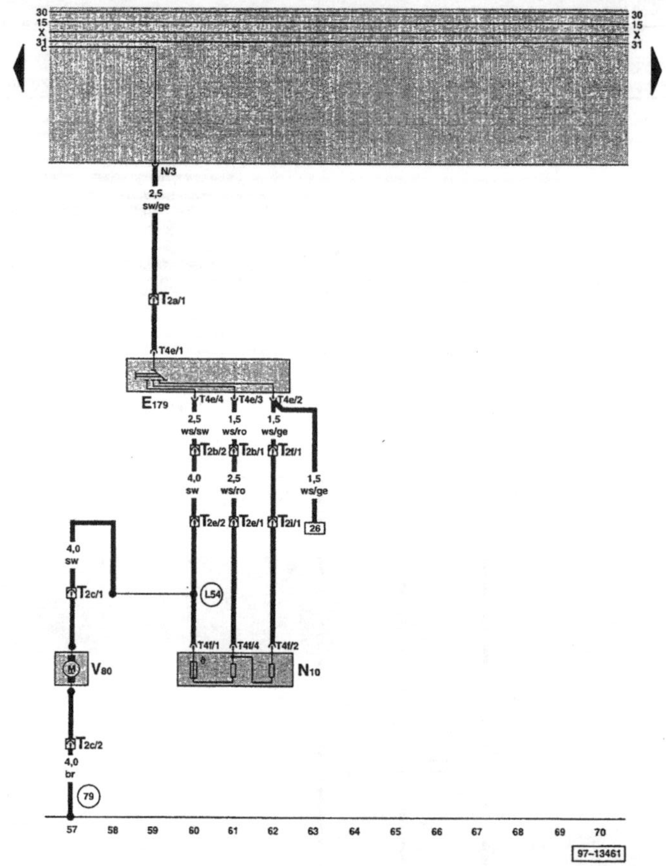

ws	=	white
sw	=	black
ro	=	red
br	=	brown
gn	=	green
bl	=	blue
gr	=	grey
li	=	violet
ge	=	yellow

E179 – Switch for Freshair blower-rear
N10 – Series Resistance for fresh air blower-rear
T2a – Double Connector, behind fuse/relay panel
T2b – Double Connector, behind instrument panel, right
T2c – Double Connector, behind trim D-pillar right, lower
T2e – Double Connector, behind trim D-pillar right, lower
T2f – Double Connector
T2i – Double Connector
T4e – 4-Pin Connector
T4f – 4-Pin Connector
V80 – Fresh air blower rear

(79) – Ground connection, near D-pillar, lower part
(L54) – Plus connector (Blower stage 3) in wiring harness

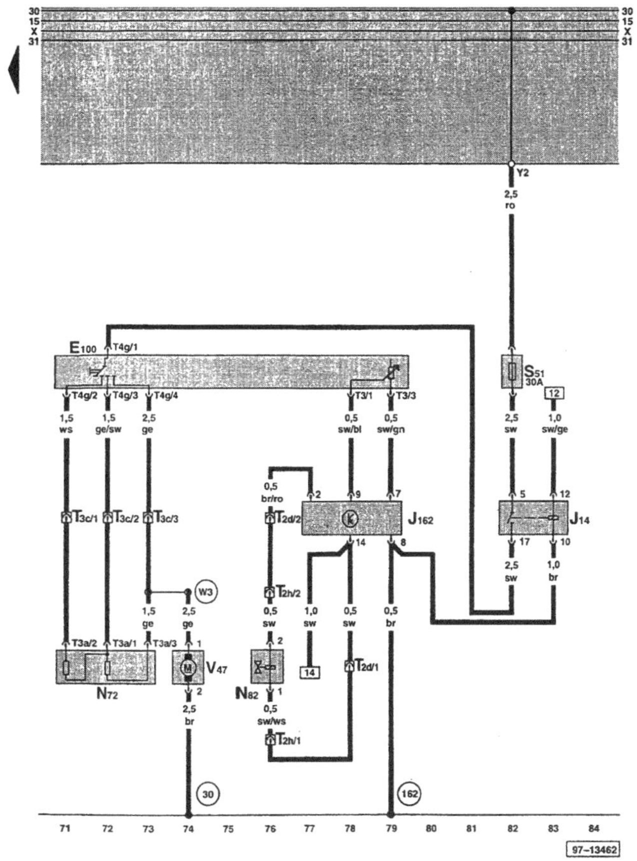

ws = white
sw = black
ro = red
br = brown
gn = green
bl = blue
gr = grey
li = violet
ge = yellow

E100 – Warm Air Blower Switch
J14 – Heater Air Blower Relay, control number (53), behind instrument panel, center
J162 – Heater Control Module, control number (152), behind instrument panel, center
N72 – Heater Blower Series Resistance
N82 – Coolant Cut-off Valve
S51 – Fuse for 2. heater core, below fuse/relay panel
T2d – Double Connector, behind instrument panel, left
T2h – Double Connector, in engine compartment, left front
T3 – 3-Pin Connector
T3a – 3-Pin Connector
T3c – 3-Pin Connector, behind fuse/relay panel
T4g – 4-Pin Connector

V47 – Rear Warm Air Fan

(30) – Ground connection, -1-, next to fuse/relay panel
(162) – Ground connection, in blower motor wiring harness
(W3) – Wire connection, in rear wiring harness

THIS PAGE INTENTIONALLY LEFT **BLANK**

Air-Conditioning,

from May1996

● 2.4 L-Diesel Engine/57 kW, code AAB, with manual transmission

Deviate relay location and fuseplacements as well as the locations of multiple connectors- see section "component locations".

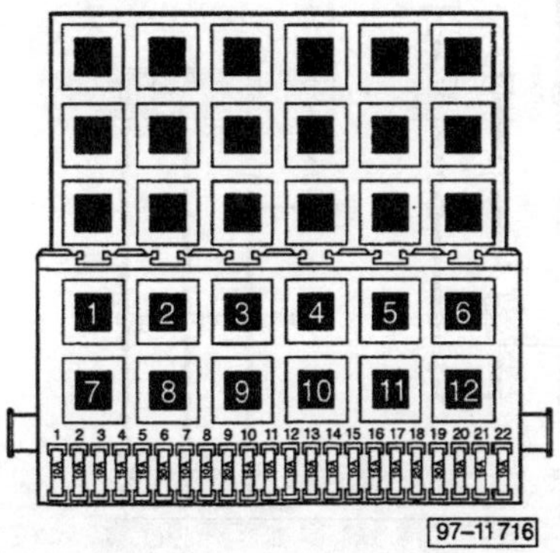

1 – A/C Relay (140)

Note:

Number in parentheses indicates production control number stamped on relay housing.

Fuse colors

30 A – green
25 A – white
20 A – yellow
15 A – blue
10 A – red
7,5 A – brown
5 A – beige
3 A – violet

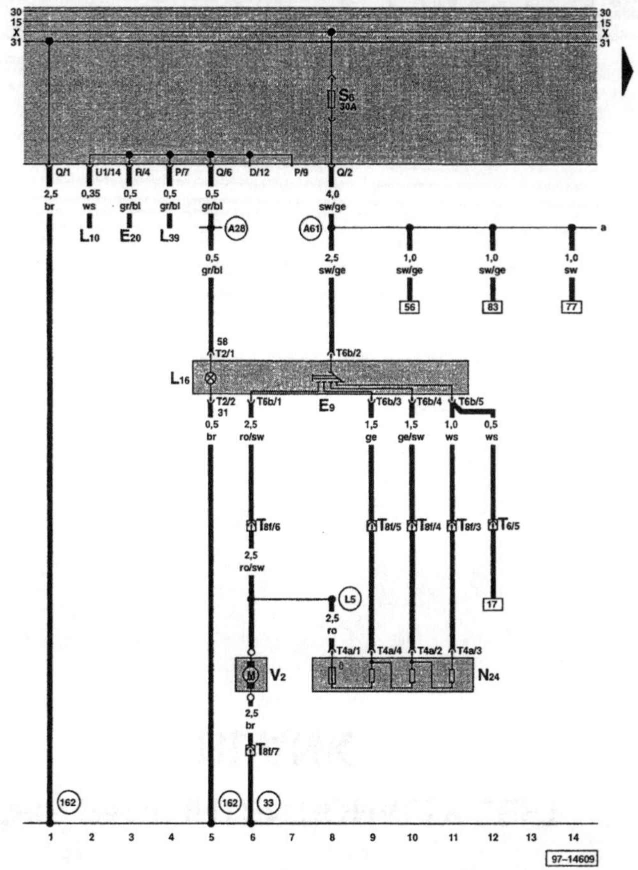

ws = white
sw = black
ro = red
br = brown
gn = green
bl = blue
gr = grey
li = violet
ge = yellow

E9 – Fresh Air Blower Switch
E20 – Instrument Panel Light Dimmer Switch
L10 – Instrument Cluster Light
L16 – Fresh Air Control Lever Light
L39 – Rear Window Defogger Switch Light
N24 – Fresh Air Blower Series Resistance with Fuse
T2 – Double Connector, on Fresh Air Blower Switch
T4a – 4-Pin Connector, on Resistance
T6 – 6-Pin Connector, behind instrument panel, center
T6b – 6-Pin Connector
T8f – 8-Pin Connector, behind instrument panel, right
V2 – Fresh Air Blower

(33) – Ground connection, behind instrument panel, right
(162) – Ground connection, in blower motor wiring harness
(A28) – Plus connection (58b), in fresh air blower Wiring harness
(A61) – Plus Connection (X), in Instrument cluster Wiring harness
(L5) – Wire connection -1-, in evaporator housing wiring harness

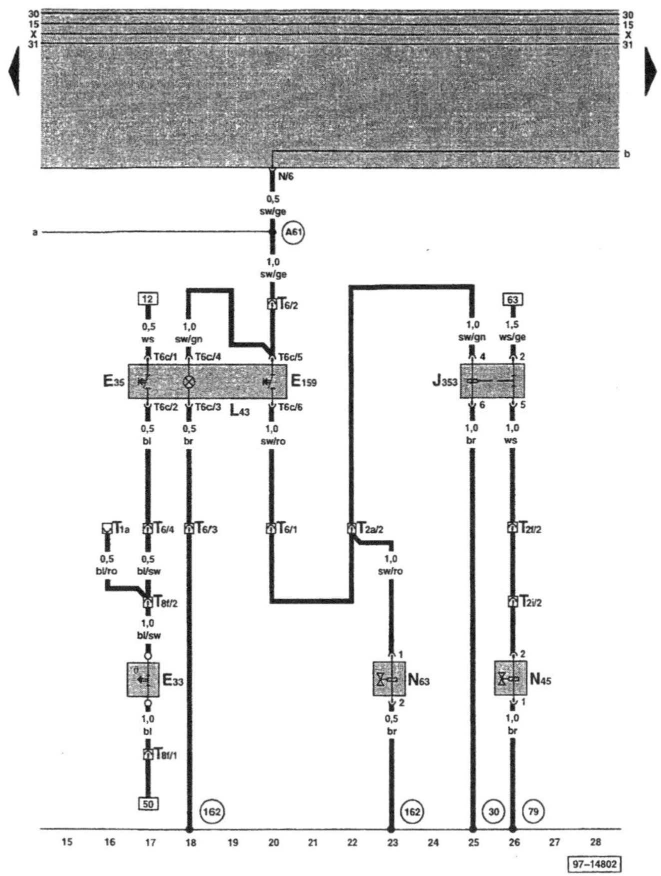

ws =	white
sw =	black
ro =	red
br =	brown
gn =	green
bl =	blue
gr =	grey
li =	violet
ge =	yellow

E33 – A/C Evaporator Temperature Switch
E35 – A/C Switch
E159 – Fresh Air/Recirculating Flap Switch
J353 – Relay for Fresh-air and recirculating air flap rear, control number (94), behind instrument panel, center
L43 – A/C Switch Illumination Light
N45 – Two-way valve for Fresh-and recirculation air flap-rear
N63 – Fresh Air/Recirculating Flap Two-Way Valve
T1a – Single Connector, behind fuse/relay panel, ("Stand by")-signal
T2a – Double Connector, behind fuse/relay panel
T2f – Double Connector
T2i – Double Connector
T6 – 6-Pin Connector, behind instrument panel, center
T6c – 6-Pin Connector
T8f – 8-Pin Connector, behind instrument panel, right

(30) – Ground connection, -1-, next to fuse/relay panel
(79) – Ground connection, near D-pillar, lower part
(162) – Ground connection, in blower motor wiring harness
(A61) – Plus Connection (X), in Instrument cluster wiring harness

Control for recirculating flap front and rear

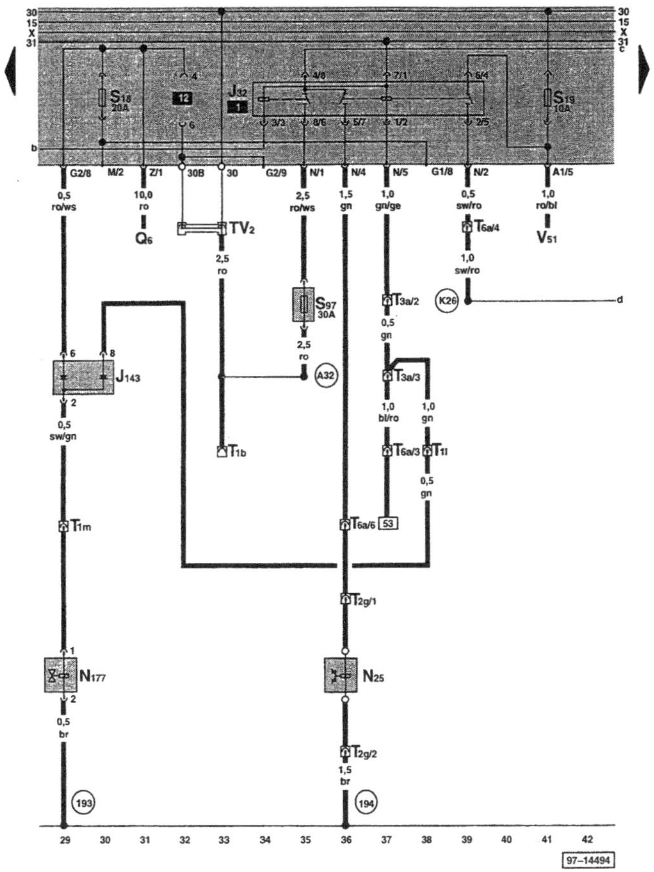

ws =	white
sw =	black
ro =	red
br =	brown
gn =	green
bl =	blue
gr =	grey
li =	violet
ge =	yellow

J32 – A/C Relay, control number (140), switch deceleration time 1/2 second
J143 – Idle Air Control (IAC) Control Module, control number (153), above fuse/relay panel
N25 – A/C Clutch
N117 – Throttle Position (TP) Actuator Breather Valve
Q6 – Spark plugs (engine)
S97 – Fresh Air Blower Fuse 1, below fuse/relay panel
T1b – Single Connector, behind fuse/relay panel
T1l – Single Connecto, behind fuse/relay panel
T1m – Single Connector, behind instrument panel, left
T2g – Double Connector, near compressor
T3a – 3-Pin Connector, behind fuse/relay panel
T6a – 6-Pin Connector, behind instrument panel, left
TV2 – Terminal 30 (B+) Wire Connector
V51 – After-Run Coolant Pump

(193) – Ground connection -1-, in coolant fan wiring harness
(194) – Ground connection -2-, in coolant fan wiring harness
(A32) – Plus connection (30), in instrument panel wiring harness
(K26) – Wire connection -5-, in coolant fan wiring harness

Magnetic clutch, idle air control (IAC)

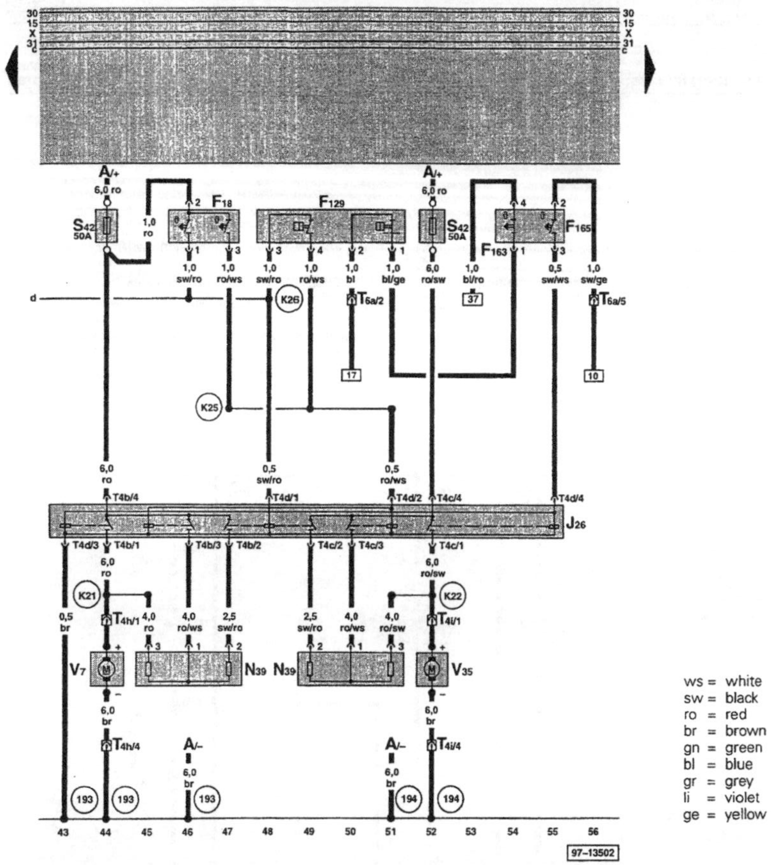

ws = white
sw = black
ro = red
br = brown
gn = green
bl = blue
gr = grey
li = violet
ge = yellow

A – Battery
F18 – Coolant Fan Control (FC) Thermal Switch
F129 – A/C Pressure Switch
F163 – A/C Cut-Out Thermal Switch
F165 – Third Speed Coolant Fan Control (FC) Thermal Switch
J26 – Coolant Fan control (FC) Relay, in engine compartment, left front
N39 – Coolant Fan Control (FC) Series Resistance
S42 – Coolant Fan Fuse, in engine compartment, left front
T4b – 4-Pin Connector
T4c – 4-Pin Connector
T4d – 4-Pin Connector
T4h – 4-Pin Connector
T4i – 4-Pin Connector
T6a – 6-Pin Connector, behind instrument panel, left

V7 – Coolant Fan (450 W)
V35 – Right Coolant Fan (450 W)

(193) – Ground connection -1-, in coolant fan wiring harness
(194) – Ground connection -2-, in coolant fan wiring harness
(K21) – wire connection -1-, in coolant fan wiring harness
(K22) – Wire connection -2-, in coolant fan wiring harness
(K25) – Wire connection -4-, in coolant fan wiring harness
(K26) – Wire connection -5-, in coolant fan wiring harness

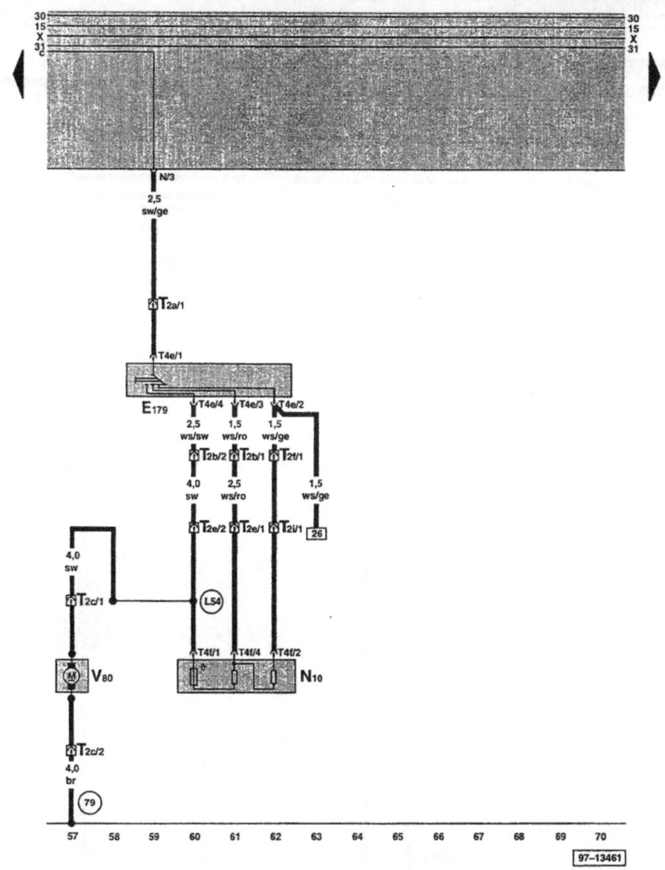

ws = white
sw = black
ro = red
br = brown
gn = green
bl = blue
gr = grey
li = violet
ge = yellow

E179 – Switch for Freshair blower-rear
N10 – Series Resistance for fresh air blower-rear
T2a – Double Connector,
T2b – Double Connector, behind instrument panel, right
T2c – Double Connector, behind trim D-pillar right, lower
T2e – Double Connector, behind trim D-pillar right, lower
T2f – Double Connector
T2i – Double Connector
T4e – 4-Pin Connector
T4f – 4-Pin Connector
V80 – Fresh air blower rear

(79) – Ground connection, near D-pillar, lower part
(L54) – Plus connector (Blower stage 3) in wiring harness

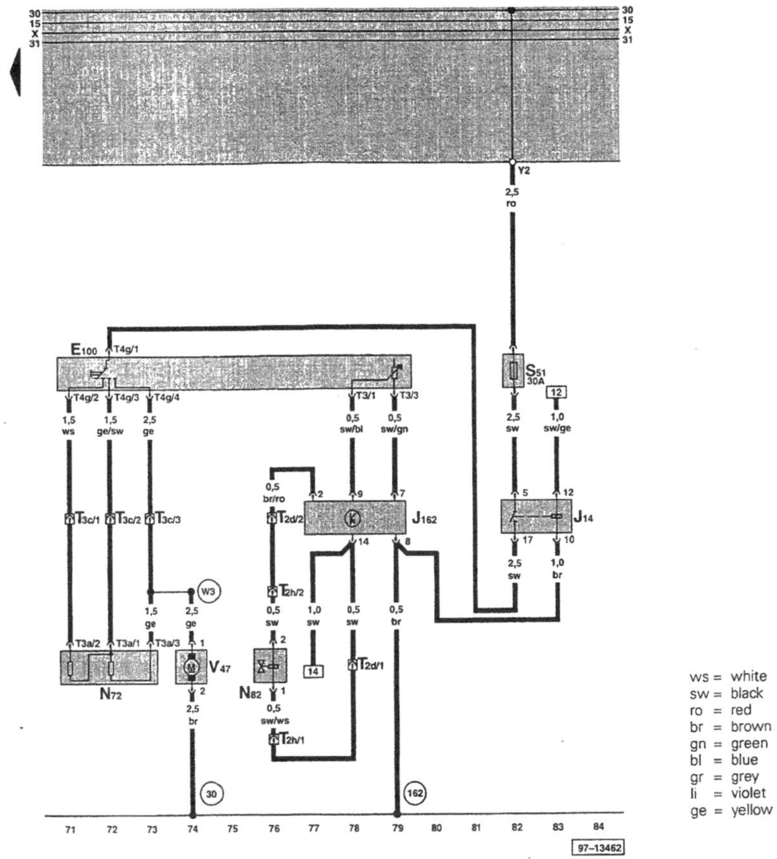

ws = white
sw = black
ro = red
br = brown
gn = green
bl = blue
gr = grey
li = violet
ge = yellow

E100 – Warm Air Blower Switch
J14 – Heater Air Blower Relay, control number (53), behind instrument panel, center
J162 – Heater Control Module, control number (152), behind instrument panel, center
N72 – Heater Blower Series Resistance
N82 – Coolant Cut-off Valve
S51 – Fuse for 2. heater core, below fuse/relay panel
T2d – Double Connector, behind instrument panel, left
T2h – Double Connector, in engine compartment, left front
T3 – 3-Pin Connector
T3a – 3-Pin Connector
T3c – 3-Pin Connector, behind fuse/relay panel
T4g – 4-Pin Connector
V47 – Rear Warm Air Fan

(30) – Ground connection, -1-, next to fuse/relay panel
(162) – Ground connection, in blower motor wiring harness
(W3) – Wire connection, in rear wiring harness

THIS PAGE INTENTIONALLY LEFT
BLANK

Climatronic,
from May 1996

- **2.8 L- Engine - Motronic Multiport Fuel Injection (MFI)/103 kW, code AES, with automatic transmission**

Deviate relay location and fuseplacements as well as the locations of multiple connectors- see section "component locations".

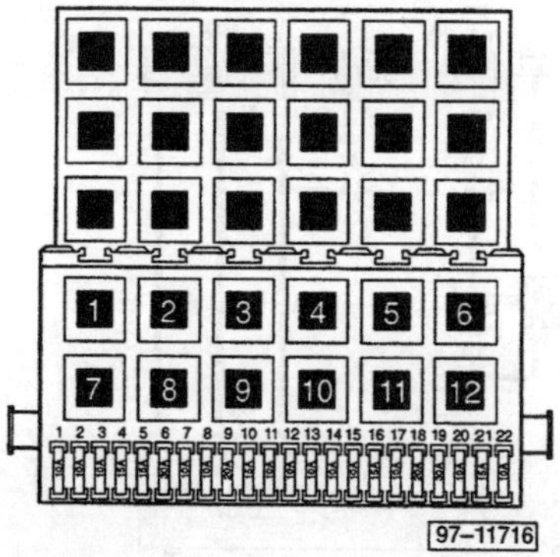

1 – Relay for Climatronic (140)

Note:

Number in parentheses indicates production control number stamped on relay housing.

Fuse colors

30 A – green
25 A – white
20 A – yellow
15 A – blue
10 A – red
7,5 A – brown
5 A – beige
3 A – violet

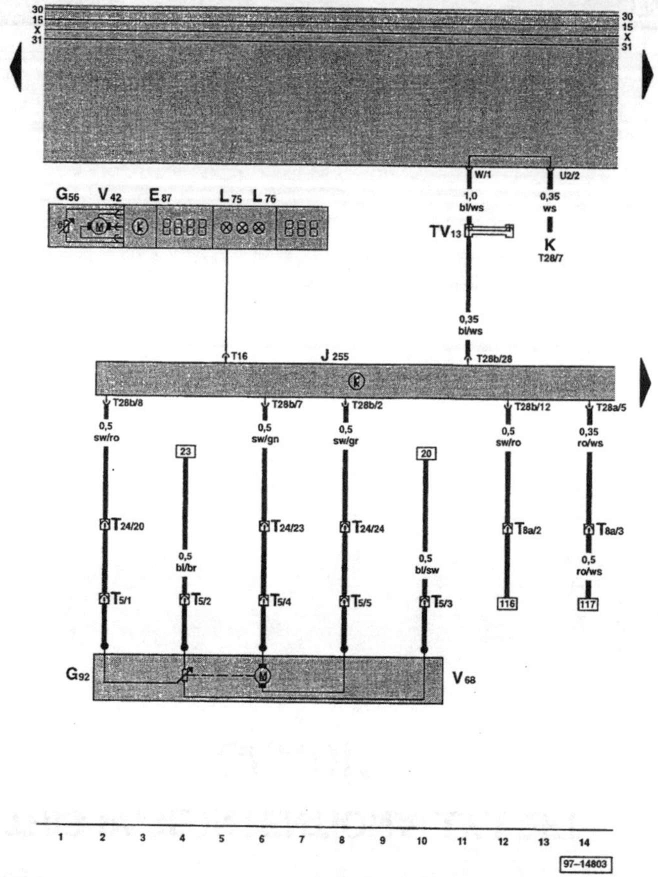

ws = white
sw = black
ro = red
br = brown
gn = green
bl = blue
gr = grey
li = violet
ge = yellow

E87 – A/C Control Head
G56 – Instrument Panel Interior Temperature Sensor, in the A/C Control Head
G92 – Temperature Regulator Flap Motor Position Sensor
J255 – Climatronic Control Module
K – Instrument Cluster
L75 – Digital Display Light
L76 – Push Button Light
T5 – 5-Pin Connector, behind instrument panel, center
T8a – 8-Pin Connector, behind fuse/relay panel
T16 – 16-Pin Connector
T24 – 24-Pin Connector, behind instrument panel, center
T28 – 28-Pin Connector, on instrument cluster
T28a – 28-Pin Connector
T28b – 28-Pin Connector
TV13 – Vehicle Speed Signal Wire Connector
V42 – Interior Temperature Sensor Fan, in the A/C Control Head
V68 – Temperature Regulator Flap Motor

A/C control head with instrument panel interior temperature sensor, temperature regulator flap motor

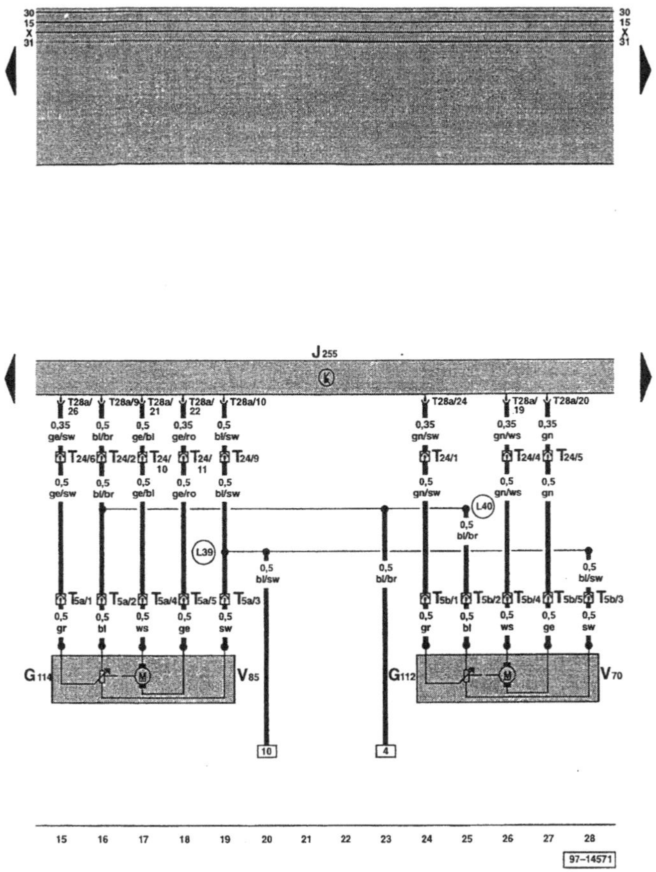

ws = white
sw = black
ro = red
br = brown
gn = green
bl = blue
gr = grey
li = violet
ge = yellow

G112 – Central Flap Motor Position Sensor
G114 – Footwell/Defrost Flap Motor Position Sensor
J255 – Climatronic Control Module
T5a – 5-Pin Connector, behind instrument panel, center
T5b – 5-Pin Connector, behind instrument panel, center
T24 – 24-Pin Connector, behind instrument panel, center
T28a – 28-Pin Connector
V70 – Central Air Flap Motor
V85 – Footwell/Defroster Flap Motor

(L39) – Wire Connection (A/C Pressure Switch) -1-, in Climatronic wiring harness
(L40) – Wire Connection (A/C Pressure Switch) -2-, in Climatronic wiring harness

Central air flap motor for footwell/defroster flap

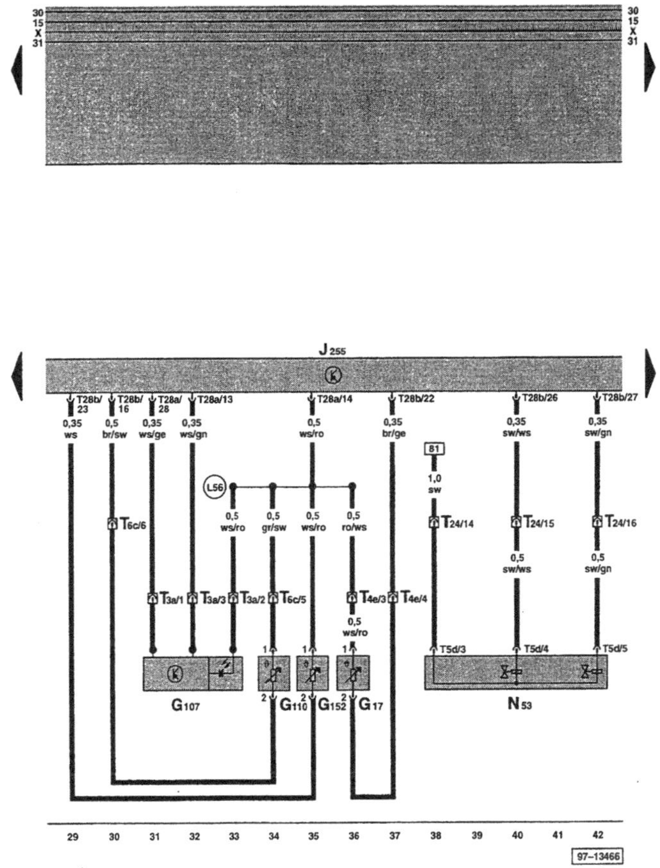

ws = white
sw = black
ro = red
br = brown
gn = green
bl = blue
gr = grey
li = violet
ge = yellow

G17 – Outside Air Temperature Sensor,
G107 – Sunlight Photo Sensor, on instrument panel, left
G110 – A/C Engine Coolant Temperature (ECT) Sensor in engine compartment, left
G152 – Front Vent Temperature Sensor, on console for footwell, left vent side
J255 – Climatronic Control Module
N53 – Vacuum Valve Rail, behind instrument panel, center
T3a – 3-Pin Connector, behind instrument panel, left
T4e – 4-Pin Connector, behind fuse/relay panel
T5d – 5-Pin Connector
T6c – 6-Pin Connector, behind fuse/relay panel
T24 – 24-Pin Connector, behind instrument panel, center
T28a – 28-Pin Connector
T28b – 28-Pin Connector

(L56) – Connector (Sensor), in wiring harness Climatronic

Sunlight photo sensor, engine coolant temperature sensor, vent temperature sensor

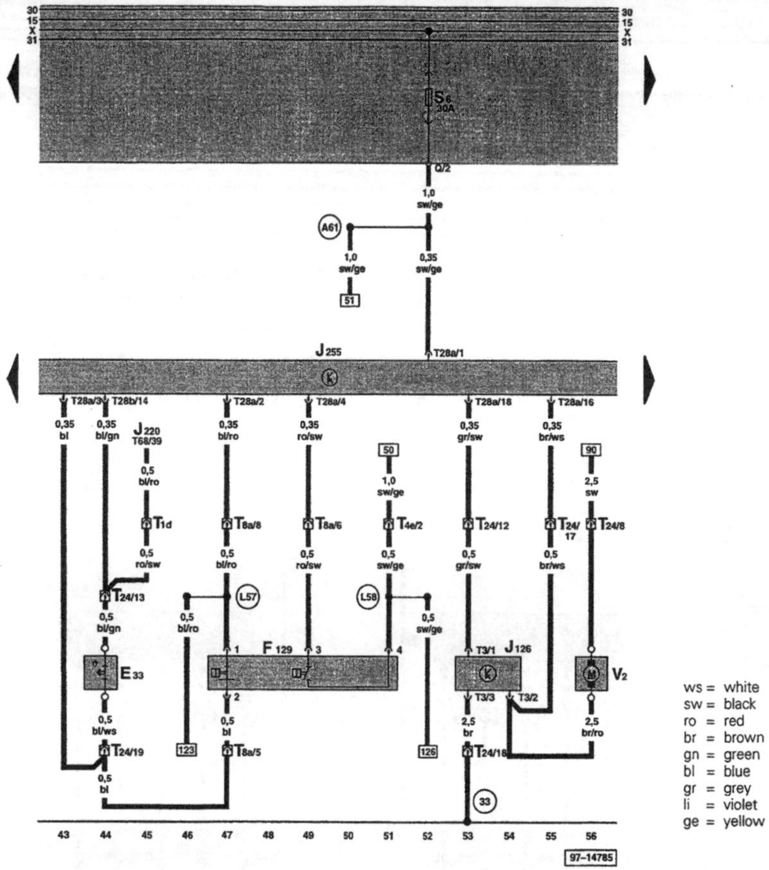

ws = white
sw = black
ro = red
br = brown
gn = green
bl = blue
gr = grey
li = violet
ge = yellow

E33 – A/C Evaporator Temperature Switch
F129 – A/C Pressure Switch
J126 – Control module for fresh air blower, behind instrument panel, right
J220 – Motronic Engine Control Module (ECM)
J255 – Climatronic Control Module
T1d – Single Connector, behind fuse/relay panel A/C "Stand by"-signal
T3 – 3-Pin Connector
T4e – 4-Pin Connector, behind fuse/relay panel
T8a – 8-Pin Connector, behind fuse/relay panel
T24 – 24-Pin Connector, behind instrument panel, center
T28a – 28-Pin Connector
T28b – 28-Pin Connector
T68 – 68-Pin Connector
V2 – Fresh Air Blower

(33) – Ground connection, behind instrument panel, right
(A61) – Plus Connection (X), in Instrument cluster wiring harness
(L57) – Connector (Pressure switch for A/C) -1- in wiring harness Climatronic-Fan for codant
(L58) – Connector (Pressure switch for A/C) -2- in wiring harness Climatronic - Fan for coolant

A/C evaporator temperature switch, pressure switch for A/C, fresh air-blower

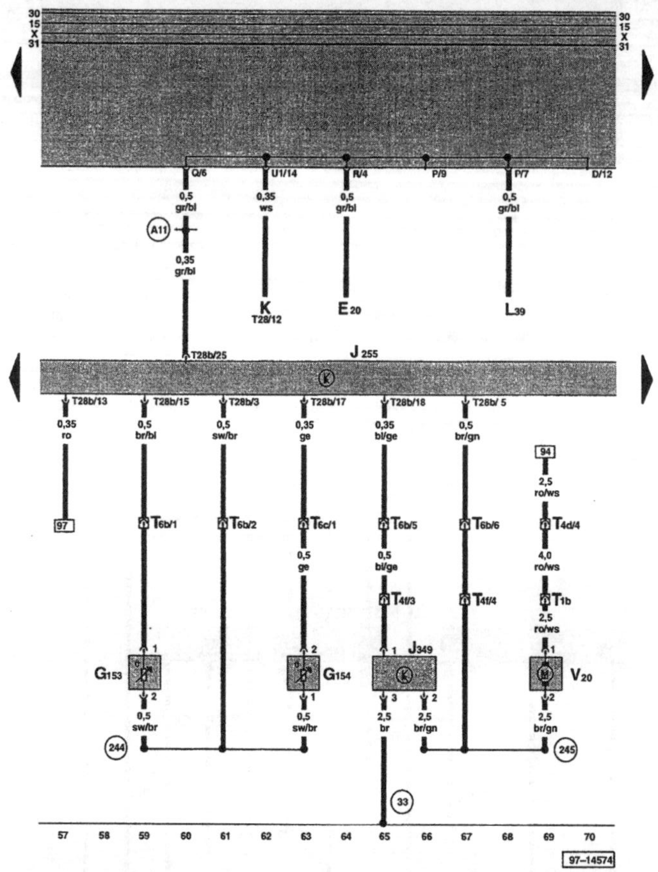

ws = white
sw = black
ro = red
br = brown
gn = green
bl = blue
gr = grey
li = violet
ge = yellow

E20 – Instrument Panel Light Dimmer Switch
G153 – Rear Evaporator Temperature Sensor
G154 – Rear Heater Core Temperature Sensor
J255 – Climatronic Control Module
J349 – Control module for blower for evaporator
K – Instrument Cluster
L39 – Rear Window Defogger Switch Light
T1b – Single Connector, behind trim D-pillar right, lower
T4d – 4-Pin Connector, behind fuse/relay panel
T4f – 4-Pin Connector, behind trim D-pillar right, lower
T6b – 6-Pin Connector, behind fuse/relay panel
T6c – 6-Pin Connector, behind fuse/relay panel
T28 – 28-Pin Connector, on instrument cluster
T28b – 28-Pin Connector
V20 – Rear Evaporator Fan

(33) – Ground connection, behind instrument panel, right
(244) – Ground connection (Sensor ground), in wiring harness Climatronic
(245) – Ground connection in wiring harness blower for 2nd Evaporatur
(A11) – Plus connection (58b) in instrument panel wiring harness

Rear evaporator temperature sensor, rear heater core temperature sensor, rear evaporator fan

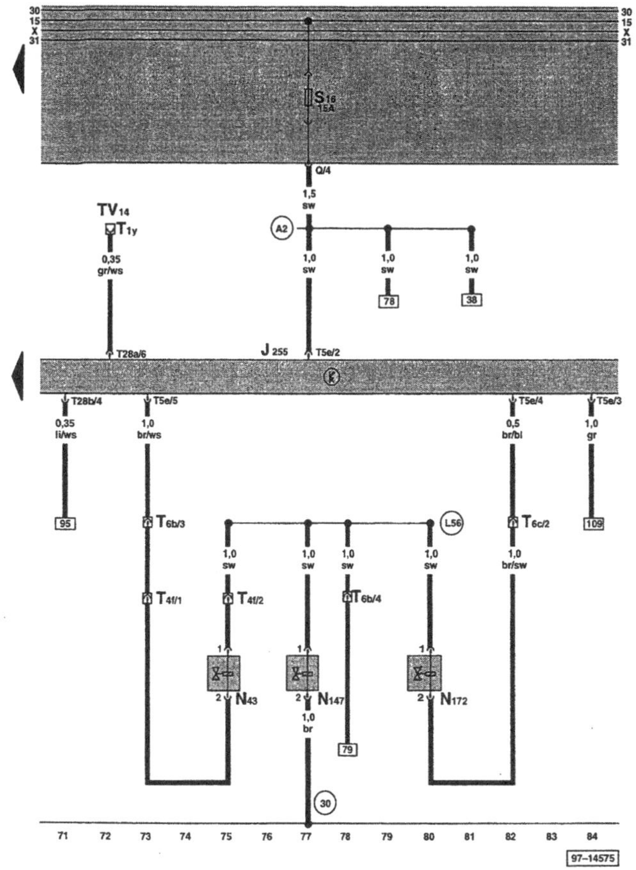

ws = white
sw = black
ro = red
br = brown
gn = green
bl = blue
gr = grey
li = violet
ge = yellow

J255 – Climatronic Control Module
N43 – A/C Refrigerant Shut-Off Valve, behind trim D-pillar right, lower
N147 – Engine Coolant Two-way Vacuum Valve, on firewall, left
N172 – Auxiliary Heater Valve, on firewall, left
T1y – Single Connector
T4f – 4-Pin Connector, behind trim D-pillar right, lower
T5e – 5-Pin Connector
T6b – 6-Pin Connector, behind fuse/relay panel
T6c – 6-Pin Connector, behind fuse/relay panel
T28a – 28-Pin Connector
T28b – 28-Pin Connector
TV14 – Data Link Connector (DLC) wire Connector, above fuse/relay panel

(30) – Ground connection, -1-, next to fuse/relay panel
(A2) – Plus connection (15), in instrument panel wiring harness
(L56) – Connector (Sensor), in wiring harness Climatronic

A/C refrigerant shut-off valve, engine coolant two-way vacuum valve, auxiliary heater valve, terminal for data link

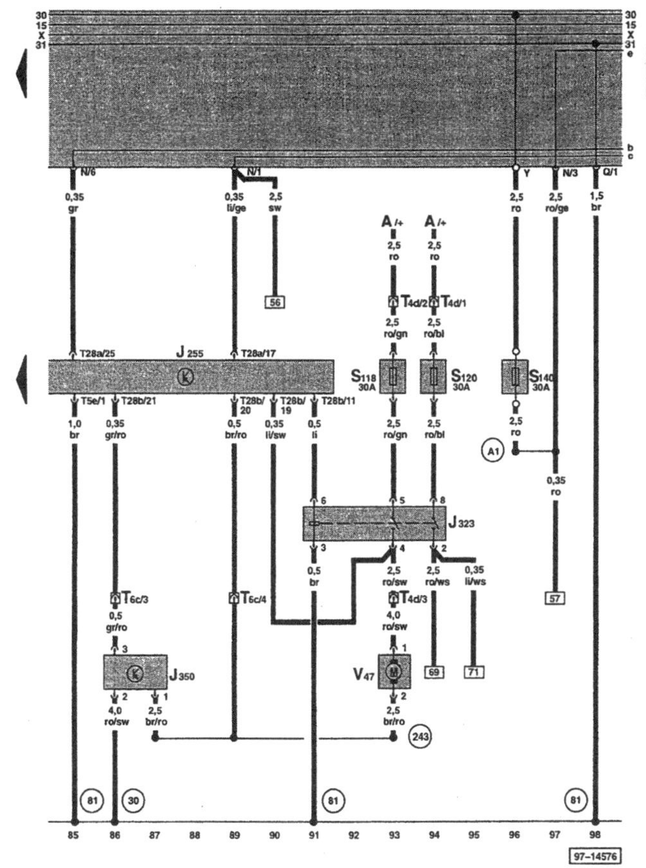

ws = white
sw = black
ro = red
br = brown
gn = green
bl = blue
gr = grey
li = violet
ge = yellow

A – Battery
J255 – Climatronic Control Module
J323 – Fan Relay, control number (114), behind instrument panel, center
J350 – Control module for warm air blower
S118 – Fuse for warmair blower, below fuse/relay panel
S120 – Fuse for Blower for Evaporator, below fuse/relay panel
S140 – Fuse for Climatronic, front, below fuse/relay panel
T4d – 4-Pin Connector, behind instrument panel
T5e – 5-Pin Connector
T6c – 6-Pin Connector, behind fuse/relay panel
T28a – 28-Pin Connector
T28b – 28-Pin Connector
V47 – Rear Warm Air Fan

(30) – Ground connection, -1-, next to fuse/relay panel
(81) – Ground connection -1-, in instrument panel wiring harness
(243) – Ground connection -1- in wiring harness, Climatronic
(A1) – Plus connection (30a), in instrument panel wiring harness

Fuse for warmair blower, fan relay, rear warm air fan

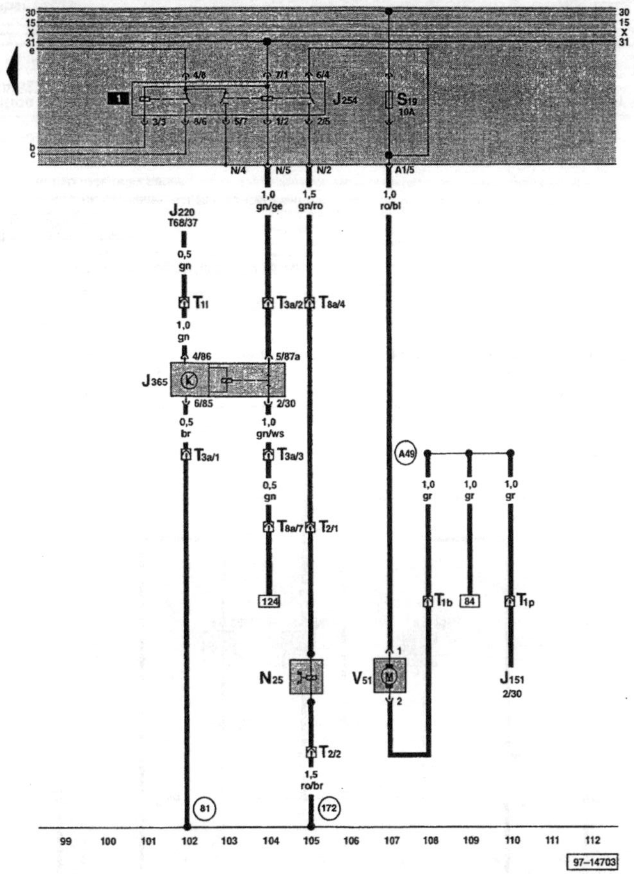

ws = white
sw = black
ro = red
br = brown
gn = green
bl = blue
gr = grey
li = violet
ge = yellow

J151 – Coolant Circulation Pump Relay
J220 – Motronic Engine Control Module (ECM)
J254 – Relay for Climatronic, control number (140), switch delay time 1/2 second
J365 – Relay for A/C shutoff, control number (147), above fuse/relay panel
N25 – A/C Clutch
T1b – Single Connector, behind fuse/relay panel
T1l – Single Connector, behind fuse/relay panel
T1p – Single Connector, behind fuse/relay panel
T2 – Double Connector, in engine compartment, right front
T3a – 3-Pin Connector, behind fuse/relay panel
T8a – 8-Pin Connector, behind fuse/relay panel
T68 – 68-Pin Connector, on Motronic Engine Control Module (ECM)
V51 – After-Run Coolant Pump

(81) – Ground connection -1-, in instrument panel wiring harness
(172) – Ground connection, in engine compartment climatronic wiring harness
(A49) – Wire connection -1-, in instrument panel wiring harness

Relay for climatronic, magnetic clutch, coolant after-run, relay for A/C shut-off

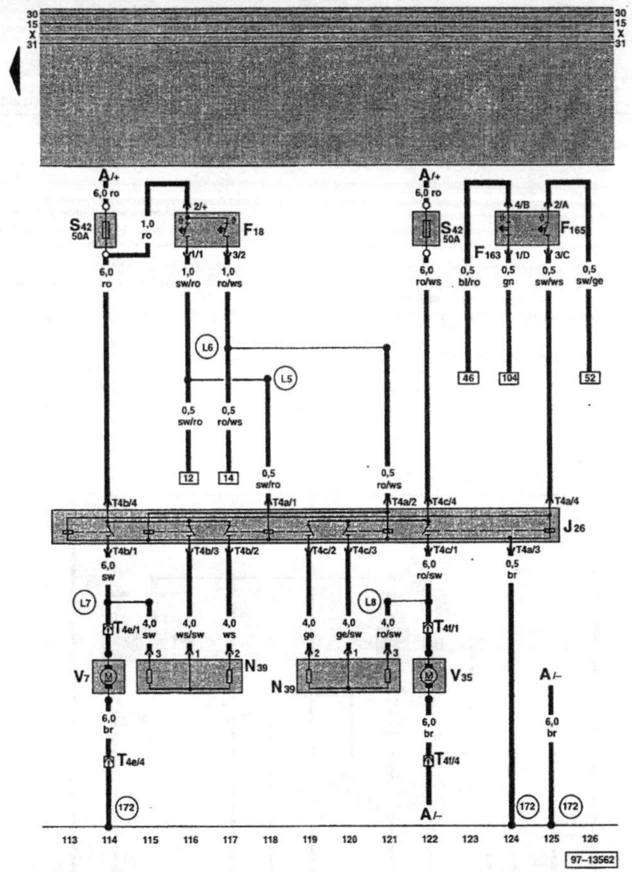

ws = white
sw = black
ro = red
br = brown
gn = green
bl = blue
gr = grey
li = violet
ge = yellow

A – Battery
F18 – Coolant Fan Control (FC) Thermal Switch
F163 – A/C Cut-Out Thermal Switch, in engine compartment, front
F165 – Third Speed Coolant Fan Control (FC) Thermal Switch, in engine compartment, front
J26 – Coolant Fan control (FC) Relay, in engine compartment, left front
N39 – Coolant Fan Control (FC) Series Resistance, behind headlight, left
S42 – Coolant Fan Fuse, in engine compartment, left front
T4a – 4-Pin Connector
T4b – 4-Pin Connector
T4c – 4-Pin Connector
T4e – 4-Pin Connector, for 450 W coolant fan
T4f – 4-Pin Connector, for 450 W coolant fan

V7 – Left Coolant Fan (450 W)
V35 – Right Coolant Fan (450 W)
(172) – Ground connection, in engine compartment climatronic wiring harness
(L5) – Wire connection -1-, in evaporator housing wiring harness
(L6) – Wire connection -2-, in evaporator housing wiring harness
(L7) – Wire connection -3-, in evaporator housing wiring harness
(L8) – Wire connection -4-, in evaporator housing wiring harness

Coolant fan control

Climatronic,
from May 1996

- **2.8 L- Engine - Motronic Multiport Fuel Injection (MFI)/103 kW, code AES, with manual transmission**

Deviate relay location and fuseplacements as well as the locations of multiple connectors- see section "component locations".

1 – Relay for Climatronic (140)

Note:

Number in parentheses indicates production control number stamped on relay housing.

Fuse colors

30 A – green
25 A – white
20 A – yellow
15 A – blue
10 A – red
7,5 A – brown
5 A – beige
3 A – violet

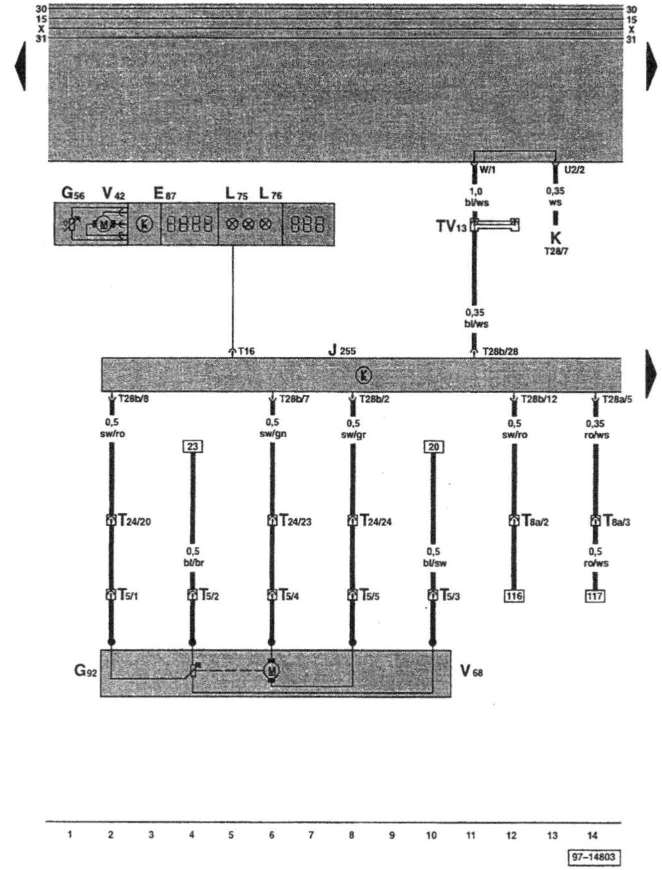

ws = white
sw = black
ro = red
br = brown
gn = green
bl = blue
gr = grey
li = violet
ge = yellow

E87 – A/C Control Head
G56 – Instrument Panel Interior Temperature Sensor, in the A/C Control Head
G92 – Temperature Regulator Flap Motor Position Sensor
J255 – Climatronic Control Module
K – Instrument Cluster
L75 – Digital Display Light
L76 – Push Button Light
T5 – 5-Pin Connector, behind instrument panel, center
T8a – 8-Pin Connector, behind fuse/relay panel
T16 – 16-Pin Connector
T24 – 24-Pin Connector, behind instrument panel, center
T28 – 28-Pin Connector, on instrument cluster
T28a – 28-Pin Connector
T28b – 28-Pin Connector
TV13 – Vehicle Speed Signal Wire Connector
V42 – Interior Temperature Sensor Fan, in the A/C Control Head
V68 – Temperature Regulator Flap Motor

A/C control head with instrument panel interior temperature sensor, temperature regulator flap motor

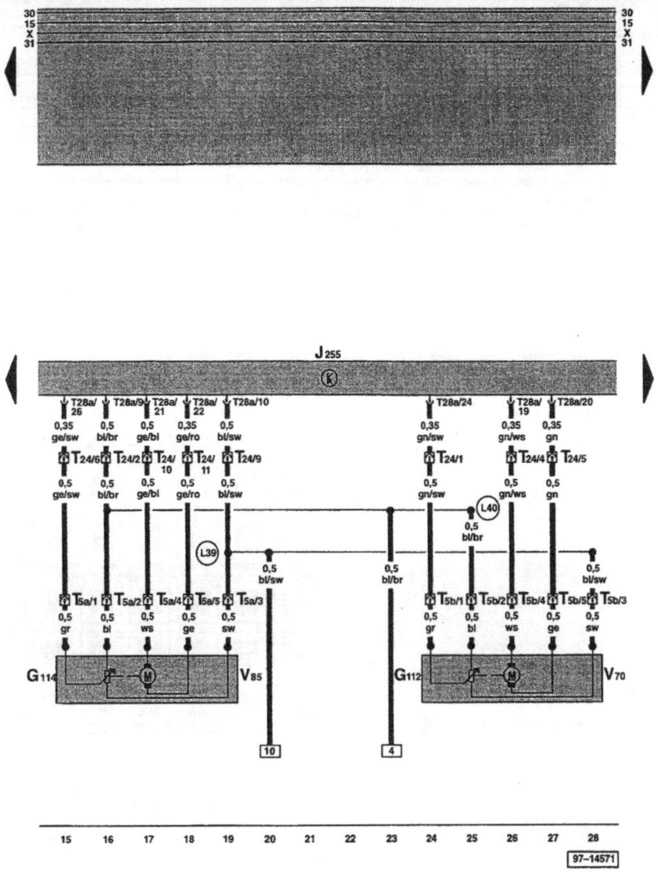

ws = white
sw = black
ro = red
br = brown
gn = green
bl = blue
gr = grey
li = violet
ge = yellow

G112 – Central Flap Motor Position Sensor
G114 – Footwell/Defrost Flap Motor Position Sensor
J255 – Climatronic Control Module
T5a – 5-Pin Connector, behind instrument panel, center
T5b – 5-Pin Connector, behind instrument panel, center
T24 – 24-Pin Connector, behind instrument panel, center
T28a – 28-Pin Connector
V70 – Central Air Flap Motor
V85 – Footwell/Defroster Flap Motor

(L39) – Wire Connection (A/C Pressure Switch) -1-, in Climatronic wiring harness
(L40) – Wire Connection (A/C Pressure Switch) -2-, in Climatronic wiring harness

Central air flap motor for footwell/defroster flap

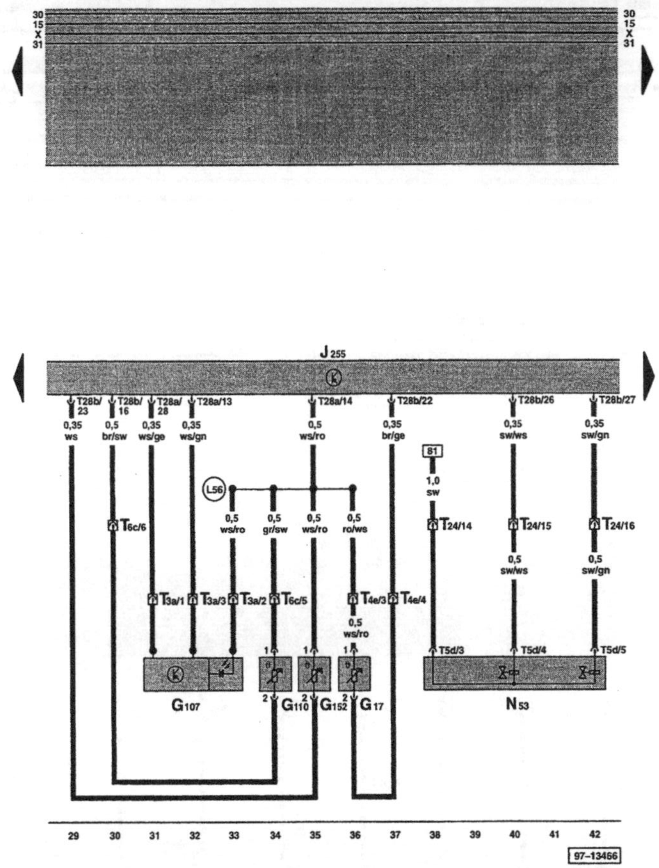

ws = white
sw = black
ro = red
br = brown
gn = green
bl = blue
gr = grey
li = violet
ge = yellow

G17 – Outside Air Temperature Sensor,
G107 – Sunlight Photo Sensor, on instrument panel, left
G110 – A/C Engine Coolant Temperature (ECT) Sensor in engine compartment, left
G152 – Front Vent Temperature Sensor, on console for footwell, left vent side
J255 – Climatronic Control Module
N53 – Vacuum Valve Rail, behind instrument panel, center
T3a – 3-Pin Connector, behind instrument panel, left
T4e – 4-Pin Connector, behind fuse/relay panel
T5d – 5-Pin Connector
T6c – 6-Pin Connector, behind fuse/relay panel
T24 – 24-Pin Connector, behind instrument panel, center
T28a – 28-Pin Connector
T28b – 28-Pin Connector

(L56) – Connector (Sensor), in wiring harness Climatronic

Sunlight photo sensor, engine coolant temperature sensor, vent temperature sensor

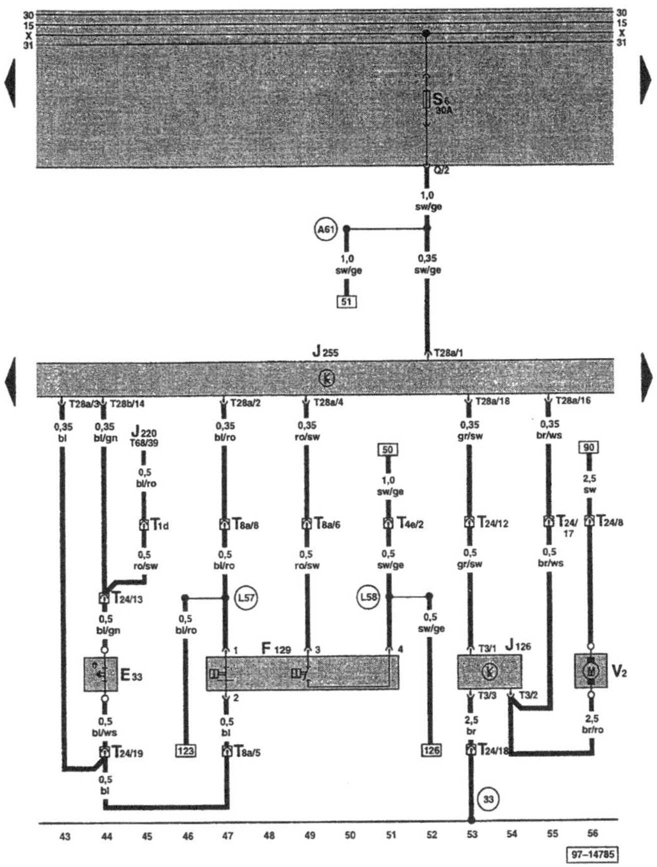

ws = white
sw = black
ro = red
br = brown
gn = green
bl = blue
gr = grey
li = violet
ge = yellow

- E33 – A/C Evaporator Temperature Switch
- F129 – A/C Pressure Switch
- J126 – Control module for fresh air blower, behind instrument panel, right
- J220 – Motronic Engine Control Module (ECM)
- J255 – Climatronic Control Module
- T1d – Single Connector, behind fuse/relay panel A/C "Stand by"-signal
- T3 – 3-Pin Connector
- T4e – 4-Pin Connector, behind fuse/relay panel
- T8a – 8-Pin Connector, behind fuse/relay panel
- T24 – 24-Pin Connector, behind instrument panel, center
- T28a – 28-Pin Connector
- T28b – 28-Pin Connector
- T68 – 68-Pin Connector
- V2 – Fresh Air Blower

- (33) – Ground connection, behind instrument panel, right
- (A61) – Plus Connection (X), in Instrument cluster wiring harness
- (L57) – Connector (Pressure switch for A/C) -1- in wiring harness Climatronic-Fan for codant
- (L58) – Connector (Pressure switch for A/C) -2- in wiring harness Climatronic - Fan for coolant

A/C evaporator temperature switch, pressure switch for A/C, fresh air-blower

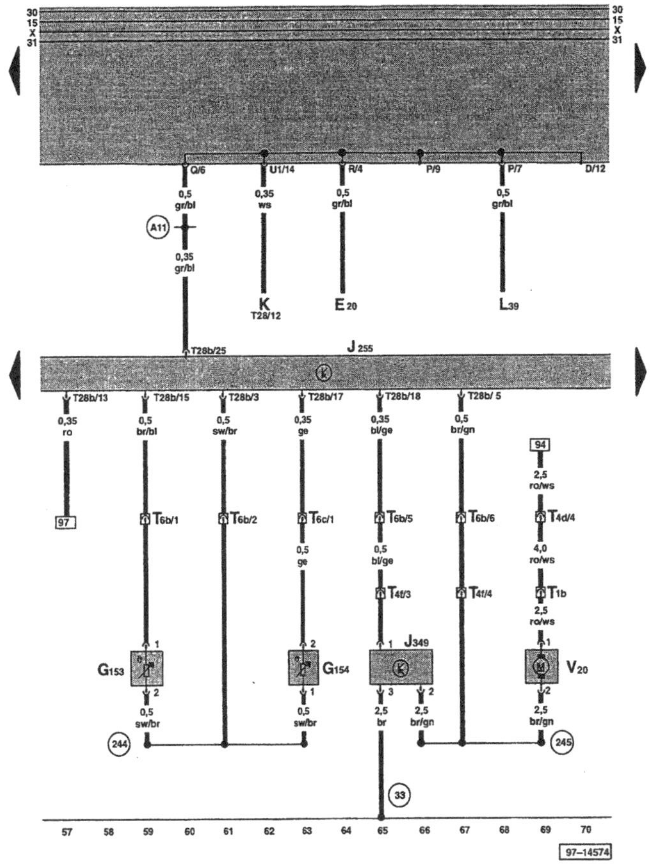

ws = white
sw = black
ro = red
br = brown
gn = green
bl = blue
gr = grey
li = violet
ge = yellow

- E20 – Instrument Panel Light Dimmer Switch
- G153 – Rear Evaporator Temperature Sensor
- G154 – Rear Heater Core Temperature Sensor
- J255 – Climatronic Control Module
- J349 – Control module for blower for evaparator
- K – Instrument Cluster
- L39 – Rear Window Defogger Switch Light
- T1b – Single Connector, behind trim D-pillar right, lower
- T4d – 4-Pin Connector, behind fuse/relay panel
- T4f – 4-Pin Connector,behind trim D-pillar right, lower
- T6b – 6-Pin Connector, behind fuse/relay panel
- T6c – 6-Pin Connector, behind fuse/relay panel
- T28 – 28-Pin Connector, on instrument cluster
- T28b – 28-Pin Connector
- V20 – Rear Evaporator Fan

- (33) – Ground connection, behind instrument panel, right
- (244) – Ground connection (Sensor ground), in wiring harness Climatronic
- (245) – Ground connection in wiring harness blower for 2nd Evaporatur
- (A11) – Plus connection (58b) in instrument panel wiring harness

Rear evaporator temperature sensor, rear heater core temperature sensor, rear evaporator fan

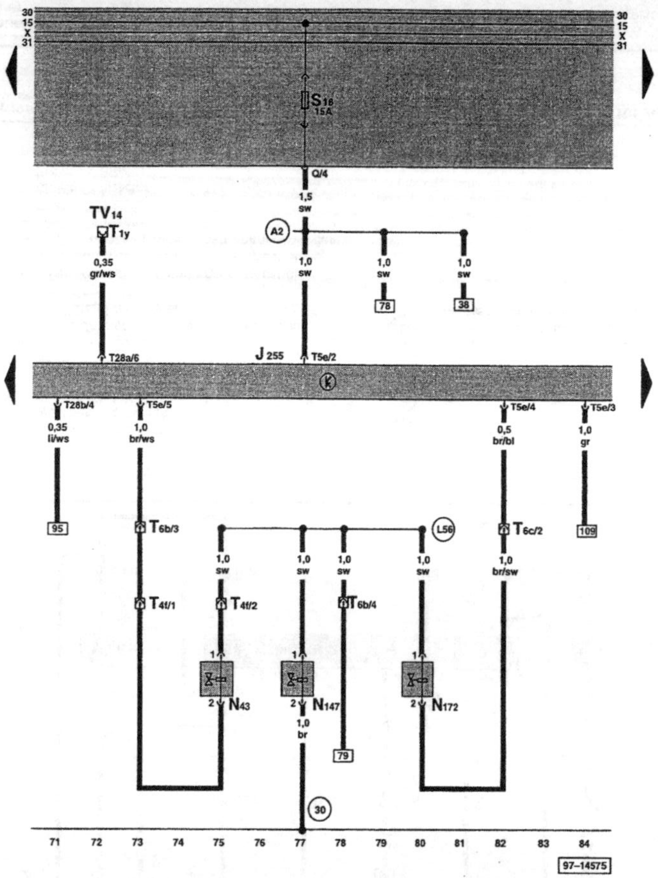

ws = white
sw = black
ro = red
br = brown
gn = green
bl = blue
gr = grey
li = violet
ge = yellow

J255 – Climatronic Control Module
N43 – A/C Refrigerant Shut-Off Valve, behind trim D-pillar right, lower
N147 – Engine Coolant Two-way Vacuum Valve, on firewall, left
N172 – Auxiliary Heater Valve, on firewall, left
T1y – Single Connector
T4f – 4-Pin Connector, behind trim D-pillar right, lower
T5e – 5-Pin Connector
T6b – 6-Pin Connector, behind fuse/relay panel
T6c – 6-Pin Connector, behind fuse/relay panel
T28a – 28-Pin Connector
T28b – 28-Pin Connector
TV14 – Data Link Connector (DLC) wire Connector, above fuse/relay panel

(30) – Ground connection, -1-, next to fuse/relay panel
(A2) – Plus connection (15), in instrument panel wiring harness
(L56) – Connector (Sensor), in wiring harness Climatronic

Edition 12/96
USA.5272.08.21

A/C refrigerant shut-off valve, engine coolant two-way vacuum valve, auxiliary heater valve, terminal for data link

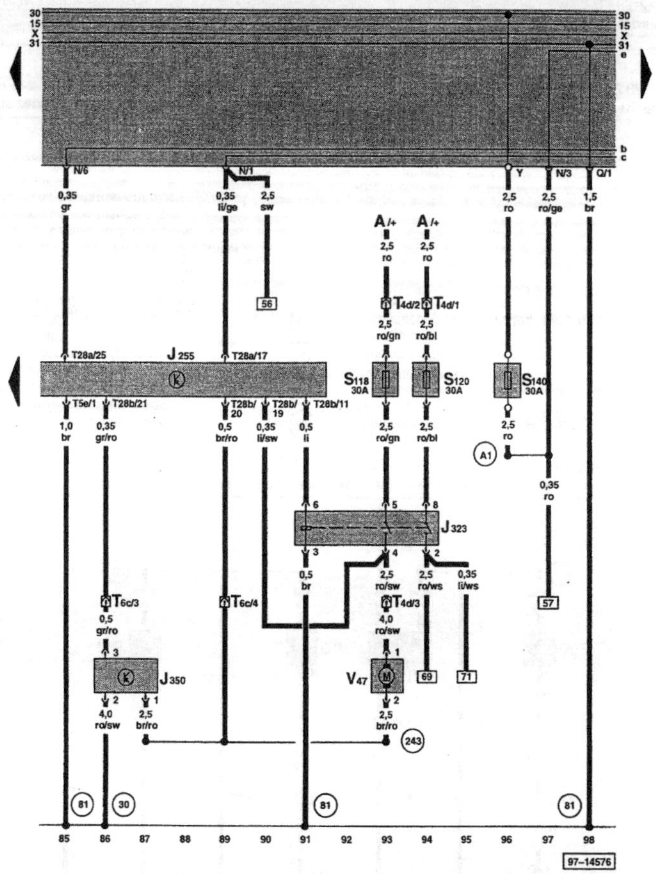

ws = white
sw = black
ro = red
br = brown
gn = green
bl = blue
gr = grey
li = violet
ge = yellow

A – Battery
J255 – Climatronic Control Module
J323 – Fan Relay, control number (114), behind instrument panel, center
J350 – Control module for warm air blower
S118 – Fuse for warmair blower, below fuse/relay panel
S120 – Fuse for Blower for Evaporator, below fuse/relay panel
S140 – Fuse for Climatronic, front, below fuse/relay panel
T4d – 4-Pin Connector, behind instrument panel
T5e – 5-Pin Connector
T6c – 6-Pin Connector, behind fuse/relay panel
T28a – 28-Pin Connector
T28b – 28-Pin Connector
V47 – Rear Warm Air Fan

(30) – Ground connection, -1-, next to fuse/relay panel
(81) – Ground connection -1-, in instrument panel wiring harness
(243) – Ground connection -1- in wiring harness, Climatronic
(A1) – Plus connection (30a), in instrument panel wiring harness

Fuse for warmair blower, fan relay, rear warm air fan

Edition 12/96
USA.5272.08.21

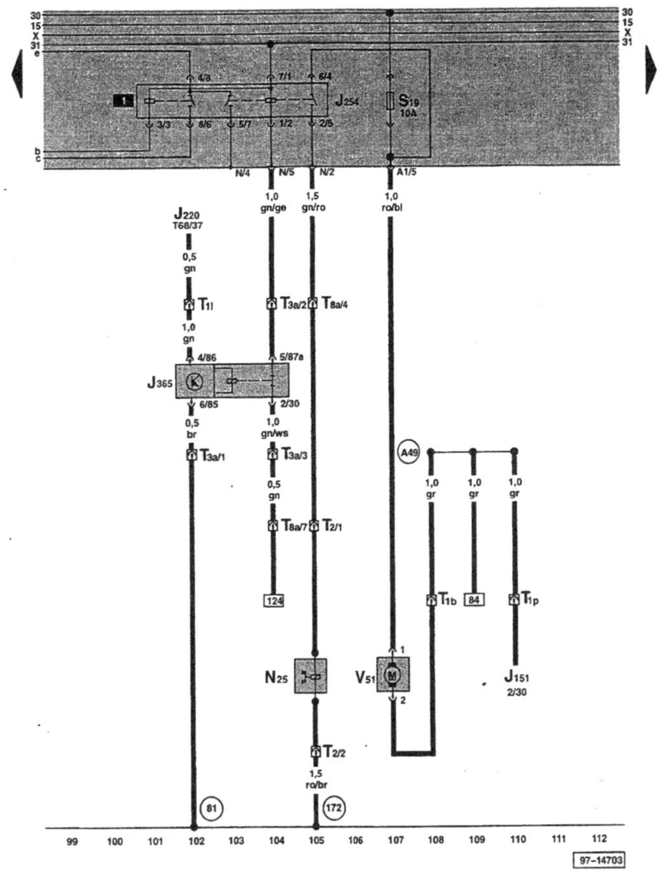

ws = white
sw = black
ro = red
br = brown
gn = green
bl = blue
gr = grey
li = violet
ge = yellow

J151 – Coolant Circulation Pump Relay
J220 – Motronic Engine Control Module (ECM)
J254 – Relay for Climatronic, control number (140), switch delay time 1/2 second
J365 – Relay for A/C shutoff, control number (147), above fuse/relay panel
N25 – A/C Clutch
T1b – Single Connector, behind fuse/relay panel
T1l – Single Connector, behind fuse/relay panel
T1p – Single Connector, behind fuse/relay panel
T2 – Double Connector, in engine compartment, right front
T3a – 3-Pin Connector, behind fuse/relay panel
T8a – 8-Pin Connector, behind fuse/relay panel
T68 – 68-Pin Connector, on Motronic Engine Control Module (ECM)
V51 – After-Run Coolant Pump

(81) – Ground connection -1-, in instrument panel wiring harness
(172) – Ground connection, in engine compartment climatronic wiring harness
(A49) – Wire connection -1-, in instrument panel wiring harness

Relay for climatronic, magnetic clutch, coolant after-run, relay for A/C shut-off

ws = white
sw = black
ro = red
br = brown
gn = green
bl = blue
gr = grey
li = violet
ge = yellow

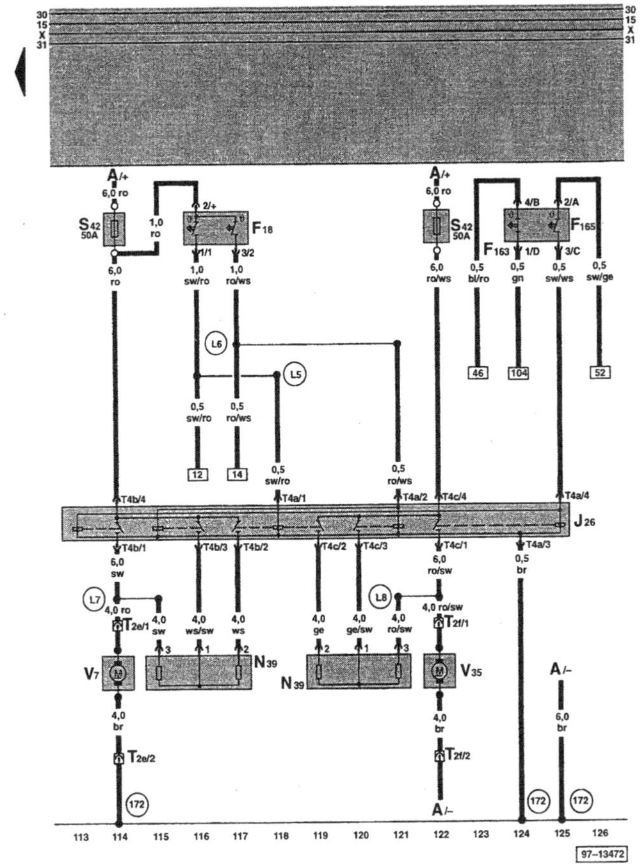

A – Battery
F18 – Coolant Fan Control (FC) Thermal Switch
F163 – A/C Cut-Out Thermal Switch, in engine compartment, front
F165 – Third Speed Coolant Fan Control (FC) Thermal Switch, in engine compartment, front
J26 – Coolant Fan control (FC) Relay, in engine compartment, left front
N39 – Coolant Fan Control (FC) Series Resistance, behind headlight, left
S42 – Coolant Fan Fuse, in engine compartment, left front
T2e – Double Connector, for 350 W coolant fan
T2f – Double Connector, for 350 W coolant fan
T4a – 4-Pin Connector
T4b – 4-Pin Connector
T4c – 4-Pin Connector

V7 – Coolant Fan (350 W)
V35 – Right Coolant Fan (350 W)

(172) – Ground connection, in engine compartment climatronic wiring harness
(L5) – Wire connection -1-, in evaporator housing wiring harness
(L6) – Wire connection -2-, in evaporator housing wiring harness
(L7) – Wire connection -3-, in evaporator housing wiring harness
(L8) – Wire connection -4-, in evaporator housing wiring harness

Coolant fan control

Climatronic,

from May 1996

- **2.4 L-Diesel Engine/57 kW, code AAB, with manual transmission**

Deviate relay location and fuseplacements as well as the locations of multiple connectors- see section "component locations".

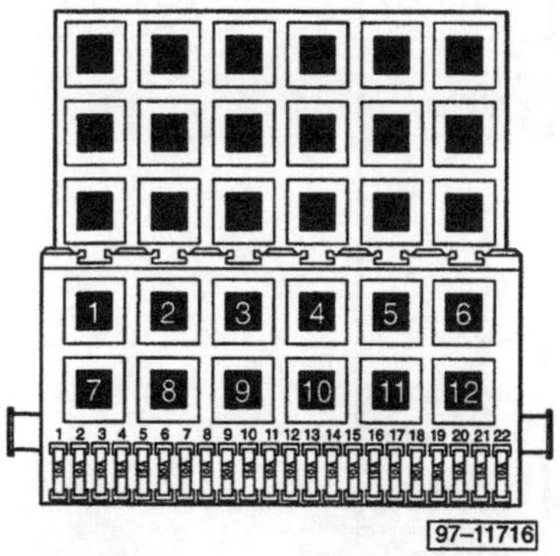

1 – Relay for Climatronic (140)

Note:

Number in parentheses indicates production control number stamped on relay housing.

Fuse colors

30 A – green
25 A – white
20 A – yellow
15 A – blue
10 A – red
7,5 A – brown
5 A – beige
3 A – violet

Edition 12/96
USA.5272.08.21

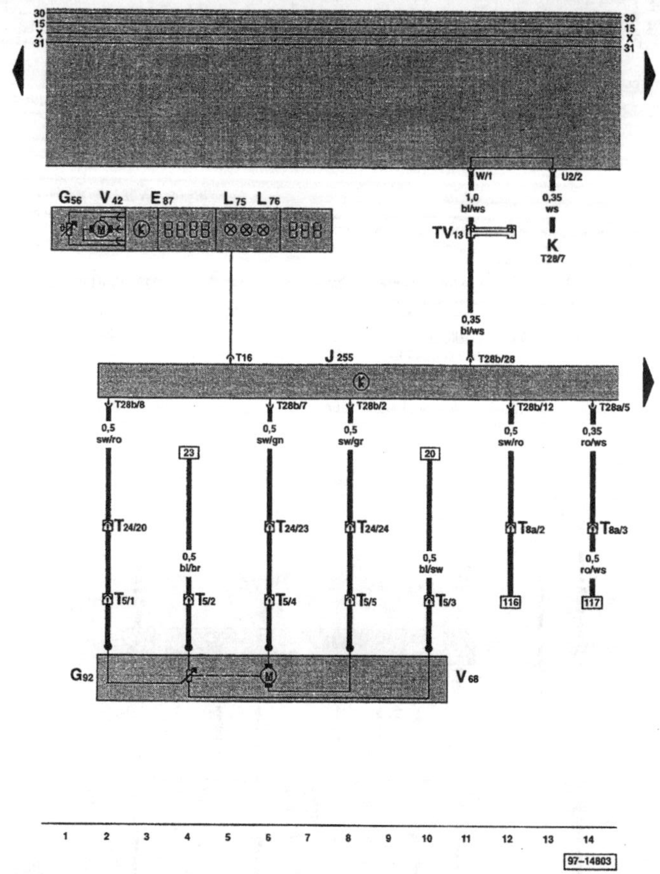

ws = white
sw = black
ro = red
br = brown
gn = green
bl = blue
gr = grey
li = violet
ge = yellow

E87 – A/C Control Head
G56 – Instrument Panel Interior Temperature Sensor, in the A/C Control Head
G92 – Temperature Regulator Flap Motor Position Sensor
J255 – Climatronic Control Module
K – Instrument Cluster
L75 – Digital Display Light
L76 – Push Button Light
T5 – 5-Pin Connector, behind instrument panel, center
T8a – 8-Pin Connector, behind fuse/relay panel
T16 – 16-Pin Connector
T24 – 24-Pin Connector, behind instrument panel, center
T28 – 28-Pin Connector, on instrument cluster
T28a – 28-Pin Connector
T28b – 28-Pin Connector
TV13 – Vehicle Speed Signal Wire Connector
V42 – Interior Temperature Sensor Fan, in the A/C Control Head
V68 – Temperature Regulator Flap Motor

A/C control head with instrument panel interior temperature sensor, temperature regulator flap motor

Edition 12.96
USA.5272.08.21

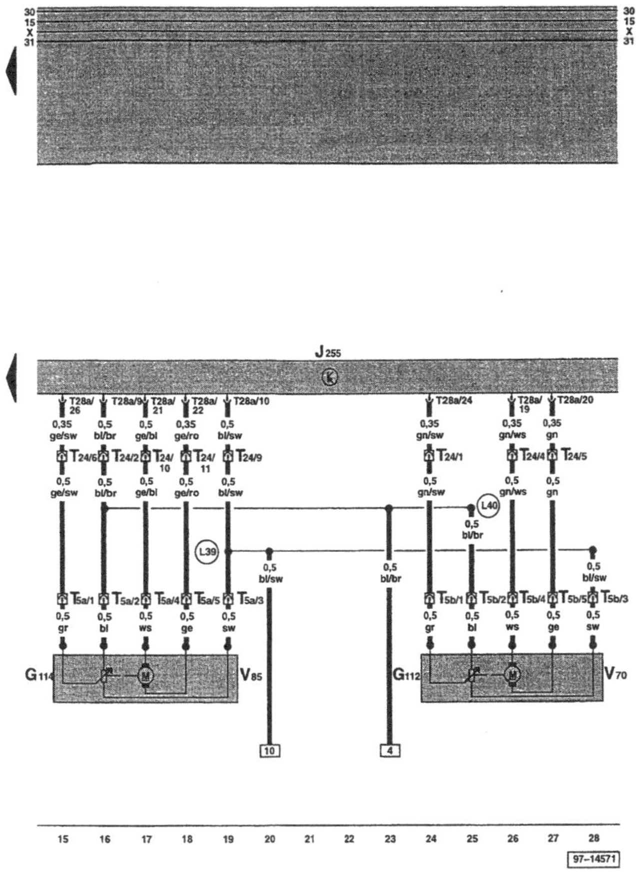

ws = white
sw = black
ro = red
br = brown
gn = green
bl = blue
gr = grey
li = violet
ge = yellow

G112 – Central Flap Motor Position Sensor
G114 – Footwell/Defrost Flap Motor Position Sensor
J255 – Climatronic Control Module
T5a – 5-Pin Connector, behind instrument panel, center
T5b – 5-Pin Connector, behind instrument panel, center
T24 – 24-Pin Connector, behind instrument panel, center
T28a – 28-Pin Connector
V70 – Central Air Flap Motor
V85 – Footwell/Defroster Flap Motor

(L39) – Wire Connection (A/C Pressure Switch) -1-, in Climatronic wiring harness
(L40) – Wire Connection (A/C Pressure Switch) -2-, in Climatronic wiring harness

Central air flap motor for footwell/defroster flap

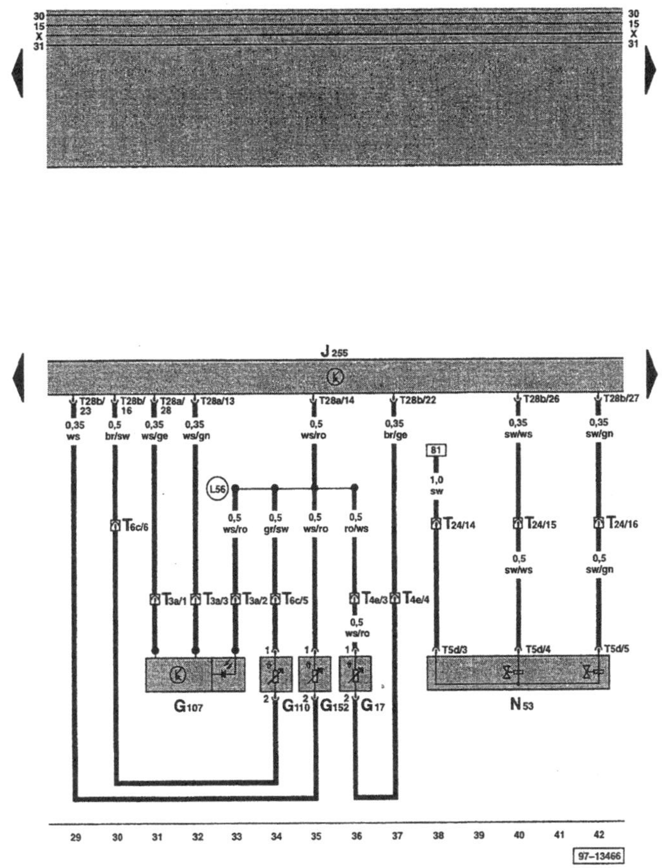

ws = white
sw = black
ro = red
br = brown
gn = green
bl = blue
gr = grey
li = violet
ge = yellow

G17 – Outside Air Temperature Sensor,
G107 – Sunlight Photo Sensor, on instrument panel, left
G110 – A/C Engine Coolant Temperature (ECT) Sensor in engine compartment, left
G152 – Front Vent Temperature Sensor, on console for footwell, left vent side
J255 – Climatronic Control Module
N53 – Vacuum Valve Rail, behind instrument panel, center
T3a – 3-Pin Connector, behind instrument panel, left
T4e – 4-Pin Connector, behind fuse/relay panel
T5d – 5-Pin Connector
T6c – 6-Pin Connector, behind fuse/relay panel
T24 – 24-Pin Connector, behind instrument panel, center
T28a – 28-Pin Connector
T28b – 28-Pin Connector

(L56) – Connector (Sensor), in wiring harness Climatronic

Sunlight photo sensor, engine coolant temperature sensor, vent temperature sensor

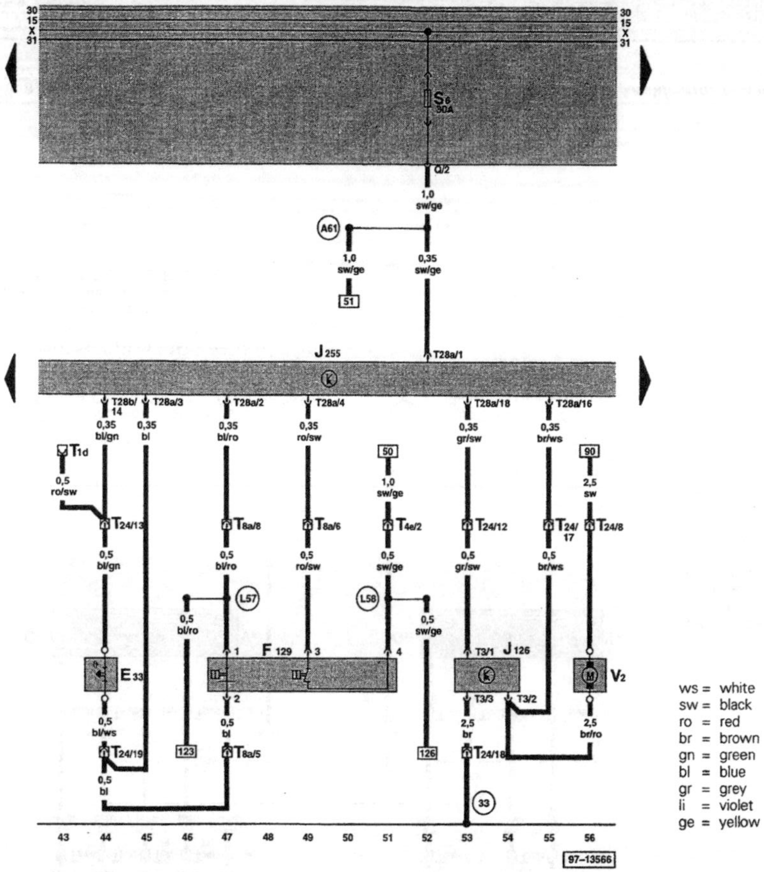

ws = white
sw = black
ro = red
br = brown
gn = green
bl = blue
gr = grey
li = violet
ge = yellow

E33 – A/C Evaporator Temperature Switch
F129 – A/C Pressure Switch
J126 – Control module for fresh air blower, behind instrument panel, right
J255 – Climatronic Control Module
T1d – Single Connector, behind fuse/relay panel, ("Stand by")-signal, Motorcode AAB not connected
T3 – 3-Pin Connector
T4e – 4-Pin Connector, behind fuse/relay panel
T8a – 8-Pin Connector, behind fuse/relay panel
T24 – 24-Pin Connector,behind instrument panel, center
T28a – 28-Pin Connector
T28b – 28-Pin Connector
V2 – Fresh Air Blower

(33) – Ground connection, behind instrument panel, right
(A61) – Plus Connection (X), in Instrument cluster wiring harness
(L57) – Connector (Pressure switch for A/C-1- in wiring harness Climatronic-Fan for coolant
(L58) – Connector (Pressure switch for A/C) -2- in wiring harness Climatronic - Fan for coolant

A/C evaporator temperature switch, A/C pressure switch, fresh air blower

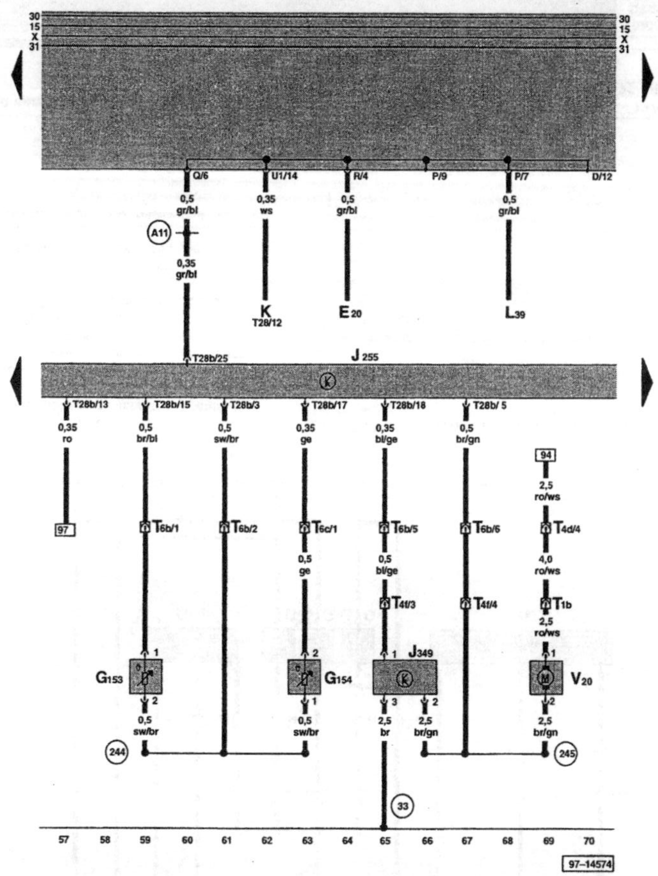

ws = white
sw = black
ro = red
br = brown
gn = green
bl = blue
gr = grey
li = violet
ge = yellow

E20 – Instrument Panel Light Dimmer Switch
G153 – Rear Evaporator Temperature Sensor
G154 – Rear Heater Core Temperature Sensor
J255 – Climatronic Control Module
J349 – Control module for blower for evaporator
K – Instrument Cluster
L39 – Rear Window Defogger Switch Light
T1b – Single Connector, behind trim D-pillar right, lower
T4d – 4-Pin Connector, behind fuse/relay panel
T4f – 4-Pin Connector,behind trim D-pillar right, lower
T6b – 6-Pin Connector, behind fuse/relay panel
T6c – 6-Pin Connector, behind fuse/relay panel
T28 – 28-Pin Connector, on instrument cluster
T28b – 28-Pin Connector
V20 – Rear Evaporator Fan

(33) – Ground connection, behind instrument panel, right
(244) – Ground connection (Sensor ground), in wiring harness Climatronic
(245) – Ground connection in wiring harness blower for 2nd Evaporatur
(A11) – Plus connection (58b) in instrument panel wiring harness

Rear evaporator temperature sensor, rear heater core temperature sensor, rear evaporator fan

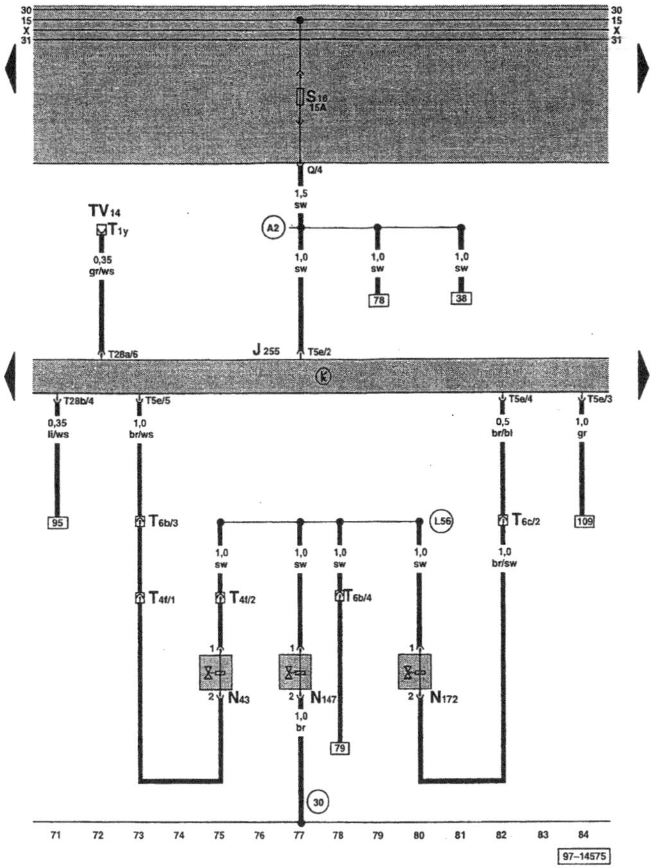

ws = white
sw = black
ro = red
br = brown
gn = green
bl = blue
gr = grey
li = violet
ge = yellow

J255 – Climatronic Control Module
N43 – A/C Refrigerant Shut-Off Valve, behind trim D-pillar right, lower
N147 – Engine Coolant Two-way Vacuum Valve, on firewall, left
N172 – Auxiliary Heater Valve, on firewall, left
T1y – Single Connector
T4f – 4-Pin Connector, behind trim D-pillar right, lower
T5e – 5-Pin Connector
T6b – 6-Pin Connector, behind fuse/relay panel
T6c – 6-Pin Connector, behind fuse/relay panel
T28a – 28-Pin Connector
T28b – 28-Pin Connector
TV14 – Data Link Connector (DLC) wire Connector, above fuse/relay panel

(30) – Ground connection, -1-, next to fuse/relay panel
(A2) – Plus connection (15), in instrument panel wiring harness
(L56) – Connector (Sensor), in wiring harness Climatronic

A/C refrigerant shut-off valve, engine coolant two-way vacuum valve, auxiliary heater valve, terminal for data link

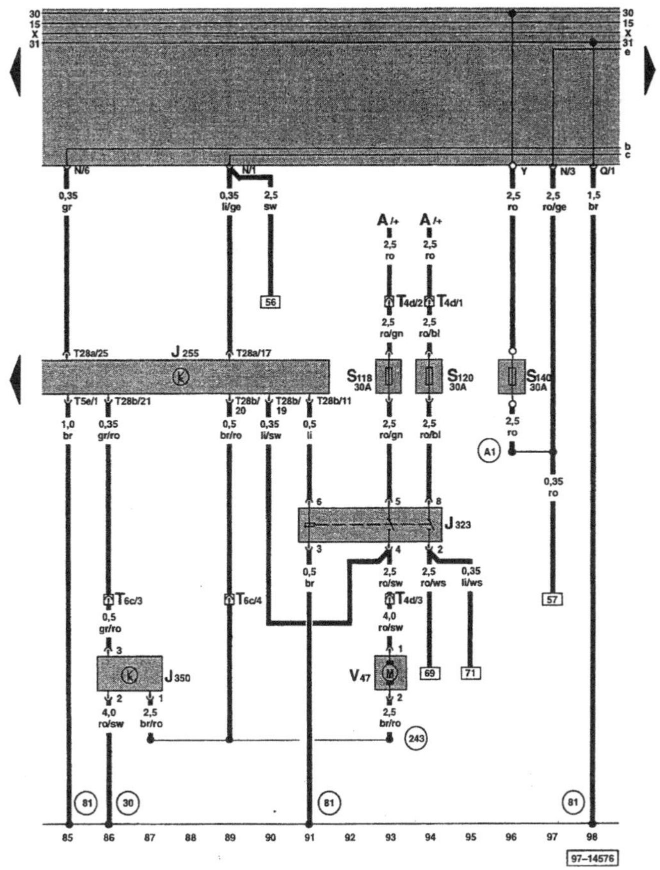

ws = white
sw = black
ro = red
br = brown
gn = green
bl = blue
gr = grey
li = violet
ge = yellow

A – Battery
J255 – Climatronic Control Module
J323 – Fan Relay, control number (114), behind instrument panel, center
J350 – Control module for warm air blower
S118 – Fuse for warmair blower, below fuse/relay panel
S120 – Fuse for Blower for Evaporator, below fuse/relay panel
S140 – Fuse for Climatronic, front, below fuse/relay panel
T4d – 4-Pin Connector, behind instrument panel
T5e – 5-Pin Connector
T6c – 6-Pin Connector, behind fuse/relay panel
T28a – 28-Pin Connector
T28b – 28-Pin Connector
V47 – Rear Warm Air Fan

(30) – Ground connection, -1-, next to fuse/relay panel
(81) – Ground connection -1-, in instrument panel wiring harness
(243) – Ground connection -1- in wiring harness, Climatronic
(A1) – Plus connection (30a), in instrument panel wiring harness

Fuse for warmair blower, fan relay, rear warm air fan

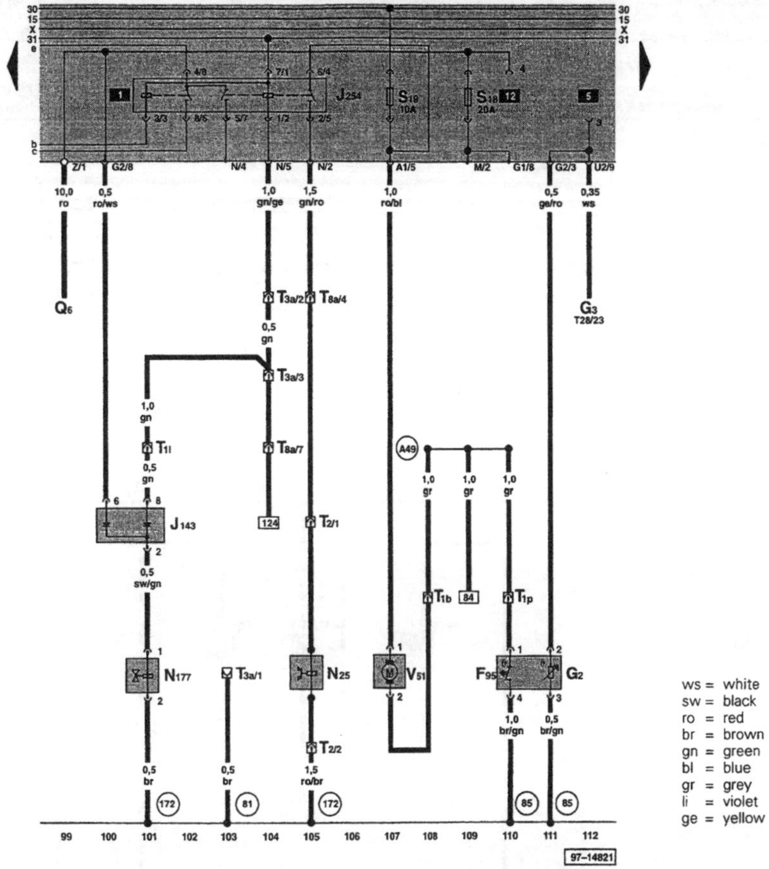

ws = white
sw = black
ro = red
br = brown
gn = green
bl = blue
gr = grey
li = violet
ge = yellow

F95 – After-Run Coolant Thermal Switch
G2 – Engine Coolant Temperature (ECT) Sensor
G3 – Engine Coolant Temperature (ECT) Gauge
J143 – Idle Air Control (IAC) Control Module, control number (153)
J254 – Relay for Climatronic, control number (140), switch delay time 1/2 second
N25 – A/C Clutch
N177 – Valve for idle rem advance
Q6 – Spark plugs (engine)
T1b – Single Connector, behind fuse/relay panel
T1l – Single Connector, behind fuse/relay panel
T1p – Single Connector, behind fuse/relay panel
T2 – Double Connector, in engine compartment, right front
T3a – 3-Pin Connector, behind fuse/relay panel
T8a – 8-Pin Connector, behind fuse/relay panel
T28 – 28-Pin Connector, on instrument cluster
V51 – After-Run Coolant Pump

(81) – Ground connection -1-, in instrument panel wiring harness
(85) – Ground connection -1-, in engine compartment wiring harness
(172) – Ground connection, in engine compartment climatronic wiring harness
(A49) – Wire connection -1-, in instrument panel wiring harness

Relay for climatronic, magnetic clutch, after-run coolant

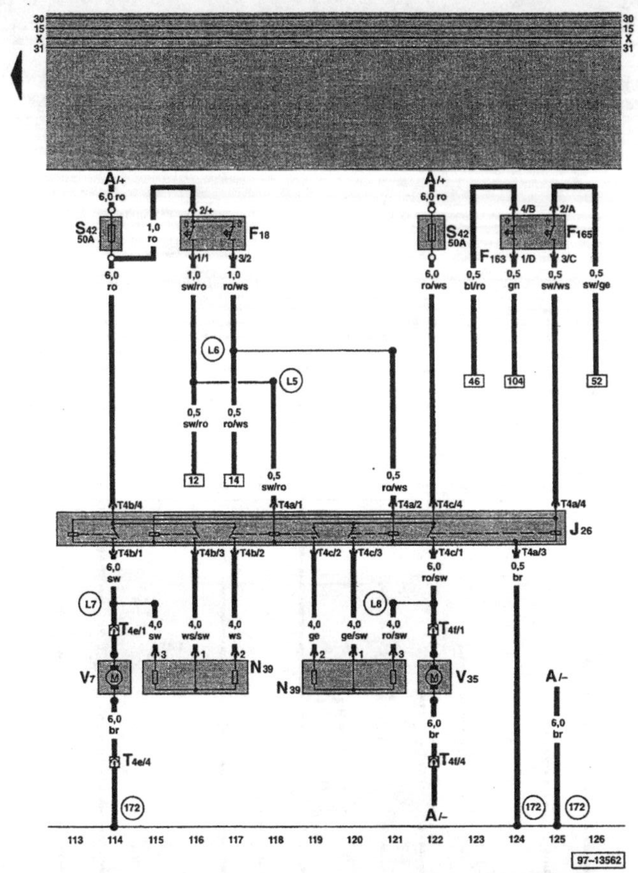

ws = white
sw = black
ro = red
br = brown
gn = green
bl = blue
gr = grey
li = violet
ge = yellow

A – Battery
F18 – Coolant Fan Control (FC) Thermal Switch
F163 – A/C Cut-Out Thermal Switch, in engine compartment, front
F165 – Third Speed Coolant Fan Control (FC) Thermal Switch, in engine compartment, front
J26 – Coolant Fan control (FC) Relay, in engine compartment, left front
N39 – Coolant Fan Control (FC) Series Resistance behind headlight, left
S42 – Coolant Fan Fuse, in engine compartment, left front
T4a – 4-Pin Connector
T4b – 4-Pin Connector
T4c – 4-Pin Connector
T4e – 4-Pin Connector, for 450 W coolant fan
T4f – 4-Pin Connector, for 450 W coolant fan
V7 – Coolant Fan (450 W)
V35 – Right Coolant Fan (450 W)

(172) – Ground connection, in engine compartment climatronic wiring harness
(L5) – Wire connection -1-, in evaporator housing wiring harness
(L6) – Wire connection -2-, in evaporator housing wiring harness
(L7) – Wire connection -3-, in evaporator housing wiring harness
(L8) – Wire connection -4-, in evaporator housing wiring harness

Coolant fan

Air-Conditioning,

from May1996

- **2.8 L- Engine - Motronic Multiport Fuel Injection (MFI)/103 kW, code AES, (automatic transmission)**

Deviate relay location and fuseplacements as well as the locations of multiple connectors- see section "component locations".

1 – A/C Relay (140)

Note:

Number in parentheses indicates production control number stamped on relay housing.

Fuse colors

30 A – green
25 A – white
20 A – yellow
15 A – blue
10 A – red
7,5 A – brown
5 A – beige
3 A – violet

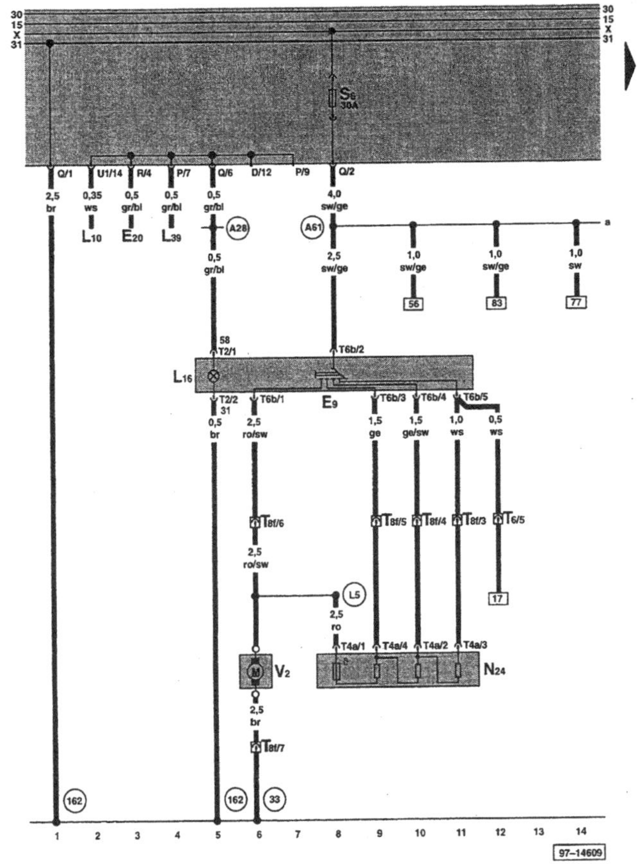

ws = white
sw = black
ro = red
br = brown
gn = green
bl = blue
gr = grey
li = violet
ge = yellow

E9 – Fresh Air Blower Switch
E20 – Instrument Panel Light Dimmer Switch
L10 – Instrument Cluster Light
L16 – Fresh Air Control Lever Light
L39 – Rear Window Defogger Switch Light
N24 – Fresh Air Blower Series Resistance with Fuse
T2 – Double Connector, on Fresh Air Blower Switch
T4a – 4-Pin Connector, on Resistance
T6 – 6-Pin Connector, behind instrument panel, center
T6b – 6-Pin Connector
T8f – 8-Pin Connector, behind instrument panel, right
V2 – Fresh Air Blower

(33) – Ground connection, behind instrument panel, right
(162) – Ground connection, in blower motor wiring harness
(A28) – Plus connection (58b), in fresh air blower Wiring harness
(A61) – Plus Connection (X), in Instrument cluster Wiring harness
(L5) – Wire connection -1-, in evaporator housing wiring harness

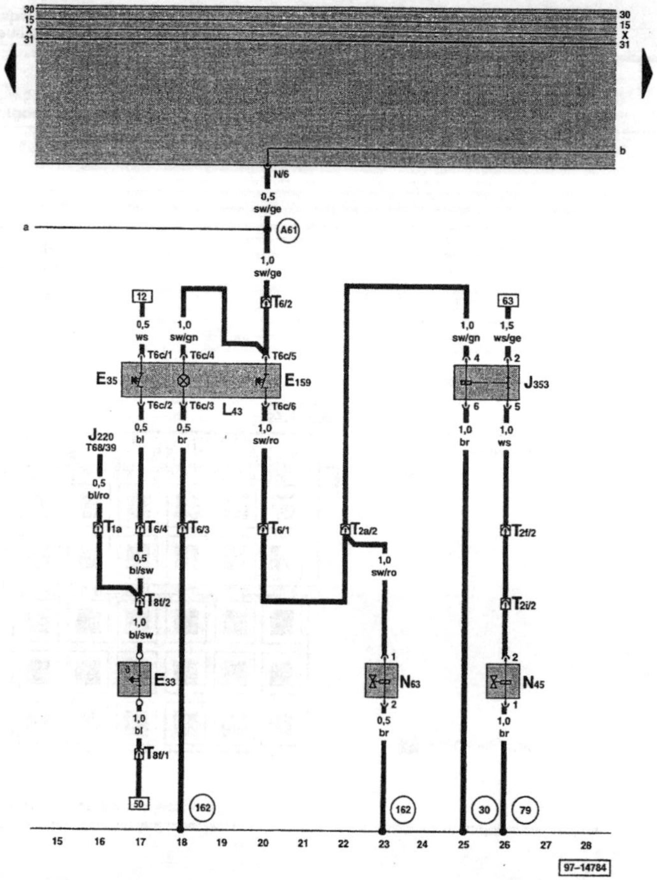

ws = white
sw = black
ro = red
br = brown
gn = green
bl = blue
gr = grey
li = violet
ge = yellow

- E33 – A/C Evaporator Temperature Switch
- E35 – A/C Switch
- E159 – Fresh Air/Recirculating Flap Switch
- J220 – Motronic Engine Control Module (ECM), ("Stand by")-signal
- J353 – Relay for Fresh-air and recirculating air flap-rear, control number (94), behind instrument panel, center
- L43 – A/C Switch Illumination Light
- N45 – Two-way valve for Fresh-and recirculation air flap-rear
- N63 – Fresh Air/Recirculating Flap Two-Way Valve
- T1a – Single Connector, behind fuse/relay panel
- T2a – Double Connector, behind fuse/relay panel
- T2f – Double Connector
- T2i – Double Connector
- T6 – 6-Pin Connector, behind instrument panel, center
- T6c – 6-Pin Connector
- T8f – 8-Pin Connector, behind instrument panel, right
- T68 – 68-Pin Connector, on Motronic Engine Control Module (ECM)
- (30) – Ground connection, -1-, next to fuse/relay panel
- (79) – Ground connection, near D-pillar, lower part
- (162) – Ground connection, in blower motor wiring harness
- (A61) – Plus Connection (X), in Instrument cluster wiring harness

Control for recirculating flap front and rear

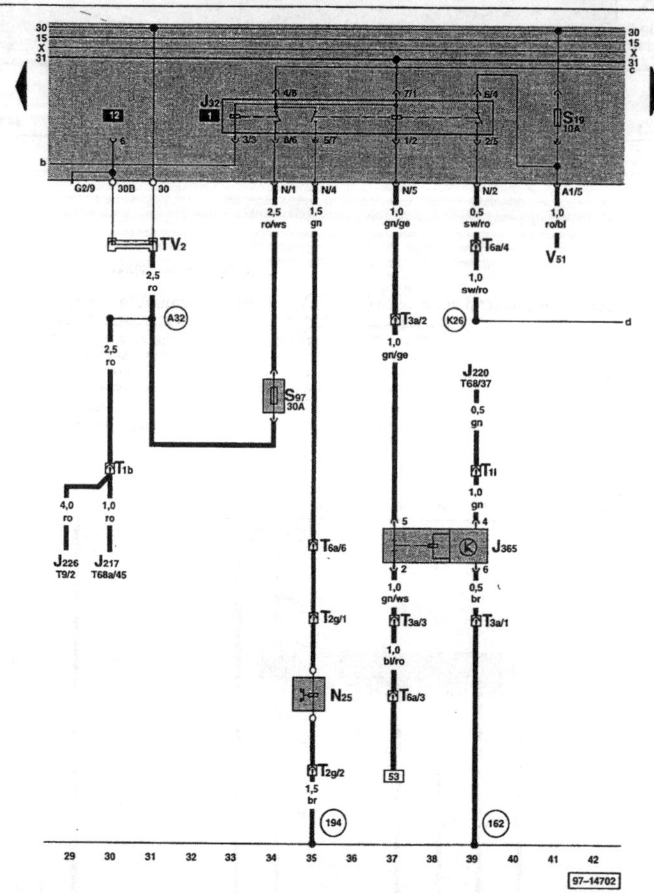

ws = white
sw = black
ro = red
br = brown
gn = green
bl = blue
gr = grey
li = violet
ge = yellow

- J32 – A/C Relay, control number (140), switch delay time 1/2 second
- J217 – Transmission Control Module (TCM)
- J220 – Motronic Engine Control Module (ECM), ("on/off",-signal)
- J226 – Park/Neutral Position (PNP) Relay
- J365 – Relay for A/C shutoff, control number (147), above fuse/relay panel
- N25 – A/C Clutch
- S97 – Fresh Air Blower Fuse 1, below fuse/relay panel
- T1b – Single Connector, behind fuse/relay panel
- T1l – Single Connector, behind fuse/relay panel
- T2g – Double Connector, near kompressor
- T3a – 3-Pin Connector, behind fuse/relay panel
- T6a – 6-Pin Connector, behind instrument panel, left
- T9 – 9-Pin Connector
- T68 – 68-Pin Connector, on Engine Control Module (ECM)
- T68a – 68-Pin Connector
- TV2 – Terminal 30 (B+) Wire Connector
- V51 – After-Run Coolant Pump
- (162) – Ground connection, in blower motor wiring harness
- (194) – Ground connection -2-, in coolant fan wiring harness
- (A32) – Plus connection (30), in instrument panel wiring harness
- (K26) – Wire connection -5-, in coolant fan wiring harness

Magnetic clutch, A/C shut off

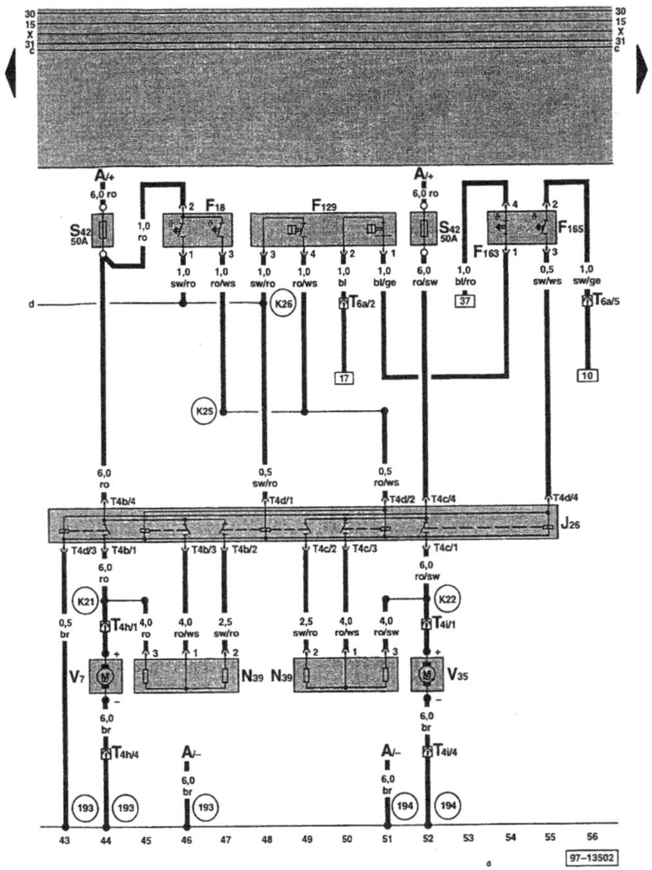

ws = white
sw = black
ro = red
br = brown
gn = green
bl = blue
gr = grey
li = violet
ge = yellow

- A – Battery
- F18 – Coolant Fan Control (FC) Thermal Switch
- F129 – A/C Pressure Switch
- F163 – A/C Cut-Out Thermal Switch
- F165 – Third Speed Coolant Fan Control (FC) Thermal Switch
- J26 – Coolant Fan control (FC) Relay, in engine compartment, left front
- N39 – Coolant Fan Control (FC) Series Resistance
- S42 – Coolant Fan Fuse, in engine compartment, left front
- T4b – 4-Pin Connector
- T4c – 4-Pin Connector
- T4d – 4-Pin Connector
- T4h – 4-Pin Connector
- T4i – 4-Pin Connector
- T6a – 6-Pin Connector, behind instrument panel, left
- V7 – Coolant Fan (450 W)
- V35 – Right Coolant Fan (450 W)
- (193) – Ground connection -1-, in coolant fan wiring harness
- (194) – Ground connection -2-, in coolant fan wiring harness
- (K21) – Wire connection -1-, in coolant fan wiring harness
- (K22) – Wire connection -2-, in coolant fan wiring harness
- (K25) – Wire connection -4-, in coolant fan wiring harness
- (K26) – Wire connection -5-, in coolant fan wiring harness

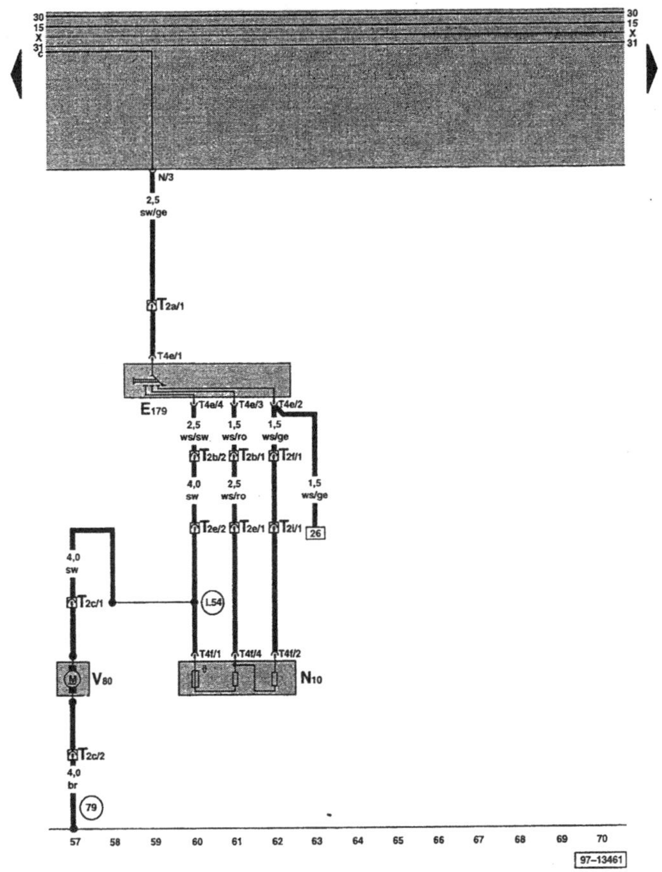

ws = white
sw = black
ro = red
br = brown
gn = green
bl = blue
gr = grey
li = violet
ge = yellow

- E179 – Switch for Freshair blower-rear
- N10 – Series Resistance for fresh air blower-rear
- T2a – Double Connector, behind fuse/relay panel
- T2b – Double Connector, behind instrument panel, right
- T2c – Double Connector, behind trim D-pillar right, lower
- T2e – Double Connector, behind trim D-pillar right, lower
- T2f – Double Connector
- T2i – Double Connector
- T4e – 4-Pin Connector
- T4f – 4-Pin Connector
- V80 – Fresh air blower rear
- (79) – Ground connection, near D-pillar, lower part
- (L54) – Plus connector (Blower stage 3) in wiring harness

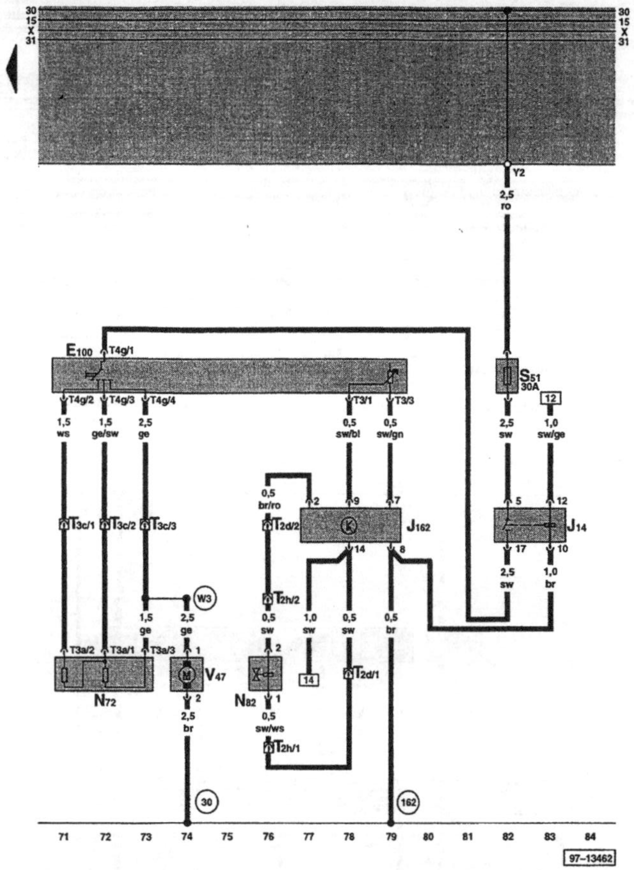

ws = white
sw = black
ro = red
br = brown
gn = green
bl = blue
gr = grey
li = violet
ge = yellow

E100 – Warm Air Blower Switch
J14 – Heater Air Blower Relay, control number (53), behind instrument panel, center
J162 – Heater Control Module, control number (152), behind instrument panel, center
N72 – Heater Blower Series Resistance
N82 – Coolant Cut-off Valve
S51 – Fuse for 2. heater core, below fuse/relay panel
T2d – Double Connector, behind instrument panel, left
T2h – Double Connector, in engine compartment, left front
T3 – 3-Pin Connector
T3a – 3-Pin Connector
T3c – 3-Pin Connector, behind fuse/relay panel
T4g – 4-Pin Connector

V47 – Rear Warm Air Fan

(30) – Ground connection, -1-, next to fuse/relay panel
(162) – Ground connection, in blower motor wiring harness
(W3) – Wire connection, in rear wiring harness

THIS PAGE INTENTIONALLY LEFT
BLANK

Auxiliary Heater B3LC/D3LC,
from May 1996

Deviate relay location and fuseplacements as well as the locations of multiple connectors-see section "component locations".

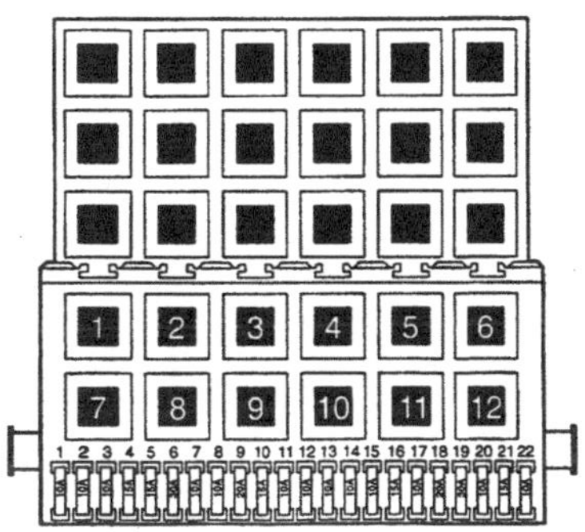

Fuse colors

30 A – green
25 A – white
20 A – yellow
15 A – blue
10 A – red
7,5 A – brown
5 A – beige
3 A – violet

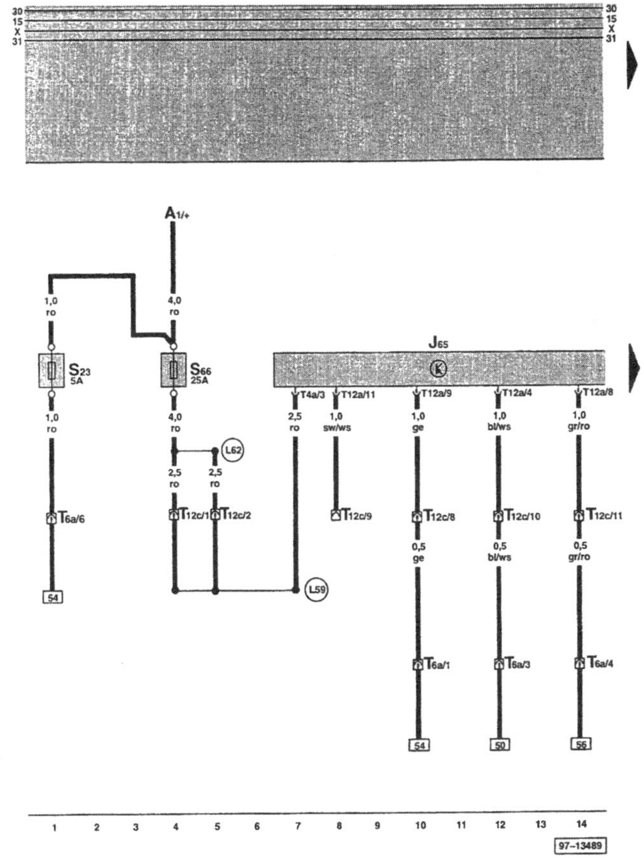

ws = white
sw = black
ro = red
br = brown
gn = green
bl = blue
gr = grey
li = violet
ge = yellow

A1 – Auxiliary Battery
J65 – Heater Control Modul, in intake channel
S23 – Heater Fuse, below driver's seat
S66 – Auxiliary Heater Glow Plug Fuse, below driver's seat
T4a – 4-Pin Connector
T6a – 6-Pin Connector, below driver's seat
T12a – 12-Pin Conenctor
T12c – 12-Pin Conenctor, to intake channel

(L59) – Plus connector (30) in wiring harness heater

(L62) – Plus connector (30a), in wiring harness auxiliary heater

Heater control module, auxiliary heater fuse

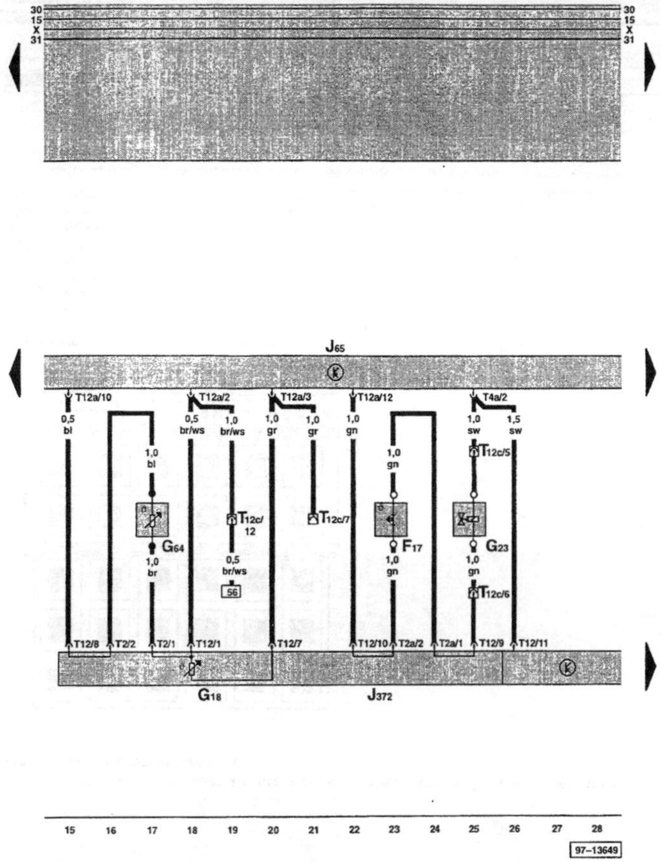

ws = white
sw = black
ro = red
br = brown
gn = green
bl = blue
gr = grey
li = violet
ge = yellow

F17 – Heater Overheat Control Switch, in heater unit
G18 – Temperature Sensor II, in heater unit
G23 – Transfer Fuel Pump (FP)
G64 – Flame Sensor in heater unit
J65 – Heater Control Module, in intake channel
J372 – Circuit board for temperature sensor, control module for blower motor and glow plug
T2 – Double Connector
T2a – Double Connector
T4a – 4-Pin Connector
T12 – 12-Pin Conenctor
T12a – 12-Pin Conenctor
T12c – 12-Pin Conenctor, to intake channel

Heater controle module, heater overheat control switch, temperature sensor, transfer fuel pump (FP), flame sensor

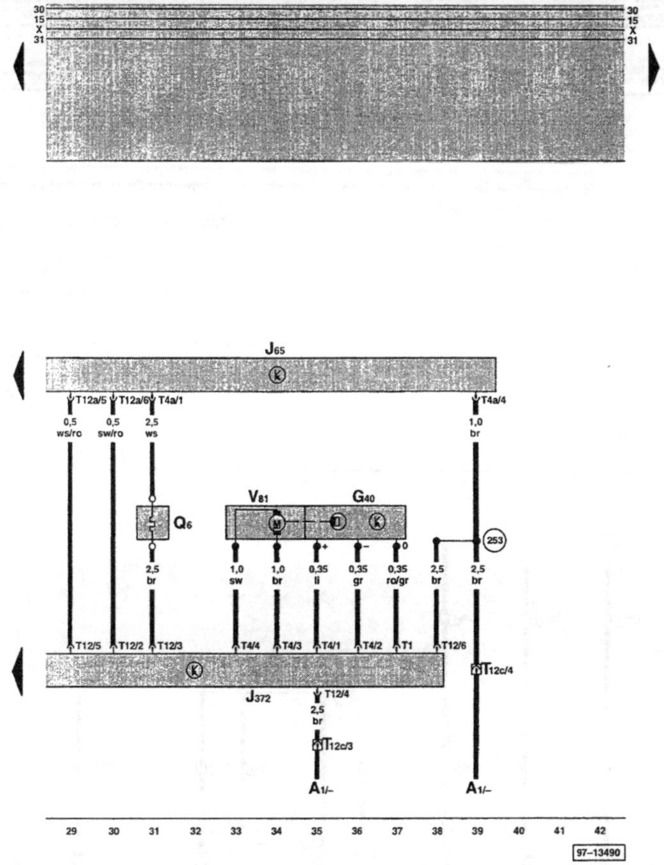

ws = white
sw = black
ro = red
br = brown
gn = green
bl = blue
gr = grey
li = violet
ge = yellow

A1 – Auxiliary Battery
G40 – Camshaft Position (CMP) Sensor
J65 – Heater Control Module
J372 – Circuit board for temperature sensor, control module for blower motor and glow plug
Q6 – glow plug, in heater unit
T1 – Single Connector
T4 – 4-Pin Connector
T4a – 4-Pin Connector
T12 – 12-Pin Conenctor
T12a – 12-Pin Conenctor
T12c – 12-Pin Conenctor, to intake channel
V81 – Fresh Air/Heater blower, in heater unit

(253) – Ground connection, in wiring harness heater unit

Heater control module, glow plug, fresh air/heater blower

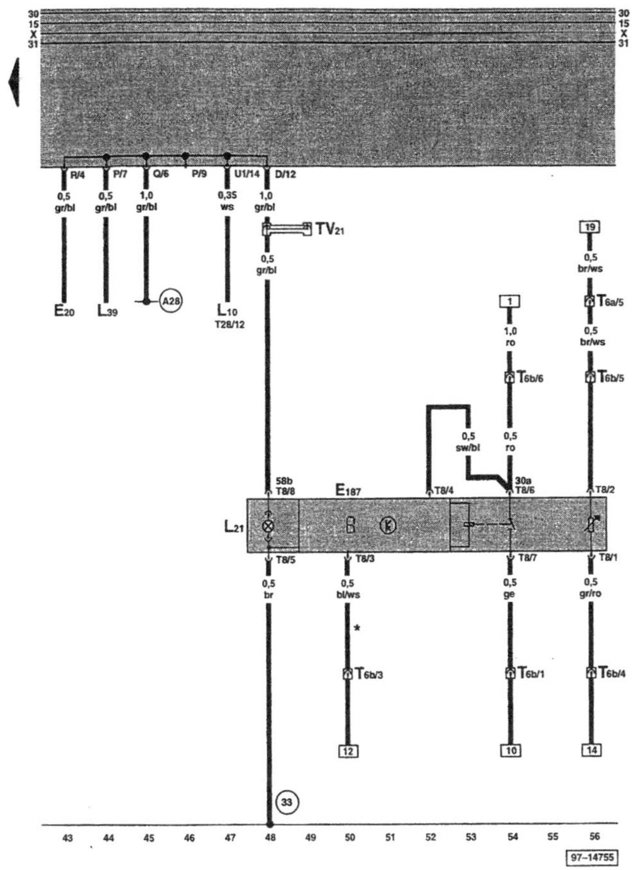

ws = white
sw = black
ro = red
br = brown
gn = green
bl = blue
gr = grey
li = violet
ge = yellow

E20 – Instrument Panel Light Dimmer Switch
E187 – Programable timer with temperature control
L10 – Instrument Cluster Light
L21 – Heater Control Lever Light
L39 – Rear Window Defogger Switch Light
T6a – 6-Pin Connector, below driver's seat
T6b – 6-Pin Connector, behind instrument panel, left
T8 – 8-Pin Connector
T28 – 28-Pin Connector, on instrument cluster
TV21 – Wire junction -2- for terminal 58b, above fuse/relay panel

(33) – Ground connection, behind instrument panel, right
(A28) – Plus connection (58b), in fresh air blower wiring harness

* – Data link wire

THIS PAGE INTENTIONALLY LEFT
BLANK

Fresh Air Blower front and rear with auxiliary heater core,

from May 1996

Deviate relay location and fuseplacements as well as the locations of multiple connectors- see section "component locations".

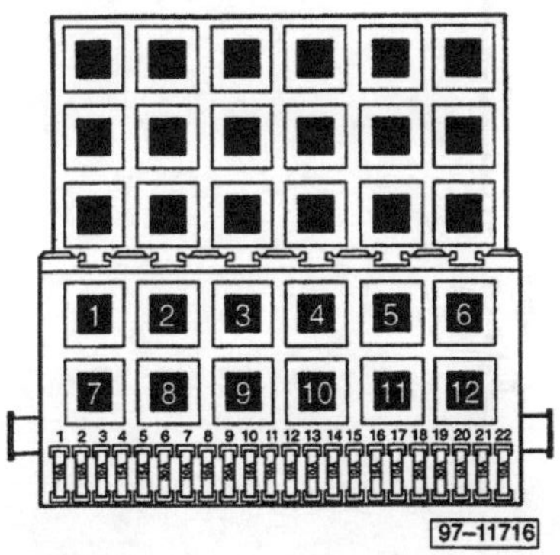

1 – Fresh Air Blower Relay (105)

Note:

Number in parentheses indicates production control number stamped on relay housing.

Fuse colors

30 A – green
25 A – white
20 A – yellow
15 A – blue
10 A – red
7,5 A – brown
5 A – beige
3 A – violet

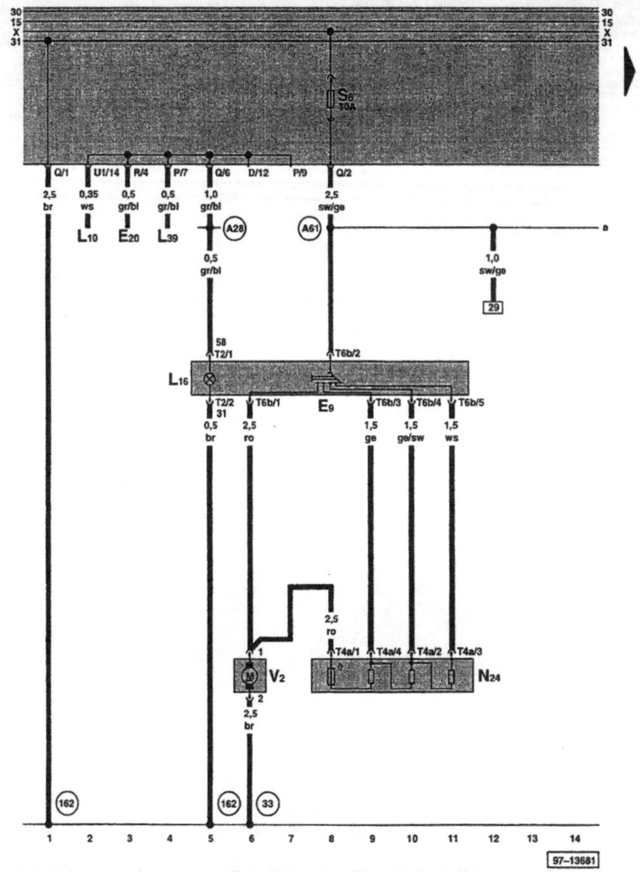

ws = white
sw = black
ro = red
br = brown
gn = green
bl = blue
gr = grey
li = violet
ge = yellow

E9 – Fresh Air Blower Switch
E20 – Instrument Panel Light Dimmer Switch
L10 – Instrument Cluster Light
L16 – Fresh Air Control Lever Light
L39 – Rear Window Defogger Switch Light
N24 – Fresh Air Blower Series Resistance with Fuse
T2 – Double Connector
T4a – 4-Pin Connector
T6b – 6-Pin Connector
V2 – Fresh Air Blower

(33) – Ground connection, behind instrument panel, right
(162) – Ground connection, in blower motor wiring harness

(A28) – Plus connection (58b), in fresh air blower wiring harness
(A61) – Plus Connection (X), in Instrument cluster wiring harness

Fresh air blower - front

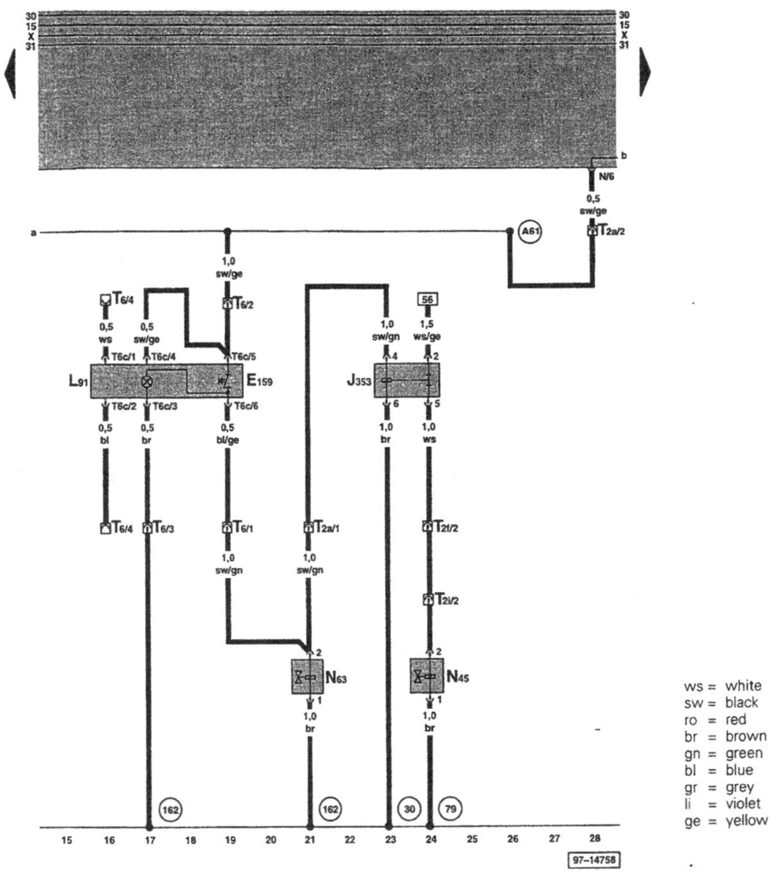

ws = white
sw = black
ro = red
br = brown
gn = green
bl = blue
gr = grey
li = violet
ge = yellow

E159 – Fresh Air/Recirculating Flap Switch
J353 – Relay for Fresh-air and recirculating air flap - rear, control number (94), behind instrument panel, center
L91 – Illumination for Switch for Fresh-air blower - rear
N45 – Two-way valve for Fresh-end recirculation air flap - rear
N63 – Fresh Air/Recirculating Flap Two-Way Valve
T2a – Double Connector, behind fuse/relay panel
T2f – Double Connector
T2i – Double Connector
T6 – 6-Pin Connector, behind instrument panel, center
T6c – 6-Pin Connector

(30) – Ground connection, -1-, beside fuse/relay panel

(79) – Ground connection, near D-pillar right, lower

(162) – Ground connection, in blower motor wiring harness

(A61) – Plus connection (X), in instrument cluster wiring harness

(L61) – Connector, in wiring harness 2. heater core

Fresh air/recirculating flap switch, Fresh air/recirculating two-way valve

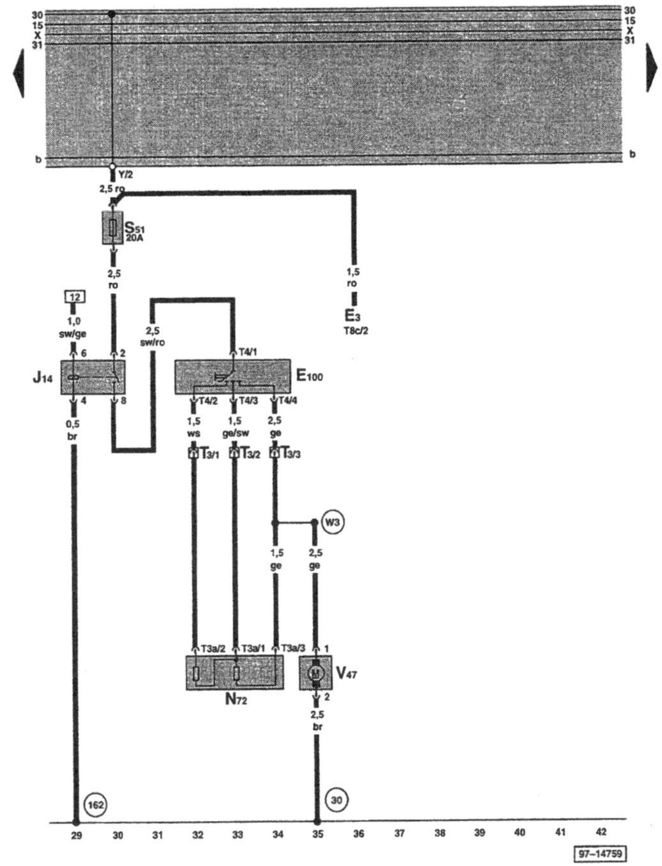

ws = white
sw = black
ro = red
br = brown
gn = green
bl = blue
gr = grey
li = violet
ge = yellow

E3 – Emergency Flasher Switch
E100 – Warm Air Blower Switch, in instrument panel
J14 – Heater Air Blower Relay, control number (53), behind instrument panel, center
N72 – Heater Blower Series Resistance
S51 – Rear Warm Air Fan Fuse
T3 – 3-Pin Connector, behind instrument panel, left
T3a – 3-Pin Connector
T4 – 4-Pin Connector
T8c – 8-Pin Connector
V47 – Rear Warm Air Fan

(30) – Ground connection, -1-, beside fuse/relay panel

(162) – Ground connection, in blower motor wiring harness

(W3) – Wire connection, in rear wiring harness

Auxiliary heater core for passenger's room

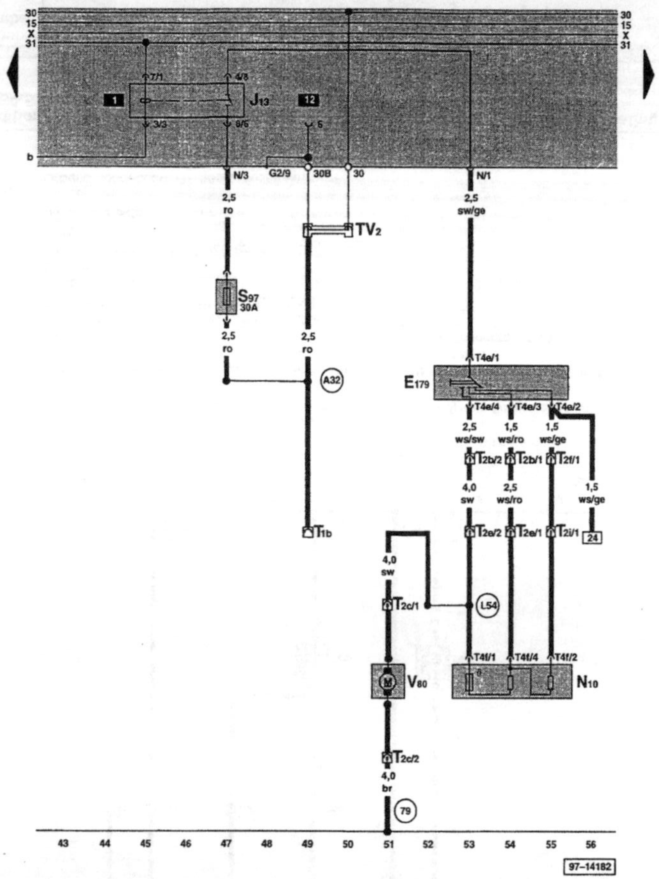

ws = white
sw = black
ro = red
br = brown
gn = green
bl = blue
gr = grey
li = violet
ge = yellow

- E179 – Switch for Freshair blower - rear
- J13 – Fresh Air Blower Relay, control number (105)
- N10 – Series Resistance for fresh air blower - rear
- S97 – Fresh Air Blower Fuse 1
- T1b – Single Connector, behind fuse/relay panel
- T2b – Double Connector, behind instrument panel, right
- T2c – Double Connector, behind trim D-pillar right, lower
- T2e – Double Connector
- T2f – Double Connector
- T2i – Double Connector
- T4e – 4-Pin Connector
- T4f – 4-Pin Connector
- TV2 – Terminal 30 (B+) Wire Connector
- V80 – Fresh air blower rear

- (79) – Ground connection, near D-pillar right, lower
- (A32) – Plus connection (30), in instrument panel wiring harness
- (L54) – Plus connector (Blower stage 3) in wiring harness blower - Passenger ventilation

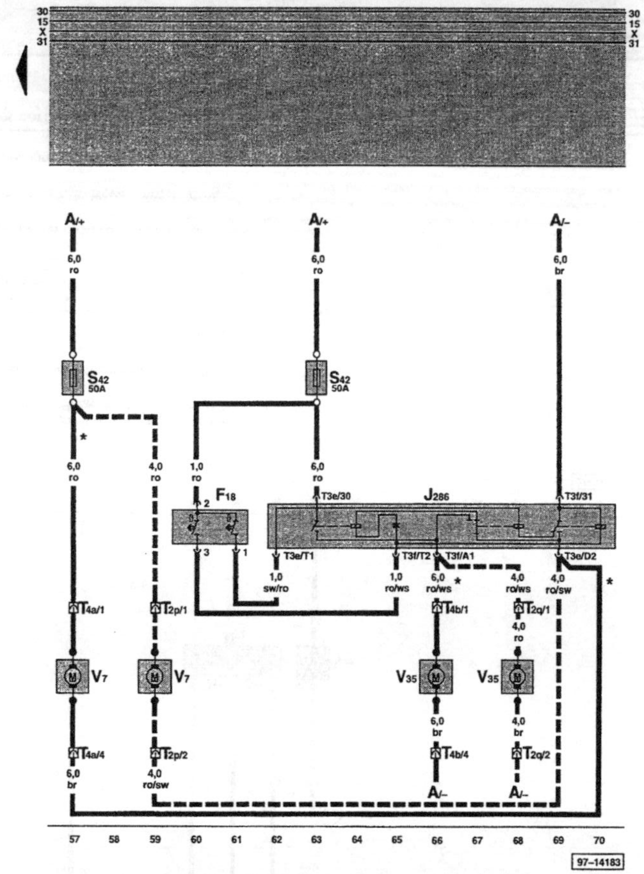

ws = white
sw = black
ro = red
br = brown
gn = green
bl = blue
gr = grey
li = violet
ge = yellow

- A – Battery
- F18 – Coolant Fan Control (FC) Thermal Switch
- J286 – Series/Parallel Coolant Fan Control (FC) Relay (V7, V35)
- S42 – Coolant Fan Fuse
- T2p – Double Connector, in engine compartment, front, only Coolant Fan (350 W)
- T2q – Double Connector, in engine compartment, front, only Coolant Fan (350 W)
- T3e – 3-Pin Connector
- T3f – 3-Pin Connector
- T4a – 4-Pin Connector, in engine compartment, front, only Coolant Fan (450 W)
- T4b – 4-Pin Connector, in engine compartment, front, only Coolant Fan (450 W)
- V7 – Left Coolant Fan
- V35 – Right Coolant Fan

- - - - only coolant fans (350 W)
- * only coolant fans (450 W)

Cruise Control (automatic transmission),

from May 1996

Deviate relay location and fuseplacements as well as the locations of multiple connectors-see section "component locations".

Fuse colors

- 30 A – green
- 25 A – white
- 20 A – yellow
- 15 A – blue
- 10 A – red
- 7,5 A – brown
- 5 A – beige
- 3 A – violet

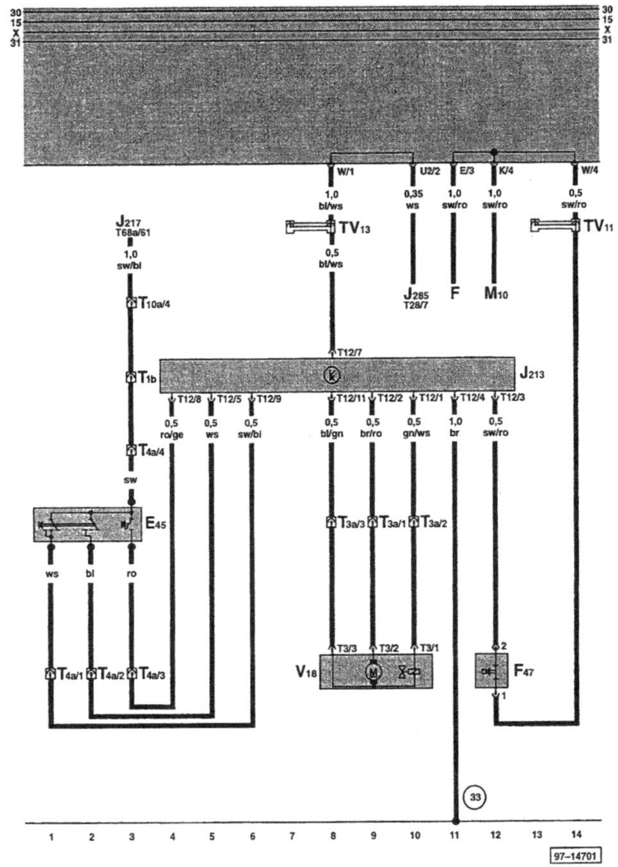

- ws = white
- sw = black
- ro = red
- br = brown
- gn = green
- bl = blue
- gr = grey
- li = violet
- ge = yellow

E45 – Cruise Control Switch
F – Brake Light Switch
F47 – Brake Vacuum Vent Valve Switch for cruise control
J213 – Cruise Control, Control Module, behind instrument panel, right
J217 – Transmission Control Module (TCM), on left A-pillar, lower part
J285 – Control module with indicator unit in instrument panel insert
M10 – Right Brake Light
T1b – Single Connector, behind fuse/relay panel
T3 – 3-Pin Connector
T3a – 3-Pin Connector, behind fuse/relay panel
T4a – 4-Pin Connector, behind steering column switch cover
T10a – 10-Pin Connector, behind fuse/relay panel
T12 – 12-Pin Connector
T28 – 28-Pin Connector, on instrument cluster
T68a – 68-Pin Connector, on Transmission Control Module (TCM)
TV11 – Terminal 54 Wire Connector, above fuse/relay panel
TV13 – Vehicle Speed Signal Wire Connector, above fuse/relay panel
V18 – Cruise Control Vacuum Pump
(33) – Ground connection, behind instrument panel, right

Automatic transmission,

from May 1996

- **2.8 L- Engine - Motronic Multiport Fuel Injection (MFI)/103 kW, code AES**

Deviate relay location and fuseplacements as well as the locations of multiple connectors- see section "component locations".

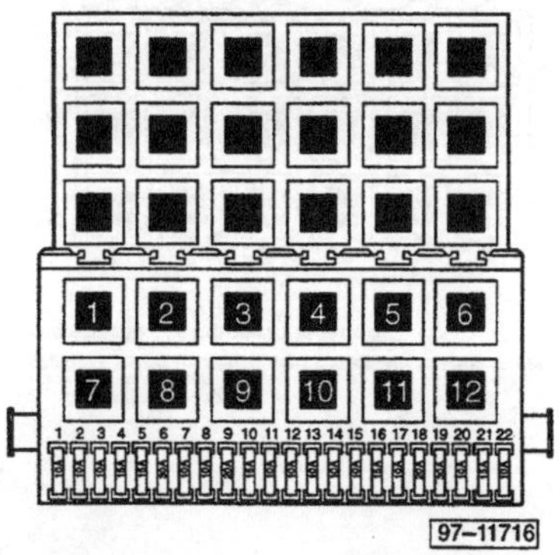

12 – Fuel Pump (FP) Relay (167)

Note:

Number in parentheses indicates production control number stamped on relay housing.

Fuse colors

30 A – green
25 A – white
20 A – yellow
15 A – blue
10 A – red
7,5 A – brown
5 A – beige
3 A – violet

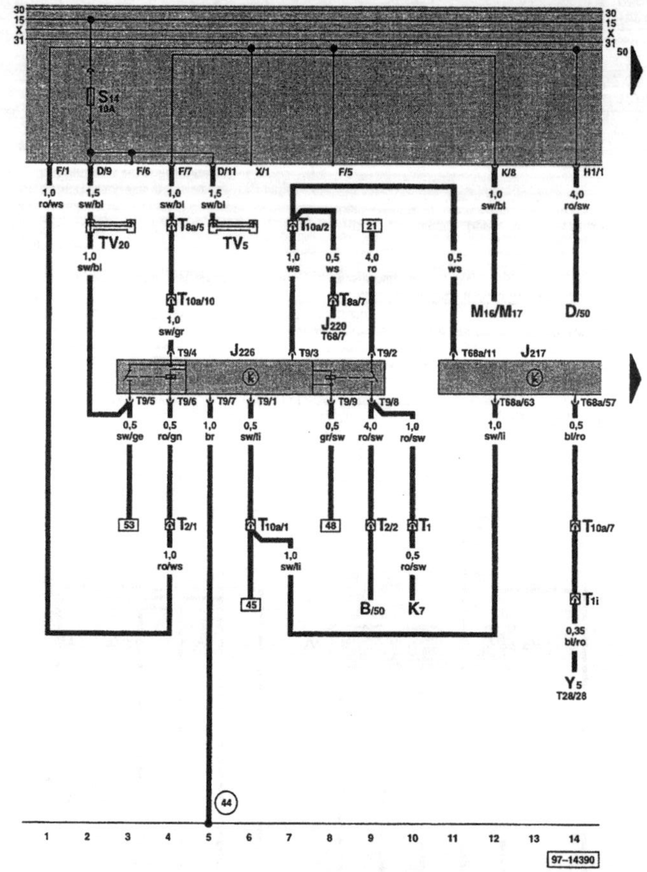

ws = white
sw = black
ro = red
br = brown
gn = green
bl = blue
gr = grey
li = violet
ge = yellow

B – Starter
D – Ignition/Starter Switch
J217 – Transmission Control Module (TCM), on left A-pillar, lower part
J220 – Motronic Engine Control Module (ECM), in engine compartment, left
J226 – Park/Neutral Position (PNP) Relay, behind instrument panel, center
K7 – Brake and Parking Brake Warning Light
M16 – Left Back-Up Light
M17 – Right Back-Up Light
T1 – Single Connector, behind fuse/relay panel
T1i – Single Connector, behind fuse/relay panel
T2 – Double Connector, behind fuse/relay panel
T8a – 8-Pin Connector, behind fuse/relay panel
T9 – 9-Pin Connector
T10a – 10-Pin Connector, behind fuse/relay panel
T28 – 28-Pin Connector, on instrument cluster
T68 – 68-Pin Connector, on Motronic Engine Control Module (ECM)
T68a – 68-Pin Connector
TV5 – Terminal 15a Wire Connector, above fuse/relay panel
TV20 – Wire junktion -2- for terminal 15a, above fuse/relay panel
Y5 – Transmission Range (TR) Selector Lever Display, in instrument cluster
(44) – Ground connection, on left A-pillar, lower part

Park neutral position (PNP) relay, transmission control module (TCM)

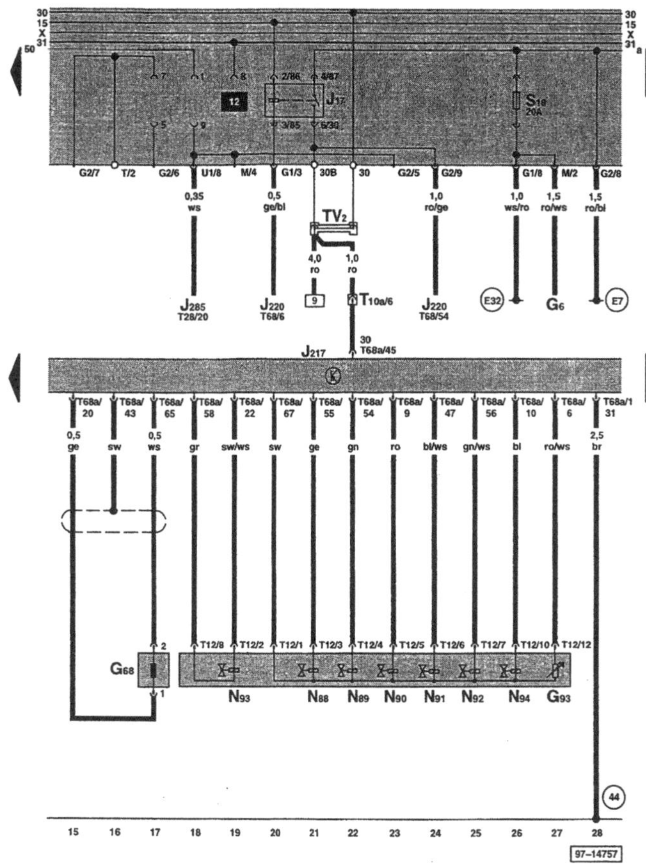

ws = white
sw = black
ro = red
br = brown
gn = green
bl = blue
gr = grey
li = violet
ge = yellow

G6 – Fuel Pump (FP)
G68 – Vehicle Speed Sensor (HO2S)
G93 – Transmission Fluid Temperature Sensor
J17 – Fuel Pump (FP) Relay
J217 – Transmission Control Module (TCM), on left A-pillar, lower part
J220 – Motronic Engine Control Module (ECM), in engine compartment, left
J285 – Control module with indicator unit in instrument panel insert
N88 – Solenoid valve 1
N89 – Solenoid valve 2
N90 – Solenoid valve 3
N91 – Solenoid valve 4
N92 – Solenoid valve 5
N93 – Solenoid valve 6
N94 – Solenoid valve 7

T10a – 10-Pin Connector, behind fuse/relay panel
T12 – 12-Pin Connector
T28 – 28-Pin Connector, on instrument cluster
T68 – 68-Pin Connector, on Motronic Engine Control Module (ECM)
T68a – 68-Pin Connector
TV2 – Terminal 30 (B+) Wire Connector

(44) – Ground connection, on left A-pillar, lower part
(E7) – Plus connection (87a), in Motronic Multiport Fuel Injection (MFI) wiring harness
(E32) – Connector (87), in wiring harness Motronic

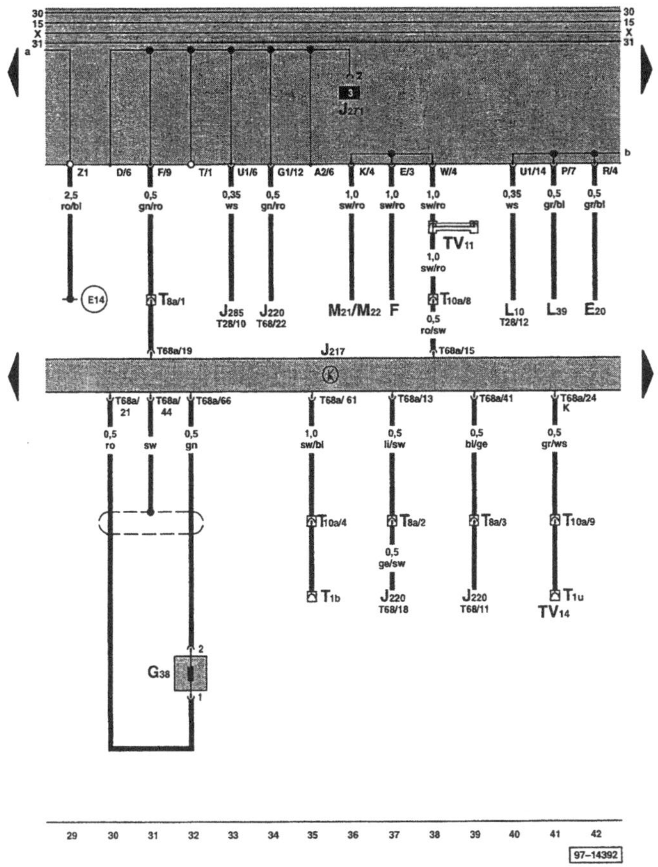

ws = white
sw = black
ro = red
br = brown
gn = green
bl = blue
gr = grey
li = violet
ge = yellow

E20 – Instrument Panel Light Dimmer Switch
F – Brake Light Switch
G38 – Transmission Vehicle Speed Sensor (VSS)
J217 – Transmission Control Module (TCM), on left A-pillar, lower part
J220 – Motronic Engine Control Module (ECM), in engine compartment, left
J271 – Motronic Engine Control Module (ECM) Power Supply Relay
J285 – Control module with indicator unit in instrument panel insert
L10 – Instrument Cluster Light
L39 – Rear Window Defogger Switch Light
M21 – Left Brake/Tail Light
M22 – Right Brake/Tail Light
T1b – Single Connector, behind fuse/relay panel

T1u – Single Connector
T8a – 8-Pin Connector, behind fuse/relay panel
T10a – 10-Pin Connector, behind fuse/relay panel
T28 – 28-Pin Connector, on instrument cluster
T68 – 68-Pin Connector, on Motronic Engine Control Module (ECM)
T68a – 68-Pin Connector
TV11 – Terminal 54 Wire Connector, above fuse/relay panel
TV14 – Data Link Connector (DLC) wire Connector

(E14) – Wire connection -1-, in Motronic Multiport Fuel Injection (MFI) wiring harness

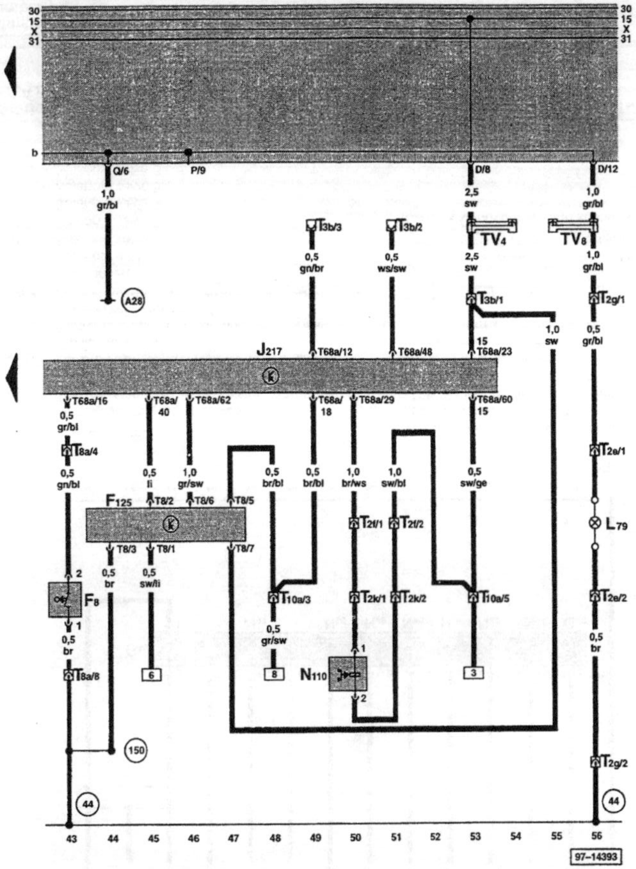

ws = white
sw = black
ro = red
br = brown
gn = green
bl = blue
gr = grey
li = violet
ge = yellow

F8 – Kick Down Switch
F125 – Multi-Function Transmission Range (TR) Switch
J217 – Transmission Control Module (TCM), on left A-pillar, lower part
L79 – Selector Lever Light
N110 – Shift Lock Solenoid
T2e – Double Connector, near selector lever
T2f – Double Connector, behind fuse/relay panel
T2g – Double Connector, behind fuse/relay panel
T2k – Double Connector, near selector lever
T3b – 3-Pin Connector, behind fuse/relay panel
T8 – 8-Pin Connector
T8a – 8-Pin Connector, behind fuse/relay panel
T10a – 10-Pin Connector, behind fuse/relay panel
T68a – 68-Pin Connector

TV4 – Terminal 15 Wire Connector, above fuse/relay panel
TV8 – Terminal 58b Wire Connector, above fuse/relay panel
(44) – Ground connection, on left A-pillar, lower part
(150) – Ground connection, in automatic transmission wiring harness
(A28) – Plus connection (58b), in fresh air blower wiring harness

Transmission control module (TCM), multi-function transmission range (TR) switch, kick down switch, shift lock solenoid

THIS PAGE INTENTIONALLY LEFT
BLANK

Daytime running lights (DRL) and headlight washer system

from May 1996

Deviate relay location and fuseplacements as well as the locations of multiple connectors- see section "component locations".

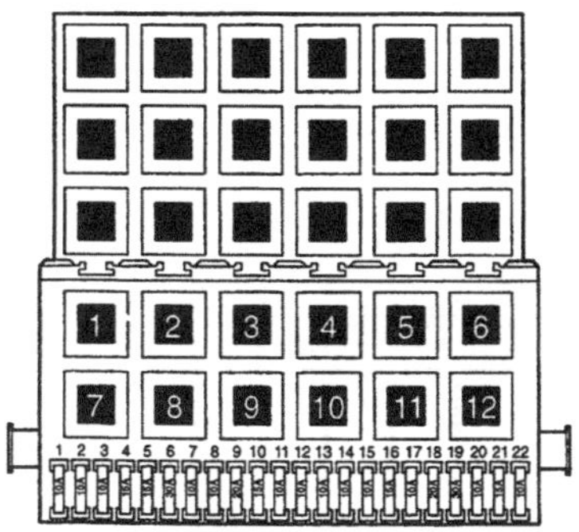

97–11716

6 – Emergency Flasher Relay (21)

7 – Headlight Washer System Relay (95)

8 – Program. Washer/Wiper Intermittent Relay (99)

Note:

Number in parentheses indicates production control number stamped on relay housing.

Fuse colors

30 A – green
25 A – white
20 A – yellow
15 A – blue
10 A – red
7,5 A – brown
5 A – beige
3 A – violet

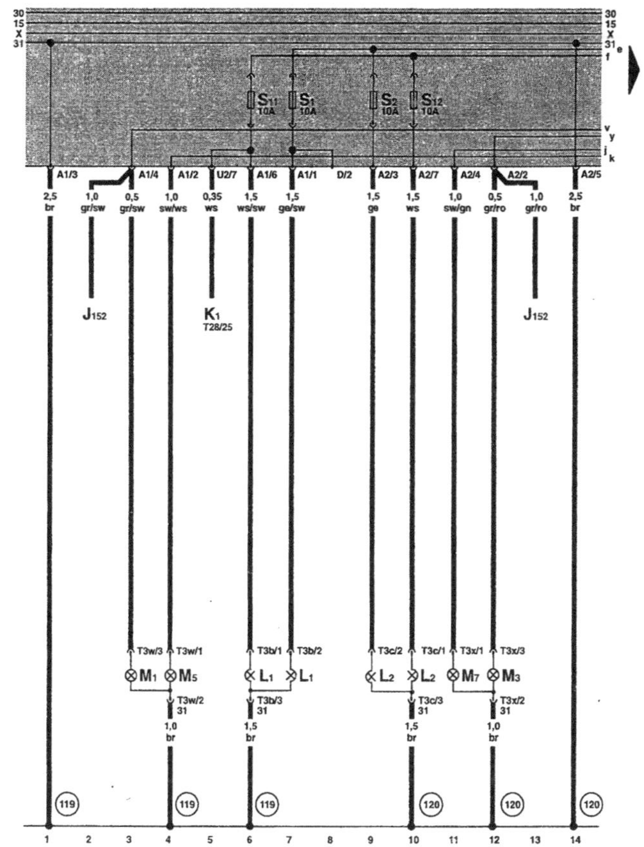

97–14670

ws = white
sw = black
ro = red
br = brown
gn = green
bl = blue
gr = grey
li = violet
ge = yellow

J152 – Parking Light Warning Buzzer
K1 – Headlight High Beam Indicator Light
L1 – Left Headlight
L2 – Right Headlight
M1 – Left Parking Light
M3 – Right Parking Light
M5 – Left Front Turn Signal Light
M7 – Right Front Turn Signal Light
T3b – 3-Pin Connector, behind headlight, left
T3c – 3-Pin Connector, behind headlight, right
T3w – 3-Pin Connector, behind headlight, left
T3x – 3-Pin Connector, behind headlight, right
T28 – 28-Pin Connector

(119) – Ground connection -1-, in headlight wiring harness
(120) – Ground connection -2-, in headlight wiring harness

Headlights, parking light, front turn signal lights

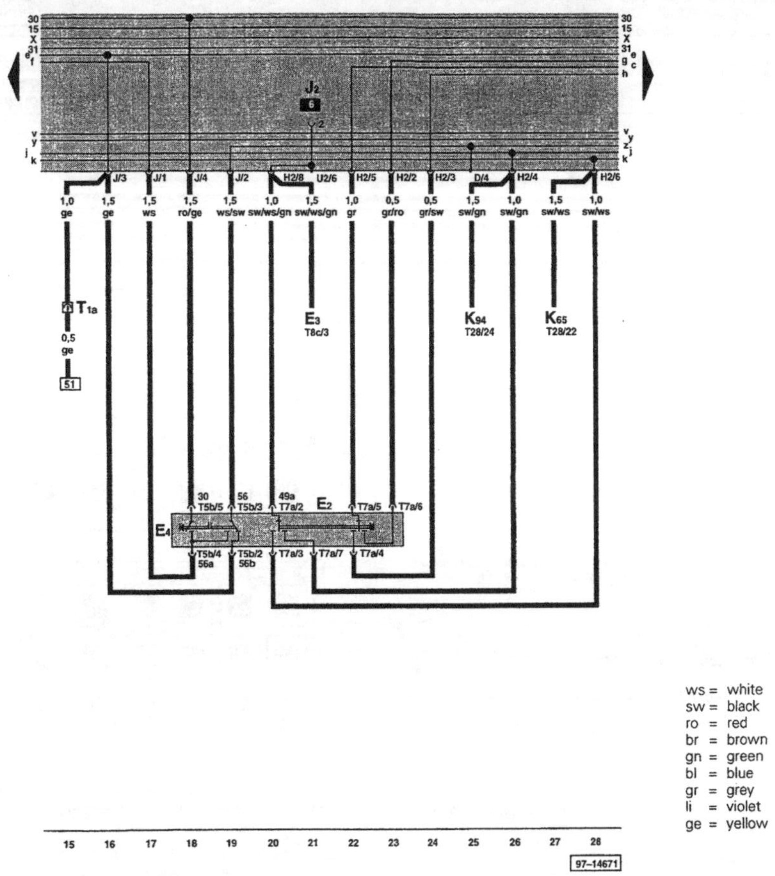

ws = white
sw = black
ro = red
br = brown
gn = green
bl = blue
gr = grey
li = violet
ge = yellow

- E2 – Turn Signal Switch
- E3 – Emergency Flasher Switch
- E4 – Headlight Dimmer/Flasher Switch
- J2 – Emergency Flasher Relay
- K65 – Left Turn Signal Indicator Light
- K94 – Right Turn Signal Indicator Light
- T1a – Single Connector, behind instrument panel, left
- T5b – 5-Pin Connector, behind steering column switch cover
- T7a – 7-Pin Connector, behind steering column switch cover
- T8c – 8-Pin Connector
- T28 – 28-Pin Connector

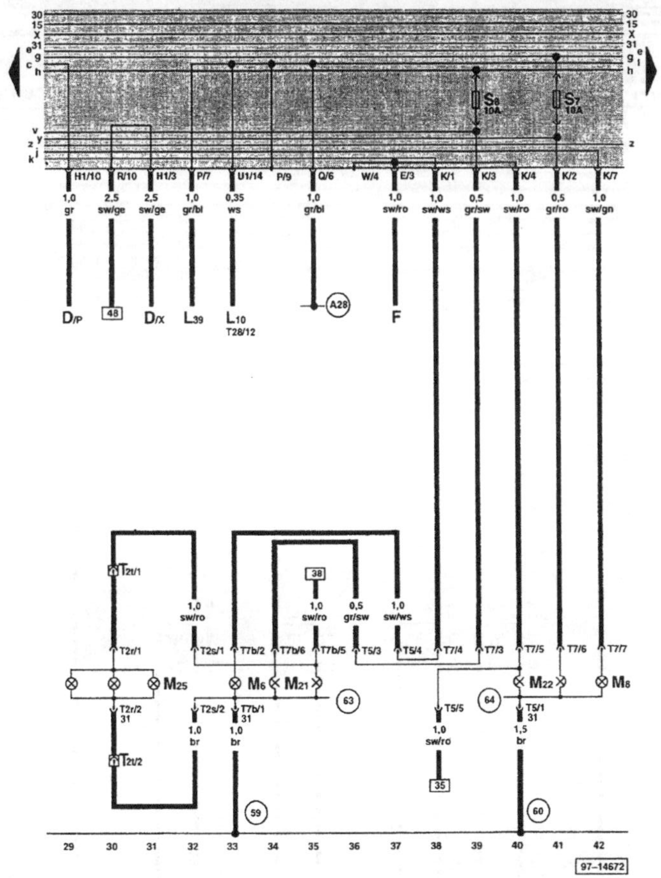

ws = white
sw = black
ro = red
br = brown
gn = green
bl = blue
gr = grey
li = violet
ge = yellow

- D – Ignition/Starter Switch
- F – Brake Light Switch
- L10 – Instrument Cluster Light
- L39 – Rear Window Defogger Indicator Light
- M6 – Left Rear Turn Signal Light
- M8 – Right Rear Turn Signal Light
- M21 – Left Brake/Tail Light
- M22 – Right Brake/Tail Light
- M25 – High-mount Brake Light
- T2r – Double Connector, in rear lid
- T2s – Double Connector, on taillight, left
- T2t – Double Connector, in rear lid
- T5 – 5-Pin Connector, on taillight, right
- T7 – 7-Pin Connector, on taillight, right
- T7b – 7-Pin Connector, on taillight, left
- T28 – 28-Pin Connector

- (59) – Ground connection, near taillight, left
- (60) – Ground point -2- near taillamp assembly, right
- (63) – Ground connection, on left taillight bulb holder
- (64) – Ground connection, on right taillight bulb holder
- (A28) – Plus connection (58b), in fresh air blower wiring harness

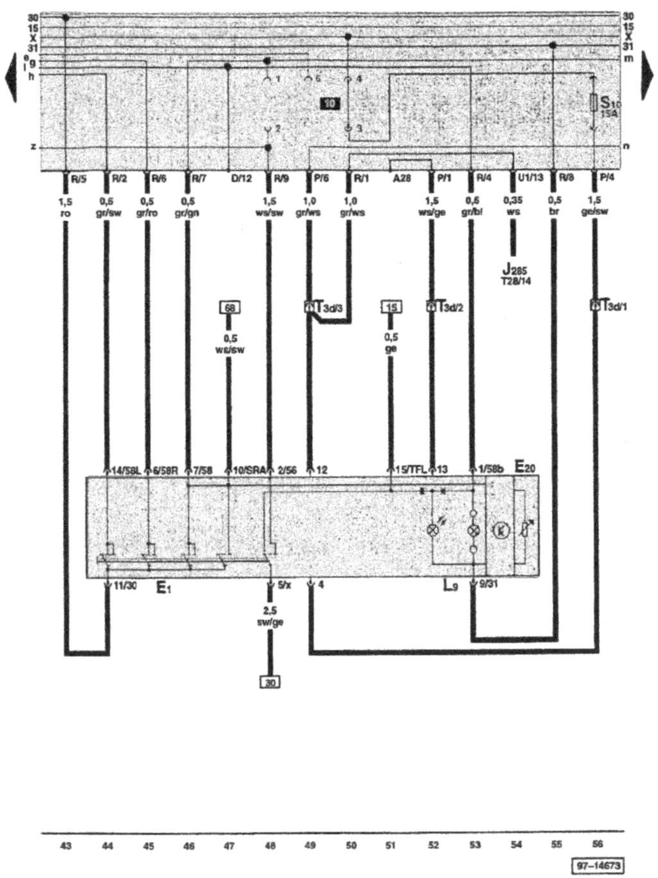

ws = white
sw = black
ro = red
br = brown
gn = green
bl = blue
gr = grey
li = violet
ge = yellow

E1 – Light switch
E20 – Instrument Panel Light Dimmer Switch
J285 – Control module with indicator unit in instrument panel insert
L9 – Headlight Switch Light
T3d – 3-Pin Connector, behind fuse/relay panel
T28 – 28-Pin Connector

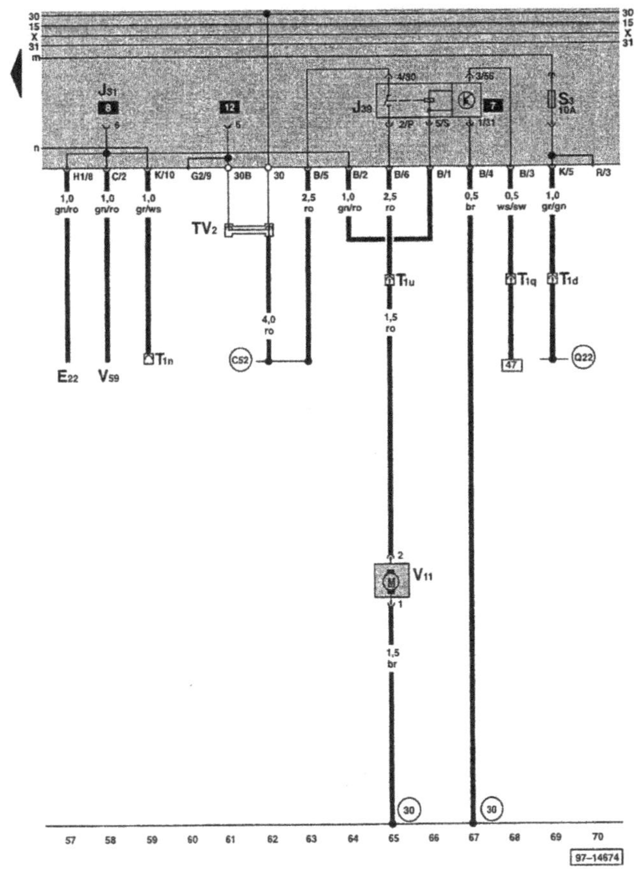

ws = white
sw = black
ro = red
br = brown
gn = green
bl = blue
gr = grey
li = violet
ge = yellow

E22 – Windshield Wiper Intermittent Switch
J31 – Wiper/Washer Intermittent Relay
J39 – Headlight Washer System Relay
T1d – Single Connector, near tail lamp assembly
T1n – Single Connector, on taillight, right
T1q – Single Connector, behind fuse/relay panel
T1u – Single Connector, behind fuse/relay panel
TV2 – Terminal 30 (B+) Wire Connector
V11 – Headlight Washer Pump
V59 – Windshield and Rear Window Washer Pump

(30) – Ground connection, -1-, next to fuse/relay panel
(C52) – Plus connector (30), in wiring harness Head washer assembly
(Q22) – Wire connection -1-, in rear lid wiring harness

Heated seats,

from May 1996

Deviate relay location and fuseplacements as well as the locations of multiple connectors- see section "component locations".

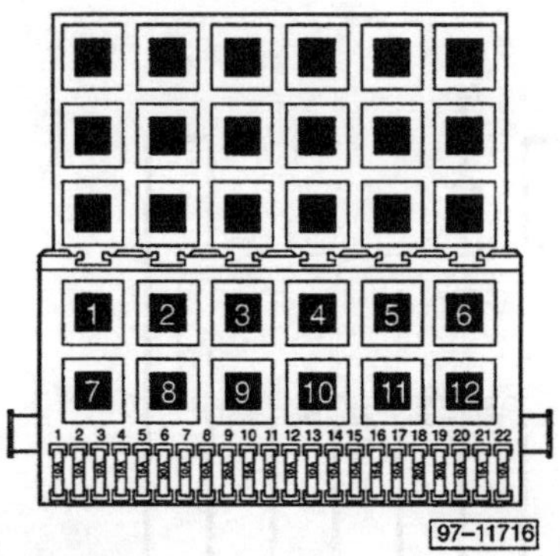

Fuse colors

30 A - green
25 A - white
20 A - yellow
15 A - blue
10 A - red
7,5 A - brown
5 A - beige
3 A - violet

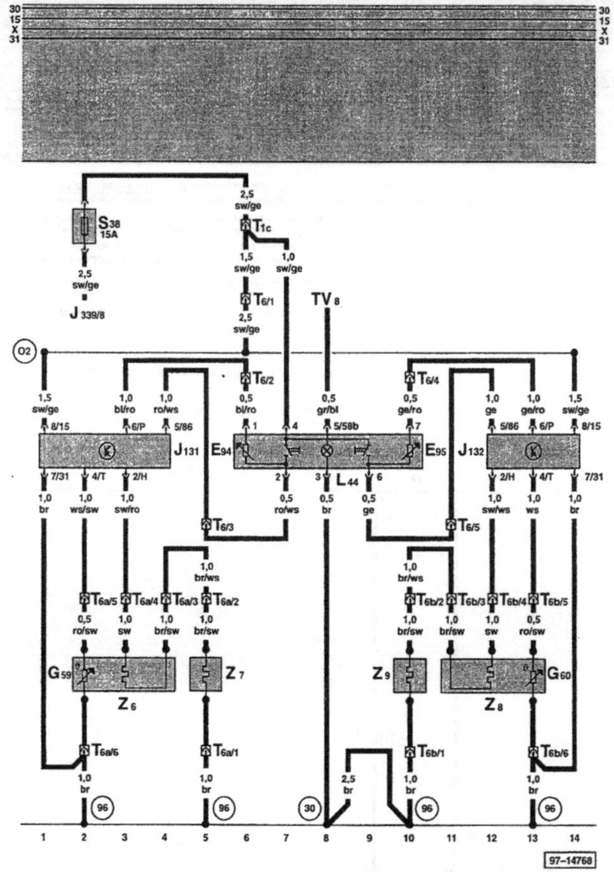

ws = white
sw = black
ro = red
br = brown
gn = green
bl = blue
gr = grey
li = violet
ge = yellow

E94 – Driver's Seat Heat Regulating Switch
E95 – Passenger's Seat Heat Regulating Switch
G59 – Driver Seat Heater Temperature Sensor
G60 – Passenger Seat Heater Temperature Sensor
J131 – Driver's Seat Heater Control Module, below driver's seat
J132 – Passenger's Seat Heater Control Module, below driver's seat
J339 – Load Reduction Relay II, above fuse/relay panel
L44 – Seat Heater Switch Illumination Light
S38 – Heated Seats Fuse, below fuse/relay panel
T1c – Single Connector, behind fuse/relay panel
T6 – 6-Pin Connector, behind fuse/relay panel
T6a – 6-Pin Connector, below driver's seat
T6b – 6-Pin Connector, below passenger's seat
TV8 – Terminal 58b Wire Connector, above fuse/relay panel
Z6 – Driver's Seat Heat Element
Z7 – Driver's Backrest Heat Element
Z8 – Passenger's Seat Heat Element
Z9 – Passenger's Backrest Heat Element
(30) – Ground connection, beside fuse/relay panel
(96) – Ground connection -1-, in heated seats wiring harness
(02) – Plus connection (15), in heated seats wiring harness

2.8 Liter engine (code AES) from Sept. 1997 to Aug. 1998

2.4 Liter engine (code AAB) from May 1996

1998 EuroVan additional electrical wiring diagrams

THIS PAGE INTENTIONALLY LEFT

BLANK

Control modules and relays

Table of contents:

Climatronic Control Module

Component location: Behind the A/C Control Head

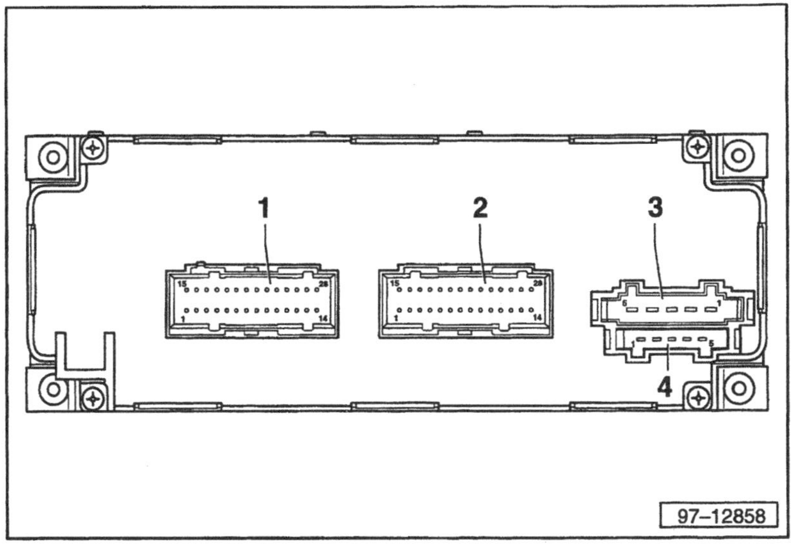

Connectors:

1	=	T28a – 28-Pin Connector, for red Connector
2	=	T28b – 28-Pin Connector, for white Connector
3	=	T5e – 5-Pin Connector
4	=	Not used

Front Power Window Control Module

Component location: In driver's and passenger's door

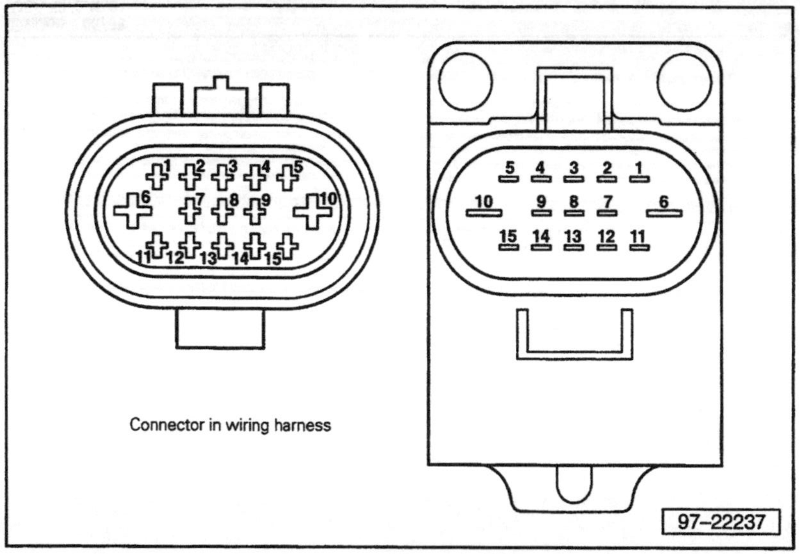

Power Sunroof Control Module

Component location: Behind Interior light front

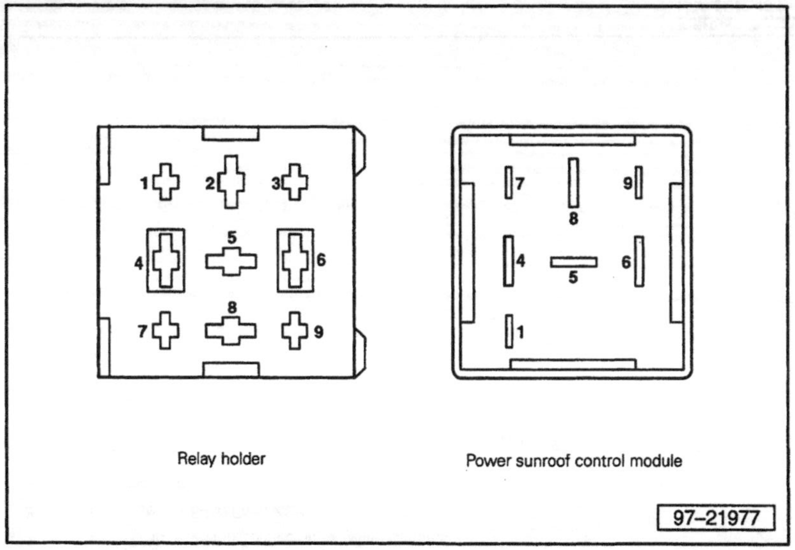

ABS Control Module (w/EDL), 31-Pin Connector

Component location: In engine compartment, left front

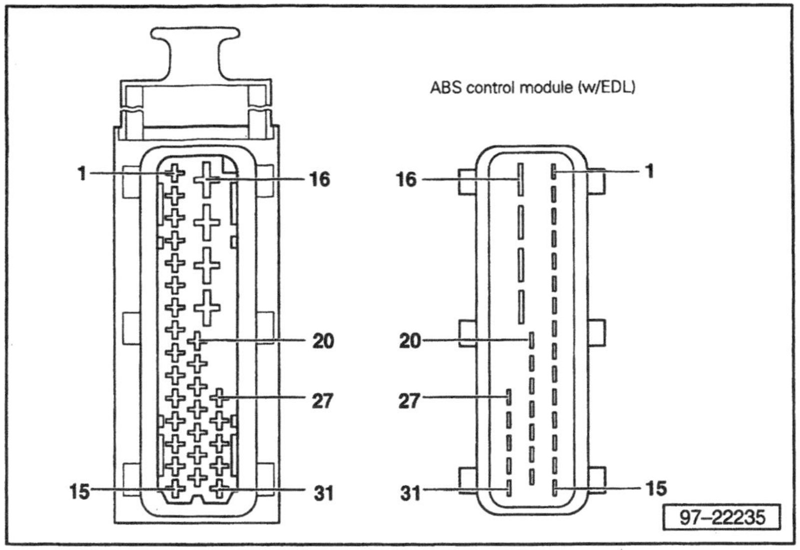

Coolant Fan Control (FC) Relay, for vehicles with Air conditioning or Climatronic

Component location: In engine compartment, left front

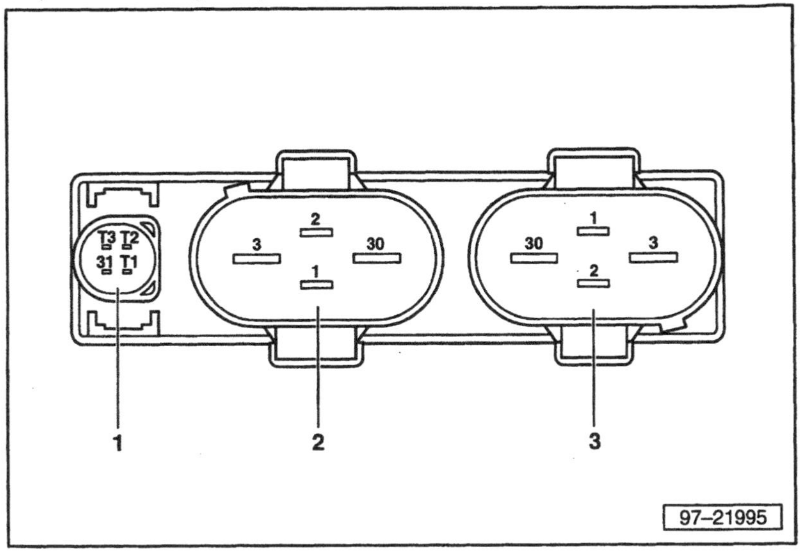

Connectors for vehicles with Air conditioning:

Connector 3:			Connector 2:			Connector 1:		
3	=	T4d/1	3	=	T4b/1	T1	=	T4c/1
2	=	T4d/2	2	=	T4b/2	T2	=	T4c/2
1	=	T4d/3	1	=	T4b/3	31	=	T4c/3
30	=	T4d/4	30	=	T4b/4	T3	=	T4c/4

Connectors for vehicles with Climatronic:

Connector 3:			Connector 2:			Connector 1:		
3	=	T4a/1	3	=	T4b/1	T1	=	T4c/1
2	=	T4a/2	2	=	T4b/2	T2	=	T4c/2
1	=	T4a/3	1	=	T4b/3	31	=	T4c/3
30	=	T4a/4	30	=	T4b/4	T3	=	T4c/4

Cruise Control Control Module

Component location: Behind instrument panel, right

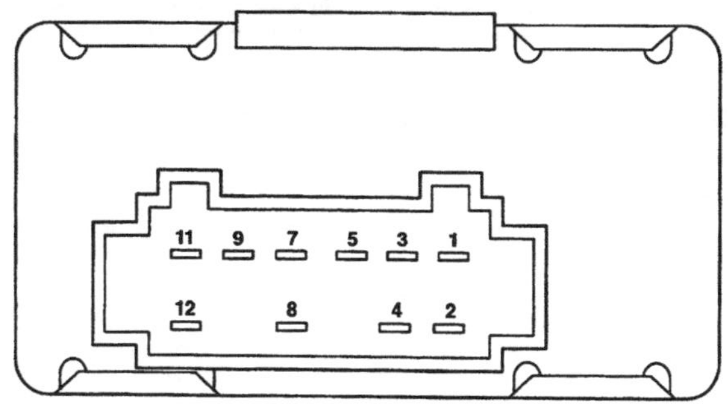

THIS PAGE INTENTIONALLY LEFT

BLANK

Edition 02/98
USA.5272.10.21

Multiple-pin connectors

Table of contents:

Automatic Transmission

Solenoid valves and Transmission Fluid Temperature Sensor

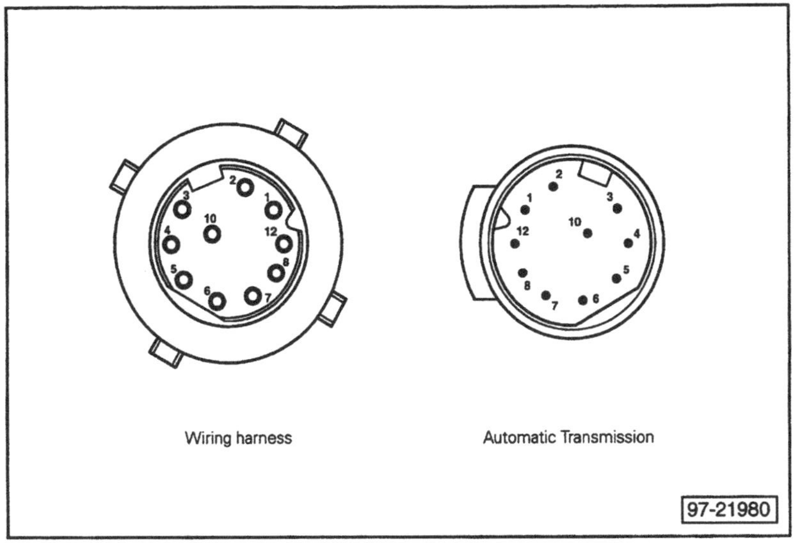

Multi-Function Transmission Range (TR) Switch

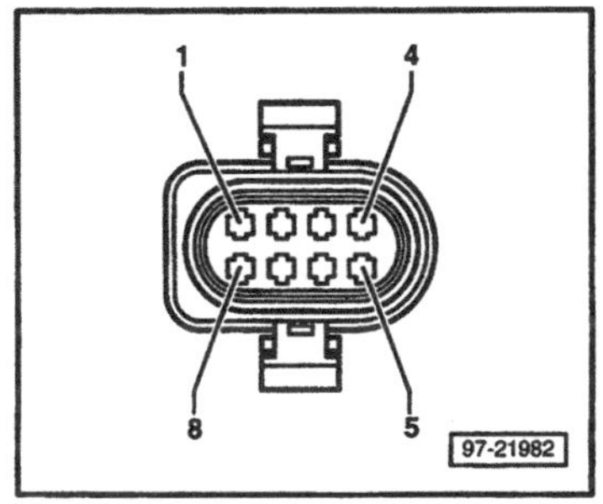

Interior lights, switches, Anti-theft protection

Light switch

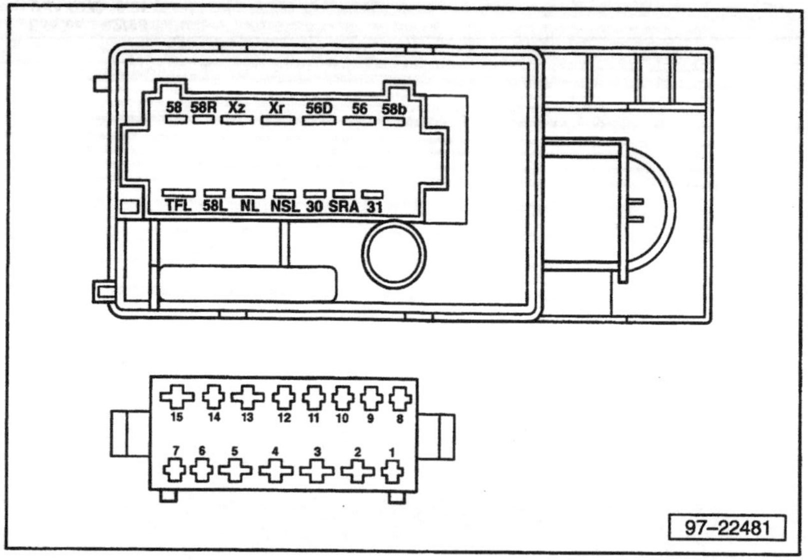

Mirror Adjustment Switch and Mirror Selector Switch

A – Wiring harness

B – Mirror adjustment switch and mirror selector switch

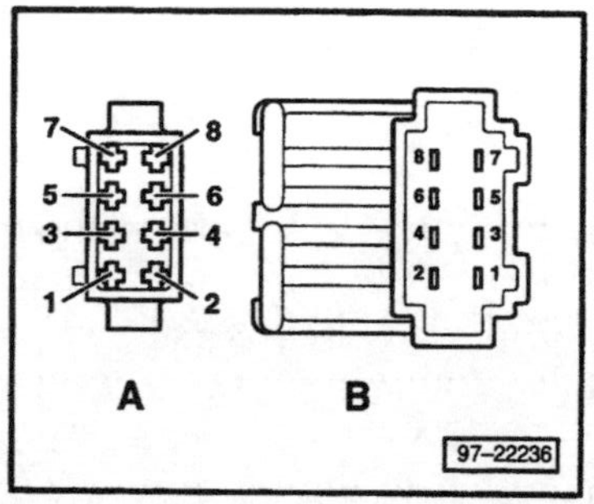

Climatronic

24-Pin Connector

Component location: Behind instrument panel

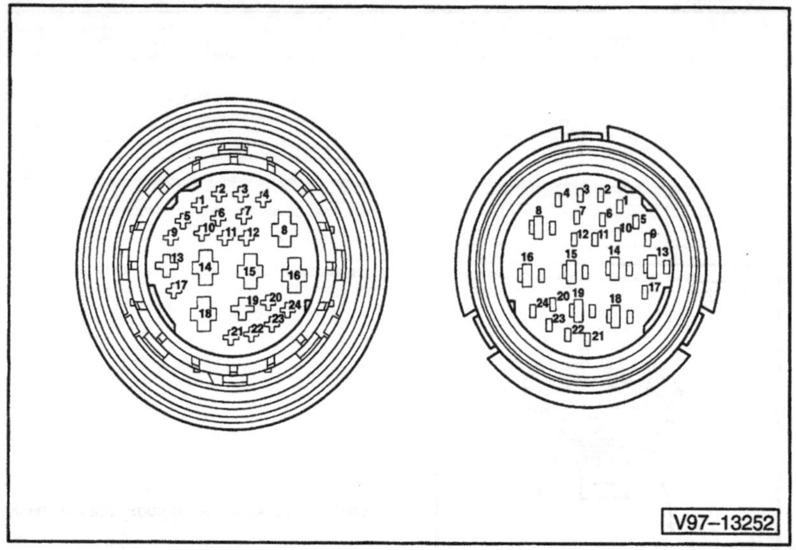

On Board Diagnostic

Data Link Connector (DLC) 16-Pin Connector

Component location: In instrument panel, center

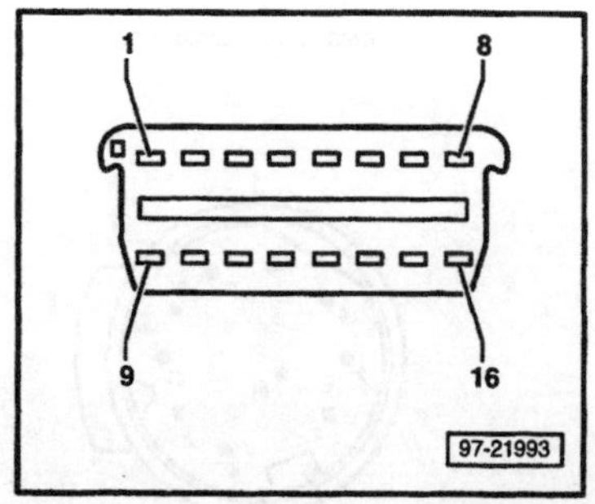

Heated seats,

from September 1997

Devite relay location and fuseplacements as well as locations of multiple connectors - see section "component locations".

Fuse colors

- 30 A – green
- 25 A – white
- 20 A – yellow
- 15 A – blue
- 10 A – red
- 7,5 A – brown
- 5 A – beige
- 3 A – violet

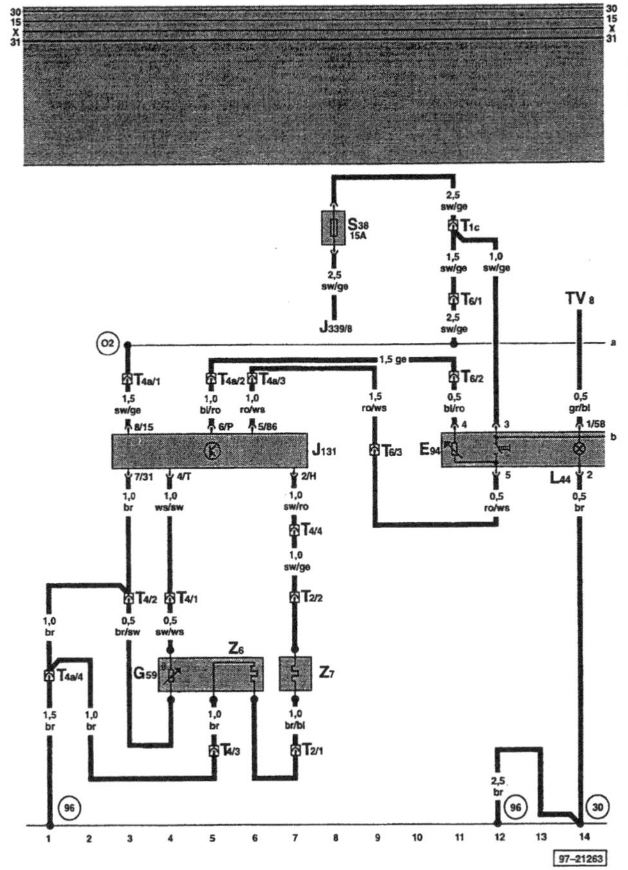

ws = white
sw = black
ro = red
br = brown
gn = green
bl = blue
gr = grey
li = violet
ge = yellow

- E94 – Adjuster for heated driver's seat
- G59 – Driver's Heated Seat Temperature Element
- J131 – Driver's Heated Seat Controle Module, below driver's seat
- J339 – Load Reduction relay II
- L44 – Heated Switch Illumination Light
- S38 – Heated Seats Fuse, above fuse/relay panel
- T1c – Single Connector, behind fuse/relay panel
- T2 – Double Connector
- T4 – 4-Pin Connector, below driver's seat
- T4a – 4-Pin Connector
- T6 – 4-Pin Connector, behind fuse/relay panel
- TV8 – Terminal 58b Wire Connector, above fuse/relay panel
- Z6 – Heated Driver's Seat
- Z7 – Heated Driver's Backrest

- (30) – Ground connection, -1-, beside fuse/relay panel
- (96) – Ground connection, -1-, in heated seats wiring harness
- (O2) – plus connection (15), in heated seats wiring harness

Driver's heated seat control module, adjuster for heated driver's seat, driver's seat heating element, driver's backrest heating element

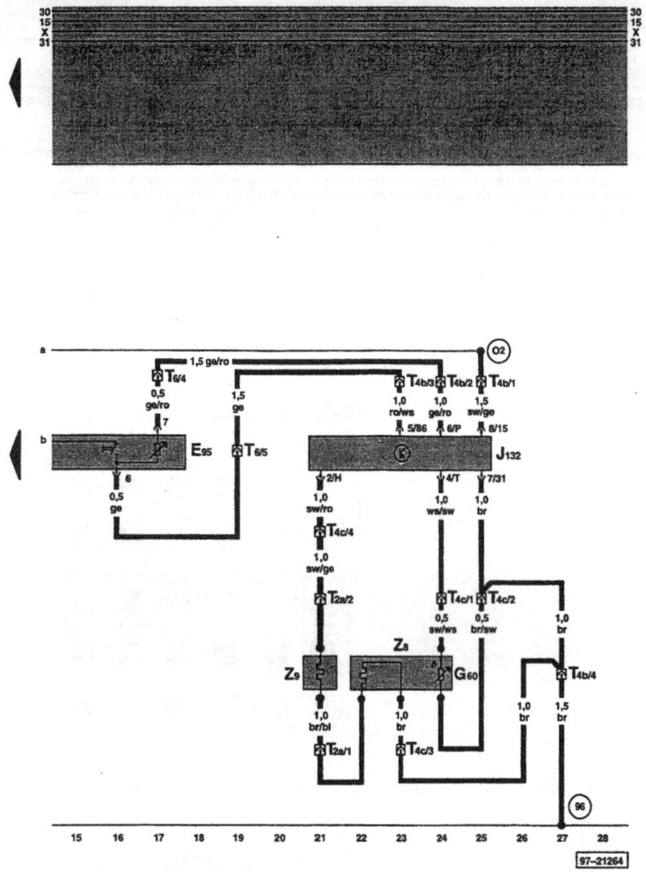

ws = white
sw = black
ro = red
br = brown
gn = green
bl = blue
gr = grey
li = violet
ge = yellow

E95 – Adjuster for heated passenger's seat
G60 – Passenger's Heated Seat Temperature Element
J132 – Passenger's Heated Seat Controle Module, below fuse/relay panel
T2a – Double Connector
T4b – 4-Pin Connector
T4c – 4-Pin Connector, below passenger's seat
T6 – 6-Pin Connector, behind fuse/relay panel
Z8 – Heated Passenger's Seat
Z9 – Heated Passenger's Backrest

(96) – Ground connection, -1-, in heated seats wiring harness
(O2) – Plus connection, in heated seats wiring harness

Passenger's heated seat control module, adjuster for heated passenger's seat, passenger's seat heating element, passenger's backrest heating element

THIS PAGE INTENTIONALLY LEFT
BLANK

Swivel seat lock,

from September 1997

Devite relay location and fuseplacements as well as locations of multiple connectors - see section "component locations".

Fuse colors

30 A – green
25 A – white
20 A – yellow
15 A – blue
10 A – red
7,5 A – brown
5 A – beige
3 A – violet

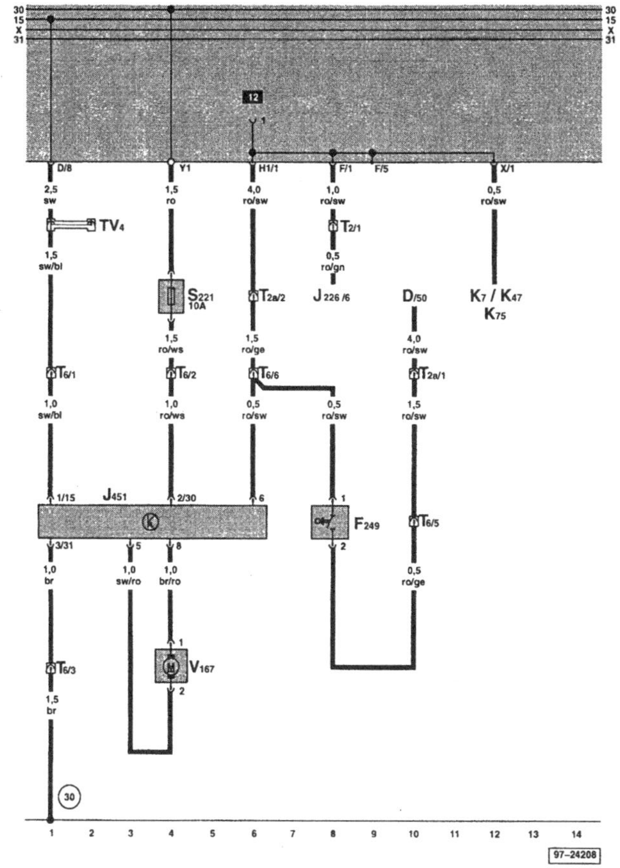

ws = white
sw = black
ro = red
br = brown
gn = green
bl = blue
gr = grey
li = violet
ge = yellow

D – Ignition/Starter Switch
F249 – End switch for swivel seat lock, below passenger's seat
J226 – Park/Neutral Position (PNP) Relay, behind instrument panel, center
J451 – Control module for swivel seat lock, below passenger's seat
K7 – Brake and Parking Brake Warning Light
K47 – ABS Warning Light
K75 – Airbag Malfunction Indicator Lamp (MIL)
S221 – Fuse for Swivel Seat lock, below fuse/relay panel
T2 – Double Connector, behind fuse/relay panel
T2a – Double Connector, behind steering column switch cover
T6 – 6-Pin Connector, below passenger's seat
TV4 – Terminal 15 Wire Connector, above fuse/relay panel
V167 – Motor for swivel seat lock, below passenger's seat

(30) – Ground connection, -1-, beside fuse/relay panel

Control module for swivel seat lock, end switch for swivel seat lock, motor for swivel seat lock

Rear window wiper,

from September 1997

Devite relay location and fuseplacements as well as locations of multiple connectors - see section "component locations".

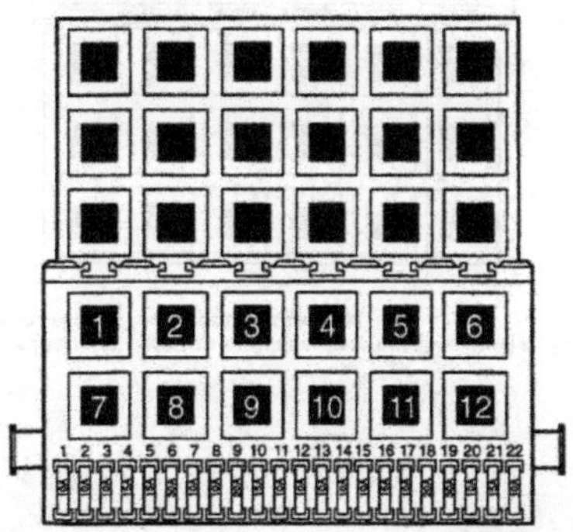

97-11716

8 Program. Washer/Wiper Intermittent Relay (99)

Note: Number in parentheses indicates production control number stamped on relay housing.

Fuse colors

30 A – green
25 A – white
20 A – yellow
15 A – blue
10 A – red
7,5 A – brown
5 A – beige
3 A – violet

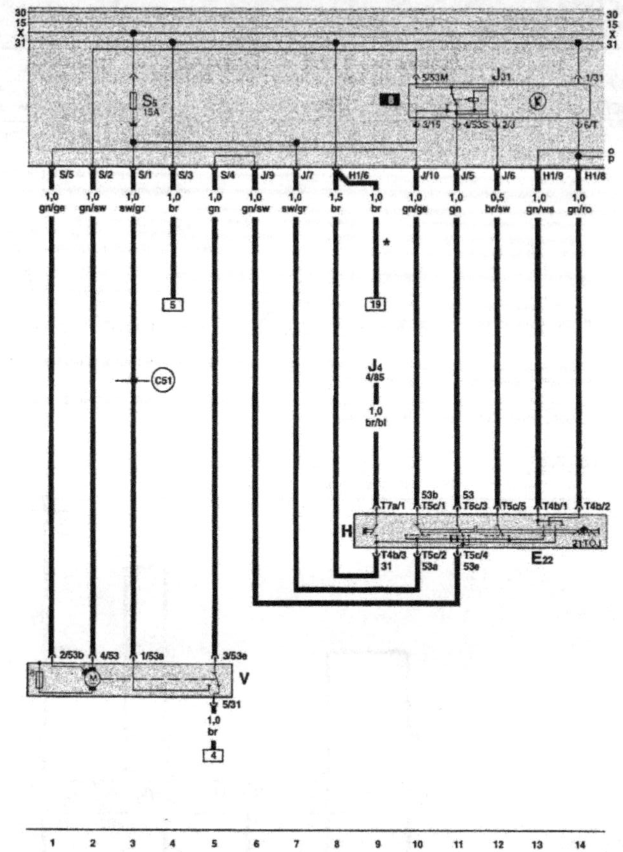

97-20538

ws = white
sw = black
ro = red
br = brown
gn = green
bl = blue
gr = grey
li = violet
ge = yellow

E22 – Windshield Wiper/Washer Switch
H – Horn Button
J4 – Dual Horn Relay
J31 – Wiper/Washer Intermittent Relay
T4b – 4-Pin Connector, behind steering column switch cover
T5c – 5-Pin Connector, behind steering column switch cover
T7a – 7-Pin Connector, behind steering column switch cover
V – Windshield Wiper Motor
(C51) – Plus connection (X), in wiper wiring harness
* – not in vehicles with Multifunction Indicator (see also wiring diagram for Multifunction Indicator)

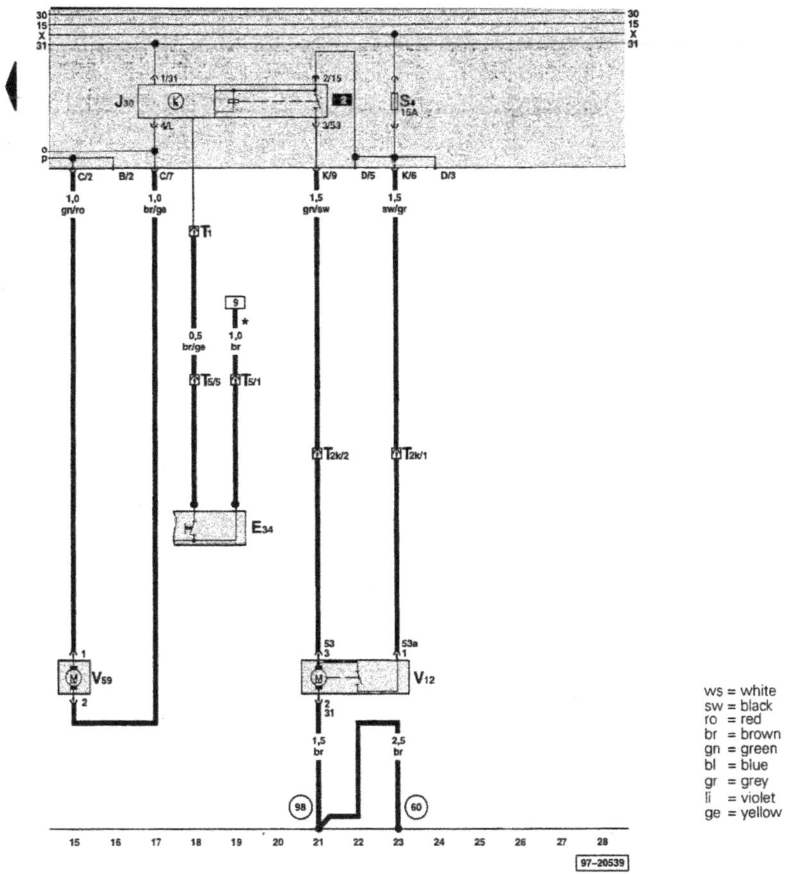

ws = white
sw = black
ro = red
br = brown
gn = green
bl = blue
gr = grey
li = violet
ge = yellow

- E34 – Rear Window Wiper Switch (interval)
- J30 – Rear Window Wiper/Washer Relay
- T1 – Sinlge Connector, on Rear Window Wiper/Washer Relay
- T2k – Double Connector, Near tail lamp assembly
- T5 – 5-Pin Connector, behind steering column switch cover
- V12 – Rear Window Wiper Motor
- V59 – Windshield and Rear Window Washer Pump
- (60) – Ground point -2- near tail lamp assembly
- (98) – Ground connection, in rear lid wiring harness

* – not in vehicles with Multifunction Indicator (see also wiring diagram for Multifunction Indicator)

THIS PAGE INTENTIONALLY LEFT
BLANK

12V outlets,

from September 1997

Devite relay location and fuseplacements as well as locations of multiple connectors - see section "component locations".

Fuse colors

30 A – green
25 A – white
20 A – yellow
15 A – blue
10 A – red
7,5 A – brown
5 A – beige
3 A – violet

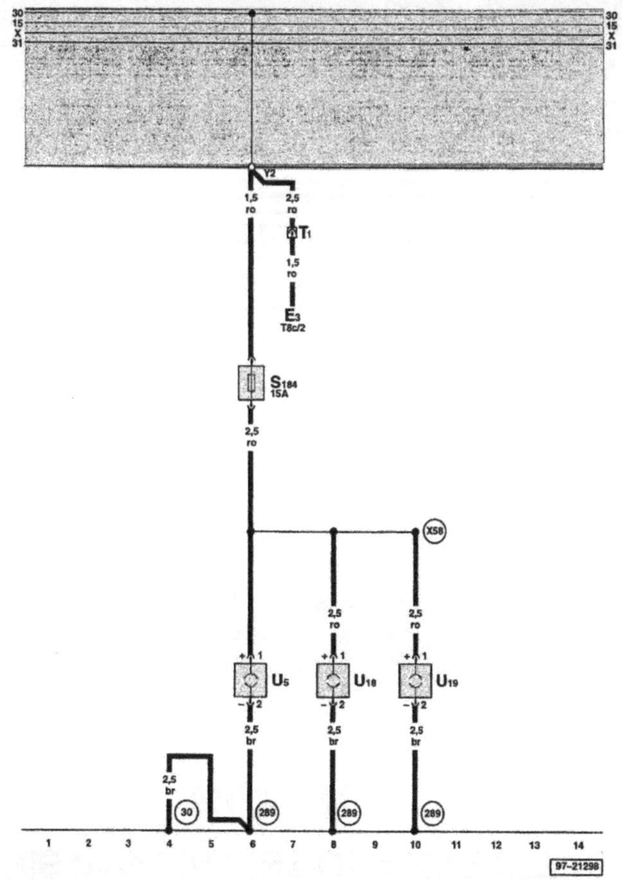

ws = white
sw = black
ro = red
br = brown
gn = green
bl = blue
gr = grey
li = violet
ge = yellow

E3 – Emergency Flasher Switch
S184 – Fuse for 12V outlets, below fuse/relay panel
T1 – Single Connector, behind fuse/relay panel
T8c – 8-Pin Connector, on Emergency Flasher Switch
U5 – 12V outlet -1-, in passenger compartment left
U18 – 12V outlet -2-, in passenger compartment right
U19 – 12V outlet -3-, in passenger compartment right

(30) – Ground Connection, -1-, beside fuse/relay panel

(289) – Ground Connection in plug wiring harness

(X58) – Positive Connection (30), in plug wiring harness

12V outlets, in passenger compartment

Anti-lock brake system (ABS) with electronic differential lock (EDL),

from September 1997

Devite relay location and fuseplacements as well as locations of multiple connectors - see section "component locations".

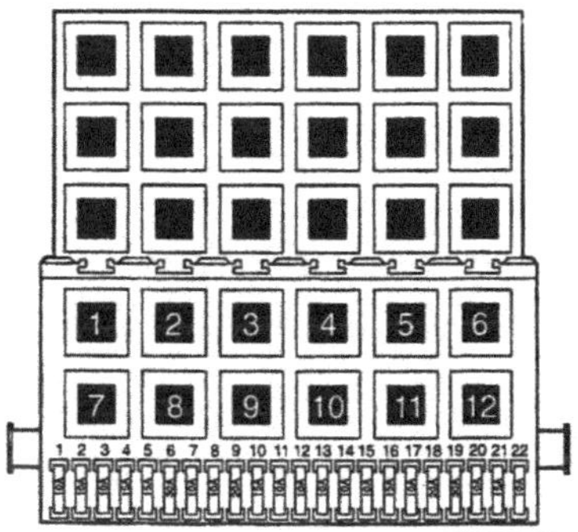

4 Load Reduction Relay (18)

Note: Number in parentheses indicates production control number stamped on relay housing.

Fuse colors

30 A – green
25 A – white
20 A – yellow
15 A – blue
10 A – red
7,5 A – brown
5 A – beige
3 A – violet

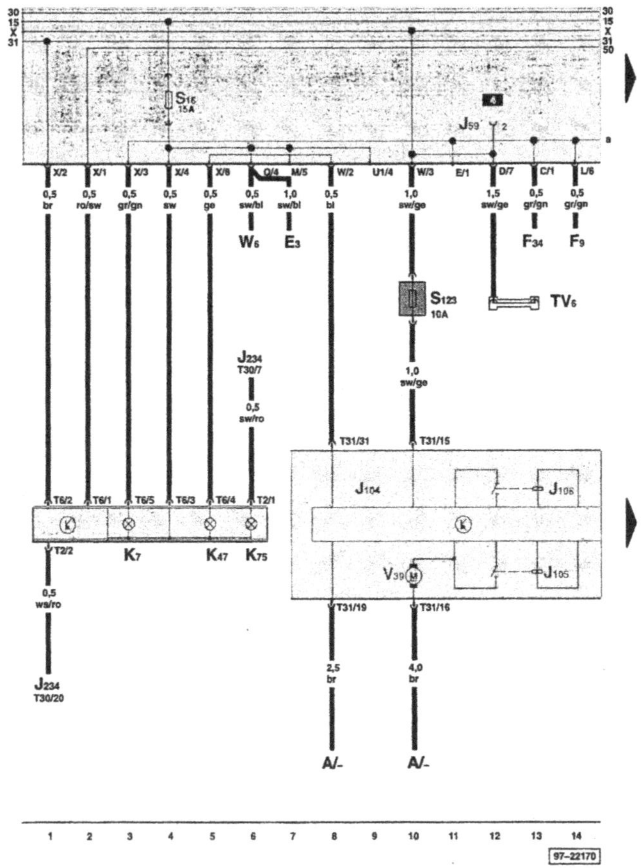

ws = white
sw = black
ro = red
br = brown
gn = green
bl = blue
gr = grey
li = violet
ge = yellow

A – Battery
E3 – Emergency Flasher Switch
F9 – Parking Brake Warning Light Switch
F34 – Brake Fluid Level Warning Switch
J59 – Load Reduction Relay
J104 – ABS Control Module (w/EDL), in engine compartment, left
J105 – ABS Return Flow Pump Relay
J106 – ABS Solenoid Valve Relay
J234 – Airbag Control Module
K7 – Brake and Parking Brake Warning Light
K47 – ABS Warning Light
K75 – Airbag Malfunction Indicator Lamp (MIL)
S123 – Fuse for control module - ABS, No. 5 in Multiple fuse-adapter
T2 – Double Connector
T6 – 6-Pin Connector
T30 – 30-Pin Connector, on Airbag Control Module
T31 – 31-Pin Connector
TV6 – Terminal X Wire Connector, above fuse/relay panel
V39 – ABS Return Flow Pump
W6 – Glove Compartment Light

ABS control module (w/EDL), ABS return flow pump, ABS return flow pump relay, ABS solenoid valve relay, ABS warning light

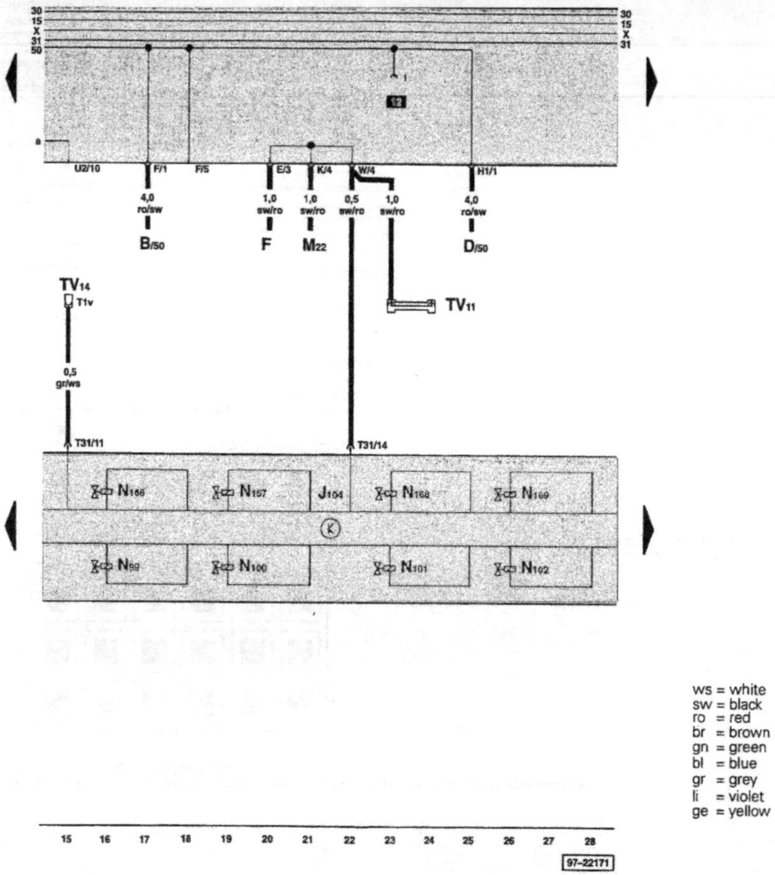

ws = white
sw = black
ro = red
br = brown
gn = green
bl = blue
gr = grey
li = violet
ge = yellow

- B – Starter
- D – Ignition/Starter Switch
- F – Brake Light Switch
- J104 – ABS Control Module (w/EDL), in engine compartment, left
- M22 – Right Brake/Tail Light
- N99 – Right Front ABS Inlet Valve
- N100 – Right Front ABS Outlet Valve
- N101 – Left Front ABS Inlet Valve
- N102 – Left Front ABS Outlet Valve
- N166 – Right Front Traction Control Switch-Over Valve
- N167 – Right Front Traction Control Outlet Valve
- N168 – Left Front Traction Control Switch-Over Valve
- N169 – Left Front Traction Control Outlet Valve
- T1v – Single Connector, on Data Link Connector (DLC) wire Connector
- T31 – 31-Pin Connector
- TV11 – Terminal 54 Wire Connector, above fuse/relay panel
- TV14 – Data Link Connector (DLC) wire Connector, above fuse/relay panel

ABS control module (w/EDL), valves for ABS front, valves for differential lock

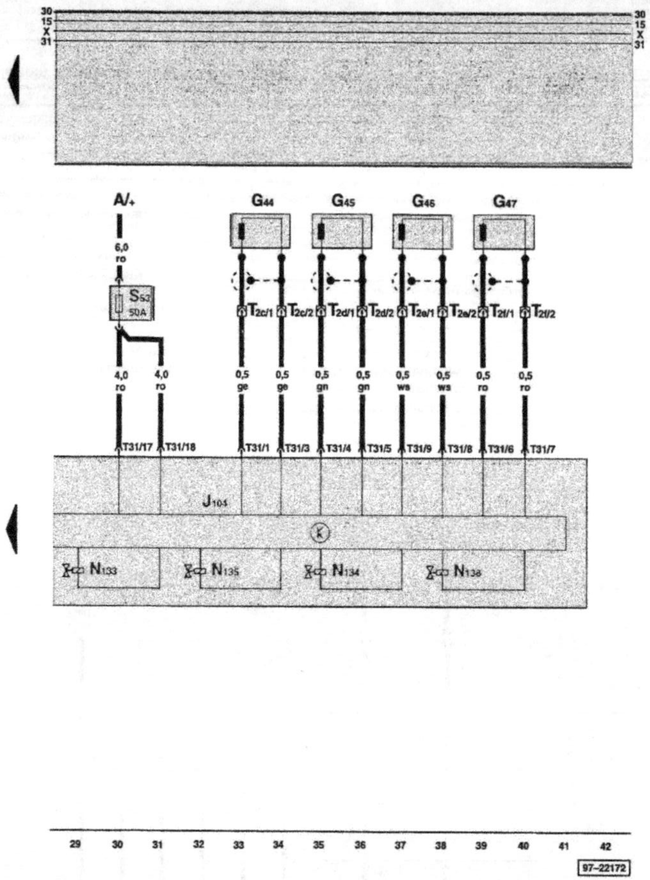

ws = white
sw = black
ro = red
br = brown
gn = green
bl = blue
gr = grey
li = violet
ge = yellow

- A – Battery
- G44 – Right Rear ABS Wheel Speed Sensor
- G45 – Right Front ABS Wheel Speed Sensor
- G46 – Left Rear ABS Wheel Speed Sensor
- G47 – Left Front ABS Wheel Speed Sensor
- J104 – ABS Control Module (w/EDL), in engine compartment, left
- N133 – Right Rear ABS Inlet Valve
- N134 – Left Rear ABS Inlet Valve
- N135 – Right Rear ABS Outlet Valve
- N136 – Left Rear ABS Outlet Valve
- S53 – ABS Hydraulic Pump/Valves Fuse, in engine compartment, left
- T2c – Double Connector
- T2d – Double Connector, in front wheel casing, right
- T2e – Double Connector
- T2f – Double Connector, in front wheel casing, left
- T31 – 31-Pin Connector

ABS control module (w/EDL), valves for ABS rear, ABS hydraulic pump/valves fuse

2.8 L- Engine - Motronic Multiport Fuel Injection (MFI)/103 kW, code AES,

from September 1997

Deviate relay location and fuseplacements as well as the locations of multiple connectors see section "component locations".

12 Fuel Pump (FP) Relay (167)

Note: Number in parentheses indicates production control number stamped on relay housing.

Fuse colors

- 30 A – green
- 25 A – white
- 20 A – yellow
- 15 A – blue
- 10 A – red
- 7,5 A – brown
- 5 A – beige
- 3 A – violet

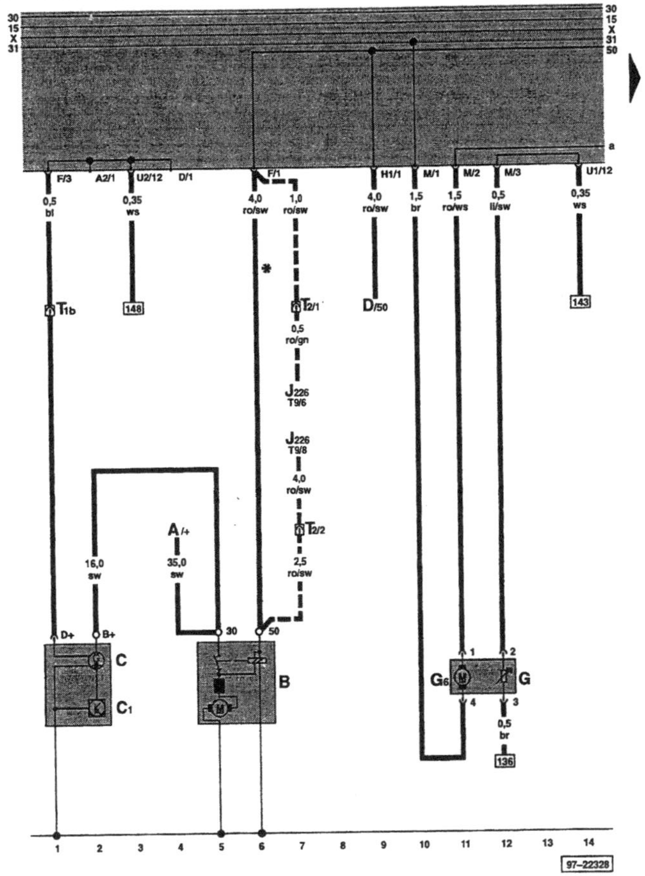

- ws = white
- sw = black
- ro = red
- br = brown
- gn = green
- bl = blue
- gr = grey
- li = violet
- ge = yellow

- A – Battery
- B – Starter
- C – Generator (GEN)
- C1 – Voltage Regulator (VR)
- D – Ignition/Starter Switch
- G – Fuel Level Sensor
- G6 – Fuel Pump (FP)
- J226 – Park/Neutral Position (PNP) Relay
- T1b – Single Connector, near starter
- T2 – Double Connector
- T9 – 9-Pin Connector, on Park/Neutral Position (PNP) Relay

- - - automatic transmission only

* – manual transmission only

Generator (GEN), starter, fuel pump (FP), fuel level sensor

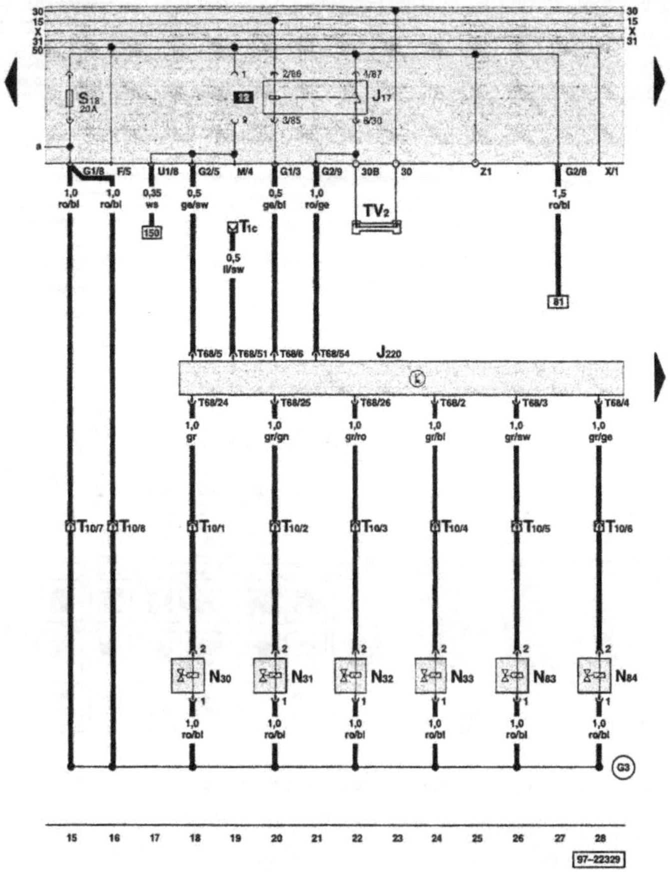

ws = white
sw = black
ro = red
br = brown
gn = green
bl = blue
gr = grey
li = violet
ge = yellow

- J17 – Fuel Pump (FP) Relay
- J220 – Motronic Engine Control Module (ECM), in engine compartment, left
- N30 – Cylinder 1 Fuel Injector
- N31 – Cylinder 2 Fuel Injector
- N32 – Cylinder 3 Fuel Injector
- N33 – Cylinder 4 Fuel Injector
- N83 – Cylinder 5 Fuel Injector
- N84 – Cylinder 6 Fuel Injector
- T1c – Single Connector, behind fuse/relay panel
- T10 – 10-Pin Connector, near cylinder head cover
- T68 – 68-Pin Connector
- TV2 – Terminal 30 (B+) Wire Connector

(G3) – Plus connection, in injector wiring harness

Motronic engine control module (ECM), fuel injectors

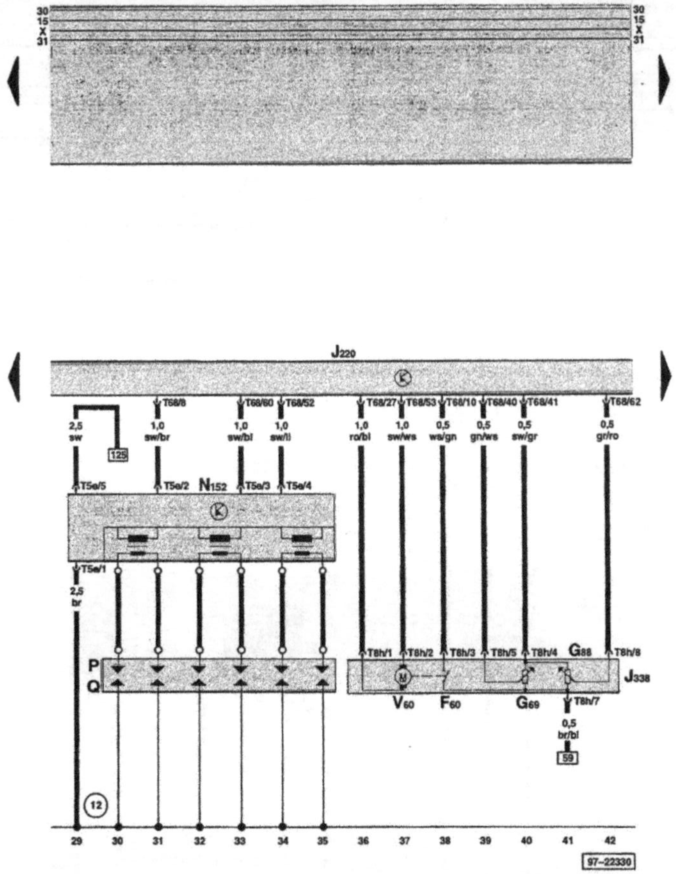

ws = white
sw = black
ro = red
br = brown
gn = green
bl = blue
gr = grey
li = violet
ge = yellow

- F60 – Closed Throttle Position (CTP) Switch
- G69 – Throttle Position (TP) Sensor
- G88 – Throttle Position (TP) Sensor
- J220 – Motronic Engine Control Module (ECM), in engine compartment, left
- J338 – Throttle Valve Control Module
- N152 – Ignition Coil
- P – Spark Plug Connectors
- Q – Spark Plugs
- T5e – 5-Pin Connector
- T8h – 8-Pin Connector
- T68 – 68-Pin Connector
- V60 – Throttle Position (TP) Actuator

(12) – Ground connection, in engine compartment, left

Motronic engine control module (ECM), ignition system, throttle valve control module

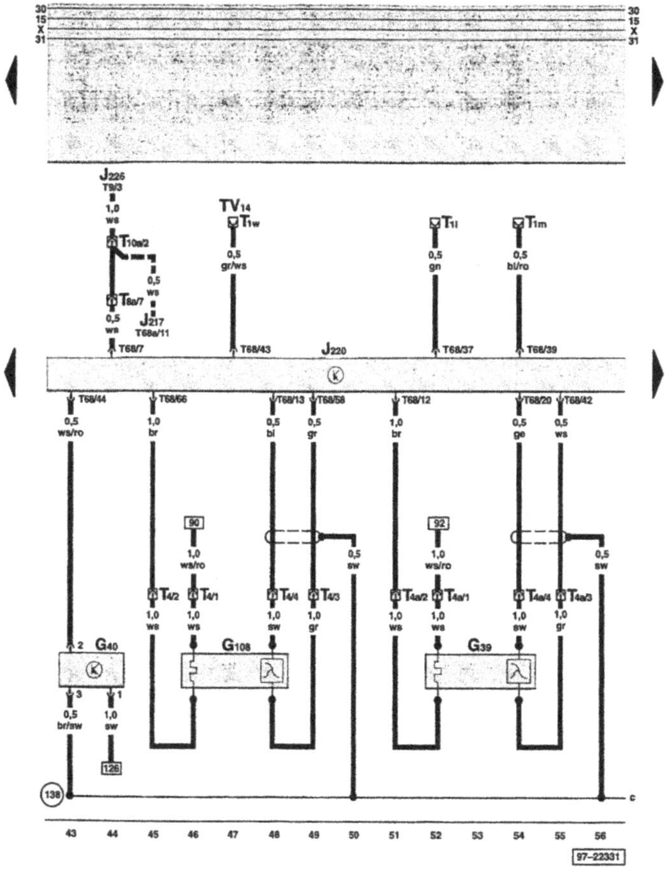

G39 – Heated Oxygen Sensor (HO2S)
G40 – Camshaft Position (CMP) Sensor
G108 – Heated Oxygen Sensor (HO2S) 2
J217 – Transmission Control Module (TCM)
J220 – Motronic Engine Control Module (ECM), in engine compartment, left
J226 – Park/Neutral Position (PNP) Relay
T1l – Single Connector (Connection A/C)
T1m – Single Connector (Connection A/C)
T1w – Single Connector
T4 – 4-Pin Connector, near Heated Oxygen Sensor (HO2S)
T4a – 4-Pin Connector, near Heated Oxygen Sensor (HO2S)
T8a – 8-Pin Connector, behind fuse/relay panel
T9 – 9-Pin Connector, on Park/Neutral Position (PNP) Relay
T10a – 10-Pin Connector, behind fuse/relay panel
T68 – 68-Pin Connector
T68a – 68-Pin Connector, on Transmission Control Module (TCM)
TV14 – Data Link Connector (DLC) wire Connector, above fuse/relay panel
(138) – Ground connection (control module), in Motronic Multiport Fuel Injection (MFI) wiring harness
- - - automatic transmission only

ws = white
sw = black
ro = red
br = brown
gn = green
bl = blue
gr = grey
li = violet
ge = yellow

Motronic engine control module (ECM), heated oxygen sensors, camshaft position (CMP) sensor

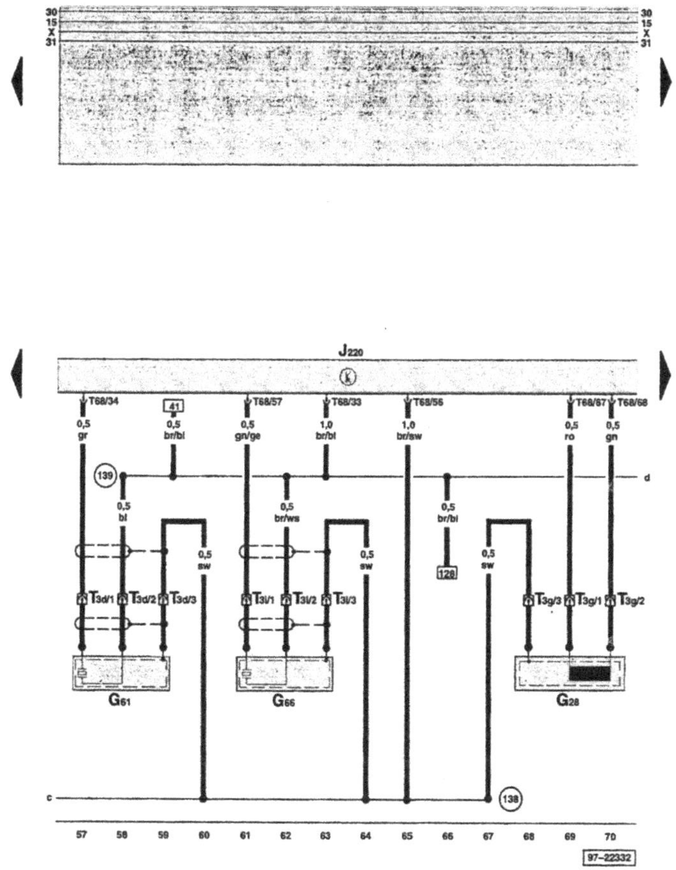

ws = white
sw = black
ro = red
br = brown
gn = green
bl = blue
gr = grey
li = violet
ge = yellow

G28 – Engine Speed (RPM) Sensor
G61 – Knock Sensor (KS) 1
G66 – Knock Sensor (KS) 2
J220 – Motronic Engine Control Module (ECM), in engine compartment, left
T3d – 3-Pin Connector, near cylinder head cover
T3g – 3-Pin Connector, near starter
T3i – 3-Pin Connector, near cylinder head cover
T68 – 68-Pin Connector
(138) – Ground connection (control module), in Motronic Multiport Fuel Injection (MFI) wiring harness
(139) – Ground connection (sensor Ground), in Motronic Multiport Fuel Injection (MFI) wiring harness

Motronic engine control module (ECM), knock sensors, engine speed (RPM) sensor

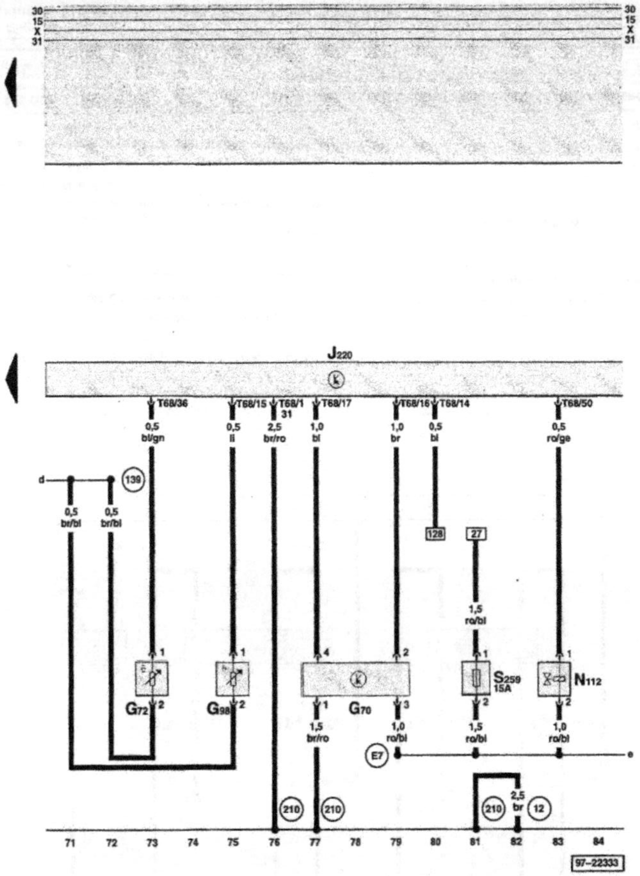

ws = white
sw = black
ro = red
br = brown
gn = green
bl = blue
gr = grey
li = violet
ge = yellow

G70 – Mass Air Flow (MAF) Sensor
G72 – Intake Air Temperature (IAT) Sensor
G98 – EGR Temperature Sensor
J220 – Motronic Engine Control Module (ECM), in engine compartment, left
S259 – Fuel system control fuse, above fuse/relay panel
T68 – 68-Pin Connector
N112 – Secondary Air Injection (AIR) Solenoid Valve

(12) – Ground connection, in engine compartment, left
(139) – Ground connection (sensor Ground), in Motronic Multiport Fuel Injection (MFI) wiring harness

(210) – Ground connection -1-, in Motronic Multiport Fuel Injection (MFI) wiring harness
(E7) – Plus connection (87a), in Motronic Multiport Fuel Injection (MFI) wiring harness

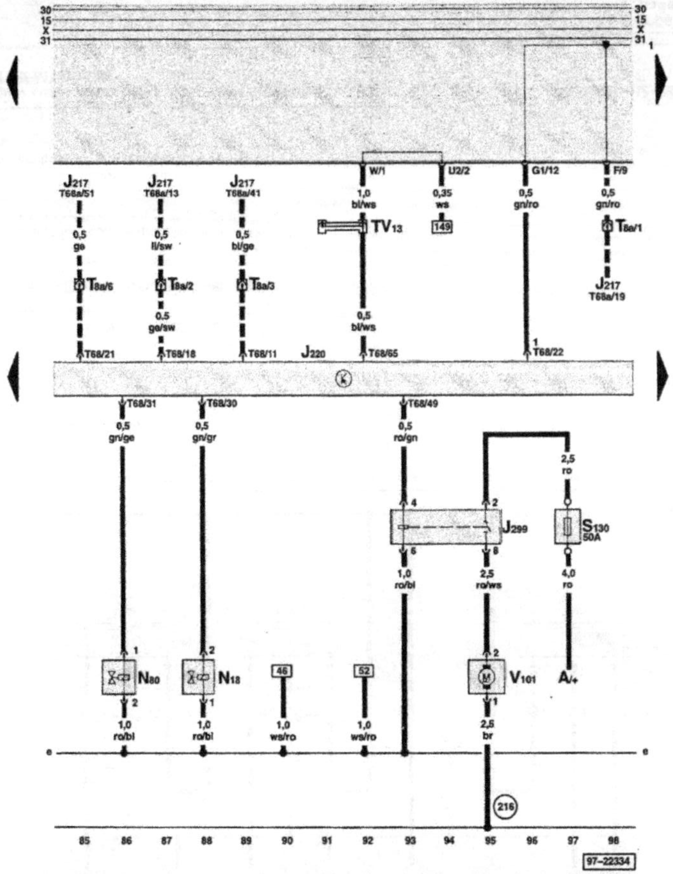

ws = white
sw = black
ro = red
br = brown
gn = green
bl = blue
gr = grey
li = violet
ge = yellow

A – Battery
J217 – Transmission Control Module (TCM)
J220 – Motronic Engine Control Module (ECM), in engine compartment, left
J299 – Secondary Air Injection (AIR) Pump Relay, above fuse/relay panel
N18 – EGR Vacuum Regulator Solenoid Valve
N80 – Evaporative Emission (EVAP) Canister Purge Regulator Valve
S130 – Fuse for secondary air pump, in engine compartment, left
T8a – 8-Pin Connector, behind fuse/relay panel
T68 – 68-Pin Connector
T68a – 68-Pin Connector, on Transmission Control Module (TCM)
TV13 – Vehicle Speed Signal Wire Connector, above fuse/relay panel

V101 – Secondary Air Injection (AIR) Pump Motor

(216) – Ground connection -2-, in Motronic Multiport Fuel Injection (MFI) wiring harness

--- automatic transmission only

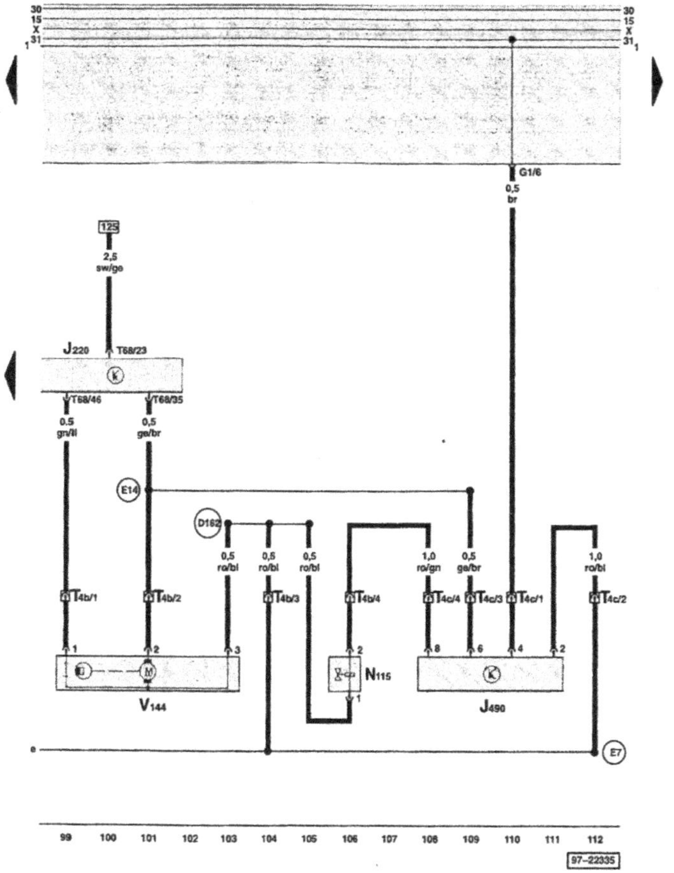

ws = white
sw = black
ro = red
br = brown
gn = green
bl = blue
gr = grey
li = violet
ge = yellow

J220 – Motronic Engine Control Module (ECM), in engine compartment, left
J490 – Fuel system diagnostic control module, production control number (194), behind instrument panel, center
N115 – Evaporative Emission (EVAP) Canister Purge Solenoid Valve
T4b – 4-Pin Connector
T4c – 4-Pin Connector
T68 – 68-Pin Connector
V144 – Leak detection pump (LDP)

(D162) – Wire connection, leak detection pump, in engine wiring harness
(E7) – Plus connection (87a), in Motronic Multiport Fuel Injection (MFI) wiring harness
(E14) – Wire connection -1-, in Motronic Multiport Fuel Injection (MFI) wiring harness

Motronic Engine Control Module (ECM), leak detection pump (LDP)

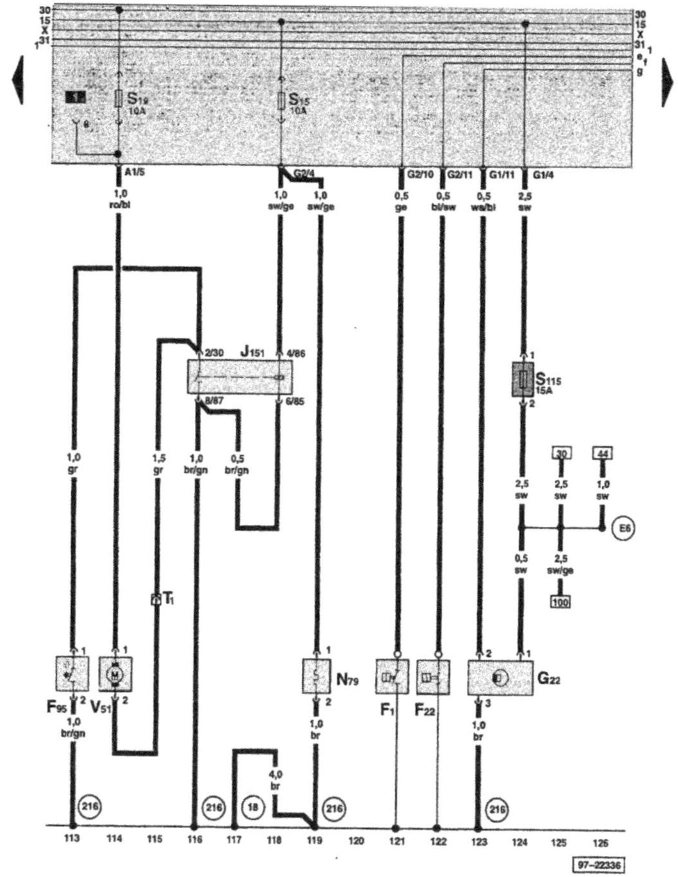

ws = white
sw = black
ro = red
br = brown
gn = green
bl = blue
gr = grey
li = violet
ge = yellow

F1 – Oil Pressure Switch
F22 – 0.3 bar Oil Pressure Switch
F95 – After-Run Coolant Thermal Switch
G22 – Speedometer Vehicle Speed Sensor (VSS)
J151 – Coolant Circulation Pump Relay, above fuse/relay panel
N79 – Positive Crankcase Ventilation (PCV) Heating Element
S115 – Fuse for ignition coil terminal 15, above fuse/relay panel
T1 – Single Connector
V51 – After-Run Coolant Pump

(18) – Ground connection, on engine block
(216) – Ground connection -2-, in Motronic Multiport Fuel Injection (MFI) wiring harness
(E6) – Plus connection -1- (15), in Motronic Multiport Fuel Injection (MFI) wiring harness

After run coolant circulation, oil pressure switch, speedometer vehicle speed sensor (VSS)

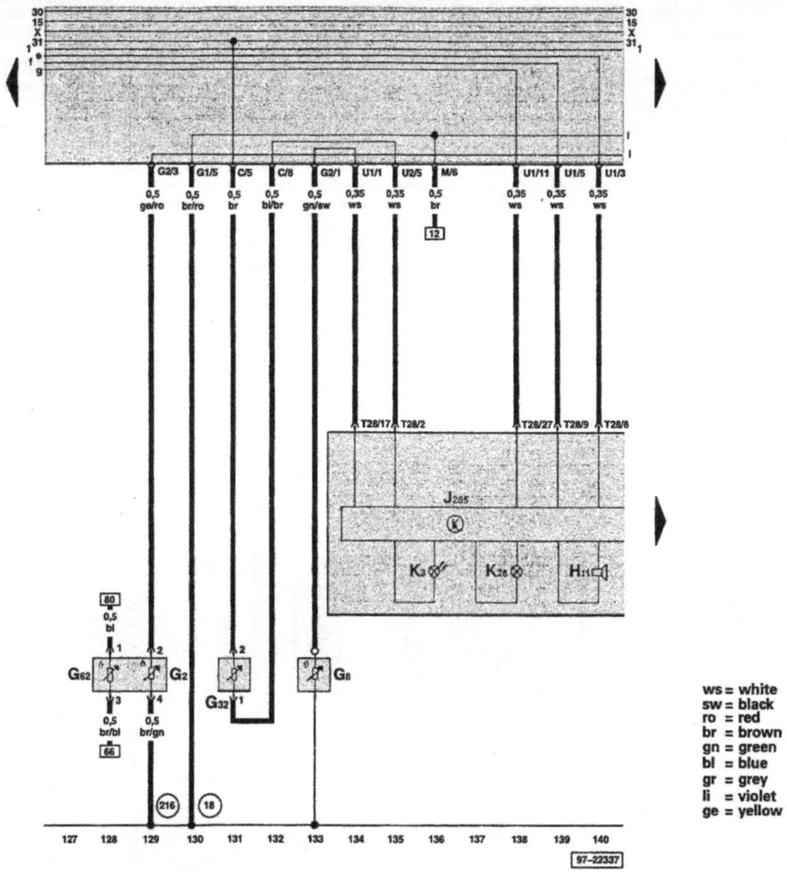

ws = white
sw = black
ro = red
br = brown
gn = green
bl = blue
gr = grey
li = violet
ge = yellow

- G2 – Engine Coolant Temperature (ECT) Sensor
- G8 – Engine Oil Temperature Sensor
- G32 – Engine Coolant Level (ECL) Sensor
- G62 – Engine Coolant Temperature (ECT) Sensor
- H11 – Oil Pressure Warning Buzzer
- J285 – Control module with indicator unit in instrument panel insert
- K3 – Oil Pressure Warning Light
- K28 – Engine Coolant Level/Temperature (ECL/ECT) Warning Light
- T28 – 28-Pin Connector
- (18) – Ground connection, on engine block
- (216) – Ground connection -2-, in Motronic Multiport Fuel Injection (MFI) wiring harness

Edition 02/98
USA.5272.10.21

Instrument cluster, visual acoustical pressure oil control, engine coolant level (ECL) sensor, engine coolant temperature (ECT) sensor

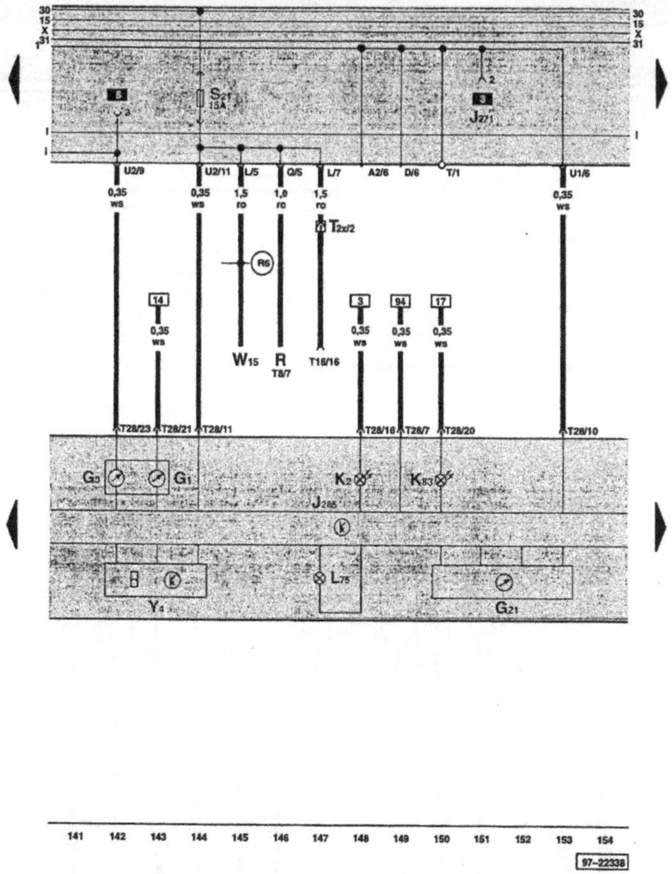

ws = white
sw = black
ro = red
br = brown
gn = green
bl = blue
gr = grey
li = violet
ge = yellow

- G1 – Fuel Gauge
- G3 – Engine Coolant Temperature (ECT) Gauge
- G21 – Speedometer
- J271 – Motronic Engine Control Module (ECM) Power Supply Relay
- J285 – Control module with indicator unit in instrument panel insert
- K2 – Generator (GEN) Warning Light
- K83 – Malfunction Indicator Lamp (MIL)
- L75 – Digital Display Light
- R – Radio
- T2x – Double Connector, behind fuse/relay panel
- T8 – 8-Pin Connector, on Radio
- T16 – 16-Pin Connector, On Board Diagnostic, in instrument panel, center
- T28 – 28-Pin Connector
- W15 – Interior Light with Delay Switch
- Y4 – Odometer Display
- (R6) – Plus connection -1-, in interior light wiring harness

Instrument cluster, fuel gauge, engine coolant temperature (ECT) gauge, speedometer, odometer display

Edition 02/98
USA.5272.10.21

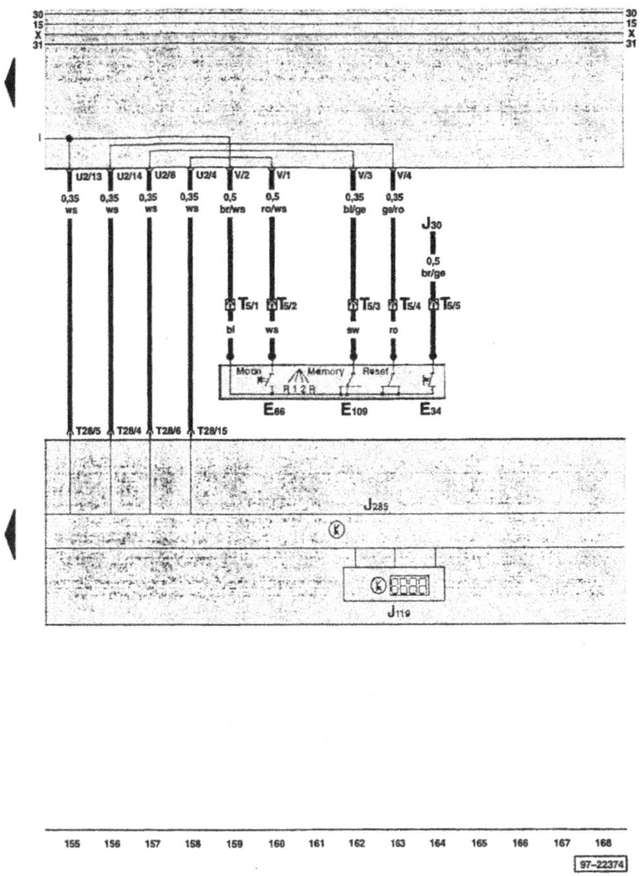

ws = white
sw = black
ro = red
br = brown
gn = green
bl = blue
gr = grey
li = violet
ge = yellow

- E34 – Rear Window Wiper Switch (interval)
- E86 – Multi-Function Indicator Mode Select Switch
- E109 – Multi-Function Indicator Memory Switch
- J30 – Rear Window Wiper/Washer Relay
- J119 – Multi-Function Indicator (MFA)
- J285 – Control module with indicator unit in instrument panel insert
- T5 – 5-Pin Connector, behind steering column switch cover
- T28 – 28-Pin Connector

THIS PAGE INTENTIONALLY LEFT BLANK

Power Sunroof (driver's cabin),

from September 1997

Deviate relay location and fuseplacements as well as the locations of multiple connectors see section "component locations".

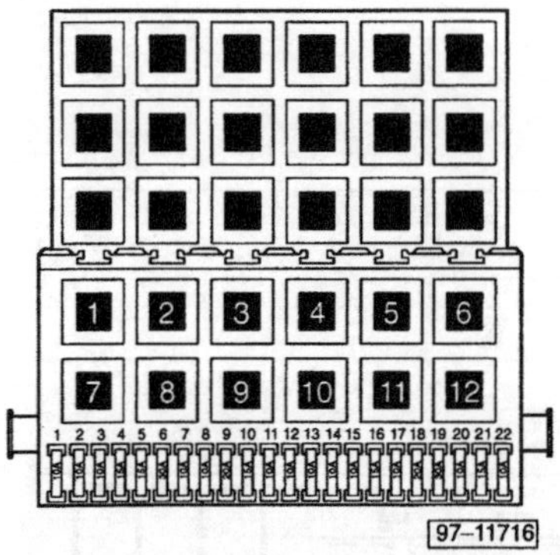

Fuse colors

30 A – green
25 A – white
20 A – yellow
15 A – blue
10 A – red
7,5 A – brown
5 A – beige
3 A – violet

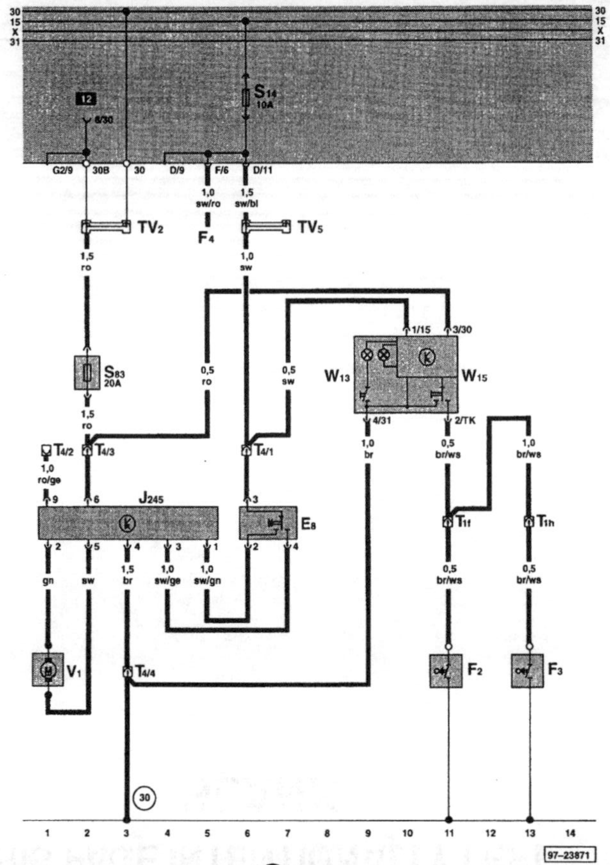

ws = white
sw = black
ro = red
br = brown
gn = green
bl = blue
gr = grey
li = violet
ge = yellow

E8 – Sunroof switch
F2 – Door contact Switch – Driver's side
F3 – Door contact Switch – Passenger side
F4 – Back-Up Light Switch
J245 – Power Sunroof Control Module
S83 – Power Sunroof Fuse, fuse No. 3 in Multiple fuse-adapter
T1f – Single Connector, behind fuse/relay-panel
T1h – Single Connector, behind instrument-panel, right
T4 – 4-Pin Connector, near Interior Light
TV2 – Terminal 30 (B+) Wire Connector
TV5 – Terminal 15a Wire Connector
V1 – Sunroof Motor
W13 – Right Front Map/Reading Light
W15 – Interior Light With Delay Switch

(30) – Ground connection, -1-, beside fuse/relay panel

Power Sunroof, Interior Light, Front Map/Reading Light

1999-2000 Electrical Wiring Supplement

GENERAL

This section contains wiring diagrams for 1999-2000 model year EuroVan. Also included here are the fuse and relay positions.

> ***WARNING —***
> *Special airbag precautions apply to any electrical system testing or repair. the airbag unit is an explosive device and must be handled with extreme care. Before starting any work on the vehicle, refer to the warnings and cautions in* **69 Seatbelts, Airbags**.

> ***CAUTION —***
> - *Before working on the electrical system always switch the ignition off and disconnect battery ground strap from battery negative (–) terminal.*
> - *Disconnecting the negative (–) battery cable may erase fault codes and basic settings in the engine management and automatic transmission control modules. Some driveability problems may be noticed until the system re-adapts to operating conditions. OBD II readiness codes, which may be required for emissions testing, may also be erased.*
> - *Be sure to have the anti-theft radio code on hand before disconnecting the battery.*
> - *Always switch test meter to the appropriate function and range before making test connections.*

Relay panel with fuse holder and auxiliary relay panel

from September 1998

Index

Fuse arrangement

Note:

Fuses for optional equipment not listed are indicated in the respective wiring diagram.

Fuse colors

30 A	–	green	15 A	–	blue	5 A	–	beige
25 A	–	white	10 A	–	red	3 A	–	violet
20 A	–	yellow	7,5 A	–	brown			

Fuses in fuse holder on relay panel

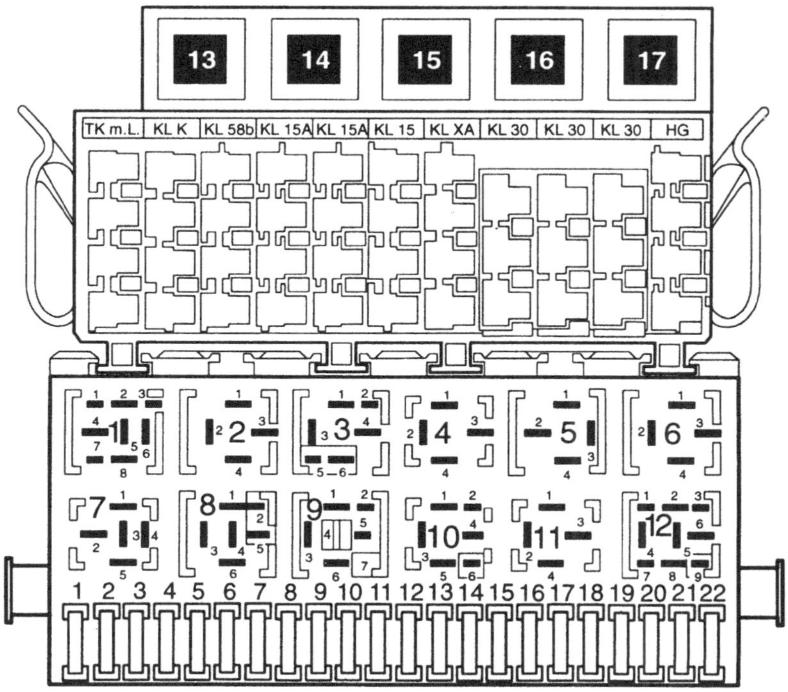

97–23226

Pos.		Function/components	Terminal	Note	Nominal value
1	–	Left headlight	56b		10 A
2	–	Right headlight	56b		10 A
3	–	License plate light	58		10 A
4	–	Rear window wiper, load reduction relay II, optional equipment	X		15 A
5	–	Windshield wiper/washer system, heated washer nozzles	X		15 A
6	–	Fresh air blower, airconditioning	X		30 A

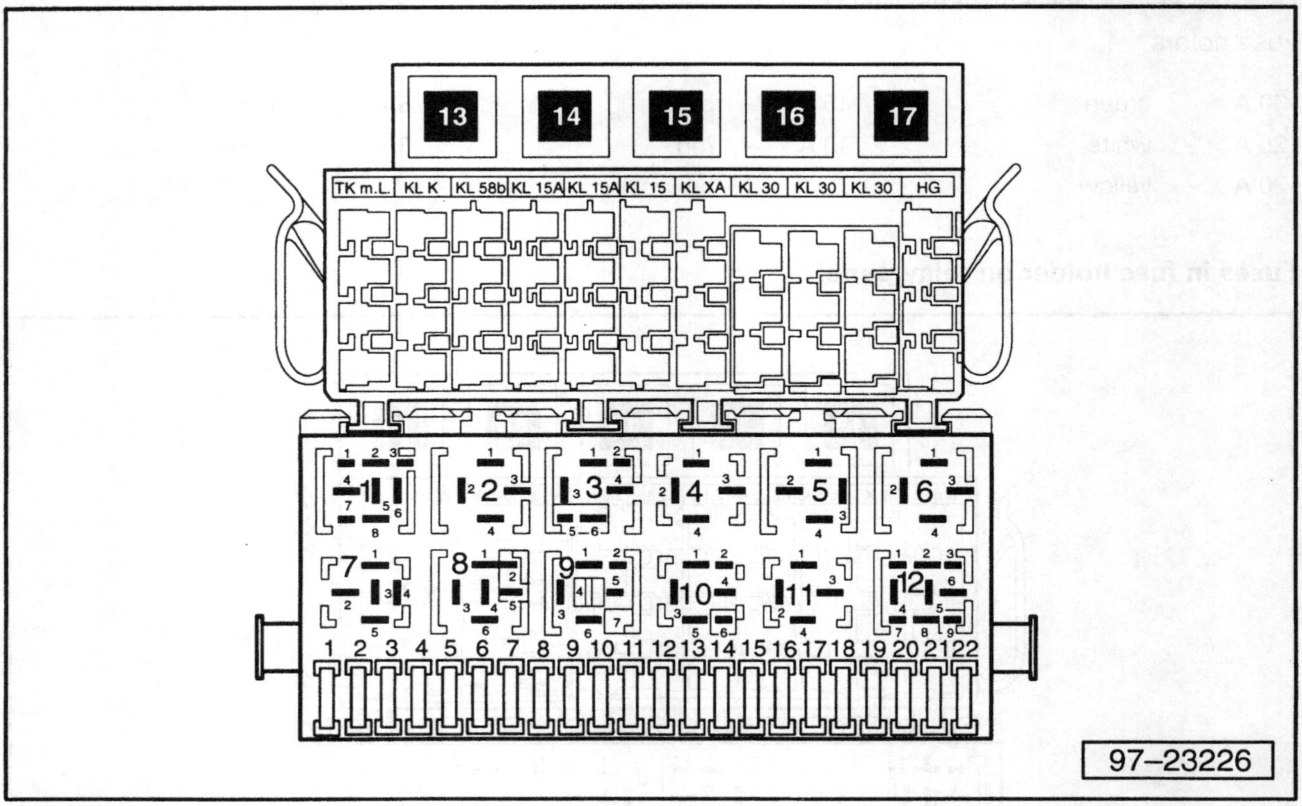

Pos.		Function/components	Terminal	Note	Nominal value
7	–	Right parking and tail light	58R		10 A
8	–	Left parking and tail light	58L		10 A
9	–	Open	X		
10	–	Open	X		
11	–	Left headlight, high beam	56a		10 A
12	–	Right headlight, high beam	56a		10 A
13	–	Dual horns	15		10 A
14	–	Back-up light, automatic transmission, optional equipment, cruise control	15		10 A
15	–	Positive crankcase ventilation (PCV) heating element	15		10 A

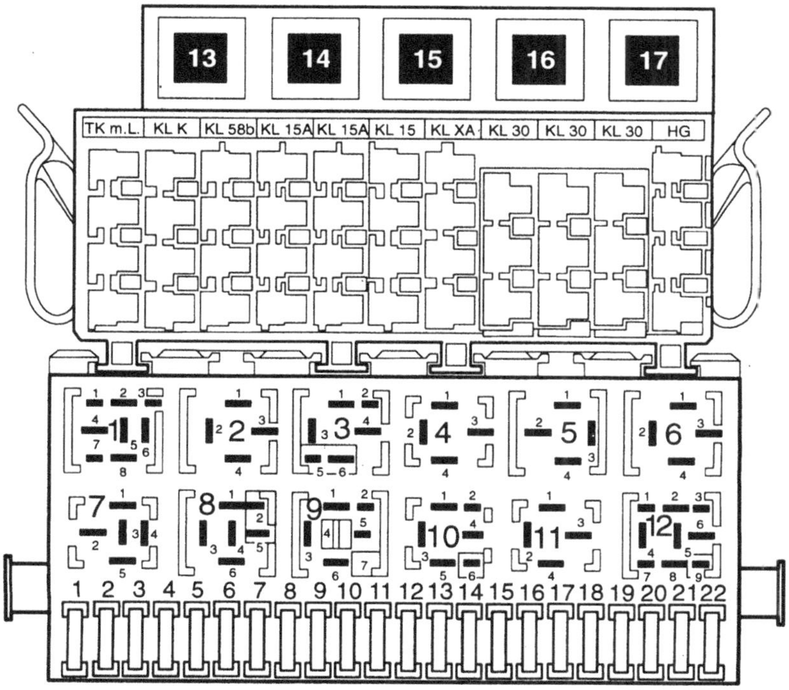

Pos.		Function/components	Terminal	Note	Nominal value
16	–	Turn signal light system, glove compartment light	15		15 A
17	–	Open	15		
18	–	Fuel pump (FP), engine electric			20 A
19	–	Coolant pump after-run, coolant fan control	30		10 A
20	–	Brake light	30		10 A
21	–	Instrument cluster, interior lights, radio, B+ (battery positiv voltage) for on board diagnostic (OBD)	30		15 A
22	–	Cigarette lighter	30		10 A

Fuses in fuse holder -B- below relay panel

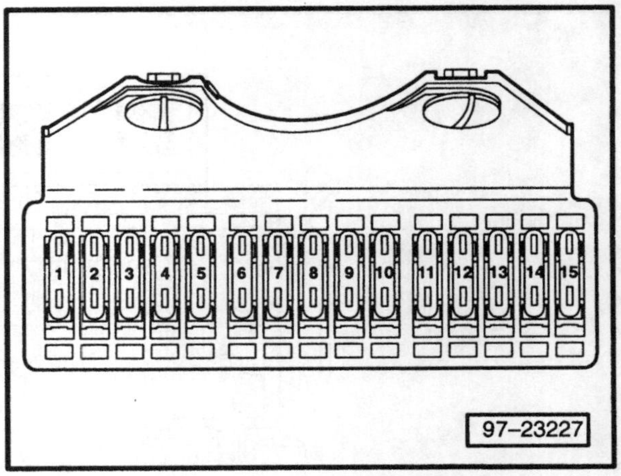

Pos.		Function/components	Terminal	Note	Nominal value
1	–	12V outlets		without 2nd battery	15 A
2	–	Swivel seat lock			10 A
3	–	Power sunroof			20 A
4	–	Open			
5	–	ABS		Control Module	10 A
6	–	Open			
7	–	Telephone			3 A
8	–	Open			
9	–	Open			
10	–	Central locking system			15 A
11	–	Heated seats			15 A
12	–	Rear window defogger and power mirrows			30 A
13	–	Climatronic		2nd evaporator	30 A
14	–	Climatronic		Control Module, Relay for Climatronic	30 A
	–	Air-Conditioning		Fresh air blower rear	
	–	Fresh air blower rear with auxiliary heater core		Fresh air blower rear	
15	–	Climatronic		2nd heater core	30 A
	–	Air-Conditioning		2nd heater core	
	–	Fresh air blower rear with auxiliary heater core		2nd heater core	20 A

Fuses in auxiliary fuse holder above relay panel

	Function/components	Terminal	Note	Nominal value
S43 –	Power window circuit breaker			30 A
S102 –	Engine control module (ECM) fuse			15 A
S117 –	Fuse for control elements			15 A

Fuses in auxiliary fuse holder below driver's seat

	Function/components	Terminal	Note	Nominal value
S23 –	Heater fuse		Auxiliary Heater B3L	5 A
S66 –	Auxiliary heater glow plug fuse		Auxiliary Heater B3L	25 A

Fuses in fuse holder -A- on battery

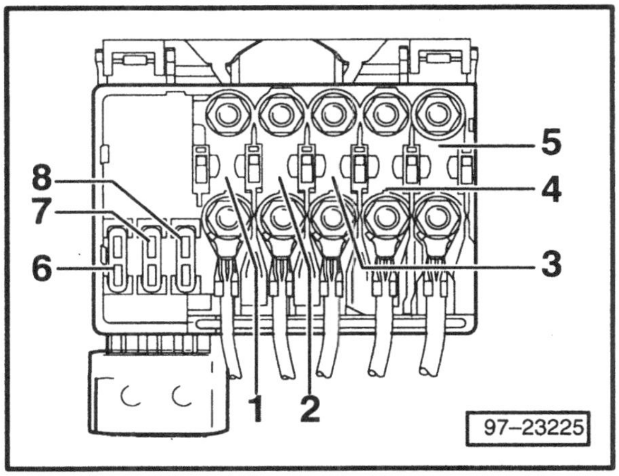

Pos.	Function/components	Terminal	Note	Nominal value
1 –	Secondary air injection			50 A
2 –	Coolant fan			50 A
3 –	Coolant fan			50 A
4 –	ABS		ABS return flow pump, ABS valves	50 A
5 –	Generator (GEN) (90A)			110 A
–	Generator (GEN) (120A)			150 A
–	Generator (GEN) (150A)			175 A
6 –	Open			
7 –	Open			
8 –	Open			

Relay arrangement

Relays on relay panel

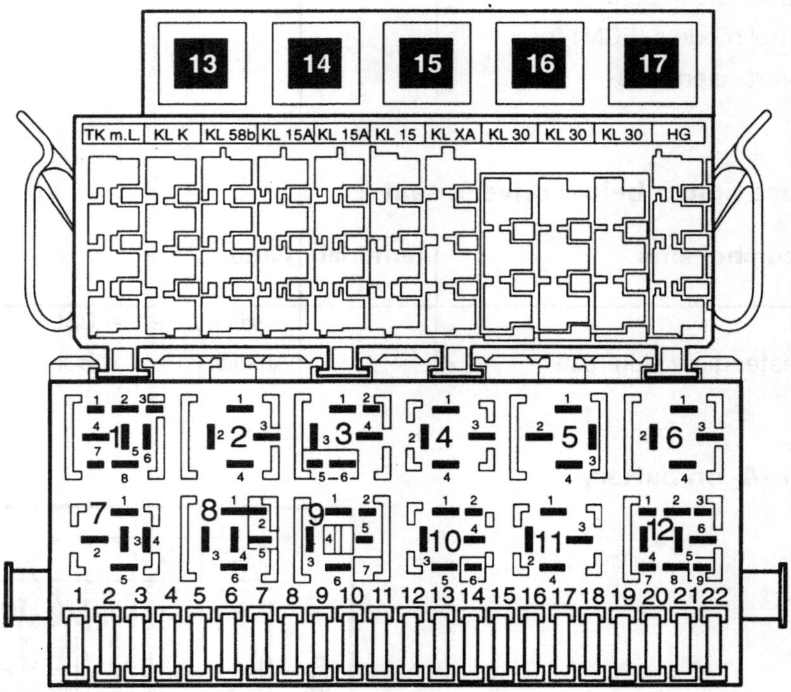

Components/function	Production control number	Note
1 – A/C Relay	140	
– Fresh Air Blower Relay	105	
2 – Rear Window Wiper/Washer Relay	174	
3 – Not used		
4 – Load Reduction Relay	18	
5 – Not used		
6 – Not used		
7 – Headlight Washer System Relay	95	
8 – Wiper/Washer Intermittent Relay (programable)	99	
9 – Not used		
10 – Busbar		
11 – Busbar for Horn		
– Dual Horn Relay	53	
12 – Fuel Pump (FP) Relay	167	

Relays above relay panel

Note:

Relay and fuse locations may vary on vehicles without power windows or air-conditioning.

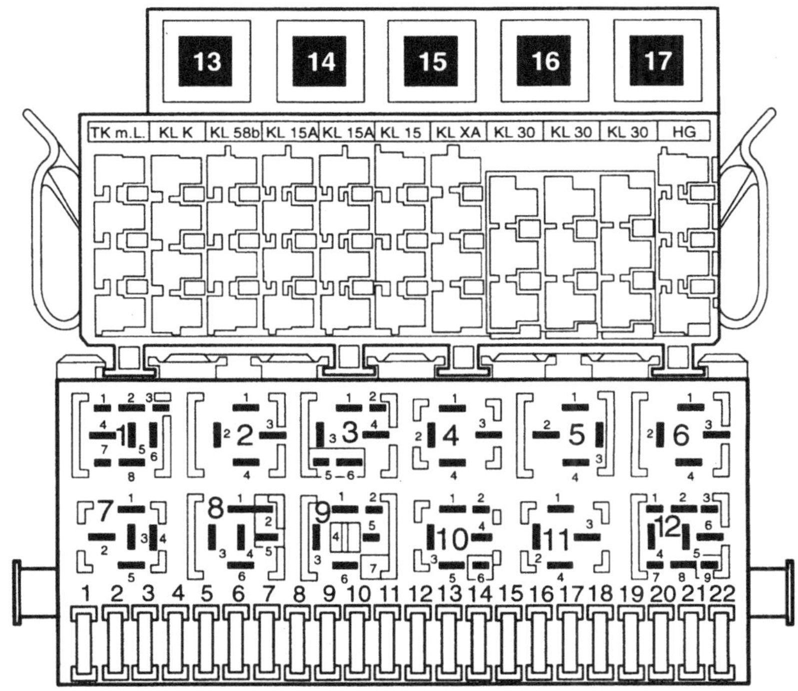

97–23226

Components/function	Production control number	Note
13 – Coolant Circulation Pump Relay	53	
14 – Engine control module (ECM) fuse		
– Power window circuit breaker		
15 – Secondary Air Injection (AIR) Pump Relay	53	
– Fuse for control elements		
16 – Relay for A/C shutoff	147	
17 – Load Reduction Relay II	53	

Relays behind instrument cluster, center

Components/function	Production control number	Note
– Daytime Running Lights Change-over Relay	94	
– Heater Air Blower Relay	53	2nd heater core
– Fan Relay	114	Climatronic
– Fuel system Diagnostic control module	194	
– Heater Control Module	152	2nd heater core
– Relay for Fresh-air and recirculating air flap - rear	38 or 94	
– Park/Neutral Position (PNP) Relay	175	Automatic transmission

Relays below driver's seat

Components/function	Production control number	Note
– Battery Cut Out Relay	214	2nd battery

Wire connectors

Front of connector unit

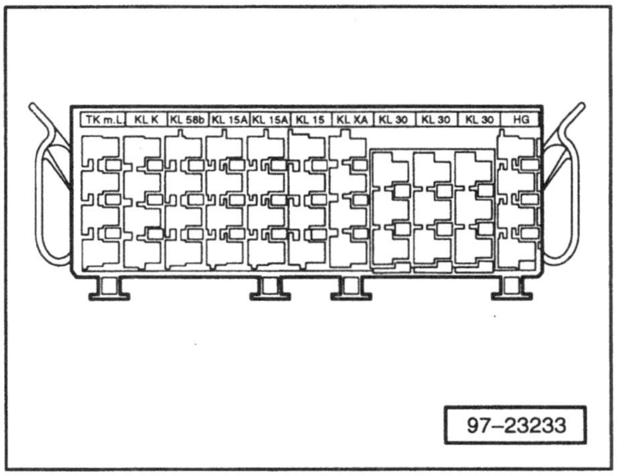

Components/function	Terminal	Color code of connectors
– Door Contact Wire Connector (with load)	TK m.L.	brown
– Data Link Connector (DLC) Wire Connector for on-board diagnostic (OBD) system	KL K	white
– Terminal 58b Wire Connector	KL 58b	grey
– Terminal 15a Wire Connector	KL 15A	green
– Terminal 15 Wire Connector	KL 15	black
– Terminal Xa Wire Connector	KL XA	violet
– Terminal 30 Wire Connector	KL 30	red
– Vehicle Speed Signal Wire Connector	HG	blue

Back of connector unit

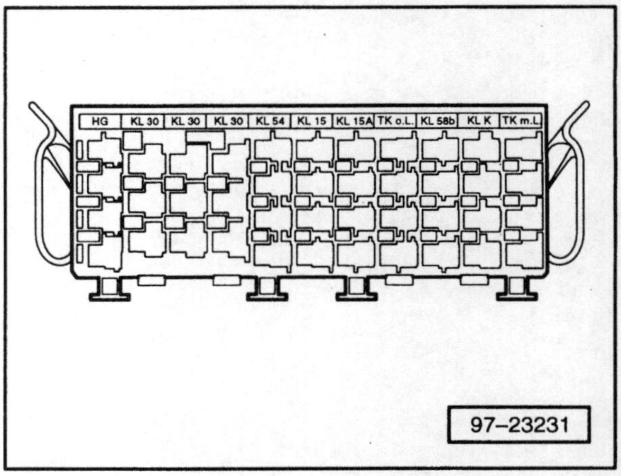

Components/function	Terminal	Color code of connectors
– Vehicle Speed Signal Wire Connector	HG	blue
– Terminal 30 Wire Connector	KL 30	red
– Terminal 54 Wire Connector	KL 54	orange
– Terminal 15 Wire Connector	KL 15	black
– Terminal 15A Wire Connector	KL 15A	green
– Door Contact Wire Connector (without load)	TK o.L.	pink
– Terminal 58b Wire Connector	KL 58b	grey
– Data Link Connector (DLC) Wire Connector for on-board diagnostic (OBD) system	KL K	white
– Door Contact Wire Connector (with load)	TK m.L.	brown

Multiple connector arrangement

Multiple connectors behind relay panel

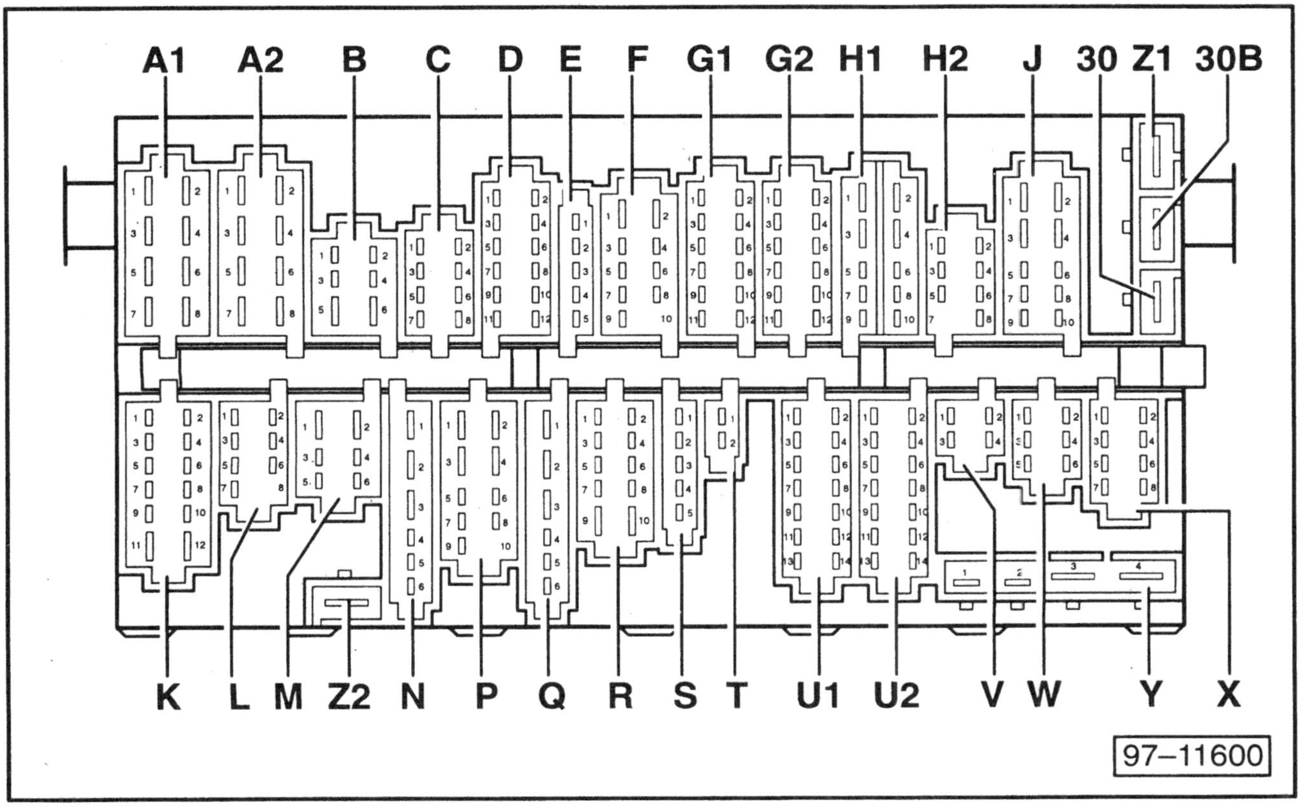

A1 – Headlight wiring harness (yellow), 8-Pin

A2 – Headlight wiring harness (yellow), 8-Pin

B – Headlight washer system wiring harness (green), 6-Pin

C – Engine compartment wiring harness (yellow), 8-Pin

D – Optional equipment wiring harness (green), 12-Pin

E – Brake light switch wiring harness (green), 5-Pin

F – Engine compartment wiring harness (white), 10-Pin

G1 – Engine compartment wiring harness (white), 12-Pin

G2 – Engine compartment wiring harness (white), 12-Pin

H1 – Steering column switch wiring harness (red), 10-Pin

H2 – Steering column switch wiring harness (red), 8-Pin

J – Steering column switch wiring harness (red), 10-Pin

K – Rear wiring harness (black), 12-Pin

L – Parking brake wiring harness (black), 8-Pin

M – Fuel tank wiring harness (black), 6-Pin

N – A/C wiring harness (green), 6-Pin

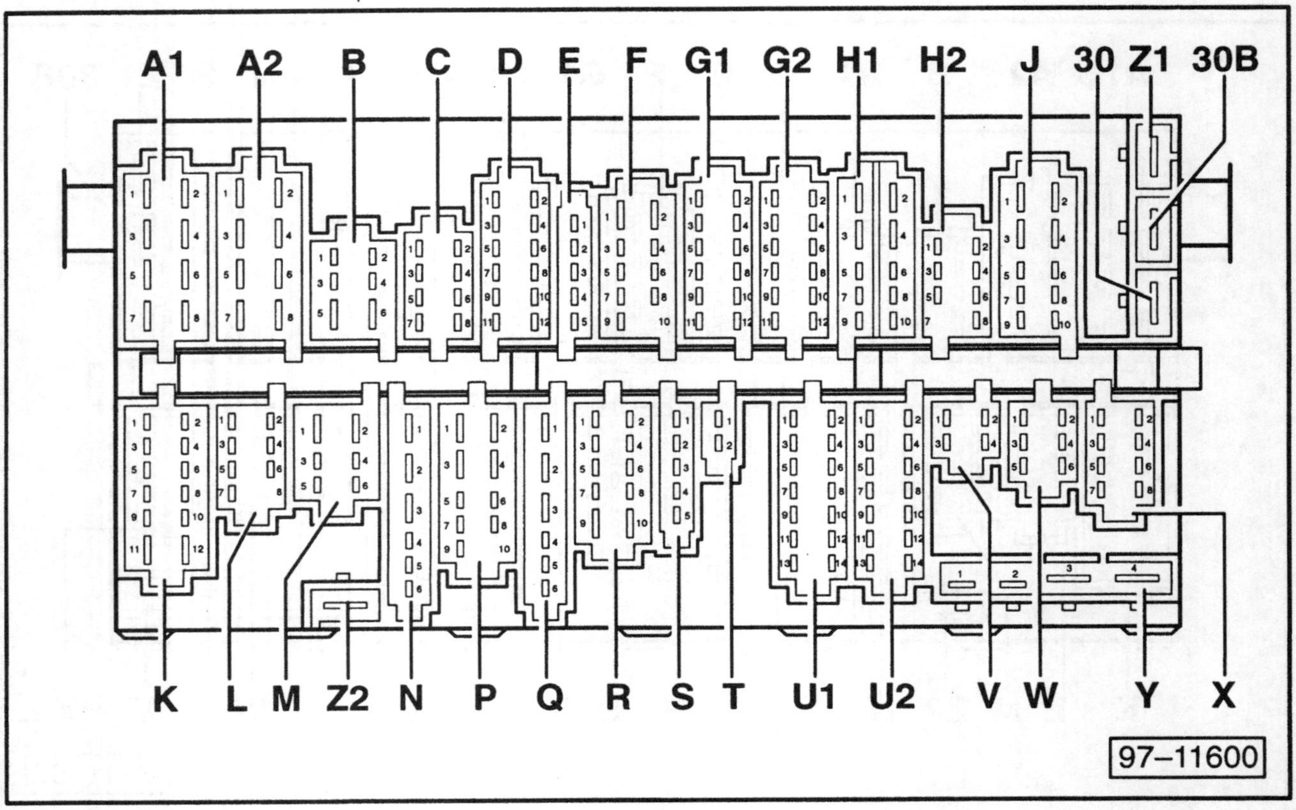

P – Rear window defogger switch and fog light switch wiring harness (blue), 10-Pin

Q – Fresh air blower wiring harness (blue), 6-Pin

R – Light switch wiring harness (blue), 10-Pin

S – Wiper motor wiring harness (white), 5-Pin

T – Double connector (green)

U1 – Instrument cluster wiring harness (blue), 14-Pin

U2 – Instrument cluster wiring harness (blue), 14-Pin

V – Steering column switch wiring harness multi-function-display (green), 4-Pin

W – Optional equipment wiring harness (green), 6-Pin

X – Warning lamps wiring harness (green), 8-Pin

Y – Single connector, terminal 30

Z1 – Single connector

Z2 – Single connector, terminal 31, battery minus

30 – Single connector, terminal 30

30B – Single connector

EuroVan – Standard Equipment, only for cars with blue illuminated instrument cluster

from September 1998

Deviate relay location and fuseplacements as well as the locations of multiple connectors see section "component locations".

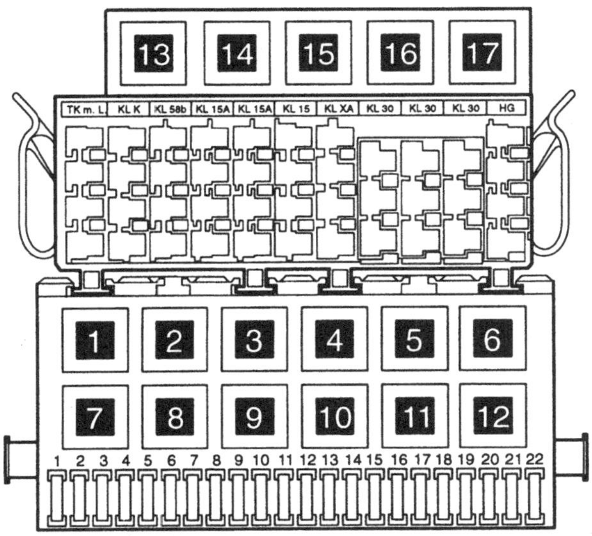

97–23232

- **2** Rear Window Wiper/Washer Relay (174)
- **4** Load Reduction Relay (18)
- **8** Wiper/Washer Intermittent Relay (programable) (99)
- **11** Dual Horn Relay (53)
- **17** Load Reduction Relay II (53)

Note: Number in parentheses indicates production control number stamped on relay housing.

Fuse colors

30 A – green
25 A – white
20 A – yellow
15 A – blue
10 A – red
7,5 A – brown
5 A – beige
3 A – violet

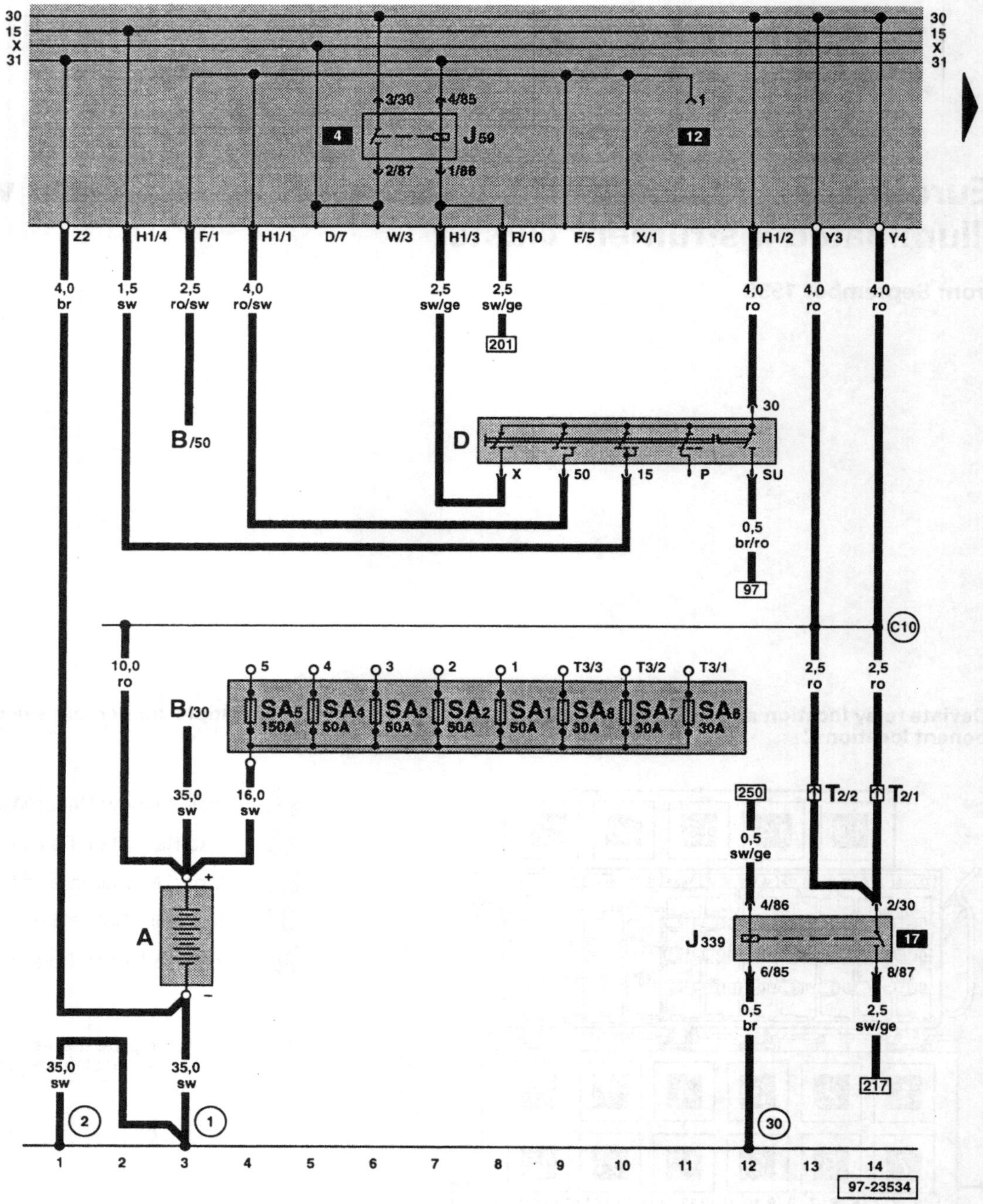

ws = white
sw = black
ro = red
br = brown
gn = green
bl = blue
gr = grey
li = violet
ge = yellow

A – Battery
B – Starter
D – Ignition/Starter Switch
J59 – Load Reduction Relay
J339 – Load Reduction Relay II, above fuse/relay panel
SA1 – Fuse -1- in fuse bracket/battery
SA2 – Fuse -2- in fuse bracket/battery
SA3 – Fuse -3- in fuse bracket/battery
SA4 – Fuse -4- in fuse bracket/battery
SA5 – Fuse -5- in fuse bracket/battery
SA6 – Fuse -6- in fuse bracket/battery
SA7 – Fuse -7- in fuse bracket/battery
SA8 – Fuse -8- in fuse bracket/battery
T2 – Double Connector, behind fuse/relay panel
T3 – 3-Pin Connector

(1) – Ground strap, battery to body
(2) – Ground strap, transmission to body
(30) – Ground connection, -1-, beside fuse/relay panel
(C10) – Plus connection -1- (30), in headlight wiring harness

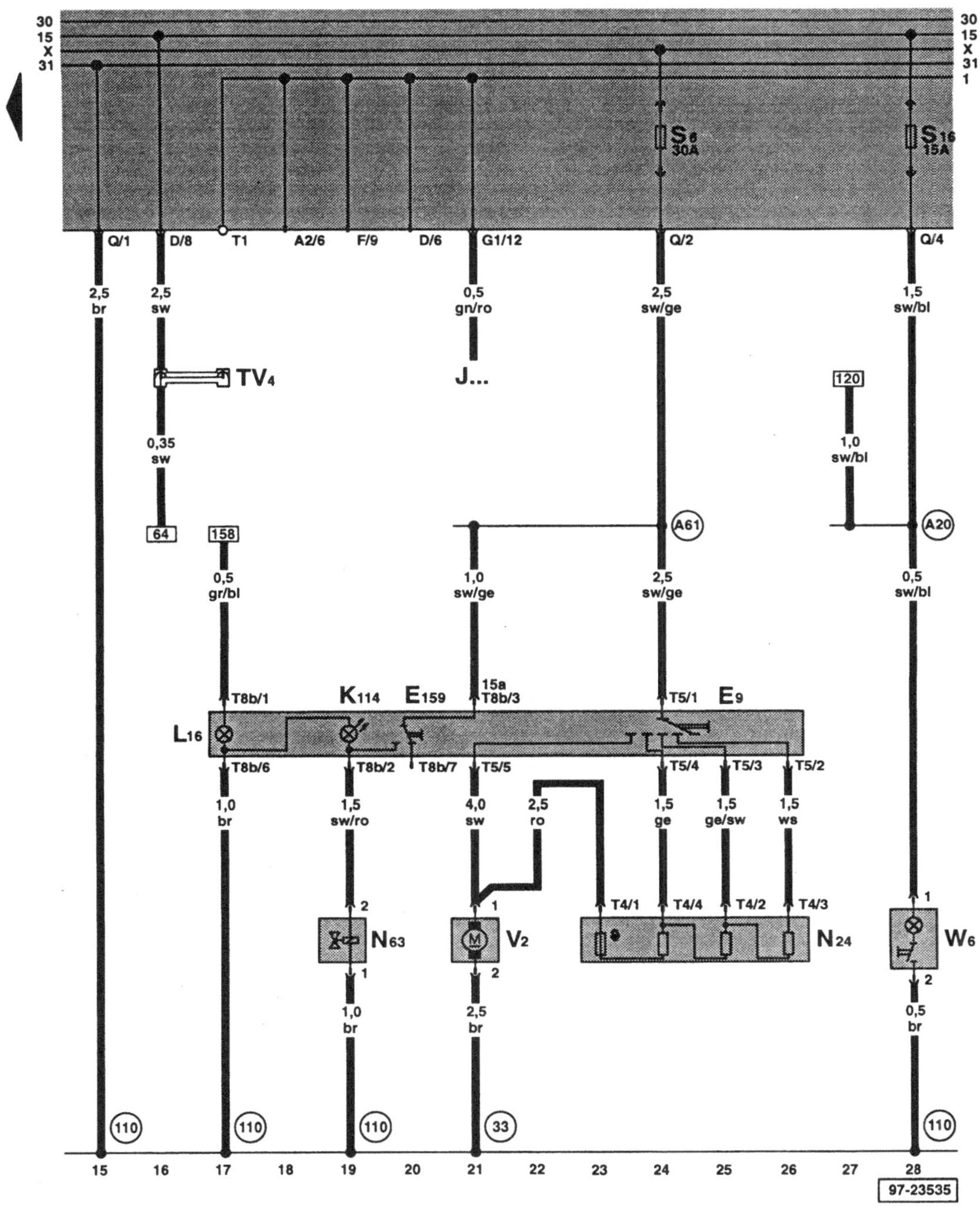

ws = white
sw = black
ro = red
br = brown
gn = green
bl = blue
gr = grey
li = violet
ge = yellow

E9 – Fresh Air Blower Switch
E159 – Fresh Air/Recirculating Flap Switch
J... – Engine Control Module (ECM)
K114 – Fresh Air and Recirculating Air Mode Indicator Light
L16 – Fresh Air Control Lever Light
N24 – Fresh Air Blower Series Resistance with Fuse
N63 – Fresh Air/Recirculating Flap Two-Way Valve
T4 – 4-Pin Connector
T5 – 5-Pin Connector
T8b – 8-Pin Connector
TV4 – Terminal 15 Wire Connector, black
V2 – Fresh Air Blower
W6 – Glove Compartment Light

(33) – Ground connection, behind instrument panel right
(110) – Ground connection -2-, in instrument panel wiring harness
(A20) – Wire connection (15a), in instrument panel wiring harness
(A61) – Plus connection (X), in Instrument cluster wiring harness

Fresh air blower switch, fresh air/recirculating flap switch, fresh air blower, fresh air/recirculating flap two-way valve, glove compartment light

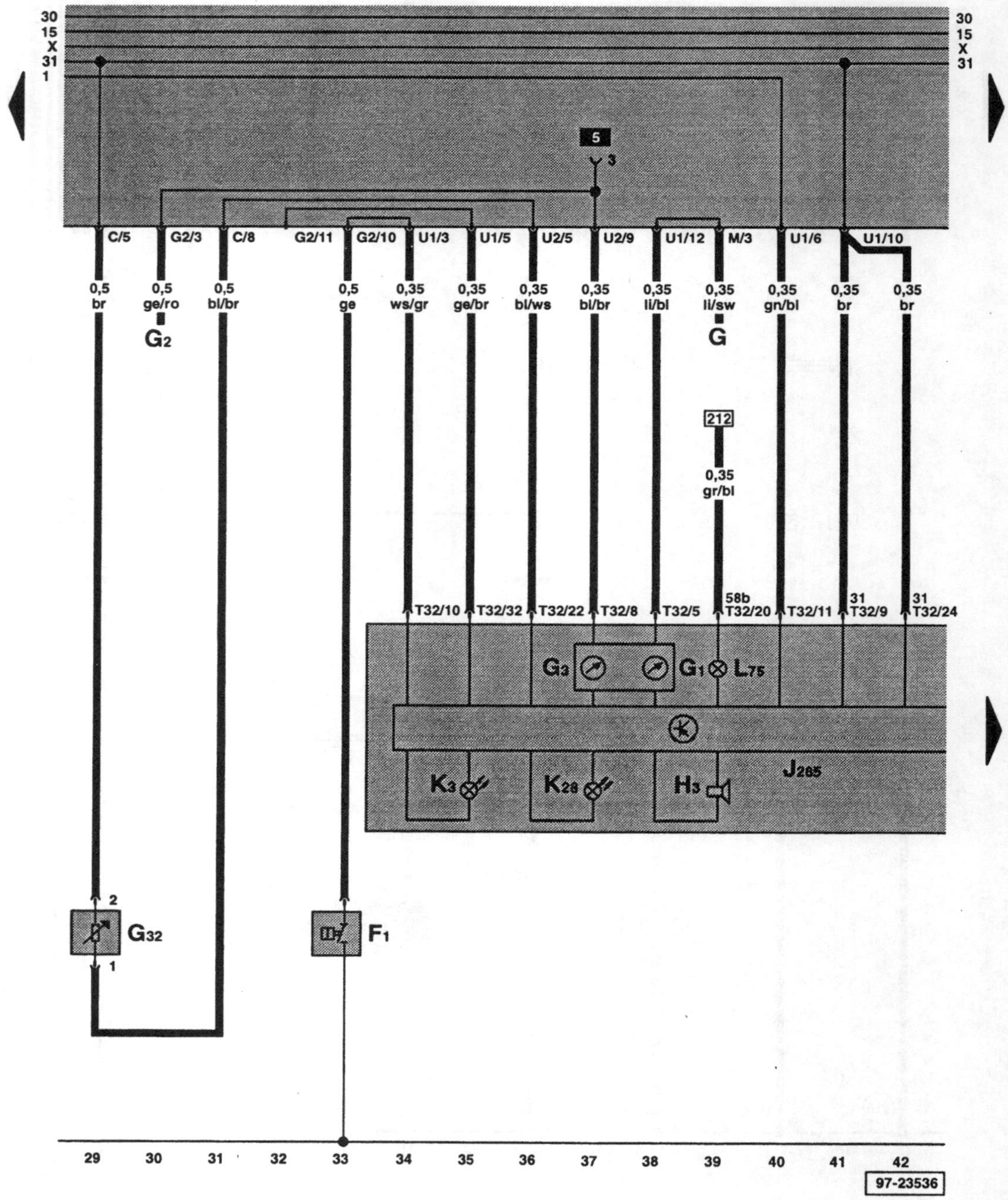

ws = white
sw = black
ro = red
br = brown
gn = green
bl = blue
gr = grey
li = violet
ge = yellow

F1 – Oil Pressure Switch
G – Fuel Level Sensor
G1 – Fuel Gauge
G2 – Engine Coolant Temperature (ECT) Sensor
G3 – Engine Coolant Temperature (ECT) Gauge
G32 – Engine Coolant Level (ECL) Sensor
H3 – Warning Buzzer
J285 – Control module with indicator unit in instrument panel insert
K3 – Oil Pressure Warning Light
K28 – Engine Coolant Level/Temperature (ECL/ECT) Warning Light
L75 – Digital Display Light
T32 – 32-Pin Connector, blue

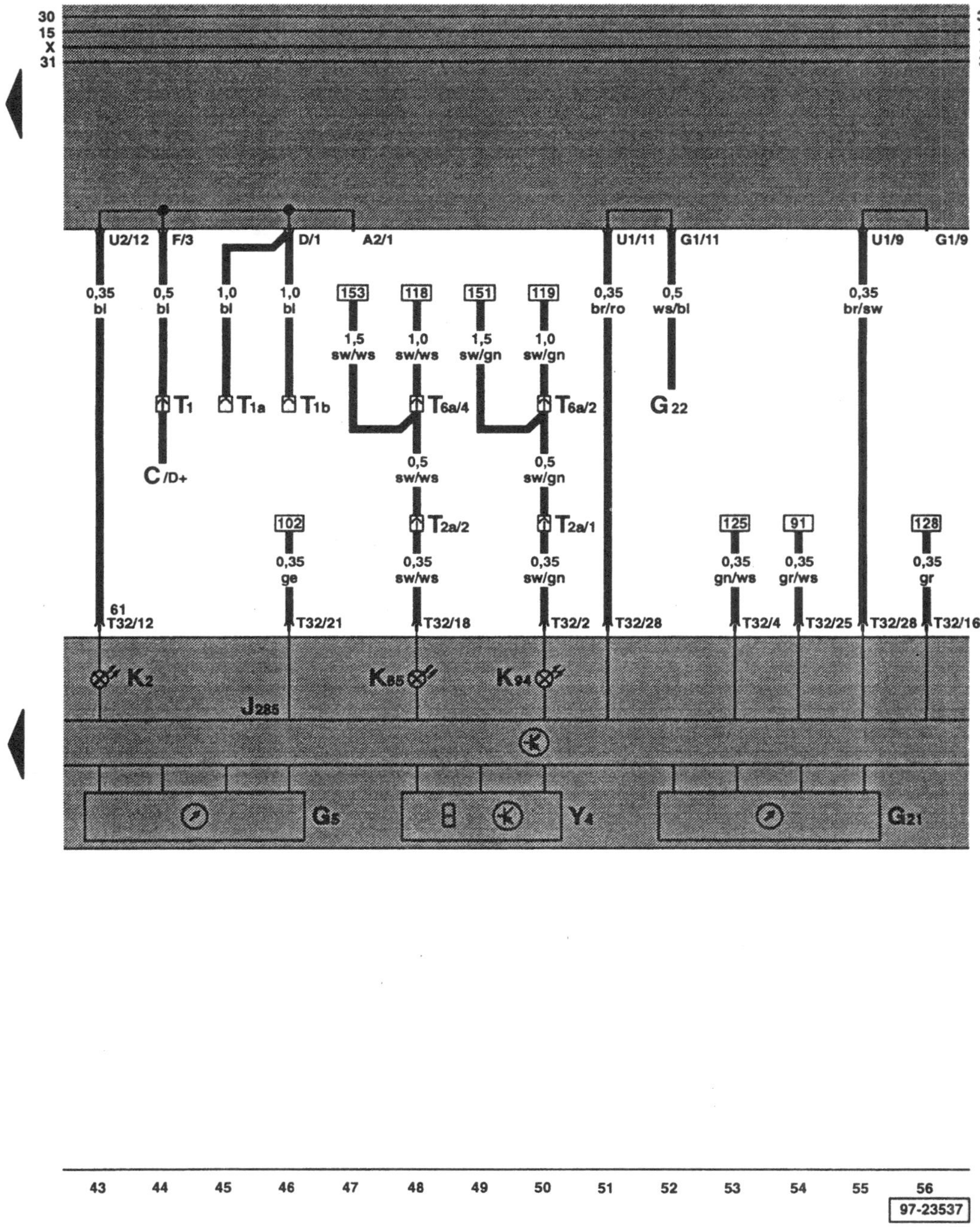

ws = white
sw = black
ro = red
br = brown
gn = green
bl = blue
gr = grey
li = violet
ge = yellow

C – Generator (GEN)
G5 – Tachometer
G21 – Speedometer
G22 – Speedometer Vehicle Speed Sensor (VSS)
J285 – Control module with indicator unit in instrument panel insert
K2 – Generator (GEN) Warning Light
K65 – Left Turn Signal Indicator Light
K94 – Right Turn Signal Indicator Light
T1 – Single Connector, near starter
T1a – Single Connector, behind fuse/relay panel
T1b – Single Connector, behind fuse/relay panel
T2a – Double Connector, behind fuse/relay panel
T6a – 6-Pin Connector, behind fuse/relay panel
T32 – 32-Pin Connector, blue
Y4 – Odometer Display

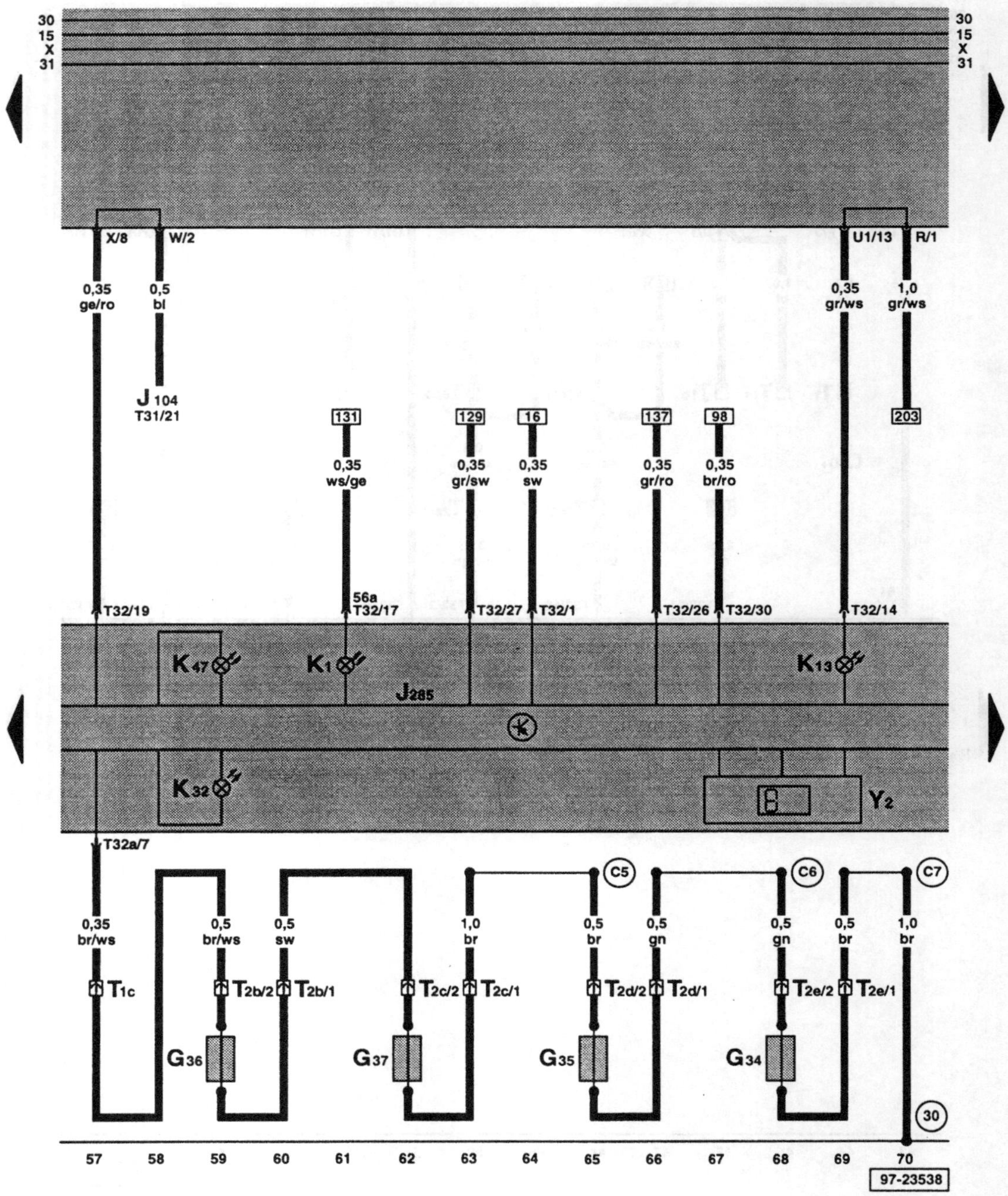

ws = white
sw = black
ro = red
br = brown
gn = green
bl = blue
gr = grey
li = violet
ge = yellow

G34 – Left Front Brake Pad Wear Sensor
G35 – Right Front Brake Pad Wear Sensor
G36 – Left Rear Brake Pad Wear Sensor
G37 – Right Rear Brake Pad Wear Sensor
J104 – ABS Control Module (w/EDL), in engine compartment, left
J285 – Control module with indicator unit in instrument panel insert
K1 – Headlight High Beam Indicator Light
K13 – Rear Fog Light Indicator Light
K32 – Brake Pad Wear Indicator Light
K47 – ABS Warning Light
T1c – Single Connector, behind fuse/relay panel
T2b – Double Connector
T2c – Double Connector
T2d – Double Connector
T2e – Double Connector
T31 – 31-Pin Connector, on ABS Control Module (w/EDL)
T32 – 32-Pin Connector, blue
T32a – 32-Pin Connector, green
Y2 – Digital Clock
(30) – Ground connection, -1-, beside fuse/relay panel
(C5) – Wire connection -1- (brake pad wear monitor), in left front wiring harness
(C6) – Wire connection -2- (brake pad wear monitor), in left front wiring harness
(C7) – Wire connection -3- (brake pad wear monitor), in left front wiring harness

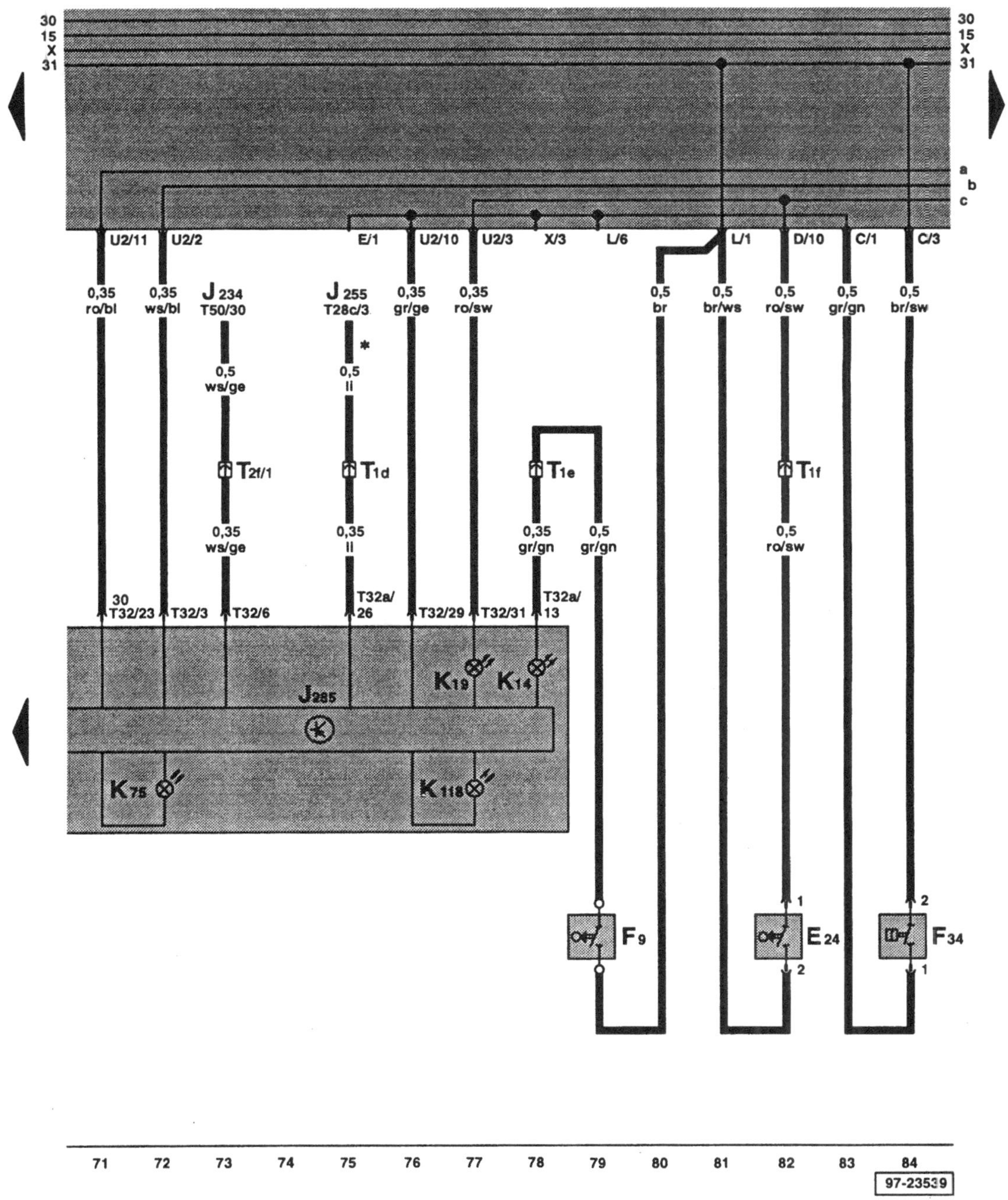

ws = white
sw = black
ro = red
br = brown
gn = green
bl = blue
gr = grey
li = violet
ge = yellow

E24 – Left Seat Belt Switch
F9 – Parking Brake Warning Light Switch
F34 – Brake Fluid Level Warning Switch
J234 – Airbag Control Module
J255 – Climatronic Control Module
J285 – Control module with indicator unit in instrument panel insert
K14 – Parking Brake Indicator Light
K19 – Seat Belt Warning Light
K75 – Airbag Malfunction Indicator Lamp (MIL)
K118 – Warning light for brake system
T1d – Single Connector, behind fuse/relay panel
T1e – Single Connector, behind fuse/relay panel
T1f – Single Connector, behind fuse/relay panel
T2f – Double Connector, behind fuse/relay panel
T28c – 28-Pin Connector, on climatronic control module
T32 – 32-Pin Connector, blue
T32a – 32-Pin Connector, green
T50 – 50-Pin Connector, on airbag control module

* – Vehicles with Climatronic only

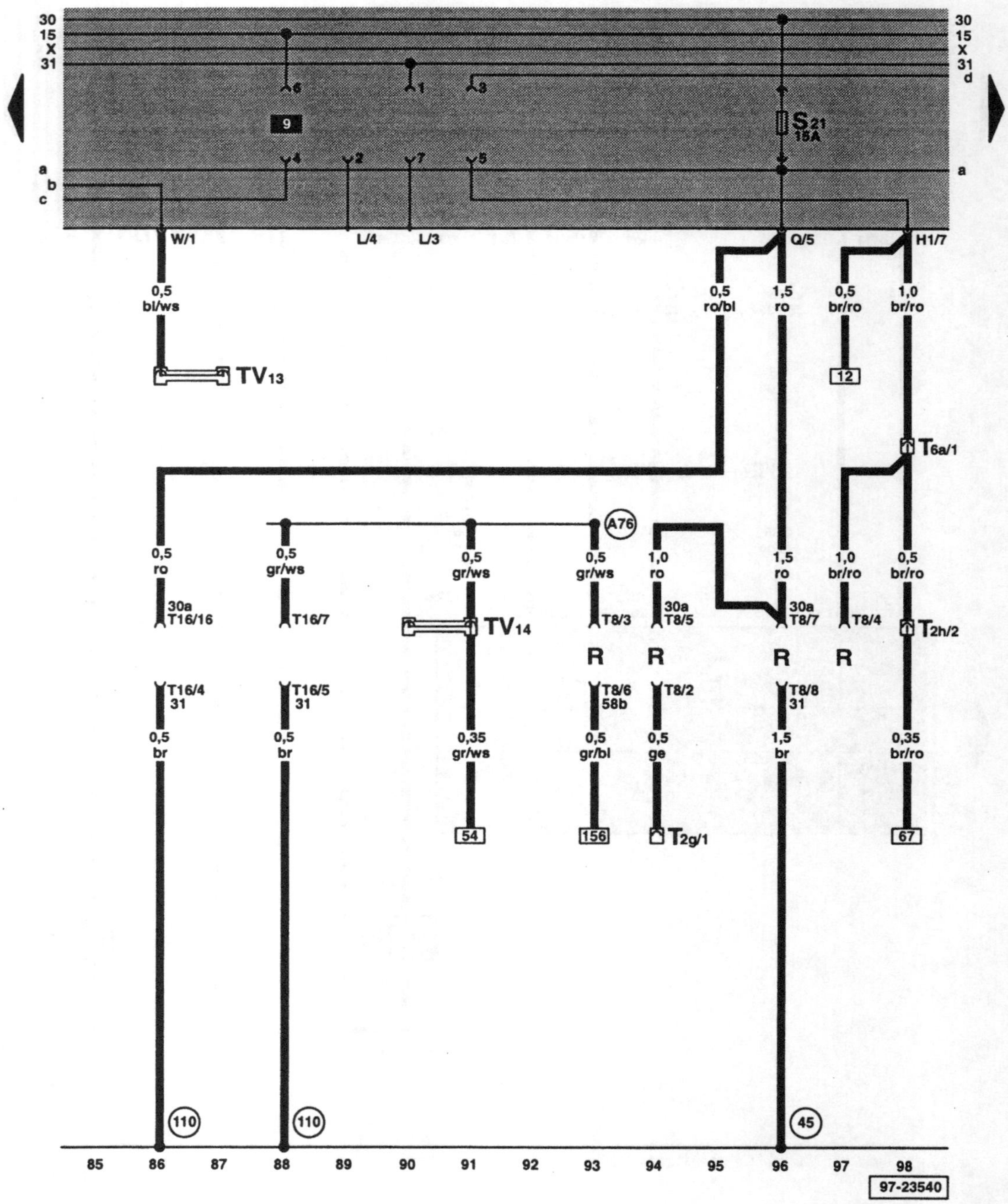

ws = white
sw = black
ro = red
br = brown
gn = green
bl = blue
gr = grey
li = violet
ge = yellow

R – Radio

T2g – Double Connector

T2h – Double Connector, behind fuse/relay panel

T6a – 6-Pin Connector, behind fuse/relay panel

T8 – 8-Pin Connector

T16 – Data Link Connector (DLC), behind steering column cover

TV13 – Vehicle Speed Signal Wire Connector, blue

TV14 – Data Link Connector (DLC) wire Connector, white

(45) – Ground connection, behind instrument panel, center

(110) – Ground connection -2-, in instrument panel wiring harness

(A76) – Connector (K-diagnosis wire) in instrument panel wiring harness

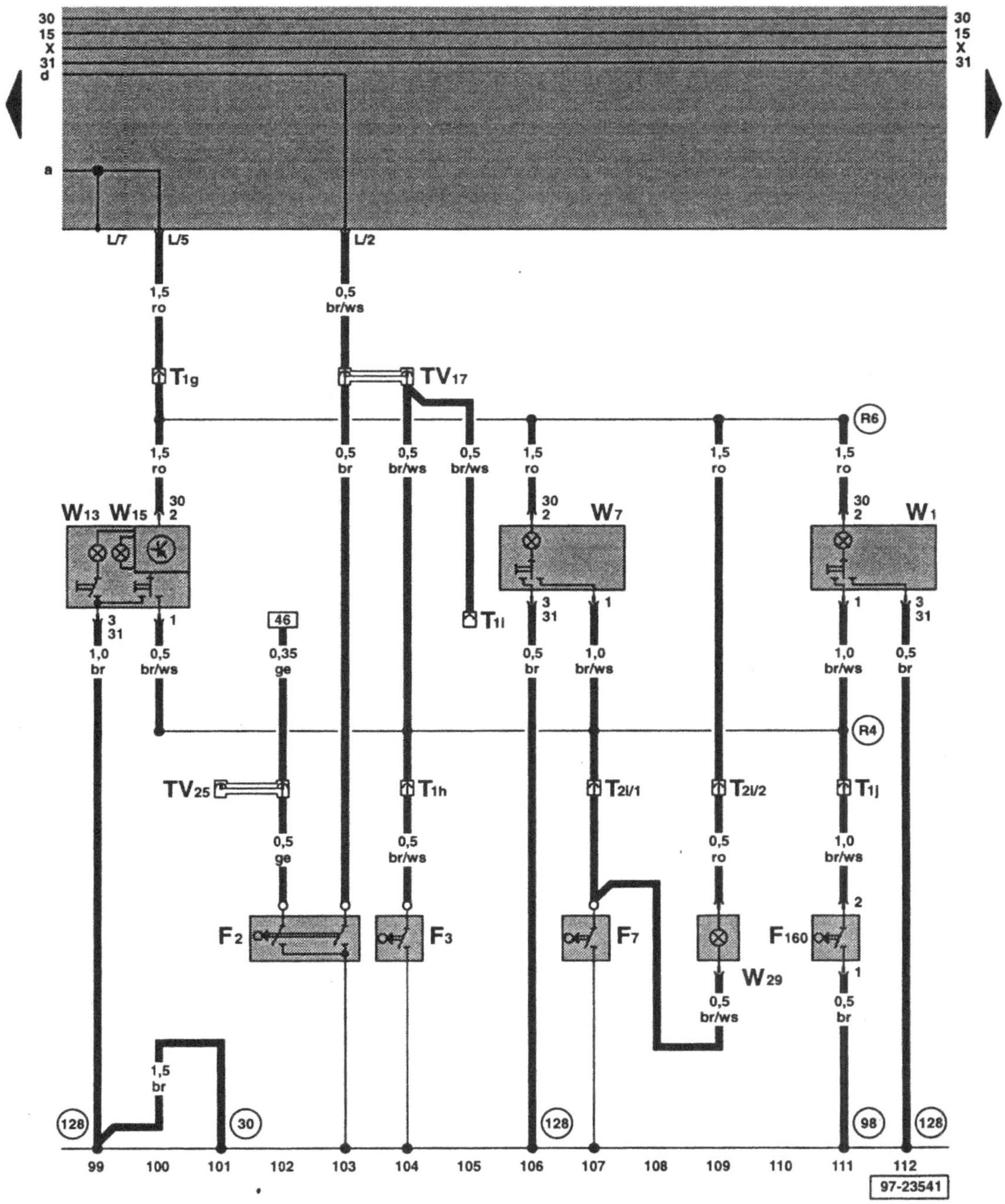

ws = white
sw = black
ro = red
br = brown
gn = green
bl = blue
gr = grey
li = violet
ge = yellow

F2 – Door contact Switch - Driver's side
F3 – Door contact switch-Passenger side
F7 – Right Rear Side/Sliding Door Contact Switch
F160 – Rear Lid Contact Switch
T1g – Single Connector, behind fuse/relay panel
T1h – Single Connector, behind instrument panel, right
T1i – Single Connector, behind fuse/relay panel
T1j – Single Connector, near taillamp assembly
T2i – Double Connector, near interior light, center
TV17 – Door Contact Switch Wire Connector, brown
TV25 – Door Contact Switch Wire Connector, pink
W1 – Interior light, front
W7 – Center Interior Light
W13 – Right Front Map/Reading Light
W15 – Interior Light with Delay Switch
W29 – Running board light

(30) – Ground connection, -1-, beside fuse/relay panel
(98) – Ground connection, in rear lid wiring harness
(128) – Ground connection -1-, in interior light wiring harness
(R4) – Wire connection -1-, in interior light/door contact switch wiring harness
(R6) – Plus connection -1-, in interior light wiring harness

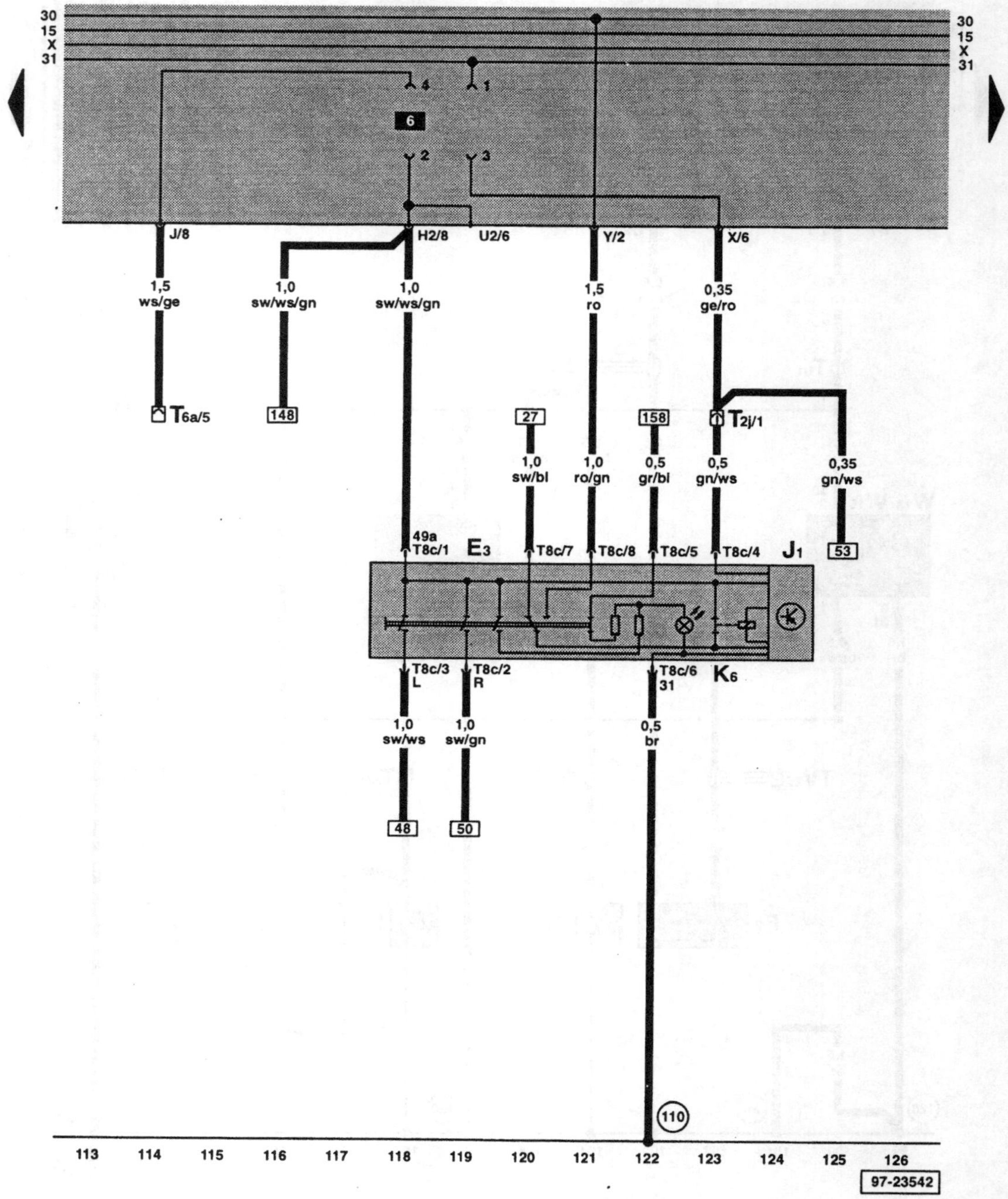

ws = white
sw = black
ro = red
br = brown
gn = green
bl = blue
gr = grey
li = violet
ge = yellow

E3 – Emergency Flasher Switch
J1 – Turn Signal Relay
K6 – Indicator light for emergency flasher system
T2j – Double Connector, behind fuse/relay panel
T6a – 6-Pin Connector, behind fuse/relay panel
T8c – 8-Pin Connector

(110) – Ground connection -2-, in instrument panel wiring harness

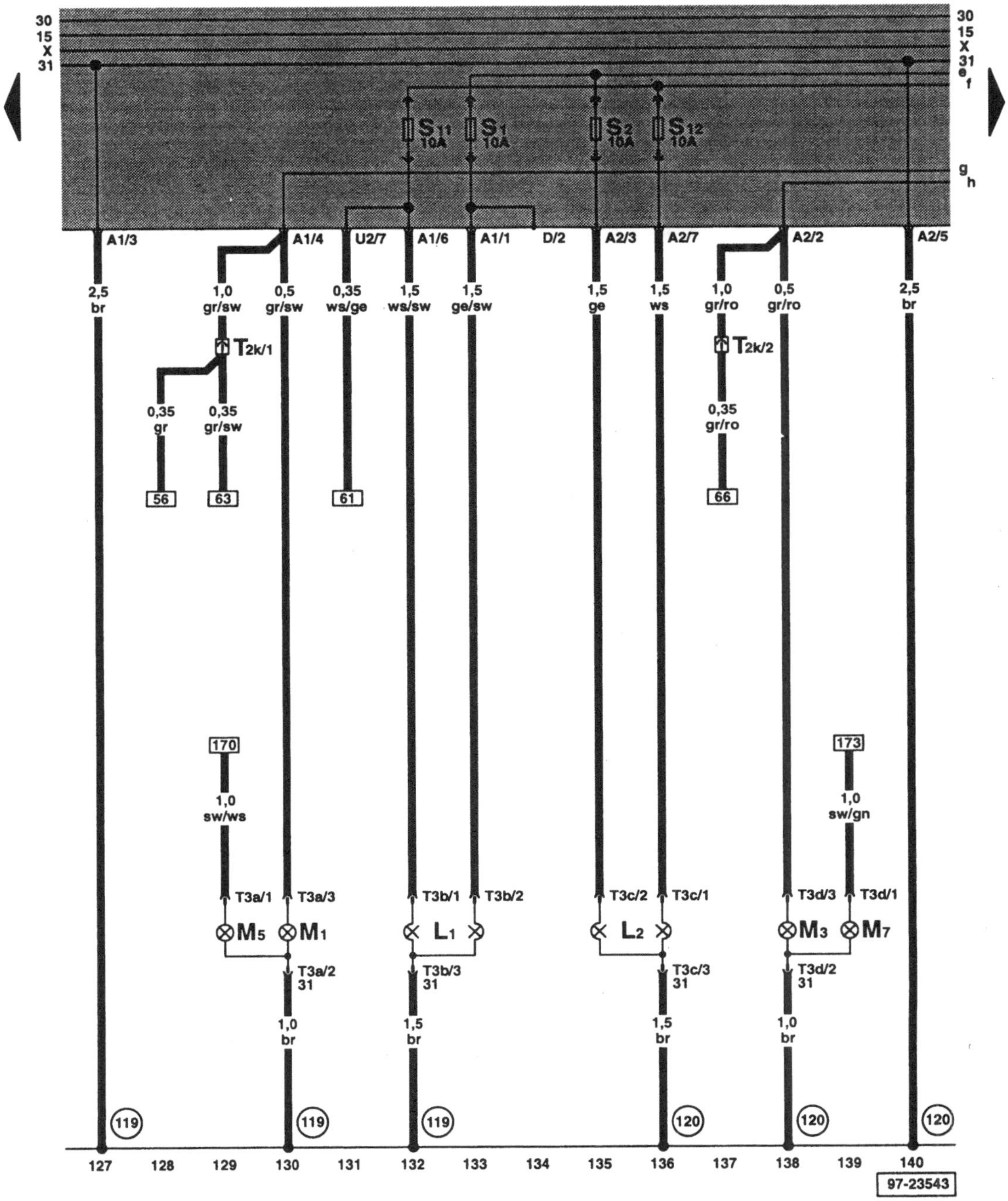

ws = white
sw = black
ro = red
br = brown
gn = green
bl = blue
gr = grey
li = violet
ge = yellow

L1 – Left Headlight
L2 – Right Headlight
M1 – Left Parking Light
M3 – Right Parking Light
M5 – Left Front Turn Signal Light
M7 – Right Front Turn Signal Light
T2k – Double Connector, behind fuse/relay panel
T3a – 3-Pin Connector
T3b – 3-Pin Connector
T3c – 3-Pin Connector
T3d – 3-Pin Connector

(119) – Ground connection -1-, in headlight wiring harness
(120) – Ground connection -2-, in headlight wiring harness

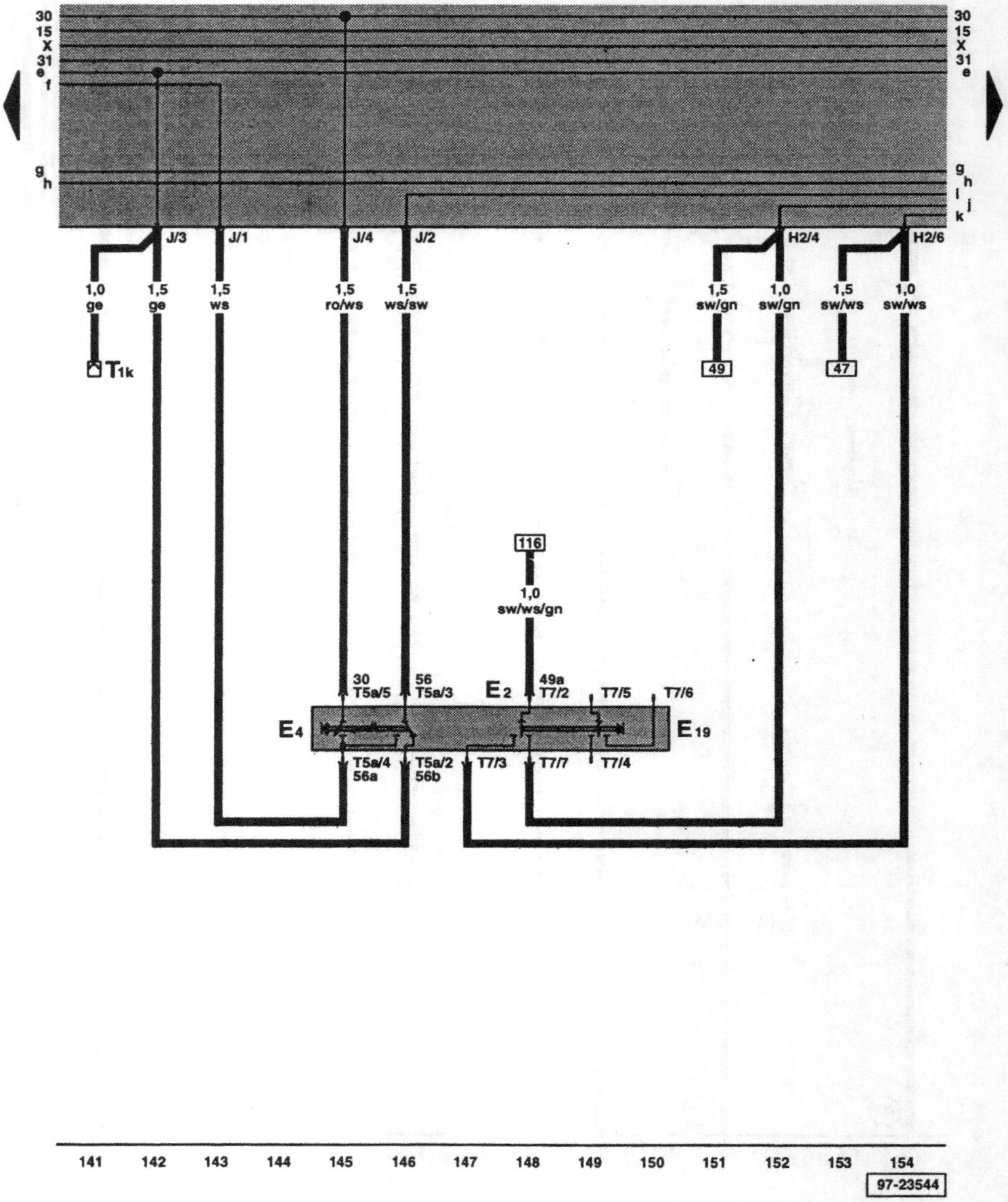

ws = white
sw = black
ro = red
br = brown
gn = green
bl = blue
gr = grey
li = violet
ge = yellow

- E2 – Turn signal switch
- E4 – Headlight Dimmer/Flasher Switch
- E19 – Park Light Switch
- T1k – Single Connector, behind fuse/relay panel
- T5a – 5-Pin Connector, behind steering column switch cover
- T7 – 7-Pin Connector, behind steering column switch cover

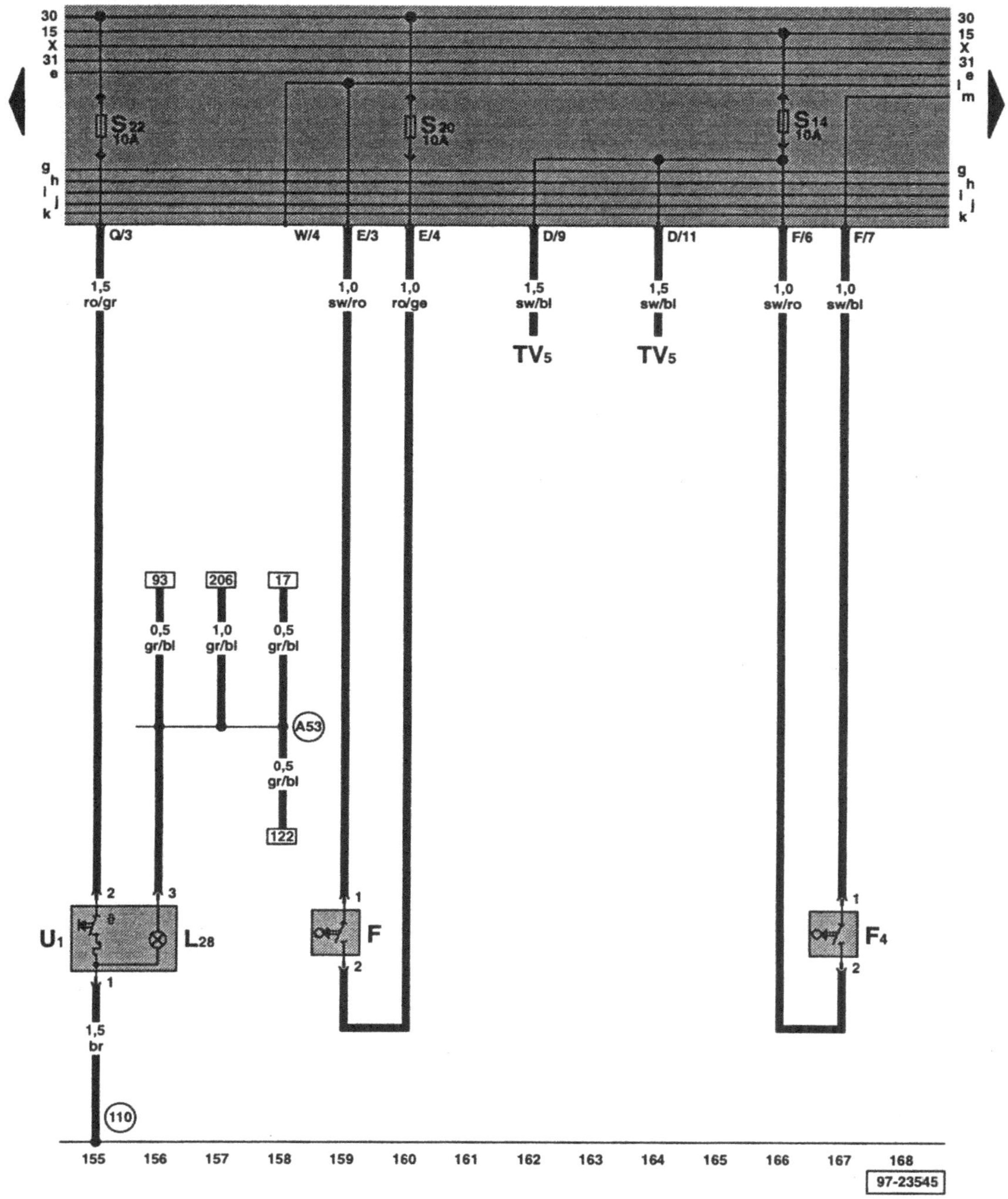

ws = white
sw = black
ro = red
br = brown
gn = green
bl = blue
gr = grey
li = violet
ge = yellow

F – Brake Light Switch
F4 – Back-Up Light Switch
L28 – Cigarette Lighter Light
TV5 – Terminal 15a Wire Connector, green, 4-Pin and 8-Pin
U1 – Cigarette Lighter

(110) – Ground connection -2-, in instrument panel wiring harness
(A53) – Plus connection (58b), in instrument panel wiring harness

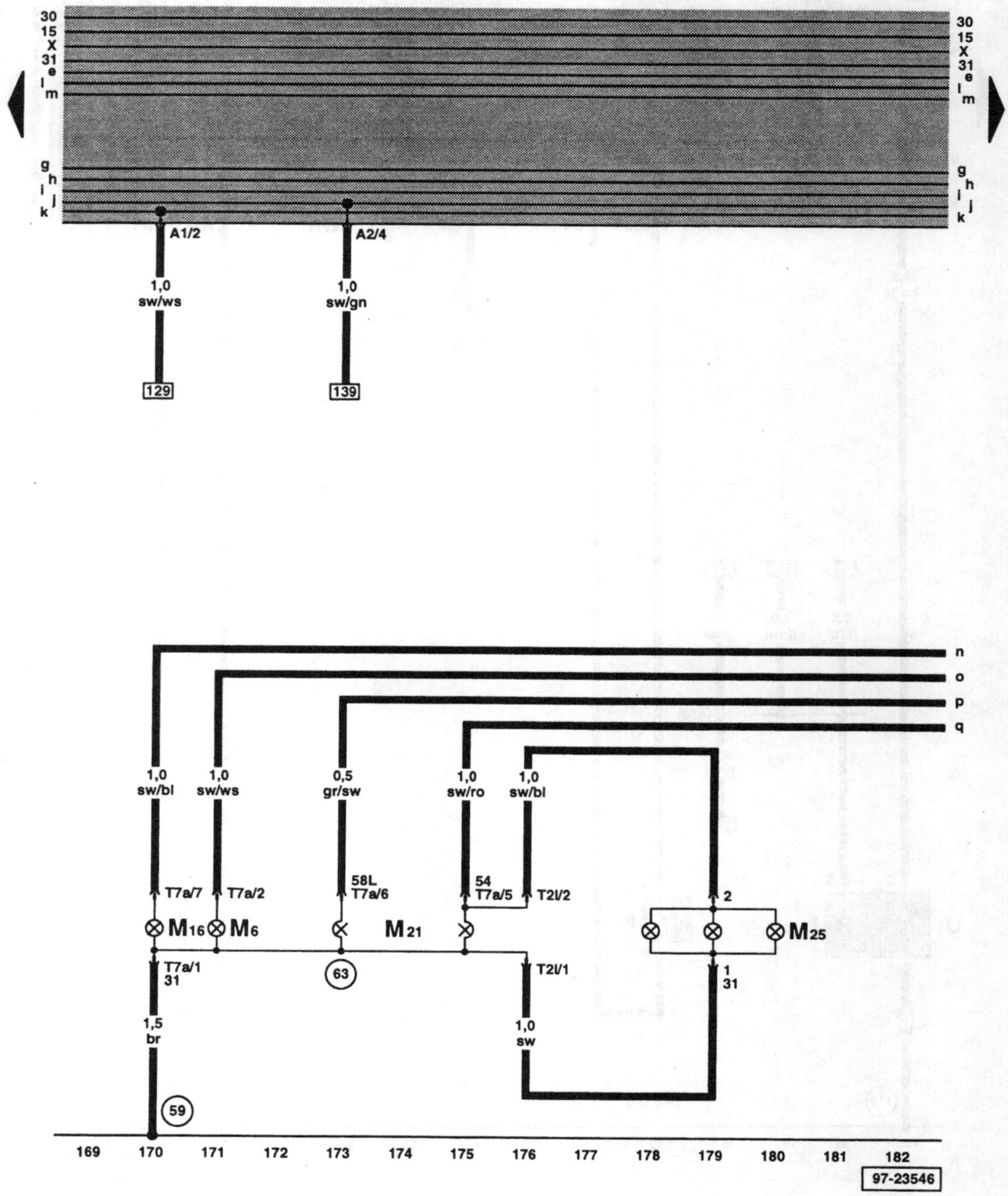

ws = white
sw = black
ro = red
br = brown
gn = green
bl = blue
gr = grey
li = violet
ge = yellow

M6 – Left Rear Turn Signal Light
M16 – Left Back-Up Light
M21 – Left Brake/Tail Light
M25 – High-mount Brake Light
T2l – Double Connector
T7a – 7-Pin Connector

(59) – Ground connection, near taillight, left

(63) – Ground connection, on left taillight bulb holder

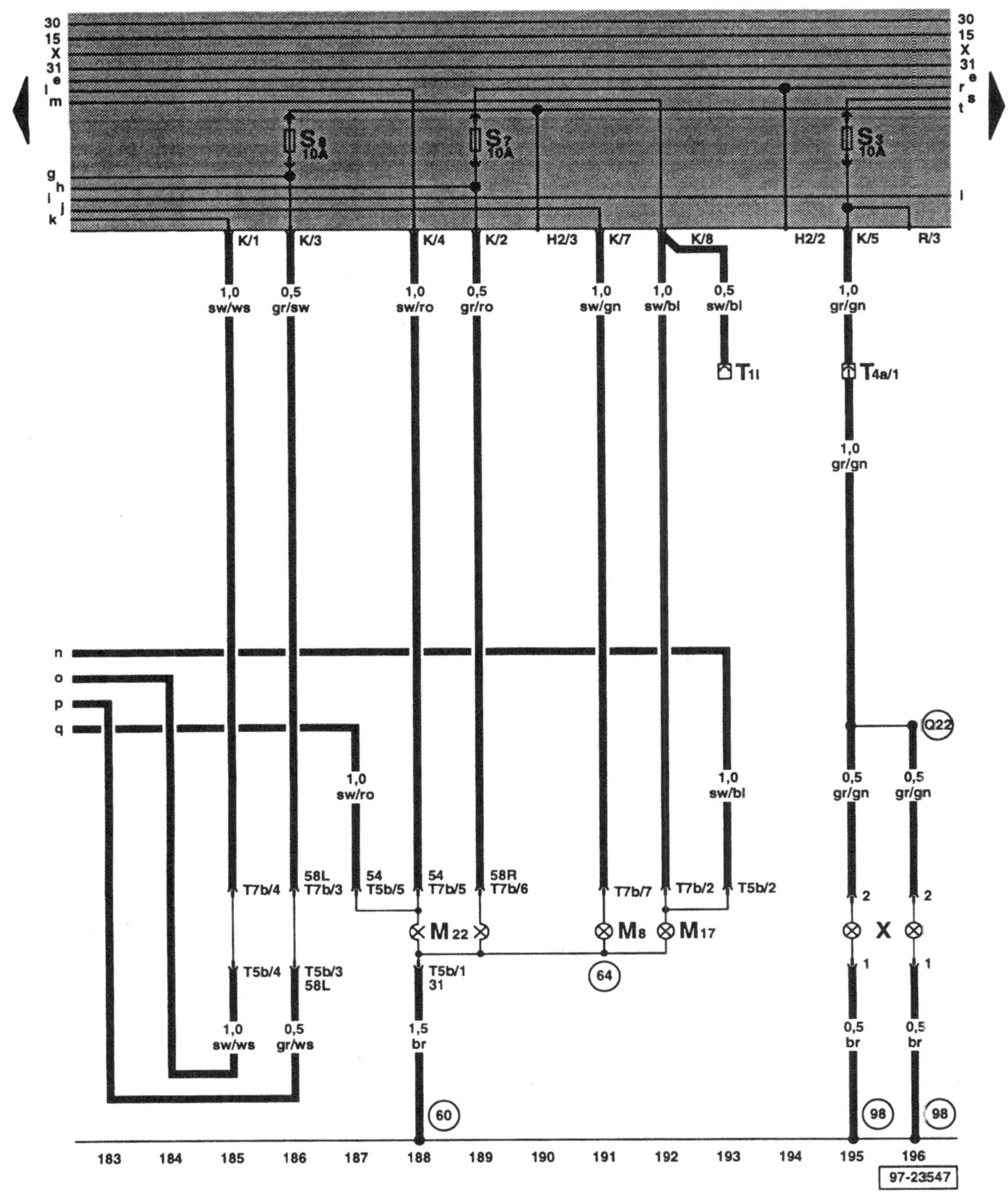

ws = white
sw = black
ro = red
br = brown
gn = green
bl = blue
gr = grey
li = violet
ge = yellow

M8 – Right Rear Turn Signal Light
M17 – Right Back-Up Light
M22 – Right Brake/Tail Light
T1l – Single Connector, behind fuse/relay panel
T4a – 4-Pin Connector, near taillamp assembly
T5b – 5-Pin Connector
T7b – 7-Pin Connector
X – License Plate Light

(60) – Ground point -2- near taillamp assembly, right

(64) – Ground connection, on right taillight bulb holder

(98) – Ground connection, in rear lid wiring harness

(Q22) – Wire connection -1-, in rear lid wiring harness

Right back-up light, right brake/tail light, right rear turn signal light, license plate light

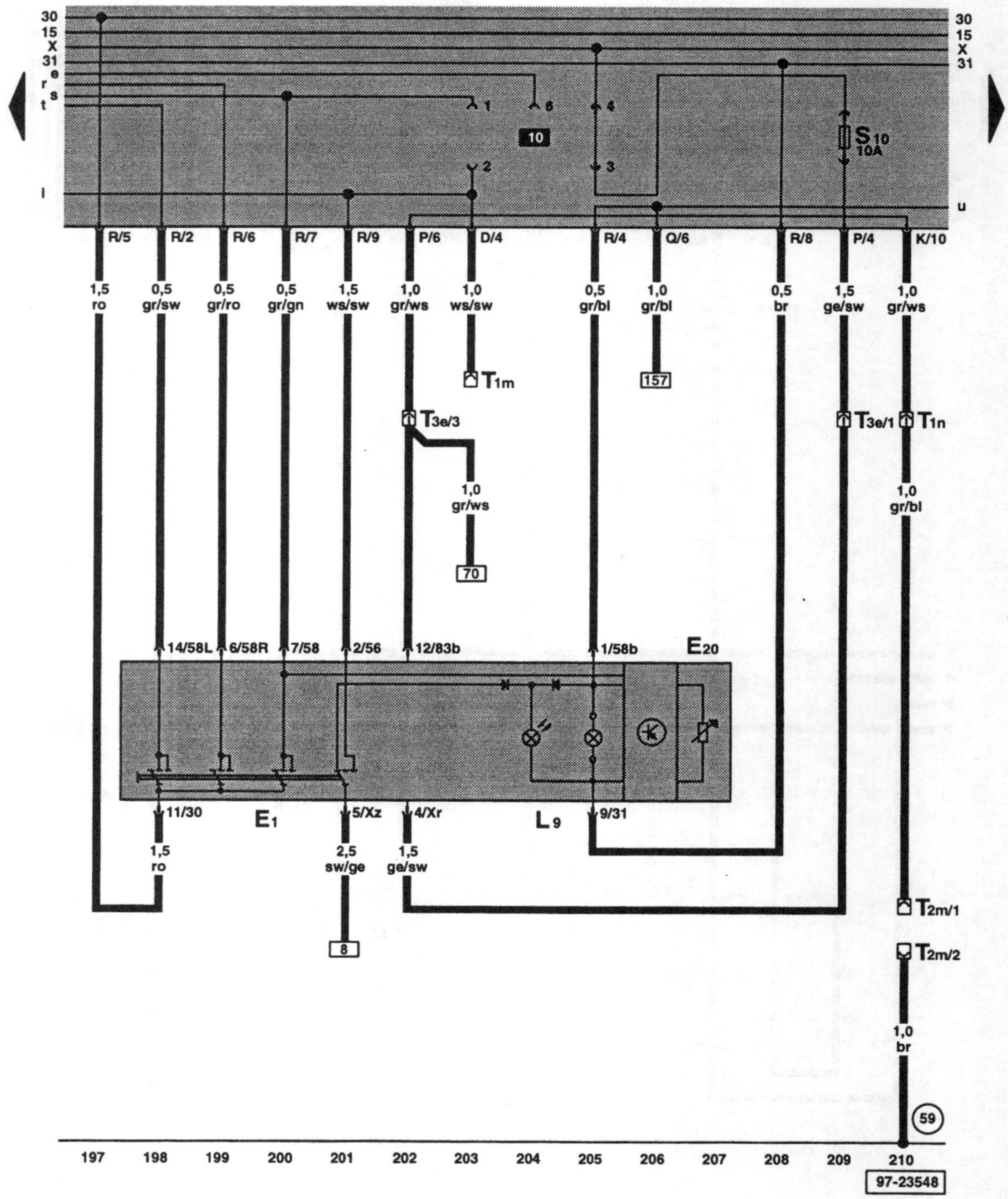

ws = white
sw = black
ro = red
br = brown
gn = green
bl = blue
gr = grey
li = violet
ge = yellow

E1 – Light switch
E20 – Instrument Panel Light Dimmer Switch
L9 – Headlight Switch Light
T1m – Single Connector, behind fuse/relay panel
T1n – Single Connector, near taillamp assembly
T2m – Double Connector
T3e – 3-Pin Connector, behind fuse/relay panel

(59) – Ground connection, near taillight, left

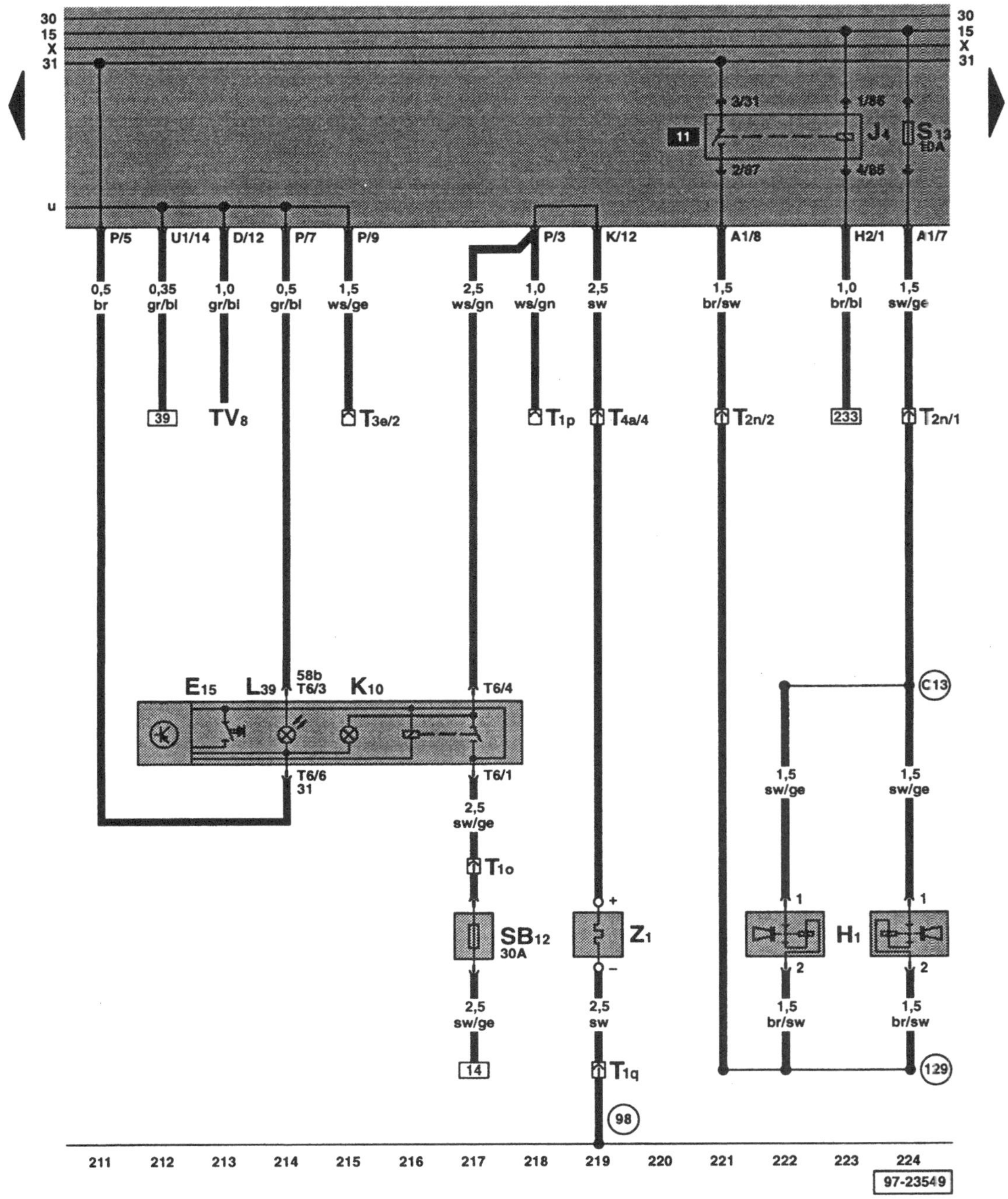

ws = white
sw = black
ro = red
br = brown
gn = green
bl = blue
gr = grey
li = violet
ge = yellow

- E15 – Rear window defogger switch
- H1 – Dual tone horn
- J4 – Dual Horn Relay
- K10 – Indicator light for heated rear windshield
- L39 – Rear Window Defogger Switch Light
- SB12 – Fuse 12 in fuse holder, below relay panel
- T1o – Single Connector, behind fuse/relay panel
- T1p – Single Connector, behind fuse/relay panel
- T1q – Single Connector, in rear lid
- T2n – Double Connector, in engine compartment, left front
- T3e – 3-Pin Connector, behind fuse/relay panel
- T4a – 4-Pin Connector, near taillamp assembly
- T6 – 6-Pin Connector
- TV8 – Terminal 58b Wire Connector, grey
- Z1 – Heated rear window

- (98) – Ground connection, in rear lid wiring harness
- (129) – Ground connection, in dual-tone horn wiring harness
- (C13) – Plus connection, in dual tone horn wiring harness

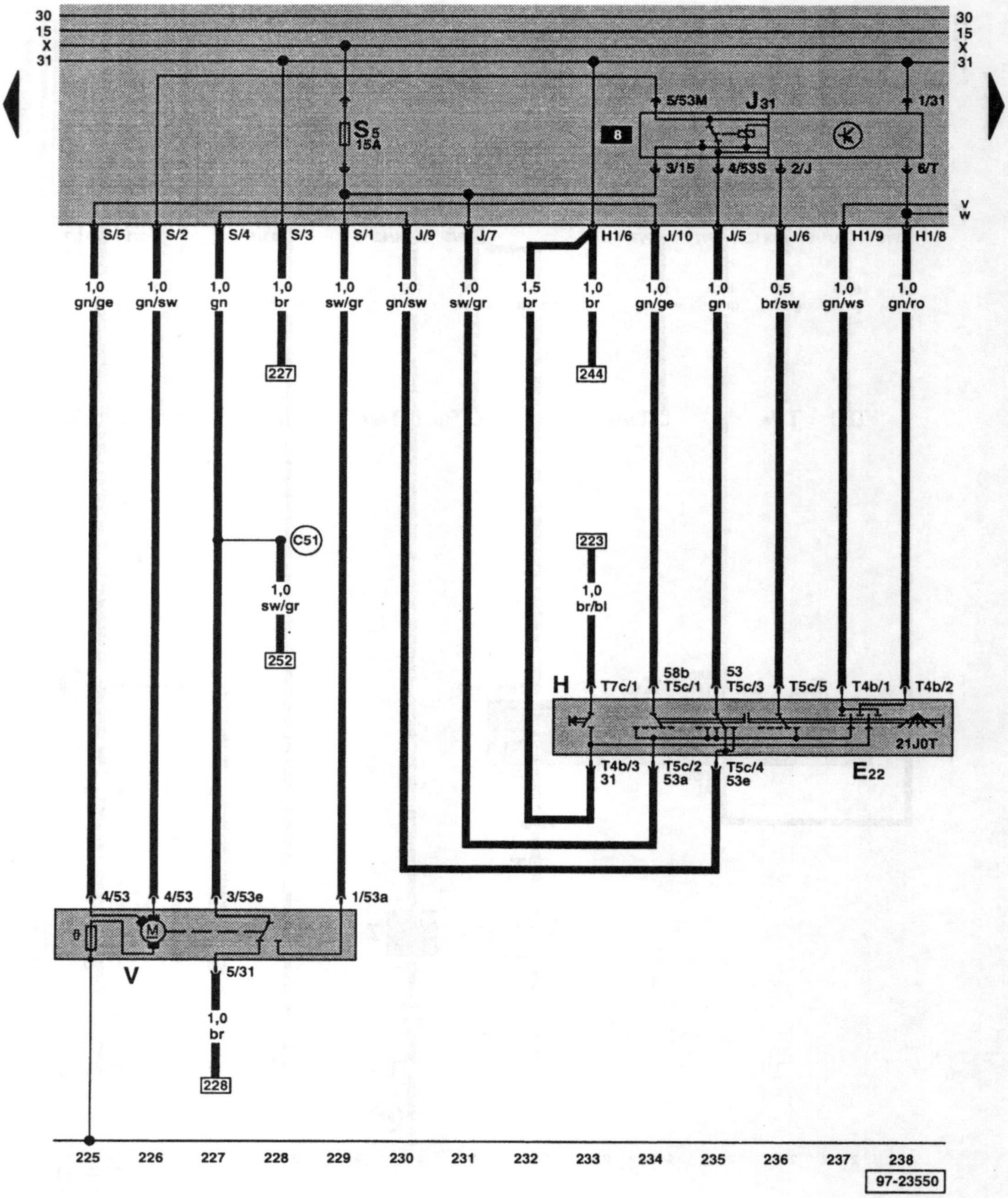

ws = white
sw = black
ro = red
br = brown
gn = green
bl = blue
gr = grey
li = violet
ge = yellow

E22 – Windshield Wiper/Washer Switch

H – Signal horn activation

J31 – Wiper/Washer Intermittent Relay (programable)

T4b – 4-Pin Connector, behind steering column switch cover

T5c – 5-Pin Connector, behind steering column switch cover

T7c – 7-Pin Connector, behind steering column switch cover

V – Windshield Wiper Motor

(C51) – Plus connection (x), in wiper motor wiring harness

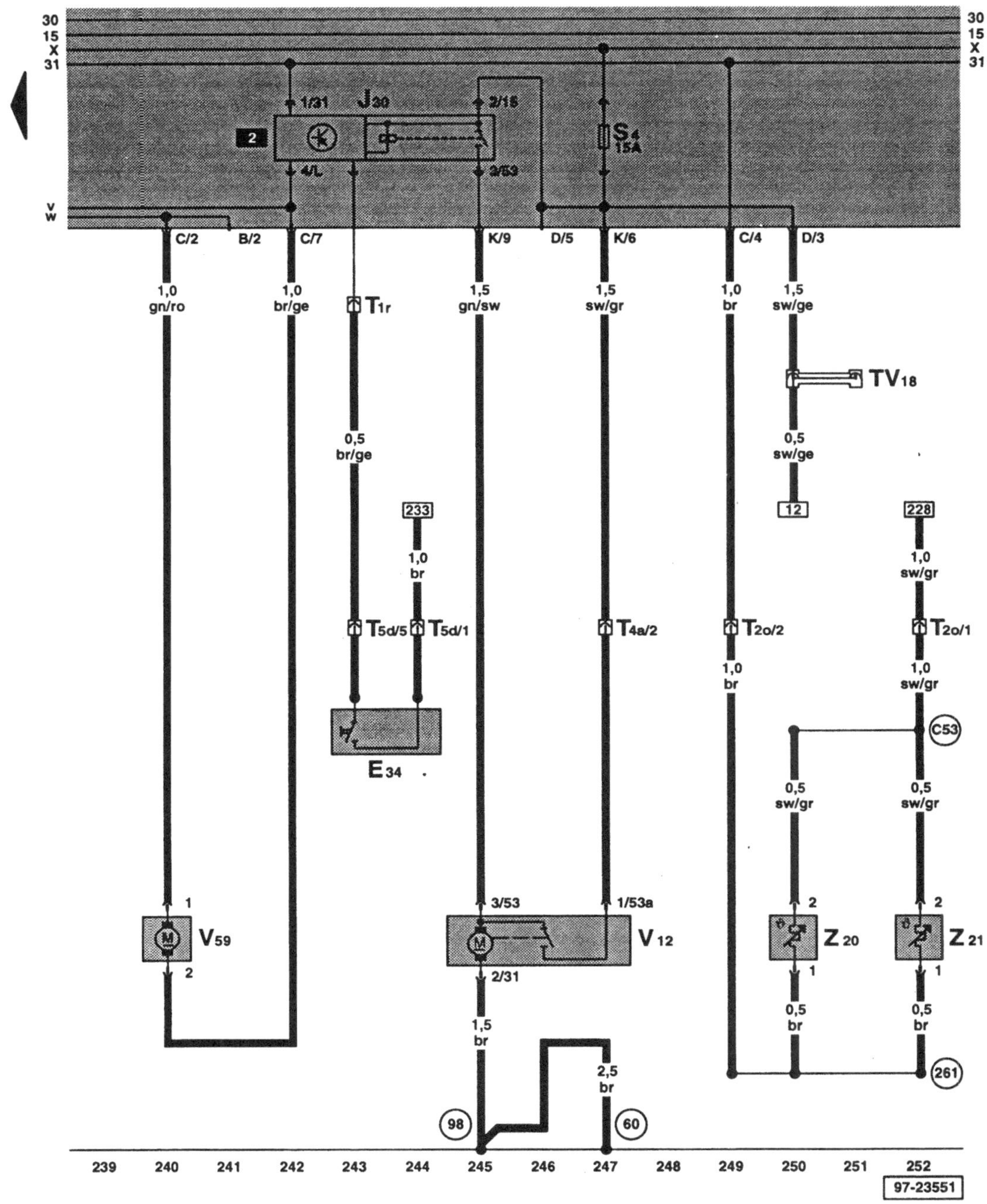

ws = white
sw = black
ro = red
br = brown
gn = green
bl = blue
gr = grey
li = violet
ge = yellow

E34 – Rear Window Wiper Switch
J30 – Rear Window Wiper/Washer Relay
T1r – Single Connector, on rear window wiper/washer relay
T2o – Double Connector, on firewall, right
T4a – 4-Pin Connector, near taillamp assembly
T5d – 5-Pin Connector, behind steering column switch cover
TV18 – Terminal Xa Wire Connector, violet
V12 – Motor for rear windshield wiper
V59 – Windshield and Rear Window Washer Pump
Z20 – Left Washer Nozzle Heater
Z21 – Right Washer Nozzle Heater

(60) – Ground point -2- near taillamp assembly, right

(98) – Ground connection, in rear lid wiring harness

(261) – Ground connection, in wiring harness heated spray jet

(C53) – Plus connection (x) in wiring harness heated spray jet

Rear window wiper switch, motor for rear windshield wiper, windshield and rear window washer pump, washer nozzle heater

THIS PAGE INTENTIONALLY LEFT
BLANK

2.8 L- Engine - Motronic Multiport Fuel Injection (MFI)/103 kW, code AES,

from September 1998

Deviate relay location and fuseplacements as well as the locations of multiple connectors see section "component locations".

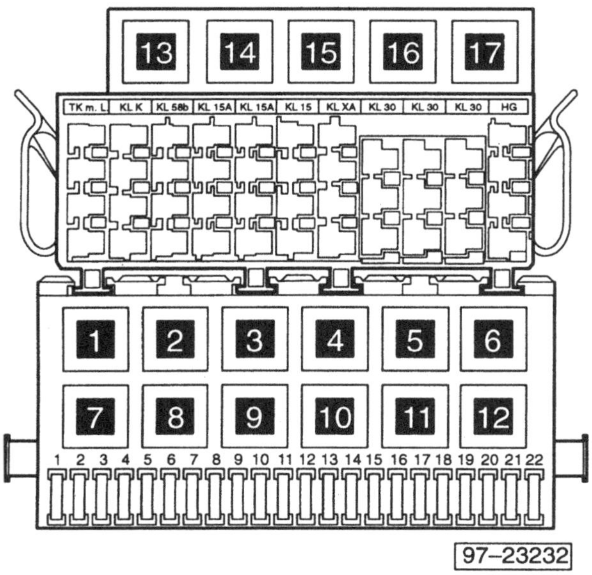

- **12** Fuel Pump (FP) Relay (167)
- **13** Coolant Circulation Pump Relay (53)
- **15** Secondary Air Injection (AIR) Pump Relay (53)

Note: Number in parentheses indicates production control number stamped on relay housing.

Fuse colors

30 A – green
25 A – white
20 A – yellow
15 A – blue
10 A – red
7,5 A – brown
5 A – beige
3 A – violet

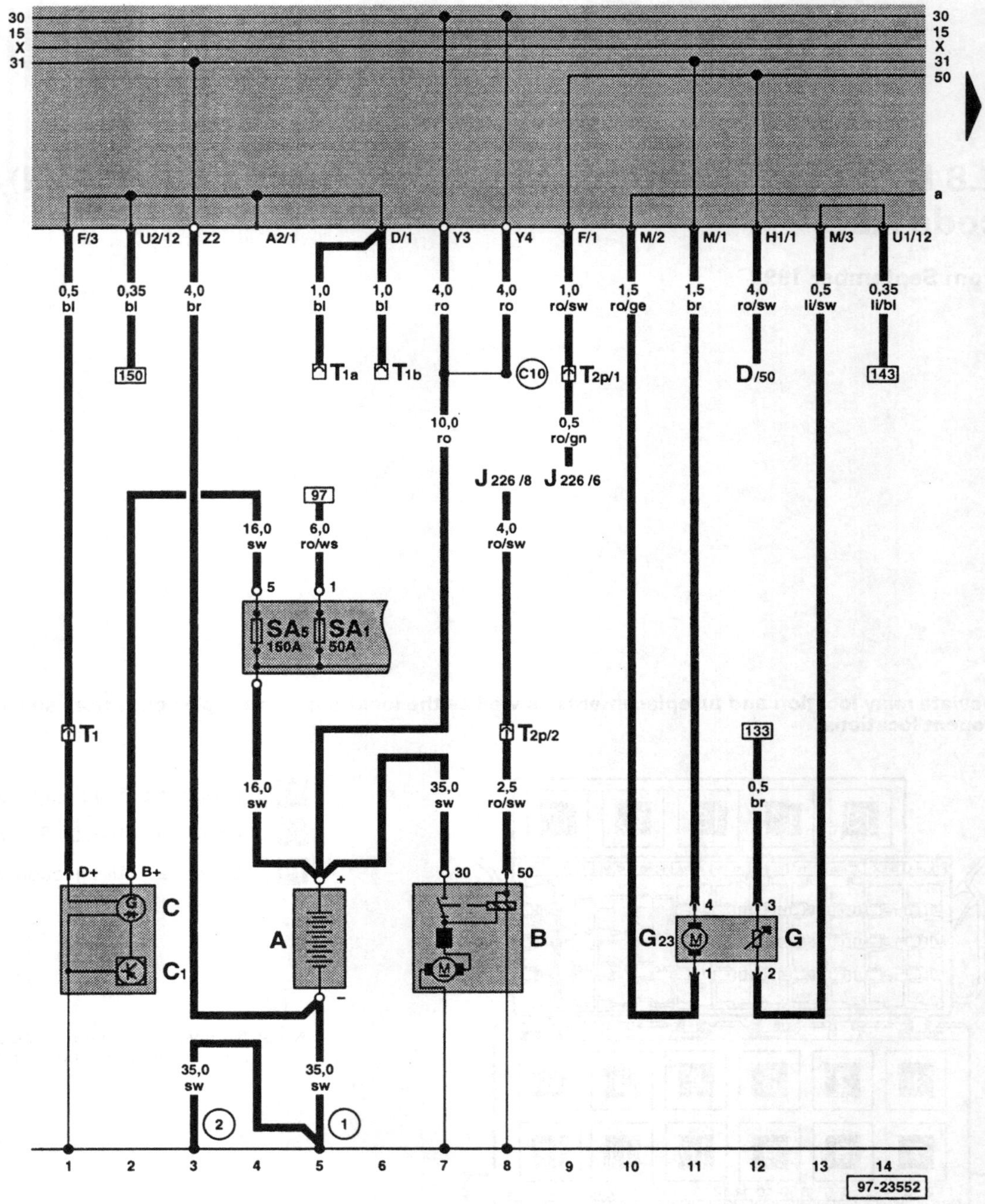

ws = white
sw = black
ro = red
br = brown
gn = green
bl = blue
gr = grey
li = violet
ge = yellow

A – Battery
B – Starter
C – Generator (GEN)
C1 – Voltage Regulator (VR)
D – Ignition/Starter Switch
G – Fuel Level Sensor
G23 – Transfer Fuel Pump (FP)
J226 – Park/Neutral Position (PNP) Relay, behind instrument panel, center
SA1 – Fuse -1- in fuse bracket/battery
SA5 – Fuse -5- in fuse bracket/battery
T1 – Single Connector, near starter
T1a – Single Connector, behind fuse/relay panel
T1b – Single Connector, behind fuse/relay panel
T2p – Double Connector, behind fuse/relay panel

(1) – Ground strap, battery to body
(2) – Ground strap, transmission to body
(C10) – Plus connection -1- (30), in headlight wiring harness

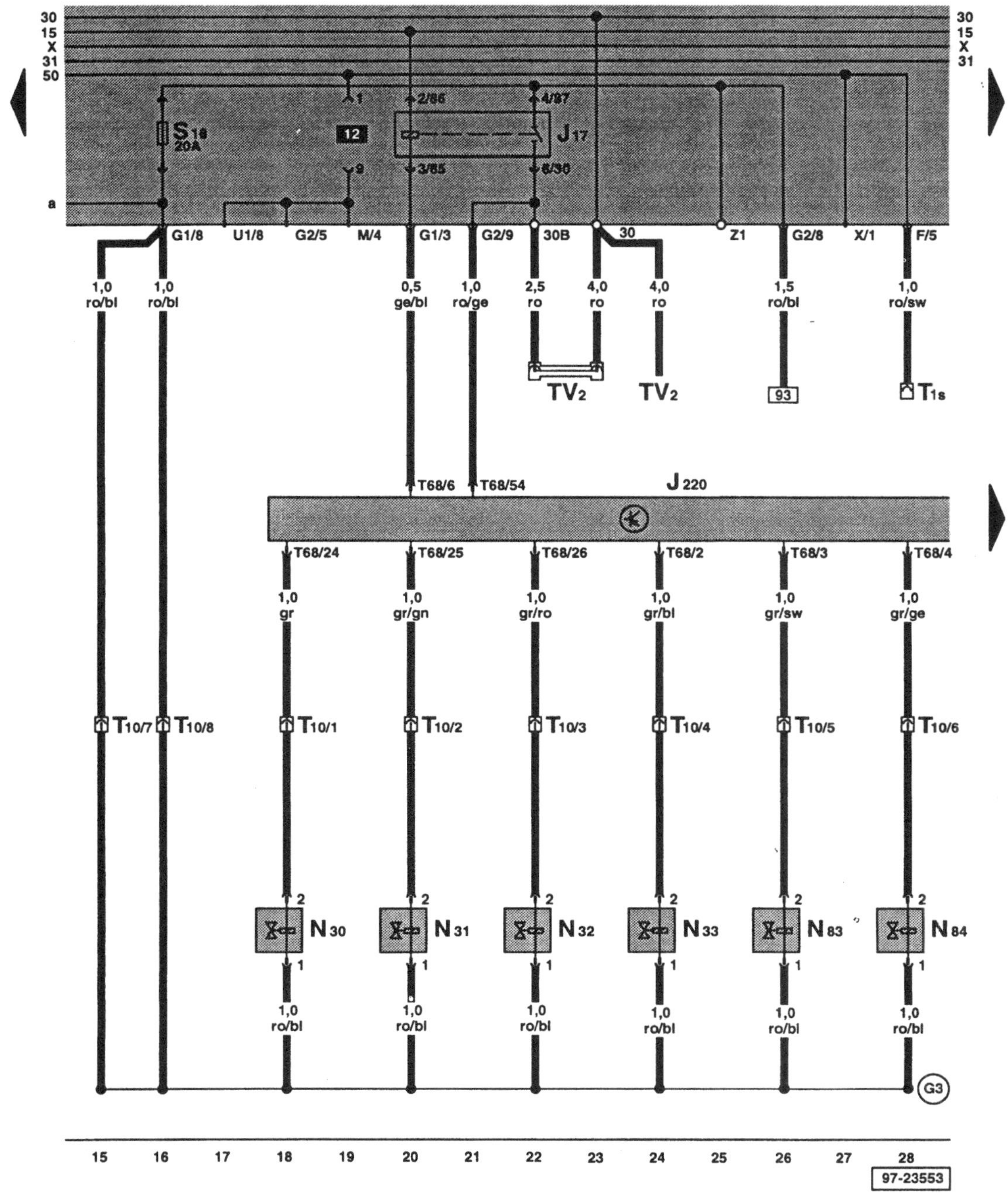

ws = white
sw = black
ro = red
br = brown
gn = green
bl = blue
gr = grey
li = violet
ge = yellow

J17 – Fuel Pump (FP) Relay
J220 – Motronic Engine Control Module (ECM), in engine compartment, left
N30 – Cylinder 1 Fuel Injector
N31 – Cylinder 2 Fuel Injector
N32 – Cylinder 3 Fuel Injector
N33 – Cylinder 4 Fuel Injector
N83 – Cylinder 5 Fuel Injector
N84 – Cylinder 6 Fuel Injector
T1s – Single Connector, behind fuse/relay panel
T10 – 10-Pin Connector, near cylinder head cover
T68 – 68-Pin Connector
TV2 – Terminal 30 (B+) Wire Connector, red, each 6 points

(G3) – Plus connection, in injector wiring harness

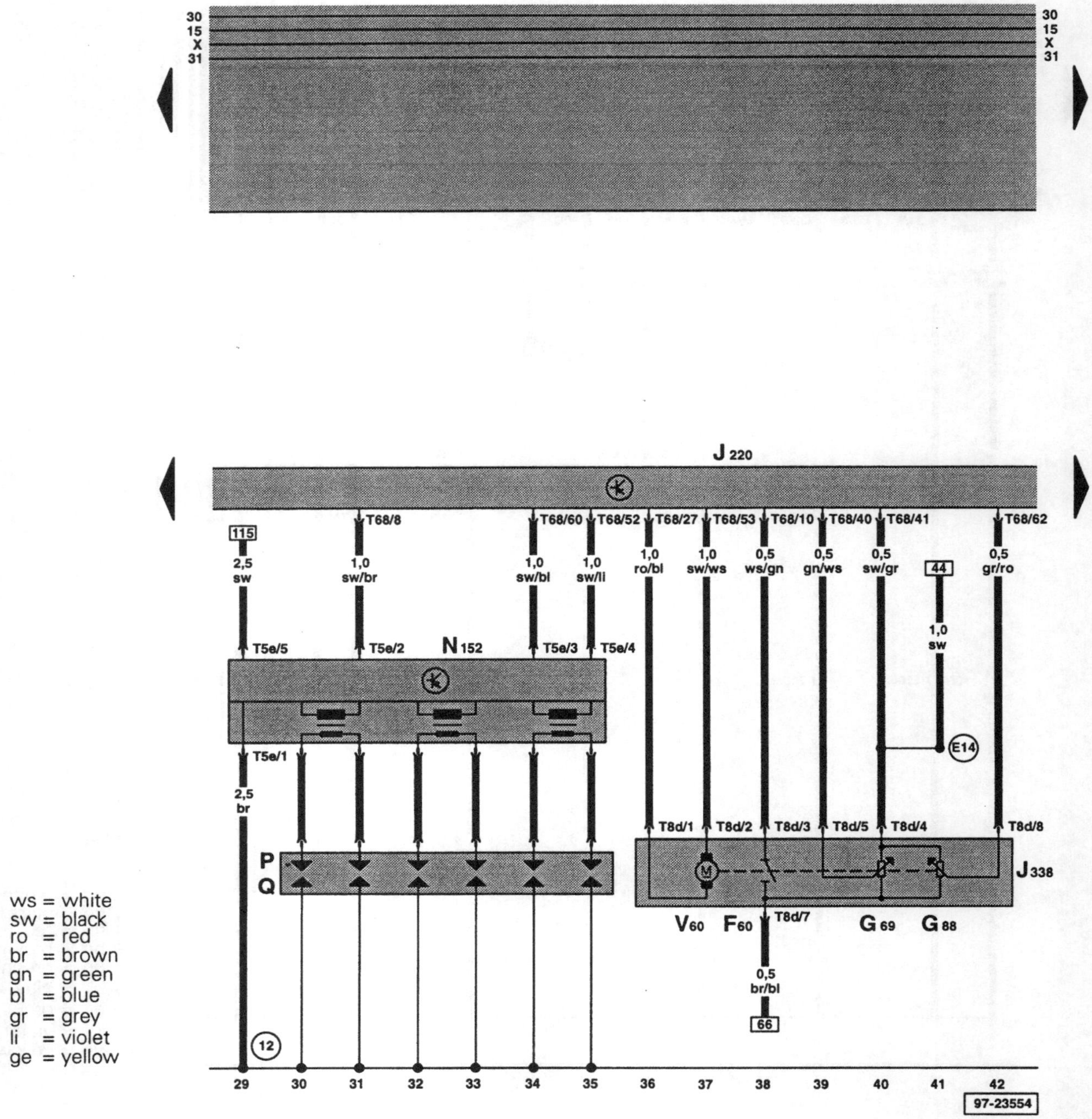

F60 – Closed Throttle Position (CTP) Switch
G69 – Throttle Position (TP) Sensor
G88 – Throttle Position (TP) Sensor
J220 – Motronic Engine Control Module (ECM), in engine compartment, left
J338 – Throttle Valve Control Module
N152 – Ignition Coil
P – Spark Plug Connectors
Q – Spark Plugs
T5e – 5-Pin Connector
T8d – 8-Pin Connector
T68 – 68-Pin Connector
V60 – Throttle Position (TP) Actuator

(12) – Ground connection, in engine compartment, left

(E14) – Wire connection -1-, in Motronic Multiport Fuel Injection (MFI) wiring harness

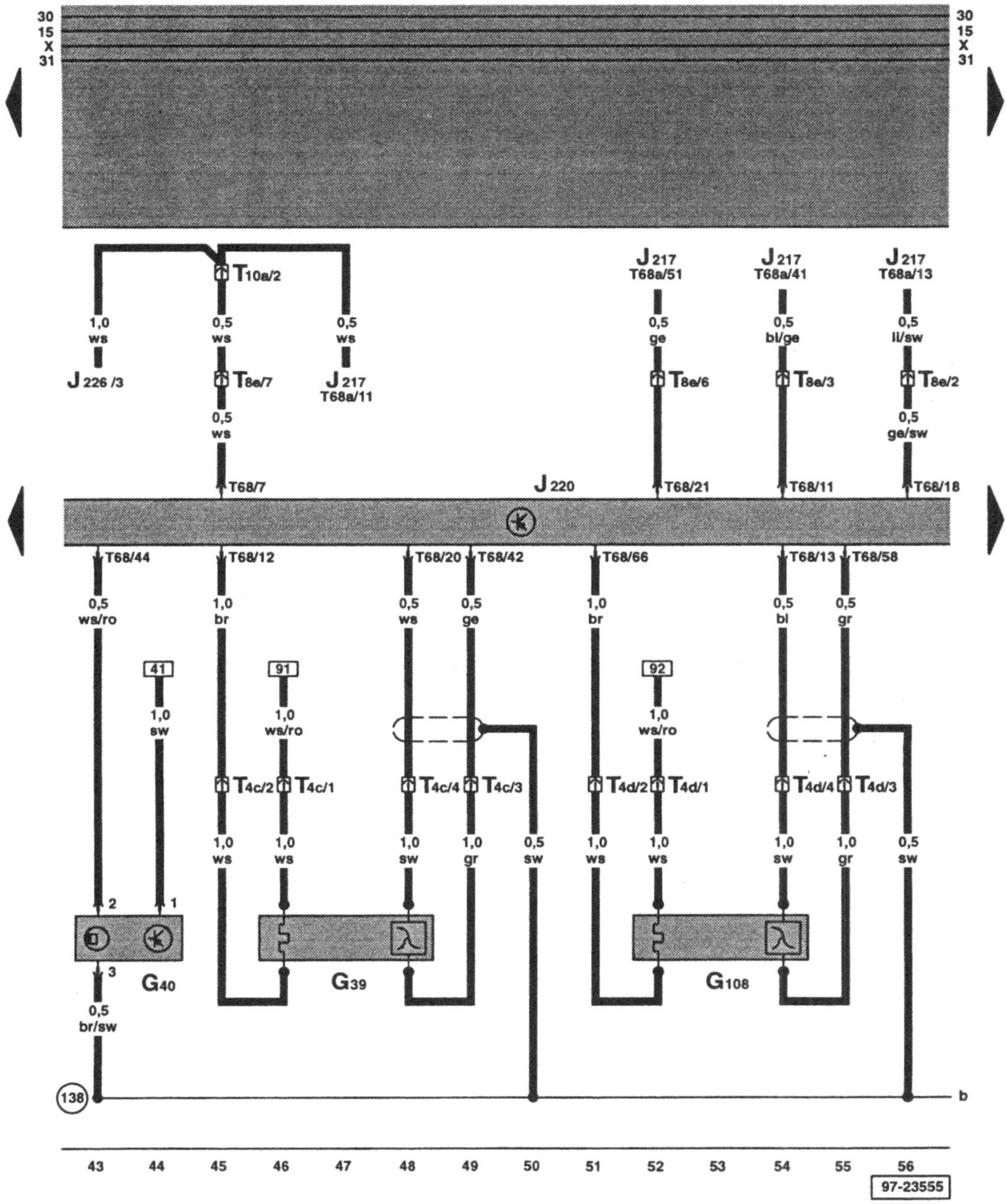

ws = white
sw = black
ro = red
br = brown
gn = green
bl = blue
gr = grey
li = violet
ge = yellow

G39 – Heated Oxygen Sensor (HO2S)
G40 – Camshaft Position (CMP) Sensor
G108 – Heated Oxygen Sensor (HO2S) 2
J217 – Transmission Control Module (TCM), on left A-pillar, lower part
J220 – Motronic Engine Control Module (ECM), in engine compartment, left
J226 – Park/Neutral Position (PNP) Relay, behind instrument panel, center
T4c – 4-Pin Connector, near Heated Oxygen Sensor (HO2S)
T4d – 4-Pin Connector, near Heated Oxygen Sensor (HO2S)
T8e – 8-Pin Connector, behind fuse/relay panel
T10a – 10-Pin Connector, behind fuse/relay panel
T68 – 68-Pin Connector
T68a – 68-Pin Connector, on transmission control module (TCM)
(138) – Ground connection (control module), in Motronic Multiport Fuel Injection (MFI) wiring harness

Motronic engine control module (ECM), heated oxygen sensors, camshaft position (CMP) sensor

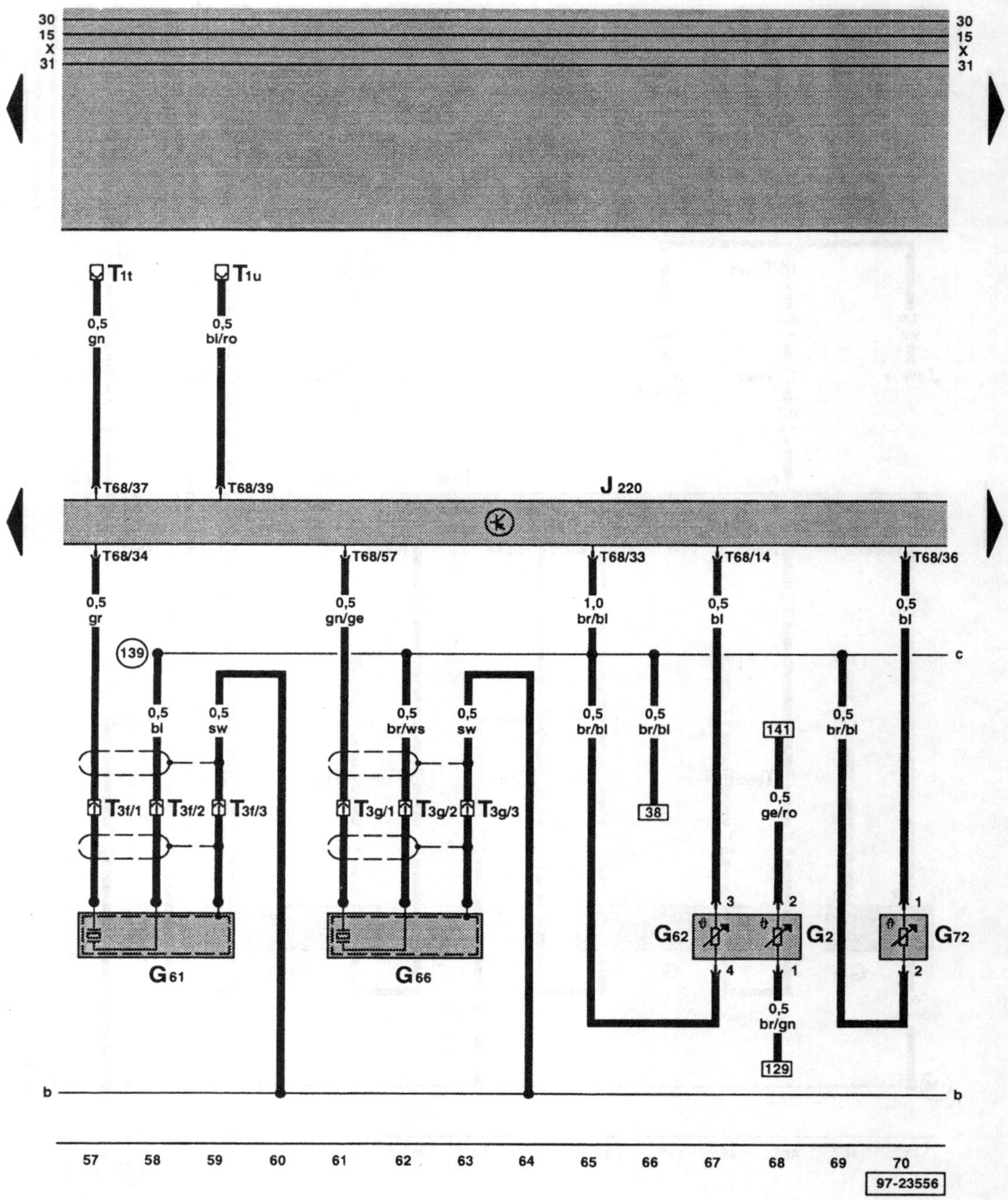

ws = white
sw = black
ro = red
br = brown
gn = green
bl = blue
gr = grey
li = violet
ge = yellow

G2 – Engine Coolant Temperature (ECT) Sensor
G61 – Knock Sensor (KS) 1
G62 – Engine Coolant Temperature (ECT) Sensor
G66 – Knock Sensor (KS) 2
G72 – Intake Air Temperature (IAT) Sensor
J220 – Motronic Engine Control Module (ECM), in engine compartment, left
T1t – Single Connector (Connection A/C), behind fuse/relay panel
T1u – Single Connector (Connection A/C), behind fuse/relay panel
T3f – 3-Pin Connector, near cylinder head cover
T3g – 3-Pin Connector, near cylinder head cover
T68 – 68-Pin Connector

(139) – Ground connection (sensor Ground), in Motronic Multiport Fuel Injection (MFI) wiring harness

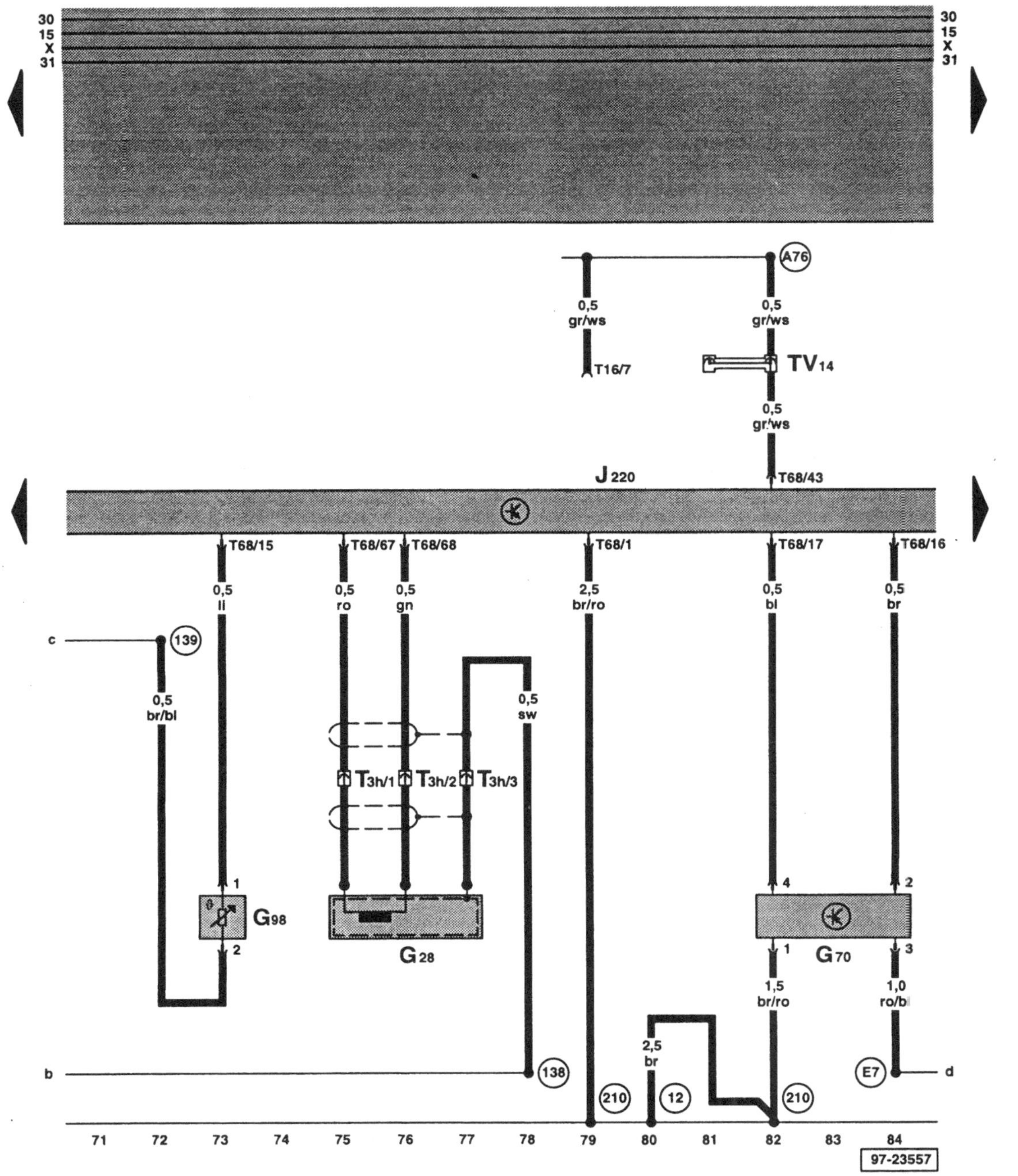

ws = white
sw = black
ro = red
br = brown
gn = green
bl = blue
gr = grey
li = violet
ge = yellow

G28 – Engine Speed (RPM) Sensor
G70 – Mass Air Flow (MAF) Sensor
G98 – EGR Temperature Sensor
J220 – Motronic Engine Control Module (ECM), in engine compartment, left
T3h – 3-Pin Connector, near starter
T16 – Data Link Connector (DLC), behind steering column cover
T68 – 68-Pin Connector
TV14 – Data Link Connector (DLC) wire Connector, white

(12) – Ground connection, in engine compartment, left

(138) – Ground connection (control module), in Motronic Multiport Fuel Injection (MFI) wiring harness
(139) – Ground connection (sensor Ground), in Motronic Multiport Fuel Injection (MFI) wiring harness
(210) – Ground connection -1-, in Motronic Multiport Fuel Injection (MFI) wiring harness
(A76) – Connector (K-diagnosis wire), in instrument panel wiring harness
(E7) – Plus connection (87a), in Motronic Multiport Fuel Injection (MFI) wiring harness

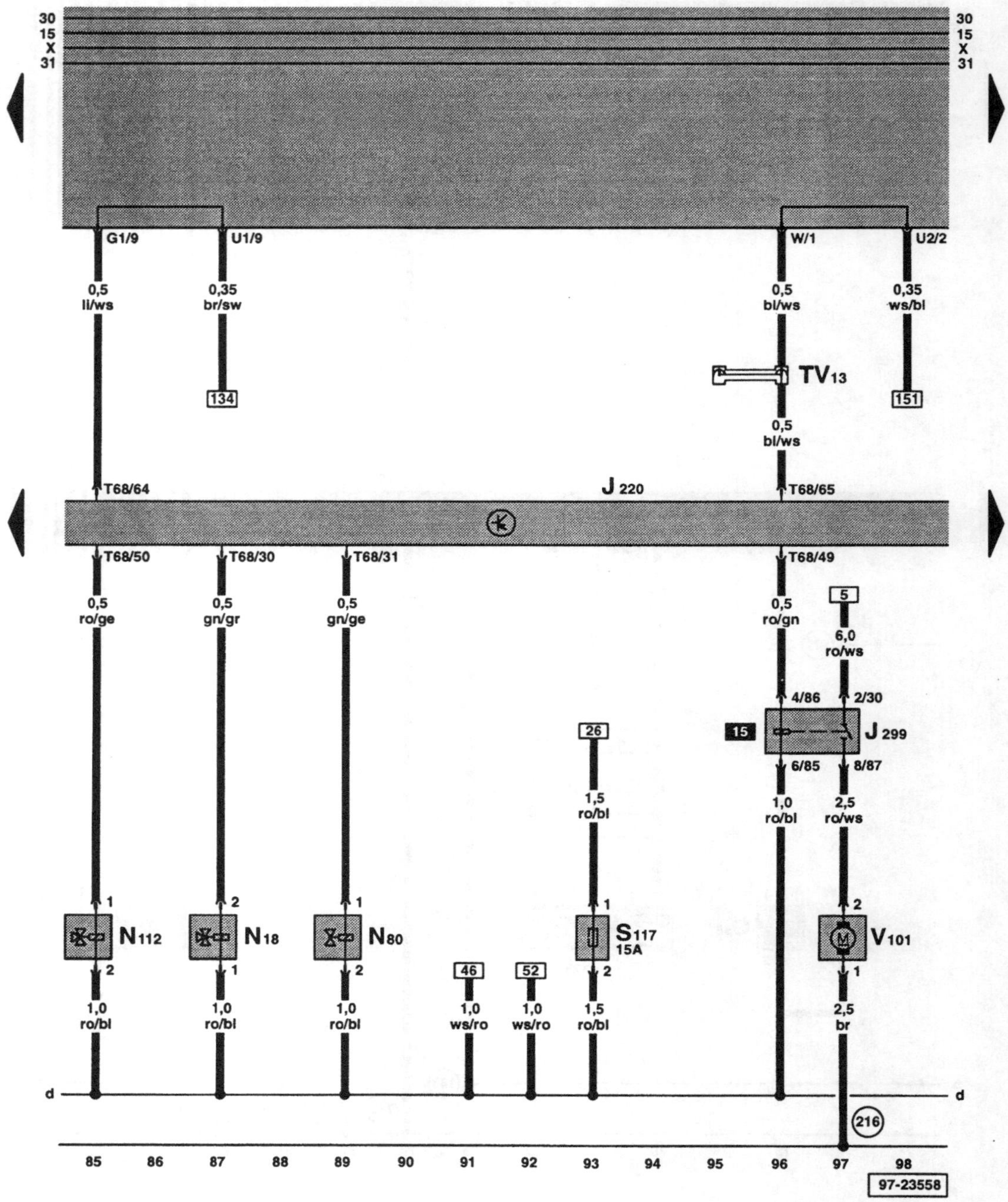

ws = white
sw = black
ro = red
br = brown
gn = green
bl = blue
gr = grey
li = violet
ge = yellow

J220 – Motronic Engine Control Module (ECM), in engine compartment, left

J299 – Secondary Air Injection (AIR) Pump Relay, above fuse/relay panel

N18 – EGR Vacuum Regulator Solenoid Valve

N80 – Evaporative Emission (EVAP) Canister Purge Regulator Valve

N112 – Secondary Air Injection (AIR) Solenoid Valve

S117 – Fuse for control elements, above fuse/relay panel

T68 – 68-Pin Connector

TV13 – Vehicle Speed Signal Wire Connector, blue

V101 – Secondary Air Injection (AIR) Pump Motor

(216) – Ground connection -2-, in Motronic Multiport Fuel Injection (MFI) wiring harness

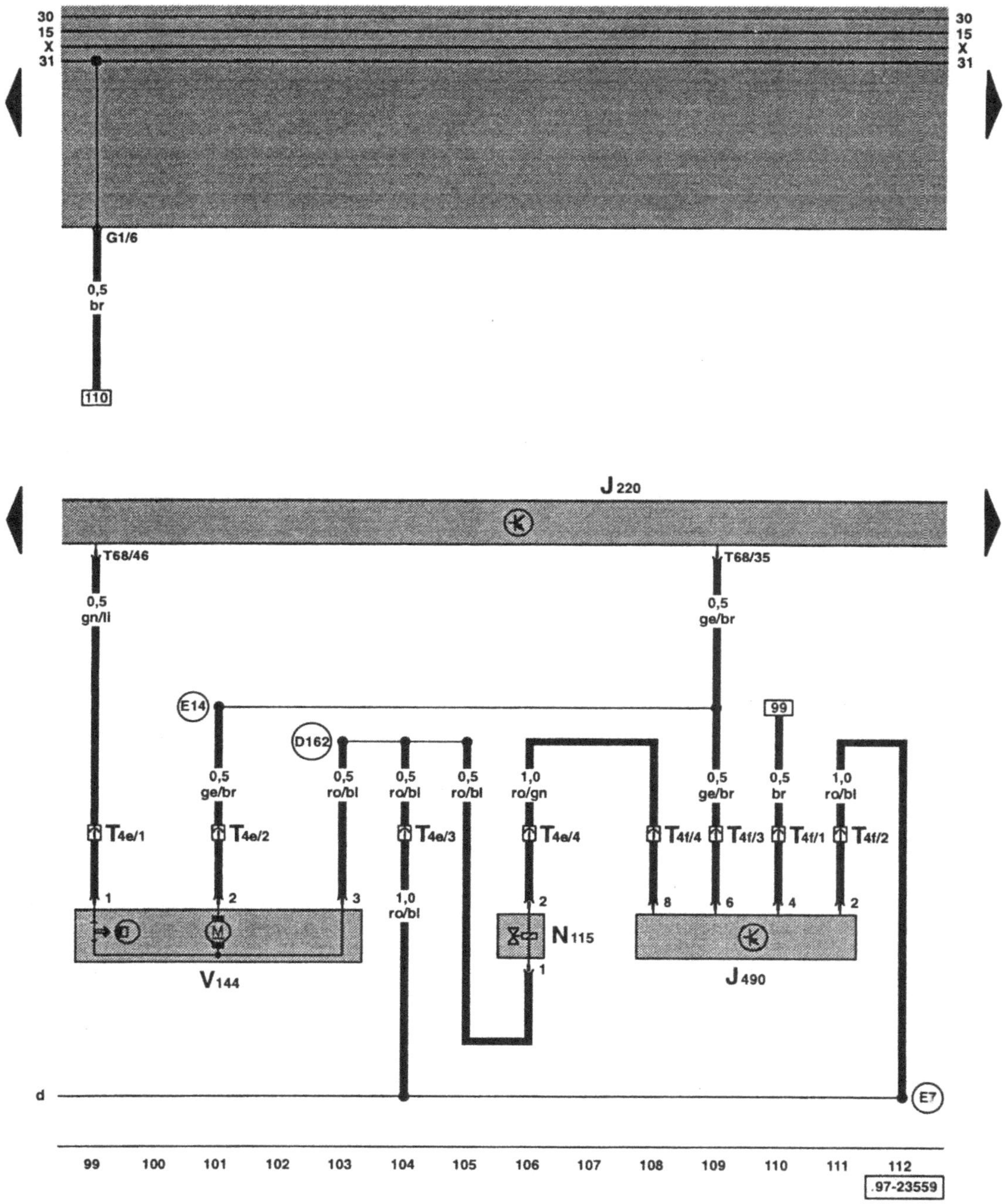

ws = white
sw = black
ro = red
br = brown
gn = green
bl = blue
gr = grey
li = violet
ge = yellow

J220 – Motronic Engine Control Module (ECM), in engine compartment, left

J490 – Fuel system Diagnostic control module, production control number (194), behind instrument panel, center

N115 – Evaporative Emission (EVAP) Canister Purge Solenoid Valve

T4e – 4-Pin Connector, in engine compartment, left

T4f – 4-Pin Connector, behind instrument panel, center

T68 – 68-Pin Connector

V144 – Leak detection pump (LDP)

(D162) – Wire connection, leak detection pump, in engine wiring harness

(E7) – Plus connection (87a), in Motronic Multiport Fuel Injection (MFI) wiring harness

(E14) – Wire connection -1-, in Motronic Multiport Fuel Injection (MFI) wiring harness

Motronic engine control module (ECM), leak detection pump (LDP), fuel system diagnostic control module, evaporative emission (EVAP) canister purge solenoid valve

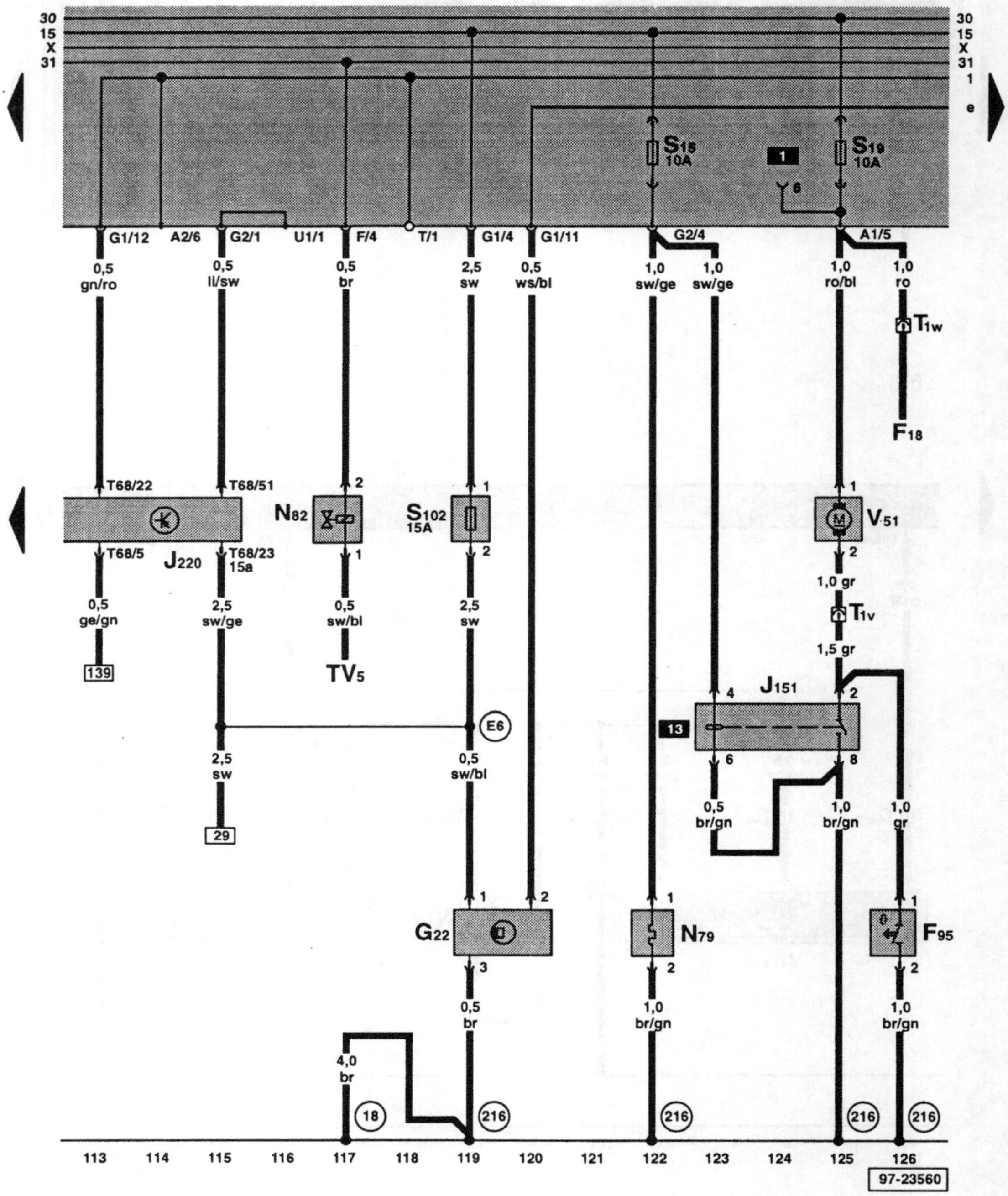

ws = white
sw = black
ro = red
br = brown
gn = green
bl = blue
gr = grey
li = violet
ge = yellow

- F18 – Coolant Fan Control (FC) Thermal Switch
- F95 – After-Run Coolant Thermal Switch
- G22 – Speedometer Vehicle Speed Sensor
- J151 – Coolant Circulation Pump Relay, above fuse/relay panel
- J220 – Motronic Engine Control Module (ECM), in engine compartment, left
- N79 – Positive Crankcase Ventilation (PCV) Heating Element
- N82 – Coolant Cut-off Valve
- S102 – Engine Control Module (ECM) Fuse, above fuse/relay panel
- T1v – Single Connector, behind fuse/relay panel
- T1w – Single Connector, behind fuse/relay panel
- T68 – 68-Pin Connector
- TV5 – Terminal 15a Wire Connector, green, 4-Pin and 8-Pin
- V51 – After-Run Coolant Pump
- (18) – Ground connection, on engine block
- (216) – Ground connection -2-, in Motronic Multiport Fuel Injection (MFI) wiring harness
- (E6) – Plus connection -1- (15), in Motronic Multiport Fuel Injection (MFI) wiring harness

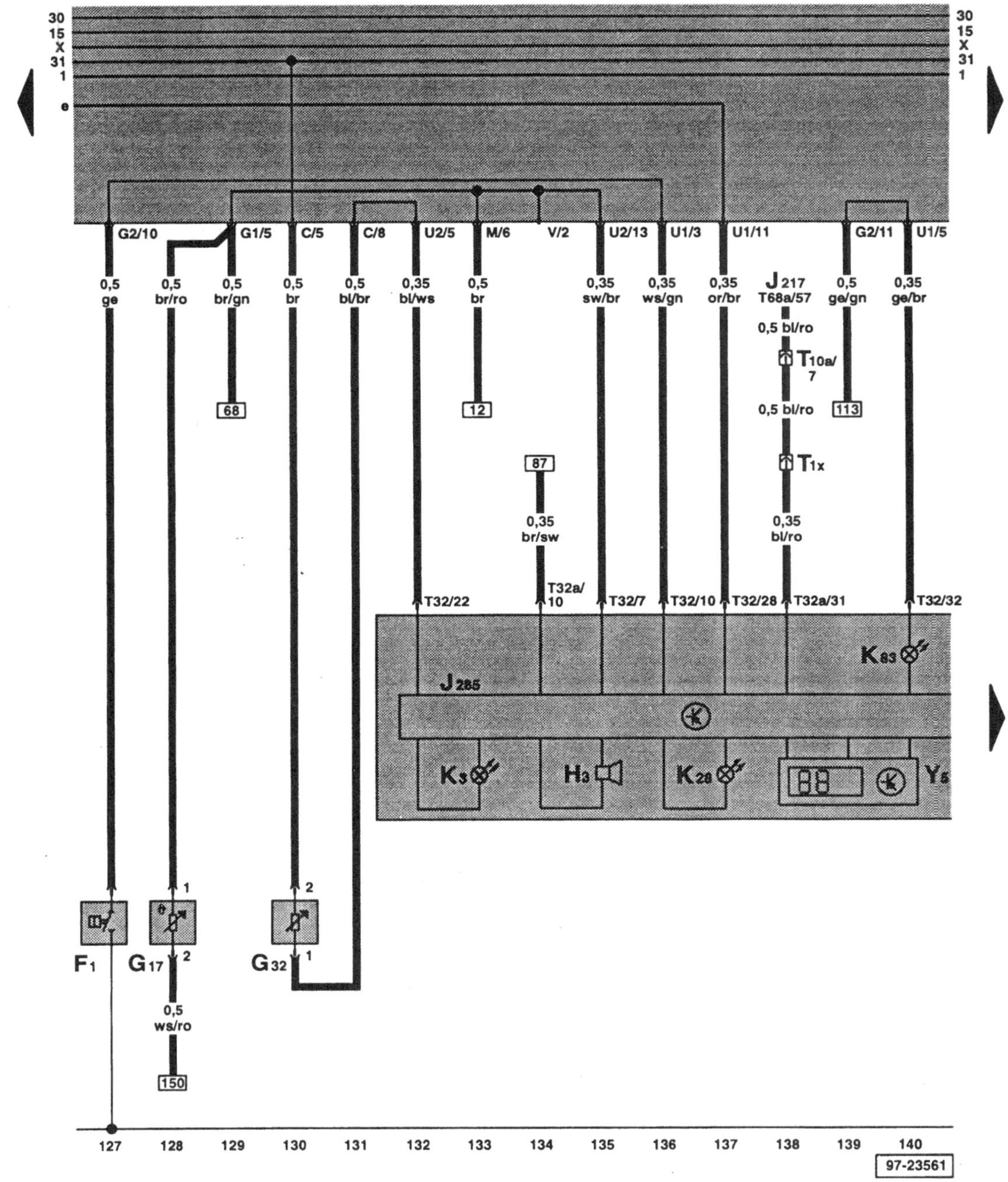

ws = white
sw = black
ro = red
br = brown
gn = green
bl = blue
gr = grey
li = violet
ge = yellow

F1 – Oil Pressure Switch
G17 – Outside Air Temperature Sensor
G32 – Engine Coolant Level (ECL) Sensor
H3 – Warning Buzzer
J217 – Transmission Control Module (TCM), on left A-pillar, lower part
J285 – Control module with indicator unit in instrument panel insert
K3 – Oil Pressure Warning Light
K28 – Engine Coolant Level/Temperature (ECL/ECT Warning Light)
K83 – Malfunction Indicator Lamp (MIL)
T1x – Single Connector, behind fuse/relay panel
T10a – 10-Pin Connector, behind fuse/relay panel
T32 – 32-Pin Connector, blue
T32a – 32-Pin Connector, green
T68a – 68-Pin Connector, on Transmission Control Module (TCM)
Y5 – Transmission Range (TR) Selector Lever Display

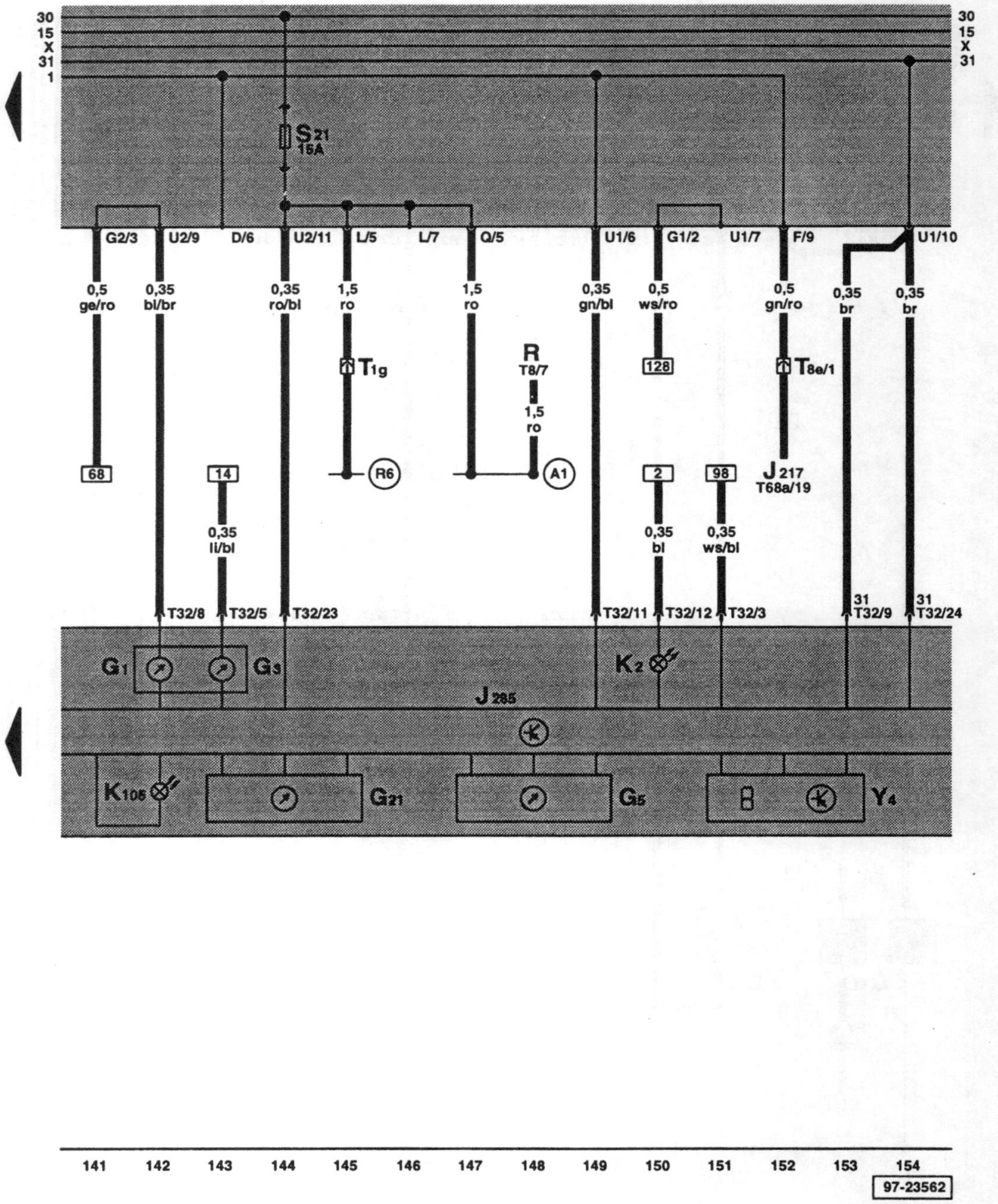

ws = white
sw = black
ro = red
br = brown
gn = green
bl = blue
gr = grey
li = violet
ge = yellow

G1 – Fuel gauge
G3 – Engine Coolant Temperature (ECT) Gauge
G5 – Tachometer
G21 – Speedometer
J217 – Transmission Control Module (TCM), on left A-pillar, lower part
J285 – Control module with indicator unit in instrument panel insert
K2 – Generator (GEN) Warning Light
K105 – Low Fuel Level Warning Light
R – Radio
T1g – Single Connector, behind fuse/relay panel
T8 – 8-Pin Connector, on radio
T8e – 8-Pin Connector, behind fuse/relay panel
T32 – 32-Pin Connector, blue
T68a – 68-Pin Connector, on transmission control module (TCM)
Y4 – Odometer Display

(A1) – Plus connection (30a), in instrument panel wiring harness
(R6) – Plus connection -1-, in interior light wiring harness

Automatic transmission

from September 1998

● 2.8L - Engine - Motronic Multiport Fuel Injection (MFI)/103 kW, code AES

Deviate relay location and fuseplacements as well as the locations of multiple connectors see section "component locations".

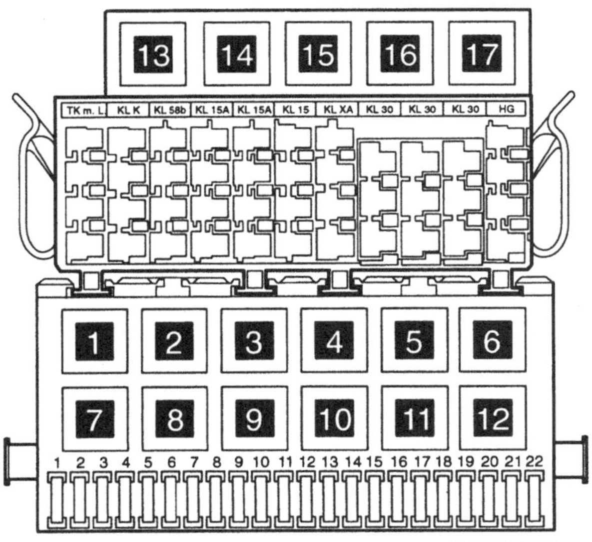

97–23232

12 Fuel Pump (FP) Relay (167)

Note: Number in parentheses indicates production control number stamped on relay housing.

Fuse colors

30 A – green
25 A – white
20 A – yellow
15 A – blue
10 A – red
7,5 A – brown
5 A – beige
3 A – violet

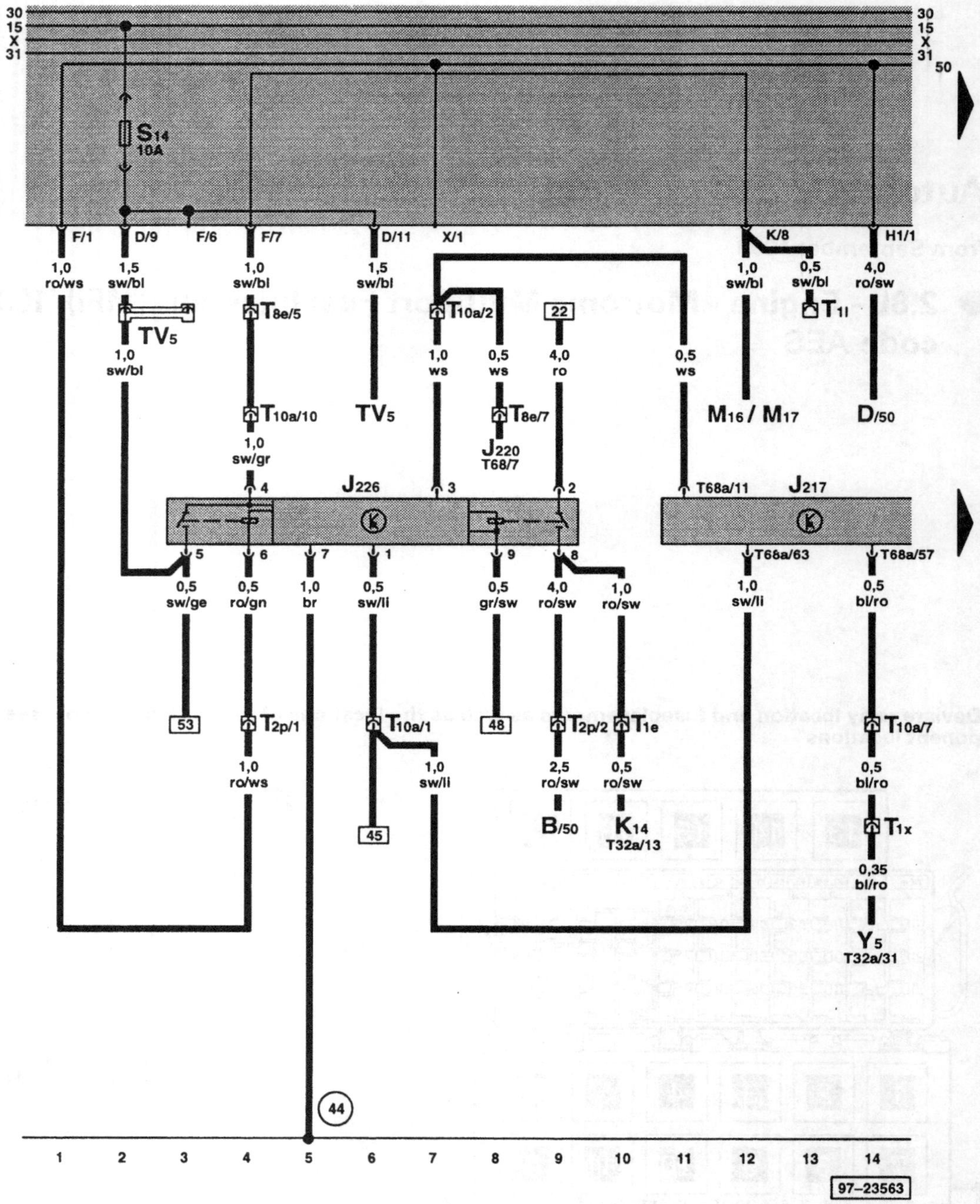

ws = white
sw = black
ro = red
br = brown
gn = green
bl = blue
gr = grey
li = violet
ge = yellow

- B – Starter
- D – Ignition/Starter Switch
- J217 – Transmission Control Module (TCM), on left A-pillar, lower part
- J220 – Motronic Engine Control Module (ECM), in engine compartment, left
- J226 – Park/Neutral Position (PNP) Relay, behind instrument panel, center
- K14 – Parking Brake Indicator Light
- M16 – Left Back-Up Light
- M17 – Right Back-Up Light
- T1e – Single Connector, behind fuse/relay panel
- T1l – Single Connector, behind fuse/relay panel
- T1x – Single Connector, behind fuse/relay panel
- T2p – Double Connector, behind fuse/relay panel
- T8e – 8-Pin Connector, behind fuse/relay panel
- T10a – 10-Pin Connector, behind fuse/relay panel
- T32a – 32-Pin Connector, green, on Instrument cluster
- T68 – 68-Pin Connector, on Motronic Engine Control Module (ECM)
- T68a – 68-Pin Connector
- TV5 – Terminal 15a Wire Connector, green, 4-Pin and 8-Pin
- Y5 – Transmission Range (TR) Selector Lever Display
- (44) – Ground connection, on left A-pillar, lower part

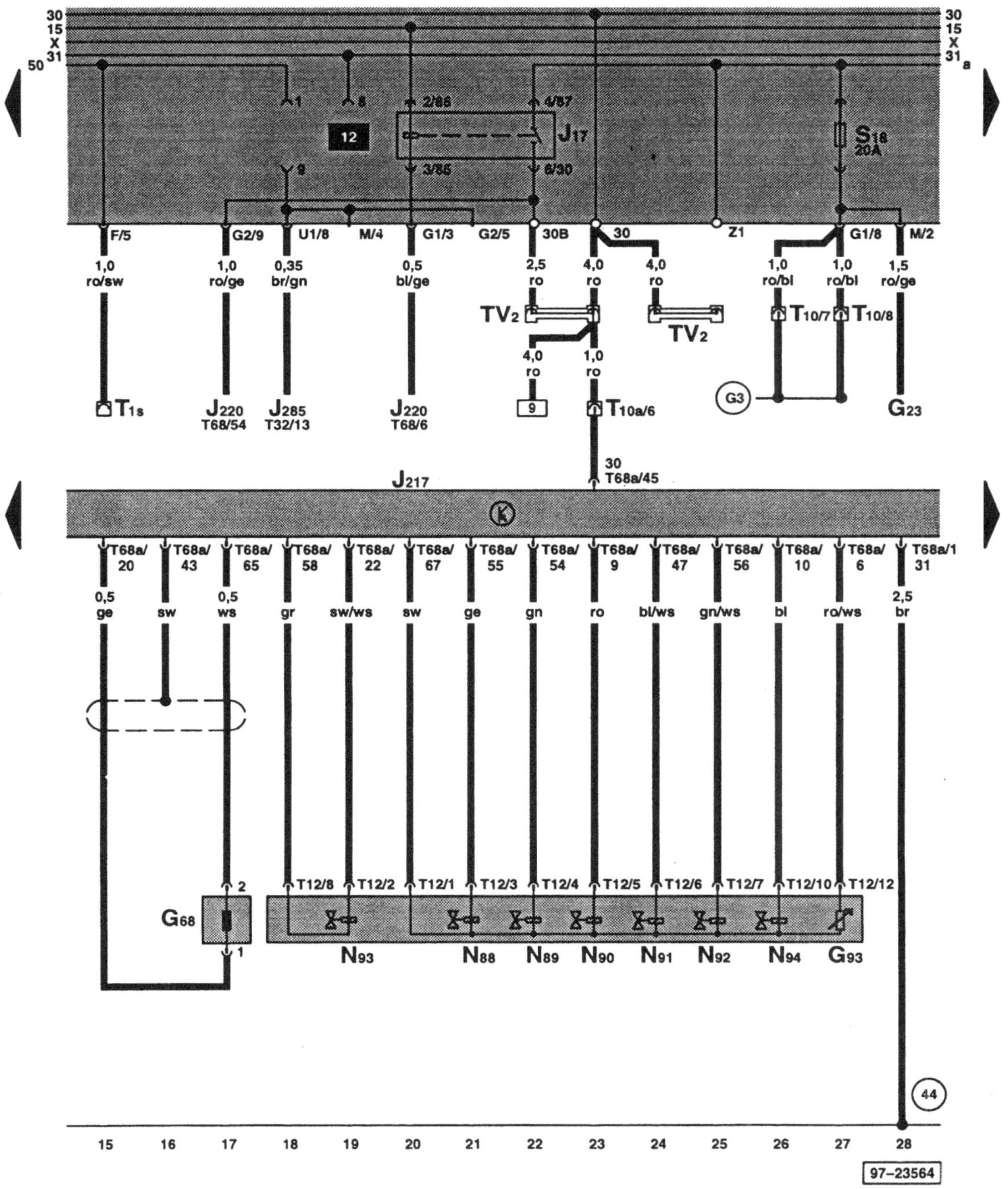

ws = white
sw = black
ro = red
br = brown
gn = green
bl = blue
gr = grey
li = violet
ge = yellow

G23 – Transfer Fuel Pump
G68 – Vehicle Speed Sensor (VSS)
G93 – Transmission Fluid Temperature Sensor
J17 – Fuel Pump (FP) Relay
J217 – Transmission Control Module (TCM), on left A-pillar, lower part
J220 – Motronic Engine Control Module (ECM), in engine compartment, left
J285 – Control module with indicator unit in instrument panel insert
N88 – Solenoid Valve 1
N89 – Solenoid Valve 2
N90 – Solenoid Valve 3
N91 – Solenoid Valve 4
N92 – Solenoid Valve 5
N93 – Solenoid Valve 6
N94 – Solenoid Valve 7
T1s – Single Connector, behind fuse/relay panel
T10 – 10-Pin Connector, near cylinder head cover
T10a – 10-Pin Connector, behind fuse/relay panel
T12 – 12-Pin Connector
T32 – 32-Pin Connector, blue, on Instrument cluster
T68 – 68-Pin Connector, on Motronic Engine Control Module (ECM)
T68a – 68-Pin Connector
TV2 – Terminal 30 (B+) Wire Connector, red, each 6 points

(44) – Ground connection, on left A-pillar, lower part

(G3) – Plus connection, in injector wiring harness

Transmission control module (TCM), vehicle speed sensor (VSS), transmission fluid temperature sensor, solenoid valves

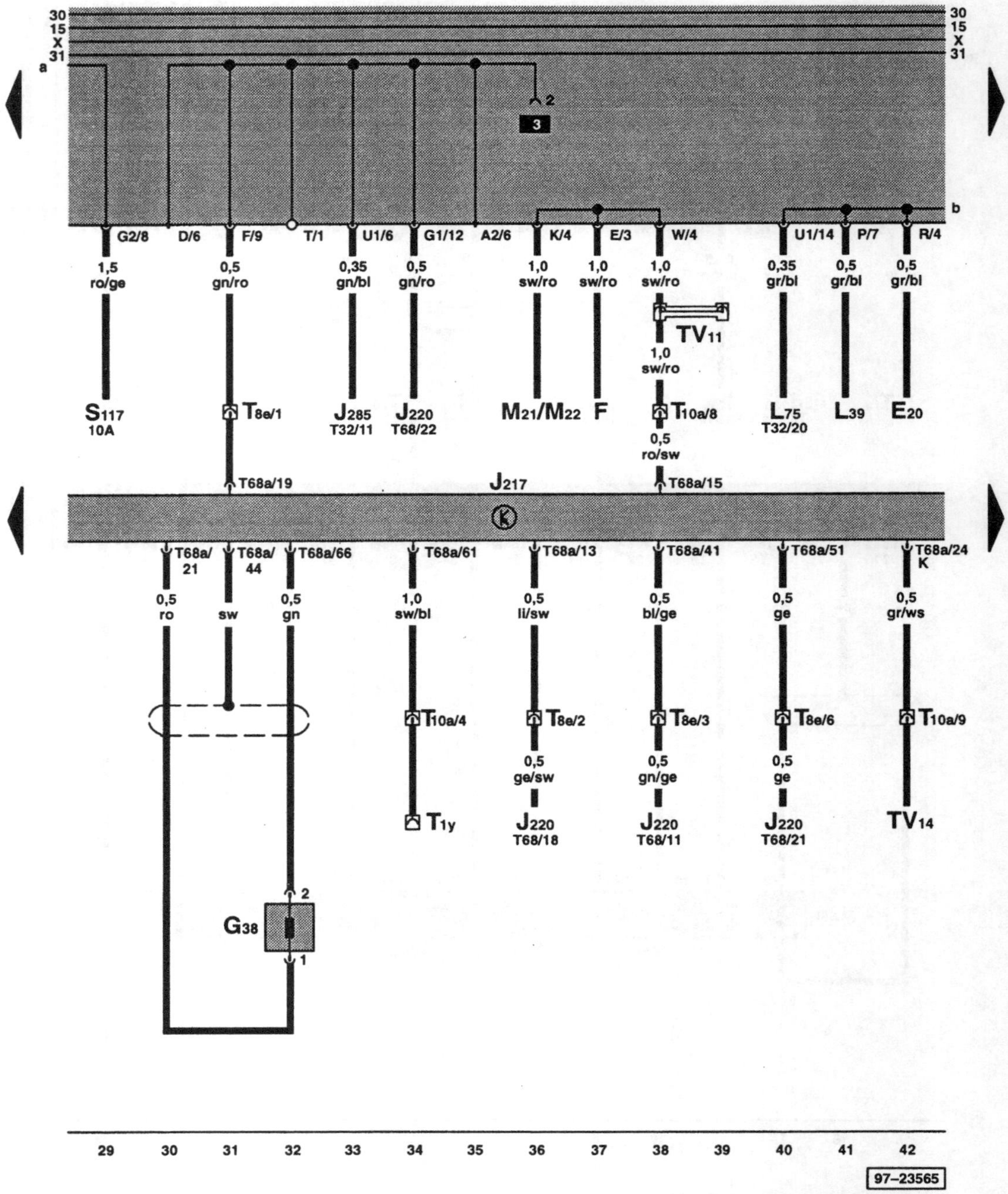

ws = white
sw = black
ro = red
br = brown
gn = green
bl = blue
gr = grey
li = violet
ge = yellow

E20 – Instrument Panel Light Dimmer Switch
F – Brake Light Switch
G38 – Transmission Vehicle Speed Sensor (VSS)
J217 – Transmission Control Module (TCM), on left A-pillar, lower part
J220 – Motronic Engine Control Module (ECM), in engine compartment, left
J285 – Control module with indicator unit in instrument panel insert
L39 – Rear Window Defogger Switch Light
L75 – Digital Display Light
M21 – Left Brake/Tail Light
M22 – Right Brake/Tail Light
S117 – Fuse for control elements, above fuse/relay panel
T1y – Single Connector, behind fuse/relay panel
T8e – 8-Pin Connector, behind fuse/relay panel
T10a – 10-Pin Connector, behind fuse/relay panel
T32 – 32-Pin Connector, blue, on Instrument cluster
T68 – 68-Pin Connector, on Motronic Engine Control Module (ECM)
T68a – 68-Pin Connector
TV11 – Terminal 54 Wire Connector, orange
TV14 – Data Link Connector (DLC) wire Connector, white

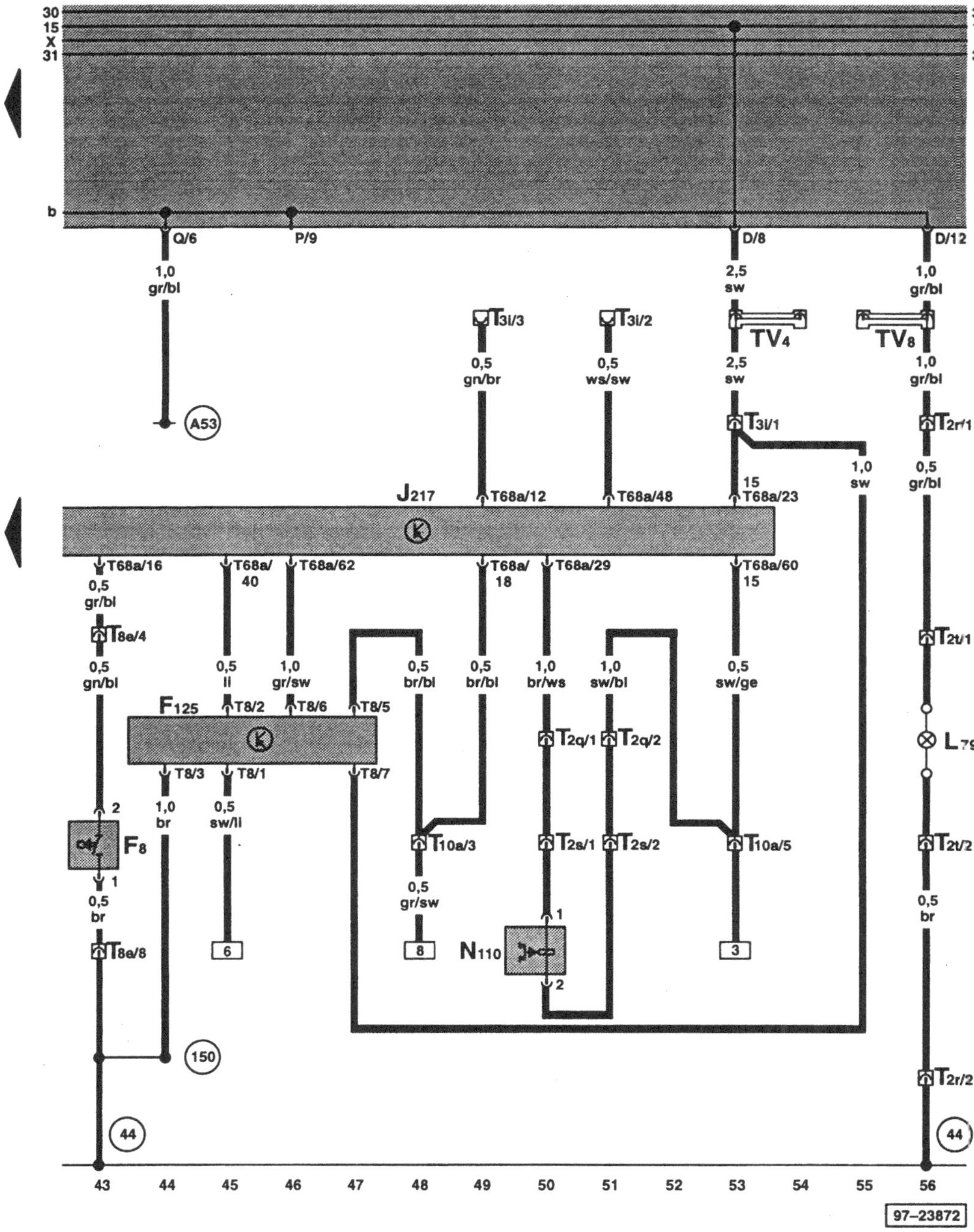

ws = white
sw = black
ro = red
br = brown
gn = green
bl = blue
gr = grey
li = violet
ge = yellow

- F8 – Kick Down Switch
- F125 – Multi-Function Transmission Range (TR) Switch
- J217 – Transmission Control Module (TCM), on left A-pillar, lower part
- L79 – Selector Lever Light
- N110 – Shift Lock Solenoid
- T2q – Double Connector, behind fuse/relay panel
- T2r – Double Connector, behind fuse/relay panel
- T2s – Double Connector, near selector lever
- T2t – Double Connector, near selector lever
- T3i – 3-Pin Connector, behind fuse/relay panel
- T8 – 8-Pin Connector
- T8e – 8-Pin Connector, behind fuse/relay panel
- T10a – 10-Pin Connector, behind fuse/relay panel
- T68a – 68-Pin Connector
- TV4 – Terminal 15 Wire Connector, black
- TV8 – Terminal 58b Wire Connector, grey
- (44) – Ground connection, on left A-pillar, lower part
- (150) – Ground connection, in automatic transmission wiring harness
- (A53) – Plus connection (58b), in instrument panel wiring harness

THIS PAGE INTENTIONALLY LEFT BLANK

Cruise Control (automatic transmission),

from September 1998

Deviate relay location and fuseplacements as well as the locations of multiple connectors see section "component locations".

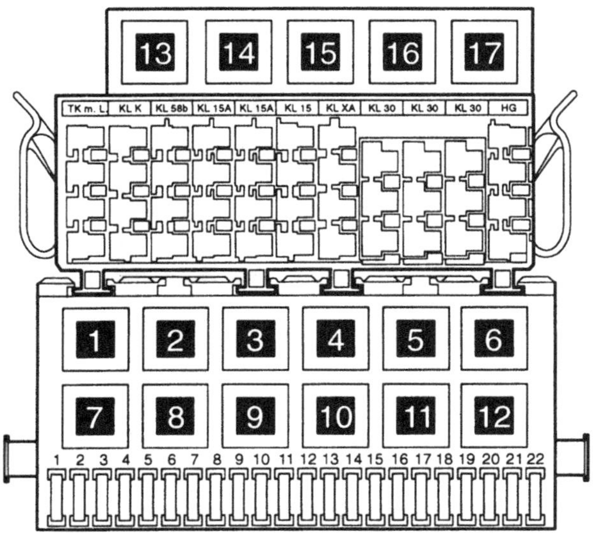

97–23232

Fuse colors

30 A – green
25 A – white
20 A – yellow
15 A – blue
10 A – red
7,5 A – brown
5 A – beige
3 A – violet

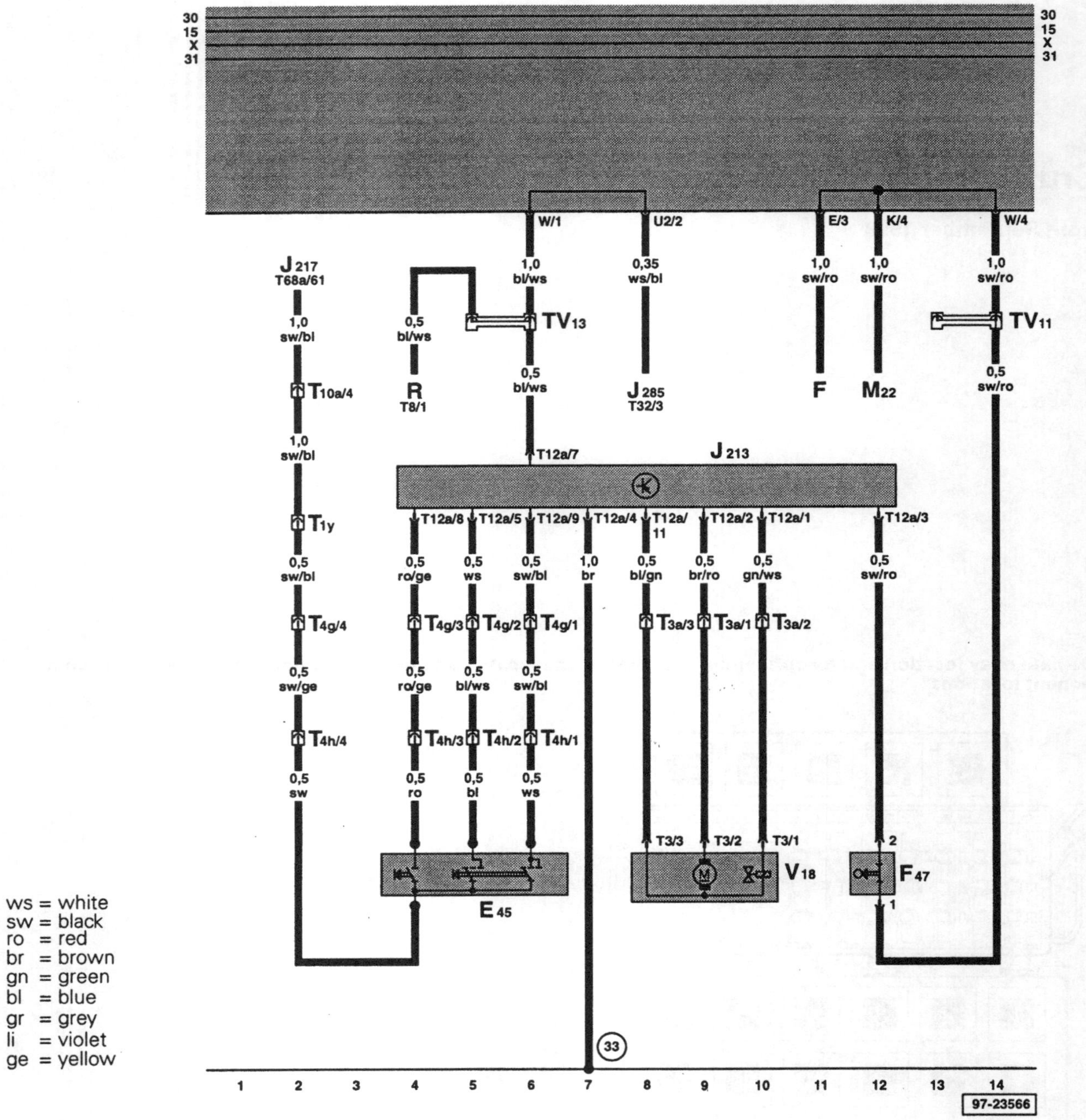

E45 – Cruise Control Switch
F – Brake Light Switch
F47 – Brake Vacuum Vent Valve Switch for cruise control
J213 – Cruise Control, Control Module, behind instrument panel, right
J217 – Transmission Control Module (TCM), on left A-pillar, lower part
J285 – Control module with indicator unit in instrument panel insert
M22 – Right Brake/Tail Light
R – Radio
T1y – Single Connector, behind fuse/relay panel
T3 – 3-Pin Connector
T3a – 3-Pin Connector, behind fuse/relay panel
T4g – 4-Pin Connector, behind fuse/relay panel
T4h – 4-Pin Connector, behind steering column switch cover
T8 – 8-Pin Connector, on Radio
T10a – 10-Pin Connector, behind fuse/relay panel
T12a – 12-Pin Connector
T32 – 32-Pin Connector, blue, on Instrument cluster
T68a – 68-Pin Connector
TV11 – Terminal 54 Wire Connector, orange
TV13 – Vehicle Speed Signal Wire Connector, blue
V18 – Cruise Control Vacuum Pump

(33) – Ground connection, behind instrument panel, right

Anti-lock brake system (ABS) with electronic differential lock (EDL),

from September 1998

Deviate relay location and fuseplacements as well as the locations of multiple connectors see section "component locations".

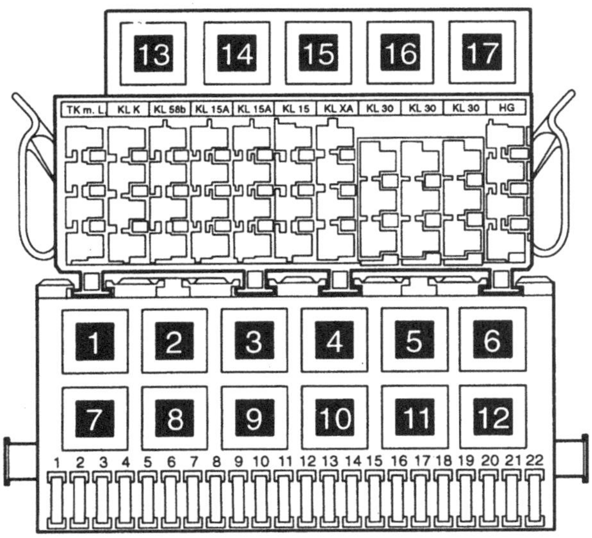

97–23232

4 Load Reduction Relay (18)

Note: Number in parentheses indicates production control number stamped on relay housing.

Fuse colors

30 A – green
25 A – white
20 A – yellow
15 A – blue
10 A – red
7,5 A – brown
5 A – beige
3 A – violet

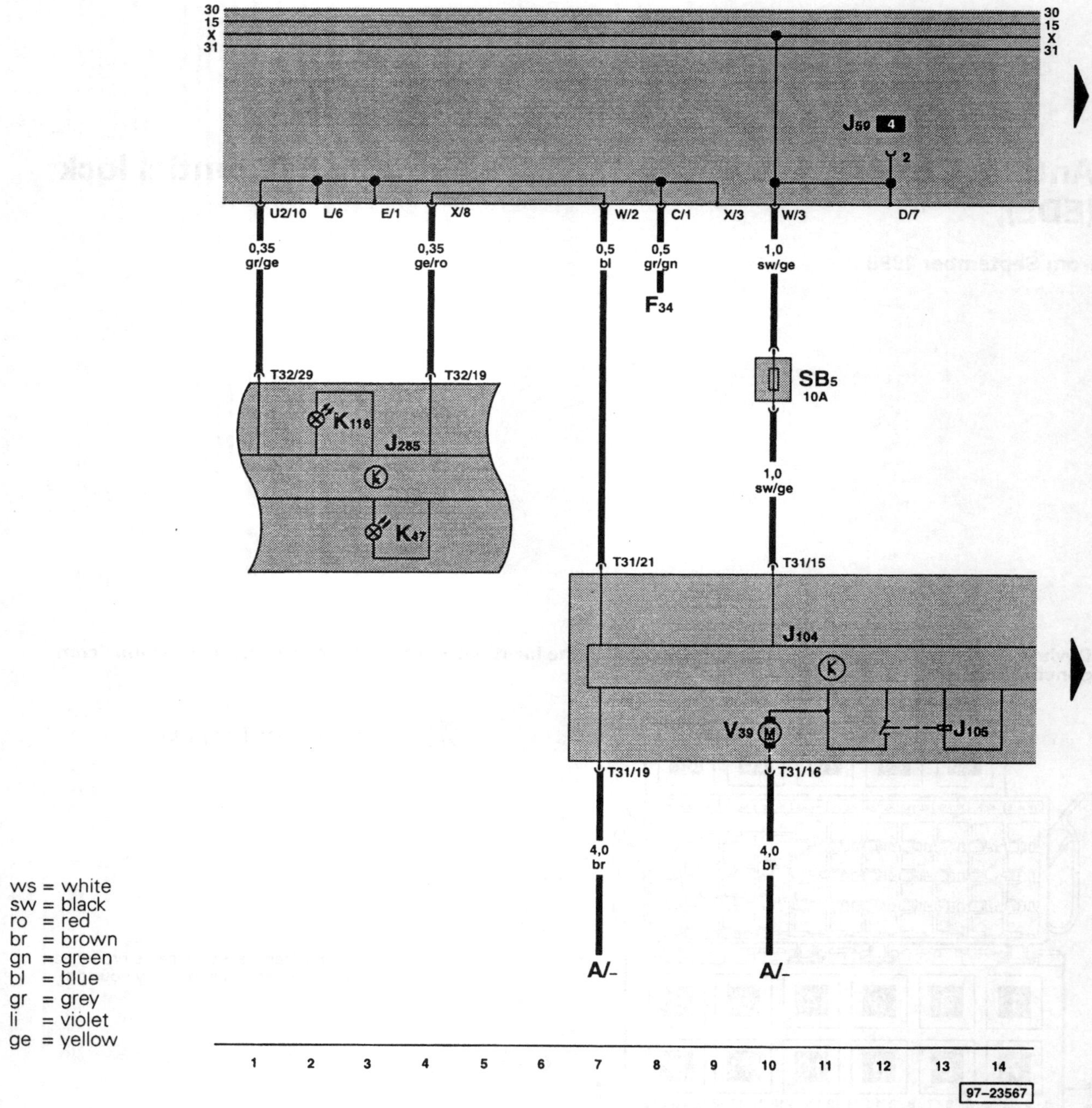

A – Battery
F34 – Brake Fluid Level Warning Switch
J59 – Load Reduction Relay
J104 – ABS Control Module (w/EDL), in engine compartment, left
J105 – ABS Return Flow Pump Relay
J285 – Control module with indicator unit in instrument panel insert
K47 – ABS Warning Light
K118 – Warning light for brake system
SB5 – Fuse 5 in fuse holder, below fuse/relay panel
T31 – 31-Pin Connector
T32 – 32-Pin Connector, blue
V39 – ABS Hydraulic Pump

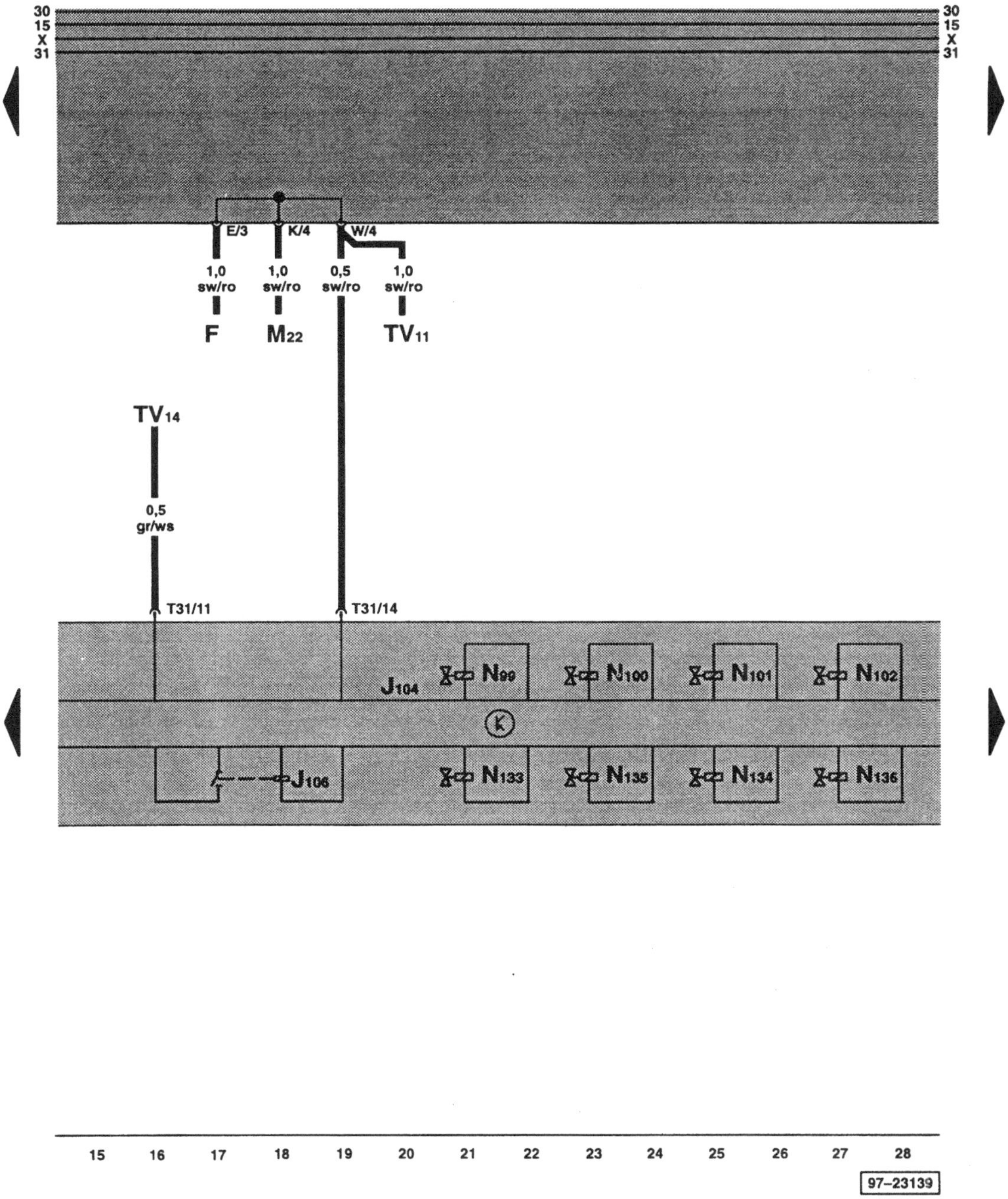

ws = white
sw = black
ro = red
br = brown
gn = green
bl = blue
gr = grey
li = violet
ge = yellow

F – Brake Light Switch
J104 – ABS Control Module (w/EDL), in engine compartment, left
J106 – ABS Solenoid Valve Relay
M22 – Right Brake/Tail Light
N99 – Right Front ABS Inlet Valve
N100 – Right Front ABS Outlet Valve
N101 – Left Front ABS Inlet Valve
N102 – Left Front ABS Outlet Valve
N133 – Right Rear ABS Inlet Valve
N134 – Left Rear ABS Inlet Valve
N135 – Right Rear ABS Outlet Valve
N136 – Left Rear ABS Outlet Valve
T31 – 31-Pin Connector
TV11 – Terminal 54 Wire Connector, orange
TV14 – Data Link Connector (DLC) wire Connector, white

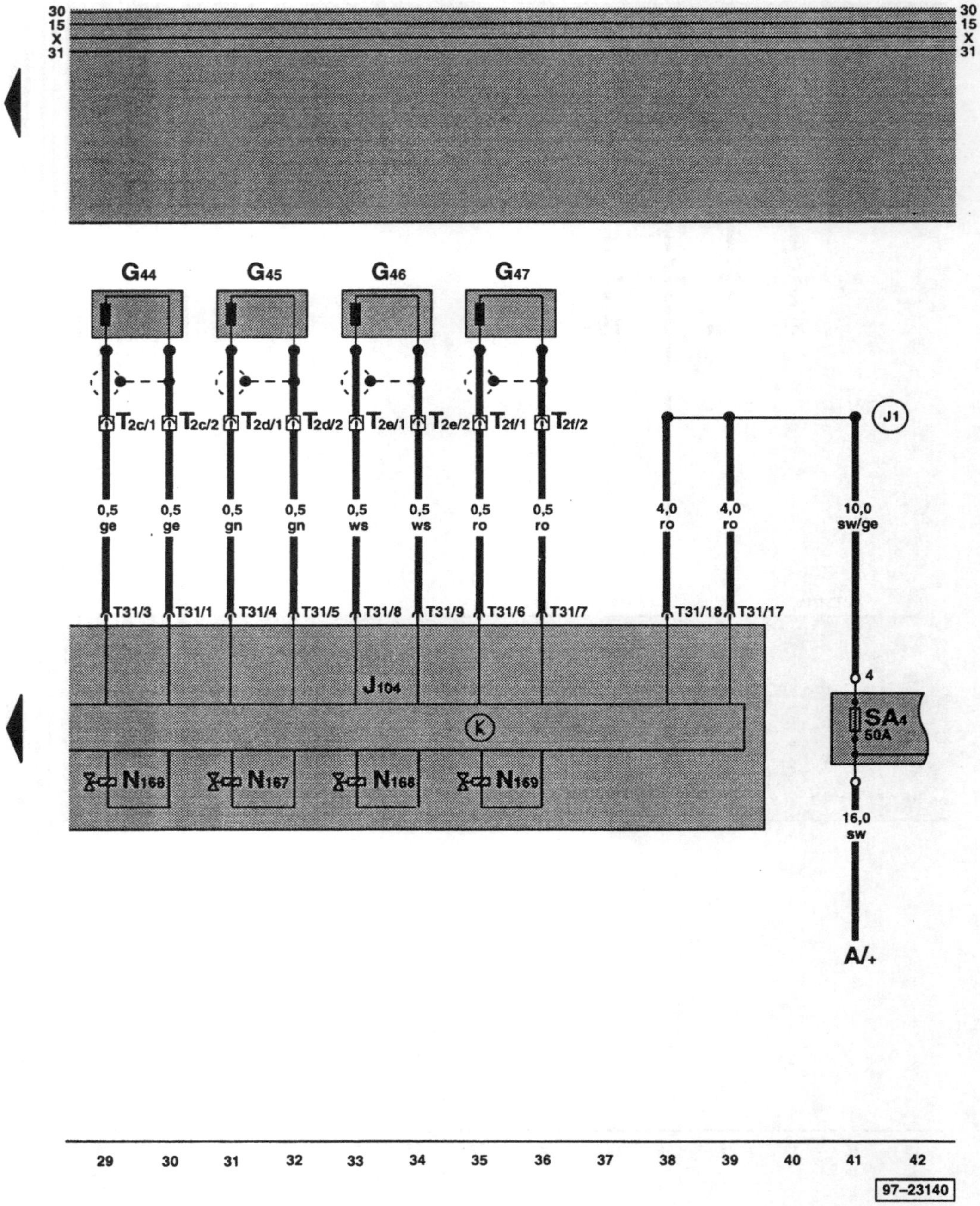

ws = white
sw = black
ro = red
br = brown
gn = green
bl = blue
gr = grey
li = violet
ge = yellow

A – Battery
G44 – Right Rear ABS Wheel Speed Sensor
G45 – Right Front ABS Wheel Speed Sensor
G46 – Left Rear ABS Wheel Speed Sensor
G47 – Left Front ABS Wheel Speed Sensor
J104 – ABS Control Module (w/EDL), in engine compartment, left
N166 – Right Front Traction Control Switch-Over Valve
N167 – Right Front Traction Control Outlet Valve
N168 – Left Front Traction Control Switch-Over Valve
N169 – Left Front Traction Control Outlet Valve
SA4 – Fuse -4- in fuse bracket/battery
T2c – Double Connector, on chassis side member, right rear
T2d – Double Connector, in front wheel house, right
T2e – Double Connector, on chassis side member, left rear
T2f – Double Connector, in front wheel house, left
T31 – 31-Pin Connector

(J1) – Plus connection (30), in ABS wiring harness

Airbag systems,

from September 1998

Deviate relay location and fuseplacements as well as the locations of multiple connectors see section "component locations".

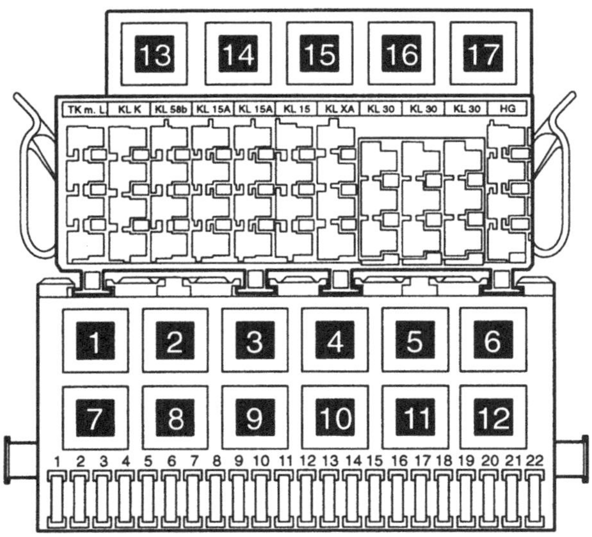

97–23232

Fuse colors

30 A – green

25 A – white

20 A – yellow

15 A – blue

10 A – red

7,5 A – brown

5 A – beige

3 A – violet

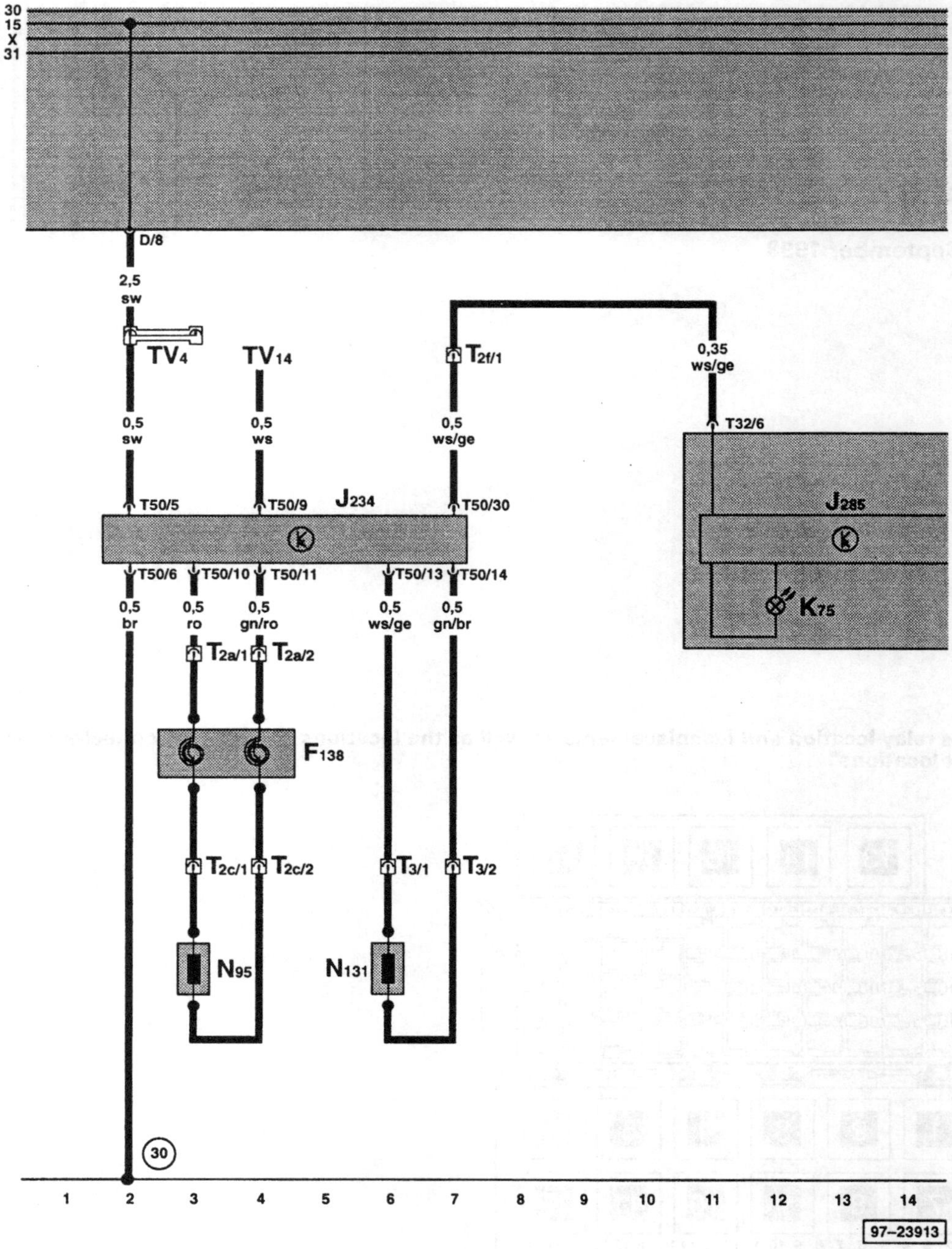

ws = white
sw = black
ro = red
br = brown
gn = green
bl = blue
gr = grey
li = violet
ge = yellow

F138 – Airbag Spiral Spring
J234 – Airbag Control Module, behind instrument panel, center
J285 – Control module with indicator unit in instrument panel insert
K75 – Airbag Malfunction Indicator Lamp (MIL)
N95 – Driver's Side Airbag Igniter
N131 – Passenger's Side Airbag Igniter 1
T2a – Double Connector, behind steering column switch cover
T2c – Double Connector, in steering wheel
T2f – Double Connector, behind fuse/relay panel
T3 – 3-Pin Connector, behind instrument panel, rigth
T32 – 32-Pin Connector, blue
T50 – 50-Pin Connector
TV4 – Terminal 15 Wire Connector, black
TV14 – Data Link Connector (DLC) wire Connector, white

(30) – Ground connection, -1-, beside fuse/relay panel

Coolant Fans (2 x 450 W),

from September 1998

Deviate relay location and fuseplacements as well as the locations of multiple connectors see section "component locations".

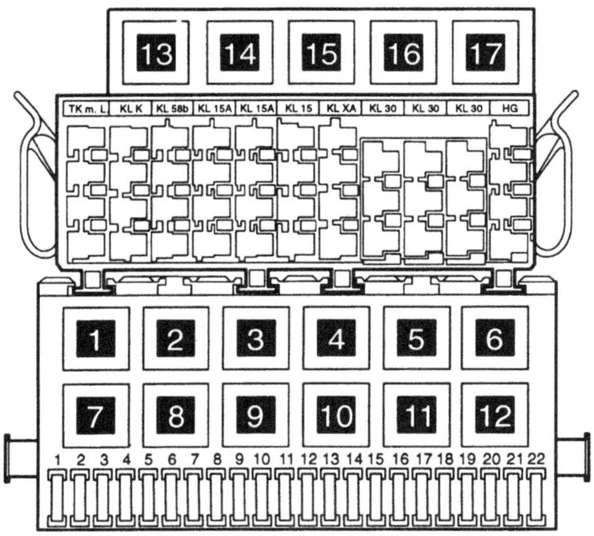

97–23232

Fuse colors

30 A – green
25 A – white
20 A – yellow
15 A – blue
10 A – red
7,5 A – brown
5 A – beige
3 A – violet

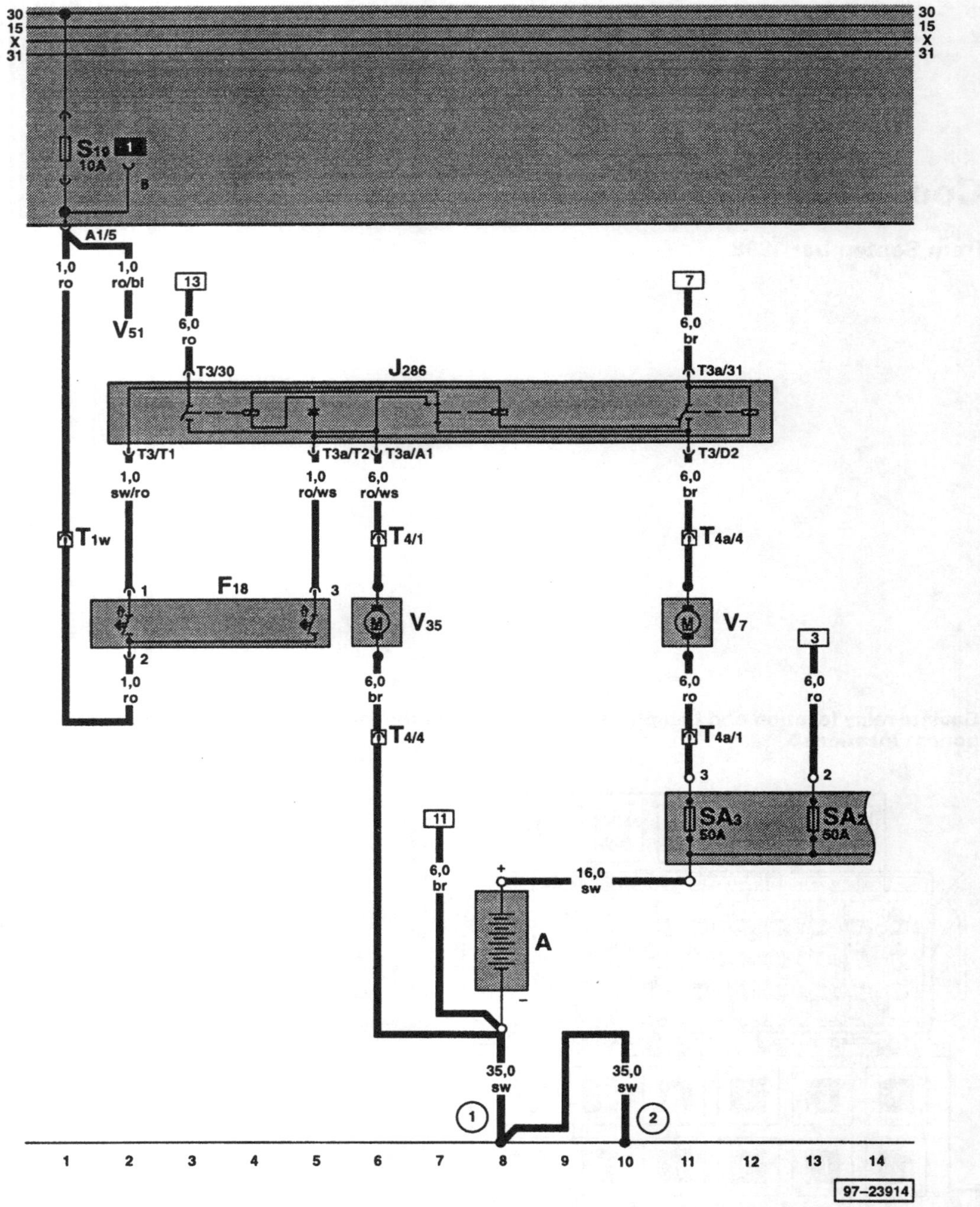

ws = white
sw = black
ro = red
br = brown
gn = green
bl = blue
gr = grey
li = violet
ge = yellow

A – Battery
F18 – Coolant Fan Control (FC) Thermal Switch
J286 – Series/Parallel Coolant Fan Control (FC) Relay (V7, V35), in engine compartment, left front
SA2 – Fuse -2- in fuse bracket/battery
SA3 – Fuse -3- in fuse bracket/battery
T1w – Single Connector, behind fuse/relay panel
T3 – 3-Pin Connector
T3a – 3-Pin Connector
T4 – 4-Pin Connector, in engine compartment, front
T4a – 4-Pin Connector, in engine compartment, front
V7 – Left Coolant Fan
V35 – Right Coolant Fan
V51 – After-Run Coolant Pump

(1) – Ground strap, battery to body
(2) – Ground strap, transmission to body

Air-Conditioning,

from September 1998

- **2.8 L- Engine - Motronic Multiport Fuel Injection (MFI)/103 kW, code AES, with automatic transmission**

Deviate relay location and fuseplacements as well as the locations of multiple connectors see section "component locations".

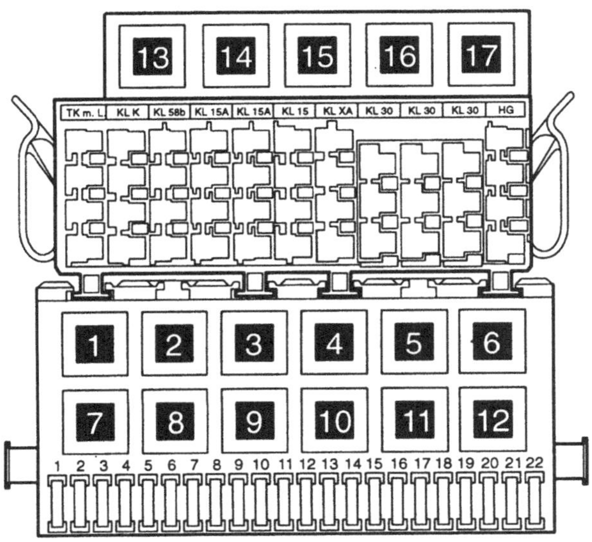

97-23232

1 A/C Relay (140)

16 Relay for A/C shutoff (147)

Note: Number in parentheses indicates production control number stamped on relay housing.

Fuse colors

30 A – green
25 A – white
20 A – yellow
15 A – blue
10 A – red
7,5 A – brown
5 A – beige
3 A – violet

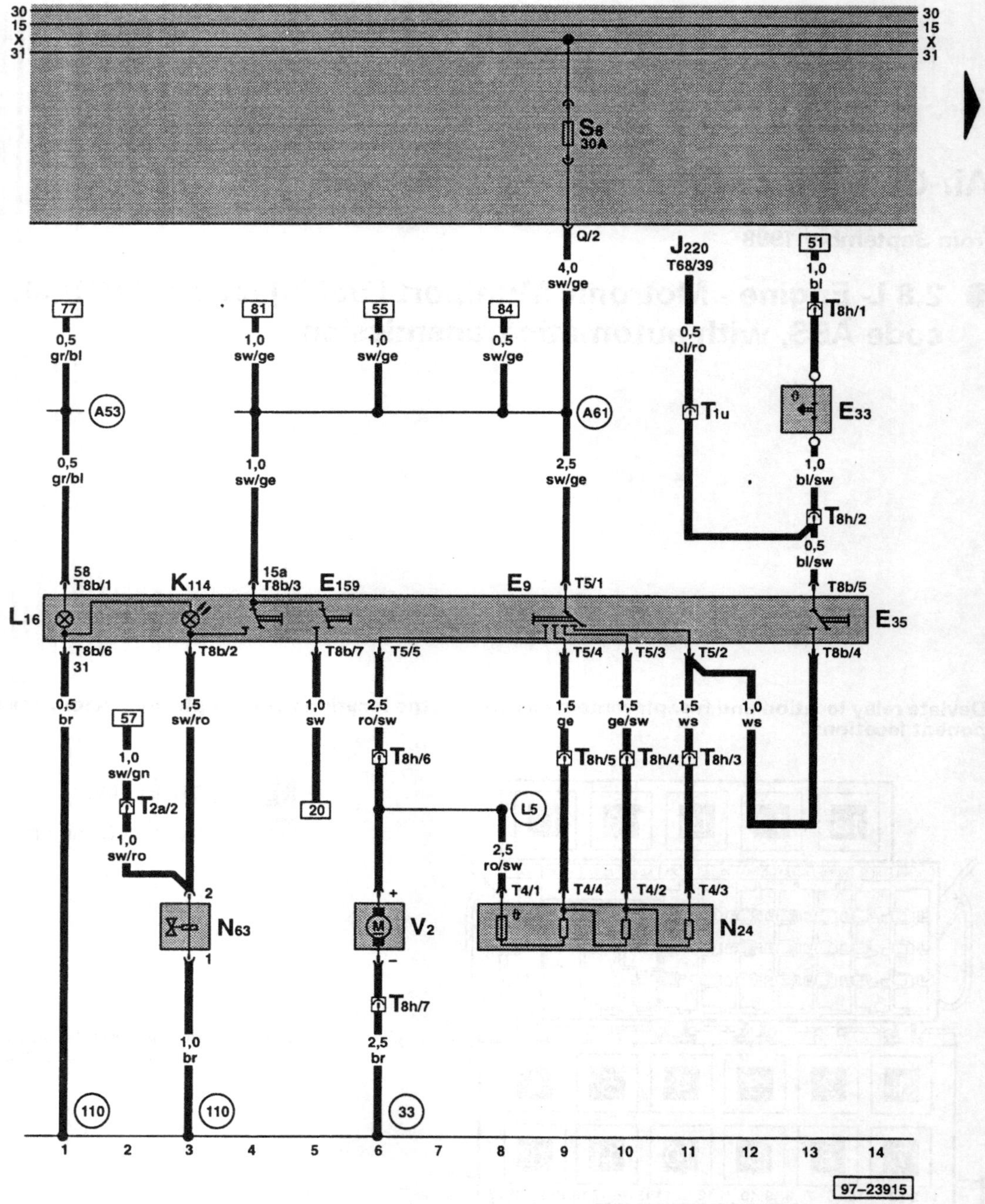

ws = white
sw = black
ro = red
br = brown
gn = green
bl = blue
gr = grey
li = violet
ge = yellow

E9 – Fresh Air Blower Switch
E33 – A/C Evaporator Temperature Switch
E35 – A/C Switch
E159 – Fresh Air/Recirculating Flap Switch
J220 – Motronic Engine Control Module (ECM), in engine compartment, left
K114 – Fresh Air and Recirculating Air Mode Indicator Light
L16 – Fresh Air Control Lever Light
N24 – Fresh Air Blower Series Resistance with Fuse
N63 – Fresh Air/Recirculating Flap Two-Way Valve
T1u – Single Connector, behind fuse/relay panel
T2a – Double Connector, behind fuse/relay panel
T4 – 4-Pin Connector
T5 – 5-Pin Connector
T8b – 8-Pin Connector
T8h – 8-Pin Connector, behind instrument panel, center
T68 – 68-Pin Connector, on Motronic Engine Control Module (ECM)
V2 – Fresh Air Blower
(33) – Ground connection, behind instrument panel right
(110) – Ground connection -2-, in instrument panel wiring harness
(A53) – Plus connection (58b), in instrument panel wiring harness
(A61) – Plus Connection (X), in Instrument cluster wiring harness
(L5) – Wire connection -1-, in evaporator housing wiring harness

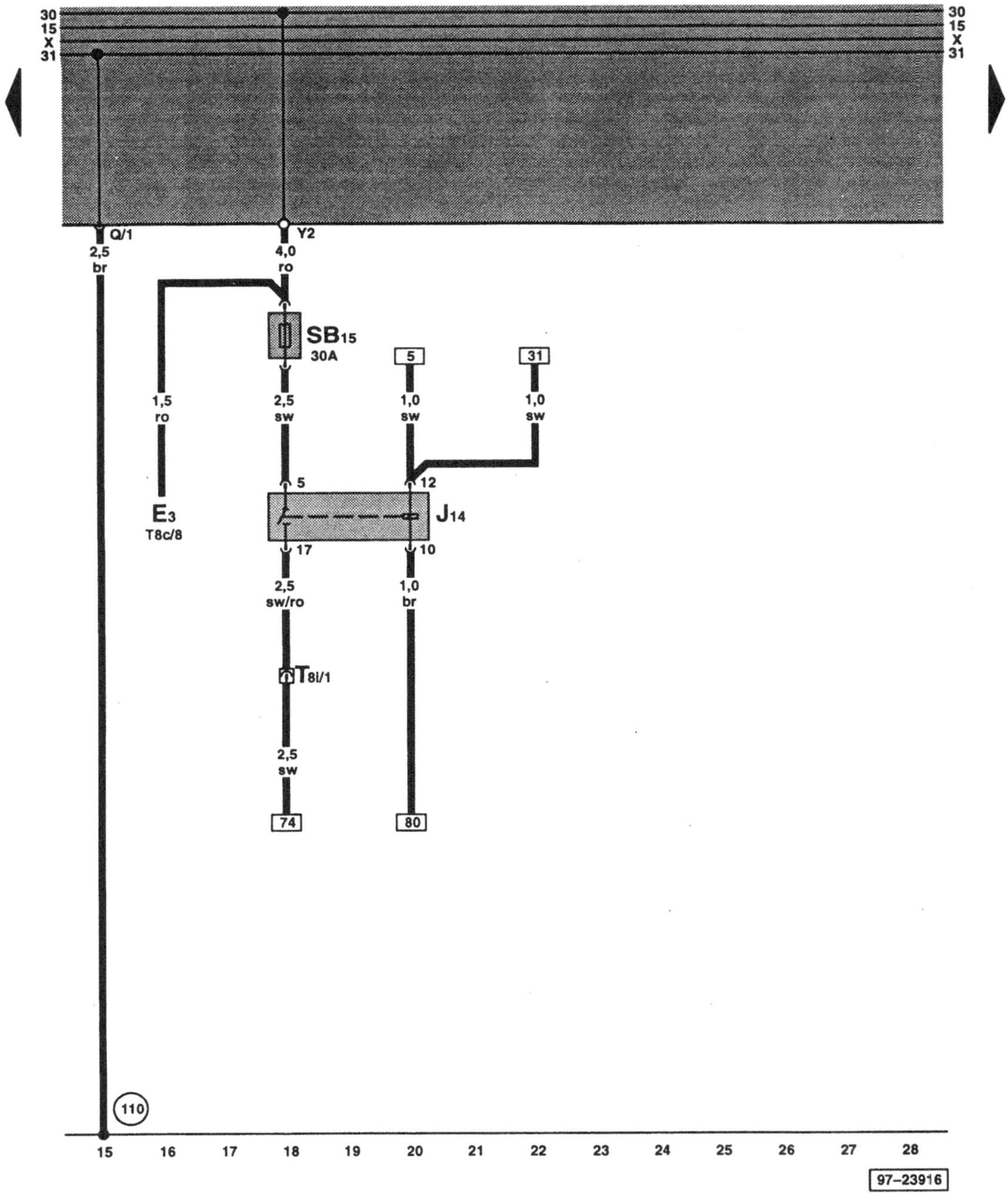

ws = white
sw = black
ro = red
br = brown
gn = green
bl = blue
gr = grey
li = violet
ge = yellow

E3 – Emergency Flasher Switch
J14 – Heater Air Blower Relay, control number (53), behind instrument panel center
SB15 – Fuse 15 in fuse holder, below relay panel
T8c – 8-Pin Connector, on Emergency Flasher Switch
T8i – 8-Pin Connector, behind fuse/relay panel

(110) – Ground connection -2-, in instrument panel wiring harness

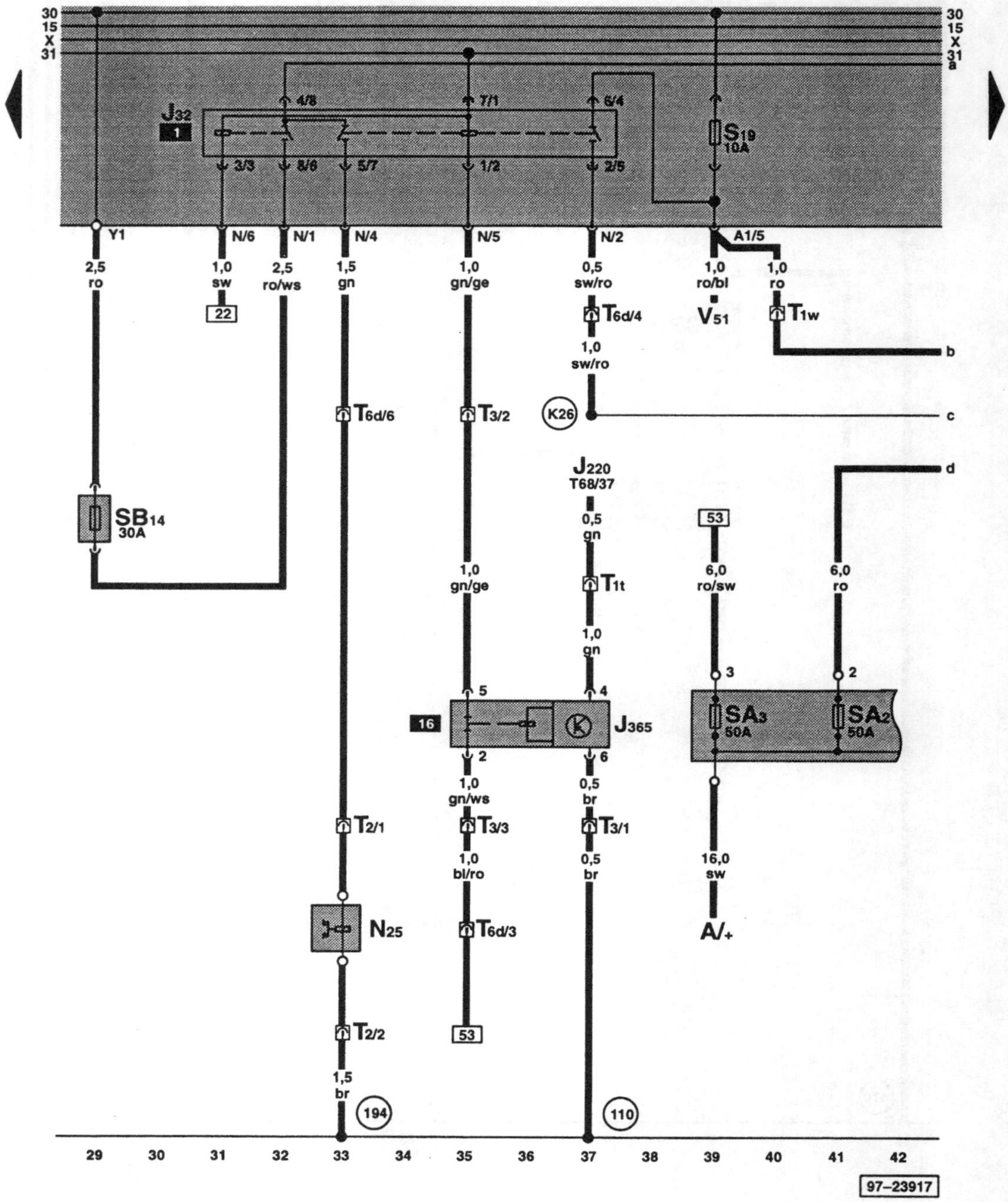

ws = white
sw = black
ro = red
br = brown
gn = green
bl = blue
gr = grey
li = violet
ge = yellow

A – Battery
J32 – A/C Relay, control number (140), switch delay time 1/2 second
J220 – Motronic Engine Control Module (ECM), in engine compartment, left
J365 – Relay for A/C shutoff, control number (147)
N25 – A/C Clutch
SA2 – Fuse -2- in fuse bracket/battery
SA3 – Fuse -3- in fuse bracket/battery
SB14 – Fuse 14 in fuse holder, below relay panel
T1t – Single Connector, behind fuse/relay panel
T1w – Single Connector, behind fuse/relay panel
T2 – Double Connector, near A/C Clutch
T3 – 3-Pin Connector, behind fuse/relay panel
T6d – 6-Pin Connector, behind fuse/relay panel
T68 – 68-Pin Connector, on Motronic Engine Control Module (ECM)
V51 – After-Run Coolant Pump
(110) – Ground connection -2-, in instrument panel wiring harness
(194) – Ground connection -2-, in coolant fan wiring harness
(K26) – Wire connection -5-, in coolant fan wiring harness

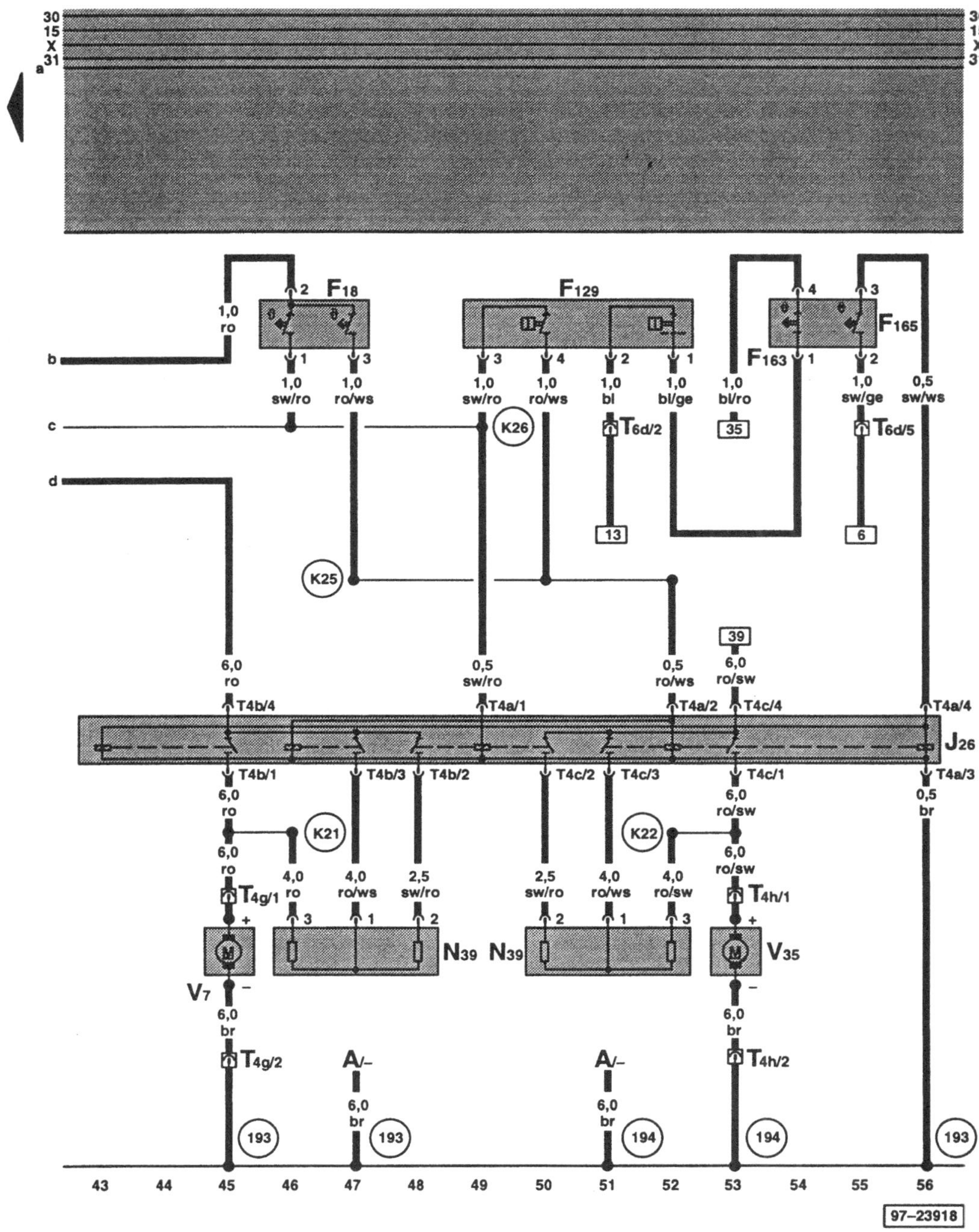

ws = white
sw = black
ro = red
br = brown
gn = green
bl = blue
gr = grey
li = violet
ge = yellow

A – Battery
F18 – Coolant Fan Control (FC) Thermal Switch
F129 – A/C Pressure Switch
F163 – A/C Cut-Out Thermal Switch
F165 – Third Speed Coolant Fan Control (FC) Thermal Switch, in engine compartment, front
J26 – Coolant Fan control (FC) Relay, in engine compartment, left front
N39 – Coolant Fan Control (FC) Series Resistance
T4a – 4-Pin Connector
T4b – 4-Pin Connector
T4c – 4-Pin Connector
T4g – 4-Pin Connector, near coolant fan
T4h – 4-Pin Connector, near coolant fan
T6d – 6-Pin Connector, behind fuse/relay panel
V7 – Left Coolant Fan (450 W)
V35 – Right Coolant Fan (450 W)
(193) – Ground connection -1-, in coolant fan wiring harness
(194) – Ground connection -2-, in coolant fan wiring harness
(K21) – Wire connection -1-, in coolant fan wiring harness
(K22) – Wire connection -2-, in coolant fan wiring harness
(K25) – Wire connection -4-, in coolant fan wiring harness
(K26) – Wire connection -5-, in coolant fan wiring harness

 Coolant fan control (FC) relay, coolant fan control (FC) thermal switch, A/C cut-out thermal switch, third speed coolant fan control (FC) thermal switch, coolant fan, A/C pressure switch

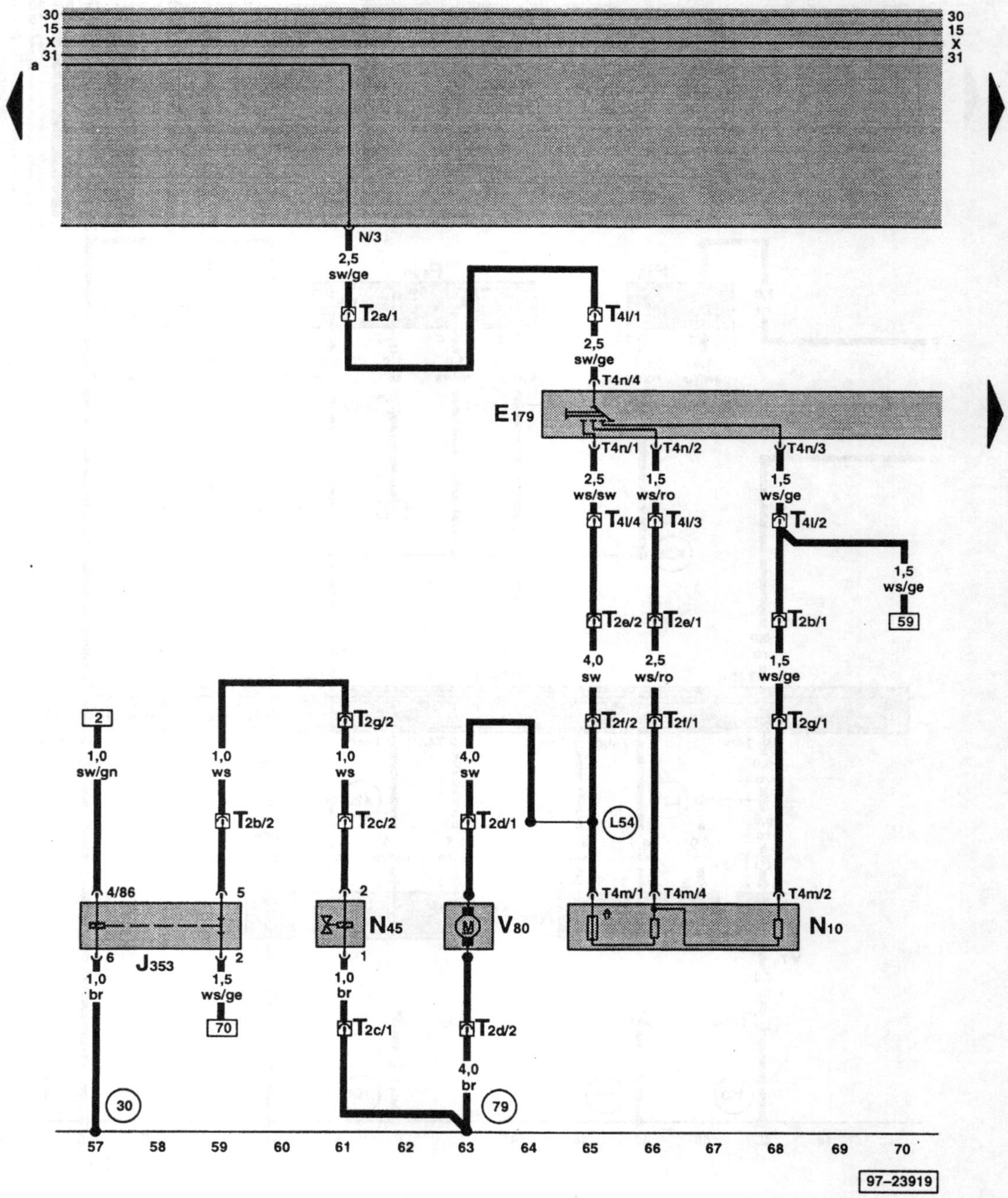

ws = white
sw = black
ro = red
br = brown
gn = green
bl = blue
gr = grey
li = violet
ge = yellow

E179 – Switch for Freshair blower-rear
J353 – Relay for Fresh-air and recirculating air flap-rear, control number (94), behind instrument panel center
N10 – Series Resistance for fresh air blower-rear
N45 – Two-way valve for Fresh- and recirculation air flap-rear
T2a – Double Connector, behind fuse/relay panel
T2b – Double Connector, behind instrument panel, right
T2c – Double Connector, behind instrument panel center
T2d – Double Connector, behind trim D-pillar right, lower
T2e – Double Connector, behind instrument panel, right
T2f – Double Connector, behind instrument panel, center
T2g – Double Connector, behind instrument panel, center
T4l – 4-Pin Connector, behind fuse/relay panel
T4m – 4-Pin Connector
T4n – 4-Pin Connector
V80 – Fresh air blower rear
(30) – Ground connection, -1-, beside fuse/relay panel
(79) – Ground connection, near right D-pillar, lower part
(L54) – Plus connector (Blower stage 3) in wiring harness blower-Passenger ventilation

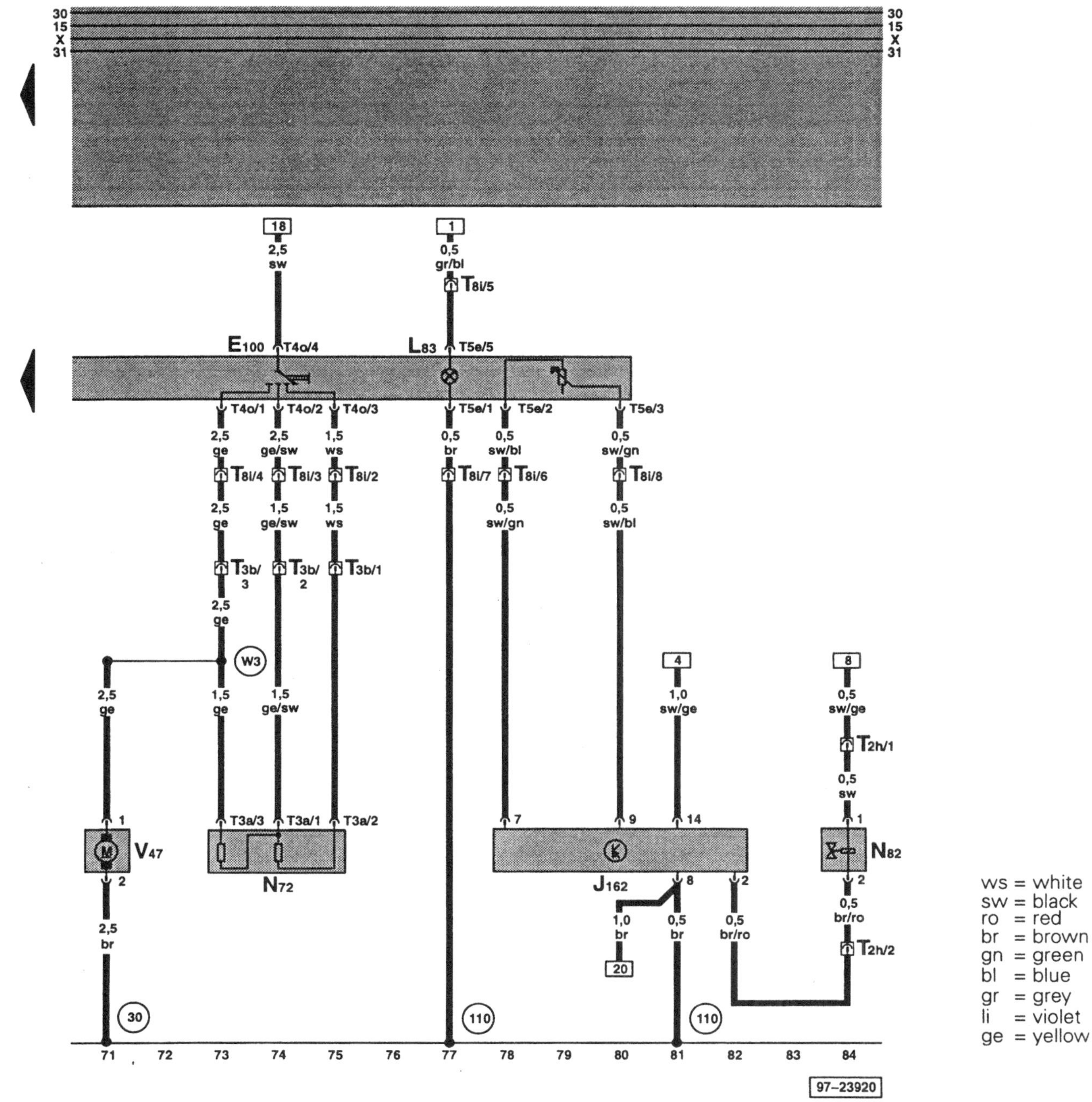

E100 – Warm Air Blower Switch
J162 – Heater Control Module, control number (152), behind instrument panel center
L83 – Warm Air Blower Switch Light
N72 – Heater Blower Series Resistance
N82 – Coolant Cut-off Valve
T2h – Double Connector, behind fuse/relay panel
T3a – 3-Pin Connector
T3b – 3-Pin Connector, behind fuse/relay panel
T4o – 4-Pin Connector
T5e – 5-Pin Connector
T8i – 8-Pin Connector, behind fuse/relay panel
V47 – Rear Warm Air Fan

(30) – Ground connection, -1-, beside fuse/relay panel

(110) – Ground connection -2-, in instrument panel wiring harness
(W3) – Wire connection, in rear wiring harness

THIS PAGE INTENTIONALLY LEFT BLANK

Climatronic,

from September 1998

● 2.8 L- Engine - Motronic Multiport Fuel Injection (MFI)/103 kW, code AES, with automatic transmission

Deviate relay location and fuseplacements as well as the locations of multiple connectors see section "component locations".

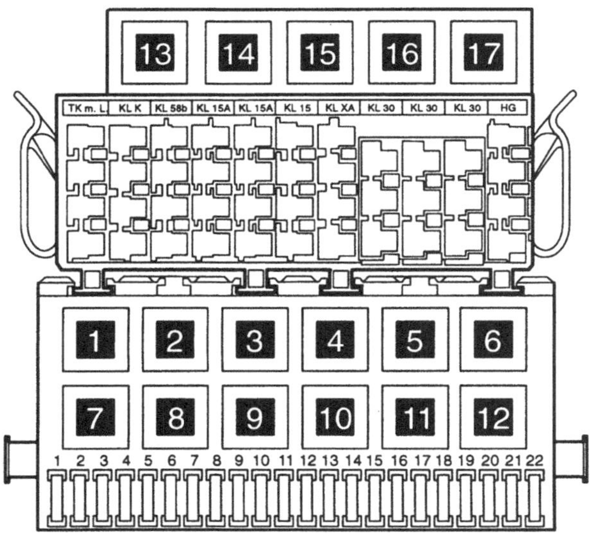

97–23232

1 Relay for Climatronic (140)

Note: Number in parentheses indicates production control number stamped on relay housing.

Fuse colors

30 A – green
25 A – white
20 A – yellow
15 A – blue
10 A – red
7,5 A – brown
5 A – beige
3 A – violet

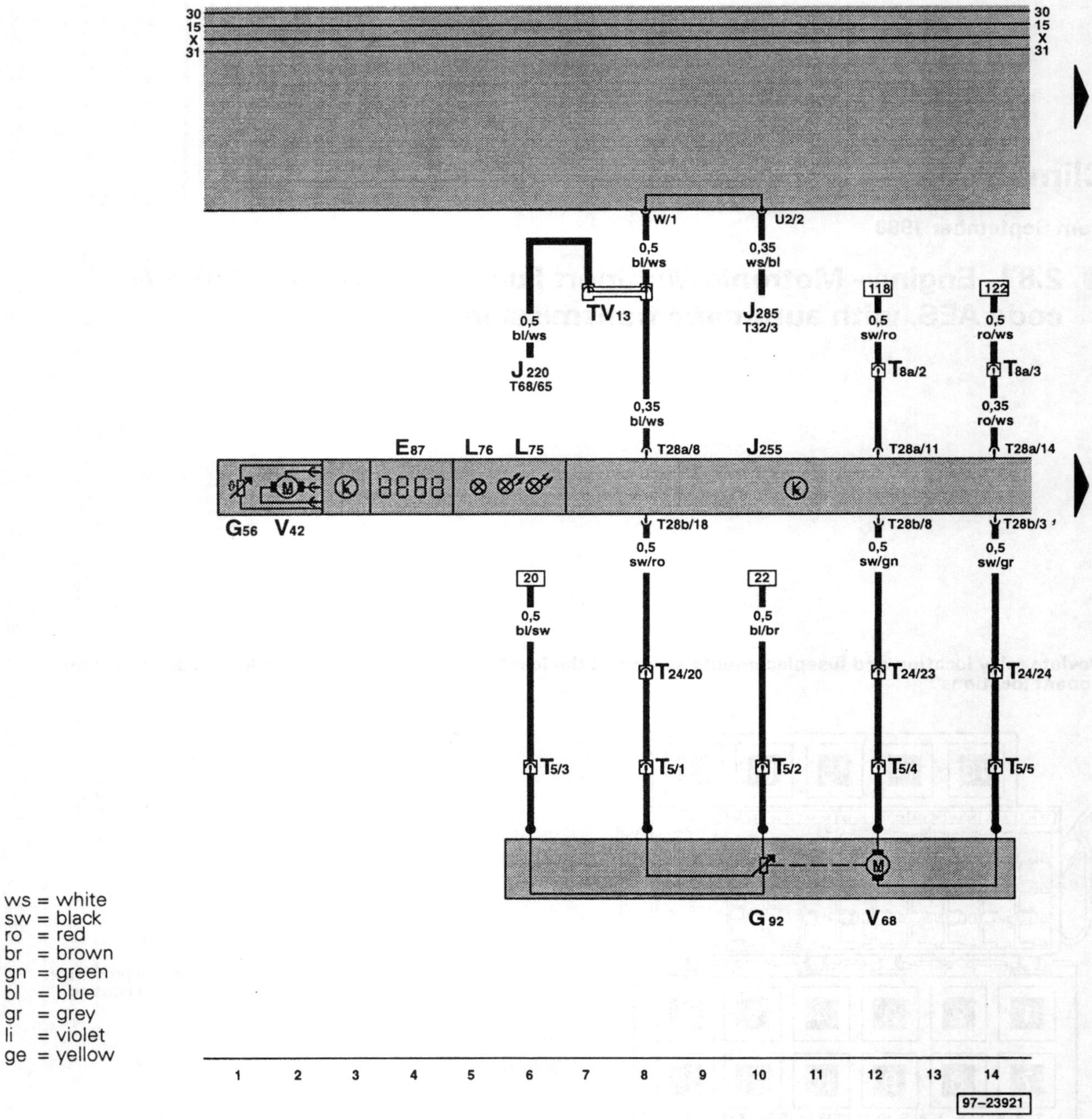

E87 – A/C Control Head
G56 – Instrument Panel Interior Temperature Sensor
G92 – Temperature Regulator Flap Motor Position Sensor
J220 – Motronic Engine Control Module (ECM), in engine compartment, left
J255 – Climatronic Control Module
J285 – Control module with indicator unit in instrument panel insert
L75 – Digital Display Light
L76 – Push Button Light
T5 – 5-Pin Connector, behind instrument panel, center
T8a – 8-Pin Connector, behind fuse/relay panel
T24 – 24-Pin Connector, behind instrument panel, left
T28a – 28-Pin Connector, red
T28b – 28-Pin Connector, blue
T32 – 32-Pin Connector, blue
T68 – 68-Pin Connector, on Motronic Engine Control Module (ECM)
TV13 – Vehicle Speed Signal Wire Connector, blue
V42 – Interior Temperature Sensor Fan
V68 – Temperature Regulator Flap Motor

Climatronic control module, A/C control head, temperature regulator flap motor, temperature regulator flap motor position sensor

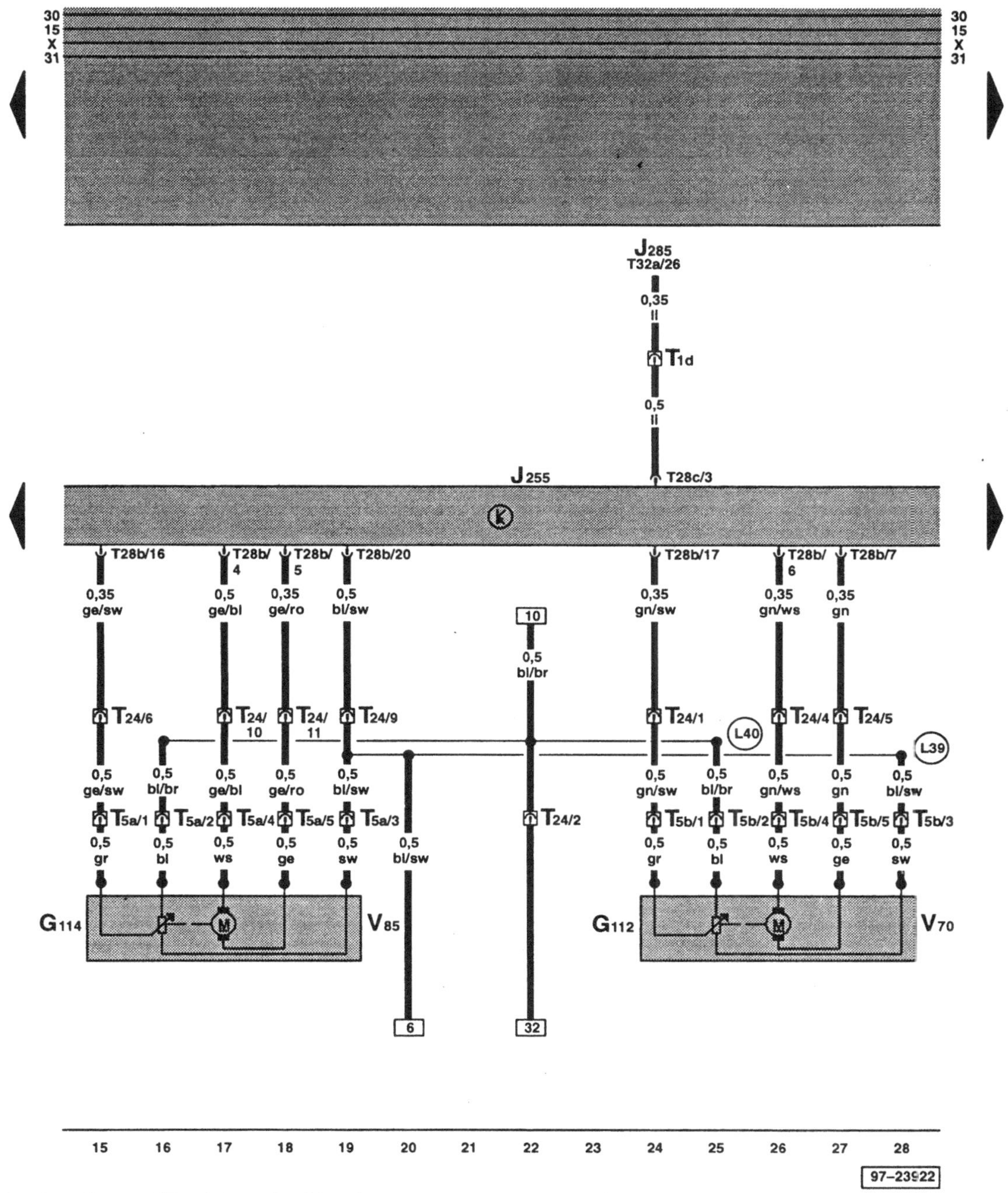

ws = white
sw = black
ro = red
br = brown
gn = green
bl = blue
gr = grey
li = violet
ge = yellow

G112 – Central Flap Motor Position Sensor
G114 – Footwell/Defrost Flap Motor Position Sensor
J255 – Climatronic Control Module
J285 – Control module with indicator unit in instrument panel insert
T1d – Single Connector, behind fuse/relay panel
T5a – 5-Pin Connector, behind instrument panel, center
T5b – 5-Pin Connector, behind instrument panel, center
T24 – 24-Pin Connector, behind instrument panel, right
T28b – 28-Pin Connector, blue
T28c – 28-Pin Connector, white
T32a – 32-Pin Connector, green
V70 – Central Air Flap Motor
V85 – Footwell/Defroster Flap Motor
(L39) – Wire connection (A/C Pressure Switch) -1-, in Climatronic wiring harness
(L40) – Wire connection (A/C Pressure Switch) -2-, in Climatronic wiring harness

Climatronic control module, central air flap motor, central flap motor position sensor, footwell/defroster flap motor, footwell/defrost flap motor position sensor

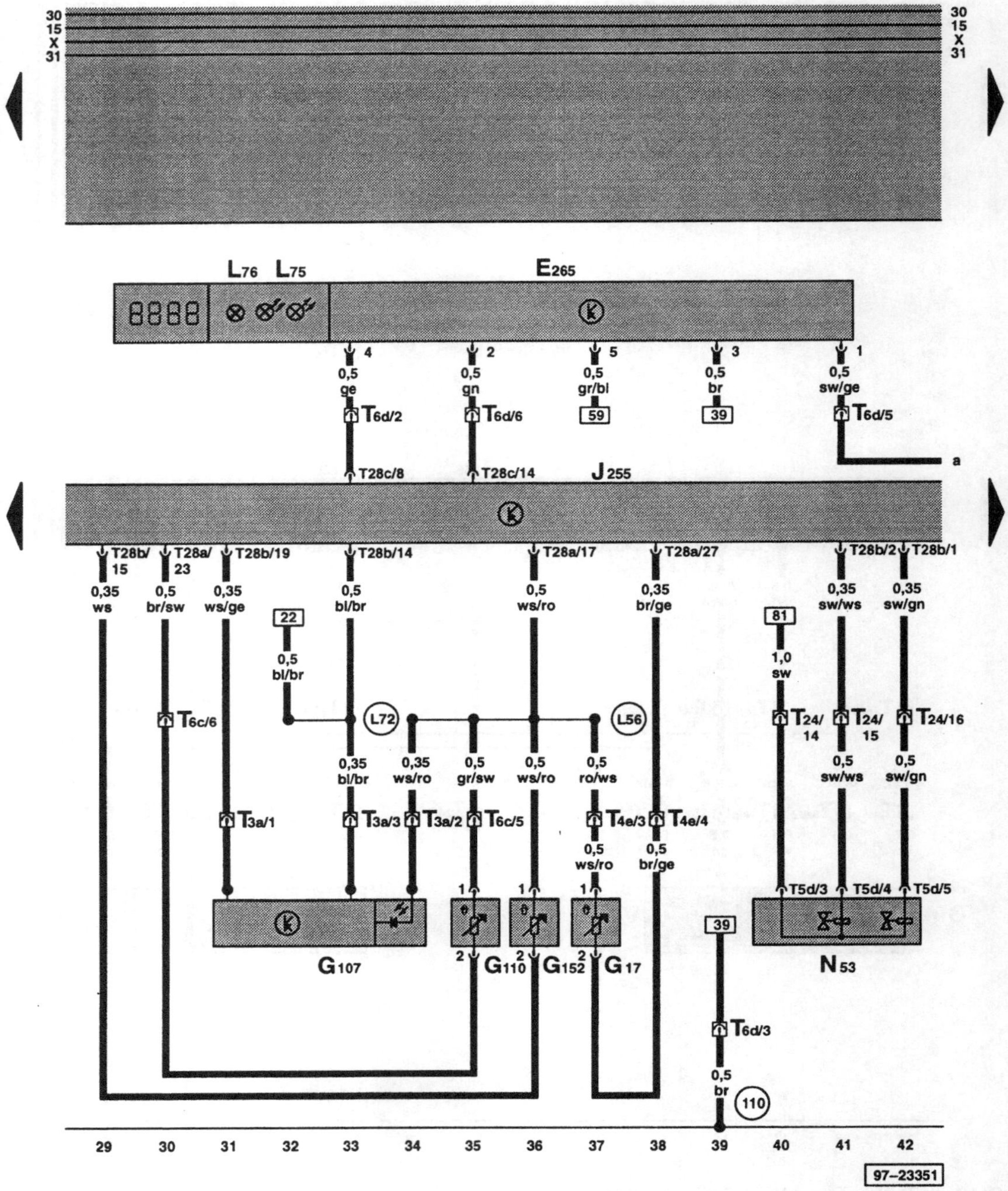

ws = white
sw = black
ro = red
br = brown
gn = green
bl = blue
gr = grey
li = violet
ge = yellow

E265 – A/C Control Head rear
G17 – Outside Air Temperature Sensor, on front bumper left
G107 – Sunlight Photo Sensor, on instrument panel, left
G110 – A/C Engine Coolant Temperature (ECT) Sensor, in engine compartment, left
G152 – Front Vent Temperature Sensor, on console for footwell, left vent side
J255 – Climatronic Control Module
L75 – Digital Display Light
L76 – Push Button Light
N53 – Vacuum Valve Rail
T3a – 3-Pin Connector, behind instrument panel, left
T4e – 4-Pin Connector, behind fuse/relay panel
T5d – 5-Pin Connector
T6c – 6-Pin Connector, behind fuse/relay panel
T6d – 6-Pin Connector, behind fuse/relay panel
T24 – 24-Pin Connector, behind instrument panel, right
T28a – 28-Pin Connector, red
T28b – 28-Pin Connector, blue
T28c – 28-Pin Connector, white

(110) – Ground connection -2-, in instrument panel wiring harness
(L56) – Connector (Sensor), in wiring harness Climatronic
(L72) – Connector (Photo Sensor), in wiring harness Climatronic

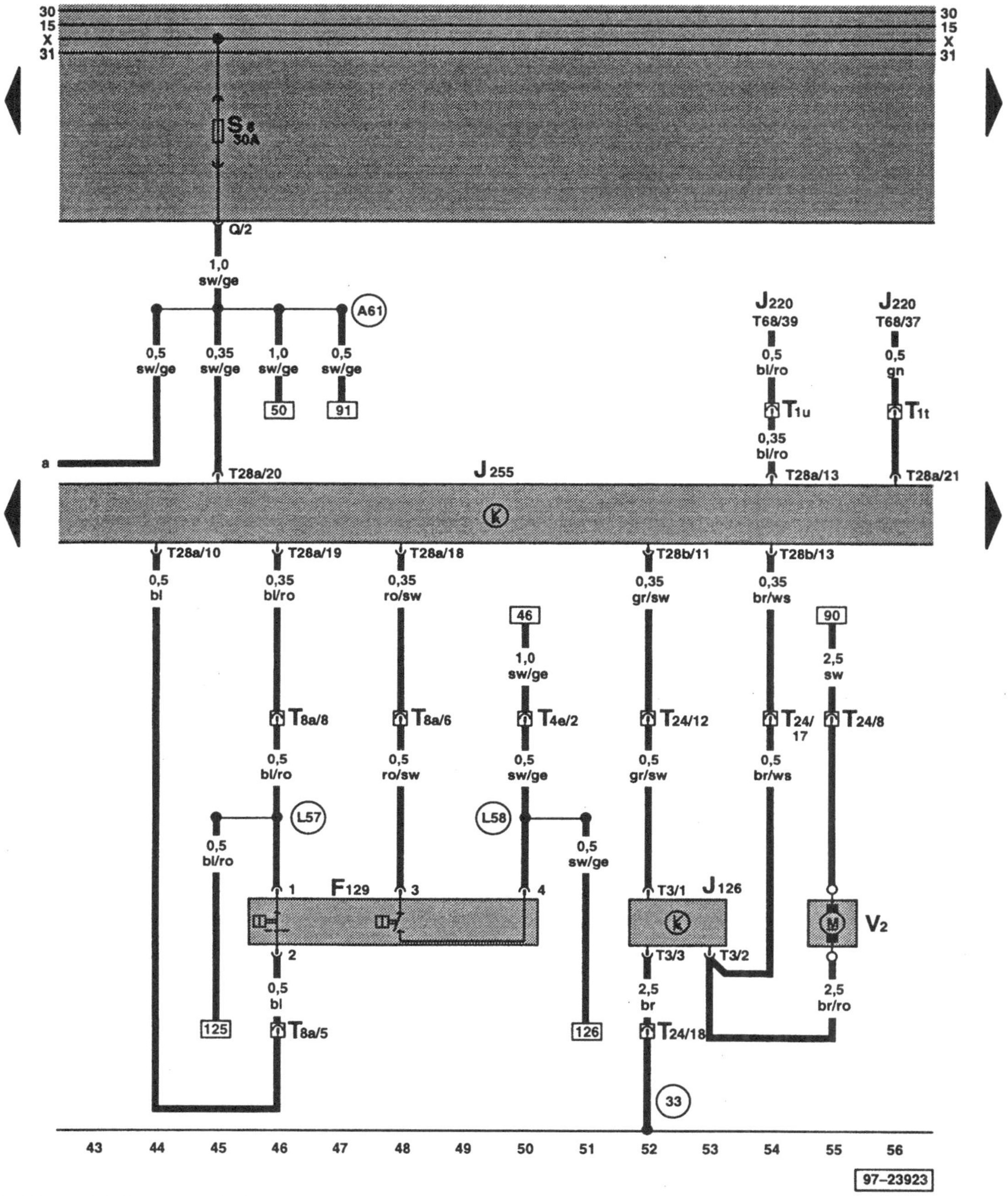

ws = white
sw = black
ro = red
br = brown
gn = green
bl = blue
gr = grey
li = violet
ge = yellow

F129 – A/C Pressure Switch
J126 – Control module for fresh air blower
J220 – Motronic Engine Control Module (ECM), in engine compartment, left
J255 – Climatronic Control Module
J285 – Control module with indicator unit in instrument panel insert
T1t – Single Connector (Connection A/C), behind fuse/relay panel
T1u – Single Connector (Connection A/C), behind fuse/relay panel
T3 – 3-Pin Connector
T4e – 4-Pin Connector, behind fuse/relay panel
T8a – 8-Pin Connector, behind fuse/relay panel
T24 – 24-Pin Connector, behind instrument panel, right
T28a – 28-Pin Connector, red
T28b – 28-Pin Connector, blue
T68 – 68-Pin Connector
V2 – Fresh Air Blower

(33) – Ground connection, behind instrument panel, right
(A61) – Plus Connection (X), in Instrument cluster wiring harness
(L57) – Connector (Pressure switch for A/C) -1- in wiring harness Climatronic - Fan for coolant
(L58) – Connector (Pressure switch for A/C) -2- in wiring harness Climatronic - Fan for coolant

Climatronic control module, A/C pressure switch, control module for fresh air blower, fresh air blower

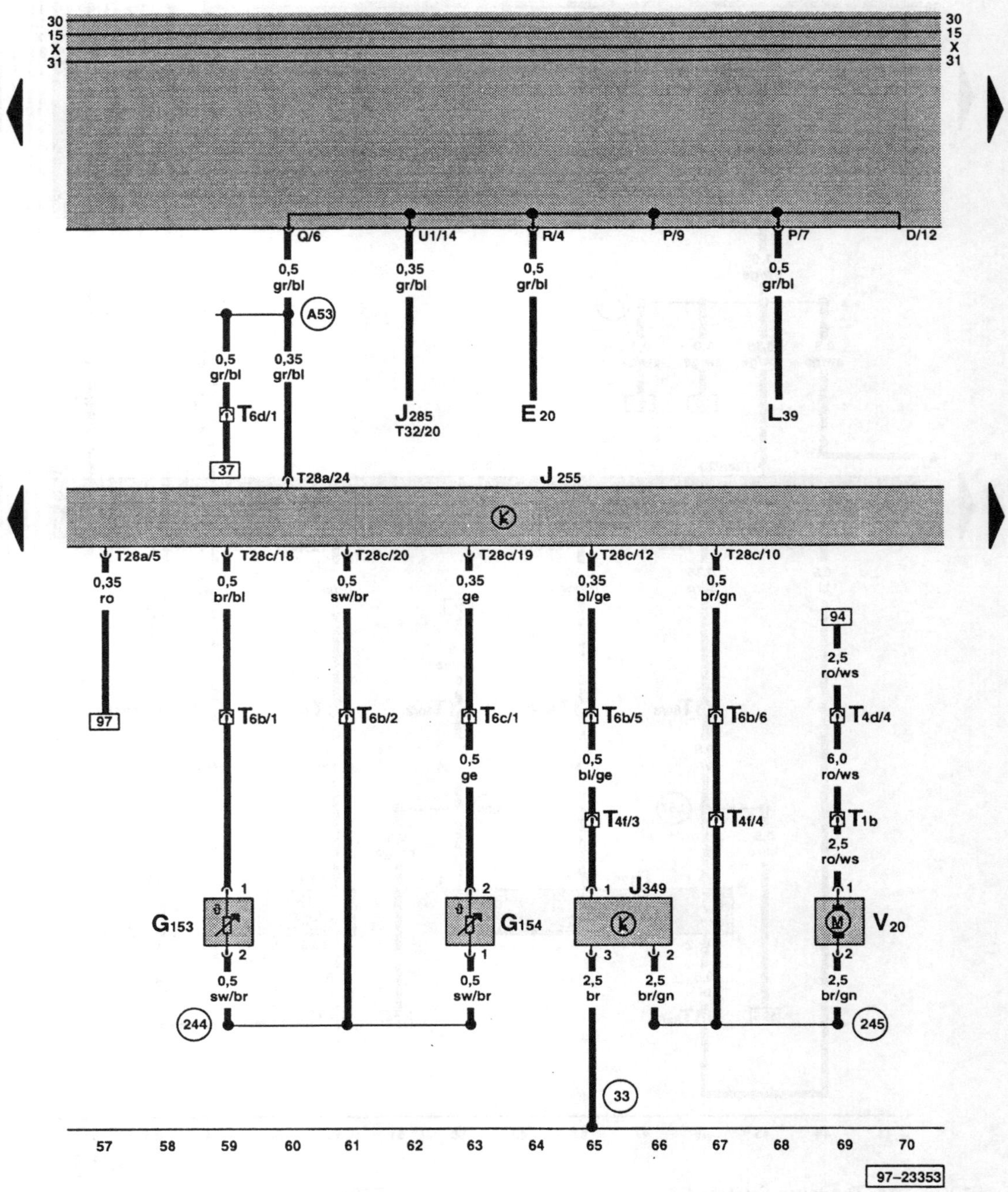

ws = white
sw = black
ro = red
br = brown
gn = green
bl = blue
gr = grey
li = violet
ge = yellow

E20 – Instrument Panel Light Dimmer Switch
G153 – Rear Evaporator Temperature Sensor
G154 – Rear Heater Core Temperature Sensor
J255 – Climatronic Control Module
J285 – Control module with indicator unit in instrument panel insert
J349 – Control module for blower for evaparator
L39 – Rear Window Defogger Switch Light
T1b – Single Connector, behind trim D-pillar right, lower
T4d – 4-Pin Connector, behind fuse/relay panel
T4f – 4-Pin Connector, behind trim D-pillar right, lower
T6b – 6-Pin Connector, behind fuse/relay panel
T6c – 6-Pin Connector, behind fuse/relay panel
T6d – 6-Pin Connector, behind fuse/relay panel
T28a – 28-Pin Connector, red
T28c – 28-Pin Connector, white
T32 – 32-Pin Connector, blue
V20 – Rear Evaporator Fan
(33) – Ground connection, behind instrument panel, right
(244) – Ground connection (Sensor ground), in wiring harness Climatronic
(245) – Ground connection in wiring harness blower for 2nd Evaporatur
(A53) – Plus connection (58b), in instrument panel wiring harness

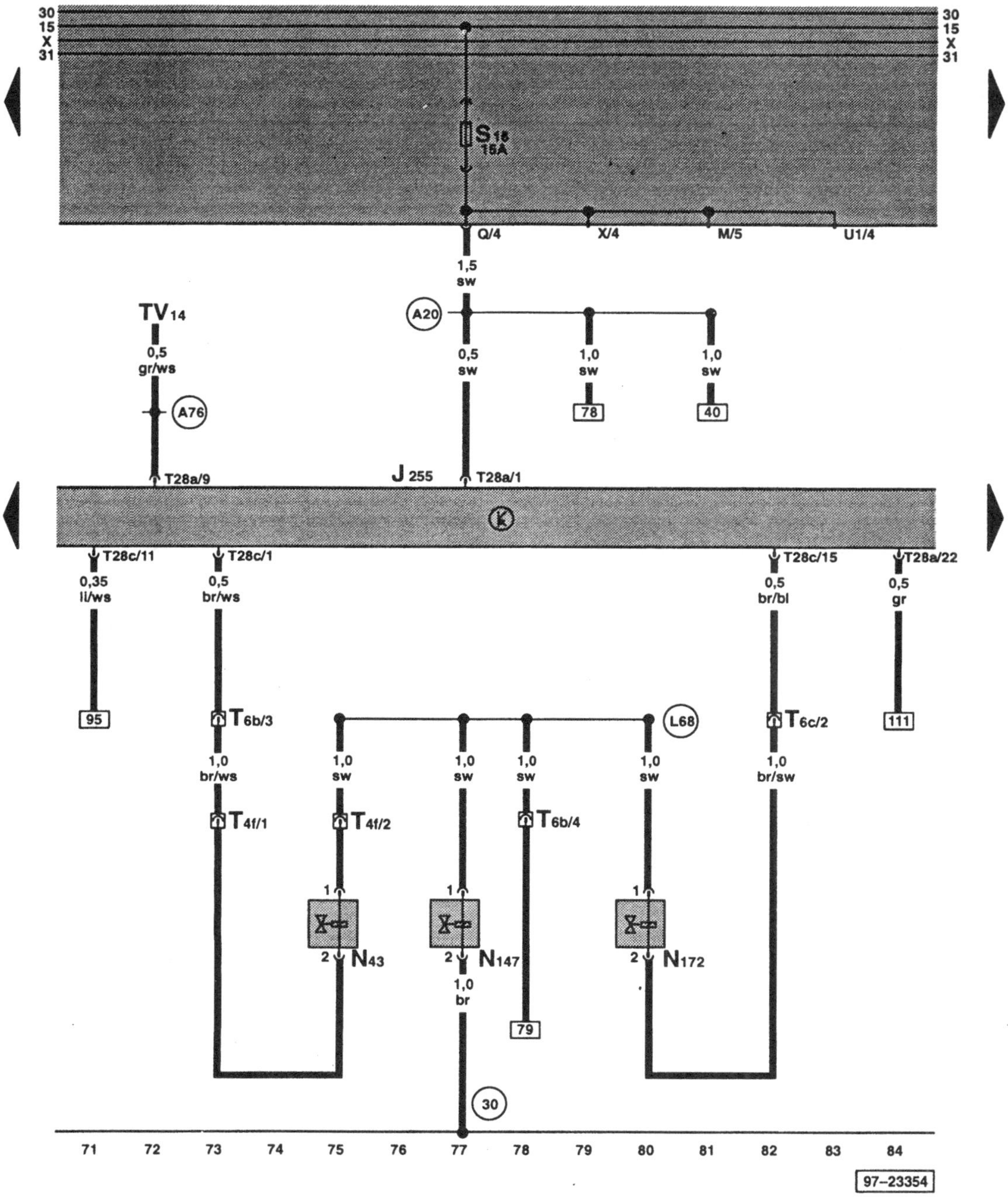

ws = white
sw = black
ro = red
br = brown
gn = green
bl = blue
gr = grey
li = violet
ge = yellow

J255 – Climatronic Control Module
N43 – A/C Refrigerant Shut-Off Valve, behind trim D-pillar right, lower
N147 – Engine Coolant Two-way Vacuum Valve, on firewall, left
N172 – Auxiliary Heater Valve, on firewall, left
T4f – 4-Pin Connector, behind trim D-pillar right, lower
T6b – 6-Pin Connector, behind fuse/relay panel
T6c – 6-Pin Connector, behind fuse/relay panel
T28a – 28-Pin Connector, red
T28c – 28-Pin Connector, white
TV14 – Data Link Connector (DLC) wire Connector, white

(30) – Ground connection, -1-, beside fuse/relay panel
(A20) – Wire connection (15a), in instrument panel wiring harness
(A76) – Connector (K-diagnosis wire), in instrument panel wiring harness
(L68) – Connection 2 (sensor) in Climatronic wiring harness

Climatronic control module, A/C refrigerant shut-Off valve, engine coolant two-way vacuum valve, auxiliary heater valve

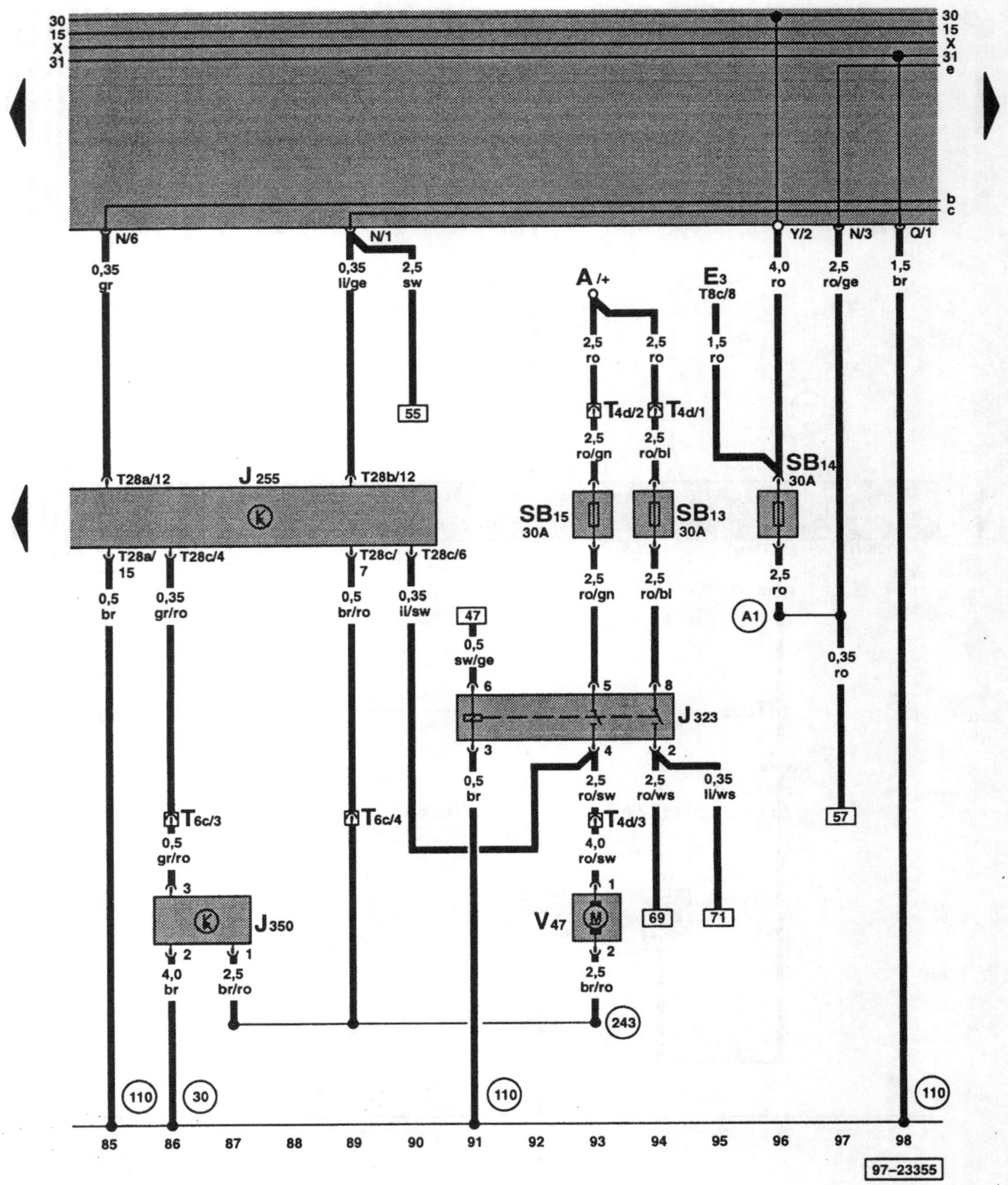

ws = white
sw = black
ro = red
br = brown
gn = green
bl = blue
gr = grey
li = violet
ge = yellow

A – Battery
E3 – Emergency Flasher Switch
J255 – Climatronic Control Module
J323 – Fan Relay, control number (114), behind instrument panel, center
J350 – Control module for warm air blower
SB13 – Fuse 13 in fuse holder, below relay panel
SB14 – Fuse 14 in fuse holder, below relay panel
SB15 – Fuse 15 in fuse holder, below relay panel
T4d – 4-Pin Connector, behind fuse/relay panel
T6c – 6-Pin Connector, behind fuse/relay panel
T8c – 8-Pin Connector, on Emergency Flasher Switch
T28a – 28-Pin Connector, red
T28b – 28-Pin Connector, blue
T28c – 28-Pin Connector, white
V47 – Rear Warm Air Fan

(30) – Ground connection, -1-, beside fuse/relay panel
(110) – Ground connection -2-, in instrument panel wiring harness
(243) – Ground connection -1- in wiring harness, Climatronic
(A1) – Plus connection (30a), in instrument panel wiring harness

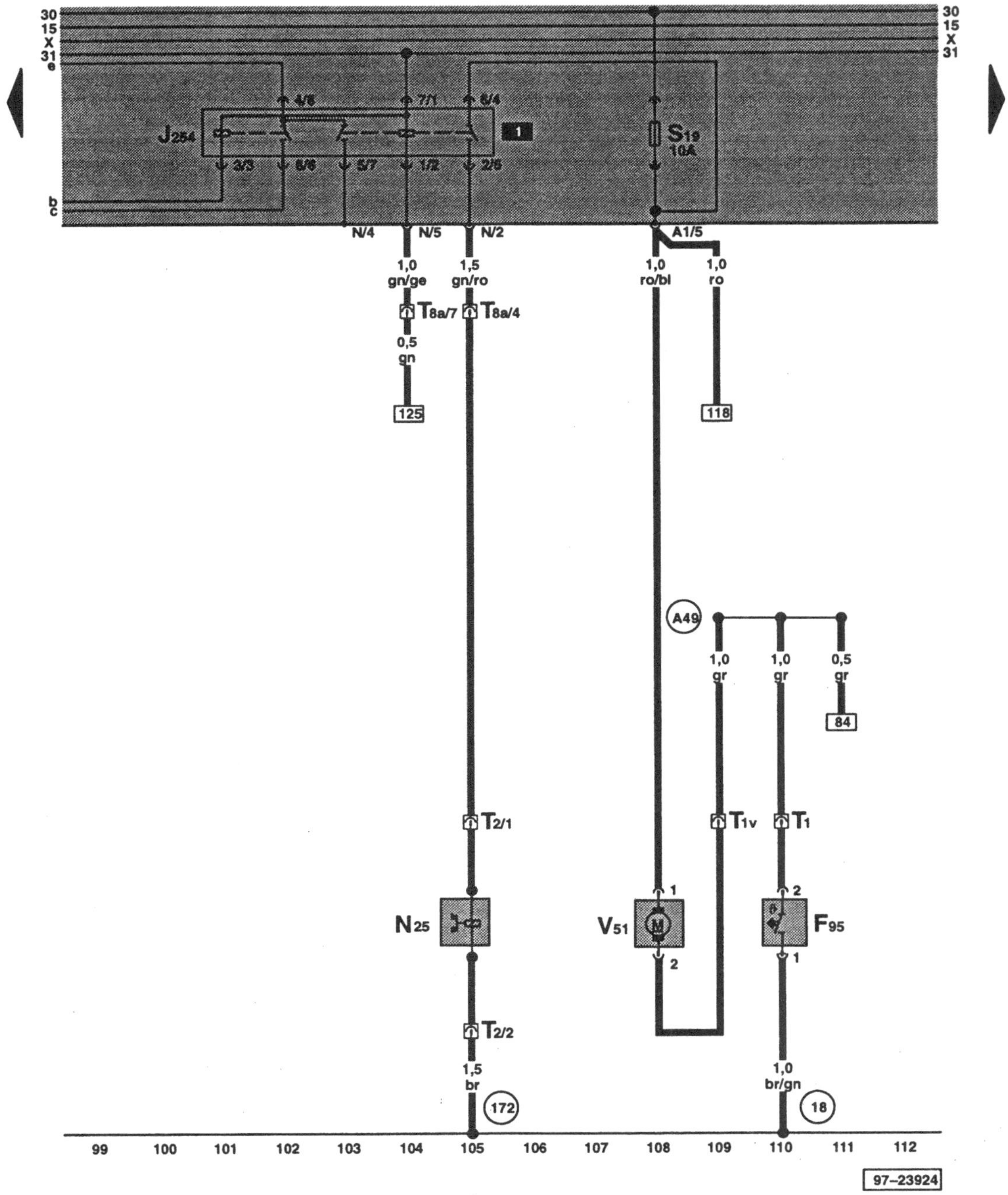

ws = white
sw = black
ro = red
br = brown
gn = green
bl = blue
gr = grey
li = violet
ge = yellow

F95 – After-Run Coolant Thermal Switch
J254 – Relay for Climatronic, control number (140), switch delay time 1/2 second
N25 – A/C Clutch
T1 – Single Connector, behind fuse/relay panel
T1v – Single Connector, behind fuse/relay panel
T2 – Double Connector, in engine compartment, right front
T8a – 8-Pin Connector, behind fuse/relay panel
V51 – After-Run Coolant Pump

(18) – Ground connection, on engine block

(172) – Ground connection, in engine compartment climatronic wiring harness

(A49) – Wire connection -1-, in instrument panel wiring harness

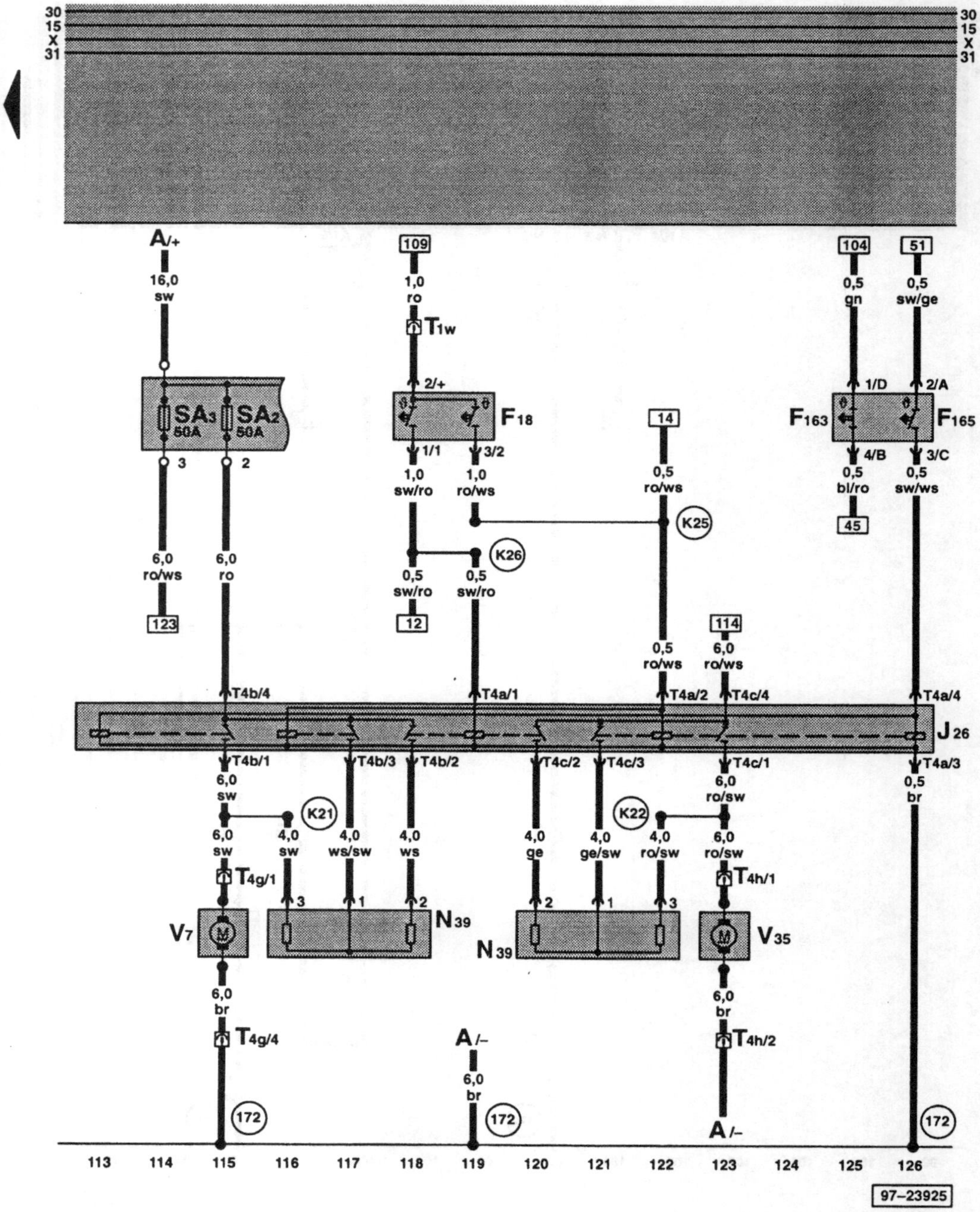

ws = white
sw = black
ro = red
br = brown
gn = green
bl = blue
gr = grey
li = violet
ge = yellow

A – Battery
F18 – Coolant Fan Control (FC) Thermal Switch
F163 – A/C Cut-Out Thermal Switch
F165 – Third Speed Coolant Fan Control (FC) Thermal Switch, in engine compartment, front
J26 – Coolant Fan control (FC) Relay, in engine compartment, left front
N39 – Coolant Fan Control (FC) Series Resistance
SA2 – Fuse -2- in fuse bracket/battery
SA3 – Fuse -3- in fuse bracket/battery
T1w – Single Connector, behind fuse/relay panel
T4a – 4-Pin Connector
T4b – 4-Pin Connector
T4c – 4-Pin Connector
T4g – 4-Pin Connector, near coolant fan
T4h – 4-Pin Connector, near coolant fan

V7 – Left Coolant Fan (450 W)
V35 – Right Coolant Fan (450 W)

(172) – Ground connection, in engine compartment climatronic wiring harness
(K21) – Wire connection -1-, in coolant fan wiring harness
(K22) – Wire connection -2-, in coolant fan wiring harness
(K25) – Wire connection -4-, in coolant fan wiring harness
(K26) – Wire connection -5-, in coolant fan wiring harness

Fresh Air Blower front and rear with auxiliary heater core,

from September 1998

Deviate relay location and fuseplacements as well as the locations of multiple connectors see section "component locations".

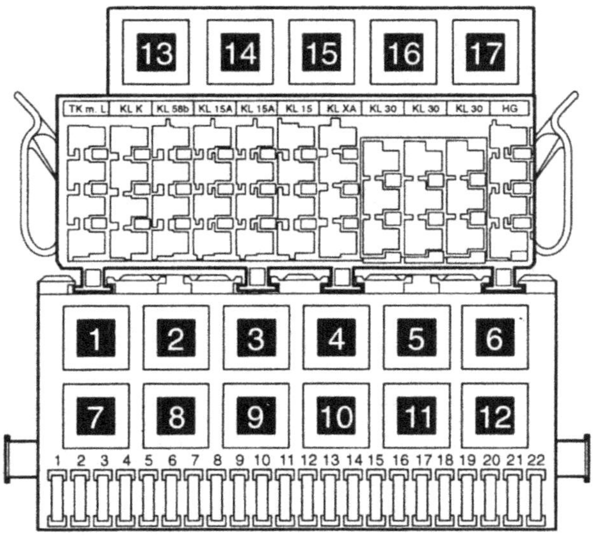

97–23232

1 Fresh Air Blower Relay (105)

Note: Number in parentheses indicates production control number stamped on relay housing.

Fuse colors

30 A – green
25 A – white
20 A – yellow
15 A – blue
10 A – red
7,5 A – brown
5 A – beige
3 A – violet

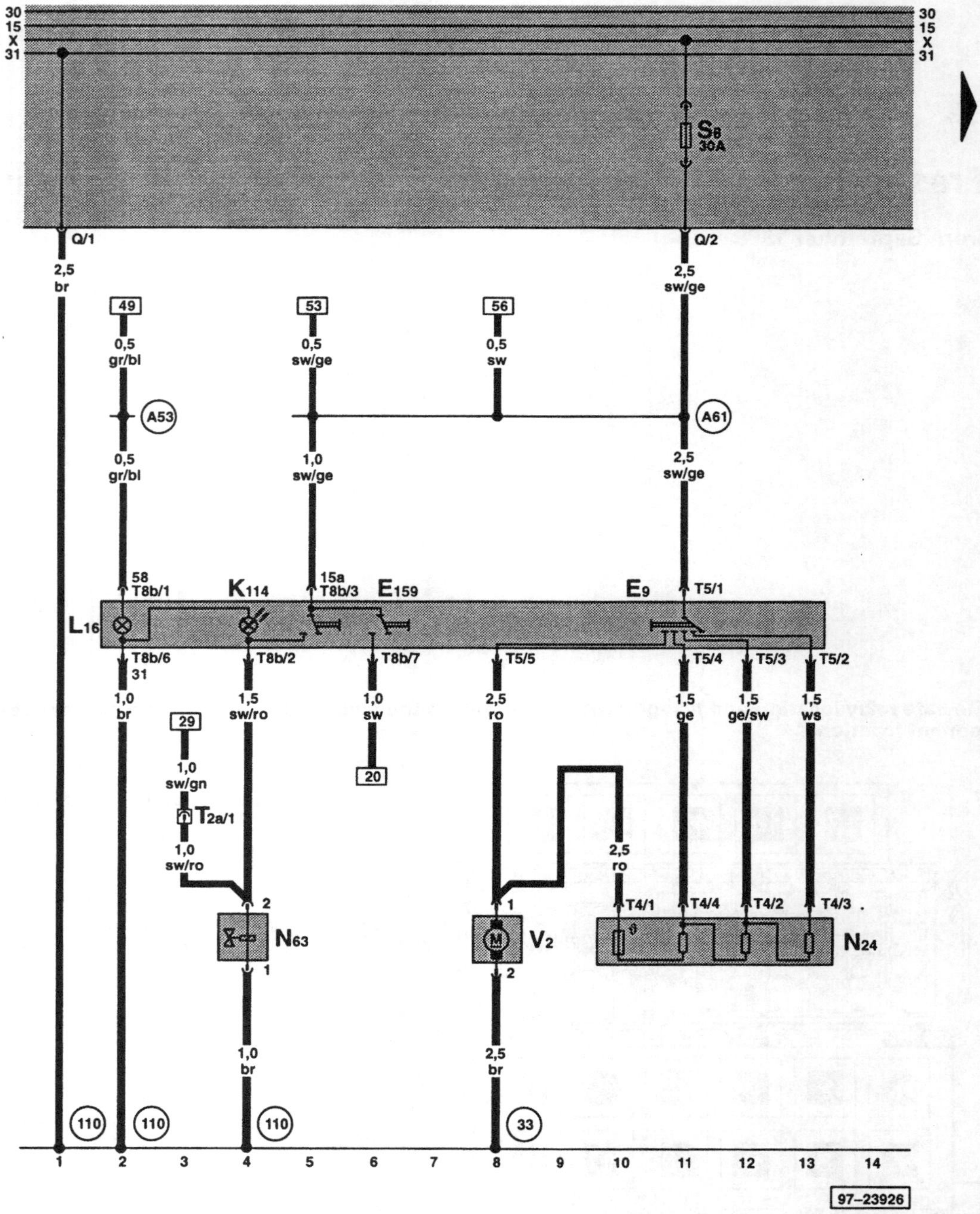

ws = white
sw = black
ro = red
br = brown
gn = green
bl = blue
gr = grey
li = violet
ge = yellow

- E9 – Fresh Air Blower Switch
- E159 – Fresh Air/Recirculating Flap Switch
- K114 – Fresh Air and Recirculating Air Mode Indicator Light
- L16 – Fresh Air Control Lever Light
- N24 – Fresh Air Blower Series Resistance with Fuse
- N63 – Fresh Air/Recirculating Flap Two-Way Valve
- T2a – Double Connector, behind fuse/relay panel
- T4 – 4-Pin Connector
- T5 – 5-Pin Connector
- T8b – 8-Pin Connector
- V2 – Fresh Air Blower

- (33) – Ground connection, behind instrument panel right
- (110) – Ground connection -2-, in instrument panel wiring harness
- (A53) – Plus connection (58b), in instrument panel wiring harness
- (A61) – Plus Connection (X), in Instrument cluster wiring harness

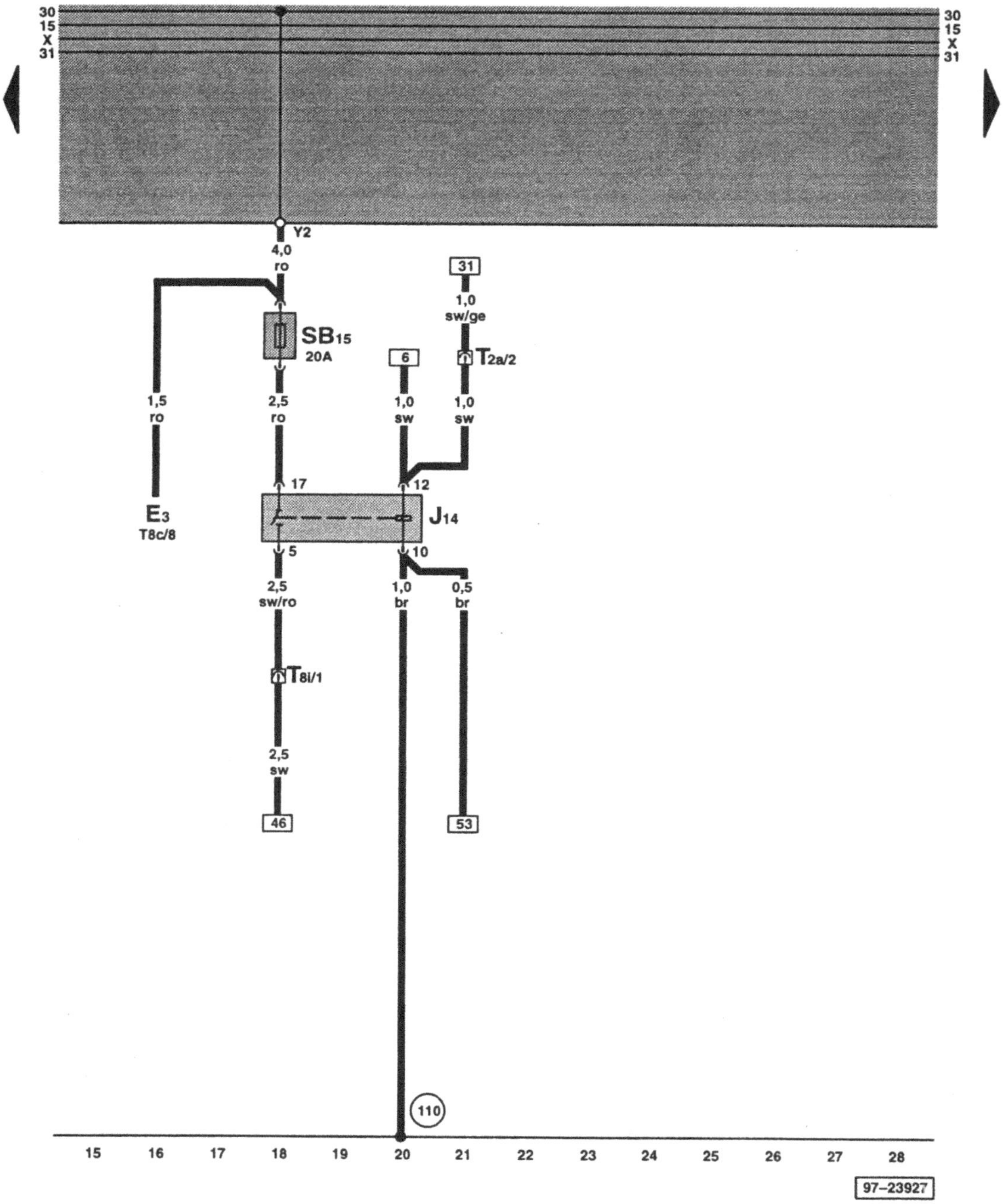

ws = white
sw = black
ro = red
br = brown
gn = green
bl = blue
gr = grey
li = violet
ge = yellow

E3 – Emergency Flasher Switch
J14 – Heater Air Blower Relay, control number (53), behind instrument panel center
SB15 – Fuse 15 in fuse holder, below relay panel
T2a – Double Connector, behind fuse/relay panel
T8c – 8-Pin Connector, on Emergency Flasher Switch
T8i – 8-Pin Connector, behind fuse/relay panel

(110) – Ground connection -2-, in instrument panel wiring harness

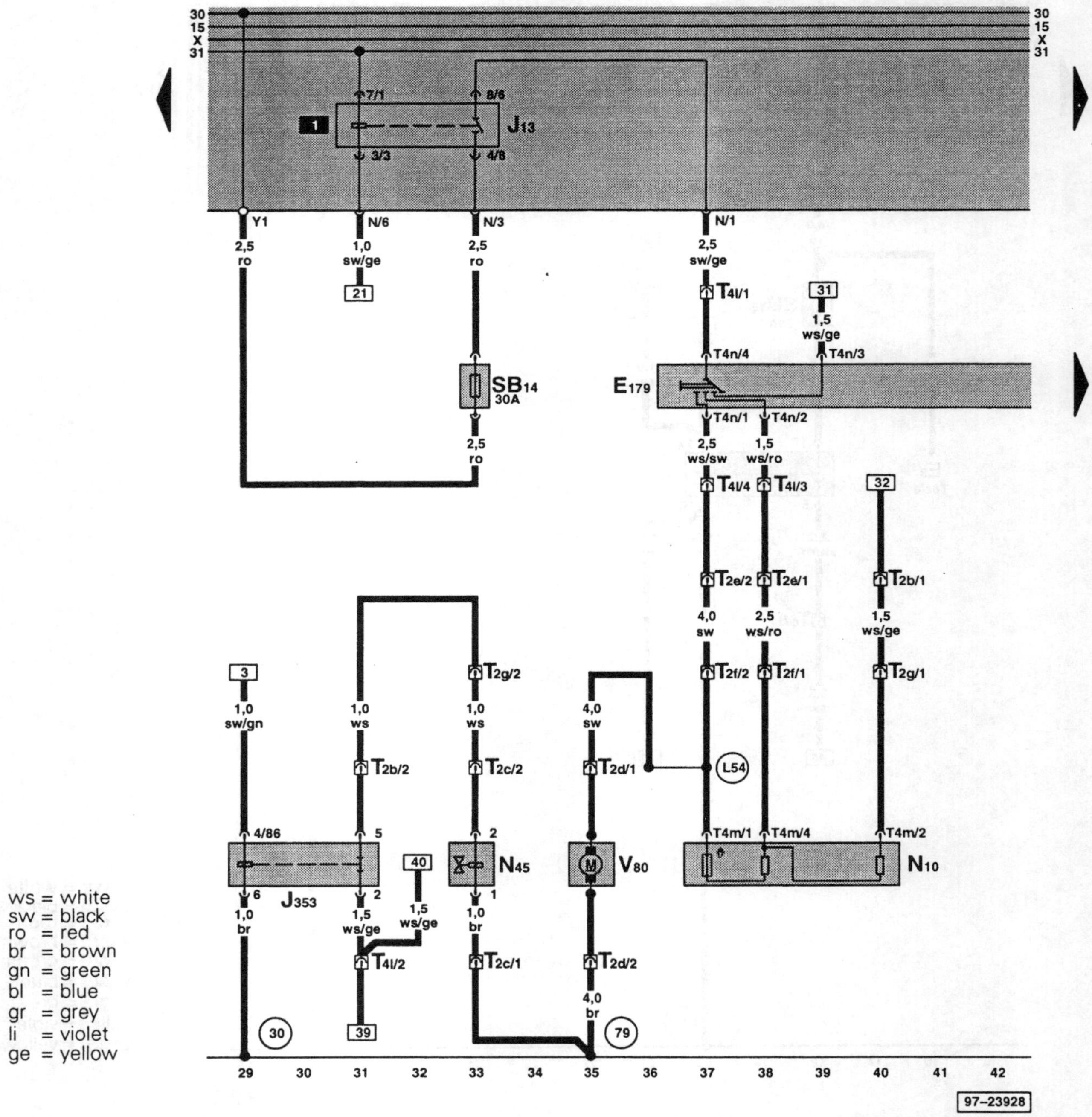

E179 – Switch for Freshair blower-rear
J13 – Fresh Air Blower Relay, control number (105)
J353 – Relay for Fresh-air and recirculating air flap-rear, control number (94), behind instrument panel center
N10 – Series Resistance for fresh air blower-rear
N45 – Two-way valve for Fresh- and recirculation air flap-rear
SB14 – Fuse 14 in fuse holder, below relay panel
T2b – Double Connector, behind instrument panel, right
T2c – Double Connector, behind right D-pillar trim, lower part
T2d – Double Connector, behind right D-pillar trim, lower part
T2e – Double Connector, behind instrument panel, right
T2f – Double Connector behind right D-pillar trim, lower part
T2g – Double Connector, behind right D-pillar trim, lower part
T4l – 4-Pin Connector, behind fuse/relay panel
T4m – 4-Pin Connector
T4n – 4-Pin Connector
V80 – Fresh air blower rear
(30) – Ground connection, -1-, beside fuse/relay panel
(79) – Ground connection, near right D-pillar, lower part
(L54) – Plus connector (Blower stage 3) in wiring harness blower-Passenger ventilation

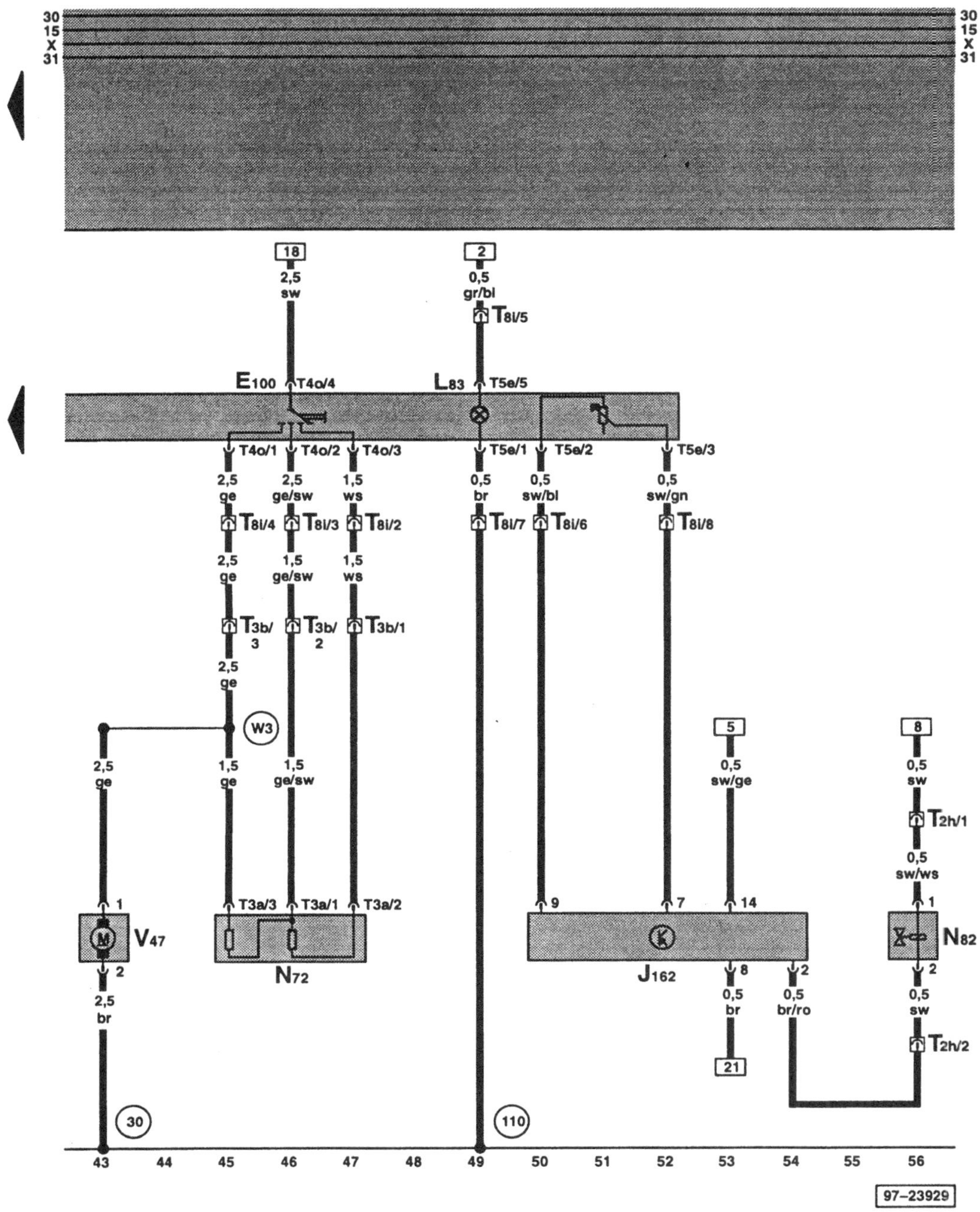

ws = white
sw = black
ro = red
br = brown
gn = green
bl = blue
gr = grey
li = violet
ge = yellow

- E100 – Warm Air Blower Switch
- J162 – Heater Control Module, control number (152), behind instrument panel center
- L83 – Warm Air Blower Switch Light
- N72 – Heater Blower Series Resistance
- N82 – Coolant Cut-off Valve
- T2h – Double Connector, behind fuse/relay panel
- T3a – 3-Pin Connector
- T3b – 3-Pin Connector, behind fuse/relay panel
- T4o – 4-Pin Connector
- T5e – 5-Pin Connector
- T8i – 8-Pin Connector, behind fuse/relay panel
- V47 – Rear Warm Air Fan

- (30) – Ground connection, -1-, beside fuse/relay panel
- (110) – Ground connection -2-, in instrument panel wiring harness
- (W3) – Wire connection, in rear wiring harness

THIS PAGE INTENTIONALLY LEFT
BLANK

Auxiliary Heater B3LC/D3LC,

from September 1998

Deviate relay location and fuseplacements as well as the locations of multiple connectors see section "component locations".

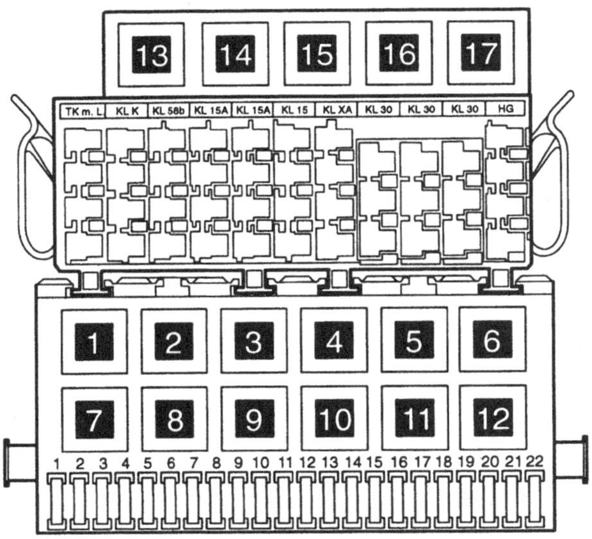

97–23232

Fuse colors

30 A – green
25 A – white
20 A – yellow
15 A – blue
10 A – red
7,5 A – brown
5 A – beige
3 A – violet

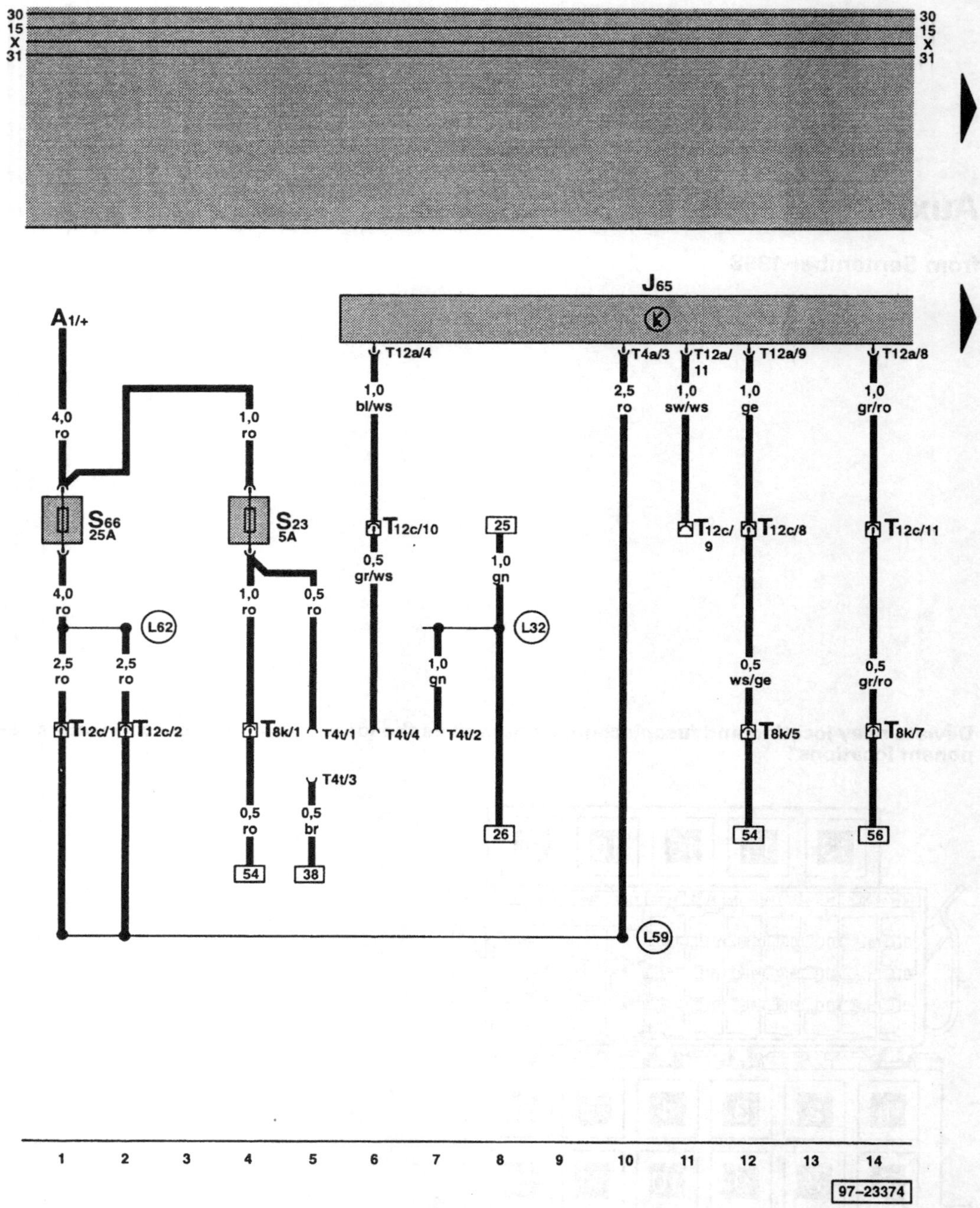

ws = white
sw = black
ro = red
br = brown
gn = green
bl = blue
gr = grey
li = violet
ge = yellow

A1 – Auxiliary Battery, below driver's seat
J65 – Heater Control Modul, in intake channel
S23 – Heater Fuse, below driver's seat
S66 – Auxiliary Heater Glow Plug Fuse, below driver's seat
T4a – 4-Pin Connector
T4t – 4-Pin Connector, below driver's seat, data link Connection
T8k – 8-Pin Connector, below driver's seat
T12a – 12-Pin Connector
T12c – 12-Pin Conenctor, to intake channel

(L32) – Wire connection -1-, in heater wiring harness

(L59) – Plus connector (30) in wiring harness heater

(L62) – Plus connector (30a), in wiring harness auxiliary heater

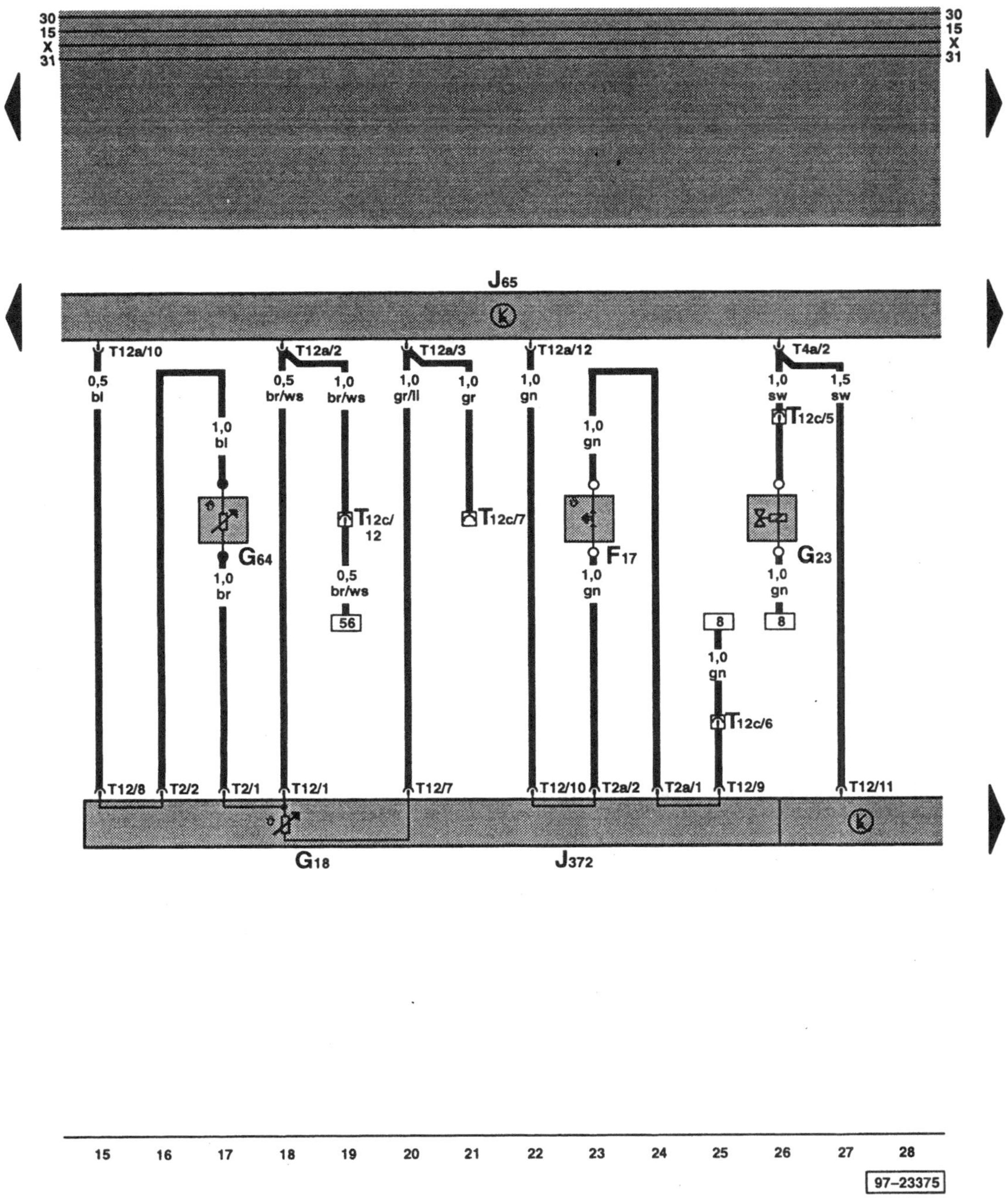

ws = white
sw = black
ro = red
br = brown
gn = green
bl = blue
gr = grey
li = violet
ge = yellow

F17 – Heater Overheat Control Switch, in heater unit
G18 – Temperature Sensor II, in heater unit
G23 – Transfer Fuel Pump (FP)
G64 – Flame Sensor in heater unit
J65 – Heater Control Module, in intake channel
J372 – Circuit board for temperature sensor, control module for blower motor and glow plug
T2 – Double Connector
T2a – Double Connector
T4a – 4-Pin Connector
T12 – 12-Pin Conenctor
T12a – 12-Pin Conenctor
T12c – 12-Pin Conenctor, to intake channel

Heater controle module, heater overheat control switch, temperature sensor, transfer fuel pump (FP), flame sensor

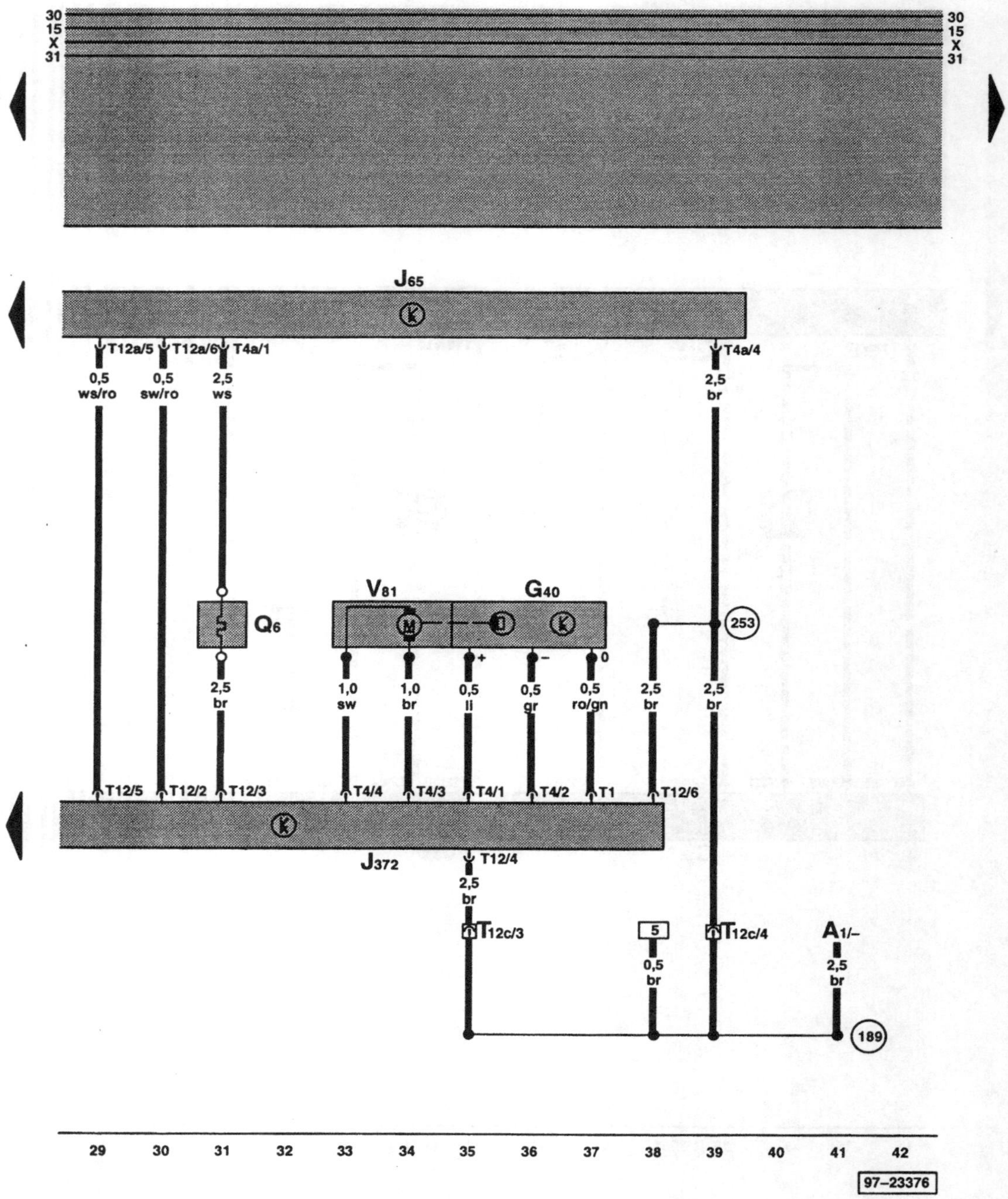

ws = white
sw = black
ro = red
br = brown
gn = green
bl = blue
gr = grey
li = violet
ge = yellow

A1 – Auxiliary Battery
G40 – Camshaft Position (CMP) Sensor
J65 – Heater Control Module, in intake channel
J372 – Circuit board for temperature sensor, control module for blower motor and glow plug
Q6 – Glow plug, in heater unit
T1 – Single Connector
T4 – 4-Pin Connector
T4a – 4-Pin Connector
T12 – 12-Pin Conenctor
T12a – 12-Pin Conenctor
T12c – 12-Pin Conenctor, to intake channel
V81 – Fresh Air/Heater blower, in heater unit

(189) – Ground connection -1-, in heater wiring harness

(253) – Ground connection, in wiring harness heater unit

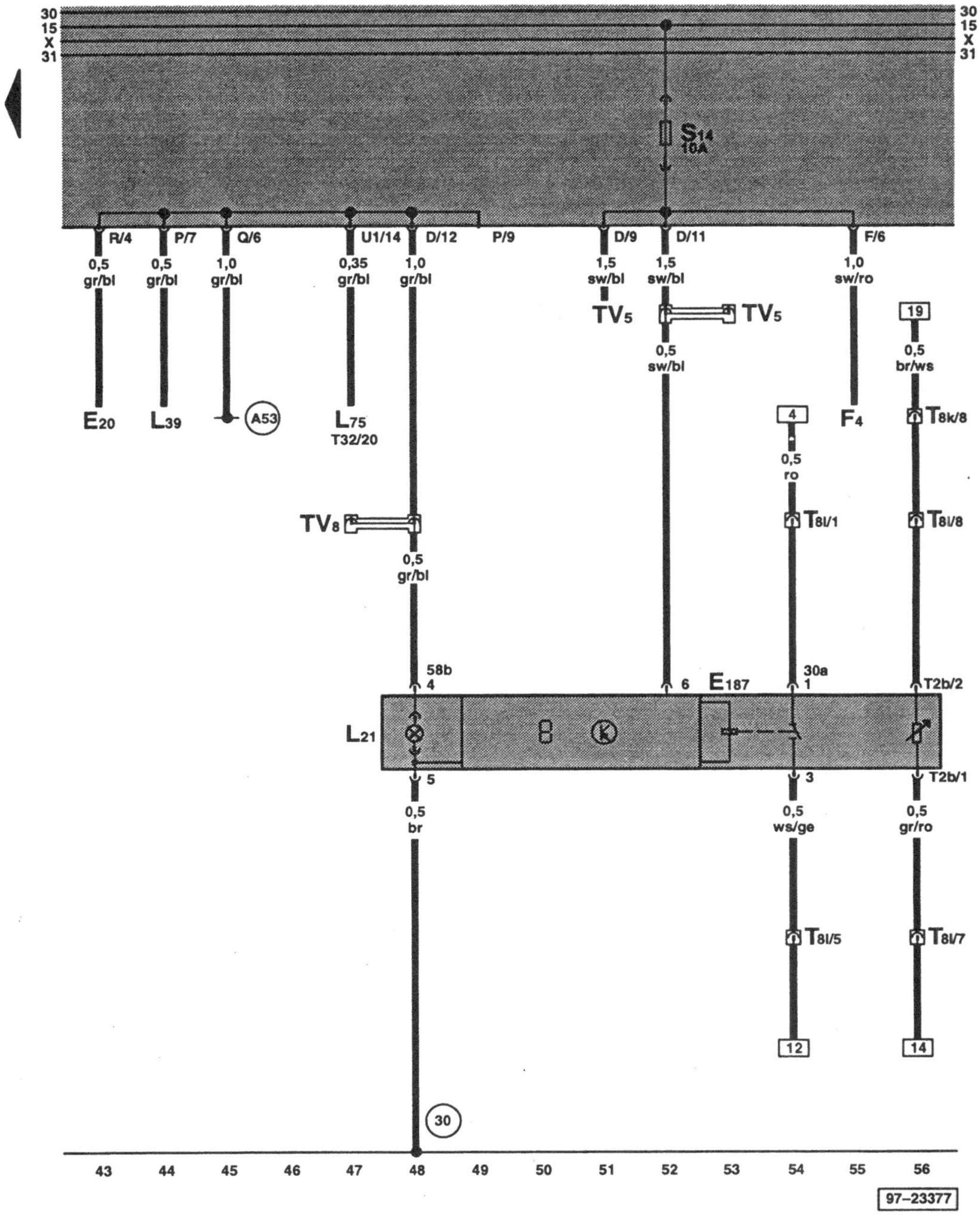

ws = white
sw = black
ro = red
br = brown
gn = green
bl = blue
gr = grey
li = violet
ge = yellow

E20 – Instrument Panel Light Dimmer Switch
E187 – Programable timer with temperature control
F4 – Back-Up Light Switch
L21 – Heater Control Lever Light
L39 – Rear Window Defogger Switch Light
L75 – Digital Display Light
T2b – Double Connector
T8k – 8-Pin Connector, below driver's seat
T8l – 8-Pin Connector, behind fuse/relay panel
T32 – 32-Pin Connector, blue, on instrument cluster
TV5 – Terminal 15a Wire Connector, green, 4-Pin and 8-Pin
TV8 – Terminal 58b Wire Connector, grey

(30) – Ground connection, -1-, beside fuse/relay panel

(A53) – Plus connection (58b), in instrument panel wiring harness

THIS PAGE INTENTIONALLY LEFT
BLANK

Front Power Windows,

from September 1998

Deviate relay location and fuseplacements as well as the locations of multiple connectors see section "component locations".

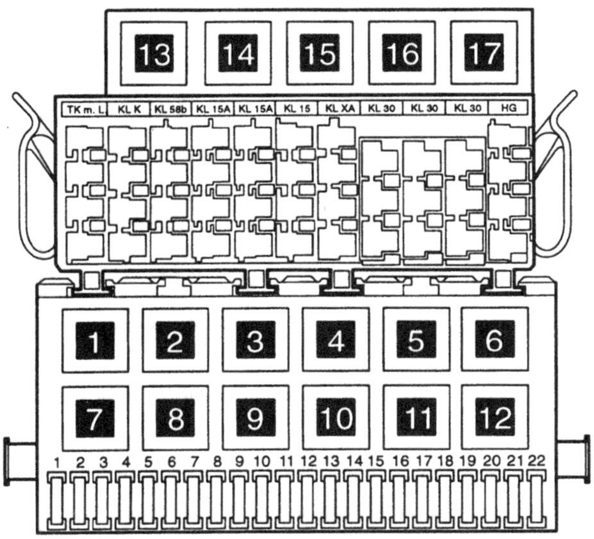

97–23232

Fuse colors

30 A – green
25 A – white
20 A – yellow
15 A – blue
10 A – red
7,5 A – brown
5 A – beige
3 A – violet

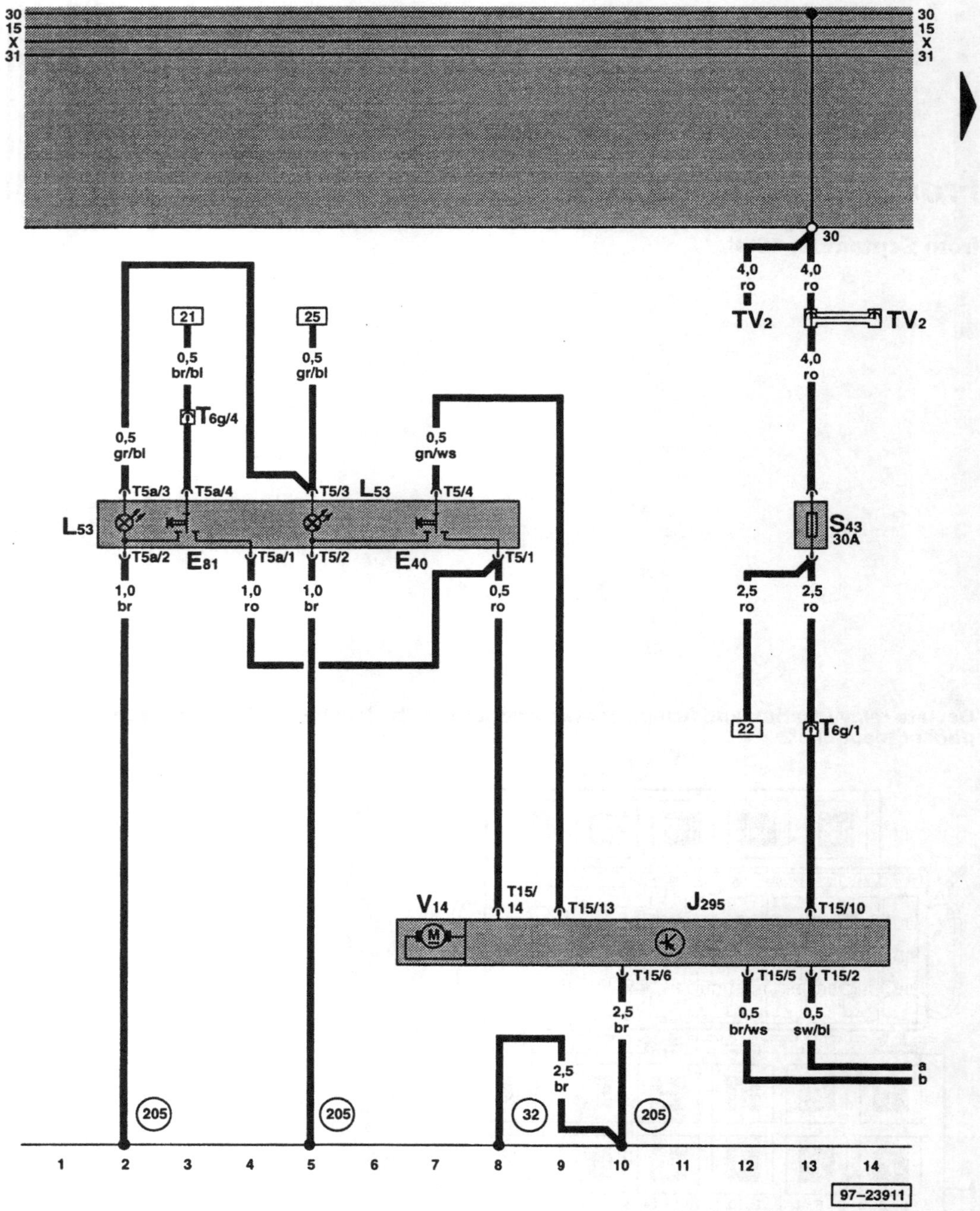

ws = white
sw = black
ro = red
br = brown
gn = green
bl = blue
gr = grey
li = violet
ge = yellow

- E40 – Switch for window regulator front left
- E81 – Switch for window regulator front right, driver
- J295 – Left Front Power Window Control Module
- L53 – Power Window Switch Light
- S43 – Power Window Circuit Breaker, above fuse/relay panel
- T5 – 5-Pin Connector
- T5a – 5-Pin Connector
- T6g – 6-Pin Connector, behind instrument panel, left
- T15 – 15-Pin Connector
- TV2 – Terminal 30 (B+) Wire Connector, red, each 6 points
- V14 – Left Window Motor
- (32) – Ground connection, behind instrument panel, left
- (205) – Ground connection, in driver's door wiring harness

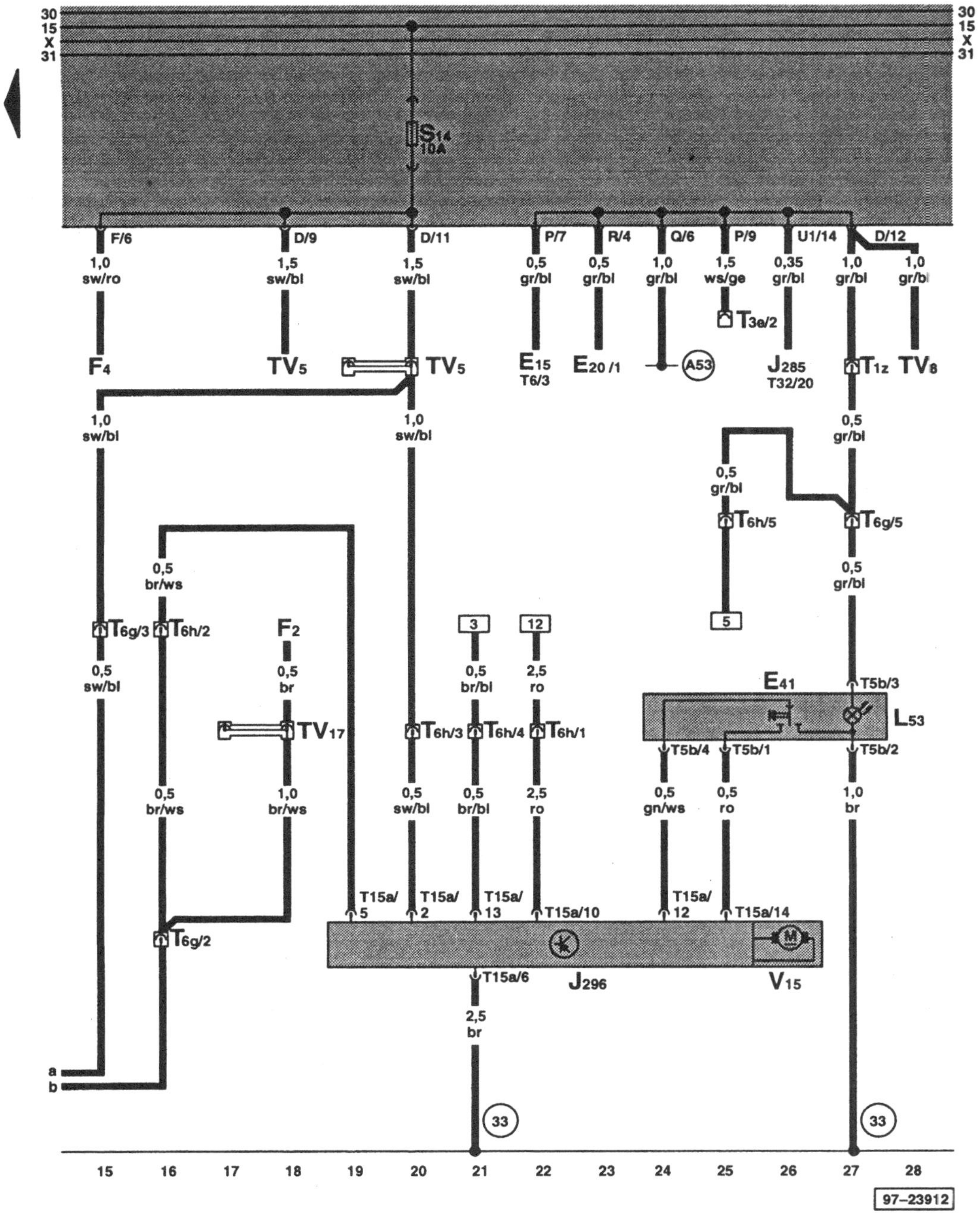

ws = white
sw = black
ro = red
br = brown
gn = green
bl = blue
gr = grey
li = violet
ge = yellow

E15 – Rear window defogger switch
E20 – Instrument Panel Light Dimmer Switch
E41 – Right Front Window Switch, in front passenger's door
F2 – Door contact Switch - Driver's side
F4 – Back-Up Light Switch
J285 – Control module with indicator unit in instrument panel insert
J296 – Right Front Power Window Control Module
L53 – Power Window Switch Light
T1z – Single Connector, behind fuse/relay panel
T3e – 3-Pin Connector, behind fuse/relay panel
T5b – 5-Pin Connector
T6 – 6-Pin Connector
T6g – 6-Pin Connector, behind instrument panel, left
T6h – 6-Pin Connector, behind instrument panel, right

T15a – 15-Pin Connector
T32 – 32-Pin Connector, blue, on instrument cluster
TV5 – Terminal 15a Wire Connector, green, 4-Pin and 8-Pin
TV8 – Terminal 58b Wire Connector, grey
TV17 – Door Contact Switch Wire Connector, brown
V15 – Right Window Motor

(33) – Ground connection, behind instrument panel, right
(A53) – Plus connection (58b), in instrument panel wiring harness

THIS PAGE INTENTIONALLY LEFT BLANK

Heated power mirrors,

from September 1998

Deviate relay location and fuseplacements as well as the locations of multiple connectors see section "component locations".

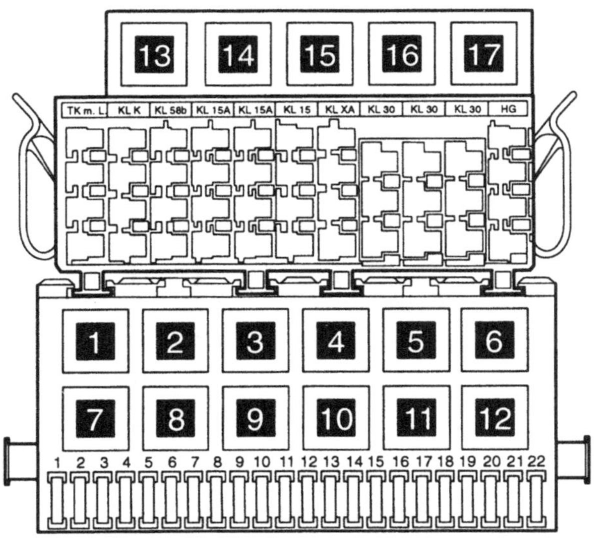

97–23232

17 Load Reduction Relay II (53)

Note: Number in parentheses indicates production control number stamped on relay housing.

Fuse colors

30 A – green
25 A – white
20 A – yellow
15 A – blue
10 A – red
7,5 A – brown
5 A – beige
3 A – violet

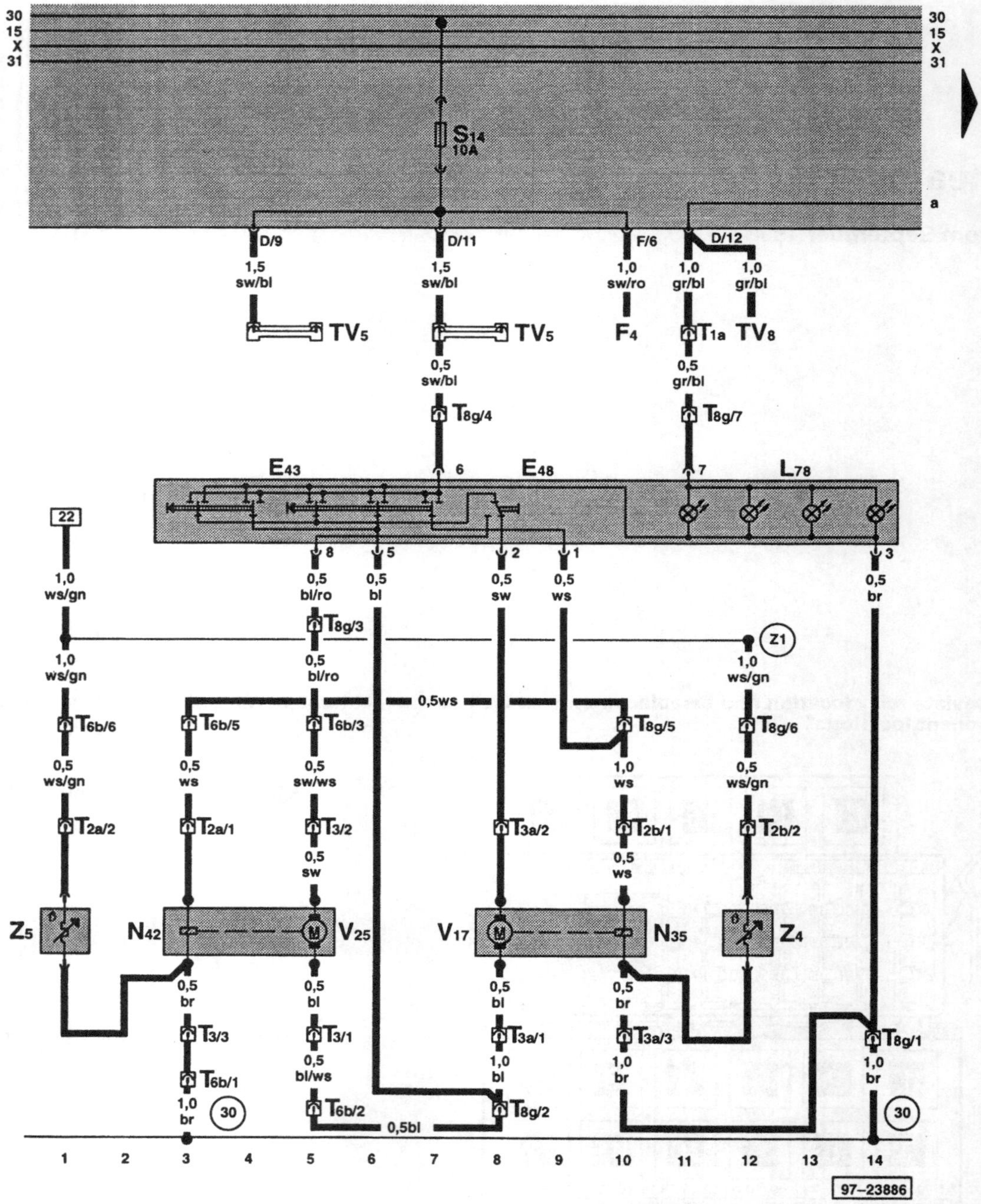

ws = white
sw = black
ro = red
br = brown
gn = green
bl = blue
gr = grey
li = violet
ge = yellow

E43 – Mirror Adjustment Switch
E48 – Mirror Selector Switch
F4 – Back-Up Light Switch
L78 – Mirror Adjusting Switch Light
N35 – Left Mirror Magnetic Clutch
N42 – Right Mirror Magnetic Clutch
T1a – Single Connector, behind fuse/relay panel
T2a – Double Connector, in front passenger's door
T2b – Double Connector, in driver's door
T3 – 3-Pin Connector, in front passenger's door
T3a – 3-Pin Connector, in driver's door
T6b – 6-Pin Connector, on right A-pillar
T8g – 8-Pin Connector, on left A-pillar
TV5 – Terminal 15a Wire Connector, green, 4-Pin and 8-Pin
TV8 – Terminal 58b Wire Connector, grey

V17 – Driver's Side Mirror Adjustment Motor
V25 – Passenger's Side Mirror Adjustment Motor
Z4 – Heated outside mirror, driver side
Z5 – Heated outside mirror, passenger side

(30) – Ground connection, -1-, beside fuse/relay panel

(Z1) – Wire connection -1-, in power heated mirror wiring harness

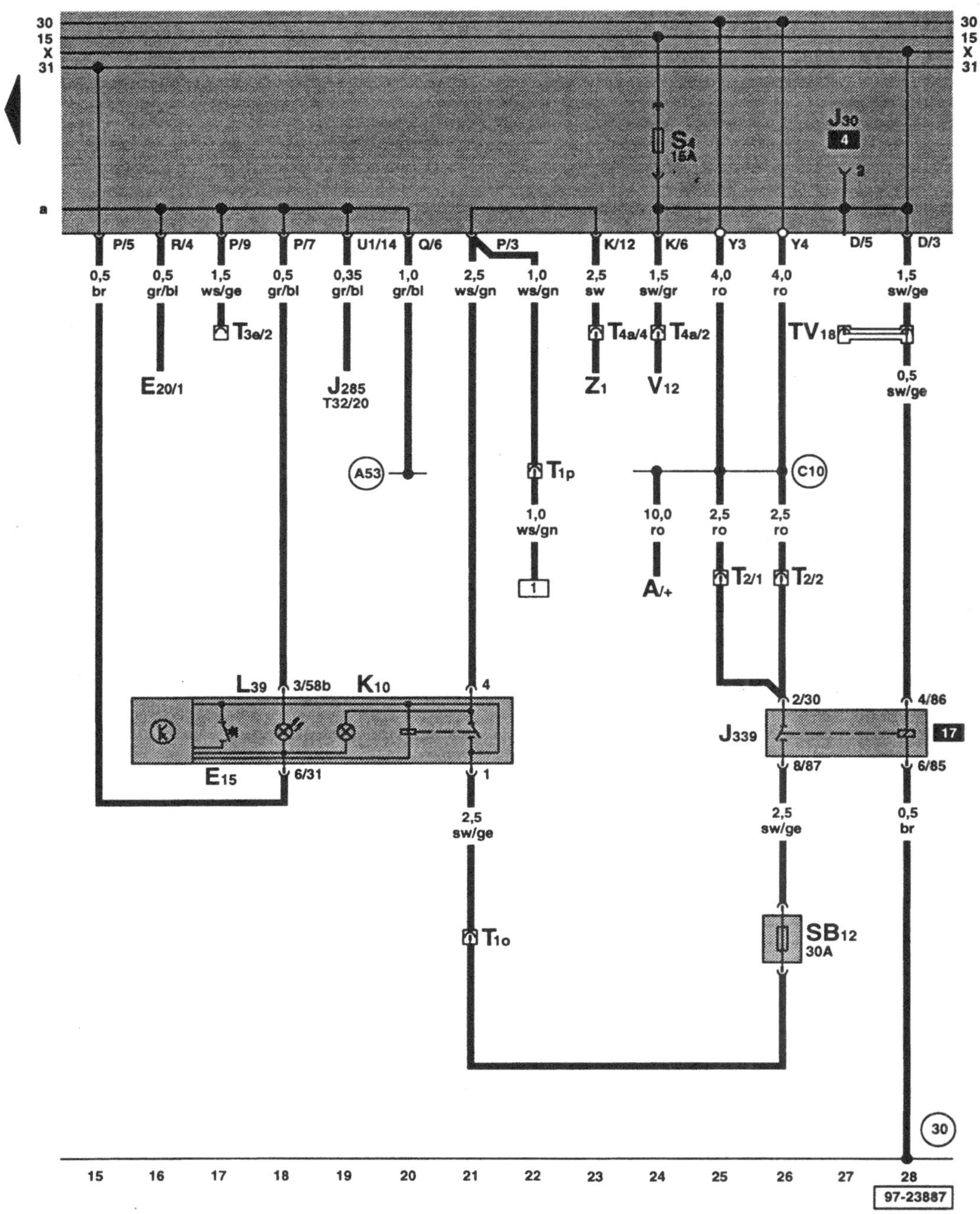

ws = white
sw = black
ro = red
br = brown
gn = green
bl = blue
gr = grey
li = violet
ge = yellow

A – Battery
E15 – Rear Window Defogger Switch
E20 – Instrument Panel Light Dimmer Switch
J30 – Rear Window Wiper/Washer Relay
J285 – Control module with indicator unit in instrument panel insert
J339 – Load Reduction Relay II, above fuse/relay panel
K10 – Rear Window Defogger Indicator Light
L39 – Rear Window Defogger Switch Light
SB12 – Fuse 12 in fuse holder, below fuse/relay panel
T1o – Single Connector, behind fuse/relay panel
T1p – Single Connector, behind fuse/relay panel
T2 – Double Connector, behind fuse/relay panel
T3e – 3-Pin Connector, behind fuse/relay panel
T4a – 4-Pin Connector, near taillamp assembly
T32 – 32-Pin Connector, blue, on instrument cluster
TV18 – Terminal Xa Wire Connector, lilac
V12 – Motor for rear windshield wiper
Z1 – Rear Window Defogger/Heat Element

(30) – Ground connection, -1-, next to fuse/relay panel
(A53) – Plus connection (58b), in instrument panel wiring harness
(C10) – Plus connection -1- (30), in headlight wiring harness

THIS PAGE INTENTIONALLY LEFT BLANK

Heated seats,

from September 1998

Deviate relay location and fuseplacements as well as the locations of multiple connectors see section "component locations".

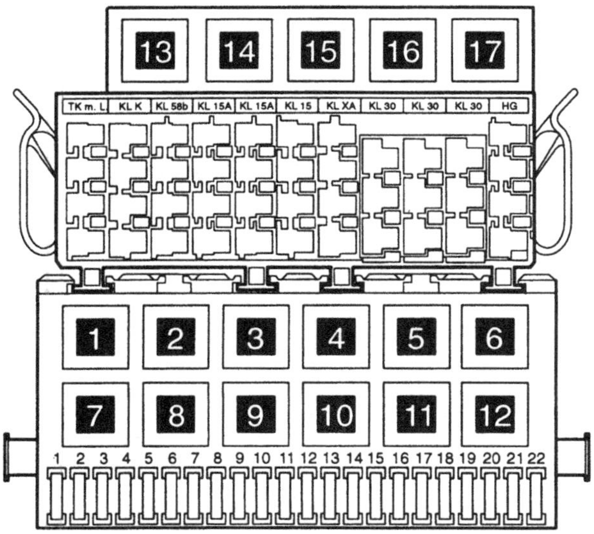

97-23232

17 Load Reduction Relay II (53)

Note: Number in parentheses indicates production control number stamped on relay housing.

Fuse colors

30 A – green

25 A – white

20 A – yellow

15 A – blue

10 A – red

7,5 A – brown

5 A – beige

3 A – violet

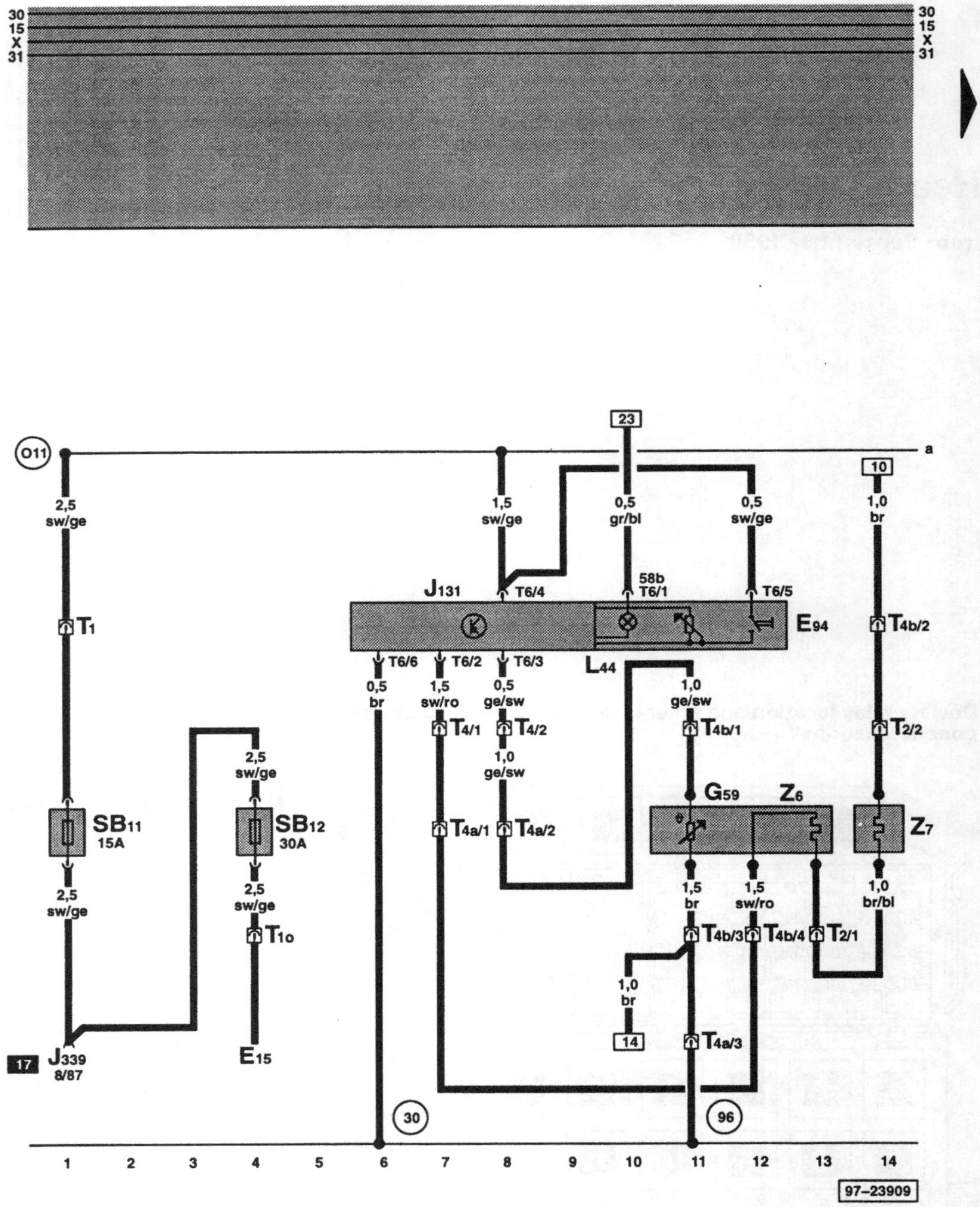

ws = white
sw = black
ro = red
br = brown
gn = green
bl = blue
gr = grey
li = violet
ge = yellow

E15 – Rear Window Defogger Switch
E94 – Adjuster for heated driver's seat
G59 – Driver's Heated Seat Temperature Element
J131 – Driver's Heated Seat Control Module
J339 – Load Reduction relay II, above fuse/relay panel
L44 – Heated Switch Illumination Light
SB11 – Fuse 11 in fuse holder, below fuse/relay panel
SB12 – Fuse 12 in fuse holder, below fuse/relay panel
T1 – Single Connector, behind fuse/relay panel
T1o – Single Connector, behind fuse/relay panel
T2 – Double Connector, below driver's seat
T4 – 4-Pin Connector, behind fuse/relay panel
T4a – 4-Pin Connector, below driver's seat
T4b – 4-Pin Connector, below driver's seat
T6 – 6-Pin Connector
Z6 – Heated Driver's Seat
Z7 – Heated Driver's Backrest
(30) – Ground connection, -1-, beside fuse/relay panel
(96) – Ground connection, -1-, in heated seats wiring harness
(O11) – Plus connection (Xa), in heated seats wiring harness

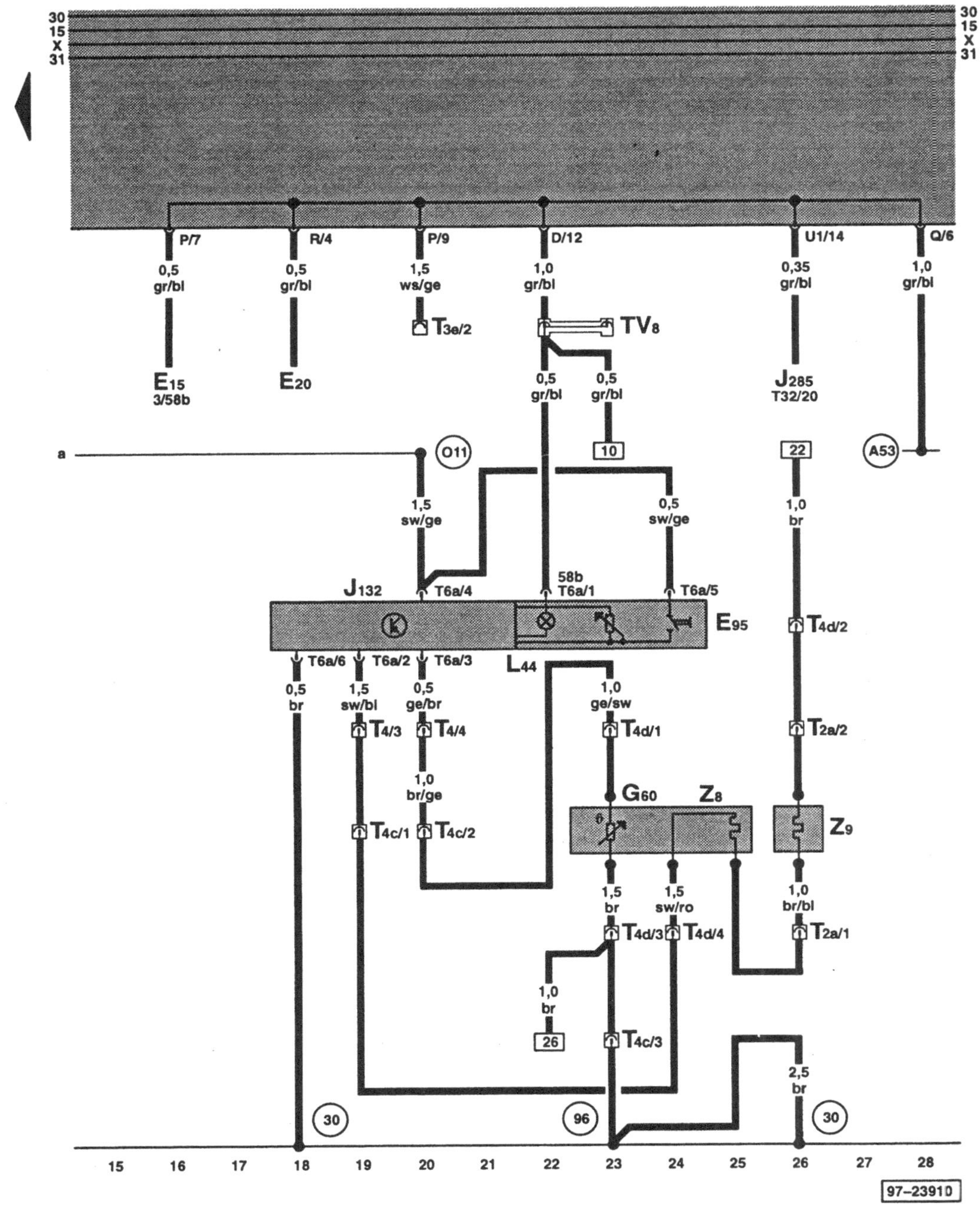

ws = white
sw = black
ro = red
br = brown
gn = green
bl = blue
gr = grey
li = violet
ge = yellow

E15 – Rear Window Defogger Switch
E20 – Instrument Panel Light Dimmer Switch
E95 – Adjuster for heated passenger's seat
G60 – Passenger's Heated Seat Temperature Element
J132 – Passenger's Heated Seat Control Module
J285 – Control module with indicator unit in instrument panel insert
L44 – Heated Switch Illumination Light
T2a – Double Connector, below passenger's seat
T3e – 3-Pin Connector, behind fuse/relay panel
T4 – 4-Pin Connector, behind fuse/relay panel
T4c – 4-Pin Connector, below passenger's seat
T4d – 4-Pin Connector, below passenger's seat
T6a – 6-Pin Connector
T32 – 32-Pin Connector, blue, on instrument cluster
TV8 – Terminal 58b Wire Connector, grey

Z8 – Heated Passenger's Seat
Z9 – Heated Passenger's Backrest

(30) – Ground connection, -1-, beside fuse/relay panel
(96) – Ground connection, -1-, in heated seats wiring harness
(A53) – Plus connection (58b), in instrument panel wiring harness
(O11) – Plus connection (Xa), in heated seats wiring harness

Adjuster for heated passenger's seat, passenger's heated seat temperature element, heated passenger's seat, heated passenger's backrest

THIS PAGE INTENTIONALLY LEFT BLANK

Swivel seat lock,

from September 1998

Deviate relay location and fuseplacements as well as the locations of multiple connectors see section "component locations".

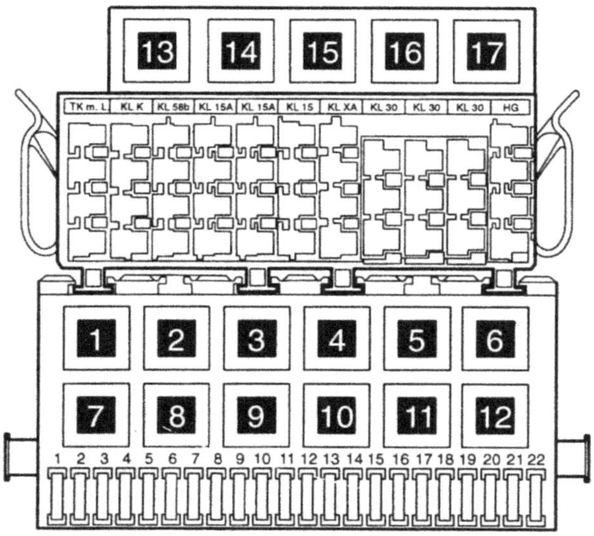

97–23232

Fuse colors

30 A – green
25 A – white
20 A – yellow
15 A – blue
10 A – red
7,5 A – brown
5 A – beige
3 A – violet

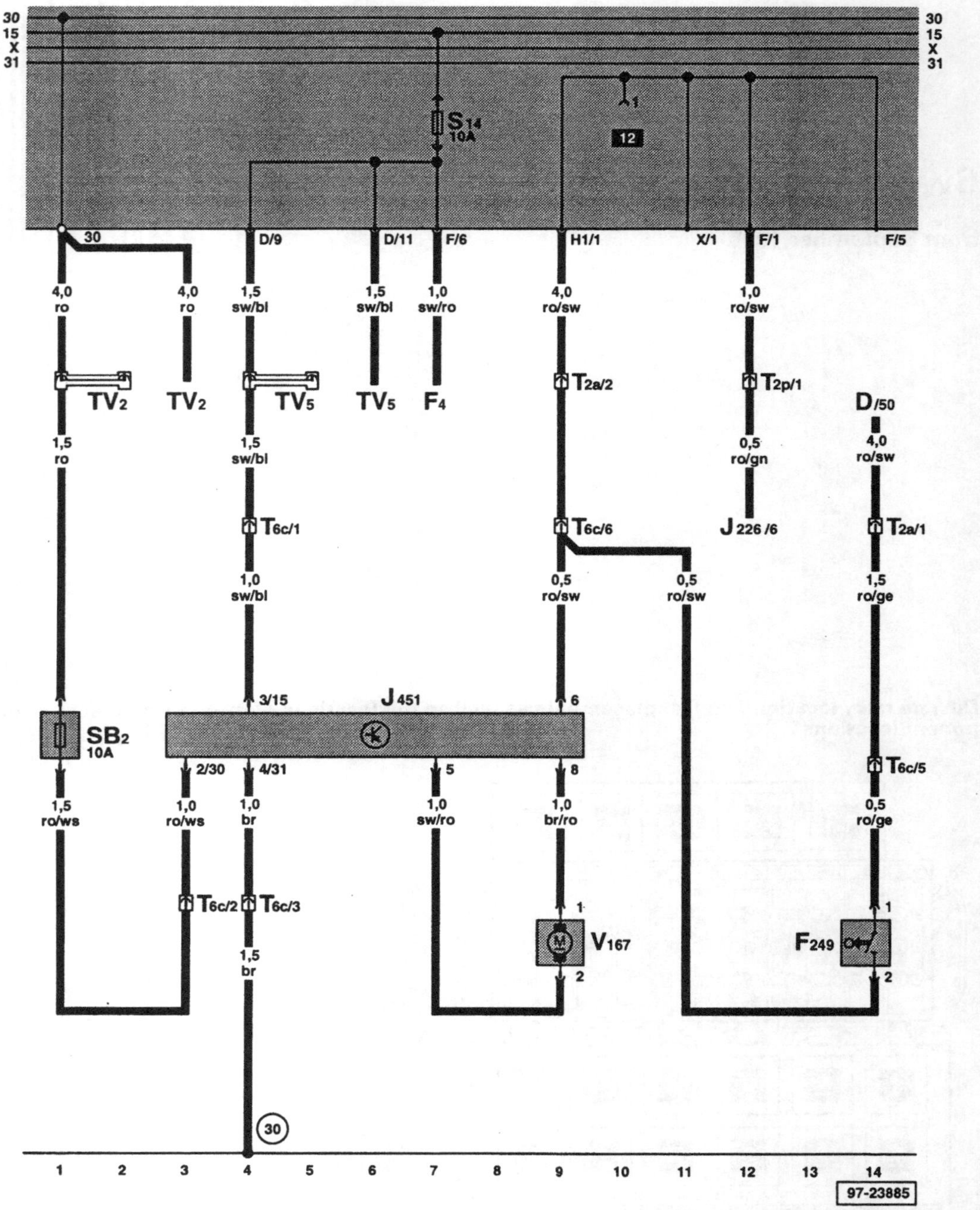

ws = white
sw = black
ro = red
br = brown
gn = green
bl = blue
gr = grey
li = violet
ge = yellow

D – Ignition/Starter Switch

F4 – Back-Up Light Switch

F249 – End switch for swivel seat lock, below passenger's seat

J226 – Park/Neutral Position (PNP) Relay, behind instrument panel, center

J451 – Control module for swivel seat lock, below passenger's seat

SB2 – Fuse 2 in fuse holder, below fuse/relay panel

T2a – Double Connector, behind steering column switch cover

T2p – Double Connector, behind fuse/relay panel

T6c – 6-Pin Connector, below passenger's seat

TV2 – Terminal 30 (B+) Wire Connector, red, each 6 points

TV5 – Terminal 15a Wire Connector, green, 4-Pin and 8-Pin

V167 – Motor for swivel seat lock, below passenger's seat

(30) – Ground connection, -1-, beside fuse/relay panel

Power sunroof (driver's cabin),

from September 1998

Deviate relay location and fuseplacements as well as the locations of multiple connectors see section "component locations".

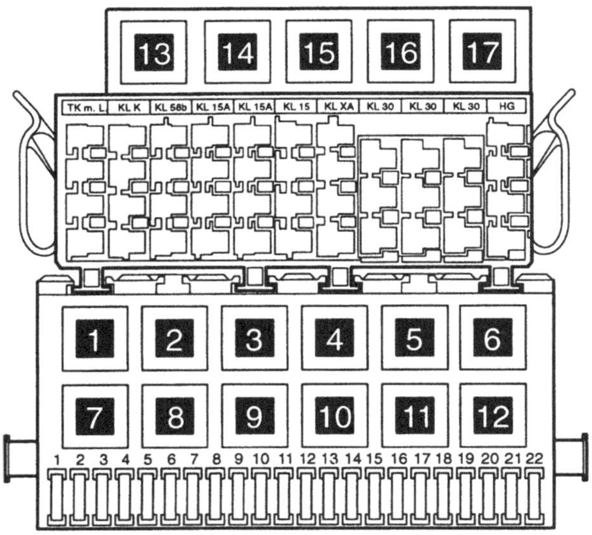

97–23232

Fuse colors

30 A – green
25 A – white
20 A – yellow
15 A – blue
10 A – red
7,5 A – brown
5 A – beige
3 A – violet

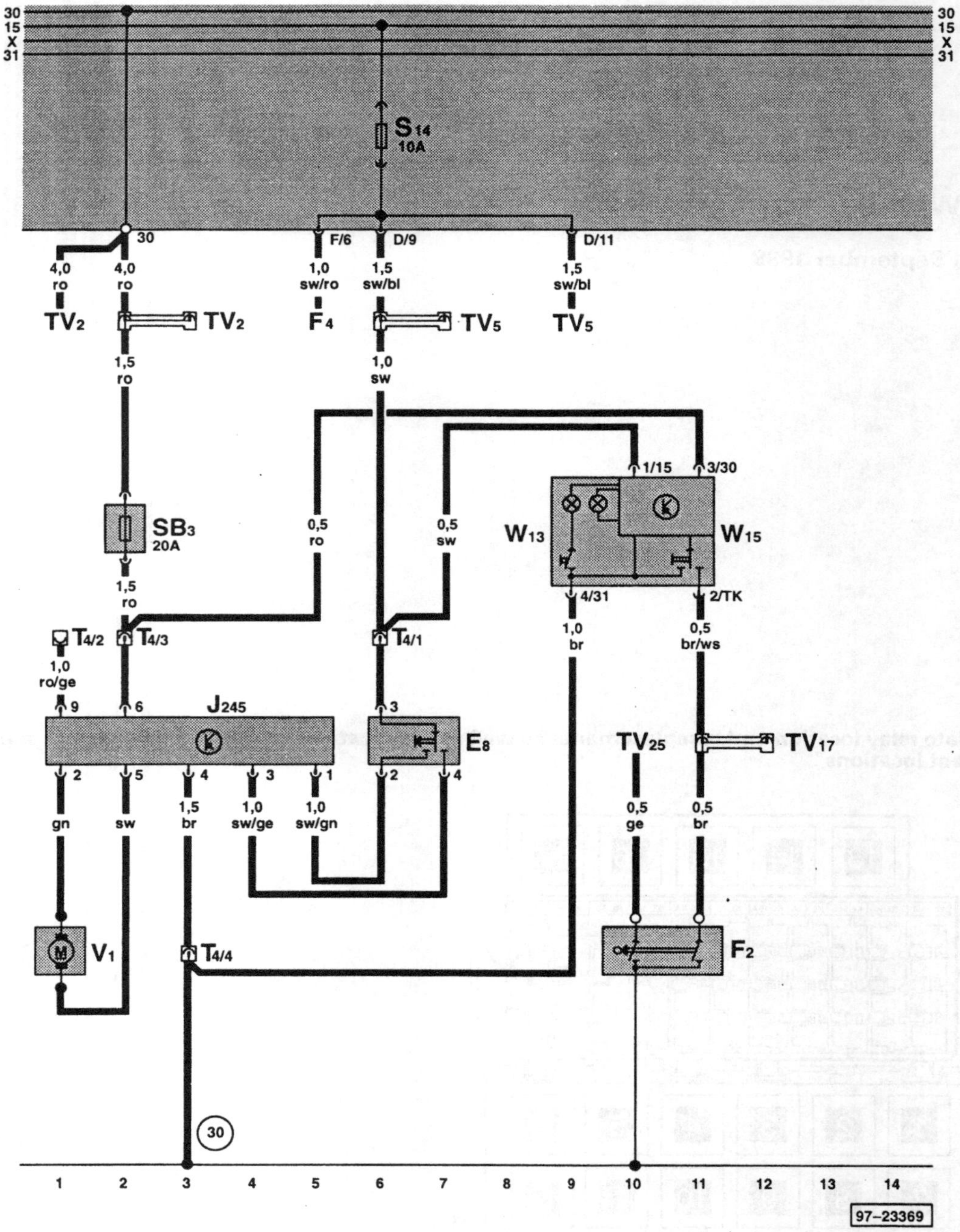

ws = white
sw = black
ro = red
br = brown
gn = green
bl = blue
gr = grey
li = violet
ge = yellow

- E8 – Sunroof switch
- F2 – Door contact Switch - Driver's side
- F4 – Back-Up Light Switch
- J245 – Power Sunroof Control Module
- SB3 – Fuse 3 in fuse holder, below fuse/relay panel
- T4 – 4-Pin Connector, near interior light
- TV2 – Terminal 30 (B+) Wire Connector, red, each 6 points
- TV5 – Terminal 15a Wire Connector, green, 4-Pin and 8-Pin
- TV17 – Door Contact Switch Wire Connector, brown
- TV25 – Door Contact Switch Wire Connector, pink
- V1 – Sunroof Motor
- W13 – Right Front Map/Reading Light
- W15 – Interior Light with Delay Switch

(30) – Ground connection, -1-, beside fuse/relay panel

Power sunroof (passenger compartment),

from September 1998

Deviate relay location and fuseplacements as well as the locations of multiple connectors see section "component locations".

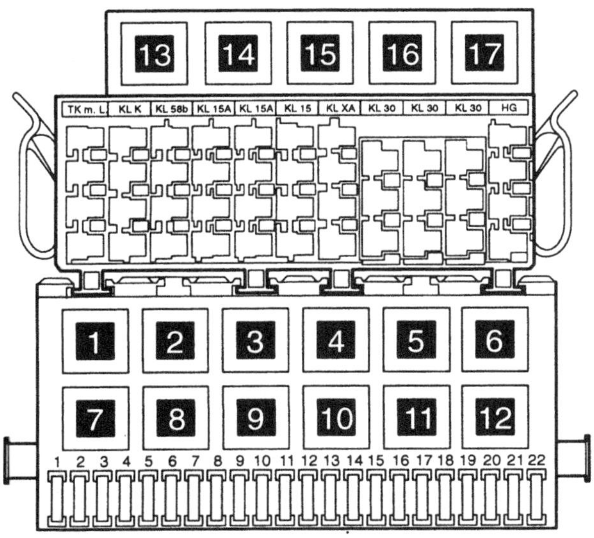

97–23232

Fuse colors

30 A – green
25 A – white
20 A – yellow
15 A – blue
10 A – red
7,5 A – brown
5 A – beige
3 A – violet

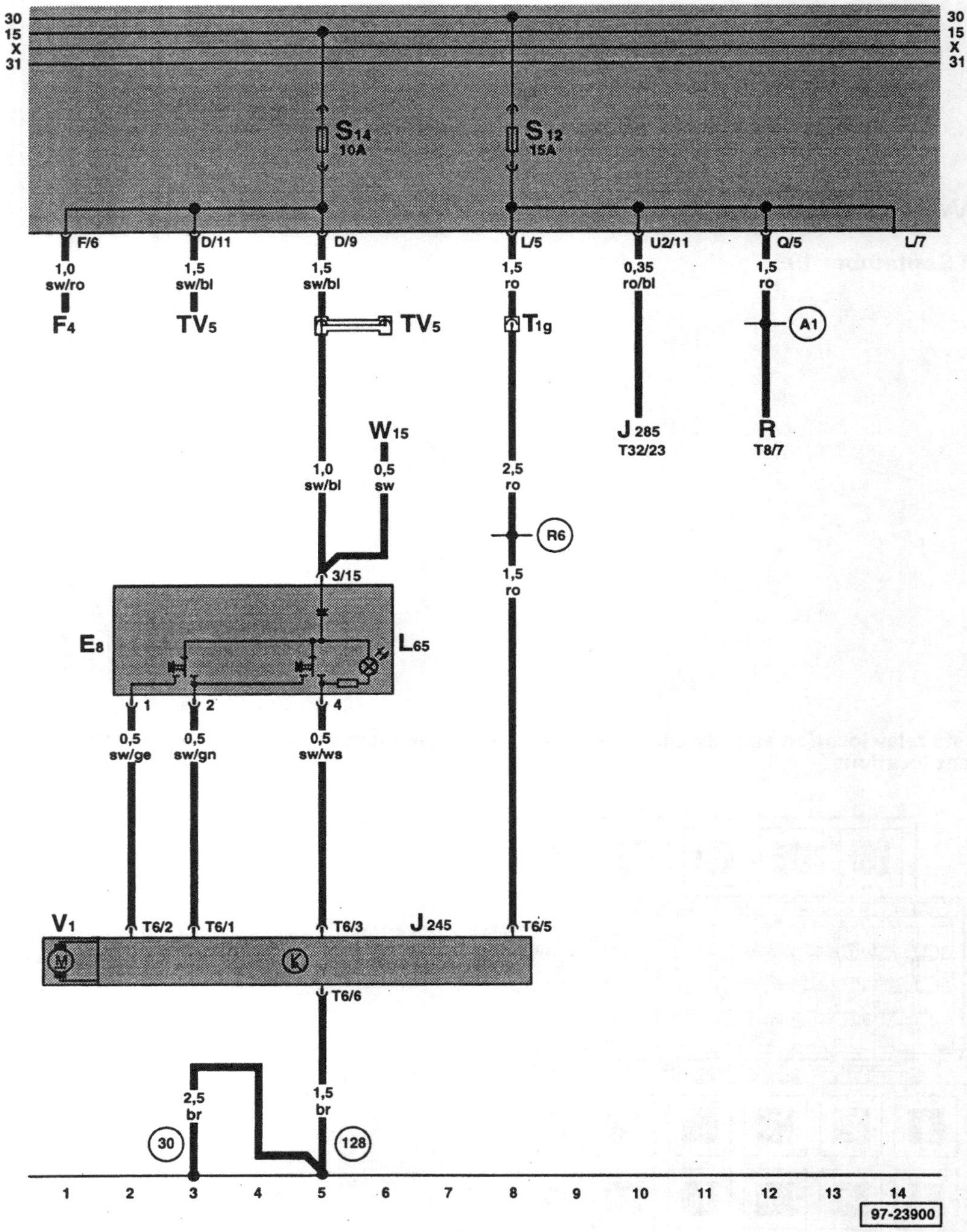

ws = white
sw = black
ro = red
br = brown
gn = green
bl = blue
gr = grey
li = violet
ge = yellow

E8 – Sunroof switch
F4 – Back-Up Light Switch
J245 – Power Sunroof Control Module
J285 – Control module with indicator unit in instrument panel insert
L65 – Sunroof Switch Light
R – Radio
T1g – Single Connector, behind fuse/relay panel
T6 – 6-Pin Connector
T8 – 8-Pin Connector, on Radio
T32 – 32-Pin Connector, blue
TV5 – Terminal 15a Wire Connector, green, 4-Pin and 8-Pin
V1 – Sunroof Motor
W15 – Interior Light With Delay Switch

(30) – Ground connection, -1-, beside fuse/relay panel

(128) – Ground connection -1-, in interior light wiring harness

(A1) – Plus connection (30a), in instrument panel wiring harness

(R6) – Plus connection -1-, in interior light wiring harness

Central locking system with remote control,

from September 1998

● Vehicles with right sliding door and tailgate or rear swinging doors

Deviate relay location and fuseplacements as well as the locations of multiple connectors see section "component locations".

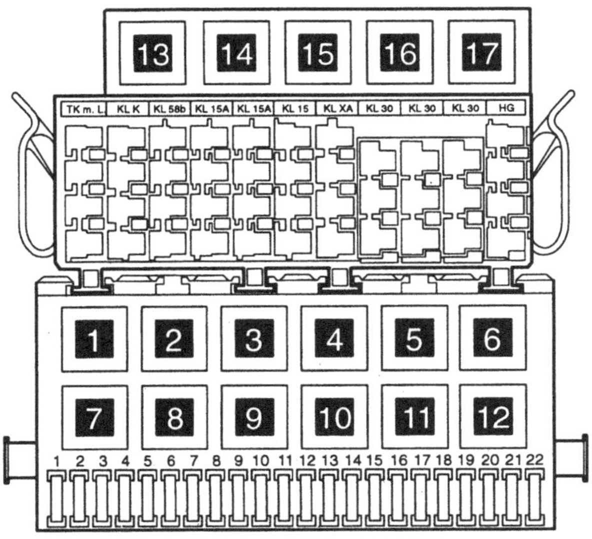

97–23232

Fuse colors

30 A – green
25 A – white
20 A – yellow
15 A – blue
10 A – red
7,5 A – brown
5 A – beige
3 A – violet

Edition 10/98
USA.5272.11.21

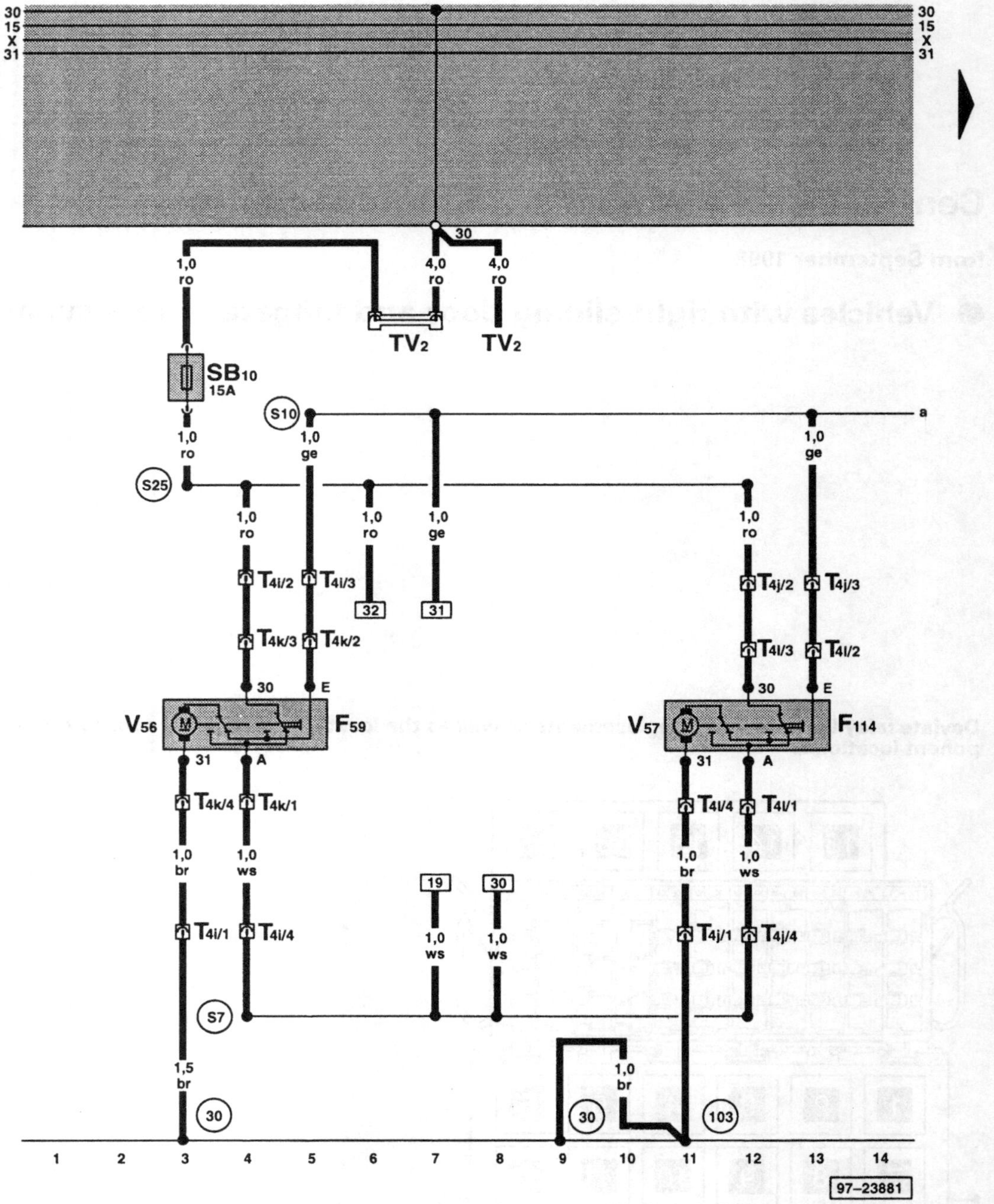

ws = white
sw = black
ro = red
br = brown
gn = green
bl = blue
gr = grey
li = violet
ge = yellow

F59 – Driver's Door Central Locking System Switch
F114 – Passenger's Door Central Locking System Switch
SB10 – Fuse 10 in fuse holder, below relay panel
T4i – 4-Pin Connector, behind fuse/relay panel
T4j – 4-Pin Connector, behind instrument panel, right
T4k – 4-Pin Connector, in driver's door
T4l – 4-Pin Connector, in front passenger's door
TV2 – Terminal 30 (B+) Wire Connector, red, each 6 points
V56 – Driver's Door Central Locking System Motor
V57 – Passenger's Door Central Locking System Motor

(30) – Ground connection, -1-, beside fuse/relay panel

(103) – Ground connection -2-, in power window/ central locking system/ door contact switch wiring harness
(S7) – Wire connection (A) -2-, in central locking system wiring harness
(S10) – Wire connection (E) -2-, in central locking system wiring harness
(S25) – Connector (Sensor/Control Module) -1- in wiring harness, central locking

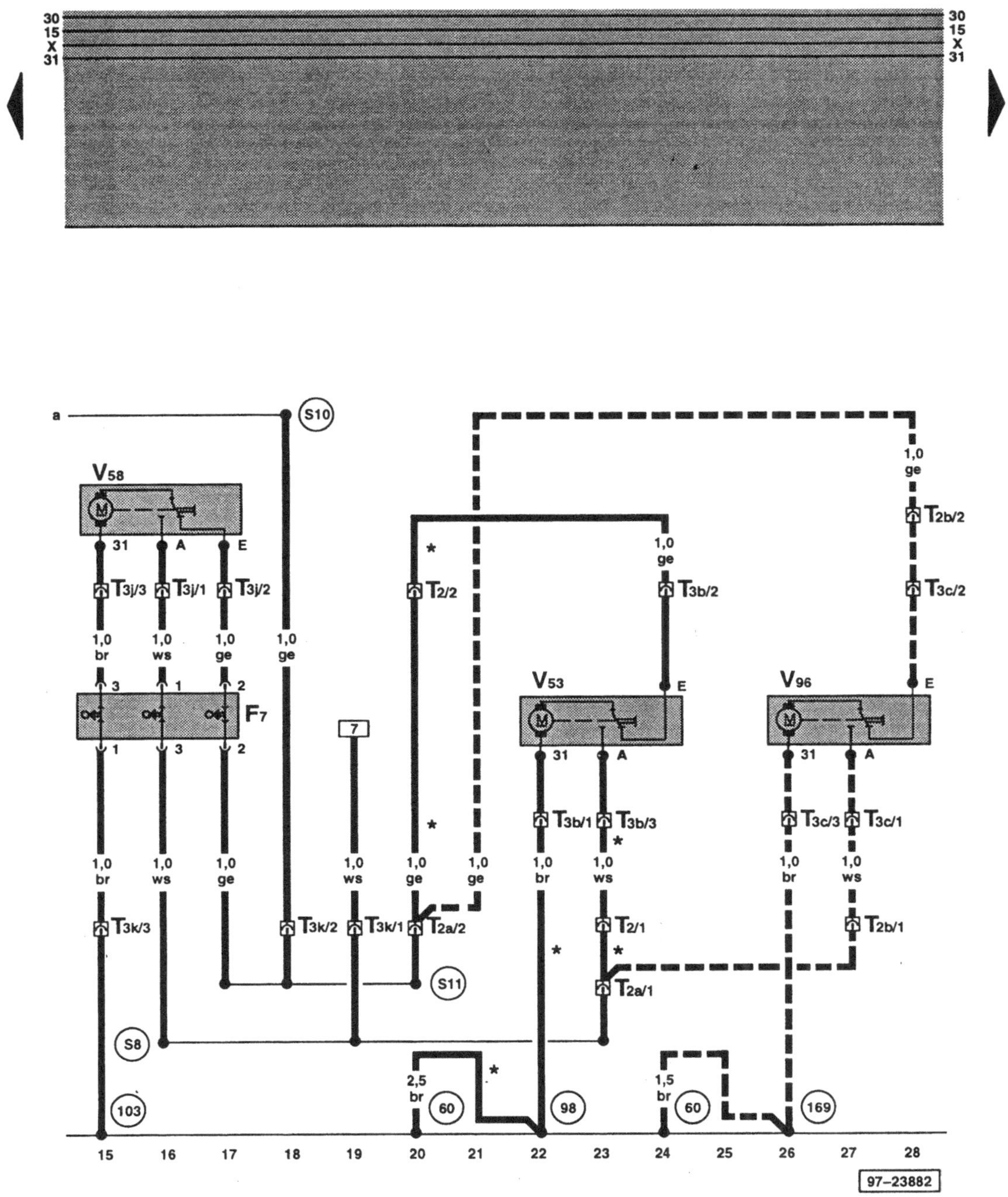

ws = white
sw = black
ro = red
br = brown
gn = green
bl = blue
gr = grey
li = violet
ge = yellow

- F7 – Right Rear Side/Sliding Door Contact Switch
- T2 – Double Connector, on right D-pillar, upper part
- T2a – Double Connector, on right D-pillar, upper part
- T2b – Double Connector, on right D-pillar, upper part
- T3b – 3-Pin Connector, in rear lid
- T3c – 3-Pin Connector, in rear swinging door,
- T3j – 3-Pin Connector, in sliding door, right
- T3k – 3-Pin Connector, behind instrument panel, right
- V53 – Decklid Central Locking System Motor
- V58 – Sliding Door Central Locking System Motor
- V96 – Left Rear Door Central Locking System Motor
- (60) – Ground point -2- near taillame assembly, right
- (98) – Ground connection, in rear lid wiring harness
- (103) – Ground connection -2-, in power window/ central locking system/ door contact switch wiring harness
- (169) – Ground connection, in right swinging door wiring harness
- (S8) – Wire connection (A) -3-, in central locking system/passenger compartment wiring harness
- (S10) – Connection (E) -2-, in central locking system wiring harness
- (S11) – Wire connection (E) -3-, in central locking system passenger compartment wiring harness
- * – Vehicles with tailgate only
- – – – Vehicles with rear swinging doors only

Decklid central locking system motor, sliding door central locking system motor, left rear door central locking system motor

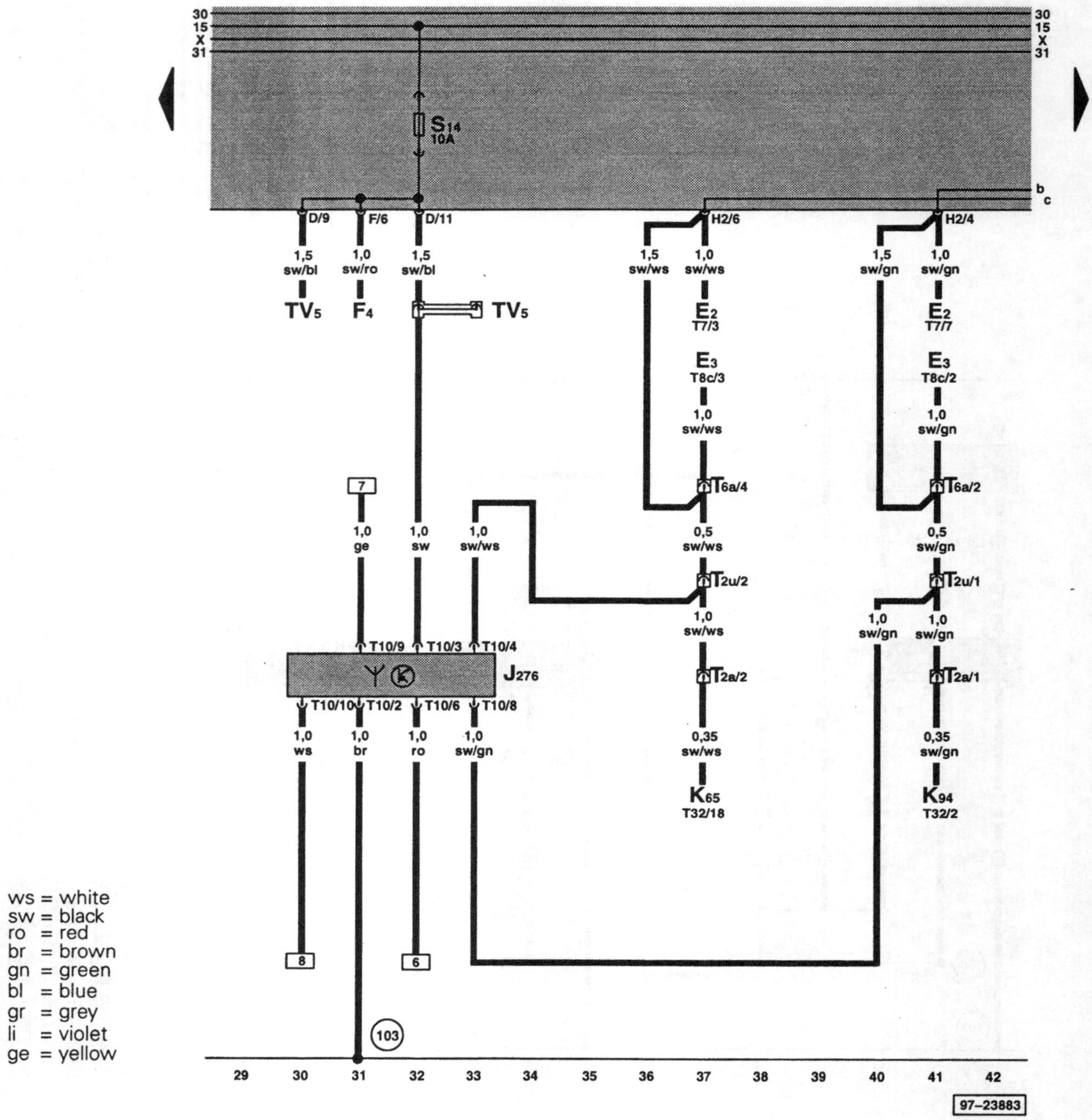

E2 – Turn signal switch
E3 – Emergency Flasher Switch
F4 – Back-Up Light Switch
J276 – Central Locking System Control Module (Remote Control), behind instrument cluster
K65 – Left Turn Signal Indicator Light
K94 – Right Turn Signal Indicator Light
T2a – Double Connector, behind fuse/relay panel
T2u – Double Connector, behind fuse/relay panel
T6a – 6-Pin Connector, behind fuse/relay panel
T7 – 7-Pin Connector, behind steering column switch cover
T8c – 8-Pin Connector, on Emergency Flasher Switch
T10 – 10-Pin Connector
T32 – 32-Pin Connector, blue

TV5 – Terminal 15a Wire Connector, green, 4-Pin and 8-Pin

(103) – Ground connection -2-, in power window/ central locking system/ door contact switch wiring harness

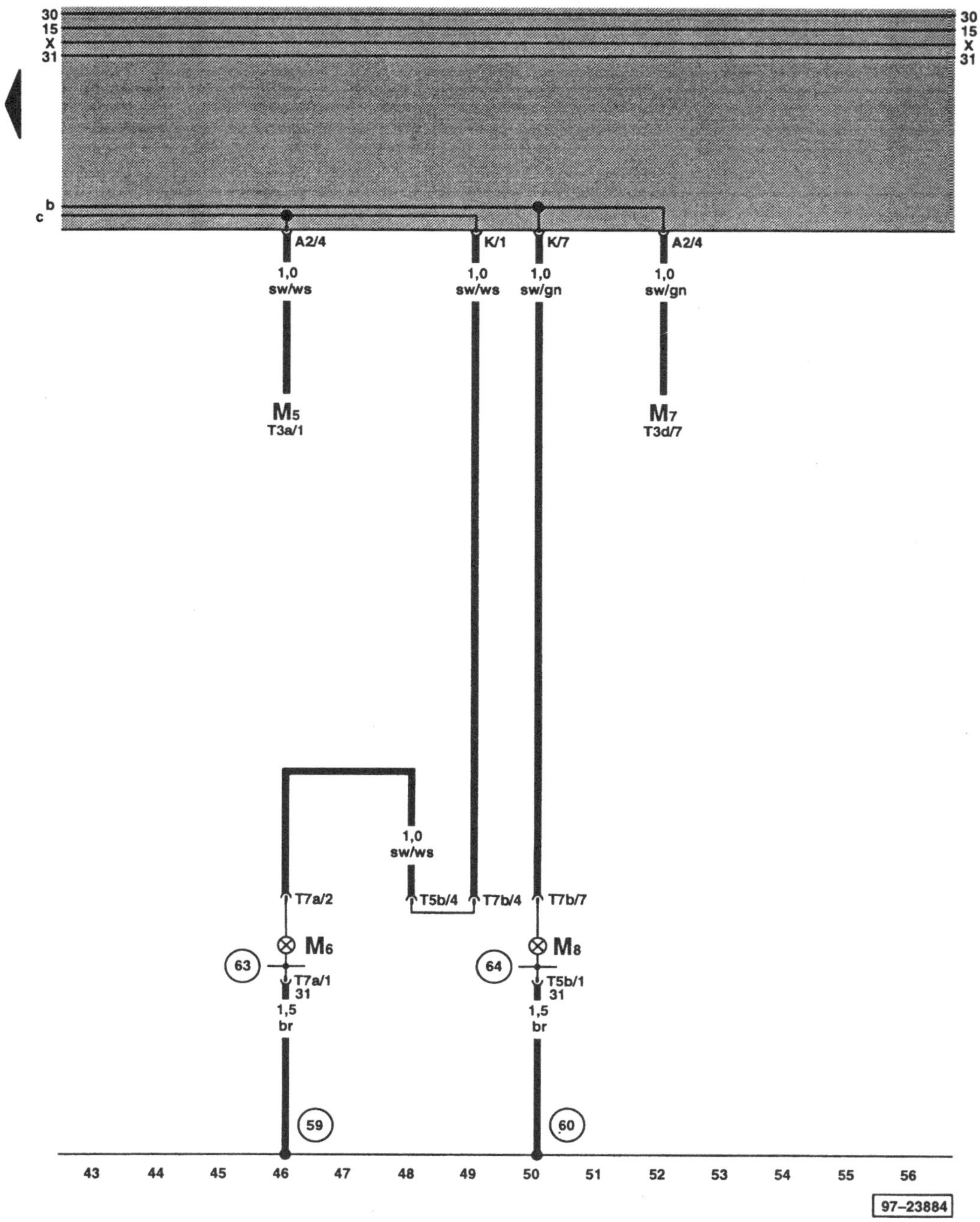

ws = white
sw = black
ro = red
br = brown
gn = green
bl = blue
gr = grey
li = violet
ge = yellow

- M5 – Left Front Turn Signal Light
- M6 – Left Rear Turn Signal Light
- M7 – Right Front Turn Signal Light
- M8 – Right Rear Turn Signal Light
- T3a – 3-Pin Connector
- T3d – 3-Pin Connector
- T5b – 5-Pin Connector
- T7a – 7-Pin Connector
- T7b – 7-Pin Connector
- (59) – Ground connection, near taillight, left
- (60) – Ground point -2- near taillamp assembly, right
- (63) – Ground connection, on left taillight bulb holder
- (64) – Ground connection, on right taillight bulb holder

THIS PAGE INTENTIONALLY LEFT BLANK

Central locking system with remote control,

from September 1998

● Vehicles with two sliding doors

Deviate relay location and fuseplacements as well as the locations of multiple connectors see section "component locations".

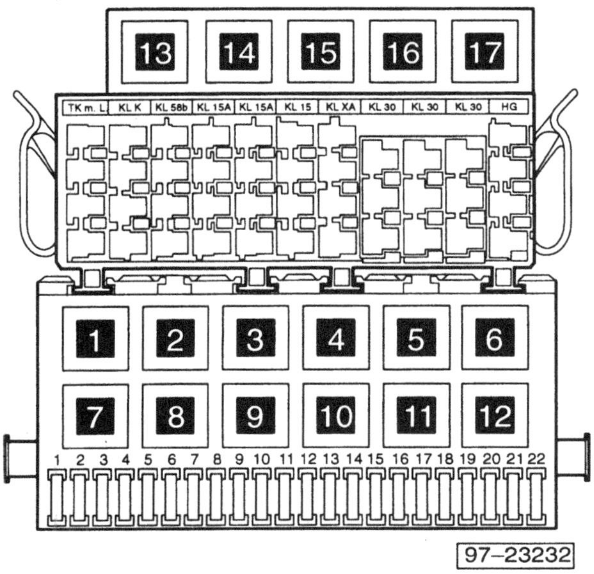

Fuse colors

30 A – green

25 A – white

20 A – yellow

15 A – blue

10 A – red

7,5 A – brown

5 A – beige

3 A – violet

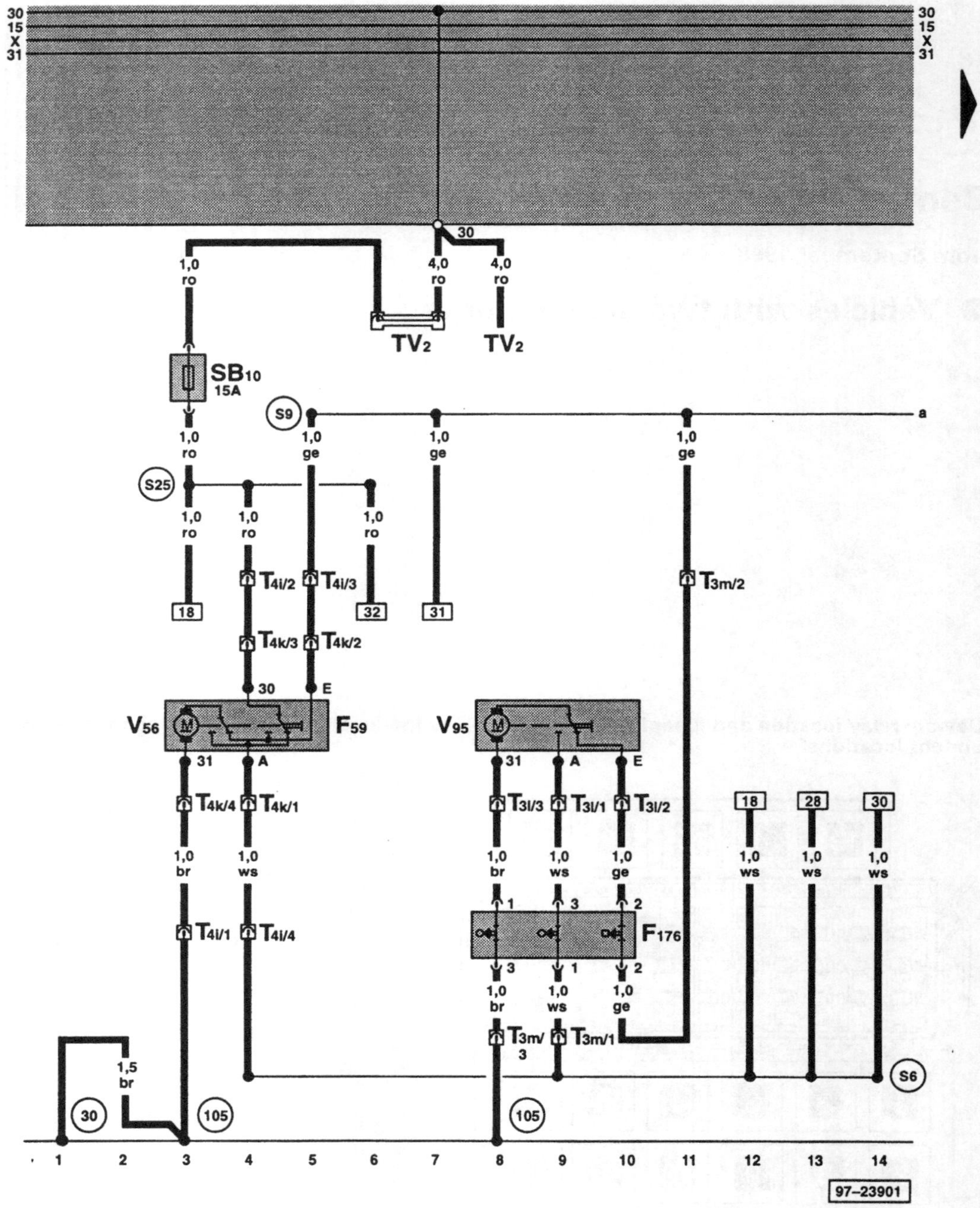

ws = white
sw = black
ro = red
br = brown
gn = green
bl = blue
gr = grey
li = violet
ge = yellow

- F59 – Driver's Door Central Locking System Switch
- F176 – Left Rear Side/Sliding Door Contact Switch
- SB10 – Fuse 10 in fuse holder, below relay panel
- T3l – 3-Pin Connector, in sliding door, left
- T3m – 3-Pin Connector, behind fuse/relay panel
- T4i – 4-Pin Connector, behind fuse/relay panel
- T4k – 4-Pin Connector, in driver's door
- TV2 – Terminal 30 (B+) Wire Connector, red, each 6 points
- V56 – Driver's Door Central Locking System Motor
- V95 – Left Sliding Door Central Locking System Motor
- (30) – Ground connection, -1-, beside fuse/relay panel
- (105) – Ground connection -1-, in power window/ central locking system/ door contact switch wiring harness
- (S6) – Wire connection (A) -1-, in central locking system wiring harness
- (S9) – Wire connection (E) -1-, in central locking system wiring harness
- (S25) – Connector (Sensor/Control Module) -1- in wiring harness, central locking

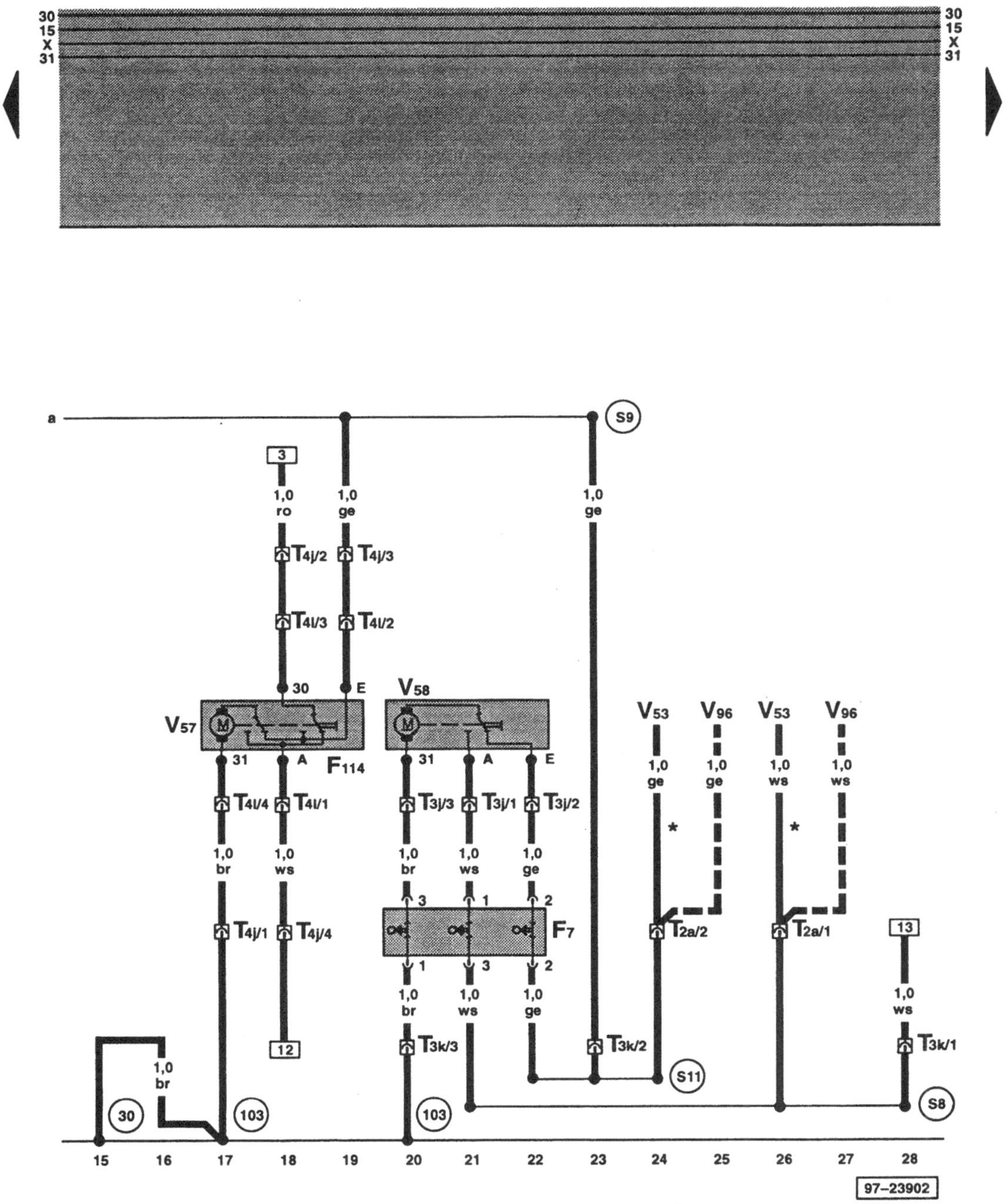

ws = white
sw = black
ro = red
br = brown
gn = green
bl = blue
gr = grey
li = violet
ge = yellow

F7 – Right Rear Side/Sliding Door Contact Switch
F114 – Passenger's Door Central Locking System Switch
T2a – Double Connector, on right D-pillar, upper part
T3j – 3-Pin Connector, in sliding door, right
T3k – 3-Pin Connector, behind instrument panel, right
T4j – 4-Pin Connector, behind instrument panel, right
T4l – 4-Pin Connector, in front passenger's door
V53 – Decklid Central Locking System Motor
V57 – Passenger's Door Central Locking System Motor
V58 – Sliding Door Central Locking System Motor
V96 – Left Rear Door Central Locking System Motor

(30) – Ground connection, -1-, beside fuse/relay panel

(103) – Ground connection -2-, in power window/ central locking system/ door contact switch wiring harness
(S8) – Wire connection (A) -3-, in central locking system/passenger compartment wiring harness
(S9) – Connection (E) -1-, in central locking system wiring harness
(S11) – Wire connection (E) -3-, in central locking system passenger compartment wiring harness

* – Vehicles with tailgate only
- - - Vehicles with rear swinging doors only

Passenger's door central locking system motor, right sliding door central locking system motor

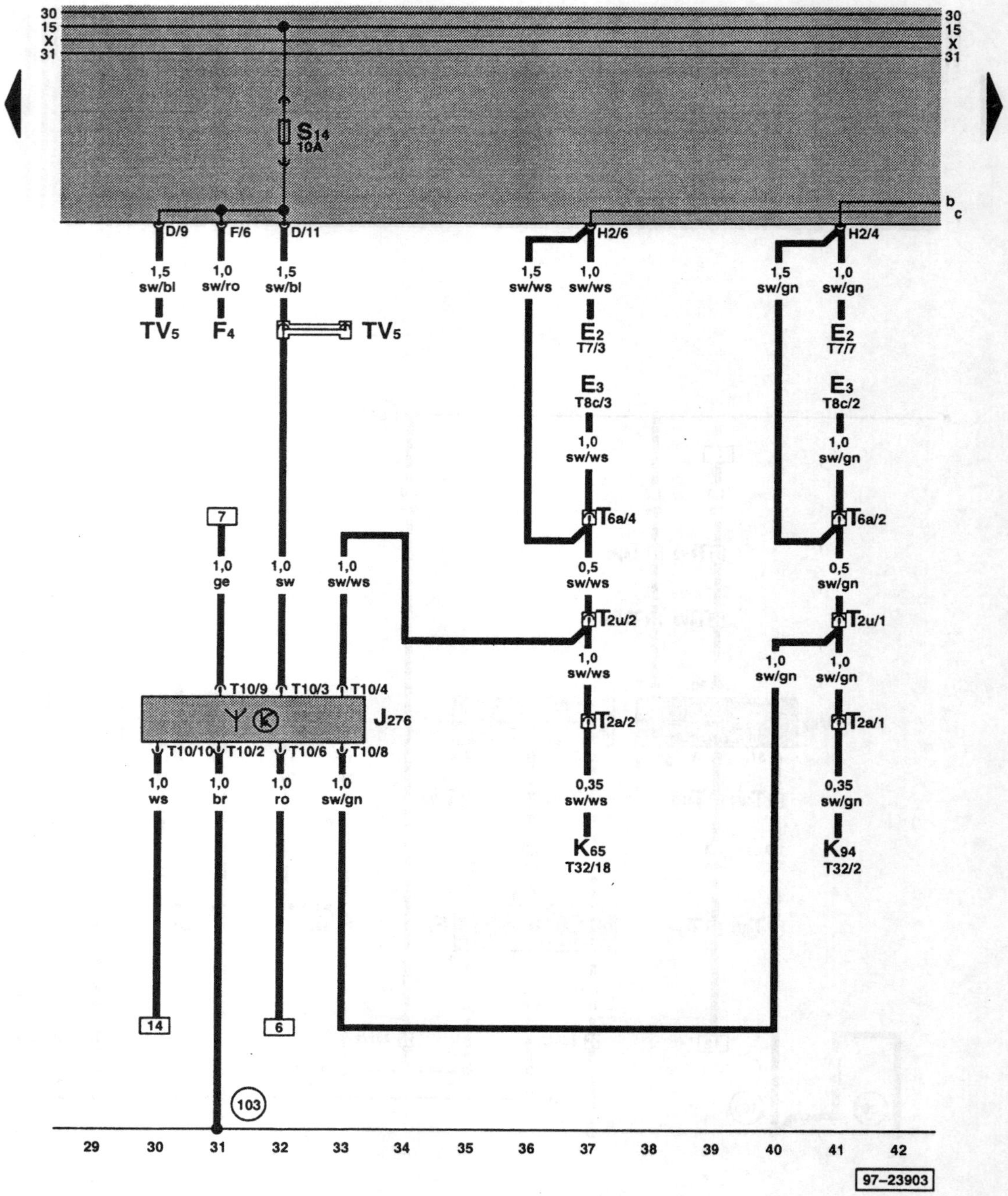

ws = white
sw = black
ro = red
br = brown
gn = green
bl = blue
gr = grey
li = violet
ge = yellow

E2 – Turn signal switch
E3 – Emergency Flasher Switch
F4 – Back-Up Light Switch
J276 – Central Locking System Control Module (Remote Control), behind instrument cluster
K65 – Left Turn Signal Indicator Light
K94 – Right Turn Signal Indicator Light
T2a – Double Connector, behind fuse/relay panel
T2u – Double Connector, behind fuse/relay panel
T6a – 6-Pin Connector, behind fuse/relay panel
T7 – 7-Pin Connector, behind steering column switch cover
T8c – 8-Pin Connector, on Emergency Flasher Switch
T10 – 10-Pin Connector
T32 – 32-Pin Connector, blue, on instrument cluster

TV5 – Terminal 15a Wire Connector, green, 4-Pin and 8-Pin

(103) – Ground connection -2-, in power window/ central locking system/ door contact switch wiring harness

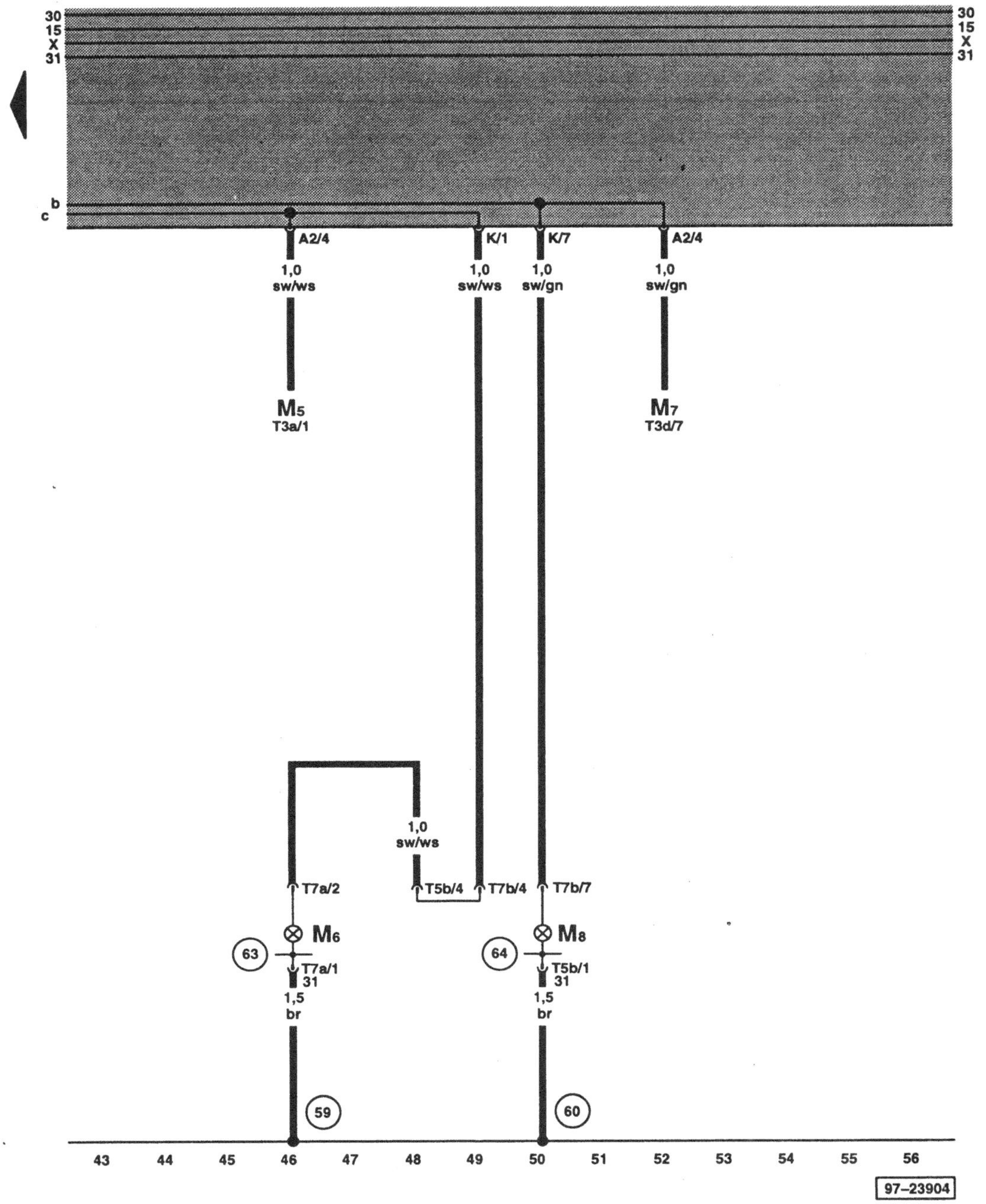

ws = white
sw = black
ro = red
br = brown
gn = green
bl = blue
gr = grey
li = violet
ge = yellow

M5 – Left Front Turn Signal Light
M6 – Left Rear Turn Signal Light
M7 – Right Front Turn Signal Light
M8 – Right Rear Turn Signal Light
T3a – 3-Pin Connector
T3d – 3-Pin Connector
T5b – 5-Pin Connector
T7a – 7-Pin Connector
T7b – 7-Pin Connector

(59) – Ground connection, near taillight, left

(60) – Ground point -2- near taillamp assembly, right

(63) – Ground connection, on left taillight bulb holder

(64) – Ground connection, on right taillight bulb holder

THIS PAGE INTENTIONALLY LEFT BLANK

Central locking system with remote control,

from September 1998

● Vehicles with double cabin

Deviate relay location and fuseplacements as well as the locations of multiple connectors see section "component locations".

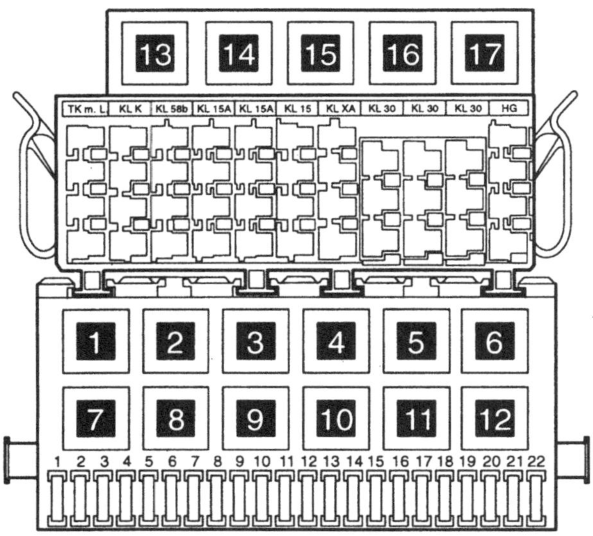

97–23232

Fuse colors

30 A – green
25 A – white
20 A – yellow
15 A – blue
10 A – red
7,5 A – brown
5 A – beige
3 A – violet

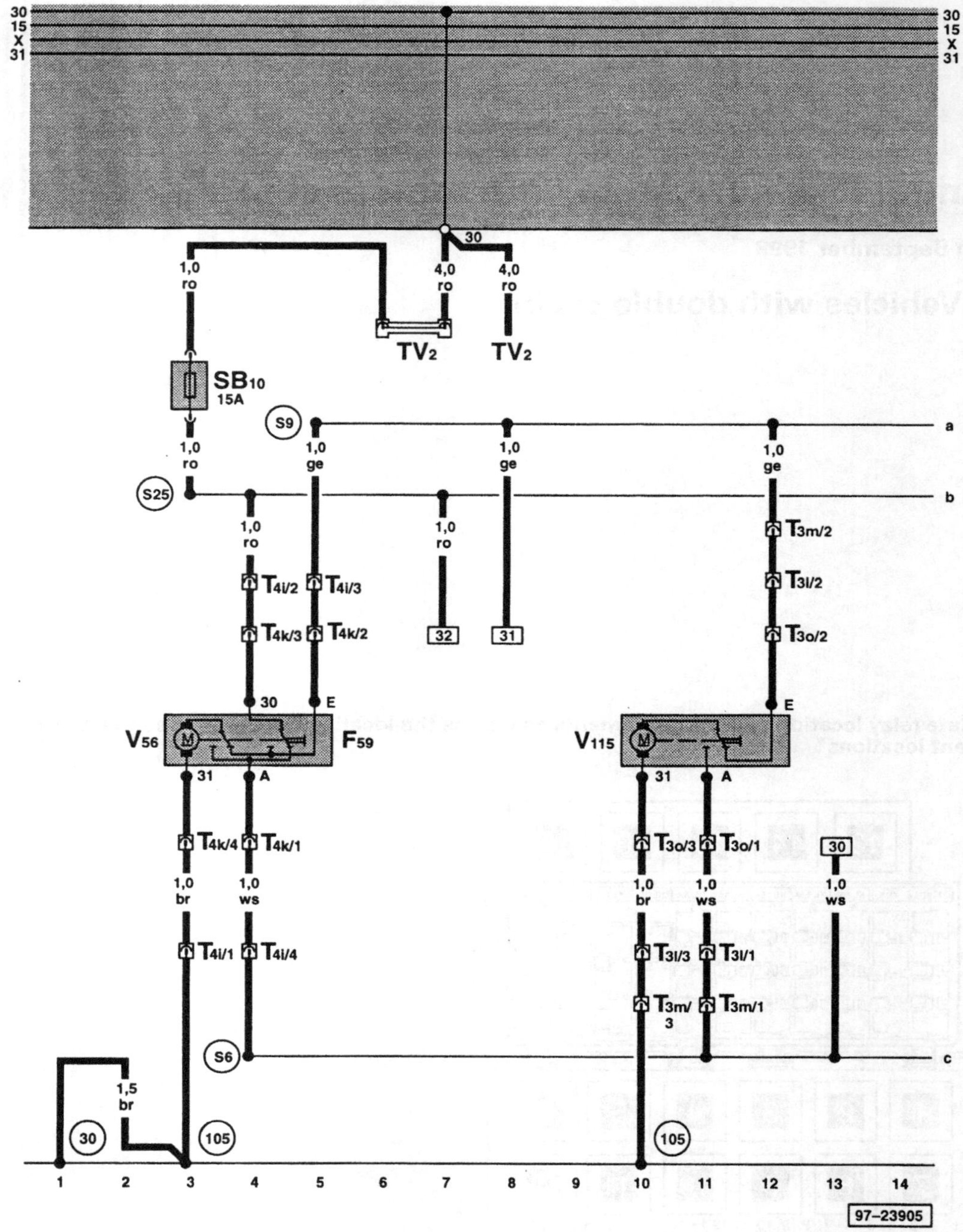

ws = white
sw = black
ro = red
br = brown
gn = green
bl = blue
gr = grey
li = violet
ge = yellow

F59 – Driver's Door Central Locking System Switch
SB10 – Fuse 10 in fuse holder, below relay panel
T3l – 3-Pin Connector, on B-pillar, left
T3m – 3-Pin Connector, behind instrument panel, left
T3o – 3-Pin Connector, in left rear door
T4i – 4-Pin Connector, behind fuse/relay panel
T4k – 4-Pin Connector, in driver's door
TV2 – Terminal 30 (B+) Wire Connector, red, each 6 points
V56 – Driver's Door Central Locking System Motor
V115 – Left Rear Central Locking System Motor

(30) – Ground connection, -1-, beside fuse/relay panel

(105) – Ground connection -1-, in power window/ central locking system/ door contact switch wiring harness
(S6) – Wire connection (A) -1-, in central locking system wiring harness
(S9) – Wire connection (E) -1-, in central locking system wiring harness
(S25) – Connector (Sensor/Control Module) -1- in wiring harness, central locking

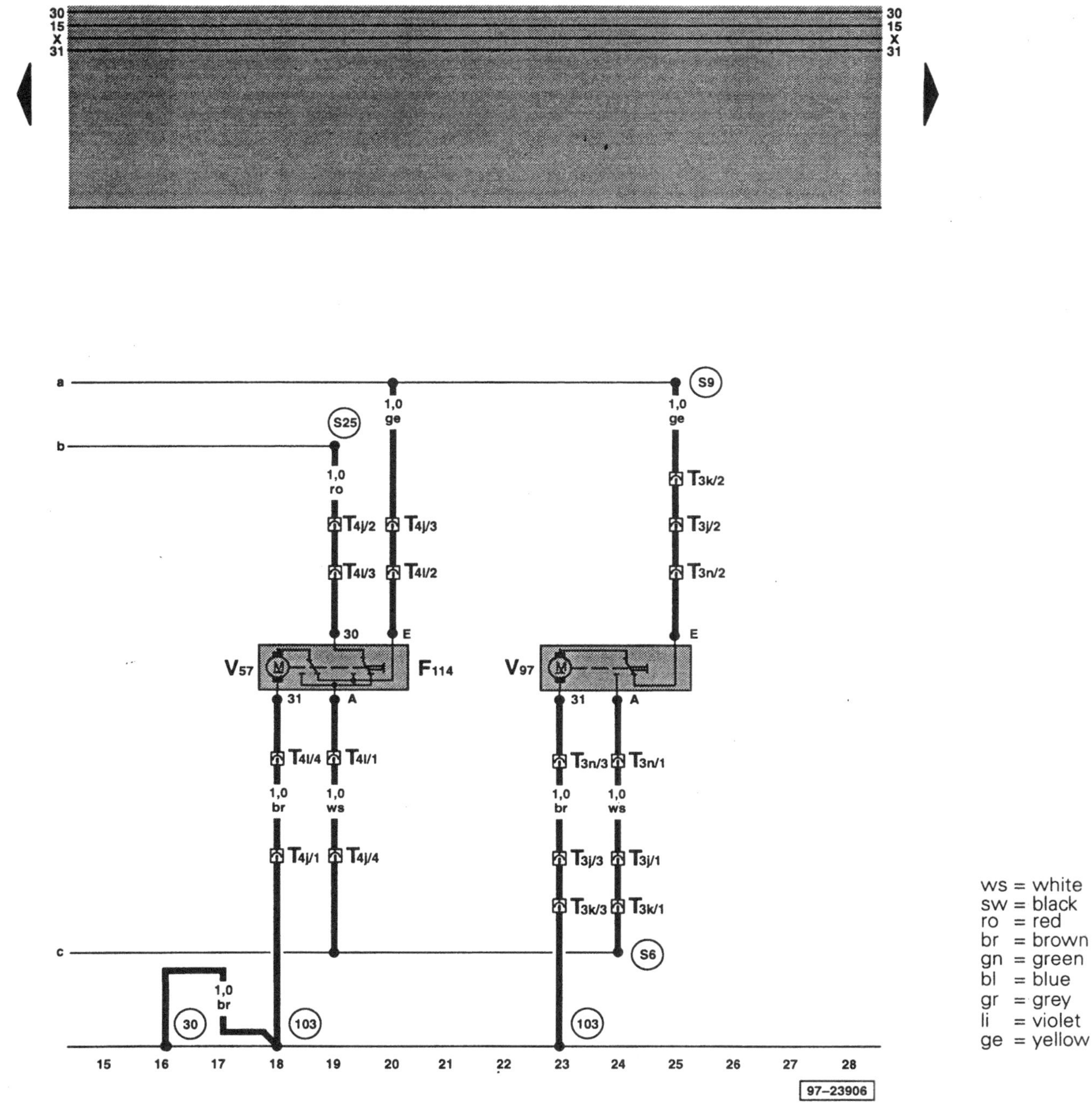

ws = white
sw = black
ro = red
br = brown
gn = green
bl = blue
gr = grey
li = violet
ge = yellow

F114 – Passenger's Door Central Locking System Switch

T3j – 3-Pin Connector, on B-pillar, right

T3k – 3-Pin Connector, behind instrument panel, right

T3n – 3-Pin Connector, in right rear door

T4j – 4-Pin Connector, behind instrument panel, right

T4l – 4-Pin Connector, in front passenger's door

V57 – Passenger's Door Central Locking System Motor

V97 – Double Cab Right Rear Door Central Locking System Motor

(30) – Ground connection, -1-, beside fuse/relay panel

(103) – Ground connection -2-, in power window/ central locking system/ door contact switch wiring harness

(S6) – Wire connection (A) -1-, in central locking system wiring harness

(S9) – Wire connection (E) -1-, in central locking system wiring harness

(S25) – Connector (Sensor/Control Module) -1- in wiring harness, central locking

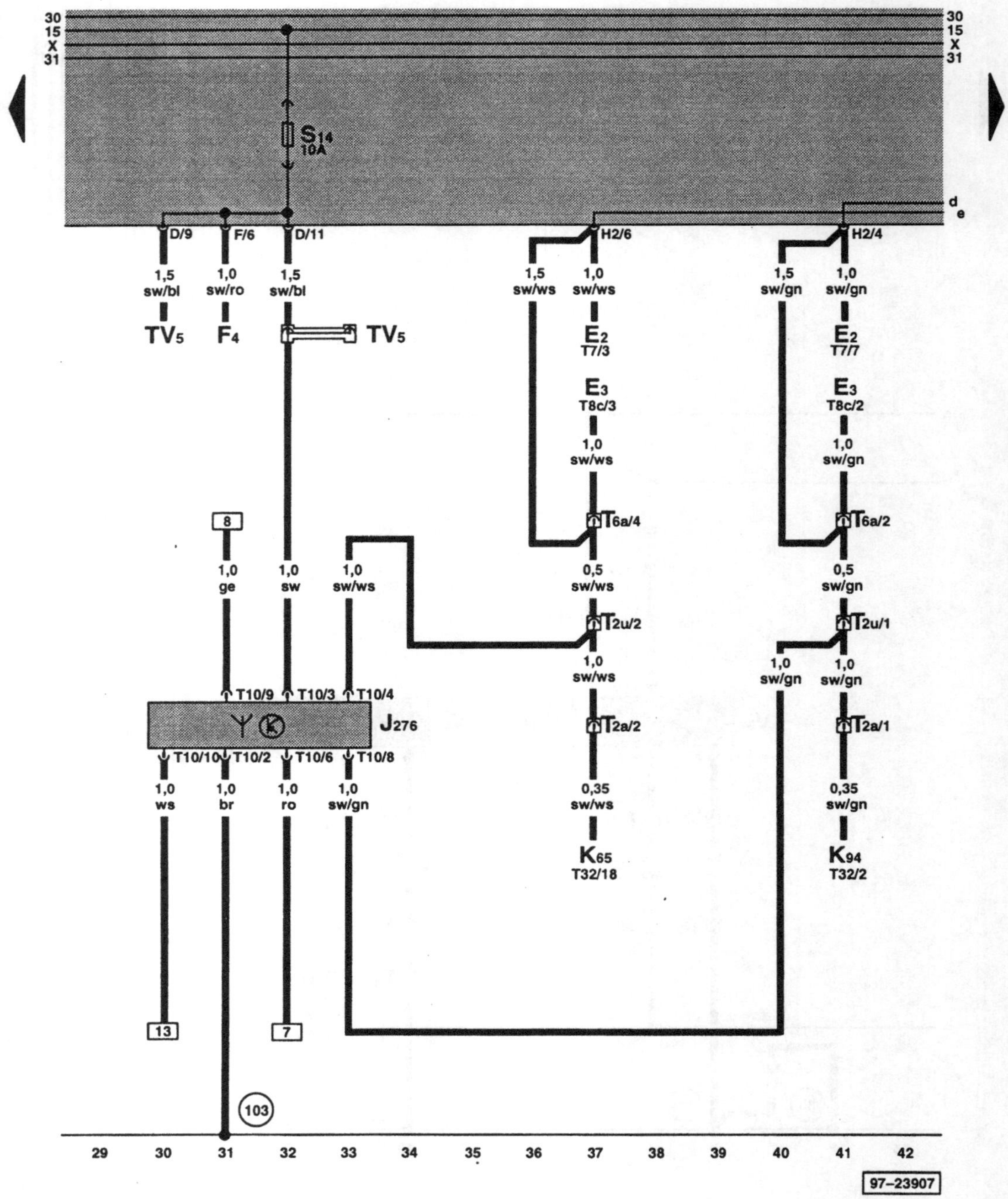

ws = white
sw = black
ro = red
br = brown
gn = green
bl = blue
gr = grey
li = violet
ge = yellow

E2 – Turn signal switch
E3 – Emergency Flasher Switch
F4 – Back-Up Light Switch
J276 – Central Locking System Control Module (Remote Control), behind instrument cluster
K65 – Left Turn Signal Indicator Light
K94 – Right Turn Signal Indicator Light
T2a – Double Connector, behind fuse/relay panel
T2u – Double Connector, behind fuse/relay panel
T6a – 6-Pin Connector, behind fuse/relay panel
T7 – 7-Pin Connector, behind steering column switch cover
T8c – 8-Pin Connector, on Emergency Flasher Switch
T10 – 10-Pin Connector
T32 – 32-Pin Connector, blue, on instrument cluster

TV5 – Terminal 15a Wire Connector, green, 4-Pin and 8-Pin

(103) – Ground connection -2-, in power window/ central locking system/ door contact switch wiring harness

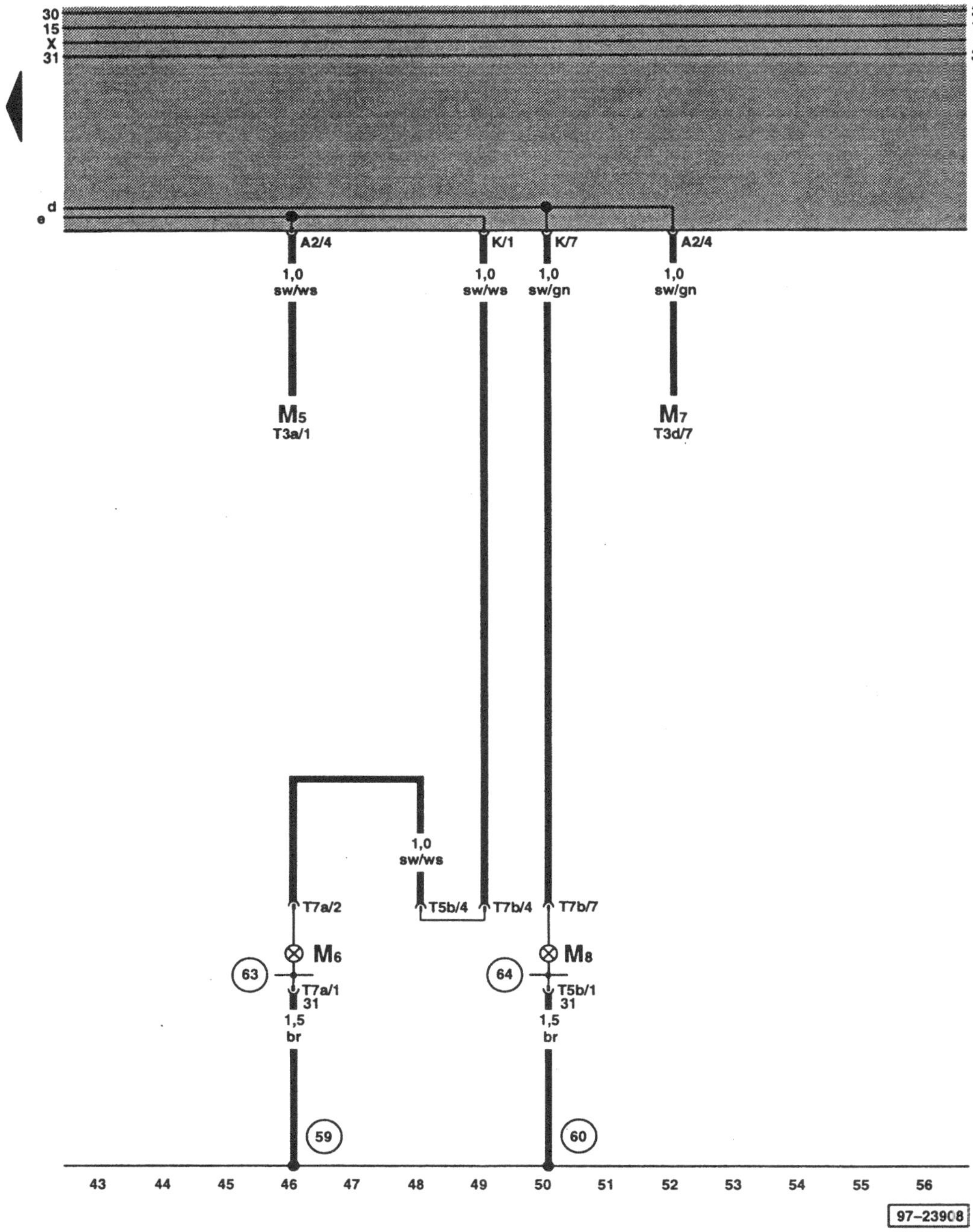

ws = white
sw = black
ro = red
br = brown
gn = green
bl = blue
gr = grey
li = violet
ge = yellow

- M5 – Left Front Turn Signal Light
- M6 – Left Rear Turn Signal Light
- M7 – Right Front Turn Signal Light
- M8 – Right Rear Turn Signal Light
- T3a – 3-Pin Connector
- T3d – 3-Pin Connector
- T5b – 5-Pin Connector
- T7a – 7-Pin Connector
- T7b – 7-Pin Connector

- (59) – Ground connection, near taillight, left
- (60) – Ground point -2- near taillamp assembly, right
- (63) – Ground connection, on left taillight bulb holder
- (64) – Ground connection, on right taillight bulb holder

THIS PAGE INTENTIONALLY LEFT BLANK

Daytime running lights (DRL),

from September 1998

Deviate relay location and fuseplacements as well as the locations of multiple connectors see section "component locations".

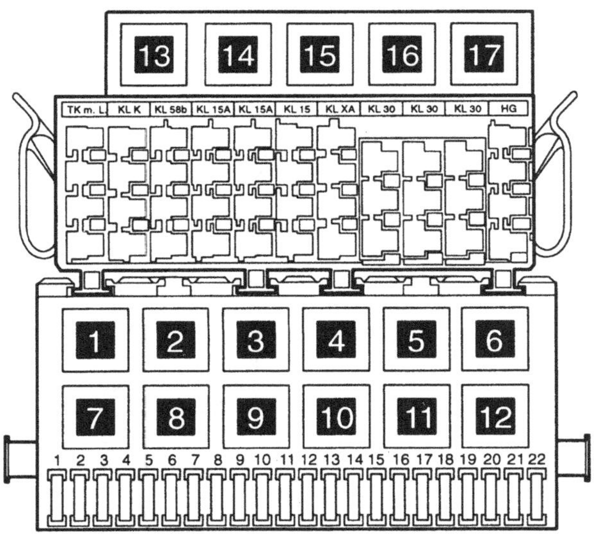

97–23232

4 Load Reduction Relay (18)

Note: Number in parentheses indicates production control number stamped on relay housing.

Fuse colors

30 A – green
25 A – white
20 A – yellow
15 A – blue
10 A – red
7,5 A – brown
5 A – beige
3 A – violet

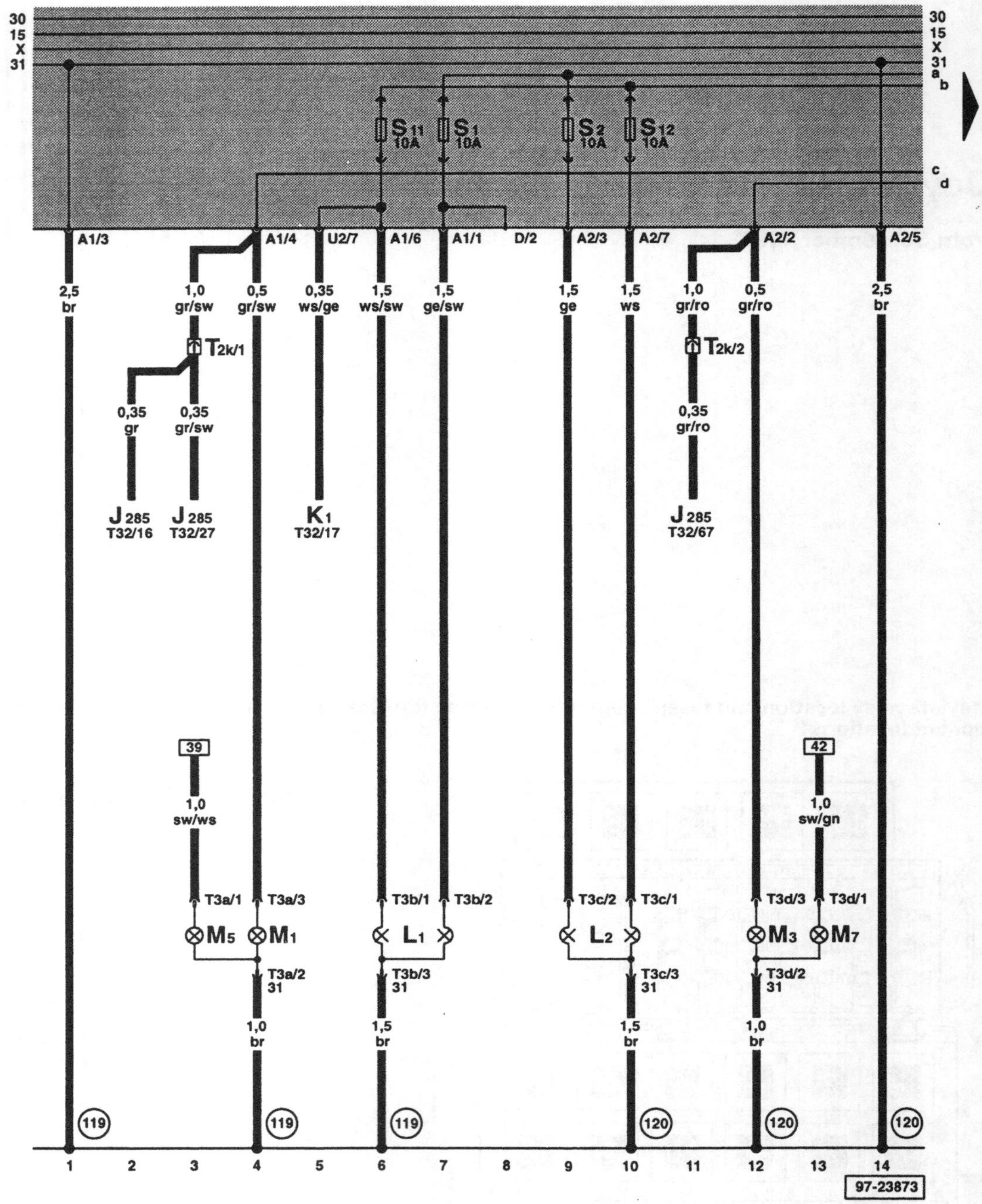

ws = white
sw = black
ro = red
br = brown
gn = green
bl = blue
gr = grey
li = violet
ge = yellow

J285 – Control module with indicator unit in instrument panel insert
K1 – Headlight High Beam Indicator Light
L1 – Left Headlight
L2 – Right Headlight
M1 – Left Parking Light
M3 – Right Parking Light
M5 – Left Front Turn Signal Light
M7 – Right Front Turn Signal Light
T2k – Double Connector, behind fuse/relay panel
T3a – 3-Pin Connector
T3b – 3-Pin Connector
T3c – 3-Pin Connector
T3d – 3-Pin Connector
T32 – 32-Pin Connector, blue, on Instrument cluster

(119) – Ground connection -1-, in headlight wiring harness
(120) – Ground connection -2-, in headlight wiring harness

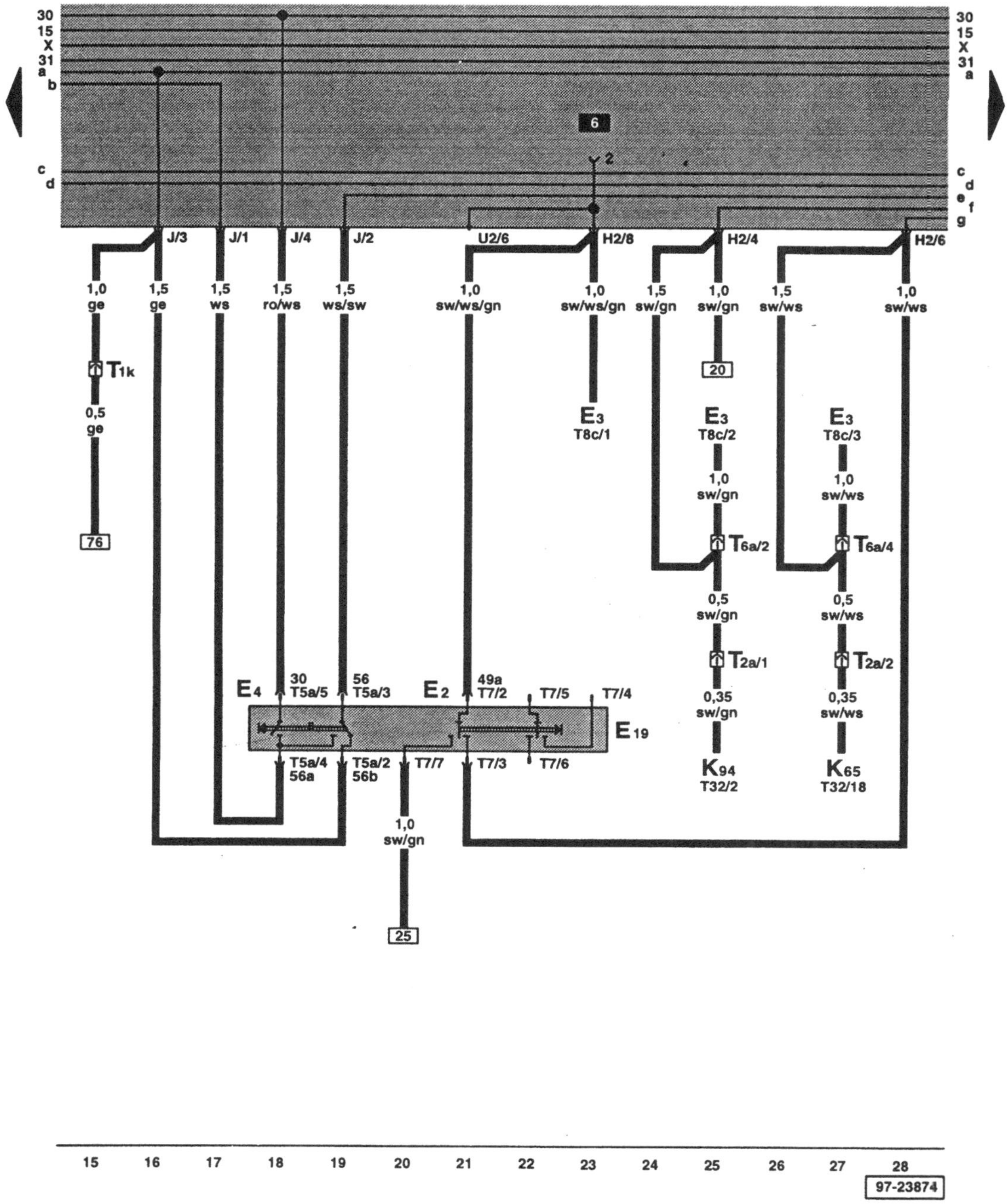

ws = white
sw = black
ro = red
br = brown
gn = green
bl = blue
gr = grey
li = violet
ge = yellow

E2 – Turn signal switch
E3 – Emergency Flasher Switch
E4 – Headlight Dimmer/Flasher Switch
E19 – Park Light Switch
K65 – Left Turn Signal Indicator Light
K94 – Right Turn Signal Indicator Light
T1k – Single Connector, behind fuse/relay panel
T2a – Double Connector, behind fuse/relay panel
T5a – 5-Pin Connector, behind steering column switch cover
T6a – 6-Pin Connector, behind fuse/relay panel
T7 – 7-Pin Connector, behind steering column switch cover
T8c – 8-Pin Connector
T32 – 32-Pin Connector, blue, on Instrument cluster

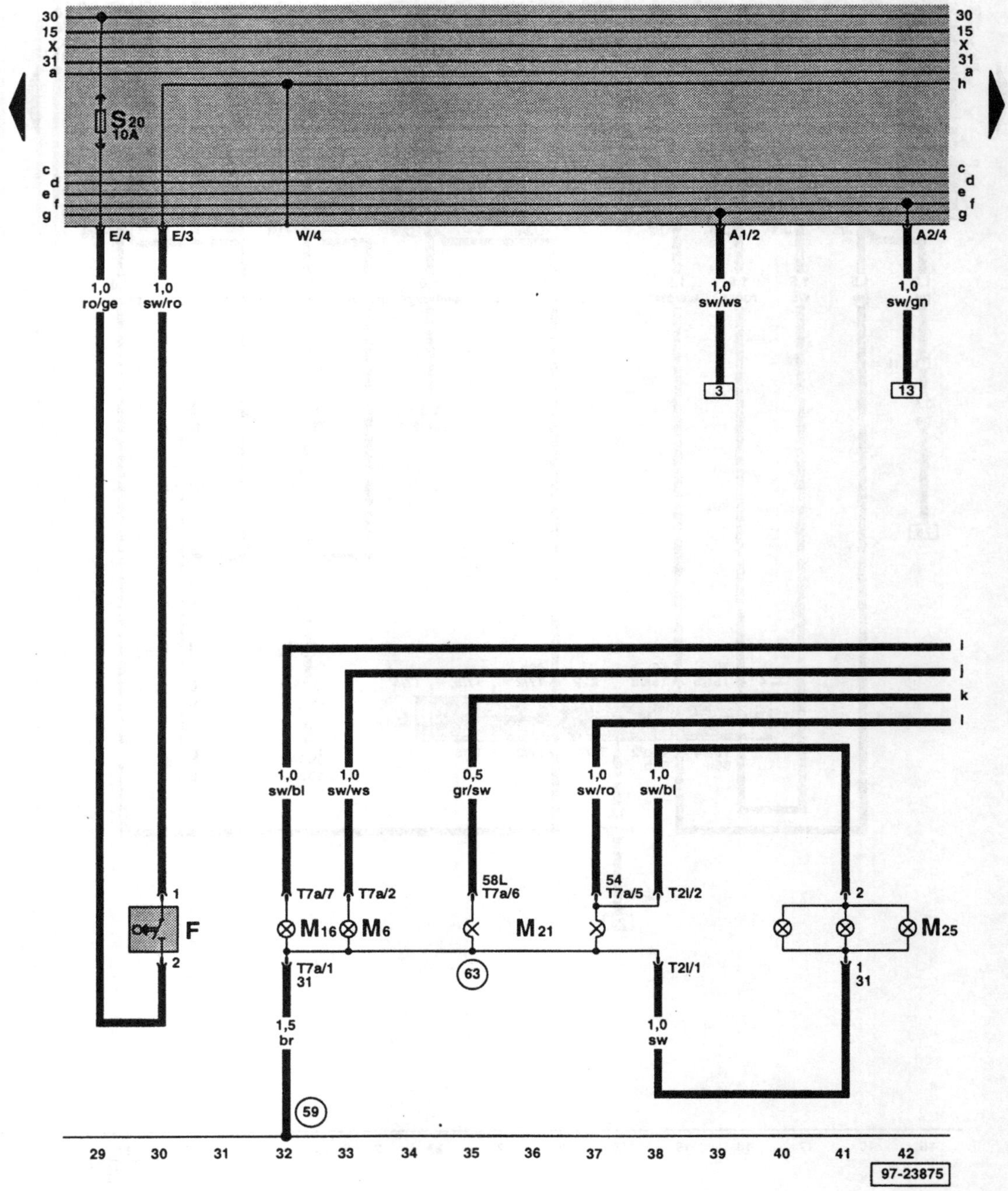

ws = white
sw = black
ro = red
br = brown
gn = green
bl = blue
gr = grey
li = violet
ge = yellow

F – Brake Light Switch
M6 – Left Rear Turn Signal Light
M16 – Left Back-Up Light
M21 – Left Brake/Tail Light
M25 – High-mount Brake Light
T2l – Double Connector
T7a – 7-Pin Connector

(59) – Ground connection, near taillight, left

(63) – Ground connection, on left taillight bulb holder

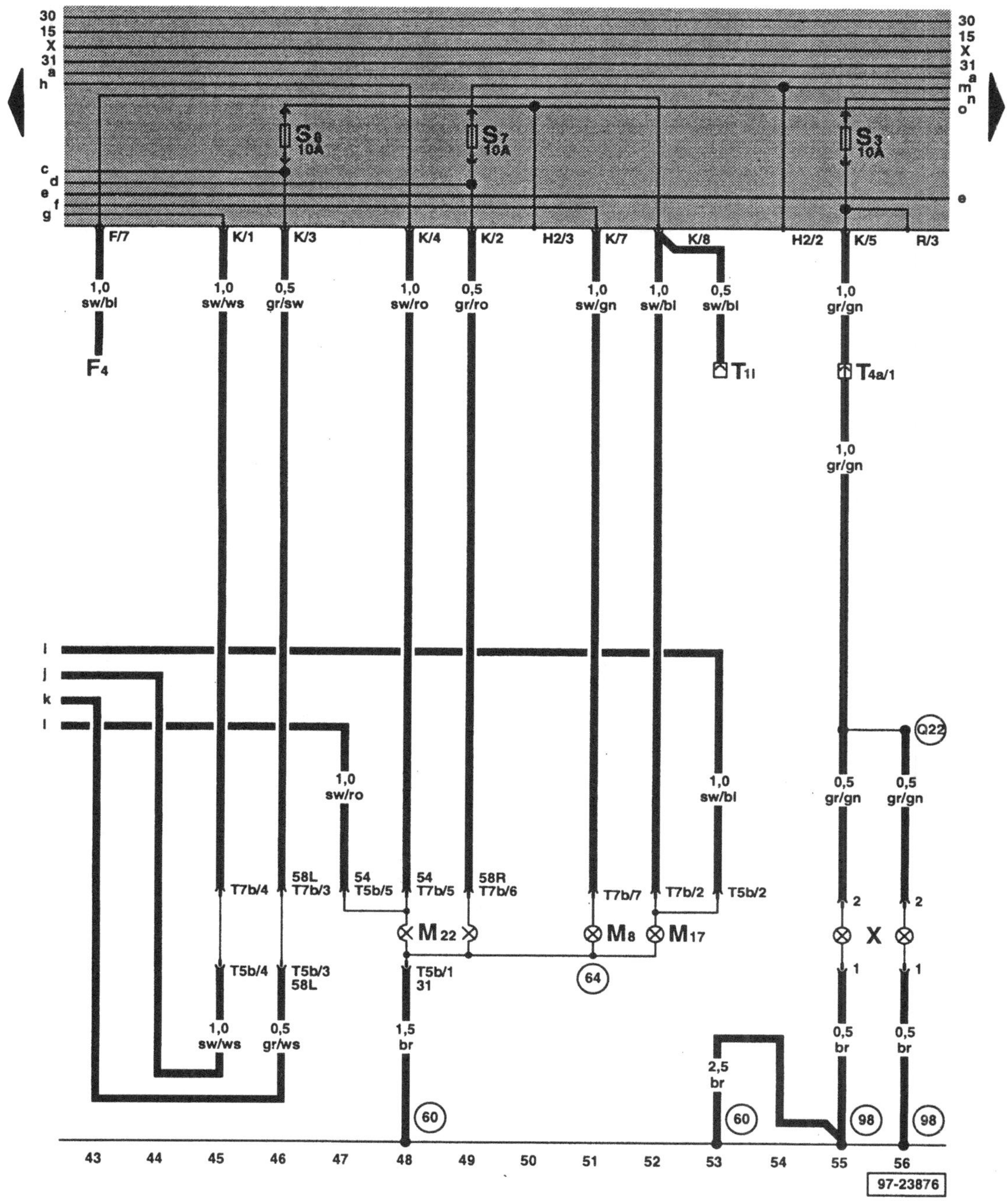

ws = white
sw = black
ro = red
br = brown
gn = green
bl = blue
gr = grey
li = violet
ge = yellow

F4 – Back-Up Light Switch
M8 – Right Rear Turn Signal Light
M17 – Right Back-Up Light
M22 – Right Brake/Tail Light
T1l – Single Connector, behind fuse/relay panel
T4a – 4-Pin Connector, near taillamp assembly
T5b – 5-Pin Connector
T7b – 7-Pin Connector
X – License Plate Light

(Q22) – Wire connection -1-, in rear lid wiring harness

(60) – Ground point -2- near taillamp assembly, right
(64) – Ground connection, on right taillight bulb holder
(98) – Ground connection, in rear lid wiring harness

Right back-up light, right brake/tail light, right rear turn signal light, license plate light

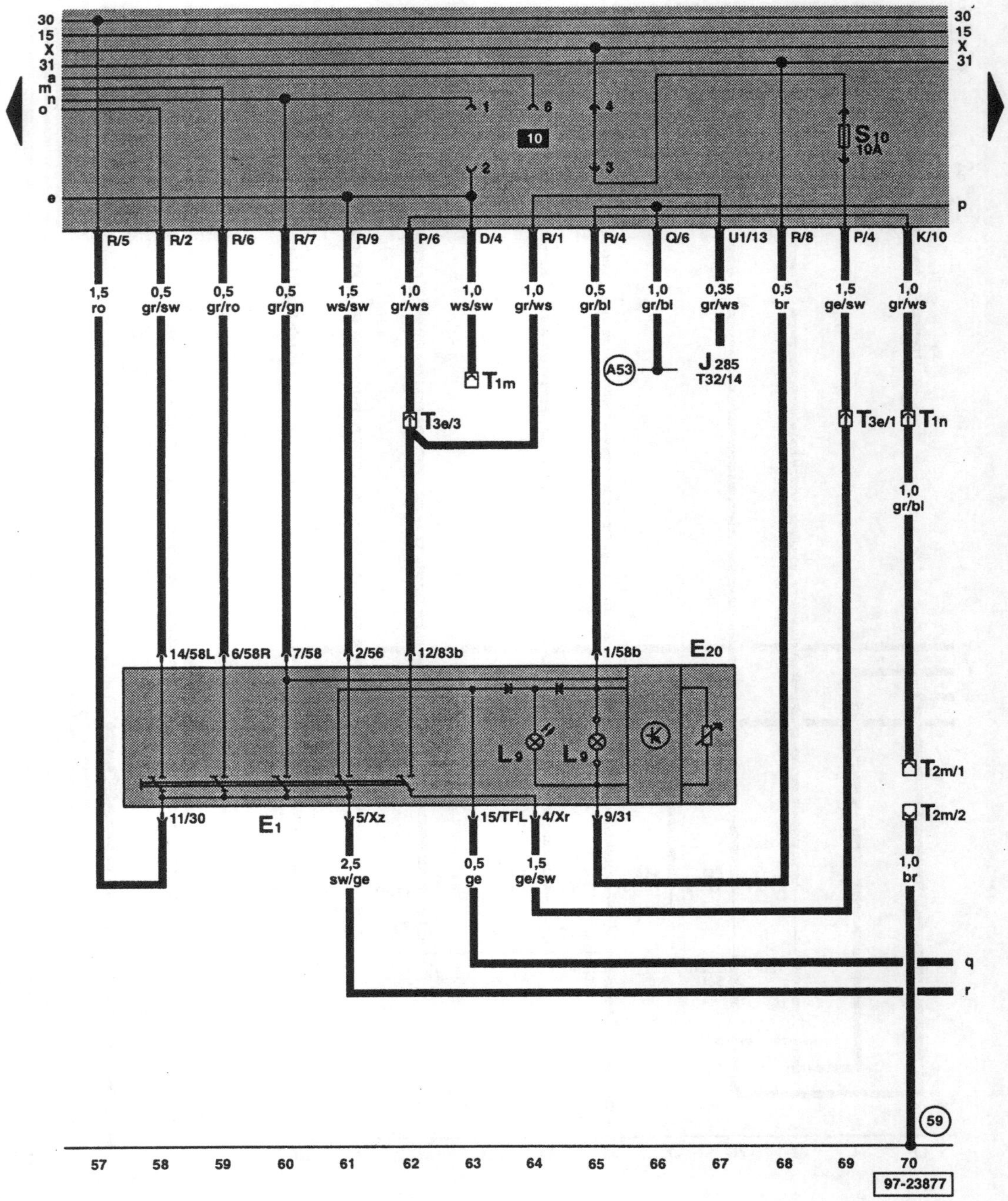

ws = white
sw = black
ro = red
br = brown
gn = green
bl = blue
gr = grey
li = violet
ge = yellow

E1 – Light switch
E20 – Instrument Panel Light Dimmer Switch
J285 – Control module with indicator unit in instrument panel insert
L9 – Headlight Switch Light
T1m – Single Connector, behind fuse/relay panel
T1n – Single Connector, near tail lamp assembly
T2m – Double Connector
T3e – 3-Pin Connector, behind fuse/relay panel
T32 – 32-Pin Connector, blue, on Instrument cluster

(59) – Ground connection, near taillight, left

(A53) – Plus connection (58b), in instrument panel wiring harness

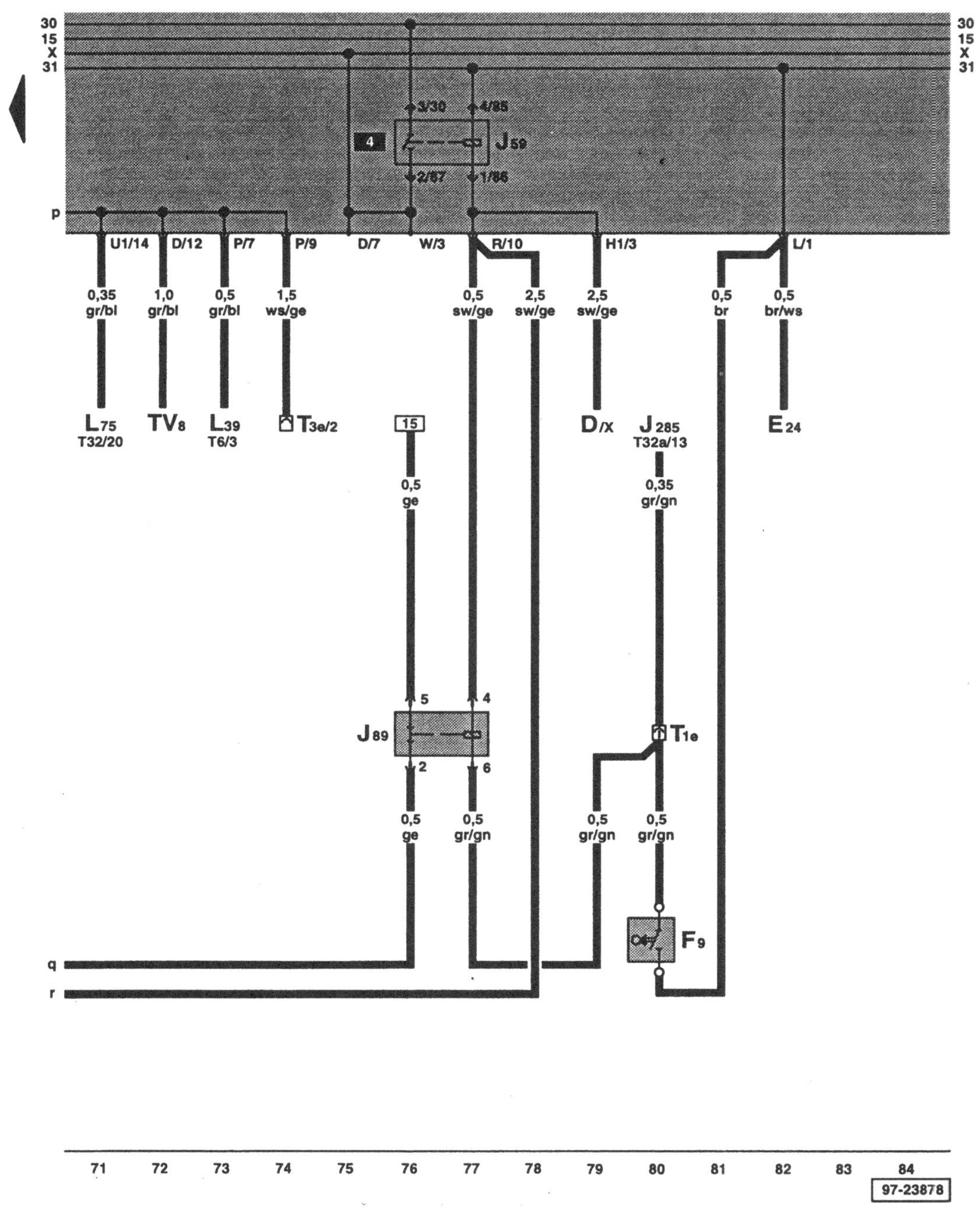

ws = white
sw = black
ro = red
br = brown
gn = green
bl = blue
gr = grey
li = violet
ge = yellow

- D – Ignition/Starter Switch
- E24 – Left Seat Belt Switch
- F9 – Parking Brake Warning Light Switch
- J59 – Load Reduction Relay
- J89 – Daytime Running Lights Change-over Relay, production control number (94), behind instrument panel, center
- J285 – Control module with indicator unit in instrument panel insert
- L39 – Rear Window Defogger Switch Light
- L75 – Digital Display Light
- T1e – Single Connector, behind fuse/relay panel
- T3e – 3-Pin Connector, behind fuse/relay panel
- T6 – 6-Pin Connector, on rear window defogger switch
- T32 – 32-Pin Connector, blue, on Instrument cluster
- T32a – 32-Pin Connector, green, on Instrument cluster
- TV8 – Terminal 58b Wire Connector, grey

THIS PAGE INTENTIONALLY LEFT

BLANK

Headlight washer system,

from September 1998

Deviate relay location and fuseplacements as well as the locations of multiple connectors see section "component locations".

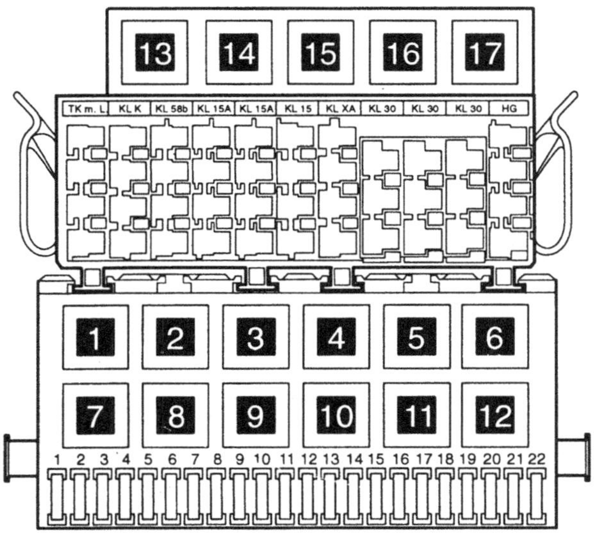

97–23232

Fuse colors

30 A – green
25 A – white
20 A – yellow
15 A – blue
10 A – red
7,5 A – brown
5 A – beige
3 A – violet

7 Relay for headlamp cleaning system (95)

Note: Number in parentheses indicates production control number stamped on relay housing.

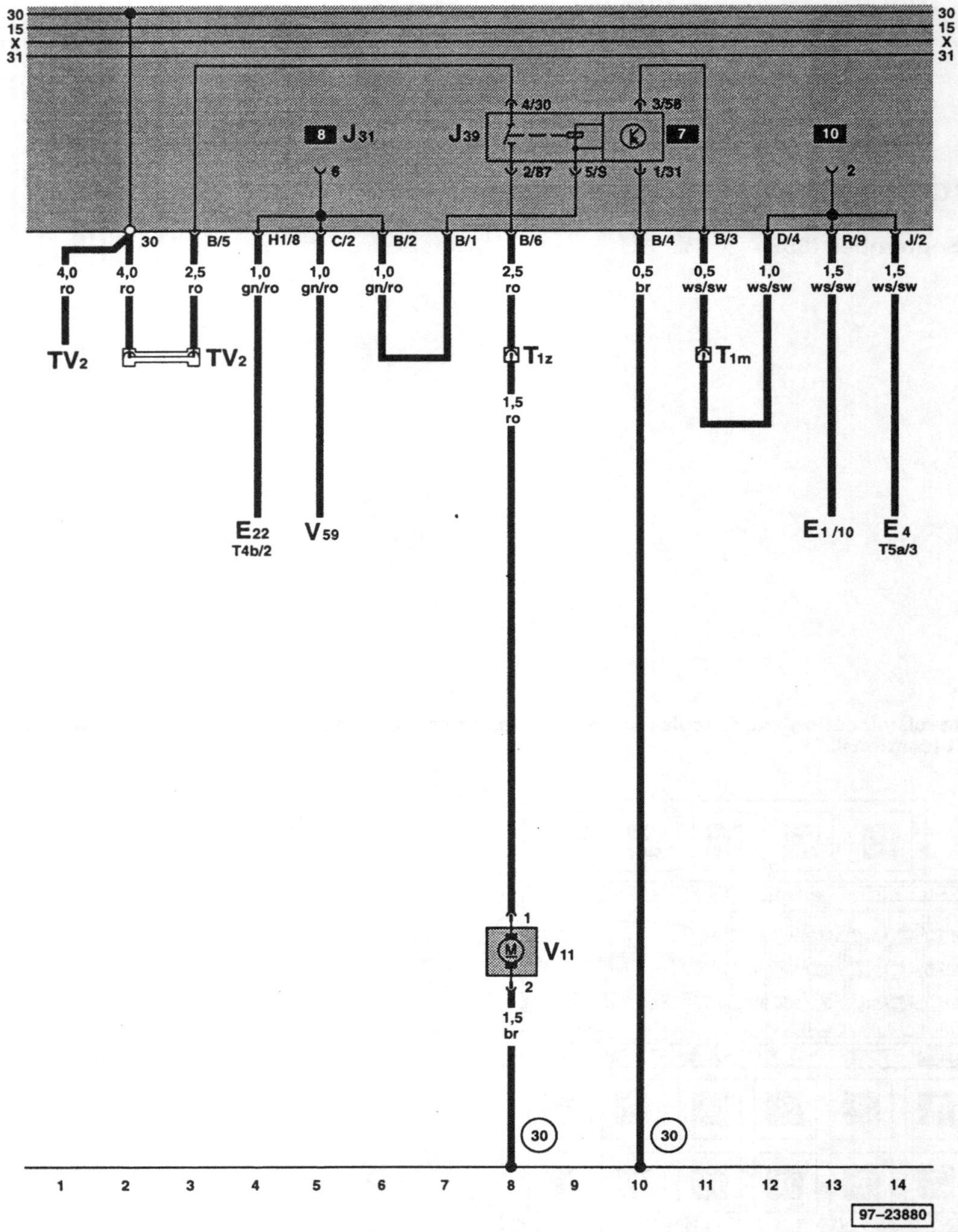

ws = white
sw = black
ro = red
br = brown
gn = green
bl = blue
gr = grey
li = violet
ge = yellow

- E1 – Light switch
- E4 – Headlight Dimmer/Flasher Switch
- E22 – Windshield Wiper/Washer Switch
- J31 – Wiper/Washer Intermittent Relay, on relais panel
- J39 – Relay for headlamp cleaning system
- T1m – Single Connector, behind fuse/relay panel
- T1z – Single Connector, behind fuse/relay panel
- T4b – 4-Pin Connector, behind steering column switch cover
- T5a – 5-Pin Connector, behind steering column switch cover
- TV2 – Terminal 30 (B+) Wire Connector, red, each 6 points
- V11 – Headlight Washer Pump
- V59 – Windshield and Rear Window Washer Pump

(30) – Ground connection, -1-, beside fuse/relay panel

12V outlets,

from September 1998

Deviate relay location and fuseplacements as well as the locations of multiple connectors see section "component locations".

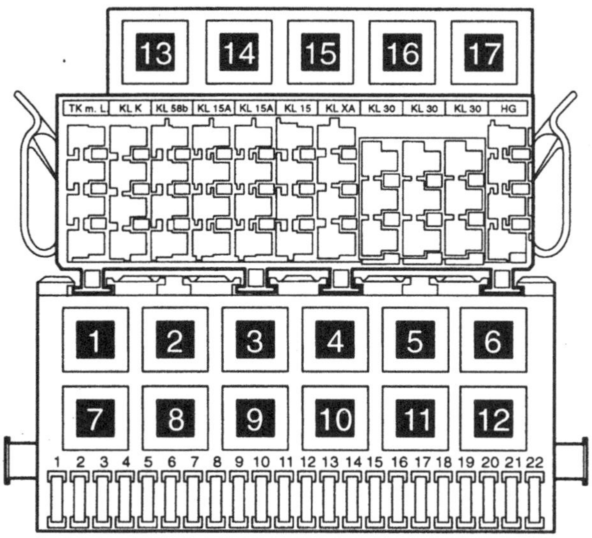

97–23232

Fuse colors

30 A – green
25 A – white
20 A – yellow
15 A – blue
10 A – red
7,5 A – brown
5 A – beige
3 A – violet

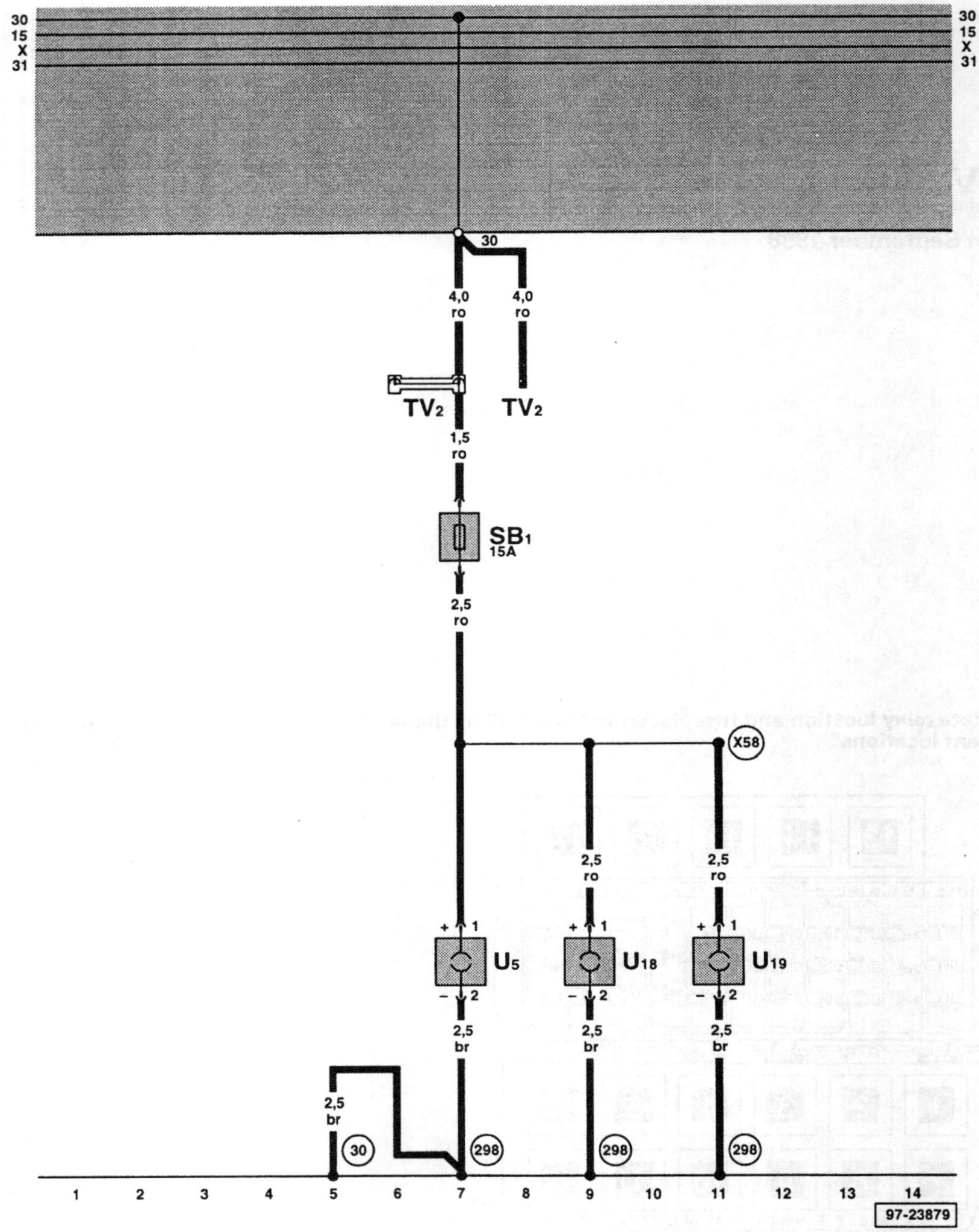

ws = white
sw = black
ro = red
br = brown
gn = green
bl = blue
gr = grey
li = violet
ge = yellow

SB1 – Fuse 1 in fuse holder, below relay panel
TV2 – Terminal 30 (B+) Wire Connector, red, each 6 points
U5 – 12V outlet -1-, in passenger compartment left
U18 – 12V outlet -2-, in passenger compartment right
U19 – 12V outlet -3-, in passenger compartment right

(30) – Ground connection, -1-, beside fuse/relay panel

(289) – Ground Connection in plug wiring harness

(X58) – Positive Connection (30), in plug wiring harness

2001-2003 Electrical Wiring Supplement

*Applicable to m.y. 2001-2003, see previous (1992-2000 m.y.) electrical wiring diagrams

Detailed Index

Control modules and relays

Table of contents:

Climatronic Control Module, from May 1996 to August 1998

Component location: Behind the A/C Control Head

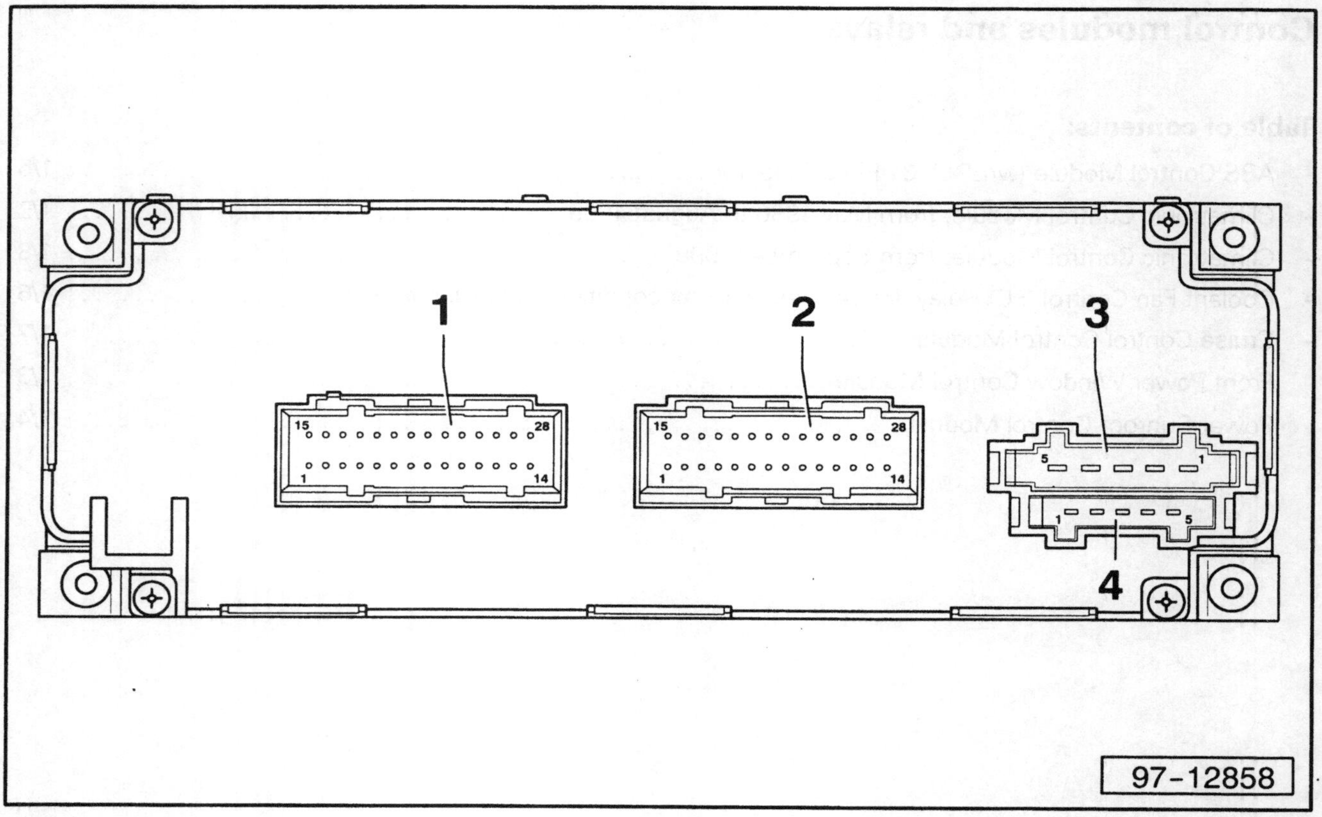

Connectors:

1	=	T28a	-	28-Pin Connector, for red Connector
2	=	T28b	-	28-Pin Connector, for white Connector
3	=	T5e	-	5-Pin Connector
4	=	Not used		

Front Power Window Control Module

Component location: In driver's and passenger's door

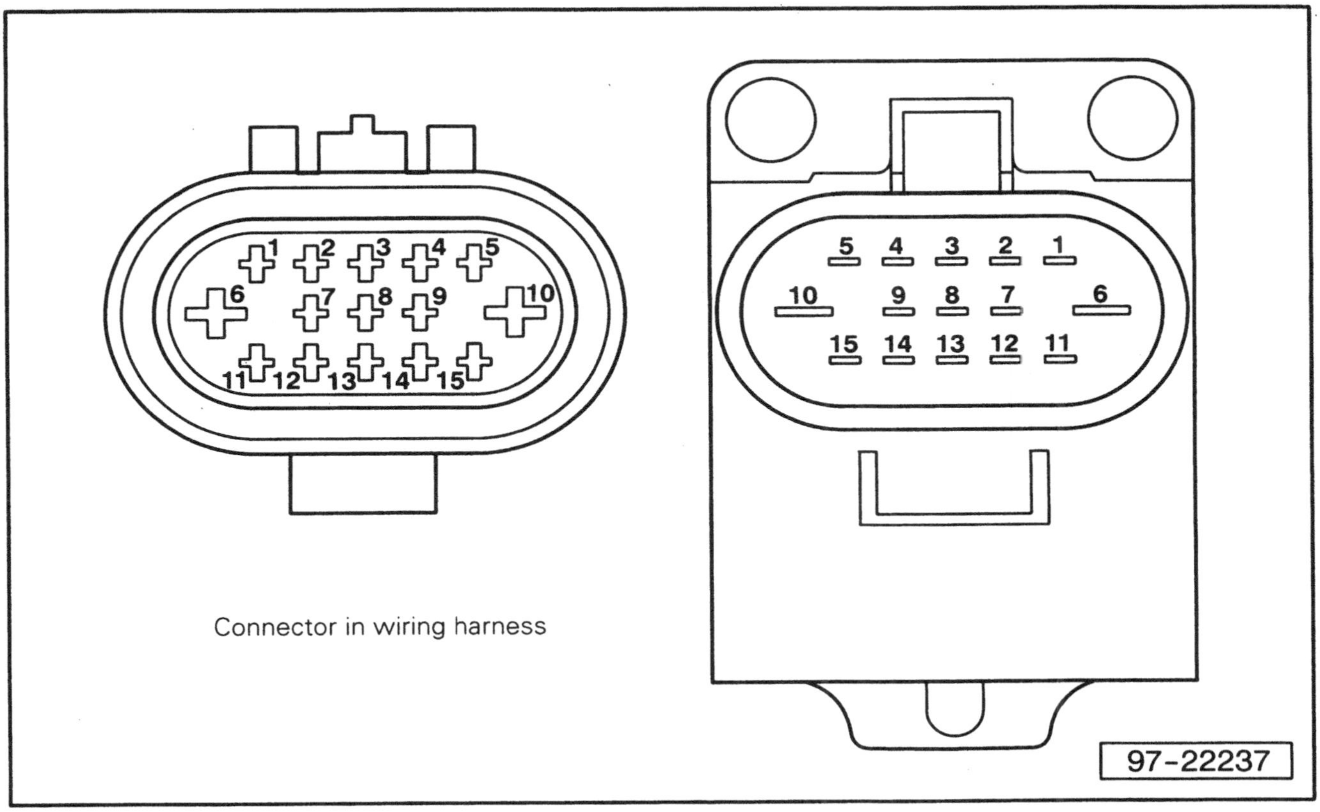

Power Sunroof Control Module

Component location: Behind Interior light front

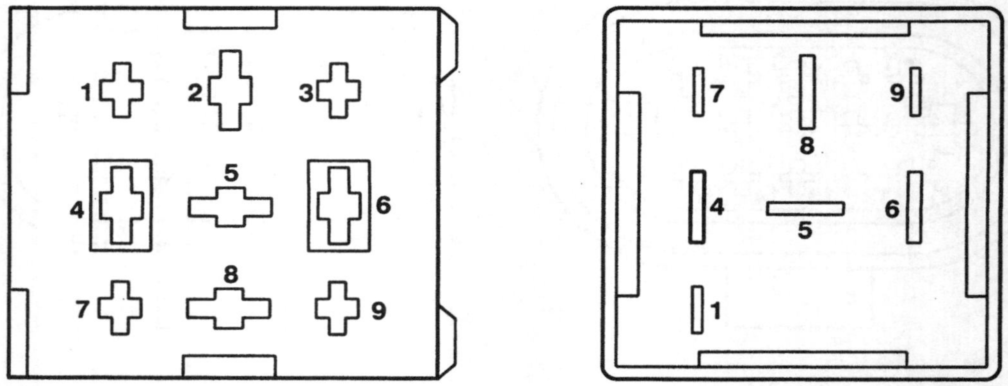

Relay holder

Power sunroof control module

97-21977

ABS Control Module (w/EDL), 31-Pin Connector

Component location: In engine compartment, left front

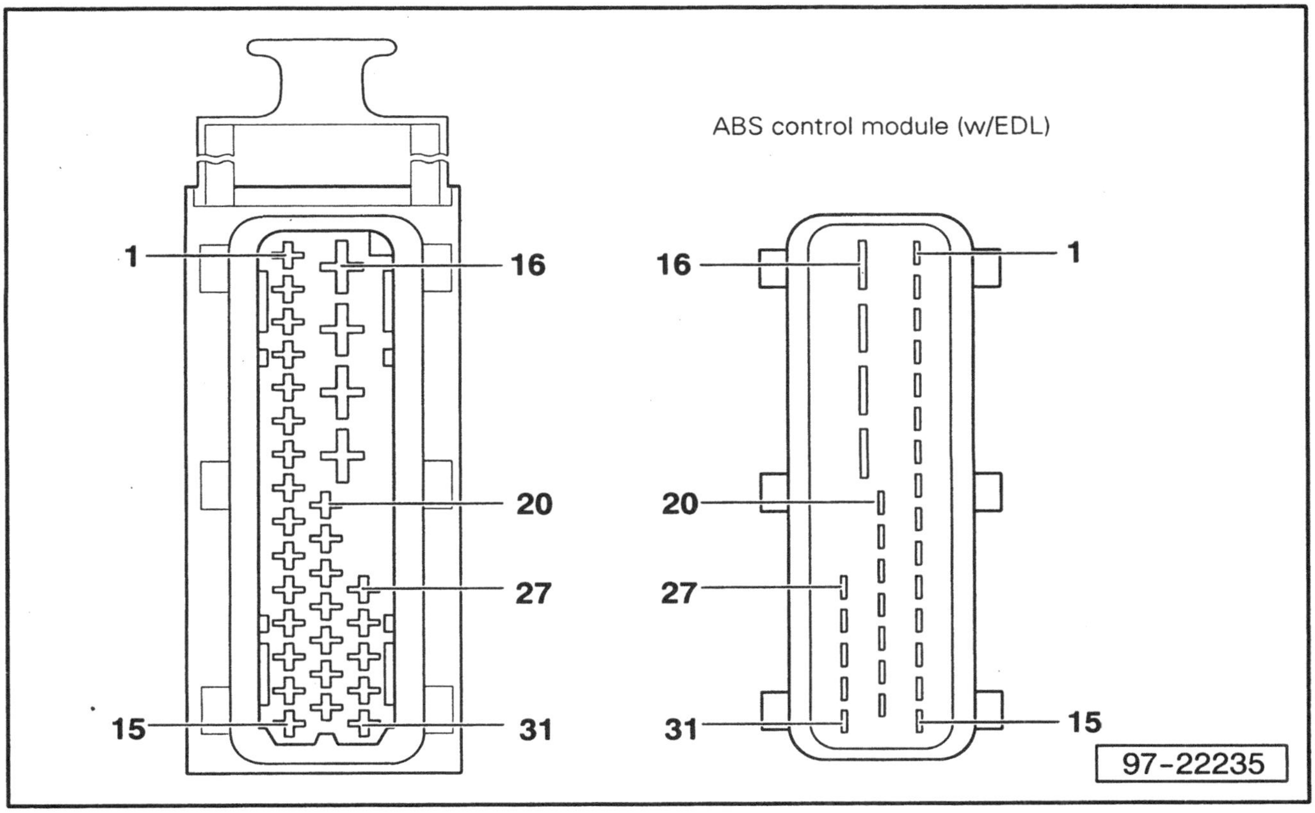

Coolant Fan Control (FC) Relay, for vehicles with Air conditioning or Climatronic

Component location: In engine compartment, left front

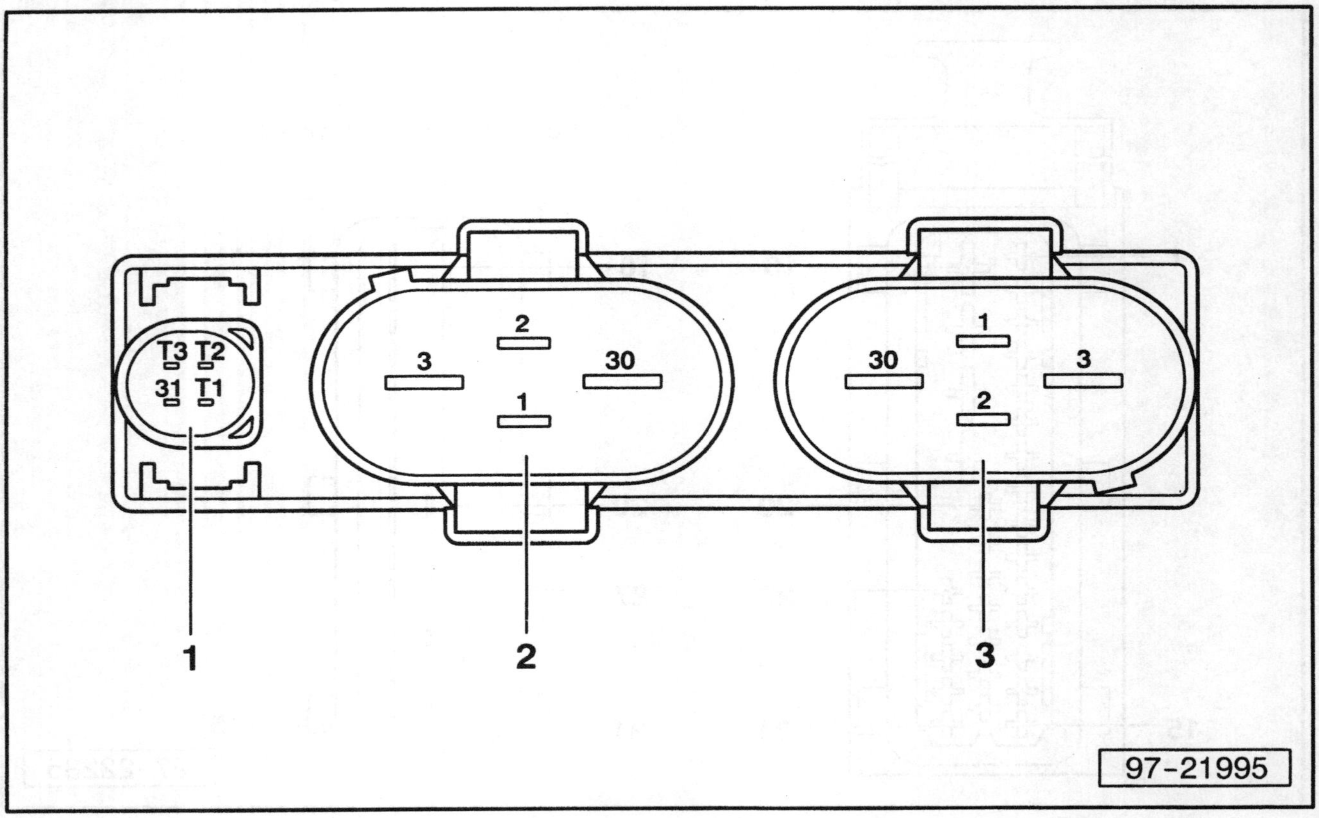

Connectors:

Connector 3:			Connector 2:			Connector 1:		
3	=	T4c/1	3	=	T4b/1	T1	=	T4a/1
2	=	T4c/2	2	=	T4b/2	T2	=	T4a/2
1	=	T4c/3	1	=	T4b/3	31	=	T4a/3
30	=	T4c/4	30	=	T4b/4	T3	=	T4a/4

Note:

The letter suffixes used in "T" connector component codes are for reference only. When a given connector appears in more than one wiring diagram, the same suffix may or may not be used.

Cruise Control Control Module

Component location: Behind instrument panel, right

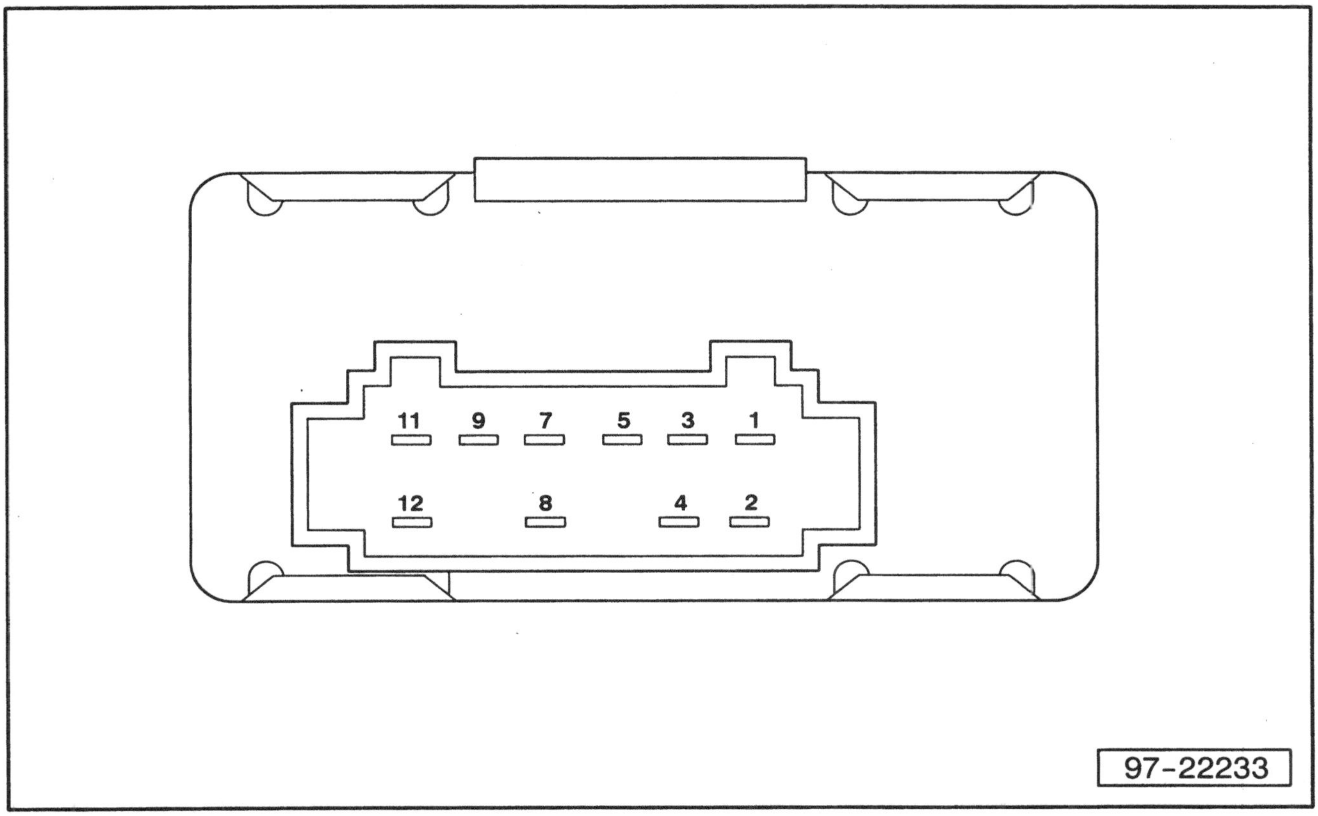

Climatronic Control Module, from September 1998

Component location: Behind the A/C Control Head

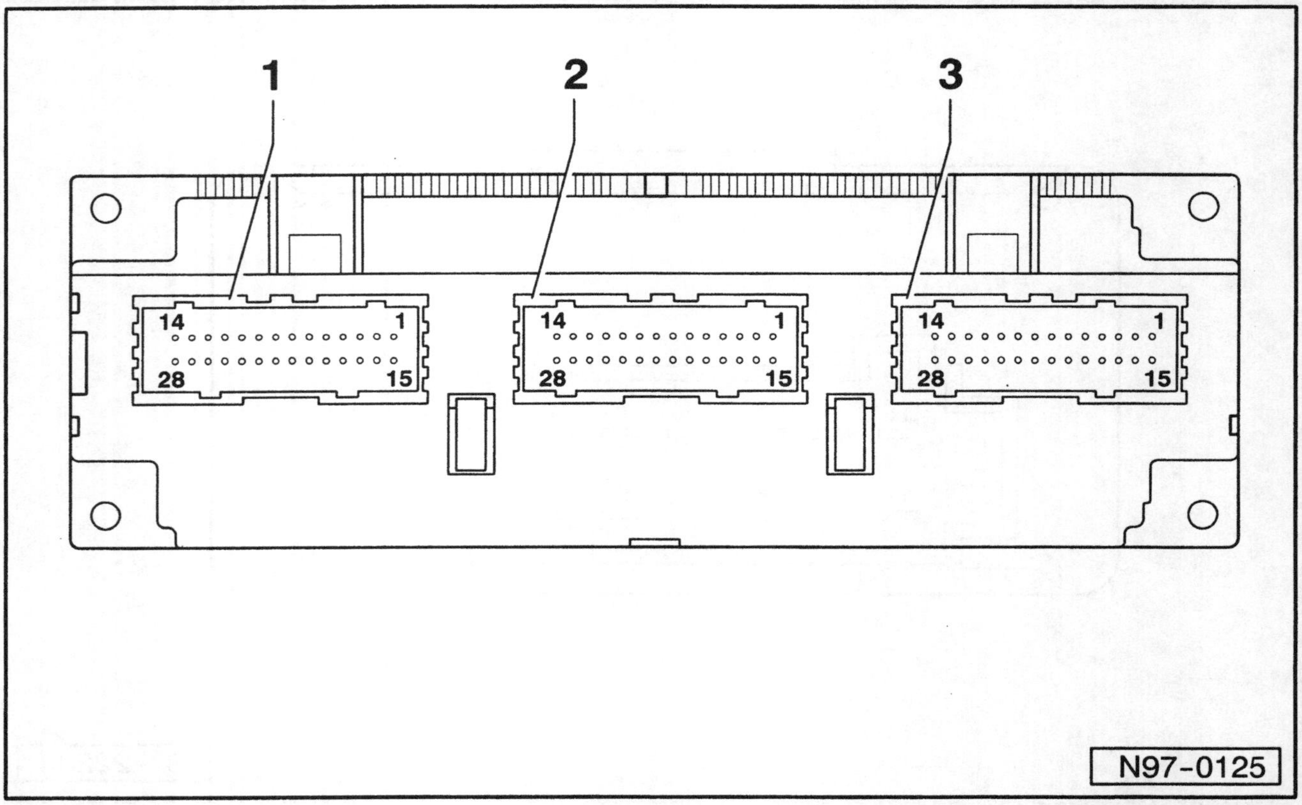

Connectors:

1	=	T28c	-	28-Pin Connector, for white Connector
2	=	T28b	-	28-Pin Connector, for blue Connector
3	=	T28a	-	28-Pin Connector, for red Connector

Multiple-pin connectors

Table of contents:

Automatic Transmission

Solenoid valves and Transmission Fluid Temperature Sensor

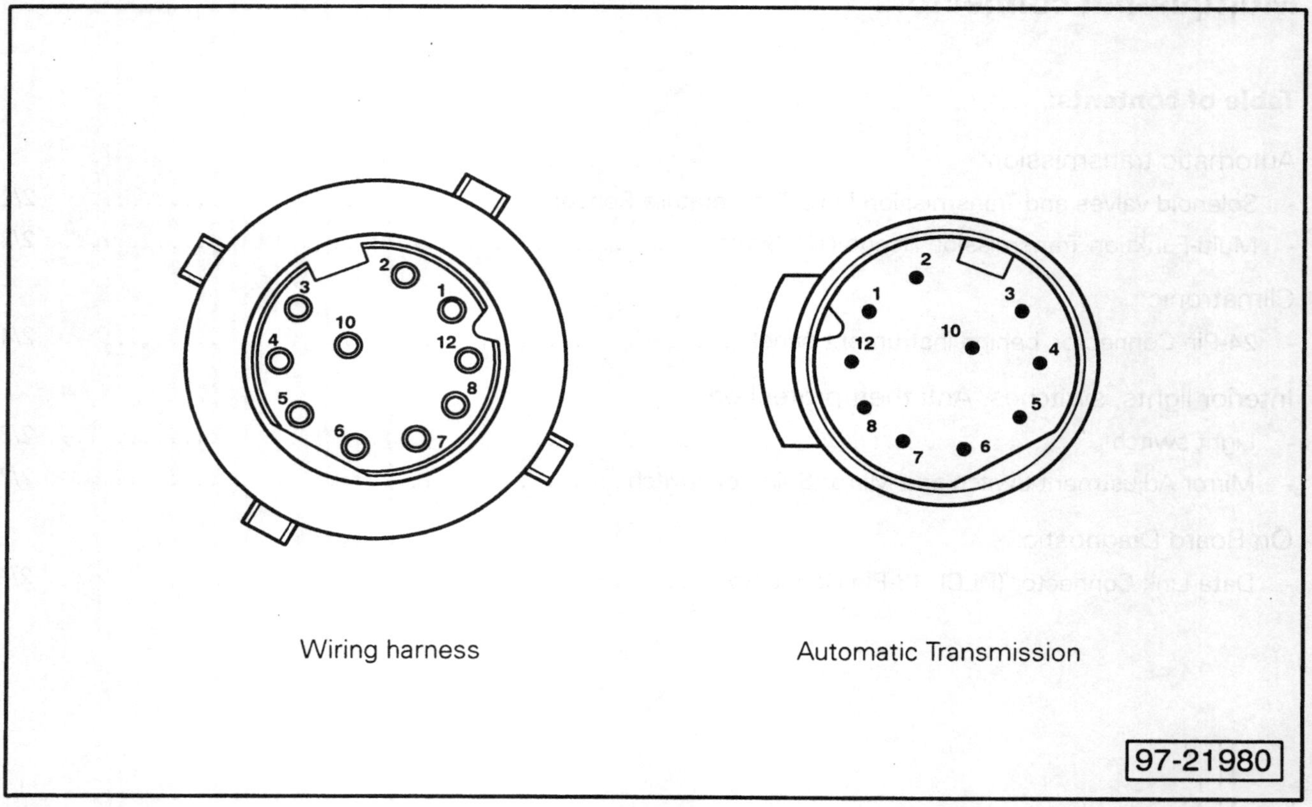

Multi-Function Transmission Range (TR) Switch

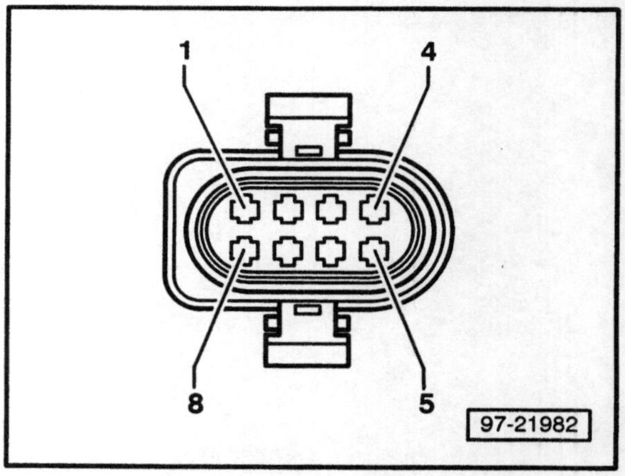

Interior lights, switches, Anti-theft protection

Light switch

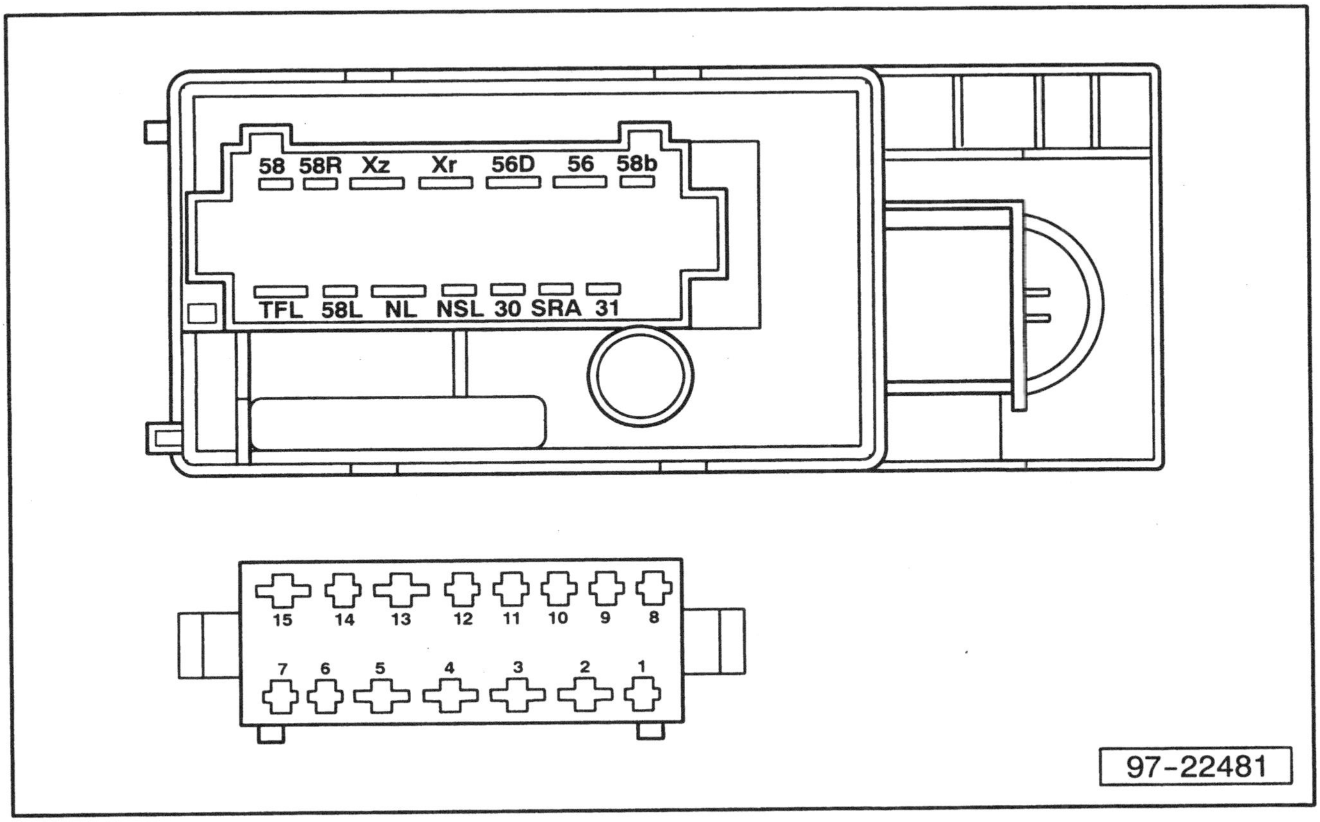

Mirror Adjustment Switch and Mirror Selector Switch

A - Wiring harness

B - Mirror adjustment switch and mirror selector switch

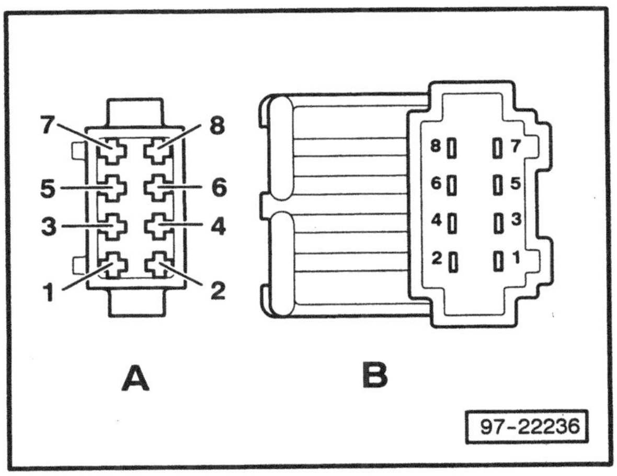

Climatronic

24-Pin Connector

Component location: Behind instrument panel

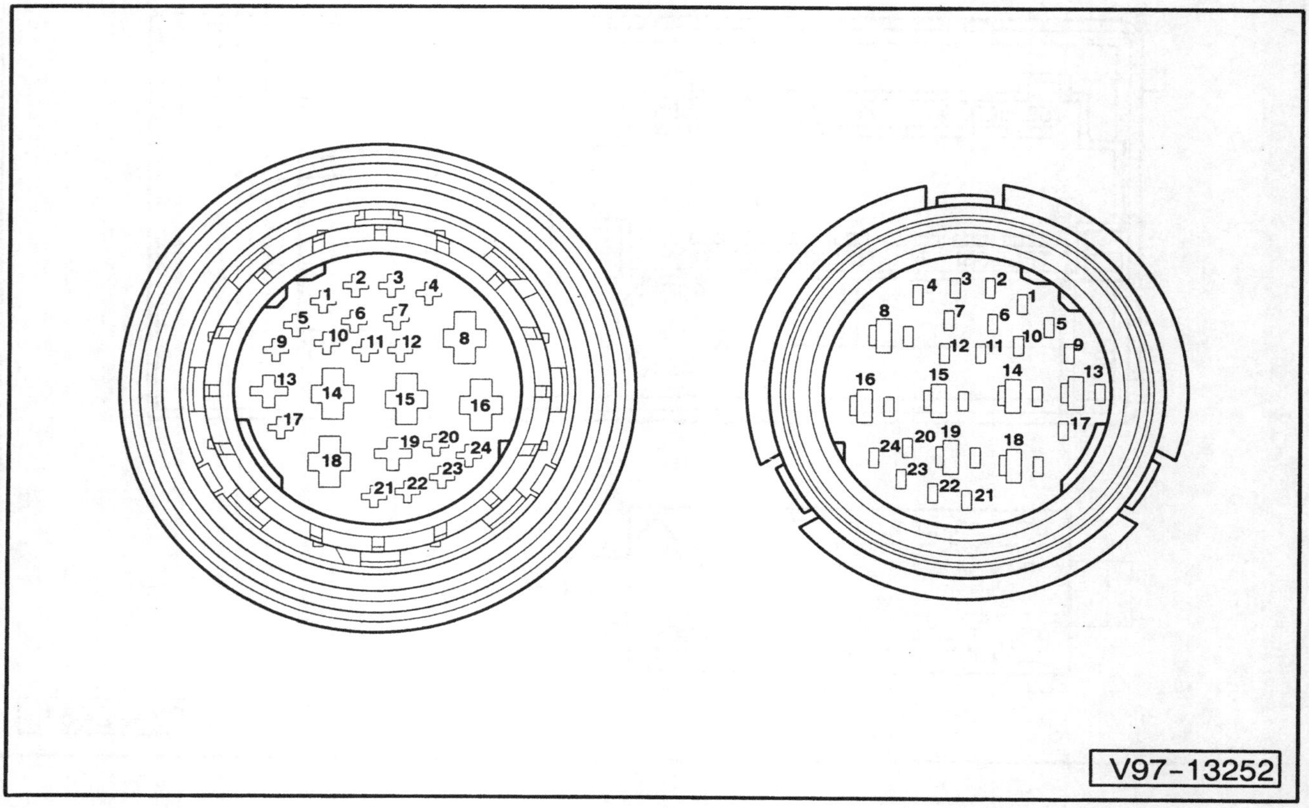

On Board Diagnostic

Data Link Connector (DLC) 16-Pin Connector

⇒ Repair Manual, *Electrical Equipment,* Repair Group 01

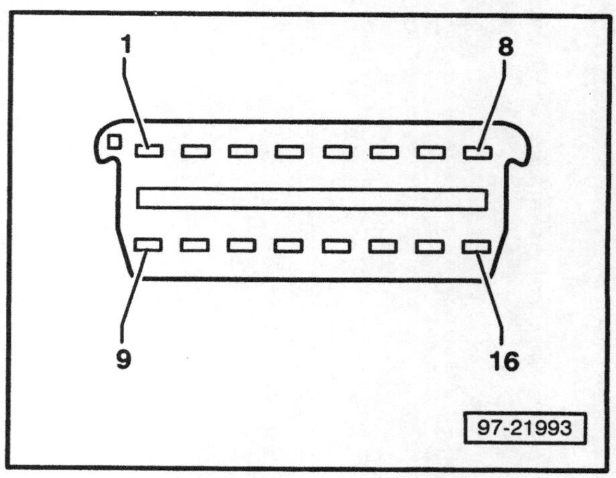

Component locations:

Vehicles ➤ 08.98 - in instrument panel, center

Vehicles 08.98 ➤ - under steering coulm

Relay panel with fuse holder and auxiliary relay panel

from September 1998

Index

Fuse arrangement

Note:

Fuses for optional equipment not listed are indicated in the respective wiring diagram.

Fuse colors

30 A	-	green	15 A	-	blue	5 A	-	beige
25 A	-	white	10 A	-	red	3 A	-	violet
20 A	-	yellow	7,5 A	-	brown			

Fuses in fuse holder on relay panel

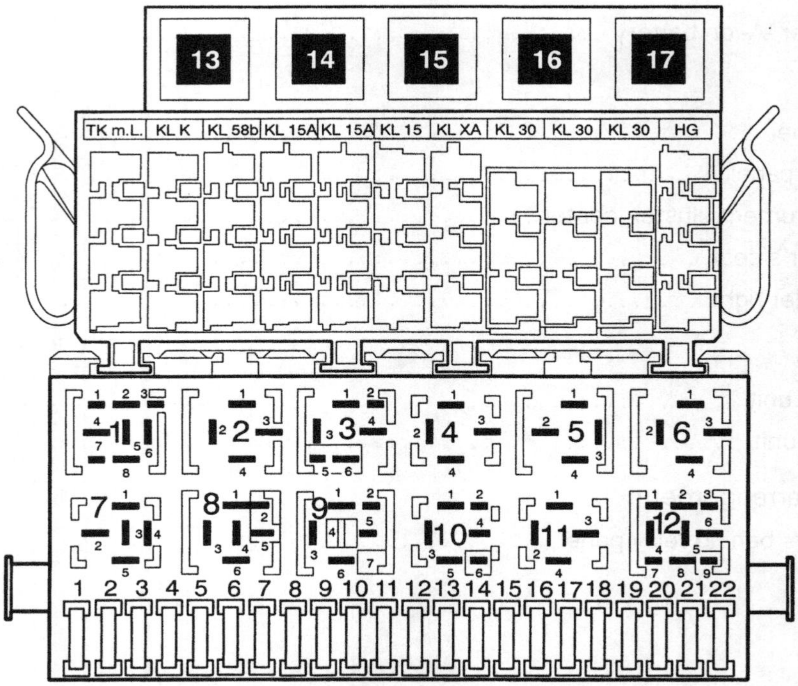

97-23226

Pos.		Function/components	Terminal	Note	Nominal value
1	-	Left headlight	56b		10 A
2	-	Right headlight	56b		10 A
3	-	License plate light	58		10 A
4	-	Rear window wiper, load reduction relay II, optional equipment	X		15 A
5	-	Windshield wiper/washer system, heated washer nozzles	X		15 A
6	-	Fresh air blower, airconditioning	X		30 A

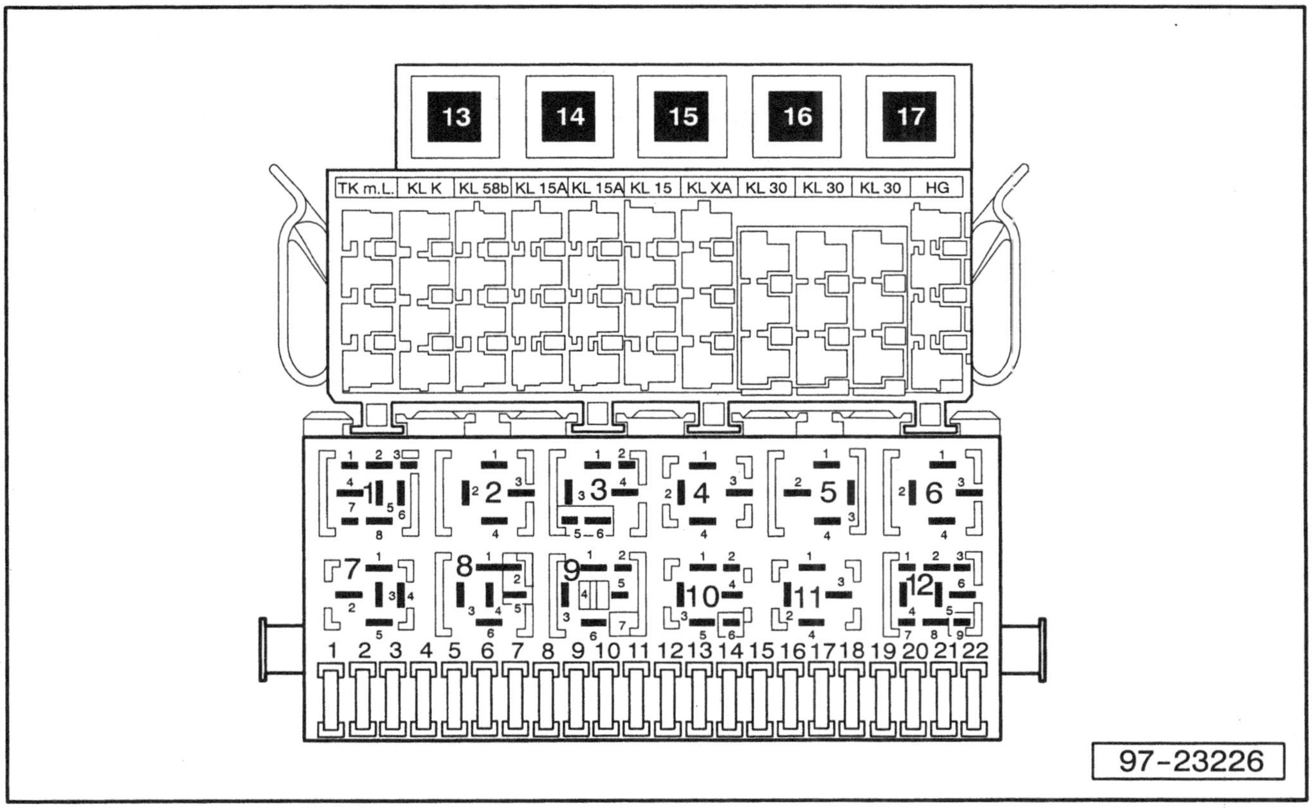

Pos.		Function/components	Terminal	Note	Nominal value
7	-	Right parking and tail light	58R		10 A
8	-	Left parking and tail light	58L		10 A
9	-	Open	X		
10	-	Fog Lights	X		15 A
11	-	Left headlight, high beam	56a		10 A
12	-	Right headlight, high beam	56a		10 A
13	-	Dual horns	15		10 A
14	-	Back-up light, automatic transmission, optional equipment, cruise control	15		10 A
15	-	Positive crankcase ventilation (PCV) heating element	15		10 A

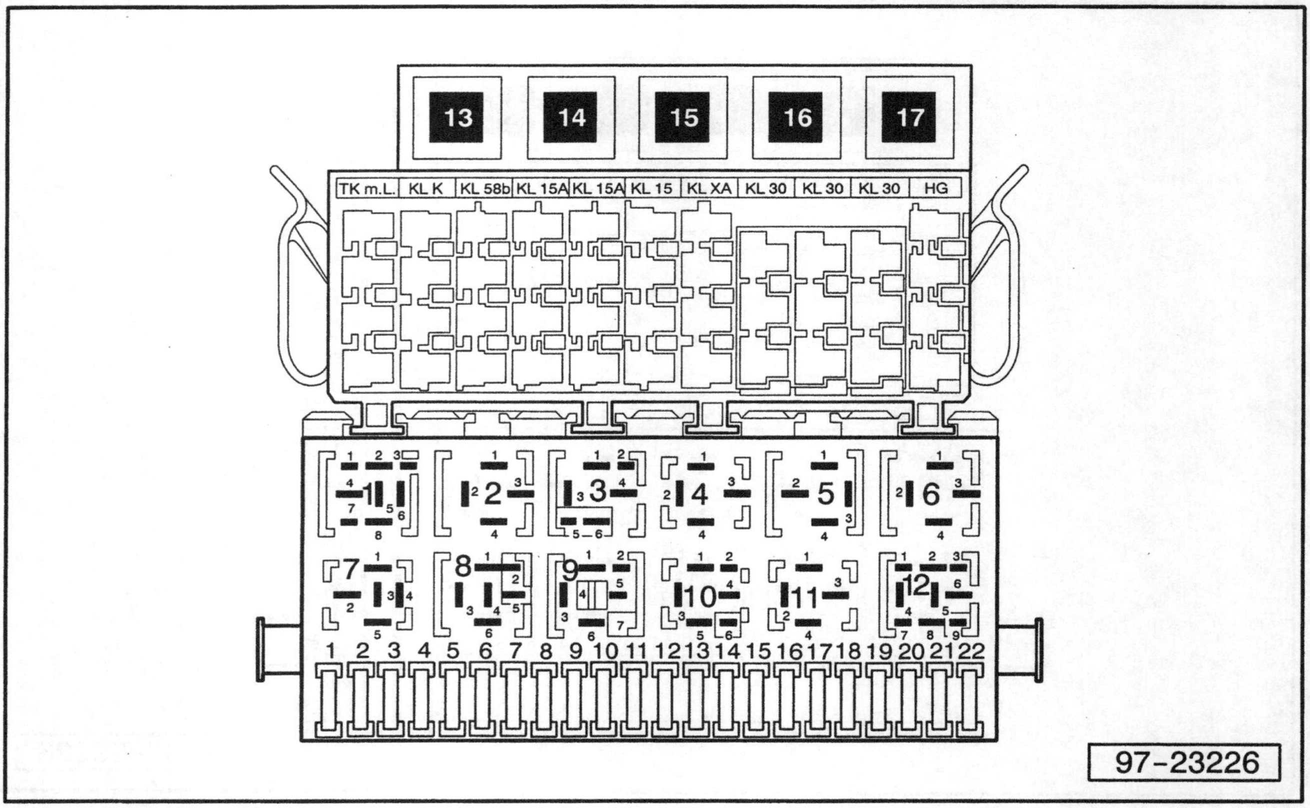

Pos.		Function/components	Terminal	Note	Nominal value
16	-	Turn signal light system, glove compartment light	15		15 A
17	-	Open	15		
18	-	Fuel pump (FP), engine electric			20 A
19	-	Coolant pump after-run, coolant fan control	30		10 A
20	-	Brake light	30		10 A
21	-	Instrument cluster, interior lights, radio, B+ (battery positiv voltage) for on board diagnostic (OBD)	30		15 A
22	-	Cigarette lighter	30		10 A

Fuses in fuse holder -B- below relay panel

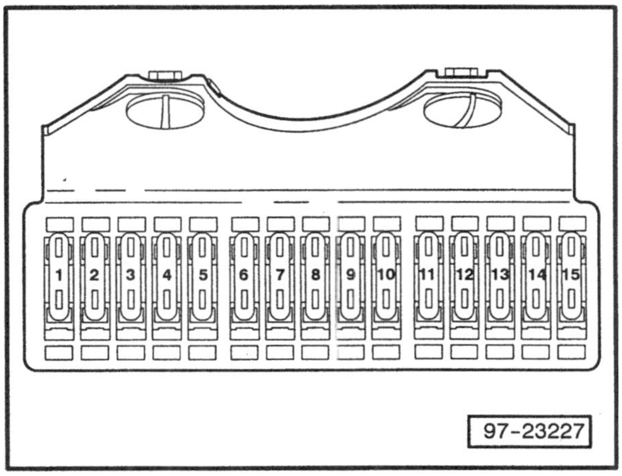

Pos.		Function/components	Terminal	Note	Nominal value
1	-	12V outlets		without 2nd battery	15 A
2	-	Swivel seat lock			10 A
3	-	Power sunroof			20 A
4	-	Open			
5	-	ABS		Control Module	10 A
6	-	Open			
7	-	Telephone			3 A
8	-	Open			
9	-	Open			
10	-	Central locking system			15 A
11	-	Heated seats			15 A
12	-	Rear window defogger and power mirrors			30 A
13	-	Climatronic		2nd evaporator	30 A
14	-	Climatronic		Control Module, Relay for Climatronic	30 A
	-	Air-Conditioning		Fresh air blower rear	
	-	Fresh air blower rear with auxiliary heater core		Fresh air blower rear	
15	-	Climatronic		2nd heater core	30 A
	-	Air-Conditioning		2nd heater core	
	-	Fresh air blower rear with auxiliary heater core		2nd heater core	20 A

Fuses in auxiliary fuse holder above relay panel

		Function/components	Terminal	Note	Nominal value
S43	-	Power window circuit breaker			30 A
S102	-	Engine control module (ECM) fuse	15	2.8L-Engine - code AES only	15 A
			15	2.8L-Engine - code AXK only	20 A
S117	-	Fuse for control elements			15 A

Fuses in auxiliary fuse holder below driver's seat

		Function/components	Terminal	Note	Nominal value
S23	-	Heater fuse		Auxiliary Heater B3L	5 A
S66	-	Auxiliary heater glow plug fuse		Auxiliary Heater B3L	25 A

Fuses in fuse holder -A- on battery

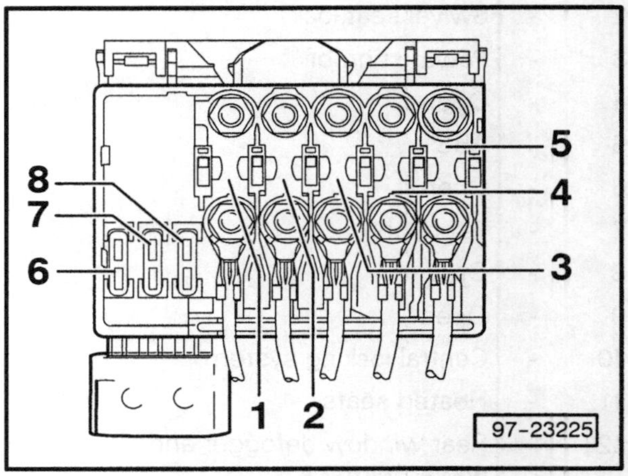

Pos.		Function/components	Terminal	Note	Nominal value
1	-	Secondary air injection			50 A
2	-	Coolant fan			50 A
3	-	Coolant fan			50 A
4	-	ABS		ABS return flow pump, ABS valves	50 A
5	-	Generator (GEN) (90A)			110 A
	-	Generator (GEN) (120A)			150 A
	-	Generator (GEN) (150A)			175 A
6	-	ABS Control Module with EDL/ASC/ESP		Vehicles with ESP only	30 A
7	-	Inlet and outlet valves for ABS		Vehicles with ESP only	30 A
8	-	Steering Angle Sensor		Vehicles with ESP only	5 A

Relay arrangement

Relays on relay panel

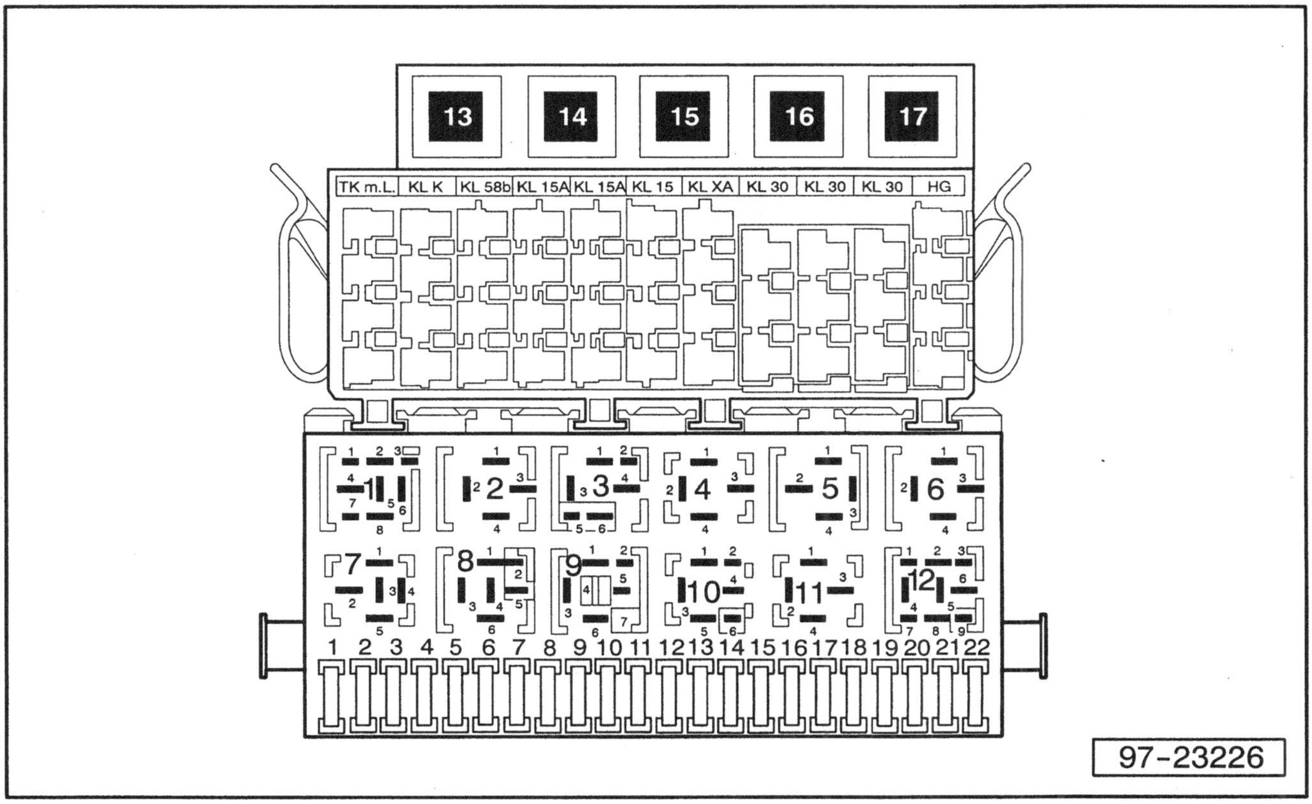

Components/function			Production control number	Note
1	-	A/C Relay	140	
	-	Fresh Air Blower Relay	105	
2	-	Rear Window Wiper/Washer Relay	174	
3	-	Open		
4	-	Load Reduction Relay	18	
5	-	Wire distributor for Data-Bus		
6	-	Open		
7	-	Headlight Washer System Relay	95	
8	-	Wiper/Washer Intermittent Relay (programable)	99	
9	-	Open		
10	-	Busbar		
	-	Fog Light Relay	53	
11	-	Busbar for Horn		
	-	Dual Horn Relay	53	
12	-	Fuel Pump (FP) Relay	167	

Relays above relay panel

Note:

Relay and fuse locations may vary on vehicles without power windows or air-conditioning.

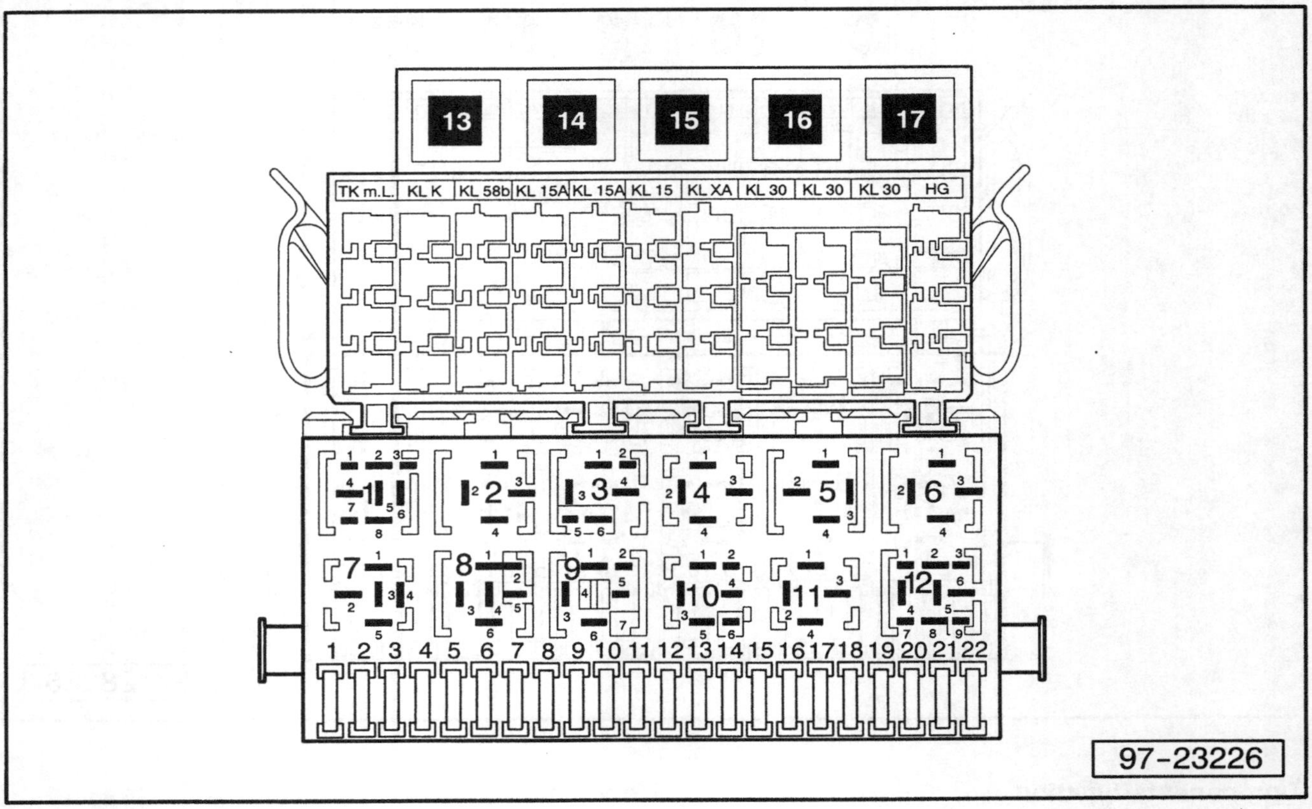

Components/function	Production control number	Note
13 - Coolant Circulation Pump Relay	53	
14 - Engine control module (ECM) fuse		
- Power window circuit breaker		
15 - Secondary Air Injection (AIR) Pump Relay	53	
- Fuse for control elements		
16 - Relay for A/C shutoff	147	
17 - Load Reduction Relay II	53	

Relays behind instrument cluster, center

Components/function	Production control number	Note
- Daytime Running Lights Change-over Relay	94	
- Heater Air Blower Relay	53	2nd heater core
- Fan Relay	114	Climatronic
- Fuel system Diagnostic control module	194	
- Heater Control Module	152	2nd heater core
- Relay for Fresh-air and recirculating air flap - rear	38 or 94	
- Park/Neutral Position (PNP) Relay	175	Automatic transmission

Relays below driver's seat

Components/function	Production control number	Note
- Battery Cut Out Relay	214	2nd battery (through August 2000)
- Battery Cut Out Relay	426	2nd battery (from December 2000)

Relays near air outlet right

Components/function	Production control number	Note
- ABS Return Flow Pump Relay	214	Vehicles with ESP only
- ABS Solenoid Valve Relay	214	Vehicles with ESP only

Wire connectors

Front of connector unit

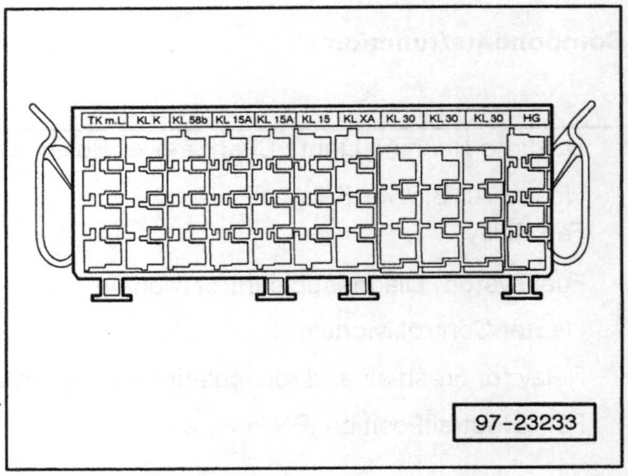

Components/function	Terminal	Color code of connectors
- Door Contact Wire Connector (with load)	TK m.L.	brown
- Data Link Connector (DLC) Wire Connector for on-board diagnostic (OBD) system	KL K	white
- Terminal 58b Wire Connector	KL 58b	grey
- Terminal 15a Wire Connector	KL 15A	green
- Terminal 15 Wire Connector	KL 15	black
- Terminal Xa Wire Connector	KL XA	violet
- Terminal 30 Wire Connector	KL 30	red
- Vehicle Speed Signal Wire Connector	HG	blue

EuroVan - Standard Equipment

from December 2000

Deviate relay location and fuseplacements as well as the locations of multiple connectors see section "component locations".

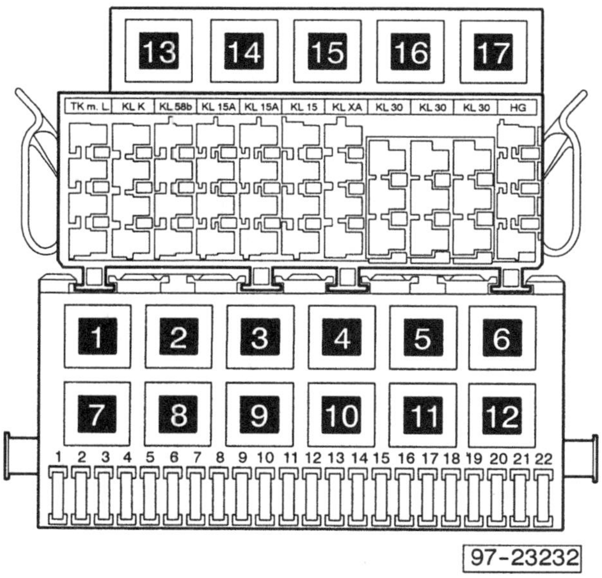

- **2** Rear Window Wiper/Washer Relay (174)
- **4** Load Reduction Relay (18)
- **5** Wire distributor for Data-Bus
- **8** Wiper/Washer Intermittent Relay (programable) (99
- **11** Dual Horn Relay (53)
- **17** Load Reduction Relay II (53)

Note: Number in parentheses indicates production control number stamped on relay housing.

Fuse colors

30 A - green
25 A - white
20 A - yellow
15 A - blue
10 A - red
7,5 A - brown
5 A - beige
3 A - violet

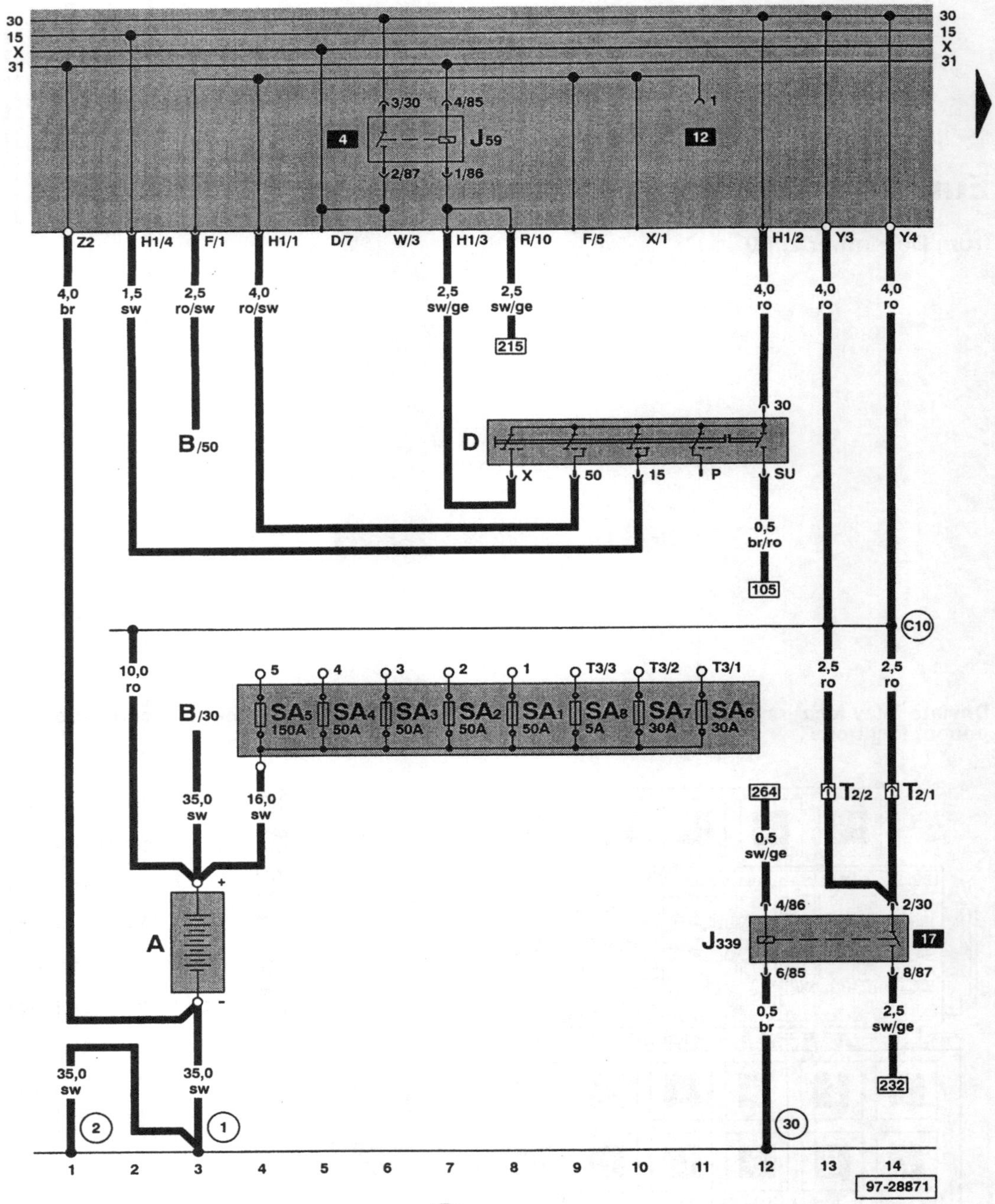

ws = white
sw = black
ro = red
br = brown
gn = green
bl = blue
gr = grey
li = lilac
ge = yellow
or = orange

A - Battery
B - Starter
D - Ignition/Starter Switch
J59 - Load Reduction Relay
J339 - Load Reduction Relay II, above fuse/relay panel
SA1 - Fuse -1- in fuse bracket/battery
SA2 - Fuse -2- in fuse bracket/battery
SA3 - Fuse -3- in fuse bracket/battery
SA4 - Fuse -4- in fuse bracket/battery
SA5 - Fuse -5- in fuse bracket/battery
SA6 - Fuse -6- in fuse bracket/battery
SA7 - Fuse -7- in fuse bracket/battery
SA8 - Fuse -8- in fuse bracket/battery
T2 - Double Connector, behind fuse/relay panel
T3 - 3-Pin Connector

(1) - Ground strap, battery to body
(2) - Ground strap, transmission to body
(30) - Ground connection, -1-, beside fuse/relay panel
(C10) - Plus connection -1- (30), in headlight wiring harness

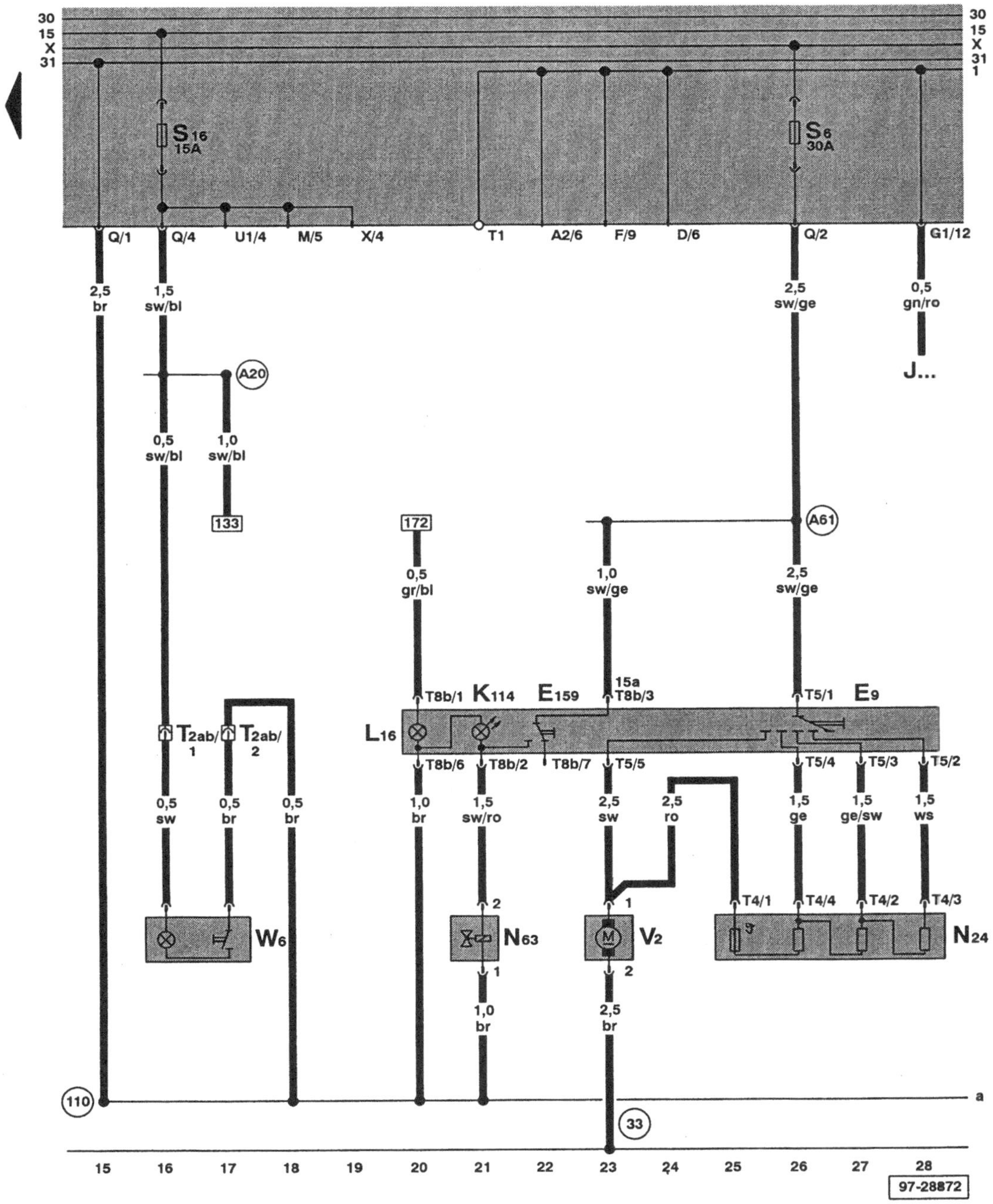

ws = white
sw = black
ro = red
br = brown
gn = green
bl = blue
gr = grey
li = lilac
ge = yellow
or = orange

- E9 - Fresh Air Blower Switch
- E159 - Fresh Air/Recirculating Flap Switch
- J... - Engine Control Module (ECM)
- K114 - Fresh Air and Recirculating Air Mode Indicator Light
- L16 - Fresh Air Control Lever Light
- N24 - Fresh Air Blower Series Resistance with Fuse
- N63 - Fresh Air/Recirculating Flap Two-Way Valve
- T2ab - Double Connector, behind instrument panel, left
- T4 - 4-Pin Connector
- T5 - 5-Pin Connector
- T8b - 8-Pin Connector
- V2 - Fresh Air Blower
- W6 - Glove Compartment Light

- (33) - Ground connection, behind instrument panel right
- (110) - Ground connection -2-, in instrument panel wiring harness
- (A20) - Wire connection (15a), in instrument panel wiring harness
- (A61) - Plus connection (X), in Instrument cluster wiring harness

Fresh air blower switch, fresh air/recirculating flap switch, fresh air blower, fresh air/recirculating flap two-way valve, glove compartment light

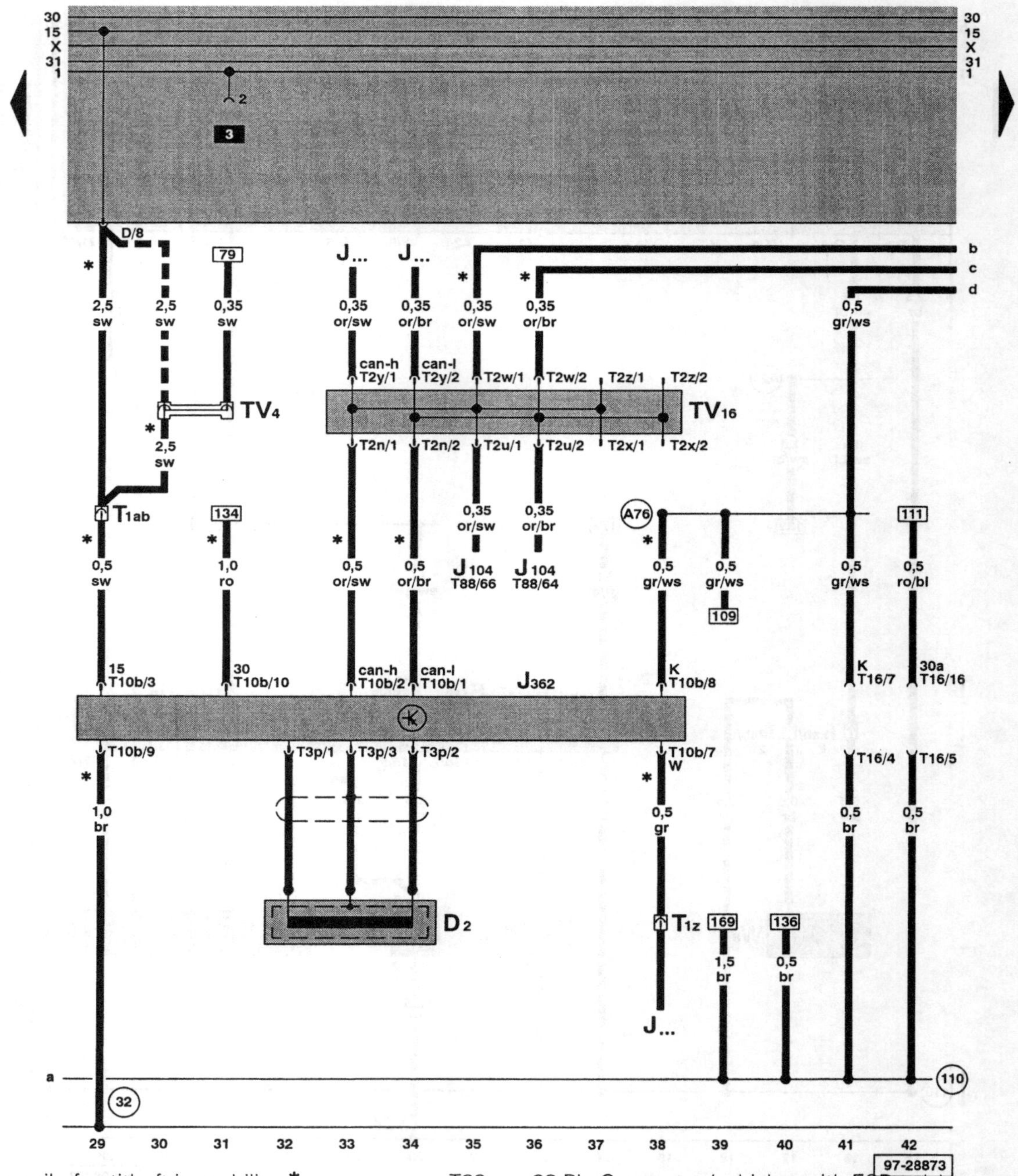

ws = white
sw = black
ro = red
br = brown
gn = green
bl = blue
gr = grey
li = lilac
ge = yellow
or = orange

D2 - Induction coil of antitheft immobilizer*
J... - Engine Control Module (ECM)
J104 - ABS Control Module (with EDL)
J362 - Control module for anti-theft immobilizer, behind instrument cluster, right*
T1ab - Single Connector, behind fuse/relay panel
T1z - Single Connector, behind fuse/relay panel
T2n - Double Connector
T2u - Double Connector
T2w - Double Connector
T2x - Double Connector
T2y - Double Connector
T2z - Double Connector
T3p - 3-Pin Connector
T10b - 10-Pin Connector
T16 - Data Link Connector (DLC), behind steering column cover
T88 - 88-Pin Connector (vehicles with ESP only)
TV4 - Terminal 15 Wire Connector, black
TV16 - Wire distributor for Data-Bus, on relay panel No. 5

(32) - Ground connection, behind instrument panel, left
(110) - Ground connection -2-, in instrument panel wiring harness
(A76) - Connector (K-diagnosis wire) in instrument panel wiring harness

* - Vehicles with antitheft immobilizer
--- - Vehicles without antitheft immobilizer

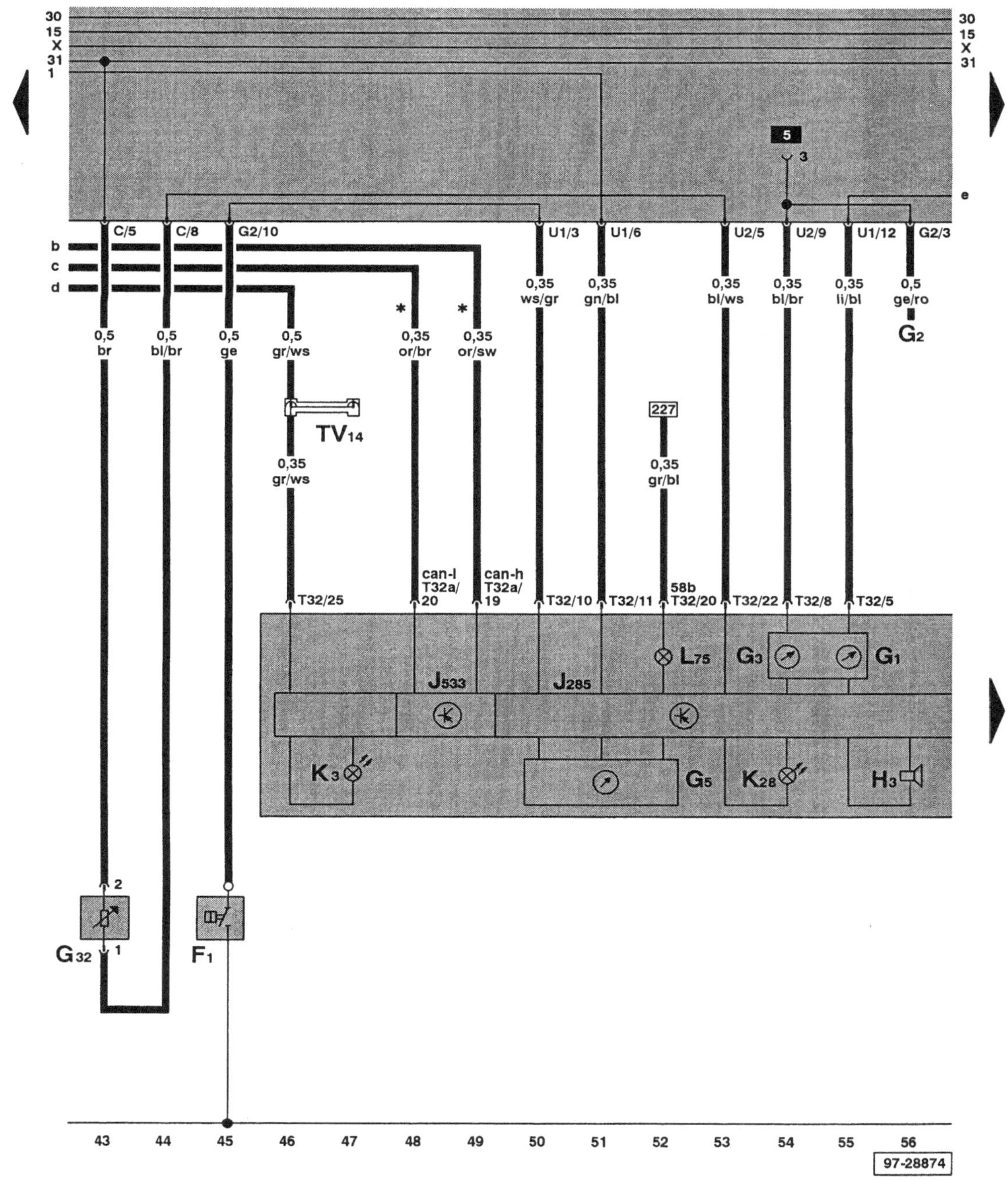

ws = white
sw = black
ro = red
br = brown
gn = green
bl = blue
gr = grey
li = lilac
ge = yellow
or = orange

F1 - Oil Pressure Switch
G1 - Fuel Gauge
G2 - Engine Coolant Temperature (ECT) Sensor
G3 - Engine Coolant Temperature (ECT) Gauge
G5 - Tachometer
G32 - Engine Coolant Level (ECL) Sensor
H3 - Warning Buzzer
J285 - Control module with indicator unit in instrument panel insert
J533 - Data Bus On Board Diagnostic Interface
K3 - Oil Pressure Warning Light
K28 - Engine Coolant Level/Temperature (ECL/ECT) Warning Light
L75 - Digital Display Light
T32 - 32-Pin Connector, blue
T32a - 32-Pin Connector, green
TV14 - Data Link Connector (DLC) wire Connector, white

* - Vehicles with antitheft immobilizer

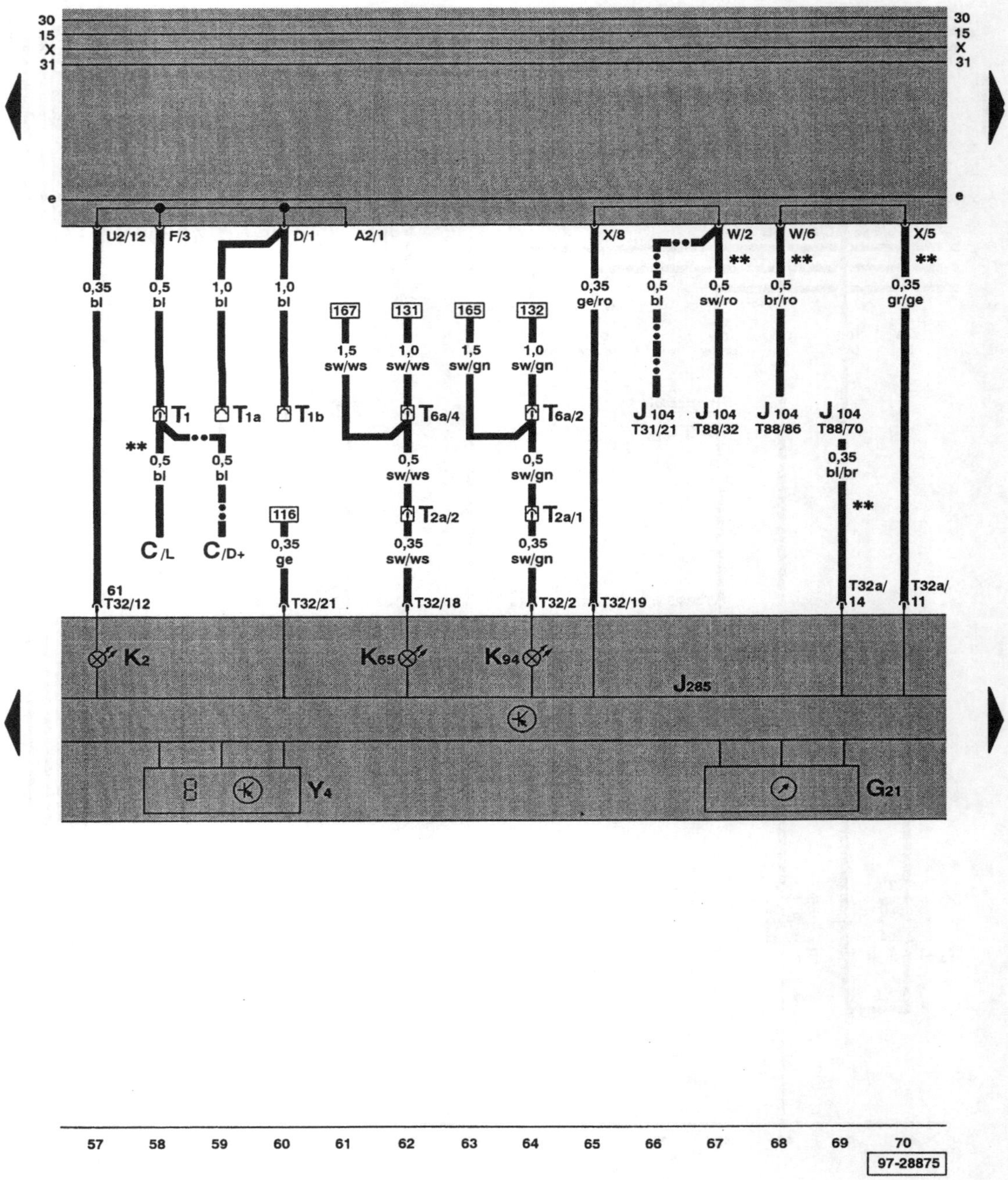

ws = white
sw = black
ro = red
br = brown
gn = green
bl = blue
gr = grey
li = lilac
ge = yellow
or = orange

C - Generator (GEN)
G21 - Speedometer
J104 - ABS Control Module (w/EDL), in engine compartment, left (through November 2000)
J104 - ABS Control Module (with EDL), below instrument panel, right**
J285 - Control module with indicator unit in instrument panel insert
K2 - Generator (GEN) Warning Light
K65 - Left Turn Signal Indicator Light
K94 - Right Turn Signal Indicator Light
T1 - Single Connector, near starter
T1a - Single Connector, behind fuse/relay panel
T1b - Single Connector, behind fuse/relay panel
T2a - Double Connector, behind fuse/relay panel
T6a - 6-Pin Connector, behind fuse/relay panel
T31 - 31-Pin Connector, on ABS Control Module (w/EDL) (through November 2000)
T32 - 32-Pin Connector, blue
T32a - 32-Pin Connector, green
T88 - 88-Pin Connector at -J104-
Y4 - Odometer Display

** - Beginning December 2000
-••- - Through November 2000

Instrument cluster, speedometer, turn signal indicator lights, odometer display, generator (GEN) warning light

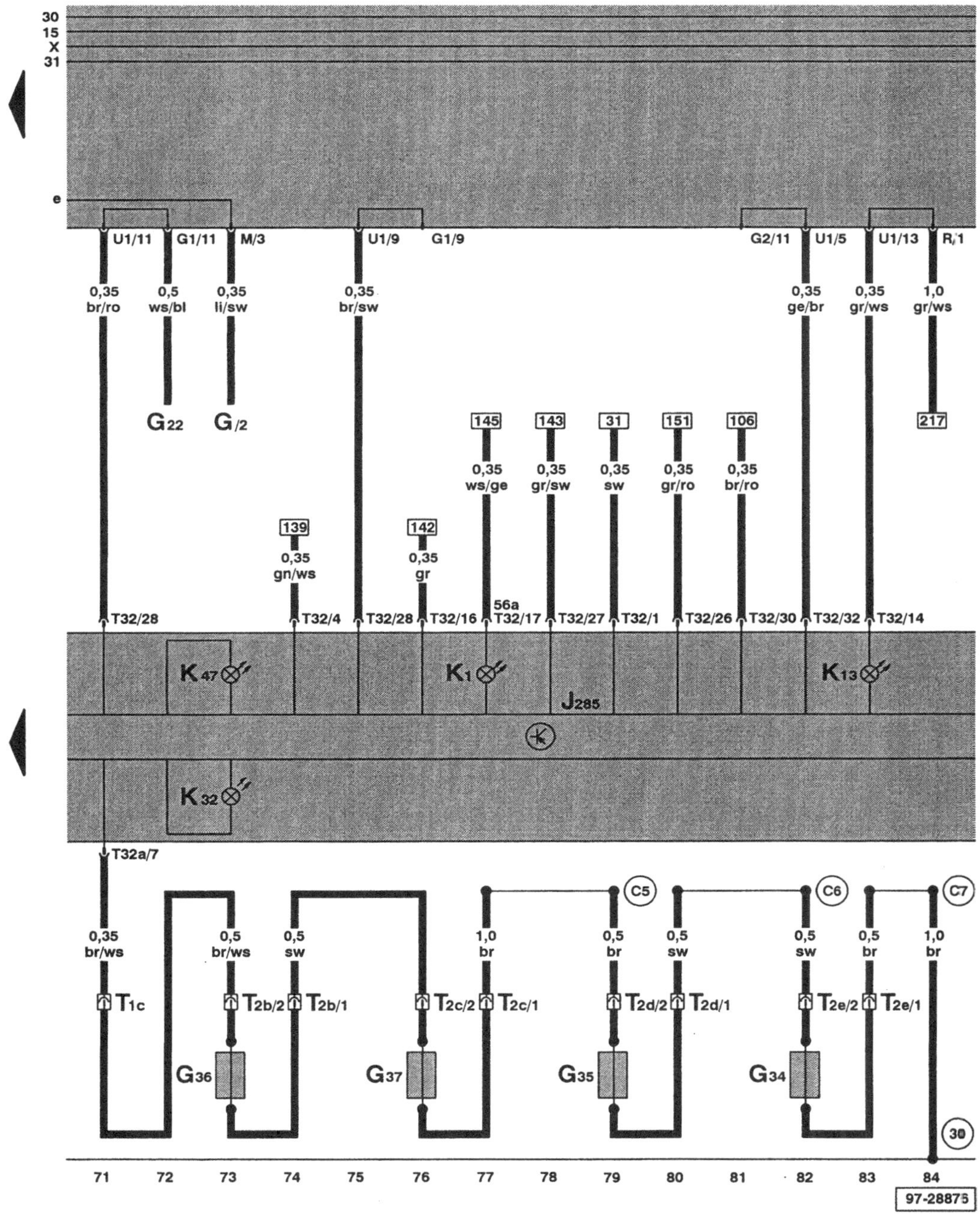

ws = white
sw = black
ro = red
br = brown
gn = green
bl = blue
gr = grey
li = lilac
ge = yellow
or = orange

G - Fuel Level Sensor
G22 - Speedometer Vehicle Speed Sensor (VSS)
G34 - Left Front Brake Pad Wear Sensor
G35 - Right Front Brake Pad Wear Sensor
G36 - Left Rear Brake Pad Wear Sensor
G37 - Right Rear Brake Pad Wear Sensor
J285 - Control module with indicator unit in instrument panel insert
K1 - Headlight High Beam Indicator Light
K13 - Rear Fog Light Indicator Light
K32 - Brake Pad Wear Indicator Light
K47 - ABS Warning Light
T1c - Single Connector, behind fuse/relay panel
T2b - Double Connector, near left rear brake pad wear sensor
T2c - Double Connector, near right rear brake pad wear sensor
T2d - Double Connector, near right front brake pad wear sensor
T2e - Double Connector, near left front brake pad wear sensor
T32 - 32-Pin Connector, blue
T32a - 32-Pin Connector, green
(30) - Ground connection, -1-, beside fuse/relay panel
(C5) - Wire connection -1- (brake pad wear monitor), in left front wiring harness
(C6) - Wire connection -2- (brake pad wear monitor), in left front wiring harness
(C7) - Wire connection -3- (brake pad wear monitor), in left front wiring harness

Instrument cluster, brake pad wear sensors, brake pad wear indicator light, headlight high beam indicator light

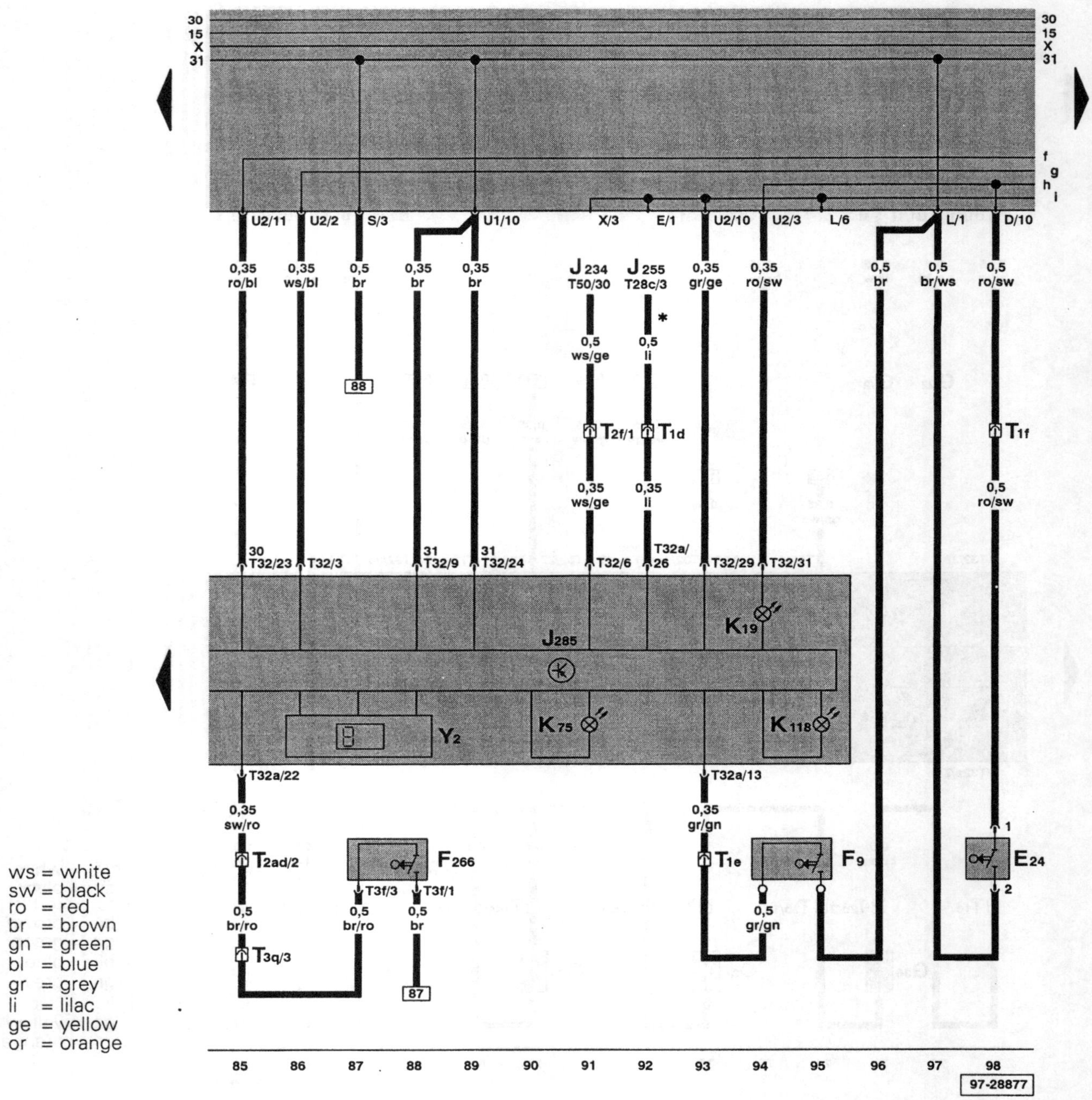

E24 - Left Seat Belt Switch
F9 - Parking Brake Warning Light Switch
F266 - Front Hood Switch
J234 - Airbag Control Module
J255 - Climatronic Control Module
J285 - Control module with indicator unit in instrument panel insert
K19 - Seat Belt Warning Light
K75 - Airbag Malfunction Indicator Lamp (MIL)
K118 - Warning light for brake system
T1d - Single Connector, behind fuse/relay panel
T1e - Single Connector, behind fuse/relay panel
T1f - Single Connector, behind fuse/relay panel
T2ad - Double Connector, behind fuse/relay panel
T2f - Double Connector, behind fuse/relay panel
T3f - 3-Pin Connector
T3q - 3-Pin Connector, behind fuse/relay panel
T28c - 28-Pin Connector, on climatronic control module
T32 - 32-Pin Connector, blue
T32a - 32-Pin Connector, green
T50 - 50-Pin Connector, on airbag control module
Y2 - Digital Clock

* - Vehicles with Climatronic only

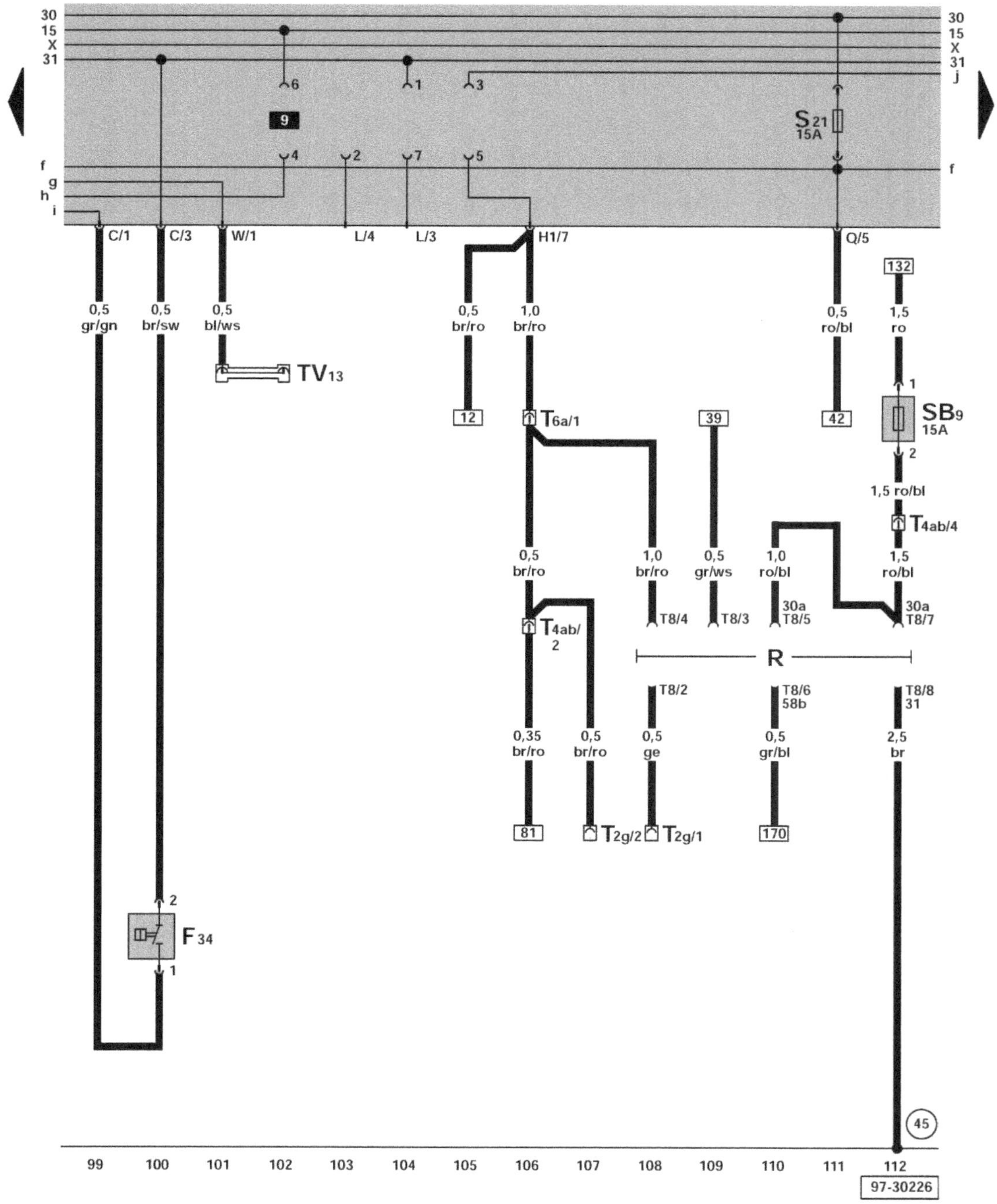

ws = white
sw = black
ro = red
br = brown
gn = green
bl = blue
gr = grey
li = lilac
ge = yellow
or = orange

F34 - Brake Fluid Level Warning Switch
R - Radio
SB9 - Fuse 9 in fuse holder, below relay panel
T2g - Double Connector
T4ab - 4-PinConnector, behind fuse/relay panel
T6a - 6-Pin Connector, behind fuse/relay panel
T8 - 8-Pin Connector
TV13 - Vehicle Speed Signal Wire Connector, blue

(45) - Ground connection, behind instrument panel, center

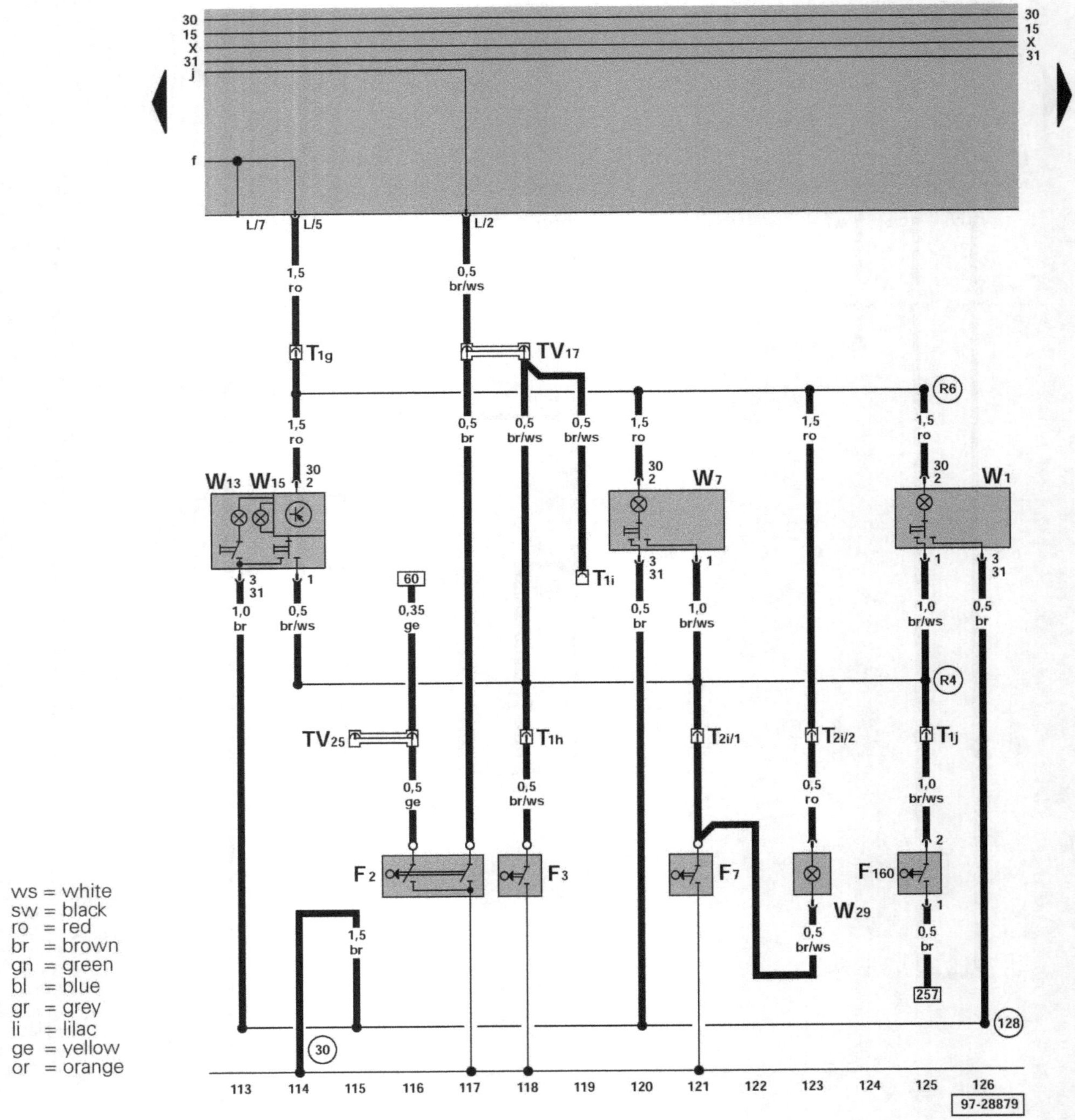

F2 - Door contact Switch - Driver's side

F3 - Door contact switch-Passenger side

F7 - Right Rear Side/Sliding Door Contact Switch

F160 - Rear Lid Contact Switch

T1g - Single Connector, behind fuse/relay panel

T1h - Single Connector, behind instrument panel, right

T1i - Single Connector, behind fuse/relay panel

T1j - Single Connector, near taillamp assembly

T2i - Double Connector, near interior light, center

TV17 - Door Contact Switch Wire Connector, brown

TV25 - Door Contact Switch Wire Connector, pink

W1 - Interior light, front

W7 - Center Interior Light

W13 - Right Front Map/Reading Light

W15 - Interior Light with Delay Switch

W29 - Running board light

(30) - Ground connection, -1-, beside fuse/relay panel

(128) - Ground connection -1-, in interior light wiring harness

(R4) - Wire connection -1-, in interior light/door contact switch wiring harness

(R6) - Plus connection -1-, in interior light wiring harness

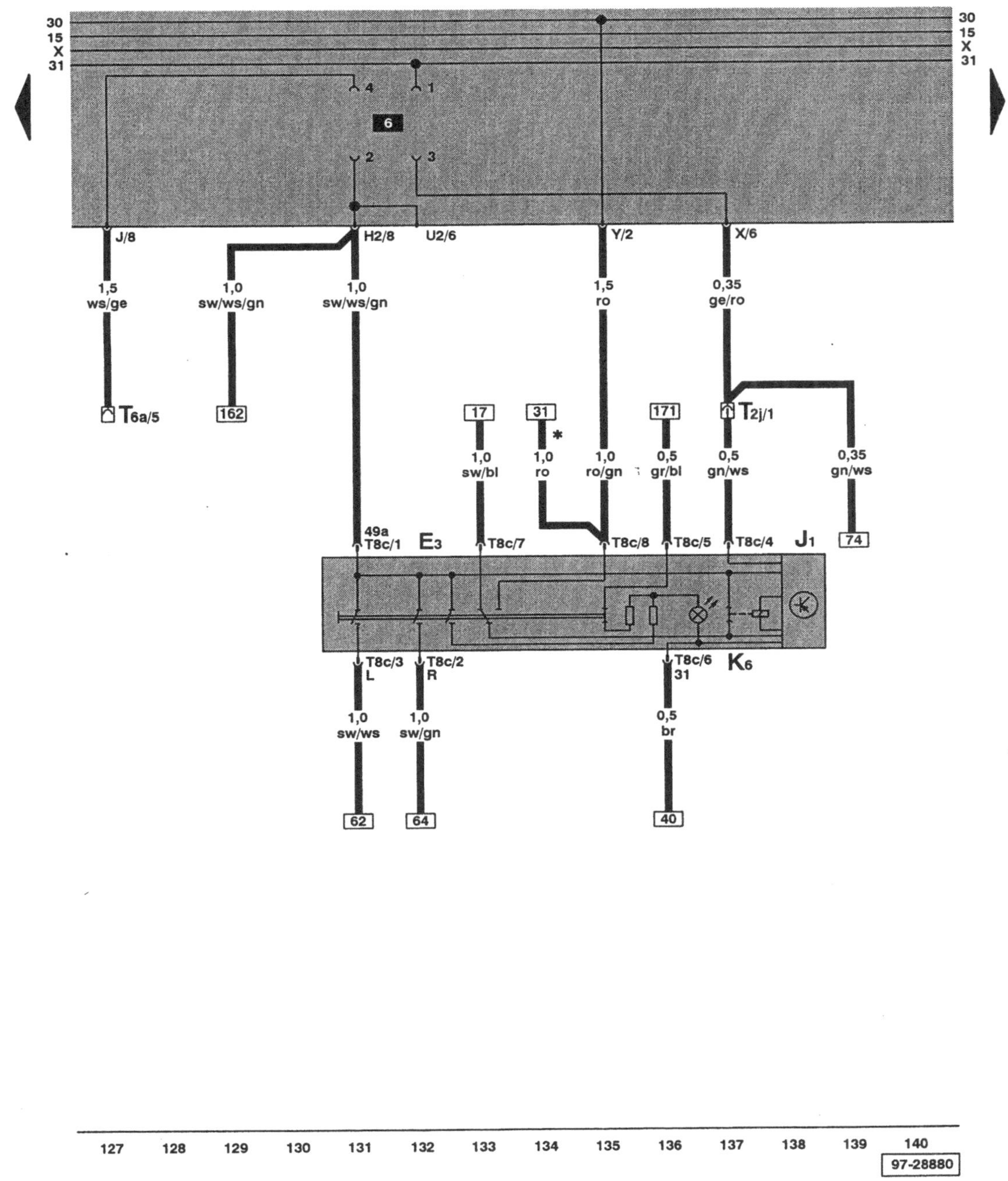

ws = white
sw = black
ro = red
br = brown
gn = green
bl = blue
gr = grey
li = lilac
ge = yellow
or = orange

E3 - Emergency Flasher Switch
J1 - Turn Signal Relay
K6 - Indicator light for emergency flasher system
T2j - Double Connector, behind fuse/relay panel
T6a - 6-Pin Connector, behind fuse/relay panel
T8c - 8-Pin Connector

* - Vehicles with antitheft immobilizer

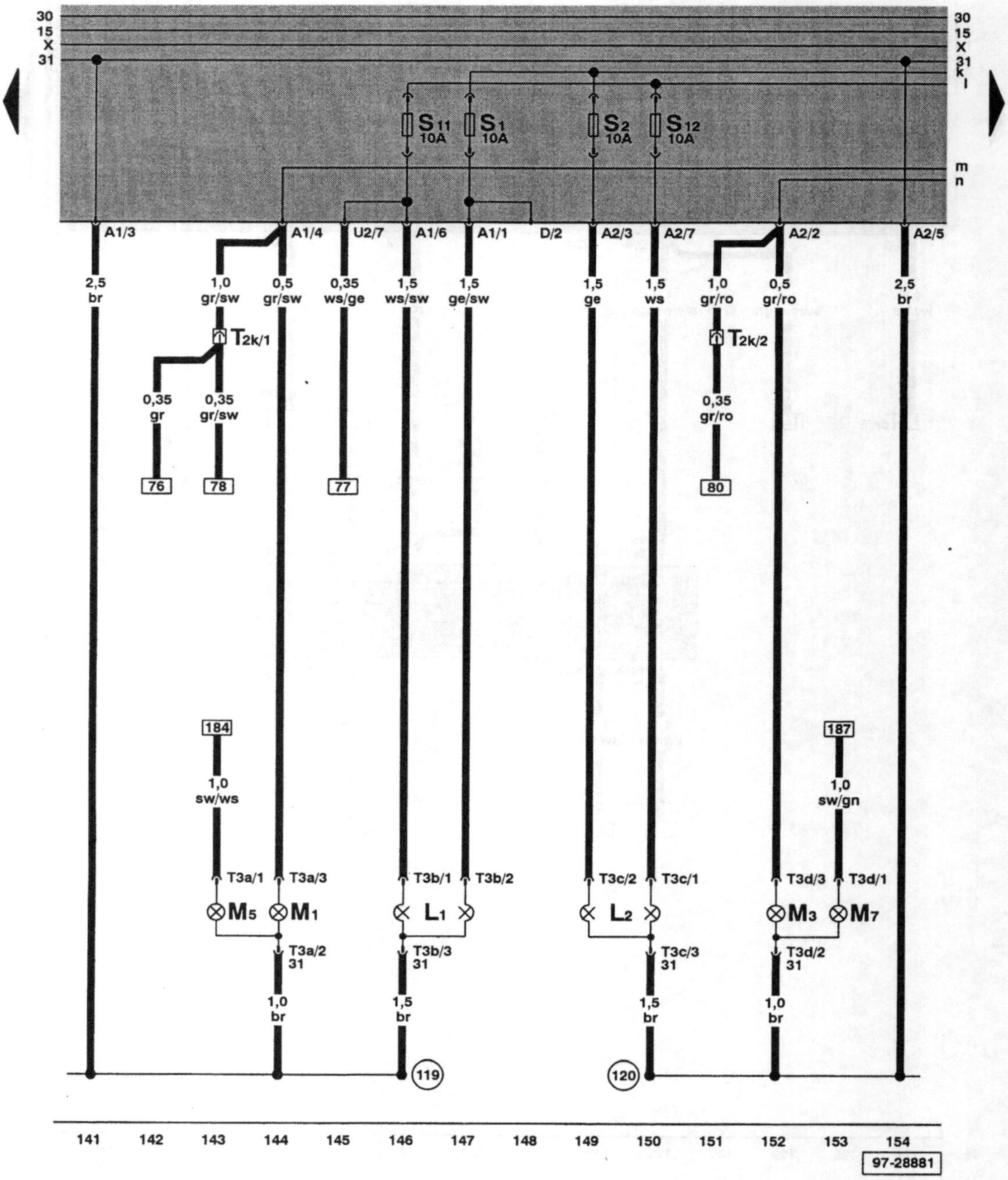

ws = white
sw = black
ro = red
br = brown
gn = green
bl = blue
gr = grey
li = lilac
ge = yellow
or = orange

L1 - Left Headlight
L2 - Right Headlight
M1 - Left Parking Light
M3 - Right Parking Light
M5 - Left Front Turn Signal Light
M7 - Right Front Turn Signal Light
T2k - Double Connector, behind fuse/relay panel
T3a - 3-Pin Connector
T3b - 3-Pin Connector
T3c - 3-Pin Connector
T3d - 3-Pin Connector

(119) - Ground connection -1-, in headlight wiring harness
(120) - Ground connection -2-, in headlight wiring harness

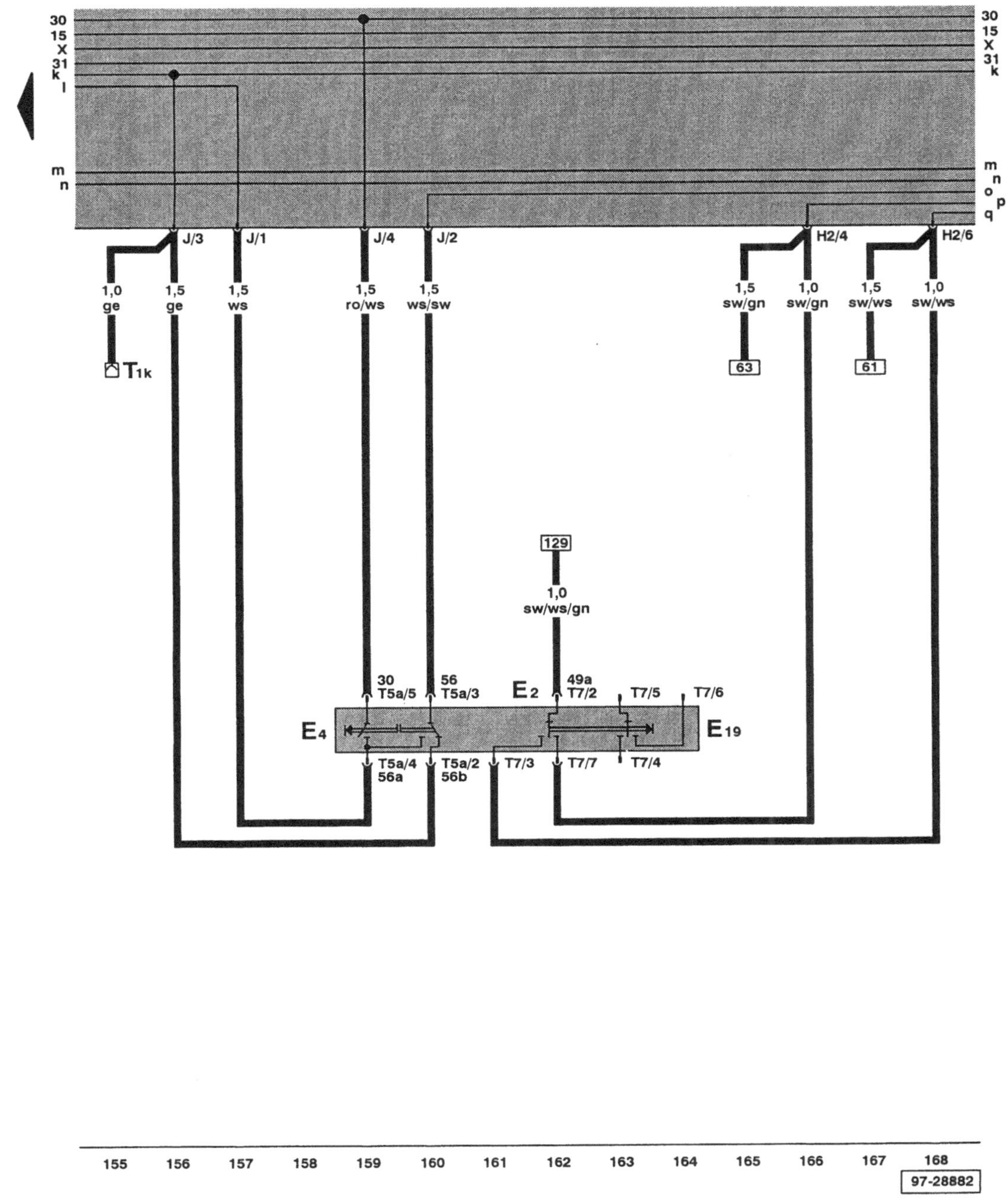

ws = white
sw = black
ro = red
br = brown
gn = green
bl = blue
gr = grey
li = lilac
ge = yellow
or = orange

E2 - Turn signal switch
E4 - Headlight Dimmer/Flasher Switch
E19 - Park Light Switch
T1k - Single Connector, behind fuse/relay panel
T5a - 5-Pin Connector, behind steering column switch cover
T7 - 7-Pin Connector, behind steering column switch cover

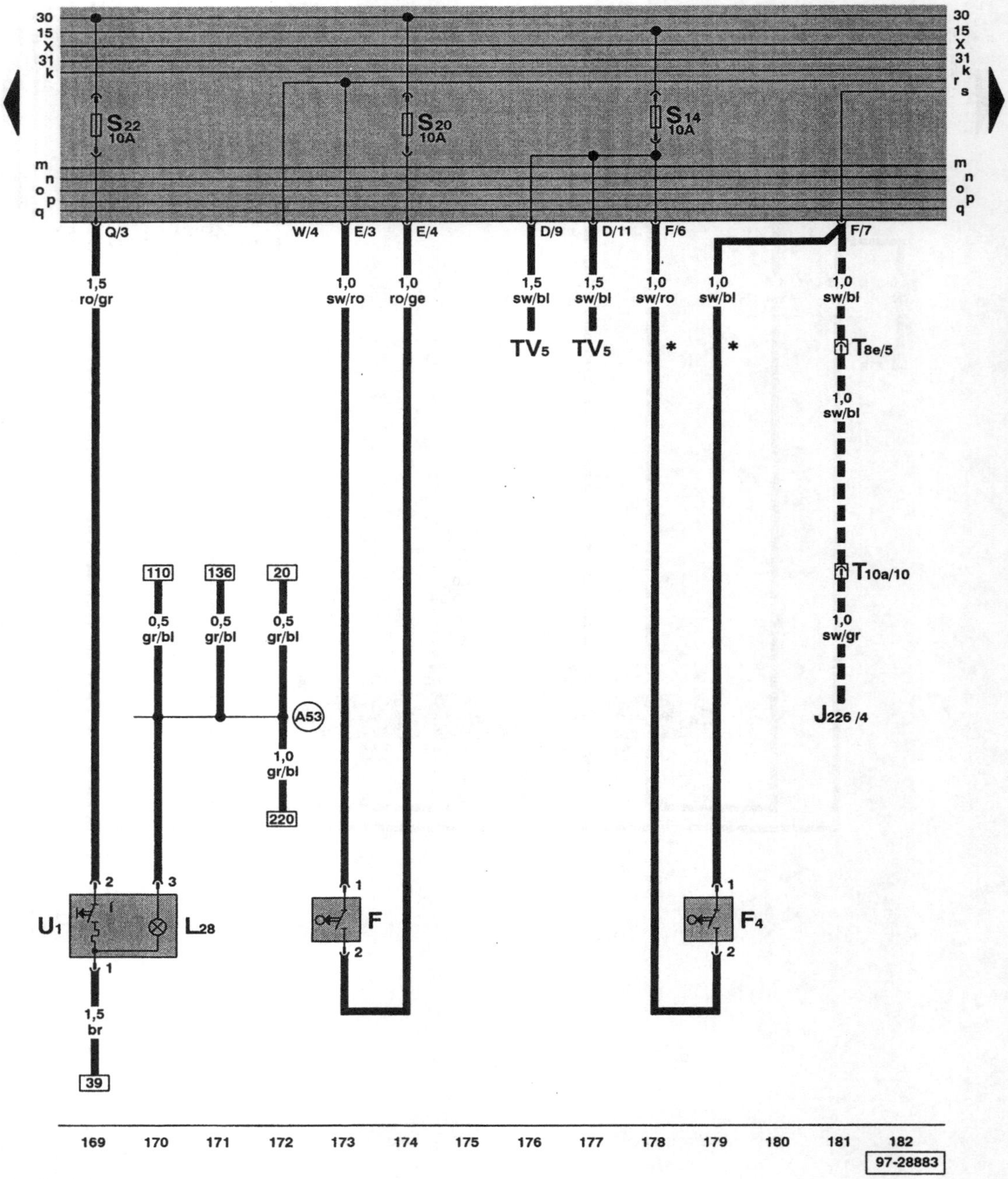

ws = white
sw = black
ro = red
br = brown
gn = green
bl = blue
gr = grey
li = lilac
ge = yellow
or = orange

F - Brake Light Switch
F4 - Back-Up Light Switch*
J226 - Park/Neutral Position (PNP) Relay
L28 - Cigarette Lighter Light
T8e - 8-Pin Connector, behind fuse/relay panel
T10a - 10-Pin Connector, behind fuse/relay panel
TV5 - Terminal 15a Wire Connector, green, 4-Pin and 8-Pin
U1 - Cigarette Lighter

(A53) - Plus connection (58b), in instrument panel wiring harness

* - Manual transmission only

--- - Automatic transmission only

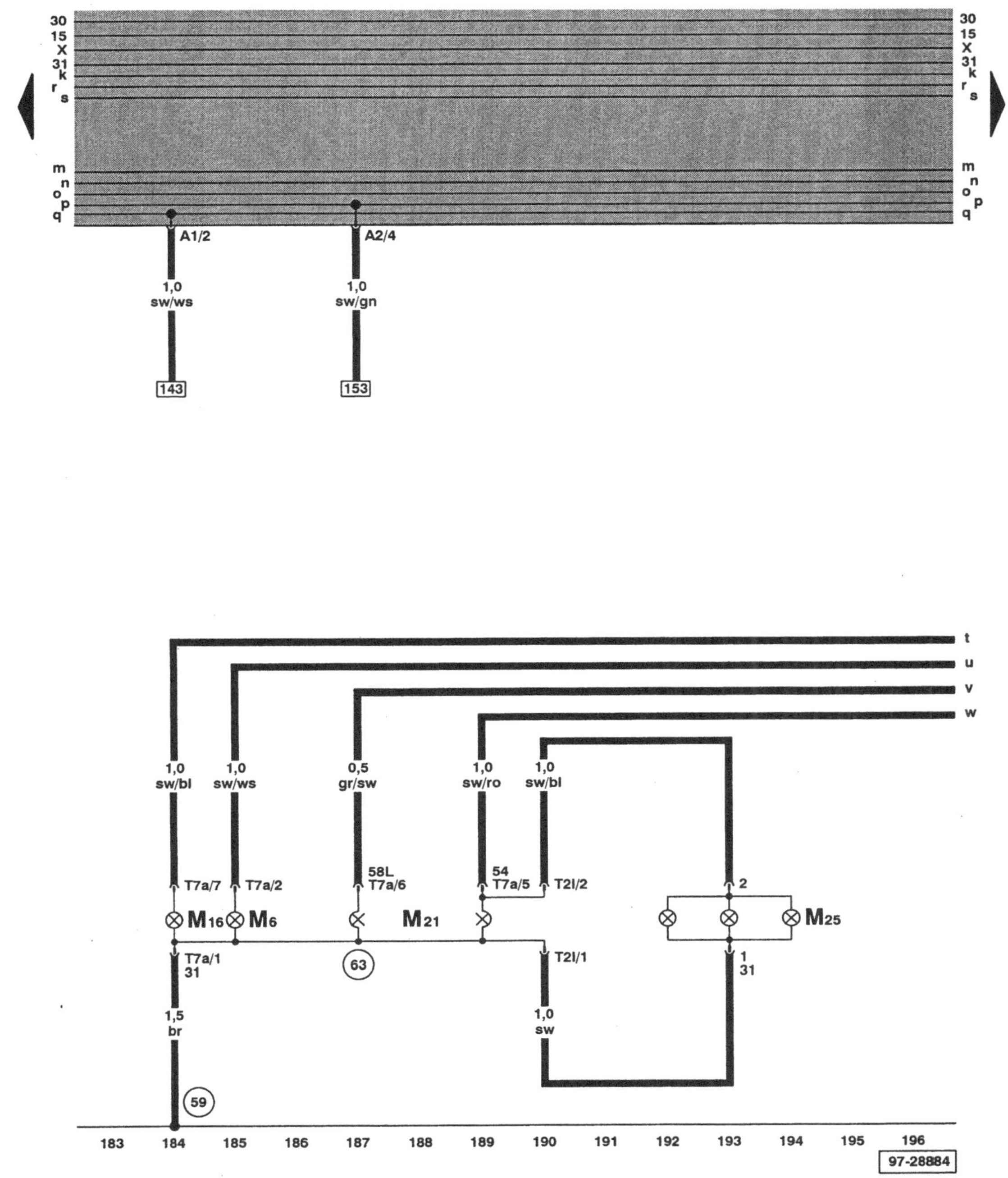

ws = white
sw = black
ro = red
br = brown
gn = green
bl = blue
gr = grey
li = lilac
ge = yellow
or = orange

M6 - Left Rear Turn Signal Light
M16 - Left Back-Up Light
M21 - Left Brake/Tail Light
M25 - High-mount Brake Light
T2l - Double Connector
T7a - 7-Pin Connector

(59) - Ground connection, near taillight, left

(63) - Ground connection, on left taillight bulb holder

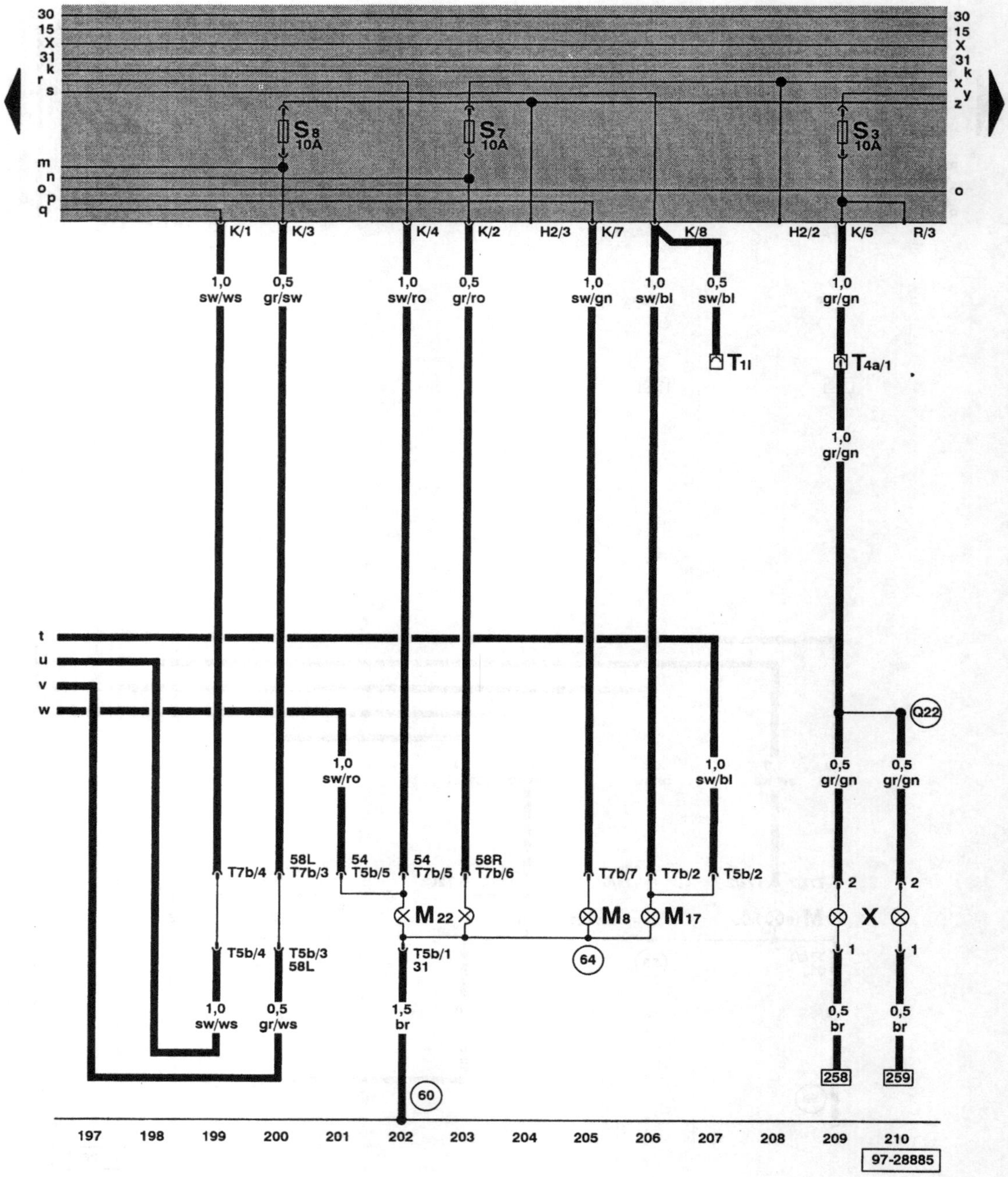

ws = white
sw = black
ro = red
br = brown
gn = green
bl = blue
gr = grey
li = lilac
ge = yellow
or = orange

M8 - Right Rear Turn Signal Light
M17 - Right Back-Up Light
M22 - Right Brake/Tail Light
T1l - Single Connector, behind fuse/relay panel
T4a - 4-Pin Connector, near taillamp assembly
T5b - 5-Pin Connector
T7b - 7-Pin Connector
X - License Plate Light

(60) - Ground point -2- near taillamp assembly, right

(64) - Ground connection, on right taillight bulb holder

(Q22) - Wire connection -1-, in rear lid wiring harness

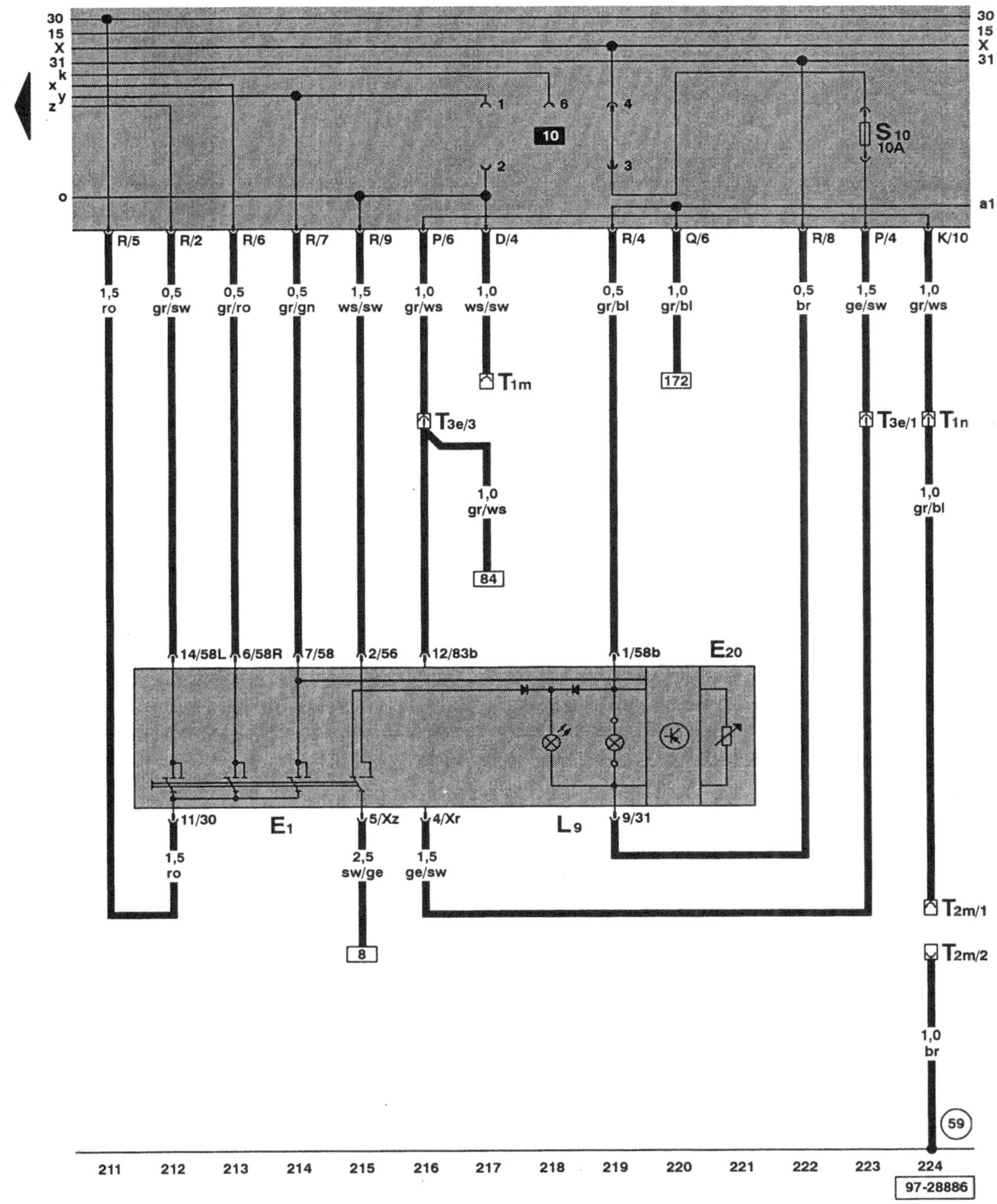

ws = white
sw = black
ro = red
br = brown
gn = green
bl = blue
gr = grey
li = lilac
ge = yellow
or = orange

E1 - Light switch
E20 - Instrument Panel Light Dimmer Switch
L9 - Headlight Switch Light
T1m - Single Connector, behind fuse/relay panel
T1n - Single Connector, near taillamp assembly
T2m - Double Connector
T3e - 3-Pin Connector, behind fuse/relay panel

(59) - Ground connection, near taillight, left

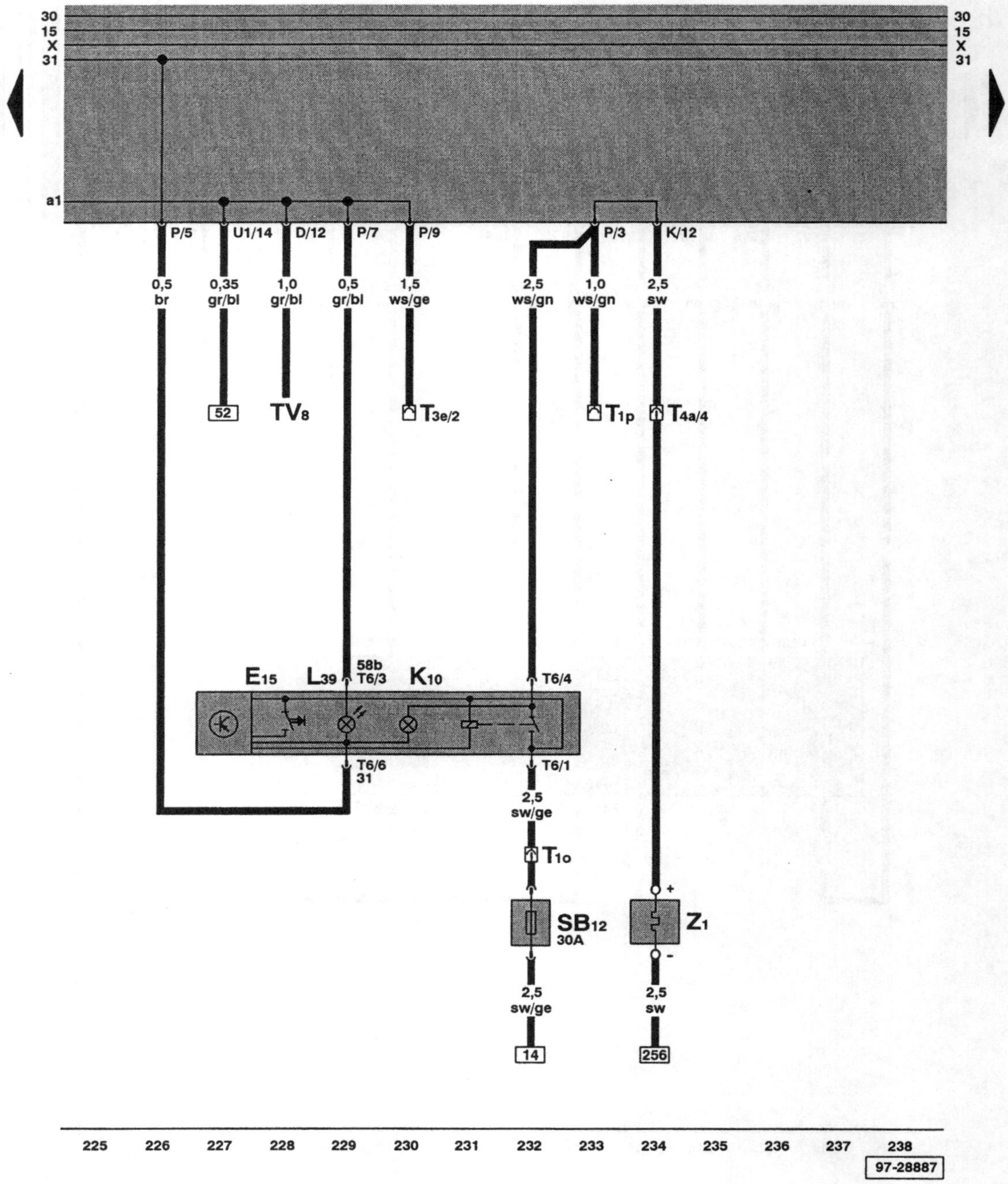

ws = white
sw = black
ro = red
br = brown
gn = green
bl = blue
gr = grey
li = lilac
ge = yellow
or = orange

E15 - Rear window defogger switch
K10 - Indicator light for heated rear windshield
L39 - Rear Window Defogger Switch Light
SB12 - Fuse 12 in fuse holder, below relay panel
T1o - Single Connector, behind fuse/relay panel
T1p - Single Connector, behind fuse/relay panel
T3e - 3-Pin Connector, behind fuse/relay panel
T4a - 4-Pin Connector, near taillamp assembly
T6 - 6-Pin Connector
TV8 - Terminal 58b Wire Connector, grey
Z1 - Heated rear window

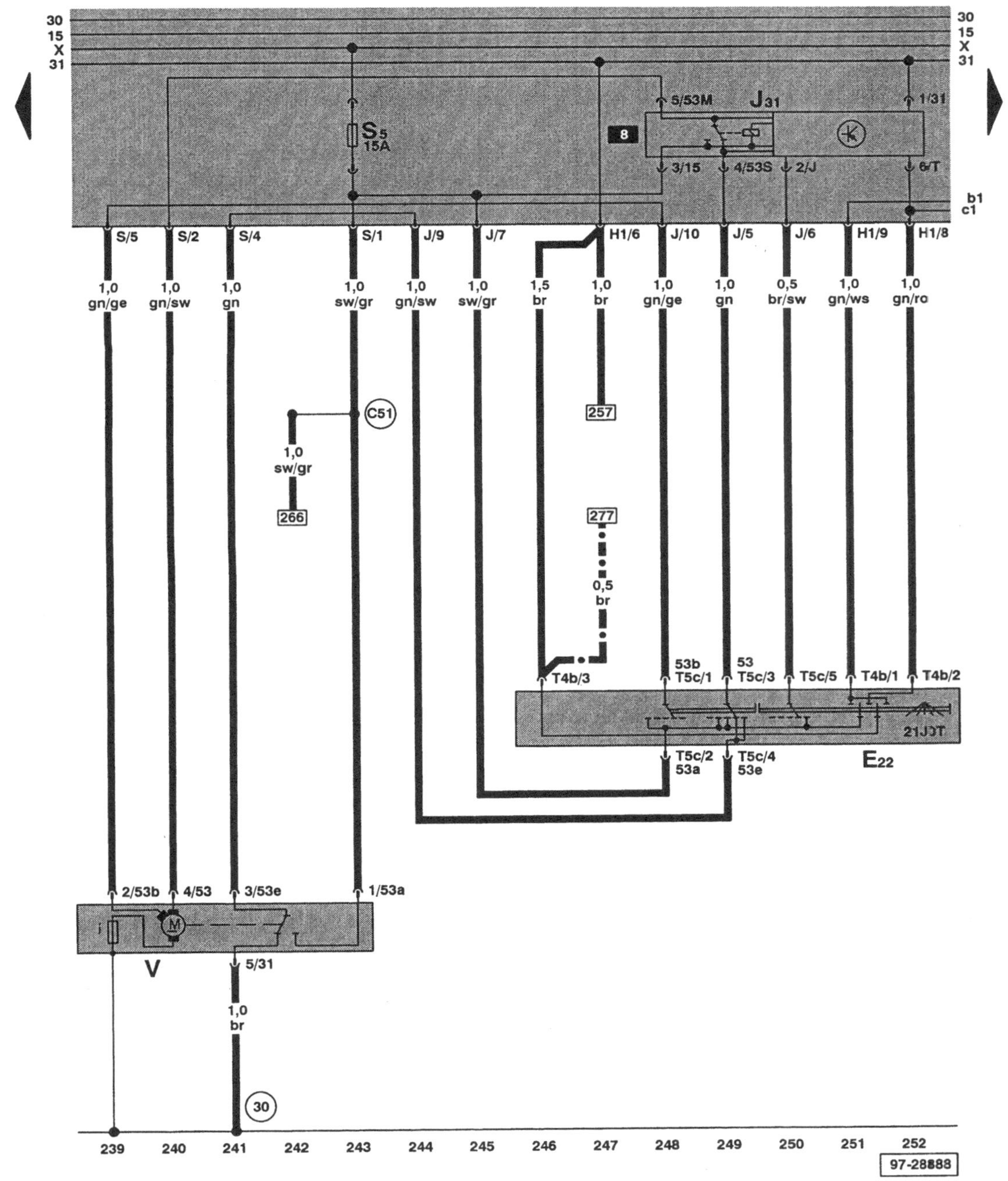

ws = white
sw = black
ro = red
br = brown
gn = green
bl = blue
gr = grey
li = lilac
ge = yellow
or = orange

E22 - Windshield Wiper/Washer Switch
J31 - Wiper/Washer Intermittent Relay (programable)
T4b - 4-Pin Connector, behind steering column switch cover
T5c - 5-Pin Connector, behind steering column switch cover
V - Windshield Wiper Motor

(30) - Ground connection, -1-, beside fuse/relay panel

(C51) - Plus connection (x), in wiper motor wiring harness

-•- - Vehicles with electronic stability program (ESP) only

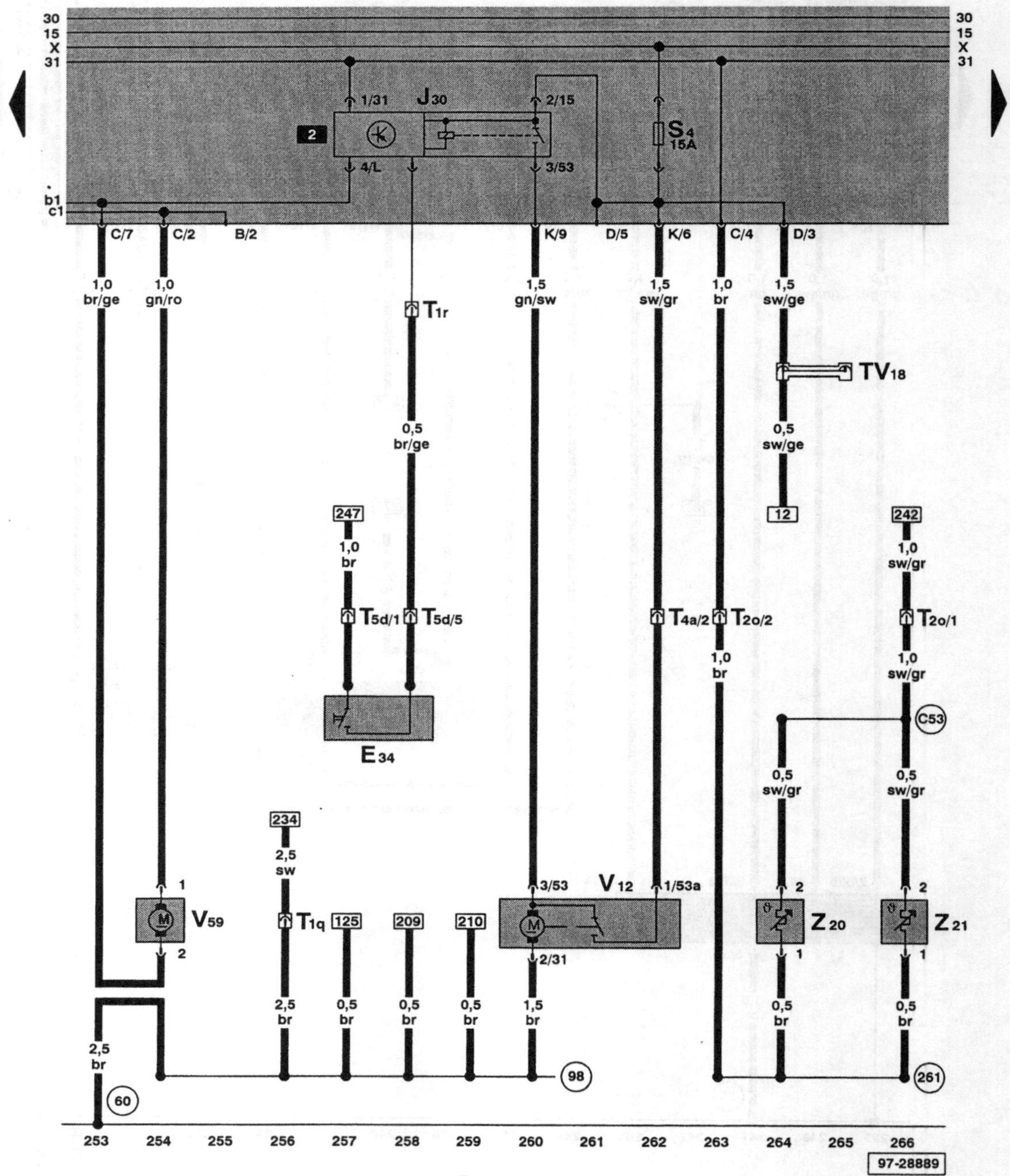

ws = white
sw = black
ro = red
br = brown
gn = green
bl = blue
gr = grey
li = lilac
ge = yellow
or = orange

- E34 - Rear Window Wiper Switch
- J30 - Rear Window Wiper/Washer Relay
- T1q - Single Connector, in rear lid
- T1r - Single Connector, on rear window wiper/washer relay
- T2o - Double Connector, on firewall, right
- T4a - 4-Pin Connector, near taillamp assembly
- T5d - 5-Pin Connector, behind steering column switch cover
- TV18 - Terminal Xa Wire Connector, violet
- V12 - Motor for rear windshield wiper
- V59 - Windshield and Rear Window Washer Pump
- Z20 - Left Washer Nozzle Heater
- Z21 - Right Washer Nozzle Heater

- (60) - Ground point -2- near taillamp assembly, right
- (98) - Ground connection, in rear lid wiring harness
- (261) - Ground connection, in wiring harness heated spray jet
- (C53) - Plus connection (x) in wiring harness heated spray jet

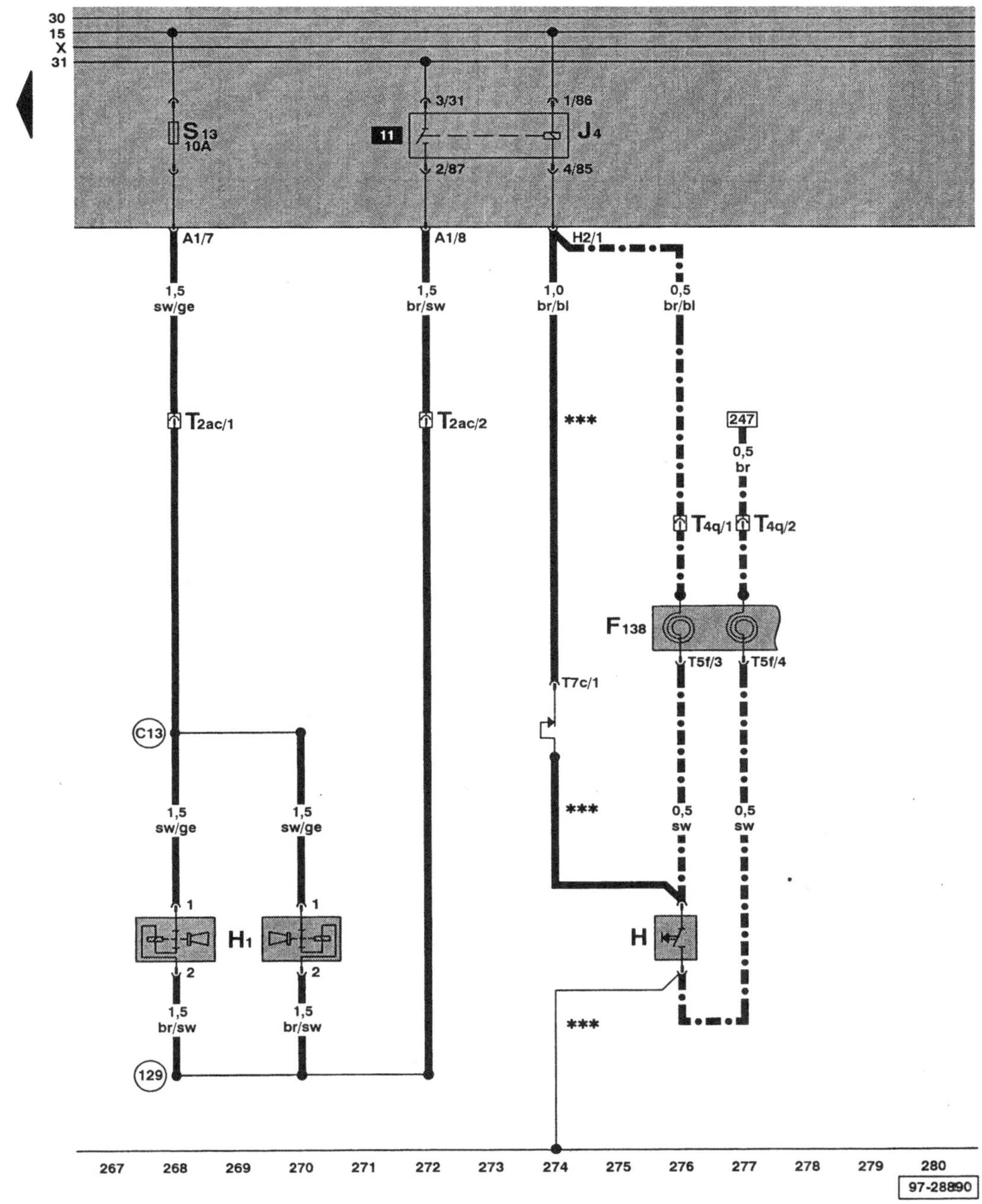

ws = white
sw = black
ro = red
br = brown
gn = green
bl = blue
gr = grey
li = lilac
ge = yellow
or = orange

F138 - Airbag Spiral Spring/Return Spring with Slip Ring
H - Signal horn activation
H1 - Dual tone horn
J4 - Dual Horn Relay
T2ac - Double Connector, in engine compartment, left front
T4q - 4-Pin Connector, behind steering column switch cover
T5f - 5-Pin Connector
T7c - 7-Pin Connector, behind steering column switch cover

(129) - Ground connection, in dual-tone horn wiring harness

(C13) - Plus connection, in dual tone horn wiring harness

*** - Vehicles without electronic stability program (ESP) only

-•- - Vehicles with electronic stability program (ESP) only

THIS PAGE INTENTIONALLY LEFT BLANK

Cruise Control (manual transmission),

from May 1999

- **2.8L - Engine - Motronic Multiport Fuel Injection (MFI)/103 kW, code AES**

Deviate relay location and fuseplacements as well as the locations of multiple connectors see section "component locations".

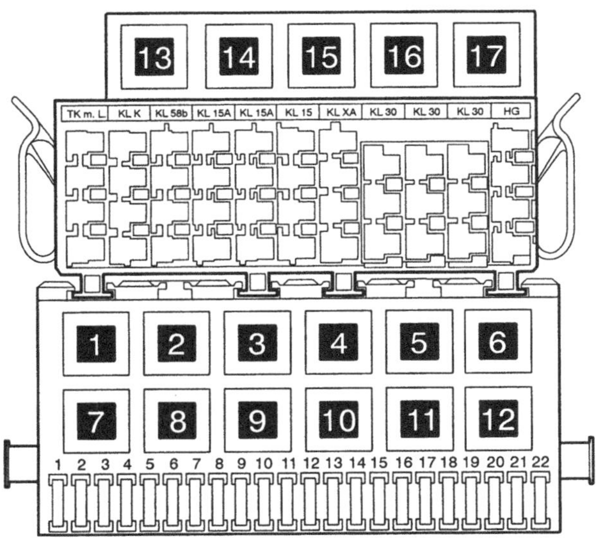

97-23232

Fuse colors

30 A - green
25 A - white
20 A - yellow
15 A - blue
10 A - red
7,5 A - brown
5 A - beige
3 A - violet

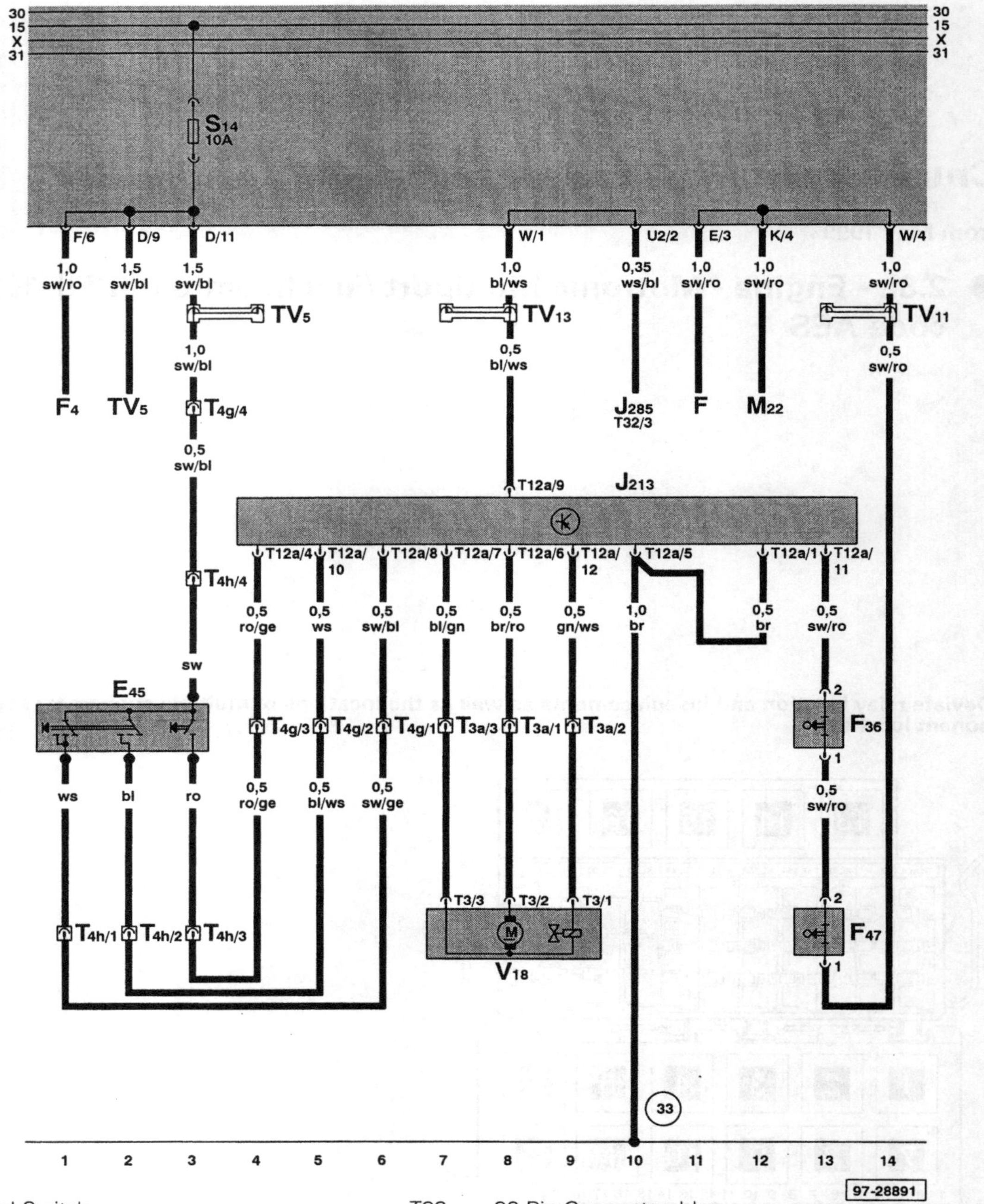

ws = white
sw = black
ro = red
br = brown
gn = green
bl = blue
gr = grey
li = lilac
ge = yellow
or = orange

E45 - Cruise Control Switch
F - Brake Light Switch
F4 - Back-Up Light Switch
F36 - Clutch Vacuum Vent Valve Switch
F47 - Brake Vacuum Vent Valve Switch for cruise control
J213 - Cruise Control, Control Module, behind instrument panel, right
J285 - Control module with indicator unit in instrument panel insert
M22 - Right Brake/Tail Light
T3 - 3-Pin Connector
T3a - 3-Pin Connector, behind fuse/relay panel
T4g - 4-Pin Connector, behind fuse/relay panel
T4h - 4-Pin Connector, behind steering column switch cover
T12a - 12-Pin Connector
T32 - 32-Pin Connector, blue
TV5 - Terminal 15a Wire Connector, green, 4-Pin and 8-Pin
TV11 - Terminal 54 Wire Connector, orange
TV13 - Vehicle Speed Signal Wire Connector
V18 - Cruise Control Vacuum Pump
(33) - Ground connection, behind instrument panel, right

Cruise Control (automatic transmission),

from May 1999

- **2.8L - Engine - Motronic Multiport Fuel Injection (MFI)/103 kW, code AES**

Deviate relay location and fuseplacements as well as the locations of multiple connectors see section "component locations".

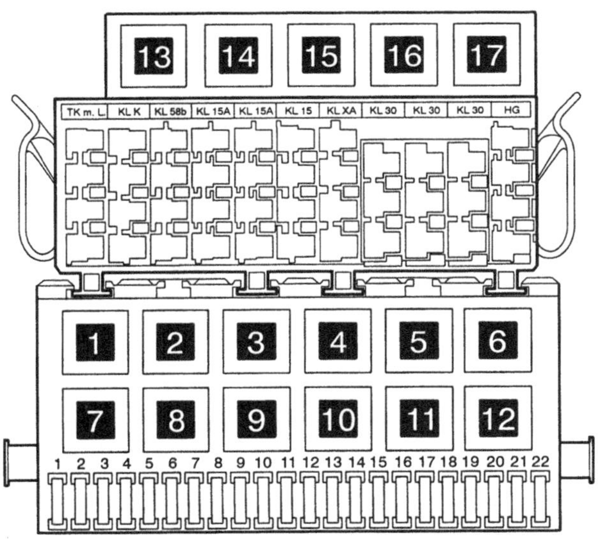

97-23232

Fuse colors

30 A - green
25 A - white
20 A - yellow
15 A - blue
10 A - red
7,5 A - brown
5 A - beige
3 A - violet

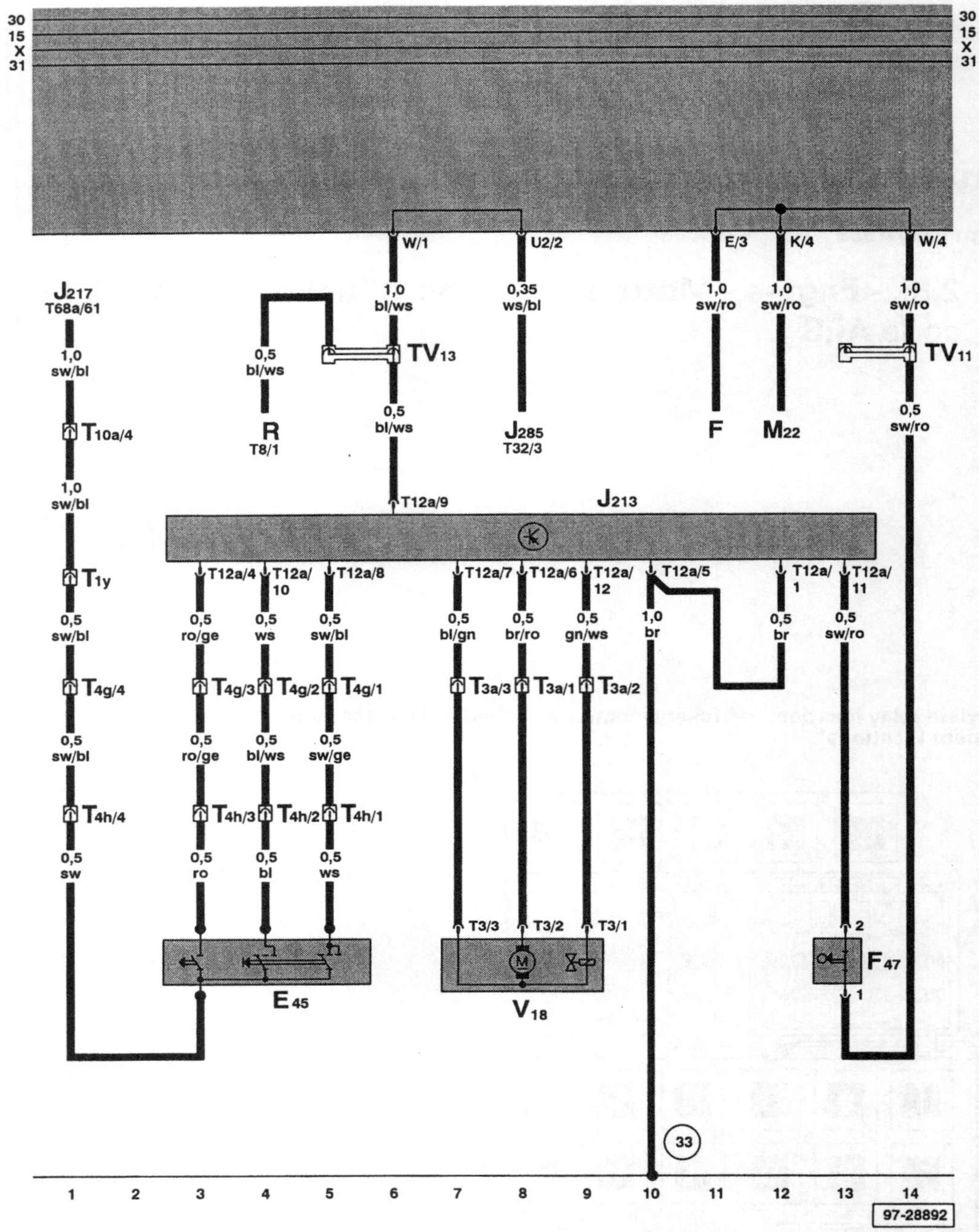

ws = white
sw = black
ro = red
br = brown
gn = green
bl = blue
gr = grey
li = lilac
ge = yellow
or = orange

E45 - Cruise Control Switch
F - Brake Light Switch
F47 - Brake Vacuum Vent Valve Switch for cruise control
J213 - Cruise Control, Control Module, behind instrument panel, right
J217 - Transmission Control Module (TCM), on left A-pillar, lower part
J285 - Control module with indicator unit in instrument panel insert
M22 - Right Brake/Tail Light
R - Radio
T1y - Single Connector, behind fuse/relay panel
T3 - 3-Pin Connector
T3a - 3-Pin Connector, behind fuse/relay panel
T4g - 4-Pin Connector, behind fuse/relay panel
T4h - 4-Pin Connector, behind steering column switch cover
T8 - 8-Pin Connector, on Radio
T10a - 10-Pin Connector, behind fuse/relay panel
T12a - 12-Pin Connector
T32 - 32-Pin Connector, blue, on Instrument cluster
T68a - 68-Pin Connector
TV11 - Terminal 54 Wire Connector, orange
TV13 - Vehicle Speed Signal Wire Connector, blue
V18 - Cruise Control Vacuum Pump
(33) - Ground connection, behind instrument panel, right

Central locking system with remote control,

from December 2000

- **Vehicles with right sliding door and tailgate**
- **Vehicles with right sliding door and swinging doors**

Deviate relay location and fuseplacements as well as the locations of multiple connectors see section "component locations".

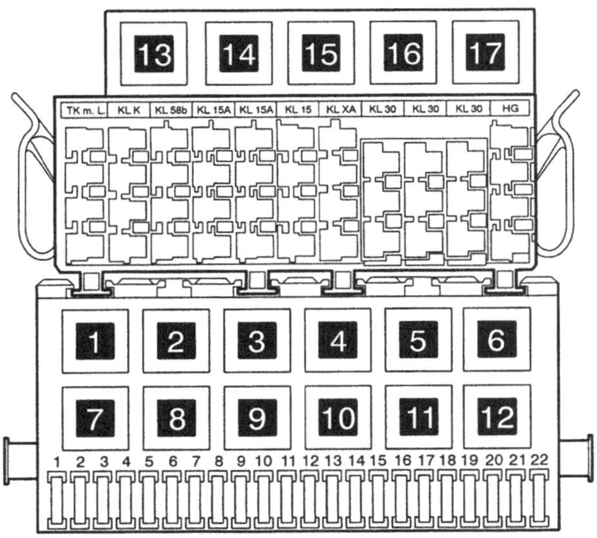

97-23232

Fuse colors

30 A - green
25 A - white
20 A - yellow
15 A - blue
10 A - red
7,5 A - brown
5 A - beige
3 A - violet

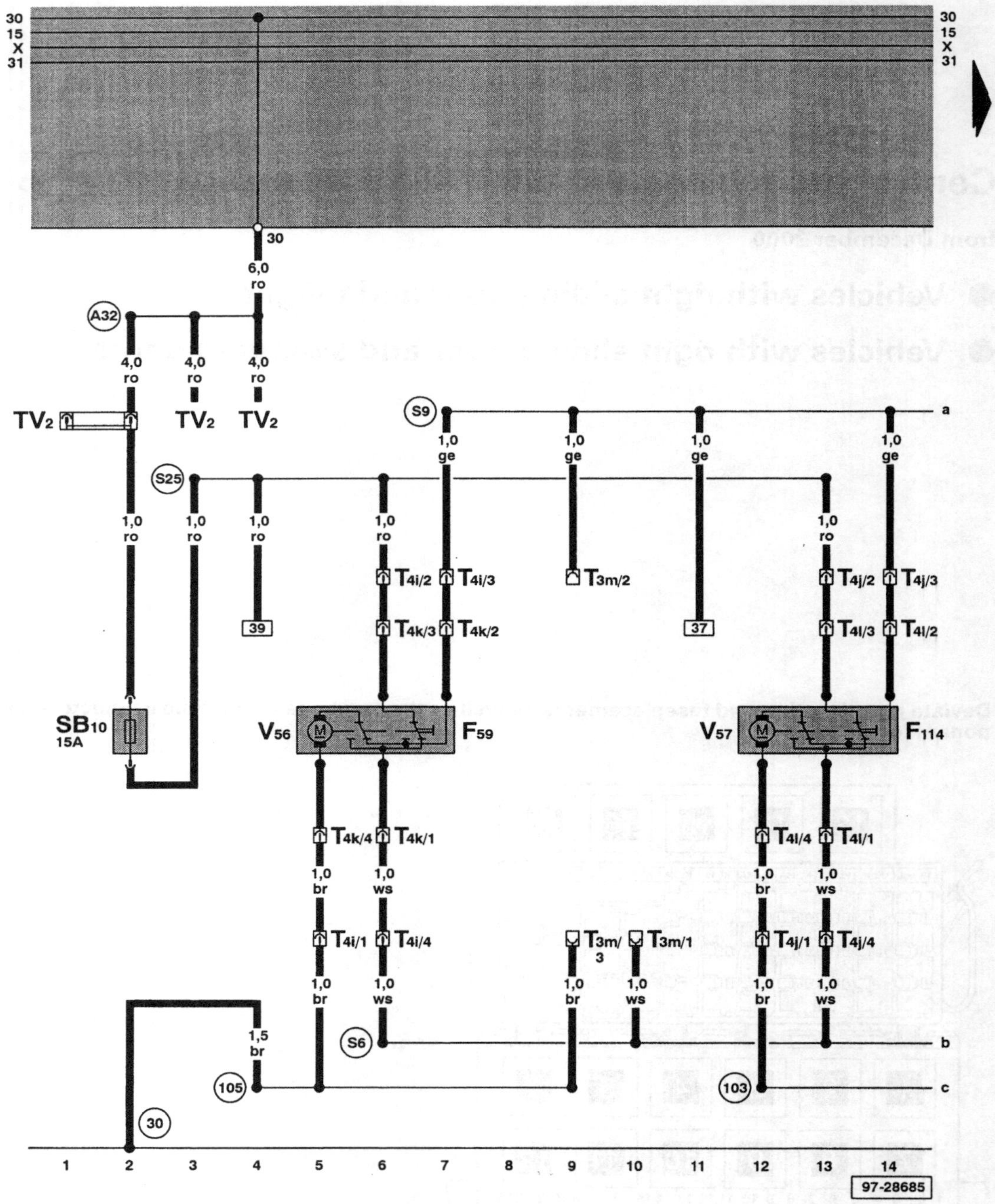

ws = white
sw = black
ro = red
br = brown
gn = green
bl = blue
gr = grey
li = lilac
ge = yellow
or = orange

- F59 - Driver's Door Central Locking System Switch
- F114 - Passenger's Door Central Locking System Switch
- SB10 - Fuse 10 in fuse holder, below relay panel
- T3m - 3-Pin Connector, behind instrument panel, left
- T4i - 4-Pin Connector, behind fuse/relay panel
- T4j - 4-Pin Connector, behind, instrument panel, right
- T4k - 4-Pin Connector, in driver's door
- T4l - 4-Pin Connector, in front passenger's door
- TV2 - Terminal 30 (B+) Wire Connector, red, each 6 points
- V56 - Driver's Door Central Locking System Motor
- V57 - Passenger's Door Central Locking System Motor
- (30) - Ground connection, -1-, beside fuse/relay panel
- (103) - Ground connection -2-, in central locking system wiring harness
- (105) - Ground connection -1-, in central locking system wiring harness
- (A32) - Plus connection (30), in instrument panel wiring harness
- (S6) - Wire connection (A) -1-, in central locking system wiring harness
- (S9) - Wire connection (E) -1-, in central locking system wiring harness
- (S25) - Connector (Sensor/Control Module) -1- in wiring harness, central locking

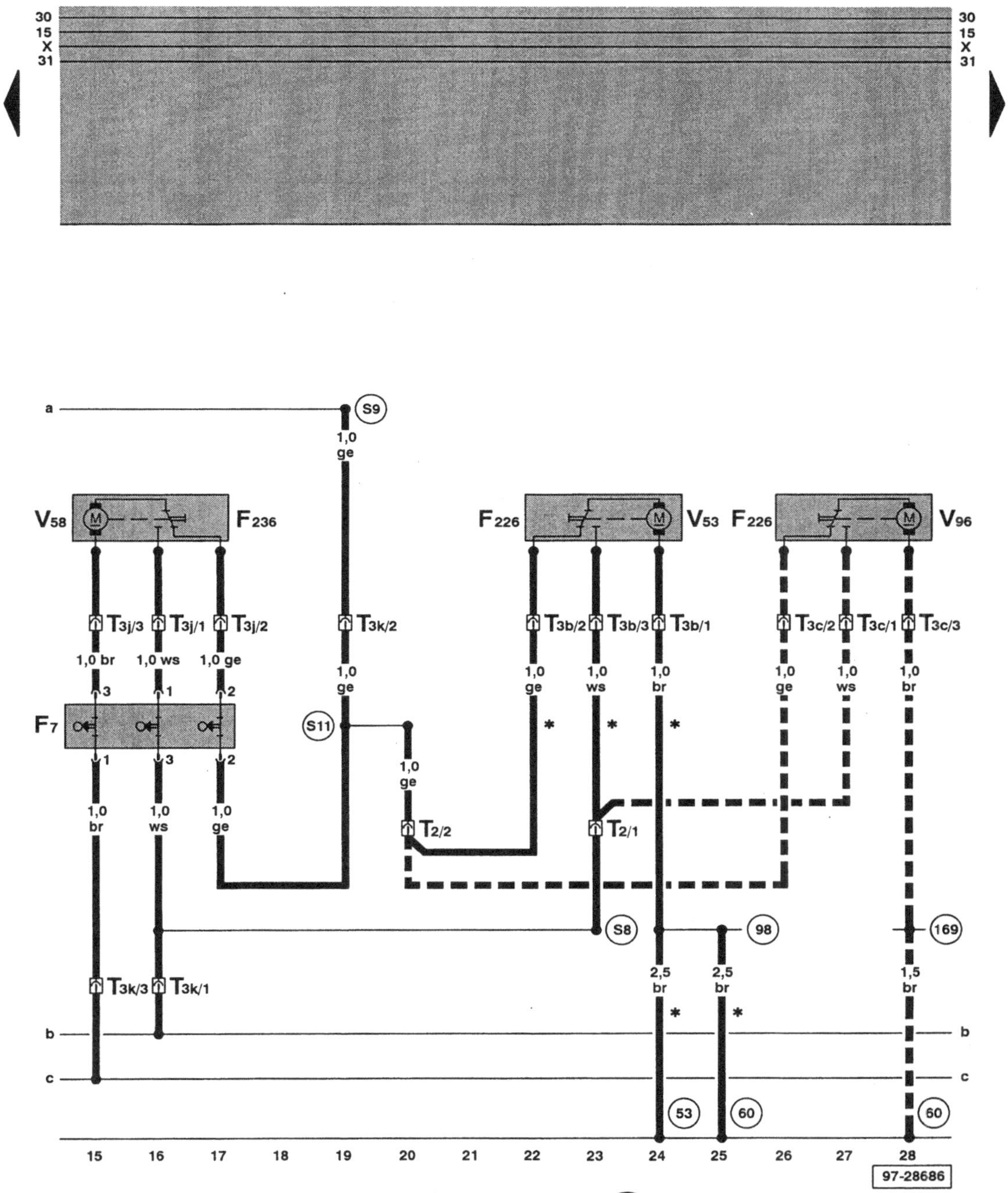

ws = white
sw = black
ro = red
br = brown
gn = green
bl = blue
gr = grey
li = lilac
ge = yellow
or = orange

- F7 - Right Rear Sliding Door Contact Switch
- F226 - Right rear door/swinging door contact switch
- F236 - Door contact switch for central locking (right sliding door)
- T2 - Double Connector, on right D-pillar, upper part
- T3b - 3-Pin Connector, in rear lid
- T3c - 3-Pin Connector, in rear swinging door
- T3j - 3-Pin Connector, in sliding door, right
- T3k - 3-Pin Connector, behind instrument panel, right
- V53 - Decklid Central Locking System Motor
- V58 - Sliding Door Central Locking System Motor
- V96 - Left Rear Door Central Locking System Motor
- (53) - Ground connection, in rear lid, right
- (60) - Ground point -2- near taillamp assembly, right
- (98) - Ground connection, in rear lid wiring harness
- (169) - Ground connection, in right sliding door wiring harness
- (S8) - Wire connection (A) -3-, in central locking system/passenger compartment wiring harness
- (S9) - Wire connection (E) -1-, in central locking system wiring harness
- (S11) - Wire connection (E) -3-, in central locking system passenger compartment wiring harness
- * - Vehicles with rear lid only
- - - - - Vehicles with rear swinging doors only

Decklid central locking system motor, sliding door central locking system motor, left rear door central locking system motor

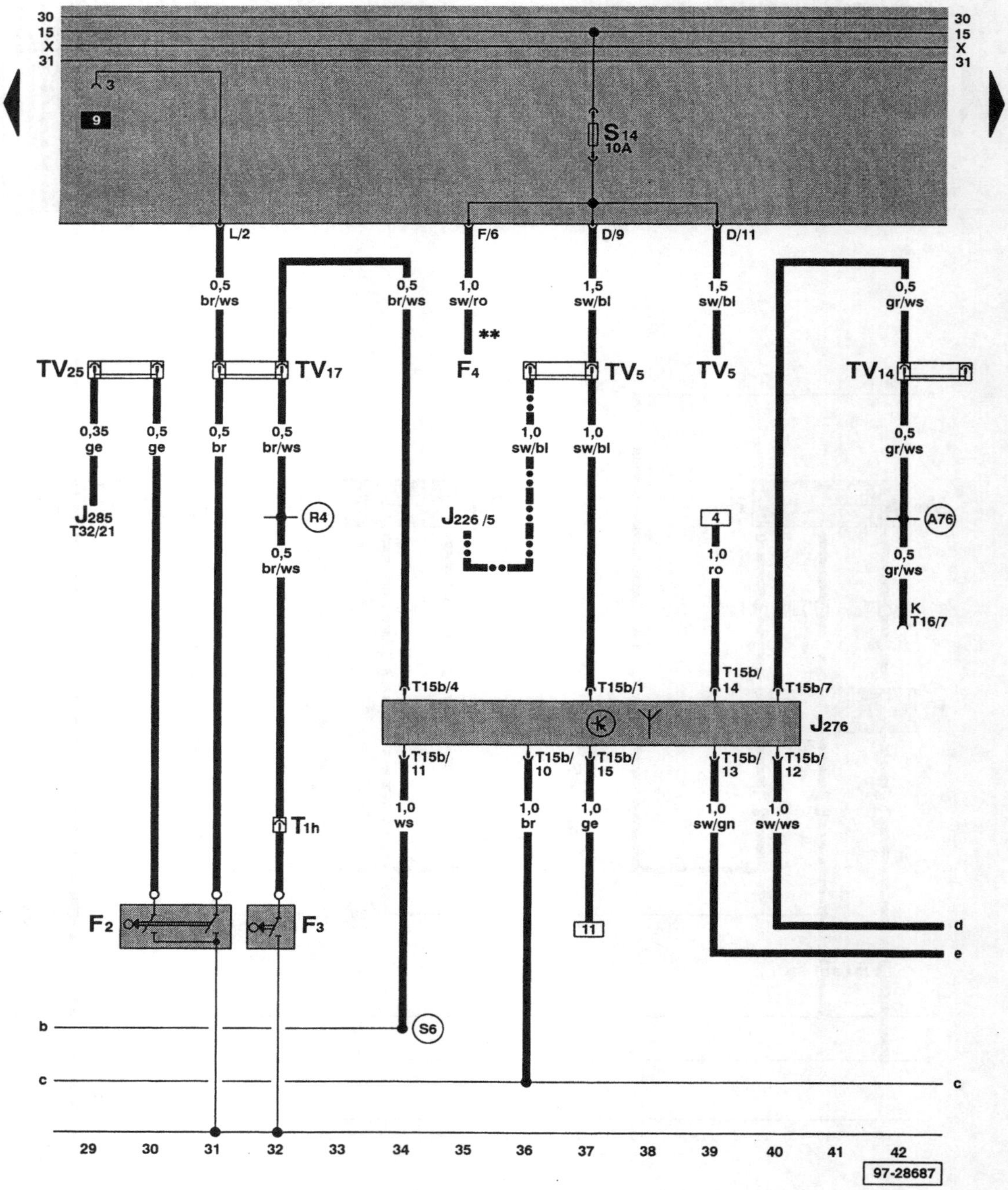

ws = white
sw = black
ro = red
br = brown
gn = green
bl = blue
gr = grey
li = lilac
ge = yellow
or = orange

- F2 - Door contact Switch - Driver's side
- F3 - Door contact Switch-Passenger side
- F4 - Back-Up Light Switch**
- J226 - Park/Neutral Position (PNP) Relay
- J276 - Central Locking System Control Module (Remote Control), behind instrument cluster
- J285 - Control module with indicator unit in instrument panel insert
- T1h - Single Connector, behind instrument panel, right
- T15b - 15-Pin Connector
- T16 - Data Link Connector (DLC), behind steering column cover
- T32 - 32-Pin Connector, blue, on instrument cluster
- TV5 - Terminal 15a Wire Connector, green, 4-Pin and 8-Pin
- TV14 - Data Link Connector (DLC) wire Connector, white
- TV17 - Door Contact Switch Wire Connector, brown
- TV25 - Door Contact Switch Wire Connector, pink
- (A76) - Connector (K-diagnosis wire) in instrument panel wiring harness
- (R4) - Wire connection -1-, in interior light/door contact switch wiring harness
- (S6) - Wire connection (A) -1-, in central locking system wiring harness
- ** - Manual transmission only
- -•••- Automatic transmission only

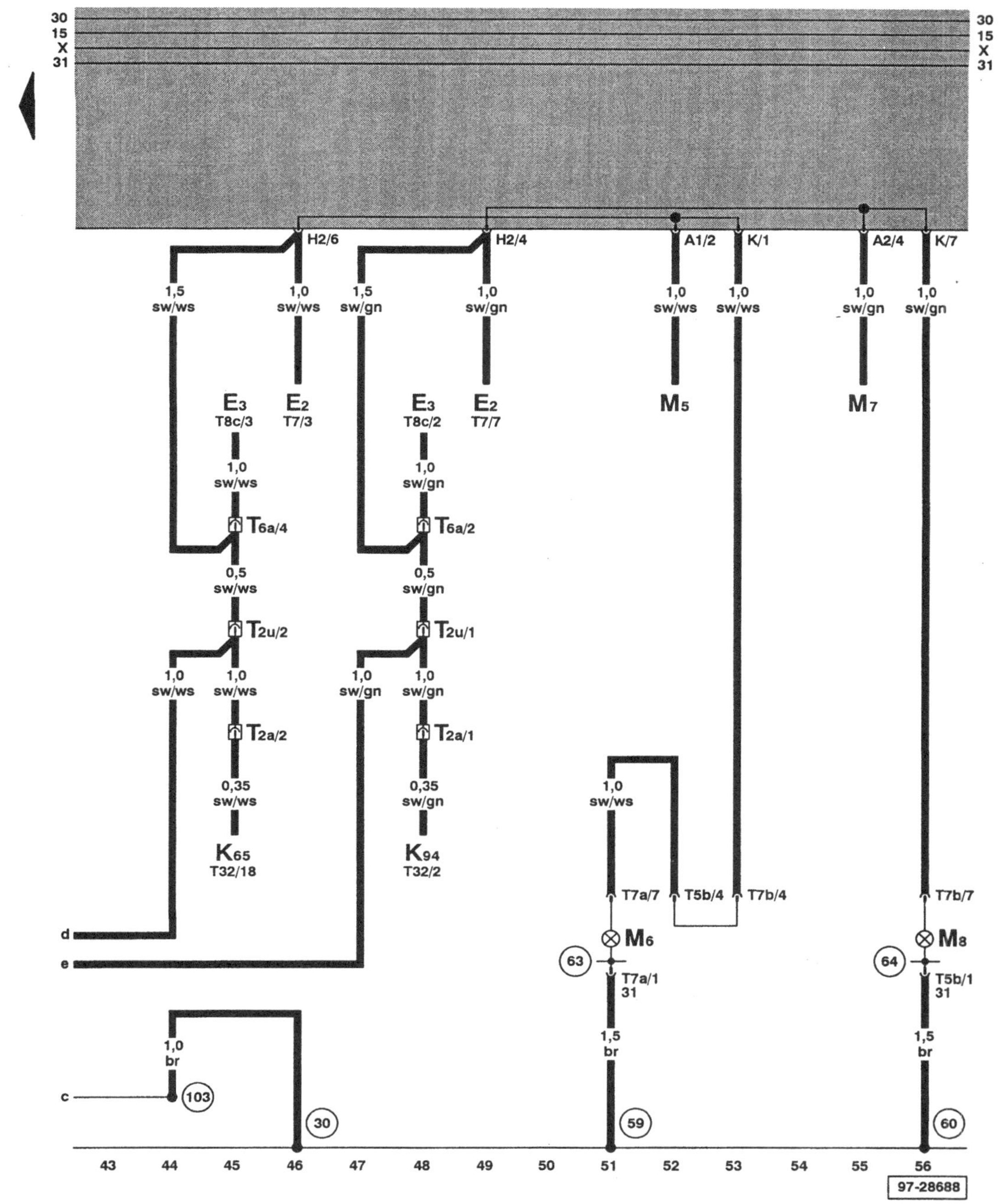

ws = white
sw = black
ro = red
br = brown
gn = green
bl = blue
gr = grey
li = lilac
ge = yellow
or = orange

E2 - Turn signal switch
E3 - Emergency Flasher Switch
K65 - Left Turn Signal Indicator Light
K94 - Right Turn Signal Indicator Light
M5 - Left Front Turn Signal Light
M6 - Left Rear Turn Signal Light
M7 - Right Front Turn Signal Light
M8 - Right Rear Turn Signal Light
T2a - Double Connector, behind fuse/relay panel
T2u - Double Connector, behind fuse/relay panel
T5b - 5-Pin Connector, on right tail light
T6a - 6-Pin Connector, behind fuse/relay panel
T7 - 7-Pin Connector, behind steering column switch cover
T7a - 7-Pin Connector, on left tail light
T7b - 7-Pin Connector, on right tail light
T8c - 8-Pin Connector, on Emergency Flasher Switch
T32 - 32-Pin Connector, blue, on instrument cluster

(30) - Ground connection, -1-, beside fuse/relay panel
(59) - Ground connection, near taillight, left
(60) - Ground point -2- near taillamp assembly, right
(63) - Ground connection, on left taillight bulb holder
(64) - Ground connection, on right taillight bulb holder
(103) - Ground connection -2-, in central locking system wiring harness

THIS PAGE INTENTIONALLY LEFT

BLANK

Fog lights,

from May 2000

Deviate relay location and fuseplacements as well as the locations of multiple connectors see section "component locations".

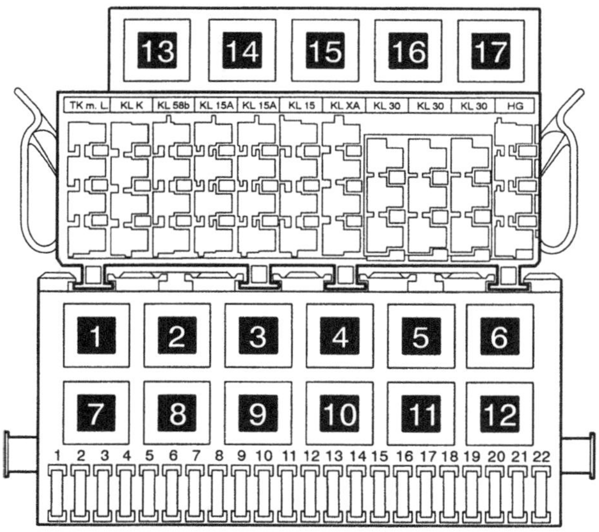

97-23232

Fuse colors

30 A - green
25 A - white
20 A - yellow
15 A - blue
10 A - red
7,5 A - brown
5 A - beige
3 A - violet

4 Load Reduction Relay (18)

10 Fog Light Relay (53)

Note: Number in parentheses indicates production control number stamped on relay housing.

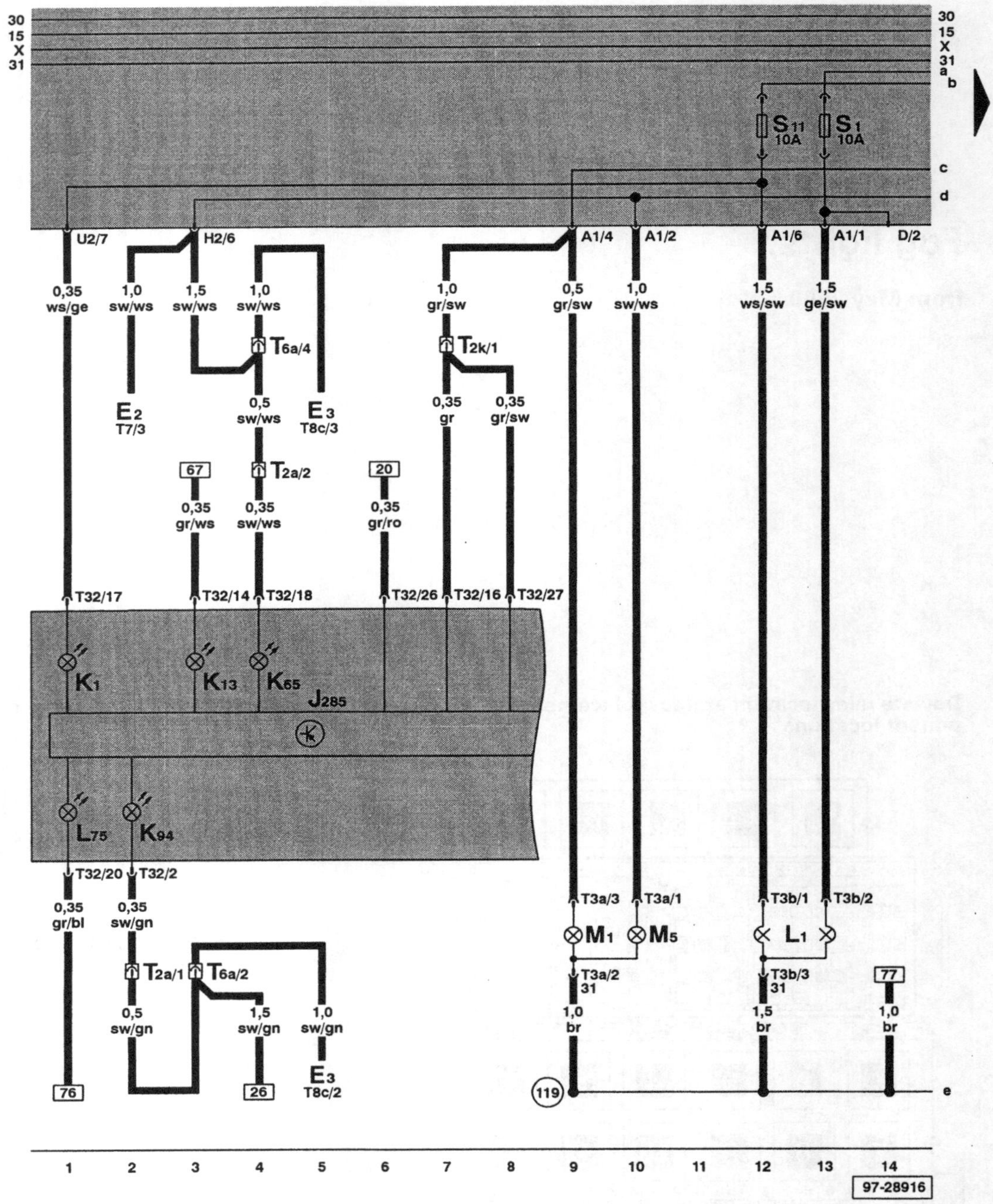

ws = white
sw = black
ro = red
br = brown
gn = green
bl = blue
gr = grey
li = lilac
ge = yellow
or = orange

E2 - Turn signal switch
E3 - Emergency Flasher Switch
J285 - Control module with indicator unit in instrument panel insert
K1 - Headlight High Beam Indicator Light
K13 - Rear Fog Light Indicator Light
K65 - Left Turn Signal Indicator Light
K94 - Right Turn Signal Indicator Light
L1 - Left Headlight
L75 - Digital Display Light
M1 - Left Parking Light
M5 - Left Front Turn Signal Light
T2a - Double Connector, behind fuse/relay panel
T2k - Double Connector, behind fuse/relay panel
T3a - 3-Pin Connector
T3b - 3-Pin Connector
T6a - 6-Pin Connector, behind fuse/relay panel
T7 - 7-Pin Connector, behind steering column switch cover
T8c - 8-Pin Connector
T32 - 32-Pin Connector, blue

(119) - Ground connection -1-, in headlight wiring harness

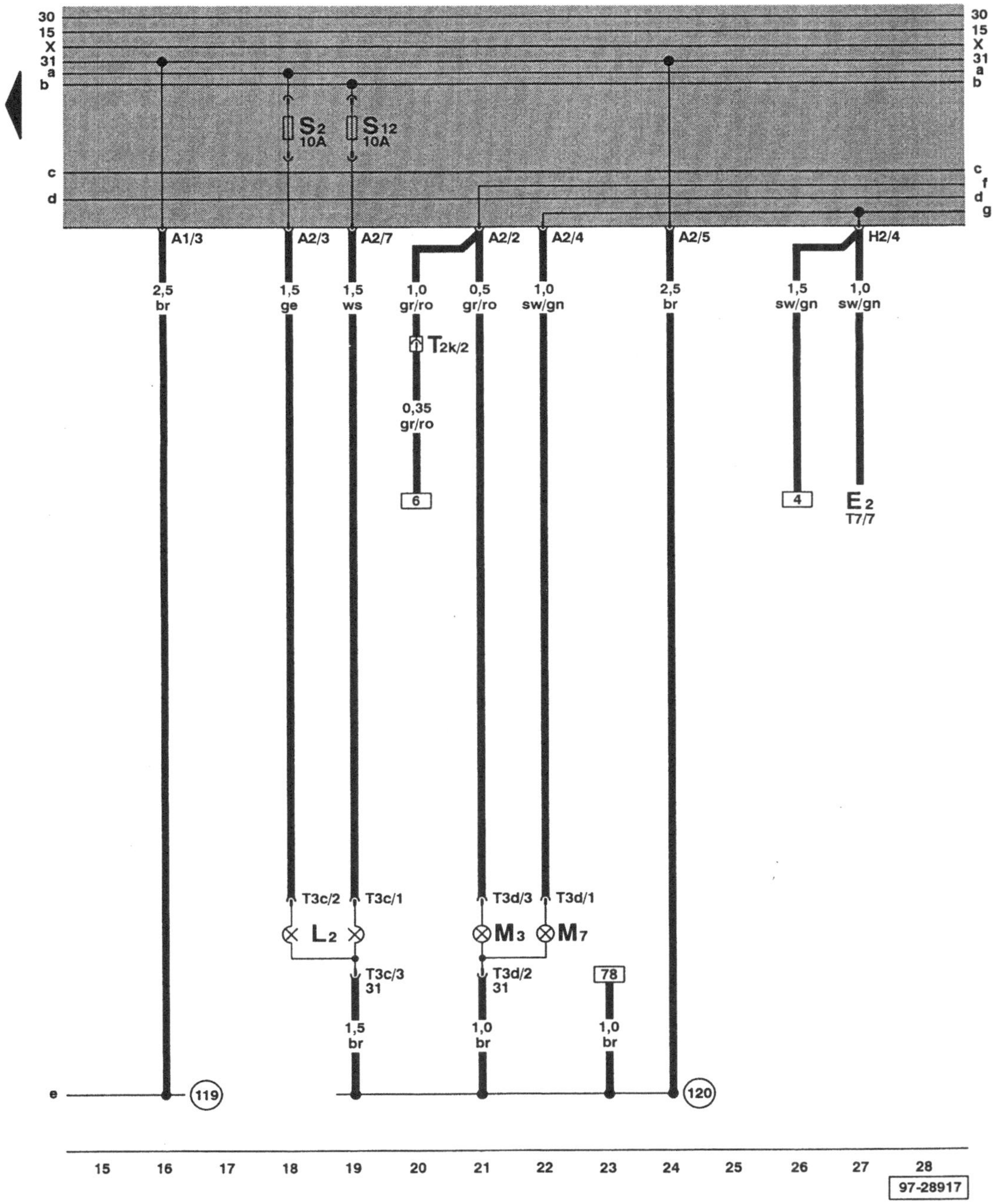

ws = white
sw = black
ro = red
br = brown
gn = green
bl = blue
gr = grey
li = lilac
ge = yellow
or = orange

E2 - Turn signal switch
L2 - Right Headlight
M3 - Right Parking Light
M7 - Right Front Turn Signal Light
T2k - Double Connector, behind fuse/relay panel
T3c - 3-Pin Connector
T3d - 3-Pin Connector
T7 - 7-Pin Connector, behind steering column switch cover

(119) - Ground connection -1-, in headlight wiring harness
(120) - Ground connection -2-, in headlight wiring harness

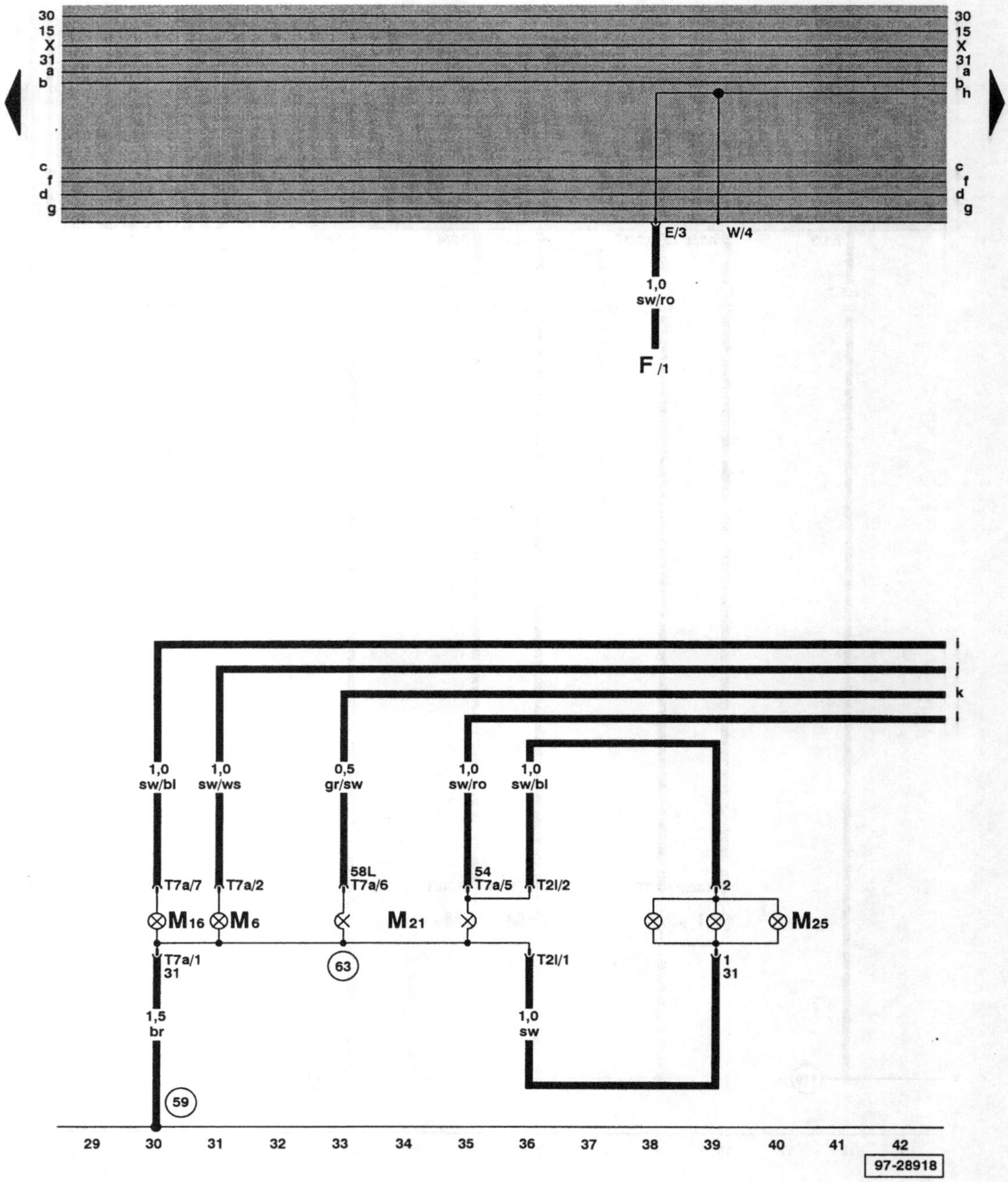

ws = white
sw = black
ro = red
br = brown
gn = green
bl = blue
gr = grey
li = lilac
ge = yellow
or = orange

F - Brake Light Switch
M6 - Left Rear Turn Signal Light
M16 - Left Back-Up Light
M21 - Left Brake/Tail Light
M25 - High-mount Brake Light
T2l - Double Connector
T7a - 7-Pin Connector

(59) - Ground connection, near taillight, left

(63) - Ground connection, on left taillight bulb holder

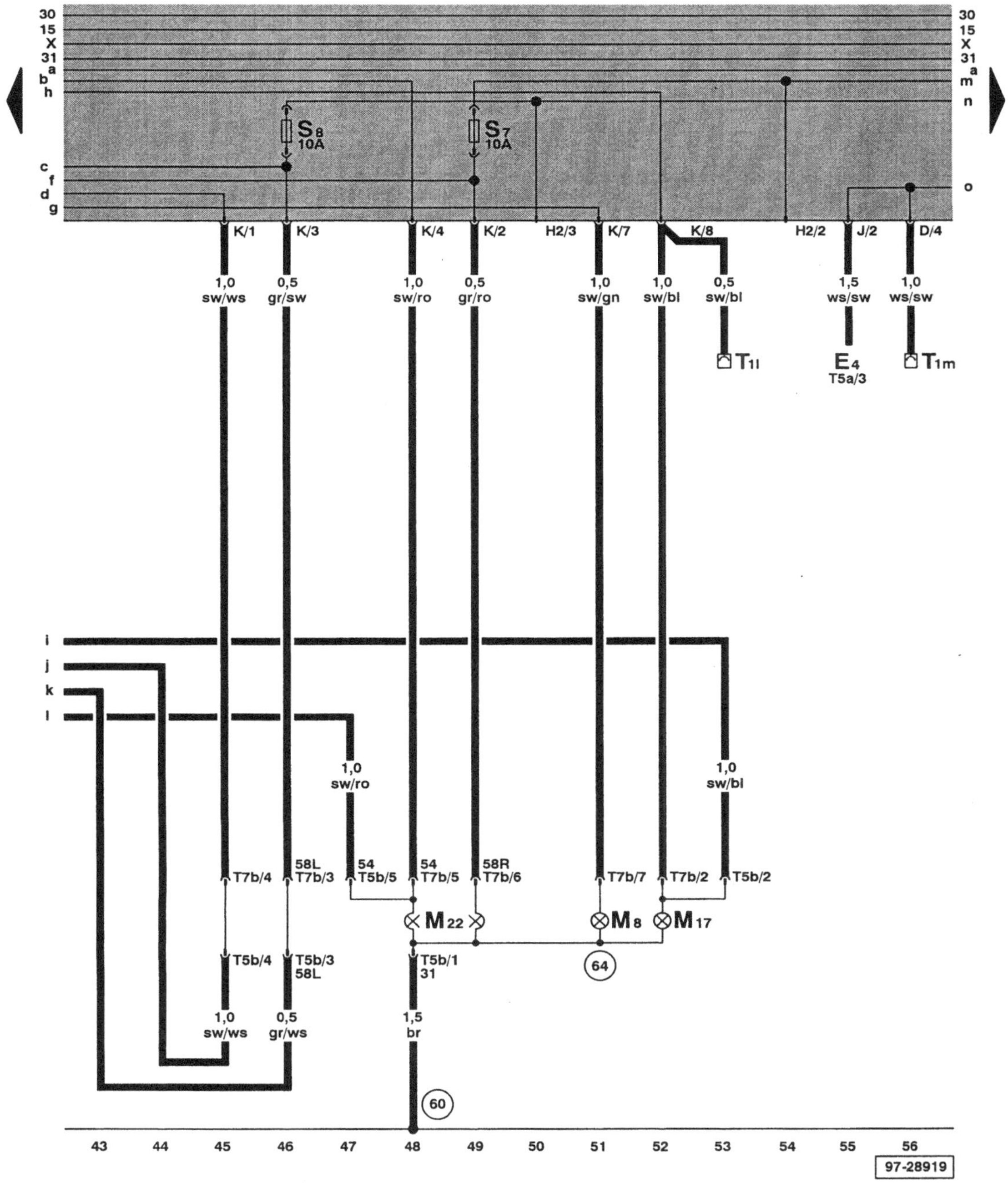

ws = white
sw = black
ro = red
br = brown
gn = green
bl = blue
gr = grey
li = lilac
ge = yellow
or = orange

E4 - Headlight Dimmer/Flasher Switch
M8 - Right Rear Turn Signal Light
M17 - Right Back-Up Light
M22 - Right Brake/Tail Light
T1l - Single Connector, behind fuse/relay panel
T1m - Single Connector, behind fuse/relay panel
T5a - 5-Pin Connector, behind steering column switch cover
T5b - 5-Pin Connector
T7b - 7-Pin Connector

(60) - Ground point -2- near taillamp assembly, right

(64) - Ground connection, on right taillight bulb holder

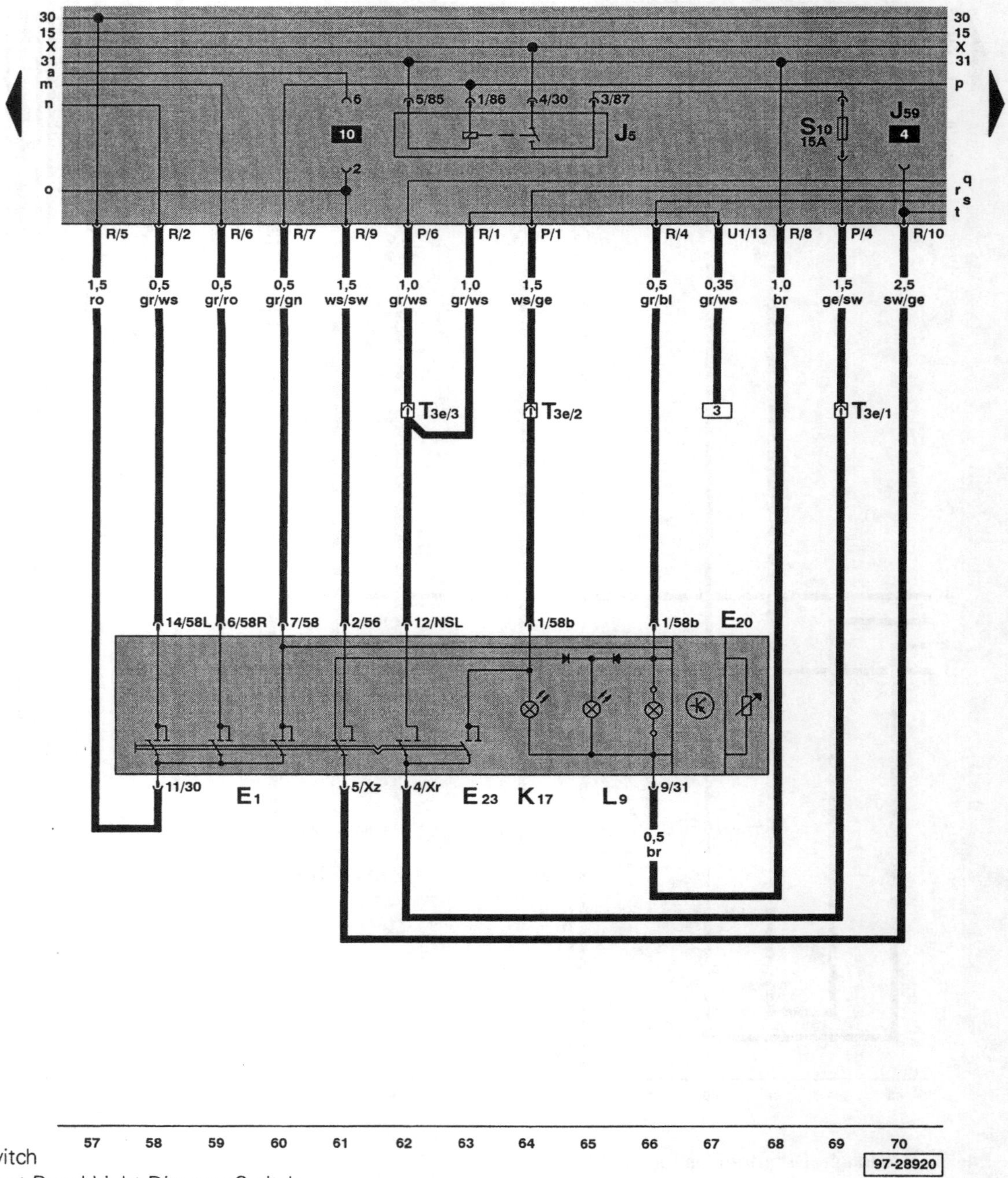

ws = white
sw = black
ro = red
br = brown
gn = green
bl = blue
gr = grey
li = lilac
ge = yellow
or = orange

E1 - Light switch
E20 - Instrument Panel Light Dimmer Switch
E23 - Fog Light Switch
J5 - Fog Light Relay
J59 - Load Reduction Relay
K17 - Fog Light Indicator Light
L9 - Headlight Switch Light
T3e - 3-Pin Connector, behind fuse/relay panel

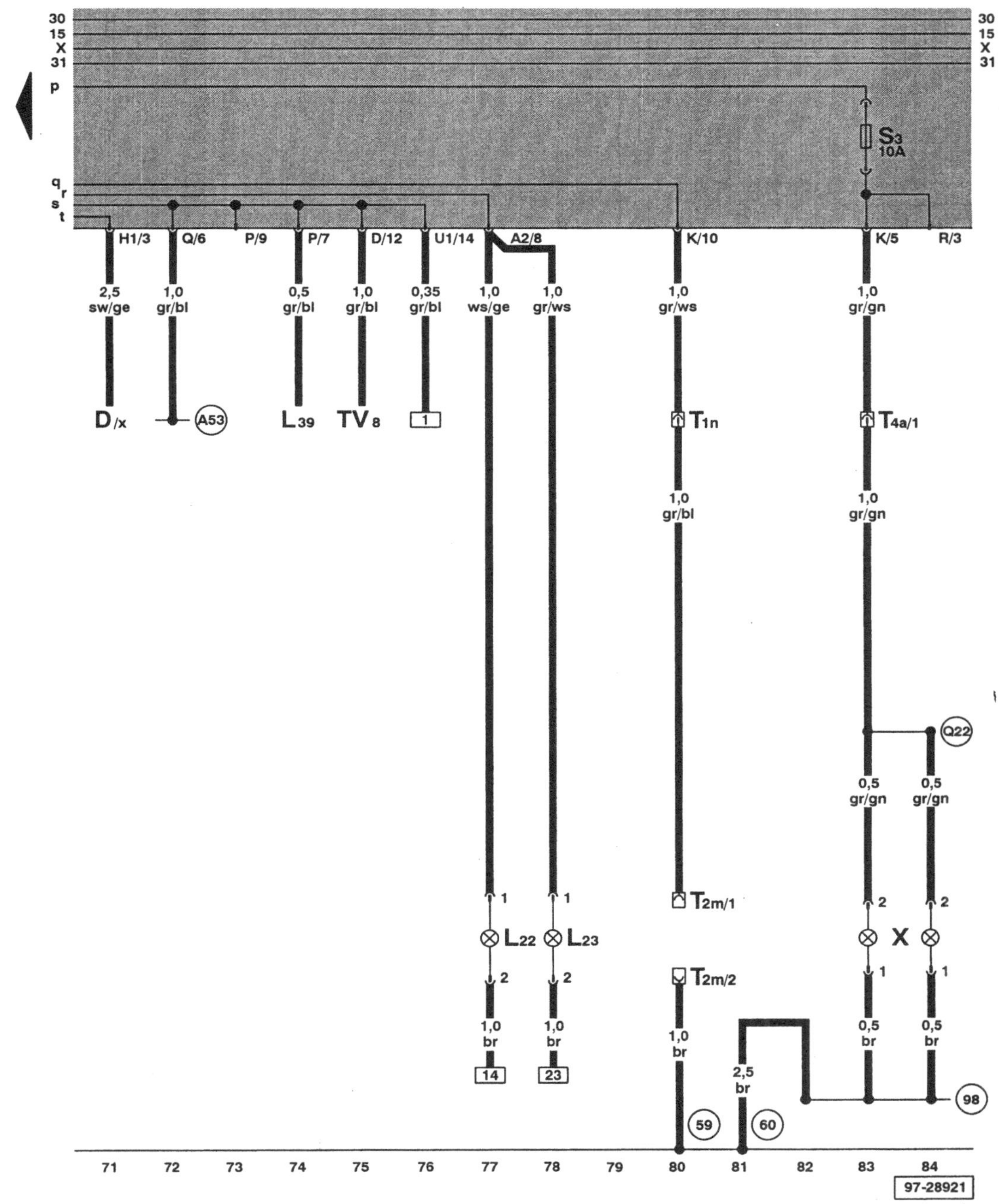

ws = white
sw = black
ro = red
br = brown
gn = green
bl = blue
gr = grey
li = lilac
ge = yellow
or = orange

D - Ignition/Starter Switch
L22 - Left Front Fog Light
L23 - Right Front Fog Light
L39 - Rear Window Defogger Switch Light
T1n - Single Connector, near taillamp assembly
T2m - Double Connector
T4a - 4-Pin Connector, near taillamp assembly
TV8 - Terminal 58b Wire Connector, grey
X - License Plate Light

(59) - Ground connection, near taillight, left

(60) - Ground point -2- near taillamp assembly, right

(98) - Ground connection, in rear lid wiring harness

(A53) - Plus connection (58b), in instrument panel wiring harness

(Q22) - Wire connection -1-, in rear lid wiring harness

THIS PAGE INTENTIONALLY LEFT BLANK

Anti-lock brake system (ABS) with electronic differential lock (EDL), anti-slip control (ASC) and electronic stability program (ESP),

from December 2000

Deviate relay location and fuseplacements as well as the locations of multiple connectors see section "component locations".

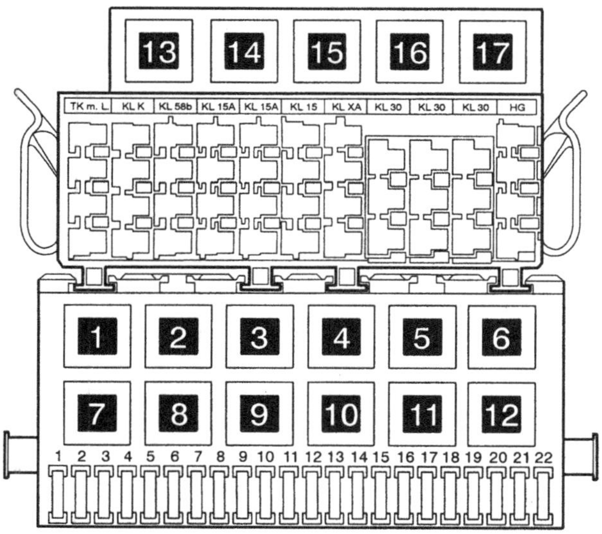

97-23232

5 Wire distributor for Data-Bus

Note: Number in parentheses indicates production control number stamped on relay housing.

Fuse colors

30 A - green
25 A - white
20 A - yellow
15 A - blue
10 A - red
7,5 A - brown
5 A - beige
3 A - violet

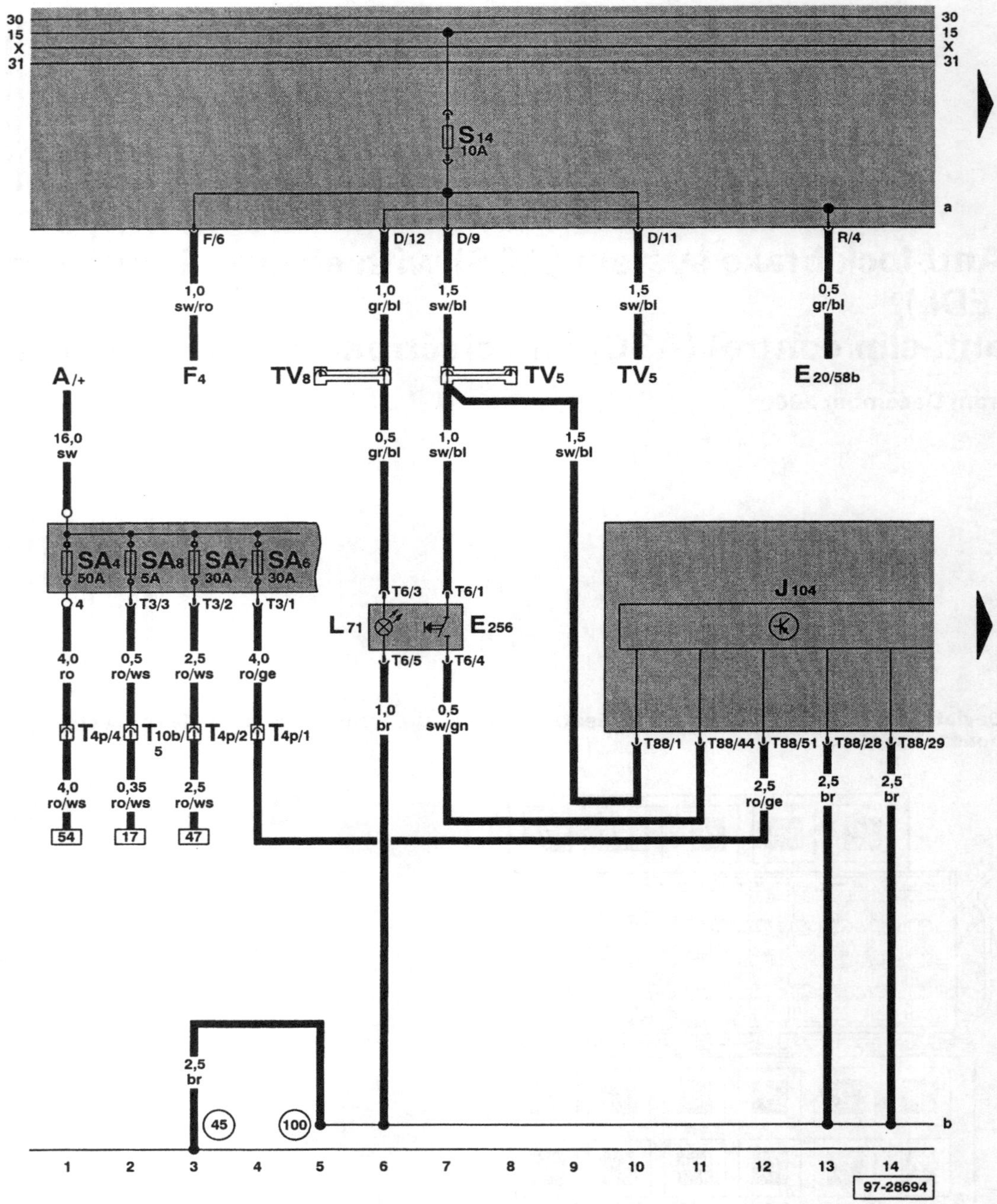

ws = white
sw = black
ro = red
br = brown
gn = green
bl = blue
gr = grey
li = lilac
ge = yellow
or = orange

A - Battery
E20 - Instrument Panel Light Dimmer Switch
E256 - ASR/ESP Button
F4 - Back-Up Light Switch
J104 - ABS Control Module (with EDL), below instrument panel, right
L71 - Anti-slip Control/ESP Switch Illumination
SA4 - Fuse -4- in fuse bracket/battery
SA6 - Fuse -6- in fuse bracket/battery
SA7 - Fuse -7- in fuse bracket/battery
SA8 - Fuse -8- in fuse bracket/battery
T3 - 3-Pin Connector
T4p - 4-Pin Connector, behind fuse/relay panel
T6 - 6-Pin Connector
T10b - 10-Pin Connector, brown, behind fuse/relay panel
T88 - 88-Pin Connector
TV5 - Terminal 15a Wire Connector, green, 4-Pin and 8-Pin
TV8 - Terminal 58b Wire Connector, grey
(45) - Ground connection, behind instrument panel, center
(100) - Ground connection -1-, in ABS wiring harness

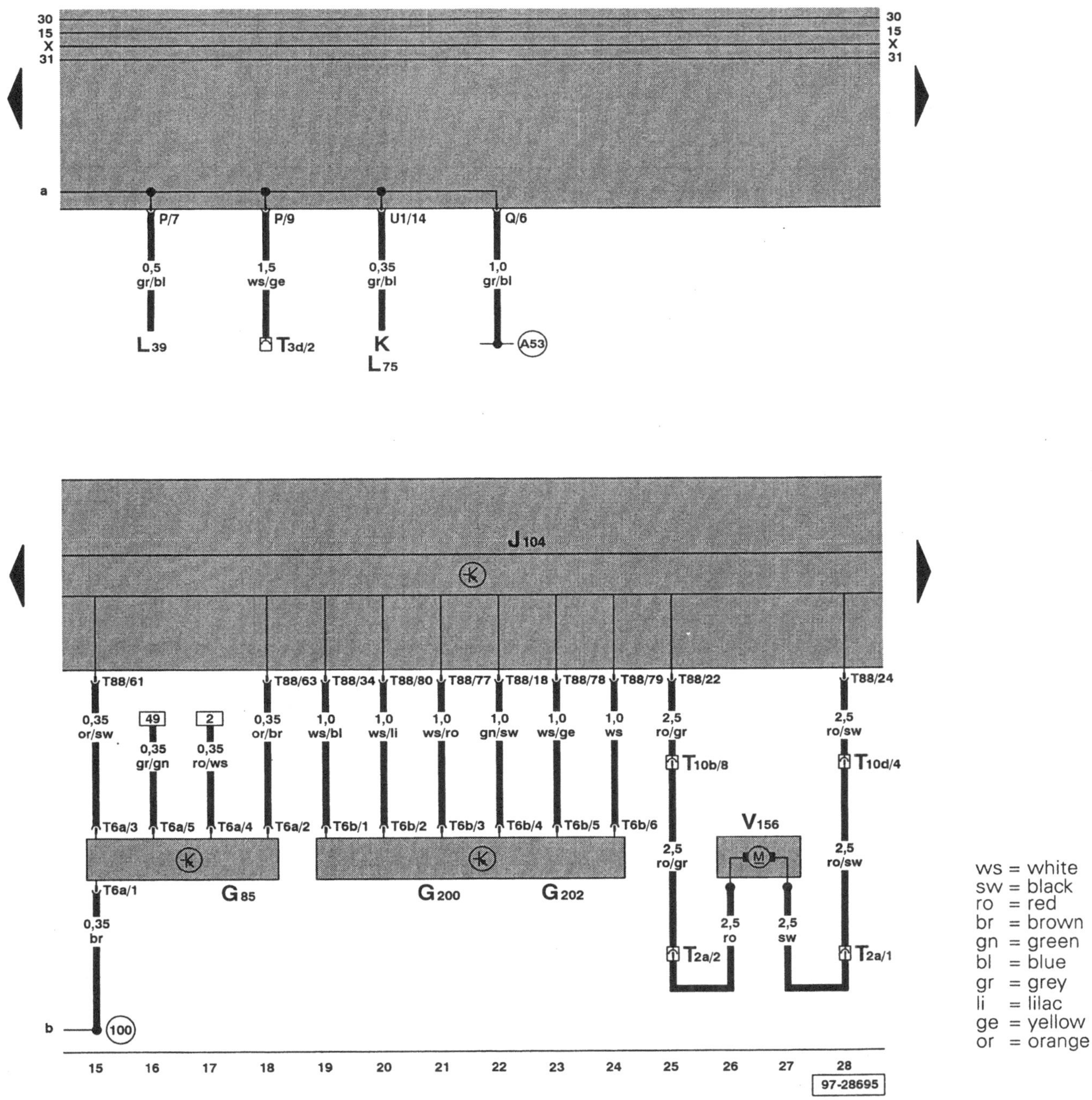

G85 - Steering Angle Sensor
G200 - Sensor for transverse acceleration
G202 - Sender for rotation rate
J104 - ABS Control Module (with EDL), below instrument panel, right
K - Instrument Cluster
L39 - Rear Window Defogger Switch Light
L75 - Digital Display Light
T2a - Double Connector, near headlight, left
T3d - 3-Pin Connector, behind fuse/relay panel
T6a - 6-Pin Connector
T6b - 6-Pin Connector
T10b - 10-Pin Connector, brown, behind fuse/relay panel
T10d - 10-Pin Connector, grey, behind fuse/relay panel
T88 - 88-Pin Connector

V156 - Hydraulic pump for traction control

(100) - Ground connection -1-, in ABS wiring harness

(A53) - Plus connection (58b), in instrument panel wiring harness

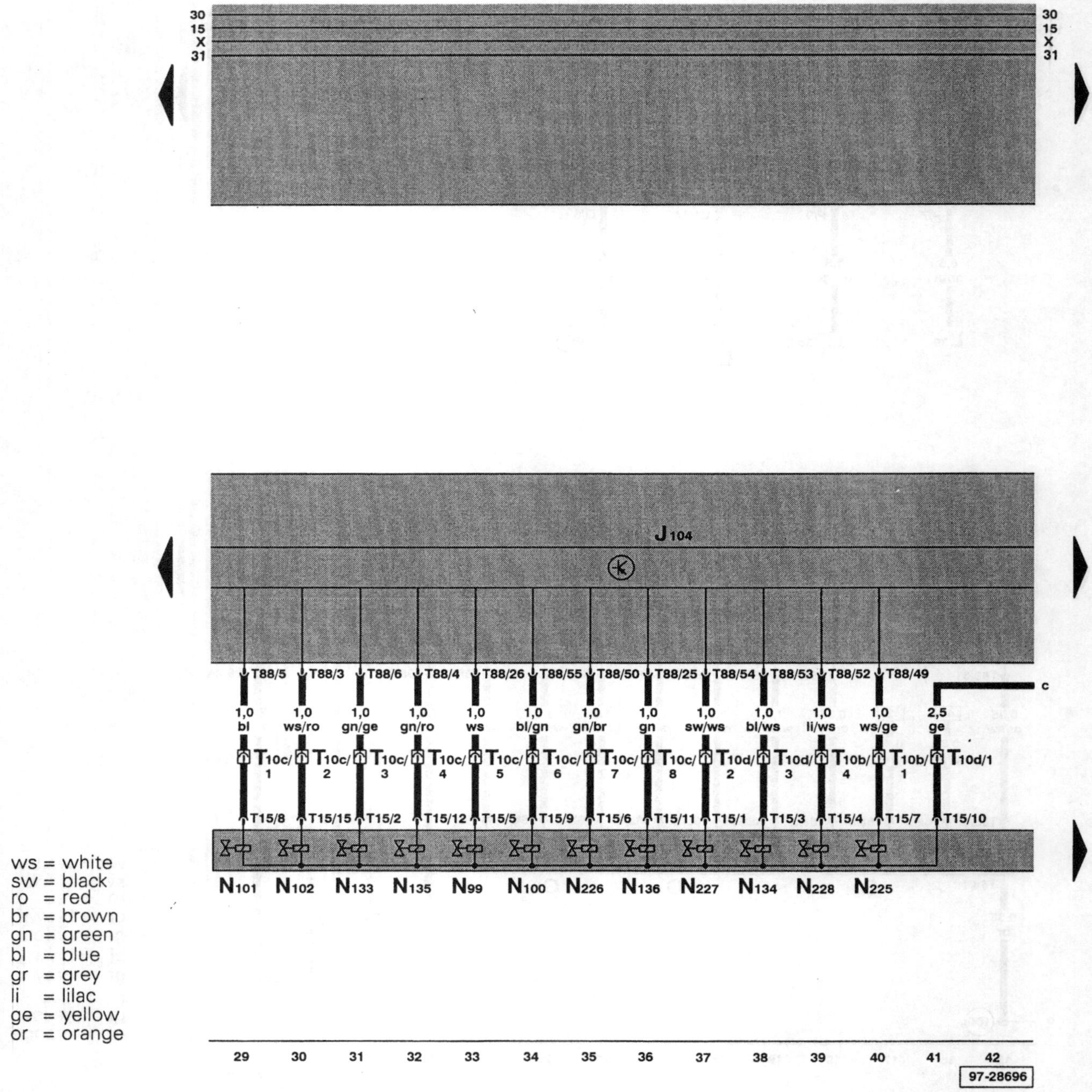

J104 - ABS Control Module (with EDL), below instrument panel, right
N99 - Right Front ABS Inlet Valve
N100 - Right Front ABS Outlet Valve
N101 - Left Front ABS Inlet Valve
N102 - Left Front ABS Outlet Valve
N133 - Right Rear ABS Inlet Valve
N134 - Left Rear ABS Inlet Valve
N135 - Right Rear ABS Outlet Valve
N136 - Left Rear ABS Outlet Valve
N225 - Pilot valve -1- traction control
N226 - Pilot valve -2- traction control
N227 - High pressure switch valve -1- traction control
N228 - High pressure switch valve -2- traction control
T10b - 10-Pin Connector, brown, behind fuse/relay panel
T10c - 10-Pin Connector, black, behind fuse/relay panel
T10d - 10-Pin Connector, grey, behind fuse/relay panel
T15 - 15-Pin Connector
T88 - 88-Pin Connector

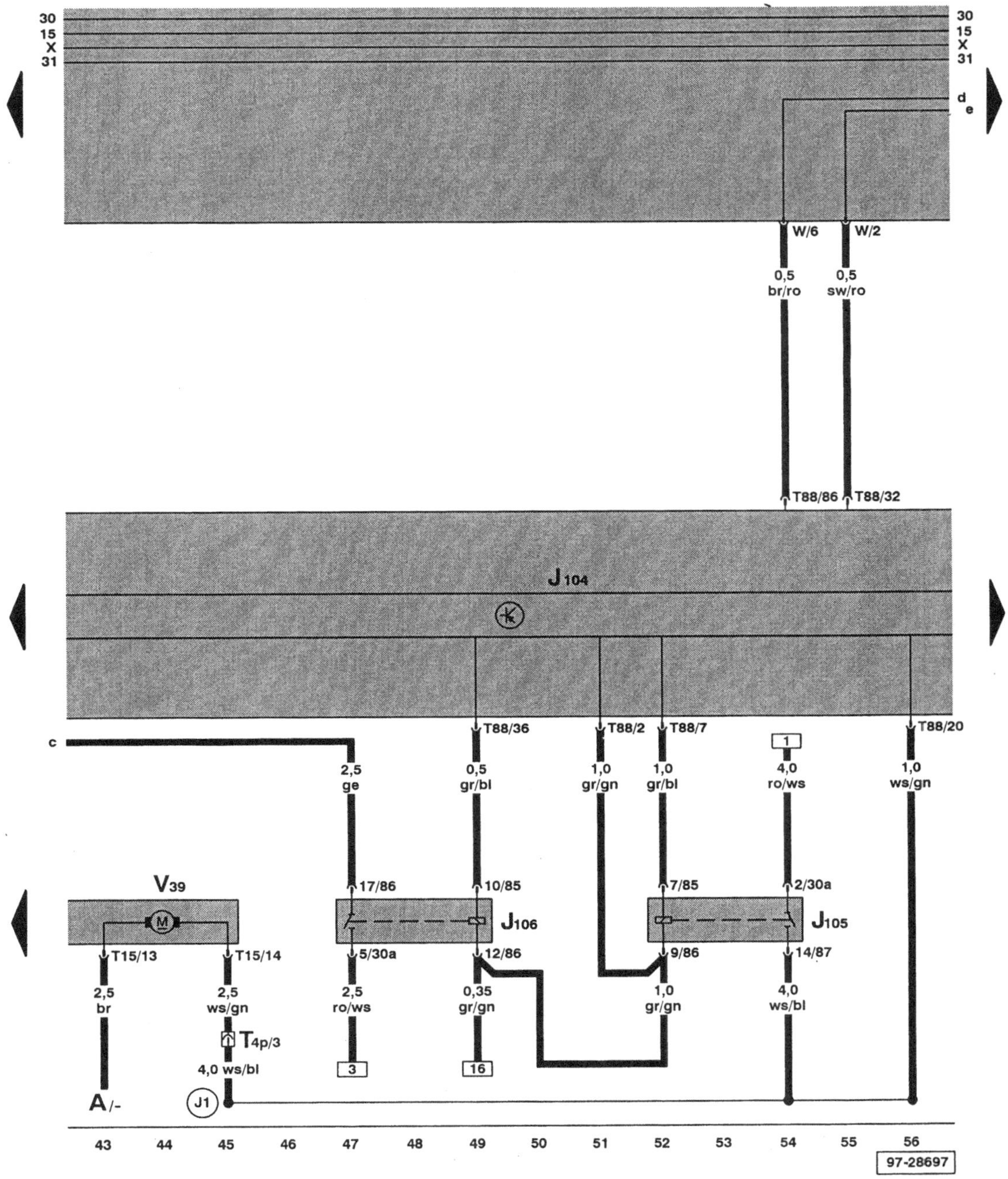

ws = white
sw = black
ro = red
br = brown
gn = green
bl = blue
gr = grey
li = lilac
ge = yellow
or = orange

A - Battery
J104 - ABS Control Module (with EDL), below instrument panel, right
J105 - ABS Return Flow Pump Relay, near air outlet right
J106 - ABS Solenoid Valve Relay, near air outlet right
T4p - 4-Pin Connector, behind fuse/relay panel
T15 - 15-Pin Connector
T88 - 88-Pin Connector
V39 - ABS Return Flow Pump

(J1) - Plus connection (30), in ABS wiring harness

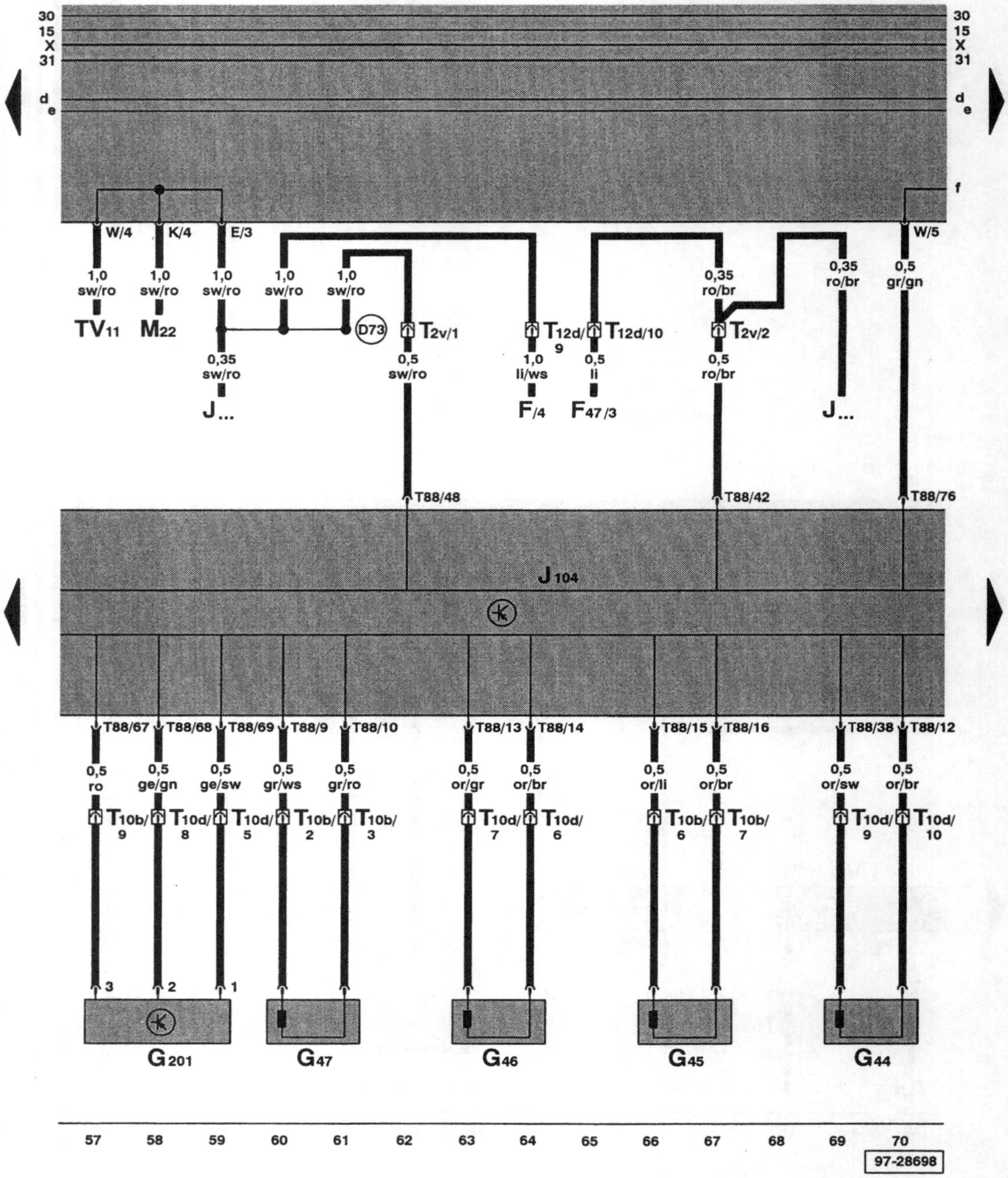

ws = white
sw = black
ro = red
br = brown
gn = green
bl = blue
gr = grey
li = lilac
ge = yellow
or = orange

- F - Brake Light Switch
- F47 - Brake Vacuum Vent Valve Switch for cruise control
- G44 - Right Rear ABS Wheel Speed Sensor
- G45 - Right Front ABS Wheel Speed Sensor
- G46 - Left Rear ABS Wheel Speed Sensor
- G47 - Left Front ABS Wheel Speed Sensor
- G201 - Sender 1 for brake booster
- J104 - ABS Control Module (with EDL), below instrument panel, right
- J... - Engine Control Module (ECM)
- M22 - Right Brake/Tail Light
- T2v - Double Connector, behind fuse/relay panel
- T10b - 10-Pin Connector, brown, behind fuse/relay panel
- T10d - 10-Pin Connector, grey, behind fuse/relay panel
- T12d - 12-Pin Connector, behind fuse/relay panel
- T88 - 88-Pin Connector
- TV11 - Terminal 54 Wire Connector, orange
- (D73) - Plus connection (54), in engine compartment wiring harness

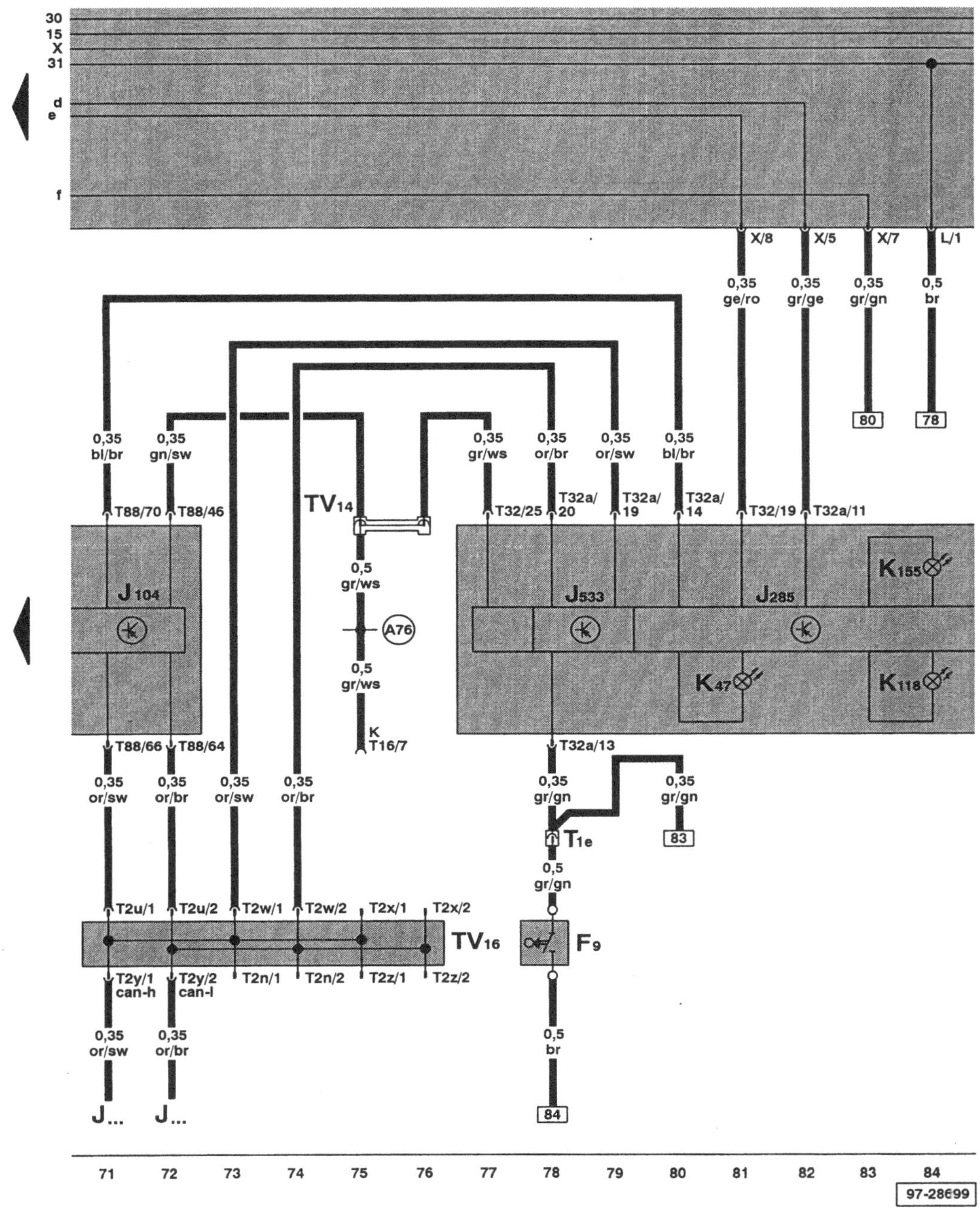

ws = white
sw = black
ro = red
br = brown
gn = green
bl = blue
gr = grey
li = lilac
ge = yellow
or = orange

F9 - Parking Brake Warning Light Switch
J104 - ABS Control Module (with EDL) below instrument panel, right
J285 - Control module with indicator unit in instrument panel insert
J533 - Data Bus On Board Diagnostic Interface
J... - Engine Control Module (ECM)
K47 - ABS Warning Light
K118 - Warning light for brake system
K155 - ESP Control Lamp
T1e - Single Connector, behind fuse/relay panel
T2n - Double Connector
T2u - Double Connector
T2w - Double Connector
T2x - Double Connector
T2y - Double Connector
T2z - Double Connector
T16 - 16-Pin Data Link Connector (DLC), behind steering column cover
T32 - 32-Pin Connector, blue, on instrument cluster
T32a - 32-Pin Connector, green, on instrument cluster
T88 - 88-Pin Connector
TV14 - Data Link Connector (DLC) wire Connector, white
TV16 - Wire distributor for Data-Bus, on relay panel No. 5

(A76) - Connector (K-diagnosis wire) in instrument panel wiring harness

THIS PAGE INTENTIONALLY LEFT BLANK

Automatic transmission,

from December 2000

● 2.8L - Engine - Motronic Multiport Fuel Injection (MFI)/150 kW, code AXK

Deviate relay location and fuseplacements as well as the locations of multiple connectors see section "component locations".

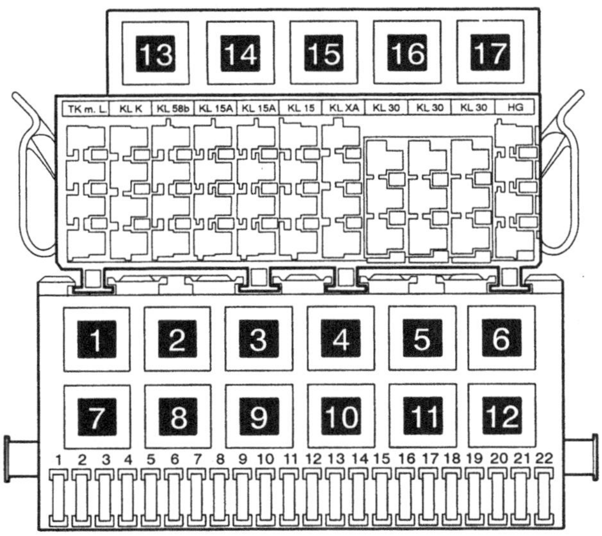

97-23232

Fuse colors

30 A - green

25 A - white

20 A - yellow

15 A - blue

10 A - red

7,5 A - brown

5 A - beige

3 A - violet

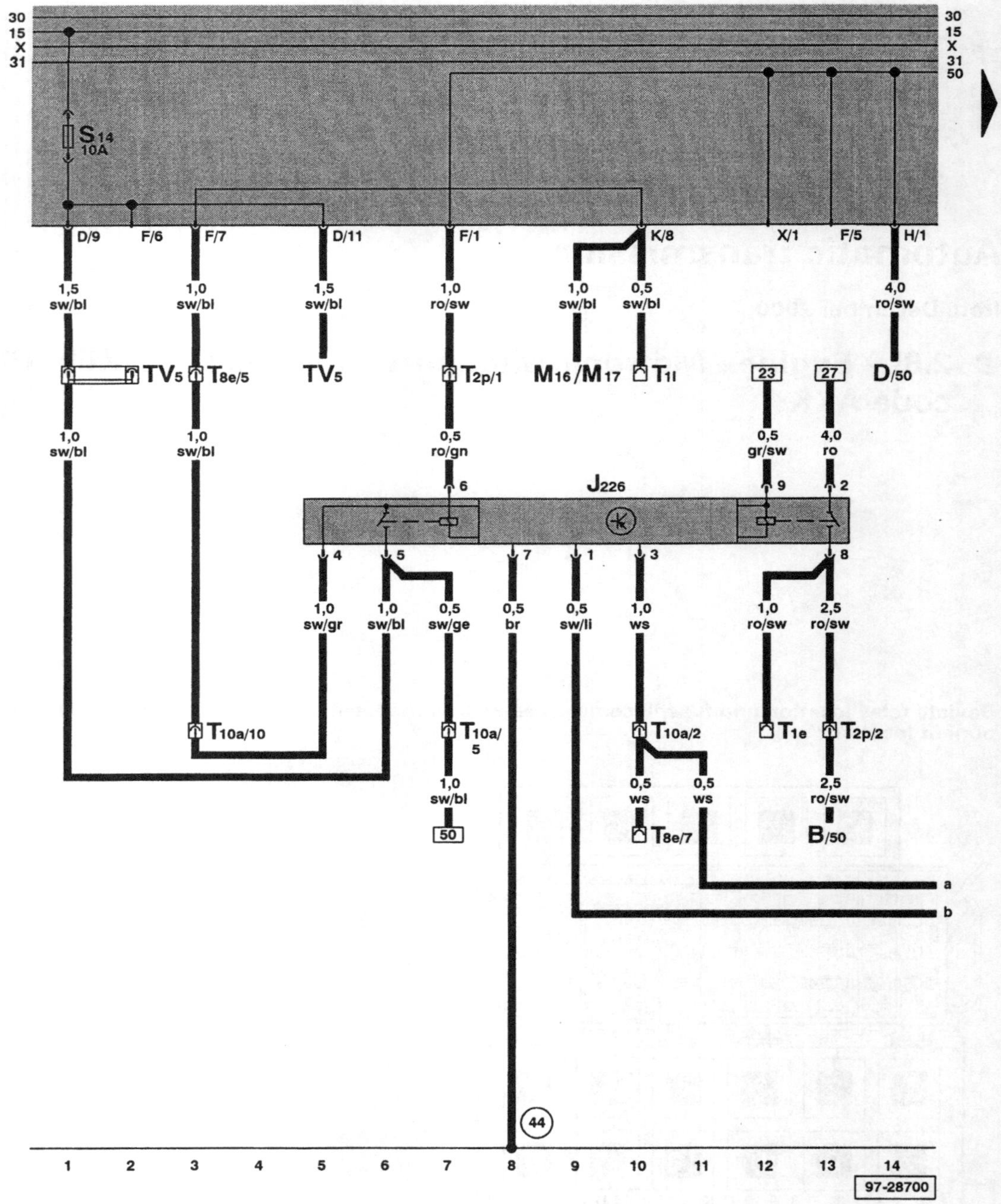

ws = white
sw = black
ro = red
br = brown
gn = green
bl = blue
gr = grey
li = lilac
ge = yellow
or = orange

B - Starter
D - Ignition/Starter Switch
J226 - Park/Neutral Position (PNP) Relay, behind instrument panel, center
M16 - Left Back-Up Light
M17 - Right Back-Up Light
T1e - Single Connector, behind fuse/relay panel
T1l - Single Connector, behind fuse/relay panel
T2p - Double Connector, behind fuse/relay panel
T8e - 8-Pin Connector, behind fuse/relay panel
T10a - 10-Pin Connector, behind fuse/relay panel
TV5 - Terminal 15a Wire Connector, green, 4-Pin and 8-Pin

(44) - Ground connection, on left A-pillar, lower part

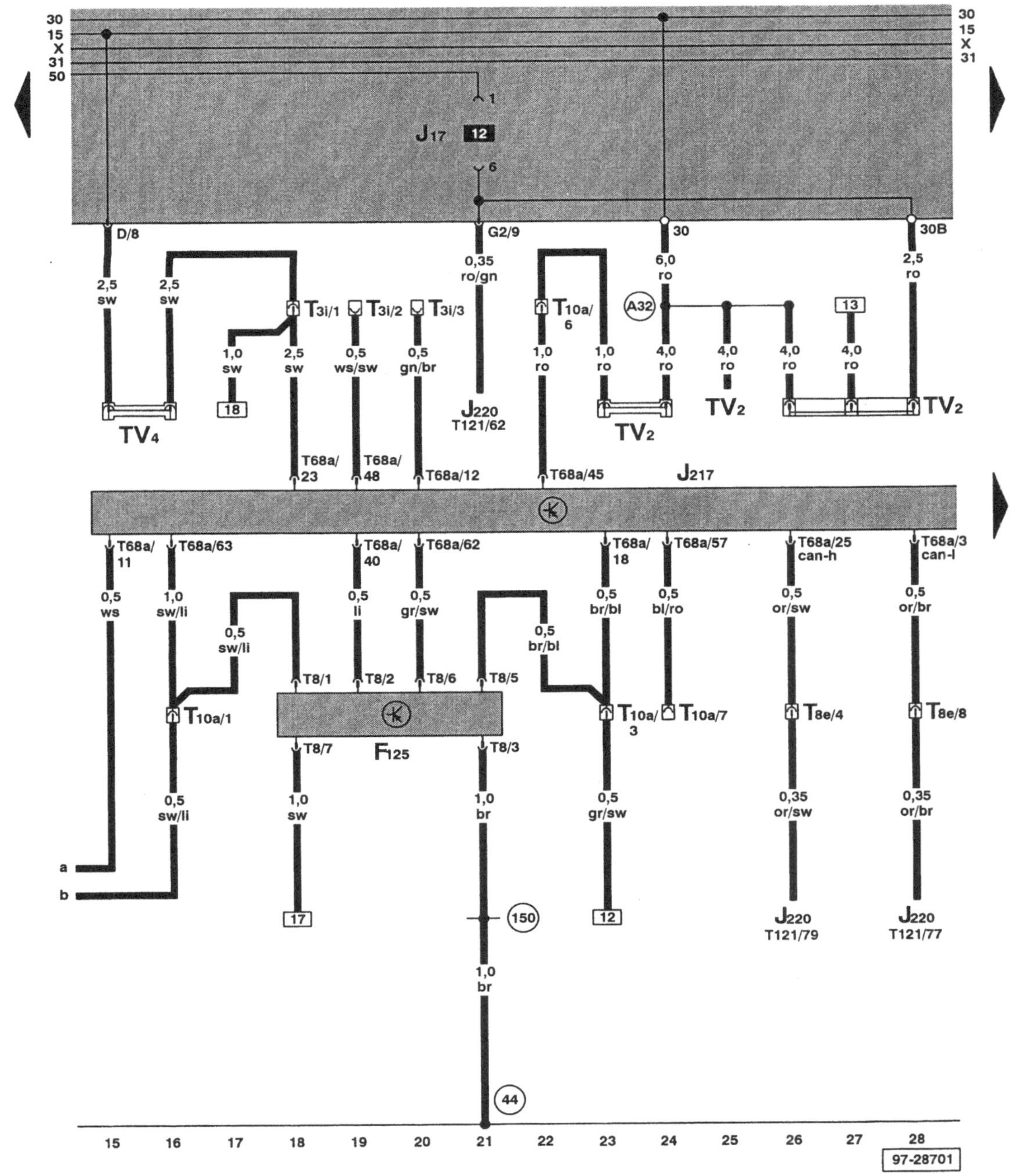

ws = white
sw = black
ro = red
br = brown
gn = green
bl = blue
gr = grey
li = lilac
ge = yellow
or = orange

F125 - Multi-Function Transmission Range (TR) Switch
J17 - Fuel Pump (FP) Relay
J217 - Transmission Control Module (TCM), on left A-pillar, lower part
J220 - Motronic Engine Control Module (ECM)
T3i - 3-Pin Connector, behind fuse/relay panel
T8 - 8-Pin Connector
T8e - 8-Pin Connector, behind fuse/relay panel
T10a - 10-Pin Connector, behind fuse/relay panel
T68a - 68-Pin Connector
T121 - 121-Pin Connector
TV2 - Terminal 30 (B+) Wire Connector, red, each 6 points
TV4 - Terminal 15 Wire Connector, black

(44) - Ground connection, on left A-pillar, lower part
(150) - Ground connection, in automatic transmission wiring harness
(A32) - Plus connection (30), in instrument panel wiring harness

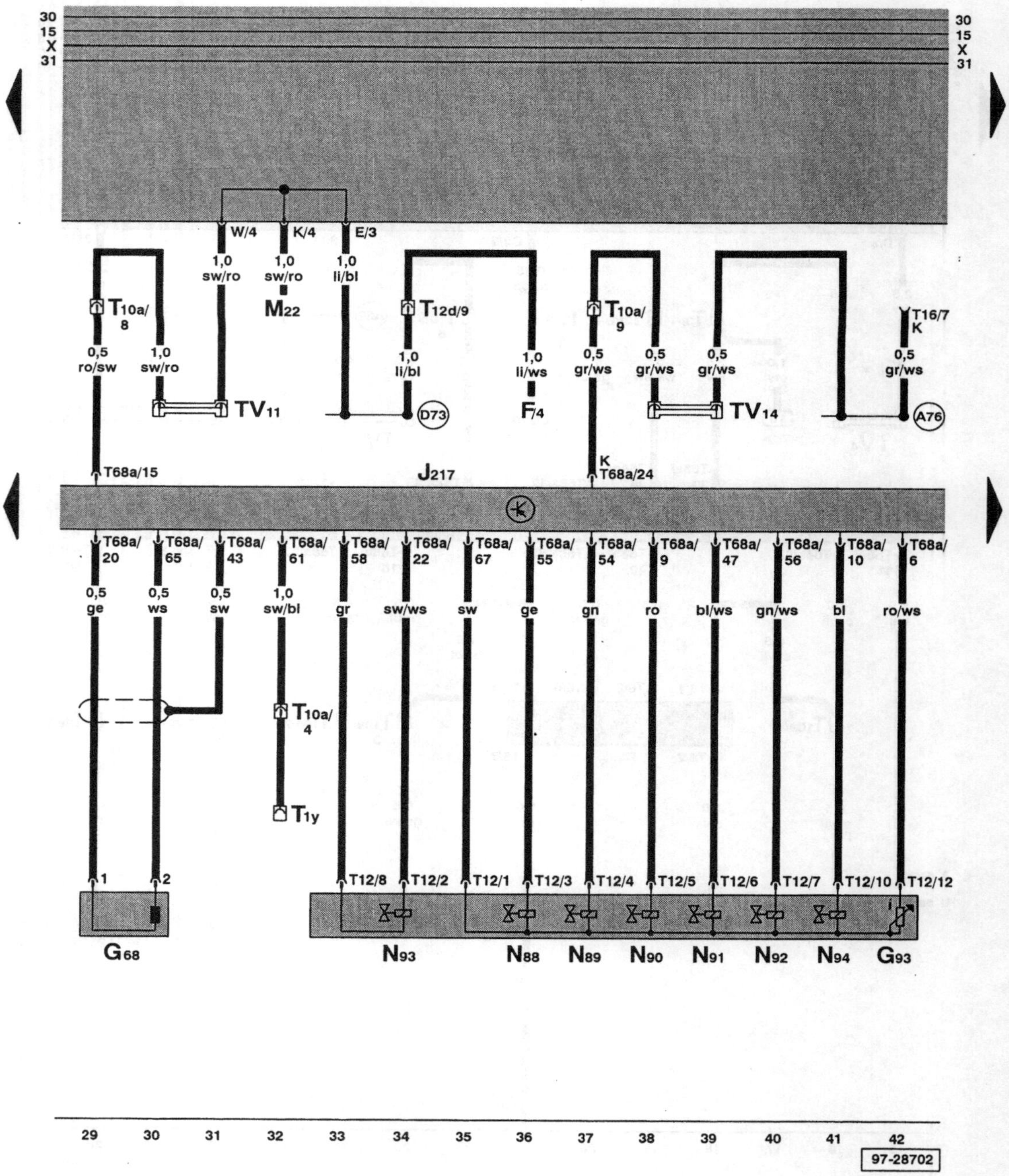

ws = white
sw = black
ro = red
br = brown
gn = green
bl = blue
gr = grey
li = lilac
ge = yellow
or = orange

F - Brake Light Switch
G68 - Vehicle Speed Sensor (VSS)
G93 - Transmission Fluid Temperature Sensor
J217 - Transmission Control Module (TCM), on left A-pillar, lower part
M22 - Right Brake/Tail Light
N88 - Solenoid Valve 1
N89 - Solenoid Valve 2
N90 - Solenoid Valve 3
N91 - Solenoid Valve 4
N92 - Solenoid Valve 5
N93 - Solenoid Valve 6
N94 - Solenoid Valve 7
T1y - Single Connector, behind fuse/relay panel
T10a - 10-Pin Connector, behind fuse/relay panel
T12 - 12-Pin Connector
T12d - 12-Pin Connector, behind fuse/relay panel
T16 - Data Link Connector (DLC), behind steering column cover
T68a - 68-Pin Connector
TV11 - Terminal 54 Wire Connector, orange
TV14 - Data Link Connector (DLC) wire Connector, white

A76 - Connector (K-diagnosis wire) in instrument panel wiring harness
D73 - Plus connection (54), in engine compartment wiring harness

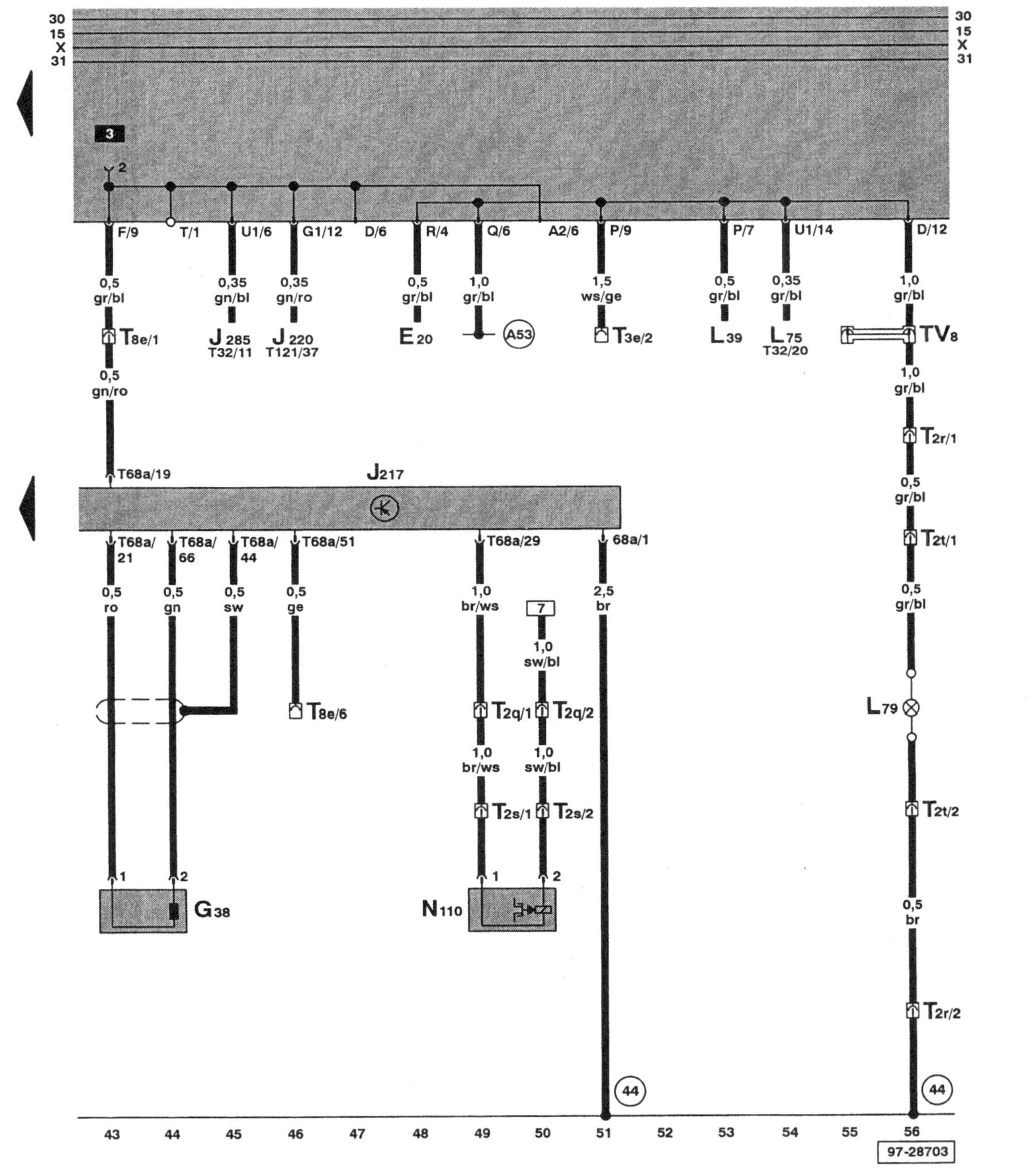

ws = white
sw = black
ro = red
br = brown
gn = green
bl = blue
gr = grey
li = lilac
ge = yellow
or = orange

E20 - Instrument Panel Light Dimmer Switch
G38 - Transmission Vehicle Speed Sensor (VSS)
J217 - Transmission Control Module (TCM), on left A-pillar, lower part
J220 - Motronic Engine Control Module (ECM
J285 - Control module with indicator unit in instrument panel insert
L39 - Rear Window Defogger Switch Light
L75 - Digital Display Light
L79 - Selector Lever Light
N110 - Shift Lock Solenoid
T2q - Double Connector, behind fuse/relay panel
T2r - Double Connector, behind fuse/relay panel
T2s - Double Connector, near selector lever
T2t - Double Connector, near selector lever
T3e - 3-Pin Connector, behind fuse/relay panel
T8e - 8-Pin Connector, behind fuse/relay panel
T32 - 32-Pin Connector, blue, on Instrument cluster
T68a - 68-Pin Connector
T121 - 121-Pin Connector
TV8 - Terminal 58b Wire Connector, grey

(44) - Ground connection, on left A-pillar, lower part

(A53) - Plus connection (58b), in instrument panel wiring harness

Transmission control module (TCM), shift lock solenoid, transmission vehicle speed sensor (VSS), selector lever light

THIS PAGE INTENTIONALLY LEFT

BLANK

2.8L - Engine - Motronic Multiport Fuel Injection (MFI)/150 kW, code AXK,

from December 2000

Deviate relay location and fuseplacements as well as the locations of multiple connectors see section "component locations".

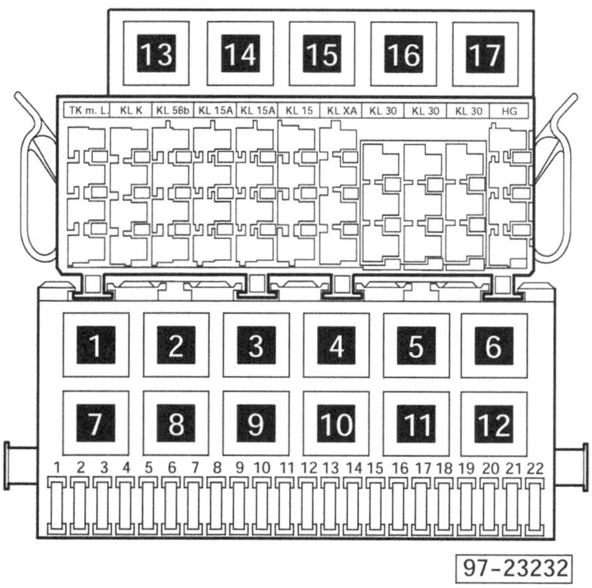

5	Wire distributor for Data-Bus
12	Fuel Pump (FP) Relay (167)
13	Coolant Circulation Pump Relay (53)
15	Secondary Air Injection (AIR) Pump Relay(53)

Note: Number in parentheses indicates production control number stamped on relay housing.

Fuse colors

30 A - green
25 A - white
20 A - yellow
15 A - blue
10 A - red
7,5 A - brown
5 A - beige
3 A - violet

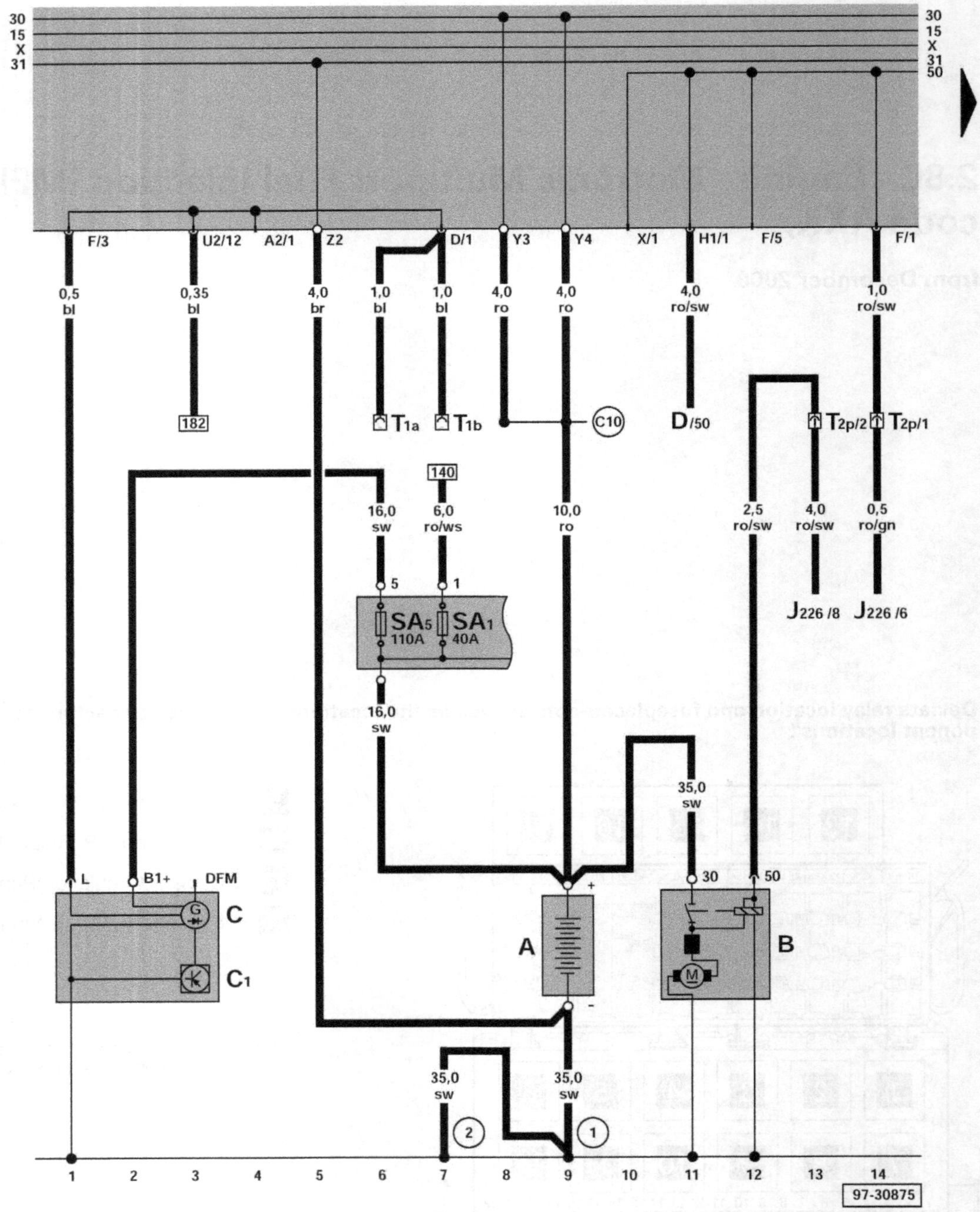

ws = white
sw = black
ro = red
br = brown
gn = green
bl = blue
gr = grey
li = lilac
ge = yellow
or = orange

A - Battery
B - Starter
C - Generator
C1 - Voltage Regulator
D - Ignition/Starter Switch
J226 - Park/Neutral Position (PNP) Relay
SA1 - Fuse 1 (in fuse holder/battery)
SA5 - Fuse 5 (in fuse holder/battery)
T1a - Single Connector, behind fuse/relay panel
T1b - Single Connector, behind fuse/relay panel
T2p - Double Connector, behind fuse/relay panel

(1) - Ground strap, battery to body

(2) - Ground strap, transmission to body

(C10) - Plus connection -1- (30), in headlight wiring harness

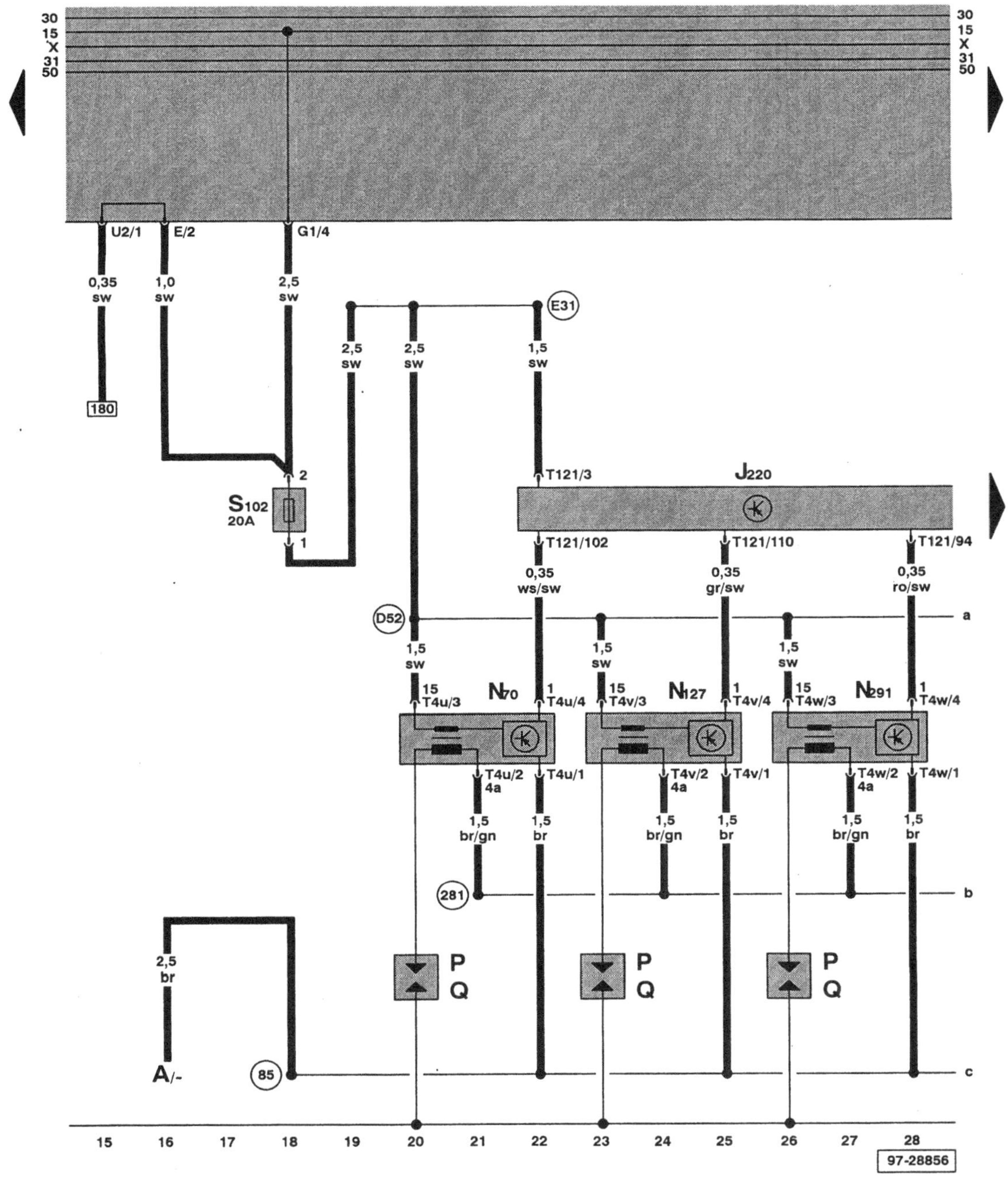

ws = white
sw = black
ro = red
br = brown
gn = green
bl = blue
gr = grey
li = lilac
ge = yellow
or = orange

A - Battery
J220 - Motronic Engine Control Module (ECM), in engine compartment, left
N70 - Ignition Coil 1 with Power Output Stage
N127 - Ignition Coil 2 with Power Output Stage
N291 - Ignition Coil 3 with Power Output Stage
P - Spark Plug Connectors
Q - Spark Plugs
S102 - Engine Control Module (ECM) Fuse
T4u - 4-Pin Connector
T4v - 4-Pin Connector
T4w - 4-Pin Connector
T121 - 121-Pin Connector

(85) - Ground connection -1-, in power seat wiring harness

(281) - Ground connector -1-, in wiring harness engine pre-wiring
(D52) - plus connection (15a), in engine compartment wiring harness
(E31) - Connector (15), in wiring harness engine

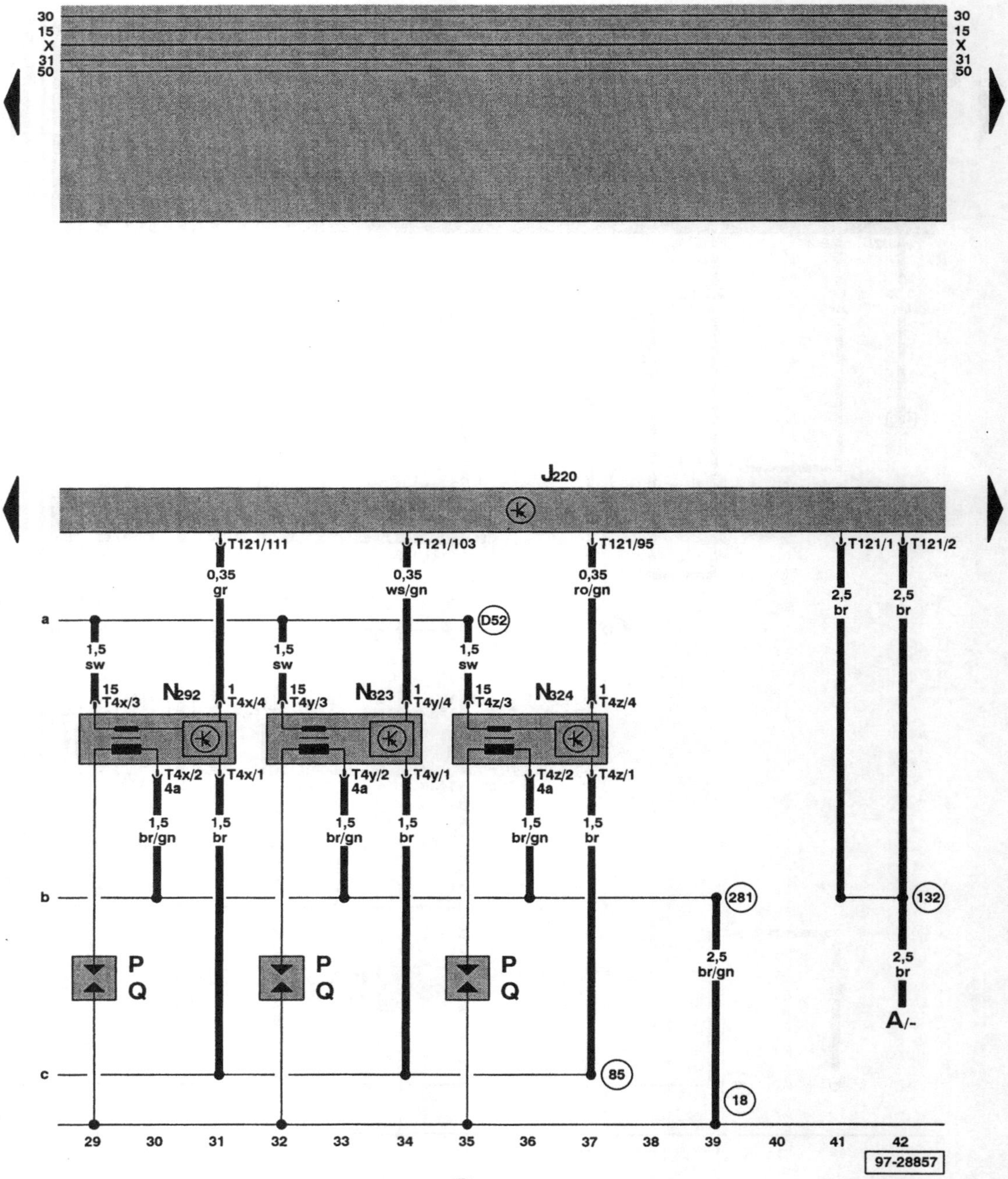

ws = white
sw = black
ro = red
br = brown
gn = green
bl = blue
gr = grey
li = lilac
ge = yellow
or = orange

A - Battery
J220 - Motronic Engine Control Module (ECM), in engine compartment, left
N292 - Ignition Coil 4 with Power Output Stage
N323 - Ignition Coil 5 with Power Output Stage
N324 - Ignition Coil 6 with Power Output Stage
P - Spark Plug Connectors
Q - Spark Plugs
T4x - 4-Pin Connector
T4y - 4-Pin Connector
T4z - 4-Pin Connector
T121 - 121-Pin Connector

(18) - Ground connection, on engine block, left

(85) - Ground connection -1-, in engine compartment wiring harness
(132) - Ground connection -3-, in engine compartment wiring harness
(281) - Ground connector -1-, in wiring harness engine pre-wiring
(D52) - plus connection (15a), in engine compartment wiring harness

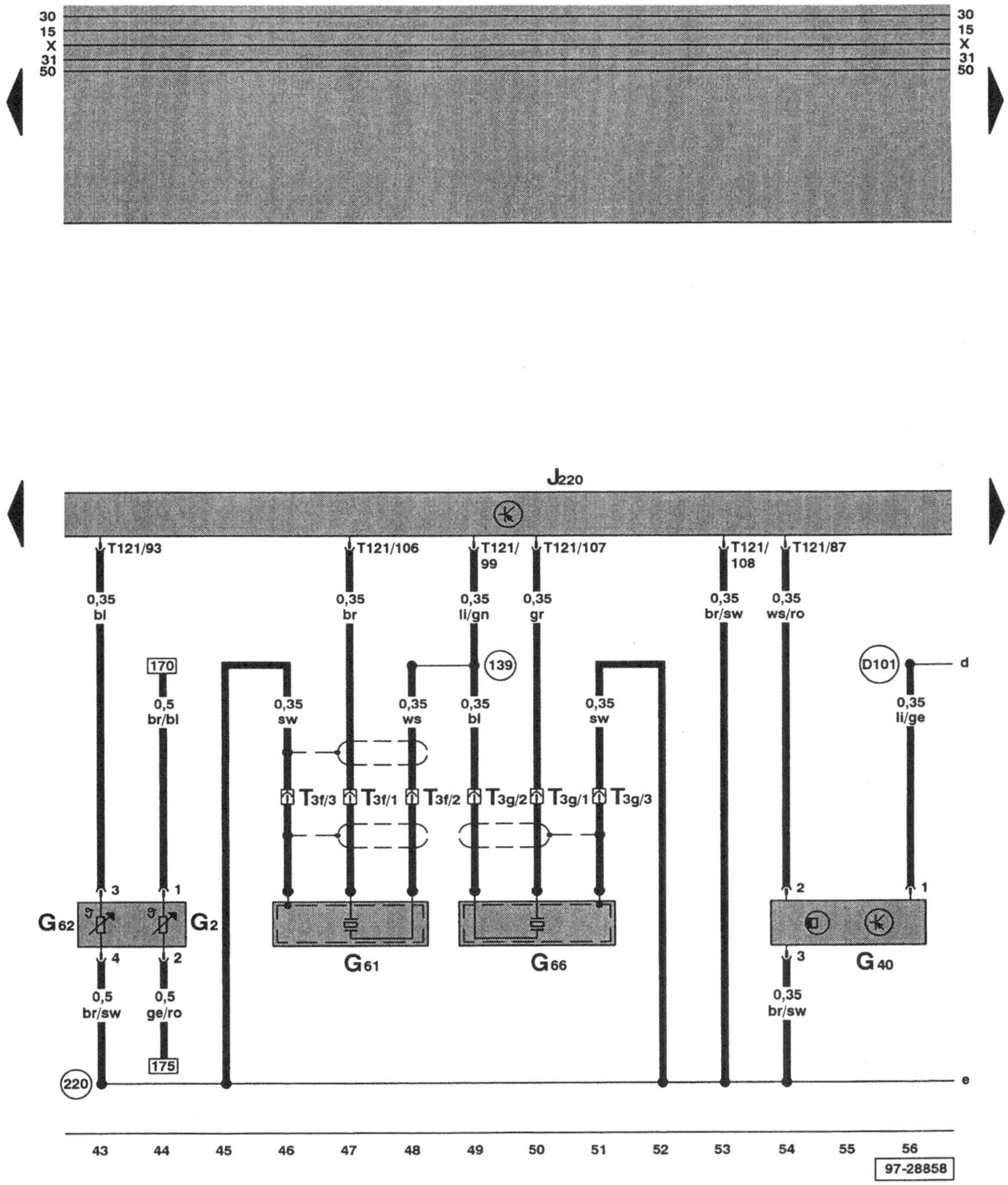

ws = white
sw = black
ro = red
br = brown
gn = green
bl = blue
gr = grey
li = lilac
ge = yellow
or = orange

G2 - Engine Coolant Temperature (ECT) Sensor
G40 - Camshaft Position (CMP) Sensor
G61 - Knock Sensor (KS) 1
G62 - Engine Coolant Temperature (ECT) Sensor
G66 - Knock Sensor (KS) 2
J220 - Motronic Engine Control Module (ECM), in engine compartment, left
T3f - 3-Pin Connector, near cylinder head cover
T3g - 3-Pin Connector, near cylinder head cover
T121 - 121-Pin Connector

(139) - Ground connection (sensor Ground), in Motronic Multiport Fuel Injection
(220) - Ground connection (sensor ground), in engine compartment wiring harness

(D101) - Wire connection -1-, in engine compartment wiring harness

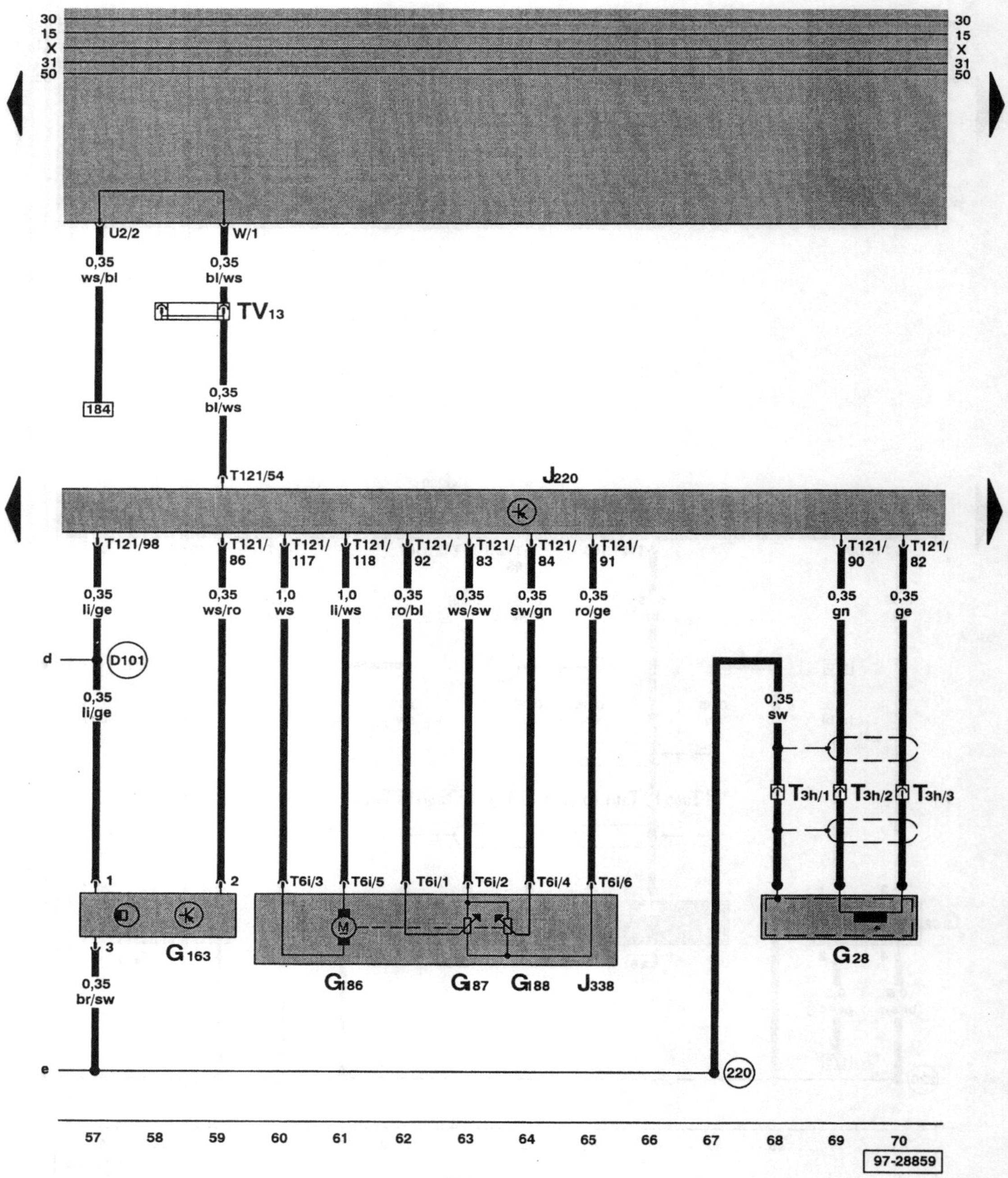

ws = white
sw = black
ro = red
br = brown
gn = green
bl = blue
gr = grey
li = lilac
ge = yellow
or = orange

G28 - Engine Speed (RPM) Sensor
G163 - Camshaft Position (CMP) Sensor 2
G186 - Throttle drive (power accelerator actuation)
G187 - Angle sensor -1- for throttle drive (power accelerator actuation)
G188 - Angle sensor -2- for throttle drive (power accelerator actuation)
J220 - Motronic Engine Control Module (ECM), in engine compartment, left
J338 - Throttle Valve Control Module
T3h - 3-Pin Connector, near starter
T6i - 6-Pin Connector
T121 - 121-Pin Connector
TV13 - Vehicle Speed Signal Wire Connector

(220) - Ground connection (sensor ground), in engine compartment wiring harness
(D101) - Wire connection -1-, in engine compartment wiring harness

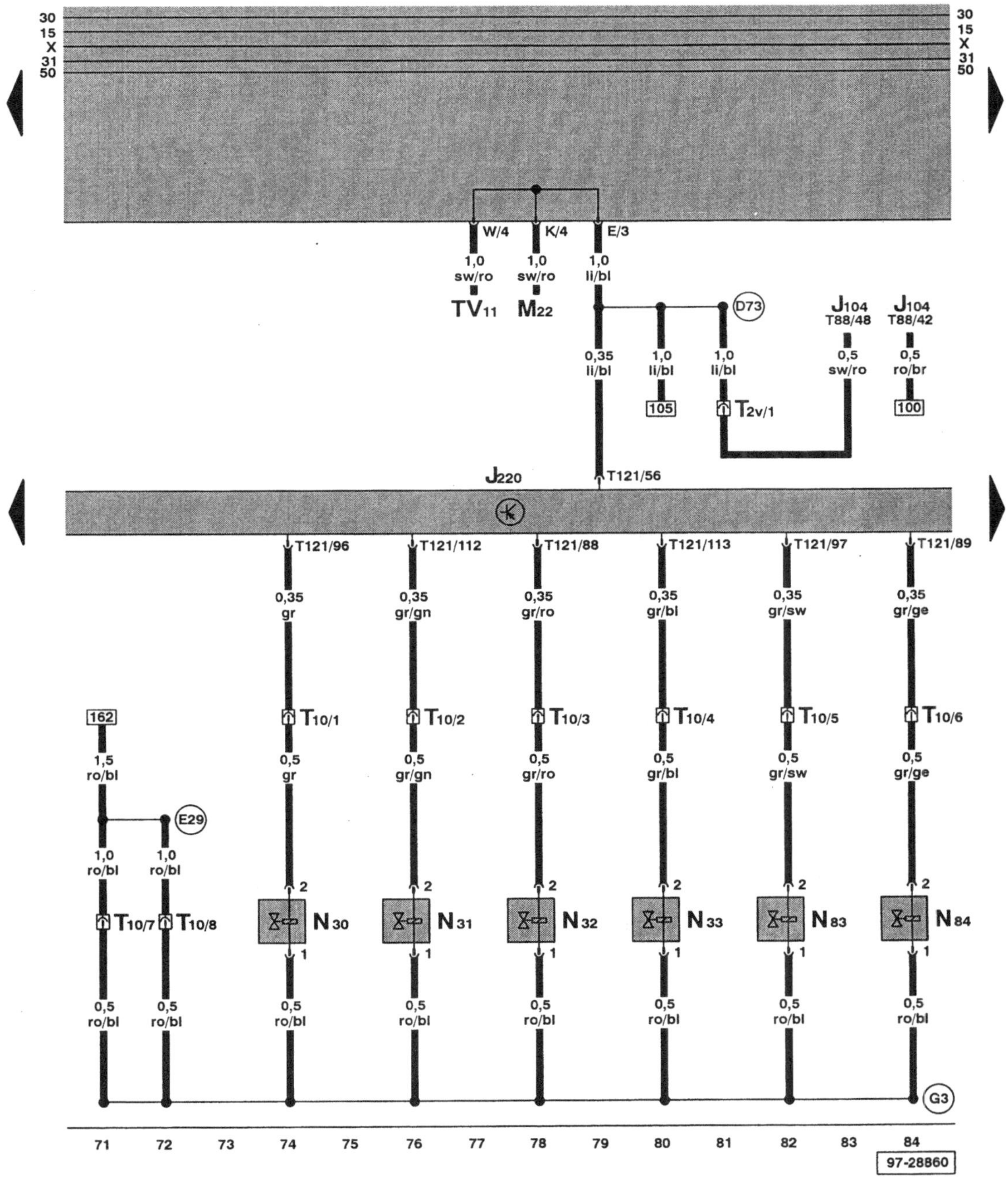

ws = white
sw = black
ro = red
br = brown
gn = green
bl = blue
gr = grey
li = lilac
ge = yellow
or = orange

J104 - ABS Control Module (with EDL), below instrument panel, right
J220 - Motronic Engine Control Module (ECM), in engine compartment, left
M22 - Right Brake/Tail Light
N30 - Cylinder 1 Fuel Injector
N31 - Cylinder 2 Fuel Injector
N32 - Cylinder 3 Fuel Injector
N33 - Cylinder 4 Fuel Injector
N83 - Cylinder 5 Fuel Injector
N84 - Cylinder 6 Fuel Injector
T2v - Double Connector, behind fuse/relay panel
T10 - 10-Pin Connector, near cylinder head cover
T88 - 88-Pin Connector
T121 - 121-Pin Connector
TV11 - Terminal 54 Wire Connector, orange

(D73) - Plus connection (54), in engine compartment wiring harness
(E29) - Connector, in wiring harness engine
(G3) - Plus connection, in injector wiring harness

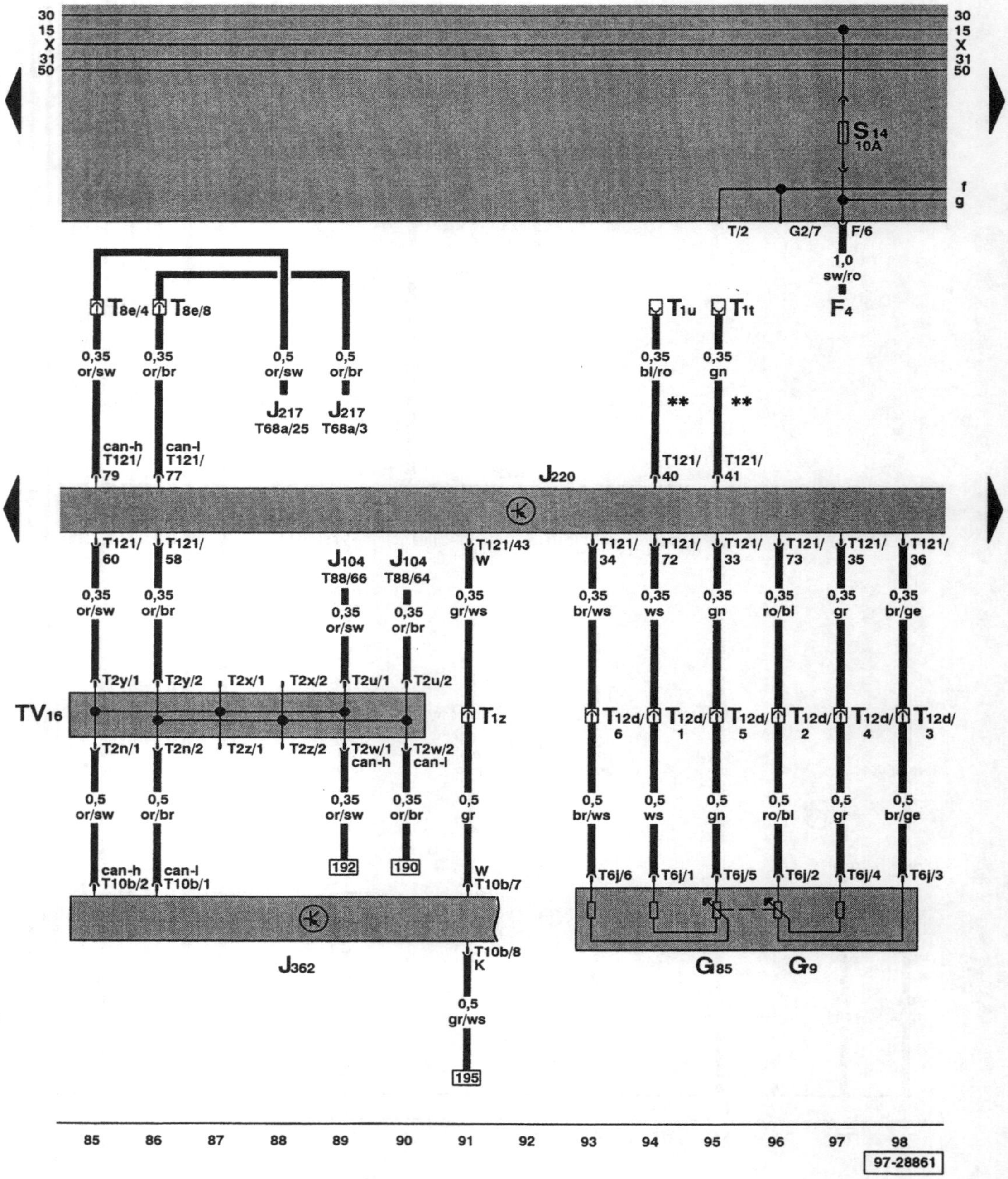

ws = white
sw = black
ro = red
br = brown
gn = green
bl = blue
gr = grey
li = lilac
ge = yellow
or = orange

F4 - Back-Up Light Switch
G79 - Throttle Position (TP) Sensor
G185 - Sender -2- for accelerator pedal position
J104 - ABS Control Module (with EDL), below instrument panel, right
J220 - Motronic Engine Control Module (ECM), in engine compartment, left
J217 - Transmission Control Module (TCM)
J362 - Control module for anti-theft immobilizer
T1t - Single Connector, behind fuse/relay panel
T1u - Single Connector, behind fuse/relay panel
T1z - Single Connector, behind fuse/relay panel
T2n - Double Connector
T2u - Double Connector
T2w - Double Connector
T2x - Double Connector
T2y - Double Connector
T2z - Double Connector
T6j - 6-Pin Connector
T8e - 8-Pin Connector, behind fuse/relay panel
T10b - 10-Pin Connector
T12d - 12-Pin Connector, behind fuse/relay panel
T68a - 68-Pin Connector
T88 - 88-Pin Connector
T121 - 121-Pin Connector
TV16 - Wire distributor for Data-Bus, on relay panel No. 5

** - A/C connection

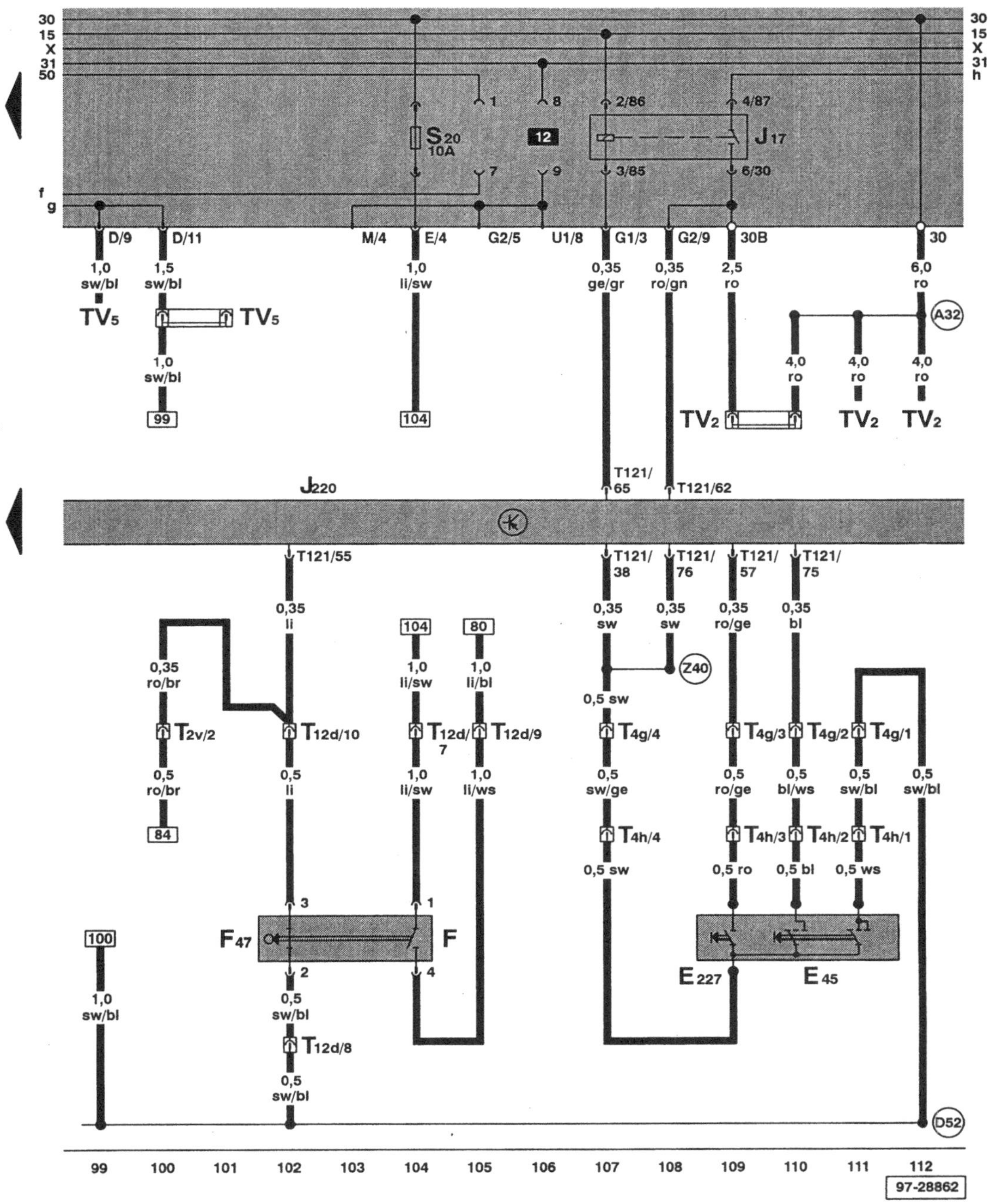

ws = white
sw = black
ro = red
br = brown
gn = green
bl = blue
gr = grey
li = lilac
ge = yellow
or = orange

E45 - Cruise Control Switch
E227 - Button for cruise control (set)
F - Brake Light Switch
F47 - Brake Vacuum Vent Valve Switch for cruise control
J17 - Fuel Pump (FP) Relay
J220 - Motronic Engine Control Module (ECM), in engine compartment, left
T2v - Double Connector, behind fuse/relay panel
T4g - 4-Pin Connector, behind fuse/relay panel
T4h - 4-Pin Connector, near steering column
T12d - 12-Pin Connector, behind fuse/relay panel
T121 - 121-Pin Connector
TV2 - Terminal 30 (B+) Wire Connector, red, each 6 points
TV5 - Terminal 15a Wire Connector, green, 4-Pin and 8-Pin

(A32) - Plus connection (30), in instrument panel wiring harness
(D52) - Plus connection (15a), in engine compartment wiring harness
(Z40) - Plus connection (15), in cruise control wiring harness

Motronic engine control module (ECM), cruise control switch, brake vacuum vent valve switch for cruise control, fuel pump (FP) relay

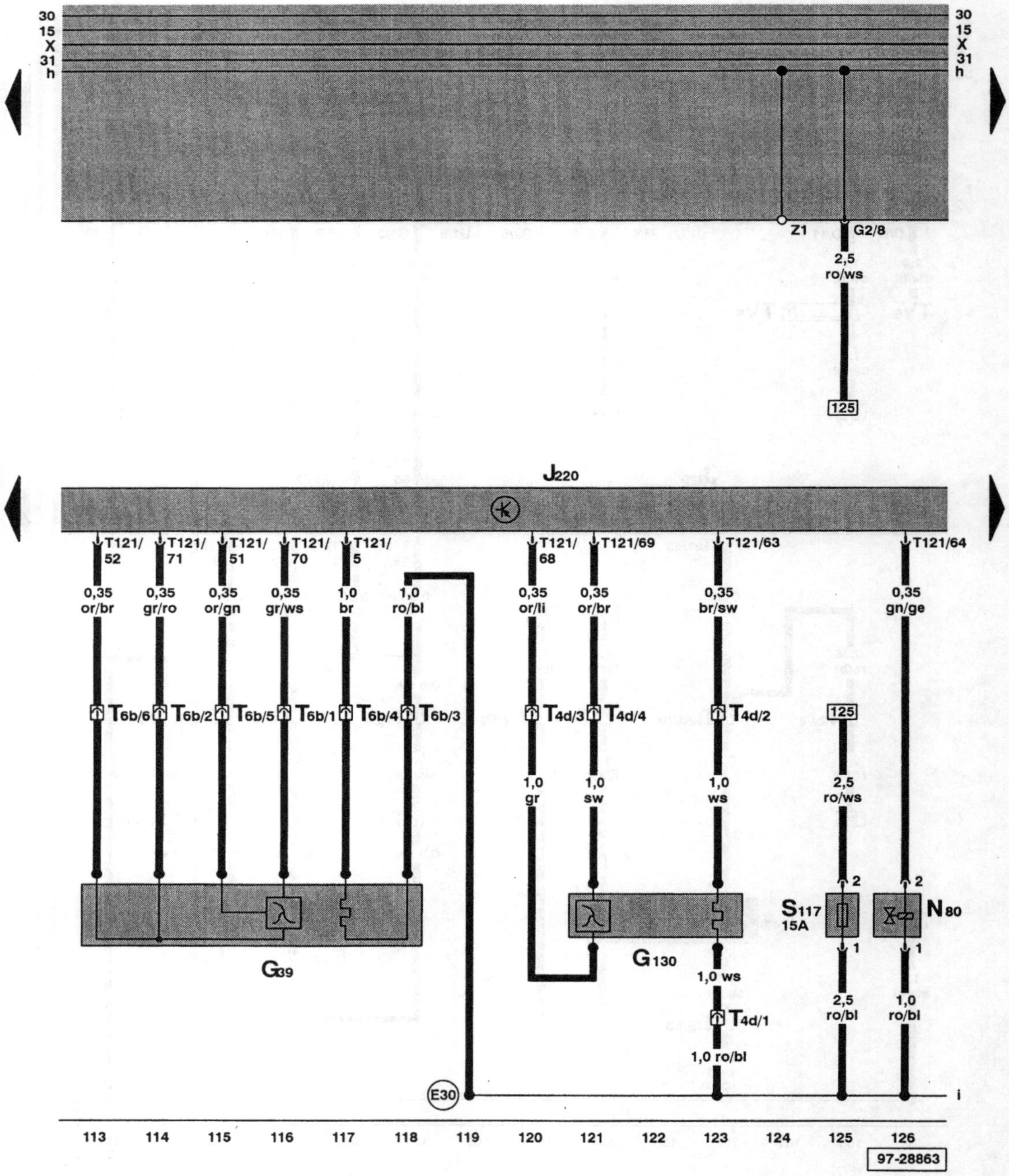

ws = white
sw = black
ro = red
br = brown
gn = green
bl = blue
gr = grey
li = lilac
ge = yellow
or = orange

G39 - Heated Oxygen Sensor (HO2S)
G130 - Oxygen Sensor (O2S) Behind Three Way Catalytic Converter (TWC)
J220 - Motronic Engine Control Module (ECM), in engine compartment, left
N80 - Evaporative Emission (EVAP) Canister Purge Regulator Valve
S117 - Fuse for control elements
T4d - 4-Pin Connector, near catalytic converter
T6b - 4-Pin Connector, near catalytic converter
T121 - 121-Pin Connector

(E30) - Connector (87a), in wiring harness engine

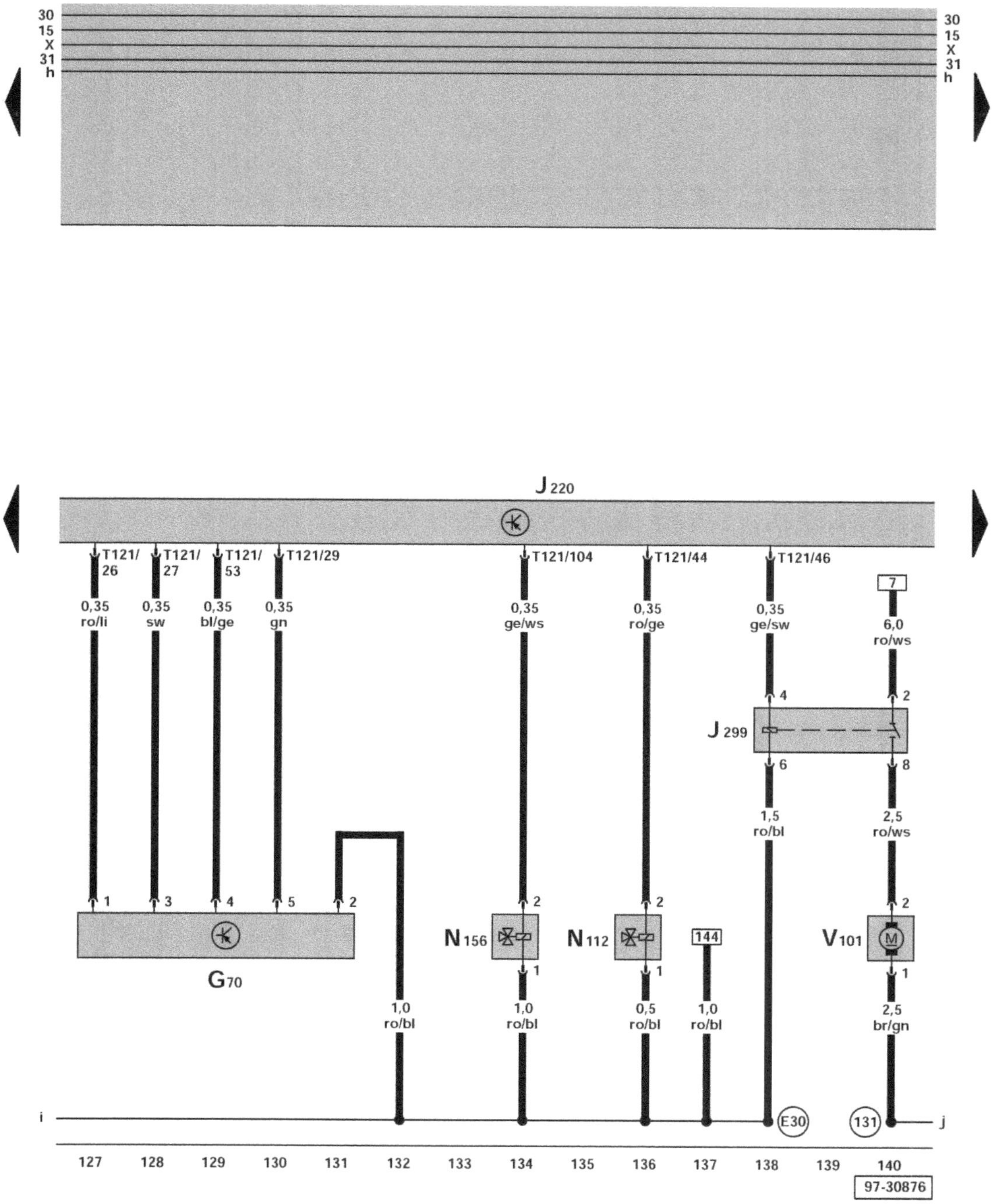

ws = white
sw = black
ro = red
br = brown
gn = green
bl = blue
gr = grey
li = lilac
ge = yellow
or = orange

G70 - Mass Air Flow (MAF) Sensor
J220 - Motronic Engine Control Module (ECM), in engine compartment, left
J299 - Secondary Air Injection (AIR) Pump Relay, above fuse/relay panel
N112 - Secondary Air Injection (AIR) Solenoid Valve
N156 - Intake Manifold Change-Over Valve
T121 - 121-Pin Connector
V101 - Secondary Air Injection (AIR) Pump Motor

(131) - Ground connection -2-, in engine compartment wiring harness
(E30) - Connector (87a), in wiring harness engine

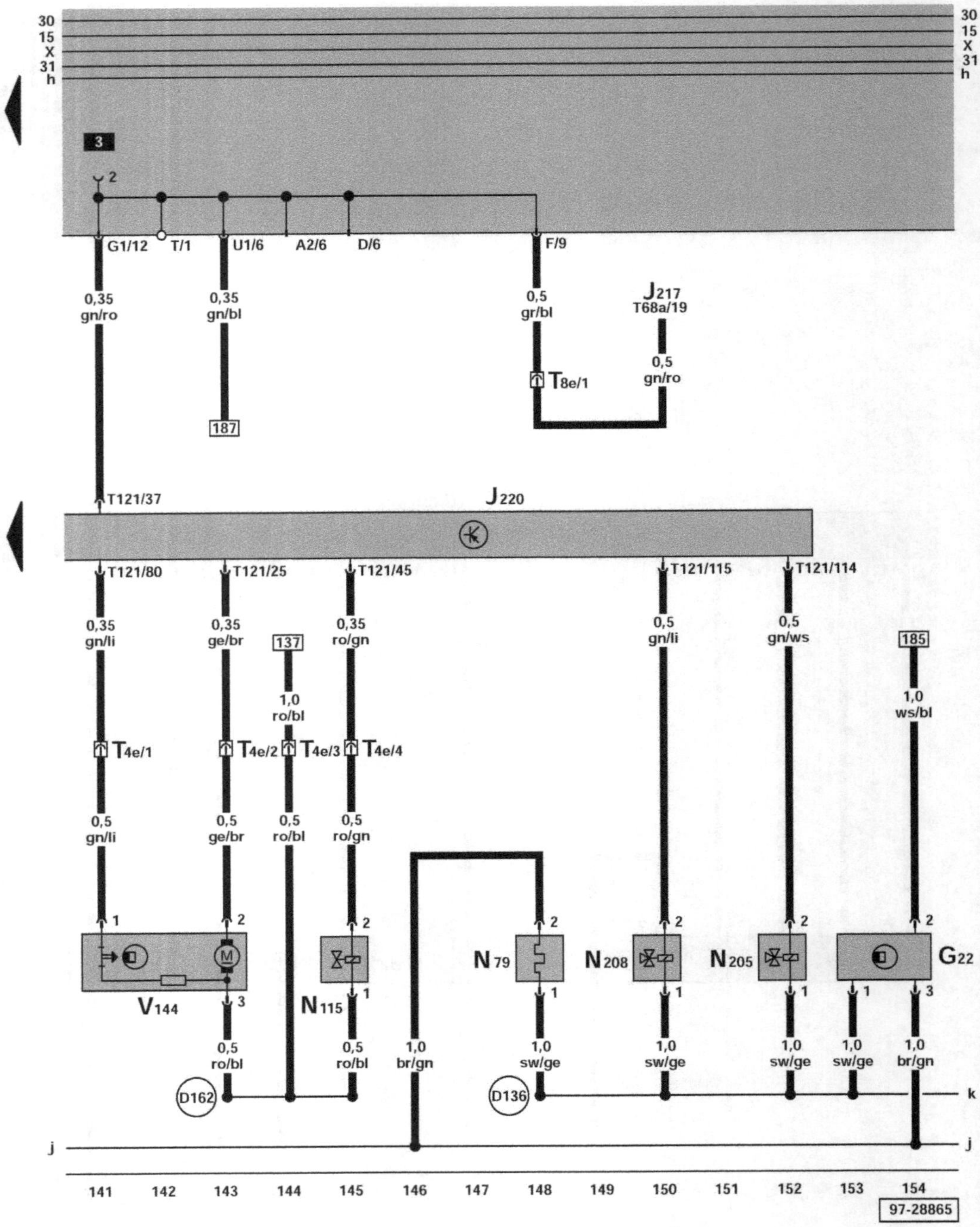

ws = white
sw = black
ro = red
br = brown
gn = green
bl = blue
gr = grey
li = lilac
ge = yellow
or = orange

G22 - Speedometer Vehicle Speed Sensor (VSS)
J217 - Transmission Control Module (TCM)
J220 - Motronic Engine Control Module (ECM), in engine compartment, left
N79 - Positive Crankcase Ventilation (PCV) Heating Element
N115 - Evaporative Emission (EVAP) Canister Purge Solenoid Valve
N205 - Valve -1- for camshaft adjustment
N208 - Valve -2- for camshaft adjustment
T4e - 4-Pin Connector, in engine compartment, left
T8e - 8-Pin Connector, behind fuse/relay panel
T68a - 68-Pin Connector
T121 - 121-Pin Connector
V144 - Leak detection pump (LDP)

(D136) - Plus connection -2- (15a) in wiring harness engine compartment
(D162) - Wire connection, leak detection pump, in engine wiring harness

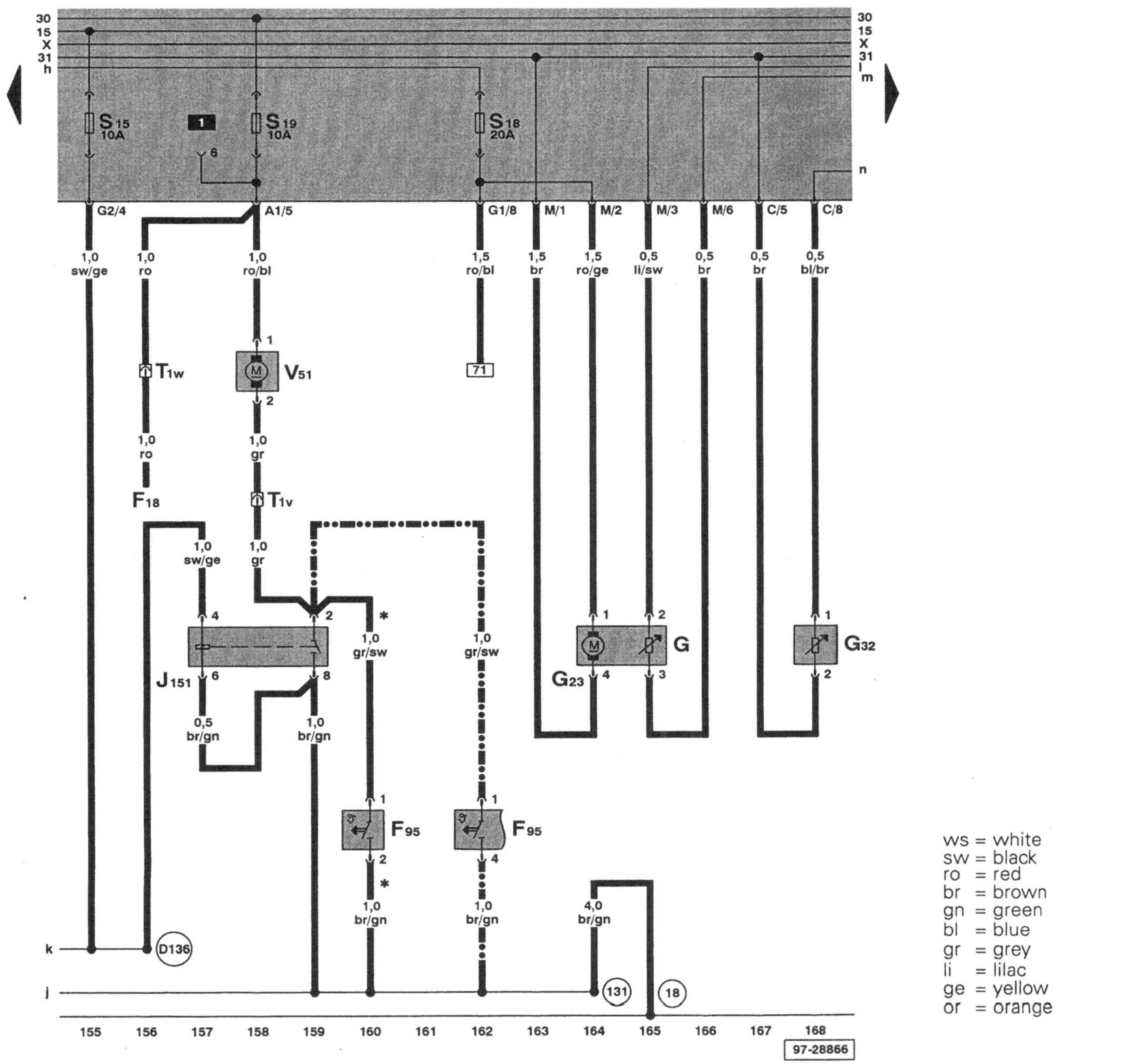

F18 - Coolant Fan Control (FC) Thermal Switch
F95 - After-Run Coolant Thermal Switch
G - Sender for fuel gauge
G23 - Transfer Fuel Pump (FP)
G32 - Engine Coolant Level (ECL) Sensor
J151 - Coolant Circulation Pump Relay, above fuse/relay panel
T1v - Single Connector, behind fuse/relay panel
T1w - Single Connector, behind fuse/relay panel
V51 - After-Run Coolant Pump

(18) - Ground connection, on engine block, left

(131) - Ground connection -2-, in engine compartment wiring harness

(D136) - Plus connection -2- (15a) in wiring harness engine compartment

* - Vehicles without Climatronic

-•••- - Vehicles with Climatronic

Coolant circulation pump relay, after-run coolant thermal switch, after-run coolant pump, transfer fuel pump (FP), sender for fuel gauge, engine coolant level (ECL) sensor

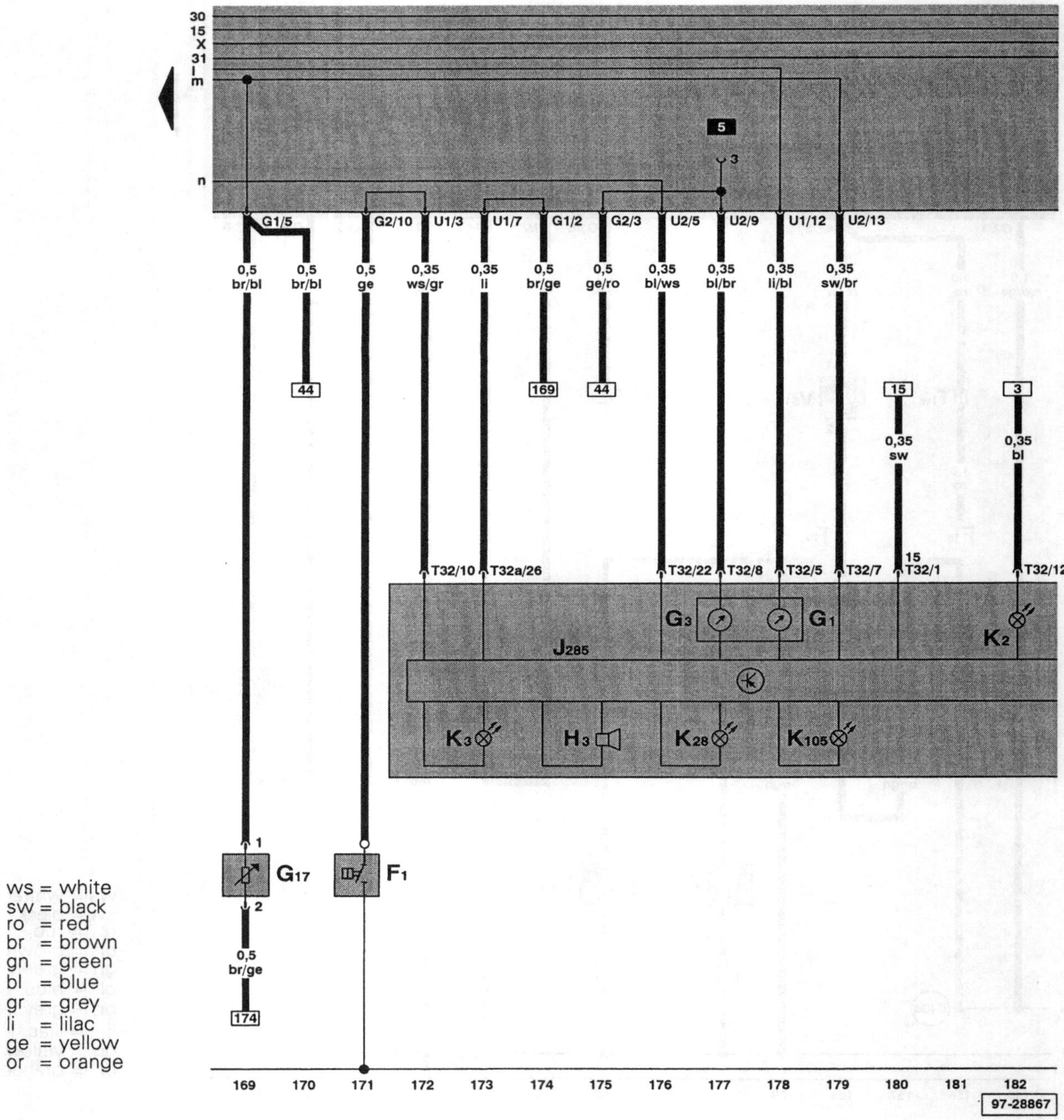

F1 - Oil Pressure Switch
G1 - Fuel Gauge
G3 - Engine Coolant Temperature (ECT) Gauge
G17 - Outside Air Temperature Sensor
H3 - Warning Buzzer
J285 - Control module with indicator unit in instrument panel insert
K2 - Generator (GEN) Warning Light
K3 - Oil Pressure Warning Light
K28 - Engine Coolant Level/Temperature (ECL/ECT) Warning Light
K105 - Low Fuel Level Warning Light
T32 - 32-Pin Connector, blue
T32a - 32-Pin Connector, green

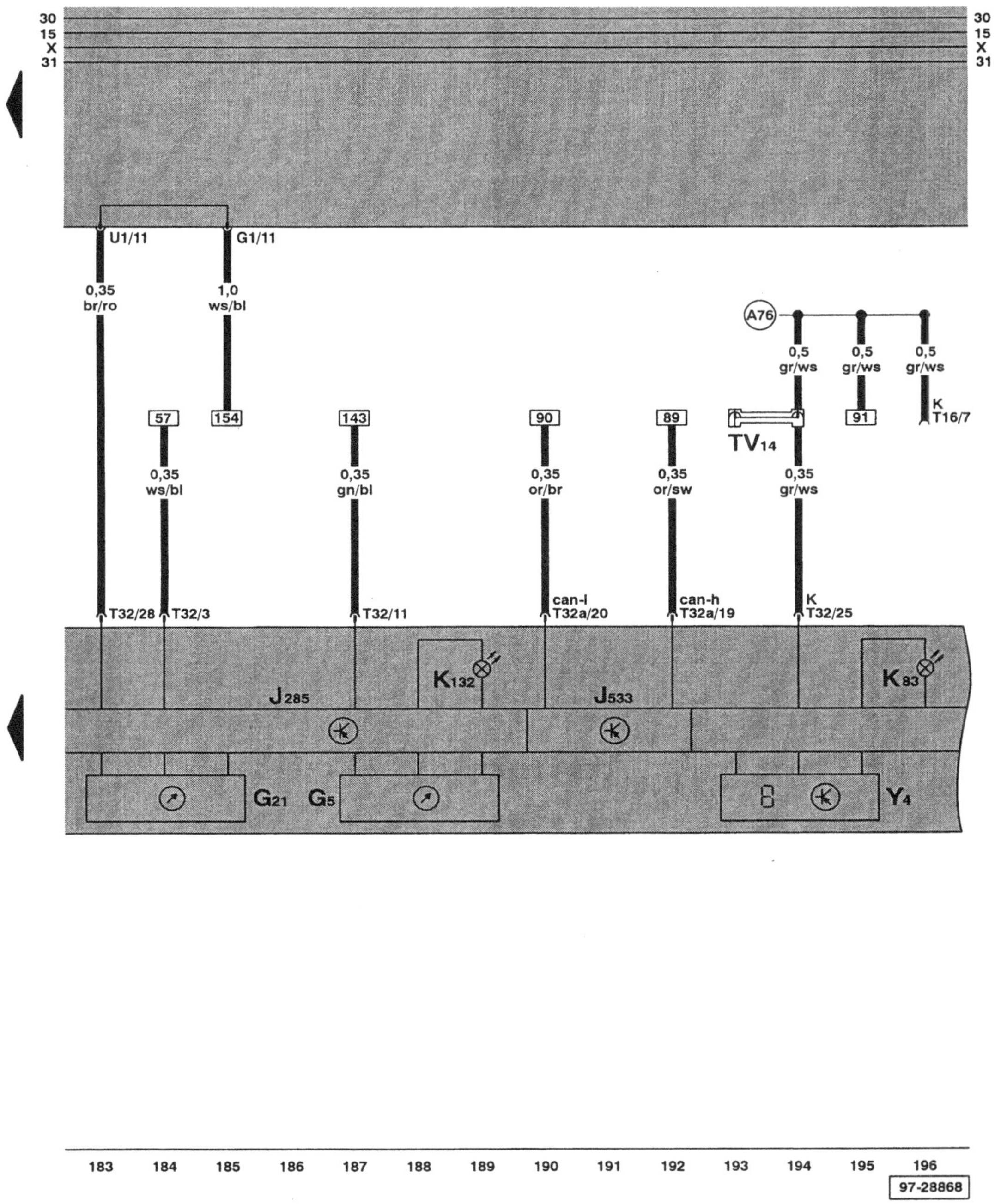

ws = white
sw = black
ro = red
br = brown
gn = green
bl = blue
gr = grey
li = lilac
ge = yellow
or = orange

G5 - Tachometer
G21 - Speedometer
J285 - Control module with indicator unit in instrument panel insert
J533 - Data Bus On Board Diagnostic Interface
K83 - Malfunction Indicator Lamp (MIL)
K132 - Electronic Power Control (EPC) Warning Lamp
T16 - Data Link Connector (DLC), behind steering column cover
T32 - 32-Pin Connector, blue
T32a - 32-Pin Connector, green
TV14 - Data Link Connector (DLC) wire Connector
Y4 - Odometer Display

(A76) - Connector (K-diagnosis wire) in instrument panel wiring harness

Instrument cluster, malfunction indicator lamp (MIL), electronic power control (EPC) warning lamp, speedometer, tachometer, data link connector (DLC)

THIS PAGE INTENTIONALLY LEFT
BLANK

Auxiliary Battery,

from December 2000

Deviate relay location and fuseplacements as well as the locations of multiple connectors see section "component locations".

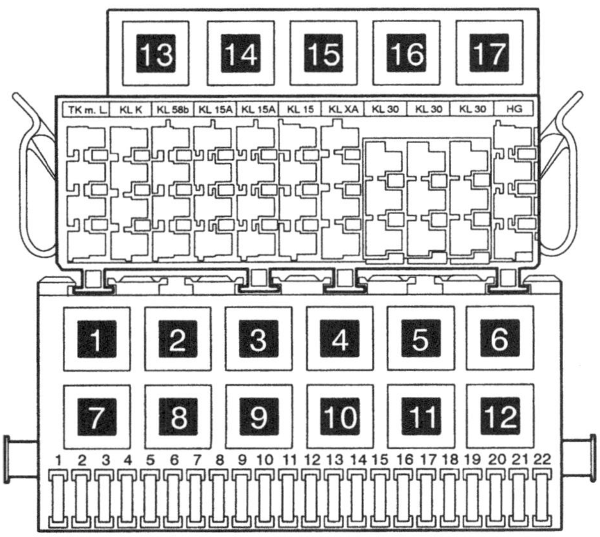

97-23232

Fuse colors

30 A - green

25 A - white

20 A - yellow

15 A - blue

10 A - red

7,5 A - brown

5 A - beige

3 A - violet

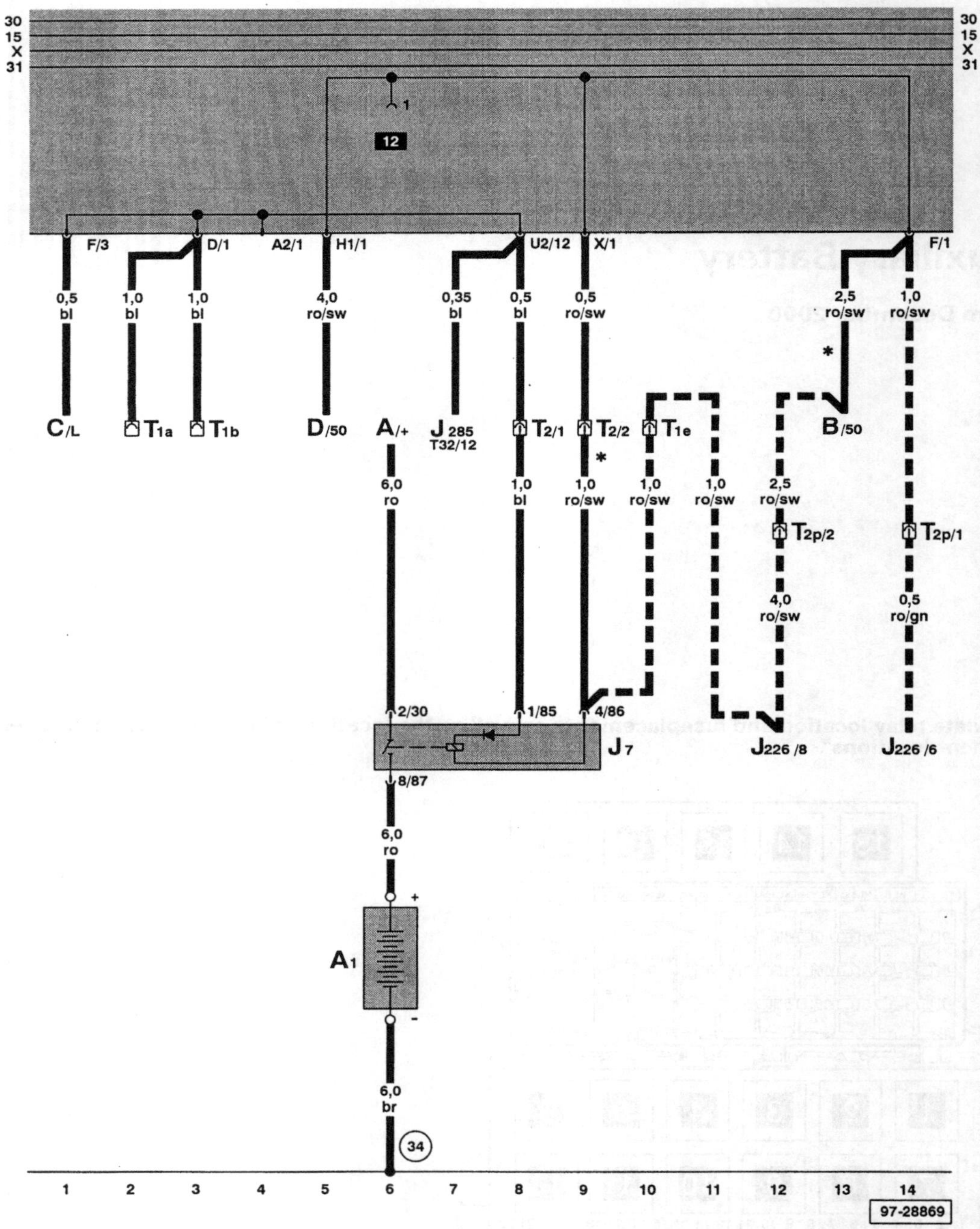

ws = white
sw = black
ro = red
br = brown
gn = green
bl = blue
gr = grey
li = lilac
ge = yellow
or = orange

A - Battery
A1 - Auxiliary Battery
B - Starter
C - Generator
D - Ignition/Starter Switch
J7 - Battery Cut Out Relay (production control number 426), below driver's seat
J226 - Park/Neutral Position (PNP) Relay
J285 - Control module with indicator unit in instrument panel insert
T1a - Single Connector, behind fuse/relay panel
T1b - Single Connector, behind fuse/relay panel
T1e - Single Connector, behind fuse/relay panel
T2 - Double Connector, behind fuse/relay panel
T2p - Double Connector, behind fuse/relay panel
T32 - 32-Pin Connector, blue, on Instrument cluster

(34) - Ground connection, below driver's seat

* - Manual transmission only

- - - - Automatic transmission only

Air-Conditioning,

from December 2000

● 2.8L - Engine - Motronic Multiport Fuel Injection (MFI)/150 kW, code AXK

Deviate relay location and fuseplacements as well as the locations of multiple connectors see section "component locations".

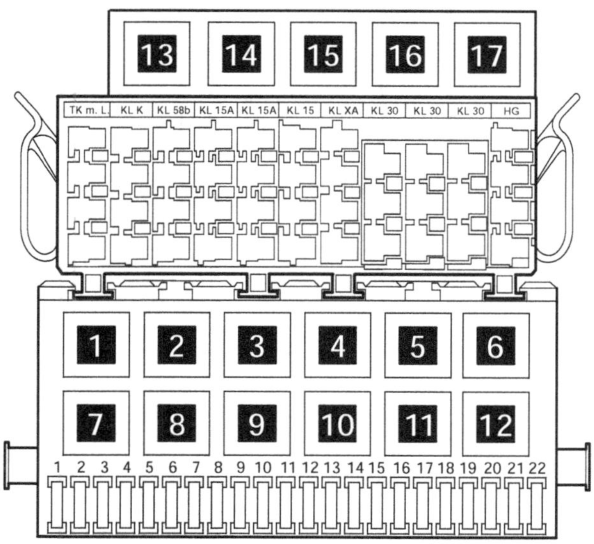

97-23232

1 A/C Relay (140)

16 Relay for A/C shutoff (147)

Note: Number in parentheses indicates production control number stamped on relay housing.

Fuse colors

30 A - green

25 A - white

20 A - yellow

15 A - blue

10 A - red

7,5 A - brown

5 A - beige

3 A - violet

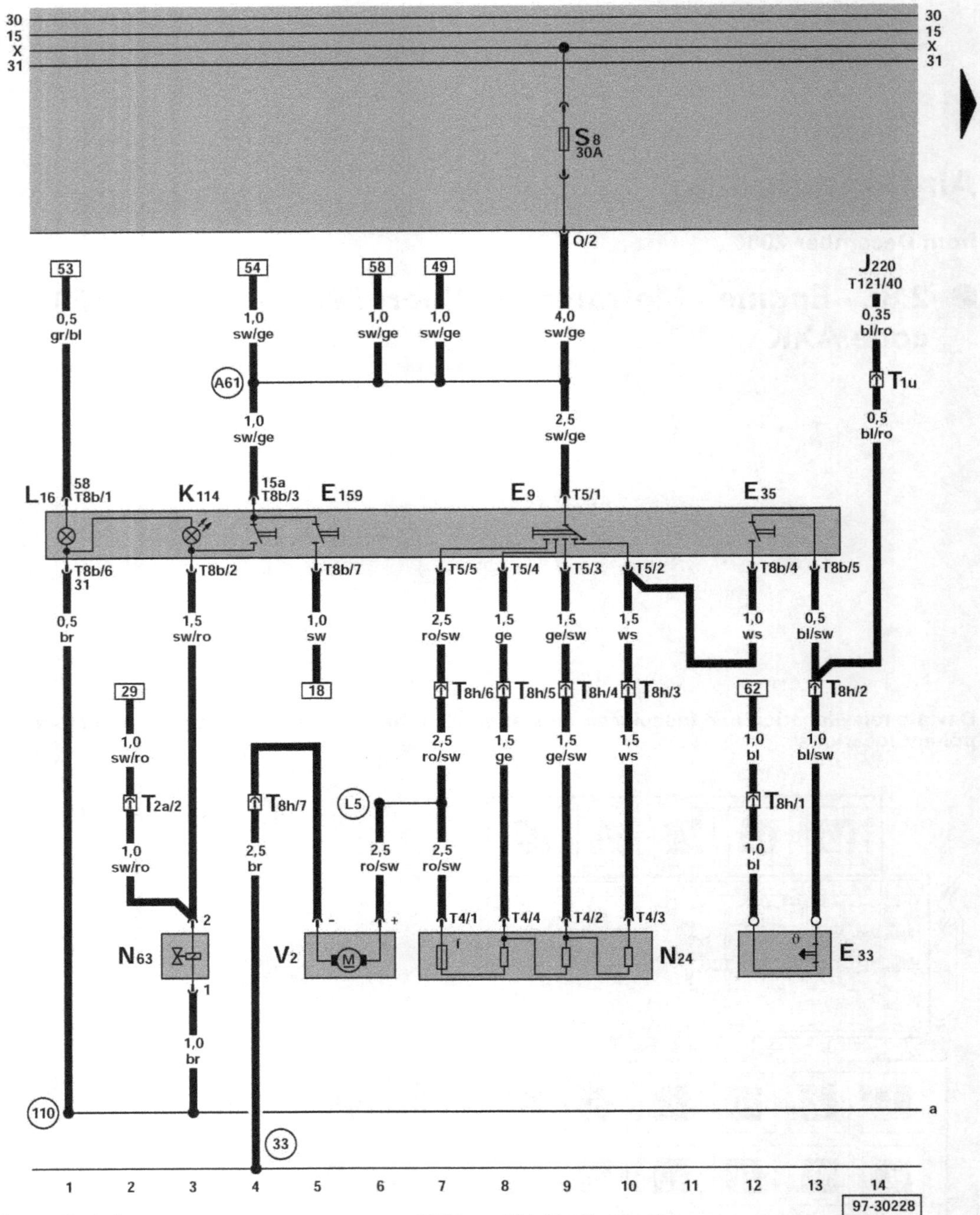

ws = white
sw = black
ro = red
br = brown
gn = green
bl = blue
gr = grey
li = lilac
ge = yellow
or = orange

E9 - Fresh Air Blower Switch
E33 - A/C Evaporator Temperature Switch
E35 - A/C Switch
E159 - Fresh Air/Recirculating Flap Switch
J220 - Motronic Engine Control Module (ECM)
K114 - Fresh Air and Recirculating Air Mode Indicator Light
L16 - Fresh Air Control Lever Light
N24 - Fresh Air Blower Series Resistance with Fuse
N63 - Fresh Air/Recirculating Flap Two-Way Valve
T1u - Single Connector, behind fuse/relay panel
T2a - Double Connector, behind fuse/relay panel
T4 - 4-Pin Connector
T5 - 5-Pin Connector
T8b - 8-Pin Connector
T8h - 8-Pin Connector, behind instrument panel, center
T121 - 121-Pin Connector
V2 - Fresh Air Blower

(33) - Ground connection, behind instrument panel right
(110) - Ground connection -2-, in instrument panel wiring harness
(A61) - Plus Connection (X), in Instrument cluster wiring harness
(L5) - Wire connection -1-, in evaporator housing wiring harness

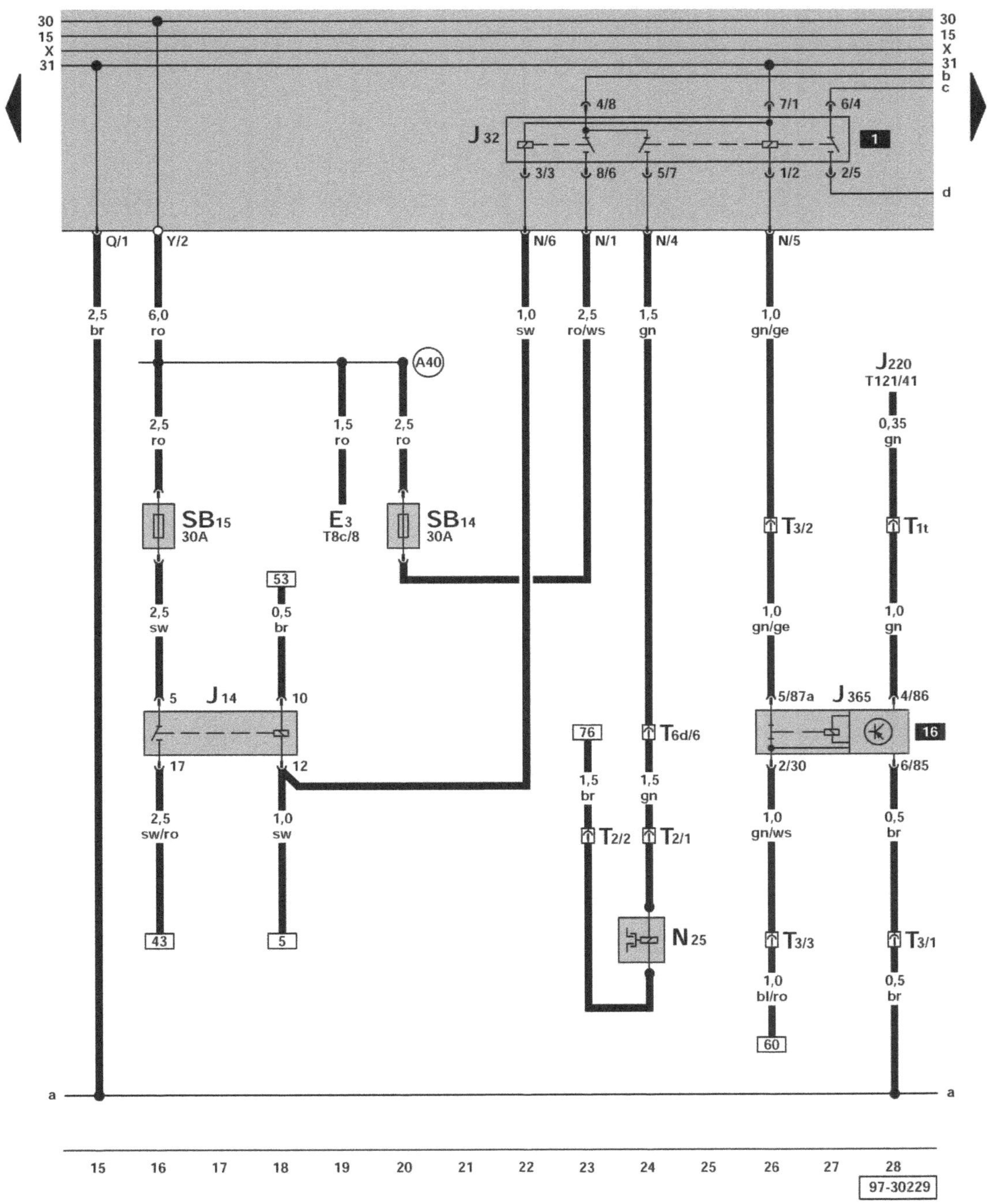

ws = white
sw = black
ro = red
br = brown
gn = green
bl = blue
gr = grey
li = lilac
ge = yellow
or = orange

E3 - Emergency Flasher Switch
J14 - Heater Air Blower Relay, control number (53), behind instrument panel center
J32 - A/C Relay, control number (140), switch delay time 1/2 second
J220 - Motronic Engine Control Module (ECM)
J365 - Relay for A/C shutoff, control number (147), above relay panel
N25 - A/C Clutch
SB14 - Fuse 14 in fuse holder, below relay panel
SB15 - Fuse 15 in fuse holder, below relay panel
T1t - Single Connector, behind fuse/relay panel
T2 - Double Connector, near A/C Clutch
T3 - 3-Pin Connector, behind fuse/relay panel
T6d - 6-Pin Connector, behind fuse/relay panel
T8c - 8-Pin Connector, on Emergency Flasher Switch
T121 - 121-Pin Connector
(A40) - Plus connection -1- (30), in instrument cluster wiring harness

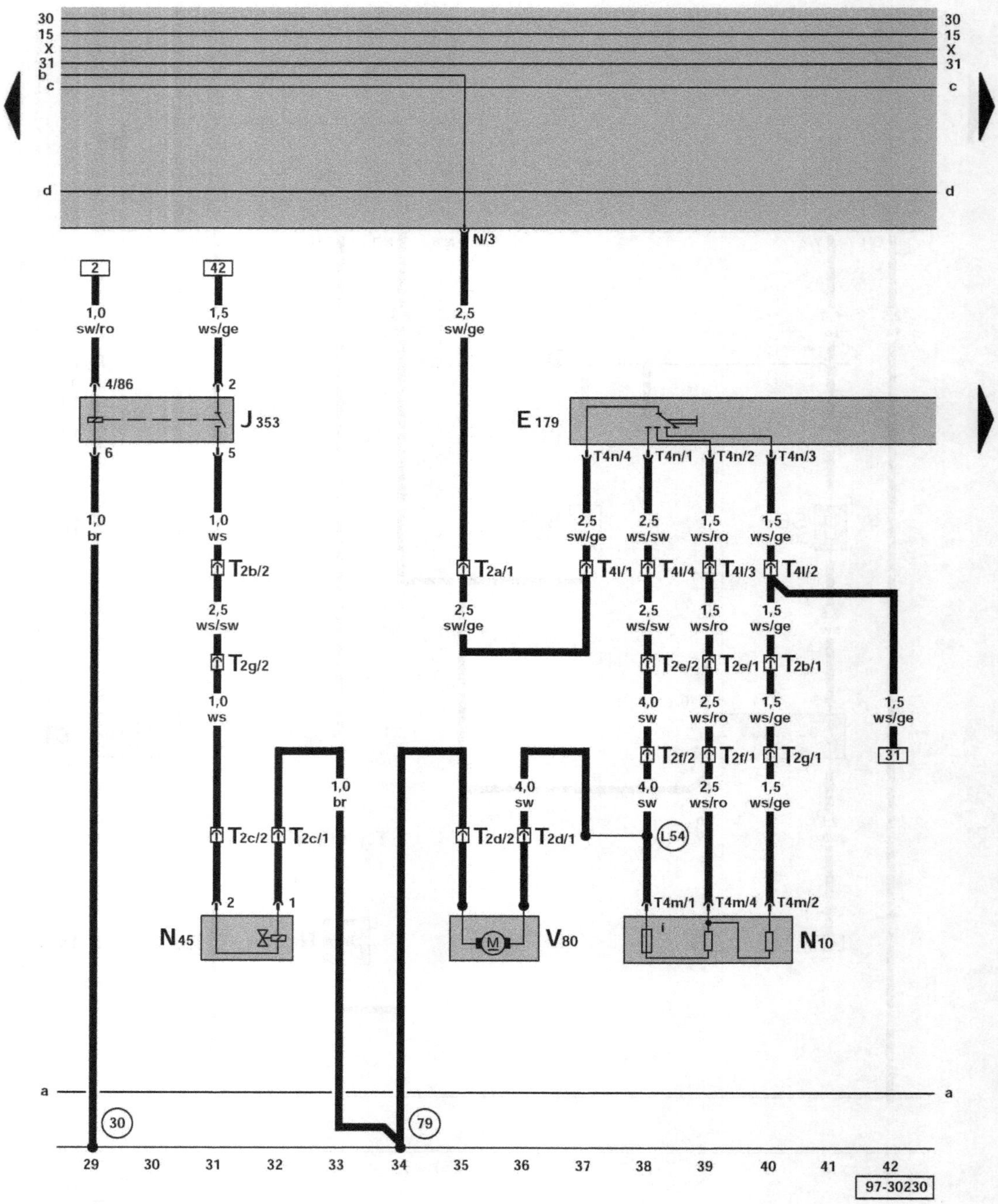

ws = white
sw = black
ro = red
br = brown
gn = green
bl = blue
gr = grey
li = lilac
ge = yellow
or = orange

E179 - Switch for Freshair blower-rear
J353 - Relay for Fresh-air and recirculating air flap-rear, control number (94), behind instrument panel center
N10 - Series Resistance for fresh air blower-rear
N45 - Two-way valve for Fresh- and recirculation air flap-rear
T2a - Double Connector, behind fuse/relay panel
T2b - Double Connector, behind instrument panel, right
T2c - Double Connector, behind instrument panel center
T2d - Double Connector, behind trim D-pillar right, lower
T2e - Double Connector, behind instrument panel, right
T2f - Double Connector, behind instrument panel, center
T2g - Double Connector, behind instrument panel, center
T4l - 4-Pin Connector, behind fuse/relay panel
T4m - 4-Pin Connector
T4n - 4-Pin Connector
V80 - Fresh air blower rear
(30) - Ground connection, -1-, beside fuse/relay panel
(79) - Ground connection, near right D-pillar, lower part
(L54) - Plus connector (Blower stage 3) in wiring harness blower-Passenger ventilation

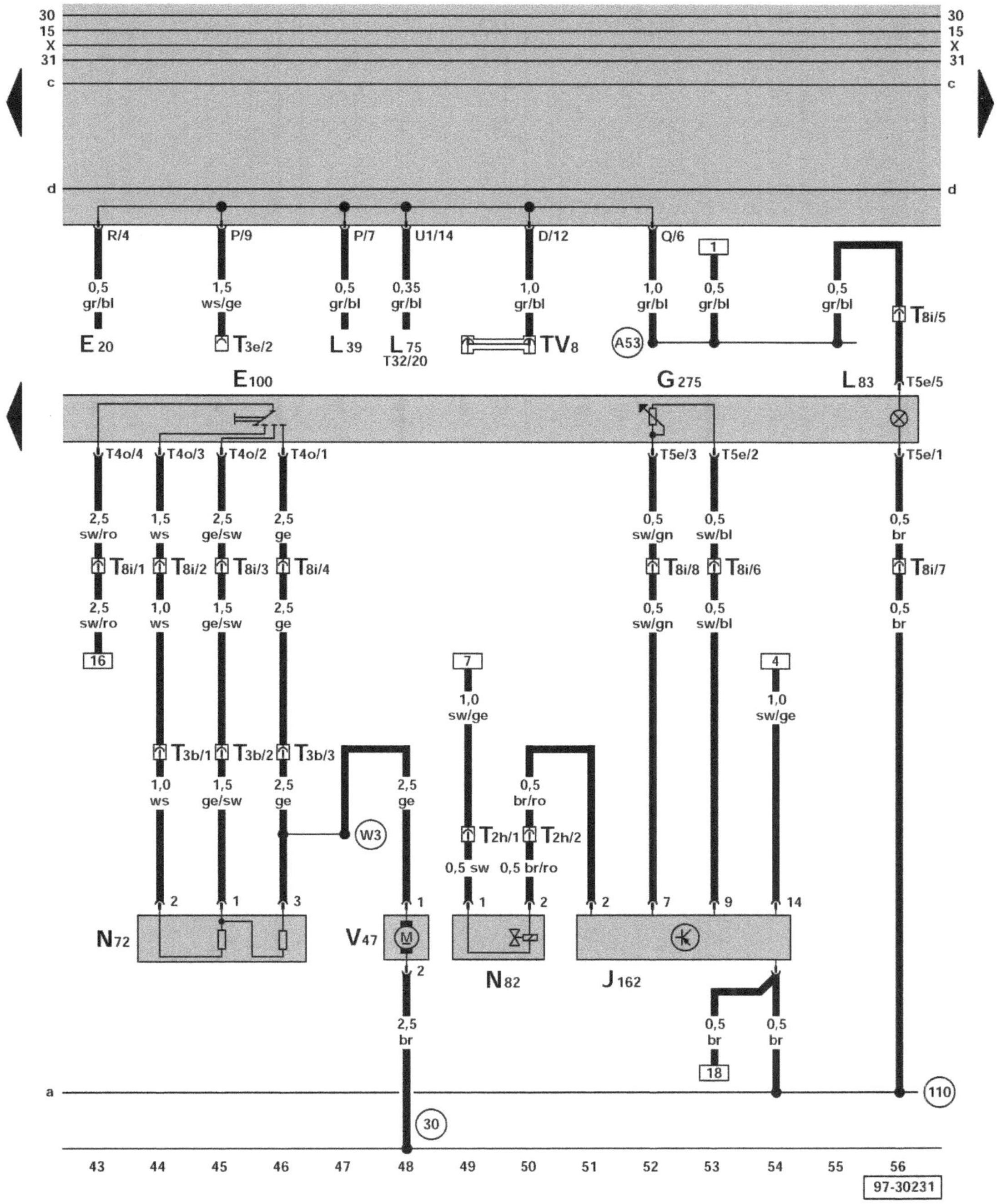

ws = white
sw = black
ro = red
br = brown
gn = green
bl = blue
gr = grey
li = lilac
ge = yellow
or = orange

E20 - Instrument Panel Light Dimmer Switch
E100 - Warm Air Blower Switch
G275 - Rear Interior Temperature Potentiometer
J162 - Heater Control Module, control number (152), behind instrument panel center
L39 - Rear Window Defogger Switch Light
L75 - Digital Display Light
L83 - Warm Air Blower Switch Light
N72 - Heater Blower Series Resistance
N82 - Coolant Cut-off Valve
T2h - Double Connector, behind fuse/relay panel
T3b - 3-Pin Connector, behind fuse/relay panel
T3e - 3-Pin Connector, behind fuse/relay panel
T4o - 4-Pin Connector
T5e - 5-Pin Connector
T8i - 8-Pin Connector, behind fuse/relay panel
T32 - 32-Pin Connector, blue
TV8 - Terminal 58b Wire Connector, grey
V47 - Rear Warm Air Fan

(30) - Ground connection, -1-, beside fuse/relay panel

(110) - Ground connection -2-, in instrument panel wiring harness

(A53) - Plus connection (58b), in instrument panel wiring harness

(W3) - Wire connection, in rear wiring harness

Warm air blower switch, rear interior temperature potentiometer, heater control module, coolant cut-off valve, rear warm air fan

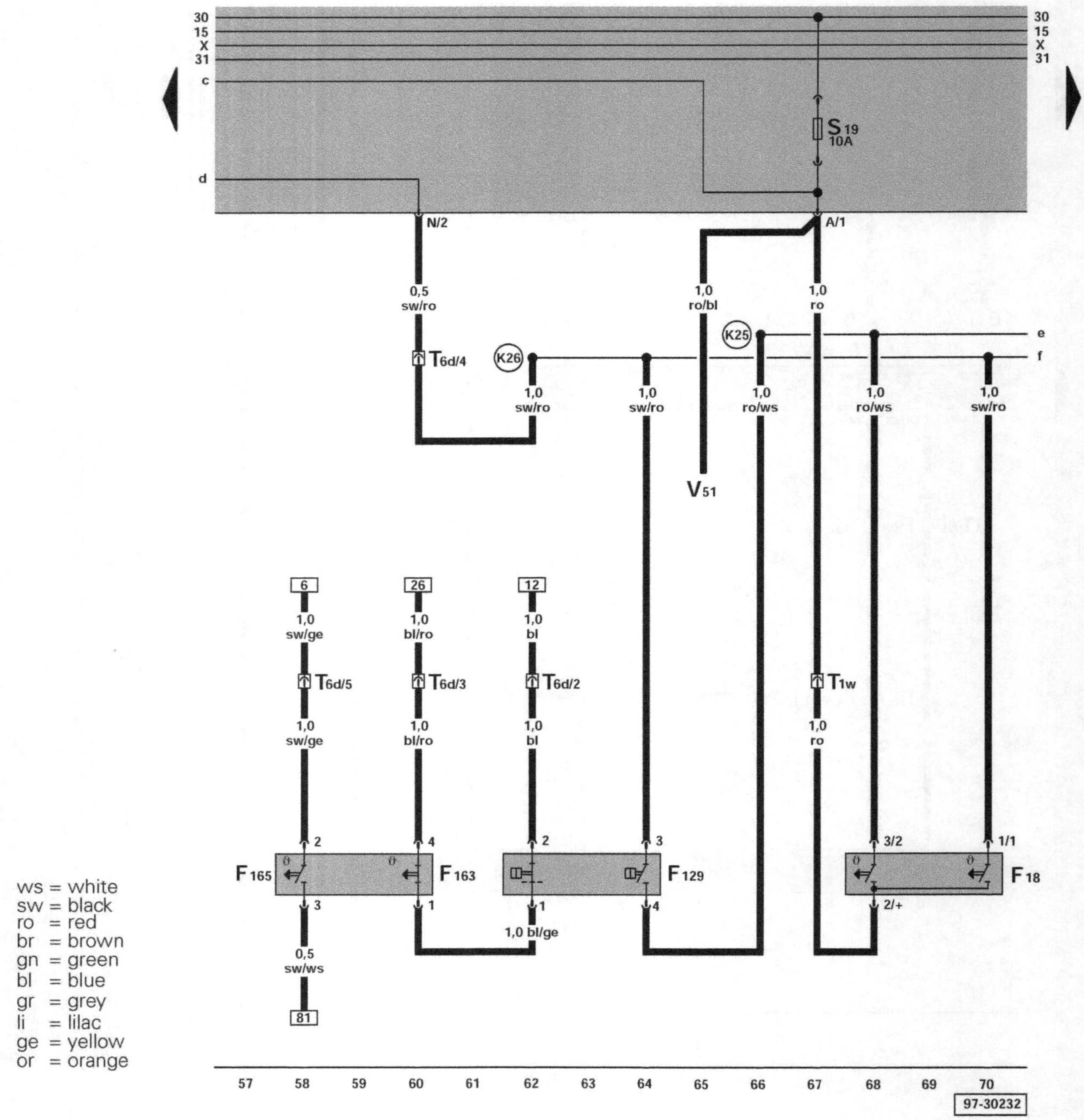

F18 - Coolant Fan Control (FC) Thermal Switch
F129 - A/C Pressure Switch
F163 - A/C Cut-Out Thermal Switch
F165 - Third Speed Coolant Fan Control (FC) Thermal Switch, in engine compartment, front
T1w - Single Connector, behind fuse/relay panel
T6d - 6-Pin Connector, behind fuse/relay panel
V51 - After-Run Coolant Pump

(K25) - Wire connection -4-, in coolant fan wiring harness
(K26) - Wire connection -5-, in coolant fan wiring harness

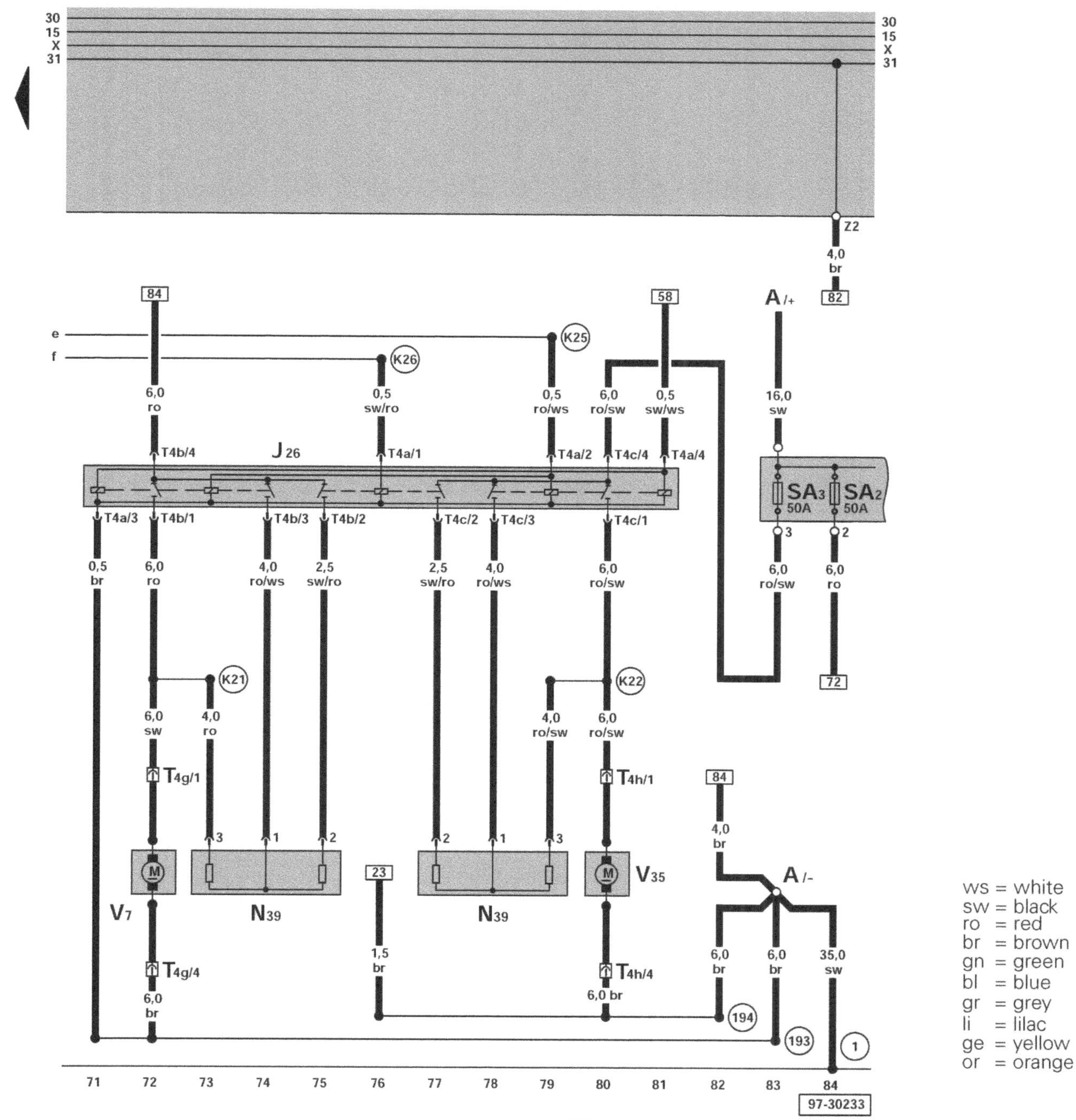

A - Battery
J26 - Coolant Fan Control (FC) Relay, in engine compartment, left front
N39 - Coolant Fan Control (FC) Series Resistance
SA2 - Fuse -2- in fuse bracket/battery
SA3 - Fuse -3- in fuse bracket/battery
T4a - 4-Pin Connector
T4b - 4-Pin Connector
T4c - 4-Pin Connector
T4g - 4-Pin Connector, near coolant fan
T4h - 4-Pin Connector, near coolant fan
V7 - Left Coolant Fan (450 W)
V35 - Right Coolant Fan (450 W)

(1) - Ground strap, battery to body

(193) - Ground connection -1-, in coolant fan wiring harness
(194) - Ground connection -2-, in coolant fan wiring harness
(K21) - Wire connection -1-, in coolant fan wiring harness
(K22) - Wire connection -2-, in coolant fan wiring harness
(K25) - Wire connection -4-, in coolant fan wiring harness
(K26) - Wire connection -5-, in coolant fan wiring harness

THIS PAGE INTENTIONALLY LEFT
BLANK

Climatronic,

from December 2000

- **2.8L - Engine - Motronic Multiport Fuel Injection (MFI)/150 kW, code AXK**

Deviate relay location and fuseplacements as well as the locations of multiple connectors see section "component locations".

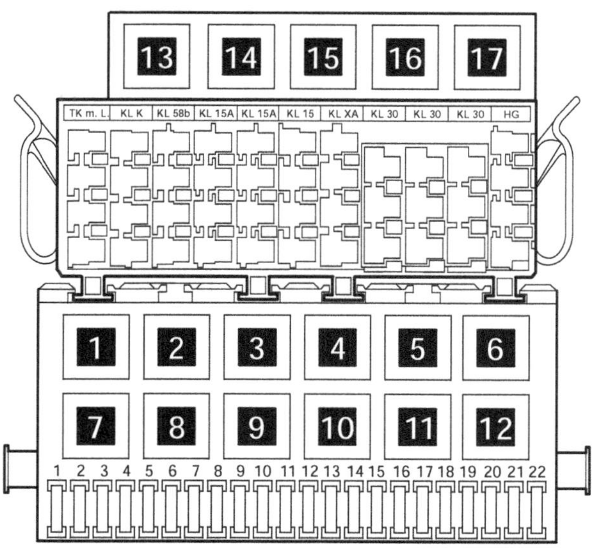

97-23232

1 Relay for Climatronic (140)

13 Coolant Circulation Pump Relay (53)

Note: Number in parentheses indicates production control number stamped on relay housing.

Fuse colors

30 A - green
25 A - white
20 A - yellow
15 A - blue
10 A - red
7,5 A - brown
5 A - beige
3 A - violet

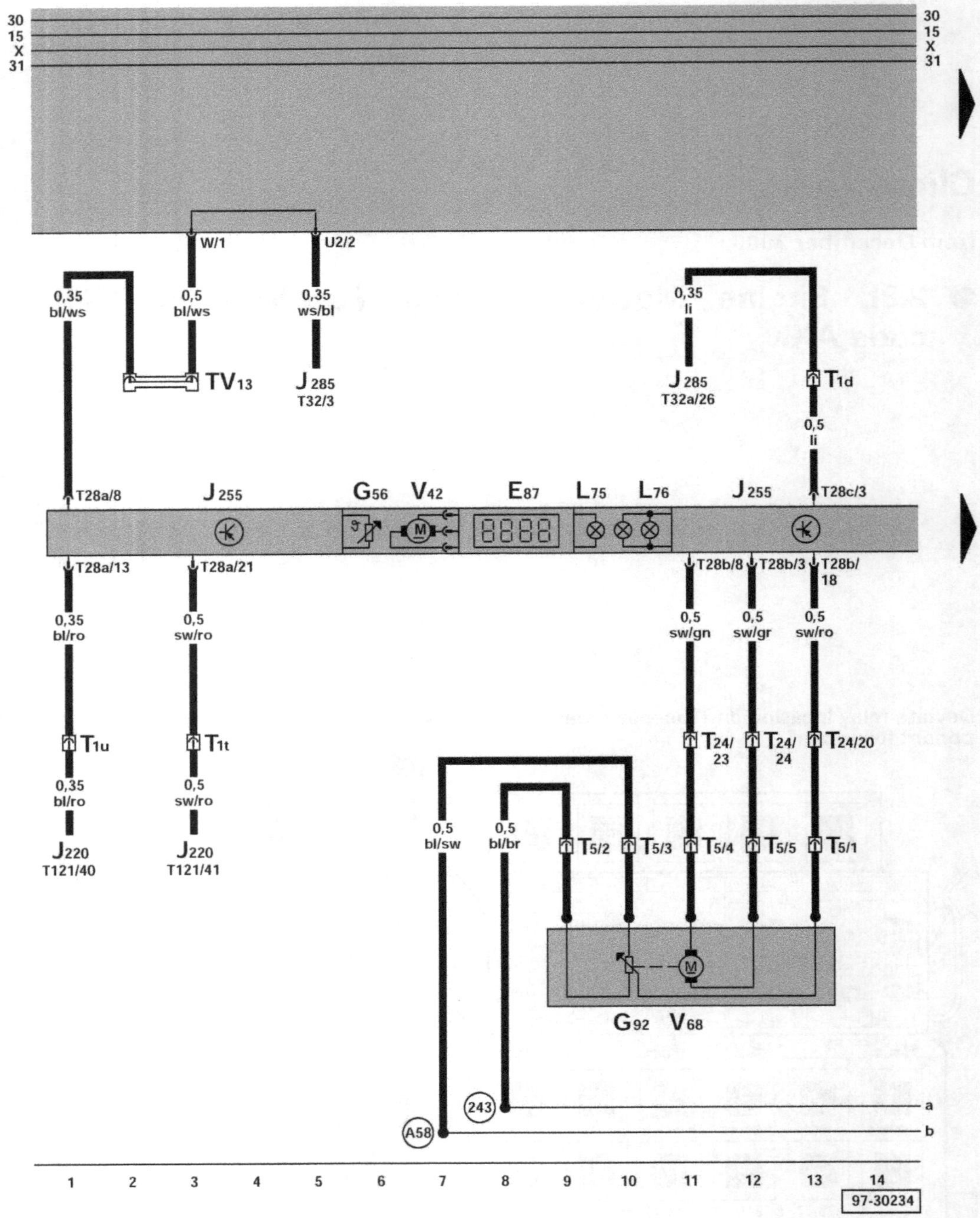

ws = white
sw = black
ro = red
br = brown
gn = green
bl = blue
gr = grey
li = lilac
ge = yellow
or = orange

E87 - A/C Control Head
G56 - Instrument Panel Interior Temperature Sensor
G92 - Temperature Regulator Flap Motor Position Sensor
J220 - Motronic Engine Control Module (ECM)
J255 - Climatronic Control Module
J285 - Control module with indicator unit in instrument panel insert
L75 - Digital Display Light
L76 - Push Button Light
T1d - Single Connector, behind fuse/relay panel
T1t - Single Connector, behind fuse/relay panel
T1u - Single Connector, behind fuse/relay panel
T5 - 5-Pin Connector, behind instrument panel, center
T24 - 24-Pin Connector, behind instrument panel, left
T28a - 28-Pin Connector, red
T28b - 28-Pin Connector, blue
T28c - 28-Pin Connector, white
T32 - 32-Pin Connector, blue
T32a - 32-Pin Connector, green
T121 - 121-Pin Connector
TV13 - Vehicle Speed Signal Wire Connector, blue
V42 - Interior Temperature Sensor Fan
V68 - Temperature Regulator Flap Motor

(243) - Ground connection -1- in wiring harness, Climatronic
(A58) - Wire connection (5 V), in instrument cluster wiring harness

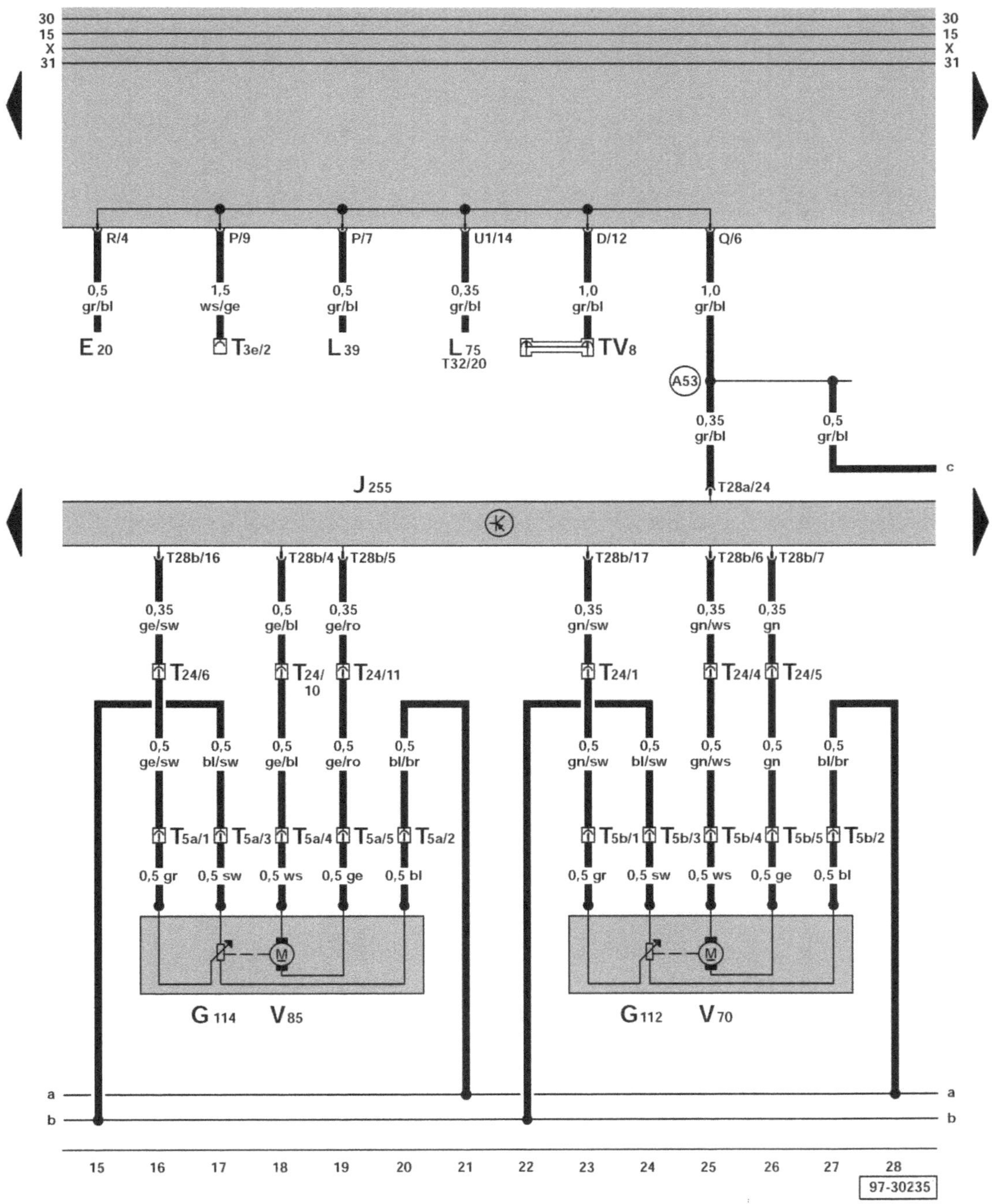

ws = white
sw = black
ro = red
br = brown
gn = green
bl = blue
gr = grey
li = lilac
ge = yellow
or = orange

E20 - Instrument Panel Light Dimmer Switch
G112 - Central Flap Motor Position Sensor
G114 - Footwell/Defrost Flap Motor Position Sensor
J255 - Climatronic Control Module
L39 - Rear Window Defogger Switch Light
L75 - Digital Display Light
T3e - 3-Pin Connector, behind fuse/relay panel
T5a - 5-Pin Connector, behind instrument panel, center
T5b - 5-Pin Connector, behind instrument panel, center
T24 - 24-Pin Connector, behind instrument panel, right
T28a - 28-Pin Connector, red
T28b - 28-Pin Connector, blue
T32 - 32-Pin Connector, blue

TV8 - Terminal 58b Wire Connector, grey
V70 - Central Air Flap Motor
V85 - Footwell/Defroster Flap Motor

(A53) - Plus connection (58b), in instrument panel wiring harness

Climatronic control module, central air flap motor, central flap motor position sensor, footwell/defroster flap motor, footwell/defrost flap motor position sensor

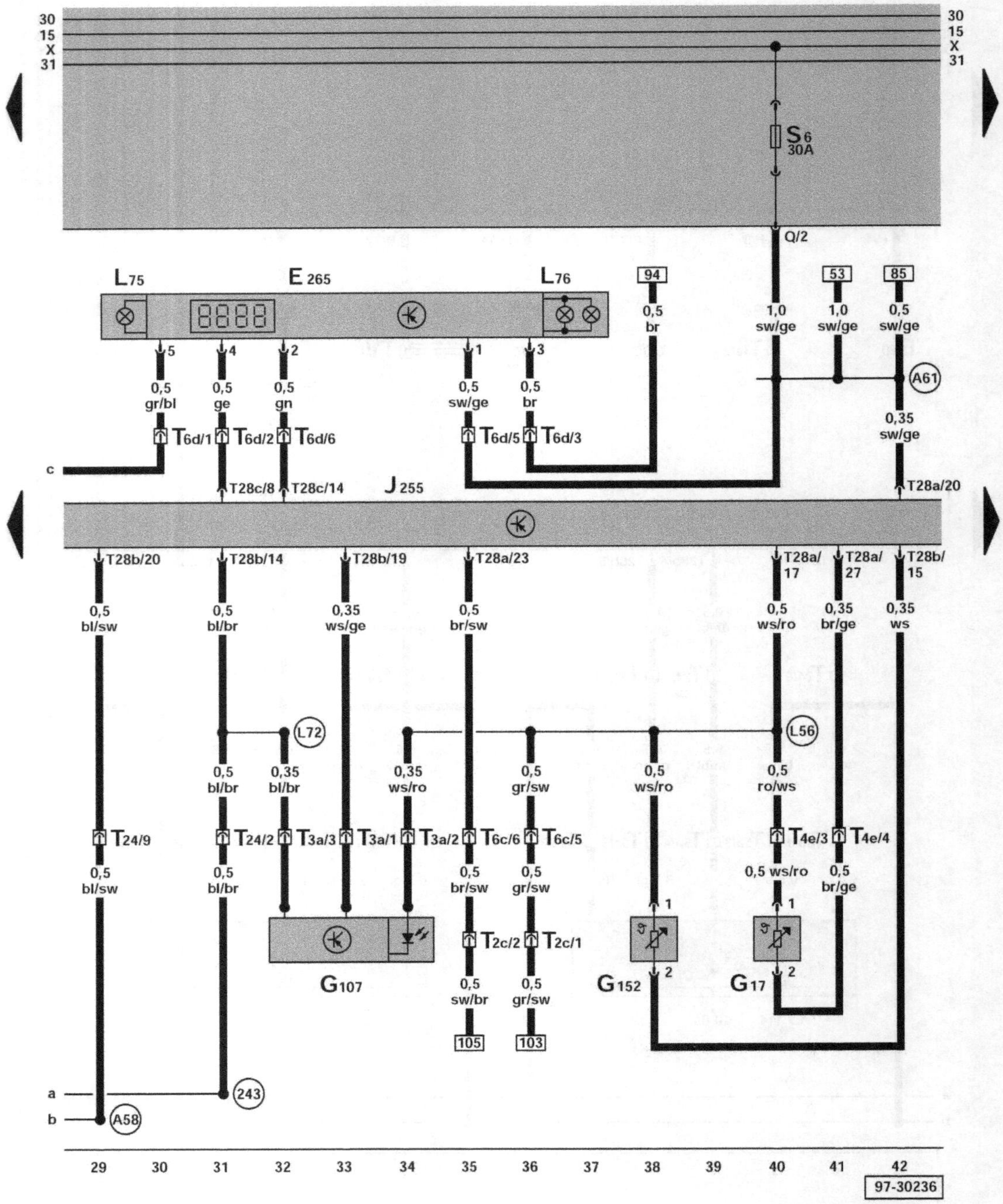

ws = white
sw = black
ro = red
br = brown
gn = green
bl = blue
gr = grey
li = lilac
ge = yellow
or = orange

E265 - A/C Control Head rear
G17 - Outside Air Temperature Sensor, on front bumper left
G107 - Sunlight Photo Sensor, on instrument panel, left
G152 - Front Vent Temperature Sensor, on console for footwell, left vent side
J255 - Climatronic Control Module
L75 - Digital Display Light
L76 - Push Button Light
T2c - Double Connector, in engine compartment, left
T3a - 3-Pin Connector, behind instrument panel, left
T4e - 4-Pin Connector, behind fuse/relay panel
T6c - 6-Pin Connector, behind fuse/relay panel
T6d - 6-Pin Connector, behind fuse/relay panel
T24 - 24-Pin Connector, behind instrument panel, right
T28a - 28-Pin Connector, red
T28b - 28-Pin Connector, blue
T28c - 28-Pin Connector, white

(243) - Ground connection -1- in wiring harness, Climatronic
(A58) - Wire connection (5 V), in instrument cluster wiring harness
(A61) - Plus Connection (X), in Instrument cluster wiring harness
(L56) - Connector (Sensor), in wiring harness Climatronic
(L72) - Connector (Photo Sensor), in wiring harness Climatronic

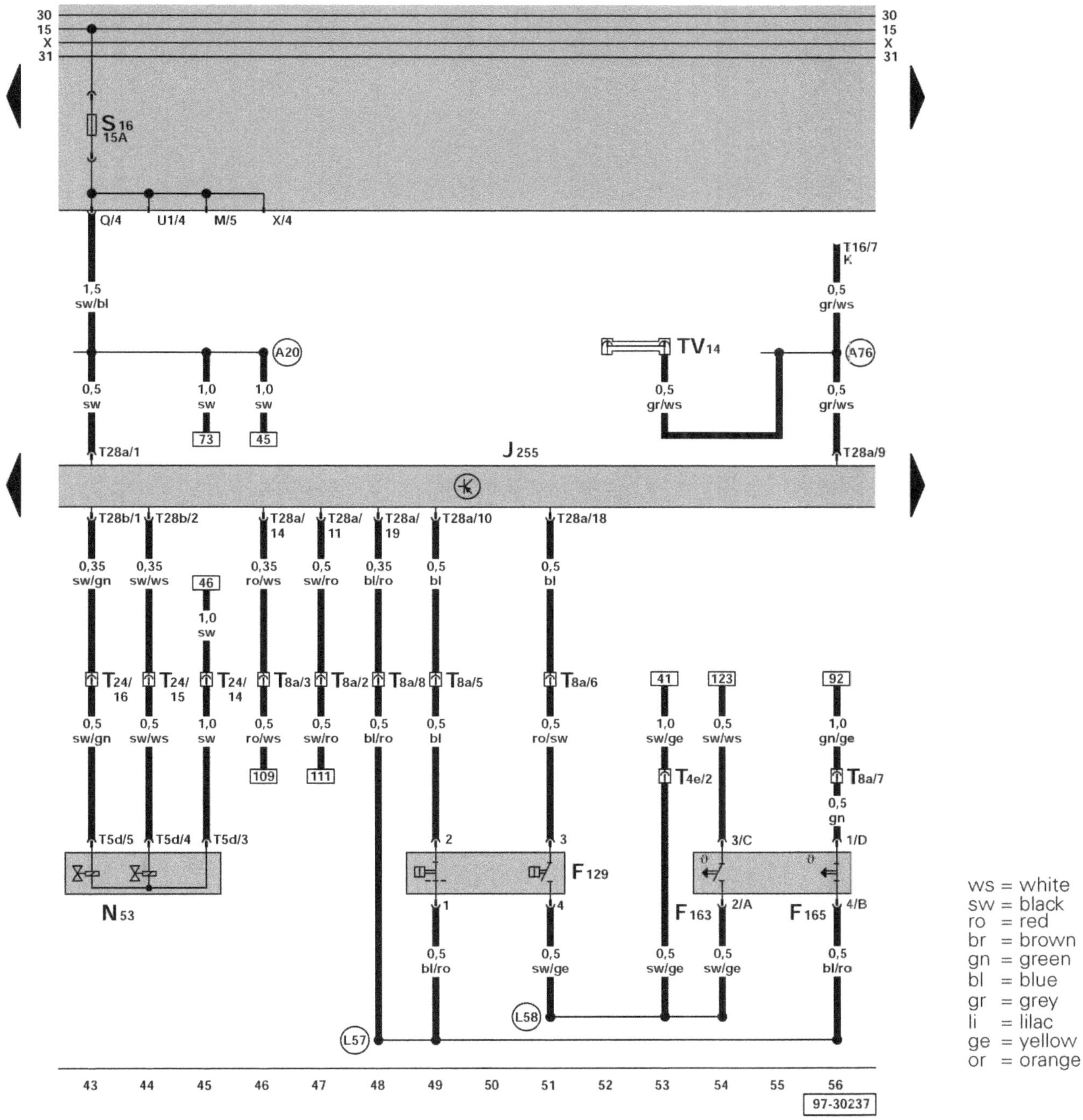

F129 - A/C Pressure Switch
F163 - A/C Cut-Out Thermal Switch
F165 - Third Speed Coolant Fan Control (FC) Thermal Switch, in engine compartment, front
J255 - Climatronic Control Module
N53 - Vacuum Valve Rail
T4e - 4-Pin Connector, behind fuse/relay panel
T5d - 5-Pin Connector
T8a - 8-Pin Connector, behind fuse/relay panel
T16 - Data Link Connector (DLC), behind steering column cover
T24 - 24-Pin Connector, behind instrument panel, right
T28a - 28-Pin Connector, red
T28b - 28-Pin Connector, blue

TV14 - Data Link Connector (DLC) wire Connector, white

(A20) - Wire connection (15a), in instrument panel wiring harness
(A76) - Connector (K-diagnosis wire) in instrument panel wiring harness
(L57) - Connector (Pressure switch for A/C) -1- in wiring harness Climatronic - Fan for coolant
(L58) - Connector (Pressure switch for A/C) -2- in wiring harness Climatronic - Fan for coolant

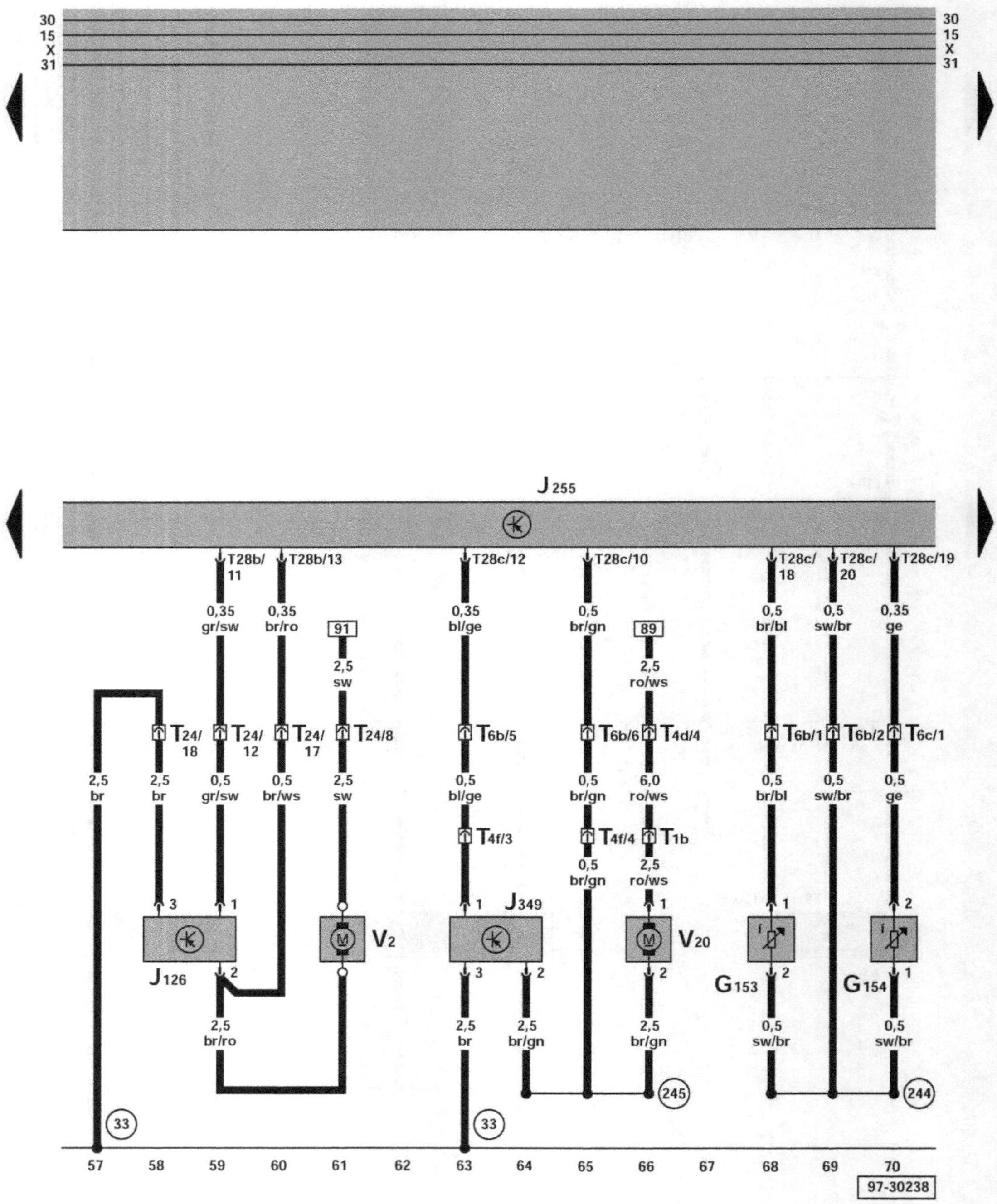

ws = white
sw = black
ro = red
br = brown
gn = green
bl = blue
gr = grey
li = lilac
ge = yellow
or = orange

G153 - Rear Evaporator Temperature Sensor
G154 - Rear Heater Core Temperature Sensor
J126 - Control module for fresh air blower
J255 - Climatronic Control Module
J349 - Control module for blower for evaparator
T1b - Single Connector, behind trim D-pillar right, lower
T4d - 4-Pin Connector, behind fuse/relay panel
T4f - 4-Pin Connector, behind trim D-pillar right, lower
T6b - 6-Pin Connector, behind fuse/relay panel
T6c - 6-Pin Connector, behind fuse/relay panel
T24 - 24-Pin Connector, behind instrument panel, right
T28b - 28-Pin Connector, blue
T28c - 28-Pin Connector, white

V2 - Fresh Air Blower
V20 - Rear Evaporator Fan

(33) - Ground connection, behind instrument panel, right
(244) - Ground connection (sensor ground), in wiring harness Climatronic
(245) - Ground connection in wiring harness blower for 2nd Evaporatur

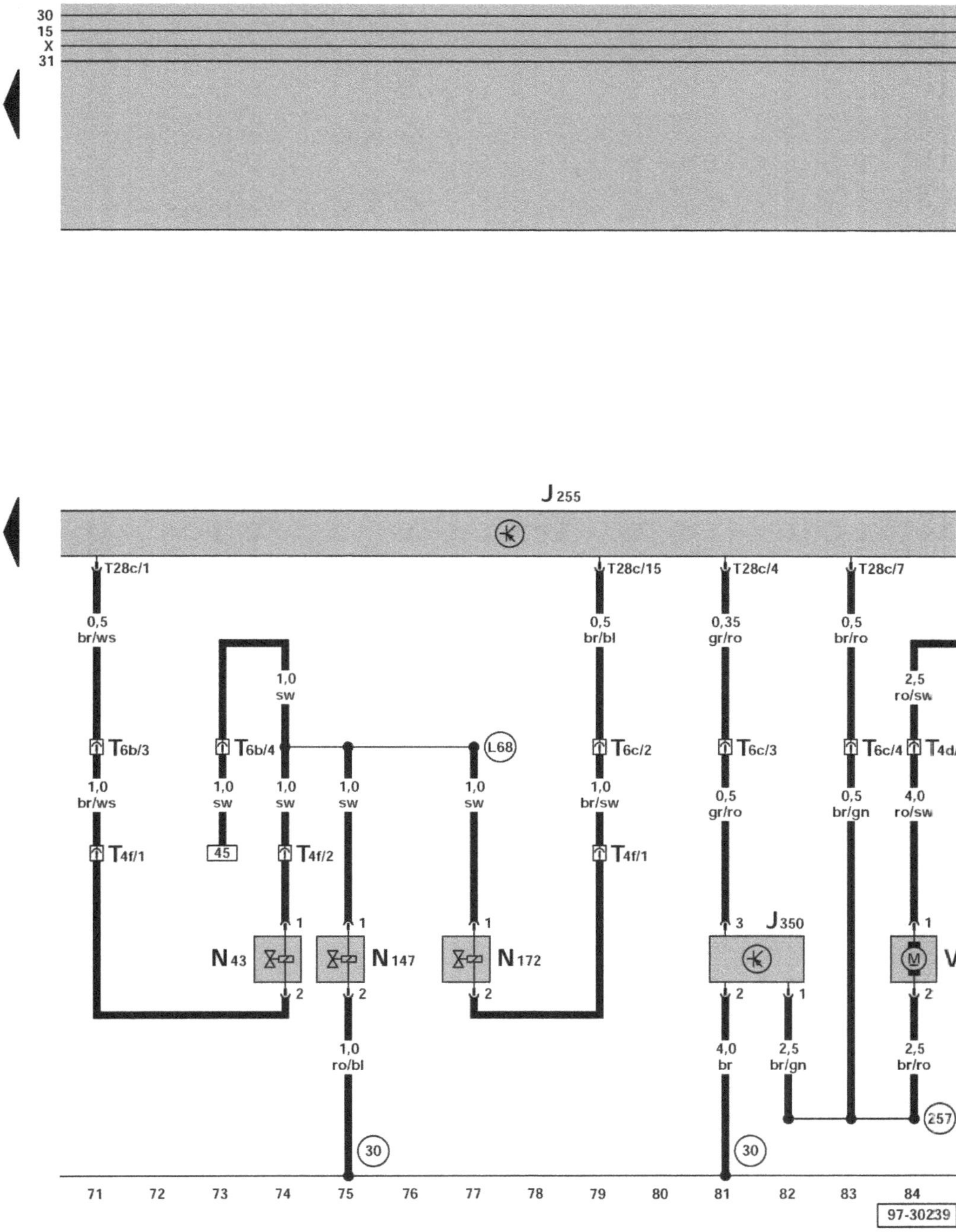

ws = white
sw = black
ro = red
br = brown
gn = green
bl = blue
gr = grey
li = lilac
ge = yellow
or = orange

J255 - Climatronic Control Module
J350 - Control module for warm air blower
N43 - A/C Refrigerant Shut-Off Valve, behind trim D-pillar right, lower
N147 - Engine Coolant Two-way Vacuum Valve, on firewall, left
N172 - Auxiliary Heater Valve, on firewall, left
T4d - 4-Pin Connector, behind fuse/relay panel
T4f - 4-Pin Connector, behind trim D-pillar right, lower
T6b - 6-Pin Connector, behind fuse/relay panel
T6c - 6-Pin Connector, behind fuse/relay panel
T28c - 28-Pin Connector, white
V47 - Rear Warm Air Fan

(30) - Ground connection, -1-, beside fuse/relay panel
(257) - Ground Connection -2-, in wirning harness Climatronic
(L68) - Connection 2 (sensor) in Climatronic wiring harness

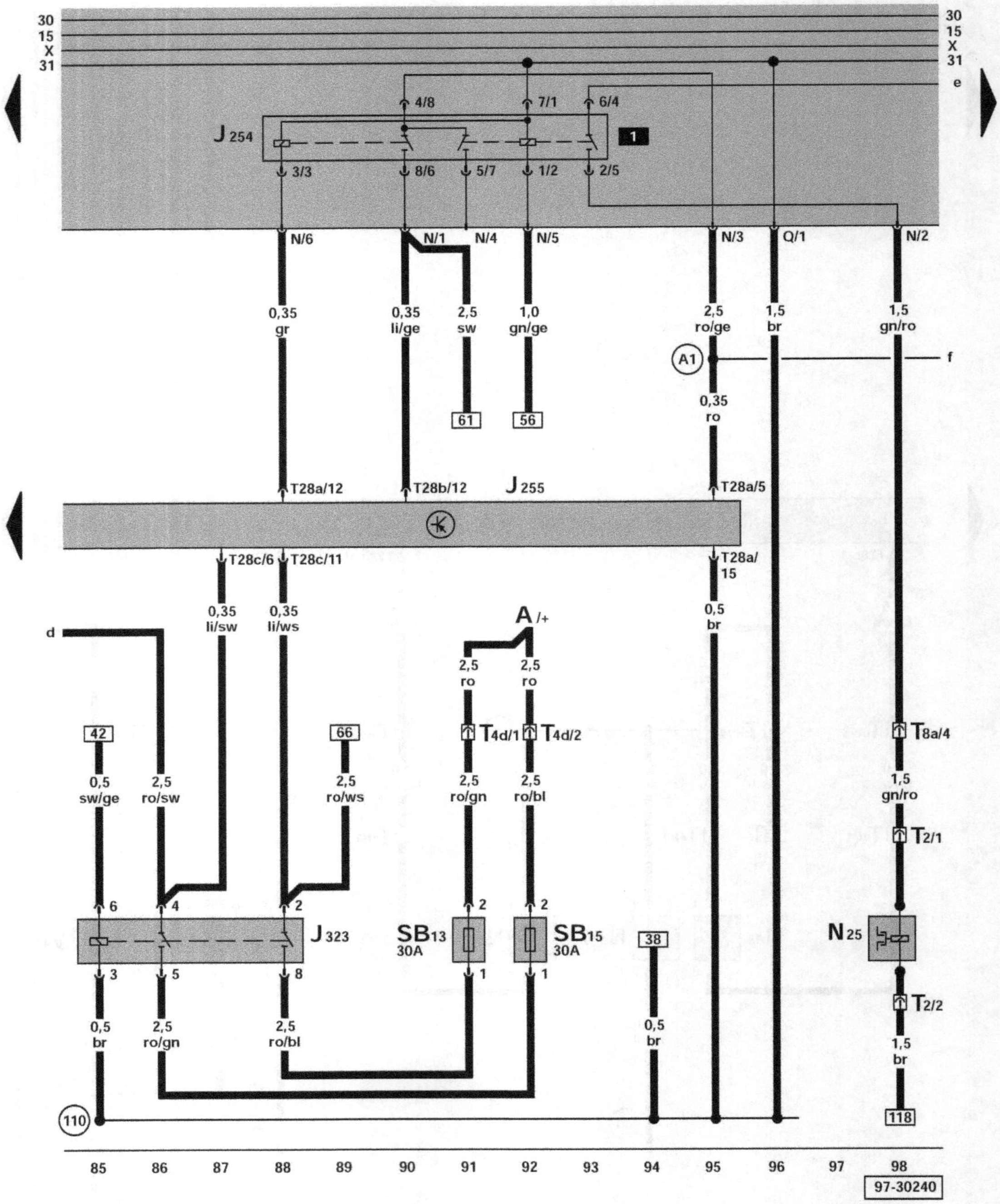

ws = white
sw = black
ro = red
br = brown
gn = green
bl = blue
gr = grey
li = lilac
ge = yellow
or = orange

- A - Battery
- J254 - Relay for Climatronic, control number (140), switch delay time 1/2 second
- J255 - Climatronic Control Module
- J323 - Fan Relay, control number (114), behind instrument panel, center
- N25 - A/C Clutch
- SB13 - Fuse 13 in fuse holder, below relay panel
- SB15 - Fuse 15 in fuse holder, below relay panel
- T2 - Double Connector, in engine compartment, right front
- T4d - 4-Pin Connector, behind fuse/relay panel
- T8a - 8-Pin Connector, behind fuse/relay panel
- T28a - 28-Pin Connector, red
- T28b - 28-Pin Connector, blue
- T28c - 28-Pin Connector, white

- (110) - Ground connection -2-, in instrument panel wiring harness
- (A1) - Plus connection (30a), in instrument panel wiring harness

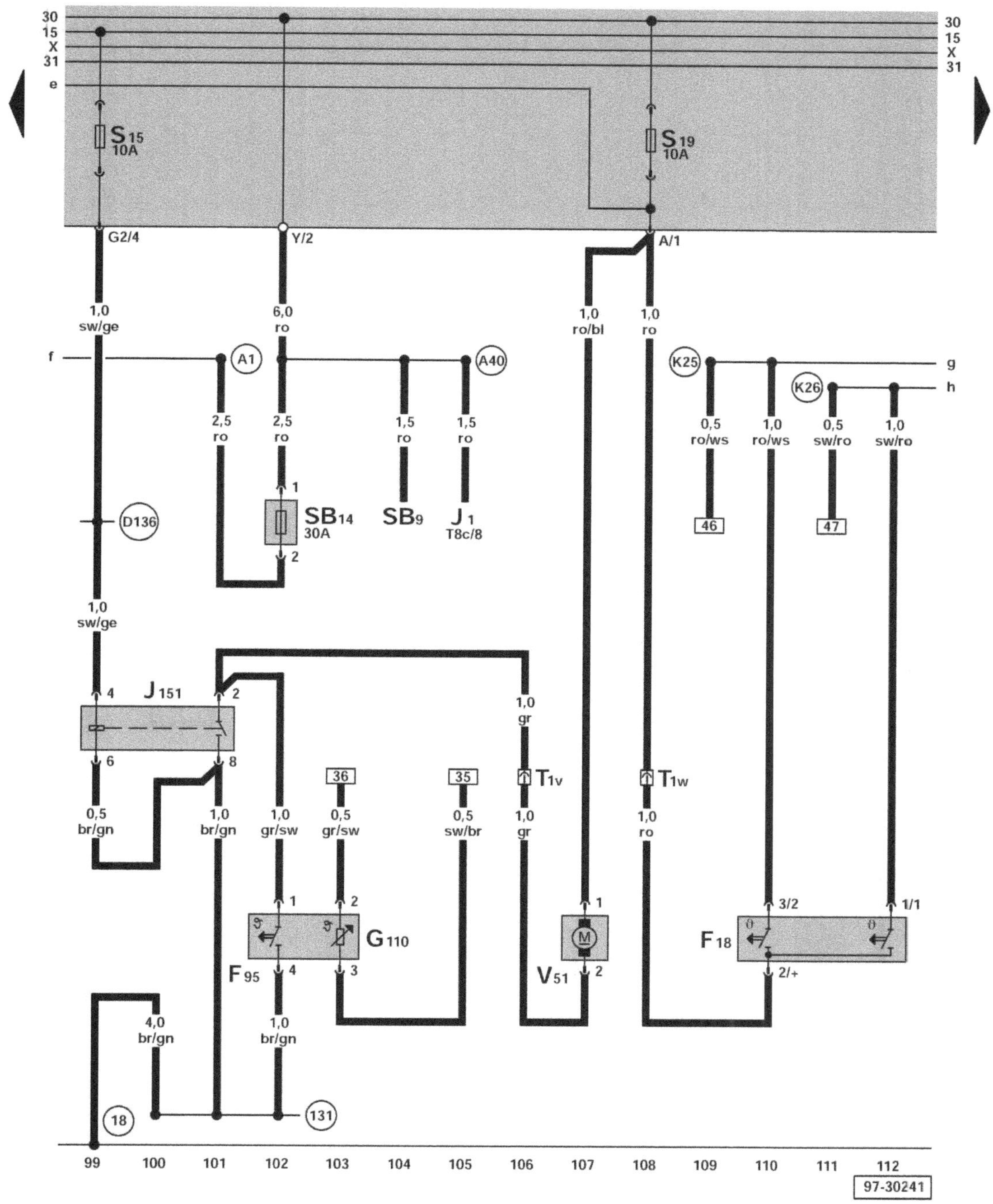

ws = white
sw = black
ro = red
br = brown
gn = green
bl = blue
gr = grey
li = lilac
ge = yellow
or = orange

F18 - Coolant Fan Control (FC) Thermal Switch
F95 - After-Run Coolant Thermal Switch
G110 - A/C Engine Coolant Temperature (ECT) Sensor, in engine compartment, left
J1 - Turn Signal Relay
J151 - Coolant Circulation Pump Relay, above fuse/relay panel
SB9 - Fuse 9 in fuse holder, below relay panel
SB14 - Fuse 14 in fuse holder, below relay panel
T1v - Single Connector, behind fuse/relay panel
T1w - Single Connector, behind fuse/relay panel
T8c - 8-Pin Connector
V51 - After-Run Coolant Pump

(18) - Ground connection, on engine block, left

(131) - Ground connection -2-, in engine compartment wiring harness
(A1) - Plus connection (30a), in instrument panel wiring harness
(A40) - Plus connection -1- (30), in instrument cluster wiring harness
(D136) - Plus connection -2- (15a) in wiring harness engine compartment
(K25) - Wire connection -4-, in coolant fan wiring harness
(K26) - Wire connection -5-, in coolant fan wiring harness

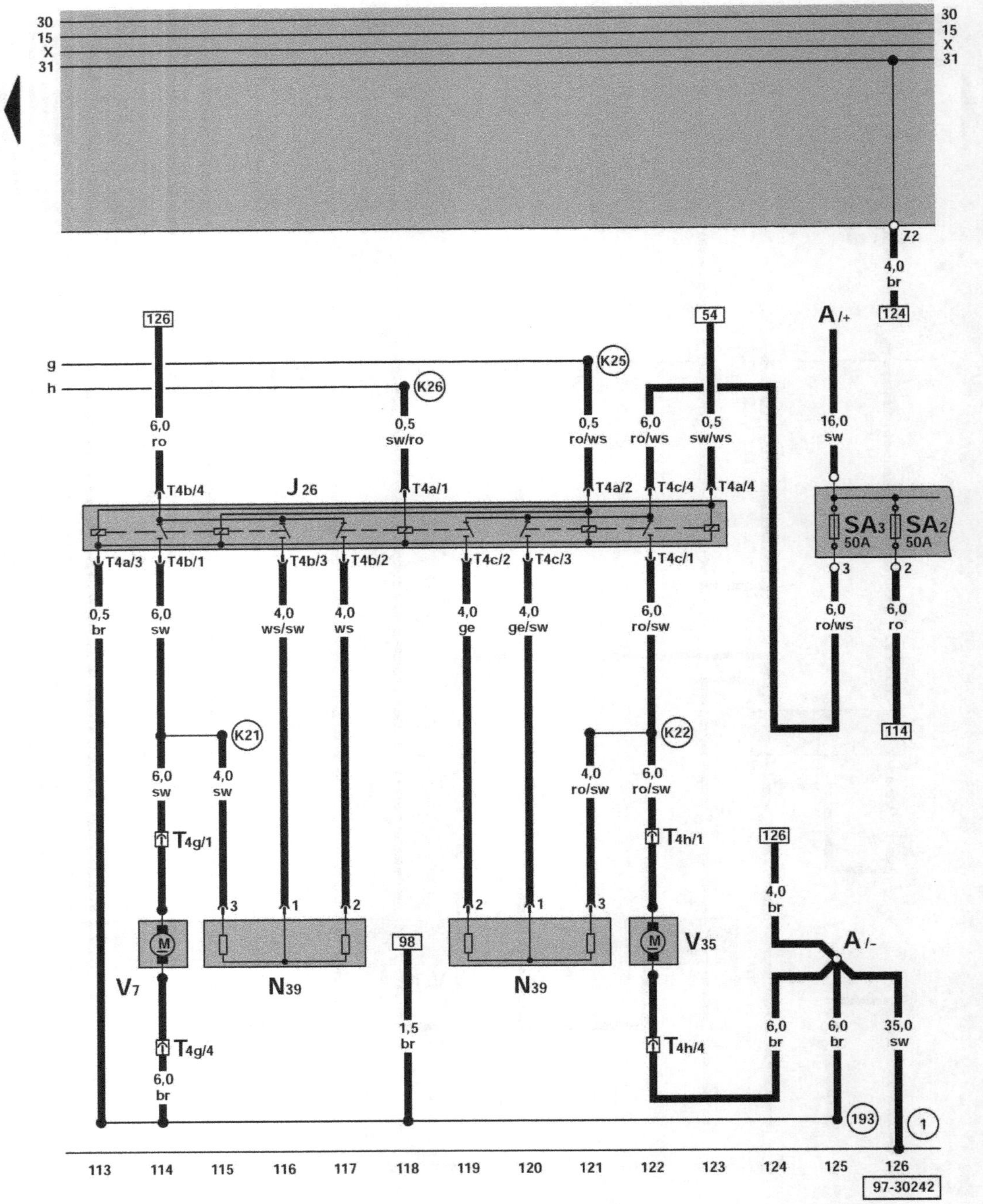

ws = white
sw = black
ro = red
br = brown
gn = green
bl = blue
gr = grey
li = lilac
ge = yellow
or = orange

A - Battery
J26 - Coolant Fan Control (FC) Relay, in engine compartment, left front
N39 - Coolant Fan Control (FC) Series Resistance
SA2 - Fuse -2- in fuse bracket/battery
SA3 - Fuse -3- in fuse bracket/battery
T4a - 4-Pin Connector
T4b - 4-Pin Connector
T4c - 4-Pin Connector
T4g - 4-Pin Connector, near coolant fan
T4h - 4-Pin Connector, near coolant fan
V7 - Left Coolant Fan (450 W)
V35 - Right Coolant Fan (450 W)

(1) - Ground strap, battery to body

(193) - Ground connection -1-, in coolant fan wiring harness
(K21) - Wire connection -1-, in coolant fan wiring harness
(K22) - Wire connection -2-, in coolant fan wiring harness
(K25) - Wire connection -4-, in coolant fan wiring harness
(K26) - Wire connection -5-, in coolant fan wiring harness

Central locking system with remote control (Camper),

from December 2000

- **Vehicles with right sliding door and tailgate**
- **Vehicles with right sliding door and swinging doors**

Deviate relay location and fuseplacements as well as the locations of multiple connectors see section "component locations".

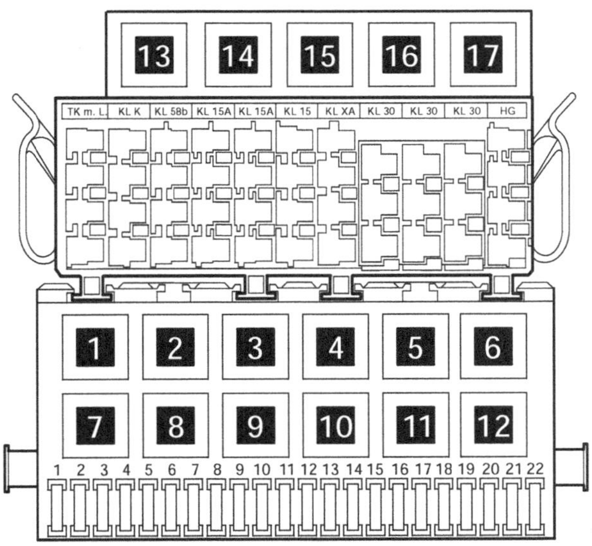

97-23232

Fuse colors

30 A - green
25 A - white
20 A - yellow
15 A - blue
10 A - red
7,5 A - brown
5 A - beige
3 A - violet

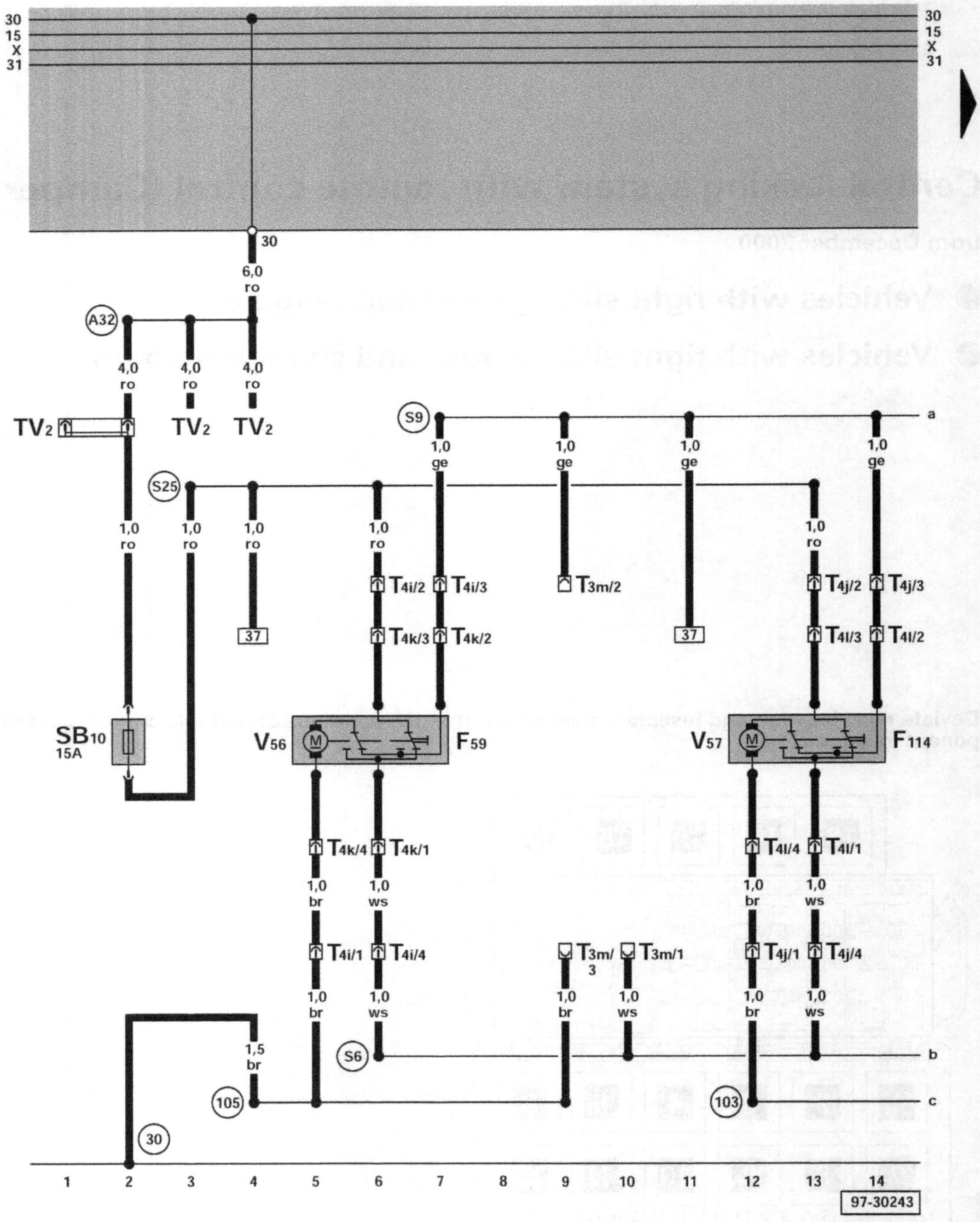

ws = white
sw = black
ro = red
br = brown
gn = green
bl = blue
gr = grey
li = lilac
ge = yellow
or = orange

F59 - Driver's Door Central Locking System Switch
F114 - Passenger's Door Central Locking System Switch
SB10 - Fuse 10 in fuse holder, below relay panel
T3m - 3-Pin Connector, behind instrument panel, left
T4i - 4-Pin Connector, behind fuse/relay panel
T4j - 4-Pin Connector, behind, instrument panel, right
T4k - 4-Pin Connector, in driver's door
T4l - 4-Pin Connector, in front passenger's door
TV2 - Terminal 30 (B+) Wire Connector, red, each 6 points
V56 - Driver's Door Central Locking System Motor
V57 - Passenger's Door Central Locking System Motor
(30) - Ground connection, -1-, beside fuse/relay panel
(103) - Ground connection -2-, in central locking system wiring harness
(105) - Ground connection -1-, in central locking system wiring harness
(A32) - Plus connection (30), in instrument panel wiring harness
(S6) - Wire connection (A) -1-, in central locking system wiring harness
(S9) - Wire connection (E) -1-, in central locking system wiring harness
(S25) - Connector (Sensor/Control Module) -1- in wiring harness, central locking

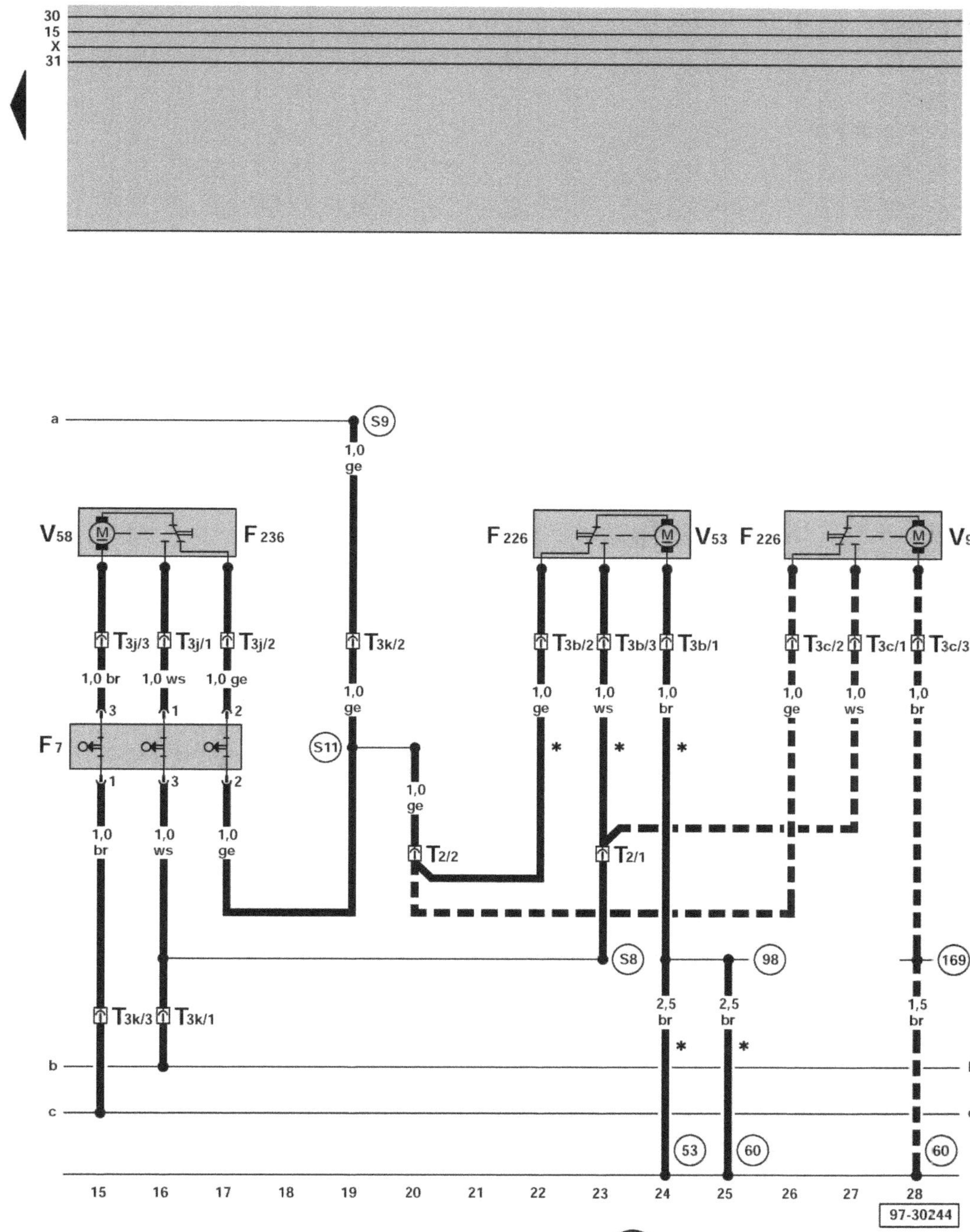

ws = white
sw = black
ro = red
br = brown
gn = green
bl = blue
gr = grey
li = lilac
ge = yellow
or = orange

F7 - Right Rear Sliding Door Contact Switch
F226 - Right rear door/swinging door contact switch
F236 - Door contact switch for central locking (right sliding door)
T2 - Double Connector, on right D-pillar, upper part
T3b - 3-Pin Connector, in rear lid
T3c - 3-Pin Connector, in rear swinging door
T3j - 3-Pin Connector, in sliding door, right
T3k - 3-Pin Connector, behind instrument panel, right
V53 - Decklid Central Locking System Motor
V58 - Sliding Door Central Locking System Motor
V96 - Left Rear Door Central Locking System Motor

(53) - Ground connection, in rear lid, right

(60) - Ground point -2- near taillamp assembly, right

(98) - Ground connection, in rear lid wiring harness

(169) - Ground connection, in right sliding door wiring harness

(S8) - Wire connection (A) -3-, in central locking system/passenger compartment wiring harness

(S9) - Wire connection (E) -1-, in central locking system wiring harness

(S11) - Wire connection (E) -3-, in central locking system passenger compartment wiring harness

* - Vehicles with rear lid only

--- -Vehicles with rear swinging doors only

Decklid central locking system motor, sliding door central locking system motor, left rear door central locking system motor

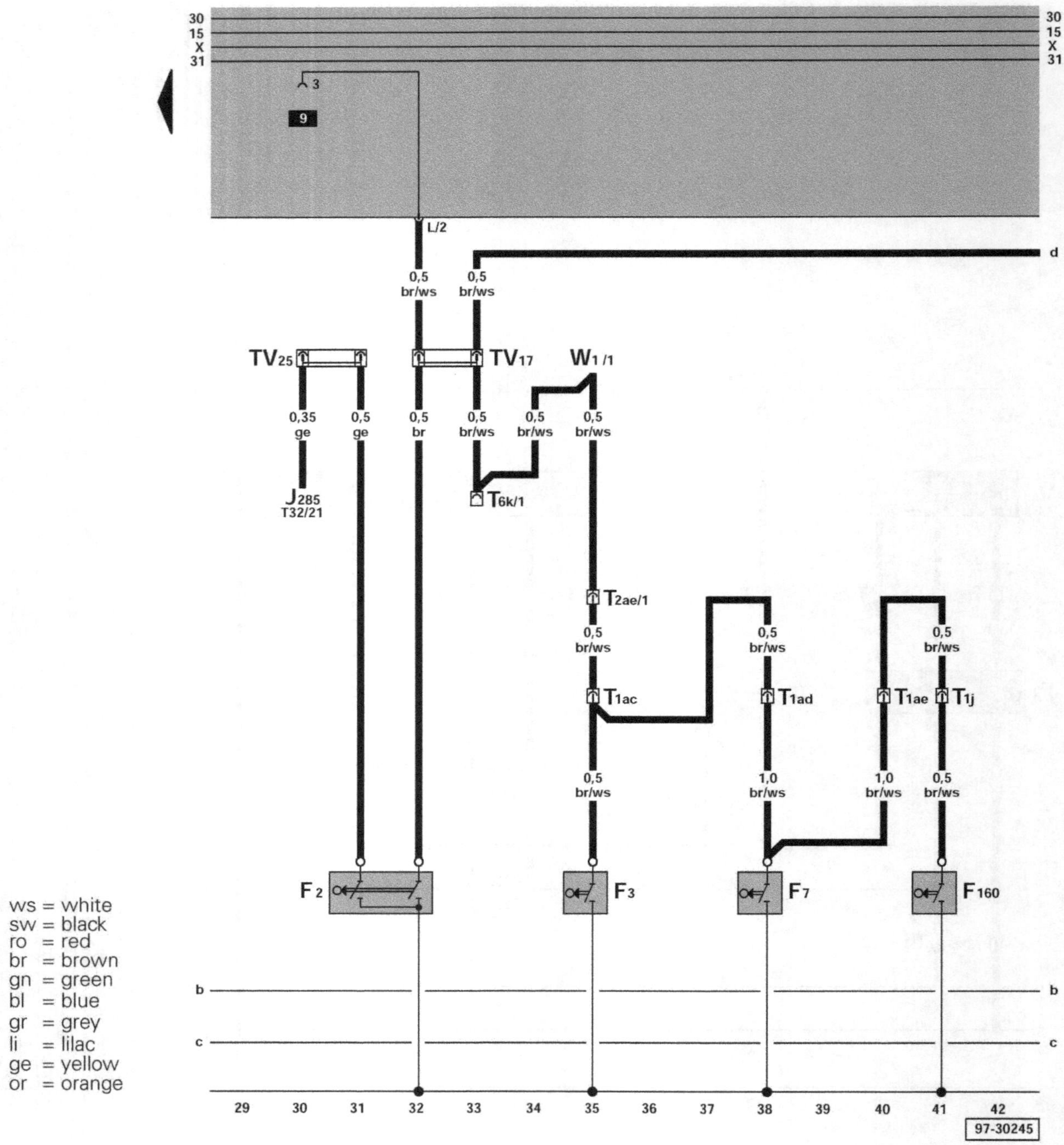

F2 - Door contact Switch - Driver's side
F3 - Door contact Switch-Passenger side
F7 - Right Rear Side/Sliding Door Contact Switch
F160 - Rear Lid Contact Switch
J285 - Control module with indicator unit in instrument panel insert
T1ac - Single Connector, near right B-pillar, roof
T1ad - Single Connector, near Right Rear Side/Sliding Door Contact Switch
T1ae - Single Connector, near Right Rear Side/Sliding Door Contact Switch
T1j - Single Connector, near taillamp assembly
T2ae - Double Connector, near right B-pillar, roof
T6k - 6-Pin Connector, behind instrument panel, right
T32 - 32-Pin Connector, blue, on instrument cluster
TV17 - Door Contact Switch Wire Connector, brown
TV25 - Door Contact Switch Wire Connector, pink
W1 - Interior light, front

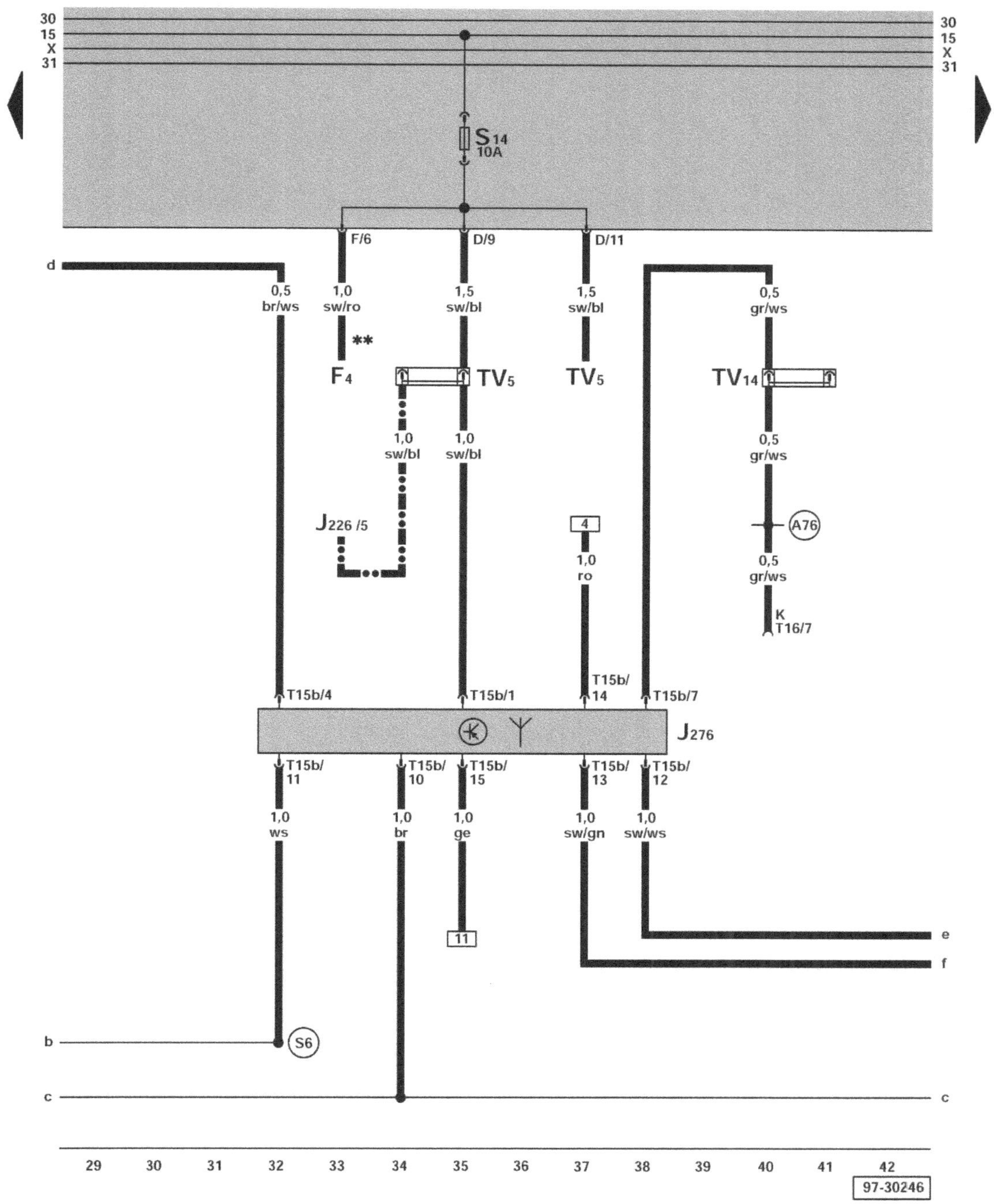

ws = white
sw = black
ro = red
br = brown
gn = green
bl = blue
gr = grey
li = lilac
ge = yellow
or = orange

F4 - Back-Up Light Switch**

J226 - Park/Neutral Position (PNP) Relay

J276 - Central Locking System Control Module (Remote Control), behind instrument cluster

T1h - Single Connector, behind instrument panel, right

T15b - 15-Pin Connector

T16 - Data Link Connector (DLC), behind steering column cover

TV5 - Terminal 15a Wire Connector, green, 4-Pin and 8-Pin

TV14 - Data Link Connector (DLC) wire Connector, white

(A76) - Connector (K-diagnosis wire) in instrument panel wiring harness

(S6) - Wire connection (A) -1-, in central locking system wiring harness

** - Manual transmission only

-•••- Automatic transmission only

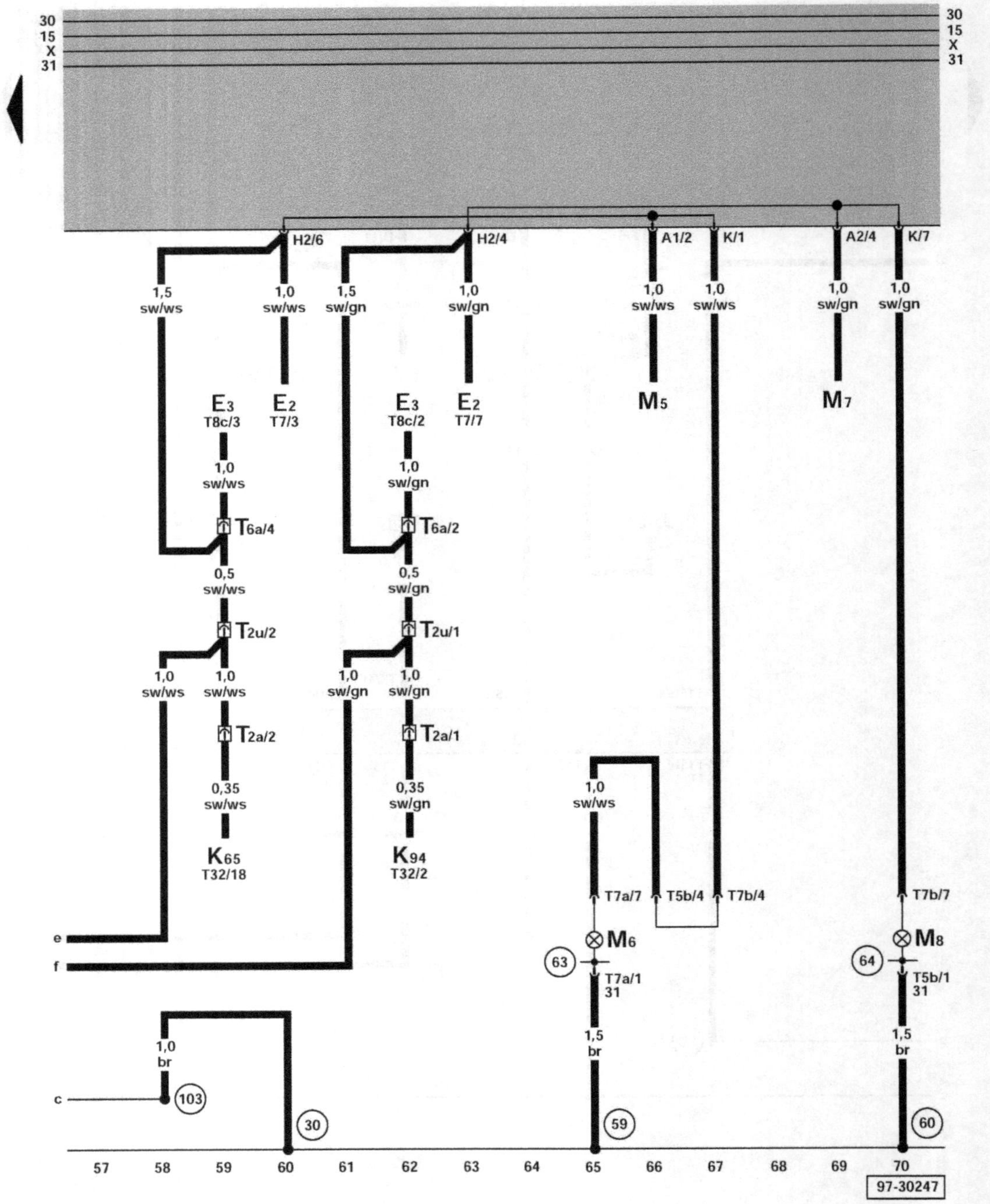

ws = white
sw = black
ro = red
br = brown
gn = green
bl = blue
gr = grey
li = lilac
ge = yellow
or = orange

E2	-	Turn signal switch
E3	-	Emergency Flasher Switch
K65	-	Left Turn Signal Indicator Light
K94	-	Right Turn Signal Indicator Light
M5	-	Left Front Turn Signal Light
M6	-	Left Rear Turn Signal Light
M7	-	Right Front Turn Signal Light
M8	-	Right Rear Turn Signal Light
T2a	-	Double Connector, behind fuse/relay panel
T2u	-	Double Connector, behind fuse/relay panel
T5b	-	5-Pin Connector, on right tail light
T6a	-	6-Pin Connector, behind fuse/relay panel
T7	-	7-Pin Connector, behind steering column switch cover
T7a	-	7-Pin Connector, on left tail light
T7b	-	7-Pin Connector, on right tail light
T8c	-	8-Pin Connector, on Emergency Flasher Switch
T32	-	32-Pin Connector, blue, on instrument cluster
30	-	Ground connection, -1-, beside fuse/relay panel
59	-	Ground connection, near taillight, left
60	-	Ground point -2- near taillamp assembly, right
63	-	Ground connection, on left taillight bulb holder
64	-	Ground connection, on right taillight bulb holder
103	-	Ground connection -2-, in central locking system wiring harness

Fog lights,

from September 2001

Deviate relay location and fuseplacements as well as the locations of multiple connectors see section "component locations".

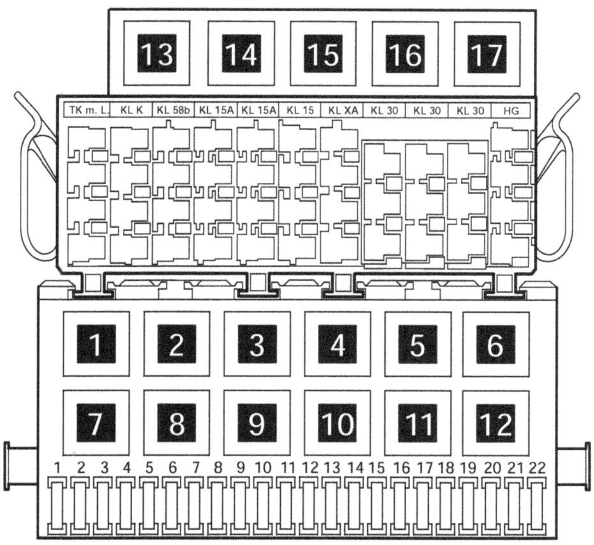

97-23232

Fuse colors

30 A - green

25 A - white

20 A - yellow

15 A - blue

10 A - red

7,5 A - brown

5 A - beige

3 A - violet

4 Load Reduction Relay (18)

Note: Number in parentheses indicates production control number stamped on relay housing.

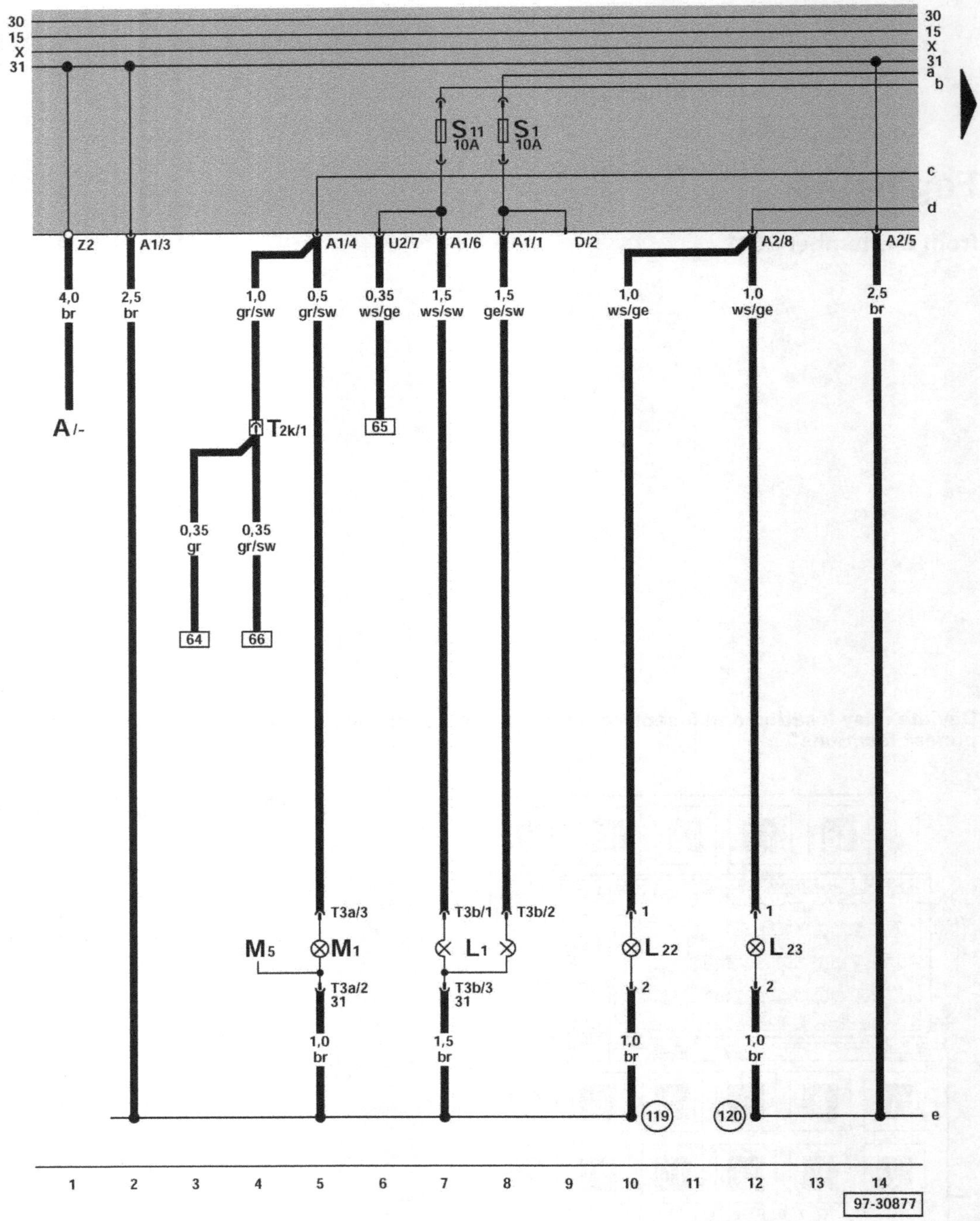

ws = white
sw = black
ro = red
br = brown
gn = green
bl = blue
gr = grey
li = lilac
ge = yellow
or = orange

A - Battery
L1 - Left Headlight
L22 - Left Front Fog Light
L23 - Right Front Fog Light
M1 - Left Parking Light
M5 - Left Front Turn Signal Light
T2k - Double Connector, behind fuse/relay panel
T3a - 3-Pin Connector
T3b - 3-Pin Connector

(119) - Ground connection -1-, in headlight wiring harness
(120) - Ground connection -2-, in headlight wiring harness

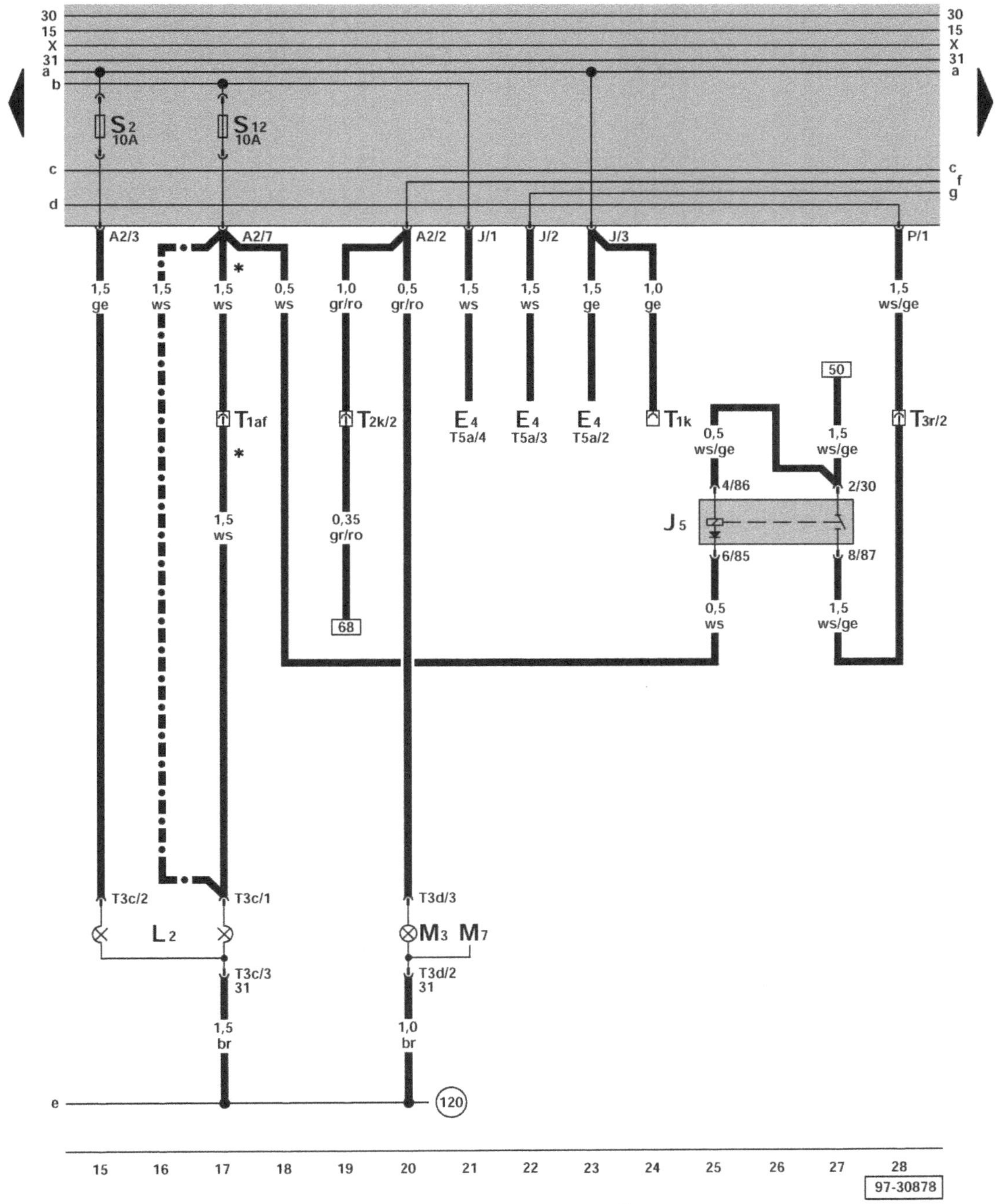

ws = white
sw = black
ro = red
br = brown
gn = green
bl = blue
gr = grey
li = lilac
ge = yellow
or = orange

E4 - Headlight Dimmer/Flasher Switch
J5 - Fog Light Relay (production control number 381)**
L2 - Right Headlight
M3 - Right Parking Light
M7 - Right Front Turn Signal Light
T1af - Single Connector, behind fuse/relay panel
T1k - Single Connector, behind fuse/relay panel
T2k - Double Connector, behind fuse/relay panel
T3c - 3-Pin Connector
T3d - 3-Pin Connector
T3r - 3-Pin Connector, behind fuse/relay panel
T5a - 5-Pin Connector, behind steering column switch cover

(120) - Ground connection -2-, in headlight wiring harness

* - Vehicles with standard equipment only

-•-•- Vehicles with overlay wiring harness only

** - Vehicles with standard equipment behind instrument panel, center.
Vehicles with overlay wiring harness, above fuse/relay panel.

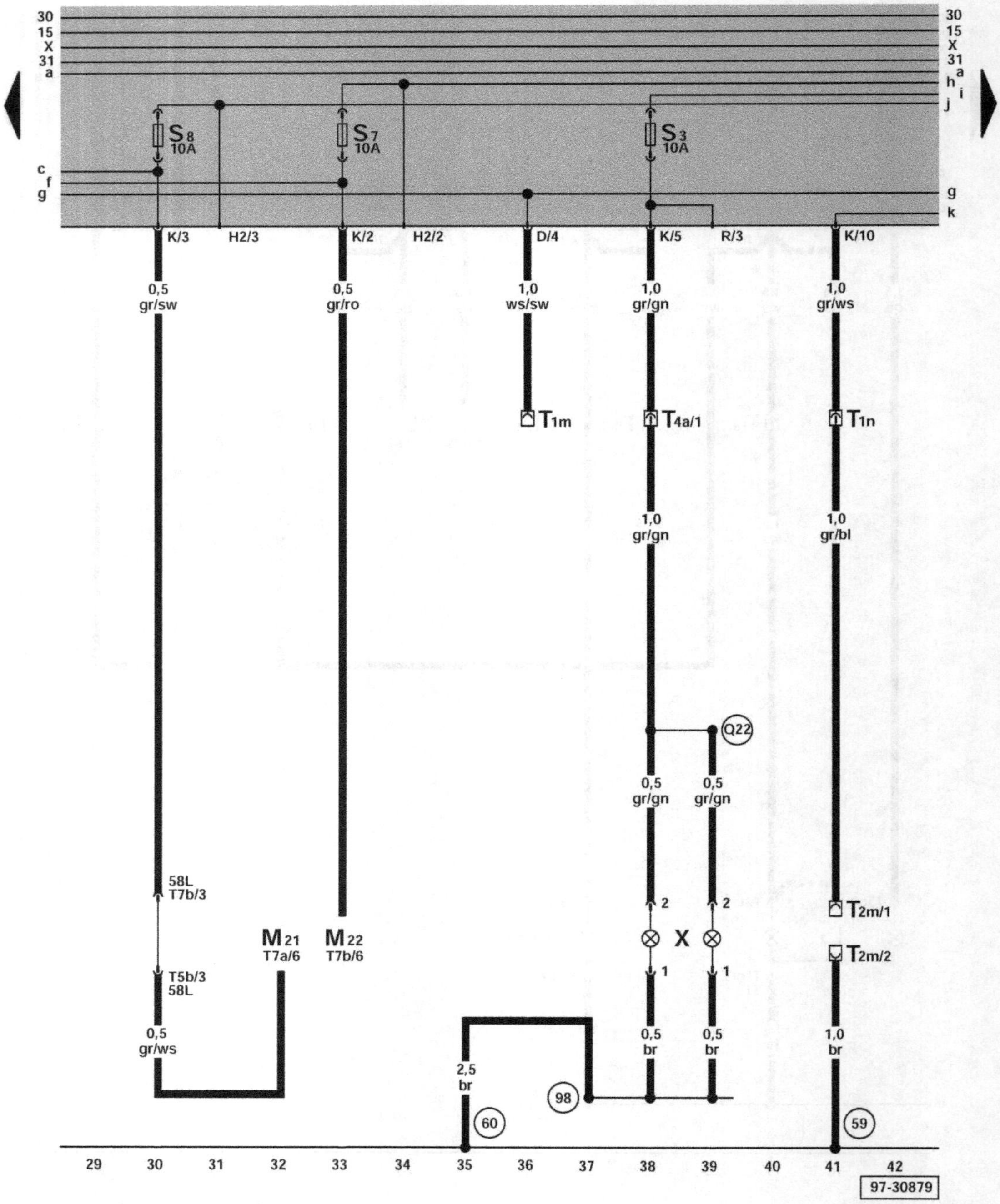

ws = white
sw = black
ro = red
br = brown
gn = green
bl = blue
gr = grey
li = lilac
ge = yellow
or = orange

M21 - Left Brake/Tail Light
M22 - Right Brake/Tail Light
T1m - Single Connector, behind fuse/relay panel
T1n - Single Connector, near taillamp assembly
T2m - Double Connector, on taillight, left
T4a - 4-Pin Connector, near taillamp assembly
T7a - 7-Pin Connector
T5b - 5-Pin Connector
T7b - 7-Pin Connector
X - License Plate Light

(59) - Ground connection, near taillight, left

(60) - Ground point -2- near taillamp assembly, right

(98) - Ground connection, in rear lid wiring harness

(Q22) - Wire connection -1-, in rear lid wiring harness

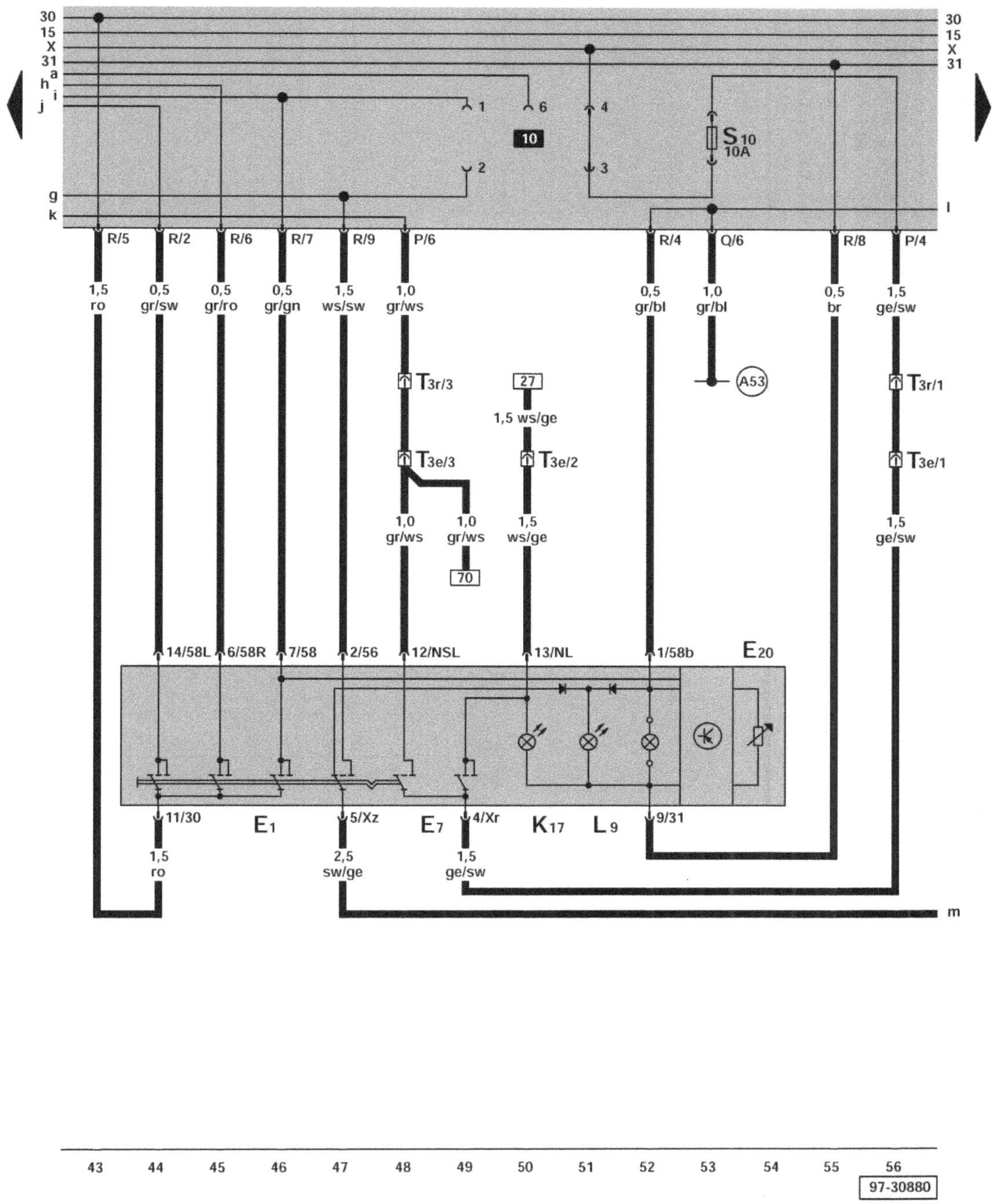

ws = white
sw = black
ro = red
br = brown
gn = green
bl = blue
gr = grey
li = lilac
ge = yellow
or = orange

E7 - Fog Light Switch
E1 - Light switch
E20 - Instrument Panel Light Dimmer Switch
K17 - Fog Light Indicator Light
L9 - Headlight Switch Light
T3e - 3-Pin Connector, behind fuse/relay panel
T3r - 3-Pin Connector, behind fuse/relay panel

(A53) - Plus connection (58b), in instrument panel wiring harness

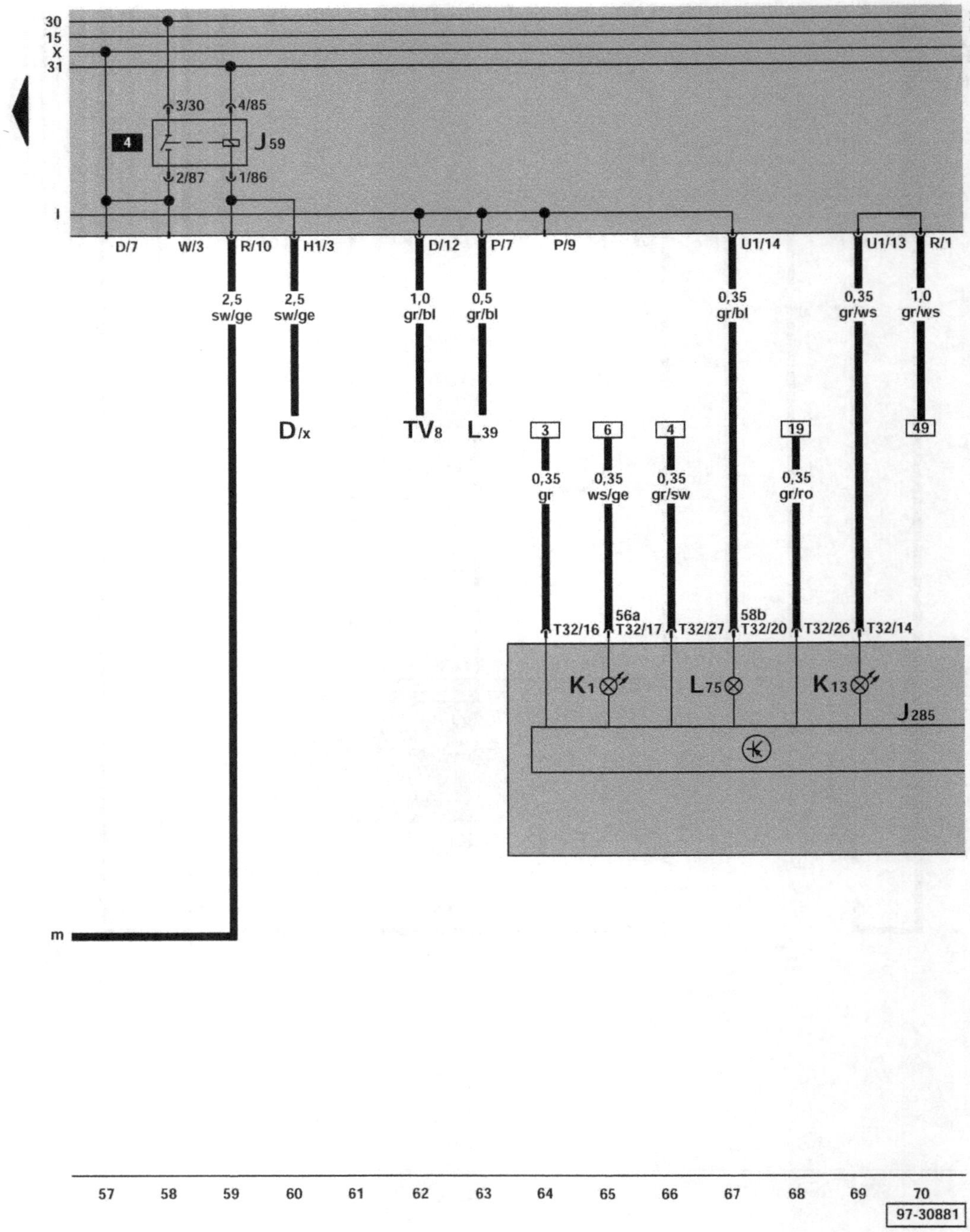

ws = white
sw = black
ro = red
br = brown
gn = green
bl = blue
gr = grey
li = lilac
ge = yellow
or = orange

- D - Ignition/Starter Switch
- J59 - Load Reduction Relay
- J285 - Control module with indicator unit in instrument panel insert
- K1 - Headlight High Beam Indicator Light
- K13 - Rear Fog Light Indicator Light
- L39 - Rear Window Defogger Switch Light
- L75 - Digital Display Light
- T32 - 32-Pin Connector, blue
- TV8 - Terminal 58b Wire Connector, grey

Generator (150 A), with pre-exciter relay

from May 2000 through April 2001

Deviate relay location and fuseplacements as well as the locations of multiple connectors see section "component locations".

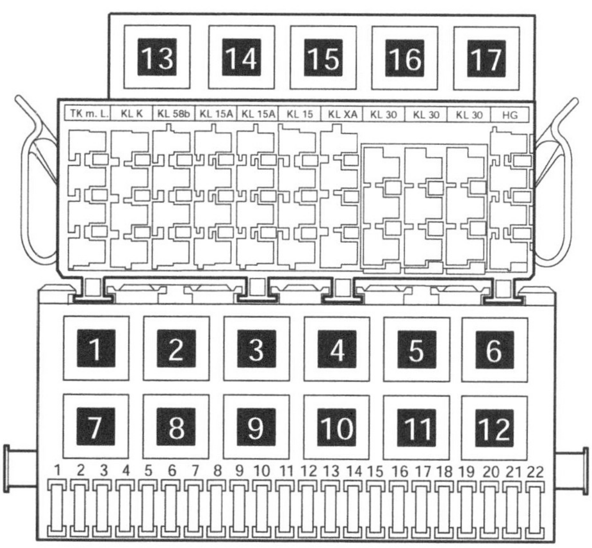

97-23232

Fuse colors

30 A - green
25 A - white
20 A - yellow
15 A - blue
10 A - red
7,5 A - brown
5 A - beige
3 A - violet

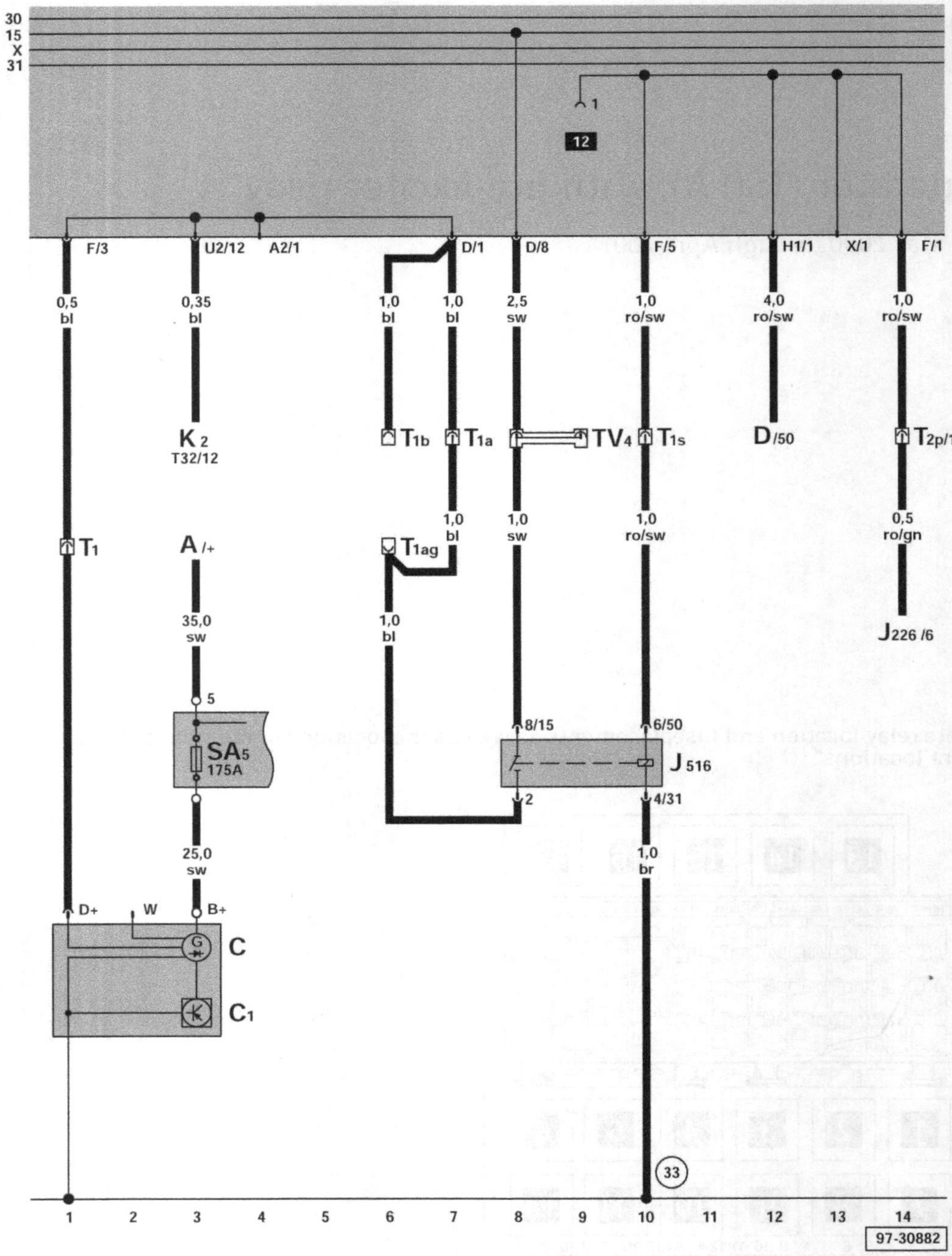

ws = white
sw = black
ro = red
br = brown
gn = green
bl = blue
gr = grey
li = lilac
ge = yellow
or = orange

- A - Battery
- C - Generator
- C1 - Voltage Regulator
- D - Ignition/Starter Switch
- J226 - Park/Neutral Position (PNP) Relay
- J516 - Pre-Excited Generator (GEN) Relay, (control number 53) behind instrument panel, center
- K2 - Generator (GEN) Warning Light
- SA5 - Fuse 5 (in fuse holder/battery)
- T1 - Single Connector, near starter
- T1a - Single Connector, behind fuse/relay panel
- T1ag - Single Connector, behind fuse/relay panel
- T1b - Single Connector, behind fuse/relay panel
- T1s - Single Connector, behind fuse/relay panel
- T2p - Double Connector, behind fuse/relay panel
- T32 - 32-Pin Connector, blue, on instrument cluster
- TV4 - Terminal 15 Wire Connector, black
- (33) - Ground connection, behind instrument panel right